Table of Atomic Weights (Relative Atomic Masses)

Name	Symbol	Atomic Number	Atomic Weight
Actinium*	Ac	89	(227.028)
Aluminum	Al	13	26.981539(5)
Americium*	Am	95	(243)
Antimony (Stibium)	Sb	51	121.757(3)
Argon	Ar	18	39.948(1)
Arsenic	As	33	74.92159(2)
Astatine*	At	85	(210)
Barium	Ba	56	137.327(7)
Berkelium*	Bk	97	(247)
Beryllium	Be	4	9.012182(3)
Bismuth	Bi	83	208.98037(3)
Bohrium*	Bh	107	(262)
Boron	B	5	10.811(5)
Bromine	Br	35	79.904(1)
Cadmium	Cd	48	112.411(8)
Cesium	Cs	55	132.90543(5)
Calcium	Ca	20	40.078(4)
Californium*	Cf	98	(251)
Carbon	C	6	12.011(1)
Cerium	Ce	58	140.115(4)
Chlorine	Cl	17	35.4527(9)
Chromium	Cr	24	51.9961(6)
Cobalt	Co	27	58.93320(1)
Copper	Cu	29	63.546(3)
Curium*	Cm	96	(247)
Dubnium*	Db	105	(262)
Dysprosium	Dy	66	162.50(3)
Einsteinium*	Es	99	(254)
Erbium	Er	68	167.26(3)
Europium	Eu	63	151.965(9)
Fermium*	Fm	100	(257)
Fluorine	F	9	18.9984032(9)
Francium*	Fr	87	(223)
Gadolinium	Gd	64	157.25(3)
Gallium	Ga	31	69.723(1)
Germanium	Ge	32	72.61(2)
Gold	Au	79	196.96654(3)
Hafnium	Hf	72	178.49(2)
Hassium*	Hs	108	(265)
Helium	He	2	4.002602(2)
Holmium	Ho	67	164.93032(3)
Hydrogen	H	1	1.00794(7)
Indium	In	49	114.818(3)
Iodine	I	53	126.90447(3)
Iridium	Ir	77	192.22(3)
Iron	Fe	26	55.847(3)
Krypton	Kr	36	83.80(1)
Lanthanum	La	57	138.9055(2)
Lawrencium*	Lr	103	(262)
Lead	Pb	82	207.2(1)
Lithium	Li	3	6.941(2)
Lutetium	Lu	71	174.967(1)
Magnesium	Mg	12	24.3050(6)
Manganese	Mn	25	54.93805(1)
Meitnerium*	Mt	109	(266)
Mendelevium*	Md	101	(258)
Mercury	Hg	80	200.59(2)
Molybdenum	Mo	42	95.94(1)
Neodymium	Nd	60	144.24(3)
Neon	Ne	10	20.1797(6)
Neptunium*	Np	93	(237.048)
Nickel	Ni	28	58.6934(2)
Niobium	Nb	41	92.90638(2)
Nitrogen	N	7	14.00674(7)
Nobelium*	No	102	(259)
Osmium	Os	76	190.23(3)
Oxygen	O	8	15.9994(3)
Palladium	Pd	46	106.42(1)
Phosphorus	P	15	30.973762(4)
Platinum	Pt	78	195.08(3)
Plutonium*	Pu	94	(244)
Polonium*	Po	84	(209)
Potassium (Kalium)	K	19	39.0983(1)
Praseodymium	Pr	59	140.90765(3)
Promethium*	Pm	61	(145)
Protactinium*	Pa	91	231.03588(2)
Radium*	Ra	88	(226.025)
Radon*	Rn	86	(222)
Rhenium	Re	75	186.207(1)
Rhodium	Rh	45	102.90550(3)
Rubidium	Rb	37	85.4678(3)
Ruthenium	Ru	44	101.07(2)
Rutherfordium*	Rf	104	(261)
Samarium	Sm	62	150.36(3)
Scandium	Sc	21	44.955910(9)
Seaborgium*	Sg	106	(263)
Selenium	Se	34	78.96(3)
Silicon	Si	14	28.0855(3)
Silver	Ag	47	107.8682(2)
Sodium (Natrium)	Na	11	22.989768(6)
Strontium	Sr	38	87.62(1)
Sulfur	S	16	32.066(6)
Tantalum	Ta	73	180.9479(1)
Technetium*	Tc	43	(98)
Tellurium	Te	52	127.60(3)
Terbium	Tb	65	158.92534(3)
Thallium	Tl	81	204.3833(2)
Thorium*	Th	90	232.0381(1)
Thulium	Tm	69	168.93421(3)
Tin	Sn	50	118.710(7)
Titanium	Ti	22	47.88(3)
Tungsten (Wolfram)	W	74	183.84(1)
Ununbium*	Uub	112	(277)
Ununhexium	Uuh	116	(289)
Ununnilium*	Uun	110	(269)
Ununoctium	Uuo	118	(293)
Ununquadium	Uuq	114	(285)
Unununium*	Uuu	111	(272)
Uranium*	U	92	238.0289(1)
Vanadium	V	23	50.9415(1)
Xenon	Xe	54	131.29(2)
Ytterbium	Yb	70	173.04(3)
Yttrium	Y	39	88.90585(2)
Zinc	Zn	30	65.39(2)
Zirconium	Zr	40	91.224(2)

*The elements marked with an asterisk have no stable isotopes.

Recommended by the IUPAC Commission on Atomic Weights and Atomic Abundances; further details are to be found in "Atomic weights of the elements 1991," *Pure and Applied Chemistry, 64,* 1519–1534(1992). The values are reliable to ± the figure given in parentheses, applicable to the last digit. Numbers for atomic weights in parentheses are estimates.

Physical Chemistry

PHYSICAL CHEMISTRY

FOURTH EDITION

KEITH J. LAIDLER
University of Ottawa, Emeritus

JOHN H. MEISER
Ball State University

BRYAN C. SANCTUARY
McGill University

HOUGHTON MIFFLIN COMPANY BOSTON NEW YORK

Vice President and Publisher: Charles Hartford
Executive Editor: Richard Stratton
Editorial Associate: Marisa Papile
Senior Project Editor: Kathryn Dinovo
Cover Design Manager: Diana Coe
Manufacturing Manager: Florence Cadran
Senior Marketing Manager: Katherine Greig

Cover image: © Greene, Dickinson, and Sadeghpour and the Department of Physics & JILA, University of Colorado, Boulder; Sharon Donahue, Photo Editor

On the cover is a trilobite-like long-range Rydberg rubidium dimer showing a cylindrical coordinate surface plot of the electronic probability density for the lowest Born-Oppenheimer state. The equilibrium internuclear distance is R = 1232 a. u. for the $^3\Sigma$ perturbed hydrogenic state, with $n = 30$. The Rb(5s) atom is beneath the towers to the right and the Rb^+ ion is represented as a small blue sphere to the left. From the work of C. H. Greene, A. S. Dickinson, and H. R. Sadeghpour, *Phys. Rev. Lett.,* 85, 2458-2461 (2000).

Photo Credits
Robert Boyle, page 11: Courtesy of the Edgar Fahs Smith Collection, University of Pennsylvania Library
Rudolf Julius Emmanuel Clausius, page 95: Courtesy of the Edgar Fahs Smith Collection, University of Pennsylvania Library
Jacobus Henricus van't Hoff, page 152: Courtesy of the Edgar Fahs Smith Collection, University of Pennsylvania Library
Michael Faraday, page 268: Courtesy of the Edgar Fahs Smith Collection, University of Pennsylvania Library
Svante August Arrhenius, page 273: Courtesy of the Edgar Fahs Smith Collection, University of Pennsylvania Library
Henry Eyring, page 391: Courtesy of Special Collections Department, J. Willard Marriott Library, University of Utah
Gilbert Newton Lewis, page 579: Courtesy of MIT Museum
Gerhard Herzberg, page 649: Bettmann/CORBIS
Ludwig Boltzmann, page 788: AIP Emilio Segre Visual Archives, Physics Today Collection
Dorothy Crowfoot Hodgkin, page 862: Hulton-Deutsch Collection/CORBIS
Agnes Pockels, page 953: Reproduced courtesy of the Library and Information Centre, Royal Society of Chemistry

Printed in the U.S.A.

Library of Congress Control Number: 2001051624

ISBN: 0-618-12341-5

123456789-DOW-06 05 04 03 02

CONTENTS

PREFACE

This fourth edition of *Physical Chemistry,* like its predecessors, has been written in such a way as to be a suitable introduction for students who intend to become chemists, and also for the many others who find physical chemistry essential in their careers. The field of physical chemistry has now become so broad that it has invaded all of the sciences. Physicists, engineers, biologists, and workers in the medical sciences—all find a knowledge of physical chemistry to be important in their work.

The students for which this book is intended are assumed to have a basic knowledge of chemistry such as they usually gain in their first year at a North American university. (In the British system, where the science degree is usually gained after three years, this basic material is taught in the high schools.) This book is intended primarily for the conventional full-year course at a university. However, it covers a good deal more than can be included in a one-year course. It may therefore also be useful in more advanced courses and as a general reference book for those working in fields that require a basic knowledge of the subject.

Changes in This Edition

In this fourth edition of our book we have preserved much of the material of the former editions, making changes only to improve understanding of the concepts or include some of the latest discoveries in physical chemistry. Many chapters have new sections and the coverage of several chapters has been greatly expanded. Unfortunately, in order to save space, we had to delete Chapter 20, Macromolecules. Because of the importance of some ideas in that chapter to other areas of physical chemistry, we have, however, transferred that material to other appropriate chapters.

Most of the numerical values for fundamental properties had to be adjusted in the light of recent data. A major new addition to thermodynamic data has been made in Appendix D; in addition, a table of CODATA thermodynamic data has been added that includes values for $H^\circ(298.15\ \text{K}) - H^\circ(0)$. The data in Appendix D relate to data at 1 bar pressure.

A number of new problems are included in the fourth edition, giving instructors ample choices for their students. The sets of problems cover a wide range of subject matter and difficulty. This new edition, as well as its accompanying *Solutions Manual,* has been thoroughly checked for accuracy.

With this edition, we are pleased to introduce a particularly useful way to visualize physical processes. The CD included with this text allows the student to utilize many interactive graphs of physical relationships. The user can ascertain the effect of changing a variable in what might be a complex relationship. In addition, many figures are animated and give a clear understanding of difficult concepts. Included dialogue, textual information, and text links give the student a well rounded way to learn.

As always, we welcome receiving student and other user comments and suggestions for future editions. We look forward to your input.

Special Features

We have deliberately given a distinctive historical flavor to the book, in part because the history of the subject is of special interest to many students. More importantly, we are convinced that many scientific topics are more comprehensible if they are introduced with some regard to the way in which they originally came to be understood. For example, attempts to present the laws of thermodynamics as postulates are in our opinion unsatisfactory from the pedagogical point of view. A presentation in terms of how the laws of thermodynamics were deduced from the experimental evidence is, we think, much easier for students to understand. In addition, by seeing the historical development of a subject we learn more about the scientific method than we can learn in any other way.

We realize that an historical approach may be dubbed "old-fashioned," but fashion must surely give way to effectiveness. We have also included eleven short biographies of scientists, chosen not because we think their work more important than that of others (for who is able to make such a judgment?), but because we find their lives and careers to be of particular interest.

Several special aids are provided for the student in this book. New to this edition is the *Objectives* section listing key ideas or techniques that the student should have mastered after finishing the chapter. The Preview of each chapter describes the material to be presented in a brief narrative that gives a sense of unity to the material of the chapter. All new terms are in *italics* or in **boldface** type. Particular attention should be paid to these terms as well as to the equations that are boxed for special emphasis. Key equations that appear in the chapter occur in a concise listing at the end of each chapter. The mathematical relationships provided in Appendix C should prove useful as a handy reference.

Organization and Flexibility

The order in which we have treated the various branches of physical chemistry is a matter of personal preference; other teachers may prefer a different order. The book has been written with flexibility in mind. The subject matter may be grouped into the following topics:

A. Chapters 1–6: General properties of gases, liquids, and solutions; thermodynamics; physical and chemical equilibrium

B. Chapters 7–8: Electrochemistry

C. Chapters 9–10: Chemical kinetics

D. Chapters 11–15: Quantum chemistry; spectroscopy; statistical mechanics

E. Chapters 16–19: Some special topics: solids, liquids, surfaces, transport properties

Our sequence has the advantage that the more difficult topics of Chapters 11–15 can come at the beginning of the second half of the course. The book also lends itself without difficulty to various alternative sequences, such as the following:

A	B	C
Chapters 1–6	Chapters 1–6	Chapters 1–6
Chapters 9–10	Chapters 11–15	Chapters 11–15
Chapters 7–8	Chapters 7–8	Chapters 9–10
Chapters 11–15	Chapters 9–10	Chapters 7–8
Chapters 16–19	Chapters 16–19	Chapters 16–19

Aside from this, the order of topics in some of the chapters, particularly those in Chapters 16–19, can be varied.

End-of-Chapter Material

The Key Equation section lists equations with which the student should become thoroughly familiar. This listing should not be construed as the only equations that are important but rather as foundation expressions that are widely applicable to chemical problems. The Problems have been organized according to subject matter, and the more difficult problems are indicated with an asterisk. Answers to all problems are included at the back of the book, with detailed solutions provided in a separate *Solutions Manual for Physical Chemistry*.

Units and Symbols

We have adhered to the Système International d'Unités (SI) and to the recommendations of the International Union of Pure and Applied Chemistry (IUPAC), as presented in the IUPAC "Green Book"; the reader is referred to Appendix A for an outline of these units and recommendations. The essential feature of these recommendations is that the methods of *quantity algebra* (often called "quantity calculus") are used; a symbol represents a physical quantity, which is the product of a pure number (the value of the quantity) and a unit. Sometimes, as in taking a logarithm or making a plot, one needs the *value* of a quantity, which is simply the quantity divided by the unit quantity. The IUPAC Green Book has made no recommendation in this regard, and we have made the innovation of using in the earlier chapters a superscript u (for unitless) to denote such a value. We felt it unnecessary to continue the practice in the later chapters, as the point would be soon appreciated.

Acknowledgments

We are particularly grateful to a number of colleagues for their stimulating conversations, help, and advice over many years, in particular: from the National Research Council of Canada, Drs. R. Norman Jones and D. A. Ramsay (spectroscopy); from the University of Ottawa, Dr. Glenn Facey (NMR spectroscopy), Dr. Brian E. Conway (electrochemistry), and Dr. Robert A. Smith (quantum mechanics); from the University of South Dakota, Dr. Donald Abraham (physics); from Beloit College, Dr. David A. Dobson (physics); from Argonne National Laboratory, Dr. Mark A. Beno (X-ray spectroscopy), Drs. Michael J. Pellin and Stephen L. Dieckman (spectroscopy), and Dr Victor A. Maroni (solid state and superconductors); from John Carroll University, Dr. Michael J. Setter (electrochemistry); from Ball State University, Dr. Jason W. Ribblett (spectroscopy and quantum mechanics); from the University of York, Drs. Graham Doggett, Tom Halstead, Ron Hester, and Robin

Perutz; from McGill University, Drs. John Harrod, Anne-Marie Lebuis (X-ray spectroscopy), David Ronis, Frederick Morin, Zhicheng (Paul) Xia (NMR spectroscopy), and Nadim Saade (mass spectroscopy).

Special acknowledgment is due to those who have contributed to the multimedia component of this work: Dr. Tom Halstead, University of York; Adam Halstead, Emily Cranston, and Jürgen Karir, MCH Multimedia Inc.; M. S. Krishnan, Institute of Technology, Madras, India; J. Anantha Krishnan, Pronexus Infoworld, Animations.

Thanks are also due our reviewers for this edition, including:

Edmund Tisko, University of Nebraska
Gordon Atkinson, University of Oklahoma
Bernard Laurenzi, University of Albany
Stephen Kelty, Seton Hall University
Pedro Muino, Saint Francis College
Ruben Parra, University of Nebraska, Lincoln
Jonathan Kenny, Tufts University
James Whitten, University of Massachusetts, Lowell
Therese Michels, Dana College
Renee Cole, Missouri State University
Robert Brown, Douglas College
Phillip Pacey, Dalhousie University
Christine Lamont, University of Huddersfield
Curt Wentrup, University of Queensland
N.K. Singh, University of New South Wales
David Hawkes, Lambuth University

In addition, we would like to thank the following reviewers for their suggestions in the previous edition: William R. Brennen (University of Pennsylvania), John W. Coutts (Lake Forest College), Nordulf Debye (Towson State University), D. J. Donaldson (University of Toronto), Walter Drost-Hansen (University of Miami), David E. Draper (Johns Hopkins University), Darrel D. Ebbing (Wayne State University), Brian G. Gowenlock (University of Exeter), Robert A. Jacobson (Iowa State University), Gerald M. Korenowski (Rensselaer Polytechnic Institute), Craig C. Martens (University of California, Irvine), Noel L. Owen (Brigham Young University), John Parson (The Ohio State University), David W. Pratt (University of Pittsburgh), Lee Pederson (University of North Carolina at Chapel Hill), Richard A. Pethrick (University of Strathclyde), Mark A. Smith (University of Arizona), Charles A. Trapp (University of Louisville), Gene A. Westenbarger (Ohio University), Max Wolfsberg (University of California, Irvine), John D. Vaughan (Colorado State University), Josef W. Zwanziger (Indiana University).

We would be amiss if we did not acknowledge the careful work of our project editor, Gina J. Linko. Finally, we would like to especially note the contribution of B. Ramu Ramachandran, Louisiana Tech University, whose work on the end-of-chapter problems and on the *Solutions Manual* has been an important part of this edition.

Keith J. Laidler
John H. Meiser
Bryan C. Sanctuary

PHYSICAL CHEMISTRY

The Nature of Physical Chemistry and the Kinetic Theory of Gases

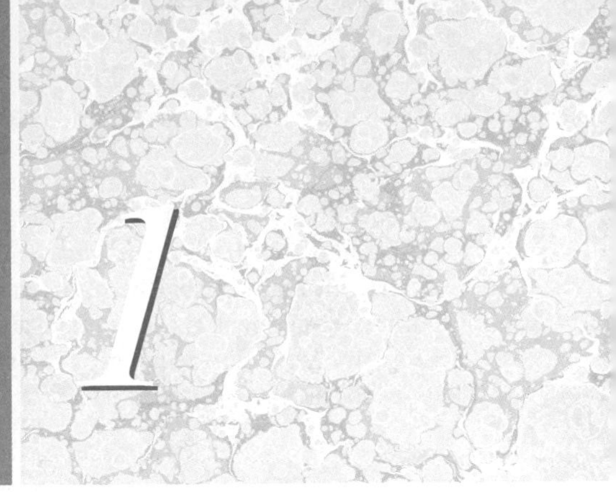

PREVIEW

In each Preview we focus on the highlights of the chapter topics and attempt to draw attention to their importance. As you begin to learn the language of physical chemistry, pay particular attention to definitions or special terms, which in this book are printed in **boldface** or *italic* type.

Physical chemistry is the application of the methods of physics to chemical problems. It can be organized into *thermodynamics, kinetic theory, electrochemistry, quantum mechanics, chemical kinetics,* and *statistical thermodynamics.* Basic concepts of physics, including classical mechanics, are important to these areas. We begin by developing the relation between *work* and *kinetic energy*. Our main interest is in the *system* and its *surroundings*.

Gases are easier to treat than liquids or solids, so we treat gases first. Following are two experimentally derived equations relating to a fixed amount of gas:

$$\textit{Boyle's Law: } PV = \text{constant}_1, \quad \text{(at constant } T \text{ and } n\text{)}$$

$$\textit{Gay-Lussac's Law: } \frac{V}{T} = \text{constant}_2, \quad \text{(at constant } P \text{ and } n\text{)}$$

These expressions combine, with the use of *Avogadro's hypothesis* that the *amount of substance n* (SI unit: mole) is proportional to the volume at a fixed T and P, to give the *ideal gas law:*

$$PV = nRT$$

where R is the *gas constant.* A gas that obeys this equation is called an *ideal gas.*

Experimental observations as embodied in these laws are important but so too is the development of a theoretical explanation for these observations. An important development in this regard is the calculation of the *pressure of a gas* from the kinetic-molecular theory. The relation of the mean molecular kinetic energy to temperature, namely,

$$\bar{\epsilon}_k = \tfrac{3}{2}k_BT \qquad (k_B = \text{Boltzmann constant})$$

allows a theoretical derivation of the ideal gas law and of laws found experimentally.

Molecular collisions between gas molecules play an important role in many concepts. *Collision densities,* often called *collision numbers,* tell us how often collisions occur in unit volume between like or unlike molecules in unit time. Related to collisions is the idea of *mean free path,* which is the average distance gas molecules travel between collisions.

Real gases differ in their behavior from ideal gases, and this difference can be expressed using the *compression factor* $Z = PV/nRT$ where $Z = 1$ if the real gas behavior is identical to that of an ideal gas. Values of Z above or below unity indicate deviations from ideal behavior. Real gases also show *critical phenomena* and *liquefaction,* phenomena that are impossible for an ideal gas. Study of critical phenomena, in particular *supercritical fluids*, has led to development of industrial processes as well as analytical techniques. The concept that there is complete *continuity of states* in the transformation from the gas to the liquid state is important in the treatment of the condensation of gas. The *van der Waals equation,* in which the pressure of the ideal gas is modified to account for attractive forces between gas particles and in which the ideal volume is reduced to allow for the actual size of the gas particles, is an important expression for describing real gases. This equation and others led to a greater understanding of gas behavior, and also provided the means to predict the behavior of chemical processes involving gases.

OBJECTIVES

The purpose of this section is to give a *minimum* listing of knowledge or computational skills that should be mastered from each chapter. This section is not meant to be all inclusive since the true understanding of physical chemistry should allow the application of the principles presented to situations and cases not covered here. Some instructors may emphasize additional areas of study.

In this chapter and each succeeding chapter, the student must be able to define and understand all of the **boldface** terms and should be familiar with, as well as able to use, all the key equations at the end of each chapter.

After studying this chapter, the student should be able to:

- Show the relationship between work and force and calculate the work under various force conditions.
- Calculate kinetic and potential energies. Identify systems and states and be able to determine equilibrium conditions.
- Calculate the temperature using different fluids.
- Determine *P, V, n, T,* or *R* relationships under conditions for Boyle's law, Gay-Lussac's (Charles's) law, and the ideal gas law.
- Understand the concept of absolute zero and the use of the Kelvin temperature.
- Develop the mole concept and link it with Avogadro's hypothesis.
- Clearly define the conditions of the kinetic-molecular theory and be able to calculate the pressure of an ideal gas from its premises.
- Calculate the mean-square speed of molecules.
- Determine the partial pressure and the rate of effusion of gases.
- Calculate the number of molecular collisions under given conditions, the related collision diameters, and the frequency, density, and mean free path.
- Be able to derive the barometric distribution law and to work through the Maxwell-Boltzmann distribution law.
- Explain and use the compression factor, critical point, critical temperature, and critical volume, and relate these to a supercritical fluid.
- Be able to work the problems related to the various equations of state, including the law of corresponding states and the virial equation.

Humans are exceedingly complex creatures, and they live in a very complicated universe. In searching for a place in their environment, they have developed a number of intellectual disciplines through which they have gained some insight into themselves and their surroundings. They are not content merely to acquire the means of putting their environment to practical use, but they also have an insatiable desire to discover the basic principles that govern the behavior of all matter. These endeavors have led to the development of bodies of knowledge that were formerly known as *natural philosophy,* but that are now generally known as *science.*

1.1 The Nature of Physical Chemistry

In this book, we are concerned with the branch of science known as *physical chemistry*. Physical chemistry is the application of the methods of physics to chemical problems. It includes the qualitative and quantitative study, both experimental and theoretical, of the general principles determining the behavior of matter, particularly the transformation of one substance into another. Although physical chemists use many of the methods of the physicist, they apply the methods to chemical structures and chemical processes. Physical chemistry is not so much concerned with the description of chemical substances and their reactions—this is the concern of organic and inorganic chemistry—as it is with theoretical principles and with quantitative problems.

Microscopic Properties

Two approaches are possible in a physicochemical study. In what might be called a *systemic* approach, the investigation begins with the very basic constituents of matter—the fundamental particles—and proceeds conceptually to construct larger systems from them. The adjective *microscopic* (Greek *micros,* small) is used to refer to these tiny constituents. In this way, increasingly complex phenomena can be interpreted on the basis of the elementary particles and their interactions.

Macroscopic Properties

In the second approach, the study starts with investigations of *macroscopic* material (Greek *macros,* large), such as a sample of liquid or solid that can be easily observed with the eye. Measurements are made of macroscopic properties such as pressure, temperature, and volume. In this *phenomenological* approach, more detailed studies of microscopic behavior are made only insofar as they are needed to understand the macroscopic behavior in terms of the microscopic.

In the early development of physical chemistry, the more traditional macroscopic approach predominated. The development of thermodynamics is a clear example of this. In the late nineteenth century, a small number of experiments in physics that were difficult to explain on the basis of classical theory led to a revolution in thought. Growing out of this development, quantum mechanics, statistical thermodynamics, and new spectroscopic methods have caused physical chemistry to take on a more microscopic flavor, particularly throughout the latter part of the twentieth century.

Physical chemistry encompasses the structure of matter at equilibrium as well as the processes of chemical change. Its three principal subject areas are thermodynamics, quantum chemistry, and chemical kinetics; other topics, such as electrochemistry, have aspects that lie in all three of these categories. *Thermodynamics,* as applied to chemical problems, is primarily concerned with the position of chemical equilibrium, with the direction of chemical change, and with the associated changes in energy. *Quantum chemistry* theoretically describes bonding at a molecular level. In its exact treatments, it deals only with the simplest of atomic and molecular systems, but it can be extended in an approximate way to deal with bonding in much more complex molecular structures. *Chemical kinetics* is concerned with the rates and mechanisms with which processes occur as equilibrium is approached.

An intermediate area, known as *statistical thermodynamics,* links the three main areas of thermodynamics, quantum chemistry, and kinetics and also, through computer simulations, provides a basic relationship between the microscopic and macroscopic worlds. Related to this area is *nonequilibrium statistical mechanics,* which is becoming an increasingly important part of modern physical chemistry. This field includes problems in such areas as the theory of dynamics in liquids, and light scattering.

1.2 Some Concepts from Classical Mechanics

We will often calculate the work done or a change in energy when a chemical process takes place. It is important to know how these are related; therefore, our study of physical chemistry begins with some fundamental macroscopic principles in mechanics.

Work

Work can take many forms, but any type of work can be resolved through dimensional analysis as the application of a force through a distance. If a force $\boldsymbol{F}$ (a vector indicated by boldface type) acts through an infinitesimal distance $d\boldsymbol{l}$ ($\boldsymbol{l}$ is the position vector), the work is

$$dw = \boldsymbol{F} \cdot d\boldsymbol{l} \tag{1.1}$$

If the applied force is not in the direction of motion but makes an angle θ with this direction (as shown in Figure 1.1), the work is the component $F \cos \theta$ in the direction of the motion multiplied by the distance traveled, $d\boldsymbol{l}$:

$$dw = F \cos \theta \, d\boldsymbol{l} \tag{1.2}$$

Equation 1.2 can then be integrated to determine the work in a single direction. The force $\boldsymbol{F}$ can also be resolved into three components, F_x, F_y, F_z, one along each of the three-dimensional axes. For instance, for a constant force F_x in the X-direction,

$$w = \int_{x_0}^{x} F_x \, dx = F_x(x - x_0) \qquad (x_0 = \text{initial value of } x) \tag{1.3}$$

Hooke's Law

Several important cases exist where the force does not remain constant, including gravitation, electrical charges, and springs. As an example, **Hooke's law** states that for an idealized spring

$$F = -k_h x \tag{1.4}$$

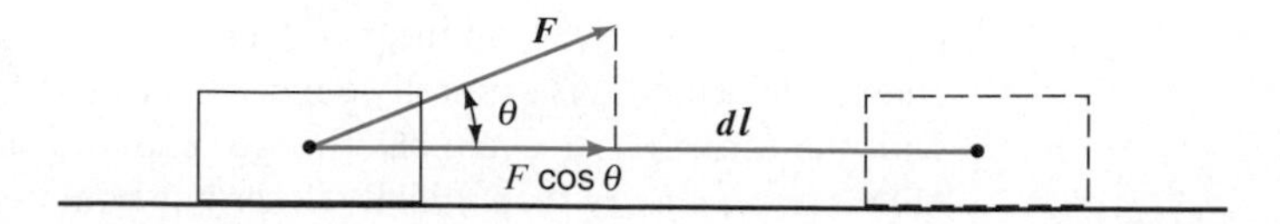

FIGURE 1.1
Work is the applied force in the direction of motion multiplied by *dl*.

*Hook's Law: Calibrate a spring balance and weigh masses on different planets.

where x is the displacement from a position ($x_0 = 0$) at which F is initially zero, and k_h (known as a *force constant*) relates the displacement to the force. See Figure 1.2. The work done on the spring to extend it is found from Eq. 1.3:

$$w = \int_0^x -k_h x\,dx = -\frac{k_h}{2}x^2 \tag{1.5}$$

Harmonic Oscillator

A particle vibrating under the influence of a restoring force that obeys Hooke's law is called a **harmonic oscillator.** These relationships apply fairly well to vibrational variations in bond lengths and consequently to the stretching of a chemical bond.

Kinetic and Potential Energy[1]

Kinetic Energy

Energy: Calculate the kinetic energy of a bullet.

The energy possessed by a moving body by virtue of its motion is called its **kinetic energy** and can be expressed as

$$E_k = \frac{1}{2}mu^2 \tag{1.6}$$

where $\boldsymbol{u}$ (= $d\boldsymbol{l}/dt$) is the velocity (i.e., the instantaneous rate of change of the position vector $\boldsymbol{l}$ with respect to time) and m is the mass. An important relation between work and kinetic energy for a point mass can be demonstrated by casting Eq. 1.1 into an integral over time:

$$w = \int_{l_0}^{l} \boldsymbol{F}(\boldsymbol{l}) \cdot d\boldsymbol{l} = \int_{t_0}^{t} \boldsymbol{F}(\boldsymbol{l}) \cdot \frac{d\boldsymbol{l}}{dt}\,dt = \int_{t_0}^{t} \boldsymbol{F}(\boldsymbol{l}) \cdot \boldsymbol{u}\,dt \tag{1.7}$$

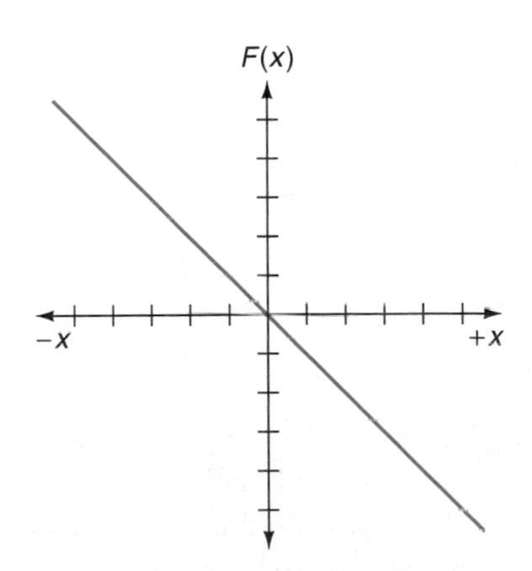

FIGURE 1.2
Plot of Hooke's law for arbitrary k_h. The force, F, is plotted against displacement, x, from an equilibrium position at $x = 0$. The force constant k_h is normally constant only over short ranges and then deviates, causing a departure from the straight-line condition shown.

Substitution of Newton's second law,

$$\boldsymbol{F} = m\boldsymbol{a} = m\frac{d\boldsymbol{u}}{dt} \tag{1.8}$$

where $\boldsymbol{a}$ is the acceleration, yields

$$w = \int_{t_0}^{t} m\frac{d\boldsymbol{u}}{dt} \cdot \boldsymbol{u}\,dt = m\int_{u_0}^{u} \boldsymbol{u} \cdot d\boldsymbol{u} \tag{1.9}$$

After integration and substitution of the limits if $\boldsymbol{u}$ is in the same direction as $d\boldsymbol{u}$ ($\cos\theta = 1$), the expression becomes, with the definition of kinetic energy (Eq. 1.6),

$$w = \int_{l_0}^{l} \boldsymbol{F}(\boldsymbol{l}) \cdot d\boldsymbol{l} = \frac{1}{2}mu_1^2 - \frac{1}{2}mu_0^2 = E_{k_1} - E_{k_0} \tag{1.10}$$

Thus, we find that the difference in kinetic energy between the initial and final states of a point body is the work performed in the process.

Another useful expression is found if we assume that the force is conservative. Since the integral in Eq. 1.10 is a function of l alone, it can be used to define a new function of l that can be written as

$$\boldsymbol{F}(\boldsymbol{l}) \cdot d\boldsymbol{l} = -dE_p(\boldsymbol{l}) \tag{1.11}$$

[1]Some mathematical relationships are found in Appendix C.

*For CD references, insert CD and click on the chapter you wish to study. Locate the desired topic on the pull-down menu.

Potential Energy

Energy: Find the potential energy of a spring.

This new function $E_p(l)$ is the **potential energy,** which is the energy a body possesses by virtue of its position.

For the case of a system that obeys Hooke's law, the potential energy for a mass in position x is usually defined as the work done against a force in moving the mass to the position from one at which the potential energy is arbitrarily taken as zero:

$$E_p = \int_0^x -F\,dx = \int_0^x k_h x\,dx = \tfrac{1}{2}k_h x^2 \tag{1.12}$$

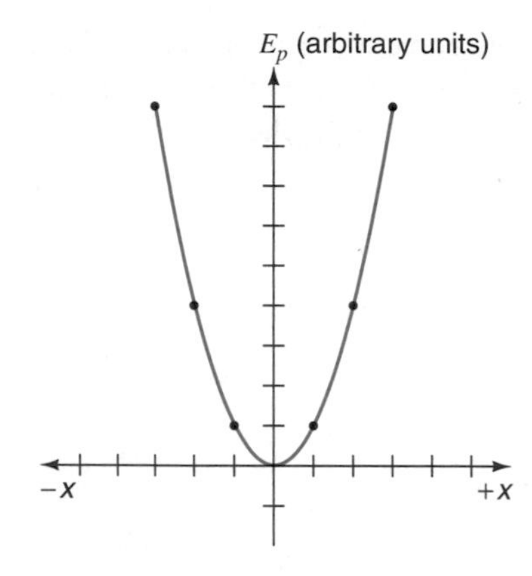

FIGURE 1.3
Plot of $E_p = -\int_0^x F\,dx = -\frac{1}{2}k_h x^2$ for the case of a system that obeys Hooke's law.

Thus, the potential energy rises parabolically on either side of the equilibrium position. See Figure 1.3. There is no naturally defined zero of potential energy. This means that absolute potential energy values cannot be given but only values that relate to an arbitrarily defined zero energy.

An expression similar to Eq. 1.10 but now involving potential energy can be obtained by substituting Eq. 1.11 into Eq. 1.10:

$$w = \int_{l_0}^{l} \mathbf{F}(\mathbf{l}) \cdot d\mathbf{l} = E_{p_0} - E_{p_1} = E_{k_1} - E_{k_0} \tag{1.13}$$

Rearrangement gives

$$E_{p_0} + E_{k_0} = E_{p_1} + E_{k_1} \tag{1.14}$$

which states that the sum of the potential and kinetic energies, $E_p + E_k$, remains constant in a transformation. Although Eq. 1.14 was derived for a body moving between two locations, it is easy to extend the idea to two colliding particles. We then find that the sum of the kinetic energy of translation of two or more bodies in an *elastic collision* (no energy lost to internal motion of the bodies) is equal to the sum after impact. This is equivalent to saying that there is no potential energy change of interaction between the bodies in collision. Expressions such as Eq. 1.14 are known as *conservation laws* and are important in the development of kinetic theory.

Elastic Collision

Energy: Follow E_k and E_p of the arrow of an archer.

1.3 Systems, States, and Equilibrium

Systems, states, and equilibrium: Example of open, closed, and isolated systems.

Physical chemists attempt to define very precisely the object of their study, which is called the *system.* It may be solid, liquid, gaseous, or any combination of these. The study may be concerned with a large number of individual components that comprise a macroscopic system. Alternatively, if the study focuses on individual atoms and molecules, a microscopic system is involved. We may summarize by saying that the system is a particular segment of the world (with definite boundaries) on which we focus our attention. Outside the system are the *surroundings,* and the system plus the surroundings compose a *universe.* In an *open system* there can be transfer of heat and also material. If no material can pass between the system and the surroundings, but there can be transfer of heat, the system is said to be a *closed system.* Finally, a system is said to be *isolated* if neither matter nor heat is permitted to exchange across the boundary. This could be accomplished by surrounding the system with an insulating container. These three possibilities are illustrated in Figure 1.4.

Physical chemists generally concern themselves with measuring the properties of a system, properties such as pressure, temperature, and volume. These properties

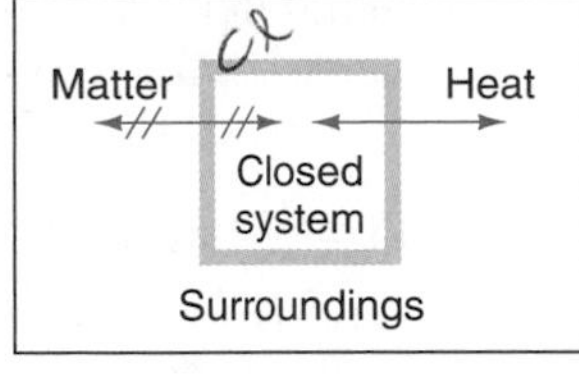

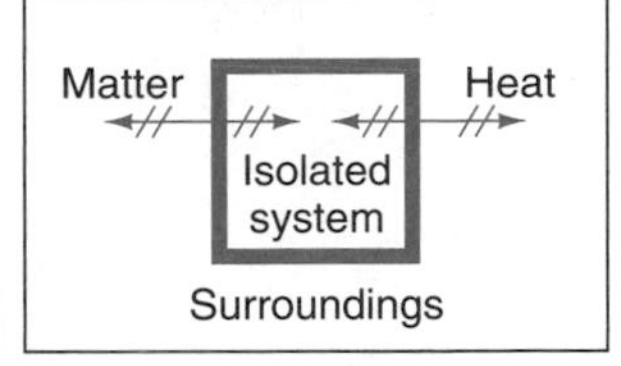

FIGURE 1.4 Relationship of heat and matter flow in open, closed, and isolated systems.

Intensive and Extensive Properties

may be of two types. If the value of the property *does not change with the quantity of matter present* (i.e., if it does not change when the system is subdivided), we say that the property is an *intensive* property. Examples are pressure, temperature, and refractive index. If the property *does change with the quantity of matter present,* the property is called an *extensive* property. Volume and mass are extensive. The ratio of two extensive properties is an intensive property. There is a familiar example of this; the density of a sample is an intensive quantity obtained by the division of mass by volume, two extensive properties.

Compare and contrast stable and unstable equilibrium states of mechanical systems. An oscillatory reaction fails to reach equilibrium.

A certain minimum number of properties have to be measured in order to determine the condition or state of a macroscopic system completely. For a given amount of material it is then usually possible to write an equation describing the state in terms of intensive variables. This equation is known as an *equation of state* and is our attempt to relate empirical data that are summarized in terms of experimentally defined variables. For example, if our system consists of gas, we normally could describe its state by specifying properties such as amount of substance, temperature, and pressure. The volume of gas is another property that will change as temperature and pressure are altered, but this fourth variable is fixed by an equation of state that connects these four properties. In some cases it is important to specify the shape or extent of the surface. Therefore, we cannot state unequivocally that a predetermined number of independent variables will always be sufficient to specify the state of an arbitrary system. However, if the variables that specify the state of the system do not change with time, then we say the system is in **equilibrium.** Thus, a state of equilibrium exists when there is no change with time in any of the system's macroscopic properties.

Equilibrium

1.4 Thermal Equilibrium

Zeroth Law of Thermodynamics: Animation of bodies with the same temperature at equlibrium.

It is common experience that when two objects at different temperatures are placed in contact with each other for a long enough period of time, their temperatures will become equal; they are then in equilibrium with respect to temperature. The concept of *heat* as a form of energy enters here. We observe that the flow of heat from a warmer body serves to increase the temperature of a colder body. However, heat is not temperature.

Zeroth Law of Thermodynamics

We extend the concept of equilibrium by considering two bodies *A* and *B* that are in thermal equilibrium with each other; at the same time an additional body *C* is in equilibrium with *B*. Experimentally we find that *A* and *C* also are in equilibrium with each other. This is a statement of the **zeroth law of thermodynamics:** Two bodies in thermal equilibrium with a third are in equilibrium with each other. This then leads to a way to measure temperature.

The Concept of Temperature and Its Measurement

The physiological sensations that we accept as indications of whether an object is hot or cold cannot serve us quantitatively; they are relative and qualitative. The first thermometer using the freezing point and boiling point of water as references was introduced by the Danish astronomer Olaus Rømer (1644–1710). On the old centigrade scale [Latin *centum,* hundred; *gradus,* step; also called the *Celsius scale,* named in honor of the Swedish astronomer Anders Celsius (1701–1744)] the freezing point of water at 1 atmosphere (atm) pressure was fixed at exactly 0 °C, and the boiling point at exactly 100 °C. We shall see later that the Celsius scale is now defined somewhat differently.

Celsius Scale

Temperature: Construct a thermometer with a cold and hot reference; Explore different temperature scales.

The construction of many thermometers is based on the fact that a column of mercury changes its length when its temperature is changed. The length of a solid metal rod or the volume of a gas at constant pressure is also used in some thermometers. Indeed, for any thermometric property, whether a length change is involved or not, the old centigrade temperature θ was related to two defined temperatures. In the case of the mercury column, we assign its length the value l_{100} when it is at thermal equilibrium with boiling water vapor at 1 atm pressure. The achievement of equilibrium with melting ice exposed to 1 atm pressure establishes the value of l_0 for this length. Assuming a linear relationship between the temperature θ and the thermometric property (length, in this case), and assuming 100 divisions between the fixed marks, allows us to write

$$\theta = \frac{(l - l_0)}{(l_{100} - l_0)}(100°) \tag{1.15}$$

Temperature: Use of Eq. 1.15.

where l is the length at temperature θ, and l_0 and l_{100} are the lengths at the freezing and boiling water temperatures, respectively. Some thermometric properties do not depend on a length, such as in a quartz thermometer where the resonance frequency response of quartz crystal is used as the thermometric property. An equation of the form of Eq. 1.15 still applies, however. Thermometric properties of actual materials generally deviate from exact linearity, even over short ranges, because of the atomic or molecular interactions within the specific material, thus reducing the value of that substance to function as a thermometric material over large temperature ranges.

1.5 Pressure and Boyle's Law

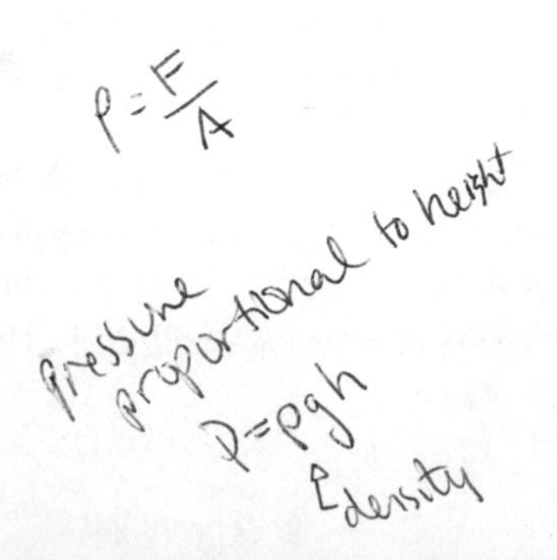

The measurement of pressure, like temperature, is a comparatively recent development. Pressure is the force per unit area. Perhaps one of the most familiar forms of pressure is that exerted by the atmosphere. Atmospheric pressure is often measured as a difference in height, h, of a mercury column trapped in an inverted tube suspended in a pool of mercury. The pressure is proportional to h, where $P = \rho gh$, ρ is the density, and g is the acceleration of gravity. The Italian physicist Evangelista

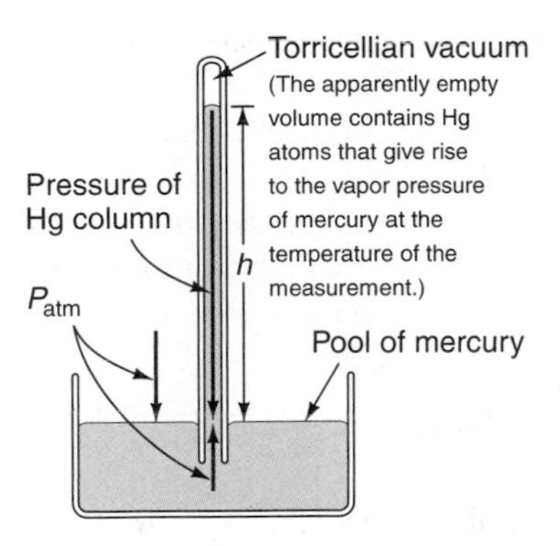

FIGURE 1.5
A barometer. The height, h, of the mercury column exerts a pressure at the surface of the mercury pool. This pressure is exactly counterbalanced by the pressure of the atmosphere, P_{atm}, on the surface of the mercury pool.

Torricelli (1608–1647) used such a device called a **barometer.** (See Figure 1.5.) In the past the standard atmosphere has been defined as the pressure exerted by a column of mercury 760 mm high at 0 °C. In SI units,[2] the standard atmospheric pressure (1 atm) is defined as exactly 101 325 Pa, where the abbreviation Pa stands for the SI unit of pressure, the pascal (kg m^{-1} s^{-2} = N m^{-2}). In this system, 133.322 Pa is equal to the pressure produced by a column of mercury exactly 1 millimeter (mm) in height. Since the pascal is inconveniently small for many uses, the unit **torr** (named after Torricelli) is defined so that 1 atm = 760 Torr exactly. Thus, the torr is almost exactly equal to 1 mmHg. (See Problem 1.7.) Another unit of pressure commonly in use is the **bar:**

$$1 \text{ bar} = 100 \text{ kPa} = 0.986923 \text{ atm}$$

compared to

$$1 \text{ atm} = 101.325 \text{ kPa}.$$

Today, 1 bar is the standard pressure used to report thermodynamic data (Section 2.5).

On page 8 we implied that the temperature of a gaseous system could be determined through its relationship to pressure and volume. To see how this can be done, consider the pressure-volume relation at constant temperature. A gas contained within a closed vessel exerts a force on the walls of the vessel. This force is related to the pressure P of gas (the force F divided by the wall area A) and is a scalar quantity; that is, it is independent of direction.

The pressure of a gas contained in a closed vessel may be measured using a manometer. Two versions are in common use (Figure 1.6). Both consist of a U-tube filled with a liquid of low volatility (such as mercury or silicone oil). In both, the top of one leg of the U-tube is attached to the sample in its container. In the closed-end manometer, the sample pressure is directly proportional to the height difference of the two columns. In the open-end manometer, the difference in height of the two

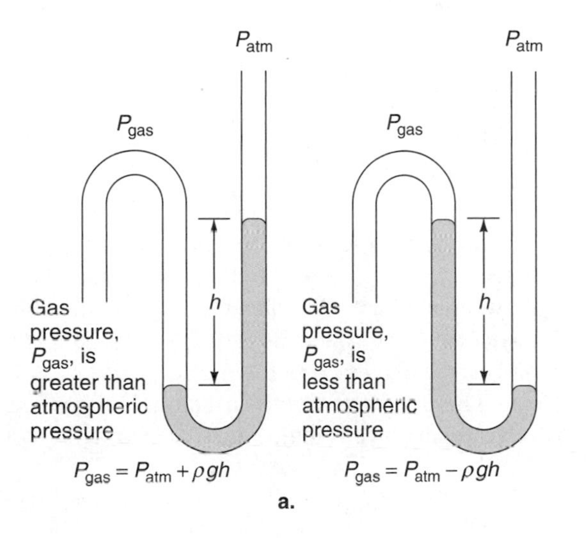

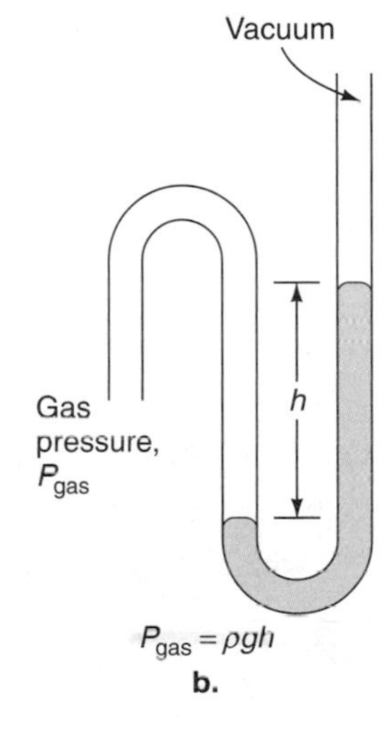

FIGURE 1.6
Two types of manometer. In (a), depicting an open-end manometer, h is added to or subtracted from the atmospheric pressure in order to find the pressure of gas. In (b), depicting a closed-end manometer, the height difference h is directly proportional to the pressure of the sample, ρgh, where ρ is the density of the manometric liquid.

Pressure: The use of the barometer.

[2] See Appendix A for a discussion of SI units and the recommendations of the International Union of Pure and Applied Chemistry.

columns is proportional to the difference in pressure between the sample and the atmospheric pressure.

Barometers and manometers fall into a class of pressure measurement devices, which depend on the measurement of the height of a liquid column. These are used for only moderate pressures. A second class of devices involves the measurement of the distortion of an elastic pressure chamber. These devices include Bourdon-tube gauges for high pressures and diaphragm gauges for more moderate pressures. The third class of devices is based on electrical sensors. Strain gauges are used for moderate pressures into the vacuum range. For the measurement of still lower vacuum, Pirani gauges or thermocouple gauges are used down to 10^{-3} Torr. Below this pressure more sophisticated gauges are used, such as thermionic ionization gauges, cold cathode gauges, and Baynard-Alpert ion gauges. Discussion of these devices is beyond the scope of this book.

EXAMPLE 1.1 Compare the length of a column of mercury to that of a column of water required to produce a pressure of 1.000 bar. The densities of mercury and water at 0.00 °C are 13.596 g cm^{-3} and 0.99987 g cm^{-3}, respectively.

Solution The pressure exerted by both liquids is given by $P = \rho gh$. Since the length of both liquids must exert the same pressure, we can set the pressures equal with the subscripts Hg and w, denoting mercury and water, respectively.

$$\rho_{\mathrm{Hg}} g h_{\mathrm{Hg}} = \rho_{\mathrm{w}} g h_{\mathrm{w}}$$

The height of mercury column required to produce 1 bar pressure in mm is found by using what are called *unit conversions* derived from the definitions. Thus from the definition: 1 bar = 0.986923 atm, a unit conversion factor may be written as 0.986923 atm bar^{-1}. The value of this result is unity. Multiplication by 760 mm atm^{-1} (another unit conversion) merely provides one more unit conversion. Thus 0.986923 atm bar^{-1} × 760 mm atm^{-1} = 750.06 mm bar^{-1}, or 1 bar = 750.06 mmHg. Substitution of this into the rearranged earlier equation gives

$$h_w = 750.06 \frac{\text{mm}}{\text{bar}} \times \frac{13.596 \text{ g cm}^{-3}}{0.99987 \text{ g cm}^{-3}} = 10\,199 \frac{\text{mm}}{\text{bar}} \quad \text{or} \quad 10.199 \frac{\text{m}}{\text{bar}}$$

Thus, 1 bar = 10.199 m water.

Boyle's Law

Ideal gases: Plot isotherms of Boyle's Law.

In the middle of the seventeenth century, Robert Boyle (1627–1691) and his assistant Robert Hooke (1635–1703) made many investigations of the relationship between pressure and volume of a gas. They did not actually discover the law that has come to be called **Boyle's law,**[3] but it was first announced by Boyle in 1662, and can be expressed as follows:

The pressure of a fixed amount of gas varies inversely with the volume if the temperature is maintained constant.

[3]The law was discovered by the amateur investigator Richard Towneley (1629–1668) and his family physician Henry Power (1623–1668). The law was communicated to Boyle who with Hooke confirmed the relationship in numerous experiments. The first publication of the law was in the second edition of Boyle's *Experiments physico-mechanical, touching the Spring of the Air,* which appeared in 1662. Boyle never claimed to have discovered the law himself, his work being of a more qualitative kind.

ROBERT BOYLE (1627–1691)

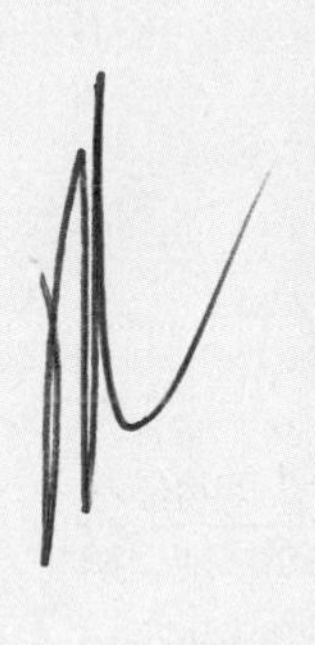

Robert Boyle was born in Lismore Castle, County Waterford, in the south of Ireland. His parents were of English rather than Irish descent but lived for a period in Ireland. In 1620 his father, Richard Boyle, a successful and wealthy businessman, was made the Earl of Cork. Later, after his elder brother had become the Earl of Cork, Robert Boyle was often humorously described as the "Father of Chemistry and Brother of the Earl of Cork." Boyle inherited a substantial income, and had no need to work for a living. He was privately educated for the most part. He never attended a university but he developed a considerable interest in philosophy and science.

In or about 1655 Boyle lodged in a house on High Street in Oxford where he carried out many scientific investigations on combustion, respiration, and the properties of gases (see Section 1.5), ably assisted by Robert Hooke (1638–1703), who designed the famous air pump that was so effective in the study of gases. In 1668 he moved to London where in the Pall Mall house of his sister Katherine, Lady Ranelagh, he established a laboratory. He worked on scientific problems, such as the properties of acids and alkalies and the purity of salts. After the discovery of phosphorus in about 1669, Boyle established many of its properties, particularly its reaction with air accompanied by light emission.

Boyle did not discover, but confirmed and publicized, "Boyle's law," as he made clear in his own writings. He was one of the first to apply the inductive methods that scientists today take for granted. Boyle almost single-handedly transformed chemistry into a respectable branch of science. Boyle can be reasonably regarded as the first physical chemist.

Boyle was a generous and charismatic man with a wide circle of friends who were devoted to him. He exerted a powerful influence on the scientific work of others, including Newton, who was fifteen years his junior. Boyle's investigations were carried out in collaboration with a team of technicians and research assistants. He was also deeply interested in other matters, such as world religions and languages. He was a member, or "adventurer" as it was called, of the Hudson's Bay Company, in order to learn about the effects of low temperatures. He was active in charitable organizations and he became Governor of the New England Company. Funds identified as the "Hon. Robert Boyle's Trust Fund" are still used to aid native peoples in parts of Canada and the West Indies.

Boyle died on December 31, 1691, and was buried in the chancel of the Church of St. Martin's-in-the-Fields in London.

References: Marie Boas Hall, *Dictionary of Scientific Biography, 1*, 1970, pp. 377–382.

J. B. Conant, "Robert Boyle's experiments in pneumatics," in *Harvard Case Histories in Experimental Science* (Ed. J. B. Conant), Cambridge: Harvard University Press, 1957.

K. J. Laidler, *The World of Physical Chemistry*, New York: Oxford University Press, 1993; this book contains biographies and further scientific information about many of the scientists mentioned in this book.

R. E. W. Maddison, *The Life of the Honourable Robert Boyle, F. R. S.*, London: Taylor and Francis, 1969.

Mathematically, Boyle's law can be stated as

$$P \propto 1/V, \qquad 1/P \propto V, \qquad P = \text{constant}/V, \qquad \text{or}$$
$$\mathbf{PV = constant} \qquad \textbf{(valid at constant } \boldsymbol{T} \textbf{ and } \boldsymbol{n}\textbf{)} \tag{1.16}$$

A plot of $1/P$ against V for some of Boyle's original data is shown in Figure 1.7a. The advantage of this plot over a P against V plot is that the linear relationship makes it easier to see deviations from the law. Boyle's law is surprisingly accurate

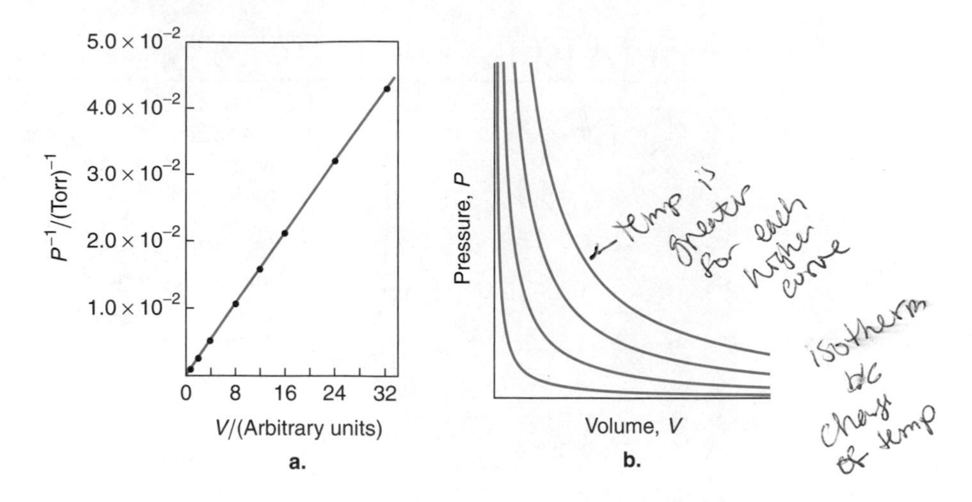

FIGURE 1.7
(a) A plot of $1/P$ against V for Boyle's original data. This linear plot, passing through the origin, shows that PV = constant.
(b) Plot of PV = constant at several different constant temperatures. The temperature is greater for each higher curve.

for many gases at moderate pressures. In Figure 1.7(b), we plot P against V for a gas at several different temperatures. Each curve of PV = constant is a hyperbola, and since it represents a change at constant temperature, the curve is called an *isotherm.*

1.6 Gay-Lussac's (Charles's) Law

Ideal gases: Plot isobars of Charles's Law.

The French physicist Guillaume Amontons (1663–1705) measured the influence of temperature on the pressure of a fixed volume of a number of different gases and predicted that as the air cooled, the pressure should become zero at some low temperature, which he estimated to be −240 °C. He thus anticipated the work of Jacques Alexandre Charles (1746–1823), who a century later independently derived the direct proportionality between the volume of a gas and the temperature. Since Charles never published his work, it was left to the French chemist Joseph Louis Gay-Lussac (1778–1850), proceeding independently, to make a more careful study using mercury to confine the gas and to report that all gases showed the same dependence of V on θ. He developed the idea of an absolute zero of temperature and calculated its value to be −273 °C. Thus, for a particular value of the temperature θ and a fixed volume of gas V_0 and 0 °C, we have the linear relation

$$V = V_0(1 + \alpha\theta) \tag{1.17}$$

where α is the *cubic expansion coefficient.* The modern value of α is 1/273.15. Plots of the volume against temperature for several different gases are shown in Figure 1.8. It can be seen that the curves of the experimentally determined region can be extrapolated to zero volume where θ is −273.15 °C. This fact immediately suggests that the addition of 273.15° to the Celsius temperature would serve to define a new temperature scale T that would not have negative numbers. The relationship between the two scales is best expressed as

$$\frac{T}{\text{K}} = \frac{\theta}{^\circ\text{C}} + 273.15 \tag{1.18}$$

That is, the value of the absolute temperature (i.e., the temperature divided by its unit) is obtained simply by adding 273.15 to the *value* of the Celsius temperature.

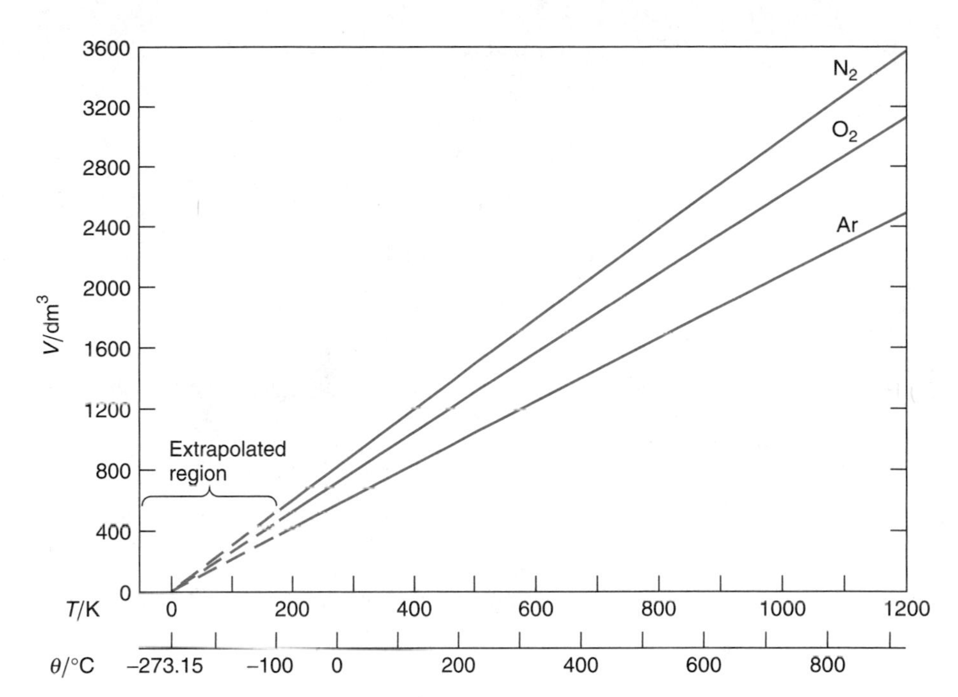

FIGURE 1.8
A plot of volume against temperature for argon, nitrogen, and oxygen. The individual curves show the effect of a change in molar mass for the three gases. In each case one kilogram of gas is used at 1.00 atmosphere.

On the new scale, 100 °C is therefore 373.15 K. Note that temperature intervals remain the same as on the Celsius scale.

Kelvin Temperature

This new scale is called the *absolute Kelvin temperature* scale or the **Kelvin temperature** scale, after William Thomson, Lord Kelvin of Largs (1824–1907), who as will be discussed in Section 3.1 suggested such a scale on the basis of a thermodynamic engine.

Gay-Lussac's law may thus be written in a convenient form in terms of the absolute temperature

$$V \propto T, \qquad V = \text{constant} \times T, \qquad \text{or}$$

$$\frac{V}{T} = \text{constant} \qquad \text{(valid at constant } P \text{ and } n\text{)} \tag{1.19}$$

The behavior of many gases near atmospheric pressure is approximated quite well by this law at moderate to high temperatures.

For all gases that obey the three laws just considered, there exists a surface in a P, V, T diagram that represents the only states (conditions of P, V, and T) of the gas that may exist. Figure 1.9 shows lines of constant P (isobar) and constant PV (isotherm) on the smooth surface.

1.7 The Ideal Gas Thermometer

The work of Gay-Lussac provided an important advance in the development of science, but temperature was still somewhat dependent on the working substance used in its determination. Since molecular interactions are responsible for the nonlinear

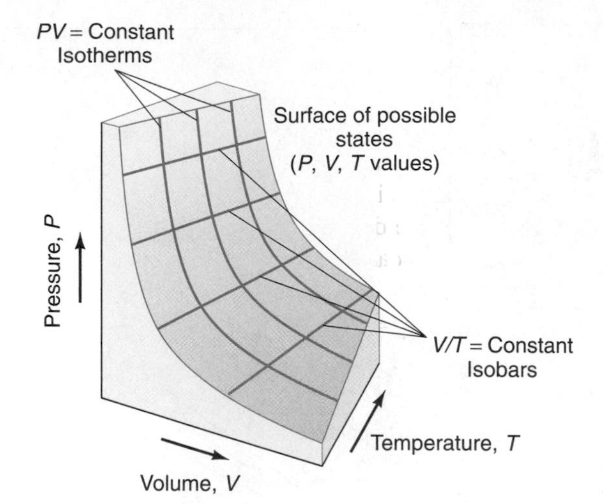

FIGURE 1.9
A three-dimensional plot of *P, V,* and *T* for a gas that obeys Boyle's and Gay-Lussac's laws. Isobars and isotherms are shown on the surface of the figure.

behavior found in materials used for their thermometric properties, we now look for a means to measure temperature that does not depend on the properties of any one substance.

Experimentally one finds that Eq. 1.19 holds true for real gases at moderate to high temperatures only as the pressure is reduced to zero. Thus, we could rewrite Gay-Lussac's law, Eq. 1.19, as

$$\lim_{P \to 0} V = CT \tag{1.20}$$

where C is a constant. This expression may serve as the basis of a new temperature scale. Thus, if the temperature and volume of a fixed amount of gas held at some low pressure are T_1 and V_1, respectively, before the addition of heat, the ratio of the temperature T_2 to T_1 after the addition of heat is given by the ratio of the initial volume and the final volume, V_2, of the gas. Thus,

$$\frac{T_2}{T_1} = \frac{\lim_{P \to 0} V_2}{\lim_{P \to 0} V_1} \tag{1.21}$$

However, the low-pressure limit of a gas volume is infinite, and this relationship is therefore impractical. Instead, we can cast the expression into the form

$$\frac{T_2}{T_1} = \frac{\lim_{P \to 0}(PV)_2}{\lim_{P \to 0}(PV)_1} \tag{1.22}$$

Ideal Gases

We thus have a gas thermometer that will work equally well for any gas. If a gas can be imagined to obey Eq. 1.19 or Eq. 1.22 exactly for all values of P, then we have defined an **ideal gas;** the thermometer using such a gas is known as the *ideal gas thermometer*. Such a concept is particularly useful for work at low temperatures.

Although considerable work went into defining a two-reference-point temperature scale as described earlier, low-temperature work has demonstrated the need for a temperature scale based only on one experimental point along with the absolute zero. In 1954, the decision was made to redefine the absolute scale and the Celsius scale. The absolute zero is zero kelvin, written as 0 K. Since the most careful measurements of the ice point (water-ice equilibrium at 1 atm pressure) varied over

Triple Point

several hundredths of a kelvin, the **triple point** *of water* (water-ice-water vapor equilibrium) *is defined as* 273.16 K *exactly*. The freezing point is then almost exactly 273.15 K and the boiling point is simply another temperature to be measured experimentally; it is almost exactly 373.15 K or 100 °C. *The value of the temperature in kelvins is by definition obtained by adding exactly* 273.15 *to the value of the temperature in degrees Celsius*. As far as temperature *intervals* are concerned, the degree Celsius is the same as the kelvin.

Using the defined value of the triple point for T_1, a working definition for the new Celsius scale becomes

$$T_2 = 273.16° \times \frac{\lim_{P\to 0}(PV)_{T_2}}{\lim_{P\to 0}(PV)_{\text{triple point}}} \qquad \text{(constant } P \text{ and } n) \tag{1.23}$$

where $\lim_{P\to 0}(PV)_T = 0$ when $T_2 = 0$.

1.8 The Equation of State for an Ideal Gas

The Gas Constant and the Mole Concept

In Sections 1.5 and 1.6, we found that pressure, volume, and temperature are related. Experimentally, these three properties of a gas cannot be arbitrarily chosen to describe the state of a fixed amount of gas. (The effects of gravitational, electric, and magnetic fields are neglected in this treatment.) For a fixed amount of gas these basic properties are related by an equation of state. However, one finds experimentally that the linear relationship of Boyle's law (Eq. 1.16) for *P-V* data is attained only at very low pressures. Hence, in the limit of zero pressure, all gases should obey Boyle's law to the same degree of accuracy. Thus, Boyle's law could be better written as

$$\lim_{P\to 0}(PV) = C' \tag{1.24}$$

Ideal gases: Plot Avogadro's Law.

In the same manner, Gay-Lussac's law can be written in the form seen in Eq. 1.20. On combining these two expressions, we obtain:

$$\lim_{P\to 0}(PV) = C''T \tag{1.25}$$

In order to determine the value of the constant C'', we can utilize an important hypothesis proposed in 1811 by the Italian physicist Amedeo Avogadro (1776–1856) that a given volume of any gas (at a fixed temperature and pressure) must contain the same number of independent particles. Furthermore, he specified that the particles of gas could be atoms or combinations of atoms, coining the word *molecule* for the latter case. One may state **Avogadro's hypothesis** as

Avogadro's Hypothesis

$$V \propto n \quad \text{or} \quad V = C'''n \qquad \text{(valid at constant } P \text{ and } T) \tag{1.26}$$

where n is the amount of substance, the SI unit for which is the *mole*.

Definition of the Mole

One **mole** is the amount of any substance containing the same number of elementary entities (atoms, molecules, ions, and so forth) as there are in exactly 0.012 kg of carbon 12 (see also Appendix A). The number of elementary entities is related to the amount of substance by the Avogadro constant; this is given the symbol L [after Joseph Loschmidt (1821–1895), who first measured its magnitude] and has the value $6.022\ 137 \times 10^{23}\ \text{mol}^{-1}$. The *numerical value* of the Avogadro constant, $6.022\ 137 \times 10^{23}$, is the *number* of elementary entities in a mole.

TABLE 1.1 Numerical Values of *R* in Various Units

8.314 51 J K^{-1} mol^{-1} (SI unit)
0.082 057 atm dm^3 K^{-1} mol^{-1}, commonly listed as 0.082 057 L atm K^{-1} mol^{-1}*
$8.314\,51 \times 10^7$ erg K^{-1} mol^{-1}
1.987 19 cal K^{-1} mol^{-1}
0.083 145 10 bar dm^3 K^{-1} mol^{-1}

*In SI the volume is expressed in cubic decimeters (dm^3). The more familiar unit, the liter, is now defined as 1 dm^3.

Since Eqs. 1.20, 1.24, and 1.26 show that the volume of a gas depends on T, $1/P$, and n, respectively, these three expressions may be combined as

$$\lim_{P\to 0}(PV) = nRT \tag{1.27}$$

If we keep in mind the limitations imposed by the $\lim_{P\to 0}$ requirement, the equation may be approximated by

$$PV = nRT \tag{1.28}$$

where R is the *universal gas constant*. The value of R will depend on the units of measurements of P and V. In terms of the variables involved, R must be of the form

$$R = \frac{PV}{nT} = \frac{(\text{force/area}) \cdot \text{volume}}{\text{mol} \cdot \text{K}} = \text{force} \cdot \text{length} \cdot \text{K}^{-1} \cdot \text{mol}^{-1}$$
$$= \text{energy} \cdot \text{K}^{-1} \cdot \text{mol}^{-1}$$

Various units of R are required for different purposes, and some of the most often used values of R are given in Table 1.1.

Equation of State of an Ideal Gas

Equation 1.28, the **equation of state of an ideal gas,** is one of the most important expressions in physical chemistry. What it states is that in any sample of gas behaving ideally, if one of the four variables (amount, pressure, volume, or temperature) is allowed to change, the values of the other three variables will always be such that a constant value of R is maintained.

A useful relation can be obtained by rearranging Eq. 1.28 into the form

$$P = \frac{n}{V}RT = \frac{m/M}{V}RT = \frac{\rho}{M}RT \tag{1.29}$$

or

$$M = \rho\frac{RT}{P}$$

where n, the amount of substance, is the mass m divided by the *molar mass*[4] M, and m/V is the density ρ of the gas.

EXAMPLE 1.2 Calculate the average molar mass of air at sea level and 0 °C if the density of air is 1.29 kg m^{-3}.

Solution At sea level the pressure may be taken equal to 1 atm or 101 325 Pa. Using Eq. 1.29,

$$M = \frac{\rho RT}{P} = \frac{1.29 \text{ kg m}^{-3} \times 8.3145 \text{ J K}^{-1} \text{ mol}^{-1} \times 273.15 \text{ K}}{101\,325 \text{ N m}^{-2} \text{ (or Pa)}}$$

$$= \frac{1.29 \text{ kg m}^{-3} \times 8.3145 \text{ kg m}^2 \text{ s}^{-2} \text{ K}^{-1} \text{ mol}^{-1} \times 273.15 \text{ K}}{101\,325 \text{ kg m s}^{-2} \text{ m}^{-2}}$$

$$= 0.0289 \text{ kg mol}^{-1} = 28.9 \text{ g mol}^{-1}$$

[4]The molar mass is mass divided by the amount of substance. The relative molecular mass of a substance M_r is usually called the *molecular weight* and is dimensionless.

1.9 The Kinetic-Molecular Theory of Ideal Gases

An experimental study of the behavior of a gas, such as that carried out by Boyle, cannot determine the nature of the gas or why the gas obeys particular laws. In order to understand gases, one could first propose some hypothesis about the nature of gases. Such hypotheses are often referred to as constituting a "model" for a gas. The properties of the gas that are deduced from this model are then compared to the experimental properties of the gas. The validity of the model is reflected in its ability to predict the behavior of gases.

We will simply state three postulates of the kinetic-molecular model and show how this model leads to the ideal gas laws. This model of an idealized gas will be seen to fit the behavior of many real gases.

1. The gas is assumed to be composed of individual particles (atoms or molecules) whose actual dimensions are small in comparison to the distances between them.
2. These particles are in constant motion and therefore have kinetic energy.
3. Neither attractive nor repulsive forces exist between the particles.

In order to see how this model predicts the observed behavior quantitatively, we must arrive at an equation relating the pressure and volume of a gas to the important characteristics of the gas, namely the number of particles present, their mass, and the velocity with which they move. We will first focus our attention on a single molecule of mass m confined in an otherwise empty container of volume V. The particle traverses the container with velocity $\boldsymbol{u}$ (a vector quantity indicating both speed and direction). Because the particle's X-component of velocity (i.e., the component $\boldsymbol{u}_x$ as shown in Figure 1.10) is normal to the YZ wall, the molecule will traverse the container of length x in the X-direction, collide with the wall YZ, and then rebound. In the impact, the molecule exerts a force F_w on the wall. This force is exactly counterbalanced by the force F exerted on the molecule by the

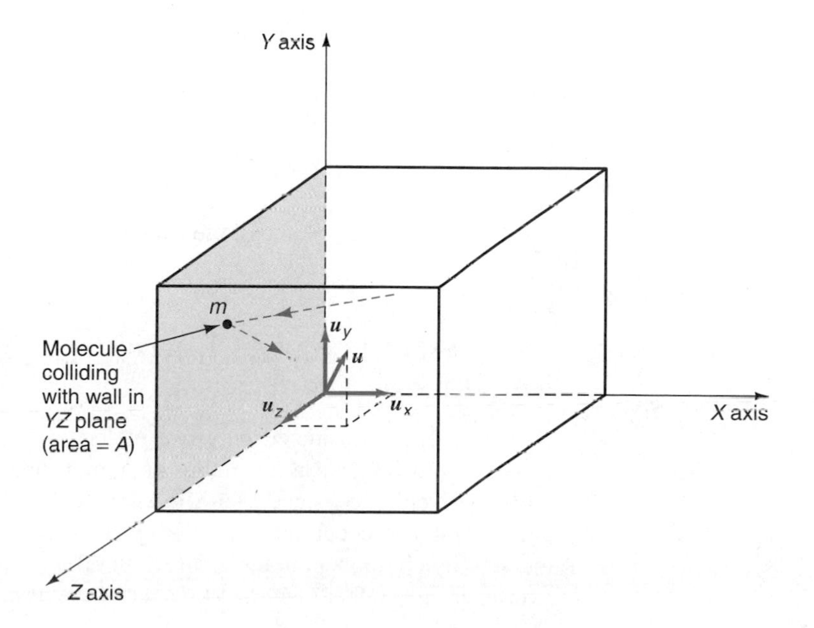

FIGURE 1.10
Container showing coordinates and velocity components for gas particle of mass *m*.

wall. The force F is equal to the change of momentum $\boldsymbol{p}$ of the molecule in the given direction per unit time, in agreement with Newton's second law of motion,

$$F = \frac{d\boldsymbol{p}}{dt} = m\boldsymbol{a} = \frac{d(m\boldsymbol{u})}{dt} = m\frac{d\boldsymbol{u}}{dt} \tag{1.30}$$

The molecule's momentum in the X-direction when it strikes the wall is mu_x. Since we assume that the collision is perfectly elastic, the molecule bounces off the wall with a velocity $-u_x$ in the opposite direction.

The change in velocity of the molecule on each collision is

$$\begin{aligned} \Delta u_x &= [-u_x \text{ (after collision)}] - [u_x \text{ (before collision)}] \\ &= -2u_x \end{aligned} \tag{1.31}$$

We drop the vector notation since we are only interested in the magnitude of $\boldsymbol{u}$. The corresponding change of momentum is $-2mu_x$. Since each collision with this wall occurs only after the molecule travels a distance of $2x$ (i.e., one round trip), the number of collisions a molecule makes in unit time may be calculated by dividing the distance u_x the molecule travels in unit time by $2x$. The result is

$$\text{number of collisions in unit time} = \frac{u_x}{2x} \tag{1.32}$$

The change of momentum per unit time is thus

$$F = m\frac{d\boldsymbol{u}}{dt} = (-2u_x m)\left(\frac{u_x}{2x}\right) = -\frac{mu_x^2}{x} \tag{1.33}$$

The force F_w exerted on the wall by the particle is exactly equal in magnitude to this but with opposite sign:

$$F_w = -F = \frac{mu_x^2}{x} \tag{1.34}$$

Since the pressure is force per unit area and the area A is yz, we may write the pressure in the X-direction, P_x, as

$$\begin{aligned} P_x &= \frac{F_w}{A} = \frac{F_w}{yz} \\ &= \frac{mu_x^2}{xyz} = \frac{mu_x^2}{V} \end{aligned} \tag{1.35}$$

since $xyz = V$ is the volume of the container.

The Pressure of a Gas Derived from Kinetic Theory

So far we have considered only one molecule that has been assumed to travel at constant velocity. For an assembly of N molecules there will be a distribution of molecular velocities, since even if the molecules all began with the same velocity, collisions would occur altering the original velocity. If we define u_i^2 as the square of the velocity component in the X-direction of molecule i and take the average of this over molecules rather than the sum over u_i^2, we have

$$\overline{u_x^2} = \frac{u_1^2 + u_2^2 + u_3^2 + \cdots + u_N^2}{N} = \frac{\sum_{i=1}^{N} u_i^2}{N} \tag{1.36}$$

where $\overline{u_x^2}$ is the mean of the squares of the normal component of velocity in the X-direction. The pressure expressed in Eq. 1.35 should therefore be written as

$$P_x = \frac{Nm\overline{u_x^2}}{V} \tag{1.37}$$

Pressure of a One-Dimensional Gas

This is the equation for the pressure of a one-dimensional gas. For the components of velocity in the Y-direction and in the Z-direction, we would obtain expressions similar to Eq. 1.37 but now involving $\overline{u_y^2}$ and $\overline{u_z^2}$, respectively. It is more convenient to write these expressions in terms of the magnitude of the velocity $\boldsymbol{u}$ rather than in terms of the squares of the velocity components. The word *speed* is used for the magnitude of the velocity; speed is defined as the positive square root of $\boldsymbol{u}^2$ and is given by

$$|u| = \sqrt{\boldsymbol{u}^2} = \sqrt{u_x^2 + u_y^2 + u_z^2} \tag{1.38}$$

If we average over all molecules, we obtain

$$\overline{u^2} = \overline{u_x^2} + \overline{u_y^2} + \overline{u_z^2} \tag{1.39}$$

Since there is no reason for one direction to be favored over the others, the mean of the u_x^2 values will be the same as the mean of the u_y^2 values and the mean of the u_z^2 values. Hence, the sum of the means is equal to $\overline{u^2}$ and each mean is equal to $\frac{1}{3}\overline{u^2}$; that is,

$$\overline{u_x^2} = \overline{u_y^2} = \overline{u_z^2} = \tfrac{1}{3}\overline{u^2} \tag{1.40}$$

Substituting Eq. 1.40 into Eq. 1.37 gives the final expression for the pressure on any wall:

Pressure of a Gas in Terms of Mean-Square Speed

$$P = \frac{Nm\overline{u^2}}{3V} \quad \text{or} \quad PV = \tfrac{1}{3}Nm\overline{u^2} \tag{1.41}$$

This is the fundamental equation as derived from the simple kinetic theory of gases. We see that Eq. 1.41 is in the form of Boyle's law and is consistent with Charles's law if $m\overline{u^2}$ is directly proportional to the absolute temperature. In order to make this relation exact, we use Eq. 1.28 and substitute nRT for PV in Eq. 1.41:

$$nRT = \tfrac{1}{3}Nm\overline{u^2} \tag{1.42}$$

or

$$\tfrac{1}{3}Lm\overline{u^2} = RT \quad \text{or} \quad \overline{u^2} = \frac{3RT}{M} \tag{1.43}$$

since N/n is equal to the Avogadro constant L, and $mL = M$, the molar mass.

Kinetic Energy and Temperature

We have seen how the kinetic molecular theory of gases may be used to explain the experimental form of two gas laws. It can also shed light on the nature of kinetic energy. In order to determine the exact relation between $\overline{u^2}$ and T, the mechanical variable u of Eq. 1.41 must be related to the temperature, which is not a mechanical variable. For our purpose of determining the relation between kinetic energy and temperature, Eq. 1.41 may be converted into another useful form by recognizing that the average kinetic energy $\bar{\epsilon}_k$ per molecule is

$$\bar{\epsilon}_k = \frac{1}{2}m\overline{u^2} \tag{1.44}$$

Substitution of this expression into Eq. 1.41 gives

$$PV = \frac{1}{3}N \cdot 2\bar{\epsilon}_k = \frac{2}{3}N\bar{\epsilon}_k \tag{1.45}$$

Thus, at constant pressure, the volume of a gas is proportional to the number of gas molecules and the average kinetic energy of the molecules. Since

$$N = nL \tag{1.46}$$

substitution into Eq. 1.45 yields

$$PV = \frac{2}{3}nL\bar{\epsilon}_k \tag{1.47}$$

Since $L\bar{\epsilon}_k$ is the total kinetic energy E_k per mole of gas,

$$PV = \frac{2}{3}nE_k \tag{1.48}$$

The connection between E_k and the temperature is provided by the empirical ideal gas law, $PV = nRT$. By equating the right-hand sides of the last two equations we obtain

$$\frac{2}{3}nE_k = nRT \quad \text{or} \quad E_k = \frac{3}{2}RT \tag{1.49}$$

The average kinetic energy per molecule $\bar{\epsilon}_k$ is obtained by dividing both sides by the Avogadro constant L:

$$\bar{\epsilon}_k = \frac{3}{2}k_{\mathrm{B}}T \tag{1.50}$$

Boltzmann Constant

where $k_{\mathrm{B}} = R/L$. Named after the Austrian physicist Ludwig Edward Boltzmann (1844–1906), the **Boltzmann constant**[5] k_{B} is the gas constant per molecule. Thus the average kinetic energy of the molecules is proportional to the absolute temperature. This provides an alternate basis for a definition of temperature. Since $\bar{\epsilon}_k$ in Eq. 1.50 is independent of the kind of substance, the average molecular kinetic energy of all substances is the same at a fixed temperature.

An interesting aspect of this fact is seen when we consider a number of different gases all at the same temperature and pressure. Then Eq. 1.41 may be written as

$$\frac{N_1 m_1 \overline{u_1^2}}{3V_1} = \frac{N_2 m_2 \overline{u_2^2}}{3V_2} = \cdots = \frac{N_i m_i \overline{u_i^2}}{3V_i} \tag{1.51}$$

[5] $k_{\mathrm{B}} = 1.380\,622 \times 10^{-23}\ \mathrm{J\ K^{-1}}$.

or

$$\frac{N_1}{3V_1} = \frac{N_2}{3V_2} = \cdots = \frac{N_i}{3V_i} \tag{1.52}$$

Thus $N_1 = N_2 = \cdots = N_i$ when the volumes are equal. In other words, equal volumes of gases at the same pressure and temperature contain equal numbers of molecules. This is just a statement of Avogadro's hypothesis already seen in Eq. 1.26.

Dalton's Law of Partial Pressures

The studies of the English chemist John Dalton (1766–1844) showed in 1801 that the *total pressure* observed for a mixture of gases is equal to the *sum of the pressures that each individual component gas would exert* had it alone occupied the container at the same temperature. This is known as **Dalton's law of partial pressures**. Of course, in order for the law to be obeyed, no chemical reactions between component gases may occur.

Partial Pressure

The term **partial pressure,** P_i, is used to express the *pressure exerted by one component of the gas mixture* and is defined as

$$P_i = x_i P_t$$

where x_i is the mole fraction of the gas(i). This avoids the necessity for the gas(i) to behave ideally. Thus

$$\begin{aligned} P_t &= P_1 + P_2 + P_3 + \cdots + P_i \\ &= x_1 P_t + x_2 P_t + x_3 P_t + \cdots + x_i P_t \end{aligned} \tag{1.53}$$

where P_t is the total pressure. Then, using a form of Eq. 1.28, we may write

$$\begin{aligned} P_t &= \frac{n_1 RT}{V} + \frac{n_2 RT}{V} + \cdots + \frac{n_i RT}{V} \\ &= (n_1 + n_2 + \cdots + n_i)\frac{RT}{V} \end{aligned} \tag{1.54}$$

where the P_i's are the partial pressures and the n_i's are the amounts of the individual gases. Dalton's law can then be predicted by the simple kinetic molecular theory (Eq. 1.41) by writing expressions of the form $P_i = N_i m_i \overline{u_i^2}/3V$ for each gas. Thus

$$P_t = \sum_1^i P_i = \frac{N_1 m_1 \overline{u_1^2}}{3V} + \frac{N_2 m_2 \overline{u_2^2}}{3V} + \cdots + \frac{N_i m_i \overline{u_i^2}}{3V} \tag{1.55}$$

Dalton's law immediately follows from the kinetic theory of gases since the average kinetic energy from Eq. 1.49 is $\frac{3}{2}RT$ and is the same for all gases at a fixed temperature.

Application of Dalton's law is particularly useful when a gas is generated and subsequently collected over water. The total gas pressure consists of the pressure of the water vapor present in addition to the pressure of the gas that is generated. The vapor pressure of water is in the order of 20 Torr near room temperature and can be a significant correction to the total pressure of the gas.

The pressure of water vapor, P_{water} in Torr, is approximately equal to θ in °C near room temperature.

EXAMPLE 1.3 The composition in volume percent of standard "1976" dry air at sea level is approximately 78.084% N_2, 20.948% O_2, 0.934% Ar, and 0.031% CO_2. All other gases, including (in decreasing order) Ne, He, Kr, Xe, CH_4, and H_2, constitute only 2.7×10^{-3}% of the atmosphere. What is the partial pressure of each of the first four gases listed at a total pressure of 1 atmosphere? What are the partial pressures (in units of kPa) if the total pressure is 1 bar?

Solution To use the expressions relating to Dalton's law of partial pressure, the volume percent, as generally listed in tables, must be converted to mass percent. This is accomplished by assuming 1 unit volume and multiplying the fractional volume of each component (the volume percentage divided by 100%) by the molecular weight or atomic weight. For N_2 this is

$$0.78084 \times 28.0134 \text{ g} = 21.8740 \text{ g}$$

The corresponding values for the other gases are as follows: O_2, 6.7030 g; Ar, 0.3731 g; CO_2, 0.0138 g; all other gases using the molecular weight of the most abundant of these gases, Ne, 5.43×10^{-4}. The sum of these values is 28.9644 g. The weight percents are found by dividing each mass in the unit volume by the total mass and multiplying by 100%. For N_2: (21.8740 g/28.9644 g) $\times$ 100% = 75.520%. The results in weight percent for the others are as follows: O_2, 23.142%; Ar, 1.288%; CO_2, 4.76×10^{-2}%; all others (approximate), 1.87×10^{-3}%.

The next step is to calculate the moles of each gas in a sample. For convenience, assume a sample mass of 100.00 g. Since moles of any component A equal the mass of A divided by its molar mass, we have for N_2

$$n_{N_2} = \frac{100.00 \text{ g} \times 0.75520}{28.0134 \text{ g mol}^{-1}} = 2.6958 \text{ mol}$$

For the other components, we have: O_2, 0.72232 mol; Ar, 0.03224 mol; CO_2, 0.00108 mol; for all others (approximate), 0.00926 mol. The sum of these is $\sum n_i = 3.4607$ mol.

The mole fractions are found from n_i/n_{total}. For N_2 the value is 2.6958/3.4607 = 0.77898. The other gases have the following mole fractions: O_2; 0.20872; Ar, 0.009316; CO_2, 0.000312.

Now the partial pressures can be determined from the definition. For N_2,

$$x_i P_{total} = P_i = 0.77898 \times 1 \text{ atm} = 0.77898 \text{ atm}$$

The other gas pressures are: O_2, 0.280872 atm; Ar, 0.009316 atm; CO_2, 0.000312 atm.

Since 1 bar = 100 kPa, at 1 bar total pressure the partial pressure of N_2 is found from

$$x_i P_{total} = 0.77898 \times 100 \text{ kPa} = 77.898 \text{ kPa}$$

The other gas pressures are as follows: O_2, 28.0872 kPa; Ar, 0.9316 kPa; CO_2, 0.0312 kPa.

Graham's Law of Effusion

Another confirmation of the kinetic theory of gases was provided by the work of the Scottish physical chemist Thomas Graham (1805–1869). He measured the movement of gases through plaster of Paris plugs, fine tubes, and small orifices in plates where the passages for the gas are small as compared with the average distance that the gas molecules travel between collisions (see the following section on Molecular Collisions). Such movement is known as **effusion.** This is sometimes mistakenly referred to as *diffusion*, but in that process there are no small passageways to inhibit the moving particles. See Section 19.2. In 1831 Graham showed that the *rate of effusion* of a gas was *inversely proportional to the square root of its density* ρ. Later, in 1848, he showed that the *rate of effusion was inversely proportional to the square root of the molar mass M.* This is known as **Graham's law of effusion.** Thus, in the case of the gases oxygen and hydrogen at equal pressures, oxygen molecules are 32/2 = 16 times more dense than those of hydrogen. Therefore, hydrogen diffuses 4 times as fast as oxygen:

Graham's Law of Effusion

Effusion: Compare and contrast diffusion and effusion.

$$\frac{\text{rate}(H_2)}{\text{rate}(O_2)} = \sqrt{\frac{\rho(O_2)}{\rho(H_2)}} = \sqrt{\frac{M(O_2)}{M(H_2)}} = \sqrt{\frac{32}{2}} = \frac{4}{1}$$

Graham's law also can be explained on the basis of the simple kinetic molecular theory. Our starting point is Eq. 1.43, which shows that the **root-mean-square velocity** $\sqrt{\overline{u^2}}$ is proportional to the square root of the absolute temperature. However, the speed with which gases effuse is proportional to the mean velocity $\bar{u}\ \left(= \sqrt{(8RT/\pi M)}\right)$ and not $\sqrt{\overline{u^2}}$. The mean velocity is the average velocity at which the bulk of the molecules move. The mean velocity differs from the root-mean-square velocity by a constant factor of 0.92 (see Section 1.11); consequently, proportionality relationships deduced for $\sqrt{\overline{u^2}}$ are equally valid for $\bar{u}$. Again, we can see from Eq. 1.43 that, at constant temperature, the quantity $m\overline{u^2}$ is constant and the mean-square velocity is therefore inversely proportional to the molecular mass,

$$\overline{u^2} = \frac{\text{constant}_1}{m} \tag{1.56}$$

The mean-square velocity is also inversely proportional to the molar mass M, which is simply mL:

$$\overline{u^2} = \frac{\text{constant}_2}{M} \tag{1.57}$$

Consequently, the mean velocity $\bar{u}$ is inversely proportional to the square root of the molar mass:

$$\bar{u} = \frac{\text{constant}}{M^{1/2}} \tag{1.58}$$

This, therefore, provides a theoretical explanation for Graham's law of effusion. Table 1.2 gives some average velocities of gases at 293.15 K.

These results from simple kinetic molecular theory, being confirmed through empirical observation, give additional support to the original postulates of the theory.

TABLE 1.2 Average Velocities of Gas Molecules at 293.15 K

Gas	Mean Velocity, $\bar{u}$/m s^{-1}
Ammonia, NH_3	582.7
Argon, Ar	380.8
Carbon dioxide, CO_2	454.5
Chlorine, Cl_2	285.6
Helium, He	1204.0
Hydrogen, H_2	1692.0
Oxygen, O_2	425.1
Water, H_2O	566.5

As a comparison, the speed of sound in dry air of density 1.29 g L^{-1} is 346.2 m s^{-1} at 298.15 K.

Molecular Collisions

The ability to relate pressure and temperature to molecular quantities in Eqs. 1.41 and 1.50 is a major accomplishment of the kinetic theory of gases. We now apply the kinetic theory to the investigation of the collisions of molecules in order to have

The speed of sound waves in a gas originates in longitudinal contractions and rarefactions.

a clearer understanding of the interactions in a gas. We will be interested in three aspects of molecular interaction: the number of collisions experienced by a molecule per unit time, the total number of collisions in a unit volume per unit time, and how far the molecules travel between collisions.

In this development we consider that the molecules behave as rigid spheres in elastic collisions and that there are two kinds of molecules, A and B, with diameters d_A and d_B. Suppose that a molecule of A travels with an average speed of $\bar{u}_A$ in a container that contains both A and B molecules. We will first assume that the molecules of B are stationary and later remove this restriction. As shown in Figure 1.11, a collision will occur each time the distance between the center of a molecule A and that of a molecule B becomes equal to $d_{AB} = (d_A + d_B)/2$, where d_{AB} is called the **collision diameter.** A convenient way to visualize this is to construct around the center of A an imaginary sphere of radius d_{AB}, which is the sum of the two radii. In unit time this imaginary sphere, represented by the dashed circle in Figure 1.11, will sweep out a volume of $\pi d_{AB}^2 \bar{u}_A$. If the center of a B molecule is in this volume, there will be a collision. If N_B is the total number of B molecules in the system, the number per unit volume is N_B/V, and the number of centers of B molecules in the volume swept out is $\pi d_{AB}^2 \bar{u}_A N_B/V$. This number is the number Z_A of collisions experienced by the one molecule of A in unit time:

Collision Diameter

$$Z_A = \frac{\pi d^2{}_{AB} \bar{u}_A N_B}{V} \qquad \text{(SI unit: s}^{-1}\text{)} \tag{1.59}$$

Collision Frequency

This number is known as the **collision frequency**. If there is a total of N_A/V molecules of A per unit volume in addition to the B molecules, the total number Z_{AB} of A-B collisions per unit volume per unit time is Z_A multiplied by N_A/V:

$$Z_{AB} = \frac{\pi d_{AB}^2 \bar{u}_A N_A N_B}{V^2} \qquad \text{(SI unit: m}^{-3}\text{ s}^{-1}\text{)} \tag{1.60}$$

Collision Density

This quantity, the number of collisions divided by volume and by time, is known as the **collision density,** often called the **collision number.**

If only A molecules are present, we are concerned with the collision density Z_{AA} for the collisions of A molecules with one another. The expression for this resembles that for Z_{AB} (Eq. 1.60), but a factor $\frac{1}{2}$ has to be introduced:

$$Z_{AA} = \frac{\frac{1}{2}\pi d_A^2 \bar{u}_A N_A^2}{V^2} \qquad \text{(SI unit: m}^{-3}\text{ s}^{-1}\text{)} \tag{1.61}$$

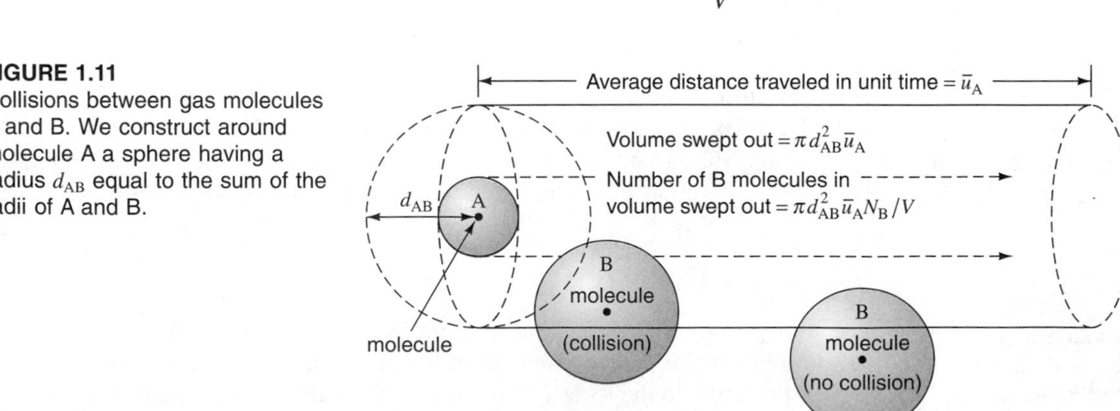

FIGURE 1.11
Collisions between gas molecules A and B. We construct around molecule A a sphere having a radius d_{AB} equal to the sum of the radii of A and B.

The reason for the factor $\frac{1}{2}$ is that otherwise we would count each collision twice. For example, we would count a collision of A_1 with A_2 and A_2 with A_1 as two separate collisions instead of one.

An error in this treatment of collisions is that we have only considered the average speed $\bar{u}_A$ of the A molecules. More correctly, we should consider the relative speeds $\bar{u}_{AB}$ or $\bar{u}_{AA}$ of the molecules.

When we have a mixture containing two kinds of molecules A and B of different masses, the $\bar{u}_A$ and $\bar{u}_B$ values are different. The average relative speed $\bar{u}_{AB}$ is equal to $(\bar{u}_A^2 + \bar{u}_B^2)^{1/2}$, and we modify Eqs. 1.59 and 1.60 by replacing $\bar{u}_A$ by $\bar{u}_{AB}$:

$$Z_A = \frac{\pi d_{AB}^2(\bar{u}_A^2 + \bar{u}_B^2)^{1/2} N_B}{V} \quad \text{(SI unit: s}^{-1}\text{)} \tag{1.62}$$

$$Z_{AB} = \frac{\pi d_{AB}^2(\bar{u}_A^2 + \bar{u}_B^2)^{1/2} N_A N_B}{V^2} \quad \text{(SI unit: m}^{-3}\text{ s}^{-1}\text{)} \tag{1.63}$$

If we have only A molecules, the average relative speed of two A molecules is

$$\bar{u}_{AA} = \sqrt{2}\bar{u}_A \tag{1.64}$$

We must therefore replace $\bar{u}_A$ in Eq. 1.61 by $\bar{u}_{AA}$ or $\sqrt{2}\bar{u}_A$, and we obtain for the collision density the more correct expression

$$mZ_{AA} = \frac{\sqrt{2}\pi d_A^2 \bar{u}_A N_A^2}{2V^2} \quad \text{(SI unit: m}^{-3}\text{ s}^{-1}\text{)} \tag{1.65}$$

The corresponding expression for Z_A, the number of collisions of 1 molecule of A in 1 s, when only A molecules are present, is

$$Z_A = \frac{\sqrt{2}\pi d_A^2 \bar{u}_A N_A}{V} \quad \text{(SI unit: s}^{-1}\text{)} \tag{1.66}$$

EXAMPLE 1.4 Nitrogen and oxygen are held in a 1.00 m^3 container maintained at 300 K at partial pressures of P_{N_2} = 80 kPa and P_{O_2} = 21 kPa. If the collision diameters are

$$d_{N_2} = 3.74 \times 10^{-10}\text{ m} \quad \text{and} \quad d_{O_2} = 3.57 \times 10^{-10}\text{ m}$$

calculate Z_A, the average number of collisions experienced in unit time by one molecule of nitrogen and by one molecule of oxygen. Also calculate Z_{AB}, the average number of collisions per unit volume per unit time. Do this last calculation both at 300 K and 3000 K on the assumption that the values for d and N do not change. At 300 K, $(u_{N_2}^2 + u_{O_2}^2)^{1/2}$ is 625 m s^{-1}; at 3000 K, it is 2062 m s^{-1}.

Solution The values of N_{N_2} and N_{O_2} are calculated from the ideal gas law:

$$PV = nRT = \frac{NRT}{L}$$

For N_2

$$N_{N_2} = \frac{LPV}{RT}$$

$$N_{N_2} = \frac{6.022 \times 10^{23}(\text{mol}^{-1}) \times 80\ 000(\text{Pa}) \times 1(\text{m}^3)}{8.314(\text{J K}^{-1}\ \text{mol}^{-1}) \times 300(\text{K})}$$
$$= 1.93 \times 10^{25} \text{ at } 300 \text{ K}$$

For O_2,

$$N_{O_2} = 5.07 \times 10^{24} \text{ at } 300 \text{ K}$$

The total number of collisions with unlike molecules is given by Eq. 1.62:

$$Z_{N_2} = \pi\left[\left(\frac{3.74 + 3.57}{2}\right) \times 10^{-10}(\text{m})\right]^2$$
$$\times 625(\text{m s}^{-1}) \times 5.07 \times 10^{24} \times 1(\text{m}^{-3})$$
$$= 1.33 \times 10^9 \text{ s}^{-1}$$

$$Z_{O_2} = \pi[3.66 \times 10^{-10}(\text{m})]^2 \times 625(\text{m s}^{-1}) \times 1.93 \times 10^{25} \times 1(\text{m}^{-3})$$
$$= 5.08 \times 10^9 \text{ s}^{-1}$$

Using Eq. 1.63, we have for the total number of collisions per cubic metre per second

$$Z_{N_2,O_2} = \pi(3.66 \times 10^{-10})^2(625)(1.93 \times 10^{25})(5.07 \times 10^{24})$$
$$= 2.57 \times 10^{34} \text{ m}^{-3} \text{ s}^{-1}$$

At 3000 K, $Z_{N_2,O_2} = 8.49 \times 10^{34}$ m^{-3} s^{-1}. From this example it can be seen that the effect of T on Z is not large, since, from Eq. 1.43 and the later discussion, T enters as $\sqrt{T}$. The effect of d is much more pronounced since it enters as d^2.

Mean Free Path

A concept of particular significance in the treatment of certain properties is the **mean free path** λ. This is the average distance that a molecule travels between two successive collisions. We have seen that for a single gas the number of collisions that a molecule makes in unit time, Z_A, is $\sqrt{2}\pi d_A^2 \bar{u}_A N_A/V$ (Eq. 1.66). In unit time the molecule has traveled, on the average, a distance of $\bar{u}_A$. The mean free path is therefore

$$\lambda = \frac{\text{distance traveled in unit time}}{\text{number of collisions in unit time}} = \frac{\bar{u}_A}{\sqrt{2}\pi d_A^2 \bar{u}_A N_A/V} \quad (1.67)$$

$$= \frac{V}{\sqrt{2}\pi d_A^2 N_A} \quad (1.68)$$

The magnitude of the d_A's is obviously very important to the kinetic theory of gases, since it is the only molecular property that we need to know in order to calculate collision numbers and the mean free path.

EXAMPLE 1.5 Molecular oxygen has a collision diameter of 3.57×10^{-10} m. Calculate λ for oxygen at 300 K and 101.325 kPa.

Solution Since $PV = nRT = (N/L)RT$, λ may be written as

$$\lambda = \frac{RT}{\sqrt{2}\pi d^2 LP}$$

$$= \frac{8.3145(\text{J K}^{-1}\ \text{mol}^{-1}) \times 300(\text{K})}{\sqrt{2}\pi[3.57 \times 10^{-10}(\text{m})]^2 \times 6.022 \times 10^{23}(\text{mol}^{-1}) \times 101\ 325(\text{Pa})}$$

$$= 7.22 \times 10^{-8}\ \text{m}$$

since J/Pa = m^3 (see Appendix A).

1.10 The Barometric Distribution Law

We have mentioned that a change in the properties of a system can be brought about by an applied potential field. In this section we will consider the effect of a gravitational field. In a laboratory experiment the effect of the gravitational field can usually be ignored, except in cases involving surface and interfacial tensions. However, for a large-scale system such as our atmosphere or an ocean, gravity can cause appreciable variation in properties. One example is the large increase in hydrostatic pressure at greater ocean depths. (See Example 1.1, p. 10.) A study of this simple system leads to the development of the Boltzmann distribution law.

The effect of gravity on the pressure can be determined by considering a column of fluid (either liquid or gas) at constant temperature as shown in Figure 1.12. The column has cross-sectional area A and is subjected to a gravitational field that gives the individual particles an acceleration g in the downward direction. Because of the field, the particles will experience different forces, and as a result the pressure will vary with height.

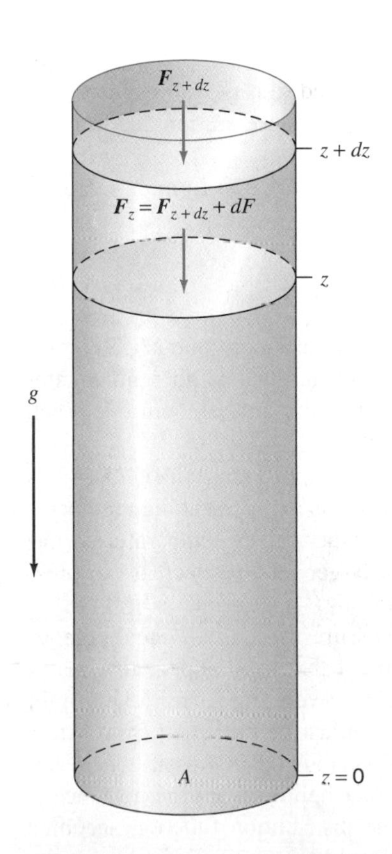

FIGURE 1.12
Distribution of a gas in a gravity field.

The force at level z due to the weight of fluid above z is F_z. At level $z + dz$ the force is F_{z+dz}, as shown in Figure 1.12. The force due to the weight within the volume element $A\ dz$ is dF. Thus, we may write

$$F_z = dF + F_{z+dz} \tag{1.69}$$

Equation 1.69 may be written in terms of the pressure P, since $F_z = PA$ and $F_{z+dz} = (P + dP)A$. After eliminating terms we obtain

$$A\ dP = -dF \tag{1.70}$$

The force dF is the mass in the volume element $A\ dz$ multiplied by the standard gravitational acceleration g (9.806 65 m s^{-2}). If the density (i.e., the mass per unit volume) is ρ, the mass of the element is $\rho A\ dz$ and its weight is $\rho gA\ dz$. Thus, on substitution for dF,

$$A\ dP = -\rho gA\ dz \tag{1.71}$$

or

$$dP = -\rho g\ dz \tag{1.72}$$

The change in pressure is thus proportional to the length of the column, and since dz is positive, the pressure decreases with an increase in height.

In general, the density depends on the pressure. For liquids, however, the density is practically independent of pressure, that is, the compressibility, κ $(= -1/V\ dV/dP)$, of the liquid is very small compared to that of a gas. For liquids, Eq. 1.72 can be integrated at once:

$$\int_{P_0}^{P} dP = P - P_0 = -\rho g \int_0^z dz \tag{1.73}$$

where P_0 is the reference pressure at the base of the column and P is the pressure at height z. This quantity $P - P_0$ is the familiar *hydrostatic pressure* in liquids.

In a vessel of usual laboratory size, the effect of gravity on the pressure of a gas is negligibly small. On a larger scale, however, such as in our atmosphere, there is a marked variation in pressure, and we must now consider the effect of pressure on the density of the gas. For an ideal gas, from Eq. 1.29, ρ is equal to PM/RT, and substitution into Eq. 1.72 gives

$$\frac{dP}{P} = -\left(\frac{Mg}{RT}\right) dz \tag{1.74}$$

Integration of this expression, with the boundary condition that $P = P_0$ when $z = 0$, gives

$$\ln \frac{P}{P_0} = -\frac{Mgz}{RT} \tag{1.75}$$

or

$$P = P_0 e^{-Mgz/RT} \tag{1.76}$$

Barometric Distribution Law

This expression describes the distribution of gas molecules in the atmosphere as a function of their molar mass, height, temperature, and the acceleration due to gravity. It is known as the **barometric distribution law.**

The distribution function in Eq. 1.74, which is in differential form, is more informative when the sign is transposed to the dP term:

$$-\frac{dP}{P} = \frac{Mg\,dz}{RT} \tag{1.77}$$

Here $-dP/P$ represents a relative decrease in pressure; it is a constant, Mg/RT, multiplied by the differential increase in height. This means that it does not matter where the origin is chosen; the function will decrease the same amount over each increment of height.

The fact that the relative decrease in pressure is proportional to Mg/RT shows that, for a given gas, a smaller relative pressure change is expected at high temperatures than at low temperatures. In a similar manner, at a given temperature a gas having a higher molar mass is expected to have a larger relative decrease in pressure than a gas with a lower molar mass.

Equation 1.76 applies equally to the partial pressures of the individual components in a gas. It follows from the previous treatment that in the upper reaches of the earth's atmosphere the partial pressure will be relatively higher for a very light gas such as helium. This fact explains why helium must be extracted from a few helium-producing natural-gas wells in the United States and in Russia where helium occurs underground. It is impractical to extract helium from the air since it tends to concentrate in the upper atmosphere. The distribution function accounts satisfactorily for the gross details of the atmosphere, although winds and temperature variations lead to nonequilibrium conditions.

Since Mgz is the gravitational potential energy, E_p, Eq. 1.76 can be written as

$$P = P_0 e^{-E_p/RT} \tag{1.78}$$

Since the density ρ is directly proportional to the pressure, we also may write

$$\rho = \rho_0 e^{-E_p/RT} \tag{1.79}$$

where ρ_0 represents the density at the reference state height of $z = 0$.

Boltzmann Distribution Law

These equations, in which the property varies exponentially with $-E_p/RT$, are special cases of the **Boltzmann distribution law**. We deal with this law in more detail in Section 15.2, where we will encounter a number of its important applications in physical chemistry. Before we proceed, note that we have considered only the average velocity of gas particles without investigating the range of their velocities. To treat this problem we will now deal with another special case of the Boltzmann distribution law, the Maxwell distribution of molecular speeds.

1.11 The Maxwell Distribution of Molecular Speeds and Translational Energies

Maxwell's famous equation for the distribution of molecular speeds in gases, developed in the early 1860s, inspired Boltzmann to produce a more general equation that dealt with energies of any kind. We will now obtain Maxwell's equation, and we will derive Boltzmann's equation later in Chapter 15, Section 15.2.

The Distribution of Speeds

In his derivation of the distribution of molecular speeds, Maxwell paid no attention to the mechanics of collisions between molecules. Instead he based his arguments on probability theory, such as the theory of errors that had been given by Pierre-Simon Laplace (1749–1827). It is interesting to note that although Maxwell had obtained an important result, he himself was not convinced that his work was of real significance. The reason for his doubt was that he had not been able to explain the specific heats of gases on the basis of his kinetic theory; we now realize that those results require the quantum theory for their explanation.

The treatment we will describe is essentially that given in 1860 by Maxwell. The velocity u of a molecule can be resolved into its three components along the X, Y, and Z axes. If u_x is the component along the X axis, the corresponding kinetic energy is $\frac{1}{2}mu_x^2$, where m is the mass of the molecule. We can then ask: What is the probability dP_x that the molecule has a speed component along the X axis between u_x and $u_x + du_x$? We must, of course, consider a range of speeds, since otherwise the probability is zero. This probability is proportional to the range du_x, and Maxwell deduced on the basis of probability theory that it is proportional to $e^{-\beta\epsilon_x}$, or $e^{-mu_x^2\beta/2}$, where β is a constant. We can therefore write

$$dP_x = Be^{-mu_x^2\beta/2}\, du_x \tag{1.80}$$

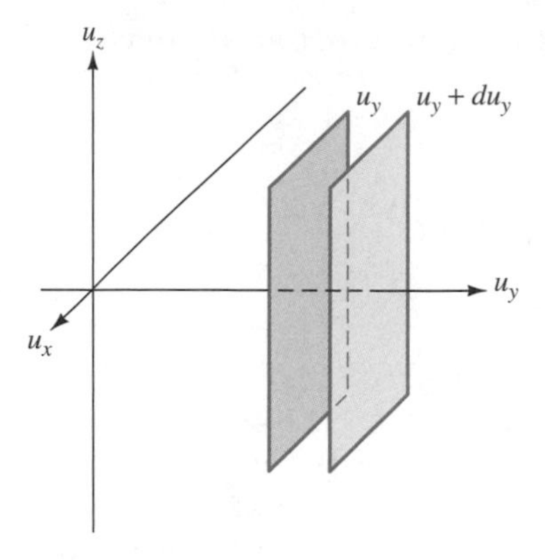

FIGURE 1.13
Plot showing volume element du_y between two planes.

where B is the proportionality constant. We also have similar expressions for the other components:

$$dP_y = Be^{-mu_y^2\beta/2}\,du_y \quad \text{and} \quad dP_z = Be^{-mu_z^2\beta/2}\,du_z \tag{1.81}$$

Each of these expressions gives the probability of finding the velocity between planes such as u_y and $u_y + du_y$, as shown in Figure 1.13.

The product of these three expressions is

$$dP_x\,dP_y\,dP_z = B^3e^{-mu_x^2\beta/2}e^{-mu_y^2\beta/2}\,e^{-mu_z^2\beta/2}\,du_x\,du_y\,du_z \tag{1.82}$$

Since $u^2 = u_x^2 + u_y^2 + u_z^2$, where u is the speed,

$$dP_x\,dP_y\,dP_z = B^3e^{-mu^2\beta/2}\,du_x\,du_y\,du_z \tag{1.83}$$

This expression is the probability that the three components of speed have values between u_x and $u_x + du_x$, u_y and $u_y + du_y$, and u_z and $u_z + du_z$. However, what we are really interested in is the probability that the actual speed of the molecule lies between u and $u + du$, and this can be achieved by a variety of combinations of the three components of velocity. In order to convert this notation into a suitable volume element, we transform the Cartesian coordinates into spherical polar coordinates. The result of this change is shown in Figure 1.14, where the relations

$$z = r\sin\theta\,\cos\varphi,\ y = r\sin\theta\,\sin\varphi,\ \text{and}\ z = r\cos\theta$$

are used to convert the Cartesian coordinates to the spherical polar coordinate system. This transformation is generally done in physics or calculus texts. We then have a spherical shell of radius u and thickness du. What we want is the probability of having particles that lie within the entire shell. The total volume of the spherical

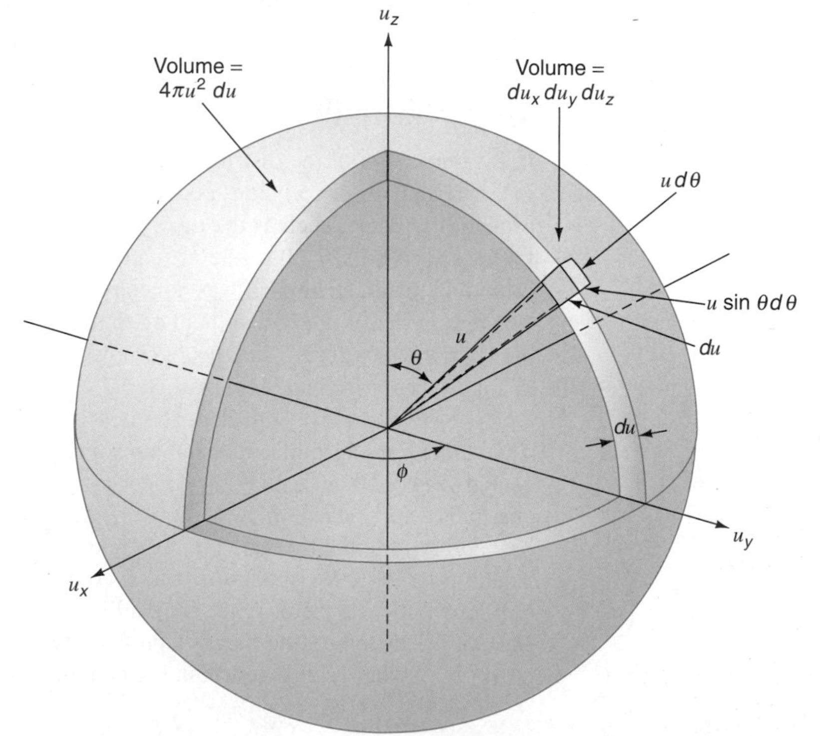

FIGURE 1.14
A diagram with axes u_x, u_y, and u_z. Equation 1.83 gives the probability that the speed is represented by the cube of volume $du_x\,du_y\,du_z$. We are interested in the speed corresponding to the volume $4\pi u^2\,du$ that lies between the concentric spherical surfaces.

shell is $4\pi u^2 du$. In order to obtain the probability dP that the speed lies between u and $u + du$, we must therefore replace $du_x\, du_y\, du_z$ in Eq. 1.83 by $4\pi u^2\, du$:

$$dP = 4\pi B^3 e^{-mu^2\beta/2} u^2\, du \tag{1.84}$$

The fraction dN/N of the N molecules that have speeds between u and $u + du$ is given by the ratio of this expression to its value integrated from $u = 0$ to $u = \infty$, since that is the range of possible speeds. Thus

$$\frac{dN}{N} = \frac{4\pi B^3 e^{-mu^2\beta/2} u^2\, du}{4\pi B^3 \int_0^\infty e^{-mu^2\beta/2} u^2\, du} \tag{1.85}$$

The integral in the denominator is a standard one,[6] and when we evaluate it, we obtain

$$\frac{dN}{N} = 4\pi \left(\frac{\beta m}{2\pi}\right)^{3/2} e^{-mu^2\beta/2} u^2\, du \tag{1.86}$$

We now consider what the mean-square speed is on the basis of this treatment. The point of doing this is that we already know the mean-square speed from kinetic theory. Thus, from Eq. 1.43 we have

$$\overline{u^2} = \frac{3RT}{Lm} \tag{1.87}$$

This quantity is obtained from Eq. 1.86 by multiplying u^2 by the fraction dN/N and integrating:

$$\overline{u^2} = \int_0^\infty u^2 \frac{dN}{N} = 4\pi \left(\frac{\beta m}{2\pi}\right)^{3/2} \int_0^\infty e^{-mu^2\beta/2} u^4\, du \tag{1.88}$$

Maxwell distribution: Vary the temperature and follow the Maxwell distribution of He, O_2 and Cl_2.

The integral is again a standard one (see the appendix to this chapter, p. 43), and evaluating it leads to

$$\overline{u^2} = \frac{3}{m\beta} \tag{1.89}$$

Comparison of the expressions in Eqs. 1.87 and 1.89 then gives the important result that

$$\beta = \frac{L}{RT} = \frac{1}{k_B T} \tag{1.90}$$

Boltzmann Constant

where R/L has been written as k_B, which is called the **Boltzmann constant.** Equation 1.85 may thus be written as

$$\frac{dN}{N} = 4\pi \left(\frac{m}{2\pi k_B T}\right)^{3/2} e^{-mu^2/2k_B T} u^2\, du \tag{1.91}$$

Maxwell Distribution Law

This is usually known as the **Maxwell distribution law.**

Figure 1.15 shows a plot of $(1/N)(dN/du)$ for oxygen gas at two temperatures. Near the origin the curves are parabolic because of the dominance of the u^2 term in the equation, but at higher speeds the exponential term is more important. Note that

[6] The appendix to this chapter (p. 43) gives a number of the integrals that are useful in distribution problems.

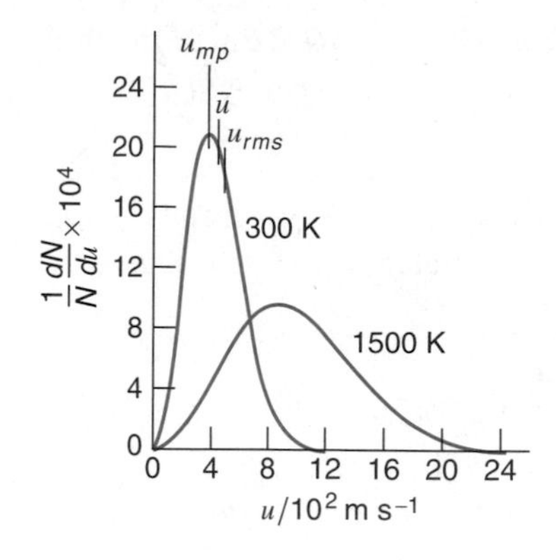

FIGURE 1.15
The Maxwell distribution of molecular speeds for oxygen at 300 K and 1500 K. Note that the areas below the two curves are identical since the total number of molecules is fixed.

the curve becomes much flatter as the temperature is raised. Indicated on the curve for 300 K are the root-mean-square speed $\sqrt{\overline{u^2}}$, the average speed $\bar{u}$, and the most probable speed u_{mp}. The mean-square speed is given in Eq. 1.89, and insertion of $1/k_BT$ for β gives the expression in Table 1.3; the root-mean-square speed is its square root. The mean speed or average speed is given by

$$\bar{u} = \int_0^{\infty} u \frac{dN}{N} \tag{1.92}$$

Insertion of Eq. 1.91 and integration leads to the expression in Table 1.3. The most probable speed is the speed at the maximum of the curve and is obtained by differentiating $(1/N)(dN/du)$ with respect to u and setting the result equal to zero. The resulting expression for this quantity is also given in Table 1.3.

The Distribution of Translational Energy

We can convert Eq. 1.91 into an equation for the energy distribution. We start with

$$\epsilon = \frac{1}{2}mu^2 \tag{1.93}$$

and differentiation gives

$$\frac{d\epsilon}{du} = mu = m\left(\frac{2\epsilon}{m}\right)^{1/2} = (2\epsilon m)^{1/2} \tag{1.94}$$

We can therefore replace du in Eq. 1.91 by $d\epsilon/(2\epsilon m)^{1/2}$ and u^2 by $2\epsilon/m$, and we obtain

$$\frac{dN}{N} = 4\pi\left(\frac{m}{2\pi k_BT}\right)^{3/2} e^{-\epsilon/k_BT} \frac{2\epsilon}{m(2\epsilon m)^{1/2}} d\epsilon \tag{1.95}$$

$$= \frac{2\pi}{(\pi k_BT)^{3/2}} e^{-\epsilon/k_BT} \epsilon^{1/2} d\epsilon \tag{1.96}$$

Maxwell-Boltzmann Distribution Law

This is often known as the **Maxwell-Boltzmann distribution law.**

TABLE 1.3 Quantities Relating to the Maxwell Distribution of Molecular Speeds

Mean-square speed	$\overline{u^2} = \frac{3k_BT}{m}$
Square root of the mean-square speed	$\sqrt{\overline{u^2}} = \sqrt{\frac{3k_BT}{m}}$
Average speed	$\bar{u} = \sqrt{\frac{8k_BT}{\pi m}}$
Most probable speed	$u_{mp} = \sqrt{\frac{2k_BT}{m}}$
Average translational energy	$\bar{\epsilon} = \frac{3}{2}k_BT$

An expression for the average translational energy is obtained by evaluating the integral

$$\bar{\epsilon} = \int_0^\infty \epsilon \frac{dN}{N} \tag{1.97}$$

The result is $\frac{3}{2}k_B T$, which we already found, in another way, in Section 1.9.

1.12 Real Gases

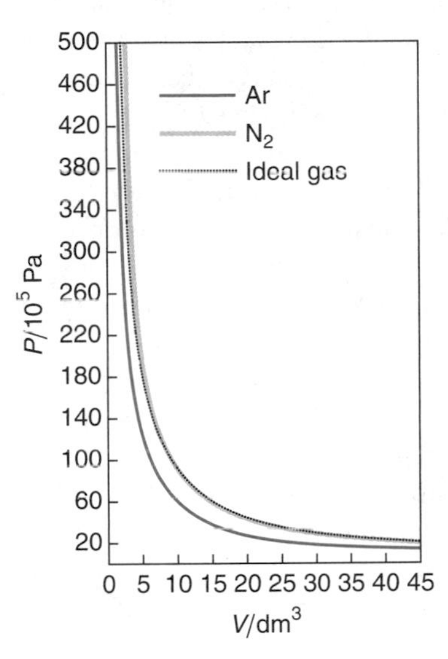

FIGURE 1.16
Plots of pressure against volume for 1.000 kg each of nitrogen and argon at 300 K. Nitrogen follows the ideal gas law very closely, but argon shows significant deviations.

The Compression Factor

The gas laws that we have treated in the preceding sections hold fairly well for most gases over a limited range of pressures and temperatures. However, when the range and the accuracy of experimental measurements were extended and improved, real gases were found to deviate from the expected behavior of an ideal gas. For instance, the PV_m product does not have the same value for all gases nor is the pressure dependence the same for different gases. (V_m represents the **molar volume,** the volume occupied by 1 mol of gas.) Figure 1.16 shows the deviations of N_2 and Ar from the expected behavior of an ideal gas under particular *isothermal* (constant temperature) conditions. It is difficult, however, to determine the relative deviation from ideal conditions from a graph of this sort or even from a P against $1/V$ plot. A more convenient technique often used to show the deviation from ideal behavior involves the use of graphs or tables of the **compression** (or **compressibility) factor** Z, defined by

$$Z = \frac{PV}{nRT} = \frac{PV_m}{RT} \tag{1.98}$$

and normally presented as a function of pressure or volume.

For the ideal gas, $Z = 1$. Therefore, departures from the value of unity indicate nonideal behavior. Since each gas will have different interactions between its molecules, the behavior of Z can be expected to be quite varied. In Figure 1.17, a plot of Z versus P for several gases shows the variation of deviations typically found. The shapes of the initial negative slopes of some of these curves ($Z < 1$) can be related to attractive forces prevalent among the molecules in the gas, whereas initial positive slopes ($Z > 1$) indicate that repulsive forces between molecules predominate. The initial negative slope for CH_4 turns positive at about $Z = 0.25$. All gases at sufficiently high pressure will have $Z > 1$ since repulsive forces are dominant. Note that as $P \to 0$, the curves for all gases will approach $Z = 1$, although with different slopes.

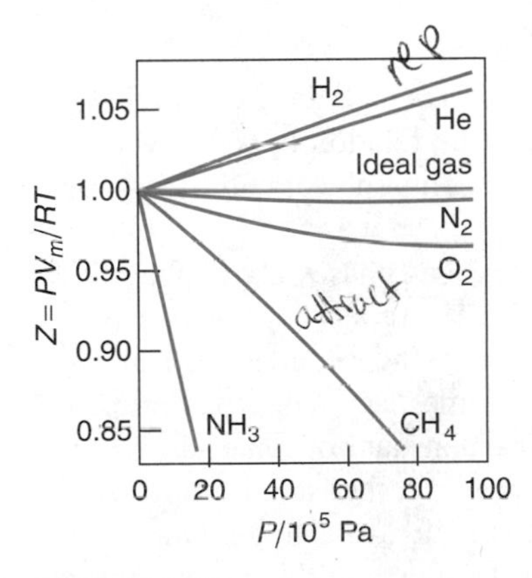

FIGURE 1.17
Plot of the compression factor, Z, against pressure for several gases at 273 K.

Condensation of Gases: The Critical Point

The Irish physical chemist Thomas Andrews (1813–1885) studied the behavior of carbon dioxide under pressure at varying temperatures. Using a sample of liquid carbon dioxide (CO_2 can be liquefied at room temperature using sufficiently high pressure), he gradually raised its temperature while maintaining the pressure constant. To his surprise, the boundary between the gas and liquid regions disappeared at 31.1 °C. Further increase in pressure could not bring about a return to the liquid

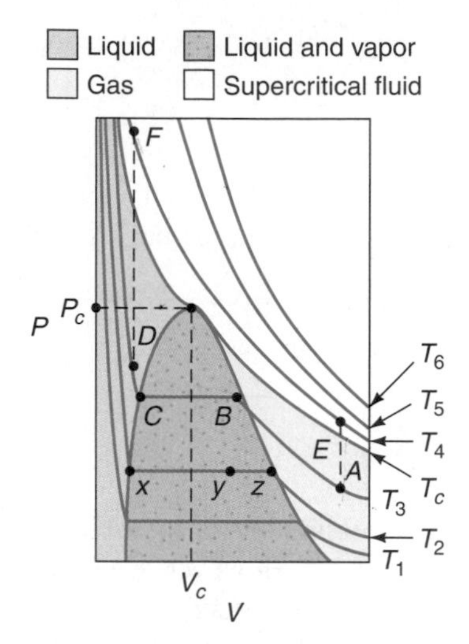

FIGURE 1.18
Isotherms for a typical real gas.

state. This experiment led Andrews to suggest that a **critical temperature** T_c existed for each gas. Above this temperature, pressure alone could not liquefy the gas. In other words, T_c is the highest temperature at which a liquid can exist as a distinct phase or region. He introduced the word *vapor* for the gas that is in equilibrium with its liquid below the critical temperature.

Investigation of other real gases showed similar behavior. In Figure 1.18 the isotherms (lines of constant temperature) are shown for a typical real gas. For the higher temperatures T_5 and T_6, the appearance of the isotherms is much like the hyperbolic curves expected of an ideal gas. However, below T_c the appearance of the curves is quite different. The horizontal portions along T_1, T_2, and T_3 are called *tie lines* and contain a new feature not explained by the ideal gas law. (Also see p. 37 for a discussion of tie lines.) The significance of the tie lines can be seen by considering a point such as y on the T_2 isotherm. Since y lies in the stippled region representing the coexistence of liquid and vapor, any variation in volume of the system between the values x and z serves merely to change the liquid-vapor ratio. The pressure of the system corresponding to the tie line pressure is the saturated vapor pressure of the liquefied gas. The right endpoint z of the tie line represents the molar volume in the gas phase; the left endpoint x represents the molar volume in the liquid phase. The distance between the endpoints and y is related to the percentage of each phase present.

If a sample of gas is compressed along isotherm T_3, starting at point A, the PV curve is approximately the Boyle's law isotherm until point B is reached. As soon as we move into the stippled region, liquid and vapor coexist as pointed out previously. Liquefaction begins at point B and ends at point C, when all the gas is converted to liquid. As we pass out of the two-phase region, only liquid is present along CD. The steepness of the isotherm indicates the rather low ability of the liquid to be compressed compared to that of the gas.

Critical Point

As the isotherms approach that of T_c, the tie lines become successively shorter until at the **critical point** they cease to exist and only one phase is present. The pressure and volume of the substance corresponding to this critical point are called the **critical pressure** P_c and **critical volume** V_c, respectively. The critical point is a point of inflection; therefore, *the equations*

$$\left(\frac{\partial P}{\partial V}\right)_{T_c} = 0 \quad \text{and} \quad \left(\frac{\partial^2 P}{\partial V^2}\right)_{T_c} = 0 \tag{1.99}$$

serve to define the critical point. The notation ∂ indicates partial differentiation and merely means that another variable that influences the relation is being held constant as indicated by the subscript on the parentheses. This subject will be discussed more fully later. Also see Appendix C.

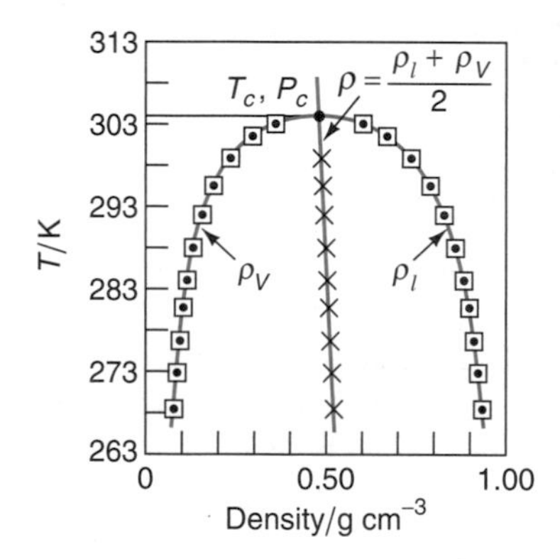

FIGURE 1.19
A schematic representation of the density of the liquid and vapor states of CO_2 near the critical point. Because it is difficult to precisely determine the critical point visually, the law of rectilinear diameters is used. At different temperatures the densities of the liquid and vapor phases are plotted and a straight line is obtained from their average. (Reprinted with permission from *Industrial & Engineering Chemistry, 38.* Copyright 1946 American Chemical Society.)

Since there is no distinction between liquid and gas phases above the critical point and no second phase is formed regardless of the pressure of the system, the term *supercritical fluid* is used instead of *liquid* or *vapor*. Indeed, at the critical point, the gas and liquid densities, the respective molar volumes, and indexes of refraction are identical. See Figure 1.19. Table 1.4 gives critical constants of some gases.

Furthermore, since gas and liquid can coexist only in the isolated dark region, it must be possible to pass from a single-phase gas region to a single-phase liquid region without noticing a phase change. In order to illustrate this point, consider 1 mol of liquid contained in a sealed vessel at the condition represented by point D in Figure 1.18; the temperature is T_3. We now heat the liquid above the critical temperature to point F. With the vessel thermostatted at T_5, the volume is allowed to

TABLE 1.4 Critical Constants for Some Gases

Substance	T_c / K	P_c / bar	V_c / $dm^3\ mol^{-1}$
H_2	33.2	12.97	0.0650
He	5.3	2.29	0.0577
N_2	126.0	33.9	0.0900
O_2	154.3	50.4	0.0744
Cl_2	417	77.1	0.123
Ar	151	48.6	0.0752
Kr	210.6	54.9	0.092
CO	134	35.5	0.040
NO	183	65.9	0.058
CO_2	304.16	73.9	0.0956
HCl	325	82.7	0.0862
SO_2	430	78.7	0.123
H_2O	647.1	220.6	0.0450
NH_3	405.5	113.0	0.0723
CH_4	190.6	46.4	0.0988
CCl_2F_2	385.1	41.1	0.217
C_5H_{12}	469.8	33.7	
C_6H_{14}	507.4	30.3	

increase to E. There is no change of phase during these processes. Then the temperature is dropped to the isotherm T_3 at point A. Again, no phase change is involved. The system now is in the gaseous state without ever having undergone a discontinuous change of phase. There is thus a complete *continuity of states* in which the transformation from the gas to the liquid state occurs continuously. In the light of the previous discussion, the distinction between the liquid and gaseous states under these conditions can be made only when two phases coexist.

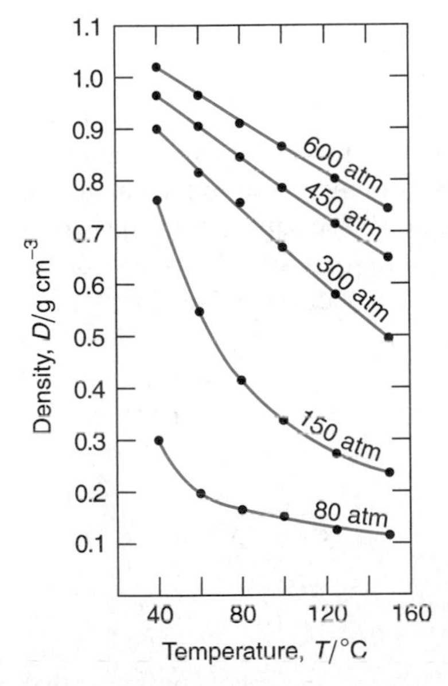

FIGURE 1.20
Variation of the density of CO_2 with temperature and total pressure. Total atmospheric pressure listed is gauge pressure. (Data taken from J. J. Langenfeld et al., *Anal. Chem., 64*, 2265(1992).)

Uses of Supercritical Fluids

Gases and liquids in the supercritical region are called *supercritical fluids*. They are formed by heating a substance above its critical temperature, and they have properties that may be significantly different from the properties in the normal state. For instance, the densities of gases as supercritical fluids may rise over 1.0 g cm^{-3}. See Figure 1.20 in which the variation of the density of CO_2 is plotted against temperature at various pressures. This high density is associated with the ability of some supercritical fluids to dissolve large nonvolatile molecules. This allows the development of interesting industrial applications, such as use of supercritical carbon dioxide as a solvent for automotive paint and for the removal of caffeine from coffee beans (thus avoiding the use of environmentally harmful solvents or chlorinated hydrocarbons).

Supercritical-fluid (SCF) chromatography is a developing area of analysis in which supercritical ammonia, carbon dioxide, heptane, and hexane have all been used as the mobile phase in a hybrid of gas and liquid chromatography. SCF combines the best features of each and, in general, considerably speeds the separation of the components in a mixture. SCF can also be used for separation of ions such as Cr^{3+} and Cr^{6+}.

1.13 Equations of State

J. D. van der Waals received the Nobel Prize in physics in 1910 for his development of the simple model that predicts the physics of gas imperfections and condensations.

The van der Waals Equation of State

The Dutch physicist Johannes Diderik van der Waals (1837–1923) found that two simple modifications in the ideal gas equation could account for the two-phase equilibrium when a gas is liquefied. The first of these modifications involves the volume of the molecules used in the ideal gas expression. In simple kinetic theory, the molecules are point particles; that is, they occupy no space. Actually, however, individual molecules do have a finite size and do occupy space. If V is the volume available for the ideal gas to move within its container, then the measured or observed volume of the container must be reduced by a volume b, called the *covolume*, which is approximately four times the volume occupied by the individual molecules. This volume will be proportional to the amount of substance n and may be written as nb. A term of the form $V - nb$ should therefore replace V_{ideal} in the ideal gas law.

In the same manner that the ideal volume is substituted by the actual volume, the pressure term can be modified by considering that in real gases there are intermolecular attractive forces not accounted for in the simple theory. (Repulsive forces are neglected in this treatment. See Chapter 17 for a fuller discussion of intermolecular forces.) Although the attractive forces are relatively small, they account for the ultimate liquefaction of gases when they are cooled sufficiently. With a force of attraction present, part of the pressure expected from the ideal gas law calculation is reduced in overcoming the force of intermolecular attraction. Thus, the observed pressure is less than the ideal pressure, and a correction term must be added to the observed pressure. This attractive force is found to be inversely proportional to the seventh power of the separation distance between nonpolar molecules and is proportional to the square of the density of the gaseous molecules. The density is proportional to the amount of substance n and inversely proportional to the volume V. The square of the density is thus proportional to n^2/V^2, and van der Waals therefore suggested that a term of the form an^2/V^2 be added to the observed pressure. His total equation then becomes:

Van der Waals Equation

$$\left(P + \frac{an^2}{V^2}\right)(V - nb) = nRT$$

or for one mole of gas (1.100)

$$\left(P + \frac{a}{V_m^2}\right)(V_m - b) = RT$$

where a and b are the *van der Waals constants*. They are *empirical* constants; that is, their values are chosen to give the best agreement between the points experimentally observed and the points calculated from the van der Waals equation.

The van der Waals equation may be solved for P giving

$$P = \frac{RT}{V_m - b} - \frac{a}{V_m^2} \tag{1.101}$$

Substitution of constants a and b (values of a and b for several gases are given in Table 1.5) allows the determination of the volume for a particular isotherm. Figure 1.21 was obtained using a = 0.680 3 Pa m^6 mol^{-2} and b = 0.056 4 × 10^{-3} m^3 mol^{-1}, the van der Waals constants for SO_2. In this figure, the gas-liquid equilibrium region for SO_2 has been superimposed on the van der Waals isotherms as a dashed line.

Below T_c there are three values of V for each value of P. To explain this, Eq. 1.100 may be multiplied by V^2 and expanded as a cubic equation in V:

$$PV^3 - (nbP + nRT)V^2 + n^2aV - n^3ab = 0 \tag{1.102}$$

Mathematically, a cubic equation may have three real roots, or one real and two complex roots. From the dashed line in the inset in Figure 1.21 we see that there are three real roots for pressure below T_c and only one above T_c. The three real roots give an oscillatory behavior to the curve; this is at variance with the normal experimental fact that the pressure remains constant along a tie line. However, some physical significance can be attached to two of the regions in the S-shaped part of the curve. The regions of the kind marked A correspond to situations in which a higher vapor pressure occurs than the liquefaction pressure. This is known as *supersaturation* and may be achieved experimentally if the vapor is entirely dust free. This is exploited in cloud chambers, used in physics. The ionizing particles produce a streak of nucleation centers, which result in localized condensation. The condensation makes visible the path of the ionizing particles. The region marked B corresponds to having the liquid under less pressure than its vapor pressure. Here bubbles of vapor should form spontaneously to offset the difference in pressures. Both metastable effects are due to the surface tension of the liquid (see Section 18.7).

C. T. R. Wilson developed cloud chambers, for which he won the Nobel Prize for physics in 1927.

Note that negative pressures may be achieved, as seen from the lowest isotherms in Figure 1.21. This is equivalent to saying that the liquid is under tension. These last three phenomena have all been experimentally demonstrated.

TABLE 1.5 Van der Waals Constants for Some Gases

Substance	a Pa m^6 mol^{-2}	b m^3 mol^{-1} × 10^{-3}
H_2	0.0248	0.0266
He	0.0034	0.0237
N_2	0.1408	0.0391
O_2	0.1378	0.0318
Cl_2	0.6579	0.0562
Ar	0.1355	0.0322
Kr	0.2349	0.0398
CO	0.1505	0.0399
NO	0.1358	0.0279
CO_2	0.3640	0.0427
HCl	0.3716	0.0408
SO_2	0.6803	0.0564
H_2O	0.5536	0.0305
NH_3	0.4225	0.0371
CH_4	0.2283	0.0428
CCl_2F_2	0.1066	0.0973

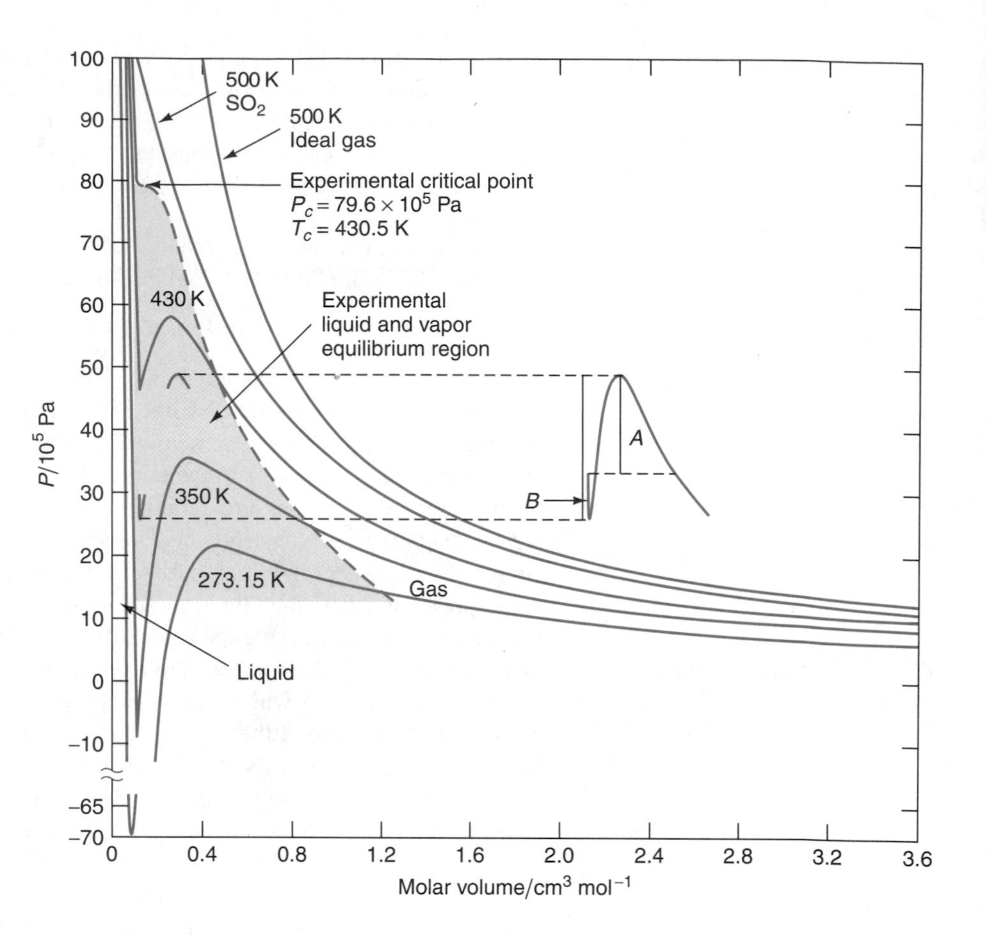

FIGURE 1.21
Isotherms for SO_2 showing the behavior predicted by the van der Waals equation. The inset shows the 400 K isotherm marked to indicate regions of supersaturation (*A*) and of less than expected pressure (*B*).

In spite of its simplicity and the easily understood significance of its constants, the van der Waals equation can be used to treat critical phenomena as well as a wide range of properties. It has therefore become one of the most widely used approximate expressions for work with gases.

The Law of Corresponding States

In the last section, on the van der Waals equation, we showed that both gases and liquids may be characterized by the critical constants T_c, P_c, and V_c. A useful rule is that the normal boiling point of a liquid is usually about two-thirds of its critical temperature. The relationship between critical constants of two *different* substances is found from the equations linking the van der Waals constants a and b to the critical constants. Equation 1.102 may be rewritten for 1 mol of gas ($n = 1$) as

$$V_m^3 - \left(b + \frac{RT}{P}\right)V_m^2 + \frac{a}{P}V_m - \frac{ab}{P} = 0 \tag{1.103}$$

At T_c the volume has three real roots that are all identical. This may be expressed as

$$(V_m - V_c)^3 = 0 \tag{1.104}$$

or

$$V_m^3 - 3V_cV_m^2 + 3V_c^2V_m - V_c^3 = 0 \tag{1.105}$$

Since Eqs. 1.103 and 1.105 describe the same condition when P and T are replaced by P_c and T_c in Eq. 1.103, we may equate coefficients of like powers of V_m. From the coefficients of V_m^2 we have

$$3V_c = b + \frac{RT_c}{P_c} \tag{1.106}$$

From terms in V_m,

$$3V_c^2 = \frac{a}{P_c} \tag{1.107}$$

and finally, from the constant terms,

$$V_c^3 = \frac{ab}{P_c} \tag{1.108}$$

From these last three equations, we obtain

$$a = 3P_cV_c^2, \qquad b = \frac{V_c}{3}, \qquad R = \frac{8P_cV_c}{3T_c} \tag{1.109}$$

Although the van der Waals constants may be evaluated from these equations, the method of choice is to determine a and b empirically from experimental PVT data.

Alternatively, the same results may be determined using the expressions in Eq. 1.99. Application of the mathematical conditions in these equations to the van der Waals equation will eventually yield Eq. 1.109. See Problem 1.57 for a similar application.

If the expressions obtained in Eq. 1.109 are inserted into the van der Waals equation for 1 mol of gas, we obtain

$$\left(\frac{P}{P_c} + \frac{3V_c^2}{V_m^2}\right)\left(\frac{V_m}{V_c} - \frac{1}{3}\right) = \frac{8T}{3T_c} \tag{1.110}$$

It is convenient at this point to replace each of the ratios P/P_c, V/V_c, and T/T_c by P_r, V_r, and T_r, respectively; these represent the *reduced pressure* P_r, *reduced volume* V_r, and *reduced temperature* T_r and are dimensionless variables. Equation 1.110 then takes the form

$$\left(P_r + \frac{3}{V_r^2}\right)\left(V_r - \frac{1}{3}\right) = \frac{8}{3}T_r \tag{1.111}$$

It is thus seen that all gases obey the same equation of state within the accuracy of the van der Waals relation when there are no arbitrary constants specific to the individual gas. This is a statement of the **law of corresponding states.**

Law of Corresponding States

As an illustration, two gases having the same reduced temperature and reduced pressure are in corresponding states and should occupy the same reduced volume. Thus, if 1 mol of helium is held at 3.43×10^5 kPa and 15.75 K, and 1 mol of carbon dioxide is held at 110.95×10^5 kPa and 912 K, they are in corresponding states (in both cases, $P/P_c = 1.5$ and $T/T_c = 3$) and hence should occupy the same reduced volume. The law's usefulness lies particularly in engineering where its range of validity is sufficient for many applications. The ability of the law to predict

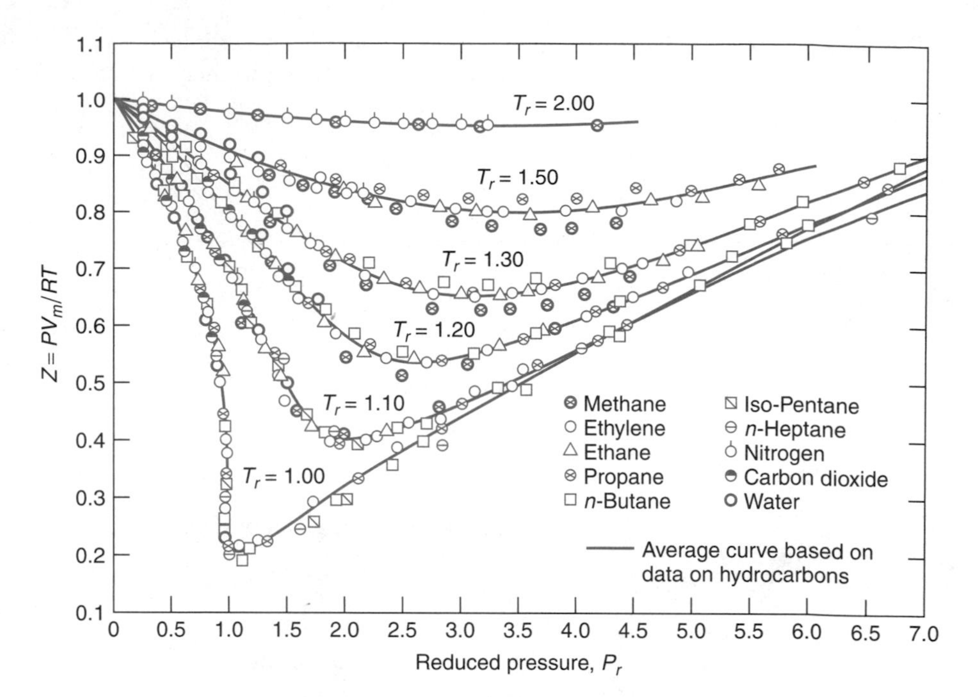

FIGURE 1.22
Compression factor versus reduced pressure for 10 gases. (Reprinted with permission from *Industrial and Engineering Chemistry,* Vol. 38. Copyright 1946 American Chemical Society.)

experimental behavior is nicely shown in Figure 1.22, where the reduced pressure is plotted against the compression factor for 10 different gases at various reduced temperatures.

Other Equations of State

There are two other major equations of state in common use. P. A. Daniel Berthelot (1865–1927) developed the equation

Berthelot Equation

$$\left(P + \frac{a}{V_m^2 T}\right)(V_m - B) = RT \tag{1.112}$$

which is the van der Waals equation modified for the temperature dependence of the attractive term. It may also be slightly modified in terms of the reduced variables as

$$P = \left(\frac{RT}{V_m}\right)\left(1 + \frac{9}{128T_r} - \frac{27}{64T_r^3}P_r\right) \tag{1.113}$$

where it provides high accuracy at low pressures and temperatures.

Another major equation of state is the one introduced in 1899 by C. Dieterici. The Dieterici equation involves the transcendental number, e (the base of the natural logarithms); however, it gives a better representation than the other expressions near the critical point. It may be written as

Dieterici Equation

$$(Pe^{a/V_mRT})(V_m - b) = RT \tag{1.114}$$

where a and b are constants not necessarily equal to the van der Waals constants. In reduced form, Eq. 1.114 becomes

$$P_r = \frac{T_r}{2V_r - 1} \exp\left(2 - \frac{2}{T_r V_r}\right) \tag{1.115}$$

1.14 The Virial Equation

The advantage of the equations discussed in the last sections is that the constants are kept to a minimum and relate to theoretically defined parameters. Another technique is to use a large number of constants to fit the behavior of a gas almost exactly, but the resulting equation is then less practical for general use and particularly for thermodynamic applications. Furthermore, as the number of constants is increased, it becomes more difficult to correlate the constants with physical parameters. However, two such expressions are of such general usefulness that they are discussed here.

H. Kamerlingh Onnes received the Nobel Prize in physics in 1913 for liquefying helium (1908) and for the discovery of superconductivity.

The Dutch physicist Heike Kamerlingh Onnes (1853–1926) suggested in 1901 that a power series, called a **virial equation,** be used to account for the deviations from linearity shown by real gases. The general form of the power series for Z as a function of P is

Virial Equation

$$Z(P, T) = \frac{PV_m}{RT} = 1 + B'(T)P + C'(T)P^2 + D'(T)P^3 + \cdots \tag{1.116}$$

However, this does not represent the data as well as a series in $1/V_m$, where the odd powers greater than unity are omitted. Thus the form of the equation of state of real gases presented by Kamerlingh Onnes is

$$\frac{PV}{nRT} = 1 + \frac{B(T)n}{V} + \frac{C(T)n^2}{V^2} + \frac{D(T)n^4}{V^4} + \cdots \tag{1.117}$$

where the coefficients $B'(T)$, $C'(T)$, $D'(T)$ and $B(T)$, $C(T)$, $D(T)$ are called the second, third, and fourth *virial coefficients,* respectively, and the notation indicates that they are functions of temperature. When Eq. 1.117 is multiplied through by R, the first term on the right is R; sometimes, therefore, R is called the *first virial coefficient.* For mixtures the coefficients are functions of both temperature and composition, and they are found experimentally from low-pressure PVT data by a graphical procedure. To illustrate how this is done, Eq. 1.117 is rewritten in terms of the molar density $\rho_m \equiv n/V$.

$$\frac{1}{\rho_m}\left(\frac{P}{\rho_m RT} - 1\right) = B(T) + C(T)\rho_m + \cdots \tag{1.118}$$

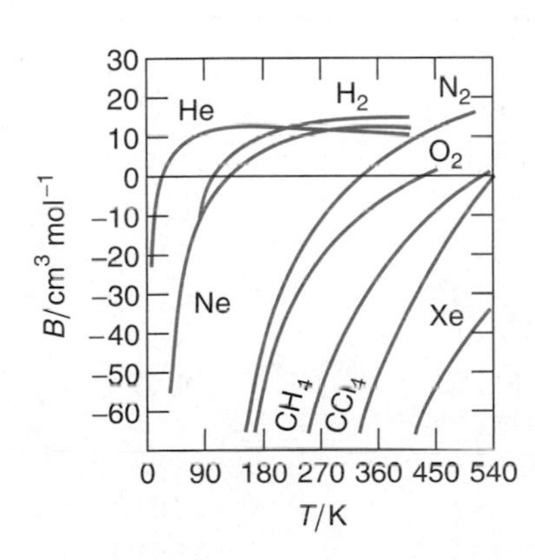

FIGURE 1.23
Plot of the second virial coefficient, $B(T)$, against T for several gases.

The left-hand side of Eq. 1.118 is plotted against ρ for fixed T, yielding a value of B at the intercept where $\rho = 0$. At the **Boyle temperature,** $B(T) = 0$ and, as a first approximation, the attractive forces balance the repulsive forces, allowing the gas to behave nearly ideally. Another way of stating this is that the partial derivative $[\partial(PV)/\partial P]_T$ becomes zero as $P \to 0$. (See Appendix C for a discussion of partial derivatives.) The slope of the curve at $\rho = 0$ gives the value of C for that particular temperature. In Figure 1.23 a plot of the second virial coefficient $B(T)$ is made showing the dependence of B on temperature for several gases.

EXAMPLE 1.6 Evaluate the Boyle temperature in terms of the known constants A, b, and R for a gas having the equation of state

$$PV_m = RT + (b - A/RT^{2/3})P$$

Solution The Boyle temperature occurs when the second virial coefficient $B(T) = 0$. This condition is fulfilled when the quantity $(b - A/RT^{2/3}) = 0$. Under this condition $T = T_B$. Solving for T_B

$$b = A/RT_B^{2/3}; \qquad T_B^{2/3} = A/bR; \qquad T_B = (A/bR)^{3/2}$$

The importance of the virial coefficients lies in the fact that, through the methods of statistical mechanics, an equation of state of a real gas may be developed in the virial form. The empirically derived coefficients can thus be related to their theoretical counterparts, which (it turns out) are the intermolecular potential energies. In this interpretation, the second virial coefficients, for instance, are due to molecular pair interactions; the other coefficients are due to higher order interactions.

The virial equation is not particularly useful at high pressures or near the critical point because the power series does not rapidly converge under conditions of higher order interactions. Furthermore, if one is to proceed on a theoretical basis rather than an empirical one, the calculation of the constants from statistical mechanics is difficult because the potential functions are not well known and the evaluation of the multiple integrals involved is very difficult.

The final expression to be considered here is the equation proposed in 1927–1928 by the American chemists James Alexander Beattie and Oscar C. Bridgeman.[7]

Beattie-Bridgeman Equation

$$P = \frac{RT[1 - (c/V_m T^3)]}{V_m^2}(V_m + B) - \frac{A}{V_m^2} \qquad (1.119)$$

TABLE 1.6 Constants for Use in the Beattie-Bridgeman Equation with R = 8.3145 J K^{-1} mol^{-1}

Gas	A_0 / Pa m^6 mol^{-2}	a / 10^{-6} m^3 mol^{-1}	B_0 / 10^{-6} m^3 mol^{-1}	b / 10^{-6} m^3 mol^{-1}	c / 10 m^3 K^3 mol^{-1}
He	0.00219	59.84	14.00	0.0	0.0040
Ne	0.02153	21.96	20.60	0.0	0.101
Ar	0.13078	23.28	39.31	0.0	5.99
H_2	0.02001	−5.06	20.96	−43.59	0.0504
N_2	0.1362	26.17	50.46	−6.91	4.20
O_2	0.1511	25.62	46.24	4.208	4.80
Air	0.13184	19.31	46.11	−11.01	4.34
CO_2	0.50728	71.32	104.76	72.35	66.00
CH_4	0.23071	18.55	55.87	−15.87	12.83
$(C_2H_5)_2O$	3.1692	124.26	454.46	119.54	33.33

[7]J. A. Beattie and O. C. Bridgeman, *J. Am. Chem. Soc., 49,* 1665(1927); *50,* 3133, 3151(1928).

where

$$A = A_0\left(1 - \frac{a}{V_m}\right) \qquad B = B_0\left(1 - \frac{b}{V_m}\right)$$

and a, b, A_0, B_0, and c are empirically determined constants. The *Beattie-Bridgeman equation* uses five constants in addition to R and is well suited for precise work, especially in the high-pressure range. Table 1.6 gives the Beattie-Bridgeman[8] constants for 10 gases.

Appendix: Some Definite and Indefinite Integrals Often Used in Physical Chemistry

$$\int \sin x\, dx = -\cos x$$

$$\int \cos x\, dx = \sin x$$

$$\int \sin^2 x\, dx = \frac{1}{2}x - \frac{1}{4}x \sin^2 x$$

$$\int x^n\, dx = \frac{x^{n+1}}{n+1} \text{ (except } n \neq -1)$$

$$\int \frac{1}{x}\, dx = \ln x$$

$$\int \mathrm{e}^x\, dx = \mathrm{e}^x$$

$$\int \ln(x)\, dx = x \ln x - x$$

$$\int \frac{dx}{a^2 + x^2} = \frac{1}{a}\tan^{-1}\left(\frac{x}{a}\right)$$

$$\int \frac{dx}{\sqrt{a^2 - x^2}} = \sin^{-1}\left(\frac{x}{a}\right)$$

$$\int_0^\infty \mathrm{e}^{-ax^2}\, dx = \frac{1}{2}\left(\frac{\pi}{a}\right)^{1/2}$$

$$\int_0^\infty \mathrm{e}^{-ax^2} x\, dx = \frac{1}{2a}$$

$$\int_0^\infty \mathrm{e}^{-ax^2} x^2\, dx = \frac{1}{4}\left(\frac{\pi}{a^3}\right)^{1/2}$$

$$\int_0^\infty \mathrm{e}^{-ax^2} x^3\, dx = \frac{1}{2a^2}$$

$$\int_0^\infty \mathrm{e}^{-ax^2} x^4\, dx = \frac{3}{8}\left(\frac{\pi}{a^5}\right)^{1/2}$$

$$\int_0^\infty \mathrm{e}^{-ax}\, dx = \frac{1}{a}, \ (a > 0)$$

$$\int_0^\infty \mathrm{e}^{-ax} x^{1/2}\, dx = \frac{1}{2a}\left(\frac{\pi}{a}\right)^{1/2}$$

$$\int_0^\infty \mathrm{e}^{-ax} x\, dx = \frac{1}{a^2}$$

$$\int_0^\infty \mathrm{e}^{-ax} x^2\, dx = \frac{2}{a^3}$$

$$\int_0^\infty \mathrm{e}^{\ ax} x^n\, dx = \frac{n!}{a^{n+1}}, \ (a > 0, n \text{ a positive integer})$$

[8] J. A. Beattie and O. C. Bridgeman, *Proc. Am. Acad. Arts Sci.*, *63*, 229(1928).

KEY EQUATIONS

Definition of *kinetic energy:*

$$E_k = \frac{1}{2}mu^2$$

Potential energy for a body obeying Hooke's law:

$$E_p = \frac{1}{2}k_h x^2$$

Boyle's law:

$$P \propto \frac{1}{V} \quad \text{or} \quad PV = \text{constant} \qquad \text{(at constant } T)$$

Gay-Lussac's (Charles's) law:

$$V \propto T \quad \text{or} \quad \frac{V}{T} = \text{constant} \qquad \text{(at constant } P)$$

Equation of state of an ideal gas:

$$PV = nRT$$

Pressure of a gas derived from the *kinetic-molecular theory:*

$$P = \frac{Nm\overline{u^2}}{3V}$$

where $\overline{u^2}$ is the mean-square speed.

Relation of *kinetic energy* to *temperature:*

$$\bar{\epsilon}_k = \frac{3}{2}k_BT$$

where k_B is the *Boltzmann constant.*

Dalton's law of partial pressures:

$$P_t = (n_1 + n_2 + \cdots + n_i)\frac{RT}{V}$$

Graham's law of effusion:

$$\frac{\text{rate (gas}_1)}{\text{rate (gas}_2)} = \sqrt{\frac{\rho\ (\text{gas}_2)}{\rho\ (\text{gas}_1)}} = \sqrt{\frac{M\ (\text{gas}_2)}{M\ (\text{gas}_1)}}$$

where ρ is the density and M is the molar mass.

Collision frequency (SI unit: s^{-1}):

$$Z_A = \frac{\sqrt{2}\pi d_A^2\bar{u}_A N_A}{V}$$

Collision density (SI unit: $m^{-3}\ s^{-1}$):

$$Z_{AA} = \frac{\sqrt{2}\pi d_A^2\bar{u}_A N_A^2}{V^2} \quad \text{and}$$

$$Z_{AB} = \frac{\pi d_{AB}^2(\bar{u}_A^2 + \bar{u}_B^2)^{1/2}N_A N_B}{V^2}$$

Mean free path:

$$\lambda = \frac{V}{\sqrt{2}\pi d_A^2 N_A}$$

Average speed:

$$\bar{u} = \sqrt{\frac{8RT}{\pi M}}$$

Mean-square speed:

$$\overline{u^2} = \frac{3RT}{M}$$

Barometric distribution law:

$$P = P_0 e^{-E_p/RT}$$

Compression factor:

$$Z = \frac{PV}{nRT} = \frac{PV_m}{RT}$$

Van der Waals equation:

$$\left(P + \frac{an^2}{V^2}\right)(V - nb) = nRT$$

where a and b are the *van der Waals constants.*

PROBLEMS[9]

(Problems marked with an asterisk are more demanding.)

Classical Mechanics and Thermal Equilibrium

1.1. Calculate the amount of work required to accelerate a 1000-kg car (typical of a Honda Civic) to 88 km hr^{-1} (55 miles hr^{-1}). Compare this value to the amount of work required for a 1600-kg car (typical of a Ford Taurus) under the same conditions.

1.2. Assume that a rod of copper is used to determine the temperature of some system. The rod's length at 0 °C is 27.5 cm, and at the temperature of the system it is 28.1 cm. What is the temperature of the system? The linear expansion of copper is given by an equation of the form $l_t = l_0(1 + \alpha t + \beta t^2)$ where $\alpha = 0.160 \times 10^{-4}\ K^{-1}$, $\beta = 0.10 \times 10^{-7}\ K^{-2}$, l_0 is the length at 0 °C, and l_t is the length at t °C.

1.3. Atoms can transfer kinetic energy in a collision. If an atom has a mass of 1×10^{-24} g and travels with a velocity of 500 m s^{-1}, what is the maximum kinetic energy that can be transferred from the moving atom in a head-on elastic collision to the stationary atom of mass 1×10^{-23} g?

1.4. Power is defined as the rate at which work is done. The unit of power is the watt (W = 1 J s^{-1}). What is the

[9] In all problems in this book, temperatures and other quantities given as whole numbers (e.g., 25 °C, 300 K, 2 g, 5 dm^3) may be assumed to be exact to two decimal places. Other quantities are to be considered exact to the number of decimal places specified.

power that a man can expend if all his food consumption of 8000 kJ a day ($\approx$ 2000 kcal) is his only source of energy and it is used entirely for work?

1.5. State whether the following properties are intensive or extensive: (a) mass; (b) density; (c) temperature; (d) gravitational field.

Gas Laws and Temperature

1.6. The mercury level in the left arm of the J-shaped tube in Fig. 1.6a is attached to a thermostatted gas-containing bulb. The left arm is 10.83 cm and the right arm is 34.71 cm above the bottom of the manometer. If the barometric pressure reads 738.4 Torr, what is the pressure of the gas? Assume that temperature-induced changes in the reading of the barometer and J tube are small enough to neglect.

1.7. Vacuum technology has become increasingly more important in many scientific and industrial applications. The unit torr, defined as 1/760 atm, is commonly used in the measurement of low pressures.

a. Find the relation between the older unit mmHg and the torr. The density of mercury is 13.5951 g cm^{-3} at 0.0 °C. The standard acceleration of gravity is defined as 9.806 65 m s^{-2}.
b. Calculate at 298.15 K the number of molecules present in 1.00 m^3 at 1.00×10^{-6} Torr and at 1.00×10^{-15} Torr (approximately the best vacuum obtainable).

1.8. The standard atmosphere of pressure is the force per unit area exerted by a 760-mm column of mercury, the density of which is 13.595 11 g cm^{-3} at 0 °C. If the gravitational acceleration is 9.806 65 m s^{-2}, calculate the pressure of 1 atm in kPa.

1.9. Dibutyl phthalate is often used as a manometer fluid. Its density is 1.047 g cm^{-3}. What is the relationship between 1.000 mm in height of this fluid and the pressure in torr?

1.10. The volume of a vacuum manifold used to transfer gases is calibrated using Boyle's law. A 0.251-dm^3 flask at a pressure of 697 Torr is attached, and after system pumpdown, the manifold is at 10.4 mTorr. The stopcock between the manifold and flask is opened and the system reaches an equilibrium pressure of 287 Torr. Assuming isothermal conditions, what is the volume of the manifold?

1.11. An ideal gas occupies a volume of 0.300 dm^3 at a pressure of 1.80×10^5 Pa. What is the new volume of the gas maintained at the same temperature if the pressure is reduced to 1.15×10^5 Pa?

1.12. If the gas in Problem 1.11 were initially at 330 K, what will be the final volume if the temperature were raised to 550 K at constant pressure?

1.13. Calculate the concentration in mol dm^{-3} of an ideal gas at 298.15 K and at (a) 101.325 kPa (1 atm), and (b) 1.00×10^{-4} Pa (= 10^{-9} atm). In each case, determine the number of molecules in 1.00 dm^3.

***1.14.** A J-shaped tube is filled with air at 760 Torr and 22 °C. The long arm is closed off at the top and is 100.0 cm long; the short arm is 40.00 cm high. Mercury is poured through a funnel into the open end. When the mercury spills over the top of the short arm, what is the pressure on the trapped air? Let h be the length of mercury in the long arm.

1.15. A Dumas experiment to determine molar mass is conducted in which a gas sample's P, θ, and V are determined. If a 1.08-g sample is held in 0.250 dm^3 at 303 K and 101.3 kPa:

a. What would the sample's volume be at 273.15 K, at constant pressure?
b. What is the molar mass of the sample?

1.16. A gas that behaves ideally has a density of 1.92 g dm^{-3} at 150 kPa and 298 K. What is the molar mass of the sample?

1.17. The density of air at 101.325 kPa and 298.15 K is 1.159 g dm^{-3}. Assuming that air behaves as an ideal gas, calculate its molar mass.

1.18. A 0.200-dm^3 sample of H_2 is collected over water at a temperature of 298.15 K and at a pressure of 99.99 kPa. What is the pressure of hydrogen in the dry state at 298.15 K? The vapor pressure of water at 298.15 K is 3.17 kPa.

1.19. What are the mole fractions and partial pressures of each gas in a 2.50-L container into which 100.00 g of nitrogen and 100.00 g of carbon dioxide are added at 25 °C? What is the total pressure?

1.20. The decomposition of $KClO_3$ produces 27.8 cm^3 of O_2 collected over water at 27.5 °C. The vapor pressure of water at this temperature is 27.5 Torr. If the barometer reads 751.4 Torr, find the volume the dry gas would occupy at 25.0 °C and 1.00 bar.

1.21. Balloons now are used to move huge trees from their cutting place on mountain slopes to conventional transportation. Calculate the volume of a balloon needed if it is desired to have a lifting force of 1000 kg when the temperature is 290 K at 0.940 atm. The balloon is to be filled with helium. Assume that air is 80 mol % N_2 and 20 mol % O_2. Ignore the mass of the superstructure and propulsion engines of the balloon.

***1.22.** A gas mixture containing 5 mol % butane and 95 mol % argon (such as is used in Geiger-Müller counter tubes) is to be prepared by allowing gaseous butane to fill an evacuated cylinder at 1 atm pressure. The 40.0-dm^3 cylinder is then weighed. Calculate the mass of argon that gives the desired composition if the temperature is maintained at 25.0 °C. Calculate the total pressure of the final mixture. The molar mass of argon is 39.9 g mol^{-1}.

1.23. The gravitational constant g decreases by 0.010 m s^{-2} km^{-1} of altitude.

a. Modify the barometric equation to take this variation into account. Assume that the temperature remains constant.
b. Calculate the pressure of nitrogen at an altitude of 100 km assuming that sea-level pressure is exactly 1 atm and that the temperature of 298.15 K is constant.

1.24. Suppose that on another planet where the atmosphere is ammonia that the pressure on the surface, at $h = 0$, is 400 Torr at 250 K. Calculate the pressure of ammonia at a height of 8000 metres. The planet has the same g value as the earth.

1.25. Pilots are well aware that in the lower part of the atmosphere the temperature decreases linearly with altitude. This dependency may be written as $T = T_0 - az$, where a is a proportionality constant, z is the altitude, and T_0 and T are the temperatures at ground level and at altitude z, respectively. Derive an expression for the barometric equation that takes this into account. Work to a form involving $\ln P/P_0$.

1.26. An ideal gas thermometer and a mercury thermometer are calibrated at 0 °C and at 100 °C. The thermal expansion coefficient for mercury is

$$\alpha = \frac{1}{V_0}(\partial V/\partial T)_P$$

$$= 1.817 \times 10^{-4} + 5.90 \times 10^{-9}\theta + 3.45 \times 10^{-10}\theta^2$$

where θ is the value of the Celsius temperature and $V_0 = V$ at $\theta = 0$. What temperature would appear on the mercury scale when the ideal gas scale reads 50 °C?

Graham's Law, Molecular Collisions, and Kinetic Theory

1.27. It takes gas A 2.3 times as long to effuse through an orifice as the same amount of nitrogen. What is the molar mass of gas A?

1.28. Exactly 1 dm^3 of nitrogen, under a pressure of 1 bar, takes 5.80 minutes to effuse through an orifice. How long will it take for helium to effuse under the same conditions?

1.29. What is the total kinetic energy of 0.50 mol of an ideal monatomic gas confined to 8.0 dm^3 at 200 kPa?

1.30. Nitrogen gas is maintained at 152 kPa in a 2.00-dm^3 vessel at 298.15 K. If its molar mass is 28.0134 g mol^{-1} calculate:

a. The amount of N_2 present.
b. The number of molecules present.
c. The root-mean-square speed of the molecules.
d. The average translational kinetic energy of each molecule.
e. The total translational kinetic energy in the system.

1.31. By what factor are the root-mean-square speeds changed if a gas is heated from 300 K to 400 K?

***1.32.** The collision diameter of N_2 is 3.74×10^{-10} m at 298.15 K and 101.325 kPa. Its average speed is 474.6 m s^{-1}. Calculate the mean free path, the average number of collisions Z_A experienced by one molecule in unit time, and the average number of collisions Z_{AA} per unit volume per unit time for N_2.

***1.33.** Express the mean free path of a gas in terms of the variables pressure and temperature, which are more easily measured than the volume.

1.34. Calculate Z_A and Z_{AA} for argon at 25 °C and a pressure of 1.00 bar using $d = 3.84 \times 10^{-10}$ m obtained from X-ray crystallographic measurements.

1.35. Calculate the mean free path of Ar at 20 °C and 1.00 bar. The collision diameter $d = 3.84 \times 10^{-10}$ m.

1.36. Hydrogen gas has a molecular collision diameter of 0.258 nm. Calculate the mean free path of hydrogen at 298.15 K and (a) 133.32 Pa, (b) 101.325 k Pa, and (c) 1.0×10^8 Pa.

1.37. In interstellar space it is estimated that atomic hydrogen exists at a concentration of one particle per cubic meter. If the collision diameter is 2.5×10^{-10} m, calculate the mean free path λ. The temperature of interstellar space is 2.7 K.

***1.38.** Calculate the value of Avogadro's constant from a study made by Perrin [*Ann. Chem. Phys., 18,* 1(1909)] in which he measured as a function of height the distribution of bright yellow colloidal gamboge (a gum resin) particles suspended in water. Some data at 15 °C are

height, $z/10^{-6}$	5	35
N, relative number of gamboge particles at height z	100	47

$\rho_{gamboge} = 1.206$ g cm^{-3}
$\rho_{water} = 0.999$ g cm^{-3}
radius of gamboge particles, $r = 0.212 \times 10^{-6}$ m

(*Hint:* Consider the particles to be gas molecules in a column of air and that the number of particles is proportional to the pressure.)

Distributions of Speeds and Energies[10]

1.39. Refer to Table 1.3 (p. 32) and write expressions and values for (a) the ratio $\sqrt{\overline{u^2}}/\bar{u}$, and (b) the ratio $\bar{u}/u_{mp}$. Note that these ratios are independent of the mass and the temperature. How do the *differences* between them depend on these quantities?

1.40. The speed that a body of any mass must have to escape from the earth is 1.07×10^4 m s^{-1}. At what temperature would the average speed of (a) an H_2

[10]Refer to the integrals given in the appendix to this chapter, p. 43.

molecule, and (b) an O_2 molecule be equal to this escape speed?

1.41. a. For H_2 gas at 25 °C, calculate the ratio of the fraction of molecules that have a speed $2u$ to the fraction that have the average speed $\bar{u}$. How does this ratio depend on the mass of the molecules and the temperature?
b. Calculate the ratio of the fraction of the molecules that have the average speed $\bar{u}_{100\ °C}$ at 100 °C to the fraction that have the average speed $\bar{u}_{25\ °C}$ at 25 °C. How does this ratio depend on the mass?

1.42. Suppose that two ideal gases are heated to different temperatures such that their pressures and vapor densities are the same. What is the relationship between their average molecular speeds?

1.43. a. If $\bar{u}_{25\ °C}$ is the average speed of the molecules in a gas at 25 °C, calculate the ratio of the fraction that will have the speed $\bar{u}_{25\ °C}$ at 100° to the fraction that will have the same speed at 25 °C.
b. Repeat this calculation for a speed of 10 $\bar{u}_{25\ °C}$.

1.44. On the basis of Eq. 1.80 with $\beta = 1/k_BT$, derive an expression for the fraction of molecules in a one-dimensional gas having speeds between u_x and $u_x + du_x$. What is the most probable speed?

***1.45.** Derive an expression for the fraction of molecules in a one-dimensional gas having energies between ϵ_x and $\epsilon_x + d\epsilon_x$. Also, obtain an expression for the average energy $\bar{\epsilon}_x$.

***1.46.** Derive an expression for the fraction of molecules in a two-dimensional gas having speeds between u and $u + du$. (*Hint*: Proceed by analogy with the derivation of Eq. 1.91.) Then obtain the expression for the fraction having energies between and $\epsilon + d\epsilon$. What fraction will have energies in excess of ϵ^*?

Real Gases

1.47. In Section 1.13 it was stated that the van der Waals constant b is approximately four times the volume occupied by the molecules themselves. Justify this relationship for a gas composed of spherical molecules.

1.48. Draw the van der Waals PV isotherm over the same range of P and V as in Figure 1.21 at 350 K and 450 K for Cl_2 using the values in Table 1.4.

1.49. Compare the pressures predicted for 0.8 dm^3 of Cl_2 weighing 17.5 g at 273.15 K using (a) the ideal gas equation and (b) the van der Waals equation.

1.50. A particular mass of N_2 occupies a volume of 1.00 L at −50 °C and 800 bar. Determine the volume occupied by the same mass of N_2 at 100 °C and 200 bar using the compressibility factor for N_2. At −50 °C and 800 bar it is 1.95; at 100 °C and 200 bar it is 1.10. Compare this value to that obtained from the ideal gas law.

1.51. A gas is found to obey the equation of state

$$P = \frac{RT}{V - b} - \frac{a}{V}$$

where a and b are constants not equal to zero. Determine whether this gas has a critical point; if it does, express the critical constants in terms of a and b. If it does not, explain how you determined this and the implications for the statement of the problem.

1.52. Ethylene (C_2H_4) has a critical pressure of P_c = 61.659 atm and a critical temperature of T_c = 308.6 K. Calculate the molar volume of the gas at T = 97.2 °C and 90.0 atm using Figure 1.22. Compare the value so found with that calculated from the ideal gas equation.

1.53. Assuming that methane is a perfectly spherical molecule, find the radius of one methane molecule using the value of b listed in Table 1.5.

1.54. Determine the Boyle temperature in terms of constants for the equation of state:

$$PV_m = RT\{1 + 8/57(P/P_c)(T_c/T)[1 - 4(T_c/T)^2]\}$$

R, P_c, and T_c are constants.

1.55. Establish the relationships between van der Waals parameters a and b and the virial coefficients B and C of Eq. 1.117 by performing the following steps:
a. Starting with Eq. 1.101, show that

$$\frac{PV_m}{RT} = \frac{V_m}{V_m - b} - \frac{a}{RT}\frac{1}{V_m}.$$

b. Since $V_m/(V_m - b) = (1 - b/V_m)^{-1}$, and $(1 - x)^{-1} = 1 + x + x^2 + \cdots$, expand $(1 - b/V_m)^{-1}$ to the quadratic term and substitute into the result of part (a).
c. Group terms containing the same power of V_m and compare to Eq. 1.117 for the case $n = 1$.
d. What is the expression for the Boyle temperature in terms of van der Waals parameters?

***1.56.** Determine the Boyle temperature of a van der Waals gas in terms of the constants a, b, and R.

1.57. The critical temperature T_c of nitrous oxide (N_2O) is 36.5 °C, and its critical pressure P_c is 71.7 atm. Suppose that 1 mol of N_2O is compressed to 54.0 atm at 356 K. Calculate the reduced temperature and pressure, and use Figure 1.22, interpolating as necessary, to estimate the volume occupied by 1 mol of the gas at 54.0 atm and 356 K.

1.58. At what temperature and pressure will H_2 be in a corresponding state with CH_4 at 500.0 K and 2.00 bar pressure? Given T_c = 33.2 K for H_2, 190.6 K for CH_4; P_c = 13.0 bar for H_2, 46.0 bar for CH_4.

***1.59.** For the Dieterici equation, derive the relationship of a and b to the critical volume and temperature. [*Hint:* Remember that at the critical point $(\partial P/\partial V)_T = 0$ and $(\partial^2 P/\partial V^2)_T = 0$.]

1.60. In Eq. 1.103 a cubic equation has to be solved in order to find the volume of a van der Waals gas. However, reasonably accurate estimates of volumes can be made by deriving an expression for the compression factor Z in terms of P from the result of the previous problem. One simply substitutes for the terms V_m on the right-hand side in terms of the ideal gas law expression $V_m = RT/P$. Derive this expression and use it to find the volume of CCl_2F_2 at 30.0 °C and 5.00 bar pressure. What will be the molar volume computed using the ideal gas law under the same conditions?

***1.61.** A general requirement of all equations of state for gases is that they reduce to the ideal gas equation (Eq. 1.28) in the limit of low pressures. Show that this is true for the van der Waals equation.

1.62. The van der Waals constants for C_2H_6 in the older literature are found to be

$$a = 5.49 \text{ atm L}^2 \text{ mol}^{-2} \quad \text{and} \quad b = 0.0638 \text{ L mol}^{-1}$$

Express these constants in SI units (L = liter = dm^3).

***1.63.** Compare the values obtained for the pressure of 3.00 mol CO_2 at 298.15 K held in a 8.25-dm^3 bulb using the ideal gas, van der Waals, Dieterici, and Beattie-Bridgeman equations. For CO_2 the Dieterici equation constants are

$$a = 0.462 \text{ Pa m}^6 \text{ mol}^{-2},$$

$$b = 4.63 \times 10^{-5} \text{ m}^3 \text{ mol}^{-1}$$

***1.64.** A gas obeys the van der Waals equation with $P_c = 3.040 \times 10^6$ Pa (= 30 atm) and $T_c = 473$ K. Calculate the value of the van der Waals constant b for this gas.

***1.65.** Expand the Dieterici equation in powers of V_m^{-1} in order to cast it into the virial form. Find the second and third virial coefficients. Then show that at low densities the Dieterici and van der Waals equations give essentially the same result for P.

Essay Questions

1.66. In light of the van der Waals equation, explain the liquefaction of gases.

1.67. State the postulates of the kinetic molecular theory of gases.

1.68. Eq. 1.22 defines the ideal-gas thermometer. Describe how an actual measurement would be made using such a thermometer starting with a fixed quantity of gas at a pressure of 150 Torr.

SUGGESTED READING

The references at the end of each chapter are to specialized books or papers where more information can be obtained on the particular subjects of the chapter.

For an authoritative account of the discovery of Boyle's law, see I. Bernard Cohen, "Newton, Hooke, and 'Boyle's Law' (discovered by Power and Towneley)," *Nature, 204*, 1964, pp. 618–621.

For interesting accounts of the lives of early chemists, see H. A. Boorse and L. Motz (Eds.), *The World of the Atom,* New York: Basic Books, 1966.

For a discussion of pressure gauges, see R. J. Sime, *Physical Chemistry: Methods, Techniques, Experiments,* Philadelphia: Founder's College Publishing, 1990, pp. 314–335.

For problems and worked solutions covering many of the types of problems encountered in this book, see B. Murphy, C. Murphy, and B. J. Hathaway, *A Working Method Approach for Introductory Physical Chemistry Calculations,* New York: Springer-Verlag (Royal Society of Chemistry Paperback), 1997.

For an in-depth account of virial coefficients, see J. H. Dymond and E. B. Smith, *The Virial Coefficients of Gases, A Critical Compilation,* Oxford: Clarendon Press, 1969.

For much more on the use of equations of state, see O. A. Hougen, K. M. Watson, and R. A. Ragatz, *Chemical Process Principles: Part II, Thermodynamics* (2nd ed.), New York: Wiley, 1959, Chapter 14.

For more on supercritical fluid use in chromatography, see M. J. Schoenmaker, L. G. M. Uunk, and H-G Janseen, *Journal of Chromatography, 506*, 1990, pp. 563–578.

For accounts of the kinetic theory of gases, see:

J. H. Jeans, *Introduction to the Kinetic Theory of Gases,* Cambridge: University Press, 1959.

L. B. Loeb, *Kinetic Theory of Gases,* New York: Dover, 1961. This is a particularly useful account, with many applications.

N. G. Parsonage, *The Gaseous State,* Oxford: Pergamon Press, 1966.

M. J. Pilling (Ed.), *Modern Gas Kinetics, Theory, Experiments, and Applications,* New York: Blackwell Scientific, 1987.

R. D. Present, *Kinetic Theory of Gases,* New York: McGraw-Hill, 1960.

For an account of the development of physical chemistry, with biographies of many of the scientists mentioned in this book, see K. J. Laidler, *The World of Physical Chemistry,* Oxford University Press, 1993.

The First Law of Thermodynamics

PREVIEW

According to the first law of thermodynamics, energy cannot be created or destroyed but can only be converted into other forms. Heat and work are forms of energy, and the law can be expressed by stating that the change in the internal energy of a system ΔU is the sum of the heat q supplied to the system and the work w done on it:

$$\Delta U = q + w$$

In this chapter the nature of *work* is considered, and the important concept of thermodynamically reversible work will be introduced and explained.

It will be shown that the internal energy U is an important property for processes that occur at constant volume. For processes at constant pressure the fundamental property is the enthalpy H, which is the internal energy plus the product of the pressure and the volume:

$$H = U + PV$$

This chapter is also concerned with *thermochemistry* (Section 2.5), a field that is based on the first law and is concerned primarily with heat changes when a process occurs under precisely defined conditions. The concept of *standard states* is introduced to define these conditions. An important consequence of the first law is that we can write balanced equations for chemical reactions, together with their enthalpy changes, and we will show how they can be manipulated algebraically so as to obtain enthalpy changes for other reactions. A vast amount of thermochemical information is summarized in the form of *standard enthalpies of formation,* which are the enthalpy changes when compounds are formed from the elements in certain defined states which we call *standard states.*

In Section 2.6 a number of processes involving ideal gases are considered. It is shown that for an ideal gas the internal energy and the enthalpy depend only on the temperature and not on the pressure or volume. This is not true for real gases, with which Section 2.7 is concerned.

OBJECTIVES

After studying this chapter, the student should be able to:

- Understand the nature of heat and the first law of thermodynamics.
- Appreciate the difference between state functions and those that are not state functions and be able to cite several examples of each.

- Recognize the criteria for equilibrium and what constitutes a reversible system.
- Separate the concepts of energy, heat, and work and define the first law in terms of these quantities.
- Apply the conditions for reversible and irreversible work to determine the work changes under different conditions.
- Define *enthalpy* and understand its use in terms of endothermic and exothermic processes.
- Apply the definition of heat capacity under constant pressure and constant volume conditions.
- Use the concept of extent of reaction to determine stoichiometric coefficients for equations written in different forms.
- Define the standard states under all of the different conditions treated.
- Apply Hess's law to calculate enthalpy changes for chemical processes whose enthalpies are not given.
- Calculate internal energy values when necessary and through calorimetry show the relationship of ΔU and ΔH.
- Determine the ΔH when the temperature of the system changes.
- State the standard enthalpies for the various states of matter and use them to calculate ΔH for chemical reactions.
- Use conventional standard enthalpies of formation and bond energies to calculate bond dissociation enthalpies.
- Compute the different work obtained in reversible and irreversible processes under constant pressure, constant volume, and isothermal conditions where possible.
- Derive equations relating to an ideal gas in an adiabatic process and express the gas laws in terms of that behavior.
- Explain the significance of the Joule-Thomson experiment and the coefficient derived from it. Also be able to work with the partial derivatives leading to Eq. 2.119.
- Determine the internal pressure of the van der Waals equation and be able to work problems relating to the equation.

Chapter 1 was mainly concerned with macroscopic properties such as pressure, volume, and temperature. We have seen how some of the relationships between these properties of ideal and real gases can be interpreted in terms of the behavior of molecules, that is, of the *microscopic* properties. This kinetic-molecular theory is of great value, but it is possible to interpret many of the relationships between macroscopic properties without any reference to the behavior of molecules, or even without assuming that molecules exist. This is what is done in the most formal treatments of the science of thermodynamics, which is concerned with the general relationships between the various forms of energy, including heat. In this book we shall present thermodynamics in a less formal way, and from time to time we shall clarify some of the basic ideas by showing how they relate to molecular behavior.

At first it might appear that the formal study of thermodynamics, with no regard to molecular behavior, would not lead us very far. The contrary, however, is true. It has proved possible to develop some very far-reaching conclusions on the basis of purely thermodynamic arguments, and these conclusions are all the more convincing because they do not depend on the truth or falsity of any theories of atomic and molecular behavior. Pure thermodynamics starts with a small number of assumptions that are based on very well-established experimental results and makes logical deductions from them, finally giving a set of relationships that are bound to be true provided that the original premises are true.

2.1 Origins of the First Law

The Nature of Heat

There are three laws of thermodynamics (aside from the zeroth law, which was mentioned in Section 1.4). The first law is essentially the principle of conservation of energy. Before the first law could be formulated it was necessary for the nature of heat to be understood. We know today that heat is a form of energy, but it took a long time for this to be realized and become generally accepted. In the seventeenth century Robert Boyle, Isaac Newton, and others proceeded on the correct assumption that heat is a manifestation of the motions of the particles of which matter is composed, but the evidence was not then compelling. In the next century investigators such as Joseph Black and Antoine Lavoisier, both of whom did important experiments on heat, were convinced that heat is a substance, and Lavoisier even listed it as one of the chemical elements.

Firm evidence that heat is related to motion and is, therefore, a form of energy came only toward the end of the eighteenth century. Experiments were carried out which showed that expenditure of work causes the production of heat. The first quantitative experiments along these lines were performed by the Massachusetts-born Benjamin Thompson (1753–1814), who had a colorful career in Europe and was created Count of Rumford while serving in the Bavarian army. During his supervision of the boring of cannon at the Munich Arsenal, he became interested in the generation of heat during the process. He suggested in 1798 that the heat arose from the work expended, and he obtained a numerical value for the amount of heat generated by a given amount of work.

A more persuasive contribution was made by the German physician Julius Robert Mayer (1814–1878). His medical observations led him to conclude that work performed by humans is derived from the food they eat, and in 1842 he made the important suggestion that the total energy is conserved. At the same time, and independently, precise experiments on the interconversion of work and heat, under a variety of conditions, were carried out by the English scientist James Prescott Joule (1818–1887), and in his honor the modern unit of energy, work, and heat is called the joule (J).

The First Law

The experiments of Joule in particular led to the conclusion that **the energy of the universe remains constant,** which is a compact statement of the first law of thermodynamics. Both work and heat are quantities that describe the transfer of energy from one system to another.

If two systems are at different temperatures, heat can pass from one to the other directly, and there can also be transfer of matter from one to the other. Energy can also be transferred from one place to another in the form of work, the nature of which is considered later (Section 2.4). No matter how these transfers occur, the total energy of the universe remains the same.

2.2 States and State Functions

Section 1.3 emphasized the important distinction between a system and its surroundings. We also explained the differences between open systems, closed systems, and isolated systems. The distinction between a system and its surroundings is particularly important in thermodynamics, since we are constantly concerned with transfer of heat between a system and its surroundings. We are also concerned with the work done by the system on its surroundings or by the surroundings on the system. In all cases the system must be carefully defined.

Certain of the macroscopic properties have fixed values for a particular state of the system, whereas others do not. Suppose, for example, that we maintain 1 g of water in a vessel at 25 °C and a pressure of 10^5 Pa (1 bar); it will have a volume of close to 1 cm^3. These quantities, 1 g of H_2O, 25 °C, 10^5 Pa, and 1 cm^3, all specify the state of the system. Whenever we satisfy these four conditions, we have the water in precisely the same state, and this means that the total amount of energy in the molecules is the same. As long as the system is in this state, it will have these particular specifications. These macroscopic properties that we have mentioned (mass, pressure, temperature, and volume) are known as **state functions** or **state variables.**

One very important characteristic of a state function is that once we have specified the state of a system by giving the values of *some* of the state functions, the values of all other state functions are fixed. Thus, in the example just given, once we have specified the mass, temperature, and pressure of the water, the volume is fixed. So, too, is the total energy in the molecules that make up the system, and energy is therefore another state function. The pressure and temperature, in fact, depend on the molecular motion in the system.

Cool and heat a drink; States and state functions; Change of state.

Another important characteristic of a state function is that when the state of a system is changed, the change in any state function depends only on the initial and final states of the system, and not on the path followed in making the change. For example, if we heat the water from 25 °C to 26 °C, the change in temperature is equal to the difference between the initial and final temperatures:

$$\Delta T = T_{\text{final}} - T_{\text{initial}} = 1\ ^\circ\text{C} \tag{2.1}$$

The way in which the temperature change is brought about has no effect on this result.

Drop an ice cube in the ocean; Reservoirs; States and state functions.

This example may seem trivial, but it is to be emphasized that not all functions have this characteristic. For example, raising the temperature of water from 25 °C to 26 °C can be done in various ways; the simplest is to add heat. Alternatively, we could stir the water vigorously with a paddle until the desired temperature rise had been achieved; this means that we are doing work on the system. We could also add some heat and do some work in addition. This shows that heat and work are not state functions.

Choose two paths to take a man up the mountain; Different paths; States and state functions.

In the meantime, it is useful to consider an analogy. Suppose that there is a point A on the earth's surface that is 1000 m above sea level and another point B that is 4000 m above sea level. The difference, 3000 m, is the height of B with respect to A. In other words, the difference in height can be expressed as

$$\Delta h = h_B - h_A \tag{2.2}$$

where h_A and h_B are the heights of A and B above sea level. Height above sea level is thus a state function, the difference Δh being in no way dependent on the path chosen. However, the distance we have to travel in order to go from A to B is dependent on the path; we can go by the shortest route or take a longer route. Distance traveled is therefore not a state function.

2.3 Equilibrium States and Reversibility

Choose the external pressure of a piston and see how the gas pressure responds; Equilibrium states; States and state functions.

Thermodynamics is directly concerned only with equilibrium states, in which the state functions have constant values throughout the system. It provides us with information about the circumstances under which nonequilibrium states will move toward equilibrium, but by itself it tells us nothing about the nonequilibrium states.

Suppose, for example, that we have a gas confined in a cylinder having a frictionless movable piston (Figure 2.1). If the piston is motionless, the state of the gas can be specified by giving the values of pressure, volume, and temperature. However, if the gas is compressed very rapidly, it passes through states for which pressure and temperature cannot be specified, there being a variation of these properties throughout the gas; the gas near to the piston is at first more compressed and heated than the gas at the far end of the cylinder. The gas then would be said to be in a **nonequilibrium** state. Pure thermodynamics could not deal with such a state, although it could tell us what kind of a change would spontaneously occur in order for equilibrium to be attained.

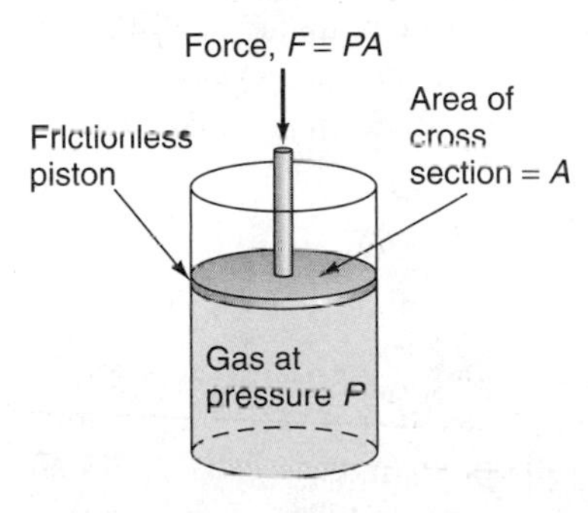

FIGURE 2.1
A gas at pressure P maintained at equilibrium by an external force, F, equal to PA, where A is the area of cross section of the piston. The force applied depends on the mass of the piston, and any masses that are placed on it.

The criteria for equilibrium are very important. The mechanical properties, the chemical properties, and the temperature must be uniform throughout the system and constant in time. The force acting on the system must be exactly balanced by the force exerted by the system, as otherwise the volume would be changing. If we consider the system illustrated in Figure 2.1, we see that for the system to be at equilibrium the force F exerted on the piston must exactly balance the pressure P of the gas; if A is the area of the piston,

$$PA = F \tag{2.3}$$

If we increase the force, for example, by adding mass to the piston, the gas will be compressed; if we decrease it, by removing mass, the gas will expand.

Suppose that we increase the force F by an infinitesimal amount dF. The pressure that we are exerting on the gas will now be infinitesimally greater than the pressure of the gas (i.e., it will be $P + dP$). The gas will therefore be compressed. We can make dP as small as we like, and at all stages during the infinitely slow compression we are therefore maintaining the gas in a state of equilibrium. We refer to a process of this kind as a **reversible** process. If we reduce the pressure to $P - dP$, the gas will expand infinitely slowly, that is, reversibly. Reversible processes play very important roles in thermodynamic arguments. All processes that actually occur are, however, irreversible; since they do not occur infinitely slowly, there is necessarily some departure from equilibrium.

Reversible Processes

2.4 Energy, Heat, and Work

We come now to a statement of the first law of thermodynamics, according to which the total amount of energy in the universe is conserved. Suppose that we add heat q to a system such as a gas confined in a cylinder (Figure 2.1). If nothing else is done to the system, the internal energy U increases by an amount that is exactly equal to the heat supplied:

Internal Energy

$$\Delta U = q \qquad \text{(with no work done)} \tag{2.4}$$

This increase in internal energy is the increase in the energy of the molecules which comprise the system.

Suppose instead that no heat is transferred to the system but that by the addition of mass to the piston an amount of work w is performed on it; the details of this are considered later (Eqs. 2.7–2.14). The system then gains internal energy by an amount equal to the work done[1]:

$$\Delta U = w \qquad \text{(with no transfer of heat)} \tag{2.5}$$

In general, if heat q is supplied to the system, and an amount of work w is also performed on the system, the increase in internal energy is given by

$$\Delta U = q \text{ (heat absorbed by the system)} + w \text{ (work done on the system)} \tag{2.6}$$

This is a statement of the first law of thermodynamics. We can understand the law by noting that a collection of molecules, on absorbing some heat, can store some of it internally, and can do some work on the surroundings. According to the IUPAC convention the work done by the system is $-w$, so that

Heat and work sign convention
- $q > 0$, heat is added to the system
- $q < 0$, heat flows out of the system
- $w > 0$, work is done on the system
- $w < 0$, work is done by the system

$$q \text{ (heat absorbed by the system)} = \Delta U - w \tag{2.6a}$$

which is equivalent to Eq. 2.6.

In applying Eq. 2.6 it is, of course, necessary to employ the same units for U, q, and w. The SI unit of energy is the joule (J = kg m^2 s^{-2}); it is the energy corresponding to a force of one newton (N = kg m s^{-2}) operating over a distance of one metre. In this book we shall use joules entirely, although many thermodynamic values have been reported in calories, one thermochemical calorie[2] being 4.184 J.

[1]The IUPAC recommendation is to use the symbol w for the work done on the system. The reader is warned that many older treatments use the symbol w for the work done *by* the system.

[2]There are other calories: The "15° calorie" is ≈ 4.1855 J; the "international calorie" is ≈ 4.1868 J.

We should note that Eq. 2.6 leaves the absolute value of the internal energy U indefinite, in that we are dealing with only the energy change ΔU, and for most practical purposes this is adequate. Absolute values can in principle be calculated, although they must always be referred to some arbitrary zero of energy. Thermodynamics is concerned almost entirely with energy changes.

State Functions

See how the energy that flows from a hot to a cold reservoir can be harnessed; Heat engines; Energy, heat, and work.

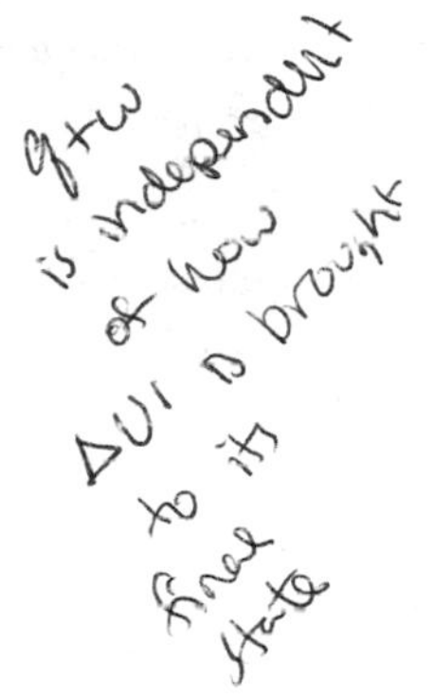

The internal energy U is a state function of the system; that is, it depends only on the state of the system and not on how the system achieved its particular state. Earlier we saw that a change from one state to another, such as from 25 °C to 26 °C, can be achieved by adding heat, by doing work, or by a combination of the two. It is found experimentally that however we bring about the temperature rise, the sum $q + w$ is always the same. In other words, for a particular change in state, the quantity ΔU, equal to $q + w$, is independent of the way in which the change is brought about. This behavior is characteristic of a state function. This example demonstrates that heat q and work w are not state functions since the change can be brought about by various divisions of the energy between heat and work; only the sum $q + w$ is fixed.

The distinction between state functions such as U and quantities such as q and w that are not state functions may be considered from another point of view. Whether or not a property is a state function is related to the mathematical concept of exact and inexact differentials. The definite integral of a state function such as U,

$$\int_{U_2}^{U_1} dU$$

is an exact differential because it has a value of $U_2 - U_1 = \Delta U$, which is independent of the path by which the process occurs. On the other hand, heat and work are the integrals of inexact differentials, symbolized by $đq$ or $đw$, the slashes indicating

$$\int_1^2 đq \quad \text{and} \quad \int_1^2 đw$$

that the integrals are inexact. These integrals are not fixed but depend on the process by which the change from state 1 to state 2 occurs. It would therefore be wrong to write $q_2 - q_1 = \Delta q$ or $w_2 - w_1 = \Delta w$; the quantities Δq and Δw have no meaning. Heat and work make themselves evident only during a change from one state to another and have no significance when a system remains in a particular state; they are properties of the path and not of the state. A state function such as the internal energy U, on the other hand, has a significance in relation to a particular state. A way to identify whether a function has an exact or inexact differential is to use Euler's **criterion for exactness** as described in Appendix C. Problem 2.12 treats this matter.

Study the parts of a heat engine; Heat engine: the parts; Energy, heat, and work.

Apply the first law and understand the efficiency of a heat engine; Heat engine efficiency; Energy, heat, and work.

If U were not a state function, we could have violations of the principle of conservation of energy, something that has never been observed. To see how a violation could occur, consider two states A and B, and suppose that there are two alternative paths between them. Suppose that for one of these paths U is 10 J; and for the other, 30 J:

$$\Delta U_1 = 10 \text{ J} \qquad \Delta U_2 = 30 \text{ J}$$

We could go from A to B by the first path and expend 10 J of heat. If we then returned from B to A by the second path, we would gain 30 J. We would then have the system in its original state, and would have a net gain of 20 J. Energy would therefore have been created from nothing. The process could be continued indefinitely, with a gain of energy at each completion of the cycle. Many attempts have

Reverse a heat engine to get a heat pump; Heat pump; Energy, heat, and work.

been made to create energy in this way, by the construction of *perpetual motion machines of the first kind,* but all have ended in failure—patent offices are constantly rejecting devices that would only work if the first law of thermodynamics were violated! The inability to make perpetual motion machines provides convincing evidence that energy cannot be created or destroyed.

Nature of Internal Energy

In purely thermodynamic studies it is not necessary to consider what internal energy really consists of; however, most of us like to have some answer to this question, in terms of molecular energies. There are contributions to the internal energy of a substance from

1. the kinetic energy of motion of the individual molecules,
2. the potential energy that arises from interactions between molecules,
3. the kinetic and potential energy of the nuclei and electrons within the individual molecules.

The precise treatment of these factors is somewhat complicated, and it is a great strength of thermodynamics that we can make use of the concept of internal energy without having to deal with it on a detailed molecular basis.

The Nature of Work

Electrical Work

Chemical Work

Mechanical Work

There are various ways in which a system may do work, or by which work may be done on a system. For example, if we pass a current through a solution and electrolyze it, we are performing one form of work—electrical work. Conversely, an electrochemical cell may *perform* work. Other forms of work are chemical work, osmotic work, and mechanical work. *Chemical* work is usually, but not quite always, involved when larger molecules are synthesized from smaller ones, as in living organisms. *Osmotic* work is the work required to transport and concentrate chemical substances. It is involved, for example, when seawater is purified by reverse osmosis (p. 218) and in the formation of the gastric juice, where the acid concentration is much higher than that of the surroundings. *Mechanical* work is performed, for example, when a weight is lifted.

Sign of PV work; PV work; Energy, heat, and work.

One simple way in which work is done is when an external force brings about a compression of a system. Suppose, for example, that we have an arrangement in which a gas or liquid is maintained at constant pressure P, which it exerts against a movable piston (Figure 2.2). In order for the system to be at equilibrium we must apply a force P to the piston, the force being related to the pressure by the relationship

$$F = PA \tag{2.7}$$

where A is the area of the piston. Suppose that the force is increased by an infinitesimal amount dF, so that the piston moves infinitely slowly, the process being reversible. If the piston moves to the left a distance l, the reversible work w_{rev} done on the system is

$$w_{rev} = Fl = PAl \tag{2.8}$$

However, Al is the volume swept *out* by the movement of the piston, that is, the decrease in volume of the gas, which is $-\Delta V$. The work done on the system is thus

$$w_{rev} = -P\Delta V \tag{2.9}$$

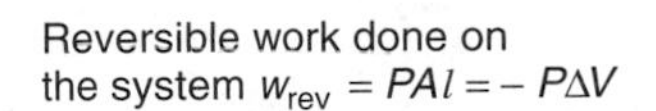

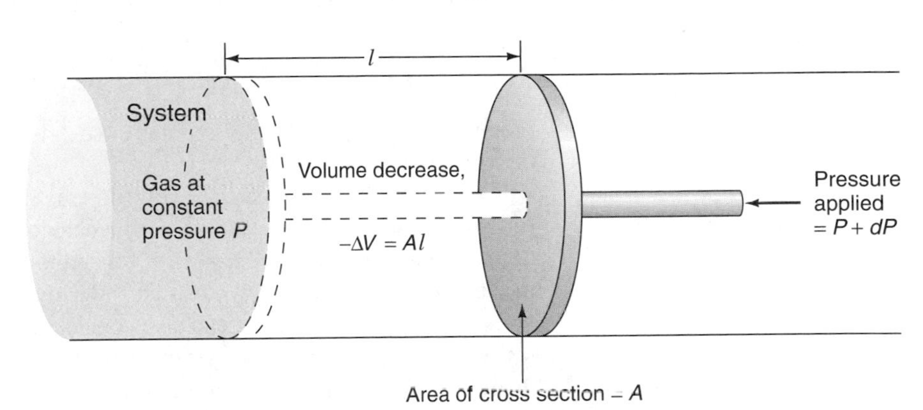

FIGURE 2.2
The reversible work done by a constant pressure P moving a piston. A simple way for a gas to be at constant pressure is to have a vapor in equilibrium with its liquid.

In our example this is a positive quantity since we have compressed the gas and ΔV is negative. If the gas had expanded, ΔV would have been positive and the work done on the system would have been negative; that is, the gas would have done a positive amount of work on the surroundings.

EXAMPLE 2.1 Suppose that a chemical reaction is caused to occur in a bulb to which is attached a capillary tube having a cross-sectional area of 2.50 mm^2. The tube is open to the atmosphere (pressure = 101.325 kPa), and during the course of the reaction the rise in the capillary is 2.40 cm. Calculate the work done by the reaction system.

Solution The volume increase is

$$2.50 \times 10^{-6}\ \text{m}^2 \times 2.40 \times 10^{-2}\ \text{m} = 6.00 \times 10^{-8}\ \text{m}^3$$

The work done by the system, which following the IUPAC convention must be written as $-w$, is $P\Delta V$:

$$-w = P\Delta V = 1.01325 \times 10^5\ \text{Pa} \times 6.00 \times 10^{-8}\ \text{m}^3$$
$$= 6.08 \times 10^{-3}\ \text{J} \qquad [\text{Pa} = \text{N m}^{-2};\ \text{N m} = \text{J}]$$

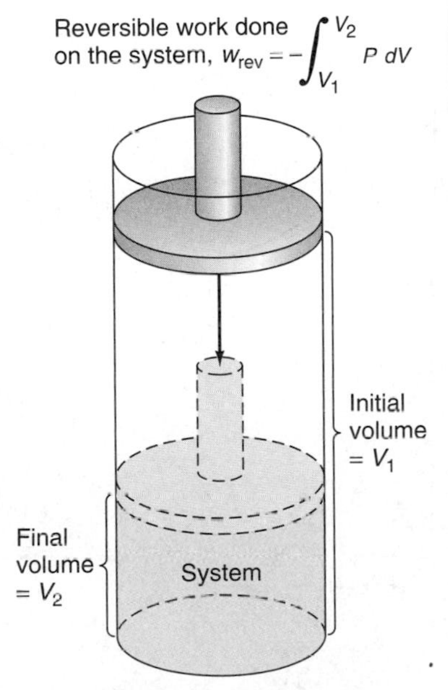

FIGURE 2.3
The reversible work performed when there is a volume decrease from V_1 to V_2.

If the pressure P varies during a volume change, we must obtain the work done by a process of integration. The work done on the system while an external pressure P moves the piston so that the volume of the gas changes by an infinitesimal volume dV is

$$dw_{rev} = -P\,dV \tag{2.10}$$

If, as illustrated in Figure 2.3, the volume changes from a value V_1 to a value V_2, the **reversible work done on the system** is

$$w_{rev} = -\int_{V_1}^{V_2} P\,dV \tag{2.11}$$

In the example shown in Figure 2.3, $V_1 > V_2$ (i.e., we have compressed the gas) and this work is positive. Only if P is constant is it permissible to integrate this directly to give

$$w_{\text{rev}} = -P \int_{V_1}^{V_2} dV = -P(V_2 - V_1) = -P\Delta V \tag{2.12}$$

(compare Eq. 2.9). If P is not constant, we must express it as a function of V before performing the integration.

Follow reversible and irreversible paths of an ideal gas PV change; Ideal gas work; Energy, heat, and work.

We have already noted that work done is not a state function, and this may be further stressed with reference to the mechanical work of expansion. The previous derivation has shown that the work is related to the *process* carried out rather than to the initial and final states. We can consider the *reversible expansion* of a gas from volume V_1 to volume V_2, and can also consider an *irreversible* process, in which case *less* work will be done *by* the system. This is illustrated in Figure 2.4. The diagram to the left shows the *expansion* of a gas, in which the pressure is falling as the volume increases. The reversible work done *by* the system is given by the integral

$$-w_{\text{rev}} = \int_{V_1}^{V_2} P\, dV \tag{2.13}$$

which is represented by the shaded area in Figure 2.4a. Suppose instead that we performed the process irreversibly, by instantaneously dropping the external pressure to the final pressure P_2. The work done by the system is now against this pressure P_2 throughout the whole expansion, and is given by

$$-w_{\text{irr}} = P_2(V_2 - V_1) \tag{2.14}$$

This work done by the system is represented by the shaded area in Figure 2.4b and is *less* than the reversible work. Thus, although in both processes the state of the system has changed from A to B, the work done is different.

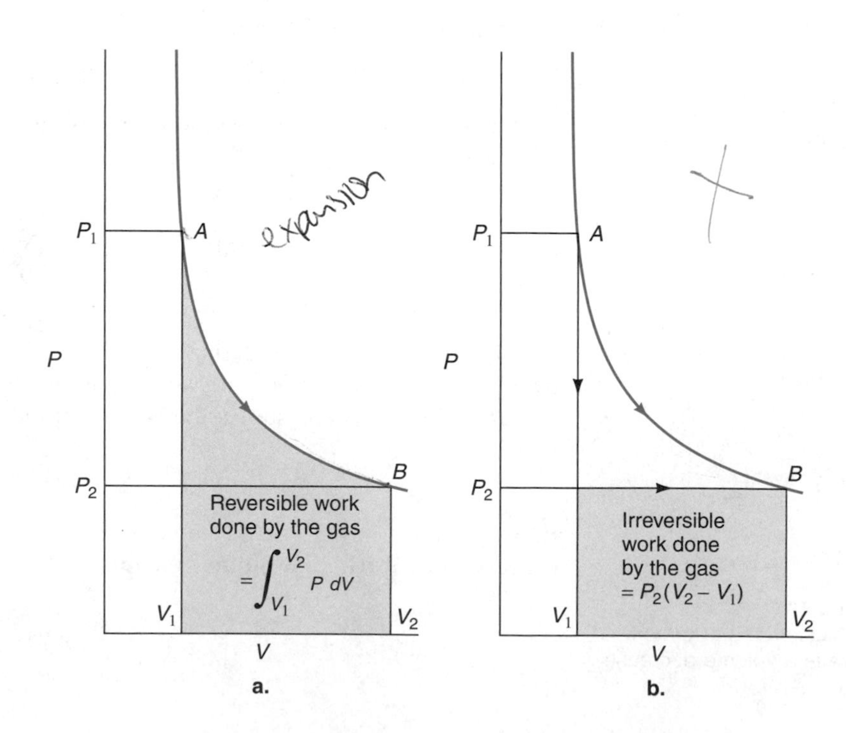

FIGURE 2.4
The left-hand diagram (a) illustrates the reversible work of expansion from V_1 to V_2. The right-hand diagram (b) shows the irreversible work that would be performed by the system if the external pressure were suddenly dropped to the final value P_2.

Maximum Work

This argument leads us to another important point. **The work done by the system in a *reversible* expansion from *A* to *B* represents the *maximum work that the system can perform* in changing from *A* to *B*.**

EXAMPLE 2.2 Suppose that water at its boiling point is maintained in a cylinder that has a frictionless piston. For equilibrium to be established, the pressure that must be applied to the piston is 1 atm (101.325 kPa). Suppose that we now reduce the external pressure by an infinitesimal amount in order to have a reversible expansion. If the piston sweeps out a volume of 2.00 dm^3, what is the work done by the system?

Solution The external pressure remains constant at 101.325 kPa, and, therefore, the reversible work done by the system is

$$-w_{rev} = P\Delta V = 101\ 325 \text{ Pa} \times 2.00 \text{ dm}^3 = 202.65 \text{ Pa m}^3$$

Since Pa = kg m^{-1} s^{-2} (see Appendix A), the units are kg m^2 s^{-2} = J; thus the work done by the system is 202.65 J.

For many purposes it is convenient to express the first law of thermodynamics with respect to an infinitesimal change. In place of Eq. 2.6, we have

$$dU = đq + đw \tag{2.15}$$

where again the symbol $đ$ denotes an inexact differential. However, if only PV work is involved and P is a constant, $đw$ may be written as $-P\,dV$, where dV is the infinitesimal increase in volume; thus,

$$dU = đq - P\,dV \tag{2.16}$$

Processes at Constant Volume

It follows from this equation that if an infinitesimal process occurs at constant volume, and only PV work is involved,

$$dU = dq_V \tag{2.17}$$

where the subscript V indicates that the heat is supplied at constant volume. (Note that under these circumstances dq_V is an exact differential, so that the d has lost its slash.) This equation integrates to

$$\Delta U = U_2 - U_1 = q_V \tag{2.18}$$

The increase of internal energy of a system in a rigid container (i.e., at constant volume) is thus equal to the heat q_V that is supplied to it.

Processes at Constant Pressure: Enthalpy

In most chemical systems we are concerned with processes occurring in open vessels, which means that they occur at constant pressure rather than at constant volume. The relationships valid for constant-pressure processes may readily be

See that the enthalpy is the heat at constant pressure; Heat of reaction; Thermochemistry.

deduced from Eq. 2.16. For an infinitesimal process at constant pressure the heat absorbed dq_P is given by

$$dq_P = dU + P\,dV \tag{2.19}$$

provided that no work other than PV work is performed. If the process involves a change from state 1 to state 2, this equation integrates as follows:

$$q_P = \int_{U_1}^{U_2} dU + \int_{V_1}^{V_2} P\,dV \tag{2.20}$$

Since P is constant,

$$q_P = \int_{U_1}^{U_2} dU + P\int_{V_1}^{V_2} dV \tag{2.21}$$

$$= (U_2 - U_1) + P(V_2 - V_1) = (U_2 + PV_2) - (U_1 + PV_1) \tag{2.22}$$

Definition of *Enthalpy*

This relationship suggests that it would be convenient to give a name to the quantity $U + PV$, and it is known as the **enthalpy,**[3] with the symbol H:

$$H \equiv U + PV \tag{2.23}$$

We thus have

$$q_P = H_2 - H_1 = \Delta H \tag{2.24}$$

This equation is valid only if the work is all PV work. Under these circumstances the increase in enthalpy ΔH of a system is equal to the heat q_P that is supplied to it at constant pressure. Since U, P, and V are all state functions, it follows from Eq. 2.23 that *enthalpy is also a state function.*

Endothermic Process

Exothermic Process

A chemical process occurring at constant pressure for which q_P and ΔH are *positive* is one in which a positive amount of heat is *absorbed* by the system. Such processes are known as **endothermic processes** (Greek *endo,* inside; *therme,* heat). Conversely, processes in which heat is *evolved* (q_P and ΔH are *negative*) are known as **exothermic processes** (Greek *exo,* outside).

Heat Capacity

The amount of heat required to raise the temperature of any substance by 1 K (which of course is the same as 1 °C) is known as its **heat capacity,** and is given the symbol C; its SI unit is J K^{-1}. The word *specific* before the name of any extensive physical quantity refers to the quantity per unit mass. The term *specific heat capacity* is thus the amount of heat required to raise the temperature of unit mass of a material by 1 K; if the unit mass is 1 kg, the unit is J K^{-1} kg^{-1}, which is the SI unit for specific heat capacity. The word *molar* before the name of a quantity refers to the quantity divided by the amount of substance. The SI unit for the molar heat capacity is J K^{-1} mol^{-1}.

Since heat is not a state function, neither is the heat capacity. It is therefore always necessary, when stating a heat capacity, to specify the process by which the temperature is raised by 1 K. Two heat capacities are of special importance:

[3]In the older scientific literature it is known as the **heat content,** but this term can be misleading.

Isochoric Process

1. The heat capacity related to a process occurring at *constant volume* (an **isochoric process**); this is denoted by C_V and is defined by

$$C_V \equiv \frac{dq_V}{dT} \tag{2.25}$$

where q_V is the heat supplied at constant volume. Since q_V is equal to ΔU, it follows that

Heat Capacity at Constant Volume

$$C_V = \left(\frac{\partial U}{\partial T}\right)_V \tag{2.26}$$

If we are working with 1 mol of the substance, this heat capacity is the *molar heat capacity at constant volume*, and is represented by the symbol $C_{V,m}$.[4]

Isobaric Process

2. The heat capacity related to a process occurring at *constant pressure* (an **isobaric process**) is C_P and is defined by

Heat Capacity at Constant Pressure

$$C_P \equiv \frac{dq_P}{dT} = \left(\frac{\partial H}{\partial T}\right)_P \tag{2.27}$$

The molar quantity is represented by the symbol $C_{P,m}$.

The heat required to raise the temperature of 1 mol of material from T_1 to T_2 at *constant volume* is

$$q_{V,m} = \int_{T_1}^{T_2} C_{V,m}\, dT \tag{2.28}$$

If $C_{V,m}$ is independent of temperature, this integrates to

$$q_{V,m} = C_{V,m}(T_2 - T_1) = \Delta U_m \tag{2.29}$$

Similarly, for a process at *constant pressure*

$$q_{P,m} = \int_{T_1}^{T_2} C_{P,m}\, dT \tag{2.30}$$

This integrates to

$$q_{P,m} = C_{P,m}(T_2 - T_1) = \Delta H_m \tag{2.31}$$

if C_P is independent of temperature. The expressions in Eqs. 2.29 and 2.31 represent ΔU_m and ΔH_m, respectively, per mole of material.

For liquids and solids, ΔU_m and ΔH_m are very close to one another. Consequently, $C_{V,m}$ and $C_{P,m}$ are essentially the same for solids and liquids. For gases, however, the $\Delta(PV)$ term is appreciable, and there is a significant difference between $C_{V,m}$ and $C_{P,m}$. For an ideal gas, which obeys the equation

$$PV = nRT \tag{2.32}$$

the relationship between C_V and C_P can be derived as follows. For 1 mol of gas

$$H_m = U_m + PV_m = U_m + RT \tag{2.33}$$

[4]The subscript m may be omitted when there is no danger of ambiguity.

and therefore

$$\frac{dH_m}{dT} = \frac{dU_m}{dT} + \frac{d(RT)}{dT} \tag{2.34}$$

Thus,

$$C_{P,m} = C_{V,m} + \frac{d(RT)}{dT} \tag{2.35}$$

or

$$C_{P,m} = C_{V,m} + R \tag{2.36}$$

The general relationship between $C_{P,m}$ and $C_{V,m}$ for a gas that is not necessarily ideal is obtained in Section 2.7; see Eq. 2.117.

2.5 Thermochemistry

We have seen that the heat supplied to a system at constant pressure is equal to the enthalpy increase. The study of enthalpy changes in chemical processes is known as *thermochemistry.*

Extent of Reaction

In dealing with enthalpy changes in chemical processes it is very convenient to make use of a quantity known as the **extent of reaction;** it is given the symbol ξ. This quantity was introduced in 1922 by the Belgian thermodynamicist T. de Donder, and IUPAC recommends that enthalpy changes be considered with reference to it. The extent of reaction *must be related to a specified stoichiometric equation for a reaction.* A chemical reaction can be written in general as

$$a\text{A} + b\text{B} + \cdots \rightarrow \cdots + y\text{Y} + z\text{Z}$$

It can also be written as

$$(-\nu_\text{A})\text{A} + (-\nu_\text{B})\text{B} + \cdots \rightarrow \cdots + \nu_\text{Y}\text{Y} + \nu_\text{Z}\text{Z}$$

Stoichiometric Coefficients

where ν_A, ν_B, ν_Y, and ν_Z are known as *stoichiometric coefficients.* By definition the stoichiometric coefficient is *positive for a product and negative for a reactant.*

The extent of reaction is then defined by

$$\xi \equiv \frac{n_i - n_{i,0}}{\nu_i} \tag{2.37}$$

where $n_{i,0}$ is the initial amount of the substance i and n_i is the amount at any time. What makes the extent of reaction so useful is that it is *the same for every reactant and product.* Thus the extent of reaction is the amount of any product formed divided by its stoichiometric coefficient:

$$\xi = \frac{\Delta n_\text{Y}}{\nu_\text{Y}} = \frac{\Delta n_\text{Z}}{\nu_\text{Z}} \tag{2.38}$$

It is also the change in the amount of any reactant (a negative quantity), divided by its stoichiometric coefficient (also a negative quantity):

$$\xi = \frac{\Delta n_A}{\nu_A} = \frac{\Delta n_B}{\nu_B} \tag{2.39}$$

These quantities are all equal.

EXAMPLE 2.3 When 10 mol of nitrogen and 20 mol of hydrogen are passed through a catalytic converter, after a certain time 5 mol of ammonia are produced. Calculate the amounts of nitrogen and hydrogen that remain unreacted. Calculate also the extent of reaction:

a. on the basis of the stoichiometric equation

$$N_2 + 3H_2 \rightarrow 2NH_3$$

b. on the basis of the stoichiometric equation

$$\tfrac{1}{2}N_2 + \tfrac{3}{2}H_2 \rightarrow NH_3$$

Solution The amounts are

	N_2	H_2	NH_3	
Initially	10	20	0	mol
Finally	7.5	12.5	5	mol

a. The extent of reaction is the amount of ammonia formed, 5 mol, divided by the stoichiometric coefficient for NH_3:

$$\xi = \tfrac{5}{2} = 2.5 \text{ mol}$$

The same answer is obtained if we divide the amounts of N_2 and H_2 consumed, 2.5 mol and 7.5 mol, by the respective stoichiometric coefficients:

$$\xi = \tfrac{2.5}{1} = \tfrac{7.5}{3} = 2.5 \text{ mol}$$

b. The extent of reaction is now doubled, since the stoichiometric coefficients are halved:

$$\xi = \underset{N_2}{\tfrac{2.5}{0.5}} = \underset{H_2}{\tfrac{7.5}{1.5}} = \underset{NH_3}{\tfrac{5}{1}} = 5.0 \text{ mol}$$

The SI unit for the extent of reaction is the mole, and the mole referred to relates to the stoichiometric equation. For example, if the equation is specified to be

$$N_2 + 3H_2 \rightarrow 2NH_3$$

the mole relates to N_2, $3H_2$, or $2NH_3$. If, therefore, the ΔH for this reaction is stated to be -46.0 kJ mol^{-1}, it is to be understood that this value refers to the removal of 1 mol of N_2 and 1 mol of $3H_2$, which is the same as 3 mol of H_2. It also refers to the formation of 1 mol of $2NH_3$, which is the same as 2 mol of NH_3. In other words, the ΔH value relates to the reaction *as written in the stoichiometric equation.*

This recommended IUPAC procedure, which we shall use throughout this book, avoids the necessity of saying, for example, -46.0 kJ per mol of nitrogen, or

−23.0 kJ per mol of ammonia. It cannot be emphasized too strongly that when this IUPAC procedure is used, the *stoichiometric equation must be specified.*

Standard States

Enthalpy is a state function, and the enthalpy change that occurs in a chemical process depends on the states of the reactants and products. Consider, for example, the complete combustion of ethanol, in which 1 mol is oxidized to carbon dioxide and water:

$$C_2H_5OH + 3O_2 \rightarrow 2CO_2 + 3H_2O$$

The enthalpy change in this reaction depends on whether we start with liquid ethanol or with ethanol in the vapor phase. It also depends on whether liquid or gaseous water is produced in the reaction. Another factor is the pressure of the reactants and products. Also, the enthalpy change in a reaction varies with the temperature at which the process occurs. In giving a value for an enthalpy change it is therefore necessary to specify (1) the state of matter of the reactants and products (gaseous, liquid, or solid; if the last, the allotropic form), (2) the pressure, and (3) the temperature. If the reaction occurs in solution, the concentrations must also be specified.

It has proved convenient in thermodynamic work to define certain **standard states** and to quote data for reactions involving these standard states. By general agreement the standard state of a substance is the form in which it is most stable at 25.00 °C (298.15 K) and 1 bar (10^5 Pa) pressure. For example, the standard state of oxygen is the gas, and we specify this by writing $O_2(g)$. Since mercury, water, and ethanol are liquids at 25 °C, their standard states are Hg(l), $H_2O(l)$, and $C_2H_5OH(l)$. The standard state of carbon is graphite. These standard states should be specified if there is any ambiguity; for example,

$$C_2H_5OH(l) + 3O_2(g, 1\ \text{bar}) \rightarrow 2CO_2(g, 1\ \text{bar}) + 3H_2O(l).$$

It is quite legitimate, of course, to consider an enthalpy change for a process not involving standard states; for example,

$$C_2H_5OH(g, 1\ \text{atm}) + 3O_2(g, 1\ \text{atm}) \rightarrow 2CO_2(g, 1\ \text{atm}) + 3H_2O(g, 1\ \text{atm})$$

If a reaction involves species in solution, their standard state is 1 mol kg^{-1} (1 *m*)[5]; for example,

$$H^+(1\ m) + OH^-(1\ m) \rightarrow H_2O(l)$$

Enthalpy changes depend somewhat on the temperature at which the process occurs. Standard thermodynamic data are commonly quoted for a temperature of 25.00 °C (298.15 K), and this can be given as a subscript or in parentheses; thus

$$C_2H_5OH(l) + 3O_2(g) \rightarrow 2CO_2(g) + 3H_2O(l)$$

$$\Delta_c H°(298\ \text{K}) = 1357.7\ \text{kJ mol}^{-1}$$

The superscript ° on the $\Delta H°$ specifies that we are dealing with standard states, so that a pressure of 1 bar is assumed and need not be stated. The subscript c on the Δ

[5]A solution having 1 mol of solute in 1 kg of solvent is known as a 1-molal (1 *m*) solution: The **molality** of a solution is the *amount of substance per kilogram of solvent.*

is commonly used to indicate complete combustion, and the modern practice is to attach such subscripts to the Δ and not to the H. As emphasized in our discussion of extent of reaction, the value 1357.7 kJ mol^{-1} relates to the combustion of 1 mol of ethanol, since that is what appears in the equation.

Standard thermodynamic values can be given for a temperature other than 25 °C; for example, we could give a value for $\Delta H°$ (100 °C), and it would be understood that the pressure was again 1 bar and that reactants and products were in their standard states but at 100 °C.

Measurement of Enthalpy Changes

Take various paths from reactants to products to see how Hess's Law works; Hess's Law; Thermochemistry.

The enthalpy changes occurring in chemical processes may be measured by three main methods:

1. *Direct Calorimetry.* Some reactions occur to completion and without side reactions, and it is therefore possible to measure their $\Delta H°$ values by causing the reactions to occur in a calorimeter. The neutralization of an aqueous solution of a strong acid by a solution of a strong base is an example of such a process, the reaction that occurs being

$$H^+(aq) + OH^-(aq) \rightarrow H_2O(l)$$

Combustion processes also frequently occur to completion with simple stoichiometry. When an organic compound is burnt in excess of oxygen, the carbon is practically all converted into CO_2 and the hydrogen into H_2O, while the nitrogen is usually present as N_2 in the final products. Often such combustions of organic compounds occur cleanly, and much thermochemical information has been obtained by burning organic compounds in calorimeters.

Hess's Law

2. *Indirect Calorimetry. Use of Hess's Law.* Few reactions occur in a simple manner, following a simple chemical equation, with the result that the enthalpy changes corresponding to a simple chemical equation often cannot be measured directly. For many of these, the enthalpy changes can be calculated from the values for other reactions by making use of **Hess's law,** named after Germain Henri Hess (1802–1850). According to this law, it is permissible to write stoichiometric equations, together with the enthalpy changes, and to treat them as mathematical equations, thereby obtaining a thermochemically valid result. For example, suppose that a substance A reacts with B according to the equation

   1. $A + B \rightarrow X \qquad \Delta H_1 = -10 \text{ kJ mol}^{-1}$

   Suppose that X reacts with an additional molecule of A to give another product Y:

   2. $A + X \rightarrow Y \qquad \Delta H_2 = -20 \text{ kJ mol}^{-1}$

   According to Hess's law, it is permissible to add these two equations and obtain:

   3. $2A + B \rightarrow Y \qquad \Delta H_3 = \Delta H_1 + \Delta H_2 = -30 \text{ kJ mol}^{-1}$

The law follows at once from the principle of conservation of energy and from the fact that enthalpy is a state function. Thus, if reactions 1 and 2 occur, there is a net *evolution* of 30 kJ when 1 mol of Y is produced. In principle we could reconvert Y into 2A + B by the reverse of reaction 3. If the heat required to do this were different from 30 kJ, we should have obtained the starting materials with a net gain or loss of heat, and this would violate the principle of conservation of energy.

EXAMPLE 2.4 The enthalpy changes in the complete combustion of crystalline α-D-glucose and maltose at 298 K, with the formation of gaseous CO_2 and liquid H_2O, are:

	$\Delta_c H°$/kJ mol^{-1}
α-D-Glucose, $C_6H_{12}O_6$(c)	−2809.1
Maltose, $C_{12}H_{22}O_{11}$(c)	−5645.5

Calculate the enthalpy change accompanying the conversion of 1 mol of crystalline glucose into crystalline maltose.

Solution The enthalpy changes given relate to the processes

1. $C_6H_{12}O_6(c) + 6O_2(g) \rightarrow 6CO_2(g) + 6H_2O(l)$
$$\Delta_c H° = -2809.1 \text{ kJ mol}^{-1}$$

2. $C_{12}H_{22}O_{11}(c) + 12O_2(g) \rightarrow 12CO_2(g) + 11H_2O(l)$
$$\Delta_c H° = -5645.5 \text{ kJ mol}^{-1}$$

We are asked to convert 1 mol of glucose into maltose; the reaction is

$$C_6H_{12}O_6(c) \rightarrow \tfrac{1}{2}C_{12}H_{22}O_{11}(c) + \tfrac{1}{2}H_2O(l)$$

Reaction 2 can be written in reverse and divided by 2:

2′. $6CO_2(g) + \tfrac{11}{2}H_2O(l) \rightarrow \tfrac{1}{2}C_{12}H_{22}O_{11}(c) + 6O_2(g)$

$$\Delta H° = \frac{5645.5}{2} = 2822.8 \text{ kJ mol}^{-1}$$

If we add reactions **1** and **2′**, we obtain the required equation, with

$$\Delta H° = -2809.1 + 2822.8 = 13.7 \text{ kJ mol}^{-1}$$

3. *Variation of Equilibrium Constant with Temperature.* A third general method of measuring $\Delta H°$ will be mentioned here only very briefly, since it is based on the second law of thermodynamics and is considered in Section 4.8. This method is based on the equation for the variation of the equilibrium constant K with the temperature:

$$\frac{d \ln K}{d(1/T)} = -\frac{\Delta H°}{R} = -\frac{\Delta H°/\text{J mol}^{-1}}{8.3145} \tag{2.40}$$

If, therefore, we measure K at a series of temperatures and plot ln K against $1/T$, the slope of the line at any temperature will be $-\Delta H°/8.3145$ J mol^{-1}, and hence $\Delta H°$ can be calculated. Whenever an equilibrium constant for a reaction can be measured satisfactorily at various temperatures, this method thus provides a very useful way of obtaining $\Delta H°$. The method cannot be used for reactions that go essentially to completion, in which case a reliable K cannot be obtained, or for reactions that are complicated by side reactions.

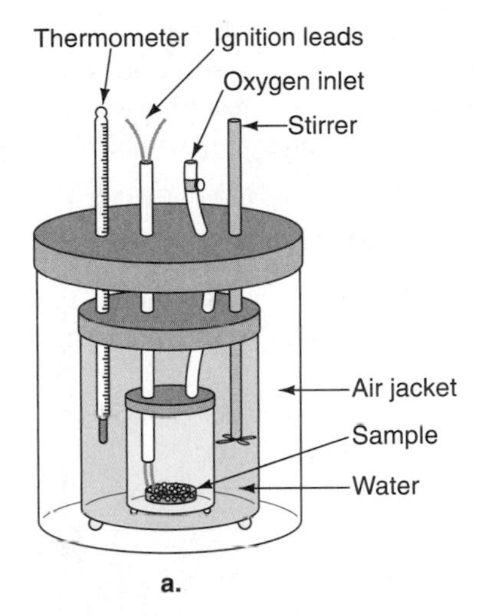

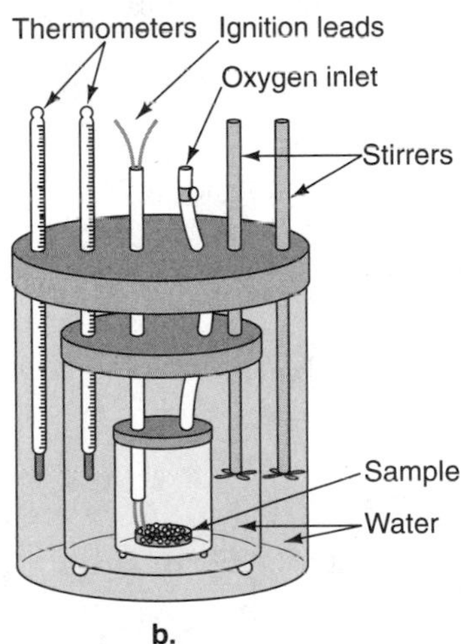

FIGURE 2.5
Schematic diagrams of two types of bomb calorimeter: (a) a conventional calorimeter, (b) an adiabatic calorimeter.

Calorimetry

The heats evolved in combustion processes are determined in *bomb calorimeters,* two types of which are shown in Figure 2.5. A weighed sample of the material to be burnt is placed in the cup supported in the reaction vessel, or bomb, which is designed to withstand high pressures. The heavy-walled steel bomb of about 400 mL volume is filled with oxygen at a pressure of perhaps 25 atm, this being more than enough to cause complete combustion. The reaction is initiated by passing an electric current through the ignition wire. Heat is evolved rapidly in the combustion process and is determined in two different ways in the two types of calorimeter.

In the type of calorimeter shown in Figure 2.5a, the bomb is surrounded by a water jacket that is insulated as much as possible from the surroundings. The water in the jacket is stirred, and the rise in temperature brought about by the combustion is measured. From the thermal characteristics of the apparatus the heat evolved can be calculated. A correction is made for the heat produced in the ignition wire, and it is customary to calibrate the apparatus by burning a sample having a known heat of combustion.

Adiabatic Calorimeter; Microcalorimeter

The type of calorimeter illustrated in Figure 2.5b is known as an **adiabatic calorimeter.** The word *adiabatic* comes from the Greek word *adiabatos,* meaning impassable, which in turn is derived from the Greek prefix *a-*, not, and the words *dia,* through, and *bainen,* to go. An adiabatic process is thus one in which there is no flow of heat. In the adiabatic calorimeter this is achieved by surrounding the inner water jacket by an outer water jacket which by means of a heating coil is maintained at the same temperature as the inner jacket. When this is done the amount of heat supplied to the outer jacket just cancels the heat loss to the surroundings. This allows a simpler determination of the temperature rise due to the combustion; the measured ΔT ($T_{\text{final}} - T_{\text{initial}}$) is directly related to the amount of heat evolved in the combustion.

By the use of calorimeters of these types, heats of combustion can be measured with an accuracy of better than 0.01%. Such high precision is necessary, since heats evolved in combustion are large, and sometimes we are more interested in the differences between the values for two compounds than in the absolute values.

Many other experimental techniques have been developed for the measurement of heats of reactions. Sometimes the heat changes occurring in chemical reactions are exceedingly small, and it is then necessary to employ very sensitive calorimeters. Such instruments are known as **microcalorimeters**. The prefix *micro* refers to the amount of heat and not to the physical dimensions of the instrument; some microcalorimeters are extremely large.

Another type of microcalorimeter is the *continuous flow calorimeter,* which permits two reactant solutions to be thermally equilibrated during passage through separate platinum tubes and then brought together in a mixing chamber; the heat change in the reaction is measured.

Relationship between ΔU and ΔH

Calibrate a calorimeter and measure the heat of a sample; Calorimetry; Thermochemistry.

Bomb calorimeters and other calorimeters in which the volume is constant give the internal energy change ΔU. Other calorimeters operate at constant pressure and therefore give ΔH values. Whether ΔU or ΔH is determined, the other quantity is easily calculated from the stoichiometric equation for the reaction. From Eq. 2.23 we see that ΔH and ΔU are related by

$$\Delta H = \Delta U + \Delta(PV) \tag{2.41}$$

If all reactants and products are solids or liquids, the change in volume if a reaction occurs at constant pressure is quite small. Usually 1 mol of a solid or liquid has a volume of less than 1 dm^3, and the volume change in a reaction will always be less than 1% (i.e., less than 0.01 dm^3). At 1 bar pressure, with $\Delta V =$ 0.01 dm^3,

$$\Delta(PV) = 100\,000 \text{ Pa} \times 10^{-5} \text{ m}^3 \text{ mol}^{-1} = 1.000 \text{ J mol}^{-1}$$

This is quite negligible compared with most heats of reaction, which are of the order of kilojoules, and is much less than the experimental error of most determinations.

If gases are involved in the reaction, however, either as reactants or products, ΔU and ΔH may differ significantly, as illustrated in the following example.

EXAMPLE 2.5 For the complete combustion of ethanol,

$$C_2H_5OH(l) + 3O_2(g) \rightarrow 2CO_2(g) + 3H_2O(l)$$

the amount of heat produced, as measured in a bomb calorimeter, is 1364.47 kJ mol^{-1} at 25 °C. Calculate $\Delta_c H$ for the reaction.

Solution Since the bomb calorimeter operates at constant volume, $\Delta_c U =$ -1364.47 kJ mol^{-1}. The product of reaction contains 2 mol of gas and the reactants, 3 mol; the change Δn is therefore -1 mol. Assuming the ideal gas law to apply, $\Delta(PV)$ is equal to ΔnRT, and therefore to

$$(-1)RT = -8.3145 \times 298.15 \text{ J mol}^{-1} = -2.48 \text{ kJ mol}^{-1}$$

Therefore,

$$\Delta_c H = -1364.47 + (-2.48) = -1366.95 \text{ kJ mol}^{-1}$$

The difference between ΔU and ΔH is now large enough to be experimentally significant.

Temperature Dependence of Enthalpies of Reaction

Enthalpy changes are commonly tabulated at 25 °C, and it is frequently necessary to have the values at other temperatures. These can be calculated from the heat capacities of the reactants and products. The enthalpy change in a reaction can be written as

$$\Delta H = H\,(\text{products}) - H\,(\text{reactants}) \tag{2.42}$$

Partial differentiation with respect to temperature at constant pressure gives[6]

$$\left(\frac{\partial \Delta H}{\partial T}\right)_P = \left[\left(\frac{\partial H\,(\text{products})}{\partial T}\right)_P - \left(\frac{\partial H(\text{reactants})}{\partial T}\right)_P\right] \tag{2.43}$$

$$= C_P\,(\text{products}) - C_P\,(\text{reactants}) = \Delta C_P \tag{2.44}$$

[6]These relationships (Eqs. 2.43, 2.44) were first deduced in 1858 by the German physicist Gustav Robert Kirchhoff (1824–1887).

Similarly,

$$\left(\frac{\partial \Delta U}{\partial T}\right)_V = \Delta C_V \tag{2.45}$$

For small changes in temperature the heat capacities, and hence ΔC_P and ΔC_V, may be taken as constant. In that case Eq. 2.44 can be integrated between two temperatures T_1 and T_2 to give

$$\Delta(\Delta H) = \Delta H_2 - \Delta H_1 = \Delta C_P(T_2 - T_1) \tag{2.46}$$

If there is a large difference between the temperatures T_1 and T_2, this procedure is not satisfactory, and it is necessary to take into account the variation of C_P with temperature. This is often done by expressing the molar value $C_{P,m}$ as a power series:

$$C_{P,m} = a + bT + cT^2 + \cdots \tag{2.47}$$

To a good approximation the values can be calculated over a wide temperature range by using only the first three terms of this expansion. For hydrogen, for example, the $C_{P,m}$ values are fitted to within 0.5% over the temperature range from 273 K to 1500 K if the following constants are used:

$$a = 29.07 \text{ J K}^{-1} \text{ mol}^{-1} \qquad b = -0.836 \times 10^{-3} \text{ J K}^{-2} \text{ mol}^{-1}$$
$$c = 20.1 \times 10^{-7} \text{ J K}^{-3} \text{ mol}^{-1}$$

These values lead to

$$C_{P,m} = 28.99 \text{ J K}^{-1} \text{ mol}^{-1} \text{ at 273 K} \quad \text{and} \quad C_{P,m} = 32.34 \text{ J K}^{-1} \text{ mol}^{-1} \text{ at 1500 K}$$

Alternatively, and somewhat more satisfactorily, we can use an equation of the form

$$C_{P,m} = d + eT + fT^{-2} \tag{2.48}$$

Some values of d, e, and f are given in Table 2.1.

If each of the $C_{P,m}$ values for products and reactants is written in the form of Eq. 2.48, the $\Delta C_{P,m}$ for the reaction will have the same form:

$$\Delta C_{P,m} = \Delta d + \Delta eT + \Delta fT^{-2} \tag{2.49}$$

TABLE 2.1 Parameters for the Equation $C_{P,m} = d + eT + fT^{-2}$

Substance	State	d / J K^{-1} mol^{-1}	e / J K^{-2} mol^{-1}	f / J K mol^{-1}
He, Ne, Ar, Kr, Xe	Gas	20.79	0	0
H_2	Gas	27.28	3.26×10^{-3}	5.0×10^4
O_2	Gas	29.96	4.18×10^{-3}	-1.67×10^5
N_2	Gas	28.58	3.76×10^{-3}	-5.0×10^4
CO	Gas	28.41	4.10×10^{-3}	-4.6×10^4
CO_2	Gas	44.22	8.79×10^{-3}	-8.62×10^5
H_2O	Vapor	30.54	10.29×10^{-3}	0
H_2O	Liquid	75.48	0	0
C (graphite)	Solid	16.86	4.77×10^{-3}	-8.54×10^5
NaCl	Solid	45.94	16.32×10^{-3}	0

Integration of Eq. 2.44 between the limits T_1 and T_2 leads to

$$\Delta H_m(T_2) - \Delta H_m(T_1) = \int_{T_1}^{T_2} \Delta C_P \, dT \tag{2.50}$$

If $\Delta H_m(T_1)$ is known for $T_1 = 25$ °C, the Δ value of $H_m(T_2)$ at any temperature T_2 is thus given by this equation, and substitution of Eq. 2.49 leads to

$$\Delta H_m(T_2) = \Delta H_m(T_1) + \int_{T_1}^{T_2} (\Delta d + \Delta e T + \Delta f T^{-2}) \, dT \tag{2.51}$$

$$= \Delta H_m(T_1) + \Delta d(T_2 - T_1) + \frac{1}{2}\Delta e(T_2^2 - T_1^2) - \Delta f\left(\frac{1}{T_2} - \frac{1}{T_1}\right) \tag{2.52}$$

EXAMPLE 2.6 Consider the gas-phase reaction

$$2CO(g) + O_2(g) \rightarrow 2CO_2(g)$$

A bomb-calorimetric study of this reaction at 25 °C leads to $\Delta H° = -565.98$ kJ mol^{-1}. Calculate $\Delta H°$ for this reaction at 2000 K.

Solution From the values in Table 2.1 we obtain

$$\Delta d = d(\text{products}) - d(\text{reactants})$$
$$= (2 \times 44.22) - (2 \times 28.41) - 29.96 = 1.66 \text{ J K}^{-1} \text{ mol}^{-1}$$

$$\Delta e = e(\text{products}) - e(\text{reactants})$$
$$= (2 \times 8.79 \times 10^{-3}) - (2 \times 4.10 \times 10^{-3}) - 4.18 \times 10^{-3}$$
$$= 5.29 \times 10^{-3} \text{ J K}^{-2} \text{ mol}^{-1}$$

$$\Delta f = f(\text{products}) - f(\text{reactants})$$
$$= [2 \times (-8.62 \times 10^5)] + (2 \times 0.46 \times 10^5) + 1.67 \times 10^5$$
$$= -14.65 \times 10^5 \text{ J K mol}^{-1}$$

Then, from Eq. 2.52,

$$\Delta H°(2000 \text{ K})/\text{J mol}^{-1} = -565\,980 + 1.66(2000 - 298)$$
$$+ \left(\frac{1}{2}\right)5.20 \times 10^{-3}(2000^2 - 298^2)$$
$$+ 14.65 \times 10^5\left(\frac{1}{2000} - \frac{1}{298}\right)$$
$$= -565\,980 + 2825 + 10169 - 4183$$
$$\Delta H°(2000 \text{ K}) = -557\,169 \text{ J mol}^{-1} = -557.17 \text{ kJ mol}^{-1}$$

Note that when numerical values are given, it is permissible to drop the subscript m from $\Delta H°$, since the unit kJ mol^{-1} avoids ambiguity. Remember that the mole referred to always relates to the reaction as written.

Enthalpies of Formation

The total number of known chemical reactions is enormous, and it would be very inconvenient if one had to tabulate enthalpies of reaction for all of them. We can avoid having to do this by tabulating *molar enthalpies of formation* of chemical compounds, which are the enthalpy changes associated with the formation of 1 mol of the substance from the elements in their standard states. From these enthalpies of formation it is possible to calculate enthalpy changes in chemical reactions.

We have seen that the standard state of each element and compound is taken to be the most stable form in which it occurs at 1 bar pressure and at 25 °C. Suppose that we form methane, at 1 bar and 25 °C, from C(graphite) and $H_2(g)$, which are the standard states; the stoichiometric equation is

$$\text{C(graphite)} + 2\text{H}_2(\text{g}) \rightarrow \text{CH}_4(\text{g})$$

It does not matter that we cannot make this reaction occur cleanly and, therefore, that we cannot directly measure its enthalpy change; as seen previously, indirect methods can be used. In such ways it is found that $\Delta H°$ for this reaction is -74.81 kJ mol^{-1}, and this quantity is known as the *standard molar enthalpy of formation* $\Delta_f H°$ of methane at 25 °C (298.15 K). The term **standard enthalpy of formation** refers to the enthalpy change when the compound in its standard state is formed from the elements in their standard states; it must not be used in any other sense. Obviously, the standard enthalpy of formation of any *element* in its standard state is zero.

Standard Enthalpy of Formation

Enthalpies of formation of organic compounds are commonly obtained from their enthalpies of combustion, by application of Hess's law. When, for example, 1 mol of methane is burned in an excess of oxygen, 802.37 kJ of heat is evolved, and we can therefore write

1. $\text{CH}_4(\text{g}) + 2\text{O}_2(\text{g}) \rightarrow \text{CO}_2(\text{g}) + 2\text{H}_2\text{O}(\text{g}) \qquad \Delta_c H° = -802.37 \text{ kJ mol}^{-1}$

In addition, we have the following data:

2. $\text{C(graphite)} + \text{O}_2(\text{g}) \rightarrow \text{CO}_2(\text{g}) \qquad \Delta H° = -393.50 \text{ kJ mol}^{-1}$
3. $2\text{H}_2(\text{g}) + \text{O}_2(\text{g}) \rightarrow 2\text{H}_2\text{O}(\text{g}) \qquad \Delta H° = 2(-241.83)\text{kJ mol}^{-1}$

If we add reactions 2 and 3 and subtract reaction 1, the result is

$$\text{C(graphite)} + 2\text{H}_2(\text{g}) \rightarrow \text{CH}_4(\text{g})$$

$$\Delta_f H°(\text{CH}_4) = 2(-241.84) - 393.50 - (-802.37) = -74.80 \text{ kJ mol}^{-1}$$

Enthalpies of formation of many other compounds can be deduced in a similar way.

Appendix D gives some enthalpies of formation.[7] The values, of course, depend on the state in which the substance occurs, and this is indicated in the table; the value for liquid ethanol, for example, is a little different from that for ethanol in aqueous solution.

[7]The table in Appendix D also includes, for convenience, values of Gibbs energies of formation; these are considered in Chapter 3.

Included in Appendix D are enthalpies of formation of individual ions. There is an arbitrariness about these values, because thermodynamic quantities can never be determined experimentally for individual ions; it is always necessary to work with assemblies of positive and negative ions having a net charge of zero. For example, $\Delta_f H°$ for HCl in aqueous solution is -167.15 kJ mol^{-1}, but there is no way that one can make experimental determinations on the individual H^+ and Cl^- ions. The convention is to take $\Delta_f H°$ for the aqueous ion H^+ in its standard state (1 mol kg^{-1}) to be zero; it then follows that, on this basis, the value of $\Delta_f H°$ for the Cl^- ion is -167.15 kJ mol^{-1}. Then, since the $\Delta_f H°$ value for the NaCl in aqueous solution is -407.27 kJ mol^{-1}, we have

$$\Delta_f H°(Na^+) = \Delta H°(NaCl, aq) - \Delta_f H°(Cl^-) = -407.27 + 167.15$$
$$= -240.12 \text{ kJ mol}^{-1}$$

Conventional Standard Enthalpies of Formation

In this way a whole set of values can be built up. Such values are often known as *conventional standard enthalpies of formation;* the word *conventional* refers to the value of zero for the aqueous proton. In spite of the arbitrariness of the procedure, correct values are always obtained when one uses these conventional values in making calculations for reactions; this follows from the fact that there is always a balancing of charges in a chemical reaction.

Enthalpies of formation allow us to calculate enthalpies of any reaction, provided that we know the $\Delta_f H°$ values for all the reactants and products. The $\Delta H°$ for any reaction is the difference between the sum of the $\Delta_f H°$ values for all the products and the sum of the $\Delta_f H°$ values for all the reactants:

$$\Delta H° = \sum \Delta_f H° \text{ (products)} - \sum \Delta_f H° \text{ (reactants)} \qquad (2.53)$$

EXAMPLE 2.7 Calculate, from the data in Appendix D, $\Delta H°$ for the hydrolysis of urea to give carbon dioxide and ammonia in aqueous solution:

$$H_2NCONH_2(aq) + H_2O(l) \rightarrow CO_2(aq) + 2NH_3(aq)$$

Solution From the data in Appendix D we have

1. $C(\text{graphite}) + 2H_2(g) + \frac{1}{2}O_2(g) + N_2(g) \rightarrow H_2NCONH_2(aq)$ $\quad \Delta_f H° = -317.77$ kJ mol^{-1}

2. $H_2(g) + \frac{1}{2}O_2(g) \rightarrow H_2O(l)$ $\quad \Delta_f H° = -285.85$ kJ mol^{-1}

3. $C(\text{graphite}) + O_2(g) \rightarrow CO_2(aq)$ $\quad \Delta_f H° = -413.80$ kJ mol^{-1}

4. $\frac{1}{2}N_2(g) + \frac{3}{2}H_2(g) \rightarrow NH_3(aq)$ $\quad \Delta_f H° = -80.71$ kJ mol^{-1}

4′. $N_2(g) + 3H_2(g) \rightarrow 2NH_3(aq)$ $\quad \Delta_f H° = 2(-80.71)$ kJ mol^{-1}

Subtraction of reactions 1 + 2 from reactions 3 + 4′ then leads to the desired equation, and the enthalpy change in the reaction is thus

$$\Delta H° = -413.80 + 2 \times (-80.71) + 285.85 + 317.77 = 28.32 \text{ kJ mol}^{-1}$$

It is not necessary, of course, to write out the reactions; Eq. 2.53 may be used directly.

Bond Enthalpies

One important aspect of thermochemistry relates to the enthalpies of different chemical bonds. As a very simple example, consider the case of methane, CH_4. The standard enthalpy of formation of methane is -74.81 kJ mol^{-1}:

See the difference between bond energy and bond dissociation energy for CH_4; Bond enthalpies; Thermochemistry.

1. C(graphite) + $2H_2$(g) → CH_4(g) $\quad \Delta_f H° = -74.81$ kJ mol^{-1}

We also know the following thermochemical values:

2. C(graphite) → C(gaseous atoms) $\quad \Delta H° = 716.7$ kJ mol^{-1}
3. $\frac{1}{2}H_2$(g) → H(gaseous atoms) $\quad \Delta H° = 218.0$ kJ mol^{-1}

The former is the enthalpy of sublimation of graphite, and the latter is one-half the enthalpy of dissociation of hydrogen. We may now apply Hess's law in the following manner:

1′. CH_4(g) → C(graphite) + $2H_2$(g) $\quad \Delta H° = 74.8$ kJ mol^{-1}
2. C(graphite) → C(gaseous atoms) $\quad \Delta H° = 716.7$ kJ mol^{-1}
4. (4 × 3.) $2H_2$(g) → 4H(gaseous atoms) $\quad \Delta H° = 872.0$ kJ mol^{-1}

If these three equations are added:

$$CH_4(g) \rightarrow C + 4H(\text{gaseous atoms}) \qquad \Delta H° = 1663.5 \text{ kJ mol}^{-1}$$

This quantity, 1663.5 kJ mol^{-1}, is known as the *enthalpy of atomization* of methane; it is the heat that has to be supplied to 1 mol of methane at constant pressure in order to dissociate all the molecules into gaseous atoms. Since each CH_4 molecule has four C—H bonds, we can divide 1663.5 by 4, obtaining 415.9 kJ mol^{-1}, and we call this the C—H **bond enthalpy;** it is an average quantity, and is commonly called the *bond strength.*

A similar procedure with ethane, C_2H_6, leads to an enthalpy of atomization of 2829.2 kJ mol^{-1}. This molecule contains one C—C bond and six C—H bonds. If we subtract 6 × 415.9 = 2495.4 kJ mol^{-1} as the contribution of the C—H bonds, we are left with 333.8 kJ mol^{-1} as the C—C bond enthalpy.

However, if we calculate the enthalpies of atomization of the higher paraffin hydrocarbons using these values, we find that the agreement with experiment is by no means perfect. In other words, there is not a strict *additivity* of bond enthalpies. The reason for this is that chemical bonds in a given molecule have different environments. On the whole, enthalpies of atomization are more satisfactorily predicted if we use the following bond enthalpies rather than the ones deduced from the data for CH_4 and C_2H_6:

C—H 413 kJ mol^{-1}

C—C 348 kJ mol^{-1}

TABLE 2.2 Bond Enthalpies

Bond	Bond Enthalpy/kJ mol^{-1}
H—H	436
C—C	348
C—H	413
C=C	682
C≡C	962
N—H	391
O—H	463
C—O	351
C=O	732

These values are for gaseous molecules; adjustments must be made for substances in condensed phases.

By the use of similar procedures for molecules containing different kinds of bonds, it is possible to arrive at a set of bond enthalpies that will allow us to make approximate estimates of enthalpies of atomization and enthalpies of formation. Such a set is shown in Table 2.2. Values of this kind have proved very useful in deducing approximate thermochemical information when the experimental enthalpies of formation are not available. These simple additive procedures can be improved in various ways.

Bond Dissociation Enthalpies

It is important to distinguish clearly between bond enthalpies, the *additive* quantities we have just considered, and **bond dissociation enthalpies.** The distinction can be illustrated by the case of methane. We can consider the successive removal of hydrogen atoms from methane in the gas phase:

1. $CH_4 \rightarrow CH_3 + H \qquad \Delta H_1^\circ = 431.8 \text{ kJ mol}^{-1}$
2. $CH_3 \rightarrow CH_2 + H \qquad \Delta H_2^\circ = 471.1 \text{ kJ mol}^{-1}$
3. $CH_2 \rightarrow CH + H \qquad \Delta H_3^\circ = 421.7 \text{ kJ mol}^{-1}$
4. $CH \rightarrow C + H \qquad \Delta H_4^\circ = 338.8 \text{ kJ mol}^{-1}$

The first and last of these bond dissociation enthalpies are known with some certainty; the values for the second and third reactions are less reliable. What is certain is that the sum of the four values must be 1663.5 kJ mol^{-1}, the enthalpy of atomization of CH_4. Although all four C—H bonds in methane are the same, the four dissociation enthalpies are not all the same, because there are adjustments in the electron distributions as each successive hydrogen atom is removed. The additive bond strength is the average of these four dissociation enthalpies.

For the gaseous water molecule we have similarly

$$1.\ HOH \rightarrow H + OH \qquad \Delta H^\circ = 498.7 \text{ kJ mol}^{-1}$$

$$2.\ OH \rightarrow H + O \qquad \Delta H^\circ = 428.2 \text{ kJ mol}^{-1}$$

$$1. + 2.\ H_2O \rightarrow 2H + O \qquad \Delta H^\circ = 926.9 \text{ kJ mol}^{-1}$$

Note that it is easier to remove the second hydrogen atom than the first. The bond enthalpy is 926.9/2 = 463.5 kJ mol^{-1}, which is the mean of the two dissociation enthalpies.

Only in the case of diatomic molecules can bond enthalpy be identified with the bond dissociation enthalpy; for example,

$$H_2 \rightarrow 2H \qquad \Delta H^\circ = 435.9 \text{ kJ mol}^{-1}$$

which is both the dissociation enthalpy and the bond enthalpy.

2.6 Ideal Gas Relationships

The various transformations that can be brought about on ideal gases have played a very important part in the development of thermodynamics. There are good reasons for devoting careful study to ideal gases. In the first place, ideal gases are the simplest systems to deal with, and they therefore provide us with valuable and not too difficult exercises for testing our understanding of the subject. In addition, some of the simple conclusions that we can draw for ideal gases can readily be adapted to more complicated systems, such as solutions. A direct application of thermodynamics to solutions would be difficult if we did not have the ideal gas equations to guide our way.

Reversible Compression at Constant Pressure

As a first example, we will consider reducing the volume of an ideal gas by lowering its temperature at constant pressure. We are going to find how much work is done on the system during this process, how much heat is lost, and what the changes are in internal energy and enthalpy. Suppose that we have 1 mol of an

Details of reversible and irreversible isobaric processes; Isobaric calculations; Ideal gas relationships.

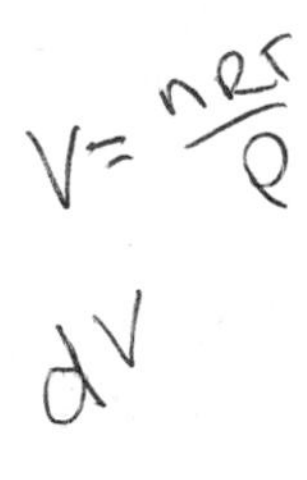

ideal gas confined in a cylinder with a piston, at a pressure P_1, a molar volume $V_{m,1}$, and an absolute temperature T_1. The **isotherm** (i.e., the PV relationship) for this temperature is shown in the upper curve in Figure 2.6a and the initial state is represented by point A. We now remove heat from the system reversibly, at the constant pressure P_1, until the volume has fallen to $V_{m,2}$ (point B). This could be done by lowering the temperature of the surroundings by infinitesimal amounts, until the temperature of the system is T_2; the isotherm for T_2 is the lower curve in Figure 2.6a.

As far as work and heat changes are concerned, we could not obtain values unless we had specified the path taken, since work and heat are not state functions. In this example, we have specified that the compression is reversible and is occurring at constant pressure. The work done on the system is

$$w_{\text{rev}} = -\int_{V_{m,1}}^{V_{m,2}} P_1\, dV = P_1(V_{m,1} - V_{m,2}) \tag{2.54}$$

This is true whether the gas is ideal or not. If the gas is ideal, use of the gas law $PV_m = RT$ (for 1 mol of gas) leads to the expression:

$$w_{\text{rev}} = P_1\left(\frac{RT_1}{P_1} - \frac{RT_2}{P_1}\right) \tag{2.55}$$

$$= R(T_1 - T_2) \qquad \text{(ideal gases only)} \tag{2.56}$$

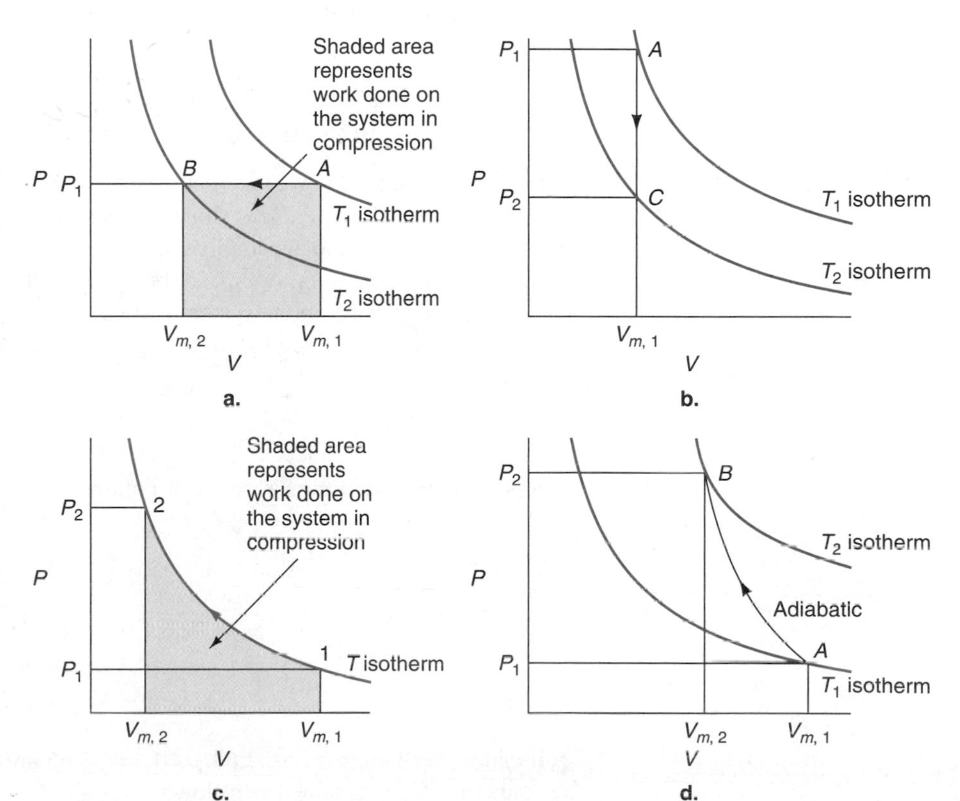

FIGURE 2.6
Pressure-volume relationships for an ideal gas. (a) Isotherms at two temperatures, T_1 and T_2. The gas is compressed reversibly at constant pressure from state A to state B. (b) Isotherms at two temperatures, T_1 and T_2. The gas at state A is cooled at constant volume to bring it to state C. (c) An isotherm, showing a reversible isothermal compression from state 1 to state 2. (d) Isotherms and an adiabatic. The gas at state A is compressed reversibly and adiabatically to state B.

Since $V_1 > V_2$, and $T_1 > T_2$, a positive amount of work has been done on the system. This work is represented by the shaded area in Figure 2.6a. If the system had expanded isothermally and at constant pressure, from state B to state A, the shaded area would represent the work done by the system.

The heat absorbed by the system during the process $A \rightarrow B$ is given by

$$q_{P,m} = \int_{T_1}^{T_2} C_{P,m}\, dT \tag{2.57}$$

since the process occurs at constant pressure. For an ideal gas, $C_{P,m}$ is independent of temperature and this expression therefore integrates to

$$q_{P,m} = C_{P,m}(T_2 - T_1) \tag{2.58}$$

Since $T_1 > T_2$, a negative amount of heat is absorbed (i.e., heat is released by the system). This amount of heat $q_{P,m}$, being absorbed at constant pressure, is the molar enthalpy change, which is also negative:

$$\Delta H_m = C_{P,m}(T_2 - T_1) \tag{2.59}$$

The molar internal energy change ΔU_m (also negative for this process) is obtained by use of the first law:

$$\Delta U_m = q + w = C_{P,m}(T_2 - T_1) + R(T_1 - T_2) \tag{2.60}$$

$$= (C_{P,m} - R)(T_2 - T_1) = C_{V,m}(T_2 - T_1) \tag{2.61}$$

using Eq. 2.36. It can easily be verified that these expressions for ΔH_m and ΔU_m are consistent with the relationship

$$\Delta H_m = \Delta U_m + \Delta(PV) \tag{2.62}$$

Reversible Pressure Change at Constant Volume

Details of constant volume reversible and irreversible processes; Isochoric processes; Ideal gas relationships.

Suppose, instead, that we take 1 mol of ideal gas from the initial state P_1, $V_{m,1}$, T_1 to the final state P_2, $V_{m,1}$, T_2 as shown in Figure 2.6b. The pressure P_1 is taken to be higher than P_2, and to accomplish this at constant volume we must remove heat until the temperature is T_2. Again, we bring about the change reversibly.

The work done on the system is the area below the line AC in Figure 2.6b and is zero. This is confirmed by considering the integral

$$w_{\text{rev}} = -\int_{V_{m,1}}^{V_{m,1}} P\, dV = 0 \tag{2.63}$$

Since the process occurs at constant volume, the heat absorbed is given by

$$q_{V,m} = \int_{T_1}^{T_2} C_{V,m}\, dT = C_{V,m}(T_2 - T_1) \tag{2.64}$$

which is negative since $T_1 > T_2$. This expression is also ΔU_m:

$$\Delta U_m = C_{V,m}(T_2 - T_1) \tag{2.65}$$

It can be verified that Eqs. 2.63, 2.64, and 2.65 are consistent with the first law. The value of ΔH_m is obtained as follows:

$$\Delta H_m = \Delta U_m + \Delta(PV_m) = \Delta U_m + \Delta(RT) \tag{2.66}$$
$$= C_{V,m}(T_2 - T_1) + R(T_2 - T_1) = (C_{V,m} + R)(T_2 - T_1) \tag{2.67}$$

$$\Delta H_m = C_{P,m}(T_2 - T_1) \tag{2.68}$$

It is interesting to compare the changes that occur in going from A to B (volume decrease at constant pressure, Figure 2.6a) with those when we go from A to C (pressure decrease at constant volume, Figure 2.6b). The work and heat values are different in the two cases. However, the ΔU_m and ΔH_m values are the same. This means that the internal energy is the same at point B on the T_2 isotherm as it is at point C; the same is true of the enthalpy. This result can be proved for any two points on an isotherm. We thus reach the very important conclusion that the *internal energy and enthalpy of an ideal gas depend only on the temperature and remain constant under isothermal conditions.*

Reversible Isothermal Compression

Details of reversible and irreversible isothermal processes; Isothermal calculations; Ideal gas relationships.

Another process of very great importance is that of compression of an ideal gas along an isotherm (i.e., at constant temperature). Such a process is illustrated in Figure 2.6c, the temperature being written simply as T. The initial conditions are P_1, $V_{m,1}$ and the final are P_2, $V_{m,2}$ with $V_{m,1} > V_{m,2}$. We have just seen that for an isothermal process in an ideal gas

$$\Delta U_m = 0 \quad \text{and} \quad \Delta H_m = 0 \tag{2.69}$$

The work done on the system in a reversible compression is

$$w_{\text{rev}} = -\int_{V_{m,1}}^{V_{m,2}} P\,dV \tag{2.70}$$

Since P is varying, we must express it in terms of V_m by use of the ideal gas equation, which for 1 mol is $PV_m = RT$; thus

$$w_{\text{rev}} = -\int \frac{RT}{V_m}\,dV = -RT \ln V \Big|_{V_{m,1}}^{V_{m,2}} = -RT \ln \frac{V_{m,2}}{V_{m,1}} \tag{2.71}$$

Since $V_{m,1} > V_{m,2}$, this is a positive quantity.

$$w_{\text{rev}} = RT \ln \frac{V_{m,1}}{V_{m,2}} \tag{2.72}$$

The heat absorbed is found by use of the equation for the first law:

$$\Delta U_m = q_{\text{rev}} + w_{\text{rev}} \tag{2.73}$$

Therefore, from Eq. 2.71,

$$q_{\text{rev}} = \Delta U_m - w_{\text{rev}} = 0 - w_{\text{rev}} = RT \ln \frac{V_{m,2}}{V_{m,1}} \tag{2.74}$$

This is negative; that is, heat is evolved during the compression. When we compress a gas, we do work on it and supply energy to it; if the temperature is to remain constant, heat must be evolved.

If 1 mol of gas has a volume $V_{m,1}$, the concentration is

$$c_1 = \frac{1}{V_{m,1}} \tag{2.75}$$

Similarly,

$$c_2 = \frac{1}{V_{m,2}} \tag{2.76}$$

The ratio of volumes is therefore the inverse ratio of the concentrations:

$$\frac{V_{m,2}}{V_{m,1}} = \frac{c_1}{c_2} \tag{2.77}$$

Equation 2.72 for the work done in the isothermal reversible expansion of 1 mol of an ideal gas can therefore be written alternatively as

$$w_{\text{rev},m} = RT \ln \frac{c_2}{c_1} \tag{2.78}$$

For n mol,

$$w_{\text{rev}} = nRT \ln \frac{c_2}{c_1} \tag{2.79}$$

This is a useful form of the equation, because it will be seen that certain types of solutions, known as *ideal* solutions, obey exactly the same relationship.

EXAMPLE 2.8 Calculate the work done by the system when 6.00 mol of an ideal gas at 25.0 °C are allowed to expand isothermally and reversibly from an initial volume of 5.00 dm^3 to a final volume of 15.00 dm^3.

Solution From Eq. 2.71 for 6.00 mol of gas, the work done (on the system) is

$$w_{\text{rev}} = -6RT \ln \tfrac{15}{5}$$

Since $R = 8.3145$ J K^{-1} mol^{-1},

$$w_{\text{rev}} = -6 \times 8.3145 \times 298.15 \text{ (J)} \ln 3 = -16\,338 \text{ J} = -16.3 \text{ kJ}$$

This is negative, so that 16.3 kJ of work has been done by the system.

EXAMPLE 2.9 Gastric juice in humans has an acid concentration of about[8] 1.00×10^{-1} M (pH $\approx$ 1) and it is formed from other body fluids, such as blood, which have an acid concentration of about 4.00×10^{-8} M (pH $\approx$ 7.4). On the average, about 3.00 dm^3 of gastric juice are produced per day. Calculate the minimum work required to produce this quantity at 37 °C, assuming the behavior to be ideal.

Solution Equation 2.79 gives us the reversible work required to produce n mol of acid, and 3 dm^3 of 10^{-1} M acid contains 0.300 mol, so that the reversible work is

[8]The symbol M (molar) represents mol dm^{-3}.

Details of reversible and irreversible adiabatic processes; Adiabatic calculations; Ideal gas relationships.

$$w_{rev} = 0.300 \times 8.3145 \times 310.15 \ln \frac{1.00 \times 10^{-1}}{4.00 \times 10^{-8}} = 11\ 396 \text{ J} = 11.4 \text{ kJ}$$

The actual work required will, of course, be greater than this because the process, being a natural one, does not occur reversibly.

Reversible Adiabatic Compression

Adiabatic Process

The final process that we will consider is the compression of an ideal gas contained in a vessel whose walls are perfectly insulating, so that no heat can pass through them. Such processes are said to be **adiabatic.**

The pressure-volume diagram for the process is shown in Figure 2.6d. Since work is performed on the gas in order to compress it, and no heat can leave the system, the final temperature T_2 must be higher than the initial temperature T_1. The figure shows the T_1 and T_2 isotherms, as well as the adiabatic curve AB. We will now consider n mol of the ideal gas.

We first need the equation for the adiabatic AB. According to the first law

$$dU = dq - P\,dV \tag{2.80}$$

Since the process is adiabatic, $dq = 0$, and therefore

$$dU + P\,dV = 0 \tag{2.81}$$

Also, for n mol, $dU = nC_{V,m}\,dT$, and thus

$$nC_{V,m}\,dT + P\,dV = 0 \tag{2.82}$$

This is true whether the gas is ideal or not. For n mol of an ideal gas

$$PV = nRT \tag{2.83}$$

Elimination of P between Eqs. 2.82 and 2.83 leads to

$$C_{V,m}\frac{dT}{T} + R\frac{dV}{V} = 0 \tag{2.84}$$

For the adiabatic AB we integrate this equation between the temperatures T_1 and T_2 and the volumes V_1 and V_2, with the assumption that $C_{V,m}$ is a constant:

$$C_{V,m}\int_{T_1}^{T_2}\frac{dT}{T} + R\int_{V_1}^{V_2}\frac{dV}{V} = 0 \tag{2.85}$$

and, therefore,

$$C_{V,m}\ln\frac{T_2}{T_1} + R\ln\frac{V_2}{V_1} = 0 \tag{2.86}$$

Since $R = C_{P,m} - C_{V,m}$, this equation may be written as[9]

$$\ln\frac{T_2}{T_1} + \frac{(C_{P,m} - C_{V,m})}{C_{V,m}}\ln\frac{V_2}{V_1} = 0 \tag{2.87}$$

[9]For a discussion of the magnitudes of $C_{P,m}$ and $C_{V,m}$, see Section 15.1.

The ratio of $C_{P,m}$ to $C_{V,m}$ is often written as γ:

$$\gamma \equiv \frac{C_{P,m}}{C_{V,m}} \tag{2.88}$$

Eq. 2.87 can therefore be written as

$$\ln \frac{T_2}{T_1} + (\gamma - 1) \ln \frac{V_2}{V_1} = 0 \tag{2.89}$$

or

$$\frac{T_2}{T_1} = \left(\frac{V_1}{V_2}\right)^{\gamma-1} \tag{2.90}$$

We can eliminate the temperature by making use of the ideal gas relationship

$$\frac{T_2}{T_1} = \frac{P_2V_2}{P_1V_1} \tag{2.91}$$

Equating the right-hand sides of Eqs. 2.90 and 2.91 gives

$$\frac{P_2}{P_1} = \left(\frac{V_1}{V_2}\right)^{\gamma} \tag{2.92}$$

or

Ideal Gas Law for the Adiabatic Case

$$P_1V_1^{\gamma} = P_2V_2^{\gamma} \tag{2.93}$$

This is to be contrasted with the Boyle's law equation $P_1V_1 = P_2V_2$ for the isothermal process. For a perfect monatomic gas, $\gamma = \frac{5}{3}$, and for all gases, $\gamma > 1$. Because γ is necessarily greater than unity, the adiabatic is steeper than the isotherm, as shown in Figure 2.6d. The rationale for this is that in the adiabatic process, no flow of heat is permitted, and the expansion occurs at the expense of a drop in temperature. In an isothermal expansion, heat flows into the system, thereby maintaining a constant temperature. Consequently, the pressure will not drop as rapidly as in the adiabatic process.

We now consider the various changes in the thermodynamic quantities when the process $A \rightarrow B$ in Figure 2.6d is undergone by an ideal gas. Since the process is adiabatic,

$$q = 0 \tag{2.94}$$

Both U and H remain unchanged as we move along the T_1 isothermal, and the same is true of the T_2 isothermal. The changes are (Eqs. 2.29 and 2.31)

$$\Delta U = C_V(T_2 - T_1) \tag{2.95}$$

and

$$\Delta H = C_P(T_2 - T_1) \tag{2.96}$$

Since $\Delta U = q + w$, and $q = 0$, the work done on the system during the adiabatic compression is

$$w = C_V(T_2 - T_1) = nC_{V,m}(T_2 - T_1) \tag{2.97}$$

2.7 Real Gases[10]

We have seen that the internal energy U of an ideal gas is a function of temperature but remains constant as we move along any isotherm. For an ideal gas, therefore, the following conditions hold:

1. $PV = nRT$ (2.98)
2. $\left(\frac{\partial U}{\partial V}\right)_T = 0$ (2.99)

The second of these conditions follows from the first. For a nonideal gas, neither of these conditions is satisfied.

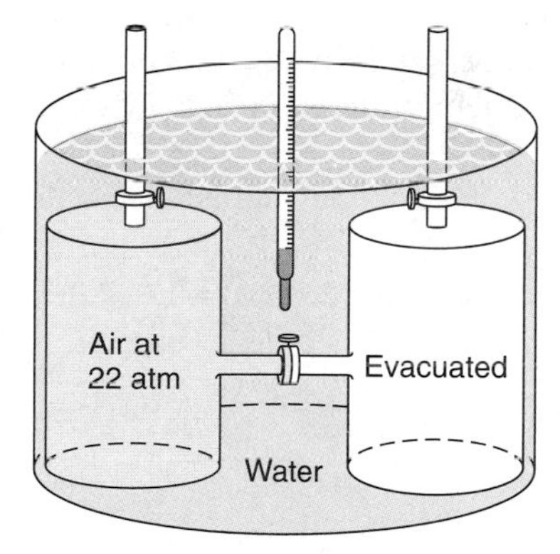

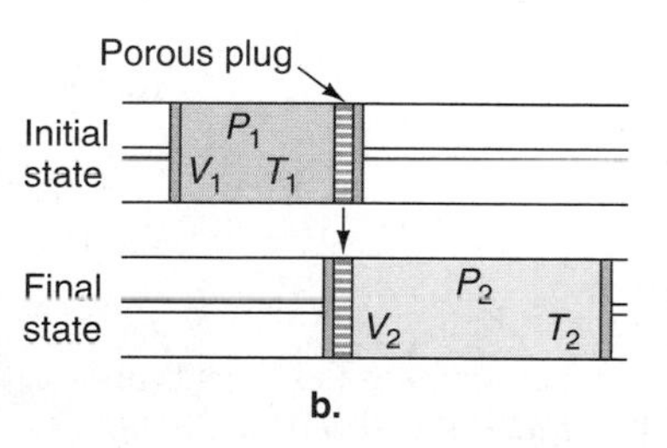

FIGURE 2.7
(a) Joule's apparatus for attempting to measure a temperature change on the free expansion of a gas.
(b) The Joule-Thomson experiment: The gas is forced through a membrane and the temperature change measured.

The Joule-Thomson Experiment

In 1845 Joule published the results of an experiment that was designed to determine whether $(\partial U/\partial V)_T = 0$ for a real gas. His apparatus, which is shown in Figure 2.7a, consisted of two containers, connected through a stopcock, the whole apparatus being placed in a water bath, the temperature of which could be measured very accurately. He filled one container with air at 22 bar pressure and evacuated the other. On opening the stopcock connecting the two containers, he was unable to detect any change in temperature.

The expansion in Joule's experiment was an irreversible process. The gas underwent no detectable change in internal energy, since no work was done and there was no detectable exchange of heat with the surroundings. The change dU can be expressed in terms of how the internal energy changes with respect to volume or temperature times the respective incremental changes in volume or temperature. This is mathematically expressed in terms of differentials and the total differential (see Appendix C) as

$$dU = \left(\frac{\partial U}{\partial V}\right)_T dV + \left(\frac{\partial U}{\partial T}\right)_V dT = \left(\frac{\partial U}{\partial V}\right)_T dV + C_V\,dT \tag{2.100}$$

Joule concluded that $dU = 0$ and, therefore, that

$$\left(\frac{\partial U}{\partial V}\right)_T = -\left(\frac{\partial U}{\partial T}\right)_V\left(\frac{\partial T}{\partial V}\right)_U \tag{2.101}$$

$$= -C_V\left(\frac{\partial T}{\partial V}\right)_U \tag{2.102}$$

[10]This section could be omitted on first reading; for a mathematical review related to this section see Appendix C.

by Eq. 2.26. Joule's experiment indicated no temperature change, [i.e., $(\partial T/\partial V)_U = 0$] and, therefore,

$$\left(\frac{\partial U}{\partial V}\right)_T = 0 \quad (2.103)$$

As far as Joule could detect, therefore, the gas was behaving ideally.

His experiment, however, was not very satisfactory, because the heat capacity of the water was extremely large compared with the heat capacity of the gas, so that any temperature rise would be very small and hard to detect. William Thomson (Lord Kelvin) suggested a better procedure and, with Joule, carried out a series of experiments between 1852 and 1862. Their apparatus, shown schematically in Figure 2.7b, consisted essentially of a cylinder containing a porous plug. By means of a piston, the gas on one side of the plug was forced completely through the plug at constant pressure P_1, the second piston moving back and maintaining the gas at constant pressure P_2, which is lower than P_1. Since the whole system is thermally isolated, the process is adiabatic (i.e., $q = 0$). The temperatures T_1 and T_2 were measured on the two sides of the plug.

Suppose that the initial volume is V_1. The work done in moving the piston to the porous plug, thereby forcing the gas through the plug, is P_1V_1. If the piston in the right-hand chamber starts at the plug, and if the final volume is V_2, the work done by the gas in expanding is P_2V_2. The net work done on the gas is thus

$$w = P_1V_1 - P_2V_2 \quad (2.104)$$

Since $q = 0$, the change in internal energy is

$$\Delta U = U_2 - U_1 = q + w = w = P_1V_1 - P_2V_2 \quad (2.105)$$

Thus

$$U_2 + P_2V_2 = U_1 + P_1V_1 \quad (2.106)$$

or, from the definition of enthalpy (Eq. 2.23),

$$H_2 = H_1 \quad (2.107)$$

The Joule-Thomson expansion thus occurs at constant enthalpy: $\Delta H = 0$.

TABLE 2.3 Joule-Thomson Coefficient μ_{JT} and Inversion Temperature T_i/(K atm^{-1}) for Several Gases at 1 atm and 298 K

Gas	μ_{JT}/(K atm^{-1})	T_i/K
Air	0.189 at 50 °C	603
He Helium	−0.062	40
H_2 Hydrogen	−0.03	193
N_2 Nitrogen	0.27	621
O_2 Oxygen	0.31	764

Joule-Thomson Coefficient

Joule and Thomson were able to detect a temperature change ΔT as the gas undergoes a pressure change P through the porous plug. The *Joule-Thomson coefficient* μ is defined as

$$\mu = \left(\frac{\partial T}{\partial P}\right)_H \approx \frac{\Delta T}{\Delta P} \quad (2.108)$$

Joule-Thomson Inversion Temperature

This is zero for an ideal gas, but may be positive or negative for a real gas. When there is an expansion, ΔP is negative; if there is a cooling on expansion (i.e., ΔT is negative when ΔP is negative), the Joule-Thomson coefficient μ is positive. This effect is easily seen in the use of ammonia as the working gas in some industrial refrigerators. Conversely, a negative μ corresponds to a rise in temperature on expansion. Most gases at ordinary temperatures cool on expansion. Hydrogen is exceptional, in that its Joule-Thomson coefficient is negative above 193 K (i.e., it warms on expansion); below 193 K, the coefficient is positive. The temperature 193 K, at which $\mu = 0$, is known as the *Joule-Thomson inversion temperature* for hydrogen. Some values of μ_{JT} are given in Table 2.3.

Since the Joule-Thomson expansion occurs at constant enthalpy, the total differential is

$$dH = \left(\frac{\partial H}{\partial P}\right)_T dP + \left(\frac{\partial H}{\partial T}\right)_P dT = 0 \tag{2.109}$$

from which it follows that

$$\left(\frac{\partial H}{\partial P}\right)_T = -\left(\frac{\partial H}{\partial T}\right)_P\left(\frac{\partial T}{\partial P}\right)_H = -C_P\mu \tag{2.110}$$

For an ideal gas, $\mu = 0$ and the enthalpy is therefore independent of pressure, as we have seen previously. For a real gas, μ is in general different from zero, and H shows some variation with P.

An equation for the difference between the heat capacities C_P and C_V, and applicable to gases, liquids, and solids, is obtained as follows:

$$C_P - C_V = \left(\frac{\partial H}{\partial T}\right)_P - \left(\frac{\partial U}{\partial T}\right)_V \tag{2.111}$$

$$= \left(\frac{\partial U}{\partial T}\right)_P + P\left(\frac{\partial V}{\partial T}\right)_P - \left(\frac{\partial U}{\partial T}\right)_V \tag{2.112}$$

We also have the following relationships from the total differentials of U and V:

$$dU = \left(\frac{\partial U}{\partial V}\right)_T dV + \left(\frac{\partial U}{\partial T}\right)_V dT \tag{2.113}$$

and

$$dV = \left(\frac{\partial V}{\partial T}\right)_P dT + \left(\frac{\partial V}{\partial P}\right)_T dP \tag{2.114}$$

Substitution of this expression for dV into Eq. 2.113 gives

$$dU = \left(\frac{\partial U}{\partial V}\right)_T\left(\frac{\partial V}{\partial T}\right)_P dT + \left(\frac{\partial U}{\partial V}\right)_T\left(\frac{\partial V}{\partial P}\right)_T dP + \left(\frac{\partial U}{\partial T}\right)_V dT \tag{2.115}$$

At constant pressure, we may eliminate the second term on the right-hand side:

$$\left(\frac{\partial U}{\partial T}\right)_P = \left(\frac{\partial U}{\partial V}\right)_T\left(\frac{\partial V}{\partial T}\right)_P + \left(\frac{\partial U}{\partial T}\right)_V \tag{2.116}$$

Substitution of this expression into Eq. 2.112 gives

$$C_P - C_V = \left[P + \left(\frac{\partial U}{\partial V}\right)_T\right]\left(\frac{\partial V}{\partial T}\right)_P \tag{2.117}$$

For an ideal gas, $(\partial U/\partial V)_T = 0$ and thus

$$C_P - C_V = P\left(\frac{\partial V}{\partial T}\right)_P \tag{2.118}$$

For 1 mol of an ideal gas, $PV_m = RT$ and, therefore, $(\partial V_m/\partial T)_P = R/P$; hence,

$$C_{P,m} - C_{V,m} = R \tag{2.119}$$

as we proved earlier in Eq. 2.36.

In general, the term $P(\partial V/\partial T)_P$ in Eq. 2.117 represents the contribution to C_P caused by the change in volume of the system acting against the *external* pressure P. The other term, $(\partial U/\partial V)_T(\partial V/\partial T)_P$, is an additional contribution due to $(\partial U/\partial V)_T$, which is an effective pressure arising from the attractive or repulsive forces between the molecules; the quantity $(\partial U/\partial V)_T$ is known as the **internal pressure.** In an ideal gas the internal pressure is zero, and in real gases the term is usually small compared to the external pressure P. Liquids and solids, on the other hand, have strong attractive or *cohesive* forces, and the internal pressure $(\partial U/\partial V)_T$ may therefore be large compared to the external pressure.

Internal Pressure

Van der Waals Gases

We have seen in Chapter 1 that for many real gases the van der Waals equation (Eq. 1.100) is obeyed. In this equation the volume V occupied by n mol of gas is replaced by $V - nb$, where b is a correction to allow for the volume actually occupied by the molecules. The pressure P becomes

$$P + \frac{an^2}{V^2}$$

where the term an^2/V^2 is due to the attractions between the molecules. The van der Waals equation is thus

Van der Waals Equation

$$\left(P + \frac{an^2}{V^2}\right)(V - nb) = nRT \quad \text{or} \quad \left(P + \frac{a}{V_m^2}\right)(V_m - b) = RT \qquad (2.120)$$

where V_m, equal to V/n, is the molar volume. It is evident that the two relations that hold for an ideal gas, Eqs. 2.98 and 2.99, do not hold for a van der Waals gas. Instead of Eq. 2.98 we have Eq. 2.120, and instead of Eq. 2.99 the equation

$$\left(\frac{\partial U}{\partial V}\right)_T = \frac{n^2 a}{V^2} = \frac{a}{V_m^2} \qquad (2.121)$$

applies. This relationship is derived in Section 3.8 (p. 132, Eq. 3.144).

EXAMPLE 2.10 Suppose that a gas obeys the modified van der Waals equation

$$P(V_m - b) = RT$$

and that b has a value of 0.0200 $dm^3\ mol^{-1}$. If 0.500 mol of the gas is reversibly compressed from an initial volume of 2.00 dm^3 to a final volume of 0.500 dm^3, how much work is done on the system? How much work would have been done if the gas were ideal? Account for the difference between the two values.

Solution The equation that applies to n mol of gas is

$$P(V - nb) = nRT$$

By Eq. 2.11, the work done on the system is

$$w = -\int_{V_1}^{V_2} P\, dV = -nRT \int_{V_1}^{V_2} \frac{dV}{V - nb}$$

The integration is performed by putting $V - nb = x$. Since $d(V - nb) = dx$

$$w = -nRT \int_{x_1}^{x_2} dx/x$$

The solution is

$$\begin{aligned} w_{rev} &= nRT \ln\left(\frac{V_2 - nb}{V_1 - nb}\right) \\ &= -0.500 \times 8.3145 \times 300(\text{J}) \ln[(0.500 - 0.01)/(2.00 - 0.01)] \\ &= 1247(\text{J}) \ln(0.49/1.99) = 1247(\text{J}) \times 1.401 \\ &= 1.75\ \text{kJ} \end{aligned}$$

If the gas had been ideal, the work done would have been

$$\begin{aligned} w_{rev} &= nRT \ln(2.00/0.500) \\ &= 1247(\text{J}) \times 1.386 = 1.73\ \text{kJ} \end{aligned}$$

The work is greater for the real gas because the ratio of the free volumes $V - nb$ is greater than for the ideal gas. Note that the difference is small.

The calculation of thermodynamic quantities such as w_{rev}, ΔU, and ΔH for processes occurring in a gas obeying the van der Waals equation is a little more complicated than for an ideal gas, and only a few cases will be considered in order to illustrate the procedures used. Suppose that a van der Waals gas is compressed reversibly and isothermally from a volume V_1 to a volume V_2.

The reversible work is

$$w_{rev} = -\int_{V_1}^{V_2} P\,dV = -\int_{V_1}^{V_2} \left(\frac{nRT}{V - nb} - \frac{n^2a}{V^2}\right) dV \tag{2.122}$$

$$= -nRT \ln\left(\frac{V_2 - nb}{V_1 - nb}\right) - n^2a\left(\frac{1}{V_2} - \frac{1}{V_1}\right) \tag{2.123}$$

The change in internal energy is obtained by starting with the general relationship 2.100, which at constant temperature reduces to

$$dU = \left(\frac{\partial U}{\partial V}\right)_T dV = \frac{n^2a}{V^2} dV \tag{2.124}$$

Then

$$\Delta U = \int_{V_1}^{V_2} \frac{n^2a}{V_2} dV = n^2a\left(-\frac{1}{V}\Big|_{V_1}^{V_2}\right) \tag{2.125}$$

$$= n^2a\left(\frac{1}{V_1} - \frac{1}{V_2}\right) \tag{2.126}$$

The value of ΔH may be obtained by obtaining $\Delta(PV)$, and then using the relationship

$$\Delta H = \Delta U + \Delta(PV) \tag{2.127}$$

KEY EQUATIONS

The first law of thermodynamics:

$$\Delta U = q + w$$

where ΔU = change in internal energy; q = heat given to system; w = work done on system.

Reversible PV work done on a gas:

$$w_{rev} = -\int_{V_1}^{V_2} P\,dV$$

where V_1 = initial volume; V_2 = final volume.

Definition of enthalpy H:

$$H \equiv U + PV$$

Heat capacities:

$$\text{At constant volume, } C_V = \frac{dq_V}{dT} = \left(\frac{\partial U}{\partial T}\right)_V$$

$$\text{At constant pressure, } C_P = \frac{dq_P}{dT} = \left(\frac{\partial H}{\partial T}\right)_P$$

For an ideal gas:

$$PV = nRT \quad \text{and} \quad \left(\frac{\partial U}{\partial V}\right)_T = 0$$

Isothermal reversible compression of 1 mol of an ideal gas:

$$w_{rev} = -q_{rev} = RT\ln\frac{V_1}{V_2}$$

For any system:

$$dU = C_V\,dT + \left(\frac{\partial U}{\partial V}\right)_T dV$$

For a reversible adiabatic process in an ideal gas:

$$P_1V_1^{\gamma} = P_2V_2^{\gamma}$$

PROBLEMS

Energy, Heat, and Work

2.1. A bird weighing 1.5 kg leaves the ground and flies to a height of 75 metres, where it attains a velocity of 20 m s^{-1}. What change in energy is involved in the process? (Acceleration of gravity = 9.81 m s^{-2}.)

2.2. The densities of ice and water at 0 °C are 0.9168 and 0.9998 g cm^{-3}, respectively. If ΔH for the fusion process at atmospheric pressure is 6.025 kJ mol^{-1}, what is ΔU? How much work is done on the system?

2.3. The density of liquid water at 100 °C is 0.9584 g cm^{-3}, and that of steam at the same temperature is 0.000 596 g cm^{-3}. If the enthalpy of evaporation of water at atmospheric pressure is 40.63 kJ mol^{-1}, what is ΔU? How much work is done by the system during the evaporation process?

2.4. The latent heat of fusion of water at 0 °C is 6.025 kJ mol^{-1} and the molar heat capacities ($C_{P,m}$) of water and ice are 75.3 and 37.7 J K^{-1} mol^{-1}, respectively. The C_P values can be taken to be independent of temperature. Calculate ΔH for the freezing of 1 mol of supercooled water at −10.0 °C.

2.5. A sample of liquid acetone weighing 0.700 g was burned in a bomb calorimeter for which the heat capacity (including the sample) is 6937 J K^{-1}. The observed temperature rise was from 25.00 °C to 26.69 °C.

a. Calculate ΔU for the combustion of 1 mol of acetone.
b. Calculate ΔH for the combustion of 1 mol of acetone.

2.6. An average man weighs about 70 kg and produces about 10 460 kJ of heat per day.

a. Suppose that a man were an isolated system and that his heat capacity were 4.18 J K^{-1} g^{-1}; if his temperature were 37 °C at a given time, what would be his temperature 24 h later?
b. A man is in fact an open system, and the main mechanism for maintaining his temperature constant is evaporation of water. If the enthalpy of vaporization of water at 37 °C is 43.4 kJ mol^{-1}, how much water needs to be evaporated per day to keep the temperature constant?

2.7. In an open beaker at 25 °C and 1 atm pressure, 100 g of zinc are caused to react with dilute sulfuric acid. Calculate the work done by the liberated hydrogen gas, assuming it behaves ideally. What would be the work done if the reaction took place in a sealed vessel?

2.8. A balloon is 0.50 m in diameter and contains air at 25 °C and 1 bar pressure. It is then filled with air isothermally and reversibly until the pressure reaches 5 bar. Assume that the pressure is proportional to the diameter of the balloon and calculate (a) the final diameter of the balloon and (b) the work done in the process.

2.9. When 1 cal of heat is given to 1 g of water at 14.5 °C, the temperature rises to 15.5 °C. Calculate the molar heat capacity of water at 15 °C.

2.10. A vessel containing 1 kg of water at 25 °C is heated until it boils. How much heat is supplied? How long would it take a one-kilowatt heater to supply this amount of heat? Assume the heat capacity calculated in Problem 2.9 to apply over the temperature range.

2.11. A nonporous ceramic of volume V m^3 and mass M kg is immersed in a liquid of density d kg m^{-3}. What is the work done on the ceramic if it is slowly raised a height h m through the liquid? Neglect any resistance caused by viscosity. What is the change in the potential energy of the ceramic?

2.12. Show that the differential dP of the pressure of an ideal gas is an exact differential.

2.13. Determine whether $dU = xy^2dx + x^2ydy$ is an exact differential. If it is find the function U of which dU is the differential. Do this by integrating over suitable paths. In a plot of y against x, show a plot of the paths that you chose.

Thermochemistry

2.14. Using the data given in Table 2.1 and Appendix D, find the enthalpy change for the reaction $2H_2(g) + O_2(g) \rightarrow 2H_2O(g)$ at 800 K.

2.15. A sample of liquid benzene weighing 0.633 g is burned in a bomb calorimeter at 25 °C, and 26.54 kJ of heat are evolved.

a. Calculate ΔU per mole of benzene.
b. Calculate ΔH per mole of benzene.

2.16. Deduce the standard enthalpy change for the process

$$2CH_4(g) \rightarrow C_2H_6(g) + H_2(g)$$

from the data in Appendix D.

2.17. A sample of liquid methanol weighing 5.27 g was burned in a bomb calorimeter at 25 °C, and 119.50 kJ of heat was evolved (after correction for standard conditions).

a. Calculate $\Delta_c H°$ for the combustion of 1 mol of methanol.
b. Use this value and the data in Appendix D for $H_2O(l)$ and $CO_2(g)$ to obtain a value for $\Delta_f H°(CH_3OH,l)$, and compare with the value given in the table.
c. If the enthalpy of vaporization of methanol is 35.27 kJ mol^{-1}, calculate $\Delta_f H°$ for $CH_3OH(g)$.

2.18. Calculate the heat of combustion ($\Delta_c H°$) of ethane from the data given in Appendix D.

2.19. The model used to describe the temperature dependence of heat capacities (Eq. 2.48; Table 2.1) cannot remain valid as the temperature approaches absolute zero because of the $1/T^2$ term. In some cases, the model starts to break down at temperatures significantly higher than absolute zero. The following data for nickel are taken from a very old textbook (*Numerical Problems in Advanced Physical Chemistry,* J. H. Wolfenden, London: Oxford, 1938, p. 45). Fit these data to the model and find the optimum values of the parameters.

T/K	15.05	25.20	47.10	67.13
C_P/J K^{-1} mol^{-1}	0.1943	0.5987	3.5333	7.6360

82.11	133.4	204.05	256.5	283.0
10.0953	17.8780	22.7202	24.8038	26.0833

Examine the behavior of the fit in the range $10 \leq T \leq 25$ and comment on this.

2.20. Suggest a practicable method for determining the enthalpy of formation $\Delta_f H°$ of gaseous carbon monoxide at 25 °C. (*Note:* Burning graphite in a limited supply of oxygen is not satisfactory, since the product will be a mixture of unburned graphite, CO, and CO_2.)

2.21. If the enthalpy of combustion $\Delta_c H°$ of gaseous cyclopropane, C_3H_6, is -2091.2 kJ mol^{-1} at 25 °C, calculate the standard enthalpy of formation $\Delta_f H°$.

2.22. The parameters for expressing the temperature dependence of molar heat capacities for various substances listed in Table 2.1 are obtained by fitting the model $C_{P,m} = d + eT + fT^2$ to experimental data at various temperatures and finding the values of the parameters d, e, and f that yield the best fit. Several mathematical software packages (Mathematica, Mathcad, Macsyma, etc.) and several scientific plotting packages (Axum, Origin, PSIPlot, etc.) can perform these fits very quickly. Fit the following data given the temperature dependence of $C_{P,m}$ for n-butane to the model and obtain the optimum values of the parameters.

T/K	220	250	275	300
C_P/J K^{-1} mol^{-1}	0.642	0.759	0.861	0.952

325	350	380	400
1.025	1.085	1.142	1.177

2.23. From the data in Appendix D, calculate $\Delta H°$ for the reaction

$$C_2H_4(g) + H_2O(l) \rightarrow C_2H_5OH(l)$$

at 25 °C.

2.24. The bacterium *Acetobacter suboxydans* obtains energy for growth by oxidizing ethanol in two stages, as follows:

a. $C_2H_5OH(l) + \frac{1}{2}O_2(g) \rightarrow CH_3CHO(l) + H_2O(l)$
b. $CH_3CHO(l) + \frac{1}{2}O_2(g) \rightarrow CH_3COOH(l)$

The enthalpy increases in the complete combustion (to CO_2 and liquid H_2O) of the three compounds are

	$\Delta_c H°$/kJ mol^{-1}
Ethanol (l)	−1370.7
Acetaldehyde (l)	−1167.3
Acetic acid (l)	−876.1

Calculate the $\Delta H°$ values for reactions (a) and (b).

2.25. The enthalpy of combustion of acrylonitrile (C_3H_3N) at 25 °C and 1 atm pressure is -1760.9 kJ mol^{-1} [Stamm, Halverson, and Whalen, *J. Chem. Phys., 17,* 105(1949)]. Under the same conditions, the heats of formation of HCN(g) and C_2H_2(g) from the elements are 135.1 and 226.73 kJ mol^{-1}, respectively [*The NBS Tables of Chemical and Thermodynamic Properties,* Supp. 2 to Vol. 11 of *J. Phys. Chem. Ref. Data*]. Combining these data with the standard enthalpies of formation of CO_2(g) and H_2O(g), calculate the enthalpy change in the reaction HCN(g) + C_2H_2(g) → $H_2C{=}CH{-}CN$(g). [*Notes:* (a) Assume that the nitrogen present in acrylonitrile is converted into nitrogen gas during combustion. (b) Assume that all substances except for graphite (for the formation of CO_2) are gases, i.e., ignore the fact that acrylonitrile and water will be liquids under the conditions given here.]

2.26. Calculate ΔH for the reaction

$$C_2H_5OH(l) + O_2(g) \rightarrow CH_3COOH(l) + H_2O(l)$$

making use of the enthalpies of formation given in Appendix D. Is the result consistent with the results obtained for Problem 2.24?

2.27. The disaccharide α-maltose can be hydrolyzed to glucose according to the equation

$$C_{12}H_{22}O_{11}(aq) + H_2O(l) \rightarrow 2C_6H_{12}O_6(aq)$$

Using data in Appendix D and the following values, calculate the standard enthalpy change in this reaction:

	$\Delta_f H°$/kJ mol^{-1}
$C_6H_{12}O_6$(aq)	−1263.1
$C_{12}H_{22}O_{11}$(aq)	−2238.3

2.28. The standard enthalpy of formation of the fumarate ion is -777.4 kJ mol^{-1}. If the standard enthalpy change of the reaction

$$\text{fumarate}^{2-}(aq) + H_2(g) \rightarrow \text{succinate}^{2-}(aq)$$

is 131.4 kJ mol^{-1}, calculate the enthalpy of formation of the succinate ion.

2.29. The $\Delta H°$ for the mutarotation of glucose in aqueous solution,

$$\alpha\text{-D-glucose}(aq) \rightarrow \beta\text{-D-glucose}(aq)$$

has been measured in a microcalorimeter and found to be -1.16 kJ mol^{-1}. The enthalpies of solution of the two forms of glucose have been determined to be

$$\alpha\text{-D-glucose}(s) \rightarrow \alpha\text{-D-glucose}(aq) \quad \Delta H° = 10.72 \text{ kJ mol}^{-1}$$

$$\beta\text{-D-glucose}(s) \rightarrow \beta\text{-D-glucose}(aq) \quad \Delta H° = 4.68 \text{ kJ mol}^{-1}$$

Calculate $\Delta H°$ for the mutarotation of solid α-D-glucose to solid β-D-glucose.

2.30. Use the data in Appendix D to calculate $\Delta H°$ for the hydrolysis of urea into carbon dioxide and ammonia at 25 °C.

2.31. Here is a problem with a chemical engineering flavor: Ethanol is oxidized to acetic acid in a catalyst chamber at 25 °C. Calculate the rate at which heat will have to be removed (in J h^{-1}) from the chamber in order to maintain the reaction chamber at 25 °C, if the feed rate is 45.00 kg h^{-1} of ethanol and the conversion rate is 42 mole % of ethanol. Excess oxygen is assumed to be available.

2.32. a. An ice cube at 0 °C weighing 100 g is dropped into 1 kg of water at 20 °C. Does all of the ice melt? If not, how much of it remains? What is the final temperature? The latent heat of fusion of ice at 0 °C is 6.025 kJ mol^{-1}, and the molar heat capacity of water, $C_{P,m}$ is 75.3 J K^{-1} mol^{-1}. **b.** Perform the same calculations with 10 ice cubes of the same size dropped into the water. (See Problem 3.33 of Chapter 3 for the calculation of the corresponding entropy changes.)

***2.33.** From the data in Table 2.1 and Appendix D, calculate the enthalpy change in the reaction

$$C(\text{graphite}) + O_2(g) \rightarrow CO_2(g)$$

at 1000 K.

2.34. From the bond strengths in Table 2.2, estimate the enthalpy of formation of gaseous propane, C_3H_8, using the following additional data:

	$\Delta H°$/kJ mol^{-1}
C(graphite) → C(g)	716.7
H_2(g) → 2H(g)	436.0

2.35. A sample of sucrose, $C_{12}H_{22}O_{11}$ weighing 0.1328 g, was burned to completion in a bomb calorimeter at 25 °C, and the heat evolved was measured to be 2186.0 J.

a. Calculate $\Delta_c U_m$ and $\Delta_c H_m$ for the combustion of sucrose. **b.** Use data in Appendix D to calculate $\Delta_f H_m$ for the formation of sucrose.

2.36. The value of $\Delta H°$ for the reaction

$$CO(g) + \tfrac{1}{2}O_2(g) \rightarrow CO_2(g)$$

is -282.97 kJ mol^{-1} at 298 K. Calculate $\Delta U°$ for the reaction.

Ideal Gases

2.37. One mole of an ideal gas initially at 10.00 bar and 298.0 K is allowed to expand against a constant external pressure of 2.000 bar to a final pressure of 2.000 bar. During this process, the temperature of the gas falls to 253.2 K. We wish to construct a reversible path connecting these initial and final steps as a combination of a reversible isothermal expansion followed by a reversible adiabatic expansion. To what volume should we allow the gas to expand isothermally so that subsequent adiabatic expansion is guaranteed to take the gas to the final state? Assume that $C_{V,m} = \frac{3}{2}R$.

2.38. Two moles of oxygen gas, which can be regarded as ideal with C_P = 29.4 J K^{-1} mol^{-1} (independent of temperature), are maintained at 273 K in a volume of 11.35 dm^3.

a. What is the pressure of the gas?
b. What is PV?
c. What is C_V?

2.39. Suppose that the gas in Problem 2.38 is heated reversibly to 373 K at constant volume:

a. How much work is done on the system?
b. What is the increase in internal energy, ΔU?
c. How much heat was added to the system?
d. What is the final pressure?
e. What is the final value of PV?
f. What is the increase in enthalpy, ΔH?

2.40. Suppose that the gas in Problem 2.38 is heated reversibly to 373 K at constant pressure.

a. What is the final volume?
b. How much work is done on the system?
c. How much heat is supplied to the system?
d. What is the increase in enthalpy?
e. What is the increase in internal energy?

2.41. Suppose that the gas in Problem 2.38 is reversibly compressed to half its volume at constant temperature (273 K).

a. What is the change in U?
b. What is the final pressure?
c. How much work is done on the system?
d. How much heat flows out of the system?
e. What is the change in H?

2.42. With the temperature maintained at 0 °C, 2 mol of an ideal gas are allowed to expand against a piston that supports 2 bar pressure. The initial pressure of the gas is 10 bar and the final pressure 2 bar.

a. How much energy is transferred to the surroundings during the expansion?
b. What is the change in the internal energy and the enthalpy of the gas?
c. How much heat has been absorbed by the gas?

2.43. Suppose that the gas in Problem 2.42 is allowed to expand *reversibly* and *isothermally* from the initial pressure of 10 bar to the final pressure of 2 bar.

a. How much work is done by the gas?
b. What are ΔU and ΔH?
c. How much heat is absorbed by the gas?

2.44. A sample of hydrogen gas, which may be assumed to be ideal, is initially at 3.0 bar pressure and a temperature of 25.0 °C, and has a volume of 1.5 dm^3. It is expanded reversibly and adiabatically until the volume is 5.0 dm^3. The heat capacity C_P of H_2 is 28.80 J K^{-1} mol^{-1} and may be assumed to be independent of temperature.

a. Calculate the final pressure and temperature after the expansion.
b. Calculate ΔU and ΔH for the process.

***2.45.** Initially 0.1 mol of methane is at 1 bar pressure and 80 °C. The gas behaves ideally and the value of C_P/C_V is 1.31. The gas is allowed to expand reversibly and adiabatically to a pressure of 0.1 bar.

a. What are the initial and final volumes of the gas?
b. What is the final temperature?
c. Calculate ΔU and ΔH for the process.

2.46. A gas behaves ideally and its C_V is given by

$$C_V/\text{J K}^{-1}\text{ mol}^{-1} = 21.52 + 8.2 \times 10^{-3}T/\text{K}$$

a. What is $C_{P,m}$ as a function of T?
b. A sample of this gas is initially at T_1 = 300 K, P_1 = 10 bar, and V_1 = 1 dm^3. It is allowed to expand until P_2 = 1 bar and V_2 = 10 dm^3. What are ΔU and ΔH for this process? Could the process be carried out adiabatically?

2.47. Prove that for an ideal gas two reversible adiabatic curves on a P-V diagram cannot intersect.

2.48. An ideal gas is defined as one that obeys the relationship $PV = nRT$. We showed in Section 2.7 that for such gases

$$(\partial U/\partial V)_T = 0 \quad \text{and} \quad (\partial H/\partial P)_T = 0$$

Prove that for an ideal gas C_V and C_P are independent of volume and pressure.

2.49. One mole of an ideal gas underwent a reversible isothermal expansion until its volume was doubled. If the gas performed 1 kJ of work, what was its temperature?

2.50. A gas that behaves ideally was allowed to expand reversibly and adiabatically to twice its volume. Its initial temperature was 25.00 °C, and $C_{V,m} = (5/2)R$. Calculate ΔU_m and ΔH_m for the expansion process.

2.51. With $C_{V,m} = (3/2)R$, 1 mol of an ideal monatomic gas undergoes a reversible process in which the volume is doubled and in which 1 kJ of heat is absorbed by the gas. The initial pressure is 1 bar and the initial temperature is 300 K. The enthalpy change is 1.50 kJ.

a. Calculate the final pressure and temperature.
b. Calculate ΔU and w for the process.

***2.52.** Prove that

$$C_V = -\left(\frac{\partial U}{\partial V}\right)_T \left(\frac{\partial V}{\partial T}\right)_U$$

***2.53.** Prove that for an ideal gas the rate of change of the pressure dP/dt is related to the rates of change of the volume and temperature by

$$\frac{1}{P}\frac{dP}{dt} = -\frac{1}{V}\frac{dV}{dt} + \frac{1}{T}\frac{dT}{dt}$$

***2.54.** Initially 5 mol of nitrogen are at a temperature of 25 °C and a pressure of 10 bar. The gas may be assumed to be ideal; $C_{V,m} = 20.8$ J K^{-1} mol^{-1} and is independent of temperature. Suppose that the pressure is *suddenly* dropped to 1 bar; calculate the final temperature, ΔU, and ΔH.

2.55. A chemical reaction occurs at 300 K in a gas mixture that behaves ideally, and the total amount of gas increases by 0.27 mol. If $\Delta U = 9.4$ kJ, what is ΔH?

2.56. Suppose that 1.00 mol of an ideal monatomic gas ($C_V = (3/2)R$) at 1 bar is adiabatically and reversibly compressed starting at 25.0 °C from 0.1000 m^3 to 0.0100 m^3. Calculate q, w, ΔU, and ΔH.

2.57. Suppose that an ideal gas undergoes an irreversible isobaric adiabatic process. Derive expressions for q, w, ΔU, and ΔH and the final temperature of the gas undergoing the process.

2.58. Exactly one mole of an ideal monatomic gas at 25.0 °C is cooled and allowed to expand from 1.00 dm^3 to 10.00 dm^3 against an external pressure of 1.00 bar. Calculate the final temperature, and q, w, ΔU, and ΔH.

2.59. A balloon 15 m in diameter is inflated with helium at 20 °C.

a. What is the mass of helium in the balloon, assuming the gas to be ideal?
b. How much work is done by the balloon during the process of inflation against an external pressure of 1 atm (101.315 kPa), from an initial volume of zero to the final volume?

2.60. a. Calculate the work done when 1 mol of an ideal gas at 2 bar pressure and 300 K is expanded isothermally to a volume of 1.5 L, with the external pressure held constant at 1.5 bar.
b. Suppose instead that the gas is expanded isothermally and *reversibly* to the same final volume; calculate the work done.

2.61. The heat capacity difference can be determined experimentally in terms of the two variables α and β in the equation for an ideal gas. Determine the value of C_P and C_V for an ideal gas in the equation $C_P - C_V = TV\alpha^2/\beta$ where

$$\alpha = \frac{1}{V}\left(\frac{\partial V}{\partial T}\right)_P \text{ and } \beta = -\frac{1}{V}\left(\frac{\partial V}{\partial P}\right)_T.$$

Real Gases

2.62. For an ideal gas, $PV_m = RT$ and therefore $(dT/dP)_V = V_m/R$. Derive the corresponding relationship for a van der Waals gas.

***2.63.** One mole of a gas at 300 K is compressed isothermally and reversibly from an initial volume of 10 dm^3 to a final volume of 0.2 dm^3. Calculate the work done on the system if

a. the gas is ideal.
b. the equation of state of the gas is $P(V_m - b) = RT$, with $b = 0.03$ dm^3 mol^{-1}.

Explain the difference between the two values.

***2.64.** One mole of a gas at 100 K is compressed isothermally from an initial volume of 20 dm^3 to a final volume of 5 dm^3. Calculate the work done on the system if

a. the gas is ideal.
b. the equation of state is

$$\left(P + \frac{a}{V_m^2}\right)V_m = RT \quad \text{where} \quad a = 0.384 \text{ m}^6 \text{ Pa mol}^{-1}$$

[This equation is obeyed approximately at low temperatures, whereas $P(V_m - b) = RT$ (see Problem 2.63) is obeyed more closely at higher temperatures.] Account for the difference between the values in (a) and (b).

2.65. Derive the expression

$$dP = \frac{P\,dV_m}{V_m - b} - \frac{ab}{V_m^3(V_m - b)}dV_m + \frac{P\,dT}{T} + \frac{a\,dT}{V_m^2 T}$$

for 1 mol of a van der Waals gas.

2.66. If a substance is burned at constant volume with no heat loss, so that the heat evolved is all used to heat the product gases, the temperature attained is known as the *adiabatic flame temperature.* Calculate this quantity for methane burned at 25 °C in the amount of oxygen required to give complete combustion to CO_2 and H_2O. Use the data in Appendix D and the following approximate expressions for the heat capacities:

$$C_{P,m}(CO_2)/\text{J K}^{-1}\text{ mol}^{-1} = 44.22 + 8.79 \times 10^{-3}T/\text{K}$$

$$C_{P,m}(H_2O)/\text{J K}^{-1}\text{ mol}^{-1} = 30.54 + 1.03 \times 10^{-2}T/\text{K}$$

***2.67.** Two moles of a gas are compressed isothermally and reversibly, at 300 K, from an initial volume of 10 dm^3 to a final volume of 1 dm^3. If the equation of state of the gas is $P(V_m - b) = RT$, with $b = 0.04$ dm^3 mol^{-1}, calculate the work done on the system, ΔU, and ΔH.

***2.68.** Three moles of a gas are compressed isothermally and reversibly, at 300 K, from an initial volume of 20 dm^3 to a final volume of 1 dm^3. If the equation of state of the gas is

$$\left(P + \frac{n^2 a}{V_m^2}\right)V_m = nRT$$

with $a = 0.55$ Pa m^6 mol^{-1}, calculate the work done, ΔU, and ΔH.

***2.69.** One mole of a van der Waals gas at 300 K is compressed isothermally and reversibly from 60 dm^3 to 20 dm^3. If the constants in the van der Waals equation are

$$a = 0.556 \text{ Pa m}^6 \text{ mol}^{-1} \quad \text{and} \quad b = 0.064 \text{ dm}^3 \text{ mol}^{-1}$$

calculate w_{rev}, ΔU, and ΔH.

***2.70.** Show that the Joule-Thomson coefficient μ can be written as

$$\mu = -\frac{1}{C_P}\left(\frac{\partial H}{\partial P}\right)_T$$

Then, for a van der Waals gas for which μ can be written as

$$\mu = \frac{2a/RT - b}{C_P}$$

calculate ΔH for the isothermal compression of 1.00 mol of the gas at 300 K from 1 bar to 100 bar.

Essay Questions

2.71. Explain clearly what is meant by a thermodynamically reversible process. Why is the reversible work done by a system the maximum work?

2.72. Explain the thermodynamic meaning of a system, distinguishing between open, closed, and isolated systems. Which one of these systems is (a) a fish swimming in the sea or (b) an egg?

SUGGESTED READING

E. F. Caldin, *An Introduction to Chemical Thermodynamics,* Oxford: Clarendon Press, 1961.

J. de Heer, *Phenomenological Thermodynamics,* Englewood Cliffs, NJ: Prentice Hall, 1986. This textbook gives a thorough account of the fundamental principles.

I. M. Klotz and R. M. Rosenberg, *Chemical Thermodynamics,* New York: Benjamin, 1972, 4th edition, 1986; New York: John Wiley, 5th edition, 1994.

B. H. Mahan, *Elementary Chemical Thermodynamics,* New York: W. A. Benjamin, 1963.

M. L. McGlashan, *Chemical Thermodynamics,* New York: Academic Press, 1979. This is a somewhat more advanced account of the subject.

C. T. Mortimer, *Reaction Heats and Bond Strengths,* Reading, MA: Addison-Wesley, 1962.

E. B. Smith, *Basic Chemical Thermodynamics*, Oxford: Clarendon Press, paperback, 1990.

For a discussion of the thermodynamic meaning of the word 'adiabatic,' and the different meaning that it acquired by mistake in chemical kinetics, see:

K. J. Laidler, "The meaning of 'adiabatic,'" *Canadian Journal of Chemistry, 72,* 936–938(1994).

The Second and Third Laws of Thermodynamics

3

PREVIEW

The second law of thermodynamics deals with whether a chemical or physical process occurs spontaneously (i.e., naturally). Many processes, such as the flow of heat from a lower to a higher temperature, do not violate the first law, but nevertheless they do not occur naturally. The second law defines the conditions under which processes take place.

The property that is of fundamental importance in this connection is the *entropy.* If a process occurs reversibly from state A to state B, with infinitesimal amounts of heat dq_{rev} absorbed at each stage, the entropy change is defined as

$$\Delta S_{A \to B} = \int_A^B \frac{dq_{rev}}{T}$$

This relationship provides a way of determining entropy changes from thermal data.

The great importance of the entropy is that if a process $A \to B$ is to occur spontaneously, the total entropy of the system plus the surroundings must *increase.* Entropy provides a measure of *probability,* in that a process that occurs spontaneously must involve a change from a state of lower probability to one of higher probability.

The *third law of thermodynamics* is concerned with the absolute values of entropies, which are based on the values at the absolute zero. This law is particularly important for chemical and physical studies near the absolute zero. It makes possible the tabulation of absolute entropies (Table 3.2).

It is usually inconvenient to consider the entropy of the surroundings as well as the entropy of the system itself. It has proved possible to define thermodynamic functions that are related only to the *system* but which nevertheless provide information about spontaneity of reaction. Such a function is the *Gibbs energy G,* defined as equal to $H - TS$, where H is the enthalpy. For a system *at constant temperature and pressure* this function has a minimum value when the system is at equilibrium. For a reaction to proceed spontaneously, there must be a *decrease* in G. The decrease in G during a process at constant T and P is the amount of work, other than PV work, that can be performed.

For processes occurring at constant temperature and volume, use is made of the *Helmholtz energy A*, equal to $U - TS$, where U is the internal energy. A process *at constant temperature and volume* tends toward a state of lower A, and at equilibrium A has a minimum value.

An important thermodynamic relationship is the *Gibbs-Helmholtz equation,* which is concerned with the effect of *temperature* on the Gibbs energy. In Chapter 4 we will see that this relationship leads to a way of obtaining enthalpies of reaction from the temperature dependence of the equilibrium constant.

OBJECTIVES

After studying this chapter, the student should be able to:

- Understand the Carnot cycle and its significance, including the reversible isothermal and adiabatic expansions and contractions.
- Calculate internal energy changes and reversible heat changes, particularly those relating to the Carnot cycle.
- Obtain an expression for the efficiency of a Carnot cycle and of other types of cycles.
- Understand the proof of Carnot's theorem and link this idea to the thermodynamic scale of temperature.
- Apply the generalized cycle to understand the concepts of state functions and the definition of *entropy*.
- Understand the molecular interpretation of entropy and the entropy changes during changes of state at constant temperature and the other entropic processes listed in Section 3.4.
- Calculate the entropy changes in the various transitions listed and, for an ideal gas, determine the ΔS at constant T and for changing T and V.
- Determine the entropy of mixing in terms of mole fractions and volumes and understand all examples.
- Apply the concept of entropy to understand Nernst's heat theorem. Understand the various ways of stating the theorem and the application of cryogenics to reaching absolute zero. Define and use the absolute entropies given at 25 °C and 1 bar pressure.
- List the conditions for equilibrium in terms of the Gibbs and Helmholtz energies.
- Calculate changes in enthalpy and correctly link the terms *exothermic* and *endothermic* to the respective changes in enthalpy and the terms *exergonic* and *endergonic* to the respective changes in Gibbs energy. Also calculate standard Gibbs energies of formation for chemical processes as well as determine the reversible and nonreversible work in a process.
- Understand the Maxwell relations and Euler's reciprocity theorem and apply them to derive the thermodynamic equation of state.
- Work through the Joule-Thomson treatment and be able to apply it to the cubic expansion coefficient and internal pressure.
- Understand the concepts of fugacity and activity and their use.
- Understand the Gibbs-Helmholtz equation and its application.
- Apply thermodynamic limitations to energy conversion in order to calculate first and second law efficiencies for heating and cooling.

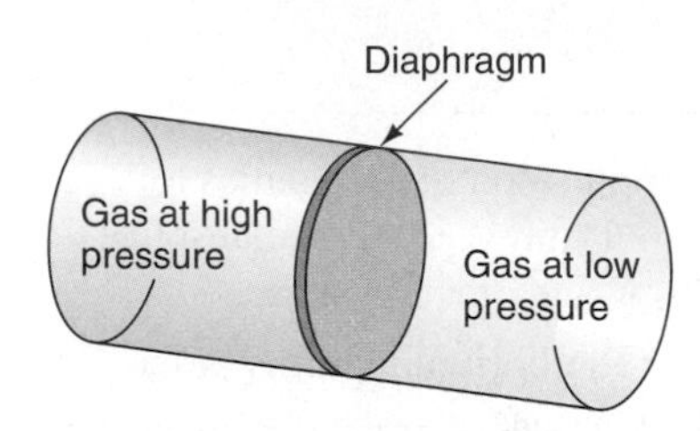

FIGURE 3.1
A cylinder separated into two compartments, with gases at two different pressures on the two sides of the partition.

Natural (Spontaneous) Processes

We have seen in the last chapter that the first law of thermodynamics is concerned with the conservation of energy and with the interrelationship of work and heat. A second important problem with which thermodynamics deals is *whether a chemical or physical change can take place spontaneously.* This particular aspect is the concern of the **second law of thermodynamics.**

There are several well-known examples of processes which do not violate the first law but which do not occur naturally. Suppose, for example, that a cylinder is separated into two compartments by means of a diaphragm (Figure 3.1), with a gas at high pressure on one side and a gas at low pressure on the other. If the diaphragm is ruptured, there will be an *equalization of pressure.* However, the reverse of this process does not occur; if we start with gas at uniform pressure, it is highly unlikely that we will obtain gas at high pressure on one side and gas at low pressure on the other side. The first process, in which the gas pressures equalize, is known as a *natural* or *spontaneous* process. The reverse process, which does not occur but can only be imagined, is known as an *unnatural process.* Note that the natural, spontaneous process is *irreversible* in the sense in which we used the term in Section 2.4. We could instead equalize gas pressures by reversibly expanding the gas at the higher pressure, in which case it would perform work. The reverse process could then be carried out, but only at the expense of work that we would have to perform on the system. Such a process would not be a spontaneous one.

Another example of a process that occurs naturally and spontaneously in one direction but not in the other is a mixing process. Suppose with reference to Figure 3.1 that we have oxygen on one side of the partition and nitrogen at the same pressure on the other side. If the partition is removed, the gases will mix. Once two gases have mixed, however, we cannot unmix them simply by waiting for them to do so; we can only unmix them by carrying out a process which involves the performance of work on the system. Unmixing does not involve a violation of the first law, but it would violate the second law.

A third example is the *equalization of temperature.* Suppose that we bring together a hot solid and a cold one. Heat will pass from the hot to the cold solid until the temperatures are equal; this is a spontaneous process. We know from common experience that heat will not flow in the opposite direction, from a cold to a hot body. There would be no violation of the first law if this occurred, but there would be a violation of the second law.

A fourth example relates to a *chemical reaction.* There are many reactions that will go spontaneously in one direction but not in the other. One of them is the reaction between hydrogen and oxygen to form water:

$$2H_2 + O_2 \rightarrow 2H_2O$$

If we simply bring together two parts of hydrogen and one of oxygen, they react together so slowly that there is no observable change. However, we can readily cause reaction to occur essentially to completion, for example by passing a spark through the system. The reaction is accompanied by the evolution of considerable heat (ΔH° is negative). There would be no violation of the first law if we were to return this heat to water and reconvert it into hydrogen and oxygen, but in practice this cannot be done unless we employ some process in which we perform work on the system. The reaction from left to right is spontaneous; that from right to left does not occur naturally.

It is obviously a matter of great importance to understand the factors that determine *the direction in which a process can occur spontaneously.* This amounts

RUDOLF JULIUS EMMANUEL CLAUSIUS (1822–1888)

Clausius was born in Köslin, Prussia (now part of Poland) and educated at the Universities of Berlin and Halle, obtaining his doctorate at Halle in 1847. After appointments in Zürich and Würzburg, he moved to the University of Bonn in 1869 and remained there the rest of his life, serving as Rector in his later years.

Clausius was a theoretical scientist, but much of his work had practical implications. He investigated many topics, but his most important contributions were made in thermodynamics and the kinetic theory of gases. His main contribution to thermodynamics was a detailed analysis of the Carnot cycle and the introduction, in 1854–1865, of the concept of entropy (Section 3.1). In 1865–1867 he published an important book, *Die mechanische Wärmetheorie,* which was later translated into English as *The Mechanical Theory of Heat.* His most important contribution to kinetic theory was introducing the concept of the mean free path (Section 1.9). A weakness of his kinetic theory work was that he dealt with average velocities without using Maxwell's distribution equations (Section 1.11). Also, he never recognized that the second law of thermodynamics is a statistical law; he believed it was a consequence of the principles of mechanics.

From time to time he became interested in electricity and magnetism and electrokinetic effects. In 1857, he developed the theory that electrolyte molecules in solution are constantly interchanging, and that the effect of an imposed electromotive force is to influence the interchange and not to cause it. This has been referred to as the Williamson-Clausius hypothesis, but the expression does not seem justified as Williamson did not, in fact, consider electrolytes. After 1875 Clausius's main work was on electrodynamic theory, but it was not very successful.

Because Clausius did not write with great clarity, general acceptance of his idea of entropy was slow. Lord Kelvin, to the end of his life in 1907, never believed that the concept was of any help to the understanding of the second law of thermodynamics. Clausius was an influential man of fine character and personality, and was exceptionally patient with those who did not accept his ideas. He received many honors at home and abroad.

References: E. E. Daub, *Dictionary of Scientific Biography, 3,* 303–311(1971).

S. G. Brush, *The Kind of Motion that We Call Heat,* Amsterdam: North Holland Publishing Co., 1976.

M. J. Klein, "Gibbs on Clausius," *Historical Studies in the Physical Sciences, 1,* 127–149(1969).

to asking what factors determine the *position of equilibrium,* because a system will move spontaneously toward the state of equilibrium (although it may do so exceedingly slowly). These matters are the concern of the second law of thermodynamics.

The second law of thermodynamics was deduced in 1850–1851 independently and in slightly different forms by the German physicist Rudolf Clausius (1822–1888) and the British physicist William Thomson (1824–1907), who is usually known by his later title of Lord Kelvin. Both of them deduced the law from the Carnot cycle, to be considered in some detail in the next section. The law can be expressed as follows:

A Statement of the Second Law

It is impossible for an engine to perform work by cooling a portion of matter to a temperature below that of the coldest part of the surroundings.

For example, a ship could not be driven by abstracting heat from the surrounding water and converting it into work. Other violations of the second law can be related to this particular expression of the law. As we will see, Clausius later introduced the concept of *entropy* to interpret the second law, and this brought about a great clarification of the law.

Some textbooks, beginning with E. A. Guggenheim's *Modern Thermodynamics* published in 1933, introduce the concept of entropy as a postulate and then proceed to consider its applications. This shortcut may appear to have some advantages, but it has the serious shortcoming of not revealing the second law as a direct deduction from experimental observations. We consider that approaching the second law by way of the Carnot cycle, considered old-fashioned by some (although a practice begun in 1933 might also be branded old-fashioned!), gives a much deeper understanding of the law.

3.1 The Carnot Cycle

Watt's Steam Engine

Study the 4 strokes of the Carnot cycle; The strokes; The Carnot cycle.

The first great step in the direction of the second law of thermodynamics was taken in 1825 by the young French engineer Sadi Carnot (1796–1832) in his book *La puissance motrice du feu* (The motive power of heat). Carnot presented an interpretation of a particular type of steam engine that had been invented by the Scottish engineer James Watt (1736–1819). The earliest steam engines had involved a single cylinder with a piston; steam was introduced into the cylinder, causing the piston to move, and the cylinder was then cooled, causing the piston to move in the opposite direction. There was great wastage of heat, and such engines had a very low efficiency, converting only about 1% of the heat into work. Watt's great innovation was to use two cylinders connected together; one was kept at the temperature of steam, and the other (the "separate condenser") was kept cooled. Early Watt engines had an efficiency of about 8%, and his further innovations led to engines having an efficiency of about 19%.

Carnot's great theoretical contribution was to realize that the production of work by an engine of this type depends on a flow of heat from a higher temperature T_h (h standing for hotter) to a lower temperature T_c (c standing for colder). He considered an ideal type of engine, involving an ideal gas, in which all processes occurred reversibly, and showed that such an engine would have the maximum possible efficiency for any engine working between the same two temperatures. Carnot's treatment was greatly clarified in 1834 by the French engineer Benoit Clapeyron (1799–1864), who restated Carnot's ideas in the language of calculus, and made use of pressure-volume diagrams, as will now be done in our presentation of the Carnot cycle.

Suppose that 1 mol of an ideal gas is contained in a cylinder with a piston, at an initial pressure of P_1, an initial volume of V_1, and an initial temperature of T_h. We refer to this gas as being in state A. Figure 3.2(a) is a pressure-volume diagram indicating the initial state A and paths. Figure 3.2(b) plots the Carnot cycle for 1 mole of an ideal gas using $C_P/C_V = \gamma = 1.40$, the value for N_2 gas.

Now we will bring about four reversible changes in the system, which will eventually bring it back to the initial state A. We first bring about an *isothermal* expansion $A \rightarrow B$, the pressure and volume changing to P_2 and V_2 and the temperature remaining at T_h; we can imagine the cylinder to be immersed in a bath of liquid at temperature T_h. In the second step we bring about an adiabatic expansion (i.e., one in which no heat is allowed to leave or enter the system); we could accomplish this

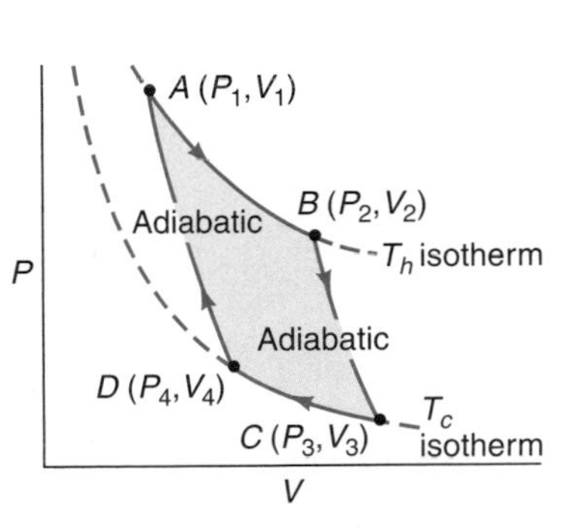

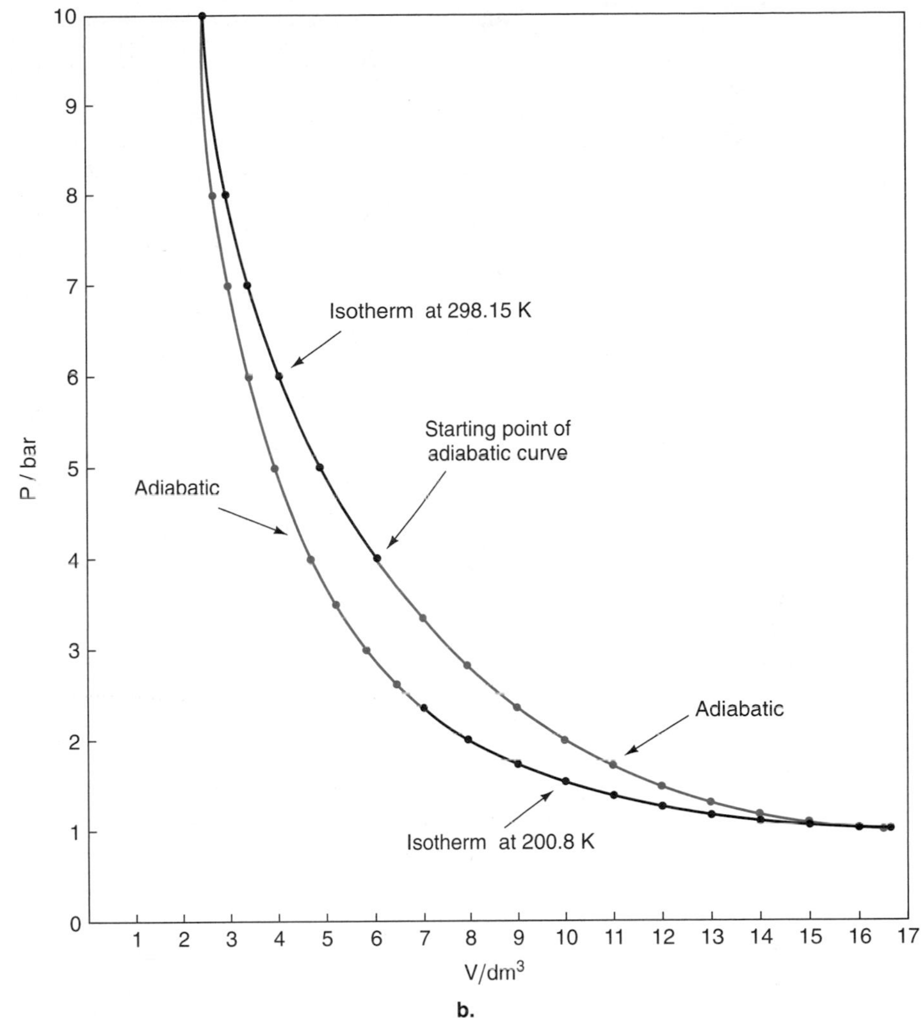

FIGURE 3.2
(a) Pressure-volume diagram for the Carnot cycle; *AB* and *CD* are isotherms, and *BC* and *DA* are adiabatics (no heat transfer). (b) The Carnot cycle for 1 mole of an ideal gas that starts at the top at 10.0 bar and 298.15 K. The value of $C_P/C_V = \gamma$ used is that for nitrogen gas, 1.40.

by surrounding the cylinder with insulating material. Since the gas does work during expansion, and no heat is supplied, the temperature must fall; we call the final temperature T_c and the pressure and volume P_3 and V_3, respectively. Third, we compress the gas isothermally (at temperature T_c) until the pressure and volume are P_4 and V_4. Finally, the gas is compressed adiabatically until it returns to its original state *A* (P_1, V_1, T_h). The performance of work on the system, with no heat transfer permitted, raises the temperature from T_c to T_h.

Let us consider these four steps in further detail. In particular we want to know the ΔU values for each step, the amounts of heat absorbed (q), and the work done (w). The expressions for these quantities are summarized in Table 3.1.

Reversible Isothermal Expansion

1. *Step* $A \rightarrow B$ is the reversible isothermal expansion at T_h. In Section 2.6, pp. 77, we proved that for the isothermal expansion of an ideal gas there is no change of internal energy:

$$\Delta U_{A \rightarrow B} = 0 \tag{3.1}$$

TABLE 3.1 Values of ΔU, q, and w for the Four Reversible Steps in the Carnot Cycle, for 1 mol of Ideal Gas

Step	ΔU	q_{rev}	w_{rev}
$A \rightarrow B$	0	$RT_h \ln \frac{V_2}{V_1}$	$RT_h \ln \frac{V_1}{V_2}$
$B \rightarrow C$	$C_V(T_c - T_h)$	0	$C_V(T_c - T_h)$
$C \rightarrow D$	0	$RT_c \ln \frac{V_4}{V_3}$	$RT_c \ln \frac{V_3}{V_4}$
$D \rightarrow A$	$C_V(T_h - T_c)$	0	$C_V(T_h - T_c)$
Net	0	$R(T_h - T_c) \ln \frac{V_2}{V_1}$	$R(T_h - T_c) \ln \frac{V_1}{V_2}$ (since $\frac{V_1}{V_2} = \frac{V_4}{V_3}$; see Eq. 3.16)

Understand how each stroke of the carnot cycle is evaluated; The parts; The Carnot cycle.

We also showed (Eq. 2.72) that the work done on the system in an isothermal reversible process is $RT \ln(V_{initial}/V_{final})$:

$$w_{A \rightarrow B} = RT_h \ln \frac{V_1}{V_2} \quad \text{(for 1 mol)} \tag{3.2}$$

Since by the first law

$$\Delta U_{A \rightarrow B} = q_{A \rightarrow B} + w_{A \rightarrow B} \tag{3.3}$$

it follows that

$$q_{A \rightarrow B} = RT_h \ln \frac{V_2}{V_1} \tag{3.4}$$

Reversible Adiabatic Expansion

2. *Step* $B \rightarrow C$ involves surrounding the cylinder with an insulating jacket and allowing the system to expand reversibly and adiabatically to a volume of V_3. Since the process is adiabatic,

$$q_{B \rightarrow C} = 0 \tag{3.5}$$

We saw in Eq. 2.29 that ΔU for an adiabatic process involving 1 mol of gas is

$$\Delta U = C_V(T_{final} - T_{initial}) \tag{3.6}$$

so that for the process $B \rightarrow C$

$$\Delta U_{B \rightarrow C} = C_V(T_c - T_h) \tag{3.7}$$

Application of the first law then gives

$$w_{B \rightarrow C} = C_V(T_c - T_h) \tag{3.8}$$

Reversible Isothermal Compression

3. *Step* $C \rightarrow D$ involves placing the cylinder in a heat bath at temperature T_c and compressing the gas reversibly until the volume and pressure are V_4 and P_4, respectively. The state D must lie on the adiabatic that passes through A (Figure 3.2(a)). Since process $C \rightarrow D$ is isothermal,

$$\Delta U_{C \rightarrow D} = 0 \tag{3.9}$$

The work done on the system is

$$w_{C \rightarrow D} = RT_c \ln \frac{V_3}{V_4} \tag{3.10}$$

and this is a positive quantity since $V_3 > V_4$ (i.e., we must do work to compress the gas). By the first law,

$$q_{C\to D} = RT_c \ln \frac{V_4}{V_3} \tag{3.11}$$

which means that a negative amount of heat is absorbed (i.e., heat is actually rejected).

Reversible Adiabatic Compression

4. *Step D → A*. The gas is finally compressed reversibly and adiabatically from D to A. The heat absorbed is zero:

$$q_{D\to A} = 0 \tag{3.12}$$

The $\Delta U_{D\to A}$ value is

$$\Delta U_{D\to A} = C_V(T_h - T_c) \tag{3.13}$$

By the first law,

$$w_{D\to A} = C_V(T_h - T_c) \tag{3.14}$$

Table 3.1 gives, as well as the individual contributions, the net contributions for the entire cycle. We see that ΔU for the cycle is zero; the contributions for the isotherms are zero, whereas those for the adiabatics are equal and opposite to each other. This result that ΔU is zero for the entire cycle is necessary in view of the fact that the internal energy is a state function; in completing the cycle the system is returned to its original state, and the internal energy is therefore unchanged.

Equation 2.90 applies to an adiabatic process, and if we apply this equation to the two processes $B \to C$ and $D \to A$ in Figure 3.2(a), we have

$$\frac{T_h}{T_c} = \left(\frac{V_4}{V_1}\right)^{\gamma-1} \quad \text{and} \quad \frac{T_h}{T_c} = \left(\frac{V_3}{V_2}\right)^{\gamma-1} \tag{3.15}$$

and therefore

$$\frac{V_4}{V_1} = \frac{V_3}{V_2} \quad \text{or} \quad \frac{V_4}{V_3} = \frac{V_1}{V_2} \tag{3.16}$$

The net q_{rev} value (see Table 3.1) is

$$q_{rev} = RT_h \ln \frac{V_2}{V_1} + RT_c \ln \frac{V_4}{V_3} \tag{3.17}$$

and since $V_4/V_3 = V_1/V_2$, this becomes

$$q_{rev} = R(T_h - T_c) \ln \frac{V_2}{V_1} \tag{3.18}$$

which is a positive quantity. Since $q = -w$, by the first law, it follows that

$$w_{rev} = R(T_h - T_c) \ln \frac{V_1}{V_2} \tag{3.19}$$

This is a negative quantity (i.e., a positive amount of work has been done by the system).

Note that the net work done *by* the system, and hence the net heat absorbed, is represented by the area within the Carnot diagram. This is illustrated in Figure 3.3. Diagram (a) shows the processes $A \to B$ and $B \to C$; both are expansions, and the work done by the system is represented by the area below the lines, which is shaded. Diagram (b) shows the processes $C \to D$ and $D \to A$, in which work is

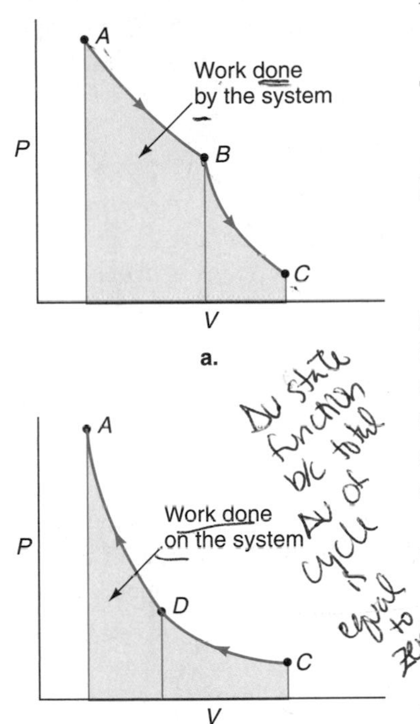

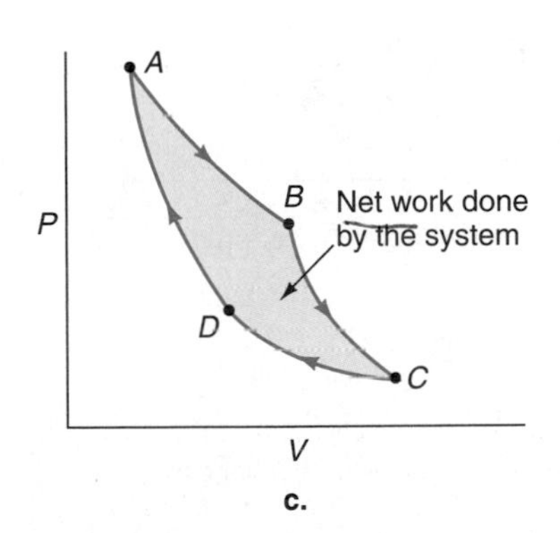

FIGURE 3.3
Diagram (a) shows the work done *by* the system in going from $A \to B$ and then from $B \to C$; (b) shows the work done *on* the system in the return process, via D. The *net* work, obtained by subtracting the shaded area in (b) from that in (a), is thus the area enclosed by the cycle (c).

done on the system in the amount shown by the shaded area. The *net* work done by the system is thus represented by the area in (a) minus the area in (b) and is thus the area within the cycle shown in (c).

The important thing to note about the Carnot cycle is that we have returned the system to its original state by processes in the course of which a net amount of work has been done by the system. This work has been performed at the expense of heat absorbed, as required by the first law. Since work and heat are not state functions, net work can be done even though the system returns to its original state.

Efficiency of a Reversible Carnot Engine

The *efficiency* of the reversible Carnot engine can be defined as *the work done by the system* during the cycle, *divided by the work that would have been done if all the heat absorbed at the higher temperature had been converted into work.* Thus

Efficiency of a Reversible Heat Engine

$$\text{efficiency} = \frac{w}{q_h} = \frac{R(T_h - T_c)\ln(V_2/V_1)}{RT_h \ln(V_2/V_1)} \tag{3.20}$$

$$= \frac{T_h - T_c}{T_h} \tag{3.21}$$

This efficiency is unity (i.e., 100%) only if the lower temperature T_c is zero (i.e., if the heat is rejected at the absolute zero). This result provides a definition of the absolute zero. The efficiency is also the net heat absorbed, $q_h + q_c$, divided by the heat absorbed at the higher temperature, q_h:

$$\text{efficiency} = \frac{q_h + q_c}{q_h} \quad \text{(Note that } q_c \text{ is in fact negative.)} \tag{3.22}$$

From Eqs. 3.21 and 3.22 it follows that, for the reversible engine,

$$\frac{T_h - T_c}{T_h} = \frac{q_h + q_c}{q_h} \quad \text{or} \quad -\frac{T_h}{T_c} = \frac{q_h}{q_c} \tag{3.23}$$

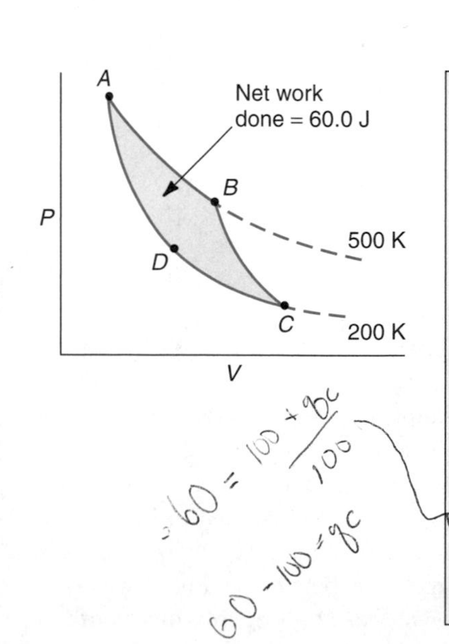

EXAMPLE 3.1 The accompanying diagram represents a reversible Carnot cycle for an ideal gas:

a. What is the thermodynamic efficiency of the engine?
b. How much heat is absorbed at 500 K?
c. How much heat is rejected at 200 K?
d. In order for the engine to perform 1.00 kJ of work, how much heat must be absorbed?

Solution

a. By Eq. 3.21, the efficiency = $(T_h - T_c)/T_h = 0.600 = 60.0\%$.
b. Since the work done is 60.0 J, and the efficiency is 0.600, the heat absorbed at 500 K = 60.0/0.600 = 100 J.
c. The heat rejected at 200 K = 100 − 60 = 40 J.
d. Since the efficiency is 0.600, the heat required to produce 1000 J of work is 1000/0.600 = 1667 J = 1.67 kJ.

Carnot's Theorem

Carnot's cycle was discussed previously for an ideal gas. Similar reversible cycles can be performed with other materials, including solids and liquids, and the efficiencies of these cycles determined. The importance of the reversible cycle for the ideal gas is that it gives us an extremely simple expression for the efficiency, namely $(T_h - T_c)/T_h$. We will now show, by a theorem due to Carnot, that *the efficiency of all reversible cycles operating between the temperatures T_h and T_c is the same,* namely $(T_h - T_c)/T_h$. This result leads to important quantitative formulations of the second law of thermodynamics.

Proof of Carnot's Theorem

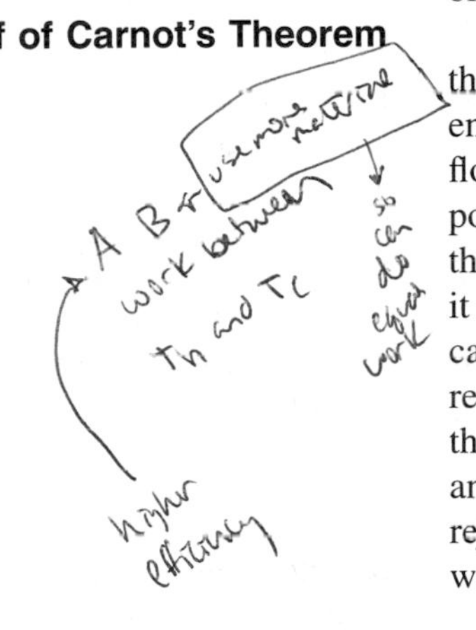

Carnot's theorem employs the method of *reductio ad absurdum.* It supposes that there are two reversible engines that operate between T_h and T_c but have different efficiencies, and then it shows that this would lead to the possibility of heat flowing from a lower to a higher temperature, which is contrary to experience. Suppose that there are two engines *A* and *B* operating reversibly between T_h and T_c and that *A* has a higher efficiency than *B*. By making *B* use a larger amount of material, it is possible to arrange for it to do exactly the same amount of work in a cycle. We can then imagine the engines to be coupled together so that *A* forces *B* to work in reverse, so that it rejects heat at T_h and absorbs it at T_c. During each complete cycle the more efficient engine *A* draws a quantity of heat q_h from the heat reservoir at T_h and rejects $-q_c$ at T_c. The less efficient engine *B*, which is being driven backward, rejects q'_h at T_h and absorbs $-q'_c$ at T_c. (If it were operating in its normal manner, it would absorb q'_h at T_h and reject $-q'_c$ at T_c.)

Since the engines were adjusted to perform equal amounts of work in a cycle, engine *A* only just operates *B* in reverse, with no extra energy available. The work performed *by A* is $q_h + q_c$, while that performed *on B* is $q'_h + q'_c$, and these quantities are equal:

$$q_h + q_c = q'_h + q'_c \tag{3.24}$$

Since engine *A* is more efficient than *B*,

$$\frac{q_h + q_c}{q_h} > \frac{q'_h + q'_c}{q'_h} \tag{3.25}$$

From Eqs. 3.24 and 3.25 it follows that

$$q'_h > q_h \tag{3.26}$$

and

$$q_c > q'_c \tag{3.27}$$

During the operation of the cycle, in which *A* has forced *B* to work backward, *A* has absorbed q_h at T_h and *B* has rejected q'_h at T_h; the combined system $A + B$ has therefore absorbed $q_h - q'_h$ at T_h, and since q'_h is greater than q_h (Eq. 3.26), this is a negative quantity (i.e., the system has rejected a positive amount of heat at the higher temperature). At the lower temperature T_c, *A* has absorbed q_c while *B* has absorbed $-q'_c$; the combined system has therefore absorbed $q_c - q'_c$ at this temperature, and this according to Eq. 3.27 is a positive quantity (it is in fact equal to $q'_h - q_h$ by Eq. 3.24). The $A + B$ system has thus, in performing the cycle, absorbed heat at a lower temperature and rejected it at a higher temperature. It is contrary to experience that heat can flow uphill in this way, during a complete cycle of operations in which the system ends up in its initial state. It must therefore be concluded that the original postulate is invalid; there cannot be two reversible engines, A and B, operating reversibly between two fixed temperatures and having different efficiencies. Thus, the efficiencies of all reversible engines must be the same as that for the ideal gas reversible engine, namely,

$$\frac{T_h - T_c}{T_h}$$

This conclusion is not a necessary consequence of the *first* law of thermodynamics. If engine A were more efficient than engine B, energy would not have been created or destroyed if engine A drove engine B backward, because the net work done would be equivalent to the heat extracted from the reservoir. However, the removal of heat from a reservoir and its conversion into work without any other changes in the system have never been observed. If they could occur, it would be possible, for example, for a ship to propel itself by removing heat from the surrounding water and converting it into work; no fuel would be needed. Such a continuous extraction of work from the environment has been called *perpetual motion of the second kind.* The first law forbids *perpetual motion of the first kind,* which involves the creation of energy from nothing. The second law forbids the operation of engines in which energy is continuously extracted from an environment which is at the same temperature as the system or at a lower temperature. We shall see later, when we consider *heat pumps* (Section 3.10), that it is possible to extract some heat from a cooler environment provided that some work is done by the system at the same time.

The Thermodynamic Scale of Temperature

In 1848 William Thomson (Lord Kelvin) proposed constructing an absolute temperature scale based on the Carnot cycle. We saw in Section 1.6 that such a scale could be constructed on the basis of the law of expansion of gases, but one weakness of this procedure is that it depends on the behavior of particular substances. Another weakness is that the procedure involves the assumption that the true temperature is reliably measured by the variation of the length of a column of mercury in a thermometer. Even if the assumption of a linear variation of volume of mercury with temperature is valid at ordinary temperatures, it may be questioned whether the linear dependence is valid down to the absolute zero.

Accounts of Kelvin's absolute temperature scale usually state that he based his scale on the efficiency of a reversible Carnot engine, but this is not really the case. For one thing, the concept of the efficiency of a Carnot engine had not been appreciated in his time. Also, a reversible engine must operate infinitely slowly; one would have to work with engines that are not quite reversible, and it would be quite impractical to measure their efficiencies in a reliable way. What in fact Kelvin did was to analyze data obtained mainly by the French physical chemist Henri Victor Regnault (1810–1878) on the physical properties of steam at various temperatures. From this analysis he was able to conclude that the work done in an isothermal reversible process is correctly given by equations such as Eq. 3.2, using as the temperature that given by the law of expansion of gases, with approximately 273° being added to the Celsius temperatures. He did not really create a scale; he merely confirmed that the scale based on the gas law was acceptable.

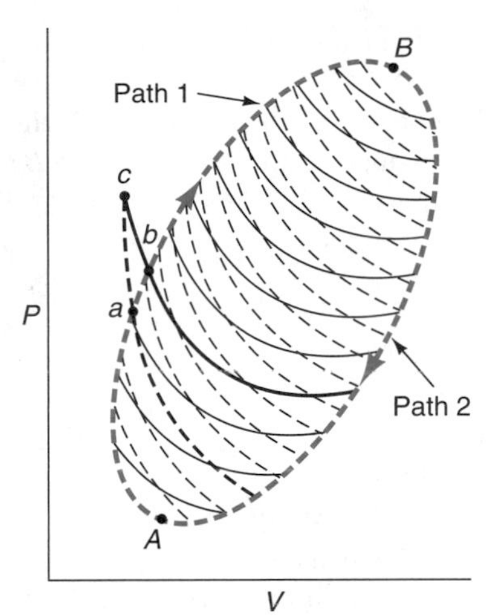

FIGURE 3.4
A generalized cycle *ABA*. The cycle can be traversed via infinitesimal adiabatics and isothermals; thus we can go from *a* to *b* by the adiabatic *ac* followed by the isothermal *cb*.

The Generalized Cycle: The Concept of Entropy

From Eq. 3.23 for the efficiency of a reversible Carnot cycle, it follows that

$$\frac{q_h}{T_h} + \frac{q_c}{T_c} = 0 \tag{3.28}$$

This equation applies to any reversible cycle that has distinct isothermal and adiabatic parts, and it can be put into a more general form to apply to any reversible cycle. Consider the cycle represented by *ABA* in Figure 3.4. During the operation of the engine from *A* to *B* and back to *A*, there is heat exchange between the system and its environment at various temperatures. The whole cycle may be split up into elements such as *ab,* shown in the figure. The distance between *a* and *b* should be infinitesimally small, but it is enlarged in the diagram. During the change from *a* to *b*, the pressure, volume, and temperature have all increased, and a quantity of heat has been absorbed. Let the temperature corresponding to *a* be T and that corresponding to *b* be $T + dT$. The isothermal corresponding to $T + dT$ is shown as *bc* and the adiabatic at *a* as *ac*; the two intersect at *c*. The change from *a* to *b* may therefore be carried out by means of an adiabatic change *ac*, followed by an isothermal change *cb*. During the isothermal process an amount of heat dq has been absorbed by the system. If the whole cycle is completed in this manner, quantities of heat dq_1, dq_2, etc., will have been absorbed by the system during isothermal changes carried out at the temperatures T_1, T_2, etc. Equation 3.28 is therefore replaced by

$$\sum_i \frac{dq_i}{T_i} = \frac{dq_1}{T_1} + \frac{dq_2}{T_2} + \frac{dq_3}{T_3} + \cdots = 0 \tag{3.29}$$

See how irreversible processes can be carried out by a series of reversible Carnot cycles; Carnot grid; Cycles are natural.

the summation being made round the entire cycle. Since the *ab* elements are all infinitesimal, the cycle consisting of reversible isothermal and adiabatic steps is equivalent to the original cycle $A \rightarrow B \rightarrow A$. It follows that for the original cycle

$$\oint \frac{dq_{rev}}{T} = 0 \tag{3.30}$$

where the symbol $\oint$ denotes integration over a complete cycle.

State Functions

This result that the integral of dq_{rev}/T over the entire cycle is equal to zero is a very important one. We have seen that certain functions are *state functions,* which means that their value is a true property of the system; the change in a state function when we pass from state *A* to state *B* is independent of the path, and therefore a state function will not change when we traverse a complete cycle. For example, pressure P, volume V, temperature T, and internal energy U are state functions; thus we can write

$$\oint dP = 0; \quad \oint dV = 0; \quad \oint dT = 0; \quad \oint dU = 0 \tag{3.31}$$

The relationship expressed in Eq. 3.30 is therefore an important one, and it is convenient to write dq_{rev}/T as dS, so that

Definition of Entropy

$$\oint \frac{dq_{rev}}{T} \equiv \oint dS = 0 \tag{3.32}$$

The property S is known as the *entropy* of the system, and it is a state function. The word **entropy** was coined in 1854 by the German physicist Rudolf Julius

Emmanuel Clausius (1822–1888); the word literally means *to give a direction* (Greek *en,* in; *trope,* change, transformation).

Since entropy is a state function, its value is independent of the path by which the state is reached. Thus if we consider a reversible change from state A to state B and back again (Figure 3.4), it follows from Eq. 3.32 that

$$\int_A^B dS + \int_B^A dS = 0 \tag{3.33}$$

or

$$\Delta S_{A\to B}^{(1)} + \Delta S_{B\to A}^{(2)} = 0 \tag{3.34}$$

where $\Delta S_{A\to B}^{(1)}$ denotes the change of entropy in going from A to B by path (1) and $\Delta S_{B\to A}^{(2)}$ is the change in going from B to A by path (2). The change of entropy in going from B to A by path (2) is the negative of the change in going from A to B by this path,

$$\Delta S_{B\to A}^{(2)} = -\Delta S_{A\to B}^{(2)} \tag{3.35}$$

It therefore follows from Eqs. 3.34 and 3.35 that

$$\Delta S_{A\to B}^{(1)} = \Delta S_{A\to B}^{(2)} \tag{3.36}$$

That is, the change of entropy is the same whatever path is followed.

3.2 Irreversible Processes

The treatment of thermodynamically reversible processes is of great importance in connection with the second law. However, in practice we are concerned with thermodynamically irreversible processes, because these are the processes that occur naturally. It is therefore important to consider the relationships that apply to irreversible processes.

Heat Transfer

A simple example of an irreversible process is the transfer of heat from a warmer to a colder body. Suppose that we have two reservoirs, a warm one at a temperature T_h and a cooler one at a temperature T_c. We might imagine connecting these together by a metal rod, as shown in Figure 3.5a, and waiting until an amount of heat q has flowed from the hotter to the cooler reservoir. To simplify the argument, let us suppose that the reservoirs are so large that the transfer of heat does not change their temperatures appreciably.

In order to calculate the entropy changes in the two reservoirs after this irreversible process has occurred, we must devise a way of transferring the heat reversibly. We can make use of an ideal gas to carry out the heat transfer process, as shown in Figure 3.5b. The gas is contained in a cylinder with a piston, and we first place it in the warm reservoir, at temperature T_h, and expand it reversibly and isothermally until it has taken up heat equal to q. The gas is then removed from the hot reservoir, placed in an insulating container, and allowed to expand reversibly and adiabatically until its temperature has fallen to T_c. Finally, the gas is placed in contact with the colder reservoir at T_c and compressed isothermally until it has given up heat equal to q.

The entropy changes that have occurred in the two reservoirs and in the gas are shown in Figure 3.5. We see that the two reservoirs, i.e., the surroundings, have experienced a net entropy change of

Understand the heat engine; Heat engines; Energy, heat, and work. (See CD-Chapter 2).

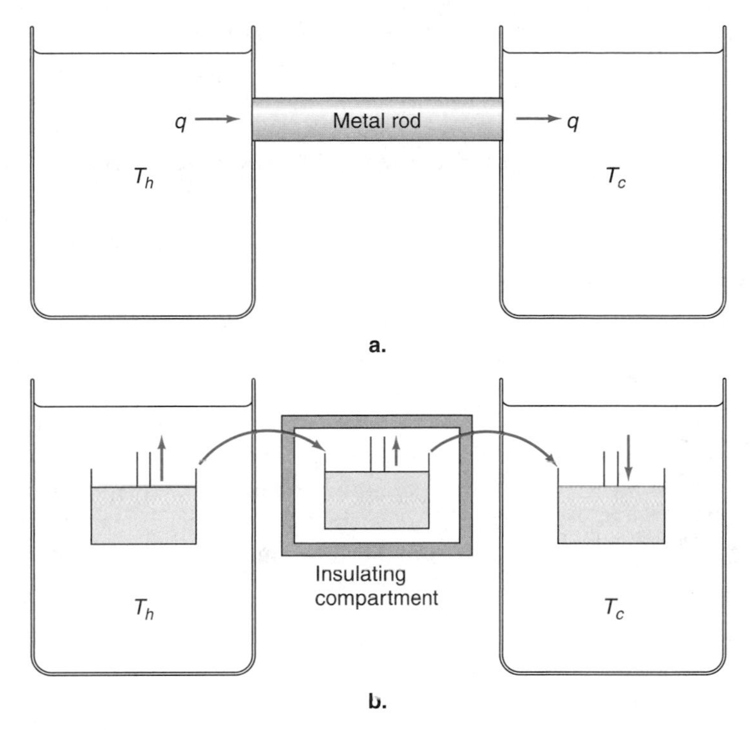

Isothermal reversible expansion at T_h:

$\Delta S_{\text{reservoir}} = -\frac{q_{\text{rev}}}{T_h}$

$\Delta S_{\text{gas}} = \frac{q_{\text{rev}}}{T_h}$

Reversible adiabatic expansion:

$T_h \rightarrow T_c$

$q_{\text{rev}} = 0$

$\Delta S = 0$

Isothermal reversible compression at T_c:

$\Delta S_{\text{reservoir}} = \frac{q_{\text{rev}}}{T_c}$

$\Delta S_{\text{gas}} = -\frac{q_{\text{rev}}}{T_c}$

FIGURE 3.5
The transfer of heat q from a hot reservoir at temperature T_h to a cooler one at temperature T_c: (a) irreversible transfer through a metal rod; (b) reversible transfer by use of an ideal gas.

$$\Delta S_{\text{reservoirs}} = -\frac{q}{T_h} + \frac{q}{T_c} \tag{3.37}$$

Study heat pumps and perpetual motion: Heat pump; Heat engines, pumps, and efficiency.

and this is a positive quantity since $T_h > T_c$. The gas, i.e., the system, has experienced an exactly equal and opposite entropy change:

$$\Delta S_{\text{gas}} = \frac{q}{T_h} - \frac{q}{T_c} \tag{3.38}$$

There is thus no overall entropy change, as is necessarily the case for reversible changes in an isolated system.

On the other hand, for the irreversible change in which the reservoirs are in thermal contact (Figure 3.5a) there is no compensating entropy decrease in the expansion of an ideal gas. The entropy increase in the two reservoirs is the same as for the reversible process (Eq. 3.37), and this is the overall entropy increase.

Proof That Total Entropy Must Increase in a Natural Process

This result, that a spontaneous (and therefore irreversible) process occurs with an overall increase of entropy in the system plus its surroundings, is universally true. The proof of this is based on the fact that the efficiency of a Carnot cycle in which some of the steps are irreversible must be less than that of a purely reversible cycle, since the maximum work is performed by systems that are undergoing reversible processes. Thus, in place of Eq. 3.23 we have, for an irreversible cycle,

$$\frac{q_h^{\text{irr}} + q_c^{\text{irr}}}{q_h^{\text{irr}}} < \frac{T_h - T_c}{T_h} \tag{3.39}$$

System isolated:
process $A \rightarrow B$ irreversible

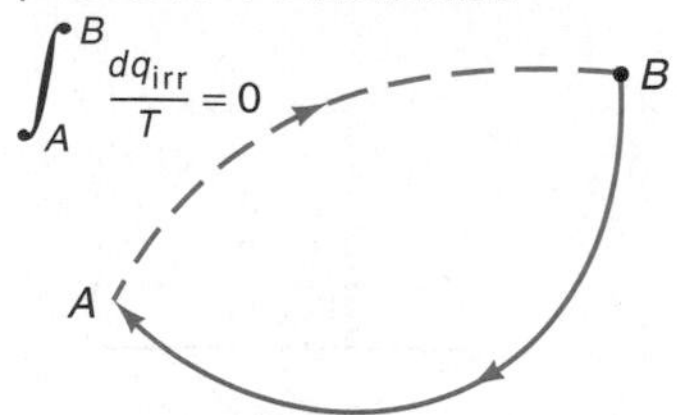

System not isolated:
process $B \rightarrow A$ reversible

$\Delta S_{B \rightarrow A} = \int_B^A \frac{dq_{rev}}{T} < 0$

$\Delta S_{A \rightarrow B} > 0$

FIGURE 3.6
A cycle process in two stages: (1) The system is isolated and changes its state from A to B by an irreversible process (dashed line); (2) The system is not isolated and changes its state by a reversible process (solid line).

This relationship reduces to

$$\frac{q_h^{\text{irr}}}{T_h} + \frac{q_c^{\text{irr}}}{T_c} < 0 \tag{3.40}$$

so that in general, for any cycle that is not completely reversible,

$$\oint \frac{dq_{\text{irr}}}{T} < 0 \tag{3.41}$$

This is known as the **inequality of Clausius,** after Rudolf Clausius, who suggested this relationship in 1854.

Consider an irreversible change from state A to state B in an *isolated* system (Figure 1.4c), as represented by the dashed line in Figure 3.6. Suppose that the conditions are then changed; the system is no longer isolated. It is finally returned to its initial state A by a reversible path represented by the solid line in Figure 3.6. During this reversible process the system is not isolated and can exchange heat and work with the environment. Since the entire cycle $A \rightarrow B \rightarrow A$ is in part reversible, Eq. 3.41 applies, which means that

$$\int_A^B \frac{dq_{\text{irr}}}{T} + \int_B^A \frac{dq_{\text{rev}}}{T} < 0 \tag{3.42}$$

The first integral is equal to zero since the system was isolated during the irreversible process, so that any heat change in one part of the system is exactly compensated by an equal and opposite change in another part. The second integral is the entropy change when the process $B \rightarrow A$ occurs, so that

$$\Delta S_{B \rightarrow A} < 0 \tag{3.43}$$

Follow the steps and understand the Clausius inequality; Irreversible processes

It thus follows that

$$\Delta S_{A \rightarrow B} > 0 \tag{3.44}$$

The entropy of the final state B is thus always greater than that of the initial state A if the process $A \rightarrow B$ occurs irreversibly in an isolated system.

Any change that occurs in nature is spontaneous and is therefore accompanied by a net increase in entropy. This conclusion led Clausius to his famous concise statement of the laws of thermodynamics:

> **The energy of the universe is a constant; the entropy of the universe tends always towards a maximum.**

3.3 Molecular Interpretation of Entropy

It has been emphasized earlier that thermodynamics is a branch of science that can be developed without any regard to the molecular nature of matter. The logical arguments employed do not require any knowledge of molecules, but many of us find it helpful, in the understanding of thermodynamics, to interpret its principles in the light of molecular structure.

In specifying a thermodynamic state, we ignore the question of the positions and velocities of individual atoms and molecules. However, any macroscopic property is in reality a consequence of the position and motion of these particles. At any instant we

Expand a gas into a bulb; Disorder; Molecular interpretation.

could *in principle* define the *microscopic* state of a system, which means that we would specify the position and momentum of each atom. An instant later, even though the system might remain in the same *macroscopic* state, the *microscopic* state would be completely different, since at ordinary temperatures molecules are changing their positions at speeds of the order of 10^3 m s^{-1}. A system at equilibrium thus remains in the same macroscopic state, even though its microscopic state is changing rapidly.

There is an enormous number of microscopic states that are consistent with any given macroscopic state. This concept leads at once to a molecular interpretation of entropy: It is *a measure of how many different microscopic states are consistent with a given macroscopic state.* When a system moves spontaneously from one state to another, it goes to a state in which there are more microscopic states. We can express this differently by saying that when a spontaneous change takes place, there is an increase in *disorder.* In other words, *entropy is a measure of disorder;* an *increase in entropy means an increase in disorder.*

Entropy as a Measure of Disorder

Deck of Cards Analogy

Roll 2, 3, 4, 6 and 6.02 × 10^{23} dice; Rolling dice.

A deck of playing cards provides us with a useful analogy.[1] A deck of 52 cards may be arranged in a particular specified order (suits separate, cards arranged from ace to king) or in a completely shuffled and disordered state. There are many sequences (analogous to microscopic states) that correspond to the shuffled and disordered (macroscopic) state, whereas there is only one microscopic state of the specified order, in which there is less disorder. Thus, the shuffled state has higher entropy than the unshuffled. If we start with an ordered deck and shuffle it, the deck moves toward a state of greater randomness or disorder; the entropy increases. The reason the random state is approached when the ordered deck is shuffled is simply that there are many microscopic states consistent with the shuffled condition, and only one consistent with the ordered condition. The chance of producing an ordered deck by shuffling a disordered one is obviously very small. This is true for 52 cards; when we are dealing with a much larger number of molecules (e.g., 6.022 × 10^{23} in a mole), the likelihood of a net decrease in entropy is obviously much more remote. (See Sections 15.1, 15.3, and p. 113.)

See several different types of disorder; Molecular interpretation.

In the light of these ideas, it is easy to predict what kinds of entropy changes will occur when various processes take place. If, for example, we raise the temperature of a gas, the range of molecular speeds becomes more extended; a larger proportion of the molecules have speeds that differ from the most probable value. There is thus *more disorder at a high temperature,* and the *entropy is greater.*

Entropy of Fusion

Entropy also increases if a solid melts. The entropy change on melting is the enthalpy of fusion $\Delta_{fus}H$ divided by the melting point T_m, and since $\Delta_{fus}H$ must be positive, there is always an entropy increase on melting. This is understandable on a molecular basis, since in the solid the molecules occupy fixed sites, while in a liquid there is much less restriction as to position. Similarly, for the evaporation of a liquid at the boiling point T_b, the entropy change $\Delta_{vap}H/T_b$ must be positive since the latent heat of vaporization $\Delta_{vap}H$ must be positive.[2] In the liquid the attractive forces between molecules are much greater than in the vapor, and there is a large increase in disorder in going from the liquid to the vapor state; there are many more microscopic states for the gas compared with the liquid. The conversion of a solid into a gas is also accompanied by an entropy increase, for the same reason.

[1] See the subsection on Informational or Configurational Entropy (p. 113) for a discussion of this analogy.

[2] For many liquids, $\Delta_{vap}S$ is around 88 J K^{-1} mol^{-1}. This is the basis of Trouton's rule (Eq. 5.18), according to which $\Delta_{vap}S$/(calories per mole) is about 88 J K^{-1} mol^{-1}/4.184 J cal^{-1} = 21 times the boiling point in kelvins.

Entropy changes in chemical reactions can also be understood on a molecular basis. Consider, for example, the process

Entropy Changes in Chemical Reactions

$$H_2 \rightarrow 2H$$

If we convert 1 mol of hydrogen molecules into 2 mol of hydrogen atoms, there is a considerable increase in entropy. The reason for this is that there are more microscopic states (more disorder) associated with the separated hydrogen atoms than with the molecules, in which the atoms are paired together. Again, an analogy is provided by a deck of cards. The hydrogen atoms are like a completely shuffled deck, while the molecular system is like a deck in which aces, twos, etc., are paired. The latter restriction means fewer permissible states and, therefore, a lower entropy.

In general, for a gaseous chemical reaction there is an increase of entropy in the system if there is an increase in the number of molecules. The dissociation of ammonia, for example,

$$2NH_3 \rightarrow N_2 + 3H_2$$

is accompanied by an entropy increase, because we are imposing a smaller restriction on the system by pairing the atoms as N_2 and H_2, as compared with organizing them as NH_3 molecules.

Processes Involving Ions in Solution

The situation with reactions in solution is, however, a good deal more complicated. It might be thought, for example, that a process of the type

$$MX \rightarrow M^+ + X^-$$

occurring in aqueous solution, would be accompanied by an entropy increase, by analogy with the dissociation of H_2 into 2H. However, there is now an additional factor, arising from the fact that ions interact with surrounding water molecules, which tend to orient themselves in such a way that there is electrostatic attraction between the ion and the dipolar water molecules. This effect is known as *electrostriction* or more simply as the *binding* of water molecules.[3] This electrostriction leads to a considerable reduction in entropy, since the bound water molecules have a restricted freedom of motion. As a result, ionization processes always involve an entropy *decrease.*

An interesting example is provided by the attachment of adenosine triphosphate (ATP) to myosin, a protein that is an important constituent of muscle and that plays an important role in muscular contraction. Myosin is an extended protein that bears a number of positive charges, whereas ATP under normal physiological conditions bears four negative charges. These charges attract and bind water molecules, thus bringing about a lowering of the entropy of the water molecules; when the ATP and myosin molecules come together, there is some charge neutralization and a consequent increase in entropy because of the release of water molecules. This entropy increase associated with the binding of ATP to myosin plays a significant role in connection with the mechanism of muscular contraction.

Contraction of Muscle or Rubber

The entropy change that occurs on the contraction of a muscle is also of interest. A stretched strip of muscle, or a stretched piece of rubber, contracts spontaneously. When stretched, muscle or rubber is in a state of lower entropy than in the contracted state. Both muscle or rubber consist of very long molecules. If a long molecule is stretched as far as possible without breaking bonds, there are few conformations available to it. However, if the ends of the molecule are brought

[3]This matter is discussed in more detail in Section 7.9; see especially Figure 7.17.

closer together, the molecules can then assume a large number of conformations, and the entropy will therefore be higher. In 1913 the British physiologist Archibald Vivian Hill made accurate measurements of the heat produced when a muscle contracts and found it to be extremely small. The same is true of a piece of rubber. When muscle or rubber contracts, there is therefore very little entropy change in the surroundings, so that the overall entropy change is essentially that of the material itself, which is positive. We can therefore understand why muscular contraction, or the contraction of a stretched piece of rubber, occurs spontaneously. Processes of this kind that are largely controlled by the entropy change in the system are referred to as *entropic processes.*

Entropic Processes

3.4 The Calculation of Entropy Changes

We have seen that entropy is a state function, which means that an entropy change $\Delta S_{A \to B}$ when a system changes from state A to state B is independent of the path. This entropy change is given by

$$\Delta S_{A \to B} = \int_A^B \frac{dq_{\text{rev}}}{T} \tag{3.45}$$

for the transition $A \to B$ by a reversible path. This integral is the negative of the second integral in Eq. 3.42, so that we have

$$\int_A^B \frac{dq_{\text{irr}}}{T} < \Delta S_{A \to B} \tag{3.46}$$

That is, whereas $\Delta S_{A \to B}$ is independent of the path, the integral

$$\int_A^B \frac{dq}{T}$$

may be *equal to or less than* the entropy change: **It is equal to the entropy change if the process is reversible and less than the entropy change if the process is irreversible.**

It follows that if a system changes from state A to state B by an irreversible process, we cannot calculate the entropy change from the heat transfers that occur. Instead **we must contrive a way of bringing about the change by purely reversible processes.** Some examples of how this is done will now be considered.

Changes of State of Aggregation

The melting of a solid and the vaporization of a liquid are examples of changes of state of aggregation. If we keep the pressure constant, a solid will melt at a fixed temperature, the melting point T_m at which solid and liquid are at equilibrium. As long as both solid and liquid are present, heat can be added to the system without changing the temperature; the heat absorbed is known as the *latent heat of melting*[4] (or *fusion*)

[4]The word *latent* is from the Latin word *latere,* meaning hidden or concealed. Latent heat refers to the heat required to cause a transition without a change in temperature (for example, to melt ice to liquid water).

$\Delta_{fus}H$ of the solid. Since the change occurs at constant pressure, this heat is an enthalpy change and is the difference in enthalpy between liquid and solid. Thus,

$$\Delta_{fus}H = H_{liquid} - H_{solid} \tag{3.47}$$

It is easy to heat a solid sufficiently slowly at its melting point that the equilibrium between liquid and solid is hardly disturbed. The process is therefore reversible, since it follows a path of successive equilibrium states, and the latent heat of melting is thus a reversible heat. Since the temperature remains constant, the integral becomes simply the heat divided by the temperature:

$$\int_A^B \frac{dq_{rev}}{T} = \frac{q_{rev}^{(A \rightarrow B)}}{T} \tag{3.48}$$

The entropy of melting or fusion is thus

Entropy of Fusion

$$\Delta_{fus}S = \frac{\Delta_{fus}H}{T_{fus}} \tag{3.49}$$

For example, $\Delta_{fus}H$ for ice is 6.02 kJ mol^{-1} and the melting point is 273.15 K, so that

$$\Delta_{fus}S = \frac{6020 \text{ J mol}^{-1}}{273.15 \text{ K}} = 22.0 \text{ J K}^{-1} \text{ mol}^{-1}$$

Entropy of Vaporization

The entropy of vaporization can be dealt with in the same way, since when a liquid is vaporized without any rise in temperature, the equilibrium between liquid and vapor remains undisturbed. Thus for water at 100 °C,

$$\Delta_{vap}S = \frac{\Delta_{vap}H}{T_{vap}} = \frac{40\,600 \text{ J mol}^{-1}}{373.15 \text{ K}} = 108.8 \text{ J K}^{-1} \text{ mol}^{-1}$$

The same procedure can be used for a transition from one allotropic form to another, provided that the process occurs at a temperature and pressure at which the two forms are in equilibrium. Gray tin and white tin, for example, are in equilibrium at 1 atm pressure and 286.0 K, and $\Delta_{trs}H = 2.09$ kJ mol^{-1}. The entropy change is thus

Entropy Change in a Transition

$$\Delta_{trs}S = \frac{\Delta_{trs}H}{T_{trs}} = \frac{2090 \text{ J mol}^{-1}}{286.0 \text{ K}} = 7.31 \text{ J K}^{-1} \text{ mol}^{-1}$$

Ideal Gases

A particularly simple process is the isothermal expansion of an ideal gas. Suppose that an ideal gas changes its volume from V_1 to V_2 at constant temperature. In order to calculate the entropy change we must consider the reversible expansion; since entropy is a state function, ΔS is the same however the isothermal expansion from V_1 to V_2 occurs.

See some solutions of idea gas entropy calculations; Entropy calculations.

We have seen (Eq. 2.74) that if n mol of an ideal gas undergoes a reversible isothermal expansion, at temperature T, from volume V_1 to volume V_2, the heat absorbed is

$$q_{rev} = nRT \ln \frac{V_2}{V_1} \tag{3.50}$$

Since the temperature is constant, ΔS is simply the reversible heat absorbed divided by the temperature:

ΔS at Constant Temperature

$$\Delta S = nR \ln \frac{V_2}{V_1} \tag{3.51}$$

If a volume change occurs in an ideal gas with a change in temperature, we proceed as follows. Suppose that the volume changes from V_1 to V_2 and that the temperature changes from T_1 to T_2. Again, we imagine a reversible change, knowing that ΔS will be the same whether the change is reversible or not:

$$dq_{\text{rev}} = dU + P\,dV \tag{3.52}$$

$$= C_V\,dT + \frac{nRT\,dV}{V} \tag{3.53}$$

$$= nC_{V,m}\,dT + \frac{nRT\,dV}{V} \tag{3.54}$$

Then

$$dS = \frac{dq_{\text{rev}}}{T} = nC_{V,m}\frac{dT}{T} + nR\frac{dV}{V} \tag{3.55}$$

Integration leads to

$$\Delta S = S_2 - S_1 = n\int_{T_1}^{T_2} C_{V,m}\,dT/T + nR\int_{V_1}^{V_2} dV/V \tag{3.56}$$

If $C_{V,m}$ is independent of temperature,

ΔS for Changing T and V

$$\Delta S = nC_{V,m}\ln\frac{T_2}{T_1} + nR\ln\frac{V_2}{V_1} \tag{3.57}$$

If the temperature of a gas is changed by a substantial amount, it will be necessary to take into account the variation of the heat capacity with the temperature. The procedure is best illustrated by an example.

EXAMPLE 3.2 A mole of hydrogen gas is heated from 300 K to 1000 K at constant volume. The gas may be treated as ideal with

$$C_{P,m}/\text{J K}^{-1}\,\text{mol}^{-1} = 27.28 + 3.26 \times 10^{-3}T/\text{K} + 5.0 \times 10^{4}(T/\text{K})^{-2}$$

Calculate the entropy change.

Solution From Eq. 3.55, with V constant and $n = 1$,

$$\Delta S_m = \int_{300}^{1000} \frac{C_{V,m}}{T}\,dT$$

But

$$C_{V,m} = C_{P,m} - R$$
$$= (18.97 + 3.26 \times 10^{-3}T/\text{K} + 5.0 \times 10^{4}(T/\text{K})^{-2}\ \text{J K}^{-1}\,\text{mol}^{-1}$$

Therefore

$$\Delta S/\text{J K}^{-1}\,\text{mol}^{-1} = \int_{300}^{1000}\left(\frac{18.97}{T/\text{K}} + 3.26 \times 10^{-3} + 5.0 \times 10^{4}T^{-3}/\text{K}^{-3}\right)dT$$

$$= 18.97\ln\frac{1000}{300} + (3.26 \times 10^{-3} \times 700) - 2.5$$

$$\times 10^{4} \times \left(\frac{1}{1000^2} - \frac{1}{300^2}\right)$$

$$\therefore \Delta S = 22.84 + 2.28 + 0.25 = 25.37\ \text{J K}^{-1}\,\text{mol}^{-1}$$

Entropy of Mixing

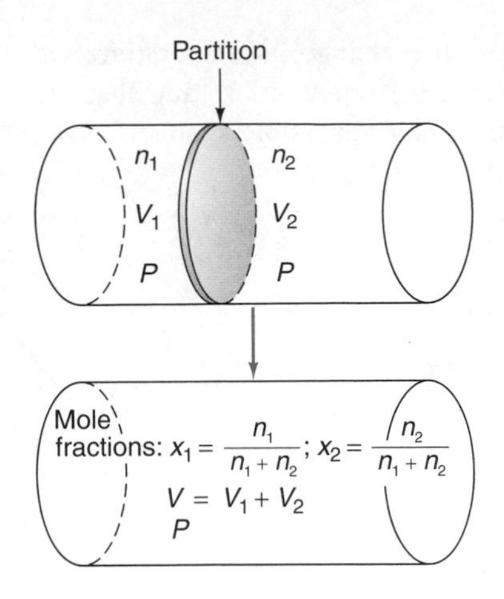

FIGURE 3.7
The mixing of ideal gases at equal pressures and temperatures.

Suppose that we have two ideal gases at equal pressures, separated by a partition as shown in Figure 3.7. If the partition is removed, the gases will mix with no change of temperature. If we want to calculate the entropy change, we must imagine the mixing to occur reversibly. This can be done by allowing the first gas to expand reversibly from its initial volume V_1 to the final volume $V_1 + V_2$; the entropy change is

$$\Delta S_1 = n_1 R \ln \frac{V_1 + V_2}{V_1} \tag{3.58}$$

Similarly, for the second gas

$$\Delta S_2 = n_2 R \ln \frac{V_1 + V_2}{V_2} \tag{3.59}$$

Since the pressures and temperatures are the same,

$$\frac{n_1}{V_1} = \frac{n_2}{V_2} \tag{3.60}$$

and the mole fractions x_1 and x_2 of the two gases in the final mixture are

$$x_1 = \frac{n_1}{n_1 + n_2} = \frac{V_1}{V_1 + V_2} \tag{3.61}$$

and

$$x_2 = \frac{n_2}{n_1 + n_2} = \frac{V_2}{V_1 + V_2} \tag{3.62}$$

The total entropy change $\Delta S_1 + \Delta S_2$ is thus

Entropy of Mixing in Terms of Mole Fractions

$$\Delta S = n_1 R \ln \frac{1}{x_1} + n_2 R \ln \frac{1}{x_2} \tag{3.63}$$

$$= -(n_1 + n_2)R(x_1 \ln x_1 + x_2 \ln x_2) \tag{3.64}$$

The entropy change per mole of mixture is thus

$$\Delta S = -R(x_1 \ln x_1 + x_2 \ln x_2) \tag{3.65}$$

For any number of gases, initially at the same pressure, the entropy change per mole of mixture is

$$\Delta S = -R(x_1 \ln x_1 + x_2 \ln x_2 + x_3 \ln x_3 + \cdots) \tag{3.66}$$

Entropy of Mixing in Terms of Volumes

If the initial pressures of two ideal gases are not the same, Eqs. 3.58 and 3.59 are still applicable, and the entropy increase is

$$\Delta S = n_1 R \ln\left(\frac{V_1 + V_2}{V_1}\right) + n_2 R \ln\left(\frac{V_1 + V_2}{V_2}\right) \tag{3.67}$$

Note that the mixing of two gases each at volume $V_1 + V_2$, to give a mixture of volume $V_1 + V_2$, involves no change in entropy.

Equation 3.67 also applies to the mixing of ideal solutions, as in the following example.

> ***EXAMPLE 3.3*** Exactly one liter of a 0.100 *M* solution of a substance A is added to 3.00 liters of a 0.050 *M* solution of a substance B. Assume ideal behavior and calculate the entropy of mixing.
>
> ***Solution*** 0.100 mol of substance A is present and the volume increases by a factor of 4:
>
> $$\Delta S(\text{A}) = 0.100 \times 8.3145 \times \ln 4.00$$
> $$= 1.153 \text{ J K}^{-1}$$
>
> 0.150 mol of B is present and the volume increases by a factor of 4/3:
>
> $$\Delta S(\text{B}) = 0.150 \times 8.3145 \times \ln(4/3)$$
> $$= 0.359 \text{ J K}^{-1}$$
>
> The net ΔS is therefore
>
> $$1.153 + 0.359 = 1.512 \text{ J K}^{-1}$$

Informational or Configurational Entropy

This treatment of entropy of mixing brings out an important point that is often misunderstood. Consider a simple example relating to the mixing of ideal gases. Suppose that we have half a mole of an ideal gas, separated by a partition from another half mole at the same volume, pressure, and temperature; if we remove the partition, what is the entropy change? The important point to appreciate is that this cannot be answered unless we have further information. If the gases are identical, there is obviously no entropy change. There is, however, an entropy of mixing if the gases are different. As just explained (Eqs. 3.58 and 3.59), we calculate an entropy of mixing by expanding each gas reversibly and isothermally to twice its volume; the heat absorbed in each case is $\frac{1}{2}RT \ln 2$ so that the total entropy change is $R \ln 2$, which is equal to 5.76 J K^{-1}. There is no entropy change when we mix the two expanded gases at the same volume; the entropy of mixing $R \ln 2$ is really an informational or configurational entropy, and its value depends on the information we possess.

For the shuffling of 52 cards, discussed on p. 107, we can say nothing about the entropy change unless we have certain information. If the cards have been printed so that all are identical, there will be no entropy change on shuffling. If, however, the cards have been printed normally, there is an increase in entropy when an ordered deck is shuffled. This is entirely a configurational entropy, arising from the fact that for an ordered deck we accept only one arrangement of cards out of the large number (52!, approximately 8.07×10^{67}) of possible arrangements. The calculated configurational entropy change for shuffling is $52k_B \ln 52! = 1.12 \times 10^{19}$ J K^{-1}. This, of course, is much smaller than the change for the mixing of two gases, because the number of cards is much smaller than the number of molecules.

Entropy is a measure of the number of different states associated with a specified state of the system. In counting these microstates, we must include the configurations associated with the state. In the shuffling of cards there is no change in the number of states arising from electronic, vibrational, and rotational levels, but we must include the configurational (informational) states. Some scientists have argued that there is no entropy change for macroscopic processes like the shuffling of cards, but this is a mistaken view, arising from the neglect of the configurational states.

Solids and Liquids

Entropy changes when solids or liquids are heated or cooled can readily be calculated provided that we know the relevant heat capacities and also that we know the latent heats for any phase transitions that occur. The entropy contributions arising from volume changes, which are small in these cases, can usually be neglected. The methods will now be illustrated by means of a number of examples.

EXAMPLE 3.4 Two moles of water at 50 °C are placed in a refrigerator which is maintained at 5 °C. Taking the heat capacity of water as 75.3 J K^{-1} mol^{-1} and independent of temperature, calculate the entropy change for the cooling of the water to 5 °C.

Also calculate the entropy change in the refrigerator, and the net entropy change.

Solution For the cooling of the water,

$$\Delta S = 2 \times 75.3 \ln \frac{278.15}{323.15}$$

$$= -22.59 \text{ J K}^{-1}$$

The heat gained by the refrigerator is

$$2 \times 75.3 \times (50 - 5) = 6777 \text{ J}$$

The entropy change in the refrigerator, at 5 °C, is

$$\frac{6777}{278.15} = 24.38 \text{ J K}^{-1}$$

The net entropy change is $24.38 - 22.57 = 1.79$ J K^{-1}. Note that, as is necessary, this net entropy change is positive.

EXAMPLE 3.5 Calculate the entropy change when 1 mol of ice is heated from 250 K to 300 K. Take the heat capacities ($C_{P,m}$) of water and ice to be constant at 75.3 and 37.7 J K^{-1} mol^{-1}, respectively, and the latent heat of fusion of ice as 6.02 kJ mol^{-1}.

Solution The entropy change when 1 mol of ice is heated from 250 K to 273.15 K is

$$\Delta S_1/\text{J K}^{-1} \text{ mol}^{-1} = \int_{250}^{273} \frac{37.7}{T}\, dT = 37.7 \ln \frac{273.15}{250}$$

$$\Delta S_1 = 3.34 \text{ J K}^{-1} \text{ mol}^{-1}$$

For the melting at 273.15 K,

$$\Delta S_2 = \frac{6020 \text{ J mol}^{-1}}{273.15 \text{ K}} = 22.04 \text{ J K}^{-1} \text{ mol}^{-1}$$

For the heating from 273.15 K to 300 K,

$$\Delta S_3 = \int_{273.15}^{300} \frac{75.3}{T} dT = 75.3 \ln \frac{300}{273.15}$$
$$= 7.06 \text{ J K}^{-1} \text{ mol}^{-1}$$

The total entropy change is

$$\Delta S = (3.34 + 22.04 + 7.06) \text{ J K}^{-1} \text{ mol}^{-1} = 32.44 \text{ J K}^{-1} \text{ mol}^{-1}$$

EXAMPLE 3.6 One mole of supercooled water at −10 °C and 1 atm pressure turns into ice. Calculate the entropy change in the system and in the surroundings and the net entropy change, using the data given in the previous example.

Solution Supercooled water at −10 °C and the surroundings at −10 °C are not at equilibrium, and the freezing is therefore not reversible. To calculate the entropy change we must devise a series of reversible processes by which supercooled water at −10 °C is converted into ice at −10 °C. We can (a) heat the water reversibly to 0 °C, (b) bring about the reversible freezing, and then (c) cool the ice reversibly to −10 °C.

a. The entropy change in heating the supercooled water from 263.15 K to 273.15 K is

$$\Delta S_1/\text{J K}^{-1} \text{ mol}^{-1} = \int_{263.15}^{273.15} \frac{75.3}{T} dT = 75.3 \ln \frac{273.15}{263.15}$$
$$\Delta S_1 = 2.81 \text{ J K}^{-1} \text{ mol}^{-1}$$

b. The entropy change in the reversible freezing of water at 0 °C is

$$\Delta S_2 = -\frac{q_{\text{fusion}}}{T} = -\frac{6020 \text{ J mol}^{-1}}{273.15 \text{ K}} = -22.04 \text{ J K}^{-1} \text{ mol}^{-1}$$

c. The entropy change in cooling the ice from 273.15 K to 263.15 K is

$$\Delta S_3 = 37.7 \int_{273.15}^{263.15} \frac{dT}{T} = 37.7 \ln \frac{263.15}{273.15}$$
$$= -1.41 \text{ J K}^{-1} \text{ mol}^{-1}$$

The entropy change in the water when it froze at −10 °C is therefore

$$\Delta S_{\text{syst}} = \Delta S_1 + \Delta S_2 + \Delta S_3$$
$$= 2.81 - 22.04 - 1.41$$
$$= -20.64 \text{ J K}^{-1} \text{ mol}^{-1}$$

The entropy increase in the surroundings is obtained by first calculating the net amount of heat that has been transferred to the surroundings. This is the sum of three terms corresponding to the three steps:

Step 1: Heat lost by the surroundings in heating the water from -10 °C is $10\ \text{K} \times 75.3\ \text{J K}^{-1}\ \text{mol}^{-1} = 753\ \text{J mol}^{-1}$; the surroundings thus gain $-753\ \text{J mol}^{-1}$.

Step 2: Heat gained by the surroundings when the water freezes at 0 °C is $6020\ \text{J mol}^{-1}$.

Step 3: Heat gained by the surroundings when the ice is cooled to -10 °C is $10\ \text{K} \times 37.7\ \text{J K}^{-1} = 377\ \text{J mol}^{-1}$.

The net heat transferred to the surroundings when the water freezes at -10 °C is therefore

$$-753 + 6020 + 377 = 5644\ \text{J mol}^{-1}$$

This heat is taken up by the surroundings at the constant temperature of -10 °C, and the entropy increase is

$$\Delta S_{\text{surr}} = \frac{5644\ \text{J mol}^{-1}}{263.15\ \text{K}} = 21.45\ \text{J K}^{-1}\ \text{mol}^{-1}$$

The overall entropy change in the system and the surroundings is therefore

$$\Delta S_{\text{overall}} = \Delta S_{\text{syst}} + \Delta S_{\text{surr}} = -20.64 + 21.45 = 0.81\ \text{J K}^{-1}\ \text{mol}^{-1}$$

A net entropy increase in the system and surroundings is, of course, what we expect for an irreversible process.

EXAMPLE 3.7 One mole of ice at -10 °C is placed in a room at a temperature of 10 °C. Using the heat capacity data given in the previous examples, calculate the entropy change in the system and in the surroundings.

Solution The ice can be converted into water at 10 °C by the following reversible steps:

Step 1: Convert the ice at -10 °C into ice at 0 °C:

$$\Delta S_1 = 37.7\ (\text{J K}^{-1})\ \ln \frac{273.15}{263.15} = 1.406\ \text{J K}^{-1}$$

Step 2: Melt the ice at 0 °C:

$$\Delta S_2 = \frac{6020}{273.15} = 22.04\ \text{J K}^{-1}$$

Step 3: Heat the water from 0 °C to 10 °C:

$$\Delta S_3 = 75.3\ (\text{J K}^{-1})\ \ln \frac{283.15}{273.15} = 2.71\ \text{J K}^{-1}$$

The net entropy change in the system $= 1.404 + 22.04 + 2.71 = 26.16\ \text{J K}^{-1}$. To calculate the entropy change in the environment we first calculate the heat lost by the environment:

Step 1: $37.7 \times 10 = 377$ J

Step 2: 6020 J

> Step 3: $75.3 \times 10 = 753$ J
> Total heat lost by the environment at 10 °C = 7150 J
> Entropy change in the environment = $-7150/283.15 = -25.25$ J K^{-1}
> Net entropy change = $26.16 - 25.25 = 0.91$ J K^{-1}

When liquids are mixed, the entropy change is sometimes given by Eq. 3.66, which we derived for ideal gases. The condition for this equation to apply to liquids is that the intermolecular forces between the different components must all be equal. We have seen (Section 2.7) that for a gas to behave ideally there must be a complete absence of forces between the molecules; the internal pressure $(\partial U/\partial V)_T$ must be zero. The cohesive forces between molecules in a liquid can never be zero, but if the various intermolecular forces are all equal, Eq. 3.66 will be obeyed. For a two-component system, for example, the intermolecular forces between A and A, B and B, and A and B, must be the same.

Further details about the thermodynamics of solutions will be considered in later chapters.

3.5 The Third Law of Thermodynamics

W. H. Nernst received the 1920 Nobel Prize in chemistry in recognition of his work in thermochemistry.

See a summary of the four laws of thermodynamics including the third law; Summary of the laws.

What has come to be called Nernst's heat theorem or the third law of thermodynamics was first formulated in 1906 by the German physical chemist Walther Hermann Nernst (1861–1941). The theorem was developed as a result of a need to calculate entropy changes for chemical reactions. We shall see in Chapter 4 that equilibrium constants are related to changes in both enthalpy and entropy. Enthalpy changes can be calculated from thermochemical data, as discussed in Section 2.5. Entropy changes, however, present a special problem. As noted in Section 3.3, we can calculate the entropy of a substance in a state B with reference to that in a state A by means of the relationship

$$\Delta S = \int_A^B \frac{dq_{\text{rev}}}{T} \tag{3.68}$$

We can thus calculate the entropy of a substance at temperature T with respect to the value at the absolute zero if we have the necessary thermal data. However, for a chemical reaction, which usually we cannot cause to occur reversibly, we cannot know the entropy change at temperature T unless we know it at the absolute zero.

Cryogenics: The Approach to Absolute Zero

The problem therefore relates to the experimental study of the behavior of matter at very low temperatures, a subject known as cryogenics (Greek *kryos,* frost; *genes,* become). Nernst carried out many such investigations; his first publication on this topic, concerned in particular with the calculation of equilibrium constants, appeared in 1906. His original conclusion can be summarized by the statement that

Entropy changes become zero at the absolute zero.

Nernst himself, however, always disliked stating his law in terms of entropy changes, preferring a different way of looking at the problem.

However, the results of some of the low-temperature experiments were inconsistent with this formulation of the heat theorem, and it became necessary to modify it. It was found that the reason for the deviations is that at very low temperatures substances are often not in a state of true equilibrium, since at these temperatures equilibrium is established exceedingly slowly. For example, many substances are frozen into metastable glassy states as the temperature is lowered, and such states persist for long periods of time. The original formulation of the heat theorem therefore has to be qualified by saying that it applies only if there is true equilibrium. A correct formulation of the theorem is therefore: **Entropy changes become zero at the absolute zero provided that the states of the system are in thermodynamic equilibrium.**

Statements of Nernst's Heat Theorem

Alternatively, we can express Nernst's heat theorem or the third law of thermodynamics as: **The entropies of all perfectly crystalline substances must be the same at the absolute zero.**

Various techniques are used for producing low temperatures. The most familiar one, used in commercial refrigerators, is based on the fact that under certain circumstances gases become cooler when they expand, as a result of the work done in overcoming the mutual attraction of the molecules. This is an application of the Joule-Thomson effect that we considered in Section 2.7. Liquid nitrogen, which boils at 77 K, is manufactured commercially by the application of this principle, a cascade process being employed; further details are given in Section 3.10 (Figure 3.14). By performing successive expansions first with nitrogen, then with hydrogen, and finally with helium, the Dutch physicist Heike Kamerlingh Onnes (1853–1926) liquefied helium, the last gas to be liquefied, in 1908. He thus opened a temperature region to somewhat below 1 K for study and eventual exploitation.

The attainment of still lower temperatures requires the use of another principle. It was suggested independently in 1926 by the American chemist William Francis Giauque (1895–1982) and the Dutch chemist Peter Joseph Wilhelm Debye (1884–1966) that one can make use of the temperature changes occurring during magnetization and demagnetization procedures. Certain salts, such as those of the rare earths, have high paramagnetic susceptibilities. The cations act as little magnets, which line up when a magnetic field is applied, and the substance is then in a state of lower entropy. When the magnetic field is decreased, the magnets adopt a more random arrangement, and the entropy increases. That this can occur was demonstrated in 1933 by Giauque.

Figure 3.8 illustrates a procedure that can be employed to achieve a low temperature. A paramagnetic salt, such as gadolinium sulfate octahydrate, is placed between the poles of an electromagnet and is cooled to about 1 K, which can be done by the expansion techniques mentioned previously. The magnetic field is then applied, and the heat produced is allowed to flow into the surrounding liquid helium (Step 1). In Step 2 the system is then isolated and the magnetic field removed;

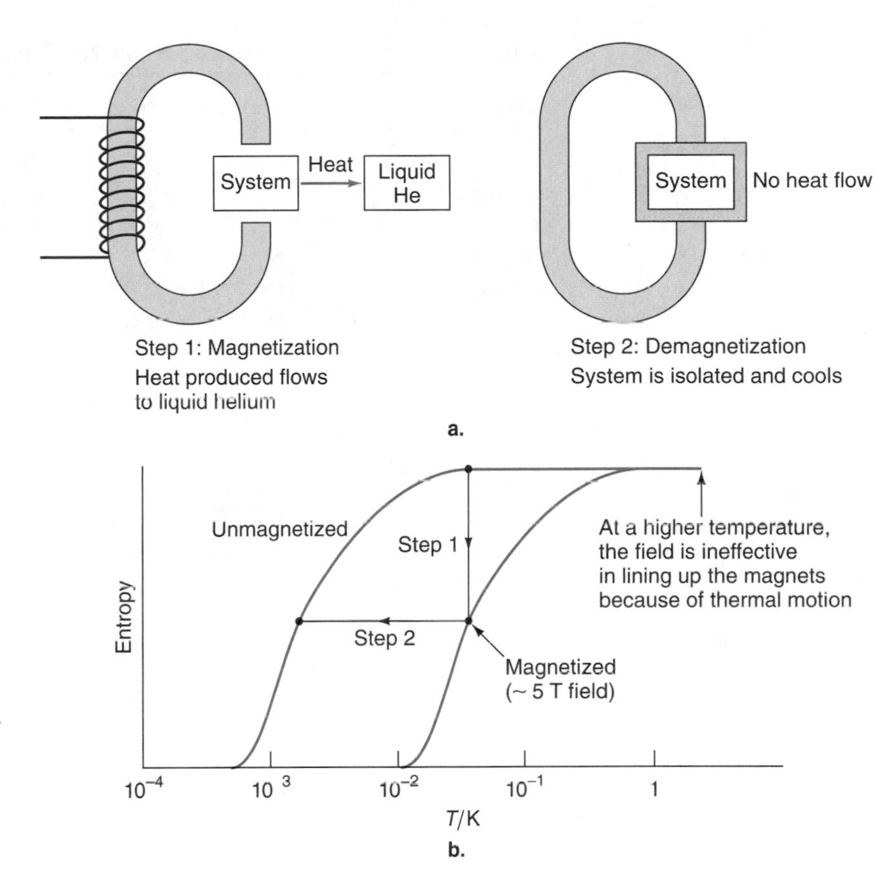

FIGURE 3.8
The production of very low temperatures by adiabatic demagnetization: (a) the magnetization and demagnetization steps; (b) the variation of entropy with temperature for the magnetized and demagnetized material.

this *adiabatic relaxation* process leads to a cooling. Temperatures of about 0.005 K are produced in this way.

The attainment of still lower temperatures, down to about 10^{-6} K, is achieved by making use of nuclear magnetic properties (see Chapter 14). The nuclear magnets are about 2000 times smaller than the electron magnet in a paramagnetic substance such as gadolinium sulfate, but there is a significant difference between the nuclear entropies even at temperatures as low as 10^{-6} K.

Absolute Entropies

In view of the Nernst heat theorem it is convenient to adopt the convention of assigning a value of zero to the entropy of every crystalline substance at the absolute zero. Entropies can then be determined at other temperatures by considering a series of reversible processes by which the temperature is raised from the absolute zero to the temperature in question. Table 3.2 lists some absolute entropies obtained in this way.

If the absolute entropies of all the substances in a chemical reaction are known, it is then a simple matter to calculate the entropy change in the reaction; the relationship is

$$\Delta S = \Sigma\, S\,(\text{products}) - \Sigma\, S\,(\text{reactants}) \tag{3.69}$$

TABLE 3.2 Absolute Entropies, $S°$, at 25 °C and 1 bar Pressure

Substance	Formula	State	$S°$/J K^{-1} mol^{-1}
Carbon	C	s(graphite*)	5.74
Hydrogen	H_2	g*	130.68
Oxygen	O_2	g*	205.14
Nitrogen	N_2	g*	191.61
Carbon dioxide	CO_2	g	213.60
Water	H_2O	l	69.91
Ammonia	NH_3	g	192.45
Ethane	C_2H_6	g	229.60
Ethene	C_2H_4	g	219.56
Methanol	CH_3OH	l	126.80
Ethanol	C_2H_5OH	l	160.70
Acetic acid	CH_3COOH	l	159.80
Acetaldehyde	CH_3CHO	g	250.30
Urea	NH_2CONH_2	s	104.60

*These are the standard states of the elements.

These data are from *The NBS Tables of Chemical Thermodynamic Properties in SI Units: Selected Values for Inorganic and C1 and C2 Organic Substances,* New York: American Chemical Society and the American Institute of Physics, for the National Bureau of Standards, 1982. For other recent thermodynamic data, see Appendix D.

3.6 Conditions for Equilibrium

The second law, in stating that any spontaneous process must be accompanied by an increase in the total entropy, gives us at once a condition for equilibrium, since a system at equilibrium cannot undergo any spontaneous change. Suppose, with reference to Figure 3.9a, that a system is at equilibrium in state A and that an infinitesimal change takes it to another state B where it is still at equilibrium. The change $A \rightarrow B$ cannot involve a total entropy increase, since otherwise the change would be spontaneous and equilibrium could not exist at A; by the same argument the change $B \rightarrow A$ cannot involve a total entropy increase. It follows that the states A and B must have the same total entropies. The *condition for equilibrium* is therefore

$$dS^{\text{total}} = dS^{\text{syst}} + dS^{\text{surr}} = 0 \tag{3.70}$$

where S^{syst} is the entropy of the system and S^{surr} is the entropy of the surroundings. This is shown in Figure 3.9a, and we see that the *position of equilibrium must correspond to a state of maximum total entropy,* since the total entropy increases in any spontaneous process.

It is more convenient to define equilibrium with reference to *changes in the system only,* without explicitly considering the environment. Suppose that the system and the surroundings are at the same temperature,

$$T^{\text{syst}} = T^{\text{surr}} \tag{3.71}$$

Suppose that a process occurs spontaneously in the system and that an amount of heat dq leaves the system and enters the surroundings. This amount entering the

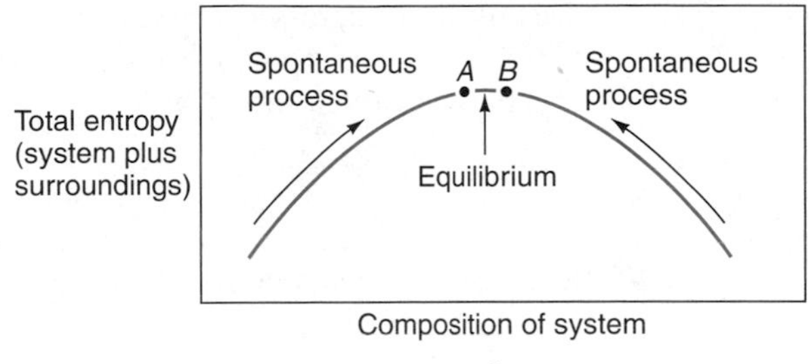

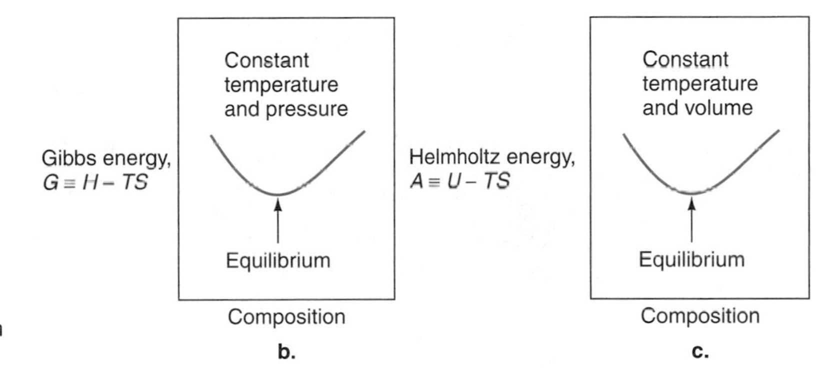

FIGURE 3.9
Conditions for chemical equilibrium: (a) the system moves toward a state of maximum total entropy; (b) at constant T and P, the system moves toward a state of minimum Gibbs energy; (c) at constant T and V, the system moves toward a state of minimum Helmholtz energy.

surroundings may be written as dq^{surr}, and it is equal to the heat change $-dq^{syst}$ in the system:

$$dq^{surr} = -dq^{syst} \tag{3.72}$$

We now come to a very important point about dq^{surr}, which arises on account of the vastness of the surroundings. As a result of this, the surroundings experience no volume change when heat is transferred to them, and dq^{surr} is therefore equal to the change dU^{surr} in the internal energy, which is a state function. It therefore does not matter whether the heat dq^{surr} enters the surroundings reversibly or irreversibly; the heat dq_{irrev}^{surr} is still equal to the increase in internal energy of the surroundings:

$$dq_{irrev}^{surr} - dq_{rev}^{surr} = dU^{surr} \tag{3.73}$$

The entropy change in the surroundings is

$$dS^{surr} = \frac{dq^{surr}}{T^{surr}} \tag{3.74}$$

But $T^{surr} = T^{syst}$ and $dq^{surr} = -dq^{syst}$; therefore

$$dS^{surr} = -\frac{dq^{syst}}{T^{syst}} \tag{3.75}$$

Instead of Eq. 3.70 we can therefore write the equilibrium condition as

$$dS^{total} = dS^{syst} - \frac{dq^{syst}}{T^{syst}} = 0 \tag{3.76}$$

We now have expressed everything in terms of the system, and to simplify the notation from now on we will drop the superscript "syst." The condition for equilibrium will thus be written simply as

$$dS - \frac{dq}{T} = 0 \tag{3.77}$$

with the understanding that it is *the system we are referring to.* Alternatively, we can write

$$dq - T\,dS = 0 \tag{3.78}$$

as the condition for equilibrium.

Constant Temperature and Pressure: The Gibbs Energy

Chemical processes commonly occur in open vessels at constant pressure, in which case dq can be equated to dH, the enthalpy change. Equation 3.78 therefore becomes

$$dH - T\,dS = 0 \tag{3.79}$$

In view of this relationship the American physicist Josiah Willard Gibbs (1839–1903) defined a new thermodynamic function that is now known as the *Gibbs function* or **Gibbs energy**[5] and is given the symbol G:

Definition of Gibbs Energy

$$G \equiv H - TS \tag{3.80}$$

At constant temperature

$$dG = dH - T\,dS \tag{3.81}$$

and it follows from Eq. 3.79 that the *condition for equilibrium at constant T and P* is

$$dG = 0 \tag{3.82}$$

Condition for Equilibrium at Constant T and P

Since G is composed of H, T, and S, which are state functions, it is also a state function.

This condition for equilibrium is represented in Figure 3.9b. We see that systems tend to move toward a state of *minimum* Gibbs energy, and this is easily understood if we follow through the preceding arguments beginning with the inequality

$$dS^{\text{total}} = dS^{\text{syst}} + dS^{\text{surr}} > 0 \tag{3.83}$$

which applies to the irreversible case. Instead of Eq. 3.82 we then find that

$$dG < 0 \tag{3.84}$$

In other words, *in spontaneous processes at constant T and P, systems move toward a state of minimum Gibbs energy.*

[5]It has long been known as the Gibbs *free* energy. However, IUPAC has recommended that the *free* be dropped and that we call it the *Gibbs function* or the *Gibbs energy.* The same recommendation applies to the Helmholtz energy.

Constant Temperature and Volume: The Helmholtz Energy

The argument for constant volume conditions is very similar; the quantity dq in Eq. 3.78 now is equated to dU:

$$dU - T\,dS = 0 \tag{3.85}$$

The quantity $U - TS$, also a state function, is called the *Helmholtz function* or **Helmholtz energy,** after the German physiologist and physicist Ludwig Ferdinand von Helmholtz (1821–1894), and is given the symbol A:

Definition of Helmholtz Energy

$$A \equiv U - TS \tag{3.86}$$

Equation 3.85 therefore can be written as

Condition for Equilibrium at Constant T and V

$$dA = 0 \tag{3.87}$$

and this is the condition for equilibrium at *constant T and V.* Under these conditions, systems tend to move toward a state of *minimum Helmholtz energy,* as shown in Figure 3.9c.

3.7 The Gibbs Energy

Molecular Interpretation

Although, as previously emphasized, thermodynamic arguments can be developed without any reference to the existence and behavior of atoms and molecules, it is nevertheless instructive to interpret the arguments in terms of molecular structure.

Consider first the dissociation of hydrogen molecules into hydrogen atoms,

$$H_2 \rightleftharpoons 2H$$

This process from left to right occurs only to a very slight extent at ordinary temperatures, but if we start with hydrogen atoms, the combination will occur spontaneously. We have seen that at constant temperature and pressure a natural or spontaneous process is one in which there is a decrease in Gibbs energy; the system approaches an equilibrium state in which the Gibbs energy is at a minimum. Therefore, if the process

$$2H \rightarrow H_2$$

occurs at ordinary temperatures, ΔG is negative. Let us now consider how this can be interpreted in terms of enthalpy and entropy changes, in the light of the molecular structures.

We know that when hydrogen atoms are brought together and combine, there is evolution of heat, which means that the enthalpy goes to a lower level; that is

$$\Delta H\,(2H \rightarrow H_2) < 0$$

The entropy change is also negative when hydrogen atoms combine, because the atoms have a less ordered arrangement than the molecules:

$$\Delta S(2H \rightarrow H_2) < 0$$

The Gibbs energy change for the combination process at constant temperature is made up as follows:

$$\Delta G = \underset{<0}{\Delta H} - \underset{<0}{T\Delta S} \tag{3.88}$$

If T is small enough, ΔG will be negative. This is the situation at room temperature; indeed up to quite high temperatures the negative ΔH term dominates the situation and ΔG is negative, which means that the process occurs spontaneously.

If, however, we go to very high temperatures,[6] the $T\Delta S$ term will become dominant; since ΔS is negative and $T\Delta S$ is *subtracted* from ΔH, the net value of ΔG becomes positive when T is large enough. Therefore, we predict that at very high temperatures hydrogen atoms will not spontaneously combine; instead, hydrogen molecules will spontaneously dissociate into atoms. This is indeed the case.

In this example the ΔH and $T\Delta S$ terms (except at very high temperatures) work in opposite directions; both are negative. In almost all reactions, ΔH and $T\Delta S$ work against each other. In the reaction

$$2H_2 + O_2 \rightarrow 2H_2O$$

ΔH is negative (the reaction is exothermic) and $T\Delta S$ is negative; there is a decrease in the number of molecules and an increase of order. Thus, at a fixed temperature,

$$\Delta G = \underset{<0}{\Delta H} - \underset{<0}{T\Delta S} \tag{3.89}$$

At ordinary temperatures, $T\Delta S$ is negligible compared with ΔH; ΔG is therefore negative and reaction occurs spontaneously from left to right. As the temperature is raised, $T\Delta S$ becomes more negative, and at sufficiently high temperatures $\Delta H - T\Delta S$ becomes positive. The spontaneous reaction is then from right to left.

It follows from the relationship

$$\Delta G = \Delta H - T\Delta S \tag{3.90}$$

that temperature is a weighting factor that determines the relative importance of enthalpy and entropy. At the absolute zero, $\Delta G = \Delta H$, and the direction of spontaneous change is determined solely by the enthalpy change. At very high temperatures, on the other hand, the entropy is the driving force that determines the direction of spontaneous change.

EXAMPLE 3.8

a. Liquid water at 100 °C is in equilibrium with water vapor at 1 atm pressure. If the enthalpy change associated with the vaporization of liquid water at 100 °C is 40.60 kJ mol^{-1}, what are ΔG and ΔS?

b. Suppose that water at 100 °C is in contact with water vapor at 0.900 atm. Calculate ΔG and ΔS for the vaporization process.

[6]It must be emphasized that ΔH and ΔS are functions of temperature, as reflected in the temperature dependence of C_P, so that this discussion is oversimplified.

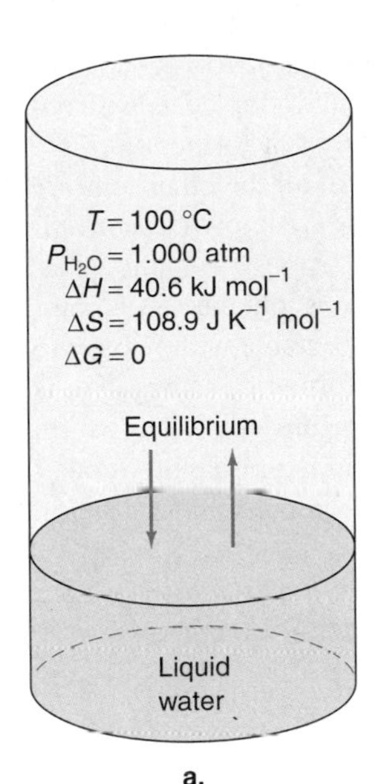

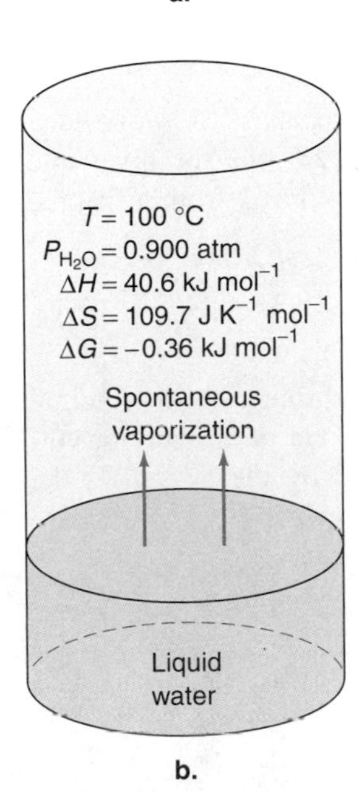

FIGURE 3.10
The vaporization of water at 100 °C. In (a), liquid water at 100 °C is in equilibrium with water vapor at 1 atm pressure. In (b), liquid water at 100 °C is in contact with water vapor at 0.9 atm pressure, and there is spontaneous vaporization.

Solution (See Figure 3.10.)

a. Since liquid water at 100 °C is in equilibrium with water vapor at 1 atm pressure,

$$\Delta G = 0$$

Since $\Delta H = 40.60$ kJ mol^{-1}, and

$$\Delta G = \Delta H - T\Delta S$$

it follows that

$$\Delta S = \frac{40\ 600\ \text{J mol}^{-1}}{373.15\ \text{K}} = 108.9\ \text{J K}^{-1}\ \text{mol}^{-1}$$

b. The entropy increase for the expansion of 1 mol of gas from 1 atm pressure to 0.900 atm is

$$\Delta S = R \ln \frac{V_2}{V_1} = R \ln \frac{P_1}{P_2}$$

$$\Delta S/\text{J K}^{-1}\ \text{mol}^{-1} = 8.3145 \ln \frac{1.00}{0.900}$$

$$\Delta S = 0.876\ \text{J K}^{-1}\ \text{mol}^{-1}$$

The entropy increase when 1 mol of liquid water evaporates to give vapor at 0.900 atm pressure is thus

$$\Delta S = 108.9 + 0.876 = 109.7\ \text{J K}^{-1}\ \text{mol}^{-1}$$

The value of $T\Delta S$ is

$$109.7 \times 373.15 = 40.96\ \text{kJ mol}^{-1}$$

The value of ΔH has not changed in this process, and the value of the Gibbs energy change is thus

$$\Delta G = \Delta H - T\Delta S = 40.60 - 40.96 = -0.36\ \text{kJ mol}^{-1}$$

Since this is a negative quantity, the vaporization process is spontaneous.

Gibbs Energies of Formation

In Section 2.5 we dealt with enthalpy changes in chemical reactions and found that it was very convenient to tabulate enthalpies of formation of compounds.

The same procedure is followed with Gibbs energies. The *standard Gibbs energy of formation* of any compound is then simply the Gibbs energy change $\Delta_f G°$ that accompanies the formation of the compound in its standard state from its elements in their standard states. We can then calculate the standard Gibbs energy change for any reaction, $\Delta G°$, by adding the Gibbs energies of formation of all the products and subtracting the sum of the Gibbs energies of formation of all the reactants:

$$\Delta G° = \Sigma\, \Delta_f G°(\text{products}) - \Sigma\, \Delta_f G°(\text{reactants}) \tag{3.91}$$

Appendix D lists Gibbs energies of formation of a number of compounds, and of ions in aqueous solution. It is impossible to measure Gibbs energies of formation of individual ions since experiments are always done with systems involving ions of opposite signs. To overcome this difficulty the same procedure is adopted as with enthalpies; the arbitrary assumption is made that the Gibbs energy of formation of the proton in water is zero, and the Gibbs energies of formation of all the other ions are calculated on that basis. The ionic values obtained in this way are known as *conventional* Gibbs energies of formation.

A negative ΔG° for a reaction means that the process is spontaneous; a compound having a negative $\Delta_f G^\circ$ is therefore *thermodynamically* stable with respect to its elements. A compound whose standard Gibbs energy of formation is negative is known as an *exergonic* (Greek *ergon,* work) compound (compare *exothermic,* for a compound formed with a negative $\Delta_f H^\circ$). Conversely, a compound having a positive $\Delta_f G^\circ$ value is known as an *endergonic* compound (compare *endothermic*). Most compounds are exergonic.

Exergonic and Endergonic

The terms *exergonic* and *endergonic* (also called *exoergic* and *endoergic*) are also employed with respect to other processes. Thus any reaction having a negative ΔG° value (i.e., accompanied by a liberation of Gibbs energy) is said to be *exergonic.* A reaction having a positive ΔG° is said to be *endergonic.*

Gibbs Energy and Reversible Work

When an ideal gas is reversibly compressed at constant temperature, the work done is the increase in Gibbs energy. We have seen in Section 2.6 that the reversible work done in compressing n mol of an ideal gas, at temperature T, from a volume V_1 to a volume V_2 is

$$w_{\text{rev}} = nRT \ln \frac{V_1}{V_2} = nRT \ln \frac{P_2}{P_1} \tag{3.92}$$

During this isothermal process there is no change in internal energy; the internal energy of an ideal gas is a function of temperature only and not of pressure or volume. It follows from the first law that the heat absorbed by the system is the negative of the work done on the system:

$$q_{\text{rev}} = nRT \ln \frac{V_2}{V_1} \tag{3.93}$$

The entropy change (numerically, a decrease since $V_1 > V_2$) is therefore

$$\Delta S = \frac{q_{\text{rev}}}{T} = nR \ln \frac{V_2}{V_1} \tag{3.94}$$

There is no change in enthalpy; the enthalpy for an ideal gas is a function only of temperature. The change in Gibbs energy is thus

$$\Delta G = \Delta H - T\Delta S \tag{3.95}$$

$$= nRT \ln \frac{V_1}{V_2} = nRT \ln \frac{P_2}{P_1} \tag{3.96}$$

The reversible work done on the system (Eq. 3.92) is thus the change in its Gibbs energy.

An even more important relationship between Gibbs energy and work arises for processes occurring at constant temperature and pressure. Work can be classified into two types: work that arises from a volume change that occurs when a process takes place, and any other type of work. For example, when an electrochemical cell operates, there may be a small volume change; work may be performed on the surroundings, or the surroundings may do work on the system. Of much more interest and practical importance, however, is the electrical work that results from the operation of the cell. We will call the work arising from the volume change the *PV* work and give it the symbol w_{PV}. Any other kind of work we will call the non-*PV* work, $w_{\text{non-}PV}$. The total work is thus

$$w = w_{PV} + w_{\text{non-}PV} \quad (3.97)$$

Another kind of non-*PV* work is osmotic work. Non-*PV* work is sometimes called the *available work.*

Available (non-*PV*) Work

We will now derive the important result that the *non-PV work is equal to the change in Gibbs energy for a reversible process occurring at constant temperature and pressure.* We start with the definition of Gibbs energy (Eq. 3.80):

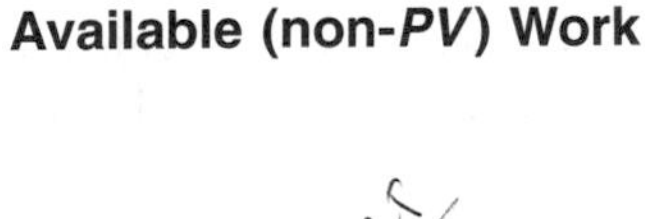

$$G \equiv H - TS \equiv U + PV - TS \quad (3.98)$$

For any change,

$$dG = dU + P\,dV + V\,dP - T\,dS - S\,dT \quad (3.99)$$

At constant *T* and *P*,

$$dG = dU + P\,dV - T\,dS \quad (3.100)$$

From the first law, dU is equal to $dq_P + dw$:

$$dG = dq_P + dw + P\,dV - T\,dS \quad (3.101)$$

For a process in which the system undergoes a volume change dV, the *PV* work is $-P\,dV$ and the total work is thus

$$dw = dw_{PV} + dw_{\text{non-}PV} = -P\,dV + dw_{\text{non-}PV} \quad (3.102)$$

Together with Eq. 3.101 this gives

$$dG = dq_P + dw_{\text{non-}PV} - T\,dS \quad (3.103)$$

However, since the process is reversible, $dq_P = T\,dS$ and therefore

$$dG = dw_{\text{non-}PV} \quad \text{or} \quad \Delta G = w_{\text{non-}PV} \text{ at constant } T \text{ and } P \quad (3.104)$$

This result has many important applications in physical chemistry. In Section 8.3 we shall use it to derive the emf of a reversible electrochemical cell.

Another important result, which we will leave for the reader to derive (see Problem 3.65), is that at a given temperature the *change in Helmholtz energy, for all processes irrespective of changes in P and V, is equal to the total work* (*PV* + *non-PV*). Because of this relationship the Helmholtz function has often been called the *work function.*

3.8 Some Thermodynamic Relationships

On the basis of the principles so far developed, it is possible to derive a number of relationships between different thermodynamic quantities. Some of the most important of these will now be obtained.

Maxwell Relations

For an infinitesimal process involving only PV work, we can combine the first and second laws in the equation

$$dU = dw + dq = -P\,dV + T\,dS \tag{3.105}$$

For dH we have similarly

$$dH = d(U + PV) = dU + d(PV) = dU + P\,dV + V\,dP \tag{3.106}$$

$$= -P\,dV + T\,dS + P\,dV + V\,dP = V\,dP + T\,dS \tag{3.107}$$

In making use of the relationship $d(PV) = P\,dV + V\,dP$, we have performed a *Legendre transformation,* named after the French mathematician Adrien Marie Legendre (1752–1853).

Similarly, for the Helmholtz and Gibbs energies, we have

$$dA = d(U - TS) = dU - d(TS) = dU - T\,dS - S\,dT \tag{3.108}$$

$$= -P\,dV + T\,dS - T\,dS - S\,dT = -P\,dV - S\,dT \tag{3.109}$$

and

$$dG = d(H - TS) = dH - T\,dS - S\,dT \tag{3.110}$$

$$= V\,dP + T\,dS - T\,dS - S\,dT = V\,dP - S\,dT \tag{3.111}$$

We can now combine the expressions we have obtained for dU, dH, dA, and dG with general relationships from differential calculus:

$$dU = -P\,dV + T\,dS = \left(\frac{\partial U}{\partial V}\right)_S dV + \left(\frac{\partial U}{\partial S}\right)_V dS \tag{3.112}$$

$$dH = V\,dP + T\,dS = \left(\frac{\partial H}{\partial P}\right)_S dP + \left(\frac{\partial H}{\partial S}\right)_P dS \tag{3.113}$$

$$dA = -P\,dV - S\,dT = \left(\frac{\partial A}{\partial V}\right)_T dV + \left(\frac{\partial A}{\partial T}\right)_V dT \tag{3.114}$$

$$dG = V\,dP - S\,dT = \left(\frac{\partial G}{\partial P}\right)_T dP + \left(\frac{\partial G}{\partial T}\right)_P dT \tag{3.115}$$

Important relationships are now obtained by equating coefficients; thus from Eq. 3.112 we have

$$\left(\frac{\partial U}{\partial V}\right)_S = -P \qquad \left(\frac{\partial U}{\partial S}\right)_V = T \tag{3.116}$$

Similarly, from Eqs. 3.113, 3.114, and 3.115,

$$\left(\frac{\partial H}{\partial P}\right)_S = V \qquad \left(\frac{\partial H}{\partial S}\right)_P = T \tag{3.117}$$

$$\left(\frac{\partial A}{\partial V}\right)_T = -P \qquad \left(\frac{\partial A}{\partial T}\right)_V = -S \tag{3.118}$$

$$\left(\frac{\partial G}{\partial P}\right)_T = V \qquad \left(\frac{\partial G}{\partial T}\right)_P = -S \tag{3.119}$$

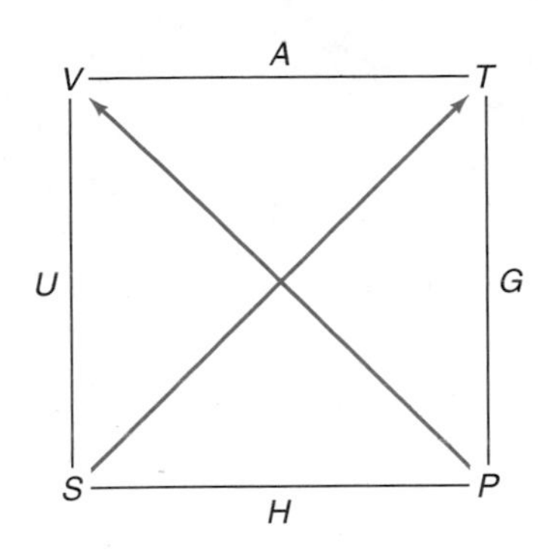

FIGURE 3.11
A mnemonic device for obtaining Eqs. 3.116–3.119 and the Maxwell relations 3.122–3.125. Each of the four thermodynamic potentials U, A, G, and H is flanked by two properties (e.g., U by V and S) to which it has a special relationship. The direction of the arrow indicates the signs of the equation. (a) For equations 3.116–3.119, any thermodynamic potential can be differentiated with respect to one of its neighboring properties, the other being held constant. The result is obtained by following the arrow. For example,

$$\left(\frac{\partial U}{\partial S}\right)_V = T \quad \text{and} \quad \left(\frac{\partial U}{\partial V}\right)_S = -P$$

(b) For Eqs. 3.122 and 3.125, any partial derivative of a property with respect to a neighboring property [e.g., $(\partial V/\partial T)_P$] is related to the corresponding derivative at the other side of the square [e.g., $(\partial S/\partial P)_T$], the arrows indicating the signs (in this example, negative).

A mnemonic device for obtaining these eight relationships is illustrated in Figure 3.11.

According to a theorem due to the Swiss mathematician Leonhard Euler (1707–1783), the order of differentiation does not matter. Thus, if f is a function of the variables x and y,

$$\frac{\partial}{\partial x}\left(\frac{\partial f}{\partial y}\right)_x = \frac{\partial}{\partial y}\left(\frac{\partial f}{\partial x}\right)_y \tag{3.120}$$

Application of this *Euler reciprocity theorem* to Eqs. 3.108 to 3.111 and use of Eqs. 3.112 to 3.115 lead to a number of useful relationships. For example, in Eq. 3.112, U is a function of V and S, so that by Euler's theorem

$$\frac{\partial}{\partial V}\left(\frac{\partial U}{\partial S}\right)_V = \frac{\partial}{\partial S}\left(\frac{\partial U}{\partial V}\right)_S \tag{3.121}$$

and introduction of Eq. 3.116 gives

$$\left(\frac{\partial T}{\partial V}\right)_S = -\left(\frac{\partial P}{\partial S}\right)_V \tag{3.122}$$

Similarly, from Eqs. 3.113 to 3.115 we obtain

$$\left(\frac{\partial T}{\partial P}\right)_S = \left(\frac{\partial V}{\partial S}\right)_P \tag{3.123}$$

$$\left(\frac{\partial P}{\partial T}\right)_V = \left(\frac{\partial S}{\partial V}\right)_T \tag{3.124}$$

$$\left(\frac{\partial V}{\partial T}\right)_P = -\left(\frac{\partial S}{\partial P}\right)_T \tag{3.125}$$

These are known as **Maxwell relations,** after the Scottish physicist James Clerk Maxwell (1831–1879) who presented them in his book *Theory of Heat,* which first appeared in 1870 and ran to 11 editions. The relations are particularly useful for obtaining quantities that are not easily measurable directly. For example, it would be difficult to measure $(\partial S/\partial P)_T$, but $(\partial V/\partial T)_P$ is easily obtained.

A useful device for obtaining these four Maxwell relations is included in Figure 3.11.

Thermodynamic Equations of State

We can also derive equations that give U and H in terms of P, V, and T and that are therefore called *thermodynamic equations of state.* From the definition of A, which equals $U - TS$, we have

$$\left(\frac{\partial U}{\partial V}\right)_T = \left[\frac{\partial(A + TS)}{\partial V}\right]_T \tag{3.126}$$

$$= \left(\frac{\partial A}{\partial V}\right)_T + T\left(\frac{\partial S}{\partial V}\right)_T \tag{3.127}$$

Then, from Eqs. 3.118 and 3.124 the thermodynamic equation of state for U is

$$\left(\frac{\partial U}{\partial V}\right)_T = -P + T\left(\frac{\partial P}{\partial T}\right)_V \tag{3.128}$$

The corresponding thermodynamic equation of state for H is obtained as follows:

$$\left(\frac{\partial H}{\partial P}\right)_T = \left[\frac{\partial(G + TS)}{P}\right]_T = \left(\frac{\partial G}{\partial P}\right)_T + T\left(\frac{\partial S}{\partial P}\right)_T \tag{3.129}$$

and by use of Eqs. 3.119 and 3.125 we have the thermodynamic equation of state for H

Thermodynamic Equation of State

$$\left(\frac{\partial H}{\partial P}\right)_T = V - T\left(\frac{\partial V}{\partial T}\right)_P \tag{3.130}$$

Some Applications of the Thermodynamic Relationships

A great many further relationships between the thermodynamic quantities can be obtained on the basis of the equations we have derived in the last few pages. Only a few examples can be included here.

One useful application is related to the theory of the Joule-Thomson effect, which we considered in Section 2.7. The Joule-Thomson coefficient μ is defined by Eq. 2.108:

$$\mu \equiv \left(\frac{\partial T}{\partial P}\right)_H \tag{3.131}$$

and C_P is defined by Eq. 2.27:

$$C_P \equiv \left(\frac{\partial H}{\partial T}\right)_P \tag{3.132}$$

But, from the theory of partial derivatives (Appendix C),

$$\left(\frac{\partial T}{\partial P}\right)_H = -\left(\frac{\partial H}{\partial P}\right)_T\left(\frac{\partial T}{\partial H}\right)_P \tag{3.133}$$

so that

$$\mu = -\frac{1}{C_P}\left(\frac{\partial H}{\partial P}\right)_T \tag{3.134}$$

Use of the thermodynamic equation of state, Eq. 3.130, then gives

$$\mu = \left(\frac{T(\partial V/\partial T)_P - V}{C_P}\right) \tag{3.135}$$

For an ideal gas

$$V = \frac{nRT}{P} \tag{3.136}$$

and

$$\left(\frac{\partial V}{\partial T}\right)_P = \frac{nR}{P} = \frac{V}{T} \tag{3.137}$$

so that the numerator in Eq. 3.135 is equal to zero. In the Joule-Thomson experiment there is therefore no temperature change for an ideal gas. For real gases, the numerator in Eq. 3.135 is in general other than zero. At the inversion temperature for a gas $\mu = 0$, and the condition for the inversion temperature is, therefore,

$$T\left(\frac{\partial V}{\partial T}\right)_P = V \tag{3.138}$$

The **cubic expansion coefficient** (formerly called the *thermal expansivity*) of a substance is defined as

Cubic Expansion Coefficient

$$\alpha \equiv \frac{1}{V}\left(\frac{\partial V}{\partial T}\right)_P \tag{3.139}$$

so that Eq. 3.135 can alternatively be written as

$$\mu = \frac{\alpha VT - V}{C_P} = \frac{V(\alpha T - 1)}{C_P} \tag{3.140}$$

The condition for the inversion temperature is therefore that

$$\alpha = T^{-1} \tag{3.141}$$

A similar expression that has wide utility is the **isothermal compressibility,** κ, defined as

$$\kappa \equiv \frac{1}{V}\left(\frac{\partial V}{\partial P}\right)_T \tag{3.142}$$

This is another partial derivative and, along with α, μ_{JT}, and the heat capacities, is an easily measurable quantity. They can often be related to calculate otherwise immeasurable values.

Another example of the application of the thermodynamic relationships is concerned with the van der Waals equation of state for 1 mol of gas:

$$\left(P + \frac{a}{V_m^2}\right)(V_m - b) = RT \tag{3.143}$$

Internal Pressure

We saw in Section 2.7 (p. 84) that the *internal pressure* of a gas is $(\partial U/\partial V)_T$, and we will now prove that if a gas obeys the van der Waals equation, the internal pressure is a/V_m^2. In other words, we will prove that

$$\left(\frac{\partial U}{\partial V}\right)_T = \frac{a}{V_m^2} \tag{3.144}$$

We do this by starting with the thermodynamic equation of state, Eq. 3.128, for the internal energy:

$$\left(\frac{\partial U}{\partial V}\right)_T = -P + T\left(\frac{\partial P}{\partial T}\right)_V \tag{3.145}$$

Equation 3.143 can be written as

$$P = \frac{RT}{V_m - b} - \frac{a}{V_m^2} \tag{3.146}$$

and therefore

$$\left(\frac{\partial P}{\partial T}\right)_V = \frac{R}{V_m - b} = \frac{1}{T}\left(P + \frac{a}{V_m^2}\right) \tag{3.147}$$

Substitution of this expression into Eq. 3.145 gives

$$\left(\frac{\partial U}{\partial V}\right)_T = \frac{a}{V_m^2} \tag{3.148}$$

and this is the internal pressure.

Fugacity and Activity

We have seen (Eq. 3.96) that for an isothermal process involving 1 mol of an ideal gas,

$$\Delta G_m = RT \ln \frac{V_1}{V_2} = RT \ln \frac{P_2}{P_1} \tag{3.149}$$

If the initial pressure P_1 is 1 bar, the gas is in a standard state and we write its molar Gibbs energy as G_m°; the Gibbs energy at any pressure P is then

$$G_m = G_m^\circ + RT \ln (P/\text{bar}) \tag{3.150}$$

If the gas is not ideal, this expression no longer applies. In order to have a parallel treatment of real gases, the American chemist Gilbert Newton Lewis (1875–1946) introduced a new function known as the **fugacity** (Latin *fugare,* to fly) and given the symbol f. The fugacity is *the pressure adjusted for lack of ideality;* if a gas is behaving ideally, the fugacity is equal to the pressure.

The fugacity is such that, to parallel Eq. 3.149,

$$\Delta G_m = RT \ln \frac{f_2}{f_1} \tag{3.151}$$

For the ideal gas we took the standard state to correspond to 1 bar pressure and so obtained Eq. 3.150. Similarly, for the nonideal gas we define the standard state to correspond to *unit fugacity* (i.e., $f_1 = 1$ bar) so that we obtain

$$G_m = G_m^\circ + RT \ln (f/\text{bar}) \tag{3.152}$$

The ratio of the fugacity of a substance in any state to the fugacity in the standard state is known as the **activity** and given the symbol a. Since for a gas the fugacity in the standard state is by definition 1 bar (10^5 Pa), the *fugacity of a gas is numerically equal to its activity.* Equation 3.152 can thus be written as

$$G_m = G_m^\circ + RT \ln a \tag{3.153}$$

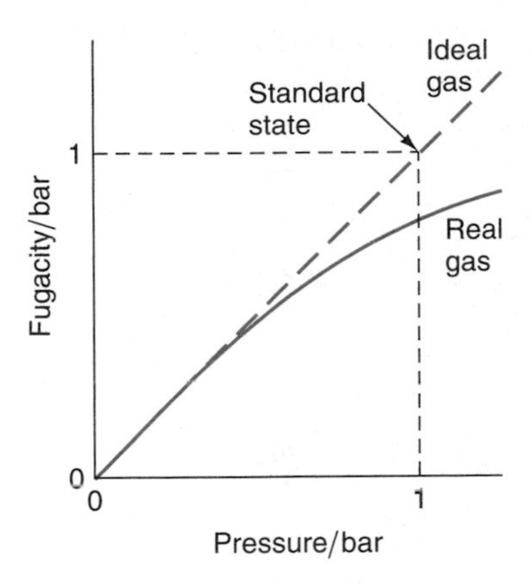

FIGURE 3.12
Fugacity of a real gas as a function of pressure. The standard state is the state at which the fugacity would be equal to 1 bar if the gas remained ideal from low pressures to 1 bar pressure.

Figure 3.12 shows schematically how the fugacity of a pure gas may vary with its pressure. At sufficiently low pressure, every gas behaves ideally, because the intermolecular forces are negligible and because the effective volume of the molecules is insignificant compared with the total volume. Therefore, as shown in Figure 3.12, we can draw a tangent to the curve at low pressures, and this line will represent the behavior of the gas if it were ideal. This line has a slope of unity; the fugacity of an ideal gas is 1 bar when the pressure is 1 bar.

It might be thought that the best procedure would have been to choose the standard state of a nonideal gas to correspond to some low pressure, such as 10^{-6} bar, at which the behavior is ideal. It has proved more convenient, however, to make use of the hypothetical ideal gas line in Figure 3.12 and to choose the point corresponding to 1 bar as the standard state. The true fugacity at 1 bar pressure is different from 1 bar (it is less in Figure 3.12), but we choose the **standard state** as the **state at which the fugacity would be equal to 1 bar if the gas remained ideal from low pressures to 1 bar pressure.**

The fugacity of a pure gas or a gas in a mixture can be evaluated if adequate *P-V-T* data are available. For 1 mol of gas at constant temperature,

$$dG = V_m\, dP \tag{3.154}$$

so that, for two states 1 and 2,

$$\Delta G_m = G_{m,2} - G_{m,1} = \int_{P_1}^{P_2} V_m\, dP \tag{3.155}$$

The quantity *RT/P* can be added to and subtracted from the integrand to give

$$\Delta G_m = \int_{P_1}^{P_2} \left[\frac{RT}{P} + \left(V_m - \frac{RT}{P}\right)\right] dP \tag{3.156}$$

$$= RT \ln \frac{P_2}{P_1} + \int_{P_1}^{P_2} \left(V_m - \frac{RT}{P}\right) dP \tag{3.157}$$

With Eq. 3.151 we have

$$RT \ln \frac{f_2}{f_1} = RT \ln \frac{P_2}{P_1} + \int_{P_1}^{P_2} \left(V_m - \frac{RT}{P}\right) dP \tag{3.158}$$

and therefore

$$RT \ln \frac{f_2/P_2}{f_1/P_1} = \int_{P_1}^{P_2} \left(V_m - \frac{RT}{P}\right) dP \tag{3.159}$$

If P_1 is a sufficiently low pressure, f_1/P_1 is equal to unity, so that

$$RT \ln \frac{f_2}{P_2} = \int_{P_1}^{P_2} \left(V_m - \frac{RT}{P}\right) dP \tag{3.160}$$

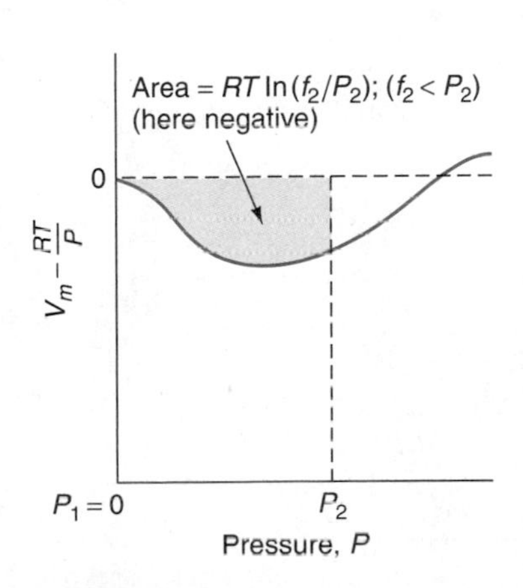

FIGURE 3.13
Schematic plot of $V_m - RT/P$ for 1 mol of a real gas.

Suppose, for example, that we have reliable pressure-volume data, covering a wide range of pressures and going down to low pressures, at a particular temperature *T*. We can then calculate ($V_m - RT/P$) and plot this quantity against *P*, as shown schematically in Figure 3.13. We can then obtain the value of the shaded area in Figure 3.13, from $P_1 = 0$ to any value P_2. This area is the integral on the right-hand side of Eq. 3.159, and the fugacity f_2 can therefore be calculated. Various analytical procedures have been used for evaluating these integrals. Use is sometimes made of the law of corresponding states (Section 1.13); to the extent that this law is valid, all gases fit a single set of curves, and for a given gas all that is necessary is to know its critical constants.

EXAMPLE 3.9 Oxygen at pressures that are not too high obeys the equation

$$P(V_m - b) = RT$$

where $b = 0.0211\ \text{dm}^3\ \text{mol}^{-1}$.

a. Calculate the fugacity of oxygen gas at 25 °C and 1 bar pressure.
b. At what pressure is the fugacity equal to 1 bar?

Solution For this equation of state

$$V_m - \frac{RT}{P} = b$$

so that

$$\int_{P_1}^{P_2} \left(V_m - \frac{RT}{P}\right) dP = \int_{P_1}^{P_2} b\, dP = b(P_2 - P_1)$$

a. Equation 3.160 then gives, with $P_1 = 0$, $P_2 = 1$ bar, and $b = 0.0211\ \text{dm}^3\ \text{mol}^{-1}$,

$$RT \ln f_2 = 0.0211\ \text{bar dm}^3\ \text{mol}^{-1}$$
$$= 0.0211 \times 10^5 \times 10^{-3}\ \text{Pa m}^3\ \text{mol}^{-1}$$
$$\ln f_2 = \frac{0.0211 \times 10^2(\text{Pa m}^3\ \text{mol}^{-1})}{8.3145(\text{J K}^{-1}\ \text{mol}^{-1}) \times 298.15\ \text{K}}$$
$$= 8.51 \times 10^{-4}$$
$$f_2 = 1.0009\ \text{bar}$$

b. $\ln f = \ln P + \dfrac{bP}{RT}$

The pressure at which $f = 1$ bar ($\ln f = 0$) is thus given by

$$\ln P = -\frac{bP}{RT} \quad \text{or} \quad P = e^{-bP/RT}$$

Since bP/RT is very small, we can expand the exponential and keep only the first term:[7]

$$P = 1 - \frac{bP}{RT}$$
$$P = \frac{RT}{RT + b}$$
$$= \frac{0.0831 \times 298.15}{0.0831 \times 298.15 + 0.0211}$$
$$= 0.9991\ \text{bar}$$

[7] $R = 8.3145\ \text{J K}^{-1}\ \text{mol}^{-1} = 0.083092\ \text{bar dm}^3\ \text{K}^{-1}\ \text{mol}^{-1}$; the latter value is often convenient when pressures are given in bars.

3.9 The Gibbs-Helmholtz Equation

The variation with temperature of the Gibbs energy change in a chemical process is a matter of great importance, and we shall see in Chapter 4 that it leads to an extremely useful way of determining the enthalpy change in a reaction.

Equation 3.119 gives the variation of G with T:

$$\left(\frac{\partial G}{\partial T}\right)_P = -S \tag{3.161}$$

For a change from one state to another, therefore,

$$\left(\frac{\partial \Delta G}{\partial T}\right)_P = -\Delta S \tag{3.162}$$

However, since

$$\Delta G = \Delta H - T\Delta S \tag{3.163}$$

$$\Delta S = \frac{\Delta H - \Delta G}{T} \tag{3.164}$$

so that, with Eq. 3.162,

$$\left(\frac{\partial \Delta G}{\partial T}\right)_P - \frac{\Delta G}{T} = -\frac{\Delta H}{T} \tag{3.165}$$

From differential calculus (Appendix C),

$$\frac{\partial}{\partial T}\left(\frac{\Delta G}{T}\right) = \frac{1}{T}\frac{\partial \Delta G}{\partial T} - \frac{\Delta G}{T^2} \tag{3.166}$$

so that the left-hand side of Eq. 3.165 can be written as

$$T\left[\frac{\partial}{\partial T}\left(\frac{\Delta G}{T}\right)\right]_P$$

Thus

$$T\left[\frac{\partial}{\partial T}\left(\frac{\Delta G}{T}\right)\right]_P = -\frac{\Delta H}{T} \tag{3.167}$$

or

$$\left[\frac{\partial}{\partial T}\left(\frac{\Delta G}{T}\right)\right]_P = -\frac{\Delta H}{T^2} \tag{3.168}$$

This important thermodynamic relationship is known as the **Gibbs-Helmholtz equation.** If the reactants and products are in their standard states, the equation takes the form

$$\left[\frac{\partial}{\partial T}\left(\frac{\Delta G^\circ}{T}\right)\right]_P = -\frac{\Delta H^\circ}{T^2} \tag{3.169}$$

We will find this equation to be particularly important when we come to consider the temperature dependence of equilibrium constants (Section 4.8).

3.10 Thermodynamic Limitations to Energy Conversion

The laws of thermodynamics have many practical applications to the interconversion of the various forms of energy, problems that are becoming of increasing technical and economic importance. Both the first law and the second law place limits on how much useful energy or work can be obtained from a given source.

First Law Efficiencies

The first law is merely a statement of the principle of conservation of energy, and its application is very straightforward; any energy that does not serve the purpose intended must be subtracted from the total in order to obtain the amount of useful energy. The efficiency with which the energy contained in any fuel is converted into useful energy varies very widely.[8] When wood or coal is burned in an open fireplace, about 80% of the heat escapes up the chimney; only about 20% enters the room. A good home furnace, on the other hand, can convert about 75% of the energy in the fuel into useful heat. Recent high-efficiency furnaces may even reach 90% conversion. However, many home furnaces operate at lower efficiencies of perhaps 50 or 55%. These low efficiencies are simply due to the fact that much of the heat produced in the combustion of the fuel passes to the outside of the building.

Second Law Efficiencies

An entirely different type of problem is encountered with energy-conversion devices for which there is a Carnot, or second law, limitation. We have seen that for a *reversible* engine operating between two temperatures T_h and T_c the efficiency is

$$\frac{T_h - T_c}{T_h}$$

In practice, since the behavior cannot be reversible, a lower efficiency will be obtained. The higher temperature T_h in a modern steam turbine, which uses steam at high pressure, may be 811 K (1000 °F) and the lower temperature T_c may be 311 K (100 °F). The Carnot efficiency is thus

$$\frac{811 - 311}{811} = 0.62 = 62\%$$

However, because the two temperatures cannot be held constant and because the behavior is not reversible, the efficiency actually obtained is more like 35%. This is considerably greater than the efficiency of the old steam engines, which operated at a much lower T_h; their efficiencies were often less than 10%.

If we want to calculate the overall efficiency for the conversion of the energy of a fuel into electricity, we must consider the efficiencies of the three processes involved:

[8] Typical efficiency values are given in an article by C. M. Summers, *Scientific American, 225,* 149(September 1971).

1. Conversion of the energy of a fuel into heat; in a modern boiler this efficiency is typically about 88%.
2. Conversion of the heat into mechanical energy; efficiency is 35% (as noted previously).
3. Conversion of the mechanical energy into electricity; modern generators have a very high efficiency of about 99%.

The overall efficiency is thus

$$0.88 \times 0.35 \times 0.99 = 0.30 = 30\%$$

Nuclear power plants operate at lower efficiencies, largely because of the lower Carnot efficiency. Nuclear reactors are usually run at lower temperatures than boilers burning fossil fuel. A typical value of T_h is 623 K, and if T_c is 311 K, the Carnot efficiency is

$$\frac{623 - 311}{623} = 0.50 = 50\%$$

In practice this is further reduced to about 25% because of irreversibility.

A somewhat different way of looking at efficiencies is to consider power, the rate of doing work, rather than work itself. We have seen that for an engine operating reversibly between 811 K and 311 K the Carnot efficiency $1 - (T_c/T_h)$ is 62%. If, however, an engine is acting reversibly, everything occurs infinitely slowly and the power output is zero since the work is done over an infinite period of time. A more useful question to ask is, If we maximize the power instead of the work, what will the efficiency be? Engineers have looked at this problem using a number of different models, and one of these leads to the simple result that the efficiency is instead $1 - (T_c/T_h)^{1/2}$. We can refer to this quantity as the **maximum power efficiency.** For the temperatures 811 K and 311 K, the maximum power efficiency works out to be 38%. This is obviously a more realistic way of estimating the efficiencies of heat engines.

Maximum Power Efficiency

Refrigeration and Liquefaction

The Carnot limitation to efficiency also becomes important when we consider a refrigerator or a device for liquefying gases. The principle of operation of a refrigerator is shown schematically in Figure 3.14a. A refrigerator consists essentially of a compressor, which can pump vapor out of an evaporator and pump it into a condenser, where it liquefies. The fluid employed is one that has a high latent heat of evaporation. For many years, ammonia, carbon dioxide, and sulfur dioxide were used, but fluorinated hydrocarbons are commonly used at the present time. Evaporation of the liquid produces cooling, and condensation leads to release of heat. The work done by the compressor thus transfers heat from the evaporator to the condenser.

The thermodynamics of refrigeration were worked out in 1895 by the German chemist Karl von Linde (1842–1934). The essential principles are illustrated in Figure 3.14b. Suppose that the refrigerator is operating reversibly and that the condenser is at a temperature of 20 °C and the evaporator at a temperature of 0 °C. We thus have a Carnot engine operating in reverse, and the ratio of the absolute

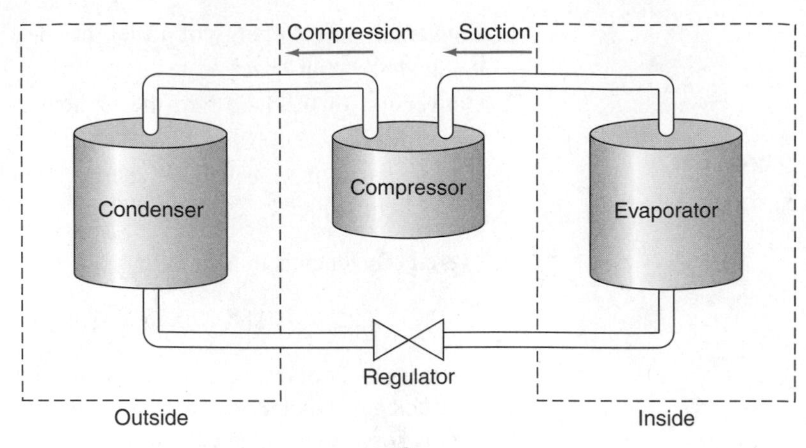

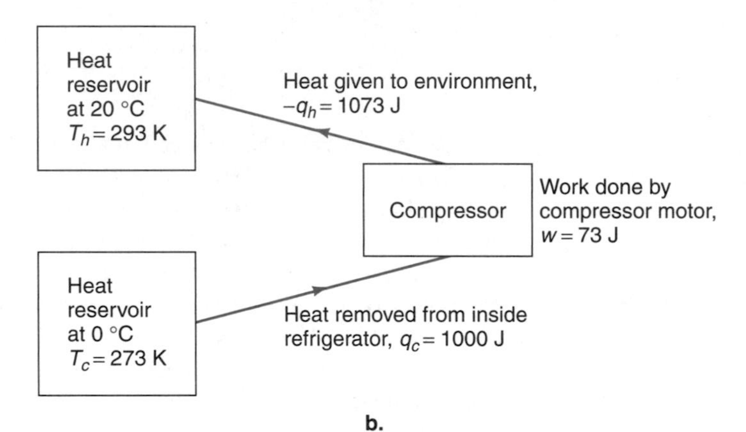

FIGURE 3.14
(a) Schematic diagram of a domestic refrigerator. (b) Analysis of the operation of a domestic refrigerator, on the assumption of reversible behavior.

Study a heat pump and understand efficiency; Heat engines, pumps, and efficiency.

temperature, 293/273, is the ratio of the heat liberated at the higher temperature to the heat absorbed at the lower temperature (see Eq. 3.23):

$$\frac{293}{273} = -\frac{q_h}{q_c}$$

Thus if the evaporator removes 1000 J from inside the refrigerator, at 273 K, the amount of heat $-q_h$ discharged by the condenser into its environment at 293 K is

$$-q_h = \frac{293}{273} \times 1000 = 1073 \text{ J}$$

Performance factor

The difference, 73 J, is the work that has to be done by the compressor. We can define the *performance factor* of the refrigerator as the heat removed from the environment at the lower temperature divided by the work done by the compressor; the performance factor in this example, for reversible behavior, is thus

$$\frac{1000}{73} = 13.7$$

In general, the maximum performance factor for a refrigerator is given by

$$\text{max performance factor} = \frac{T_c}{T_h - T_c} \tag{3.170}$$

In practice the performance is considerably less than this since the cycle does not operate reversibly; in order to maintain the inside of the refrigerator at 0 °C the temperature of the evaporator will have to be significantly less than 0 °C, and the condenser will be significantly warmer than its surroundings.

Cascade processes

When much lower temperatures are required, it is necessary to employ *cascade* processes. The first practical way of liquefying air, based on work done earlier by Linde, was devised in 1877 by Raoul Pierre Pictet (1846–1929), a Swiss chemist and refrigeration engineer. The critical temperature of air (approximately 20% O_2, 79% N_2, and 1% Ar) is about −141°C, and this temperature must be reached before pressure will bring about liquefaction. Ammonia has a critical temperature of 132.9 °C, and it can therefore be liquefied by the same type of expansion and compression cycle as in Figure 3.14; this process provides a vessel with liquid ammonia at −34 °C. A stream of ethylene compressed to 19 atm is passed through this bath and cooled to about −31 °C, which is below its critical point (9.6 °C); on being throttled through a valve, two-thirds of it is liquefied and collects in a bath. When it boils away at atmospheric pressure, it cools to −104 °C. The next stage employs methane at 25 atm, which is cooled by the liquid ethylene and produces, on liquefaction and subsequent evaporation at 1 atm, liquid methane at −161 °C in the bath. This suffices to liquefy the nitrogen after the latter has been compressed by 18.6 atm.

Heat Pumps

The heat pump works on exactly the same principle as the refrigerator, but the purpose is to bring about heating instead of cooling. Figure 3.15a shows schematically the type of arrangement; the building is being maintained at 30 °C with an external temperature of −15 °C. The temperature ratio is 303/258 = 1.17; therefore, in order for 1000 J of heat to leave the condenser (Figure 3.15b),

$$\frac{1000}{1.17} = 855 \text{ J}$$

must be extracted from the external environment. The difference, 145 J, is the work that must be done by the compressor, assuming all processes to be reversible. The performance factor can be defined as the heat provided to the building divided by the work done by the compressor, and in this example it is thus

$$\frac{1000}{145} = 6.9$$

In general, the performance factor of a heat pump, operating reversibly, is given by

$$\text{max performance factor} = \frac{T_h}{T_h - T_c} \qquad (3.171)$$

In practice, the performance factor will be much less, since the behavior will be far from reversible.

The principle of the heat pump has been known for over a century, but for many years little practical application was made of it. The example just given shows that use of the heat pump leads to a considerable saving of energy, in that an expenditure of 145 J of energy led to heating which would have required 1000 J if carried out directly. This estimate was made for an outside temperature of −15 °C; if it is greater than this the performance factor is larger, and the saving of energy is greater.

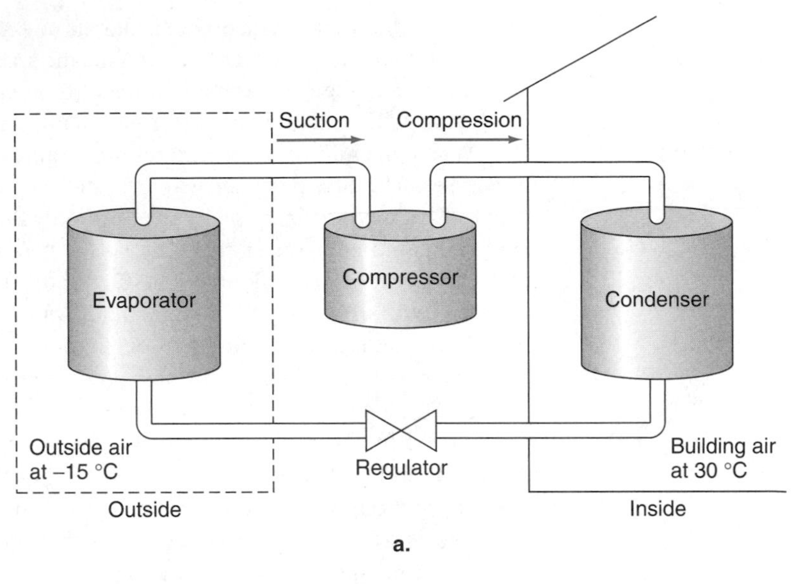

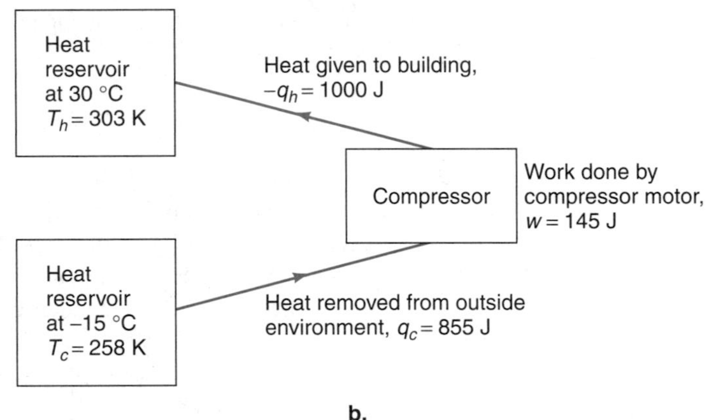

FIGURE 3.15
(a) Schematic diagram of a heat pump used for heating a building to 30 °C, with an external temperature of −15 °C. (b) Analysis of the operation of the heat pump.

However, there are practical problems which tend to reduce the performance factors to below the ideal values calculated on the basis of the second law. If the evaporator is maintained outside the building, the air near it undergoes cooling, and the performance factor is lowered; agitation of the outside air is therefore important. Over the years there have been technical improvements which bring performance factors closer to the theoretical ones. If a building is near a rapidly flowing river, the heat sink could be in the river where the external temperature remains fairly constant, and then higher practical efficiencies can be achieved.

In modern domestic applications it is common for a heat pump and air conditioner to be combined in the same unit; as has been noted, a heat pump is essentially an air conditioner operating in reverse. This combination of the two functions is effective in practice, but design parameters tend to reduce the performance factors to some extent.

Chemical Conversion

The thermodynamics of the conversion of the chemical energy of a fuel into heat and work is best considered with reference to the energy and enthalpy changes and the Gibbs and Helmholtz energy changes involved. Consider, for example, the combustion of 1 mol of isooctane (2,2,4-trimethylpentane), an important constituent of gasoline:

$$C_8H_{18}(g) + 12\tfrac{1}{2}O_2(g) \rightarrow 8CO_2(g) + 9H_2O(g)$$

Measurement of the heat of this reaction, in a calorimeter at 25 °C and at constant volume, leads to

$$\Delta U = -5109 \text{ kJ mol}^{-1}$$

The stoichiometric sum[9] $\Sigma\nu$ for the process is $8 + 9 - (12\tfrac{1}{2} + 1) = 3.5$, with the result that Eq. 2.23 becomes

$$\begin{aligned}\Delta H &= \Delta U + \Sigma\nu RT \\ &= -5\,109\,000 \text{ J mol}^{-1} + 3.5 \times 8.314 \text{ J K}^{-1} \text{ mol}^{-1} \times 298.15 \text{ K} \\ &= -5\,100\,000 \text{ J mol}^{-1} = -5100 \text{ kJ mol}^{-1}\end{aligned}$$

From the absolute entropies of the reactants and products it is found that, at 298 K,

$$\Delta S = 422 \text{ J K}^{-1} \text{ mol}^{-1}$$

We thus obtain

$$\begin{aligned}\Delta A = \Delta U - T\Delta S &= -5\,109\,000 \text{ J mol}^{-1} - 422 \text{ J K}^{-1} \text{ mol}^{-1} \times 298.15 \text{ K} \\ &= -5\,235\,000 \text{ J mol}^{-1} = -5235 \text{ kJ mol}^{-1}\end{aligned}$$

and

$$\begin{aligned}\Delta G = \Delta H - T\Delta S &= -5\,100\,000 \text{ J mol}^{-1} - 422 \text{ J K}^{-1} \text{ mol}^{-1} \times 298.15 \text{ K} \\ &= -5\,226\,000 \text{ J mol}^{-1} = -5226 \text{ kJ mol}^{-1}\end{aligned}$$

These values, with the negative signs removed, are the maximum amounts of work that could be obtained by the combustion of 1 mol of isooctane, at constant volume and constant pressure, respectively.

Note that the maximum amount of work that could be obtained from the oxidation of 1 mol of isooctane at constant volume, 5235 kJ, is actually greater than the heat liberated at constant volume, 5109 kJ. There is, of course, no violation of the first law; the system is not isolated and heat will enter from the environment; there is an entropy increase in the system and an entropy decrease in the environment. In practice, however, it will never be possible to obtain anything like 5235 kJ of work from 1 mol of isooctane. If the fuel is simply burned in a calorimeter, all the energy is released as heat, and no work is done. If it is burned in an internal combustion engine, a good deal of heat would again be released; a typical efficiency for an automobile engine is 25%, so that it might be possible to obtain about 1300 kJ of work from 1 mol. To obtain more work we must devise a process in which less heat is evolved. A possible arrangement is a fuel cell (Section 8.6), in which the isooctane is catalytically oxidized at the surface of an electrode, with the production of an electric potential. Much research is going on at the present time with the object of increasing the efficiencies of such devices.

[9] The *stoichiometric sum* $\Sigma\nu$ is the sum of the stoichiometric coefficients (Section 2.5) in a reaction. Since stoichiometric coefficients are negative for reactants and positive for products, the stoichiometric sum is the change in the number of molecules for the reaction as written; e.g., for $A + B \rightarrow Z$, $\Sigma\nu = -1$.

Another example is provided by the reaction between hydrogen and oxygen:

$$2H_2(g) + O_2(g) \rightarrow 2H_2O(l)$$

The thermodynamic data are

$$\Delta U = -562.86 \text{ kJ mol}^{-1} \qquad \Delta H = -570.30 \text{ kJ mol}^{-1}$$
$$\Delta A = -466.94 \text{ kJ mol}^{-1} \qquad \Delta G = -474.38 \text{ kJ mol}^{-1}$$

Note that now the maximum work per mole obtainable at constant volume, 466.94 kJ, is *less* than the energy released; this is because there is an entropy *loss* on the formation of liquid water from gaseous hydrogen and oxygen. In a fuel cell at constant pressure it would be theoretically possible to obtain 474.38 kJ mol^{-1} of electrical work, somewhat less than the heat evolved (570.30 kJ mol^{-1}) in the combustion at constant pressure. In practice the efficiency of a modern fuel cell is about 60%, so that about 280 kJ of work might be produced from the reaction of 2 mol of hydrogen with 1 mol of oxygen.

KEY EQUATIONS

Definition of entropy change ΔS:

$$\Delta S_{A\rightarrow B} \equiv \int_A^B \frac{dq_{\text{rev}}}{T}$$

For any completely reversible cycle,

$$\oint \frac{dq_{\text{rev}}}{T} \equiv \oint dS = 0$$

If any part of the cycle is irreversible,

$$\oint \frac{dq_{\text{irr}}}{T} < 0 \qquad \text{(the inequality of Clausius)}$$

Entropy of mixing of ideal gases, per mole of mixture:

$$\Delta S_{\text{mix}} = -R(x_1 \ln x_1 + x_2 \ln x_2)$$

where x_1 and x_2 are the mole fractions.

Definition of Helmholtz energy A: $A \equiv U - TS$

Definition of Gibbs energy G: $G \equiv H - TS$

Conditions for equilibrium:
Constant V and T: $dA = 0$
Constant P and T: $dG = 0$

At constant P and T:

$$dG = w_{\text{non-}PV} \qquad (w_{\text{non-}PV} = \text{non-}PV \text{ work})$$

Important relationships:

$$\left(\frac{\partial U}{\partial V}\right)_S = -P \qquad \left(\frac{\partial U}{\partial S}\right)_V = T$$

$$\left(\frac{\partial H}{\partial P}\right)_S = V \qquad \left(\frac{\partial H}{\partial S}\right)_P = T$$

$$\left(\frac{\partial A}{\partial V}\right)_T = -P \qquad \left(\frac{\partial A}{\partial T}\right)_V = -S$$

$$\left(\frac{\partial G}{\partial P}\right)_T = V \qquad \left(\frac{\partial G}{\partial T}\right)_P = -S$$

Maxwell's relations:

$$\left(\frac{\partial T}{\partial V}\right)_S = -\left(\frac{\partial P}{\partial S}\right)_V \qquad \left(\frac{\partial T}{\partial P}\right)_S = \left(\frac{\partial V}{\partial S}\right)_P$$

$$\left(\frac{\partial P}{\partial T}\right)_V = \left(\frac{\partial S}{\partial V}\right)_T \qquad \left(\frac{\partial V}{\partial T}\right)_P = -\left(\frac{\partial S}{\partial P}\right)_T$$

The *Gibbs-Helmholtz equation:*

$$\left[\frac{\partial}{\partial T}\left(\frac{\Delta G}{T}\right)\right]_P = -\frac{\Delta H}{T^2}$$

The *cubic expansion coefficient* α:

$$\alpha \equiv \frac{1}{V}\left(\frac{\partial V}{\partial T}\right)_P$$

The *isothermal compressibility* κ:

$$\kappa \equiv \frac{1}{V}\left(\frac{\partial V}{\partial P}\right)_T$$

The *Joule-Thomson coefficient* μ:

$$\mu \equiv \left(\frac{\partial T}{\partial P}\right)_H$$

PROBLEMS

The Carnot Cycle (see also Section 3.1)

3.1. The accompanying diagram represents a reversible Carnot cycle for an ideal gas:

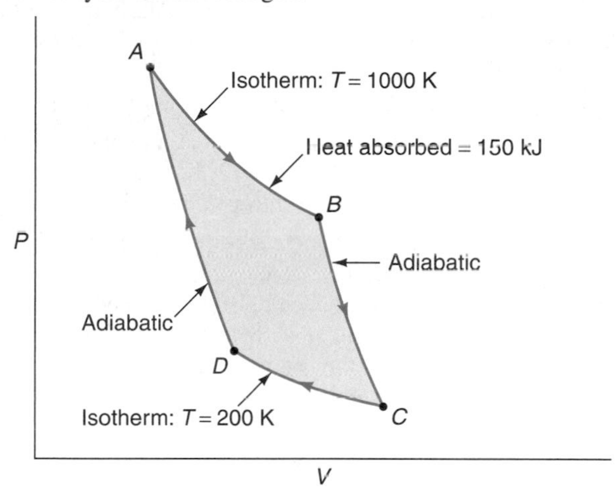

a. What is the thermodynamic efficiency of the engine?
b. How much heat is rejected at the lower temperature, 200 K, during the isothermal compression?
c. What is the entropy increase during the isothermal expansion at 1000 K?
d. What is the entropy decrease during the isothermal compression at 200 K?
e. What is the overall entropy change for the entire cycle?
f. What is the increase in Gibbs energy during the process $A \rightarrow B$?

3.2. An engine operates between 125 °C and 40 °C. What is the minimum amount of heat that must be withdrawn from the reservoir to obtain 1500 J of work?

3.3. a. Figure 3.2 shows a Carnot cycle in the form of a pressure-volume diagram. Sketch the corresponding entropy-temperature diagram, labeling the individual steps $A \rightarrow B$ (isotherm at T_h), $B \rightarrow C$ (adiabatic), $C \rightarrow D$ (isotherm at T_c), and $D \rightarrow A$ (adiabatic).
b. Suppose that a reversible Carnot engine operates between 300 K and a higher temperature T_h. If the engine produces 10 kJ of work per cycle and the entropy change in the isothermal expansion at T_h is 100 J K^{-1}, what are q_h, q_c, and T_h?

3.4. The following diagram represents a reversible Carnot cycle for an ideal gas:

a. What is the thermodynamic efficiency of the engine?
b. How much heat is absorbed at 400 K?
c. How much heat is rejected at 300 K?
d. What is the entropy change in the process $A \rightarrow B$?
e. What is the entropy change in the entire cycle?

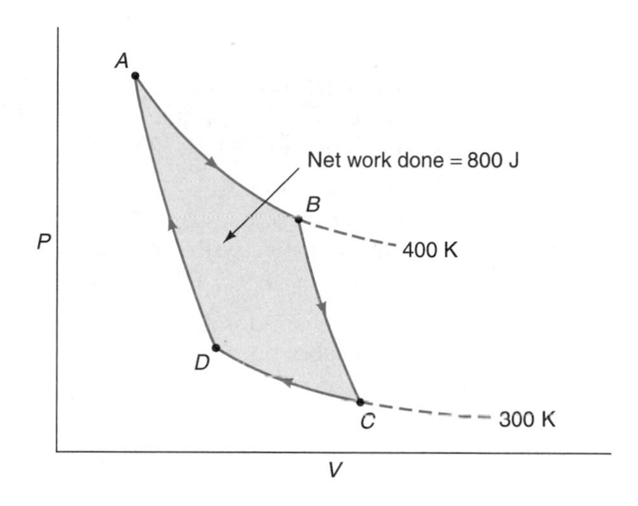

f. What is the Gibbs energy change in the process $A \rightarrow B$?
g. In order for the engine to perform 2 kJ of work, how much heat must be absorbed?

3.5. Suppose that an iceberg weighing 10^9 kg were to drift into a part of the ocean where the temperature is 20 °C. What is the maximum amount of work that could be generated while the iceberg is melting? Assume the temperature of the iceberg to be 0 °C. The latent heat of fusion of ice is 6.025 kJ mol^{-1}.

If the process occurred in one day, what would be the power produced?

3.6. Show that the change in the internal energy of an ideal gas during an isothermal expansion is zero, i.e., $\left(\frac{\partial U}{\partial V}\right)_T = 0$. Compare this result to Eq. 3.148 for a van der Waals gas.

Entropy Changes

3.7. Calculate the entropies of vaporization in J K^{-1} mol^{-1} of the following substances, from their boiling points and enthalpies of vaporization:

	Boiling Point/K	$\Delta_{vap}H$/kJ mol^{-1}
C_6H_6	353	30.8
$CHCl_3$	334	29.4
H_2O	373	40.6
C_2H_5OH	351	38.5

In terms of the structures of the liquids, suggest reasons for the higher values observed for H_2O and C_2H_5OH.

3.8. Calculate the standard entropies of formation of (a) liquid methanol and (b) solid urea, making use of the absolute entropies listed in Table 3.2 (p. 120).

3.9. Calculate the standard entropies for the following reactions at 25 °C:

a. $N_2(g) + 3H_2(g) \rightarrow 2NH_3(g)$
b. $N_2O_4(g) \rightarrow 2NO_2(g)$

3.10. Calculate the standard entropy for the dissociation of $H_2(g)$ into atomic hydrogen 2[H(g)] at 298.15 K and 1273.15 K. C_P°/J K^{-1} mol^{-1}: $H_2(g)$, 28.824; H(g), 20.784.

3.11. One mole of an ideal gas, with $C_{V,m} = \frac{3}{2}R$, is heated (a) at constant pressure and (b) at constant volume, from 298 K to 353 K. Calculate ΔS for the system in each case.

3.12. One mole each of N_2 and O_2 and $\frac{1}{2}$ mol of H_2, at 25 °C and 1 atm pressure, are mixed isothermally; the final total pressure is 1 atm. Calculate ΔS, on the assumption of ideal behavior.

3.13. Initially 1 mol of O_2 is contained in a 1-liter vessel, and 5 mol of N_2 are in a 2-liter vessel; the two vessels are connected by a tube with a stopcock. If the stopcock is opened and the gases mix, what is the entropy change?

3.14. Calculate the entropy of mixing per mole of air, taking the composition by volume to be 79% N_2, 20% O_2, and 1% Ar.

3.15. From the data given in Table 3.2 (p. 120), calculate the standard entropy of formation $\Delta_f S^\circ$ of liquid ethanol at 25 °C.

3.16. a. One mole of an ideal gas at 25 °C is allowed to expand reversibly and isothermally from 1 dm^3 to 10 dm^3. What is ΔS for the gas, and what is ΔS for its surroundings?

b. The same gas is expanded adiabatically and irreversibly from 1 dm^3 to 10 dm^3 with no work done. What is the final temperature of the gas? What is ΔS for the gas, and what is ΔS for the surroundings? What is the net ΔS?

3.17. One mole of liquid water at 0.00 °C and 1 atm pressure is turned into steam at 100.0 °C and 1 atm pressure by the following two paths:

a. Heated at constant pressure to 100.0°C, and allowed to boil into steam ($\Delta_{vap}H^\circ$ = 40.67 J mol^{-1} at this temperature).

b. Pressure lowered to 0.006 02 atm so that water evaporates to steam at 0 °C ($\Delta_{vap}H^\circ$ = 44.92 J mol^{-1} at this temperature), heated at the constant pressure of 0.006 02 atm to 100.0 °C, and compressed at 100.0 °C to 1 atm pressure.

Calculate the entropy change along each path and verify that they are the same, thus proving that ΔS° is a state property. The $C_{P,m}$ for liquid water and water vapor can be found in Table 2.1. [The paths and the enthalpies of vaporization are adapted from Table 6.1, Gordon M. Barrow, *Physical Chemistry*, 5th Ed., New York: McGraw-Hill, 1988.]

3.18. Predict the signs of the entropy changes in the following reactions when they occur in aqueous solution.

a. Hydrolysis of urea: $H_2NCONH_2 + H_2O \rightarrow CO_2 + 2NH_3$
b. $H^+ + OH^- \rightarrow H_2O$
c. $CH_3COOH \rightarrow CH_3COO^- + H^+$
d. $CH_2BrCOOCH_3 + S_2O_3^{2-} \rightarrow CH_2(S_2O_3^-)COOCH_3 + Br^-$

3.19. Obtain a general expression, in terms of the molar heat capacity $C_{P,m}$ and temperature T_1 and T_2, for the entropy increase of n mol of a gas (not necessarily ideal) that is heated at constant pressure so that its temperature changes from T_1 to T_2. To what does your expression reduce if the gas is ideal?

3.20. Initially 5 mol of an ideal gas, with $C_{V,m}$ = 12.5 J K^{-1} mol^{-1}, are at a volume of 5 dm^3 and a temperature of 300 K. If the gas is heated to 373 K and the volume changed to 10 dm^3, what is the entropy change?

***3.21.** At 100 °C 200 g of mercury are added to 80 g of water at 20 °C in a vessel that has a water equivalent of 20 g. The specific heat capacities of water and mercury may be taken as constant at 4.18 and 0.140 J K^{-1} g^{-1}, respectively. Calculate the entropy change of (a) the mercury; (b) the water and vessel; (c) the mercury, water, and vessel together.

***3.22.** At 0 °C 20 g of ice are added to 50 g of water at 30°C in a vessel that has a water equivalent of 20 g. Calculate the entropy changes in the system and in the surroundings. The heat of fusion of ice at 0 °C is 6.02 kJ mol^{-1}, and the specific heat capacities of water and ice may be taken as constant at 4.184 and 2.094 J K^{-1} g^{-1}, respectively, and independent of temperature.

***3.23.** Calculate the increase in entropy of 1 mol of nitrogen if it is heated from 300 K to 1000 K at a constant pressure of 1 atm; use the C_P data in Table 2.1.

***3.24.** The entropy change for the isothermal expansion of an ideal gas at 300 K from a particular state A to a state B is 50 J K^{-1}. When an expansion was performed, the work done by the system was 6 kJ. Was the process reversible or irreversible? If the latter, calculate the *degree of irreversibility* (i.e., the ratio of the work done to the reversible work).

3.25. One mole of water is placed in surroundings at −3 °C, but at first it does not freeze (it remains as supercooled water). Suddenly it freezes. Calculate the entropy change in the system during the freezing, making use of the following data:

$$C_{P,m}(\text{water}) = 75.3 \text{ J K}^{-1} \text{ mol}^{-1}$$
$$C_{P,m}(\text{ice}) = 37.7 \text{ J K}^{-1} \text{ mol}^{-1}$$
$$\Delta_f H(\text{ice} \rightarrow \text{water}) = 6.02 \text{ kJ mol}^{-1} \text{ at } 0\ ^\circ\text{C}$$

The two C_P values can be assumed to be independent of temperature. Also, calculate the entropy change in the surroundings, and the net entropy change in the system and surroundings.

3.26. 200 cm^3 of a 0.5 m solution of sucrose is diluted to 1 dm^3 by the addition of 800 cm^3 of water. Assume ideal behavior and calculate the entropy change.

3.27. One liter of a 0.1 *M* solution of a substance A is added to 3 liters of a 0.05 *M* solution of a substance B. Assume ideal behavior and calculate the entropy of mixing.

3.28. Ten moles of water at 60 °C are mixed with an equal amount of water at 20 °C. Neglect any heat exchange with the surroundings and calculate the entropy change. The heat capacity of water may be taken to be 75.3 J K^{-1} mol^{-1} and independent of temperature.

3.29. A vessel is divided by a partition into two compartments. One side contains 5 mol O_2 at 1 atm pressure; the other, 5 mol N_2 at 1 atm pressure. Calculate the entropy change when the partition is removed.

3.30. One mole of liquid water at 0 °C is placed in a freezer having a temperature of −12 °C. The water freezes and the ice cools to −12 °C. Making use of the data given in Problem 3.25, calculate the change in entropy in the system and in surroundings (the freezer), and the net entropy change.

3.31. One mole of liquid water at 0 °C is placed in a freezer which is maintained at −10 °C. Carry out the same calculations as for Problem 3.30.

3.32. Two moles of water at 60 °C are added to 4 mol of water at 20 °C. Calculate the entropy change, assuming that there is no loss of heat to the surroundings. The heat capacity of water is 75.3 J K^{-1} mol^{-1}.

3.33. One mole of an ideal gas is initially at 10 bar and 298 K. It is allowed to expand against a constant external pressure of 2 bar to a final pressure of 2 bar. During this process, the temperature of the gas falls to 253.2 K. Find ΔU, ΔH, ΔS, ΔS_{therm}, and ΔS_{univ} for the process. Assume that the thermal surroundings remain at 298 K throughout. Devise at least three different paths to accomplish this change and show that no matter which path is used, the desired values are the same.

3.34. Five moles of water at 50 °C are placed in a refrigerator maintained at 3 °C. Calculate ΔS for the system and for the environment, and the net entropy change, taking C_P for water at 75.3 J K^{-1} mol^{-1} and independent of temperature.

3.35. Problem 2.32 of Chapter 2 was concerned with dropping (a) one ice cube, (b) 10 ice cubes, each weighing 100 g, into 1 kg of water at 20 °C. Calculate the entropy change in each case. (ΔH_{fus} of ice at 0 °C is 6.026 kJ mol^{-1}; $C_{P,m}$ for water is 75.3 J K^{-1} mol^{-1}.)

3.36. The absolute entropy of nitrogen at its vaporization point of 77.32 K and exactly 1 bar is 151.94 J K^{-1} mol^{-1}. Using the expression for $C_{P,m}$ for nitrogen given in Table 2.1, find the entropy of the gas at 800.0 K and 1 bar.

Gibbs and Helmholtz Energies

3.37. Calculate $\Delta G°$ at 25 °C for the following fermentation reaction:

$$\underset{\text{glucose}}{C_6H_{12}O_6(aq)} \rightarrow \underset{\text{ethanol}}{2C_2H_5OH(aq)} + 2CO_2(g)$$

The standard Gibbs energies of formation of glucose, ethanol, and carbon dioxide are given in Appendix D.

Also use the data in Appendix D to calculate $\Delta S°$ for the fermentation reaction.

3.38. The latent heat of vaporization of water at 100 °C is 40.6 kJ mol^{-1} and when 1 mol of water is vaporized at 100 °C and 1 atm pressure, the volume increase is 30.19 dm^3. Calculate the work done by the system, the change in internal energy ΔU, the change in Gibbs energy ΔG and the entropy change ΔS.

3.39. On pages 115–116 we worked out the ΔS values for the freezing of water at 0 °C and at −10 °C. What are the corresponding ΔG values?

3.40. At 25 °C 1 mol of an ideal gas is expanded isothermally from 2 to 20 dm^3. Calculate ΔU, ΔH, ΔS, ΔA, and ΔG. Do the values depend on whether the process is reversible or irreversible?

3.41. The values of ΔH and ΔS for a chemical reaction are −85.2 kJ mol^{-1} and −170.2 J K^{-1} mol^{-1}, respectively, and the values can be taken to be independent of temperature.

a. Calculate ΔG for the reaction at (a) 300 K, (b) 600 K, and (c) 1000 K.

b. At what temperature would ΔG be zero?

3.42. The standard Gibbs energy for the combustion, $\Delta_c G°$, of methane has been measured as −815.04 kJ mol^{-1} at 25.0 °C and −802.57 kJ mol^{-1} at 75.0 °C. Assuming that Eq. 3.169 applies and that $\Delta_c G°$ changes linearly with temperature in this range, estimate the enthalpy of combustion at the midpoint of this temperature range, i.e., 50.0 °C.

3.43. The heat of vaporization of water at 25 °C is 44.01 kJ mol^{-1}, and the equilibrium vapor pressure at that temperature is 0.0313 atm. Calculate ΔS, ΔH, and ΔG when 1 mol of liquid water at 25 °C is converted into vapor at 25 °C and a pressure of 10^{-5} atm, assuming the vapor to behave ideally.

3.44. For each of the following processes, state which of the quantities ΔU, ΔH, ΔS, ΔA, and ΔG are equal to zero:

a. Isothermal reversible expansion of an ideal gas.

b. Adiabatic reversible expansion of a nonideal gas.

c. Adiabatic expansion of an ideal gas through a throttling valve.

d. Adiabatic expansion of a nonideal gas through a throttling valve.

e. Vaporization of liquid water at 80 °C and 1 bar pressure.
f. Vaporization of liquid water at 100 °C and 1 bar pressure.
g. Reaction between H_2 and O_2 in a thermally insulated bomb.
h. Reaction between H_2SO_4 and NaOH in dilute aqueous solution at constant temperature and pressure.

3.45. Calculate the change ΔG_m in the Gibbs energy of 1 mol of liquid mercury initially at 1 bar pressure if a pressure of 1000 bar is applied to it. The process occurs at the constant temperature of 25 °C, and the mercury may be assumed to be incompressible and to have a density of 13.5 g cm^{-3}.

3.46. The entropy of argon is given to a good approximation by the expression

$$S_m/\text{J K}^{-1}\text{ mol}^{-1} = 36.36 + 20.79 \ln(T/\text{K})$$

Calculate the change in Gibbs energy of 1 mol of argon if it is heated at constant pressure from 25 °C to 50 °C.

3.47. Calculate the absolute entropy of SO_2(g) at 300.0 K and 1 bar given the following information: $S°(15.0\text{ K}) = 1.26$ J K^{-1} mol^{-1}, $C_{P,m}(\text{s}) = 32.65$ J K^{-1} mol^{-1}, $T_{\text{fus}} = 197.64$ K, $\Delta_{\text{fus}}H° = 7\,402$ J mol^{-1}, $C_{P,m}(\text{l}) = 87.20$ J K^{-1} mol^{-1}, $T_{\text{vap}} =$ 263.08 K, $\Delta_{\text{vap}}H° = 24\,937$ J mol^{-1}, $C_{P,m}(\text{g}) = 39.88$ J K^{-1} mol^{-1}.

3.48. Initially at 300 K and 1 bar pressure, 1 mol of an ideal gas undergoes an irreversible isothermal expansion in which its volume is doubled, and the work it performs is 500 J mol^{-1}. What are the values of q, ΔU, ΔH, ΔG, and ΔS? What would q and w be if the expansion occurred reversibly?

***3.49.** At 100 °C 1 mol of liquid water is allowed to expand isothermally into an evacuated vessel of such a volume that the final pressure is 0.5 atm. The amount of heat absorbed in the process was found to be 30 kJ mol^{-1}. What are w, ΔU, ΔH, ΔS, and ΔG?

***3.50.** Water vapor can be maintained at 100 °C and 2 atm pressure for a time, but it is in a state of metastable equilibrium and is said to be supersaturated. Such a system will undergo spontaneous condensation; the process is

$$H_2O(\text{g, 100 °C, 2 atm}) \rightarrow H_2O(\text{l, 100 °C, 2 atm})$$

Calculate ΔH_m, ΔS_m, and ΔG_m. The molar enthalpy of vaporization $\Delta_{\text{vap}}H_m$ is 40.60 kJ mol^{-1}; assume the vapor to behave ideally and liquid water to be incompressible.

***3.51.** Initially at 300 K and 10 atm pressure, 1 mol of a gas is allowed to expand adiabatically against a constant pressure of 4 atm until equilibrium is reached. Assume the gas to be ideal with

$$C_{P,m}/\text{J K}^{-1}\text{ mol}^{-1} = 28.58 + 1.76 \times 10^{-2}\, T/\text{K}$$

and calculate ΔU, ΔH, and ΔS.

3.52. Calculate $\Delta H°$, $\Delta G°$, and $\Delta S°$ for the reaction

$$CH_4(\text{g}) + 2O_2(\text{g}) \rightarrow CO_2(\text{g}) + 2H_2O(\text{l})$$

making use of the data in Appendix D.

3.53. The following is a set of special conditions:

a. True only for an ideal gas.
b. True only for a reversible process.
c. True only if S is the total entropy (system + surroundings).
d. True only for an isothermal process occurring at constant pressure.
e. True only for an isothermal process occurring at constant volume.

Consider each of the following statements, and indicate which of the above conditions must apply in order for the statement to be true:

a. $\Delta U = 0$ for an isothermal process.
b. $\Delta H = 0$ for an isothermal process.
c. The total $\Delta S = 0$ for an adiabatic process.
d. $\Delta S > 0$ for a spontaneous process.
e. $\Delta G < 0$ for a spontaneous process.

3.54. Calculate the entropy and Gibbs energy changes for the conversion of 1 mol of liquid water at 100 °C and 1 bar pressure into vapor at the same temperature and a pressure of 0.1 bar. Assume ideal behavior. The heat of vaporization of water at 100 °C is 40.6 kJ mol^{-1}.

3.55. In the bacterium *nitrobacter* the following reaction occurs:

$$NO_2^- + \tfrac{1}{2}O_2 \rightarrow NO_3^-$$

Use the data in Appendix D to calculate $\Delta H°$, $\Delta G°$, and $\Delta S°$ for the reaction.

Energy Conversion

3.56. At 100 atm pressure water boils at 312 °C, while at 5 atm it boils at 152 °C. Compare the Carnot efficiencies of 100-atm and 5-atm steam engines, if T_c is 30 °C.

3.57. A cooling system is designed to maintain a refrigerator at −4 °C in a room at 20 °C. If 10^4 J of heat leaks into the refrigerator each minute, and the system works at 40% of its maximum thermodynamic efficiency, what is the power requirement in watts? [1 watt (W) = 1 J s^{-1}.]

3.58. A heat pump is employed to maintain the temperature of a house at 25 °C. Calculate the maximum performance factor of the pump when the external temperature is (a) 20 °C, (b) 0 °C, and (c) −20 °C.

3.59. A typical automobile engine works with a cylinder temperature of 2000 °C and an exit temperature of 800 °C. A typical octane fuel (molar mass = 114.2 g mol^{-1}) has an enthalpy of combustion of −5500 kJ mol^{-1} and 1 dm^3

(0.264 U.S. gal) has a mass of 0.80 kg. Calculate the maximum amount of work that could be performed by the combustion of 10 dm^3 of the fuel.

3.60. The temperature of a building is maintained at 20 °C by means of a heat pump, and on a particular day the external temperature is 10 °C. The work is supplied to the heat pump by a heat engine that burns fuel at 1000 °C and operates at 20 °C. Calculate the performance factor for the system (i.e., the ratio of the heat delivered to the building to the heat produced by the fuel in the heat engine). Assume perfect efficiencies of the pump and the engine.

3.61. Suppose that a refrigerator cools to 0 °C, discharges heat at 25 °C, and operates with 40% efficiency.

a. How much work would be required to freeze 1 kg of water ($\Delta_f H = -6.02$ kJ mol^{-1})?
b. How much heat would be discharged during the process?

Thermodynamic Relationships

3.62. Show that (a) $\left(\frac{\partial U}{\partial V}\right)_T = \frac{\alpha T - \kappa P}{\kappa}$, and (b) $\left(\frac{\partial U}{\partial P}\right)_T = V(\kappa P - \alpha T)$, where $\kappa = \frac{-1}{V}\left(\frac{\partial V}{\partial P}\right)_T$, is called the isothermal compressibility coefficient. [Use the relationship $\left(\frac{\partial P}{\partial T}\right)_V = -\left(\frac{\partial P}{\partial V}\right)_T\left(\frac{\partial V}{\partial T}\right)_P$.]

3.63. Derive an equation of state from

$$dH = T\,dS + V\,dP$$

by taking the partial derivative with respect to P at constant temperature. Then use the appropriate Maxwell relation and the definition of α to express the partial in terms of easily measured quantities.

3.64. Derive expressions for (a) α and (b) κ for an ideal gas.

***3.65.** Suppose that a gas obeys the van der Waals equation

$$\left(P + \frac{a}{V_m^2}\right)(V_m - b) = RT$$

Prove that

$$\left(\frac{\partial U}{\partial V_m}\right)_T = \frac{a}{V_m^2}$$

***3.66.** Obtain an expression for the Joule-Thomson coefficient for a gas obeying the equation of state

$$P(V_m - b) = RT$$

in terms of R, T, P, V_m, and $C_{P,m}$.

***3.67.** Derive the following equations:

a. $C_P = -T\left(\frac{\partial^2 G}{\partial T^2}\right)$

b. $\left(\frac{\partial C_P}{\partial P}\right)_T = -T\left(\frac{\partial^2 V}{\partial T^2}\right)_P$

***3.68.** Starting with the definition of the Helmholtz energy, $A = U - TS$, prove that the change in Helmholtz energy for a process at constant temperature is the total work (PV and non-PV). (This relationship holds without any restriction as to volume or pressure changes.)

***3.69.** Prove that if a gas obeys Boyle's law and if in addition $(\partial U/\partial V)_T = 0$, it must obey the equation of state $PV = \text{constant} \times T$.

***3.70.** Derive the relationship

$$\left(\frac{\partial S}{\partial V}\right)_U = \frac{P}{T}$$

and confirm that it applies to an ideal gas.

3.71. Starting from Eq. 3.160,

a. Show that $\ln\left(\frac{f_2}{P_2}\right) = \int_{P_1}^{P_2}\left(\frac{Z-1}{P}\right)dP$, where $Z = \frac{PV_m}{RT}$.

b. For a nonideal gas, the equation of state is given as $PV_m = RT + (b - A/RT^{2/3})P$ (see Example 1.6, p. 42). Derive an expression to find the fugacity of the gas at a given temperature and pressure when the constants b and A are given.

3.72. The van der Waals constants for methane in older units are $a = 2.283$ L^2 bar mol^{-2} and $b = 0.0428$ L mol^{-1}. Expressing the compression factor as (see Problem 1.52 in Chapter 1)

$$Z = 1 + \frac{1}{RT}\left(b - \frac{a}{RT}\right)P + \left(\frac{b}{RT}\right)^2 P^2,$$

find the fugacity of methane at 500 bar and 298 K.

Essay Questions

3.73. The frying of a hen's egg is a spontaneous reaction and has a negative Gibbs energy change. The process can apparently be reversed by feeding the fried egg to a hen and waiting for it to lay another egg. Does this constitute a violation of the second law? Discuss.[10]

[10]In answering this question, a student commented that a hen would never eat a fried egg. We suspect she would if she were hungry and had no alternative. In any case, let us postulate a hen sufficiently eccentric to eat a fried egg.

3.74. Consider the following statements:

a. In a reversible process there is no change in the entropy.
b. In a reversible process the entropy change is dq_{rev}/T.
How must these statements be qualified so that they are correct and not contradictory?

3.75. Consider the following statements:

a. The solution of certain salts in water involves a decrease in entropy.
b. For any process to occur spontaneously there must be an increase in entropy.

Qualify these statements so that they are correct and not contradictory, and suggest a molecular explanation for the behavior.

3.76. A phase transition, such as the melting of a solid, can occur reversibly and, therefore, $\Delta S = 0$. But it is often stated that melting involves an entropy increase. Reconcile these two statements.

SUGGESTED READING

See also Suggested Reading for Chapter 2 (p. 91). For treatments particularly of the second law see

P. W. Atkins, *The Second Law,* New York: Scientific American Books, 1984.

H. A. Bent, *The Second Law,* New York: Oxford University Press, 1965.

K. G. Denbigh, *The Principles of Chemical Equilibrium, with Applications to Chemistry and Chemical Engineering,* Cambridge University Press, 1961, 4th edition, 1981.

K. G. Denbigh and J. S. Denbigh, *Entropy in Relation to Incomplete Knowledge,* Cambridge University Press, 1985.

For details on the determination of entropies for a wide range of compounds see

A. A. Bondi, *Physical Properties of Molecular Crystals,* New York: Wiley, 1968.

For accounts of Nernst's heat theorem and low-temperature work see

D. K. C. MacDonald, *Near Zero: An Introduction to Low Temperature Physics,* Garden City, NY: Doubleday, 1961.

K. Mendelssohn, *The Quest for Absolute Zero,* New York: McGraw-Hill, 1966.

Accounts of the application of thermodynamic principles to practical problems of energy conservation are to be found in:

Efficient Use of Energy (A.I.P. Conference Proceedings No. 25), published in 1975 by the American Institute of Physics, 335 East 45th Street, New York, NY 10017.

C. M. Summers, "The Conservation of Energy," *Scientific American, 225*, 149(September 1971).

Chemical Equilibrium

PREVIEW

We saw in the last chapter that certain thermodynamic functions give conditions for equilibrium. For example, if a system is maintained at constant temperature and pressure, it settles down at an equilibrium state in which *its Gibbs energy is a minimum.* In this chapter we treat chemical equilibrium in greater detail and introduce the important idea of the *equilibrium constant.*

It is found empirically that the equilibrium condition for a chemical process depends only on the stoichiometry and not at all on the mechanism by which the process occurs. For a general reaction of stoichiometry

$$a\text{A} + b\text{B} + \cdots \rightarrow \cdots + y\text{Y} + z\text{Z}$$

involving gases that behave ideally, equilibrium is established at a given temperature when the ratio

$$\left(\frac{P_{\text{Y}}^{y} P_{\text{Z}}^{z} \cdots}{P_{\text{A}}^{a} P_{\text{B}}^{b} \cdots}\right)_{\text{eq}}$$

has a particular value. The P's are the partial pressures, and this ratio is the equilibrium constant with respect to pressure and given the symbol, K_P. The equilibrium constant can also be expressed in terms of concentrations and then has the symbol K_c. In Section 4.1 the equilibrium constant expressions are deduced on the basis of thermodynamic arguments originally given by van't Hoff.

The thermodynamic equilibrium constant K° is related to the standard Gibbs energy change for a process by

$$\Delta G^\circ = -RT \ln K^\circ$$

This thermodynamic equilibrium constant K° is a dimensionless quantity. The recommended standard state for ΔG° is 1 bar pressure, and then K° is K_P°, which is the dimensionless form of K_P in which pressures are in bars; if the standard state is 1 mol dm^{-3}, the K° is K_c°, which is the dimensionless form of K_c in which concentrations are expressed in mol dm^{-3}.

Deviations from ideal behavior are dealt with by using activities instead of concentrations. Two or more chemical reactions can be coupled, by either a common reactant or a catalyst.

The variation of an equilibrium constant with the temperature provides a useful way of obtaining enthalpy changes for chemical reactions.

OBJECTIVES

After studying this chapter, the student should be able to:

- Understand the concept of dynamical equilibrium and write the equilibrium constant, K_c.
- Write the Gibbs energy for gas phase reactions in terms of the pressure.
- Define the chemical potential and derive the expression for the Gibbs energy change written for the overall gas phase reaction.
- Define the thermodynamic equilibrium constant and write the Gibbs energy in terms of it.
- Express the equilibrium constant in terms of concentration units and use the stoichiometric sum.
- Use the thermodynamic equilibrium and practical equilibrium constant.
- Apply the equilibrium condition to nonideal gaseous systems.
- Apply the condition of chemical equilibrium to solutions.
- Express the equilibrium constant for heterogeneous equilibrium and define the activities of pure solids and liquids as unity.
- Recognize solubility product problems and know the ways to test for chemical equilibrium.
- Apply Le Chatelier's principle to reactions where temperature changes and compare it to van't Hoff's explanation of the process.
- Recognize coupled reactions and apply that concept to solving problems including systems that use catalysts.
- Derive van't Hoff's equation and apply it in either form to equilibrium problems.
- Use C_P and apply it to systems at two different temperatures to calculate changes in K or ΔH.
- Understand the relationships involving the pressure dependence of equilibrium constants.

The concept of equilibrium had its origin early in the history of chemistry. It was first thought that there was an analogy between chemical equilibrium and mechanical equilibrium, in which the forces acting on a system balance one another. However, in 1851 it was suggested by the British chemist Alexander William Williamson (1824–1904), in a study of esterifications, that when equilibrium is reached in a chemical system all reaction has not ceased. Instead, Williamson argued, reaction is still occurring in forward and reverse directions, the rates being the same in the two directions. This idea of **dynamical equilibrium** was soon accepted and is the basis of modern ideas about chemical equilibrium. Because we look at equilibrium in this way, it is now customary to write chemical equations using arrows instead of equal signs. A single arrow, $\rightarrow$ or $\leftarrow$, indicates that we are concerned particularly with reaction in one direction, while the double arrow, $\rightleftharpoons$, shows that the chemical equilibrium, and the reactions in the two opposite directions, are of interest.

Dynamical Equilibrium

The first quantitative work on chemical equilibrium was carried out in 1862 by the French chemists Pierre Eugène Marcellin Berthelot (1827–1907) and Péan de St. Gilles, who studied the equilibrium

$$CH_3COOH + C_2H_5OH \rightleftharpoons CH_3COOC_2H_5 + H_2O$$

They showed empirically that at a fixed temperature the concentration ratio

$$\frac{[CH_3COOC_2H_5]\,[H_2O]}{[CH_3COOH]\,[C_2H_5OH]}$$

is always the same after equilibrium has been reached. A vast amount of subsequent work has established that for any reaction

$$a\mathrm{A} + b\mathrm{B} + \cdots \rightleftharpoons \cdots y\mathrm{Y} + z\mathrm{Z}$$

the ratio

$$\frac{\cdots [\mathrm{Y}]^y[\mathrm{Z}]^z}{[\mathrm{A}]^a[\mathrm{B}]^b \cdots}$$

is constant at a given temperature—aside from deviations that may arise from nonideal behavior (Section 4.2). This ratio is known as the **equilibrium constant** and given the symbol K_c.

Equilibrium Constant

In 1864 the Norwegian mathematician Cato Maximillian Guldberg (1836–1902) and his brother-in-law the chemist Peter Waage (1833–1900) arrived at the correct equilibrium expression on the basis of an argument that is now known, rather paradoxically, to have no general validity. Their procedure was to write the rates in forward and reverse directions as

$$v_1 = k_1[\mathrm{A}]^a[\mathrm{B}]^b \cdots \quad \text{and} \quad v_{-1} = k_{-1}[\mathrm{Y}]^y[\mathrm{Z}]^z \cdots \tag{4.1}$$

At equilibrium, when there is no net concentration change, these rates are equal and therefore

$$\frac{\cdots [\mathrm{Y}]^y[\mathrm{Z}]^z}{[\mathrm{A}]^a[\mathrm{B}]^b \cdots} = \frac{k_1}{k_{-1}} = K_c \tag{4.2}$$

Although this procedure always gives the correct result, it is unsatisfactory, because the rate equations are not necessarily those given in Eq. 4.1. The rate equations in fact depend on the mechanism, as we will see in Chapters 9 and 10. The equilib-

Jacobus Henricus van't Hoff (1852–1911)

Van't Hoff was born in Rotterdam and studied in the Netherlands, Germany, and France. Before obtaining his Ph.D. degree in 1874 he published his now-famous pamphlet on the "tetrahedral carbon atom." After obtaining his degree, he searched for employment for a year and a half and then went to teach at the State Veterinary School in Utrecht. In 1878 he was appointed professor of chemistry, mineralogy, and geology at the University of Amsterdam, where he carried out important research on the fundamentals of solution theory, thermodynamics, and kinetics. The work is summarized in his *Études de dynamique chimique* (Studies in Chemical Dynamics), which appeared in 1884, and in its second edition, in German, published in 1896.

These investigations had a profound influence on the development of physical chemistry for many years. Van't Hoff's thermodynamics, although not as rigorous as Willard Gibbs's, was much easier to understand and more closely related to practical problems. Particularly, van't Hoff based his thermodynamics on the concept of the equilibrium constant, a property never mentioned by Gibbs in his treatments of chemical equilibrium. In kinetics van't Hoff clarified many basic principles such as the concept of order of reaction.

In 1896 van't Hoff moved to the University of Berlin, where he carried out a study of the equilibria involved in the marine salt deposits at Stassfurt, Germany, in terms of Gibbs's phase rule. This work established van't Hoff's position as the founder of the science of petrology.

In 1901 van't Hoff was awarded the first Nobel Prize in chemistry, "in recognition of the extraordinary value of his discovery of the laws of chemical dynamics and of the osmotic pressure in solutions."

Van't Hoff was a man of kindness and modesty who greatly influenced his students. He was a man of few words, sometimes remaining silent for several days in order to give complete, thorough answers to his students' questions.

References: H. P. M. Snelders, *Dictionary of Scientific Biography, 13*, 565–581(1976).

A. F. Holleman, "My reminiscences of van't Hoff," *J. Chemical Education, 29*, 379–382(1952).

K. J. Laidler, *The World of Physical Chemistry,* Oxford University Press, 1993.

W. A. E. McBride, "J. H. van't Hoff," *J. Chemical Education, 54*, 573–574(1987).

J. W. Servos, *Physical Chemistry from Ostwald to Pauling: The Making of a Science in America,* Princeton University Press, 1990.

rium expression, on the other hand, is independent of the mechanism and—aside from the matter of nonideal behavior—is correctly given by the ratio of concentrations as expressed in Eq. 4.2.

A satisfactory theoretical basis for the equilibrium equation is provided by thermodynamics. The first thermodynamic treatment of equilibrium in chemical systems was given in 1874–1878 by Willard Gibbs, but only in a very formal way, with no reference to equilibrium constants. The first treatment that was of particular value to chemists was given a few years later by the Dutch chemist Jacobus Henricus van't Hoff (1852–1911), whose discussion began with the concept of the equilibrium constant. In 1887 he presented a derivation of the equilibrium constant that is essentially that given in the next section.

Modern treatments of equilibrium are based on the idea that equilibrium is a state of **minimum free energy** (Gibbs energy or Helmholtz energy according to conditions), and that forward and reverse processes are occurring **at equal rates.**

4.1 Chemical Equilibrium Involving Ideal Gases

Consider a gas-phase reaction

$$aA + bB = yY + zZ$$

in which all the reactants and products are ideal gases. We have seen in Eq. 3.96 that when the volume of n mol of an ideal gas changes from V_1 to V_2 at constant temperature T, the Gibbs energy change is

$$\Delta G = nRT \ln \frac{V_1}{V_2} \tag{4.3}$$

Since the pressures P_1 and P_2 are inversely proportional to the volumes V_1 and V_2, we also have

$$\Delta G = nRT \ln \frac{P_2}{P_1} \tag{4.4}$$

The usual standard state for a gas is 1 bar, and if this is used, P_1 is 1 bar; Eq. 4.4 becomes

$$\Delta G = nRT \ln(P_2/\text{bar}) = nRT \ln P_2^u \tag{4.5}$$

where for convenience we have written the *value* of the pressure, a dimensionless quantity, as P_2^u.[1] The Gibbs energy at pressure P (we will now drop the subscript 2) is thus greater than that at 1 bar by $nRT \ln P^u$. If the Gibbs energy at 1 bar is denoted as $G°$, that at pressure P is thus given by

$$G = G° + nRT \ln P^u \tag{4.6}$$

For 1 mol of gas this equation becomes

$$G_m = G_m^\circ + RT \ln P^u \tag{4.7}$$

The molar Gibbs energy is known as the **chemical potential** and is given the symbol μ (mu); thus

$$\mu = \mu° + RT \ln P^u \tag{4.8}$$

[1] We realize that the device we have used—a superscript ^{u} or $°$ (the latter in the case of thermodynamic equilibrium constants)—will be felt by many to be unnecessary and perhaps irritating. This whole issue has been a matter of some controversy among physicists and physical chemists, and has been debated by IUPAC commissions. What we propose is something of a compromise, which will certainly not satisfy everyone. The use of the dimensionless equilibrium constant $K°$, called the *thermodynamic equilibrium constant*, is a recommendation of IUPAC, and appears in the Green Book (Appendix A). With regard to other quantities, such as ratios of concentrations in which the units do not cancel out, some (but by no means all) scientists feel strongly that it is not legitimate to work with their logarithms without indicating explicitly that one has made them dimensionless by dividing by the unit quantity. IUPAC has not agreed on any particular way of indicating that a quantity has been made dimensionless, and some rather clumsy procedures are often employed. We thought that the use of a superscript u was a simple way of reminding the student, in the earlier parts of this book, that the unitless quantity is being used. In later chapters we drop the procedure.

Anyone who feels that the superscript ^{u} is unnecessary is, of course, free to leave it out; we often do so ourselves!

If n_A mol of A are present in a mixture of A, B, Y, etc., the chemical potential μ_A of A is defined as

Definition of Chemical Potential

$$\mu_A \equiv \left(\frac{\partial G}{\partial n_A}\right)_{T,P,n_B,n_Y\cdots} \tag{4.9}$$

The subscripts indicate that the temperature, pressure, and amounts of all components other than A itself are held constant.

In the reaction under consideration we have a mol of A, the Gibbs energy of which is given by

$$G_A = a\mu_A = a\mu_A^\circ + aRT \ln P_A^u \tag{4.10}$$

where μ_A° is the standard chemical potential of A (i.e., the Gibbs energy of 1 mol of A at a pressure of 1 bar). Similarly, for B, Y, and Z

$$G_B = b\mu_B = b\mu_B^\circ + bRT \ln P_B^u \tag{4.11}$$

$$G_Y = y\mu_Y = y\mu_Y^\circ + yRT \ln P_Y^u \tag{4.12}$$

$$G_Z = z\mu_Z = z\mu_Z^\circ + zRT \ln P_Z^u \tag{4.13}$$

Form the pressure equilibrium constant; Thermodynamic equilibrium constant.

The increase in Gibbs energy when a mol of A (at pressure P_A) react with b mol of B (at pressure P_B) to give y mol of Y (at pressure P_Y) and z mol of Z (at pressure P_Z) is thus

$$\Delta G = y\mu_Y + z\mu_Z - a\mu_A - b\mu_B \tag{4.14}$$

$$= y\mu_X^\circ + z\mu_Z^\circ - a\mu_A^\circ - b\mu_B^\circ + RT \ln\left(\frac{P_Y^y P_Z^z}{P_A^a P_B^b}\right)^u \tag{4.15}$$

$$= \Delta G^\circ + RT \ln\left(\frac{P_Y^y P_Z^z}{P_A^a P_B^b}\right)^u \tag{4.16}$$

where ΔG°, equal to $y\mu_Y^\circ + z\mu_Z^\circ - a\mu_A^\circ - b\mu_B^\circ$, is the *standard Gibbs energy change.* The *standard state* for this ΔG° is 1 bar pressure.

If the pressures P_A, P_B, P_Y, P_Z correspond to equilibrium pressures, the Gibbs energy change ΔG is equal to zero, and, therefore,

$$\Delta G^\circ = -RT \ln\left(\frac{P_Y^y P_Z^z}{P_A^a P_B^b}\right)^u_{eq} \tag{4.17}$$

Since ΔG° is independent of pressure, the ratio of pressures in Eq. 4.17 is also a constant, and as recommended by the IUPAC Commission on Thermodynamics we call it the *thermodynamic equilibrium constant* K_P°.

Thermodynamic Equilibrium Constant

$$\left(\frac{P_Y^y P_Z^z}{P_A^a P_B^b}\right)^u_{eq} = K_P^\circ \tag{4.18}$$

Note that this equilibrium constant K_P° is a *dimensionless* quantity; it has the same *value* as the pressure equilibrium constant K_P, which in general has units and which is defined by

$$K_P = \left(\frac{P_Y^y P_Z^z}{P_A^a P_B^b}\right)_{eq} \tag{4.19}$$

From Eqs. 4.17 and 4.18 we have

$$\Delta G^\circ = -RT \ln K_P^\circ \tag{4.20}$$

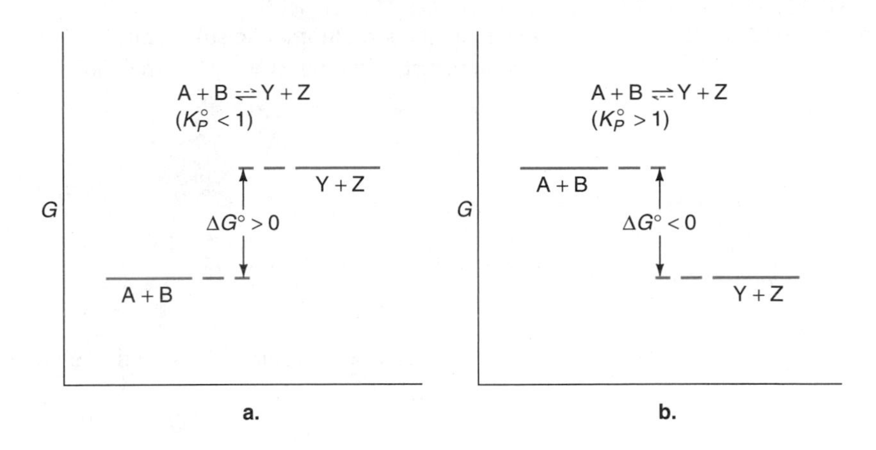

FIGURE 4.1
A standard Gibbs energy diagram, illustrating the relationship between ΔG° for a reaction and the equilibrium constant K_P.
(a) ΔG° is positive and $K_P^\circ < 1$;
(b) ΔG° is negative and $K_P^\circ > 1$;
(c) ΔG° is zero and $K_P^\circ = 1$.

If the initial and final pressures do not correspond to equilibrium, Eq. 4.16 applies and can be written as

Relate the Gibbs energy to the equilibrium constant; Thermodynamic equilibrium constant.

$$\Delta G = -RT \ln K_P^\circ + RT \ln\left(\frac{P_Y^y P_Z^z}{P_A^a P_B^b}\right)^u \tag{4.21}$$

The relationship between ΔG° and K_P°, as expressed by Eq. 4.20, is illustrated in Figure 4.1.

Equilibrium Constant in Concentration Units

It is often convenient to use concentrations instead of pressures in dealing with equilibrium problems. For an ideal gas

Convert to concentration units; Equilibrium constant in concentration units.

$$P = \frac{nRT}{V} = cRT \tag{4.22}$$

Substitution of the expressions for P_Y, P_Z, P_A, and P_B into Eq. 4.19 leads to

$$K_P = \left(\frac{c_Y^y c_Z^z}{c_A^a c_B^b}\right)_{eq} (RT)^{y+z-a-b} \tag{4.23}$$

This is more conveniently written as

$$K_P = \left(\frac{[Y]^y[Z]^z}{[A]^a[B]^b}\right)_{eq} (RT)^{\Sigma\nu} \tag{4.24}$$

Stoichiometric Sum

where $\Sigma\nu$, the **stoichiometric sum,** is the difference between the coefficients in the stoichiometric equation. If we write the equilibrium constant in terms of concentrations as

$$K_c = \left(\frac{[Y]^y[Z]^z}{[A]^a[B]^b}\right)_{eq} \tag{4.25}$$

we see that K_c and K_P are related by

$$K_P = K_c(RT)^{\Sigma\nu} \tag{4.26}$$

From K_c we can calculate a $\Delta G°$, using the relationship

$$\Delta G° = -RT \ln K_c^\circ \tag{4.27}$$

where K_c° is the numerical value of K_c and is thus the *standard* equilibrium constant in terms of concentration.

If the concentrations are in mol dm^{-3}, the standard state for $\Delta G°$ is 1 mol dm^{-3}. In general, the change of standard state will change the value of $\Delta G°$, and it is important always to indicate the standard state. By analogy with Eq. 4.16 we also have

$$\Delta G = \Delta G° + RT \ln\left(\frac{[Y]^y[Z]^z}{[A]^a[B]^b}\right)^u \tag{4.28}$$

The superscript u to the ratio again indicates that we have made it dimensionless. This is the Gibbs energy change when a mol of A, at concentration [A], reacts with b mol of B, at concentration [B], to produce y mol of Y, at concentration [Y], and z mol of Z, at concentration [Z].

Alternatively, we can work with mole fractions. The pressure P_A of substance A is related to the total pressure P by

$$P_A = x_A P \tag{4.29}$$

where x_A is the mole fraction of A; similar equations apply to B, Y, and Z. Substitution of these expressions into Eq. 4.18 gives

$$K_P = \left(\frac{x_Y^y x_Z^z}{x_A^a x_B^b}\right)_{eq} P^{y+z-a-b} = \left(\frac{x_Y^y x_Z^z}{x_A^a x_B^b}\right)_{eq} P^{\Sigma\nu} \tag{4.30}$$

Thus, if we write the equilibrium constant in terms of mole fractions as K_x, namely

$$K_x = \left(\frac{x_Y^y x_Z^z}{x_A^a x_B^b}\right)_{eq} \tag{4.31}$$

we see that K_P and K_x are related by

$$K_P = K_x P^{\Sigma\nu} \tag{4.32}$$

Since, for ideal gases, K_P is a true constant at constant temperature, it follows that K_x will be pressure dependent unless $\Sigma\nu = 0$. The ratio K_x is thus only constant at constant temperature and constant total pressure. The functions K_P and K_c, on the other hand, are (for an ideal gas) functions of temperature only.

From K_x, which is dimensionless, we can calculate a Gibbs energy change,

$$\Delta G° = -RT \ln K_x \tag{4.33}$$

the standard state for which is unit mole fraction. Also,

$$\Delta G = \Delta G^\circ + RT \ln\left(\frac{x_Y^y x_Z^z}{x_A^a x_B^b}\right) \tag{4.34}$$

This is the Gibbs energy change when a mol of A at mole fraction x_A reacts with b mol of B at mole fraction x_B to yield y mol of Y at mole fraction x_Y plus z mol of Z at mole fraction x_Z.

Units of the Equilibrium Constant

Thermodynamic and Practical Equilibrium Constants

In thermodynamics it is convenient to work with two different kinds of equilibrium constants, the **thermodynamic equilibrium constant** and the **practical equilibrium constant.** The thermodynamic equilibrium constant is dimensionless, and it is this constant that is used in equations such as Eq. 4.20, in which its logarithm is involved; it is denoted by the superscript zero. This thermodynamic equilibrium constant has the same numerical value as the practical equilibrium constant; it has simply been made dimensionless by division by the unit quantity.

The practical equilibrium constant in general has units, which depend on the type of reaction. Consider, for example, a reaction of the type

$$A + B \rightleftharpoons Z$$

in which there is a decrease, by one, in the number of molecules. The equilibrium constant for the standard state of 1 mol dm^{-3} is then

$$K_c = \frac{[Z]}{[A][B]} \tag{4.35}$$

and since the units of [A], [B], and [Z] are mol dm^{-3}, the units of K_c are dm^3 mol^{-1}. If there is no change in the number of molecules, as in a reaction of the type

$$A + B \rightleftharpoons Y + Z$$

the units cancel out, so that K_c is dimensionless. In general, if the stoichiometric sum is $\Sigma\nu$, the units are

$$\text{mol}^{\Sigma\nu}\ \text{dm}^{-3\Sigma\nu}$$

Another important point about equilibrium constants is that their value depends on how the stoichiometric equation is written. Consider, for example, the dissociation

$$1.\ A \rightleftharpoons 2Z$$

the equilibrium constant K_c for which is

$$K_c = \frac{[Z]^2}{[A]} \tag{4.36}$$

If the concentrations are expressed as mol dm^{-3}, the units of K_c are mol dm^{-3}. Alternatively, we could write the reaction as

$$2.\ \tfrac{1}{2}A \rightleftharpoons Z$$

and express the equilibrium constant as

$$K_c' = \frac{[Z]}{[A]^{1/2}} \tag{4.37}$$

This latter constant K_c' is obviously the square root of K_c:

$$K_c' = K_c^{1/2} \tag{4.38}$$

If the concentrations are in mol dm^{-3}, the units of K_c' are mol$^{1/2}$ dm$^{-3/2}$. The standard Gibbs energy change for reaction 1 is

$$\Delta G_1^\circ = -RT \ln K_c^\circ \tag{4.39}$$

while that for reaction 2 is

$$\Delta G_2^\circ = -RT \ln K_c^{\circ\prime} \tag{4.40}$$

Since

$$\ln K_c^{\circ\prime} = \tfrac{1}{2} \ln K_c^\circ \tag{4.41}$$

we see that

$$\Delta G_2^\circ = \tfrac{1}{2} \Delta G_1^\circ \tag{4.42}$$

This is as it should be, since reaction 2 to which ΔG_2° refers is for the conversion of 0.5 mol of A into X; for reaction 1 we are referring to the conversion of 1 mol of A into 2X.

EXAMPLE 4.1 The Gibbs energies of formation of NO_2(g) and N_2O_4(g) are 51.30 and 102.00 kJ mol^{-1}, respectively (standard state: 1 bar and 25 °C).

a. Assume ideal behavior and calculate, for the reaction $N_2O_4 \rightleftharpoons 2NO_2$, K_P (standard state; 1 bar) and K_c (standard state; 1 mol dm^{-3}).
b. Calculate K_x at 1 bar pressure.
c. At what pressure is N_2O_4 50% dissociated?
d. What is ΔG° if the standard state is 1 mol dm^{-3}?

Solution For the reaction

$$N_2O_4(g) \rightleftharpoons 2NO_2(g)$$

$$\Delta G^\circ(\text{standard state: 1 bar}) = 2 \times 51.30 - 102.0 = 0.60 \text{ kJ mol}^{-1}$$

a. Therefore, from Eq. 4.20,

$$\ln(K_P/\text{bar}) = -\frac{600 \text{ J mol}^{-1}}{8.3145 \text{ J K}^{-1} \text{ mol}^{-1} \times 298.15 \text{ K}} = -0.242$$

$$K_P = 0.785 \text{ bar}$$

From Eq. 4.26, since $\Sigma\nu = 1$

$$K_c = \frac{K_P}{RT} = \frac{0.785(\text{bar}) \times 10^5(\text{Pa bar}^{-1})}{8.3145(\text{J K}^{-1} \text{ mol}^{-1}) \times 298.15(\text{K})}$$

$$= 31.7 \text{ mol m}^{-3}$$

$$= 3.17 \times 10^{-2} \text{ mol dm}^{-3}$$

b. From Eq. 4.32, with $\Sigma\nu = 1$,

$$K_x = K_P P^{-1} = 0.785 \text{ at } P = 1 \text{ bar}$$

c. Suppose that we start with 1 mol of N_2O_4 and that α mol have become converted into NO_2; the amounts at equilibrium are

$$\underset{1-\alpha}{N_2O_4} \rightleftharpoons \underset{2\alpha}{2NO_2}$$

and the total amount is $(1 + \alpha)$ mol. If P is the total pressure, the partial pressures are

$$N_2O_4\text{:} \quad \frac{1-\alpha}{1+\alpha}P$$

$$NO_2\text{:} \quad \frac{2\alpha}{1+\alpha}P$$

The equilibrium equation is thus

$$K_P^\circ = K_P/\text{bar} = 0.785 = \frac{[2\alpha/(1+\alpha)]^2(P/\text{bar})^2}{[(1-\alpha)/(1+\alpha)](P/\text{bar})} = \frac{4\alpha^2 P/\text{bar}}{1-\alpha^2}$$

If $\alpha = 0.5$,

$$0.785 = \frac{4 \times 0.25 \times P/\text{bar}}{1 - 0.25} = \frac{P/\text{bar}}{0.75}$$

and, therefore,

$$P = 0.589 \text{ bar}$$

d. K_c, with the concentrations in mol dm^{-3}, is 3.17×10^{-2} mol dm^{-3}; then

$$\Delta G^\circ/\text{J mol}^{-1} = -8.3145 \times 298.15 \ln(3.17 \times 10^{-2})$$
$$\Delta G^\circ = 8555 \text{ J mol}^{-1} = 8.56 \text{ kJ mol}^{-1}$$

EXAMPLE 4.2 From the 1890s and for many years after the German physical chemist Max Bodenstein (1871–1942), who was a remarkably skillful experimentalist, carried out detailed studies on several reactions, including that between hydrogen and iodine,

$$H_2 + I_2 \rightleftharpoons 2HI$$

He studied both the equilibrium established and the rates in forward and reverse directions. In one investigation of the equilibrium constant at 731 K, he introduced 22.13 cm^3 of hydrogen and 16.18 cm^3 of iodine into a vessel. After waiting a sufficient time for equilibrium to be established, he chilled the reaction vessel rapidly so as to "freeze" the equilibrium, (i.e., prevent its shifting appreciably), and he then analyzed the products, finding that 28.98 cm^3 of hydrogen iodide had been formed. (These volumes relate to STP, which at the time meant a temperature of 0 °C and a pressure of 1 atm; see also Appendix A, p. 999)

Calculate the equilibrium constant and the Gibbs energy change for the reaction.

Solution Note first that because the reaction involves no change in the number of molecules, the volume of the vessel is irrelevant, since it cancels out in the equilibrium equation. Note also that from Eq. 4.26 and Eq. 4.32 the equilibrium constants K_P, K_c, and K_x are all the same and that they are dimensionless.

The volumes at STP of the substances present at equilibrium are:

$$\begin{aligned} &[\text{HI}]: && 28.98 \text{ cm}^3 \\ &[\text{H}_2]: && (22.13 - 14.49) = 7.64 \text{ cm}^3 \\ &[\text{I}_2]: && (16.18 - 14.49) = 1.69 \text{ cm}^3 \end{aligned}$$

The equilibrium constants K_c, K_P, and K_x are all therefore

$$K = 28.98^2/7.64 \times 1.69 = 65.0$$

The Gibbs energy change is obtained from Eq. 4.20, Eq. 4.27, or Eq. 4.33, all leading to the same value:

$$\begin{aligned} \Delta G^\circ &= -RT \ln 65.0 \\ &= -8.3145 \times 731(\text{J mol}^{-1}) \ln 65.0 \\ &= -9128 \text{ J mol}^{-1} = -9.13 \text{ kJ mol}^{-1} \end{aligned}$$

For this special case of a reaction with no change in the number of molecules, the standard state is irrelevant.

4.2 Equilibrium in Nonideal Gaseous Systems

We saw in Section 3.8 that the thermodynamics of real gases are dealt with by making use of *fugacities* or *activities,* which play the same role for real gases that pressures do for ideal gases.

The procedure for treating chemical equilibria involving nonideal gases is to follow through exactly the same treatment as for ideal gases, replacing pressures by activities that are dimensionless. Then, instead of Eq. 4.20, we have

$$\Delta G^\circ = -RT \ln K_a^\circ \tag{4.43}$$

where K_a° is the dimensionless equilibrium constant in terms of activities:

$$K_a^\circ = \left(\frac{a_X^x a_Y^y}{a_A^a a_B^b}\right)_{eq}^u \tag{4.44}$$

Similarly, in place of Eq. 4.16, we have

$$\Delta G = \Delta G^\circ + RT \ln\left(\frac{a_X^x a_Y^y}{a_A^a a_B^b}\right)^u = -RT \ln K_a^\circ + RT \ln\left(\frac{a_X^x a_Y^y}{a_A^a a_B^b}\right)^u \tag{4.45}$$

4.3 Chemical Equilibrium in Solution

We have seen that chemical equilibria in ideal gases can be formulated in terms of the molar concentrations of the reactants and products. The same relationships frequently apply to equilibria involving substances present in solution. Equations involving mole fractions also often apply to a good approximation to reactions in solution. Alternatively, molalities (mol kg^{-1}) may be used instead of concentrations. When any of these relationships apply, the behavior is said to be *ideal*.

When molalities are used, deviations from ideality are taken care of by replacing the molality m by the activity a, which is the molality multiplied by the activity coefficient γ. The Gibbs energy is then given in terms of the dimensionless activity a^u by

$$G = G^\circ + RT \ln a^u = G^\circ + RT \ln \gamma m^u \tag{4.46}$$

where G° is the value at unit activity. The standard equilibrium constant for a reaction

$$aA + bB \rightleftharpoons yY + zZ$$

is then given by

$$K_a^\circ = \left(\frac{a_Y^y a_Z^z}{a_A^a a_B^b}\right)^u = \left(\frac{m_Y^y m_Z^z}{m_A^a m_B^b}\right)^u \cdot \left(\frac{\gamma_Y^y \gamma_Z^z}{\gamma_A^a \gamma_B^b}\right)^u \tag{4.47}$$

where the superscripts ^{u} indicate that we have made the ratios dimensionless. A similar procedure is followed if we work with concentrations or with mole fractions. The symbol y is used for an activity coefficient used with concentration, while f is used for an activity coefficient used with mole fractions.

The theory of activities of species in solution and the way activities are determined experimentally are considered further in Section 5.6. For uncharged species in solution the behavior is often close to ideal, and equilibrium constants are then sometimes satisfactorily expressed in terms of molalities, concentrations, or mole fractions, without the use of activity coefficients. For ions, however, there may be serious deviations from ideality, and activities must be used. The activity coefficients of ions are considered in Sections 7.10 and 8.5.

The usual procedure in dealing with *solvent* species is to use the mole fraction. The Gibbs energy is expressed as

$$G = G^\circ + RT \ln x_1 f_1 \tag{4.48}$$

where x_1 is the mole fraction of the solvent and f_1 its activity coefficient.

EXAMPLE 4.3 The equilibrium constant K_c for the reaction

fructose-1,6-diphosphate $\rightleftharpoons$ glyceraldehyde-3-phosphate + dihydroxyacetone phosphate

is 8.9×10^{-5} *M* at 25 °C, and we can assume the behavior to be ideal.

a. Calculate ΔG° for the process (standard state: 1 *M*).
b. Suppose that we have a mixture that is initially 0.01 *M* in fructose-1,6-diphosphate and 10^{-5} *M* in both glyceraldehyde-3-phosphate and dihydroxyacetone phosphate. What is ΔG? In which direction will reaction occur?

Solution

a. $$\Delta G^\circ = -RT \ln K^\circ$$

$$\Delta G^\circ/\text{J mol}^{-1} = -8.3145 \times 298.15 \ln(8.9 \times 10^{-5})$$

$$\therefore \Delta G^\circ = 23\ 120 \text{ J mol}^{-1} = 23.12 \text{ kJ mol}^{-1}$$

b. $$\Delta G = \Delta G^\circ + RT \ln \frac{[\text{G-3-P}][\text{DHAP}]/\text{mol dm}^{-3}}{[\text{FDF}]}$$

$$\Delta G/\text{J mol}^{-1} = 23\ 120 + 8.3145 \times 298.15 \ln \frac{10^{-5} \times 10^{-5}}{10^{-2}}$$

$$\therefore \Delta G = (23\ 120 - 45\ 660)\ \text{J mol}^{-1} = -22\ 540\ \text{J mol}^{-1}$$

$$= -22.54\ \text{kJ mol}^{-1}$$

Because ΔG is negative, reaction proceeds spontaneously from left to right. If the initial concentrations had been unity, the positive ΔG° value would have forced the reaction to the left; however, the low concentrations of products have reversed the sign and the reaction proceeds to the right.

Another way of seeing that the reaction will shift to the right is to note that the concentration ratio

$$\frac{[\text{G-3-P}][\text{DHAP}]}{[\text{FDF}]} = \frac{10^{-5}10^{-5}}{10^{-2}} = 10^{-8}\ \text{mol dm}^{-3}$$

is *less* than the equilibrium constant 8.9×10^{-5} mol dm^{-3}.

4.4 Heterogeneous Equilibrium

When equilibrium involves substances in different phases, such as different states of matter, we speak of **heterogeneous equilibrium.** A simple example is provided by the dissociation of solid calcium carbonate into solid calcium oxide and gaseous carbon dioxide:

$$CaCO_3(s) \rightleftharpoons CaO(s) + CO_2(g)$$

As before, we may write

$$\frac{[CaO][CO_2]}{[CaCO_3]} = K_c' \tag{4.49}$$

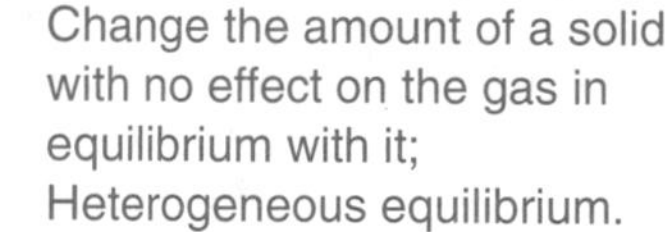
Change the amount of a solid with no effect on the gas in equilibrium with it; Heterogeneous equilibrium.

The concentration of a solid or liquid, however, has a special feature. As always, it is the amount of substance divided by the volume. However, in the case of a given solid or pure liquid, the amount contained in a given volume is a fixed quantity. The concentrations [CaO] and [CaO_3] are therefore constants, independent of the total quantity of solid present, as are their activities.

Since in Eq. 4.49 the concentrations [CaO] and [CaO_3] are constants, we can incorporate them into the constant K_c' and simply write

$$[CO_2] = K_c \tag{4.50}$$

This is the procedure adopted for pure solids and pure liquids; their concentrations are always incorporated into the equilibrium constant. Equation 4.50 can be expressed more exactly in terms of the activity of CO_2:

$$a_{CO_2} = K_c \tag{4.51}$$

Note that this procedure of incorporating the concentrations of pure solids and liquids into the equilibrium constants is equivalent to **defining the activities of pure solids and liquids as unity.**

It is important to note that if solids are present in solid solution or liquids are present mixed with other liquids, their concentrations are obviously not fixed, and they must not be incorporated into the equilibrium constant in this way.

Another example of heterogeneous equilibrium is that between a salt and its saturated solution. Suppose, for example, that the sparingly soluble salt silver chloride, AgCl, is placed in contact with water. The equilibrium between the solid and its saturated solution, in which the AgCl is completely dissociated into Ag^+ and Cl^- ions, is

$$AgCl(s) \rightleftharpoons Ag^+(aq) + Cl^-(aq)$$

The equilibrium equation could be written as

$$\frac{[Ag^+][Cl^-]}{[AgCl]} = K_c' \qquad (4.52)$$

Again, however, the concentration [AgCl] is a constant and the convention is to combine it with K_c' to give

Solubility Product

$$[Ag^+][Cl^-] = K_{sp} \qquad (4.53)$$

This constant K_{sp} is known as the **solubility product.** For greater precision, activities are used:

$$a_{Ag^+}\, a_{Cl^-} = K_{sp} \qquad (4.54)$$

EXAMPLE 4.4 The solubility of silver chloride in pure water at 25 °C is 1.265×10^{-5} mol dm^{-3}. Calculate the solubility product and $\Delta G°$ for the process

$$AgCl(s) \rightleftharpoons Ag^+(aq) + Cl^-(aq)$$

Solution The solubility product K_{sp}, equal to $[Ag^+][Cl^-]$ for this system, is the square of the solubility:

$$K_{sp} = (1.265 \times 10^{-5})^2 = 1.599 \times 10^{-10}$$

The $\Delta G°$ value is

$$-RT \ln(1.599 \times 10^{-10}) = 55\,913 \text{ J mol}^{-1}$$
$$= 55.9 \text{ kJ mol}^{-1}$$

4.5 Tests for Chemical Equilibrium

Any reaction

$$aA + bB + \cdots \rightleftharpoons \cdots yY + zZ$$

will eventually reach a state of equilibrium at which

$$\left(\frac{\cdots[Y]^y[Z]^z}{[A]^a[B]^b\cdots}\right)_{eq} = K_c \qquad (4.55)$$

It is obviously of great practical importance to determine whether a chemical system is at equilibrium.

It is not enough to establish that the composition of the system is not changing as time goes on. Reaction may be proceeding so slowly that no detectable change will occur over a long period of time. A good example is the reaction

$$2H_2 + O_2 \rightarrow 2H_2O$$

Follow the kinetics of a reaction to the equilibrium state and verify the equilibrium expression; Tests for chemical equilibrium.

Except at extremely high temperatures the equilibrium for this reaction lies almost completely to the right. However, the reaction is so slow that if we bring hydrogen and oxygen together at room temperature, there will be no detectable change even over many hundreds of years. In fact, even after 5×10^9 years (the estimated age of the solar system) only an insignificant amount of this reaction will have taken place!

Obviously, more practical tests for equilibrium are required. One of these consists of adding a substance that speeds up reaction. For example, if to the hydrogen-oxygen mixture we introduce a lighted match, or if we add certain substances known as **catalysts** (such as powdered platinum), the reaction will occur with explosive violence. This shows that the system was not at equilibrium, but it appeared to be at equilibrium because of the slowness of the reaction. The function of catalysts in speeding up reactions will be considered later (Section 10.9); here we may simply note that catalysts do not affect the position of equilibrium but merely decrease the time required for equilibrium to be attained.

A second and more fundamental test for equilibrium is as follows: If a system

$$A + B \rightleftharpoons Y + Z$$

is truly at equilibrium, the addition of a small amount of A (or B) will cause the equilibrium to shift to the right, with the formation of more Y and Z. Similarly, the addition of more Y or Z will cause a shift to the left. If the system is not in a state of true equilibrium, either these shifts will not occur at all or they will not occur in the manner predicted by the equilibrium equations.

4.6 Shifts of Equilibrium at Constant Temperature

One of the important consequences of the theory of equilibrium relates to the way in which equilibria shift as the volume of the system changes. The equilibrium constants K_P and K_c are functions of temperature only and do not change when we vary the volume. However, as a result of this constancy of the equilibrium constant, the *position* of equilibrium does shift as the volume changes. Consider the equilibrium

$$AB \rightleftharpoons A^+ + B^-$$

in which AB is dissociating into ions. The equilibrium constant K_c is

$$K_c = \frac{[A^+][B^-]}{[AB]} \tag{4.56}$$

If we add A^+ to the system, the ratio will temporarily be increased; the equilibrium will have to shift from right to left to maintain K_c constant. Similarly, addition of B^- will cause a shift to the left. Addition of AB will cause a shift to the right.

The effect of changing the *volume* can be seen by expressing each concentration in Eq. 4.56 as the ratio of the amount of each substance to the volume;

$$[AB] = \frac{n_{AB}}{V} \qquad [A^+] = \frac{n_{A^+}}{V} \qquad [B^-] = \frac{n_{B^-}}{V} \tag{4.57}$$

so that Eq. 4.56 becomes

$$\frac{(n_{A^+}/V)(n_{B^-}/V)}{n_{AB}/V} = K_c \tag{4.58}$$

or

$$\frac{n_{A^+}n_{B^-}}{n_{AB}} \cdot \frac{1}{V} = K_c \tag{4.59}$$

Suppose that we dilute the system, by adding solvent and making V larger. Since K_c must remain constant, the ratio

$$\frac{n_{A^+}n_{B^-}}{n_{AB}}$$

must become larger, in proportion to V; that is, there will be more dissociation into ions at the larger volume. This result follows from what is generally known as the **Le Chatelier principle,** after the French chemist Henri Louis Le Chatelier (1850–1936). In 1884 he suggested that

Le Chatelier Principle

> **if there is a change in the condition of a system in equilibrium, the system will adjust itself in such a way as to counteract, as far as possible, the effect of that change.**

Understand how changing the concentration and pressure shift the equilibrium; Le Chatelier's principle.

A few months earlier, however, the same principle had been proposed, and explained in terms of thermodynamics, by van't Hoff. In their formulations, both van't Hoff and Le Chatelier considered the effects of temperature changes (Section 4.8) as well as volume changes.

EXAMPLE 4.5 The equilibrium constant K_P for the dissociation of chlorine into atoms

$$Cl_2 \rightleftharpoons 2Cl$$

has been determined to be 0.570 bar at 2000 K. Suppose that 0.1 mol of chlorine is present in a volume of 5 L at that temperature. What will be the degree of dissociation (the fraction present as chlorine atoms)? If the volume is expanded to 20 L, what will then be the degree of dissociation?

Solution Since we are working with concentrations it will be convenient to convert K_P into K_c using Eq. 4.26:

$$\begin{aligned} K_c &= K_P(RT)^{-\Sigma\nu} = K_P(RT)^{-1} \\ &= \frac{0.570 \text{ bar} \times 10^5 \text{ Pa bar}^{-1}}{8.3145 \times 2000 \text{ J mol}^{-1}} \\ &= 3.43 \text{ mol m}^{-3} = 3.43 \times 10^{-3} \text{ mol dm}^{-3} \end{aligned}$$

If the degree of dissociation is α, the amounts of Cl_2 and Cl present are:

$$Cl_2\text{:} \quad 0.1(1 \quad \alpha) \text{ mol}$$
$$Cl\text{:} \quad 0.1 \times (2\alpha) = 0.2\alpha \text{ mol}$$

If the volume is 5 dm^3, the equilibrium equation is therefore

$$\frac{(0.2\alpha/5)^2 \text{ mol}^2 \text{ dm}^{-6}}{0.1(1 - \alpha)/5 \text{ mol dm}^{-3}} = 3.43 \times 10^{-3} \text{ mol dm}^{-3}$$

This reduces to the quadratic equation

$$0.04\alpha^2 + 1.715 \times 10^{-3}\alpha - 1.715 \times 10^{-3} = 0$$

the solution of which is

$$\alpha = 0.185$$

(we ignore the negative root, which is physically impossible). For a volume of 20 L the corresponding equation is

$$\frac{(0.2\alpha)^2}{0.1(1 - \alpha)20} = 3.473 \times 10^{-3}$$

or

$$0.04\alpha^2 + 6.86 \times 10^{-3}\alpha - 6.86 \times 10^{-3} = 0$$

the solution of which is

$$\alpha = 0.319$$

Note that, in accordance with the Le Chatelier principle, there is a greater degree of dissociation when the volume is increased.

4.7 Coupling of Reactions

For a chemical reaction

$$1.\ A + B \rightleftharpoons X + Y$$

removal of either of the products X or Y will lead to a shift of equilibrium to the right. Frequently a product is removed by the occurrence of another reaction; for example, X might isomerize into Z:

$$2.\ X \rightleftharpoons Z$$

We can write an equilibrium constant for reaction 1,

$$K_1 = \left(\frac{[X][Y]}{[A][B]}\right)_{eq} \tag{4.60}$$

and a corresponding standard Gibbs energy change:

$$\Delta G_1^\circ = -RT \ln K_1 \tag{4.61}$$

Similarly for the second reaction

$$K_2 = \left(\frac{[Z]}{[X]}\right)_{eq} \tag{4.62}$$

and

$$\Delta G_2^\circ = -RT \ln K_2 \tag{4.63}$$

We can add reactions 1 and 2 to obtain

$$3.\ \mathrm{A + B \rightleftharpoons Y + Z}$$

The equilibrium constant K_3 for the combined reaction 3 is

$$K_3 = \left(\frac{[\mathrm{Y}][\mathrm{Z}]}{[\mathrm{A}][\mathrm{B}]}\right)_{\mathrm{eq}} \tag{4.64}$$

The same result is obtained by multiplying K_1 and K_2:

$$K_1K_2 = \left(\frac{[\mathrm{X}][\mathrm{Y}]}{[\mathrm{A}][\mathrm{B}]}\right)_{\mathrm{eq}} \times \left(\frac{[\mathrm{Z}]}{[\mathrm{X}]}\right)_{\mathrm{eq}} = \left(\frac{[\mathrm{Y}][\mathrm{Z}]}{[\mathrm{A}][\mathrm{B}]}\right)_{\mathrm{eq}} = K_3 \tag{4.65}$$

We can take natural logarithms of both sides of this equation, and then multiply by $-RT$:

$$-RT \ln K_3 = -RT \ln K_1 - RT \ln K_2 \tag{4.66}$$

Combine three acid dissociations of a polyprotic acid into one equation; Coupling reactions.

so that

$$\Delta G_3^\circ = \Delta G_1^\circ + \Delta G_2^\circ \tag{4.67}$$

The result is easily shown to be quite general; if we add reactions, the ΔG° of the resulting reaction is the sum of the ΔG° values of the component reactions. This is an extension of Hess's law (Section 2.5) to standard Gibbs energy changes.

As a result of this **coupling** of chemical reactions, it is quite possible for a reaction to occur to a considerable extent even though it has a positive value of ΔG°. Thus, for the scheme just given, suppose that reaction 1 has a positive value:

$$\Delta G_1^\circ > 0 \tag{4.68}$$

Reaction 1 by itself will therefore not occur to a considerable extent ($K_1 < 1$). Suppose however, that reaction 2 has a negative value. It is then possible for ΔG_3° to be negative, so that $A + B$ may react to a considerable extent to give X + Y. In terms of the Le Chatelier principle, this simply means that X is constantly removed by the isomerization reaction 2, so that the equilibrium of reaction 1 is shifted to the right.

EXAMPLE 4.6 The aminotransferases, also known as transaminases, are enzymes which perform the function of transferring amino groups from one amino acid to another, as for example in the reaction

$$\alpha\text{-ketoglutarate} + \text{alanine} \rightleftharpoons \text{glutamine} + \text{pyruvate}$$

At 25 °C the ΔG° for this reaction is 251 J mol^{-1}. Calculate the equilibrium constant for the process. Suppose that the pyruvate formed is oxidized in a process for which ΔG° is -258.6 kJ mol^{-1}; what effect does this have on the transamination process?

Solution The equilibrium constant for the simple process is

$$\begin{aligned} K_c &= e^{-\Delta G^\circ/RT} \\ &= e^{-251/8.3145 \times 298.15} \\ &= e^{-0.101} = 0.90 \end{aligned}$$

When the reaction is combined with the oxidation process, the overall $\Delta G°$ is 251 − 258 600 = −258 350 J mol^{-1}. The equilibrium constant is

$$K_c = e^{258\ 350/8.3145 \times 298.15}$$
$$= e^{104.2} = 1.8 \times 10^{45}$$

The oxidation of the pyruvate has greatly enhanced the formation of glutamine in the process, effectively taking it to completion.

Coupling by a Catalyst

There is another way in which chemical reactions can be coupled. In the cases we have considered, the product of one reaction is removed in a second reaction. Alternatively, two reactions might be coupled by a *catalyst.* Suppose, for example, that we have two independent reactions, having no common reactant or product:

$$1.\ A + B \rightleftharpoons X + Y$$
$$2.\ C \rightleftharpoons Z$$

There might exist some catalyst that brings about the reaction

$$3.\ A + B + C \rightleftharpoons X + Y + Z$$

and this catalyst would couple the reactions 1 and 2. This type of coupling of reactions is quite common in biological systems.

EXAMPLE 4.7 A good example is to be found in the synthesis of glutamine in living systems, by the process

$$\text{glutamate} + NH_4^+ \rightleftharpoons \text{glutamine}$$

for which $\Delta G°$ is 15.7 kJ mol^{-1} at 37 °C (standard state, 1 mol dm^{-1}).

a. Calculate the equilibrium constant for this reaction.
b. Another reaction is the hydrolysis of adenosine triphosphate (ATP) into adenosine diphosphate (ADP) and phosphate (P):

$$\text{ATP} \rightleftharpoons \text{ADP} + \text{P}$$

for which the $\Delta G°$ value may be taken as −31.0 kJ mol^{-1} (the value varies considerably with the ions present in the solution). Since these two reactants have no reactants or products in common they can only be coupled together by a catalyst. The enzyme glutamine synthetase in fact catalyzes the reaction

$$\text{glutamate} + NH_4^+ + \text{ATP} \rightleftharpoons \text{glutamine} + \text{ADP} + \text{P}$$

and therefore couples the reactions together. Calculate the equilibrium constant for this combined reaction.

Solution

a. The equilibrium constant for the first reaction is

$$K_c = e^{-15\ 700/8.3145 \times 310.15}$$
$$= e^{-6.09} = 2.27 \times 10^{-3}\ \text{dm}^3\ \text{mol}^{-1}$$

b. For the coupled reaction

$$\Delta G^\circ = 15.7 - 31.0 = -15.3 \text{ kJ mol}^{-1}$$

The equilibrium constant for the coupled reaction is therefore

$$K_c = e^{15\,300/8.3145 \times 310.15}$$
$$= e^{5.93} = 3.8 \times 10^2$$

The occurrence of the second reaction, coupled to the first, has therefore greatly enhanced the glutamine formation.

4.8 Temperature Dependence of Equilibrium Constants

In Section 3.9 on p. 135 we derived the Gibbs-Helmholtz equation, Eq. 3.169:

$$\left[\frac{\partial}{\partial T}\left(\frac{\Delta G^\circ}{T}\right)\right]_P = -\frac{\Delta H^\circ}{T^2} \tag{4.69}$$

Since $\Delta G^\circ = -RT \ln K_P^\circ$ (Eq. 4.20), this at once leads to an equation that gives the temperature dependence of K_P°; thus

$$\left[\frac{\partial}{\partial T}(-R \ln K_P^\circ)\right]_P = -\frac{\Delta H^\circ}{T^2} \tag{4.70}$$

or

$$\left(\frac{\partial \ln K_P^\circ}{\partial T}\right)_P = \frac{\Delta H^\circ}{RT^2} \tag{4.71}$$

The equilibrium constant K_P is affected very little by the pressure (see Section 4.9), and the equation is usually written as

$$\frac{d \ln K_P^\circ}{dT} = \frac{\Delta H^\circ}{RT^2} \tag{4.72}$$

van't Hoff Equation

This is known as the **van't Hoff Equation.** Since $d(1/T) = -dT/T^2$, the equation can also be written as

$$\frac{d \ln K_P^\circ}{d(1/T)} = -\frac{\Delta H^\circ}{R} \tag{4.73}$$

This equation tells us that a plot of $\ln K_P^\circ$ against $1/T$ will have a slope equal to $\Delta H^\circ/R$. In Figure 4.2a the slope is varying with temperature so that ΔH° is a function of temperature. Frequently, however, plots of $\ln K^\circ$ against $1/T$ are linear, as in Figure 4.2b, which means that ΔH° is independent of temperature over the range investigated. When this is so, we can integrate Eq. 4.73 as follows:

$$\int d \ln K_P^\circ = -\int \frac{\Delta H^\circ}{R}\, d\left(\frac{1}{T}\right) \tag{4.74}$$

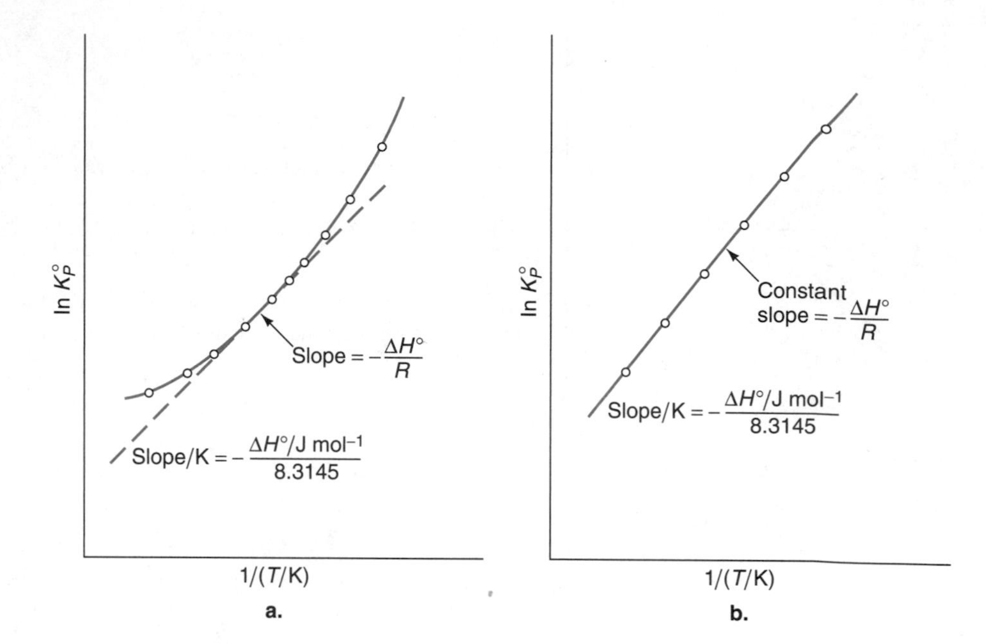

FIGURE 4.2
Schematic plots of ln K_P° against $1/T$. (a) ΔH° is temperature dependent; (b) ΔH° is independent of temperature, in which case Eq. 4.76 applies.

and thus

$$\ln K_P^\circ = -\frac{\Delta H^\circ}{RT} + I \tag{4.75}$$

where I is the constant of integration. However, from Eq. 4.20

$$\ln K_P^\circ = -\frac{\Delta G^\circ}{RT} = -\frac{\Delta H^\circ}{RT} + \frac{\Delta S^\circ}{R} \tag{4.76}$$

The intercept on the ln K_P° axis, at $1/T = 0$, is therefore $\Delta S^\circ/R$.

Equation 4.72 gives the temperature dependence of K_P°, and we can obtain analogous equations for the temperature dependence of K_x, the equilibrium constant in terms of mole fractions, and of K_c°, the equilibrium constant in terms of concentrations. Equation 4.32 relates K_P and K_x,

$$K_P = K_x P^{\Sigma\nu} \tag{4.77}$$

and from Eq. 4.71 it follows that

$$\left(\frac{\partial \ln K_x^\circ}{\partial T}\right)_P = \frac{\Delta H^\circ}{RT^2} \tag{4.78}$$

Since in general K_x is pressure dependent, it is now necessary to express the condition that the pressure must be held constant.

For ideal gases the concentration equilibrium constant K_c is related to K_P by Eq. 4.26:

$$K_P = K_c(RT)^{\Sigma\nu} \tag{4.79}$$

and therefore

$$\ln K_P^\circ = \ln K_c^\circ + \Sigma\nu \ln R^u + \Sigma\nu \ln T^u \tag{4.80}$$

where R^u and T^u are the numerical values of R and T, respectively. Partial differentiation at constant pressure gives

$$\left(\frac{\partial \ln K_P^\circ}{\partial T}\right)_P = \left(\frac{\partial \ln K_c^\circ}{\partial T}\right)_P + \frac{\Sigma\nu}{T} \tag{4.81}$$

By Eq. 4.71 the left-hand side of this equation is $\Delta H^\circ/RT^2$, and therefore

$$\left(\frac{\partial \ln K_c^\circ}{\partial T}\right)_P = \frac{\Delta H^\circ}{RT^2} - \frac{\Sigma\nu}{T} = \frac{\Delta H^\circ - \Sigma\nu RT}{RT^2} \tag{4.82}$$

However, $\Delta H^\circ - \Sigma\nu RT$ for ideal gases is equal to ΔU° and therefore

$$\frac{d \ln K_c^\circ}{dT} = \frac{\Delta U^\circ}{RT^2} \quad \text{or} \quad \frac{d \ln K_c^\circ}{d(1/T)} = -\frac{\Delta U^\circ}{R} \tag{4.83}$$

Again, since K_c° changes only slightly with pressure, it is usually not necessary to specify constant pressure conditions. For reactions in solution, Eq. 4.83 is the one usually employed, but ΔU° and ΔH° are then very close to each other.

EXAMPLE 4.8 The equilibrium constant for an association reaction

$$A + B \rightleftharpoons AB$$

is 1.80×10^3 dm^3 mol^{-1} at 25 °C and 3.45×10^3 dm^3 mol^{-1} at 40 °C. Assuming ΔH° to be independent of temperature, calculate ΔH° and ΔS°.

Solution If the lower temperature, 298.15 K, is written as T_1 and the corresponding dimensionless equilibrium constant as K_1°,

a. $$\ln K_1^\circ = -\frac{\Delta H^\circ}{RT_1} + \frac{\Delta S^\circ}{R}$$

Similarly, for the higher temperature, 313.15 K,

b. $$\ln K_2^\circ = -\frac{\Delta H^\circ}{RT_2} + \frac{\Delta S^\circ}{R}$$

Thus,

$$\ln \frac{K_2^\circ}{K_1^\circ} = \frac{\Delta H^\circ}{R} \cdot \frac{T_2 - T_1}{T_1 T_2}$$

In our particular example,

$$\ln \frac{3.45}{1.80} = \frac{(\Delta H^\circ/\text{J mol}^{-1})}{8.3145} \cdot \frac{15.00}{298.15 \times 313.15}$$

from which it follows that

$$\Delta H^\circ = 33.67 \text{ kJ mol}^{-1}$$

The entropy change ΔS° can be calculated by inserting the values of K°, ΔH°, and T into Eq. a. For example,

$$\ln(3.45 \times 10^3) = 8.146 = -\frac{33\,670}{8.3145 \times 313.15} + \frac{\Delta S^\circ/\text{J K}^{-1}\text{ mol}^{-1}}{8.3145}$$

so that

$$\Delta S^\circ = 175.2 \text{ J K}^{-1} \text{ mol}^{-1}$$

The procedure is somewhat more complicated when $\Delta H°$ is a function of temperature. We have seen in Section 2.5 that for a gas the heat capacity at constant pressure can often be expressed as

$$C_P = d + eT + fT^{-2} \tag{4.84}$$

and that as a result the enthalpy change at a temperature T_2 is related to that at T_1 by an equation of the form (see Eq. 2.52 on p. 70)

$$\Delta H_m(T_2) = \Delta H_m(T_1) + \Delta d(T_2 - T_1) + \frac{1}{2}\Delta e(T_2^2 - T_1^2) - \Delta f\left(\frac{1}{T_2} - \frac{1}{T_1}\right) \tag{4.85}$$

If T_1 is 25 °C and ΔH_m° (25 °C) refers to specified standard states, ΔH_m° at any temperature T is

$$\Delta H_m^\circ(T) = \Delta H_m^\circ(25\text{ °C}) + \Delta d(T - 298.15\text{ K}) + \frac{1}{2}\Delta e[T^2 - (298.15\text{ K})^2] - \Delta f\left(\frac{1}{T} - \frac{1}{298.15\text{ K}}\right) \tag{4.86}$$

Substitution of this into Eq. 4.71, and integration, gives

$$\ln K_P^\circ = \frac{1}{R}\int\left[\frac{\Delta H_m^\circ}{T^2} + \frac{\Delta d}{T} - \frac{(298.15\text{ K})\Delta d}{T^2} + \frac{\Delta e}{2} - \frac{(298.15\text{ K})^2\Delta e}{2T^2} - \frac{\Delta f}{T^3} + \frac{\Delta f}{(298.15\text{ K})T^2}\right]dT + I \tag{4.87}$$

where I is the constant of integration. This equation is readily integrated, so that if ΔH_m°, Δd, Δe, and Δf are known for a particular system, and K is known at one temperature (thus allowing I to be determined), K can be calculated at any other temperature.

4.9 Pressure Dependence of Equilibrium Constants

See how pressure can change the equilibrium concentrations; Le Chatelier's principle; Pressure changes.

We have seen that for equilibria involving ideal gases the equilibrium constants K_P and K_c are independent of the total pressure. A shift in equilibrium is brought about by a change in pressure when $\Sigma\nu$ is other than zero, but the equilibrium constant itself does not change. For nonideal systems, however—and this includes nonideal gases and all solids and liquids—there is in general a change in the equilibrium constant itself as the pressure is varied.

The theory of this can be deduced from Eq. 3.119 on p. 129:

$$\left(\frac{\partial G}{\partial P}\right)_T = V \tag{4.88}$$

The change $\Delta G°$ for a chemical reaction, in which the reactants in their standard states are transformed into the products in their standard states, is thus

$$\left(\frac{\partial \Delta G^\circ}{\partial P}\right)_T = \Delta V^\circ \tag{4.89}$$

where $\Delta V°$ is the corresponding volume change. Substitution of $-RT \ln K°$ for $\Delta G°$ then leads to

$$\left(\frac{\partial \ln K^\circ}{\partial P}\right)_T = -\frac{\Delta V^\circ}{RT} \tag{4.90}$$

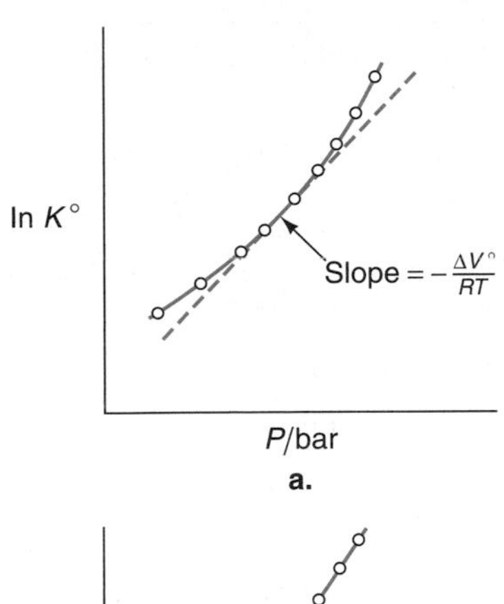

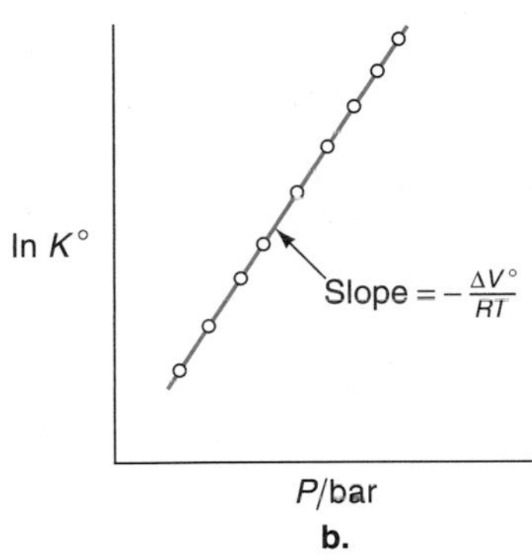

FIGURE 4.3
Plots of ln K° against pressure, P. (a) ΔV° dependent on pressure; (b) ΔV° independent of pressure, in which case Eq. 4.92 applies.

Schematic plots of ln K° against P are shown in Figure 4.3. In Figure 4.3a the volume change ΔV° is itself a function of pressure, and the slope of the plot varies as the pressure changes. In Figure 4.3b the plot is linear, which means that ΔV° is independent of pressure. In that case, Eq. 4.90 can be integrated as follows:

$$\int d \ln K^\circ = -\int \frac{\Delta V^\circ}{RT} dP \tag{4.91}$$

so that

$$\ln K^\circ = -\frac{\Delta V^\circ}{RT} P + \text{constant} \tag{4.92}$$

In practice, the change of an equilibrium constant with pressure is fairly small, and quite high pressures have to be used in order to produce an observable effect. Suppose, for example, that ΔV° for a reaction were 10 $cm^3\ mol^{-1}$ and independent of pressure. The value of $\Delta V^\circ/RT$ at 25 °C is then

$$\frac{10^{-5}\ \text{m}^3\ \text{mol}^{-1} \times 10^5(\text{Pa bar}^{-1})}{8.3145(\text{J K}^{-1}\ \text{mol}^{-1}) \times 298.15(\text{K})} = 4.03 \times 10^{-4}\ \text{bar}^{-1}$$

A pressure of $1/4.03 \times 10^{-4} = 2480$ bar will therefore change ln K° by one unit (i.e., will change K by a factor of 2.718).

EXAMPLE 4.9 The equilibrium constant for the reaction

$$2NO_2 \rightleftharpoons N_2O_4$$

in carbon tetrachloride solution at 22 °C is increased by a factor of 3.77 when the pressure is increased from 1 bar to 1500 bar. Calculate ΔV°, on the assumption that the equilibrium constant is independent of pressure.

Solution An increase by a factor of 3.77 means that ln K° has increased by 1.327 as P increases by $1500 - 1 = 1499$ bar. The slope of a plot of ln K° versus P is thus

$$\frac{1.327}{1499}\ \text{bar}^{-1}$$

and this is equal to $-\Delta V^\circ/RT$. Thus

$$\Delta V^\circ = -\frac{(1.327/1499)(\text{bar}^{-1}) \times 8.3145 \times 293.15(\text{J mol}^{-1})}{1 \times 10^5(\text{Pa bar}^{-1})}$$
$$= -2.16 \times 10^{-5}\ \text{m}^3\ \text{mol}^{-1} = -21.6\ \text{cm}^3\ \text{mol}^{-1}$$

KEY EQUATIONS

For a reaction $aA + bB + \cdots \rightleftharpoons \cdots yY + zZ$:

$$\left(\frac{\cdots[Y]^y[Z]^z}{[A]^a[B]^b\cdots}\right)_{eq} = K_c$$

where K_c is the equilibrium constant (concentration basis). Other equilibrium constants are:

Activity basis: $K_a = \left(\frac{\cdots a_Y^y a_Z^z}{a_A^a a_B^b \cdots}\right)_{eq}$

Pressure basis: $K_P = K_c(RT)^{\Sigma\nu}$

Mole fraction basis: $K_x = K_P P^{-\Sigma\nu}$

Definition of *chemical potential* for species A:

$$\mu_A \equiv \left(\frac{\partial G}{\partial n_A}\right)_{T,P,n_B,n_Y,\ldots}$$

Relationship between *standard Gibbs energy change and equilibrium constant:*

$$\Delta G^\circ = -RT \ln K^\circ$$

$$\Delta G = \Delta G^\circ + RT \ln\left(\frac{\cdots[Y]^y[Z]^z}{[A]^a[B]^b\cdots}\right)^u$$

Temperature dependence of equilibrium constants:

$$\frac{d \ln K_P^\circ}{d(1/T)} = -\frac{\Delta H^\circ}{R}$$

$$\frac{d \ln K_c^\circ}{d(1/T)} = -\frac{\Delta U^\circ}{R}$$

PROBLEMS

Equilibrium Constants

4.1. A reaction occurs according to the equation

$$2A \rightleftharpoons Y + 2Z$$

If in a volume of 5 dm^3 we start with 4 mol of pure A and find that 1 mol of A remains at equilibrium, what is the equilibrium constant K_c?

4.2. The equilibrium constant for a reaction

$$A + B \rightleftharpoons Y + Z$$

is 0.1. What amount of A must be mixed with 3 mol of B to yield, at equilibrium, 2 mol of Y?

4.3. The equilibrium constant for the reaction

$$A + 2B \rightleftharpoons Z$$

is 0.25 dm^6 mol^{-2}. In a volume of 5 dm^3, what amount of A must be mixed with 4 mol of B to yield 1 mol of Z at equilibrium?

4.4. The equilibrium constant K_c for the reaction

$$2SO_3(g) \rightleftharpoons 2SO_2(g) + O_2(g)$$

is 0.0271 mol dm^{-3} at 1100 K. Calculate K_P at that temperature.

4.5. When gaseous iodine is heated, dissociation occurs:

$$I_2 \rightleftharpoons 2I$$

It was found that when 0.0061 mol of iodine was placed in a volume of 0.5 dm^3 at 900 K, the degree of dissociation (the fraction of the iodine that is dissociated) was 0.0274. Calculate K_c and K_P at that temperature.

4.6. It has been observed with the ammonia equilibrium

$$N_2 + 3H_2 \rightleftharpoons 2NH_3$$

that under certain conditions the addition of nitrogen to an equilibrium mixture, *with the temperature and pressure held constant,* causes further dissociation of ammonia. Explain how this is possible. Under what particular conditions would you expect this to occur? Would it be possible for added hydrogen to produce the same effect?

4.7. Nitrogen dioxide, NO_2, exists in equilibrium with dinitrogen tetroxide, N_2O_4:

$$N_2O_4(g) \rightleftharpoons 2NO_2(g)$$

At 25.0 °C and a pressure of 0.597 bar the density of the gas is 1.477 g dm^{-3}. Calculate the degree of dissociation under those conditions, and the equilibrium constants K_c, K_P, and K_x. What shift in equilibrium would occur if the pressure were increased by the addition of helium gas?

4.8. At 25.0 °C the equilibrium

$$2NOBr(g) \rightleftharpoons 2NO(g) + Br_2(g)$$

is rapidly established. When 1.10 g of NOBr is present in a 1.0-dm^3 vessel at 25.0 °C the pressure is 0.355 bar. Calculate the equilibrium constants K_P, K_c, and K_x.

4.9. At 100 °C and 2 bar pressure the degree of dissociation of phosgene is 6.30×10^{-5}. Calculate K_P, K_c, and K_x for the dissociation

$$COCl_2(g) \rightleftharpoons CO(g) + Cl_2(g)$$

4.10. In a study of the equilibrium

$$H_2 + I_2 \rightleftharpoons 2HI$$

1 mol of H_2 and 3 mol of I_2 gave rise at equilibrium to x mol of HI. Addition of a further 2 mol of H_2 gave an additional x mol of HI. What is x? What is K at the temperature of the experiment?

***4.11.** The equilibrium constant for the reaction

$$H_2(g) + I_2(g) \rightleftharpoons 2HI(g)$$

is 20.0 at 40.0 °C, and the vapor pressure of solid iodine is 0.10 bar at that temperature. If 12.7 g of solid iodine are placed in a 10-dm^3 vessel at 40.0 °C, what is the minimum amount of hydrogen gas that must be introduced in order to remove all the solid iodine?

4.12. The degree of dissociation α of $N_2O_4(g)$ is 0.483 at 0.597 bar and 0.174 at 6.18 bar. The temperature is 298 K for both measurements. Calculate K_P, K_c, and K_x in each case. (See Example 4.1.)

4.13. One mole of HCl mixed with oxygen is brought into contact with a catalyst until the following equilibrium has been established:

$$4HCl(g) + O_2(g) \rightleftharpoons 2Cl_2(g) + 2H_2O(g).$$

If y mol of HCl is formed, derive an expression for K_P in terms of y and the partial pressure of oxygen. [*Hint:* First develop expressions for the ratios x_{Cl_2}/x_{HCl} and x_{H_2O}/x_{Cl_2} in terms of y and P_{O_2}.]

4.14. Using the result of Problem 4.13, evaluate K_P for an experiment in which 49% HCl and 51% O_2 are brought into contact with a catalyst until the reaction is complete at 1 bar and 480 °C. The fraction of HCl converted per mole is found to be 0.76.

4.15. 10.0 g of HI is introduced into an evacuated vessel at 731 K and allowed to reach equilibrium. Find the mole fractions of H_2, I_2, and HI present at equilibrium. $K_P = K_c = K_x = 65.0$ for the reaction $H_2(g) + I_2(g) \rightleftharpoons 2HI(g)$ (see Example 4.2).

Equilibrium Constants and Gibbs Energy Changes

4.16. The equilibrium constant for the reaction

$$(C_6H_5COOH)_2 \rightleftharpoons 2C_6H_5COOH$$

in benzene solution at 10 °C is 2.19×10^{-3} mol dm^{-3}.

a. Calculate $\Delta G°$ for the dissociation of the dimer.
b. If 0.1 mol of benzoic acid is present in 1 dm^3 of benzene at 10 °C, what are the concentrations of the monomer and of the dimer?

4.17. At 3000 K the equilibrium partial pressures of CO_2, CO, and O_2 are 0.6, 0.4, and 0.2 atm, respectively. Calculate $\Delta G°$ at 3000 K for the reaction

$$2CO_2(g) \rightleftharpoons 2CO(g) + O_2(g)$$

4.18. The conversion of malate into fumarate

$$\text{1. malate(aq)} \rightleftharpoons \text{fumarate(aq)} + H_2O(l)$$

is endergonic at body temperature, 37 °C; $\Delta G°$ is 2.93 kJ mol^{-1}. In metabolism the reaction is coupled with

$$\text{2. fumarate(aq)} \rightleftharpoons \text{aspartate(aq)}$$

for which $\Delta G°$ is -15.5 kJ mol^{-1} at 37 °C.

a. Calculate K_c for reaction 1.
b. Calculate K_c for reaction 2.
c. Calculate K_c and $\Delta G°$ for the coupled reaction 1 + 2.

4.19. From the data in Appendix D, deduce the $\Delta G°$ and K_P values for the following reactions at 25.0 °C:

a. $N_2(g) + 3H_2(g) \rightleftharpoons 2NH_3(g)$
b. $2H_2(g) + C_2H_2(g) \rightleftharpoons C_2H_6(g)$
c. $H_2(g) + C_2H_4(g) \rightleftharpoons C_2H_6(g)$
d. $2CH_4(g) \rightleftharpoons C_2H_6(g) + H_2(g)$

4.20. Calculate K_c and K_x for each of the reactions in Problem 4.19 assuming total pressures of 1 bar in each case.

4.21. At 25.0 °C the equilibrium constant for the reaction

$$CO(g) + H_2O(g) \rightleftharpoons CO_2(g) + H_2(g)$$

is 1.00×10^{-5}, and $\Delta S°$ is 41.8 J K^{-1} mol^{-1}.

a. Calculate $\Delta G°$ and $\Delta H°$ at 25.0 °C.
b. Suppose that 2 mol of CO and 2 mol of H_2O are introduced into a 10-dm^3 vessel at 25.0 °C. What are the amounts of CO, H_2O, CO_2, and H_2 at equilibrium?

4.22. Suppose that there is a biological reaction

$$\text{1. A + B} \rightleftharpoons \text{Z}$$

for which the $\Delta G°$ value at 37.0 °C is 23.8 kJ mol^{-1}. (Standard state = 1 mol dm^{-3}.) Suppose that an enzyme couples this reaction with

$$\text{2. ATP} \rightleftharpoons \text{ADP + phosphate}$$

for which $\Delta G° = -31.0$ kJ mol^{-1}. Calculate the equilibrium constant at 37.0 °C for these two reactions and for the coupled reaction

$$\text{3. A + B + ATP} \rightleftharpoons \text{Z + ADP + phosphate}$$

4.23. The equilibrium between citrate and isocitrate involves *cis*-aconitate as an intermediate:

$$\text{citrate} \rightleftharpoons \textit{cis}\text{-aconitate} + H_2O \rightleftharpoons \text{isocitrate}$$

At 25 °C and pH 7.4 it was found that the molar composition of the mixture was

90.9% citrate

2.9% *cis*-aconitate

6.2% isocitrate

Calculate the equilibrium constants for the individual reactions, and for the overall reaction, and $\Delta G°$ for the citrate-isocitrate system.

4.24. The solubility product of $Cr(OH)_3$ is 3.0×10^{-29} mol^4 dm^{-12} at 25 °C. What is the solubility of $Cr(OH)_3$ in water at this temperature?

Temperature Dependence of Equilibrium Constants

4.25. A gas reaction

$$A \rightleftharpoons B + C$$

is endothermic and its equilibrium constant K_P is 1 bar at 25 °C.

a. What is $\Delta G°$ at 25 °C (standard state: 1 bar)?
b. Is $\Delta S°$, with the same standard state, positive or negative?
c. For the standard state of 1 *M*, what are K_c and $\Delta G°$?
d. Will K_P at 40 °C be greater than or less than 1 bar?
e. Will $\Delta G°$ at 40 °C (standard state: 1 bar) be positive or negative?

4.26. A solution reaction

$$A + B \rightleftharpoons X + Y$$

is endothermic, and K_c at 25 °C is 10.

a. Is the formation of X + Y exergonic at 25 °C?
b. Will raising the temperature increase the equilibrium yield of X + Y?
c. Is $\Delta S°$ positive or negative?

4.27. From the data given in Appendix D, for the reaction

$$C_2H_4(g) + H_2(g) \rightleftharpoons C_2H_6(g)$$

calculate the following:

a. $\Delta G°$, $\Delta H°$, and $\Delta S°$ at 25 °C; what is the standard state?
b. K_P at 25 °C.
c. K_c at 25 °C (standard state: 1 *M*).
d. $\Delta G°$ at 25 °C (standard state: 1 *M*).
e. $\Delta S°$ at 25 °C (standard state: 1 *M*).
f. K_P at 100 °C, on the assumption that $\Delta H°$ and $\Delta S°$ are temperature independent.

4.28. From the data in Appendix D, for the reaction

$$2H_2(g) + O_2(g) \rightleftharpoons 2H_2O(g)$$

calculate the following:

a. $\Delta G°$, $\Delta H°$, and $\Delta S°$ at 25 °C (standard state: 1 bar).
b. K_P at 25 °C.
c. $\Delta G°$ and K_P at 2000 °C, on the assumption that $\Delta H°$ and $\Delta S°$ are temperature independent.

4.29. Calculate the equilibrium constant at 400 K for the reaction

$$3O_2(g) \rightarrow 2O_3(g).$$

where $\Delta_f G°(O_3, g) = 163.2$ kJ mol^{-1}.

4.30. The hydrolysis of adenosine triphosphate to give adenosine diphosphate and phosphate can be represented by

$$ATP \rightleftharpoons ADP + P$$

The following values have been obtained for the reaction at 37 °C (standard state: 1 *M*):

$$\Delta G° = -31.0 \text{ kJ mol}^{-1}$$
$$\Delta H° = -20.1 \text{ kJ mol}^{-1}$$

a. Calculate $\Delta S°$.
b. Calculate K_c at 37 °C.
c. On the assumption that $\Delta H°$ and $\Delta S°$ are temperature independent, calculate $\Delta G°$ and K_c at 25 °C.

4.31. Thermodynamic data for *n*-pentane(g) and neopentane(g) (standard state: 1 bar and 25 °C) are as follows

Compound	Enthalpy of Formation, $\Delta H_f°$ kJ mol^{-1}	Entropy, $S°$ J K^{-1} mol^{-1}
n-Pentane(g)	−146.44	349.0
Neopentane(g)	−165.98	306.4

a. Calculate $\Delta G°$ for *n*-pentane → neopentane.
b. Pure *n*-pentane is in a vessel at 1 bar and 25 °C, and a catalyst is added to bring about the equilibrium between *n*-pentane and neopentane. Calculate the final partial pressures of the two isomers.

4.32. a. An equilibrium constant K_c is increased by a factor of 3 when the temperature is raised from 25.0 °C to 40.0 °C. Calculate the standard enthalpy change.
b. What is the standard enthalpy change if instead K_c is *decreased* by a factor of 3 under the same conditions?

4.33. a. The ionic product $[H^+][OH^-]$, which is the equilibrium constant for the dissociation of water,

$$H_2O \rightleftharpoons H^+ + OH^-$$

is 1.00×10^{-14} mol^2 dm^{-6} at 25.0 °C and 1.45×10^{-14} mol^2 dm^{-6} at 30.0 °C. Deduce $\Delta H°$ and $\Delta S°$ for the process.
b. Calculate the value of the ionic product at body temperature (37 °C).

4.34. The equilibrium constant K_P for the reaction $I_2(g)$ + cyclopentane(g) $\rightleftharpoons$ 2 HI(g) + cyclopentadiene(g) varies with temperatures according to the equation

$$\log_{10}(K_P/\text{bar}) = 7.55 - 4844/(T/\text{K})$$

a. Calculate K_P, $\Delta G°$, $\Delta H°$, $\Delta S°$ (standard state: 1 bar) at 400 °C.
b. Calculate K_c and $\Delta G°$ (standard state: 1 M) at 400 °C.
c. If I_2 and cyclopentane are initially at 400 °C and at concentrations of 0.1 M, calculate the final equilibrium concentrations of I_2, cyclopentane, HI, and cyclopentadiene.

4.35. From the data in Appendix D, for the synthesis of methanol,

$$CO(g) + 2H_2(g) \rightleftharpoons CH_3OH(l)$$

calculate $\Delta H°$, $\Delta G°$, and $\Delta S°$ and the equilibrium constant at 25 °C.

4.36. The bacterium *nitrobacter* plays an important role in the "nitrogen cycle" by oxidizing nitrite to nitrate. It obtains the energy it requires for growth from the reaction

$$NO_2^-(aq) + \tfrac{1}{2}O_2(g) \rightarrow NO_3^-(aq)$$

Calculate $\Delta H°$, $\Delta G°$, and $\Delta S°$ for this reaction from the following data, at 25 °C:

Ion	$\Delta_f H°$ / kJ mol^{-1}	$\Delta_f G°$ / kJ mol^{-1}
NO_2^-	−104.6	−37.2
NO_3^-	−207.4	−111.3

4.37. When the reaction

$$\text{glucose-1-phosphate(aq)} \rightleftharpoons \text{glucose-6-phosphate(aq)}$$

is at equilibrium at 25 °C, the amount of glucose-6-phosphate present is 95% of the total.
a. Calculate $\Delta G°$ at 25 °C.
b. Calculate ΔG for reaction in the presence of 10^{-2} M glucose-1-phosphate and 10^{-4} M glucose-6-phosphate. In which direction does reaction occur under these conditions?

4.38. From the data in Appendix D, for the reaction

$$CO_2(g) + H_2(g) \rightleftharpoons CO(g) + H_2O(g)$$

calculate the following:
a. $\Delta H°$, $\Delta G°$, and $\Delta S°$ (standard state: 1 bar and 25 °C).
b. The equilibrium constant at 25 °C.
c. From the heat capacity data in Table 2.1, obtain an expression for $\Delta H°$ as a function of temperature.
d. Obtain an expression for ln K_P as a function of temperature.
e. Calculate K_P at 1000 K.

4.39. Irving Langmuir [*J. Amer. Chem. Soc.*, *28*, 1357 (1906)] studied the dissociation of CO_2 into CO and O_2 by bringing the gas at 1 atm pressure into contact with a heated platinum wire. He obtained the following results:

T/K	Percent Dissociation
1395	0.0140
1443	0.0250
1498	0.0471

Calculate K_P for $2CO_2(g) = 2CO(g) + O_2(g)$ at each temperature, and estimate $\Delta H°$, $\Delta G°$, and $\Delta S°$ at 1395 K.

4.40. G. Stark and M. Bodenstein [*Z. Electrochem.*, *16*, 961(1910)] carried out experiments in which they sealed iodine in a glass bulb and measured the vapor pressure. The following are some of the results they obtained:

volume of bulb = 249.8 cm^3

amount of iodine = 1.958 mmol

Temperature/°C	Pressure/Torr
800	558.0
1000	748.0
1200	1019.2

a. Calculate the degree of dissociation at each temperature.
b. Calculate K_c at each temperature, for the process $I_2 \rightleftharpoons 2I$.
c. Calculate K_P at each temperature.
d. Obtain values for $\Delta H°$ and $\Delta U°$ at 1000 °C.
e. Calculate $\Delta G°$ and $\Delta S°$ at 1000 °C.

4.41. The following diagram shows the variation with temperature of the equilibrium constant K_c for a reaction. Calculate $\Delta G°$, $\Delta H°$, and $\Delta S°$ at 300 K.

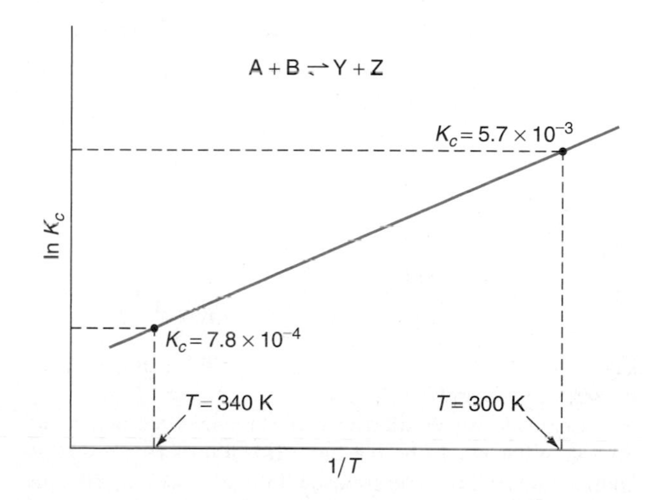

4.42. The following values apply to a chemical reaction A $\rightleftharpoons$ Z:

$$\Delta H° = -85.2 \text{ kJ mol}^{-1}$$
$$\Delta S° = -170.2 \text{ J K}^{-1} \text{ mol}^{-1}$$

Assuming these values to be temperature independent, calculate the equilibrium constant for the reaction at 300 K. At what temperature is the equilibrium constant equal to unity?

4.43. The equilibrium constant K_c for the hydrolysis of adenosine triphosphate (ATP) to adenosine diphosphate (ADP) and phosphate is 1.66×10^5 mol dm^{-3} at 37 °C, and $\Delta H°$ is -20.1 kJ mol^{-1}. Calculate $\Delta S°$ for the hydrolysis at 37 °C. On the assumption that $\Delta H°$ and $\Delta S°$ are temperature independent, calculate K_c at 25 °C.

4.44. A dissociation $A_2 \rightleftharpoons 2A$ has an equilibrium constant of 7.2×10^{-5} mol dm^{-3} at 300 K, and a $\Delta H°$ value of 40.0 kJ mol^{-1}. Calculate the standard entropy change for the reaction at 300 K. (What is its standard state?) If the $\Delta H°$ and $\Delta S°$ values for this reaction are temperature independent, at what temperature is the equilibrium constant equal to unity?

4.45. A reaction $A + B \rightleftharpoons Z$ has an equilibrium constant of 4.5×10^4 dm^3 mol^{-1} at 300 K, and a $\Delta H°$ value of -40.2 kJ mol^{-1}. Calculate the entropy change for the reaction at 300 K. If the $\Delta H°$ and $\Delta S°$ values are temperature independent, at what temperature is the equilibrium constant equal to unity?

4.46. At 1 bar pressure liquid bromine boils at 58.2 °C, and at 9.3 °C its vapor pressure is 0.1334 bar. Assuming $\Delta H°$ and $\Delta S°$ to be temperature independent, calculate their values, and calculate the vapor pressure and $\Delta G°$ at 25 °C.

4.47. The standard Gibbs energy of formation of gaseous ozone at 25 °C, $\Delta_f G°$, is 162.3 kJ mol^{-1}, for a standard state of 1 bar. Calculate the equilibrium constants K_P, K_c, and K_x for the process

$$3O_2(g) \rightleftharpoons 2O_3(g)$$

What is the mole fraction of O_3 present at 25 °C at 2 bar pressure?

4.48. For the equilibrium

$$H_2(g) + I_2(g) \rightleftharpoons 2HI(g)$$

the following data apply:

$$\Delta H° \text{ (300 K)} = -9.6 \text{ kJ mol}^{-1}$$

$$\Delta S° \text{ (300 K)} = 22.18 \text{ J K}^{-1} \text{ mol}^{-1}$$

$$\Delta C_p \text{ (500 K)} = -7.11 \text{ J K}^{-1} \text{ mol}^{-1}$$

The latter value can be taken to be the average value between 300 K and 500 K.

Calculate the equilibrium constants K_P, K_c, and K_x at 500 K. What would be the mole fraction of HI present at equilibrium if HI is introduced into a vessel at 10 atm pressure; how would the mole fraction change with pressure?

***4.49.** Protein denaturations are usually irreversible but may be reversible under a narrow range of conditions. At pH 2.0, at temperatures ranging from about 40 °C to 50 °C, there is an equilibrium between the active form P and the deactivated form D of the enzyme trypsin:

$$P \rightleftharpoons D$$

Thermodynamic values are $\Delta H° = 283$ kJ mol^{-1} and $\Delta S° = 891$ J K^{-1} mol^{-1}. Assume these values to be temperature independent over this narrow range, and calculate $\Delta G°$ and K_c values at 40.0 °C, 42.0 °C, 44.0 °C, 46.0 °C, 48.0 °C, and 50.0 °C. At what temperature will there be equal concentrations of P and D?

Note that the high thermodynamic values lead to a considerable change in K over this 10 °C range.

Binding to Protein Molecules

***4.50.** Suppose that a large molecule, such as a protein, contains n sites to which a molecule A (a ligand) can become attached. Assume that the sites are equivalent and independent, so that the reactions M + A = MA, MA + A = MA_2, etc., all have the same equilibrium constant K_s. Show that the average number of occupied sites per molecule is

$$\bar{\nu} = \frac{nK_s[A]}{1 + K_s[A]}$$

***4.51.** Modify the derivation in Problem 4.50 so as to deal with sites that are not all equivalent; the equilibrium constants for the attachments of successive ligands are each different:

$$M + A \rightleftharpoons MA \qquad K_1 = \frac{[MA]}{[M][A]}$$

$$MA + A \rightleftharpoons MA_2 \qquad K_2 = \frac{[MA_2]}{[MA][A]}$$

$$MA_{n-1} + A \rightleftharpoons MA_n \qquad K_n = \frac{[MA_n]}{[MA_{n-1}][A]}$$

Show that the average number of molecules of A bound per molecule M is

$$\bar{\nu} = \frac{K_1[A] + 2K_1K_2[A]^2 + \cdots + n(K_1K_2K_3 \cdots K_n)[A]^n}{1 + K_1[A] + K_1K_2[A]^2 + \cdots + (K_1K_2K_3 \cdots K_n)[A]^n}$$

This equation is important in biology and biochemistry and is often called the *Adair equation,* after the British biophysical chemist G. S. Adair.

***4.52.** Now show that the Adair equation, derived in Problem 4.51, reduces to the equation obtained in Problem 4.50 when the sites are equivalent and independent. [It is not correct simply to put $K_1 = K_2 = K_3 \cdots = K_n$; certain statistical factors must be introduced. Thus, if K_s is the equilibrium constant for the binding at a given site, $K_1 = nK_s$, since there are n ways for A to become attached to a given molecule and one way for it to come off. Similarly $K_2 = (n - 1)K_s/2$; $n - 1$ ways on and 2 ways off. Continue this argument and develop an expression for ν that will factorize into $nK_s[A]/(1 + K_s[A])$. Suggest a method of testing the equilibrium obtained and arriving at a value of n from experimental data.]

***4.53.** Another special case of the equation derived in Problem 4.51 is if the binding on one site affects that on another. An extreme case is highly cooperative binding, in which the binding of A on one site influences the other sites so that they fill up immediately. This means that K_n is much greater than K_1, K_2, etc. Show that now

$$\bar{\nu} = \frac{nK[A]^n}{1 + K[A]^n}$$

where K is the product of $K_1, K_2, \cdots K_n$. The British physiologist A. V. Hill suggested that binding problems can be treated by plotting

$$\ln \frac{\theta}{1 - \theta} \quad \text{against} \quad \ln[A]$$

where θ is the fraction of sites that are occupied. Consider the significance of such *Hill plots,* especially their shapes and slopes, with reference to the equations obtained in Problems 4.50 to 4.53.

Essay Questions

4.54. Give an account of the effect of temperature on equilibrium constants, and explain how such experimental studies lead to thermodynamic data.

4.55. Give an account of the effect of pressure on (a) the position of equilibrium and (b) the equilibrium constant.

4.56. Explain what experimental studies might be made to decide whether a chemical system is at equilibrium or not.

4.57. Give an account of the coupling of chemical reactions.

4.58. State the Le Chatelier principle, and give several examples.

SUGGESTED READING

See the listing at the end of Chapter 3. For a discussion of binding problems relating to Problems 4.50–4.53 see

K. J. Laidler, *Physical Chemistry with Biological Applications,* Menlo Park, California: Benjamin/Cummings, 1978; especially Section 11.2, "Multiple Equilibria."

J. Steinhart, and J. A. Reynolds, *Multiple Equilibria in Proteins,* New York: Academic Press, 1969, especially Chapter 2, "Thermodynamics and Model Systems."

C. Tanford, *Physical Chemistry of Macromolecules,* New York: Wiley, 1961, especially Chapter 8, "Multiple Equilibria."

Phases and Solutions

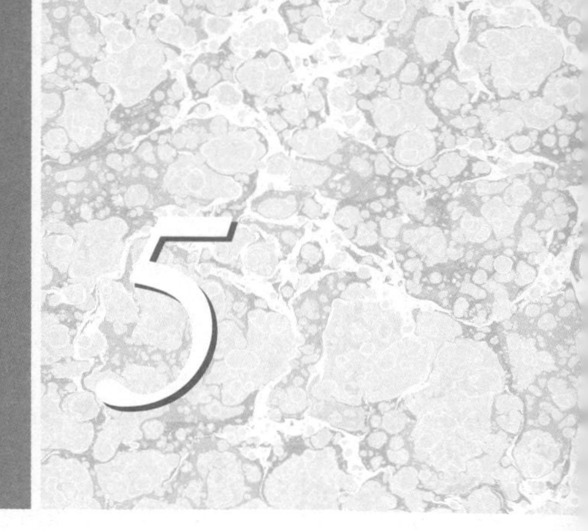

PREVIEW

Most substances can exist in more than one *phase* or *state of aggregation.* This chapter examines the conditions for equilibrium at various temperatures and pressures. The criterion for equilibrium between two phases is that the Gibbs energy is the same in the two phases. This condition leads to insight into the thermodynamics of vapor pressure. In particular, the *Clapeyron equation* and the *Clausius-Clapeyron* equation, which are concerned with the variation of vapor pressure with temperature, are two important expressions for treating liquid-vapor systems.

It is useful to deal with ideal solutions first. Such solutions obey *Raoult's law,* according to which the vapor pressure of each component of a solution is proportional to its mole fraction. For real solutions Raoult's law applies to the *solvent* when the concentration of solute is small, but deviations occur at higher concentrations. At low solute concentrations the *solute* obeys *Henry's law,* according to which the solute vapor pressure is proportional to the solute concentration.

The contribution of each component in a solution is described in terms of *partial molar quantities.* Any extensive thermodynamic property $X = X(T, P, n_i)$, has the partial molar value X_i defined as

$$X_i \equiv \left(\frac{\partial X}{\partial n_i}\right)_{T,P,n_j}$$

All the thermodynamic equations apply to these partial molar quantities. The partial molar Gibbs energy is known as the *chemical potential* μ_i, and for multicomponent systems the chemical potential plays a thermodynamic role equivalent to that of the Gibbs energy.

The concepts of *activity* a and *activity coefficient* f or γ are introduced to handle nonideal solutions. In the *rational system,* activities for the solvent are defined so that $a_i = 1$ and $f_i \to 1$ as $x_i \to 1$. In the *practical system,* activities for the solute are defined so that $a_i = 1$ and $\gamma_i \to 1$ as $n_i \to 0$.

Colligative properties, which depend only on the number of molecules of solute present, are a consequence of vapor pressure lowering as expressed by Raoult's law. Aside from *vapor pressure lowering* the colligative properties are *boiling point elevation, freezing point depression,* and *osmotic pressure.*

In Chapter 4 we have been concerned primarily with the study of equilibrium in reacting systems. Up to this point little has been said concerning the nature of the equilibria that can exist within a pure material, or in a solution between a solute and its solvent. No formal chemical reaction need occur in such equilibria. For instance, the melting of ice, the boiling of water, and the dissolving of sugar in coffee are all examples of **phase** (Greek *phasis,* appearance) **changes,** or changes in the state of aggregation. For each of these changes an equilibrium relation defines the behavior of the system, in much the same way as the equilibrium constant does for chemically reacting systems. In this chapter we will investigate what constitutes a phase, and the criteria of equilibrium as applied to phases, for both pure substances and solutions. Then the thermodynamics needed to treat the behavior of solutions will be explored. In Chapter 6, the ideas developed in this chapter will be applied to a systematic study in a less mathematical vein of systems consisting of many components and/or phases.

OBJECTIVES

After studying this chapter, the student should be able to:

- Distinguish between homogeneous and heterogeneous solutions.
- Discuss the phase diagram of water and explain why the separate phases occur when they do.
- Explain what a metastable state is.
- Understand why the Gibbs energy is so important when explaining equilibrium conditions.
- Understand and apply the Clapeyron and Clausius-Clapeyron equations.
- Explain the reason for Trouton's focus and why associated liquids give different results.
- Classify first- and second-order phase transitions.
- Define and work problems involving Raoult's and Henry's laws.
- Develop partial molar quantities to the point of using the Gibbs-Duhem equation.
- Show how the chemical potential, μ, is the partial of G with respect to n_i, holding T, P, and n_j, constant. Then show how μ can equally be defined in terms of the other thermodynamic state functions.
- Define and use the activity in both the practical and rational systems.
- Explain the meaning of the various thermodynamic functions when the symbol "*" is used.
- Understand how ΔG and ΔS influence the mixing of solutions.
- Explain the use of activity and activity coefficients.
- Solve problems using each of the colligative properties.
- Explain osmotic pressure and how a semipermeable membrane works.
- State the importance of van't Hoff's equation and give an example of how it is used.

5.1 Phase Recognition

Homogeneous and Heterogeneous

The vaporization of a liquid into its vapor state is an example of a phase change. For pure water we say that both the liquid and the gaseous states are single phases. A single phase is uniform throughout both in chemical composition and physical state, and it is said to be *homogeneous.* Note that subdivision does not produce new phases; a block of ice reduced to crushed ice still consists of only one phase. Mixtures of compounds or elements may also appear as a single phase as long as no boundaries exist that allow one region to be distinguished from another region.

In contrast to this, a *heterogeneous* system consists of more than one phase; the phases are distinguished from each other through separation by distinct boundaries. A familiar example is provided by ice cubes in water; there two phases coexist, one solid and one liquid, even though the chemical composition is the same.

Two solids of different substances may be mixed, and if each retains its characteristic boundary, the mixture will consist of two phases. Even if two solids are melted together and then cooled to give the outward appearance of a single uniform solid, more than one phase would exist if the phase boundaries in the solid can be discerned by using a microscope.

Mixtures of liquids also can occur in one or more phases. For example, carbon tetrachloride forms a separate layer when mixed with water. In this system, two phases coexist; whereas, for example, only one phase exists in the ethanol-water system. Although systems with many liquid phases are known, generally only one or two liquid phases are present, larger numbers of liquid phases in contact being rather unusual.

Although solid and liquid mixtures may consist of a number of phases, gases can exist in only one phase at normal pressures since gases mix in all proportions to give a uniform mixture.

Phase Distinctions in the Water System

We will first investigate the equilibria present in the ice, liquid water, water-vapor system. The variables associated with the equilibrium criterion of the Gibbs energy are pressure and temperature, and it is therefore natural to depict the phase equilibria on a pressure-temperature diagram. This is done in Figure 5.1.

The areas marked *solid, liquid,* and *vapor* are regions where only one phase may exist. Where these single-phase regions are indicated, arbitrary values of P and T may exist within the lines defining the phase limits.

Metastable State

The solid lines on the diagram give the conditions under which the two adjoining phases are in equilibrium. Thus, the curve TT_c gives the vapor pressure of water up to the critical point T_c. This line therefore defines the pressures and temperatures at which the gas and liquid phases can coexist at equilibrium. The extension of this line to B gives the equilibrium conditions for supercooled water, an example of a substance in a **metastable state** (i.e., not the thermodynamically most stable state). (See p. 229 for further discussion.) This state can be achieved under certain conditions because, in this case, the rate of formation of ice has been reduced by using a very clean sample of water, thereby reducing the number of nucleation sites. Formation of metastable states often occurs when the entropy change during the phase transition is negative, as in this case. (See the section on liquid crystals, pp. 183–184.)

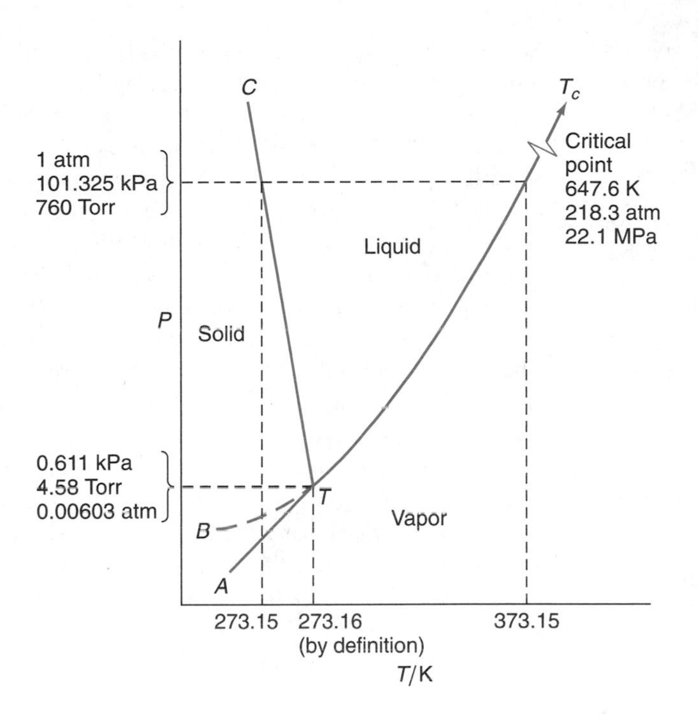

FIGURE 5.1
Phase diagram for water (not drawn to scale).

In a similar manner, the line *TC* gives the conditions under which solid and liquid coexist at equilibrium; that is, it represents the melting point of ice at different pressures of the solid. The final line *TA* represents the equilibrium between the vapor and solid and shows the vapor pressure of the solid as a function of temperature. Above approximately 2×10^5 kPa (2000 bar) different crystalline forms of ice may exist, giving rise to a phenomenon known as **polymorphism** (Greek *pollor,* many; *morphe,* shape, form).

Polymorphism

It should be emphasized that water is not truly representative of most materials, in that the slope of the line *TC* is negative. Only a few materials behave as water does in this respect, among them bismuth and gallium.

Triple Point
Invariant Point

At the **triple point** *T*, all three phases (liquid, solid, vapor) coexist at the same vapor pressure, 0.611 kPa or 4.58/760 atm. This is an **invariant** point for the water system, which means that only under this set of conditions can the three phases coexist. The triple-point temperature is a defining point in the modern Celsius scale and, by definition, is exactly 273.16 K or 0.01 °C.

Phase Changes in Liquid Crystals

From our everyday experiences, we are familiar with the phase distinctions in water just discussed. Transitions between the solid and liquid states of many other substances are also familiar to us. However, another class of materials shows phase properties under certain conditions that are quite different. The term **liquid crystal** is used to describe an intermediate phase that has something of the properties of both solids and liquids. Besides cholesterol (see marginal note on p. 184), many other examples of liquid crystals have been discovered. The molecules of substances forming liquid crystals are known as **mesogens** and fall into two main classes.

The discovery of liquid crystals is usually attributed to the Austrian botanist Friedrich Reinitzer (1857–1927), whose main interest was the function of cholesterol in plants, the structure of which was then not known. In 1888, working with a substance related to cholesterol, Reinitzer noticed that it first melted to a cloudy liquid and then, at a temperature of about 30 °C higher, suddenly turned into a clear liquid. He sent a sample of his material to the German crystallographer and physicist Otto Lehmann (1855–1922), who in 1889 introduced the term **liquid crystal** to describe the intermediate phase lying between the solid phase and the normal liquid phase.

Some of them are long and fairly rigid, while others are disk shaped. The mesogen units can form ordered structures having long-range order, with the long axes of the mesogenic groups oriented in one preferred direction. The liquid-like properties arise from the fact that these structures can flow past one another quite readily. The solid-like properties arise because the structures themselves are not disturbed when the sliding occurs. Since the latent heat of the transition from solid to liquid crystal is always greater than that of the transition from liquid crystal to liquid, it can be said that the liquid crystal is more a liquid than a solid. However, as previously noted, polymer liquid crystals can have a high rigidity.

Structural work has shown that liquid crystals can be of two general classes. These are illustrated schematically in Figure 5.2, which compares the structures with those of the solid and liquid states. In a **nematic** liquid crystal the molecules are oriented, as in a solid, but there is no periodicity; the word *nematic* comes from the Greek word *nematos* meaning a thread. In a **smectic** liquid crystal there is also orientation, but a certain periodicity in that the molecules are arranged in layers. The word *smectic* comes from the Greek word for soap, and a familiar example of a smectic liquid crystal is the "goo" commonly found at the bottom of a soap dish.

Details of the behavior of liquid crystals have been investigated by computer simulation, on the basis of models in which the dispersion and other intermolecular forces are taken into account.

Phase Equilibria in a One-Component System: Water

The pressure-temperature diagram for water is well known, but what dictates the dependence of the equilibrium pressure on temperature for different phases? Only a thermodynamic analysis can give a quantitative expression for this relationship. In a reacting system, equilibrium is established when the change in Gibbs energy is zero (Section 3.6). In like manner, for two phases of the same pure substance to be in equilibrium under given conditions of temperature and pressure, the Gibbs energy of one phase must be equal to that of the other. In other words, the change in Gibbs energy between the two phases is zero. Specifically, when equilibrium occurs between ice and water

$$G(\mathrm{s}) = G(\mathrm{l}) \tag{5.1}$$

Solid state: orientation and periodicity

a.

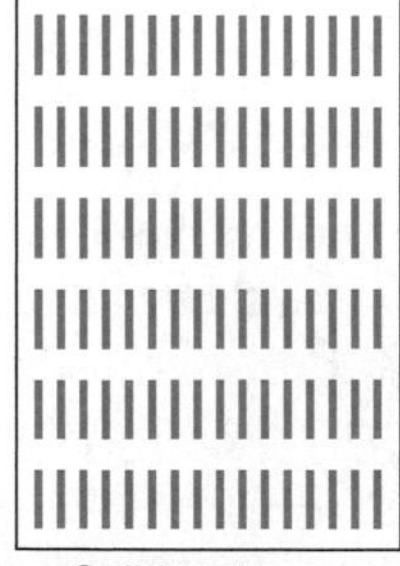
Liquid state: disorientation

b.

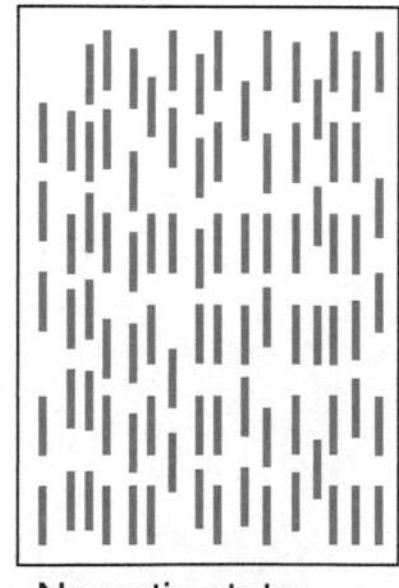
Nematic state: orientation without periodicity

c.

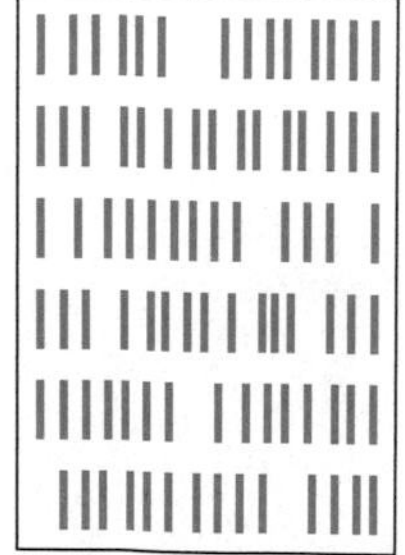
Smectic state: orientation with periodicity

d.

FIGURE 5.2
Two liquid-crystal states, shown schematically in comparison with the solid state (a) and the liquid state (b). The nematic state is shown in (c); the molecules are oriented in one direction, but there is no periodicity in their positions. In the smectic state (d) there is orientation, and also periodicity along the direction of orientation of the molecules.

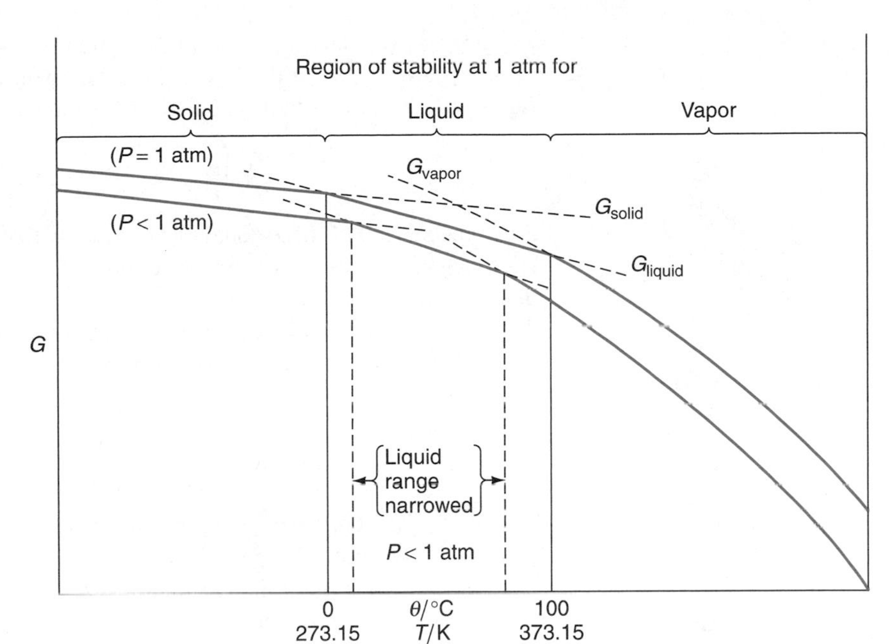

FIGURE 5.3
The variation of G with T for water. The upper solid line represents the variation of G at 1 atm pressure, and the lower line represents the variation of G at a reduced pressure. The figure shows the narrowing of the liquid range with a decrease in pressure.

and when there is equilibrium between water and steam

$$G(\mathrm{l}) = G(\mathrm{g}) \tag{5.2}$$

where s stands for solid, l for liquid, and g for vapor. At the triple point, both of these conditions apply. If only one phase is present under the particular P-T conditions, it will be the phase with the lowest value of G.

With Eqs. 5.1 and 5.2 as our criteria of equilibrium we now examine the equilibrium existing at 1 atm in the water system. Moving horizontally at 1 atm it is seen from Figure 5.1 that any one of three phases may be present, depending on the temperature of the system. This may be depicted in another way by plotting the variation of G with T for water at 1 atm, as shown in Figure 5.3. The curvatures for the ice, liquid, and vapor regions are all different, but the ice-liquid curves intersect at 0 °C, the normal melting point, and the liquid-vapor curves intersect at 100 °C, the normal boiling point. At these intersections Eqs. 5.1 and 5.2 apply and the respective phases are in equilibrium.

EXAMPLE 5.1 From the position of the curves representing G_{solid}, G_{liquid}, and G_{vapor}, predict the phases that are present (more stable) at 1 atm pressure as the temperature increases in Figure 5.3.

Solution At temperatures less than 0 °C, the curve representing G_{solid} has the lowest value of Gibbs energy. The solid is the most stable because this phase has the lowest Gibbs energy in this temperature range. Between 0 °C and 100 °C, the curve for G_{liquid} is lowest and the liquid is the most stable phase; and above 100 °C, the curve for G_{vapor} is lowest and the vapor is the most stable phase.

For an explanation of the slopes of the lines depicting individual phase equilibria in Figure 5.3, we utilize two of the Maxwell relations from Chapter 3, namely Eq. 3.119 (p. 129) written for 1 mol of substance:

$$\left(\frac{\partial G_m}{\partial P}\right)_T = V_m \quad \text{and} \quad \left(\frac{\partial G_m}{\partial T}\right)_P = -S_m \tag{5.3}$$

Since the second of these equations has a negative sign and the curves in Figure 5.3 have negative slopes, the entropy is positive. Moreover, the relative slopes show that $(S_m)_g > (S_m)_l > (S_m)_s$. Furthermore, the curves depicting the variation of G with T for each phase do not have the same curvature because $C_{P,m}$ for each phase is different. Then from the second equality in Eq. 5.3, take the second derivative of G with respect to T,

$$\left(\frac{\partial^2 G_m}{\partial T^2}\right)_P = -\left(\frac{\partial S_m}{\partial T}\right)_P = -\left(\frac{C_{P,m}}{T}\right) \tag{5.4}$$

we see that the curvature given by the second derivative increases as T increases since $C_{P,m}$ does not vary greatly with T.

From an analysis of the left-hand expression in Eq. 5.3, G must decrease as the pressure is decreased, since V_m is always positive. The effect of reducing the pressure from 1 atm is indicated by a narrowing of the liquid range as shown by the vertical dashed lines in Figure 5.3. Notice that the decrease in Gibbs energy for the gas phase is much greater than for the other two phases. This is because the molar volume of the gas V_m(g) is much larger compared to that of the liquid or solid. Note too that the reduction of pressure reduces the boiling point of all liquids. This is the basic reason for vacuum distillation of organic compounds: to cause the distillation in a temperature range over which the liquid is stable. Also in Figure 5.3 observe that there is an increase in the melting point for water as the pressure is reduced. The situation described here for water is different for most other substances, for which a reduction in pressure normally decreases the melting point of the substance.

Now consider an increase in pressure; we expect the Gibbs energy of a phase to increase. In a normal substance, $V_m(\text{l}) > V_m(\text{s})$, and thus the increase in pressure will cause a larger increase in Gibbs energy for the liquid than for the solid. In order to have equilibrium between the two phases at a higher pressure, ΔG must equal 0 and, to accomplish this, the melting point must shift to a higher value as shown in Figure 5.4a.

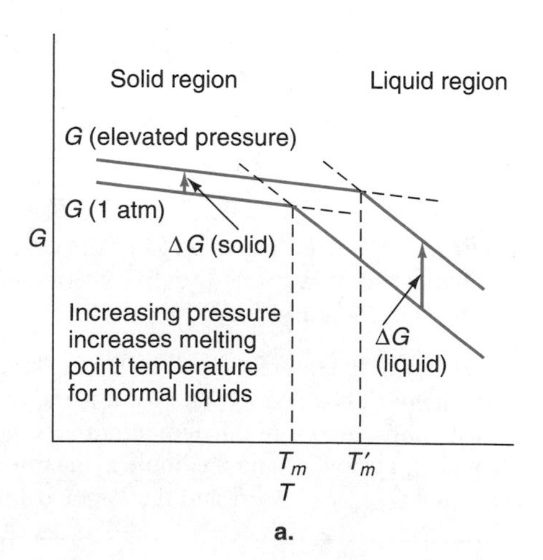

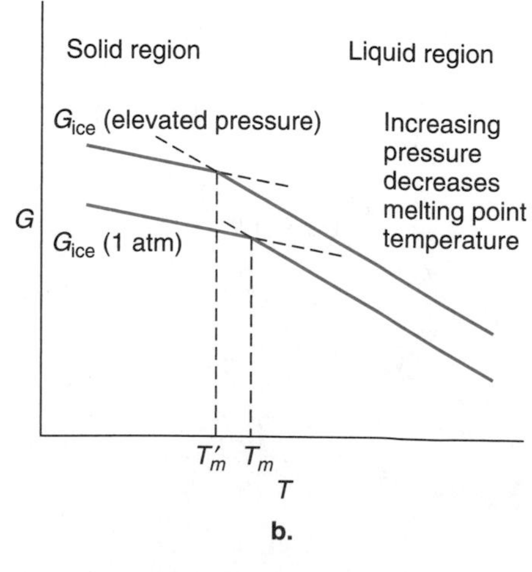

FIGURE 5.4
The effect of increased pressure on solid-liquid phase transitions. (a) Normal solid-liquid transitions; (b) ice-water transition. Dashed lines indicate hypothetical value of G of phases at temperatures where the phases are not stable. The phase with the lowest value of G is the most stable at a particular temperature.

(The phase with the lowest Gibbs energy at a particular temperature is the stable phase.) Just the opposite result is obtained for water. Since the molar volume of ice is larger than that of liquid water, the increase in G will be greater for the solid form; thus the melting point is decreased as shown in Figure 5.4b.

If the pressure is reduced far enough below that shown in Figure 5.3, the system will reach the triple point and all three phases will be at equilibrium. Below this pressure the curve for the gas will intersect that for the solid phase and the liquid phase is bypassed. The practical result is that the solid will pass directly into the gaseous state, a process called **sublimation.** Again the Gibbs energies of the two phases must be equal for equilibrium to exist.

Sublimation

5.2 Vaporization and Vapor Pressure

Thermodynamics of Vapor Pressure: The Clapeyron Equation

In Section 5.1 we found that the criterion for two phases of a pure substance to coexist is that their Gibbs energies must be equal at a given temperature and pressure. However, if either T or P is varied, with the other held constant, one of the phases will disappear. Thus, if we are at a particular point on one of the phase-equilibria lines on a P-T diagram (gas-liquid, liquid-solid, solid-gas, or even solid-solid), our problem becomes how to vary P and T while maintaining equilibrium. In 1834 Benoit Clapeyron (1799–1864) published a solution for the case of a liquid and its vapor. We investigate the vapor pressure because, since the vapor generally behaves as an ideal gas, it provides, through thermodynamics, one of the most important means of determining the properties of either a solid or a liquid.

We begin with the statement of phase equilibrium, Eq. 5.2, written for a single pure substance in the liquid and vapor states. If the pressure and temperature are changed infinitesimally in such a way that equilibrium is maintained,

$$dG_{\text{v}} = dG_{\text{l}} \tag{5.5}$$

Since this expression depends only on T and P, the total derivatives may be written as

$$\left(\frac{\partial G_{\text{v}}}{\partial P}\right)_T dP + \left(\frac{\partial G_{\text{v}}}{\partial T}\right)_P dT = \left(\frac{\partial G_{\text{l}}}{\partial P}\right)_T dP + \left(\frac{\partial G_{\text{l}}}{\partial T}\right)_P dT \tag{5.6}$$

From Eq. 5.3 we may substitute for these partials, obtaining

$$V_m(\text{v})\, dP - S_m(\text{v})\, dT = V_m(\text{l})\, dP - S_m(\text{l})\, dT \tag{5.7}$$

or

$$\frac{dP}{dT} = \frac{S_m(\text{v}) - S_m(\text{l})}{V_m(\text{v}) - V_m(\text{l})} = \frac{\Delta S_m}{\Delta V_m} \tag{5.8}$$

If the molar enthalpy change at constant pressure for the phase transformation is ΔH_m, the term ΔS_m may be written as $\Delta H_m/T$ and Eq. 5.8 becomes

$$\frac{dP}{dT} = \frac{\Delta H_m}{T\Delta V_m} \tag{5.9}$$

Clapeyron Equation

The general form of Eq. 5.9 is known as the **Clapeyron equation** and may be applied to vaporization, sublimation, fusion, or solid phase transitions of a pure substance. Its derivation involves no assumptions and is thus valid for the general process of equilibrium between any two phases of the same substance.

In order to use the Clapeyron equation the molar enthalpy of the process must be known along with the equilibrium vapor pressure. If the two phases are condensed phases (solid or liquid), then the pressure is the mechanical pressure established. To integrate this expression exactly, both ΔH_m and ΔV_m must be known as functions of temperature and pressure; however, they may be considered constant over short temperature ranges. For solid-to-liquid transitions, ΔH_m is almost invariably positive; ΔV_m is usually positive except in a few cases, such as H_2O, Bi, and Ga, where it is negative.

It should be pointed out that the enthalpies of sublimation, fusion, and vaporization are related at constant temperature by the expression

$$\Delta_{\text{sub}}H_m = \Delta_{\text{fus}}H_m + \Delta_{\text{vap}}H_m \tag{5.10}$$

This is as expected since enthalpy is a state function and the same amount of heat is required to vaporize a solid directly as is required to go through an intermediate melting stage to reach the final vapor state.

EXAMPLE 5.2 What is the expected boiling point of water at 98.7 kPa (approximately 740 Torr, a typical barometric pressure at 275 m altitude)? The heat of vaporization is 2258 J g^{-1}, the molar volume of liquid water is 18.78 cm^3 mol^{-1}, and the molar volume of steam is 30.199 dm^3 mol^{-1}, all values referring to 373.1 K and 101.325 kPa (1 atm).

Solution Apply Eq. 5.9 to find the change in boiling point for 1 Pa and then multiply by the difference between the given pressure and the standard atmosphere:

$$\frac{dP}{dT} = \frac{\Delta_{\text{vap}}H_m}{T_b[V_m(\text{v}) - V_m(\text{l})]}$$

where T_b refers to the temperature at boiling.

$$\begin{aligned}\frac{dP}{dT} &= \frac{2258(\text{J g}^{-1}) \times 18.01(\text{g mol}^{-1})}{373.1(\text{K}) \times [30.199(\text{dm}^3\ \text{mol}^{-1}) - 0.019(\text{dm}^3\ \text{mol}^{-1})]}\\ &= 3.614\ \text{J K}^{-1}\ \text{dm}^{-3} = 3.614 \times 10^3\ \text{J m}^{-3}\ \text{K}^{-1}\\ &= 3.614 \times 10^3\ \text{Pa K}^{-1} = 3.614\ \text{kPa K}^{-1}\end{aligned}$$

or

$$\frac{dT}{dP} = 2.767 \times 10^{-4}\ \text{K Pa}^{-1}$$

For a decrease of 101.325 kPa − 98.7 kPa = 2.625 kPa, there is a decrease in temperature of

$$2.767 \times 10^{-4}\ \text{K Pa}^{-1} \times 2625\ \text{Pa} = 0.73\ \text{K}$$

Therefore the new boiling point is

$$373.15\ \text{K} - 0.73\ \text{K} = 372.42\ \text{K}$$

EXAMPLE 5.3 Determine the change in the freezing point of ice with increasing pressure. The molar volume of water is 18.02 cm^3 mol^{-1} and the molar volume of ice is 19.63 cm^3 mol^{-1} at 273.15 K. The molar heat of fusion $\Delta_{fus}H_m = 6.009 \times 10^3$ J mol^{-1}.

Solution Equation 5.9 applies to all phase equilibria. For a fusion process we have

$$\frac{dP}{dT} = \frac{\Delta_{fus}H_m}{T_m \Delta_{fus}V_m}$$

where the subscript fus refers to the value of the variables at the melting point. Here, since the pressure is not the equilibrium value, P refers to the applied pressure maintained mechanically or through an inert gas. Also $\Delta_{fus}V_m$ is the molar volume difference $V_m(\text{v}) - V_m(\text{l})$ and is assumed to be approximately constant over a moderate pressure range.

$$\frac{dP}{dT} = \frac{6.009 \times 10^3(\text{J mol}^{-1})}{273.15(\text{K}) \times [0.018\,02(\text{dm}^3\ \text{mol}^{-1}) - 0.019\,65(\text{dm}^3\ \text{mol}^{-1})]}$$

$$= -13\,498\ \text{J K}^{-1}\ \text{dm}^{-3} = -13\,498\ \text{kPa K}^{-1}$$

$$\frac{dT}{dP} = -7.408 \times 10^{-5}\ \text{K kPa}^{-1} = -7.408 \times 10^{-3}\ \text{K bar}^{-1}$$

This is a 0.74 K decrease in temperature per 100 bar increase in pressure. Students are advised to check the units conversion above.

The Clausius-Clapeyron Equation

Some 30 years after Clapeyron introduced this equation, Clausius introduced a modification that improved the versatility of the expression. When one of the phases in equilibrium is a vapor phase, we assume that $V_m(\text{v})$ is so much larger than $V_m(\text{l})$ that we may neglect $V_m(\text{l})$ in comparison to $V_m(\text{v})$ when the pressure is near 1 bar. (For water, the volume of vapor is at least a thousand times that of a liquid. See Example 5.2.) The second assumption is to replace $V_m(\text{v})$ by its equivalent from the ideal gas law RT/P. Equation 5.9 then becomes

$$\frac{dP}{dT} = \frac{\Delta_{vap}H_m P}{RT^2} \tag{5.11}$$

or if the standard pressure is 1 atm,[1]

$$\frac{dP}{P\,dT} = \frac{d\ln(P/\text{atm})}{dT} = \frac{\Delta_{vap}H_m}{RT^2} \tag{5.12}$$

Clausius-Clapeyron Equation

This expression is known as the **Clausius-Clapeyron equation.** For pressure expressed in other units, P^u is used to indicate the numerical value of the pressure. In

[1]The bar is recommended as the reference pressure for reporting thermodynamic data. The atmosphere (atm) is still used when reporting normal boiling and freezing points.

order to make use of Eq. 5.12, integration is performed assuming $\Delta_{vap}H_m$ to be independent of temperature and pressure. We thus obtain

$$\int d \ln P^u = \frac{\Delta_{vap}H_m}{R} \int T^{-2}\, dT \tag{5.13}$$

$$\ln P^u = -\frac{\Delta_{vap}H_m}{RT} + C \tag{5.14}$$

where C is a constant of integration. A plot of ln P^u against $1/T$ should be linear. The slope of the line is $-\Delta_{vap}H_m/R$. A plot of R ln P^u against $1/T$ could also be made in which case the intercept on the R ln P^u axis (i.e., where the temperature is infinitely high) will give the value of $\Delta_{vap}S_m$. This is shown in Figure 5.5 for five liquids, the dotted lines representing the extrapolation of data. As might be expected over a wide range of temperature, deviations from linearity occur as a result of the temperature variation of $\Delta_{vap}H_m$ and of nonideal gas behavior of the vapor.

Another useful form of Eq. 5.14 may be obtained by integrating between specific limits:

$$\int_{P_1}^{P_2} d \ln P^u = \frac{\Delta_{vap}H_m}{R} \int_{T_1}^{T_2} T^{-2}\, dT \tag{5.15}$$

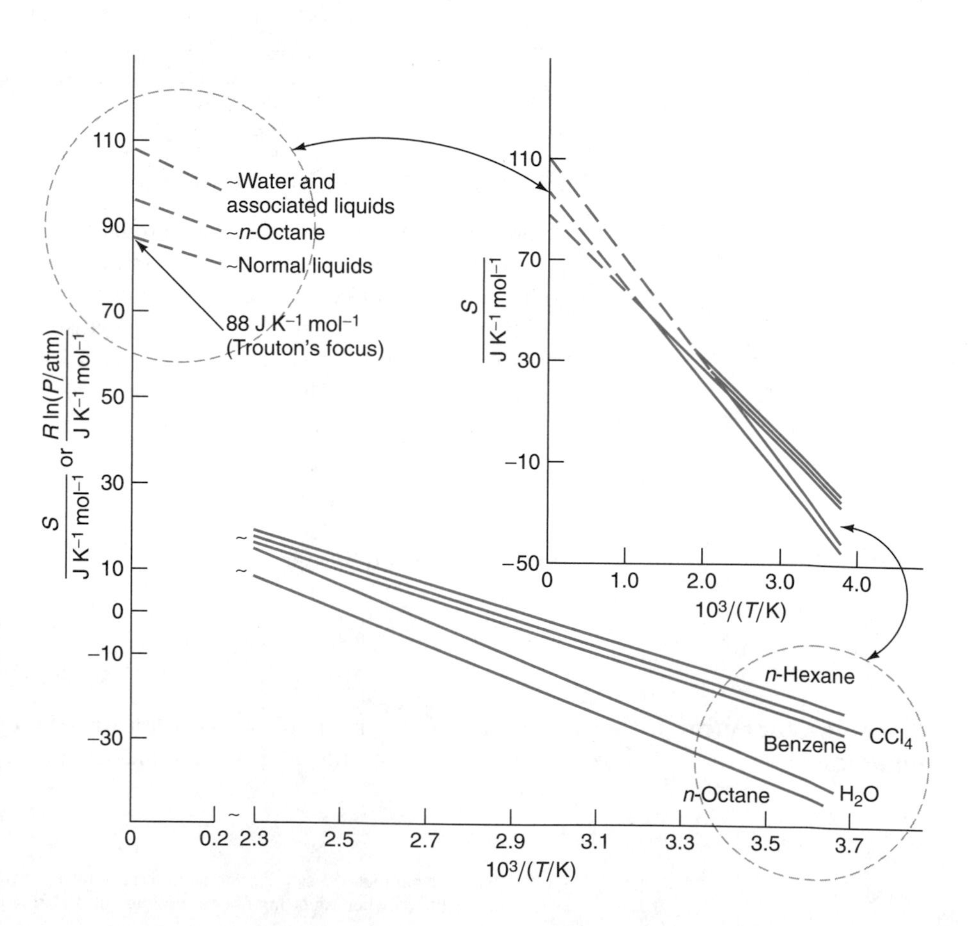

FIGURE 5.5
Plot of R ln(P/atm) against $1/T$ for five liquids. The right-hand portion gives experimental points where the slope is $-\Delta_{vap}H_m$. The left-hand portion shows the extrapolation of the data to Trouton's focus, discussed on page 191. For normal liquids, the intercept is about 88 J K^{-1} mol^{-1}. The inset shows the overall plot in perspective.

$$\ln \frac{P_2}{P_1} = \frac{\Delta_{\text{vap}} H_m}{R}\left(\frac{1}{T_1} - \frac{1}{T_2}\right) \tag{5.16}$$

It is immediately evident that one may calculate $\Delta_{\text{vap}}H_m$ (or $\Delta_{\text{sub}}H_m$ for sublimation from a similar expression) by measuring the vapor pressure of a substance at two different temperatures. It is generally convenient to let $P_1 = 1$ atm so that $T_1 = T_b$. Even if curvature exists in the plot, the enthalpy change may be calculated by drawing a tangent to the curve at the temperature of interest. Fairly good results are obtained if the equilibrium vapor density is not too high.

EXAMPLE 5.4 Benzene has a normal boiling point at 760 Torr of 353.25 K and $\Delta_{\text{vap}}H_m = 30.76$ kJ mol^{-1}. If benzene is to be boiled at 30.00 °C in a vacuum distillation, to what value of P must the pressure be lowered?

Solution In order to boil benzene its vapor pressure must equal the pressure on the system. The problem is thus a matter of finding the vapor pressure of benzene at 30.00 °C. We may use the Clausius-Clapeyron equation considering $\Delta_{\text{vap}}H_m$ constant over the temperature range. $T_2/\text{K} = 30.00\ °\text{C}/°\text{C} + 273.15 = 303.15$.

Using Eq. 5.16, we have

$$\ln \frac{P_2}{760.0} = \frac{\Delta_{\text{vap}}H}{8.3145\ \text{J mol}^{-1}\ \text{K}^{-1}}\left(\frac{1}{353.25} - \frac{1}{303.15}\right)$$

$$= -1.7309$$

$$\ln P_2 = 4.9024; \quad P_2 = 134.6\ \text{Torr}$$

Enthalpy and Entropy of Vaporization: Trouton's Rule

In Section 3.4, p. 191, we saw that entropy values may be obtained from the expression

$$\Delta_{\text{trs}}S = \frac{\Delta_{\text{trs}}H}{T_{\text{trs}}} \tag{5.17}$$

where trs specifics a particular transition. There are no easy generalizations to suggest even approximate values of the three quantities appearing in Eq. 5.17 when the transition process is fusion, although the melting point is easily determined experimentally. However, the entropies of vaporization $\Delta_{\text{vap}}S_m$ of most non-hydrogen-bonded compounds have values of $\Delta_{\text{vap}}S_m$ in the neighborhood of 88 J K^{-1} mol^{-1}. This generalization is known as **Trouton's rule** and was pointed out in 1884:

$$\frac{\Delta_{\text{vap}}H_m}{T_b} = \Delta_{\text{vap}}S_m \approx 88\ \text{J K}^{-1}\ \text{mol}^{-1} \tag{5.18}$$

The intersection of lines on the $R \ln(P/\text{atm})$ axis in Figure 5.5 at approximately 88 J mol^{-1} K^{-1} is known as *Trouton's focus*. Table 5.1 lists for a number

Trouton's Rule

Frederick Thomas Trouton (1863–1922), an Irish physicist born in Dublin, held academic positions there and in London. His main work was on electricity and on Maxwell's theory of electromagnetic radiation. He somewhat deprecated his work on Trouton's rule, saying that he had done it in an afternoon.

TABLE 5.1 Enthalpies and Entropies of Vaporization and Fusion

	Liquid ⇌ Vapor			Solid ⇌ Liquid		
Substance	T_b K	$\Delta_{vap}H_m$ kJ mol^{-1}	$\Delta_{vap}S_m$ J K^{-1} mol^{-1}	T_m K	$\Delta_{fus}H_m$ kJ mol^{-1}	$\Delta_{fus}S_m$ J K^{-1} mol^{-1}
He	4.20	0.084	19.66	3.45	0.021	6.28
H_2	20.38	0.904	44.35	13.95	0.117	8.37
N_2	77.33	1.777	72.13	63.14	0.720	11.38
O_2	90.18	6.820	75.60	54.39	0.444	8.16
H_2O	373.15	40.656	108.951	273.15	6.009	22.096
SO_2	263.13	24.916	94.68	197.48	7.401	37.45
CH_4	111.16	8.180	73.26	190.67	.941	10.38
C_2H_6	184.52	14.715	79.75	89.88	2.858	31.80
CH_3OH	337.85	35.27	104.39	175.25	3.167	18.07
C_2H_5OH	351.65	38.58	109.70	158.55	5.021	31.67
n-C_4H_{10}	272.65	22.40	82.13	134.80	4.661	34.572
C_6H_6	353.25	30.76	87.07	278.68	10.590	35.296
C_7H_8 (toluene)	383.77	33.48	87.19			
CH_3COOH	391.45	24.35	61.92	289.76	11.72	40.42

of substances the values of the enthalpy and entropy of both fusion and vaporization. The average of $\Delta_{vap}S_m$ for a large number of liquids that are not appreciably hydrogen-bonded bears out the value given in Eq. 5.18.

The effect of hydrogen bonding is seen in the case of water. The abnormally low value for acetic acid may be explained by its appreciable association in the vapor state. Allowing for its apparent molecular weight of about 100 in the vapor state, acetic acid will have a value of $\Delta_{vap}S_m$ of approximately 100 J K^{-1} mol^{-1}, in line with other associated liquids.

Hildebrand Rule

An alternative rule, known as the **Hildebrand rule** (1915, 1918) and named after the American physical chemist Joel Henry Hildebrand (1881–1983), states that the entropies of vaporization of unassociated liquids are equal, not at their boiling points, but at temperatures at which the vapors occupy equal volumes.

The accuracy of the Hildebrand rule indicates that, under the conditions of comparison, all liquids with fairly symmetrical molecules possess equal amounts of configurational entropy. This means that, on the molecular level, maximum molecular disorder exists in a liquid that adheres to the rule.

EXAMPLE 5.5 Estimate the enthalpy of vaporization of CS_2 if its boiling point is 319.40 K.

Solution From Eq. 5.18,

$$\Delta_{vap}H_m = \Delta_{vap}S_m \times T_b$$

$$\Delta_{vap}H_m = 88 \text{ J K}^{-1} \text{ mol}^{-1} \times 319.40 \text{ K} = 28.11 \text{ kJ mol}^{-1}$$

The experimental value is 28.40 kJ mol^{-1}.

Normally, boiling points are recorded at 101.325 kPa (1 atm), whereas they seldom are obtained experimentally under this exact pressure. A useful expression for correcting the boiling point to the standard pressure was derived in 1887 by James Mason Crafts (1839–1917) by combining Eq. 5.11 with Trouton's rule. Assuming that dP/dT is approximately $\Delta P/\Delta T$, we obtain

Crafts' Rule

$$\frac{\Delta P}{\Delta T} = \frac{\Delta_{\text{vap}} H_m}{T_b} \times \frac{P}{RT_b} \tag{5.19}$$

where T_b is the normal boiling point. For ordinary liquids we may substitute $\Delta_{\text{vap}} S_m = 88 \text{ J K}^{-1} \text{ mol}^{-1}$ for $\Delta_{\text{vap}} H_m / T_b$:

$$\frac{\Delta P}{\Delta T} \approx \frac{88(\text{J K}^{-1} \text{ mol}^{-1}) \times 101.3(\text{kPa})}{8.3145(\text{J K}^{-1} \text{ mol}^{-1}) T_b} = \frac{1072}{T_b} \text{ kPa} \tag{5.20}$$

or

$$\Delta T \approx 9.3 \times 10^{-4} T_b \Delta P/\text{kPa}$$

For associated liquids, a numerical coefficient of 7.5×10^{-4} gives better results than 9.3×10^{-4}, which is used for normal liquids.

Variation of Vapor Pressure with External Pressure

In the preceding sections the saturated vapor pressure of a pure liquid was considered as a function of temperature alone. However, if an external pressure is applied in addition to the saturated vapor pressure, the vapor pressure becomes a function of pressure as well as of temperature. The increased pressure may be applied by adding an inert gas that is insoluble in the liquid or through the action of a piston that is permeable to the gas.

Gibbs Equation

For a closed system, with T constant and the total pressure P_t equal to the pressure exerted on the liquid, Eq. 5.7 may be written in a form known as the **Gibbs equation:**

$$V_m(\text{v})\, dP = V_m(\text{l})\, dP_t \quad \text{or} \quad \frac{dP}{dP_t} = \frac{V_m(\text{l})}{V_m(\text{v})} \tag{5.21}$$

where P is the pressure of the vapor under the total pressure P_t. If we assume that the vapor phase behaves ideally, we may substitute $V_m(\text{v}) - RT/P$, obtaining

$$\frac{d \ln P^u}{dP_t} = \frac{V_m(\text{l})}{RT} \tag{5.22}$$

Since the molar volume of a liquid does not change significantly with pressure, we may assume $V_m(\text{l})$ to be constant through the range of pressure and obtain, by integrating,

$$\ln \frac{P}{P_\text{v}} = \frac{V_m(\text{l})}{RT}(P_t - P_\text{v}) \tag{5.23}$$

where P_v is the saturated vapor pressure of the pure liquid without an external pressure. Note that in this case the vapor pressure is now dependent on the total pressure as well as on the temperature. This is because the two phases exist under different pressures.

EXAMPLE 5.6 The vapor pressure of water without the presence of any other gas, such as air, at 25 °C is 3167 Pa. Calculate the vapor pressure of water when the *head space,* that is, the enclosed volume above the water, already contains an insoluble gas at a pressure of 1 bar. This is an important problem in the bottling industry in the design of beverage containers.

Solution Apply Eq. 5.23, where P_v is 3167 Pa, and P_t is 1.000×10^5 Pa (1 bar). The molar volume of liquid water can be found from its density at 25.0 °C, 0.9971 g cm^{-3}. With a molar mass of 18.02 g mol^{-1},

$$\text{Molar volume of water} = \frac{18.02 \text{ g mol}^{-1}}{0.9971 \text{ g cm}^{-3}} = 18.07 \text{ cm}^3 \text{ mol}^{-1}$$
$$= 18.07 \times 10^{-6} \text{ m}^3 \text{ mol}^{-1}$$

Then

$$\ln\left(\frac{P/\text{Pa}}{3167}\right) = \frac{18.07 \times 10^{-6} \text{ m}^3 \text{ mol}^{-1} \times 1.000 \times 10^5 \text{ Pa}}{8.3145 \text{ J mol}^{-1} \text{ K}^{-1} \times 298.15 \text{ K}}$$

$$\ln P/\text{Pa} = 7.290 \times 10^{-4} + \ln 3167 = 7.29 \times 10^{-4} + 8.06054$$
$$= 8.0613$$
$$P = 3169 \text{ Pa}$$

This shows that the presence of the insoluble (inert) gas has very little effect on the water vapor pressure. Students should convince themselves that the units cancel in the expression for ln(P/Pa).

5.3 Classification of Transitions in Single-Component Systems

We have seen in some detail the nature of transitions involving vaporization, fusion, and sublimation. In each of these the transition occurs without a change in temperature as the enthalpy changes. Other types of phase transitions also occur, such as polymorphic transitions in solids (i.e., changes from one crystalline structure to another). The character of the enthalpy and other changes is quite different in these transformations.

First-Order Phase Transition

In order to classify the transitions, a scheme was devised by Paul Ehrenfest (1880–1933). According to his classification, phase transitions are said to be **first-order phase transitions** if the molar Gibbs energy is continuous from one phase to another, but the derivatives of the Gibbs energy

$$S_m = -(\partial G_m/\partial T)_P \quad \text{and} \quad V_m = (\partial G_m/\partial P)_T \tag{5.24}$$

are discontinuous. For example, at the melting point of ice, even though the Gibbs energies of the liquid water and ice are equal, as required for equilibrium, there is a change in the slope of the G against T curve as seen in Figure 5.3. There is consequently a break in the S against T curve and also in the V against T curve, since

$$(\partial G_{\text{water}}/\partial T)_P - (\partial G_{\text{ice}}/\partial T)_P = -S_{\text{water}} + S_{\text{ice}} = -\Delta S$$

and

$$(\partial G_{\text{water}}/\partial P)_T - (\partial G_{\text{ice}}/\partial P)_T = V_{\text{water}} - V_{\text{ice}} = \Delta V.$$

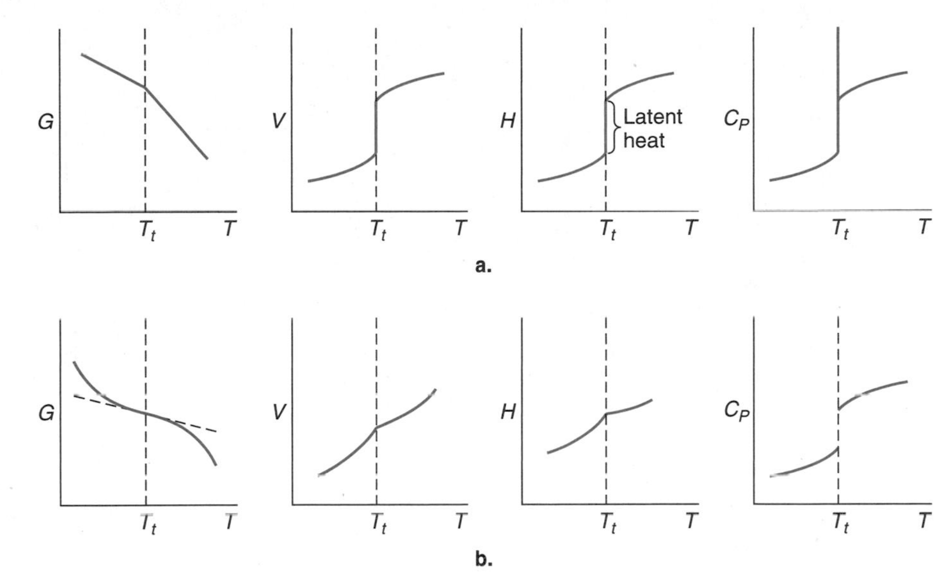

FIGURE 5.6
(a) Schematic plots for G, V, H, and C_P for a first-order transition such as melting and vaporization. (b) Schematic plots for G, V, H, and C_P for a second-order transition.

This is plotted in the general case in Figure 5.6 along with plots of ΔH, and C_P against T for first-order and second-order transitions. The infinite value of C_P at 273.15 K for the transition in water is related to the change in enthalpy for the transition. There is an increase in heat, while the change in temperature is zero. Consequently, since $C_P = (\partial H/\partial T)_P$, its value is infinity at the transition temperature. There is quite different behavior in higher order transitions.

Although the Ehrenfest classification, involving finite jumps in entropy and volume, works well for first-order transitions such as vaporization, most higher order phase transitions have discontinuities at which the second or higher derivatives are infinite. The more general theory developed by Laszlo Tisza (b. 1907) is applicable to such transitions. According to Tisza's theory, a **second-order phase transition** is one in which the discontinuity appears in the thermodynamic properties that are expressed as the second derivative of the Gibbs energy. For second-order transitions, there is no latent heat although there is a finite discontinuity in the heat capacity. In addition the enthalpy and entropy are continuous functions, although their first derivatives, $C_P = (\partial H/\partial T)_P$ and $C_P = T(\partial S/\partial T)_P$, are not. Thus, C_P shows a finite change at the transition temperature and fulfills Tisza's definition since

Second-Order Phase Transition

$$C_P = T(\partial S/\partial T)_P = -T(\partial^2 G/\partial T^2)_P \tag{5.25}$$

An example of this type of transition occurs in the change from the conducting to superconducting state (Section 16.6) in some metals at temperatures near 20 K.

Another common transition, called the *lambda transition*, is one in which ΔS and ΔV are zero but the heat capacity becomes infinite in the form of the Greek letter lambda (λ). This is illustrated in Figure 5.7a. A plot of the actual data for the entropy of a system exhibiting a lambda transition is shown in Figure 5.7b. Unlike the case in a first-order transition, the heat capacity can be measured as close to the transition temperature as possible and is found still to be rising. The enthalpy is continuous but has a vertical inflection point. The transition between He I and He II is an example of a lambda transition that occurs at about 2.2 K. He II has very unusual properties including a practically zero viscosity, leading it to be called a *superfluid*.

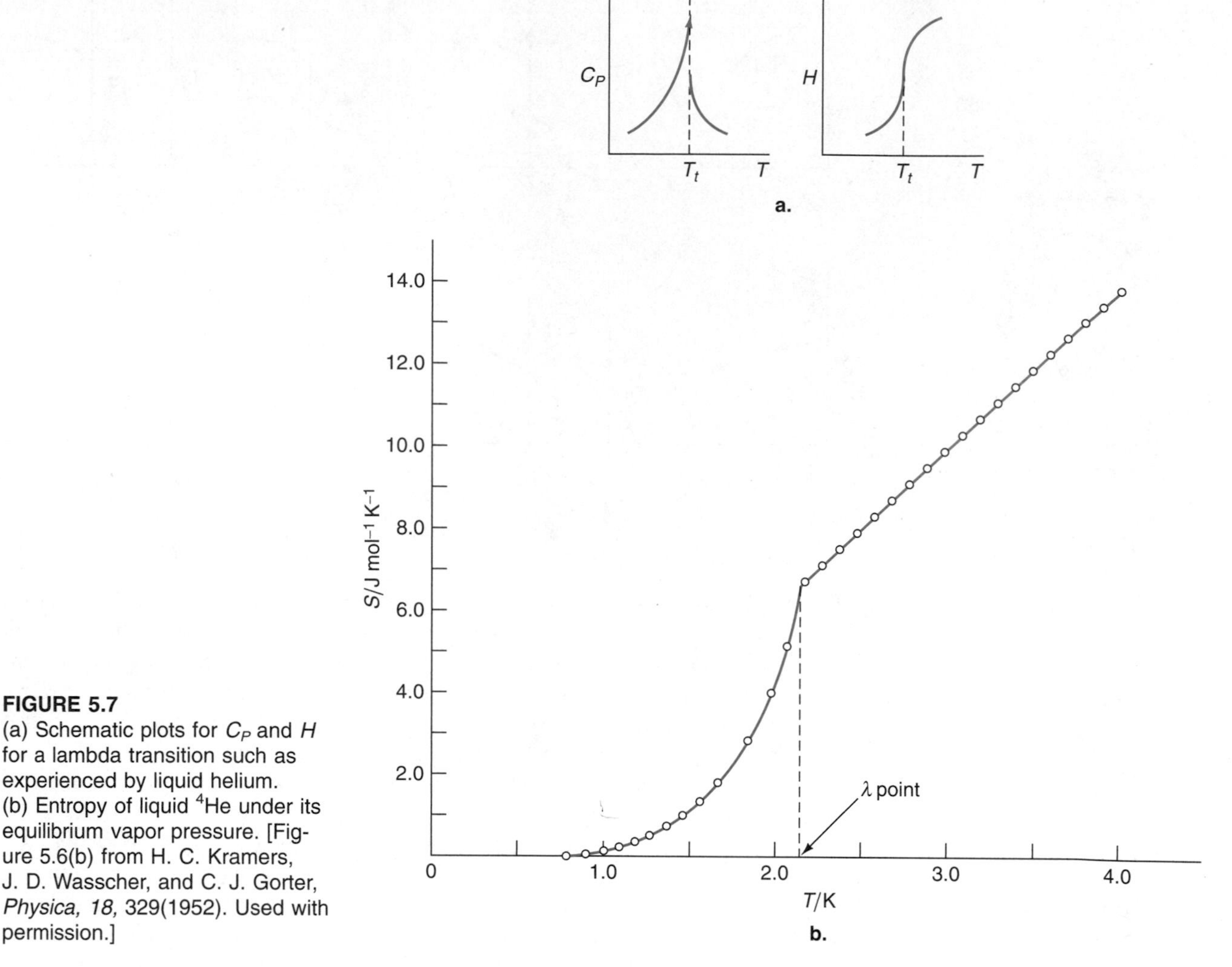

FIGURE 5.7
(a) Schematic plots for C_P and H for a lambda transition such as experienced by liquid helium.
(b) Entropy of liquid ^{4}He under its equilibrium vapor pressure. [Figure 5.6(b) from H. C. Kramers, J. D. Wasscher, and C. J. Gorter, *Physica, 18,* 329(1952). Used with permission.]

5.4 Ideal Solutions: Raoult's and Henry's Laws

In the earlier sections of this chapter we have been primarily concerned with phase equilibria of one-component systems. When we consider a two-component system with variable composition, a way must be found to represent the composition variable. Since three-dimensional plots generally are difficult to work with, most variable composition equilibria are represented either at constant temperature, with pressure and composition as the variables, or at constant pressure, with temperature and composition as the variables.

We generally speak of solutions when discussing systems of variable composition. A *solution* is any homogeneous phase that contains more than one component. Although there is no fundamental difference between components in a solution, we call the component that constitutes the larger proportion of the solution the *solvent.* The component in lesser proportion is called the *solute.*

In Chapter 1 we studied the behavior of gases and characterized them as ideal or nonideal on the basis of whether or not they obeyed a particularly simple mathematical expression, the ideal gas law. When examining the behavior of liquid-vapor equilibria of solutions, we are naturally inclined to ask the question: Is there an ideal solution? We look for what we may call an ideal solution with the hope that the concept may lead us to a better understanding of solutions in general, just as the concept of the ideal gas has helped in understanding real gases.

Begin by imagining a solution of molecules of A and B. If the molecular sizes are the same, and the intermolecular attractions of A for A and of B for B are the same as the attraction of A for B, then we may expect the most simple behavior possible from a solution. Thus the solution is considered to be ideal when there is a complete uniformity of intermolecular forces, arising from similarity in molecular size and structure; a thermodynamic definition will be given later.

The partial vapor pressures of the individual components within the solution are a good measure of the behavior of the individual components. This is because the partial vapor pressure measures the escaping tendency of a molecule from solution, which is in turn a measure of the cohesive forces present in solution. Thus, measurements of the vapor pressure of each component as a function of pressure, temperature, and mole fraction lead to a good understanding of the system.

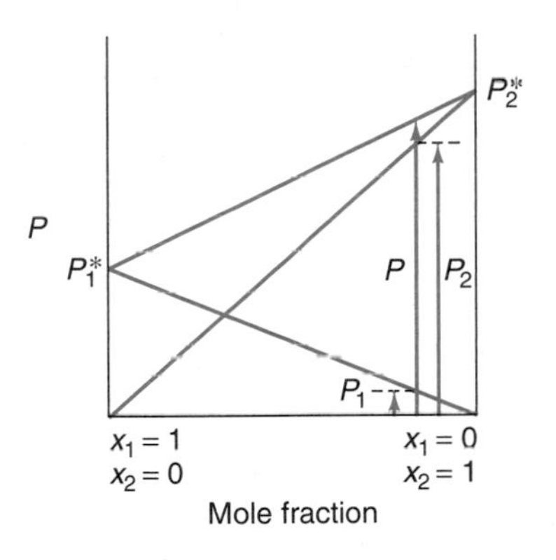

FIGURE 5.8
Vapor pressure of two liquids obeying Raoult's law. The total pressure at constant temperature is obtained from $P = x_1P_1^* + (1 - x_1)P_2^*$.

François Marie Raoult (1830–1916) formed a generalization for an ideal solution known as **Raoult's law.** According to Raoult's law, the vapor pressure P_1 of solvent 1 is equal to its mole fraction in the solution multiplied by the vapor pressure P_1^* of the pure solvent. The superscript * indicates that the substance is at the temperature of the solution but not necessarily at the normal reference temperature. A similar expression holds for substance 2 in the mixture. Mathematically this may be stated as

$$P_1 = x_1P_1^*; \qquad P_2 = x_2P_2^* \tag{5.26}$$

If the solution has partial vapor pressures that follow Eq. 5.26, we say that the solution obeys Raoult's law and behaves ideally.

This behavior is shown in Figure 5.8. A number of pairs of liquids obey Raoult's law over a wide range of compositions. Among these solutions are benzene–toluene, ethylene bromide–ethylene chloride, carbon tetrachloride–trichloroethylene, and acetic acid–isobornyl acetate. It should be noted that air is generally present above the solution at a pressure that makes up the difference between the total vapor pressure P_t and atmospheric pressure. The fact that air is present can generally be ignored in the vapor phase.

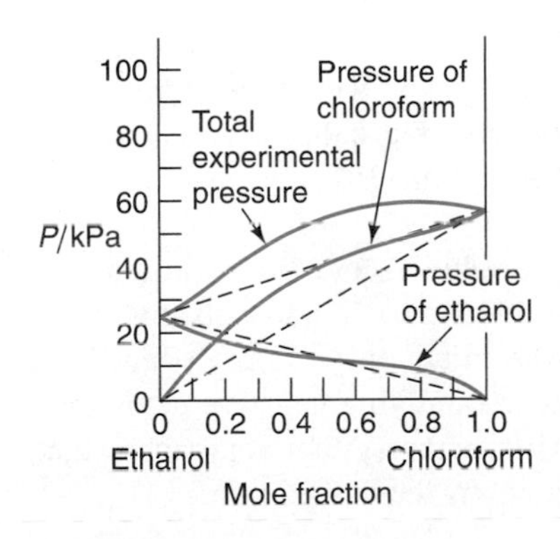

FIGURE 5.9
Vapor pressure of the system ethanol–chloroform at 318 K showing positive deviations from Raoult's law. Dashed lines show Raoult's law behavior.

Deviations from Raoult's law do occur and may be explained if we consider again the interaction between molecules 1 and 2. If the strength of the interaction between like molecules, 1–1 or 2–2, is greater than that between 1 and 2, the tendency will be to force both components into the vapor phase. This increases the pressure above what is predicted by Raoult's law and is known as a *positive deviation.* As an illustration, consider a drop of water initially dispersed in a container of oil. The attraction of oil molecules and of water molecules for their own kind is great enough that the dispersed water is re-formed into a droplet and is excluded from solution. In the same manner, when 1–1 or 2–2 interactions are strong, both 1 and 2 are excluded from solutions and enter the vapor state. Figure 5.9 illustrates positive deviations from Raoult's law for a binary system of chloroform–ethanol at 318 K. Many other binary systems consisting of dissimilar liquids show positive

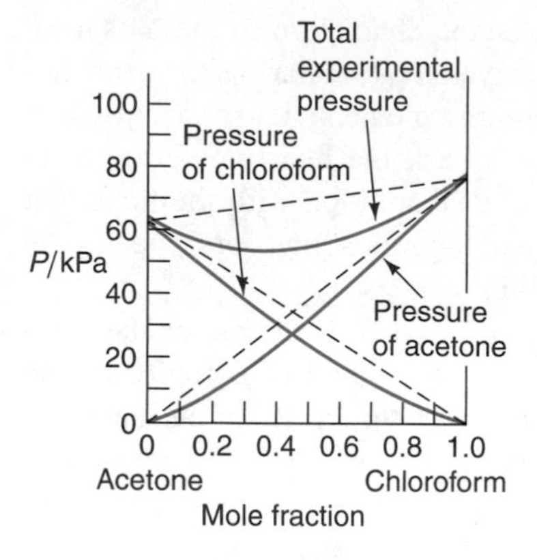

FIGURE 5.10
Vapor pressure of the system acetone–chloroform at 328 K showing negative deviations from Raoult's law. Dashed lines show Raoult's law behavior.

deviations. In addition to the aforementioned, negative deviations occur when the attractions between components 1 and 2 are strong. This may be visualized as a holding back of molecules that would otherwise go into the vapor state. Figure 5.10 shows one example of *negative deviation* from Raoult's law.

In examining Figures 5.9 and 5.10 we notice that, in the regions where the solutions are dilute, the vapor pressure of the solvent in greatest concentration approaches ideal behavior, which is shown by the dashed lines. Therefore, in the limit of infinite dilution the vapor pressure of the *solvent* obeys Raoult's law.

Henry's Law

Another property of binary systems was discovered by the English physical chemist William Henry (1774–1836). He found that the mass of gas m_2 dissolved by a given volume of solvent at constant temperature is proportional to the pressure of the gas in equilibrium with the solution. Mathematically stated,

$$m_2 = k_2 P_2 \tag{5.27}$$

where generally the subscript 2 is used to refer to the solute (the subscript 1 is used to refer to the solvent) and k is the *Henry's law constant.* Most gases obey Henry's law when the temperatures are not too low and the pressures are moderate.

If several gases from a mixture of gases dissolve in a solution, Henry's law applies to each gas independently, regardless of the pressure of the other gases present in the mixture.

Although Eq. 5.27 is the historical form of Henry's law, since mass per unit volume is equal to a concentration term, we may write Eq. 5.27 as

$$P_2 = k'x_2 \quad \text{or} \quad P_2 = k''c_2 \tag{5.28}$$

where for dilute solutions the concentration c_2 is proportional to the mole fraction of dissolved substance. Equation 5.28, in either form, is now often referred to as **Henry's law.**

Since Raoult's law may be applied to the volatile solute as well as to the volatile solvent, the pressure P_2 of the volatile solute in equilibrium with a solution in which the solute has a mole fraction of x_2 is given by

$$P_2 = x_2 P_2^* \tag{5.29}$$

In this case, P_2^* is the vapor pressure of the liquified gas at the temperature of the solution. From the last two equations it is easily seen that either a dissolved gas may be viewed in terms of its solubility under the pressure P_2, or P_2 may be taken as the vapor pressure of the volatile solute.

Thus Henry's law may also be applied to dilute solutions of a binary liquid system. It is found that in the limit of infinite dilution most liquid solvents obey Raoult's law but that under the same conditions the solute obeys Henry's law. Figure 5.11 shows a hypothetical binary liquid system exhibiting negative deviations from Raoult's law. The regions where the behavior of the system approaches that predicted by Henry's law and Raoult's law are shown by the dashed lines.

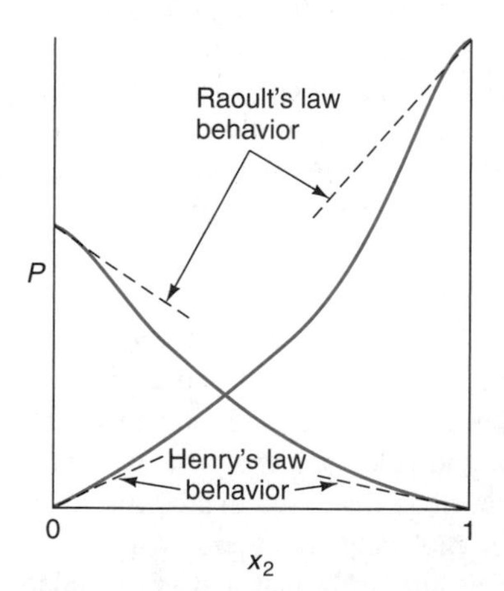

FIGURE 5.11
A hypothetical binary liquid mixture exhibiting negative deviation from Raoult's law, showing regions where Henry's and Raoult's laws are obeyed.

EXAMPLE 5.7 Dry air contains 78.084 mol % N_2 and 20.946 mol % O_2. Calculate the relative proportion of N_2 and O_2 dissolved in water under a total pressure of 1.000 bar. Henry's law constants, k', for N_2 and O_2 are 6.51×10^7 Torr and 3.30×10^7 Torr, respectively, at 25 °C.

Solution First find the partial pressures and from these, the mole fractions. The partial pressures are determined from Dalton's law.

$$P_{N_2} = 0.78084 \times 750.06 \text{ Torr} = 585.7 \text{ Torr}$$

$$P_{O_2} = 0.20946 \times 750.06 \text{ Torr} = 157.1 \text{ Torr}$$

From Eq. 5.28 (an application of Henry's law),

$$x_{N_2} = \frac{P_{N_2}}{P^*_{N_2}} = \frac{585.7 \text{ Torr}}{6.51 \times 10^7 \text{ Torr}} = 9.00 \times 10^{-6}$$

$$x_{O_2} = \frac{P_{O_2}}{P^*_{O_2}} = \frac{157.1 \text{ Torr}}{3.30 \times 10^7 \text{ Torr}} = 4.76 \times 10^{-6}$$

The relative proportions are

$$\text{for } N_2 \ \frac{9.00 \times 10^{-6}}{9.00 \times 10^{-6} + 4.76 \times 10^{-6}} = \frac{9.00 \times 10^{-6}}{1.376 \times 10^{-6}} = 0.654$$

$$\text{for } O_2 \ \frac{4.76 \times 10^{-6}}{9.00 \times 10^{-6} + 4.76 \times 10^{-6}} = 0.346$$

This is approximately a ratio of 2 N_2 to 1 O_2.

5.5 Partial Molar Quantities

Previously, in Chapter 4, we examined chemical system equilibrium, each component being considered to be fixed in composition. This corresponds to a closed system, as discussed in Section 1.3. We made the tacit assumption that the system under investigation is an ideal one in which each component has the same thermodynamic properties as it would have if it were the only reagent present. The thermodynamics developed thus far is adequate to deal with such systems. But, after examining Raoult's and Henry's laws and the deviations therefrom, we must develop a thermodynamic treatment capable of handling situations in which the thermodynamic quantities not only vary with the composition of the system but also allow for nonideal behavior. In order to consider the changing composition of a solution, it generally is found convenient to consider the changes that occur when different amounts of any component are added to or subtracted from the solution.

The type of problem that we are confronted with in dealing with solutions may best be considered by utilizing what might be the most easily visualized property of a solution, its volume. For example, it is well known that upon mixing 50 cm^3 of water with 50 cm^3 of methanol the total volume of solution is approximately 95 cm^3 instead of the expected 100 cm^3. Although the molar volume of both water and methanol may be uniquely and unambiguously determined by experimentation, there exists no unique and unambiguous method to determine the molar volume of either in solution. This is in spite of the fact that the total volume is easily determined. If we were to assign a fraction of the total volume to each component, there is no way to determine what part of the contraction that occurs upon mixing is due to the water and what part to the alcohol. Similar difficulties are present for all the other thermodynamic properties.

In our search for a method of handling variable compositions, we avoid the problem just stated by the invention of partial molar quantities. The following treatment is applicable to any extensive thermodynamic quantity such as the enthalpy, internal energy, or the Gibbs energy. Each of these extensive functions depends on the amounts of every component in the system, along with the two natural independent variables for the particular function. Explicitly, for the Gibbs energy, $G = G(T, P, n_1, n_2,..., n_i)$. Initially we will proceed in our development using the volume as the extensive function and then later focus on the Gibbs energy.

Let us imagine that we start with a large volume of solution and to it add an infinitesimal amount of component 1, dn_1, without changing the amounts of the other components. If this is done at constant temperature and pressure, the corresponding increment in the volume V is dV. Since $V = V(n_1, n_2, n_3,\ldots)$, at constant temperature and pressure, the increase in V for the general case is given by

$$dV = \left(\frac{\partial V}{\partial n_1}\right)_{T,P,n_2,n_3,\ldots} dn_1 + \left(\frac{\partial V}{\partial n_2}\right)_{T,P,n_1,n_3,\ldots} dn_2 + \cdots \tag{5.30}$$

where the second and following terms on the right are zero when only dn_1 is changed.

Partial Molar Volume

The increase in volume per mole of component 1 is known as the **partial molar volume** of component 1. It is given the symbol V_1 and is written as

$$V_1 \equiv \left(\frac{\partial V}{\partial n_1}\right)_{T,P,n_2,n_3\cdots} \tag{5.31}$$

Analogous definitions apply to the other components. Note that the same symbol V_1 is used both for the molar volume of pure component 1 and for the partial molar volume. This is logical since the latter becomes the molar volume when the component is pure. If there is any danger of confusion, the molar volume of pure component 1 will be designated as V_1^*.

We may also interpret the partial molar volume as the increase in V at the specified composition, temperature, and pressure when 1 mol of component 1 is added to such a large amount of solution that the addition does not appreciably change the concentration.

In either case, the definition for the partial molar volume, Eq. 5.31, may be used to rewrite Eq. 5.30 as

$$dV = V_1\, dn_1 + V_2\, dn_2 + \cdots \tag{5.32}$$

This expression may be integrated under the condition of constant composition so that the V_i's are constant. We can visualize the physical process of this integration as the addition of infinitesimals of each component in the same proportions in which they exist in the final solution. We obtain

$$V = n_1V_1 + n_2V_2 + \cdots \tag{5.33}$$

Although this expression was obtained under the restriction of constant composition, it is generally applicable, since each term in it is a function of state, which in turn is independent of the way in which the solution is formed.

If the total differential of Eq. 5.33 is now taken at constant temperature and pressure, we obtain

$$dV = n_1\,dV_1 + n_2\,dV_2 + \cdots + V_1\,dn_1 + V_2\,dn_2 + \cdots \tag{5.34}$$

Upon subtracting Eq. 5.32, we have

$$n_1\,dV_1 + n_2\,dV_2 + \cdots = 0 \tag{5.35}$$

If we now divide by the total amount of the components, $n_1 + n_2 + \cdots + n_i$, Eq. 5.35 may be expressed in terms of the mole fractions, x_i:

$$x_1\,dV_1 + x_2\,dV_2 + \cdots = 0 \tag{5.36}$$

Gibbs-Duhem Equation

This expression is one form of the **Gibbs-Duhem equation.** For a two-component system, Eqs. 5.35 and 5.36 may be put into the form

$$dV_1 = -\frac{n_2}{n_1}\,dV_2 \quad \text{or} \quad dV_1 = -\frac{x_2}{x_1}\,dV_2 = \frac{x_2}{x_2 - 1}\,dV_2 \tag{5.37}$$

If the variation of either V_1 or V_2 with concentration is known, the other may be calculated from the last equations. For example, by integration we have

$$\int dV_1 = \int \frac{x_2}{x_2 - 1}\,dV_2 \tag{5.38}$$

A plot of $x_2/(x_2 - 1)$ against V_2 gives the change in V_1 between the limits of integration. The molar volume V_1^* of pure component 1 may be used as the starting point of the integration to some final concentration.

Although these expressions are written in terms of volume, they apply equally well to any partial molar quantity.

Relation of Partial Molar Quantities to Normal Thermodynamic Properties[2]

The introduction of partial molar quantities is designed to handle open systems. We now inquire whether or not the thermodynamic equations developed in the earlier chapters are applicable to partial molar quantities.

We may begin by differentiating the definition of the Gibbs energy, $G = H - TS$, with respect to the amounts n_i of the various components:

$$\left(\frac{\partial G}{\partial n_1}\right)_{T,P,n_j} = \left(\frac{\partial H}{\partial n_1}\right)_{T,P,n_j} - T\left(\frac{\partial S}{\partial n_1}\right)_{T,P,n_j} \tag{5.39}$$

Since each of these partials is taken with respect to n_1, with T, P, and n_j held constant, they are in the generalized form of the expression for partial molar quantities. We may write therefore

$$G_1 = H_1 - TS_1 \tag{5.40}$$

For the partial molar enthalpy we obtain in a similar manner

$$H_1 = U_1 + PV_1 \tag{5.41}$$

Expressions could be written for each of the other extensive functions, resulting in similar equations. Since Eqs. 5.40 and 5.41 are of the same form as those for constant composition systems, our task is to show that this is a general result.

[2]This section may be omitted on first reading.

In order to show this, we combine the first and second laws of thermodynamics in terms of the Gibbs energy. From Eq. 3.111, p. 128, we have

$$dG = V\,dP - S\,dT \tag{5.42}$$

which is valid for all substances of constant composition, including solutions. It immediately follows that

$$\left(\frac{\partial G}{\partial P}\right)_{T,n_i} = V \quad \text{and} \quad \left(\frac{\partial G}{\partial T}\right)_{P,n_i} = -S \tag{5.43}$$

We now differentiate the expression on the right with respect to n_1:

$$\left[\frac{\partial}{\partial n_1}\left(\frac{\partial G}{\partial T}\right)_{P,n_i}\right]_{T,P,n_j} = -\left(\frac{\partial S}{\partial n_1}\right)_{T,P,n_j} \equiv -S_1 \tag{5.44}$$

Since the order of differentiation is immaterial, the left-hand side of Eq. 5.44 may be written as

$$\left[\frac{\partial}{\partial T}\left(\frac{\partial G}{\partial n_1}\right)_{T,P,n_j}\right]_{P,n_i} = \left(\frac{\partial G_1}{\partial T}\right)_{P,n_i} = -S_1 \tag{5.45}$$

Similarly, the left-hand expression of Eq. 5.43 leads to

$$\left(\frac{\partial G_1}{\partial P}\right)_{T,n_i} = V_1 \tag{5.46}$$

The partial molar Gibbs energy is still a state function and may be represented as a function of temperature and pressure at constant composition. Therefore, upon differentiation,

$$dG_1 = \left(\frac{\partial G_1}{\partial P}\right)_{T,n_i} dP + \left(\frac{\partial G_1}{\partial T}\right)_{P,n_i} dT \tag{5.47}$$

or upon substitution of Eqs. 5.45 and 5.46 we have

$$dG_1 = V_1\,dP - S_1\,dT \tag{5.48}$$

which is identical in form with the combined statement of the first and second laws, Eq. 5.42.

If we now add Eqs. 5.40 and 5.41

$$G_1 = U_1 + PV_1 - TS_1 \tag{5.49}$$

and take the total differential, we have

$$dG_1 = dU_1 + P\,dV_1 + V_1\,dP - T\,dS_1 - S_1\,dT \tag{5.50}$$

Subtracting Eq. 5.48 leaves

$$dU_1 = T\,dS_1 - P\,dV_1 \tag{5.51}$$

which is the partial molar internal energy analog of the original combined expression of the first and second laws for systems of fixed composition. All the earlier thermodynamic reactions for systems of fixed composition were derived from it, along with the definitions for H, A, and G. Therefore, since Eq. 5.51 and Eqs. 5.40 and 5.41 are identical in form to their counterparts previously derived, every relation developed for a system of fixed composition expressed in molar quantities is applicable to each component of the system expressed in terms of the partial molar quantities.

5.6 The Chemical Potential

The thermodynamic functions such as V, U, H, and G are extensive properties, since their values depend on the amount of substance present. Thus a total differential of the Gibbs function $G = G(T, P, n_i)$ is written as

$$dG = \left(\frac{\partial G}{\partial T}\right)_{P,n_i} dT + \left(\frac{\partial G}{\partial P}\right)_{T,n_i} dP + \sum_i \left(\frac{\partial G}{\partial n_i}\right)_{T,P,n_j} dn_i \tag{5.52}$$

Substitution of Eq. 5.13 into Eq. 5.52 gives

$$dG = -S\,dT + V\,dP + \sum_i \left(\frac{\partial G}{\partial n_i}\right)_{T,P,n_j} dn_i \tag{5.53}$$

We will now use a symbolism introduced by Willard Gibbs, in which the coefficient of the form $(\partial G/\partial n_i)_{T,P,n_j}$ (the partial molar Gibbs energy) is called the **chemical potential** μ_i, for the ith component. Therefore, Eq. 5.53 may be written in the form

$$dG = -S\,dT + V\,dP + \sum_i \mu_i\,dn_i \tag{5.54}$$

In a similar manner, the total differential of the internal energy may be written as

$$dU = \left(\frac{\partial U}{\partial S}\right)_{V,n_j} dS + \left(\frac{\partial U}{\partial V}\right)_{S,n_j} dV + \sum_i \left(\frac{\partial U}{\partial n_i}\right)_{S,V,n_j} dn_i \tag{5.55}$$

Again, substitution for the first two partials (Eq. 3.116 on p. 128) gives

$$dU = T\,dS - P\,dV + \sum_i \left(\frac{\partial U}{\partial n_i}\right)_{S,V,n_j} dn_i \tag{5.56}$$

However, since $G \equiv U + PV - TS$ for any closed system, we may differentiate this expression,

$$dG = dU + P\,dV + V\,dP - T\,dS - S\,dT \tag{5.57}$$

and substitute Eq. 5.56 into it:

$$dG = -S\,dT + V\,dP + \sum_i \left(\frac{\partial U}{\partial n_i}\right)_{S,V,n_j} dn_i \tag{5.58}$$

Upon comparison of Eqs. 5.53 and 5.58 we see that

$$\mu_i = \left(\frac{\partial G}{\partial n_i}\right)_{T,P,n_j} = \left(\frac{\partial U}{\partial n_i}\right)_{S,V,n_j} \tag{5.59}$$

A similar treatment of the other thermodynamic functions shows that

$$\begin{aligned}\mu_i &= \left(\frac{\partial G}{\partial n_i}\right)_{T,P,n_j} = \left(\frac{\partial U}{\partial n_i}\right)_{S,V,n_j} = \left(\frac{\partial H}{\partial n_i}\right)_{S,P,n_j} \\ &= \left(\frac{\partial A}{\partial n_i}\right)_{T,V,n_j} = -T\left(\frac{\partial S}{\partial n_i}\right)_{U,V,n_j}\end{aligned} \tag{5.60}$$

Note that the constant quantities are the natural variables for each function.

We now see why a special name is warranted for μ_i rather than identifying it with a particular thermodynamic function. It should be pointed out, however, that the chemical potential is most commonly associated with the Gibbs energy because we most often work with systems at constant temperature and pressure.

One may wonder why the word *potential* is used here. If we were to integrate Eq. 5.54 under the condition of constant temperature, pressure, and composition, the μ_i would remain constant. Hence, the total Gibbs energy increase would be proportional to the size of the system. The chemical potential may be viewed as the potential for moving matter. This is analogous to the electrical concept of a fixed voltage being a *potential* or *capacity factor*.

One of the most common uses of the chemical potential is as the criterion of equilibrium for a component distributed between two or more phases. Let us investigate this use. From Eq. 5.54 we have, under conditions of constant temperature and pressure,

$$dG = \sum_i \mu_i \, dn_i \tag{5.61}$$

This expression allows the calculation of the Gibbs energy for the change in both the amount of substance present in a phase and also the number of the phase's components. Therefore, this equation applies to phases that are open to the transport of matter. If a single phase is closed, and no matter is transferred across its boundary ($dG = 0$ for a closed system),

$$\sum_i \mu_i \, dn_i = 0 \tag{5.62}$$

However, if a system consisting of several phases in contact is closed but matter is transferred between phases, the condition for equilibrium at constant T and P becomes

$$dG = dG^\alpha + dG^\beta + dG^\gamma + \cdots = 0 \tag{5.63}$$

where α, β, and γ refer to the different phases that are in contact. Since an expression of the form of Eq. 5.62 exists for each individual term, we may write

$$dG = \sum_i \mu_i^\alpha dn_i^\alpha + \sum_i \mu_i^\beta dn_i^\beta + \sum_i \mu_i^\gamma dn_i^\gamma + \cdots = 0 \tag{5.64}$$

Now suppose that dn_i mol of component i are transferred from phase α to phase β in the closed system, without any mass crossing the system boundaries. The equilibrium condition would require that

$$dG = -\mu_i^\alpha \, dn_i + \mu_i^\beta \, dn_i = 0 \tag{5.65}$$

so that

$$\mu_i^\alpha = \mu_i^\beta \tag{5.66}$$

This can be generalized by stating that the equilibrium condition for transport of matter between phases, as well as chemical equilibrium between phases, requires that the value of the chemical potential μ_i for each component i be the same in every phase at constant T and P. Thus, for a one-component system, the requirement for equilibrium is that the chemical potential of the substance i is the same in the two phases.

To further emphasize the importance of Eq. 5.65 for a nonequilibrium situation, we write it in the form

$$dG = (\mu_i^\beta - \mu_i^\alpha) \, dn_i \tag{5.67}$$

If μ_i^β is less than μ_i^α, dG is negative and a transfer of matter occurs with a decrease of the Gibbs energy of the system. The transfer occurs spontaneously by the flow of substance i from a region of high μ_i to a region of low μ_i. The flow of matter continues until μ_i is constant throughout the system, that is, until $dG = 0$. This is the driving force for certain processes. Diffusion is an example; see Chapter 19.

It might be emphasized that the conditions of constant T and P are required to maintain thermal and mechanical equilibrium. Thus if all phases were not at the same temperature, heat could flow from one phase to another, and equilibrium would not exist. In the same manner, if one phase were under a pressure different from the rest, it could change its volume, thus destroying the condition of equilibrium.

5.7 Thermodynamics of Solutions

Raoult's Law Revisited

We may approach Raoult's law in a thermodynamic vein through Eq. 5.66. For any component i of a solution in equilibrium with its vapor, we may write

$$\mu_{i,\text{sol}} = \mu_{i,\text{vap}} \tag{5.68}$$

If the vapor behaves ideally, the Gibbs energy for each component is given by Eq. 3.96 on p. 126:

$$G_i = G_i^\circ + n_i RT \ln P_i^u \tag{5.69}$$

Since $\mu_i = G_i/n_i$, we may write

$$\mu_{i,\text{vap}} = \mu_{i,\text{vap}}^\circ + RT \ln P_i^u \tag{5.70}$$

where $\mu_{i,\text{vap}}^\circ$ is the chemical potential of the vapor when $P_i = 1$ bar at the temperature T of the system. But at equilibrium, since Eq. 5.68 holds, substitution gives

$$\mu_{i,\text{sol}} = \mu_{i,\text{vap}}^\circ + RT \ln P_i^u \tag{5.71}$$

However, for pure liquid in equilibrium with its vapor,

$$\mu_i^* = \mu_{i,\text{vap}}^\circ + RT \ln(P_i^*)^u \tag{5.72}$$

where the superscript * represents the value for the pure material. By subtraction of the last two equations, we obtain an expression for the difference between the chemical potentials of the solution and the pure material:

$$\mu_{i,\text{sol}} - \mu_i^* = RT \ln \frac{P_i}{P_i^*} \tag{5.73}$$

The fugacity of a gas is smaller than the pressure when the temperature is below the Boyle temperature, and larger when it is above the Boyle temperature. (See Section 1.14, p. 41.)

The fugacity of a nonideal gas was defined (Eq. 3.152) as numerically equal to its activity, since the standard state of fugacity is by definition 1 bar. Thus the chemical potential may be written in terms of a dimensionless activity a:

$$\mu_i = \mu_i^* + RT \ln f_i^u = \mu_i^* + RT \ln a_i \tag{5.74}$$

If the gas behaves ideally, as is required for exact adherence to Raoult's law, the partial pressure is equal to its fugacity. Then, by comparison with Eq. 5.73, we may substitute the pressure for the fugacity when the gas behaves ideally,

$$a_i = \frac{f_i}{f_i^\circ} = \frac{P_i}{P_i^*} \tag{5.75}$$

where f_i° is the fugacity in the standard state and is 1 bar, and P_i^* is the reference state for pure component i. The relative activity a_i of a component in solution is just the ratio of the partial pressure of component i above its solution compared to the vapor pressure of pure component i at the temperature of the system. By comparison with Raoult's law written in the form

$$x_i = \frac{P_i}{P_i^*} \tag{5.76}$$

it is seen that

$$a_i = x_i \tag{5.77}$$

Definition of an Ideal Solution

Equation 5.77 may be used as a definition of an ideal solution.

EXAMPLE 5.8 Calculate the activities and activity coefficients for an acetone-chloroform solution in which $x_2 = 0.6$. The vapor pressure of pure chloroform at 323 K is $P_2^* = 98.6$ kPa and the vapor pressure above the solution is $P_2 = 53.3$ kPa. For acetone, the corresponding values are $P_1^* = 84.0$ kPa and $P_1 = 26.6$ kPa.

Solution The activities, from Eq. 5.75, are

$$a_1 = \frac{26.6}{84.0} = 0.317, \qquad a_2 = \frac{53.3}{98.6} = 0.540$$

The activity coefficients are determined from the definition $f_i = a_i/x_i$ where f_i is the activity coefficient. (See Section 4.3.) Therefore,

$$f_1 = \frac{0.317}{0.400} = 0.792, \qquad f_2 = \frac{0.540}{0.600} = 0.900$$

Note that these values are for a solution exhibiting negative behavior from Raoult's law.

An interesting extension of Eq. 5.76 results in the original historical form of Raoult's law. If the addition of component 2 lowers the vapor pressure of component 1, then the difference $P^* - P$ is the lowering of the vapor pressure. Dividing by P^* gives a relative vapor pressure lowering, which is equal to the mole fraction of the solute. The mathematical form of this statement may be shown to be derived directly from Eq. 5.76, by subtracting unity from both sides. Assuming a two-component system, we have

$$\left(\frac{P_1}{P_1^*}\right) - 1 = x_1 - 1, \qquad \frac{P_1^* - P_1}{P_1^*} = x_2 \tag{5.78}$$

This form of Raoult's law is especially useful for solutions of relatively involatile solutes in a volatile solvent.

Equation 5.78 may be written so that the molar mass of the solute 2 may be determined. Since

$$x_2 = \frac{n_2}{n_1 + n_2} \quad \text{and} \quad n_i = \frac{W_i}{M_i}$$

where W_i is the mass and M_i is the molar mass, we write

$$\frac{P_1^* - P_1}{P_1^*} = \frac{n_2}{n_1 + n_2} = \frac{W_2/M_2}{(W_1/M_1) + (W_2/M_2)} \tag{5.79}$$

For a dilute solution, n_2 may be neglected in the denominator and we obtain

$$\frac{P_1^* - P_1}{P_1^*} = \frac{n_2}{n_1} = \frac{W_2}{M_2} \cdot \frac{M_1}{W_1} \tag{5.80}$$

Ideal Solutions

In view of our foregoing discussion in which Eq. 5.71 leads to a statement of Raoult's law, we may inquire as to the values of other thermodynamic functions for mixtures.

The value of the change in volume upon mixing is our first consideration. We have already seen from Eq. 5.46 and the fact that $\mu = G_1$ that

$$\left(\frac{\partial \mu}{\partial P}\right)_{T,n_i} = V_1 \tag{5.81}$$

If we look at the difference in chemical potentials upon mixing, namely $\mu_i - \mu_i^*$, and substitute into the last expression, we have

$$\left(\frac{\partial(\mu_i - \mu_i^*)}{\partial P}\right)_{T,n_i} = V_i - V_i^* \tag{5.82}$$

Then using the definition of the activity, Eq. 5.74, we have

$$RT\left(\frac{\partial \ln a_i}{\partial P}\right)_{T,n_i} = V_i - V_i^* \tag{5.83}$$

Inserting the definition of an ideal solution from Eq. 5.77, namely $a_i = x_i$, the partial derivative in Eq. 5.83 is found to be zero. Consequently,

$$V_i = V_i^* \tag{5.84}$$

Thus, the partial molar volume of a component in an ideal solution is equal to the molar volume of the pure component. There is therefore no change in volume on mixing ($\Delta_{mix}V^{id} = 0$); that is, the volume of the solution is equal to the sum of the molar volumes of the pure components.

The enthalpy of formation of an ideal solution may be investigated in a similar manner starting from the Gibbs-Helmholtz equation (Eq. 3.167 on p. 135). Again inserting our definition for activity, $\mu_i - \mu_i^* = RT \ln a_i$, we have

$$R\left(\frac{\partial \ln a_i}{\partial(1/T)}\right)_{P,n_i} = H_i - H_i^* \tag{5.85}$$

But if $a_i = x_i$ as required for an ideal solution, the partial derivative is zero and

$$\Delta_{mix}H^{id} = \sum_i (H_i - H_i^*) = 0 \tag{5.86}$$

or

$$H_i = H_i^* \tag{5.87}$$

As with the volume, the change in enthalpy of mixing is the same as though the components in an ideal solution were pure. Thus there is no heat of solution ($\Delta_{mix}H^{id} = 0$) when an ideal solution is formed from its components.

The value of the Gibbs energy change for mixing cannot be zero as with the two previous functions since the formation of a mixture from a solution is a spontaneous process and must show a decrease.

The Gibbs energy of mixing $\Delta_{mix}G$ for any solution is

$$\Delta_{mix}G = G - G^* = \sum_i n_i\mu_i - \sum_i n_i\mu_i^* \tag{5.88}$$

From the definition of activity in Eq. 5.74 and after multiplication by n_i, we obtain

$$n_i\mu_i - n_i\mu_i^* = n_iRT \ln a_i \tag{5.89}$$

Combining this with Eq. 5.88 gives

$$\Delta_{mix}G = RT(n_1 \ln a_1 + n_2 \ln a_2 + \cdots) \tag{5.90}$$

If the solution is ideal, then $a_i = x_i$. We have, therefore,

$$\Delta_{mix}G^{id} = RT(n_1 \ln x_1 + n_2 \ln x_2 + \cdots) \tag{5.91}$$

This can be put into a more convenient form by using the substitution

$$n_i = x_i \sum_i n_i \quad \text{where} \quad \sum_i n_i = n_1 + n_2 + \cdots$$

is the total amount of material in the mixture and x_i is the mole fraction of component i. Then,

$$\Delta_{mix}G^{id} = \sum_i n_iRT\,(x_1 \ln x_1 + x_2 \ln x_2 + \cdots) = \sum_i n_iRT \sum_i x_i \ln x_i \tag{5.92}$$

which is the Gibbs energy of mixing in terms of the mole fractions of the components. Note that each term on the right-hand side is negative since the logarithm of a fraction is negative. Thus the sum is always negative.

For a two-component mixture, Eq. 5.92 may be written as

$$\Delta_{mix}G^{id} = \sum_i n_iRT[x_1 \ln x_1 + (1 - x_1)\ln(1 - x_1)] \tag{5.93}$$

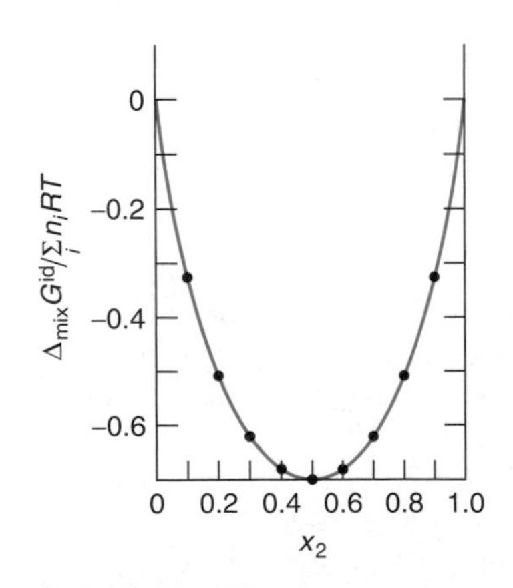

FIGURE 5.12
$\Delta_{mix}G/\Sigma_i\, n_iRT$ against x for a binary ideal mixture.

A plot of this function is shown in Figure 5.12, where it is seen that the curve is symmetrical about $x_1 = 0.5$. For more complex systems, the lowest value will occur when the mole fraction is equal to $1/n$ where n is the number of components. These equations apply to mixtures of liquids and gases.

The final thermodynamic function to be considered here is the entropy. Using Eq. 5.43, differentiate $\Delta_{mix}G$ with respect to temperature and obtain $\Delta_{mix}S$ directly:

$$\left(\frac{\partial \Delta_{mix}G}{\partial T}\right)_{P,n_i} = \left(\frac{\partial G}{\partial T}\right)_{P,n_i} - \frac{\partial}{\partial T}(n_1\mu_1^* + n_2\mu_2^* + \cdots)_{P,n_i}$$
$$= -S + n_1S_1^* + n_2S_2^* + \cdots = -\Delta_{mix}S \tag{5.94}$$

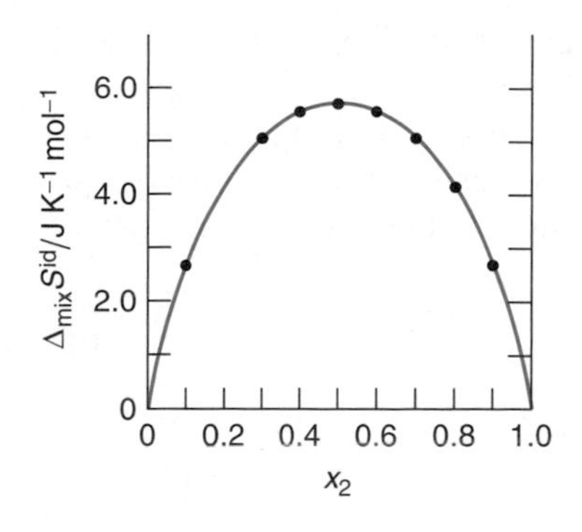

FIGURE 5.13
$\Delta_{mix}S$ of 1 mol of an ideal solution from the pure components.

Differentiating the right-hand side of Eq. 5.92 with respect to T and substituting it into the last expression gives, for an ideal mixture,

$$-\Delta_{mix}S^{id} = \sum_i n_i R \sum_i x_i \ln x_i \tag{5.95}$$

Thus the entropy of mixing for an ideal solution is independent of temperature and pressure. Figure 5.13 is a plot of this function for a binary system.

Another interesting aspect of mixing for an ideal solution can be seen by writing

$$\Delta_{mix}G = \Delta_{mix}H - T\Delta_{mix}S \tag{5.96}$$

Since we have seen that $\Delta_{mix}H^{id} = 0$,

$$-\Delta_{mix}G^{id} = T\Delta_{mix}S^{id} \tag{5.97}$$

Thus the driving force for mixing is purely an entropy effect. Since the mixed state is a more random state, it is therefore a more probable state. However, this need not mean that the driving force is large. If $\Delta_{mix}G$ were to be positive, the material would stay as distinct phases; this explains the lack of miscibility found in some liquid systems.

Nonideal Solutions; Activity and Activity Coefficients

We now approach the problem of how solutions are treated when they are not ideal. We have already seen from Eq. 3.153 on p. 132 and the discussion of the last section that

$$\mu_i = \mu_i^\circ + RT \ln a_i \tag{5.98}$$

There are two ways of defining μ_i° that lead to different systems of activities. In both systems the activity remains a measure of the chemical potential.

In the *rational system,* μ_i° is identified with the chemical potential of the pure liquid μ_i^*:

$$\mu_i = \mu_i^* + RT \ln a_i \tag{5.99}$$

As $x_i \rightarrow 1$ and the system approaches pure i, μ_i must approach μ_i^*. Hence, $\ln a_i = 0$ as $x_i \rightarrow 1$. In other words,

$$a_i \rightarrow 1 \quad \text{as} \quad x_i \rightarrow 1 \tag{5.100}$$

and the activity of the pure liquid, a dimensionless quantity, is equal to unity. For an ideal liquid solution a_i may be replaced by x_i, as seen in Eq. 5.77:

$$\mu_{i,\text{id}} = \mu_i^* + RT \ln x_i \tag{5.101}$$

Subtraction of Eq. 5.101 from 5.99 gives the difference between a nonideal and an ideal solution:

$$\mu_i - \mu_{i,\text{id}} = RT \ln \frac{a_i}{x_i} \tag{5.102}$$

We may now define the *rational activity coefficient* of i, f_i, as

$$f_i = \frac{a_i}{x_i} \quad \text{and} \quad f_i \rightarrow 1 \quad \text{as} \quad x_i \rightarrow 1 \tag{5.103}$$

from which it is seen that f_i is a measure of the extent of deviation from ideal behavior. This is basically the system employed earlier.

For a solution, it is most useful to use the rational system for the solvent when the solvent has a mole fraction near unity. Although this same system may be used for solutes, the *practical system* may be used for the solute (if the solute is present in small amounts). In this system

$$\mu_{i,\text{id}} = \mu_i^{\ominus} + RT \ln m_i^u \tag{5.104}$$

where m_i^u is the value of the molality (i.e., m_i/mol kg^{-1}) and μ_i° in Eq. 5.98 is identified with $\mu_i^{\ominus}$ (the value μ_i° would have in a hypothetical state of unit molality if the solution behaved as an ideal dilute solution, i.e., obeyed Henry's law, over the range of $m_i = 0$ to 1). This identification defines the practical system of activities. In the same manner in which we obtained Eq. 5.101, setting $\mu_i^{\ominus} = \mu_i^{\circ}$ in Eq. 5.98 gives us

$$\mu_i = \mu_i^{\ominus} + RT \ln a_i \tag{5.105}$$

and then, subtracting Eq. 5.104, we have

$$\mu_i - \mu_{i,\text{id}} = RT \ln \frac{a_i}{m_i} \tag{5.106}$$

From this, the *practical activity coefficient* is defined by

$$\gamma_i = \frac{a_i}{m_i} \quad \text{and} \quad \gamma_i \to 1 \quad \text{as} \quad m_i \to 0 \tag{5.107}$$

Thus γ_i like f_i is a measure of the departure of substance i from the behavior expected in an ideal solution.

The practical system of activities defined by Eq. 5.105 measures the chemical potential of the solute relative to the chemical potential of the hypothetical ideal solution of unit molality. Equation 5.106 applies to both volatile and nonvolatile solutes.

EXAMPLE 5.9 The solubility of oxygen in water at 1 atm pressure and 298.15 K is 0.001 15 mol kg^{-1} of water. Under these conditions, calculate μ_i° for a saturated solution of oxygen in water.

Solution Since O_2(g, 1 atm) $\rightleftharpoons$ O_2(aq, 0.001 15 m), the chemical potentials of the two sides are equal. Since $\mu_{O_2}(g)$ for formation equals zero, $\mu_{O_2}(aq) = 0$. Substitution of this into Eq. 5.104 gives

$$\mu_{O_2}(g) = 0 = \mu_{O_2}^{\circ}(aq) + RT \ln m_i^u$$

$$\mu_{O_2}^{\circ}(aq) = -8.3145(298.15) \ln 0.001\,15 = 16.78 \text{ kJ mol}^{-1}$$

This is the hypothetical chemical potential of O_2 if the solution behaved according to Henry's law. The hypothetical chemical potential is useful in biological reactions for establishing the position of equilibrium when oxygen, dissolved in the solution, is also involved in the reaction.

5.8 The Colligative Properties

The properties of dilute solutions that depend on only the number of solute molecules and not on the type of species present are called **colligative properties** (Latin *colligatus,* bound together). All colligative properties (such as *boiling point elevation, freezing point depression,* and *osmotic pressure*) ultimately can be related to a lowering of the vapor pressure $P^* - P$ for dilute solutions of nonvolatile solutes.

Freezing Point Depression

The boiling point elevation and freezing point depression are related to the vapor pressure through the chemical potential. The situation is simplified by recognizing that the nonvolatile solute does not appear in the gas phase, and consequently the curve of μ for the gas phase is unchanged. An additional simplifying assumption is that the solid phase contains only the solid solvent, since when freezing occurs, only the solvent usually becomes solid. This leaves the curve for μ for the solid unchanged. From the last section we combine Eqs. 5.73 and 5.76 and use the convention that the subscript 1 refers to the solvent. Thus we have

$$\mu_1^{\mathrm{l}} - \mu_1^{*,\mathrm{l}} = RT \ln x_1 \tag{5.108}$$

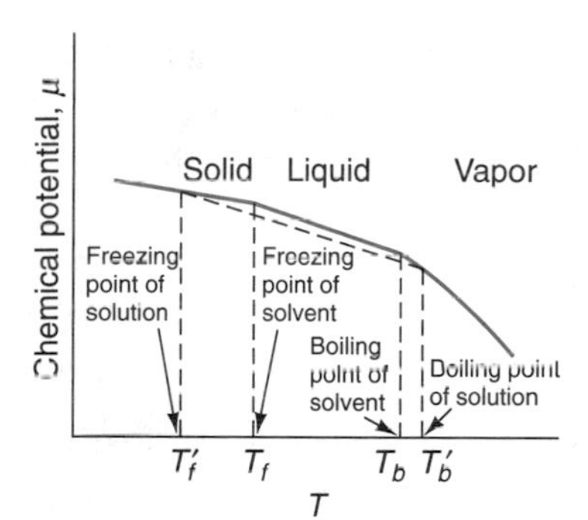

FIGURE 5.14
Variation of the chemical potential. Note the enhanced decrease in freezing point from the normal value compared to the increase in boiling point.

where μ_1^{l} now refers to the chemical potential of the liquid solvent in solution, and x_1 is its mole fraction. As a result, the freezing point of the solution is different from that of the pure solvent, as shown in Figure 5.14. Drawn for constant pressure, the figure is similar to Figure 5.3. The solid lines show normal behavior of the pure solvent. In view of Eq. 5.108, the superimposed dashed curve represents the effects of lowering the vapor pressure, as a result of the term $RT \ln x_1$. The size of the phenomenon is larger for the freezing point depression than for the boiling point elevation in a solution of the same concentration.

Another way to illustrate the change is to use the ordinary phase diagram plot of P against T for water shown in Figure 5.15. This plot takes the form of Figure 5.1 but again has dashed lines superimposed to indicate the effects of vapor pressure lowering. The intersections of the horizontal line at the reference point of 1 atm with the solid and dashed vertical lines give the freezing points and boiling points.

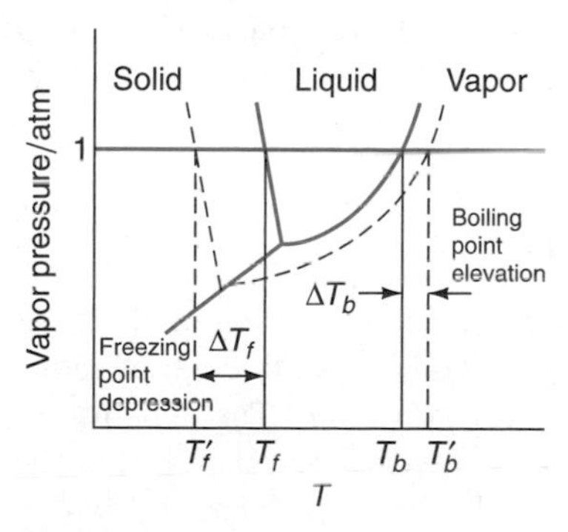

FIGURE 5.15
Phase diagram (P against T) of a solvent (water)-nonvolatile solute system showing depression of freezing point and elevation of boiling point (not drawn to scale).

Now consider a solution in equilibrium with pure solid solvent. Our aim is to discover how T depends on x_1. The chemical potential is a function of both T and P and will be written to emphasize this point. Since by our previous hypothesis the solid is pure, the chemical potential of the pure solid $\mu_1^{\mathrm{s}}(T, P)$ is independent of the mole fraction of the solute. On the other hand, the chemical potential of the liquid solvent $\mu_1^{\mathrm{l}}(T, P, x_1)$ is a function of the mole fraction of solvent, x_1. The equilibrium condition is

$$\mu_1^{\mathrm{l}}(T, P, x_1) = \mu_1^{\mathrm{s}}(T, P) \tag{5.109}$$

For an ideal solution, we may substitute from Eq. 5.108 for μ_1^{l} and obtain

$$\mu_1^{*,\mathrm{l}}(T, P) + RT \ln x_1 = \mu_1^{\mathrm{s}}(T, P) \tag{5.110}$$

or

$$\ln x_1 = \frac{-(\mu_1^{*,\mathrm{l}}(T, P) - \mu_1^{\mathrm{s}}(T, P))}{RT} \tag{5.111}$$

The difference in chemical potentials for pure liquid and pure solid expressed in the numerator is simply $\Delta_{\text{fus}}G_m$ (i.e., the molar Gibbs energy of fusion of the solvent at T and 1 bar of pressure). Thus we may write

$$\ln x_1 = \frac{-\Delta_{\text{fus}}G_m}{RT} \tag{5.112}$$

Differentiation with respect to T gives

$$\frac{d \ln x_1}{dT} = -\frac{1}{R}\left[\frac{\partial(\Delta_{\text{fus}}G_m/T)}{\partial T}\right]_P \tag{5.113}$$

Substitution of the Gibbs-Helmholtz equation, Eq. 3.168 on p. 135, yields

$$\frac{d \ln x_1}{dT} = \frac{\Delta_{\text{fus}}H_m}{RT^2} \tag{5.114}$$

If $\Delta_{\text{fus}}H_m$, the heat of fusion for pure solvent, is independent of T over a moderate range of temperature, we may integrate Eq. 5.114 from T_f^*, the freezing point of pure solvent at $x_1 = 1$, to T, the temperature at which solid solvent is in equilibrium with liquid solvent of mole fraction x_1. The result is

$$\ln x_1 = \frac{\Delta_{\text{fus}}H_m}{R}\left(\frac{1}{T_f^*} - \frac{1}{T}\right) \tag{5.115}$$

This expression gives the mole fraction of the solvent in relation to the freezing point of an ideal solution and to the freezing point of the pure solvent.

This relation may be simplified by expressing x_1 in terms of x_2, the mole fraction of the solute:

$$\ln(1 - x_2) = \frac{\Delta_{\text{fus}}H_m}{R}\left(\frac{T - T_f^*}{TT_f^*}\right) \tag{5.116}$$

The term $\ln(1 - x_2)$ may be expanded in a power series, and if x_2 is small (corresponding to a dilute solution), only the first term need be kept:

$$\ln(1 - x_2) = -x_2 - \tfrac{1}{2}x_2^2 - \tfrac{1}{3}x_2^3 - \cdots = -x_2 \tag{5.117}$$

in which only the first term is kept. The freezing point depression is $T_f^* - T = \Delta_{\text{fus}}T$. Since $\Delta_{\text{fus}}T$ is small in comparison to T_f^*, we may set the product $TT_f^* \approx T_f^{*2}$. These changes convert Eq. 5.116 to

$$x_2 = \frac{\Delta_{\text{fus}}H_m}{R} \cdot \frac{\Delta_{\text{fus}}T}{T_f^{*2}} \tag{5.118}$$

Since the most important application of this expression is for determining molar masses of dissolved solutes, an alternate form is useful. The mole fraction of solute is $x_2 \equiv n_2/(n_1 + n_2)$, and in dilute solution is approximately n_2/n_1. The molality m_2 is the amount n_2 of solute divided by the mass W_1 of solvent: $m_2 = n_2/W_1$. For the solvent, the amount present is its mass W_1 divided by its molar mass M_1: $m_1 = W_1/M_1$. Then $x_2 \approx m_2W_1/W_1/M_1 = m_2M_1$. Rearrangement of Eq. 5.118 and substitution for x_2 yields

$$\Delta_{\text{fus}}T \approx \frac{M_1RT_f^{*2}}{\Delta_{\text{fus}}H_m} \cdot m_2 \tag{5.119}$$

This can be further simplified by introducing the *freezing point depression* or *cryoscopic constant* K_f, defined as

Freezing Point Depression Constant

$$K_f = \frac{M_1 R T_f^{*2}}{\Delta_{\text{fus}} H_m} \tag{5.120}$$

EXAMPLE 5.10 Find the value of K_f for the solvent 1,4-dichlorobenzene from the following data:

$$M = 147.01 \text{ g mol}^{-1}; \qquad T_f^* = 326.28 \text{ K}; \qquad \Delta_{\text{fus}} H_m = 17.88 \text{ kJ mol}^{-1}$$

Solution The value of K_f depends only on the properties of the pure solvent. The value of R is expressed in joules, which have the units kg m^2 s^{-1}. Consequently, M must be expressed in units of kg mol^{-1}. Therefore,

$$K_f = \frac{0.147\,01(\text{kg mol}^{-1}) \times 8.3145(\text{J K}^{-1} \text{ mol}^{-1}) \times 326.28^2(\text{K}^2)}{17\,880(\text{J mol}^{-1})}$$

$$= 7.28 \text{ K kg mol}^{-1}$$

With the definition of K_f, Eq. 5.119 may be expressed as

$$\Delta_{\text{fus}} T = K_f m_2 \tag{5.121}$$

This is a particularly simple relation between the freezing point depression and the solute's molality in a dilute ideal solution. Table 5.2 lists K_f for several substances. If the depression constant is known, the molar mass of the dissolved substance may be found. Since $m_2 = W_2/M_2W_1$, we have after rearrangement

$$M_2 = \frac{K_f W_2}{\Delta_{\text{fus}} T W_1} \tag{5.122}$$

TABLE 5.2 Freezing Point Depression Constants K_f

Solvent	K_f / K kg mol^{-1}	Melting Point, T_m/K
Water	1.86	273.15
Acetic acid	3.90	289.75
Dioxane	4.71	284.85
Benzene	4.90	278.63
Phenol	7.40	316.15
Camphor	37.7	451.55

EXAMPLE 5.11 A solution contains 1.50 g of solute in 30.0 g of benzene and its freezing point is 3.74 °C. The freezing point of pure benzene is 5.48 °C. Calculate the molar mass of the solute.

Solution From Table 5.2, K_f for benzene is 4.90 K kg mol^{-1}. The value of $\Delta_{\text{fus}}T = 5.48 - 3.74 = 1.74$ K. Substitution into Eq. 5.122 gives

$$M_2 = \frac{4.90(\text{K kg mol}^{-1}) \times 1.50(\text{g})}{1.74(\text{K}) \times 30.0(\text{g})}$$
$$= 0.147 \text{ kg mol}^{-1} = 147 \text{ g mol}^{-1}$$

It should be pointed out that the foregoing treatment does not apply to a solid solution in which a homogeneous solid containing both solute and solvent solidifies.

Ideal Solubility and the Freezing Point Depression

We may digress from our main development briefly to investigate some implications of Eq. 5.115. This expression gives the temperature variation of the solubility x_1 of a pure solid in an ideal solution. This expression applies to the equilibrium

$$\text{pure solid 1} \rightleftharpoons \text{ideal solution of 1 and 2} \qquad (5.123)$$

Thus, in one sense the system could be viewed as a saturated solution of solid 1 in solvent 2 with some excess solid 1 present. In this perspective, solid component 1 is now thought of as the solute in the solvent compound 2. Furthermore T_f^* is now simply the temperature at which the solubility is measured. Thus, Eq. 5.115 can equally well be written for solutes where x_2 for the solute replaces x_1, with the temperature and the enthalpy now referring to the solute x_2. In that form, Eq. 5.115 is known as the *ideal solubility equation.* The basic idea is that the mole fraction solubility of a substance in an ideal solution depends only on the properties of that substance. Stated differently, the mole fraction solubility of a substance in an ideal solution is the same in all solvents.

In Table 5.3 are listed the mole fraction solubilities of several solutes. In the row marked *Ideal solution* are the solubilities as calculated from Eq. 5.115.

EXAMPLE 5.12 Calculate the ideal solubility of phenanthrene at 298.1 K in a solvent in which it forms an ideal solution.

Solution The heat of fusion of phenanthrene is 18.64 kJ mol^{-1} at the melting point of 369.4 K. Using Eq. 5.115 we find

$$\ln x_1 = \frac{18\,640}{8.3145}\left(\frac{1}{369.4} - \frac{1}{298.1}\right) = -1.45$$
$$x_1 = 0.234$$

Comparisons are available in Table 5.3.

Two observations may be made with respect to the solubility of all solutes from Eq. 5.115. First, the *solubility of all solutes should increase as the temperature is increased.* Second, the solubility in a series of similar solutes will decrease

TABLE 5.3 Mole-Fraction Solubilities of Several Solutes in Different Solvents

	Solute	Naphthalene	Phenanthrene	*p*-Dibromobenzene
Solvent	**Temperature**	**298.1 K**	**298.1 K**	**298.1 K**
Benzene		0.241	0.207	0.217
Hexane		0.090	0.042	0.086
Carbon tetrachloride		0.205	0.086	0.193
Toluene		0.224	0.255	0.224
Ideal solution		0.268	0.234	0.248

as the melting point increases. The first prediction is universally true, whereas the second applies if the heats of fusion lie fairly close to each other. Some deviations do occur, and generally a low melting point and a low heat of fusion will favor increased solubility. Hydrogen bonding also plays an important role in causing deviations. In spite of the deviations from ideal solubility values in some cases, techniques have been developed that allow accurate prediction of values. However, further discussion is beyond the scope of this book.[3]

Boiling Point Elevation

The physical changes occurring on the addition of a nonvolatile solute to a boiling solution have already been shown in Figures 5.14 and 5.15. The condition for equilibrium of a solution with the vapor of the pure solvent is

$$\mu_1^{l}(T, P, x_1) = \mu_1^{v}(T, P) \tag{5.124}$$

The development is the same as in the section on freezing point depression. Following the same development through Eq. 5.115 we have

$$\ln x_1 = \frac{\Delta_{vap}H_m}{R}\left(\frac{1}{T} - \frac{1}{T_b^*}\right) \tag{5.125}$$

where T is the boiling point of the solution, T_b^* is now the normal boiling point of pure solvent, and $\Delta_{vap}H_m$ is the heat of vaporization. The development may be continued in the same manner as before and we obtain

Boiling Point Elevation Constant

$$\Delta_{vap}T = K_b m_2 \tag{5.126}$$

where

$$K_b = M_1 R T_b^{*2}/\Delta_{vap}H_m \quad \text{and} \quad \Delta_{vap}T \quad \text{is} \quad T - T_b^*.$$

In Table 5.4 are listed several *boiling point elevation* or *ebullioscopic constants,* K_b. Generally, the higher the molar mass of the solute, the larger the value of K_b. These data may be used to determine molar mass in the same manner as for freezing point depression constants. However, since the boiling point is a function of the total pressure, K_b is also a function of temperature. Its variation may be calculated using the Clausius-Clapeyron equation, and it is generally small. If the solution does not behave ideally, the mole fraction can be written in terms of the activity.

[3]The interested reader should consult the Hildebrand and Temperley books in the Suggested Reading.

TABLE 5.4 Boiling Point Elevation Constants K_b

Solvent	K_b / K kg mol^{-1}	Boiling Point, T_b/K
Water	0.51	373.15
Acetic acid	3.07	391.45
Acetone	1.71	329.25
Benzene	2.53	353.35
Ethanol	1.22	351.65

Osmotic Pressure

In 1748 the French physicist Jean Antoine Nollet (1700–1770) placed wine in an animal bladder and immersed the sealed bladder in pure water. Water entered the bladder and the bladder expanded and eventually burst. The bladder acted somewhat as a **semipermeable membrane.** Such a membrane permits the solvent molecules, but not the solute molecules, to pass through it. The concentration of the solvent molecules is greater on the pure solvent side of the membrane than on the solution side, since, on the solution side, some of the volume is occupied by solute molecules. There is therefore a tendency for molecules of the solvent to pass through the membrane. This tendency increases with increasing concentration of the solute. Although there is a flow of solvent molecules in both directions through the membrane, there is a more rapid flow from the pure solvent into the solution than from the solution into the pure solvent. The overall effect is a net flow of solvent into the solution, thus diluting its concentration and expanding its original volume.

Semipermeable Membrane

Much later, in 1887, Wilhelm Friedrick Philipp Pfeffer (1845–1920) utilized colloidal cupric ferrocyanide, supported in the pores of unglazed porcelain, as the semipermeable membrane to study this effect quantitatively.

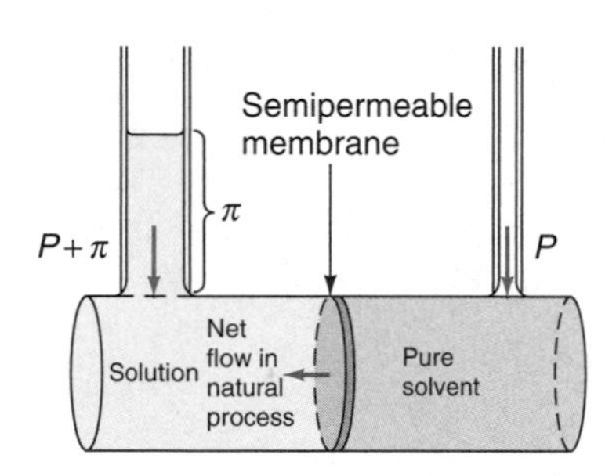

FIGURE 5.16
Schematic representation of apparatus for measuring osmotic pressure. At equilibrium the natural process of flow of pure solvent into solution is just stopped by the osmotic pressure π due to the mass of the column of fluid. In actual osmotic pressure determinations an external pressure π is applied to stop the flow.

This phenomenon may be investigated through use of the apparatus shown in Figure 5.16. The solution and the pure solvent are separated from each other by a semipermeable membrane. The natural flow of solvent molecules can be stopped by applying a hydrostatic pressure to the solution side. The effect of the pressure is to increase the tendency of the solvent molecules to flow from solution into pure solvent. The particular pressure that causes the net flow to be reduced to zero is known as the **osmotic pressure** π of the solution.

The equilibrium requirement for the system in Figure 5.16 is that the chemical potentials on the two sides of the membrane be equal. This can only be achieved through a pressure difference on the two sides of the membrane. If the pure solvent is under atmospheric pressure P, the solution is under the pressure $P + \pi$. The chemical potentials of the two sides are

$$\mu(T, P + \pi, x_1) = \mu^*(T, P) \tag{5.127}$$

where x_1 is the mole fraction of the solvent. Replacing the left-hand side with the explicit dependency on the mole fraction (Eq. 5.101), we have

$$\mu^*(T, P + \pi) + RT \ln x_1 = \mu^*(T, P) \tag{5.128}$$

We must now express the variation of μ as a function of pressure. One of the Maxwell relations is $(\partial G_1/\partial P)_{T,n_i} = V_1$ and since $G_1 = \mu$, it follows that $d\mu^* = V_1^* \, dP$. Integrating between the limits of P and $P + \pi$ gives

$$\mu^*(T, P + \pi) - \mu^*(T, P) = \int_P^{P+\pi} V_1^* \, dP \tag{5.129}$$

where V_1^* is the molar volume of the pure solvent. Substitution into Eq. 5.128 yields

$$\int_P^{P+\pi} V_1^* \, dP + RT \ln x_1 = 0 \tag{5.130}$$

On the assumption that the solvent is incompressible, V_1^* is independent of pressure and our expression, after integration, reduces to

$$V_1^* \pi + RT \ln x_1 = 0 \tag{5.131}$$

We need to express this relationship in terms of the solute concentration. Since $\ln x_1 = \ln(1 - x_2)$, and if the solution is dilute, $x_2 \ll 1$, the logarithm may be expanded in a series. Only the first term of the series $-x_2$ makes an important contribution. From its definition, x_2 may be expressed in terms of the amount of solute n_2 and solvent n_1 present. We have, therefore,

$$\ln(1 - x_2) \approx -x_2 = -\frac{n_2}{n_1 + n_2} \approx -\frac{n_2}{n_1} \tag{5.132}$$

since $n_2 \ll n_1$ in dilute solution. Equation 5.131 then becomes

$$\pi = \frac{n_2 RT}{n_1 V_1^*} \tag{5.133}$$

If a solution is dilute, as required in Eq. 5.133, the contribution of the molar volume of n_2 is small so that $n_1 V_1^* \approx V$, the volume of the solution. Using this, our expression for π becomes

$$\pi = \frac{n_2 RT}{V} \quad \text{or} \quad \pi = cRT \tag{5.134}$$

Van't Hoff's Equation

where $c = n_2/V$ is the concentration of the solute. This is **van't Hoff's equation** for **osmotic pressure** and may be used for molar mass determinations.

EXAMPLE 5.13 In a study of the osmotic pressure of hemoglobin at 276.15 K, the pressure was found to be equal to that of a column of water 3.51 cm in height. The concentration was 1 g per 0.100 dm^3. Calculate the molar mass.

Solution The pressure must first be converted to Pa. This can be done using 13.59 as the relative density of mercury with reference to water. Also, 1 mmHg = 133.32 Pa. Therefore,

$$P \text{ of } 3.51 \text{ cmH}_2\text{O} = \frac{3.51(\text{cmH}_2\text{O}) \times 1333.2(\text{Pa/cmHg})}{13.59(\text{cmH}_2\text{O/cmHg})} = 344.3 \text{ Pa}$$

TABLE 5.5 Average Composition of Seawater* for Selected Elements in Order of Decreasing Mass

Element	Concentration ($mol\ L^{-1}$)
Cl	0.536
Na	0.457
Mg	0.0555
S	0.0276
Ca	0.010
K	0.0097
Br	0.00081
C	0.0023

*There is a natural variability in concentrations of seawater at different locations around the world.

From Eq. 5.134, and since $J = kg\ m^2\ s^{-2}$ and $Pa = kg\ m^{-1}\ s^{-2}$,

$$c = \frac{\pi}{RT} = \frac{344.3(\text{Pa})}{8.3145(\text{J K}^{-1}\ \text{mol}^{-1}) \times 276.1(\text{K})} = 0.150\ \text{mol m}^{-3}$$
$$= 1.50 \times 10^{-4}\ \text{mol dm}^{-3}$$

Since the concentration of the solution is 10.0 g dm^{-3}, 10.0 g is 1.50×10^{-4} mol. Consequently,

$$\text{molar mass} = \frac{10.0\ \text{g}}{1.50 \times 10^{-4}\ \text{mol}} = 66\ 700\ \text{g mol}^{-1}$$

Besides being a valuable technique for high-molecular-mass materials, osmotic pressure effects are important in physiological systems. Another area for large-scale industrial application is in water desalination. Seawater contains approximately 35 000 parts per million (ppm) of dissolved salts consisting mostly of sodium chloride. See Table 5.5.

Reverse Osmosis

The osmotic pressure of seawater due to these salts is approximately 3.0 MPa (30.0 atm). Individually, all other elements not listed occur at less than 9 mg L^{-1} in seawater. A hydrostatic pressure in excess of 3.0 MPa applied to seawater will separate pure water through a suitable semipermeable membrane. This process is known as *reverse osmosis.* Unfortunately, a very slight amount of salt may still pass through such a membrane. When the water is used for agricultural pruposes, over time this small amount of salt can cause salt buildup. With time this buildup may force a shift in the crops planted to those that are resistant to salt. In the extreme, the buildup may cause the land to be lost to agriculture. Work continues to improve present membranes.

KEY EQUATIONS

Clapeyron equation:

$$\frac{dP}{dT} = \frac{\Delta H_m}{T\,\Delta V_m}$$

Clausius-Clapeyron equation:

$$\frac{d \ln P}{dT} = \frac{\Delta_{\text{vap}} H_m}{RT^2}$$

Trouton's rule:

$$\frac{\Delta_{\text{vap}} H_m}{T_b} = \Delta_{\text{vap}} S_m \approx 88\ \text{J K}^{-1}\ \text{mol}^{-1}$$

Raoult's law:

$$P_1 = x_1 P_1^* \qquad P_2 = x_2 P_2^*$$

Henry's law:

$$P_2 = k' x_2 \quad \text{or} \quad P_2 = k'' c_2$$

Definition of *partial molar quantity:*

$$X_i \equiv \left(\frac{\partial X}{\partial n_i}\right)_{T,P,n_j}$$

Definition of *chemical potential:*

$$\mu_i \equiv \left(\frac{\partial G}{\partial n_i}\right)_{T,P,n_j}$$

Condition for equilibrium between phases:

$$\mu_i^\alpha = \mu_i^\beta$$

Depression of the freezing point:

$$\Delta_{\text{fus}} T \approx \frac{M_1 R T_f^{*2}}{\Delta_{\text{fus}} H_m} \cdot m_2 = K_f m_2$$

where K_f is the *freezing point depression constant.*

Elevation of the boiling point:

$$\Delta_{\text{vap}} T \approx \frac{M_1 R T_b^{*2}}{\Delta_{\text{vap}} H_m} \cdot m_2 = K_b m_2$$

Osmotic pressure:

$$\pi = \frac{n_2 RT}{n_1 V_1^*} \approx cRT$$

PROBLEMS

Thermodynamics of Vapor Pressure

5.1. Diamonds have successfully been prepared by submitting graphite to high pressure. Calculate the approximate minimum pressure needed using $\Delta_f G = 0$ for graphite and $\Delta_f G = 2.90 \times 10^3$ J mol^{-1} for diamond. The densities of the two forms may be taken as independent of pressure and are 2.25 and 3.51 g cm^{-3}, respectively.

5.2. The molar entropy of vaporization of water is 108.72 J K^{-1} at 760 Torr. The corresponding densities of liquid water and water vapor are 0.958 kg dm^{-3} and 5.98×10^{-4} kg dm^{-3}, respectively. Calculate the change of pressure for a one-degree change in temperature.

5.3. Calculate the heat of vaporization of water at 373.15 K and 101.325 kPa using the Clausius-Clapeyron equation. The vapor pressure of water is 3.17 kPa at 298.15 K. Compare your answer to the CRC Handbook[4] value.

5.4. Liquid water and vapor are in equilibrium at the triple point of water (0.00603 atm and 273.16 K). Assuming that the enthalpy of vaporization of water does not change over the temperature range considered, calculate the equilibrium vapor pressure of water at 373.15 K. Comment on the assumption made here. ($\Delta_{vap}H^{\circ}$ = 40 656 J mol^{-1} at 1 atm.)

5.5. Estimate the vapor pressure of iodine under an external pressure of 101.3×10^6 Pa at 313.15 K. The density of iodine is 4.93 g cm^{-3}. The vapor pressure at 101.3 kPa is 133 Pa.

5.6. The cubic expansion coefficient is given by $\alpha = 1/V (\partial V/\partial T)_P$. According to Ehrenfest's or Tisza's theory, find the order of the transition. Suggest what a plot of α against T would look like near the transition point.

5.7. The vapor pressure of n-propanol is 1.94 kPa at 293 K and 31.86 kPa at 343 K. What is the enthalpy of vaporization?

5.8. The compound 2-hydroxybiphenyl (o-phenylphenol) boils at 286 °C under 101.325 kPa and at 145 °C under a reduced pressure of 14 Torr. Calculate the value of the molar enthalpy of vaporization. Compare this value to that given in the CRC Handbook.

5.9. Using Trouton's rule, estimate the molar enthalpy of vaporization of n-hexane, the normal boiling point of which is 342.10 K. Compare the value obtained to the value 31.912 kJ mol^{-1} obtained in vapor pressure studies.

5.10. The normal boiling point of toluene is 110.62 °C. Estimate its vapor pressure at 80.00 °C assuming that toluene obeys Trouton's rule.

[4] *Handbook of Chemistry and Physics*, 82nd ed., D. R. Lide, Ed., Boca Raton, FL: CRC Press, 2001.

5.11. 2-Propanone (acetone) boils at 329.35 K at 1 atm of pressure. Estimate its boiling point at 98.5 kPa using Crafts' rule.

5.12. The variation of the equilibrium vapor pressure with temperature for liquid and solid chlorine in the vicinity of the triple point is given by

$$\ln P_l = \frac{-2661}{T} + 22.76,$$

$$\ln P_s = \frac{-3755}{T} + 26.88.$$

Use P/pascal in the equations. Calculate the triple point pressure and temperature.

5.13. The boiling point of water at 102.7 kPa is 373.52 K. Calculate the value at 101.325 kPa (1 atm) using Crafts' rule.

5.14. The vapor pressure of water at 27.5 °C, a calibration temperature for glassware used in warmer climates, is 27.536 Torr under its own vapor pressure. Calculate the vapor pressure of water under an air pressure of 1.00 atm. Assume that air is inert. The density of water at 27.5 °C is 996.374 g dm^{-3}.

5.15. Following the derivation of the expression for ΔS in terms of ΔG in Eq. 3.161, derive an expression for ΔV, the volume change accompanying a transition from one state to another, in terms of ΔG starting with the definition given in Eq. 3.115.

5.16. Derive an equation for the temperature dependence of the vapor pressure of a liquid (analogous to the integrated form of the Clausius-Clapeyron equation) assuming that the vapor has the equation of state $PV = RT + M$ where M is a constant.

5.17. Calculate the vapor pressure above liquid ethanol at 35.0 °C when Ar is added until the total pressure is 100 bar. The density of liquid ethanol at this temperature is 0.7767 kg dm^{-3} and the true vapor pressure is 100.0 Torr.

5.18. A solid exists in two forms, A and B, whose densities are 3.5155 g cm^{-3} and 2.2670 g cm^{-3}, respectively. If the standard Gibbs energy change for the reaction A $\rightleftharpoons$ B is 240 kJ kg^{-1}, find the pressure at which the two forms of the solid are in equilibrium at 25 °C. Assume that the volume change in going from A to B is independent of the pressure.

5.19. What are the partial pressures of toluene (0.60 mole fraction) and benzene (mole fraction) in a solution at 60 °C? What is the total pressure in the vapor? The vapor pressures of the pure substances at 60 °C are as follows: toluene, 0.185 bar; benzene, 0.513 bar.

5.20. The normal boiling point of ethylene glycol ($C_2H_6O_2$) is 197 °C; its enthalpy of vaporization is 801 J mol^{-1}. Estimate the temperature at which ethylene glycol will boil in a vacuum distillation if the system were maintained at 50 Torr.

Raoult's Law, Equivalence of Units, and Partial Molar Quantities

5.21. Benzene and toluene form nearly ideal solutions. If, at 300 K, P^* (toluene) = 3.572 kPa and P^* (benzene) = 9.657 kPa, compute the vapor pressure of a solution containing 0.60 mol fraction of toluene. What is the mole fraction of toluene in the vapor over this liquid?

5.22. Often it is important to express one unit of concentration in terms of another. Derive a general expression to find the mole fraction x_2 in a two-component system where the molality is given as m_2.

5.23. Assuming that commercially available automotive antifreeze is pure ethylene glycol (it actually also contains relatively small amounts of added rust inhibitors and a fluorescent dye that helps to differentiate a radiator leak from condensation from the air conditioner), in what ratio *by volume* will antifreeze and water have to be mixed in order to have a solution that freezes at −20.0 °C? What will be the boiling point of this solution at 1 atm pressure? (MW = 62.02 g mol^{-1}, density = 1.1088 g cm^{-3}.)

***5.24.** The familiar term *molarity* is now discouraged by IUPAC because of the danger of confusion with molality. In its place, *concentration* is defined as the amount of substance 2, n_2, dissolved in unit volume of solution. Derive a general relation to find x_2 from the concentration c_2. Let the solution density be ρ.

5.25. Show that if a solute follows Henry's law in the form of $P_2 = k'x_2$, then the solvent must follow Raoult's law. (*Hint:* The use of the Gibbs-Duhem equation might prove useful.)

5.26. A 1.0 *m* solution of NaCl in water produces a freezing point depression of approximately 3.7 K. How can we account for this observation?

5.27. Derive a general expression to relate the molality m to concentration c_2.

5.28. An amalgam of 1.152 g of a metal dissolved in 100.0 g of mercury is heated to boiling. The partial pressure of mercury vapor over the boiling mixture is 754.1 Torr and the total pressure is 768.8 Torr. Find the atomic weight of the metal and, therefore, its identity.

***5.29.** The volume of a solution of NaCl in water is given by the expression

$$V/\text{cm}^3 = 1002.874 + 17.8213\, m + 0.873\,91\, m^2 - 0.047\,225\, m^3$$

where m is the molality. Assume that $m \propto n_{NaCl}$ and that $n_{H_2O} = 55.508$ mol, where $V^*_{H_2O} = 18.068$ cm^3. Derive an analytical expression for the partial molar volume of H_2O in the solution.

***5.30.** The partial molar volume of component 2 in a solution may be written as

$$V_2 = \left(\frac{\partial V}{\partial n_2}\right)_{n_1} = \frac{M_2}{\rho} - (M_1 n_1 + M_2 n_2)\frac{1}{\rho^2}\left(\frac{\partial \rho}{\partial n_2}\right)_{n_1}$$

where n_1 and M_1 are amount and molar mass of component 1 and n_2 and M_2 represent the same quantities for component 2. The density is ρ. Rewrite the expression in terms of the mole fractions x_1 and x_2.

***5.31.** Mikhail and Kimel, *J. Chem. Eng. Data, 6,* 533(1961), give the density of a water-methanol solution in g cm^{-3} at 298 K related to the mole fraction x_2 of the methanol through the equation

$$\rho/\text{g cm}^{-3} = 0.9971 - 0.289\,30x_2 + 0.299\,07x_2^2 - 0.608\,76x_2^3 + 0.594\,38x_2^4 - 0.205\,81x_2^5$$

Using the equation developed in Problem 5.30, calculate V_2 at 298 K when $x_2 = 0.100$.

5.32. Beckmann and Faust [*Z. Physik. Chemie, 89,* 235(1915)] found that a solution of chloroform in acetone in which the mole fraction of the latter is 0.713 has a total vapor pressure of 220.5 Torr at 28.15 °C. The mole fraction of acetone in the vapor is 0.818. The vapor pressure of pure chloroform at this temperature is 221.8 Torr. Assuming that the vapor behaves ideally, calculate the activity and the activity coefficient of chloroform.

5.33. When 12.5 g of A, a nonvolatile compound, is dissolved in 520.8 g of ethanol, the vapor pressure of the pure solvent, 56.18 Torr, is reduced to 55.24 Torr. Calculate the molar mass of compound A.

5.34. The following data are for mixtures of isopropanol (I) in benzene (B) at 25 °C.

x_I	0	0.059	0.146	0.362	0.521	0.700	0.836	0.924	1.0
P_I (Torr)	0	12.9	22.4	27.6	30.5	36.4	39.5	42.2	44.0
P_{tot}	94.4	104.5	109.0	108.4	105.8	99.8	84.0	66.4	44.0

Does this solution exhibit positive or negative deviation from Raoult's law? From a pressure-composition plot, estimate the activities a_I and a_B and activity coefficients f_I and f_B at x_I = 0.20, 0.50, and 0.80. [Data from Olsen and Washburn, *J. Phys. Chem., 41,* 457(1937).]

5.35. The vapor pressure of pure ethylene dibromide is 172 Torr and that of pure propylene dibromide is 128 Torr both at 358 K and 1 atm pressure. If these two components follow Raoult's law, estimate the total vapor pressure in kPa and the vapor composition in equilibrium with a solution that is 0.600 mol fraction propylene dibromide.

5.36. Calculate Henry's law constant and the vapor pressure of pure liquid A (molar mass = 89.5 g mol^{-1}) and that of 75.0 g of liquid A in solution with 1000 g of liquid B. Liquid B (molar mass = 185 g mol^{-1}) has a pressure in this solution of 430 Torr and the total solution pressure is 520 Torr.

***5.37.** Henry's law constants k' for N_2 and O_2 in water at 20.0 °C and 1 atm pressure are 7.58×10^4 atm and 3.88×10^4 atm, respectively. If the density of water at 20.0 °C is 0.9982 g cm^{-3}, calculate (a) the equilibrium mole fraction and (b) the concentration of N_2 and O_2 in water exposed to air at 20.0 °C and 1 atm total pressure. Assume in this case that air is 80.0 mol % N_2 and 20.0 mol % O_2.

5.38. Methane dissolves in benzene with a Henry's law constant of 4.27×10^5 Torr. Calculate methane's molal solubility in benzene at 25 °C if the pressure above benzene is 750 Torr. The vapor pressure of benzene is 94.6 Torr at 25 °C.

Thermodynamics of Solutions

5.39. In a molar mass determination, 18.04 g of the sugar mannitol was dissolved in 100.0 g of water. The vapor pressure of the solution at 298 K was 2.291 kPa, having been lowered by 0.0410 kPa from the value for pure water. Calculate the molar mass of mannitol.

5.40. A liquid has a vapor pressure of 40.00 kPa at 298.15 K. When 0.080 kg of an involatile solute is dissolved in 1 mol of the liquid, the new vapor pressure is 26.66 kPa. What is the molar mass of the solute? Assume that the solution is ideal.

***5.41.** Components 1 and 2 form an ideal solution. The pressure of pure component 1 is 13.3 kPa at 298 K, and the corresponding vapor pressure of component 2 is approximately zero. If the addition of 1.00 g of component 2 to 10.00 g of component 1 reduces the total vapor pressure to 12.6 kPa, find the ratio of the molar mass of component 2 to that of component 1.

5.42. Pure naphthalene has a melting point of 353.35 K. Estimate the purity of a sample of naphthalene in mol %, if its freezing point is 351.85 K ($K_f = 7.0$ K kg mol^{-1}).

5.43. Calculate the activity and activity coefficients for 0.330 mol fraction toluene in benzene. The vapor pressure of pure benzene is 9.657 kPa at 298 K. $P_2^* = 3.572$ kPa for toluene. The vapor pressure for benzene above the solution is $P_1 = 6.677$ kPa and for toluene $P_2 = 1.214$ kPa.

5.44. Calculate the mole fraction, activity, and activity coefficients for water when 11.5 g NaCl are dissolved in 100 g water at 298 K. The vapor pressure is 95.325 kPa.

5.45. Determine the range for the Gibbs energy of mixing for an ideal 50/50 mixture at 300 K. How does this value limit $\Delta_{mix}H$?

5.46. The mole fraction of a nonvolatile solute dissolved in water is 0.010. If the vapor pressure of pure water at 293 K is 2.339 kPa and that of the solution is 2.269 kPa, calculate the activity and activity coefficient of water.

***5.47.** A nonideal solution contains n_A of substance A and n_B of substance B and the mole fractions of A and B are x_A and x_B. The Gibbs energy of the solution is given by the equation

$$G = n_A\mu_A^\circ + n_B\mu_B^\circ + RT(n_A \ln x_A + n_B \ln x_B) + Cn_An_B/(n_A + n_B)$$

where C is a constant and describes the pair interaction.

a. Derive an equation for μ_A in the solution in terms of the quantities on the right-hand side. {*Hint:* $(\partial \ln x_A/\partial n_A)_{n_B} = (1/n_A) - [1/(n_A + n_B)]$.}

b. Derive a similar expression for the activity coefficient of A. Specify the conditions when the activity coefficient equals unity.

Colligative Properties

5.48. Calculate the mole fraction solubility of naphthalene at 25 °C in a liquid with which it forms an ideal solution. The $\Delta_{fus}H = 19.0$ kJ mol^{-1} for naphthalene at 25 °C. Its normal melting point is 80.2 °C.

5.49. Using Henry's law, determine the difference between the freezing point of pure water and water saturated with air at 1 atm. For N_2 at 298.15 K,

$$(k'')^{-1} = 2.17 \times 10^{-8} \text{ mol dm}^{-3} \text{ Pa}^{-1}$$

For O_2 at 298.15 K,

$$(k'')^{-1} = 1.02 \times 10^{-8} \text{ mol dm}^{-3} \text{ Pa}^{-1}$$

5.50. Using van't Hoff's equation, calculate the osmotic pressure developed if 6.00 g of urea, $(NH_2)_2CO$, is dissolved in 1.00 dm^3 of solution at 27 °C.

5.51. The apparent value of K_f in 1.50-molal aqueous sucrose ($C_{12}H_{22}O_{11}$) solution is 2.17 K kg mol^{-1}. The solution does not behave ideally; calculate its activity and activity coefficient ($\Delta_{fus}H^\circ = 6009.5$ J mol^{-1}).

5.52. A 0.85-g sample is dissolved in 0.150 kg of bromobenzene. Determine the molar mass of the solute if the solution boils at 429.0 K at 1 atm pressure. The normal boiling point of bromobenzene is 428.1 K and the boiling point elevation constant is 6.26 K kg mol^{-1}.

***5.53.** If in a colligative properties experiment a solute dissociates, a term i known as van't Hoff's factor, which is the total concentration of ions divided by the nominal concentration, must be included as a factor. Thus, for the lowering of the freezing point, $\Delta_{fus}T = imK_f$. Derive an expression that relates to the degree of dissociation α and to ν, the number of particles that would be produced if the solute were completely dissociated. Then calculate van't Hoff's i factor and α for a 0.010-m solution of HCl that freezes at 273.114 K.

5.54. In an osmotic pressure experiment to determine the molar mass of a sugar, the following data were taken at 20 °C:

π/atm	2.59	5.06	7.61	12.75	18.13	23.72
$m_2 V^{-1}$/g dm^{-3}	33.5	65.7	96.5	155	209	259

Estimate the molar mass of the sugar. If the sugar is sucrose, what is the percentage error and why?

5.55. When 3.78 g of a nonvolatile solute is dissolved in 300.0 g of water, the freezing point depression is 0.646 °C. Calculate the molar mass of the compound. $K_f = 1.856$ K kg mol^{-1}.

5.56. Calculate the elevation in the boiling point of water if 6.09 g of a nonvolatile compound with molar mass of 187.4 g mol^{-1} is dissolved in 250.0 g of water. Compare the values obtained using Eq. 5.125 and Eq. 5.126. The value of $K_b = 0.541$ K kg mol^{-1}; $\Delta_{vap}H = 40.66$ kJ mol^{-1}.

5.57. Suppose that you find in the older literature the vapor pressure P of a liquid with molar mass of 63.9×10^{-3} kg mol^{-1} listed with P in mmHg as

$$\log P = 5.4672 - 1427.3\, T^{-1} - 3169.3\, T^{-2}$$

The densities of the liquid and vapor phases are 0.819 kg dm^{-3} and 3.15×10^{-4} kg dm^{-3}, respectively. Calculate the $\Delta_{vap}H$ at the normal boiling point, 398.4 K. How do you handle the fact that P is listed in mmHg?

5.58. Calculate the osmotic pressure of seawater using the data of Table 5.5. Assume a temperature of 298 K and that the concentration of the additional salts not listed does not substantially contribute to the osmotic pressure.

Essay Questions

5.59. Describe the form of a typical $P\theta$ diagram and how the Gibbs-energy diagram may be generated for a one-component system. What is the requirement of stability for each region in the $P\theta$ diagram?

5.60. Detail the steps in going from the Clapeyron equation to the Clausius-Clapeyron equation. What specific assumptions are made?

5.61. Explain why Trouton's rule, according to which the entropy of vaporization is 88 J K^{-1} mol^{-1}, holds fairly closely for normal liquids.

5.62. Describe three colligative properties and comment on their relative merits for the determination of molar masses of proteins.

5.63. Show mathematically how the chemical potential is the driving force of diffusion for component A between two phases α and β.

5.64. Why do positive and negative deviations from Raoult's law occur?

SUGGESTED READING

J. R. Anderson, "Determination of boiling and condensation temperatures." In *Techniques of Chemistry,* A. Weissberger and B. W. Rossiter, Eds., *V,* 199, New York: Wiley-Interscience, 1971.

A. A. Bondi, *Physical Properties of Molecular Crystals, Liquids and Gases,* New York: Wiley, 1968.

D. Eisenberg and W. Kauzmann, *The Structure and Properties of Water,* New York: Oxford University Press, 1969.

A. Findlay, *Phase Rule* (9th ed., revised and enlarged by A. N. Campbell and N. O. Smith), New York: Dover Pub., 1951.

J. H. Hildebrand, J. M. Prausnitz, and R. L. Scott, *Regular and Related Solutions, The Solubility of Gases, Liquids, and Solids,* New York: Van Nostrand Reinhold Co., 1970.

I. M. Klotz and R. M. Rosenberg, *Chemical Thermodynamics: Basic Theory and Methods* (5th ed.), New York: Wiley, 1994.

R. G. Laughlin, *The Aqueous Phase Behavior of Surfactants,* San Diego: Academic Press, 1994.

E. L. Skau and J. C. Arthur, "Determination of melting and freezing temperatures." In *Techniques of Chemistry,* A. Weissberger and B. W. Rossiter, Eds., *V,* 199, New York: Wiley-Interscience, 1971.

R. A. Swalin, *Thermodynamics of Solids* (2nd ed.), New York: Wiley, 1972.

H. N. V. Temperley, *Changes of State,* London: Cleaver-Hune, 1956.

R. H. Wagner and L. D. Moore, "Determination of Osmotic Pressure," in A. Weissberger (Ed.), *Physical Methods of Organic Chemistry,* Part 1 (3rd ed.), New York: Interscience Pub., 1959.

Phase Equilibria

PREVIEW

This chapter is concerned mainly with the application of the famous **phase rule** of J. Willard Gibbs to a number of examples. The phase rule is important when one is concerned with equilibria between different phases. It provides an expression for the *number of degrees of freedom,* which is the number of intensive variables, such as temperature, pressure, and concentration, that can be independently varied without changing the number of phases.

Whenever one is concerned with systems in which there is more than one phase, the phase rule is of great value. For example, it is of paramount importance in the field of petrology, particularly in connection with mineral deposits, and in the field of metallurgy, where modifying physical properties depends upon its successful application.

In binary-liquid systems, i.e., systems involving two liquids, there is a finite temperature and pressure range over which liquid and vapor coexist. One can represent the total pressure P of the vapor above the liquids in terms of the individual vapor pressures of the pure liquids and the mole fractions in the vapor of one of the components. When P is plotted against this mole fraction, the curve is called the *vapor curve* (for an example, see Figure 6.2). The curve representing the liquid and liquid-vapor boundary lies above the vapor curve and is called the *liquid curve.*

The existence of the region between the two curves allows *distillation* to be carried out. A *tie line* between the liquid and vapor curves determines the ratio of liquid to vapor through the use of the *lever rule,* which is explained in Section 6.3. Both positive and negative deviations from Raoult's law occur, and there is the possibility of *azeotropes,* which are mixtures that correspond to a maximum or minimum in the boiling point curve. When there are azeotropes, it is impossible by normal distillation to separate both pure components of a liquid mixture.

It is often convenient to ignore the vapor in a mixture and consider only the liquid-solid equilibrium. Often there is a single composition that has the lowest melting point in the phase diagram. This is called the *eutectic composition,* and the temperature of melting is the *eutectic temperature* (Section 6.4). The method of *thermal analysis* is useful for constructing such phase diagrams. Sometimes a chemical reaction occurs when there is a phase change; such reactions are known as *phase reactions* or as *peritectic reactions.* The point in the phase diagram at which they occur is known as a *peritectic point.*

This chapter also deals with equilibria involving three components (i.e., ternary systems).

OBJECTIVES

After studying this chapter, the student should be able to:

- Determine the number of components and phases in a system.
- Use the phase rule to determine the number of degrees of freedom, f, or the variance of a system.
- Recognize the features, such as the presence of allotropes, in a one-component system.
- Calculate the pressure in the vapor curve of a two-component system.
- Discuss the liquid-vapor equilibrium of binary liquid systems and determine the concentrations present by using the lever rule.
- Describe liquid-vapor systems not obeying Raoult's law and the process of isothermal distillation.
- Plot temperature-composition diagrams for boiling point curves and describe the process of distillation in detail.
- Describe in detail the different types of azeotropes and the result of boiling them.
- Explain how steam distillation works and calculate the ratio of the two components in the distillation.
- Discuss condensed binary systems, including conjugate solutions, isopleths, upper and lower consolute temperatures, and eutectics.
- Show how thermal analysis is done to determine a phase diagram.
- Discuss the salt-water phase diagram, solid solution, partial miscibility, and compound formation.
- Describe the Krafft boundary and items related to systems having liquid crystal, colloid, or gels present.
- Determine compositions from ternary systems and describe the behavior of the types of systems described.
- Plot three-dimensional ternary phase diagrams and read the information on them.

Chapter 5 was mainly concerned with general principles related to equilibria involving vapors, liquids, solids, and solutions. In this chapter we apply these principles to a number of additional examples, and we will see that they lead to useful ways of presenting and classifying experimental data. The approach we take has been found useful in a wide range of scientific and technical problems in such fields as chemistry, engineering, geology, metallurgy, and physics.

6.1 Equilibrium Between Phases

Number of Components

In Section 5.1 we considered what constituted a phase (p), which is easily distinguished since it is a homogeneous and distinct part of a system separated by definite boundaries from the rest of the system. A brief review of those concepts and an extension to multiple solids is provided by the following example.

EXAMPLE 6.1 How many phases are present in the following system at equilibrium?

$$CaCO_3(s) \rightleftharpoons CaO(s) + CO_2(g)$$

Solution Although there are two solids present, each has its own structure and is separated by distinct boundaries. Therefore, we have two solid phases and one gas phase, a total of three phases.

We have not concerned ourselves with discussing what constitutes the **number of components** c in a phase diagram, since we have dealt with at most only two nonreacting constituents. However, this idea needs careful development. In the sense used here the number of components is the smallest number of independent chemical *constituents* needed to fix the composition of *every* phase in the system. If we consider water, it consists of only one component even though it may exist in three different phases: solid, liquid, and vapor. However, if sodium chloride is now dissolved in the water, the system becomes a two-component system and remains so even though under some conditions the salt may separate and form a pure solid phase within the system. A person might conjecture that since the salt dissociates into ions, additional components are present. However, this is not the case. Even though there are two more constituents, namely Na^+ and Cl^- ions, there is a material balance since the reduction in amount of NaCl dissociated must equal the amounts of cations or anions formed. There is in addition a requirement of electroneutrality since the number of Na^+ ions must equal the number of Cl^- ions. Consequently, the material balance and electroneutrality conditions reduce the number of constituents, H_2O, NaCl, Na^+, and Cl^-, to two; that is, $c = 4 - 1 - 1 = 2$. The same argument may be made for the small amount of H_3O^+ and OH^- formed from the dissociation of water but will not otherwise be considered since there again exist material balance and electroneutrality expressions. Thus, although there is no unique set of components from the possible constituents, the *number* of components is unique. (See a further discussion of the water-phase diagram with a non-volatile solute in Section 6.6.)

The number of components is unique.

If a chemical reaction can take place between constituents of a solution, the number of components is reduced by the number of equilibrium conditions. Considering the system

$$PCl_5 \rightleftharpoons PCl_3 + Cl_2 \tag{6.1}$$

we recognize three distinct chemical species. This number of species is reduced by the independent equilibrium condition and, therefore, $c = 3 - 1 = 2$. Another reduction in the number of components is possible if we start with pure PCl_5, in which case $[PCl_3] = [Cl_2]$. In this case the number of components is unity because of the additional mathematical relation.

EXAMPLE 6.2 How many components are present when ethanol and acetic acid are mixed assuming the reaction to occur to equilibrium?

Solution At first sight we might predict two components because there are two constituents, HOAc and EtOH. However, these constituents react and ethyl acetate and water are also present at equilibrium. This raises the number of constituents from 2 to 4.

$$\text{HOAc} + \text{EtOH} \rightleftharpoons \text{EtOAc} + \text{HOH}$$

But now the equilibrium condition is applied and reduces by 1 the number of components (i.e., the number of constituents whose value is required to specify the system). Furthermore, since EtOAc and HOH must be formed in equal amounts, another mathematical condition exists, reducing the number of components back to 2; thus, $c = 4 - 2 = 2$.

There are many systems in which it might be difficult to decide how to apply the equilibrium condition to arrive at the correct number of components. For example, we know that an equilibrium condition can be written for water, hydrogen, and oxygen. If these three constituents are present in the absence of a catalyst, we have a three-component system because the equilibrium that could reduce the number of constituents, $H_2 + \frac{1}{2} O_2 \rightleftharpoons H_2O$, for all practical purposes is not achieved. There are other instances where our criterion of establishing equilibrium is not so clear-cut as this. For example, a mixture of water, hydrogen, and oxygen is a three-component system at room temperature in the absence of a catalyst, but at higher temperatures, because the equilibrium becomes established, it consists of two components. Where does the shift from a three-component system to a two-component system occur? To resolve this problem we must consider the time required to make the measurement of the specific property of the system. If the equilibrium shifts with a change in the measured variable in a time very short in comparison to the time required for the measurement, the equilibrium condition is applicable. In our example the number of components at high temperature is two. However, if the time required to make the measurement of the property is very short in comparison to the time required for equilibrium to be reestablished (that will be the case at low temperatures), then effectively there is no equilibrium in the H_2O–H_2–O_2 system and it remains a three-component system.

Degrees of Freedom

Another important idea for understanding phase diagrams is the concept of the *number of* **degrees of freedom** f or the *variance* of a system. The number of degrees of freedom is the number of intensive variables, such as temperature, pressure, and concentration, that can be independently varied without changing the number of phases. Since some variables are fixed by the values chosen for the independent variables and by the requirement of equilibrium, it is only the number of those that remain unencumbered that is referred to as the number of degrees of freedom. Stated differently, the number of degrees of freedom is the number of variables that must be fixed in order for the condition of a system at equilibrium to be completely specified.

If the system has one degree of freedom, we say it is *univariant.* If a system has two degrees of freedom, it is a *bivariant* system, etc. Thus pure water is univariant since, at any given temperature, the pressure of vapor in equilibrium with liquid is fixed (only the one variable, temperature, may be varied independently).

The Phase Rule

The Phase Rule

The application to a system of the thermodynamic techniques developed in Chapter 5 can be laborious if each system must be treated as an individual case. However, the general conditions for equilibrium between phases, the **phase rule,** was deduced theoretically by J. Willard Gibbs in 1875–1876. We assume in this development that the equilibrium between phases is not influenced by gravity, electrical forces, magnetic forces, or surface forces but only by pressure, temperature, and composition. Under these conditions the phase rule is given by the expression

$$f = c - p + 2 \tag{6.2}$$

where the term 2 is for the two variables, temperature and pressure.

This result may be derived by expanding upon the development in Section 5.6. We showed there that, for equilibrium to exist throughout a system consisting of more than one phase, the chemical potential of any given component must have the same value in every phase. From Eq. 5.66 on p. 204 we have that $\mu_i^\alpha = \mu_i^\beta$ at constant T and P. Extending this to a system consisting of p phases, we have

$$\mu_1^\alpha = \mu_1^\beta = \mu_1^\gamma = \cdots = \mu_1^p \tag{6.3}$$

There will be an equation similar to Eq. 6.3 for each component present in the system. This leads to an array of information needed to specify the system, along with the temperature and pressure:

$$\begin{aligned} \mu_1^\alpha &= \mu_1^\beta = \mu_1^\gamma = \cdots = \mu_1^p \\ \mu_2^\alpha &= \mu_2^\beta = \mu_2^\gamma = \cdots = \mu_2^p \\ \mu_3^\alpha &= \mu_3^\beta = \mu_3^\gamma = \cdots = \mu_3^p \\ &\vdots \qquad \vdots \qquad \vdots \qquad\qquad \vdots \\ \mu_c^\alpha &= \mu_c^\beta = \mu_c^\gamma = \cdots = \mu_c^p \end{aligned} \tag{6.4}$$

Our task now is to reduce the information of this array to the bare minimum required to describe the system at constant T and P. In any phase containing c components, the composition is completely specified by $c - 1$ concentration terms if they are expressed in *mole fractions* or *weight percent.* This is the case since the last term is automatically determined from the remaining mole fraction or remaining weight percent. Thus, for a system of p phases, $p(c - 1)$ concentration terms are required to define the composition completely, and an additional two terms are required for the temperature and pressure of the system. Thus, $p(c - 1) + 2$ terms are required. But from Eq. 6.4 we see that each line contains $p - 1$ independent equations specifying the state of the system. Each equality sign is a condition imposed on the system that reduces by one the pieces of information required. Since there are $c(p - 1)$ of these expressions defining $c(p - 1)$ independent variables, the total number of variables left to be defined (i.e., the number of degrees of freedom) is

$$f = [p(c - 1) + 2] - [c(p - 1)] = c - p + 2 \quad (6.5)$$

This is of course the phase rule given in Eq. 6.2.

6.2 One-Component Systems

In Section 5.1 we saw an example of phase equilibrium in a one-component system, water. We will now apply the phase rule to another one-component system, sulfur. The general features of the phase diagram of sulfur are shown in Figure 6.1, with P plotted against T. There are several features in this figure not present in Figure 5.1 for water. The stable form of sulfur at 1 atm pressure and room temperature is a crystalline form called *(ortho)rhombic* sulfur. (See Section 16.2, p. 850, for the details on the arrangement of atoms in this form of sulfur.) As rhombic sulfur is heated slowly at 1 atm, it transforms to a different crystalline form called *monoclinic* sulfur at 368.55 K. This is an example of polymorphism (see Section 5.1).

Allotropes

The name **allotrope** (Greek *allos,* other; *tropia,* turning) refers to each of the crystalline forms when this type of transformation occurs in elements.

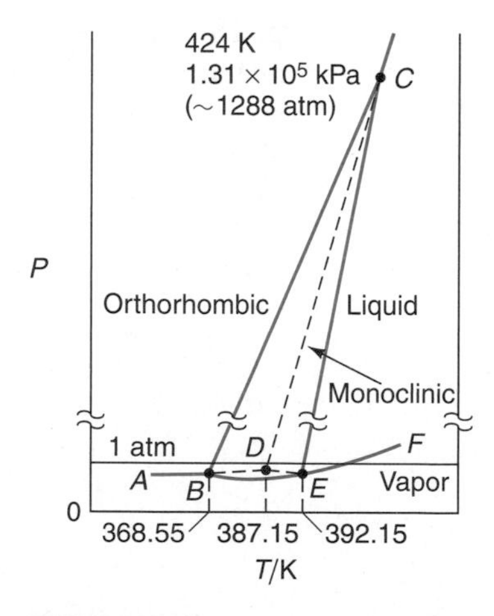

FIGURE 6.1
Phase diagram for sulfur. Temperature along abscissa refers to transitions at 1 atm.

At any particular pressure along line BC there is a definite temperature, called the *transition point,* at which one form will change reversibly to the other. In sulfur the transition point at pressures less than 1 atm (101.325 kPa) lies below the melting point of the solid. Each crystalline form possesses a definite range of stable existence. *Enantiotropy* (Greek *enantios,* opposite; *tropia,* turning) is the name given to this phenomenon and the crystalline forms are said to be *enantiotropic.*

Monoclinic sulfur melts to the liquid along line CE. However, if the rhombic form is rapidly heated at 1 atm pressure, the transformation temperature of 368.55 K is bypassed and the rhombic form melts directly to liquid sulfur at 387 K. When the rhombic sulfur is in equilibrium with liquid (dashed lines in Figure 6.1), we have an example of a metastable equilibrium since this equilibrium position lies in the region of the more thermodynamically stable monoclinic form of sulfur. This is not a true equilibrium, but it appears to be one because the process of change into the more favored form is slow. In this region both the rhombic and liquid forms of sulfur can decrease their Gibbs energy by converting to the favored monoclinic form.

The intersection of the lines BC and CE is brought about by the difference in density of the two crystalline forms. The density of rhombic sulfur is greater than that of the monoclinic form, which has a density greater than that of the liquid.

Therefore the lines BC and EC have positive slopes. The variation of vapor pressure with temperature may be handled using the Clapeyron equation. There are three triple points of stable equilibrium in this system:

Point B at 368.55 K: $S_{rhombic} \rightleftharpoons S_{monoclinic} \rightleftharpoons S_{vapor}$

Point E at 392.15 K: $S_{monoclinic} \rightleftharpoons S_{liquid} \rightleftharpoons S_{vapor}$

Point C at 424 K: $S_{rhombic} \rightleftharpoons S_{monoclinic} \rightleftharpoons S_{liquid}$

At these three points (B, C, E), there is one component, $c = 1$, and there are three phases, $p = 3$. By the phase rule then, $f = c - p + 2$, and we have

$$f = 1 - 3 + 2 = 0 \tag{6.6}$$

For an invariant system, no change is possible unless there is a phase change.

From this we see that there are no degrees of freedom; the system is *invariant*. This means that the equilibrium of the three phases automatically fixes both the temperature and pressure of the system, and no variable may be changed without reducing the number of phases present.

The curves AB, BC, BE, CE, and EF describe the equilibria between two phases. The phase rule requires that along these curves

$$f = c - p + 2 = 1 - 2 + 2 = 1 \tag{6.7}$$

The system is thus *univariant*. Thus under conditions of two phases present either temperature or pressure may be varied, but once one of them is fixed along the respective equilibrium line, the final state of the system is completely defined.

In the four regions where single phases exist (namely rhombic, monoclinic, liquid, or vapor) we have $f = 1 - 1 + 2 = 2$ and the system is *bivariant*. To define the state of the system completely the two variables temperature and pressure must both be specified.

Point D is a metastable triple point with two-phase equilibrium occurring along lines BD, CD, and DE. The phase rule predicts the number of degrees of freedom regardless of the fact that the system is metastable, since the metastable system, within the time scale of the measurements, behaves as if it were at equilibrium. Thus we find invariant and univariant metastable systems as we did earlier for the corresponding stable equilibria.

Supercooling

We met another case of metastable equilibrium earlier, on p. 182, where we saw that it occurs with liquid water. It has long been known that water can be cooled to as much as 9.4 K below its normal freezing point without solidification occurring, a process called **supercooling** discovered in 1724 by Fahrenheit (1686–1736). Super-cooled water is thus another example of a metastable system, and it becomes *unstable* in the presence of solid ice.

6.3 Binary Systems Involving Vapor

Liquid-Vapor Equilibria of Two-Component Systems

Raoult's law (Section 5.4) may be applied to binary solutions of volatile components to predict their total pressure. A curve such as Figure 5.8 in which the vapor pressure is plotted against x (the mole fraction of one of the liquid components) is incapable of describing states of the system that are completely gaseous at lower pressures. In order to describe the boundary between the liquid-vapor and pure vapor regions, we introduce the mole fraction of component 1 in the vapor, y_1. By Dalton's law this is just the ratio of the partial pressure of component 1 to the total pressure,

$$y_1 = \frac{P_1}{P} \tag{6.8}$$

Our task then is to describe the total pressure in terms of the mole fraction of component 1 in the vapor state and the individual pressures of the pure components.

Raoult's law, $P_1 = x_1 P_1^*$, and the expression for P

$$P = P_1 + P_2 = x_1 P_1^* + (1 - x_1)P_2^* = P_2^* + (P_1^* - P_2^*)x_1 \tag{6.9}$$

may be substituted into Eq. 6.8 and the resulting equation solved for x_1:

$$x_1 = \frac{y_1 P_2^*}{P_1^* + (P_2^* - P_1^*)y_1} \tag{6.10}$$

This expression may now be substituted back into Eq. 6.9 with the elimination of x_1. The result is

$$P = \frac{P_1^* P_2^*}{P_1^* + (P_2^* - P_1^*)y_1} \tag{6.11}$$

When P is plotted against y_1, or y_2 as in the vapor curve shown in Figure 6.2, the curve is concave upward; P becomes P_1^* when $y_1 = 1$ and P_2^* when $y_1 = 0$.

In Figure 6.2 the pressure-composition curve for the mixture isobutyl alcohol and isoamyl alcohol is plotted. This mixture behaves almost ideally (i.e., almost as predicted from Raoult's law). The upper curve is a straight line giving the boundary

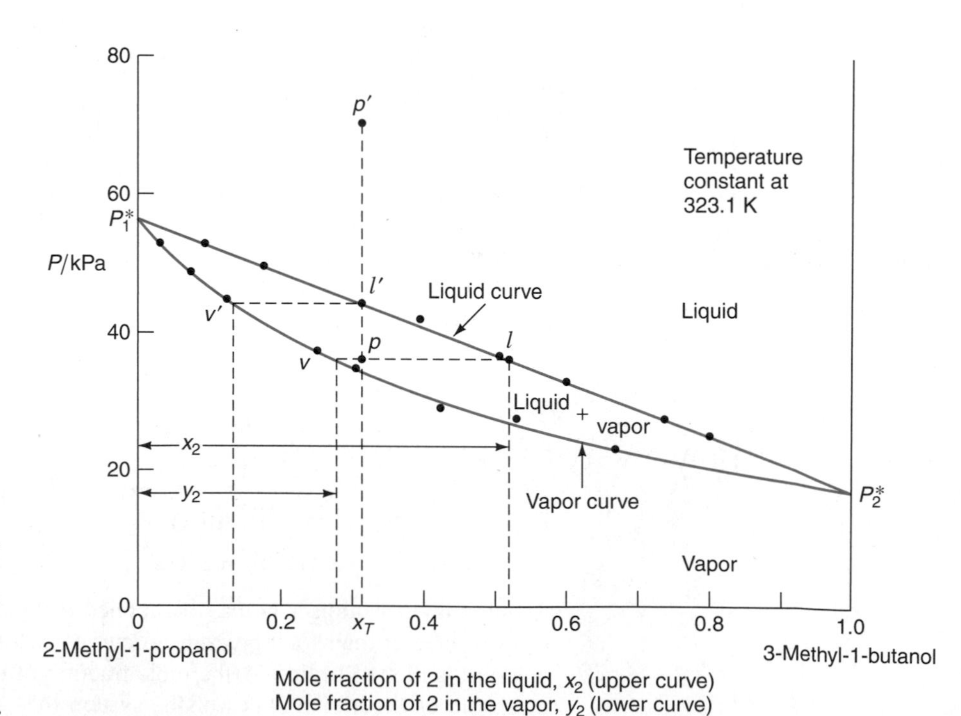

FIGURE 6.2
Pressure-composition diagram for two volatile components showing liquid-vapor equilibrium as a function of vapor pressure and mole fraction for the system isobutyl alcohol (component 1)–isoamyl alcohol (component 2) at 323.1 K.

between the pure liquid and liquid-vapor regions. This line is called the *liquid curve* and is merely a plot of the total vapor pressure for the liquid mixture against x_2. Above this line, vapor will condense to a liquid.

The lower curve is a plot of P against y_2 from Eq. 6.11 and is called the *vapor curve.* It gives the phase boundary between the liquid-vapor and pure vapor regions. Below this line, liquid cannot exist in equilibrium.

In the region marked Liquid, only one phase exists. There are two components so that from the phase rule, $f = 2 - 1 + 2 = 3$. Since Figure 6.2 is drawn at a specific temperature, one degree of freedom is already determined. Only two more variables are needed to specify the state of the system completely. A choice of P and x_1 then completes the description. In the region marked Vapor, specific values of P and y_1 will define the system.

In the region marked Liquid + vapor, there are two phases present and the phase rule requires specification of only one variable since T is fixed. This variable may be P, x_1, or y_1. To demonstrate this, consider point p in Figure 6.2. The value x_T at point p corresponds to the mole fraction of component 2 in the entire system. To find the compositions of both liquid and vapor in equilibrium at this pressure, we draw a horizontal line through the point p corresponding to constant pressure. This line, called a **tie line,** intersects the liquid curve at l and the vapor curve at v. The compositions x_2 and y_2 corresponding to l and v give the mole fractions of component 2 in the liquid phase and vapor phase, respectively. The intersection of the vertical composition line representing x_2 or y_2 with the respective liquid or vapor curves gives the pressure. The tie line at that pressure gives the other composition variable. Thus the system is fully defined by specifying only P or x_1, or y_1 in the two-phase region, since $x_1 = 1 - x_2$.

It is often necessary to determine the relative amounts of the components in the two phases at equilibrium. When point p in Figure 6.2 lies close to point v, the amount of substance in the vapor state is large in comparison to the amount in the liquid state. Conversely, if p were close to l, the liquid phase would predominate. To make this observation quantitative, we will obtain an expression for the ratio of the amount of substance in the liquid state, n_l, to the amount in the vapor state, n_v.

The total amount present is given by the mass balance

$$n = n_l + n_\mathrm{v} \tag{6.12}$$

A similar expression can be written for each component; thus

$$n_1 = n_{l,1} + n_{\mathrm{v},1} \tag{6.13}$$

In terms of mole fractions (see Figure 6.2 for x_T)

$$x_T = \frac{n_1}{n}, \qquad x_1 = \frac{n_{l,1}}{n_l}, \qquad y_1 = \frac{n_{\mathrm{v},1}}{n_\mathrm{v}} \tag{6.14}$$

Substituting these expressions into Eq. 6.13 gives

$$nx_T = n_l x_1 + n_\mathrm{v} y_1 \tag{6.15}$$

Substitution for n from Eq. 6.12 gives, after rearrangement,

$$\frac{n_l}{n_\mathrm{v}} = \frac{y_1 - x_T}{x_T - x_1} = \frac{\overline{p\mathrm{v}}}{\overline{lp}} \tag{6.16}$$

The Lever Rule

where $\overline{pv}$ is the length of the line segment between p and v, and $\overline{lp}$ is the length of the line segment between l and p. This is a statement of the **lever rule.**

EXAMPLE 6.3 Determine the mass percentage of carbon tetrachloride CCl_4 ($P_1^* = 114.5$ Torr) in the vapor phase at equilibrium in a 1:1 mole ideal solution with trichloromethane $CHCl_3$ ($P_2^* = 199.1$ Torr) at 25 °C.

Solution Eq. 6.16 provides a way to determine the relative amount of the liquid phase and vapor phase present at equilibrium. In this example, the relative amounts of the two components in the vapor state are required. Let the mole fraction of CCl_4 in the vapor state be y_1 and that of $CHCl_3$ be y_2 above the solution having mole fraction $x_1 = 0.5$ CCl_4. Then

$$y_1 = \frac{P_1}{P} \quad \text{and} \quad y_2 = \frac{P_2}{P}$$

The ratio is

$$\frac{y_1}{y_2} = \frac{P_1}{P_2} = \frac{P_1^* x_1}{P_2^* x_2}$$

Since the mole fraction is proportional to the mass, W_1, divided by the molar mass, M_1, we can write

$$\frac{y_1 M_1}{y_2 M_2} = \frac{W_1}{W_2} = \frac{114.5 \text{ Torr} \times 0.500 \times 53.823 \text{ g mol}^{-1}}{199.1 \text{ Torr} \times 0.500 \times 119.378 \text{ g mol}^{-1}}$$

$$\frac{W_{CCl_4}}{W_{CHCl_3}} = 0.741$$

Thus the ratio is 0.741 to 1. So the mass percent of carbon tetrachloride in the vapor is

$$\frac{0.741}{1.741} \times 100\% = 42.6\%.$$

The vapor contains relatively more of the more volatile component than does the liquid.

An interesting feature of curves similar to those in Figure 6.2 is that the vapor always contains relatively more of the more volatile component than does the liquid. This applies to all liquid mixtures and is useful in interpreting even those systems not obeying Raoult's law. Assuming that the vapors behave ideally, we have from Eq. 6.8 and Dalton's law

$$y_1 = \frac{P_1}{P_1 + P_2} \tag{6.17}$$

If the liquid mixture in contact with the vapor follows Raoult's law, we may express the mole fraction in the vapor phase in terms of the mole fractions in the liquid state. Thus

$$y_1 = \frac{x_1 P_1^*}{x_1 P_1^* + x_2 P_2^*} \tag{6.18}$$

Taking the ratio of y_1 to x_1 we have, from Eq. 6.18,

$$\frac{\text{mole fraction of component 1 in vapor}}{\text{mole fraction of component 1 in liquid}} = \frac{y_1}{x_1} = \frac{1}{x_1 + x_2(P_2^*/P_1^*)} \tag{6.19}$$

In a binary solution, $x_1 + x_2 = 1$. Therefore the ratio of the mole fraction in the vapor state to that in the liquid state can be unity only if the pressure $P_1^* = P_2^*$. If $P_2^* > P_1^*$, the denominator is greater than 1 and therefore $y_1 < x_1$. Thus, if the liquid of pure component 2 has a higher vapor pressure than that of pure component 1 ($P_2^* > P_1^*$), the vapor will contain relatively more of component 2 than does the liquid that is in equilibrium with it.

Isothermal Distillation

Application of the aforementioned principle is found in **isothermal distillation.** In this process the temperature of a liquid mixture is held constant and the vapor is progressively removed. As the process proceeds, the composition of the liquid is altered. As an illustration, consider a mixture whose composition is represented by point p' in Figure 6.2. As the pressure on the liquid is reduced by removing vapor, the solution does not reach equilibrium with the vapor until the pressure at point l' is reached. Then the vapor has composition v'. This vapor is much richer in the more volatile component than is the liquid. Hence, as this vapor is removed, the remaining solution will become concentrated in the less volatile component, changing composition along line $\overline{l'\,l}$. This method is not very commonly used for separation, but the method is particularly useful if the mixture were to decompose upon heating when distilled at atmospheric pressure. It also is useful when one of the components is much more volatile than the other. For instance, in the preparation of freeze-dried coffee, water is removed from the frozen coffee solution by the sublimation of ice, leaving behind a rigid, porous structure that is readily soluble.

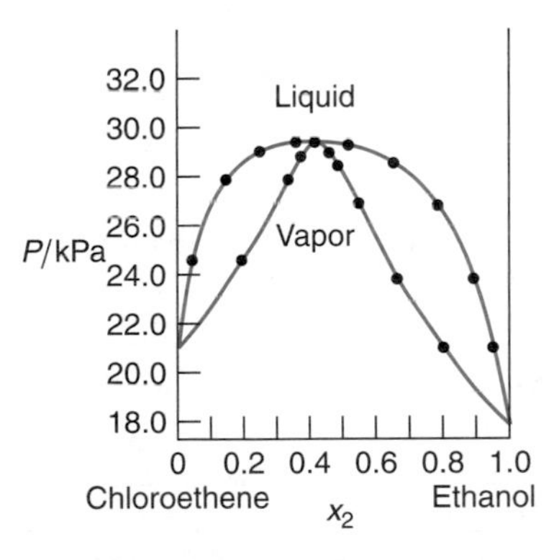

FIGURE 6.3
Liquid-vapor equilibrium as a function of vapor pressure and mole fraction for the system chloroethene–ethanol at 313.1 K. Dots represent data points.

Liquid-Vapor Equilibrium in Systems Not Obeying Raoult's Law

When Raoult's law holds reasonably well for completely miscible liquid pairs, the total vapor pressure in a binary solution will be *intermediate* between the vapor pressures of the two pure components. The pressure-composition diagram will be similar to that in Figure 6.2.

When there are large, positive deviations from Raoult's law (Section 5.4), there will be a *maximum* in the total vapor pressure curve. The liquid-vapor equilibrium will then be established, as shown in Figure 6.3 for the chloroethene–ethanol system. This curve shows a maximum at 38 mol % ethanol at a constant temperature of 313.1 K.

In a similar manner, there may be a *minimum* in the total vapor-pressure curve for *negative deviations* from Raoult's law. The liquid-vapor equilibrium is established as shown in Figure 6.4 for the acetone–chloroform system. Here a minimum occurs at approximately 58 mol % chloroform at 308 K.

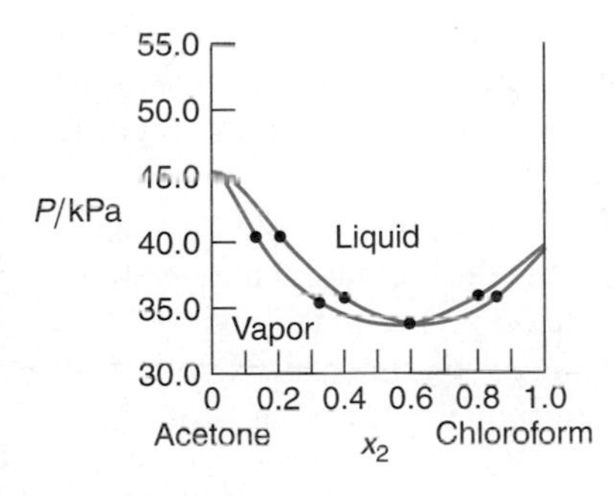

FIGURE 6.4
Liquid-vapor equilibrium as a function of vapor pressure and mole fraction for the system acetone–chloroform at 308 K. Dots represent data points.

Temperature-Composition Diagrams: Boiling Point Curves

The plots of vapor pressure against composition in the last two subsections have been made at constant temperature. It is more common to work at constant pressure, and temperature-composition diagrams are then required to detail the behavior of the liquid-vapor equilibrium. We first consider a full representation of a binary

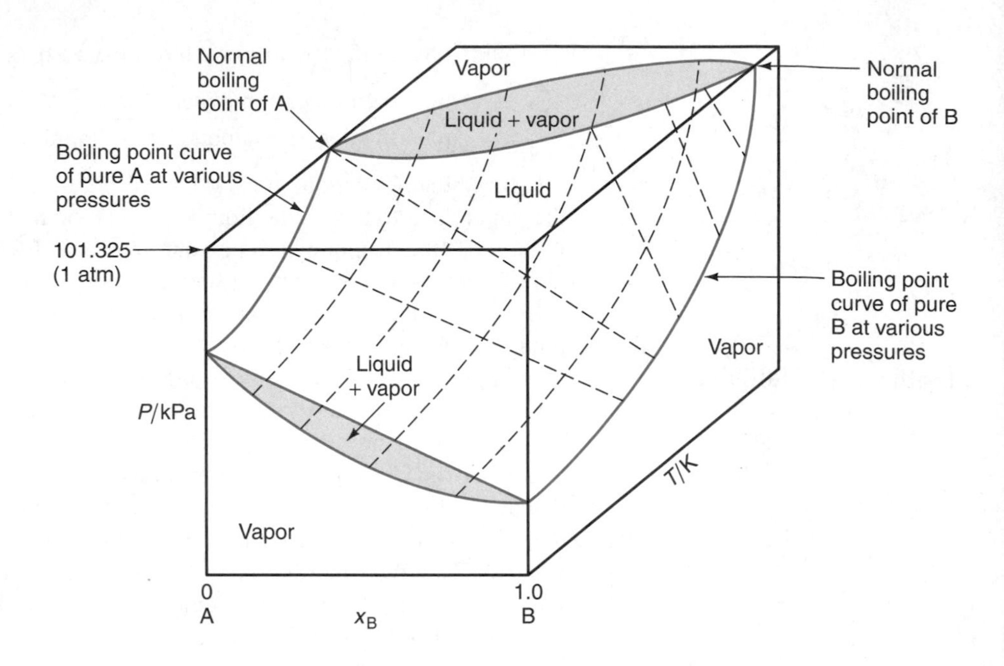

FIGURE 6.5
Typical three-dimensional phase diagram for a binary system obeying Raoult's law.

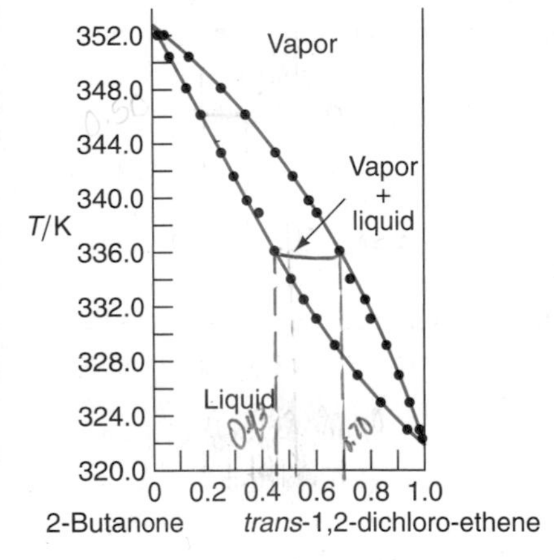

FIGURE 6.6
Temperature-vs.-composition diagram for 2-butanone and *trans*-1, 2-dichloroethene at 101.325 kPa (1 atm). Dots represent data points.

system with respect to T, P, and composition. Such a description requires a three-dimensional figure.

For a typical, completely miscible binary system, the temperature and composition are represented along the two horizontal axes, and pressure along the vertical axis. A solid figure will be formed from the volume representing the region of liquid-vapor equilibrium. It lies between a concave and convex surface meeting in edges lying in the two faces of the figure representing the T-P planes at $x = 0$ and $x = 1$. This is shown in Figure 6.5, where the right and left faces of the figure are the respective vapor-pressure curves of the pure liquids. Note that, in the region above the envelope of liquid-vapor equilibrium, only liquid exists; below the envelope, only vapor exists.

A typical temperature-composition plane (similar to the isobaric section at 1 atm pressure in Figure 6.5) is shown in Figure 6.6 for a binary system consisting of 2-butanone and *trans*-1,2-dichloroethene. In comparison to the simple double curve-shaped or lens form in Figure 6.2, the temperature-composition curve will have the component with the higher vapor pressure at the low end of the lens in the T-x plot. The vapor-composition curve is now the upper curve because the vapor is more stable at high temperatures.

Although the binary system may behave ideally, the straight line from a P against x plot for the total pressure of the liquids is not a straight line in the temperature-composition plot. (The vapor pressure does not increase proportionally with T. See Eq. 5.12 on p. 189.) This curve is normally determined experimentally by measuring the boiling points and vapor compositions corresponding to a range of liquid mixtures.

Distillation

Changes in state brought about by an increase in temperature may be utilized to separate the components in a binary mixture. To see how this can be accomplished,

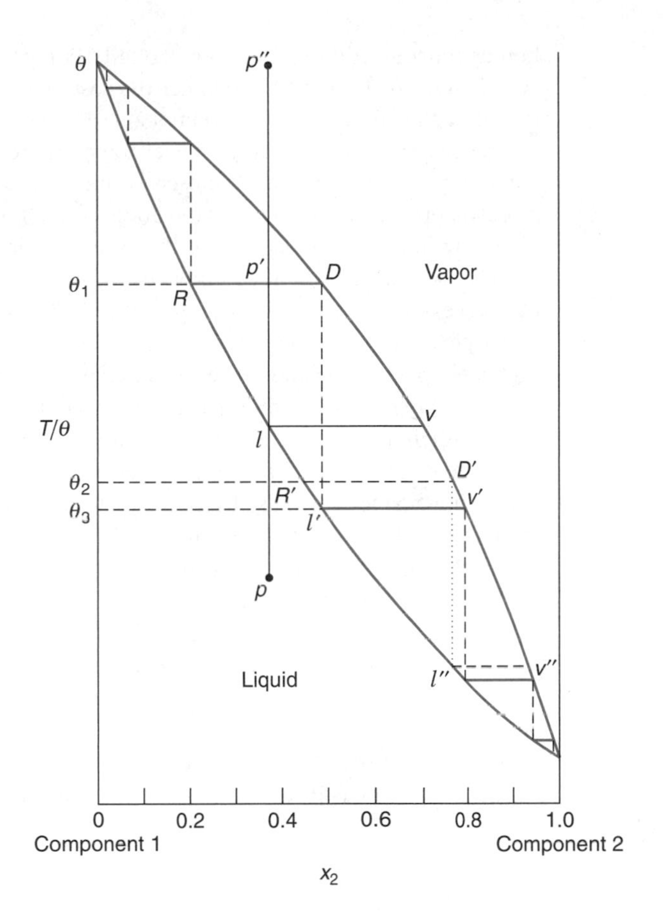

FIGURE 6.7
Temperature-composition diagram showing the technique for separation of a mixture into its pure components.

examine Figure 6.7. The boiling points of all the mixtures are intermediate between the boiling points of the pure components. Point p in Figure 6.7 describes the mixture in the liquid state, and we will consider what happens as the temperature is raised. The liquid system does not reach equilibrium with its vapor until point l is reached on the liquid curve. At this point the vapor has the composition v, which is much richer in the more volatile component than the original liquid. As vapor is removed, the composition of the remaining liquid has a lower mole fraction of component 2. Boiling will not continue (i.e., equilibrium will not be maintained) unless the temperature is raised. As the temperature is raised to θ_1 (point p'), the composition of the liquid will have moved from l to R along line $\overline{lR}$. In a similar way the composition of the vapor will have moved from v to D along line $\overline{vD}$. The lever rule may be used to determine the amounts of the two phases present. Finally, at any position higher than the vapor curve, say at p'', the entire state consists of vapor.

With composition and pressure already fixed, the phase rule allows only one degree of freedom, namely temperature, in the single-phase regions such as at point p. As we have seen on page 229, the system is invariant in the two-phase region when one degree of freedom is taken by the constant pressure and the other by either temperature, or by either of the composition variables.

Within a closed system, the movement of the system variables from p along the line to p' leaves the composition of the overall system unchanged. The amount of vapor formed at p' is given by the lever rule and the vapor has composition D. The vapor D may be removed and condensed to point l'. This liquid l' will have a much higher concentration of the more volatile component than the original mixture of composition p. In a similar manner the residual liquid R, boiling at θ_1, will be enriched in the higher boiling component. The condensate from D, now at point l', can be evaporated again at the lower temperature θ_3. The resulting composition of the vapor at v' will be much closer to that of the pure component than before. As more vapor is collected, the composition of the higher boiling material will increase along the line $v'D'$. Assume that the composition of the vapor reaches point D' at θ_2. The D' distillate may be collected and condensed to l'' and reevaporated, and the process repeated as shown in Figure 6.7 until the pure component of higher vapor pressure is achieved. Using the same technique, the component boiling at the higher temperature also may be separated as the residual liquid.

Fractionating Column

The *batch-type* operation just described is obviously a very time-consuming process. In practice, a *fractionating column* is used and the process is automated using a variety of techniques. The fundamental requirement of the column is to provide the contacting surface for mass transfer between the liquid and vapor at a desired rate. Plate columns are common in which liquid counterflows downward through the same orifice through which the vapor rises. Sieve plates (a perforated plate and simplest of plates used), valve plates, and bubble-cap plates fall into this category. Perhaps the easiest of these to picture is the use of a *bubble-cap* variety, shown in Figure 6.8.

In this device preheated liquid mixture enters the column about halfway up. The less volatile components drop to the bottom boiler or still, A, where they are reheated. As the vapor ascends the column, the higher-boiling components begin to condense while the lower-boiling materials proceed to higher stages. Thus a temperature gradient is established, with the highest temperature at the bottom of the

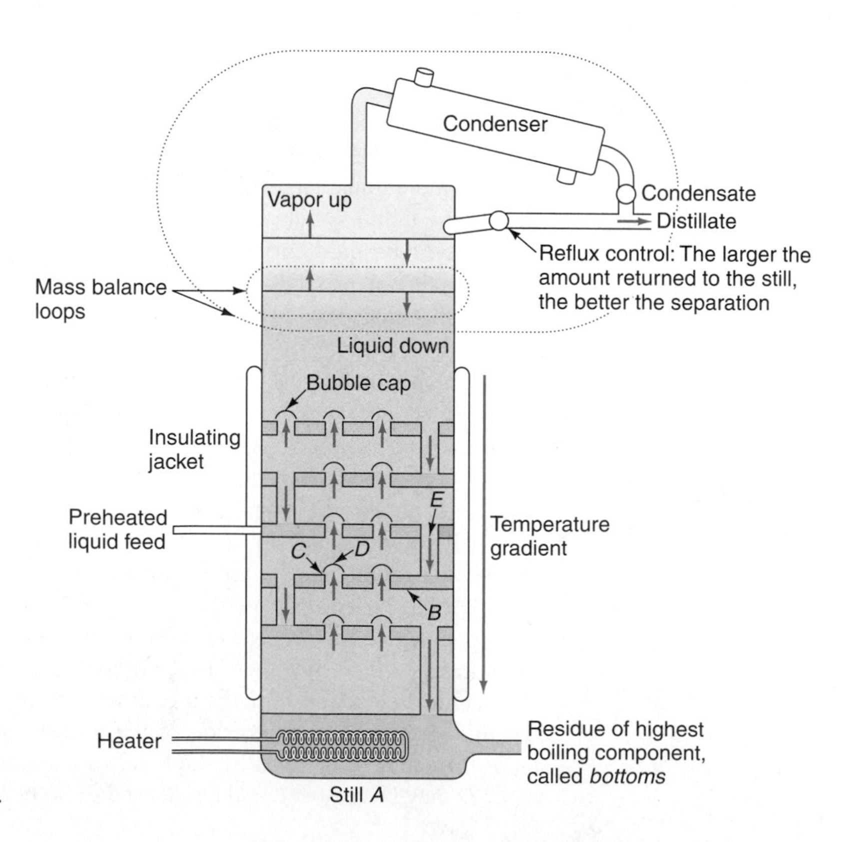

FIGURE 6.8
A bubble-cap fractionating column.

column and the lowest at the top from which the lowest-boiling solution may be removed. At each level in the column, such as *B*, vapor from the level or *plate* below bubbles through a thin film of liquid *C*, at a temperature slightly lower than that of the vapor coming through the bubble cap, *D*. Partial condensation of the vapor occurs. The lower-boiling mixture remains as vapor and moves to the next plate. Thus the vapor leaving each plate is enriched in the more volatile component compared to the entering vapor from the plate below. The action of the vapor condensing and then reevaporating is the same as described for the behavior shown in Figure 6.7. Excess liquid at each plate is returned to the plates below via overflow tubes, *E*. The reflux control allows part of the condensate to return to the column, where it continues to undergo further separation.

The intimate contact required between liquid and vapor to achieve the equilibrium at each plate may be achieved in other ways. In the *Hempel column,* glass beads or other packing is used to increase the surface area of the liquid in the column. The vapor must then pass through and over more liquid on its path to the top of the column. Direct flow of the liquid back to the boiler must be avoided because this type of pathway does not provide the intimate contact needed between liquid and vapor to approximate equilibrium.

The efficiency of the column is generally measured in terms of the number of *theoretical plates* required to separate two components. A plate is a level, such as *RD* or $l'v'$, of liquid-vapor equilibrium. Then the number required is the number of plates, such as *RD* and $l'v'$, in the stair-step form linking the two pure components, minus one. We describe the height equivalent of a theoretical plate (HETP), h, as the distance between the vapor in the column and the liquid with which it would be exactly in equilibrium. This separation comes about in practice because equilibrium is not established at every position.

If two components obey Raoult's law, the closeness of their boiling points dictates how many theoretical plates are needed for separation. For widely separated boiling points a few plates will suffice, whereas if the boiling points are close, many theoretical plates will be needed. Although the word *equilibrium* is sometimes used to refer to an operating column, it is more correct to say that the column is in a **steady state.** In a thermodynamic sense, true equilibrium is not established throughout since a uniformity of temperature does not exist and because the counterflow of liquid and vapor in itself is another nonequilibrium condition.

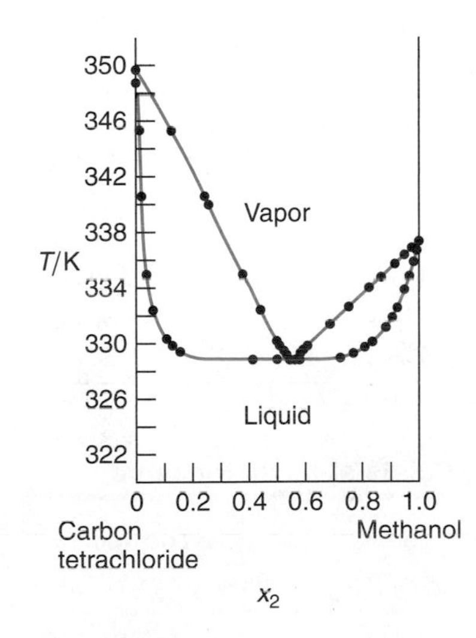

FIGURE 6.9
Temperature-composition diagram for carbon tetrachloride–methanol at 101.325 kPa (1 atm). Dots represent data points.

Azeotropes

The separation of two liquids just described relates to liquids conforming to Raoult's law. If a plot of vapor pressure against composition shows a maximum, the boiling point curve will show a minimum. An example of this is shown in Figure 6.9 for carbon tetrachloride–methanol at 101.325 kPa (1 atm). This should be compared with Figure 5.9 on p. 197, which is the vapor pressure curve for two compounds showing a maximum in the vapor pressure. In the same way, a minimum in the vapor pressure curve will result in a maximum in the boiling point curve. This behavior is shown in Figure 6.10 for the system tetrahydrofuran *cis*-1,2-dichloroethene at 101.325 kPa (1 atm). Mixtures corresponding to either a minimum or maximum in the boiling point curves are called **azeotropes** (Greek *a*, without; *zein,* to boil; *trope,* change).

Separation of azeotropic mixtures into two pure components by direct distillation is impossible. However, one pure component may be separated as well as the

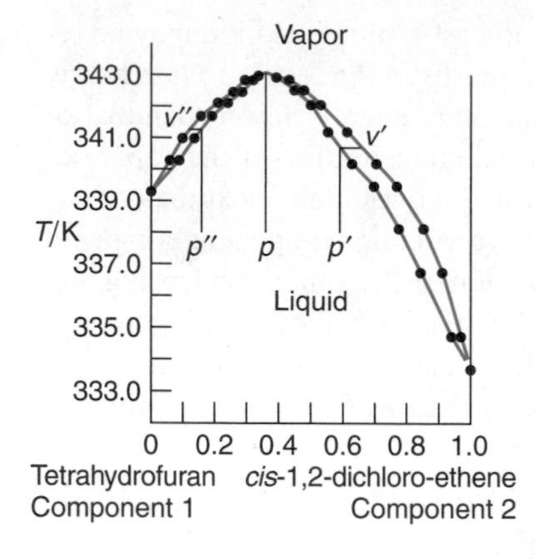

FIGURE 6.10
Temperature against composition diagram for the system tetrahydrofuran–*cis*-1,2-dichloroethene at 101.325 kPa (1 atm). Dots represent data points.

azeotropic composition. As an example, consider Figure 6.10. If we heat a mixture having the azeotropic composition p, the mixture will boil unchanged; no separation is accomplished. If we now boil a mixture having composition p', the first vapor to appear will have composition v' and is richer in *cis*-1,2-dichloroethene, component 2. Continued fractionation, as described before, will eventually produce pure component 2 in the distillate with azeotrope as the residual liquid in the pot. Boiling a mixture at point p'' results in vapor having composition v'', which is much richer in tetrahydrofuran, component 1, than before. Fractionation can eventually yield pure component 1 at the head of the column and leave the azeotropic mixture in the pot.

Similar results are obtained with a minimum boiling azeotrope except that the azeotropic composition is obtained at the head of the column and pure component remains behind in the still. Examples of minimum boiling azeotropes are far more numerous than those exhibiting a maximum in their curves.

In Tables 6.1 and 6.2 are listed a number of minimum and maximum boiling azeotropes at 101.325 kPa (1 atm). Azeotropes have sometimes been mistaken for pure compounds because they boil at a constant temperature. However, for an azeotrope a variation in pressure changes not only the boiling temperature but also the composition of the mixture, and this easily distinguishes it from a true compound. This is demonstrated for HCl in Table 6.3.

One of the more useful applications of azeotropes in binary systems is the preparation of a constant composition mixture. In the case of H_2O and HCl, the constancy of the composition of this azeotrope allows its use as a standard solution of known composition.

TABLE 6.1 Azeotropes with Minimum Boiling Points at 1 atm Pressure

				Azeotrope	
Component 1	T_b/K	Component 2	T_b/K	Wt % 1	T_b/K
H_2O	373.15	Ethanol	351.45	4.0	351.32
H_2O	373.15	2-Propanol	355.65	12.0	353.25
H_2O	373.15	1-Chlorohexane	407.65	29.7	364.95
H_2O	373.15	Acetophenone	474.75	81.5	372.25
Carbon disulfide	319.35	Iodomethane	315.7	18.6	314.35
Methanol	337.85	Pentane	309.3	7	304.00
Acetic acid	391.25	Heptane	371.4	33	364.87
Ethanol	351.45	Benzene	353.25	31.7	341.05

TABLE 6.2 Azeotropes with Maximum Boiling Points at 1 atm Pressure

				Azeotrope	
Component 1	T_b/K	Component 2	T_b/K	Wt % 1	T_b/K
H_2O	373.15	HNO_3	359.15	85.6	393.85
Chloroform	334.35	Methyl acetate	330.25	64.35	337.89
Acetic acid	391.25	Butanol	390.25	43	393.45
Ethanol	351.45	1-Aminobutane	350.95	82.2	355.35

TABLE 6.3 Influence of Pressure on Azeotropic Temperature and Composition

Pressure/ kPa	Wt % HCl in H_2O	T_b/K
66.660	20.916	370.728
93.324	20.360	379.574
101.325	20.222	381.734
106.656	20.155	383.157

Distillation of Immiscible Liquids: Steam Distillation

We have just considered the distillation of miscible liquids. Now consider the behavior of two liquids whose mutual solubility is so small that they may be considered immiscible. In this case each liquid exerts the same pressure as though it were the only liquid present. Thus the total pressure above the mixture at a particular temperature is simply the sum of the vapor pressures of the two components and remains so until one of the components disappears. This fact makes possible a distillation quite different from the type discussed previously.

Since the two vapor pressures are added together, any total pressure is reached at a much lower temperature than the boiling point of either component. As the distillation proceeds, both components are distilled in a definite ratio by weight. Thus if P_T is the total pressure, and P_A^* and P_B^* are the vapor pressures of pure liquids A and B, respectively, then

$$P_T = P_A^* + P_B^* \tag{6.20}$$

If n_A and n_B are the amounts of each component present in the vapor, the composition of the vapor is

$$\frac{n_A}{n_B} = \frac{P_A^*}{P_B^*} \tag{6.21}$$

Since the ratio of partial pressures is a constant at a particular temperature, the ratio n_A/n_B must also be a constant. Thus the distillate is of constant composition as long as both liquids are present, and it boils at a constant temperature.

Steam Distillation

Water is often one component when this type of distillation is used for purifying organic compounds. This process, called **steam distillation,** is frequently used for substances that would decompose when boiled at atmospheric pressure. What makes this process attractive is the high yield of organic materials brought about by the low molar mass of water and its convenient boiling point, in contrast to the relatively high molar masses of most organic substances.

EXAMPLE 6.4 Toluene (methylbenzene) and water are immiscible. If boiled together under an atmospheric pressure of 755 Torr at 83 °C, what is the ratio of toluene to water in the distillate? The vapor pressure of pure toluene and water at 83 °C are 322 Torr and 400.6 Torr, respectively.

Solution Since $n = W/M$ where M is the molar mass and W is the mass in a given volume of vapor, using Eq. 6.21, we may write

$$\frac{n_{\text{toluene}}}{n_{\text{water}}} = \frac{P^*_{\text{toluene}}}{P^*_{\text{water}}} = \frac{W_{\text{toluene}} M_{\text{water}}}{W_{\text{water}} M_{\text{toluene}}}$$

Rearranging and substituting gives

$$\frac{W_{\text{toluene}}}{W_{\text{water}}} = \frac{322 \text{ Torr} \times 92.15}{400.6 \text{ Torr} \times 18.02} = 4.11$$

If a liquid is partially miscible with water, it too may be distilled as long as the solubility is not too great. However, one important difference arises when the composition of distillate is calculated. The actual partial pressures must replace P_A^* and P_B^* in Eq. 6.21. The relative molecular mass of an unknown substance may also be estimated if the masses and vapor pressures of the two components are known.

Distillation of Partially Miscible Liquids

Many pairs of liquids are miscible only to a limited extent; an example is the water–butanol system. Generally partial miscibility at low temperatures is caused by large positive deviations from Raoult's law, in which case we expect to find a minimum in the boiling point–composition curve. As the pressure on the system is reduced, the boiling point curve generally intersects the liquid-liquid equilibrium curve, resulting in the curve for a typical system shown in Figure 6.11.

Any composition in the range from 0 to x_a and from x_c to 1 will show the same behavior upon boiling as already demonstrated for minimum boiling azeotropes, with one exception. Two layers are formed if liquid at point p is evaporated and its vapor v condensed. The two liquids at p' are L_1 and L_2, in amounts given by the lever rule. L_1 has the composition given by f and L_2 that given by g. If solution of overall composition p' is boiled at T_e, three phases will be in equilibrium: liquid phase L_1 having composition x_a, liquid phase L_2 having composition x_c, and vapor having composition y_b. From the phase rule the system is invariant at the specified pressure. Thus, as vapor of composition y_b is removed, the composition of the two liquid phases does not change, only the relative amounts of the two layers. In this particular case, continued distillation will cause all the L_2 layer to be consumed. When it is exhausted, liquid L_1 at a and vapor at b are left. The temperature may be

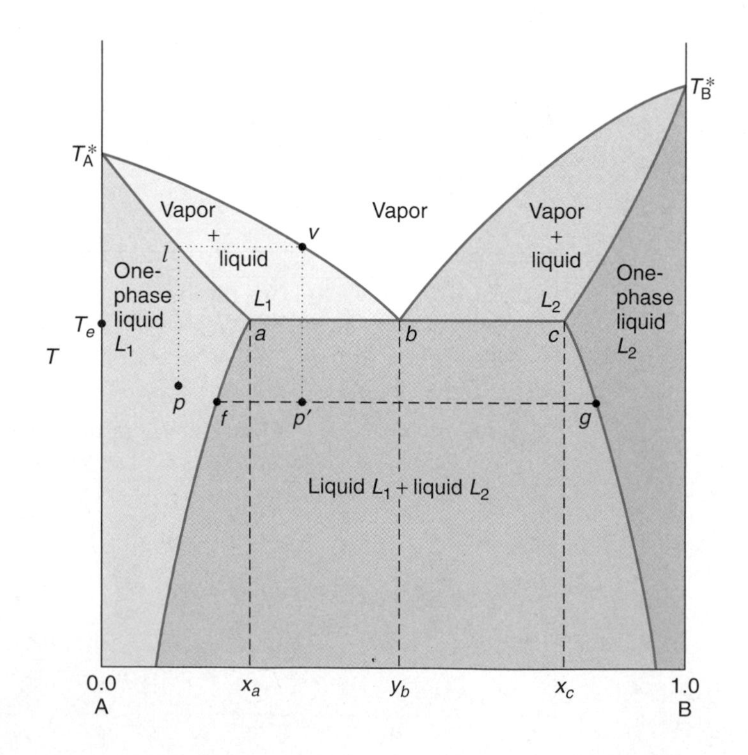

FIGURE 6.11
Liquid-vapor equilibrium for a system with partial miscibility in the liquid phase.

increased at this same pressure; then the liquid composition changes along curve $\overline{al}$ and the vapor along $\overline{bv}$. The last drop of liquid disappears when l and v are reached, and only vapor remains since this is the original composition of liquid from which the liquid p' was produced.

If the composition of the liquid layers lies in the range b to c, then as distillation occurs, vapor at b will be formed at the expense of the L_1 layer. The rest of the distillation will be similar to that already described.

6.4 Condensed Binary Systems

Two-Liquid Components

The vapor phase in an equilibrium system is often of little or no interest compared to the interaction in the liquid and solid phases. It is customary, therefore, in these cases to disregard the vapor phase and to fix a total pressure, normally 1 atm. The system may then be studied in open vessels. When only solid and liquid phases are considered, we speak of a *condensed system.* There is a minor drawback in this procedure in that the pressure is generally not the equilibrium value, and the system is consequently not in true thermodynamic equilibrium. However, for liquid-solid systems this does not have much effect on the behavior.

Since we have just discussed the distillation of two immiscible liquids, it seems appropriate to investigate the types of behavior exhibited by a condensed system consisting of two liquids that are only partially miscible.

The water–aniline system provides a simple example of partial miscibility (see Figure 6.12). If a small amount of aniline is added to pure water at any temperature below 441 K, the aniline dissolves in the water. If we work at a constant temperature of 363 K, pure water is present at point a and only one phase is present as aniline is added. However, as more aniline is added, point b on the solubility curve is reached and then, in addition to phase L_1 of composition b, a slight amount of a second liquid

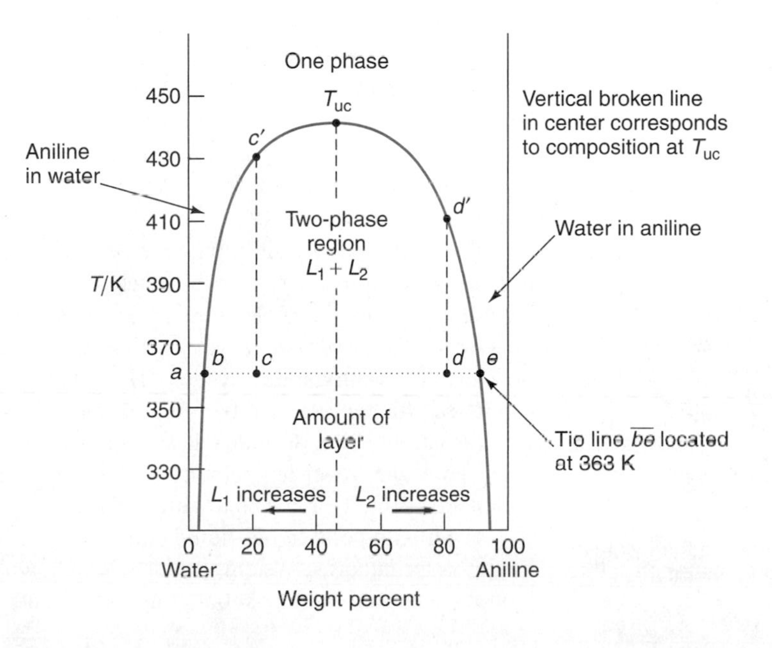

FIGURE 6.12
Solubility of water and aniline as a function of temperature. The **upper consolute** temperature, T_{uc}, is 441 K. The lines $\overline{cc'}$ and $\overline{dd'}$ are isopleths at 20 and 80 wt % aniline, respectively.

phase L_2 appears having composition e. The composition of the L_1 layer is a solution of aniline in water and that of the L_2 layer is a solution of water in aniline. As more aniline is added, the second liquid layer L_2 becomes more evident and continually increases with the addition of aniline until the composition is given by point e. Beyond point e, only one phase is present. The same type of behavior is observed as water is added to pure aniline. The composition of any point in the two-phase region along the tie line between points b and e is composed of varying proportions of solution L_1 and of solution L_2. These solutions are called **conjugate solutions.** The composition of these layers depends on the temperature. The amount of the individual layers present may be determined using the lever rule (Section 6.3).

Conjugate Solutions

EXAMPLE 6.5 Calculate the ratio of the mass of the water-rich layer to that of the aniline-rich layer, for a 20-wt % water–aniline mixture at 363 K.

Solution The compositions along the tie line $\overline{be}$ are maintained at 363 K. The composition at c is 20%; for L_1 at b, it is 8%; and for L_2 at e, it is 90% (see Figure 6.12).

Using the lever rule, we have

$$\frac{\text{mass of water-rich layer, } L_1}{\text{mass of aniline-rich layer, } L_2} = \frac{\overline{ce}}{\overline{bc}} = \frac{90 - 20}{20 - 8} = 5.83$$

Isopleths

Constant composition lines, vertical on the diagram in Figure 6.12, are known as **isopleths** (Greek *iso,* the same; *plethora,* fullness). As the temperature is increased from point c along the isopleth cc' or indeed from any point left of the vertical dashed line joining T_{uc}, the solubility of the aniline in water layer, L_1, grows as does the solubility of the water in aniline layer, L_2. As a result of the change in solubility, the predominant layer L_1 increases at the expense of the L_2 layer. Similar behavior is observed for the L_2 layer when the temperature is increased from point d. A different behavior is observed with the *critical composition,* which is the composition corresponding to the highest temperature T_{uc} at which two layers may coexist. (This temperature is known as the *upper consolute temperature* or *critical solution temperature.*) If the curve is symmetrical, the relative size of the layers remains constant as the temperature is raised along the dotted line. Above T_{uc}, only one phase exists.

The increased solubility with temperature can be explained by the fact that the forces holding different types of molecules apart are counteracted by the thermal kinetic energy of the molecules. It is curious, therefore, that for some systems a *lower consolute temperature* exists; Figure 6.13 shows an example of this behavior in the water–triethylamine system. The lower consolute temperature is 291.65 K, and above this temperature two immiscible layers exist. In this case the large positive deviations from Raoult's law responsible for the immiscibility may be just balanced at the lower temperature by large negative deviations from Raoult's law, which are normally associated with compound formation.

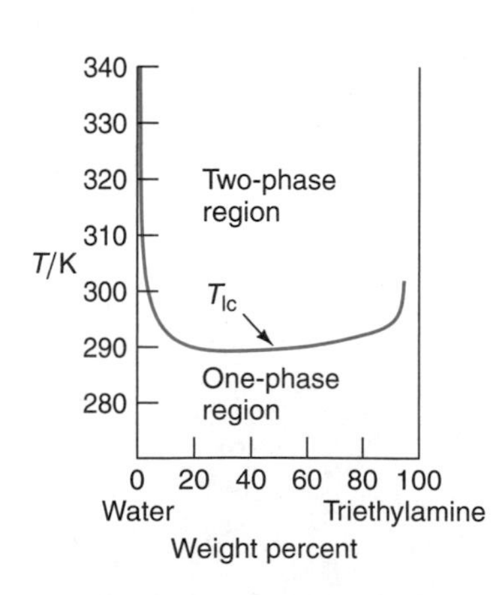

FIGURE 6.13
Solubility of water and triethylamine as a function of temperature (T_{lc} = 291.6 K).

The final type of liquid-liquid equilibrium, called a *miscibility gap,* is exhibited by the water–nicotine system. In this case, shown in Figure 6.14, the two-phase region is enclosed and has both an upper and lower consolute temperature at

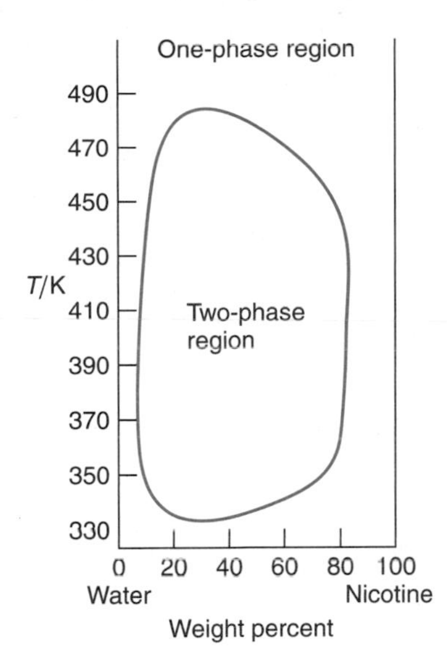

FIGURE 6.14
Solubility of water and nicotine as a function of temperature.

atmospheric pressure. This may be called a *closed miscibility loop*. It has been shown for this system that an increase of pressure can cause total solubility.

An interesting example of liquid-liquid solubility in two materials normally solid, and its application to a practical problem, is afforded by the Pb–Zn system. Figure 6.15 gives the phase diagram for this system to 1178 K, above which boiling occurs. Ignoring the details in the right- and left-hand extremes of the diagram, we see that miscibility occurs above the upper consolute temperature at 1071.1 K. The rather limited solubility of zinc in lead may be used in the metallurgical separation of dissolved silver in lead. The zinc is added to the melted lead, which has an economically recoverable amount of silver dissolved in it. The melt is agitated to effect thorough mixing. The zinc is then allowed to rise to the surface of the lead and is skimmed off. Because of the much higher solubility of silver in Zn than in Pb, most of the silver will now be in the zinc. The zinc may be boiled off from this liquid to give the desired silver.

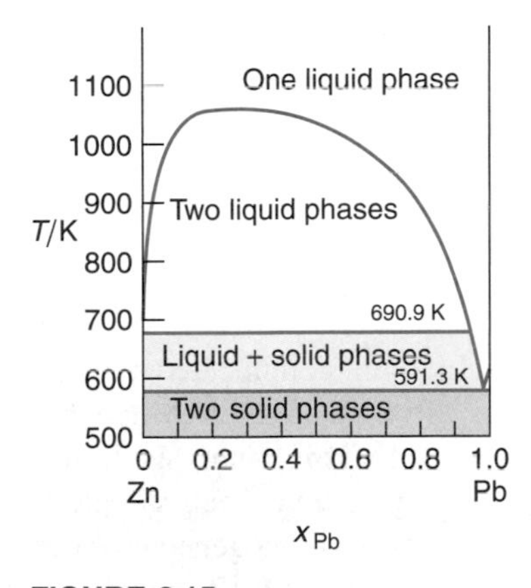

FIGURE 6.15
The zinc–lead system. This is an example of a two-phase liquid system.

Solid-Liquid Equilibrium: Simple Eutectic Phase Diagrams

When a single-liquid melt formed from two immiscible solids is cooled sufficiently, a solid is formed. The temperature at which the solid is first formed is the freezing point of the solution, and this is dependent on the composition. Such a case is provided by the gold–silicon system shown in Figure 6.16. A curve that represents the boundary between liquid only and the liquid plus solid phase is known as a **liquidus** curve. If we begin with pure gold, the liquidus curve will start at $x_{Au} = 1$ and drop toward the center of the figure. The curve from the silicon side behaves in a similar manner. The temperature of intersection of the two curves, T_e, is called the **eutectic** (Greek *eu,* easily; *tecktos,* molten) temperature. The eutectic composition x_e has the lowest melting point in the phase diagram.

At the eutectic point, three phases are in equilibrium: solid Au, solid Si, and liquid. At fixed pressure the eutectic point is invariant. This means that the temperature is fixed until one of the phases disappears. The relationship between phases is easy to follow if we isobarically cool the liquid represented by point p in a single-liquid-phase region. As the temperature is lowered, the first solid appears at l at the same temperature as the liquid. Since this is an almost completely immiscible system, the solid is practically pure gold. As the temperature is dropped further, more crystals of pure gold form. The *composition of the liquid* follows the line $\overline{le}$, and the overall composition between liquid and solid is given by the lever rule. In this region of liquid plus solid, either the temperature or the composition may be varied. When point s is reached, three phases are in equilibrium: solid Au, solid Si, and liquid of composition x_e. Since the system is invariant, the liquid phase is entirely converted into the two solids before the temperature may drop lower. Finally, at point p', only two solids, Au and Si, exist. The right-hand side of the diagram may be treated in the same way.

Eutectic Temperature

A mixture of eutectic composition will melt sharply at the eutectic temperature to form a liquid of the same composition. Materials of eutectic composition are generally fine grained and uniformly dispersed, as revealed by high-magnification microscopy. Microscopic observation of a noneutectic melt cooled to solidification reveals individual crystals of either solid dissolved in a uniform matrix of eutectic composition.

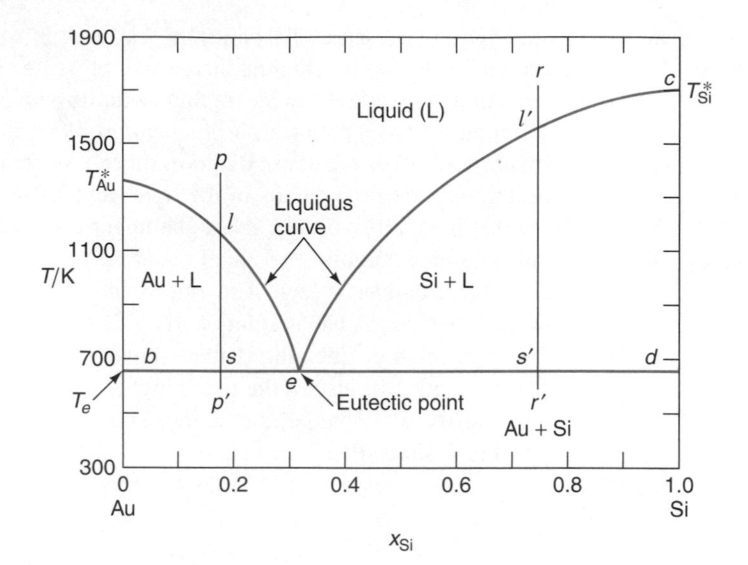

FIGURE 6.16
The gold–silicon system. This is an example of a simple eutectic system without miscibility. T_e is the eutectic temperature.

The mole fraction of silicon in the region near pure silicon varies with temperature according to Eq. 5.115 on p. 212:

$$\ln x_{Si} = -\frac{\Delta_{fus} H_{Si}}{R}\left(\frac{1}{T} - \frac{1}{T_0}\right) \tag{6.22}$$

where T_0 = melting point of Si.

6.5 Thermal Analysis

The careful determination of phase boundaries, particularly in complicated metallic systems, is quite difficult and requires considerable effort. One method that has proved useful for phase determination is the technique of **thermal analysis.** In this technique a series of mixtures of known composition is prepared. Each sample is heated above its melting point and, where possible, is made homogeneous. Then the rate of cooling of each sample is followed very closely. Figure 6.17 shows a se-

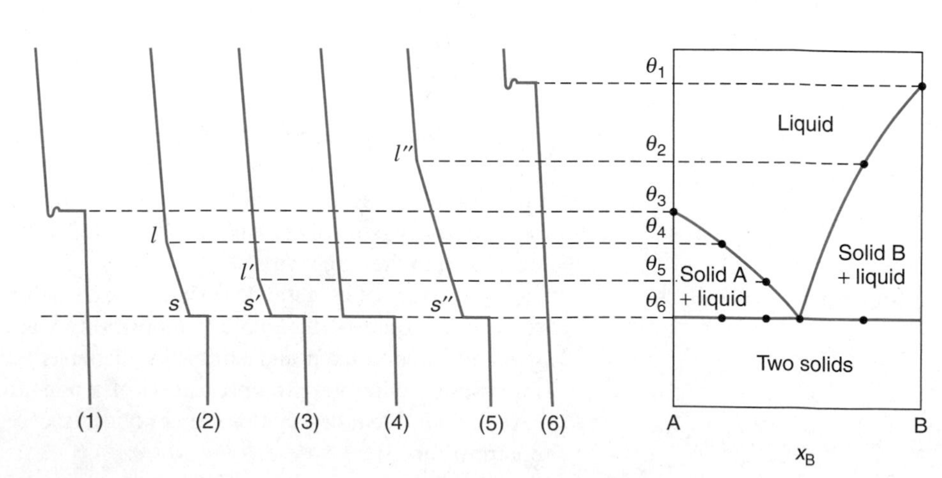

FIGURE 6.17
Use of thermal analysis. Demonstration of how thermal analysis can be used to determine a phase diagram. Cooling curves (1) and (6) represent behavior for pure A and B, respectively.

ries of cooling curves in a plot of T against time and shows how individual points are used to form a phase diagram similar to Figure 6.16. For the two pure materials the rate of cooling of the liquid melt is fairly rapid. When the melting temperature is reached, there is generally a little supercooling that is evidenced by a slight jog in the curve. This is shown in curve 1. The curve returns to the melting point and remains there until all the liquid is converted to solid. The temperature then drops more rapidly for the solid than for the liquid since the heat capacity of a solid is generally lower than that of a liquid. Thus it requires less heat removal to cool the sample a fixed number of degrees.

Curve 2 represents a mixture of some B in A. The mixture cools rapidly until point l is reached. This point appears on the liquidus curve. Liquid and solid are in equilibrium as the mixture cools more slowly along line $\overline{ls}$. This is because of the heat that is released on solidification. At point s, a horizontal region appears called the *eutectic halt.* The liquid still present in the system must completely solidify before the temperature can drop farther. Once all the liquid is converted to solid, the temperature will drop. As with the pure material, the cooling of two solids is much more rapid than if liquid were present. The descriptions of curves 3 and 5 are the same as for curve 2 except for the lengths of the lines $\overline{l's'}$ and $\overline{l''s''}$ and for the time that the system stays at the eutectic halt. This period at the eutectic halt provides a means to establish the eutectic temperature. In general the eutectic halt will lengthen, and lines like $\overline{l's'}$ will shorten, as the composition approaches the eutectic composition. The reason for this is that for the eutectic composition the cooling is rapid until the eutectic temperature is reached. After all the liquid is converted to solid, the mixture can then cool further. The temperature for each composition at which a change occurs in the cooling curve is then used to establish a point on the phase diagram, as is shown in the right-hand portion of Figure 6.17.

Eutectic Halt

6.6 More Complicated Binary Systems

It may be beneficial to focus on several useful ideas before proceeding. Curved lines in a phase diagram normally indicate the limits of miscibility and as such are phase boundaries. A one-phase region always exists to one side of the boundary, and a two-phase region exists to the other side. Such regions are always labeled so that the number of phases present and their structures can be read from the diagram.

Isothermal discontinuities as well as discontinuities in isopleths show changes in phase behavior that occur at a particular temperature or composition. Isothermal discontinuities are most easily seen along isoplethal lines that fall within the composition span of the discontinuity.

With these ideas in mind, let us examine the phase diagram of the sodium chloride–water system, as shown in Figure 6.18. Perhaps the most obvious isothermal discontinuity occurs at −21.1 °C, which is the eutectic temperature at a concentration of 23.3 wt %. A second discontinuity is a peritectic reaction at 0.1 °C and 61.7%, where there is exactly a 2:1 mole ratio of water to Na^+Cl^- ion pairs. This combination is sometimes referred to as a *crystal dihydrate* and more generally is termed a *phase compound.* Such mixtures have a precise integral mole ratio defining the stoichiometry. Most elementary chemistry students are already familiar with sodium sulfate decahydrate, which is another phase compound.

Isoplethal discontinuities are easily seen along the 0% and 100% compositions, which reflect the phase behavior of the pure components. For water, this shows the

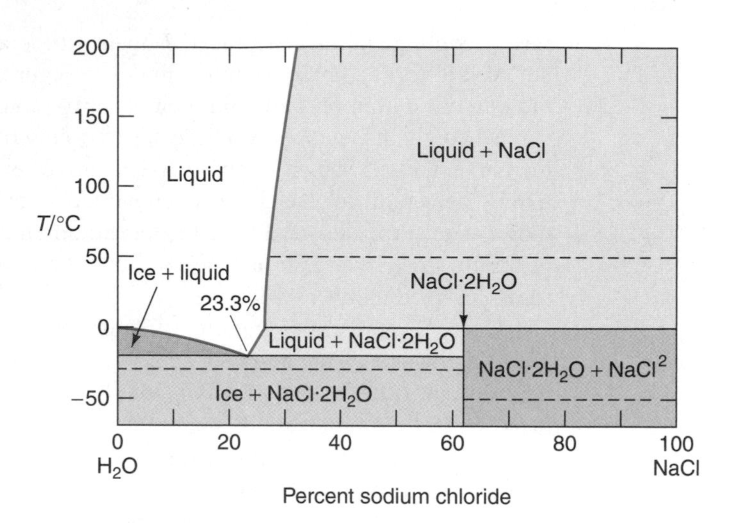

FIGURE 6.18
The phase diagram of the sodium chloride–water system. Dashed lines are tie lines that indicate the composition of coexisting phases. NaCl represents the dry crystal (molecular formula NaCl) and $NaCl \cdot 2H_2O$ indicates the dihydrate crystals. This type of diagram is typical of the salt-water interactions given in Table 6.4.

transition from ice (crystal) to liquid phase at 0 °C (273.15 K). Since salt melts at 801 °C (1074 K), there is no discontinuity in the salt crystals in the temperature range shown. However, another isoplethal discontinuity occurs below 0.1 °C at 61.9%, where there is exactly a 2:1 ratio of water to Na^+Cl^- ion pairs. This is a *crystal dihydrate,* as just mentioned.

The behavior of NaCl in water is an example of phase behavior that is much more complex than might have been anticipated. As a few crystals of NaCl are dissolved in liquid water at room temperature, a liquid phase is formed that contains both water and the ions Na^+ and Cl^-. More added salt will dissolve up to a limit called the "solubility limit," shown along $\overline{EB}$ in Figure 6.18. Up to this limit the salt and water are completely miscible. At this limit the solution is said to be "saturated" with salt. As still more salt is added, the salt concentration in the liquid phase does not change, and any crystals added beyond the solubility limit remain seemingly unchanged in contact with the liquid.

Now suppose in another experiment that sufficient salt crystals are mechanically mixed with ice crystals to form a 40 wt % mixture of NaCl in water maintained at −10 °C. A liquid phase is quickly formed, dissolving salt crystals up to the solubility limit. This freezing point curve for water is represented by the curve $\overline{AE}$. In this case, however, the crystals in equilibrium with the saturated liquid are not the same as those formed above 0 °C. These latter crystals have the formula $NaCl \cdot 2H_2O$. They differ in composition and structure from dry sodium chloride.

One use of eutectic systems such as just described is to prepare a constant-temperature bath at some temperature below that of melting ice. If NaCl is added to ice, the ice melts. Indeed, if this is done in an insulated container, ice continues to melt with the addition of NaCl until 252.0 K is reached. Then the temperature of the system and any vessel immersed in the insulated container will remain invariant until all the ice has been melted by heat from an outside source.

The eutectic point is again seen to be associated with two solid phases (ice and salt hydrate), similar to the two solid phases at the eutectic point in the metallic systems that we have considered. In Table 6.4 several eutectic compositions involving different salts and water are presented.

TABLE 6.4 Eutectics Involving Salts and Ice

Salt	Eutectic Temperature/K	Eutectic Composition/ Wt % Salt
Ammonium chloride	257.7	19.7
Potassium chloride	262.4	19.7
Sodium bromide	245.1	40.3
Sodium chloride	252.0	23.3
Sodium iodide	241.6	39.0
Sodium nitrate	257.7	44.8
Sodium sulfate	272.0	3.84

Mixed Melting Points

Organic compounds can also form simple eutectic mixtures. The *method of mixed melting points* is seen as an application of eutectics in organic chemistry. To confirm the identity of a substance, the sample is mixed with a pure sample of the compound thought to be identical to it. If the two substances are the same, the melting point will not change. If they are not the same, the temperature will drop, since the sample lowers the freezing point, as shown in Figure 6.17.

Solid Solutions

Only one type of situation is known in which the mixture of two different substances may result in an increase of melting point. This is the case in which the two substances are **isomorphous** (Greek *iso,* the same; *morphos,* form). In terms of metallic alloys this behavior is a result of the complete mutual solubility of the binary components. This can occur when the sizes of the two atoms of the two components are about the same. Then atoms of one type may replace the atoms of the other type and form a *substitutional alloy.* An example of this behavior is found in the Mo–V system, the phase diagram for which is shown in Figure 6.19. Addition of molybdenum to vanadium will raise the melting point. The Cu–Ni system also forms a solid solution. Copper melts at 1356 K, and addition of nickel raises the temperature until for $x_{Ni} = 1$ the temperature reaches 1726 K. An alloy known as *constantan,*

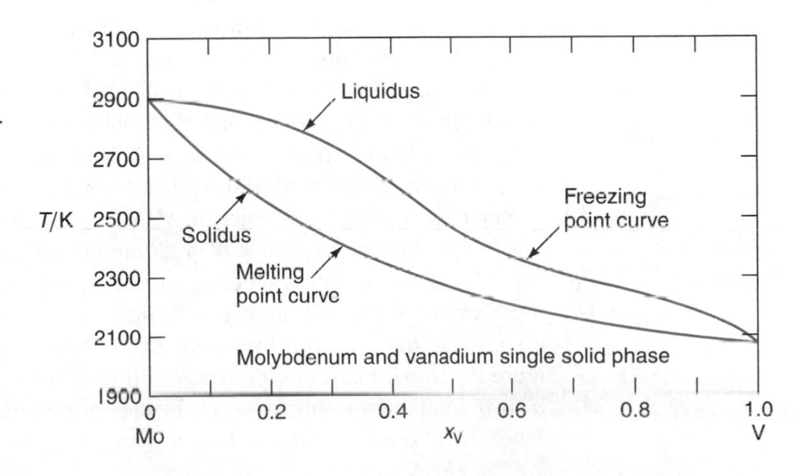

FIGURE 6.19 The molybdenum–vanadium system. This is an example of a solid solution. The **solidus** curve establishes the equilibrium between solid and solid + liquid regions.

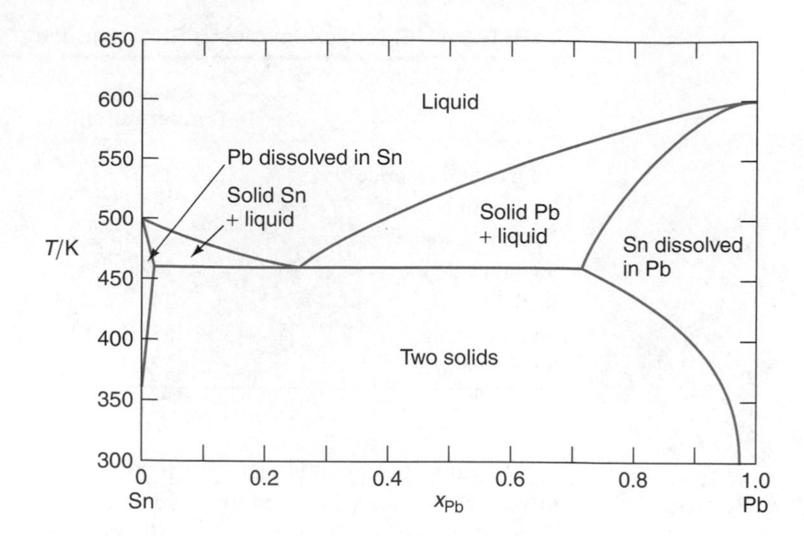

FIGURE 6.20
The tin–lead system. An example of a simple eutectic system with limited miscibility of the components.

consisting of 60 wt % Cu and 40 wt % Ni, has special interest since it is useful as one component of a thermocouple for the determination of temperature.

Partial Miscibility

The two previous sections covered two extremes, namely complete immiscibility of components and complete solubility of components. Most actual cases show limited solubility of one component in the other. Figure 6.20 is an example of this limited solubility for both components. Tin dissolves lead to a maximum of approximately 2.5 mol % or 1.45 wt %. Tin is more soluble in lead, dissolving to a maximum of 29 mol % at 466 K. The two-phase solid region is composed of these two solid alloys in proportions dictated by the lever rule. The situation is analogous to that represented in Figure 6.11 concerning liquid-vapor equilibrium where partial miscibility occurs in the liquid phase.

This is an important system from a practical standpoint, because in the electrical industry and in plumbing the low-temperature eutectic is used to make commercial solders. Actually the tin content should be about 20 mol % rather than the eutectic composition, to improve "wiping" characteristics for solder application.

A second type of system in which partial miscibility occurs involves a transition point. An example is provided by the Mn_2O_3–Al_2O_3 system, the phase diagram for which is shown in Figure 6.21. This system gives rise to one series of *spinels* (Latin *spina,* thorn), which are of variable composition with the general formula AB_2O_4 in which A may be magnesium, iron, zinc, manganese, or nickel and B may be aluminum, chromium, or iron. These form octahedral crystals. (The gem ruby is a magnesium aluminum spinel.) Also formed in this system are the *corundums,* which are abrasive materials of high aluminum oxide content that form hexagonal crystals. The gem sapphire belongs to this group of substances. A transition temperature exists along *abc* at which spinel, corundum, and liquid of composition *a* coexist and the system is invariant. At any temperature above the transition temperature the spinel phase disappears. Cooling of the liquid and corundum phases in the *a* to *b* composition range results in the formation of the spinel phase and coexistence of two solid phases. Cooling in the *b* to *c* range initially results in the

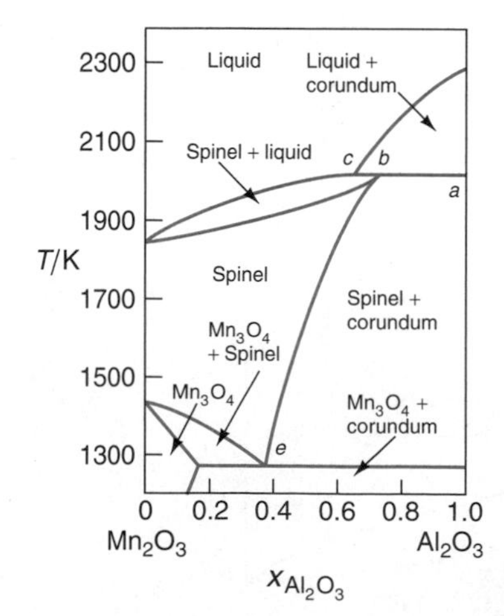

FIGURE 6.21
The system Mn_2O_3–Al_2O_3 exposed to air.

disappearance of corundum and formation of spinel along with the liquid. Further cooling results in solid spinel only. However, as the temperature falls further, corundum makes an appearance again.

Another feature of this system deserves mention. The region near 1300 K and 0.4 mol fraction Al_2O_3 appears to be similar to what has been described as a eutectic point. However, where liquid would be expected in a normal eutectic system, this region is entirely solid. An invariant point such as *e* surrounded solely by crystalline phases is called a *eutectoid.* At the eutectoid, phase reactions occur on change of heat resulting in a change in proportions of the solid phases exactly analogous to that at a eutectic point.

Compound Formation

Somctimes there are such strong interactions between components that an actual compound is formed. Two types of behavior can then be found. In the first type, the compound formed melts into liquid having the same composition as the compound. This process is called **congruent melting.** In the second type, when the compound melts, the liquid does not contain melt of the same composition as the compound. This process is called **incongruent melting.**

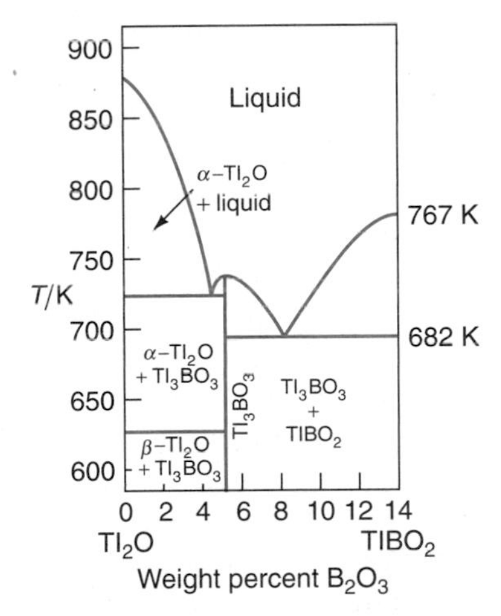

FIGURE 6.22
The Tl_2O–$TlBO_2$ system.

In Figure 6.22 for the system Tl_2O–$TlBO_2$ the compound Tl_3BO_3 is formed and melts congruently at 725 K. A eutectic occurs at 8.2 wt % B_2O_3. Note that the *X* axis is plotted as weight percent B_2O_3 over a very limited range. The easiest way to interpret this figure is to mentally cut it in half along the line representing pure Tl_3BO_3. Then each half is treated as in Figure 6.16. The left-hand portion introduces a new feature that is often found not only in ceramic systems such as this but also in metallic systems. At 627 K a reversible transformation occurs in the crystalline structure of Tl_2O. A form called β is stable below 627 K and the second form α is stable above the transition temperature all the way to the melting point.

It appears that there are now two points in this system that were previously described as eutectic points. However, only the lowest one is referred to as the eutectic point; others are called **monotectic points.**

In contrast to congruent melting, incongruent melting occurs for each of the compounds in the Au–K system shown in Figure 6.23. The composition of each compound formed is given by the formula alongside the line representing that compound. The composition in the region around K_2Au is given in detail. From this example, and the following discussion, the details in the other regions may be supplied.

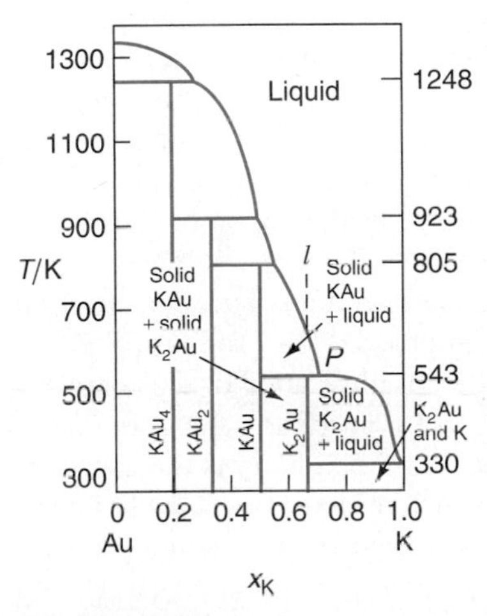

FIGURE 6.23
The Au–K system.

If we examine the compound K_2Au as it is heated, we find that liquid of composition *P* is formed at 543 K:

$$K_2Au(s) \rightleftharpoons KAu(s) + \text{liquid(composition } P)$$

Since the liquid is richer in potassium than is solid KAu, some solid KAu will remain as solid. Thus the reaction is known as a **phase reaction** or, more commonly, a **peritectic reaction** (Greek *peri,* around; *tektos,* melting). The point *P* is known as the **peritectic point.** This reaction is reversible if liquid of the same total composition as K_2Au is cooled. Solid KAu begins to separate at *l*. More solid KAu forms until the temperature of 543 K is reached. As heat is removed, the reverse of the peritectic reaction just shown occurs. From the lever rule, approximately 25% of the material initially exists as particles of solid KAu surrounded by liquid of composition *P*. Thus the KAu is consumed as the reaction proceeds. As the last trace of liquid and KAu disappears, the temperature is free to drop and only K_2Au is present.

If the starting melt has a composition that is slightly rich in Au compared to K_2Au (i.e., lies to the left of K_2Au), cooling into the two-solid region will produce crystals of KAu surrounded by the compound K_2Au. It is difficult to establish equilibrium in these systems unless a long annealing period is used to homogenize the sample.

6.7 Crystal Solubility: The Krafft Boundary and Krafft Eutectic

In Section 5.1 we introduced liquid crystals and discussed their phase implications. Colloidal and gel structures have somewhat similar properties and are discussed more fully in Section 18.9. So far we have not represented these structures in phase diagrams. Many surfactants used in almost all consumer products, such as shampoos, lotions, etc., may fall into one of these classes. They generally are highly soluble in liquid water at high temperatures. At lower temperatures they may separate from solution as a crystal-like phase. So far we have discussed defining the state of the system in terms of T, P, components, and phases. Phase structure is the highest level of structure that is fully defined by thermodynamics. It is defined as the manner in which molecules are arranged in space within a phase. Phase structure is a fundamental property of state, but this is not true of still higher levels of structure, such as that possessed by liquid crystals, colloidal, or gel structures. All of these may be considered as forms of conformational structure. Assuming equilibrium, once a phase structure is defined, conformational structure is also fixed.

Conformational structure may exist within both single- and multi-phase mixtures and may persist for long periods of time, particularly within phases that are both highly structured and very viscous. The concentration of defects that generally exists within liquid crystal phases gives even uniform samples a characteristic turbid appearance. In defining colloidal structure, the concern is with structure within mixtures that are thermodynamically heterogeneous because they consist of two (or more) phases. It is how these phases are arranged in space that defines colloidal structure. Great care must be exercised, for if a biphasic mixture having colloidal structure is described as a single phase, the conclusions made from the phase rule with be incorrect.

The Krafft boundary and discontinuity are named in honor of the German organic and physical chemist Wilhelm Ludwig Friedrich Emil Krafft (1852–1923). Friedrich Krafft discovered that the crystal solubility of soaps displayed a sudden and dramatic increase over a narrow range of temperature. He was one of the pioneers of surfactant phase science. His work attracted the interest of the Canadian physical chemist James William McBain (1882–1953), who then led the way in developing this field into a science.

Two broad categories of colloidal structure exist. In *reversible colloidal structure* there exist micellar liquid solutions of surfactants. Such solutions have a variance, f, of three, are thermodynamically homogeneous, and are a single phase. *Irreversible colloidal structures* are found in emulsions of a crystal phase within a fluid phase. Such systems change with time and are dependent upon the history of the system.

Many surfactants are quite soluble at elevated temperatures but separate from solution at lower temperatures into a crystal phase. The sodium dodecyl sulfate–water phase diagram in Figure 6.24 provides an illustration of the complexity that is often found. The important aspects for our purposes are shown in Figure 6.25 and should be matched with Figure 6.24. The *Krafft boundary* is the name applied to the crystal solubility boundary. It has a distinctive shape (see *knee* in Figure 6.25) and at its upper temperature limit terminates at a eutectic discontinuity known as the *Krafft discontinuity* or *Krafft eutectic*.

The saturating phase at solubilities and temperatures above the Krafft eutectic is usually (but not always) a liquid crystal phase. At temperatures far below the

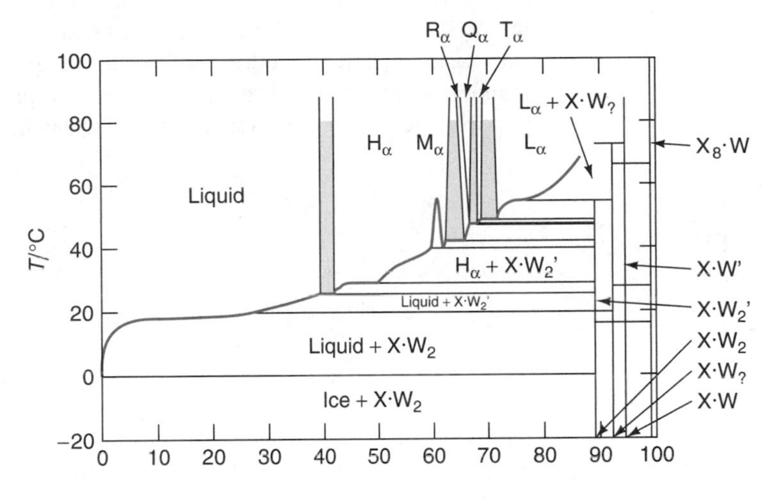

FIGURE 6.24
Phase diagram in simplified form of the anionic surfactant, sodium dodecyl sulfate, in water. This diagram shows a rather large difference between the temperature of the knee (see text) and the Krafft eutectic. Also shown is the large array of liquid crystal phases that exist in this system. The terms in the form of X · W indicate composition of sodium dodecyl as X with water as W in varying combinations and with different structures indicated by apostrophes. The long, narrow regions near the top center of the figure are regions of stability of different liquid crystals. Cusps are shown at the bottom of M_α, R_α, and T_α.

Krafft eutectic, the solubility of surfactants is small and the crystal solubility boundary is steep. This indicates that in this region there is a low temperature coefficient of solubility. As the temperature is increased, a turnover or *knee* develops, and almost immediately a leveling off or *plateau* exists. In this region the temperature coefficient of solubility becames very large. This plateau ends at the Krafft eutectic, as shown in Figure 6.25. At this point the crystal boundary intersects the solubility boundary that exists at still higher temperatures. At the intersection of these two solubility boundaries is a liquid phase that is the most dilute of the three phases that can exist at the Krafft eutectic.

Along the liquid boundary is a cusp at the intersection of the two other solubility boundaries. This indicates a discontinuity of state within the coexisting phase. Finding such a cusp demands further investigation of the system because often a peritectic decomposition of a crystal hydrate or a polymorhpic phase transition in

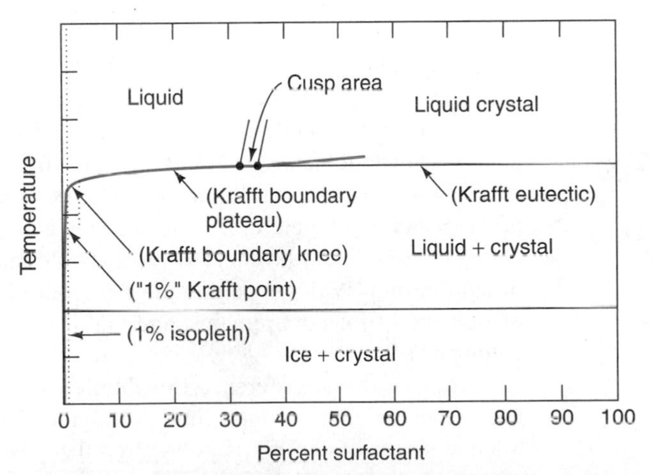

FIGURE 6.25
Details of the Krafft phase boundary. The lower near-vertical portion, the knee, and the plateau region are shown. The boundary ends at the Krafft eutectic. Also shown are some other terms used in phase work.

the crystal has occurred. For instance, crystal hydrates in dodecylammonium chloride–water and in dimethyl sulfone diimine–water systems were found in this manner. Indeed, many of the advances in the products that we use today are made possible because of our knowledge of phase behavior.

6.8 Ternary Systems

When we consider a three-component, one-phase system, the phase rule allows for $f = c - p + 2 = 3 - 1 + 2 = 4$ degrees of freedom. These four independent variables are generally taken as pressure, temperature, and two composition variables, since only two mole fractions are necessary to define the composition. Thus the composition of a three-component system can be represented in two dimensions with T and P constant.

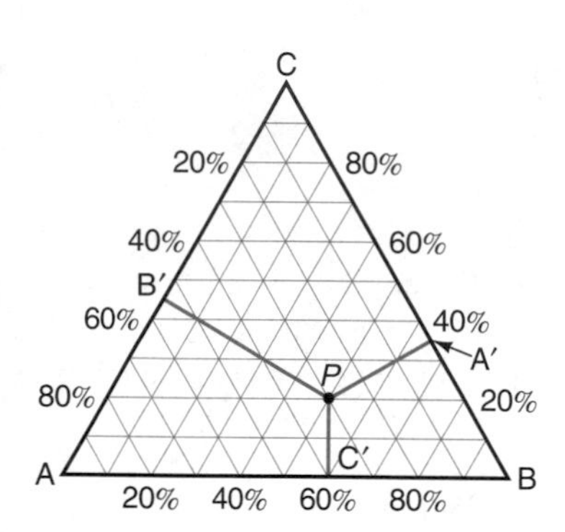

FIGURE 6.26
The triangular diagram for representing a three-component system at constant temperature and pressure.

The most convenient technique for plotting such a phase diagram is due to Hendrik William Bakhuis Roozeboom (1854–1907). The composition is determined using the fact that, from any point within an equilateral triangle, *the sum of the distances perpendicular to each side is equal to the height of the triangle.* The height is set equal to 100% and is divided into 10 equal parts. A network of small equilateral triangles is formed by drawing lines parallel to the three sides through the 10 equal divisions. Each apex of the equilateral triangle in Figure 6.26 represents one of the three pure components, namely 100% A, 100% B, or 100% C. The three sides of the triangle represent the three possible binary systems and 0% of the third component. Thus any point on the line $\overline{BC}$ represents 0% A. A line parallel to $\overline{BC}$ through P represents all possible compositions of B and of C in combination with 30% A. Here the percentage of A is read from the value of the length of the line $\overline{A'P}$. Compositions along the other two sides are read in a similar manner. Since the distance perpendicular to a given side of the triangle represents the percentage of the component in the opposite apex, the compositions at P are 30% A, 50% B, and 20% C. Any composition of a ternary system can thus be represented within the equilateral triangle.

Liquid-Liquid Ternary Equilibrium

A simple example for demonstrating the behavior of a three-component liquid system is the system toluene–water–acetic acid. In this system, toluene and acetic acid are completely miscible in all proportions. The same is true for water and acetic acid. However, toluene and water are only slightly soluble in each other. Their limited solubility causes two liquids to form, as shown along the base of the triangle at points p and q in Figure 6.27. Added acetic acid will dissolve, distributing itself between the two liquid layers. Therefore, two conjugate ternary solutions are formed in equilibrium. With temperature and pressure fixed in the two-phase region, only one degree of freedom remains, and that is given by the composition of one of the conjugate solutions.

Because of the difference in solubility of acetic acid in the two layers, however, the tie lines connecting conjugate solutions are not parallel to the toluene–water base. This is shown for the tie lines $\overline{p'q'}$, $\overline{p''q''}$, etc. This type of curve is called *binodal.* The relative amounts of the two liquids are given by the lever rule. As the two liquid solutions become more nearly the same, the tie lines

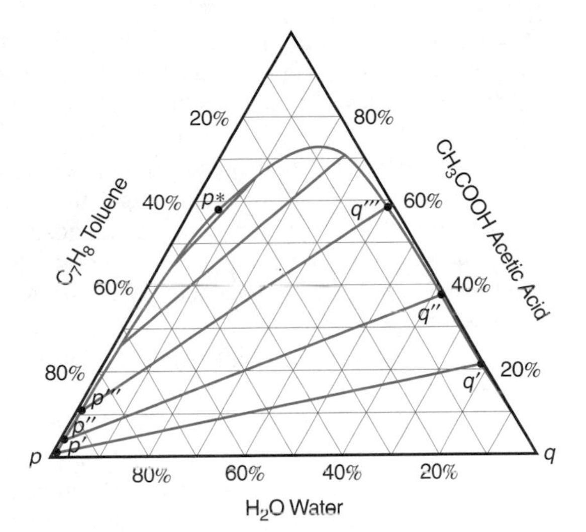

FIGURE 6.27
The ternary system acetic acid–toluene–water at a fixed temperature and pressure. The formation and representation of two conjugate solutions are shown.

become shorter, finally reducing in length to a point. This point generally does not occur at the top of the solubility curve and is called an *isothermal critical* point or *plait* point, p^* in the diagram. At p^* both layers are present in approximately the same proportion, whereas at p'' only a trace of water remains in the toluene layer. This curve becomes more complicated if the other sets of components are only partially miscible.

Solid-Liquid Equilibrium in Three-Component Systems

Common-ion Effect

The *common-ion effect* may be explained by use of phase diagrams. Water and two salts with an ion in common form a three-component system. A typical phase diagram for such a system is shown in Figure 6.28. Such systems as NaCl–KCl–H_2O and NH_4Cl–$(NH_4)_2SO_4$–H_2O give this type of equilibrium diagram. We now will see how each salt influences the solubility of the other and how one salt may actually be separated.

In Figure 6.28, A, B, and C represent nonreactive pure components with C being the liquid. Point *a* gives the maximum solubility of A in C when B is not present. Point *c* gives the maximum solubility of B in C in the absence of A. Points along the line $\overline{Aa}$ represent various amounts of solid A in equilibrium with saturated solution *a*. Solutions having composition between *a* and C are unsaturated solutions of *a* in C. When B is added to a mixture of A and C, the solubility of A usually decreases along the line $\overline{ab}$. In like manner, addition of A to a solution B in C usually decreases the solubility along the line $\overline{cb}$. This is the effect normally called the *common-ion effect.* The meeting of these curves at *b* represents a solution that is saturated with respect to both salts. In the region A*b*B, three phases coexist, the two pure solids A and B and a saturated solution of composition *b*. In the tie line regions, the pure solid and saturated solution are in equilibrium. Thus, if point *d* gives the composition of the mixture, the amount of solid phase present is given by the length of the line $\overline{de}$ and the amount of saturated solution is given by the length of the line $\overline{dA}$. Considering the regions of the three-component systems with one solid phase present in contact with liquid, we find a bivariant system (no vapor phase

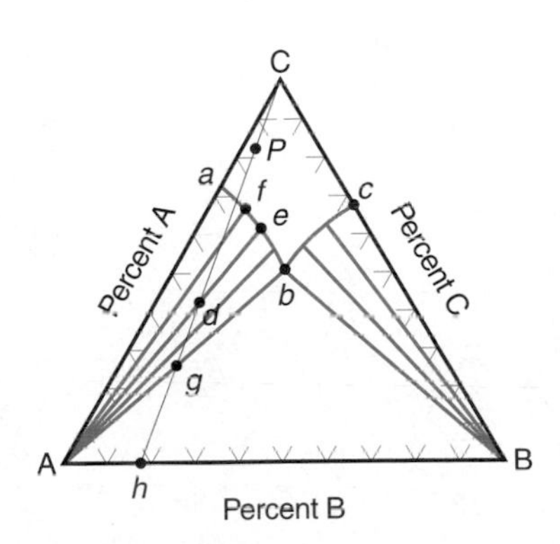

FIGURE 6.28
Phase equilibrium for two solids and a liquid showing the common-ion effect.

present and the system at constant pressure). Thus, at any given temperature the concentration of the solution may be changed. However, at the point *b*, two solid phases are present and the system is isothermally invariant and must have a definite composition.

Now consider what happens to an unsaturated solution *P* as it is isothermally evaporated. The overall or state composition moves along the line $\overline{Ch}$. Pure A begins to crystallize at point *f*, with the composition of the solution moving along $\overline{fb}$. At point *g* the solution composition is *b* and B begins to crystallize. As evaporation and hence removal of C continue, both solid A and solid B are deposited until at *h* all the solution is gone.

The process of recrystallization[1] can also be interpreted from Figure 6.28. Let A be the solid to be purified from the only soluble impurity B. If the original composition of the solid mixture of A and B is *h*, water is added to achieve the overall composition *d*. The mixture is heated, thus changing the state to an unsaturated liquid and effecting complete solubility. (Complete solubility can usually be brought about by increasing the temperature sufficiently.) When the liquid is cooled, the impurity B stays in the liquid phase as pure A crystallizes out along the line equivalent to $\overline{ab}$ at the higher temperature. Thus, when the solution returns to room temperature, the crystals of pure A may be filtered off.

Another type of ternary system to be considered involves one solid and two liquids. There are many examples in organic chemistry in which the addition of a salt to a mixture of an organic liquid such as alcohol and water results in the separation of two liquid layers, one rich in the organic liquid, one rich in water. With the addition of enough salt, a third layer, this time a solid, also appears. By separating the organic layer, a separation or *salting out* has been achieved.

A typical salt–alcohol–water system is given in Figure 6.29 for purposes of illustration. This diagram is characterized by the formation of a two-liquid region *bfc*. The solubility of salt in water is given by point *a*. The nearness of point *d* to 100% alcohol indicates a rather low solubility of the salt in alcohol. As before, the solubility of salt in water is changed by the addition of alcohol along the line $\overline{ab}$. In a similar manner, water increases the solubility of salt in alcohol along $\overline{dc}$. These

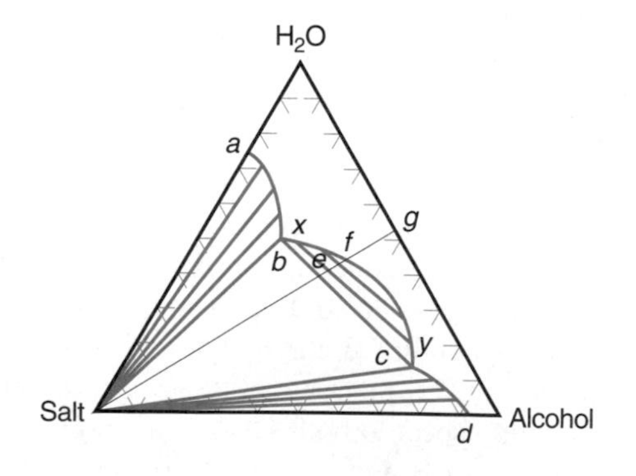

FIGURE 6.29
Typical salt–alcohol–water ternary diagram—basis of "salting out" effect.

[1]For an interesting application of the use of three-component phase diagrams in the extraction of tantalum from its ore, see G. B. Alexander, *J. Chem. Educ., 46,* 157(1969).

solutions, *b* and *c*, are immiscible and form the ends of the binodal curve *bfc*. This area enclosed by *bfc* consists of two conjugate liquids having one degree of freedom. The greatest solubility of the two liquids occurs at the plait point *f*, where the solutions are saturated with salt. Any composition lying within the triangle salt, *b*, *c* must yield salt and two liquid layers *b* and *c*. The "salting out" effect is easily demonstrated by starting with solution *g* composed of only water and alcohol. As salt is added point *e* is reached, where two liquids *x* and *y* are formed. Layer *y* containing the high proportion of alcohol can now be separated.

Representation of Temperature in Ternary Systems

If we consider three-component systems when temperature is a variable, it is obvious that a third dimension must be added to the equilateral triangle for representation of the system. The temperatures of the features being represented may be indicated by the length of lines perpendicular to the plane of the composition triangle. A solid diagram for the system LiF–NaF–CaF_2 is shown in Figure 6.30. The three uppermost peaks represent the melting points of the three pure components. The liquidus surface is depicted in this case by three intersecting curved surfaces representing the *primary phase fields* of the three compounds. A primary phase field of a congruently melting compound is a domed surface, the peak of which represents the melting point of the compound. For the three components at the apices of the triangle, the dome is truncated. The fields of adjacent compounds intersect in a sloping valley or boundary line. These valleys are shown in the figure along lines $\overline{e_1E}$, $\overline{e_2E}$, and $\overline{e_3E}$.

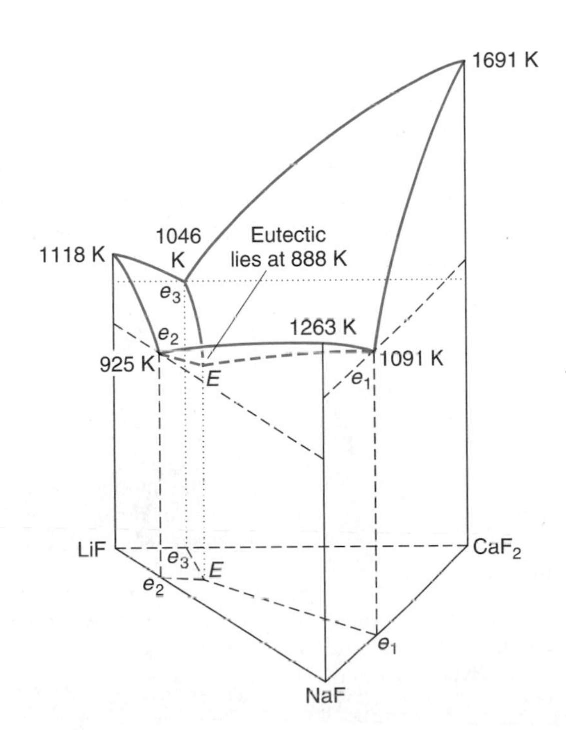

FIGURE 6.30 Perspective drawing of a three-dimensional phase diagram for the system LiF–NaF–CaF_2. The troughs from the binary eutectics to the ternary eutectic are projected to the triangular base. The triangular base for this system is shown in Figure 6.31.

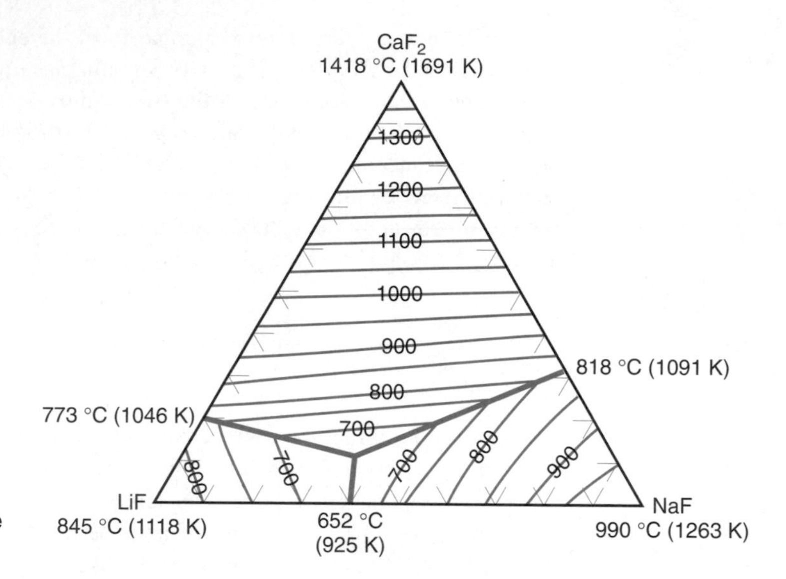

FIGURE 6.31
Representation of temperature profiles in the system LiF–NaF–CaF_2. Temperatures internal to the triangle are in degrees C.

Generally it is more convenient to construct isotherms at uniform temperature intervals on one of the thermal surfaces. Thus the isotherms of the liquidus surface may be represented in two dimensions, much like the elevation contours on topographic maps. An example of this is shown in Figure 6.31 for the same system as in Figure 6.30.

We immediately see the correspondence to the melting points of the pure compounds. The sloping valleys are indicated by the heavy lines from the binary eutectics at 1046 K, 1091 K, and 925 K to the system eutectic at 888 K. The spacing of the isothermal lines gives an idea of the steepness of the surface (closely spaced lines indicate a steep surface). The shapes of the contours indicate whether the surface is highly curved or almost planar.

Many other features, such as compound formation, may be represented on triangular diagrams. Graphical representation of four-component systems may be made using a square. The procedures for use of these techniques are outside the scope of this book. The information available from such phase diagrams has made possible advances in many fields.

KEY EQUATIONS

The phase rule:

$$f = c - p + 2$$

where the quantities are precisely defined as follows:

f is the number of degrees of freedom, which is the number of intensive variables, such as temperature, pressure, and concentrations, which can be independently varied without changing the number of phases.

c is the number of components, which is the smallest number of independent chemical constituents needed to fix the composition of every phase in the system.

p is the number of phases, which are homogeneous and distinct regions separated by definite boundaries from the rest of the system.

PROBLEMS

Number of Components and Degrees of Freedom

6.1. In Figure 6.1, in the region marked *orthorhombic,* how many degrees of freedom exist? How many components are present? How many phases? How many phases exist in the region marked *monoclinic?*

6.2. What is the composition of the two-phase region in Figure 6.14? How many degrees of freedom exist in this region?

6.3. Determine the number of degrees of freedom for the following systems:

a. A solution of potassium chloride in water at the equilibrium pressure.
b. A solution of potassium chloride and sodium chloride at 298 K at 1 atm pressure.
c. Ice in a solution of water and alcohol.

6.4. How many components are present in a water solution of sodium acetate?

6.5. How many components are present in the system $CaCO_3$–CaO–CO_2?

6.6. How many components are present in the following system?

$$CO(g) + 3H_2(g) \rightleftharpoons CH_4(g) + H_2O(g)$$

6.7. A certain substance exists in two solid phases A and B and also in the liquid and gaseous states. Construct a *P-T* phase diagram indicating the regions of stable existence for each phase from the following triple-point data:

T/K	*P*/kPa	Phases in Equilibrium
200	100	A, B, gas
300	300	A, B, liquid
400	400	B, liquid, gas

Use of the Lever Rule; Distillation

6.8. Answer the following questions, using the accompanying figure.

a. A liquid mixture consists of 33 g of component A and 99 g of component B. At what temperature would the mixture begin to boil?
b. Under the conditions in (a), what is the composition of the vapor when boiling first occurs?
c. If the distillation is continued until the boiling point is raised by 5.0 °C, what would be the composition of the liquid left in the still?

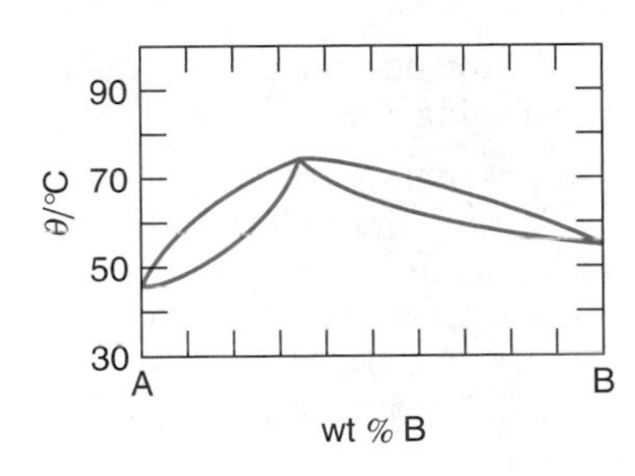

d. Under the conditions in (c), what are the composition and mass of the two components collected over the initial 5.0 °C interval?

6.9. From the data of Figure 6.14, calculate the ratio of the mass of the water-rich layer to that of the nicotine-rich layer, for a 40 wt % water–nicotine mixture at 350 K.

6.10. The ratio of the mass of chlorobenzene to that of water collected in a steam distillation is 1.93 when the mixture was boiled at 343.85 K and 56.434 kPa. If the vapor pressure of water at this temperature is 43.102 kPa, calculate the molar mass of chlorobenzene.

6.11. a. Do the actual derivation of Eq. 6.11 from Eq. 6.8.
b. From Eq. 6.8 derive an expression that gives you the ratio of the mass of two volatile components, 1 and 2, in terms of their mole fractions in the vapor and their molar masses.

6.12. Obtain an expression for the ratio of masses of the materials distilled in a steam distillation in terms of the molar masses and the partial pressures of the two components.

6.13. Under atmospheric pressure 1 kg of pure naphthalene is to be prepared by steam distillation at 372.4 K. What mass of steam is required to perform this purification? The vapor pressure of pure water at 372.4 K is 98.805 kPa.

6.14. The vapor pressure of water at 343.85 K is 43.102 kPa. A certain mixture of chlorobenzene and water boils at 343.85 K under a reduced pressure of 56.434 kPa. What is the composition of the distillate?

6.15. Calculate the composition of the vapor in equilibrium at 323 K with a liquid solution of 0.600 mol fraction 2-methyl-1-propanol (isobutyl alcohol) and 0.400 mol fraction 3-methyl-1-butanol (isoamyl alcohol). The vapor pressure of pure isobutyl alcohol is 7.46 kPa and that of pure isoamyl alcohol is 2.33 kPa both at 323 K.

6.16. The thermal expansion coefficient $\alpha = (1/V)(\partial V/\partial T)_P$ is often used when predicting changes in vapor pressure induced by temperature changes. From the relation $\rho = m/V$, show that $\alpha = -(\partial \ln \rho/\partial T)_P$.

6.17. At 293.15 K the density of water is 0.998 234 g cm^{-3} and at 294.15 K it is 0.998 022 g cm^{-3} under 1 atm of pressure. Estimate the value of α for water at 1 atm.

6.18. How many theoretical plates are required to separate the mixture shown in Fig. 6.7?

***6.19.** A sealed reaction vessel is completely filled with liquid water at 293.15 K and 1.00 atm. If the temperature is raised exactly 6 K and the walls of the vessel remain rigid, what is the pressure in the container if the average value of $\alpha = 2.85 \times 10^{-4}$ K^{-1} and the compressibility coefficient κ [$= -(1/V)(\partial V/\partial P)_T$] is 4.49×10^{-5} atm^{-1}?

Construction of Phase Diagrams from Physical Data

6.20. In Figure 6.16, a solution having composition p is cooled to just above the eutectic temperature (point s is about 0.18 x_{Si}, and x_e is 0.31 x_{Si}); calculate the composition of the solid that separates and that of the liquid that remains.

***6.21.** The melting points and heats of fusion of gold and silicon are

	Au	Si
T/K	1337	1683
$\Delta_{fus}H$/J mol^{-1}	12 677.5	39 622.5

For the data, calculate the solid-liquid equilibrium lines and estimate the eutectic composition graphically. Compare the result with the values given by Figure 6.16.

6.22. Use the following data to construct a phase diagram of the phenol–water system and answer the following questions (the compositions are given in grams of phenol in 100 grams of solution):

t/°C	20	25	30	35	40	45	50	55	60	65	68.8
Aqueous layer	8.40	8.71	8.92	9.34	9.78	10.62	12.08	13.88	17.10	22.26	35.90
Phenol layer	72.24	71.38	69.95	68.28	66.81	65.02	62.83	60.18	56.10	49.34	35.90

a. What will be the compositions of the layers formed from a solution of 30 g phenol and 70 g water maintained at 30 °C?

b. A solution of 20 g phenol and 80 g water is prepared at 70 °C. How many phases will be present?

c. At what temperature will two phases appear if the solution in part (b) is cooled gradually? What will be the compositions of the two phases?

6.23. The following information is obtained from cooling curve data on the partial system Fe_2O_3–Y_2O_3 [J. W. Nielsen and E. F. Dearborn, *Phys. Chem. Solids, 5,* 203(1958)]:

Composition of Melt/ mol % Y_2O_3	Temperature of Break/°C	Temperature of Halt/°C
0		1550
5	1540	1440
10	1515	1440
15	1450	1440
20	1520	1440
25	1560	1440
30	1620	1575/1440
40	1705	1575
50		1720

Sketch the simplest melting point diagram consistent with these data. Label the phase regions and give the composition of any compounds formed.

6.24. The study of cooling curves for the thallium–gold system yields the following data. Construct the phase diagram and identify the eutectic composition and temperature. Pure gold melts at 1063 °C and pure thallium melts at 302 °C. In each region, identify the number of phases and the solid that separates out, if any.

Wt % Au	10	20	30	40	60	80	90
First break (°C)	272	204	200	400	686	910	998
Eutectic halt (°C)	128	128	128	128	128	128	128

[Data adapted from A. C. K. Smith, *Applied Physical Chemistry Problems,* London: McGraw-Hill, 1968, p. 13.]

6.25. a. From the following information, draw the binary phase diagram for the system FeO (mp. 1370 °C)–MnO (mp. 1785 °C). A peritectic reaction occurs at 1430 °C between α solid solution containing 30 mass % MnO and solid solution containing 60 mass % MnO. These are in equilibrium with melt that contains 15 mass % MnO. At 1200 °C the composition of α and β solution is 28 mass % and 63 mass %, respectively.

b. Describe what happens as a liquid containing 28 mass % MnO is cooled to 1200 °C.

6.26. The following data for the magnesium–copper system is the result of analyzing cooling curves. Pure copper melts at 1085 °C while pure magnesium melts at 659 °C. Two compounds are formed, one at 16.05 wt % Mg with a melting point of 800 °C, and the other at 43.44 wt % Mg with a melting point of 583 °C, respectively. Construct the phase diagram from this information and identify the compositions of the eutectics. [Data adapted from A. C. K. Smith, *Applied Physical Chemistry Problems,* London: McGraw-Hill, 1968, p. 14.]

Wt % Mg	5	10	15	20	30	35
First break (°C)	900	702	785	765	636	565
Eutectic halt (°C)	680	680	680	560	560	560

40	45	50	60	70	80	90
581	575	546	448	423	525	600
560	360	360	360	360	360	360

6.27. What are the empirical formulae of the compounds represented by the vertical lines formed in the magnesium–copper system described in Problem 6.26?

6.28. A preliminary thermal analysis of the Fe–Au system showed two solid phases of composition 8.1 mol % Au and 25.5 mol % Au in equilibrium at 1168 °C with liquid of composition 43 mol % Au. Construct the simplest melting point diagram consistent with this information and label all the phase regions. Sketch the cooling curves for the composition 10 mol % Au, 30 mol % A, and 60 mol % Au, and make them consistent with the fact that there is an α-γ phase transition in iron at 903 °C and that the γ-phase field extends to 45 mol % Au at this temperature. Iron melts at 1536 °C and gold at 1063 °C.

6.29. The aluminum–selenium system was determined from thermal analysis. Al_2Se_3 melts congruently at approximately 950 °C and forms a eutectic both with aluminum and with selenium at a very low concentration of the alloying element and at a temperature close to the melting point of the base elements. Draw a diagram from this information and give the composition of the phases. Aluminum melts at 659.7 °C and selenium melts at approximately 217 °C.

***6.30.** The metals Al and Ca form the compounds Al_4Ca and Al_2Ca. The solids Al, Ca, Al_4Ca, and Al_2Ca essentially are immiscible in each other but are completely miscible as liquids. Maximum Ca solubility in Al is about 2% and occurs at 616 °C. Al melts at 659.7 °C and Ca melts at 848 °C. Compound Al_2Ca melts congruently at 1079 °C and gives a simple eutectic with Ca at 545°C. Compound Al_4Ca decomposes at 700 °C to give Al_2Ca and a melt, the peritectic lying at 10 mol %. A monotectic exists at 616 °C. At approximately 450 °C a transition occurs between α-Ca and β-Ca.

a. Draw the simplest phase diagram consistent with this information and label all phase regions.
b. Sketch cooling curves for melts of composition 15 mol % Ca and 80 mol % Ca.

***6.31.** The extent of dehydration of a salt such as $CuSO_4$ can often be followed by measuring the vapor pressure over the hydrated salt. The system H_2O–$CuSO_4$ is shown in the accompanying figure as an example of such a system. Label the areas as to the phase(s) present. Then describe the sequence of phase changes if a dilute solution of copper sulfate is dehydrated at 275 K, ending with anhydrous copper sulfate. What would a vacuum gauge read starting with pure water during the dehydration process at 298.15 K? Sketch a plot of P/Torr against $CuSO_4$/wt %. Relevant data are:

The Vapor Pressure of $CuSO_4$–H_2O at 298.15 K	P/Torr
Vapor + saturated solution + $CuSO_4 \cdot 5H_2O$	16
Vapor $CuSO_4 \cdot 5H_2O$ + $CuSO_4 \cdot 3H_2O$	7.85
Vapor + $CuSO_4 \cdot 3H_2O$ + $CuSO_4 \cdot H_2O$	4.32
Vapor + $CuSO_4 \cdot H_2O$ + $CuSO_4$	0.017
Vapor pressure of water	23.8

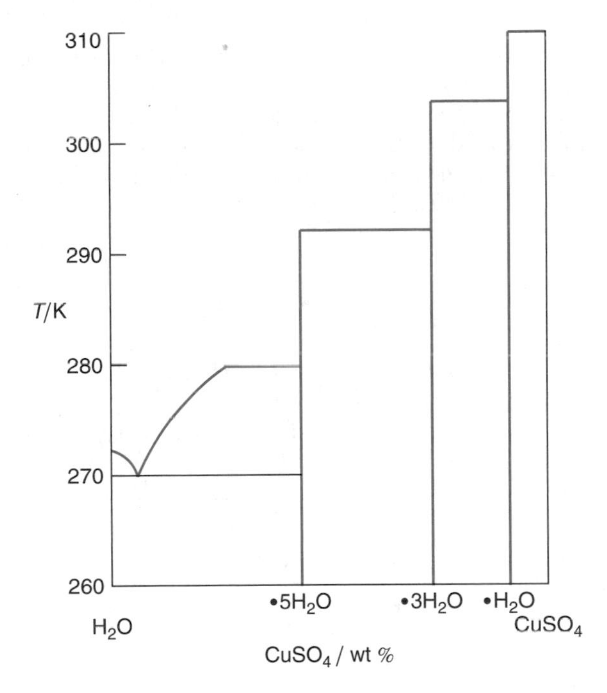

***6.32.** The data in the accompanying table are approximate for the isobaric-isothermal system SnO_2–CaO–MgO at 298.15 K and 1 atm. Sketch a reasonable phase diagram in mol % with SnO_2 at the apex of the triangle. Label all phase regions; the results are known as composition triangles.

Material	In Equilibrium with Solid Phases
SnO_2	$(MgO)_2SnO_2$, $CaOSnO_2$
$(MgO)_2SnO_2$	SnO_2, $(CaO)SnO_2$, MgO
MgO	CaO, $(CaO)_2SnO_2$
CaO	MgO, $(CaO)_2SnO_2$
$(CaO)_2SnO_2$	CaO, MgO, $CaOSnO_2$
$CaOSnO_2$	$(CaO)_2SnO_2$, MgO, $(MgO)_2SnO_2$, SnO_2

6.33. Sketch the *P* against *T* diagram for phosphorous from the following information. White phosphorous melts at 311 K and 0.2 Torr; red phosphorous melts at 763 K and 43 atm. The white form is more dense than the liquid and the red form is less dense than the liquid. The vapor pressure of the white form is everywhere greater than that of the red form. Label each area on the plot, and explain which triple point(s) is (are) stable or metastable.

6.34. Giguère and Turrell, *J. Am. Chem. Soc., 102,* 5476(1980), describe three ionic hydrates formed between HF and H_2O. Sketch the H_2O-HF phase diagram in mol % HF from the following information. HF · H_2O melts at −35.2 °C, 2HF · H_2O decomposes by a peritectic reaction at −75 °C, and 4HF · H_2O melts at −98.2 °C. HF melts at −83.1 °C. Label the composition of all regions. The eutectic occurs at −111 °C with monotectics at −71 °C, −77 °C, and −102 °C.

6.35. In the system A–B a line of three-phase equilibrium occurs at 900 K as determined by thermal analysis. A second three-phase equilibrium occurs at 500 K. Only one halt is observed for any one cooling curve. The compound AB_2 is known and melts at 600 K. If A melts at 1200 K and B at 700 K, sketch the simplest phase diagram consistent with the given data. Label each region.

Data Derived from Phase Diagrams of Condensed Systems

6.36. The following questions refer to Figure 6.28:

a. If liquid C were added to the system, what changes would occur if the system originally contained 80% salt A and 20% salt B?

b. What changes would occur if the system originally contained 50% salt A and 50% salt B upon the addition of liquid?

c. If liquid is added to an unsaturated solution of salt A and salt B in solution of composition lying at *e*, what changes would occur?

6.37. In the accompanying diagram, due to B. S. R. Sastry and F. A. Hammel, *J. Am. Ceramic Soc., 42;* 218(1959), identify the composition of all the areas. Identify the phenomenon associated with each lettered position.

6.38. Describe what happens within the system Mn_2O_3–Al_2O_3 in Fig. 6.21 when a liquid of $x_{Al_2O_3} = 0.2$ is cooled from 2100 K to 1200 K.

6.39. The isobaric solubility diagram for the system acetic acid–toluene–water is shown in Figure 6.27. What phase(s) and their composition(s) will be present if 0.2 mol of toluene is added to a system consisting of 0.5 mol of water and 0.3 mol of acetic acid? Give the relative amounts of each phase.

6.40. A fictitious ternary system composed of liquids A, B, and C was constructed by adding the component B to various binary A–C mixtures and noting the point at which complete miscibility occurred. The following are the mole-percents of A and B at which complete miscibility was observed. Construct the phase diagram on a triangular graph paper.

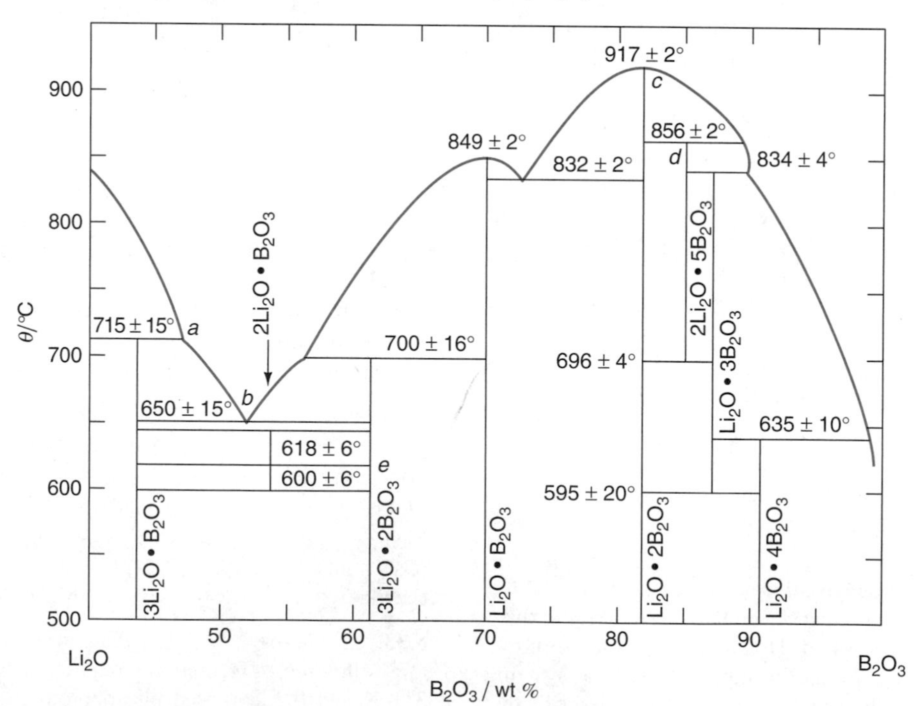

Diagram for Problem 6.37. Reprinted with permission of The American Ceramic Society, Post Office Box 6136, Westerville, OH 43086-6136,

$x_A(\%)$	10.0	20.0	30.0	40.0	50.0	60.0	70.0	80.0	90.0
$x_B(\%)$	20.0	27.0	30.0	28.0	26.0	22.0	17.0	12.0	7.0

Comment on the variation of the mutual solubility of A and C as B is added.

6.41. In organic chemistry it is a common procedure to separate a mixture of an organic liquid in water by adding a salt to it. This is known as "salting out." The ternary system K_2CO_3–H_2O–CH_3OH is typical. The system is distinguished by the appearance of the two-liquid region *abc*.

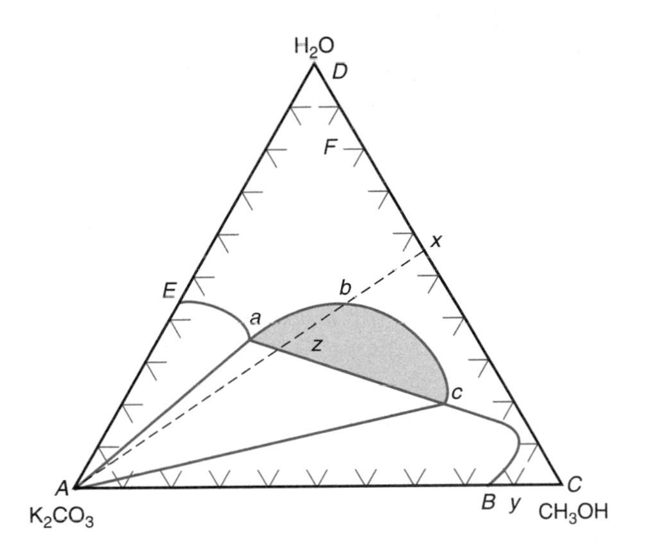

a. Describe the phase(s) present in each region of the diagram.
b. What would occur as solid K_2CO_3 is added to a solution of H_2O and CH_3OH of composition *x*?
c. How can the organic-rich phase in (b) be separated?
d. How can K_2CO_3 be precipitated from a solution having composition *y*?
e. Describe in detail the sequence of events when a solution of composition *F* is evaporated.

Essay Questions

6.42. How is thermal analysis used to determine the liquid-solid equilibria and the eutectic temperature?

6.43. Explain what is meant by a *metastable system.*

6.44. Outline how isothermal distillation may be used to prepare a pure sample.

6.45. Detail the process by which a pure sample is obtained using a fractionating column.

6.46. What is the difference on a molecular level between a maximum and minimum boiling azeotrope? How do the plots of *P* against *x* and *T* against *x* differ?

6.47. How would you distinguish between an azeotrope and a pure compound?

6.48. A synthetic chemist has prepared several zwitterionic compounds in a homogeneous series. With each compound a reproducible melting point is determined using different samples from a fresh batch of material. If, however, the same sample is used in repeating the determination, a progressively lower melting-point temperature is obtained. Explain what is happening.

SUGGESTED READING

A. Findlay, *Phase Rule* (revised and enlarged 9th ed., by A. N. Campbell and N. O. Smith), New York: Dover, 1951.

R. J. Forbes, *A Short History of the Art of Distillation,* Leiden: E. J. Brill, 1970.

J. H. Hildebrand, J. M. Prausnits, and R. L. Scott, *Regular and Related Solutions,* New York: Van Nostrand Reinhold, 1970.

W. Hume-Rothery, R. E. Smallman, and C. W. Haworth, *The Structure of Metals and Alloys,* The Metals and Metallurgy Trust of the Institute of Metals and the Institution of Metallurgists, London, 1969.

R. G. Laughlin, *The Aqueous Phase Behavior of Surfactants,* New York: Academic Press, 1994.

C. S. Robinson and E. R. Gilliland, *Fractional Distillation,* New York: McGraw-Hill, 1950.

B. D. Smith, *Design of Equilibrium Stage Processes,* New York: McGraw-Hill, 1963.

Specifically on miscibility:

J. S. Walker and C. A. Vance. *Scientific American,* May 1987, p. 98.

Much of the literature has been reviewed and compiled in several areas. The following are convenient sources of much of the work.

Azeotropic Data, Advances in Chemistry Series No. 35, American Chemical Society, Washington, DC, 1962.

R. P. Elliott, *Constitution of Binary Alloys* (1st Suppl.), New York: McGraw-Hill, 1965.

M. Hansen, *Constitution of Binary Alloys* (2nd ed.), New York: McGraw-Hill, 1958.

M. Hirata, S. One, and K. Nagahama, *Computer Aided Data Book of Vapor-Liquid Equilibria,* New York: Kodansha Limited, Elsevier, Scientific Publishing Co., 1990.

E. M. Levin, R. Robbins, and H. F. McMurdie, *Phase Diagrams for Ceramists,* The American Ceramic Society, Inc., 1964; 1969 Supplement (Figures 2067–4149); E. M. Levin and H. F. McMurdie, 1975 Supplement (Figures 4150–4999). (Series through 1992.)

T. B. Massalski, Ed. *Binary Alloy Phase Diagrams,* ASM/NIST Data Program for Alloy Phase Diagrams, ASM International, Materials Park, Ohio 44073. 2nd Ed., 1990, in three volumes.

W. G. Moffatt, Ed., *The Handbook of Binary Phase Diagrams,* Genium Publishing Corporation, Schenectady, NY: 1994. (In five volumes.)

G. Petzow and G. Effenberg, Eds., *Ternary Alloys, A Comprehensive Compendium of Evaluated Constitutional Data and Phase Diagrams,* Materials Science; International Services GmbH and the Max Planck-Institut fur Metallforschung, Stuttgart, VCH Verlagsgesellschaft mbH, Weinheim, Germany, 1992. (In five volumes.)

F. A. Shunk, *Constitution of Binary Alloys* (2nd Suppl.), New York: McGraw-Hill, 1969.

I. Wichterle, J. Linek, and E. Hala, *Vapor-Liquid Equilibrium Data Bibliography,* New York: Elsevier Science Publishers, 1985. Covers the literature with four supplements to 1985.

Solutions of Electrolytes

7

PREVIEW

Electrochemistry is concerned with the properties of solutions of electrolytes and with processes that occur at electrodes. A useful starting point is provided by *Faraday's laws of electrolysis,* which relate the mass of a substance deposited at an electrode to the quantity of electricity (current × time) passed through the solution, and to the relative atomic or molecular mass of the substance.

Important properties of solutions of electrolytes are the resistance, the conductance, and the electrolytic conductivity. To compare solutions of different concentrations, it is convenient to introduce the *molar conductivity* Λ (formerly called the equivalent conductivity), which is the conductivity when one mole of the electrolyte is in between two electrodes that are a standard distance apart.

Different theories are required for solutions of weak electrolytes, which are only slightly dissociated, and for strong electrolytes, which are completely dissociated. The former are satisfactorily treated by a theory due to Arrhenius, which considers how the dissociation equilibrium shifts as the concentration of the electrolyte is varied. The theory finds its quantitative expression in *Ostwald's dilution law,* which relates the molar conductivity Λ at any concentration to that at infinite dilution Λ°. The ratio Λ/Λ° is the degree of dissociation α.

This theory does not apply to strong electrolytes, which are dealt with by an entirely different treatment due mainly to Debye and Hückel. This theory focuses attention on the distribution of positive and negative ions in solution as a result of the electrostatic forces. In the near neighborhood of each ion there are more ions of opposite sign than of the same sign, and it is convenient to speak of an *ionic atmosphere* as existing around each ion. The thickness of the ionic atmosphere decreases as the concentration is raised. When an ion moves through the solution as a result of an applied potential field, the atmosphere lags behind and causes a retardation of the motion; this is known as the *relaxation* or the *asymmetry* effect. The moving ionic atmosphere drags solvent molecules with it, and as a result there is an additional retardation; this is known as the *electrophoretic* effect.

Kohlrausch's law of independent migration of ions states that the molar conductivity of an electrolyte is the sum of the individual ionic conductivities. The *mobility* of an ion is the speed with which it moves in a unit potential gradient, and it is proportional to the ionic conductivity. The *transport number* of an ion is the fraction of the current carried by that ion. Transport numbers can be measured experimentally, and they allow the molar conductivity to be split into the individual ionic conductivities.

This chapter is also concerned with the thermodynamics of ions in solution, and several important matters are involved: the enthalpies and entropies of hydration of ions and the activity coefficients of ions. The latter are treated in an approximate manner by the *Debye-Hückel limiting law,* according to which the logarithm of the activity coefficient γ_i of an ion of the *i*th type is proportional to the square root of the *ionic strength* of the solution.

Another important matter, particularly in biological systems, is the distribution of electrolytes across membranes that are permeable to some ions but not to others. This problem is treated by the theory of the *Donnan equilibrium.*

OBJECTIVES

After studying this chapter, the student should be able to:

- Understand Faraday's laws of electrolysis.
- Define *molar conductivity.*
- Explain the Arrhenius theory of strong electrolytes and derive Ostwald's dilution law.
- Describe the ionic atmosphere and how it affects the conductivity of a solution of an electrolyte (the Debye-Hückel theory).
- Understand the concept of a transport number and how it is measured.
- Explain the properties of ions in solution, including the special properties of the hydrogen ion in aqueous solution.
- Understand how the Debye-Hückel theory interprets the activity coefficients of ions in solution.
- Explain the concept of the ionic product.
- Understand solubility products and the theory of the Donnan equilibrium.

At the atomic level, electrical factors determine the structures and properties of chemical substances. The important branch of physical chemistry known as electrochemistry is particularly concerned with the properties of solutions of electrolytes and with processes occurring when electrodes are immersed in these solutions. This chapter deals with the electrochemistry of solutions, a topic that is of special interest since studies in this area, in the late nineteenth century, played an important role in the development of physical chemistry.

Much can be learned about the behavior of ions in water and other solvents by investigations of electrical effects. Thus, measurements of the conductivities of aqueous solutions at various concentrations have led to an understanding of the extent to which substances are ionized in water, of the association of ions with surrounding water molecules, and of the way in which ions move in water.

It is convenient to begin our study of electrochemistry with a brief survey of the development of concepts related to electricity. It has been known since ancient times that amber, when rubbed, acquires the property of attracting light objects such as small pieces of paper or pitch. In 1600 Sir William Gilbert (1544–1603) coined the word *electric,* from the Greek word for amber (*elektron*), to describe substances that acquired this power to attract. It was subsequently found that materials such as glass, when rubbed with silk, exerted forces opposed to those from amber. Two types of electricity were thus distinguished: resinous (from substances like amber) and vitreous (from substances like glass).

In 1747 the great American scientist and statesman Benjamin Franklin (1706–1790) proposed a "one-fluid" theory of electricity. According to him, if bodies such as amber and fur are rubbed together, one of them acquires a surplus of "electric fluid" and the other acquires a deficit of "electric fluid." Quite arbitrarily, Franklin established the convention, still used today, that the type of electricity produced on glass rubbed with silk is positive; conversely, the resinous type of electricity is negative. The positive electricity was supposed to involve an excess of electric fluid; the negative electricity, a deficit. We now know that negative electricity corresponds to a surplus of electrons, whereas positive electricity involves a deficit. Franklin's convention, however, is now so well entrenched that it would be futile to try to change it.

Until 1791 the study of electricity was entirely concerned with static electricity, in the form of charges developed by friction. In that year Luigi Galvani (1737–1798) brought the nerve of a frog's leg into contact with an electrostatic machine and observed a sharp convulsion of the leg muscles. Later he found that the twitching could be produced simply by bringing the nerve ending and the end of the leg into contact by means of a metal strip. Galvani believed that this effect could only be produced by living tissues. Then in 1794 the Italian physicist Alessandro Volta (1745–1827) discovered that the same type of electricity could be produced from inanimate objects. In 1800 he constructed his famous "Voltaic pile,"

Voltaic Pile

which consisted of a number of consecutive plates of silver and zinc, separated by cloth soaked in salt solution. From the terminals of the pile he produced the shocks and sparks that had previously been observed only with frictional machines. With the experiments of Galvani and more particularly of Volta, the subject of electricity entered into a new era in which electric currents played predominant roles.

Very shortly after Volta carried out his experiments, the British scientific writer William Nicholson (1753–1815), in collaboration with the surgeon Anthony Carlisle (1768–1842), decomposed water by means of an electric current and observed that

Electrolysis

oxygen appeared at one pole and hydrogen at the other. The word **electrolysis** (Greek *lysis,* setting free) was later used by the English scientist Michael Faraday (1791–1867) to describe such splitting by an electric current. Faraday gave the name **cathode** (Greek *cathodos,* way down) to the electrode at which the hydrogen collects and **anode** (Greek *ana,* up; *hodos,* way) to the other electrode. The important studies of Faraday, which put the subject of electrolysis on a quantitative basis, are considered later.

In an electrolytic cell, the negative electrode is called the *cathode,* and the positive electrode the *anode;* these terms were introduced by Faraday in 1834. In a cell that generates electricity (Chapter 8), the opposite is true; the *cathode* is the positive electrode and the *anode* the negative electrode. To avoid confusion it is best to specify the charges, plus or minus.

Electrical Units

The *electrostatic force* F between two charges Q_1 and Q_2 separated by a distance r in a vacuum is

$$F = \frac{Q_1 Q_2}{4\pi\epsilon_0 r^2} \tag{7.1}$$

Dielectric Constant

The constant ϵ_0 is the *permittivity of a vacuum* and has the value 8.854×10^{-12} C^2 J^{-1} m^{-1}. The factor 4π is introduced into this expression in order that the Gauss and Poisson equations (see pp. 278–279) are free of this factor.[1] If the charges are in a medium having a *relative permittivity,* or *dielectric constant,* of ϵ, the equation for the force is

$$F = \frac{Q_1 Q_2}{4\pi\epsilon_0 \epsilon r^2} \tag{7.2}$$

For example, water at 25 °C has a dielectric constant of about 78, with the result that the electrostatic forces between ions are reduced by this factor. The SI unit of charge Q is the coulomb, C, and that of distance r is the metre, m. The unit of force F is the newton, N, which is joule per metre, J m^{-1}.

The *electric field* E at any point is the force exerted on a unit charge (1 C) at that point. The field strength at a distance r from a charge Q, in a medium of dielectric constant ϵ, is thus

$$E = \frac{Q}{4\pi\epsilon_0 \epsilon r^2} \tag{7.3}$$

The SI unit of field strength is therefore newton per coulomb (N C^{-1} = J C^{-1} m^{-1}). However, since joule = volt coulomb = newton metre (see Appendix A), the SI unit of field strength is usually expressed as volt per metre (V m^{-1}).

Just as a mechanical force is the negative gradient of a potential, the *electric field strength* is the negative gradient of an *electric potential* ϕ. Thus the field strength E at a distance r from a charge Q is given by

$$E = -\frac{d\phi}{dr} \tag{7.4}$$

and the electric potential is thus

[1]See Appendix A for a further discussion of the term $4\pi\epsilon_0$.

$$\phi = -\int E\, dr = \frac{Q}{4\pi\epsilon_0\epsilon r} \tag{7.5}$$

The SI unit of the electric potential is the volt (V).[2]

Further details about electrical units, including a comparison of SI units with the electrostatic units, are to be found in Appendix A.

7.1 Faraday's Laws of Electrolysis

During the years 1833 and 1834 Michael Faraday published the results of an extended series of investigations on the relationships between the quantity of electricity passing through a solution and the amount of material liberated at the electrodes. He formulated **Faraday's laws of electrolysis,** which may be summarized as follows:

1. **The mass of an element produced at an electrode is proportional to the *quantity of electricity* Q passed through the liquid; the SI unit of Q is the coulomb (C). The quantity of electricity is defined as equal to the current I (SI unit = ampere, A) multiplied by the time t (SI unit = second, s):**

$$Q = It \tag{7.6}$$

2. **The mass of an element liberated at an electrode is proportional to the *equivalent weight* of the element.**

The SI recommendation, supported by the International Union of Pure and Applied Chemistry and other international scientific bodies, is to abandon the use of the word *equivalent* and to refer to *moles* instead. However, the concept of the equivalent is still employed, even though the name has been dropped. Suppose, for example, that we pass 1 A of electricity for 1 h through a dilute sulfuric acid solution and through solutions of silver nitrate, $AgNO_3$, and cupric sulfate, $CuSO_4$.[3] The masses liberated at the respective cathodes are

0.038 g of H_2 (relative atomic mass, $A_r = 1.008$; relative molecular mass, $M_r = 2.016$)

4.025 g of Ag ($A_r = 107.9$)

1.186 g of Cu ($A_r = 63.6$)

These masses 0.038, 4.025, and 1.186 are approximately in the ratio 1.008/107.9/31.8. Therefore, the amount of electricity that liberates 1 mol of Ag liberates 0.5 mol of H_2 and 0.5 mol of Cu. These quantities were formerly referred to as 1 equivalent, but the modern practice is to speak instead of 1 mol of $\frac{1}{2}H_2$ and 1 mol of $\frac{1}{2}Cu$. These are the quantities liberated by 1 mol of electrons:

$$e^- + H^+ \rightarrow \tfrac{1}{2}H_2 \quad \text{and} \quad e^- + \tfrac{1}{2}Cu^{2+} \rightarrow \tfrac{1}{2}Cu$$

In what follows, when we refer to 1 mol we may mean 1 mol of a fraction of an entity (e.g., to $\frac{1}{2}Cu$ or $\frac{1}{2}SO_4^{2-}$).

[2] In terms of base SI units, volt = kg m^2 s^{-3} A^{-1}.

[3] The symbol for hour is h.

MICHAEL FARADAY (1791–1867)

Faraday was born in Newington, Surrey, which is not far from London. His father was a blacksmith and the family was of very limited means. At the age of 14 Faraday was apprenticed to a bookbinder, and he took the advantage of reading many of the books he bound. His interest in science was particularly aroused by Jane Marcet's famous *Conversations on Chemistry,* which first appeared in 1809.

Faraday was fortunate to obtain the position of laboratory assistant to Sir Humphry Davy (1778–1829), the Director of the Royal Institution. In 1825 he succeeded Davy as director of the Royal Institution laboratories, and in 1833 he also became Fullerian Professor of Chemistry, a post that was created for him by John Fuller, a Member of Parliament. Fuller had attended many of the public lectures at the Royal Institution, and had often slept through them; his endowment was "in gratitude for the peaceful hours thus snatched from an otherwise restless life."

After 1821 much of Faraday's work was on electricity and magnetism, which at the time was considered to be chemistry rather than physics. In 1831 he discovered the phenomenon of electromagnetic induction, and in the following year he began to investigate whether the electricities produced in various ways were identical. It is interesting to note that he did not obtain his two laws of electrolysis empirically; he deduced them from what he discovered about the nature of electricity, and proceeded to verify them.

Faraday was ignorant of mathematics beyond simple arithmetic, but he developed qualitative theories. In Faraday's view an electric current is due to a buildup and breakdown of strain within the particles of matter through which the current flows. His ideas about electricity and magnetism led to the establishment of modern field theory. In the 1850s and 1860s Clerk Maxwell (1831–1879) formulated Faraday's ideas into a mathematical theory of electromagnetic radiation.

Faraday was deeply religious, being a loyal member of the Sandemanian Church. He modestly declined all honors and avoided scientific controversy. He preferred to work alone and without an assistant, and as a result he founded no school of research. He became famous for his Friday Evening Discourses, which he founded at the Royal Institution and of which he gave over a hundred himself. One of them was his well-known Christmas lecture on the "Chemical History of a Candle."

References: L. Pierce Williams, *Dictionary of Scientific Biography, 4,* 527–540(1971).

D. K. C. Macdonald, *Faraday, Maxwell and Kelvin,* New York: Doubleday & Co., 1964.

Sir George Porter, "Michael Faraday—Chemist" (Faraday Lecture of the Chemical Society), *Proc. Roy. Institution, 53,* 90–99(1981).

L. Pierce Williams, *Michael Faraday, A Biography,* New York: Basic Books, 1965.

The proportionality factor that relates the amount of substance deposited to the quantity of electricity passed through the solution is known as the **Faraday constant** and given the symbol F. In modern terms, the charge carried by 1 mol of ions bearing z unit charges is zF. According to the latest measurements, the Faraday constant F is equal to 96 485.309 coulombs per mole (C mol^{-1}); in this discussion the figure will be rounded to 96 485 C mol^{-1}. In other words, 96 485 C will liberate 1 mol of Ag, 1 mol of $\frac{1}{2}H_2$ (i.e., 0.5 mol of H_2), etc. If a constant current I is passed for a period of time t, the amount of substance deposited is $It/(96\,485 \text{ C mol}^{-1})$.

It follows that the (negative) charge on 1 mol of electrons is 96 485 C. The charge on one electron is this quantity divided by the Avogadro constant L and is thus

$$\frac{96\,485 \text{ C mol}^{-1}}{6.022 \times 10^{23} \text{ mol}^{-1}} = 1.602 \times 10^{-19} \text{ C}$$

EXAMPLE 7.1 An aqueous solution of gold (III) nitrate, $Au(NO_3)_3$, was electrolyzed with a current of 0.0250 A until 1.200 g of Au (atomic weight 197.0) had been deposited at the cathode. Calculate (a) the quantity of electricity passed, (b) the duration of the experiment, and (c) the volume of O_2 (at NSTP[4]) liberated at the anode. The reaction at the cathode is

$$3e^- + Au^{3+} \rightarrow Au \quad \text{or} \quad e^- + \tfrac{1}{3}Au^{3+} \rightarrow \tfrac{1}{3}Au$$

The equation for the liberation of O_2 is

$$2H_2O \rightarrow O_2 + 4H^+ + 4e^- \quad \text{or} \quad \tfrac{1}{2}H_2O \rightarrow \tfrac{1}{4}O_2 + H^+ + e^-$$

Solution a. From the cathode reaction, 96 485 C of electricity liberates 1 mol of $\frac{1}{3}$Au (i.e., 197.0/3 g). The quantity of electricity required to produce 1.200 g of Au is thus

$$\frac{96\,485 \times 1.200}{197.0/3} = 1.76 \times 10^3 \text{ C}$$

b. Since the current was 0.0250 A, the time required was

$$\frac{1.76 \times 10^3 (\text{A s})}{0.025(\text{A})} = 7.4 \times 10^4 \text{ s}$$

c. From the anode reaction, 96 485 C liberates $\frac{1}{4}$ mol O_2, so that 1.76×10^3 C produces

$$\frac{1.76 \times 10^3}{96\,485 \times 4} = 4.56 \times 10^{-3} \text{ mol } O_2$$

The volume of this at NSTP is

$$4.56 \times 10^{-3} \text{ (mol)} \times 24.8(\text{dm}^3 \text{ mol}^{-1}) = 0.113 \text{ dm}^3$$

Faraday's work on electrolysis was of great importance in that it was the first to suggest a relationship between matter and electricity. His work indicated that *atoms,* recently postulated by Dalton, might contain electrically charged particles. Later work led to the conclusion that an electric current is a stream of electrons and that electrons are universal components of atoms.

7.2 Molar Conductivity

Important information about the nature of solutions has been provided by measurements of their conductivities. A solution of sucrose in water does not dissociate into ions and has the same electrical conductivity as water itself. Such substances are known as *nonelectrolytes.* A solution of sodium chloride or acetic acid, on the other hand, forms ions in solution and has a much higher conductivity. These substances are known as *electrolytes.*

[4]New standard temperature and pressure, viz. 25.0 °C and 1 bar; at NSTP 1 mol of gas occupies 24.8 dm^3.

Some electrolytes in aqueous solutions of the same molality conduct better than others. This suggests that some electrolytes exist as ions in solution to a greater extent than do others. More detailed investigations have indicated that certain substances such as the salts sodium chloride and cupric sulfate, as well as some acids such as hydrochloric acid, occur almost entirely as ions when they are in aqueous solution. Such substances are known as **strong electrolytes.** Other substances, including acetic acid and ammonia, are present only partially as ions (e.g., acetic acid exists in solution partly as CH_3COOH and partly as $CH_3COO^- + H^+$). These substances are known as **weak electrolytes.**

Strong and Weak Electrolytes

Further insight into these matters is provided by studies of the way in which the electrical conductivities of solutions vary with the concentration of solute.

According to **Ohm's law,** the resistance R of a slab of material is equal to the electric potential difference V divided by the electric current I:

$$R = \frac{V}{I} \tag{7.7}$$

The SI unit of potential is the volt (V) and that for current is the ampere (A). The unit of electrical resistance is the ohm,[5] given the symbol Ω (omega). The reciprocal of the resistance is the *electrical conductance* G, the SI unit of which is the *siemens* ($S \equiv \Omega^{-1}$). The electrical conductance of material of length l and cross-sectional area A is proportional to A and inversely proportional to l:

$$G\ (\text{conductance}) = \kappa \frac{A}{l} \tag{7.8}$$

Electrolytic Conductivity

The proportionality constant κ, which is the conductance of a unit cube (see Figure 7.1a), is known as the *conductivity* (formerly as the specific conductivity); for the particular case of a solution of an electrolyte, it is known as the **electrolytic conductivity.** Its SI unit is $\Omega^{-1}\ m^{-1}$ ($\equiv S\ m^{-1}$), but the unit more commonly used is $\Omega^{-1}\ cm^{-1}$ ($\equiv S\ cm^{-1}$).

Molar Conductivity

The electrolytic conductivity is not a suitable quantity for comparing the conductivities of different solutions. If a solution of one electrolyte is much more concentrated than another, it may have a higher conductivity simply because it contains more ions. What is needed instead is a property in which there has been some compensation for the difference in concentrations. In the 1880s the German physicist Friedrich Wilhelm Georg Kohlrausch (1840–1910) clarified the early theory by employing the concept of what is now called the **molar conductivity**[6] and given the symbol Λ (lambda). It is defined as the electrolytic conductivity κ divided by concentration c,

$$\Lambda = \frac{\kappa}{c} \tag{7.9}$$

The meaning of Λ can be visualized as follows. Suppose that we construct a cell having parallel plates a unit distance apart, the plates being of such an area that for

[5]In terms of base SI units, ohm $\equiv$ kg $m^2\ s^{-3}\ A^{-2}$.

[6]It was formerly called the equivalent conductivity, but (as already noted) IUPAC no longer recommends the use of the word "equivalent," which is sometimes ambiguous: An entirely different equivalent is involved if a substance acts as an oxidizing or reducing agent. Instead of speaking of an equivalent of $CuSO_4$, for example, which requires us to know that we mean half a mole of $CuSO_4$, we speak of a mole of $\frac{1}{2}CuSO_4$.

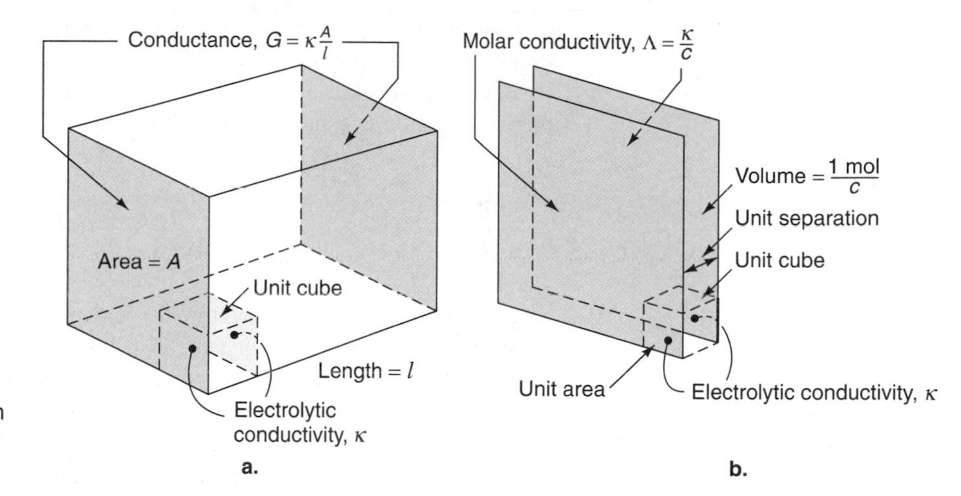

FIGURE 7.1
(a) The relationship between conductance and electrolytic conductivity. (b) The relationship between molar conductivity and electrolytic conductivity.

a particular solution 1 mol of electrolyte (e.g., HCl, $\frac{1}{2}CuSO_4$) is present in it (see Figure 7.1b). The molar conductivity Λ is the conductance (i.e., the reciprocal of the resistance) across the plates. We need not actually construct cells for each solution for which we need to know the molar conductivity, since we can calculate it from the electrolytic conductivity κ. If the concentration of the solution is c, the volume of the hypothetical molar conductivity cell must be (1 mol)/c in order for 1 mol to be between the plates (Figure 7.1b). The molar conductivity Λ is therefore the electrolytic conductivity κ multiplied by $1/c$. In using Eq. 7.9, care must be taken with units, as shown by Example 7.2.

EXAMPLE 7.2 The electrolytic conductivity of a 0.1 M solution of acetic acid (corrected for the conductivity of the water) was found to be $5.3 \times 10^{-4}\ \Omega^{-1}$ cm^{-1}. Calculate the molar conductivity.

Solution From Eq. 7.9,

$$\Lambda = \frac{\kappa}{c} = \frac{5.3 \times 10^{-4}(\Omega^{-1}\ \text{cm}^{-1})}{0.1(\text{mol dm}^{-3})} = \frac{5.3 \times 10^{-4}(\Omega^{-1}\ \text{cm}^{-1})}{10^{-4}(\text{mol cm}^{-3})}$$
$$= 5.3\ \Omega^{-1}\ \text{cm}^2\ \text{mol}^{-1}$$

This result can also be written as 5.3 S cm^2 mol^{-1}, where S stands for siemens, the unit of conductance, which is equivalent to the reciprocal ohm.

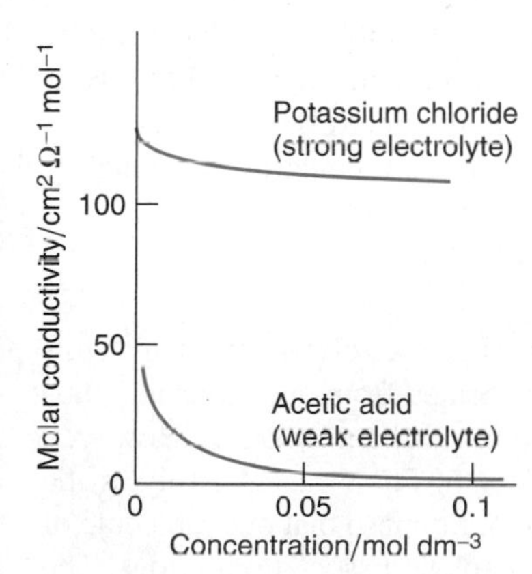

FIGURE 7.2
The variations of molar conductivity Λ with concentration for strong and weak electrolytes.

The importance of the molar conductivity is that it gives information about the conductivity of the ions produced in solution by 1 mol of a substance. In all cases the molar conductivity diminishes as the concentration is raised, and two patterns of behavior can be distinguished. From Figure 7.2 we see that for strong electrolytes the molar conductivity falls only slightly as the concentration is raised. On the other hand, the weak electrolytes produce fewer ions and exhibit a much more pronounced fall of Λ with increasing concentration, as shown by the lower curve in Figure 7.2.

We can extrapolate the curves back to zero concentration and obtain a quantity known as Λ°, the *molar conductivity at infinite dilution,* or zero concentration. With weak electrolytes this extrapolation is unreliable, and an indirect method, explained on pp. 291–292, is usually employed for its calculation.

7.3 Weak Electrolytes: The Arrhenius Theory

The conductivity data of Kohlrausch, and the fact that the heat of neutralization of a strong acid by a strong base in dilute aqueous solution was practically the same for all strong acids and strong bases (about 54.7 kJ mol^{-1} at 25 °C), led the Swedish scientist Svante August Arrhenius (1859–1927) to propose in 1887 a new theory of electrolyte solutions. According to this theory, there exists an equilibrium in solution between undissociated molecules AB and the ions A^+ and B^-:

$$AB \rightleftharpoons A^+ + B^- \tag{7.10}$$

At very low concentrations this equilibrium lies over to the right, and the molar conductivity is close to Λ°. As the concentration is increased, this equilibrium shifts to the left and the molar conductivity decreases from Λ° to a lower value Λ. The

Degree of Dissociation

degree of dissociation, that is, the fraction of AB in the form $A^+ + B^-$, is Λ/Λ°, which is denoted by the symbol α:

$$\alpha = \frac{\Lambda}{\Lambda^\circ} \tag{7.11}$$

At high concentrations there is less dissociation, while at infinite dilution there is complete dissociation and the molar conductivity then is Λ°.

This theory explains the constant heat of neutralization of all strong acids and bases, which are dissociated to a considerable extent, the neutralization involving simply the reaction $H^+ + OH^- \rightarrow H_2O$. However, as we shall see (Section 7.4), Arrhenius's explanation of the decrease in Λ with increasing concentration is valid only for weak electrolytes; another explanation is required for strong electrolytes.

Soon after Arrhenius's theory was proposed, van't Hoff found that osmotic pressure measurements that he had made gave it considerable support. Van't Hoff had found that the osmotic pressures of solutions of electrolytes were always considerably higher than predicted by the osmotic pressure equation for nonelectrolytes (Eq. 5.134). He proposed the modified equation

$$\pi = icRT \tag{7.12}$$

Van't Hoff Factor

where i is known as the **van't Hoff factor.** For strong electrolytes, the van't Hoff factor is approximately equal to the number of ions formed from one molecule; thus for NaCl, HCl, etc., $i = 2$; for Na_2SO_4, $BaCl_2$, etc., $i = 3$; and so forth. However, the value of i decreases with increasing ionic strength. For weak electrolytes the van't Hoff factor i involves the degree of dissociation. Suppose that one molecule of a weak electrolyte would produce, if there were complete dissociation, ν ions. The number of ions actually produced is thus $\nu\alpha$, and the number of undissociated molecules is $1 - \alpha$, so that the total number of particles produced from 1 molecule is

$$i = 1 - \alpha + \nu\alpha \tag{7.13}$$

Svante August Arrhenius (1859–1927)

Arrhenius was born in Vik (Wijk), which is near Uppsala, Sweden. He was a student at the University of Uppsala but carried out his doctoral research in Stockholm under the physicist Erik Edlund. His thesis, completed in 1884, was on the conductivities of electrolytic solutions. It did not suggest the dissociation of electrolytes into ions, and received a poor rating from the examiners. Arrhenius then had discussions with Ostwald, van't Hoff, and others, and as a result postulated his theory of electrolytic dissociation in 1887.

After holding some teaching positions, in 1895 Arrhenius was appointed professor of physics at the University of Stockholm; he served as its Rector from 1897 to 1902. From 1905 until his death he was director of physical chemistry at the Nobel Institute in Stockholm.

Arrhenius continued for many years to work on electrolytic solutions, but to the end of his life never accepted the fact that strong electrolytes are completely dissociated, and that their behavior is strongly influenced by interionic forces (Section 7.4). He worked in a variety of other fields, and made important contributions to immunochemistry, cosmology, the origin of life, and the causes of the ice age. It was he who first discussed what is now called the "greenhouse effect." He published many papers and a number of books, including *Lehrbuch der kosmischen Physik* (1903), *Immunochemistry* (1907), and *Quantitative Laws in Biological Chemistry* (1915).

Arrhenius received the 1903 Nobel Prize in chemistry, the citation making reference to the "special value of his theory of electrolytic dissociation in the development of chemistry."

References: H. A. M. Snelders, *Dictionary of Scientific Biography, 1*, 296–302(1970).
J. W. Servos, *Physical Chemistry from Ostwald to Pauling: The Making of a Science in America*, Princeton University Press, 1990.
K. J. Laidler, *The World of Physical Chemistry*, New York: Oxford University Press, 1993.
Elisabeth Crawford, *Arrhenius: From Ionic Theory to the Greenhouse Effect*, Science History Publications, Canton, Ohio, 1998.

Therefore

$$\alpha = \frac{i-1}{\nu-1} \tag{7.14}$$

It is thus possible to calculate α from osmotic pressure data. For weak electrolytes the values so obtained are in good agreement with Λ/Λ°, and the Arrhenius theory therefore applies satisfactorily to such systems.

Ostwald's Dilution Law

Wilhelm Ostwald received the 1909 Nobel Prize in chemistry for his wide-ranging research on catalysis, chemical equilibrium, and rates of chemical reactions. He was very concerned with energy waste, writing "Dissipate no energy, but strive to use energy by converting it into more useful forms."

In 1888 the ideas of Arrhenius were expressed quantitatively by Friedrich Wilhelm Ostwald (1853–1932) in terms of a *dilution law*. Consider an electrolyte AB that exists in solution partly as the undissociated species AB and partly as the ions A^+ and B^-:

$$AB \rightleftharpoons A^+ + B^-$$

The equilibrium constant, on the assumption of ideal behavior, is

$$K_c = \frac{[A^+][B^-]}{[AB]} \tag{7.15}$$

Suppose that an amount n of the electrolyte is present in a volume V and that the fraction dissociated is α; the fraction not dissociated is $1 - \alpha$. The amounts of the three species present at equilibrium, and the corresponding concentrations, are therefore

	AB	$\rightleftharpoons$	A^+	+	B^-
Amounts present at equilibrium:	$n(1-\alpha)$		$n\alpha$		$n\alpha$
Concentrations at equilibrium:	$\frac{n(1-\alpha)}{V}$		$\frac{n\alpha}{V}$		$\frac{n\alpha}{V}$

The equilibrium constant is

$$K_c = \frac{(n\alpha/V)^2}{[n(1-\alpha)]/V} = \frac{n\alpha^2}{V(1-\alpha)} \tag{7.16}$$

Therefore, for a given amount of substance the degree of dissociation α must vary with the volume V as follows:

$$\frac{\alpha^2}{1-\alpha} = \text{constant} \times V \tag{7.17}$$

Alternatively, since $n/V = c$, we can write

$$\frac{c\alpha^2}{1-\alpha} = K \tag{7.18}$$

The larger the V, the lower the concentration c and the larger the degree of dissociation. Thus, if we start with a solution containing 1 mol of electrolyte and dilute it, the degree of dissociation increases and the amounts of the ionized species increase. As V becomes very large (the concentration approaching zero), the degree of dissociation α approaches unity; that is, dissociation approaches 100% as infinite dilution is approached. The experimental value of Λ° corresponds to complete dissociation; at finite concentrations the molar conductivity Λ is therefore lower by the factor $\alpha(= \Lambda/\Lambda^\circ)$. The dilution law can thus be expressed as

$$KV = \frac{n(\Lambda/\Lambda^\circ)^2}{1 - (\Lambda/\Lambda^\circ)} \tag{7.19}$$

or

$$K = \frac{c(\Lambda/\Lambda^\circ)^2}{1 - (\Lambda/\Lambda^\circ)} \tag{7.20}$$

Equation 7.20 has been found to give a satisfactory interpretation of the variation of Λ with c for a number of weak electrolytes, but for strong electrolytes there are serious deviations, as discussed in the next section.

7.4 Strong Electrolytes

Arrhenius maintained all his life that his theory applied without any modification to strong as well as to weak electrolytes. However, some of the results for strong electrolytes were soon found to be seriously inconsistent with the theory and with Ostwald's dilution law. For example, the "constants" K calculated from Λ/Λ° values on the basis of Ostwald's dilution law (Eq. 7.20), were found for stronger acids and bases to vary, sometimes by several powers of 10, as the concentration was varied. Furthermore, values for the conductivity ratio Λ/Λ° were sometimes significantly different from those obtained from the van't Hoff factor (Eq. 7.12).

We saw earlier that the constancy of heats of neutralization of strong acids and bases was originally regarded as good evidence for the Arrhenius theory. But the heats were *too* constant to be consistent with the theory, because at a given concentration there should have been small but significant differences in the degrees of ionization of acids such as HCl, HNO_3, and H_2SO_4, and these differences should have given measurable differences between the heats of neutralization. These differences could not be observed, a result that suggests that these acids are completely dissociated at all concentrations.

Strong support for this idea was obtained between 1909 and 1916 by the Danish physical chemist Niels Janniksen Bjerrum (1879–1958). He examined the absorption spectra of solutions of a number of strong electrolytes and found no evidence for the existence of undissociated molecules. The fall in Λ with increasing concentration must therefore, for strong electrolytes, be attributed to some cause other than a decrease in the degree of dissociation.

These anomalies led a number of scientists to reject Arrhenius's theory completely, and to conclude that ions are not present free in solution. One of the most vigorous opponents of electrolytic dissociation was the British scientist Henry Edward Armstrong (1848–1937), who even went so far as to lampoon the idea in two fairy tales, entitled *A Dream of Fair Hydrone* and *The Thirst of Salt Water.* His main objection to the theory was that it appeared to give the solvent merely a passive role—in his expressive phrase, water was "merely a dance floor for ions." He preferred a solvation theory, in which the water played a more active role. There was indeed some substance to his criticism, but Armstrong and others were wrong to try to overthrow completely the idea that substances can be dissociated into ions in solution. What was needed instead was modification of the idea. Later treatments allowed the solvent to play a more active role, and it was realized that solvent molecules are sometimes bound quite strongly to ions, as will be considered in Section 7.9. The importance of the electrostatic forces between ions was also taken into account, as will now be discussed.

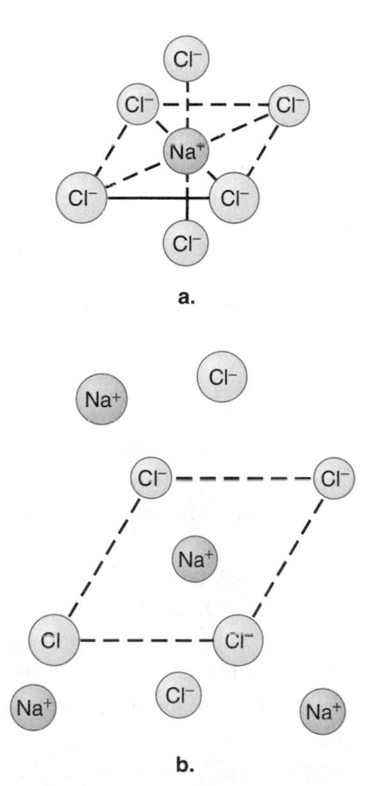

FIGURE 7.3
The distribution of chloride ions around a sodium ion (a) in the crystal lattice and (b) in a solution of sodium chloride. In the solution the interionic distances are greater, and the distribution is not so regular; but near to the sodium ion there are more chloride ions than sodium ions.

Debye-Hückel Theory

In 1923, the Dutch physical chemist Peter Debye (1884–1966) and the German physicist Erich Hückel (1896–1980) published a very important mathematical treatment of the problem of the conductivities of strong electrolytes. The decrease in the molar conductivity of a strong electrolyte was attributed to the mutual interference of the ions, which becomes more pronounced as the concentration increases. Because of the strong attractive forces between ions of opposite signs, the arrangement of ions in solution is not completely random. In the immediate neighborhood of any positive ion, there tend to be more negative than positive ions, whereas for a negative ion there are more positive than negative ions. This is shown schematically in Figure 7.3 for a sodium chloride solution. In solid sodium chloride there is a regular array of sodium and chloride ions. As seen in Figure 7.3a, each sodium ion has six chloride ions as its nearest neighbors, and each chloride ion has six sodium ions. When the sodium chloride is dissolved in water, this ordering is still preserved to a very slight extent (Figure 7.3b). The ions are much farther apart than in the solid; the electrical attractions are therefore much smaller and the thermal motions cause irregularity. The small amount of ordering that does exist, however, is sufficient to exert an important effect on the conductivity of the solution.

The way in which this ionic distribution affects the conductivity is in brief as follows; the matter is discussed in more detail later (see Figure 7.6). If an electric potential is applied, a positive ion will move toward the negative electrode and must drag along with it an entourage of negative ions. The more concentrated the solution, the closer these negative ions are to the positive ion under consideration, and the greater is the drag. The ionic "atmosphere" around a moving ion is therefore not symmetrical; the charge density behind is greater than that in front, and this will result in a retardation in the motion of the ion. This influence on the speed of an ion is called the **relaxation,** or **asymmetry,** effect.

Relaxation Effect

A second factor that retards the motion of an ion in solution is the tendency of the applied potential to move the ionic atmosphere itself. This in turn will tend to drag solvent molecules, because of the attractive forces between ions and solvent molecules. As a result, the ion at the center of the ionic atmosphere is required to move upstream, and this is an additional retarding influence. This effect is known as the **electrophoretic** effect.

Electrophoretic Effect

The Ionic Atmosphere

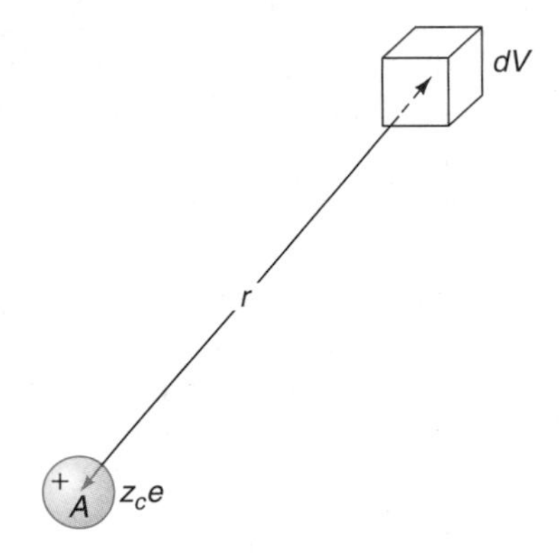

FIGURE 7.4
An ion in solution of charge z_ce, with an element of volume dV situated at a distance r from it.

The Debye-Hückel theory is very complicated mathematically and a detailed treatment is outside the scope of this book. We will give a brief account of the theoretical treatment of the ionic atmosphere, followed by a qualitative discussion of the way in which the theory of the ionic atmosphere explains conductivity behavior.

Figure 7.4 shows a positive ion, of charge $z_c e$, situated at a point A; e is the unit positive charge (the negative of the electronic charge) and z_c is the valence of the ion. As a result of thermal motion there will sometimes be an excess of positive ions in the volume element and sometimes an excess of negative ions. On the average, the charge density will be negative because of the electrostatic forces. In other words, the probability that there is a negative ion in the volume element is greater than the probability that there is a positive ion. The negative charge density will be greater if r is small than if it is large, and we will see later that it is possible to define an effective thickness of the ionic atmosphere.

Suppose that the average electric potential in the volume element dV is ϕ, and let z_+ and z_- be the positive numerical values of the ionic valencies; for example, for Na^+, $z_+ = 1$; for Cl^-, $z_- = 1$. The work required to bring a positive ion of charge z_+e from infinity up to this volume element is $z_+e\phi$, a positive quantity. Because the central positive ion attracts the negative ion, the work required to bring a negative ion of charge $-z_-e$ is $-z_-e\phi$, a negative quantity.

The time-average numbers of the positive and negative ions present in the volume element are given by the Boltzmann principle (see Eq. 1.78, p. 29; the principle is dealt with in detail in Section 1.11):[7]

$$dN_+ = N_+\mathrm{e}^{-z_+e\phi/k_BT}\,dV \tag{7.21}$$

and

$$dN_- = N_-\mathrm{e}^{-(-z_-e\phi)/k_BT}\,dV \tag{7.22}$$

[7]We use roman e here for the exponential function to distinguish it from the charge on the electron e.

where N_+ and N_- are the total numbers of positive and negative ions, respectively, per unit volume of solution; k_B is the Boltzmann constant; and T is the absolute temperature. The charge density in the volume element (i.e., the net charge per unit volume) is given by

$$\rho = \frac{e(z_+ \, dN_+ - z_- \, dN_-)}{dV} = e(N_+ z_+ e^{-z_+ e\phi/k_B T} - N_- z_- e^{z_- e\phi/k_B T}) \tag{7.23}$$

For a univalent electrolyte (one in which both ions are univalent) z_+ and z_- are unity, and N_+ and N_- must be equal because the entire solution is neutral; Eq. 7.23 then becomes

$$\rho = Ne(e^{-e\phi/k_B T} - e^{e\phi/k_B T}) \tag{7.24}$$

where $N = N_+ = N_-$. If $e\phi/k_B T$ is sufficiently small (which requires that the volume element is not too close to the central ion, so that ϕ is small), the exponential terms are given by[8]

$$e^{-e\phi/k_B T} \approx 1 - \frac{e\phi}{k_B T} \quad \text{and} \quad e^{e\phi/k_B T} \approx 1 + \frac{e\phi}{k_B T} \tag{7.25}$$

Equation 7.24 thus reduces to

$$\rho = -\frac{2Ne^2\phi}{k_B T} \tag{7.26}$$

In the more general case, in which there are a number of different types of ions, the expression is

$$\rho = -\frac{e^2\phi}{k_B T} \sum_i N_i z_i \tag{7.27}$$

where N_i and z_i represent the number per unit volume and the positive value of the valence of the ions of the ith type. The summation is taken over all types of ions present.

Equation 7.27 relates the charge density to the average potential ϕ. In order to obtain these quantities separately it is necessary to have another relationship between ρ and ϕ.

This is provided in the following way.[9] From Eq. 7.3 we have the magnitude E of the electric field at a point P, a distance r from the origin. The direction of the vector field due to a positive charge at the origin is the same as the direction of the vector from the origin to the point P (see Figure 7.5a). To make the connection we will be interested in determining the flux of the electric field (i.e., the number of lines of force) through an element of area on a surface perpendicular to the electric field vector. Consider a closed surface S surrounding a charge Q (Figure 7.5b). At

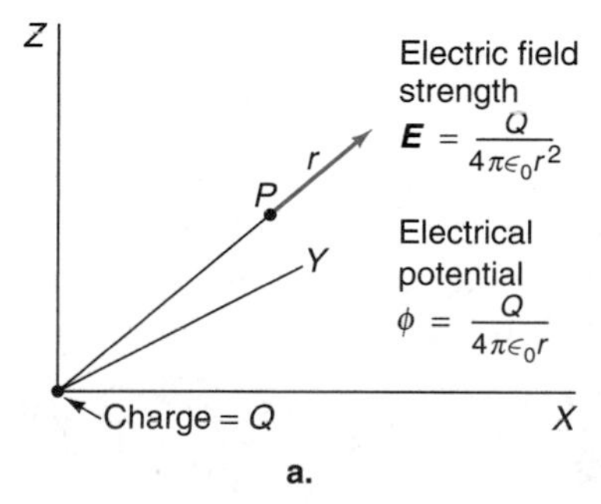

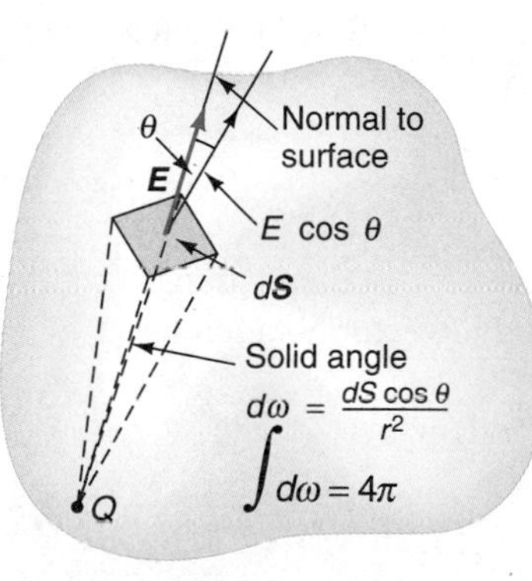

FIGURE 7.5
(a) The electric field strength E and the electric potential ϕ, at a distance r from a charge Q.
(b) A diagram to illustrate Gauss's theorem. A charge Q is surrounded by a surface S.

---

[8] The expansion of e^x is

$$e^x = 1 + x + \frac{x^2}{2!} + \frac{x^3}{3!} + \frac{x^4}{4!} + \cdots$$

and when x is sufficiently small, we can neglect terms beyond x; thus $e^x \approx 1 + x$ and $e^{-x} \approx 1 - x$. The condition is that $e\phi$ must be much smaller than $k_B T$.

[9] The reader may wish to skip this mathematical treatment, and resume the argument at Eq. 7.42.

any point on the surface the field strength is $\boldsymbol{E}$, and for an element of area dS the scalar product $\boldsymbol{E} \cdot dS$ is defined as

$$\boldsymbol{E} \cdot dS = \boldsymbol{E} \cos \theta \, dS \tag{7.28}$$

where θ is the angle between $\boldsymbol{E}$ and the normal to dS. This scalar product is the flux of the electric field through dS. The element of solid angle $d\omega$ subtended by dS at the charge Q is $dS(\cos \theta)/r^2$, and the total solid angle $d\omega$ subtended by the entire closed surface S at any point within the surface is 4π. These considerations lead to an expression known as **Gauss's theorem:**

Gauss's Theorem

$$\int \boldsymbol{E} \cdot dS = \frac{Q}{\epsilon_0} \tag{7.29}$$

In our case the distribution of charge can vary, so that Q is replaced by the charge density ρ integrated over the volume enclosed by the surface; that is,

$$\int \boldsymbol{E} \cdot dS = \frac{1}{\epsilon_0} \int \rho \, dV \tag{7.30}$$

It is known from vector analysis that the divergence (div) of a vector $\boldsymbol{E}$

Divergence of *E*

$$\text{div } \boldsymbol{E} = \nabla \boldsymbol{E} = \frac{\partial \boldsymbol{E}_x}{\partial x} + \frac{\partial \boldsymbol{E}_y}{\partial y} + \frac{\partial \boldsymbol{E}_z}{\partial z} \tag{7.31}$$

Del Operator

is equal to the flux of $\boldsymbol{E}$ from dV per unit volume. Here ∇ is known as the **del operator** and when applied to a vector it represents the operation $\partial/\partial x + \partial/\partial y + \partial/\partial z$ on its components in Cartesian coordinates. If Eq. 7.31 is integrated over the system of charges enclosed by the surface and equated to Eq. 7.30, we have

$$\int \text{div } \boldsymbol{E} \, dV = \int \boldsymbol{E} \cdot dS = \frac{1}{\epsilon_0} \int \rho \, dV \tag{7.32}$$

Divergence Theorem

The equality on the left is known as the *divergence theorem.* Since the integrals are equal for any volume,

$$\text{div } \boldsymbol{E} = \frac{\rho}{\epsilon_0} \tag{7.33}$$

We now relate $\boldsymbol{E}$ to the potential ϕ. From Eq. 7.4 the electric field strength can be represented as the gradient of an electric potential and in vector notation

$$\boldsymbol{E} = -\text{grad } \phi = -\nabla \phi \tag{7.34}$$

Substitution into Eq. 7.33 gives

$$\text{div } \boldsymbol{E} = -\text{div grad } \phi = \frac{\rho}{\epsilon_0} \tag{7.35}$$

The operator div grad is often written as ∇^2 and read as "del squared"; it is known as the *Laplacian operator.* Equation 7.35 can thus be written as

Poisson Equation

$$\nabla^2 \phi = -\frac{\rho}{\epsilon_0} \tag{7.36}$$

This is the **Poisson equation;** the Laplacian operator ∇^2 in it is given by

Laplacian or Del-Squared Operator

$$\nabla^2 = \frac{\partial^2}{\partial x^2} + \frac{\partial^2}{\partial y^2} + \frac{\partial^2}{\partial z^2} \tag{7.37}$$

For a spherically symmetrical field, such as that produced by an isolated charge, the Laplacian operator may be written in terms of spherical polar coordinates $[f(r, \theta, \phi)]$. The Laplacian operator applied to Eq. 7.36 then gives

$$\frac{1}{r^2}\frac{\partial}{\partial r}\left(r^2\frac{\partial \phi}{\partial r}\right) = -\frac{\rho}{\epsilon_0} \tag{7.38}$$

because no dependency on the angles θ and ϕ is found.

The preceding equations are for a vacuum, which has a permittivity ϵ_0 equal to 8.854×10^{-12} C^2 J^{-1} m^{-1}. If the potential ϕ is in a medium having a relative permittivity, or dielectric constant, of ϵ, the absolute permittivity is $\epsilon_0\epsilon$ and must replace ϵ_0 in the preceding equations.

Equation 7.38 is the needed relation to determine ρ and ϕ in conjunction with Eq. 7.27. Insertion of the expression for ρ gives

$$\frac{1}{r^2}\frac{\partial}{\partial r}\left(r^2\frac{\partial \phi}{\partial r}\right) = \kappa^2\phi \tag{7.39}$$

where the quantity κ (not to be confused with electrolytic conductivity) is given by

$$\kappa^2 = \frac{e^2}{\epsilon_0\epsilon k_B T}\sum_i N_i z_i^2 \tag{7.40}$$

The general solution of Eq. 7.39 is

$$\phi = \frac{Ae^{-\kappa r}}{r} + \frac{Be^{\kappa r}}{r} \tag{7.41}$$

where A and B are constants. Since ϕ must approach zero as r becomes very large, and $e^{\kappa r}/r$ becomes very large as r becomes large, the multiplying factor B must be zero, and Eq. 7.41 thus becomes

$$\phi = \frac{Ae^{-\kappa r}}{r} \tag{7.42}$$

The constant A can be evaluated by considering the situation when the solution is infinitely dilute, when the central ion can be considered to be isolated. In such a solution $\Sigma_i N_i z_i^2$ approaches zero, so that κ also approaches zero and the potential is therefore

$$\phi = \frac{A}{r} \tag{7.43}$$

In these circumstances, however, the potential at a distance r will simply be the potential due to the ion itself, since there is no interference by the atmosphere. The potential at a distance r due to an ion of charge z_c is

$$\phi = \frac{z_c e}{4\pi\epsilon_0\epsilon r} \tag{7.44}$$

Equating these two expressions for ϕ leads to

$$A = \frac{z_c e}{4\pi\epsilon_0\epsilon} \tag{7.45}$$

and insertion of this in Eq. 7.42 gives

$$\phi = \frac{z_c e e^{-\kappa r}}{4\pi\epsilon_0\epsilon r} \tag{7.46}$$

If κ is sufficiently small (i.e., if the solution is sufficiently dilute), the exponential $e^{-\kappa r}$ is approximately $1 - \kappa r$, and Eq. 7.46 therefore becomes

$$\phi = \frac{z_c e}{4\pi\epsilon_0\epsilon r} - \frac{z_c e \kappa}{4\pi\epsilon_0\epsilon} \tag{7.47}$$

The first term on the right-hand side of this equation is the potential at a distance r, due to the central ion itself. The second term is therefore the potential produced by the ionic atmosphere:

$$\phi_a = -\frac{z_c e \kappa}{4\pi\epsilon_0\epsilon} \tag{7.48}$$

There is now no dependence on r; this potential is therefore uniform and exists at the central ion. If the ionic atmosphere were replaced by a charge $-z_c e$ situated at a distance $1/\kappa$ from the central ion, the effect due to it at the central ion would be exactly the same as that produced by the ionic atmosphere. The distance $1/\kappa$ is therefore referred to as the **thickness of the ionic atmosphere.**

The quantity κ is given by Eq. 7.40; the thickness of the ionic atmosphere is thus

Thickness of the Ionic Atmosphere

$$\frac{1}{\kappa} = \left(\frac{\epsilon_0\epsilon k_B T}{e^2 \sum_i N_i z_i^2}\right)^{1/2} \tag{7.49}$$

Instead of the number of ions N_i per unit volume, it is usually more convenient to use the concentration c_i; $N_i = c_i L$, where L is the Avogadro constant, and thus

$$\frac{1}{\kappa} = \left(\frac{\epsilon_0\epsilon k_B T}{e^2 \sum_i c_i z_i^2 L}\right)^{1/2} \tag{7.50}$$

This equation allows values of the thickness of the ionic atmosphere to be estimated.

Table 7.1 shows values of $1/\kappa$ for various types of electrolytes at different molar concentrations in aqueous solution. We see from Eq. 7.50 that the thickness of the ionic atmosphere is inversely proportional to the square root of the concentration; the atmosphere moves further from the central ion as the solution is diluted.

TABLE 7.1 Thickness of Ionic Atmospheres in Water at 25 °C

	Molar Concentration		
Type of Electrolyte	**0.10 *M***	**0.01 *M***	**0.001 *M***
Uni-univalent	0.962 nm	3.03 nm	9.62 nm
Uni-bivalent and bi-univalent	0.556 nm	1.76 nm	5.53 nm
Bi-bivalent	0.481 nm	1.52 nm	4.81 nm
Uni-tervalent and ter-univalent	0.392 nm	1.24 nm	3.92 nm

EXAMPLE 7.3 Estimate the thickness of the ionic atmosphere for a solution of (a) 0.01 *M* NaCl and (b) 0.001 *M* $ZnCl_2$, both in water at 25 °C, with $\epsilon = 78$.

Solution

a. The summation in Eq. 7.50 is

$$\sum_i c_i z_i^2 = 0.01 \times 1^2 + 0.01 \times 1^2 = 0.02 \text{ mol dm}^{-3}$$

Insertion of the appropriate values into Eq. 7.50 then gives

$$\frac{1}{\kappa} = \left[\frac{8.854 \times 10^{-12}(\text{C}^2\,\text{N}^{-1}\,\text{m}^{-2}) \times 78 \times 1.381 \times 10^{-23}(\text{J K}^{-1}) \times 298.15(\text{K})}{(1.602 \times 10^{-19})^2(\text{C}^2) \times 0.02(\text{mol dm}^{-3}) \times 10^3(\text{dm}^3\,\text{m}^{-3}) \times 6.022 \times 10^{23}(\text{mol}^{-1})} \right]^{1/2}$$

$$= 3.03 \times 10^{-9}\ (\text{J N}^{-1}\ \text{m})^{1/2}$$

Since $\text{J} = \text{kg m}^2\ \text{s}^{-2}$ and $\text{N} = \text{kg m s}^{-2}$ or J = Nm, the units are metres; the thickness is thus 3.03 nm (nanometres).

b. The summation is now

$$\sum_i c_i z_i^2 = 0.001 \times 2^2 + 0.002 \times 1^2 = 0.006 \text{ mol dm}^{-3}$$

and the result is

$$\frac{1}{\kappa} = 5.53 \times 10^{-9}\ \text{m} = 5.53\ \text{nm}$$

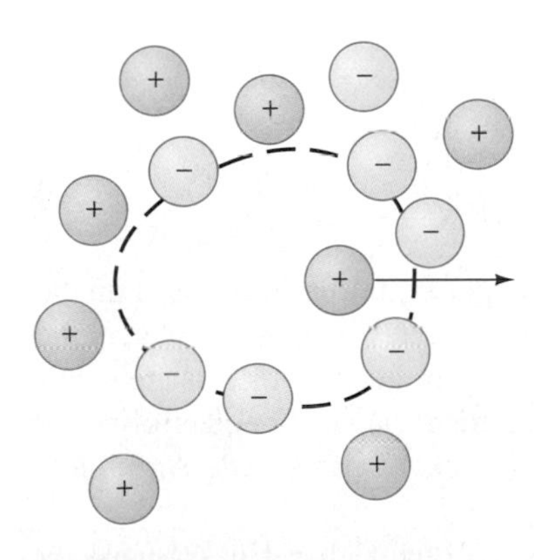

FIGURE 7.6
(a) A stationary central positive ion with a spherically symmetrical ion atmosphere. (b) A positive ion moving to the right. The ion atmosphere behind it is relaxing, while that in front is building up. The distribution of negative ions around the positive ion is now asymmetric.

Mechanism of Conductivity

The mathematical treatment of conductivity is somewhat complicated and only a very general account of the main ideas can be given here.

The effect of the ionic atmosphere is to exert a drag on the movement of a given ion. If the ion is stationary, the atmosphere is arranged symmetrically about it and does not tend to move it in either direction (see Figure 7.6a). However, if a potential that tends to move the ion to the right is applied, the atmosphere will decay to some extent on the left of the ion and build up more on the right (Figure 7.6b). Since it takes time for these *relaxation* processes to occur, there will be an excess of ionic atmosphere to the left of the ion (i.e., behind it) and a deficit to the right (in front of it). This asymmetry of the atmosphere will have the effect of dragging the central ion back. This is the **relaxation** or **asymmetry** effect.

There is a second reason why the existence of the ionic atmosphere impedes the motion of an ion. Ions are attracted to solvent molecules mainly by ion-dipole forces; therefore, when they move, they drag solvent along with them. The ionic atmosphere, having a charge opposite to that of the central ion, moves in the opposite direction to it and therefore drags solvent in the opposite direction. This means that the central ion has to travel upstream, and it therefore travels more slowly than if there were no effect of this kind. This is the **electrophoretic** effect.

Debye and Hückel carried through a theoretical treatment that led them to expressions for the forces exerted upon the central ion by the relaxation effect. They supposed the ions to travel through the solution in straight lines, neglecting the

Lars Onsager received the Nobel Prize in chemistry in 1968 for his discovery of the Onsager reciprocal relations, mathematical equations that are fundamental for the thermodynamic theory of irreversible processes.

zigzag Brownian motion brought about by the collisions of surrounding solvent molecules. This theory was improved in 1926 by the Norwegian-American physical chemist Lars Onsager (1903–1976), who took Brownian motion into account and whose expression for the *relaxation force* f_r was

Relaxation Force

$$f_r = \frac{e^2 z_i \kappa}{24\pi\epsilon_0 \epsilon k_B T} w V' \tag{7.51}$$

where V' is the applied potential gradient and w is a number whose magnitude depends on the type of electrolyte; for a uni-univalent electrolyte, w is $2 - \sqrt{2} = 0.586$.

Electrophoretic Force

The *electrophoretic force* f_e was given by Debye and Hückel as

$$f_e = \frac{e z_i \kappa}{6\pi\eta} K_c V' \tag{7.52}$$

where K_c is the coefficient of frictional resistance of the central ion with reference to the solvent and η is the viscosity of the medium. The viscosity enters into this expression because we are concerned with the motion of the ion past the solvent molecules, which depends on the viscosity of the solvent.

The final expression obtained on this basis for the molar conductivity Λ of an electrolyte solution of concentration c is

Debye-Hückel-Onsager Equation

$$\Lambda = \Lambda^\circ - (P + Q\Lambda^\circ)\sqrt{c} \tag{7.53}$$

where P and Q can be expressed in terms of various constants and properties of the system. For the particular case of a *symmetrical* electrolyte [i.e., one for which the two ions have equal and opposite signs ($z_+ = z_- = z$)], P and Q are given by

$$P = \frac{zeF}{3\pi\eta}\left(\frac{2z^2e^2L}{\epsilon_0\epsilon k_B T}\right)^{1/2} \tag{7.54}$$

and

$$Q = \frac{z^2e^2w}{24\pi\epsilon_0\epsilon k_B T}\left(\frac{2z^2e^2L}{\pi\epsilon_0\epsilon k_B T}\right)^{1/2} \tag{7.55}$$

Equation 7.53 is known as the **Debye-Hückel-Onsager equation.** It is based on the assumption of complete dissociation. With weak electrolytes such as acetic acid the ions are sufficiently far apart that the relaxation and electrophoretic effects are negligible; the Arrhenius-Ostwald treatment (Section 7.3) then applies. For electrolytes of intermediate strength a combination of the Debye-Hückel-Onsager and Arrhenius-Ostwald theories must be used.

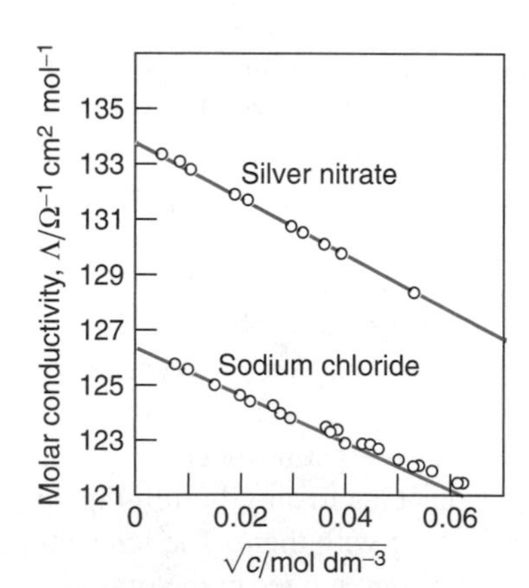

FIGURE 7.7
Plots of molar conductivity Λ against the square root of the concentration of the solution. The plots provide a test of the Debye-Hückel-Onsager equation, Eq. 7.53.

A number of experimental tests have been made of the Debye-Hückel-Onsager equation. Equation 7.53 can be tested by seeing whether a plot of Λ against $\sqrt{c}$ is linear and whether the slopes and intercepts of the lines are consistent with Eqs. 7.54 and 7.55. An example of such a plot is shown in Figure 7.7. For aqueous solutions of uni-univalent electrolytes, the theoretical equation is found to be obeyed very satisfactorily up to a concentration of about 2×10^{-3} mol dm^{-3}; at higher concentrations, deviations are found. The corresponding equations for other types of electrolytes in water are also obeyed satisfactorily at very low concentrations, but deviations are found at lower concentrations than with uni-univalent electrolytes.

Various attempts have been made to improve the Debye-Hückel treatment. One of these attempts involved taking ion association into account, a matter that is dealt with in the next subsection. Another attempt, made by Gronwall, La Mer, and coworkers in 1928, was based on including higher terms in the expansions of the exponentials in Eq. 7.24. This led to improvement for ions of higher valence, but unfortunately the resulting equations are complicated and difficult to apply to the experimental results.

Another weakness of the Debye-Hückel theory is that the ions are treated as point charges; no allowance is made for the fact they occupy a finite volume and cannot come close to each other. This difficulty was dealt with in 1952 by E. Wicke and the German chemist Manfred Eigen (b. 1927) in a manner similar to that of introducing the excluded volume b into the van der Waals equation for a real gas (Section 1.13).

Manfred Eigen shared the 1967 Nobel Prize in chemistry for his development of relaxation spectrometry.

Ion Association

A common type of deviation from the Debye-Hückel-Onsager conductivity equation is for the negative slopes of the Λ versus c plots to be greater than predicted by the theory; that is, the experimental conductivities are lower than the theory predicts. This result, and also certain anomalies found with ion activities (Section 7.10), led to the suggestion that the assumption of complete dissociation sometimes needs qualification for strong electrolytes. In the case of an electrolyte such as $ZnSO_4$ there are strong electrostatic attractions between the Zn^{2+} and SO_4^{2-} ions, and it is possible for pairs of ions to become associated in solution. In this example, in which the ions have equal charges, the ion pairs have no net charge and therefore make no contribution to the conductivity, thus accounting for the anomalously low conductivity. It is important to distinguish clearly between ion association and bond formation; the latter is more permanent, whereas in ion association there is a constant interchange between the various ions in the solution.

The term *ion association* was first employed by Niels Bjerrum, who in 1926 developed a theory of it. The basis of his theory is that ions of opposite signs separated in solution by a distance r^* form an associated ion pair held together by coulombic forces. The calculation of Bjerrum's distance r^* is based on the Boltzmann equations (Eqs. 7.21 and 7.22) and the final result is that, with charge numbers z_c and z_i,

$$r^* = \frac{z_c z_i e^2}{8\pi\epsilon_0 \epsilon k_B T} \tag{7.56}$$

For the case of a *uni-univalent* electrolyte ($z_c z_i = 1$) the value of r^* at 25 °C, with $\epsilon = 78.3$, is equal to 3.58×10^{-10} m $= 0.358$ nm. The expression for the electrostatic potential energy of interaction between two univalent ions separated by a distance r is

$$E = \frac{e^2}{4\pi\epsilon_0 \epsilon k_B T r} \tag{7.57}$$

Substitution of r^* for r (Eq. 7.56) into this expression leads to

$$E^* = 2k_B T \tag{7.58}$$

The electrostatic potential energy at this distance r^* is thus four times the mean kinetic energy per degree of freedom. The energy at this distance is therefore sufficient

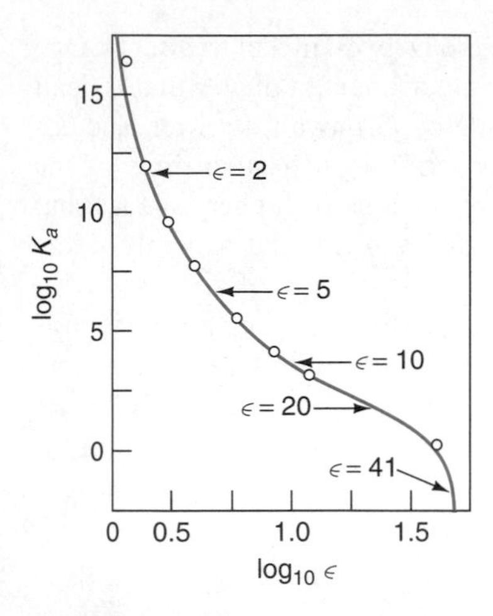

FIGURE 7.8
A double-logarithmic plot of K_a, the association constant for ion-pair formation with tetraisoamylammonium nitrate, against the dielectric constant of the solvent (data of R. M. Fuoss and C. A. Kraus, 1923).

for there to be significant ion association, which will be dynamic in character in that there will be a rapid exchange with the surrounding ions. At distances smaller than 0.358 nm the probability of ionic association increases rapidly. Therefore, to a good approximation one can say that if the ions are closer than 0.358 nm, they can be considered to be "undissociated" and to make no contribution to the conductivity.

However, in aqueous solution univalent ions can rarely approach one another as closely as 0.358 nm, and ion association is therefore of little importance for such ions. In solvents having lower dielectric constants, however, there can be substantial ion-pair formation even for uni-univalent electrolytes. This is strikingly illustrated by results obtained in 1923 by the American electrochemists Raymond M. Fuoss and Charles A. Kraus. For tetraisoamylammonium nitrate in a series of solvents having dielectric constants ranging from 2.2 to 78.6, they made measurements of conductances, and from the results they calculated the association constants K_a for ion association. As shown in Figure 7.8, the formation of ion pairs was negligible in solutions of high dielectric constant, but very considerable in solutions of low dielectric constant. Dioxane has a dielectric constant of 2.2 at 25 °C, and the r^* value for a uni-univalent electrolyte is 12.7 nm; ion association is therefore expected to be important.

Much more ion association is found with ions of higher valence. With salts having ions of different valences, such as Na_2SO_4, ion association will lead to the formation of species such as $Na^+SO_4^{2-}$. Again there will be a reduction in conductivity, since these species will carry less current than the free ions. Bjerrum's theory has been extended to deal with such unsymmetrical electrolytes and also with the formation of triple ions.

Conductivity at High Frequencies and Potentials

An important consequence of the existence of the ionic atmosphere is that the conductivity should depend on the frequency if an alternating potential is applied to the solution. Suppose that the alternating potential is of sufficiently high frequency that the time of oscillation is small compared with the time it takes for the ionic atmosphere to relax. There will then not be time for the atmosphere to relax behind the ion and to form in front of it; the ion will be virtually stationary and its ionic atmosphere will remain symmetrical. Therefore, as the frequency of the potential increases, the relaxation and electrophoretic effects will become less and less important, and there will be an increase in the molar conductivity.

A related effect is observed if conductivities are measured at very high potential gradients. For example, if the applied potential is 20 000 V cm^{-1}, an ion will move at a speed of about 1 m s^{-1} and will travel several times the thickness of the effective ionic atmosphere in the time of relaxation of the atmosphere. Consequently, the moving ion is essentially free from the effect of the ionic atmosphere, which does not have time to build up around it to any extent. Therefore, at sufficiently high voltages, the relaxation and electrophoretic effects will diminish and eventually disappear and the molar conductivity will increase. This effect was first observed experimentally by the German physicist Max Carl Wien (1866–1938) in 1927 and is known as the **Wien effect.** However, molar conductivities of weak electrolytes at high potentials are anomalously large, and it appears that very high potentials bring about a dissociation of the molecules into ions. This phenomenon is known as the **dissociation field effect.**

Wien Effect

Dissociation Field Effect

7.5 Independent Migration of Ions

In principle the plots of Λ against concentration (Figure 7.2) can be extrapolated back to zero concentration to give the Λ° value. In practice this extrapolation can only satisfactorily be made with strong electrolytes. With weak electrolytes there is a strong dependence of Λ on c at low concentrations and therefore the extrapolations do not lead to reliable Λ° values. An indirect method for obtaining Λ° for weak electrolytes is considered later.

In 1875 Kohlrausch made a number of determinations of the Λ° values and observed that they exhibited certain regularities. Some values are given in Table 7.2 for corresponding sodium and potassium salts. We see that the difference between the molar conductivities of a potassium and a sodium salt of the same anion is independent of the nature of the anion. Similar results were obtained for a variety of pairs of salts with common cations or anions, in both aqueous and nonaqueous solvents.

This behavior was explained in terms of **Kohlrausch's law of independent migration of ions.** Each ion is assumed to make its own contribution to the molar conductivity, irrespective of the nature of the other ion with which it is associated. In other words,

$$\Lambda^\circ = \lambda_+^\circ + \lambda_-^\circ \tag{7.59}$$

where λ_+° and λ_-° are the ion conductivities of cation and anion, respectively, at infinite dilution. Thus, for potassium chloride (see Table 7.2)

$$\Lambda^\circ(\text{KCl}) = \lambda_{\text{K}^+}^\circ + \lambda_{\text{Cl}^-}^\circ = 149.9 \text{ S cm}^2 \text{ mol}^{-1} \tag{7.60}$$

and for sodium chloride

$$\Lambda^\circ(\text{NaCl}) = \lambda_{\text{Na}^+}^\circ + \lambda_{\text{Cl}^-}^\circ = 126.5 \text{ S cm}^2 \text{ mol}^{-1} \tag{7.61}$$

The difference, 23.4 S cm^2 mol^{-1}, is thus the difference between the λ_+° values for K^+ and Na^+:

$$\lambda_{\text{K}^+}^\circ - \lambda_{\text{Na}^+}^\circ = 23.4 \text{ S cm}^2 \text{ mol}^{-1} \tag{7.62}$$

and this will be the same whatever the nature of the anion.

TABLE 7.2 Molar Conductivities at Infinite Dilution for Various Sodium and Potassium Salts in Aqueous Solution at 25 °C

Electrolyte	Λ° / S cm^2 mol^{-1}	Electrolyte	Λ° / S cm^2 mol^{-1}	Difference
KCl	149.79	NaCl	126.39	23.4
KI	150.31	NaI	126.88	23.4
$\frac{1}{2}K_2SO_4$	153.48	$\frac{1}{2}Na_2SO_4$	130.1	23.4

Note that these molar conductivities are what were formerly called "equivalent conductivities," the latter term being no longer recommended by IUPAC. The units S cm^2 mol^{-1} are equivalent to Ω^{-1} cm^2 mol^{-1}.

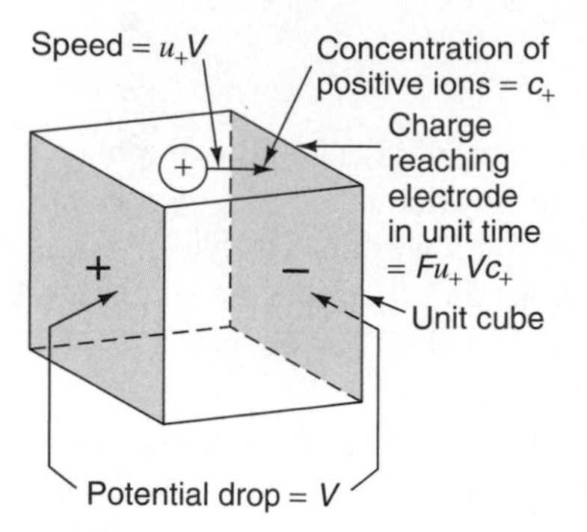

FIGURE 7.9
A unit cube containing a solution in which the concentration of positive ions is c_+, and where there is a potential drop of V between the opposite faces.

Ionic Mobilities

It is not possible from the $\Lambda°$ values alone to determine the individual λ_+° and λ_-° values; however, once one λ_+° or λ_-° value (such as $\lambda_{\text{Na+}}^\circ$) has been determined, the rest can be calculated. The λ_+° and λ_-° values are proportional to the speeds with which the ions move under standard conditions, and this is the basis of the methods by which the individual conductivities are determined. The **mobility** of an ion, u, is defined as the speed with which the ion moves under a unit potential gradient. Suppose that the potential drop across the opposite faces of a unit cube is V and that the concentration of univalent positive ions in the cube is c_+ (see Figure 7.9). If the mobility of the positive ions is u_+, the speed of the ions is u_+V. All positive ions within a distance of u_+V of the negative plate will therefore reach that plate in unit time, and the number of such ions is u_+Vc_+. The charge they carry is Fu_+Vc_+, and since this is carried in unit time, the current is Fu_+Vc_+. The electrolytic conductivity due to the positive ions, κ_+, is thus

$$\kappa_+ = \frac{\text{current}}{\text{potential drop}} = \frac{Fu_+Vc_+}{V} = Fu_+c_+ \tag{7.63}$$

The molar conductivity due to these positive ions is therefore

$$\lambda_+^\circ = \frac{\kappa_+}{c_+} = Fu_+ \tag{7.64}$$

The SI unit of mobility is m s^{-1}/V m^{-1} = m^2 V^{-1} s^{-1}, but it is more common to use the unit cm^2 V^{-1} s^{-1}.

EXAMPLE 7.4 The mobility of a sodium ion in water at 25 °C is 5.19×10^{-4} cm^2 V^{-1} s^{-1}. Calculate the molar conductivity of the sodium ion.

Solution From Eq. 7.64,

$$\lambda_{\text{Na+}}^\circ = 96\,485 \text{ A s mol}^{-1} \times 5.19 \times 10^{-4} \text{ cm}^2 \text{ V}^{-1} \text{ s}^{-1}$$
$$= 50.1 \text{ A V}^{-1} \text{ cm}^2 \text{ mol}^{-1}$$

A V^{-1} = Ω^{-1} = S; therefore

$$\lambda_{\text{Na+}}^\circ = 50.1 \text{ S cm}^2 \text{ mol}^{-1}$$

7.6 Transport Numbers

In order to split the $\Lambda°$ values into the λ_+° and λ_-° values for the individual ions, we make use of a property known variously as the **transport number,** the *transference number,* or the *migration number.* It is the *fraction of the current carried by each ion* present in solution.

Consider an electrolyte of formula A_aB_b, which ionizes as follows:

$$A_aB_b \rightleftharpoons aA^{z_++} + bB^{z_--}$$

The quantity of electricity carried by a mol of the ions A^{z_++}, whose charge number is z_+, is aFz_+, and the quantity of electricity crossing a given cross-sectional area in unit time is aFz_+u_+, where u_+ is the mobility. This quantity is the current carried by the positive ions when 1 mol of the electrolyte is present in the solution. Similarly, the current carried by b mol of the negative ions is bFz_-u_-. The fraction of the current carried by the positive ions is therefore

$$t_+ = \frac{aFz_+u_+}{aFz_+u_+ + bFz_-u_-} = \frac{az_+u_+}{az_+u_+ + bz_-u_-} \tag{7.65}$$

This is the transport number of the positive ions. Similarly, the transport number of the negative ions is

$$t_- = \frac{bz_-u_-}{az_+u_+ + bz_-u_-} \tag{7.66}$$

In order for the solution to be electrically neutral it is necessary that

$$az_+ = bz_- \tag{7.67}$$

and Eq. 7.65 and 7.66 then reduce to

Transport Numbers

$$t_+ = \frac{u_+}{u_+ + u_-} \quad \text{and} \quad t_- = \frac{u_-}{u_+ + u_-} \tag{7.68}$$

The individual molar ion conductivities are given by

$$\lambda(\mathrm{A}^{z_+ +}) = Fz_+u_+ \quad \text{and} \quad \lambda(\mathrm{B}^{z_- -}) = Fz_-u_- \tag{7.69}$$

The molar conductivity of the electrolyte is

$$\Lambda(\mathrm{A}_a\mathrm{B}_b) = F(az_+u_+ + bz_-u_-) \tag{7.70}$$

It then follows that the transport number t_+ is related to the individual ion conductivities by

$$t_+ = \frac{\lambda(\mathrm{A}^{z_+ +})/z_+}{\lambda(\mathrm{A}^{z_+ +})/z_+ + \lambda(\mathrm{B}^{z_- -})/z_-} \tag{7.71}$$

Together with Eq. 7.67 this gives

$$t_+ = \frac{a\lambda(\mathrm{A}^{z_+ +})}{a\lambda(\mathrm{A}^{z_+ +}) + b\lambda(\mathrm{B}^{z_- -})} = \frac{a\lambda(\mathrm{A}^{z_+ +})}{\Lambda(\mathrm{A}_a\mathrm{B}_b)} \tag{7.72}$$

Similarly, for t_-,

$$t_- = \frac{b\lambda(\mathrm{B}^{z_- -})}{a\lambda(\mathrm{A}^{z_+ +}) + b\lambda(\mathrm{B}^{z_- -})} = \frac{b\lambda(\mathrm{B}^{z_- -})}{\Lambda(\mathrm{A}_a\mathrm{B}_b)} \tag{7.73}$$

If t_+ and t_- can be measured over a range of concentrations, the values at zero concentration, t_+° and t_-°, can be obtained by extrapolation. These values allow $\Lambda^\circ(\mathrm{A}_a\mathrm{B}_b)$ to be split into the individual ion conductivities $\lambda^\circ(\mathrm{A}^{z_+ +})$ and $\lambda^\circ(\mathrm{B}^{z_- -})$ by use of Eqs. 7.72 and 7.73.

At first sight it may appear surprising that Faraday's laws, according to which equivalent quantities of different ions are liberated at the two electrodes, can be reconciled with the fact that the ions are moving at different speeds toward the electrodes. Figure 7.10, however, shows how these two facts can be reconciled. The diagram shows in a very schematic way an electrolysis cell in which there are equal numbers of positive and negative ions of unit charge. The situation before

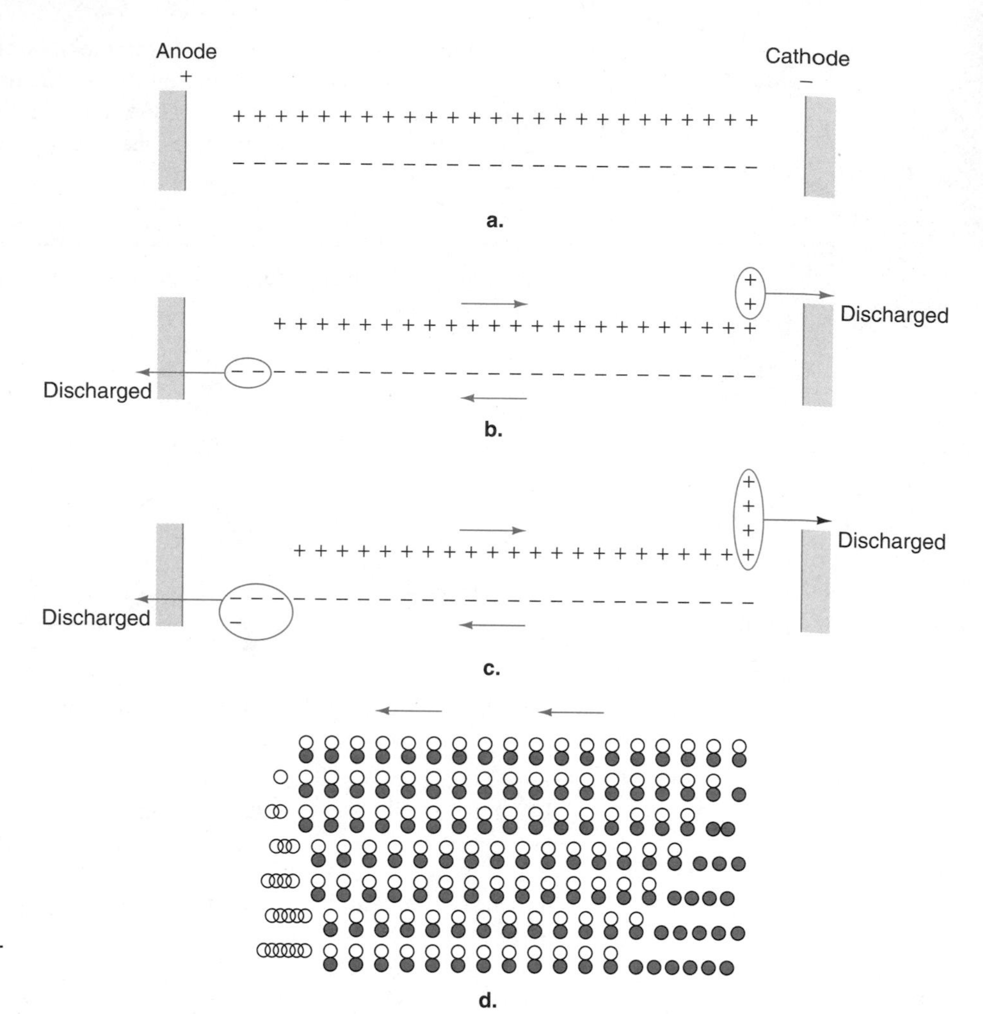

FIGURE 7.10
Schematic representation of the movement of ions during an electrolysis experiment, showing that equivalent amounts are neutralized at the electrodes in spite of differences in ionic velocities. Diagram (d) is reproduced from an original publication by Hittorf.

electrolysis occurs is shown in Figure 7.10a. Suppose that the cations only were able to move, the anions having zero mobility; after some motion has occurred, the situation will be as represented in Figure 7.10b. At each electrode two ions remain unpaired and are discharged, two electrons at the same time traveling in the outer circuit from the anode to the cathode. Thus, although only the cations have moved through the bulk of the solution, equivalent amounts have been discharged at the two electrodes. It is easy to extend this argument to the case in which both ions are moving but at different speeds. Thus Figure 7.10c shows the situation when the speed of the cation is three times that of the anion, four ions being discharged at each electrode.

This problem was considered by Hittorf, and Figure 7.10d is reproduced from one of his publications.

Hittorf Method

Two experimental methods have been employed for the determination of transport numbers. One of them, developed by the German physicist Johann Wilhelm Hittorf (1824–1914) in 1853, involves measuring the changes of concentration in the vicin-

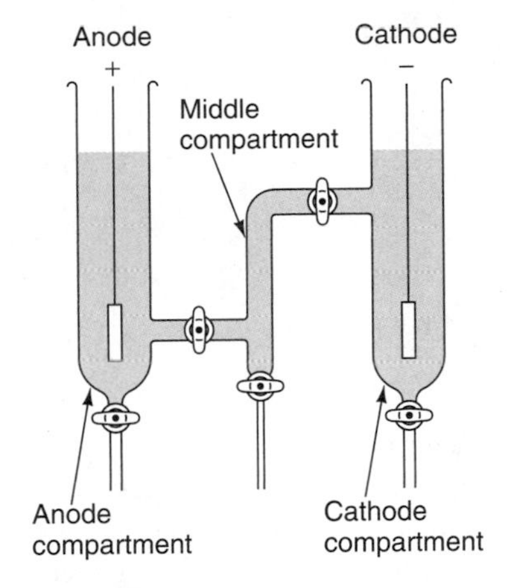

FIGURE 7.11
Simple type of apparatus used for measuring transference numbers by the Hittorf method.

ity of the electrodes. In the second, the **moving boundary method,** a study is made of the rate of movement, under the influence of a current, of the boundary between two solutions. This method is described on p. 290. A third method, involving the measurement of the electromotive force of certain cells, is considered in Section 8.5.

A very simple type of apparatus used in the **Hittorf method** is shown in Figure 7.11. The solution to be electrolyzed is placed in the cell, and a small current is passed between the electrodes for a period of time. Solution then is run out through the stopcocks, and the samples are analyzed for concentration changes.

The theory of the method is as follows. Suppose that the solution contains the ions M^+ and A^-, which are not necessarily univalent but are denoted as such for simplicity. The fraction of the total current carried by the cations is t_+, and that carried by the anions is t_-. Thus, when 96 485 C of electricity passes through the solution, 96 485 t_+C are carried in one direction by t_+ mol of M^+ ions, and 96 485 t_-C are carried in the other direction by t_- mol of A^- ions. At the same time 1 mol of each ion is discharged at an electrode.

Suppose that, as represented in Figure 7.12, the cell containing the electrolyte is divided into three compartments by hypothetical partitions. One is a compartment near the cathode, one is near the anode, and the middle compartment is one in which no concentration change occurs. It is supposed that the two ions are discharged at the electrodes, which are unchanged in the process. The changes that occur when 96 485 C is passed through the solution are shown in Figure 7.10. The results are

At the cathode: A loss of $1 - t_+ = t_-$ mol of M^+, and a loss of t_- mol of A^- (i.e., a loss of t_- mol of the electrolyte MA).

In the middle compartment: No concentration change.

At the anode: Loss of $1 - t_- = t_+$ mol of A^-; loss of t_+ mol of M^+; the net loss is therefore t_+ mol of MA.

It follows that

$$\frac{\text{amount lost from anode compartment}}{\text{amount lost from cathode compartment}} = \frac{t_+}{t_-} \tag{7.74}$$

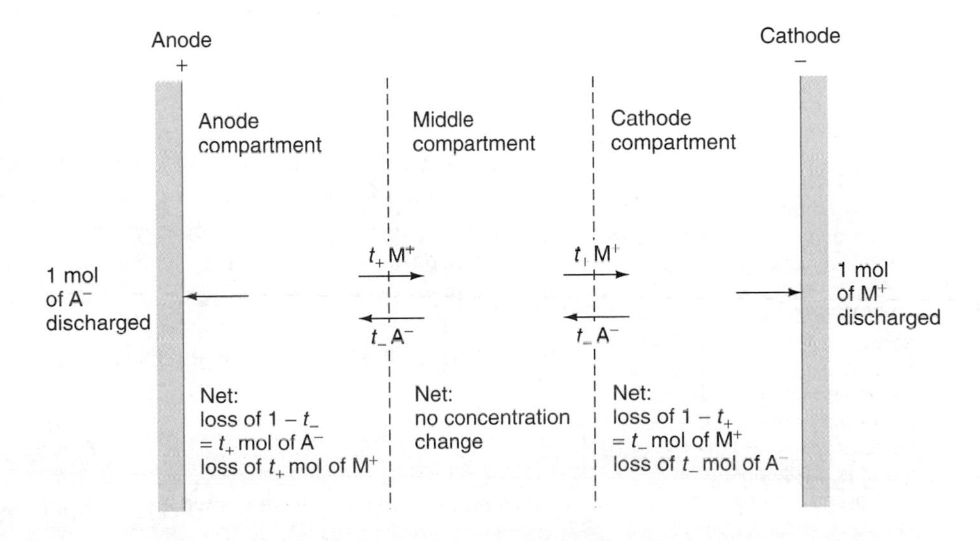

FIGURE 7.12
The concentration changes in three compartments of a conductivity cell (Hittorf method).

Then, since $t_+ + t_- = 1$, the values of t_+ and t_- can be calculated from the experimental results. Alternatively,

$$\frac{\text{amount lost from anode compartment}}{\text{amount deposited}} = t_+ \tag{7.75}$$

and

$$\frac{\text{amount lost from cathode compartment}}{\text{amount deposited}} = t_- \tag{7.76}$$

Any of these three equations allows t_+ and t_- to be determined.

In the preceding discussion we have assumed that the electrodes are inert (i.e., are not attacked as electrolysis proceeds) and that the ions M^+ and A^- are deposited. If instead, for example, the anode were to pass into solution during electrolysis, the concentration changes would be correspondingly different; the treatment can readily be modified for this and other situations.

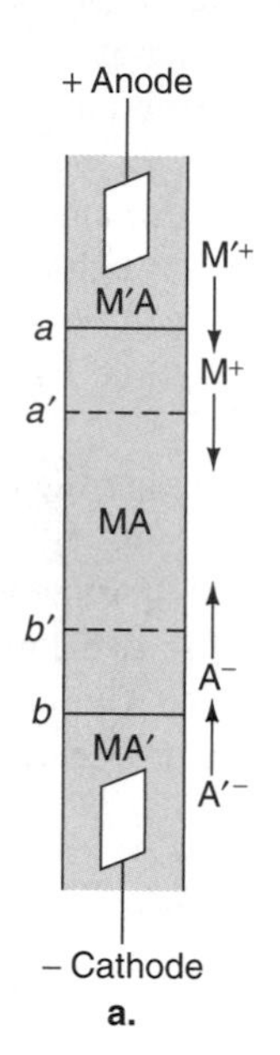

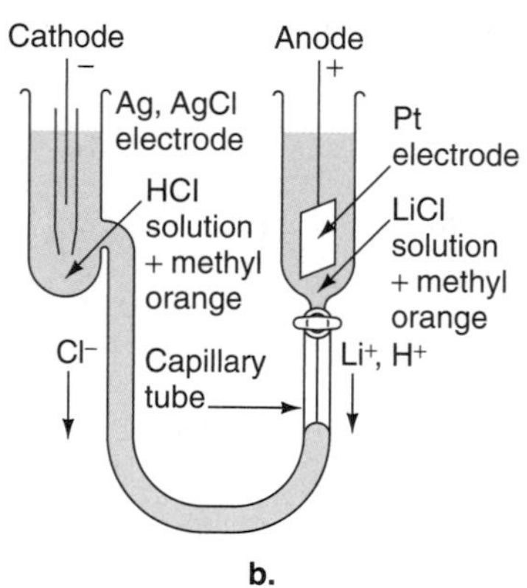

FIGURE 7.13
The determination of transport numbers by the moving boundary method. (a) Schematic diagram showing the movement of the ions and the boundaries. (b) Simple apparatus for measuring the transport number of H^+, using Li^+ as the following cation. A clear boundary is formed between the receding acid solution and the LiCl solution and is easily observed if methyl orange is present.

Moving Boundary Method

The moving boundary method was developed in 1886 by the British physicist Sir Oliver Joseph Lodge (1851–1940) and in 1893 by the British physicist Sir William Cecil Dampier (formerly Whetham) (1867–1952).

The method is illustrated in Figure 7.13. Suppose that it is necessary to measure the transport numbers of the ions in the electrolyte MA. Two other electrolytes M′A and MA′ are selected as "indicators"; each has an ion in common with MA, and the electrolytes are such that M'^+ moves more slowly than M^+ and A'^- moves more slowly than A^-. The solution of MA is placed in the electrolysis tube with the solution of M′A on one side of it and that of MA′ on the other; the electrode in M′A is the anode and that in MA′ is the cathode. As the current flows, the boundaries remain distinct, since the M'^+ ions cannot overtake the M^+ ions at the boundary a, and the A'^- ions cannot overtake the A^- ions at boundary b. The slower ions M'^+ and A'^- are known as the *following* ions. The boundaries a and b move, as shown in Figure 7.13a, to positions a' and b'. The distances aa' and bb' are proportional to the ionic velocities, and therefore

$$\frac{aa'}{aa' + bb'} = t_+ \quad \text{and} \quad \frac{bb'}{aa' + bb'} = t_- \tag{7.77}$$

Alternatively, the movement of only one boundary may be followed, as shown by the example in Figure 7.13b. If Q is the quantity of electricity passed through the system, t_+Q is the quantity of electricity carried by the positive ions, and the amount of positive ions to which this corresponds is t_+Q/F. If the concentration of positive ions is c, the amount t_+Q/F occupies a volume of t_+Q/Fc. This volume is equal to the area of cross section of the tube A multiplied by the distance aa' through which the boundary moves. Thus it follows that

$$aa' = \frac{t_+Q}{FcA} \tag{7.78}$$

Since aa', Q, F, c, and A are known, the transport number t_+ can be calculated.

7.7 Ion Conductivities

With the use of the transport numbers the $\Lambda°$ values can be split into λ_+° and λ_-°, the contributions for the individual ions; the transport numbers extrapolated to infinite dilution are

$$t_+ = \frac{\lambda_+^\circ}{\Lambda^\circ} \quad \text{and} \quad t_- = \frac{\lambda_-^\circ}{\Lambda^\circ} \tag{7.79}$$

Once a single λ_+° or λ_-° value has been determined, a complete set can be calculated from the available $\Lambda°$ values. For example, suppose that a transport number study on NaCl led to λ_{Na+}° and λ_{Cl-}°. If $\Lambda°$ for KCl is known, the value of λ_{K+}° [$= \Lambda°(KCl) - \lambda_{Cl-}^\circ$] can be obtained; from λ_{K+}° and $\Lambda°(KBr)$ the value of λ_{Br-}° can be calculated, and so on. A set of values obtained in this way is given in Table 7.3. Symbols such as $\frac{1}{2}Ca^{2+}$, $\frac{1}{2}SO_4^{2-}$ are used in the case of the polyvalent ions.

An important use for individual ion conductivities is in determining the $\Lambda°$ values for weak electrolytes. We have seen (see Figure 7.2) that for a weak electrolyte the extrapolation to zero concentration is rather a long one, with the result that a reliable figure cannot be obtained. For acetic acid, however, $\Lambda°$ is $\lambda_{H^+}^\circ + \lambda_{CH_3COO^-}^\circ$; from the values in Table 7.3 the value of $\Lambda°$ is thus 349.8 + 40.9 = 390.7 S cm^2 mol^{-1}.

TABLE 7.3 Individual Molar Ionic Conductivities in Water at 25 °C

Cation	λ_+° / S cm^2 mol^{-1}	Anion	λ_-° / S cm^2 mol^{-1}
H^+	349.65	OH^-	198
Li^+	38.66	F^-	55.5
Na^+	50.08	Cl^-	76.31
K^+	73.48	Br^-	78.1
Rb^+	77.8	I^-	76.8
Cs^+	77.2	CH_3COO^-	40.9
Ag^+	61.9	$\frac{1}{2}SO_4^{2-}$	80.0
Tl^+	74.7	$\frac{1}{2}CO_3^{2-}$	69.3
$\frac{1}{2}Mg^{2+}$	53.0		
$\frac{1}{2}Ca^{2+}$	59.47		
$\frac{1}{2}Sr^{2+}$	59.4		
$\frac{1}{2}Ba^{2+}$	63.6		
$\frac{1}{2}Cu^{2+}$	56.6		
$\frac{1}{2}Zn^{2+}$	52.8		
$\frac{1}{3}La^{3+}$	69.7		

Again, these molar conductivities are what were formerly called equivalent conductivities.

To calculate the mobility u, divide the $\lambda°$ by the Faraday constant 96 485 C mol^{-1}; the units of u are cm^2 V^{-1} s^{-1} (see Eq. 7.64 and Example 7.4 on p. 311).

An equivalent but alternative procedure avoids the necessity of splitting the Λ° values into the individual ionic contributions. The molar conductivity Λ° of any electrolyte MA can be expressed, for example, as

$$\Lambda^\circ(\text{MA}) = \Lambda^\circ(\text{MCl}) + \Lambda^\circ(\text{NaA}) - \Lambda^\circ(\text{NaCl}) \tag{7.80}$$

If MCl, NaA, and NaCl are all strong electrolytes, their Λ° values are readily obtained by extrapolation and therefore permit the calculation of Λ° for MA, which may be a weak electrolyte. The following example illustrates the use of this method for acetic acid:

$$\begin{aligned}\Lambda^\circ(\text{CH}_3\text{COOH}) &= \Lambda^\circ(\text{HCl}) + \Lambda^\circ(\text{CH}_3\text{COONa}) - \Lambda^\circ(\text{NaCl}) \qquad (7.81)\\ &= 426.2 + 91.0 - 126.5 = 390.7 \times \text{S cm}^2\ \text{mol}^{-1} \quad \text{at } 25\ ^\circ\text{C}\end{aligned}$$

This procedure is also useful for a highly insoluble salt, for which direct determinations of Λ° values might be impractical.

Ionic Solvation

The magnitudes of the individual ion conductivities (Table 7.3) are of considerable interest. There is no simple dependence of the conductivity on the size of the ion. It might be thought that the smallest ions, such as Li^+, would be the fastest moving, but the values for Li^+, K^+, Rb^+, and Cs^+ show that this is by no means true; K^+ moves faster than either Li^+ or Na^+. The explanation is that Li^+, because of its small size, becomes strongly attached to about four surrounding water molecules by ion-dipole and other forces, so that when the current passes, it is $Li(H_2O)_4^+$ and not Li^+ that moves. With Na^+ the binding of water molecules is less strong, and with K^+ it is still weaker. This matter of the solvation of ions is discussed further in Section 7.9.

Mobilities of Hydrogen and Hydroxide Ions

Of particular interest is the very high conductivity of the hydrogen ion. This ion has an abnormally high conductivity in a number of hydroxylic solvents such as water, methanol, and ethanol, but it behaves more normally in nonhydroxylic solvents such as nitrobenzene and liquid ammonia. At first sight the high values might appear to be due to the small size of the proton. However, there is a powerful electrostatic attraction between a water molecule and the proton, which because of its small size can come very close to the water molecule. As a consequence the equilibrium

$$H^+ + H_2O \rightleftharpoons H_3O^+$$

lies very much to the right. In other words, there are very few free protons in water; the ions exist as H_3O^+ ions, which are hydrated by other molecules (see Figure 7.21, p. 299). Thus there remains a difficulty in explaining the very high conductivity and mobility of the hydrogen ion. By virtue of its size it would be expected to move about as fast as the Na^+ ion, which in fact is the case in the nonhydroxylic solvents.

In order to explain the high mobilities of hydrogen ions in hydroxylic solvents such as water, a special mechanism must be invoked. As well as moving through the solution in the way that other ions do, the H_3O^+ ion can also transfer its proton to a neighboring water molecule:

$$H_3O^+ + H_2O \rightarrow H_2O + H_3O^+$$

The resulting H_3O^+ ion can now transfer a proton to another H_2O molecule. In other words, protons, although not free in solution, can be passed from one water molecule to another. Calculations from the known structure of water show that the proton must on the average jump a distance of 86 pm from an H_3O^+ ion to a water molecule but that as a result the proton moves effectively through a distance of 310 pm. The conductivity by this mechanism therefore will be much greater than by the normal mechanism. The proton transfer must be accompanied by some rotation of H_3O^+ and H_2O molecules in order for them to be positioned correctly for the next proton transfer. Similar types of mechanisms have been proposed for the hydroxylic solvents. In methyl alcohol, for example, the process is

$$\begin{array}{c} CH_3 \\ | \\ O^+ \\ H \diagup \quad \diagdown H \end{array} + \begin{array}{c} CH_3 \\ | \\ O \\ \quad \diagdown H \end{array} \rightarrow \begin{array}{c} CH_3 \\ | \\ O \\ H \diagup \quad \end{array} + \begin{array}{c} CH_3 \\ | \\ O^+ \\ H \diagup \quad \diagdown H \end{array}$$

Mechanisms of this type bear some resemblance to a mechanism that was suggested in 1805 by the German physicist Christian Johann Dietrich von Grotthuss (1785–1822) as a general explanation of conductance and are frequently referred to as **Grotthuss mechanisms.**

Grotthuss Mechanisms

In non-hydroxylic solvents, in which such mechanisms cannot be involved, the mobilities of hydrogen and hydroxide ions are no longer abnormally large, as would be expected.

Ionic Mobilities and Diffusion Coefficients

Ions also migrate in the absence of an electric field, if there is a concentration gradient. The migration of a substance when there is a concentration gradient is known as **diffusion,** a topic that will be considered in further detail in Chapter 19. It will there be seen that the tendency of a substance to move in a concentration gradient is measured in terms of a *diffusion coefficient* D. In 1888 the German physicist Walther Herman Nernst (1864–1941) showed that the relationship between the diffusion coefficient and the mobility u of an ion is

$$D = \frac{k_B T}{Q} u \tag{7.82}$$

where Q is the charge on the ion. This equation is derived in Section 19.2. Since the molar ionic conductivity λ is equal to Fu and the charge Q is equal to $F|z|/L$, this equation can be written as

$$D = \frac{L k_B T \lambda}{F^2 |z|} = \frac{RT\lambda}{F^2 |z|} \tag{7.83}$$

These relationships show that diffusion experiments, as well as conductivity measurements, can provide information about mobilities and hence about molar conductivities. Equations 7.82 and 7.83 relate to a single ionic species but, in practice, experiments must involve at least two types of ions. Diffusion experiments with a uni-univalent electrolyte A^+B^-, such as NaCl, will be concerned with the diffusion of the two ions A^+ and B^-. The overall diffusion constant D must then relate to the individual diffusion constants D_{A+} and D_{B-}, and Nernst showed that for uni-univalent electrolytes the proper average is

$$D = \frac{2D_{A+}D_{B-}}{D_{A+} + D_{B-}} \tag{7.84}$$

Walden's Rule

An important relationship involving molar conductivity was discovered in 1906 by Paul Walden (1863–1957), who made measurements of the molar conductivities of tetramethylammonium iodide in various solvents having different viscosities. The viscosity of a liquid, which is treated in more detail in Section 19.1, is a measure of the ease with which a liquid flows. Walden noticed that the product of the molar conductivity, Λ, and the viscosity η of the solvent was approximately constant:

$$\Lambda\eta = \text{constant} \tag{7.85}$$

and this is known as **Walden's rule.** In Section 19.2 we will meet the *Stokes-Einstein equation,* Eq. 19.77, according to which the diffusion coefficient of a particle is inversely proportional to the viscosity of the solvent. Since ionic conductivities and diffusion coefficients are proportional to one another (Eq. 7.83), Walden's rule is easy to understand.

7.8 Thermodynamics of Ions

In thermodynamic work it is customary to obtain values of standard enthalpies and Gibbs energies of species. These quantities relate to the formation of 1 mol of the substance from the elements in their standard states and usually refer to 25.0 °C. In the case of entropies it is common to obtain absolute values based on the third law of thermodynamics.

It is a comparatively straightforward matter to determine standard enthalpies and Gibbs energies of formation and absolute entropies for *pairs* of ions in solution, for example, for a solution of sodium chloride. We cannot, however, carry out experiments on ions of one kind. The conventional procedure is to set the value for H^+ as zero. This allows complete sets of values to be built up, and one then speaks of *conventional standard enthalpies of formation,* of *conventional Gibbs energies of formation,* and of *conventional absolute entropies.*

A few such values for ions are included in Appendix D (p. 1019); more complete compilations are given in Table 7.4. This table includes values of enthalpies, Gibbs energies, and entropies of hydration; the subscript *hyd* is used to identify these quantities. These values relate the thermodynamic properties of the ions in water to their properties in the gas phase and are therefore important in leading to an understanding of the effect of the surrounding water molecules. The Gibbs energy of hydration is the change in Gibbs energy when an ion is transferred from the gas phase into aqueous solution. Again, the convention is that the values for the hydrogen ion H^+ are all zero.

Whereas absolute thermodynamic values for individual ions cannot be measured directly, they can be estimated on the basis of theory. Unfortunately, owing to the large number of interactions involved, the theoretical treatment of an ion in

TABLE 7.4 Conventional Standard Enthalpies, Gibbs Energies, and Entropies of Formation and Hydration of Individual Ions at 25 °C

Ion	$\Delta_f H^\circ$ / kJ mol^{-1}	$\Delta_{hyd} H^\circ$ / kJ mol^{-1}	$\Delta_f G^\circ$ / kJ mol^{-1}	$\Delta_{hyd} G^\circ$ / kJ mol^{-1}	$\Delta_f S^\circ$ / J K^{-1} mol^{-1}	$\Delta_{hyd} S^\circ$ / J K^{-1} mol^{-1}
H^+	0	0	0	0	0	0
Li^+	−278.7	576.1	−293.7	579.1	14.2	10.0
Na^+	−239.7	685.3	−261.9	679.1	60.2	21.3
K^+	−251.0	769.9	−282.4	751.4	102.5	56.9
Rb^+	—	—	−282.4	774.0	124.3	69.0
Cs^+	—	—	−282.0	806.3	133.1	72.0
Ag^+	105.9	615.5	77.1	610.9	77.0	15.5
Mg^{2+}	—	—	−456.1	274.1	−118.0	−49.0
Ca^{2+}	−543.1	589.1	−553.1	586.6	−55.2	7.5
Sr^{2+}	—	—	−557.3	732.6	−39.2	14.2
Ba^{2+}	−538.5	879.1	−560.7	861.5	—	—
Al^{3+}	—	—	−481.2	−1346.4	−313.4	−136.8
F^-	—	—	−276.6	−1524.2	−9.6	−264.0
Cl^-	−167.4	−1469.0	−131.4	−1407.1	55.2	−207.1
Br^-	−120.9	−1454.8	−102.9	−1393.3	80.8	−191.6
I^-	−56.1	−1397.0	−51.9	−1346.8	109.2	−168.6
OH^-	−230.1	−1552.3	—	—	—	—

See Appendix D for sources of thermochemical data. Standard enthalpies of formation and standard Gibbs energies of formation are from a National Bureau of Standards compilation, "Selected Values of Chemical Thermodynamical Properties," *N. B. S. Circular 500* (1952); heats of hydration, Gibbs energies of hydration, and entropies of hydration are from D.R. Rossinsky, *Chem. Rev., 65,* 467(1965).

aqueous solution is very difficult. The following absolute values are generally agreed to be approximately correct, for the proton:

$$\Delta_{hyd}H^\circ(\text{absolute}) = -1090.8 \text{ kJ mol}^{-1}$$
$$\Delta_{hyd}S^\circ(\text{absolute}) = -131.8 \text{ J K}^{-1} \text{ mol}^{-1}$$
$$\Delta_{hyd}G^\circ(\text{absolute}) = -1051.4 \text{ kJ mol}^{-1}$$

Once these values are accepted, absolute values for the other ions can be calculated from their conventional values.

EXAMPLE 7.5 Calculate the absolute enthalpies of hydration of Li^+, I^-, and Ca^{2+} on the basis of a value of -1090.8 kJ mol^{-1} for the proton, using the values listed in Table 7.4.

Solution The enthalpy of hydration of a uni-univalent electrolyte M^+X^- is independent of whether conventional or absolute ionic values are used. Thus, for H^+I^-, if we lower the value of H^+ by 1090.8, we must raise that for I^- by the same amount. Thus the absolute value for I^- will be

$$-1397.0 + 1090.8 = -306.2 \text{ kJ mol}^{-1}$$

Similarly, the value for Li^+ must be lowered by 1090.8; the absolute value is thus

$$576.1 - 1090.8 = -514.7 \text{ kJ mol}^{-1}$$

To obtain the value for Ca^{2+}, we may consider a salt like CuI_2 ($Cu^{2+} + 2I^-$). Since we have raised the value for *each* I^- ion by 1090.8, we must lower the Ca^{2+} value by 2×1090.8; thus, $589.1 - (2 \times 1090.8) = -1592.5$ kJ mol^{-1}.

7.9 Theories of Ions in Solution

Many theoretical treatments of ions in solution, especially in aqueous solution, have been put forward. Some of the treatments are quantitative, attempts being made to obtain expressions for the strength of the interactions between ions and the solvent. Other treatments are qualitative, but based on the mathematical treatments. Another classification of theories of ions in solution is based on whether the solvent is treated as if it were a continuum, or whether the molecular nature of the solvent is taken into account. The latter theories are obviously more realistic, but unfortunately they are difficult to develop in any detail. The continuum theories, such as the electrostriction treatment of Nernst and Drude and the treatment of Max Born, are much simpler, and in spite of their lack of realism they give a surprisingly useful interpretation of ionic behavior in solution.

Drude and Nernst's Electrostriction Model

As early as 1894 Paul Drude (1863–1906) and Walther Nernst published a paper of great importance on what they called the **electrostriction** of solvent molecules by ions. They pointed out that the strong electrostatic field due to an ion causes neighboring polar solvent molecules, such as water molecules, to become closely packed and to occupy less volume. In their treatment, as a matter of mathematical convenience, they regarded the solvent as if it were a continuous dielectric. It was noted in Section 7.7 that this electrostriction plays an important role in connection with the mobilities of ions, and in Section 3.4 that it greatly affects the entropies of ions in solution.

It was commented earlier (Section 7.4) that one of the criticisms leveled by some chemists at the theory of electrolytic dissociation was that it appeared to take no account of the role of the solvent. The recognition by Drude and Nernst that ions bring about electrostriction of ions played an important role in leading to a general acceptance of the theory.

Born's Model

A simple interpretation of the thermodynamic quantities for ions in solution was suggested in 1920 by the physicist Max Born (1882–1970). In Born's model the solvent is assumed to be a continuous dielectric and the ion a conducting sphere. Born obtained an expression for the work of charging such a sphere, which is the Gibbs energy change during the charging process. The total reversible work in transporting increments until the sphere has a charge of ze is, according to Born's model,

Max Born did important work in a number of fields. He did pioneering work on quantum mechanics in the 1920s, and was awarded the 1954 Nobel Prize in physics for his work on the quantum mechanics of collision processes.

$$w_{\text{rev}} = \frac{z^2e^2}{8\pi\epsilon_0\epsilon r} \tag{7.86}$$

This work, being non-PV work, is the electrostatic contribution to the Gibbs energy of the ion:

$$G^\circ_{\text{es}} = \frac{z^2e^2}{8\pi\epsilon_0\epsilon r} \tag{7.87}$$

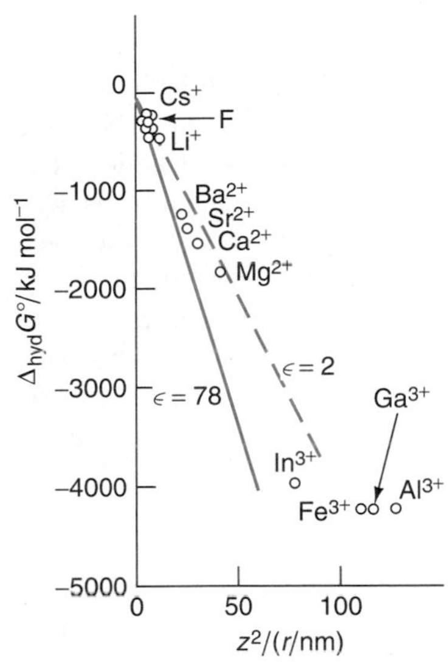

FIGURE 7.14
A plot of the absolute Gibbs energy of hydration of ions against z^2/r. The solid line is the theoretical line for a dielectric constant of 78; the dashed line is for a dielectric constant of 2.

If the same charging process is carried out in vacuo ($\epsilon = 1$), the electrostatic Gibbs energy is

$$G^\circ_{es}(\text{vacuum}) = \frac{z^2e^2}{8\pi\epsilon_0 r} \tag{7.88}$$

The electrostatic Gibbs energy of hydration is therefore

$$\Delta_{hyd}G^\circ = G^\circ_{es} - G^\circ_{es}(\text{vacuum}) = \frac{z^2e^2}{8\pi\epsilon_0 r}\left(\frac{1}{\epsilon} - 1\right) \tag{7.89}$$

For water at 25 °C, $\epsilon = 78$, and Eq. 7.89 leads to

$$\Delta_{hyd}G^\circ = -68.6\frac{z^2}{r/\text{nm}}\ \text{J mol}^{-1} \tag{7.90}$$

Figure 7.14 shows a plot of absolute $\Delta_{hyd}G^\circ$ values against z^2/r. In view of the simplistic nature of the model, the agreement with the predictions of the treatments is not unsatisfactory; the theory accounts quite well for the main effects. Note, however, that the agreement with the theory is worse the higher the charge and the smaller the radius. Figure 7.14 also shows the prediction for $\epsilon = 2$, which is approximately the dielectric constant of water when it is subjected to an intense field. The line for $\epsilon = 2$ is closer to the points for ions of high charge and small radius. The significance of this will be discussed later.

The Born equation also leads to a simple interpretation of entropies of hydration and of absolute entropies of ions. The thermodynamic relationship between entropy and Gibbs energy (see Eq. 3.119) is

$$S = -\left(\frac{\partial G}{\partial T}\right)_P \tag{7.91}$$

The only quantity in Eq. 7.89 that is temperature dependent is the dielectric constant ϵ, so that

$$S^\circ_{es} = \Delta_{hyd}S^\circ = \frac{z^2e^2}{8\pi\epsilon_0\epsilon r}\frac{\partial \ln \epsilon}{\partial T} \tag{7.92}$$

The reason that the theory leads to the same expression for the absolute entropy of the ion and for its entropy of hydration is that it gives zero entropy for the ion in the gas phase, since the Gibbs energy expression in Eq. 7.88 contains no temperature-dependent terms. For water at 25 °C, ϵ is approximately 78 and $(\partial \ln \epsilon)/\partial T$ is about $-0.0046\ \text{K}^{-1}$ over a considerable temperature range. Insertion of these values into Eq. 7.92 leads to

$$S^\circ_{es} = \Delta_{hyd}S^\circ = -4.10\frac{z^2}{r/\text{nm}}\ \text{J K}^{-1}\ \text{mol}^{-1} \tag{7.93}$$

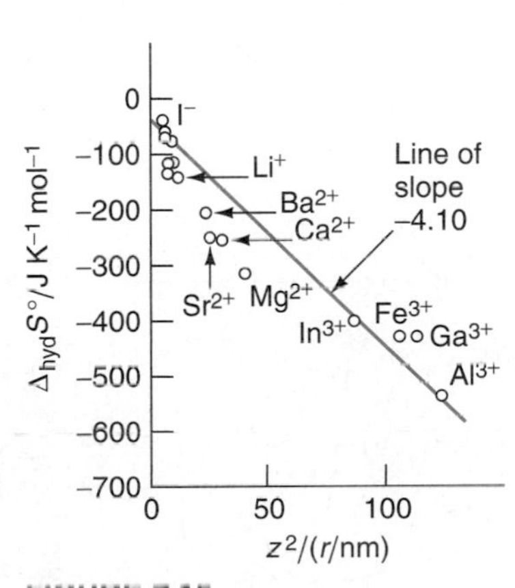

FIGURE 7.15
Plot of entropy of hydration against $z^2/(r/\text{nm})$.

Figure 7.15 shows a plot of the experimental $\Delta_{hyd}S^\circ$ values, together with a line of slope −4.10. The line has been drawn through a value of $-42\ \text{J K}^{-1}\ \text{mol}^{-1}$ at $z^2/r = 0$, since it is estimated that there will be a $\Delta_{hyd}S^\circ$ value of about $-42\ \text{J K}^{-1}\ \text{mol}^{-1}$ due to nonelectrostatic effects. In particular, there is an entropy loss resulting from the fact that the ion is confined in a small solvent "cage," an effect that is ignored in the electrostatic treatment.

Again, the simple model of Born accounts satisfactorily for the main effects. The deviations are primarily for small ions.

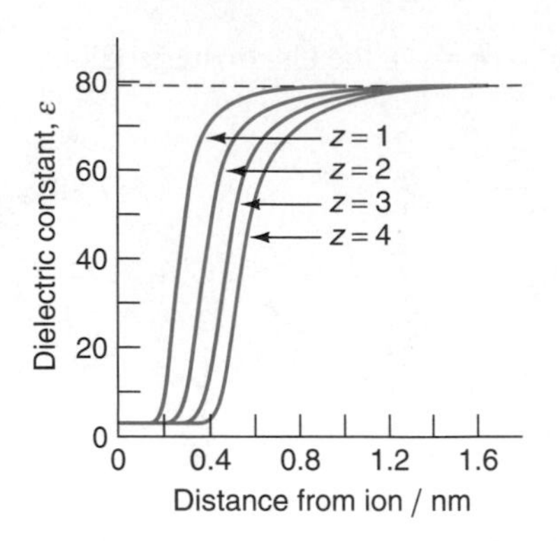

FIGURE 7.16
The variation of dielectric constant of water in the neighborhood of ions.

More Advanced Theories

Various attempts have been made to improve Born's simple treatment while still regarding the solvent as continuous. One way of doing this is to consider how the effective dielectric constant of a solvent varies in the neighborhood of an ion. The dielectric behavior of a liquid is related to the tendency of ions to orient themselves in an electric field. At very high field strengths the molecules become fully aligned in the field and there can be no further orientation. The dielectric constant then falls to a very low value of about 2, an effect that is known as *dielectric saturation.*

It has been estimated that the effective dielectric constant in the neighborhood of ions of various types will be as shown in Figure 7.16. For example, for a ferric (Fe^{3+}) ion the dielectric constant has a value of about 1.78 up to a distance of about 0.3 nm from the center of the ion, owing to the high field produced by the ion. At greater distances, where the field is less, the effective dielectric constant rises toward its limiting value of 78.

When the Born treatment of Gibbs energies and entropies of hydration of ions are modified by taking these dielectric saturation effects into account, the agreement with experiment is considerably better. The experimental values in Figure 7.12 are consistent with a dielectric constant of somewhat less than the true value of 78, and this can be understood in terms of the reduction in dielectric constant arising from dielectric saturation.

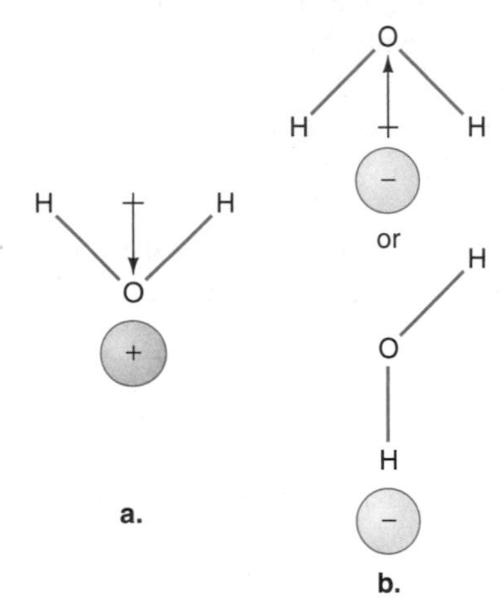

FIGURE 7.17
Orientation of a water molecule.
(a) Close to a positive ion.
(b) Close to a negative ion.

Qualitative Treatments

On the basis of these quantitative theories it is possible to arrive at useful qualitative pictures of ions in solution. The main attractive force between an ion and a surrounding water molecule is the ion-dipole attraction. The way in which a water molecule is expected to orient itself in the neighborhood of a positive or negative ion is shown in Figure 7.17. There are two possible orientations for a negative ion, and it appears that both may play a role. Quite strong binding can result from ion-dipole forces; the binding energy for a Na^+ ion and a neighboring water molecule, for example, is about 80 kJ mol^{-1}.

The smaller the ion, the greater the binding energy between the ion and a water molecule, because the water molecule can approach the ion more closely. Thus, the binding energy is particularly strong for the Li^+ ion, and calculations suggest that a Li^+ ion in water is surrounded tetrahedrally by four water molecules ori-

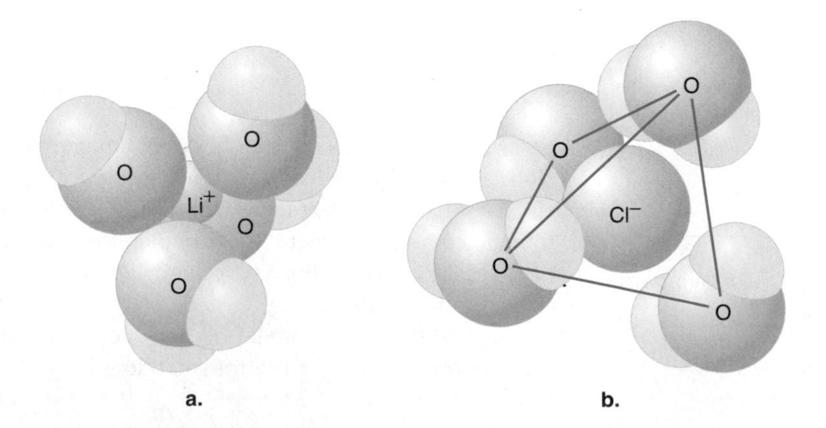

FIGURE 7.18
The tetrahedral arrangement of four water molecules. (a) Around a Li^+ ion. (b) Around a Cl^- ion that can probably also have five or six water molecules around it.

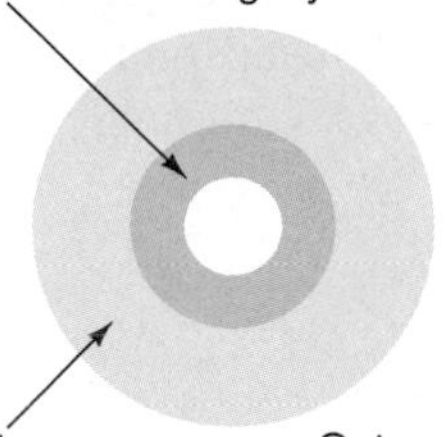

Ordered structure of water molecules bound tightly to the ion

Disorder zone in which structure-making forces oppose each other

Outer zone consisting of ordinary water

FIGURE 7.19
Frank's concept of the structure of water in the neighborhood of an ion.

ented as in Figure 7.17a; this is shown in Figure 7.18. Thus we can say that Li^+ has a **hydration number** of 4. The four water molecules are sufficiently strongly attached to the Li^+ ion that during electrolysis they are dragged along with it. For this reason (Table 7.3) Li^+ has a lower mobility than Na^+ and K^+; the latter ions, being larger, have a smaller tendency to drag water molecules. However, the situation is not so clear-cut as just implied. A moving Li^+ ion drags not only its four neighboring water molecules but also some of the water molecules that are further away from it.

A very useful treatment of ions in solution has been developed by the American physical chemist Henry Sorg Frank (1902–1990) and his coworkers. They have concluded that for ions in aqueous solution it is possible to distinguish three different zones in the neighborhood of the solute molecule, as shown in Figure 7.19. In the immediate vicinity of the ion there is a shell of water molecules that are more or less immobilized by the very high field due to the ion. These water molecules can be described as constituting the *inner hydration shell* and are sometimes described as an *iceberg,* since their structure has some icelike characteristics. However, this analogy should not be pressed too far: Ice is less dense than liquid water, whereas the iceberg around the ion is more compressed than normal liquid water.

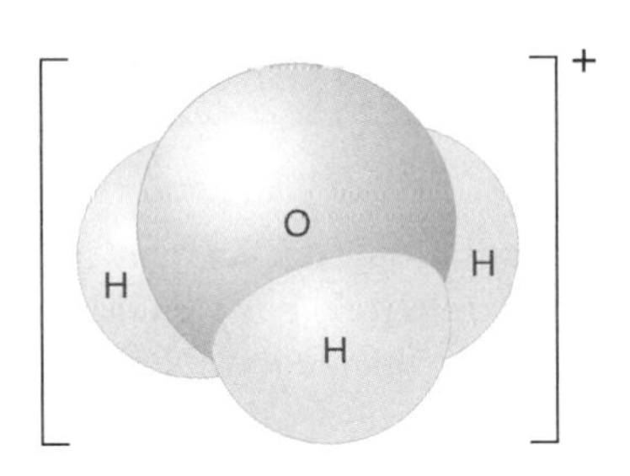

FIGURE 7.20
The structure of the oxonium (H_3O^+) ion.

Surrounding this iceberg there is a second region, which Frank and his coworkers refer to as a region of *structure breaking.* Here the water molecules are oriented more randomly than in ordinary water, where there is considerable ordering due to hydrogen bonding. The occurrence of this structure-breaking region results from two competing orienting influences on each water molecule. One of these is the normal structural effect of the neighboring water molecules; the other is the orienting influence upon the dipole of the spherically symmetrical ionic field.

The third region around the ion comprises all the water sufficiently far from the ion that its effect is not felt.

The behavior of the proton H^+ in water deserves some special discussion. Because of its small size, the proton attaches itself very strongly to a water molecule,

$$H^+ + H_2O \rightleftharpoons H_3O^+$$

The equilibrium for this process, which is strongly exothermic, lies very far to the right. Thus, the hydrogen ion is frequently regarded as existing as the H_3O^+ ion, which is known as the **oxonium ion.** Experimental studies and theoretical calculations have indicated that the angle between the O–H bonds in H_3O^+ is about 115°. The ion is thus almost, but not quite, flat, as shown in Figure 7.20.

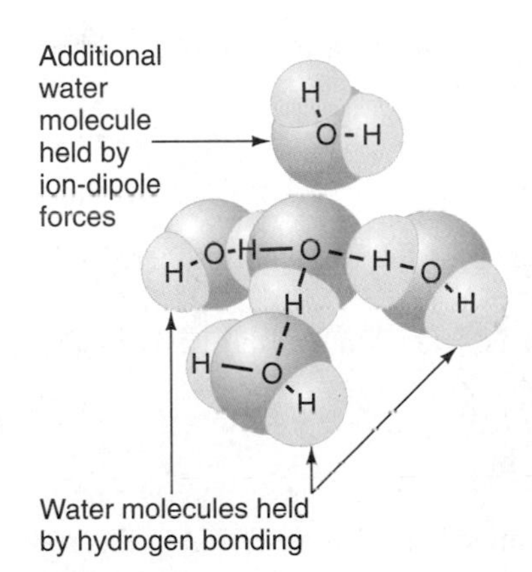

FIGURE 7.21
A possible structure for the hydrated proton. The hydrated H_3O^+ ion (e.g., $H_9O_4^+$) is called the *hydronium ion.*

However, the state of the hydrogen ion in water is more complicated than implied by this description. Because of ion-dipole attractions, other molecules are held quite closely to the H_3O^+ species. The most recent calculations tend to support the view that three water molecules are held particularly strongly, in the manner shown in Figure 7.21. These three molecules are held by hydrogen bonds involving three hydrogen atoms of the H_3O^+ species. The hydrated proton may thus be written as $H_3O^+ \cdot 3H_2O$, or as $H_9O_4^+$. Alternatively, some think that the species $H_9O_4^+$ has the same kind of structure as the hydrated Li^+ ion (Figure 7.17a), in which the proton is surrounded tetrahedrally by four water molecules. Other water molecules will also be held to $H_9O_4^+$ but not so strongly. Thus in Figure 7.21 there is shown an additional H_2O molecule held by ion-dipole forces but not by hydrogen bonding, and it therefore is not held so strongly as the three other water molecules. IUPAC has recommended that the term **hydronium ion** be employed for these hydrated H_3O^+ ions, in contrast to *oxonium ion* for H_3O^+ itself.

The situation is hardly clear-cut; we can write the hydrated proton as H^+, H_3O^+, $H_9O_4^+$, or $H_{11}O_5^+$. It is satisfactory to write it simply as H^+, as long as we remember that the proton is strongly hydrated.

7.10 Activity Coefficients

The electrostatic interactions between ions, besides having an important effect on the conductivities of solutions of strong electrolytes, have an effect on the thermodynamic properties of ions. This matter is most conveniently dealt with in terms of **activity coefficients.** Several experimental methods are available for determining the activity coefficients of ions—see Sections 7.11 and 8.5 (emf measurements). First we will apply the Debye-Hückel treatment to ionic activity coefficients, as was first done in 1922 by the Danish physical chemist Johannes Nicolaus Brønsted (1879–1947) and the American physical chemist Victor K. La Mer (1895–1966).

Debye-Hückel Limiting Law

There are various reasons why a solution shows deviations from ideality, and the matter is quite complicated. Here we are concerned with a relatively simple reason for non-ideality in an ionic solution, namely the electrostatic interactions between ions, as interpreted by the Debye-Hückel theory.

If the behavior of a single ion of type i were ideal, its Gibbs energy could be expressed by the relationship

$$G_i = G_i^\circ + k_B T \ln c_i \tag{7.94}$$

where c_i is the concentration of the ion. Note that in this equation we have used k_B, the Boltzmann constant, instead of the gas constant R, since we are concerned with single ions instead of a mole of ions.

To take into account deviations from ideality we write instead, for the Gibbs energy,

$$G_i = G_i^\circ + k_B T \ln c_i \gamma_i \tag{7.95}$$

$$= G_i^\circ + k_B T \ln c_i + k_B T \ln \gamma_i \tag{7.96}$$

The additional term $k_B T \ln \gamma_i$ is due to the presence of the ionic atmosphere; γ is the activity coefficient.

We have seen that the work of charging an isolated ion, on the basis of the Born model, Eq. 7.86, is

$$w = \frac{z_i^2 e^2}{8\pi\epsilon_0 \epsilon r} \tag{7.97}$$

We also need the work of charging the ionic atmosphere, because this is the required correction to the Gibbs energy; this work is equal to $k_B T \ln \gamma_i$. If ϕ is the potential due to the ionic atmosphere, the work of transporting a charge dQ to the ion is $\phi\, dQ$; the net work of charging the ionic atmosphere is thus

$$w = \int_0^{z_i e} \phi\, dQ \tag{7.98}$$

The potential ϕ corresponding to the charge Q is given by Eq. 7.48 as $-Q\kappa/4\pi\epsilon_0\epsilon$, where $1/\kappa$ is the radius of the ionic atmosphere. The work is thus

$$w = -\int_0^{z_ie} \frac{Q\kappa}{4\pi\epsilon_0\epsilon}\,dQ = -\frac{Q^2\kappa}{8\pi\epsilon_0\epsilon}\Bigg|_0^{z_ie} = -\frac{z_i^2e^2\kappa}{8\pi\epsilon_0\epsilon} \tag{7.99}$$

Note that Eq. 7.99 can be obtained from Eq. 7.97 by replacing r by $1/\kappa$ and changing the sign. The reciprocal $1/\kappa$ plays the same role for the atmosphere as does r for the ion, and the change of sign is required because the net charge on the atmosphere is opposite to that on the ion.

The expression for the activity coefficient γ_i is therefore

$$k_BT \ln \gamma_i = -\frac{z_i^2e^2\kappa}{8\pi\epsilon_0\epsilon} \tag{7.100}$$

whence

$$\ln \gamma_i = -\frac{z_i^2e^2\kappa}{8\pi\epsilon_0\epsilon k_BT} \tag{7.101}$$

or[10]

$$\log_{10} \gamma_i = -\frac{z_i^2e^2\kappa}{8 \times 2.303\pi\epsilon_0\epsilon k_BT} \tag{7.102}$$

Equation 7.40 shows that κ is proportional to the square root of the quantity $\Sigma_i N_iz_i^2$, where N_i is the number of ions of the ith type per unit volume and z_i is its valency.

The **ionic strength** I of a solution is defined as

Ionic Strength

$$I = \frac{1}{2}\sum_i c_iz_i^2 \tag{7.103}$$

where c_i is the *molar* concentration of the ions of type i.[11] The ionic strength is proportional to $\Sigma_i N_iz_i^2$, and the reciprocal of the radius of the ionic atmosphere, κ, is thus proportional to $\sqrt{I}$ (see Eq. 7.49). Equation 7.102 may thus be written as

$$\log_{10} \gamma_i = -z_i^2B\sqrt{I} \tag{7.104}$$

where B is a quantity that depends on properties such as ϵ and T. When water is the solvent at 25 °C, the value of B is 0.51 mol$^{-1/2}$ dm$^{3/2}$.

Experimentally we cannot measure the activity coefficient or indeed any thermodynamic property of a simple ion, since at least two types of ions must be present in any solution. To circumvent this difficulty we define a **mean activity coefficient** $\gamma_\pm$ in terms of the individual values for γ_+ and γ_- by the relationship

Mean Activity Coefficient

$$\gamma_\pm^{\nu_+ + \nu_-} = \gamma_+^{\nu_+}\gamma_-^{\nu_-} \tag{7.105}$$

where ν_+ and ν_- are the numbers of ions of the two kinds produced by the electrolyte. For example, for $ZnCl_2$, $\nu_+ = 1$ and $\nu_- = 2$. For a uni-univalent electrolyte

[10] It is convenient to use common logarithms in this treatment, as the resulting constant for water at 25 °C is then easy to remember (see Eq. 7.111).

[11] For a uni-univalent electrolyte such as NaCl the ionic strength is equal to the molar concentration. Thus for 1 M solution $c_+ = 1$ and $c_- = 1$, $z_+ = 1$ and $z_- = -1$; hence

$$\frac{1}{2}\sum_i c_iz_i^2 = \frac{1}{2}(1 + 1) = 1\ M$$

For a 1 M solution of a uni-bivalent electrolyte such as K_2SO_4, $c_+ = 2$, $c_- = 1$, $z_+ = 1$, and $z_- = -2$; hence the ionic strength is $\frac{1}{2}(2 \times 1 + 1 \times 4) = 3\ M$. Similarly, for a 1 M solution of a uni-trivalent electrolyte such as Na_3PO_4, the ionic strength is 6 M.

TABLE 7.5 Mean Activity Coefficients of Electrolytes as a Function of Concentration

m/mol kg^{-1}	NaCl	$NaNO_3$	Na_2HPO_4
0.001	0.965	0.965	0.887
0.002	0.952	0.951	0.848
0.005	0.928	0.926	0.780
0.010	0.903	0.900	0.717
0.020	0.872	0.866	0.644
0.050	0.822	0.810	0.539
0.100	0.779	0.759	0.456
0.200	0.734	0.701	0.373
0.500	0.681	0.617	0.266
1.000	0.657	0.550	0.191
2.000	0.668	0.480	0.133
5.000	0.874	0.388	—

($\nu_+ = \nu_- = 1$) the mean activity coefficient is the geometric mean $(\gamma_+\gamma_-)^{1/2}$ of the individual values. Table 7.5 shows the effect of differences in type of ions within a compound on the mean activity coefficient as the concentration changes.

In order to express $\gamma_\pm$ in terms of the ionic strength, we proceed as follows. From Eq. 7.105

$$(\nu_+ + \nu_-) \log_{10} \gamma_\pm = \nu_+ \log_{10} \gamma_+ + \nu_- \log_{10} \gamma_- \tag{7.106}$$

Insertion of the expression in Eq. 7.104 for log γ_+ and log γ_- gives

$$(\nu_+ + \nu_-) \log_{10} \gamma_\pm = -(\nu_+ z_+^2 + \nu_- z_-^2) B\sqrt{I} \tag{7.107}$$

For electrical neutrality[12]

$$\nu_+ z_+ = \nu_- |z_-| \quad \text{or} \quad \nu_+^2 z_+^2 = \nu_-^2 z_-^2 \tag{7.108}$$

and therefore

$$(\nu_+ + \nu_-) \log_{10} \gamma_\pm = -\nu_+^2 z_+^2 \left(\frac{1}{\nu_+} + \frac{1}{\nu_-}\right) B\sqrt{I} \tag{7.109}$$

Thus

$$\log_{10} \gamma_\pm = -\frac{\nu_+ z_+^2}{\nu_-} B\sqrt{I} = -z_+|z_-| B\sqrt{I} \tag{7.110}$$

For aqueous solutions at 25 °C,

$$\log_{10} \gamma_\pm = -0.51 z_+|z_-| \sqrt{I/\text{mol dm}^{-3}} \tag{7.111}$$

Equation 7.111 is known as the **Debye-Hückel limiting law** (DHLL).

[12]Note that in this treatment we use the positive sign for the valency of negative ions (e.g., $|z_{Cl^-}| = +1$).

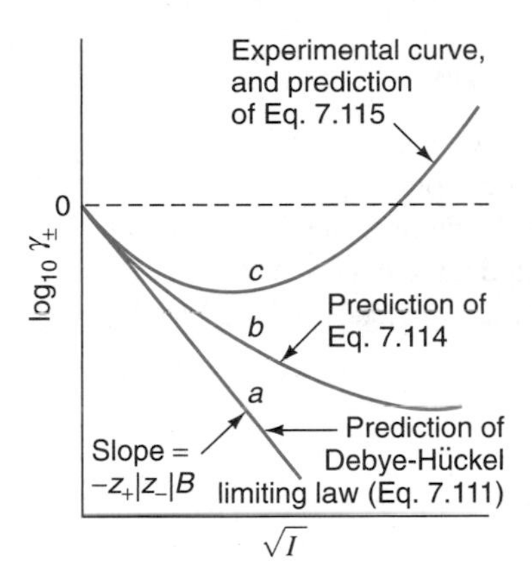

FIGURE 7.22
Variation of $\log_{10} \gamma_\pm$ with the square root of the ionic strength.

Deviations from the Debye-Hückel Limiting Law

Experimentally the DHLL is found to apply satisfactorily only at extremely low concentrations[13]; at higher concentrations there are very significant deviations, as shown schematically in Figure 7.22. Equation 7.110 predicts that a plot of $\log_{10} \gamma_\pm$ against $\sqrt{I}$ will have a negative slope, the magnitude of the slope being $-z_+|z_-|B$. The results with a number of electrolytes have shown that at very low I values $\log_{10} \gamma_\pm$ does indeed fall linearly with I, with the correct slope; for this reason Eq. 7.111 is very satisfactory as a *limiting* law. At higher $\sqrt{I}$ values, however, the value of $\log_{10} \gamma_\pm$ becomes significantly less negative than predicted by the law and, at sufficiently high ionic strengths, may actually attain positive values. The significance of this in connection with the solubilities of salts is considered in Section 7.11.

Various theories have been put forward to explain these deviations from the DHLL, but here they can be considered only briefly. One factor that has been neglected in the development of the equations is the fact that the ions occupy space; thus far they have been treated as point charges with no restrictions on how closely they can come together. If the theory is modified in such a way that the centers of the ions cannot approach one another closer than the distance a, the expression for the activity coefficient of an individual ion (compare Eq. 7.101) becomes

$$\ln \gamma_i = -\frac{z_c^2 e^2 \kappa}{8\pi\epsilon_0\epsilon k_B T} \cdot \frac{1}{1 + \kappa a} \tag{7.112}$$

Since κ is proportional to $\sqrt{I}$, this equation can be written as

$$\log_{10} \gamma_i = -\frac{z_c^2 B\sqrt{I}}{1 + aB'\sqrt{I}} \tag{7.113}$$

(compare Eq. 7.104), where B is the same constant as used previously and B' is a new constant. The corresponding equation for the mean activity coefficient $\gamma_\pm$ is

$$\log_{10} \gamma_\pm = -\frac{z_+|z_-|B\sqrt{I}}{1 + aB'\sqrt{I}} \tag{7.114}$$

(compare Eq. 7.110). Equation 7.114 leads to the curve shown as b in Figure 7.22; however, it cannot explain the positive values of $\log_{10} \gamma_\pm$ that are obtained at high ionic strengths.

In order to explain these positive values it is necessary to take account of the orientation of solvent molecules by the ionic atmosphere. This was considered by Hückel, who showed that it gives a term linear in I in the expression for $\log_{10} \gamma_\pm$:

$$\log_{10} \gamma_\pm = \frac{Bz_+|z_-|\sqrt{I}}{1 + aB'\sqrt{I}} + CI \tag{7.115}$$

where C is a constant. At sufficiently high ionic strengths the last term (CI) predominates, and $\log_{10} \gamma_\pm$ is approximately linear in I, as found experimentally. The CI term is often known as the "salting-out" term since, as seen in Section 7.11, it accounts for the lowered solubilities of salts at high ionic strengths.

[13]For example, for a uni-univalent electrolyte such as KCl, the law is obeyed satisfactorily up to a concentration of about 0.01 M; for other types of electrolytes, deviations appear at even lower concentrations.

7.11 Ionic Equilibria

Equilibrium is usually established very rapidly between ionic species in solution. In this section an account will be given of some of the more important ways in which activity coefficients can be determined when equilibria are established in solution. Only some special topics are presented here; for a detailed account of how calculations can be made for fairly complicated systems, the reader is referred to the book by J. N. Butler in *Suggested Reading* (p. 304).

Activity Coefficients from Equilibrium Constant Measurements

Equilibrium constant determinations can provide values of activity coefficients. The procedure may be illustrated with reference to the dissociation of acetic acid,

$$CH_3COOH \rightleftharpoons H^+ + CH_3COO^-$$

The practical equilibrium constant is

$$K_a = \frac{[H^+][CH_3COO^-]}{[CH_3COOH]} \cdot \frac{\gamma_+ \gamma_-}{\gamma_u} \tag{7.116}$$

where γ_+ and γ_- are the activity coefficients of the ions and γ_u is that of the undissociated acid. In reasonably dilute solution the undissociated acid will behave ideally ($\gamma_u = 1$), but γ_+ and γ_- may be significantly different from unity because of the electrostatic interactions. Replacement of $\gamma_+\gamma_-$ by $\gamma_\pm^2$ and taking logarithms of Eq. 7.116 leads to[14]

$$\log_{10}\left(\frac{[H^+][CH_3COO^-]}{[CH_3COOH]}\right)^u = \log_{10} K^\circ - 2 \log_{10} \gamma_\pm \tag{7.117}$$

The left-hand side can be written as

$$\log_{10}\left(\frac{c\alpha^2}{1-\alpha}\right)^u = \log_{10} K^\circ - 2 \log_{10} \gamma_\pm \tag{7.118}$$

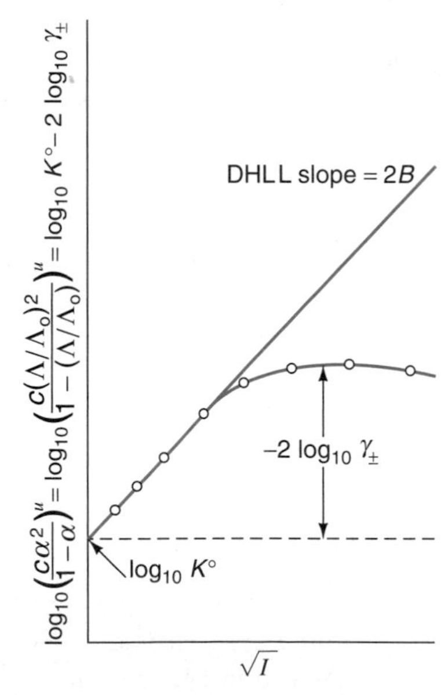

FIGURE 7.23
A schematic plot against $\sqrt{I}$ of $\log_{10}[c\alpha^2/(1-\alpha)]^u$, where α, the degree of dissociation, may be obtained from conductivity measurements.

where c is the concentration and α is the degree of dissociation, which can be determined from conductivity measurements. (See p. 274.) Values of the left-hand side of this equation can therefore be calculated for a variety of concentrations, and these values are equal to $\log_{10} K^\circ - 2 \log_{10} \gamma_\pm$.

If the Debye-Hückel limiting law applies, $\log_{10} \gamma_\pm$ is given by Eq. 7.111. If the solution contains only acetic acid, the ionic strength I is given by

$$I = \tfrac{1}{2}[(c\alpha)(1)^2 + (c\alpha)(-1)^2] = c\alpha \tag{7.119}$$

If other ions are present, their contributions must be added. It is convenient to plot the left-hand side of Eq. 7.118 against the ionic strength, as shown schematically in Figure 7.23. If there were exact agreement with the DHLL, the points would lie on

[14]In Eqs. 7.117 and 7.118 we again use the superscript u to indicate the numerical value of the quantity. K° is the thermodynamic equilibrium constant, i.e., the dimensionless form of the practical constant K_a.

a line of slope $2B$ (i.e., 1.02 for water at 25 °C). Extrapolation to zero ionic strength gives the value of K. Once this value has been obtained, subtraction of $\log_{10} K°$ from the experimental curve at any ionic strength gives $-\log_{10} \gamma_{\pm}$, and $\gamma_{\pm}$ can therefore be calculated at that ionic strength. The individual γ_{+} and γ_{-} values cannot, of course, be determined experimentally.

Solubility Products

When we write solubility products (Section 4.4), we often ignore activity coefficients; the solubility products are expressed as products of concentrations instead of activities. The solubility product for silver chloride should more accurately be written as

$$K_s = [Ag^+][Cl^-]\gamma_+\gamma_- \tag{7.120}$$

where γ_+ and γ_- are the activity coefficients of Ag^+ and Cl^-, respectively. The product $\gamma_+\gamma_-$ is equal to $\gamma_{\pm}^2$, where $\gamma_{\pm}$ is the mean activity coefficient, and therefore

$$K_s = [Ag^+][Cl^-]\gamma_{\pm}^2 \tag{7.121}$$

One matter of interest that can be understood in terms of this equation is the effect of inert electrolytes on solubilities. An inert electrolyte is one that does not contain a common ion (Ag^+ or Cl^- in this instance) and also does not contain any ion that will complicate the situation by forming a precipitate with either the Ag^+ or the Cl^- ions. In other words, the added inert electrolyte does not bring about a chemical effect; its influence arises only because of its ionic strength.

The influence of ionic strength I on the activity coefficient of an ion is given according to the DHLL by the equation

$$\log_{10} \gamma_{\pm} = -z_+|z_-|B\sqrt{I} \tag{7.122}$$

Figure 7.22 shows that this equation satisfactorily accounts for the drop in $\gamma_{\pm}$ that occurs at very low ionic strengths but that considerable deviations occur at higher ones; the value of $\log_{10} \gamma_{\pm}$ becomes positive (i.e., $\gamma_{\pm}$ is greater than unity) at sufficiently high values of I.

It follows as a result of this behavior that there are two qualitatively different ionic-strength effects on solubilities, one arising at low I values when the $\gamma_{\pm}$ falls with increasing I and the other being found when $\gamma_{\pm}$ increases with increasing I. Thus, at low ionic strengths the product $[Ag^+][Cl^-]$ will increase with increasing I, because the product $[Ag^+][Cl^-]\gamma_{\pm}^2$ remains constant and $\gamma_{\pm}$ decreases. Under these conditions, added salt increases solubility, and we speak of *salting in*.

Salting In

At higher ionic strengths, however, $\gamma_{\pm}$ rises as I increases, and $[Ag^+][Cl^-]$ therefore diminishes. Thus there is a decrease in solubility, and we speak of *salting out*. Of particular interest are the salting-in and salting-out effects found with protein molecules, a matter of considerable practical importance since proteins are conveniently classified in terms of their solubility behavior.

Salting Out

EXAMPLE 7.6 The solubility product of $BaSO_4$ is 9.2×10^{-11} mol^2 dm^{-6}. Calculate the mean activity coefficient of the Ba^{2+} and SO_4^{2-} ions in a solution that is 0.05 *M* in KNO_3 and 0.05 *M* in KCl, assuming the Debye-Hückel limiting law to apply. What is the solubility of $BaSO_4$ in that solution, and in pure water?

Solution The ionic strength of the solution is

$$I = \frac{1}{2}(0.05 + 0.05 + 0.05 + 0.05) = 0.1\ M$$

According to the DHLL,

$$\log_{10} \gamma_+ = -2^2 \times 0.51\sqrt{0.1}$$
$$= -0.645$$
$$\gamma_+ = 0.226$$

If the solubility in the solution is s,

$$K_s = s^2\gamma_\pm^2 = s^2 \times (0.226)^2$$
$$9.2 \times 10^{-11} = s^2 \times 0.05126$$

Therefore

$$s = 4.24 \times 10^{-5}\ M$$

In pure water the activity coefficients are taken to be unity so that the solubility product is simply the square of the solubility. The solubility in pure water is therefore

$$(9.2 \times 10^{-11}\ \text{mol}^2\ \text{dm}^{-6})^{\frac{1}{2}} = 9.6 \times 10^{-6}\ \text{mol dm}^{-3}$$

We note that the solubility is higher in the salt solution than in water ("salting in"), because the activity coefficients have been lowered.

We saw earlier that measurements of equilibrium constants over a range of ionic strengths allow activity coefficients to be obtained. The same can be done with measurements of solubility. We will outline the method for a sparingly soluble uni-univalent salt AB, for which the solubility equilibrium is

$$AB(s) \rightleftharpoons A^+ + B^-$$

The solubility product is

$$K_s = a_{A+}a_{B-} = [A^+][B^-]\gamma_\pm^2 \tag{7.123}$$

Thus

$$\log_{10}([A^+][B^-])^u = \log_{10} K_s^\circ - 2\log_{10}\gamma_\pm \tag{7.124}$$

For a solution in which no common ions are present, the solubility is

$$s = [A^+] = [B^-] \tag{7.125}$$

and therefore

$$\log_{10}(s^2)^u = \log K_s^\circ - 2\log_{10}\gamma_\pm \tag{7.126}$$

$$\log_{10} s^u = \frac{1}{2}\log_{10} K_s^u - \log_{10}\gamma_\pm \tag{7.127}$$

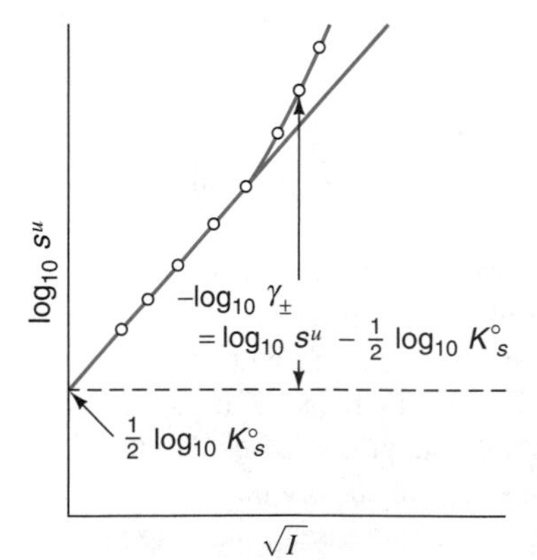

FIGURE 7.24
A schematic plot of $\log_{10} s^u$ against the square root of the ionic strength, showing how K_s° and $\gamma_\pm$ are obtained.

Insofar as the DHLL is obeyed, a plot of $\log_{10} s^u$ against $\sqrt{I/\text{mol dm}^{-3}}$ will therefore be a straight line of slope $B = 0.51$ in water at 25 °C.

Figure 7.24 shows the type of plot that is obtained in this way. At sufficiently low ionic strengths the points lie on a line of slope B. Extrapolation to zero ionic strength, where $\log_{10} \gamma_\pm = 0$, therefore gives $\frac{1}{2} \log_{10} K_s^\circ$, from which the true thermodynamic solubility product K_s is obtained. The value of $\log \gamma_\pm$ is then given by the difference between $\frac{1}{2} \log_{10} K_s^\circ$ and $\log_{10} s^u$ at that ionic strength, as shown in the figure.

7.12 Ionization of Water

Although water itself ionizes only to a small extent, the fact that it does so at all plays a vital role in many aspects of physical chemistry. The equilibrium constant for the ionic dissociation

$$H_2O \rightleftharpoons H^+ + OH^-$$

is

$$\frac{[H^+][OH^-]}{[H_2O]} = K \tag{7.128}$$

Ionic Product

However, because the concentration of water hardly varies from solution to solution, it is usual to incorporate this concentration into the equilibrium constant and to work in terms of the **ionic product** of water, defined as

$$K_w = [H^+][OH^-] \tag{7.129}$$

At 25 °C the value of the ionic product is almost exactly 10^{-14} mol^2 dm^{-6}.

In chemically pure water the concentrations of hydrogen and hydroxyl ions are equal and at 25 °C are 10^{-7} mol dm^{-3}. A solution in which the hydrogen ion concentration is greater than this is said to be acidic; one with a lower concentration is said to be basic or alkaline.

Acids and bases have been defined in various ways. According to Brønsted, an acid is a substance that can donate a proton, and a base a substance that can accept one. The ionization of an acid HA can be written as

$$HA + H_2O \rightleftharpoons H_3O^+ + A^-$$

and that of a base B as

$$B + H_2O \rightleftharpoons BH^+ + OH^-$$

According to Brønsted's definition the species HA, H_3O^+, and BH^+ are acids, while B, OH^-, and A^- are bases. In the first process the water is acting as a base, in the second as an acid. Substances like water that can be both acids and bases are said to be **amphoteric.** The species A^- is said to be the conjugate base of the acid HA, while BH^+ is the conjugate acid of the base B.

The acid dissociation constant of HA is defined as

$$K_a = \frac{[H_3O^+][A^-]}{[HA]} \tag{7.130}$$

It is convenient to work with pK_a values, defined as the negative common logarithm of the K_a value. Thus the dissociation constant of acetic acid is 1.8×10^{-5} M, and the pK_a value is 4.75.

Of special interest are polybasic acids, which undergo successive ionizations. An example is phosphoric acid, for which the ionizations and the corresponding pK values are

$$H_3PO_4 \overset{pK_1=2.1}{\rightleftharpoons} H_2PO_4^- \overset{pK_2=7.2}{\rightleftharpoons} HPO_4^{2-} \overset{pK_3=12.3}{\rightleftharpoons} PO_4^{3-}$$

The preponderant species in solution at various pH values can easily be deduced if the pK values are known (see Problem 7.43).

Amphoteric Electrolytes

Also of special interest are **amphoteric electrolytes,** or ampholytes. Glycine, for example, ionizes as follows:

$$H_3N^+CH_2COOH \rightleftharpoons H_3N^+CH_2COO^- + H^+ \rightleftharpoons H_2NCH_2COO^- + 2H^+$$

The fact that the intermediate species is the **zwitterion** (double ion) $H_3N^+CH_2COO^-$ rather than H_2NCH_2COOH can be shown by an argument such as that in Problem 7.42.

7.13 The Donnan Equilibrium

Of considerable importance are the ionic equilibria that are established when two solutions are separated by a membrane. Complications arise with ions that are too large to diffuse through the membrane, and the diffusible ions then reach a special type of equilibrium, known as the **Donnan equilibrium.** Its theory was first worked out in 1911 by the British physical chemist Frederick George Donnan (1870–1956).

Consider first the equilibrium established when all ions can diffuse through the membrane. Suppose that solutions of sodium chloride, of volume 1 dm^3, are separated by the membrane, as shown in Figure 7.25a. At equilibrium the concentrations are $[Na^+]_1$ and $[Cl^-]_1$ on the left-hand side and $[Na^+]_2$ and $[Cl^-]_2$ on the right-hand side. Intuitively, we know that in this case these concentrations must be all the same, and in thermodynamic terms we can arrive at this conclusion by noting that at equilibrium

$$\Delta G = \Delta G_{Na^+} + \Delta G_{Cl^-} = 0 \tag{7.131}$$

where the terms are the Gibbs energy differences across the membrane (e.g., ΔG is the change in Gibbs energy in going from left to right). The expressions for the individual molar ionic Gibbs energy terms are

$$\Delta G_{Na^+} = RT \ln \frac{[Na^+]_2}{[Na^+]_1} \quad \text{and} \quad \Delta G_{Cl^-} = RT \ln \frac{[Cl^-]_2}{[Cl^-]_1} \tag{7.132}$$

Thus

$$RT \ln \frac{[Na^+]_2}{[Na^+]_1} + RT \ln \frac{[Cl^-]_2}{[Cl^-]_1} = 0 \tag{7.133}$$

from which it follows that

$$\frac{[Na^+]_2[Cl^-]_2}{[Na^+]_1[Cl^-]_1} = 1 \tag{7.134}$$

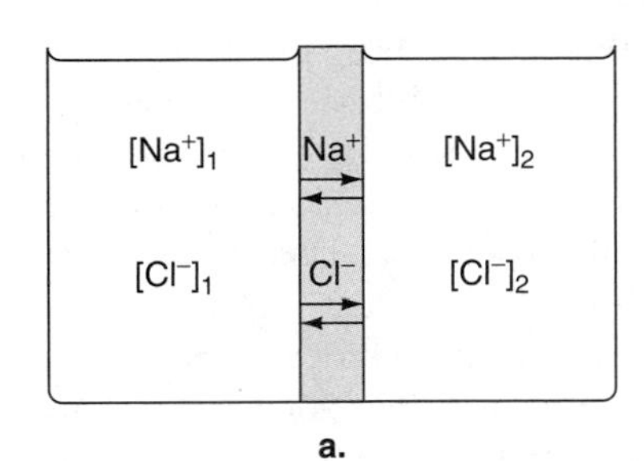

FIGURE 7.25
(a) Sodium and chloride ions separated by a membrane.
(b) Na^+P^- and Na^+Cl^- separated by a membrane: initial conditions.
(c) The final Donnan equilibrium conditions arising from (b).

Since, for electrical neutrality, $[Na^+]_1 = [Cl^-]_1$ and $[Na^+]_2 = [Cl^-]_2$, the result is that

$$[Na^+]_1 = [Na^+]_2 = [Cl^-]_1 = [Cl^-]_2 \tag{7.135}$$

In other words, at equilibrium we have equal concentrations of the electrolyte on each side of the membrane.

This rather trivial case is useful as an introduction to the Donnan equilibrium, since Eq. 7.131 is still obeyed even if the system contains a nondiffusible ion in addition to sodium and chloride ions. For example, suppose that we initially have the situation represented in Figure 7.25b. On the left-hand side, there are sodium ions and nondiffusible anions P^-; on the right-hand side there are sodium and chloride ions. Since there are no chloride ions on the left-hand side, spontaneous diffusion of chloride ions from right to left will occur. Since there must always be electrical neutrality on each side of the membrane, an equal number of sodium ions must also diffuse from right to left. Figure 7.25c shows the situation at equilibrium; x mol dm^{-3} of $[Na^+]$ and $[Cl^-]$ have diffused from right to left; the initial concentrations on the two sides are c_1 and c_2.

Application of Eq. 7.134 then leads to

$$(c_2 - x)^2 = (c_1 + x)x \tag{7.136}$$

whence

$$x = \frac{c_2^2}{c_1 + 2c_2} \tag{7.137}$$

EXAMPLE 7.7 Suppose that a liter of a solution of sodium palmitate at concentration $c_1 = 0.01\ M$ is separated by a membrane from a liter of solution of sodium chloride at concentration $c_2 = 0.05\ M$. If the membrane is permeable to sodium and chloride ions, but not to palmitate ions, what are the final concentrations after the Donnan equilibrium has become established?

Solution Suppose that x mol of sodium ions and x mol of chloride ions cross the membrane from right to left. The concentrations are then:

Left side: $[Na^+] = (0.10 + x)\ M$; $[Cl^-] = x\ M$
Right side: $[Na^+] = [Cl^-] = (0.05 - x)\ M$

The products of the concentrations of Na^+ and Cl^- ions must be the same on the two sides, and therefore

$$(0.01 + x)x = (0.05 - x)^2$$

The solution of this quadratic equation is $x = 0.023$. The final concentrations are therefore:

Left side: $[Na^+] = 0.33\ M$; $[Cl^{-1}] = 0.023\ M$
Right side: $[Na^+] = [Cl^-] = 0.027\ M$

It is easy to check this solution by confirming that the products of the concentrations of the diffusible ions are the same on each side of the membrane.

Equations for more complicated Donnan equilibria (e.g., those involving divalent ions) can be worked out using the same principles. Equilibria of the Donnan type are relevant to many types of biological systems; the theory is particularly important with reference to the passage of ions across the membranes of nerve fibers. However, under physiological conditions the significance of the Donnan effect is not easily assessed, because of the complication of *active transport,* a phenomenon in which ions are transported against concentration gradients by processes requiring the expenditure of energy. One straightforward example of the Donnan effect in which there is little or no active transport is found with the erythrocytes (red blood cells). Here the concentration of chloride ions within the cells is significantly smaller than the concentration in the plasma surrounding the cells. This effect can be attributed to the much higher concentration of protein anions retained within the erythrocyte; hemoglobin accounts for a third of the dry weight of the cell.

Active Transport

Under certain circumstances, the establishment of the Donnan equilibrium can lead to other effects, such as changes in pH. Suppose, for example, that an electrolyte NaP (where P is a large anion) is on one side of a membrane, with pure water on the other. The Na^+ ions will tend to cross the membrane and, to restore the electrostatic balance, H^+ ions will cross in the other direction, leaving an excess of OH^- ions. Dissociation of water molecules will occur as required. There will therefore be a lowering of pH on the NaP side of the membrane and a raising on the other side.

The Donnan equilibrium is associated with Nernst potentials, as will be considered in Section 8.3.

KEY EQUATIONS

Definition of molar conductivity Λ*:*

$$\Lambda \equiv \frac{\kappa}{c}$$

where

κ = electrolytic conductivity;

c = concentration.

Ostwald's dilution law:

$$\frac{c(\Lambda/\Lambda^\circ)^2}{1-(\Lambda/\Lambda^\circ)} = K$$

where

Λ° = molar conductivity at infinite dilution;

Λ/Λ° = degree of dissociation;

K = equilibrium constant.

Law of independent migration of ions:

$$\Lambda = \lambda_+ + \lambda_-$$

where λ_+ and λ_- are the individual ion conductivities.

Definitions of transport numbers:

$$t_+ = \frac{u_+}{u_+ + u_-} \qquad t_- = \frac{u_-}{u_+ + u_-}$$

where u_+ and u_- are the ion mobilities.

Definition of ionic strength I:

$$I \equiv \frac{1}{2}\sum_i c_i z_i^2$$

where

c_i = concentration of ion of ith type;

z_i = its charge number.

Debye-Hückel limiting law for activity coefficient γ_i:

$$\log_{10}\gamma_i = -z_i^2 B\sqrt{I}$$

For the mean activity coefficient $\gamma_\pm$,

$$\log_{10}\gamma_\pm = -z_+|z_-|B\sqrt{I}$$
$$= -0.51 z_+|z_-|\sqrt{I/\text{mol dm}^{-3}}$$

for water at 25 °C.

PROBLEMS

Faraday's Laws, Molar Conductivity, and Weak Electrolytes

7.1. A constant current was passed through a solution of cupric sulfate, $CuSO_4$, for 1 h, and 0.040 g of copper was deposited. Calculate the current (atomic weight of Cu = 63.5).

7.2. After passage of a constant current for 45 min, 7.19 mg of silver (atomic weight = 107.9) was deposited from a solution of silver nitrate. Calculate the current.

7.3. Electrolysis of molten KBr generates bromine gas, which can be used in industrial bromination processes. How long will it take to convert a 500.00-kg batch of phenol (C_6H_5OH) to monobromophenol using a current of 20 000 A?

7.4. The following are the molar conductivities Λ of chloroacetic acid in aqueous solution at 25 °C and at various concentrations c:

$\frac{c}{10^{-4}\,M}$	625	312.5	156.3	78.1	39.1	19.6	9.8
$\frac{\Lambda}{\Omega^{-1}\text{ cm}^2\text{ mol}^{-1}}$	53.1	72.4	96.8	127.7	164.0	205.8	249.2

Plot Λ against c. If $\Lambda° = 362\ \Omega^{-1}\text{ cm}^2\text{ mol}^{-1}$, are these values in accord with the Ostwald dilution law? What is the value of the dissociation constant? (See also Problem 7.11.)

7.5. The electrolytic conductivity of a saturated solution of silver chloride, AgCl, in pure water at 25 °C is $1.26 \times 10^{-6}\ \Omega^{-1}\text{ cm}^{-1}$ higher than that for the water used. Calculate the solubility of AgCl in water if the molar ionic conductivities are Ag^+, 61.9 $\Omega^{-1}\text{ cm}^2\text{ mol}^{-1}$; Cl^-, 76.4 $\Omega^{-1}\text{ cm}^2\text{ mol}^{-1}$.

***7.6.** The electrolytic conductivity of a 0.001 M solution of Na_2SO_4 is $2.6 \times 10^{-4}\ \Omega^{-1}\text{ cm}^{-1}$. If the solution is saturated with $CaSO_4$, the conductivity becomes $7.0 \times 10^{-4}\ \Omega^{-1}\text{ cm}^{-1}$. Calculate the solubility product for $CaSO_4$ using the following molar conductivities at these concentrations: $\lambda(Na^+) = 50.1\ \Omega^{-1}\text{ cm}^2\text{ mol}^{-1}$; $\lambda(\frac{1}{2}Ca^{2+}) = 59.5\ \Omega^{-1}\text{ cm}^2\text{ mol}^{-1}$.

7.7. The quantity l/A of a conductance cell (see Eq. 7.8) is called the cell constant. Find the cell constant for a conductance cell in which the conductance, G, of a 0.100 M KCl solution is 0.01178 S at 25 °C. The equivalent conductance for 0.100 M KCl at 25 °C is 128.96 S cm^2 mol^{-1}. If a 0.0500 M solution of an electrolyte has a measured conductance of 0.00824 S using this cell, what is the equivalent conductance of the electrolyte?

***7.8.** A conductivity cell when standardized with 0.01 M KCl was found to have a resistance of 189 Ω. With 0.01 M ammonia solution the resistance was 2460 Ω. Calculate the base dissociation constant of ammonia, given the following molar conductivities at these concentrations: $\lambda(K^+) = 73.5\ \Omega^{-1}\text{ cm}^2\text{ mol}^{-1}$; $\lambda(Cl^-) = 76.4\ \Omega^{-1}\text{ cm}^2\text{ mol}^{-1}$; $\lambda(NH_4^+) = 73.4\ \Omega^{-1}\text{ cm}^2\text{ mol}^{-1}$; $\lambda(OH^-) = 198.6\ \Omega^{-1}\text{ cm}^2\text{ mol}^{-1}$.

7.9. The conductivity of a 0.0312 M solution of a weak base is 1.53×10^{-4} S cm^{-1}. If the sum of the limiting ionic conductances for BH^+ and OH^- is 237.0 S cm^2 mol^{-1}, what is the value of the base constant K_b?

7.10. The equivalent conductance of KBr solutions as a function of concentration at 25 °C is given in the following table. By a linear regression analysis of suitable variables, find the value of $\Lambda°$ for KBr.

$c/10^{-3}\ M$	0.25	0.36	0.50	0.75	1.00	1.60	2.00	5.00	10.00
Λ/S cm^2 mol^{-1}	150.16	149.87	149.55	149.12	148.78	148.02	147.64	145.47	143.15

7.11. Equation 7.20 is one form of Ostwald's dilution law. Show how it can be linearized (i.e., convert it into a form that will allow experimental values of Λ at various concentrations to be tested by means of a straight-line plot). Explain how $\Lambda°$ and K can be obtained from the plot.

Kraus and Callis, *J. Amer. Chem. Soc., 45,* 2624(1923), obtained the following electrolytic conductivities κ for the dissociation of tetramethyl tin chloride, $(CH_3)_4SnCl$, in ethyl alcohol solution at 25.0 °C and at various concentrations c:

$c/10^{-4}$ mol dm^{-3}	1.566	2.600	6.219	10.441
$\kappa/10^{-6}\ \Omega^{-1}\text{ cm}^{-1}$	1.788	2.418	4.009	5.336

By the use of the linear plot you have devised, determine $\Lambda°$ and K.

7.12. A certain chemical company wishes to dispose of its acetic acid waste into a local river by first diluting it with water to meet the regulation that the total acetic acid concentration cannot exceed 1500 ppm by weight. You are asked to design a system using conductance to continuously monitor the acid concentration in the water and trigger an alarm if the 1500 ppm limit is exceeded. What is the maximum conductance at which the system should trigger an alarm at a constant temperature of 25 °C? (Assume that the cell constant is 1.0 cm^{-1} and that the density of 1500 ppm acetic acid solution is not appreciably different from that of pure water. The $\Lambda°$ for acetic acid is 390.7 S cm^2 mol^{-1} and $K_a = 1.81 \times 10^{-5}$ mol dm^{-3} at 25 °C. Ignore the conductance of water.)

7.13. How far can the conductivity of water at 25 °C be lowered *in theory* by removing impurities? The $\Lambda°$ (in S cm^2 mol^{-1}) for KOH, HCl, and KCl are, respectively, 274.4, 426.04, and 149.86. $K_w = 1.008 \times 10^{-14}$. Compare your answer to the *experimental* value of 5.8×10^{-8} S cm^{-1} obtained by Kohlrausch and Heydweiller, *Z. phys. Chem. 14*, 317(1894).

Debye-Hückel Theory and Transport of Electrolytes

7.14. The radius of the ionic atmosphere ($1/\kappa$) for a univalent electrolyte is 0.964 nm at a concentration of 0.10 *M* in water at 25 °C ($\epsilon = 78$). Estimate the radius (a) in water at a concentration of 0.0001 *M* and (b) in a solvent of $\epsilon = 38$ at a concentration of 0.10 *M*.

7.15. The molar conductivities of 0.001 *M* solutions of potassium chloride, sodium chloride, and potassium sulfate $\left(\frac{1}{2}K_2SO_4\right)$ are 149.9, 126.5, and 153.3 Ω^{-1} cm^2 mol^{-1}, respectively. Calculate an approximate value for the molar conductivity of a solution of sodium sulfate of the same concentration.

7.16. The molar conductivity at 18 °C of a 0.0100 *M* aqueous solution of ammonia is 9.6 Ω^{-1} cm^2 mol^{-1}. For NH_4Cl, $\Lambda° =$ 129.8 Ω^{-1} cm^2 mol^{-1} and the molar ionic conductivities of OH^- and Cl^- are 174.0 and 65.6 Ω^{-1} cm^2 mol^{-1}, respectively. Calculate $\Lambda°$ for NH_3 and the degree of ionization in 0.01 *M* solution.

7.17. A solution of LiCl was electrolyzed in a Hittorf cell. After a current of 0.79 A had been passed for 2 h, the mass of LiCl in the anode compartment had decreased by 0.793 g.

a. Calculate the transport numbers of the Li^+ and Cl^- ions.
b. If $\Lambda°$ (LiCl) is 115.0 Ω^{-1} cm^2 mol^{-1}, what are the molar ionic conductivities and the ionic mobilities?

7.18. A solution of cadmium iodide, CdI_2, having a molality of 7.545×10^{-3} mol kg^{-1}, was electrolyzed in a Hittorf cell. The mass of cadmium deposited at the cathode was 0.03462 g. Solution weighing 152.64 g was withdrawn from the anode compartment and was found to contain 0.3718 g of cadmium iodide. Calculate the transport numbers of Cd^{2+} and I^-.

7.19. The transport numbers for HCl at infinite dilution are estimated to be $t^+ = 0.821$ and $t^- = 0.179$ and the molar conductivity is 426.16 Ω^{-1} cm^2 mol^{-1}. Calculate the mobilities of the hydrogen and chloride ions.

7.20. If a potential gradient of 100 V cm^{-1} is applied to a 0.01 *M* solution of NaCl, what are the speeds of the Na^+ and Cl^- ions? Take the ionic conductivities to be those listed in Table 7.3 on p. 291.

***7.21.** A solution of LiCl at a concentration of 0.01 *M* is contained in a tube having a cross-sectional area of 5 cm^2. Calculate the speeds of the Li^+ and Cl^- ions if a current of 1 A is passed. Use the ion conductivities listed in Table 7.3.

7.22. What is the work required to separate in vacuum two particles, one with the charge of the proton, from another particle with the same charge of opposite sign? Carry out the calculations for an initial distance of (a) 1.0 nm to an infinite distance apart and (b) from 1.0 mm to an infinite distance apart. (c) In (a) how much work would be required if the charge is moved to a distance of 0.1 m? The charge on a proton is 1.6×10^{-19} C.

***7.23.** According to Bjerrum's theory of ion association, the number of ions of type *i* present in a spherical shell of thickness *dr* and distance *r* from a central ion is

$$dN_i = N_i \exp(-z_i z_c e^2/4\pi\epsilon_0 \epsilon r k_B T)\, 4\pi r^2\, dr$$

where z_i and z_c are the charge numbers of the ion of type *i* and of the central ion and e, ϵ_0, ϵ, and k_B have their usual significance. Plot the exponential in this expression and also $4\pi r^2$ against *r* for a uni-univalent electrolyte in water at 25.0 °C ($\epsilon = 78.3$). Allow *r* to have values from 0 to 1 nm. Plot also the product of these functions, which is $(dN_1/N_1)dr$ and is the probability of finding an ion of type *i* at a distance between *r* and $r + dr$ of the central ion.

By differentiation, obtain a value r^* for which the probability is a minimum, and calculate the value for water at 25.0 °C. The electrostatic potential is given to a good approximation by the first term in Eq. 7.47 on p. 280. Obtain an expression, in terms of k_BT, for the electrostatic energy between the two univalent ions at this minimum distance, and evaluate this energy at 25 °C.

Thermodynamics of Ions

7.24. The following are some conventional standard enthalpies of ions in aqueous solution at 25 °C:

Ion	$\Delta_f H°$/kJ mol^{-1}
H^+	0
Na^+	−239.7
Ca^{2+}	−543.1
Zn^{2+}	−152.3
Cl^-	−167.4
Br^-	−120.9

Calculate the enthalpy of formation in aqueous solution of 1 mol of NaCl, $CaCl_2$, and $ZnBr_2$, assuming complete dissociation.

7.25. One estimate for the absolute Gibbs energy of hydration of the H^+ ion in aqueous solution is −1051.4 kJ mol^{-1}. On this basis, calculate the absolute Gibbs energies of hydration of the following ions, whose conventional standard Gibbs energies of hydration are as follows:

Ion	$\Delta_{hyd} G°$ kJ mol^{-1}
H^+	0
Na^+	679.1
Mg^{2+}	274.1
Al^{3+}	−1346.4
Cl^-	−1407.1
Br^-	−1393.3

7.26. Calculate the ionic strengths of 0.1 *M* solutions of KNO_3, K_2SO_4, $ZnSO_4$, $ZnCl_2$, and $K_4Fe(CN)_6$; assume complete dissociation and neglect hydrolysis.

7.27. Calculate the mean activity coefficient $\gamma_\pm$ for the Ba^{2+} and SO_4^{2-} ions in a saturated solution of $BaSO_4$ ($K_{sp} = 9.2 \times 10^{-11}$ mol^2 dm^{-6}) in 0.2 *M* K_2SO_4, assuming the Debye-Hückel limiting law to apply.

7.28. The solubility of AgCl in water at 25 °C is 1.274×10^{-5} mol dm^{-3}. On the assumption that the Debye-Hückel limiting law applies,

a. Calculate $\Delta G°$ for the process $AgCl(s) \rightarrow Ag^+(aq) + Cl^-(aq)$.
b. Calculate the solubility of AgCl in an 0.005 *M* solution of K_2SO_4.

7.29. Employ Eq. 7.114 to make plots of log $\gamma_\pm$ against $\sqrt{I}$ for a uni-univalent electrolyte in water at 25 °C, with B = 0.51 mol^{-1} dm$^{3/2}$ and $B' = 0.33 \times 10^{10}$ mol^{-1} dm$^{3/2}$ m^{-1}, and for the following values of the interionic distance a:

$$a = 0, \quad 0.1, \quad 0.2, \quad 0.4, \quad \text{and} \quad 0.8 \text{ nm}$$

7.30. Estimate the change in Gibbs energy ΔG when 1 mol of K^+ ions (radius 0.133 nm) is transported from aqueous solution ($\epsilon = 78$) to the lipid environment of a cell membrane ($\epsilon = 4$) at 25 °C.

7.31. At 18 °C the electrolytic conductivity of a saturated solution of CaF_2 is 3.86×10^{-5} Ω^{-1} cm^{-1}, and that of pure water is 1.5×10^{-6} Ω^{-1} cm^{-1}. The molar ionic conductivities of $\frac{1}{2}Ca^{2+}$ and F^- are 51.1 Ω^{-1} cm^2 mol^{-1} and 47.0 Ω^{-1} cm^2 mol^{-1}, respectively. Calculate the solubility of CaF_2 in pure water at 18 °C and the solubility product.

7.32. What concentrations of the following have the same ionic strength as 0.1 *M* NaCl?

$$CuSO_4, \quad Ni(NO_3)_2, \quad Al_2(SO_4)_3, \quad Na_3PO_4$$

Assume complete dissociation and neglect hydrolysis.

7.33. The solubility product of PbF_2 at 25.0 °C is 4.0×10^{-9} mol^3 dm^{-9}. Assuming the Debye-Hückel limiting law to apply, calculate the solubility of PbF_2 in (a) pure water and (b) 0.01 *M* NaF.

7.34. Calculate the solubility of silver acetate in water at 25 °C, assuming the DHLL to apply; the solubility product is 4.0×10^{-3} mol^2 dm^{-6}.

***7.35.** Problem 7.30 was concerned with the Gibbs energy change when 1 mol of K^+ ions are transported from water to a lipid. Estimate the electrostatic contribution to the entropy change when this occurs, assuming the dielectric constant of the lipid to be temperature independent, and the following values for water at 25 °C: $\epsilon = 78$, $\partial \ln \epsilon/\partial T = -0.0046$ K^{-1}. Suggest a qualitative explanation for the sign of the value you obtain.

***7.36.** Assuming the Born equation (Eq. 7.86) to apply, make an estimate of the reversible work of charging 1 mol of Na^+Cl^- in aqueous solution at 25 °C ($\epsilon = 78$), under the following conditions:

a. The electrolyte is present at infinite dilution.
b. The electrolyte is present at such a concentration that the mean activity coefficient is 0.70. The ionic radii are 95 pm for Na^+ and 181 pm for Cl^-.

7.37. If the solubility product of barium sulfate is 9.2×10^{-11} mol^2 dm^{-6}, calculate the solubility of $BaSO_4$ in a solution that is 0.10 *M* in $NaNO_3$ and 0.20 *M* in $Zn(NO_3)_2$; assume the DHLL to apply.

7.38. Silver chloride, AgCl, is found to have a solubility of 1.561×10^{-5} *M* in a solution that is 0.01 *M* in K_2SO_4. Assume the DHLL to apply and calculate the solubility in pure water.

7.39. The enthalpy of neutralization of a strong acid by a strong base, corresponding to the process

$$H^+(aq) + OH^-(aq) \rightarrow H_2O$$

is -55.90 kJ mol^{-1}. The enthalpy of neutralization of HCN by NaOH is -12.13 kJ mol^{-1}. Make an estimate of the enthalpy of dissociation of HCN.

7.40. Make use of the Debye-Hückel limiting law to estimate the activity coefficients of the ions in an aqueous 0.004 *M* solution of sodium sulfate at 298 K. Estimate also the mean activity coefficient.

Ionic Equilibria

7.41. A 0.1 *M* solution of sodium palmitate, $C_{15}H_{31}COONa$, is separated from a 0.2 *M* solution of sodium chloride by a membrane that is permeable to Na^+ and Cl^- ions but not to palmitate ions. Calculate the concentrations of Na^+ and Cl^- ions on the two sides of the membrane after equilibrium has become established. (For a calculation of the Nernst potential, see Problem 8.18.)

7.42. Consider the ionizations

$$H^+ + H_3N^+CH_2COO^- \rightleftharpoons H_3N^+CH_2COOH \rightleftharpoons H_2NCH_2COOH + H^+$$

Assume that the following acid dissociation constants apply to the ionizations:

$$-NH_3^+ \rightleftharpoons -NH_2 + H^+; \; K_a = 1.5 \times 10^{-10} \; M$$
$$-COOH \rightleftharpoons -COO^- + H^+; \; K_a = 4.0 \times 10^{-3} \; M$$

Estimate a value for the equilibrium constant for the process

$$H_3N^+CH_3COO^- \rightleftharpoons H_2NCH_2COOH$$

7.43. The p*K* values for the successive ionizations of phosphoric acid are given on p. 308. Which of the four species is predominant at the following values of the hydrogen or hydroxide concentration?

a. $[H^+] = 0.1$ *M*.
b. $[H^+] = 2 \times 10^{-3}$ *M*.
c. $[H^+] = 5 \times 10^{-5}$ *M*.
d. $[OH^-] = 2 \times 10^{-3}$ *M*.
e. $[OH^-] = 1$ *M*.

7.44. Two solutions of equal volume are separated by a membrane which is permeable to K^+ and Cl^- ions but not to P^- ions. The initial concentrations are as shown below.

$[K^+] = 0.05\ M$	$[K^+] = 0.15\ M$
$[Cl^-] = 0.05\ M$	$[P^-] = 0.15\ M$

Calculate the concentrations on each side of the membrane after equilibrium has become established. (See Problem 8.26 in Chapter 8 for the calculation of the Nernst potential for this system.)

Essay Questions

7.45. State Faraday's two laws of electrolysis and discuss their significance in connection with the electrical nature of matter.

7.46. Discuss the main ideas that lie behind the Debye-Hückel theory, as applied to the conductivities of solutions of strong electrolytes.

7.47. Outline two important methods for determining transport numbers of ions.

7.48. Explain why Li^+ has a lower ionic conductivity than Na^+ and why the value for H^+ is so much higher than the values for both of these ions.

7.49. Describe briefly the type of hydration found with the following ions in aqueous solution: Li^+, Br^-, H^+, OH^-.

7.50. What modifications to the Debye-Hückel limiting law are required to explain the influence of ionic strength on solubilities?

SUGGESTED READING

For general accounts of electrochemistry, see

J. M. G. Barthal, K. Krienke, and W. Kunz, *Physical Chemistry of Electrolyte Solutions: Modern Aspects,* New York: Springer-Verlag, 1998.

J. O'M. Bockris and A. K. N. Reddy, *Modern Electrochemistry,* 2 volumes, New York: Plenum Press, 1973. Volume 1 deals with ionic solutions, Volume 2 with the kinetics of electrode processes.

C. M. A. Brett and A. M. O. Brett, *Electrochemistry: Principles, Methods, and Applications,* Oxford: University Press, 1993.

S. Glasstone, *An Introduction to Electrochemistry,* New York: Van Nostrand Co., 1942. This old book, long out of print but still to be found in libraries, gives an extremely clear account of the subject.

G. Kortüm, *Treatise on Electrochemistry,* Amsterdam: Elsevier Publishing Corporation, 1965.

J. T. Stock and M. V. Orna (Eds.), *Electrochemistry, Past and Present,* American Chemical Society, Washington, D.C., 1989.

The following publications deal with special aspects of electrolyte solutions.

J. N. Butler, *Ionic Equilibrium: A Mathematical Approach,* Reading, MA: Addison-Wesley, 1964.

B. E. Conway, *Ionic Hydration in Chemistry and Biophysics,* Amsterdam: Elsevier North Holland, 1980.

B. E. Conway and R. G. Barradas, (Eds.), *Chemical Physics of Ionic Solutions,* New York: Wiley, 1966.

C. W. Davies, *Ion Associations,* London: Butterworth, 1962.

D. Eisenberg and W. J. Kauzmann, *The Structure and Properties of Water,* Oxford: Clarendon Press, 1969.

R. M. Fuoss and F. Accascina, *Electrolytic Conductance,* New York: Interscience, 1959.

H. S. Harned and B. B. Owen, *The Physical Chemistry of Electrolytic Solutions,* New York: Reinhold, 1950 (3rd ed., 1958).

H. S. Harned and R. A. Robinson, *Multicomponent Electrolyte Solutions,* Oxford and New York: Pergamon Press, 1968.

J. L. Kavanau, *Water and Solute-Water Interactions,* San Francisco: Holden-Day, 1964.

L. Onsager, "The Motion of Ions: Principles and Concepts," *Science, 166,* 1359(1969).

J. Robbins, *Ions in Solution,* Oxford: Clarendon Press, 1972.

R. A. Robinson and R. H. Stokes, *Electrolyte Solutions* (2nd ed.), London: Butterworth, 1959.

Electrochemical Cells

PREVIEW

This chapter is concerned with galvanic cells, in which a chemical reaction produces an electric potential difference between two electrodes. An example is the Daniell cell, in which zinc and copper electrodes are immersed in solutions of Zn^{2+} and Cu^{2+} ions separated by a porous membrane, which prevents bulk mixing. The chemical reactions occurring at the two electrodes cause a flow of electrons in the outer circuit.

It is impossible to measure the electromotive force (emf) of a single electrode, and the convention is to use a standard hydrogen electrode as the left-hand electrode in a cell. With another standard electrode on the right-hand side, the cell emf is then taken to be the standard electrode potential for the right-hand electrode. Such standard electrode potentials are useful for predicting the direction of a chemical reaction. The thermodynamic relationship between the Gibbs energy change and the emf of a galvanic cell leads to the determination of equilibrium constants. An equation due to Nernst relates the cell emf to the activities of solutions in the cell. The **Nernst potential** is the potential difference established across a membrane when two different solutions are on opposite sides of it.

An important type of cell is a redox cell, in which oxidation-reduction processes occur in a special way. It is often convenient to relate the emfs of such cells to a standard pH value, usually taken to be 7.

Practical applications of emf measurements include the determination of pH, activity coefficients, equilibrium constants, solubility products, and transport numbers. The technique of polarography has become an important analytical tool. Much valuable information has also been derived from studies of the kinetics of electrode processes, particularly with reference to the phenomenon of **overvoltage.**

Certain electrochemical cells are useful devices for the generation of electric power. In a fuel cell, for example, the reacting substances are continuously fed into the system. Photogalvanic cells are electrochemical cells in which radiation induces a chemical process that gives rise to an electric current. Among batteries discussed, lithium ion batteries give a high energy per unit density. Consequently, they will be used in hybrid automobiles and other applications requiring high energy drain and light weight.

OBJECTIVES

After studying this chapter, the student should be able to:

- Explain how electrochemical cells, such as the Daniell cell, function.
- Understand the concept of the standard hydrogen electrode and describe other standard electrodes.
- Apply the principles of thermodynamics to electrochemical cells, including the derivation of the Nernst equation.
- Obtain activity coefficients and equilibrium constants from emf measurements.
- Describe fuel cells and photogalvanic cells.
- Understand the use of the terms *battery* and *cells*.
- Describe the different batteries discussed, especially the chemistry of the lithium ion battery.

Anode and cathode: Review REDOX reactions and ionic bonds.

An *electrochemical cell,* also called a *voltaic cell* or a *galvanic cell,* is a device in which a chemical reaction occurs with the production of an electric potential difference between two electrodes. If the electrodes are connected to an external circuit there results a flow of current, which can lead to the performance of mechanical work so that the electrochemical cell transforms chemical energy into work. Besides being of considerable practical importance, electrochemical cells are valuable laboratory instruments, because they provide some extremely useful scientific data. For example, they lead to thermodynamic quantities such as enthalpies and Gibbs energies, and they allow us to determine transport numbers and activity coefficients for ions in solution. This chapter deals with the general principles of electrochemical cells and with some of their more important applications.

8.1 The Daniell Cell

The cell originally developed by the English chemist John Frederic Daniell (1790–1845) consisted of a zinc electrode immersed in dilute sulfuric acid and a copper electrode immersed in a cupric sulfate solution. It was later found that the cell gave a more stable voltage if the sulfuric acid was replaced by zinc sulfate solution, and the expression "Daniell cell" today usually refers to such an arrangement, which is shown schematically in Figure 8.1. The voltage produced by the cell depends on the activities of the Zn^{2+} and Cu^{2+} ions in the two solutions. If the molalities of the two solutions are 1 mol kg^{-1} (1 *m*), the cell is called a **standard Daniell cell.**

Anode and cathode examples; Electrolytes.

When a Daniell cell is set up, there is a flow of electrons from the zinc to the copper electrode in the outer circuit. This means that positive current is moving from left to right in the cell itself. By convention, a **potential difference corresponding to an external flow of electrons from the left-hand electrode to the right-hand electrode is said to be a positive potential difference.**

The processes that occur when this cell operates are shown in Figure 8.1. Since positive electricity in the form of positive ions moves from left to right within the cell, zinc metal dissolves to form Zn^{2+} ions,

$$Zn \rightarrow Zn^{2+} + 2e^-$$

Some of these zinc ions pass through the membrane into the right-hand solution, and at the right-hand electrode cupric ions interact with electrons to form metallic copper:

$$Cu^{2+} + 2e^- \rightarrow Cu$$

Every time a zinc atom dissolves and a copper atom is deposited, two electrons travel round the outer circuit.

Cells of this kind can be made to behave in a reversible fashion, by balancing their potentials by an external potential so that no current flows. This can be done by means of a potentiometer, the principle of which is illustrated in Figure 8.2a. A current is passed through a uniform slide wire *AB*, along which there is a linear potential drop, the magnitude of which can be adjusted by a rheostat. The cell under investigation is connected through a tap-key switch and a galvanometer to one end (*A*) of the slide wire and to a movable contact *C*. The contact is moved along the slide wire until, when the switch is depressed, no current passes through the galvanometer. The potential difference produced by the cell is then exactly balanced

REDOX reactions: Zinc reduces copper; Copper oxidizes zinc.

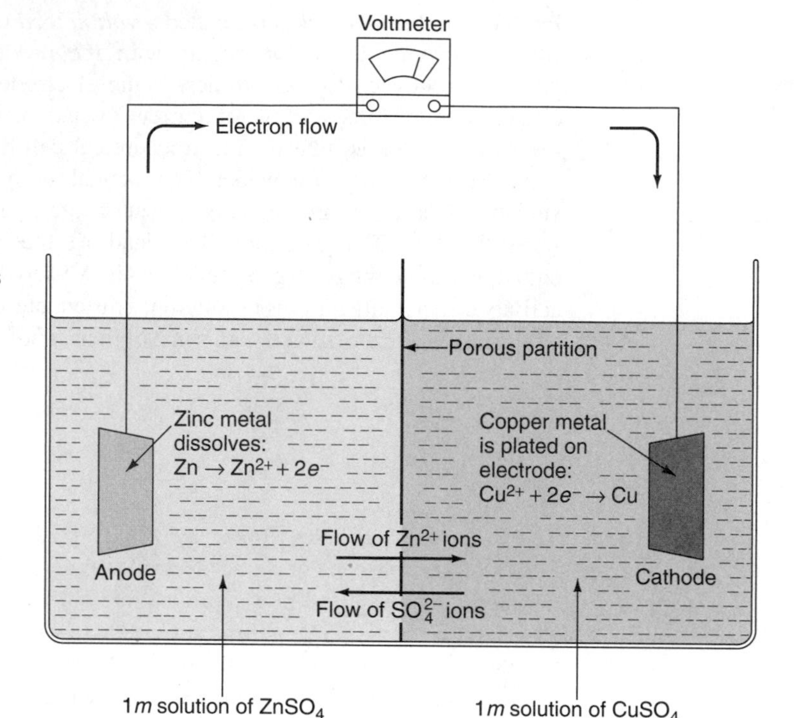

FIGURE 8.1
The standard Daniell cell.

Anode and cathode: See cations migrate to the cathode and anions migrate to the anode.

by the potential difference between points A and C. The potentiometer wire can be calibrated by use of a standard cell, such as the Weston cell (Figure 8.2b).

When the electric potential of a cell is exactly balanced in this way, the cell is operating reversibly and its potential is then referred to as the *electromotive force* (*emf*) of the cell. If the counter-potential in the slide wire is slightly less than the emf of the cell, there is a small flow of electrons from left to right in the external circuit. If the counter-potential is adjusted to be slightly greater than the cell emf, the cell is forced to operate in reverse; zinc is deposited at the left-hand electrode,

$$Zn^{2+} + 2e^- \rightarrow Zn$$

and copper dissolves at the right,

$$Cu \rightarrow Cu^{2+} + 2e^-$$

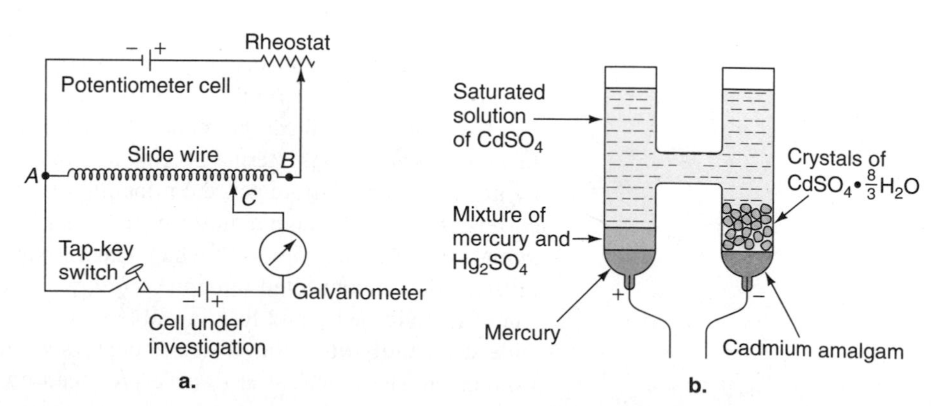

FIGURE 8.2
(a) A potentiometer circuit used for determining the reversible emf of a cell. (b) The Weston standard cell, which is the most widely used standard cell. At 25 °C its emf is 1.018 32 V and it has a very small temperature coefficient (see Problem 8.33).

The fact that the electrons normally flow from the zinc to the copper electrode indicates that the tendency for $Zn \rightarrow Zn^{2+} + 2e^-$ to occur is greater than for the reaction $Cu \rightarrow Cu^{2+} + 2e^-$, which is forced to occur in the reverse direction. The magnitude of the emf developed is a measure of the relative tendencies of the two processes. The emf varies with the activities of the Zn^{2+} and Cu^{2+} ions in the two solutions. Thus the tendency for $Zn \rightarrow Zn^{2+} + 2e^-$ to occur is smaller when the concentration of Zn^{2+} is large, while the tendency for $2e^- + Cu^{2+} \rightarrow Cu$ to occur increases when the concentration of Cu^{2+} is increased. The relationship between emf and the activities will be considered in Section 8.5.

8.2 Standard Electrode Potentials

It would be very convenient if we could measure the potential of a single electrode, such as the right-hand electrode in Figure 8.1, which we will write as

$$Cu^{2+}|Cu$$

The potential of such an electrode would be a measure of the tendency of the process

$$Cu^{2+} + 2e^- \rightarrow Cu$$

to occur. However, there is no way to measure the emf of a single electrode, since in order to obtain an emf there must be two electrodes, with an emf associated with each one. The procedure used is to choose one electrode as a standard and to measure emf values of other electrodes with reference to that standard.

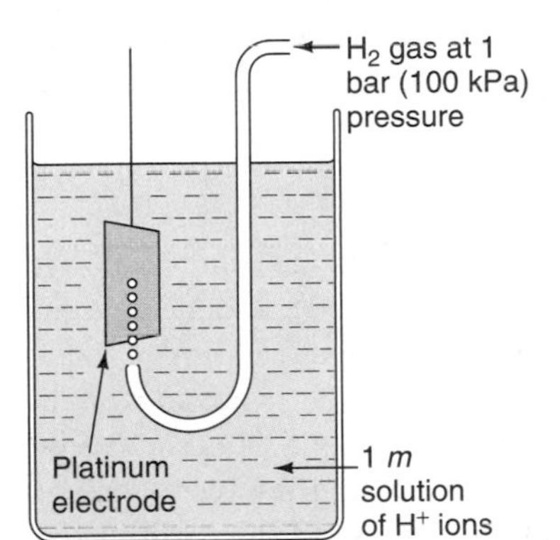

FIGURE 8.3
The standard hydrogen electrode.

The Standard Hydrogen Electrode

The electrode chosen as the ultimate standard is the standard hydrogen electrode, which is illustrated in Figure 8.3. It consists of a platinum electrode immersed in a 1 *m* solution of hydrogen ions maintained at 25 °C and 1 bar pressure. Hydrogen gas is bubbled over the electrode and passes into solution, forming hydrogen ions and electrons:

$$H_2 \rightarrow 2H^+ + 2e^-$$

The emf corresponding to this electrode is arbitrarily assigned the value of zero, and this electrode is used as a standard for other electrodes.

Standard Electrode Potentials

There are two conventions in common use, and the student should be aware of both methods of procedure. The standard hydrogen electrode can be either the left-hand or the right-hand electrode. In the convention adopted by the International Union of Pure and Applied Chemistry (IUPAC) the hydrogen electrode is placed on the left-hand side, and the emf of the other electrode is taken to be that of the cell. Such emf values, under standard conditions, are known as **standard electrode potentials** or *standard reduction potentials* and are given the symbol E°. Alternatively, the standard hydrogen electrode may be placed on the right-hand side; the potential so obtained is known as the *standard oxidation potential.* The latter potentials are the standard electrode potentials with the signs reversed; the only difference is that the cells have been turned around.

To illustrate the standard electrode (reduction) potentials (IUPAC convention), consider the cell shown in Figure 8.4. The left-hand electrode is the standard

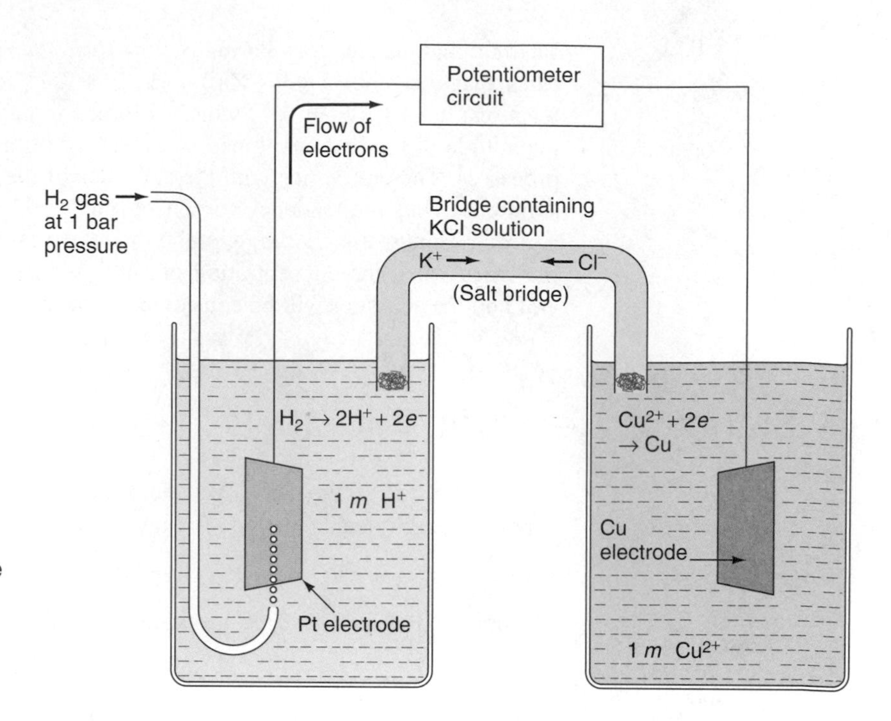

FIGURE 8.4
A voltaic cell in which a standard hydrogen electrode has been combined with a copper electrode immersed in a 1 *m* solution of cupric ions; the two solutions are connected by a potassium chloride bridge.

A single vertical line represents a phase separation.
Double vertical dashed lines represent a salt bridge.

REDOX reactions: Galvanic cell; Watch the electrons flow through the external circuit and anions and cations pass through the salt bridge.

hydrogen electrode, with a hydrogen gas pressure of 1 bar, and the acid solution is 1 *m* in H^+ ions. The right-hand electrode is the $Cu^{2+}|Cu$ electrode, and the concentration of Cu^{2+} ions is 1 *m*. The two solutions are connected by a "salt bridge" such as a potassium chloride solution, which conducts electricity but does not allow bulk mixing of the two solutions. Alternatively, an agar gel containing KCl is commonly used. This procedure is somewhat more reliable than separating the two solutions by a porous partition, which itself sets up a small emf; the salt bridge minimizes this effect.

The voltaic cell shown in Figure 8.4 can be represented as follows:[1]

$$\text{Pt}, H_2(1\ \text{bar})|H^+(1\ m) \;\vdots\vdots\; Cu^{2+}(1\ m)|Cu$$

where the double vertical dashed lines represent the salt bridge. The observed emf is +0.34 V; the sign is positive by convention since electrons flow from left to right in the outer circuit. There is therefore a greater tendency for the process

$$Cu^{2+} + 2e^- \rightarrow Cu$$

to occur than for

$$2H^+ + 2e^- \rightarrow H_2$$

to occur; the latter process is forced to go in the reverse direction. By the IUPAC convention, the $Cu^{2+}|Cu$ electrode is on the right, and the standard electrode poten-

[1]When a gas is a reactant in an electrochemical cell, its pressure must be known. Throughout this discussion, the gas pressure, if not stated, is assumed to be 1 bar. Values of the emf in many tables relate to 1 atm pressure. The differences in values are generally within the margins of error of the measurement.

REDOX agents: Distinguish between what is reduced and what is oxidized.

tial $E°$ of this electrode is $+0.34$ V. This is a measure of the tendency for the cupric ions to be reduced by the process

$$Cu^{2+} + 2e^{-} \rightarrow Cu$$

It is for this reason that these electrode potentials are also known as *standard reduction potentials.*

In this book, standard electrode (reduction) potentials will always be employed, as recommended by the IUPAC. Table 8.1 lists such potentials; the reactions are written as reduction processes (e.g., $Cu^{2+} + 2e^{-} \rightarrow Cu$). A table of standard oxidation potentials, on the other hand, would have all the signs reversed and the corresponding reactions would be oxidations; for example,

$$Cu \rightarrow Cu^{2+} + 2e^{-}$$

REDOX reactions: Electrolytic cells; Compare galvanic and electrolytic cells.

By combining the standard electrode potentials for two electrodes we can deduce the emf of a cell involving the two electrodes, neglecting the hydrogen electrode. Consider, for example, the following items in Table 8.1:

$$Cu^{2+} + 2e^{-} \rightarrow Cu \quad E° = 0.34 \text{ V}$$
$$Zn^{2+} + 2e^{-} \rightarrow Zn \quad E° = -0.76 \text{ V}$$

These values are the emf values for the following cells, in which the electrode processes are shown:

$$\text{Pt, } H_2 \text{ (1 bar)} |H^{+}(1\ m) \,\vdots\vdots\, Cu^{2+}(1\ m)|Cu \qquad E° = 0.34 \text{ V}$$
$$H_2 \rightarrow 2H^{+} + 2e^{-} \qquad Cu^{2+} + 2e^{-} \rightarrow Cu$$
$$\text{Pt, } H_2 \text{ (1 bar)} |H^{+}(1\ m) \,\vdots\vdots\, Zn^{2+}(1\ m)|Zn \qquad E° = -0.76 \text{ V}$$
$$H_2 \rightarrow 2H^{+} + 2e^{-} \qquad Zn^{2+} + 2e^{-} \rightarrow Zn$$

Standard reduction potentials: Example; Learn the strongest to weakest oxidizing and reducing agents.

We could connect the two cells together as follows:

$$Zn|Zn^{2+}(1\ m) \,\vdots\vdots\, H^{+}(1\ m)|H_2 \text{ (1 bar), Pt—Pt, } H_2 \text{ (1 bar)} |H^{+}(1\ m) \,\vdots\vdots\, Cu^{2+}(1\ m)|Cu$$

and the emf would then be $E° = 0.34 - (-0.76) = 1.10$ V (i.e., the right-hand electrode potential minus the left-hand potential). We could also eliminate the hydrogen electrodes altogether and set up the cell

$$Zn|Zn^{2+}(1\ m) \,\vdots\vdots\, Cu^{2+}(1\ m)|Cu$$

The emf of this would be the same, 1.10 V, since we have merely eliminated two identical hydrogen electrodes working in opposition to each other. This cell is, of course, just the standard Daniell cell (see Figure 8.1).

The potential for a half-reaction is an intensive property.

Note that in writing the individual cell reactions it makes no difference whether they are written with one or with more electrons. Thus the hydrogen electrode reaction can be written as either

$$2H^{+} + 2e^{-} \rightarrow H_2 \quad \text{or} \quad H^{+} + e^{-} \rightarrow \tfrac{1}{2}H_2$$

Standard reduction potentials: Reactivity and stoichiometry; Take quiz on standard reduction potentials.

However, in considering the overall process we must obviously balance out the electrons. Thus, for the cell

$$\text{Pt, } H_2|H^{+} \,\vdots\vdots\, Cu^{2+}|Cu$$

the individual reactions can be written as

$$H_2 \rightarrow 2H^{+} + 2e^{-} \quad \text{and} \quad 2e^{-} + Cu^{2+} \rightarrow Cu$$

Oxidation states; Summarize the oxidation states of elements by use of the periodic table.

Oxidation states; Examples; Test your ability to assign oxidation states.

Standard reduction potential; Cell EMF; Confirm what is oxidized and what is reduced.

TABLE 8.1 Standard Electrode (Reduction) Potentials, T = 25 °C

Half-Reaction	Standard Electrode Potential, $E°$/V
$F_2 + 2e^- \rightarrow 2F^-$	2.866
$H_2O_2 + 2H^+ + 2e^- \rightarrow 2H_2O$	1.776
$Ce^{4+} + e^- \rightarrow Ce^{3+}$	1.72
$Au^+ + e^- \rightarrow Au$	1.692
$MnO_4^- + 8H^+ + 5e^- \rightarrow Mn^{2+} + 4H_2O$	1.52
$Ce^{4+} + e^- \rightarrow Ce^{3+}$	1.72
$Cl_2 + 2e^- \rightarrow 2Cl^-$	1.35827
$Cr_2O_7^{2-} + 14H^+ + 6e^- \rightarrow 2Cr^{3+} + 7H_2O$	1.232
$MnO_2 + 4H^+ + 2e^- \rightarrow Mn^{2+} + 2H_2O$	1.224
$O_2 + 4H^+ + 4e^- \rightarrow 2H_2O$	1.229
$Pt^{2+} + 2e^- \rightarrow Pt$	1.18
$Br_2 + 2e^- \rightarrow 2Br^-$	1.0873
$Hg^{2+} + 2e^- \rightarrow Hg$	0.851
$Ag^+ + e^- \rightarrow Ag$	0.7996
$Hg_2^{2+} + 2e^- \rightarrow 2Hg$	0.7973
$Fe^{3+} + e^- \rightarrow Fe^{2+}$	0.771
$O_2 + 2H^+ + 2e^- \rightarrow H_2O_2$	0.695
$I_2 + 2e^- \rightarrow 2I^-$	0.5355
$Cu^{2+} + 2e^- \rightarrow Cu$	0.3419
$Hg_2Cl_2 + 2e^- \rightarrow 2Hg + 2Cl^-$, 0.1 m HCl	0.3337
$Hg_2Cl_2 + 2e^- \rightarrow 2Hg + 2Cl^-$, saturated KCl	0.2412
$AgCl(s) + e^- \rightarrow Ag + Cl^-$	0.22233
$Cu^{2+} + e^- \rightarrow Cu^+$	0.153
$Sn^{4+} + 2e^- \rightarrow Sn^{2+}$	0.151
$HgO + H_2O + 2e^- \rightarrow Hg + 2OH^-$	0.0977
$2H^+ + 2e^- \rightarrow H_2$	0.00 (by definition)
$Pb^{2+} + 2e^- \rightarrow Pb$	−0.1262
$Sn^{2+} + 2e^- \rightarrow Sn$	−0.1375
$Ni^{2+} + 2e^- \rightarrow Ni$	−0.257
$PbCl_2 + 2e^- \rightarrow Pb + 2Cl^-$	−0.2675
$Co^{2+} + 2e^- \rightarrow Co$	−0.280
$Fe^{2+} + 2e^- \rightarrow Fe$	−0.447
$Cr^{3+} + 3e^- \rightarrow Cr$	−0.744
$Zn^{2+} + 2e^- \rightarrow Zn$	−0.7618
$Al^{3+} + 3e^- \rightarrow Al$	−1.662
$Mg^{2+} + 2e^- \rightarrow Mg$	−2.372
$Na^+ + e^- \rightarrow Na$	−2.71
$Ca^{2+} + 2e^- \rightarrow Ca$	−2.868
$K^+ + e^- \rightarrow K$	−2.931
$Li^+ + e^- \rightarrow Li$	−3.0401

The standard oxidation potentials are the negatives of the values given here: The reactions are written in the opposite direction.

Some of the values in this table were determined indirectly from other experimental results, since hypothetical electrodes (e.g., $Li|Li^+$) are impossible to set up.

The values are referenced to 1 bar pressure. See footnote on p. 320

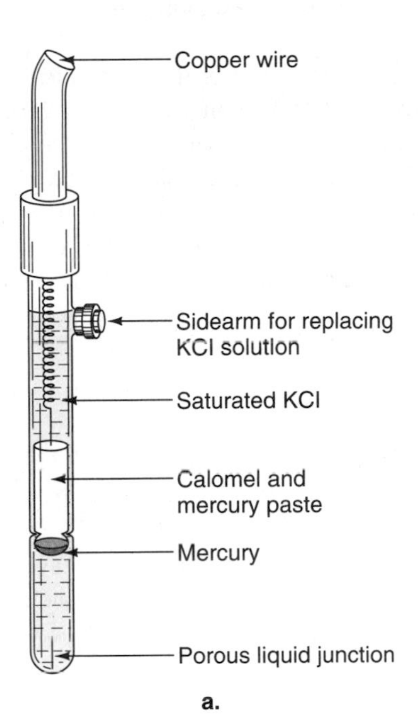

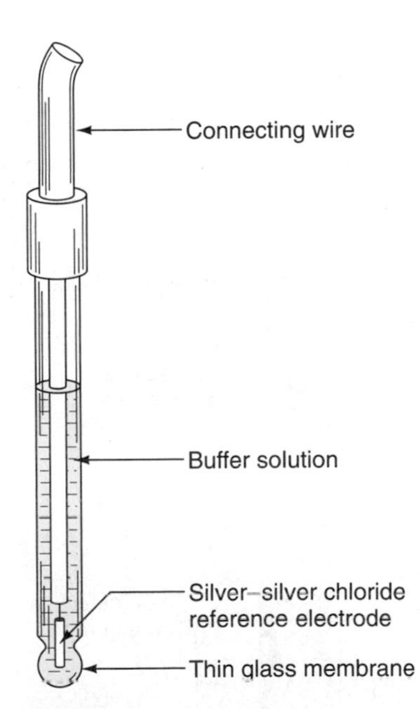

FIGURE 8.5
The calomel electrode (a), and the glass electrode (b). The pH meter, commonly used in chemical and biological laboratories, often employs a glass electrode that is immersed in the unknown solution and is used with a reference calomel electrode.

so that the overall process is

$$H_2 + Cu^{2+} \rightarrow Cu + 2H^+$$

This process is accompanied by the passage of *two* electrons around the outer circuit. We could equally well write the reactions as

$$\frac{1}{2}H_2 \rightarrow H^+ + e^-$$
$$\frac{1}{2}Cu^{2+} + e^- \rightarrow \frac{1}{2}Cu$$
$$\overline{\frac{1}{2}H_2 + \frac{1}{2}Cu^{2+} \rightarrow \frac{1}{2}Cu + H^+}$$

This tells us that every time 0.5 mol of Cu^{2+} disappears and 0.5 mol of Cu appears, 1 mol of electrons passes from the left-hand electrode to the right-hand electrode.

Other Standard Electrodes

The standard hydrogen electrode is not the most convenient electrode because of the necessity of bubbling hydrogen over the platinum electrode. Several other electrodes are commonly used as secondary standard electrodes. One of these is the standard silver–silver chloride electrode, in which a silver electrode is in contact with solid silver chloride, which is a highly insoluble salt. The whole is immersed in potassium chloride solution in which the chloride-ion concentration is 1 *m*. This electrode can be represented as

$$Ag|AgCl(s)|Cl^-(1\ m)$$

We can set up a cell involving this electrode and the hydrogen electrode,

$$Pt, H_2\ (1\ bar)\ |H^+(1\ m) \ \vdots\vdots\ Cl^-(1\ m)|AgCl(s)|Ag$$

with a salt bridge connecting the two solutions. The emf at 25 °C is found to be 0.22233 V. The individual reactions are

$$\frac{1}{2}H_2 \rightarrow H^+ + e^-$$
$$e^- + AgCl \rightarrow Ag + Cl^-$$

and the overall process is

$$\frac{1}{2}H_2 + AgCl \rightarrow H^+ + Cl^- + Ag$$

The standard electrode potential for the silver–silver chloride electrode is thus 0.22233 V.

Another commonly used electrode is the **calomel electrode,** illustrated in Figure 8.5a. In this, mercury is in contact with mercurous chloride (calomel, Hg_2Cl_2) immersed either in a 0.1 *m* solution of potassium chloride or in a saturated solution of potassium chloride. If the cell

$$Pt, H_2(1\ bar)|H^+(1\ m) \ \vdots\vdots\ Cl^-(0.1\ m)|Hg_2Cl_2(s)|Hg$$

is set up, the individual reactions are

$$\frac{1}{2}H_2 \rightarrow H^+ + e^-$$
$$e^- + \frac{1}{2}Hg_2Cl_2 \rightarrow Hg + Cl^-$$

and the overall process is

$$\frac{1}{2}H_2 + \frac{1}{2}Hg_2Cl_2 \rightarrow H^+ + Cl^- + Hg$$

The emf at 25 °C is 0.3337 V, which is the standard electrode potential $E°$. If a saturated solution of KCl is used with the calomel electrode, the standard electrode potential is 0.2412 V.

Another electrode commonly used as a secondary standard is the **glass electrode,** illustrated in Figure 8.5b. In its simplest form this consists of a tube terminating in a thin-walled glass bulb; the glass is reasonably permeable to hydrogen ions. The glass bulb contains a 0.1 *m* hydrochloric acid solution and a tiny silver–silver chloride electrode. The theory of the glass electrode is somewhat complicated, but when the bulb is inserted into an acid solution, it behaves like a hydrogen electrode. This electrode is particularly convenient for making pH determinations.

Ion-Selective Electrodes

The glass electrode was devised in 1906 by the German biologist M. Cremer and was the prototype of a considerable number of membrane electrodes that have been developed subsequently. The importance of membrane electrodes is that some of them are highly selective to particular ions. For example, membrane electrodes have been constructed that are 10^4 times as responsive to Na^+ as to K^+ ions, and such electrodes are of great value for chemical analysis. Electrodes that are selective for more than 50 ions are now available, and most of them are membrane electrodes.

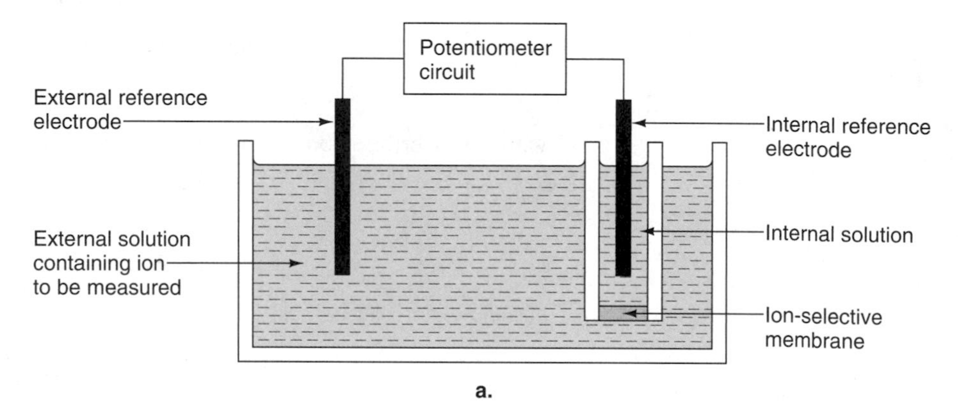

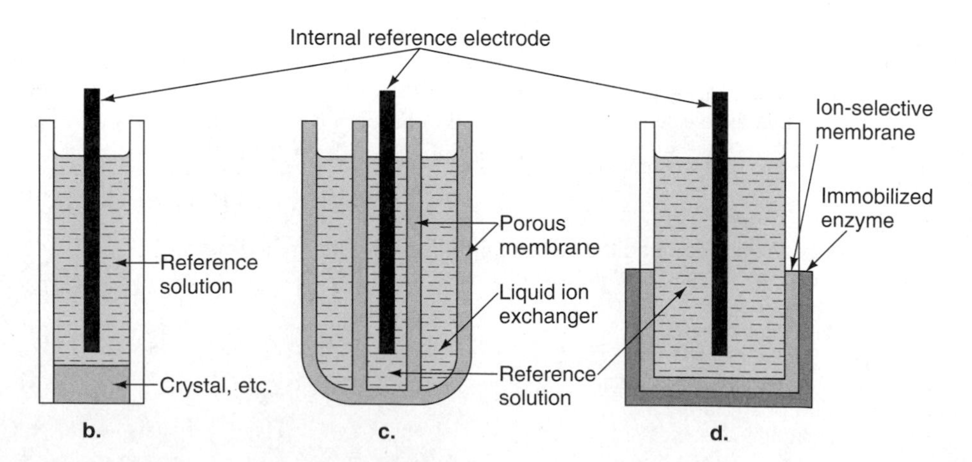

FIGURE 8.6
(a) The principle of the membrane electrode. (b) A membrane electrode in which the membrane is a single crystal, or a mixed crystal, or a matrix impregnated with a precipitate. (c) A membrane electrode in which the membrane is a liquid ion exchanger. (d) An enzyme-substrate electrode, which makes use of the ability of an enzyme to react selectively with an organic ion.

The principle of the membrane electrode is illustrated in Figure 8.6a. The sample solution is separated from an internal solution by an ion-selective membrane, and an internal reference electrode is placed within the internal solution. An external reference electrode, such as a silver–silver chloride electrode, is also immersed in the sample solution, and a measurement is made of the reversible emf of the assembly. The glass electrode is illustrated in Figure 8.5b, and some other assemblies are shown in Figure 8.6b–d.

The theory of membrane electrodes is quite complicated and is different in detail for each type of electrode. It is not necessary for the ion to which the electrode is sensitive actually to be transported through the membrane. What occurs at the membrane is a combination of an ion-exchange process at the solution-membrane interface and the movement of various cations at the interface. It is not necessary for the ion to be measured to be especially mobile or for the same ion to be present in the membrane. A complete theoretical treatment of a membrane electrode requires a consideration of the Donnan equilibrium that is established (Section 7.13), of the Nernst potential (Section 8.3), and of complications arising from deviations from equilibrium. Since so many factors are involved, the development of ion-selective electrodes is necessarily done on the basis of a good deal of empiricism.

8.3 Thermodynamics of Electrochemical Cells

During the last century studies were made of the relationship between the emf of a cell and the thermodynamics of the chemical reactions occurring in the cell. These studies made important contributions to the understanding of the basic principles of thermodynamics. An early contribution was made by Joule who, with very simple apparatus but with accurate temperature and current measurements, found in 1840 that

> The calorific effects of equal quantities of transmitted electricity are proportional to the resistance opposed to its passage, whatever may be the length, thickness, shape, or kind of metal which closes the circuit; and also that, *caeteris paribus,* these effects are in the duplicate ratio of the quantities of transmitted electricity, and, consequently, also in the duplicate ratio of the velocity of transmission.

By "quantity of transmitted electricity," Joule meant the current; by "duplicate ratio," the square. His conclusion was therefore that the heat produced was proportional to the square of the current I^2 and to the resistance R. Since it is also proportional to the time t, Joule had shown that the heat is proportional to

$$I^2Rt$$

Since the resistance R is the potential drop V divided by the current I (Ohm's law), it follows that the heat is proportional to

$$IVt$$

These conclusions have been confirmed by many later investigations.

The SI unit of heat is the joule, that of current is the ampere, and that of electric potential the volt; in these units the proportionality factor relating heat to IVt is unity:

$$q = IVt \tag{8.1}$$

This is readily seen by expressing the joule and the volt in terms of the base SI units (see Appendix A, Table A-1); thus

$$J \equiv kg\ m^2\ s^{-2}$$
$$V \equiv kg\ m^2\ s^{-3}\ A^{-1}$$

The product IVt is

$$A \times kg\ m^2\ s^{-3}\ A^{-1} \times s \equiv kg\ m^2\ s^{-2} \equiv J$$

Joule's conclusion that the heat generated in a wire is IVt is quite correct, but later he and others went wrong. In 1852 he concluded that there is a correspondence between the heat of reaction of a cell and the electrical work. This error was also made by Helmholtz and by William Thomson (later Lord Kelvin). Thomson's conclusion appeared to be supported by his calculation of the emf of the Daniell cell from the heat of the reaction; his value, 1.074 V, is practically the measured value, but this agreement is accidental.[2]

It remained for Willard Gibbs to draw the correct conclusion, in 1878, that the work done in an electrochemical cell is equal to the decrease in what is now known as the Gibbs energy. This is an example of the deduction we have already made, in Section 3.7, that non-*PV* work (i.e., available work) is equal to the decrease in Gibbs energy.

This may be illustrated for the standard cell

$$Pt, H_2\ (1\ bar)\ |H^+(1\ m)\ \vdots\vdots\ Cu^{2+}(1\ m)|Cu$$

for which the emf ($E°$) is 0.3419 V. The overall reaction is

$$H_2 + Cu^{2+} \rightarrow 2H^+ + Cu$$

Every time 1 mol of H_2 reacts with 1 mol of Cu^{2+}, 2 mol of electrons pass through the outer circuit. According to Faraday's laws, this means the transfer of 2 × 96 485 C of electricity. The emf developed is +0.3419 V, and the passage of 2 × 96 485 C across this potential drop means that

$$2 \times 96\,485 \times 0.3419\ C\ V = 6.598 \times 10^4\ J$$

of work has been done by the system. Thus, for this cell process,

$$\Delta G° = -6.598 \times 10^4\ J$$

In general, for any standard-cell reaction associated with the passage of z electrons and an emf of $E°$, the change in Gibbs energy is

Gibbs Energy Change in a Cell

$$\Delta G° = -zFE° \tag{8.2}$$

Since this Gibbs energy change is calculated from the $E°$ value, which relates to a cell in which the molalities are unity, it is a *standard* Gibbs energy change, as indicated by the superscript °.

The same argument applies to any cell; if the emf is E, the Gibbs energy change is

$$\Delta G = -zFE \tag{8.3}$$

[2]It results from the fact that there is only a small entropy change in the Daniell cell ($\partial E/\partial T = 0$, see Eq. 8.23), so that $\Delta H \approx \Delta G$.

The ΔG is the change in Gibbs energy when the reaction occurs with the concentrations having the values employed in the cell. Note that if E is positive, ΔG is negative; a positive E means that the cell is operating spontaneously with the reactions occurring in the forward direction (e.g., $H_2 + Cu^{2+} \rightarrow 2H^+ + Cu$), and this requires ΔG to be negative.

For any reaction

$$aA + bB + \cdots \rightarrow \cdots + yY + zZ$$

the Gibbs energy change that occurs when a mol of A at a concentration [A] reacts with b mol of B at [B], etc., is given by[3]

$$\Delta G = -RT\left[\ln K^\circ - \ln\left(\frac{\cdots [Y]^y[Z]^z}{[A]^a[B]^b \cdots}\right)^u\right] \tag{8.4}$$

If the initial and final concentrations are unity, ΔG is ΔG° and is given by

$$\Delta G^\circ = -RT \ln K^\circ \tag{8.5}$$

For any cell involving standard electrodes, such as the standard Daniell cell

$$Zn|Zn^{2+}(1\ m) ⁞⁞ Cu^{2+}(1\ m)|Cu$$

Eq. 8.2 applies and therefore

$$E^\circ = \frac{RT}{zF} \ln K^\circ \tag{8.6}$$

At 25 °C this becomes

$$E^\circ/V = \frac{0.0257}{z} \ln K^\circ \tag{8.7}$$

These equations provide a very important method for calculating Gibbs energy changes and equilibrium constants. The extension of the method to the calculation of ΔH° and ΔS° values is considered on p. 335.

EXAMPLE 8.1 Calculate the equilibrium constant at 25 °C for the reaction occurring in the Daniell cell, if the standard emf is 1.10 V.

Solution The reaction is

$$Zn + Cu^{2+} \rightleftharpoons Zn^{2+} + Cu$$

and $z = 2$. From Eq. 8.7,

$$\ln K^\circ = \frac{zE^\circ}{0.0257} = \frac{2 \times 1.10}{0.0257} = 85.6$$

and thus

$$K^\circ = 1.50 \times 10^{37}$$

[3] This is the approximate relationship in which concentrations rather than activities are used. In Eq. 8.4 and subsequent equations we are again using the superscript ^{u} to indicate the *numerical value* of the quantities such as the ratio $[Y]^y[Z]^z/[A]^a[B]^b$.

EXAMPLE 8.2 Using the data in Table 8.1, calculate the equilibrium constant for the reaction

$$H_2 + 2Fe^{3+} \rightleftharpoons 2H^+ + 2Fe^{2+}$$

Solution From Table 8.1, the standard electrode potentials are

$$2H^+ + 2e^- \rightarrow H_2 \qquad E° = 0$$

$$Fe^{3+} + e^- \rightarrow Fe^{2+} \qquad E° = 0.771 \text{ V}$$

The $E°$ value for the process

$$H_2 + 2Fe^{3+} \rightarrow 2H^+ + 2Fe^{2+}$$

for which $z = 2$ is thus 0.771 V. Therefore

$$0.771 = \frac{0.0257}{2} \ln K°$$

and

$$K = 1.14 \times 10^{26} \text{ mol}^2 \text{ dm}^{-6}$$

It should be emphasized that the emf is an intensive property, so that in making this calculation *we must not* multiply the value of 0.771 V by 2; the emf of 0.771 V applies equally well to the process

$$2Fe^{3+} + 2e^- \rightarrow 2Fe^{2+}$$

If the problem had been to calculate the equilibrium constant K' for the process

$$\tfrac{1}{2}H_2 + Fe^{3+} \rightleftharpoons H^+ + Fe^{2+}$$

the reactions would have been written as

$$H^+ + e^- \rightarrow \tfrac{1}{2}H_2 \qquad E° = 0$$
$$Fe^{3+} + e^- \rightarrow Fe^{2+} \qquad E° = 0.771 \text{ V}$$

and again $E° = 0.771$ V. In this case $z = 1$ and

$$0.771 = 0.0257 \ln K°' \quad \text{or} \quad \ln K°' = \frac{0.771}{0.0257} = 30.01$$

and therefore

$$K°' = 1.1 \times 10^{13} \quad \text{or} \quad K' = 1.1 \times 10^{13} \text{ mol dm}^{-3}$$

K' is, of course, the square root of K.

In general, a more complicated procedure must be used to combine half-cell reactions in order to find an $E°$ for a desired half-cell reaction, as is shown in the following example.

EXAMPLE 8.3 Calculate $E°$ for the process

$$Cu^+ + e^- \rightarrow Cu$$

making use of the following $E°$ values:

1. $Cu^{2+} + e^- \rightarrow Cu^+ \quad E_1^\circ = 0.153$ V
2. $Cu^{2+} + 2e^- \rightarrow Cu \quad E_2^\circ = 0.337$ V

Solution The $\Delta G°$ values for these two reactions are

$$Cu^{2+} + e^- \rightarrow Cu^+ \quad \Delta G_1^\circ = -zE_1^\circ F = -1 \times 0.153 \times 96\,485 \text{ J mol}^{-1}$$

$$Cu^{2+} + 2e^- \rightarrow Cu \quad \Delta G_2^\circ = -zE_2^\circ F = -2 \times 0.341 \times 96\,485 \text{ J mol}^{-1}$$

The reaction $Cu^+ + e^- \rightarrow Cu$ is obtained by subtracting reaction 1 from reaction 2 and the $\Delta G°$ value for $Cu^+ + e^- \rightarrow Cu$ is therefore obtained by subtracting ΔG_1° from ΔG_2°:

$$\begin{aligned}\Delta G° &= [-2 \times 0.341 \times 96\,485 - (-1 \times 0.153 \times 96\,485)] \text{ J mol}^{-1} \\ &= (0.153 - 0.682)\, 96\,485 \text{ J mol}^{-1} \\ &= -0.529 \times 96\,485 \text{ J mol}^{-1}\end{aligned}$$

Since, for $Cu^+ + e^- \rightarrow Cu$, $z = 1$, it follows that

$$E° = 0.529 \text{ V}$$

Note that it is incorrect, in working the previous example, which is for a single electrode *half cell,* simply to combine the $E°$ values directly.

In view of this the student may wonder why it is legitimate to calculate $E°$ values for overall cell reactions by simply combining the $E°$ values for individual electrodes. Consider, for example, the following $E°$ values involving both a one-electron and a two-electron process as in the last example:

$$Fe^{3+} + e^- \rightarrow Fe^{2+} \quad E° = 0.771 \text{ V}$$

$$I_2 + 2e^- \rightarrow 2I^- \quad E° = 0.536 \text{ V}$$

The procedure we have been adopting is to combine these two $E°$ values:

$$2Fe^{3+} + 2I^- \rightarrow 2Fe^{2+} + I_2 \quad E° = 0.771 - 0.536 = 0.235 \text{ V}$$

The fact that this is justified can be seen by writing the $\Delta G°$ values:

$$Fe^{3+} + e^- \rightarrow Fe^{2+} \quad \Delta G° = -1 \times 0.771 \times 96\,485 \text{ J mol}^{-1}$$

$$I_2 + 2e^- \rightarrow 2I^- \quad \Delta G° = -2 \times 0.536 \times 96\,485 \text{ J mol}^{-1}$$

We combine the two equations by multiplying the first by 2 and subtracting the second:

$$2Fe^{3+} + 2I^- \rightarrow 2Fe^{2+} + I_2$$

$$\begin{aligned}\Delta G° &= [2(-1 \times 0.771 \times 96\,485) - (-2 \times 0.536 \times 96\,485)] \text{ J mol}^{-1} \\ &= -2 \times 0.235 \times 96\,485 \text{ J mol}^{-1}\end{aligned}$$

Thus $E° = 0.235$ V, and this is simply $0.771 - 0.536$. We are justified in simply subtracting $E°$ values to find $E°$ for an overall reaction, in which there are no electrons left over. However, to obtain an $E°$ for a half-reaction (as in Example 8.3), we can in general *not* simply combine E° values but must calculate the ΔG° values as just done.

The Nernst Equation

So far we have limited our discussion to standard electrode potentials E° and to E° values for cells in which the active species are present at 1 *m* concentrations. The corresponding standard Gibbs energies have been written as ΔG°.

Let us now remove this restriction and consider cells in which the concentrations are other than unity. Consider, for example, the cell

The Nernst equation; Follow its derivation.

$$\text{Pt, H}_2\text{ (1 bar)}\,|\text{H}^+\text{(aq)} \,\vdots\vdots\, \text{Cu}^{2+}\text{ (aq)}|\text{Cu}$$

in which a hydrogen electrode has been combined with a copper electrode immersed in a Cu^{2+} solution, the concentration of which is other than unity. The overall cell reaction is

$$\text{H}_2 + \text{Cu}^{2+} \rightarrow 2\text{H}^+ + \text{Cu}$$

and the Gibbs energy change is (see Eq. 8.4)[4]

$$\Delta G = -RT\left[\ln K^\circ - \ln\left(\frac{[\text{H}^+]^2}{[\text{Cu}^{2+}]}\right)^u\right] \tag{8.8}$$

However, $\Delta G^\circ = -RT \ln K^\circ = -zE^\circ F$, and therefore

$$\Delta G = -zE^\circ F + RT\ln\left(\frac{[\text{H}^+]^2}{[\text{Cu}^{2+}]}\right)^u \tag{8.9}$$

Since $\Delta G = -zEF$, where E is the emf of this cell, we obtain

$$-zEF = -zE^\circ F + RT\ln\left(\frac{[\text{H}^+]^2}{[\text{Cu}^{2+}]}\right)^u \tag{8.10}$$

and therefore

$$E = E^\circ - \frac{RT}{zF}\ln\left(\frac{[\text{H}^+]^2}{[\text{Cu}^{2+}]}\right)^u \tag{8.11}$$

In general, we may consider any cell for which the overall reaction has the general form

$$a\text{A} + b\text{B} + \cdots \rightarrow \cdots + y\text{Y} + z\text{Z}$$

ΔG is given by Eq. 8.3 and $\Delta G^\circ = -RT \ln K^\circ = -zE^\circ F$, and therefore

$$\Delta G = -zE^\circ F + RT\ln\left(\frac{\cdots[\text{Y}]^y[\text{Z}]^z}{[\text{A}]^a[\text{B}]^b\cdots}\right)^u \tag{8.12}$$

Note that, as in an equilibrium constant, products are in the numerator and reactants in the denominator. Since $\Delta G = -zEF$, Eq. 8.12 leads to the following general expression for the emf:

Nernst Equation

$$E = E^\circ - \frac{RT}{zF}\ln\left(\frac{\cdots[\text{Y}]^y[\text{Z}]^z}{[\text{A}]^a[\text{B}]^b\cdots}\right)^u \tag{8.13}$$

[4]Again, for simplicity, we use concentrations instead of activities. See Section 8.5, where equations involving activities are employed.

This relationship was first given in 1889 by Nernst and is known as the **Nernst equation.**

Suppose that we apply the Nernst equation to the cell

$$Zn|Zn^{2+}||Ni^{2+}|Ni$$

for which the overall reaction is

$$Zn + Ni^{2+} \rightarrow Zn^{2+} + Ni$$

Nernst equations; Simplify and restrict it to 298 K.

The standard electrode potentials (see Table 8.1) are

$$Ni^{2+} + 2e^- \rightarrow Ni \quad E° = -0.257 \text{ V}$$
$$Zn^{2+} + 2e^- \rightarrow Zn \quad E° = -0.762 \text{ V}$$

and $E°$ for the overall process is $-0.257 - (-0.762) = 0.505$ V. The Nernst equation is thus

$$E = 0.505(\text{V}) - \frac{RT}{zF} \ln \frac{[Zn^{2+}]}{[Ni^{2+}]} \tag{8.14}$$

(As always, the concentrations of solid species such as Zn and Ni are incorporated into the equilibrium constant and are therefore not included explicitly in the equation.) At 25 °C this equation becomes

$$E/\text{V} = 0.505 - \frac{0.0257}{2} \ln \frac{[Zn^{2+}]}{[Ni^{2+}]} \tag{8.15}$$

since $z = 2$. We see from this equation that increasing the ratio $[Zn^{2+}]/[Ni^{2+}]$ decreases the cell emf; this is understandable in view of the fact that a positive emf means that the cell is producing Zn^{2+} and that Ni^{2+} ions are being removed.

EXAMPLE 8.4 Calculate the emf of the cell

$$Co|Co^{2+}||Ni^{2+}|Ni$$

if the concentrations are

a. $[Ni^{2+}] = 1.0\ m$ and $[Co^{2+}] = 0.10\ m$
b. $[Ni^{2+}] = 0.010\ m$ and $[Co^{2+}] = 1.0\ m$

Solution The cell reaction is

$$Co + Ni^{2+} \rightarrow Co^{2+} + Ni$$

and from Table 8.1 the standard electrode potentials are

$$Ni^{2+} + 2e^- \rightarrow Ni \quad E° = -0.257 \text{ V}$$
$$Co^{2+} + 2e^- \rightarrow Co \quad E° = -0.280 \text{ V}$$

The standard emf is thus $-0.257 - (-0.280) = 0.023$ V, and $z = 2$. The cell emf at the first concentrations specified (a) is

$$E = 0.023 - \frac{0.0257}{2} \ln \frac{[Co^{2+}]}{[Ni^{2+}]}$$
$$= 0.023 - \frac{0.0257}{2} \ln 0.10 = 0.023 + 0.030 = 0.053 \text{ V}$$

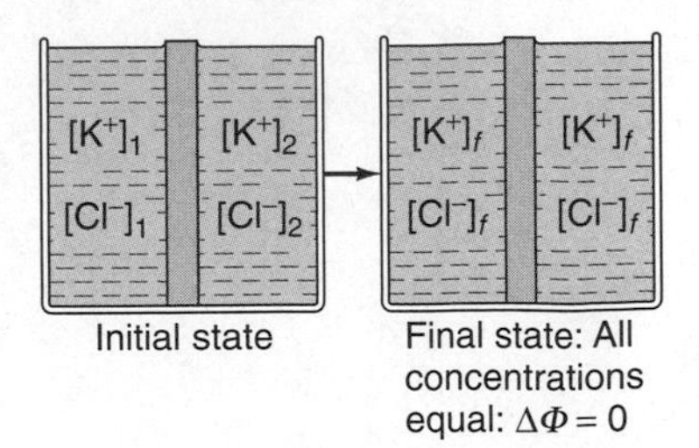

a. Membrane permeable to both ions

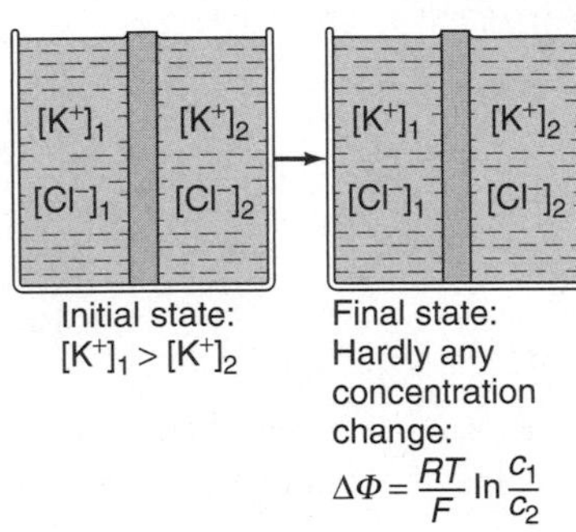

b. Membrane permeable to K^+ only

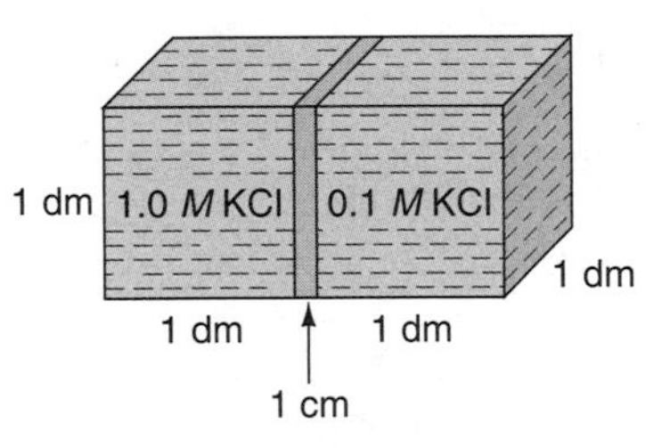

c. Two cubic cells separated by a membrane

FIGURE 8.7
(a) Solutions of KCl separated by a membrane that is permeable to both ions. (b) Solutions of KCl separated by a membrane that is permeable only to K^+. (c) Two 1-dm^3 cubes separated by a membrane of area 1 dm^2 and thickness 1 cm.

In b,

$$E = 0.023 - \frac{0.0257}{2} \ln \frac{1.0}{0.010}$$
$$= 0.023 - 0.059 = -0.036 \text{ V}$$

We see that the cell operates in opposite directions in the two cases.

Nernst Potentials

If two electrolyte solutions of different concentrations are separated by a membrane, in general there will be an electric potential difference across the membrane. Various situations are possible. Suppose, for example, that solutions of potassium chloride were separated by a membrane, as shown in Figure 8.7. If both ions could cross the membrane, the concentrations would eventually become equal and there would be no potential difference (Figure 8.7a). If, however, only the potassium ions can cross, and the membrane is not permeable to the solvent,[5] there will be very little change in the concentrations on the two sides of the membrane. Some K^+ ions will cross from the more concentrated side (the left-hand side in Figure 8.7b), and as a result the left-hand side will have a negative potential with respect to the right-hand side. The effect of this will be to prevent more K^+ ions from crossing. An equilibrium will therefore be established at which the electric potential will exactly balance the tendency of the concentrations to become equal. This potential is known as the **Nernst potential.**

The Gibbs energy difference ΔG_e arising from the potential difference $\Delta\Phi$ is

$$\Delta G_e = zF\,\Delta\Phi \tag{8.16}$$

where z is the charge on the permeable ions ($z = 1$ in this case). In our particular example (Figure 8.7b), the potential is higher on the right-hand side ($\Delta\Phi > 0$), because a few K^+ ions have crossed, and there will thus be a higher Gibbs energy, as far as K^+ ions are concerned, on the right-hand side ($\Delta G_e > 0$). The Gibbs energy difference arising from the concentration difference is

$$\Delta G_c = RT \ln \frac{c_2}{c_1} \tag{8.17}$$

At equilibrium there is no net ΔG across the membrane and therefore

$$\Delta G_e + \Delta G_c = zF\,\Delta\Phi + RT \ln \frac{c_2}{c_1} = 0 \tag{8.18}$$

Thus

$$\Delta\Phi = \frac{RT}{zF} \ln \frac{c_1}{c_2} \tag{8.19}$$

It is important to realize that in situations of this kind the Nernst potential is due to the transfer of only an exceedingly small fraction of the diffusible ions, so

[5]This restriction is made so that there will be no complications due to osmotic effects.

By convention we take the ratio of the concentration of ions outside the cell to the concentration of ions inside the cell. The potential obtained is that inside the cell.

EXAMPLE 8.5 Mammalian muscle cells are freely permeable to K^+ ions but much less permeable to Na^+ and Cl^- ions. Typical concentrations of K^+ ions are

Inside the cell: $[K^+] = 155\ mM$

Outside the cell: $[K^+] = 4\ mM$

Calculate the Nernst potential at 310 K (37 °C) on the assumption that the membrane is impermeable to Na^+ and Cl^-.

Solution From Eq. 8.19 with $z = 1$,

$$\Delta\Phi = \frac{8.3145(\text{J K}^{-1}\text{ mol}^{-1}) \times 310(\text{K})}{96\,485(\text{C mol}^{-1})} \ln \frac{4}{155}$$

$$= 0.0267 \ln \frac{4}{155} = -0.0977\text{ V} = -97.7\text{ mV}$$

The potential is negative inside the cell and positive outside. In reality, such potentials are more like −85 mV, because there is a certain amount of diffusion of Na^+ and Cl^-, and there is also a biological pumping mechanism.

that there is no detectable concentration change. Suppose, for example, that two solutions 1 dm^3 in volume are separated by a membrane 1 dm^2 in area (A) and 1 cm in thickness (l), and suppose that 0.1 M and 1.0 M solutions of KCl are present on the two sides (Figure 8.7c), the membrane again being permeable only to the K^+ ions. The capacitance C of the membrane is

$$C = \frac{\epsilon_0 \epsilon A}{l} \tag{8.20}$$

where ϵ, the dielectric constant of the membrane, will be taken to have a value of 3 (this is typical of biological and other organic membranes). The capacitance in this example is therefore

$$C = \frac{8.854 \times 10^{-12}(\text{C}^2\text{ N}^{-1}\text{ m}^{-2}) \times 3 \times 10^{-2}(\text{m}^2)}{10^{-2}(\text{m})}$$

$$= 2.66 \times 10^{-11}\text{ C}^2\text{ N}^{-1}\text{ m}^{-1} = 2.66 \times 10^{-11}\text{ F}$$

where F is the *farad*, the unit of electric capacitance. In terms of base SI units:

$$\text{F} \equiv \text{C V}^{-1} \equiv \text{A s (kg m}^2\text{ s}^{-3}\text{ A}^{-1})^{-1} \equiv \text{A}^2\text{ s}^4\text{ kg}^{-1}\text{ m}^{-2}$$
$$\text{C}^2\text{ N}^{-1}\text{ m}^{-1} \equiv (\text{A s})^2(\text{kg m s}^{-2})^{-1}\text{ m}^{-1} \equiv \text{A}^2\text{ s}^4\text{ kg}^{-1}\text{ m}^{-2} \equiv \text{F}$$

The Nernst potential at 25 °C is

$$\Delta\Phi = \frac{8.3145 \times 298.15}{96\,485} \ln 10 = 0.059\text{ V}$$

With a capacitance of 2.66×10^{-11} F, the net charge on each side of the wall, required to maintain this potential, is

$$Q = 2.66 \times 10^{-11}\text{ F} \times 0.0591\text{ V}$$
$$= 1.57 \times 10^{-12}\text{ C}$$

The number of K^+ ions required to produce this charge is

$$\frac{Q}{e} = \frac{1.57 \times 10^{-12}\ \text{C}}{1.602 \times 10^{-19}\ \text{C}} = 9.80 \times 10^6$$

However, 1 dm^3 of the 0.1 *M* solution contains $0.1 \times 6.022 \times 10^{23} = 6.022 \times 10^{22}$ ions. The fraction involved in establishing the potential difference is therefore exceedingly small ($<10^{-14}$) and no concentration change could be detected. The fraction is larger if the system is smaller, but even for a biological cell it is no more than 10^{-6} (see Problem 8.22).

The situation that we have described relates to a membrane permeable to only one ion. If more than one ion can pass through the membrane, the situation is more complex. The distribution of K^+, Na^+, and Cl^- ions across a biological membrane provides an interesting example. In Example 8.5 on p. 333 we saw that mainly as a result of the K^+ ions, which are freely permeable, the Nernst potential is about -85 mV. The Cl^- ions are also somewhat permeable, and typical concentrations are

Inside the cell: $[Cl^-] = 4.3$ m*M*

Outside the cell: $[Cl^-] = 104$ m*M*

The Nernst potential corresponding to this distribution is, since $z = -1$,

$$\Delta\Phi = -0.0267 \ln \frac{104}{4.3} = -0.0851\ \text{V} = -85.1\ \text{mV}$$

which is close to the observed potential. The Cl^- ions are therefore more or less at equilibrium. Typical concentrations for Na^+ ions are

Inside the cell: $[Na^+] = 12$ m*M*

Outside the cell: $[Na^+] = 145$ m*M*

If these ions were permeable, their Nernst potential would be

$$\Delta\Phi = 0.0267 \ln \frac{145}{12} = 66.5\ \text{mV}$$

This is of the opposite sign to the true potential, and this distribution of Na^+ ions only arises because of their inability to cross the membrane. These ionic distributions are of great importance in connection with nerve impulses, which are activated by changes in permeability of the nerve membranes.

Potential differences across membranes are set up when a Donnan equilibrium (Section 7.13) is established. The procedure for calculating them is best illustrated by an example.

EXAMPLE 8.6 A 0.10 *M* solution of sodium palmitate is separated from an equal volume of a 0.20 *M* solution of sodium chloride by a membrane that is permeable to Na^+ and Cl^- but not to palmitate ions. Calculate the final concentrations and the Nernst potential at 25 °C, assuming ideal behavior.

Solution Suppose that x mol dm^{-3} of Na^+ and Cl^- ions move to the palmitate side; the final concentrations are

Palmitate side: $[Na^+] = (0.1 + x)\ M$; $[Cl^-] = x\ M$

Other side: $[Na^+] = (0.2 - x)\ M$; $[Cl^-] = (0.2 - x)\ M$

At equilibrium

$$(0.2 - x)^2 = x(0.1 + x)$$
$$x = 0.08\ M$$

The final concentrations are therefore

Palmitate side: $[Na^+] = 0.18\ M$; $[Cl^-] = 0.08\ M$
Other side: $[Na^+] = [Cl^-] = 0.12\ M$

The Nernst potential arising from the distribution of Na^+ ions is, at 25 °C,

$$\Delta\Phi = \frac{8.3145 \times 298.15}{96\ 485} \ln \frac{0.18}{0.12} = 0.0104 \text{ V} = 10.4 \text{ mV}$$

The palmitate side, having the higher concentration of Na^+, is the negative side, since Na^+ ions will tend to cross from that side. The Nernst potential calculated from the distribution of Cl^- ions is exactly the same:

$$\Delta\Phi = \frac{8.3145 \times 298.15}{96\ 485} \ln \frac{0.12}{0.08} = 10.4 \text{ mV}$$

Strictly speaking, $\Delta\Phi$ values measured by placing platinum electrodes in two solutions separated by a membrane are not precisely the same as these calculated Nernst potentials. The measured values are for the potential drop between the Pt electrodes, but the potential drop between the electrode and the solution is not quite the same for the two solutions. However, the error is about the same as the error in the calculations, and in electrophysiological and other studies it is usually assumed that the measured values can be equated to those obtained in the calculations.

Temperature Coefficients of Cell emfs

Since a Gibbs energy change can be obtained from the standard emf of a reversible cell (Eq. 8.2), the $\Delta S°$ and $\Delta H°$ values can be calculated if emf measurements are made over a range of temperature.

The basic relationship is Eq. 3.119:

$$S = -\left(\frac{\partial G}{\partial T}\right)_P \tag{8.21}$$

and for an overall reaction

$$\Delta S = -\left(\frac{\partial \Delta G}{\partial T}\right)_P \tag{8.22}$$

Introduction of Eq. 8.3 gives

$$\Delta S = zF\left(\frac{\partial E}{\partial T}\right)_P \tag{8.23}$$

The enthalpy change is thus

$$\Delta H = \Delta G + T\Delta S \tag{8.24}$$

$$\Delta H = -zF\left(E - T\frac{\partial E}{\partial T}\right) \tag{8.25}$$

The measurement of emf values at various temperatures provides a very convenient method of obtaining thermodynamic values for chemical reactions and has frequently been employed. For the results to be reliable the temperature coefficients should be known to three significant figures, and this requires careful temperature and emf measurements.

EXAMPLE 8.7 The emf of the cell

$$\text{Pt, H}_2\text{(1 bar)}|\text{HCl(0.01 } m)|\text{AgCl(s)}|\text{Ag}$$

is 0.2002 V at 25 °C; $\partial E/\partial T$ is -8.665×10^{-5} V K^{-1}. Write the cell reaction and calculate ΔG, ΔS, and ΔH at 25 °C.

Solution The electrode reactions are

$$\tfrac{1}{2}\text{H}_2 \rightarrow \text{H}^+ + e^- \quad \text{and} \quad e^- + \text{AgCl(s)} \rightarrow \text{Ag} + \text{Cl}^-$$

The cell reaction is

$$\tfrac{1}{2}\text{H}_2 + \text{AgCl(s)} \rightarrow \text{Ag} + \text{H}^+ + \text{Cl}^-$$

The Gibbs energy change is

$$\Delta G = -96\,485 \times 0.2002 = -19\,320 \text{ J mol}^{-1}$$

The entropy change is obtained by use of Eq. 8.23:

$$\Delta S = 96\,485 \times -8.665 \times 10^{-5} = -8.360 \text{ J K}^{-1} \text{ mol}^{-1}$$

The enthalpy change can be calculated by use of Eq. 8.25 or more easily from the ΔG and ΔS values:

$$\begin{aligned}\Delta H &= \Delta G + T\,\Delta S \\ &= -1.932 \times 10^4 + (-8.360 \times 298.15) \\ &= -2.181 \times 10^4 \text{ J mol}^{-1}\end{aligned}$$

8.4 Types of Electrochemical Cells

In the cells we have considered so far there is a net chemical change. Such electrochemical cells are classified as **chemical cells.** There are also cells in which the driving force, instead of being a chemical reaction taking place between electrodes and solute ions, is a dilution process. Such cells are known as **concentration cells.** The changes in concentration can occur either in the electrolyte or at the electrodes. Examples of concentration changes at electrodes are found with electrodes made of amalgams or consisting of alloys and with gas electrodes (e.g., the Pt, H_2 electrode) when there are different gas pressures at the two electrodes.

Figure 8.8 shows a classification of electrochemical cells. A subclassification of chemical and concentration cells relates to whether or not there is a boundary between two solutions. If there is not, as in the cell

$$\text{Pt, H}_2|\text{HCl(aq)}|\text{AgCl(s)}|\text{Ag}$$

we have a *cell without transference.* If there is a boundary, as in the Daniell cell (Figure 8.1), the cell is known as a *cell with transference.* In the latter case there

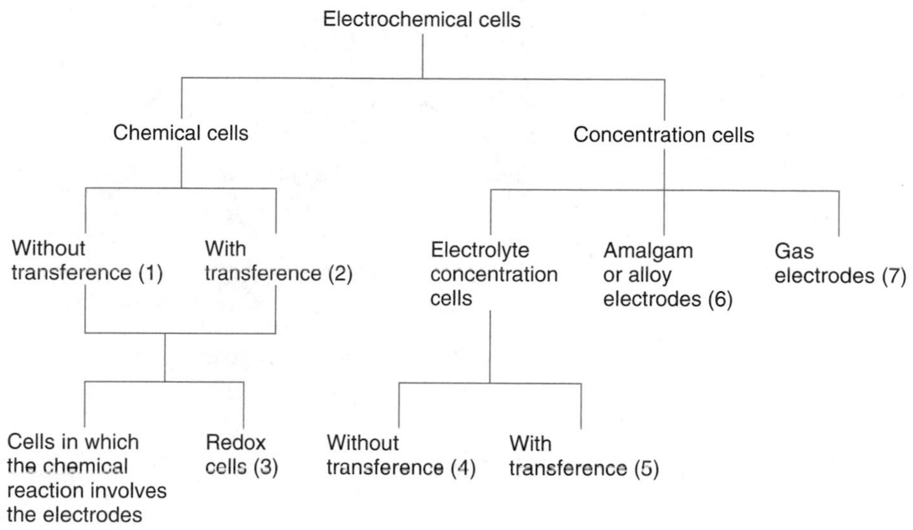

Electrochemical cells

Chemical cells

Concentration cells

Without transference (1)

With transference (2)

Electrolyte concentration cells

Amalgam or alloy electrodes (6)

Gas electrodes (7)

Cells in which the chemical reaction involves the electrodes

Redox cells (3)

Without transference (4)

With transference (5)

FIGURE 8.8
Classification of electrochemical cells.
1. Pt, $H_2|HCl|AgCl|Ag$
2. $Zn|Zn^{2+} \,\vdots\vdots\, Cu^{2+}|Cu$
3. Pt, $H_2|H^+$, Fe^{2+}, $Fe^{3+}|Pt$
4. Pt, $H_2|HCl(m_1)|AgCl|Ag$ – $AgCl|HCl(m_2)|Pt$, H_2
5. Pt, $H_2|H^+(m_1) \,\vdots\vdots\, H^+(m_2)|Pt$, H_2
6. Na in Hg at $c_1|Na^+|$Na in Hg at c_2
7. Pt, $H_2(P_1)|H^+|Pt$, $H_2(P_2)$.

is a potential difference between the solutions—which can be minimized by use of a salt bridge—and there are irreversible changes in the two solutions as the cell is operated.

Concentration cells; Watch a reaction in a concentration cell driven by a concentration difference.

Concentration Cells

A simple example of a concentration cell is obtained by connecting two hydrogen electrodes by means of a salt bridge:

$$\text{Pt, } H_2|HCl(m_1) \,\vdots\vdots\, HCl(m_2)|H_2\text{, Pt}$$

The salt bridge could be a tube containing saturated potassium chloride solution. The reaction at the left-hand electrode is

$$\tfrac{1}{2}H_2 \rightarrow H^+(m_1) + e^-$$

while that at the right-hand electrode is

$$H^+(m_2) + e^- \rightarrow \tfrac{1}{2}H_2$$

The net process is therefore

$$H^+(m_2) \rightarrow H^+(m_1)$$

and is simply the transfer of hydrogen ions from a solution of molality m_2 to one of molality m_1. If m_2 is greater than m_1, the process will actually occur in this direction, and a positive emf is produced; if m_2 is less than m_1, the emf is negative and electrons flow from the right-hand to the left-hand electrode.

The Gibbs energy change associated with the transfer of H^+ ions from a molality m_2 to a molality m_1 is

$$\Delta G = RT \ln \frac{m_1}{m_2} \tag{8.26}$$

Since $z = 1$, the emf produced is

$$E = \frac{RT}{F} \ln \frac{m_2}{m_1} \tag{8.27}$$

and is positive when $m_2 > m_1$.

EXAMPLE 8.8 Calculate the emf at 25 °C of a concentration cell of this type in which the molalities are 0.2 *m* and 3.0 *m*.

Solution The emf is given by

$$E = \frac{RT}{F} \ln \frac{m_2}{m_1} = 0.0257 \ln \frac{3.0}{0.2} = 0.0696 \text{ V}$$

Redox Cells

Since, when cells operate, there is an electron transfer at the electrodes, oxidations and reductions are occurring. In all the cells considered so far these oxidations and reductions have involved the electrodes themselves—for example, the H_2 gas in the hydrogen electrode. There is also an important class of oxidation-reduction cells, known as **redox cells,** in which both the oxidized and reduced species are in solution; their interconversion is effected by an inert electrode such as one of platinum.

Consider, for example, the cell

$$\text{Pt, H}_2|\text{H}^+(1\ m) \,\vdots\vdots\, \text{Fe}^{2+}, \text{Fe}^{3+}|\text{Pt}$$

The left-hand electrode is the standard hydrogen electrode. The right-hand electrode consists simply of a platinum electrode immersed in a solution containing both Fe^{2+} and Fe^{3+} ions. The platinum electrode is able to catalyze the interconversion of these ions, and the reaction at this electrode is

$$e^- + \text{Fe}^{3+} \rightarrow \text{Fe}^{2+}$$

Since the reaction at the hydrogen electrode is

$$\tfrac{1}{2}\text{H}_2 \rightarrow \text{H}^+ + e^-$$

the overall process is

$$\text{Fe}^{3+} + \tfrac{1}{2}\text{H}_2 \rightarrow \text{Fe}^{2+} + \text{H}^+$$

The emf of this cell represents the ease with which the Fe^{3+} is reduced to Fe^{2+}. The emf of the cell is

$$E = E^\circ - \frac{RT}{F} \ln \frac{[\text{Fe}^{2+}]}{[\text{Fe}^{3+}]} \tag{8.28}$$

(Note that the E° relates to $P_{H_2} = 1$ bar and $[H^+] = [Fe^{2+}] = [Fe^{3+}] = 1\ m$.) The E° for this system is +0.771 V (Table 8.1).

The interconversion of oxidized and reduced forms frequently involves also the participation of hydrogen ions. Thus, the half-reaction for the reduction of fumarate ions to succinate ions is

$$\begin{array}{lcl} \mathrm{CHCOO^-} + 2\mathrm{H^+} + 2e^- & \rightarrow & \mathrm{CH_2COO^-} \\ \| & & | \\ \mathrm{CHCOO^-} & & \mathrm{CH_2COO^-} \\ \text{fumarate} & & \text{succinate} \end{array}$$

If we wished to study this system, we could set up the following cell:

$$\mathrm{Pt, H_2|H^+(1\ \mathit{m})} \,\vdots\vdots\, \mathrm{F^{2-}, S^{2-}, H^+(\mathit{m}_{H^+})|Pt}$$

where F^{2-} and S^{2-} represent fumarate and succinate, respectively. The hydrogen-ion molality in the right-hand solution is not necessarily 1 m; it will here be denoted as m_{H^+}. If we combine the standard hydrogen electrode,

$$\mathrm{H_2} \rightarrow 2\mathrm{H^+}(1\ m) + 2e^-$$

we obtain, for the overall cell reaction,

$$\mathrm{F^{2-}} + 2\mathrm{H^+}(m_{\mathrm{H^+}}) + \mathrm{H_2} \rightarrow \mathrm{S^{2-}} + 2\mathrm{H^+}(1\ m)$$

The equation for the emf is thus, since $z = 2$,

$$E = E^\circ - \frac{RT}{2F} \ln\left(\frac{[\mathrm{S^{2-}}]}{[\mathrm{F^{2-}}]m_{\mathrm{H^+}}^2}\right)^u \tag{8.29}$$

The E° for this system is related to a standard Gibbs energy change ΔG° by the usual equation

$$\Delta G^\circ = -zFE^\circ \tag{8.30}$$

This standard Gibbs energy change is related to the equilibrium constant:

$$K = \frac{[\mathrm{S^{2-}}]}{[\mathrm{F^{2-}}]m_{\mathrm{H^+}}^2} \tag{8.31}$$

However, it is frequently convenient to deal with the modified equilibrium constant

$$K' = \frac{[\mathrm{S^{2-}}]}{[\mathrm{F^{2-}}]} \tag{8.32}$$

at some specified hydrogen-ion concentration. Often this standard concentration is taken to be $10^{-7}\,M$, corresponding to a pH of 7. In that case

$$K' = (10^{-7})^2 K = 10^{-14} K \tag{8.33}$$

In many cases the K' value corresponds to a fairly well-balanced equilibrium at pH 7, whereas K will be larger by the factor 10^{14}; the K' value and the corresponding $\Delta G^{\circ\prime}$ at pH 7 therefore give a clearer indication of the situation at that pH.

Equation 8.29 for the emf can be written as

$$E = E^\circ - \frac{RT}{2F}\ln\frac{[\mathrm{S^{2-}}]}{[\mathrm{F^{2-}}]} + \frac{RT}{F}\ln m_{\mathrm{H^+}} \tag{8.34}$$

$$= E^\circ - \frac{RT}{2F}\ln\frac{[\mathrm{S^{2-}}]}{[\mathrm{F^{2-}}]} + 0.0257\ (\mathrm{V}) \ln m_{\mathrm{H^+}} \tag{8.35}$$

$$= E^\circ - \frac{RT}{2F}\ln\frac{[\mathrm{S^{2-}}]}{[\mathrm{F^{2-}}]} - 0.05916\ (\mathrm{V})\ \mathrm{pH} \tag{8.36}$$

where pH is the pH of the solution in which the $S^{2-}:F^{2-}$ system is maintained (pH $= -\log_{10} m_{H^+} = -2.303 \ln m_{H^+}$). Then, if we define a modified standard potential $E^{\circ\prime}$ by

$$E = E^{\circ\prime} - \frac{RT}{2F} \ln \frac{[S^{2-}]}{[F^{2-}]} \tag{8.37}$$

it follows that

$$E^{\circ\prime} = E^\circ - 0.05916\ (V)\ pH \tag{8.38}$$

Different relationships apply to other reaction types.

Redox systems are particularly important in biological systems, as illustrated by the following example.

EXAMPLE 8.9 The enzyme glycollate oxidase is a catalyst for the reduction of cytochrome c in its oxidized form—denoted as cytochrome c (Fe^{3+})—by glycollate ions. The relevant standard electrode potentials $E^{\circ\prime}$, relating to 25 °C and pH 7, are as follows:

	$E^{\circ\prime}/V$
cytochrome c (Fe^{3+}) + e^- → cytochrome c (Fe^{2+})	0.250
glyoxylate$^-$ + $2H^+$ + $2e^-$ → glycollate$^-$	−0.085

a. Calculate $\Delta G^{\circ\prime}$ for the reduction, at pH 7 and 25 °C.
b. Then calculate the equilibrium ratio

$$\frac{[\text{cytochrome } c\ (Fe^{2+})]^2[\text{glyoxylate}^-]}{[\text{cytochrome } c\ (Fe^{3+})]^2[\text{glycollate}^-]}$$

at pH 7 and 25 °C.
c. What is the equilibrium ratio at pH 7.5 and 25 °C?

Solution a. The balanced reaction is

$$2\ \text{cytochrome } c\ (Fe^{3+}) + \text{glycollate}^- \rightarrow \text{glyoxylate}^- + 2\ \text{cytochrome } c\ (Fe^{2+}) + 2H^+$$

and this equation corresponds to $z = 2$.

$$E^\circ = 0.250 - (-0.085) = 0.335\ V$$
$$\Delta G^{\circ\prime} = -2 \times 0.335 \times 96\,485$$
$$= -64\,645\ J\ mol^{-1} = -64.6\ kJ\ mol^{-1}$$

b. The equilibrium ratio at pH 7 is

$$K' = \exp\,[64\,645/(8.3145 \times 298.15)]$$
$$= 2.115 \times 10^{11}$$

c. The equilibrium ratio at pH 7.5 is conveniently calculated in terms of the true (pH-independent) equilibrium constant K_{true}. If K'' is the equilibrium ratio at pH 7.5,

$$K_{true} = K' \times (10^{-7})^2 = K'' \times (10^{-7.5})^2$$
$$K'' = 2.115 \times 10^{11} \times (10^{0.5})^2$$
$$= 2.115 \times 10^{12}$$

8.5 Applications of emf Measurements

A number of physical measurements are conveniently made by setting up appropriate electrochemical cells. Since emf values can be determined very accurately, such techniques are frequently employed. A few of them will be mentioned briefly.

pH Determinations

Since the emf of a cell such as that shown in Figure 8.4 depends on the hydrogen-ion concentration, pH values can be determined by dipping hydrogen electrodes into solutions and measuring the emf with reference to another electrode. In commercial pH meters the electrode immersed in the unknown solution is often a glass electrode, and the other electrode may be the silver–silver chloride or the calomel electrode. In accordance with the Nernst equation the emf varies logarithmically with the hydrogen-ion concentration and therefore varies linearly with the pH. Commercial instruments are calibrated so as to give a direct reading of the pH.

Activity Coefficients

Up to now we have expressed the Gibbs energy changes and emf values of cells in terms of molalities. This is an approximation and the errors become more serious as concentrations are increased. For a correct formulation, *activities* must be employed, and the emf measurements over a range of concentrations lead to values for the activity coefficients.

Consider the cell

$$\text{Pt, H}_2(1\text{ bar})|\text{HCl(aq)}|\text{AgCl(s)}|\text{Ag}$$

The overall process is

$$\tfrac{1}{2}\text{H}_2 + \text{AgCl} \rightarrow \text{Ag} + \text{H}^+ + \text{Cl}^-$$

and the Gibbs energy change is

$$\Delta G = \Delta G^\circ + RT \ln[a_+ a_-]^u \tag{8.39}$$

where a_+ and a_- are the activities of the H^+ and Cl^- ions, and ΔG° is the standard Gibbs energy change, when the activities are unity. The emf is, since $z = 1$,

$$E = E^\circ - \frac{RT}{F} \ln[a_+ a_-]^u \tag{8.40}$$

Since each of the activities a_+ and a_- is the molality m multiplied by the activity coefficient γ_+ or γ_-, we can write

$$E = E^\circ - \frac{RT}{F} \ln[m^u]^2 - \frac{RT}{F} \ln \gamma_+ \gamma_- \tag{8.41}$$

$$= E^\circ - \frac{2RT}{F} \ln m^u - \frac{2RT}{F} \ln \gamma_\pm \tag{8.42}$$

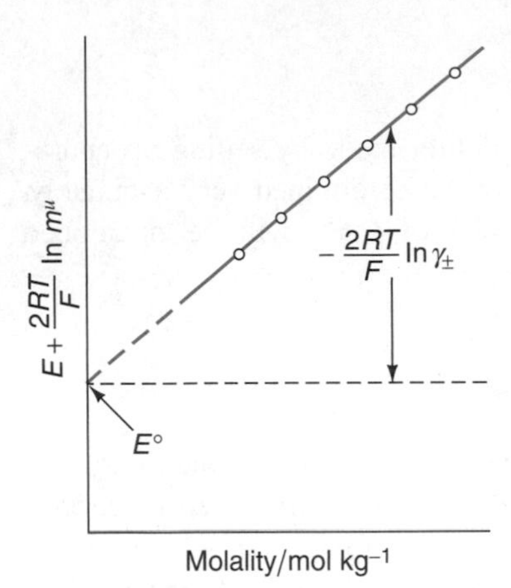

FIGURE 8.9
A plot that will provide the value of E° for a cell and the mean activity coefficients at various concentrations.

where $\gamma_\pm$, equal to $\sqrt{\gamma_+\gamma_-}$, is the **mean activity coefficient.** This equation can be written as

$$E + \frac{2RT}{F}\ln m^u = E^\circ - \frac{2RT}{F}\ln\gamma_\pm \tag{8.43}$$

and if E is measured over a range of molalities of HCl, the quantity on the left-hand side can be calculated at various molalities. If this quantity is plotted against m, as shown schematically in Figure 8.9, the value extrapolated to zero m gives E°, since at zero m the activity coefficient $\gamma_\pm$ is unity so that the final term vanishes. At any molality the ordinate minus E° then yields

$$-\frac{2RT}{F}\ln\gamma_\pm$$

from which the activity coefficient $\gamma_\pm$ can be calculated.

Equilibrium Constants

The emf determination of equilibrium constants may be illustrated with reference to the measurement of the dissociation constant of an electrolyte HA such as acetic acid. Suppose that the following cell is set up:

$$\text{Pt, H}_2(1\text{ bar})|\text{HA}(m_1),\ \text{NaA}(m_2),\ \text{NaCl}(m_3)|\text{AgCl(s)}|\text{Ag}$$

The essential feature of this cell is that the solution contains NaA, which being a salt is essentially completely dissociated and therefore provides a known concentration of A^- ions. The emf of the cell provides a measure of the hydrogen-ion concentration (since the hydrogen electrode is used); since A^- and H^+ are known, and the total amount of HA is known, the amount of undissociated HA is known, and hence the dissociation constant can be obtained.

In more detail, the result is obtained as follows. The reactions at the two electrodes and the overall reaction are

$$\frac{1}{2}\text{H}_2 \rightarrow \text{H}^+ + e^-$$
$$e^- + \text{AgCl} \rightarrow \text{Ag} + \text{Cl}^-$$
$$\overline{\frac{1}{2}\text{H}_2 + \text{AgCl} \rightarrow \text{H}^+ + \text{Cl}^- + \text{Ag}}$$

The emf of the cell is

$$E = E^\circ - \frac{RT}{F}\ln[a_{\text{H}^+}a_{\text{Cl}^-}]^u \tag{8.44}$$

$$= E^\circ - \frac{RT}{F}\ln[m_{\text{H}^+}m_{\text{Cl}^-}]^u - \frac{RT}{F}\ln\gamma_{\text{H}^+}\gamma_{\text{Cl}^-} \tag{8.45}$$

E° is the standard electrode potential of the silver–silver chloride electrode (0.2224 V at 25 °C).

The dissociation constant for the acid HA is

$$K_a = \frac{m_{\text{H}^+}m_{\text{A}^-}}{m_{\text{HA}}}\cdot\frac{\gamma_{\text{H}^+}\gamma_{\text{A}^-}}{\gamma_{\text{HA}}} \tag{8.46}$$

and Eq. 8.45 can be written as

$$E = E^\circ - \frac{RT}{F}\ln\left[\frac{m_{\text{HA}}m_{\text{Cl}^-}}{m_{\text{A}^-}}\right]^u - \frac{RT}{F}\ln\frac{\gamma_{\text{HA}}\gamma_{\text{Cl}^-}}{\gamma_{\text{A}^-}} - \frac{RT}{F}\ln K_a^\circ \tag{8.47}$$

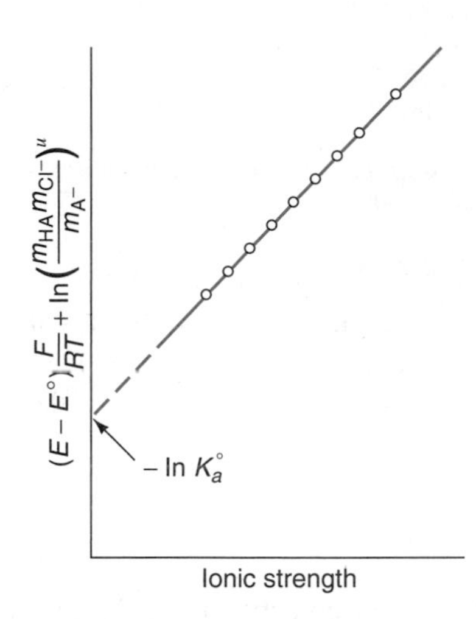

FIGURE 8.10
Plot of the function shown on the left-hand side of Eq. 8.48 against the ionic strength.

or as

$$(E - E^\circ)\frac{F}{RT} + \ln\left[\frac{m_{HA}m_{Cl^-}}{m_{A^-}}\right]^u = -\ln\frac{\gamma_{HA}\gamma_{Cl^-}}{\gamma_{A^-}} - \ln K_a^\circ \tag{8.48}$$

The molality m_{Cl^-} is equal to m_3, the molality of the NaCl solution. The value of m_{HA} is equal to $m_1 - m_{H^+}$; and the value of m_{A^-} is $m_2 + m_{H^+}$. The molality m_{H^+} can be sufficiently well estimated from an approximate value of the dissociation constant; it is usually much less than m_1 and m_2 so that not much error arises from a rough estimate. The quantities on the left-hand side of Eq. 8.48 are therefore known at various values of the molalities m_1, m_2, and m_3, and the left-hand side can be plotted against the ionic strength I, as shown schematically in Figure 8.10. At zero ionic strength the activity coefficients become unity, and the first term on the right-hand side of Eq. 8.48 is therefore zero; extrapolation to $I = 0$ thus gives $-\ln K_a^\circ$.

Similar methods can be employed for the determination of the dissociation constants of bases and of polybasic acids.

Solubility Products

The determination of solubility products by emf measurements may be exemplified by the use of the following cell, which gives the solubility product of silver chloride:

$$Cl_2(1\ \text{bar})|HCl(aq)|AgCl(s)|Ag$$

The electrode processes are

$$Cl^- \rightarrow \tfrac{1}{2}Cl_2 + e^- \quad \text{and} \quad e^- + AgCl(s) \rightarrow Ag + Cl^-$$

and the overall process is

$$AgCl(s) \rightarrow Ag + \tfrac{1}{2}Cl_2$$

However, the AgCl(s) is in equilibrium with Ag^+ and Cl^- ions present in solution, and we can write the overall process as

$$Ag^+ + Cl^- \rightarrow Ag + \tfrac{1}{2}Cl_2$$

The emf corresponding to this process is[6]

$$E = E^\circ + \frac{RT}{F}\ln[a_{Ag^+}a_{Cl^-}]^u \tag{8.49}$$

$$= E^\circ + \frac{RT}{F}\ln K_{sp}^\circ \tag{8.50}$$

where E° is given by

$$E^\circ = E^\circ(Ag^+ + e^- \rightarrow Ag) - E^\circ(\tfrac{1}{2}Cl_2 + e^- \rightarrow Cl^-) \tag{8.51}$$

(see Table 8.1). At 25 °C the value of E° is 0.7996 − 1.35827 = −0.5587 V, and the measured emf of the cell is −1.140 V; thus

$$-1.140 = -0.5587 + \frac{RT}{F}\ln K_{sp}^\circ$$

[6]It is of interest that E does not depend on the concentration of HCl used in the cell, since if a_{Cl^-} is large, a_{Ag^+} is correspondingly small.

and

$$K_{sp} = 1.57 \times 10^{-10} \text{ mol}^2 \text{ dm}^{-6}$$

Similar cells can be devised for other electrochemical reactions. The general principle, for a salt AB, is to use the following type of cell:

$$\text{B|soluble salt of B}^- \text{ ions|AB(s)|A}$$

The emf method is a valuable one for measuring solubility products for salts of very low solubility, for which direct solubility measurements cannot be made with high accuracy. A practical difficulty with the emf method is that sometimes the electrodes do not operate reversibly.

Potentiometric Titrations

One of the most important practical applications of electrode potentials is to titrations. The procedure will be briefly explained with reference to a titration involving a precipitation, but it can also be applied to acid-base and oxidation-reduction titrations.

Suppose that a solution of sodium chloride is titrated with a solution of silver nitrate; silver chloride, which is only slightly soluble, is precipitated, and at the end point of the titration the concentration of the silver ions in the solution rises sharply. The titration can therefore be followed by inserting a clean silver sheet or wire into the solution, and connecting it, through a salt bridge, with a reference electrode, such as a calomel electrode. Typical results of such a potentiometric titration are shown in Figure 8.11a, in which the electromotive force, E, is plotted against the volume V of silver nitrate solution added to a solution of sodium chloride. To obtain a precise value of the end point of the titration, $\partial E/\partial V$ can be determined and plotted against V; as shown in Figure 8.11b, the end point is given by the maximum value. Modern instrumentation allows this differentiation to be obtained electronically.

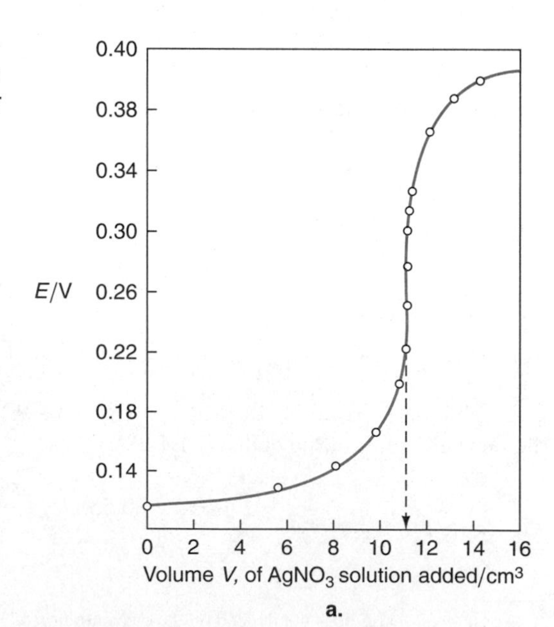

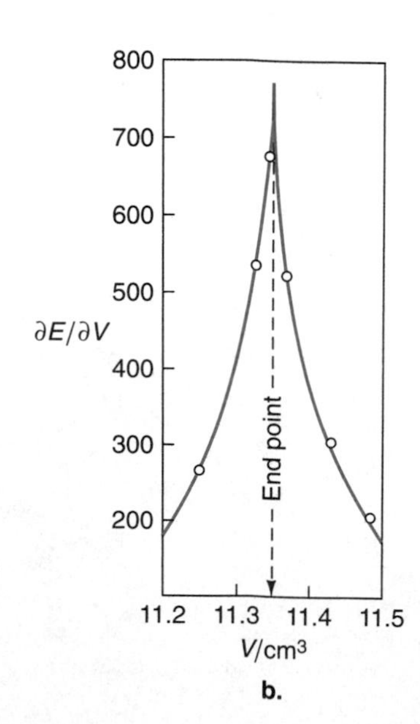

FIGURE 8.11
Typical curves for a potentiometric titration. The example shown is for the titration of a solution of sodium chloride with one of silver nitrate. At the end point there is a sharp change in the electromotive force E (a), and a maximum in $\partial E/\partial V$ (b).

To carry out an acid-base titration potentiometrically, one uses any convenient form of hydrogen electrode, such as a glass electrode. At the end point of the titration there is a sharp change in the concentration of hydrogen ions, which is reflected in a sharp change in the electromotive force. To carry out a titration of an oxidizing agent against a reducing agent one measures the potential of the redox system, which again varies sharply at the end point. In this case an inert electrode, such as a platinum electrode, is employed.

8.6 Fuel Cells

The first fuel cell was constructed in 1839 by the British physicist and lawyer Sir William Robert Grove (1811–1896); it had platinum electrodes with hydrogen bubbled over one electrode and oxygen over the other. For many years little was done to develop fuel cells for commercial purposes, but since the 1960s there has been a considerable revival of interest in this problem, particularly in view of present energy shortages. Fuel cells have been used as sources of auxiliary power in spacecraft, and major research efforts are under way to develop their use in automobiles in order to minimize air pollution and noise.

Fuel cells employ the same electrochemical principles as conventional cells. Their distinguishing feature is that the reacting substances are continuously fed into the system, so that fuel cells, unlike conventional cells, do not have to be discarded when the chemicals are consumed. The simplest type of fuel cell uses hydrogen and oxygen as fuel. Figure 8.12 is a schematic representation of a hydrogen-oxygen fuel

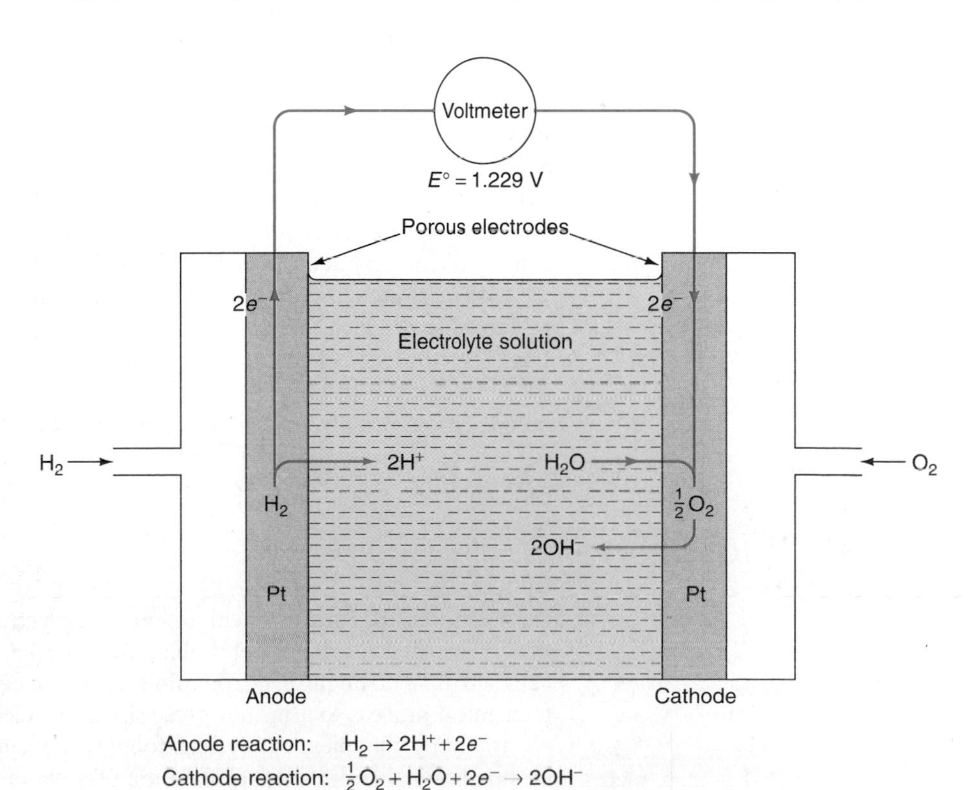

FIGURE 8.12
Diagrammatic representation of a hydrogen-oxygen fuel cell, showing the reactions occurring at the two electrodes and the overall reaction.

cell and shows the reactions occurring at the two electrodes. Various electrolyte solutions, such as sulfuric acid, phosphoric acid, and potassium hydroxide solutions, have been employed in such cells.

The overall reaction in this type of fuel cell is

$$H_2(g) + \frac{1}{2}O_2(g) \rightarrow H_2O(l)$$

This process corresponds to the transfer of two electrons. The standard Gibbs energy change in this reaction [i.e., the standard Gibbs energy of formation of $H_2O(l)$] is -237.18 kJ mol^{-1} at 25 °C, and the theoretical reversible emf, from Eq. 8.2, is

$$E° = \frac{237\,180}{2 \times 96\,485} = 1.23 \text{ V}$$

In practice voltages are less than this, because of deviations from reversible behavior. The extent of these deviations depends on the materials used as electrodes. Much present research is devoted to the development of improved electrodes. If the difficulties can be overcome, fuel cells will offer a considerable advantage over other methods of obtaining energy from fuels. As discussed in Section 3.10, furnaces in which fuels are burned suffer from second-law limitations to their efficiency. There are no such Carnot limitations to the efficiencies of fuel cells.

Because of the hazards of using hydrogen, and the cost of preparing it from other fuels, attention is being given to the development of fuel cells using other materials, such as hydrocarbons and alcohols. Much modern research has been done along these lines, some of it focusing on specially designed solid oxide electrodes that operate at 1000 °C. The focus is now on cells that operate at much lower temperatures of 600 °C and below. These cells consist of a number of units, each one of which is a ceramic device consisting of an electrode which uses air as the oxidant, the electrolyte, and a fuel electrode. These sets are arranged in stacked layers to provide the desired voltage, the oxidant and fuel being able to flow between the electrodes. Such solid oxide fuel cells are capable of operating at high energy conversion efficiency and with low pollutant emissions, but a number of practical problems still remain. Other types of fuel cells that operate at much lower temperatures include those based on polymer electrolytes, phosphoric acid, or alkaline systems, and are well under development.

Attempts are also being made to develop biochemical fuel cells that would use as fuels the products of biological processes. For example, at the bottom of the Black Sea, bacteria obtain oxygen from sulfates and produce large amounts of hydrogen sulfide that could be used as anodic fuel. It will probably be many years before these speculative possibilities are turned into practical devices.

8.7 Photogalvanic Cells

Certain materials, such as selenium, produce electricity directly when they are irradiated, and devices that employ this effect are known as *photovoltaic cells.* Such cells are to be distinguished from **photogalvanic cells,** in which irradiation induces a chemical process that in turn gives rise to an electric current. The generation of electricity by the chemical system follows the same principles as in an ordinary electrochemical cell; the special feature of a photogalvanic cell is that the reaction is brought about photochemically.

A number of devices of this kind have been set up on the laboratory scale, but more remains to be achieved in the direction of producing photogalvanic cells that will be of practical use in the utilization of solar energy. One reaction that has been studied in the laboratory is the light-induced reaction between the purple dye thionine and Fe^{2+} ions in aqueous solution. If we represent the thionine ion as T^+, and its reduced colorless form as TH^+, the overall reaction is

$$\underset{\text{purple}}{T^+} + Fe^{2+} + H^+ \rightarrow \underset{\text{colorless}}{TH^+} + Fe^{3+}$$

In the dark the equilibrium for this reaction lies over to the left. Irradiation with visible light causes this equilibrium to shift considerably to the right, and the solution becomes colorless. The most effective wavelength for bringing about this equilibrium shift is 478 nm, corresponding to the yellow-green region of the spectrum. In the dark the reaction has a positive standard Gibbs energy change, but absorption of photons by the thionine molecules provides energy for displacement of the equilibrium.

A cell that utilizes this equilibrium shift is shown schematically in Figure 8.13. There are two compartments separated by a membrane that is impermeable to thionine but permeable to Fe^{2+} and Fe^{3+}. The left-hand compartment, which contains thionine as well as Fe^{2+} and Fe^{3+}, is irradiated. The right-hand compartment contains no thionine and is kept in darkness. The irradiation of the

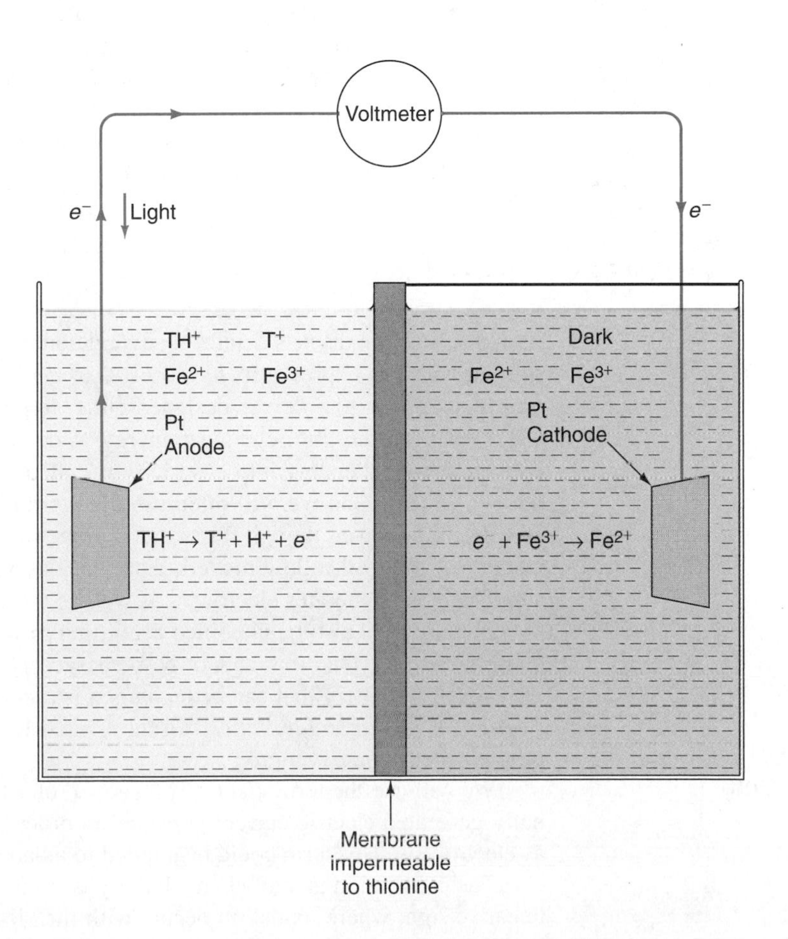

FIGURE 8.13
Schematic representation of a photogalvanic cell.

left-hand compartment causes the ratio $[TH^+]/[T^+]$ to be abnormally large, and the process

$$TH^+ \rightarrow T^+ + H^+ + e^-$$

therefore occurs at the electrode. In the dark right-hand solution, the process

$$e^- + Fe^{3+} \rightarrow Fe^{2+}$$

occurs at the electrode. The overall reaction giving rise to the emf is thus

$$TH^+ + Fe^{3+} \rightarrow T^+ + Fe^{2+} + H^+$$

and is the reverse of the reaction that is occurring as a result of the irradiation. As long as the light shines on the thionine solution, the current flows; but when it is turned off, the reaction rapidly reverts to equilibrium and the current ceases. The theoretical standard emf for the system is 0.47 V. The reversible potential obtained depends on the concentrations, in accordance with the Nernst equation, and these concentrations depend on the initial concentrations of the materials and also on the intensity of the radiation. In practice the voltages obtained are only 2 to 3% of the theoretical values because of deviations from reversibility.

There are formidable problems that make it difficult to make effective use of solar radiation with photogalvanic cells. The most important of these relates to the small proportion of light that is absorbed. With thionine the maximum absorption is at a wavelength of 478 nm, and even when monochromatic light of this wavelength is used only about 0.1% of the light is absorbed in a typical experiment. With sunlight, covering a wide range of wavelengths, the fraction absorbed is reduced by a further factor of 10 or more, since most of the visible and near-ultraviolet radiation lies outside the absorption region of the dye.

Photogalvanic cells thus represent an interesting laboratory device, but only if completely new systems are used will they become of practical importance.

8.8 Batteries, Old and New

In this section we discuss details of several commercial batteries, but first we define some terms. The expression *electrochemical cell,* or more simply, *cell,* is a general term that applies to a device designed either to convert chemical energy into electricity or electricity into chemical energy. A device that is designed specifically to convert chemical energy into electricity is called a *voltaic cell,* a *galvanic cell,* or a *battery.* In this book we will often use the word *battery* for any device that generates electricity, even though this word is sometimes regarded as applying only to a single cell; we will follow modern dictionary usage and use the word *battery* for a single cell that generates electricity, as well as for a battery of such cells.

Voltaic Cell

Batteries can also be classified according to whether or not they are capable of being recharged. One that cannot be recharged is known as a *primary* cell or battery. A cell or battery that can be recharged is commonly referred to as a *secondary, storage,* or *rechargeable battery.* In the United Kingdom a storage battery is commonly called an *accumulator.*

Electrolytic Cell

We will use the term *electrolytic cell* to refer to a cell through which an externally generated electric current is passed in order to produce chemical action, such as electrolysis. This term could be applied to a battery that is being recharged.

The convention is that when a battery is in the discharge mode the electrode of negative sign, where oxidation occurs with the electrode providing electrons to the

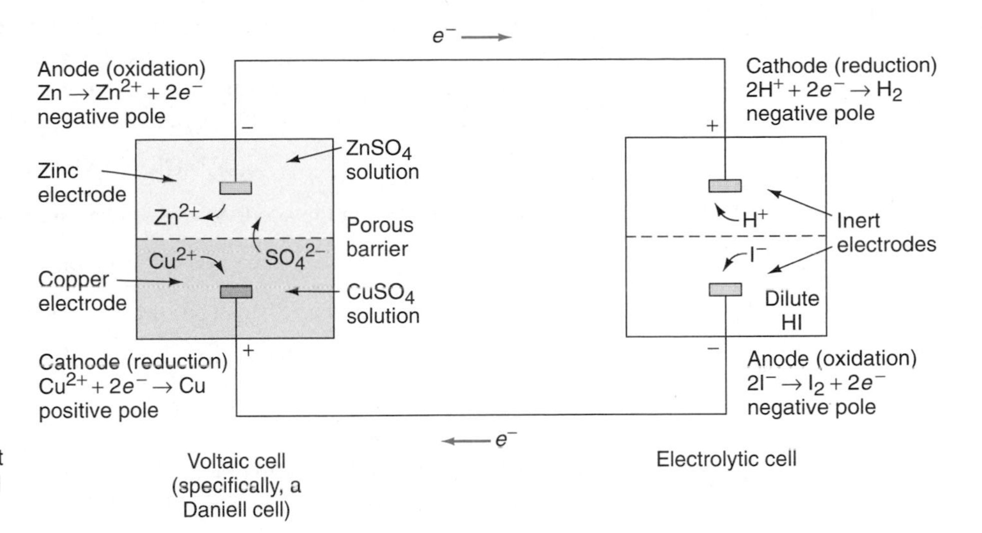

FIGURE 8.14
Relationship between voltaic and electrolytic cells. The voltaic cell on the left provides the electrical current to operate the cell on the right. The electrode signs are different for the two types of cell, but oxidation occurs at the anode and reduction occurs at the cathode.

external circuit, is called the *anode.* In Section 8.1 we discussed the Daniell cell or battery, where the anode is the left-hand electrode. The potential in the external circuit, as measured with a voltmeter, is positive. Figure 8.14 shows the relationship between batteries and electrolytic cells.

The reason a primary battery cannot be recharged is that the electrodes and electrolytes are consumed and cannot be restored to their original state by reversing the flow of electrical current. The Daniell cell (Figure 8.1) is an example of a primary battery. A secondary battery on the other hand may be assembled in a discharged state and be charged or recharged when its original reactants have become depleted. Passing current from an outside source through the cell in the direction opposite to the discharge process can reverse the spontaneous discharge process. The battery therefore functions as an electrolytic cell in the recharge process.

Most modern primary batteries are based on aqueous electrolyte systems, but free liquid in the battery introduces the possibility of leaks. Thus almost all commercial primary batteries have the electrolyte immobilized in a gel or incorporated into a microporous separator that prevents direct interaction between the electrodes. Although some water is present in the gel such batteries are referred to as *dry cells,* where the word *cell* now refers to the smallest physical unit having the components necessary to produce a voltage. Several cells may be linked in series (anode of 1 to cathode of 2, and so forth) to make a battery capable of producing a higher voltage. Thus six cells each with a potential of 1.5 V are needed to form a 9 V battery. We now consider some primary batteries.

The Original and Modified Leclanché Cell

The modern Leclanché cell, first proposed by George Leclanché in 1866, has evolved into the modern "heavy duty" battery consisting of a zinc anode in the form of a cylindrical vessel that forms the outside container. Inside it is placed a central inert cylindrical carbon (graphite) rod, C(gr), acting as the electron collector, with a paste of manganese dioxide as cathode surrounding the graphite rod. This latter combination is often referred to as the *electrode,* although only the actual cathode (or anode) material undergoes change when the discharge occurs. The electrolyte consists of an aqueous solution of zinc chloride, in gel form or incorporated into an inert filler that occupies the space between anode and cathode. A porous liner separates the zinc can from the cathode, thereby preventing internal discharge (an electrical short). The cell is represented as follows:

$$Zn(s)|ZnCl_2(aq)|MnO_2(s),\ C(gr)$$

The half-reaction at the anode where oxidation occurs is

$$Zn(s) \rightarrow Zn^{2+}(aq) + 2e^-$$

and the overall cell reaction can be summarized as

$$4Zn(s) + 8MnO_2(s) + ZnCl_2(aq) + 8H_2O(l) \rightarrow 8MnO{\cdot}OH(s) + ZnCl_2{\cdot}4Zn(OH)_2(s)$$

The potential of this cell is about 1.5 V. This cell was modified from the original Leclanché cell, which contained an ammonium and zinc chloride electrolyte. One advantage over the original cell is that it has less polarization (caused by formation of a layer of ammonia molecules on the cathode) and thus avoids the subsequent reduction of voltage under heavy drain. This change allows an almost twofold increase in battery life over that of the original battery, and better service capacity under high current drain and continuous discharge.

When naming batteries it is common to place the anode name first, followed by the cathode name. However, deviations from this naming method will sometimes be found. Thus we have the zinc-carbon cell with the bottom of the Zn case in a typical "D"-cell battery being designated negative, and with the cathode center top being positive.

Alkaline Manganese Cells

In alkaline batteries the electrolyte is a concentrated aqueous solution of ($\approx$ 30 wt. %) potassium hydroxide that has been partially converted to potassium zincate by the addition of zinc oxide. The cathode is a compressed mixture of MnO_2 and graphite, moistened with electrolyte in contact with a carbon rod that acts as a current collector. The anode is a cylindrical zinc vessel in contact with the other current collector, an external steel-can casing being used to prevent leakage through the case. For most batteries, increasing the rate of current drawn from the battery decreases the voltage; however, this system gives a relatively constant voltage even at high discharge rates. The cell reaction in this battery is

$$Zn(s) + 2MnO_2(s) + H_2O(l) \rightarrow 2MnO{\cdot}OH(s) + ZnO(s)$$

Reactions involving MnO_2 are much more complex than given in this equation because of possible further reduction of the manganese. Also, because of impurities in the raw material, the source of MnO_2 plays a role in the reactions that are observed. The cell potential for the alkaline cell is 1.55 V at room temperature.

Zinc–Mercuric Oxide and Zinc–Silver Oxide Batteries

Both of these batteries are familiarly known as "button" cells. Both cells have a high volumetric capacity (that is, they can supply a high amount of current from a particular volume), and the zinc–silver oxide cell has the added advantage of having an almost constant voltage even under conditions of high discharge. The zinc–mercuric oxide cell or "mercury cell" used in watches is of environmental concern because, in its discharged state, it can release toxic elemental mercury. Only the chemistry of the zinc–silver oxide battery will be discussed here because the chemical reactions in both are quite similar. The center cap of the battery is the negative pole formed by a steel electron collector in contact with the zinc anode. This is insulated from the rest of the positive steel case that contains the silver oxide cathode in a KOH electrolyte. The internal contact is made through a porous separator. The zinc–silver oxide cell may be written as

$$Zn(s)|ZnO(s)|KOH(aq)|Ag_2O(s), C(s)$$

The overall cell reaction is

$$Zn(s) + Ag_2O(s) \rightarrow ZnO(s) + 2Ag(s)$$

where the anode (oxidation) reaction is

$$Zn(s) + 2OH^-(aq) \rightarrow ZnO(s) + H_2O(l) + 2e^-$$

and the cathode (reduction) reaction is

$$Ag_2O(s) + H_2O(l) + 2e^- \rightarrow 2Ag(s) + 2OH^-(aq)$$

The E_{cell} for this battery is 1.60 V, whereas the overall reaction for the mercury battery is

$$Zn(s) + HgO(s) \rightarrow ZnO(s) + Hg(l)$$

and its E_{cell} is 1.357 V.

Metal-Air Batteries

The primary objective of miniature battery design is to maximize the energy density or cell capacity in a small container. What better way is there to accomplish this than to cause one of the necessary components to flow continuously into the cell? In metal-air batteries the cathodic reactant is the oxygen in the air, which need not be stored within the confines of the battery. In essence, this is similar to one-half of a fuel cell. (See Section 8.6.) Aluminum, lithium, magnesium, and zinc have been examined extensively for use as the anode in these systems. The first three suffer from severe corrosion problems; only zinc is satisfactory and is used in hearing-aid batteries. The zinc anode is separated from the cathode by a separator disc that allows oxygen to pass but prevents liquid from escaping. Many of the techniques used are proprietary, that is, trade secrets. The oxygen-electrode reaction is complex but in basic solution it may be considered as a two-stage process, written as

$$O_2(aq) + H_2O(l) + 2e^- \rightleftharpoons HO_2^-(aq) + OH^-(aq)$$

followed by

$$HO_2^-(aq) + H_2O(l) + 2e^- \rightarrow 3OH^-(aq)$$

The hydroperoxide ion may form a metal-oxygen bond that may be reduced or the hydroperoxide ion may decompose to re-form oxygen. The anode reaction may be written as

$$Zn(s) + 4OH^-(aq) - 2e^- \rightarrow ZnO_2^{2-}(aq) + 2H_2O(l)$$

When the solution is saturated with zincate ion, zinc oxide is formed

$$Zn(s) + 2OH^-(aq) - 2e^- \rightarrow ZnO(s) + H_2O(l)$$

Preventing zinc-oxide passivation of the cell (in which the anode fails to function properly) is a major design problem. The cell voltage is about 1.4 V, which drops to about 1.25 V when half of the anode material is consumed. The volumetric energy density, however, is about 30% better than that of the closest competitor, the zinc–silver oxide cell.

Nickel-Cadmium (Ni-cad) Secondary Battery

The Nicad battery should be called the cadmium-nickel battery in keeping with the convention that the anode should be named first. Because of its rechargeability, this battery is popular in portable hand tools, but it presents a disposal problem because of the toxicity of cadmium. In this battery the anode reaction is

$$Cd(s) + 2OH^-(aq) \rightarrow Cd(OH)_2(s) + 2e^-$$

The cathode reaction is

$$2NiO(OH)(s) + 2H_2O(l) + 2e^- \rightarrow 2Ni(OH)_2(s) + 2OH^-(aq)$$

Combining these gives the overall cell reaction

$$Cd(s) + 2NiO(OH)(s) + 2H_2O(l) \underset{\text{charge}}{\overset{\text{discharge}}{\rightleftharpoons}} 2Ni(OH)_2(s) + Cd(OH)_2(s)$$

The $E_{cell} = 1.4$ V and the reaction can be reversed upon recharging as indicated by the double arrow.

The Lead-Acid Storage Battery

The familiar lead-acid battery was invented by Gaston Planté in 1859; it should be named the lead–lead oxide battery to be consistent with practice. Today more than a third of the world's output of lead goes into constructing these batteries. The typical lead-acid battery anode has a negative lead grid support filled with porous lead (known as lead sponge) where oxidation occurs, and a positive lead grid support cathode with a filler of lead dioxide, PbO_2. Both anode and cathode electrodes are submerged into a solution of sulfuric acid with a porous separator sheet between them to keep the plates from touching. The cell is represented by

$$Pb(s)|PbSO_4(s), H_2SO_4(aq)|PbSO_4(s)|PbO_2(s)|Pb(s)$$

The overall process may be written as

$$Pb(s) + PbO_2(s) + 2H_2SO_4(aq) \rightleftharpoons 2PbSO_4(s) + 2H_2O(l)$$

The anode (oxidation) negative electrode is

$$Pb(s) + SO_4^{2-}(aq) \underset{\text{charge}}{\overset{\text{discharge}}{\rightleftharpoons}} PbSO_4(s) + 2e^-$$

In the discharge process lead sulfate coats the lead electrode, and charging causes regeneration of the lead. The cathode (reduction) electrode reaction is

$$PbO_2(s) + 4H^+(aq) + SO_4^{2-}(aq) + 2e^- \underset{\text{charge}}{\overset{\text{discharge}}{\rightleftharpoons}} PbSO_4(s) + 2H_2O(l)$$

Again, lead sulfate forms on the lead dioxide plates. As the cell is discharged, sulfuric acid is consumed and water is formed so that the state of charge of the battery may be determined from the density of the electrolyte. At full charge, each cell produces 2.15 V at 25 °C. This value is different from that calculated at standard-state conditions and is typical for all batteries of this type. At full discharge the voltage drops to 1.98 V. Thus, the charged state of the battery may be followed by measuring the specific gravity of the sulfuric acid. When the specific gravity of the sulfuric acid falls below 1.2, the battery needs to be recharged. The open-circuit voltage may be predicted thermodynamically from the sulfuric acid and water activities and the temperature. For automotive use, six cells are placed in series, giving a voltage of 12.6 V. Complex chemistry and careful design engineering are involved in making such batteries function for five or more years, and the reader is referred to the references.

Lithium Ion Batteries

Familiar in cellular phones, camcorders, and computers, lithium ion batteries offer distinct advantages. They are rechargeable and have a high volumetric energy density of 300 W h/dm^3, where W h is the watt hour, and with a gravimetric energy density of 125 W h/kg, far surpassing most batteries. However, their development required understanding of a more complicated chemistry than that previously discussed. The materials used as electrodes in lithium ion batteries were first patented in 1981. Curiously, these batteries do not contain lithium as isolated metal. The chemistry involves a process called **intercalation,** in which a lithium ion is inserted as Li metal into a host without damaging its structure. See Figure 8.15, where, in the charging process, graphite intercalates lithium. The process is

$$Li^+ + e^- + C_6 \rightarrow LiC_6$$

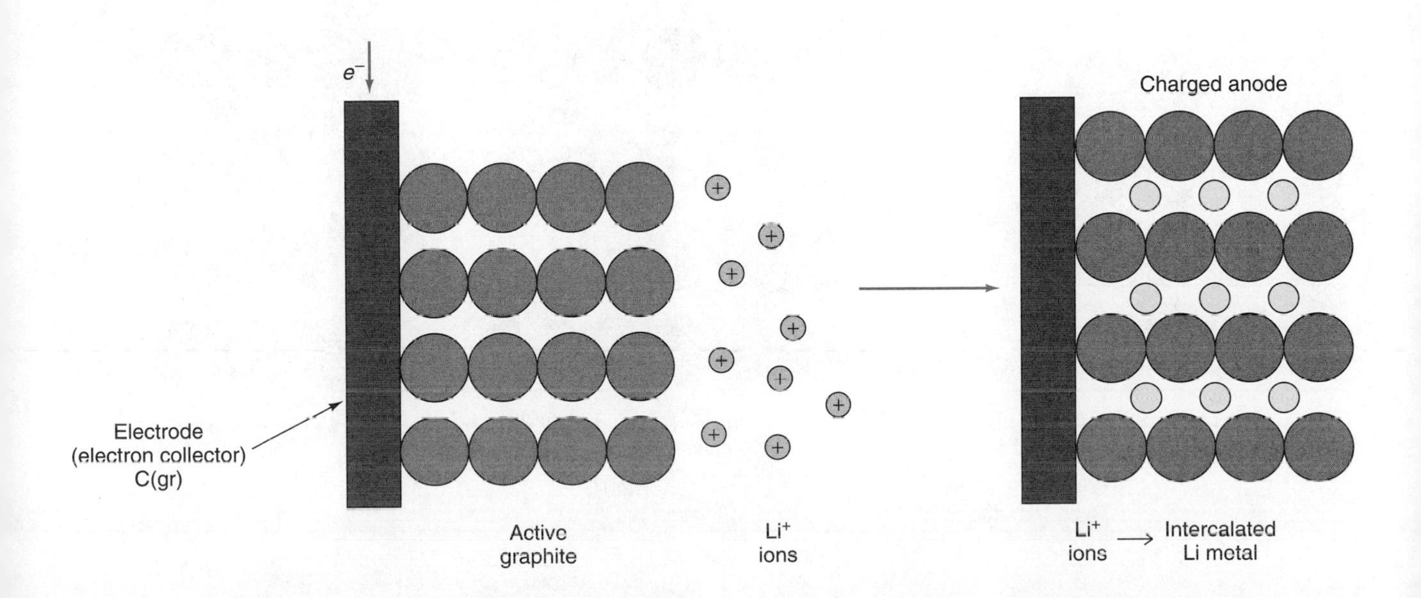

FIGURE 8.15
Intercalation. In the charging process electrons from an outside source are supplied along the conductor to the electrode, which will be the anode in the discharge process. In the process of intercalation shown, ions of Li^+ are inserted into the structure of the graphite and, in the process, each gains an electron from the conductor without damage to the original structure of the host. The process is reversed in the discharge mode.

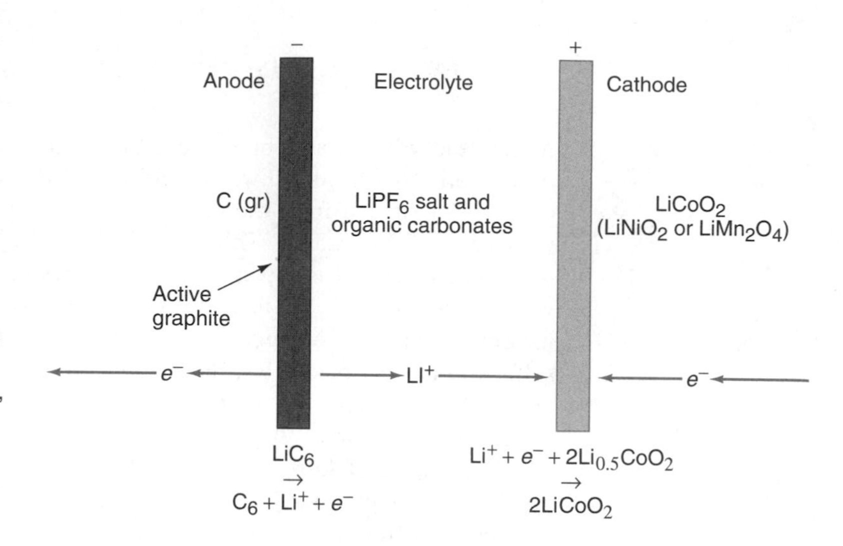

FIGURE 8.16
Discharge process for the lithium ion battery. Electrons leave the anode conductor via the external circuit. Li^+ ions move through the electrolyte ($LiPF_6$ salt and organic carbonate in the case shown) to the cathode in the $Li_{0.5}CoO_2$ state, where electrons from the external circuit enter the cathode from the conductor and the Li^+ fully loads the cathode with Li metal.

where external electrons are supplied via a carrier to the active material, graphite. On discharge, the process is reversed and the Li-loaded graphite anode de-intercalates lithium as Li^+ with the electrons going into the external circuit. Several materials ($LiCoO_2$, $LiNiO_2$, or $LiMn_2O_4$) may be used as the cathode material. The selection of the actual material depends upon the particular characteristics desired from the cell. We will use $LiCoO_2$ as an example throughout this discussion. At the cathode, $LiCoO_2$ de-intercalates lithium during charging. Thus the processes at both electrodes are somewhat similar.

$$2LiCoO_2 \rightarrow 2Li_{0.5}CoO_2 + Li^+ + e^-$$

During discharge, $Li_{0.5}CoO_2$ intercalates lithium ions. The overall reaction for the discharge process is shown in Figure 8.16, where the salt $LiPF_6$ and an organic carbonate serve as the electrolyte. In one design of the battery, an organic

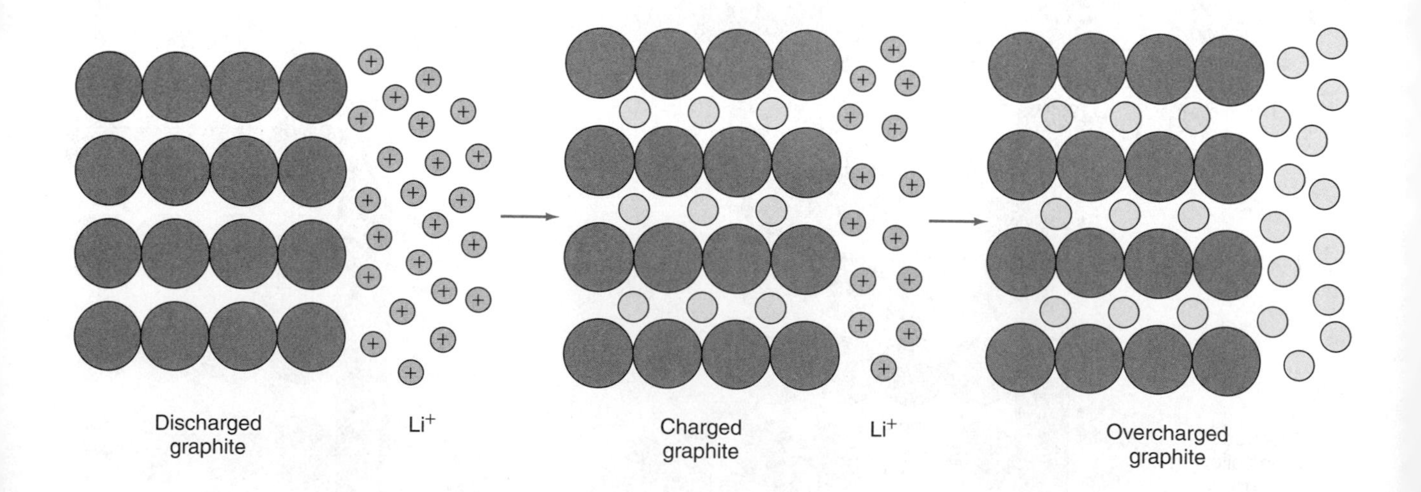

FIGURE 8.17
(a) $LiCoO_2$ in its discharged state is fully loaded with Li. (b) Charging of $LiCoO_2$ removes Li from the structure, leaving Li^+ now in the electrolyte. (c) Overcharging the $LiCoO_2$ removes too much lithium, causing its structure to collapse, which releases oxygen to the atmosphere and thereby shortens battery life.

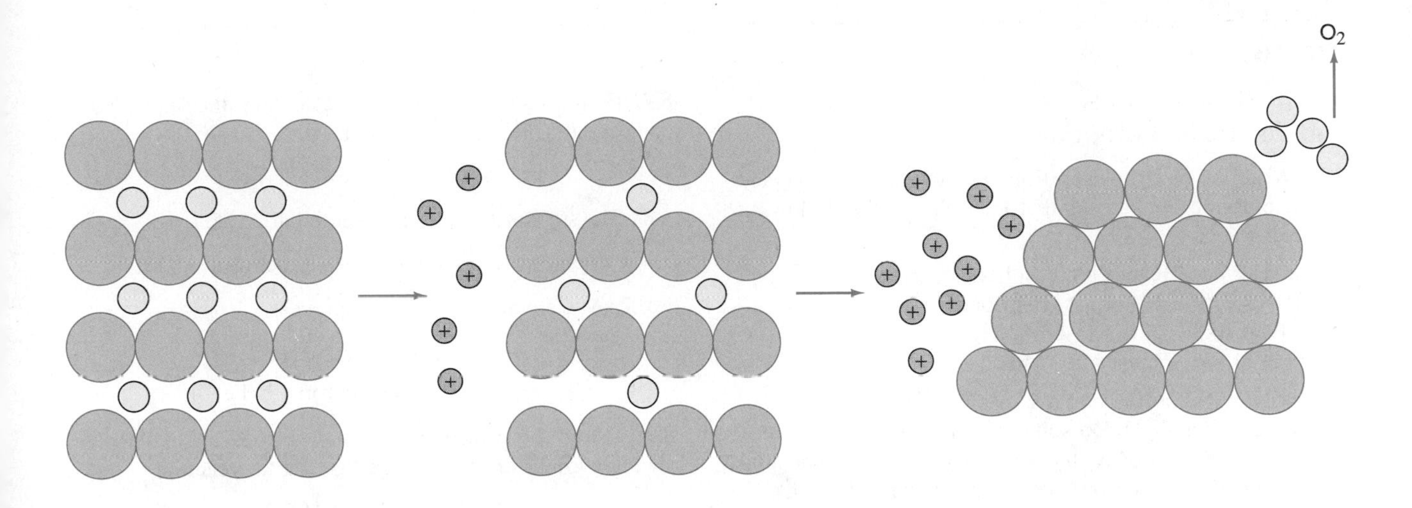

FIGURE 8.18
Discharged graphite is loaded with Li in the charging process. Overcharging of the graphite allows the excess lithium metal to plate as metal on the surface of the graphite, thereby shortening battery life.

fluoropolymer is used as the electrolyte to transfer ions, but not the electrons. Since charging involves just the reverse processes, it is instructive to consider what happens when $LiCoO_2$ and graphite are overcharged. In Figure 8.17, $LiCoO_2$ is charged and then overcharged. When overcharged, the structure of $LiCoO_2$ is destroyed and O_2 is released, thereby decreasing cycle life, which is the number of times the battery can be discharged and then recharged. At the same time electrolyte oxidation adds Li^+ to the system. In the overcharging of graphite, shown in Figure 8.18, the excess lithium can begin to plate as a metallic film by electrolytic reduction on the surface of the graphite. This can lead to an electrical short as well as decreased cycle life. To prevent overcharging, the complete characteristics of the charging process must be determined under a variety of conditions. In the case cited, a constant voltage of 4.1 V leads to complete charge and avoids overcharge of the anode and the cathode. Chemists working on the development of such new batteries require a broad range of knowledge, encompassing electrical measurements and material science besides basic chemistry.

Another popular design of a lithium ion battery incorporates the Li anode and a transition metal oxide or sulfide as the cathode. The latter includes MnO_2, V_6O_{13}, and TiS_2. The electrolyte is a polymer that allows the flow of Li ions but not of electrons.

KEY EQUATIONS

Thermodynamics of an electrochemical cell:

$$\Delta G = -zEF$$

where z = charge number of cell reaction; E = emf; F = Faraday constant.

Nernst equation:

$$E = E^\circ - \frac{RT}{zF}\ln\left(\frac{[\mathrm{Y}]^y[\mathrm{Z}]^z}{[\mathrm{A}]^a[\mathrm{B}]^b}\right)^u$$

for a reaction $a\mathrm{A} + b\mathrm{B} \rightleftharpoons y\mathrm{Y} + z\mathrm{Z}$.

Nernst potential:

$$\Delta\Phi = \frac{RT}{zF}\ln\frac{c_1}{c_2}$$

Emf of a concentration cell:

$$E = \frac{RT}{zF}\ln\frac{m_2}{m_1}$$

PROBLEMS

Electrode Reactions and Electrode Potentials

8.1. Write the electrode reactions, the overall reaction, and the expression for the emf for each of the following reversible cells.

a. Pt, H_2(1 bar)|HCl(aq)|Pt, Cl_2(1 bar)
b. Hg|Hg_2Cl_2(s)|HCl(aq)|Pt, H_2(1 bar)
c. Ag|AgCl(s)|KCl(aq)|Hg_2Cl_2(s)|Hg
d. Pt, H_2(1 bar)|HI(aq)|AuI(s)|Au
e. Ag|AgCl(s)|KCl(c_1) ¦¦ KCl(c_2)|AgCl(s)|Ag

8.2. At 25 °C and pH 7, a solution containing compound A and its reduced form AH_2 has a standard electrode potential of −0.60 V. A solution containing B and BH_2 has a standard potential of −0.16 V. If a cell were constructed with these systems as half-cells,

a. Would AH_2 be oxidized by B or BH_2 oxidized by A under standard conditions?
b. What would be the reversible emf of the cell?
c. What would be the effect of pH on the equilibrium ratio $[B][AH_2]/[A][BH_2]$?

8.3. Calculate the standard electrode potential for the reaction $Cr^{2+} + 2e^- \rightarrow Cr$ at 298 K. The necessary $E°$ values are

a. $Cr^{3+} + 3e^- \rightarrow Cr \quad E° = -0.74$ V
b. $Cr^{3+} + e^- \rightarrow Cr^{2+} \quad E° = -0.41$ V

8.4. Write the individual electrode reactions and the overall cell reaction for the following cell:

$$\text{Pt, } H_2|H^+(1\ m) \ ⁞⁞\ F^{2-}, S^{2-}, H^+(aq)|\text{Pt}$$

where F^{2-} represents the fumarate ion and S^{2-} the succinate ion. Write the expression for the emf of the cell.

8.5. Design electrochemical cells in which each of the following reactions occurs:

a. $Ce^{4+}(aq) + Fe^{2+}(aq) \rightarrow Ce^{3+}(aq) + Fe^{3+}(aq)$
b. $Ag^+(aq) + Cl^-(aq) \rightarrow AgCl(s)$
c. $HgO(s) + H_2(g) \rightarrow Hg(l) + H_2O(l)$

In each case, write the representation of the cell and the reactions at the two electrodes.

Thermodynamics of Electrochemical Cells

8.6. Calculate the equilibrium constant at 25 °C for the reaction

$$2Fe^{3+}(aq) + 2I^-(aq) \rightarrow 2Fe^{2+}(aq) + I_2(s)$$

using the standard electrode potentials given in Table 8.1.

8.7. From data in Table 8.1, calculate the equilibrium constant at 25 °C for the reaction

$$Sn + Fe^{2+} \rightarrow Sn^{2+} + Fe$$

8.8. The standard electrode potential at 25 °C for

$$\text{cytochrome } c\ (Fe^{3+}) + e^- \rightarrow \text{cytochrome } c\ (Fe^{2+})$$

is 0.25 V. Calculate $\Delta G°$ for the process

$$\tfrac{1}{2}H_2(g) + \text{cytochrome } c\ (Fe^{3+}) \rightarrow H^+ + \text{cytochrome } c\ (Fe^{2+})$$

8.9. Using the values given in Table 8.1, calculate the standard Gibbs energy change $\Delta G°$ for the reaction

$$H_2 + \tfrac{1}{2}O_2 \rightarrow H_2O$$

***8.10.** From the data in Table 8.1, calculate the equilibrium constant at 25 °C for the reaction

$$2Cu^+ \rightarrow Cu^{2+} + Cu$$

What will be produced if Cu_2O is dissolved in dilute H_2SO_4?

8.11. For the reaction $3H_2$(g, 1 atm) + Sb_2O_3(s, cubic) → 2Sb(s) + $3H_2O$(l), $\Delta G° = -83.7$ kJ [Roberts and Fenwick, *J. Amer. Chem. Soc.*, *50*, 2146(1928)]. Calculate the potential developed by the cell

$$\text{Pt}|H_2(\text{g, 1 atm})|H^+|Sb_2O_3(\text{s, cubic})|\text{Sb(s)}$$

Which electrode will be positive?

Nernst Equation and Nernst Potentials

8.12. Calculate the emf for the following cell at 25 °C:

$$\text{Pt, } H_2(1\ \text{bar})|\text{HCl}(0.5\ m) \ ⁞⁞\ \text{HCl}(1.0\ m)|\text{Pt, } H_2(1\ \text{bar})$$

8.13. The pyruvate-lactate system has an $E°'$ value of −0.185 V at 25 °C and pH 7.0. What will be the potential of this system if the oxidation has gone to 90% completion?

8.14. a. From the data in Table 8.1, calculate the standard electrode potential for the half-reaction

$$Fe^{3+} + 3e^- \rightarrow Fe$$

b. Calculate the emf at 25 °C of the cell

$$\text{Pt}|Sn^{2+}(0.1\ m), Sn^{4+}(0.01\ m) \ ⁞⁞\ Fe^{3+}(0.5\ m)|\text{Fe}$$

8.15. The cell Pt|H_2(1 bar), H^+ ¦¦ KCl(saturated)|Hg_2Cl_2|Hg was used to measure the pH of a solution of 0.010 *M* acetic acid in 0.0358 *M* sodium acetate. Calculate the cell potential expected at 25 °C [$K_a = 1.81 \times 10^{-5}$ for acetic acid].

8.16. The voltage required to electrolyze certain solutions changes as the electrolysis proceeds because the concentrations in the solution are changing. In an experiment, 500 dm^3 of a 0.0500 M solution of copper (II) bromide was electrolyzed until 2.872 g Cu was deposited. Calculate the theoretical minimum voltage required to sustain the electrolysis reaction at the beginning and at the end of the experiment.

8.17. Calculate the concentration of I_3^- in a standard solution of iodine in 0.5 M KI, making use of the following standard electrode potentials:

$$I_2 + 2e^- \rightarrow 2I^- \quad E^\circ = 0.5355 \text{ V}$$
$$I_3^- + 2e^- \rightarrow 3I^- \quad E^\circ = 0.5365 \text{ V}$$

The molality of I^- in the standard solution can be assumed to be 0.5 m.

8.18. Calculate the Nernst potential at 25 °C arising from the equilibrium established in Problem 7.41.

8.19. It might seem plausible to separate lead and gold by making use of the great difference between their standard electrode potentials (Table 8.1). In order to test this idea, one might electrolyze a solution containing 0.0100 M $AuNO_3$ and 0.0100 M $Pb(NO_3)_2$ in a well-stirred tank using platinum electrodes at low current density. As the potential difference is slowly increased from zero, which metal will be deposited first? What will be the concentration of this metal ion in solution when the second metal begins to be deposited? Do you think this is an acceptable method of separating the two metals?

8.20. Calculate the emf of the cell

$$\text{Pt, } H_2(1 \text{ bar})|HCl(0.1\ m) \,\vdots\vdots\, HCl(0.2\ m)|\text{Pt, } H_2\ (10 \text{ bar})$$

***8.21.** Suppose that the cell in Problem 8.20 is set up but that the two solutions are separated by a membrane that is permeable to H^+ ions but impermeable to Cl^- ions. What will be the emf of the cell at 25 °C?

***8.22.** A typical biological cell has a volume of 10^{-9} cm^3, a surface area of 10^{-6} cm^2, and a membrane thickness of 10^{-6} cm; the dielectric constant of the membrane may be taken as 3. Suppose that the concentration of K^+ ions inside the cell is 0.155 M and that the Nernst potential across the cell wall is 0.085 V.

a. Calculate the net charge on either side of the wall, and
b. Calculate the fraction of the K^+ ions in the cell that are required to produce this charge.

***8.23.** Calculate the emf at 25 °C of the cell

$$\text{Pt, } H_2(1 \text{ bar})|H_2SO_4(0.001\ m)|CrSO_4(s)|Cr$$

given the following standard electrode potential:

$$CrSO_4(s) + 2e^- \rightarrow Cr + SO_4^{2-} \quad E^\circ = -0.40 \text{ V}$$

a. First make the calculation neglecting activity coefficient corrections.
b. Then make the calculation using activity coefficients estimated on the basis of the Debye-Hückel limiting law.

***8.24.** Write the individual electrode reactions and the overall reaction for

$$Cu|CuCl_2(aq)|AgCl(s)|Ag$$

If the emf of the cell is 0.191 V when the concentration of $CuCl_2$ is 1.0×10^{-4} M and is -0.074 V when the concentration is 0.20 M, make an estimate of the mean activity coefficient in the latter solution.

***8.25. a.** Write both electrode reactions and the overall reaction for the cell

$$Tl|TlCl(s)|CdCl_2(0.01\ m)|Cd$$

b. Calculate E and E° for this cell at 25 °C from the following information:

$$Tl^+ + e^- \rightarrow Tl \quad E^\circ = -0.34 \text{ V}$$
$$Cd^{2+} + 2e^- \rightarrow Cd \quad E^\circ = -0.40 \text{ V}$$

The solubility product for TlCl is 1.6×10^{-3} $mol^2\ dm^{-6}$ at 25 °C.

8.26. Problem 7.44 involved calculating the concentrations on each side of a membrane after a Donnan equilibrium had become established. Which side of the membrane is positively charged? Calculate the Nernst potential across the membrane if the temperature is 37 °C.

8.27. The oxidation of lactate to pyruvate by the oxidized form of cytochrome c—represented as cytochrome c (Fe^{3+})—is an important biological reaction. The following are the relevant $E^{\circ\prime}$ values, relating to pH 7 and 25 °C:

	$E^{\circ\prime}$/V
$\text{pyruvate}^- + 2H^+ + 2e^- \rightarrow \text{lactate}^-$	-0.185
cytochrome c (Fe^{3+}) + e^- → cytochrome c (Fe^{2+})	0.254

Calculate the equilibrium ratio

$$\frac{[\text{cytochrome } c\ (Fe^{2+})]^2[\text{pyruvate}^-]}{[\text{cytochrome } c\ (Fe^{3+})]^2[\text{lactate}^-]}$$

at pH 7 and 25 °C. Also calculate the ratio at pH 6.

8.28. Suppose that the cell

$$Ag|AgCl(s)|HCl(0.10\ m) \,\vdots\vdots\, HCl(0.01\ m)|AgCl(s)|Ag$$

is set up and that the membrane separating the two solutions is permeable only to H^+ ions. What is the emf of the cell at 25 °C?

8.29. a. Consider the cell

$$\text{Pt, } H_2(1 \text{ bar})|HCl(m_1) \,\vdots\vdots\, HCl(m_2)|\text{Pt, } H_2(1 \text{ bar})$$

in which the solutions are separated by a partition that is permeable to both H^+ and Cl^-. The ratio of the speeds with which these ions pass through the membrane is the ratio of their transport numbers t_+ and t_-. Derive an expression for the emf of this cell.

b. If when $m_1 = 0.01\ m$ and $m_2 = 0.01\ m$ the emf is 0.0190 V, what are the transport numbers of the H^+ and Cl^- ions?

8.30. The metal M forms a soluble nitrate and a very slightly soluble chloride. The cell

$$M|M^+(0.1\ m), HNO_3(0.2\ m)|H_2(1\ \text{bar}), Pt$$

has a measured $E = -0.40$ V at 298.15 K. When sufficient solid KCl is added to make the solution of the cell 0.20 m in K^+, the emf changes to -0.15 V at 298.15 K as MCl precipitates. Calculate the K_{sp} of MCl, taking all activity coefficients to be unity.

8.31. The substance nicotinamide adenine dinucleotide (NAD^+) plays an important role in biological systems; under the action of certain enzymes it can react with a reducing agent and release a proton to the solution to form its reduced form NADH. With pyruvate the reduced form NADH undergoes the reaction

$$\text{NADH} + \text{pyruvate}^- + H^+ \rightleftharpoons \text{NAD}^+ + \text{lactate}^-$$

The appropriate $E^{\circ\prime}$ values, relating to 25 °C and pH 7, are

$$\text{pyruvate}^- + 2H^+ + 2e^- \rightarrow \text{lactate}^- \quad E^{\circ\prime} = -0.19\ \text{V}$$

$$\text{NAD}^+ + H^+ + 2e^- \rightarrow \text{NADH} \quad E^{\circ\prime} = -0.34\ \text{V}$$

Use these values to calculate $\Delta G^{\circ\prime}$ for the reaction, and also the equilibrium ratio

$$\frac{[\text{lactate}^-][\text{NAD}^+]}{[\text{pyruvate}^-][\text{NADH}]}$$

(a) at pH 7, and (b) at pH 8.

Temperature Dependence of Cell emfs

8.32. a. Calculate the standard emf E° for the reaction

$$\text{fumarate}^{2-} + \text{lactate}^- \rightarrow \text{succinate}^{2-} + \text{pyruvate}^-$$

on the basis of the following information:

$$\text{fumarate}^{2-} + 2H^+ + 2e^- \rightarrow \text{succinate}^{2-} \quad E^{\circ\prime} = 0.031\ \text{V}$$

$$\text{pyruvate}^- + 2H^+ + 2e^- \rightarrow \text{lactate}^- \quad E^{\circ\prime} = -0.185\ \text{V}$$

The $E^{\circ\prime}$ values relate to pH 7. The temperature coefficient $\partial E/\partial T$ for this cell is 2.18×10^{-5} V K^{-1}.

b. Calculate ΔG°, ΔH°, and ΔS° at 25 °C.

8.33. The Weston standard cell (see Figure 8.2b) is

$$\text{Cd amalgam}|CdSO_4 \cdot \tfrac{8}{3}H_2O(s)|Hg_2SO_4(s), Hg$$
(saturated solution)

a. Write the cell reaction.

b. At 25 °C, the emf is 1.018 32 V and $\partial E^\circ/\partial T = -5.00 \times 10^{-5}$ V K^{-1}. Calculate ΔG°, ΔH°, and ΔS°.

8.34. Salstrom and Hildebrand [*J. Amer. Chem. Soc., 52,* 4650(1930)] reported the following data for the cell

$$Ag(s)|AgBr(s)|HBr(aq)|Br_2(g, 1\ \text{atm})|Pt$$

t/°C	442.3	456.0	490.9	521.4	538.3	556.2
E/V	0.8031	0.7989	0.7887	0.7803	0.7751	0.7702

Find the temperature coefficient for this cell assuming a linear dependence of the cell potential with temperature. What is the entropy change for the cell reaction?

8.35. The reaction taking place in the cell $Mg(s)|Mg^{2+}(aq), Cl^-(aq)|Cl_2(g,1\ \text{atm})|Pt$ is found to have an entropy change of -337.3 J K^{-1} mol^{-1} under standard conditions. What is the temperature coefficient for the cell?

***8.36. a.** Estimate the Gibbs energy of formation of the fumarate ion, using data in Problem 8.32 and the following values.

$$\Delta_f G^\circ(\text{succinate, aq}) = -690.44\ \text{kJ mol}^{-1}$$
$$\Delta_f G^\circ(\text{acetaldehyde, aq}) = 139.08\ \text{kJ mol}^{-1}$$
$$\Delta_f G^\circ(\text{ethanol, aq}) = -181.75\ \text{kJ mol}^{-1}$$
$$\text{acetaldehyde} + 2H^+ + 2e^- \rightarrow \text{ethanol} \quad E^{\circ\prime} = -0.197\ \text{V}$$

b. If the $\partial E^\circ/\partial T$ value for the process

$$\text{fumarate}^{2-} + \text{ethanol} \rightarrow \text{succinate}^{2-} + \text{acetaldehyde}$$

is 1.45×10^{-4} V K^{-1}, estimate the enthalpy of formation of the fumarate ion from the following values.

$$\Delta_f H^\circ(\text{succinate, aq}) = -908.68\ \text{kJ mol}^{-1}$$
$$\Delta_f H^\circ(\text{acetaldehyde, aq}) = -210.66\ \text{kJ mol}^{-1}$$
$$\Delta_f H^\circ(\text{ethanol, aq}) = -287.02\ \text{kJ mol}^{-1}$$

***8.37. a.** Calculate the emf at 298.15 K for the cell

$$Tl|TlBr|HBr\ (\text{unit activity})|H_2(1\ \text{bar}), Pt$$

b. Calculate ΔH for the cell reaction in the following cell.

$$Tl|Tl^+\ (\text{unit activity}), H^+\ (\text{unit activity})|H_2(1\ \text{bar}), Pt$$

For the half-cell

$$Tl^+ + e^- \rightarrow Tl \quad E^\circ = 0.34\ \text{V}$$
$$\partial E/\partial T = -0.003\ \text{V/K} \quad \text{and}$$
$$K_{sp}(TlBr) = 10^{-4}\ \text{mol}^2\ \text{dm}^{-6}$$

Applications of emf Measurements

8.38. Calculate the solubility product and the solubility of AgBr at 25 °C on the basis of the following standard electrode potentials:

$$AgBr(s) + e^- \rightarrow Ag + Br^- \qquad E° = 0.0713 \text{ V}$$
$$Ag^+ + e^- \rightarrow Ag \qquad E° = 0.7996 \text{ V}$$

8.39. The emf of a cell

$$Pt, H_2(1 \text{ bar})|HCl(aq)|AgCl(s)|Ag$$

was found to be 0.517 V at 25 °C. Calculate the pH of the HCl solution.

8.40. The emf of the cell

$$Ag|AgI(s)I^-(aq)\vdots\vdots Ag^+(ag)|Ag$$

is 0.9509 V at 25 °C. Calculate the solubility and the solubility product of AgI at that temperature.

8.41. An electrochemical cell M(s)|MCl(aq, 1.0 *m*)|AgCl(s)|Ag(s), where MCl is the chloride salt of the metal electrode M, yields a cell potential of 0.2053 V at 25 °C. What is the mean activity coefficient $\gamma_\pm$ of the electrolyte MCl? $E°$ for the M(s)|M^+ electrode is 0.0254 V.

8.42. The following thermodynamic data apply to the complete oxidation of butane at 25 °C.

$$C_4H_{10}(g) + (13/2)O_2(g) \rightarrow 4CO_2(g) + 5H_2O(l)$$
$$\Delta H° = -2877 \text{ kJ mol}^{-1}$$
$$\Delta S° = -432.7 \text{ J K}^{-1} \text{ mol}^{-1}$$

Suppose that a completely efficient fuel cell could be set up utilizing this reaction. Calculate (a) the maximum electrical work and (b) the maximum total work that could be obtained at 25 °C.

***8.43.** At 298 K the emf of the cell

$$Cd, Hg|CdCl_2(aq, 0.01\ m), AgCl(s)|Ag$$

is 0.7585 V. The standard emf of the cell is 0.5732 V.

a. Calculate the mean activity coefficient for the Cd^{2+} and Cl^- ions.
b. Compare the value with that estimated from the Debye-Hückel limiting law, and comment on any difference.

***8.44.** The following emf values were obtained by H. S. Harned and Copson [*J. Amer. Chem. Soc., 55,* 2206(1933)] at 25 °C for the cell

$$Pt,H_2(1 \text{ bar})|LiOH(0.01\ m), LiCl(m)|AgCl(s)|Ag$$

at various molalities *m* of LiCl:

m/mol kg^{-1}	0.01	0.02	0.05	0.10	0.20
E/V	1.0498	1.0318	1.0076	1.9888	0.9696

Obtain from these data the ionic product of water.

Essay Questions

8.45. Explain how emf measurements can be used to obtain $\Delta G°$, $\Delta H°$, and $\Delta S°$ for a reaction.

8.46. Suggest an additional example, giving details, for each of the electrochemical cells listed in Figure 8.8.

SUGGESTED READING

See Suggested Reading for Chapter 7, and also:

B. E. Conway, *Electrochemical Data,* Amsterdam: Elsevier Pub. Co., 1952.

A. J. de Bethune and N. A. S. Loud, *Standard Aqueous Electrode Potentials and Temperature Coefficients at 25 °C,* Skokie, Ill.: C. A. Hampel, 1964.

D. J. G. Ives and G. J. Janz (Eds.), *Reference Electrodes; Theory and Practice,* New York: Academic Press, 1961.

C. A. Vincent, with F. Bonino, M. Lazzari, and B. Scrosati, *Modern Batteries: An Introduction to Electrochemical Power Sources,* London: Edward Arnold, 1984.

For further details on ion-selective electrodes, see

Brenda B. Shaw, "Modification of solid electrodes in electro-analytical chemistry, 1978–1988," in *Electrochemistry, Past and Present,* J. T. Stock and Mary Virginia Orna (Eds.), American Chemical Society, 1989, pp. 318–338.

Jiri Vanata, *Principles of Chemical Sensors,* New York: Plenum Press, 1989.

For further details about fuel cells, see

J. O'M. Bockris and S. Srinivasan, *Fuel Cells: Their Electrochemistry,* New York: McGraw-Hill, 1969.

A. K. Vijh, "Electrochemical principles in a fuel cell," *J. Chem. Ed., 47,* 680–686(1970).

Photogalvanic cells are treated in

J. R. Bolton (Ed.), *Solar Power and Fuels,* New York: Academic Press, 1977.

W. D. K. Clark and J. A. Eckert, *Solar Energy, 17,* 147(1975).

For treatments of the kinetics of electrode processes, see

J. Albery, *Electrode Kinetics,* Oxford: Clarendon Press, 1975.

J. O'M. Bockris and S. U. M. Khan, *Surface Electrochemistry, A Molecular Level Approach,* New York: Plenum, 1993.

J. O'M. Bockris and A. K. N. Reddy, *Modern Electrochemistry,* Volume 2, New York: Plenum Press, 1970. Vol. 2 deals with electrode kinetics.

B. E. Conway, *Theory and Principles of Electrode Processes,* New York: Ronald Press, 1964.

P. Delahay, *Double Layer and Electrode Kinetics,* New York: Interscience, 1965.

K. J. Laidler, "The kinetics of electrode processes," *J. Chemical Education, 47,* 600(1970).

K. J. Vetter, *Electrochemical Kinetics,* New York: Academic Press, 1967.

Chemical Kinetics I. The Basic Ideas

9

PREVIEW

The subject of chemical kinetics is concerned with the rates and mechanisms of chemical reactions. This chapter deals with some of the more basic ideas that are important for an understanding of the subject. Much, but not quite all, of the chapter is concerned with *elementary* reactions, which proceed in a single step, and we will learn something about the energetics and dynamics of such reactions.

For a reaction of stoichiometry

$$aA + bB \rightarrow yY + zZ$$

the rate of consumption of A is defined as $-d[A]/dt$, and the rate of formation of Y is defined as $d[Y]/dt$. These quantities are not in general the same for different reactants and products, and it is convenient to have a quantity, simply called the *rate of reaction*, which is the same for all reactants and products. This is done by making use of the concept of extent of reaction ξ, as explained in Section 9.2.

For some reactions the rate is related to reactant concentrations by an equation of the form

$$v = k[A]^{\alpha}[B]^{\beta} \cdots$$

where α and β are known as *partial orders:* α is the order with respect to A, and β the order with respect to B. The sum of all the partial orders is known as the *overall order.* The coefficient k that appears in this equation is known as the *rate constant* or the *rate coefficient.*

The effect of temperature T on a rate constant is expressed by the *Arrhenius equation*

$$k = Ae^{-E_a/RT}$$

where R is the *gas constant;* E_a is the *activation energy;* and A is the *preexponential factor.* In principle, activation energies can be calculated by the methods of quantum mechanics, but it is difficult to obtain reliable results except for the simplest of reactions. The course of a reaction is conveniently mapped by means of a *potential-energy surface,* in which energy is plotted against suitable parameters which describe the reacting system.

The preexponential factor A can be interpreted on the basis of *transition state theory,* which focuses attention on the activated complex, a reaction intermediate that lies at the highest point of the energy profile connecting the initial and the final states. This theory leads to the concepts of the entropy of activation $\Delta^{\ddagger}S^{\circ}$ and the enthalpy of activation $\Delta^{\ddagger}H^{\circ}$. These are the changes in entropy and enthalpy that occur when the activated complexes are formed from the reactants.

This chapter also discusses the influence of solvent on the rate of reaction. It also has a little to say about the new field of reaction dynamics, which is concerned with the molecular details of what occurs when a chemical process takes place.

OBJECTIVES

After studying this chapter, the student should be able to:

- Define the rate of a reaction.
- Understand the concept of the order of a reaction with respect to one reactant and to the overall order.
- Understand the definition of a rate constant and a rate coefficient.
- Integrate the rate equations for simpler systems.
- Understand the half-life of a reaction and its significance.
- Explain some of the techniques used for the study of fast reactions, including stopped flow, flash photolysis, and temperature jump.
- Use the Arrhenius equation for the temperature dependence of a chemical reaction and understand its significance.
- Determine an activation energy from the relevant data.
- Understand the concept of a potential-energy surface.
- Understand the basic ideas involved in transition-state theory.
- Understand something of quantum-mechanical tunneling and kinetic-isotope effects.
- Discuss solvent effects on reaction rates, including ionic-strength effects.
- Discuss work that has been done in the field of reaction dynamics.

Thermodynamics is concerned with the direction in which a process will occur, but in itself can tell us nothing about its rate. Chemical kinetics is the branch of physical chemistry that deals with the rates of chemical reactions and with the factors on which the rates depend. One reason for the importance of this subject is that it provides some of the information needed to arrive at the mechanism of a reaction.

A vast amount of work has been done on reaction rates. It is now realized that some reactions go in a single step; they are known as *elementary reactions* and are dealt with in this chapter. Other reactions go in more than one step and are said to be *composite, stepwise,* or *complex;* such reactions are treated in Chapter 10.

9.1 Rates of Consumption and Formation

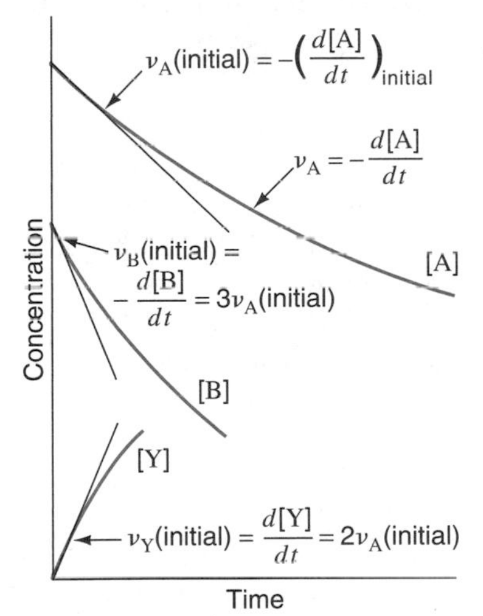

FIGURE 9.1
The variations with time of the concentrations of A, B, and Y for a reaction of the type A + 3B → 2Y.

Vary the rate constants and take slopes of an exponential growth and exponential decay; Rates of consumption and formation.

Kinetic investigations are concerned with rates of change of concentrations of reactants and products. Consider, for example, a reaction

$$A + 3B \rightarrow 2Y$$

Figure 9.1 shows schematically the variations in concentrations of A, B, and Y if we start the kinetic experiment with a mixture of A and B but no Y. At any time t we can draw a tangent to the curve representing the consumption of A, and the rate of consumption of A at that time is

$$v_A = -\frac{d[A]}{dt} \tag{9.1}$$

As a special case, we may draw a tangent at $t = 0$, corresponding to the beginning of the reaction; the negative of the slope is the initial rate of consumption of A. The rate of formation of Y, v_Y, is given by

$$v_Y = \frac{d[Y]}{dt} \tag{9.2}$$

In this particular reaction the stoichiometric coefficients are different for the three species, A, B, and Y, and the rates of change of their concentrations are correspondingly different. Thus, the rate of consumption of B, v_B, is three times the rate of consumption of A, v_A, and the rate of formation of Y, v_Y, is twice the rate of consumption of A:

$$v_A = \frac{1}{3}v_B = \frac{1}{2}v_Y \tag{9.3}$$

Individually v_A, v_B, or v_Y would be ambiguous as the rate of reaction.

9.2 Rate of Reaction

In Section 2.5 we introduced the concept of extent of reaction ξ, and this forms the basis of the definition of rate of reaction. The rate of reaction is now defined as the time derivative of the extent of reaction divided by the volume:

$$v = \frac{1}{V}\frac{d\xi}{dt} = \frac{1}{V\nu_i}\frac{dn_i}{dt} \tag{9.4}$$

If the volume is constant, dn_i/V can be replaced by the concentration change dc_i, and therefore

$$v = \frac{1}{\nu_i}\frac{dc_i}{dt} \tag{9.5}$$

Vary the stoichiometric coefficients and find the relative rates; Rates of reaction; The progress variable.

This quantity is independent of which reactant or product species is chosen. For a reaction

$$a\text{A} + b\text{B} \rightarrow y\text{Y} + z\text{Z}$$

occurring at constant volume, the rate of reaction is

$$v = -\frac{1}{a}\frac{d[\text{A}]}{dt} = -\frac{1}{b}\frac{d[\text{B}]}{dt} = \frac{1}{y}\frac{d[\text{Y}]}{dt} = \frac{1}{z}\frac{d[\text{Z}]}{dt} \tag{9.6}$$

The time derivatives of the concentrations (with a negative sign for reactants) are the rates of consumption or formation (see Figure 9.1) and thus

$$v = \frac{v_\text{A}}{a} = \frac{v_\text{B}}{b} = \frac{v_\text{Y}}{y} = \frac{v_\text{Z}}{z} \tag{9.7}$$

A distinction must be made between v without lettered subscript, meaning rate of reaction, and v with lettered subscript (e.g., v_A), meaning rate of consumption or formation. Since the stoichiometric coefficients and therefore extents of reaction depend on the way in which the reaction is written (e.g., $H_2 + Br_2 \rightarrow 2HBr$ or $\frac{1}{2}H_2 + \frac{1}{2}Br_2 \rightarrow HBr$), whenever rates of reaction are given the stoichiometric equation must be stated.

9.3 Empirical Rate Equations

For some reactions the rate of consumption or formation can be expressed empirically by an equation of the form

$$v_\text{A} = k_\text{A}[\text{A}]^\alpha[\text{B}]^\beta \tag{9.8}$$

where k_A, α, and β are independent of concentration and of time. Similarly, for a product Z, where k_Z is not necessarily the same as k_A,

$$v_\text{Z} = k_\text{Z}[\text{A}]^\alpha[\text{B}]^\beta \tag{9.9}$$

When these equations apply, the rate of reaction must also be given by an equation of the same form:

$$v = k[\text{A}]^\alpha[\text{B}]^\beta \tag{9.10}$$

The stoichiometric coefficient of the species i is ν_i (see Section 2.5, p. 62).

In these equations, k_A, k_Z, and k are not necessarily the same, being related by stoichiometric coefficients; thus if the stoichiometric equation is

$$\text{A} + 2\text{B} \rightarrow 3\text{Z}$$

$$k = k_\text{A} = \frac{1}{2}k_\text{B} = \frac{1}{3}k_\text{Z} \tag{9.11}$$

Order of Reaction

Partial Order

Overall Order

The exponent α in the previous equations is known as the order of reaction with respect to A and can be referred to as a *partial order*. Similarly, the partial order β is the order with respect to B. These orders are purely experimental quantities and are not necessarily integral. The sum of all the partial orders, $\alpha + \beta + \cdots$, is referred to as the *overall order* and is usually given the symbol n.

A very simple case is when the rate is proportional to the first power of the concentration of a single reactant:

$$v = k[\mathrm{A}] \tag{9.12}$$

Plot zeroth-, first-, and second-order reactions and vary the rate constants; Empirical rate equations.

Such a reaction is said to be of the first order. An example is the conversion of cyclopropane into propylene:

$$\underset{\mathrm{H_2C-CH_2}}{\overset{\mathrm{CH_2}}{/\ \backslash}} \rightarrow \mathrm{CH_3-CH=CH_2}$$

Over a wide range of pressure the rate of this reaction is proportional to the first power of the cyclopropane concentration.

There are many examples of second-order reactions. The reaction

$$\mathrm{H_2 + I_2 \rightleftarrows 2HI}$$

is second order in both directions. For this reaction the rate from left to right is proportional to the product of the concentrations of H_2 and I_2:

$$v_1 = k_1[\mathrm{H_2}][\mathrm{I_2}] \tag{9.13}$$

where k_1 is a constant at a given temperature. The reaction from left to right is said to be first order in H_2 and first order in I_2, and its overall order is two. The reverse reaction is also second order; the rate from right to left is proportional to the square of the concentration of hydrogen iodide:

$$v_{-1} = k_{-1}[\mathrm{HI}]^2 \tag{9.14}$$

The rate of a reaction must be proportional to the product of two concentrations [A] and [B] if the reaction simply involves collisions between A and B molecules. Similarly, the kinetics must be third order if a reaction proceeds in one stage and involves collisions between three molecules, A, B, and C. There are a few reactions of the third order, but reactions of higher order are unknown. Collisions in which three or more molecules all come together at the same time are very unlikely; reaction will instead proceed more rapidly by a composite mechanism involving two or more elementary processes, each of which is only first or second order.

There is no simple connection between the stoichiometric equation for a reaction and the order of the reaction. An example that illustrates this is the decomposition of gaseous ethanal (acetaldehyde), for which the equation is

$$\mathrm{CH_3CHO \rightarrow CH_4 + CO}$$

We might think that because there is one molecule on the left-hand side of this equation, the reaction should be first order. In fact, the order is three-halves:

$$v = k[\mathrm{CH_3CHO}]^{3/2} \tag{9.15}$$

We will see in Section 10.5 that this reaction occurs by a composite free-radical mechanism of a particular type that leads to three-halves-order behavior.

Reactions Having No Order

Not all reactions behave in the manner described by Eq. 9.8, and the term *order* should not be used for those that do not. For example, as we will discuss in Section 10.9, reactions catalyzed by enzymes frequently follow a law of the form

$$v = \frac{V[\mathrm{S}]}{K_m + [\mathrm{S}]} \tag{9.16}$$

where V and K_m are constants; and [S] is the concentration of the substance, known as the *substrate,* that is undergoing catalyzed reaction. This equation does not correspond to a simple order, but under two limiting conditions an order may be assigned. Thus, if the substrate concentration is sufficiently low that $[\mathrm{S}] \ll K_m$, Eq. 9.16 becomes

$$v = \frac{V}{K_m}[\mathrm{S}] \tag{9.17}$$

The reaction is then first order with respect to S. Also, when [S] is sufficiently large that $[\mathrm{S}] \gg K_m$, the equation reduces to

$$v = V \tag{9.18}$$

The rate is then independent of [S] (i.e., is proportional to [S] to the zero power) and the reaction is said to be *zero order.*

Rate Constants and Rate Coefficients

The constant k that appears in rate equations that are special cases of Eq. 9.8 is known as the **rate constant** or the *rate coefficient.* Some scientists use the former term when the reaction is believed to be elementary and the latter term when the reaction is known to occur in more than one stage.

The units of the rate constant or coefficient vary with the order of the reaction. Suppose, for example, that a reaction is of the first order; that is,

$$v = k[\mathrm{A}] \tag{9.19}$$

If the units of v are mol dm^{-3} s^{-1}, and those of [A] are mol dm^{-3}, the unit of the rate constant k is s^{-1}. For a second-order reaction, for which

$$v = k[\mathrm{A}]^2 \quad \text{or} \quad k[\mathrm{A}][\mathrm{B}] \tag{9.20}$$

the units of k will be

$$\frac{\text{mol dm}^{-3}\ \text{s}^{-1}}{(\text{mol dm}^{-3})^2} = \text{dm}^3\ \text{mol}^{-1}\ \text{s}^{-1}$$

The units corresponding to other orders can easily be worked out (see Table 9.1).

We have seen that the rate of change of a concentration in general depends on the reactant or product with which we are concerned. The rate constant also re-

TABLE 9.1 Rate Equations and Half-Lives

Order	Rate Equation: Differential Form	Rate Equation: Integrated Form	Units of Rate Constant	Half-Life $t_{1/2}$
0	$\frac{dx}{dt} = k$	$k = \frac{x}{t}$	mol dm^{-3} s^{-1}	$\frac{a_0}{2k}$
1	$\frac{dx}{dt} = k(a_0 - x)$	$k = \frac{1}{t} \ln \frac{a_0}{a_0 - x}$	s^{-1}	$\frac{\ln 2}{k}$
2	$\frac{dx}{dt} = k(a_0 - x)^2$	$k = \frac{1}{t} \frac{x}{a_0(a_0 - x)}$	dm^3 mol^{-1} s^{-1}	$\frac{1}{ka_0}$
2	$\frac{dx}{dt} = k(a_0 - x)(b_0 - x)$ (reactants at different concentrations)	$k = \frac{1}{t(a_0 - b_0)} \ln \frac{b_0(a_0 - x)}{a_0(b_0 - x)}$	dm^3 mol^{-1} s^{-1}	—
n	$\frac{dx}{dt} = k(a_0 - x)^n$	$k = \frac{1}{t(n-1)}\left[\frac{1}{(a_0 - x)^{n-1}} - \frac{1}{a_0^{n-1}}\right]$	mol^{1-n} dm^{3n-3} s^{-1}	$\frac{2^{n-1} - 1}{k(n - 1)a_0^{n-1}}$

flects this dependence. For example, for the dissociation of ethane into methyl radicals,

$$C_2H_6 \rightarrow 2CH_3$$

the rate of formation of methyl radicals, v_{CH_3}, is twice the rate of consumption of ethane, $v_{C_2H_6}$:

$$v_{C_2H_6}\left(= -\frac{d[C_2H_6]}{dt}\right) = \frac{1}{2} v_{CH_3}\left(= \frac{1}{2}\frac{d[CH_3]}{dt}\right) \tag{9.21}$$

The reaction is first order in ethane under certain conditions, and the rate coefficient for the consumption of ethane, $k_{C_2H_6}$, is one-half of that for the appearance of methyl radicals:

$$k_{C_2H_6} = \tfrac{1}{2} k_{CH_3} \tag{9.22}$$

In cases such as this it is obviously important to specify the species to which a rate constant applies.

If a reaction shows time-independent stoichiometry, it is possible to evaluate a rate of reaction (Eq. 9.7). For the dissociation of ethane into methyl radicals, the rate of reaction v is equal to the rate of consumption of C_2H_6 and is one-half the rate of formation of methyl radicals. Whenever possible, the rate constant should be evaluated with respect to the rate of reaction.

9.4 Analysis of Kinetic Results

The first task in the kinetic investigation of a chemical reaction is to measure rates under a variety of experimental conditions and to determine how the rates are affected by the concentrations of reactants, products of reaction, and other substances (e.g., inhibitors) that may affect the rate.

Method of Integration

There are two main methods for dealing with such problems; they are the method of integration and the differential method. In the **method of integration** we start with a rate equation that we think may be applicable. For example, if the reaction is believed to be a first-order reaction, we start with

$$-\frac{dc}{dt} = kc \tag{9.23}$$

where c is the concentration of reactant. By integration we convert this into an equation giving c as a function of t and then compare this with the experimental variation of c with t. If there is a good fit, we can then, by simple graphical procedures, determine the value of the rate constant. If the fit is not good, we must try another rate equation and go through the same procedure until the fit is satisfactory. The method involves trial and error, but is nevertheless very useful, especially when no special complications arise.

Differential Method

The second method, the **differential method,** employs the rate equation in its differential, unintegrated, form. Values of dc/dt are obtained from a plot of c against t by taking slopes, and these are directly compared with the rate equation. The main difficulty with this method is that slopes cannot always be obtained very accurately. In spite of this drawback the method is on the whole the more reliable one, and unlike the integration method it does not lead to any particular difficulties when there are complexities in the kinetic behavior.

These two methods will now be considered in further detail.

Method of Integration

A *first-order reaction* may be of the type

$$A \rightarrow Z$$

Suppose that at the beginning of the reaction ($t = 0$) the concentration of A is a_0, and that of Z is zero. If after time t the concentration of Z is x, that of A is $a_0 - x$. The rate of change of the concentration of Z is dx/dt, and thus for a first-order reaction

First-order Reaction Rate

$$\frac{dx}{dt} = k(a_0 - x) \tag{9.24}$$

Separation of the variables gives

$$\frac{dx}{a_0 - x} = k\,dt \tag{9.25}$$

and integration gives

$$-\ln(a_0 - x) = kt + I \tag{9.26}$$

where I is the constant of integration. This constant may be evaluated using the boundary condition that $x = 0$ when $t = 0$; hence

$$-\ln a_0 = I \tag{9.27}$$

and insertion of this into Eq. 9.26 leads to

$$\ln \frac{a_0}{a_0 - x} = kt \tag{9.28}$$

This equation can also be written as

$$x = a_0(1 - e^{-kt}) \tag{9.29}$$

and as

$$a_0 - x = a_0 e^{-kt} \tag{9.30}$$

Equation 9.30 shows that the concentration of reactant, $a_0 - x$, decreases exponentially with time, from an initial value of a_0 to a final value of zero.

The first-order equations can be tested and the constant evaluated using graphical procedures, two of which are illustrated in Figure 9.2.

Second-order reactions can be treated in a similar fashion. There are now two possibilities: The rate may be proportional to the product of two equal concentrations or to the product of two different ones. The first case must occur when a single reactant is involved, as in the process

$$2A \rightarrow Z$$

It may also be found in reactions between two different substances,

$$A + B \rightarrow Z$$

provided that their initial concentrations are the same.

In such situations the rate may be expressed as

Second-order Reaction Rate

$$\frac{dx}{dt} = k(a_0 - x)^2 \tag{9.31}$$

where x is the amount of A that has reacted in unit volume at time t and a_0 is the initial amount of A. Separation of the variables leads to

$$\frac{dx}{(a_0 - x)^2} = k\,dt \tag{9.32}$$

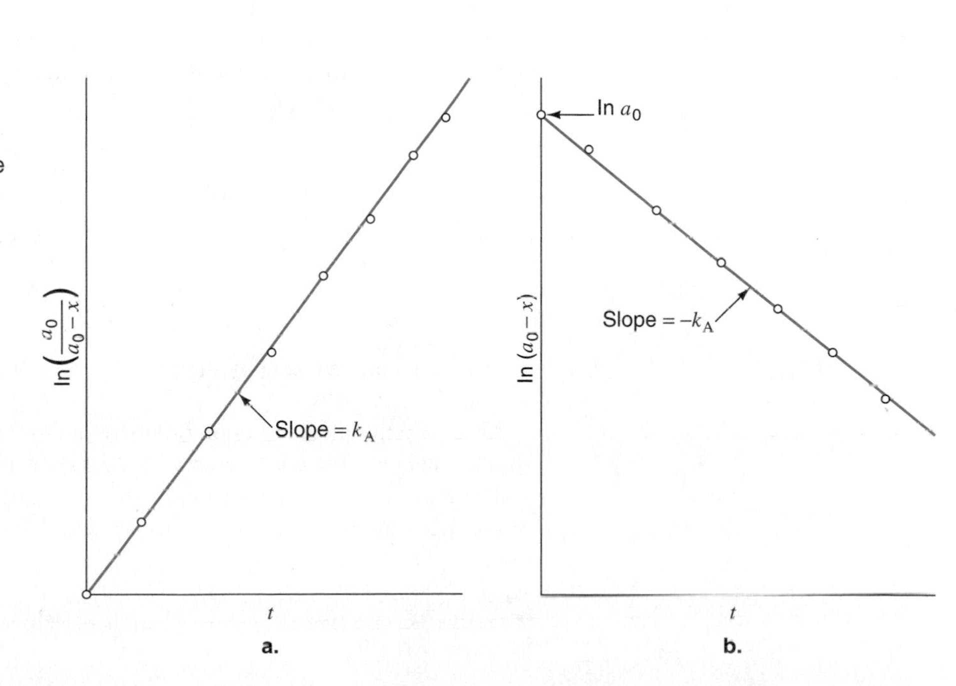

FIGURE 9.2
Two alternative ways of plotting results to see if a reaction is first order and, if it is, to determine the rate constant k.

Study Figure 9.2 by varying the rate constant and initial concentration; Take slopes; Analysis of kinetic results; First-order rate.

Study Figure 9.3 by varying the rate constant and initial concentration; Analysis of kinetic results; second-order rate.

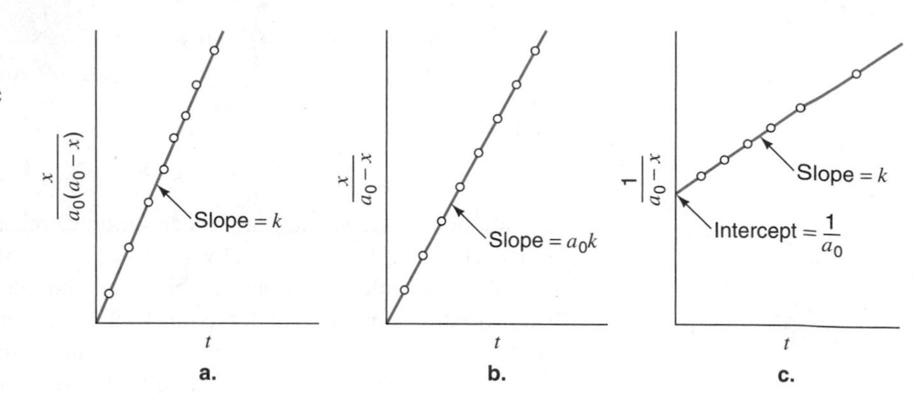

FIGURE 9.3
Three alternative ways of plotting results to see if a reaction is second order and, if it is, to determine the rate constant k.

which integrates to

$$\frac{1}{a_0 - x} = kt + I \tag{9.33}$$

where I is the constant of integration. The boundary condition is that $x = 0$ when $t = 0$; therefore

$$I = \frac{1}{a_0} \tag{9.34}$$

Hence

$$\frac{x}{a_0(a_0 - x)} = kt \tag{9.35}$$

The variation of x with t is no longer an exponential one.

Graphical methods can again be employed to test this equation and to obtain the rate constant k. Three possible plots are shown in Figure 9.3.

Vary the rate constant and initial concentrations for the second-order rate law (Eq. 9.36); Second-order Type AB.

If the rate is proportional to the concentrations of two different reactants, and these concentrations are not initially the same, the integration proceeds differently. Suppose that the initial concentrations are a_0 and b_0; the rate after an amount x (per unit volume) has reacted is

$$\frac{dx}{dt} = k(a_0 - x)(b_0 - x) \tag{9.36}$$

The result of the integration, with the boundary condition $x = 0$ when $t = 0$, is

$$\frac{1}{a_0 - b_0} \ln \frac{b_0(a_0 - x)}{a_0(b_0 - x)} = kt \tag{9.37}$$

This equation can be tested by plotting the left-hand side against t; if a straight line is obtained, its slope is k.

These are the most common orders; reactions of other orders can be treated in a similar manner.[1] Table 9.1 summarizes some of the results.

The main disadvantage of the method of integration is that the integrated expressions, giving the variation of x with t, are often quite similar for different types

[1]Solutions for many systems are given by C. Capellos and B. H. J. Bielski, *Kinetic Systems,* New York: Wiley-Interscience, 1972.

of reactions. For example, the time course of a simple second-order reaction is closely similar to that of a first-order reaction inhibited by products; unless the experiments are done very accurately, there is danger of confusion. We can test for inhibition by-products by measuring the rate after the deliberate introduction of products of reaction.

Study zeroth-order half-lives; Zeroth-order half-lives.

Half-Life

For a given reaction the **half-life** $t_{1/2}$ of a particular reactant is the time required for its concentration to reach a value that is halfway between its initial and final values. The value of the half-life is always inversely proportional to the rate constant and in general depends on reactant concentrations. For a first-order reaction the rate equation is Eq. 9.28, and the half-life is obtained by putting x equal to $a_0/2$:

$$\ln \frac{a_0}{a_0 - a_0/2} = kt_{1/2} \tag{9.38}$$

and therefore

Half-life of a First-order Reaction

$$t_{1/2} = \frac{\ln 2}{k} \tag{9.39}$$

In this case, the half-life is independent of the initial concentration. Since, for a first-order reaction, there is only one reactant, the half-life of the reactant can also be called the half-life of the *reaction*.

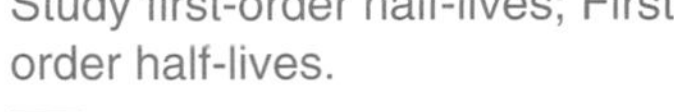

Study first-order half-lives; First-order half-lives.

For a second-order reaction involving a single reactant or two reactants of equal initial concentrations, the rate equation is Eq. 9.35, and setting $x = a_0/2$ leads to

Half-life of a Second-order Reaction

$$t_{1/2} = \frac{1}{a_0 k} \tag{9.40}$$

The half-life is now inversely proportional to the concentration of the reactant.

In the general case of a reactant of the nth order involving equal initial reactant concentrations a_0, the half-life, as seen in Table 9.1, is inversely proportional to a_0^{n-1}. For reactions of order other than unity the half-life is the same for different reactants only if they are initially present in their stoichiometric ratios; only in that case is it permissible to speak of the half-life of the *reaction*.

The order of a reaction can be determined by determining half-lives at two different initial concentrations a_1 and a_2. The half-lives are related by

Study second-order half-lives; Second-order half-lives.

$$\frac{t_{1/2}(1)}{t_{1/2}(2)} = \left(\frac{a_2}{a_1}\right)^{n-1} \tag{9.41}$$

$$n = 1 + \frac{\log[t_{1/2}(1)/t_{1/2}(2)]}{\log(a_2/a_1)} \tag{9.42}$$

and n can readily be calculated. This method can give misleading results if the reaction is not of simple order or if there are complications such as inhibition by products.

Since the half-lives of all reactions have the same units, they provide a useful way of comparing the rates of reactions of different orders. Rate constants, as we have seen, have different units for different reaction orders. Thus, if two reactions have different orders, we cannot draw conclusions about their relative rates from their rate constants.

Radioactive Disintegrations

Radioactive disintegrations follow first-order kinetics and therefore have half-lives that are independent of the amount of radioactive substance present. An atom whose nucleus has a specified number of protons and neutrons is known as a *nuclide;* if it is radioactive, it is referred to as a *radionuclide.* A radionuclide disintegrates at a rate that is a function only of the constitution of the nucleus; unlike ordinary chemical processes, radioactive disintegration cannot be influenced by any chemical or physical means, such as changing the temperature. The first-order rate constant for a radionuclide is known as its *decay constant.*

EXAMPLE 9.1 The half-life of radium, $^{226}_{88}\text{Ra}$, is 1600 years. How many disintegrations per second would be undergone by 1 g of radium?

Solution The half-life in seconds is

$$1600 \text{ yr} \times 365.25 \frac{\text{d}}{\text{yr}} \times 24 \frac{\text{h}}{\text{d}} \times 60 \frac{\text{min}}{\text{h}} \times 60 \frac{\text{s}}{\text{min}} = 5.049 \times 10^{10} \text{ s}$$

The decay constant is thus, from Eq. 9.39,

$$k = \frac{\ln 2}{t_{1/2}} = \frac{0.693}{5.049 \times 10^{10} \text{ s}} = 1.37 \times 10^{-11} \text{ s}^{-1}$$

The number of nuclei present in 1 g of radium is

$$\frac{6.022 \times 10^{23} \text{ mol}^{-1}}{226 \text{ g mol}^{-1}} = 2.666 \times 10^{21} \text{ g}^{-1}$$

The number of disintegrations is therefore

$$1.37 \times 10^{-11} \text{ s}^{-1} \times 2.67 \times 10^{21} \text{ g}^{-1} = 3.66 \times 10^{10} \text{ g}^{-1} \text{ s}^{-1}$$

The kinetic characteristics of radionuclides are used to specify their amounts. The old unit for the amount of a radionuclide was the curie (Ci), which is defined as the amount producing exactly 3.7×10^{10} disintegrations per second; we have seen in the preceding example that this is approximately the number of disintegrations produced per second by 1 g of radium. In a 1975 IUPAC recommendation the curie was replaced by the becquerel (Bq), named after the French physicist Antoine Henri Becquerel (1852–1908). The becquerel is defined as the amount of radioactive substance giving one disintegration per second; thus

$$1 \text{ Ci} \equiv 3.7 \times 10^{10} \text{ Bq}$$

Differential Method

In the differential method, which was first suggested in 1884 by van't Hoff, the procedure is to determine rates directly by measuring tangents to the experimental concentration-time curves and to introduce these into the equations in their differential forms.

The theory of the method is as follows. The instantaneous rate of a reaction of the nth order involving only one reacting substance is proportional to the nth power of its concentration:

$$v = -\frac{da}{dt} = ka^n \tag{9.43}$$

Therefore

$$\ln v = \ln k + n \ln a \tag{9.44}$$

A plot of $\ln v$ against $\ln a$ therefore will give a straight line if the reaction is of simple order; the slope is the order n, and the intercept is $\ln k$. However, an accurate value of k cannot be obtained from such a plot.

Reactions Having No Simple Order

Many reactions do not admit to the assignment of an order. Although such reactions can be treated by the method of integration, this procedure is seldom entirely satisfactory, owing to the difficulty of distinguishing between various possibilities. In such cases the best method is usually the differential method; rates of change of concentrations are measured accurately in the initial stages of the reaction, and runs are carried out at a series of initial concentrations. In this way, a plot of rate against concentration can be prepared and the dependence of rate on concentration can then be determined by various methods; one that is frequently used for enzyme reactions will be considered in Section 10.9.

Opposing Reactions

One complication is that a reaction may proceed to a state of equilibrium that differs appreciably from completion. The simplest case is when both forward and reverse reactions are of the first order:

Opposing First-order Reactions

Vary the forward and reverse rate constants, and vary the initial conditions to plot opposing reactions; Opposing first-order reactions.

$$\mathrm{A} \underset{k_{-1}}{\overset{k_1}{\rightleftarrows}} \mathrm{Z}$$

If the experiment is started using pure A, of concentration a_0, and if after time t the concentration of Z is x, that of A is $a_0 - x$. The rate of reaction 1 if it occurred in isolation is then equal to $k_1(a_0 - x)$, while the rate of the reverse reaction is $k_{-1}x$; the net rate of change of concentration of Z is thus

$$\frac{dx}{dt} = k_1(a_0 - x) - k_{-1}x \tag{9.45}$$

If x_e is the concentration of Z at equilibrium, when the net rate is zero,

$$k_1(a_0 - x_e) - k_{-1}x_e = 0 \tag{9.46}$$

It follows at once from this equation that the **equilibrium constant** K_c, equal to $x_e/(a_0 - x_e)$, is **equal to the ratio of the rate constants,** k_1/k_{-1}.

Elimination of k_{-1} between Eqs. 9.45 and 9.46 gives rise to

$$\frac{dx}{dt} = \frac{k_1 a_0}{x_e}(x_e - x) \tag{9.47}$$

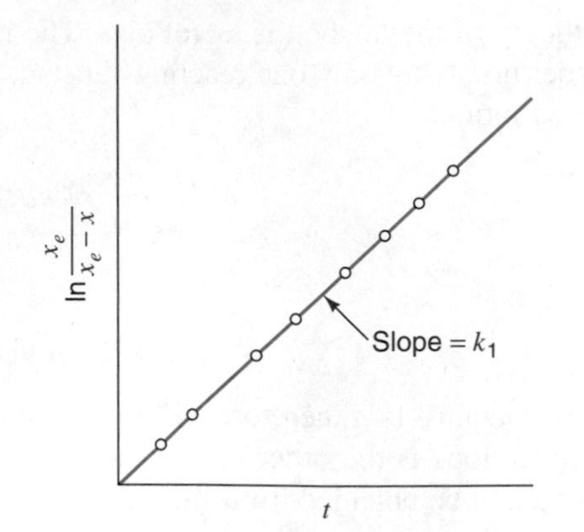

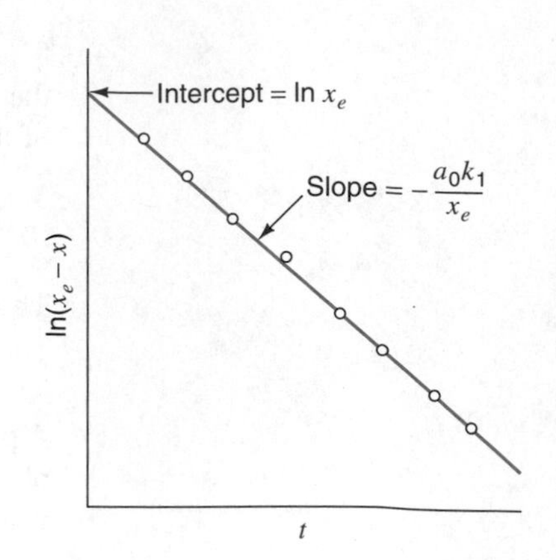

FIGURE 9.4
Two alternative plots for obtaining the rate constant k_1 when there are opposing first-order reactions.

Integration of this equation, subject to the boundary condition that $x = 0$ when $t = 0$, gives

$$k_1 t = \frac{x_e}{a_0} \ln \frac{x_e}{x_e - x} \quad \text{or} \quad \frac{a_0 k_1}{x_e} t = \ln x_e - \ln(x_e - x) \tag{9.48}$$

The amount of x present at equilibrium, x_e, can be measured directly. If values of x are determined at various values of t, the rate constant k_1 can be obtained by graphical means, as shown in Figure 9.4. Since the equilibrium constant $K_c = k_1/k_{-1}$, the rate constant k_{-1} is easily calculated.

9.5 Techniques for Very Fast Reactions

The essential feature of all methods of studying the kinetics of a reaction is to determine the time dependence of concentrations of reactants or products. Some reactions are so fast that special techniques have to be employed in order for this to be possible. There are two reasons why conventional techniques are not suitable for very rapid reactions:

1. The time that it usually takes to mix reactants, or to bring them to a specified temperature, may be significant in comparison with the half-life of the reaction. An appreciable error therefore will be made, since the initial time cannot be determined accurately.
2. The time that it takes to make a measurement of concentration is significant compared with the half-life of the reaction.

The methods that are used in the study of rapid reactions fall into two classes: **flow** methods and **pulse** methods. Some of the more important techniques for studying fast reactions are indicated in the lower part of Figure 9.5. The upper part of the diagram shows for comparison the half-lives or relaxation times for various chemical and physical processes.

Flow Methods

The difficulty regarding the mixing of reactants can sometimes be overcome by using special techniques for bringing the reactants very rapidly into the reaction

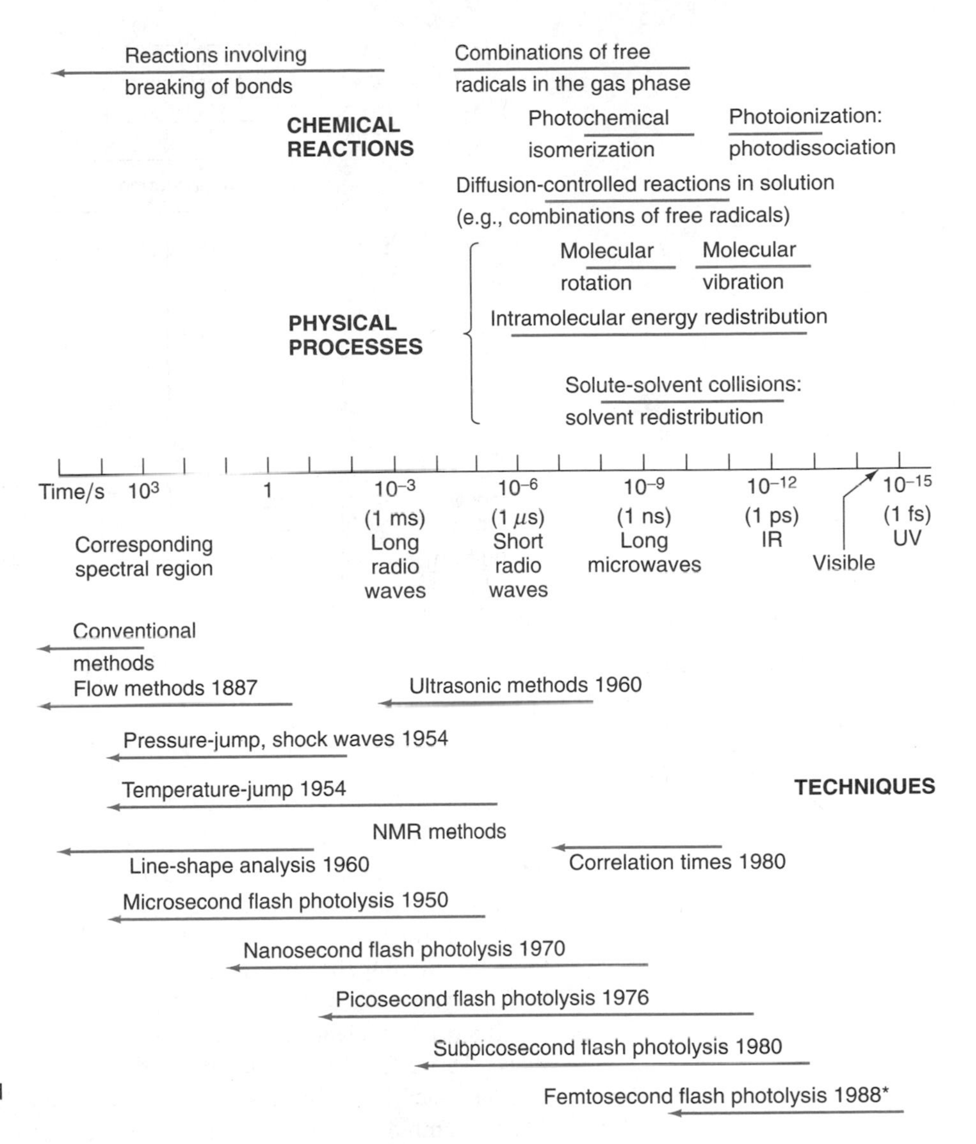

FIGURE 9.5
The range of times available to various experimental techniques for kinetic studies, and the range of half-lives for some chemical and physical processes. The approximate years in which the various techniques were first used are indicated.

vessel and for mixing them very rapidly. With the use of conventional techniques, it takes from several seconds to a minute to bring solutions into a reaction vessel and to have them completely mixed and at the temperature of the surroundings. This time can be greatly reduced by using a **flow technique.** Two gases or two solutions can be introduced into a mixing chamber, and the resulting mixture caused to pass rapidly along a tube. Concentrations of reactants or products can be determined at various positions along the tube, corresponding to various times. Reactions having half-lives down to about 10^{-2} seconds can be studied by such techniques.

Stopped-flow Technique

One useful modification is the **stopped-flow technique,** shown schematically in Figure 9.6. This particular apparatus is designed for the study of a reaction between two substances in solution. A solution of one of the substances is maintained initially in the syringe *A*, and a solution of the other is in syringe *B*. The plungers of the syringes can be forced down rapidly and a rapid stream of two solutions passes

*See pp. 442–443 and Suggested Reading, p. 414.

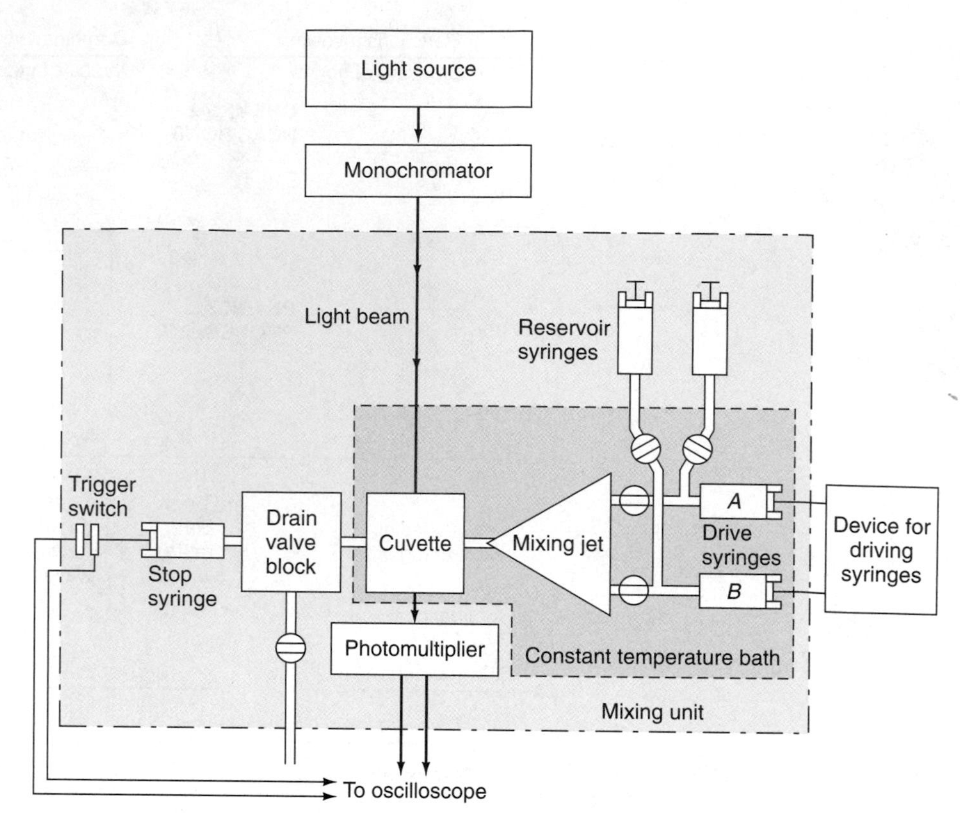

FIGURE 9.6
Schematic diagram of stopped-flow apparatus. Solutions in the two drive syringes *A* and *B* are rapidly forced through a mixing jet into a cuvette. When the flow is stopped, the oscilloscope is triggered and records light absorption as a function of time.

into the mixing system. This is designed in such a way that jets of the two solutions impinge on one another and give very rapid mixing; with a suitable design of the mixing chamber it is possible for mixing to be essentially complete in 0.001 s. From this mixing chamber the solution passes at once into the reaction cuvette; alternatively, the two may be combined in the reaction chamber.

If a reaction is rapid, it is not possible to carry out chemical analyses at various stages. This difficulty must be resolved by using techniques that allow properties to be determined instantaneously. For reactions in solution, spectrophotometric methods are commonly employed. If the products absorb differently from the reactants at a particular wavelength, we can pass monochromatic light of this wavelength through the reaction vessel, and analyze the results on the basis of the Lambert-Beer law (Section 13.1). Fluorescence, electrical conductivity, and optical rotation are also convenient properties to measure in such high-speed studies.

Pulse Methods

The flow techniques just described are limited by the speed with which it is possible to mix solutions. There is no difficulty, using optical or other techniques, in following the course of a very rapid reaction, but for hydrodynamic reasons it is impossible to mix two solutions in less than about 10^{-3} s. If the half-life is less than this, the reaction will be largely completed by the time that it takes for mixing to be achieved; any rate measurement made will be of the rate of mixing, not of the rate of reaction. The neutralization of an acid by a base, that is, the reaction

$$H^+ + OH^- \rightarrow H_2O$$

under ordinary conditions has a half-life of 10^{-6} s or less (see Problem 9.6), and its rate therefore cannot be measured by any technique involving the mixing of solutions.

These technical problems were overcome by the development of a group of methods known as **pulse methods.** The first of these to be developed, in 1949, the **flash-photolysis methods** due to Norrish and Porter, are more conveniently dealt with later in Section 10.6. Another class of pulse methods comprises the **relaxation methods,** developed in the 1950s by the German physical chemist Manfred Eigen (b. 1927). The relaxation methods differ fundamentally from conventional kinetic methods in that we start with the system at equilibrium under a given set of conditions. We then change these conditions very rapidly; the system is no longer at equilibrium, and it *relaxes* to a new state of equilibrium. The speed with which it relaxes can be measured, usually by spectrophotometry, and we can then calculate the rate constants.

Relaxation Methods

Eigen shared with Norrish and Porter the 1967 Nobel Prize in chemistry. He developed relaxation methods to pursue his interest in (1) ionic reactions, (2) proton transfer reactions, and (3) multistage processes involving metal complexes or biochemical processes.

There are various ways in which the conditions are disturbed. One is by changing the hydrostatic pressure. Another, the most common technique, is to increase the temperature suddenly, usually by the rapid discharge of a capacitor; this method is called the **temperature-jump** or **T-jump method.** It is possible to raise the temperature of a tiny cell containing a reaction mixture by a few degrees in less than 10^{-7} s, which is sufficiently rapid to permit the investigation of many fast processes. Some are too fast for this technique, and then flash photolysis (Section 10.6) must be employed.

The principle of the method is illustrated in Figure 9.7. Suppose that the reaction is of the simple type

$$\mathrm{A} \underset{k_{-1}}{\overset{k_1}{\rightleftarrows}} \mathrm{Z}$$

At the initial state of equilibrium, the product Z is at a certain concentration, and it stays at this concentration until the temperature jump occurs, when the concentration

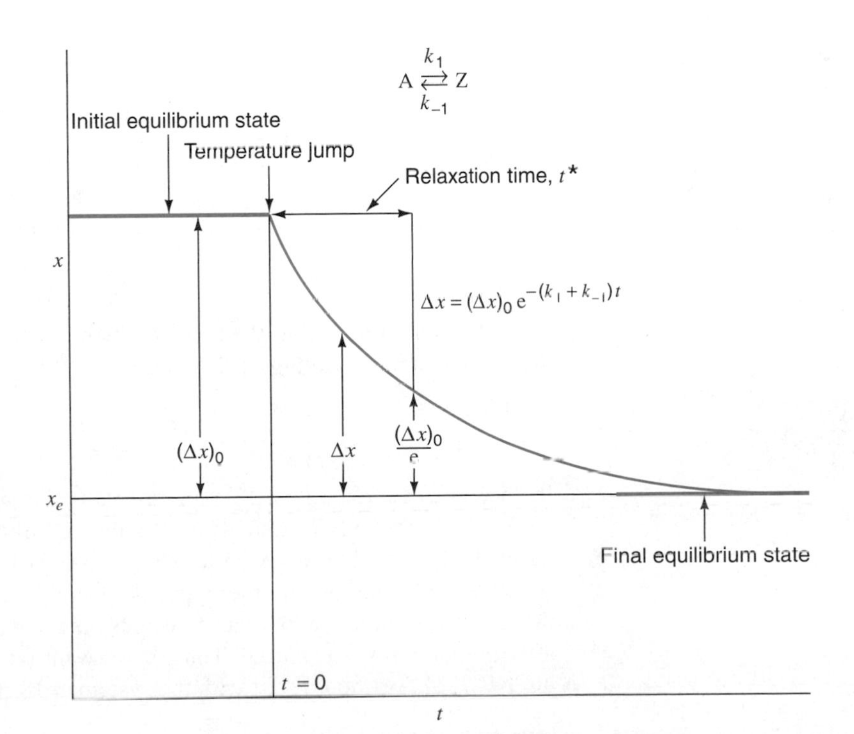

FIGURE 9.7
Principle of the temperature-jump technique.

changes to another value that will be higher or lower than the initial value according to the sign of $\Delta H°$ for the reaction. From the shape of the curve during the relaxation phase, we can obtain the sum of the rate constants, $k_1 + k_{-1}$, as shown by the following treatment.

Let a_0 be the sum of the concentrations of A and Z, and let x be the concentration of Z at any time; the concentration of A is $a_0 - x$. The kinetic equation is thus

$$\frac{dx}{dt} = k_1(a_0 - x) - k_{-1}x \tag{9.49}$$

If x_e is the concentration of Z at equilibrium,

$$k_1(a_0 - x_e) - k_{-1}x_e = 0 \tag{9.50}$$

The deviation of x from equilibrium Δx is equal to $x - x_e$, and therefore

$$\frac{d\,\Delta x}{dt} = \frac{dx}{dt} = k_1(a_0 - x) - k_{-1}x \tag{9.51}$$

Subtraction of the expression in Eq. 9.50 gives

$$\frac{d\,\Delta x}{dt} = k_1(x_e - x) - k_{-1}(x - x_e) \tag{9.52}$$

$$= -(k_1 + k_{-1})\Delta x \tag{9.53}$$

The quantity Δx thus varies with time in the same manner as does the concentration of a reactant in a first-order reaction. Integration of Eq. 9.53, subject to the boundary condition that $\Delta x = (\Delta x)_0$ when $t = 0$, leads to

$$\ln\frac{(\Delta x)_0}{\Delta x} = (k_1 + k_{-1})t \tag{9.54}$$

We can define a relaxation time t^* as the time corresponding to

$$\frac{(\Delta x)_0}{\Delta x} = \mathrm{e} \tag{9.55}$$

or to

$$\ln\frac{(\Delta x)_0}{\Delta x} = 1 \tag{9.56}$$

The relaxation time is thus the time at which the distance from equilibrium is 1/e of the initial distance (see Figure 9.7). From Eq. 9.54 we see that

$$t^* = \frac{1}{k_1 + k_{-1}} \tag{9.57}$$

If, therefore, we determine t^* experimentally for such a system, we can calculate $k_1 + k_{-1}$. However, the ratio k_1/k_{-1} is the equilibrium constant and it can be determined directly; hence, the individual constants k_1 and k_{-1} can be calculated.

Relaxation curves for other types of reactions can similarly be analyzed to give rate constants. Note that the rate constants obtained refer to the second temperature, after the T-jump has occurred. Thus, if we want rate constants at 25 °C, and the T-jump is 7 °C, we should start with the system at 18 °C.

During recent years a considerable number of investigations have been made using this technique. An important reaction studied in this way is the dissociation of water and combination of hydrogen and hydroxide ions:

$$2H_2O \rightarrow H_3O^+ + OH^-$$

The reaction was followed by measuring conductivity; after the T-jump the conductivity increased with time, with a relaxation half-life of 3.7×10^{-5} s at 23 °C. From this value it can be calculated that the second-order rate constant for the combination of H_3O^+ and OH^- ions is 1.3×10^{11} dm^3 mol^{-1} s^{-1}, which is a remarkably high value. The rate constant for the reverse dissociation, which is very small, can be calculated from this value and the equilibrium constant. A number of other hydrogen-ion transfer processes have also been studied using the same technique.

Ultrasonic techniques are also being used for the study of rapid reactions, especially in solution. The method is useful for processes occurring in the microsecond and nanosecond time ranges, such as proton transfers and conformational changes.

Nuclear magnetic resonance (NMR) spectroscopy, to be considered in Section 14.6, has also found application in high-speed kinetics. The analysis of line-widths in NMR spectra has provided valuable information about processes having half-lives of 10^{-5} s. Another technique, involving the study of correlation times, allows one to study processes occurring in times of 10^{-8} to 10^{-12} s. Similar investigations to those done with NMR can be done with electron spin resonance spectroscopy (Section 14.7), but this technique is limited to reactions involving paramagnetic species such as free radicals.

9.6 Molecular Kinetics

An important aspect of chemical kinetics is concerned with how rates depend on temperature. The conclusions from such studies lead to considerable insight into the molecular nature of chemical reactions. In this chapter we will be concerned mainly with temperature effects on **elementary reactions,** which are reactions that occur in a single stage, with no identifiable reaction intermediates. Elementary reactions are to be contrasted with **composite reactions,** which occur in more than one stage.

Molecularity and Order

Once a process has been identified as elementary, an important question that arises is: How many molecules enter into reaction? This number is referred to as the **molecularity** of the reaction. We have seen that from the variation of rate with concentration we can frequently determine a reaction order. This number, a purely *experimental* one, should be sharply distinguished from molecularity, which represents a deduction as to the *number of molecules* taking part in the reaction. It is permissible to speak of the order of a composite reaction, provided that the rate is proportional simply to concentrations raised to certain powers. It is meaningless, on the other hand, to speak of the molecularity if the mechanism is a composite one.

With certain exceptions, discussed later, we can assume that the order of an *elementary* reaction indicates the number of molecules that enter into reaction (i.e., that the order and the molecularity are the same). For example, if an elementary reaction is first order with respect to a reactant A and first order with respect to another substance B, we often find that the reaction is bimolecular, a molecule of A and a molecule of B entering into the reaction.

Sometimes, however, the kinetic order, which is determined experimentally, does not correspond to the molecularity, which is postulated on the basis of the individual elementary reactions for a proposed mechanism. Suppose, for example, that a reaction is bimolecular but that one reactant is present in large excess, so that its concentration does not change appreciably as the reaction proceeds; moreover (for example, if it is the solvent), its concentration may be the same in different kinetic runs. If this is the case, the kinetic investigation will not reveal any dependence of the rate on the concentration of this substance, which would therefore not be considered to be entering into reaction. This situation is frequently found in reactions in solution where the solvent may be a reactant. For example, in hydrolysis reactions in aqueous solution, a water molecule may undergo reaction with a solute molecule. Unless special procedures are employed, the kinetic results will not reveal the participation of the solvent. However, if it appears in the stoichiometric equation, it must participate in the reaction. This will be discussed in some detail in Chapter 10.

Catalyst

Another case in which the kinetic study may not reveal that a substance enters into reaction is when a catalyst is involved. A **catalyst,** by definition, is a substance that influences the rate of reaction without itself being used up; it may be regarded as being both a reactant and a product of reaction (see Section 10.9). The concentration of a catalyst therefore remains constant during reaction, and the kinetic analysis of a single run will not reveal the participation of the catalyst in the reaction. However, the fact that it does enter into reaction may be shown by measuring the rate at a variety of catalyst concentrations; generally a linear dependence is found.

Unimolecular gas reactions are also reactions in which the molecularity does not always correspond to the order. At sufficiently high pressures they do obey first-order kinetics, but as the pressure is lowered they approach second-order kinetics. The study of unimolecular gas reactions is an important aspect of chemical kinetics, but since an adequate treatment of them requires rather a lot of space, and since most readers of this book will probably never be concerned with them, we decided to omit this topic. There are various excellent treatments of these reactions, some of them listed at the end of this chapter.

It follows that the decision about the molecularity of an elementary reaction must involve not only a careful kinetic study in which as many factors as possible are varied, but also a consideration of other aspects of the reaction.

9.7 The Arrhenius Equation

It is found empirically for a vast number of reactions that the rate constant k is related to the absolute temperature T by the equation

$$k = Ae^{-B/T} \tag{9.58}$$

where A and B are constants. This relationship was expressed by van't Hoff and Arrhenius in the form

$$k = Ae^{-E/RT} \tag{9.59}$$

where R is the gas constant, equal to 8.3145 J K^{-1} mol^{-1}, and E is known as the *activation energy.* The equation was arrived at in 1884 by van't Hoff, who argued on the basis of the variation of the equilibrium constant with the temperature, and pointed out that a similar relationship should hold for the rate constant of a reaction. This idea was extended by Arrhenius, who successfully applied it to a large number of reactions and, as a result, Eq. 9.59 is generally referred to as the **Arrhenius equation.**

The arguments of van't Hoff are briefly as follows. The temperature dependence of a standard equilibrium constant K_c° is given by Eq. 4.83:

$$\frac{d \ln K_c^\circ}{dT} = \frac{\Delta U^\circ}{RT^2} \tag{9.60}$$

where ΔU° is the standard internal energy change in the reaction. For a reaction such as

$$A + B \underset{k_{-1}}{\overset{k_1}{\rightleftharpoons}} Y + Z$$

Visualize the effect of changing the temperature for exothermic and endothermic reactions; Arrhenius equation; Effects of temperature.

the equilibrium constant K_c is equal to the ratio of the rate constants k_1 and k_{-1},

$$K_c = \frac{k_1}{k_{-1}} \tag{9.61}$$

Equation 9.60 can therefore be written as[2]

$$\frac{d \ln k_1}{dT} - \frac{d \ln k_{-1}}{dT} = \frac{\Delta U^\circ}{RT^2} \tag{9.62}$$

and this may be split into the two equations

$$\frac{d \ln k_1}{dT} = \text{constant} + \frac{E_1}{RT^2} \tag{9.63}$$

$$\frac{d \ln k_{-1}}{dT} = \text{constant} + \frac{E_{-1}}{RT^2} \tag{9.64}$$

where $E_1 - E_{-1}$ is equal to ΔU°. Experimentally, it is found that the constants appearing in Eqs. 9.63 and 9.64 can be set equal to zero, and integration of these equations gives rise to

$$k_1 = A_1 e^{-E_1/RT} \tag{9.65}$$

$$k_{-1} = A_{-1} e^{-E_{-1}/RT} \tag{9.66}$$

[2]In this chapter, in order to simplify notation we drop the device of using superscripts u to indicate the value of a quantity. It is to be understood by the reader that one takes the logarithm of the *value* of the quantity rather than the quantity itself.

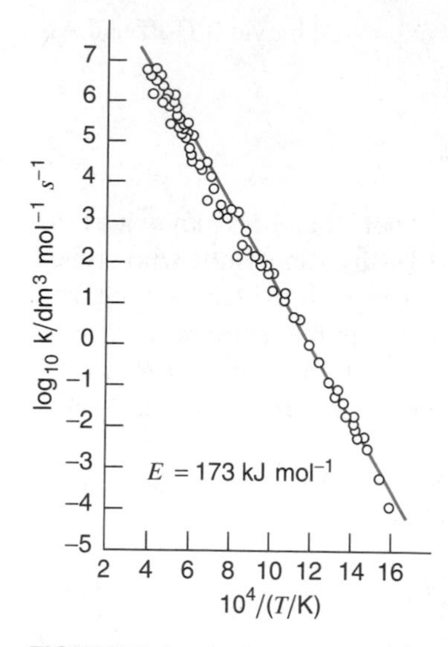

FIGURE 9.8
An Arrhenius plot for the thermal decomposition of ethyne, a second-order reaction. This plot covers a very wide temperature range, from 600 K to 2500 K, and the rate constants range over a factor of more than 10^{10}. The rates were measured by 11 different investigators, using a variety of experimental techniques. (Reproduced by permission of the National Research Council of Canada, Margaret H. Back, *Canadian Journal of Chemistry*, Vol. 49, pages 2199–2204, 1971.)

The quantities A_1 and A_{-1} are best known as the **preexponential factors**[3] of the reactions, and E_1 and E_{-1} are known as the **activation energies,** or **energies of activation.**

Arrhenius's approach to the law was a little different from that of van't Hoff. He pointed out that for ordinary chemical reactions the majority of collisions between the reactant molecules are ineffective; the energy is insufficient. The orientation of the reacting species is also a factor and is treated in the discussion of the preexponential factor in Section 9.9, p. 380–390. In a small fraction of the collisions, however, the energy is great enough to allow reaction to occur. According to the Boltzmann principle (see Section 15.2) the fraction of collisions in which the energy is in excess of a particular value E is

$$e^{-E/RT}$$

This fraction is larger the higher the temperature T and the lower the energy E. The rate constant should therefore be proportional to this fraction.

In order to test the Arrhenius equation (Eq. 9.59), we first take logarithms of both sides:

$$\ln k = \ln A - \frac{E}{RT} \tag{9.67}$$

If the Arrhenius law applies, a plot of ln k against $1/T$ will be a straight line, and the slope (which has the unit of K) will be $-E/R$. Alternatively, we can take common logarithms:

$$\log_{10} k = \log_{10} A - \frac{E}{2.303RT} \tag{9.68}$$

and the slope of a plot of $\log_{10} k$ against $1/T$ is $-E/2.303R$, or

$$\frac{\text{slope}}{\text{K}} = -\frac{E/\text{J mol}^{-1}}{19.14} \tag{9.69}$$

An example of such an Arrhenius plot is shown in Figure 9.8. It follows from Eq. 9.59 that the preexponential factor has the same units as the rate constant itself (e.g., s^{-1} for a first-order reaction, $dm^3\ mol^{-1}\ s^{-1}$ for a second-order reaction). The SI unit of activation energy is J mol^{-1}, and kJ mol^{-1} is very commonly used, but many values in the scientific literature are in cal mol^{-1} or kcal mol^{-1}.

The Arrhenius equation has a surprisingly wide applicability. It is obeyed not only by the rate constants of elementary reactions but frequently also by the rates of much more complicated processes. For example, the law is obeyed by the chirping of crickets (see Figure 9.9), the creeping of ants, the flashing of fireflies, the α brainwave rhythm, the rate of aging, and even by psychological processes such as the rates of counting and forgetting.[4] The reason for the applicability of the law to such processes is that they are controlled by chemical reactions.

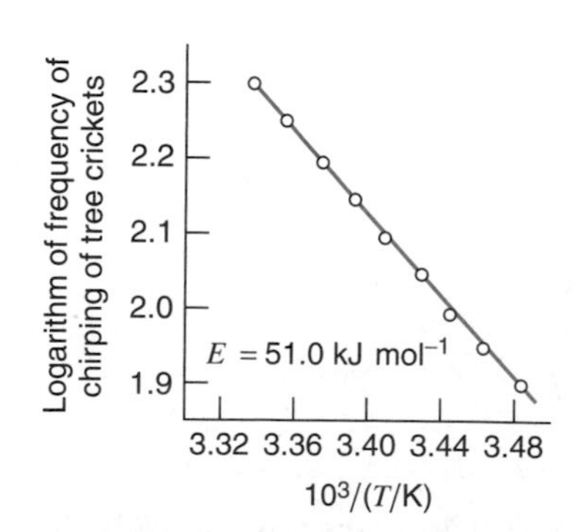

FIGURE 9.9
Arrhenius plot for the chirping of crickets.

[3]The term *frequency factor* is very commonly used, but is not recommended; A does not in general have the dimensions of frequency.

[4]For details, with Arrhenius plots, of these processes see K. J. Laidler, *J. Chem. Education, 49,* 343(1972).

EXAMPLE 9.2 A second-order reaction in solution has a rate constant of 5.7×10^{-5} dm^3 mol^{-1} s^{-1} at 25 °C and of 1.64×10^{-4} dm^3 mol^{-1} s^{-1} at 40 °C. Calculate the activation energy and the preexponential factor, assuming the Arrhenius equation to apply.

Solution To solve this type of problem it is convenient (but not necessary) to sketch an Arrhenius plot; this is shown in Figure 9.10, in which natural logarithms are plotted against the reciprocals of the rates. It is to be emphasized that, in taking the logarithms of the rate constants and the reciprocals of the rates, it is necessary to use enough significant figures because we are dealing with relatively small differences between the values.

The slope of the line in Figure 9.10 is

$$\frac{-9.772 - (-8.716)}{(3.3540 - 3.1934) \times 10^{-3}} = -\frac{1.056}{0.1606} \times 10^3\ \text{K}$$
$$= -6.576 \times 10^3\ \text{K}$$

The energy of activation is the slope multiplied by -8.314 J K^{-1} mol^{-1}:

$$E_a = 8.3145 \times 6.576 \times 10^3 = 54.68 \times 10^3\ \text{J mol}^{-1}$$
$$= 54.7\ \text{kJ mol}^{-1}$$

The preexponential factor A can be obtained from Eq. 9.67. The rate constant at either temperature can be used to evaluate A. If we use 25 °C, we obtain (see Figure 9.10)

$$-9.772 = \ln A - \frac{54\,680}{8.3145 \times 298.15} \quad \text{or} \quad -9.772 = \ln A - 22.059$$

and thus

$$\ln A = 12.287$$
$$A = 2.17 \times 10^5\ \text{dm}^3\ \text{mol}^{-1}\ \text{s}^{-1}$$

The units of A are the same as those of the rate constant.

Activation Energy

The activation energy is looked upon differently in the theories of van't Hoff and Arrhenius. Van't Hoff placed emphasis on the energy levels of the reactants and products and of species occurring during the course of reactions. Figure 9.11 shows a potential-energy diagram for a reaction

$$\text{A} + \text{B} \rightleftharpoons \text{Y} + \text{Z}$$

In this diagram the internal energy for the products Y + Z is larger than that for the reactants A + B, by an amount ΔU. In general there is not much difference between energy and enthalpy changes, so that ΔH will also be positive, that is, the reaction is endothermic.

During the course of the reaction between a molecule of A and a molecule of B the potential energy very often passes through a maximum, as shown in Figure 9.11. The height of this maximum plays a very important role in all theories of the rates of reactions. The molecular species having the maximum energy is called the

Vary the activation energy to change the barrier height; Plot Eq. 9.67; Arrhenius equation; Activation energy.

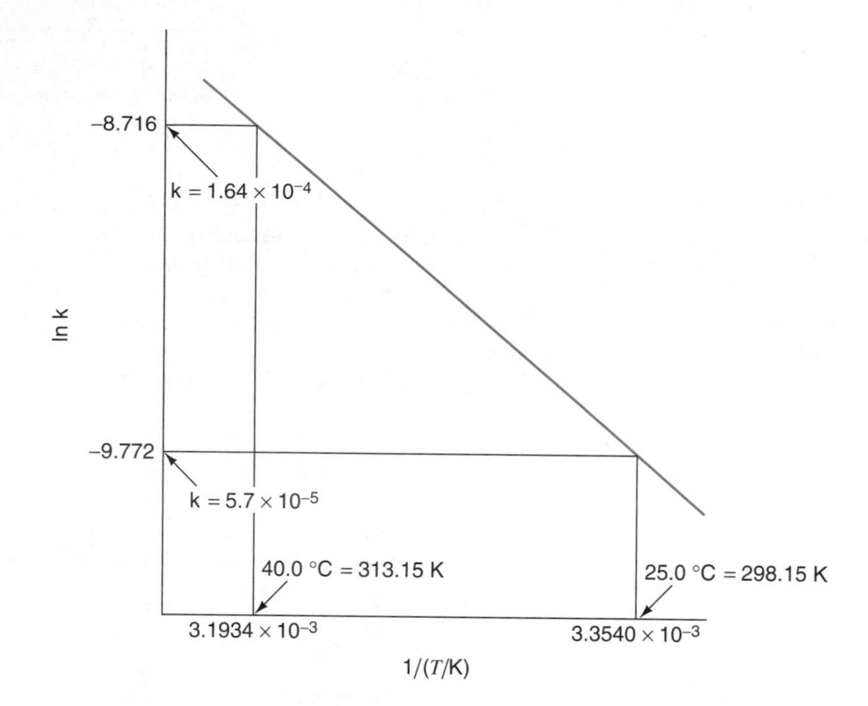

FIGURE 9.10
Schematic Arrhenius plot for Example 9.2.

activated complex, and its *state* is called the **transition state** and is usually denoted by the symbol ‡. The energy E_1 of this complex with respect to the energy of A + B is the activation energy for the reaction in the left-to-right direction. The energy E_{-1} of the activated complex with respect to X + Y is the activation energy for the reverse reaction. We see from Figure 9.11 that

$$E_1 - E_{-1} = \Delta U \tag{9.70}$$

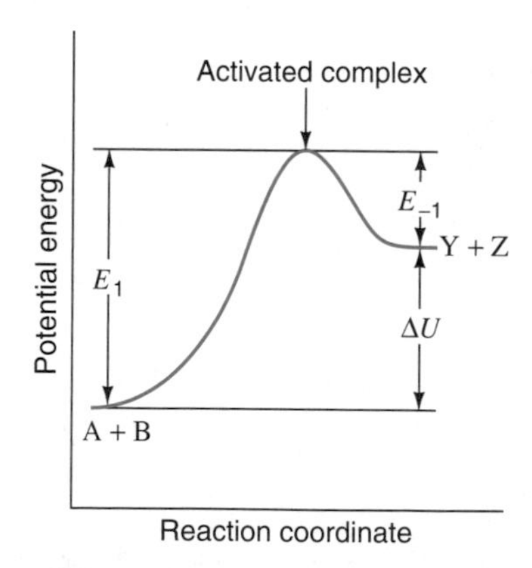

FIGURE 9.11
Potential-energy diagram for a chemical reaction A + B ⇌ Y + Z.

This equation is very useful. Sometimes we can measure both E_1 and E_{-1} and then have a value for ΔU, from which ΔH can easily be obtained by making a small correction (see Eq. 2.41). More often we know ΔH and hence ΔU and can measure one activation energy conveniently; the other is then obtained from Eq. 9.70.

This concept of an energy barrier for a reaction is a very useful and important one. It follows that for an endothermic reaction the activation energy must be at least equal to the endothermicity; for our example, E_1 must be at least equal to $\Delta U°$, since E_{-1} cannot be negative. Arrhenius's discussion of the activation energy is closely related to this concept; for reaction between A and B to occur, the molecules must come together with at least the energy E_1, in order to surmount the barrier.

Why is there usually an energy barrier to reaction? Why does not the curve in Figure 9.11 rise smoothly from one level to the other, without passing through a maximum? An important clue is provided by the fact that certain special types of reactions do occur with zero activation energy. These are reactions in which there is simply a pairing of electrons, without the breaking of any chemical bond. A free methyl radical, for example, has an odd electron; its Lewis structure is

$$\begin{array}{c} \text{H} \\ \text{H}:\ddot{\text{C}}\cdot \\ \ddot{\text{H}} \end{array}$$

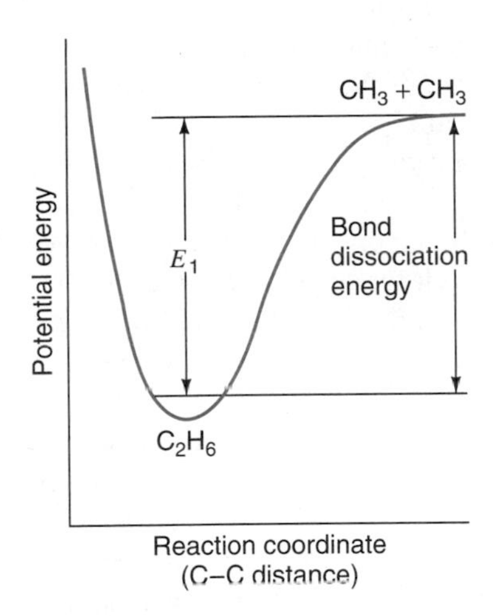

FIGURE 9.12
Potential-energy diagram for the $2CH_3 \rightarrow C_2H_6$ system.

When methyl radicals combine to form ethane,

$$2CH_3 \rightarrow C_2H_6$$

they do so with zero activation energy, and the energy diagram is as shown in Figure 9.12. The activation energy for the reverse reaction, the dissociation of C_2H_6, is equal to the dissociation energy of the bond. The activation energy is also zero for some ion-ion reactions, and reactions involving the attachment of electrons to molecules. Such reactions, although having zero activation energy, often have an observable temperature coefficient because of a temperature dependence of the pre-exponential factor.

However, in most chemical reactions at least one chemical bond is broken, and at least one bond is formed. For example, in the hydrolysis of an alkyl halide reaction in solution

$$RX + H_2O - ROH + HX$$

the C—X bond is being broken during reaction at the same time that the O—C bond is being formed. Even though the overall process is exothermic, the energy released in the formation of the O—C bond is not transferred completely efficiently into the energy required to break the C—X bond, and there is an energy barrier.

9.8 Potential-Energy Surfaces

View movies and animations of the potential energy surface for the reaction $H + HBr \rightarrow H_2 + Br$; Potential energy surfaces.

The theoretical treatment of the dynamics of a chemical reaction is a matter of considerable difficulty, even for the simplest of systems. A very important advance was made in 1931 by the American physical chemist Henry Eyring (1901–1981) and the Hungarian-British physical chemist (later philosopher and sociologist) Michael Polanyi (1891–1976), who developed the method of **potential-energy surfaces,** which resemble maps of the reaction system. Their calculations were carried out for one of the simplest of all chemical reactions, the reaction between a hydrogen atom and a hydrogen molecule. It is possible to measure the rate of a reaction of this type by labeling one or more of the atoms; a deuterium atom, for example, may be caused to react with a hydrogen molecule, in which case the products will be DH + H. Otherwise, it is possible to make atomic hydrogen react with pure *para*-hydrogen[5], in which case there is conversion of the *para*-hydrogen into the equilibrium mixture of the *ortho* and *para* forms. In either case we may represent the reaction as follows:

$$H^{\alpha} + H^{\beta}\text{—}H^{\gamma} \rightarrow H^{\alpha}\text{—}H^{\beta} + H^{\gamma}$$

This process is the simplest reaction in which there is breaking of one bond and making of another. Many reactions are of this type, and a study of this very simple process has provided great insight into the mechanisms of much more complicated processes.

It is found from experiment that the activation energy of this reaction is about 40 kJ mol^{-1} (≈ 9 kcal mol^{-1}). The energy levels of the reactants and the products

[5]*Para*-hydrogen is a form of H_2 in which the nuclear spins are opposed; in *ortho*-hydrogen the spins are parallel. *Para*-hydrogen can be prepared pure and has physical properties that differ slightly from the *ortho* form. Ordinary hydrogen is a 1:3 mixture of the *ortho* and *para* forms.

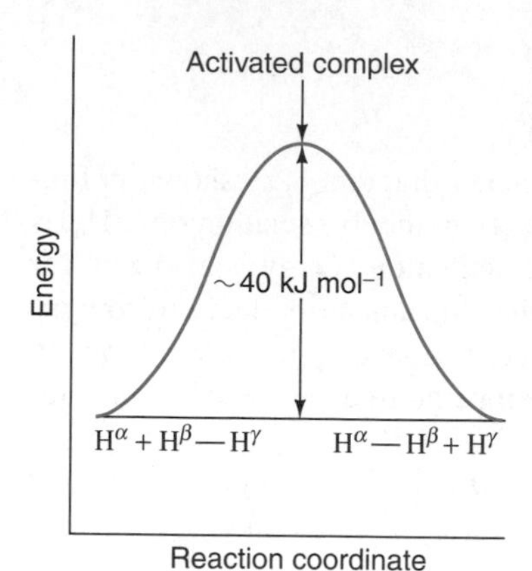

FIGURE 9.13
The energy barrier for the reaction $H + H_2 \rightarrow H_2 + H$.

are the same, but the activation energy of 40 kJ mol^{-1} implies that an energy barrier of this height must be crossed during the passage of the system from the initial to the final state. This is represented schematically in Figure 9.13, in which the energy of the system is plotted against a *reaction coordinate* that represents in some way the extent to which the individual reaction process has occurred.

Eyring's and Polanyi's treatment in 1929 was based on the following ideas. We can imagine a series of steps in which H^α is brought closer and closer to the H^β—H^γ molecule. We suppose that the atom H^α and the molecule H^β—H^γ possess sufficient energy between them to come close together and to give rise to reaction. As the atom approaches the molecule, there is an electronic interaction between them, and the potential energy of the system at first increases and later decreases. Since this particular system is symmetrical, the maximum energy corresponds to a symmetrical species in which the distance between H^α and H^β is equal to the distance between H^β and H^γ. After this complex has formed, the energy gradually decreases as the system approaches the state corresponding to the molecule H^α—H^β and the separated atom H^γ. This species, which lies at the maximum along the reaction path, is known as an *activated complex.*

The most stable complex formed by the interaction between a hydrogen atom and a hydrogen molecule is linear. We will now consider how the energy of such a linear complex varies with the distances between the atoms. The distance between H^α and H^β is designated as r_1, and the distance between H^β and H^γ as r_2; since the complex is linear, the distance between H^α and H^γ is $r_1 + r_2$. The energy of such a linear system of atoms can be represented in a three-dimensional diagram in which the energy is plotted against r_1 and r_2. Such a diagram is shown schematically in Figure 9.14. On the left-hand face of this diagram the distance r_2 may be considered to be sufficiently great that we are dealing simply with the diatomic molecule H^α—H^β. Similarly, on the right-hand face of the diagram there is a curve for the diatomic molecule H^β—H^γ, the distance r_1 now being large. The course of reaction is shown by the arrows in the diagram.

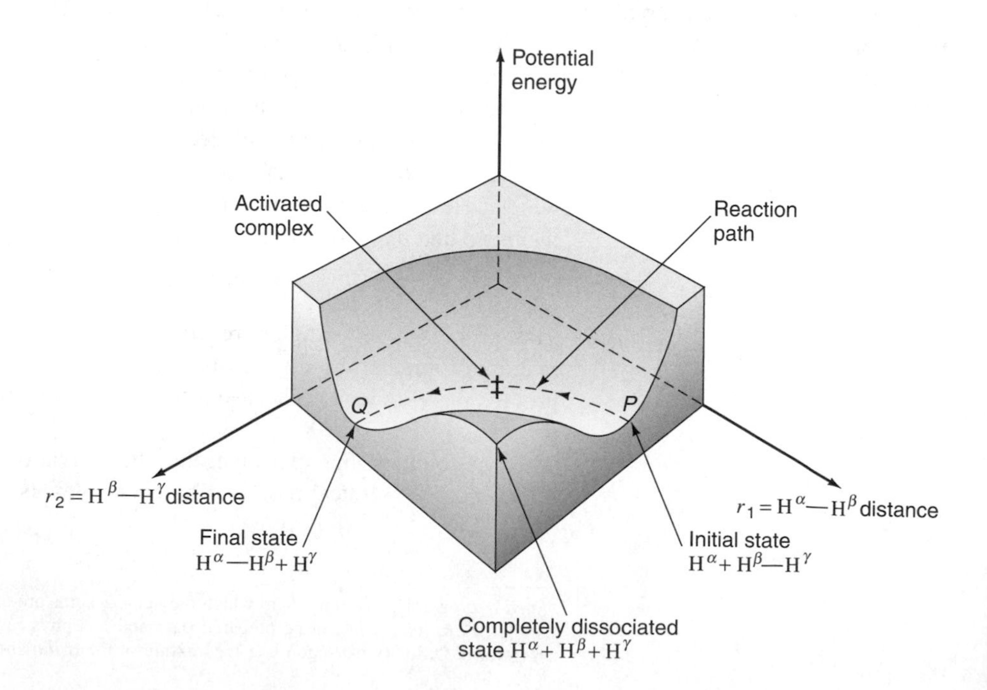

FIGURE 9.14
Schematic potential-energy surface for the reaction H^α + H^β—$H^\gamma \rightarrow H^\alpha$—$H^\beta$ + H^γ.

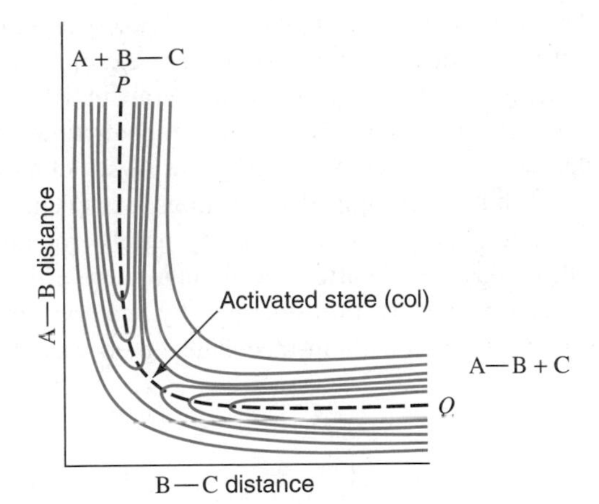

FIGURE 9.15
Schematic potential-energy surface for a reaction A + B—C → A—B + C, showing contour lines connecting positions of equal energy.

To determine the course of such a reaction it is necessary to make quantum-mechanical calculations corresponding to a number of points in the interior of the diagram. The calculations that were made in 1929 by Eyring and Polanyi were necessarily rather approximate, but they nevertheless revealed the correct general form of the potential-energy surface for the reaction, as has been revealed by later and more accurate calculations. These calculations have all shown that, running in from the points P and Q on the diagram, there are two valleys that meet in the interior of the diagram at a *col,* or *saddle point.* This result may be represented in a different type of diagram, such as that in Figure 9.15, where the energy levels are shown by means of contour lines. To go from point P to the point Q using the minimum amount of energy, the system will travel along the valley, over the col, and down into the second valley. This reaction path is represented by the dashed line shown in the contour diagram, and for energetic reasons the majority of the reaction systems will follow this path. The energy corresponding to the col is from one point of view a maximum energy and from another point of view a minimum energy. It is a minimum energy in the sense that the system cannot use less energy in going from P to Q; it is a maximum energy in the sense that, as the system travels along its most economical path, the col represents the highest point in the path—the energies are higher on each side of this path. The height of this col represents the activation energy of the system.

When the reaction is not a symmetrical one but involves three atoms that are not identical, the results are similar, but the potential-energy surface is no longer symmetrical with respect to the two axes.

Note that the mechanism of this reaction is very different from one corresponding to the complete dissociation of the molecule H^{β}—H^{γ}, followed in a separate stage by the combination of H^{α} with H^{β}. The activation energy that would be required for this process is the energy of dissociation of the hydrogen molecule, which is 431.8 kJ mol^{-1} (103.2 kcal mol^{-1}). It is evident that by moving along the valleys the system can achieve reaction at the expense of very much less energy than would be required if the mechanism involved complete dissociation. The physical explanation of this is that the energy released by the making of the second bond (the bond between H^{α} and H^{β}) continuously contributes toward the energy requirement for the breaking of the first bond, the bond between H^{β} and H^{γ}.

The calculation of potential-energy surfaces, by the use of quantum mechanics, is still a matter of considerable difficulty even for such a simple system as $H + H_2$.

Even with the latest computers it still takes a great deal of computer time to obtain accurate energies for this system. For example, an accurate calculation of the barrier height for this reaction, carried out in 1992 by D. D. Diedrich and James B. Anderson[6], required 80 days of computer time! The calculated energy at the col, relative to the energy of $H + H_2$, was 40.21 kJ mol^{-1} (9.61 kcal mol^{-1}).

With more complicated reactions it is even more difficult to make reliable calculations of activation energies by purely quantum-mechanical methods. Because of this, considerable effort has gone into the estimation of potential-energy surfaces and hence of activation energies by methods in which empirical information is used, often with quantum-mechanical theory used in addition.

9.9 The Preexponential Factor

According to the Arrhenius equation (Eq. 9.59), the rate of a reaction at a given temperature is controlled entirely by the two quantities E and A. We saw in the last section that the calculation of the activation energy E presents very considerable difficulty. The situation regarding the preexponential factor A is somewhat more encouraging, in that it is possible to make reasonable estimates of the value of this quantity, taking due account of the type of reaction and of the various factors involved.

Hard-Sphere Collision Theory

Collision Density

The first attempt to calculate the preexponential factor was based on the kinetic theory of collisions, the assumption being that the molecules are hard spheres. For a single gas A, the **collision density** (i.e., the total number of collisions per unit time per unit volume) is (see Section 1.9)

$$Z_{AA} = \frac{1}{2}\sqrt{2}\pi d^2 \bar{u} N_A^2 \tag{9.71}$$

where d is the molecular diameter, N_A is the number of molecules per unit volume, and $\bar{u}$ is the mean molecular velocity, given in Table 1.3:

$$\bar{u} = \sqrt{\frac{8k_BT}{\pi m}} \tag{9.72}$$

The collision density is therefore

$$Z_{AA} = 2d^2N_A^2\sqrt{\frac{\pi k_BT}{m}} \tag{9.73}$$

Its SI unit is m^{-3} s^{-1}. The corresponding expression for the collision density Z_{AB} for collisions between two unlike molecules A and B, of masses m_A and m_B, is

$$Z_{AB} = N_AN_Bd_{AB}^2\left(8\pi k_BT\frac{m_A + m_B}{m_Am_B}\right)^{1/2} \tag{9.74}$$

Collision Cross Section

Here d_{AB} is the average of the diameters, or the sum of the radii. The quantity d_{AB}^2 is known as the **collision cross section** and given the symbol σ (Greek lowercase letter sigma).

[6] D. D. Diedrich and James B. Anderson, "An accurate quantum Monte Carlo calculation of the barrier height for the reaction $H + H_2 \rightarrow H_2 + H$," *Science, 258,* 786–788(1992).

According to the hard-sphere kinetic theory of reactions, these collision numbers multiplied by the Arrhenius factor $e^{-E/RT}$ give the rate of formation of the products of reaction, in molecules per unit volume per unit time; thus, for reaction between A and B,

$$v_{AB} = N_A N_B d_{AB}^2 \left(8\pi k_B T \frac{m_A + m_B}{m_A m_B} \right)^{1/2} e^{-E/RT} \tag{9.75}$$

Division by $N_A N_B$ gives a rate constant in molecular units (SI unit: $m^3\ s^{-1}$); it can be put into molar units by multiplication by the Avogadro constant L:

$$k = L d_{AB}^2 \left(8\pi k_B T \frac{m_A + m_B}{m_A m_B} \right)^{1/2} e^{-E/RT} \tag{9.76}$$

Collision Frequency Factor

The preexponential factor in this expression is called the **collision frequency factor** and given the symbol Z_{AB} (or Z_{AA} if there is only one type of molecule):

$$Z_{AB} = L d_{AB}^2 \left(8\pi k_B T \frac{m_A + m_B}{m_A m_B} \right)^{1/2} \tag{9.77}$$

The development of this treatment of reactions was first made by the German chemist Max Trautz (1880–1961) in 1916 and independently in 1918 by the British chemist William Cudmore McCullagh Lewis (1855–1956). Lewis applied this treatment to the reaction

$$2HI \rightarrow H_2 + I_2$$

and calculated a preexponential factor of $3.5 \times 10^{-7}\ dm^3\ mol^{-1}\ s^{-1}$ at 25 °C, which was in excellent agreement with the experimental value of $3.52 \times 10^{-7}\ dm^3\ mol^{-1}\ s^{-1}$.

Paradoxically, this good agreement was unfortunate, because it led to undue confidence in the theory and delayed the development of the subject for many years. It later became evident that there are many reactions for which there are large discrepancies between the observed preexponential factors and the collision frequency factors. For example, it is often found that when the reactant molecules are of some complexity, the observed preexponential factors are lower by several powers of 10 than the factors calculated by the simple collision theory. Deviations from theory are also encountered with solution reactions involving ions or dipolar substances as reactants.

The source of these discrepancies lies in the use of the hard-sphere kinetic theory of gases in order to evaluate the frequency of collisions. This theory is quite satisfactory for the treatment of energy transfer processes such as occur in viscous flow, but for a theory of chemical reactivity a more precise definition of a collision is needed. If two molecules are to undergo a chemical reaction, they not only must collide with sufficient mutual energy but must come together with such a mutual orientation that the required bonds can be broken and made. The kinetic theory, by treating the reacting molecules as hard spheres, counts every collision as an effective one; if the molecules are complex, on the other hand, in only a small fraction of the collisions will the molecules come together in the right way for reaction to occur.

Some workers remedied the situation by retaining the kinetic theory of gases but introducing into the preexponential factor a steric factor P that was supposed to represent the fraction of the total number of collisions that is effective from the point of view of the mutual orientation of the molecules. The rate constant would then be written as

$$k = PZ_{AB}e^{-E/RT} \tag{9.78}$$

This procedure certainly introduces some improvement, but the evaluation of P cannot be done in an entirely satisfactory way. Moreover, certain factors other than orientation enter into the magnitude of the preexponential factor, and these cannot easily be estimated.

Transition-State Theory

Since orientation and other effects cannot be treated satisfactorily by a modification of hard-sphere collision theory, an alternative approach to the problem is necessary. A much more satisfactory treatment was presented in 1935 by Henry Eyring, and independently by M. G. Evans (1904–1952) and Michael Polanyi; their theory has become known as *activated complex theory* or as **transition-state theory.** A useful way of approaching this theory is to regard it as a logical extension of the argument of van't Hoff with regard to the energy of activation. We have seen in Section 9.7 that van't Hoff regarded the energy change of a reaction as being the difference between two terms, one of which exercises the sole control over the reaction in the forward direction and the other in the reverse direction. We can extend this argument to the standard Gibbs energy change $\Delta G°$ in a chemical reaction. This quantity can be broken up into two terms, as shown in Figure 9.16. In going to the final state a reaction system must pass over a Gibbs energy barrier, and once at the top of this barrier the reaction can proceed without the expenditure of any additional Gibbs energy. The reaction from left to right will, therefore, depend solely on $\Delta^{\ddagger}G_1^\circ$; that from right to left, solely on $\Delta^{\ddagger}G_{-1}^\circ$.

This type of theory may be formulated as follows, by analogy with the van't Hoff approach. The thermodynamic (i.e., dimensionless) equilibrium constant K_c° of the reaction is related to the Gibbs energy change by

$$\ln K_c^\circ = -\frac{\Delta G^\circ}{RT} \tag{9.79}$$

This equilibrium constant K_c° is the ratio of k_1/k_{-1} of the rate constants and therefore

$$\ln k_1 - \ln k_{-1} = -\frac{\Delta G^\circ}{RT} \tag{9.80}$$

$$= -\frac{\Delta^{\ddagger}G_1^\circ}{RT} + \frac{\Delta^{\ddagger}G_{-1}^\circ}{RT} \tag{9.81}$$

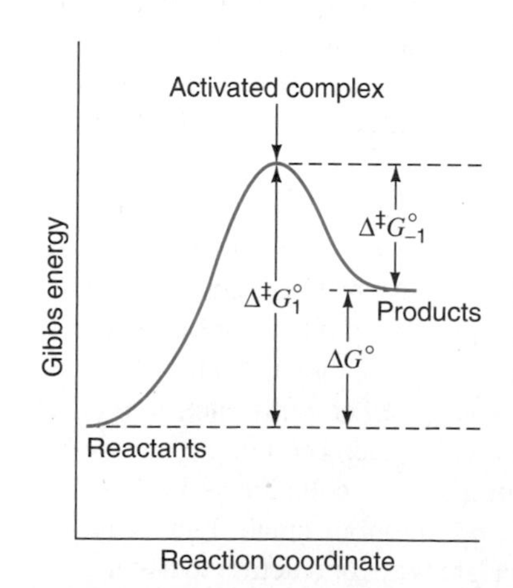

FIGURE 9.16
Gibbs energy diagram for a reaction.

In the latter equation, $\Delta G°$ has been split into the two terms $\Delta^{\ddagger}G_1^\circ$ and $\Delta^{\ddagger}G_{-1}^\circ$. This equation may be split into the two equations

$$\ln k_1 = \ln \nu - \frac{\Delta^{\ddagger}G_1^\circ}{RT} \tag{9.82}$$

$$\ln k_{-1} = \ln \nu - \frac{\Delta^{\ddagger}G_{-1}^\circ}{RT} \tag{9.83}$$

Here $\ln \nu$ is a quantity that must appear in both equations. Equation 9.82 may be written as

$$k_1 = \nu e^{-\Delta^{\ddagger}G_1^\circ/RT} \tag{9.84}$$

Henry Eyring (1901–1981)

Henry Eyring was born in Colonia Juarez, Mexico, of American parents. He studied mining engineering at the University of Arizona, and then went to the University of California at Berkeley, obtaining a Ph.D. degree in physical chemistry in 1927. He taught for a period at the University of Wisconsin, and spent a year (1929–1930) in Berlin collaborating with Michael Polanyi on the construction of the first potential-energy surface for a chemical reaction. In 1931 he was appointed professor of chemistry at Princeton, and became a naturalized American citizen in 1935. In 1946 he went to the University of Utah as Dean of Graduate Studies and professor of chemistry, remaining there until the end of his life.

Much of Eyring's research was in chemical kinetics, but he also worked on the kinetics of physical processes such as diffusion and on the structures of solids and liquids. Aside from some early experimental work at the University of Wisconsin, Eyring's research was entirely theoretical, always closely related to experimental results.

Perhaps his most important contribution was his formulation, in 1935, of transition-state theory (Sections 9.9 and 15.8). Subsequently Eyring applied this theory to many chemical and physical processes. He also did much work on extending the theory. He wrote a number of books, including *The Theory of Rate Processes,* coauthored in 1941 with S. Glasstone and K. J. Laidler, which for the first time presented a systematic account of transition-state theory and many of its applications, and *The Theory of Rate Processes in Biology and Medicine,* published in 1974 with F. H. Johnston and Betsy J. Stover.

Eyring had a friendly disposition and was always happy to discuss his scientific work with anyone who would listen. He was always full of ideas, many of them wrong, but he always welcomed criticism, and his suggestions could usually be turned into sound scientific treatments.

Eyring was a devout and active member of the Mormon Church, on which he published a number of articles. As in the case of G. N. Lewis, his failure to obtain a Nobel Prize has been a matter of surprise among many physical chemists. His transition-state theory after over half a century continues to help us understand the rates of chemical and physical processes.

References: K. J. Laidler, *Dictionary of Scientific Biography,* Supplement 2, *17,* 279–284(1990).

S. H. Heath, "The making of a physical chemist: the education and early researches of Henry Eyring," *Journal of Chemical Education, 62,* 93–98(1985).

J. O. Hirschfelder, "Henry Eyring, 1901–1981," in *American Philosophical Society Year Book 1982,* Philadelphia.

K. J. Laidler, *The World of Physical Chemistry,* Oxford University Press, 1993.

D. W. Urry, "Henry Eyring (1901–1981): A twentieth-century physical chemist and his models," *Mathematical Modelling, 3,* 503–522(1982).

Since the Gibbs energy changes may be written in terms of entropy and enthalpy changes,

$$\Delta G = \Delta H - T\Delta S \tag{9.85}$$

Eq. 9.84 may be written as

$$k_1 = \nu e^{\Delta^{\ddagger}S_1^{\circ}/R} e^{-\Delta^{\ddagger}H_1^{\circ}/RT} \tag{9.86}$$

Thermodynamic Parameters for Activation

The quantity $\Delta G_1^{\ddagger}$ appearing in Eq. 9.84 is known as the **Gibbs energy of activation** (formerly as the free energy of activation), and $\Delta^{\ddagger}S_1^{\circ}$ and $\Delta^{\ddagger}H_1^{\circ}$ are known as the **entropy of activation** and as the **enthalpy of activation,** respectively.

If we compare Eq. 9.86 with Eq. 9.78, we see that there is a close similarity as far as the heat and energy terms are concerned; E in Eq. 9.78 is the energy required to reach the activated state and $\Delta^{\ddagger}H°_1$ is the corresponding enthalpy change, and these are very similar in magnitude. The expressions differ, however, in that instead of the PZ of Eq. 9.78, the new equation involves an entropy term. It is theoretically superior to the kinetic theory in this respect, because the kinetic theory fails to give any interpretation of the fact that the ratio of the rate constants of two opposing reactions, being an equilibrium constant, must involve an entropy term. Moreover, transition-state theory gives a satisfactory interpretation of orientation and solvent effects, since it interprets these in terms of entropy changes when the reactant molecules come together.

We will defer a derivation of the transition-state theory equation until we have studied some statistical mechanics; the derivation, with an example of its use to calculate a rate constant by use of transition-state theory, is to be found in Section 15.8. Here we will anticipate the result of the derivation and state that the factor ν in Eq. 9.84 and 9.86 is k_BT/h, where k_B is the Boltzmann constant and h is the Planck constant. The Eyring equation is therefore

Eyring Equation

$$k = \frac{k_BT}{h} e^{\Delta^{\ddagger}S°/R} e^{-\Delta^{\ddagger}H°/RT} = \frac{k_BT}{h} e^{-\Delta^{\ddagger}G°/RT} \tag{9.87}$$

Another convenient form of the equation is

$$k = \frac{k_BT}{h} K^{\ddagger} \tag{9.88}$$

where $K^{\ddagger}$ is the equilibrium constant for the process

$$A + B \rightarrow X^{\ddagger}$$

in which activated complexes are formed from the reactants. An important feature of transition state theory is that the activated complexes $X^{\ddagger}$ are assumed to be in a special type of equilibrium with the reactants A and B. This assumption is shown in Section 15.8 to be satisfactory.

Equation 9.87 may be expressed in a form that involves the experimental energy of activation E_a instead of the heat of activation $\Delta^{\ddagger}H°$. Since $K^{\ddagger}$ is a concentration equilibrium constant, its variation with temperature is given by Eq. 4.83:

$$\frac{d \ln K^{\ddagger}}{dT} = \frac{\Delta^{\ddagger}U°}{RT^2} \tag{9.89}$$

where $\Delta^{\ddagger}U°$ is the increase in internal energy in passing from the initial state to the activated state. Differentiation of the logarithmic form of Eq. 9.88 gives

$$\frac{d \ln k}{dT} = \frac{1}{T} + \frac{d \ln K^{\ddagger}}{dT} \tag{9.90}$$

and together with Eq. 9.89 this gives

$$\frac{d \ln k}{dT} = \frac{1}{T} + \frac{\Delta^{\ddagger}U°}{RT^2} = \frac{RT + \Delta^{\ddagger}U°}{RT^2} \tag{9.91}$$

The experimental energy of activation E_a is thus given by

$$E_a = RT + \Delta^{\ddagger}U^{\circ} \tag{9.92}$$

The relationship between $\Delta^{\ddagger}U^{\circ}$ and $\Delta^{\ddagger}H^{\circ}$ is

$$\Delta^{\ddagger}H^{\circ} = \Delta^{\ddagger}U^{\circ} + P\,\Delta^{\ddagger}V^{\circ} \tag{9.93}$$

where $\Delta^{\ddagger}V^{\circ}$ is the increase in volume in going from the initial state to the activated state. Substitution of this in Eq. 9.92 gives

$$E_a = \Delta^{\ddagger}H^{\circ} - P\,\Delta^{\ddagger}V^{\circ} + RT \tag{9.94}$$

For unimolecular gas reactions there is no change in the number of molecules as the activated complex is formed, and $\Delta^{\ddagger}V^{\circ}$ is therefore zero; $\Delta^{\ddagger}V^{\circ}$ is also small for reactions in solution. In these cases

$$E_a = \Delta^{\ddagger}H^{\circ} + RT \tag{9.95}$$

and the rate equation may therefore be written as

$$k = \frac{k_{\mathrm{B}}T}{h} \mathrm{e}^{\Delta^{\ddagger}S^{\circ}/R} \mathrm{e}^{-(E_a - RT)/RT} \tag{9.96}$$

or as

$$k = \mathrm{e}\left(\frac{k_{\mathrm{B}}T}{h}\right) \mathrm{e}^{\Delta^{\ddagger}S^{\circ}/R} \mathrm{e}^{-E_a/RT} \tag{9.97}$$

Thus, for reactions in solution and unimolecular gas reactions the preexponential factor A is given by

$$A = \mathrm{e}\left(\frac{k_{\mathrm{B}}T}{h}\right) \mathrm{e}^{\Delta^{\ddagger}S^{\circ}/R} \tag{9.98}$$

For gas reactions the general relationship is

$$P\,\Delta^{\ddagger}V^{\circ} = \Sigma \nu RT \tag{9.99}$$

where $\Sigma\nu$ is the change in the number of molecules when the activated complex is formed from the reactants (i.e., is the stoichiometric sum for the process $\mathrm{A} + \mathrm{B} + \cdots \rightarrow \mathrm{X}^{\ddagger}$). In a bimolecular reaction two molecules become one, so that $\Sigma\nu$ is equal to -1; in this case the experimental energy of activation is related to the enthalpy of activation by the relationship

$$E_a = \Delta^{\ddagger}H^{\circ} + 2RT \tag{9.100}$$

From this it follows that the rate constant may be written as

$$k = e^2\left(\frac{k_{\mathrm{B}}T}{h}\right) \mathrm{e}^{\Delta^{\ddagger}S^{\circ}/R} \mathrm{e}^{-E_a/RT} \tag{9.101}$$

For agreement with the simple hard-sphere theory of collisions the entropy of activation, relative to the units of $\mathrm{dm^3\ mol^{-1}\ s^{-1}}$, must be about $-60\ \mathrm{J\ K^{-1}\ mol^{-1}}$ (see Problem 9.39). If, on the other hand, the standard state is $1\ \mathrm{mol\ cm^{-3}}$ (i.e., if the rate constant units are $\mathrm{cm^3\ mol^{-1}\ s^{-1}}$), the entropy of activation must be approximately zero for agreement with the collision theory.

EXAMPLE 9.3 From the data given in Example 9.2 on p. 383 calculate the Gibbs energy of activation at 25 °C, the entropy of activation, and the enthalpy of activation.

Solution The Gibbs energy of activation is related to the rate constant by Eq. 9.87, which in logarithmic form is

$$\ln k = \ln \frac{k_B T}{h} - \frac{\Delta^{\ddagger}G^{\circ}/\text{J mol}^{-1}}{8.3145 T/\text{K}}$$

At 25 °C the value of $k_B T/h$ is 6.214×10^{12} s^{-1}. Then, inserting values into the last equation,

$$-9.772 = \ln(6.214 \times 10^{12}) - \frac{\Delta^{\ddagger}G^{\circ}/\text{J mol}^{-1}}{8.3145 \times 298.15}$$

or

$$-9.772 = 29.458 - \frac{\Delta^{\ddagger}G^{\circ}/\text{J mol}^{-1}}{2479}$$

and thus

$$\Delta^{\ddagger}G^{\circ} = 97\ 250 \text{ J mol}^{-1} = 97.2 \text{ kJ mol}^{-1}$$

The entropy of activation can be calculated from the preexponential factor, using Eq. 9.98, which in logarithmic form is

$$\ln A = \ln \text{e} + \ln \frac{k_B T}{h} + \frac{\Delta^{\ddagger}S^{\circ}/\text{J K}^{-1}\text{ mol}^{-1}}{8.3145}$$

The value of $\ln A$ was calculated to be 12.287; thus

$$12.287 = \ln 2.718 + 29.458 + \frac{\Delta^{\ddagger}S^{\circ}}{8.3145 \text{ J K}^{-1}\text{ mol}^{-1}}$$

$$= 1.000 + 29.458 + \frac{\Delta^{\ddagger}S^{\circ}}{8.3145 \text{ J K}^{-1}\text{ mol}^{-1}}$$

and thus

$$\Delta^{\ddagger}S^{\circ} = -151.1 \text{ J K}^{-1}\text{ mol}^{-1}$$

From Eq. 9.95

$$\Delta^{\ddagger}H^{\circ} = E_a - RT = 54\ 700 - 8.3145 \times 298.15$$

$$= 52\ 220 \text{ J mol}^{-1} = 52.2 \text{ kJ mol}^{-1}$$

Alternatively, $\Delta H^{\ddagger}$ could have been calculated from $\Delta G^{\ddagger}$ and $\Delta S^{\ddagger}$:

$$\Delta^{\ddagger}H^{\circ} = \Delta^{\ddagger}G^{\circ} + T\,\Delta^{\ddagger}S^{\circ}$$

$$= 97\ 240 - 151.1 \times 298.15$$

$$= 52\ 189 \text{ J mol}^{-1} = 52.2 \text{ kJ mol}^{-1}$$

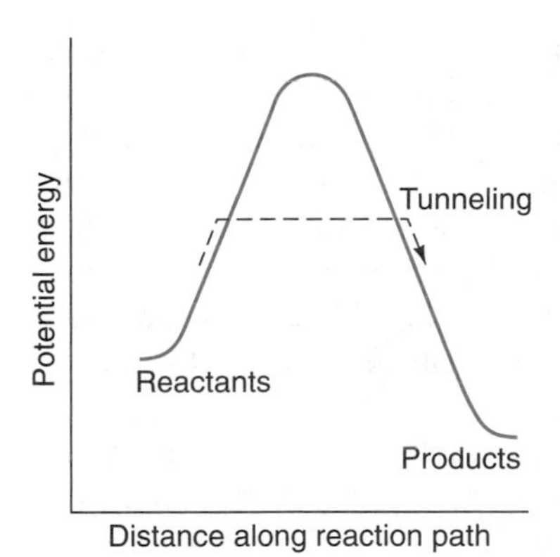

FIGURE 9.17
Schematic representation of quantum-mechanical tunneling through a potential-energy barrier.

Quantum-Mechanical Tunneling

So far, in considering transition-state theory, we have assumed that the system behaves classically, by which we mean that quantum-mechanical behavior need not be taken into account. In particular, we have assumed that when a chemical reaction occurs the system moves over the top of the potential-energy barrier, such as the one illustrated in Figure 9.13. However, according to quantum mechanics (dealt with in detail in Chapter 11) a system may under certain circumstances tunnel through an energy barrier, as shown schematically in Figure 9.17. For tunneling to occur the barrier must be sufficiently thin. Another condition is that the transfer that occurs must involve a light particle. Thus quantum-mechanical tunneling occurs readily with an electron-transfer process such as

$$Ce^{4+} + Ag^{+} \rightarrow Ce^{3+} + Ag^{2+}$$

It also occurs readily with reactions such as

$$D + H_2 \rightarrow DH + H$$

in which there is transfer of a hydrogen atom. Tunneling is also important in the transfer of a proton H^+ or a hydride ion H^-. However, even a deuterium atom is sufficiently heavy for tunneling to be negligible, so that there is little tunneling in the reaction

$$D + DH \rightarrow D_2 + H$$

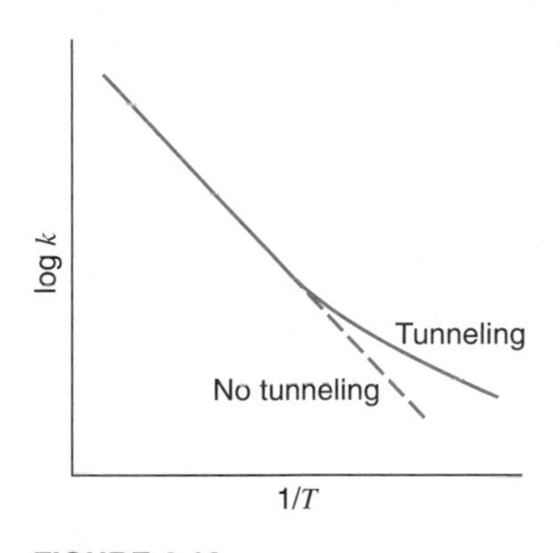

FIGURE 9.18
Schematic Arrhenius plots for a reaction with no quantum-mechanical tunneling and with tunneling. The tunneling rates are independent of temperature, and therefore become relatively more important than the non-tunneling processes when the temperature is lowered. The rates are therefore higher than predicted by the Arrhenius equation. Put differently, the apparent activation energies are lower at lower temperatures. Such curved Arrhenius plots, suggestive of tunneling, have been observed for a number of chemical reactions involving transfer of hydrogen atoms (for which tunneling is more likely), but have not yet been detected for electrochemical processes. (See p. 480.)

For reasonably simple systems it is possible to make fairly reliable quantum-mechanical calculations to predict when tunneling is to be expected. Experimental indications that tunneling is occurring are obtained in several different ways. One way is to carry out experiments down to low temperatures. At a sufficiently low temperature a chemical reaction involving an energy barrier will effectively cease, but tunneling, having a low temperature coefficient, will continue to occur. The result is that an Arrhenius plot (of the logarithm of the rate against the reciprocal of the absolute temperature), instead of being linear, will show curvature, the rates at the lower temperatures being abnormally high (Figure 9.18). By such techniques, tunneling of electrons, hydrogen atoms, and protons has been confirmed for certain ordinary chemical reactions. An example is the reaction

$$D + H_2 \rightarrow DH + H$$

A number of reactions in solution involving electron and proton transfer have also exhibited curved Arrhenius plots and are believed to involve tunneling.

A second way of obtaining evidence for tunneling is related to the first; preexponential factors will be abnormally small. We can see from Figure 9.18 that since $\ln A$ is $\ln k$ extrapolated to a zero value of $1/T$, the preexponential factor at low temperatures is smaller than it would be if there were no tunneling. Such evidence for tunneling has been obtained for a number of reactions.

Two other lines of evidence for tunneling involve kinetic-isotope effects, which will now be considered.

Kinetic-Isotope Effects

When an atom in a reactant molecule is replaced by one of its isotopes (e.g., H is replaced by D), both the equilibrium constant and the rate constant are altered.

TABLE 9.2 Typical Kinetic-Isotope Ratios at 25 °C

Isotopic Forms	Ratio
H,D	6*
H,T	15*
^{12}C, ^{13}C	1.04
^{12}C, ^{14}C	1.08
^{14}N, ^{15}N	1.04
^{15}O, ^{16}O	1.04

With elementary reactions, the lighter isotopes give the higher rates. The numbers are the ratios when there is tunneling.

*Greater if there is tunneling.

These isotope effects are due entirely to the atomic masses so that they are particularly large when an ordinary hydrogen atom H is replaced by a deuterium (D) or a tritium (T) atom, the ratio of the masses being close to 1:2:3. The theory of isotope effects is simple in principle but some of the details are a little complicated; the theory of equilibrium isotope effects is touched on in Section 15.7, where the effects will be seen to be largely due to the influences of the atomic masses on the zero-point energies (Section 11.6). The difficulty of interpreting kinetic-isotope effects springs mainly from the fact that they depend rather critically on the structure assumed for the activated complex.

Some typical kinetic-isotope ratios are shown in Table 9.2. Note that the values are particularly large for the H,D and H,T substitutions because of the much larger mass ratios. If the reaction is such that the H atom can tunnel, these ratios are even larger, since the D and T atoms are too heavy to tunnel to any appreciable extent.

Abnormally large H/D or H/T kinetic-isotope effects thus provide additional evidence for quantum-mechanical tunneling of the H atom. A value of k_H/k_D greater than 10 is good evidence for tunneling; values over 20 have been observed experimentally for some reactions.

Further evidence for tunneling is obtained if experiments are done with tritium (T) as well as with H and D. From the theory of kinetic-isotope effects, according to which the effects are largely due to the effects of the masses on the zero-point levels, the relationship between the rate constants (if there is no tunneling) should be approximately

$$\ln \frac{k_H}{k_T} = 1.44 \ln \frac{k_H}{k_D} \tag{9.102}$$

However, if there is tunneling for H the k_H value will be abnormally high, with the result that the factor on the right-hand side will be smaller than 1.44. Low values for this factor thus provide evidence for tunneling.

9.10 Reactions in Solution

We will now consider some of the factors that determine the magnitudes of the rates of reactions in solution. This topic can be divided into two parts, one concerned with the activation energy and the other with the entropy of activation. As mentioned earlier, very little can be said, from a fundamental standpoint, about the magnitudes of activation energies, and this is especially true of reactions in solution.

Rather more can be said about entropies of activation (or preexponential factors). No exact treatment is possible at the present owing to the inherent complexity of the problem of reactions in solution. A number of important generalizations are, however, well recognized, and it is now possible to make a fairly reliable estimate of the entropy of activation of a solution reaction, and hence of the preexponential factor, in terms of the various factors involved in the reaction. Conversely, it is possible, from the knowledge of values of entropies of activation, to draw some inferences about the nature of a reaction (such as its ionic character) in cases where this is not already known.

Reactions in solution are of a variety of types. There are certain reactions, involving nonpolar molecules, in which it seems that the solvent plays a relatively subsidiary role. Some reactions of this kind occur in the gas phase as well as in solution, and when they do so, they usually occur with much the same rates as in solution and with similar entropies and energies of activation. In such reactions the interactions between reactant molecules and solvent molecules are not of great importance, and the solvent can be regarded as merely filling up space between the reactant molecules. Theory and experiments with mechanical models suggest that in such cases the number of effective collisions between reactant molecules is hardly affected by the presence of solvent. If the reactant molecules are fairly simple, the orientation effect is small, and the hard-sphere collision theory is not far from the truth. The preexponential factor may then be described as a "normal" one; its magnitude for a bimolecular reaction is of the order 10^9 to 10^{11} dm^3 mol^{-1} s^{-1}. If the reacting molecules are more complex, there will be a loss of entropy in forming the activated complex and a lower preexponential factor.

More often, however, the solvent does not act as an inert space filler but is involved in a significant way in the reaction itself. This is particularly the case when the reactants are ions or neutral (but perhaps polar) molecules and when the solvent also is polar. Changes in polarity during the course of reaction will then cause a reorientation of solvent molecules and will have important effects on the entropies of activation.

For reactions between ions the electrostatic effects are very important, and the preexponential factors of such reactions depend on the ionic charges. For reactions between ions of opposite signs, the preexponential factors are much higher than for reactions between neutral molecules. If the ions are of the same sign, the preexponential factors are abnormally low. In terms of simple collision theory these effects can be explained in terms of electrostatic forces. The frequency of collisions between reactants of opposite signs is greater than for neutral molecules, whereas the frequency is smaller for ions of the same sign. These effects can be explained more satisfactorily in terms of the entropy changes that occur when the activated complexes are formed. We will now consider this matter by a treatment that will lead to a useful prediction of the influence of the solvent dielectric constant on the rates of ionic reactions.

Influence of Solvent Dielectric Constant

We can make estimates of the Gibbs energy change when two reacting ions form an activated complex and then obtain the rate constant by the use of Eq. 9.87. In reactions between ions the electrostatic interactions make the most important contribution to the Gibbs energy of activation. The simplest treatment of electrostatic interactions makes the same assumptions as the Born model for ions in solution (Section 7.9). The charged ions are considered to be conducting spheres, and the solvent is regarded as a continuous dielectric, having a fixed dielectric constant ϵ. Such a treatment represents a gross oversimplification, but it has proved surprisingly useful in that it leads to conclusions that are semiquantitatively correct.

A very simple model for a reaction between two ions in solution is represented in Figure 9.19. The reacting molecules are regarded as conducting spheres; the radii are r_A and r_B and the charges are $z_A e$ and $z_B e$; e is the electronic charge and z_A and z_B are whole numbers (positive or negative) that indicate the numbers of charges on the ion. Initially the ions are at an infinite distance from each other, and in the transition state they are considered to be intact (i.e., there is no "smearing" of charge

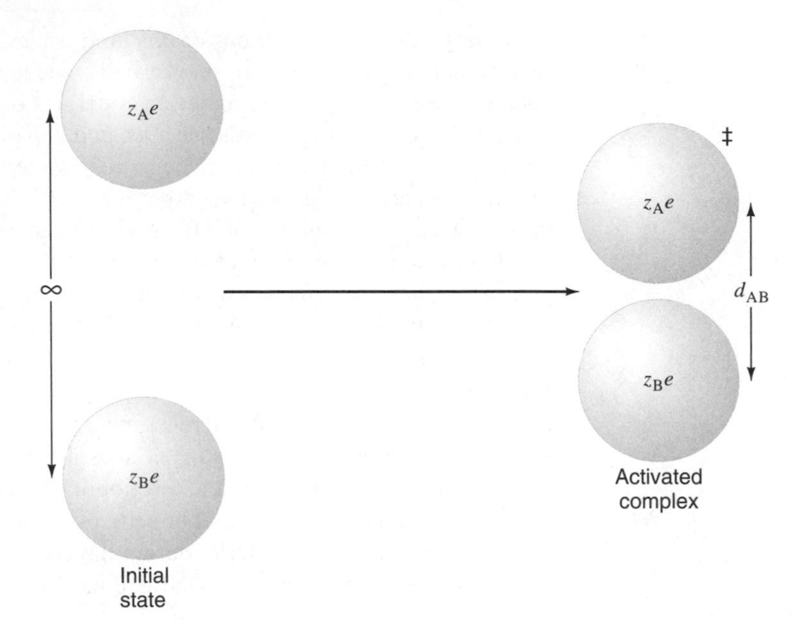

FIGURE 9.19
A simple model for a reaction between two ions, of charges z_Ae and z_Be, in a medium of dielectric constant ϵ. This is known as the "double-sphere" model.

and they are at a distance d_{AB} apart). This model is frequently referred to as the *double-sphere model.* When the ions are at a distance x apart, the force acting between them is equal to

$$f = \frac{z_A z_B e^2}{4\pi\epsilon_0\epsilon x^2} \tag{9.103}$$

The work that must be done in moving them together a distance dx is

$$dw = -\frac{z_A z_B e^2}{4\pi\epsilon_0\epsilon x^2}\,dx \tag{9.104}$$

(The negative sign is used because x decreases by dx when the ions move together by a distance dx.) The work that is done in moving the ions from an initial distance of infinity to a final distance of d_{AB} is therefore

$$w = -\int_{\infty}^{d_{AB}} \frac{z_A z_B e^2}{4\pi\epsilon_0\epsilon x^2}\,dx \tag{9.105}$$

$$= \frac{z_A z_B e^2}{4\pi\epsilon_0\epsilon\, d_{AB}} \tag{9.106}$$

If the signs on the ions are the same, this work is positive; if they are different, it is negative. This work is equal to the electrostatic contribution to the Gibbs energy increase as the ions are moved up to each other.

There is also a nonelectrostatic term $\Delta^{\ddagger}G^{\circ}_{nes}$. The Gibbs energy of activation per mole may therefore be written as

$$\Delta^{\ddagger}G^{\circ} = \Delta^{\ddagger}G^{\circ}_{nes} + \frac{L z_A z_B e^2}{4\pi\epsilon_0\epsilon d_{AB}} \tag{9.107}$$

The last term has been multiplied by the Avogadro constant L so as to give the value per mole. Introduction of this equation into Eq. 9.87 gives

$$k = \frac{k_B T}{h} \exp\left(-\frac{\Delta^{\ddagger}G^{\circ}_{nes}}{RT}\right) \exp\left(-\frac{z_A z_B e^2}{4\pi\epsilon_0 \epsilon d_{AB} k_B T}\right) \tag{9.108}$$

since $R/L = k_B$. Taking natural logarithms

$$\ln k = \ln \frac{k_B T}{h} - \frac{\Delta^{\ddagger}G^{\circ}_{nes}}{RT} - \frac{z_A z_B e^2}{4\pi\epsilon_0 \epsilon d_{AB} k_B T} \tag{9.109}$$

which may be written as

$$\ln k = \ln k_0 - \frac{z_A z_B e^2}{4\pi\epsilon_0 \epsilon d_{AB} k_B T} \tag{9.110}$$

The rate constant k_0 is the value of k in a medium of infinite dielectric constant, when the final term in Eq. 9.110 becomes zero (i.e., when there are no electrostatic forces).

Equation 9.110 leads to the prediction that the logarithm of the rate constant of a reaction between ions should vary linearly with the reciprocal of the dielectric constant. An example of a test of Eq. 9.110 is shown in Figure 9.20. Deviations from linearity can be explained as due to failure of the simple approximations involved in deriving Eq. 9.110 and in some cases to a change in reaction mechanism as the solvent is varied.

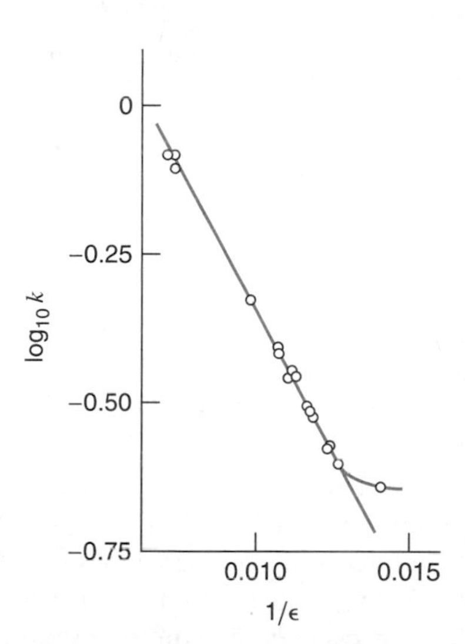

FIGURE 9.20
A plot of $\log_{10} k$ against the reciprocal of the dielectric constant, for the reaction between bromoacetate and thiosulfate ions in aqueous solution.

The slope of the line obtained by plotting $\ln k$ against $1/\epsilon$ is given by Eq. 9.110 as $z_A z_B e^2/4\pi\epsilon_0 d_{AB} k_B T$. Since everything in this expression is known except d_{AB}, it is possible to calculate d_{AB} from the experimental slope. This has been done in a number of cases, and the values obtained, of the order of a few tenths of a nanometre, have always been entirely reasonable; for the data shown in Figure 9.20 the value of d_{AB} is 0.51 nm.

We can extend this treatment to give an interpretation of the magnitudes of the preexponential factors of ionic reactions. We have seen that the electrostatic contribution to the molar Gibbs energy of activation is

$$\Delta^{\ddagger}G^{\circ}_{es} = \frac{L z_A z_B e^2}{4\pi\epsilon_0 \epsilon d_{AB}} \tag{9.111}$$

The thermodynamic relationship between entropy and Gibbs energy is given by Eq. 3.119:

$$S = -\left(\frac{\partial G}{\partial T}\right)_P \tag{9.112}$$

and the electrostatic contribution to the entropy of activation is

$$\Delta^{\ddagger}S^{\circ}_{es} = -\left(\frac{\partial \Delta^{\ddagger}G^{\circ}_{es}}{\partial T}\right)_P \tag{9.113}$$

The only quantity in Eq. 9.111 that is temperature dependent is ϵ and it therefore follows that

$$\Delta^{\ddagger}S^{\circ}_{es} = \frac{L z_A z_B e^2}{4\pi\epsilon_0 \epsilon^2 d_{AB}} \left(\frac{\partial \epsilon}{\partial T}\right)_P \tag{9.114}$$

$$= \frac{L z_A z_B e^2}{4\pi\epsilon_0 \epsilon d_{AB}} \left(\frac{\partial \ln \epsilon}{\partial T}\right)_P \tag{9.115}$$

TABLE 9.3 Some Observed and Predicted Preexponential Factors and Entropies of Activation

	Experimental Values		Estimated Values	
Reactants	A dm^3 mol^{-1} s^{-1}	$\Delta^{\ddagger}S^{\circ}$ J K^{-1} mol^{-1}	A dm^3 mol^{-1} s^{-1}	$\Delta^{\ddagger}S^{\circ}$ J K^{-1} mol^{-1}
$[Cr(H_2O)_6]^{3+} + CNS^-$	$\sim10^{19}$	~126	10^{19}	126
$Co(NH_3)_5Br^{2+} + OH^-$	5×10^{17}	92	10^{17}	84
$CH_2BrCOOCH_3 + S_2O_3^{2-}$	1×10^{14}	25	10^{13}	0
$CH_2ClCOO^- + OH^-$	6×10^{10}	−50	10^{11}	−42
$ClO^- + ClO_2^-$	9×10^{8}	−84	10^{11}	−42
$CH_2BrCOO^- + S_2O_3^{2-}$	1×10^{9}	−71	10^{9}	−84
$Co(NH_3)_5Br^{2+} + Hg^{2+}$	1×10^{8}	−100	10^{5}	−167
$S_2O_4^{2-} + S_2O_4^{2-}$	2×10^{4}	−167	10^{5}	−167
$S_2O_3^{2-} + SO_3^{2-}$	2×10^{6}	−126	10^{5}	−167

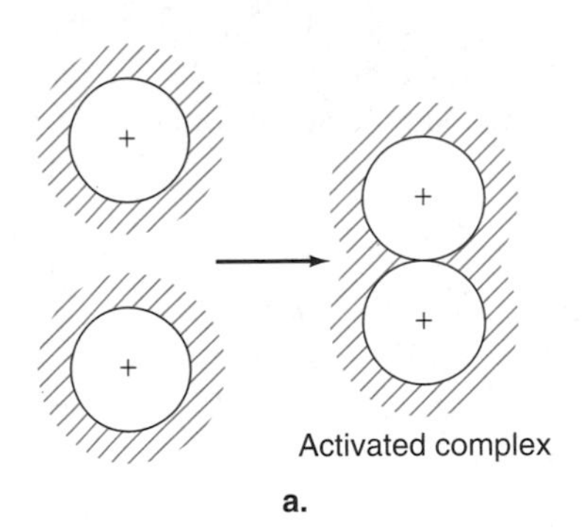

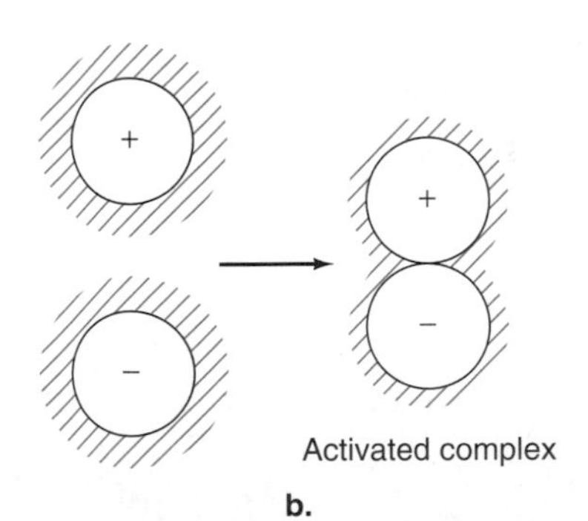

FIGURE 9.21
An interpretation of entropies of activation in terms of the electrostriction. In (a) there is more electrostriction in the activated complex; there is therefore a decrease in entropy. In (b) there is less electrostriction in the activated complex.

In aqueous solution ϵ is about 80 and $(\partial \ln \epsilon/\partial T)_P$ remains constant at -0.0046 K^{-1} over a considerable temperature range. If d_{AB} is taken as equal to 0.2 nm, it follows from Eq. 9.115 that the entropy of activation is given by

$$\Delta^{\ddagger}S^{\circ}_{es} \approx -42 z_A z_B \text{ J K}^{-1}\text{ mol}^{-1} \quad (9.116)$$

The entropy of activation in aqueous solution should thus decrease by about 42 J K^{-1} mol^{-1} for each unit of $z_A z_B$. Moreover, since the preexponential factor is proportional to $e^{\Delta S^{\ddagger}/R}$, which is equal to $10^{\Delta S^{\ddagger}/2.303R}$ or $10^{\Delta S^{\ddagger}/19.12}$, it follows that the factor should decrease by a factor of $10^{42/19.12}$ (i.e., by about one-hundredfold) for each unit of $z_A z_B$. Table 9.3 shows that these relationships are obeyed in a very approximate manner. The treatment is too crude to allow detailed predictions to be made, but it is evidently along the right lines.

The physical model that lies behind these relationships is represented schematically in Figure 9.21. In Figure 9.21a the ions are shown as having single positive charges, and the activated complex therefore bears a double positive charge. The frequency of collision between the two ions will be reduced on account of the electrostatic repulsions between the like charges, but this is by no means the entire effect. When the ions of like sign come together, they form a complex of higher charge, and the activated complex therefore exerts more electrostriction[7] on the surrounding water molecules than is exerted by the reactant ions. Since electrostriction leads to entropy loss, the result is that a reaction between like charges has a negative entropy of activation and a low preexponential factor.

If the reaction is such that oppositely charged ions come together (Figure 9.21b), there is less charge on the activated complex than on the reactants. There is then a decrease in electrostriction when the activated complex is formed, a positive entropy of activation, and a high preexponential factor.

Similar effects are found, but to a lesser extent, when reactions involve the approach of dipolar molecules.

[7]Electrostriction has been discussed in Sections 3.3 and 7.9.

Influence of Ionic Strength

The influence of ionic strength on the rates of reaction between ions may be considered with reference to a reaction of the general type.

$$A + B \rightarrow X^{\ddagger} \rightarrow \text{products}$$

The theory of ionic-strength effects was first worked out in 1924 by J. N. Brønsted. We will give a more modern version, in terms of the concept of the activated complex.

The basis of the treatment is that the rate of a reaction is proportional to the *concentration* of the activated complexes $X^{\ddagger}$ and not to their activity:

$$v = k'[X^{\ddagger}] \tag{9.117}$$

The equilibrium between the activated complexes and the reactants A and B may be expressed as

$$K^{\ddagger} = \frac{a^{\ddagger}}{a_A a_B} = \frac{[X^{\ddagger}]}{[A][B]} \frac{\gamma^{\ddagger}}{\gamma_A \gamma_B} \tag{9.118}$$

where the a's are the activities and the γ's are the activity coefficients. Introduction of Eq. 9.118 into Eq. 9.117 gives rise to

$$v = k[A][B] = k_0[A][B]\frac{\gamma_A \gamma_B}{\gamma^{\ddagger}} \tag{9.119}$$

where k is the second-order rate constant and $k_0 = k'K^{\ddagger}$. Taking logarithms,

$$\log_{10} k = \log_{10} k_0 + \log_{10} \frac{\gamma_A \gamma_B}{\gamma^{\ddagger}} \tag{9.120}$$

According to the Debye-Hückel limiting law (Section 7.10) the activity coefficient of an ion is related to its valency z and the ionic strength I by Eq. 7.104:

$$\log_{10} \gamma = -Bz^2\sqrt{I} \tag{9.121}$$

Introduction of Eq. 9.121 into the rate equation (Eq. 9.120) gives

$$\log_{10} k = \log_{10} k_0 + \log_{10} \gamma_A + \log_{10} \gamma_B - \log_{10} \gamma^{\ddagger} \tag{9.122}$$

$$= \log_{10} k_0 - B[z_A^2 + z_B^2 - (z_A + z_B)^2]\sqrt{I} \tag{9.123}$$

$$= \log_{10} k_0 + 2Bz_A z_B\sqrt{I} \tag{9.124}$$

The value of B is approximately 0.51 $dm^{-3/2}$ $mol^{-1/2}$ for aqueous solutions at 25 °C; Eq. 9.124 may therefore be written as

$$\log_{10} k = \log_{10} k_0 + 1.02\, z_A z_B \sqrt{I/\text{mol dm}^{-3}} \tag{9.125}$$

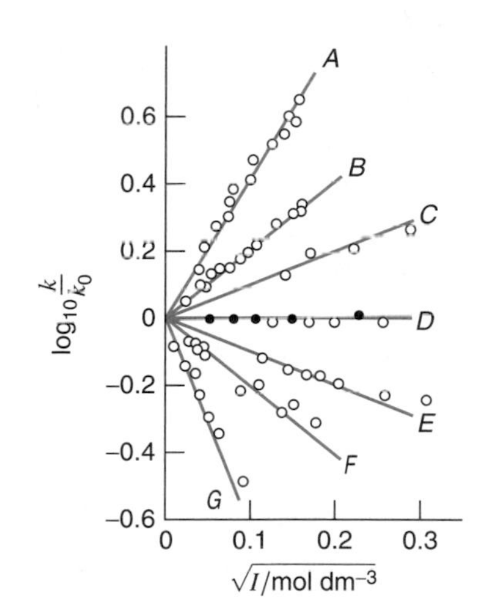

FIGURE 9.22
Plot of $\log_{10}(k/k_0)$ against the square root of the ionic strength, for ionic reactions of various types. The lines are drawn with slopes equal to $z_A z_B$. The reactions follow:

- A: $Co(NH_3)_5Br^{2+} + Hg^{2+}$ ($z_A z_B = 4$)
- B: $S_2O_8^{2-} + I^-$ ($z_A z_B = 2$)
- C: $CO(OC_2H_5)N : NO_2^- + OH^-$ ($z_A z_B = 1$)
- D: $[Cr(urea)_6]^{3+} + H_2O$ (open circles) ($z_A z_B = 0$) $CH_3COOC_2H_5 + OH^-$ (closed circles) ($z_A z_B = 0$)
- E: $H^+ + Br^- + H_2O_2$ ($z_A z_B = -1$)
- F: $Co(NH_3)_5Br^{2+} + OH^-$ ($z_A z_B = -2$)
- G: $Fe^{2+} + Co(C_2O_4)_3^{3-}$ ($z_A z_B = -6$)

This equation has been tested a considerable number of times. The procedure has usually been to measure the rates of ionic reactions in media of varying ionic strength; according to Eq. 9.125, a plot of $\log_{10} k$ against $\sqrt{I/\text{mol dm}^{-3}}$ will give a straight line of slope $1.02z_A z_B$. Figure 9.22 shows a plot of results for reactions of various types; the lines drawn are those with theoretical slopes, and the points lie close to them. If one of the reactants is a neutral molecule, $z_A z_B$ is zero, and the rate constant is expected to be independent of the ionic strength; this is true, for example, for the base-catalyzed hydrolysis of ethyl acetate, shown in Figure 9.22.

The investigations of ionic-strength effects have provided valuable support for the applicability of the Debye-Hückel limiting law and for the validity of the assumption that rate is proportional to the concentration of activated complexes, not to their activities (see Problem 9.48). They have also sometimes been found useful in determining charges on chemical species. For example, in 1960 muonium was discovered by V. W. Hughes and coworkers, and was found to have a mass 1/8.83 that of hydrogen. At first it was not known whether muonium bore an electric charge, and this question was answered by kinetic studies at different ionic strengths. In the reactions of muonium with Cu^{2+} and other ionic species the ionic strength effect was negligible (see Problem 9.49), and this is good evidence that muonium bears no charge. Muonium can be regarded as a light isotope of hydrogen; it is an electron paired with a muon, in this case a positively charged particle, μ^+.

EXAMPLE 9.4 An ionic species of unknown charge undergoes a second-order reaction with a species B^+ that has a charge number of +1. The rate constants for the reaction at 25 °C and at two different ionic strengths are as follows:

I/mol dm^{-3}	k/mol dm^{-3} s^{-1}
0.1	1.224×10^{-1}
0.2	7.21×10^{-2}

Assume the Debye-Hückel limiting law to apply and estimate the charge number z_A of the species A.

Solution A plot of $\log_{10} k$ against $\sqrt{I}$ gives a line of slope −2.04. According to Eq. 9.125 this slope is equal to $1.02z_Az_B$. It follows that z_Az_B equals −2, and since z_B is +1, z_A is −2.

Influence of Hydrostatic Pressure

Studies of the effect of hydrostatic pressure on reactions in solution provide important information about the detailed mechanisms of reactions. The theory of pressure effects on rates was first formulated in 1901 by van't Hoff, who started with Eq. 4.90 for the effect of pressure on equilibrium constants:

$$\left(\frac{\partial \ln K^\circ}{\partial P}\right)_T = -\frac{\Delta V^\circ}{RT} \tag{9.126}$$

He employed an argument very similar to the one he used to arrive at the Arrhenius law, pointing out that since K is the ratio of rate constants, the latter must vary with pressure in a similar way.

An alternative derivation is in terms of transition state theory. If we apply Eq. 9.126 to the equilibrium constant $K^\ddagger$ for the equilibrium between reactants and activated complexes, the result is

$$\left(\frac{\partial \ln K^\ddagger}{\partial P}\right)_T = -\frac{\Delta^\ddagger V^\circ}{RT} \tag{9.127}$$

Volume of Activation

Here $\Delta^{\ddagger}V^{\circ}$ is the volume change in going from the initial state A + B to the activated state $X^{\ddagger}$; this volume change is known as the **volume of activation.** According to Eq. 9.88 the constant k is proportional to $K^{\ddagger}$, and therefore

$$\left(\frac{\partial \ln k}{\partial P}\right)_T = -\frac{\Delta^{\ddagger}V^{\circ}}{RT} \tag{9.128}$$

A rate constant will therefore increase with increasing pressure if $\Delta^{\ddagger}V^{\circ}$ is negative (i.e., if the activated complex has a smaller volume than the reactants). Conversely, pressure has an adverse effect on rates if there is a volume increase when the activated complex is formed. By use of Eq. 9.128, values of $\Delta^{\ddagger}V^{\circ}$ can be determined from experimental determinations of rates at different pressures. In practice, it is usually necessary to use fairly high pressures for this purpose (at least 100 bar, or 10^4 kPa) since otherwise the changes of rate are too small for accurate $\Delta V^{\ddagger}$ values to be obtained.

According to Eq. 9.128, if ln k is plotted against pressure, the slope at any pressure is $-\Delta^{\ddagger}V^{\circ}/RT$. In some cases the plots are linear, which means that $\Delta^{\ddagger}V^{\circ}$ is independent of pressure. If this is so, Eq. 9.128 can be integrated to give

$$\ln k = \ln k_0 - \frac{\Delta^{\ddagger}V^{\circ}}{RT}P \tag{9.129}$$

where k_0 is the rate constant at zero pressure (this is always very close to the value at atmospheric pressure). Examples of such linear plots are shown in Figure 9.23.

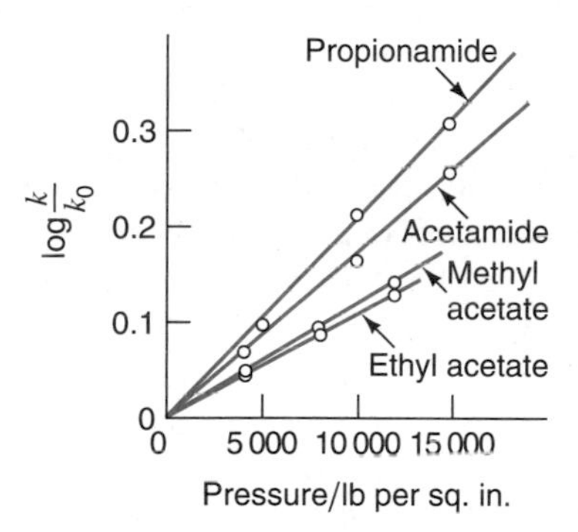

FIGURE 9.23
Plots of $\log_{10}(k/k_0)$ against the hydrostatic pressure, for the alkaline hydrolyses of some esters and amides at 25 °C. (Reprinted with permission from *Transactions of the Faraday Society,* Vol. 54, No. 1020, England, copyright © 1958. The Royal Society of Chemistry is a Signatory to the S.T.M. Guidelines on Permissions.)

EXAMPLE 9.5 From Figure 19.23, estimate an approximate value for $\Delta^{\ddagger}V^{\circ}$ for the hydrolysis of acetamide at 25 °C.

Solution The value of $\log_{10}(k/k_0)$ when the pressure is 15 000 pounds per square inch is about 0.25. The slope of a plot in which natural logarithms were used would therefore be

$$\frac{2.303 \times 0.25}{15\,000 \text{ psi}}$$

1 psi = 6.89×10^3 Pa and the slope is therefore

$$\frac{2.303 \times 0.25}{15\,000 \times 6.89 \times 10^3 \text{ Pa}} = 5.57 \times 10^{-9} \text{ Pa}^{-1}$$

The slope is $\Delta^{\ddagger}V^{\circ}/RT$, and therefore, since J Pa^{-1} = m^3,

$$\begin{aligned}\Delta^{\ddagger}V^{\circ} &= 8.3145 \times 298.15 \text{ (J mol}^{-1}) \times 5.57 \times 10^{-9} \text{ Pa}^{-1} \\ &= 1.38 \times 10^{-5} \text{ m}^3 \text{ mol}^{-1} \\ &= 13.8 \text{ cm}^3 \text{ mol}^{-1}\end{aligned}$$

The interpretation of volumes of activation for reactions in solutions involves the consideration of several factors, for example, the structural changes in the reactant molecules as they form the activated complex. Often, however, even more profound volume changes occur in the solvent itself. If, for example, a reaction occurs between like charges in aqueous solution (Figure 9.21a), there is an increase in electrostriction when the activated complex is formed. Since the bound water molecules occupy less volume than water molecules in ordinary water, there will be a negative $\Delta^{\ddagger}V^{\circ}$ for a reaction of this type, which also has a negative $\Delta^{\ddagger}S^{\circ}$. Conversely, for a reaction between oppositely charged ions, there is a decrease in electrostriction leading to a positive $\Delta^{\ddagger}V^{\circ}$ and a positive $\Delta^{\ddagger}S^{\circ}$. There is, in fact, quite a good correlation between $\Delta^{\ddagger}V^{\circ}$ and $\Delta^{\ddagger}S^{\circ}$ values for reactions in which electrostatic effects are important.

Diffusion-Controlled Reactions

Mixing Control

If a rapid bimolecular reaction in solution is initiated by mixing solutions of the reactants, the observed rate may depend on the rate with which the solutions mix. This effect is known as **mixing control.**

Microscopic Diffusion Control

We have seen in Section 9.5 that the effect of mixing can be eliminated by the use of relaxation methods. However, even when this is done, the rate of reaction may be influenced by the rate with which the reactant molecules diffuse toward each other. This effect is known as **microscopic diffusion control** or as *encounter control.* If the rate we measure is almost exactly equal to the rate of diffusion, we speak of *full microscopic diffusion control* or of *full encounter control.* An example is provided by the combination of H^+ and OH^- ions in solution, which we saw in Section 9.5 to have a rate constant of 1.3×10^{11} dm^3 mol^{-1} s^{-1} at 23 °C. For some reactions the rates of chemical reaction and of diffusion are similar to each other, and we then speak of *partial microscopic diffusion control* or of *partial encounter control.*

A mathematical theory of diffusion was proposed in 1917 by the Polish scientist Marian von Smolochowski (1872–1917), and some aspects of his work are considered in Section 19.2.

Linear Gibbs Energy Relationships

A number of relationships have been suggested in connection with the rate constants of reactions in solution. One of the most useful of these is an equation proposed by the American chemist Louis P. Hammett (1894–1987), which relates equilibrium and rate constants for the reactions of *meta-* and *para-*substituted benzene derivatives; for example, one might compare the rate constants for two types of reactions involving a given set of substituents in the *meta* position. According to Hammett's relationship, a rate or equilibrium constant for the reaction of one compound is related to that for the unsubstituted ("parent") compound in terms of two parameters ρ and σ. For rate constants the relationship is

Hammett Relationship

$$\log_{10} k = \log_{10} k_0 + \sigma\rho \qquad (9.130)$$

where k_0 is the rate constant for the parent compound. For equilibrium constants

$$\log_{10} K = \log_{10} K_0 + \sigma\rho \qquad (9.131)$$

where K_0 is for the parent compound. Of the two parameters, σ depends only on the *substituent,* whereas ρ is a *reaction constant,* varying with the nature of the reaction and the external conditions such as the solvent. A value of unity is arbitrarily chosen for ρ for the ionization equilibrium constant of benzoic acid in aqueous solution and for the substituted benzoic acids. It follows that σ is the logarithm of the ratio of the dissociation constant of a substituted benzoic acid to that of benzoic acid itself.

The Hammett relationships imply linear relationships between the Gibbs energies, of reaction or of activation, for different series of reactions. We have seen (Eq. 9.87) that the rate constant is related to the Gibbs energy of activation by

$$k = \frac{k_B T}{h} e^{-\Delta^{\ddagger} G^{\circ}/RT} \tag{9.132}$$

and therefore

$$\log_{10} k = \log_{10} \frac{k_B T}{h} - \frac{\Delta^{\ddagger} G^{\circ}}{2.303RT} \tag{9.133}$$

Equation 9.130 may therefore be written as

$$\Delta^{\ddagger} G^{\circ} = \Delta^{\ddagger} G_0^{\circ} - 2.303RT\rho\sigma \tag{9.134}$$

This equation, with a particular value of ρ, applies to any reaction involving a reactant having a series of substituents. For a second series of homologous reactions, with the reaction constant ρ',

$$\Delta^{\ddagger} G^{\circ\prime} = \Delta^{\ddagger} G_0^{\circ\prime} - 2.303RT\rho'\sigma \tag{9.135}$$

Equations 9.134 and 9.135 may be written as

$$\frac{\Delta^{\ddagger} G^{\circ}}{\rho} = \frac{\Delta^{\ddagger} G_0^{\circ}}{\rho} - 2.303RT\sigma \tag{9.136}$$

and

$$\frac{\Delta^{\ddagger} G^{\circ\prime}}{\rho'} = \frac{\Delta^{\ddagger} G_0^{\circ\prime}}{\rho'} - 2.303RT\sigma \tag{9.137}$$

Subtraction gives

$$\frac{\Delta^{\ddagger} G^{\circ}}{\rho} - \frac{\Delta^{\ddagger} G^{\circ\prime}}{\rho'} = \frac{\Delta^{\ddagger} G_0^{\circ}}{\rho} - \frac{\Delta^{\ddagger} G_0^{\circ\prime}}{\rho'}$$

or

$$\Delta^{\ddagger} G^{\circ} - \frac{\rho}{\rho'} \Delta^{\ddagger} G^{\circ\prime} = \text{constant} \tag{9.138}$$

There is thus a linear relationship between the Gibbs energies of activation for one homologous series of reactions and those for another. An equivalent relationship is easily derived for the Gibbs energies of overall reactions.

9.11 Reaction Dynamics

During recent years there has been increasing interest in what might be called the "fine structure" of reaction rates. Conventional kinetic studies, such as we have considered up to now, provide information relating to the overall rates of formation of products of reaction. They cannot reveal the details of how the system progresses over the potential-energy surface or of how the energy released in the reaction is distributed among the product molecules. There have been developed two important experimental techniques, concerned with **crossed molecular beams** and with **chemiluminescence,** which do provide evidence about energy distributions in reaction products. In order to interpret the results of such experiments it is necessary also to carry out calculations of the actual passage of systems over potential-energy surfaces. These experimental and theoretical studies are conveniently referred to as being in the area of **reaction dynamics** or of **molecular dynamics.**

Molecular Beams

The essential feature of a molecular-beam experiment, illustrated in Figure 9.24, is that narrow beams of atoms or molecules are caused to cross one another. Movable detectors determine the direction taken by the product molecules and by the unreacted molecules. Analysis of the experimental results yields information about the distribution of energy and angular momentum among the reaction products, the dependence of the total reaction probability on the molecular energies, the lifetime of the collision complex, the quantum states of the products, and a number of other important features. In recent molecular beam work, particularly that done by the American Dudley R. Herschbach (b. 1932) and by the Taiwanese-American Yuan Tseh Lee (b. 1936), great success has been achieved in putting reactant molecules into desired vibrational and rotational states, by laser excitation. The expression **state-to-state dynamics** is often applied to work of this kind, concerned with specific quantum states of the reactant and product molecules.

One important result that has emerged from molecular-beam experiments is that there are two types of reaction mechanisms, **stripping mechanisms** and **rebound mechanisms.** Figure 9.25a shows what occurs in a stripping mechanism of the type $A + BC \rightarrow AB + C$. The reactant atom, A, after it has undergone reaction and has combined with B, travels in much the same direction as previously; similarly the direction of motion of C is not greatly altered. In other words, A strips B away from C, which is hardly affected and is said to act as a "spectator"; the term **spectator stripping** is also applied to these mechanisms. An example of a reaction of this type is

$$K + HBr \rightarrow KBr + H$$

Stripping reactions are all characterized by the fact that reaction can occur when the reactants are farther apart than the sum of their kinetic theory radii. On account of

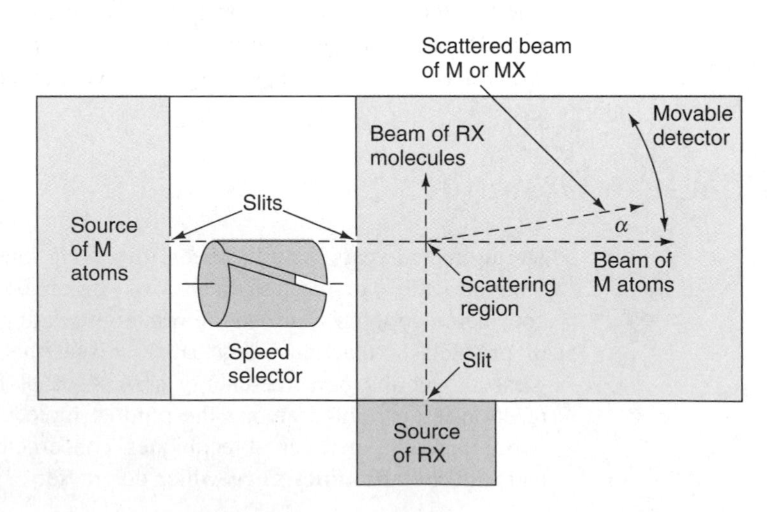

FIGURE 9.24
Schematic diagram of a molecular beam apparatus for the study of a reaction of the type $M + RX \rightarrow MX + R$.

Harpooning Mechanisms

this feature, Michael Polanyi used the expressive phrase **harpooning mechanisms,** the idea being that one reactant (e.g., the K atom) throws out an electron and "harpoons" an atom on the other reactant. In these stripping reactions there is a considerable amount of vibrational energy in the product of reaction (e.g., in KBr in the example just given).

When dynamical calculations are made with various shapes of potential-energy surfaces, it becomes clear that the type of surface that gives rise to a stripping mechanism is the type shown in Figure 9.25b. The characteristic feature of this surface is that the activated complex corresponds to a rather large A—B distance, a feature that accounts for the large reaction cross sections. After the activated complex is formed, the atom A continues to approach B, with at first little change in the B—C distance; after the activated complex is formed, the potential energy falls so that there is attraction between A and BC. Because of this feature the surface is known as an **attractive surface** or as an **early-downhill surface.** If we constructed such a surface and projected a ball along the upper valley, as indicated by the dashed line, it would travel along the lower, right-hand valley with considerable lateral motion. This lateral motion corresponds to vibrational energy in the product A—B, and we therefore have an explanation of why these attractive surfaces lead to a large amount of vibrational energy in the products of reaction. Dynamical calculations confirm this general conclusion.

Attractive Surface

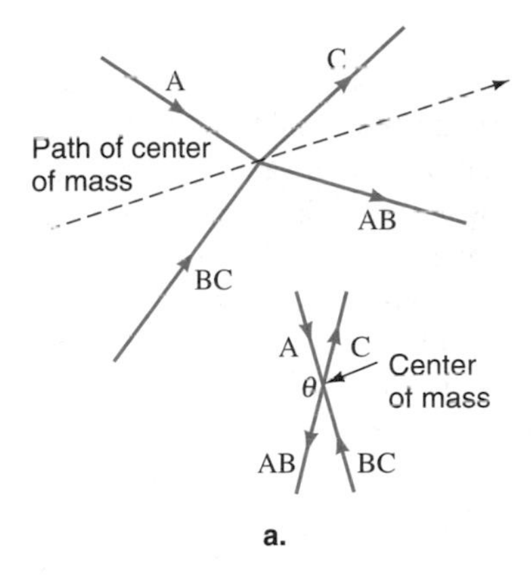

Other reactions studied in molecular beams occur by a rebound mechanism, the features of which are shown in Figure 9.26a. In such reactions the reactant A, after abstracting B from BC, does not continue on its way but rebounds, remaining on the same side of the center of mass of the system. Reactions of this type tend to have small collision cross sections, and little energy goes into the vibration of the product. An example of a reaction occurring by a rebound mechanism is

$$K + CH_3I \rightarrow KI + CH_3$$

for which the cross section is about 0.10 nm^2, which means that reaction requires the approach of the reactants to $\sqrt{0.10}$ nm = 0.3 nm.

Repulsive Surface

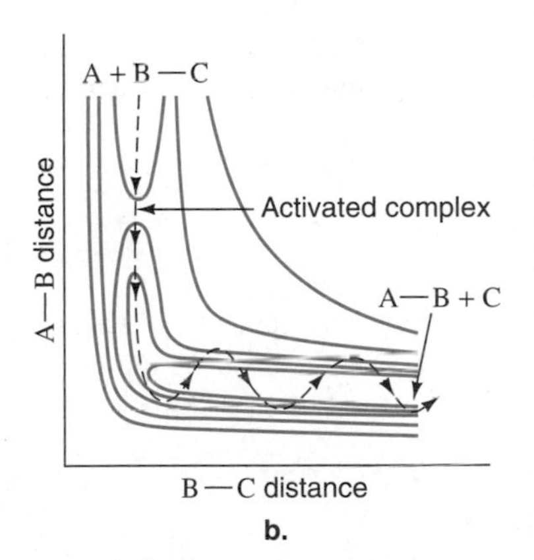

FIGURE 9.25
(a) Directions of motion of reactants and products, for a stripping mechanism. (b) An attractive potential-energy surface, which leads to high-reaction cross sections, stripping, and a large proportion of vibrational energy in the reaction product.

Dynamical calculations show that the type of potential-energy surface leading to a rebound mechanism is as shown in Figure 9.26b. This is known as a **repulsive surface** or late-downhill surface. On such a surface the system cannot reach the activated state without there being a significant extension of the B—C bond. A ball projected on a repulsive surface would follow a path such as that shown in Figure 9.26b; it would leave the lower valley without any lateral motion corresponding to the vibration of the A—B bond. Dynamical calculations on a number of repulsive surfaces have confirmed this general conclusion.

Chemiluminescence

The vibrational states of reaction products can be studied by making measurements of the radiation they emit. The chemiluminescence produced in a reaction may be in various regions of the spectrum, including the visible. Many studies have been made of infrared chemiluminescence, which provides information about the vibrational and rotational states of the product molecules.

The pioneer in this field of investigation was Michael Polanyi, who in the 1920s and 1930s studied the emission of sodium D radiation in systems in which sodium was reacting with halogens. For the reaction

$$Na_2 + Cl \rightarrow Na + NaCl$$

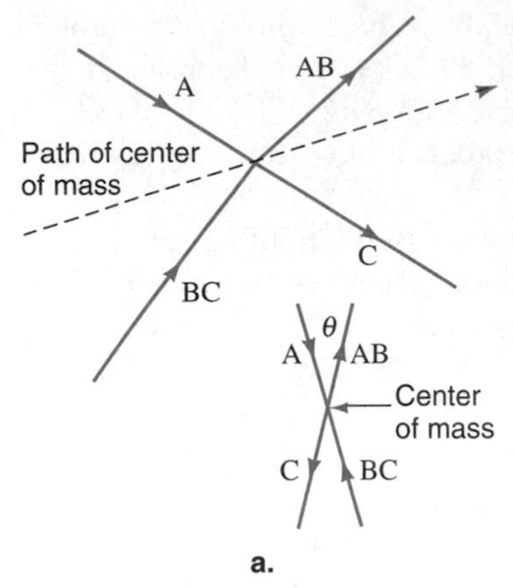

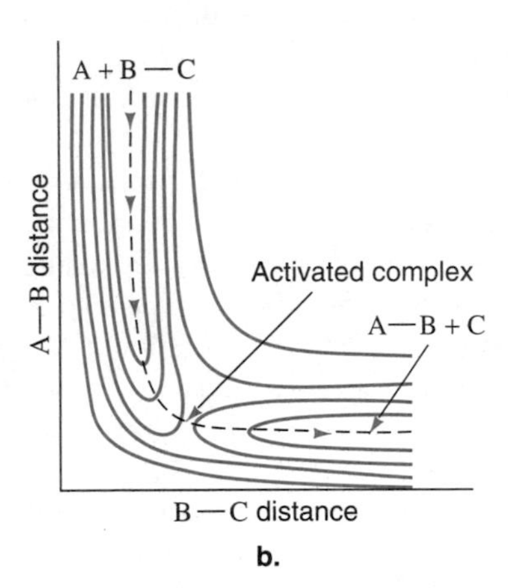

FIGURE 9.26
(a) Directions of motion of reactants and products, for a rebound mechanism. (b) A repulsive potential-energy surface, which leads to low-reaction cross sections, a rebound mechanism, and a smaller proportion of vibrational energy in the reaction product.

he found that a substantial fraction of the energy released remained as vibrational energy in the NaCl molecule, which in a subsequent collision with a sodium atom gave rise to electronically excited Na, which produced the yellow sodium D line.

More recently the infrared emissions from the products of a number of reactions have been extensively investigated, particularly by Michael Polanyi's son, the Canadian physical chemist John Polanyi (b. 1929), who has used infrared spectroscopy to study the vibrational states of reactants and products. His investigations have explored the relationship between the type of energy transfer that takes place and the shape of the potential-energy surface for the reaction. The 1986 Nobel Prize for chemistry was awarded jointly to Herschbach, Lee, and John Polanyi for their contributions to reaction dynamics.

Dynamical Calculations

A number of calculations have been made of the passage of systems over potential-energy surfaces. Such calculations are useful for testing the validity of transition-state theory, which is based on the assumption of equilibrium between reactants and activated complexes. They are also important in leading to certain information, such as the vibrational states of the reaction products and whether a stripping or a rebound mechanism is involved, which is not revealed by transition-state theory.

Many of the dynamical calculations have been made for the $H + H_2$ reaction, since reliable potential-energy surfaces have not been constructed for other systems. Some of the dynamical calculations employ quantum-mechanical methods; others use classical mechanics. It has been concluded that the rates calculated by the dynamical treatments usually agree satisfactorily with those given by transition-state theory. The treatments also show—and this is again consistent with the transition-state theory—that the reaction system usually does not linger at the activated state and perform a number of vibrations; instead it passes directly through the activated state. Calculations with other potential-energy surfaces have shown that systems perform vibrations at the activated state only if there is a depression (a "basin") at the intersection of the two valleys in the potential-energy surface.

The Detection of Transition Species

An important development has been the detection of transition species, which are molecular entities having configurations intermediate between those of the reactants and products. The first successful detection was reported in 1980 by John Polanyi and coworkers for the reaction

$$F + Na_2 \rightarrow F \cdots Na \cdots Na \rightarrow NaF + Na^*$$

The reaction was caused to occur in a molecular beam. The product Na* is in an electronically excited state and emits the yellow D line. On both sides of this line "wing" emission was observed, resulting from the transition species F ··· Na ··· Na. Similar evidence has subsequently been obtained for other reactions.

Transition species are now being actively investigated by the technique of femtosecond flash photolysis (Section 10.6), particularly by Philip R. Brooks (b. 1938) and Ahmed H. Zewail (b. 1946), Nobel Prize for chemistry, 1999.

KEY EQUATIONS

Definition of *rate of reaction:*

$$v = \frac{1}{V}\frac{d\xi}{dt}$$

where ξ = extent of reaction (Sections 2.5 and 9.2).

For an *elementary reaction*

$$v = k[A]^{\alpha}[B]^{\beta}$$

where α and β are partial orders, $\alpha + \beta = n$ is the overall order, and k is the rate constant.

Influence of temperature (Arrhenius equation):

$$k = Ae^{-E/RT}$$

Transition-state theory:

$$k = \frac{k_B T}{h}K^{\ddagger} = \frac{k_B T}{h}e^{-\Delta^{\ddagger}G^{\circ}/RT} = \frac{k_B T}{h}e^{\Delta^{\ddagger}S^{\circ}/R}e^{-\Delta^{\ddagger}H^{\circ}/RT}$$

$$= e\frac{k_B T}{h}e^{\Delta S^{\ddagger}/R}e^{-E/RT} \text{ for reactions in solution.}$$

Influence of *ionic strength I* for reaction in aqueous solution at 25 °C:

$$\log_{10} k = \log_{10} k_0 + 1.02\, z_A z_B \sqrt{I/\text{mol dm}^{-3}}$$

Influence of *hydrostatic pressure P:*

$$\left(\frac{\partial \ln k}{\partial P}\right)_T = -\frac{\Delta^{\ddagger}V^{\circ}}{RT}$$

PROBLEMS

Rate Constants and Order of Reaction

9.1. The stoichiometric equation for the oxidation of bromide ions by hydrogen peroxide in acid solution is

$$2Br^- + H_2O_2 + 2H^+ \rightarrow Br_2 + 2H_2O$$

Since the reaction does not occur in one stage, the rate equation does not correspond to this stoichiometric equation but is

$$v = k[H_2O_2][H^+][Br^-]$$

a. If the concentration of H_2O_2 is increased by a factor of 3, by what factor is the rate of consumption of Br^- ions increased?
b. If the rate of consumption of Br^- ions is 7.2×10^{-3} mol dm^{-3} s^{-1}, what is the rate of consumption of hydrogen peroxide? What is the rate of formation of bromine?
c. What is the effect on the rate constant k of increasing the concentration of bromide ions?
d. If by the addition of water to the reaction mixture the total volume were doubled, what would be the effect on the rate of change of the concentration of Br^-? What would be the effect on the rate constant k?

9.2. A reaction obeys the stoichiometric equation

$$A + 2B \rightarrow 2Z$$

Rates of formation of Z at various concentrations of A and B are as follows:

[A]/mol dm^{-3}	[B]/mol dm^{-3}	Rate/mol dm^3 s^{-1}
3.5×10^{-2}	2.3×10^{-2}	5.0×10^{-7}
7.0×10^{-2}	4.6×10^{-2}	2.0×10^{-6}
7.0×10^{-2}	9.2×10^{-2}	4.0×10^{-6}

What are α and β in the rate equation

$$v = k[A]^{\alpha}[B]^{\beta}$$

and what is the rate constant k?

9.3. Some results for the rate of a reaction between two substances A and B are shown in the following table. Deduce the order α with respect to A, the order β with respect to B, and the rate constant.

[A]/mol dm^{-3}	[B]/mol dm^{-3}	Rate/mol dm^3 s^{-1}
1.4×10^{-2}	2.3×10^{-2}	7.40×10^{-9}
2.8×10^{-2}	4.6×10^{-2}	5.92×10^{-8}
2.8×10^{-1}	4.6×10^{-2}	5.92×10^{-6}

9.4. A substance decomposes at 600 K with a rate constant of 3.72×10^{-5} s^{-1}.

a. Calculate the half-life of the reaction.
b. What fraction will remain undecomposed if the substance is heated for 3 h at 600 K?

9.5. How does the time required for a first-order reaction to go to 99% completion relate to the half-life of the reaction?

9.6. The rate constant for the reaction $H^+ + OH^- \rightarrow H_2O$ is 1.3×10^{11} dm^3 mol^{-1} s^{-1}. Calculate the half-life for the neutralization process if (a) $[H]^+ = [OH^-] = 10^{-1}$ M and (b) $[H^+] = [OH^-] = 10^{-4}$ M.

9.7. The isotope ^{90}Sr emits radiation by a first-order process (as is always the case with radioactive decay) and has a half-life of 28.1 years. When ingested by mammals it becomes permanently incorporated in bone tissue. If 1 μg is absorbed at birth, how much of this isotope remains after (a) 25 years, (b) 50 years, (c) 70 years?

9.8. The first-order decomposition of nitramide in the presence of bases, $NH_2NO_2 \rightarrow N_2O(g) + H_2O(l)$, is conveniently analyzed by collecting the gas evolved during the reaction. During an experiment, 50.0 mg of nitramide was allowed to decompose at 15 °C. The volume of dry gas evolved after 70.0 min. was measured to be 6.59 cm^3 at 1 bar pressure. Find the rate constant and the half-life for nitramide decomposition.

9.9. The reaction

$$2NO(g) + Cl_2(g) \rightarrow 2NOCl(g)$$

is second order in NO and first order in Cl_2. In a volume of 2 dm^3, 5 mol of nitric oxide and 2 mol of Cl_2 were brought together, and the initial rate was 2.4×10^{-3} mol dm^{-3} s^{-1}. What will be the rate when one-half of the chlorine has reacted?

9.10. Measuring the total pressure is a convenient way to monitor the gas phase reaction $2NOCl(g) \rightarrow 2NO(g) + Cl_2(g)$. However, the rate depends on the concentration of the reactant, which is proportional to the partial pressure of the reactant. Derive an expression relating the rate of this reaction to the initial pressure, P_0, and the total pressure, P_t, at time t. Assume that the reaction follows second-order kinetics.

9.11. The following results were obtained for the rate of decomposition of acetaldehyde:

% decomposed	0	5	10	15	20	25	30	35	40	45	50
Rate/Torr min^{-1}	8.53	7.49	6.74	5.90	5.14	4.69	4.31	3.75	3.11	2.67	2.29

Employ van't Hoff's differential method to obtain the order of reaction.

9.12. The isotope $^{32}_{15}P$ emits radiation and has a half-life of 14.3 days. Calculate the decay constant in s^{-1}. What percentage of the initial activity remains after (a) 10 days, (b) 20 days, (c) 100 days?

9.13. The following counts per minute were recorded on a counter for the isotope $^{35}_{16}S$ at various times:

Time/d	Counts/min
0	4280
1	4245
2	4212
3	4179
4	4146
5	4113
10	3952
15	3798

Determine the half-life in days and the decay constant in s^{-1}. How many counts per minute would be expected after (a) 60 days and (b) 365 days?

9.14. The reaction

$$cis\text{-}Cr(en)_2(OH)_2^+ \rightleftarrows trans\text{-}Cr(en)_2(OH)_2^+$$

is first order in both directions. At 25 °C the equilibrium constant is 0.16 and the rate constant k_1 is 3.3×10^{-4} s^{-1}. In an experiment starting with the pure *cis* form, how long would it take for half the equilibrium amount of the *trans* isomer to be formed?

9.15. Suppose that a gas phase reaction $2A(g) \rightarrow 2B(g) + C(g)$ follows second-order kinetics and goes to completion. If the reaction is allowed to proceed in a constant volume vessel at an initial pressure of 2 bar (only A is initially present), what will be the partial pressures of A, B, and C and the total pressure at $t = t_{1/2}$, $2t_{1/2}$, $3t_{1/2}$, and infinity?

9.16. Derive the following relationship for the half-life $t_{1/2}$ of a reaction of order n, with all reactants having an initial concentration a_0:

$$t_{1/2} = \frac{2^{n-1} - 1}{ka_0^{n-1}(n-1)}$$

9.17. Vaughan [*J. Am. Chem. Soc. 54*, 3867(1932)] reported the following pressure measurements as a function of time for the dimerization of 1,3-butadiene (C_4H_6) under constant volume conditions at 326 °C:

t/min	3.25	12.18	24.55	42.50	68.05
P/Torr	618.5	584.2	546.8	509.3	474.6

The initial amount of butadiene taken would have exerted a pressure of 632.0 Torr. Find whether the reaction follows first- or second-order kinetics and evaluate the rate constant.

9.18. A drug administered to a patient is usually consumed by a first-order process. Suppose that a drug is administered in equal amounts at regular intervals and that the interval between successive doses is equal to the $(1/n)$-life for the disappearance process (i.e., to the time that it takes for the fraction $1/n$ to disappear). Prove that the limiting concentration of the drug in the patient's body is equal to n times the concentration produced by an individual dose.

9.19. Equation 9.45 applies to a second-order reaction of stoichiometry A + B → Z. Derive the corresponding equation for a second-order reaction of stoichiometry 2A + B → Z.

9.20. Derive the integrated rate equation for an irreversible reaction of stoichiometry 2A + B → Z, the rate being proportional to $[A]^2[B]$ and the reactants present in stoichiometric proportions; take the initial concentration of A as $2a_0$ and that of B as a_0. Obtain an expression for the half-life of the reaction.

9.21. Prove that for two simultaneous (parallel) reactions

$$A \xrightarrow{k_1} Y \qquad A \xrightarrow{k_2} Z$$

$$\frac{[Y]}{[Z]} = \frac{k_1}{k_2} \text{ at all times.}$$

***9.22.** Prove that for two consecutive first-order reactions

$$A \rightarrow B \rightarrow C$$

the rate of formation of C is given by

$$[C] = [A]_0\left(1 + \frac{k_2 e^{-k_1 t} - k_1 e^{-k_2 t}}{k_1 - k_2}\right)$$

where $[A]_0$ is the initial concentration of A. *Hint:* The solution of the differential equation

$$\frac{dx}{dt} = abe^{-bt} - cx$$

where a, b, and c are constants, is

$$x = \frac{ab}{c - b}(e^{-bt} - e^{-ct}) + I$$

***9.23. a.** Derive the integrated rate equation for a reversible reaction of stoichiometry

$$A \underset{k_{-1}}{\overset{k_1}{\rightleftarrows}} Y + Z$$

The reaction is first order from left to right and second order from right to left. Take the initial concentration of A as a_0 and the concentration at time t as $a_0 - x$.
b. Obtain the integrated equation in terms of k, and the equilibrium constant $K = k_1/k_{-1}$.
c. A reaction to which this rate equation applies is the hydrolysis of methyl acetate. Newling and Hinshelwood, *J. Chem. Soc., 1936,* 1357(1936), obtained the following results for the hydrolysis of 0.05 M ester at 80.2 °C in the presence of 0.05 M HCl, which catalyzes the reaction:

Time, s	1350	2070	3060	5340	7740	∞
Percent hydrolysis	21.2	30.7	43.4	59.5	73.45	90.0

Obtain values for the rate constants k_1 and k_{-1}.

***9.24.** The dissociation of a weak acid

$$HA + H_2O \rightleftharpoons H_3O^+ + A^-$$

can be represented as

$$A \rightleftarrows Y + Z$$

The rate constants k_1 and k_{-1} cannot be measured by conventional methods but can be measured by the T-jump technique (Section 9.5). Prove that the relaxation time is given by

$$t^* = \frac{1}{k_1 + 2k_{-1}x_e}$$

where the concentration of the ions (Y and Z) is at equilibrium.

Temperature Dependence

9.25. The rate constant for a reaction at 30 °C is found to be exactly twice the value at 20 °C. Calculate the activation energy.

9.26. The rate constant for a reaction at 230 °C is found to be exactly twice the value at 220 °C. Calculate the activation energy.

9.27. The following data for a first-order decomposition reaction in aqueous medium was reported by E. O. Wiig [*J. Phys. Chem. 34,* 596(1930)].

t/°C	0	20	40	60
$k/10^{-5}$ min^{-1}	2.46	43.5	576	5480

Find the activation energy and the preexponential factor.

9.28. Two second-order reactions have identical preexponential factors and activation energies differing by 20.0 kJ mol^{-1}. Calculate the ratio of their rate constants (a) at 0 °C and (b) at 1000 °C.

9.29. The gas-phase reaction between nitric oxide and oxygen is third order. The following rate constants have been measured:

T/K	80.0	143.0	228.0	300.0	413.0	564.0
$k \times 10^9$/cm^6 mol^{-1} s^{-1}	41.8	20.2	10.1	7.1	4.0	2.8

The behavior is interpreted in terms of a temperature-dependent preexponential factor; the rate equation is of the form

$$k = aT^n e^{-E/RT}$$

where a and n are constants. Assume the activation energy to be zero and determine n to the nearest half-integer.

9.30. The definition of activation energy E_a is generally considered to be given by an extension of Eq. 9.91:

$$E_a = RT^2\left(\frac{d\ln k}{dT}\right)$$

Problem 9.29 shows that for certain reactions, the temperature dependence of the reaction rate constant is better described by an expression of the type

$$k = aT^n e^{-E/RT}$$

Using the definition for E_a given here, derive an expression for the activation energy from this expression.

9.31. The water flea *Daphnia* performs a constant number of heartbeats and then dies. The flea lives twice as long at 15 °C as at 25 °C. Calculate the activation energy for the reaction that controls the rate of its heartbeat.

9.32. A sample of milk kept at 25 °C is found to sour 40 times as rapidly as when it is kept at 4 °C. Estimate the activation energy for the souring process.

***9.33.** Experimentally, the rate constant for the $O(^3P)$ + HCl reaction in the gas phase is found to have a temperature dependence given by

$$k\ (\text{cm}^3\ \text{molecule}^{-1}\ \text{s}^{-1}) = 5.6 \times 10^{-21}T^{2.87}\ e^{-1766\ \text{K}/T}$$

in the range 350 − 1480 K [Mahmud, Kim, and Fontijn, *J. Phys. Chem. 94,* 2994(1990)].

a. Using the results of Problem 9.30, find the value of E_a at 900 K, which is approximately the middle of this range.
b. Using variational transition-state theory (an extension of the transition-state theory described in Section 9.9), the theoretical rate constant for this reaction is found to behave according to the equation

$$k\ (\text{cm}^3\ \text{molecule}^{-1}\ \text{s}^{-1}) = 6.9 \times 10^{-20}T^{2.60}e^{-2454\ \text{K}/T}$$

in the same temperature range [T. C. Allison, B. Ramachandran, J. Senekowitsch, D. G. Truhlar, and R. E. Wyatt, *J. Mol. Structure, Theochem,* **454**, 307, 1998.] Compare the experimental and theoretical rate constants at 900 K.

9.34. The activation energy for the reaction

$$H + CH_4 \rightarrow H_2 + CH_3$$

has been measured to be 49.8 kJ mol^{-1}. Some estimates of enthalpies of formation, $\Delta_f H°$, are:

H	218.0 kJ mol^{-1}
CH_4	−74.8 kJ mol^{-1}
CH_3	139.5 kJ mol^{-1}

Estimate a value for the activation energy of the reverse reaction.

***9.35.** By a treatment similar to that given for relaxation methods for the case A ⇌ Z, derive the rate equations for analyzing the reaction A + B ⇌ Z by carrying out the steps below.

a. Show that at equilibrium, $k_1 a_e b_e = k_{-1} z_e$, where the subscript e indicates equilibrium concentrations.
b. Show that $\frac{dx}{dt} = k_1(a_e - x)(b_e - x) - k_{-1}(z_e + x)$, where x represents a change from equilibrium.
c. Show that for small x, $\frac{dx}{dt} = -[k_1(a_e + b_e) + k_{-1}]x$. [*Hint:* Use the result of part (a).]
d. The displacement from equilibrium x always follows the first-order process $x = x_0 \exp(-t/t^*)$, where t^* is the relaxation time. Show that $dx/dt = -x/t^*$.
e. Comparing the results of parts (c) and (d), show that $\frac{1}{t^*} = 2k_1 a_e + k_{-1}$ if $a_e = b_e$.
f. For the case $a_e = b_e$, show that $\frac{1}{t^*} = 2\sqrt{k_1 k_{-1} z_e} + k_{-1}$. [*Hint:* Use the result of part (e) and the fact that $\frac{k_1}{k_{-1}} = \frac{z_e}{a_e b_e}$.]

9.36. A reaction of the type A + B ⇌ Z has been studied by relaxation methods. Some of the available data relating equilibrium concentrations of the product to the relaxation times are given below.

z_e/M	0.001	0.002	0.005	0.010	0.025	0.05	0.10
t^*/ms	4.08	3.74	2.63	1.84	1.31	0.88	0.67

Determine k_1, k_{-1}, and $K = k_1/k_{-1}$.

9.37. The equilibrium $H_2O \rightleftharpoons H^- + OH$ has a relaxation time of about 40 μs at 25 °C. Find the values of the forward and reverse rate constants. $K_w = [H^+][OH^-] = 10^{-14}$. [*Hint:* For this case, using steps similar to those of Problem 9.25, it can be shown that $\frac{1}{t^*} = k_1 + k_{-1}([H^+]_e + [OH^-]_e)$.]

Collision Theory and Transition-State Theory

9.38. Two reactions of the same order have identical activation energies and their entropies of activation differ by 50 J K^{-1} mol^{-1}. Calculate the ratio of their rate constants at any temperature.

9.39. The gas-phase reaction

$$H_2 + I_2 \rightarrow 2HI$$

is second order. Its rate constant at 400 °C is 2.34 × 10^{-2} dm^3 mol^{-1} s^{-1}, and its activation energy is 150 kJ mol^{-1}. Calculate $\Delta^{\ddagger}H°$, $\Delta^{\ddagger}S°$, and $\Delta^{\ddagger}G°$ at 400 °C, and the preexponential factor.

9.40. A substance decomposes according to first-order kinetics; the rate constants at various temperatures are as follows:

Temperature/°C	Rate constant, k/s^{-1}
15.0	4.18×10^{-6}
20.0	7.62×10^{-6}
25.0	1.37×10^{-5}
30.0	2.41×10^{-5}
37.0	5.15×10^{-5}

Calculate the activation energy. Calculate also, at 25 °C, the enthalpy of activation, the Gibbs energy of activation, the preexponential factor, and the entropy of activation.

9.41. The following data have been obtained for the hydrolysis of adenosine triphosphate, catalyzed by hydrogen ions:

Temperature/°C	Rate constant, k/s^{-1}
39.9	4.67×10^{-6}
43.8	7.22×10^{-6}
47.1	10.0×10^{-6}
50.2	13.9×10^{-6}

Calculate, at 40 °C, the Gibbs energy of activation, the energy of activation, the enthalpy of activation, the preexponential factor, and the entropy of activation.

9.42. The half-life of the thermal denaturation of hemoglobin, a first-order process, has been found to be 3460 s at 60 °C and 530 s at 65 °C. Calculate the enthalpy of activation and entropy of activation at 60 °C, assuming the Arrhenius equation to apply.

***9.43. a.** Using Eq. 9.73, calculate the collision density for 6.022×10^{23} molecules of hydrogen iodide present in a volume 1 m^3 at 300 K. Take $d_{AA} = 0.35$ nm.
b. If the activation energy for the decomposition of HI is 184 kJ mol^{-1}, what rate constant does kinetic theory predict at 300 °C? To what entropy of activation does this result correspond?

9.44. The rate constant for a first-order reaction is 7.40×10^{-9} s^{-1} at 25 °C, and the activation energy is 112.0 kJ mol^{-1}. Calculate, at 25 °C, the preexponential factor A, the enthalpy of activation $\Delta^{\ddagger}H°$, the Gibbs energy of activation $\Delta^{\ddagger}G°$, and the entropy of activation $\Delta^{\ddagger}S°$.

9.45. The rate constant for a second-order reaction in solution is 3.95×10^{-4} dm^3 mol^{-1} s^{-1} at 25 °C, and the activation energy is 120.0 kJ mol^{-1}. Calculate, at 25 °C, the preexponential factor A, the enthalpy of activation $\Delta^{\ddagger}H°$, the Gibbs energy of activation $\Delta^{\ddagger}G°$, and the entropy of activation $\Delta^{\ddagger}S°$.

Ionic-Strength Effects

9.46. The rate constant k for the reaction between persulfate ions and iodide ions varies with ionic strength I as follows:

$I/10^{-3}$ mol dm^{-3}	2.45	3.65	4.45	6.45	8.45	12.45
k/dm^3 mol^{-1} s^{-1}	1.05	1.12	1.16	1.18	1.26	1.39

Estimate the value of $z_A z_B$.

9.47. The following constants were obtained by Brønsted and Livingstone [*J. Amer. Chem. Soc., 49,* 435(1927)] for the reaction

$$[CoBr(NH_3)_5]^{2+} + OH^- \rightarrow [Co(NH_3)_5OH]^{2+} + Br^-$$

under the following conditions:

Concentration/mol dm^{-1}			k
$[CoBr(NH_3)_5]^{2+}$	NaOH	NaCl	dm^3 mol^{-1} s^{-1}
5.0×10^{-4}	7.95×10^{-4}	0	1.52
5.96×10^{-4}	1.004×10^{-3}	0	1.45
6.00×10^{-4}	0.696×10^{-3}	0.005	1.23
6.00×10^{-4}	0.696×10^{-3}	0.020	0.97
6.00×10^{-4}	0.691×10^{-3}	0.030	0.91

Make an estimate of the rate constant of the reaction at zero ionic strength. Are the results consistent with $z_A z_B = 2$?

9.48. Suppose that the rates of ionic reactions in solution were proportional to the activity rather than the concentration of activated complexes. Derive an equation relating the logarithm of the rate constant to the ionic strength and the charge numbers of the ions and contrast it with Eq. 9.124. Can the results in Figure 9.22 be reconciled with the equation you have derived?

9.49. When the subatomic species muonium (Mu) was first discovered in 1960, it was not known whether it bore an electric charge. The answer was provided by a kinetic study of the ionic strength effect on the reaction Mu + Cu^{2+} in aqueous solution. The following rate constants were measured at two ionic strengths:

$$I = 0 \qquad k = 6.50 \times 10^9 \text{ dm}^3 \text{ mol}^{-1} \text{ s}^{-1}$$
$$I = 0.9\ M \qquad k = 6.35 \times 10^9 \text{ dm}^3 \text{ mol}^{-1} \text{ s}^{-1}$$

Suppose that muonium had a single negative charge; what would k be expected to be at an ionic strength of 0.9 M? What do you deduce about the actual charge on muonium?

9.50. The rate constants of a second-order reaction in aqueous solution at 25 °C had the following values at two ionic strengths:

I/mol dm^{-3}	k/dm^3 mol^{-1} s^{-1}
2.5×10^{-3}	1.40×10^{-3}
2.5×10^{-2}	2.35×10^{-3}

Make an estimate of the value of $z_A z_B$, the product of the charge numbers.

9.51. A reaction of the type

$$A^+ + B^{2-} \rightarrow \text{products}$$

was found at 25 °C to have a rate constant of 2.8×10^{-4} dm^3 mol^{-1} s^{-1} at an ionic strength of 1.0×10^{-3} *M*. Assume the Debye-Hückel limiting law to apply and estimate the rate constant at zero ionic strength.

Pressure Effects

9.52. The rate of a reaction at 300 K is doubled when the pressure is increased from 1 bar to 2000 bar. Calculate $\Delta^{\ddagger}V^{\circ}$, assuming it to be independent of pressure.

9.53. The following results were obtained for the solvolysis of benzyl chloride in an acetone-water solution at 25 °C:

$P/10^2$ kPa	1.00	345	689	1033
$k/10^{-6}$ s^{-1}	7.18	9.58	12.2	15.8

Make an appropriate plot and estimate $\Delta^{\ddagger}V^{\circ}$.

9.54. The fading of bromphenol blue in alkaline solution is a second-order reaction between hydroxide ions and the quinoid form of the dye:

$$\text{quinoid form (blue)} + OH^- \rightarrow \text{carbinol form (colorless)}$$

The following results show the variation of the second-order rate constant *k* with the hydrostatic pressure *P* at 25 °C:

$P/10^4$ kPa	101.3	2.76	5.51	8.27	11.02
$k/10^{-4}$ M^{-1} s^{-1}	9.30	11.13	13.1	15.3	17.9

Estimate $\Delta^{\ddagger}V^{\circ}$.

9.55. Use Figure 9.23 to make approximate estimates of the volumes of activation for the alkaline hydrolyses of methyl acetate, ethyl acetate, and propionamide, at 25 °C.

Essay Questions

9.56. Explain clearly the difference between the order and the molecularity of a reaction.

9.57. Give an account of experimental methods that might be used to study the kinetics of (a) a reaction having a half-life of about 10^{-1} s and (b) a reaction having a half-life of about 10^{-7} s.

9.58. Predict the effects of (a) increasing the dielectric constant of the solvent, (b) increasing the ionic strength, and (c) increasing the pressure on the reactions of the following types:

$$A^{2+} + B^- \rightarrow X^+$$

$$A^+ + B^{2+} \rightarrow X^{3+}$$

$$A + B \rightarrow A^+B^-$$

Give a clear explanation in each case. What can you say about the entropy of activation to be expected in each case?

9.59. Van't Hoff's differential method can be applied to kinetic data in two different ways:

1. Rates can be determined at various stages of a single reaction.
2. Initial rates can be measured at a variety of initial concentrations, the reaction being run a number of times. In each case $\log_{10}$ (rate) can be plotted against $\log_{10}$ (concentration of a reactant). Can you suggest why a different order of reaction might be obtained when these two different procedures are used?

SUGGESTED READING

Chemical kinetics is a vast subject, and to avoid making this book too large a number of important topics had to be omitted altogether, or given only a brief treatment. Only a brief mention, for example, was made of unimolecular reactions in the gas phase. Much more on topics omitted, or treated too briefly, is to be found in the references listed below.

For general and more detailed accounts of chemical kinetics, see

S. W. Benson, *The Foundations of Chemical Kinetics,* New York: McGraw-Hill, 1965.

J. H. Espenson, *Chemical Kinetics and Reaction Mechanisms* (2nd ed.), New York: McGraw-Hill, 1995.

P. C. Jordan, *Chemical Kinetics and Transport,* New York: Plenum, 1979.

K. J. Laidler, *Chemical Kinetics* (3rd ed.), New York: Harper & Row, 1987.

M. J. Pilling and P. W. Seakins, *Reaction Kinetics,* New York: Oxford University Press, 1995.

The many volumes of *Comprehensive Chemical Kinetics* (Amsterdam: Elsevier), which have been appearing since 1969, cover the subject of kinetics in great detail.

For treatments of some special aspects of kinetics, see

M. H. Back and K. J. Laidler, *Selected Readings in Chemical Kinetics,* Oxford: Pergamon Press, 1967.

S. W. Benson, *Thermochemical Kinetics* (2nd ed.), New York: Wiley, 1976.

C. Capellos and B. H. J. Bielski, *Kinetic Systems,* New York: Wiley-Interscience, 1972. This book gives mathematical solutions for a number of kinetic systems.

R. G. Gilbert and S. C. Smith, *Theory of Unimolecular and Recombination Reactions,* Oxford: Blackwell, 1990.

K. A. Holbrook, M. J. Pilling, and S. H. Robertson, *Unimolecular Reactions.* London: Wiley, 1996.

K. J. Laidler, "Glossary of terms used in chemical kinetics, including reaction dynamics: IUPAC recommendations." *Pure and Applied Chemistry,* Vol. 68, pp. 149–192 (1996).

K. J. Laidler, "Just what is a transition state?" *J. Chem. Ed., 65,* 540–542(1988).

For accounts of work on the kinetics of rapid reactions, see

C. F. Bernasconi, *Relaxation Kinetics,* New York: Academic Press, 1976.

P. R. Brooks, "Spectroscopy of transition region species." *Chemical Reviews, 88,* 407–428(1988).

E. F. Caldin, *Fast Reactions in Solution,* Oxford: Blackwell, Scientific Publications, 1964.

G. R. Fleming, *Chemical Applications of Ultrafast Spectroscopy,* New York: Oxford University Press, 1986.

Raoul Hoffman, "Pulse, Pump and Probe," *American Scientist, 87,* 308–311 (July-August, 1999).

George Porter, *Chemistry in Microtime: Selected Writings on Flash Photolysis Free Radicals and the Excited State,* River Edge, NJ: World Scientific, 1997.

A. H. Zewail, *Femtochemistry: Ultrafast Dynamics of the Chemical Bond,* River Edge, NJ: World Scientific, 1994.

A. H. Zewail, "Femtochemistry: Recent progress in studies of dynamics and control of reactions and their transition states," *Journal of Physical Chemistry, 100,* 12701–12724 (1996).

For accounts of reaction dynamics, see the previously cited articles by Brooks and by Zewail, and also

R. B. Bernstein, *Chemical Dynamics via Molecular Beam and Laser Techniques,* Oxford: Clarendon Press, 1982.

J. M. Haile, *Molecular Dynamics Simulation: Elementary Methods,* New York: Wiley, 1992.

D. M. Hirst, *Potential-Energy Surfaces: Molecular Structure and Reaction Dynamics,* London: Taylor and Francis, 1985.

R. D. Levine and R. B. Bernstein, *Molecular Reaction Dynamics and Chemical Reactivity,* Oxford University Press, 1987.

W. H. Miller, "Recent advances in quantum-mechanical reactive scattering theory, including comparison of recent experiments with rigorous calculations for state-to-state cross sections for the $H/D + H_2 \rightarrow H_2/HD + H$ reactions," *Annual Review of Physical Chemistry, 41,* 245–281(1990).

J. N. Murrell and S. D. Bosanac, *Introduction to the Theory of Molecular Collisions,* New York: Wiley, 1989.

J. I. Steinfeld, J. S. Francisco, and W. L. Hase, *Chemical Kinetics and Dynamics,* Englewood Cliffs, NJ: Prentice Hall, 1989.

Chemical Kinetics II. Composite Mechanisms

10

PREVIEW

Composite reactions are reactions that occur in more than one elementary step; they are also commonly referred to as *complex* reactions or as *stepwise* reactions. There are several different types of composite reactions, which may involve *simultaneous elementary* reactions and *consecutive elementary* reactions.

The analysis of a composite mechanism involves explaining the overall behavior in terms of the kinetics of the elementary steps. This can be difficult mathematically, but it is greatly simplified if reaction intermediates are present in much smaller amounts than the reactants themselves. When this is the case, we can apply the *steady-state treatment,* according to which the rate of change of concentration of the reaction intermediate can, to a good approximation, be set equal to zero. This treatment allows the rate expression for the overall reaction to be obtained in terms of the reactant concentrations and the rate constants for the elementary steps.

A particularly important class of composite reactions comprises the *chain reactions.* An essential feature of a chain reaction is that there must be a closed sequence, or cycle, of reactions such that certain active intermediates are consumed in one step and are regenerated in another. It is also an essential feature that on the average the cycle is repeated more than once. Elementary reactions that are involved in such cycles are known as *chain-propagating steps.*

Photochemical and radiation-chemical reactions often occur by chain mechanisms. In a particular subgroup of photochemical reactions, *photosensitized* reactions, light energy is absorbed by an atom or molecule which transfers the energy to the substance undergoing reaction. An important technique, valuable for the study of the fastest of chemical and physical processes, is *flash photolysis.*

Many explosive reactions occur by a special type of chain reaction in which there are *branching chains.* These are chain-propagation steps in which one active intermediate, usually a free radical, gives rise to two or more active intermediates.

Catalysis can occur by a number of types of composite mechanisms. Catalysis by acids and bases and by enzymes occurs by nonchain mechanisms in which a reaction intermediate is first formed from the catalyst and the substance undergoing reaction (the substrate); this intermediate subsequently breaks down into the reaction products and the catalyst. Certain other catalyzed reactions occur by chain mechanisms.

When composite reactions occur in solution, there is an important distinction between collisions and encounters that occur between reactant molecules. An *encounter* is a group of collisions brought about by the fact that the reactant molecules are caged in by the surrounding solvent molecules.

OBJECTIVES

After studying this chapter, the student should be able to:

- Explain the criteria for determining if a reaction occurs by a composite mechanism.
- List the methods for arriving at a plausible mechanism on the basis of kinetic information.
- Obtain rate equations for composite mechanisms.
- Understand the steady-state hypothesis and how to apply it in order to obtain a rate equation.
- Discuss the concept of the rate-determining step and the pitfalls involved in applying it to particular mechanisms.
- Understand what is meant by a chain reaction, and explain the free-radical mechanisms for a number of chemical reactions, such as the hydrogen-bromine reaction and the decomposition of some organic compounds.
- Explain the basic principles involved in photochemical and radiation-chemical reactions.
- Understand the nature of explosions, especially in gases.
- Discuss the principles of catalysis, especially catalysis by acids and bases and by enzymes.
- Explain how oscillations can occur in chemical reactions, and why the outcome of a chemical reaction is sometimes impossible to predict.
- Apply the principles of kinetics to electrode processes and mechanisms of polymerization.

Composite Reactions

Chapter 9 dealt with the more important basic principles of chemical kinetics. To a considerable extent, but not entirely, it dealt with elementary reactions that occur in a single step; the reactant molecules form an activated complex, which passes directly into products. However, the majority of reactions with which the chemist deals are not elementary; instead, they involve two or more elementary steps and then are said to be **composite,** stepwise, or complex. This chapter is concerned with how the rates of composite reactions are related to the characteristics of the elementary steps.

There are various types of composite reactions. In some of them, relatively stable molecules occur as intermediates. A simple example is the reaction between hydrogen and iodine monochloride.

$$H_2 + 2ICl \rightarrow I_2 + 2HCl$$

If this reaction occurred in a single elementary step, it would be third order—first order in hydrogen and second order in ICl—since a molecule of hydrogen and two molecules of ICl would come together to form an activated complex. In fact, the reaction is experimentally observed to be second order, being first order in hydrogen and first order in iodine monochloride:

$$v = k[H_2][ICl] \tag{10.1}$$

This can be explained if there is initially a slow reaction between one molecule of H_2 and one of ICl,

$$1.\ H_2 + ICl \longrightarrow HI + HCl \qquad \text{(slow)}^1$$

followed by a rapid reaction between the HI formed in this step and an additional molecule of ICl,

$$2.\ HI + ICl \longrightarrow HCl + I_2 \qquad \text{(rapid)}$$

Addition of these two reactions gives the overall equation. The HI produced in the first reaction is removed as rapidly as it is formed. The rate of the second process has no effect on the overall rate, which is therefore that of the first step; the kinetic behavior is thus explained. In this scheme the first step is said to be the *rate-determining* or *rate-controlling* step, a matter that is considered further in Section 10.3.

Another reaction that involves a fairly stable intermediate, and which has a rate-determining step, is the oxidation of bromide ions by hydrogen peroxide in aqueous acid solution,

$$2Br^- + H_2O_2 + 2H^+ \rightarrow Br_2 + 2H_2O$$

The experimental rates vary with concentration according to the expression

$$v = k[H_2O_2][H^+][Br^-] \tag{10.2}$$

This result is explained if the reaction occurs in the following three stages:

1. $H^+ + H_2O_2 \rightleftharpoons H_2O^+\text{—}OH$ (rapid equilibrium)
2. $H_2O^+\text{—}OH + Br^- \longrightarrow HOBr + H_2O$ (slow)
3. $HOBr + H^+ + Br^- \longrightarrow Br_2 + H_2O$ (fast)

---

[1]The filled-in arrow ⟶ is used to emphasize that a reaction is believed to be elementary.

The concentration of H_2O^+—OH, which is a protonated hydrogen peroxide molecule, is proportional to $[H^+][H_2O_2]$, and the rate of reaction 2 is proportional to $[H_2O^+—OH]$ and therefore to $[H_2O_2][H^+]$ $[Br^-]$. The kinetic equation, Eq. 10.2, is thus explained if reaction 2 is the slow and rate-controlling step.

This explanation is supported by the fact that if solutions of HOBr and Br^- are mixed, and the mixture is acidified, bromine is produced very rapidly; reaction 3 is therefore fast. On the basis of arguments of this kind, we draw conclusions about the elementary reactions that occur in composite mechanisms.

Several other examples will be given to illustrate the methods used in arriving at a reaction mechanism. It cannot be emphasized too strongly that

kinetic evidence—or indeed any other kind of evidence—can never prove a reaction mechanism, although the evidence may disprove a mechanism!

The point is that one may suggest a mechanism that is consistent with all the kinetic evidence available at any time, but one can always find another mechanism, perhaps a more complicated one, that explains the experimental results equally well. It has often been found that a kinetic mechanism that has been accepted for many years is proved by later evidence to be quite wrong.

Of course, it is true of all science that one can never prove any theory with absolute certainty: One can say only that the theory is consistent with all the known facts, but one has to admit that newly discovered facts may disprove the theory entirely. Nowhere is this more true than in the search for the mechanism of a chemical reaction. We will now consider a few more examples.

EXAMPLE 10.1 Suppose that a reaction of stoichiometry

$$A + B \rightarrow Y + Z$$

is zero order in the reactant A, and first order in B. Suggest a mechanism that is consistent with this result.

Solution The obvious suggestion is that B forms some intermediate X by a slow step, and that X then reacts rapidly with A:

$$B \rightarrow X \qquad \text{(slow)}$$
$$X + A \rightarrow Y + Z \qquad \text{(fast)}$$

To avoid complications for the time being we will suppose that the first reaction is highly exergonic (i.e., the equilibrium lies far to the right), so that we do not have to worry about the back reaction, which would complicate the situation. (Such matters are considered later, especially when we come to consider the steady-state hypothesis in Section 10.3.)

In the mechanism proposed for this reaction, the rate of the overall process is controlled by the rate of the first step; as soon as X is formed it reacts with A, and the amount of A does not affect the rate. Of course, there must be enough A present for this to happen; if there is no A there will be no reaction, and if there is very little A the rate will depend on the amount of A. Again, these matters are covered in Section 10.3.

EXAMPLE 10.2 The kinetics of the reaction

$$2Fe(CN)_6^{3-} + 2I^- \rightarrow 2Fe(CN)_6^{4-} + I_2$$

has been studied and the rate equation found to be

$$v = k[Fe(CN)_6^{3-}]^2[I^-]^2[Fe(CN)_6^{4-}]^{-1}[I_2]^0$$

Suggest a possible mechanism to explain this behavior.

Solution What we notice at once about this rate equation is that the concentration of a product of the reaction appears in it, and to a power of -1. This can be explained if we assume that the product $Fe(CH)_6^{4-}$ is involved in a pre-equilibrium. The following is a possible scheme:

$$Fe(CN)_6^{3-} + 2I^- \rightleftharpoons Fe(CN)_6^{4-} + I_2^-$$

This may occur rapidly, and the fact that the concentration of $Fe(CN)_6^{3-}$ occurs to the second power suggests that this pre-equilibrium is followed by a slow reaction of I_2^- with another molecule of $Fe(CN)_6^{3-}$

$$I_2^- + Fe(CN)_6^{3-} \rightarrow Fe(CN)_6^{4-} + I_2$$

The equation for the pre-equilibrium, which being fast is undisturbed by the subsequent reaction, is

$$K = \frac{[Fe(CN)_6^{4-}]\,[I_2^-]}{[Fe(CN)_6^{3-}][I^-]^2}$$

The rate of formation of the products is

$$v = k_2[I_2^-][Fe(CN)_6^{3-}]$$

and by use of the equilibrium equation we obtain

$$v = k_2K[Fe(CN)_6^{3-}]^2[I^-]^2[Fe(CN)_6^{4-}]^{-1}$$

EXAMPLE 10.3 An investigation was made by M. J. Haugh and D. R. Dalton [*J. Amer. Chem. Soc., 97,* 5674(1975)] of the reaction of hydrogen chloride with propene at high pressures. They found that under some circumstances the reaction was first order in propene and third order in hydrogen chloride:

$$v = k[\text{propene}][HCl]^3$$

Suggest a mechanism that is consistent with this result.

Solution One thing we can be sure of is that the reaction does *not* involve three molecules of HCl interacting with one molecule of propene in one step. No case is known of four molecules coming together and reacting; the probability of even three molecules coming together is low.

To explain the third-order kinetics in HCl we can suppose instead that two molecules first come together and form a complex $(HCl)_2$, and that another molecule forms a complex with propene:

$$2HCl \rightleftharpoons (HCl)_2 \qquad K_1$$

$$HCl + \text{propene} \rightleftharpoons HCl\cdot\text{propene} \qquad K_2$$

There could then be a slow reaction between $(HCl)_2$ and the HCl·propene complex:

$$(HCl)_2 + HCl\cdot\text{propene} \rightarrow CH_3CHClCH_3 + 2HCl$$

The rate of this reaction is

$$v = k_3[(HCl)_2][HCl\cdot\text{propene}]$$

The expressions for $[(HCl)_2]$ and [HCl·propene] can then be written in terms of the equilibrium constants K_1 and K_2:

$$[(HCl)_2] = K_1[HCl]^2$$

$$[HCl\cdot\text{propene}] = K_2[HCl][\text{propene}]$$

so that the rate of formation of the chlorinated propene is

$$v = k_3K_1K_2[HCl]^3[\text{propene}]$$

It cannot be emphasized too strongly that in the three examples just considered we must not assume that the mechanisms proposed are the ones that really occur. All we have done was to propose plausible mechanisms which give rise to rate equations consistent with the observed kinetics. To have reasonable confidence in a mechanism (although we can never be sure of it) we would have to have much additional evidence. For example, in the case of the chlorination of propene we would want to know if any spectroscopic evidence could be obtained for the existence of $(HCl)_2$ and the HCl·propene complex in the reaction system.

So far we have been considering only rather simple situations in which the individual reactions have not interfered with each other. In the last two examples, there was a slow rate-controlling step which was assumed to have no effect on the equilibria established. Later we will meet mechanisms in which there is no such simplification, and we will see how to proceed in such cases.

10.1 Evidence for a Composite Mechanism

Watch animations of the composite mechanism for the formation of HBr; Composite mechanism; HBr.

There are various indications that a reaction is occurring by a composite mechanism. An obvious piece of evidence is when the kinetic equation does not correspond to the stoichiometric equation; several examples of this have just been noted. In other cases the kinetic equation is more complicated, sometimes involving concentrations raised to nonintegral powers and with reactant concentrations in the denominator. For example, the gas-phase reaction between hydrogen and bromine

$$H_2 + Br_2 = 2HBr$$

would be first order in hydrogen and first order in bromine if it were an elementary reaction. In fact, the reaction follows an equation of the form

$$v = \frac{k[H_2][Br_2]^{1/2}}{1 + [HBr]/m[Br_2]} \tag{10.3}$$

where k and m are constants. The mechanism of this reaction is considered in Section 10.5, where this rate equation will be derived in terms of the rate constants for the individual steps.

Another indication of a composite mechanism is the detection, by chemical or other means, of reaction intermediates. When this can be done, a kinetic scheme must be developed that will account for the existence of these intermediates. Sometimes these intermediates are relatively stable substances; in other cases they are labile substances such as atoms and free radicals. Free radicals can sometimes be observed by spectroscopic and other methods, and evidence for their existence may also be obtained by causing them to undergo specific reactions that less active substances cannot bring about. In particular, substances that react rapidly with intermediates bring about inhibition of the overall reaction.

When the nature of the reaction intermediates has been determined, the next step is to devise a reaction scheme that will involve these intermediates and account for the kinetic features of the reaction. If such a scheme fits the data satisfactorily, it can be tentatively assumed that the mechanism may be the correct one. It should be emphasized, however, that additional kinetic work frequently leads to the overthrow of schemes that have previously been supposed to be established firmly.

10.2 Types of Composite Reactions

Composite reaction mechanisms can have a number of features. Reactions occurring in parallel, such as

Simultaneous Reactions

$$A \rightarrow Y$$
$$A \rightarrow Z$$

Vary rate constants and initial concentrations and plot simultaneous reactions.

are called *simultaneous* reactions. When there are simultaneous reactions, there is sometimes *competition,* as in the scheme

$$A + B \rightarrow Y$$
$$A + C \rightarrow Z$$

Opposing Reactions

where B and C compete with one another for A.

Reactions occurring in forward and reverse directions are called *opposing,* for example;

Vary constants and plot opposing reactions; Opposing first order.

$$A + B \rightleftarrows Z$$

Reactions occurring in sequence, such as

$$A \rightarrow X \rightarrow Y \rightarrow Z$$

Consecutive Reactions

are known as *consecutive* reactions, and the overall process is said to occur by consecutive steps. Reactions are said to exhibit *feedback* if a substance formed in one step affects the rate of a previous step. For example, in the scheme

$$A \rightarrow X \rightarrow Y \rightarrow Z \quad (\text{Y feeds back to step } A \rightarrow X)$$

the intermediate Y may catalyze reaction 1 (*positive* feedback) or inhibit reaction 1 (*negative* feedback). A final product as well as an intermediate may bring about feedback.

10.3 Rate Equations for Composite Mechanisms

In dealing with composite mechanisms it is necessary to consider the rates of elementary reactions as if they occurred in isolation in one direction. For example, for the composite mechanism

$$A \underset{k_{-1}}{\overset{k_1}{\rightleftarrows}} X \underset{k_{-2}}{\overset{k_2}{\rightleftarrows}} Z$$

there are four elementary reactions. The rate for reaction 1 can be denoted by v_1 and is the rate of $A \rightarrow X$ if no other reaction were occurring; it is given by

$$v_1 = k_1[A] \tag{10.4}$$

Similarly,

$$v_{-1} = k_{-1}[X] \tag{10.5}$$

$$v_2 = k_2[X] \tag{10.6}$$

$$v_{-2} = k_{-2}[Z] \tag{10.7}$$

For any species X the *total rate into X*, Σv_x, is the sum of the rates of all reactions that produce X. The total rate out of X, Σv_{-x}, is the sum of the rates of all the reactions that remove X. For example, for the scheme just considered the total rate into X is

$$\Sigma v_x = v_1 + v_{-2} = k_1[A] + k_{-2}[Z] \tag{10.8}$$

and the total rate out of X is

$$\Sigma v_{-x} = v_{-1} + v_2 = (k_{-1} + k_2)[X] \tag{10.9}$$

For a system at complete equilibrium the total rate into each species is equal to the total rate out of it.

Consecutive Reactions

Vary the rate constants and plot consecutive reactions.

The simplest consecutive mechanism is

$$A \xrightarrow{k_1} X \xrightarrow{k_2} Z$$

A reaction of this type was first investigated in 1865 by the chemist Augustus George Vernon Harcourt (1834–1919), and the kinetic equations were solved by the mathematician William Esson (1839–1916). If the initial concentration of A is $[A]_0$ and its concentration at any time t is [A], the rate equation for A is

$$v_1 = \frac{d[A]}{dt} = k_1[A] \tag{10.10}$$

Integration of this equation, subject to the boundary condition that $[A] = [A]_0$ when $t = 0$, gives (see Eq. 9.30)

$$[A] = [A]_0 e^{-k_1 t} \tag{10.11}$$

The net rate of formation of X is

$$\frac{d[X]}{dt} = k_1[A] - k_2[X] \tag{10.12}$$

which, with Eq. 10.11, is

$$\frac{d[\mathrm{X}]}{dt} = k_1[\mathrm{A}]_0 e^{-k_1 t} - k_2[\mathrm{X}] \tag{10.13}$$

This contains only the variables [X] and t, and integration gives

$$[\mathrm{X}] = [\mathrm{A}]_0 \frac{k_1}{k_2 - k_1}(e^{-k_1 t} - e^{-k_2 t}) \tag{10.14}$$

The equation for the variation of [Z] is most easily obtained by noting that

$$[\mathrm{A}] + [\mathrm{X}] + [\mathrm{Z}] = [\mathrm{A}]_0 \tag{10.15}$$

so that

$$[\mathrm{Z}] = [\mathrm{A}]_0 - [\mathrm{A}] - [\mathrm{X}] \tag{10.16}$$

Insertion of the expressions for [A] and [X] into Eq. 10.16 leads to

$$[\mathrm{Z}] = \frac{[\mathrm{A}]_0}{k_2 - k_1}[k_2(1 - e^{-k_1 t}) - k_1(1 - e^{-k_2 t})] \tag{10.17}$$

Induction Period

Figure 10.1a shows the time variations in the concentrations of A, X, and Z as given by these equations. We see that [A] falls exponentially, while [X] goes through a maximum. Since the rate of formation of Z is proportional to the concentration of X, the rate is initially zero and is a maximum when [X] reaches its maximum value. For an initial period of time it may be impossible to detect any of the product Z, and the reaction is said to have an *induction period.* Such induction periods are commonly observed for reactions occurring by composite mechanisms.

Plot Figure 10.1 by changing the rate constants; Consecutive reactions.

Kinetic equations like Eqs. 10.14 and 10.17 are frequently obeyed by nuclides undergoing radioactive decay (see also Section 9.4), but there are not many examples of chemical reactions that show consecutive first-order behavior. Two good examples are the thermal isomerizations of 1, 1-dicyclopropylene and 1-cyclopropylcyclopentene.

Two limiting cases are of special interest. Suppose first that the rate constant k_1 is very large and that k_2 is very small. The reactant A is then rapidly converted into the intermediate X, which slowly forms Z. Figure 10.1b shows plots of the

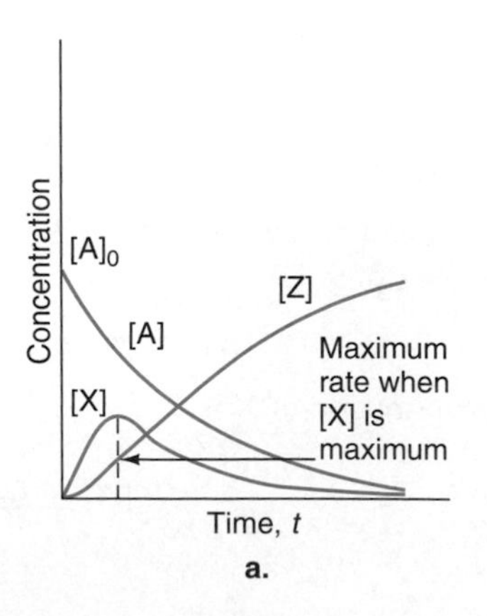

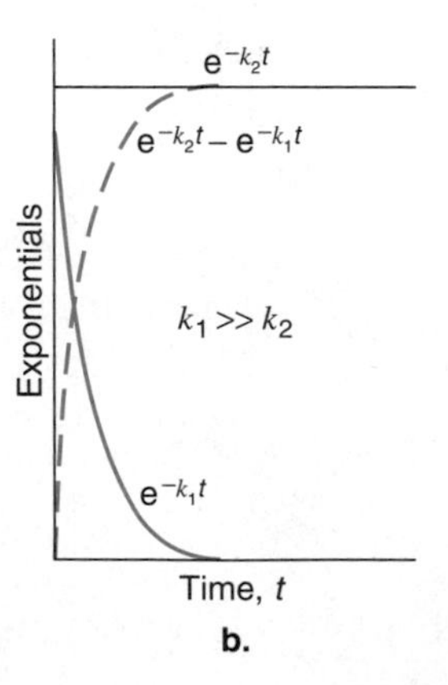

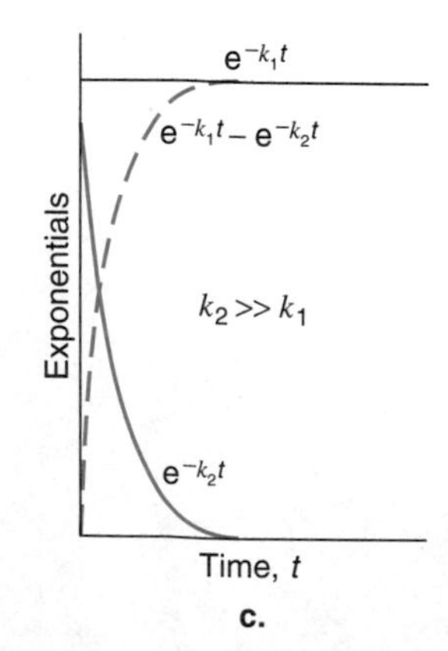

FIGURE 10.1
(a) Variations in the concentrations of A, X, and Z, for a reaction occurring by the mechanism A → X → Z. (b) Variations with time of the exponentials, when $k_1 \gg k_2$. (c) Variations of the exponentials when $k_2 \gg k_1$.

exponentials $e^{-k_1 t}$ and $e^{-k_2 t}$ and of their difference. Since k_2 is small, the exponential $e^{-k_2 t}$ shows a very slow decay, while $e^{-k_1 t}$ shows a rapid fall. The difference

$$e^{-k_2 t} - e^{-k_1 t}$$

is shown by the dashed line in Figure 10.1b. The rate of change of the concentration of X is, by Eq. 10.14, equal to this difference multiplied by $[A]_0$ (since $k_1 \gg k_2$), and [X] therefore rises rapidly to the value $[A]_0$ and then slowly declines. The rise in [Z] then follows approximately the simple first-order law.

The converse case, when $k_2 \gg k_1$, is a particularly interesting one, since it leads us to the concept of the **steady state** and will now be discussed.

Steady-State Treatment

When k_1 is small and k_2 is large, the exponentials change with time in the manner shown in Figure 10.1c. The situation is the same as in Figure 10.1b, with k_1 and k_2 interchanged. Since $k_2 \gg k_1$, the concentration of X is now given by (see Eq. 10.14)

$$[X] = [A]_0 \frac{k_1}{k_2}(e^{-k_1 t} - e^{-k_2 t}) \tag{10.18}$$

At $t = 0$, $[X] = 0$, but after a very short time, relative to the duration of the reaction, the difference

$$e^{-k_1 t} - e^{-k_2 t}$$

has attained the value of unity, and the concentration of X is then $[A]_0 k_1/k_2$, which is much less than $[A]_0$. After this short induction period the concentration of X remains practically constant, so that to a good approximation

$$\frac{d[X]}{dt} = 0 \tag{10.19}$$

Steady-state treatment requires the concentrations of the intermediates to be always much smaller than the reactant concentrations.

This is the basis of the **steady-state treatment,** which is very commonly applied to reaction mechanisms. What we have proved for the very simple scheme of two consecutive first-order reactions is that if the conditions are such that the concentration of the intermediate X is always much smaller than the reactant concentration, the concentration of X rapidly reaches a value that remains practically constant during the course of the reaction. It is not possible to give a formal proof of this hypothesis, applicable to any reaction mechanism, because the rate equations for more complicated mechanisms are often impossible to solve. However, the derivation we have given for the two-stage system of first-order reactions leads us to an important general conclusion. The rate of change of the concentration of an intermediate can, to a good approximation, be set equal to zero whenever the intermediate is formed slowly and disappears rapidly. In other words, whenever an intermediate X is such that it is always present in amounts much smaller than those of the reactants, the total rate into X, Σv_x, is nearly the same as the total rate out of X, Σv_{-x}

$$d[X]/dt = \Sigma v_x - \Sigma v_{-x} = 0 \tag{10.20}$$

The steady-state treatment is of great importance in the analysis of composite mechanisms, since often there are mathematical difficulties that make it impossible to obtain an exact solution of the rate equations for the reaction. Consider, for example, the mechanism

$$A + B \underset{k_{-1}}{\overset{k_1}{\rightleftarrows}} X$$

$$X \xrightarrow{k_2} Z$$

The differential rate equations that apply to this set of reactions are

$$-\frac{d[A]}{dt} = -\frac{d[B]}{dt} = k_1[A][B] - k_{-1}[X] \tag{10.21}$$

$$\frac{d[X]}{dt} = k_1[A][B] - k_{-1}[X] - k_2[X] \tag{10.22}$$

$$\frac{d[Z]}{dt} = k_2[X] \tag{10.23}$$

In order to treat this problem exactly it would be necessary to eliminate [X] and to solve the resulting differential equation to find [Z] as a function of t. Unfortunately, however, in spite of the simplicity of the kinetic scheme, it is not possible to obtain an explicit solution.

The steady-state treatment, which is valid provided that the concentration of X is always small, involves using Eq. 10.19 so that, from Eq. 10.22,

$$k_1[A][B] - k_{-1}[X] - k_2[X] = 0 \tag{10.24}$$

The concentration of [X] is therefore given by

$$[X] = \frac{k_1[A][B]}{k_{-1} + k_2} \tag{10.25}$$

and insertion of this into Eq. 10.23 gives

$$v = v_Z = \frac{d[Z]}{dt} = \frac{k_1 k_2[A][B]}{k_{-1} + k_2} \tag{10.26}$$

Rate-Controlling (Rate-Determining) Steps

Previously in this chapter slow reaction steps followed by rapid ones have been said to be rate-determining or rate-controlling,[2] the rate of the overall reaction being the same as the rate of the initial step. A **rate-controlling step** may be defined as a step which has a strong influence on the overall rate of the reaction. A reaction step is also often said to be rate controlling if there are *rapid pre-equilibria followed by a slow step.* Both types of rate-controlling steps are exemplified by the reaction scheme just considered, as will now be explained.

Suppose that in that reaction scheme the intermediate X is converted very rapidly into Z, much more rapidly than it can go back into A + B. In that case the rate of the overall reaction is the rate of formation of X from A + B; that is,

$$v = k_1[A][B] \tag{10.27}$$

[2]A distinction is sometimes made between rate-determining and rate-controlling steps; the former is used when the overall rate is equal to the rate of the *first* step in a consecutive mechanism, the latter when a later reaction is involved, the overall rate also involving equilibrium constants for pre-equilibria (as in Eq. 10.30). However, this distinction is not universally recognized, and will not be made in this book; those who make the distinction should therefore emphasize that they are doing so.

since as soon as X is formed, it is transformed into Z. The initial step is therefore the rate-controlling step or the rate-determining step. The exact condition, for this two-step mechanism, is

$$k_2 \gg k_{-1}$$

and we see that the steady-state rate equation (Eq. 10.26) becomes Eq. 10.27 if this inequality is satisfied.

Alternatively, suppose that the rate constant for the second reaction, X $\rightarrow$ Z, is very small compared to that for the reverse of the first reaction; that is,

$$k_2 \ll k_{-1}$$

The overall rate is

$$v = k_2[\mathrm{X}] \tag{10.28}$$

and since reaction 2 is too slow to disturb the equilibrium A + B $\rightleftharpoons$ X,

$$[\mathrm{X}] = \frac{k_1}{k_{-1}}[\mathrm{A}][\mathrm{B}] \tag{10.29}$$

Insertion of Eq. 10.29 into Eq. 10.28 gives

$$v = \frac{k_1 k_2}{k_{-1}}[\mathrm{A}][\mathrm{B}] = k_2 K_1[\mathrm{A}][\mathrm{B}] \tag{10.30}$$

where K_1 ($= k_1/k_{-1}$) is the equilibrium constant for the pre-equilibrium. Again, this is the expression to which Eq. 10.26 reduces if the inequality $k_{-1} \gg k_2$ is satisfied. Since the rate coefficient k_1k_2/k_{-1} is the product of the rate constant k_2 and the equilibrium constant k_1/k_{-1}, reaction 2 is often regarded as the rate-controlling step. It is important to recognize, however, that k_1 and k_2 have an equal effect on the rate constant.

It is essential to consider each kinetic scheme carefully, as recognizing a rate-controlling step is often far from straightforward. A helpful analysis of more complicated situations, in terms of Gibbs-energy diagrams, was given in 1981 by J. R. Murdoch [*J. Chem. Ed.*, *58,* 32(1981)].

EXAMPLE 10.4 The reaction between iodide ions and the cobalt complex $Co(CN)_5OH_2^{2-}$, for which the stoichiometric equation is

$$Co(CN)_5OH_2^{2} + I^- \rightarrow Co(CN)_5I^{3-} + H_2O,$$

is believed to go by the mechanism

$$Co(CN)_5OH_2^{2-} \underset{k_{-1}}{\overset{k_1}{\rightleftarrows}} Co(CN)_5^{2-} + H_2O$$

$$Co(CN)_5^{2-} + I^- \xrightarrow{k_2} Co(CN)_5I^{3-}$$

Assume that the intermediate exists in a steady state, and derive the general rate equation. Write the rate equation for the special cases of low and high iodide concentrations, and decide which is the rate-controlling step in each case.

Solution The steady-state equation for $Co(CN)_5^{2-}$ is

$$k_1[Co(CN)_5OH_2^{2-}] - k_{-1}[Co(CN)_5^{2-}] - k_2[Co(CN)_5^{2}\ ][I^-] = 0$$

Note that $[H_2O]$ is included in the value of k_{-1} because, as solvent, its value is essentially constant. Therefore

$$[Co(CN)_5^{2-}] = \frac{k_1[Co(CN)_5OH_2^{2-}]}{k_{-1} + k_2[I^-]}$$

The general expression for the rate is therefore

$$v = k_2[I^-][Co(CN)_5^{2-}] = \frac{k_1k_2[Co(CN)_5OH_2^{2-}][I^-]}{k_{-1} + k_2[I^-]}$$

If the concentration of iodide ions is sufficiently low the condition $k_{-1} \gg k_2[I^-]$ will apply, and the rate is then

$$v = \frac{k_1k_2}{k_{-1}}[Co(CN)_5OH_2^{2-}][I^-]$$

The rate is thus proportional to the rate constant k_2 for the second step, and to the equilibrium constant k_1/k_{-1} for the pre-equilibrium. In these circumstances reaction 2, of rate constant k_2, is commonly called the rate-controlling step.

If, on the other hand, the iodide concentration is high enough, the term k_{-1} in the rate equation can be neglected in comparison with $k_2[I^-]$, and the rate equation is then

$$v = k_1[Co(CN)_5OH_2^{2-}]$$

The rate is now simply the rate of the first step, which is the rate-controlling step.

Great care must be taken in identifying a rate-controlling step, as it is easy to make mistakes in the case of more complicated systems. Unless the situation is straightforward, it is better to avoid the concept of the rate-determining step altogether, as it is never essential to identify such a step. The following points about rate-controlling steps should be noted in particular:

1. **It is wrong to define a rate-controlling step, as is sometimes done, as the slowest step in the reaction scheme.** Sometimes it is, but often it is not. If the reaction proceeds by a chain mechanism (Section 10.5), the chain-propagating steps are proceeding at the same rate, but one of them may be rate controlling.
2. **It is impossible to decide on a rate-controlling step if one has no information about the relative values of the rate constants.** Students sometimes think that they must decide on a rate-controlling step before they can work out a rate expression on the basis of the steady-state hypothesis. The schemes just considered show that this is not the case: One decides on a rate-controlling step *after* one has analyzed the kinetics.
3. **In general, the rate-controlling step depends on concentrations of reactants.** In the example just considered the concentration of iodide ions played a critical role.

Rate-controlling steps are related to kinetic isotope effects, which we considered in Section 9.9. The kinetic isotope effect observed in a reaction occurring by a composite mechanism depends on how the rate is related to the rate constants for the individual steps as well as on how these rate constants are affected by an isotopic substitution. Suppose that a reaction occurs by a mechanism to which Eq. 10.30 applies. If either of the rate constants k_1 or k_2 is decreased by the substitution of a hydrogen atom for a deuterium atom, the overall rate will be decreased. If on the other hand the rate constant k_{-1} is decreased by the substitution, the overall rate will be increased. In the latter case we say that we have a *negative* kinetic isotope effect, the rate constant for the overall rate being greater for the heavier isotope.

Suppose that a particular step in a reaction is rate controlling. If an isotope substitution is made at a bond that is broken or formed in that step, the change in the rate constant of that step will be reflected by a change in the overall rate. It is important to realize, however, that a kinetic isotope study can never prove a particular step to be rate controlling; it can merely be concluded that the rate constant for that step appears to be involved in the rate equation for the overall reaction. A kinetic isotope study that leads to no effect on the rate of the overall reaction does, on the other hand, provide good evidence that the bond involved in the isotope substitution is playing no part in a rate-controlling step.

If the isotopic substitution generates a substantial effect on the rate, as with H, D, and T substitutions, the rates may be measured in separate kinetic experiments, and we then speak of *absolute* kinetic effects. If, on the other hand, the isotope effect is only a few percent, as with carbon substitutions (Table 9.2), the differences cannot be measured reliably in separate kinetic experiments. In such cases the reactions can be made to occur in the same reaction system, the different isotopic forms of the reactant being present together. Account must then be taken of possible complications due to interactions between the two reactions; there is competition between the two isotopic forms, and we then speak of *competitive* isotopic effects. An example will be seen later (Example 10.10 in Section 10.9) in connection with an enzyme system.

10.4 Rate Constants, Rate Coefficients, and Equilibrium Constants

For an *elementary* reaction it is easy to show that the equilibrium constant must be the ratio of the rate constants in forward and reverse directions. Thus, consider the process

$$\mathrm{A + B} \underset{k_{-1}}{\overset{k_1}{\rightleftarrows}} \mathrm{Y + Z}$$

in which the reactions in forward and reverse directions, as indicated by the filled-in arrows, are elementary. Their rates are

$$v_1 = k_1[\mathrm{A}][\mathrm{B}] \tag{10.31}$$

$$v_{-1} = k_{-1}[\mathrm{Y}][\mathrm{Z}] \tag{10.32}$$

If the system is at equilibrium, these rates are equal; hence

$$\frac{k_1}{k_{-1}} = \left(\frac{[\mathrm{Y}][\mathrm{Z}]}{[\mathrm{A}][\mathrm{B}]}\right)_{\mathrm{eq}} = K_c \tag{10.33}$$

where K_c is the equilibrium constant.

This argument can be extended to a reaction that occurs in two or more stages. Consider, for example, the reaction

$$H_2 + 2ICl \rightarrow I_2 + 2HCl$$

which we saw at the beginning of this chapter to occur in two stages, which at equilibrium will be occurring in both directions:

$$H_2 + ICl \underset{k_{-1}}{\overset{k_1}{\rightleftarrows}} HI + HCl$$

$$HI + ICl \underset{k_{-2}}{\overset{k_2}{\rightleftarrows}} HCl + I_2$$

At equilibrium the rate of each elementary reaction and its reverse must be the same:

$$k_1[H_2][ICl] = k_{-1}[HI][HCl] \tag{10.34}$$

$$k_2[HI][ICl] = k_{-2}[HCl][I_2] \tag{10.35}$$

The equilibrium constant for each reaction is thus

$$K_1 = \frac{k_1}{k_{-1}} = \left(\frac{[HI][HCl]}{[H_2][ICl]}\right)_{eq} \tag{10.36}$$

$$K_2 = \frac{k_2}{k_{-2}} = \left(\frac{[HCl][I_2]}{[HI][ICl]}\right)_{eq} \tag{10.37}$$

The product of these two equilibrium constants is

$$K_1K_2 = \frac{k_1k_2}{k_{-1}k_{-2}} = \left(\frac{[I_2][HCl]^2}{[H_2][ICl]^2}\right) = K_c \tag{10.38}$$

where K_c is the equilibrium constant for the overall reaction. It is easy to prove that for any mechanism, involving any number of *elementary* steps, the overall equilibrium constant is the product of the equilibrium constants for the individual steps and is the product of the rate constants for the reactions in the forward direction divided by the product of those for the reverse reactions:

$$K_c = K_1K_2K_3 \cdots = \frac{k_1k_2k_3 \cdots}{k_{-1}k_{-2}k_{-3} \cdots} \tag{10.39}$$

Principle of Microscopic Reversibility

This conclusion is related to an important principle, the **principle of microscopic reversibility,** which was stated in 1938 by the American physical chemist Richard Chase Tolman (1881–1948) in his important book *The Principles of Statistical Mechanics.* According to this principle,

> . . . in a system at equilibrium, any molecular process and the reverse of that process occur, on the average, at the same rate.

A closely related principle, the **principle of detailed balance at equilibrium,** was put forward in 1936 by the British theoretical physicist Ralph Howard Fowler (1884–1950) in his book *Statistical Mechanics.* This principle relates particularly to collisions between molecules, and it states that

> . . . in a system at equilibrium, each collision has its exact counterpart in the reverse direction, so that the rate of every chemical process is exactly balanced by that of the reverse process.

It is important to appreciate the fact that these principles relate to a system at equilibrium, when processes are occurring at equal rates in forward and reverse directions. Serious errors can arise if we try to apply the principles to systems not at equilibrium. If a reaction is elementary it is true to say that the ratio of rate constants in the forward and reverse directions is equal to the equilibrium constant. However, if a reaction occurs by a composite mechanism, and we measure a rate coefficient[3] k_1 for the overall reaction from left to right and also measure a rate coefficient k_{-1} for the overall reaction from right to left, at the same temperature, the ratio k_1/k_{-1} is not necessarily the equilibrium constant for the overall reaction. The reason for this is that rate equations for composite reactions change with the experimental conditions, such as reactant concentrations, and the rate coefficients also change. The ratio of the rate coefficients k_1 and k_{-1} that apply *when the system is at equilibrium* is equal to the equilibrium constant, but rate coefficients determined away from equilibrium are not necessarily the same as those at equilibrium, and their ratio is not necessarily equal to K_c. Great caution should therefore be used in deducing rate coefficients and rate laws for reactions from the equilibrium constant and the rate coefficient for the reverse reaction.

10.5 Free-Radical Reactions

Reactions frequently occur by a series of reactions in which free radicals play a part. The important distinction between a free radical and an ion may be illustrated by comparison of the hydroxyl radical and the hydroxide ion. The oxygen-hydrogen bond in the water molecule may be split *homolytically;* that is, one of the electrons goes with one fragment and the other with the other:

$$\underset{\text{water molecule}}{\overset{\text{H}}{:\ddot{\underset{\cdot\cdot}{\text{O}}}:\text{H}}} \rightarrow \underset{\text{hydroxyl radical}}{\overset{\text{H}}{:\ddot{\underset{\cdot\cdot}{\text{O}}}\cdot}} + \underset{\text{hydrogen atom}}{\cdot\text{H}}$$

In this process, two electrically neutral free radicals are produced; an atom is a special case of a radical. The hydrogen atom consists of a proton and an electron, and the hydroxyl radical consists of nine protons (one in the nucleus of the hydrogen atom and eight protons in the oxygen nucleus) and nine electrons (seven valence electrons plus two 1s electrons). The hydrogen atom and the hydroxyl radical are both one electron short of the noble gas structures and, therefore, are very reactive species. Radicals combine with one another with very low or zero activation energies, and their reactions with stable molecules occur with quite low activation energies.

In the ionization of water, on the other hand, a bond is split *heterolytically* and the electron pair remains with the oxygen atom:

$$\underset{\text{water molecule}}{\overset{\text{H}}{:\ddot{\underset{\cdot\cdot}{\text{O}}}:\text{H}}} \rightarrow \underset{\text{hydroxide ion}}{\left[\overset{\text{H}}{:\ddot{\underset{\cdot\cdot}{\text{O}}}:}\right]^-} + \underset{\text{hydrogen ion (proton)}}{[\text{H}]^+}$$

The hydroxide ion is negatively charged; it has nine protons and ten electrons. It has the same electronic configuration as neon and therefore is chemically stable and

[3]We use the term *coefficient* rather than *constant* when the reaction is not elementary.

unreactive. Whereas hydroxyl radicals cannot be stored, solutions containing hydroxide ions can remain intact for long periods of time.

Chain Reactions

Ions play little part in ordinary gas-phase reactions, owing to the difficulty with which they are formed in the absence of an ionizing solvent. They do play a role in radiolytic reactions (Section 10.7), where high energies are involved. Atoms and free radicals are produced more easily in the gas phase and, because they enter readily into further reaction, they are important intermediates in reactions. For example, consider the reaction between hydrogen and bromine, for which Eq. 10.3 is the rate equation. This rate equation can be explained by the mechanism[4]

Visualize the steps in the formation of HBr; Composite mechanism; HBr.

$$
\begin{array}{llll}
1. & Br_2 \xrightarrow{k_1} 2Br & & \text{initiation} \\
2. & Br + H_2 \xrightarrow{k_2} HBr + H & \rbrace & \text{chain propagation} \\
3. & H + Br_2 \xrightarrow{k_3} HBr + Br & \rbrace & \\
4(-2). & H + HBr \xrightarrow{k_4} H_2 + Br & & \\
-1. & 2Br \xrightarrow{k_{-1}} Br_2 & & \text{termination or chain ending}
\end{array}
$$

Initiation Reaction

Chain Propagation Step

Termination Reaction

The first reaction, the production of bromine atoms from a bromine molecule, is known as the **initiation reaction,** since it starts the whole process. Reactions 2 and 3, the so-called **chain-propagation steps,** play a very important role in reactions of this type. Bromine atoms disappear in reaction 2 and reappear in reaction 3; hydrogen atoms disappear in reaction 3 and come back again in reaction 2. Because of this feature a small number of Br atoms, produced in reaction 1, can bring about a considerable amount of reaction, since after producing two molecules of hydrogen bromide, one in reaction 2 and one in reaction 3, a bromine atom is regenerated. Reaction 4 accounts for the fact that in the rate equation (Eq. 10.3), HBr appears in the denominator; HBr reduces the rate by removing H atoms. If it were not for reaction -1, a single pair of bromine atoms could bring about the reaction of all the H_2 and Br_2 present. Because of the **termination reaction** -1, however, only a limited amount of reaction is brought about each time a pair of bromine atoms is produced. Bromine atoms are continuously formed by reaction 1, and this keeps the reaction going.

A reaction of this type is known as a **chain reaction.** One essential feature of a chain reaction is that there must be a *closed sequence,* or *cycle,* of reactions such that certain active intermediates are consumed in one step and are regenerated in another; these active intermediates may be atoms, free radicals, or ions. It is also an essential feature that the sequence is, on the average, repeated more than once.

The way in which the chain-reaction mechanism for the hydrogen-bromine reaction explains the experimental rate equation 10.3 can be shown as follows. The net rate of increase of the concentration of hydrogen bromide is equal to

$$v_{HBr} = k_2[Br][H_2] + k_3[H][Br_2] - k_4[H][HBr] \tag{10.40}$$

[4]The reason that the reverse of reaction 3 is not included is that it is quite endothermic and therefore has a high activation energy; it is too slow to play an important role.

The concentration of bromine atoms can be obtained by use of the steady-state method, which must now be applied to the two unstable intermediates H and Br. The steady-state equation for H is

$$\frac{d[\mathrm{H}]}{dt} = k_2[\mathrm{Br}][\mathrm{H}_2] - k_3[\mathrm{H}][\mathrm{Br}_2] - k_4[\mathrm{H}][\mathrm{HBr}] = 0 \qquad (10.41)$$

and that for Br is

$$\frac{d[\mathrm{Br}]}{dt} = k_1[\mathrm{Br}_2] - k_2[\mathrm{Br}][\mathrm{H}_2] + k_3[\mathrm{H}][\mathrm{Br}_2] + k_4[\mathrm{H}][\mathrm{HBr}] - k_{-1}[\mathrm{Br}]^2 = 0 \qquad (10.42)$$

We thus have two equations in the two unknowns [H] and [Br], and we can solve for both of these concentrations. A solution for [Br] is quickly obtained if we add Eqs. 10.41 and 10.42:

$$k_1[\mathrm{Br}_2] - k_{-1}[\mathrm{Br}]^2 = 0 \qquad (10.43)$$

and thus

$$[\mathrm{Br}] = \left(\frac{k_1}{k_{-1}}\right)^{1/2}[\mathrm{Br}_2]^{1/2} \qquad (10.44)$$

Note that this is the equilibrium concentration of Br atoms. It is by no means always the case that atoms and free radicals in reaction systems are present at their equilibrium concentrations; indeed, in the present example the H atoms are present at much higher concentrations than their equilibrium concentrations.

We can insert this expression for [Br] into either Eq. 10.41 or Eq. 10.42 and obtain an expression for [H]. Insertion into Eq. 10.41 leads to

$$k_2\left(\frac{k_1}{k_{-1}}\right)^{1/2}[\mathrm{Br}_2]^{1/2}[\mathrm{H}_2] - k_3[\mathrm{H}][\mathrm{Br}_2] - k_4[\mathrm{H}][\mathrm{HBr}] = 0 \qquad (10.45)$$

so that

$$[\mathrm{H}] = \frac{k_2(k_1/k_{-1})^{1/2}[\mathrm{H}_2][\mathrm{Br}_2]^{1/2}}{k_3[\mathrm{Br}_2] + k_4[\mathrm{HBr}]} \qquad (10.46)$$

If we subtract Eq. 10.41 from Eq. 10.40, we obtain

$$v_{\mathrm{HBr}} = 2k_3[\mathrm{H}][\mathrm{Br}_2] \qquad (10.47)$$

and insertion of the expression for [H] leads to

$$v_{\mathrm{HBr}} = \frac{2k_2k_3(k_1/k_{-1})^{1/2}[\mathrm{H}_2][\mathrm{Br}_2]^{3/2}}{k_3[\mathrm{Br}_2] + k_4[\mathrm{HBr}]} \qquad (10.48)$$

$$= \frac{2k_2(k_1/k_{-1})^{1/2}[\mathrm{H}_2][\mathrm{Br}_2]^{1/2}}{1 + (k_4/k_3)([\mathrm{HBr}]/[\mathrm{Br}_2])} \qquad (10.49)$$

This equation is of the same form as the empirical Eq. 10.3, and we note that the empirical k is equal to $2k_2(k_1/k_{-1})^{1/2}$ and that the empirical m is equal to k_3/k_4.

We can see that the reason the term in $[\mathrm{HBr}]/[\mathrm{Br}_2]$ appears in the denominator of the rate equation is that HBr inhibits the reaction, by undergoing reaction 4, and that Br_2 reduces the amount of inhibition, since, in reactions 3 and 4, Br_2 and HBr are competing with one another for H atoms.

Organic Decompositions

A typical organic free-radical chain reaction is the decomposition of ethane,

$$C_2H_6 \rightarrow C_2H_4 + H_2$$

Under most conditions this is a simple first-order reaction, and originally it was thought that it occurred in one stage by what is called a *molecular* reaction; that is, that a small fraction of the molecules have sufficient energy for two C—H bonds to be ruptured, so that a hydrogen molecule is liberated:

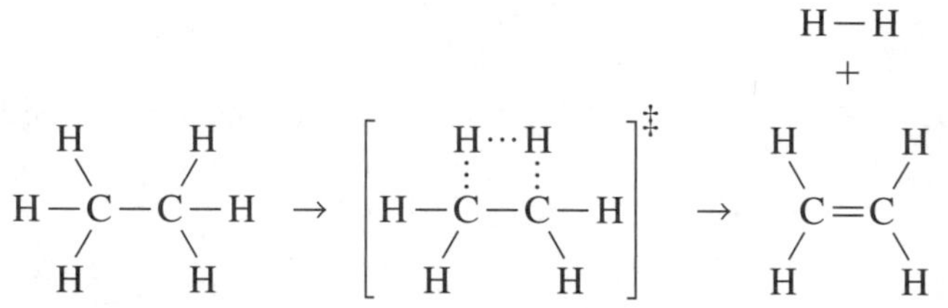

```
                                                        H—H
                                                         +
 H       H          ┌    H···H    ┐‡          H       H
  \     /           │    :   :    │            \     /
H—C—C—H    →        │H—C—C—H      │     →       C=C
  /     \           │   /   \     │            /     \
 H       H          └  H     H    ┘           H       H
```

However, the evidence indicates that such a mechanism plays an unimportant role and that practically all the decomposition occurs by the following chain mechanism:

$$\begin{array}{lrcll}
1. & C_2H_6 & \rightarrow & 2CH_3 & \\
2. & CH_3 + C_2H_6 & \rightarrow & CH_4 + C_2H_5 & \text{initiation} \\
3. & C_2H_5 & \rightarrow & C_2H_4 + H & \\
4. & H + C_2H_6 & \rightarrow & H_2 + C_2H_5 & \text{chain propagation} \\
5. & 2C_2H_5 & \rightarrow & C_4H_{10} & \text{termination}
\end{array}$$

The initiation process involves the breaking of a C—C bond, which is the weakest bond in the molecule. Reaction 2 is in a sense part of the initiation reaction; it converts CH_3 into a radical C_2H_5, which can be involved in propagation; reaction 2 is not a propagation reaction since CH_3 is not regenerated from C_2H_5. Reactions 3 and 4 are chain-propagating steps: C_2H_5 disappears in reaction 3 and appears in reaction 4, while H disappears in reaction 4 and appears in reaction 3. The main products of the reaction, C_2H_4 and H_2, are formed in these propagation steps. The termination step, reaction 5, forms butane, C_4H_{10}, which can be detected as a minor product of the reaction. Methane, formed in reaction 2, has also been observed as a minor product of the reaction.

EXAMPLE 10.5 Work out the expression for the overall rate of the ethane decomposition according to this mechanism, on the assumption that the steady-state hypothesis applies to the free radicals CH_3, C_2H_5, and H.

Solution The steady-state equations are

for CH_3: $k_1[C_2H_6] - k_2[CH_3][C_2H_6] = 0$

for C_2H_5: $k_2[CH_3][C_2H_6] - k_3[C_2H_5] + k_4[H][C_2H_6] - k_5[C_2H_5]^2 = 0$

for H: $k_3[C_2H_5] - k_4[H][C_2H_6] = 0$

Addition of all three equations gives

$$k_1[C_2H_6] - k_5[C_2H_5]^2 = 0$$

so that

$$[C_2H_5] = (k_1/k_5)^{1/2}[C_2H_6]^{1/2}$$

It is convenient to choose to calculate the rate of formation of ethylene, since it is formed in reaction 3 from the ethyl radical, the concentration of which we now have. The rate of formation of ethylene is

$$v = k_3[C_2H_5] = k_3(k_1/k_5)^{1/2}[C_2H_6]^{1/2}$$

(We might also have taken the rate of reaction to be the rate of formation of H_2, and therefore to be the rate of reaction 4. More algebra would then be involved, since we would first have to solve for the hydrogen atom concentration. The final answer would be the same. It always saves a little trouble to look for the solution requiring the least algebra.)

The steady-state solution for this mechanism has led to a rate equation in which the rate is proportional to the square root of the ethane concentration. However, most experiments have shown the reaction to be of the first order! For many years this was regarded as an insuperable objection to this mechanism, but in 1963 the key to the dilemma was ingeniously provided by the British physical chemist C. P. Quinn. He pointed out that reaction 3, a unimolecular reaction, is not necessarily a first-order reaction; in fact under the conditions usually employed in studying the reaction, the rate of reaction 3 was approximately given by

$$v = k_3[C_2H_5][C_2H_6]^{1/2}$$

With this modification the scheme correctly leads to first-order kinetics. This example shows how complicated and confusing the search for a reaction mechanism can sometimes be. It should again be emphasized that we can never be sure of a mechanism; all we can say is that all of the extensive evidence so far gathered on the ethane decomposition is consistent with this mechanism, as modified by Quinn.

EXAMPLE 10.6 The mechanism originally proposed in 1934 by F. O. Rice and K. F. Herzfeld for the ethane decomposition was

1. $C_2H_6 \rightarrow 2CH_3$
2. $CH_3 + C_2H_6 \rightarrow CH_4 + C_2H_5$
3. $C_2H_5 \rightarrow C_2H_4 + H$
4. $H + C_2H_6 \rightarrow H_2 + C_2H_5$
5. $H + C_2H_5 \rightarrow C_2H_4$

(Note that this differs from the previous scheme only in the chain-ending step.) Derive the rate equation corresponding to this mechanism, assuming the reaction orders to correspond to the molecularities.

Solution The steady-state equations are

for CH_3: $k_1[C_2H_6] - k_2[CH_3][C_2H_6] = 0$

for C_2H_5: $k_2[CH_3][C_2H_6] - k_3[C_2H_5] + k_4[H][C_2H_6] - k_5[H][C_2H_5] = 0$

for H: $k_3[C_2H_5] - k_4[H][C_2H_6] - k_5[H][C_2H_5] = 0$

Addition of all three equations gives

$$k_1[C_2H_6] - k_5[H][C_2H_5] = 0$$

so that

$$[H] = k_1[C_2H_6]/k_5[C_2H_5]$$

Insertion of this expression into the steady-state equation for H gives, after some rearrangement,

$$k_3k_5[C_2H_5]^2 - k_1k_5[C_2H_6][C_2H_5] - k_1k_4[C_2H_6]^2 = 0$$

The general solution of this quadratic equation is

$$[C_2H_5] = \left\{\frac{k_1}{2k_3} + \left[\left(\frac{k_1}{2k_3}\right)^2 + \left(\frac{k_1k_4}{k_3k_5}\right)\right]^{1/2}\right\}[C_2H_6]$$

A satisfactory approximation to the solution can be obtained by noting that the rate constant k_1 is very small, since it involves the largest activation energy. The terms involving $k_1/2k_3$ can therefore be neglected in comparison with k_1k_4/k_3k_5, and thus

$$[C_2H_5] = (k_1k_4/k_3k_5)^{1/2}[C_2H_6]$$

The rate of formation of ethylene is therefore

$$v = k_3[C_2H_5] = k_3(k_1k_4/k_3k_5)^{1/2}[C_2H_6]$$
$$= (k_1k_3k_4/k_5)^{1/2}[C_2H_6]$$

The Rice-Herzfeld mechanism thus gives first-order kinetics, in agreement with experiment. Nevertheless it turned out not to be the correct mechanism. One difficulty is that it was found that the ethyl radical concentration is much higher than the hydrogen atom concentration, so that the termination process $C_2H_5 + C_2H_5$ must be more important than $C_2H_5 + H$.

When ethanal (acetaldehyde) decomposes thermally the main products are methane and carbon monoxide, and under usual conditions the order of reaction is 1.5. A variety of experimental evidence has shown that the reaction occurs to a large extent by the mechanism

1. $CH_3CHO \xrightarrow{k_1} CH_3 + CHO$
2. $CH_3 + CH_3CHO \xrightarrow{k_2} CH_4 + CH_3CO$
3. $CH_3CO \xrightarrow{k_3} CH_3 + CO$
4. $CH_3 + CH_3 \xrightarrow{k_4} C_2H_6$

To simplify the steady-state treatment we will neglect the subsequent reactions of CHO. The steady-state equation for CH_3 is

$$k_1[CH_3CHO] - k_2[CH_3]\,[CH_3CHO] + k_3[CH_3CO] - k_4[CH_3]^2 = 0 \quad (10.50)$$

and the steady-state equation for CH_3CO is

$$k_2[CH_3][CH_3CHO] - k_3[CH_3CO] = 0 \quad (10.51)$$

Addition of these two equations gives

$$k_1[CH_3CHO] - k_4[CH_3]^2 = 0 \quad (10.52)$$

and therefore

$$[CH_3] = \left(\frac{k_1}{k_4}\right)^{1/2} [CH_3CHO]^{1/2} \quad (10.53)$$

The rate of change of the concentration of methane, which is approximately the rate of change of the concentration of acetaldehyde, is

$$v = k_2[CH_3][CH_3CHO] \quad (10.54)$$

and insertion of Eq. 10.53 gives

$$v = k_2\left(\frac{k_1}{k_4}\right)^{1/2} [CH_3CHO]^{3/2} \quad (10.55)$$

This agrees with the experimental fact that the order of the acetaldehyde decomposition is three-halves.

10.6 Photochemical Reactions

The reactions we have considered in Section 10.5 are known as **thermal reactions;** the energy needed for the activation barriers to be surmounted is provided by the thermal motions of the molecules and radicals. It is also possible for reactions to be brought about by electromagnetic radiation. For example, if a mixture of dry hydrogen and chlorine is irradiated by visible light, the reaction

$$H_2 + Cl_2 \rightarrow 2HCl$$

occurs with explosive violence. The reaction between hydrogen and bromine can also be brought about by irradiating the mixture with light of suitable wavelength, and the formation of hydrogen bromide is found to follow the rate equation

$$v_{HBr} = \frac{k'[H_2]I^{1/2}}{1 + [HBr]/m'[Br_2]} \quad (10.56)$$

where k' and m' are constants (compare Eq. 10.3) and I is the intensity of the light that is absorbed.

Later (p. 441) we will discuss the mechanism of this reaction, but first we will consider some important principles that apply to the interaction between molecules and radiation. Radiation is of two kinds, electromagnetic and particle, and examples of each follow, in order of increasing energy (see also Figure 11.1).

Electromagnetic Radiation	Particle Radiation
Infrared radiation	α particles (He nuclei)
Visible light	β particles (electrons)
Ultraviolet radiation	Cathode rays (electrons)
X rays	Beams of electrons, protons,
γ rays	Deuterons, etc., produced in an accelerator

Chemical reactions brought about by any of these radiations are known as either *photochemical* or *radiation-chemical reactions.* The distinction between these two classes of reactions is not a sharp one, but it has been found to be useful. In the case of electromagnetic radiation from the mid-ultraviolet range and beyond, and with high-energy particle radiation, ions can be detected in the reaction system, and in that case we speak of a **radiation-chemical reaction;** such reactions are considered in Section 10.7. Electromagnetic energy of higher wavelengths, such as visible and near-ultraviolet radiation, does not possess enough energy to bring about ionization and we then speak of a **photochemical** reaction.

Various special kinds of photochemical reactions may be identified by the prefix *photo.* For example, a *photolysis* is a photochemical reaction in which there is molecular dissociation; while a *photoisomerization* is a photochemical isomerization.

A number of important principles are associated with the photochemical production of excited molecules and of atoms and radicals; some of these will be discussed in Chapter 13. For a gas there is often a fairly sharp transition from a spectral region where there is no absorption and no chemical reaction to one in which a considerable amount of chemical reaction occurs. The frequency or wavelength at this transition is known as the *photochemical threshold.* Reaction occurs on the higher-frequency, or shorter-wavelength, side of the threshold.

Photochemical Threshold

Since the lifetime of an electronically excited species is very short, often about 10^{-8} s, it is very unlikely for a molecule that has absorbed one photon to absorb another before it has become deactivated.[5] Therefore, there is usually a one-to-one relationship between the number of photons absorbed by the system and the number of excited molecules produced. For example, light may be absorbed with the production of an excited species that dissociates into two radicals, as in the process

$$CH_3COCH_3 + h\nu \longrightarrow CH_3COCH_3^* \longrightarrow CO + 2CH_3$$

where the photon has been represented by the symbol $h\nu$. (The Planck constant h multiplied by the frequency ν is the energy of the photons.) The rate of formation of methyl radicals is in this case twice the rate of absorption of photons. This principle, which is due to Johannes Stark (1874–1957) and to Albert Einstein (1879–1955), is known as the **law of photochemical equivalence.** This law has been of great value in photochemical studies, since it enables the rates of formation of radicals to be calculated from the results of optical measurements. It is convenient to speak of a mole of photons as an *einstein;* the rate of production of methyl radicals in the photolysis of acetone would then be, in mol dm^{-3} s^{-1}, twice the number of einsteins absorbed per cubic decimeter per second.

Photochemical Equivalence

[5]Exceptions to this are found when lasers are used (Section 14.1); the intensity of the radiation can then be so great that two or more photons can be absorbed.

In photochemical experiments it is often found that the number of molecules that are chemically transformed differs markedly from the number of photons absorbed. In such cases it is sometimes said the law of photochemical equivalence is not obeyed, although the violation is only an apparent one. There are two main reasons for the apparent failure of the law. In the first place, radicals that are produced initially may recombine before they can undergo reaction; as will be seen, this very commonly occurs in solution. The rate of reaction is then less than predicted by the law of photochemical equivalence. In other systems the radicals produced may initiate chain reactions, in which case the rate of reaction may be much larger than expected. Deviations from the law of photochemical equivalence obviously provide valuable information about reaction mechanisms. The ratio of the number of molecules undergoing reaction in a given time to the number of photons absorbed is known as the **photon yield,** or **quantum yield,** or **quantum efficiency,** and given the symbol Φ (capital phi).

Photon Yield

EXAMPLE 10.7 Suppose that a 60-watt lamp, operating with 100% efficiency, emits radiation only at a wavelength of 313 nm. How many photons does it emit per second? Suppose that all of the radiation emitted is absorbed by heptan-4-one (di-*n*-propyl ketone), and that one product is ethene (ethylene), which is formed with a quantum yield of 0.25. How much ethene is produced in one second?

Solution The frequency corresponding to 313 nm is the speed of light divided by the wavelength:

$$\nu = 2.998 \times 10^8 \text{ m s}^{-1}/313 \times 10^{-9} \text{ m} = 9.578 \times 10^{14} \text{ s}^{-1}$$

The energy of each photon is the frequency multiplied by Planck's constant:

$$E = 9.578 \times 10^{14} \text{ s}^{-1} \times 6.626 \times 10^{-34} \text{ J s}^{-1} = 6.346 \times 10^{-19} \text{ J}$$

The 60-watt lamp emits 60 J each second, and in one second therefore emits

$$60 \text{ J}/6.346 \times 10^{-19} \text{ J} = 9.45 \times 10^{19} \text{ photons}$$

Since the quantum yield is 0.25, the number of ethene molecules produced in a second is therefore

$$0.25 \times 9.45 \times 10^{19} = 2.36 \times 10^{19} \text{ molecules}$$
$$= 3.92 \times 10^{-5} \text{ mol}$$

Chemical Actinometer

EXAMPLE 10.8 Uranyl oxalate is commonly used as a *chemical actinometer,* to determine the amount of light emitted by a lamp in a photochemical experiment. It has been established that at a wavelength of 300 nm the quantum yield for the decomposition of uranyl oxalate is 0.570. In a particular experiment it was found that the light of that wavelength, emitted by a lamp in one hour, brought about the decomposition of 5.78×10^{-5} mol of uranyl oxalate. When acetone was irradiated with the lamp for 10 hours, in such a way that all of the light was absorbed, the amount of acetone decomposed was found to be 2.32×10^{-4} mol. What is the quantum yield for the acetone decomposition under those conditions?

Solution The number of moles of photons (einsteins) required to decompose 5.78×10^{-5} mol of uranyl oxalate is

$$5.78 \times 10^{-5} \text{ mol}/0.570 = 1.014 \times 10^{-4} \text{ mol}$$

In 10 hours, 1.014×10^{-3} einsteins will be absorbed, and since 2.32×10^{-4} mol of acetone has been decomposed, the quantum yield is

$$2.32 \times 10^{-4} \text{ mol}/1.104 \times 10^{-3} \text{ mol} = 0.21$$

Photolysis of HI

A simple example of a reaction having a quantum yield of greater than unity is provided by the photochemical decomposition of hydrogen iodide. The German photochemist Emil Gabriel Warburg (1845–1931) found for this reaction that the quantum yield is 2; that is, two molecules of hydrogen iodide are transformed into $H_2 + I_2$ when one photon is absorbed. To explain this he proposed the mechanism

$$HI + h\nu \rightarrow H + I$$

$$H + HI \rightarrow H_2 + I$$

$$I + I \rightarrow I_2$$

Addition of these three reactions gives

$$2HI + h\nu \rightarrow H_2 + I_2$$

Thus, although one photon reacts with one HI molecule, in accordance with the law of photochemical equivalence, two molecules are transformed. Note that this reaction is not a chain reaction, in that there is no cycle of reactions in which intermediates are formed in one step and are reproduced in another.

The Photochemical Hydrogen-Chlorine Reaction

When there is a chain reaction, the quantum yields may be very large. This is particularly noteworthy with the photochemical reaction between hydrogen and chlorine, to give hydrogen chloride—a reaction which provides a very spectacular laboratory demonstration, as a short flash of visible light causes rapid reaction accompanied by a sharp sound. In the 1860s the German chemist Robert Bunsen (1811–1899) and his English student Henry (later Sir Henry) Roscoe (1833–1915) estimated that the light emitted by the sun in one minute and absorbed by a hydrogen-chlorine mixture would bring about the conversion of 25×10^{12} cubic miles of the mixture at atmospheric pressure! Translated into quantum yields, this amounts to something like 10^6.

The first satisfactory proposal for a chain reaction was put forward in 1918 for this reaction by Walther Nernst. For the primary process he suggested the interaction between a photon and a molecule of chlorine:

$$Cl_2 + h\nu \rightarrow 2Cl$$

He suggested that the following chain-propagating reactions occur.

$$Cl + H_2 \rightarrow HCl + H$$

$$H + Cl_2 \rightarrow HCl + Cl$$

These steps are analogous to the chain-propagating steps in the thermal (and photochemical) hydrogen-bromine reaction. The chain-ending steps for the hydrogen-chlorine reaction are, however, somewhat different from those for the hydrogen-bromine reaction. The details of the chain-ending steps in the hydrogen-chlorine reaction are a little complicated, and will not be dealt with here; it will simply be said that the chain lengths, and therefore the quantum yields, are much greater than for the hydrogen-bromine reaction. In both cases the thermal reactions proceed by the same chain mechanisms as the photochemical reactions, the only difference being that the initiation steps are the thermal dissociations of the halogen molecules into their atoms, instead of the photochemical dissociations.

The Photochemical Hydrogen-Bromine Reaction

When light of sufficiently short wavelength is passed through a mixture of hydrogen and bromine, at room temperature, reaction occurs with the formation of hydrogen bromide, and the rate equation is Eq. 10.56. The similarity of this equation to that for the thermal reaction, Eq. 10.3, suggests that the difference is only in the initiation step. In the thermal reaction, which requires higher temperatures than the photochemical reaction, the initiation reaction is the thermal dissociation of bromine molecules. In the photochemical reaction the initiation process is the photochemical dissociation of Br_2 into bromine atoms. The mechanism is therefore

1. $Br_2 + h\nu \rightarrow 2Br$ — initiation
2. $Br + H_2 \rightarrow HBr + H$ — chain propagation
3. $H + Br_2 \rightarrow HBr + Br$ — chain propagation
4. $H + HBr \rightarrow H_2 + Br$
5. $2Br \rightarrow Br_2$ — termination or chain-ending

This scheme is more straightforward than that for the hydrogen-chlorine reaction, since termination steps other than the recombination of atoms do not play any significant role.

If I_a is the intensity of light absorbed, expressed as einstein dm^{-3} s^{-1}, the rate of change of concentration of Br atoms by reaction 1 (in the same units) is $2I_a$. The steady-state equation for Br atoms is therefore

$$2I_a - k_2[Br][H_2] + k_3[H][Br_2] + k_4[H][HBr] - k_5[Br]^2 = 0 \quad (10.57)$$

The steady-state equation for H atoms is

$$k_2[Br][H_2] - k_3[H][Br_2] - k_4[H][HBr] = 0 \quad (10.58)$$

These equations are very similar to Eqs. 10.42 and 10.41, respectively; the only differences are that $k_1[Br_2]$ has been replaced by $2I_a$ and k_{-1} has been replaced by k_5. If we make these replacements in Eq. 10.49, we see that the rate equation for the photochemical reaction is

$$v_{HBr} = \frac{2k_2(2/k_5)^{1/2}[H_2]I_a^{1/2}}{1 + k_4[HBr]/k_3[Br_2]} \quad (10.59)$$

This is of the same form as Eq. 10.56.

Photosensitization

Many molecules do not absorb radiation at a convenient wavelength. The production of hydrogen atoms from a hydrogen molecule requires 431.8 kJ mol^{-1}, which corresponds to a wavelength of 277.6 nm. However, hydrogen gas does not absorb at this wavelength and it is necessary to go to very much lower wavelengths in order to produce hydrogen atoms. A very convenient procedure for avoiding this difficulty is to introduce mercury vapor into the hydrogen. Mercury atoms are normally in a 6 $^1S_0$ state,[6] and they absorb 253.7 nm radiation to give the 6 $^3P_1$ state, which is 468.6 kJ mol^{-1} higher:

$$Hg(6\ ^1S_0) + h\nu(253.7\ \text{nm}) \longrightarrow Hg^*(6\ ^3P_1)$$

The excited mercury atom, which has a relatively long life, has more than enough energy to dissociate a hydrogen molecule, and it does so with high efficiency:

$$Hg^*(6\ ^3P_1) + H_2 \longrightarrow Hg(6\ ^1S_0) + 2H$$

Many other bonds having strengths of less than 468.6 kJ mol^{-1} can be split in this way.

Photosensitized Processes

Processes of this kind, in which the radiation is absorbed by one species and the energy is passed on to another, are known as **photosensitized processes.** Zinc and cadmium atoms have also been used in photosensitization experiments. Sometimes the photosensitizer produces an excited state rather than a dissociated state. For example, with ethylene, excited mercury induces the reaction

$$Hg^* + C_2H_4 \longrightarrow Hg + C_2H_4^*$$

as well as

$$Hg^* + C_2H_4 \longrightarrow Hg + H + C_2H_3$$

Flash Photolysis

For their work in flash photolysis, Porter and Norrish shared the 1967 Nobel Prize in chemistry with Manfred Eigen.

The subject of photochemistry entered a new phase in about 1950 with the development of flash photolysis by the English physical chemists George Porter (b. 1920) and Ronald George Wreyford Norrish (1897–1978). In this technique a flash of high intensity but short duration brings about the formation of species such as atoms and molecules, the structure and reactions of which can be studied by spectroscopic means. Methods of studying rapid reactions were considered in Section 9.5 and were classified as *flow* methods and *pulse* methods. Like the relaxation techniques of Manfred Eigen (Section 9.5), flash photolysis is a pulse method. The techniques used by Porter, Norrish, and others have been referred to as *pulse and probe* methods, the initial short pulse being followed by a probing procedure in which the structures of species produced, or the kinetics of their reactions, are investigated.

Pulse and Probe Methods

In the earliest experiments on flash photolysis, in 1948, the duration of the flash was about a millisecond (10^{-3} s), and it is a remarkable fact that during the next four decades the duration of the flash was reduced by eleven powers of ten, to about 10^{-14} s, or 10 femtoseconds. As a result, techniques are now available for studying the kinetics of the fastest chemical and physical processes (see Figure 9.5 on p. 375). The flash photolysis method is in fact the only experimental technique available for studying these extremely rapid processes.

[6]The spectroscopic notation used here is explained in Section 13.2.

By 1950 flashes of about a microsecond (10^{-6} s) duration had been produced, and were used for the study of the structures of free radicals in the gas phase, for following the kinetics of a number of fast reactions such as the combinations of atoms and free radicals, and for following the decay of species in triplet excited states. By 1966 Porter had developed a highly efficient laser flash system capable of dealing with processes occurring in the nanosecond (10^{-9} s) range. It was then possible to study many additional processes, such as the decay of species in singlet excited states.

A flash of nanosecond duration is adequate for the study of almost any purely chemical process, by which is meant a process in which there is a change in chemical identity. Accompanying chemical processes, however, there are always processes of a purely physical nature, such as energy distribution and solvent reorganization, and these processes commonly occur in the picosecond (10^{-12} s) range. Flashes of these very short durations were achieved during the 1970s at the AT&T Bell Laboratories in New Jersey, notably by C. V. Shank (b. 1943) and P. M. Rentzepis (b. 1934). Many processes have now been studied with flashes of a picosecond duration.

In recent years, many such processes have been studied with flashes of only a few femtoseconds duration. The time that it takes for a chemical bond to change by 10^{-10} metres (1 Å) is about 100 fs, so that a flash of a few femtoseconds duration, closely followed by another one of the same duration, will provide information about such tiny changes in bond lengths. The technique for causing one flash to occur a few nanoseconds after another is to route the light by a slightly longer path. Since the speed of light is so great, a path of a micrometre (10^{-6}m) causes a delay of 3×10^{-15}m or three femtoseconds, and such short-path differences are now technically feasible. A pioneer in this field is Ahmed Zewail (Nobel Prize for chemistry, 1999).

10.7 Radiation-Chemical Reactions

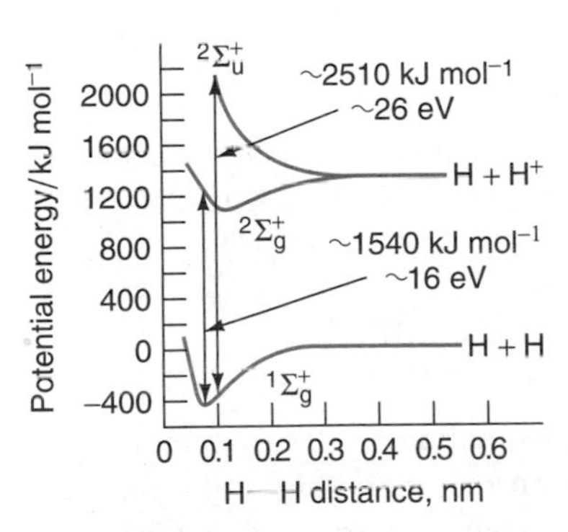

FIGURE 10.2
Potential-energy curves for H_2 and H_2^+, showing how the positive H_2^+ ion is formed by electron impact.

We saw in Section 10.6 that an arbitrary but useful distinction is often made between photochemical and radiation-chemical reactions; the latter term is usually applied only when ions can be detected in the reaction system. When molecular dissociation occurs in a radiation-chemical reaction, the process is called a *radiolysis*.

Whereas photochemical initiation reactions are fairly clear-cut, radiolytic initiation reactions frequently occur in a number of stages, and there are marked differences between different kinds of radiation. As a fairly simple example we will consider the effect of a high-energy electron beam on hydrogen molecules. Figure 10.2 shows the ground state of the hydrogen molecule and also two states of H_2^+ which dissociate into H + H^+. Electrons of energy 2.6×10^{-18} J (= 1540 kJ mol^{-1} = 16 eV) will eject an electron from H_2; the process is

$$e^- + H_2 \longrightarrow H_2^+ + 2e^-$$

When H_2^+ encounters an electron, neutralization occurs with the liberation of a considerable amount of energy, which brings about dissociation into two H atoms:

$$e^- + H_2^+ \longrightarrow H_2^* \rightarrow 2H$$

Electrons of higher energy, of about 4.2×10^{18} J (= 2510 kJ mol^{-1} = 26 eV), can also produce H_2^+ in its repulsive $^2\Sigma_u^+$ state, which at once dissociates into H + H^+:

$$e^- + H_2 \longrightarrow 2e^- + H_2^+(^2\Sigma_u^+)$$

When H^+ encounters an electron, in the presence of another molecule M (a "third body") to remove the energy, it forms H:

$$M + H^+ + e^- \longrightarrow H + M$$

The net result of these several processes is the formation of two hydrogen atoms from the hydrogen molecule. These various processes are usually referred to collectively as the *primary radiolytic process,* which is written as

Primary Radiolytic Process

$$H_2 \rightsquigarrow 2H$$

The primary process in the irradiation of hydrogen by α particles is written as

$$H_2 \overset{\alpha}{\rightsquigarrow} 2H$$

Again, this primary process occurs in a number of stages, involving the formation of ions. The first step is usually the ejection of an electron:

$$\alpha + M \longrightarrow M^+ + e^- + \alpha$$

The α particle continues on its way with little deflection and ionizes other molecules. The ejected electron frequently has sufficient energy to bring about ionization of additional molecules, by the mechanism just described.

Photons of very high energy (very low wavelengths) frequently eject electrons with the formation of the positive ion:

$$M + h\nu \longrightarrow M^+ + e^-$$

Although the details of primary radiolytic processes are often complex, the overall result is usually fairly simple. With hydrogen, for example, the initiation process with most types of radiation is largely the production of atoms:

$$H_2 \rightsquigarrow 2H$$

Similarly, with HI the main process is

$$HI \rightsquigarrow H + I$$

In addition, the formation of ions must be taken into account.

Ion-pair Yield

In investigations of overall radiation-chemical reactions two quantities are frequently quoted. The first is the *ion-pair yield,* or *ionic yield,* which is denoted as M/N and is defined by

$$\frac{M}{N} = \frac{\text{rate of formation of product molecules}}{\text{rate of formation of ion pairs}} \tag{10.60}$$

The rate of formation of ion pairs is determined by applying to the system an electrical potential sufficient to produce a limiting current (the *saturation current*); when this is done, the ions are swept out of the system as rapidly as they are produced by the radiation, and the rate of formation of the ions can be calculated from the saturation current.

G Value

The second quantity that is frequently quoted is the *G value,* which is the number of product molecules formed per 100 eV input (1 eV = 1.602×10^{-19} J = 96.47 kJ mol^{-1}). Both the ion-pair yield and the G value give some idea of the chain length.

When a gas is passed through an electric discharge, the stream of electrons usually brings about dissociation. It was first demonstrated in 1920 by the American physicist Robert W. Wood (1868–1955) that hydrogen atoms are conveniently produced in this way. Later the German physical chemist Karl Friedrich Bonhoeffer (1899–1957) modified Wood's apparatus and made a number of studies of the reactions of hydrogen atoms. This method of preparing hydrogen atoms is usually known as the Wood-Bonhoeffer method.

Wood-Bonhoeffer Method

10.8 Explosions

Detonation

An explosion is a process that occurs very rapidly and evolves much energy. There are many kinds of chemical explosions. An explosion that brings about damage to the environment, and is accompanied by the production of much noise, is called a *detonation*—a word which at first referred to the noise of an explosion, but which in the latter part of the nineteenth century became applied to the explosion itself. In a detonation a shock wave is formed, at which there is an extremely rapid rise in pressure and temperature. This shock wave moves with supersonic velocity through the explosive, where it is supported and reinforced by the chemical reaction, and continues into the unexploded environment.

A less violent explosion is called a *deflagration.* Instead of a shock wave moving at a speed of kilometers per second, there is now a flame front which moves with a speed of the order of metres per second. All detonations begin as deflagrations, which may last only a tiny fraction of a second before the detonation wave is established (Figure 10.3).

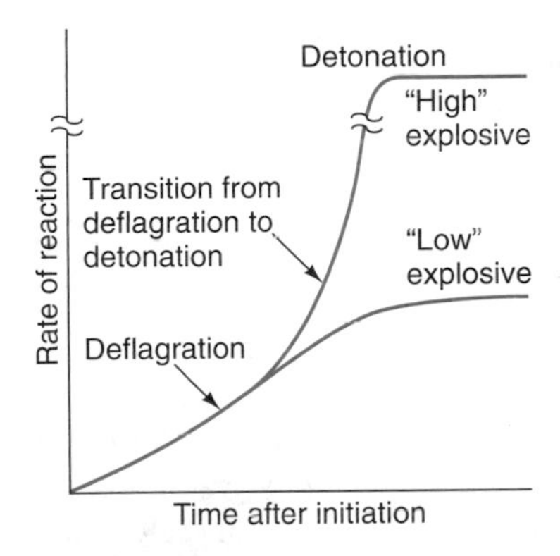

FIGURE 10.3
The transition from deflagration to detonation in a typical explosive substance or mixture.

Explosives are loosely classified as *high explosives* and *low explosives.* High explosives are those which readily give rise to detonations. Low explosives are those which under ordinary conditions merely deflagrate, although under special conditions, such as close confinement, they may detonate. One of the first explosives to be used, a low explosive, was gunpowder, also known as black powder. This is a mixture of potassium nitrate, charcoal, and sulfur, and appears to have been first mentioned in English by the Franciscan friar Roger Bacon (c. 1214–1292), who perhaps rediscovered it, although it was used earlier by the Chinese. Mixtures of hydrogen and oxygen, of hydrogen and air, and of natural gas and air are also low explosives. It should be noted that the damage caused by a low explosive can be comparable to that caused by a similar mass of high explosive. The appearance of a building that has been demolished by a natural gas explosion is superficially quite similar to that of a building on which a bomb has been dropped. Because of the confinement of the gas-air mixture, the deflagration becomes a detonation, and there can be extensive damage to surrounding buildings.

One of the first of the high explosives, dynamite was invented in 1866 by the Swedish chemist and manufacturer Alfred Nobel (1833–1896). High explosives are always solids or liquids, and other examples are trinitrotoluene (TNT) and cyclotrimethylenetrinitramine (RDX). Organic compounds such as these, which contain a high proportion of nitrogen, may be quite stable if handled with care, but explode if they are heated or if a suitable method of initiation is employed.

Explosions are sometimes quite gentle. If hydrogen and oxygen gases are brought together in their stoichiometric proportions at, say, 550 °C and at 10 Torr pressure in a glass vessel, very rapid reaction occurs, with the evolution of light, but the explosion is so gentle that it usually does not break the vessel.

The Initiation of an Explosion

There are two different ways in which an explosive substance or mixture can be caused to explode. One is by raising the temperature. A stoichiometric mixture of hydrogen and oxygen at atmospheric pressure and at room temperature is quite stable if left completely undisturbed; it can be estimated that the half-life of the reaction is greater than the age of the solar system! If, however, the temperature of the mixture is raised to about 550 °C, the mixture will explode. Any explosive substance will explode if its temperature is raised, the temperature required for explosion to occur depending on the kinetic parameters of the explosion process.

The second method by which an explosion can be made to occur, and the most usual practical one, is by the use of a suitable initiator. A stoichiometric mixture of hydrogen and oxygen at room temperature can be caused to explode by applying a lighted match, a spark, a radioactive substance, or a catalyst such as finely divided platinum. The effect of a flame, a spark, or a radioactive substance is to introduce free radicals into the reaction system, and so to initiate explosion in one small region from which it spreads to the rest of the mixture. The effect of a suitable solid surface is to bring about some dissociation of hydrogen into atoms which initiate the explosion.

The initiation of a detonation in a high explosive is usually brought about by means of a detonator, which consists of a primary explosive in which the initiation is easily effected by impact.

The Transmission of an Explosion

There are two distinctly different ways in which an explosion is transmitted, although in some situations both types of process occur together. There is *physical* transmission, involving the transfer of heat from one layer to another, and *chemical* transmission, involving reactions of atoms and free radicals. The physical processes predominate in the case of detonations of solids and liquids. As the detonation wave travels through the material the hot gases in the shock front transfer heat to the next layer of unexploded material, which itself reacts, so that the shock wave travels through the explosive.

Branching Chains

The chemical type of transmission predominates in the case of explosions in gases at low pressures, and involves the existence of *branching chains.* The idea of branching chains was first put forward in 1923, jointly by the Danish chemist Jens Anton Christiansen (1888–1969) and the Dutch physicist Hendrick Anthony Kramers (1894–1952). Their idea was that a chain reaction may involve propagating steps in which one chain carrier (atom or free radical) gives rise to more than one; an example is

$$H + O_2 \longrightarrow OH + O$$

where the atom H has produced another atom and also a free radical OH. When such chain branching occurs the number of chain carriers increases rapidly, and the rate correspondingly increases, often leading to an explosion.

The first experimental evidence for chain branching was obtained in 1927 by the Russian chemist Nikolay Nikolayevich Semenov (1896–1986). His work and the related work of the English chemist Sir Cyril Norman Hinshelwood are considered later (p. 448) in connection with explosion limits in gaseous systems.

Detonations

The theory of shock waves traveling in a chemically neutral medium was first worked out in 1870 by the Scottish engineer W. J. M. Rankine (1820–1872), whose treatment was further developed in 1887 by the French physicist P. H. Hugoniot (1851–1887).

Detonation waves arising as a result of chemical explosions in gases were first observed in 1881 by Marcellin Berthelot (1827–1907) and his assistant P. Vieille, and in the same year by H. L. Le Chatelier (1850–1936) and E. Mallard. The investigations of Le Chatelier and Mallard are remarkable for some excellent photographs of detonation fronts, obtained by allowing the detonation waves to pass along a horizontal tube, and using a photographic film which moved vertically at high speed.

Berthelot and Vieille developed a theory of detonation waves traveling in explosive gas mixtures, and also calculated the speeds of the waves in terms of the properties of the gases. Their idea was that a detonation wave has many of the same general characteristics as a sound wave, but is sustained by the energy released in the reaction occurring at the detonation front. Their theory was later modified by H. B. Dixon (1852–1930), who recognized that at the high temperatures involved in the detonation front there is some molecular dissociation, so that the specific heats are substantially larger at the higher temperatures.

Dixon's ideas were further refined and were put into mathematical form in 1899 by D. L. Chapman (1869–1958), who made calculations of the detonation speeds of a variety of gaseous mixtures, and obtained excellent agreement with experiment. A more detailed treatment, based on the same assumptions as those of Chapman, was developed in 1905 by the mathematician Emile Jouguet, and the detonation layer is now commonly known as the *Chapman-Jouguet layer.*

Chapman-Jouguet Layer

These theoretical treatments are related to explosions in gaseous mixtures. Detonations in solids and liquids are more difficult to treat theoretically because of the more complicated equations of state, and success was achieved only in the early years of World War II. The fact that these theories lead to good agreement with experiment without any consideration of chain reactions and chain branching suggests that physical transmission of the detonation wave is predominant.

Explosion Limits in Gaseous Explosions

An interesting effect, which throws light on the mechanisms of explosions in gases, is the existence of explosion limits, which occur in two sets of circumstances. An *initiated* explosion in a mixture of gases at atmospheric pressure and ordinary temperatures will occur only if the composition of the mixture lies between certain limits. For example, natural gas is largely methane, and a stoichiometric mixture consists of 2 parts of oxygen and 1 part of natural gas, or 10 parts of air and 1 of gas. If the proportion of natural gas mixed with air is less than about 5% or greater than about 15%, there can be no explosion. These limits are easy to understand; outside these explosion limits there is too much dilution of the mixture, either by the air or the fuel, to sustain the explosive wave.

The second circumstance under which explosion limits are found is when gaseous mixtures explode at higher temperatures *without any external initiation.* In 1927 Semenov studied the reaction between phosphorus vapor and oxygen at low pressures and showed that there is a critical pressure above which explosion occurs,

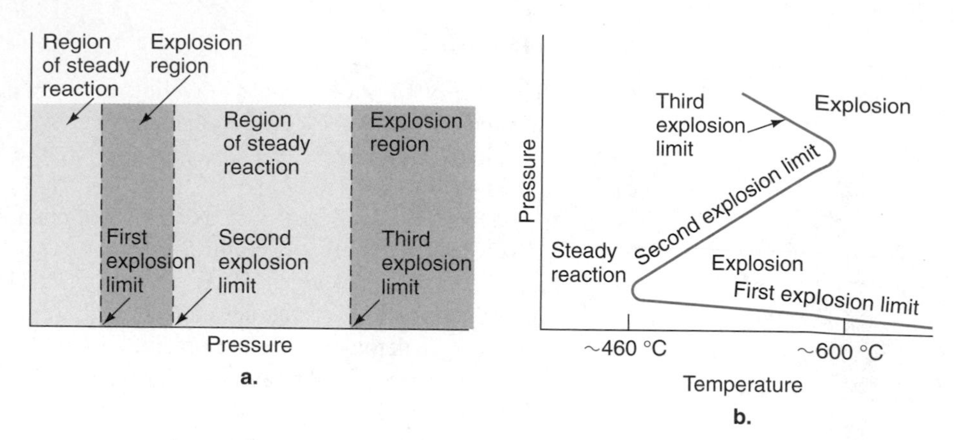

FIGURE 10.4
(a) Regions of steady reaction and explosion for a typical gaseous system, such as a mixture of hydrogen and oxygen.
(b) Variation of the explosion limits with temperature. The temperatures given are typically those for a stoichiometric $2H_2$—O_2 mixture. For such a mixture there is no explosion below about 460 °C, and above about 600 °C explosion will occur at all pressures.

but below which there is steady reaction. He related this to chain branching, and suggested that at sufficiently low pressures the chain carriers can diffuse to the walls of the reaction vessel and be removed by recombination, in this way counteracting the effect of chain branching. As the pressure is raised, however, it becomes harder for the chain carriers to reach the walls because of the molecules that impede their progress, and explosion occurs. As expected from this explanation, the lower explosion limit varies with the size of the reaction vessel, and with the nature of its surface walls.

In the same year Hinshelwood began a similar investigation of the reaction between hydrogen and oxygen. He showed that in addition to a lower explosion limit there is also a second explosion limit, above which reaction is again slow (Figure 10.4a). He attributed the second limit to removal of chain carriers in the gas phase; this becomes more likely as the pressure is raised, so that when the pressure is high enough the removal of chain carriers again counteracts the effects of chain branching. As expected from this explanation, the second limit does not vary with the size or nature of the reaction vessel. At still higher pressures there is a third explosion limit, and the explosion that occurs beyond that limit may be transmitted thermally, although some special chain processes may play a role also. Figure 10.4b shows how the explosion limits in the hydrogen-oxygen reaction change with temperature.

These contributions of Semenov and Hinshelwood played an important part in their being awarded jointly the 1956 Nobel Prize for chemistry.

Cool Flames

The gas-phase combustions of hydrocarbons show some special features. Like the hydrogen-oxygen reaction they become explosions within certain limits of temperature and pressure, but under certain other conditions a mild type of explosion occurs, with the emission of light. Since this occurs at lower temperatures than are usually found in flames, the effect is referred to as a *cool flame*. Spectroscopic studies of the flames have shown that much of the emission is due to the formation of formaldehyde in an electronically excited state.

Degenerate Branching

Hydrocarbon oxidations have been extensively investigated because of their practical interest. To explain the main features of the kinetics, and particularly the cool flames, Semenov put forward the hypothesis of *degenerate branching*. He postulated the existence of an intermediate such as formaldehyde which can break down in two different ways, one of which is a branching process and the other not; for example:

$$HCHO \begin{cases} \xrightarrow{+OH} H_2O + HCO & \text{(non-branching)} \\ \xrightarrow[+O_2]{} HCO + HO_2 & \text{(branching)} \end{cases}$$

Since the intermediate (formaldehyde, in this example) has a short life and does not invariably lead to branching, the concentration of chain-carriers increases more slowly than in a normal explosion. Under cool-flame conditions the reaction rate is therefore high, but not as high as in a normal explosion, and a cool flame has been called a *degenerate explosion.*

10.9 Catalysis

The rates of many reactions are influenced by the presence of a substance which remains unchanged at the end of the process. Examples are the conversion of starch into sugars, the rate of which is influenced by acids; the decomposition of hydrogen peroxide, influenced by ferric ions; and the formation of ammonia in the presence of spongy platinum. In 1836 such reactions were classified by the Swedish chemist Jöns Jacob Berzelius (1779–1848) under the title of **catalysis.**[7] A substance that decreases the rate of reaction is referred to as an **inhibitor.** (It was formerly sometimes called a negative catalyst, but this is not recommended in view of the mode of action of inhibitors, to be referred to later.) It is convenient to classify catalyzed reactions according to whether they occur homogeneously (in a single phase) or heterogeneously (at an interface between two phases). This section is concerned mainly with homogeneous catalysis; heterogeneous catalysis is considered in Section 10.12 (p. 465).

Various definitions of catalysis have been proposed. An early definition, suggested in 1895 by Wilhelm Ostwald (1853–1932), was that a catalyst is "any substance that alters the velocity of a chemical reaction without modification of the energy factors of the reaction." Another definition is that "a catalyst alters the velocity of a chemical reaction and is both a reactant and a product of the reaction." These definitions were intended to exclude substances that accelerated the rate of a reaction by entering into reaction, thus disturbing the position of equilibrium; such substances are reactants in the ordinary sense. Perhaps the most satisfactory definition of a catalyst is that it is

Catalyst

> a substance that increases the rate of a reaction without modifying the overall standard Gibbs energy change in the reaction.

In these definitions of catalysis there is no reference to the fact that a small amount of a catalyst has a large effect on the rate; this is frequently the case, but it is not an essential characteristic of a catalyst.

Although by definition the amount of a catalyst must be unchanged at the end of the reaction, the catalyst is invariably involved in the chemical process. In the case of a single reacting substance, a complex may be formed between this reactant

[7]This word comes from the Greek *kata,* wholly; *lyein,* to loosen.

(known as the *substrate*) and the catalyst. If there is more than one substrate, the complex may involve one or more molecules of substrate combined with the catalyst. These complexes are formed only as intermediates and decompose to give the products of the reaction, with the regeneration of the catalyst molecule. For example, when a reaction is catalyzed by hydrogen ions, an intermediate complex, involving the substrate and a hydrogen ion, is formed, and this later reacts further with the liberation of the ion and the formation of the products of the reaction.

Since the catalyst is unchanged at the end of the reaction, it gives no energy to the system; therefore it can have no influence on the position of equilibrium. It follows that since the equilibrium constant K_c is, at equilibrium, the ratio of the rate constants in the forward and reverse directions (i.e., $K_c = k_1/k_{-1}$), a catalyst must influence the forward and reverse rates in the same proportion. This conclusion has been verified experimentally in a number of instances.

An extremely small amount of a catalyst frequently causes a considerable increase in the rate of a reaction. Colloidal palladium at a concentration of 10^{-8} mol dm^{-3} has a significant catalytic effect on the decomposition of hydrogen peroxide. The effectiveness of a catalyst is sometimes expressed in terms of its *turnover number,* which is the number of molecules of substrate decomposed per minute by one molecule of the catalyst. For example, the enzyme catalase has, under certain conditions, a turnover number of 5 million for the decomposition of hydrogen peroxide ($2H_2O_2 \rightarrow 2H_2O + O_2$). Since the turnover number generally varies with the temperature and with the concentration of substrate, it is not a particularly useful quantity in kinetic work; the effectiveness of a catalyst is best measured in terms of a rate coefficient.

The rate of a catalyzed reaction is often proportional to the concentration of the catalyst,

$$v = k[\mathrm{C}] \tag{10.61}$$

where [C] represents the concentration of the catalyst, and k is a function of the concentration of the substrate. If Eq. 10.61 were obeyed exactly, the rate of reaction in the absence of the catalyst would be zero. Many examples are known for which it is necessary to introduce an additional term that is independent of the catalyst concentration,

$$v = k[\mathrm{C}] + v_0 \tag{10.62}$$

At zero concentration of the catalyst the reaction occurs with the velocity v_0.

The activation energy of a catalyzed reaction is almost always lower than that of the same reaction when it is uncatalyzed. In other words, catalysts generally work by permitting the reaction to occur by another reaction path that has a lower energy barrier. This is shown schematically in Figure 10.5. It is important to note that inhibitors do *not* work by introducing a higher reaction path; this would not reduce the rate, since the reaction would continue to occur by the alternative mechanism. Inhibitors act either by destroying catalysts already present or by removing reaction intermediates such as free radicals.

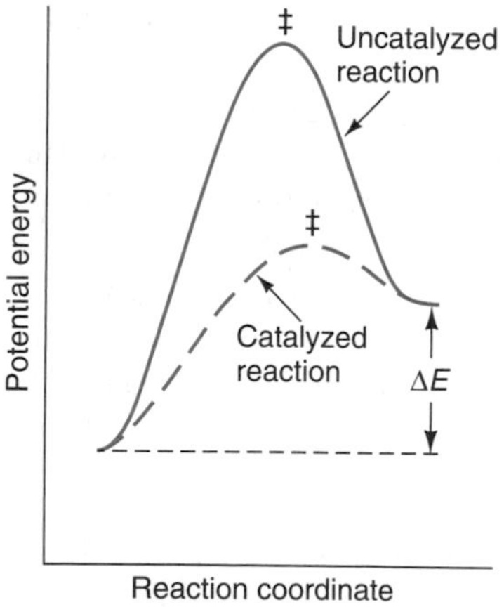

FIGURE 10.5
The lowering of the energy barrier brought about by a catalyst. Since the catalyst is not used up during the reaction, ΔE is the same for the catalyzed reaction as for the uncatalyzed, but the activation energies in both directions are lower for the catalyzed reaction.

Acid-Base Catalysis

The study of catalysis by acids and bases played a very important part in the development of chemical kinetics, since many of the reactions studied in the early days of the subject were of this type. The first investigations of the kinetics of reactions catalyzed by acids and bases were carried out at the same time that the electrolytic dissociation theory was being developed, and the kinetic studies contributed considerably to the development of that theory. The reactions considered from this point of view were chiefly the inversion of cane sugar and the hydrolysis of esters. It was first realized in 1884, by Ostwald and later by Arrhenius, that the ability of an acid to catalyze these reactions is independent of the nature of the anion but is approximately proportional to its electrical conductivity, which is a measure of an acid's strength. It was originally assumed that the hydrogen ions were the sole effective acid catalysts. Similarly, in catalysis by alkalis the rate is proportional to the concentration of the alkali but independent of the nature of the cation, suggesting that the active species is the hydroxide ion.

The idea that the only catalyzing species in reactions of this type are the hydrogen and hydroxide ions has been found to require modification in a number of instances. Many reactions do exist, however, for which only these two ions are effective catalysts; and we then say that the catalysis is **specific.**

Specific Acid-Base Catalysis

If such reactions are carried out in sufficiently strong acid solution, the concentration of hydroxide ions may be reduced to such an extent that these ions do not have any appreciable catalytic action. The hydrogen ions are then the only effective catalysts, and the rate (at least at concentrations of catalyst and substrate that are not too high) is given by an expression of the type

$$v = k_{H^+}[H^+][S] \tag{10.63}$$

where k_{H^+} is the rate constant for the hydrogen-ion-catalyzed reaction. If catalysis is effected simultaneously by hydrogen and hydroxide ions, and reaction may also occur spontaneously (i.e., without a catalyst), the rate is

$$v = k_0[S] + k_{H^+}[H^+][S] + k_{OH^-}[OH^-][S] \tag{10.64}$$

The first-order rate coefficient is therefore given by

$$k = k_0 + k_{H^+}[H^+] + k_{OH^-}[OH^-] \tag{10.65}$$

Catalytic Constants

In these equations, k_0 is the rate constant of the spontaneous reaction, and k_{H^+} and k_{OH^-} are known as the **catalytic constants** for H^+ and OH^-, respectively.

The rate equation may be expressed in a different form by making use of the fact that $[H^+][OH^-] = K_w$, where K_w is the ionic product of water. Elimination of $[OH^-]$ gives

$$k = k_0 + k_{H^+}[H^+] + \frac{k_{OH^-}K_w}{[H^+]} \tag{10.66}$$

while elimination of $[H^+]$ gives

$$k = k_0 + \frac{k_{H^+}K_w}{[OH^-]} + k_{OH^-}[OH^-] \tag{10.67}$$

In many cases one of these terms containing concentration is negligibly small compared with the other. If work is carried out with 0.1 *M* hydrochloric acid, for example, the second term in Eq. 10.66 is $k_{H^+} \times 10^{-1}$ while the third term is $k_{OH^-} \times 10^{-14}$ (since $K_w = 10^{-14}$); consequently, unless k_{OH^-} is at least 10^9 greater than k_{H^+}, the third term will be negligible compared with the second; at this acid concentration, therefore, catalysis by hydroxide ions will be negligible compared with that by hydrogen ions. Similarly, in 0.1 *M* sodium hydroxide solution catalysis by hydrogen ions will usually be unimportant compared with that by hydroxide ions. In general, there will be an upper range of hydrogen-ion concentrations at which catalysis by hydroxide ions will be unimportant and a lower range at which catalysis by hydroxide ions will predominate and catalysis by hydrogen ions will be unimportant. Within each of these ranges the rate will be a linear function of $[H^+]$ and of $[OH^-]$, respectively. In the upper range the value of the catalytic constant k_{H^+} can readily be determined from the experimental data; in the lower range k_{OH^-} can be so determined. The constants for the hydrolysis of ethyl acetate were measured in this manner by J. J. A. Wijs in 1893, and he also obtained a value for K_w, the ionic product of water, by making use of the fact that the velocity is a minimum when the second and third terms in Eq. 10.66 are equal; this gives rise to

$$[H^+]_{min} = (k_{OH^-}K_w/k_{H^+})^{1/2} \tag{10.68}$$

Thus, from the values of $[H^+]_{min}$, k_{H^+}, and k_{OH^-}, the ionic product K_w can be obtained.

A plot of the logarithm of the rate constant against the pH of the solution is shown in Figure 10.6a. There are regions of catalysis by hydrogen and hydroxide ions, separated by a region in which the amount of catalysis is unimportant in comparison with the spontaneous reaction. When the catalysis is largely by hydrogen ions, $k = k_{H^+}[H^+]$, so that

$$\log_{10} k = \log_{10} k_{H^+} + \log_{10}[H^+] \tag{10.69}$$

or

$$\log_{10} k = \log_{10} k_{H^+} - \text{pH} \tag{10.70}$$

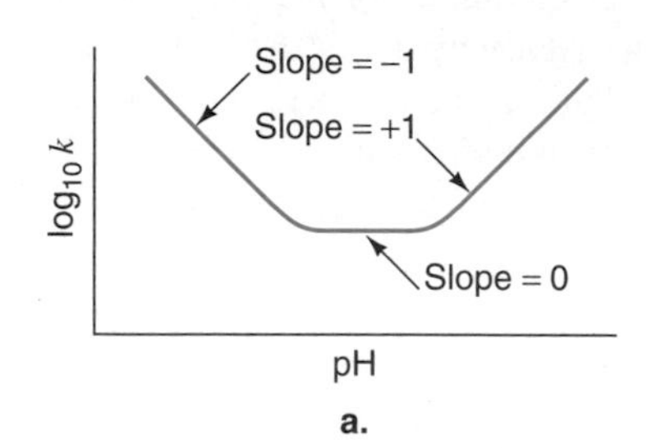

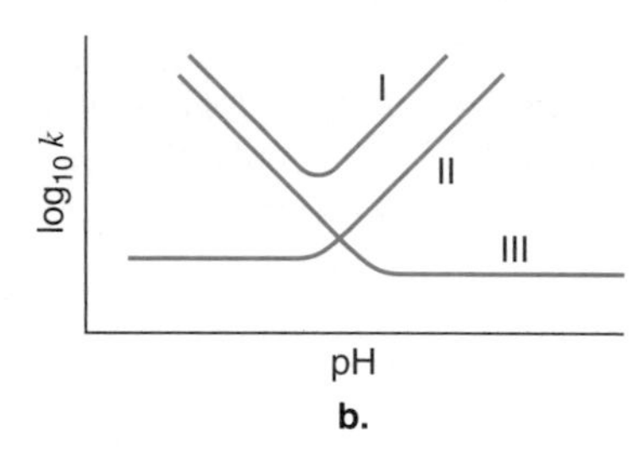

FIGURE 10.6
(a) A schematic plot of $\log_{10} k$ against pH for a reaction that is catalyzed by both hydrogen and hydroxide ions and for which the uncatalyzed reaction occurs at an appreciable rate. (b) Variants of the plot shown in (a).

The slope is therefore -1, which is the slope of the left-hand limb. The slope of the right-hand limb is similarly $+1$. The velocity in the intermediate region is equal to $k_0[S]$, and k_0 can thus be determined directly from the rate in this region. If the rate of the spontaneous reactions is sufficiently small, the horizontal part of the curve is not found, and the two limbs intersect fairly sharply (Figure 10.6b, curve I). If either k_{H^+} or k_{OH^-} is negligibly small, the corresponding sloping limb of the curve is not found (Figure 10.6b, curves II and III).

The evidence is that acid-base catalysis involves the transfer of a proton to or from the substrate molecule; it is therefore to be expected that catalysis may be effected by acids and bases other than H^+ and OH^-. This has been found in a number of instances, and **general** acid-base catalysis is then said to occur. General acid-base catalysis is to be contrasted with specific acid-base catalysis, in which one can detect catalysis only by H^+ and OH^- ions.

General Acid-Base Catalysis

An example of acid-base catalysis is the reaction between acetone and iodine in aqueous solution:

$$CH_3COCH_3 + I_2 \rightarrow CH_3COCH_2I + HI$$

The rate of this reaction is linear in the acetone concentration and in any acid species present in solution, but it is independent of the concentration of iodine; indeed, if the iodine is replaced by bromine, the corresponding bromination reaction proceeds at the same rate. This suggests that the iodine or bromine is involved in a rapid step that has no effect on the overall reaction rate. The evidence is that the rate-determining step is the conversion of the ordinary keto form of acetone into its enol form:

$$\underset{\text{keto form}}{\mathrm{CH_3\overset{\overset{\displaystyle O}{\|}}{C}CH_3}} \xrightarrow{\text{slow}} \underset{\text{enol form}}{\mathrm{CH_3\overset{\overset{\displaystyle O-H}{|}}{C}{=}CH_2}}$$

The enol form is then rapidly iodinated or brominated:

Follow in animated steps the acid and base catalyzed halogenation of ketones; Ketone halogenation.

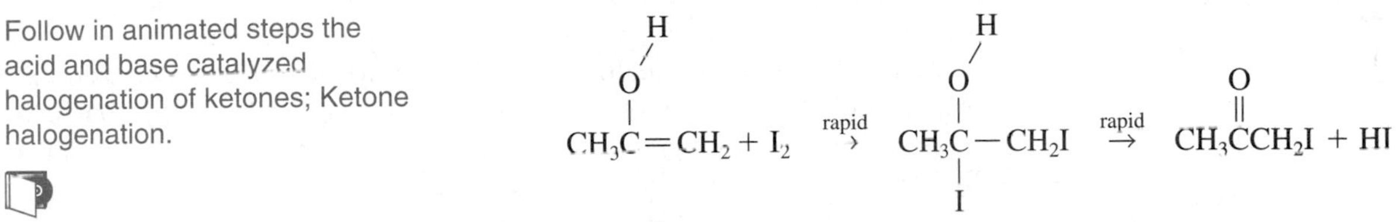

The way in which acids catalyze the conversion of the keto form into the enol form is as follows. First, the acidic species HA transfers a proton to the oxygen atom on the acetone molecule:

$$\mathrm{HA + CH_3-\overset{\overset{\displaystyle O}{\|}}{C}-CH_3 \rightarrow CH_3-\overset{\overset{\displaystyle ^{+}O-H}{\|}}{C}-CH_3 + A^-}$$

The transferred proton is bound to the oxygen atom by one of the oxygen's lone pairs of electrons. The protonated acetone then gives up one of its other hydrogen atoms to some base B present in solution (which may be water), at the same time forming the enol form of acetone:

$$\mathrm{CH_3-\overset{\overset{\displaystyle ^{+}O-H}{\|}}{C}-CH_3 + B \rightarrow CH_3-\overset{\overset{\displaystyle O-H}{|}}{C}-CH_2 + BH^+}$$

It has been demonstrated that this process is catalyzed not only by H^+ ions but by other species present in solution (i.e., there is *general* acid catalysis). There is also general basic catalysis of this reaction.

Brønsted Relationships

Since acid-base catalysis always involves the transfer of a proton from an acid catalyst or to a basic catalyst, it is natural to seek a correlation between the effectiveness of a catalyst and its strength as an acid or base; this strength is a measure of the ease with which the catalyst transfers a proton to or from a water molecule. The most satisfactory relationship between the rate constant k_a and the acid dissociation constant K_a of a monobasic acid is the equation

$$k_a = G_a K_a^{\alpha} \tag{10.71}$$

Brønsted Equation

which was proposed in 1924 by the Danish chemists J. N. Brønsted (1879–1947) and K. J. Pedersen. G_a and α are constants, the latter always being less than unity. The analogous equation for basic catalysis is

$$k_b = G_b K_b^{\beta} \tag{10.72}$$

Similarly, the relationship between the catalytic constant of a base and the acid strength of the conjugate acid may be expressed as

$$k_b = G_b'(1/K_a)^{\beta} \tag{10.73}$$

β is again always less than unity. These equations are commonly spoken of as *Brønsted relationships.*

The equations require a modification if they are to be applied to an acid that has more than one ionizable proton or to a base that can accept more than one proton. The conclusions may be generalized by means of the following relationships, given by Brønsted:

$$k_a/p = G_a(qK_a/p)^{\alpha} \tag{10.74}$$

and

$$k_b/q = G_b(p/qK_b)^{\beta} \tag{10.75}$$

Change values of variables in Michaelis-Menten equation to see the effect on Figure 10.7; Enzyme kinetics.

In Eq. 10.74, p is the number of dissociable protons bound equally strongly in the acid, whereas q is the number of equivalent positions in the conjugate base to which a proton may be attached. Similarly, in Eq. 10.75, q is the number of positions in the catalyzing base to which a proton may be attached, and p is the number of equivalent dissociable protons in the conjugate acid. Very satisfactory agreement has been obtained in all cases to which these equations have been applied.

The Brønsted relationships are special cases of linear Gibbs energy relationships, which are still often called "linear free-energy relationships" (see Problem 10.26).

Enzyme Catalysis

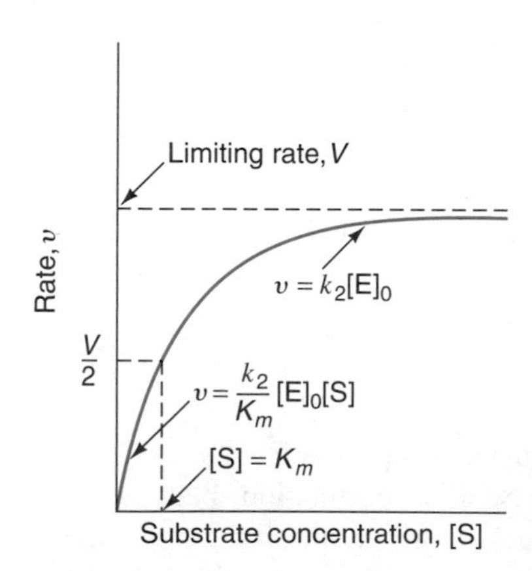

FIGURE 10.7
The variation of rate with substrate concentration for an enzyme-catalyzed reaction obeying the Michaelis-Menten equation.

Enzymes, which are proteins, are biological catalysts. Their action shows some resemblance to the catalytic action of acids and bases but is considerably more complicated. The details of the mechanisms of action of enzymes are still being worked out, and much research remains to be done. Only a brief account can be given here.

The simplest case is that of an enzyme-catalyzed reaction where there is a single substrate; an example is the hydrolysis of an ester. The dependence on substrate concentration in such cases is frequently as shown in Figure 10.7. The rate varies linearly with the substrate concentration at low concentrations (first-order kinetics) and becomes independent of substrate concentration (zero-order kinetics) at high concentrations. This type of behavior was first explained in 1913 by the German-American chemist Leonor Michaelis (1875–1949) and his Canadian colleague Maud L. Menten (1879–1960) in terms of the mechanism

$$E + S \underset{k_{-1}}{\overset{k_1}{\rightleftarrows}} ES$$

$$ES \xrightarrow{k_2} E + Z$$

Here E and S are the enzyme and substrate, Z is the product, and ES is an addition complex.

Usually the concentration of the substrate is much greater than that of the enzyme, and it is then permissible to apply the steady-state treatment in order to obtain the rate equation. The reason that this can be done is that under these conditions the concentration of the enzyme-substrate complex ES must be very much less than that of the substrate; the rate of change of its concentration is therefore much less than that of the substrate. Thus

$$\frac{d[\mathrm{ES}]}{dt} = k_1[\mathrm{E}][\mathrm{S}] - k_{-1}[\mathrm{ES}] - k_2[\mathrm{ES}] = 0 \qquad (10.76)$$

The concentration [E] that appears in this equation is the concentration of the *free* enzyme, and it may be very much less than the *total* concentration $[\mathrm{E}]_0$ of enzyme, since much of the enzyme may be in the form of ES. The total concentration is given by

$$[\mathrm{E}]_0 = [\mathrm{E}] + [\mathrm{ES}] \qquad (10.77)$$

Use of Eq. 10.76 allows [E] in this equation to be expressed in terms of [ES]:

$$[\mathrm{E}]_0 = [\mathrm{ES}][(k_{-1} + k_2)/k_1[\mathrm{S}] + 1] \qquad (10.78)$$

and therefore

$$[\mathrm{ES}] = \frac{k_1[\mathrm{E}]_0[\mathrm{S}]}{k_{-1} + k_2 + k_1[\mathrm{S}]} \qquad (10.79)$$

Vary parameters and plot the Michaelis-Menten equation as a straight line; Enzyme kinetics; Lineweaver-Burk plot.

The rate is

$$v = k_2[\mathrm{ES}] = \frac{k_1 k_2[\mathrm{E}]_0[\mathrm{S}]}{k_{-1} + k_2 + k_1[\mathrm{S}]} \qquad (10.80)$$

$$= \frac{k_2[\mathrm{E}]_0[\mathrm{S}]}{[(k_{-1} + k_2)/k_1] + [\mathrm{S}]} \qquad (10.81)$$

This equation is conveniently written as

Michaelis-Menten Equation

$$v = \frac{V[\mathrm{S}]}{K_m + [\mathrm{S}]} \qquad (10.82)$$

where $V = k_2[\mathrm{E}]_0$ and K_m, known as the **Michaelis constant,** is equal to

Michaelis Constant

$$\frac{k_{-1} + k_2}{k_1}$$

Two limiting cases of Eq. 10.82 are of interest. If $[\mathrm{S}] \gg K_m$, the rate becomes $V\ (= k_2[\mathrm{E}]_0)$, so that the kinetics are zero order in substrate. The rate V is commonly referred to as the *limiting rate.* Under these conditions the breakdown of the enzyme-substrate complex is the rate-controlling step.

If, on the other hand, $[\mathrm{S}] \ll K_m$,

$$v = \frac{V[\mathrm{S}]}{K_m} = \frac{k_2}{K_m}[\mathrm{E}]_0[\mathrm{S}] = \frac{k_1 k_2}{k_{-1} + k_2}[\mathrm{E}]_0[\mathrm{S}] \qquad (10.83)$$

and the kinetics are therefore first order in substrate. It is sometimes said that now reaction 1, the formation of the enzyme-substrate complex, is the rate-controlling step, but it is to be noted that k_1 and k_2 have equally strong effects on the overall rate, so that it is preferable to say that there is now no rate-controlling step. The second step, the breakdown of the enzyme-substrate complex, is often referred to as the *rate-limiting step,* which it is in the sense that it places a limit on the rate at a given enzyme concentration. It is important to realize, however, that this step is only the rate-controlling step in the limit of high substrate concentrations.

When [S] = K_m, Eq. 10.82 becomes

$$v = \frac{V[\mathrm{S}]}{[\mathrm{S}] + [\mathrm{S}]} = \frac{V}{2} \tag{10.84}$$

The significance of the Michaelis constant is therefore that it is the substrate concentration at which the rate is one-half the limiting rate. These relationships are illustrated in Figure 10.7.

It is important to note that adherence to the Michaelis-Menten Eq. 10.82 does not necessarily mean that the simple Michaelis-Menten mechanism applies. The mechanism

$$\mathrm{E + S} \underset{k_{-1}}{\overset{k_1}{\rightleftarrows}} \mathrm{ES} \underset{\mathrm{Y}}{\overset{k_2}{\rightarrow}} \mathrm{ES'} \overset{k_3}{\rightarrow} \mathrm{E + Z}$$

also gives the same type of rate equation (see Problem 10.42); in fact there can be any number of intermediates ES, ES′, ES″, etc., and the same dependence of v on [S] is obtained. In view of this, if the Michaelis equation applies but the mechanism is unknown, it is usual to write the equation either as Eq. 10.82 or as

$$v = \frac{k_c[\mathrm{E}]_0[\mathrm{S}]}{K_m + [\mathrm{S}]} \tag{10.85}$$

In this more general equation, k_2 in Eq. 10.81 is replaced by k_c, which is known as the *catalytic constant.*

It may well be that no enzyme-catalyzed reaction has as simple a mechanism as the one originally envisioned by Michaelis and Menten. Whenever an enzyme mechanism has been probed in detail it has invariably been found necessary to postulate modifications to the simple mechanism. It can hardly be stressed too often that one can never be certain of any proposed mechanism; the most one can say is that a mechanism is consistent with all the evidence known at a given time.

EXAMPLE 10.9 The following data apply to an enzyme-catalyzed reaction:

[S]/mol dm^{-3}	Rate, v/mol dm^{-3} s^{-1}
2.5×10^{-4}	2.3×10^{-4}
5.0×10^{-3}	7.8×10^{-4}

The concentration of the enzyme is 2 g dm^{-3} and its molecular weight is 50 000. Assume the Michaelis-Menten equation to apply and calculate the Michaelis constant K_m, the limiting rate, V, and the rate coefficient k_c.

Solution The equation to be fitted is Eq. 10.82:

$$v = \frac{V[S]}{K_m + [S]}$$

Insertion of the two sets of data gives two simultaneous equations

$$2.3 \times 10^{-4} = \frac{V \times 2.5 \times 10^{-4}}{K_m + 2.5 \times 10^{-4}} \qquad 7.8 \times 10^{-4} = \frac{V \times 5.0 \times 10^{-3}}{K_m + 5.0 \times 10^{-4}}$$

This is easily solved after cross-multiplication and leads to

$$K_m = 7.20 \times 10^{-4} \text{ mol dm}^{-3} \qquad V = 8.92 \times 10^{-4} \text{ mol dm}^{-3} \text{ s}^{-1}$$

The molar concentration of the enzyme, $[E]_0$, is

$$[E]_0 = \frac{2 \text{ g dm}^{-3}}{50\,000 \text{ g mol}^{-1}} = 4 \times 10^{-5} \text{ mol dm}^{-3}$$

Thus

$$k_c = \frac{V}{[E]_0} = 22.3 \text{ s}^{-1}$$

Many other steady-state equations have been derived for a variety of enzyme systems. Treatments have been given for enzyme reactions in which two or more substrates are involved (see Problem 10.40) and for the effects of inhibitors (Problem 10.41), pH, and temperature. It is important to remember that the steady-state hypothesis must be applied only when the concentration of enzyme is much smaller than that of the substrate; this ensures that the concentration of the enzyme-substrate complex is much smaller than that of the substrate. For some simpler systems the pre–steady state in enzyme systems has also been studied.

A remarkable feature of enzymes is that they are exceedingly effective as catalysts—much more so than other types of catalysts. Their catalytic action usually, but not always, arises from the fact that they reduce the activation energy very substantially.

EXAMPLE 10.10 Suppose that an enzyme system, involving a single substrate, proceeds by the simple Michaelis-Menten mechanism. Suppose that kinetic measurements are carried out, at a single temperature, with a particular substrate S and with a substrate S* on which an isotopic substitution has been made.

a. Suppose first that separate kinetic measurements are made with S and S*. In terms of the kinetic parameters k_c and K_m, what will be the ratio of rates at very low, and at very high, substrate concentrations?
b. Suppose instead that competitive isotope effects are studied, i.e., the two substrates S and S* are present together. What will then be the ratio of rates at the two limits?

Solution a. At low substrate concentrations the rates are given by Eq. 10.83 and a corresponding equation for S*. The ratio of rates at the same enzyme and substrate concentrations is therefore

$$k_2K_m^*/k_2^*K_m$$

At high substrate concentrations the ratio is k_2/k_2^*.
b. At first sight the answers for this case might seem to be the same, but this turns out not to be the case. The scheme is now

$$\mathrm{E + S} \underset{k_{-1}}{\overset{k_1}{\rightleftarrows}} \mathrm{ES} \xrightarrow{k_2} \mathrm{E + products}$$

$$\mathrm{E + S^*} \underset{k_{-1}^*}{\overset{k_1^*}{\rightleftarrows}} \mathrm{ES^*} \xrightarrow{k_2^*} \mathrm{E + products}$$

The total enzyme concentration is

$$[\mathrm{E}]_0 = [\mathrm{E}] + [\mathrm{ES}] + [\mathrm{ES^*}]$$

and the steady-state equations are

$$k_1[\mathrm{E}][\mathrm{S}] - (k_{-1} + k_2)[\mathrm{ES}] = 0$$
$$k_1^*[\mathrm{E}][\mathrm{S^*}] - (k_{-1}^* + k_2^*)[\mathrm{ES^*}] = 0$$

Then

$$[\mathrm{E}]_0 = [\mathrm{ES}](K_m/[\mathrm{S}] + 1 + [\mathrm{S}]\, K_m/[\mathrm{S^*}]K_m^*)$$

where $K_m = (k_{-1} + k_2)/k_1$ and $K_m^* = (k_{-1}^* + k_2^*)/k_1^*$.

The rate of formation of products from S is therefore

$$v = k_2[\mathrm{ES}] = \frac{k_2[\mathrm{E}]_0}{K_m/[\mathrm{S}] + 1 + [\mathrm{S^*}]\, K_m/[\mathrm{S}]K_m^*}$$
$$= \frac{k_2[\mathrm{E}]_0[\mathrm{S}]}{K_m(1 + [\mathrm{S}]\,/K_m + [\mathrm{S^*}]\,/K_m^*)}$$

A similar treatment of the rate of formation of the products from S* leads to

$$v^* = k_2^*[\mathrm{S}] = \frac{k_2^*[\mathrm{E}]_0[\mathrm{S^*}]}{K_m^*(1 + [\mathrm{S}]\,/K_m + [\mathrm{S^*}]\,/K_m^*)}$$

It follows that the ratio of rates at equal concentrations of S and S* is

$$v/v^* = \frac{k_2/K_m}{k_2^*/K_m^*} = k_2K_m^*/k_2^*K_m$$

Note that this is true at all substrate concentrations. This result differs from that obtained if separate experiments are made with the two isotopic forms; in that case the ratio was k_2/k_2^* at high substrate concentrations. Although the breakdown of the enzyme-substrate complex is the rate-controlling step at high substrate concentrations, competitive isotopic studies cannot reveal anything of this step.

This example illustrates how careful one must be in drawing conclusions about mechanisms from kinetic investigations. It is important not to rely on intuition, but to work out the rate equations explicitly.

10.10 Reactions in Solution: Some Special Features

Most of the general principles we have discussed in this chapter apply equally well to gas reactions and to solution reactions. In Section 9.10 we considered some aspects of solvent effects, and in this section we consider some additional matters that relate in particular to composite mechanisms.

One important difference between gas-phase and solution reactions is that solvents frequently favor mechanisms involving ions. We have seen that ordinary gas-phase reactions rarely involve ionic intermediates and usually occur by mechanisms in which atoms and free radicals are intermediates. Exceptions are reactions induced by high-energy radiation (Section 10.7), where the energy required to produce ions is provided by the radiation. For reactions in solution, on the other hand, ionic intermediates are much more common, since the ions may be stabilized by the solvent.

Collisions and Encounters

In Section 9.10 we noted that for a number of reactions the rates and the activation parameters (activation energy and preexponential factor) are sometimes the same in the gas phase and in a variety of solvents. This tends to be true if ionic effects are not important. The similarity between preexponential factors for gas-phase and solution reactions suggests that the collision frequencies are much the same.

Further investigation of this problem reveals that there are important differences between collisions in solution and collisions in the gas phase and that these differences have implications for certain composite mechanisms. In a gas the molecules are relatively far apart and move freely between collisions; after two molecules A and B collide, they separate and may not meet again for some time. In solution, on the other hand, when two solute molecules come together, they are "caged" in by surrounding solvent molecules, which hold the solute molecules together and cause them to collide a number of times before they finally separate. This effect was demonstrated in a very interesting way in 1936 by Eugene Rabinowitch and W. C. Wood, who employed a tray on which spheres were allowed to roll. Agitation of the tray caused the spheres to move around, and an electrical device was employed to record collisions between a given pair of spheres. When very few spheres were present on the tray, the collisions between a given pair occurred individually, with relatively long intervals between successive collisions (see Figure 10.8, curve I). This represents the behavior in the gas phase. The effect of solvents was studied by adding additional spheres to the tray, so that there was fairly close packing. It was then found that the distribution of collisions between a given pair was quite different. After one collision occurred, the pair tended to stay together and undergo a succession of collisions (as shown in curve II of Figure 10.8). A group of such collisions is known as an **encounter.** The effect of solvents on caging of the solute molecules is known as the **cage effect** or as the **Franck-Rabinowitch effect,** having been investigated in 1934 by James Franck (1882–1964) and Eugene Rabinowitch. Franck did important work on photochemistry and related problems and won the 1925 Nobel Prize in physics.

Cage Effect

The important result obtained in the demonstration of Rabinowitch and Wood is that although the *distribution* of collisions is quite different in the two cases, the average *frequency* of collisions is much the same. When the tray was packed with

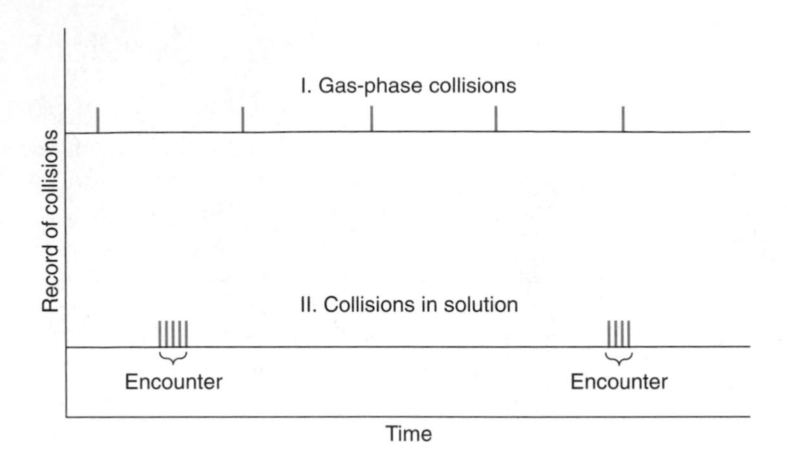

FIGURE 10.8
The distribution of collisions between solute molecules, as shown by the demonstration of Rabinowitch and Wood. Curve I shows schematically the gas-phase collisions; curve II shows the distribution of collisions when a solvent is present.

spheres, so as to represent collisions in solution, the collisions occurred in groups within an encounter, but the time between successive encounters was correspondingly long, so that the frequency of collisions was very similar.

For ordinary chemical reactions, involving an activation energy, the distribution of collisions is irrelevant; reaction may occur at any collision within the encounter, so that the rate is controlled by the frequency of collisions and not by their distribution.

On the other hand, reactions involved in composite mechanisms frequently have zero activation energies, and then the distribution of collisions becomes important. Free-radical combinations, for example, do not require an activation energy, and the process will therefore occur at the first collision within the encounter. The preexponential factor of such reactions therefore depends on the frequency of *encounters* and not on the frequency of collisions.

A related example is provided by certain photochemical reactions, where a pair of radicals may be produced initially. In the gas phase they will separate rapidly, but in solution they will be caged in by the surrounding solvent molecules and may combine with one another before they can separate. This effect is known as *primary combination,* as opposed to *secondary combination,* which occurs after the free radicals have separated from each other.

10.11 Mechanisms of Macromolecule Formation

Introduction

Staudinger won the 1953 Nobel Prize in chemistry for his work on macromolecules. He played an important role in overthrowing the view that substances like rubber, cellulose, and proteins are aggregates of smaller molecules.

The word **macromolecule** (Greek *macros,* long, large) was introduced by the German chemist Herman Staudinger (1881–1965) to refer to a substance having a molecular weight of more than 10 000. Macromolecular substances in solution show characteristic behavior with respect to properties such as viscosity, diffusion, the colligative properties, and sedimentation. Later, in Chapter 18, we will consider the properties of colloids and in so doing cover some important aspects of macromolecules. Not all colloids are macromolecules; those that are not are called *association colloids,* while the term *molecular colloids* was introduced by Staudinger to refer to colloidal particles that consist of single macromolecules.

Also used in this connection is the word **polymer** (Greek *poly,* much, many; *meros,* part). It was coined in 1830 by the Swedish chemist Jöns Jacob Berzelius (1779–1848) to refer to a molecule of general formula M_n, made up by the repetition of n identical units M, the latter being known as the **monomer** (Greek *monos,* alone). However, over the years the meaning of the word *polymer* has become extended in various directions. It is no longer necessary for the polymer to consist of exactly an integral number of monomer units. Many macromolecular substances are formed by condensation reactions between monomer molecules, with the elimination of water molecules or other small molecules. The products of such reactions are now called *step-growth polymers* even though, because of the splitting off of the small molecules, the formula of the macromolecule is not exactly M_n. Furthermore, it is no longer necessary for the polymer to be formed from *identical* monomer molecules. The word **copolymer** is used to describe a molecule composed of two different units; if it is formed by addition processes (i.e., is an *addition polymer*), its formula will be M_nN_m where M and N are the monomers and n and m are integers. The term **homopolymer** (Greek *homos,* one and the same, jointly) can be used to describe polymers composed of only one type of monomer.

Nowadays the word *polymer* is even applied to substances that are made up of a considerable variety of units. For example, protein molecules are formed by condensation reactions involving over 20 different amino acids, and they are often referred to as polymers. In some ways this extension of the meaning is unfortunate, since now there is essentially no difference between polymers and macromolecular substances, any of which can be called polymers in the broadest sense.

The technical applications of polymeric materials are well known, and need only a brief mention here. Some idea of the extent of their use is provided by the fact that in 1997 the annual production of synthetic polymers in the United States amounted to over 40 billion kilograms, with about a million workers involved in their manufacture. The volume of polymers now produced exceeds that of all metals. Furthermore, polymers can now be produced which have properties that cannot be matched by any other materials, and many technical developments have therefore depended on them.

Polymers can be classified in various ways, and a convenient way is with reference to the manner of their preparation. One class comprises the **addition polymers,** which are usually prepared by reactions involving the participation of free radicals. There are also **step-growth polymers,** in which the chains are built up step by step, usually by **condensation** reactions that occur by molecular mechanisms in which a small molecule, such as water, is split off at each step. In either case catalysts may be involved, and the polymerizations may also be brought about by radiation.

Addition Polymers

Olefinic substances, such as ethylene and styrene, usually form addition polymers. The initiation of free-radical polymerizations, leading to such polymers, can occur in various ways. For example, atoms or free radicals may be introduced to an unsaturated monomer, and they will add on to the double bond:

$$R- + CH_2{=}CHR' \rightarrow R-CH_2-CHR'-$$

This reaction has produced another radical, which in turn can add on to another molecule of monomer:

$$\mathrm{R-CH_2CHR'-} + \mathrm{CH_2{=}CHR'} \rightarrow \mathrm{-R-CH_2CHR'-CH_2-CHR'-}$$

This process can continue forming larger and larger radicals. Finally, two large radicals can combine to form a polymer molecule. The radical R that has initiated the polymerization process can be introduced into the monomer system in a number of ways. It can be produced thermally, be formed by the action of a catalyst, or be generated by photochemical or radiation-chemical processes as discussed in Sections 10.6 and 10.7. The kinetics of free-radical polymerization are briefly discussed later in Section 10.12.

Step-Growth Polymerization

An example of a step-growth polymerization occurring by a *condensation* mechanism is the reaction between ethylene glycol and succinic acid:

$$\underset{\text{ethylene glycol}}{\mathrm{HO(CH_2)_2OH}} + \underset{\text{succinic acid}}{\mathrm{HOOC(CH_2)_2COOH}} \rightarrow \mathrm{HO(CH_2)_2OCO(CH_2)_2COOH} + \mathrm{H_2O}$$

Since this product has two functional end groups, —OH and —COOH, it can react with two more monomer molecules, yielding a product that also has two functional end groups. The process can therefore continue indefinitely with the formation of a large copolymer. It can be brought about by the usual catalysts (e.g., acids and bases) for esterification reactions. Another example of a polycondensation reaction, first brought about in 1934 by the American chemist Wallace Hume Carothers (1896–1937), is the synthesis of a form of *nylon,* which is a polyamide:

$$\begin{array}{c} \mathrm{COOH} \\ | \\ \mathrm{(CH_2)_4} \\ | \\ \mathrm{COOH} \\ \text{adipic acid} \end{array} + \begin{array}{c} \mathrm{NH_2} \\ | \\ \mathrm{(CH_2)_6} \\ | \\ \mathrm{NH_2} \\ \text{hexamethylene diamine} \end{array} \rightarrow \begin{array}{c} \mathrm{COOH} \\ | \\ \mathrm{(CH_2)_4} \\ | \\ \mathrm{CO} \\ | \\ \mathrm{NH} \\ | \\ \mathrm{(CH_2)_6} \\ | \\ \mathrm{NH_2} \end{array} + \mathrm{H_2O}$$

The product has two functional end groups, —COOH and —NH_2, and the condensation reactions can continue and yield large molecules. The molecular weights of nylon molecules are typically about 15 000.

Ionic Polymerizations

Polymerization processes in solution often occur by mechanisms in which the intermediates are *ions.* Such processes are catalyzed by acidic or basic substances, and their rates vary with the dielectric constant of the solvent in the manner expected of ionic processes. Polymerizations of this type can be classified as *cationic* or *anionic,* according to whether the processes are brought about by cationic or anionic species. An example of a **cationic** process is the polymerization of a substance such as isobutene.

Cationic Process

Such polymerizations are catalyzed by a variety of Lewis acids, such as HCl, H_2SO_4, $AlCl_3$, I_2, and $AgClO_4$. All these substances, which we will write in general as MX, are electron acceptors, and in the presence of a solvent SH they form an ion pair:

$$MX + SH \rightleftharpoons H^+[SMX]^-$$

This ion pair can add on to a monomer molecule to produce another ion pair, as follows:

$$H^+[SMX]^- + CH_2{=}CHR \rightarrow CH_3{-}CHR^+[SMX]^-$$

The product of this reaction, involving a carbocation, can add on to another olefin molecule:

$$CH_3{-}CHR^+[SMX]^- + CH_2{=}CHR \rightarrow CH_3{-}CHR{-}CH_2{-}CHR^+[SMX]^-$$

The process can continue until finally a large double ion splits off $H^+[SMH]^-$ with the formation of a stable polymer molecule:

$$\begin{aligned}&CH_3{-}[CHR{-}CH_2]_n{-}CHR{-}CH_2{-}CHR^+[SMX]^- \rightarrow\\ &\quad CH_3{-}[CHR{-}CH_2]_n{-}CHR{-}CH{=}CHR + H^+[SMX]^-\end{aligned}$$

Anionic Polymerization

A well-known example of **anionic polymerization,** and in fact the first one to be achieved (by the British chemists W. C. E. Higginson and N. S. Wooding in 1952), is the polymerization of styrene initiated by potassium amide. It can be performed at 240 K in liquid ammonia, which is highly polar. Initiation involves the dissociation of the potassium amide into its ions K^+ and NH_2^- followed by addition of the anion to the monomer to produce an active chain carrier:

$$\begin{aligned}&KNH_2 \rightleftharpoons K^+ + NH_2^-\\ &NH_2^- + CH_2{=}CH\phi \rightarrow [H_2NCH_2{-}CH\phi]^-\end{aligned}$$

Here ϕ has been written for the phenyl group C_6H_5. Propagation is the addition of this carbanion to another monomer molecule:

$$[H_2N{-}CH_2{-}CH\phi]^- + CH_2{=}CH\phi \rightarrow [H_2N{-}CH_2{-}CH\phi{-}CH_2{-}CH\phi]^-$$

This process continues with the formation of larger polymer anions, and one of these may abstract a proton from the solvent, with regeneration of the amide ion:

$$\begin{aligned}&[H_2N{-}(CH_2{-}CH\phi)_n{-}CH_2{-}CH\phi]^- + NH_3 \rightarrow\\ &\quad H_2N{-}(CH_2{-}CH\phi)_n{-}CH_2{-}CH_2\phi + NH_2^-\end{aligned}$$

"Living" Polymers

An interesting feature of this reaction scheme for anionic polymerization is that there is no termination step; the reaction just written has produced an amide ion capable of initiating another chain. Thus, if one removes from the system any impurities which may react with carbanions, propagation will continue until all of the monomer has been consumed, leaving carbanions and amide ions intact and capable of undergoing further reactions. It follows that after a polymerization of this kind has gone to completion, addition of more monomer, even after a period of time, will bring about further polymerization.

Because of this unusual characteristic, systems that had undergone anionic polymerization were called "living" polymers by the British chemist Michael

Szwarc. By the use of apparatus carefully designed to exclude impurities, Szwarc was able to demonstrate the existence of these "living" polymers.

Heterogeneous Polymerization

Elastomer

Heterogeneous catalysts have proved to be very effective for bringing about polymerization. The pioneers in this field were the German chemist Karl Ziegler (1898–1973) and the Italian chemist Giulio Natta (1903–1979). They found that mixtures of $Al(C_2H_5)_3$ and $TiCl_4$ were very effective catalysts for the synthesis of polymers with specified stereoisomerism. They were successful, for example, in synthesizing a polymer that is identical to natural rubber. This natural material, which has elastic properties and is known as an **elastomer,** was studied by Michael Faraday in 1826 and is now known to be poly-*cis*-isoprene:

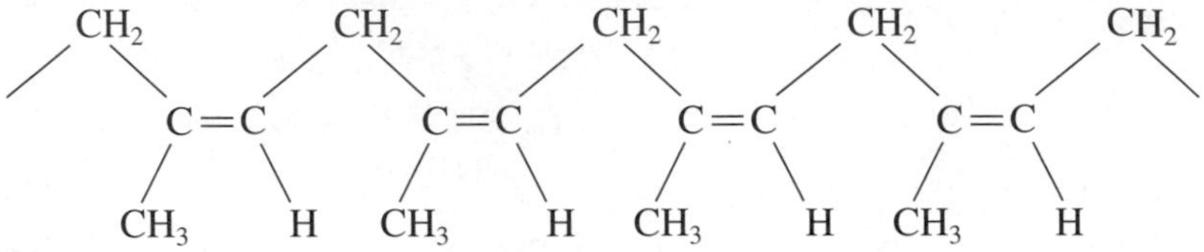

```
    CH2           CH2           CH2           CH2           CH2
   /   \         /   \         /   \         /   \         /   \
        C=C           C=C           C=C           C=C
       /   \         /   \         /   \         /   \
    CH3     H     CH3     H     CH3     H     CH3     H
```

Another naturally occurring polymer is gutta-percha, which is poly-*trans*-isoprene,

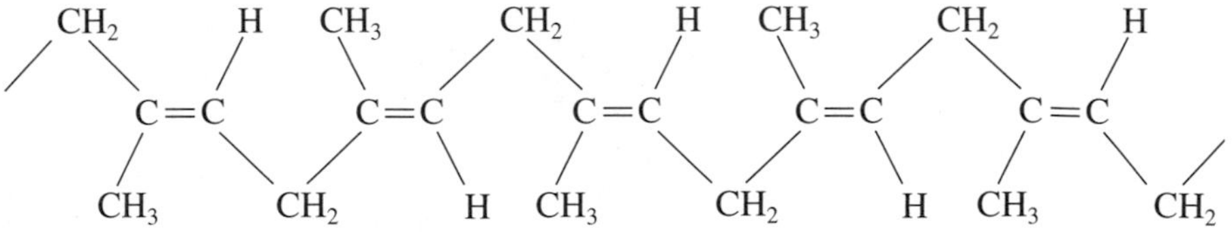

```
  CH2        H   CH3        CH2       H   CH3        CH2        H
 /   \      /       \      /   \     /       \      /   \      /
      C=C            C=C        C=C           C=C         C=C
     /   \          /   \      /   \         /   \       /   \   /
  CH3     CH2           H   CH3     CH2          H   CH3      CH2
```

Plastic

Ziegler-Natta Catalysts

Gutta-percha is **plastic** but does not have the elastic properties that rubber has. Until 1955 all processes used to polymerize isoprene gave a mixture of poly-*cis*-isoprene and poly-*trans*-isoprene, mainly the latter, and this mixture is not a suitable artificial rubber. Use of the **Ziegler-Natta catalysts,** however, led in that year to the synthesis of the separate stereoisomeric forms and therefore to important developments in artificial rubber technology. Ziegler and Natta shared the 1963 Nobel Prize for chemistry for these contributions.

Emulsion Polymerization

It is also possible, and sometimes very convenient, to bring about polymerizations in aqueous emulsions. A typical procedure that has been employed with monomers like methyl methacrylate

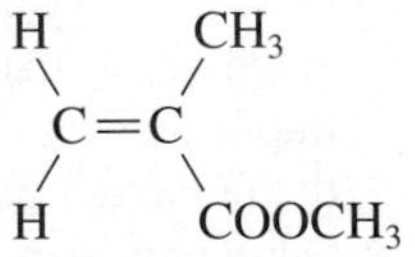

```
H      CH3
 \    /
  C=C
 /    \
H      COOCH3
```

is first to form an aqueous emulsion with the monomer, by the use of an emulsifying agent such as cetyltrimethylammonium bromide. Addition of Fenton's reagent (a solution containing Fe^{2+} ions and hydrogen peroxide) will then bring about polymerization; this reagent generates free radicals, and the polymerization occurs by a free-radical mechanism. An advantage of emulsion polymerization is that the processes proceed much more rapidly than in bulk systems.

As will be noted in Section 18.10, long-chain hydrocarbon residues having polar ends form *micelles*. When monomers such as methyl methacrylate are added to a micelle, some of it becomes *solubilized,* which means that monomer penetrates into the micelles; the remainder stays in the aqueous phase. The free radicals generated by the initiating medium (e.g., Fenton's reagent) are able to diffuse into the micelles, where they initiate polymerization. At the same time more monomer molecules enter the micelles, which increase in size and soon consist of polymer particles with the emulsifying agent adsorbed on the surface. The process goes on, and the micelle particles continue to grow until almost all the monomer has been polymerized. This theory of emulsion polymerization was first put forward in 1947 by the American physical chemist William Draper Harkins (1873–1951).

10.12 Kinetics of Polymerization

Only a few types of polymerization will be considered here, to indicate some of the kinetic features of polymerization processes.

Free-Radical Polymerization

The general reaction scheme for a free-radical polymerization can be represented as follows:

$$\begin{array}{lll} ? \rightarrow R_1 & & \text{initiation} \\ R_1 + M \rightarrow R_2 & \multirow{4}{*}{} & \\ R_2 + M \rightarrow R_3 & & \text{chain propagation} \\ \text{- - - - - - - - - -} & & \\ R_{n-1} + M \rightarrow R_n & & \\ R_n + R_m \rightarrow M_{n+m} & & \text{termination} \end{array}$$

The initiation reaction is a process in which a radical R_1 is produced; for the time being we will leave the nature of this reaction unspecified and will write the rate of formation of R_1 in this process as v_i. For simplicity we shall assume the rate constants of all the chain-propagating steps to be the same, k_p. Termination can involve any two radicals, including identical ones, and the rate constant for termination will be written as k_t.

The steady-state equation for R_1 is

$$v_i - k_p[R_1][M] - k_t[R_1]([R_1] + [R_2] + \cdots) = 0 \tag{10.86}$$

The final term is for the removal of R_1 by reaction with R_1, R_2, etc. The equation can be written as

$$v_i - k_p[R_1][M] - k_t[R_1]\sum_{n=1}^{\infty}[R_n] = 0 \tag{10.87}$$

Similarly for R_2,

$$k_p[R_1][M] - k_p[R_2][M] - k_t[R_2]\sum_{n=1}^{\infty}[R_n] = 0 \tag{10.88}$$

In general, for R_n,

$$k_p[R_{n-1}][M] - k_p[R_n][M] - k_t[R_n]\sum_{n=1}^{\infty}[R_n] = 0 \tag{10.89}$$

There is an infinite number of such equations, and the sum of all of them is

$$v_i - k_t\left(\sum_{n=1}^{\infty}[R_n]\right)^2 = 0 \tag{10.90}$$

This equation simply states that the rate of initiation is equal to the sum of the rates of all the termination processes, which must be true in the steady state.

The rate of disappearance of monomer is the sum of the rates of all the propagation reactions,

$$-\frac{d[M]}{dt} = k_p[M]\sum_{n=1}^{\infty}[R_n] \tag{10.91}$$

According to Eq. 10.90 the summation is given by

$$\sum_{n=1}^{\infty}[R_n] = \left(\frac{v_i}{k_t}\right)^{1/2} \tag{10.92}$$

and the rate of disappearance of monomer is therefore

$$-\frac{d[M]}{dt} = k_p\left(\frac{v_i}{k_t}\right)^{1/2}[M] \tag{10.93}$$

There are various special cases, according to the nature of the initiation process; we shall consider only a few of them. In a purely thermal initiation, in which the radicals are produced by heating the monomer, the initial radical formation may be second order in monomer:

$$v_i = k_i[M]^2 \tag{10.94}$$

The rate of polymerization is then

$$-\frac{d[M]}{dt} = k_p\left(\frac{k_i}{k_t}\right)^{1/2}[M]^2 \tag{10.95}$$

so that the process is second order. Alternatively, initiation might involve a second-order reaction between the monomer and a catalyst C,

$$v_i = k_i[M][C] \tag{10.96}$$

The polymerization rate is then

$$-\frac{d[M]}{dt} = k_p\left(\frac{k_i}{k_t}\right)^{1/2}[M]^{3/2}[C]^{1/2} \tag{10.97}$$

In a photochemical initiation the rate of initiation may be simply the intensity I of light absorbed, usually expressed in einsteins (mole quanta) per second (see Section 10.6),

$$v_i = I \tag{10.98}$$

The rate of polymerization is then

$$-\frac{d[\mathrm{M}]}{dt} = k_p\left(\frac{I}{k_i}\right)^{1/2}[\mathrm{M}] \tag{10.99}$$

Examples are known of all of these kinetic equations and of others we have not considered.

Rotating-Sector Technique

A technique that has frequently been used with polymerization processes is the **rotating-sector technique,** introduced in 1924 by D. L. Chapman in connection with the hydrogen-chlorine reaction (Section 10.6). In this technique, a rotating slotted disk is used in order to produce alternating periods of light and dark. The free radicals formed in a reaction are generated in the light and decay in the periods of darkness. By varying the speed of rotation of the disk, in this way varying the periods of light and darkness, and observing how the rate of reaction varies, it is possible to deduce the lifetime of the radicals.

More recently, slotted disks have been replaced by **pulsed lasers,** particularly by K. F. O'Driscoll and his collaborators. In 1989, for example, they studied the polymerization of styrene and other monomers by using laser flashes of a few nanoseconds in duration and determining the molecular weight distribution by gel permeation chromatography. This procedure allowed them to measure rate constants for the chain propagation processes.

Condensation Polymerization

The kinetic equations for condensation polymerization are most easily formulated in terms of the concentrations of the functional group (—OH, —COOH, etc.) that are undergoing reaction. Suppose that the initial concentration of these groups is c_0, and at time t the concentration is c. We shall assume that all the rate constants are the same, and therefore

$$-\frac{dc}{dt} = kc^2 \tag{10.100}$$

With the boundary condition $t = 0$, $c = c_0$, this integrates to

$$\frac{c_0 - c}{cc_0} = kt \tag{10.101}$$

(compare Eq. 9.35). The fraction f of functional groups that have disappeared at time t is

$$f = \frac{c_0 - c}{c_0} \tag{10.102}$$

Elimination of c between these equations gives

$$\frac{f}{1-f} = c_0kt \tag{10.103}$$

or

$$\frac{1}{1-f} = 1 + c_0kt \tag{10.104}$$

Plots of $f/(1-f)$ or $1/(1-f)$ should therefore be linear, and this has been verified for a number of reactions.

10.13 Induction Periods, Oscillations, and Chaos

In many chemical reactions the kinetic behavior has some special features not previously mentioned in this chapter. We saw in Section 10.3 that reactions occurring in more than one step have an induction period in which little reaction occurs. In many reactions the induction periods are so short that they easily escape detection. In some reactions, on the other hand, the induction periods are long, and the reaction seems to go in a flash after a certain time has elapsed. Such reactions are called *clock* reactions. A classic example of a clock reaction was discovered by Landolt in 1886. In this reaction an initially colorless mixture of aqueous solutions appears unchanged for a minute or so, and then a deep blue color suddenly appears. This reaction is essentially complete the instant the color appears. The mechanism of the reaction is considered later.

In some reactions, the end of the induction period does not quickly lead to a completed reaction, but oscillations occur. For example, combustion reactions often occur by a succession of small explosions. In the late 1950s the Russian chemists Boris Belousov and Anatol Zhabotinsky studied a reaction, to be discussed in detail later, involving a redox system such as Ce(III)/Ce(IV). They found that after an induction period the concentrations of intermediates, and therefore the color of the solution, oscillated in a spectacular manner.

For some chemical reactions, the final outcome is impossible to predict, and the behavior is then said to be *chaotic*. Over the past few decades a great deal of experimental and mathematical investigation has shown that unpredictable behavior is much more common than had been supposed, and that it has a perfectly rational explanation. The mathematical work has led to what is generally but rather regrettably called *chaos theory* and sometimes *catastrophe theory*. The words *chaos* and *catastrophe* are rather unfortunate in this context, since they overstate the case. By the word *chaos,* for example, we understand "utter confusion," which is the usual dictionary definition. The modern scientific meaning of *chaos* does not, however, imply utter confusion; it simply means that the final outcome is unpredictable. Before going more into this, we will give mechanisms for clock and oscillating reactions and discuss their main features.

Clock Reactions

Clock reactions show long induction periods in which nothing appears to happen, after which the reaction suddenly and rapidly occurs. The classic example discovered by Landolt involves the autocatalytic iodate-bisulfite system. The overall reaction is the reduction of the iodate ions to iodine by the bisulfite ions, which are converted into bisulfate ions:

$$2IO_3^- + 5HSO_3^- + 2H^+ \rightarrow I_2 + 5HSO_4^- + H_2O$$

If a solution at room temperature is initially 0.01 M in both iodate and bisulfite, with starch present as an indicator for iodine, there is an induction period of about 40 seconds after which the solution suddenly becomes deep blue.

The reaction is initiated by the formation of iodide ions in the process

$$IO_3^- + 3HSO_3^- \rightarrow I^- + 3HSO_4^-$$

The iodide ions produced in this reaction then react with iodate ions to form free iodine:

$$5I^- + IO_3^- + 6H^+ \rightarrow 3I_2 + 3H_2O$$

However, before the iodine can interact with the starch indicator to produce the blue color, it is reduced back to iodide by the rapid reaction

$$I_2 + HSO_3^- + H_2O \rightarrow 2I^- + HSO_4^- + 2H^+$$

Since iodide ions are formed in this process, these ions act as autocatalysts. These processes continue until the reducing bisulfite is all used up; the concentration of iodide ions and hence of iodine then rapidly rises, with the sudden formation of the blue color.

Oscillating Reactions

For oscillations to occur there must be a special mechanism by which the "clock is rewound" a number of times. Many mathematical treatments of systems leading to oscillations have been carried out and have been supplemented by extensive numerical calculations on computers. In these, various hypothetical mechanisms have been formulated and calculations made with a range of starting conditions. The general conclusion from both mathematical analysis and computer calculations is that in any system, chemical or otherwise, oscillations can result only if at least two variables are involved. This condition is of course satisfied by the vast majority of chemical reactions; there are usually two or more reacting substances so that there are two variables aside from environmental factors such as the temperature. Moreover, the differential rate equations for at least one of the processes occurring must be nonlinear, which in the case of chemical reactions means that the rate must be proportional to other than the first power of the amount of a substance (i.e., the reaction must not be a first-order reaction). Again, this condition is usually satisfied in chemical systems. Finally, a necessary condition is that the processes occurring must interact with each other in such a way that there is *positive feedback.* What is meant by feedback is that the product of one of the processes that occurs must affect another process. If it increases its rate, there is positive feedback; if it inhibits the other reaction, there is negative feedback. There is nearly always both positive and negative feedback in any chemical system undergoing oscillations.

Positive Feedback

A familiar example may help our understanding of oscillating systems. Suppose that a lawn becomes infested with beetles; it starts to deteriorate, and the beetle colony grows exponentially. Eventually the lawn becomes so bad that beetles die of starvation, and only a few remain. The lawn then starts to revive and becomes more presentable; the few beetles still alive begin to replicate, and the cycle starts over again. It is easy to see that the quality of the lawn goes through maxima and minima, the population of beetles doing the same. This type of analogy is often referred to as the *predator-prey* analogy. We see that the basic conditions are satisfied; there are two variables, the grass and the beetles, and there is positive feedback, in that the amount of grass favorably affects the growth of the beetle colony. There is also negative feedback in that the number of beetles adversely affects the growth of the grass. Positive feedback is essential for oscillations, but negative feedback is usually present in addition.

The simplest type of chemical system that leads to oscillations is one involving a rapid initial chemical process that is easily reversible and in which heat is evolved:

$$A \rightleftarrows X + \text{heat}$$

Suppose that this is followed by another process in which heat is evolved:

$$X \rightarrow \text{products} + \text{heat}$$

The heat evolved in the second reaction will cause a rise in temperature and therefore, by Le Chatelier's principle (Section 4.6), a decrease in the concentration of X (negative feedback), so that the second reaction, which leads to products, slows down. The system will then cool, resulting in an increase in the concentration of X and an increase in rate of product formation. The process can be repeated many times, and the concentration of X will therefore oscillate. Cool flames involve this general principle, although the details are somewhat more complicated.

Cool Flames

In Section 10.8 we discussed cool flames, commonly found with hydrocarbon oxidations. Aside from their other odd features, cool flames often show oscillations. For example, if propane and air are brought together under suitable conditions there may be a succession of individual ignitions, lasting for only short periods and separated perhaps by several seconds. Many spectacular demonstrations exhibit this and similar effects.

The system is a somewhat more complex example of the general mechanism just discussed, and the explanation in brief is as follows. We represent the hydrocarbon by RCH_3 (for propane, for example, $R = CH_3CH_2-$) The first step in the oxidation is the extraction of a hydrogen atom from the hydrocarbon by an oxygen molecule:

1. $$RCH_3 + O_2 \rightarrow RCH_2 + HO_2$$

The radical HO_2 (H—O—O—) plays an important role in hydrocarbon oxidations. The radical RCH_2 rapidly adds on to an oxygen molecule with the formation of a peroxide radical, which can break down into an aldehyde molecule and a hydroxyl radical:

2. $$RCH_2 + O_2 \rightleftarrows RCH_2{-}O{-}O{-} \rightarrow RCHO + OH$$

This is a non-branching reaction, only one radical having been produced, but the aldehyde can react with O_2 to give two radicals:

3. $$RCHO + O_2 \rightarrow RCO + HO_2$$

This is a branching reaction. As we discussed earlier (Section 10.8), the aldehyde can also undergo another reaction

4. $$RCHO + OH \rightarrow H_2O + RCO$$

which is non-branching. It is because of the competition between the two alternative reactions of RCHO that the explosion is a mild one, a degenerate explosion that leads to a cool flame.

The reason that this mechanism leads to oscillations is that reaction 2 quickly approaches a state of equilibrium and is exothermic:

$$RCH_2 + O_2 \rightleftarrows RCH_2{-}O{-}O{-} + \text{heat}$$

Since this process is exothermic, it is shifted to the left by a rise in temperature. After the first explosion occurs the rise in temperature leads to a reduction of the concentration of the peroxide radical (negative feedback); the rate of the combustion depends on this concentration and it therefore decreases, the explosion fading away. Then, as the temperature falls, the equilibrium shifts to the right, the reaction speeds up, and soon another mild explosion occurs. It is the shift back and forth of this equilibrium that provides the feedback required for oscillations to occur. Every explosion causes a rise in temperature, which leads to a reduction in the rate of reaction and a premature end to the explosion.

The Belousov-Zhabotinsky Reaction

The best known of the oscillating reactions in which temperature changes play no significant role is a reaction first reported in 1959 by Boris Belousov, involving the cerium-catalyzed bromination of malonic acid by bromate ions in sulfuric acid solution. The effect is obtained with a stirred reaction mixture initially containing malonic acid, bromate ions, Ce^{3+} ions, and sulfuric acid. As illustrated in Figure 10.9, the ratio $[Ce^{3+}]/[Ce^{4+}]$ and the concentration of bromide ions oscillate after an induction period; the color of the solution oscillates between red and blue. In 1967, Anatol Zhabotinsky showed that a reaction mixture that is left unstirred exhibits spatial oscillations in the concentrations of intermediates. If, for example, the reaction is carried out in a shallow dish and the initial red solution is touched with a hot needle, a blue ring propagates outwards, and eventually complex patterns evolve.

For several years this rather striking result was either ignored or treated with disbelief. Some chemists insisted that it was impossible for such oscillations to occur, arguing that the second law of thermodynamics would be violated. That objection, however, completely missed the point: it is the concentrations of intermediates that oscillate, not the concentrations of reactants and final products. It is true that a reaction can go only in the direction of increasing total entropy (of the system plus the environment) and therefore cannot reverse itself. There is, however, no thermodynamic objection to an oscillation of the concentrations of intermediates.

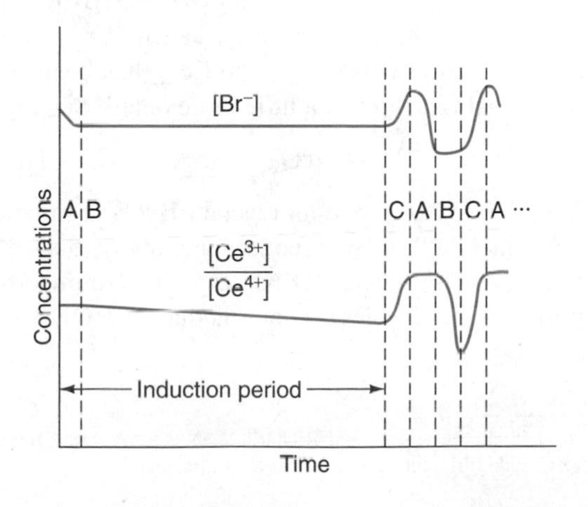

FIGURE 10.9
Schematic diagram showing oscillations in the Belousov-Zhabotinsky reaction. The characteristics of the three reaction sequences are approximately as follows, where kinetically important intermediates are underlined.

Sequence A: $[Br^-]$ declines; $[Ce^{3+}]/[Ce^{4+}]$ remains constant. This sequence begins when $[Br^-]$ is high enough and becomes unimportant when Br^- is depleted.

Sequence B: $[Br^-]$ remains constant; $[Ce^{3+}]/[Ce^{4+}]$ declines. This sequence begins when $[Br^-]$ is low and ends when $\underline{HBrO_2}$ and $\underline{Ce^{3+}}$ are depleted.

Sequence C: $[Br^-]$ increases; $[Ce^{3+}]/[Ce^{4+}]$ increases. This sequence, which resets the clock, begins after some product is formed and there is enough Ce^{4+}. It ends when Ce^{4+} is depleted, when Sequence A takes over again.

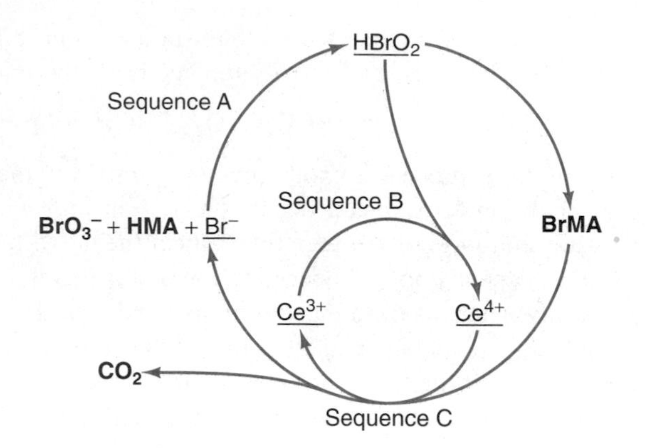

FIGURE 10.10
The three reaction sequences in the Belousov-Zhabotinsky reaction. If Sequences A and B were the only ones to occur, reaction would cease when all of the $\underline{Ce^{3+}}$ ions are converted into $\underline{Ce^{4+}}$ ions. However, the "clock is reset" by Sequence C, in which the product BrMA [brominated malonic acid, $CHBr(COOH)_2$] produces Br^- ions, which allow Sequence A to occur. It also reduces $\underline{Ce^{4+}}$ ions to $\underline{Ce^{3+}}$ ions and allows Sequence B to occur when Sequence A has produced some $\underline{HBrO_2}$. Sequence C thus shows positive feedback towards both Sequences A and B.

The overall reaction in the Belousov-Zhabotinsky reaction is approximately

$$2BrO_3^- + 3CH_2(COOH)_2 + 2H^+ \rightarrow 2CHBr(COOH)_2 + 3CO_2 + 4H_2O$$

The reaction is catalyzed by Ce^{3+} ions, which act by being oxidized to Ce^{4+} ions, which are then reduced again. During the course of the reaction the concentrations of the intermediates Ce^{3+} and Ce^{4+} oscillate, as do the concentrations of Br^- ions. The concentrations of the reactants BrO_3^- and malonic acid steadily decline, while that of the brominated acid steadily increases.

The basic ideas about the mechanism of the reaction were first worked out in 1972 by R. J. Field, E. Körös, and Richard M. Noyes.[8] The situation is quite complex, and at least 22 individual reactions have been identified as playing a part. We will not be concerned with details here, but only with the particular features of the reaction that cause the oscillations to occur. Three important reaction sequences are involved in the reaction, and these are summarized in the boxed inset. Figure 10.10 shows the three sequences in compact form.

Sequence A involves the participation of bromide ions, which are used up. Bromide ions are not normally added deliberately to the initial reaction mixture, but they are usually present in minute amounts as impurities, and Sequence A is therefore able to occur initially, but only to a limited extent since the bromide ions are soon depleted. When this occurs Sequence B takes over, carried along by the intermediate $\underline{HBrO_2}$, which is not present initially but which has been produced in small amounts by Sequence A. This intermediate is regenerated by reaction 4, so that Sequence B has the important characteristic of being autocatalytic. However, Sequence B, like Sequence A, can occur only to a limited extent, since it must come to a stop when all of the $\underline{Ce^{3+}}$ has been oxidized. We will now consider the individual reactions in a little more detail. Sequence A begins with the step

$$1'. \qquad \mathbf{BrO_3^-} + \underline{Br^-} + 2H^+ \rightarrow \underline{HBrO_2} + HOBr$$

The kinetically important reactant $\mathbf{BrO_3^-}$ is in bold type (we overlook the hydrogen ions since, although essential chemically, they do not play a particular role in the kinetics). The species $\underline{HBrO_2}$ and $\underline{Br^-}$ are underlined to indicate that they play important kinetic roles as intermediates. HOBr is not underlined since although it

[8] R. J. Field, E. Körös, and Richard M. Noyes, *J. Am. Chem. Soc., 94,* 8649 (1972); important details were added in later papers by Noyes's group and others.

SUMMARY OF REACTION SEQUENCES

Reactants and products are in bold type; kinetically important intermediates are underlined. **HMA** = $CH_2(COOH)_2$: **BrMA** = $CHBr(COOH)_2$

SEQUENCE A

1. $\mathbf{BrO_3^-} + \underline{Br^-} + 2H^+ + \mathbf{HMA} \rightarrow \underline{HBrO_2} + \mathbf{BrMA} + H_2O$
2. $\underline{HBrO_2} + \underline{Br^-} + H^+ + 2\mathbf{HMA} \rightarrow 2\mathbf{BrMA} + 2H_2O$

Overall stoichiometry for Sequence A:

$$\underline{BrO_3^-} + 2Br^- + 3\mathbf{CH_2(COOH)_2} + 3H^+ \rightarrow 3\mathbf{CHBr(COOH)_2} + 3H_2O$$

SEQUENCE B

3. $\mathbf{BrO_3^-} + \underline{HBrO_2} + H^+ \rightarrow 2\underline{BrO_2} + H_2O$
4. $\underline{BrO_2} + \underline{Ce^{3+}} + H^+ \rightarrow \underline{HBrO_2} + \underline{Ce^{4+}}$
5. $2\underline{HBrO_2} + \mathbf{HMA} \rightarrow \mathbf{BrO_3^-} + \mathbf{BrMA} + H_2O + H^+$

Overall stoichiometry for Sequence B:

$$\underline{HBrO_2} + \mathbf{CH_2(COOH)_2} + 2\underline{Ce^{3+}} + 2H^+ \rightarrow \mathbf{CHBr(COOH)_2} + 2\underline{Ce^{4+}} + 2H_2O$$

SEQUENCE C

Overall stoichiometry:

$$4Ce^{4+} + \mathbf{CHBr(COOH)_2} + 2H_2O + HOBr \rightarrow 2Br^- + 4\underline{Ce^{3+}} + 3CO_2 + 6H^+$$

There is no suggestion that the above reactions occur as elementary reactions; we can understand the oscillations without being concerned with the details of all the elementary steps involved.

plays an important part in the chemistry, it is not directly concerned with the oscil lations. It undergoes reactions such as the following, which give rise to the final product $CHBr(COOH)_2$:

$$HOBr + \underline{Br^-} + H^+ \rightarrow Br_2 + H_2O$$

$$Br_2 + \mathbf{CH_2(COOH)_2} \rightarrow \mathbf{CHBr(COOH)_2} + \underline{Br^-} + H^+$$

$$\text{Sum: } HOBr + \mathbf{CH_2(COOH)_2} \rightarrow \mathbf{CHBr(COOH)_2} + H_2O$$

Instead of reaction 1′, we can thus write

$$1. \qquad \mathbf{BrO_3^-} + \underline{Br^-} + 2H^+ + \mathbf{HMA} \rightarrow \underline{HBrO_2} + \mathbf{BrMA} + H_2O$$

where **HMA** is malonic acid and **BrMA** is brominated malonic acid.

This reaction is followed by the process $\underline{HBrO_2} + \underline{Br^-} + H^+ \rightarrow 2HOBr$, and again, in view of the boxed reactions above, we write this simply as

$$2. \qquad \underline{HBrO_2} + \underline{Br^-} + H^+ + 2HMA \rightarrow 2BrMA + 2H_2O$$

The two reactions 1 and 2 constitute a sequence of reactions in which the intermediate $\underline{HBrO_2}$ exists in a steady state; the product **BrMA** is steadily formed by this mechanism, and at the same time bromide ions are consumed. The overall stoichiometry for Sequence A is:

$$\underline{BrO_3^-} + \underline{2Br^-} + 3\mathbf{CH_2(COOH)_2} + 3H^+ \rightarrow 3\mathbf{CHBr(COOH)_2} + 3H_2O$$

After Sequence A has essentially come to a stop because of the depletion of bromide ions, Sequence B takes over. This sequence involves the intermediate $\underline{HBrO_2}$ generated in Sequence A and is

3. $$\mathbf{BrO_3^-} + H^+ + \underline{HBrO_2} \rightarrow 2\underline{BrO_2} + H_2O$$

4. $$\underline{BrO_2} + \underline{Ce^{3+}} + H^+ \rightarrow \underline{HBrO_2} + \underline{Ce^{4+}}$$

5. $$2\underline{HBrO_2} + \mathbf{HMA} \rightarrow \mathbf{BrO_3^-} + \mathbf{BrMA} + H_2O + H^+$$

In addition to $\underline{HBrO_2}$, which plays an autocatalytic role, another intermediate, $\underline{BrO_2}$, is involved in this sequence. Reaction 5 causes $\underline{HBrO_2}$ to be consumed by a second-order reaction, and when the concentrations of it and of Ce^{3+} have fallen sufficiently, Sequence B also becomes unimportant.

Note the essential differences between the mechanisms. Sequence A gives rise to bromination of malonic acid without disturbing the cerium ion ratio, and leads to depletion of bromide ions. Sequence B also leads to bromination of malonic acid and at the same time a decrease in the $\underline{Ce^{3+}}/\underline{Ce^{4+}}$ ratio. It does not require bromide ions; indeed the mechanism is inhibited by bromide ions, which compete for the $\underline{HBrO_2}$ radical by reaction 2. Thus, at high bromide ion concentrations, reaction occurs predominantly by Sequence A, while at lower bromide ion concentrations it occurs mainly by Sequence B.

At the very beginning of the reaction Sequence A predominates, the bromide ions competing successfully for the $\underline{HBrO_2}$ radicals formed in reaction 1. Later, when the bromide ion concentration has fallen to a low value, Sequence A can no longer compete for $\underline{HBrO_2}$ radicals and Sequence B takes over, giving rise to a decrease in the $\underline{Ce^{3+}}/\underline{Ce^{4+}}$ concentration ratio. The fact that both of these reactions can occur only to limited extents explains the long induction period for the reaction. Indeed, if these two sequences were all that could happen, little reaction would occur at all. The further reaction, and the oscillations, are explained by another sequence, Sequence C, which reconverts $\underline{Ce^{4+}}$ into $\underline{Ce^{3+}}$ at the expense of brominated malonic acid, $\mathbf{CHBr(COOH)_2}$. This sequence regenerates more bromide ions, which allow Sequences A and B to occur again. Sequence C is composed of reactions in which the $\underline{Ce^{4+}}$ and HOBr present bring about oxidation of the product $\mathbf{CHBr(COOH)_2}$ to carbon dioxide, water, and bromide ions. The exact mechanism of this process is complex and the details are still not entirely clear; it occurs somewhat as follows (of course, in a number of steps):

$$4\underline{Ce^{4+}} + \mathbf{CHBr(COOH)_2} + 2H_2O + HOBr \rightarrow 2Br^- + 4\underline{Ce^{3+}} + 3CO_2 + 6H^+$$

This reaction explains the long induction period in the reaction; initially there is no $\mathbf{CHBr(COOH)_2}$, and it must build up to a sufficiently high level for this reaction to occur.

We may summarize the situation as follows. Initially Sequence A occurs, to be taken over by Sequence B when the bromide concentration has become too low. The induction period ends when enough of the product $\mathbf{CHBr(COOH)_2}$ has been formed for Br^- and $[Ce^{3+}]$ ions to be regenerated by Sequence C. After this occurs, Sequence A begins again, but Sequence B takes over when the bromide concentration has fallen too low. When $[Ce^{3+}]$ has become low Sequence B is taken over by

Sequence C. The cycle then repeats itself, and there is a series of oscillations. Eventually they die down, and the reaction drifts monotonically to its final state.

Various other oscillating reactions are known. Also, a number of hypothetical kinetic schemes giving rise to oscillations have been put forward. All of these satisfy the requirement that the rate equation for at least one of the reactions is a non-linear differential equation, and that there is some positive feedback. Such systems of non-linear differential equations cannot be solved analytically. However, values can be chosen for rate constants and the equations solved numerically by the use of computers. Many hypothetical examples of oscillating reactions have been investigated in this way.

Chaos in Chemical Reactions

Theories of chaos were developed less than half a century ago, but it is now recognized that many examples of chaos had previously been known, many of them in chemistry. Organic chemists, for example, have often found that they obtained different yields of products when they repeated their experiments in what they thought to be exactly the same way. Similarly, in kinetics experiments different results have often been reported for the same reaction carried out in different laboratories, and often by the same experimenter on different days, for no ascertainable reason.

One common source of misunderstanding about modern chaos theory is that the effect is supposed to be caused by the uncertainty dealt with by Heisenberg's uncertainty principle (Section 11.3), according to which one cannot know the exact states of systems. This is not the case: The chaos we are concerned with would occur even if there were no restriction imposed by the Heisenberg principle. To avoid such misunderstanding it is helpful to use the expression *deterministic chaos.*

Deterministic Chaos

We saw earlier that the conditions for systems to oscillate are as follows:

1. At least two variables (such as the amounts of chemical substances and the temperature) must be involved.
2. The differential rate equations for at least one of the processes occurring must be non-linear.
3. There must be positive feedback.

The conditions for deterministic chaos are similar to those for oscillations, the important difference being that there must be at least *three* variables rather than two. The mathematical (including computational) work shows that if there are only two differential equations, at least one of them non-linear, the solutions may involve oscillations but will lead to only one possible final result. When there is an additional differential equation, however, the mathematical solutions may become extremely sensitive to two factors: (a) the initial amounts of the three or more substances involved in the process, and (b) the values of parameters such as rate constants. What is meant by deterministic chaos is that an imperceptibly small change in any of these quantities can have a profound effect on the *type* of behavior that occurs. Even if we start with what we believe to be exactly the same initial conditions, the tiniest of variations in them can lead to various types of behavior—to no oscillations but an unpredictable final outcome (called deterministic chaos), to oscillations followed by deterministic chaos, to oscillations followed by chaos, followed by further oscillations, and so on.

For example, consider the predator-prey analogy mentioned earlier, the infestation of a lawn by beetles, giving rise to oscillations. What will be the final outcome? One possibility is that the grass and the beetles might all be dead. Another is that the lawn might settle down into an equilibrium state in which the beetle population remains fairly low and the lawn is not too bad but not at its best. Theory tells us that if everything is held constant except the amount of grass and the number of beetles, there will be oscillations but the final outcome will be determinable from the initial conditions. This is because there are only two variables. If there is just one additional variable, such as the humidity or the temperature, it would no longer be possible to predict the outcome from the initial conditions, however carefully they were measured. There would then be deterministic chaos.

In the case of the Belousov-Zhabotinsky reaction, there are more than two variables, so that the final state is chaotic. It corresponds more or less to what is predicted by the stoichiometric equation, but there are significant variations from one experiment to another.

One of the first important mathematical treatments leading to modern chaos theory was put forward in 1963 by the American meteorologist Edward Lorenz (b. 1917). He used simplified mathematical equations to represent weather conditions, and found that trivial changes in initial conditions would lead to widely different outcomes. In 1972 he suggested what is called the "butterfly effect," with his paper entitled, "Does the Flap of a Butterfly's Wings in Brazil Set Off a Tornado in Texas?" He commented, incidentally, that just as a single flap of a wing could generate a tornado, it could also prevent a tornado. Another notable contribution to chaos theory was made in the early 1970s by the Polish-French mathematical physicist Benoit Mandelbrot (b. 1924). By-products of this work were his construction of special patterns and his introduction of the idea of fractals.

10.14 Electrochemical Dynamics

Chapters 7 and 8 dealt among other things with the fundamental principles of electrochemistry, including the thermodynamics of electrochemical cells. This section is concerned with the dynamics of some of the processes that occur when a current passes through matter. In the case of solids, such as metals and semiconductors, the current is carried by a flow of electrons, and this will be discussed in more detail in Section 16.6. In metals at the absolute zero the electrons fill a number of quantized energy levels up to a particular level known as the *Fermi level.* At higher temperatures electrons pass into higher levels, and they are then in what is called the *conduction band.* The electrons can easily pass along this conduction band. Other solids are poor conductors, and some are semiconductors or superconductors (Section 16.6).

In a solution of an electrolyte the ions, not the electrons, are the main carriers of the current, as discussed in Chapter 7. The electrons are involved in the neutralization of ions at the electrodes, but there is also the important matter of the electrode processes. If, for example, hydrogen gas is being evolved at an electrode, there is a question of eactly how the hydrogen ions in the solution are discharged, and how they become hydrogen molecules. The hydrogen ions must first pass through a layer called the "double layer" at the electrode surface, and this layer has different properties from the bulk of the solution. When the hydrogen ions reach the surface, there may be an electron transfer from the Fermi level at the surface to the ion, with the formation of a hydrogen atom. Two hydrogen atoms then recombine to form a hydrogen molecule. Several processes take place, and the rate with which the current flows usually depends largely on the rates of these electrode processes. That is to say, these electrode processes control the rate with which the current passes: The passage of the current through the wires is sufficiently easy as to have no significant effect on the overall rate.

Kinetics of Electrode Reactions

Overvoltage

Reversible Voltage

Valuable information about elecrode processes has been provided by investigations of what is known as the **overvoltage** associated with cathodes and anodes under particular circumstances. Experiments on the electrolysis of solutions in which, for example, hydrogen is evolved at a cathode and oxygen at an anode have shown that the emf at which bubbles of gas appear varies considerably with the nature of the electrodes. With electrodes consisting of "platinized platinum," which is platinum electrodeposited with a layer of platinum black, the voltage that has to be applied to an aqueous solution of an acid or base in order for bubbles of hydrogen and oxygen to appear is about 1.7 V, which is about 0.5 V greater than what is called the *equilibrium voltage,* or the *reversible voltage.* If the electrode is replaced by other metals, much higher decomposition voltages may be required. With zinc, for example, a potential of about 2.4 V must be applied in order to have an appreciable evolution of gas. The difference between the voltage required with a given *cathode,* compared with the thermodynamic decomposition voltage, is known as the *cathodic overvoltage.* Similarly, the overvoltage relating to the *anode* is called the *anodic overvoltage.* The overvoltage η is defined as

$$\eta = V - V_{rev} \tag{10.105}$$

where V is the potential at which electrolysis is actually proceeding, and V_{rev} is the reversible potential under the same conditions. Some typical values are given in Table 10.1 There are also overvoltages relating to the evolution of oxygen at the anode. The overvoltages relating to the deposition of metals and other materials at electrodes are usually quite small.

There are various reasons for the existence of overvoltages. When, for instance, hydrogen is evolved at a cathode several processes are involved: (1)

TABLE 10.1 Some Typical Overvoltages for Hydrogen Evolution

These values apply, for example, when the current density is about 1 mA (milliampere) per square centimetre of surface.

Cathode	Overvoltage/V
Platinized platinum	Small
Palladium	Small
Gold	0.02
Smooth platinum	0.09
Silver	0.15
Zinc	0.70
Mercury	0.78

diffusion of hydrogen ions from the bulk of the solution to the electrode; (2) discharge of the ions to form hydrogen atoms attached to the surface of the electrode; (3) formation of hydrogen molecules from the atoms; (4) formation of bubbles of hydrogen gas. If the diffusion of ions to the surface controls the overall rate and in this way causes the overvoltage, we speak of *concentration polarization;* there is said to be *diffusion control.* This is an important factor in some cases of electrolysis, but in the case of hydrogen overvoltage the rate is usually controlled by the chemical processes occurring at the electrode. In particular, the discharge, at the cathode, of hydrogen ions by electrons from the metal, is often rate controlling.

Concentration Polarization
Diffusion Control

The first significant work on overvoltage was done in 1905 by Julius Tafel (1862–1918), who was primarily an organic chemist, active in the field of preparative electro-organic chemistry. In the early twentieth century, he investigated the electrolysis of solutions in which hydrogen gas is evolved at the cathode. He obtained an empirical expression for the relationship between the overvoltage η and the current density i (the current per unit area of electrode):

$$\eta = a + b \ln i \tag{10.106}$$

Tafel Equation

where a and b are constants for a given system. This equation, fundamental to the study of the kinetics of electrode processes, has come to be called the **Tafel equation**.

To interpret this equation and related overvoltage behavior we must recognize that at the equilibrium or reversible voltage ($\eta = 0$), the rate of deposition of ions at an electrode is equal to the rate of reionization of the deposited substance; the net rate of deposition, and therefore the net current, is zero. We will write the equal and opposite current flowing in the forward and reverse directions as i_o; it is called the *exchange current.* When the additional voltage (the overvoltage) is applied, the rate of deposition is increased, and the rate of the reverse reaction is decreased; there is then net deposition. The greater the overvoltage, the greater the current. Suppose first that the rate-controlling step is the discharge of a hydrated hydrogen ion at the surface of the negative electrode, a process we can write as

Exchange Current

$$H_3O^+ + S^- \rightarrow H_2O + H{-}S$$

Figure 10.11 shows a schematic potential-energy diagram for this process. Curve I represents the situation where the process is occurring at a surface at which there is

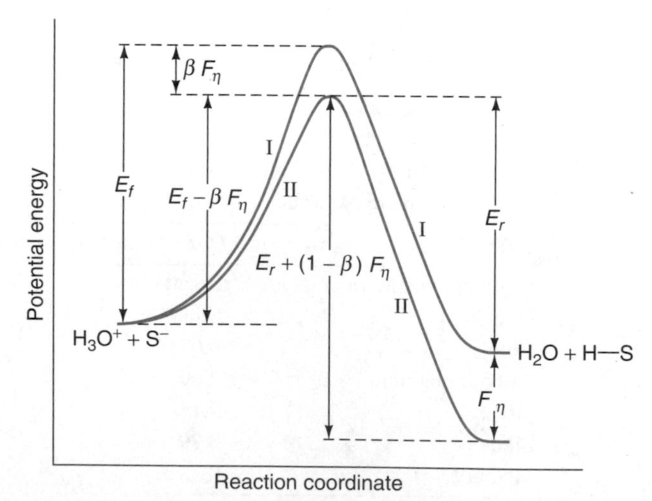

FIGURE 10.11
Schematic potential-energy curves for the discharge of an H_3O^+ ion at a negative electrode surface S. Curve I shows the case of zero overvoltage. The activation energy in the forward (discharge) direction is E_f, and that in the reverse direction E_r: there is an equal and opposite current i, flowing in each direction. Curve II shows the situation when there is an overvoltage η. This has the effect of lowering the energy of the final state by an amount $F\eta$, where F is the Faraday constant. The activation energy, however, is lowered only by a fraction β of this, i.e., by $\beta F\eta$. The activation energy of the reverse reaction is thus raised by $(1 - \beta)F\eta$.

no overvoltage. When there is an overvoltage η, the contribution of the electrical energy to the energy change in the overall reaction is $F\eta$, where F is the Faraday constant; the situation is then represented by Curve II. The activation energy for the forward reaction (deposition) is reduced, but not by the full amount $F\eta$ because the electric potential does not act over the entire reaction coordinate. Suppose that the activation energy in the forward direction (leading to hydrogen evolution) is reduced by $\beta F\eta$, where β is a fraction. According to the Arrhenius equation for reaction rates (Section 9.7), the rate of a reaction is related exponentially to the height E of the energy barrier for the reaction (the activation energy). The forward current i_f when the overvoltage η is applied can therefore be written as

$$i_f = i_0 e^{\beta F\eta/RT} \tag{10.107}$$

where i_0, the *exchange current,* is the rate when the overvoltage is zero.

For the reverse reaction the remaining fraction $1 - \beta$ of the electrical energy is involved, and it causes the activation energy to increase by $(1 - \beta)F\eta$ (Figure 10.11). The reverse current i_r can thus be written as

$$i_r = i_0 e^{-(1 - \beta)F\eta} \tag{10.108}$$

The net current that passes is therefore

$$i = i_f - i_r = i_0(e^{\beta F\eta/RT} - e^{-(1 - \beta)F\eta}) \tag{10.109}$$

When the overvoltage is zero the two exponential terms are zero, and no current flows. When the overvoltage is large the second exponential approaches zero, so that the equation becomes

$$i = i_0 e^{\beta F\eta/RT} \tag{10.110}$$

which can be written as

$$\eta = \frac{RT}{\beta F}\ln\frac{i}{i_0} = -\frac{RT}{\beta F}\ln i_0 + \frac{RT}{\beta F}\ln i \tag{10.111}$$

This is of the same form as the empirical Tafel equation, Equation 10.106, so that the empirical b in that equation is given by $RT/\beta F$.

The coefficient β plays an important role in electrochemical dynamics. Further light on its significance is thrown by showing the potential-energy diagram in the form of Figure 10.12. The left-hand curve is the hypothetical curve for the system

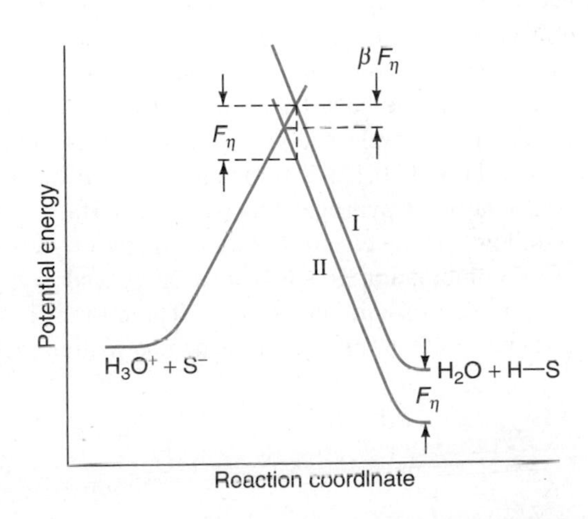

FIGURE 10.12
An alternative representation of the potential-energy variations during the discharge of an H_3O^+ ion at the surface of a negative electrode S. The left-hand curve is for the system $H_3O^+ + S^-$ in the absence of reaction. It is crossed by Curve I, which is for the system $H_2O + H-S$ in the absence of reaction and with zero overvoltage; Curve I relates to the right-hand portion of Curve I in Fig. 10.11. With an overvoltage of η Curve I is lowered at all points by the amount $F\eta$. The crossing point for the curves for the initial and final states is lowered by $\beta F\eta$. If the curves cross symmetrically, β is equal to one-half.

$H_3O^+ + S^-$ in the absence of reaction. It is crossed by Curve I, which is for the system $H_2O + H—S$ in the absence of reaction and with zero overvoltage; this Curve I relates to the Curve I of Fig. 10.11. When the overvoltage is η, Curve I is lowered at all points by the amount $F\eta$, since that is the energy provided by the overvoltage. The crossing point for the curves for the initial and final states is, however, not lowered by $F\eta$, but by a fraction of it which we write as β; it is thus lowered by $\beta F\eta$. We can see from simple geometry that the fraction β depends on the symmetry at the crossing point. If the curves cross symmetrically β is one-half, while if one curve is steeper than the other at the crossing point β is different from one-half. Because of this the coefficient β is referred to by electrochemists as the *symmetry factor.*

Symmetry Factor

For a number of electrodes over various experimental conditions the value of β, calculated from the Tafel b values, is approximately one-half, which is consistent with the hypothesis that in many cases the slow process in the evolution of hydrogen is the transfer of an electron at the surface to a hydrated proton. Other evidence is available to support this mechanism, such as studies of the influence of pH on the overvoltage. In 1936 the Russian electrochemist Alexander Naimovitch Frumkin (1895–1976) showed that under certain circumstances the overvoltage passes through a minimum when the pH of the solution is varied. This suggests that in acid solution the rate-controlling step is the neutralization of a hydrated proton at the surface. In alkaline solution, however, the hydrated protons are present only at low concentrations, and the rate-controlling step is then the transfer of a proton from a water molecule at the surface.

Quantum-mechanical treatments of proton transfers at electrodes suggest the possibility that quantum-mechanical tunneling might occur in the transfer of an electron to a proton. As discussed in Section 9.9, a good way of detecting tunneling is to carry out experiments down to low temperatures in order to see if the Arrhenius plot exhibits curvature (Figure 9.18). Another test for tunneling is to replace hydrogen by deuterium, which because of its greater mass is much less likely to tunnel; the H/D isotopic ratios will therefore be abnormal. Quantum-mechanical calculations (admittedly very approximate) have suggested that tunneling should occur in some electrochemical systems, but up to the present time, in spite of careful work done at temperatures as low as 180 K, no curvature of Arrhenius plots has been detected. Also, if tunneling occurred in overvoltage experiments there would be deviations from the Tafel equation, but none have yet been detected. There is therefore still no experimental evidence for quantum effects in overvoltage experiments.

For a few electrochemical systems the value of β calculated from the Tafel b is 2 rather than one-half. This is obviously impossible to reconcile with the idea that a fraction β of the electric potential acts between the initial and activated states (Figures 10.11 and 10.12). It is explained, however, if the slow step in the process is the combination of hydrogen atoms on the surface of the electrode. The argument goes as follows. If the electrode surface is sparsely covered by hydrogen atoms, the rate of the combination $2H \rightarrow H_2$ is proportional to n^2, where n is the number of atoms of hydrogen on any unit surface. The current i is thus kn^2, where k is a constant. The emf of the atomic electrode can be written in the form

$$V = \frac{RT}{F} \ln n + \text{constant} \tag{10.112}$$

The emf V_{rev} of the corresponding reversible electrode is

$$V_{rev} = \frac{RT}{F} \ln n_0 + \text{constant} \tag{10.113}$$

and the overvoltage η, equal to $V - V_{rev}$, is given by

$$\eta = V - V_{rev} = \frac{RT}{F} \ln \frac{n}{n_0} \tag{10.114}$$

Combining this with $i = kn^2$ leads to the expression

$$\eta = \frac{RT}{F} \ln \frac{i^{1/2}}{i_0^{1/2}} = \frac{RT}{2F} \ln \frac{i}{i_0} \tag{10.115}$$

To take into account other discharge mechanisms the practice is to generalize Eqs. 10.110 and 10.111 and write them in the form

$$i = i_0 e^{\alpha F \eta / RT} \tag{10.116}$$

and

$$\eta = \frac{RT}{\alpha F} \ln \frac{i}{i_0} = -\frac{RT}{\alpha F} \ln i_0 + \frac{RT}{\alpha F} \ln i \tag{10.117}$$

Instead of β the letter α is used when the rate-controlling step may be other than the neutralization of a hydrogen ion by an electron at the surface. This more general coefficient α is known as the *transfer coefficient.* Comparison of Eq. 10.115 with 10.117 shows that the Tafel α is equal to 2 instead of one-half. This mechanism, with the combination of atoms as the rate-controlling step, is therefore a possible one for electrode systems for which α has a value of 2. It also explains why for such systems the overvoltage is independent of the pH.

The kinetics of electrode processes are still being actively investigated. Important developments in the field have resulted from the invention of ultramicroelectrodes. Associated with every electrode is a time constant that indicates the speed with which the electrode reacts to an electric perturbation. Time constants of conventional cells are in the region of microseconds, which means that such cells are not useful for studying the kinetics of some of the more rapid processes such as solvent redistributions (see Figure 9.5). Ultramicroelectrodes have now been developed with time constants of a few nanoseconds (10^{-9} s); they are sometimes referred to as *nanoelectrodes.* They are essentially metal electrodes drawn to a sharp tip and coated with an insulating polymer in such a way that only about a nanometre of the tip is exposed to the solution.

Polarography

An important analytical technique related to electrode processes is **polarography,** which was developed in the 1920s by the Czech physical chemist Jaroslav Heyrovsky (1890–1967). Figure 10.13 illustrates the technique, which depends on the use of a **dropping mercury electrode.** The solution under investigation, kept saturated with hydrogen, is contained in a vessel through which drops of mercury constantly fall. These drops act as a cathode, the surface of which is constantly renewed; the mercury at the bottom of the flask acts as the anode.

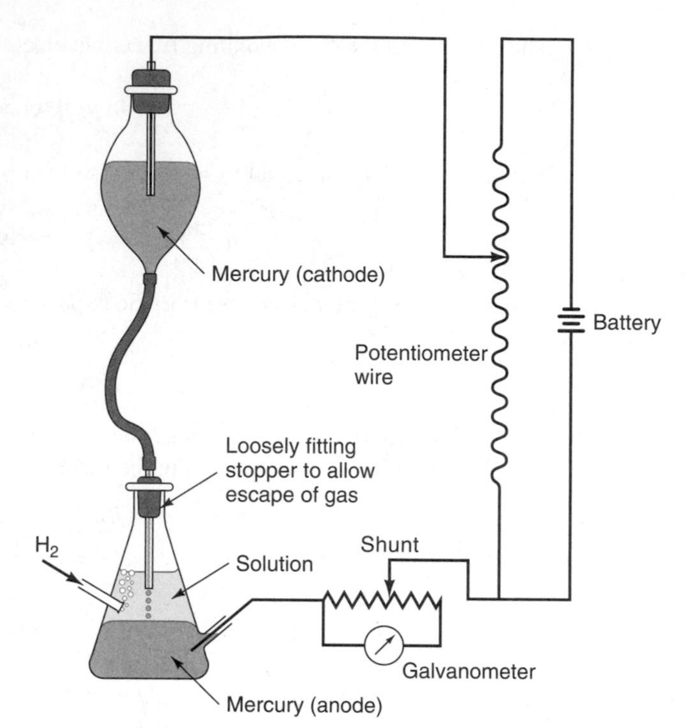

FIGURE 10.13
A simple polarographic instrument, involving the use of a dropping mercury cathode. The potential applied is controlled by the potentiometer, and the current is measured by the galvanometer. The shunt allows the sensitivity of the instrument to be varied. Modern polarographs involve much more sophisticated instrumentation, but the principle is the same.

The cathode and anode are connected to a potentiometer wire, and the current that passes is measured. Since the anode has a large area and the current is normally small (of the order of 10^{-6} A), the polarization at the electrode is negligible.

If the potential that is applied to a cell involving a dropping mercury electrode is steadily increased, the current that passes varies according to a pattern that depends on the nature of the ions present in the solution. An example is shown in Figure 10.14. At first the potential and current rise until a cathodic process, such as metal deposition, occurs (in the example shown in the figure it is the deposition of copper). The current then increases with little increase of potential until it reaches a value referred to as the *limiting current* for the particular species undergoing reduction. If the applied potential is further increased, another cathodic reduction takes place (the deposition of lead in the example shown), and another limiting current is attained. The current thus increases with increasing potential in a series of "waves," each wave corresponding to the electro-reduction of a particular cation.

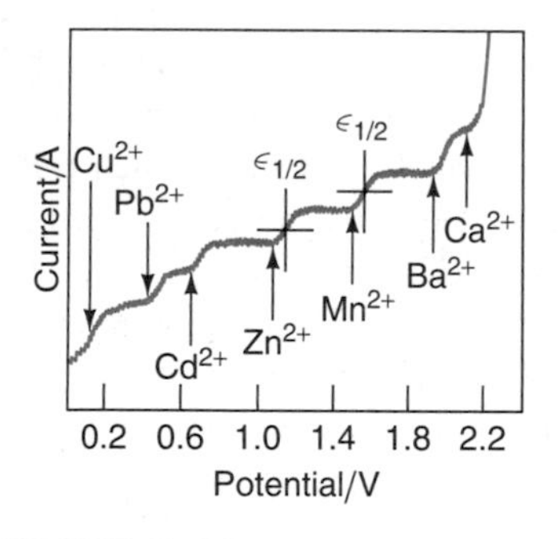

FIGURE 10.14
A typical polarogram, obtained with a solution containing various ions at concentrations of about 10^{-4} M. The crossed horizontal and vertical lines show the half-wave potentials $\epsilon_{1/2}$ for Mn^{2+} and Ba^{2+}. These half-wave potentials allow the identification of ions, and the corresponding limiting currents give the concentrations of the ions.

Patterns such as this provide valuable analytical information, as illustrated in Figure 10.14. From the half-wave potentials $\epsilon_{1/2}$, shown in the figure for Mn^{2+} and Ba^{2+}, the ions can be identified, and the corresponding limiting currents give their concentrations in the solution.

Heyrovsky was awarded the 1959 Nobel Prize for chemistry for his discovery of polarography, which has become a delicate, highly effective, and widely employed technique for chemical analysis. By his death in 1967, over 20,000 papers had been published on polarography, and work in the field continues to expand. The first commercial electronic instrument, the Mervyn Instruments' Square-Wave Polarograph, appeared in 1958, and improvements to the technique are still being made. Particularly noteworthy are the developments in pulse polarography and in the use of instruments controlled by computers and microcomputers.

A more recent technique involves the use of alternating-voltage modulation to the linear voltage sweep. The resulting alternating-current response has a maximum amplitude at the half-wave potential $\epsilon_{1/2}$. This amplitude corresponds to the situation in which the concentrations of oxidized and reduced forms at the electrode are equal at a potential equal to the half-wave potential $\epsilon_{1/2}$.

Electrokinetic Effects: The Electric Double Layer

The kinetics of electrode processes depend to a considerable extent on the nature of the interface between the electrode and the solution. This interface has another important significance; it determines the movement of particles and of solvent in an electric field. We will now consider the nature of the solid-solution interface and the electric effects associated with it.

Fixed Double Layer

The expression "electric double layer" is used to refer to the thin layer of the solution at an electrode, although the layer is a good deal more complicated than that expression would imply. The first and simplest theory of the layer was given in 1879 by Helmholtz. According to this model, which is represented in Figure 10.15a, the surface of a solid can be regarded as bearing positive or negative charges. If these are negative, as in the diagram, a unimolecular layer of positive charges will be attracted to the surface from the solution. A *fixed double layer* is therefore formed, corresponding to an electric capacitor. If the distance between the layers of positive and negative charges is a and each layer has a charge density (i.e.,

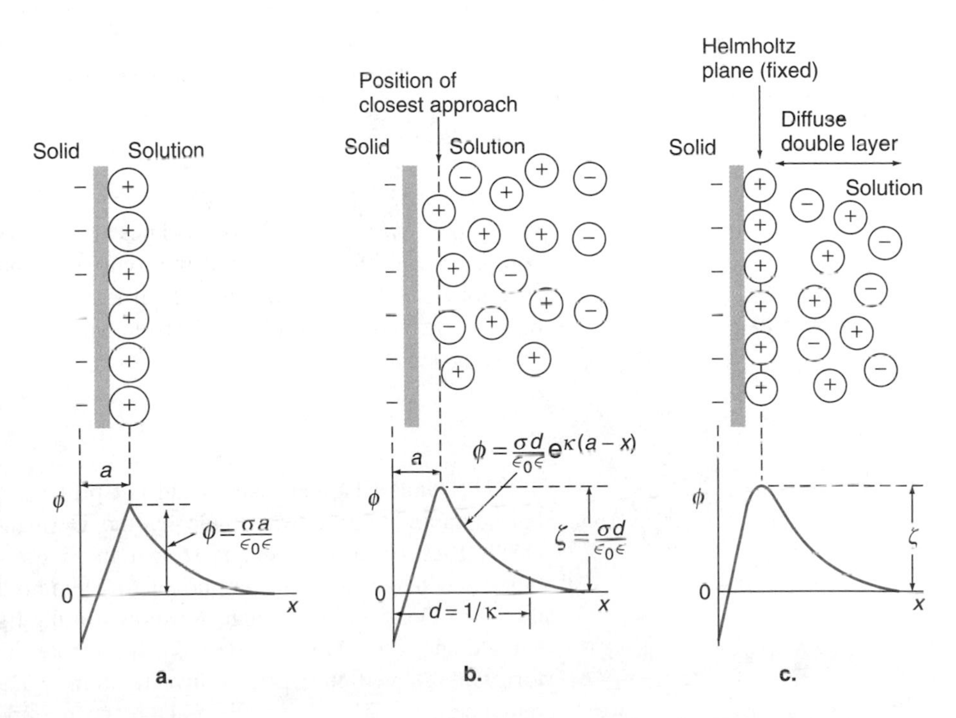

FIGURE 10.15
Three models for the structure of the electric double layer, showing the variations of electric potential ϕ with distance x from the negative charges on the surface. (a) Helmholtz model of the fixed double layer. The thickness of the fixed double layer is usually about 0.3 nm (3×10^{-10} m or 3 Å). (b) Diffuse double-layer model of Gouy and Chapman, showing Chapman's expression for the variation of ϕ with distance x and for the ζ potential. (c) Stern's model, which is a combination of (a) and (b). The thickness of the diffuse double layer roughly varies inversely with the concentration of the solution. For a 10^{-3} M solution the thickness is about 100 nm (10^{-7} m or 1000 Å).

a charge per unit area) of σ (sigma), it follows from electrostatic theory that the potential difference ϕ between the layers is

$$\phi = \frac{\sigma a}{\epsilon_0 \epsilon} \tag{10.118}$$

where ϵ is the dielectric constant of the liquid medium and ϵ_0 is the permittivity of a vacuum. The thickness of the fixed double layer is usually about 0.3 nm (3×10^{-10} m or 3 Å).

This simple idea, that a layer of ions from the solution becomes attached to the surface, was modified in 1910 by the French physicist Georges Gouy (1854–1926) and in 1913 by the British physical chemist David Leonard Chapman (1869–1959). They pointed out that the Helmholtz theory is unsatisfactory in neglecting the Boltzmann distribution of the ions. Their suggestion was that on the solution side of the interface there is not just a simple layer of ions, but an ionic distribution that extends some distance from the surface. In other words, there is a *diffuse double layer,* as shown in Figure 10.15b. Thermal agitation permits the movement of the ions present in the solution, but the distribution of the negative and positive ions is not uniform because of the electric field. In the example shown in the figure the surface is negative, so that there are more positive ions in close proximity to the surface. The thickness of the diffuse double layer depends on the concentration of the solution. For a 10^{-3} M solution the thickness is about 100 nm (10^{-7} m or 1000 Å), and roughly speaking it varies inversely with the concentration, being 1-2 nm for a 1 M solution.

Diffuse Double Layer

The idea behind the Gouy-Chapman theory is similar to that involved in the Debye-Hückel theory of the ionic atmosphere surrounding an ion. Indeed, 10 years before the formulation of the Debye-Hückel theory in 1923, Chapman had worked out the analogous treatment for the distribution of ions at a charged surface, i.e., for the two-dimensional case. The expression for the potential ϕ to which Chapman's theory leads is shown in Figure 10.15b. As x becomes very large, ϕ approaches zero; when x is equal to a (the distance of closest approach of the ions to the surface), the potential is

$$\phi = \frac{\sigma}{\epsilon_0 \epsilon \kappa} = \frac{\sigma d}{\epsilon_0 \epsilon} \tag{10.119}$$

where $1/\kappa = d$ is the effective thickness of the double layer. This quantity is to be compared with the thickness of the ionic atmosphere in the Debye-Hückel theory (Section 7.4). The potential ϕ_a, which is the potential at $x = a$ with respect to the bulk solution, is known as the electrochemical potential or ζ (zeta) potential and is given the symbol ζ; thus

Zeta Potential

$$\phi = \frac{\sigma d}{\epsilon_0 \epsilon} \tag{10.120}$$

The Gouy-Chapman theory did not prove entirely satisfactory, and in 1924 a considerable advance was made by the German-American physicist Otto Stern (1888–1969), whose model is shown in Figure 10.15c. Essentially, Stern combined the fixed double-layer model of Helmholtz with the diffuse double-layer model of Gouy and Chapman. As shown in the figure, there is a fixed layer at the surface and also a diffuse layer. On the whole this treatment has proved satisfactory, but for certain types of investigation it has been found necessary to add refinements. For example, there have been important treatments of the double layer in which account was taken of the solvation of ions and the orientation of solvent molecules.

Electroosmosis

If a membrane separates two identical liquids or solutions and a potential difference is applied across the membrane, there results a flow of liquid through the pores of the membrane. This phenomenon is known as *electroendosmosis* or as *electroosmosis*. A simplified version of the theory is as follows.

When a liquid is forced by electroosmosis through the pores of a membrane, the rate of flow is determined by two opposing factors: the force of electroosmosis, on the one hand, and the frictional force between the moving liquid layer and the wall on the other. When the forces are equal, there will be a uniform rate of flow. The situation in a single pore is represented in Figure 10.16a. On the solid side of the double layer the speed of flow is zero, whereas on the solution side it has attained the uniform speed v of the moving liquid; the velocity gradient, assumed to be uniform, is thus v/d. The force due to friction is the product of the velocity gradient and the coefficient of viscosity η of the liquid and is thus $\eta v/d$. The electric force causing electroosmosis is equal to the product of the applied electric potential gradient V and the charge density σ at the boundary at which movement occurs. Thus, in the steady state,

$$\frac{\eta v}{d} = V\sigma \tag{10.121}$$

The ζ potential is related to the distance d by eq. 10.120, and elimination of d between Eqs. 10.120 and 10.121 gives

$$v = \frac{\zeta \epsilon_0 \epsilon}{\eta} V \tag{10.122}$$

Electroosmotic Mobility

If the potential gradient V is unity the uniform speed attained, v_0, is known as the **electroosmotic mobility** and is given by

$$v_0 = \frac{\zeta \epsilon_0 \epsilon}{\eta} \tag{10.123}$$

If A is the total area of cross section of all the pores in a membrane, the volume V_l of liquid transported electroosmotically per unit time is equal to Av, so that from Eq. 10.122

$$V_l = \frac{\zeta A \epsilon_0 \epsilon}{\eta} V \tag{10.124}$$

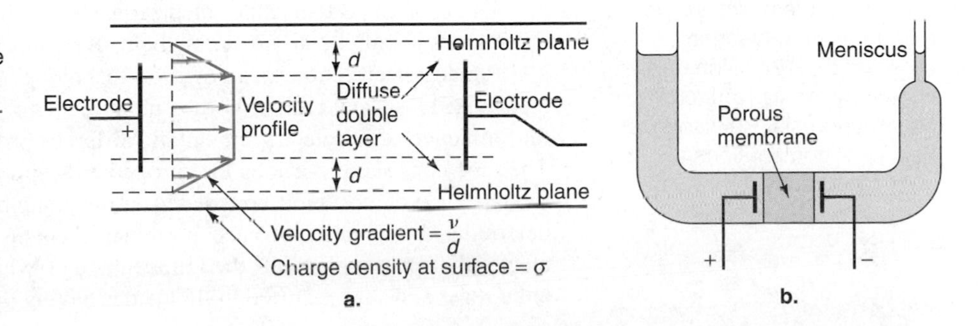

FIGURE 10.16
(a) Electroosmosis through a pore of a membrane. (b) Apparatus for studying electroosmosis; observations are made of the movement of the meniscus in a capillary tube.

Electroosmosis may alternatively be studied using a single capillary tube instead of a membrane containing many pores. The area A is then equal to πr^2, where r is the radius of the tube, and Eq. 10.124 becomes

$$V_l = \frac{\pi \zeta r^2 \epsilon_0 \epsilon}{\eta} V \tag{10.125}$$

A simple type of apparatus for studying electroosmosis through a membrane is shown in Figure 10.16b.

During recent years electroosmosis has found many practical applications. It is used, for example, in the dewatering of soils. In some parts of the world the soil is too full of water to be fertile, and the water content can be reduced by electroosmotic methods. There are also many medical and surgical applications of electroosmosis, and these are rapidly becoming more widely used. Cancerous tumors are now sometimes treated electrochemically, by a process that depends in part on the electroosmotic dewatering of the cancerous tissue.

Electrophoresis

Electrophoretic Mobility

In the derivation of Eq. 10.123 for the electroosmotic mobility under unit potential gradient, the moving liquid was regarded as a cylinder moving through a capillary tube (Figure 10.16a). The position of the liquid and wall can be reversed without affecting the argument, so that Eq. 10.123 also gives the rate of movement of a solid cylindrical particle through a liquid under the influence of an applied field of zero potential gradient. This quantity, denoted by v_0, is the **electrophoretic mobility;** thus, for a cylindrical particle moving along its axis,

$$v_0 = \frac{\zeta \epsilon_0 \epsilon}{\eta} \tag{10.126}$$

The treatment of particles of different shapes is more complicated, and various theories have been presented. In their theory of electrolytic mobility (Section 7.4), Debye and Hückel concluded in 1924 that for a spherical particle Eq. 10.126 should be replaced by

$$v_0 = \frac{2\zeta \epsilon_0 \epsilon}{3\eta} \tag{10.127}$$

This equation is valid only if the thickness of the double layer is large compared with the radius of the particle. The equation may thus be satisfactory for ions, but less so for large particles, and various improved equations have been suggested. In practice, however, electrophoresis experiments are usually carried out in an empirical manner, without reference to these equations.

Tiselius, some of whose work had been with Svedberg, received the 1948 Nobel Prize in chemistry for his work on electrophoresis. He also made important contributions to chromatography.

The first important experiments on electrophoresis were carried out by the Swedish physical chemist Arne Wilhelm Kaunin Tiselius (1902–1971), who in the 1930s made electrophoresis a powerful technique for studying mixtures of proteins. He devised a special type of U-tube along which the protein molecules move under the influence of an electric potential; different proteins move at different speeds. The tube consists of portions fitted together at ground-glass joints, so that one of a mixture of proteins could be isolated in one chamber. Optical methods are used to determine the quantity of each protein present in the mixture. The technique of electrophoresis supplements the ultracentrifuge, which separates according to molecular mass and shape. Different molecules having the same sizes and shapes behave identically in the ultracentrifuge, but may have different electrical properties and therefore can be separated by electrophoresis.

A difficulty with the use of a Tiselius tube is that a certain amount of local heating occurs, leading to convection currents that cause some mixing and disturb the separation. This problem is now usually minimized by supporting the solutions on a gel, such as a polyacrylamide gel. Another development involves the addition of detergents, such as sodium dodecyl sulfate, to solutions that are supported on gels. The dissolved molecules, such as proteins, have electrophoretic mobilities that are proportional to their molecular masses, and as a result it is possible to separate proteins that are not easily separable in the absence of the detergent.

Isoelectric Focusing

One special electrophoretic technique is *isoelectric focusing*. In this method a pH gradient is established so that the solution at the cathode is more basic and that at the anode more acidic. Molecules will migrate towards the electrode of appropriate charge, but as they do so they will be subject to changing pHs that will tend to make them change their state of ionization. This will continue until the species is concentrated at the position where the pH of the solution is the *isoelectric point* of the molecule. If several species are present, each will concentrate at a position corresponding to its isoelectric point. With suitable pH gradients it is possible to separate species with small differences in isoelectric points. Isoelectric focusing is frequently carried out with the sample supported on a gel.

The *isoelectric point* is the pH at which a molecule or other particle has no net charge. It is not quite the same as the *isoionic point* (Section 18.9) which is the pH at which, as a result of *proton ionizations*, there is an equal number of positive and negative charges. The isoelectric point takes account of other charges, such as heavy metal ions attached to the particles.

Reverse Isoelectric Effects

In the two effects just mentioned, electroosmosis and electrophoresis, the application of an electric field brings about relative motion of two phases. If, on the other hand, charged liquid layers are moved with respect to each other, their displacement produces a potential gradient in the direction of motion. For example, if a liquid is forced through the pores of a membrane or through a capillary tube, a potential difference is observed, its magnitude depending on the ζ potential. This phenomenon, the reverse of electroosmosis, is known as the **streaming potential.**

Streaming Potential

The reverse of electrophoresis occurs when small particles are allowed to fall through a liquid under the influence of gravity. A difference of potential is observed between two electrodes placed at different levels, and its magnitude again depends on the magnitude of the ζ potential. This effect is known as the **sedimentation potential** or the **Dorn effect,** after the Germany physicist Friedrich Ernst Dorn (1848–1916), who discovered it in 1880.

Dorn Effect

PROBLEMS

Composite Mechanisms and Rate Equations

10.1. Suppose that a reaction of stoichiometry A + 2B = Y + Z is believed to occur by the mechanism

$$\mathrm{A + B} \xrightarrow{k_1} \mathrm{X} \qquad \text{(very slow)}$$

$$\mathrm{X + B} \xrightarrow{k_2} \mathrm{Y + Z} \qquad \text{(very fast)}$$

where X is an intermediate. Write the expression for the rate of formation of Y.

10.2. Suppose that a reaction A + 2B = 2Y + 2Z is believed to occur according to the mechanism

$$\mathrm{A} \xrightarrow{k_1} \mathrm{2X} \qquad \text{(very rapid equilibrium)}$$

$$\mathrm{X + B} \xrightarrow{k_2} \mathrm{Y + Z} \qquad \text{(slow)}$$

Obtain an expression for the rate of formation of the product Y.

10.3. Suppose that a reaction of stoichiometry A + B = Y + Z is believed to occur according to the mechanism

$$A \underset{k_{-1}}{\overset{k_1}{\rightleftarrows}} X$$

$$X + B \xrightarrow{k_2} Y + Z$$

Apply the steady-state treatment and obtain an expression for the rate. To what expressions does the general rate equation reduce if

a. The second reaction is slow, the initial equilibrium being established very rapidly?

b. The second reaction is very rapid compared with the first reaction in either direction?

10.4. A reaction of stoichiometry

$$A + B = Y + Z$$

is found to be second order in A and zero order in B. Suggest a mechanism that is consistent with this behavior.

10.5. The rate of formation of the product of a reaction is found to give a nonlinear Arrhenius plot, the line being convex to the $1/T$ axis (i.e., the activation energy is higher at higher temperatures). Suggest a reason for this type of behavior. (*Hint:* For this and the following problem, consider the possibility of two parallel reactions and of two consecutive reactions having different activation energies.)

10.6. An Arrhenius plot is concave to the $1/T$ axis (i.e., it exhibits a lower activation energy at higher temperatures). Suggest a reason for this type of behavior.

10.7. Nitrogen pentoxide reacts with nitric oxide in the gas phase according to the stoichiometric equation

$$N_2O_5 + NO = 3NO_2$$

The following mechanism has been proposed.

$$N_2O_5 \rightarrow NO_2 + NO_3$$

$$NO_2 + NO_3 \rightarrow N_2O_5$$

$$NO + NO_3 \rightarrow 2NO_2$$

Assume that the steady-state treatment can be applied to NO_3, and derive an equation for the rate of consumption of N_2O_5.

10.8. The reaction $2NO + O_2 \rightarrow 2NO_2$ is believed to occur by the mechanism

$$2NO \xrightarrow{k_1} N_2O_2$$

$$N_2O_2 \xrightarrow{k_{-1}} 2NO$$

$$N_2O_2 + O_2 \xrightarrow{k_2} 2NO_2$$

Assume N_2O_2 to be in a steady state and derive the rate equation. Under what conditions does the rate equation reduce to second-order kinetics in NO and first-order kinetics in O_2?

***10.9.** The gas-phase reaction

$$Cl_2 + CH_4 \rightarrow CH_3Cl + HCl$$

proceeds by a free-radical chain reaction in which the chain propagators are Cl and CH_3 (but not H), and the chain-ending step is $2Cl \rightarrow Cl_2$. Write the mechanism, identify the initiation reaction and the chain-propagating steps, and obtain an expression for the rate of the overall reaction.

10.10. The following mechanism has been proposed for the thermal decomposition of pure ozone in the gas phase:

$$2O_3 \xrightarrow{k_1} O_3 + O_2 + O$$

$$O + O_3 \xrightarrow{k_2} 2O_2$$

Derive the rate equation.

***10.11.** A reaction occurs by the mechanism

$$A + B \underset{k_{-1}}{\overset{k_1}{\rightleftarrows}} X \xrightarrow{k_2} Z$$

and the concentration of X is sufficiently small compared with the concentrations of A and B that the steady-state treatment applies. Prove that the activation energy E_a at any temperature is given by

$$E_a = \frac{k_{-1}(E_1 + E_2 - E_{-1}) + k_2E_1}{k_{-1} + k_2}$$

that is, is the weighted mean of the values $E_1 + E_2 - E_{-1}$, and E_1, which apply, respectively, to the limiting cases of $k_1 \gg k_2$ and $k_2 \gg k_{-1}$.

10.12. F. A. Lindemann [*Trans. Faraday Soc., 17,* 598(1922)] proposed the following mechanism for a unimolecular gas reaction

$$A + A \underset{k_{-1}}{\overset{k_1}{\rightleftarrows}} A^* + A$$

$$A^* \xrightarrow{k_2} Y + Z$$

The species A* is an energized molecule that is present in low concentrations. Apply the steady-state treatment to A* and obtain an expression for the rate in terms of [A], k_1, k_{-1}, and k_2. Show that the mechanism predicts first-order kinetics at higher A concentrations and second-order kinetics at lower ones.

***10.13.** Certain polymerizations involve esterification reactions between −COOH groups on one molecule and −OH groups on another. Suppose that the concentration of such functional groups is c and that the rate of their removal by esterification obeys the equation

$$-\frac{dc}{dt} = kc^2$$

Obtain an equation relating the time t to the fraction f of functional groups remaining and to the initial concentration c_0 of functional groups.

***10.14.** Show that the mechanism

$$I_2 \underset{k_{-1}}{\overset{k_1}{\rightleftharpoons}} 2I \qquad \text{(fast)}$$

$$I + H_2 \underset{k_{-2}}{\overset{k_2}{\rightleftharpoons}} H_2I \qquad \text{(fast)}$$

$$H_2I + I \xrightarrow{k_3} 2HI \qquad \text{(slow)}$$

leads to the result that the rate equation for the overall reaction is $v = k[H_2][I_2]$.

10.15. Apply the steady-state treatment to the following mechanism, in which Y and Z are final products and X is a labile intermediate:

$$A + B \underset{k_{-1}}{\overset{k_1}{\rightleftharpoons}} X$$

$$X + A \xrightarrow{k_2} Y$$

$$X + B \xrightarrow{k_3} Z$$

Obtain an expression for the rate of formation of the product Z. What rate equations are obtained if (a) A and (b) B are present in great excess?

Photochemistry and Radiation Chemistry

10.16. Calculate the maximum wavelength of the radiation that will bring about dissociation of a diatomic molecule having a dissociation energy of 390.4 kJ mol^{-1}.

10.17. Hydrogen iodide undergoes decomposition into $H_2 + I_2$ when irradiated with radiation having a wavelength of 207 nm. It is found that when 1 J of energy is absorbed, 440 μg of HI is decomposed. How many molecules of HI are decomposed by 1 photon of radiation of this wavelength? Suggest a mechanism that is consistent with this result.

10.18. A 100-watt mercury-vapor lamp emits radiation of 253.7 nm wavelength and may be assumed to operate with 100% efficiency. If all the light emitted is absorbed by a substance that is decomposed with a quantum yield of unity, how long will it take for 0.01 mol to be decomposed?

10.19. Suppose that the radiation emitted by the lamp in Problem 10.18 is all absorbed by ethylene, which decomposes into $C_2H_2 + H_2$ with a quantum yield of unity. How much ethyne will be produced per hour?

10.20. A 1000-watt mercury vapor flash lamp emits radiation of 253.7 nm wavelength, and the duration of the flash is 1 μs. Suppose that all of the radiation of a single flash is absorbed by mercury vapor; how many atoms of excited mercury are formed?

***10.21.** The photochemical reaction between chlorine and chloroform in the gas phase follows the stoichiometric equation

$$CHCl_3 + Cl_2 = CCl_4 + HCl$$

It is believed to occur by the mechanism

$$Cl_2 + h\nu \xrightarrow{k_1} 2Cl$$

$$Cl + CHCl_3 \xrightarrow{k_2} HCl + CCl_3$$

$$CCl_3 + Cl_2 \xrightarrow{k_3} CCl_4 + Cl$$

$$2Cl \xrightarrow{k_4} Cl_2$$

Assume the rate of formation of Cl atoms in the initiation reaction to be $2I_a$, where I_a is the intensity of light absorbed, and obtain an expression for the overall rate in terms of I_a and $[CHCl_3]$.

***10.22.** When water vapor is irradiated with a beam of high-energy electrons, various ions such as H^+ and O^- appear. Calculate the minimum energies required for the formation of these ions, given the following thermochemical data:

$$H_2O(g) \xrightarrow{k_1} H(g) + OH(g) \qquad \Delta H^\circ = 498.7 \text{ kJ mol}^{-1}$$

$$OH(g) \xrightarrow{k_2} H(g) + O(g) \qquad \Delta H^\circ = 428.2 \text{ kJ mol}^{-1}$$

$$H(g) \xrightarrow{k_3} H^+(g) + e^-(g) \qquad \Delta H^\circ = 1312.2 \text{ kJ mol}^{-1}$$

$$O(g) + e^-(g) \xrightarrow{k_4} O^-(g) \qquad \Delta H^\circ = -213.4 \text{ kJ mol}^{-1}$$

Are the results you obtain consistent with the experimental appearance potentials of 19.5 eV for H^+ and 7.5 eV for O^-?

10.23. The mercury-photosensitized hydrogenation of ethylene in the presence of mercury vapor is first-order with respect to ethylene and half-order with respect to H_2. Its rate is proportional to the square root of the intensity of the light absorbed. The following mechanism has been suggested to account for these observations.

$$Hg + h\nu \rightarrow Hg^*$$

$$Hg^* + H_2 \xrightarrow{k_1} Hg + 2H$$

$$H + C_2H_4 \xrightarrow{k_2} C_2H_5$$

$$C_2H_5 + H_2 \xrightarrow{k_3} C_2H_6 + H$$

$$H + H \xrightarrow{k_4} H_2$$

Applying the steady-state approximation to [H] and $[C_2H_5]$, verify that the mechanism indeed supports the observations. What is the observed rate constant in terms of the rate constants of the elementary reactions?

Catalysis

10.24. The hydrolysis of a substance is specifically catalyzed by hydrogen ions, and the rate constant is given by

$$k/\text{dm}^3\ \text{mol}^{-1}\ \text{s}^{-1} = 4.7 \times 10^{-2}([\text{H}^+]/\text{mol dm}^{-3})$$

When the substance was dissolved in a 10^{-3} *M* solution of an acid HA, the rate constant was 3.2×10^{-5} $\text{dm}^3\ \text{mol}^{-1}\ \text{s}^{-1}$. Calculate the dissociation constant of HA.

***10.25.** The following is a slightly simplified version of the mechanism proposed in 1937 by G. K. Rollefson and R. F. Faull [*J. Amer. Chem. Soc., 59,* 625(1937)] to explain the iodine-catalyzed decomposition of acetaldehyde:

$$\text{I}_2 \rightarrow 2\text{I}$$
$$\text{I} + \text{CH}_3\text{CHO} \rightarrow \text{HI} + \text{CH}_3\text{CO}$$
$$\text{CH}_3\text{CO} \rightarrow \text{CH}_3 + \text{CO}$$
$$\text{CH}_3 + \text{HI} \rightarrow \text{CH}_4 + \text{I}$$
$$2\text{I} \rightarrow \text{I}_2$$

Apply the steady-state treatment to I, CH_3CO, and CH_3 and obtain an expression for the rate.

***10.26.** Suppose that a reaction is catalyzed by a series of homologous acids and that the Hammett equation (9.130) applies:

$$\log_{10} k_a = \log_{10} k_0 + \sigma\rho$$

where σ is the substituent constant and ρ is the reaction constant. Suppose that the corresponding equation for the dissociation of the acid is

$$\log_{10} K_a = \log_{10} K_0 + \sigma\rho'$$

where ρ' is the reaction constant for the dissociation; the substituent constants are the same in both equations. Prove that the Brønsted equation

$$k_a = G_a K_a^{\alpha}$$

applies. How does α relate to the reaction constants ρ and ρ'?

10.27. The hydrolysis of ethyl acetate catalyzed by hydrochloric acid obeys the rate equation

$$v = k[\text{ester}][\text{HCl}]$$

and the reaction essentially goes to completion. At 25 °C the rate constant is 2.80×10^{-5} $\text{dm}^3\ \text{mol}^{-1}\ \text{s}^{-1}$. What is the half-life of the reaction if [ester] = 0.1 *M* and [HCl] = 0.01 *M*?

10.28. The following mechanism has been proposed for the alkaline hydrolysis of $Co(NH_3)_5Cl^{2+}$.

$$\text{Co(NH}_3)_5\text{Cl}^{2+} + \text{OH}^- \rightarrow \text{Co(NH}_3)_4(\text{NH}_2)\text{Cl}^+ + \text{H}_2\text{O}$$
$$\text{Co(NH}_3)_4(\text{NH}_2)\text{Cl}^+ \rightarrow \text{Co(NH}_2)_4(\text{NH}_2)^{2+} + \text{Cl}^-$$
$$\text{Co(NH}_3)_4(\text{NH}_2)^{2+} + \text{H}_2\text{O} \rightarrow \text{Co(NH}_3)_5(\text{OH})^{2+}$$

Assume $Co(NH_3)_4(NH_2)Cl^+$ and $Co(NH_3)_4(NH_2)^{2+}$ to be in the steady state and derive an expression for the rate of reaction.

Experimentally, the rate is proportional to $[\text{Co(NH)}_5\text{Cl}^{2+}]\ [\text{OH}^-]$; does this fact tell us anything about the relative magnitudes of the rate constants?

10.29. Confirm that Eq. 10.68,

$$[\text{H}^+]_{\text{min}} = (k_{\text{OH}^-}K_w/k_{\text{H}^+})^{\frac{1}{2}}$$

follows from Eq. 10.66,

$$k = k_0 + k_{\text{H}^+}[\text{H}^+] + \frac{k_{\text{OH}^-}K_w}{[\text{H}^+]}$$

10.30. The following results have been obtained by D. B. Dahlberg and F. A. Long [*J. Amer. Chem. Soc., 95,* 3825(1973)] for the base-catalyzed enolization of 3-methyl acetone.

Catalyst	$ClCH_2COO^-$	CH_3COO^-	HPO_4^{2-}
K_a/mol dm^{-3}	1.39×10^{-3}	1.80×10^{-5}	6.25×10^{-8}
k/dm^3 mol^{-1} s^{-1}	1.41×10^{-3}	1.34×10^{-2}	0.26

Estimate the Brønsted coefficient β.

10.31. Suggest a plausible mechanism for the bromination of acetone catalyzed by hydroxide ions. As with the acid-catalyzed reaction discussed in Section 10.9, the rate is independent of the bromine concentration.

10.32. It was found by J. Halpern and coworkers [*J. Phys. Chem., 60,* 1455(1956)] that the rate equation for the oxidation of molecular hydrogen by dichromate ions ($Cr_2O_7^{2-}$) catalyzed by Cu^{2+} ions is of the form

$$v = \frac{k[\text{H}_2][\text{Cu}^{2+}]^2}{[\text{H}_2] + k'[\text{Cu}^{2+}]}$$

(Note that the rate is independent of the concentration of dichromate ions.) Suggest a mechanism consistent with this behavior, and apply the steady-state treatment to obtain the rate expression. Comment on rate-controlling steps corresponding to special cases of the mechanism.

10.33. For the oxidation of molecular hydrogen by dichromate ions catalyzed by Ag^+ ions, A. H. Webster and J. Halpern [*J. Phys. Chem., 60,* 280(1956)] obtained the rate equation

$$v = k[\text{H}_2][\text{Ag}^+]^2 + \frac{k'[\text{H}_2][\text{Ag}^+]^2}{[\text{H}_2] + k''[\text{Ag}^+]}$$

The existence of two terms suggests that two mechanisms are occurring in parallel. Suggest the two mechanisms, applying the steady-state treatment to obtain the second term in the rate equation.

10.34. The reaction

$$\text{Tl}^+ + 2\text{Ce}^{4+} \rightarrow \text{Tl}^{3+} + 2\text{Ce}^{3+}$$

is catalyzed by Ag^+ ions. Under certain conditions the rate is proportional to

$$[\text{Ce}^{4+}][\text{Tl}^+][\text{Ag}^+]/[\text{Ce}^{3+}]$$

Suggest a mechanism consistent with this behavior.

Enzyme-Catalyzed Reactions

10.35. The following rates have been obtained for an enzyme-catalyzed reaction at various substrate concentrations:

10^3[S]/mol dm^3	Rate, v/(arbitrary units)
0.4	2.41
0.6	3.33
1.0	4.78
1.5	6.17
2.0	7.41
3.0	8.70
4.0	9.52
5.0	10.5
10.0	12.5

Plot v against [S], $1/v$ against $1/$[S], and $v/$[S] against v, and from each plot estimate the Michaelis constant. Which plot appears to give the most reliable value?

10.36. The following data have been obtained for the myosin-catalyzed hydrolysis of ATP, at 25 °C and pH 7.0.

10^5[ATP]/mol dm^{-3}	$10^6 v$/mol dm^{-3} s^{-1}
7.5	0.067
12.5	0.095
20.0	0.119
32.5	0.149
62.5	0.185
155.0	0.191
320.0	0.195

Plot v against [S], $1/v$ against $1/$[S], and $v/$[S] against v, and from each plot calculate the Michaelis constant K_m and the limiting rate V.

***10.37.** The following values of V (limiting rate at high substrate concentrations) and K_m have been obtained at various temperatures for the hydrolysis of acetylcholine bromide, catalyzed by acetylcholinesterase.

T/°C	$10^6 V$/mol dm^{-3} s^{-1}	$K_m \times 10^4$/mol dm^{-3}
20.0	1.84	4.03
25.0	1.93	3.75
30.0	2.04	3.35
35.0	2.17	3.05

a. Assuming the enzyme concentration to be 1.00×10^{-11} mol dm^{-3}, calculate the energy of activation, the enthalpy of activation, the Gibbs energy of activation, and the entropy of activation for the breakdown of the enzyme-substrate complex at 25 °C.

b. Assuming K_m to be the dissociation constant k_{-1}/k_1 for the enzyme-substrate complex (ES $\underset{k_1}{\overset{k_{-1}}{\rightleftharpoons}}$ E + S), determine the following thermodynamic quantities for the *formation* of the enzyme-substrate complex at 25 °C: $\Delta G°$, $\Delta H°$, $\Delta S°$.

c. From the results obtained in parts (a) and (b), sketch a Gibbs energy diagram and an enthalpy diagram for the reaction.

***10.38.** The following data relate to an enzyme reaction.

10^3[S]/mol dm^3	$10^5 V$/mol dm^3 s^{-1}
2.0	13
4.0	20
8.0	29
12.0	33
16.0	36
20.0	38

The concentration of the enzyme is 2.0 g dm^{-3}, and its molecular weight is 50 000. Calculate K_m, the maximum rate V, and k_c.

10.39. The following data have been obtained for the myosin-catalyzed hydrolysis of ATP.

Temperature/°C	$k_c \times 10^6$/s^{-1}
39.9	4.67
43.8	7.22
47.1	10.0
50.2	13.9

Calculate, at 40 °C, the energy of activation, the enthalpy of activation, the Gibbs energy of activation, and the entropy of activation.

***10.40.** The following is a simplified version of the mechanism that has been proposed by H. Theorell and Britton Chance for certain enzyme reactions involving two substrates A and B.

$$E + A \underset{k_{-1}}{\overset{k_1}{\rightleftharpoons}} EA$$

$$EA + B \xrightarrow{k_2} EZ + Y$$

$$EZ \xrightarrow{k_3} E + Z$$

Assume that the substrates A and B are in excess of E so that the steady-state treatment can be applied to EA and EZ, and obtain an expression for the rate.

***10.41.** When an inhibitor I is added to a single-substrate enzyme system, the mechanism is sometimes

$$\begin{array}{c} S \\ + \\ I + E \underset{k_{-i}}{\overset{k_i}{\rightleftharpoons}} EI \\ k_{-1}\uparrow\downarrow k_1 \\ ES \\ \downarrow k_2 \\ E + Y \end{array}$$

This is known as a *competitive* mechanism, since S and I compete for sites on the enzyme.

a. Assume that the substrate and inhibitor are present in great excess of the enzyme, apply the steady-state treatment, and obtain the rate equation.
b. Obtain an expression for the degree of inhibition defined as

$$\epsilon = \frac{v_0 - v}{v_0}$$

where v is the rate in the presence of inhibitor and v_0 is the rate in its absence.

***10.42.** Obtain the rate equation corresponding to the mechanism

$$E + S \underset{k_{-1}}{\overset{k_1}{\rightleftarrows}} ES \overset{k_2}{\rightarrow} ES' \overset{k_3}{\rightarrow} E + Z$$
$$ES \rightarrow Y$$

Assume ES and ES′ to be in the steady state and the substrate concentration to be much higher than the enzyme concentration. Express the catalytic constant k_c and the Michaelis constant K_m in terms of k_1, k_{-1}, k_2, and k_3.

***10.43.** Enzyme-catalyzed reactions frequently follow an equation of the form of Eq. 10.85. Suppose that k_c and K_m show the following temperature dependence:

$$k_c = A_c \exp(-E_c/RT) \quad \text{and} \quad K_m = B \exp(-\Delta H_m/RT)$$

where A_c, B, E_c, and ΔH_m are temperature-independent parameters. Explain under what conditions, with [S] held constant, the rate may pass through a maximum as the temperature is raised.

10.44. Some enzyme reactions involving two substrates A and B occur by the following mechanism.

$$E + A \underset{k_{-1}}{\overset{k_1}{\rightleftarrows}} EA$$

$$EA + B \overset{k_2}{\rightarrow} EAB \overset{k_3}{\rightarrow} E + Y + Z$$

(This is known as the *ordered ternary-complex mechanism;* A must add first to E, and the resulting complex EA reacts with B; the complex EB is not formed.) The concentrations of A and B are much greater than the concentration of E. Apply the steady-state treatment and obtain an expression for the rate.

10.45. The following "ping-pong" mechanism appears sometimes to apply to an enzyme-catalyzed reaction between two substrates A and B to give the final products Y and Z:

$$E + A \underset{k_{-1}}{\overset{k_1}{\rightleftarrows}} EA \overset{k_2}{\rightarrow} EA' + Y$$

$$EA' + B \overset{k_3}{\rightarrow} EA'B \overset{k_4}{\rightarrow} E + Z$$

It can be assumed that the substrates are present in great excess of the enzyme and that steady-state conditions apply. Obtain an expression for the rate of reaction.

Polymerization

10.46. The polymerization of styrene [M] catalyzed by benzoyl peroxide [C] obeys a kinetic equation of the form

$$-\frac{d[M]}{dt} = k[M]^{3/2}[C]^{1/2}$$

Obtain an expression for the kinetic chain length, in terms of [M], [C], and the rate constants for initiation, propagation, and termination.

10.47. The polymerization of ethylene [M] photosensitized by acetone occurs by the mechanism

$$CH_3COCH_3 \overset{h\nu}{\rightarrow} CO + 2CH_3$$
$$CH_3 + C_2H_4 \overset{k_p}{\rightarrow} CH_3CH_2{-}CH_2{-}$$
$$CH_3CH_2CH_2{-} + C_2H_4 \overset{k_p}{\rightarrow} CH_3CH_2CH_2CH_2CH_2{-}$$
$$R_n + R_m \overset{k_t}{\rightarrow} M_{n+m}$$

where one quantum gives $2CH_3$.

Show that the rate equation is

$$-\frac{d[M]}{dt} = k_p\left(\frac{2I}{k_t}\right)^{1/2}[M]$$

where I is the intensity of light absorbed and k_p and k_t are the rate constants for the propagation and termination steps, respectively.

Essay Questions

10.48. Explain the essential features of a chain reaction.

10.49. Give an account of catalysis by acids and bases, distinguishing between specific and general catalysts.

10.50. Will the rate of an enzyme-catalyzed reaction usually be more sensitive to temperature than that of the same reaction when it is uncatalyzed? Discuss.

10.51. Explain how you would determine the parameters K_m and k_c for an enzyme reaction involving a single substrate.

10.52. Explain clearly the difference between collisions and encounters. What significance does this distinction have in chemical kinetics?

10.53. Explain clearly the kind of reasoning involved in deciding what might be the rate-controlling step in a chemical reaction.

10.54. Give a qualitative description of the electronic double-layer theories of Helmholtz, Gouy and Chapman, and Stern.

SUGGESTED READING

See also the suggested reading list for Chapter 9.

The following books and articles deal with special aspects of composite mechanisms.

W. P. Jencks, *Catalysis in Chemistry and Enzymology,* New York: McGraw-Hill, 1969; Dover edition, 1987.

K. J. Laidler, "Rate-controlling step: a necessary or useful concept?" *J. Chem. Ed. 65,* 250–254(1988). In this article some of the pitfalls in deciding on a rate-controlling step are pointed out. It is argued that it is not necessary to worry about identifying a rate-controlling step, and that it is better not to do so if there are any uncertainties about the identification.

K. J. Laidler and P. S. Bunting, *The Chemical Kinetics of Enzyme Action* (2nd ed.), Oxford: Clarendon Press, 1973.

G. J. Minkoff and C. F. H. Tipper, *Chemistry of Combustion Reactions,* London: Butterworth, 1962.

J. R. Murdoch, "What is the rate-limiting step of a multi-step reaction?" *J. Chem. Ed., 58,* 32–36(1981). This article gives a particularly clear analysis of the problem of deciding upon a rate-controlling step when there are a number of consecutive steps.

R. P. Wayne, *Principles and Applications of Photochemistry,* Oxford: University Press, 1988.

J. Wolfrum, H. R. Volpp, R. Rannacher, and J. Warnatzi (Eds.), *Gas Phase Chemical Reaction Systems,* Berlin: Springer, 1996.

For further details on the electrokinetic effects, see

R. C. Allen, C. A. Saravis, and H. R. Maurer, *Gel Electrophoresis and Isoelectronic Focusing of Proteins,* Berlin and New York: de Gruyter, 1984.

A. T. Andrews, *Electrophoresis,* Oxford: Clarendon Press, 1986.

A. W. Preece and K. A. Brown, "Recent trends in particle electrophoresis," in *Advances in Electrophoresis,* Vol. 3 (A. Chrambach, M. J. Dunn, and B. J. Radola, Eds.), Weinheim: Verlag Chemie, 1989.

For a detailed account of electroosmotic dewatering, see

A. K. Vijh, "Electroosmotic dewatering of clays, soils and suspensions," in *Modern Aspects of Electrochemistry* (B. E. Conway, J. O'M. Bockris, and R. E. White, Eds.), Vol. 32, New York: Kluwer Academic/Plenum, 1999.

For an advanced treatment of oscillations and chaos, see

Peter Gray and S. K. Scott, *Chemical Oscillations and Instabilities: Non-Linear Chemical Kinetics,* Oxford: University Press, 1990.

The many volumes of *Comprehensive Chemical Kinetics,* Amsterdam: Elsevier, which have appeared regularly since 1969, cover most aspects of kinetics in considerable detail.

For general accounts of macromolecular systems, see

H. R. Alcock and F. W. Lampe, *Contemporary Polymer Chemistry,* Englewood Cliffs, NJ: Prentice Hall, 1981.

F. W. Billmeyer, *Textbook of Polymer Science,* New York: Wiley, 1963; 3rd edition, 1984.

J. M. G. Cowie, *Polymers: Chemistry and Physics of Modern Materials,* Glasgow: Blackie; New York: Chapman and Hall, 2nd edition, 1991. This book gives a particularly clear account of the subject in its modern aspects, covering both the organic and physical aspects of polymers.

R. J. Hunter, *Introduction to Modern Colloid Science,* Oxford University Press, 1994.

Walter Kauzmann, "Reminiscences from a life in protein physical chemistry," *Protein Science,* 2, 671–691 (1993).

L. H. Sperling, *Introduction to Physical Polymer Science,* New York, Wiley, 1986; 2nd edition, 1992.

For treatments of special aspects of polymer systems, see

D. C. Blackley, *Synthetic Rubbers: Their Chemistry and Technology,* London, New York: Applied Science Publishers, 1983.

C. L. Brooks, M. Karplus, and B. M. Pettitt, *Proteins: A Theoretical Perspective of Dynamics, Structure, and Thermodynamics,* New York: Wiley, 1988.

J. E. Mark, A. Eisenberg, W. W. Graessley, L. Mandel Kern, E. T. Samulski, J. I. Koenig, and G. D. Wignall, *Physical Properties of Polymers,* Washington, DC: American Chemical Society, 2nd edition, 1993. This is an ACS Professional Reference Book.

J. E. Mark and B. Erman, *Rubberlike Elasticity: A Molecular Primer,* New York: Wiley, 1988.

C. Tanford, "How protein chemists learned about the hydrophobic factor," *Protein Science, 6,* 1358–1366 (1997).

An excellent resource for teachers is "Polymer Chemistry Demonstrations and Experiments" presented by POLYED National Information Center for Polymer Education, University of Wisconsin–Stevens Point, Department of Chemistry, Stevens Point, WI 54481–3897.

Quantum Mechanics and Atomic Structure

11

PREVIEW

Quantum mechanics was one of the most important scientific developments of the twentieth century. Advancements in technologies, such as embedding metal in glass and electric circuitry, led to new discoveries that defied description by either classical mechanics or classical electromagnetic theory. Quantum mechanics is the unification of the two theories, electromagnetic waves being allowed to have a particle nature and classical particles to have a wave nature. The wave-particle duality led to a startling result, which is summarized by the Heisenberg uncertainty principle. Whereas in our macroscopic world, we believe that we can improve accuracy of measurement by improving our apparatus, for quantum systems there is an inherent limit to the accuracy with which we can make measurements. This conclusion, called the Copenhagen statistical interpretation of quantum mechanics, was never fully accepted by Einstein, who said "God does not play dice." Today, however, most scientists accept the Copenhagen interpretation.

Quantum mechanics describes the wave-particle duality in terms of the four fundamental forces of nature: gravity, the electromagnetic force, the weak nuclear force, and the strong nuclear force. Quantum chemistry is concerned with the restriction of quantum mechanics to the electromagnetic force and leads to a treatment of the structure and properties of atoms and molecules.

Chemistry is essentially about bonding of atoms to form molecules. Since a chemical bond is a purely quantum phenomenon, this alone underscores the importance of quantum mechanics to chemistry. The birth of *quantum theory* occurred in 1900, but little progress was made until *quantum mechanics* was developed in the 1920s. This chapter begins by outlining what is known about *wave motion,* which played a prominent part in the development of quantum theory and of quantum mechanics. Max Planck developed his quantum theory from an analysis of radiation, and an important contribution was made in 1905 by Einstein, who suggested that light energy is quantized. The first successful application of quantum theory to a problem important in chemistry was in 1913, when Niels Bohr applied his theory to the hydrogen atom. It was soon realized, however, that this type of theory was inadequate, and that an entirely new approach was necessary.

This came in 1925 when Heisenberg first developed, in a rather abstract form involving matrices, the principles of quantum mechanics. Previously, in 1924, de Broglie had predicted that electrons and other particles exhibit *wave properties,* a prediction that was soon confirmed experimentally. This idea led Erwin Schrödinger late in 1925 to develop a system of wave mechanics, which he later showed to be mathematically equivalent to Heisenberg's *matrix mechanics.* Schrödinger made his approach more versatile by formulating quantum mechanics on the basis of a number of *postulates,* which include rules for arriving at an appropriate *quantum-mechanical operator.* When an operator is related to energy it is known as the *Hamiltonian operator* and given the symbol $\hat{H}$. The energy E is then obtained by solving the differential equation

$$\hat{H}\psi = E\psi$$

When applied to atoms and molecules this equation can only be solved for certain values of the energy E, and energy is therefore *quantized.* Only certain functions, known as *eigenfunctions* or *wave functions,* are possible solutions of the equation, so that quantization appears naturally and not as a postulate. These eigenfunctions have a special significance, since their square (or, if they are complex, the product of them and their complex conjugate) gives the *probability density* that an electron is present in a specified small region of space.

The principles of quantum mechanics can be applied to a number of problems of basic importance. One of them is the *particle in the box* and another is the *harmonic oscillator,* approximately exemplified by a *vibrating molecule.* The solution of the wave equations for these systems results in the natural appearance of quantum numbers, which in these examples can have only integral values.

The latter part of this chapter is mainly concerned with the quantum mechanics of hydrogenlike atoms or ions, which have a nucleus and a single electron. The Schrödinger equation can be solved exactly for such systems, and leads to the conclusion that the description of the orbital motion of the electron requires the use of three orbital quantum numbers. For a complete description of atoms, a fourth quantum number, the *spin quantum number,* is introduced. This can have only two values, $+\frac{1}{2}$ and $-\frac{1}{2}$. Associated with electron spin there is an *angular momentum* and a *magnetic moment.* These two properties arise from orbital motion of an electron.

This chapter also deals briefly with atoms having more than one electron, and here we have to be content with approximate solutions of the Schrödinger equation. The orbitals into which the electrons go depend in part on the order of orbital energy levels. There is also an important restriction imposed by the *Pauli exclusion principle,* according to which no two electrons in an atom can have the same set of four quantum numbers.

OBJECTIVES

After studying this chapter, the student should be able to:

- Exhibit a basic knowledge of classical electromagnetic waves.
- Appreciate how the experimental data from blackbody radiation could be described only using a quantum assumption.
- Understand that the Bohr model of the atom used the ideas of quantized angular momentum and explained emission spectroscopy.
- Recognize the failings of the Bohr model.
- Appreciate the historical development of quantum mechanics and how it led to the postulates that formulate the modern quantum theory.
- Understand that the Heisenberg uncertainty relations express a fundamental difference between measurement of classical and quantum systems.
- Recognize the importance of the non-commutation of operators in quantum theory.
- Understand the Schrödinger equation as an eigenvalue equation.
- Solve the Schrödinger equation for some simple systems.
- Formulate and solve the Schrödinger equation for the hydrogen atom.
- Understand how the hydrogen atom forms the basis for the electronic structure of the polyelectron atoms.
- Appreciate the role of symmetry in the formulation of the Pauli principle.
- Understand how Slater determinants correctly formulate the antisymmetric nature of wave function containing more than one electron.
- Appreciate that, as a result of electron-electron correlations, it is impossible to solve the Schrödinger equation for atoms that contain more than one electron.
- Understand the basic ideas of the variational method and perturbation theory as applied to quantum systems.
- Recognize the mathematical and notational advantages of the bra-ket notation due to Dirac.

So far we have developed physical chemistry with little regard to the existence of the fundamental particles that comprise all matter. This approach reveals a good deal about the properties of matter, but much more can be accomplished on the basis of the laws that govern the behavior of these fundamental particles. During the 1930s it became apparent that more progress could be made by considering the *wave properties* of particles, and we therefore begin by summarizing what is known about wave motion.

11.1 Electromagnetic Radiation and the Old Quantum Theory

Wavelength and Frequency

Visible light and many other apparently different types of radiation are all forms of *electromagnetic radiation,* and in a vacuum they all travel with the same speed of $c = 2.998 \times 10^8$ metres per second (m s^{-1}). Electromagnetic radiation is characterized by a **wavelength** λ and a **frequency** ν. These two physical quantities are related to the speed of light c by the equation

$$\lambda \nu = c \tag{11.1}$$

The SI unit of wavelength is the metre, although multiples such as nanometres (nm) are frequently used, especially in spectroscopy (Chapters 13 and 14). The SI unit of frequency is the reciprocal second (s^{-1}), which is also called the *hertz* (Hz).

Because of this relationship, all electromagnetic radiation can be classified in terms of either its wavelength or its frequency, as is done in Figure 11.1. We see that long radio waves and electric waves may have wavelengths of many kilometres and low frequencies, whereas γ (gamma) rays have high frequencies and extremely short wavelengths.

Electromagnetic radiation differs from certain other types of waves in a number of important respects. Sound waves, water waves, seismic waves, and waves in a plucked guitar string exist only by virtue of the medium in which they occur, whereas electromagnetic waves can travel in a vacuum. Sound waves traveling through a gas, for example, consist of alternating zones of compression and rarefaction, and the molecular displacements that occur are in the direction in which the wave travels. As the wave passes a certain region, the gas molecules undergo changes in energy and momentum, which then pass on to the next region. This type of wave propagation is also found with a seismic wave traveling through the earth.

An electromagnetic wave is essentially different, since it can travel through a vacuum, and the medium is not essential. However, when such a wave comes in contact with matter, important interactions affect the wave and the material. There is a coupling of the radiation with the medium, and how this occurs is best considered with reference to Figure 11.2, which shows that the wave has two components, one an electric field and the other a magnetic field. These components are in two planes at right angles to each other. A given point in space experiences a periodic disturbance in electric and magnetic field as the wave passes by. A charged particle such as an electron couples its charge with these field fluctuations and oscillates with the frequency of the wave. A useful analogy is provided by a cork floating on

Study the parts of the electromagnetic spectrum; Wavelength and frequency; Electromagnetic spectrum.

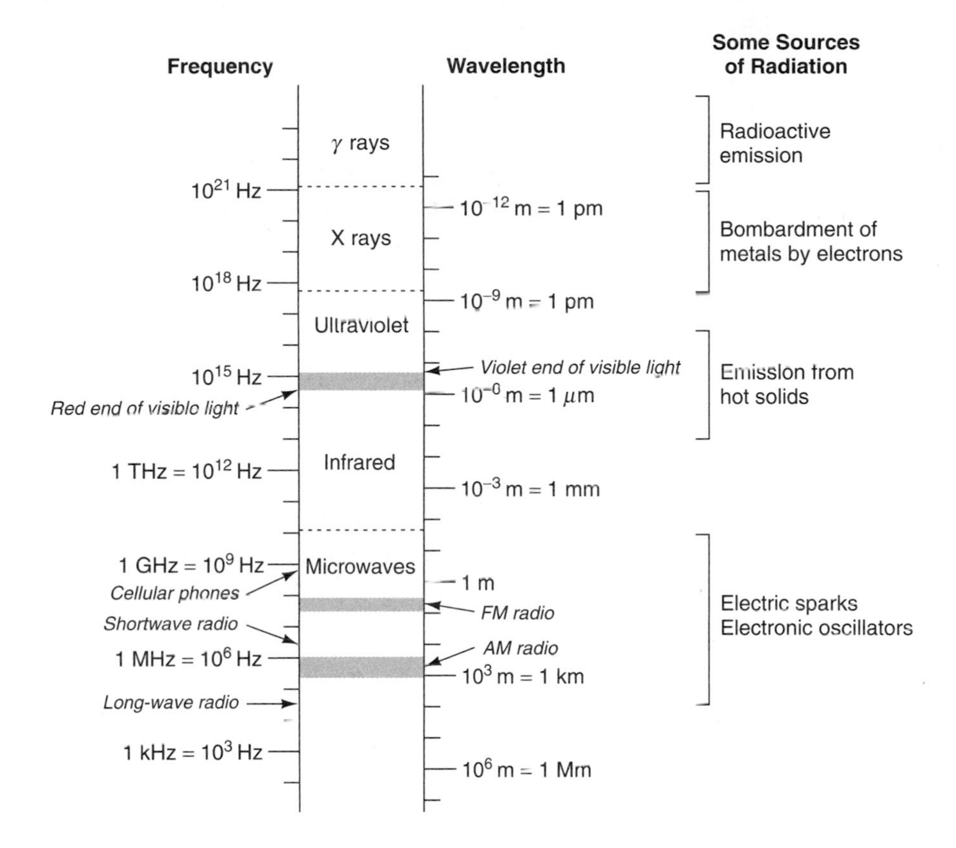

FIGURE 11.1
The electromagnetic spectrum.

still water. If a ripple passes, the cork bobs up and down; its motion is perpendicular to the direction of the wave and it has the same frequency. Similarly an electromagnetic wave causes the particle to move *transverse* (i.e., at right angles) to the direction of propagation, and it is therefore called a *transverse* wave. By contrast, a sound wave is a *longitudinal* wave since the motion of the medium is in the direction of propagation. Both longitudinal and transverse waves are possible in a given medium, a fact that is used to locate the epicenters of earthquakes.

The analogy of a cork on water may be carried further. If the cork is made to oscillate up and down on the surface, a ripple wave is generated. Similarly, an oscillating electron induces electric and magnetic fields and generates an electromagnetic wave.

FIGURE 11.2
The propagation of one type of electromagnetic wave, showing the oscillation of the electric field and the magnetic field in planes at right angles to one another, at a given time. Standing waves, and plane-polarized waves, are of this type.

Simple Harmonic Motion

It is important to understand the physics and mathematics of wave motion, since this is essential to an understanding of atomic and molecular structure. Some fundamental principles are illustrated in Figure 11.3. Suppose that point C in Figure 11.3a moves counterclockwise on the circumference of the circle; we will call this the *positive* direction. We will be particularly interested in the motion of point P, the projection of point C on the Y axis, which is plotted in Figure 11.3b. Suppose that point C is at B at zero time and moves with constant speed. During a complete revolution with the point C starting at B, the distance y between P and the origin starts from zero, becomes equal to A after a quarter revolution ($\frac{\pi}{2}$ rad), is zero again after half a revolution (π rad), is $-A$ at $\frac{3\pi}{2}$ rad, and is zero again after a complete revolution. This

Follow the time evolution of a wave; Electromagnetic waves: Harmonic oscillator.

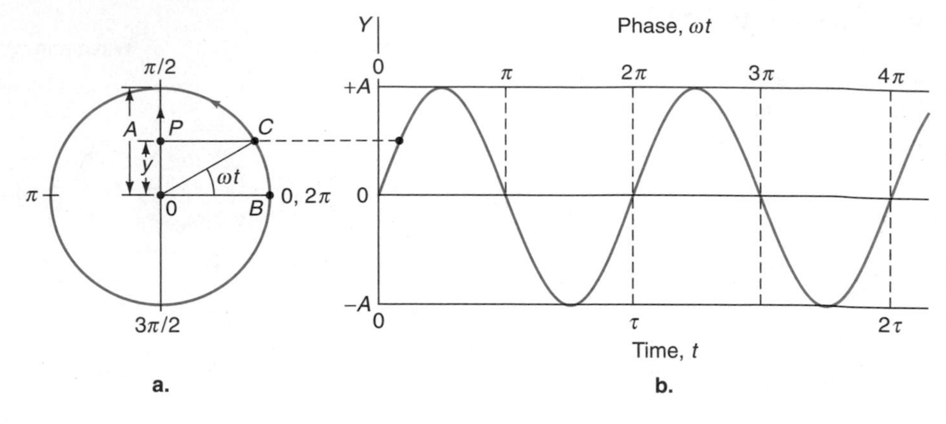

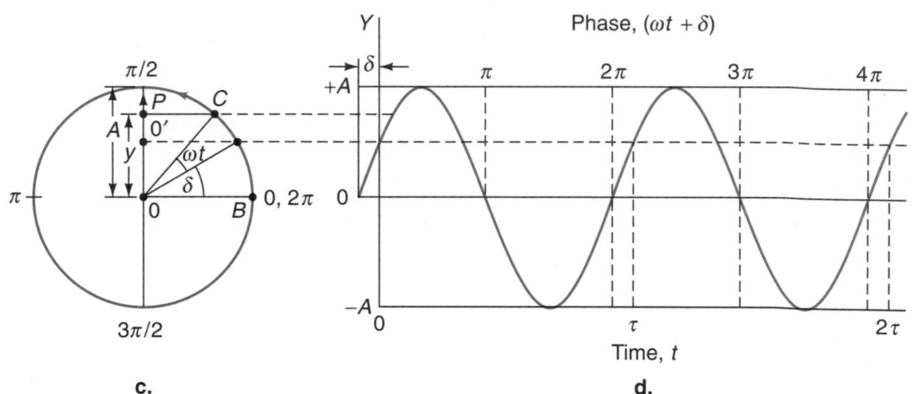

FIGURE 11.3
The generation of simple harmonic motion. (a) A point C undergoes circular motion in the anticlockwise direction. (b) The variation with time of the displacement y of point P; the projection of C on the y axis. (c) and (d) The same as (a) and (b) but with an initial phase displacement δ.

Amplitude; Period

variation of y is shown in Figure 11.3b, the same pattern then continuing indefinitely. The maximum displacement A is known as the **amplitude.** The **period** τ is the time for one revolution, and since the angular path for one revolution is 2π radians, we have[1]

$$\tau = \frac{2\pi \text{ rad}}{\omega} \tag{11.2}$$

Frequency

where ω is the angular velocity; its SI unit is rad s^{-1}. The **frequency,** ν, is the reciprocal of the period,

Identify the wavelength, frequency, and amplitude; Electromagnetic waves: Amplitude; Wavelength; Frequency.

$$\nu = \frac{1}{\tau} = \frac{\omega}{2\pi \text{ rad}} \tag{11.3}$$

and therefore

$$\omega = (2\pi \text{ rad})\nu \tag{11.4}$$

The mathematical form of the displacement y, shown in Figure 11.3b, is

$$y = A \sin(\omega t) \tag{11.5}$$

[1]A complete revolution is 2π rad. Note that the unit, the radian, must be included in Eqs. 11.2–11.4 to balance the units.

Phase

See the effects of phase; Electromagnetic waves; Phase.

where the quantity ωt (unit: radian) is called the **phase.** More generally, we can allow a displacement by the angle δ before simple harmonic motion begins, as shown in Figure 11.3c. This displacement δ is known as the **phase constant,** and this modification requires that

$$y = A \sin(\omega t + \delta) \tag{11.6}$$

The corresponding variation of y with t is shown in Figure 11.3d. Since the curves are not superimposable without a *phase shift,* they are said to be *out of phase.*

To obtain the *acceleration* of point P we differentiate this equation twice:

$$\frac{d^2y}{dt^2} = -A\omega^2 \sin(\omega t + \delta) \tag{11.7}$$

Substitution of Eq. 11.6 into Eq. 11.7 gives the equation

$$\frac{d^2y}{dt^2} = -\omega^2 y \tag{11.8}$$

Simple Harmonic Motion

This is the equation for **simple harmonic motion.**

Any motion obeying Eq. 11.8 is referred to as simple harmonic motion. Such motion is also found with a mass attached to the end of a spring in which the restoring force obeys *Hooke's law,* which means that the force is proportional to the displacement y:

Hooke's Law

$$F = -k_h y \tag{11.9}$$

Hooke's law.

where k_h is the *force constant.* According to Newton's second law of motion this force is the mass m times the acceleration d^2y/dt^2:

$$F = m\frac{d^2y}{dt^2} \tag{11.10}$$

Equating these two expressions for F gives

$$\frac{d^2y}{dt^2} = \frac{-k_h}{m}y \tag{11.11}$$

To solve this we need a function y that when differentiated twice gives the function back again. This condition is satisfied by the sine and cosine functions or by linear combinations of them. For simplicity we choose the sine function,

$$y = a \sin\left(\sqrt{\frac{k_h}{m}}t + b\right) \tag{11.12}$$

Double differentiation gives

$$\frac{d^2y}{dt^2} = -\frac{k_h}{m}a \sin\left(\sqrt{\frac{k_h}{m}}t + b\right) = -\frac{k_h}{m}y \tag{11.13}$$

Our chosen function has therefore satisfied Eq. 11.11. The choices of a and b in Eq. 11.12 are arbitrary and can be determined from initial conditions. Equation 11.12 becomes identical with Eq. 11.6 if a is the amplitude A, if b is the initial phase angle δ, and if

$$\sqrt{\frac{k_h}{m}} = \omega/\text{rad} \tag{11.14}$$

Substitution into this expression of the value of ω from Eq. 11.4 gives

$$\nu = \frac{1}{2\pi}\sqrt{\frac{k_h}{m}} \tag{11.15}$$

Natural Frequency

This frequency, known as the **natural frequency** of the simple harmonic motion, thus varies inversely with the square root of the mass.

EXAMPLE 11.1 Suppose that a hydrogen atom (mass = 1.67×10^{-27} kg) is attached to the surface of a solid by a bond having a force constant of 5.0 kg s^{-2}. Calculate the frequency of its vibration.

Solution From Eq. 11.15 it follows that

$$\nu = \frac{1}{2\pi}\sqrt{\frac{k_h}{m}} = \frac{1}{2\pi}\sqrt{\frac{5.0\ (\text{kg s}^{-2})}{1.67 \times 10^{-27}\ (\text{kg})}}$$
$$= 8.70 \times 10^{12}\ \text{s}^{-1}$$

When a body is oscillating, its kinetic energy E_k and its potential energy E_p are continuously varying, but their sum E_{total} is a constant:

$$E_{\text{total}} = E_k + E_p \tag{11.16}$$

We have seen (Eq. 1.12) that the potential energy is the work done in moving a mass from its equilibrium position to a new position; thus, for the displacement y from the equilibrium position

$$E_p = \int_0^y (-F)\,dy = \int_0^y k_h y\,dy = \frac{k_h y^2}{2} \tag{11.17}$$

The kinetic energy is $\frac{1}{2}mu^2$ where u is the velocity. We can evaluate the total energy by calculating its value when the oscillator reverses its direction; then $u = 0$ and $y = A$, the maximum amplitude. The potential energy is then $\frac{1}{2}k_h A^2$ (Eq. 11.17) and this is the total energy:

$$E_{\text{total}} = \tfrac{1}{2}k_h A^2 \tag{11.18}$$

This result, that the energy is proportional to A^2, is true for any type of simple harmonic motion. For electromagnetic wave motion in a vacuum, for example, the electrical energy is proportional to the square of the electrical field vector (see Figure 11.2), and the magnetic energy is proportional to the square of the magnetic field vector.

Plane Waves and Standing Waves

So far we have considered oscillations at right angles to the direction of propagation of the wave. Energy must also be transported along the wave. A simple type of wave is the *plane wave,* which consists of parallel waves of constant amplitude and has a planar wave front; a useful model of a plane wave is provided by a string stretched horizontally. If it is given a single vertical pulse (Figure 11.4a), the pulse

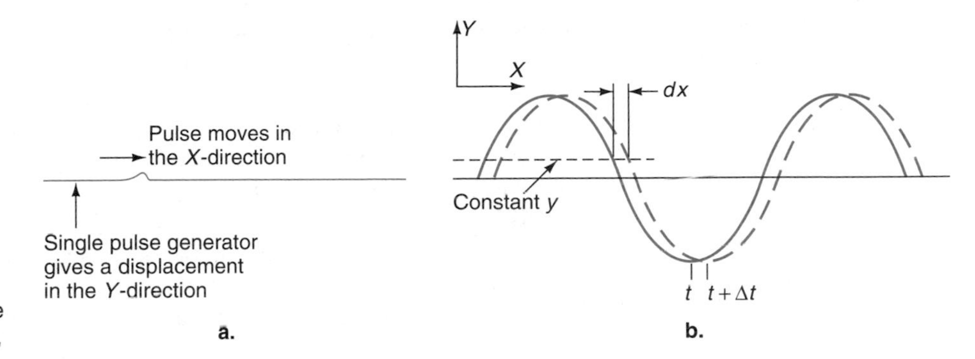

FIGURE 11.4
(a) A single wave pulse traveling along a stretched string. (b) Two profiles of a traveling wave. The solid line shows the transverse displacement y at a position x at time t. The dashed line shows the same curve at a later time $t + \Delta t$.

Traveling Wave
Watch a traveling wave; Plane and standing waves; Wave motion.

travels the length of the string. If a fixed-frequency vertical oscillation is applied, a *traveling wave* is generated. This is a wave whose amplitude at a particular position along the string changes periodically with time. Figure 11.4b shows two wave patterns captured at two different instants; these two *profiles* illustrate the fact that the transverse displacement is moving along the string.

Analysis of this type of wave motion leads to the result that the displacement y is given by the equation

$$y(x, t) = A \sin \frac{2\pi}{\lambda}(x \pm ut) \tag{11.19}$$

Phase Velocity

Watch a wave travel to the left and right; Plane and standing waves; Plane waves.

The quantity u is called the **phase velocity** because it is the velocity with which a given phase of the wave travels along the X axis. If the positive sign is taken, the wave is traveling to the right with velocity u (see Figure 11.4b); the negative sign means that it is traveling to the left. For electromagnetic radiation traveling in a vacuum the phase velocity is the speed of light, $c = 2.998 \times 10^8 \text{ m s}^{-1}$.

So far we have considered the length of the string to be infinite. A finite string vibrates in a pattern having evenly spaced **nodes** (points of zero displacement) and **antinodes** (points of maximum displacement). An example is shown in Figure 11.5,

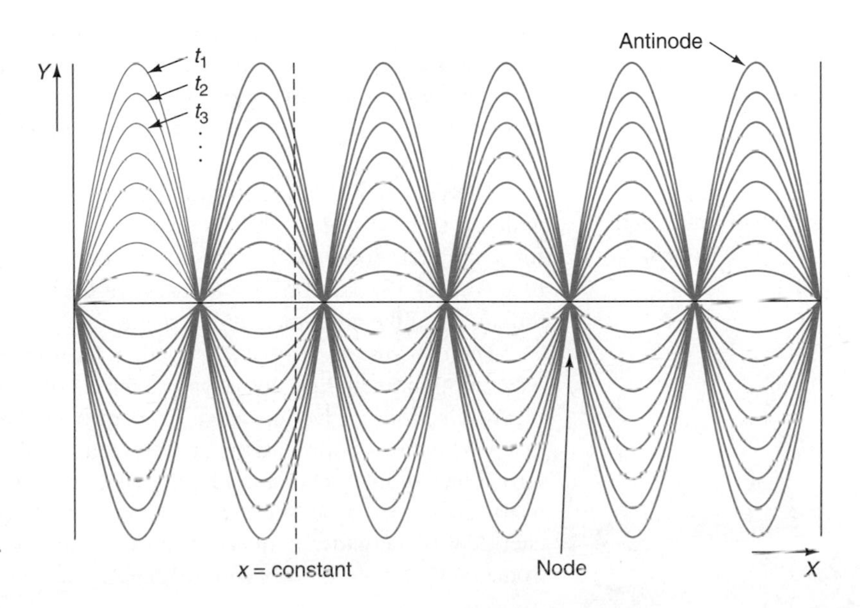

FIGURE 11.5
A typical standing wave in a stretched string fixed at both ends. This diagram shows individual profiles at different times t_1, t_2, t_3, ... Each point vibrates at the same frequency but with different amplitudes. The amplitude is zero at a node and a maximum at an antinode.

Plot Figure 11.5 and see nodes and antinodes; Plane and standing waves; Nodes and antinodes.

Create nodal lines in 2D; Plane and standing waves; Nodal lines.

See constructive and destructive interference; Plane and standing waves.

in which individual profiles are superimposed to show the patterns at various times. The fixed ends are themselves nodes and may be the only nodes; there may be any integral number of nodes as shown in Figure 11.5. This wave form is called a **standing wave** or a *stationary wave.* This situation is dealt with by applying the **principle of superposition,** according to which when two or more waves are involved, the displacement at any position is the sum of the displacements at that position for each individual wave. A special case of this is when the crest of one wave coincides with the trough of another, so that the two cancel; there is then said to be *destructive interference.* Conversely, if crests coincide, the waves reinforce each other, and we then speak of *constructive interference.* The standing wave shown in Figure 11.5 is the sum of the displacements of waves that are traveling back and forth and are reflected at the ends of the string. The magnitude of the amplitude at a given point fluctuates with time, but the positions of the maxima and minima remain fixed in space.

The equation for a standing wave, such as shown in Figure 11.5, is

$$y(x, t) = A_s \sin \frac{n\pi x}{l} \sin \sqrt{\frac{k_h}{m}}\, t \tag{11.20}$$

where A_s is the amplitude of the standing wave, l is the length of the string, and n is a constant to be determined. This equation already satisfies the left-hand boundary condition that $y = 0$ when $x = 0$. It must also satisfy the right-hand boundary condition that $y = 0$ when $x = l$. Insertion of this condition into Eq. 11.20 leads to

$$\sin n\pi = 0 \tag{11.21}$$

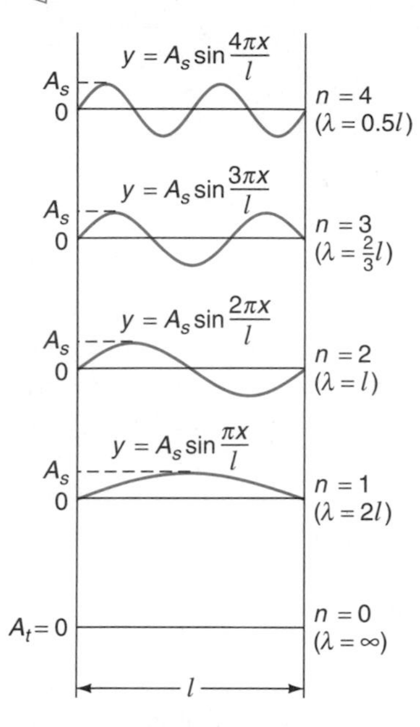

FIGURE 11.6
Some modes of vibration allowed for a standing wave in a stretched string fixed at both ends.

Since the sine is zero only for the angles 0, π, 2π, etc., it follows that *n must be an integer*.

Plot Figure 11.6 and show how standing waves fit only if $\lambda = \frac{2l}{n}$; Plane and standing waves; Create waves.

The value of n is the number of antinodes in the string, as illustrated in Figure 11.6. When $n = 0$, there is zero displacement for all values of x, and no antinodes are present; this means that there is no vibration. When $n = 1$, the value of y is zero only if $x = 0$ or $x = l$; there are only two nodes, at the ends of the string, and there is an antinode in the middle. When $n = 2$, nodes occur at $x = 0$, $x = l/2$, and $x = l$, and there are two antinodes. It is easily seen that only those wavelengths are allowed for which l is an integral multiple of one-half the wavelength. Corresponding to these allowed wavelengths are certain frequencies that correspond to what are known as **modes of vibration.** The actual vibration of a string need not correspond to a single mode but may be a linear combination of modes, formed by a superposition of modes.

Modes of Vibration

Create a variety of waves by superimposing two sine waves; Plane and standing waves; Superposition.

The energy of a standing wave depends on the distance along the wave. At a node the energy is zero, while at an antinode the energy is a maximum. We saw that for simple harmonic motion the energy is proportional to the square of the amplitude (Eq. 11.18) and for a standing wave the energy at any point is proportional to the square of the amplitude at that point.

The conclusions we have arrived at for a vibrating string apply to all kinds of wave motion, including the wave properties of atoms and molecules. We will later regard electrons as having wave properties and will consider the standing waves that are confined within atoms and molecules. A consequence of this is the natural appearance of integral quantum numbers that restrict the wave functions and lead to the quantization of measurable quantities. Finally, an understanding of wave motion is critical to the understanding of spectroscopy, which is concerned with how electromagnetic waves interact with matter.

Blackbody Radiation

Change the temperature of a blackbody; Blackbody radiation; Thermal radiation.

Classical physics was dominant until the end of the nineteenth century, when it was thought that the major problems had been solved; all that seemed to remain was to fill in some details. This view was shattered by investigations on the variation with wavelength of the intensity of radiation emitted by a hot solid.

A familiar example of how intensity and wavelength vary with temperature is provided by an element of a stove. As the element heats up, the metal first feels hot as invisible infrared radiation is emitted. At a higher temperature a red glow appears; the temperature is then about 950 K. As the temperature continues to increase, the radiation becomes bright red. This is about the limit of heating a commercial stove element, but if we could continue to heat the element, as we do in the tungsten filament of an incandescent electric light bulb, the emitted radiation would become orange, then yellow, and finally, at about 2300 K, white. A substantial portion of the radiation in the case of the stove and the light bulb, however, is still in the infrared.

Blackbody

In 1859 the German physicist Gustav Robert Kirchhoff (1824–1887) conceived of a substance that would absorb all frequencies of radiation falling on it; it would reflect no radiation and appear black. Such an idealized substance, known as a **blackbody,** would be in perfect equilibrium, such that the radiation striking it would be equal in intensity to the radiation emitted by it. No such substance really exists, but in 1894 the German physicist Wilhelm Wien (1864–1928) suggested that a close approximation is a cavity with a small hole leading to it. Any radiation passing into the cavity would have little chance of being reflected out. It was found experimentally that the intensity of radiation emitted by such a device is independent of the material of which it is constructed and depends only on the temperature.

Stefan-Boltzmann Law

Somewhat earlier, in 1879, the Austrian physicist Josef Stefan (1835–1893) found experimentally that the total energy radiated by a hot body increases with the fourth power of the absolute temperature **(Stefan's law),** and his ideas were further developed by Ludwig Boltzmann in the form of what is known as the **Stefan-Boltzmann law.** It is convenient to consider what today is called the **radiant energy density** ρ_ν, which is the energy at a frequency ν, per unit volume and per unit frequency range. Until about 1900, measurements of the frequency distribution of the radiant energy density were confined to the visible and near ultraviolet regions of the spectrum. In 1896 Wien made an attempt to explain the variation of ρ_ν with ν, but on the basis of somewhat doubtful assumptions. He obtained the expression

$$\rho_\nu = \alpha\nu^3 \exp(-\beta\nu/T) \qquad (11.22)$$

where α and β are constants. A plot of his function, which passes through a maximum, is shown in Figure 11.7.

The variation of the radiant energy density with the frequency was extensively investigated, especially by the German physicists Otto R. Lummer (1860–1925), Ernst Pringsheim (1859–1917), Heinrich Rubens (1865–1922), and Ferdinand Kurlbaum (1857–1927). At a given temperature the curves go through a maximum, as shown in Figure 11.7a, and as the temperature is raised the maximum shifts to shorter wavelengths. This result was qualitatively predicted by Wien's equation 11.22, which at first seemed to be satisfactory. In 1900, however, Rubens and Kurlbaum made measurements over a wider frequency range than previously, going further into the infrared, that is, to lower frequencies. They found that whereas Wien's formula gave excellent agreement at the higher frequencies, at the lower

Plot Figure 11.7b for any temperature; Blackbody radiation; Graphing.

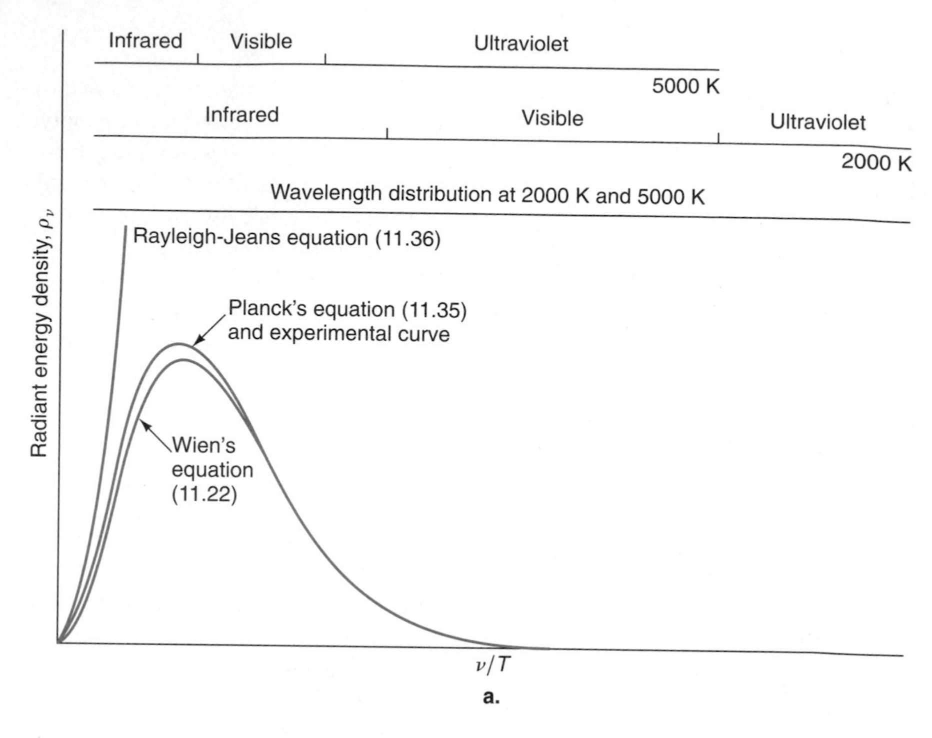

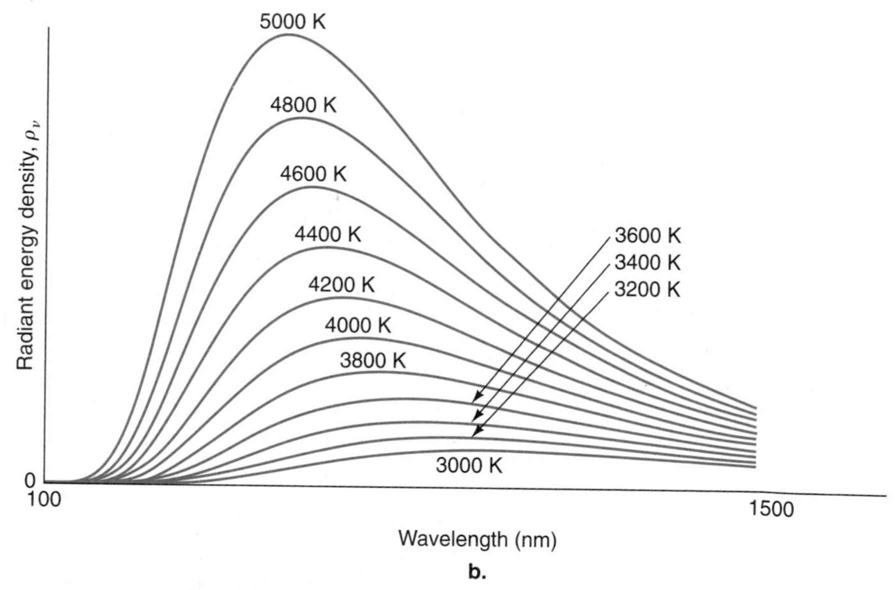

FIGURE 11.7
a) Plots of Wien's equation 11.22, Planck's equation 11.35, and the Rayleigh-Jeans equation 11.36. Since all three equations involve the ratio of the frequency ν to the temperature T, the radiant energy density ρ_ν is plotted against ν/T. The agreement with Planck's equation is within the experimental error at all temperatures measured. The approximate infrared, visible, and infrared ranges are shown at two temperatures.
b) A plot of intensity versus wavelength for a series of temperatures using Plank's expression for radiant energy density, Eq. 11.35. Note that the maximum intensity shifts to shorter wavelengths as the temperature increases.

frequencies it predicted radiant energy densities that were too low, the discrepancies becoming relatively greater as the frequency was lowered.

The solution to the problem of the discrepancies at low frequencies was given in 1900 by the German physicist Max Karl Ernst Ludwig Planck (1858–1947). He first suggested, on a purely empirical basis, the equation

$$\rho_\nu = \frac{\alpha\nu^3}{\exp(\beta\nu/T) - 1} \tag{11.23}$$

This expression differs from Wien's equation 11.22 only by the inclusion of unity in the denominator; it predicts a curve which always lies above the Wien curve (Fig. 11.7a). At higher frequencies, where unity in the denominator can be neglected, the equation reduces to Wien's equation.

When Planck found that his empirical formula fitted (within the experimental error) the experimental results over the entire frequency range, he at once began to look for a theoretical derivation that would lead to it. He based his treatment on some theoretical work on radiation that had been done by Ludwig Boltzmann (1844–1906). Purely as a mathematical convenience, Boltzmann had often treated energy as coming in small packets, but at the end of his derivations he always required the energy units to become of zero size. Planck did not, however, carry out this final step, and he did not at first appreciate the great significance of this omission. The packets of energy are called quanta, from the Latin *quantum,* for "how much?" Planck assumed that the packets are of energy $h\nu$, where the proportionality factor h has come to be called **Planck's constant.**

Planck's Constant

Planck's treatment was in terms of the vibrations of the atoms of which the solid is composed. Recall that electromagnetic radiation is generated when electric charges oscillate, and that the frequency of the radiation depends upon the frequency of the oscillating charges. Radio transmission is an example of this. In the case of blackbody radiation, as the atoms in a solid are heated, their vibration increases. Planck assumed that all the atoms have the same fundamental vibrational frequency ν, and that since the energy comes in quanta of $h\nu$ there will be

N_0 oscillators having zero energy

N_1 oscillators having energy $h\nu$

N_2 oscillators having energy $2h\nu$, and so on.

He further assumed that the distribution over these allowed energies is governed by the Boltzmann equation (Section 15.2), according to which the number of oscillators N_1 having energy $\epsilon_1(= h\nu)$ is

$$N_1 = N_0 \exp(-\epsilon_1/k_B T) = N_0 \exp(-h\nu/k_B T) \tag{11.24}$$

where N_0 is the number with zero energy and k_B, the Boltzmann constant, is R/L, the gas constant per molecule. The total number of oscillators in all states is thus

$$N = N_0 + N_0 \exp(-h\nu/k_B T) + N_0 \exp(-2h\nu/k_B T) \ldots \tag{11.25}$$

This series is of the form

$$N = N_0(1 + x + x^2 + x^3 + \cdots) \tag{11.26}$$

where x is $\exp(-h\nu/k_B T)$. The series is an expansion of $1/(1 - x)$, $x < 1$ and Eq. 11.26 thus becomes

$$N = \frac{N_0}{1 - x} = \frac{N_0}{1 - \exp(-h\nu/k_B T)} \tag{11.27}$$

The total energy E of all the oscillators is the number of oscillators in each state multiplied by the energy of that state:

$$E = N_0(0) + N_1(h\nu) + N_2(2h\nu) + \cdots \tag{11.28}$$

With the use of Eq. 11.24 applied to every level this becomes

$$E = N_0 h\nu \exp(-h\nu/k_B T) + N_0 2h\nu \exp(-2h\nu/k_B T) + \cdots \quad (11.29)$$

$$= N_0 h\nu \exp(-h\nu/k_B T)[1 + 2\exp(-h\nu/k_B T) + \cdots] \quad (11.30)$$

The series in parentheses is of the form $(1 + 2x + 3x^2 + \cdots)$, which is an expansion of $1/(1 - x)^2$, $x < 1$. Equation 11.30 may thus be written as

$$E = \frac{N_0 h\nu \exp(-h\nu/k_B T)}{[(1 - \exp(-h\nu/k_B T)]^2} \quad (11.31)$$

However, by Eq. 11.27, N_0 is equal to $N[1 - \exp(-h\nu/k_B T)]$ and therefore the average energy of an oscillator is

$$\bar{\epsilon} = \frac{E}{N} = \frac{h\nu \exp(-h\nu/k_B T)}{1 - \exp(-h\nu/k_B T)} \quad (11.32)$$

$$= \frac{h\nu}{\exp(h\nu/k_B T) - 1} \quad (11.33)$$

This is Planck's expression for the average energy. When $h\nu$ is much less than $k_B T$, the exponential is approximately $1 + h\nu/k_B T$, so that the term in the denominator is $h\nu/k_B T$, and the average energy therefore becomes

$$\bar{\epsilon} = k_B T \quad (11.34)$$

Test yourself on the units of Eq. 11.35; Blackbody radiation; Quiz.

which is the classical value. The quantum theory thus reduces to classical theory when the quanta are small or the temperature is high.

For the radiant energy density, Eq. 11.33 led Planck to the expression

$$\rho_\nu = \frac{8\pi\nu^2}{c^3} \frac{h\nu}{\exp(h\nu/k_B T) - 1} \quad (11.35)$$

This expression is of the same form as Planck's empirical equation 11.23, which gave such extremely close agreement with the experimental curves.

At about the same time that Planck made this important contribution, attempts were being made to explain blackbody radiation on the basis of classical physics, but all were complete failures. One such attempt was made in 1900 by John William Strutt, Lord Rayleigh (1842–1919), and was later slightly modified and improved by the mathematician and astronomer Sir James Hopwood Jeans (1877–1946). Their theory was based on the assumption that there is a continuous distribution of vibrational energies. It was also based on the principle of equipartition of energy between all the vibrations in the substance that is emitting the radiation, a principle that had been insisted upon by the leading theoretical physicists of the nineteenth century. The expression they obtained was

Rayleigh-Jeans Equation

$$\rho_\nu = \frac{8\pi\nu^2}{c^3} k_B T \quad (11.36)$$

According to this equation the radiant energy density would not pass through a maximum but would increase indefinitely with increase in frequency, (Figure 11.7a). That Eq. 11.36 is incorrect can be deduced from the fact that the night sky (at 3 K) would be predicted to emit a constant blue-violet glow. These conclusions are completely at variance with the experimental results, and the expression "ultraviolet catastrophe" was later used to refer to this serious discrepancy.

It is now easy to understand why the Rayleigh-Jeans theory is bound to lead to a continuous increase in radiant energy density with increasing frequency. Whereas Planck's treatment (Eq. 11.35) restricts the amount of energy in the higher levels, the Rayleigh-Jeans treatment allows every mode of vibration to have the energy k_BT, which is too large at the higher frequencies. At sufficiently low frequencies the expression $\exp(h\nu/k_BT) - 1$ in the denominator of Planck's equation 11.35 reduces to $h\nu/k_BT$, so that the equation reduces to the Rayleigh-Jeans equation.

An important contribution made by Planck in his original publication was to evaluate the constants h and k_B from the experimental data on the spectral distribution. His values are in good agreement with the best modern values. Surprisingly, Boltzmann had never stressed the great importance of his constant k_B, and had never obtained a value for it; Planck was the first to do so. Noting that k_B is the gas constant per molecule, and therefore equal to the gas constant R divided by the Avogadro constant L, Planck also obtained a value for L; his value was 6.175×10^{23} mol^{-1}, to be compared with the modern value of 6.022×10^{23} mol^{-1}. He also obtained a good value for the elementary unit of charge, e, as the ratio of the Faraday constant to the Avogadro constant. Since at the time there was doubt as to the values of some of these constants, this contribution was of importance in itself, aside from giving support to the general idea of quanta.

In spite of this, for several years little attention was paid to Planck's contribution, and most of the attention was unfavorable. In 1905 Jeans, for example, stated flatly that "the true value of h is $h = 0$"; in other words, there is no such thing as quantization, in spite of the fact that ignoring it led to the "ultraviolet catastrophe." However, in 1905, strong support was provided to the idea of quantization by Albert Einstein (1879–1955).

Einstein and the Quantization of Radiation

Planck regarded his theory as applying only to material oscillators, such as those in a solid which could emit radiation only in multiples of $h\nu$. The important contribution made by Einstein in 1905 was to suggest that electromagnetic radiation itself is quantized. Modern textbooks often refer to Einstein's paper on this topic as his paper on the **photoelectric effect,** and say that it was based on Planck's 1900 paper on the quantization of the energy of a body emitting blackbody radiation. A perusal of Einstein's paper,[2] however, shows that this is by no means the case. The paper makes only a passing reference to Planck's paper, and Planck's constant h does not appear in it explicitly; in fact, at that time Einstein was not convinced that oscillator energies are quantized, but thought it likely that radiation is quantized. The paper is based not on Planck's but on Wien's treatment of radiation, and is mainly concerned with experimental evidence relating to the absorption and emission of radiation. It discusses fluorescence and photoionization, only one of the nine sections being concerned with the photoelectric effect. Einstein later realized that his treatment, based on Wien's equation, was inconsistent with the experimental data on the spectral distribution of energies, and by 1907 he had come to accept the quantization of oscillator energies as well as of radiation.

---

[2]Einstein's paper on radiation is well worth reading today; a translation of it appears in A. B. Arons and M. B. Peppard, *Amer. J. Phys.*, *33*, 367–374(1965). The curious title of Einstein's paper, "On a heuristic viewpoint concerning the emission and transformation of light," has elicited much comment, the word *heuristic* meaning "serving to discover."

Photoelectric Effect

It is perhaps true that the **photoelectric effect** gives the most striking evidence for the quantization of radiation, and indeed Einstein's 1921 Nobel Prize was specifically for his work on the photoelectric effect. This effect was discovered in 1887 by Heinrich Rudolf Hertz (1857–1894), who found that when suitable radiation (usually in the visible or ultraviolet) falls on a metal surface, electrons are emitted from the surface. Later experiments, particularly those of Johann Elster (1854–1920), Hans Geitel (1855–1923), and Philipp Lenard (1862–1946), showed that the *frequency* of the light and *not* its *intensity* controls whether or not electrons are emitted. The *number* of electrons emitted is proportional to the intensity, and the electron emission occurs the instant the radiation strikes the metal surface.

It is impossible to explain these results on the basis of classical physics, according to which the energy of the radiation should depend only on its intensity. Whether or not electrons are emitted should therefore depend on the energy of the radiation, and even low intensities, regardless of frequency, should be effective if long periods of time were allowed for their absorption.

Einstein's suggestion was that light itself is a beam of particles each of which has energy equal to $h\nu$, where ν is the frequency of the radiation. He envisaged the photoelectric effect as a process in which the quantum of energy $h\nu$ is given to an individual electron, which may then have sufficient energy to leave the metal surface. If w is the potential energy (called the **work function**) that is required to remove the electron from the surface, then

Work Function

$$h\nu = \frac{1}{2}mu^2 + w \tag{11.37}$$

where $\frac{1}{2}mu^2$ is the kinetic energy of the electron once it has left the metal. If the energy $h\nu$ is less than w, the electron cannot leave the surface; but if $h\nu$ is greater than w, it can leave, and the excess, $h\nu - w$, will appear as kinetic energy of the electron. There is therefore a frequency ν_0, equal to w/h, which is just sufficient to remove the electron, and this is known as the **threshold frequency.**

Threshold Frequency for Electron Emission

This theory clearly explains why the emission of an electron is related to a threshold frequency of the radiation, and not to a threshold intensity. The intensity affects not whether an electron is emitted, but how many electrons are emitted. The theory also accounts for the fact that the emission is instantaneous; the energy $h\nu$ interacts with a single electron and cannot be stored in the metal.

Photon

In 1926 the American chemist Gilbert Newton Lewis (1875–1946; see p. 579 for historical notes) suggested the name **photon** (Greek *photos,* light) for these particles of radiation of energy $h\nu$. The essential conclusion from the photoelectric experiments was that there is a one-to-one relationship between a photon absorbed and an electron emitted. When a photon interacts with an electron, the electron gains the energy $h\nu$ and the photon disappears.

Test yourself on the main terms of the photoelectric effect; Photoelectric effect.

One objection raised against Einstein's theory was that the wave theory of radiation, after centuries of controversy, had become firmly entrenched, and seemed to be the only theory capable of explaining the diffraction and interference of light. Einstein anticipated these objections and answered them in his 1905 paper. He admitted that the wave theory was not to be displaced as far as certain properties are concerned. But, he pointed out, those are properties in which light interacts with matter in bulk, and therefore relate to time averages. In the photoelectric effect, and in the absorption and emission of radiation, there is a one-to-one relationship between the radiation and a particle of matter, and it is quite reasonable to suppose that the wave theory will then prove inadequate. In other words, there is need for a **dual theory of radiation.** For diffraction and interference the wave properties are

Dual Theory of Radiation

relevant; for the emission and absorption of radiation, and the photoelectric effect, one must regard the radiation as a beam of particles.

Zero-Point Energy

As the quantum theory gradually became accepted, it was applied successfully to a wider variety of problems. In 1907 Einstein made another important contribution to quantum theory, with his theory of the specific heats of solids, a matter that is considered in Section 16.5. Einstein did not consider the specific heats of gases at that time, but this was done by the Danish physical chemist Niels Janniksen Bjerrum (1879–1958), who in a series of papers from 1911 to 1914 applied the quantum theory to both spectra and the specific heats of gases, and discussed the relationship between the two properties.

In 1913 Einstein and his assistant Otto Stern (1888–1969) published a paper of great significance in which they suggested for the first time the existence of a residual energy that all oscillators have at the absolute zero. They called this residual energy the *Nullpunktsenergie,* translated as the **zero-point energy.**

They arrived at this idea from a careful analysis of the specific heat of hydrogen gas at low temperatures. Their conclusion was that the specific heats are best represented even at the absolute zero of temperature by an energy with the value $\frac{1}{2}h\nu$. More direct and convincing evidence for the zero-point energy, which is fundamentally a consequence of the Heisenberg uncertainty principle, was obtained in 1924 by the American chemical physicist Robert Sanderson Mulliken (1895–1986) from an analysis of the spectrum of boron monoxide. The theoretical interpretation of the zero-point energy had to await the development of quantum mechanics (Section 11.4).

11.2 Bohr's Atomic Theory

Another great triumph of the quantum theory, and one of particular importance to chemists, was its application to the structure of the atom. This was first done successfully by the Danish physicist Niels Henrik David Bohr (1885–1962), whose theory was based on the nuclear model of the atom proposed in 1911 by Ernest Rutherford (1871–1937). According to Rutherford, who did his Nobel Prize–winning work at McGill University in Montreal, Canada, the nucleus of an atom is much smaller than the atom itself. The electrons revolve around this dense, positively charged nucleus, and the size of the atom is determined by the size of the electronic orbits. The simplest atom, hydrogen, consists of a nucleus having a single positive charge and a single orbiting electron. However, this model presented a serious theoretical problem in terms of classical physics. An orbiting charged particle has an acceleration toward the center and according to classical physics must continuously lose energy; viewed end-on, it is the oscillating electron, which must lose energy. It would therefore fall into the nucleus, and the atom would not survive.

Stationary States

Niels Bohr overcame this dilemma by combining Rutherford's concept of the nuclear atom with the new quantum theory and with some new ideas of his own. He made the bold assumption that there are certain allowed energy states, known as **stationary states,** in which the electrons do not emit radiation. These states are characterized by discrete (quantized) values of the **angular momentum** L. Bohr further postulated that when an electronic transition occurs between two states of energies E_1 and E_2, the frequency ν of the spectral line is given by

$$h\nu = E_2 - E_1 \tag{11.38}$$

Observe the formation of different lines of the Lyman series; Bohr's atomic model; Bohr model.

Angular momentum L is (mass) × (velocity) × (orbital radius) and therefore has the dimensions (mass) (distance)^2 (time)$^{-1}$, which is energy multiplied by time, and the same is true of the Planck constant (6.626×10^{-34} J s). Bohr proposed that the angular momentum is quantized.[3] An electron of mass, m, in moving with a speed u in a circle of radius r has an angular momentum L of mur, and Bohr's suggestion was that this quantity must be a positive integer n multiplied by the Planck constant divided by 2π (often written as $\hbar$):

$$L = mur = n\frac{h}{2\pi} = n\hbar \qquad (n = 0, 1, 2, \ldots) \tag{11.39}$$

For an electron in an atomic orbital the integer n is called the **principal quantum number.**

We will consider the application of these ideas to a hydrogen-like atom; this is a species having a nucleus of charge Ze and a single orbiting electron (when $Z = 1$, we have the hydrogen atom itself, while $Z = 2$, 3, and 4 correspond to He^+, Li^{2+}, and Be^{3+}, respectively). The electron actually moves around the center of mass of the nucleus-electron system and it is its reduced mass μ (see p. 534 for definition and usage) that strictly speaking should be considered; this is close to its true mass. For the circular orbit to be stable, centripetal force $\mu u^2/r$ must be supplied by the coulombic force $Ze^2/4\pi\epsilon_0 r^2$. Thus

$$\frac{\mu u^2}{r} = \frac{Ze^2}{4\pi\epsilon_0 r^2} \tag{11.40}$$

By Eq. 11.39, u is equal to $nh/2\pi\mu r$, and insertion of this expression into Eq. 11.40 gives

$$\frac{n^2h^2}{4\pi^2\mu r^3} = \frac{Ze^2}{4\pi\epsilon_0 r^2} \tag{11.41}$$

Thus

$$r = \frac{n^2h^2\epsilon_0}{\pi\mu Ze^2} \tag{11.42}$$

It is convenient to write this as

$$r = \frac{n^2}{Z}a_0 \tag{11.43}$$

where a_0 is given by

$$a_0 = \frac{h^2\epsilon_0}{\pi\mu e^2} \tag{11.44}$$

[3]Bohr initially based his treatment on the quantization of *energy.* For a discussion see Blanca L. Haendler, "Presenting the Bohr atom," *J. Chem. Ed., 59,* 372–376(1982).

Bohr Radius

This quantity a_0 is a length and is the radius of the orbit for $n = 1$ for the hydrogen atom itself ($Z = 1$). The length a_0 is known as the **Bohr radius** and has a value of 52.92 pm (1 picometre $= 10^{-12}$ m.). It is now recognized as the atomic unit of length (Section 11.13).

The energy corresponding to the various electronic levels is the sum of the kinetic and potential energies. The kinetic energy is $\frac{1}{2}\mu u^2$ and, using Eq. 11.40, is given by

$$E_k = \frac{1}{2}\mu u^2 = \frac{Ze^2}{8\pi\epsilon_0 r} \tag{11.45}$$

With Eq. 11.43 this gives

$$E_k = \frac{Z^2e^2}{8\pi\epsilon_0 n^2 a_0} \tag{11.46}$$

The potential energy of the electron is conveniently defined with respect to infinite separation from the nucleus and is the negative of the work required to remove an electron from the distance r to infinity. It is therefore (See Eq. 11.40)

$$E_p = -\int_r^\infty \frac{Ze^2}{4\pi\epsilon_0 r^2}\,dr = -\frac{Ze^2}{4\pi\epsilon_0 r} \tag{11.47}$$

With Eq. 11.43 this gives

$$E_p = -\frac{Z^2e^2}{4\pi\epsilon_0 n^2 a_0} \tag{11.48}$$

Virial Theorem

Note that the potential energy is twice the kinetic energy with the sign changed. This is an example of the **virial theorem,** from which it follows that if a potential has the mathematical form of ar^b, then the mean kinetic and potential energies are related by $2E_k = nE_p$. In this case, the Coulombic potential has $n = -1$ and so $2E_k = -E_p$, as seen by Eqs. 11.46 and 11.48. The total energy is

$$E = E_k + E_p = -\frac{Z^2e^2}{8\pi\epsilon_0 n^2 a_0} \tag{11.49}$$

The lowest possible state for the hydrogen atom is when $n = 1$, and the value of this ground-state energy is

$$-2.179 \times 10^{-18}\ \text{J} \quad \text{or} \quad -13.60\ \text{eV}$$

Ionization Energy

This is the energy of the state relative to the state in which the electron has been completely removed; in other words, 2.179×10^{-18} J is the ground-state **ionization energy,** that is, the amount of energy required to remove the electron to infinity. This value, relating to a single electron, corresponds to 2.179×10^{-18} J $\times$ 6.022×10^{23} mol^{-1} = 1313 kJ mol^{-1} = 56.9 eV.

Spectral Series

After Bohr had developed his theory of hydrogenlike atoms he did not at first consider applying it to spectra, which he thought would be too difficult to interpret. However, a colleague pointed out to him the existence of certain regularities in atomic spectra, of which Bohr had surprisingly been unaware.

Study the hydrogen-emission spectrum; Bohr's atomic model; Hydrogen-emission spectrum.

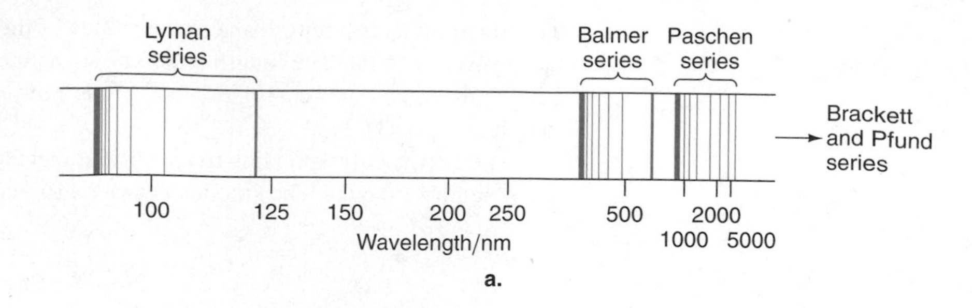

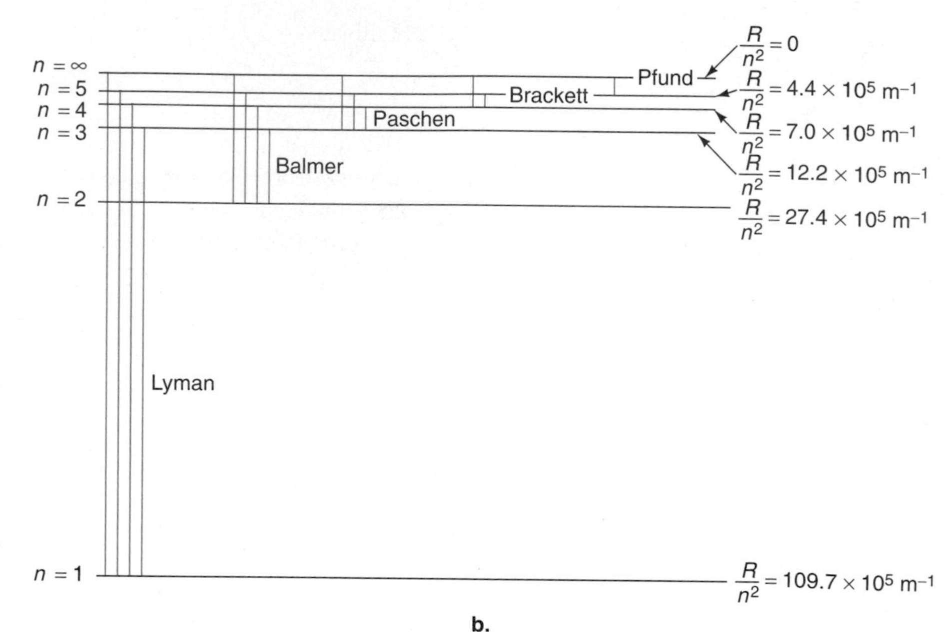

FIGURE 11.8
(a) The hydrogen spectrum in the visible and the near-ultraviolet and near-infrared regions, showing the Lyman, Balmer, and Paschen series. (b) The interpretation of the series as given by the Balmer-Rydberg-Ritz formula. The horizontal lines show R/n^2 values (the spectral terms) for various values of n. The transitions involved in the various series are shown.

The emission spectrum of hydrogen consists of sharp lines, as shown in Figure 11.8a, and some useful empirical relationships had been discovered. In 1885 Johann Jakob Balmer (1825–1898), who taught mathematics at a girls' high school in Basel, Switzerland, developed an empirical equation for those lines, now known as the Balmer series (Figure 11.8). Other similar series were later predicted by empirical equations developed by the American physicist Theodore Lyman (1874–1954) and the German physicist Friedrich Paschen (1865–1940). These equations were all generalized by the Swedish physicist Johannes Robert Rydberg (1854–1919) and by the Swiss physicist Walter Ritz (1878–1909).

The modern form of the Balmer-Rydberg-Ritz formula involves two integers, n_1 and n_2, and is

$$\tilde{\nu} = \frac{1}{\lambda} = R\left(\frac{1}{n_1^2} - \frac{1}{n_2^2}\right) \tag{11.50}$$

$$(n_1 = 1, 2, 3 \ldots; n_2 = n_1 + 1, n_1 + 2, \ldots)$$

Rydberg Constant

where $\tilde{\nu}$, the reciprocal of the wavelength λ, is known as the **wavenumber** and R is a constant known as the **Rydberg constant;** its modern value is 1.0968×10^7 m^{-1}. The wavenumbers and frequencies are thus the differences between certain *spectral* *terms,* taken in pairs. All known spectral lines in the hydrogen spectrum may be obtained by appropriate integral values of n_1 and n_2. Figure 11.8b shows the n values and the particular combinations that are involved in the various series found in the hydrogen spectrum.

However, these empirical relationships provided no real explanation for the spectral lines. Moreover, as spectroscopic instruments improved, it was found that what appeared to be single lines at low resolution were actually two or more lines; this is referred to as the *fine structure* of the lines. Theories based on classical mechanics could provide no explanation for the lines or for their fine structure.

When the existence of the spectral series had been pointed out to Bohr, he saw at once that he could derive the Balmer-Rydberg-Ritz formula. He had assumed that the atom can exist in various *stationary states* characterized by the value of the quantum number n, no radiation being emitted or absorbed as long as the electron remains in a stationary state. Equation 11.38 gives the value of the frequency ν as $(E_2 - E_1)/h$, and the wavenumber $\tilde{\nu}$ is therefore $(E_2 - E_1)/hc$. If the quantum numbers associated with E_1 and E_2 are n_1 and n_2, Eq. 11.49 leads to

$$\tilde{\nu} = \frac{Z^2e^2}{8\pi\epsilon_0 a_0 hc}\left(\frac{1}{n_1^2} - \frac{1}{n_2^2}\right) \tag{11.51}$$

This is the same form as Eq. 11.50, and when $Z = 1$, the Rydberg constant is given by

$$R = \frac{e^2}{8\pi\epsilon_0 a_0 hc} \tag{11.52}$$

The value calculated by Bohr from this relationship was extremely close to the experimental value.

This good agreement was encouraging, but serious problems still remained. The Bohr theory provided no explanation for the fine structure of the lines, and it was not found possible to extend it satisfactorily to atoms containing more than one electron. Certain attempts to modify the Bohr theory did introduce some improvement. In particular, the German physicist Arnold Johannes Wilhelm Sommerfeld (1868–1951) gave a treatment of elliptical orbits and introduced a second quantum number. The German physicist Werner Karl Heisenberg (1901–1976), in a paper published when he was 20 years old, suggested that quantum numbers could be half-integral as well as integral. However, these treatments did not prove satisfactory, and it became clear that the Bohr theory was inadequate and that a completely new approach was needed. Bohr himself again helped to point the way through his **correspondence principle,** which states that any quantum theory must reduce to the familiar classical laws under the conditions for which classical behavior is expected. We have seen an example of the correspondence principle already. Planck's blackbody radiation expression, Eq. 11.35, reduces to the classical Rayleigh-Jeans equation, Eq. 11.36, when Planck's constant, h, goes to zero.

Bohr's Correspondence Principle

11.3 The Foundations of Quantum Mechanics

The first positive suggestion for a new approach to the problem was made by the German physicist Max Born (1882–1970); his paper appeared in 1924, with the title "Zur Quantummechanik," this being the first time the term **quantum mechanics** had been used. In the spring of 1925 Werner Heisenberg (1901–1976), one of

Heisenberg, as the author of the first successful system of quantum mechanics, received the 1932 Nobel Prize in physics, although it was actually awarded to him in 1933, the year in which Dirac and Schrödinger shared their Nobel Prize.

Born's research assistants, developed the first satisfactory system of quantum mechanics, but in a rather obscure form, involving matrix mechanics (about which Heisenberg knew nothing—he had reinvented it!). Born himself initially had difficulty with Heisenberg's treatment, but was soon able to convert it into a more understandable form. Later in 1925 the English physicist Paul Adrien Maurice Dirac (1902–1984) developed a similar system of quantum mechanics, based on Hamilton's methods of classical mechanics, which he transformed by means of postulates into a versatile system of quantum mechanics.

Shortly afterwards the Austrian physicist Erwin Schrödinger (1887–1961) developed what appeared at first sight to be an entirely new system of quantum mechanics, and which, because of its use of familiar concepts related to waves, was much easier for chemists to understand than the systems of Heisenberg and Dirac. Schrödinger soon proved that his system was mathematically equivalent to the other systems. Since Schrödinger's method is easier to visualize and apply to chemical problems, it is the system often used by chemists and is dealt with in detail in following sections of this chapter.

The Wave Nature of Electrons

Schrödinger's theory, often referred to as **wave mechanics** rather than quantum mechanics, was based on the realization that elementary particles such as electrons have wave properties, just as radiation has particle properties. We have seen that Planck's treatment of blackbody radiation and Einstein's interpretation of the photoelectric effect require us to regard radiation as having particle properties (i.e., as consisting of photons), and a third phenomenon, the **Compton effect,** led to the same conclusion. The American physicist Arthur Holly Compton (1892–1962) investigated the scattering of monochromatic (Greek *mono,* one; *chroma,* color) X rays by a target such as a piece of graphite. He found in 1923 that the scattered beam consisted of radiation of two different wavelengths; one wavelength was the same as for the original beam, whereas the other was slightly longer. The results could only be interpreted on the hypothesis that photons interacted with the material and obeyed the laws of conservation of energy and momentum. The results could not be explained by treating the radiation as having wave properties. Compton's experiments finally convinced physicists to accept the wave-particle duality of electromagnetic radiation almost 20 years after Einstein proposed it.

Compton Effect

Compton's results made a particular impression on the French physicist Louis Victor, Prince de Broglie (1892–1987), who suggested in 1924 the converse hypothesis, that particles such as electrons can also have wave properties. To obtain the wavelength of the wave associated with a particle of mass m moving with speed u, de Broglie reasoned as follows. According to Einstein's special theory of relativity the energy E of a particle, its rest mass m, and its momentum p are related to each other by the equation

$$E^2 = p^2c^2 + m^2c^4 \tag{11.53}$$

where c is the velocity of light. The relativistic rest mass of the photon is zero and therefore

$$E = pc = p\lambda\nu \tag{11.54}$$

However, since $E = h\nu$, it follows that $h\nu = p\lambda\nu$ and therefore

$$\lambda = \frac{h}{p} \tag{11.55}$$

This expression is for electromagnetic radiation, and de Broglie suggested that a parallel expression would apply to a beam of particles of momentum p and velocity u:

$$\lambda = \frac{h}{p} = \frac{h}{mu} \tag{11.56}$$

where m is the mass of the particle.

EXAMPLE 11.2 An electron has a mass of 9.11×10^{-31} kg and a charge of $e = 1.602 \times 10^{-19}$ C. Calculate the de Broglie wavelength of an electron that has been accelerated by a potential of 100 V.

Solution When a charge, q, is placed in an electric potential field, V, it experiences a force depending upon its position and starts to accelerate and pick up kinetic energy. This is given by $E_k = qV$. For a potential of 100 V and a charge of one electron, this is (1 coulomb-volt = 1 joule)

$$E_k = eV = 1.602 \times 10^{-19}\ \text{C} \times 100\ \text{V} = 1.602 \times 10^{-17}\ \text{J}$$

The kinetic energy is $\frac{1}{2}mu^2$ and therefore

$$u = \sqrt{\frac{2 \times 1.602 \times 10^{-17}\ (\text{J})}{9.11 \times 10^{-31}\ (\text{kg})}} = 5.93 \times 10^6\ \text{m s}^{-1}$$

The de Broglie wavelength λ is therefore, from Eq. 11.56,

$$\begin{aligned} &\frac{6.626 \times 10^{-34}\ (\text{J s})}{9.11 \times 10^{-31}\ (\text{kg}) \times 5.93 \times 10^6\ (\text{m s}^{-1})} \\ &= 1.23 \times 10^{-10}\ \text{m} \\ &= 123\ \text{pm} \end{aligned}$$

This wavelength is of a magnitude similar to the distance between neighboring atoms or ions in a crystal, and it was therefore suggested that a beam of electrons could be diffracted by using a crystal as a diffraction grating. This prediction was confirmed experimentally in 1927 by the English physicists Sir George Paget Thomson (1892–1975) and A. Reid and by the American physicists Clinton Joseph Davisson (1881–1958) and Lester Halbert Germer. The diffraction of electrons is now commonly employed as a technique for investigating molecular structure, and is the basis for electron microscopy.

Note that J. J. Thomson won a 1906 Nobel Prize for showing that the electron is a particle, while his son G. P. Thomson shared the 1937 prize with C. J. Davisson for showing that it is a wave.

The realization that electrons have wave properties leads at once to an interpretation of Bohr's hypothesis of stationary orbits. Figure 11.9 shows the de Broglie wave associated with an electron in a Bohr orbit of radius r. If the circumference

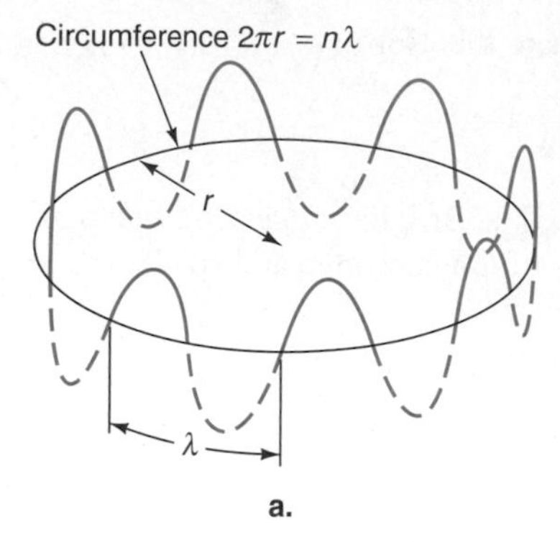

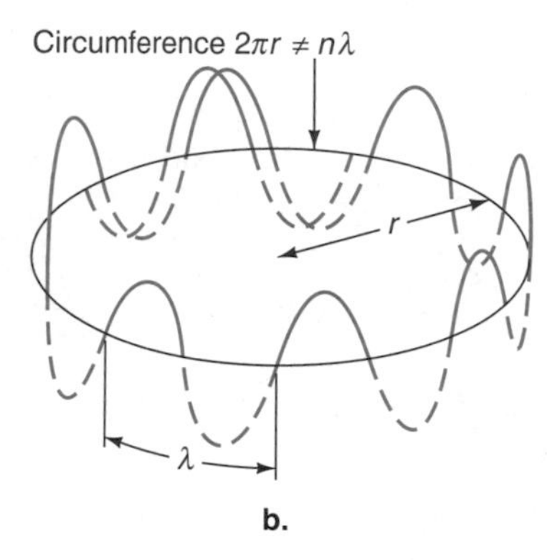

FIGURE 11.9
The de Broglie wave associated with an electron in a Bohr orbit of radius *r*. (a) Constructive interference: The wave fits into the orbit. (b) Destructive interference: The wave does not fit an integral number of times into the orbit. If in this diagram we continued the waves indefinitely, they would completely obliterate one another, the net amplitude becoming zero.

See Figure 11.9 in motion; Bohr's Atomic model; Bohr condition.

$2\pi r$ is equal to an integer n multiplied by the wavelength λ, the wave will fit into the orbit an integral number of times, and *constructive interference* will occur. The condition for this is

$$2\pi r = n\lambda \tag{11.57}$$

Destructive interference occurs with wavelengths that do not satisfy this condition, and the orbit is not a stationary one. The requirement that n must be an integer for constructive interference thus leads to a quantization condition. Substitution of λ as given by Eq. 11.56 into Eq. 11.57 gives

$$2\pi r = \frac{nh}{mu} \tag{11.58}$$

or since $\hbar = h/2\pi$

$$mur = n\hbar \tag{11.59}$$

The quantity mur is the orbital angular momentum, which Bohr had postulated to be an integral number of $\hbar$ (Eq. 11.39). We can now see that the de Broglie relationship leads at once to this quantization condition.

The Uncertainty Principle

Uncertainty Principle

Scientists have often found it useful to carry out what are called *gedanken* (German *Gedanke,* thought) experiments. These are experiments that would be difficult if not impossible to perform but that can easily be carried out in imagination with results that are quite reliable. A significant gedanken experiment was carried out in 1926 by Werner Heisenberg, who considered the procedures that would have to be followed in order to make a simultaneous measurement of the position and momentum of a small particle such as an electron. He realized that it would be impossible to make accurate measurements of both properties, since any technique for measuring one of them will necessarily disturb the system and will cause the measurement of the other one to be imprecise. Suppose, for example, that radiation is used to make an accurate determination of the position of an electron. To do so we must use radiation of short wavelength. Recall that the advantage of an electron microscope over a light microscope is that the wavelength of the electrons is much shorter than that of nomal visible light. Hence to measure the position of a tiny piece of matter, such as an electron, would require radiation of extremely short wavelength. Radiation of short wavelength, however, involves high-energy photons, and we have seen in the Compton effect that these will change the particle's momentum. If we avoid the momentum change by using radiation of long wavelength, the position of the particle will not be precisely defined. These conclusions are expressed in the **Heisenberg uncertainty principle,** or principle of indeterminacy, which may be written as

Zoom to quantum dimensions; The uncertainty principle; Quantum size.

$$\Delta q \, \Delta p \geq \frac{1}{2}\hbar \tag{11.60}$$

That is, in the direction of motion, the product of the uncertainty in the position Δq and the uncertainty in the momentum Δp can never be less than Planck constant di-

Locate a classical point in position and momentum space; The uncertainty principle; Classical point.

vided by 4π. Since the momentum p is equal to mu, the mass times the velocity, the relationship can also be written as

$$\Delta q\ \Delta u \geq \frac{\hbar}{2m} \tag{11.61}$$

Locate a ball in position and momentum space; The uncertainty principle; Classical motion.

Thus the product of the uncertainties decreases as the mass of the particle increases. The uncertainty principle is thus particularly important for light particles such as electrons; it does not apply to any significant extent to massive particles (see Problem 11.10).

Another form of the principle, involving energy and time, is also of importance. If a particle has a velocity u, the uncertainty in the time at which it exists in a given position is related to Δq by

$$\Delta t = \frac{\Delta q}{u} \tag{11.62}$$

Have a large (small) uncertainty in position with a small (large) uncertainty in momentum; The uncertainty principle; Uncertainties.

Since energy is $E = \frac{1}{2}mu^2 = p^2/2m$, and the differential $dE = pdp/m = udp$, the uncertainty in energy is given by

$$\Delta E = u\ \Delta p \tag{11.63}$$

With Eq. 11.60 these equations give

$$\Delta E\ \Delta t \geq \frac{1}{2}\hbar \tag{11.64}$$

Another uncertainty relationship exists between the angle ϕ and the angular momentum, L, that is

$$\Delta L_Z \Delta\phi \geq \frac{\hbar}{2} \tag{11.65}$$

In all three cases, the product of the uncertainties in the two variables has the units of angular momentum, J s, which are also the units of Planck's constant.

Another physical consequence of the Heisenberg uncertainty principle is that the order of measuring these quantities is important to the outcome. That is, if position is measured first and then the momentum in the same direction, the answer is different from when momentum is measured first and then position. Anticipating the results of Section 11.5, where quantum mechanical operators are introduced, we note here that the operators for position and momentum, $\hat{q}$ and $\hat{p}$, respectively, form a non-commutative algebra, $\hat{p}\hat{q} - \hat{q}\hat{p} \neq 0$. It is this fundamental property from which the relationships shown in Eqs. 11.60, 11.64, and 11.65 follow.

The circumflex (ˆ) over a symbol identifies it as an operator. Normal rules of algebra need not apply. For instance, in $\hat{p}\Psi = x\Psi$ we cannot divide both sides by Ψ and obtain $\hat{p} = x$. We must perform the mathematical operations indicated by the operator definition. Details are given in Section 11.5.

Two examples of the relationship shown in Eq. 11.64 may be mentioned. The first relates to β particles emitted in the nuclear decay of radioactive substances. The energy of such particles can be measured rather precisely, which means that the time at which the emission occurs is uncertain. Another application, considered further in Section 14.2, relates to the widths of spectral lines. If an electronic transition occurs from the ground state of an atom or molecule, the spectral line is sharp; that is, it covers only a narrow range of wavelengths. However, if an atom or molecule is in an electronically excited state, its lifetime may be short; Δt is therefore small, which means that ΔE is large. The energy of the transition therefore covers a range of values, and hence the spectral line is broadened. This is an example of lifetime broadening.

11.4 Schrödinger's Wave Mechanics

At first Heisenberg and Dirac paid little attention to de Broglie's ideas about the wave nature of electrons, and when Schrödinger's wave mechanics first appeared they were not impressed by it. However, when Schrödinger proved that his wave mechanics was mathematically equivalent to Heisenberg's and Dirac's formulations, they accepted it.

Schrödinger's equation cannot be derived but is postulated. It can be justified by plausible arguments. Schrödinger's particular interest had been in hydrodynamics and wave motion. To describe an electron moving in one dimension, the X-direction, he took as his starting point a fundamental wave equation that described an electromagnetic wave traveling in a vacuum with speed c. He thought the same equation would describe the wave nature of an electron and wrote

$$\frac{\partial^2 \Psi}{\partial x^2} = \frac{1}{c^2}\frac{\partial^2 \Psi}{\partial t^2} \tag{11.66}$$

where $\Psi(x, t)$ is a function that describes the behavior of the electron and is a function of the distance x and the time t. A solution of Eq. 11.66 is

$$\Psi(x, t) = Ce^{i\alpha} \tag{11.67}$$

where C is a constant, i is $\sqrt{-1}$, and α is the phase, given by

$$\alpha = 2\pi\left(\frac{x}{\lambda} - \nu t\right) \tag{11.68}$$

Substitution of Eq. 11.67 into Eq. 11.66 confirms that $\Psi(x, t)$ satisfies Eq. 11.66. The function $\Psi(x, t)$ is known as a **wave function,** in this case in one dimension, and it can be split into two factors as follows:

$$\Psi(x, t) = Ce^{2\pi i x/\lambda}e^{-2\pi i\nu t} \tag{11.69}$$

This may be written as

$$\Psi(x, t) = \psi(x)e^{-2\pi i\nu t} \tag{11.70}$$

where $\psi(x)$ is a function of x but not of t and $e^{-2\pi i\nu t}$ is a function of t but not of x.

The next stage in Schrödinger's argument was to replace ν by E/h and λ by h/p_x (Eq. 11.56). Equation 11.69 therefore becomes

$$\Psi(x, t) = Ce^{2\pi i x p_x/h}e^{-2\pi i E t/h} = Ce^{i x p_x/\hbar}e^{-iEt/\hbar} \tag{11.71}$$

Partial differentiation with respect to t then gives

$$\frac{\partial \Psi(x, t)}{\partial t} = -\frac{iEC}{\hbar}e^{i x p_x/\hbar}e^{-iEt/\hbar} \tag{11.72}$$

$$= -\frac{iE}{\hbar}\Psi(x, t) \tag{11.73}$$

This equation rearranges to

$$i\hbar \frac{\partial \Psi}{\partial t} = E\Psi \tag{11.74}$$

This is an *operator equation.* The operator, written as $\hat{H}$, is

$$i\hbar \frac{\partial}{\partial t}$$

Eigenfunction

which means that we take the partial derivative of the function Ψ and multiply the result by $i\hbar$. Equation 11.74 thus tells us that if we perform this operation on the function Ψ, we obtain the energy E multiplied by the function. All wave functions that satisfy this equation are called **eigenfunctions.**

We can also take the partial derivative of $\Psi(x, t)$ with respect to x. We obtain, from Eq. 11.71,

$$\frac{\partial \Psi(x, t)}{\partial x} = \frac{2\pi i p_x C}{h} e^{ixp_x/\hbar} e^{-iEt/\hbar} \tag{11.75}$$

$$= \frac{ip_x}{\hbar} \Psi(x, t) \tag{11.76}$$

This rearranges to

$$-i\hbar \frac{\partial \Psi}{\partial x} = p_x \Psi \tag{11.77}$$

The operator $\hat{p}_x$, which is

$$\hat{p}_x = -i\hbar \frac{\partial}{\partial x}$$

Linear Momentum Operator

is called a **linear momentum operator,** since when it operates on the function Ψ it gives the linear momentum multiplied by the function.

The next step in the argument is to introduce the total energy E of the system. This is the sum of the kinetic energy E_k and the potential energy E_p, and for this one-dimensional system the kinetic energy is $p_x^2/2m$. Thus

$$E = \frac{p_x^2}{2m} + E_p(x, t) \tag{11.78}$$

When energy is expressed in terms of momentum, it is said to be a *Hamiltonian,* after the Irish mathematician Sir William Rowan Hamilton (1805–1865). We can now replace p_x in his equation by the linear momentum operator (Eq. 11.77), and we obtain the operator

$$\hat{H} = -\frac{\hbar^2}{2m} \frac{\partial^2}{\partial x^2} + \hat{E}_p(x, t) \tag{11.79}$$

Hamiltonian Operator

This is known as the **Hamiltonian operator.**

If we replace the energy E in Eq. 11.74 by this operator $\hat{H}$, we obtain

$$\left[-\frac{\hbar^2}{2m} \frac{\partial^2}{\partial x^2} + \hat{E}_p(x, t) \right] \Psi = i\hbar \frac{\partial \Psi}{\partial t} \tag{11.80}$$

This equation is easily extended to three dimensions. The operator $\partial^2/\partial x^2$ is then replaced by the sum

$$\nabla^2 = \frac{\partial^2}{\partial x^2} + \frac{\partial^2}{\partial y^2} + \frac{\partial^2}{\partial z^2} \tag{11.81}$$

Del-squared or Laplacian Operator

This is the **del-squared operator** or the **Laplacian operator,** named for the French astronomer and mathematician Pierre Simon, Marquis de Laplace (1749–1827). We have seen its use earlier in Eq. 7.36 on p. 278 in the Poisson equation. The functions $\Psi(x, t)$ and $E_p(x, t)$ are now replaced by $\Psi(x, y, z, t)$ and $E_p(x, y, z, t)$, and Eq. 11.80 becomes

$$i\hbar\frac{\partial \Psi}{\partial t} = \hat{H}\Psi \tag{11.82a}$$

Time-dependent Schrödinger Equation

This is the **time-dependent Schrödinger equation,** where

$$\hat{H} = -\frac{\hbar^2}{2m}\nabla^2 + \hat{E}_p(x, y, z, t) \tag{11.82b}$$

For many problems, such as those concerned with the structures of atoms and molecules, we are not concerned with time-dependent wave functions and energies, but with *stationary states,* which implies that the potential function is time independent. The time-dependent function $\Psi(x, y, z, t)$ may be expressed as

$$\Psi(x, y, z, t) = \psi(x, y, z)e^{-iEt/\hbar} \tag{11.83}$$

where $\psi(x, y, z)$ is the time-independent wave function. (Compare Eq. 11.71 for the one-dimensional form.) Partial differentiation of $\Psi(x, y, z, t)$ with respect to t gives

$$\frac{\partial \Psi}{\partial t} = -\frac{2\pi i E}{h}\psi(x, y, z)e^{-iEt/\hbar} = -\frac{i}{\hbar}E\Psi \tag{11.84}$$

For stationary states where E_p is independent of time, Eq. 11.84 is substituted into Eq. 11.82a; and, since there is no operator in the Hamiltonian involving time, the exponential term involving time in Ψ is factored from both sides. Equation 11.82a thus becomes

$$\left[-\frac{\hbar^2}{2m}\nabla^2 + \hat{E}_p(x, y, z)\right]\psi(x, y, z) = E\psi(x, y, z) \tag{11.85}$$

Time-independent Schrödinger Equation

This is the **time-independent Schrödinger equation,** and it is frequently written in the compact form

$$\hat{H}\psi = E\psi \tag{11.86}$$

where $\hat{H}$ represents the Hamiltonian operator that appears in Eq. 11.85.

Solution of Eq. 11.86, with an appropriate expression for the potential energy $\hat{E}_p$, thus leads to stationary-state solutions, which are analogous to the stationary-state or standing-wave solutions obtained for a vibrating string (Figure 11.6). Those ψ functions which are possible solutions of the equation are called *eigenfunctions, characteristic functions,* or *wave functions.* They are analogous to the amplitude functions y of Eq. 11.20. These wave functions have the property that when the Hamiltonian operator acts on each of them, a constant (which is the energy) is formed times the original wave function. Thus, corresponding to each of these

Eigenvalues

wave functions is an allowed energy level, and these energy levels are called the *characteristic values* or the **eigenvalues** for the energy of the system. As we have just seen, eigenvalues result from the action of the operator on the eigenfunction. These energy eigenvalues are often called the allowed energies of the system.

In most of the rest of this book we will be concerned with time-independent systems, which are often called **conservative** systems.

Eigenfunctions and Normalization

A further, and very significant, interpretation of the eigenfunction ψ was suggested in 1926 by Max Born. His idea was that if the eigenfunction is real (i.e., does not involve $\sqrt{-1}$), its square ψ^2 multiplied by a small volume element $dx\,dy\,dz$ is proportional to the *probability* that the electron is present in that element.

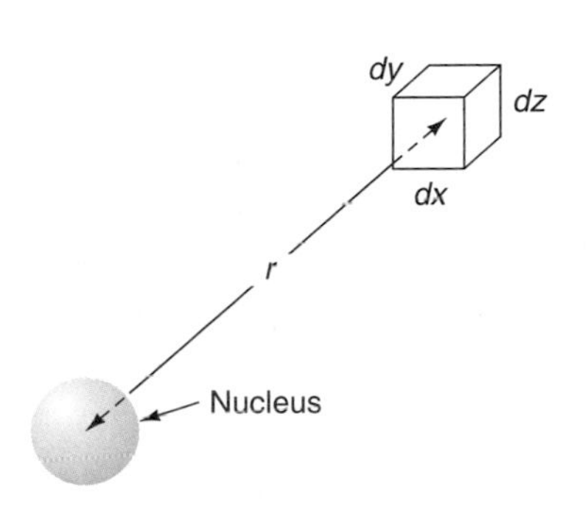

FIGURE 11.10
A volume element $dx\,dy\,dz$ at a distance r from an atomic nucleus. The probability that the electron is in the volume element is proportional to $\psi^2\,dx\,dy\,dz$ if ψ is real and to $\psi^*\psi\,dx\,dy\,dz$ if ψ is complex. If the function ψ is normalized (Eq. 11.89), the probability is *equal* to $\psi^*\psi\,dx\,dy\,dz$.

This concept is again analogous to that in classical physics, in which the intensity of radiation at any point is proportional to the square of the amplitude of the wave at that point. In other words, the greater the amplitude of the light wave in a particular region, the greater is the probability that a photon is present in that region. For an electron, Born's proposal, which is now accepted as correct, is that the probability of finding an electron within an element of volume $dx\,dy\,dz$ is proportional to

$$\psi^2(x, y, z)\,dx\,dy\,dz$$

This is illustrated in Figure 11.10. In other words, the eigenfunction has provided information about the **probability density** or *probability distribution* of the electron.

For an eigenfunction to be an acceptable one, several conditions must be satisfied. One is that it must be *single valued,* since the probability in any region can have only one value. Another is that it must be *finite* in all regions of space; otherwise the Schrödinger equation would not apply since the electron would be located at a region where ψ was infinite and would not be acting as a wave. Third, the eigenfunction must be *continuous,* since discontinuous eigenfunctions are inconsistent with Born's statistical interpretation that the square of the wave function is proportional to a probability. Finally, the eigenfunction must be *square integrable,* by which is meant that integrals such as

$$\int \psi^2(x, y, z)\,dx\,dy\,dz$$

must be finite over all space.

An eigenfunction may, however, be *complex;* that is, it may contain $i = \sqrt{-1}$. A complex function contains a real part and an imaginary part and may be written as $a + ib$. The square of this is $a^2 + 2iab - b^2$, which is also complex and cannot, therefore, represent a probability. To avoid this difficulty the probability is considered to be $\psi^*\psi$, where ψ^* is the complex conjugate of the function. Thus, if the function ψ is written as $a + ib$, the complex conjugate ψ^* is $a - ib$ and the product is $a^2 + b^2$, which is real.

Normalization

We have said that the probability of finding an electron within the element $dx\,dy\,dz$ is *proportional* to $\psi^*\psi\,dx\,dy\,dz$. However, it is possible to *normalize* an eigenfunction so that the probability is *equal* to this function. First we should note that there is an arbitrariness about a function ψ obtained by solving the Schrödinger equation, Eq. 11.86. Thus, if ψ is a solution and we multiply ψ by any number a, the function $a\psi$ is still an eigenfunction, because

$$\hat{H}(a\psi) = E(a\psi) \tag{11.87}$$

Thus, for a given Schrödinger equation there is an infinite set of eigenfunctions differing from each other by numerical factors. We need to know which particular eigenfunction is such that

$$\psi^*\psi \, dx \, dy \, dz$$

is *equal* to the probability that the electron is in the volume element of dimensions $dx\,dy\,dz$. To determine this, we note that the total probability of finding the electron anywhere at all in space must be unity. The integral over all space is

$$\int_{-\infty}^{\infty}\int_{-\infty}^{\infty}\int_{-\infty}^{\infty} \psi^*\psi \, dx \, dy \, dz = 1 \tag{11.88}$$

This may conveniently be written as

$$\int \psi^*\psi \, d\tau = 1 \tag{11.89}$$

Normalization Condition

where $d\tau$ represents $dx\,dy\,dz$ and it is understood that the integral is over all space. This is known as the **normalization condition.**

Suppose that when Eq. 11.89 is integrated, the eigenfunction ϕ obtained by solving the Schrödinger equation gives us not unity but a number b:

$$\int \phi^*\phi \, d\tau = b \tag{11.90}$$

The eigenfunction therefore requires adjustment to make the integral unity. This is done by dividing both ϕ^* and ϕ by $\sqrt{b}$, and then

$$\int \frac{\phi^*}{\sqrt{b}} \frac{\phi}{\sqrt{b}} \, d\tau = 1 \tag{11.91}$$

Normalization

The new function $\psi = \phi/\sqrt{b}$ is therefore such that $\psi^*\psi \, d\tau$ correctly represents the probability in any region. This process of adjusting an eigenfunction to satisfy Eq. 11.89 is known as **normalization,** and the resulting function is a *normalized eigenfunction* dealt with in detail later in this chapter.

11.5 Quantum-Mechanical Postulates

In Section 11.4 we have seen how quantum mechanics was developed by analogy with the wave theory of electromagnetic radiation. This approach has been fruitful, but for many purposes it is more useful to start with a series of postulates that can be shown to lead to conclusions that are verified experimentally. Different authors express the postulates in a variety of ways, and this approach leads to the treatment in Section 11.4, but it also leads to more general procedures.

Postulate I states that

> **The physical state of a particle is described as fully as possible by an appropriate wave function $\Psi(x, y, z, t)$.**

We have already considered this, and its implications, in Section 11.4.

Postulate II states that

> **The possible wave functions $\Psi(x, y, z, t)$ are obtained by solving the appropriate Schrödinger equation.**

More explicitly, the Schrödinger equation is written as Eq. 11.82b, where $\hat{E}_p(x, y, z, t)$ is the potential-energy operator that relates to the system being considered. If $\hat{E}_p$ is independent of time, the equation reduces to

$$\left[-\frac{\hbar^2}{2m}\nabla^2 + \hat{E}_p(x, y, z)\right]\psi = E\psi \tag{11.92}$$

In solving the equation, boundary conditions are important.

Postulate III states that

> **Every dynamical variable, corresponding to a physically observable quantity, can be represented by a linear operator.**[4]

Linear Operator

We represent an operator by a circumflex accent over the symbol; for example, $\hat{o}$ is an operator, and $\hat{o}\psi$ means that $\hat{o}$ is operating on the function ψ. An operator is *linear* when for any two functions ψ and ϕ

$$\hat{o}(a\psi + b\phi) = a\hat{o}\psi + b\hat{o}\phi \tag{11.93}$$

where a and b are arbitrary numbers. Multiplication and taking a derivative are examples of linear operators, but taking a square root is not; $\sqrt{\psi + \phi}$ is not in general equal to $\sqrt{\psi} + \sqrt{\phi}$. If operators were not linear, the wave functions obtained from a solution of the Schrödinger equation could not be superimposed.

The operator is always written to the left of the function on which it is to operate. The symbol $\hat{o}_2\hat{o}_1\psi$ means that we perform the operation $\hat{o}_1$ first and then perform the operation $\hat{o}_2$ on the result. Two operators are said to **commute** if for any arbitrary function

$$\hat{o}_1\hat{o}_2\psi = \hat{o}_2\hat{o}_1\psi \tag{11.94}$$

Commuting Operators

In other words, if the order of operation is irrelevant then the operators commute. We will now show that *if two operators commute, any eigenfunction of one of them is also an eigenfunction of the other.* Suppose that the function ψ is an eigenfunction of $\hat{o}_1$; this means that

$$\hat{o}_1\psi = o_1\psi \tag{11.95}$$

where o_1 is the eigenvalue, which simply multiplies ψ. If we operate on both sides of Eq. 11.95 by $\hat{o}_2$, we obtain

$$\hat{o}_2\hat{o}_1\psi = \hat{o}_2 o_1\psi \tag{11.96}$$

The right-hand side can also be written as $o_1\hat{o}_2\psi$, since multiplication by a number commutes with any operator. However, since the operators commute, the left-hand side of Eq. 11.96 is $\hat{o}_1\hat{o}_2\psi$ and therefore

$$\hat{o}_1(\hat{o}_2\psi) = o_1(\hat{o}_2\psi) \tag{11.97}$$

As indicated by the parentheses in this equation, $\hat{o}_2\psi$ is an eigenfunction for the operator $\hat{o}_1$. However, since ψ is an eigenfunction for $\hat{o}_1$, so also is $o_2\psi$; in other words

$$\hat{o}_1(o_2\psi) = o_1 o_2\psi = o_1(\hat{o}_2\psi) \tag{11.98}$$

and therefore

$$\hat{o}_2\psi = o_2\psi \tag{11.99}$$

The function ψ, an eigenfunction of $\hat{o}_1$, is therefore also an eigenfunction for $\hat{o}_2$.

[4] It will be seen later (Eq. 11.106) that it is a Hermitian operator.

TABLE 11.1 Common Quantum-Mechanical Operators as Derived from Their Classical Expressions

Classical Variable	Quantum Mechanical Operator	Operation
Position		
x	$\hat{x}$	x (multiplication)
	(similarly for the Y- and Z-directions)	
Linear momentum		
p_x (X-direction)	$\hat{p}_x$	$-i\hbar\frac{\partial}{\partial x}$
	(similarly for the Y- and Z-directions)	
Angular momentum		
L_z (rotation about the Z axis)	$\hat{L}_z$	$-i\hbar\frac{\partial}{\partial \phi}$
Kinetic energy		
E_k	$\hat{E}_k$	$-\frac{\hbar^2}{2m}\nabla^2$
Potential energy		
$E_p(x, y, z)$	$\hat{E}_p$	E_p (multiplication)
Total energy (time-dependent system)		
$E = E_k + E_p$	$\hat{H}$	$i\hbar\frac{\partial}{\partial t}$
Total energy (conservative system)		
H (Hamiltonian)	$\hat{H} = \hat{E}_k + \hat{E}_p$	$-\frac{\hbar^2}{2m}\nabla^2 + \hat{E}_p$

In the discussion following Eq. 11.60, we noted that the position and momentum operators do not commute; this is consistent with the Heisenberg uncertainty principle. If two operators commute, there simultaneously is an eigenfunction for both operators, which means that the corresponding physical properties both have precise values. When two operators do not commute, the corresponding physical properties cannot both be measured simultaneously and precisely, and this is the case with position and momentum (compare Eq. 11.60).

Postulate IV states that

Quantum mechanical operators corresponding to physical properties are obtained from the classical expressions for these quantities, using certain established procedures.

These procedures are summarized in Table 11.1. As an example of the application of the method, if we are concerned with calculating the energy for a conservative (time-independent) system, we start with the equation

$$\hat{H}\psi = E\psi \tag{11.100}$$

For the operator $\hat{H}$ we use

$$\hat{H} = -\frac{\hbar^2}{2m}\nabla^2 + \hat{E}_p \tag{11.101}$$

To form the quantum mechanical equation from a classical expression, each classical variable is replaced by its quantum mechanical operator.

and at once obtain

$$\left(-\frac{\hbar^2}{2m}\nabla^2 + \hat{E}_p\right)\psi = E\psi \tag{11.102}$$

which is Eq. 11.85. This form of the Schrödinger equation may be used in a variety of special cases by changing the potential energy to describe different systems. The postulatory approach that we are using thus has a much wider applicability than the original Schrödinger equation.

Postulate V states that

> **When the equation $\hat{F}\psi = F\psi$ has been set up, with $\hat{F}$ corresponding to a physical quantity (as in Table 11.1), the eigenvalues F that are obtained by solving the equation represent all possible values of an individual measurement of that quantity.**

For example, solution of the energy equation

$$\hat{H}\psi = E\psi \tag{11.103}$$

gives us all the possible values of energy E for a particular system.

Postulate VI states that

> **The mean value $\overline{F}$ of a physical quantity, or the expectation value $\langle F \rangle$ when measurements are made for a large number of particles, is given by the equation**

$$\overline{F} \equiv \langle F \rangle = \frac{\int_{-\infty}^{\infty} \psi^* (\hat{F}\psi)\, d\tau}{\int_{-\infty}^{\infty} \psi^*\psi\, d\tau} \tag{11.104}$$

As before, $d\tau$ stands for $dx\, dy\, dz$, and the integration is from minus infinity to plus infinity, i.e., over all space. The denominator in this expression allows for the possibility that the wave functions are not normalized; if they are normalized, only the numerator is needed.

This postulate allows us to calculate average values for quantities for which stationary-state values cannot be measured exactly. If ψ is an eigenfunction of $\hat{F}$, that is, if

$$\hat{F}\psi = F\psi \tag{11.105}$$

$\overline{F}$ is equal to F; that is, the eigenvalue is the average value.

Hermitian Operators

Measurable physical properties must be real and not complex. The operators in Table 11.1 all belong to the class of **Hermitian operators,** which means that the following relationship applies to them:

$$\int \psi_2^* \hat{F}\psi_1\, d\tau = \int \psi_1 (\hat{F}\psi_2)^*\, d\tau \tag{11.106}$$

We now show that all measurable quantities (i.e., eigenvalues corresponding to operators such as those in Table 11.1) are real. We take complex conjugates of everything in Eq. 11.104 and obtain

$$\langle F \rangle^* = \overline{F}^* = \frac{\int_{-\infty}^{\infty} \psi\, (\hat{F}\psi)^*\, d\tau}{\int_{-\infty}^{\infty} \psi\psi^*\, d\tau} \tag{11.107}$$

The denominators in Eqs. 11.104 and 11.107 are identical, and the numerators are equal because of the Hermitian condition. Thus

$$\langle F\rangle^* = \langle F\rangle \text{ or } \overline{F}^* = \overline{F} \tag{11.108}$$

which means that $\overline{F}$ is real. Hermitian operators are said to be self-adjoint, the relationship between them being shown by use of the symbol†:

$$\hat{F}^\dagger = \hat{F} \tag{11.109}$$

Orthogonality of Wave Functions

Orthogonal

Eigenfunctions for a given Hermitian operator and corresponding to different eigenvalues are necessarily **orthogonal** to one another, as will now be shown. Suppose, for example, that ψ_1 and ψ_2 are two different eigenfunctions for the Hermitian operator $\hat{F}$ and that the eigenvalues f_1 and f_2 are different from one another:

$$\hat{F}\psi_1 = f_1\psi_1 \text{ and } F\psi_2 = f_2\psi_2 \tag{11.110}$$

We will now show that ψ_1 and ψ_2 are orthogonal to each other, which means that the relationship

$$\int \psi_1^*\psi_2\, d\tau = 0 \tag{11.111}$$

is true. When we use the simplified notation of Eq. 11.111, it is to be understood that the integration is over all space.

The proof is as follows. Multiplication of both sides of Eq. 11.110 by ψ_2^* and integration over all space gives

$$\int \psi_2^*\hat{F}\psi_1\, d\tau = f_1 \int \psi_2^*\psi_1\, d\tau \tag{11.112}$$

If we take complex conjugates of both sides of Eq. 11.110, multiply by ψ_1, and integrate, we obtain

$$\int \psi_1(\hat{F}\psi_2)^*\, d\tau = f_2^* \int \psi_1\psi_2^*\, d\tau \tag{11.113}$$

The Hermitian condition (Eq. 11.106) requires that the left-hand sides of Eqs. 11.112 and 11.113 are equal so that, since $f_2^* = f_2$ (the eigenvalues must be real), we have

$$f_1 \int \psi_2^*\psi_1\, d\tau = f_2 \int \psi_1\psi_2^*\, d\tau \tag{11.114}$$

These two integrals are identical, but since by hypothesis f_1 is not equal to f_2, the integrals must be zero:

$$\int \psi_1\psi_2^*\, d\tau = 0 \tag{11.115}$$

Orthogonality Condition

This is the **orthogonality** condition.

Degeneracy

The situation is somewhat different if the eigenvalues for two eigenfunctions are equal to one another, in which case the eigenvalue is said to be **degenerate.** Suppose, for example, that two eigenfunctions of an operator $\hat{F}\psi_1$ and ψ_2 have the same eigenvalue f; in other words $\hat{F}\psi_1 = f\psi_1$ and $\hat{F}\psi_2 = f\psi_2$. It can then be seen at once that *any linear combination of* ψ_1 *and* ψ_2 *is also an eigenfunction,* because

$$\hat{F}(c_1\psi_1 + c_2\psi_2) = f(c_1\psi_1 + c_2\psi_2) \tag{11.116}$$

where c_1 and c_2 can be constants. It is not necessarily the case that ψ_1 and ψ_2 are orthogonal to each other, but it is always possible to take linear combinations of them and to obtain orthogonal functions. The way this is done is explained in books on quantum mechanics (see Suggested Reading, p. 575).

A special case of Eq. 11.116 is when the operator is the Hamiltonian operator $\hat{H}$, in which case the eigenvalue is the energy E:

$$\hat{H}(c_1\psi_1 + c_2\psi_2) = E(c_1\psi_1 + c_2\psi_2) \tag{11.117}$$

If there are two wave functions ψ_1 and ψ_2 for an energy E, the energy level is said to have *twofold degeneracy.* Again, it is often convenient to take linear combinations of ψ_1 and ψ_2 and convert them into two wave functions that are orthogonal to each other. If there are n wave functions corresponding to an energy E, the energy level is said to have *n-fold degeneracy.*

It is often useful (for example, in connection with the theory of *resonance* and when using the *variation method;* see Section 11.13) to work with functions that are linear combinations of wave functions that correspond to different energies. For example, suppose that ψ_1 and ψ_2 are normalized wave functions for a given Hamiltonian operator; we have seen that they must be orthogonal. We can construct the normalized function, which is the superposition of ψ_1 and ψ_2

Superposition Principle

$$\psi_3 = c_1\psi_1 + c_2\psi_2 \tag{11.118}$$

where c_1 and c_2 are constants. Then

$$\int \psi_3^*\psi_3\,d\tau = \int (c_1^*\psi_1^* + c_2^*\psi_2^*)(c_1\psi_1 + c_2\psi_2)\,d\tau \tag{11.119}$$

$$= c_1^*c_1 \int \psi_1^*\psi_1\,d\tau + c_2^*c_2 \int \psi_2^*\psi_2\,d\tau$$

$$+ c_1^*c_2 \int \psi_1^*\psi_2\,d\tau + c_2^*c_1 \int \psi_2^*\psi_1\,d\tau \tag{11.120}$$

The integral on the left-hand side is unity because of the normalization condition, and the same is true of the first two integrals on the right-hand side. The remaining two integrals are zero because of the orthogonality condition. It therefore follows that

$$c_1^*c_1 + c_2^*c_2 = 1 \tag{11.121}$$

This relationship is useful for the construction of suitable linear combinations (or superpositions) of wave functions. When we use such a superposition to calculate the expectation value of a quantum mechanical operator (Postulate VI), the result has a well-defined interpretation. For example, when each wave function in the superposition corresponds to a different energy, the expectation value does not give one definite energy or eigenvalue. Rather it gives the probability that a large number of particles have a particular energy eigenvalue. Alternatively, after a large number of measurements on one particle, it gives the relative probabilities of the various energy states.

Non-Commutation and the Heisenberg Uncertainty Principle

We have seen in Section 11.3 that certain quantum-mechanical observables disrupt each other so that it is not possible to measure these simultaneously to an accuracy greater than that given by the Heisenberg uncertainty principle, expressed by Eqs. 11.60, 11.64, and 11.65. In this section, we show that these relationships are a direct consequence of the non-commutation of the observable pairs: position-momentum, energy-time, and angle–angular momentum.

EXAMPLE 11.3 The operator $\hat{x}$ is the one-dimensional position operator and $\hat{p}_x = -i\hbar(\partial/\partial x)$ is the one-dimensional momentum operator in the X-direction. Calculate the commutation relation of these two operators defined by:

$$[\hat{x}, \hat{p}_x]_- \equiv \hat{x}\hat{p}_x - \hat{p}_x\hat{x}$$

Solution The commutator can be evaluated by operating it on any function of x, say $f(x)$, to give:

$$[\hat{x}, \hat{p}_x]_- f(x) = \hat{x}\hat{p}_x f(x) - \hat{p}_x\hat{x} f(x)$$

or performing the operations, which gives

$$-i\hbar[\hat{x}\frac{\partial}{\partial x}f(x) - \frac{\partial}{\partial x}\hat{x}f(x)] = -i\hbar\hat{x}\frac{\partial f(x)}{\partial x} + i\hbar\hat{x}\frac{\partial f(x)}{\partial x} + i\hbar f(x)$$

The first two terms on the right-hand side cancel each other, which leaves only the last term. Dropping the arbitrary function $f(x)$ gives the commutation relation as:

$$[\hat{x}, \hat{p}_x]_- = i\hbar$$

The Schwarz deals with a positive definite operator $\hat{A}$ and f and g that are arbitary functions where

$$\int \psi^* \hat{A}\psi\, d\tau \geqslant 0$$

If $\psi = f + \lambda g$, where λ may be any complex number, we can substitute into and expand the above inequality to give

$$(\int f^* \hat{A} f\, d\tau)(\int g^* \hat{A} g\, d\tau) > |\int f^* \hat{A} g\, d\tau|^2$$

for any positive operator.

Similar considerations show that energy-time and angle–angular momentum operators do not commute, so that (see Table 11.1)

$$[\hat{x}, \hat{p}_x]_- = [\hat{\phi}, \hat{L}_z]_- = [\hat{E}, \hat{t}\,]_- = i\hbar \tag{11.122}$$

Non-commutation of operators, and this alone, leads to the Heisenberg uncertainty relations. The variance, ΔA^2, of an operator is defined by (ψ is normalized)

$$\Delta A^2 = \int \psi^* (\hat{A} - \langle A \rangle)^2 \psi\, d\tau \tag{11.123}$$

Using Eq. 11.123, the commutation relation of two operators,

$$[\hat{A}, \hat{B}]_- = i\hat{C} \tag{11.124}$$

and the Schwarz inequality leads to the relation for the uncertainties,[5] ΔA and ΔB

$$\Delta A \Delta B \geq \left|\frac{1}{2}\langle C\rangle\right| \tag{11.125}$$

The commutation relations in Eq. 11.122 lead to the Heisenberg uncertainty relations as given by Eqs. 11.60, 11.64 and 11.65.

11.6 Quantum Mechanics of Some Simple Systems

For the trivial operator $\hat{A} = \hat{1}$, we obtain the Schwarz inequality

$$(\int |f|^2\, d\tau)(\int |g^*|^2\, d\tau) \geqslant |f^* g\, d\tau|^2.$$

We will now apply the postulates of Section 11.5 to some simple systems that are related to some important chemical problems.

[5]The derivation of the Heisenberg uncertainty relationship is given in E. Metzbacher, *Quantum Mechanics*, New York, Wiley, 2nd Ed. 1970.

The Free Particle

The simplest application is to a particle of mass m moving freely in one dimension, which we will take to be the X-direction. This is known as the *free particle problem in one dimension.* Since the particle is moving freely with no forces acting on it, the potential energy E_p is constant throughout its motion and we take it to be zero. The total energy is therefore the kinetic energy $\frac{1}{2}mu_x^2$, which is $p_x^2/2m$ where p_x is the momentum. The total energy in Hamilton's classical form is thus

$$H = \frac{p_x^2}{2m} \tag{11.126}$$

We see from Table 11.1 that p_x is to be replaced by the operator $-i\hbar(\partial/\partial x)$, so that the quantum-mechanical operator is

$$\hat{H} = -\frac{\hbar^2}{2m}\frac{\partial^2}{\partial x^2} \tag{11.127}$$

The Schrödinger equation $\hat{H}\psi = E\psi$ thus takes the form

$$-\frac{\hbar^2}{2m}\frac{\partial^2\psi}{\partial x^2} = E\psi \tag{11.128}$$

or

$$\frac{\partial^2\psi}{\partial x^2} + \frac{2mE}{\hbar^2}\psi = 0 \tag{11.129}$$

There are two solutions to this differential equation, namely

$$\psi_+ = A_+ e^{i\sqrt{2mE}x/\hbar} \tag{11.130}$$

and

$$\psi_- = A_- e^{-i\sqrt{2mE}x/\hbar} \tag{11.131}$$

where A_+ and A_- are constants. The fact that these are solutions may easily be verified by substituting them into Eq. 11.129.

The relative probability density of the particle is $|\psi|^2 = \psi^*\psi$, and for the function ψ_+ this is

$$|\psi_+|^2 = \psi_+^*\psi_+ = A_+^* e^{-i\sqrt{2mE}x/\hbar} A_+ e^{i\sqrt{2mE}x/\hbar} \tag{11.132}$$

$$= A_+^*A_+ = |A_+|^2 \tag{11.133}$$

Similarly for the function ψ_- the relative probability is $|A_-|^2$. Both of these quantities are independent of x, so that there is an equal probability of finding the particle at any distance along the X axis. The particle is therefore *nonlocalized.*

We can also use Postulate V to obtain allowed values of the momentum p_x. The momentum operator $\hat{p}_x$ is $-i\hbar(\partial/\partial x)$, and the wave equation is therefore

$$-i\hbar\frac{\partial\psi}{\partial x} = p_x\psi \tag{11.134}$$

From Eqs. 11.130 and 11.131, we obtain

$$-i\hbar\frac{\partial}{\partial x}(A_\pm e^{\pm i\sqrt{2mE}x/\hbar}) = p_x A_\pm e^{\pm i\sqrt{2mE}x/\hbar} \tag{11.135}$$

This reduces to

$$\pm\sqrt{2mE}A_{\pm}e^{\pm i\sqrt{2mE}x/\hbar} = p_x A_{\pm}e^{\pm i\sqrt{2mE}x/\hbar} \quad (11.136)$$

and therefore

$$p_x = \pm\sqrt{2mE} \quad (11.137)$$

The wave functions are

$$\psi_{\pm} = A_{\pm}e^{\pm ip_x x/\hbar} \quad (11.138)$$

These solutions correspond to the classical values; p_x is positive if the particle is moving in one direction and negative if it is moving in the other. For this particular case there is therefore no quantization; we can give the particle any energy we like.

An example of a free particle is an electron that has been separated from an atom; its energy is no longer quantized.

The Particle in a Box

Compare a classical particle and 3 quantum states; The particle in a box; Introduction and 3 quantum states.

Quantization does, however, appear if the particle is not permitted to travel an infinite distance but is confined to a certain region of space. In three dimensions this problem is referred to as the *particle in a box.* We will first consider the one-dimensional problem, with the particle moving along the X axis over a distance from 0 to a, as shown in Figure 11.11a. The potential energy E_p is taken to be zero within the box ($0 \leq x \leq a$) and infinity for $x < 0$ and $x > a$.

Within the box the wave equation is Eq. 11.128, as for the free particle, and the solutions are given by Eqs. 11.130 and 11.131. We now, however, have the additional conditions that ψ must be zero when $x = 0$ and $x = a$ and must be zero for $0 > x > a$. Neither ψ_1 nor ψ_2 as given by these equations can *individually* satisfy these *boundary conditions,* because the condition that ψ must be zero when $x = a$ requires both A_+ and A_- to be zero. However, this dilemma is avoided by taking a linear combination of ψ_1 and ψ_2, which we know is also a solution of the wave equation (compare Eq. 11.116). It is perfectly general to take $\psi_1 + \psi_2$ as the linear combination, since A_+ and A_- are in any case adjustable. Our selected wave function is therefore

$$\psi = A_+e^{i\sqrt{2mE}x/\hbar} + A_-e^{-i\sqrt{2mE}x/\hbar} \quad (11.139)$$

We will now see if this function is consistent with the boundary conditions, and if so, with what restrictions. The boundary condition that $\psi = 0$ when $x = 0$ requires that $A_+ + A_- = 0$, so that $A_- = -A_+$. Equation 11.139 can thus be rewritten as $A \equiv A_+$

$$\psi = A_+(e^{i\sqrt{2mE}x/\hbar} - e^{-i\sqrt{2mE}x/\hbar}) \quad (11.140)$$

It is now convenient to apply Euler's theorem[6], and rewrite this equation with $A \equiv A_+$.

$$\psi = 2iA\sin\left(\frac{\sqrt{2mE}}{\hbar}x\right) \quad (11.141)$$

[6]According to Euler's theorem, $e^{iy} = \cos y + i\sin y$, and therefore $e^{iy} - e^{-iy} = 2i\sin y$.

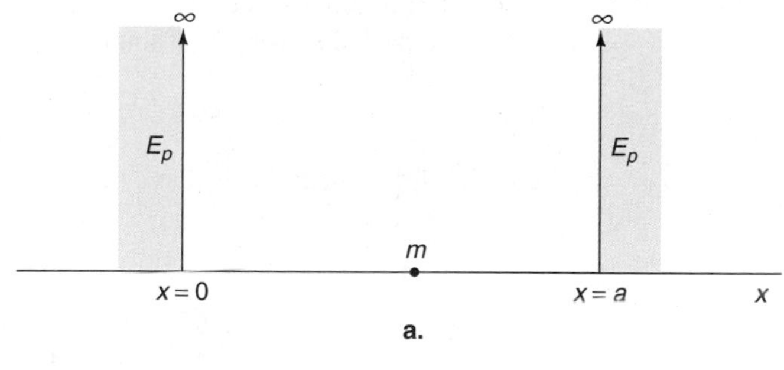

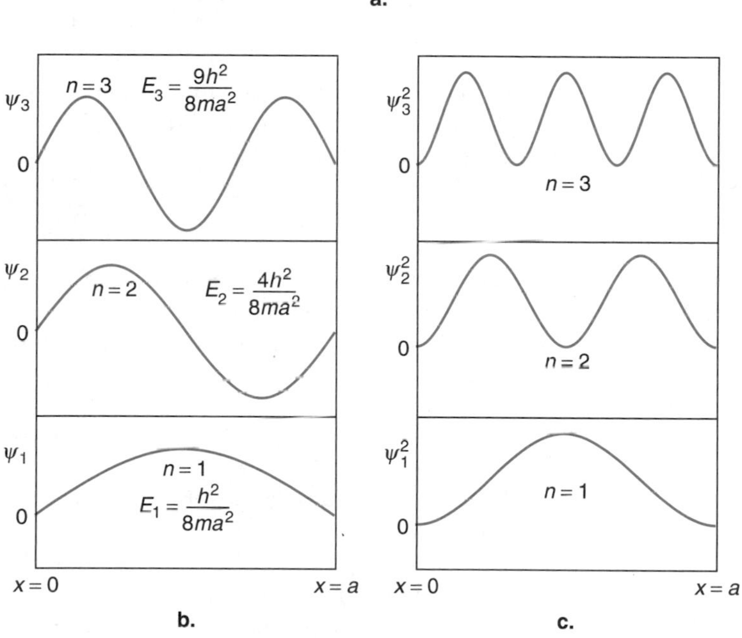

FIGURE 11.11
(a) A particle confined to a one-dimensional box of length a. The potential energy is infinite for $x < 0$ and $x > a$. (b) The form of some of the wave functions for the particle in a one-dimensional box and the corresponding energies. (c) The form of the probability densities ψ^2.

The second boundary condition, that $\psi = 0$ when $x = a$, gives

$$0 = 2iA \sin\left(\frac{\sqrt{2mE}}{\hbar} a\right) \tag{11.142}$$

The factor $2iA$ cannot be zero, but the sine of an angle is zero when the angle is an integral multiple of π. Thus

$$\frac{\sqrt{2mE_n}}{\hbar} a = \pm n\pi \tag{11.143}$$

where $n = 0, 1, 2, \ldots, \infty$. We have now written E_n instead of E since there is a different energy E_n for each value of n. The wave function (Eq. 11.141) therefore becomes

$$\psi_n = \pm 2iA \sin\frac{n\pi x}{a} \tag{11.144}$$

Note that the $\pm$ sign can go outside the sine function since $\sin(-n\pi x/a) = -\sin(n\pi x/a)$.

To determine the value of $2iA$ we use the normalization condition, Eq. 11.89, that the total probability $\psi_n^*\psi_n$ of finding the particle in the box must be unity. Thus

$$\int_0^a \psi_n^*\psi_n \, dx = 4A^2 \int_0^a \sin^2\left(\frac{n\pi x}{a}\right) dx = 1 \tag{11.145}$$

The value of the integral is $a/2$ and therefore

$$4A^2 \frac{a}{2} = 1 \tag{11.146}$$

or

$$A = \pm\sqrt{\frac{1}{2a}} \tag{11.147}$$

To make the wave function real, we introduce a phase factor $-i = e^{-i\pi}$ that multiplies the value of A. The result is $A = \mp i\sqrt{1/2a}$. Insertion of this into Eq. 11.144 gives the wave function as

$$\psi_n = \sqrt{\frac{2}{a}} \sin\left(\frac{n\pi x}{a}\right) \tag{11.148}$$

The quantum numbers can now be given as $n = 1, 2, 3, \ldots, \infty$, since $n = 0$ would mean that no particle is present.

The possible energy values are obtained by rearrangement of Eq. 11.143:

$$E_n = \frac{n^2h^2}{8ma^2} \tag{11.149}$$

These are therefore the possible stationary-state energies for a particle in a one-dimensional box.

The wave functions for various values of n and the corresponding energies are shown in Figure 11.11b, and the probability densities $\psi^*\psi$ are shown in Figure 11.11c. Note that the energy increases in proportion to n^2 and that the number of antinodes is equal to n; the number of nodes is $n - 1$ (cf. Figures 11.5 and 11.6).

The energy levels and, therefore, the separations between them are inversely proportional to ma^2. Thus, if the particles are heavy or if the box is large, the energy levels are close together and the system behaves classically, with no observable quantization. This result is consistent with Bohr's correspondence principle, according to which at certain limits quantum-mechanical behavior becomes classical behavior. At the other extreme, if m and a are of atomic dimensions, the quantization is important and can easily be detected experimentally. The following two examples illustrate the effect of mass and box size.

EXAMPLE 11.4 Calculate the energy difference between the $n = 1$ and $n = 2$ levels for an electron ($m = 9.1 \times 10^{-31}$ kg) confined to a one-dimensional box having a length of 4.0×10^{-10} m (this is the order of magnitude of an atomic diameter). What wavelength corresponds to a spectral transition between these levels?

Solution The ground-state energy is obtained from Eq. 11.149 with $n = 1$:

$$E_1 = \frac{[6.626 \times 10^{-34}\ (\text{J s})]^2}{8[9.1 \times 10^{-31}\ (\text{kg})][4.0 \times 10^{-10}\ (\text{m})]^2}$$
$$= 3.77 \times 10^{-19}\ \text{J}$$

The energy for $n = 2$ is just four times this, and the energy difference is therefore three times this value:

$$\Delta E = E_2 - E_1 = 3 \times 3.77 \times 10^{-19} \text{ J} = 1.13 \times 10^{-18} \text{ J}$$

The frequency is this energy divided by the Planck constant, and the wavelength is the velocity of light divided by the frequency:

$$\lambda = \frac{c}{\nu} = \frac{hc}{\Delta E} = \frac{6.626 \times 10^{-34} \text{ (J s)} 2.998 \times 10^{8} (\text{m s}^{-1})}{1.13 \times 10^{-18} (\text{J})}$$

$$= 1.76 \times 10^{-7} \text{ m} = 176 \text{ nm}$$

This wavelength is in the ultraviolet region of the spectrum.

EXAMPLE 11.5 Calculate the energy difference between the $n = 1$ and $n = 2$ levels for a marble of mass 1 g confined in a one-dimensional box of length 0.10 m. What wavelength corresponds to a spectral transition between these levels?

Solution The ground-state ($n = 1$) energy is

$$E_1 = \frac{[6.626 \times 10^{-34} \text{ (J s)}]^2}{8[0.001 \text{ (kg)}][0.10 (\text{m})]^2}$$

$$= 5.48 \times 10^{\ 63} \text{ J}$$

The energy difference ΔE is three times this:

$$E_2 - E_1 = 3 \times 5.48 \times 10^{-63} \text{ J} = 1.65 \times 10^{-62} \text{ J}$$

Such an energy is much too small to be measurable; it is impossible to measure an energy of less than about 10^{-25} J. The corresponding wavelength is

$$\lambda = \frac{hc}{\Delta E} = 1.20 \times 10^{37} \text{ m}$$

This wavelength is rather difficult to measure, as it is greater than the distance to any observed star! No quantization can be detected in such a system.

So far we have considered only a one-dimensional box. The extension to a three-dimensional box is quite straightforward, and we will simply state the result. For a particle confined in a box of sides a, b, and c, the allowed energy levels are

$$E = \frac{h^2}{8m}\left(\frac{n_1^2}{a^2} + \frac{n_2^2}{b^2} + \frac{n_3^2}{c^2}\right) \tag{11.150}$$

There are now three distinct quantum numbers, n_1, n_2, and n_3. If the box is a cube of sides a, the expression becomes

$$E = \frac{h^2}{8ma^2}(n_1^2 + n_2^2 + n_3^2) \tag{11.151}$$

It is now possible for three different combinations of quantum numbers to give the same energy. For example, the same energy is found for

$$n_1 = 2, \quad n_2 = 1, \quad n_3 = 1$$

as for

$$n_1 = 1, \qquad n_2 = 2, \qquad n_3 = 1$$

and for

$$n_1 = 1, \qquad n_2 = 1, \qquad n_3 = 2$$

The energy level is thus *degenerate*. In this case we speak of *threefold degeneracy,* since there are three combinations of quantum numbers that give the same energy.

Distort a cubic box to see the degenercies lift; The particle in a box; Symmetry breaking.

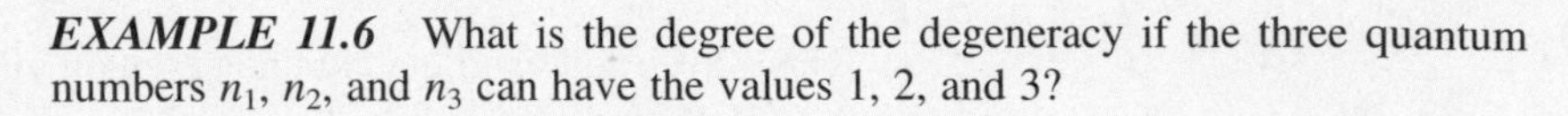

EXAMPLE 11.6 What is the degree of the degeneracy if the three quantum numbers n_1, n_2, and n_3 can have the values 1, 2, and 3?

Solution The following combinations are possible:

123 132 213
231 312 321

There is therefore *sixfold* degeneracy.

Symmetry Breaking

The three-dimensional particle in a box can be used to illustrate an important and common effect that occurs in quantum systems, namely **symmetry breaking**. In the case of a square box, the dimensions are equal, $a = b = c$. Fig. 11.12 shows a number of degenerate energy levels. If the box is stretched or squashed in various directions, the cube is distorted into a box of lower symmetry, $a \neq b \neq c$. Comparison of Eq. 11.150 with 11.151 shows that the degeneracies are lifted (removed) as the symmetry is lowered. For example, the six-fold degeneracy calculated in Exam-

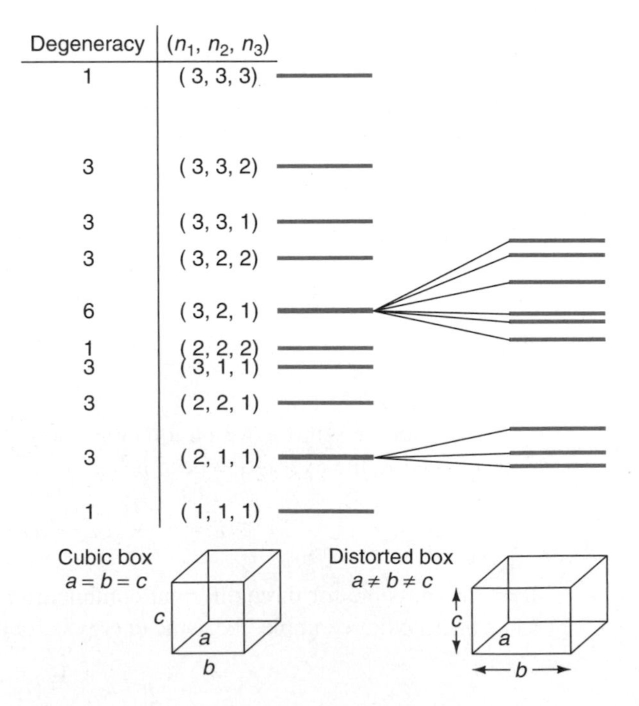

FIGURE 11.12
Symmetry breaking: On the left-hand side, a 3D cube has degenerate levels for a particle in a box. When the symmetry is lowered to $a = b = c$: $a \neq b \neq c$, the right-hand side shows how the degeneracies are removed. We also will lose the degenercy for all the other levels (with splitting into three levels) except for levels 1,1,1; 2,2,2; and 3,3,3.

ple 11.6 becomes six levels of different energy. The greater the distortion, the greater the energy differences become between the previously degenerate energy levels.

The Harmonic Oscillator

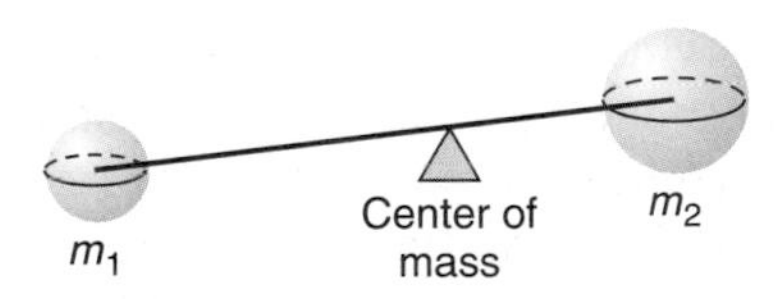

FIGURE 11.13
Two connected particles of unequal masses may be regarded as being a single mass called the *reduced mass*, μ, located at the center of mass that lies closer to the particle with greater mass. The reduced mass is given by Eq. 11.152.

Another quantum-mechanical problem of importance is molecular vibration. The simplest case to consider is a diatomic molecule in which the two atomic masses are m_1 and m_2. The potential energy is approximately symmetric and the vibration of such a molecule obeys the equations of simple harmonic motion (Eqs. 11.8 to 11.18). However, as we will discuss in more detail in Section 13.3 (Eq. 13.94), we must replace the mass m in those equations by a *reduced mass* μ, given by

$$\mu = \frac{m_1 m_2}{m_1 + m_2} \tag{11.152}$$

See Figure 11.13. If the bond is extended by a distance x, the potential energy is given by

$$E_p = \tfrac{1}{2} k_h x^2 \tag{11.153}$$

(compare Eq. 11.17), where k_h is the force constant. The kinetic energy is

$$E_k = \frac{p_x^2}{2\mu} \tag{11.154}$$

and the total energy in classical Hamiltonian form is therefore

$$H = \frac{p_x^2}{2\mu} + \frac{1}{2} k_h x^2 \tag{11.155}$$

Try to fit a parabola into a potential energy well; The harmonic oscillator.

The quantum Hamiltonian operator is then obtained by replacing p_x by $-i\hbar(\partial/\partial x)$:

$$\hat{H} = -\frac{\hbar^2}{2\mu}\frac{\partial^2}{\partial x^2} + \frac{1}{2} k_h x^2 \tag{11.156}$$

The Schrödinger equation for the system is therefore

$$\left(-\frac{\hbar^2}{2\mu}\frac{\partial^2}{\partial x^2} + \frac{1}{2} k_h x^2\right)\psi = E\psi \tag{11.157}$$

The solution of this equation involves *Hermite polynomials.* We will not give details but merely state the main conclusions. Solutions can only be obtained provided that the total energy E has a value given by the equation

$$E_v = \hbar\sqrt{\frac{k_h}{\mu}}\left(v + \frac{1}{2}\right) \tag{11.158}$$

Here v is the *vibrational quantum number,* which can have only integral values:

$$v = 0, 1, 2, 3, \ldots \tag{11.159}$$

Equation 11.158 can also be written as

$$E_v = h\nu_0\left(v + \tfrac{1}{2}\right) \tag{11.160}$$

where ν_0, equal to $(1/2\pi)\sqrt{k_h/\mu}$, is the natural frequency of the harmonic oscillator (compare Eq. 11.15).

We can see from Eq. 11.160 that the quantized energy levels for the harmonic oscillator are equidistant from one another. An important feature is that the lowest possible energy, corresponding to $v = 0$, is equal to

$$E_0 = \frac{1}{2}h\nu_0 \tag{11.161}$$

Zero-point Energy

In classical theory a bond could be undergoing no vibration, but in quantum mechanics this is not allowed; even at the absolute zero of temperature, vibration still occurs with energy $\frac{1}{2}h\nu_0$. This energy is the **zero-point energy** (p. 509). The fact that there is a zero-point energy is consistent with Heisenberg's uncertainty principle. If no vibration occurred, the position and momentum of the atoms would both have precise values, and this is impossible.

The wave functions corresponding to the various energy levels involve the Hermite polynomials, and the shapes of some of them are shown in Figure 11.14. The curves for even values of v are symmetric with respect to $x = 0$, while with v odd the curves are antisymmetric. The probability density functions $\psi^*\psi$ for several values of v are shown in Figure 11.14. Also shown on this diagram are the corresponding classical probability functions. The differences between the classical and the quantum-mechanical behavior are striking. For $v = 0$, for example, we see that

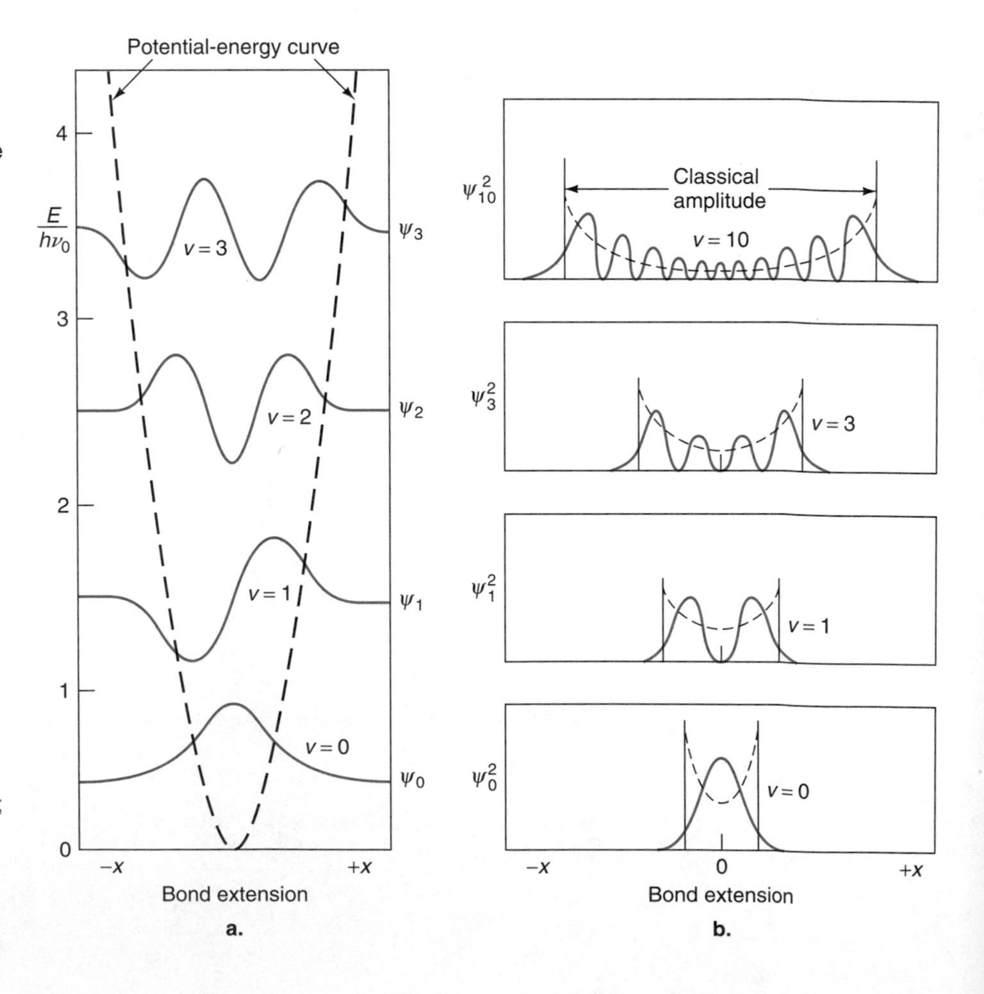

FIGURE 11.14
(a) The form of some wave functions for a harmonic oscillator.
(b) Some probability density functions for a harmonic oscillator. The corresponding classical functions are shown as dashed lines.

Animate and plot Figure 11.14; The harmonic oscillator; Potential.

Vary parameters of Figure 11.14; The harmonic oscillator; Wave functions.

the highest probability according to the quantum-mechanical treatment is when the system is passing through its equilibrium position ($x = 0$). In classical mechanics, on the other hand, the probability is *lowest* at the equilibrium position, since the system has its highest velocity at this position and therefore passes through it rapidly; the vibrational motion is slowest at the extremities of the vibration, and the probability is therefore the highest at these extremities. As the vibrational quantum number increases, however, the quantum-mechanical probabilities become closer to the classical ones, and the highest probabilities are close to the turning points of the vibrations.

Another important difference between the quantum-mechanical and classical probabilities is that the quantum-mechanical treatment gives a finite probability that the bond is extended to a greater extent than the value permitted by classical theory. In a classical vibration the extension is restricted to a certain fixed value, but the quantum-mechanical curves in Figure 11.14b show that there is a certain probability of an even greater extension. There are a number of situations where quantum mechanics permits a system to penetrate into regions that are forbidden in classical mechanics, and the effect is known as the **tunnel effect.** This effect is particularly important in chemical kinetics, and more was said about it in Section 9.10 in connection with transition-state theory (see Figures 9.17 and 9.18).

Tunneling

11.7 Quantum Mechanics of Hydrogenlike Atoms

Bohr's theory of the hydrogen atom was a great step forward, but it failed to provide a satisfactory basis for the understanding of more complex atoms. Much greater success has been achieved through quantum mechanics. We will consider treatment of a hydrogenlike atom, having a nucleus of charge Ze and a single electron. Since the mass of the electron is much smaller than that of the nucleus, the reduced mass μ, given by Eq. 11.152, is almost exactly equal to the mass of the electron.

The potential energy of the electron at a distance r from the nucleus arises entirely from the Coulombic attraction and is given by

$$E_p = -\frac{Ze^2}{4\pi\epsilon_0 r} \tag{11.162}$$

where ϵ_0 is the permittivity of a vacuum. Recall from p. 266 that the term $4\pi\epsilon_0$ enters into expressions for the electrostatic force and energy. (See also Appendix A, p. 999.) Since the energy in Eq. 11.162 is the same in all directions, we have a **symmetrical field** or a **central field**.

Central Field

The classical Hamiltonian for the system involves the components of momentum along the three axes and is

$$H = \frac{1}{2\mu}(p_x^2 + p_y^2 + p_z^2) - \frac{Ze^2}{4\pi\epsilon_0 r} \tag{11.163}$$

The quantum Hamiltonian operator is obtained by making the substitutions of Table 11.1 and is

$$\hat{H} = -\frac{\hbar^2}{2\mu}\nabla^2 - \frac{Ze^2}{4\pi\epsilon_0 r} \tag{11.164}$$

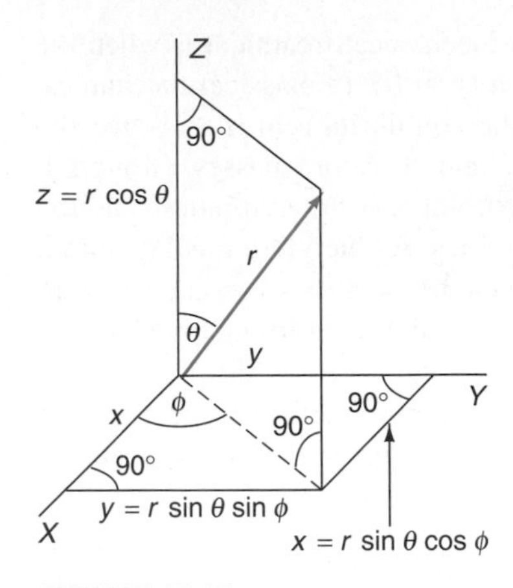

FIGURE 11.15
The relation between Cartesian and spherical polar coordinates.

Relate Cartesian and polar coordinates in 2 and 3 dimensions; Polar coordinates.

The time-independent Schrödinger equation $\hat{H}\psi = E\psi$ is therefore

$$-\frac{\hbar^2}{2\mu}\nabla^2\psi - \frac{Ze^2}{4\pi\epsilon_0 r}\psi = E\psi \tag{11.165}$$

Since the system is spherically symmetrical, it is most convenient to use spherical polar coordinates, which are related to Cartesian coordinates as shown in Figure 11.15. When the Laplacian operator ∇^2 is converted to polar coordinates, the Schrödinger equation takes the form

$$\frac{1}{r^2}\frac{\partial}{\partial r}\left(r^2\frac{\partial\psi}{\partial r}\right) + \frac{1}{r^2\sin\theta}\frac{\partial}{\partial\theta}\left(\sin\theta\frac{\partial\psi}{\partial\theta}\right) + \frac{1}{r^2\sin^2\theta}\frac{\partial^2\psi}{\partial\phi^2} + \frac{2\mu}{\hbar^2}\left(E + \frac{Ze^2}{4\pi\epsilon_0 r}\right)\psi = 0 \tag{11.166}$$

Although Eq. 11.166 looks at first more cumbersome than Eq. 11.165, it leads to a much simpler solution because it may be solved by the well-known method of separation of variables, which leads to three ordinary differential equations that can be exactly solved. Each of the three equations has only one of the variables r, θ, and ϕ. To solve these we write the wave function ψ, which is a function of r, θ, and ϕ, as the product of three functions $R(r)$, $\Theta(\theta)$, and $\Phi(\phi)$:

$$\psi(r, \theta, \phi) = R(r)\Theta(\theta)\Phi(\phi) \tag{11.167}$$

Substitution of this expression into Eq. 11.166 and division by $R\Theta\Phi$ give

$$\frac{1}{r^2R}\frac{d}{dr}\left(r^2\frac{dR}{dr}\right) + \frac{1}{r^2(\sin\theta)\Theta}\frac{d}{d\theta}\left(\sin\theta\frac{d\Theta}{d\theta}\right) + \frac{1}{r^2\sin^2\theta}\cdot\frac{1}{\Phi}\frac{d^2\Phi}{d\phi^2} + \frac{2\mu}{\hbar^2}\left(E + \frac{Ze^2}{4\pi\epsilon_0 r}\right) = 0 \tag{11.168}$$

The partial derivatives have been replaced by ordinary derivatives since each function now depends on a single variable only.

The third term in this expression contains

$$\frac{1}{\Phi}\cdot\frac{d^2\Phi}{d\phi^2}$$

and this is the only term in which Φ and ϕ appear. Since the three polar coordinates are quite independent of one another, this term must therefore be constant. We will write it as

Φ Equation

$$\frac{1}{\Phi}\cdot\frac{d^2\Phi}{d\phi^2} = -m_l^2 \tag{11.169}$$

Magnetic Quantum Number

and we will see later that m_l is the **magnetic quantum number.** We will call Eq. 11.169 the Φ *equation.*

If we make this substitution (Eq. 11.169) into Eq. 11.168 and then multiply by r^2, we obtain, after some rearrangement,

$$\frac{1}{R}\frac{d}{dr}\left(r^2\frac{dR}{dr}\right) + \frac{2\mu}{\hbar^2}\left(E + \frac{Ze^2}{4\pi\epsilon_0 r}\right)r^2 = \frac{m_l^2}{\sin^2\theta} - \frac{1}{\sin\theta}\frac{1}{\Theta}\frac{d}{d\theta}\left(\sin\theta\frac{d\Theta}{d\theta}\right) \tag{11.170}$$

The left-hand side involves R and r but not Θ and θ, while the right-hand side involves Θ and θ but not R and r. Both sides must therefore be equal to a *separation constant,* which we will write as $l(l+1)$ in anticipation of the fact that l is the **azimuthal** or **orbital angular momentum quantum number.** Equating the left-hand side of Eq. 11.170 to $l(l+1)$ gives the *radial equation,* or *R equation:*

Orbital Angular Momentum Quantum Number.

***R* Equation**

$$\frac{1}{R}\frac{d}{dr}\left(r^2\frac{dR}{dr}\right) + \frac{2\mu}{\hbar^2}\left(E + \frac{Ze^2}{4\pi\epsilon_0 r}\right)r^2 = l(l+1) \tag{11.171}$$

Similarly, from the right-hand side, with a little rearrangement, we obtain the *angular equation,* or *Θ equation:*

Θ Equation

$$\left[\frac{1}{\sin\theta}\frac{d}{d\theta}\left(\sin\theta\frac{d}{d\theta}\right) - \frac{m_l^2}{\sin^2\theta} + l(l+1)\right]\Theta = 0 \tag{11.172}$$

We have thus split the Schrödinger equation (Eq. 11.168) into three equations, one involving Φ and ϕ (Eq. 11.169), one involving R and r (Eq. 11.171), and one involving Θ and θ (Eq. 11.172). These equations must now be solved so as to eliminate the differentials. We will first solve the Φ equation so as to obtain the allowed values of m_l. These will then be used to solve the Θ equation so as to obtain the allowed l values. Finally, the l values will be used to solve the R equation.

Solution of the Φ Equation

Equation 11.169 is of a familiar form (compare Eq. 11.129) and its solution is

$$\Phi = Ae^{im_l\phi} \tag{11.173}$$

where A is a normalization constant. Since the angle ϕ is the azimuthal polar angle, the function Φ must be periodic, $\Phi(\phi) = \Phi(\phi + 2n\pi)$, because these angles correspond to the same poition; this requires that

$$m_l = 0, \pm1, \pm2, \pm3, \ldots \tag{11.174}$$

The positive and negative values relate to distinct solutions. The quantity m_l has thus become a quantum number because of the mathematical constraints on the system and not in any arbitrary way.

The value of A is obtained by applying the normalization condition, the range of ϕ being 0 to 2π:

$$\int_0^{2\pi} \Phi_{m_l}\Phi^*_{m_l}\, d\phi = A^2 \int_0^{2\pi} e^{im_l\phi}e^{-im_l\phi}\, d\phi = A^2 2\pi = 1 \tag{11.175}$$

Therefore, $A = 1/\sqrt{2\pi}$ and the solution becomes

$$\Phi_{m_l} = \frac{1}{\sqrt{2\pi}}e^{im_l\phi} \tag{11.176}$$

When $m_l = 0$, the value of Φ_0 is $1/\sqrt{2\pi}$, but for other values of m_l the solutions in volve imaginary exponents. This is awkward, and it is more usual to employ a linear combination of the functions, $\Phi_{m_l} \pm \Phi_{-m_l}$; as we have seen (Eq. 11.116), linear combinations of wave functions having the same eigenvalue are also solutions of wave equations. For example,

$$\text{for } m_l = 1, \qquad \Phi_1 = \frac{1}{\sqrt{2\pi}}e^{i\phi} \tag{11.177}$$

$$\text{for } m_l = -1, \qquad \Phi_{-1} = \frac{1}{\sqrt{2\pi}}e^{-i\phi} \tag{11.178}$$

TABLE 11.2 Solution of the Φ Equation

Value of m_l	Solution in Complex Form	Real Form
0	—	$\Phi_0 = \dfrac{1}{\sqrt{2\pi}}$
1	$\Phi_1 = \dfrac{1}{\sqrt{2\pi}}e^{i\phi}$	$\Phi_x = \dfrac{1}{\sqrt{2}}(\Phi_1 + \Phi_{-1}) = \dfrac{\cos\phi}{\sqrt{\pi}}$
−1	$\Phi_{-1} = \dfrac{1}{\sqrt{2\pi}}e^{-i\phi}$	$\Phi_y = \dfrac{1}{\sqrt{2}}(\Phi_1 - \Phi_{-1}) = \dfrac{\sin\phi}{\sqrt{\pi}}$
2	$\Phi_2 = \dfrac{1}{\sqrt{2\pi}}e^{i2\phi}$	$\Phi_{x^2-y^2} = \dfrac{1}{\sqrt{2}}(\Phi_2 + \Phi_{-2}) = \dfrac{\cos 2\phi}{\sqrt{\pi}}$
−2	$\Phi_{-2} = \dfrac{1}{\sqrt{2\pi}}e^{-i2\phi}$	$\Phi_{xy} = \dfrac{1}{\sqrt{2}}(\Phi_2 - \Phi_{-2}) = \dfrac{\sin 2\phi}{\sqrt{\pi}}$

The reasons for the $x^2 - y^2$ and xy subscripts are as follows:

1. $\cos 2\phi = \cos^2\phi - \sin^2\phi$; from Figure 11.15, $\cos^2\phi = x^2/(r^2 \sin^2\theta)$ and $\sin^2\phi = y^2/(r^2\sin^2\theta)$; thus $\cos 2\phi$ has the same dependence on ϕ as $x^2 - y^2$.
2. $\sin 2\phi = 2\sin\phi\cos\phi = xy/(r^2\sin^2\theta)$; thus $\sin 2\phi$ has the same dependence on ϕ as xy.

We can take the sum of these and divide by $\sqrt{2}$ to preserve the normalization,

$$\Phi_x = \frac{1}{\sqrt{2}}(\Phi_1 + \Phi_{-1}) = \frac{1}{2\sqrt{\pi}}(e^{i\phi} + e^{-i\phi}) = \frac{\cos\phi}{\sqrt{\pi}} \tag{11.179}$$

We designate this orbital Φ_x since $\cos\phi$ has its maximum value when $\phi = 0$, which (as we see from Figure 11.15) corresponds to the X axis. Similarly, we can take the difference $\Phi_1 - \Phi_{-1}$ divided by $\sqrt{2}$:

$$\Phi_y = \frac{1}{\sqrt{2}}(\Phi_1 - \Phi_{-1}) = \frac{1}{2\sqrt{\pi}}(e^{i\phi} - e^{-i\phi}) = \frac{i\sin\phi}{\sqrt{\pi}} \tag{11.180}$$

The imaginary number can be written as $i = \exp(i\pi)$ and as such is a phase. Since we are usually interested in probability densities $\Phi^*\Phi$, the usual phase convention is to drop i in this expression since it disappears when we take the complex conjugate. The function is therefore written as

$$\Phi_y = \frac{\sin\phi}{\sqrt{\pi}} \tag{11.181}$$

The value of this function is a maximum when $\phi = \pi/2$, which is along the Y axis.

Table 11.2 lists Φ functions for the first three values of $\pm m_l$.

Solution of the Θ Equation

The solution of the Θ equation (Eq. 11.172) is mathematically more difficult, and we will present only a brief outline, with emphasis on the main results. For details, the reader is referred to textbooks of quantum mechanics.[7]

We may introduce a transformation into Eq. 11.172 by putting

$$\xi = \cos\theta \quad \text{and} \quad P_l(\xi) = \Theta \tag{11.182}$$

[7]Donald A. McQuarrie, "Quantum Chemistry," University Science Books, Oxford University Press, 1983.

and obtain

$$(1-\xi^2)\frac{d^2P_l(\xi)}{d\xi^2} - 2\xi\frac{dP_l(\xi)}{d\xi} + \left[l(l+1) - \frac{m_l^2}{(1-\xi^2)}\right]P_l(\xi) = 0 \quad (11.183)$$

When m_l is zero, this equation is the *Legendre equation,* named after the French mathematician Adrien Marie Legendre (1752–1833). Solutions for the Legendre equation are possible only *when l is zero or has positive integral values.* These solutions are known as the *Legendre polynomials of degree l.* When m_l is not zero, a solution can be obtained only if m_l has one of the integral values $-l, -l+1, \ldots, 0, \ldots, l-1, l$. The solutions are then the *associated Legendre functions* and are usually expressed by means of a **recursion formula.** The details are omitted but can be found in the source referred to in footnote 7.

Insertion of a value into a recursion formula gives a new value back until the series of new values is naturally limited by mathematical constraints of the formula.

The conclusion is therefore that l can only be zero or have a positive integral value and that the m_l values are determined by the value of l:

$$l = 0, 1, 2, 3, \ldots \quad (11.184)$$

$$m_l = -l, -l+1, \ldots, -1, 0, 1, \ldots, l-1, l \quad (11.185)$$

We shall see that the solution of the R equation imposes an upper limit on the value of l.

Table 11.3 gives solutions of the Θ equation for l values of 0, 1, and 2 and the corresponding permitted m_l values. Because the sine and cosine functions can have positive and negative values, there are positive and negative regions of the wave functions. The functions in Table 11.3 are orthogonal to one another (as required for eigenfunctions of a Hermitian operator) and have been normalized.

Solution of the *R* Equation

The radial equation Eq. 11.171 can be cast into a form that is reminiscent of the Schrödinger equation

$$-\frac{\hbar^2}{2\mu r^2}\frac{d}{dr}\left(r^2\frac{dR}{dr}\right) - \frac{Ze^2}{4\pi\epsilon_o r}R + \frac{\hbar^2}{2\mu}\frac{l(l+1)}{r^2}R - ER \quad (11.186)$$

TABLE 11.3 Solution of the Θ Equation

l	m_l	Function
0	0	$\Theta_{00} = \frac{\sqrt{2}}{2} = \frac{1}{\sqrt{2}}$
1	0	$\Theta_{10} = \frac{\sqrt{6}}{2}\cos\theta$
1	+1, −1	$\Theta_{1\pm1} = \frac{\sqrt{3}}{2}\sin\theta$
2	0	$\Theta_{20} = \frac{\sqrt{10}}{4}(3\cos^2\theta - 1)$
2	+1, −1	$\Theta_{2\pm1} = \frac{\sqrt{15}}{2}\sin\theta\cos\theta$
2	+2, −2	$\Theta_{2\pm2} = \frac{\sqrt{15}}{4}\sin^2\theta$

Comparing this equation with Eq. 11.102, the first term can be considered to be a radial kinetic energy term for the electron. The second term is the attractive Coulombic potential. The third term is a pseudo-potential that arises from the angular momentum of the electron and can be considered to be a centrifugal distortion of the Coulombic attraction. The greater the orbital angular momentum the electron has, the greater the third term becomes. For example, when $l = 0$, this term vanishes, but for $l \neq 0$ it is repulsive and, being proportional to $1/r^2$, dominates the Coulombic attraction ($1/r$) for small r. This means that for $l \neq 0$ the electron can never reach the nucleus. This is shown in the orbital diagrams (see Figure 11.18 on p. 549), which have a node at the origin for all cases except $l = 0$ (s orbital).

Equation 11.186 is called Laguerre's equation and was studied in the nineteenth century by the French mathematician Edmond Laguerre (1834–1886). Its solution leads to the conclusion that there is a quantum number n that can have positive integral values starting with unity. The relationship between n and l is that the maximum value that l can have is one less than the value of n, hence

$$\begin{aligned} n &= 1, 2, 3, \ldots \\ l &= 0, 1, \ldots, n - 1 \end{aligned} \tag{11.187}$$

The solutions of the R equation, under these restrictions, are the *Laguerre polynomials.*

Complete Wave Functions

The complete wave functions are obtained by multiplying together the appropriate functions that are given in Tables 11.2, 11.3, and 11.4. Some examples are given in Table 11.5. The way they are constructed is shown by the following example.

EXAMPLE 11.7 Obtain a complete wave function for an electron having $n = 3$, $l = 1$, $m_l = 0$ (a 3p orbital).

Solution In order to determine an expression for ψ_{nlm}, we can multiply the radial wave function for an $n = 3$, $l = 1$ orbital from Table 11.4 (R_{31}) by the Θ function for $l = 1$, $m_l = 0$ from Table 11.3 (Θ_{10}) and then by the Φ function for $m_l = 0$ from Table 11.2 (Φ_0).

$$\left[\frac{1}{9\sqrt{6}}\left(\frac{Z}{a_0}\right)^{3/2}\left(4 - \frac{2Zr}{3a_0}\right)\frac{2Zr}{3a_0}\, e^{-Zr/3a_0}\right]\left[\frac{\sqrt{6}}{2}\cos\theta\right]\left[\frac{1}{\sqrt{2\pi}}\right]$$

The final expression is

$$\psi_{310} = \frac{1}{27\sqrt{2\pi}}\left(\frac{Z}{a_0}\right)^{5/2}\left(4 - \frac{2Zr}{3a_0}\right) r e^{-Zr/3a_0} \cos\theta$$

Note that two other equally acceptable solutions for the 3p orbitals could have been obtained by taking the products $R_{31}\Theta_{11}\Phi_x$ or $R_{31}\Theta_{1-1}\Phi_y$. These give identical orbitals to the one described except for orientation.

The wave function for hydrogenlike atoms is completely determined by the three quantum numbers n, l, and m_l. From Eq. 11.167, the wave function also depends upon the spherical coordinates r, θ, and ϕ. This is commonly expressed as.

$$\psi_{nlm_l}(r, \theta, \phi) = R_{nl}(r) Y_l^{m_l}(\theta, \phi) \tag{11.188}$$

TABLE 11.4 Solutions of the *R* Equation

Quantum Numbers *n*	*l*	Function
1	0	$R_{10} = 2\left(\frac{Z}{a_0}\right)^{3/2} e^{-Zr/a_0}$
2	0	$R_{20} = \frac{1}{2\sqrt{2}}\left(\frac{Z}{a_0}\right)^{3/2}\left(2 - \frac{Zr}{a_0}\right)e^{-Zr/2a_0}$
2	1	$R_{21} = \frac{1}{2\sqrt{6}}\left(\frac{Z}{a_0}\right)^{3/2}\frac{Zr}{a_0}e^{-Zr/2a_0}$
3	0	$R_{30} = \frac{1}{9\sqrt{3}}\left(\frac{Z}{a_0}\right)^{3/2}\left(6 - \frac{4Zr}{a_0} + \frac{4Z^2r^2}{9a_0^2}\right)e^{-Zr/3a_0}$
3	1	$R_{31} = \frac{1}{9\sqrt{6}}\left(\frac{Z}{a_0}\right)^{3/2}\left(4 - \frac{2Zr}{3a_0}\right)\frac{2Zr}{3a_0}e^{-Zr/3a_0}$
3	2	$R_{32} = \frac{1}{9\sqrt{30}}\left(\frac{Z}{a_0}\right)^{3/2}\left(\frac{2Zr}{3a_0}\right)^2 e^{-Zr/3a_0}$

The quantity a_0 is the same as the first Bohr radius, defined by Eq. 11.44.

Spherical Harmonics

where the θ, and ϕ dependence is combined into a single function $Y_l^{m_l}(\theta, \phi)$, which is called **spherical harmonics**.

$$Y_l^{m_l}(\theta, \phi) = \Theta_{lm_l}(\theta)\Phi_{m_l}(\phi) \tag{11.189}$$

Hence the Schrödinger equation is completely solved, being

$$\hat{H}\psi_{nlm_l}(r, \theta, \phi) = E_n\psi_{nlm_l}(r, \theta, \phi) \tag{11.190}$$

We can obtain the energies by substituting the various solutions for R, listed in Table 11.4, into the R equation and solving for E. When this is done, the permitted energies are given by the expression.

$$E_n = -\frac{Z^2e^2}{8\pi\epsilon_0 n^2 a_0} \tag{11.191}$$

TABLE 11.5 Selected Complete Hydrogen Atom Wave Functions $\psi_{n,l,m}$ for the Hydrogen Atom ($Z = 1$)

Quantum Numbers *n*	*l*	m_l	Function
1	0	0	$\psi_{1s} = \psi_{100} = \frac{1}{\sqrt{\pi}}\left(\frac{1}{a_0}\right)^{3/2}e^{-r/a_0}$
2	0	0	$\psi_{2s} = \psi_{200} = \frac{1}{4\sqrt{2\pi}}\left(\frac{1}{a_0}\right)^{3/2}\left(2 - \frac{r}{a_0}\right)e^{-r/2a_0}$
2	1	0	$\psi_{2p_z} = \psi_{210} = \frac{1}{4\sqrt{2\pi}}\left(\frac{1}{a_0}\right)^{3/2}\frac{r}{a_0}e^{-r/2a_0}\cos\theta$
2	1	+1	$\psi_{2p_x} = \frac{1}{4\sqrt{2\pi}}\left(\frac{1}{a_0}\right)^{3/2}\frac{r}{a_0}e^{-r/2a_0}\sin\theta\cos\phi$
2	1	−1	$\psi_{2p_y} = \frac{1}{4\sqrt{2\pi}}\left(\frac{1}{a_0}\right)^{3/2}\frac{r}{a_0}e^{-r/2a_0}\sin\theta\sin\phi$
3	0	0	$\psi_{3s} = \psi_{300} = \frac{1}{18\sqrt{3\pi}}\left(\frac{1}{a_0}\right)^{3/2}\left[6 - 6\left(\frac{2r}{3a_0}\right) + \left(\frac{2r}{3a_0}\right)^2\right]e^{-r/3a_0}$

The significance of the notation 1s, 2s, $2p_z$, etc., is considered later.

It is of interest that this is exactly the same expression that was given by Bohr's theory (Eq. 11.49). The lowest (i.e., most negative) energy occurs when $n = 1$, and this is therefore the most stable state, or ground state, of the atom. Since when $n = 1$, l must be 0, this is the 1s state.

The energy of the hydrogenlike atoms depends only on the principal quantum number n. For other atoms, with more than one electron, the energy depends upon both n and l. In the presence of a magnetic field, the energy depends upon three quantum numbers, including the magnetic quantum number, m_l, and hence its name.

Orbitals

The complete wave functions are also called **orbitals** and are commonly designated as spdf orbitals, which are alternative ways of expressing the quantum numbers (see Section 11.8). In fact, the wave functions $\psi_{nlm_l}(r,\theta,\phi)$ give the exact mathematical expressions from which these orbitals are obtained. It should be pointed out that these are exact *only* for the hydrogenlike (one-electron) atoms. For more than one electron, exact quantum mechanical solutions do not exist and hence the spdf orbitals do not exist for the many-electron atoms. In spite of this, the exact spdf . . . orbitals for hydrogen can be used as an approximation for many-electron atoms commonly used to describe the electronic configurations for the heavier atoms.

The spherical harmonics arise frequently and have convenient properties, amongst which is the orthonormality relation

$$\int_0^{\pi}\int_0^{2\pi} Y_l^{m_l}{}^*(\theta,\phi)Y_{l'}^{m'_l}(\theta,\phi)\sin\theta d\theta d\phi = \delta_{ll'}\delta_{m_l m_l} \qquad (11.192)$$

Kronecker Delta

where δ_{ij} is called the **Kronecker delta**. It is equal to 1 if $i = j$ and is equal to zero if $i \neq j$.

11.8 Physical Significance of the Orbital Quantum Numbers

The mathematical solution of the Schrödinger equation for hydrogenlike atoms has thus revealed that there are three orbital quantum numbers n, l, and m_l; their magnitudes are related to one another by Eqs. 11.184, 11.187, and 11.185, respectively. A special notation has been introduced to designate the quantum numbers n and l and the orbitals to which they correspond. The principal quantum number n is given first, followed by a letter that indicates the quantum number l, as follows:

See the spectral transitions for s, p, d, and f orbitals; Orbitals; Angular dependence.

$l = 0$	s
$l = 1$	p
$l = 2$	d
$l = 3$	f

From then on we follow the letters in alphabetical order (g, h, etc.). The letters s, p, d, and f relate to the descriptions "sharp," "principal," "diffuse," and "fundamental" that the early spectroscopists had given to series of lines in atomic spectra. As an example of the use of this notation, a 3p orbital is one for which $n = 3$ and $l = 1$. We will see later that m_l values can be indicated by the addition of subscripts to the letters, but this is frequently unnecessary.

Study the wave function for the hydrogen atom and locate the nodes of s-orbitals; Principal quantum number.

The Principal Quantum Number *n*

Inspection of Eqs. 11.169, 11.171, and 11.172 shows that the only one that contains the energy E is Eq. 11.171, the R equation. It therefore follows that the energies for a hydrogenlike atom depend only on the solutions of the R equation and not at all on the solutions of the Θ and Φ equations. In other words, the energy depends only on the principal quantum number n, which comes from the R equation, and not on l and m_l. This is true only for hydrogenlike atoms in the absence of a magnetic field.

The R equation, and therefore the value of n, determines the expectation value of the distance between the nucleus and the electron. Thus, if $n = 1$, the electron is more likely to be close to the nucleus than if $n = 2$, and so on. This may be represented quantitatively by diagrams in which the function R is plotted against the distance r. Some such diagrams are shown in Figure 11.16. Note that because of the exponential factor appearing in the expression for R (Table 11.4), the function approaches zero as r becomes large. For l equal to $n - 1$, this approach to zero is monotonic, with no maxima and minima, but for other values of l the function passes through zero at various distances and there are maxima and minima. The number of radial nodes (positions around the nucleus where $R(r) = 0$) is equal to $n - l - 1$.

Probability Density

We have seen (Figure 11.10) that the probability that an electron is in a volume element $d\tau = dx\, dy\, dz$ is given by $\psi^*\psi\, d\tau$, or $\psi^2\, d\tau$ if the wave function is real. If we divide by $d\tau$, the result ψ^2 is the probability per unit volume, or **probability density.** For an electron in the 1s state the probability density, obtained from the first item in Table 11.5, is therefore

$$\psi_{1s}^2 = \frac{e^{-2r/a_0}}{\pi a_0^3} \tag{11.193}$$

A plot of this function against r is shown in Figure 11.17a on page 547, and we see that the probability density has a maximum value at $r = 0$ and falls toward zero as r increases. It is of greater interest, however, to consider the probability that the electron is present at a distance between r and $r + dr$ from the nucleus—in other words, the probability that the electron is present between two concentric spherical surfaces; one has a radius of r and the other a radius of $r + dr$. The volume between two such concentric surfaces is $4\pi r^2\, dr$, and the probability that the 1s electron lies within the distances r and $r + dr$ is thus

$$4\pi r^2 \psi_{1s}^2\, dr$$

The corresponding probability density (which is now per unit radial distance r) is obtained by dividing by dr and is therefore

$$4\pi r^2 \psi_{1s}^2$$

Radial Distribution Function

This function is known as the **radial distribution function,** and a plot of it against r is shown in Figure 11.17a. The curve passes through a maximum when $r = a_0 = 52.92$ pm, and this distance is therefore the most probable distance between the nucleus and the electron. This emphasizes an important similarity, and also an important difference, between the Bohr theory and the wave-mechanical theory. In the Bohr theory the electron in the ground (1s) state of the hydrogen atom moves in a *precise orbit* of radius a_0, while in quantum mechanics this distance is only the *most probable distance;* the electron can be at other distances from the nucleus.

Figure 11.17b shows plots of ψ_{2s}^2 and of $4\pi r^2 \psi_{2s}^2$ for an electron in the 2s state. The latter shows two maxima. Similar plots can be given for 3s, 4s, etc., states, and

Plot Figure 11.16 for n = 1 to 6; The radial wave function.

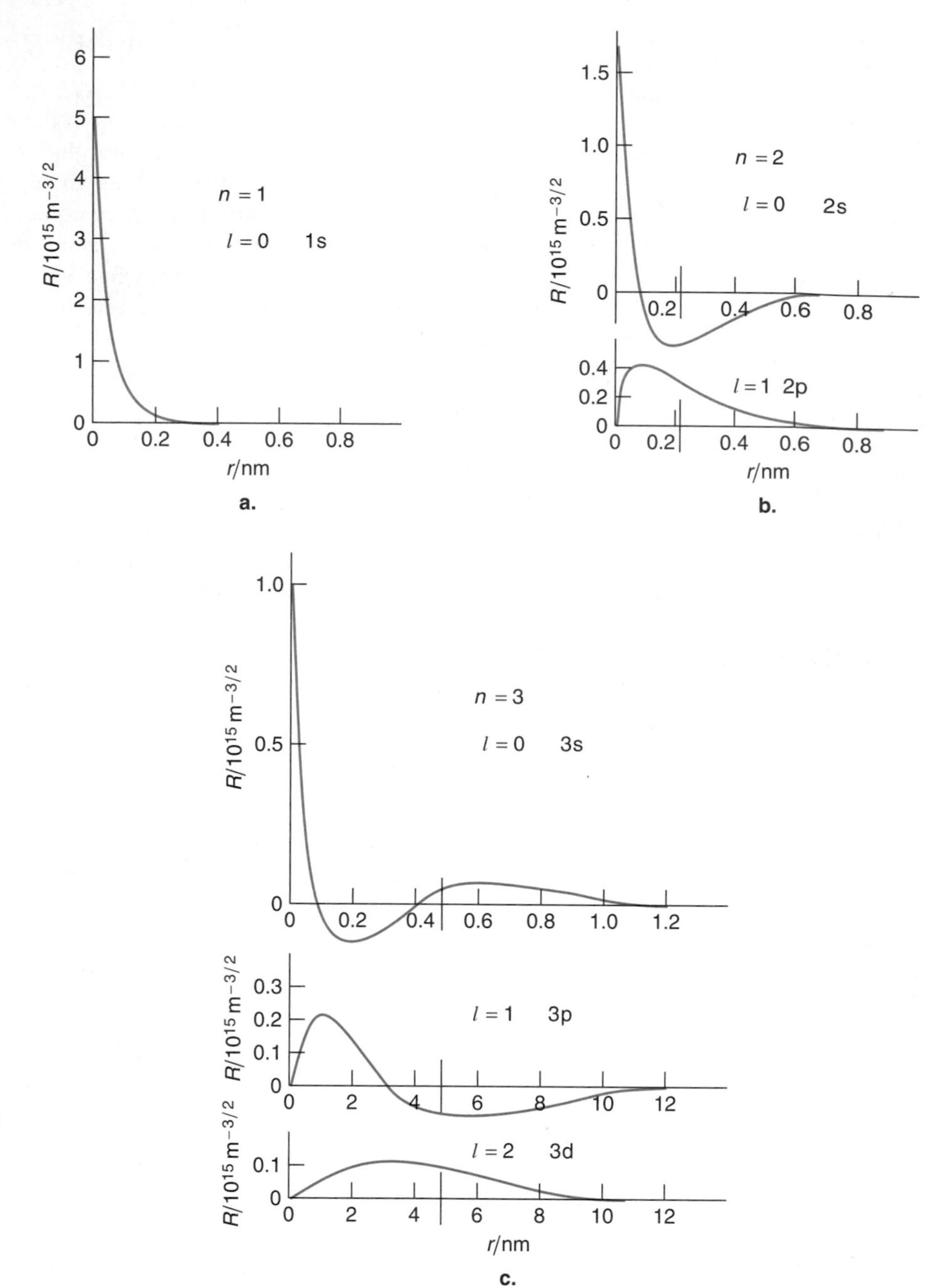

FIGURE 11.16
Radial wave functions $R(r)$ for the hydrogen atom, plotted against distance r from the nucleus. (a) The curve for $n = 1$; l must be 0. (b) The curves for $n = 2$; l can be 0 or 1. (c) The curves for $n = 3$; l can be 0, 1, or 2.

the number of maxima in the $4\pi r^2\psi^2$ plots is always equal to the principal quantum number n.

For states where l is not zero (e.g., 2p, 3d, . . . states), we have to proceed differently because there is now an angular dependence (see Table 11.5). We could make plots of ψ^2 with cos θ and sin θ taken to be unity, their maximum values. Such plots would represent probability densities *along the axis for which these have their maximum values.* However, a more usual practice is instead to make plots of R^2 against r. This is not quite the same thing, since a comparison

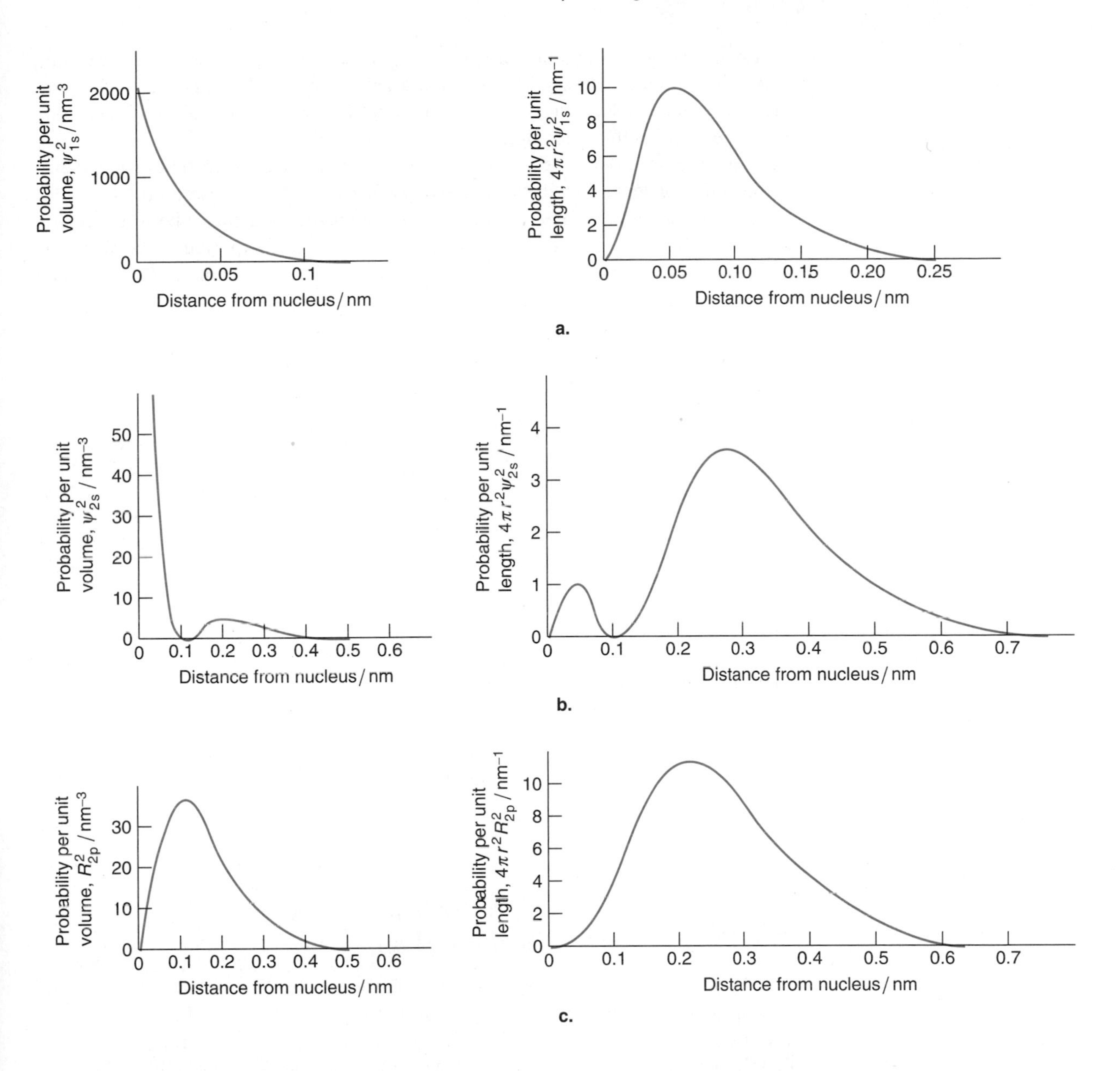

FIGURE 11.17
(a) Plots of ψ^2_{1s} and $4\pi r^2\psi^2_{1s}$ against the distance r from the nucleus, for a 1s electron in the hydrogen atom. (b) Plots of ψ^2_{2s} and $4\pi r^2\psi^2_{2s}$ against r, for a 2s electron. (c) Plots of R^2_{2p} and $4\pi r^2 R^2_{2p}$, for a 2p electron.

Plot Figure 11.17; Radial distributions.

of Tables 11.4 and 11.5 shows us that even with $Z = 1$ the numerical factors are not the same. In spite of this, however, the plots of R^2 against r do give us the *general form* of the probability density function. To obtain the function along the axis for which it is a maximum we would have to multiply by an appropriate factor. A plot of R^2 against r for the 2p state of the hydrogen atom is shown in Figure 11.17c.

We can also make plots of $4\pi r^2 R^2$ against r, and Figure 11.17c also shows such a plot for the 2p state. There are now two points to note about the significance of such plots. One is that the numerical factor in R is not the same as that in ψ. But aside from this, the plot of $4\pi r^2 R^2$ would represent the probability that the electron is in a spherical element at a distance r from the nucleus only *if the orbital had spherical symmetry.* In reality, orbitals for which l is other than zero have maximum values in certain directions. The function $4\pi r^2 R^2$ therefore does not have any simple significance in such cases, but plots of $4\pi r^2 R^2$ against r are informative as long as one recognizes this limitation.

We have seen that an electron in the 1s orbital is *most likely* to be at a distance a_0 from the nucleus. This distance, however, because of the skewed distribution of the orbitals, is not the same as the *expectation value* of the distance between the nucleus and the electron (i.e., the distance the electron is expected to be from the nucleus). We can calculate this expectation value by using Postulate VI. According to Eq. 11.104, with the wave functions real and normalized,

$$\langle r_{1s} \rangle = \int \psi_{1s} r \psi_{1s} \, d\tau = \int_0^\infty \psi_{1s} r \psi_{1s} 4\pi r^2 \, dr \tag{11.194}$$

Introduction of the expression for ψ_{1s} from Table 11.5 then gives

$$\langle r_{1s} \rangle = \frac{1}{\pi a_0^3} \int_0^\infty r e^{-2r/a_0} 4\pi r^2 \, dr \tag{11.195}$$

$$= \frac{3}{2} a_0 \tag{11.196}$$

The reason that this is greater than a_0, the most probable distance, is that the wave functions do not give a symmetrical curve when plotted against r (see Figure 11.17a); they are skewed in favor of larger r values. For s states with other values of n, the expectation values are given by

$$\langle r_{ns} \rangle = \frac{3}{2} n^2 a_0 \tag{11.197}$$

Angular Dependence of the Wave Function: The Quantum Numbers l and m_l

The Θ and Φ wave functions are best considered together, since both are concerned with the angular dependence of the orbitals. Plots of $[\Theta(\theta)\Phi(\phi)]^2$ for various l and m_l values are shown in Figure 11.18. It is to be emphasized that these plots are in no way related to distance from the nucleus but only to the variation of the wave function with the angles θ and ϕ. This is illustrated explicitly for the p_z orbital, for which the angles θ and ϕ are shown and in which the plot is shown in its three-dimensional form. For given values of θ and ϕ, the length of the line joining the origin to the surface of the solid figure is the relative probability that the electron is to be found in that direction. For this p_z orbital the *maximum* probability is along the Z axis. The other diagrams are given in a simpler form, but they are to be interpreted in the same way.

All s orbitals ($l = 0$) have spherical symmetry, as shown in Figure 11.18a. When $l = 1$ (a p orbital), the quantum number m_l can have the value -1, 0, or $+1$, and there are orbitals that have their maximum values along the X, Y, and Z axes. For the d orbitals ($l = 2$) there are five equivalent orbitals corresponding to $m_l = -2, -1, 0, 1, 2$. The subscripts for these orbitals relate to the symmetry properties.

See the orbitals in 3D; Orbitals.

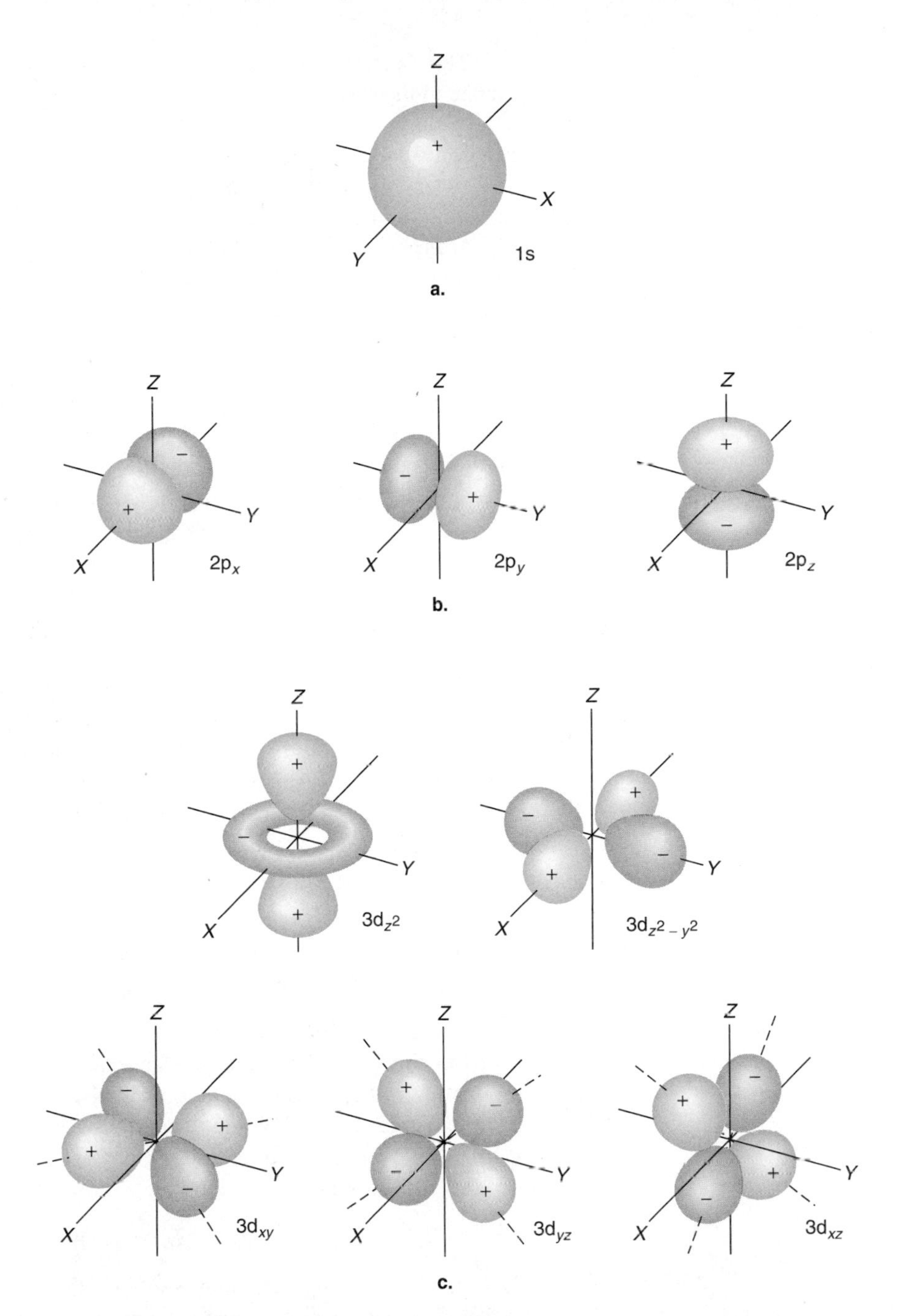

FIGURE 11.18
Plots of $[\Theta(\theta)\Phi(\phi)]^2$. (a) An s orbital ($l = 0$). (b) The three p orbitals; (c) The d_{z^2}, $d_{z^2-y^2}$, d_{xz}, d_{yz}, and d_{xy} orbitals.

For example, with $l = 2$ and $m_l = -1$ we obtain, from Tables 11.2 and 11.3, the product

$$\Phi_y\Theta_{2\pm1} = \frac{\sin\phi}{\sqrt{\pi}} \cdot \frac{\sqrt{15}}{2}\sin\theta\cos\theta = \frac{1}{2}\sqrt{\frac{15}{\pi}}\sin\theta\cos\theta\sin\phi \quad (11.198)$$

From Figure 11.15 we see that $y = r\sin\theta\sin\phi$ and that $z = r\cos\theta$. The product yz therefore has the same angular dependence as the function in Eq. 11.198, which is therefore designated d_{yz}.

The f orbitals ($l = 3$) are difficult to represent, but they bear some resemblance to the d orbitals. There are seven such orbitals, corresponding to $m_l = -3, -2, -1, 0, 1, 2, 3$. Three of them, f_{z^3}, f_{x^3}, and f_{y^3}, have *two* doughnut-shaped rings perpendicular to the appropriate axis, instead of the one found in the d_{z^2} orbital. The other four resemble the $d_{x^2-y^2}$ orbitals except that there are now eight lobes instead of four.

Unsöld's Theorem

From what has been said, it might be thought that the various orbitals corresponding to a given value of l would mean that a complete orbital diagram for an atom would look like a pincushion, with orbitals sticking out in various directions. However, a theorem due to the German physicist Albrecht Otto Johannes Unsöld (b. 1905) states that for the orbitals corresponding to a particular value of l the sum of the values of $[\Theta(\theta)\Phi(\phi)]^2$ is a constant. Such a sum therefore corresponds to completely spherical symmetry for the atom. It follows from this that the atoms of the noble gases, in which all orbitals corresponding to a given l value are filled, have complete spherical symmetry (see Problem 11.49).

11.9 Angular Momentum and Magnetic Moment

Besides determining the shape of an orbital and its orientation in space, the quantum numbers l and m_l have another important significance. They relate to the angular momentum of the electron in its orbital and also to the magnetic moment of the orbital. In this section we will see that the quantum number l determines the *magnitude* of the angular momentum and magnetic moment, and that the quantum number m_l determines the *direction in space* of the angular momentum and magnetic moment.

Angular Momentum

The classical relationship between angular momentum and linear velocity is illustrated in Figure 11.19. Suppose that a particle of mass m is undergoing circular motion in the XY plane, in an anticlockwise direction if one is looking down along the Z axis (Fig. 11.19a). When the particle is on the X axis, it has a positive velocity u_y in the Y-direction, and its linear momentum p_y is mu_y. Its angular momentum is then, by definition, positive along the Z axis and is given by

$$L_z = p_y x = mu_y x \qquad (11.199)$$

The sign of the angular momentum is determined by the *right-hand rule;* the curved fingers of the right hand are caused to point in the direction of the linear momentum vector, and the thumb then points in the direction of the angular momentum vector. Figure 11.19b shows the situation when the particle is on the Y axis; a positive momentum p_x then means that it is moving in the clockwise direction, so that the angular momentum L_z is now $-yp_x$. We can generalize these results for a particle moving in any direction (Figure 11.19c). If its momentum components are p_x and p_y and its position coordinates are x and y, the resultant component of angular momentum along the Z axis is given by

$$L_z = xp_y - yp_x \qquad (11.200)$$

Animate Figure 11.19 by varying the parameters; Angular momentum; Classical.

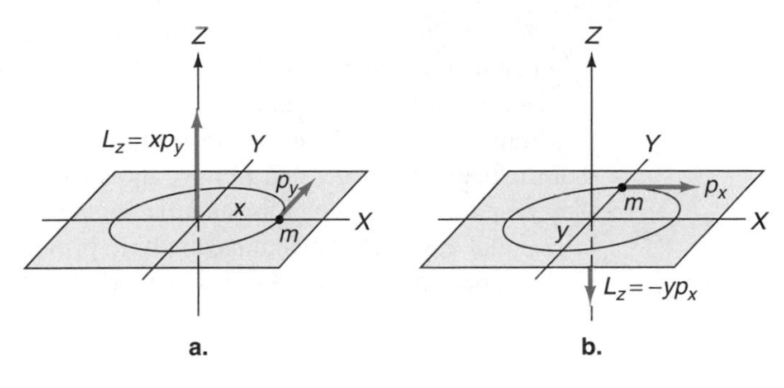

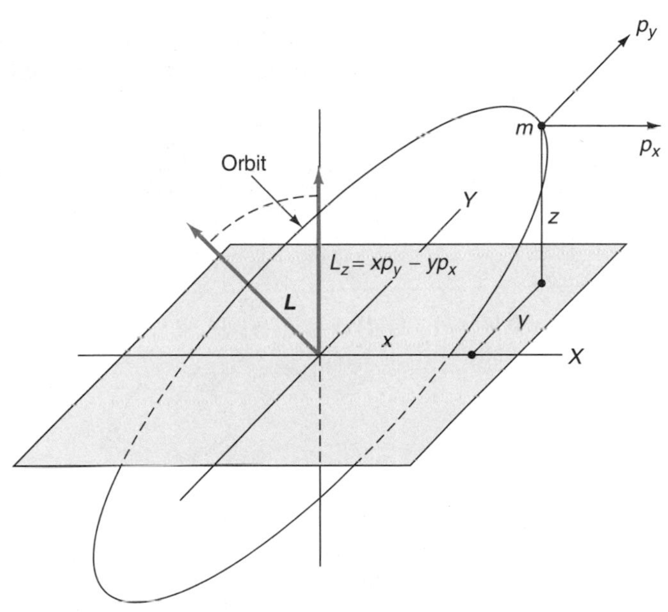

FIGURE 11.19
(a) A particle of mass m undergoing circular motion in the XY plane in an anticlockwise direction.
(b) Clockwise motion in the XY plane. (c) The general case, showing the components of velocity and of linear momentum, and the component of angular momentum along the Z axis. In vector notation, $\boldsymbol{L} = \boldsymbol{r} \times \boldsymbol{p}$.

The corresponding quantum-mechanical operator, as given in Table 11.1, is

$$\hat{L}_z = -i\hbar\left(x\frac{\partial}{\partial y} - y\frac{\partial}{\partial x}\right) \tag{11.201}$$

In polar coordinates (Figure 11.15) this becomes

Angular Momentum Operator (Z axis)

$$\hat{L}_z = -i\hbar\frac{\partial}{\partial \phi} \tag{11.202}$$

Exactly the same type of treatment can be applied to the components of angular momentum along the X and Y axes, leading to expressions for the operators $\hat{L}_x$ and $\hat{L}_y$. Of particular significance is the square of the *total* angular momentum operator, defined by

$$\hat{L}^2 = \hat{L}_x^2 + \hat{L}_y^2 + \hat{L}_z^2 \tag{11.203}$$

This is given by

Square of the Total Angular Momentum Operator

$$\hat{L}^2 = -\hbar^2\left[\frac{1}{\sin\theta}\frac{\partial}{\partial\theta}\left(\sin\theta\frac{\partial}{\partial\theta}\right) + \frac{1}{\sin^2\theta}\frac{\partial^2}{\partial\phi^2}\right] \tag{11.204}$$

We saw in Section 11.5, in our discussion of Postulate III, that if two operators are such that they have the same eigenfunctions, the corresponding physical properties can be measured simultaneously and precisely; in other words, the Heisenberg uncertainty principle places no restriction on their measurement. We will now show that the operators $\hat{H}$, $\hat{L}_z$, and $\hat{L}^2$ have the same set of eigenfunctions; this means that the energy, the component of angular momentum along the Z axis, and the square of the total angular momentum can all, in principle, be measured simultaneously.

The operator $\hat{L}_z$ involves only the angle ϕ (Eq. 11.202), and the function

$$\Phi = \frac{1}{\sqrt{2\pi}} e^{im_l\phi} \tag{11.205}$$

is the solution of the $\hat{H}$ equation that involves ϕ (Eq. 11.176). This function is therefore an eigenfunction for $\hat{H}$, and we will now see if it is an eigenfunction for $\hat{L}_z$:

$$\hat{L}_z\Phi_{m_l} = -i\hbar \frac{\partial}{\partial\phi}\left(\frac{1}{\sqrt{2\pi}} e^{im_l\phi}\right) = \hbar m_l \cdot \frac{1}{\sqrt{2\pi}} e^{im_l\phi} \tag{11.206}$$

$$= m_l\hbar\Phi_{m_l} \tag{11.207}$$

The function Φ_{m_l} is therefore an eigenfunction of $\hat{L}_z$ as well as of $\hat{H}$, and the eigenvalue of $\hat{L}_z$ is

Z Component of the Angular Momentum

$$L_z = m_l\hbar \tag{11.208}$$

This expression therefore gives the quantized values of the Z component of the angular momentum.

The operator $\hat{L}^2$ involves only θ and ϕ, and we know that the function $\Theta\Phi$ is an eigenfunction for the Hamiltonian operator $\hat{H}$; we may therefore see whether it is also an eigenfunction for $\hat{L}^2$:

$$\hat{L}^2\Theta_{lm_l}\Phi_{m_l} = -\hbar^2\left[\frac{\Phi_{m_l}}{\sin\theta}\frac{\partial}{\partial\theta}\left(\sin\theta\frac{\partial\Theta_{lm_l}}{\partial\theta}\right) + \frac{\Theta_{lm_l}}{\sin^2\theta}\frac{\partial^2\Phi_{m_l}}{\partial\phi^2}\right] \tag{11.209}$$

Comparison of this equation with the Φ_{m_l} equation (Eq. 11.169) and the Θ_{lm_l} equation (Eq. 11.172) shows that the function in brackets in Eq. 11.209 has the eigenvalues $l(l + 1)$ and may therefore be written as $l(l + 1)\Theta_{lm_l}\Phi_{ml}$; thus

$$\hat{L}^2\Theta_{lm_l}\Phi_{m_l} = l(l + 1)\hbar^2\Theta_{lm_l}\Phi_{m_l} \tag{11.210}$$

The eigenfunction for $\hat{H}$ is therefore also an eigenfunction for $\hat{L}^2$, with the eigenvalue L^2

Square of the Total Angular Momentum

$$L^2 = l(l + 1)\hbar^2 \tag{11.211}$$

We have thus proved that it is possible to make simultaneous and precise measurements of the total energy, the total angular momentum, and the component of the angular momentum along the Z axis. This arises because the operators $\hat{H}$, $\hat{L}^2$, and $\hat{L}_z$

commute (Postulate III). Of course, our choice of Z axis is entirely arbitrary; we can equally well show that $\hat{H}$, $\hat{L}^2$, and $\hat{L}_x$ commute and that $\hat{H}$, $\hat{L}^2$, and $\hat{L}_y$ commute. It is therefore possible to make simultaneous and precise measurements of the total energy, the total angular momentum, and the component of the angular momentum along the X axis *or* along the Y axis *or* along the Z axis. However, $\hat{L}_x$, $\hat{L}_y$, and $\hat{L}_z$ do *not* commute with each other, and we therefore cannot make simultaneous and precise measurements of the angular momentum along the X, Y, and Z axes.

We therefore have to choose one axis, and the convention is to choose the Z axis. For example, when an atom is placed in an electric or magnetic field, the field is taken to be along the Z axis. Similarly, when an atom is present in a diatomic molecule, we take the axis of the molecule to be the Z axis.

The commutation relations for angular momentum can easily be found from the definitions, Eq. 11.201, and Table 11.1:

$$[\hat{L}_x, \hat{L}_y]_- = i\hbar\hat{L}_z; \qquad [\hat{L}_y, \hat{L}_z]_- = i\hbar\hat{L}_x; \qquad [\hat{L}_z, \hat{L}_x]_- = i\hbar\hat{L}_y \quad (11.212)$$

A more compact way of expressing these commutation relations is in terms of the vector cross product, $\hat{\boldsymbol{L}} \times \hat{\boldsymbol{L}} = i\hbar\hat{\boldsymbol{L}}$ where $\hat{\boldsymbol{L}} = x\hat{L}_x + y\hat{L}_y + z\hat{L}_z$. At first this may seem strange since for usual vectors, the cross product of two equal vectors is zero. The vector here is the quantum-mechanical vector operator, and the cross product with itself is not zero by virtue of the commutation relations. Moreover, the angular momentum vector has another property: when all the coordinates are reversed, it does not change sign. Common vectors (called polar vectors) do change sign under coordinate reversal. From the classical definition of the angular momentum, $\boldsymbol{L} = \boldsymbol{r} \times \boldsymbol{p}$, when $\boldsymbol{r}$ is replaced by $-\boldsymbol{r}$, the vector $\boldsymbol{L}$ remains unchanged: $\boldsymbol{L}(-\boldsymbol{r}) = \boldsymbol{L}(\boldsymbol{r})$. Vectors that display this property are called **pseudo-vectors**.

Pseudo-vectors

Magnetic Moment

Watch a classical magnetic moment interact with a magnetic field; Angular momentum; Magnetic moment.

An orbiting electron, being charged, generates a magnetic field. Since the electron is negatively charged, the direction of the magnetic field is exactly opposite to that of the angular momentum, as shown in Figure 11.20a. The direction is conveniently obtained by using the *left-hand* rule; the curved fingers of the left hand point in the direction of motion of the orbiting electron, and the thumb then points toward the north pole of the electromagnet. The magnitude of the magnetic moment is proportional to that of the angular momentum, and aside from the difference of orientation the treatment of the magnetic moment is similar to that of the angular momentum. Again there is quantization, and precise and simultaneous measurements can be made of the energy, the component M_z of magnetic moment along the Z axis, and the square of the total magnetic moment. This arises because the magnetic moment operators $\hat{M}^2$ and $\hat{M}_z$ commute with $\hat{H}$, $\hat{L}^2$, and $\hat{L}_z$ and, therefore, have the same set of eigenfunctions. The same quantum numbers l and m_l therefore relate to the magnetic moment.

The magnetic moment, μ, differs in the *units* used for its measurement. We will discuss this further in Section 13.2 in connection with the Zeeman effect. Here we will simply state that the ratio of the magnetic moment to the angular momentum is known as the **gyromagnetic ratio,** or the **magnetogyric ratio.** For an electron of charge $-e$ and mass m_e this ratio is $-e/m_e$. The SI unit of magnetic moment is ampere metre^2 (A m^2), but it is usual to employ the **Bohr magneton,** which is equal to 9.274×10^{-24} A m^2 = 9.274×10^{-24} J T^{-1}, where the SI unit of magnetic field strength is the tesla, T = A^{-1} kg s^{-2} or V s m^{-2}. See Appendix A, p.999, where magnetic units are explained.

Gyromagnetic Ratio

Bohr Magneton

Animate Figure 11.20; Angular momentum; Orbital angular momentum.

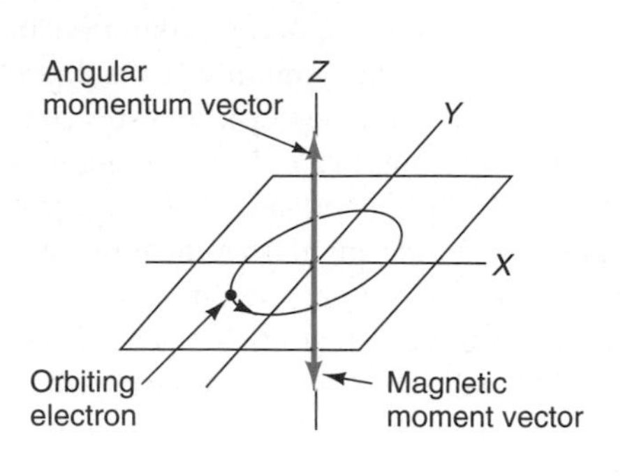

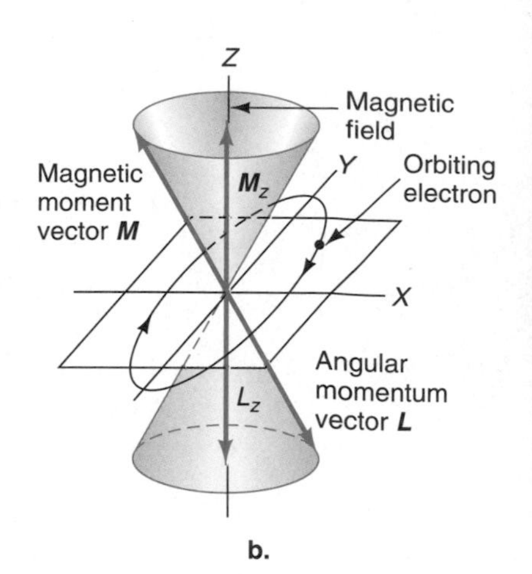

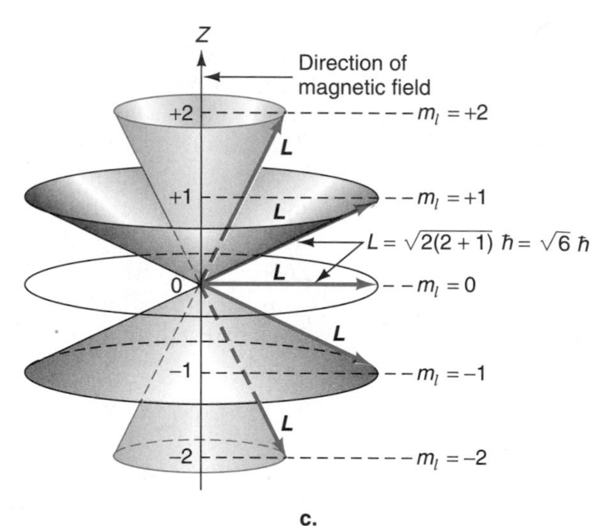

FIGURE 11.20
(a) An electron moving in an orbit in the *XY* plane, showing the angular momentum and magnetic moment vectors along the *Z* axis. (b) An orbiting electron in a magnetic field that lies along the *Z* axis, for $l = 1$. The *Z* component of the magnetic moment is shown aligned with the field and the angular momentum against it. The cones represent precession about the *Z* axis. (c) The five possible orientations of the orbital angular momentum vector, for $l = 2$, in a magnetic field. The magnetic moment lies in the opposite direction.

The importance of the magnetic moment is that it dictates the orientation of the orbital in a magnetic field, which may be that due to neighboring atoms. Figure 11.20b shows one arrangement that can arise when a p orbital ($l = 1$) is in a magnetic field along the Z axis. The Z components of the angular momentum and of the magnetic moment, opposite to one another, must have quantized values, corresponding to the magnetic quantum number m_l, along the Z axis. In the case shown in Figure 11.20b the Z component of the magnetic moment is aligned with the field, which means that the Z component of the angular momentum is aligned against the field. This is the case of $m_l = -1$, and since the magnetic moment is in the same general direction of the field, the energy has its lowest value. Another possibility is that the component L_z of angular momentum is aligned with the field ($m_l = +1$) and the magnetic moment against it; this corresponds to the highest energy. A third possibility is that L_z is at right angles to the field; this corresponds to $m_l = 0$ and to zero energy of interaction between the magnetic moment and the field.

The L_x and L_y components of the angular momentum and the M_x and M_y components of the magnetic moment do not have precise values, and the angular momentum and the magnetic moment therefore precess about the Z axis, as shown in Figure 11.20b. This motion, which corresponds to the curved surface of a cone, keeps L, L_z, M, and M_z fixed but leaves L_x, L_y, M_x, and M_y indeterminate.

Figure 11.20c shows the situation for a d orbital, for which $l = 2$ and m_l can have the values -2, -1, 0, $+1$, and $+2$. The diagram shows the *angular momentum vectors;* the magnetic moment vectors are not shown to avoid complicating the diagram, but they lie in the opposite direction. The case of $m_l = -2$ means that the angular momentum has its maximum value in the direction opposed to the field, and the magnetic moment then has its maximum value in the direction of the field. This corresponds to the lowest energy for the system. The highest energy is for $m_l = +2$, and $m_l = 0$ gives zero interaction energy.

Space Quantization

Figure 11.20 shows that for $l = 1$, only three orientations of the magnetic moment are possible, while for $l = 2$, only five orientations are possible. This means that the directions in space of the magnetic moments are quantized. This is usually referred to as **space quantization**.

11.10 The Rigid Linear Rotor

It is convenient to make a brief digression at this point and consider the rotation of a rigid linear molecule. We will see that the angular momentum of such a system obeys the same types of equations that apply to orbitals.

The classical angular momentum L of a body having a moment of inertia I (see pp. 498 and 666) and an angular velocity of ω is given by

$$L = I\omega \tag{11.213}$$

Its kinetic energy E_k is given by

$$E_k = \frac{1}{2}I\omega^2 \tag{11.214}$$

For free rotation the potential energy is zero, and the total energy E is therefore $\frac{1}{2}I\omega^2$. Elimination of ω between these two equations gives

$$E = E_k = \frac{L^2}{2I} \tag{11.215}$$

The Hamiltonian that describes the rigid rotor is thus given by

$$\hat{H} = \frac{\hat{L}^2}{2I} = \frac{\hbar^2\hat{J}^2}{2I} \tag{11.216}$$

where the rotational angular-momentum operator $\hat{\boldsymbol{J}}$ is introduced. The rotational angular-momentum operator $\hat{\boldsymbol{J}}$ has the same properties as any quantum-mechanical angular momentum. In particular, from Eqs. 11.209 and 11.210 it is shown that the square of the angular momentum has eigenvalues of $l(l + 1)\hbar^2$. Likewise the eigenvalues of $\hat{J}^2$ are $J(J + 1)$ and the solution to the Schrödinger equation is

$$\hat{H}Y_J^M(\theta, \phi) = \frac{\hbar^2 J(J + 1)}{2I} Y_J^M(\theta, \phi) \tag{11.217}$$

Note that just as the spherical harmonics are eigenfunctions of the orbital angular momentum, they are the eigenfunctions of the rigid rotor. Moreover, the energy, given by Eq. 11.215, depends only on J and hence the energy is $(2J + 1)$ degenerate, with the M quantum number spanning integral values from $-J$ to $+J$. The values of J are 0, 1, 2, . . . and the permitted energy values are then obtained from Eq. 11.217.

Energy Values for Rotation

$$E = J(J + 1)\frac{\hbar^2}{2I} \quad (11.218)$$

It is to be noted that when $J = 0$, the energy is zero; there is no zero-point energy for rotation. In this treatment the angles θ and ϕ are completely unspecified, so that one cannot assign values to the angular momenta about the X and Y axes. For a linear molecule the angular momentum about the Z axis, which is the axis of the molecule, is zero.

We shall see applications of this treatment to rotational spectroscopy in Chapter 13.

11.11 Spin Quantum Numbers

An important experiment carried out in 1921 by the physicists Otto Stern (1888–1969) and Walter Gerlach (1889–1979) showed that the three quantum numbers n, l, and m_l are insufficient to explain atomic behavior. It was found that a beam of silver atoms, each one of which has an odd electron, is split into two beams when passed through an inhomogeneous magnetic field. This result was explained in 1924 by the Austrian physicist Wolfgang Pauli (1900–1958), and in 1925 by the Dutch-American physicists George Eugene Uhlenbeck (1900–1988) and Samuel Abraham Goudsmit (1902–1978), in terms of a spin angular momentum of the electron.

As a convenient way of dealing with the situation, we often think about the electron as spinning about an axis, but the situation is in reality rather more subtle than that. If a body does spin about its axis there is indeed a resulting angular momentum. However, in the case of the electron, the spin of the electron is not revealed by quantum mechanics unless relativistic effects are taken into account. In 1928 Dirac developed a relativistic form of quantum mechanics, and it led to the conclusion that for an electron there can be two possible components of spin angular momentum about an axis.

The *spin angular momentum vector* $\boldsymbol{S}$ is exactly analogous to the orbital angular momentum $\boldsymbol{L}$ and is quantized in a similar way (compare Eqs. 11.208 and 11.211); the magnitude of the $\boldsymbol{S}$ vector is thus

$$|\boldsymbol{S}| = \sqrt{s(s + 1)}\,\hbar \quad (11.219)$$

where the quantum number s can have only the value $+\frac{1}{2}$. The component S_z of spin angular momentum along the Z axis is quantized in a similar way to L_z, with eigenvalues of

$$S_z = m_z\hbar \quad (11.220)$$

Half-Integral Quantum Numbers

but the **spin magnetic quantum number** m_s can have only the values $+\frac{1}{2}$ and $-\frac{1}{2}$.

The orbital angular momentum of an orbital electron and also the spin angular momentum are shown in Figure 11.21. The orbiting electron produces a magnetic moment vector in the direction indicated, and the electron may be considered to spin on its own axis in the magnetic field produced by the orbital motion. The magnetic moment

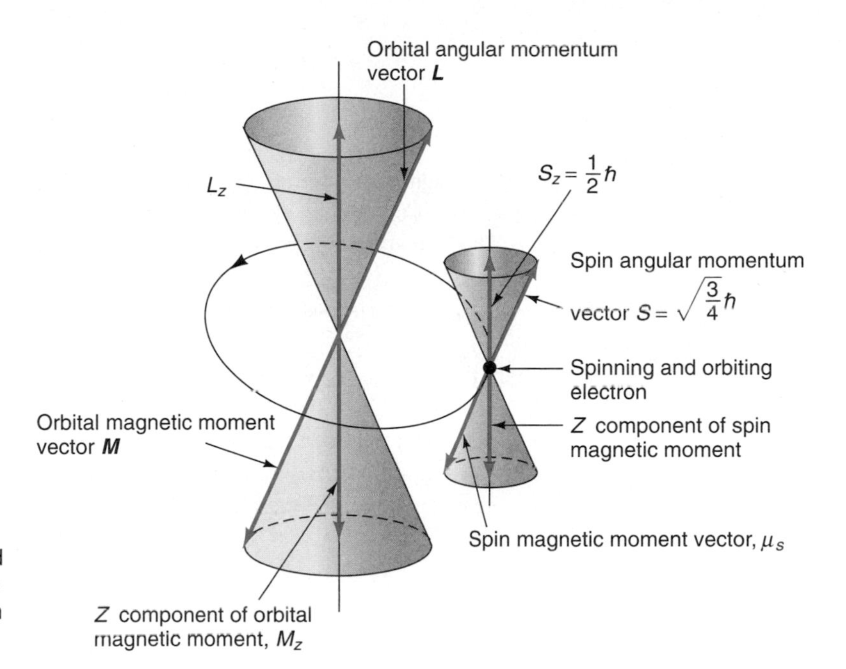

FIGURE 11.21
The spin angular momentum vector ***S*** and its relationship to the orbital angular momentum vector ***L***. The vectors are shown oriented in the same direction, which leads to a somewhat higher energy than when they are opposed.

Visualize an electron spin and produce Figure 11.21; Spin quantum number and Aufbau principle.

due to the spin may be oriented in the same direction as the magnetic field produced by the orbiting electron (as shown in the figure) or in the opposite direction. The former gives a somewhat higher energy than the latter. We shall see in Section 13.2 that the resulting splitting of energy levels gives closely spaced lines in atomic spectra.

We are now in a position to add two additional postulates to the six that we introduced in Section 11.5:

Postulate VII states that

> **There is a spin angular momentum operator $\hat{S}$ and a Z-component spin operator $\hat{S}_z$ that correspond to the angular momentum operators $\hat{L}$ and $\hat{L}_z$.**

Postulate VIII states that

> **In contrast to $\hat{L}^2$ and $\hat{L}_z$, the spin operators $\hat{S}^2$ and $\hat{S}_z$ have only two eigenfunctions α and β.**

The eigenvalue equations for $\hat{S}_z$ are

$$\hat{S}_z\alpha = \frac{\hbar}{2}\alpha \tag{11.221}$$

and

$$\hat{S}_z\beta = -\frac{\hbar}{2}\beta \tag{11.222}$$

In other words, the allowed values of the spin quantum number m_s are $+\frac{1}{2}$ and $-\frac{1}{2}$. The eigenvalue equations for $\hat{S}^2$ are

$$\hat{S}^2\alpha = \frac{1}{2}\left(\frac{1}{2} + 1\right)\hbar^2\alpha = \frac{3}{4}\hbar^2\,\alpha \tag{11.223}$$

and

$$\hat{S}^2\beta = \frac{1}{2}\left(\frac{1}{2} + 1\right)\hbar^2\beta = \frac{3}{4}\hbar^2\beta \tag{11.224}$$

It should again be emphasized that the Z axis is entirely arbitrary and has significance only if there is an external magnetic field.

The spin states are both orthogonal and normalized (*orthonormal states*)

$$\int \alpha^*\alpha\, ds = \int \beta^*\beta\, ds = 1 \qquad \text{and} \qquad \int \alpha^*\beta\, ds = \int \beta^*\alpha\, ds = 0 \tag{11.225}$$

where the integral is over spin states.

The existence of the electron spin was a complete surprise at the time of its discovery. First, it showed that non-integral quantum numbers existed. Second, it was the first quantum-mechanical observable found that did not follow the correspondence principle and had no classical analogue.

11.12 Many-Electron Atoms

For atoms containing more than one electron the form of the Schrödinger equation is more complicated, because the energy expression involves terms for the repulsive interactions between the different electrons. Even for helium, with two electrons, the Schrödinger equation is too complex to be solved in explicit form, although numerical solutions can be obtained fairly easily. For larger atoms the procedures are extremely complicated and require the use of approximation methods. We will consider some of these methods of solving the equations in Section 11.13. In this section we will outline a simple qualitative approach that assigns hydrogenlike orbital wave functions to each electron.

The Pauli Exclusion Principle

The existence of electron spin means that the wave function we obtained for the hydrogenlike atoms must be modified. Postulate VIII states that two spin eigenfunctions α and β exist and that they satisfy the orthonormal conditions, Eq. 11.225. The wave function that describes the energy and orbital angular momentum of an electron, ψ, must include spin so that the total wave function becomes

$$\psi_{\text{total}} = \psi\,\alpha \quad \text{or} \quad \psi\,\beta \tag{11.226}$$

depending on whether the electron spin is in the $+\frac{1}{2}$ or $-\frac{1}{2}$ state.

For one electron, by virtue of the orthonormality of states, the electron spin does not affect the spinless probability density,

$$|\psi_{\text{total}}|^2 = \int \psi^*\alpha^*\psi\alpha ds = \int \psi^*\beta^*\psi\beta ds = |\psi|^2 \tag{11.227}$$

where ds refers to the spin variables. The situation changes when more than one electron is present.

First, quantum mechanics cannot distinguish identical particles. If two particles are interchanged, the probability density $|\psi(1,2)|^2 = |\psi(2,1)|^2$ remains unchanged. This does not, however, preclude the sign of the total wave function changing upon the interchange of two particles. If we create an operator which permutes two identical particles, $\hat{P}_{12}$, the wave function is either symmetric or antisymmetric

$$\hat{P}_{12}\psi_{\text{total}}(1,2) = \pm\psi_{\text{total}}(2,1) \tag{11.228}$$

Fermions and Bosons

This property has fundamental consequences and divides elementary particles that possess spin into two categories, **fermions** and **bosons**. Fermions are particles with ½ integral spin (1/2, 3/2, 5/2 . . .) while bosons are particles with integral spin (0, 1, 2, . . .). The antisymmetrization principle states that when identical fermions are interchanged, the total wave function must change sign (antisymmetric), while when identical bosons are interchanged, the wave function must not change sign (symmetric).

Let us consider a two-electron system in which the two electrons have the same n, l, and m_l quantum numbers. Two spin states are available for the first, $\alpha(1)$ and $\beta(1)$. Likewise the second spin can have states $\alpha(2)$ and $\beta(2)$. We can immediately see that if the two spins are in the same state, the antisymmetrization principle is violated. That is, the total wave functions

$$\psi_{\text{total}} = \psi_{nlm_l}(1)\alpha(1)\psi_{nlm_l}(2)\alpha(2) \quad \text{or} \quad \psi_{\text{total}} = \psi_{nlm_l}(1)\beta(1)\psi_{nlm_l}(2)\beta(2) \tag{11.229}$$

are symmetric to electron interchange

$$\hat{P}_{12}\psi_{nlm_l}(1)\alpha(1)\psi_{nlm_l}(2)\alpha(2) - \psi_{nlm_l}(2)\alpha(2)\psi_{nlm_l}(1)\alpha(1) \tag{11.230}$$

or

$$\hat{P}_{12}\psi_{nlm_l}(1)\beta(1)\psi_{nlm_l}(2)\beta(2) = \psi_{nlm_l}(2)\beta(2)\psi_{nlm_l}(1)\beta(1) \tag{11.231}$$

Hence the two electrons with the same n, l, and m_l values must have different spin states, one in the α state and the other in the β state, giving a total wave function which is antisymmetric, as we will see shortly.

Pauli Exclusion Principle

This leads to the **Pauli exclusion principle**, which is an expression of the antisymmetrization principle applied to an orbital model. It can be stated as follows:

> **In an atom no two electrons can have all four quantum numbers (n, l, m_l, m_s) the same.**

Another formulation follows from the first:

> **If, in an atom, two electrons have the same three orbital quantum numbers, (n, l, and m_l), their spins must be opposed (i.e., the spin quantum number of one must be +1/2 and of the other −1/2).**

It follows at once that for a given principal quantum number n, the maximum number of electrons that can be in the various subshells is as follows:

s	2
p	6
d	10
f	14

In general, the maximum number is $2(2l + 1)$.

It is, however, not enough to define the total wave function as $\psi(1,2) = \psi_{nlm_l}(1)\alpha(1)\psi_{nlm_l}(2)\beta(2)$ because electron exchange leads to a different wave function. This means that the state as formed above, $\psi(1,2)$, is distinguishable from the other, $\psi(2,1)$. On the other hand, if we add and subtract these states, we can make them either symmetric or antisymmetric

$$\psi(1,2) \pm \psi(1,2) \tag{11.232}$$

The correct antisymmetrization of two electrons in the same orbital, normalized with a root 2, is

$$\psi_{\text{total}}(1,2) = \psi_{nlm_l}(1)\psi_{nlm_l}(2)\frac{1}{\sqrt{2}}[\alpha(1)\beta(2) - \beta(1)\alpha(2)] \tag{11.233}$$

which satisfies Eq. 11.228 and maintains $|\psi_{\text{total}}(1,2)|^2 = |\psi_{\text{total}}(2,1)|^2$.

It follows from this discussion that it is possible to construct from the four products of spin wave functions four linear combinations with differing symmetry. There is a singlet spin state that is antisymmetric to spin exchange

$$\frac{1}{\sqrt{2}}[\alpha(1)\beta(2) - \beta(1)\alpha(2)] \tag{11.234}$$

and three states, called a triplet state that is symmetric to spin exchange

$$\alpha(1)\alpha(2); \qquad \frac{1}{\sqrt{2}}[\alpha(1)\beta(2) + \beta(1)\alpha(2)]; \qquad \beta(1)\beta(2) \tag{11.235}$$

We shall see in Chapter 12 that both singlet and triplet spin states exist for molecules.

Another way of representing these antisymmetric states is to use a determinant

$$\begin{vmatrix} a & b \\ c & d \end{vmatrix} = ad - bc \tag{11.236}$$

It is a property of any determinant that the interchange of any two rows or two columns multiplies the determinant by -1. Therefore the correctly antisymmetrized state, shown in Eq. 11.233, can be written

$$\Psi_{\text{total}}(1,2) = \frac{1}{\sqrt{2}}\begin{vmatrix} \Psi_{nlm_l}(1)\alpha(1) & \Psi_{nlm_l}(1)\beta(1) \\ \Psi_{nlm_l}(2)\alpha(2) & \Psi_{nlm_l}(2)\beta(2) \end{vmatrix} \tag{11.237}$$

Slater Determinant

For two-spin systems use of such a determinant is not necessary, but it becomes invaluable when treating many-electron systems that must maintain the wave function antisymmetric. This method was developed by the American physicist John Clarke Slater (1900–1976), and the determinants are called **Slater determinants.** For example, the wave function for the ground state of an atom containing four electrons can be written immediately in terms of one or more 4×4 Slater determinants,

$$\Psi_{\text{ground state}}(1,2,3,4) = \frac{1}{\sqrt{4!}}\begin{vmatrix} 1s(1)\alpha(1) & 1s(1)\beta(1) & 2s(1)\alpha(1) & 2s(1)\beta(1) \\ 1s(2)\alpha(2) & 1s(2)\beta(2) & 2s(2)\alpha(2) & 2s(2)\beta(2) \\ 1s(3)\alpha(3) & 1s(3)\beta(3) & 2s(3)\alpha(3) & 2s(3)\beta(3) \\ 1s(4)\alpha(4) & 1s(4)\beta(4) & 2s(4)\alpha(4) & 2s(4)\beta(4) \end{vmatrix} \tag{11.238}$$

Any violation of the Pauli exclusion principle causes the wave function to vanish by virtue of the properties of determinants.

Drag electrons to create the electronic configurations for the atoms H to Ne; Spin quantum number and the Aufbau principle.

The Aufbau Principle

Only a brief account of this topic will be given, as it is assumed that the reader will already be familiar with the periodic table and with how it is interpreted in terms of adding electrons successively into the various orbitals. Use is made of the Aufbau (German *aufbau,* building up) principle, originally suggested by Pauli. The basis of the methods is that it is assumed that the orbitals obtained from the solution of the Schrödinger equation for the hydrogen atom can also be occupied in multielectron atoms. There are two important aspects to the problem: the *order* in which the different orbitals are filled and the *number of electrons* that can go into each orbital.

The first of these aspects is illustrated in Figure 11.22, which shows the order of filling of the orbitals. This order can be deduced from spectroscopic data on the various atoms and also on numerical calculations based on some of the approximation methods. There are several important points to note about this order of filling. The first relates to the influence of the azimuthal or angular momentum quantum number l on the energy levels. We have seen that for the hydrogen atom the energy is determined only by the principal quantum number n and not at all by l, so that, for example, the energy levels are the same for 2s and 2p orbitals. However, when there is more than one electron, the energy does depend upon l. The effect of the electron repulsions is to cause the 2p level to be somewhat higher than the 2s level. Similarly, when $n = 3$, the 3d level is higher than the 3p, which is higher than the 3s.

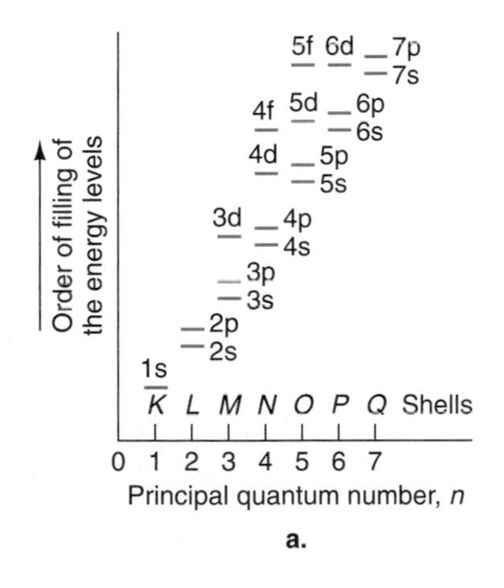

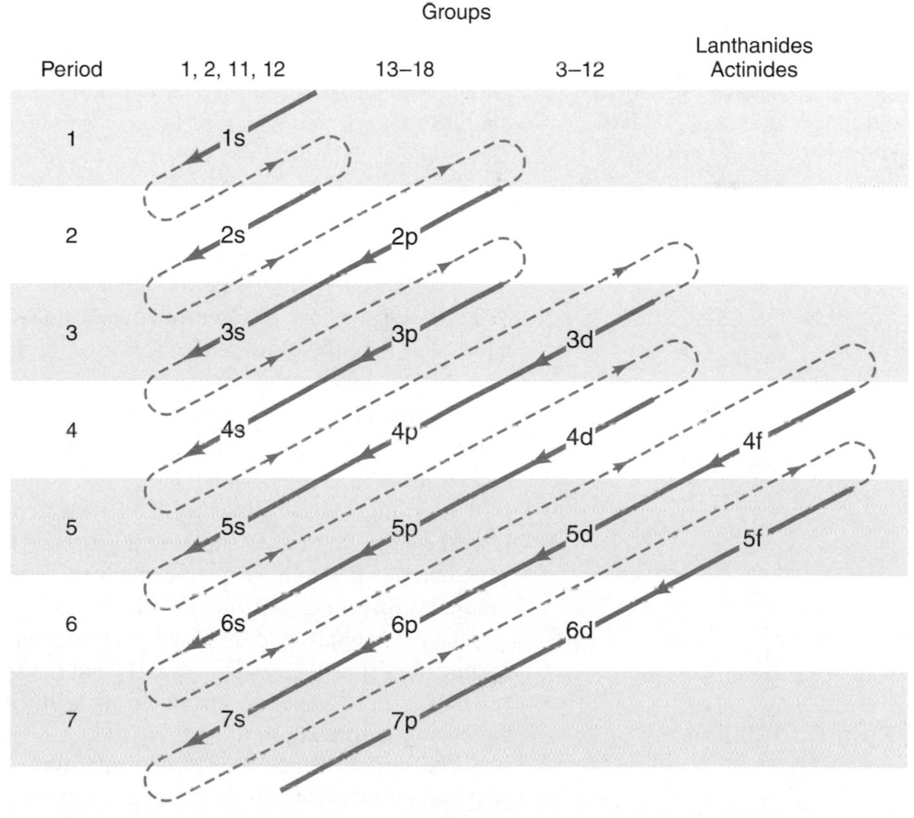

FIGURE 11.22
(a) The order of filling of the various ground-state energy levels. This is close to the order of the energy levels in an atom having a single electron, but there are some exceptions, as discussed in the text. (b) Diagram for order of orbital filling. Group numbers refer to orbitals that are filling in each energy level where the IUPAU numbering system is used.

To a first approximation the order of filling to the levels, shown in Figure 11.22, is the order of the energy levels themselves. This is certainly true for a given value of n, but there are some discrepancies for different n values. For example, the diagram shows that the 4s level is filled before the 3d level, and this is often explained by saying that the 3d orbital is higher in energy than the 4s. This statement, however, is an oversimplification, as the order of energy levels depends on what electrons are present in the atom. Calculations for a *single*-electron system show that the energy of the electron in the 3d orbital is below the energy when it is in the 4s orbital. The situation is, however, different when a number of electrons are present, because of the electron repulsions. The 3d orbital is more concentrated than the 4s, which has spherical symmetry, and there is a greater repulsion between a 3d electron and the other electrons than between a 4s electron and the other electrons. As a result, the 4s level fills before the 3d level.

Hund's Rule

The carbon atom, with six electrons, provides a simple example of the principles we have just considered. The first two electrons will go into the 1s orbital and have opposite spins. This shell is now filled and the next two go into the 2s orbital. The remaining two go into the 2p orbital, and now there are three possibilities which are conveniently represented as follows; the boxes indicate different m_l values and the arrow directions indicate the two different m_s values:

(a) [↑↓][ ][ ] (b) [↑][↑][ ] (c) [↑][↓][ ]

On the occasion of his 100th birthday, Hund said, "If I had known how long I would live, I would have made provision to work longer."

An important rule, formulated by the German physicist Friedrich Hund (1896–1997) enables us to decide which of these possibilities corresponds to the lowest energy. According to **Hund's rule,** *generally the favored configuration is the one in which the electrons have different magnetic quantum numbers m_l and the same spin quantum numbers.* Thus the arrangement b is found in the lowest energy state of carbon. The other arrangements are permitted but correspond to slightly higher energies.

The reason for the effect is rather subtle as two opposing factors are at work. When electrons have different magnetic quantum numbers, they tend to be farther apart, and they therefore repel each other less than if they had the same m_l value. This factor alone would appear to favor arrangements b and c over arrangement a. When two electrons have the same spin quantum numbers, their magnetic moments are aligned in the same direction and they therefore repel one another more than if their spins are opposed. This factor alone would appear to favor arrangements a and c over arrangement b. It might then appear that arrangement c would lead to the lowest energy. In fact, electrons in degenerate orbitals prefer to be as far away from each other as possible. When the electrons are unpaired, there is an added repulsion that helps this. In arrangement c, there is not enough repulsion to keep the electrons in separate orbitals, and on balance arrangement b is preferred. Still more detailed treatment of the atomic states reveals that the spin parallel condition is energetically favorable. This tendency for electrons with the same spin to avoid each other is called **spin correlation.**

Spin Correlation

The best evidence for Hund's rule and spin correlation comes from spectroscopic measurements of the paramagnetic properties of atoms. The paramagnetism of an atom is related to the number of unpaired electrons in orbitals where there is orbital angular momentum (i.e., all orbitals except s orbitals). If carbon has two unpaired p electrons, as predicted by Hund's rule, the paramagnetism of the carbon atom should be roughly twice that of the boron atom, which has one p electron. In fact, this is the case. If the two 2p electrons in the carbon atom were paired with each other (i.e., had opposite spins as in arrangements a and c), the atom would exhibit zero paramagnetism.

11.13 Approximation Methods in Quantum Mechanics

It is a straightforward matter to write the Schrödinger equation for any atom, but an explicit solution is possible only for the hydrogenlike atoms, which have a single electron. This is because of the terms for electron-electron repulsion, which make it impossible to separate the equation into simpler forms to which solutions can be obtained.

For many electron atoms it is therefore necessary to employ approximation methods and even then it is usually impossible to obtain explicit solutions. These approximation methods, moreover, allow numerical solutions to be obtained in a reasonable time with the use of a modern computer. As a testament to their success, today's approximation methods have moved far beyond determining the electronic structure of atoms. Software packages exist that can routinely determine the minimum energy structures of biomolecules, including small proteins and polypeptides. In this section we survey the main methods that underpin these successes, which have their origin in determining the electronic structures of many-electron atoms.

Central-field Approximation

Some of these approximation methods make use of the **central-field approximation.** The outermost electron in the atom is considered to be in the electric field that is created by the nucleus and the remaining electrons. The effect of these electrons is to shield the outermost electron from the nucleus, which is considered to have an *effective charge* $Z_{eff}e$, which is less than the true charge Ze on the nucleus. The shielded nucleus is assumed to produce a spherically symmetric field, so that the atom can in this approximate manner be treated quantum mechanically as a one-electron atom.

A simple application of this procedure is in the estimation of **ionization energies,** which are the energies required to remove an electron from its orbital to infinity. According to Bohr's theory of the hydrogenlike atom, the potential energy of the electron, relative to infinite separation, is given by Eq. 11.48:

$$E_p = -\frac{Z^2e^2}{4\pi\epsilon_0 n^2 a_0} \tag{11.239}$$

Exactly the same expression for E_p is given by the quantum-mechanical treatment. The value of this energy, with the sign dropped and with $Z = 1$ and $n = 1$, is 4.36×10^{-18} J or 27.2 eV, and it is a common practice to refer to this quantity as one **atomic unit of energy** (au). It is also known as the **hartree,** after the British physicist Douglas Rayner Hartree (1897–1958). The Bohr radius a_0, which is equal to 52.92 pm, is known as the **atomic unit of length.**

Atomic Unit of Energy

Atomic Unit of Length

The total energy of the electron, according to Bohr's theory and to quantum mechanics, is given by Eqs. 11.49 and 11.191:

$$E_n = -\frac{Z^2e^2}{8\pi\epsilon_0 n^2 a_0} \tag{11.240}$$

This is just one-half the potential energy, and the negative of its value is the ionization energy I for an electron of quantum number n. The ionization energy is therefore given by

Ionization Energy

$$I = \frac{Z^2 e^2}{8\pi\epsilon_0 n^2 a_0} = \frac{Z^2}{2n^2} \times 4.36 \times 10^{-18}\ \text{J} \tag{11.241}$$

$$= \frac{Z^2}{2n^2}\ \text{atomic units (au)} \tag{11.242}$$

To apply the central-field approximation we simply replace Z by Z_{eff}:

$$I = \frac{Z^2_{\text{eff}}}{2n^2}\ \text{au} \tag{11.243}$$

The following examples show how the value of Z_{eff} can be estimated from an experimental value of the ionization energy.

EXAMPLE 11.8 Estimate Z_{eff} for a 1s electron in He, if the first ionization energy of helium is 24.6 eV.

Solution In atomic units this ionization energy is

$$24.6/27.2 = 0.904\ \text{au}$$

For the helium atom $n = 1$ and therefore, from Eq. 11.243,

$$0.904 = Z_{\text{eff}}^2/2 \quad \text{or} \quad Z_{\text{eff}} = 1.34$$

This is significantly less than 2, the Z value of He, and shows that there is substantial screening by the other 1s electron.

EXAMPLE 11.9 Estimate the effective nuclear charge felt by the 2s electron in the lithium atom, if the ionization energy is 5.38 eV.

Solution In atomic units the ionization energy is 5.38/27.2 = 0.198 au. Then by Eq. 11.243, with $n = 2$,

$$0.198 = Z_{\text{eff}}^2/8 \quad \text{or} \quad Z_{\text{eff}} = 1.26$$

This indicates considerable screening by the 1s electrons.

The Variation Method

A powerful quantum-mechanical procedure forms the basis of most calculations that have been made on atoms and molecules. This procedure, known as the **variation method,** not only allows reliable calculations to be made on complex atoms (and, as we shall see in Chapter 12, on molecules), but it also provides considerable insight into the nature of atoms and of chemical bonds. It is most useful for obtaining the energies of atoms and molecules in their ground states.

Suppose, first, that the Schrödinger equation can be solved exactly and that we have obtained an eigenfunction ψ, which need not be normalized. In other words, for a particular system we have set up the equation

$$\hat{H}\psi = E\psi \tag{11.244}$$

and have obtained the function ψ. We can now multiply both sides by the complex conjugate ψ^* of the eigenfunction:

$$\psi^*\hat{H}\psi = \psi^*E\psi = E\psi^*\psi \tag{11.245}$$

The transformation of $\psi^*E\psi$ into $E\psi^*\psi$ is permissible because E is a constant and ψ, ψ^*, and E all commute. Integration of both sides of Eq. 11.245 over all space gives

$$\int \psi^*\hat{H}\psi\, d\tau = E\int \psi^*\psi\, d\tau \tag{11.246}$$

and therefore

$$E = \frac{\int \psi^*\hat{H}\psi\, d\tau}{\int \psi^*\psi\, d\tau} \tag{11.247}$$

The denominator of this equation is unity if the eigenfunction has been normalized.

On the other hand, suppose that ψ was not an eigenfunction of the operator $\hat{H}$ but was instead a trial function obtained in some way. We could then calculate the values of the integrals on the right-hand side of Eq. 11.247 and calculate an energy, if necessary by using numerical methods. The variation principle now tells us that the energy calculated from Eq. 11.247, with any trial function, *cannot be below* the true energy for the ground state of the system. The closer the chosen function is to the exact eigenfunction, the closer the energy is to the true (experimental) value. If the function chosen happened to be the true eigenfunction, the exact energy would be obtained. Otherwise, the calculated energy is *higher* than the true energy.

In order to calculate reliable energies by the variation method, it is best to choose trial wave functions, containing adjustable parameters, that relate as closely as possible to the system that is being studied. For a many-electron atom, for example, one might choose a linear combination of hydrogenlike wave functions and vary the coefficients, until a minimum energy is obtained. For more reliable results, a number of trial wave functions of different types may be employed.

Configuration Interaction

One procedure used with the variation method deserves special mention. It is referred to as **configuration interaction** (CI) and attempts to account for different electron configurations that can exist due to *excited electronic states*. It therefore involves adding to the trial function wave functions that correspond to these states. As has been emphasized, when the variation principle is applied, improvement is brought about by the addition of additional terms, and the addition of terms for excited states has proved to be particularly useful. With the use of supercomputers, variational calculations for atomic and molecular problems have been made with as many as a million terms in the trial eigenfunction.

EXAMPLE 11.10 Use the trial function $\psi = x(a - x)$ to estimate the energy of a particle in a box, in which the boundaries of the box are 0 and a.

Solution When using any trial function, it is important to make sure that the boundary conditions are satisfied. This function is easy to check. It vanishes at $x = 0$ and $x = a$ at the beginning and end of the box. In order to calculate the energy of this system, substitute the trial wave function into Eq. 11.247.

$$E = \frac{\int_0^a \psi^* \hat{H} \psi dx}{\int_0^a \psi^* \psi dx} = \frac{\int_0^a x(a - x)\hat{H}x(a - x)dx}{\int_0^a x^2(a - x)^2 dx}$$

The value of the integral in the denominator is $a^5/30$. The Hamiltonian used is that of the particle-in-a-box problem, which is the same as that used for a free particle, Eq. 11.127, with boundary conditions. Making this substitution into the numerator above gives

$$\int_0^a x(a - x)\hat{H}x(a - x)dx = -\frac{\hbar^2}{2m}\int_0^a x(a - x)\frac{d^2}{dx^2}x(a - x)dx$$

$$= +\frac{\hbar^2}{m}\int_0^a x(a - x)dx = \frac{\hbar^2 a^3}{6m}$$

Substitution of these two calculated values into the variational expression above gives $E = \dfrac{5h^2}{4\pi^2 ma^2}$. Comparison with the exact calculation from Eq. 11.149 for the lowest energy, $n = 1$, shows that indeed the variational energy is greater than the exact energy

$$E = \frac{5h^2}{4\pi^2 ma^2} > \frac{h^2}{8ma^2} = E_1$$

but is remarkably close, as seen from the ratio:

$$\frac{E}{E_1} = \frac{10}{\pi^2} = 1.0132.$$

Perturbation Method

Another method that has been used for dealing with interelectron repulsion energies is the *perturbation method.* The procedure is to start with an exact solution of the Schrödinger equation for a known system and then to add additional small terms to deal with the actual system. Suppose that the solution can be obtained exactly for a Hamiltonian operator $\hat{H}°$; in other words, one can solve the equation

$$\hat{H}°\psi° = \hat{E}°\psi° \tag{11.248}$$

which is called the *unperturbed equation.* In first-order perturbation theory a single term $\hat{H}'$, known as the *perturbation,* is added to the Hamiltonian $\hat{H}°$ for the unperturbed system, so that one is dealing with the operator

$$\hat{H} = \hat{H}° + \delta\hat{H}' \tag{11.249}$$

The perturbation can, for example, be related to interelectronic potential energies. Since Eq. 11.248 can be solved exactly, the eigenvalues and eigenfunctions are known. Call these $E_n^{(0)}$ and $\psi_n^{(0)}$. Successive corrections due to the perturbation, $\hat{H}'$, are usually written as

$$\psi_n = \psi_n^{(0)} + \delta\psi_n^{(1)} + \delta^2\psi_n^{(2)} + \cdots \tag{11.250}$$

$$E_n = E_n^{(0)} + \delta E_n^{(1)} + \delta^2 E_n^{(2)} + \cdots \tag{11.251}$$

where the magnitude of δ determines how many terms need to be retained to give the degree of accuracy desired. If δ is small, the first-order perturbation results are sufficient. The first-order non degenerate correction to the energy is given by

$$E_n \cong E_n^{(0)} + \delta \int \psi_n^{(0)*} \hat{H}' \psi_n^{(0)} \, d\tau \tag{11.252}$$

and the correction to the eigenfunction is

$$\psi_n \cong \psi_n^{(0)} + \delta \sum_{m \neq n} \left[\frac{\int \psi_m{}^{(0)*} \hat{H}' \psi_n{}^{(0)} \, d\tau}{E_n{}^{(0)} - E_m{}^{(0)}} \right] \psi_m{}^{(0)} \tag{11.253}$$

Notice that the first-order perturbation wave functions have contributions from all the zeroth-order eigenfunctions and eigenvalues. This works only if no degeneracies exist. When degeneracies do exist, degenerate perturbation theory must first be used to remove the degeneracies, after which non-degenerate perturbation theory can be applied.

Perturbation theory is particularly useful when the perturbation is small. The harmonic oscillator and the rigid rotor are good starting models to describe vibration and rotation of molecules. Perturbations to these motions can be included and can successfully account for anharmonic terms in vibrational motion and centrifugal distortions to rotational motion.

The Self-Consistent Field (SCF) Method

Self-consistent Field Method

An important procedure for dealing with complex atoms is the **self-consistent field (SCF) method,** which starts with the n-electron kinetic energy term and an approximate term to account for the potential. The variational principle is iteratively applied to keep improving the potential arising from the electrons, thereby generating an improved Hamiltonian with each iteration. The method was first suggested in 1928 by Hartree and it was improved by the introduction of methods earlier suggested by the Russian physicist Vladimir Alexandrovitch Fock (b. 1898). The method is therefore often known as the **Hartree-Fock method.** It is based on the construction of an approximate electronic distribution for the atom, by first considering the individual shells. Closed shells have spherical symmetry, by Unsöld's theorem (Section 11.8), and for nonfilled shells the distribution is averaged over all angles.

Hartree-Fock Method

The next step is to consider one particular electron in the potential field created by the nucleus and the remaining electrons, averaged as indicated previously. Since the averaged potential is symmetrical, the Schrödinger equation can be separated into R, Θ, and Φ equations that can be solved numerically. This procedure can be applied to all the electrons. On the basis of the wave functions obtained in this way one can then calculate a new potential-energy field that will be closer to the truth than the one originally assumed. The whole procedure is then repeated and a second set of solutions obtained. This *iterative* procedure is repeated until there is no further change in the individual electronic distributions or their potential-energy fields, which are then said to be self-consistent. This is the Hartree-Fock limit. By virtue of the variational method, the Hartree-Fock limit gives an energy which is greater than the exact ground-state energy. If configuration interaction (CI) is ignored, the difference between the Hartree-Fock energy and the exact energy is a measure of the correlation energy of the electrons. The electron densities calculated by this method have been found to give good agreement with the results of X-ray and electron-diffraction experiments.

Slater Orbitals

The SCF method is rather laborious, and a useful approximate procedure (involving analytical expressions for atomic orbitals) was suggested in 1932 by the American physicist John Clarke Slater (1900–1976). These orbitals are calculated using the effective nuclear charge that influences each electron. This effective nuclear charge is calculated from the equation

$$Z_{\text{eff}} = Z - \sigma \tag{11.254}$$

where Z is the actual charge number of the nucleus and σ is the amount of shielding, which is obtained from some semiempirical rules:

1. There is no contribution to σ from any electron with n greater than that of the electron under consideration.
2. For an s electron, the shielding σ of the other s electron is 0.30 (unless $n = 1$). For all other electrons that have the same value of n as the electron under consideration, the contribution is 0.35 for each electron.
3. If the value of n is 1 less than that of the electron of interest, σ is 0.85 for each s or p electron and is 1.00 for each d or f orbital.
4. If n is 2 or more less than the quantum number of the electron of interest, σ is 1.00 for each electron.

EXAMPLE 11.11 Calculate the effective nuclear charge for one of the 1s electrons in the helium atom.

Solution Of the two electrons in the ground state of helium, one electron contributes to the shielding of the other. Therefore, from rule 2, $\sigma = 0.30$, and $Z_{\text{eff}} = Z - \sigma = 2.00 - 0.30 = 1.70$. Note that this is somewhat higher than the value (1.34) estimated from the ionization energy. See Example 11.8 on p. 564.

EXAMPLE 11.12 Calculate the effective nuclear charge for a 2p electron in nitrogen.

Solution The electron configuration for nitrogen is $1s^2 2s^2 2p^3$ and it has a nuclear charge of 7. The shielding is $2(0.85) = 1.70$ for the 1s electrons and $(4 \times 0.35) = 1.40$ for the two 2s and the two remaining 2p electrons. The effective nuclear charge for any 2p electron is thus $Z - \sigma = 7 - (1.70 + 1.40) = 3.90$.

Slater Orbitals

The values of Z_{eff} as calculated are used to calculate *Slater orbitals,* which are of a slightly different form than the hydrogenlike functions. The Slater *ns* orbitals are defined as

$$\psi_{ns}(r) = r^{n-1}e^{-Z_{eff}r/n} \tag{11.255}$$

and the *np* orbitals as

$$\psi_{np_x}(r) = xr^{n-2}e^{-Z_{eff}r/n} \tag{11.256}$$

where x is the contribution of p(r) (a function of the variable r only) to np_x. The orbitals generated in this way are useful for order-of-magnitude calculations and can be used as a starting point in SCF calculations.

With the use of the procedures considered in this section, approximate solutions to quantum-mechanical problems can be obtained fairly easily with the use of the computers that are available at most universities and research institutes. To obtain values of the energies of atoms and molecules that are precise enough to deal with many of the practical problems of chemistry, however, it is necessary to use one of the modern supercomputers, and even then the work is often time consuming.

It is appropriate to mention that commercial programs are available which contruct wave functions with *ab initio* calculations using Gaussian functions. Such calculations can be run on personal computers with reasonable results.

Relativistic Effects in Quantum Mechanics

So far we have made only a passing reference to the application of Einstein's theory of relativity to quantum mechanics. At the beginning of Section 11.11 we mentioned that in 1928 P. A. M. Dirac deduced the fact that an electron can have two spin quantum numbers from his relativistic formulation of the equations of quantum mechanics. Interestingly, Dirac himself at the time wrote that relativistic effects would be "of no importance in the consideration of atomic and molecular structure and of ordinary chemical reactions." His reason for believing this was that relativistic effects become significant only if the electron speeds approach the speed of light, and he thought that valence electron speeds are too low. This is indeed true for light atoms, where the nuclear charges are small.

For heavy atoms, however, the nuclear charges are sufficiently large that the electrons that are close to the nucleus (particularly s and p electrons) do move at speeds which compare with the speed of light, and when such atoms are involved this has to be taken into account in detailed quantum-mechanical calculations. The modifications that have to be made when relativistic effects are introduced are always small, but sometimes significant.

Just one example, a rather striking one, will be mentioned. When quantum mechanical calculations are made on the gold atom, without taking the relativistic effects into account, the conclusion is that gold should have the same color as silver. When the relativity effects are taken into account, the electronic energy levels are altered somewhat, and the gold color of the metal can then be understood. Also see Section 16.7 on optical properties of metals.

Dirac Notation

In dealing with wave functions, a convenient notation was introduced by Dirac. His notation avoids writing the wave functions and showing integrals explicitly. In this more compact notation, a "bra" is written $\langle \,|$ and a "ket" is written $|\, \rangle$, and together they form a bracket: $\langle \,|\, \rangle$.

As an example, α^*, the complex conjugate of the spin wave function, is written as α within the bra as

$$\alpha^* \rightarrow \langle\, \alpha \,|$$

and the spin wave function α is written within the ket as

$$\alpha \rightarrow |\, \alpha\, \rangle$$

The same procedure is followed using β^* and β. Normalization over all space (Eq. 11.89) is implied when the bra and ket are put together to form the bracket $\langle \,|\, \rangle$. Thus, the normalization and orthogonal conditions that we have previously seen, for example in Eq. 11.225 are

$$\int \alpha^* \alpha \, ds = \int \beta^* \beta \, ds = 1 \quad \text{and} \quad \int \alpha^* \beta \, ds = \int \beta^* \alpha \, ds = 0 \qquad (11.257)$$

and they become

$$\langle \alpha | \alpha \rangle = \langle \beta | \beta \rangle = 1 \quad \text{and} \quad \langle \alpha | \beta \rangle = \langle \beta | \alpha \rangle = 0 \qquad (11.258)$$

This is not restricted to spin states and is quite general. The wave function ψ^* is written in the bra $\langle \psi |$ and ψ is written in the ket $| \psi \rangle$.

Furthermore there are other benefits in using this notation. If an operator $\hat{o}$ of an observable is placed in the position of c in the $\langle \text{bra} | c | \text{ket} \rangle$, this form allows the calculation of expectation values. Thus, part of Eq. 11.194 rewritten becomes

$$\langle r_{1s} \rangle = \int \psi_{1s}^* \, \hat{r} \, \psi_{1s} \, d\tau = \langle \psi_{1s} | \hat{r} | \psi_{1s} \rangle \qquad (11.259)$$

in the Dirac notation where the wave functions are normalized, and integrated over all space. Because the wave functions can have specific quantum numbers identifying the states, this specific form is a matrix element of the operator $\hat{r}$. Hence the Hermitian property also applies, which we encountered in Section 11.5. (See Eqs. 11.104 and 11.106.)

$$\int \psi_2^* \hat{F} \psi_1 d\tau = \int \psi_1 (\hat{F} \psi_2)^* d\tau \qquad (11.260)$$

can be written

$$\langle \psi_2 | \hat{F} | \psi_1 \rangle = \langle \psi_1 | \hat{F} | \psi_2 \rangle^* \qquad (11.261)$$

which expresses the invariance of transposing a matrix element and simultaneously taking its complex conjugate. Operators that obey this relationship are Hermitian operators, and such operators are said to be self-adjoint (see Eq. 11.109)

$$\hat{F}^\dagger = \hat{F} \qquad (11.262)$$

Similarly, Eq. 11.107, in Dirac notation, is

$$\langle F \rangle^* = \frac{\langle (\hat{F} \psi) | \psi \rangle}{\langle \psi | \psi \rangle} \qquad (11.263)$$

The Dirac notation is particularly useful for a finite number of states, like the two states of a spin ½ or for the $2J + 1$ rotational states, $|JM\rangle$. It is more than a convenience, however. The bras and kets form vector spaces, called Hilbert spaces. The bras are conjugate vectors to the kets and form an inner product, which is the bracket $\langle \,|\, \rangle$. The bracket is bounded (i.e., the integrals all exist and are finite) and is a complex number. The mathematics of Hilbert spaces is well established and their properties have been useful in understanding the fundamental mathematical structure of quantum mechanics.

KEY EQUATIONS

Electromagnetic radiation:

$$\lambda\nu = c$$

Natural frequency ν of simple harmonic motion:

$$\nu = \frac{1}{2\pi}\sqrt{\frac{k_h}{m}}$$

Energy quantization:

$$E = h\nu$$

Average energy of an *oscillator* of frequency ν:

$$\bar{\epsilon} = \frac{h\nu}{e^{h\nu/k_BT}} - 1$$

Wavenumbers $\tilde{\nu}$ of hydrogen atom spectral lines:

$$\tilde{\nu} \equiv \frac{1}{\lambda} = R\left(\frac{1}{n_1^2} - \frac{1}{n_2^2}\right) \qquad (R = \text{Rydberg constant})$$

Uncertainty principle:

$$\Delta q\Delta p \geq \frac{\hbar}{2}\ ;\ \Delta E\Delta t \geq \frac{\hbar}{2} \quad \text{and} \quad \Delta L\, \Delta\phi \geq \frac{\hbar}{2}$$

Schrödinger equation:

Time dependent:

$$\hat{H}\Psi = i\hbar\frac{\partial\Psi}{\partial t}$$

Time independent:

$$\left[-\frac{\hbar}{2m}\nabla^2 + \hat{E}_p(x, y, z)\right]\psi = E\psi$$

or

$$\hat{H}\psi = E\psi$$

Normalization condition:

$$\int \psi_1^*\psi_1\, d\tau = 1$$

Orthogonality condition:

$$\int \psi_1{}^*\psi_2\, d\tau = 0$$

Average value of a quantity F:

$$\bar{F} = \langle F\rangle = \frac{\int \psi^*\hat{F}\psi\, d\tau}{\int \psi^*\psi\, d\tau}$$

Energy of a particle in a box of sides a, b, and c:

$$E = \frac{h^2}{8m}\left(\frac{n_1^2}{a^2} + \frac{n_2^2}{b^2} + \frac{n_3^2}{c^2}\right)$$

Energy of a *harmonic oscillator* of reduced mass μ:

$$E_v = \hbar\sqrt{\frac{k_h}{\mu}}\left(v + \frac{1}{2}\right) = h\nu_0\left(v + \frac{1}{2}\right) \qquad v = 0, 1, 2, \ldots$$

Quantum numbers for hydrogenlike atoms:

$$n = 1, 2, 3, \ldots, \infty$$
$$l = 0, 1, 2, 3, \ldots, n - 1$$
$$m_l = -l, -l + 1, -l + 2, \ldots, -1, 0, 1, \ldots, l - 1, l$$
$$m_s = +\frac{1}{2}, -\frac{1}{2}$$

Spherical harmonics

$$Y_l^{m_l}(\theta,\phi) =$$

$$\int_0^{\pi}\int_0^{2\pi} Y_l^{m_l*}(\theta,\phi)Y_{l'}^{m_l'}(\theta,\phi)\sin\theta d\theta d\phi = \delta_{ll'}\delta_{m_lm_l'}$$

Kronecker delta function

$$\delta_{kl} = 1 \text{ if } k = l \text{ and } \delta_{kl} = 0 \text{ if } k \neq l$$

Angular momentum:

$$L_z = m_l\hbar$$
$$L^2 = l(l + 1)\hbar^2$$

Angular momentum commutation relations:

$$[\hat{L}_x, \hat{L}_y]_- = i\hbar\hat{L}_z;\ [\hat{L}_y, \hat{L}_z]_- = i\hbar\hat{L}_x;\ [\hat{L}_z, \hat{L}_x]_- = i\hbar\hat{L}_y$$

First-order non-degenerate perturbation theory:

$$E_n \simeq E_n^{(0)} + \delta\int\psi_n^{(0)*}\hat{H}'\psi_n^{(0)}\, d\tau$$

$$\psi_n \cong \psi_n^{(0)} + \delta\sum_{m\neq n}\left[\frac{\int\psi_m^{(0)*}\hat{H}'\psi_n^{(0)}d\tau}{E_n^{(0)} - E_m^{(0)}}\right]\psi_m^{(0)}$$

Dirac notation:

$$\alpha \rightarrow |\alpha\rangle \text{ and } \alpha^* \rightarrow \langle\alpha|$$
$$\langle\alpha|\alpha\rangle = \langle\beta|\beta\rangle = 1 \text{ and } \langle\alpha|\beta\rangle = \langle\beta|\alpha\rangle = 0$$

Variation method:

$$E = \frac{\int \psi^*\hat{H}\psi\, d\tau}{\int \psi^*\psi\, d\tau}$$

PROBLEMS

Electromagnetic Radiation and Wave Motion

11.1. Calculate, for light of 325 nm wavelength,

a. the frequency;
b. the wavenumber;
c. the photon energy in J, eV, and kJ mol^{-1}; and
d. the momentum of the photon.

11.2. A pulsar in the Crab Nebula, NP 0532, emits both radio pulses and optical pulses. A radio pulse is observed at 196.5 Mhz. Calculate

a. the corresponding wavelength;
b. the energy of the photon in J, eV and J mol^{-1}; and
c. the momentum of the photon.

11.3. The potassium spectrum has an intense doublet with lines at 766.494 nm and 769.901 nm. Calculate the frequency difference between these two lines.

11.4. Suppose that the position y of a particle that travels along the Y axis of a coordinate system is given by

$$y(t) = y_0 \sin\left[\frac{3\pi \text{ rad}}{5}\left(\frac{t}{s}\right) + C\right]$$

What is the frequency of the wave motion?

11.5. A mass of 0.2 kg attached to a spring has a period of vibration of 3.0 s.

a. What is the force constant of the spring?
b. If the amplitude of vibration is 0.010 m, what is the maximum velocity?

***11.6.** If the average energy associated with a standing wave of frequency ν in a cavity is

$$\overline{\varepsilon} = \frac{h\nu}{\exp(h\nu/k_BT) - 1}$$

deduce the expression for the low-frequency limit of the average energy associated with the standing wave.

Particles and Waves

11.7. A sodium lamp of 50-watt power emits yellow light at 550 nm. How many photons does it emit each second? What is the momentum of each photon?

11.8. The threshold frequency ν_0 for emission of photoelectrons from metallic sodium was found by Millikan, *Phys. Rev.*, *7*, 1916, p. 362, to be 43.9 × 10^{13} s^{-1}. Calculate the work function for sodium. A more recent value, for a carefully outgassed sample of sodium, is 5.5 × 10^{13} s^{-1}. What work function corresponds to that value?

11.9. Calculate the value of the de Broglie wavelength associated with

a. an electron moving with a speed of 6.0 × 10^7 m s^{-1} (this is the approximate velocity produced by a potential difference of 10 kV).
b. an oxygen molecule moving with a speed of 425 m s^{-1} at 0 °C.
c. an α-particle emitted by the disintegration of radium, moving at a speed of 1.5 × 10^7 m s^{-1}.
d. an electron having a speed of 2.818 × 10^8 m s^{-1}.

11.10. Consider a colloidal particle with a mass of 6 × 10^{-16} kg. Suppose that its position is measured to within 1.0 nm, which is about the resolving power of an electron microscope. Calculate the uncertainty in the velocity and comment on the significance of the result.

11.11. Calculate the velocity and the de Broglie wavelength of an electron accelerated by a potential of

a. 10 V,
b. 1 kV, and
c. 1 MV.

***11.12.** The group velocity of a wave is given by the equation

$$v_g = \frac{d\nu}{d(1/\lambda)}$$

Prove that the group velocity of a de Broglie particle wave is equal to the ordinary velocity of the particle.

11.13. Photoelectric experiments show that about 5 eV of energy are required to remove an electron from platinum.

a. What is the maximum wavelength of light that will remove an electron?
b. If light of 150 nm wavelength were used, what is the velocity of the emitted electron?

11.14. Calculate the kinetic energy of an electron that has a wavelength of (a) 10 nm, (b) 100 nm.

11.15. Calculate the de Broglie wavelength of (a) an α-particle (a helium nucleus) accelerated by a field of 100 V, and (b) a tennis ball served at 220 km h^{-1}. (An α-particle has a mass of 6.64 × 10^{-27} kg and a diameter of about 10^{-15} m. A standard tennis ball has a mass of 55.4 g and a diameter of 6.51 cm.)

Quantum-Mechanical Principles

11.16. Assume that the three real functions ψ_1, ψ_2, and ψ_3 are normalized and orthogonal. Normalize the following functions:

a. $\psi_1 + \psi_2$
b. $\psi_1 - \psi_2$
c. $\psi_1 + \psi_2 + \psi_3$
d. $\psi_1 - \frac{1}{\sqrt{2}}\psi_2 + \frac{\sqrt{3}}{\sqrt{2}}\psi_3$

11.17. Is the function Ae^{-ax} an eigenfunction of the operator d^2/dx^2? If so, what is the eigenvalue?

11.18. Prove that m_1 must be integral in order for the function

$$\Phi = \sin m_1\phi$$

to be an acceptable wave function.

11.19. The energy operator for a time-dependent system (Table 11.1) is

$$i\hbar\frac{\partial}{\partial t}$$

A possible eigenfunction for the system is

$$\Psi(x, y, z, t) = \psi(x, y, z)\exp(-2\pi iEt/h)$$

Show that $\Psi\Psi^*$, the probability density, is independent of time.

***11.20.** Prove that the momentum operator corresponding to p_x is a Hermitian operator.

11.21. Which of the following functions is an eigenfunction of the operator d/dx?

a. k
b. kx^2
c. $\sin kx$
d. $\exp(kx)$
e. $\exp(kx^2)$
f. $\exp(ikx)$

(k is a constant, and i is the square root of minus one.) Give the eigenvalue where appropriate.

11.22. Figure 11.20 shows the angular momentum vectors for 1 = 2 and for m = 2, 1, 0, −1, −2. In each case, calculate the angles the vectors make with the Z axis.

11.23. Show that the one-electron wave functions ψ_{nim} are also eigenfunctions of the operator $(\hat{L}_x^2 + \hat{L}_y^2)$. What physical property (observable) is associated with this operator?

11.24. Explain why the Heisenberg uncertainty principle would be violated if the harmonic oscillator ground-state energy were zero.

Particle in a Box

11.25. Calculate the lowest possible energy for an electron confined in a cube of sides equal to

a. 10 pm and
b. 1 fm (1 femtometre = 10^{-15} m).

The latter cube is the order of magnitude of an atomic nucleus; what do you conclude from the energy you calculate about the probability of a free electron being present in a nucleus?

11.26. A particle is moving in one dimension between $x = a$ and $x = b$. The potential energy is such that the particle cannot be outside these limits and that the wave function in between is

$$\psi = A/x$$

a. Determine the normalization constant A.
b. Calculate the average value of x.

11.27. An electron is confined in a one-dimensional box 1 nm long. How many energy levels are there with energy less than 10 eV? How many levels are there with energy between 10 and 100 eV?

11.28. Determine whether the eigenfunctions obtained in Section 11.6 for a particle in a one-dimensional box are eigenfunctions for the momentum operator. If they are, obtain the eigenvalues; if they are not, explain why.

***11.29.** Treat the three-dimensional particle in a box of sides a, b, and c by analogy with the treatment in Section 11.6. Assume the potential to be zero inside the box and infinite outside, and proceed by the following steps:

a. Write the basic differential equation that must be solved for the three-dimensional problem.
b. Separate the equation from (a) into terms involving $X(x)$, $Y(y)$, and $Z(z)$.
c. Determine the expressions for X, Y, and Z.
d. Obtain the expression (Eq. 11.150) for the total energy.

11.30. What is the quantum-mechanical probability of finding the particle in a one-dimensional "box" in the middle third of the "box"? Derive an expression that shows how this quantity depends on the quantum number n.

11.31. The classical probability for finding a particle in the region x to $x + dx$ in a one-dimensional box of length a is dx/a.

a. Derive the classical probability for finding the particle in the middle third of the box.
b. Show that as $n \to \infty$, the quantum probability obtained in the previous problem becomes identical to the classical result.

***11.32.** Problem 11.25 is concerned with the calculation of the minimum energy for an electron confined in a cube. Another approach to the problem is to consider, on the basis of the uncertainty principle (Eq. 11.60), the uncertainty in the energy if the uncertainty in the position is equal to the length of the side of the cube. Calculate ΔE for a cube of sides equal to

a. 10 pm and
b. 1 fm (10^{-15} m),

and compare the results with the minimum energies found for Problem 11.25.

***11.33.** Prove that any two wave functions for a particle in a one-dimensional box of length a are orthogonal to each other; that is, they obey the relationship

$$\int_0^a \psi_m \psi_n \, dx = 0, \text{ m} \neq \text{n}$$

11.34. Use the trial function $\Psi = x(a - x)$ and Eq. 11.247 to calculate an energy for a particle in a one-dimensional box of length a.

11.35. a. At a node, a wave function passes through zero. For the problem of the particle in a box, how many nodes are there for $n = 2$ and $n = 3$?

b. From the expression for the radial function for the 3s electron (Table 11.4), obtain expressions for the position of the radial nodes (i.e., the nodes in the solution of the radial equation) in terms of Z and a_0.

Vibration and Rotation

11.36. The vibration frequency of the N_2 molecule corresponds to a wave number of 2360 cm^{-1}. Calculate the zero-point energy and the energy corresponding to $v = 1$.

***11.37.** If a rigid body rotates in the XY plane, about the Z axis, the angular momentum operator is

$$\hat{L} = -i\hbar \frac{\partial}{\partial \phi}$$

(see Figure 11.15). If the moment of inertia is I, what is the energy operator?

(For additional problems dealing with molecular vibrations and rotations, see Chapter 13.)

The Atom

11.38. Calculate the ionization energy of the hydrogen atom on the basis of the Bohr theory.

11.39. Calculate, on the basis of the Bohr theory, the linear velocity of an electron (mass = 9.11×10^{-31} kg) in the ground state of the hydrogen atom. To what de Broglie wavelength does this velocity correspond? Deduce an equation for the de Broglie wavelength, in a Bohr orbit of quantum number n, with $Z = 1$, in terms of a_0 and n. What is the ratio of the circumference of a Bohr orbit of quantum number n to the de Broglie wavelength?

11.40. For a hydrogenlike atom (a one-electron system with a charge number of Z), find the radius of the sphere on which the probability of finding the 1s electron is a maximum. Compare the result to the expression of Eq. 11.44.

***11.41.** Calculate the reduced masses of the hydrogen and deuterium atoms, using the following masses for the particles:

Electron:	9.1095×10^{-31} kg
Proton:	1.6727×10^{-27} kg
Deuterium nucleus:	3.3434×10^{-27} kg

a. Explain qualitatively what effect the different reduced masses will have on the Bohr radii and therefore on the positions of the lines in the atomic spectra.

b. The Balmer spectrum of hydrogen has a line of wavelength 656.47 nm. Deduce the wavelength of the corresponding line in the spectrum of deuterium.

11.42. Calculate the wavelength and energy corresponding to the $n = 4$ to $n = 5$ transition in the hydrogen atom.

11.43. Calculate, in joules and in atomic units, the potential energy of an electron in the $n = 2$ orbit of the hydrogen atom.

11.44. The first ionization energy of the Li atom is 5.39 eV. Estimate an effective nuclear charge Z_{eff} for the valence electron in the Li atom.

11.45. The first ionization energy of the Na atom is 5.14 eV. Estimate the effective nuclear charge Z_{eff} for the valence electron in the Na atom.

***11.46.** Use Slater's method (Section 11.13) to determine the effective nuclear charge for

a. a 3s electron in the chlorine atom,
b. a 3p electron in the phosphorus atom, and
c. the 4s electron in the potassium atom.

***11.47.** A normalized Slater orbital for the 1s orbital in the helium atom is

$$\psi_{1s} = \frac{1}{\sqrt{\pi}}\left(\frac{Z_{eff}}{a_0}\right)^{3/2} \exp(-Z_{eff} r/a_0)$$

where Z_{eff} is the effective charge number. It leads to the following expression for the energy

$$E = \frac{e^2}{a_0}\left(Z^2_{eff} - \frac{27}{8} Z_{eff}\right).$$

Treat Z_{eff} as a variation parameter, and calculate a minimum energy in terms of e and a_0. Why is the optimum value of Z_{eff} different from the actual charge number?

***11.48.** Use the wave function for the 1s orbital of the hydrogen atom, given in Table 11.5, to obtain an expression for the probability that the electron lies between the distance r and $r + dr$ from the nucleus. (Use spherical polar coordinates, for which the volume element is $r^2 \, dr \sin\theta \, d\theta \, d\phi$.)

***11.49.** Unsöld's theorem (Section 11.8) states that, for a given value of l, the sum of the values of

$$\sum_{l,m}[\Theta_{l,m}(\theta)\Phi_m(\phi)]^2$$

is independent of θ and ϕ, i.e., is a constant. Write all these functions for the 2p orbitals (see Tables 11.2 and 11.3), and show that their sum shows no angular dependence.

Essay Questions

11.50. With emphasis on the physical significance, explain precisely what is meant by a *normalized* wave function.

11.51. Explain clearly the relationship between the Heisenberg uncertainty principle and the question of whether two operators commute.

11.52. Give an account of the main principles underlying the variation method in quantum mechanics.

11.53. Discuss the reasons for abandoning the Bohr theory of the atom.

SUGGESTED READING

For general treatments of quantum theory and quantum mechanics, see

P. W. Atkins, *Molecular Quantum Mechanics* (2nd ed.), Oxford: Clarendon Press, 1983.

P. W. Atkins, *Quanta: A Handbook of Concepts,* Oxford: Clarendon Press, 1974; 2nd edition, 1991.

J. Baggott, *The Meaning of Quantum Theory: A Guide for Students of Chemistry and Physics,* Oxford University Press, 1992. This book discusses the fundamental ideas of quantum theory and quantum mechanics in an interesting way.

S. Brandt and H. D. Dahmen, *The Picture Book of Quantum Mechanics,* 2nd ed., Springer, 1994.

J. Calais, *Quantum Chemistry Workbook,* New York: Wiley, 1994.

P. A. Cox, *Introduction to Quantum Theory and Atomic Structure.* Oxford University Press, paperback, 1995.

G. Gamow, *Mr. Tompkins in Paperback,* Cambridge University Press, 1993. This is an updated paperback edition of two classics, *Mr. Tompkins in Wonderland* (1940) and *Mr. Tompkins Explores the Atom* (1945). It provides interesting insights into atomic theory and other aspects of physics.

G. Gamow, *Thirty Years that Shook Physics,* New York: Doubleday; Dover Publications, New York, 1985. This book, written for the general reader, is entertaining as well as informative!

Werner Heisenberg, *The Physical Principles of the Quantum Theory,* London: Constable, 1930; Dover reprint, 1949. This small book gives a very clear account of the subject.

W. Heitler, *Elementary Wave Mechanics,* Oxford: Clarendon Press, 1956.

M. Karplus and R. N. Porter, *Atoms and Molecules: An Introduction for Students of Physical Chemistry,* New York: Benjamin, 1970.

I. N. Levine, *Quantum Chemistry* (4th ed.), Englewood Cliffs, NJ: Prentice Hall, 1991.

D. A. McQuarrie, *Quantum Chemistry,* University Science Books, Oxford University Press, 1983.

E. Merzbacher, *Quantum Mechanics,* New York, Wiley, 2nd ed., 1970.

L. Pauling and E. Bright Wilson, *Introduction to Quantum Mechanics, with Applications to Chemistry,* New York: McGraw-Hill, 1935; reprinted by Dover Publications, New York, 1985. Although this book was written so long ago, it is still of great value since many derivations are given explicitly and clearly.

References specifically relating to relativistic effects and quantum mechanics:

K. Balasubramanian, *Relativistic Effects in Chemistry,* New York: Wiley, 1997.

P. Pyykko, "Relativistic effects in structural chemistry," *Chem. Rev., 88,* 563(1988).

For a detailed discussion of energy levels in atomic systems, see

F. L. Pilar, "4s is always above 3d! or, How to tell the orbitals from the wavefunctions," *J. Chem. Ed., 55,* 1978, 2–6.

Some biographies are particularly instructive and interesting:

Walter J. Moore, *Schrödinger: Life and Thought,* Cambridge University Press, 1989.

A. Pais, *Niels Bohr's Times, in Physics, Philosophy and Poetry,* Oxford: Clarendon Press, 1991.

The Chemical Bond 12

PREVIEW

In this chapter we consider, particularly by the methods of quantum mechanics, how atoms join together to form chemical bonds. We will be concerned almost entirely with **covalent bonds,** which arise from the sharing of electrons between atoms. Other kinds of binding between atoms and molecules are dealt with elsewhere in this book; for example, electrostatic attractions between ions in ionic crystals are considered in Chapter 16, and a number of weaker types of bonding in Chapter 17.

The simplest of all molecules is the hydrogen molecule-ion H_2^+ in which two protons are bound together by a single electron. To gain an insight into chemical bonds, in general, it is useful to explore approximate methods in which wave functions for a molecule are constructed from the orbitals for the isolated atoms. One way of doing this is to take a **linear combination of atomic orbitals** (LCAO). Such a combination gives what is called a **molecular orbital** (MO); the variational method, considered in Section 11.13, can be used to obtain the energy of the molecule.

The hydrogen molecule H_2 is the simplest molecule having the most usual type of chemical bond, the two-electron or electron-pair bond. There are two essentially different quantum-mechanical methods for dealing with such bonds. Closely related to the familiar chemical concept of a chemical bond is the **valence-bond method,** in which a wave function for the electron-pair bond is constructed by first taking the product of two atomic wave functions, one for each electron. The other method is the **molecular-orbital method,** which focuses attention not on the electron pair but on the individual electrons. Often molecular orbitals are constructed as linear combinations of atomic orbitals.

The valence-bond method was first used in 1927 by Heitler and London, who recognized that in constructing an orbital for the molecule one must not associate a particular electron with a particular nucleus. They avoided this difficulty by adding or subtracting functions that were related to one another simply by an exchange of electrons. They found that this device leads to an important contribution to the binding energy. This *exchange* energy usually accounts for most of the binding, and it is an energy that is not explained at all by classical mechanics.

The nature of a chemical bond is conveniently considered in terms of the concept of **electronegativity.** A bond may have a dipole moment, which is related to the difference between the electronegativities of the atoms that the bond connects. Certain bonds are hybrids of different kinds of orbitals, and this concept leads to an interpretation of the shapes of molecules.

An important concept in the study of molecules is their symmetry. A consideration of planes of symmetry, centers of symmetry, and other symmetry elements allows decisions to be made as to what types of wave functions are possible for particular molecules. The systematic study of symmetry is known as **group theory,** which leads to useful rules for designating the different types of molecular wave functions.

A good general knowledge of chemical bonding is to be derived from a reading of Sections 12.1, 12.2 and 12.4. Sections 12.3 and 12.6 could be deferred for later reading. Section 12.5 covers symmetry.

OBJECTIVES

After studying this chapter, the student should be able to:

- Write the Hamiltonian operator for a molecular system.
- Appreciate the types of wave functions that are obtained for various chemical bonds.
- Understand the principle of maximum overlap and the significance of Coulombic and exchange integrals.
- Deal with the wave functions for electron spin.
- Understand electronegativity and its relationship to bond dipole moments.
- Appreciate the interpretation of chemical bonding in terms of orbital hybridization.
- Discuss the various symmetry elements and be able to apply them to molecules.
- Understand the fundamental ideas of group theory as applied to molecules.
- Understand the principles of molecular-orbital theory as applied to molecules.

Chemical bonds between atoms are due to Coulombic forces. Although a Coulombic force is a single type of force, it is manifested in various ways, giving different types of bonds. This chapter is mainly concerned with **covalent bonds,** which are bonds found in molecules and in ions containing more than one atom.

For convenience, chemists talk about different forces that hold atoms and molecules together, even though they all have their origin in electrostatic interactions. Table 12.1 gives, for comparison, information about the more important forces that hold atoms and molecules together. The forces between ions, as considered in Chapter 16, are comparable in strength to those in covalent bonds. Also of importance in chemistry are certain weaker attractive and repulsive forces. These are particularly important for an understanding of the structures of liquids, and a discussion of intermolecular forces is to be found in Section 17.3. A bond formed largely as a result of dipole-dipole attractions is the **hydrogen bond,** which is important in the structure of liquid water (see Section 17.3). The **dispersion forces,** or **van der Waals forces,** play an important role in the properties of gases and liquids, and are considered in Section 17.5.

Hydrophobic bonds are of an indirect kind; they occur when nonpolar groups are present in aqueous solution, and they result from the fact that such groups have an effect on the hydrogen-bonded structure of water. (See Section 17.3.)

The Nature of the Covalent Bond

It took considerable time to understand the nature of the covalent bond. As early as 1902 the American chemist G. N. Lewis, who had already made pioneering contributions to thermodynamics (Section 3.8), had developed an important theory of the bond, but he did not publish his ideas until 1916. The essence of his concept was that chemical bonding can take place as a result of the sharing of electrons between

Nature of the bond; Study Table 12.1 in more detail.

TABLE 12.1 The Main Types of Chemical Forces

Type of Force or Chemical Bond	Example	Equilibrium Separation/nm	Dissociation, Energy*/kJ mol^{-1}
Ionic bond (ion-ion force)	$Na^+ \cdots F^-$	0.23	670.0
Covalent bond	H—H	0.074	435.0
Ion-dipole force	$Na^+ \cdots OH_2$ (O bonded to two H)	0.24	84.0
Hydrogen bond (dipole-dipole bond)	$H_2O \cdots H—OH$ (O···H—O)	0.28	20.0
Hydrophobic bond	$CH_2 \cdots H_2C$	≈0.30	≈4.0
Van der Waals (dispersion forces)	Ne···Ne	≈0.33	≈0.25

*This is the energy that would be required per mole to dissociate the species into its units (e.g., H—H into H + H or $Na^+ \cdots OH_2$ into $Na^+ + H_2O$).

Gilbert Newton Lewis (1875–1946)

G. N. Lewis was born in West Newton, Massachusetts, and received much of his early education at home. He went to the University of Nebraska in 1893 but transferred after two years to Harvard where he obtained his B.S. degree in 1896. After a year of teaching at Phillips Academy in Andover he returned to Harvard as a graduate student, obtaining his Ph.D. degree there in 1899. He then spent some time in the laboratories of Walther Nernst at Göttingen and of Wilhelm Ostwald at Leipzig.

After teaching for a period at Harvard he was, in 1905, appointed to a position at the Massachusetts Institute of Technology where he assisted Arthur Amos Noyes (1866–1936) in establishing an important school of research in physical chemistry. In 1912 he was appointed Dean of the College of Chemistry at the University of California at Berkeley where he attracted many chemists of distinction and established an outstanding school of research.

Lewis made contributions of great importance in two branches of physical chemistry—thermodynamics and the theory of the covalent bond. He clarified and expanded Gibbs's system of thermodynamics, for the most part by employing the more explicit methods of van't Hoff. From 1900 to 1907 he introduced and developed the concepts of fugacity and activity (Sections 3.8 and 5.7).

Early in the 20th century Lewis began to develop a theory of the covalent bond (Section 12.4), which he presented to students at Harvard from 1902 on; he was disappointed, however, to receive no support for these ideas from his colleagues. His first paper on the subject did not appear until 1916, and in 1923 he presented it in an important book, *Valence and the Structure of Atoms and Molecules.* In the meantime Irving Langmuir widely publicized Lewis's ideas and somewhat extended them. To Lewis's annoyance the theory was sometimes, especially in Britain, referred to as the Langmuir theory! In addition, Lewis developed a theory of acids and bases according to which a base is an electron donor, an acid an electron acceptor.

Lewis also did important work on fluorescence and phosphorescence, and was the first to recognize that fluorescence often involves a triplet excited state.

Many physical chemists have found it surprising that Lewis never received a Nobel Prize. The reason may be that in its original form Lewis's valence theory postulated electrons stationed at the corners of cubes, which seemed naive to the physicists. However, by the time his 1923 book appeared his treatment was entirely consistent with the physicists' ideas, and was unquestionably of great significance in the later development of valence theory.

Lewis died as a result of a laboratory accident; there was an escape of hydrogen cyanide vapor in one of his experiments.

References: R. E. Kohler, *Dictionary of Scientific Biography, 8,* 289–294(1973).

K. J. Laidler, *The World of Physical Chemistry,* Oxford University Press, 1993.

J. W. Servos, *Physical Chemistry from Ostwald to Pauling: The Making of a Science in America,* Princeton University Press, 1990.

The January, February, and March, 1984, issues of the *Journal of Chemical Education* (Vol. 61) contain many articles about G. N. Lewis.

two atoms. It had previously been realized that groups of eight electrons (octets) play some role in chemical bonding, and Lewis took this idea further. In his original paper he stationed these electrons at the corners of cubes, and suggested that bonding could occur by an overlap of the edges of two cubes. He recognized that in some cases both of the bonding electrons may originally have come from one of the atoms, an arrangement later referred to as a **coordinate link** or a **semi-polar bond.** These ideas of Lewis were later somewhat extended and considerably publicized by the American chemist Irving Langmuir (1881–1957).

Lewis structures; Do an extensive review of Lewis structures.

Lewis's ideas seemed originally to imply that the electrons were held in fixed positions, instead of having mobility, but over the years his theory became suitably modified. The importance of his work was that it provided chemists with a valuable way of visualizing the electronic structures of atoms and molecules, and for practical purposes his ideas are still used today, even by those who are also concerned with the more mathematical aspects of the problem. Lewis's idea that the covalent bond is brought about by the pairing of electrons shared by two atoms has been fully justified by the quantum-mechanical treatments: Lewis's pairs of electrons are pairs of electrons with antiparallel spin. (The Lewis dot structure of bonding is discussed in all introductory chemistry courses and is not reviewed here.)

The remainder of this chapter is concerned with the quantum mechanics of the covalent bond. It is convenient to begin with the one-electron bond found in the hydrogen molecular-ion H_2^+.

12.1 The Hydrogen Molecular-Ion, H_2^+

The simplest molecular system is H_2^+, which consists of two protons and a single electron. This species forms no stable salts, but there is evidence for its existence in electrical discharges passed through hydrogen gas, and some of its properties have been studied by spectroscopic methods. The experimental potential-energy curve for H_2^+ is shown in Figure 12.1a. The minimum energy occurs at an internuclear separation of 106 pm and corresponds to an energy 269.3 kJ mol^{-1} (= 2.791 eV) below that of the separate particles H + H$^+$.

It is possible to obtain a satisfactory quantum-mechanical solution for the H_2^+ ion. The system of two protons and an electron is shown in Figure 12.1b, and we make use of confocal-elliptical coordinates in order to effect a separation of variables (see Figure 12.2). The problem is simplified by the fact that the electron moves much more rapidly than the nuclei. The assumption may therefore be made that the nuclei are held fixed in position during the period of the electron's motion. This is known as the **Born-Oppenheimer approximation,** and it permits the calculation of both the stationary-state wave functions and the energy of the electron. The energy is calculated with a fixed r_{AB}; then new fixed values of r_{AB} are chosen and the corresponding energies are determined. The potential energy E_p at different values of r_{AB}

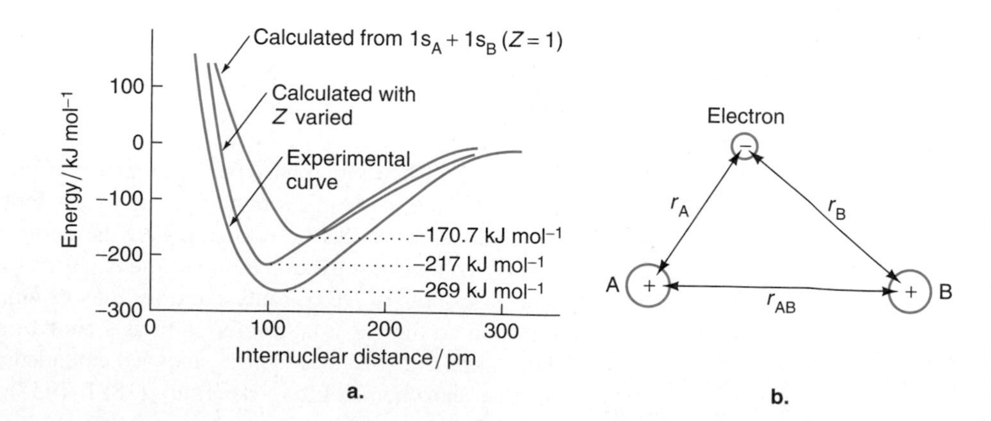

FIGURE 12.1
The hydrogen molecule-ion H_2^+. (a) Experimental and theoretical potential-energy curves for the ground electronic state. (b) The system of two protons and an electron.

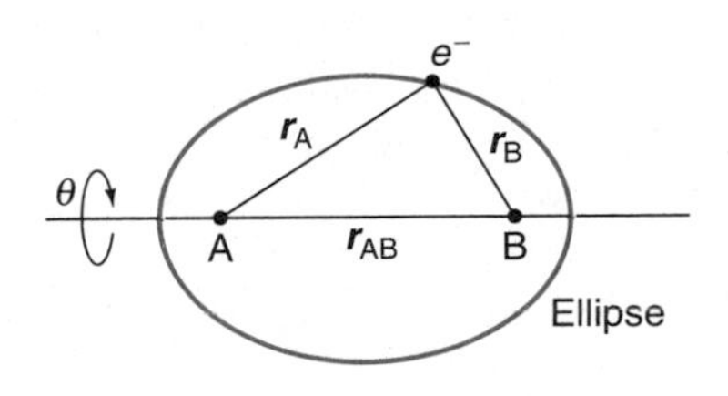

FIGURE 12.2
Confocal elliptical coordinate system for H_2^+.

is then plotted against r_{AB} to obtain the **potential-energy curve** of the system. It is possible to obtain an exact quantum-mechanical solution for this system, but instead we will consider a much simpler solution based on the variational principle.

The system shown in Figure 12.1b consists of two electron-proton attractions and a proton-proton repulsion, and its potential energy is

$$\hat{E}_p = \frac{e^2}{4\pi\epsilon_0}\left(\frac{1}{r_{AB}} - \frac{1}{r_A} - \frac{1}{r_B}\right) \tag{12.1}$$

The Hamiltonian operator for the system is, therefore, from Table 11.1,

$$\hat{H} = -\frac{\hbar^2}{2\mu}\nabla^2 + \frac{e^2}{4\pi\epsilon_0}\left(\frac{1}{r_{AB}} - \frac{1}{r_A} - \frac{1}{r_B}\right) \tag{12.2}$$

where μ is the reduced mass. If the Born-Oppenheimer approximation were not used, the kinetic energy would have two extra terms $-\frac{\hbar^2}{2\mu}(\nabla_A^2 + \nabla_B^2)$ and the internuclear separation r_{AB} would vary. Equation 12.2 treats only the motion of the one electron, and the wave function is that for one electron in H_2^+. The Schrödinger equation to be solved is therefore

$$\left[-\frac{\hbar^2}{2\mu}\nabla^2 + \frac{e^2}{4\pi\epsilon_0}\left(\frac{1}{r_{AB}} - \frac{1}{r_A} - \frac{1}{r_B}\right)\right]\psi = E\psi \tag{12.3}$$

To obtain the best energy we choose some suitable trial eigenfunction ψ that displays the generally expected shape of the "true" function. One function used was

$$\psi = e^{-b\xi}(1 + c\eta^2) \tag{12.4}$$

where ξ and η are the confocal-elliptic variables and b and c are constants adjusted to minimize the energy by the variational method. (See Figure 12.2). The energy is calculated from Eq. 11.247 and, according to the variational principle, the calculated energy cannot be below the experimental value. This function gave a dissociation energy of 268.8 kJ mol^{-1} (2.786 eV), close to the experimental value, but the internuclear distance is 206 pm, which is in considerable error.

Not many problems are simple enough for arbitrary trial variation functions to be used. A more common procedure is to construct trial functions for the molecule from the exact wave functions that apply to the atoms from which the molecule is formed. This approach is an approximation, but the atomic wave functions provide a useful starting point.

To deal with H_2^+ in this way we start with the 1s wave function for the hydrogen atom, given in Table 11.5, p. 543. We will write the wave function for the electron bound to proton A as $1s_A$, and the corresponding wave function for the electron associated with nucleus B as $1s_B$:

$$1s_A = \frac{1}{\sqrt{\pi}}\left(\frac{1}{a_0}\right)^{3/2} e^{-r_A/a_0} \quad \text{and} \quad 1s_B = \frac{1}{\sqrt{\pi}}\left(\frac{1}{a_0}\right)^{3/2} e^{-r_B/a_0} \tag{12.5}$$

These 1s functions are unsatisfactory in themselves because they do not properly describe the bonding of the molecule-ion.

We then form a linear combination of the atomic orbitals (LCAO), which when normalized gives the wave function

$$\psi = \frac{1}{\sqrt{2(1+S)}}(1s_A + 1s_B) \tag{12.6}$$

The term S is the overlap integral and its significance is developed later in Section 12.2 beginning at Eq. 12.35. A wave function such as ψ that is spread over the entire molecule is called a **molecular orbital** (MO). The distribution of the electron throughout the molecule is proportional to ψ^2 and therefore to

$$(1s_A + 1s_B)^2 = (1s_A)^2 + (1s_B)^2 + 2(1s_A)(1s_B) \tag{12.7}$$

When the electron is near proton A, the wave function resembles $1s_A$, the wave function on the isolated hydrogen atom, and the contribution from $1s_B$ is small; when the electron is near proton B, the wave function is approximately $1s_B$. Therefore, when the electron is near A or B, the distribution of electron density is similar to that in the isolated atom. In the region between the nuclei the probability of finding the electron is not simply the sum of $(1s_A)^2$ and $(1s_B)^2$ but is enhanced by the term $2(1s_A)(1s_B)$, and this accounts for the binding.

It is often stated that the accumulation of electron density between nuclei lowers the energy of the molecule and therefore accounts for its stability. This may be the case for more complicated molecules, but for the H_2^+ species detailed calculations have shown that the shifting of the electron away from its position on either nucleus into the internuclear region *raises* its potential energy. A more satisfactory explanation of the bonding is that, coincident with the electron shift into the internuclear region, the atomic orbitals shrink closer to their respective nuclei. This causes an increase in the electron-nuclear attractions and thus a lowering of the potential energy, which more than makes up for the loss in space-filling character of the orbitals. Although the kinetic energy is also changed, the overall effect is that the electron-nuclear attractions dominate.

The wave function of the bonding and lowest-energy MO in Eq. 12.6 is not spherically symmetrical like the atomic s orbital; instead it has a cylindrical symmetry about the internuclear axis. Such an MO is called a **σ (sigma) orbital** because of its similarity to the symmetry of the s atomic orbital. The bond formed by such a σ orbital is called a **σ bond.**

Sigma Bond

The variational method is used to calculate the energy of this orbital σ from Eq. 11.247:

$$E = \frac{\int \psi^* \hat{H} \psi \, d\tau}{\int \psi^* \psi \, d\tau} \tag{12.8}$$

When the integrals are evaluated and the energies calculated for various internuclear distances, the results are as shown in Figure 12.1a. As expected from the variational principle, the calculated energies are all higher than the experimental energies. The theory certainly gives a curve of the right form, but the calculated energies are not accurate. For example, the calculated dissociation energy is 170.7 kJ mol^{-1} as compared with the experimental value of 269.3 kJ mol^{-1}. The calculated internuclear distance for this minimum is 132 pm.

The function in Eq. 12.6 can then be varied to obtain better agreement with experiment. The nuclear charge may be used as a variable in order to obtain the lowest possible energy at each internuclear separation. In other words, instead of Eq. 12.6, the new trial eigenfunction is

$$\psi = \frac{1}{\sqrt{\pi}} \left(\frac{Z}{a_0}\right)^{3/2} e^{-Zr_A/a_0} + \frac{1}{\sqrt{\pi}} \left(\frac{Z}{a_0}\right)^{3/2} e^{-Zr_B/a_0} \tag{12.9}$$

and Z is varied after each new energy is calculated from Eq. 12.8. The best value for Z is 1.23, and with this value the calculated potential-energy curve is much closer to the experimental curve, as shown in Figure 12.1a. The calculated dissociation energy now is 217.0 kJ mol^{-1}, much closer to the experimental value of 269.3 kJ mol^{-1}.

With the advent of large computer facilities, it is now possible to include contributions from the complete set of atomic orbitals; each orbital has the proper weighting factor determined by the variational principle.

Although the results of this simple method are not in good agreement with the experimental energy, it does include the salient features responsible for bonding. It is also possible to improve the agreement and to give better values for molecular properties; we will later see how this can be done.

12.2 The Hydrogen Molecule

Almost every covalent bond consists of a pair of electrons that hold two nuclei together. The hydrogen molecule H_2 is the simplest molecule in which there is an *electron-pair bond.* Many calculations have been made for this molecule, which is a prototype for all covalent bonds. There are two basic quantum-mechanical treatments of the hydrogen molecule. One of them, the **valence-bond (VB) method,** considers the *two-electron (electron-pair)* bond and constructs a wave function for it by first taking the product of two atomic wave functions, one for each electron. The other treatment, the **molecular-orbital (MO) method,** is primarily concerned with the orbitals for single electrons and is based on the type of treatment we have given for the H_2^+ ion. Molecular orbitals are often constructed as linear combinations of atomic orbitals (LCAO), and electrons are fed into the corresponding orbitals with due regard to the Pauli principle.

The Heitler-London Valence-Bond Method

The valence-bond method is based on the familiar chemical ideas of **resonance** and *resonance structures.* This method is favored by some chemists because it gives a more pictorial view of bonding and is closely related to the classical structural theory of organic chemistry.

The system is represented in Figure 12.3a, where the protons are labeled A and B and the electrons are labeled 1 and 2. The potential energy is

$$\hat{E}_p = \frac{e^2}{4\pi\epsilon_0}\left(\frac{1}{r_{AB}} + \frac{1}{r_{12}} - \frac{1}{r_{A1}} - \frac{1}{r_{A2}} - \frac{1}{r_{B1}} - \frac{1}{r_{B2}}\right) \tag{12.10}$$

and the Hamiltonian operator is therefore

$$\begin{aligned}\hat{H} = \left[-\frac{\hbar^2}{2\mu}\nabla_1^2 - \frac{e^2}{4\pi\epsilon_0}\frac{1}{r_{A1}}\right] + \left[-\frac{\hbar^2}{2\mu}\nabla_2^2 - \frac{e^2}{4\pi\epsilon_0}\frac{1}{r_{B2}}\right] \\ + \frac{e^2}{4\pi\epsilon_0}\left[\frac{1}{r_{AB}} + \frac{1}{r_{12}} - \frac{1}{r_{B1}} - \frac{1}{r_{A2}}\right]\end{aligned} \tag{12.11}$$

We initially simplify the problem by ignoring the electrostatic repulsion between the electrons, which is the case if the two hydrogen atoms are infinitely far

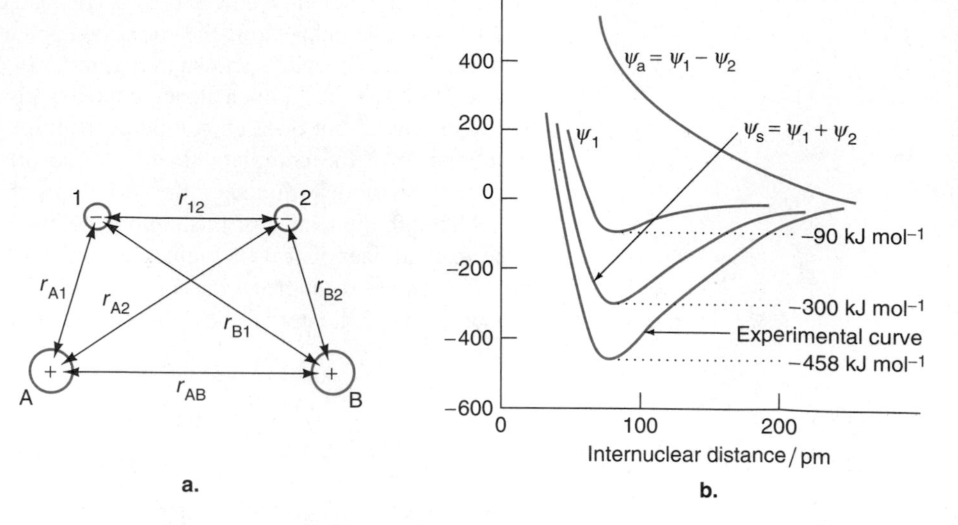

FIGURE 12.3
(a) The hydrogen molecule H_2, consisting of two protons and two electrons. (b) Experimental and theoretical potential-energy curves for the hydrogen molecule.

apart. In that case $1s_A(1)$ is an electron eigenfunction of the first term in brackets in Eq. 12.11, while$1s_B(2)$ is an electron eigenfunction of the second term in brackets. The remaining terms in the third bracket of Eq. 12.11 account for Coulombic attractions and repulsions as the two H atoms are brought closer together. When the two atoms are infinitely far apart, their energy is simply the sum of the two atomic energies, $E_A + E_B$. To meet this requirement, the Schrödinger wave function for this system must be the *product* of the individual wave functions. Thus, a possible wave function is

$$\psi_1 = 1s_A(1)1s_B(2) \tag{12.12}$$

An equally acceptable solution is obtained if electron 2 is associated with nucleus A and electron 1 with nucleus B:

$$\psi_2 = 1s_A(2)1s_B(1) \tag{12.13}$$

The curve ψ_1 in Figure 12.3b is the plot of the energy for the wave function in Eq. 12.12. These wave functions do not include electron spin.

Now, in a long example, we will go through the individual steps necessary to solve the hydrogen molecule. It is important to be able to understand and follow the individual steps, and the student is encouraged to do each calculation.

EXAMPLE 12.1 Set up the Schrödinger equation in a form to solve the hydrogen molecule using Eqs. 12.12 and 12.13 as trial functions.

Solution Using ψ_1 and ψ_2 as trial functions means that the wave function must be a linear combination of the two. Call that function ψ

$$\psi = c_1\psi_1 + c_2\psi_2$$

where c_1 and c_2 are parameters to be evaluated. By using the variational method, the energy E will be found in terms of c_1, c_2, and the energies of the atomic orbitals that form ψ_1 and ψ_2. Substitution into the Schrodinger equation, $\hat{H}\psi = E\psi$, gives

$$\hat{H}\psi = c_1\hat{H}\psi_1 + c_2\hat{H}\psi_2 = c_1E\psi_1 + c_2E\psi_2$$

Solving for E, multiplying by ψ^*, and integrating over both spin coordinates gives a form involving $\psi^*\psi$, but as long as the wave functions are real, the complex conjugate can be dropped. This gives

$$E = \frac{\int\int(c_1\psi_1 + c_2\psi_2)\hat{H}(c_1\psi_1 + c_2\psi_2)d\tau_1 d\tau_2}{\int\int(c_1\psi_1 + c_2\psi_2)^2 d\tau_1 d\tau_2}$$

$$E = \frac{\int\int(c_1\psi_1\hat{H}c_1\psi_1 + c_1\psi_1\hat{H}c_2\psi_2 + c_2\psi_2\hat{H}c_1\psi_1 + c_2\psi_2\hat{H}c_2\psi_2)d\tau_1 d\tau_2}{\int\int(c_1^2\psi_1^2 + 2c_1c_2\psi_1\psi_2 + c_2^2\psi_2^2)d\tau_1 d\tau_2}$$

Although this may look formidable, for real systems

$$\int\int\psi_1\hat{H}\psi_2 d\tau_1 d\tau_2 = \int\int\psi_2\hat{H}\psi_1 d\tau_1 d\tau_2 \qquad \text{(a Hermitian property)}$$

allows us to make several useful abbreviations, making the work comparatively simple. We define H_{ij} as

$$\hat{H}_{ij} = \int\int\psi_i\hat{H}\psi_j d\tau_i d\tau_j$$

and S_{ij} as

$$S_{ij} = \int\int\psi_i\psi_j d\tau_i d\tau_j$$

With these substitutions we have

$$E = \frac{c_1^2H_{11} + 2c_1c_2H_{12} + c_2^2H_{22}}{c_1^2S_{11} + 2c_1c_2S_{12} + c_2^2S_{22}}$$

Although this is the energy, we want the minimum energy, and that is found from the variational method (Section 11.13). First take the partial derivative of E with respect to c_1 and then of E with respect to c_2 to minimize the energy. Since the expression is a division, the calculus requires

$$\frac{\partial E}{\partial c_1} = \frac{(c_1^2S_{11} + 2c_1c_2S_{12} + c_2^2S_{22})(2c_1H_{11} + 2c_2H_{12})}{(c_1^2S_{11} + 2c_1c_2S_{12} + c_2^2S_{22})^2}$$

$$- \frac{(c_1^2H_{11} + 2c_1c_2H_{12} + c_2^2H_{22})(2c_1S_{11} + 2c_2S_{12})}{(c_1^2S_{11} + 2c_1c_2S_{12} + c_2^2S_{22})^2} = 0$$

Multiplying through by the denominator and rearrangement reduces this to

$$(2c_1H_{11} + 2c_2H_{12}) - \frac{(c_1^2H_{11} + 2c_1c_2H_{12} + c_2^2H_{22})(2c_1S_{11} + 2c_2S_{12})}{c_1^2S_{11} + 2c_1c_2S_{12} + c_2^2S_{22}}$$

Substitution of the value of E given three equations earlier and division by 2 give

$$c_1H_{11} + c_2H_{12} = E(c_1S_{11} + c_2S_{12})$$

or

$$c_1(H_{11} - ES_{11}) + c_2(H_{12} - ES_{12}) = 0$$

Similar evaluation of $\partial E/\partial c_2 = 0$ gives

$$c_1H_{21} + c_2H_{22} = E(c_1S_{21} + c_2S_{22})$$

or

$$c_1(H_{21} - ES_{21}) + c_2(H_{22} - ES_{22}) = 0$$

With $H_{12} = H_{21}$, as seen earlier, examination of the answer and its rearrangement for each partial differentiation suggests that these two coupled equations, known as **secular equations,** may be written in matrix form.

$$\begin{pmatrix} H_{11} & H_{12} \\ H_{21} & H_{22} \end{pmatrix}\begin{pmatrix} c_1 \\ c_2 \end{pmatrix} = E\begin{pmatrix} S_{11} & S_{12} \\ S_{21} & S_{22} \end{pmatrix}\begin{pmatrix} c_1 \\ c_2 \end{pmatrix}$$

From linear algebra, a set of simultaneous homogeneous linear equations (as we have from the partial differentiation) has a non-trivial solution only when the determinant of the coefficients vanishes; that is, when the secular equations (written as a determinant) is zero.

$$\begin{vmatrix} H_{11} - ES_{11} & H_{12} - ES_{12} \\ H_{21} - ES_{21} & H_{22} - ES_{22} \end{vmatrix} = 0$$

Further simplification occurs because the trial wave functions are normalized. Hence $S_{11} = S_{22} = 1$. $S_{12} = S_{21}$ and may be written individually as $\int 1s_A(1)1s_B(1)d\tau_1 \int 1s_B(2)1s_A(2)d\tau_2$. Since these integrals are equal (see Eqs. 12.12 and 12.13), we will define each integral as

$$S = \int 1s_A(1)1s_B(1)d\tau_1$$

and call S the **overlap integral.** Making these substitutions gives the **secular determinant** as

$$\begin{vmatrix} H_{11} - E & H_{12} - ES^2 \\ H_{21} - ES^2 & H_{22} - E \end{vmatrix} = 0$$

The term "secular" equation derives from the Latin *saeculum* meaning generation, age, or very long time. It relates to situations that are well defined, such as the elliptical orbit of a planet, but which are affected by small perturbations on the law of force which cause a slight change over a long period of time. Such a change is called a *secular perturbation*. The term is used here because the variational method gives the same answer as the perturbation method taken only to the first approximation.

The matrix elements in the secular determinant in Example 12.1 can be evaluated, but the work is somewhat advanced. We will merely give some salient points of the development. As seen earlier, the first two terms in the Hamiltonian Eq. 12.11 give the eigenvalues of the separated atoms. These are written as $-E_A$ and $-E_B$, respectively. The third term in the Hamiltonian Eq. 12.11 gives rise to two extra terms that become important as the atoms approach each other. The first of these terms arises for the diagonal contribution and is called the **Coulomb integral** J,

Coulomb Integral

$$J = \frac{e^2}{4\pi\epsilon_0}\iint [1s_A(1)]^2\left[\frac{1}{r_{AB}} + \frac{1}{r_{12}} - \frac{1}{r_{B1}} - \frac{1}{r_{A2}}\right][1s_B(2)]^2 d\tau_1 d\tau_2 \tag{12.14}$$

Exchange Integral

The second term arises for the off-diagonal terms and is called the **exchange integral** K,

$$K = \frac{e^2}{4\pi\epsilon_0}\iint 1s_A(1)1s_B(1)\left[\frac{1}{r_{AB}} + \frac{1}{r_{12}} - \frac{1}{r_{B1}} - \frac{1}{r_{A2}}\right]1s_A(2)1s_B(2)d\tau_1 d\tau_2 \tag{12.15}$$

Both the Coulomb and exchange integrals are negative. Substitution of these quantities into the secular determinant in Example 12.1 gives

$$\begin{vmatrix} -E_A - E_B + J - E & -S^2 + K - ES^2 \\ -S^2 + K - ES^2 & -E_A - E_B + J - E \end{vmatrix} = 0 \tag{12.16}$$

The use of atomic units leads to a more compact way to express Eqs. 12.14 to 12.17. On p. 539, the use of Eq. 11.242 shows that for hydrogen, the energy is $-\frac{1}{2}$ in atomic units. Thus in Eqs. 12.16 and 12.17, the values of $-E_A$ and $-E_B$ are each replaced by $-\frac{1}{2}$. Furthermore, by using atomic units the term $e^2/4\pi\epsilon_0$ in Eqs. 12.14 and 12.15 is no longer required.

Application of linear algebra allows us to solve the secular determinant (other examples will be given later) to obtain the eigenvalues and eigenfunctions. The results are

$$E_\pm = -E_A - E_B + \frac{J \pm K}{1 \pm S^2} \tag{12.17}$$

and

$$\psi_\pm = \frac{1}{\sqrt{2(1 \pm S^2)}}(\psi_1 \pm \psi_2) = \frac{1}{\sqrt{2(1 \pm S^2)}}\Big(1s_A(1)1s_B(2) \pm 1s_A(2)1s_B(1)\Big) \tag{12.18}$$

The purpose of this calculation is to arrive at a physical picture for the bonding in the hydrogen molecule using the simplest model. Because the first two terms in Eq. 12.17 involve only A and B, they account for the energy of the two isolated atoms at infinity. The third term accounts for the bonding energy due to the Coulomb integral and the exchange integral. Since both J and K are negative, E_+ is of lower energy than E_-, and E_+ is lower than the energy of the separate H atoms.

The above calculation was first carried out by the German physicists Walter Heinrich Heitler (1904–1981) and Fritz London (1900–1964) in 1927 (just after Schrödinger's equation appeared in 1926) and was the first calculation of molecular energies. What Heitler and London realized was that the starting equations (Eq. 12.12 and 12.13) imply that the electrons are distinguishable, whereas they must be indistinguishable.

Eq. 12.12 implies that one electron can be designated as electron 1 and is particularly associated with a nucleus designated nucleus A; a similar objection applies to the function in Eq. 12.13. Heitler and London realized that this difficulty is avoided if one uses a linear combination of these wave functions as given in Example 12.1. These account for electron repulsion and satisfy the Pauli principle. Note in Eq. 12.18 that ψ_+ is symmetric to electron exchange and ψ_- is antisymmetric. To emphasize this, we write

$$\psi_s = \psi_+ \tag{12.19}$$

and

$$\psi_a = \psi_- \tag{12.20}$$

For a stable bond the distribution of electrons must reduce the repulsion between the two protons. This is accomplished if the electrons are between the nuclei, which implies that the electrons must be close to one another. Consequently,

$$1s_A(1) \approx 1s_A(2) \quad \text{and} \quad 1s_B(1) \approx 1s_B(2)$$

Substitution of these relations into Eq. 12.18 and their addition or subtraction, respectively, lead to

$$\psi_s \approx \frac{2}{\sqrt{2(1 + S^2)}} 1s_A(1)1s_B(2) \quad \text{and} \quad \psi_a \approx 0 \tag{12.21}$$

The description of the system is given by the probabilities ψ_s^2 and ψ_a^2. The probability of finding the electrons close together is, therefore, small with the ψ_a wave function and larger with ψ_s. Plots are shown in Figure 12.3b, where the energy is calculated as a function of r_{AB}, the internuclear distance. The existence of a minimum in the ψ_s curve indicates that a stable molecule is formed; the state is *attractive*. The difference between the zero of energy and the minimum is the *classical binding energy,* or the *classical dissociation energy,* of the molecule.[1] The energy of the state ψ_a is always above the energy of two separated atoms and is therefore called an **antibonding** state; it corresponds to repulsion between the atoms. The energy difference between ψ_1 and ψ_s at the minimum in the curves is called the **resonance energy** or *resonance stabilization energy.* The energy of the electrons is decreased by their being spread out over both nuclei. There is therefore a buildup of electron density between the nuclei, and this may be thought of as an overlapping of the electron clouds originally based on the atoms. This is an example of constructive interference. This semi-quantitative rationale is the basis of the **principle of maximum overlap** first proposed by the American physical chemist Linus Pauling (1901–1994).

Antibonding

Resonance Energy

Principle of Maximum Overlap

When ψ_s is used in the variation technique to calculate the energy, the agreement with experiment is more satisfactory than when either ψ_1 or ψ_2 is used, as seen in Figure 12.3b. There is still much room for improvement, however; the calculated dissociation energy is only 66% of the experimental value. Evaluation of the integrals involving ψ_s, which occur in the variational treatment, leads to two contributions to the energy, the **Coulombic energy *J*** and the **exchange energy *K*,** which are treated in more detail later on p. 593. At the normal internuclear separation both of these energies are negative in value with respect to the energy of the separated atoms. The Coulombic energy accounts for only about 10% of the binding; the remaining 90% is exchange energy. The Coulombic energy is approximately the energy that would be calculated on the basis of electrostatic effects in a purely classical treatment, and such a treatment is therefore completely inadequate. The exchange energy, which accounts for most of the binding, is a purely quantum-mechanical contribution, arising from the interchange of electrons allowed for in the Heitler-London wave function.

Improvements to Heitler-London Treatment

The Heitler-London treatment can be improved in various ways. For example, we can add terms corresponding to ionic states, which were completely neglected in the preceding treatment. If both electrons are associated with nucleus A, we have the function

$$\psi_3 = 1s_A(1)1s_A(2) \tag{12.22}$$

but if both are associated with B, we have

$$\psi_4 = 1s_B(1)1s_B(2) \tag{12.23}$$

Therefore, a reasonable trial function is

$$\psi = c_1(\psi_1 + \psi_2) + c_2(\psi_3 + \psi_4) \tag{12.24}$$

where c_1 and c_2 are numbers that can be varied, after the integrals in Eq. 12.8 have been evaluated, to obtain the lowest energy. The number c_1 multiplies both ψ_1 and ψ_2 because these two functions are equivalent. Similarly, ψ_3 and ψ_4 have both been multiplied by c_2, because in the symmetrical H_2 molecule one ionic state cannot be favored over the other. After the variational method is applied, the function ψ leads to better energies than those obtained when only the first two terms are used. Of course, other refinements are possible to give still better agreement with experiment, but when they are made, we lose some of the clarity of the chemical description.

[1] We have seen in Sections 11.1 and 11.6 that there is a *zero-point energy* so that the true dissociation energy is somewhat less than the classical value.

Electron Spin

In the previous treatment, only the coordinate wave functions were treated and electron spin was explicitly left out. Excluding spin, it is found that when the coordinates of the electrons in ψ_s and ψ_a (Eq. 12.18) are interchanged, the wave functions are symmetric and antisymmetric, respectively; that is, when

$$1s_A(1)1s_B(2) + 1s_A(2)1s_B(1) \rightarrow 1s_A(2)1s_B(1) + 1s_A(1)1s_B(2) \quad (12.25)$$

the wave function ψ_s is *left unchanged;* it is said to be **symmetric** with respect to the interchange of electrons. The function ψ_a, however, *changes sign* upon interchange of the electrons:

$$1s_A(1)1s_B(2) - 1s_A(2)1s_B(1) \rightarrow 1s_A(2)1s_B(1) - 1s_A(1)1s_B(2) \quad (12.26)$$

This operation causes the first term on the left-hand side to be the same as the second term on the right-hand side, *with the sign changed.* Similarly, the second term on the left becomes the first term on the right, *with the sign changed.* Such behavior under this operation is said to be **antisymmetric.**

We saw in Section 11.12 that the Pauli exclusion principle places an important restriction on the possible electron spins when more than one electron is present. The most general statement of the principle is that

> **The total wave function of a system (i.e., the product of the orbital and spin wave functions) must be antisymmetric with respect to an interchange of electrons.**

Hence when the spin wave functions are included, the wave function ψ_s must be multiplied by the antisymmetric spin wave function (Eq. 11.234), while the wave function ψ_a must independently be multiplied by each of the three symmetric spin wave functions Eq. 11.235. We are thus led naturally in the first case to the formation of a **singlet wave function**

Singlet Wave Function

$$\psi_4 = \psi_s \frac{1}{\sqrt{2}} [\alpha(1)\beta(2) - \beta(1)\alpha(2)] \quad (12.27)$$

Triplet Wave Function

and in the second case to a **triplet of wave functions**

$$\begin{aligned} \psi_1 &= \psi_a \alpha(1)\alpha(2) \\ \psi_2 &= \psi_a \frac{1}{\sqrt{2}} [\alpha(1)\beta(2) + \beta(1)\alpha(2)] \\ \psi_3 &= \psi_a \beta(1)\beta(2) \end{aligned} \quad (12.28)$$

all of which satisfy the antisymmetrization principle. The three antisymmetric orbitals involve ψ_a, which we have seen leads to repulsion. The singlet state involves ψ_s, giving attraction. More is said about these states in Section 13.2.

The Molecular-Orbital Method

An alternative treatment, particularly useful for large molecules, is the molecular-orbital method. Whereas the valence-bond method was created in the single 1927 paper of Heitler and London, the method of molecular orbitals evolved over a number of years. The first publications dealing with it appeared in the 1920s and were descriptive and somewhat intuitive, being mainly concerned with the interpretation of molecular spectra. More analytical and quantitative methods were later developed, based on the mathematical methods of quantum mechanics.

The main credit for the molecular-orbital method is due to Friedrich Hund and the American chemist and physicist Robert Sanderson Mulliken (1896–1980). Their work on the method was independent in the sense that they never published together, but there was much cooperation between them. Credit must also go to John Edward Lennard-Jones (1894–1954), who in 1929 suggested the important idea of constructing a molecular orbital as a **linear combination of atomic orbitals (LCAO)**. Gerhard Herzberg (1904–1999—see a biographical sketch on page 649) also made important contributions to molecular-orbital theory through his extensive interpretations of molecular spectra (Chapter 13). Also, in 1929 he first explained chemical bonding in a simple and convincing manner in terms of **bonding** and **antibonding electrons,** an antibonding electron counteracting the effect of a bonding electron.

A simple way of looking at the molecular-orbital method is to imagine that in a diatomic molecule AB the two nuclei are brought together, resulting in a "united atom." The treatment of the united atom can then be taken as a first approximation for the AB molecule. Whereas in the valence-bond (Heitler-London) method the electrons are treated in pairs, in the method of molecular orbitals the electrons are imagined to be fed one by one into the molecule.

The molecular orbitals are formed from atomic orbitals, without consideration of the electrons they have to accommodate, and therefore the MOs for the hydrogen molecule are identical to those for the hydrogen molecule-ion. The form of the bonding orbital from Eq. 12.6 is thus $1s_A + 1s_B$, and this is the MO of lowest energy. Figure 12.4a shows the individual atomic orbitals, and Figure 12.4b shows their sum; the value of the MO is plotted along the internuclear axis A–B. This MO corresponds to a high electron density between the nuclei, and the effect is to hold the nuclei together; there is therefore bonding.

The molecular orbitals for a diatomic molecule are eigenfunctions of $\hat{L}_z$ but not of $\hat{L}^2$, and this fact is used in the designation of the orbitals. We again use the quantum numbers that relate to the eigenvalue L_z for the angular momentum along the internuclear axis. The notation is similar to that for atomic orbitals except that Greek letters are now used. For example, the quantum number L becomes λ, as follows:

	$\lambda = 0$	1	2
orbital designation:	σ	π	δ
	sigma	pi	delta

The bonding orbital is thus

$$\sigma = 1s_A + 1s_B \tag{12.29}$$

When one electron is in this orbital, we have the H_2^+ ion in which there is bonding; the energy of the H_2^+ is below that of $H + H^+$. Since the bond is due to one bonding electron, and a normal single bond requires two electrons, we say that the bond order is 0.5. A second electron, with opposite spin, can also go into this orbital, with a consequent increase in bonding; the bond order is now 1.

A second MO can be obtained by subtracting the two atomic orbitals:

$$\sigma^* = 1s_A - 1s_B \tag{12.30}$$

This corresponds to a low electron density between the nuclei, as shown in Figure 12.4c. It corresponds to $\lambda = 0$, and it is also a σ orbital; the fact that it is antibond-

Molecular orbital method; Watch LCAO make MOs.

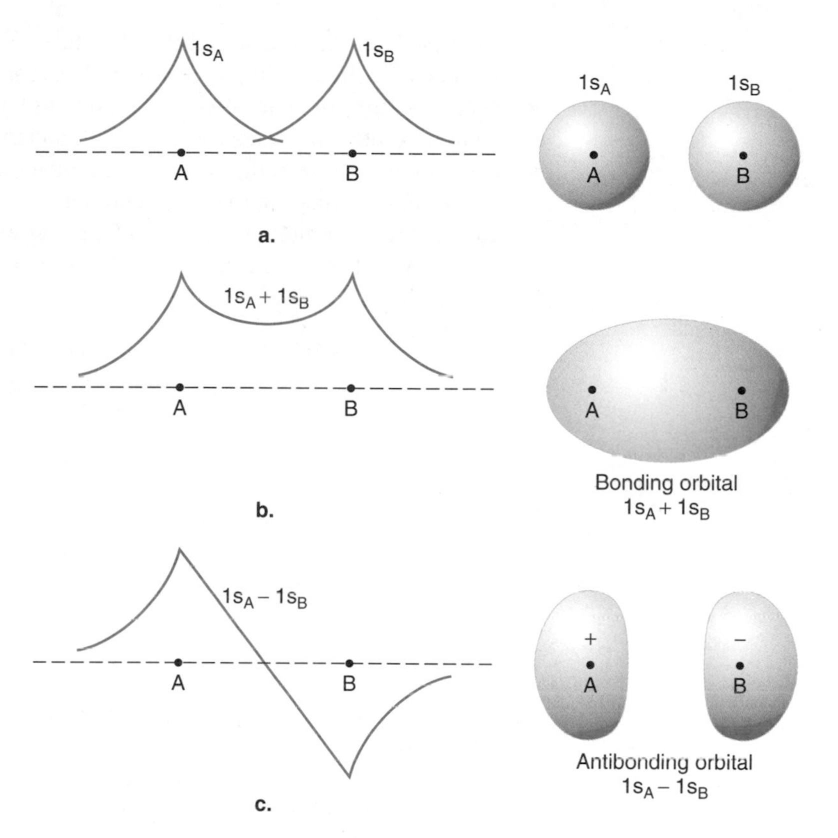

FIGURE 12.4
Wave function plots: To the right are shown contours that represent the general shapes of the orbitals. The plus and minus signs on the orbitals refer to mathematical signs on the wave functions and not to a charge. However, overlap and bond formation can only occur when the signs are the same. When the signs on the orbitals are different, an antibonding orbital occurs. (a) The individual atomic orbitals $1s_A + 1s_B$ for the hydrogen atoms. (b) The bonding orbital $1s_A + 1s_B$. (c) The antibonding orbital $1s_A - 1s_B$.

Molecular orbital method; Orbital energy; See Figure 12.4 and 12.5 animated.

ing is shown by the asterisk. The two wave functions σ and σ^* are orthogonal to each other (see Problem 12.8).

Figure 12.5 is an energy diagram showing the separated atoms and the bonding and antibonding levels. The ground state of H_2 is shown, with two electrons in the bonding orbital. If one of these is promoted into the antibonding σ^* orbital, the antibonding effect approximately overcomes the bonding effect.

The molecular orbital for the electron-pair bond in the hydrogen molecule is obtained by taking the product of the molecular orbitals for each of the two electrons, as given by Eq. 12.29. This description allows the electrons to move freely throughout the molecule. The wave function for electron 1 is written as

$$\sigma(1) = N_b[1s_A(1) + 1s_B(1)] \tag{12.31}$$

and that for electron 2 as

$$\sigma(2) = N_b[1s_A(2) + 1s_B(2)] \tag{12.32}$$

where N_b is the normalization factor. Note that Eqs. 12.31 and 12.32 show that the electrons are equally, or covalently, associated with each nucleus. The trial molecular orbital is, therefore,

$$\sigma = \sigma(1)\sigma(2) = N_b^2[1s_A(1) + 1s_B(1)][1s_A(2) + 1s_B(2)] \tag{12.33}$$

$$= N_b^2[1s_A(1)1s_B(2) + 1s_B(1)1s_A(2) + 1s_A(1)1s_A(2) + 1s_B(1)1s_B(2)] \tag{12.34}$$

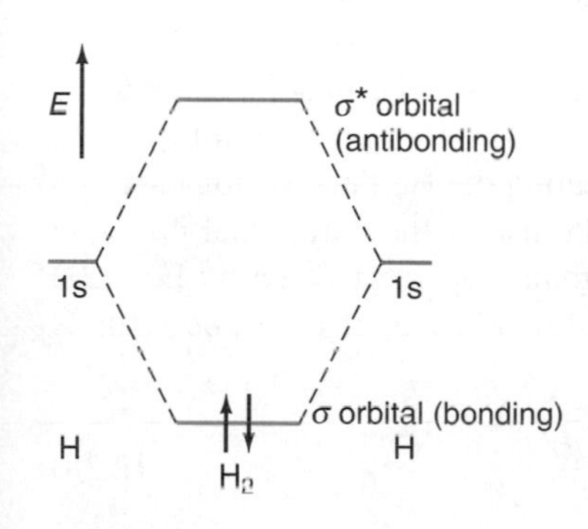

FIGURE 12.5
The energies of bonding and antibonding orbitals for the H_2 molecule. Two electrons, with opposite spins, are shown in the bonding orbital.

The first two terms correspond to the Heitler-London wave function; the other two terms describe ionic states (both electrons residing primarily around the same nucleus). These ionic structures contribute slightly, $\approx 3\%$, to the total energy of the hydrogen molecule. In this simple treatment these forms are weighted equally with the covalent structures. In more advanced treatments a weighting factor is used to reduce the contribution to a more appropriate level, and the results of the MO method then become similar to those of the valence-bond method. In fact, as refinements are made to the two methods, they approach one another.

Molecular orbital method; Bond order; See electrons in MOs.

In order to determine the energy of the system, absolute values of N_b and N_a must be known. These are found from the normalization (Eq. 11.89). The value of N_b is determined as follows:

$$\begin{aligned}\int \sigma(1)^2\, d\tau = 1 &= N_b^2 \int [1s_A(1) + 1s_B(1)]^2\, d\tau \\ &= N_b^2\Bigg[\int 1s_A(1)^2\, d\tau + \int 1s_B(1)^2\, d\tau \\ &\quad + 2\int 1s_A(1)1s_B(1)\, d\tau\Bigg]\end{aligned} \tag{12.35}$$

Since $1s_A(1)$ and $1s_B(1)$ are separately normalized, each of the first two integrals is equal to 1. Again the overlap integral, S, appears (and is defined the same way as in Example 12.1), which represents the degree of overlap of the two orbitals. This is evident from the definition.

$$S = \int 1s_A 1s_B\, d\tau \tag{12.36}$$

The integration is over the coordinates of one electron.

The wave functions extending from nucleus A and nucleus B each have a particular value at any location. If the nuclei are far apart, $1s_A$ is large at a point near A, but $1s_B$ is so small that their product is vanishingly small. At a point near B, the reverse is true for the wave functions, so that their product is again vanishingly small. In these limits, $1s_A$ and $1s_B$ are orthogonal and $S = 0$. As the nuclei approach the equilibrium bond length, S becomes larger. When S is not zero, that is, when r_{AB} is not infinite, $1s_A$ and $1s_B$ are not orthogonal and consequently are only approximations to the proper wave functions of the hydrogen molecule. The integral S is thus a measure of the interpenetration or overlapping of the electron clouds on the nuclei. Equation 12.35 with these substitutions becomes

$$N_b^2(1 + 1 + 2S) = 1 \quad \text{and} \quad N_b = \frac{1}{\sqrt{2(1 + S)}} \tag{12.37}$$

The energy of the orbitals may now be calculated for the bonding orbitals of Eqs. 12.31 and 12.32 using the variational principle. The Hamiltonian in Eq. 12.11 is used in much the same way as described for treating the Heitler-London case and is not included here. Instead, we will illustrate the use of the variational principle for the hydrogen molecular-ion using the Hamiltonian operator given by Eq. 12.2 and the bonding orbital of Eq. 12.31. (Note that Eq. 12.31 is of the same form as Eq. 12.6.) Substitution into Eq. 12.8 gives

$$E = \frac{\int \sigma(1)\hat{H}\sigma(1)\, d\tau}{\int \sigma(1)\sigma(1)\, d\tau} \tag{12.38}$$

Since $\sigma(1)$ is already normalized, we may drop the denominator in Eq. 12.38, and the energy becomes

$$E = \int \sigma(1)\hat{H}\sigma(1)\, d\tau \tag{12.39}$$

Substitution of Eq. 12.31 into Eq. 12.39 gives, for the bonding energy,

$$E_b = N_b^2 \int [1s_A(1) + 1s_B(1)]\hat{H}[1s_A(1) + 1s_B(1)]\, d\tau \tag{12.40}$$

Coulomb Integrals

Electrostatic Energy J'

$$= N_b^2 \int 1s_A(1)\hat{H}1s_A(1)\, d\tau + N_b^2 \int 1s_B(1)\hat{H}1s_B(1)\, d\tau$$

$$+2N_b^2 \int 1s_A(1)\hat{H}1s_B(1)\, d\tau \equiv N_b^2\,(2J' + 2K') \tag{12.41}$$

Resonance or Exchange Integral

Exchange Energy K'

The first two integrals are equal since $1s_A$ and $1s_B$ are identical orbitals. They are the *Coulomb integrals,* equal to the **electrostatic energy** J' contributing to the bond. The last integral is the *resonance* or *exchange integral,* which gives the **exchange energy** K'. Comparison of J' and K' with Eq. 12.14 and 12.15 shows that their definitions differ in that in the valence bond method, the integrals are over both electrons, while in Eq. 12.40 they are over one electron. Substitution of J' and K' for the integrals and the value of N_b from Eq. 12.37 gives

$$E_b = \frac{1}{2(1 + S)}(2J' + 2K') = \frac{J' + K'}{1 + S} \tag{12.42}$$

A further simple approximation is to neglect the overlap integral S; then the energy of the bonding orbital becomes

$$E_b = J' + K' \tag{12.43}$$

Since K' is a negative quantity, the energy of the molecular orbital lies *below* the energy of the individual atoms, as shown in Figure 12.5. Compare Eq. 12.42 with the valence-bond energy, Eq. 12.17.

Symmetric (g) and Antisymmetric (u)

Symmetry requires that the molecular wave function is symmetric or antisymmetric under the interchange of nuclei. For the hydrogen molecule this is equivalent to an *inversion* of spatial coordinates through a center of symmetry. Inversion means that a straight line is drawn from every element of space through the center of symmetry and is continued for the same distance beyond the center. A wave function that retains its sign after inversion is said to be **symmetric** and is labeled g (German *gerade,* even). One that changes its sign on inversion is said to be **antisymmetric** and is labeled u (German *ungerade,* odd). The bonding orbital may then be written as

$$\sigma_g(1) = N_b[1s_A(1) + 1s_B(1)] \tag{12.44}$$

For the antibonding orbital, we have

$$\sigma_u^*(1) = N_a[1s_A(1) - 1s_B(1)] \tag{12.45}$$

where $N_a = 1/\sqrt{2(1 - S)}$. The energy E_a of the antibonding orbital is thus

$$E_a = \frac{J' - K'}{1 - S} \tag{12.46}$$

Its approximate value, if S is zero, is

$$E_a = J' - K' \tag{12.47}$$

Again K' is a negative quantity, which explains why the energy of σ_u^* lies above the value of J'. As long as some overlap S exists, the difference between the energy is J' greater than the difference between J' and the energy of the bound state. We will later see that this accounts for the instability of some molecules.

12.3 Hückel Theory for More Complex Molecules

More will be said about molecular orbitals in Section 12.6, after the idea of orbital symmetry has been introduced. For more complex molecular systems the molecular-orbital method is often more convenient than the valence-bond method. Particular success with benzene and other aromatic molecules was achieved using the molecular-orbital approach by the German physicist Erich Hückel (1896–1980), who had already achieved fame with his collaboration with Peter Debye on the theory of strong electrolytes (Section 7.4). In a series of papers published from 1930 on, Hückel achieved a great clarification of the chemical behavior of more complicated chemical systems. It was he who suggested the idea of σ and π bonds, although it was Hund who suggested that particular notation.

As described by **Hückel theory,** a σ bond is a localized buildup of electron density between atoms, while in conjugated π systems, such as aromatic compounds, a π bond results from a delocalization of electrons over the whole molecule. The treatment of hydrogen as H_2^+ and H_2, as presented in Sections 12.1 and 12.2, is relatively simple since only two electrons are involved and the secular determinant is a 2×2. For more complicated cases, in particular, chains and rings of carbon-containing molecules, the dimension of the secular determinant increases. To handle such cases and to enable calculations to be more easily performed, Hückel made a number of approximations. We will first apply his approximations to H_2^+ and H_2 and then describe more complicated cases.

Hückel assumed that all terms involving the overlap integral, S_{ij}, can be simplified. If $i = j$, then for normalized atomic orbitals

$$S_{ii} = \int \psi_i \psi_i \, d\tau = \int \psi_i^2 d\tau = 1 \tag{12.48}$$

When $i \neq j$,

$$S_{ij} = \int \psi_i \psi_j \, d\tau = 0 \tag{12.49}$$

and the wave functions are orthogonal. In other words, the wave functions are independent, and being centered on widely separated atoms, there is no interaction of the electrons. However, when the atoms on which the wave functions are based become closer, the value of S_{ij} can make a contribution. The magnitude of S_{ij} is a measure of how much the orbitals on atoms i and j overlap or interact. In this sense S_{ij} is an *overlap integral*. Generally, setting $S_{ij} = 0$ introduces error, but it considerably simplifies the calculations. Hückel also simplified how the Coulomb integrals needed to be handled. He did not distinguish between the Coulomb integrals based on different atoms of the same type; that is, he set

$$H_{ii} = \int \psi_i \hat{H} \psi_i \, d\tau = \langle \psi_i | \hat{H} | \psi_i \rangle = \alpha \tag{12.50}$$

where ψ_i represents the wave function for each electron and α is a function of nuclear charge and type of orbital of the electron. Finally, he set equal to zero all exchange, or resonance, integrals except for those between adjacent atoms. That is, he set

$$H_{ij} = \int \psi_i \hat{H} \psi_j d\tau = \langle \psi_i | \hat{H} | \psi_j \rangle = \beta \tag{12.51}$$

for adjacent atoms; otherwise, he set $H_{ij} = 0$ $(i \neq j)$ for all atoms that are not at the usual bond-forming distance. Note, however, that β is also a function of the degree of overlap of the orbitals. This means that β depends on the distance and, except for s orbitals, the angles at which the orbitals are set with respect to the line between nuclei. Then, if all of these approximations are applied to the secular determinant in Example 12.1, it becomes

$$\begin{vmatrix} \alpha - E & \beta \\ \beta & \alpha - E \end{vmatrix} = 0 \tag{12.52}$$

The determinant is evaluated by multiplying the diagonal elements together and subtracting the square of the off-diagonal elements to give

$$\alpha^2 - 2\alpha E + E^2 - \beta^2 = 0 \tag{12.53}$$

This is rearranged as a quadratic equation in E and solved. The answers are either

$$E = \alpha + \beta \qquad \text{or} \qquad E = \alpha - \beta \tag{12.54}$$

which show that there are two possible energy levels.

We now must determine the wave function corresponding to each energy state. From Example 12.1 we found the secular equations to be

$$c_1(\alpha - E) + c_2\beta = 0 \qquad \text{and} \qquad c_1\beta + c_2(\alpha - E) = 0 \tag{12.55}$$

Making substitution of $E = \alpha + \beta$ in Equation 12.55 allows us to find the ratio of c_1/c_2 as

$$c_1/c_2 = -\frac{\beta}{\alpha - E} = -\frac{\beta}{-\beta} = 1 \tag{12.56}$$

Thus the original wave function for the energy level $E = \alpha + \beta$ would be $\psi = \psi_1 + \psi_2$, but to represent the magnitude of the electron density the wave function must be normalized.

$$\int \psi^2 d\tau = \int (\psi_1 + \psi_2) d\tau = \int \psi_1^2 d\tau + \int \psi_2^2 d\tau + 2\int \psi_1\psi_2 d\tau \tag{12.57}$$

Assuming that ψ_1 and ψ_2 are individually normalized and mutally orthogonal wave functions, we have

$$\int \psi^2 d\tau = 1 + 1 + 0 = 2 \tag{12.58}$$

Because this is not equal to 1, ψ is not normalized. It can be made so by multiplying by $1/\sqrt{2}$, the normalization factor. Thus the complete wave function is

$$\Psi_1 = (1/\sqrt{2})(\psi_1 + \psi_2). \tag{12.59}$$

For the other combination of c_1 and c_2, the value is

$$\Psi_2 = (1/\sqrt{2})(\psi_1 - \psi_2). \tag{12.60}$$

The electron distribution for these wave functions has already been shown in Figure 12.4, where Ψ_1 corresponds to a bonding orbital and Ψ_2 corresponds to an antibonding orbital. For H_2^+ the energy levels may be represented as in Figure 12.5 but with only one electron shown. Of course, for H_2 the representation is the same. However, note that because of the interelectronic repulsion terms, the values for α and β change, and are not the same for species with different numbers of electrons even if the molecular orbitals are similar.

The importance of the Hückel theory is that the treatment that was just used to describe H_2^+ and H_2 can be applied to organic molecules of some complexity by making the approximation that electrons in most of the bonds are "localized." This means that these electrons do not contribute greatly to the electronic character of other bonds in the molecule. First Hückel treated the localized σ electrons separately from the delocalized π electrons. For example, in ethylene (C_2H_4), the σ framework is sp^2 hybridized, leaving two $2p_z$ orbitals into which two electrons must go. In the case of ethylene we are taking each bond as a localized molecular orbital similar to the one in H_2^+ in that both π and σ bonds are treated as localized. Here we focus only on the different kinds of atomic orbitals involved. This works for ethylene and many other organic compounds, giving meaningful results, because in most reactions of simple systems the bonds are essentially independent of each other.

Ethene (Ethylene)

In ethylene (C_2H_4) we focus on the π electrons, assuming that the σ-bond framework has standard properties. To the right in Figure 12.6 a and b are shown the σ-framework and p orbitals that result in the double bond. Since there are only two electrons in this system, ethylene becomes a two-orbital problem like H_2. The mathematical operations are the same as just completed where here $\Psi_{\pi \text{ electron}} = c_1\psi_1 \pm c_2\psi_2$. The solutions are $E_1 = \alpha + \beta$ and $E_2 = \alpha - \beta$, shown on the left of Figure 12.6 a and b.

The wave functions are shown in Figure 12.6. The antibonding orbital has a node in the plane of the figure and changes sign on passing through the node. The eigenvalues are $\alpha \pm \beta$ for the two levels so that the resonance stability of the two π electrons is $2(\alpha + \beta)$ (since $\beta < 0$, this is the lowest energy and therefore each electron has energy $\alpha + \beta$), as shown in Figure 12.6c.

Butadiene

In the case of butadiene (C_4H_8), we consider the four π electrons (one based on each of the four carbons) so that π overlap may occur over four parallel 2p orbitals. The wave function is then written as

$$\Psi = c_1\psi_1 + c_2\psi_2 + c_3\psi_3 + c_4\psi_4 \tag{12.61}$$

and the individual ψs are again assumed normalized. The possible energies correspond to the roots of the secular determinant. We have already seen how to make

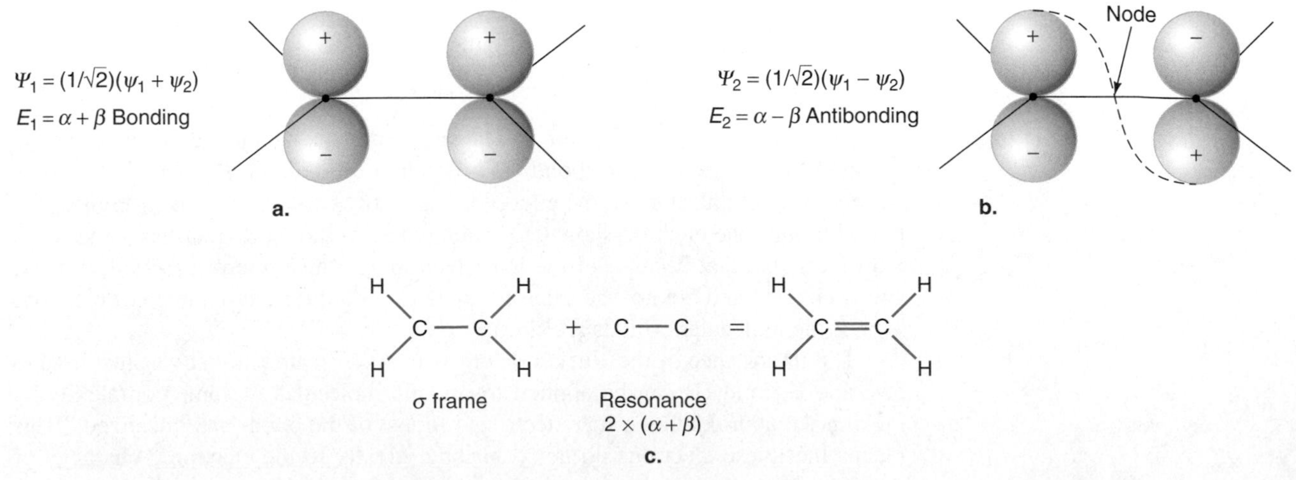

FIGURE 12.6
The energies, wave functions, and orbital representaions for ethylene, showing (a) bonding and (b) antibonding orbitals. (c) The occurrence of resonance in ethene. At the left is the σ-framework of the molecule. To the right the Cs are repeated with the resonance contribution.

the simplifying assumptions of $S_{ij}(i \neq j) = 0$ and even though the surroundings of the carbons are different, we assume that each wave function is similar to the others. Furthermore, $H_{ii} = \alpha_i$, where $\alpha_1 = \alpha_4$ and $\alpha_2 = \alpha_3$ exactly, and for expediency of calculation, we set α equal to each of the α_i s because they are similar. Furthermore, $H_{ij} = \beta_{ij}$ where for adjacent atoms $\beta_{12} = \beta_{23} = \beta_{34} = \beta$ and for nonadjacent atoms $\beta_{13} = \beta_{14} = \beta_{24} = 0$. Making these substitutions, the secular determinant to be evaluated is

$$\begin{bmatrix} \alpha - E & \beta & 0 & 0 \\ \beta & \alpha - E & \beta & 0 \\ 0 & \beta & \alpha - E & \beta \\ 0 & 0 & \beta & \alpha - E \end{bmatrix} = 0 \tag{12.62}$$

To simplify, we divide through by β and then let $(\alpha - E)/\beta = x$.

EXAMPLE 12.2 Evaluate the determinantal equation derived from Eq. 12.62.

$$\begin{vmatrix} x & 1 & 0 & 0 \\ 1 & x & 1 & 0 \\ 0 & 1 & x & 1 \\ 0 & 0 & 1 & x \end{vmatrix} = 0$$

Solution Evaluate the determinant by expanding about either the first row or the first column, whichever appears to be simpler. We'll pick the first column of 4 terms. The cofactor then is the determinant with the first column and nth row removed. Now multiply each term by the corresponding (nth) cofactor with a + sign for the product when n is odd and a − sign when n is even. The result is

$$x \begin{vmatrix} x & 1 & 0 \\ 1 & x & 1 \\ 0 & 1 & x \end{vmatrix} - 1 \begin{vmatrix} 1 & 0 & 0 \\ 1 & x & 1 \\ 0 & 1 & x \end{vmatrix} + 0 \begin{vmatrix} 1 & 0 & 0 \\ x & 1 & 0 \\ 0 & 1 & x \end{vmatrix} - 0 \begin{vmatrix} 1 & 0 & 0 \\ x & 1 & 0 \\ 1 & x & 1 \end{vmatrix} = 0$$

Now expand about the first column of each of the 3 × 3 determinants after dropping the zero terms.

$$x(x) \begin{vmatrix} x & 1 \\ 1 & \text{x} \end{vmatrix} + x(-1) \begin{vmatrix} 1 & 0 \\ 1 & x \end{vmatrix} - 1(1) \begin{vmatrix} x & 1 \\ 1 & x \end{vmatrix} - 1(-1) \begin{vmatrix} 0 & 0 \\ 1 & x \end{vmatrix} = 0$$

The two-row determinants yield

$$x^2(x^2 - 1) + x(-1)(x) - 1(x^2 - 1) + 0 = x^4 - 3x^2 + 1 = 0$$

Using the quadratic equation on the last equality yields

$$x = \pm\sqrt{\frac{3 \pm \sqrt{9 - 4}}{2}} = \pm 1.6180, \pm 0.6180$$

Thus four values are obtained as solutions to the equation, as expected for a fourth-degree equation.

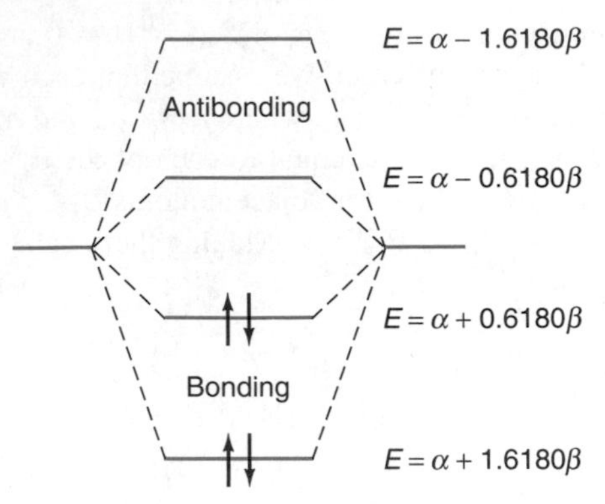

FIGURE 12.7
Energy levels of butadiene according to Hückel theory.

The energy levels are found in the form $E = \alpha \pm \beta$ from the eigenvalues in Example 12.2, along with $x = (\alpha - E)/\beta$. These are shown in Figure 12.7 along with the numerical values.

The wave functions of butadiene depend upon the energy level. We proceed to calculate them as we did for hydrogen by finding the ratio of cs. First find the ratios of c_n/c_1 using $+$ for the ratio: cofactor (n)/cofactor (1) when n is odd and a $-$ sign before the ratio when n is odd. For instance $c_3/c_1 = 1/(x^2 - 2)$ with the normalization constant of $1/\sqrt{\Sigma(c_n/c_1)^2} = 1/1.6625$. The final wave functions are:

$$
\begin{aligned}
\Psi_1 &= 0.3717\,\psi_1 + 0.6015\psi_2 + 0.6015\,\psi_3 + 0.3717\,\psi_4 \\
\Psi_2 &= 0.6015\,\psi_1 + 0.3717\psi_2 - 0.3717\,\psi_3 - 0.6015\,\psi_4 \\
\Psi_3 &= 0.6015\,\psi_1 - 0.3717\psi_2 - 0.3717\,\psi_3 + 0.6015\,\psi_4 \\
\Psi_4 &= 0.3717\,\psi_1 - 0.6015\psi_2 + 0.6015\,\psi_3 - 0.3717\,\psi_4
\end{aligned}
\tag{12.63}
$$

These functions are drawn in Figure 12.8. Notice that wherever the wave function changes sign between nuclei, a node results and the energy of the orbital increases as the number of nodes increases. The lowest energy orbital has no node and is totally bonding. Such calculations lend much to the understanding of bonding.

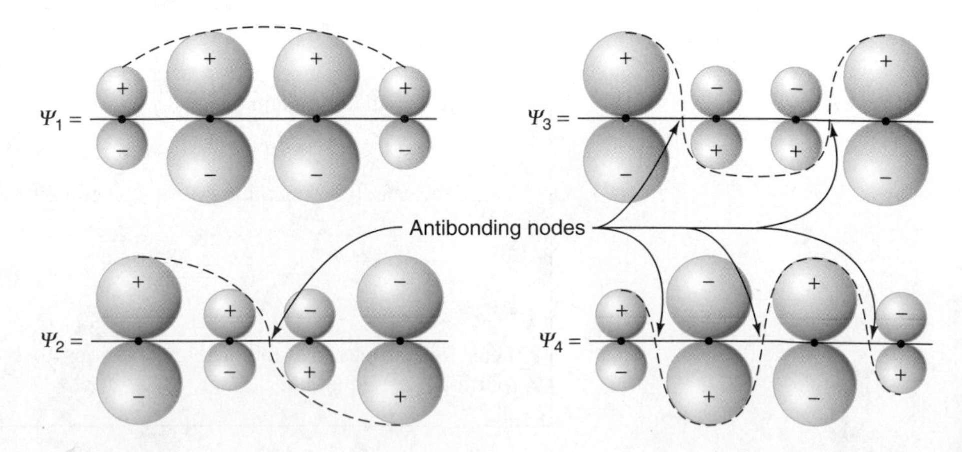

FIGURE 12.8
The four wave functions of butadiene using the Hückel approximations.

Benzene

For benzene (C_6H_6), the Hückel approximation for the six $2p_z$ orbitals is illustrated in Figure 12.9a and the secular detrminant is

$$\begin{vmatrix} \alpha - E & \beta & 0 & 0 & 0 & \beta \\ \beta & \alpha - E & \beta & 0 & 0 & 0 \\ 0 & \beta & \alpha - E & \beta & 0 & 0 \\ 0 & 0 & \beta & \alpha - E & \beta & 0 \\ 0 & 0 & 0 & \beta & \alpha - E & \beta \\ \beta & 0 & 0 & 0 & \beta & \alpha - E \end{vmatrix} - 0 \tag{12.64}$$

which shows how only the nearest-neighbor resonance integrals are retained. This 6×6 determinant has six eigenvalues. Placing the six electrons in the lowest energy levels results in a stabilization of

$$E_{\pi}\,(\text{benzene}) = 6\alpha + 8\beta. \tag{12.65}$$

On the other hand, if we consider benzene to be composed of three single and three double bonds as shown in Figure 12.9b, the secular determinant becomes

$$\begin{vmatrix} \alpha - E & \beta & 0 & 0 & 0 & 0 \\ \beta & \alpha - E & 0 & 0 & 0 & 0 \\ 0 & 0 & \alpha - E & \beta & 0 & 0 \\ 0 & 0 & \beta & \alpha - E & 0 & 0 \\ 0 & 0 & 0 & 0 & \alpha - E & \beta \\ 0 & 0 & 0 & 0 & \beta & \alpha - E \end{vmatrix} = 0 \tag{12.66}$$

Resonance Stability

which is simply the sum of three 2×2 determinants like ethylene, with triply degenerate eigenvalues of $\alpha \pm \beta$. This would give a **resonance stability** for the six π electrons of only $6(\alpha + \beta)$. Comparing the model for benzene resonance with that of three ethylene molecules, gives a difference of stability of

$$E_{\pi}(\text{benzene}) - 3E_{\pi}(\text{ethylene}) = 6\alpha + 8\beta - 6(\alpha + \beta) = 2\beta \tag{12.67}$$

which is the delocalization energy for benzene. For a number of conjugated and aromatic organic molecules, the accepted value for β is -75 kJ, mol^{-1}. The Hückel approximation therefore predicts a resonance stability for benzene, Figure 12.9a,

σ frame + Resonance $6\alpha + 8\beta$ = (a.)

σ frame + 3-ethylene resonances $3(\alpha + \beta)$ = (b.)

FIGURE 12.9
Resonance stability of benzene using Eq. 12.64. (a) The circle represents inclusion of the β terms in Eq. 12.64. (b) Benzene is considered as though it were formed from three ethylene molecules. Some resonance energy is not accounted for in this process. Compare Eq. 12.66 with Eq. 12.64.

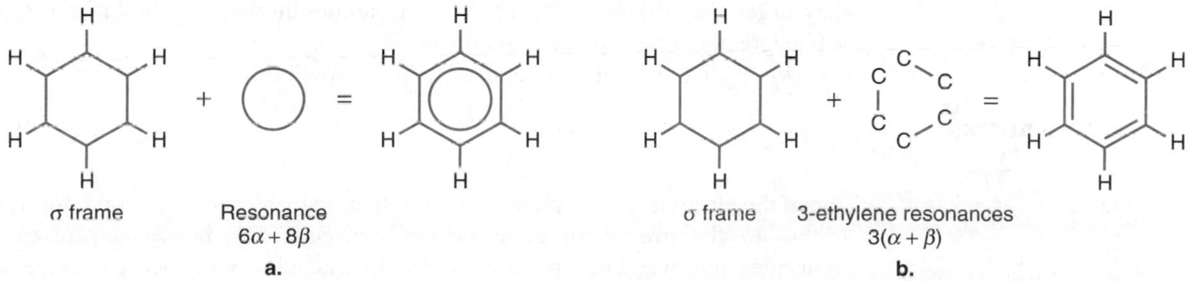

over that of three ethylene molecules, Figure 12.9b, of $2\beta = -150$ kJ mol^{-1}. The experimental value is -160 kJ mol^{-1}.

It is possible to use this procedure to calculate properties such as bond order and charge distribution, and obtain considerable insight into resonance stability limited only by the assumptions that are made.

12.4 Valence-Bond Theory for More Complex Molecules

For molecules considerably larger than hydrogen, valence-bond theory has not been as successful as MO theory in obtaining exact calculations of molecular properties. This is because of the difficulty of including the contribution to bonding from all the possible ionic structures as well as for all the possible weakly bonded structures. Nevertheless, valence-bond theory is of great importance in providing an easily visualized conceptual basis for understanding the chemical bond. Important concepts that have been developed from valence-bond theory have profoundly influenced our chemical thinking, and will now be briefly discussed.

The Covalent Bond

In general, the covalent bond between two atoms, such as H and Cl, can be described by a wave function that is similar to that for hydrogen (Eq. 12.12). Thus, for the structures marked 1 and 2,

Pure Covalent Structures

$$1.\ \mathrm{H}^{\cdot 1}\ \ {}^{2\cdot}\mathrm{Cl} \quad \psi_1 = \psi_{\mathrm{H}}(1)\psi_{\mathrm{Cl}}(2) \tag{12.68}$$

$$2.\ \mathrm{H}^{\cdot 2}\ \ {}^{1\cdot}\mathrm{Cl} \quad \psi_2 = \psi_{\mathrm{H}}(2)\psi_{\mathrm{Cl}}(1) \tag{12.69}$$

ψ_1 and ψ_2 are the valence-bond orbitals constructed from the appropriate H and Cl orbitals, and the electrons are differentiated by numbers in parentheses. The symmetric, bonding orbital is then

$$\psi_s = \frac{1}{\sqrt{2(1 + S^2)}}(\psi_1 + \psi_2) \tag{12.70}$$

This wave function ψ_s leads to a minimum in the potential-energy curve, and the resonance energy is calculated as the difference between that computed for ψ_s and that computed for ψ_1 or ψ_2. Since the resonance energy is the prime source of stability for the bond, it is important to know the factors that influence this resonance stability. In writing structures that can contribute to the resonance stability, two factors have been recognized. One is that atomic positions must be the same for all contributing forms. The second is that the forms must have the same number of unpaired electrons. In Eq. 12.70 both ψ_1 and ψ_2 have exactly the same energy since they differ only by the exchange of electron coordinates, and they therefore contribute greatly to the resonance stabilization.

When ionic structures are considered, two possibilities exist:

Pure Ionic Structures

$$\begin{array}{ll} 3.\ [\mathrm{H}]^+ & [{}^{1\cdot}_{2\cdot}\mathrm{Cl}]^- \\ 4.\ [\mathrm{H}^{1\cdot}_{2\cdot}]^- & [\mathrm{Cl}]^+ \end{array} \tag{12.71}$$

Again the electrons on the atoms are identified. In the general case, both forms contribute to the overall structure of the molecule. For heteronuclear diatomic molecules, however, one structure is normally much lower in energy than the other.

In larger molecules many ionic structural permutations may contribute to the overall structure of the molecule. In our example, structure 3 is more important than structure 4, so that the structure of HCl may be described by the wave function

$$\Psi = [\psi_1 + \psi_2]_{\text{covalent}} + \lambda[\psi_3]_{\text{ionic}} \tag{12.72}$$

The molecule is said to exist as a **resonance hybrid** between covalent and ionic forms. The coefficient λ indicates that the contribution of ψ_3 is different from that of ψ_1 and ψ_2. Calculations show that these contributions are 26% each for the covalent structures ψ_1 and ψ_2 and 48% for the ionic structure ψ_3.

Even in homonuclear molecules there is a small contribution from ionic forms. For the hydrogen molecule this contribution is approximately 3%.

Electronegativity

Electronegativity

Dipole Moment

As with the hydrogen molecule, the calculations for hydrogen chloride lead to the conclusion that there is a build-up of electron density between the nuclei. The H_2 molecule is symmetrical, so that the electron cloud lies symmetrically between the nuclei. The quantum-mechanical calculations for hydrogen chloride, on the other hand, show that the electron cloud lies more toward the chlorine atom. This effect is dealt with in terms of a property called **electronegativity,** which is the tendency of an atom to attract an electron to itself along a chemical bond within a molecule.

The consequence of this asymmetry is that the molecule has a **dipole moment.** The dipole moment of a diatomic molecule is equal to the effective charge q at the positive and negative ends multiplied by the distance between them:

$$\mu = qd \tag{12.73}$$

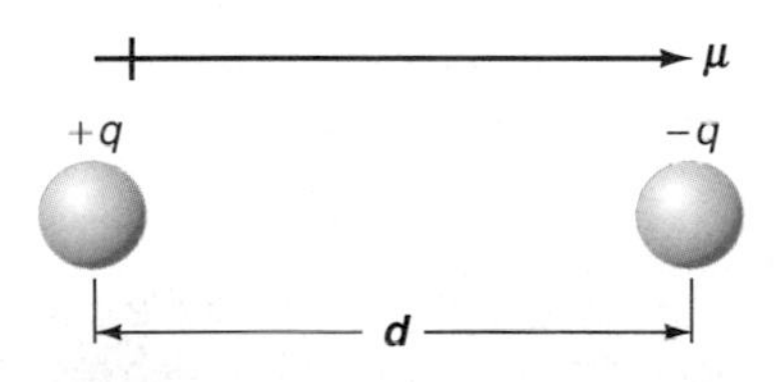

FIGURE 12.10
Two charges $+q$ and $-q$, separated by a distance d; the dipole moment is qd. The direction of the moment is often represented by an arrow ↦, as shown.

This is shown in Figure 12.10. It has been common practice to express the charge in electrostatic units (esu) and the distance in angstroms. If an electronic charge of 4.8×10^{-10} esu were separated by a distance of 1 Å (10^{-8} cm or 0.1 nm) from an equal charge of opposite sign, the dipole moment would be 4.8×10^{-18} esu cm or 4.8 debyes (D); one debye equals 10^{-18} esu cm.

The SI unit of dipole moment is Coulomb metre, C m. The elementary charge is 1.602×10^{-19} C; and if two such charges, one positive and one negative, are separated by 1 Å (1×10^{-10} m), the dipole moment is

$$1.602 \times 10^{-19}\ \text{C} \times 1 \times 10^{-10}\ \text{m} = 1.602 \times 10^{-29}\ \text{C m}$$

Since this is 4.8 D, it follows that

$$1\ \text{D} = \frac{1.602 \times 10^{-29}\ \text{C m}}{4.8} = 3.336 \times 10^{-30}\ \text{C m}$$

Electronegativity; Dipole moment; Make dipole line up in an electric field.

If we know the distance between two atoms in a diatomic molecule, we can calculate a dipole moment μ_{ionic} by making the assumption that the atoms bear a full elementary charge. The percentage ionic character of the bond can then be calculated as

$$\%\ \text{ionic character} = \frac{\mu_{\text{exp}}}{\mu_{\text{ionic}}} \times 100 \tag{12.74}$$

where μ_{exp} is the experimental dipole moment. The percent ionic character also can be calculated from valence-bond theory. Equation 12.72 is constructed as a linear combination of covalent and ionic wave functions. The lowest-energy ionic function is weighted by λ, as compared with unity for the covalent function. When the energy is calculated using the variational method, Eq. 12.8, the ratio of the contribution to the energy of the ionic wave function to that of the covalent wave function is $\lambda^2/1$. The theoretical percent ionic character is therefore

$$\% \text{ ionic character} = \frac{\lambda^2}{1 + \lambda^2} \times 100 \tag{12.75}$$

Electronegativity; Change q and d to make bonds more or less polar.

Linus Pauling made important contributions to our understanding of electronegativity and the ionic character of bonds by using both quantum-mechanical theory and experimental results. He considered a reaction such as

$$AA + BB \rightarrow 2AB$$

in which two homonuclear molecules form two heteronuclear molecules. He regarded the molecules AA and BB as purely covalent in the sense that the molecules are symmetrical and cannot have dipole moments. However, the unsymmetrical molecule AB can have a dipole moment, and there will be an ionic contribution to its energy, which will make the molecule more stable. Pauling concluded empirically that the purely covalent bond dissociation energy of AB is the geometric mean of the values for AA and BB:

$$E_{\text{covalent}} = [D(\text{AA})D(\text{BB})]^{1/2} \tag{12.76}$$

The experimental dissociation energy D(AB) will, in general, be greater than this, and the difference is taken to be the ionic energy of the bond:

$$E_{\text{ionic}} = D(\text{AB}) - [D(\text{AA})D(\text{BB})]^{1/2} \tag{12.77}$$

This quantity, therefore, can be calculated from the dissociation energies, which are usually known.

Pauling found empirically that the square roots of these ionic energies, $(E_{\text{ionic}})^{1/2}$, were additive with respect to the atoms A and B. In other words, the $(E_{\text{ionic}})^{1/2}$ values were proportional to the difference between certain numbers χ assigned to each atom:

$$(E_{\text{ionic}})^{1/2} = K|\chi_{\text{A}} - \chi_{\text{B}}| \tag{12.78}$$

Pauling chose his proportionality constant K in such a way that the difference $\chi_{\text{A}} - \chi_{\text{B}}$ also gave a reliable estimate of the dipole moment of AB measured in debyes. For energies in kJ, $K = 10$. In this way, he was able to construct a table of χ values, or **electronegativities.** A few such values are given in Table 12.2. These values are useful for making rough estimates of dipole moments. For example, hydrogen has an electronegativity of 2.20 and chlorine of 3.16; the estimated dipole moment of HCl is thus $3.16 - 2.20 = 0.96$ D (debye) with the chlorine atom being at the negative end of the dipole.

Ionization Potential, IP
$A \rightarrow A^+ + e^-$
Electron Affinity, EA
$A + e^- \rightarrow A^-$

A second but limited method of expressing electronegativities is due to Robert S. Mulliken. Mulliken considered that the attraction of an atom in a molecule for a pair of electrons in the bond is an average of the attraction of the free ion for an electron (the ionization potential IP) and the attraction of the neutral atom for an electron (the electron affinity EA). A scale factor of 5.6 is used to make coincident the values of IP and EA with χ from the Pauling scale. Thus, if IP and EA are expressed in electron volts,

$$\chi_M = \frac{(\text{IP} + \text{EA})/\text{eV}}{5.6} \tag{12.79}$$

Values of the Mulliken χ_M are listed in parentheses in Table 12.2.

TABLE 12.2 Atomic Electronegativities on the Pauling (Mulliken) Scale[2]

H						
2.20 (3.06)						
Li	Be	B	C	N	O	F
0.98 (1.28)	1.57 (1.99)	2.04 (1.83)	2.55 (2.67)	3.04 (3.08)	3.44 (3.22)	3.98 (4.44)
Na	Mg	Al	Si	P	S	Cl
0.93 (1.21)	1.31 (1.63)	1.61 (1.37)	1.90 (2.03)	2.19 (2.39)	2.58 (2.65)	3.16 (3.54)
K	Ca	Ga	Ge	As	Se	Br
0.82 (1.03)	1.00 (1.30)	1.81 (1.34)	2.01 (1.95)	2.18 (2.26)	2.55 (2.51)	2.96 (3.24)
Rb	Sr	In	Sn	Sb	Te	I
0.82 (.99)	0.95 (1.21)	1.78 (1.30)	1.96 (1.83)	2.05 (2.06)	2.10 (2.34)	2.66 (2.88)

Electronegativity; Pauling's table; See an expanded version of Table 12.2.

EXAMPLE 12.3 The internuclear distance in HF has been determined to be 92 pm. Use the Pauling electronegativities in Table 12.2 to estimate the percentage ionic character of the H—F bond. The dipole moment of HF has been measured to be 1.91 D; does this affect the estimate of the ionic character?

Solution The difference between the electronegativities gives an estimate of the dipole moment in debyes:

$$\mu = 3.98 - 2.20 = 1.78 \text{ D}.$$

Since $1 \text{ D} = 3.336 \times 10^{-30}$ C m, the dipole moment is

$$\mu = 1.78 \times 3.336 \times 10^{-30} = 5.94 \times 10^{-30} \text{ C m}$$

If the molecule were completely ionic, the dipole moment would be the product of the electronic charge and the internuclear distance:

$$\mu_{\text{ionic}} = 1.602 \times 10^{-19} \text{ C} \times 92 \times 10^{-12} \text{ m}$$
$$= 1.474 \times 10^{-29} \text{ C m}$$

The ionic character is therefore

$$\frac{5.94 \times 10^{-30} \text{ C m}}{1.474 \times 10^{-29} \text{ C m}} = 0.40 = 40\%$$

The observed dipole moment of 1.91 D is sufficiently close to the estimated value, 1.78D, to make no difference to the estimated ionic character.

Orbital Overlap

We think of the covalent bond as a pair of electrons with their spins opposed in a stable orbital based on two adjacent atoms in a molecule. The strength of the bond depends on the extent of overlap or interpenetration of the charge clouds on the two atoms. We have already seen that the overlap integral S (Eq. 12.36) is a measure of this overlap and that its value depends on the orientation of the orbitals. In addition, overlap of orbitals, and hence bond formation, can occur only in regions of like sign. Associated with the orbitals s, p, and d are + or − regions that give

[2]Values from the Mulliken scale are in parentheses and are taken from L. C. Allen, *J. Am. Chem. Soc.* *111,* 9003(1989). Pauling values are from A. L. Alfred, *J. Inorg. Nuc. Chem. 17,* 215(1961).

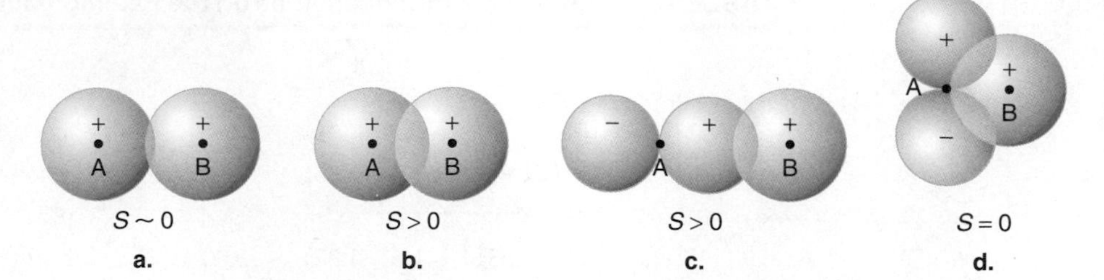

FIGURE 12.11
Orbital overlap involving s and p orbitals. (a) Slight overlap between two s orbitals. (b) Larger overlap between two s orbitals. (c) Overlap between an s orbital and a p orbital, with the s orbital interacting with the positive lobe of the p orbital; this gives significant orbital overlap and hence bonding. (d) Lateral overlap between an s orbital and a p orbital. The overlap involving the positive lobe of the p orbital is exactly canceled by that involving the negative lobe, and there is no bonding.

the algebraic sign of the wave function in its different regions. Bonding can occur only if the signs are the same for two orbitals in an overlap region.

Since the s orbital has a positive value everywhere, the overlap of s orbitals on two different atoms is independent of the direction of approach, as shown in Figure 12.11a and b; there can be more or less overlap. These bonds are σ (sigma) bonds.

A p orbital has a positive lobe and a negative lobe, and orientation is important. If the positive lobe is oriented toward an s orbital, which is positive, significant overlap may occur (Figure 12.11c); this is a σ bond. The s orbital can also be normal to the longitudinal axis of the p orbital, as shown in Figure 12.11d. In this case the integral formed from the s orbital and the negative region of the p orbital exactly counteracts that from the s orbital and the positive lobe of the p orbital; the integral S is therefore zero and there is no bonding. It is common to speak of this as a non-bonding situation. Orientations intermediate to these are also possible.

Two p orbitals can also overlap, and three cases are shown in Figure 12.12. The bond shown in Figure 12.12a is a σ bond, whereas in Figure 12.12b there is no bonding. Figure 12.12c shows lateral overlapping of p orbitals with the positive lobes coming together and the negative lobes coming together; the bond formed from such an arrangement is called a **π (pi) bond**.

π Bond

The value of S depends on the orientation of the orbitals. Values in the range $S = 0.2$ to 0.3 usually indicate an effective bond.

Orbital Hybridization

In the description of H_2 by the valence-bond method, one atomic orbital (1s) was used from each of the hydrogen atoms. Sometimes, however, this is not satisfactory and we must use two or more orbitals from a given atom.

The simplest **hybrid** to imagine is one made from one s orbital and one p orbital. The hybrid is formed by adding the wave functions for the s and p orbitals and dividing by $\sqrt{2}$. Figure 12.13 shows how these wave functions constructively and destructively interfere to form the sp hybrids.

A simple example of a more complicated hybridization is provided by the molecule methane, CH_4. In its ground state, carbon has two unpaired 2p electrons with

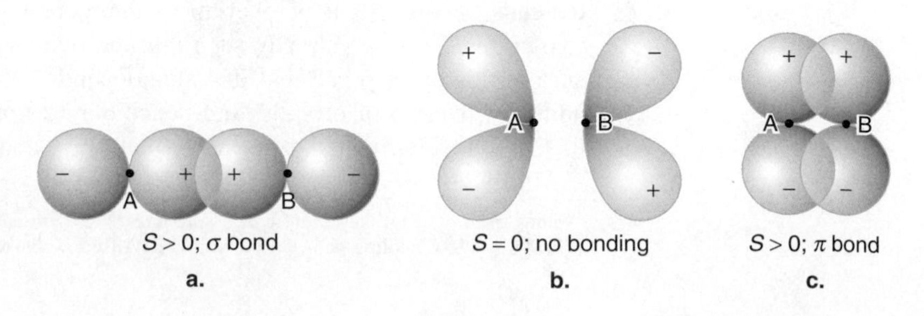

FIGURE 12.12
Orbital overlap involving two p orbitals. (a) Overlap along the internuclear axis, with lobes of the same sign coming together; this is a σ bond. (b) Lateral overlap in which lobes of opposite sign come together, giving no net bonding. (c) Lateral overlap in which lobes of the same sign come together, giving a π bond. The convention is to define the axis along which overlap occurs as the Z axis.

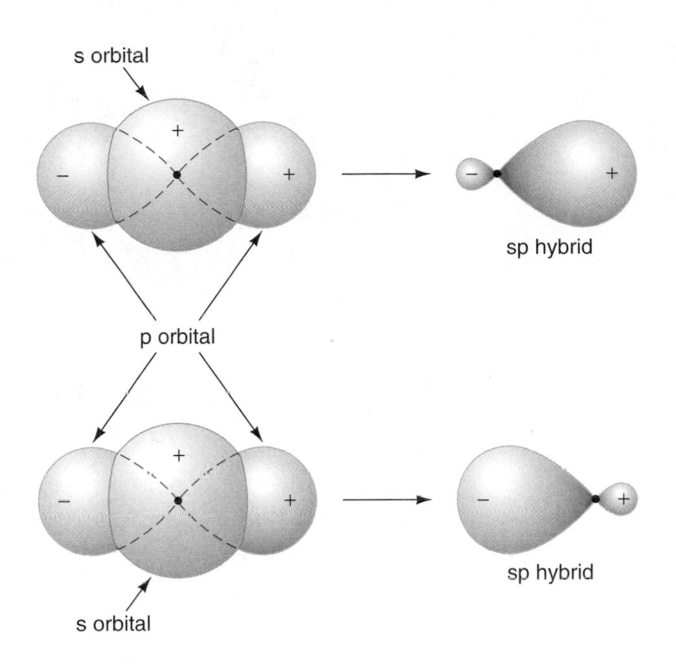

FIGURE 12.13
The formation of two sp hybrids from an s orbital and a p orbital.

the same spin. Although it might appear at first sight that carbon would form only two single bonds, carbon usually shows a valency of 4. As the four bonds are formed, it may at first be thought that one of the 2s electrons is "promoted" to the 2p state to give the configuration

Orbital hybridizations; Follow the animations of the formation of sp, sp^2, and sp^3 hybrid orbitals.

$$C^* \qquad 1s^2 2s 2p^3$$

for the excited C* carbon atom.

It might appear that, of the four bonds formed by a carbon atom, one would involve the 2s orbital and the other three would involve the three 2p orbitals. This, however, implies that one bond is different from the other three, whereas experimentally the CH_4 molecule is perfectly symmetrical; all four bonds are identical. The solution to this dilemma was given in the 1920s by Pauling, who suggested that we should use a linear combination of orbitals instead of the pure s and p orbitals of the carbon atom. On the basis of the required geometry, Pauling concluded that, from one s and three p orbitals, four hybridized orbitals, labeled t_1 to t_4, can be constructed;

sp^3 Hybridization

$$t_1 = \frac{1}{2}(s + p_x + p_y + p_z) \tag{12.80}$$

$$t_2 = \frac{1}{2}(s + p_x - p_y - p_z) \tag{12.81}$$

$$t_3 = \frac{1}{2}(s - p_x + p_y - p_z) \tag{12.82}$$

$$t_4 = \frac{1}{2}(s - p_x - p_y + p_z) \tag{12.83}$$

Principle of Maximum Overlap

These orbitals are normalized and are mutually orthogonal (see Problem 12.10). Their extension in space is as large as possible, so that when bonds are formed, the overlap is a maximum. This condition embodies Pauling's **principle of maximum overlap**. These four new orbitals are directed to the apices of a tetrahedron, and the individual bonds are formed by combining one of these hybrid orbitals and a 1s orbital from the hydrogen atom. This can be visualized as the overlapping of the electron clouds, as shown in Figure 12.14a. Obviously, the maximum overlap is obtained when the hydrogen atoms are located along the axes of the orbitals. This

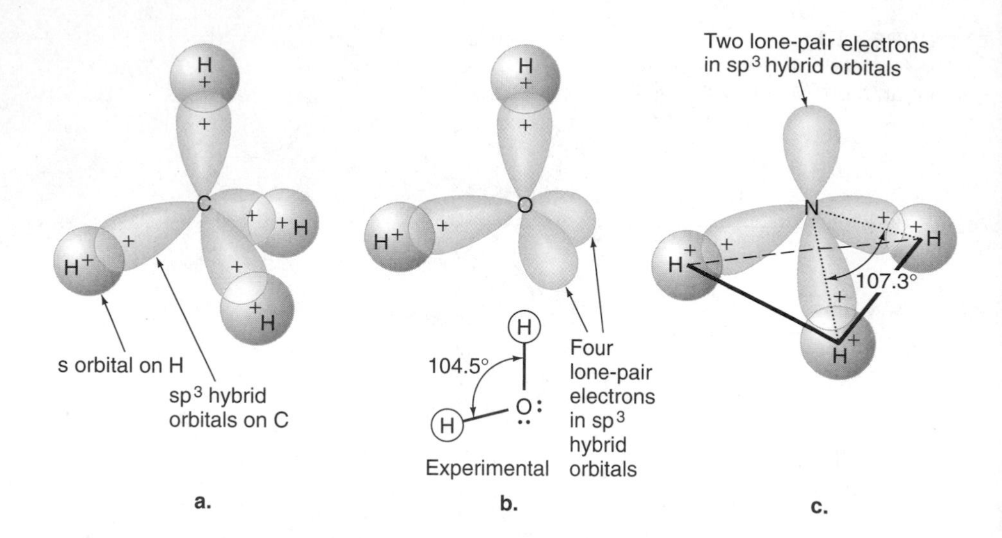

FIGURE 12.14
Hybrid orbitals in the bonding of (a) methane, (b) water, and (c) ammonia. Small negative lobes of the hybrids are not shown for clarity.

principle of maximum overlap is of great value in predicting the shapes of molecules, which follow at once from the shapes of the orbitals.

Multiple bonds; Ethene (ethylene); See animations of the hybrid orbitals of ethene and acetylene.

Hybridization also arises with the bonding in water and ammonia, although matters are not so clear as with methane. The ground-state oxygen atom configuration is $1s^2 2s^2 2p^4$. In accordance with Hund's rule (Section 11.12), there are two unpaired 2p electrons in orbitals at right angles to each other. If the H_2O molecule were made up directly from these 2p orbitals, a bond angle of 90° would be predicted from the principle of maximum overlap. Experimentally, however, the angle is about 104.5°, as shown in Figure 12.14b.

Alternatively, hybridization of the one 2s and the three 2p orbitals gives a tetrahedral sp^3 arrangement, in which the angle between the orbitals is 109.47°, the angle between bonds shown for methane in Figure 12.14a. Detailed quantum-mechanical treatments suggest that there is only partial hybridization of the orbitals.

A similar case is found with ammonia. If there were no hybridization, the bond angles would be 90°. The experimental bond angle is 107.3° (Figure 12.14c), and the fact that this is only slightly below the tetrahedral angle suggests that the sp^3 hybridization is almost complete.

VSEPR Theory

A theory that is particularly useful in making predictions about the shapes is the **valence-shell electron-pair repulsion (VSEPR) theory.** The ideas behind this theory were first suggested in 1940 by the British chemists Nevil Vincent Sidwick (1873–1952) and H. E. Powell. These concepts were later developed further by Sir Ronald Nyholm (1917–1971) and, more particularly, by the Canadian chemist Ronald James Gillespie (b. 1924).

The basis of the theory is the following fundamental rule:

> **The pairs of electrons in a valence shell adopt an arrangement that maximizes their distance apart; that is, the electron pairs behave as if they repel each other.**

For an application of this rule, consider the water molecule. This molecule has two pairs of bonding electrons and two lone pairs:

$$:\ddot{O}:H$$
$$\ddot{H}$$

VSEPR theory; Use a table to see the various structures and test yourself on the molecular shapes.

Application of the VSEPR theory suggests that the four pairs will be arranged in an approximately tetrahedral manner, as seen in Figure 12.14b. The lone-pair orbitals lie closer to the oxygen outermost shell and there will be a greater repulsion between them than between a lone-pair orbital and a bonding-pair orbital containing the H—O bond. Consequently, to minimize this repulsion, the H—O—H angle is less than the tetrahedral angle, in agreement with experiment.

Rules have been worked out for more complicated cases and are valuable in predicting the geometries of more complicated molecules. Most first-year chemistry texts cover this subject adequately. The student is referred to such texts for additional examples.

Multiple Bonds

In addition to the sp^3 hybridization that carbon exhibits, two other types must be invoked to explain molecules having double and triple bonds. Simple examples of these compounds are

```
H       H    H
 \     /      \
  C=C          C=O     H−C≡C−H
 /     \      /
H       H    H
 ethylene   formaldehyde   acetylene
```

The existence of such molecules is explained in valence-bond theory in terms of two different kinds of hybridization, sp^2 and sp, which are illustrated in Figure 12.15b and c. In sp^2 hybridization, the orthogonal bonding orbitals involve a linear combination of an s orbital and two p orbitals (e.g., p_x and p_y) as follows:

sp^2 Hybridization

$$\psi_1 = \frac{1}{\sqrt{3}}\psi_{2s} + \frac{\sqrt{2}}{\sqrt{3}}\psi_{2p_x} \tag{12.84}$$

$$\psi_2 = \frac{1}{\sqrt{3}}\psi_{2s} - \frac{1}{\sqrt{6}}\psi_{2p_x} + \frac{1}{\sqrt{2}}\psi_{2p_y} \tag{12.85}$$

$$\psi_3 = \frac{1}{\sqrt{3}}\psi_{2s} - \frac{1}{\sqrt{6}}\psi_{2p_x} - \frac{1}{\sqrt{2}}\psi_{2p_y} \tag{12.86}$$

As shown in Figure 12.15b, these orbitals lie symmetrically in the *XY* plane; the angle between orbitals is 120°. In sp hybridization, we combine the s orbital with one p orbital (e.g., p_x) as follows:

sp Hybridization

$$\psi_1 = \psi_{2s} + \psi_{p_x} \tag{12.87}$$

$$\psi_2 = \psi_{2s} - \psi_{p_x} \tag{12.88}$$

The hybrid sp orbital is linear as shown in Figure 12.15c.

In sp^2 hybridization, the p orbital that is not involved in hybridization lies above and below the plane of the hybrid bonds and is thus capable of overlapping with a similar p orbital on another atom. This accounts for one rather weak π bond. The second bond, the σ bond, is formed by overlap of two of the sp^2 hybrid orbitals. The σ bond is much stronger than the π bond. In ethylene, all the atoms lie in one plane.

An interesting feature of the double bond is the *torsional rigidity* (absence of free rotation) that occurs because of the overlap of the two p orbitals. The rotation of one CH_2 group out of the plane relative to the other CH_2 group decreases overlap

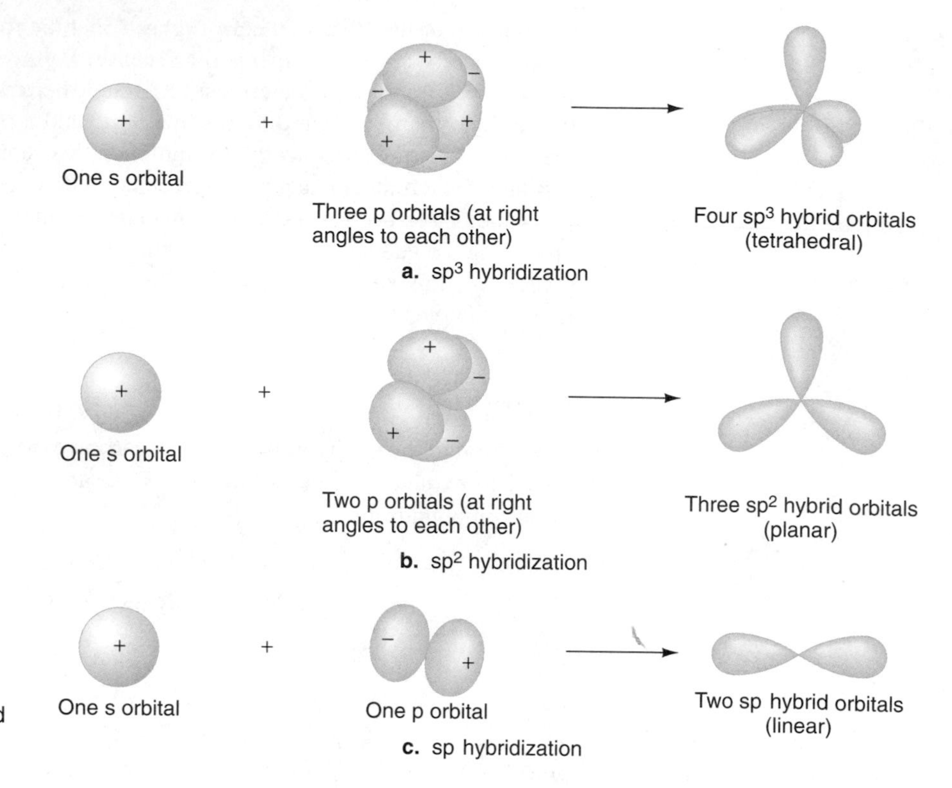

FIGURE 12.15
The three sets of hybrids formed from s and p orbitals.

of the p orbitals. This can only be accomplished by the performance of work on the molecule to distort the planar arrangement. If this happens, the π bond becomes weaker because the energy of the molecules rises. Only a small amount of energy is required to do this. Thus the reactivity of the double bond is easy to understand since replacement of the π bond by a σ bond is rather easily accomplished; the energy of the system is then lowered when this occurs.

It is often stated that carbon is to life what silicon is to rocks. Silicon lies one below carbon in the periodic table. The rich chemistry of carbon is partly due to the fact that carbon bonds are hybridized and that the unused p orbitals extend far beyond the 2s orbitals and the hybrid sp and sp^2 orbitals, thereby allowing for extensive resonance structures and π bonding. In contrast, for silicon, the vacant p orbitals do not extend far enough out to afford a similar behavior as in carbon, thus accounting for significant differences in its chemistry.

12.5 Symmetry in Chemistry

Although the valence-bond method gives good pictorial representations of molecules, it is difficult to apply it to complex molecules, particularly when hybridized orbitals are involved. As a result, there is now more emphasis on the molecular-orbital method. Even this method, however, usually requires complicated calculations, and any device that simplifies the computational procedures is welcome. Considerable aid is provided by the study of *symmetry.* In fact, symmetry plays an important role in science. We have already seen in the distorted 3-D particle in a box (Eqs. 11.150 and 11.151) how symmetry-breaking leads to the lifting of degeneracy. Many other examples exist. The mathematical basis for the study of symmetry is group theory. First we show how molecules are invariant to some symmetry operations. The relationships between the symmetry operators leads to the concept of point groups, which is summarized here.

Symmetry Elements and Symmetry Operations

A molecule, like any other geometry figure or object, may have one or more **symmetry elements.** For example, a molecule may have an *axis of symmetry* which is such that a rotation about it, through a specified angle, leads to a configuration that is superimposable on the original molecule and indistinguishable from it. Such a rotation is an example of a *symmetry operation.* We may think of an operator which, when applied to a figure or molecule, accomplishes the symmetry operation; thus, there is a *symmetry operator.* Other symmetry elements are a *plane of symmetry* and a *center of symmetry.* Associated with these three symmetry elements are four different symmetry operations, each one of which leaves the center of gravity unchanged but transforms the molecule into a configuration that is indistinguishable from the original one.

Symmetry Axis

1. *Rotation About a Symmetry Axis.* A molecule may have one or more axes about which a rotation leads to a configuration that is indistinguishable from the original one. Such an axis is a symmetry element and is called a **symmetry axis** or a **rotational axis.** A rotation about a symmetry axis is a symmetry operation. A simple example is provided by the water molecule, shown in Figure 12.16a and b. We can draw the Z axis through the oxygen atom, bisecting the angle between the two O—H bonds. The X axis can be drawn through the center of gravity of the molecule and in the plane of the molecule, and the Y axis is at right angles to these two

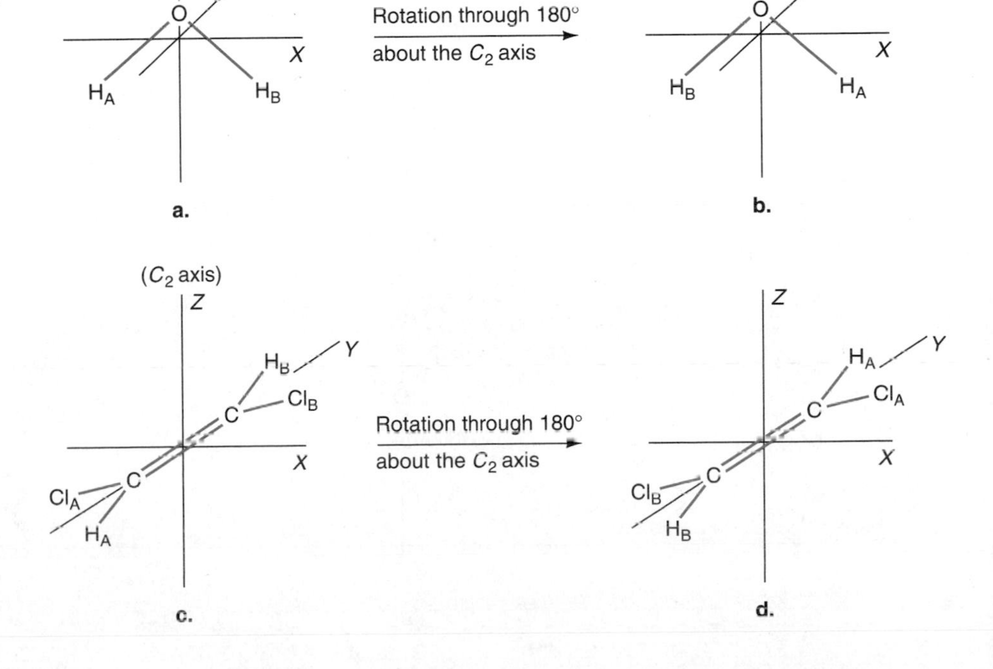

FIGURE 12.16
Two molecules having a C_2 rotational axis. (a) and (b) The water molecule. (c) and (d) *Trans*-dichloroethylene. In both cases rotation through 180° about the Z axis leads to an indistinguishable configuration.

Symmetry in chemistry; Symmetry; Try the C_2 operation on the water molecule.

axes. The X and Y axes are not symmetry axes, but the Z axis is a symmetry axis because a 180° rotation about this axis leads to an indistinguishable configuration. To show this we have labeled the two hydrogen atoms A and B. If Figure 12.16a represents the original configuration, Figure 12.16b represents the configuration after rotation through 180°. Since the labels are only mental constructs, the configuration in Figure 12.16b is indistinguishable from that in Figure 12.16a.

If a molecule is rotated through an angle θ in order to achieve the indistinguishable configuration, the axis is said to be a 360°/θ-fold rotational axis and to have an **order p** of 360°/θ; such an axis is designated C_p. In the example of water, $p = 360°/180° = 2$ and the designation is therefore C_2; the axis is said to be a *twofold* rotational axis. The planar molecule *trans*-dichloroethylene also has a twofold (C_2) axis, as shown in Figure 12.16c. All molecules remain the same if they are rotated through 360°, so that all molecules have C_1 axes, and these need not be specified.

The nonplanar ammonia molecule has a threefold (C_3) axis; its order is $360°/120° = 3$. If we label the hydrogen atoms A, B, and C, Figure 12.17a shows

FIGURE 12.17
(a), (b), and (c) The nonplanar ammonia molecule, having a C_3 axis. (d) The planar BF_3 molecule, which has a C_3 axis and three C_2 axes. (e) Benzene, having a C_6 axis and six C_2 axes. (f) The linear molecule HCN, having a C_∞ axis. (g) The linear molecule CO_2, which has a C_∞ axis and an infinite number of C_2 axes.

Symmetry in chemistry; Symmetry axis; See the C_3 axis of Figures 12.17 a-e.

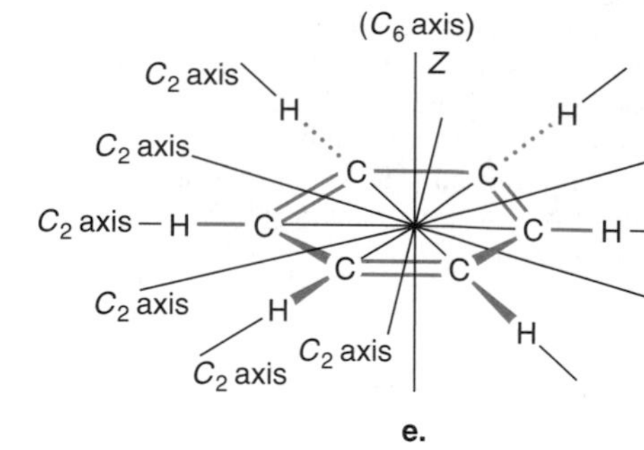

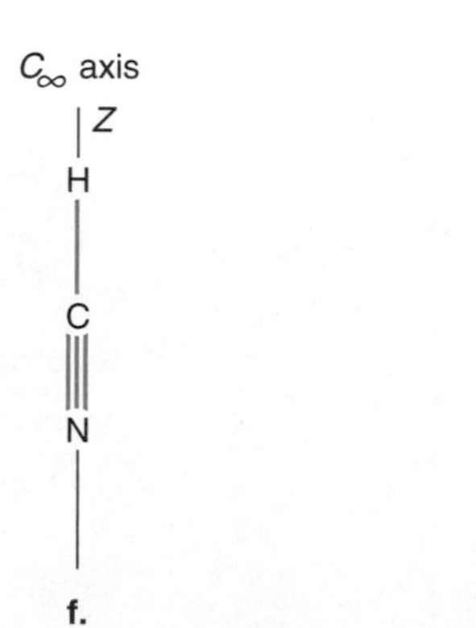

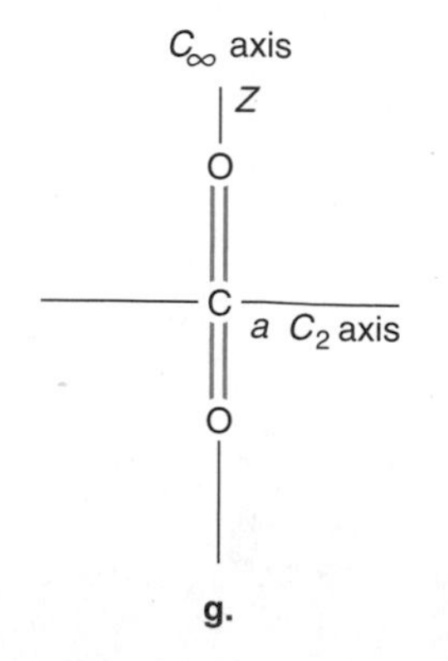

one configuration, and Figure 12.17b shows the configuration after a rotation through 120°; Figure 12.17c shows the result of a further rotation through 120°.

The planar molecule BF_3 (Figure 12.17d) has a threefold axis at right angles to the plane of the molecule and passing through the boron atom. In addition, each of the bonds lies along a twofold rotational axis; there are thus three C_2 axes as well as the one C_3 axis. When a molecule has symmetry axes of different orders, a useful convention is to take an axis of the highest order as the Z axis and to draw it vertically. The axis of highest order is often called the **principal axis.** Thus in the example of BF_3, shown in Figure 12.17d, the C_3 axis is taken to be the vertical Z axis, so that the molecule lies in a horizontal plane.

Benzene, shown in Figure 12.17e, is a molecule having a C_6 axis. It also has six C_2 axes at right angles to the C_6 axis. The C_6 axis is thus the principal axis. All linear molecules have a C_∞ rotational axis, since an infinitesimal rotation about the axis of the molecule leaves the molecule unchanged; thus $\theta \to 0$ and $p = 360°/\theta \to \infty$. An example is H—C≡N, shown in Figure 12.17f. The linear symmetrical molecule CO_2 has, in addition to the C_∞ axis, which is the principal axis, an infinite number of C_2 axes, since any axis through the carbon atom and at right angles to the C_∞ axis is a C_2 axis (see Figure 12.17g).

Center of Symmetry

Symmetry in chemistry; Center of symmetry; See this inversion for the H_2 molecule.

2. *Inversion About a Center of Symmetry.* A molecule has a **center of symmetry,** designated i (for *inversion*), if a straight line drawn from every atom through the center and extended in the same direction encounters an equivalent atom equidistant from the center. Such a center of symmetry, which can also be called a **center of inversion,** is another example of a symmetry element. Figure 12.18 shows some examples of molecules having a center of symmetry. In each case there are pairs of atoms equidistant from the center and situated in opposite directions from the center.

Plane of Symmetry

3. *Reflection through a Plane of Symmetry.* For some molecules there is a plane such that if the molecule is reflected in the plane, it is indistinguishable from the original. Such a plane, which is another symmetry element, is called a **plane of symmetry** and is denoted by the symbol σ (sigma); it is also often called a **mirror plane.** Several types of mirror planes may be distinguished according to their orientation. Consider, for example, the water molecule, shown again in Figure 12.19a.

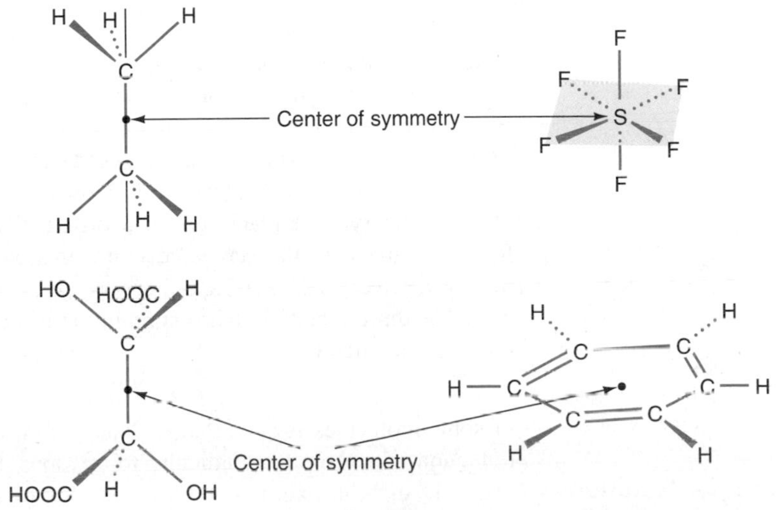

FIGURE 12.18
Four molecules having a center of symmetry. Heavy wedged lines represent bonds projecting toward the reader out of the plane of the paper. Dashed lines represent bonds projecting behind the plane of the paper.

Symmetry in chemistry; Plane of symmetry; See the actual symmetry operations of Figure 12.19.

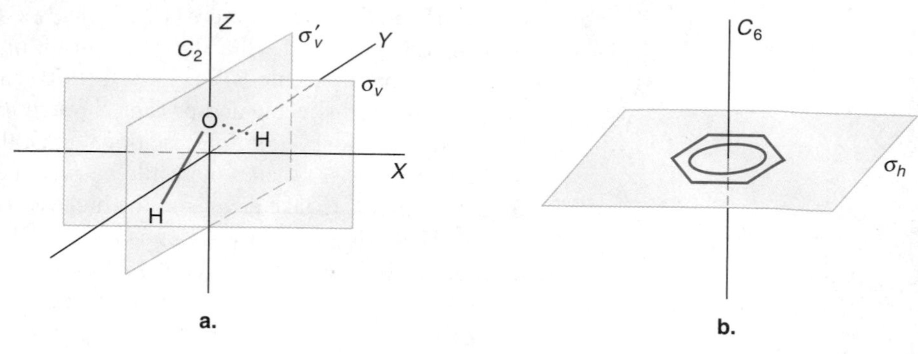

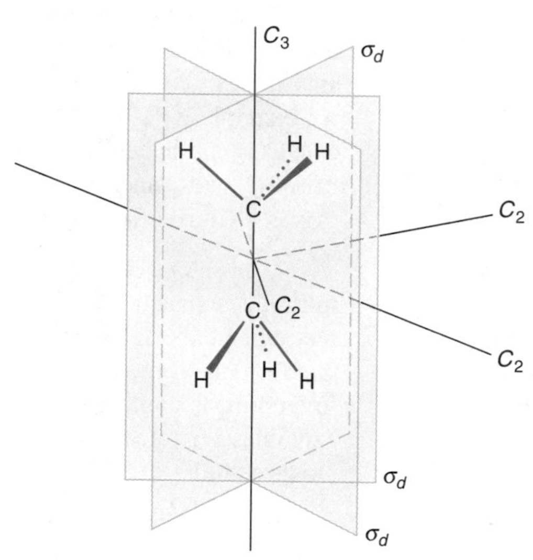

FIGURE 12.19
Some molecules having planes of symmetry. (a) H_2O: The σ_v' plane is the plane of the molecule: The σ_v plane is at right angles to the σ_v' plane. (b) Benzene, C_6H_6, which has a σ_h plane. (c) Staggered ethane, which has three dihedral (σ_d) planes; these bisect the angles between the three C_2 axes.

We have seen that there is one axis of symmetry, a C_2 axis, and that this is conventionally taken as the Z axis and drawn vertically. Both planes of symmetry pass through this axis and are therefore vertical planes. One of them is the plane of the molecule itself; we take it to be the YZ plane and denote it by the symbol σ_v'. The other plane of symmetry is at right angles to this σ_v' plane and is the XZ plane; we denote it by the symbol σ_v. In this example of water there are only two planes and both pass through the axis of symmetry. Some molecules have a plane that is perpendicular to the principal axis of symmetry. Since this axis is conventionally taken to the vertical, this type of plane is horizontal and is designated σ_h. An example of such a plane is found in benzene, as shown in Figure 12.19b.

Another type of plane, called a **dihedral** plane and given the symbol σ_d, is found in ethane in the staggered conformation (Figure 12.19c). In this molecule there is a principal C_3 axis, and there are also three C_2 axes at right angles to the C_3 axis. The three dihedral planes contain the principal axis and also bisect the angles between the three C_2 axes.

4. *Rotation About an Axis Followed by Reflection in a Plane.* The symmetry of some molecules is such that if there is first a rotation about an axis and then a reflection in a plane perpendicular to this axis, the result is superimposable on the original. Such an axis is known as an **axis of improper rotation,** or a **rotation-**

Axis of Improper Rotation

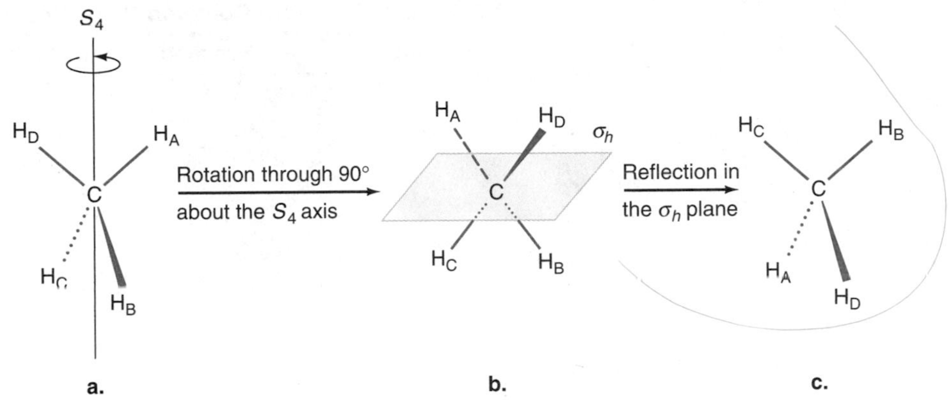

FIGURE 12.20
Methane, which has an S_4 rotation reflection axis. (a) The original form. (b) After rotation through 90° (i.e., a C_4 rotation). (c) After reflection in the horizontal plane perpendicular to the S_4 axis.

reflection axis. The designation used for such an axis is S_p; p is equal to $360°/\theta$, where θ is the angle through which the molecule is rotated in order for the reflection to give a superimposable form. Methane has an S_4 rotation-reflection axis, as shown in Figure 12.20.

Symmetry in chemistry; Improper rotation; Animation of Figure 12.20.

5. *The Identity Operation.* We have seen that all molecules have a C_1 axis of symmetry, since rotation through 360° restores anything to its original condition. The same is, of course, true if we perform no operation at all; we then say that we are performing the **identity operation.** This operation is included for mathematical completeness, and the symbol E is used for it.

If an operation $\hat{A}$ is performed and is followed by an operation $\hat{B}$, we write the combined operation as $\hat{B}\,\hat{A}$. If a C_2 operation is performed, and then another C_2 operation, the combined operation is written as C_2^2. Since C_2 rotates about 180°, C_2^2 rotates about 360° and is therefore equal to C_1 and to E.

Point Groups and Multiplication Tables

None of the operations we have considered displaces the molecule to another position. There is always one point in the molecule (at which an atom is not necessarily present) that remains unmoved by the operation. Since this is the case, we call the group of all possible symmetry operations for a given molecule a **point group.** This is in contrast to a *space group,* which is concerned with operations that move the molecule to another position in space.

Order of a Group

The number of elements in a group is known as its **order.** For a symmetry group the order is therefore the number of operations that leave the molecule unchanged. An important property of a mathematical group is that the product of any pair of operations in the group results in another operation that is also a member of the group. This result is helpful in that by taking products of known symmetry operations for a molecule we may discover other symmetry operations that were not at once apparent.

Point groups are defined with respect to the particular symmetry elements involved. Table 12.3 lists the more important point groups and gives some examples. A molecule having no symmetry at all, such as CHClBrI (Figure 12.21a), is said to belong to the C_1 point group. The hydrogen peroxide molecule, in which the two O—H bonds do not lie in the same plane (Figure 12.21b), has a twofold (C_2) axis of symmetry. Its symmetry operations are therefore C_2 and E, and it is said to belong to the C_2 point group. Since the identity operation E does nothing, EC_2 and

Group theory; Learn to use a table to determine molecular symmetry.

TABLE 12.3 The More Common Point Groups,* with Some Examples

Point Group	Symmetry Elements (besides E)	Examples
C_1	None	CHFClBr
C_2	C_2	H_2O_2
C_{2v}	C_2, $2\sigma_v$	H_2O, H_2CO
C_{3v}	C_3, $3\sigma_v$	NH_3, CH_3Cl
$C_{\infty v}$	C_∞, $\infty\sigma_v$	HCN
C_{2h}	C_2, σ_h, i	*trans*-$C_2H_2Cl_2$
D_{2h}	C_2, $2C_2$, 3σ, i	C_2H_4
D_{3h}	C_3, $3C_2$, $3\sigma_v$, σ_h	BF_3
$D_{\infty h}$	C_∞, ∞C_2, $\infty\sigma_v$, σ_h, i	H_2, O_2, CO_2
T_d	$3C_2$, $4C_3$, 6σ, $3S_4$	CH_4
O_h	$3C_4$, $4C_3$, i, $3S_4$, $8C_2$, 9σ, $4S_6$	SF_6

*Character tables can be found for many point groups of chemical interest in Appendix E, page 1028.

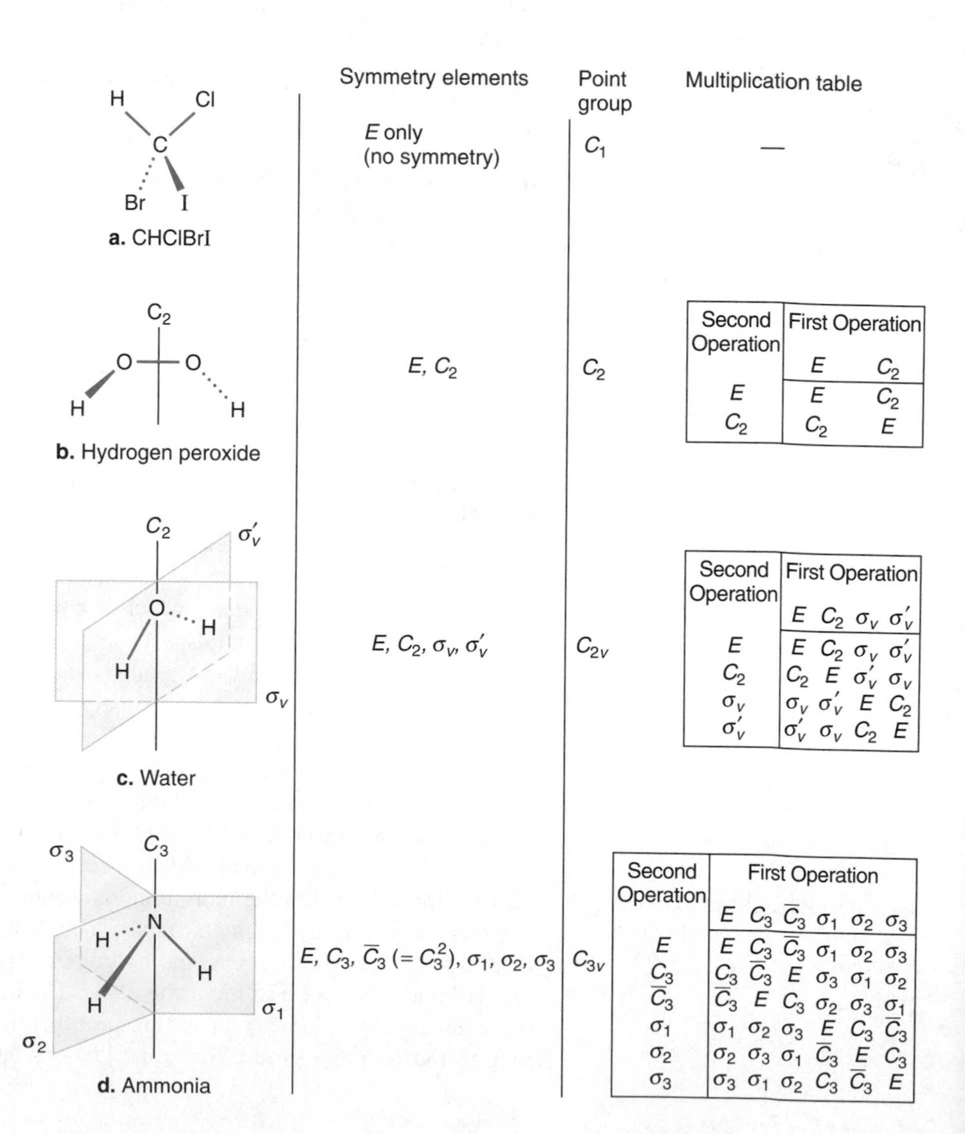

FIGURE 12.21
Examples of the point groups C_1, C_2, C_{2v}, and C_{3v}, with their multiplication tables.

Group theory; See the symmetry elements in action for Table 12.21.

C_2E are both equal to C_2, and the C_2 operation performed twice (i.e., C_2^2) is equal to E. A *multiplication table* for a point group shows all possible products of the symmetry operations and such a table is included in Figure 12.21b.

The water molecule contains the symmetry elements E, C_2, σ_v, and σ_v' and is said to belong to the point group C_{2v}. Figure 12.21c gives the multiplication table. The reader should verify the individual entries in this table. Consider, for example, the item

$$\sigma_v' C_2 = \sigma_v \tag{12.89}$$

This means that rotation by 180° about the C_2 axis followed by reflection in the σ_v' plane is equivalent to reflection in the σ_v plane.

Ammonia (Figure 12.21d) has the symmetry elements E and C_3 and three vertical planes that we may designate σ_1, σ_2, and σ_3. Its point group is called C_{3v}, and its multiplication table is shown in Figure 12.21d. One special point that arises here is the direction of rotation. If we take C_3 to mean a clockwise rotation through 120°, then C_3^2 means a clockwise rotation through 240°; this however is equivalent to an anticlockwise rotation through 120°, an operation that can be written as $\overline{C}_3$. Thus $C_3^2 = \overline{C}_3$. In the multiplication table it is therefore necessary to include both C_3 and $\overline{C}_3$.

The rest of this subsection can be omitted on first reading.

EXAMPLE 12.4 Find matrices that can perform the operations of the C_2 group, that is, E, C_2, σ_v, and σ_v' for the water molecule.

Solution We can consider the coordinates for the water molecules as represented by an array:

$$\begin{bmatrix} H_A \\ O \\ H_B \end{bmatrix}$$

and the various symmetry operations change this but leave the actual molecule indistinguishable. For example, the C_2 group operations have the following effects:

$$\begin{bmatrix} H_A \\ O \\ H_B \end{bmatrix} \xrightarrow{E} \begin{bmatrix} H_A \\ O \\ H_B \end{bmatrix}; \begin{bmatrix} H_A \\ O \\ H_B \end{bmatrix} \xrightarrow{C_2} \begin{bmatrix} -H_B \\ -O \\ -H_A \end{bmatrix}; \begin{bmatrix} H_A \\ O \\ H_B \end{bmatrix} \xrightarrow{\sigma_v} \begin{bmatrix} H_B \\ O \\ H_A \end{bmatrix}; \begin{bmatrix} H_B \\ O \\ H_B \end{bmatrix} \xrightarrow{\sigma_v'} \begin{bmatrix} -H_A \\ -O \\ -H_B \end{bmatrix}$$

By studying these and using Figure 12.19a, we can visualize that the identity E leaves the figure unchanged; the C_2 operation rotates the three atoms so we are looking at the back of them (hence the minus signs), which are indistinguishable from the front, and also interchanges the two H atoms; the σ_v operation interchanges the two H atoms, but we are still looking at the same side of all three, hence no minus signs; and the σ_v' operation simply means we are looking at the back of the three atoms. All four operations leave the water molecule indistinguishable from its starting position.

We can easily create matrices to represent these operations.

The identity:

$$\begin{bmatrix} H_A \\ O \\ H_B \end{bmatrix} = \begin{bmatrix} 1 & 0 & 0 \\ 0 & 1 & 0 \\ 0 & 0 & 1 \end{bmatrix} \begin{bmatrix} H_A \\ O \\ H_B \end{bmatrix}$$

The C_2 operation:

$$\begin{bmatrix} -H_B \\ -O \\ -H_A \end{bmatrix} = \begin{bmatrix} 0 & 0 & -1 \\ 0 & -1 & 0 \\ -1 & 0 & 0 \end{bmatrix} \begin{bmatrix} H_A \\ O \\ H_B \end{bmatrix}$$

The σ_v operation:

$$\begin{bmatrix} H_B \\ O \\ H_A \end{bmatrix} = \begin{bmatrix} 0 & 0 & 1 \\ 0 & 1 & 0 \\ 1 & 0 & 0 \end{bmatrix} \begin{bmatrix} H_A \\ O \\ H_B \end{bmatrix}$$

The σ'_v operation:

$$\begin{bmatrix} -H_A \\ -O \\ -H_B \end{bmatrix} = \begin{bmatrix} -1 & 0 & 0 \\ 0 & -1 & 0 \\ 0 & 0 & -1 \end{bmatrix} \begin{bmatrix} H_A \\ O \\ H_B \end{bmatrix}$$

Using matrix representations of the symmetry operations allows various group operations to be obtained by matrix multiplication. For example, Eq. 12.89 shows

$$\sigma'_v C_2 = \begin{bmatrix} -1 & 0 & 0 \\ 0 & -1 & 0 \\ 0 & 0 & -1 \end{bmatrix} \begin{bmatrix} 0 & 0 & -1 \\ 0 & -1 & 0 \\ -1 & 0 & 0 \end{bmatrix} = \begin{bmatrix} 0 & 0 & 1 \\ 0 & 1 & 0 \\ 1 & 0 & 0 \end{bmatrix} = \sigma_v$$

If all the matrix representations are set up in a table and multiplied out, the result is a multiplication table that must be satisfied by these group operations (see Figure 12.21c). There is, however, an arbitrariness to the way the initial array is chosen for the water molecule. For example, rather than use

$$\begin{bmatrix} H_B \\ O \\ H_A \end{bmatrix}$$

there is no difficulty in using

$$\begin{bmatrix} O \\ H_B \\ H_A \end{bmatrix}$$

The matrices in this example will be different depending upon the choice of array even though they all perform the correct operation. We call this the **basis** in which the representation is realized.

In fact, it is not necessary to actually represent the symmetry operations by matrices since the group properties are reflected in the multiplication table. Group theory gives the recipe for replacing the symmetry operations by numbers, called **characters,** that maintain the integrity of the multiplication table. For example, the following four sets of characters can be designated for the four symmetry operations that maintain the structure of the character table:

$$
\begin{array}{c|cccc}
 & E & C_2 & \sigma_V & \sigma' \\
\hline
E & E & C_2 & \sigma_V & \sigma' \\
C_2 & C_2 & E & \sigma' & \sigma_V \\
\sigma_V & \sigma_V & \sigma' & E & C_2 \\
\sigma' & \sigma' & \sigma_V & C_2 & E
\end{array}
\tag{12.90}
$$

A totally symmetric representation (denoted by the letter A_1):

$$
\begin{array}{c|cccc}
 & +1 & +1 & +1 & +1 \\
\hline
+1 & +1 & +1 & +1 & +1 \\
+1 & +1 & +1 & +1 & +1 \\
+1 & +1 & +1 & +1 & +1 \\
+1 & +1 & +1 & +1 & +1
\end{array}
\tag{12.91}
$$

A representation that has a character of $+1$ for the rotation and -1 for reflections (denoted by the letter A_2):

$$
\begin{array}{c|cccc}
 & +1 & +1 & -1 & -1 \\
\hline
+1 & +1 & +1 & -1 & -1 \\
+1 & +1 & +1 & -1 & -1 \\
-1 & -1 & -1 & +1 & +1 \\
-1 & -1 & -1 & +1 & +1
\end{array}
\tag{12.92}
$$

A representation that has a character of -1 for the rotation (denoted by the letter B_1):

$$
\begin{array}{c|cccc}
 & +1 & -1 & +1 & -1 \\
\hline
+1 & +1 & -1 & +1 & -1 \\
-1 & -1 & +1 & -1 & +1 \\
+1 & +1 & -1 & +1 & -1 \\
-1 & -1 & +1 & -1 & +1
\end{array}
\tag{12.93}
$$

Another representation that has a character of -1 for the rotation (denoted by the letter B_2):

$$
\begin{array}{c|cccc}
 & +1 & -1 & -1 & +1 \\
\hline
+1 & +1 & -1 & -1 & +1 \\
-1 & -1 & +1 & +1 & -1 \\
-1 & -1 & +1 & +1 & -1 \\
+1 & +1 & -1 & -1 & +1
\end{array}
\tag{12.94}
$$

Group Theory

The branch of mathematics that allows us to deal with symmetry operations is known as **group theory.** It is outside the scope of this book to give a detailed account of this subject. Instead we will be content with presenting some of the more important aspects, which are sufficient to allow us to appreciate the symbolism used in designating molecular states. We will also be able to understand some of the spectroscopic selection rules, discussed in Chapter 13.

Representations

The essential feature of group theory is that the symmetry operations are replaced by numbers or matrices that multiply in the same way as the operations themselves. These sets of numbers and matrices are known as **representations** or as *symmetry species,* and we will first give some examples for the point group C_{2v}.

The multiplication table for this point group is shown in Figure 12.21c. There are four operations, E, C_2, σ_v, and σ_v', and we ask what sets of numbers multiply in the same way as the actual operations. One obvious possibility is to replace all the operations by unity:

$$E \rightarrow 1, \quad C_2 \rightarrow 1, \quad \sigma_v \rightarrow 1, \quad \sigma_v' \rightarrow 1$$

Another is to replace E and C_2 by unity, and σ_v and σ_v' by -1:

$$E \rightarrow 1, \quad C_2 \rightarrow 1, \quad \sigma_v \rightarrow -1, \quad \sigma_v' \rightarrow -1$$

We can easily verify that this representation also leads to the same multiplication table as shown in Figure 12.16c; for example, $\sigma_v \sigma_v' = C_2$ and $(-1)(-1) = 1$. Two other representations are also shown in Table 12.4. The four representations have been designated A_1, A_2, B_1, and B_2 (Eq. 12.91 to Eq. 12.94). The A designation means that there is no change on rotation about the principal axis; B means that there is a change.

Irreducible Representations

It turns out that there are certain representations that are the most fundamental and useful, and these are known as **irreducible representations.** The symmetry species shown in Table 12.4 are in fact all the irreducible representations for this point group as seen from Eqs. 12.91–12.94.

Classes

For the point group C_{2v} there are four different types of operation: the identity operation (E), rotation (C_2), and two reflections (σ_v and σ_v'). We refer to these types of operation as **classes.** Sometimes (although not in the present example) there is more than one operation in a class, and we then speak of the *degeneracy* of the class. An important theorem in group theory is that *the number of irreducible representations is equal to the number of classes.* Thus, for the C_{2v} point group there are four classes, E, C_2, σ_v and σ_v' and, therefore, four irreducible representations as listed in Table 12.4.

TABLE 12.4 Representations (Symmetry Species) for the Point Group C_{2v}

Designation of Symmetry Species	E	C_2	$\sigma_v(xz)$	$\sigma_v'(yz)$
A_1	+1	+1	+1	+1
A_2	+1	+1	−1	−1
B_1	+1	−1	+1	−1
B_2	+1	−1	−1	+1

The modern convention is to take the $\sigma_v'(yz)$ plane to be the plane of the molecule. For the B_1 species there is therefore antisymmetry with respect to a reflection in the plane of the molecule.

Character Tables

Tables such as Table 12.4 are known as **character tables.** The numbers +1, −1, etc., that appear in these tables are known as **characters.**[3] In Appendix E, p. 1028, we give character tables for some of the more common point groups. In these tables we have made use of symbols such as A_1, B_2, and Σ_g^+, and these are part of a scheme proposed by Robert S. Mulliken.

The terms **symmetric** and **antisymmetric** are used to refer to the characters +1 and −1, respectively, for a particular operation. For example, suppose that a molecular orbital is such that it retains its sign when a rotation is carried out; the character is +1 and the orbital is said to be symmetric with respect to that rotation. If it changes sign, the character is −1 and the function is said to be antisymmetric. A particular notation is used if the molecule (e.g., a homonuclear diatomic molecule) has a center of symmetry, in which case one of the symmetry elements is inversion. If the symmetry species is such that inversion brings about no change of sign, the species is *symmetric with respect to inversion* and the subscript g (German *gerade*) is added to the symbol. An example is the Σ_g^+ species shown in Appendix E, p. 1028, for point group $D_{\infty h}$. If, on the other hand, inversion brings about a change of sign, the subscript is u (German *ungerade*). An example is the symmetry species Σ_u^+ found in the point group $D_{\infty h}$. The word **parity** is used with reference to inversion about a center of symmetry, and therefore relates to the g-u classification. Other features of the notation for the symmetry species of diatomic molecules are considered in Section 12.6.

Parity

Many of the applications of group theory to molecular problems depend on the fact that certain simple functions behave, with respect to the symmetry operations, in the same way as certain symmetry species. Such a function is said to be a **basis** for the particular species.

Basis for a Representation

Consider, for example, the point group C_{2v}, which is shown in Figure 12.21c. A z coordinate is unchanged by a rotation about the C_2 axis and by reflections in the σ_v and σ_v' planes. It therefore behaves in the same way as the species A_1;

E	C_2	σ_v	σ_v'
+1	+1	+1	+1

A z coordinate is therefore a basis for the A_1 representation. An x coordinate, however, changes its sign on a C_2 rotation and on reflection in the σ_v' plane; reflection in the σ_v plane, however, does not affect it. It therefore transforms according to the scheme

E	C_2	σ_v	σ_v'
+1	−1	+1	−1

and it is therefore a basis for the B_1 representation (Table 12.4). A y coordinate can similarly be shown to be a basis for the B_2 representation.

Rotations can be considered from the same point of view. Suppose that the water molecule shown in Figure 12.19a is rotating about the Z axis. The direction of rotation is unchanged by the C_2 operation, whereas it is reversed by reflections in the σ_v and σ_v' planes. The rotation R_z therefore transforms according to the scheme

E	C_2	σ_v	σ_v'
+1	+1	−1	−1

[3]They are also known as *traces*. The word *character* is then reserved for the *set* of traces that are given in a character table.

and the function R_z therefore belongs to the A_2 symmetry species. Rotation about the X axis is unchanged by the σ_v' operation but is reversed by C_2 and by σ_v, and R_x therefore transforms according to

E	C_2	σ_v	σ_v'
+1	−1	−1	+1

The function R_x therefore belongs to the B_2 symmetry species. Similarly, R_y belongs to the B_1 species.

These and other relationships are shown in the character tables given in Appendix E, page 1028. We shall see in Chapter 13 that the x, y, and z relationships are relevant to infrared vibrational spectra; R_x, R_y, and R_z to infrared rotational spectra; and x^2, y^2, z^2, xy, xz, and yz to Raman spectra.

We have seen that systems with symmetry display invariance to symmetry operations such as rotations and reflections. When a system is invariant to more symmetry operations, we say it has higher symmetry than a system that is invariant to fewer symmetry operations. The process of lowering the symmetry of the particle in a cube to a particle in a box led to the lifting of degeneracies. When a system has no symmetry, the only operation possible is the identity operation (see Figure 12.21a).

Consider a system that has a certain symmetry described by a group G. When symmetry operations are applied to the system, it is changed to a form indistinguishable from the first, and the transformed form must still be described by the same Hamiltonian operator. That is, the Hamiltonian operator must be invariant under all the group operations. Symmetry operations do, however, change wave functions. Assume a symmetry operation such as θ; it follows that under this operation a wave function becomes

$$|\psi_i\rangle \xrightarrow{\theta} |\theta\psi_i\rangle \tag{12.95}$$

EXAMPLE 12.5 Using the group G = C_{2v}, find the four symmetry adapted wave functions ψ_i.

Solution Suppose that a system displaying C_{2v} symmetry has wave functions ψ_i. For the group C_{2v}, there are four symmetry operations $\hat{E}$, $\hat{C}_2$, $\hat{\sigma}_v$, and $\hat{\sigma}'$ and these operations can change the wave functions according to Eq. 12.95. The four symmetry adapted wave functions are found from Eqs. 12.91 to 12.94 and are:

$$\psi_i^{A_1} = \frac{1}{2}\left[\psi_i + \hat{C}_2\psi_i + \hat{\sigma}_v\psi_i + \hat{\sigma}'\psi_i\right]$$

$$\psi_i^{A_2} = \frac{1}{2}\left[\psi_i + \hat{C}_2\psi_i - \hat{\sigma}_v\psi_i - \hat{\sigma}'\psi_i\right]$$

$$\psi_i^{B_1} = \frac{1}{2}\left[\psi_i - \hat{C}_2\psi_i + \hat{\sigma}_v\psi_i - \hat{\sigma}'\psi_i\right]$$

$$\psi_i^{B_2} = \frac{1}{2}\left[\psi_i - \hat{C}_2\psi_i - \hat{\sigma}_v\psi_i + \hat{\sigma}'\psi_i\right]$$

In general, if a group G has n representations, $\boldsymbol{\Gamma}_j$, the symmetrized wave functions are given by

$$\psi_i^{\boldsymbol{\Gamma}_j} = \hat{\boldsymbol{\Gamma}}_j\,\psi_i \tag{12.96}$$

The effect of the operator $\hat{\Gamma}_j$ is given for C_{2v} in Example 12.5. It should be apparent from the fact that a symmetry operation leaves the system unchanged that the Hamiltonian operator is invariant to those symmetry operations. That is, the Hamiltonian operator is exactly the same for the system after each symmetry operation has been applied to the system. This invariance of the Hamiltonian operator can be expressed by first noting that the use of symmetrized wave functions must give the same result as the unsymmetrized wave functions.

$$\langle\psi_i|\hat{H}|\psi_k\rangle = \langle\hat{\Gamma}_j\,\psi_i|\hat{H}|\hat{\Gamma}_j\psi_k\rangle = \langle\psi_i|\hat{\Gamma}_j^{-1}\hat{H}\hat{\Gamma}_j|\psi_k\rangle \tag{12.97}$$

This is true only if the Hamiltonian operator commutes with $\hat{\Gamma}_j$

$$\hat{H} = \Gamma_j^{-1}\hat{H}\hat{\Gamma}_j \qquad \text{or} \qquad \hat{\Gamma}_j\hat{H} = \hat{H}\hat{\Gamma}_j \tag{12.98}$$

or

$$[\hat{H}, \hat{\Gamma}_j] = 0 \tag{12.99}$$

As a result of the above considerations, there is a useful simplifying effect of using symmetrized wave functions, as follows. *There are no matrix elements between wave functions having different symmetry.* Stated differently, for a symmetry adapted wave function $\psi_i^{\Gamma_j}$ and one of a different symmetry, $\psi_i^{\Gamma_m}$, all matrix elements are zero. Thus

$$\langle\psi_i^{\Gamma_i}|\hat{H}|\psi_k^{\Gamma_m}\rangle = \langle\hat{\Gamma}_j\,\psi_i|\hat{H}|\hat{\Gamma}_m\psi_k\rangle = \langle\psi_i|\hat{\Gamma}_j^{-1}\hat{H}\hat{\Gamma}_m|\psi_k\rangle = 0 \text{ for } j \neq k \tag{12.100}$$

If these matrix elements were not zero, then the Hamiltonian operator would not be invariant by Eq. 12.99 and the orthogonality of the irreducible representations $\hat{\Gamma}_j$ would be lost.

The importance of the foregoing is that because there are no non-zero matrix elements between symmetrized wave functions of differing symmetry, the secular determinant block diagonalizes, making it easier to solve for the eigenfunctions.

Symmetry Breaking

When a perturbation is introduced into the Hamiltonian operator, it may cause the matrix elements in Eq. 12.100 to become non-zero. When this happens, the effect is called **symmetry breaking.**

Group theory has much greater depth than is presented here. It is a complete mathematical field in itself, but like much of abstract mathematics, has wide application and has had profound and simplifying effects on science. We have seen how the total wave functions can be symmetric or antisymmetric to electron interchange and how the Pauli principle follows. Many other symmetries exist in nature. We have mentioned parity earlier, and the symmetries that occur in molecules give us enormous insight. Examples can be found for which the parity operation causes an object to change sign (we say the object is antisymmetric with respect to parity). Some objects are antisymmetric to a time-reversal symmetry operation, and yet others are antisymmetric to changing the sign of the electric charge. In nature, however, no structure has been found that is antisymmetric to these three operations combined.

12.6 Symmetry of Molecular Orbitals

The main reason that the study of symmetry, particularly by the use of group theory, is of such importance in the quantum-mechanical treatment of molecules is that it places restrictions on the possible orbitals. The pattern of electron density around a molecule must have the same symmetry as the molecule itself. For example, the

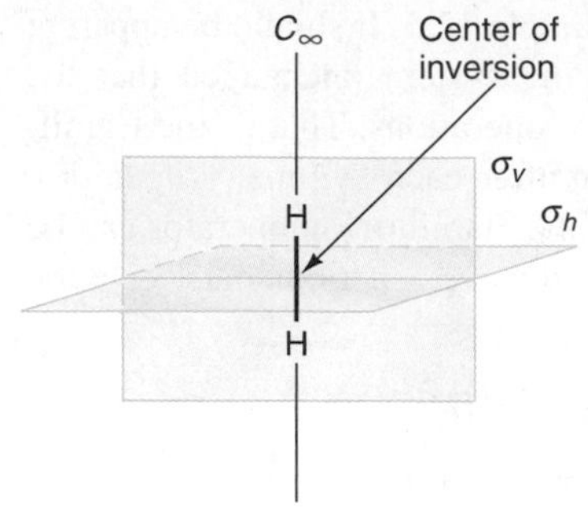

a. Point group $D_{\infty h}$

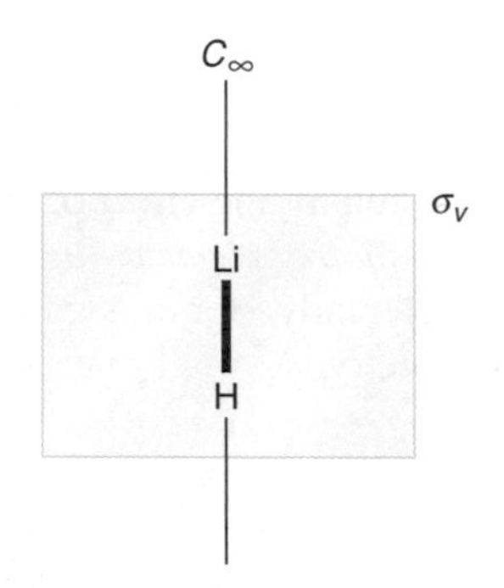

b. Point group $C_{\infty v}$

FIGURE 12.22
Examples of the symmetry elements of a (a) homonuclear and a (b) heteronuclear molecule.

Molecular orbitals; Symmetry elements; Animation of Figure 12.22.

plane of the water molecule (Figure 12.19a) is a plane of symmetry, and there cannot be a different pattern of electron density on one side of this plane than on the other. The same is true for every symmetry operation.

Since the electron density is given by ψ^2, it follows that a wave function ψ must be such that if a symmetry operation is performed, ψ either remains unchanged or simply changes its sign. For the water molecule, for example, the wave function may belong to any one of the symmetry species A_1, A_2, B_1, and B_2, but it cannot show any other behavior. This conclusion is of great help in constructing molecular orbitals from atomic orbitals, as we will now examine in more detail.

Homonuclear Diatomic Molecules

Every homonuclear diatomic molecule (Figure 12.22a) is of symmetry $D_{\infty h}$. Its axis is a C_∞ axis of rotation, and there is an infinite number of axes at right angles to the C_∞ axis and passing through the center of gravity; these are C_2 axes. Any plane passing through the molecule is a σ_v plane, and there is also a σ_h plane passing through the center of mass and at right angles to the principal C_∞ axis. There is also a center of inversion i. Examples of these symmetry elements are shown in Figure 12.22a.

The complete wave functions for homonuclear diatomic molecules must be consistent with this $D_{\infty h}$ symmetry and must therefore correspond to one of the symmetry species listed in the character table. (See Appendix E.)

For example, a molecular orbital for the H_2 molecule can be constructed by adding the 1s orbitals for the individual atoms. The resulting MO, $1s_A + 1s_B$, belongs to the Σ_g^+ symmetry species for this point group. From Appendix E for the $D_{\infty h}$ group, adding $1s_A$ and $1s_B$ is given by the first row (all 1s). The convention used for individual molecular orbitals, as opposed to the complete wave functions for the molecule, is to use lowercase Greek letters. Thus, this orbital would be represented as σ_g (the plus sign is not usually added), and to indicate that the orbital is constructed from 1s atomic orbitals we can write it as $1s\sigma_g$.

The molecular orbital obtained by subtracting the 1s atomic orbitals, namely $1s_A - 1s_B$, is of symmetry species Σ_u^+ (second row of the $D_{\infty h}$ group in Appendix E) and we write it as $1s\sigma_u^*$; the asterisk indicates that it corresponds to a repulsive state.

Bond Order

Earlier in this chapter we discussed the simplest homonuclear molecule H_2 from both the valence-bond and molecular-orbital points of view. The molecular-orbital diagram for the molecule was shown as Figure 12.5, which shows the two electrons in bonding orbitals. It is convenient to refer to one-half the difference between the number of bonding electrons and the number of antibonding electrons as the **bond order.** Since hydrogen has two bonding electrons and no antibonding electrons, its bond order is one.

The next simplest homonuclear molecule He_2 has the same MO diagram as hydrogen (Figure 12.5), since only two molecular orbitals are needed to accommodate four electrons. The first two electrons enter the lowest-lying orbital $1s\sigma_g$ with spins paired as shown in Figure 12.23. The Pauli exclusion principle forces the next electron to be placed in the antibonding $1s\sigma_u^*$ orbital. The two electrons in the $1s\sigma_g$ orbital produce a stabilization of the molecule, but the electron in the antibonding orbital tends to destabilize the structure. The fourth electron must also be placed into the antibonding orbital, with opposite spin. Thus, He_2 has two bonding electrons and two antibonding electrons, and its bond order is therefore zero. This conclusion does not, however, preclude the possibility of the existence

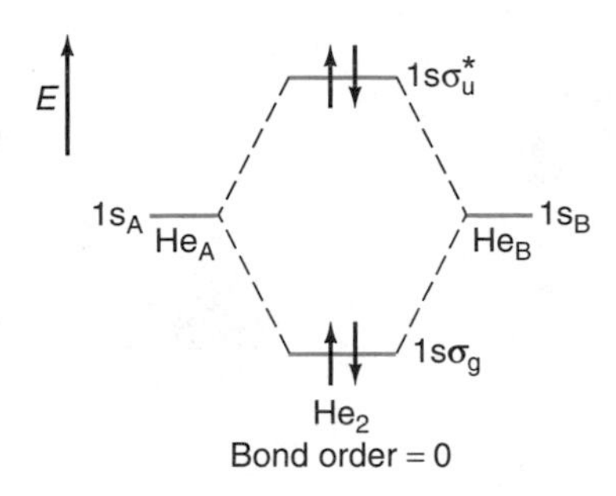

FIGURE 12.23
The energies of bonding and antibonding orbitals for the He_2 molecule.

of an He_2 dimer. This dimer has, in fact, been shown to exist, and since there are some interesting features associated with it, we will discuss it later in a separate subsection. (See pp. 626–627.)

Lithium can exist as dilithium, Li_2, and alternative molecular-orbital diagrams are shown in Figure 12.24. In Figure 12.24a the 1s orbitals remain intact, whereas in Figure 12.24b they combine to give $1s\sigma_g$ and $1s\sigma_u$ orbitals. In both representations the two 2s orbitals combine to form two molecular orbitals $2s\sigma_g$ and $2s\sigma_u^*$. Six electrons must be accommodated, and in the second representation (Figure 12.24b) the first two, with spins paired, occupy the lowest-energy $1s\sigma_g$ orbital; the next two occupy the $1s\sigma_u^*$ orbital. As a first approximation we may assume that the energy of the bonding MO is as much below the atomic 1s orbital as the energy of the antibonding orbital is above it. This is equivalent to setting the overlap integral S_{AB} equal to zero. In other words, this arrangement of four electrons is just about as stable as the separated atoms (Figure 12.24b). If S_{AB} is included, the antibonding electrons destabilize the bond, an effect known as **inner-shell repulsion;** however, S_{AB} is small because of the small size of the 1s orbitals. The stability of Li_2 therefore comes from the final two electrons that are placed in the bonding $2s\sigma_g$ molecular orbital.

In Be_2 there are two additional electrons, which must go into the $2s\sigma_u^*$ orbital. There are therefore as many antibonding electrons as bonding electrons, and the bonding order is zero; the molecule has not been detected.

To deal with diatomic molecules having more electrons than Be_2, we must construct molecular orbitals from 2p atomic orbitals. If we start with p orbitals that lie along the axis of the atomic molecule (i.e., p_z orbitals), these combine as follows:

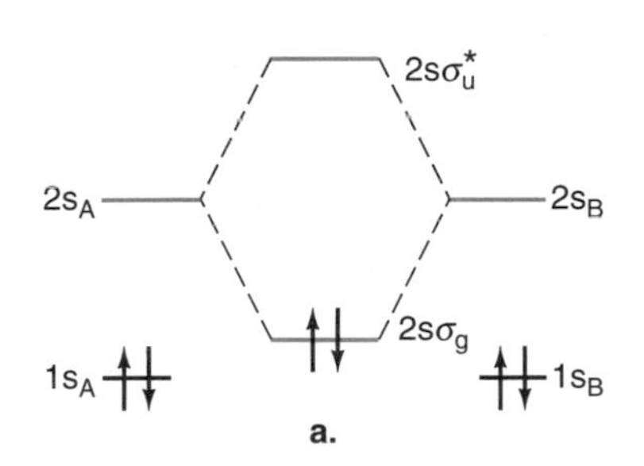

$$2p\sigma_g = 2p_{zA} + (-2p_{zB}) = 2p_{zA} - 2p_{zB} \tag{12.101}$$
$$2p\sigma_u^* = 2p_{zA} - (-2p_{zB}) = 2p_{zA} + 2p_{zB} \tag{12.102}$$

The former is bonding, and the latter is antibonding, as shown in Figure 12.25a. The original orientation of the p_z orbitals centered on the two atoms is such that the lobes of the regions signed + are both pointed in the same direction. Only orbitals of like sign can overlap and bond. Thus we reverse the direction of orientation of one p orbital by multiplication by a -1 to achieve proper orientation for bonding. The + sign in Eq. 12.101 is maintained immediately after the first equals sign to emphasize that bonding occurs because of overlap.

We can also add and subtract $2p_y$ orbitals, which lie perpendicular to the axis of the molecule, and obtain bonding and antibonding MOs:

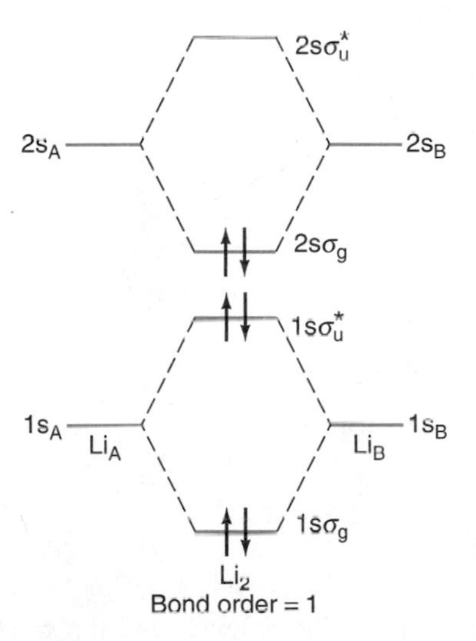

FIGURE 12.24
Molecular-orbital descriptions for Li_2. In (a) the 1s electrons are regarded as remaining in atomic orbitals. In (b) they are in σ_g and σ_u^* molecular orbitals.

$$2p\pi_u = 2p_{yA} + 2p_{yB} \tag{12.103}$$
$$2p\pi_g^* = 2p_{yA} - 2p_{yB} \tag{12.104}$$

These orbitals are represented in Figure 12.25b. Note the analogy with the π bonds formed in valence-bond theory by the sideways overlapping of p orbitals. We see from the figure, in which the + and − signs on the orbitals are shown, that the bonding orbital is now of u symmetry, whereas the antibonding orbital is of g symmetry. We can also form an equivalent set of MOs from the $2p_x$ orbitals, as shown in Figure 12.25c:

$$2p\pi_u = 2p_{xA} + 2p_{xB} \tag{12.105}$$
$$2p\pi_g^* = 2p_{xA} - 2p_{xB} \tag{12.106}$$

Molecular orbitals; p orbitals; Animation of Figure 12.25.

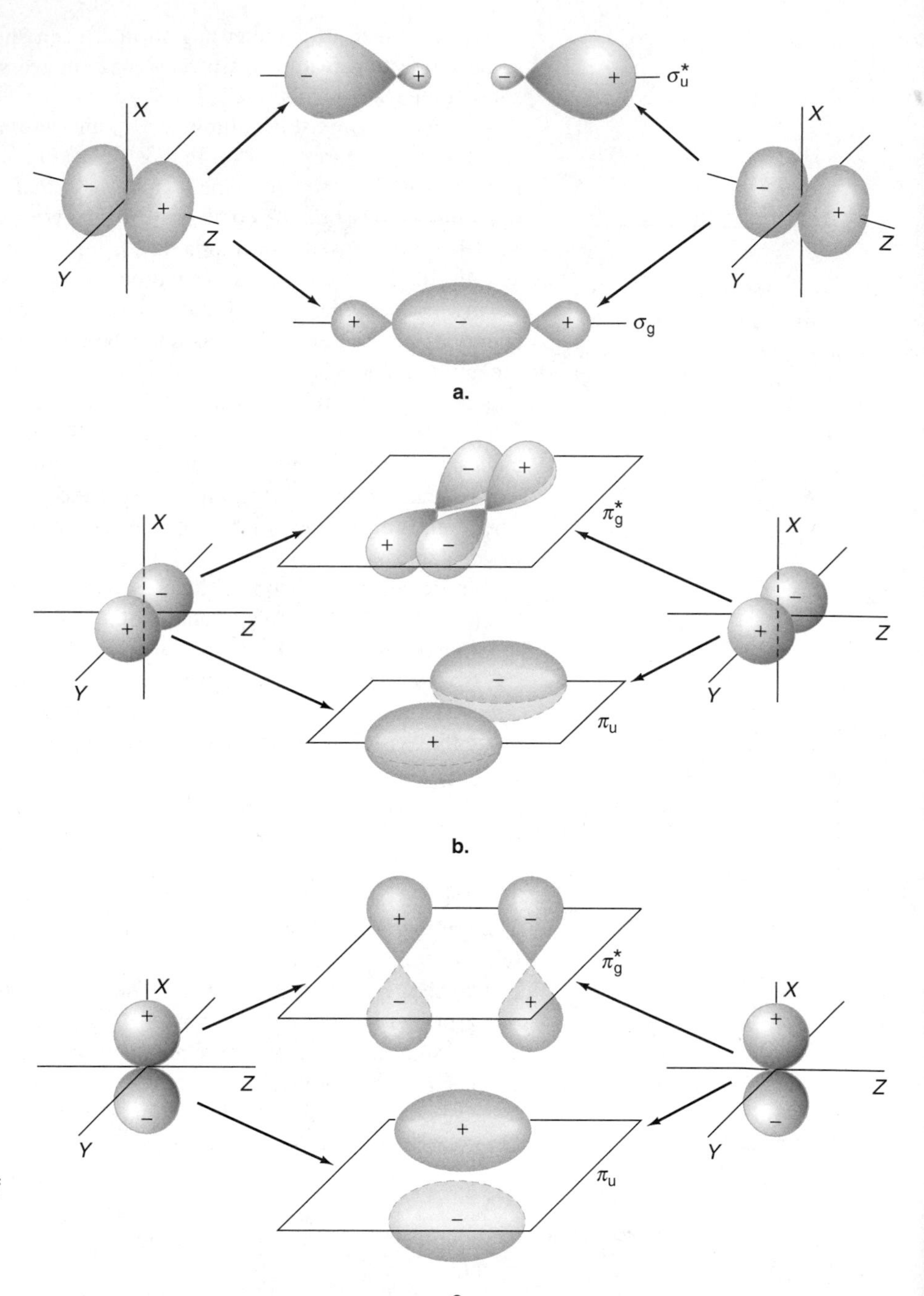

FIGURE 12.25
Combination of p atomic orbitals to give molecular orbitals. (a) σ orbitals formed from two p_z orbitals, which lie along the axis of the molecule (the Z axis). (b) π orbitals formed from p_y orbitals. (c) π orbitals formed from p_x orbitals.

Again, the bonding orbital is u and the antibonding is g. The π levels and the π^* levels are therefore each doubly *degenerate,* and each can be occupied by two pairs of electrons. Note that we cannot combine, for example, p_x and p_z orbitals, since the resulting MO would be inconsistent with the symmetry of the molecule. The molecular orbital energy diagrams become slightly modified when hybridization of the atomic orbitals is taken into account (see page 630).

Some mixing between the 2s states and the 2p levels are shown in Figure 12.26a by dotted lines. The order of filling orbitals is controlled by the separation of the 2s and 2p orbitals, which increases across the $n = 2$ level. Energies are sufficiently different at O_2 to cause the difference in order of filling.

The electronic configurations in molecules like N_2, O_2, and F_2 depend on the order of the energy levels. Calculations by the self-consistent field method have shown that the $2p\sigma_g$ and $2p\pi_u$ levels have much the same energy, but photoelectron spectroscopy has shown that for diatomics up through N_2 the order is as indicated in Figure 12.26a, which gives the assignment of electrons in N_2. The first four electrons go into the $1s\sigma_g$ and $1s\sigma_u^*$ orbitals. The high nuclear charge draws these electrons close to the nuclei, and there is virtually no overlap for the 1s AOs. The next four electrons fill the $2s\sigma_g$ and $2s\sigma_u^*$ MOs. The next level, $2p\pi_u$, shown as two lines, is degenerate and can accommodate four electrons. The remaining two electrons go into the $2p\sigma_g$ orbital. There are therefore six bonding electrons, and the bond order is three. The configuration of N_2 can be written as

$$1s\sigma_g^2 1s\sigma_u^{*2} 2s\sigma_g^2 2s\sigma_u^{*2} 2p\pi_u^4 2p\sigma_g^2$$

For O_2 and F_2, however, the order of filling is different, as shown in Figure 12.26b for O_2.

A valuable feature of MO theory is that it provides a simple interpretation of molecular magnetism. In Section 11.9 we saw that an electron in any orbital has a magnetic moment. If all such electrons in an atom or molecule are paired, there is no resultant magnetic moment, and there is no interaction with a magnetic field; the atom or molecule is then said to be **diamagnetic.** If, on the other hand, there are p, d, etc., electrons that are not paired and have the same spin (i.e., *odd* electrons), there is a resultant magnetic moment. The atom or molecule then interacts with an external magnetic field, and is said to be **paramagnetic.** The degree of paramagnetism depends on the number of unpaired electrons.

Diamagnetism

Paramagnetism

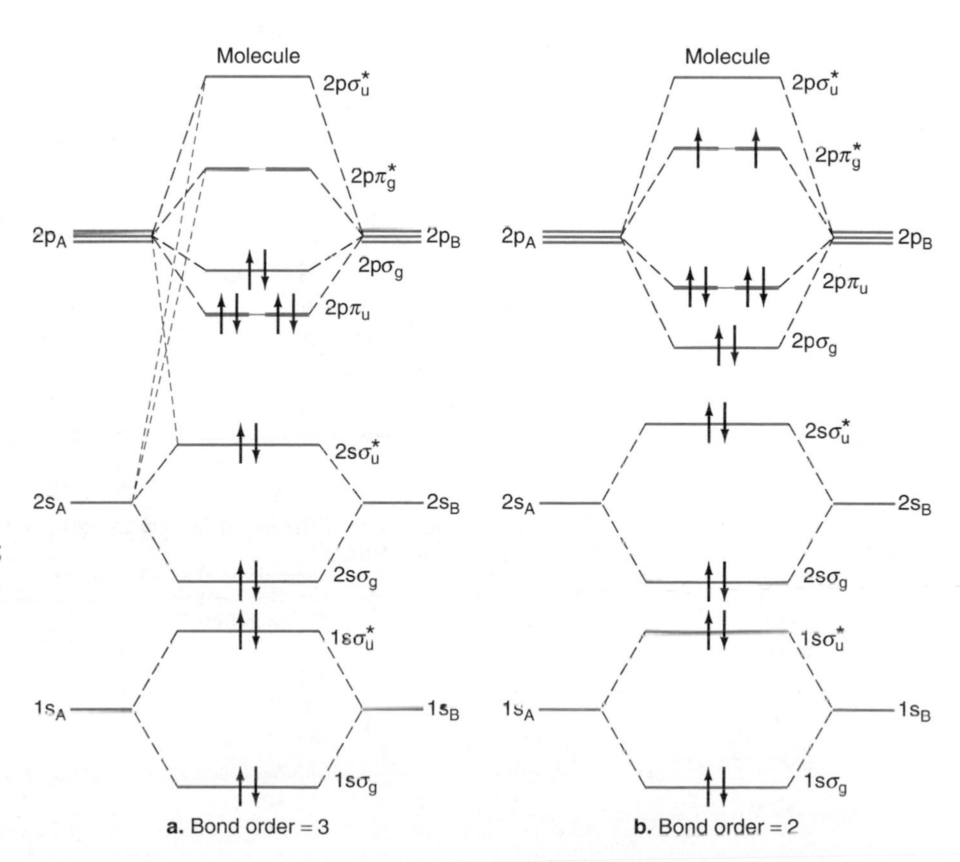

FIGURE 12.26
Molecular-orbital energy diagrams for homonuclear diatomic molecules. (a) The order of energy levels for N_2, showing the assignment of the 14 electrons; the order is the same for the elements B_2 through N_2. (b) The order of levels for O_2, and F_2, showing the electron filling for O_2.

Dashed lines between $2s_A$ and $2p_A$ on the left of Figure 12.26a show how mixing of 2s and 2p orbitals occurs to give the resulting orbitals.

Molecular orbitals; Homonuclear; Animation of Figure 12.26 for B_2, C_2, N_2, O_2, and F_2.

In the MO diagram of Figure 12.26b we see that 14 of the 16 electrons in O_2 are paired with one another. The remaining 2 electrons, on the other hand, are in the $2p\pi_g^*$ orbital, which could accommodate a total of 4 electrons, and by Hund's rule they have parallel spins. The oxygen molecule is therefore predicted to have a resultant magnetic moment (i.e., to be paramagnetic), and this is in agreement with experiment.

The MO energy diagram of F_2 is similar to that for N_2, but four additional electrons must be accommodated. These electrons go into the $2p\pi_g^*$ orbitals in Figure 12.26a. These electrons in the two antibonding orbitals cancel the bonding from the $2p\pi_u$ orbitals, leaving only one bonding orbital having its pair of electrons; the bond order is therefore one. Because the antibonding cancels the bonding, there are three lone pairs on each atom, corresponding to the conventional valence-bond structure $:\ddot{\underset{..}{F}}:\ddot{\underset{..}{F}}:$. The fluorine molecule is not paramagnetic, in agreement with this MO structure.

In the series N, O, F there is increasing nuclear charge but decreasing bond order; this is partly responsible for the decreased bonding in the order N_2, O_2, and F_2. However, the importance of the antibonding orbitals is illustrated by the decreasing bond energies of N≡N, O═O, and F—F, which are 941.4, 493.7, and 153.0 kJ mol^{-1}, respectively, and by the increasing bond distances, which are 110 pm for N_2, 121 pm for O_2, 142 pm for F_2.

There is a correlation between the concept of bond order and experimental data for the homonuclear diatomics. One might expect that actual bond strength increases with bond order and that bond lengths decrease. Table 12.5 presents data that illustrate this correlation.

The Weakest Known Bond: The Helium Dimer

From the simple molecular-orbital point of view we are left in doubt as to whether the helium dimer He_2 can exist. The molecular-orbital diagram (Figure 12.23) shows that there are two bonding electrons and two antibonding electrons. If a bond can be formed, it is obviously a very weak one. The first quantum-mechanical calculations on the dimer were made in 1928 by J. C. Slater, who concluded that the classical dissociation energy of the molecule (often known as the **well depth**) is 1.23×10^{-22} K, or 74.0 J mol^{-1}. This is smaller by a factor of more than a thousand than, for example, the dissociation energy of H_2, which is 435.0 kJ mol^{-1} (Table 12.5). Furthermore, the calculations indicated that the zero-point energy of the molecule is close to the classical energy. If it is slightly less than the classical energy, the molecule can exist, but would readily dissociate except at exceedingly low temperatures. If it is greater, the molecule cannot exist at all.

TABLE 12.5 Bond Strength, Bond Length, and Bond Order for Some Diatomic Molecules

Molecule	Bond Strength $D_{298}^0 \times$ kJ mol^{-1}	Bond Length Å	Bond Order
H_2	435.990	0.74144	1
N_2	945.33	1.09769	3
O_2	498.36	1.20752	1
F_2	158.78	1.41193	1
Br_2	192.807	2.2811	1
I_2	151.088	2.666	1

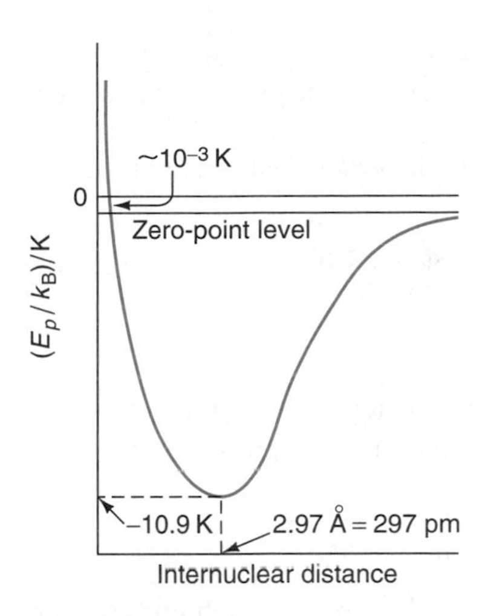

FIGURE 12.27
A schematic diagram of potential energy divided by the Boltzmann constant (E_p/k_B) for the He_2 dimer. It is not possible to draw this diagram to scale, since the zero-point level is in fact extremely close to the line corresponding to dissociation. Note that the zero-point line meets the curve at high separations, so that the dimer has a large internuclear separation.

Since it is difficult to appreciate such low energies, it is convenient to work with potential energies E_p divided by the Boltzmann constant k_B; the ratio E_p/k_B has the unit of K. In these terms, Slater's well depth corresponded to 8.9 K. The advantage of quoting energies in this way is that the thermal energy in terms of E_p/k_B is of the order of unity. Thus, when we know in terms of E_p/k_B the difference between the zero-point energy and the well depth, we know at once how low a temperature is needed for the molecule to survive.

Accurate quantum-mechanical calculations carried out by James B. Anderson and others have indicated a well depth corresponding to 10.9 K at an internuclear separation of 297 pm (2.97 Å). They show, moreover, that the zero point level lies about 10^{-3} K (1 mK) *below* the classical dissociation level. These results are shown schematically in Figure 12.27. The conclusion is that He_2 can exist, but would only survive for any appreciable time at temperatures below 1 mK.

In February 1993, C. F. Giese and W. R. Gentry and their coworkers[4] at the University of Minnesota reported that they had obtained conclusive experimental evidence for the existence of He_2. They had previously found that temperatures below 1 mK could be achieved by rapid expansion of helium gas, and in their later work they were able to detect the existence of He_2^+, which they showed convincingly could only be formed by loss of an electron from He_2. The He_2 is presumably formed by a collision between three helium atoms, one of the atoms removing the excess energy.

This molecule has by far the weakest bond of any known molecule. The other inert gas dimers are all known, and have binding energies larger by several powers of ten.

Heteronuclear Diatomic Molecules

Every heteronuclear diatomic molecule belongs to the point group $C_{\infty v}$. The axis of the molecule is a C_∞ axis, and any plane passing through the molecule is a plane of symmetry, designated σ_v (see Figure 12.22b). These are the only symmetry elements. In particular, a heteronuclear molecule lacks a center of symmetry and in this respect differs from a homonuclear diatomic molecule.

The wave functions for heteronuclear molecules must therefore correspond to one of the symmetry species for this point group. Subscripts g and u are now inappropriate, since there is no center of inversion, but for some of the representations the superscripts + and − are used, the former to indicate no change of sign on reflection in a σ_v plane and the latter to indicate a change of sign.

Lowercase Greek letters are again used to indicate the symmetry species of the individual molecular orbitals. For example, the MO formed by adding two 1s atomic orbitals is called a $1s\sigma$ orbital, while that obtained by subtracting them is a $1s\sigma^*$ orbital. Since the gerade-ungerade notation no longer applies, these two orbitals are distinguished simply by using the asterisk to indicate antibonding. If a molecular orbital is formed by adding or subtracting two different atomic orbitals, such as $1s_A + 2s_B$ or $1s_A - 2s_B$, the symbol used is simply σ or σ^*.

On this basis we may consider LiH as follows. There is a total of four electrons, and two of these may be considered to remain in the 1s atomic orbital of Li.

[4]F. Luo, G. C. McBane, G. Kim, C. F. Giese, and W. R. Gentry, "The weakest bond: experimental observation of helium dimer," *Journal of Chemical Physics, 98,* 3564–3567(1993).

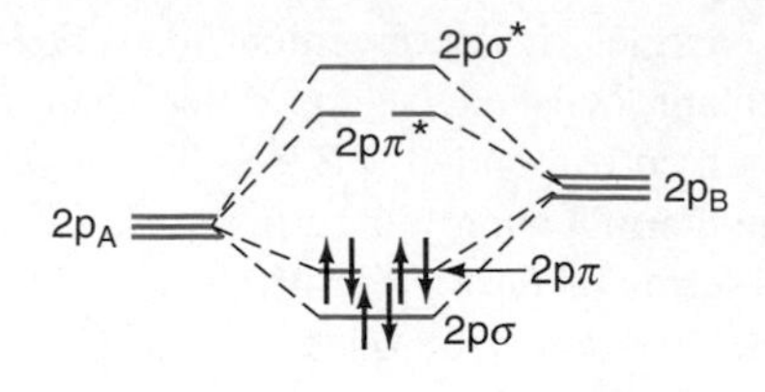

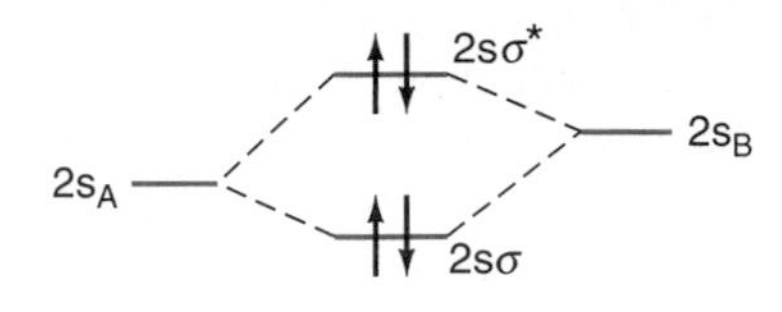

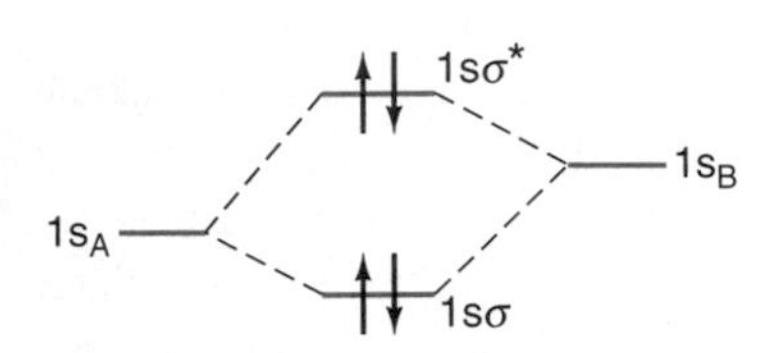

FIGURE 12.28
Molecular-orbital diagram for a heteronuclear diatomic molecule. As an example, the 14 electrons of carbon monoxide are shown as arrows.

Molecular orbitals; Heteronuclear; Vary the energy of the atomic orbitals to produce a heteronuclear MO.

The bond is formed from the valence (2s) electron of Li and the 1s electron of H. Addition of the $1s_H$ and $2s_{Li}$ orbitals will give a σ molecular orbital that will accommodate the two electrons.

The two modifications required to deal with heteronuclear diatomic molecules are:

1. The g and u suffixes are dropped, since the molecules have no center of symmetry.
2. The diagrams are no longer symmetrical.

Figure 12.28 shows a schematic MO diagram for a heteronuclear diatomic molecule. As an example, carbon monoxide has 14 electrons, and on the basis of the Aufbau principle we have

$$(1s\sigma)^2(1s\sigma^*)^2(2s\sigma)^2(2s\sigma^*)^2(2p\pi)^4(2p\sigma)^2$$

The two pairs of $\sigma - \sigma^*$ orbitals cancel and give no bonding. Bonding, however, results from the six electrons in the $2p\pi$ and $2p\sigma$ orbitals, and the bond order is three.

The Water Molecule

The procedures for constructing molecular orbitals for more complicated molecules may be exemplified by a brief treatment of the water molecule, the symmetry elements for which are shown in Figure 12.19a. The oxygen atom lies on all three of these symmetry elements, and in Figure 12.29a we show orbitals of the oxygen atom in relation to these elements. Since the 1s and 2s orbitals are spherically symmetrical, none of the symmetry operations brings about any change in them and both are therefore of A_1 symmetry. The $2p_z$ orbital has a positive lobe in one direction along the Z axis and a negative lobe in the other. The C_2, σ_v, and σ_v' operations bring about no change in this orbital, which is therefore also of A_1 symmetry.

The $2p_x$ and $2p_y$ orbitals, however, behave differently. Reflection of the $2p_x$ orbital in the σ_v' plane interchanges the + and − lobes and therefore changes the sign, and the same happens with a C_2 rotation. Reflection in the σ_v plane brings about no change. The $2p_x$ orbital is therefore of B_1 symmetry. By similar arguments we find that the $2p_y$ orbital is of B_2 symmetry.

The situation with the orbitals on the hydrogen atoms in the water molecule is a little more complicated, since these atoms do not lie on all the symmetry elements of the molecule; they lie only in the σ_v' plane. The 1s orbital of the H_A atom, which we will write as $1s_A$, therefore does not correspond to the symmetry of the molecule, and the same is true of the $1s_B$ orbital. However, from these two orbitals we can construct orbitals that do have the right symmetry. Thus the sum of the atomic orbitals, $1s_A + 1s_B$, is unchanged when any of the symmetry operations is performed, and it is therefore of A_1 symmetry. The difference, $1s_A - 1s_B$, changes its sign when the C_2 operation is performed and when the σ_v operation is performed, but it remains unchanged when σ_v' is performed. It is therefore of B_2 symmetry.

The symmetry properties of the various orbitals on the H and O atoms are summarized in Figure 12.29b. The complete molecular orbital must correspond in symmetry to one of the symmetry species of the molecule, and this is impossible if we combine atomic orbitals of different symmetries. We can, however, obtain a molecular orbital of symmetry A_1 if we make a linear combination of 2s, $2p_z$, and

Molecular orbitals; Water molecule; See all the symmetry operations shown in Figure 12.29.

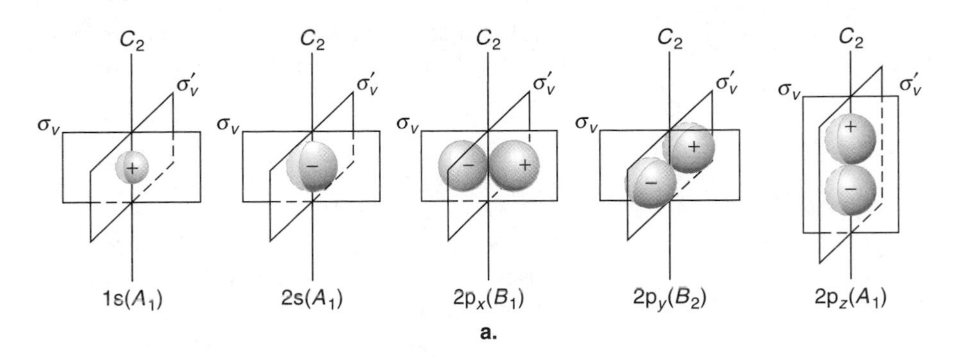

	E	C_2	σ_v	σ_v'	Atomic Orbitals
A_1	+1	+1	+1	+1	1s(O), 2s(O), $2p_z$(O), $1s_A + 1s_B$
A_2	+1	+1	−1	−1	—
B_1	+1	−1	+1	−1	$2p_x$(O)
B_2	+1	−1	−1	+1	$2p_y$(O), $1s_A - 1s_B$

b.

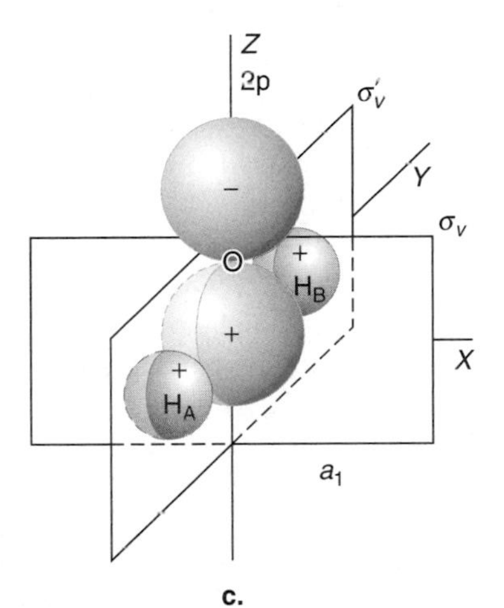

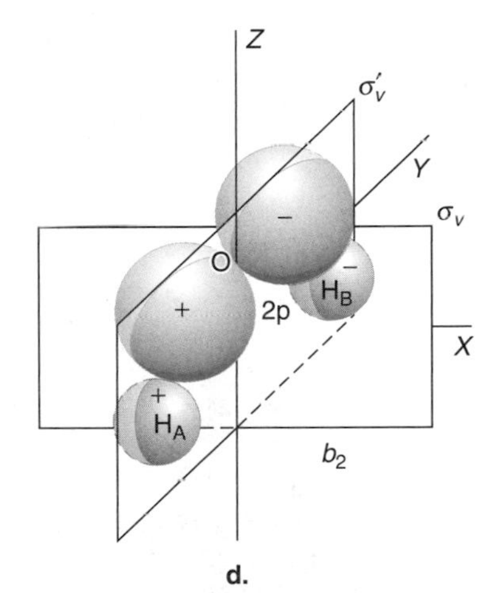

FIGURE 12.29
Orbitals for the water molecule. (a) The atomic orbitals and their symmetry species. The σ_v' plane is the plane of the molecule. (b) The C_{2v} character table showing the symmetries of the atomic orbitals and the symmetry-adapted orbitals for oxygen (indicated by an O). (c) and (d) The bonding molecular orbitals a_1 and b_2.

$1s_A + 1s_B$ orbitals, all of which are of A_1 symmetry. Such an orbital is designated a_1 (the lowercase letter is used) and is of the form

$$a_1 = 1s_A + 1s_B + \lambda_1 2s + \lambda_2 2p_z \tag{12.107}$$

where λ_1 and λ_2 are coefficients that could be determined by a variation procedure. This orbital is represented schematically in Figure 12.29c.

We can also construct the orbital

$$b_2 = 1s_A - 1s_B + \lambda 2p_y \tag{12.108}$$

in which orbitals of B_2 symmetry are combined. This orbital is represented in Figure 12.29d.

Both of these orbitals are bonding, since the atomic orbitals have been added together, and there is a piling up of charge between the nuclei. For each bonding molecular orbital there is an antibonding one, of the same symmetry, formed by subtracting the atomic orbitals. These are indicated by an asterisk. The antibonding molecular orbital corresponding to a_1 is

$$a_1^* = 1s_A + 1s_B - \lambda_1' 2s - \lambda_2' 2p_z \tag{12.109}$$

That corresponding to b_2 is

$$b_2^* = 1s_A - 1s_B - \lambda' 2p_y \tag{12.110}$$

There is also an antibonding orbital of B_1 symmetry, which consists only of $2p_x$(O) itself; this cannot combine in any way with the hydrogen atom orbitals and, therefore, gives no bonding.

These conclusions are shown in Figure 12.30 in the form of an energy diagram. The order of the energy levels is that predicted by self-consistent field calculations. The 1s atomic orbital of the oxygen atom has a much lower energy than the other atomic orbitals, and it is therefore carried over as the lowest MO; electrons in this orbital are nonbonding. The 2s atomic orbital of the oxygen atom also forms an MO, designated a_1', and electrons in this level are also nonbonding.

The assignment of the 10 electrons in the water molecule is shown in Figure 12.30. The two electrons in the 1s orbital of oxygen make no contribution to bonding. Two electrons are in each of the a_1 and b_2 orbitals, and these four electrons are responsible for the two O—H bonds. The remaining four electrons are in the a_1' and b_1 orbitals and constitute the two lone pairs.

This chapter has dealt almost entirely with the general principles relating to the nature of the chemical bond. Much of the modern work being done in this field is concerned with making accurate *ab initio* calculations, by which is meant that no empirical quantities are employed aside from the masses and charges of the electrons and nuclei, and the fundamental physical constants.

Many of the accurate calculations now being made are of energies of molecules. With the aid of supercomputers it is now possible to obtain results that agree

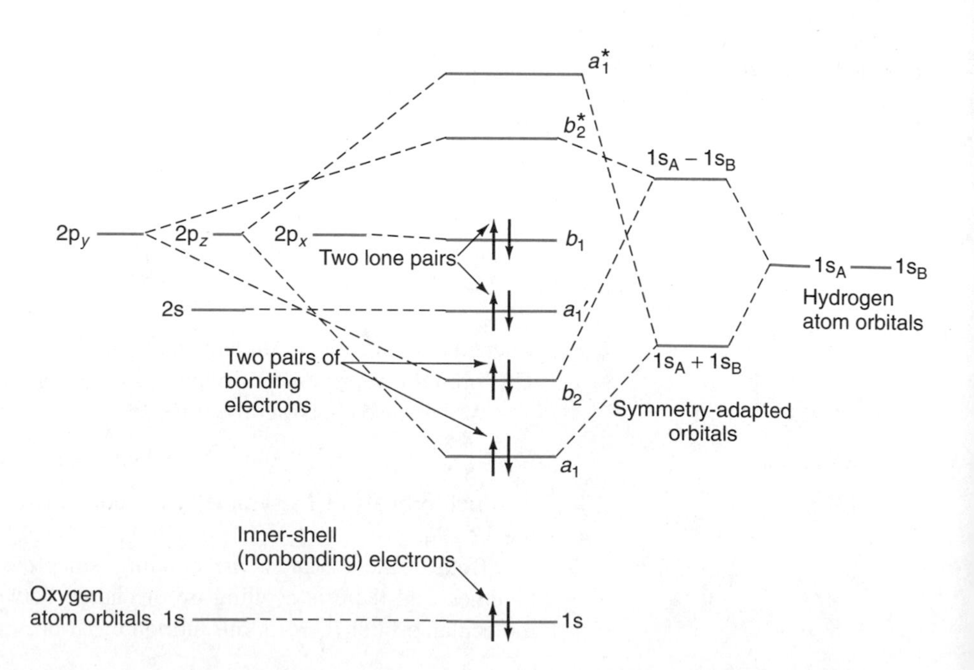

FIGURE 12.30
The energy levels corresponding to the molecular orbitals for water, showing how they are derived from the atomic orbitals of oxygen and hydrogen (not to scale).

with experiment within experimental error, even for quite complicated molecules. One example already mentioned, on the helium dimer, was particularly challenging, and anything but an extremely accurate calculation would not have led to a useful conclusion; the outcome was successful.

One application of quantum chemistry, important for pharmaceutical research, depends upon knowledge of how proteins fold into 3-D structures. With this knowledge, drugs can be tailored to intervene in biochemical processes. Currently, structures of small proteins (less than about 150 amino acid residues) can be obtained by x-ray crystallography and NMR, and no doubt in the next few years it will be possible to deal with larger proteins. An active area of research as yet unresolved, is aimed at predicting protein folding. At the time of writing, IMB is building a computer to perform the enormous number of calculations necessary to obtain protein folding data based upon theories of bonding discussed in this chapter. In addition to calculating the actual folding of proteins, it is anticipated that rules can be found from which protein folding can be determined.

Another important value of such accurate calculations is that they confirm the validity of the quantum-mechanical methods that are being used. In some cases, as with the helium dimer, the calculations have the important result of acting as guide to future experiments. Accurate quantum-mechanical calculations of potential-energy surfaces for reactions (Section 9.8) are also of great help in leading to interpretations of the mechanisms of chemical reactions.

A difficulty that remains with the accurate quantum-mechanical calculations is that even with supercomputers they still take a considerable amount of time if the molecule is of any complexity. It is necessary to use a large number of terms, perhaps a million, in the trial eigenfunctions to which the variational treatment is applied. Even though the final results may be of great importance, the methods employed do not shed significant light on the nature of the covalent bond. Our insights into the nature of the bond still have to come from the approximate treatments, developed mainly in the 1930s, that have been discussed in this chapter.

KEY EQUATIONS

Heitler-London wave functions for the H_2 molecule:

$$\psi_s = \frac{1}{\sqrt{2}}[1s_A(1)1s_B(2) + 1s_A(2)1s_B(1)]$$

$$\psi_a = \frac{1}{\sqrt{2}}[1s_A(1)1s_B(2) - 1s_A(2)1s_B(1)]$$

Hamiltonian operator for the hydrogen molecule:

$$\hat{H} = \left[-\frac{\hbar^2}{2\mu}\nabla_1^2 - \frac{e^2}{4\pi\epsilon_0}\frac{1}{r_{A1}}\right] + \left[-\frac{\hbar^2}{2\mu}\nabla_2^2 - \frac{e^2}{4\pi\epsilon_0}\frac{1}{r_{B2}}\right] + \frac{e^2}{4\pi\epsilon_0}\left[\frac{1}{r_{AB}} + \frac{1}{r_{12}} - \frac{1}{r_{B1}} - \frac{1}{r_{A2}}\right]$$

The Hamiltonian operator commutes with symmetry operators $\hat{\Gamma}$:

$$[\hat{H}, \hat{\Gamma}] = 0$$

Coulomb integral J in atomic units:

$$J = \frac{e^2}{4\pi\epsilon_o}\iint 1s_A(1)^2\left[\frac{1}{r_{AB}} + \frac{1}{r_{12}} - \frac{1}{r_{B1}} - \frac{1}{r_{A2}}\right]1s_B(2)^2\, d\tau_1 d\tau_2$$

Exchange integral K in atomic units:

$$K = \frac{e^2}{4\pi\epsilon_o}\iint 1s_A(1)1s_B(1)\left[\frac{1}{r_{AB}} + \frac{1}{r_{12}} - \frac{1}{r_{B1}} - \frac{1}{r_{A2}}\right] 1s_A(2)1s_B(2) d\tau_1 d\tau_2$$

Overlap integral S:

$$S = \int 1s_A(1)1S_B(1)d\tau_1$$

Molecular orbitals for the H_2 molecule:

$$\sigma = 1s_A + 1s_B$$

$$\sigma^* = 1s_A - 1s_B$$

PROBLEMS

Bond Energies, Shapes of Molecules, and Dipole Moments

12.1. The attractive energy between two univalent ions M^+ and A^- separated by a distance r is

$$-\frac{137.2}{r/\text{nm}}\ \text{kJ mol}^{-1}$$

Suppose that there is also a repulsive energy given by

$$\frac{0.0975}{(r/\text{nm})^6}\ \text{kJ mol}^{-1}$$

Plot the attractive and repulsive energies against r, and also plot the resultant energy, all on the same graph. By differentiating the equation for the resultant energy, calculate the equilibrium interionic distance and the net energy at that distance.

12.2. The equilibrium internuclear distance in gaseous LiI is 239 pm, and the dipole moment is 2.09×10^{-29} C m. Estimate the percentage ionic character of the bond.

12.3. The following are bond dissociation energies:

$$D(Li_2) = 113\ \text{kJ mol}^{-1}$$
$$D(H_2) = 435\ \text{kJ mol}^{-1}$$
$$D(LiH) = 243\ \text{kJ mol}^{-1}$$

a. Use Pauling's relationship (Eq. 12.78) to estimate the electronegativity difference between Li and H.
b. Estimate the percentage ionic character of the Li—H bond, given the following covalent radii:

Li: 126 pm

H: 36 pm

(*Note*: Since the Pauling relationship leads to dipole moments in debyes, it is often convenient to work problems of this type taking the electronic charge as 4.8×10^{-10} esu.)

12.4. Deduce the shapes of the following, on the basis of valence-shell electron-pair repulsion (VSEPR) theory:

$BeCl_2$, SF_6, H_3O^+, NH_4^+, PCl_6^-, AlF_6^{3-},
PO_4^{3-}, CO_2, SO_2, NH_3^{2+}, CO_3^{2-}, NO_3^-

(See a freshman chemistry text for help.)

12.5. Calculate the percentage ionic character of the HCl, HBr, HI, and CO bonds from the following data:

	HCl	HBr	HI	CO
Internuclear distance/pm	127	141	160	113
Dipole moment/10^{-30} C m	3.60	2.67	1.40	0.33

Molecular Orbitals

12.6. Use Figures 12.26 and 12.28 to construct molecular-orbital diagrams for the following:

B_2, CO, BN, BN^{2-}, BO, BF, OF, OF^-, OF^+

Deduce the bond order and paramagnetism in each case.

12.7. Sketch the molecular-orbital diagrams for the following:

N_2, O_2, C_2, F_2, CN, NO

Which of these species would you expect to become more stable if (a) an electron is added and (b) an electron is removed?

12.8. The hydrogen atom wave functions $1s_A$ and $1s_B$ are normalized. Prove that the molecular orbitals σ and σ^* given in Eqs. 12.29 and 12.30 are mutually orthogonal.

12.9. The four sp^3 hybrid orbitals are given in Eqs. 12.80–12.83. They are constructed from atomic orbitals, s, p_x, p_y, and p_z, which are normalized and mutually orthogonal.

a. The orbitals t_1, t_2, t_3, and t_4 are normalized. Prove that this is the case for t_3.
b. The orbitals t_1, t_2, t_3, and t_4 are mutually orthogonal. Prove that this is the case for t_2 and t_4.

12.10. Eigenfunctions for sp^2 hybridization, with maxima at 120° to one another, can be constructed with reference to the following diagram:

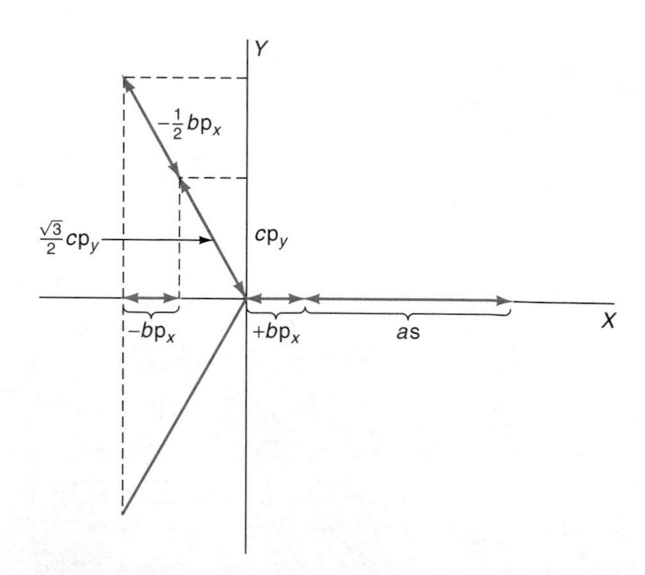

The plane is chosen as the XY plane, and there will be no contribution from the p_z orbital. An orbital along the X axis can be constructed as a linear combination of the s and p_x orbitals:

$$\psi_1 = as + bp_x$$

where a and b are numbers to be determined. The orbitals s and p_x are normalized and orthogonal. To obtain the other two orbitals, the p_x and p_y orbitals are resolved along the two directions 120° from the X axis and are combined with the s orbital:

$$\psi_2 = as - \frac{1}{2}bp_x + \frac{\sqrt{3}}{2}cp_y$$

$$\psi_3 = as - \frac{1}{2}bp_x - \frac{\sqrt{3}}{2}cp_y$$

Make use of the normalization and orthogonality conditions to determine the numbers a, b, and c.

12.11. Use the procedure of Problem 12.10 to construct two normalized and orthogonal wave functions for sp hydridization, at an angle of 180° to one another.

12.12. Find the expectation values of the $\hat{L}^2$ operator for the three sp^2 hybrid orbitals of Eqs. 12.84–12.86. (*Hint:* $L^2 = J(J + 1)\hbar^2$. Calculate $\langle L^2 \rangle$ for only $i = 1$ in the set of orbitals.)

12.13. As explained in Section 12.6, the exact energy ordering of the $2p\sigma_g$ and $2p\pi_u$ molecular orbitals depends, to some extent, on the particular molecule examined. Figure 12.26 shows the energy ordering of the molecular orbitals for the case of the nitrogen and oxygen molecules. Show for the C_2 molecule that we can determine the ordering of the $2p\sigma_g$ and $2p\pi_u$ molecular orbitals simply by checking whether the molecule is paramagnetic.

12.14. Using the LCAO MO technique, determine which of the two arrangements, linear $(H - H - H)^+$ or the triangular H_3^+, is the more stable state. *Hint*: Compute the energy of the molecular orbitals in each arrangement.

12.15. Calculate the coefficients of the occupied orbitals of butadiene with a value of $x = -1.6180$. The normalization factor is given by $1/\sqrt{(c_n/c_1)^2}$.

Group Theory

12.16. Give the point groups to which the following belong: (a) an equilateral triangle, (b) an isosceles triangle, and (c) a cylinder.

12.17. List the symmetry elements for each of the following molecules, and give the point group:

$CHCl_3$, CH_2Cl_2, naphthalene, chlorobenzene, NO_2(bent), cyclopropane, CO_3^{2-}, C_2H_2

12.18. The condition for optical activity, and for a molecule to exist in two enantiomeric forms, is that the molecule has neither a plane of symmetry nor a center of inversion. Can H_2O_2 exist in two mirror-image forms?

***12.19.** Deduce the symmetry species of the following vibrations:

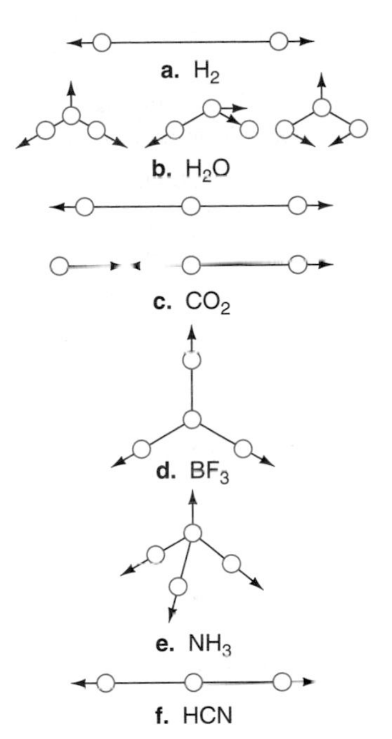

12.20. A molecule having a center of symmetry (i) or having an axis of improper rotation (S) cannot have a dipole moment; all other molecules can. Note, however, that in tables such as Table 12.3 the symmetry operation S is not always included, since it follows from other operations (for example, as seen in Figure 12.20, S_4 is the product $C_4\sigma_h$). The following are molecules that have no center of symmetry but have no dipole moment:

cyclopropane   cyclopentane

In each case, identify the axis of improper rotation, designate the operation, and relate the operation to the appropriate rotations and reflections. Which of the point groups in Table 12.3 are such that there can be a dipole moment?

***12.21.** The spectroscopic properties of several inorganic complexes can be explained in terms of the energy splitting of the five 3d orbitals as a result of the electron-electron repulsion between the 3d electrons and those on the approaching ligands. In the case of octahedral complexes like $[Fe(CN)_6]^{3+}$, it is seen that the set of five 3d orbitals on Fe give rise to three symmetry-adapted orbitals of what is known as T_{2g} symmetry (d_{xy}, d_{yz}, d_{xz}) and two of what is known as E_g symmetry (d_{z^2}, $d_{x^2-y^2}$). If the coordinate system for analyzing the energy levels of this complex is set up such that Fe is at the origin and the six ligands approach along the three Cartesian axes (from the positive and negative sides of each), explain which set of orbitals will have higher energy.

Essay Questions

12.22. Give an account of the valence-shell electron-pair repulsion (VSEPR) theory of the shapes of molecules.

12.23. Explain orbital hydridization, with special reference to sp, sp^2, and sp^3 hydridization.

12.24. Explain the principles underlying the construction of trial wave functions (a) in the valence-bond method and (b) in the molecular-orbital method.

12.25. Explain the theory underlying the estimation of bond dipole moments from electronegativities.

A considerable number of problems and solutions relating to molecular structure and spectroscopy are to be found in

T. A. Albright and J. K. Burdett, *Problems in Molecular Orbital Theory,* New York: Oxford University Press, 1992.

G. J. Bullen and D. J. Greenslade, *Problems in Molecular Structure,* London: Pion, 1983.

C. S. Johnson and L. G. Pedersen, *Problems and Solutions in Quantum Chemistry and Physics,* New York: Dover, 1986.

SUGGESTED READING

See also Suggested Reading for Chapter 11 (p. 575).

For general accounts of chemical bonding, see

P. W. Atkins, *Molecular Quantum Mechanics,* Oxford University Press, 1970.

R. F. W. Bader, *Atoms in Molecules: A Quantum Theory,* Oxford: Clarendon Press, 1990; softcover edition 1994.

C. J. Ballhausen and H. B. Gray, *Molecular Electronic Structures, An Introduction,* Reading, MA: Benjamin Cummings, 1980.

R. Carbo (Ed.), *Quantum Chemistry: Basic Aspects, Actual Trends,* Amsterdam: Elsevier, 1989.

E. Cartmell and G. W. A. Fowles, *Valency and Molecular Structure,* London: Butterworth, 1961; 4th edition, 1977.

C. A. Coulson (revised by R. McWeeny), *The Shape and Structure of Molecules,* Oxford: Clarendon Press, 1982.

C. E. Dykstra, *Ab Initio Calculations of the Structures and Properties of Molecules,* Amsterdam: Elsevier, 1988.

R. J. Gillespie, *Molecular Geometry,* London: Van Nostrand Reinhold, 1972.

W. J. Hehre, L. Radom, P. V. R. Schleyer, and J. A. Pople, *Ab Initio Molecular Orbital Theory,* New York: John Wiley, 1986.

Y. Jean and François Volatron, *An Introduction to Molecular Orbitals* (translated and edited by J. K. Burdett), New York: Oxford University Press, 1993.

D. A. McQuarrie, *Quantum Chemistry*, Mill Valley, CA: University Science Books, 1983.

R. McWeeny, *Coulson's Valence,* Oxford University Press, 3rd edition, 1979.

J. N. Murrell, S. F. A. Kettle, and J. M. Tedder, *Valence Theory,* New York: John Wiley, 1965. A less mathematical version of this book has appeared as *The Chemical Bond,* Wiley, 1978.

L. Pauling, *The Nature of the Chemical Bond,* Ithaca, NY: Cornell University Press, 1939, 3rd edition, 1960.

J. D. Roberts, *Notes on Molecular Orbital Calculations*, New York: W. A. Benjamin, Inc., 1962.

J. G. Verkade, *A Pictorial Approach to Chemical Bonding,* New York: Springer, 1986.

B. Webster, *Chemical Bonding Theory,* Oxford: Blackwell, 1990.

M. J. Winter, *Chemical Bonding,* Oxford University Press, paperback, 1994.

K. Yates, *Hückel Molecular Orbital Theory,* New York: Academic Press, 1978.

A. Zewail (Ed.), *The Chemical Bond: Structure and Dynamics,* San Diego: Academic Press, 1992. This volume, in honor of Linus Pauling's ninetieth birthday, contains articles by a number of distinguished scientists.

Some interesting special aspects of valence theory are also to be found in

R. S. Mulliken, "Spectroscopy, molecular orbitals, and chemical bonding," *Science, 157,* 13–24(1967). This is Mulliken's Nobel Prize address.

R. S. Mulliken, *Life of a Scientist,* New York: Springer Verlag, 1989. This book, edited by B. J. Ransil, includes an introduction by F. Hund, an outline of Mulliken's career, and a complete list of his publications.

R. G. Pearson Electronegativity Scales, *Accounts of Chemical Research, 23,* 1(1990).

C. A. Russell, *The History of Valency,* Leicester: The University Press, 1971.

J. W. Servos, *Physical Chemistry from Ostwald to Pauling: The Making of a Science in America,* Princeton: The University Press, 1990. This book contains interesting accounts of the work of Lewis and Pauling on the chemical bond.

The application of group theory to chemical bonding is treated in

D. M. Bishop, *Group Theory and Chemistry,* Oxford: Clarendon Press, 1973.

P. R. Bunker, *Molecular Symmetry and Spectroscopy,* New York: Academic Press, 1979.

F. A. Cotton, *Chemical Applications of Group Theory* (3rd ed.), New York: John Wiley, 1990.

J. L. Kavanau, *Symmetry: an Analytical Treatment,* Los Angeles: Science Software Systems, 1980. This is a somewhat advanced and detailed (800 pp.) treatment of the subject.

M. F. C. Ladd, *Symmetry in Molecules and Crystals,* New York: John Wiley, 1989.

Foundations of Chemical Spectroscopy

13

PREVIEW

Chapters 11 and 12 deal with atoms and molecules from the point of view of quantum mechanics. Important experimental information about atoms and molecules is provided by *spectroscopy,* which is the study of the interaction between electromagnetic radiation and matter. A fundamental relationship is the *Lambert-Beer law,* which relates the amount of radiation absorbed to the concentration of absorbing material and the distance the light passes through the absorber. Many techniques in analytical chemistry are based on this basic law.

Of more fundamental importance to the understanding of the nature of atoms and molecules is the examination of the individual lines that arise from absorption or emission of electromagnetic radiation. Data on the location of lines in the spectrum give information about the energy levels that are occupied. The Rydberg spectrum of the hydrogen atom, in which groups of lines become more closely spaced at increasing wavelength, is shown in Figure 11.8 and is characteristic of all *atomic spectra.* The simple Bohr treatment in Chapter 11 of the atomic spectrum of the hydrogen atom can be improved by the inclusion of some additional effects; in particular, *Coulombic interactions* between the different electrons in an atom, *exchange interactions,* and *spin-orbit interactions,* which result from the magnetic fields associated with electron spin and orbital motion. The *Zeeman effect* and the *anomalous Zeeman effect* are the effects of external magnetic fields on spectra.

In *molecular spectra* rotational and vibrational transitions can be superimposed on the electronic transitions. When the light quanta are of low energy (in the far infrared or microwave regions of the spectrum), only rotational transitions can occur. At higher energies (the near infrared) both rotational and vibrational transitions occur. At still higher energies (the visible and ultraviolet) electronic transitions are possible, accompanied by vibrational and rotational transitions. Analysis of the rotational transitions provides *interatomic distances* and information about molecular geometries. Analysis of the vibrational transitions yields the *force constants* that give information about the strength of bonds. Analysis of the ultraviolet spectra yields information about bonding and antibonding states.

OBJECTIVES

After studying this chapter, the student should be able to:

- Understand that chemical spectroscopy is concerned with a coupling mechanism between classical electromagnetic waves and quantum-mechanical atoms and molecules.
- Understand how absorption, spontaneous emission, and stimulated emission are described by the Einstein coefficients.
- Appreciate how a transition takes place during the absorption or emission of a photon, and understand the importance of the existence of non-zero matrix elements and of satisfying the Bohr condition for such transitions.
- Appreciate the existence of electric and magnetic moments in atoms and molecules and their role in spectroscopy.
- Understand the origin of spectroscopic selection rules and of symmetry breaking.
- Describe the principles of atomic adsorption and emission spectroscopy and the use of term symbols to describe atomic states.
- Understand the vector model description of the atom and its application to the electronic structure of molecules and to magnetic systems discussed in Chapter 14.
- Understand how spectroscopic studies in different regions of the spectrum probe different types of molecular transitions: rotational, rotational-vibrational, and electronic.
- Describe diatomic, symmetric-top, and asymmetric top molecules.
- Appreciate how in certain situations the total wave function can be approximated as a product of wave functions.
- Understand the essential difference between emission and absorption spectroscopies and Raman spectroscopy.
- Understand molecular term symbols and the vector model of simple molecules.
- Appreciate the various coupling situations that are summarized in the four Hund cases.

The subject of *spectroscopy* is concerned with the interaction of matter with electromagnetic radiation. We saw in Chapter 11 that the electromagnetic spectrum spans a vast range of frequencies, from radio waves to gamma rays. We also saw that electromagnetic radiation behaves both as a wave and as a beam of particles.

As far as spectroscopy is concerned we need to consider only the wave nature of radiation. In other words, we can deal with spectra by regarding atoms and molecules as existing in a sea of electromagnetic waves. In spite of this, in chemical spectroscopy we commonly use language that suggests the particle nature of radiation; we speak of the emission or adsorption of a quantum, or photon, of radiation.

Spectroscopic measurements, combined with theoretical interpretations, provide detailed information about chemical structure and the arrangement of electrons in atoms and molecules. Atomic spectra, for example, have been valuable in providing information about the energy levels that are available to the electrons, and the result can be compared with those predicted by quantum mechanics. Molecules often are more complex because of the variety of ways in which energy transformations can occur, and thus their spectra can be more complicated. Molecular spectra provide valuable information about molecular vibrations, bond energies, and other properties.

Spectroscopy also plays a very useful role in the determination of the concentrations of substances. Some of the experimental methods are simple and are widely used.

13.1 Emission and Absorption Spectra

The foundation stone of spectroscopy was laid in 1665–1666 when the English scientist Sir Isaac Newton (1642–1727) demonstrated that white light passed through a prism is split into a spectrum of colors ranging from red to violet. The red end of the visible spectrum corresponds to longer wavelengths (700 nm or 7000 Å) and lower frequencies. The violet end of the spectrum has shorter wavelengths (400 nm or 4000 Å) and higher frequencies. The violet end corresponds to higher photon energies, since photon energies are equal to the frequency multiplied by the Planck constant. Because of its higher energy, violet light is in general more effective than red light in bringing about chemical and biological changes.

Classical Electromagnetic Waves

In Section 11.1 we considered the electromagnetic spectrum, which is represented in Figure 11.1, and we saw that there are vast spectral regions to which the human eye is not sensitive. The classical treatment of EM radiation by James Clerk Maxwell led to the conclusion that it is composed of electric and magnetic components oscillating at right angles to each other, as seen in Figure 11.2. If we position ourselves at some point along the line of propagation, such as the X direction, the electric and magnetic components can be written respectively as

$$E_x(\omega,t) = 2E_x^0 \cos \omega t \tag{13.1}$$

and

$$H_x(\omega,t) = 2H_x^0 \cos \omega t \tag{13.2}$$

The terms E_x^0 and H_x^0 give the amplitudes of the electric and magnetic components of the waves. As written, Eqs. 13.1 and 13.2 describe one frequency component ω (called monochromatic EM radiation). EM radiation can be characterized by either a frequency ν or a wavelength λ, the product being $\lambda\nu = c$, the speed of light (see Eq. 11.1). Frequencies can be expressed in the SI unit of reciprocal seconds or Hertz (s^{-1} = Hz). The quantity omega, ω ($= 2\pi\nu$), has units of radians per second and is widely used. For practical applications of spectroscopy these units are often inconvenient, because the numbers involved are usually large. A common practice is to use the reciprocal of the wavelength instead of the frequency; that is, $1/\lambda$ instead of c/λ. The term **wavenumber** is then employed, and its usual symbol is $\tilde{\nu}$. If λ is expressed in metres, $\tilde{\nu} = 1/\lambda$ will be in reciprocal metres (m^{-1}), and multiplication by the velocity of light (2.998×10^8 m s^{-1}) gives ν in reciprocal seconds (s^{-1}). Wavenumbers are more often expressed in the units of reciprocal centimetres; 1 cm^{-1} = 100 m^{-1}. Multiplication of $\tilde{\nu}$ cm^{-1} by 2.998×10^{10} cm s^{-1} gives the frequency ν. The quantity omega, ω, is also widely used to represent frequency.

Wavenumber

Various units have been used for wavelength, the most important of these being:

$$1 \text{ nanometre (nm)} = 10^{-9} \text{ m}$$
$$1 \text{ angstrom (Å)} = 10^{-8} \text{ cm} = 10^{-10} \text{ m} = 0.1 \text{ nm}$$
$$1 \text{ micrometre } (\mu\text{m}) = 10^{-6} \text{ m}$$

Although not an SI unit, the angstrom has been approved by IUPAC for temporary use. The micrometre used to be called a micron, and the nanometre a millimicron, but these terms are now discouraged.

The Energy of Radiation in Emission and Absorption

Spectra can result from either emission or absorption of energy. In order to observe an emission spectrum, the substance must be excited in some way. This may be done for an atomic species by introducing it into a flame or by passing an electric discharge through it. If the emitted light is observed through a spectrometer, the characteristic **emission spectrum** of the substance will be seen. In Section 11.2 we considered the emission spectrum of hydrogen.

The procedure for obtaining an **absorption spectrum** is quite different. An absorption spectrum can be observed by passing continuous radiation, such as white light, through a substance in the vapor phase or in solution and observing the spectrum with a spectrometer, (see Figure 13.1). A spectrometer controls the frequency and energy density (J m^{-3} s), $w(\omega)$, of the electromagnetic radiation along the direction of the light beam into which a sample is placed. The energy density is proportional to the average of the amplitude squared, $(E_x^0)^2$, (refer to Eq. 11.18),

$$w(\omega) = \frac{1}{2}\,\epsilon_o\,[(E^0)^2]_{ave} = \frac{3}{2}\,\epsilon_o\,[(E_x^0)^2]_{ave} \tag{13.3}$$

When a sample is placed in a spectrometer, certain wavelengths are missing from the spectrum. For example, if continuous radiation of white light is used, its normal spectrum consists of colors from red to violet, which blend smoothly into one another. After the white light has passed through the substance, however, the absorption spectrum consists of black lines superimposed on the continuous spec-

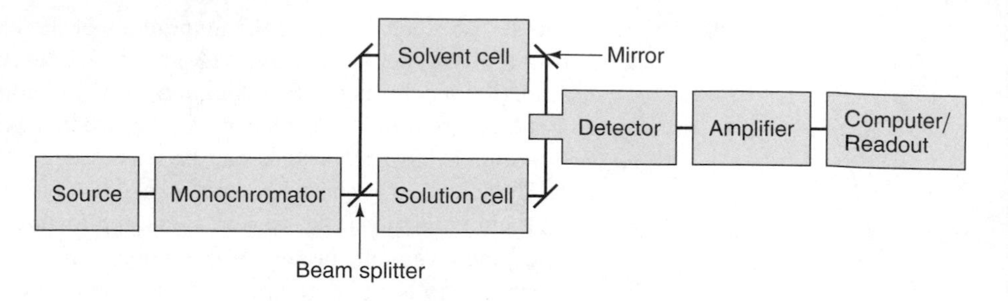

FIGURE 13.1
Schematic of a simple double-beam spectrophotometer. A chopper before the detector may break the light beams so that the detector can tell the difference between the two signals. Alternatively, two matched detectors may be used and the difference measured electronically.

The components of a spectrophotometer change depending upon the region of the spectrum being investigated. A partial listing includes:

For the visible and ultraviolet region: *Source:* tungsten-I_2 lamp or H_2 discharge lamp; *Cell:* Quartz, plastic, or glass; *Detector:* Prism or grating with photomultiplier; *Information:* Electronic spectra.

For the infrared region: *Source:* Hot ceramic source; *Cell:* Sample dispersed in KBr or Nujol, or gas in sample cell; *Detector:* Rock-salt prism or grating with thermocouple detector, or thermistor bolometer (This consists of a semiconductor material whose electrical resistance changes with temperature.) *Information:* Molecular vibration spectra.

For the microwave region: *Source:* Klystron tube or other oscillator; *Cell:* Cavity and waveguide; *Detector:* Crystal diode [W tip in contact with semiconductor (Ge, Si, GaAs)]; *Information:* Molecular rotation spectra.

trum. These black lines occur because the substance through which the light passed has removed (absorbed) light corresponding to certain wavelengths. The energy of a transition causing such a line is the frequency of the line multiplied by the Planck constant (Eq. 11.38).

It is often important to evaluate the energy associated with absorption in a specific region of the spectrum. Since the energy of a photon ϵ is equal to $h\nu$, we can calculate an energy per mole as follows:

$$\begin{aligned} E &= L\epsilon = Lh\nu \\ &= 6.022 \times 10^{23}(\text{mol}^{-1}) \times 6.626 \times 10^{-34}(\text{J s}) \\ &\quad \times \frac{2.998 \times 10^{8}(\text{m s}^{-1}) \times 10^{-3}(\text{kJ J}^{-1})}{10^{-9}(\text{m nm}^{-1}) \times \lambda/\text{nm}} \\ &= \frac{1.196 \times 10^{5}}{\lambda/\text{nm}} \text{ kJ mol}^{-1} \end{aligned} \tag{13.4}$$

EXAMPLE 13.1 Calculate the energy in kJ mol^{-1} corresponding to absorption at 400 nm.

Solution Substitution of $\lambda = 400$ nm into Eq. 13.4 gives 2.99×10^2 kJ mol^{-1}.

The amount of energy absorbed by 1 mol of a compound at a particular wavelength has been called an **einstein** and is the energy of 1 mol of photons at that wavelength. In Example 13.1, 1 einstein equals 2.99×10^2 kJ.

Time-Dependent Perturbation Theory and Spectral Transitions

Up to this stage in the treatment of atoms and molecules, we have used the time-independent Schrödinger equation, Eq. 11.86, to obtain the stationary states that characterized their quantum nature. That is, we started with the time-dependent Schrödinger equation (Eq. 11.82a), and wrote the wave function as Eq. 11.83, repeated here:

$$\Psi(x, y, z, t) = \psi(x, y, z)e^{-iEt/\hbar} \tag{11.83}$$

This gave us the time-independent Schrödinger equation, Eq. 11.86, which for a specific eigenvalue and eigenfunction is

$$\hat{H}\psi_i = E_i\psi_i \tag{11.86}$$

Let us assume that we can solve this equation and that all the eigenvalues are known. We will call these the **eigenstates** of the system. Figure 13.2 shows schematically a set of such levels. To each level, there corresponds an eigenfunction, which we will assume to be known. For simplicity, we will also assume that no degenerate energy levels exist, so that there is one and only one eigenvalue for each eigenfunction. From Eq. 11.83 each eigenfunction can be written as

$$\Psi_i^0 = \psi_i e^{-iE_i t/\hbar} \tag{13.5}$$

where, again for simplicity of notation, we suppress the position and time arguments of the wave functions.

Equation 13.5, written for all the eigenfunctions of our system, expresses that certain stationary states exist at various quantized energies E_i. Now we ask: What happens when we place this sample into an electromagnetic (EM) field of a certain amplitude and frequency? That is, how do transitions occur between the various levels? In other words, how is light absorbed or emitted from one eigenstate to another?

For interactions to occur, there always has to be some mechanism that couples them. In this particular case, the classical EM radiation must couple in some way with the molecule, which is suspended in space in one of its eigenstates. (Alternately, a large number of molecules can be in a certain distribution of eigenstates, such distribution usually being given by the Boltzmann distribution.) Recall from our study of blackbody radiation that when a blackbody is heated, EM radiation is emitted. Molecules having oscillating dipole moments (see Section 11.1) emit light. In general, atoms and molecules can possess a variety of electric and magnetic moments, and it is these that can couple with the EM radiation. For example, the electric dipole moment, μ_x, (see Figure 12.18) results from a separation of electric charge δq, by a distance x, and is given by

$$\mu_x = x\delta q \tag{13.6}$$

It is responsible for much of the spectroscopy that we observe (atomic absorption, atomic emission, rotational, vibrational, and electronic spectroscopy). Raman spectroscopy depends upon the ability of light to induce a dipole moment in a molecule. Electron spins generate a magnetic moment, and this is responsible for electron spin resonance spectroscopy (ESR). Nuclei can give rise to nuclear magnetic moments by virtue of their nuclear spin leading to nuclear magnetic resonance (NMR).

The electric moments interact with the electric components of the EM radiation (Eq. 13.1), and the magnetic moments interact with the magnetic components of the EM radiation (Eq. 13.2). We treat here only the electric moments, and these allow molecules and EM radiation to couple. The strength of this coupling is not large under normal circumstances and can be treated as a perturbation. Hence Eq. 11.82a becomes

$$[\hat{H} + \hat{H}'(t)]\Psi = i\hbar\frac{\partial \Psi}{\partial t} \tag{13.7}$$

The perturbing Hamiltonian expression is

$$\hat{H}'(t) = 2\hat{\mu}_x E_x^0 \cos \omega t \tag{13.8}$$

which is the electric dipole moment of a particle coupling with the electric component of the EM radiation, Eq. 13.1.

We have treated perturbation theory before, but this case is different because the perturbation is time dependent. The procedure is similar, but we must start with

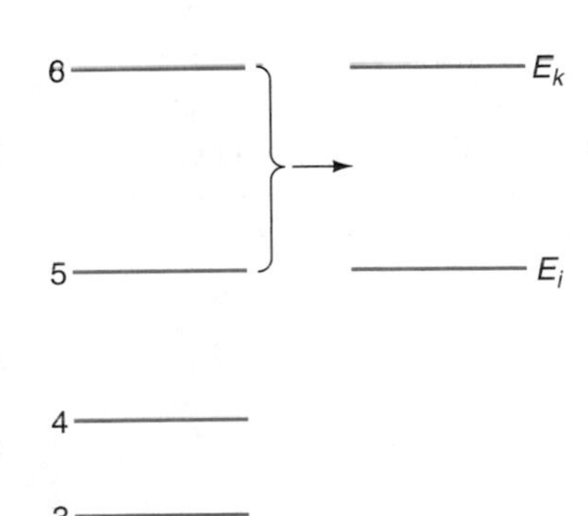

FIGURE 13.2
Typical stationary states arising from the solution of the time-independent Schrödinger equation, Eq. 11.86. Two levels, E_i and E_k, are treated as a general pair in Figures 13.3 and 13.4.

the time-dependent Schrödinger equation and take into account that the perturbing Hamiltonian is time dependent. To proceed, we assume that the EM radiation mixes the various eigenstates so that rather than having a set of pure eigenstates (Eq. 13.5), we must account for the possible mixing (transitions) between them and write

$$\Psi = \sum_i c_i\,(t)\,\Psi_i^0(t) \tag{13.9}$$

where the unperturbed eigenfunctions (Eq. 13.5) have been used and the sum is over all the eigenstates available.

Equation 13.9 expresses exactly what we want to know. In the absence of the EM radiation, $H'(t) = 0$, so that the stationary states exist (Figure 13.2). When the perturbation is non-zero, the states are mixed, or, in other words, transitions can occur between one state ψ_i and other ψ_k. If we can calculate the coefficients $c_k(t)$, then the problem is solved. In particular, due to orthogonality of the states, the probability Ψ_k^0 that the system is found in eigenstate k after a time t is

$$c_k^*(t)\,c_k(t) \tag{13.10}$$

EXAMPLE 13.2 Assume that a system is initially in one eigenstate at time $t = 0$, (perhaps the ground state) given by $\Psi_i^0\,(0)$; that is, $c_i(0) = 1$ and $c_k(0) = 0$ for $k \neq i$. Obtain the equation that determines the coefficients, $c_k(t)$, after a time-dependent perturbation is introduced.

Solution Substitute Eq. 13.9 into the right-hand side of Eq. 13.7 and use the chain rule to give.

$$[\hat{H} + \hat{H}'(t)]\Psi = i\hbar\,\frac{\partial}{\partial t}\sum_i c_i(t)\Psi_i^0 = i\hbar\sum_i \frac{dc_i(t)}{dt}\,\Psi_i^0$$

$$+\, i\hbar\sum_i c_i(t)\,\frac{\partial \Psi_i^0}{\partial t} = i\hbar\sum_i \frac{dc_i(t)}{dt}\,\Psi_i^0 + \sum_i c_i(t)\hat{H}\,\Psi_i^0$$

The last step uses Eq. 11.82, which can be simplified using Eq. 13.9 to give

$$[\hat{H} + \hat{H}'(t)]\Psi = i\hbar\sum_i \frac{dc_i(t)}{dt}\,\Psi_i^0 + \hat{H}\,\Psi \tag{13.11}$$

Note that the first term on the left-hand side and last term on the right-hand side cancel. We can cancel these, rearrange the equation a little, multiply both sides by Eq. 13.9, and use the orthogonality of the original unperturbed eigenfunctions,

$$\int \Psi_k^{0*}\,\Psi_i^0\,d\tau = \delta_{ki}$$

The result is

$$\frac{dc_k\,(t)}{dt} = -\frac{i}{\hbar}\sum_i c_i(t)\int \Psi_k^{0*}\,\hat{H}'(t)\Psi_i^0\,d\tau \tag{13.12}$$

At the beginning of this example, we assumed that the system was initially in one eigenstate with $c_i(0) = 1$ and $c_k(0) = 0$ for $k \neq i$. This removes the summation. Finally, substitute Eq. 11.83 to give our desired equation

$$\frac{dc_k\,(t)}{dt} = -\frac{i}{\hbar}\,e^{[i(E_k\,-\,E_i)t/\hbar]}\int \psi_k^*\,\hat{H}'(t)\psi_i\,d\tau = -\frac{i}{\hbar}\,e^{[i(E_k\,-\,E_i)t/\hbar]}\,\langle \psi_k \mid \hat{H}'(t) \mid \psi_i\rangle \tag{13.13}$$

Equation 13.13 can be solved further only by introducing the perturbing Hamiltonian, such as Eq. 13.1, which contains a $\cos(\omega t)$ term from the oscillating EM radiation.

> ***EXAMPLE 13.3*** Collect the time dependencies that occur on the left-hand side of Eq. 13.13.
>
> ***Solution*** The $\cos \omega t$ term from the perturbing Hamiltonian can be written as
>
> $$\cos \omega t = \frac{e^{i\omega t} + e^{-i\omega t}}{2} \tag{13.14}$$
>
> which combines to give
>
> $$f_{ik}(\omega, t) = \frac{e^{[i(E_k - E_i)t/\hbar + \omega t]} + e^{-[i(E_i - E_k)t/\hbar + \omega t]}}{2} \tag{13.15}$$

We obtain, after using the results of Example 13.3 on Equation 13.11,

$$\frac{dc_k(t)}{dt} = -\frac{i}{\hbar} f_{ik}(\omega, t)\, E_x^0 \langle \psi_k \,|\, \mu_x \,|\, \psi_i \rangle \tag{13.16}$$

where E^0 is a function of w.

This can be integrated over time to give the coefficient $c_i(t)$ from the initial condition that $c_k(0) = 0$.

Before doing this, to complete the problem, note that it is necessary that the two states i and k be coupled by, in this case, the electric dipole moment. That is, the integral

$$\langle \psi_k \,|\, \mu_x \,|\, \psi_i \rangle = \int \psi_k^* \mu_x \psi_i \, d\tau \tag{13.17}$$

must be non-zero for $i \neq k$. If this integral is zero, then the transition between such states is *forbidden*. All the selection rules for spectroscopy, in this case due to the electric dipole moment, arise from solving these integrals. We call these integrals **matrix elements**.

Matrix Elements

Bohr Condition

Not only must the matrix element be non-zero, but the frequency of the EM radiation must also be close to the **Bohr condition**, that is, in a change of energy from one energy level to another, the difference in energy is carried away by a photon of energy $\hbar\omega_{ki}$

$$E_k - E_i \cong \hbar\omega_{ki} \tag{13.18}$$

Unless this is true, the integral over time involving Eq. 13.15 gives a vanishingly small value. Application of this condition leads fairly easily to the final result for the probability of ending up in eigenstate k after starting from eigenstate i and is (see Eq. 13.12)

$$c_k^*(t)\, c_k(t) = \frac{1}{\hbar^2} [\langle \psi_i \,|\, \mu_x \,|\, \psi_k \rangle]^2 (E_x^0)^2 t \tag{13.19}$$

The Einstein Coefficients

When a sample is placed in a spectrometer, the greater the energy density $w(\omega_{ki})$, Eq. 13.3, and the longer the sample is exposed, the greater the transition probability

Einstein Coefficient

for the system to end up in state k when it started out in state i. Einstein expressed this conclusion as

$$c_k^*(t)c_k(t) = B_{ki}\,\rho(\omega_{ki})t \tag{13.20}$$

In other words, starting in an initial eigenstate i at time $t = 0$, the probability of a particle being in eigenstate k at a later time t in an EM field of energy density $\rho(\omega_{ki})$ depends upon the coefficient B_{ki}. These **Einstein coefficients** can be identified from Eqs. 13.19 and 13.20 using Eq. 13.3, and are given by

$$B_{ki} = B_{ik} = \frac{2\pi[\langle\psi_i\,|\,\mu_x\,|\,\psi_k\rangle]^2}{3\hbar^2} \tag{13.21}$$

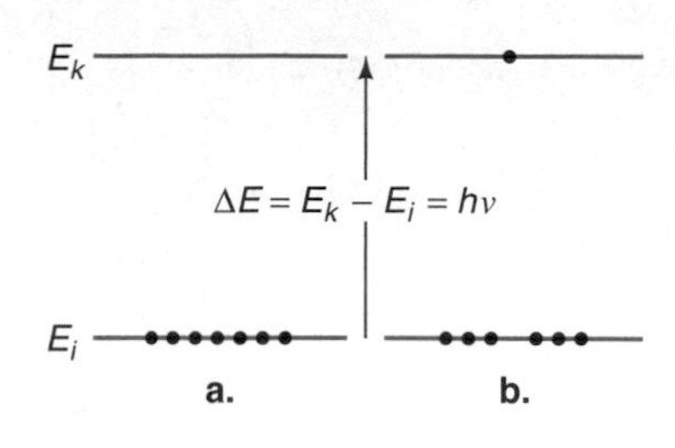

FIGURE 13.3
In (a), two energy levels are shown where only the ground state E_i is occupied. When an electron is excited by stimulated absorption in (b) the electron moves to an excited state E_k with a change of energy $\Delta E = h\nu = E_k - E_i$. The number of electrons in the k state when a large population of identical particles is present is given by the Boltzmann distribution.

If the number of particles in state i at time $t = 0$ is N_i, then the rate of transition to state k of higher energy is

$$\frac{dN_i}{dt} = -N_iB_{ki}\rho(\omega_{ki}) \tag{13.22}$$

This is the rate of absorption. Energy levels for this kind of transition are shown in Figure 13.3.

It was Einstein who pointed out that that there are two processes to consider in emission, *spontaneous emission* and *radiation-induced emission* (*stimulated emission*). Spontaneous emission does not depend on the radiation and is given by

$$\frac{dN_k}{dt} = -N_kA_{ki} \tag{13.23}$$

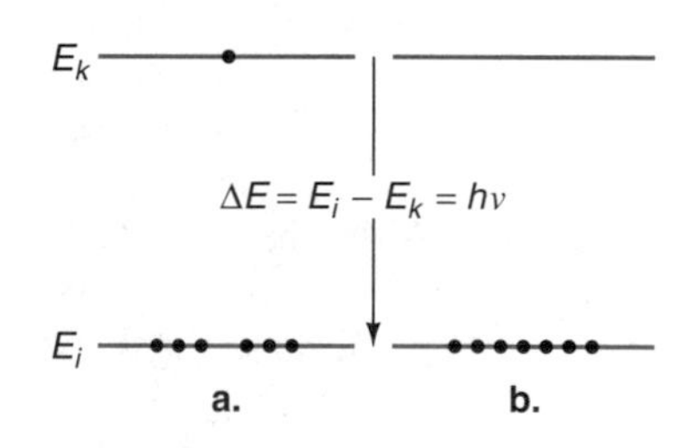

FIGURE 13.4
As shown in (a) an electron starts in an excited state E_k. In spontaneous emission, shown in (b), the electron drops back to the lower level E_i.

where A_{ki} is the **Einstein coefficient of spontaneous emission**. Figure 13.4 shows a normal emission in which an electron falls back to a ground state. An energy diagram illustrating the process of **stimulated emission** is shown in Figure 13.5. The number of emission transitions brought about by the exciting radiation is

$$\frac{dN_k}{dt} = -N_kB_{ki}\,\rho(\omega_{ki}) \tag{13.24}$$

where B_{ki} is the **Einstein coefficient of stimulated emission** since the two coefficients B_{ik} and B_{ki} are equal to one another. The total number of transitions from the higher to the lower state is thus

$$\frac{dN_k}{dt} = -N_k[A_{ki} + B_{ki}\rho(\omega_{ki})] \tag{13.25}$$

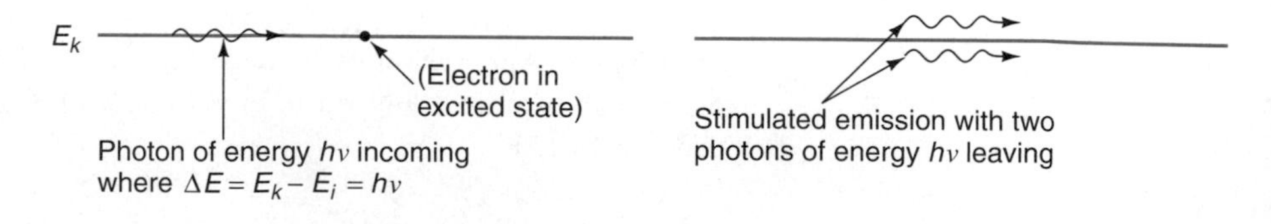

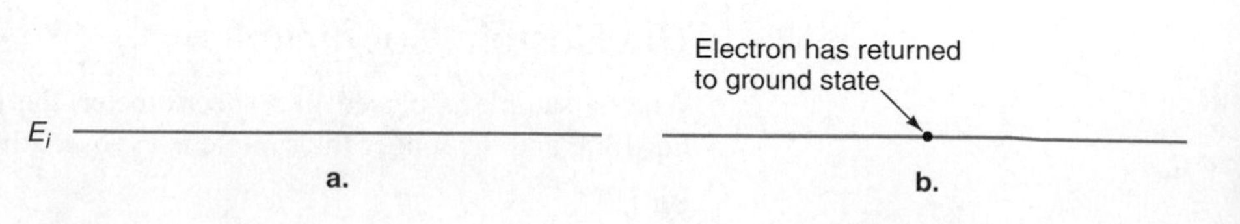

FIGURE 13.5
Stimulated emission by an incoming photon in (a) causes the electron to return to the ground state with the emission of a photon as shown in (b). Both photons have energy $\Delta E = E_n - E_m$.

A steady state will be established in which the number of molecules passing from i to k is equal to the number passing from k to i:

$$N_i B_{ki}\rho(\omega_{ki}) = N_k[A_{ik} + B_{ki}\rho(\omega_{ki})] \tag{13.26}$$

The value of A_{ik} cannot be calculated from radiation theory, but it can be deduced in the following way on the basis of the theory of the Boltzmann distribution. According to this theory, which is considered in Section 15.2, at equilibrium the ratio of populations of molecules in the two states k and i is

$$N_k/N_i = e^{-\hbar\omega ki/kBT} \tag{13.27}$$

Thus the equilibrium population of the lower-energy level i is always larger than that of the excited level k, except in the limit of infinite temperature.

Equations 13.26 and 13.27 give, since $B_{ki} = B_{ik}$,

$$e^{-\hbar\omega ki/kBT} = \frac{B_{ki}\rho(\omega_{ki})}{A_{ki} + B_{ki}\rho(\omega_{ki})} \tag{13.28}$$

The radiant energy density $\rho(\nu_{ik})$ is given by the Planck radiation law (compare Eq. 11.35, and note that $\hbar\omega = h\nu$):

$$\rho(\omega_{ki}) = \frac{h\omega_{ki}^3}{\pi^2 c^3} \frac{1}{e^{-\hbar\omega_{ki}/k_B T} - 1} \tag{13.29}$$

Since Eq. 13.28 holds at equilibrium, we may substitute Eq. 13.29 into it and obtain, after simplification,

$$\frac{A_{ki}}{B_{ki}} = \frac{8\pi h\nu^3{}_{ki}}{c^3} \tag{13.30}$$

Equations 13.26 and 13.27 also suggest that if a way were found to populate the upper state independently of the exciting radiation, large quantities of energy could be stored in the excited state. Normally radiation that excites the electrons into the higher state simultaneously induces them to emit radiation, thereby returning the electrons to the lower energy state at a rate equal to that of their production. Thus there is no net gain in energy. However, if a population difference can be maintained so that the upper state has a greater population than the lower state, then stimulated emission will dominate. If this can be maintained, we have the conditions for light amplification by the stimulated emission of radiation, or laser. See Section 14.1.

One competing process that opposes establishment of a population inversion between the excited and ground state is the spontaneous emission process governed by A_{ki}. From Eq. 13.30, the frequency dependence of A_{ki} relative to B_{ki} shows that the third power of the frequency differentiates the two. This means that spontaneous emission will increase as the energy of the excited state increases. For this reason, it is more difficult to fabricate lasers that work in the UV and higher frequencies than in the visible and lower frequencies. Lasers are considered in more detail in Section 14.1.

The Laws of Lambert and Beer

When spectroscopic studies of substances are combined with measurements of light absorbed and transmitted at various wavelengths, the term **spectrophotometry** is employed. This type of investigation is based on laws propounded by the German mathematician Johann Heinrich Lambert (1728–1777) and the German astronomer

Wilhelm Beer (1797–1850). These laws are concerned with the intensities of light absorbed or transmitted when incident light is passed through some material.

Lambert's law states that the proportion of radiation absorbed by a substance is independent of the intensity of the incident radiation. This means that each successive layer of thickness dx of the medium absorbs an equal fraction $-dI/I$ of the radiant intensity I incident upon it. In other words, the relative loss of intensity is proportional to the thickness (dx),

$$-\frac{dI}{I} = b\,dx \tag{13.31}$$

where b is a constant.[1] Integration of this equation, for passage of light through a distance l, proceeds as follows:

$$\int \frac{dI}{I} = -b \int_0^l dx \tag{13.32}$$

or

$$\ln I = -bl + g \tag{13.33}$$

where g is the constant of integration. This constant can be evaluated using the boundary condition that $I = I_0$ when $l = 0$, where I_0 is the intensity of radiation before passage through the medium. Thus,

$$g = \ln I_0 \tag{13.34}$$

and therefore

$$\ln I = -bl + \ln I_0 \tag{13.35}$$

or

$$\ln \frac{I_0}{I} = bl \tag{13.36}$$

or

$$I = I_0 e^{-bl} \tag{13.37}$$

It is more usual to employ common logarithms, and instead of Eq. 13.36 we have

$$\log_{10} \frac{I_0}{I} = \frac{bl}{2.303} = A \tag{13.38}$$

Absorbance

The quantity A is known as the *decadic absorbance,* or simply as the **absorbance.** It was formerly called the *extinction* or the *optical density,* but IUPAC now discourages these terms.

Transmittance

The **transmittance** T is the ratio of the intensities of transmitted to incident light:

$$T \equiv \frac{I}{I_0} \tag{13.39}$$

[1] The IUPAC "Green Book" (see Appendix A) gives special names to a number of the parameters that relate to the absorption of radiation. The names that relate to Napierian (natural) logarithms are not much used, but we include them in this footnote for reference. The parameter b that appears in Eqs. 13.25 and 13.37 is known as the *Napierian absorption coefficient.* The product bl which appears in Eq. 13.36 and other equations can be written as B and is known as the *Napierian absorbance.*

Thus, from Eq. 13.38,

$$\log_{10}\frac{1}{T} = A \tag{13.40}$$

The **percentage transmittance** $T\%$ is the transmittance multiplied by 100:

$$T\% = 100\,T = \frac{100\,I}{I_0} \tag{13.41}$$

Taking logarithms gives

$$\log_{10} T\% = \log_{10} 100 + \log_{10} T \tag{13.42}$$

$$\log_{10} T\% = 2 - A \tag{13.43}$$

Beer studied the influence of the concentration of a substance in solution on the absorbance, and he found the same linear relationship between absorbance and concentration as Lambert had found between absorbance and thickness (see Eq. 13.38). Thus, for a substance in solution at a concentration c, **Beer's law** states that

$$\log_{10}\frac{I_0}{I} = A = \text{constant} \times c \tag{13.44}$$

The **Lambert-Beer law** combines Eqs. 13.38 and 13.44 into a single equation

$$A = \log_{10}\frac{I_0}{I} = \epsilon c l \tag{13.45}$$

Absorption Coefficient

where ϵ is a constant known as the **absorption coefficient** and l is the length of the light path. The absorption coefficient is usually written as $\epsilon_l^c(\lambda)$ so that if the concentration units are moles per cubic decimeter, the light path is 1 cm, and the wavelength is 430 nm, the coefficient would be written as

$$\epsilon_{1\ \mathrm{cm}}^{1\,M}(430\ \mathrm{nm})$$

and would then be called the *molar absorption coefficient*. Since $\epsilon c l$ is dimensionless, the molar absorption coefficient then has the units $\mathrm{dm^3\ mol^{-1}\ cm^{-1}}$.

EXAMPLE 13.4 An aqueous solution of a purine triphosphate, at a concentration of 57.8 mg $\mathrm{dm^{-3}}$ of its trisodium dihydrate (molar mass 586 g $\mathrm{mol^{-1}}$), gave an absorbance of 1.014 with a light path of 1 cm. Calculate the molar absorption coefficient. What would be the absorbance of a 10 μM solution, and what would be the percentage of light transmittance?

Solution From Eq. 13.45

$$A = 1.014 = \epsilon c l$$

$$c = \frac{57.8 \times 10^{-3}(\mathrm{g\ dm^{-3}})}{586(\mathrm{g\ mol^{-1}})} = 9.86 \times 10^{-5}\ M$$

Thus

$$\epsilon = \frac{1.014}{9.86 \times 10^{-5}(\mathrm{mol\ dm^{-3}}) \times 1(\mathrm{cm})} = 1.028 \times 10^{4}\ \mathrm{dm^3\ mol^{-1}\ cm^{-1}}$$

For a 10 μM solution,

$$A = \epsilon cl = 1.028 \times 10^4(\text{dm}^3\ \text{mol}^{-1}\ \text{cm}^{-1}) \times 10 \times 10^{-6}(\text{mol dm}^{-3}) \times 1(\text{cm})$$
$$= 0.1028$$

The percentage light transmittance $T\%$ is given by Eq. 13.43,

$$\log_{10} T\% = 2 - A = 1.8972$$

and thus

$$T\% = 78.9\%$$

Deviations from the Lambert-Beer equation are sometimes observed, especially at high concentrations. It is then necessary to find an explanation in terms of extraneous effects such as dissociation and complex formation.

13.2 Atomic Spectra

In spectroscopy, the terms *near* and *far* used with a region of the spectrum refer to its position relative to the visible region of the spectrum. Thus, *near* infrared is the infrared region nearest the visible, approximately 7×10^{-7} m to 2×10^{-4} m. Vacuum ultraviolet refers to the region of the spectrum below about 200 nm, which is observable only if water vapor is removed because water absorbs in this region. Its removal is normally accomplished in a vacuum to exclude water vapor from the light path.

Atomic spectra may be observed in different regions of the spectrum, depending on the energies involved in the electronic transitions. The hydrogen spectrum, which we considered in Section 11.2, occurs in the vacuum ultraviolet, near-ultraviolet, visible, and infrared regions. Sometimes, on the other hand, the differences between atomic energy levels are small, and lines may be observed only in the microwave region. In this section we first present experimental spectroscopic evidence that leads to the vector model of the atom.

The characteristics of atomic spectra depend to a considerable extent on the type of spectrometer used. Simple instruments give what are called *low-resolution* spectra. If more advanced instruments are employed, a *high-resolution* spectrum is observed; what appeared on the simpler instrument to be single lines are split into two or more lines. Several effects are responsible for this splitting; in order of decreasing strength, they are

1. Coulombic interactions
2. exchange interactions
3. spin-orbit interactions.

In addition, splitting can be brought about by an external magnetic field. We shall now consider these effects in some detail.

Coulombic Interaction and Term Symbols

For a hydrogen atom the energy of the electron depends to a good approximation only on the principal quantum number, as seen from Eq. 11.46, or Eq. 11.191. Thus, an electron has the same energy whether it is in the 2s or the 2p state; the energy is the same for 3s, 3p, and 3d electrons, and so on. This is true for all hydrogenlike atoms (i.e., for atoms or ions having a single electron). When there is more than one electron, however, there is a significant difference in energy for electrons having different orbital angular momentum quantum numbers. Thus, the 2p level lies

GERHARD HERZBERG (B. 1904)

Herzberg was born on Christmas Day in 1904 in Hamburg, Germany. In 1924 he began to study at the Technische Hochschule in Darmstadt, and obtained his Ph.D. degree in 1928 for a thesis in the field of spectroscopy. From 1928 to 1929 he was a postdoctoral fellow at the University of Göttingen and carried out some research with Walter Heitler, who later became well known for his quantum-mechanical (valence-bond) treatment, with Fritz London, of the hydrogen molecule (Section 12.2). At about this time Herzberg published his suggestion that chemical bonding can be understood in terms of bonding and antibonding electrons. In 1929 Herzberg went to the University of Bristol for another postdoctoral fellowship, and carried out further research in spectroscopy and molecular-orbital theory. He returned to Darmstadt in 1930 as *privatdozent.*

By 1935 Herzberg and his wife felt it inadvisable to remain in Nazi Germany, and he accepted an appointment at the University of Saskatchewan. There, besides teaching and carrying out research, Herzberg completed the German and English versions of his *Atomic Spectra and Atomic Structure.* Some of his research in Saskatoon was concerned with the spectra of free radicals and free-radical ions, found in planets and also produced in the laboratory. He later wrote three books on molecular spectra, *Spectra of Diatomic Molecules* (1939), *Infrared and Raman Spectra* (1945), and *Electronic Spectra and Electronic Structure of Polyatomic Molecules* (1966). Later editions of these books have appeared, and collectively these publications are often referred to as the "bible" of spectroscopy. All of the books contain much original material, and they are remarkable for their comprehensiveness and reliability.

From 1945 to 1948 Herzberg was professor of spectroscopy at the Yerkes Observatory of the University of Chicago, where he continued his spectroscopic research and also did a little teaching. In 1949 he returned to Canada as director of the division of physics at the National Research Council in Ottawa. In 1969 he retired from the directorship but continued to carry out a great deal of research with the title of Distinguished Research Scientist. In 1975 the National Research Council created the Herzberg Institute of Astrophysics, in which Herzberg has continued to carry out research and revise his four monographs.

Herzberg has made many pioneering contributions in spectroscopy and molecular theory, and most of them are of particular interest to chemists; it is therefore appropriate that although he is a physicist, his 1971 Nobel Prize was for chemistry.

References: G. Herzberg, "Molecular spectroscopy: a personal history," *Annual Reviews of Physical Chemistry, 36,* 1–30(1985).

K. J. Laidler, *The World of Physical Chemistry,* Oxford University Press, 1993.

above the 2s level, and 3d above the 3p level, and so on (see Figure 11.22). These differences are due to **Coulombic interactions** between the electrons in the atom.

Term Symbols

To take account of the effect of Coulombic interactions, the atomic energy levels are described by capital letters, called **term symbols,** which indicate the magnitude of L, the total angular momentum. For a single electron the orbital angular momentum has values

$$L = \sqrt{l(l + 1)}\hbar \tag{13.46}$$

(see Eq. 11.211 and Section 11.9). For more than one electron in an atom, the resultant total angular momentum is with values

$$\boldsymbol{L} = \Sigma \boldsymbol{l}_i = \sqrt{L(L + 1)}\hbar \tag{13.47}$$

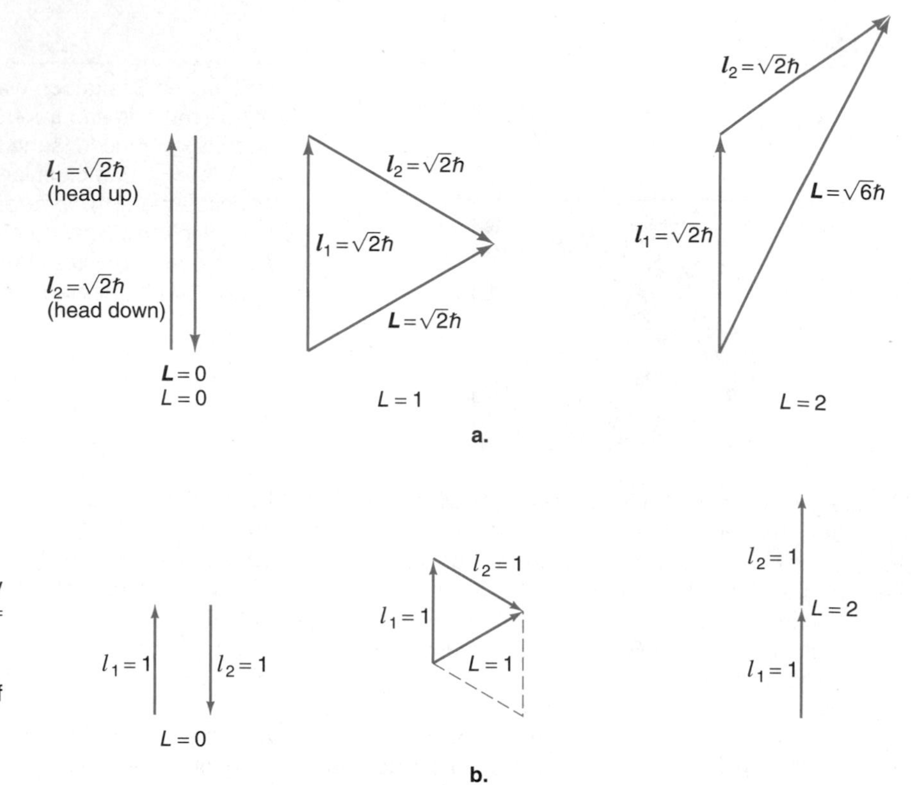

FIGURE 13.6
Vector addition of l_1 and l_2 to give the resultant values of $\boldsymbol{L}$. (a) In this case $l_1 = l_2 = 1$ and vector addition proceeds normally to give the quantum numbers $L = 0$, 1, and 2. (b) Simplified vector additions for $l_1 = l_2 = 1$ that appear in (a). Only orientations of l_1, etc., are allowed that result in values of L equal to 0 or whole numbers.

where $\boldsymbol{l}_i$ refers to the angular momentum vector of each electron in an atom and L, which has only integral values, is the total angular momentum quantum number of the atom. The significance of Eq. 13.47 is that the vector sum of the $\boldsymbol{l}_i$ is constrained to certain values. The rule for coupling angular momenta is given by the **triangular inequality**, which states that when coupling two angular momenta, l_1, and l_2, to give a third, the values of l_3, must be between the integer limits, $|l_1 - l_2| \leq l_3 \leq l_1 + l_2$. Hence when $L = 0, 1, 2, 3 \ldots$, the corresponding term symbols are S, P, D, F, . . . by analogy to the one-electron angular-momentum terms.

For hydrogen, with only one electron, $L = l$. The term symbols for hydrogen in its ground and first two excited states are therefore

$$\begin{array}{lll} n = 1 & l = 0 & \text{S} \\ n = 2 & l = 0, 1 & \text{S, P} \\ n = 3 & l = 0, 1, 2 & \text{S, P, D} \end{array} \tag{13.48}$$

For a multielectron atom, vector addition is used to find the resulting value of L. We will illustrate this process for two electrons with the same value of l but different values of n. The orbital angular momenta for the individual electrons are assumed to be coupled strongly and the vectors representing the angular momentum of the two electrons, $\boldsymbol{l}_1$ and $\boldsymbol{l}_2$, are placed head to tail as shown in Figure 13.6a. Only those angles between $\boldsymbol{l}_1$ and $\boldsymbol{l}_2$ are allowed that are consistent with the triangular inequality and that result in integral values of L. These vectors are cumbersome to use because of the square-root terms, and it is simpler to use schematic vectors that are in direct proportion to the quantum numbers. The simplified technique is demonstrated in Figure 13.6b and by the following example.

EXAMPLE 13.5 Determine the allowed values of L for two electrons in an atom where $l_1 = 2$ and $l_2 = 1$.

Solution Using the same method as in Figure 13.6b, we have

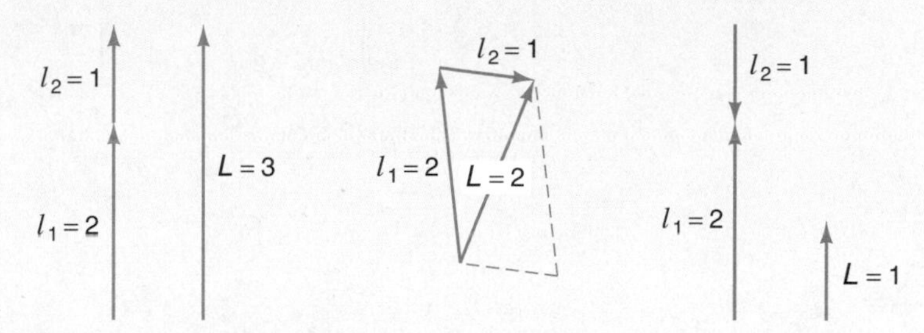

The three values allowed for L are thus 1, 2, and 3.

To summarize the results of this subsection, the Coulombic interaction between electrons gives rise to term symbols that designate the total angular momentum, L. They are S, P, D, F, respectively, for $L = 0, 1, 2, 3$.

Exchange Interaction: Multiplicity of States

Spin Correlation

After the Coulombic interaction, the greatest cause of splitting of spectral lines arises from what is known as **exchange interaction,** or perhaps better, as **spin correlation.** This effect arises from the fact that the electrons may be thought of as spinning (Section 11.11 p. 562) and that electrons with the same spin interact with one another differently from electrons with opposite spins. Consider, for example, the helium atom, in which there are two electrons. If these are both in the 1s orbital, the Pauli principle requires them to have opposite spins. Suppose, however, that one is in the 1s orbital and the other in the 2s orbital. There are then two possibilities:

1. They may both have the same spin.
2. The spins may be opposed.

According to Hund's rule (Section 11.12) for equivalent electrons (same n and l), the state in which the spins are the same is lower in energy than that in which the spins are opposed. Two different energy states therefore arise, according to whether the spins are the same or are opposed.

The resultant spin quantum number for more than one electron is represented by the symbol S (not to be confused with the term symbol). The individual spin quantum numbers are $+\frac{1}{2}$ and $-\frac{1}{2}$ (see Section 11.11, especially Eq. 11.220), and we obtain the resultant S in much the same way as we obtained the resultant orbital quantum number L from the l values for the individual electrons (Figure 13.6). The triangular inequality in this case is $|S_1 - S_2| \leq S_3 \leq S_1 + S_2$. Figure 13.7 shows a few examples. The case of a single electron is shown in Figure 13.7a, and of course the resultant spin quantum number is the spin quantum number of the individual electron; thus $m_s = \frac{1}{2}$ and $S = \frac{1}{2}$. If there are two electrons, there are two possibilities, shown in Figure 13.7b. If the spins are opposed, the resultant spin S is zero; whereas if the electrons have the same value of m_s, the resultant spin S is $\frac{1}{2} + \frac{1}{2} = 1$. The case of two electrons of opposite spins is of particular interest, because this is

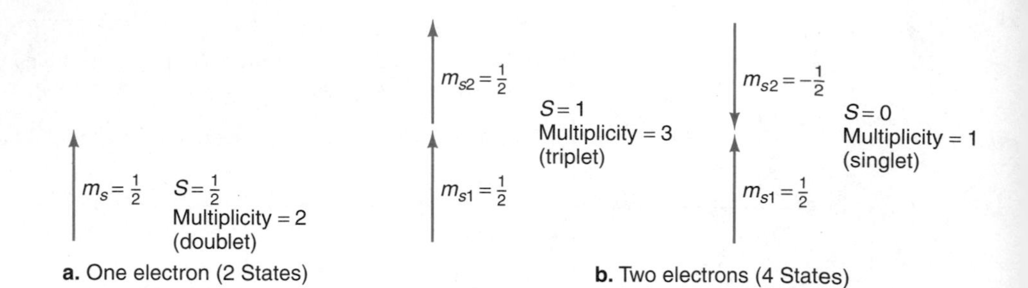

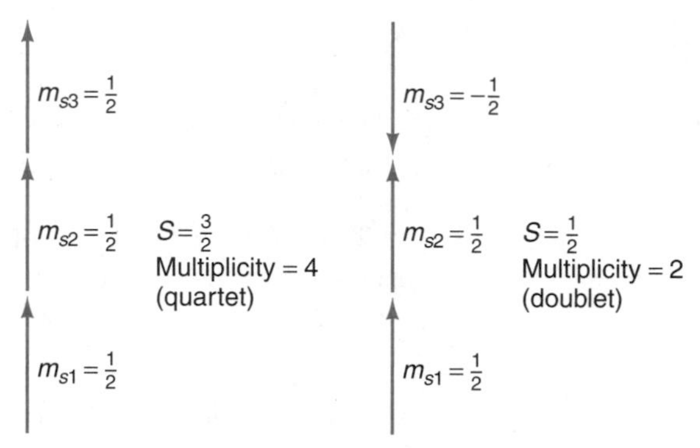

c. Three electrons (8 states: a quartet with $S = \frac{3}{7}$; two doublet states of $S = \frac{1}{2}$)

FIGURE 13.7
Vector diagrams showing how the resultant spin quantum number S is obtained from the spin quantum numbers for the individual electrons.

what arises when two electrons in an atom have the same three orbital quantum numbers, n, l, and m. By the Pauli principle (Section 11.12) they must have opposite spins, and it follows that electrons in a completed shell or in completed subshells do not contribute to the resultant spin quantum number S.

The case of two electrons that are not restricted in this way is also of special interest. We considered this problem for molecules in Section 12.2, where we saw that there are four possible wave functions (Eqs. 12.27 and 12.28), three of them symmetric and one antisymmetric:

$$\left.\begin{array}{lll} 1. & \alpha(1)\alpha(2) & m_{s1} + m_{s2} = +1 \\ 2. & \alpha(1)\beta(2) + \alpha(2)\beta(1) & m_{s1} + m_{s2} = 0 \\ 3. & \beta(1)\beta(2) & m_{s1} + m_{s2} = -1 \end{array}\right\} \quad S = 1$$

$$\begin{array}{lll} 4. & \alpha(1)\beta(2) - \alpha(2)\beta(1) & m_{s1} + m_{s2} = 0 \end{array} \qquad S = 0$$

Two electrons with spins combine to give a resultant spin S that can be 1 or 0, and the components $m_{s1} + m_{s2}$ can be 1, 0, or -1. It is obvious that the spin functions 1 and 3 must belong to $S = 1$, because only $S = 1$ can have components $+1$ or -1. Both of these functions are symmetric with respect to an interchange of electrons, as is also function 2. This function, having a spin component 0, also belongs to the group of three functions having $S = 1$. The remaining function 4, which is antisymmetric, has $S = 0$.

Multiplicity

Since there are three wave functions corresponding to two electrons having $S = 1$, we say that the state has a **multiplicity** of 3 or that it is a *triplet* state. In general the multiplicity is related to the resultant spin by the formula

$$\text{multiplicity} = 2S + 1 \tag{13.49}$$

In this particular example $S = 1$ and the multiplicity is 3; for $S = 0$ the multiplicity is 1, and we say that the state is a *singlet* state.

Other situations are also represented in Figure 13.7. If we have three electrons, there can be two resultant spin quantum numbers, $\frac{3}{2}$ and $\frac{1}{2}$, as shown in Figure 13.7c. If $S = \frac{3}{2}$, the multiplicity is 4 (a *quartet*), while if $S = \frac{1}{2}$, the multiplicity is 2 (a *doublet*).

When adding angular momenta together, whatever their origin (orbital, electron spin, nuclear spin), the rules are the same. First, the number of states between the uncoupled spins and the coupled spins must be preserved. Secondly, when vectors are added the sums range from $l_1 + l_2$ to $|l_1 - l_2|$ for both integer and half-integer angular momenta (triangular inequality).

In the case of three spins of magnitude $\frac{1}{2}$ (see Figure 13.7c), there are two states for each spin, $\pm\frac{1}{2}$, giving a total of $2^3 = 8$ states. This number must be preserved. The addition of the three spins is performed as follows:

1. Add two spins of $S = \frac{1}{2}$ to give intermediate state of $\mathrm{S} = 1$ (triplet) and $\mathrm{S} = 0$ (singlet).
2. Add the third spin to the first intermediate state $\mathrm{S} = 1$ to give two states of magnitude $\mathrm{S} = \frac{3}{2}$ (quartet) and $\mathrm{S} = \frac{1}{2}$ (doublet).
3. Add the third spin to the second intermediate state $\mathrm{S} = 0$ to give one state of magnitude $\mathrm{S} = \frac{1}{2}$ (doublet).

That is, there are two ways of producing doublet states. Hence three spins coupled together lead to one quartet (total spin $\mathrm{S} = \frac{3}{2}$) and two doublet states (total spin of $\mathrm{S} = \frac{1}{2}$) for a total of 8 states.

The multiplicity $2S + 1$ is indicated in the term symbol as a left-hand superscript. Thus, for the ground state of hydrogen with $S = \frac{1}{2}$, the multiplicity is $2(\frac{1}{2}) + 1 = 2$. The term symbol is thus $^2\mathrm{S}$ and is read *doublet S*.

EXAMPLE 13.6 Determine the possible multiplicities for an atom in the configuration $1s^2 2s^2 2p^2$.

Solution Electrons in a complete shell or in completed subshells do not contribute to the total spin angular momentum. Consequently, only the interaction of the spins of the two p electrons need be considered. The resultant values of S are 1 and 0, giving rise to two states: $2(1) + 1 = 3$, a triplet state; and $2(0) + 1 = 1$, a singlet state.

The way in which exchange interaction (spin correlation) brings about the splitting of energy levels can be illustrated by the simplest atom that shows this effect, namely helium. The term diagram of helium is shown in Figure 13.8, in which the terms are divided into two sets. At one time these were thought to arise from two distinct forms of helium, *parahelium* and *orthohelium*, but this is now known not to be the case. The first set consists of *singlets* (multiplicity = 1) and the other of *triplets* (multiplicity = 3). The ground state of helium is $1s^2$ with $L = 0$ ($l_1 = 0$, $l_2 = 0$) and $S = 0$ since the electrons must have opposite spins. Hence, its ground-state term symbol is $^1\mathrm{S}$.

If one electron is promoted to $n = 2$, two possibilities exist, $1s^1 2s^1$ and $1s^1 2p^1$. The first of these has $L = 0$ and is an S state. The second has $L = 1$ and is a P state. Since in both states the electrons may have either $m_s = +\frac{1}{2}$ or $m_s = -\frac{1}{2}$, the value

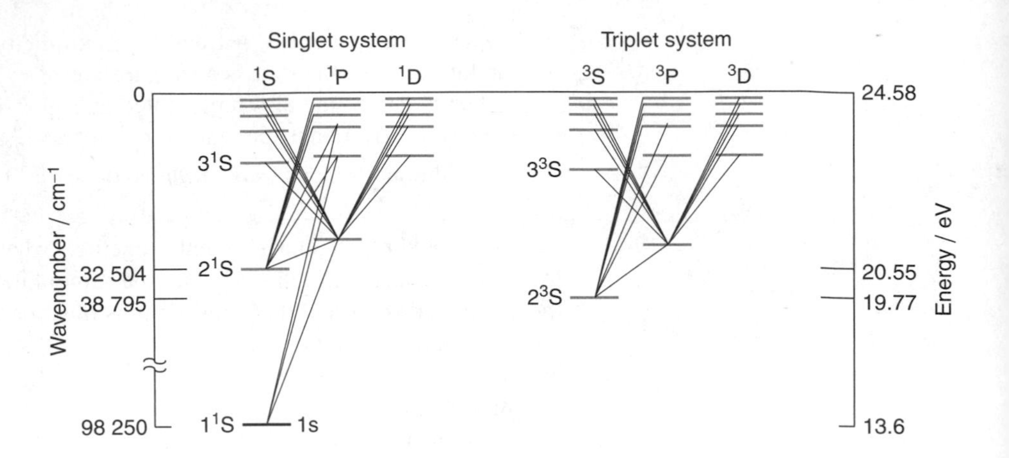

FIGURE 13.8
Simplified energy-level diagram for helium showing the singlet and triplet term systems.

of S is 0 or 1, and ^{1}S, ^{3}S, ^{1}P, and ^{3}P states are formed. The S terms lie below the P terms of the same principal quantum number, and by Hund's rule the triplet states lie below the singlets of the corresponding n and l values. The electrostatic energies of two such states may in fact be written as

$$\begin{aligned} {}^1\mathrm{S} \quad E_1 &= J + K \\ {}^3\mathrm{S} \quad E_3 &= J - K \end{aligned} \tag{13.50}$$

where J is the Coulombic integral and K is the exchange integral. These integrals are analogous to those we encountered for molecules in Section 12.4. These energy integrals are usually difficult to calculate from quantum-mechanical theory, but spectroscopic measurements can provide accurate values for them.

To summarize the results of this subsection, exchange interaction of spins leads to a multiplicity of states, which is expressed in the term symbol as $^{2S+1}$L.

Spin-Orbit Interactions

The effects that we have just considered, Coulombic interactions between electrons and exchange interactions between electrons, do not apply to the hydrogen atom, in which there is only one electron. We have seen that to a good approximation the energy levels of the hydrogen atom depend only on the principal quantum number n and not on the second quantum number l. However, if measurements are made with a spectrograph of high resolving power, the situation is not quite so simple. Some of the lines found in a low-resolution instrument are then split into component lines. For example, the H_α line of the Balmer series (Section 11.2 and Figure 11.8) corresponds to transitions between $n = 2$ and $n = 3$. In an ordinary spectrometer it appears as a single line, but in a high-resolution instrument it is split into seven lines, corresponding to transitions between the states shown in Figure 13.9a. Lines in other series, such as the Lyman series, are also split, and some of the transitions are shown in Figure 13.9b.

These splittings are smaller than those resulting from Coulombic and exchange interactions between electrons. They are due to a magnetic coupling between the magnetic moment associated with the spin of the electron and the magnetic moment associated with the orbital motion of the electron. This coupling is referred to as a **spin-orbit interaction,** and it can arise in two different ways. In hydrogen, $\boldsymbol{l}$ and $\boldsymbol{s}$

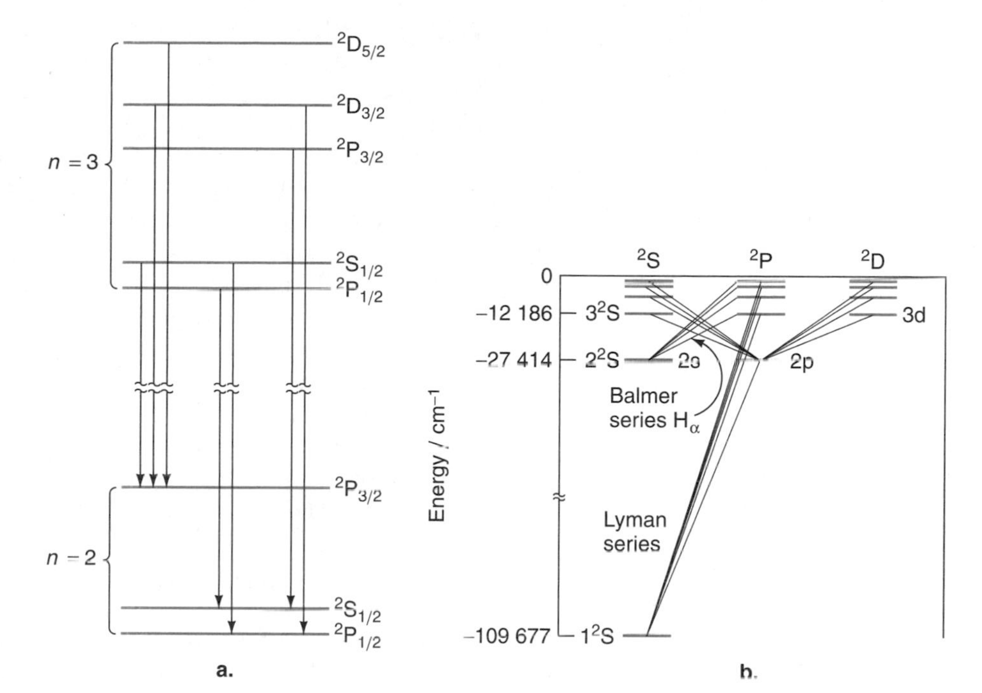

FIGURE 13.9
(a) States associated with the H_α line ($n_1 = 2 \leftrightarrow n_2 = 3$) of the Balmer series for the hydrogen atom spectrum, as detected by high-resolution spectroscopy.
(b) A simplified energy diagram for the hydrogen atom, showing various transitions. This type of diagram is known as a Grotrian diagram, named for Otto Natalies August Grotrian (1847–1921).

are coupled in such a way that there is formed a new *inner quantum number j* with the values

$$j = l \pm s \tag{13.51}$$

This quantum number j describes the allowed energy levels associated with the total angular momentum, both spin and orbital. Since m_s can only have the values $\pm\frac{1}{2}$, for a P state ($l = 1$) only two possibilities exist, namely

$$j = l + s = \frac{3}{2} \quad \text{and} \quad j = l - s = \frac{1}{2} \tag{13.52}$$

The value of j is written as a right-hand subscript on the term symbol. Corresponding to Eq. 13.48, we have

$$\begin{aligned} n &= 1: \quad {}^2S_{1/2} \\ n &= 2: \quad {}^2S_{1/2}, {}^2P_{1/2}, {}^2P_{3/2} \\ n &= 3: \quad {}^2S_{1/2}, {}^2P_{1/2}, {}^2P_{3/2}, {}^2D_{3/2}, {}^2D_{5/2} \end{aligned} \tag{13.53}$$

Only certain transitions occur and these are given by a set of **selection rules,** namely

$$\Delta n = \pm 1, \pm 2, \ldots, \quad \Delta l = \pm 1 \tag{13.54}$$

and

$$\Delta j = 0, \pm 1 \tag{13.55}$$

Transitions are allowed between S levels and P levels. The ${}^2D_{3/2}$–${}^2P_{3/2}$ ($\Delta j = 0$) transition[2] between the closest pair of these levels is lowest in frequency and the

[2]The convention employed in spectroscopy is always to write the *upper* state first, with the arrow pointing in the appropriate direction; ← represents absorption and → represents emission. If either emission or absorption is referred to, a simple dash (−) is used.

$^2D_{5/2}$–$^2P_{3/2}$ ($\Delta j = +1$) transition is next lowest. The lines for these are close together because the separation between the ^2D states is small. The next line, farther away, is for the transition $^2D_{3/2}$–$^2P_{1/2}$ ($\Delta j = \pm 1$) but the transition $^2D_{5/2}$–$^2P_{1/2}$ is not allowed since for this case $\Delta j = \pm 2$. When three lines arise from transitions between doublet levels, we speak of a *compound doublet* spectrum. Thus in the ground-state spectrum of hydrogen, the Lyman spectrum, practically every line is a doublet. This is difficult to observe experimentally with hydrogen, but it is rather easily seen in sodium where the yellow line, the strongest line in the spectrum, has two maxima, one at 589.0 nm and the other at 589.6 nm. This splitting of the yellow sodium line into two separate lines was in fact the first spectroscopic proof of the existence of electron spin.

To summarize the results of this subsection, spin-orbit interactions lead to a total angular momentum $\boldsymbol{J} = \boldsymbol{L} + \boldsymbol{S}$, which is expressed in the term symbol as $^{2S+1}L_J$. The term symbols are used to designate the various energy levels in atoms. Absorption and emission spectra result from transition between different energy levels and are governed by various selection rules.

The Vector Model of the Atom

The vector additions of angular momenta that we have considered earlier in this section are examples of a more general method for dealing with the energy levels of molecules and, therefore, of interpreting spectra. We have seen that individual electrons have orbital angular momenta, which couple together to give a resultant orbital angular momentum L. Each individual l_i precesses around the resultant vector, much as the angular momentum vector of Figure 11.20 interacts with an external magnetic field. Similar considerations apply to the spin angular momenta. These couple together to form a resultant S, which interacts with the resultant orbital angular momentum L. From the standpoint of an orbital electron the nucleus is rotating around it, just as the sun appears to us to rotate around the earth. The magnetic field generated by the nucleus causes an electron to align in one of two possible directions, and there is a coupling of the resultant spin angular momentum with the resultant orbital angular momentum.

The details of the way in which coupling occurs vary somewhat, and it is possible to distinguish two extreme situations, according to whether the nuclear charge is small (i.e., the atoms are light) or is large (the atoms are heavy). The way in which the coupling occurs for light atoms is described by **Russell-Saunders coupling,** named in recognition of the pioneering work in this area by the American astronomer Henry Norris Russell (1877–1957) and physicist Frederick Albert Saunders (1875–1963). In this scheme, the individual spin vectors of each electron s_i form a resultant spin vector S. In the same way that we treated the vector addition of the l_i, in Figure 13.6b, we may simplify the analysis by requiring that the spin quantum numbers add to give S, the total spin quantum number, with integral or half-integral values:

Russell-Saunders Coupling

$$\sum_i m_{s,i} = S \tag{13.56}$$

where vector addition is implied. For example, two spins of $+\frac{1}{2}$ can couple to yield $S = 0$ or 1, whereas three spins could give two doublets with $S = 1/2$ and one quartet with $S = 3/2$, as shown in Figure 13.7 and described on page 653.

The individual orbital angular momentum quantum numbers l_i couple to form a resultant L which is restricted to integral values, as seen above in Eq. 13.47.

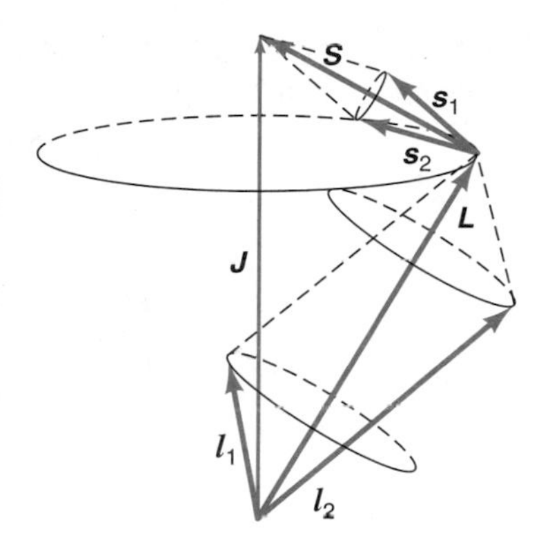

FIGURE 13.10
Precession of $\boldsymbol{L}$ and $\boldsymbol{S}$ about $\boldsymbol{J}$ for Russell-Saunders coupling. The individual $\boldsymbol{l}$ couple and $\boldsymbol{s}$ couple form resultant $\boldsymbol{L}$ and $\boldsymbol{S}$ before coupling to form $\boldsymbol{J}$.

The vectors represented by $\boldsymbol{L}$ and $\boldsymbol{S}$ exert magnetic forces on each other, and they couple to form a resultant total angular momentum vector $\boldsymbol{J}$ for the atom. Thus

$$\boldsymbol{J} = \boldsymbol{L} + \boldsymbol{S} \tag{13.57}$$

where the values of $\boldsymbol{J}$ are quantized with values

$$\sqrt{J(J+1)}\hbar \tag{13.58}$$

and the inner quantum number J may have integral or half-integral values. In the absence of an external field, $\boldsymbol{L}$ and $\boldsymbol{S}$ precess around their resultant $\boldsymbol{J}$ in order to keep the total angular momentum a constant, as shown in Figure 13.10. Because of the coupling, $\boldsymbol{J}$ precesses relatively more slowly around the field direction.

In Russell-Saunders coupling, which applies to light atoms, the coupling to form $\boldsymbol{L}$ and to form $\boldsymbol{S}$ is strong. This means that the different values of L and S represent states of considerably different energies. On the other hand, the coupling of $\boldsymbol{L}$ and $\boldsymbol{S}$ to form $\boldsymbol{J}$ is weak, resulting in J values that do not differ much in energy.

The selection rules that apply to transitions in Russell-Saunders coupling can be summarized as

$$\Delta S = 0 \tag{13.59}$$
$$\Delta L = 0, \pm 1 \tag{13.60}$$
$$\Delta J = 0, \pm 1 \tag{13.61}$$

There is a further restriction, known as the *Laporte rule,* that even terms only combine with odd terms:

$$g \not\leftrightarrow g \qquad u \not\leftrightarrow u \qquad g \leftrightarrow u \tag{13.62}$$

This is a symmetry effect. In addition

$$J = 0 \not\leftrightarrow J = 0 \tag{13.63}$$

The underlying reason behind these selection rules is that the photon carries one unit of spin angular momentum; there therefore must be a change of angular momentum when a photon is absorbed or emitted.

j-j Coupling

For heavy atoms (i.e., for atoms of high atomic number) there is another type of coupling, known as ***j-j* coupling.** The spin-orbit interactions are now stronger and cause the individual $\boldsymbol{l}$ and $\boldsymbol{s}$ vectors for each electron to couple and to form resultant $\boldsymbol{j}$ vectors. These in turn couple to form a total $\boldsymbol{J}$ for the atom. This $j\,j$ coupling may be explained in terms of enhanced spin-orbit interactions caused by the shielding of the outermost electrons by all the remaining electrons. In effect, the electrostatic potential changes rapidly as a function of distance from the nucleus as a valence electron penetrates deeply into the shielding electron cloud. As a result of an increase in the number of electrons, there is an increase in the electron velocity and in the average magnetic field felt by an electron in a heavy atom. This type of spin-orbit interaction is strong even in halogens since their valence electrons may also contribute to the increased magnetic field. Thus, for heavy atoms the quantum numbers S and L lose their significance, and the selection rules for Russell-Saunders coupling fail. Instead, single-triplet transitions ($S = 0 \leftrightarrow S = 1$) are permitted in heavier atoms, whereas they are forbidden in light atoms.

The value of J is written as a right-hand subscript to the term symbol. Since a closed shell is spherically symmetric, its contribution to the total angular momentum is zero; for a closed shell, therefore, $L = 0$, $S = 0$, and $J = 0$.

So far the discussion has been concerned with nonequivalent electrons. (The set n, l, m_l is not the same for each electron.) For two such electrons, with $L > S$, the permitted values, which must be positive, are

$$L = (l_1 + l_2), (l_1 + l_2 - 1), \ldots, |l_1 - l_2| \tag{13.64}$$

$$S = (s_1 + s_2), (s_1 - s_2) = 1, 0 \tag{13.65}$$

$$J = (L + S), (L + S - 1), \ldots, |L - S| \tag{13.66}$$

A procedure can be worked out for equivalent electrons (having the same value of n and l), but it is rather lengthy and outside the scope of this book.

The Effect of an External Magnetic Field

Zeeman Effect

P. Zeeman received the Nobel Prize in physics in 1902 for discovering the Zeeman effect, along with Hendrik Antoon Lorentz (1853–1928) for predicting the effect.

In 1896 the Dutch physicist Pieter Zeeman (1865–1943) discovered that spectral lines were split into component lines by a magnetic field. This effect is sometimes due to the interaction of the orbital magnetic moment of the atom with the magnetic field and is then called the **Zeeman effect.** A more common effect in a magnetic field is the *anomalous Zeeman effect,* in which the electron spin is also involved.

We saw in Section 11.9 that an orbiting electron has an angular momentum and also a magnetic moment which is in the opposite direction to the angular momentum. We can arrive at a simple quantitative interpretation of the normal Zeeman effect by considering an electron moving in a circular orbit of radius r; $\mu = IA$, where μ is the magnetic moment, I the current, and A the area of the loop (see Figure 13.11a). This orbital motion is a circulation of charge, and that movement must give rise to a magnetic field that is associated with a magnetic dipole source. The electric current that travels around this orbit is the charge that passes a given point per unit time; it is therefore the product of the charge $-e$ and the frequency with which the electron passes the given point. The circumference of the orbit is $2\pi r$, and if the electron moves with velocity u, it performs $u/2\pi r$ revolutions in unit time. The current is, therefore,

$$I = -\frac{eu}{2\pi r} \tag{13.67}$$

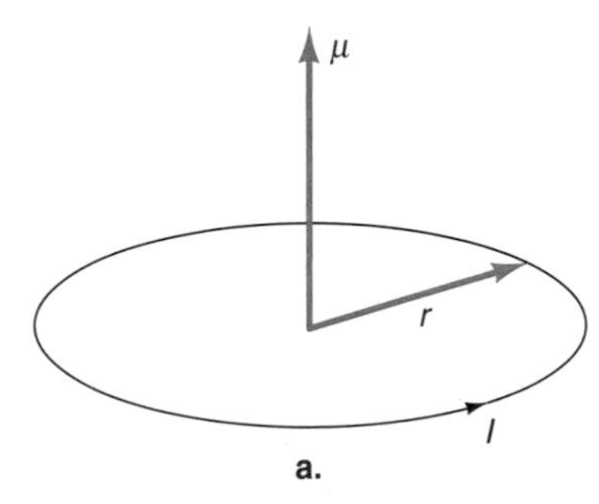

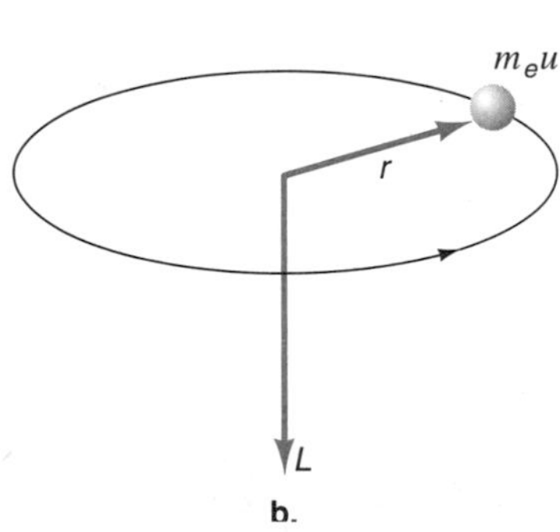

FIGURE 13.11
(a) A magnetic dipole moment produced by a circulating current *I*. (b) An orbiting electron with velocity u produces an orbital angular moment vector $\boldsymbol{L}$.

From electromagnetic theory the magnetic moment μ_l produced from this charge flowing through a circular loop or orbit is the product of the current and the area πr^2 of the orbit:

$$\mu_l = -\frac{eu}{2\pi r} \cdot \pi r^2 = -\frac{eur}{2} \tag{13.68}$$

The direction of this magnetic moment is given by the left-hand rule (Section 11.9, and especially Figure 11.20a) and is perpendicular to the plane of the orbit.

The angular momentum is in the opposite direction, and its direction is given by the right-hand rule (Figure 11.20a). Its magnitude is

$$L = m_e ur \tag{13.69}$$

where m_e is the mass of the electron. The ratio of the magnetic moment to the angular momentum is known as the **magnetogyric ratio** or in the past as the **gyromagnetic ratio** and is given the symbol γ:

Gyromagnetic Ratio

$$\gamma = \frac{\mu_l}{L} = -\frac{eur/2}{m_e ur} = -\frac{e}{2m_e} \tag{13.70}$$

More generally, the magnetic moment is a vector and the angular momentum is replaced by the quantum mechanical operator $\hat{\boldsymbol{L}}$,

$$\hat{\boldsymbol{\mu}}_l = -\frac{e}{2m_e}\hat{\boldsymbol{L}} = \gamma_e\hat{\boldsymbol{L}} \tag{13.71}$$

In Section 11.9 we considered the orientation of the orbital angular momentum in a magnetic field, and we have seen (Eq. 11.208) that the component L_z of the angular momentum along the direction of the field (the Z axis) is

$$L_z = m_l\hbar \tag{13.72}$$

where m_l is the magnetic quantum number. It then follows that the corresponding component $\mu_{l,z}$ of the magnetic moment along the Z axis is

$$\mu_{l,z} = -m_l\frac{e\hbar}{2m_e} \tag{13.73}$$

Bohr Magneton

The quantity $e\hbar/2m_e$, plays an important part in the theory; it is called the **Bohr magneton** and given the symbol μ_B. It is the basic unit for electronic magnetic dipole moments just as $\hbar$ is the basic unit for angular momentum. Insertion of the values of e, $\hbar$, and m_e leads to

$$\mu_B = \frac{e\hbar}{2m_e} = 9.274 \times 10^{-24}\ \text{A m}^2 = 9.274 \times 10^{-24}\ \text{J T}^{-1}$$

The Z component of the magnetic moment is thus

$$\mu_{l,z} = -m_l\mu_B \tag{13.74}$$

Magnetic Flux Density

The magnetic flux density B is the field strength H multiplied by the permeability μ; see Appendix A. The use of the old unit, the gauss (G), is now discouraged; 1 T $\equiv$ 10^4 G.

In order to treat the energy of a magnetic dipole in a magnetic field it is simplest to work not with the magnetic field strength H (SI unit: ampere/metre, A m^{-1}) but with the **magnetic flux density** B. The SI unit of magnetic flux density is kg s^{-2} A^{-1} ($\equiv$ V s m^{-2}), and this unit is known as the **tesla,** for which the symbol is T. Use of the magnetic flux density produces a considerable simplification, since the energy of a magnetic moment $\mu_{l,z}$ in a field of magnetic flux density B is the product $\mu_{l,z}B$:

$$E = -\mu_{l,z}B \tag{13.75}$$

Insertion of the expression for $\mu_{l,z}$ (Eq. 13.74) gives

$$E = -\mu_{l,z}B = m_l\frac{e\hbar}{2m_e}B = +m_l\mu_B B \tag{13.76}$$

More generally, the Hamiltonian operator for the static magnetic energy due to a magnetic moment (Eq. 13.73) is given by the vector dot product

$$\hat{H} = -\hat{\boldsymbol{\mu}}_l \cdot \boldsymbol{B} \tag{13.77}$$

and when the magnetic field is oriented in the Z direction, $\boldsymbol{B} = zB$. Eq. 13.76 is obtained.

Since the unit of E is the joule (J) and that of B is the tesla (T), the unit of μ_B is most conveniently written as J T^{-1}, which is equivalent to A m^2.

Since there are $2l + 1$ possible values of m_l, it follows from Eq. 13.76 that the magnetic field splits each energy level of the atom into $2l + 1$ components. The energy separation between adjacent components is

$$\Delta E = \mu_B B \tag{13.78}$$

and is independent of m_l.

> ***EXAMPLE 13.7*** Calculate the energy splitting when the hydrogen atom is placed in a magnetic field of 1 T (10^4 gauss). Calculate the wavelength splitting expected for the $n = 2$ to $n = 3$ transition in the Balmer series (the 656.2-nm line).
>
> ***Solution*** For such a field, we have from Eq. 13.78,
>
> $$\Delta E = \mu_B B = \frac{e\hbar B}{2m_e} = \frac{1.60 \times 10^{-19}(\text{C})6.626 \times 10^{-34}(\text{J s})\ 1(\text{T})}{4\pi(9.11 \times 10^{-31})(\text{kg})}$$
>
> $$= 9.274 \times 10^{-24}\ \text{J} \quad \text{or} \quad 5.79 \times 10^{-5}\ \text{eV}$$
>
> The energy of the $n = 2$ level is $(1/2^2)(13.6\ \text{eV}) = 3.40$ eV, and that for the $n = 3$ level, $(1/3^2)(13.6\ \text{eV}) = 1.51$ eV, a difference of about 1.9 eV. The fractional change in energy is
>
> $$\frac{5.8 \times 10^{-5}}{1.9} \approx 3 \times 10^{-5}$$
>
> Since $\Delta E \propto \Delta\left(\frac{1}{\lambda}\right)$, for the 656.2-nm line, we expect a splitting of
>
> $$\Delta\lambda = (656.2\ \text{nm})(3 \times 10^{-5}) = 0.02\ \text{nm}$$
>
> This is easily observable with modern grating spectrometers.

The previous treatment to determine the values of ***LS*** coupling in atoms has been done by application of the vector model. For diatomic and other molecules, another method has proven useful. In this technique, we introduce the operator associated with the required vectors. For example, we take the classical angular momentum and express it in quantum mechanical terms. The angular momentum ***L*** is defined classically as $\boldsymbol{L} = \boldsymbol{r} \times \boldsymbol{p}$ and has Cartesian components:

$\hat{L}_x = yp_z - zp_y$,

$\hat{L}_y = zp_x - xp_z$, and

$\hat{L}_z = xp_y - yp_x$.

In each expression only compatible variables are combined making it easy to

For a multielectron atom the treatment is similar, but l and Δm_l are now replaced by L and ΔM_L, the resultant values. The normal Zeeman effect consists of lines arising from transitions between the $2L + 1$ new energy levels produced by the magnetic field; however, the normal Zeeman effect applies only to transitions between singlet states, for which $S = 0$, and as a consequence $L = J$. Therefore an $^1S_0$ term with $L = J = 0$ remains unsplit, but a $^1P_1$ term with $L = J = 1$ has $M_L = +1, 0, -1$ and splits into three levels. The transition $^1P_1 \rightarrow {}^1S_0$ thus gives a splitting into three lines, as shown in Figure 13.12a. Because the energy levels are equally spaced and the selection rule $\Delta M_L = 0, \pm 1$ applies, the transition $^1D_2 \rightarrow {}^1P_1$ also gives three lines, as shown in Figure 13.12b.

In a magnetic field most atoms behave in a more complicated way than predicted on the basis of the normal Zeeman effect. The behavior, known as the **anomalous Zeeman effect,** arises when $S > 0$; the Zeeman splitting is then no longer the same in different terms of the atom. When $S > 0$, there are contributions to the magnetic moment operator from both $\hat{L}$ and $\hat{S}$

$$\hat{\mu} = -\frac{e}{2m_e}(g_L\hat{L} + g_S\hat{S}) \tag{13.79}$$

We have already found that $g_L = 1$ (Eq 13.71), while for electrons, $g_S \cong 2$. In the case of Russell-Saunders coupling (Eq. 13.57), ***L*** and ***S*** are coupled to give the resultant ***J*** and the magnetic moment becomes substitute the quantum mechanical operators from Table 11.1, p. 524. Three expressions of the form $\hat{L}_z = yp_z - xp_y = (h/i)(y\partial/\partial z - z\partial/\partial y)$ result from the substitution with $L^2_y + L^2_z$. This operator does not commute with L_x and L_y but leads again to three relations of the form $\hat{L}_x\hat{L}_y - \hat{L}_y\hat{L}_x = i\hbar\hat{L}_z$. We need to look for a variable compatible, with $\hat{L}_x$, $\hat{L}_y$, $\hat{L}_z$, much as the Hamiltonian operator is compatible with r, ψ, and ϕ, and in this case that function is the square of the angular momentum defined by $L^2 = L^2_z + L^2_y + L^2_z$ and its associated operator is $\hat{L}^2 = \hat{L}^2_x + \hat{L}^2_y + \hat{L}^2_z$.

Rather than develop specific forms of these operators in spherical-polar or confocal elliptical coordinates for use with atoms and molecules in special cases, we find that work can be saved by a different approach. The fact that certain of these operators commute assures the existence of common eigenfunctions, which are the object of our study. As an example, consider the eigenvalue of $\hat{L}_z$, which is $m\hbar$ where ψ_m is the corresponding wave function. The entire mathematical relationship can then be built which is used from Eq. 13.79 through 13.85. The interested reader is directed to more advanced books such as that by Barriol in the suggested readings.

$$\hat{\boldsymbol{\mu}} = -\frac{e}{2m_e}(g_J\hat{\boldsymbol{J}}) \tag{13.80}$$

In the presence of a magnetic field, the vector $\boldsymbol{J}$ rather than $\boldsymbol{L}$ precesses about the magnetic field. From Figure 13.10, the component of $\boldsymbol{L}$ in the direction of $\boldsymbol{J}$ is given by application of $\hat{\boldsymbol{J}} \cdot \hat{\boldsymbol{L}}$, while the component of $\boldsymbol{S}$ is $\hat{\boldsymbol{J}} \cdot \hat{\boldsymbol{S}}$; hence

$$\hat{\boldsymbol{L}} = \hat{\boldsymbol{L}} \cdot \hat{\boldsymbol{J}}\frac{\hat{\boldsymbol{J}}}{J(J+1)} \quad \text{and} \quad \hat{\boldsymbol{S}} = \hat{\boldsymbol{S}} \cdot \hat{\boldsymbol{J}}\frac{\hat{\boldsymbol{J}}}{J(J+1)} \tag{13.81}$$

giving

$$\mu = -\frac{e}{2m_e}(\hat{\boldsymbol{L}} \cdot \hat{\boldsymbol{J}} + 2\hat{\boldsymbol{S}} \cdot \hat{\boldsymbol{J}})\frac{\hat{\boldsymbol{J}}}{J(J+1)} \tag{13.82}$$

A little vector algebra shows that $\hat{S}^2 = (\hat{\boldsymbol{J}} - \hat{\boldsymbol{L}}) \cdot (\hat{\boldsymbol{J}} - \hat{\boldsymbol{L}}) = \hat{\boldsymbol{J}}^2 + \hat{\boldsymbol{L}}^2 - 2\hat{\boldsymbol{L}} \cdot \hat{\boldsymbol{J}}$ or

$$2\hat{\boldsymbol{L}} \cdot \hat{\boldsymbol{J}} = \hat{\boldsymbol{J}}^2 + \hat{\boldsymbol{L}}^2 - \hat{\boldsymbol{S}}^2 \tag{13.83}$$

and $\hat{\boldsymbol{L}}^2 = (\hat{\boldsymbol{J}} - \hat{\boldsymbol{S}}) \cdot (\hat{\boldsymbol{J}} - \hat{\boldsymbol{S}}) = \hat{\boldsymbol{J}}^2 + \hat{\boldsymbol{S}}^2 - 2\boldsymbol{S} \cdot \boldsymbol{J}$ or

$$2\hat{\boldsymbol{S}} \cdot \hat{\boldsymbol{J}} = \hat{\boldsymbol{J}}^2 + \hat{\boldsymbol{S}}^2 - \hat{\boldsymbol{L}}^2 \tag{13.84}$$

Substitution of Eq. 13.83 and Eq. 13.84 into Eq. 13.82 gives

$$\hat{\boldsymbol{\mu}} = -\frac{e}{4m_e}(3\hat{\boldsymbol{J}}^2 - \hat{\boldsymbol{L}}^2 + \hat{\boldsymbol{S}}^2)\frac{\hat{\boldsymbol{J}}}{J(J+1)} = -\frac{e}{2m_e}\left(1 + \frac{\hat{\boldsymbol{J}}^2 - \hat{\boldsymbol{L}}^2 + \hat{\boldsymbol{S}}^2}{2J(J+1)}\right)\hat{\boldsymbol{J}} \tag{13.85}$$

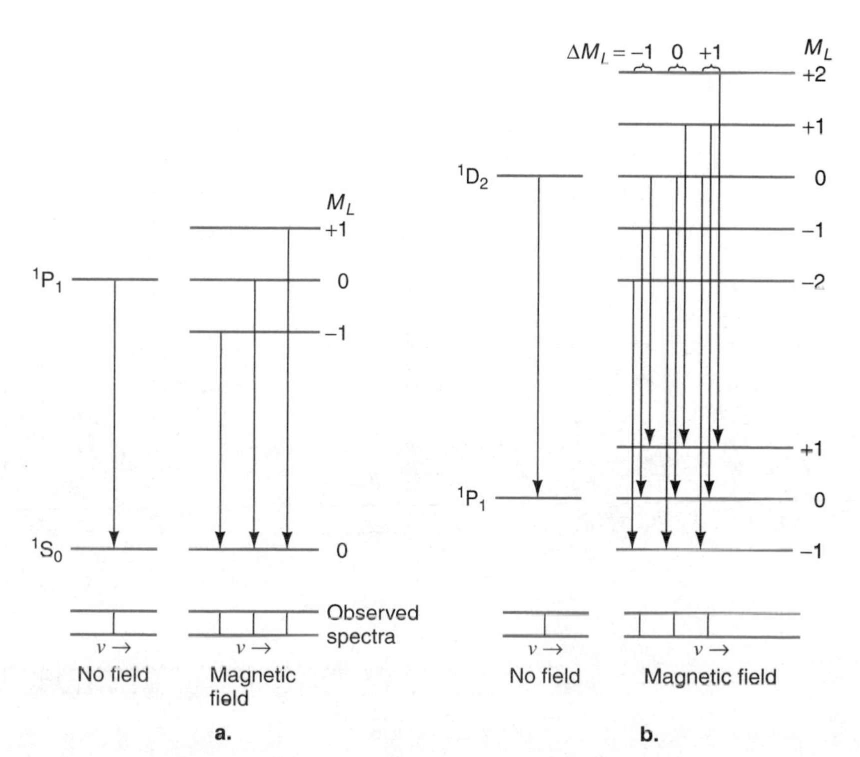

FIGURE 13.12
The normal Zeeman effect, arising from the interaction of the orbital angular momentum with a magnetic field. (a) The $^1P_1 \rightarrow {}^1S_0$ transition, where the $^1P_1$ state is split into three levels so that three Zeeman lines are observed. (b) The $^1D_2 \rightarrow {}^1P_1$ transition. Because the split levels are equally spaced, there are still only three Zeeman lines.

Further substitutions of the magnitudes $J^2 = J(J+1)$, $L^2 = L(L+1)$, and $S^2 = S(S+1)$ give an expression that is usually written as

$$\mu = -\frac{e}{2m_e} g_J J \tag{13.86}$$

where the **Landé-g-factor** is defined as

$$g_J = 1 + \frac{J(J+1) - L(L+1) + S(S+1)}{2J(J+1)} \tag{13.87}$$

If only the Z component of the operator $\hat{\boldsymbol{J}}$ is required, then this has eigenvalues $\hbar M_J$, giving a series of equally spaced energy levels in a magnetic field $\boldsymbol{B} = zB_z$:

$$E_J = g_J M_J \mu_B B_z \tag{13.88}$$

The treatment is more complicated for j-j splitting, but g_J normally has a value between 0 and 2.

For a $^3S_1$ state the experimental value of g_J is 2.0023, and the splitting of the energy levels in the anomalous Zeeman effect is therefore about twice the value for the normal Zeeman effect. For other terms the g_J values are as follows:

$$^3P_0: \quad 0$$
$$^3P_1 \quad \text{and} \quad ^3P_2: \quad \frac{3}{2}$$

Figure 13.13 shows the anomalous Zeeman effect of a $^3P_2 \rightarrow {}^3S_1$ transition.

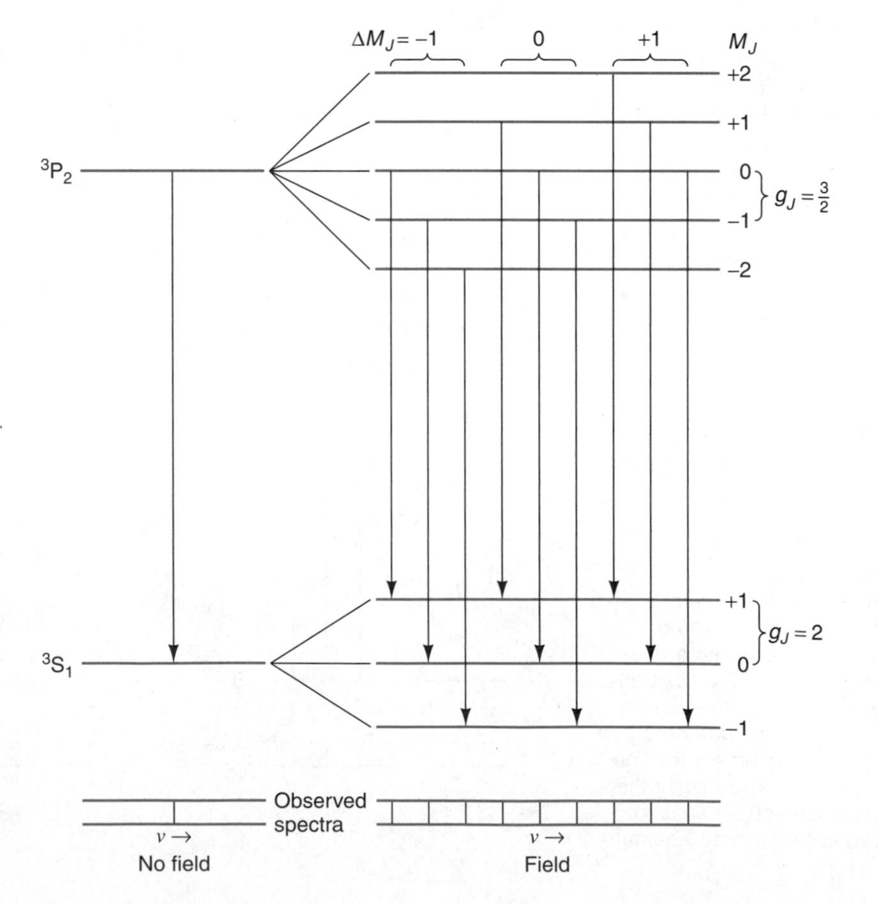

FIGURE 13.13
The anomalous Zeeman effect for the transitions $^3P_2 \rightarrow {}^3S_1$.

Matters become more complicated at high magnetic fields. If the separations between Zeeman components exceed the separation arising because of the fine structure, the magnetic coupling between $\boldsymbol{J}$ and the field exceeds that between $\boldsymbol{L}$ and $\boldsymbol{S}$. In this case, $\boldsymbol{L}$ and $\boldsymbol{S}$ are uncoupled and precess independently about the field direction. The spectral lines now revert back to the normal triplet but are split into closely spaced components. This is known as the **Paschen-Back effect,** named after the German physicists Friedrich Paschen (1865–1940) and E. Back.

Paschen-Back Effect

EXAMPLE 13.8 Into how many levels will a $^3D_2$ level split in a magnetic field? If the field is 4.0 T, calculate the separation between neighboring energy levels.

Solution For the $^3D_2$ level,

$$J = 2, \quad L = 2, \quad \text{and} \quad S = 1$$

and the M_J values are 2, 1, 0, −1, and −2; there is therefore a splitting into five levels.

The Landé-g-factor is, according to Eq. 13.87,

$$g_J = 1 + \frac{(2 \times 3) - (2 \times 3) + (1 \times 2)}{2 \times 2 \times 3} = 1.167$$

The energy levels are given by Eq. 13.88:

$$E = 1.167 \times 9.273 \times 10^{-24} (\text{J T}^{-1}) \times 4.0(\text{T}) M_J$$

where M_J can have the aforementioned values. The separation between neighboring levels is thus

$$\Delta E = 1.167 \times 9.273 \times 10^{-24} \times 4.0 \text{ J} = 4.329 \times 10^{-23} \text{ J}$$

The separation in cm^{-1} is obtained by dividing by hc with $c = 2.998 \times 10^{10} \text{ cm s}^{-1}$:

$$\frac{E}{hc} = \frac{4.329 \times 10^{-23} (\text{J})}{6.626 \times 10^{-34} (\text{J s}) \times 2.998 \times 10^{10} (\text{cm s}^{-1})}$$

$$= 2.18 \text{ cm}^{-1}$$

13.3 Pure Rotational Spectra of Molecules

The energy of an atom can only be electronic energy, and atomic spectra are due to transitions between different electronic states. Molecular spectra may also involve transitions between electronic states, but energy can also reside in molecules in the form of rotational and vibrational energy. As a result, molecular spectra are considerably more complicated than atomic spectra because in a molecule a transition may involve simultaneous changes in electronic, vibrational, and rotational energy. Some simplification, however, is brought about by the fact that the amount of energy involved in a pure rotational transition is considerably less than that in a change of vibrational state, which in turn involves less energy than a change of electronic state. An indication of the different energies and of the spectral regions where each type of transition occurs is given in Table 13.1. The relationships between the different types of motion are illustrated in Figure 13.14, in which rotational levels are superimposed on vibrational levels, and vibrational levels are superimposed on electronic levels.

TABLE 13.1 Types of Optical Spectra

Spectroscopic Region	Approximate Range: Frequency s^{-1}	Wavenumber cm^{-1}	Energy $kJ\ mol^{-1}$	Types of Molecular Energy	Information Obtained
Microwave and far infrared	10^9–10^{12}	0.03–30	4×10^{-4}–0.4	Rotation	Interatomic distance and rotational constants
Infrared	10^{12}–10^{14}	30–3000	0.4–40	Vibration and rotation	Interatomic distances and force constants of bonds
Visible and ultraviolet	10^{14}–10^{16}	3×10^3–3×10^5	40–4000	Electronic, vibration, and rotation	Electronic energy levels, bond dissociation energies, force constants of bonds, and interatomic distances

Born-Oppenheimer Approximation

In representing the situation as we do in Figure 13.14 we are making use of the **Born-Oppenheimer approximation** (see also Section 12.1). When we try to solve the exact Schrödinger equation, which takes into account different forms of energy, the energy cannot be separated into rotational, vibrational, and electronic contributions. This is discussed more fully starting at Eq. 13.147, which gives the condition under which the total energy E is the sum of the electronic energy E_e, the vibrational energy E_v, and the rotational energy E_r. This means, for example, that in dealing with electronic energy of a diatomic molecule we can take the distance between the nuclei to be the equilibrium distance, as if the molecule were not

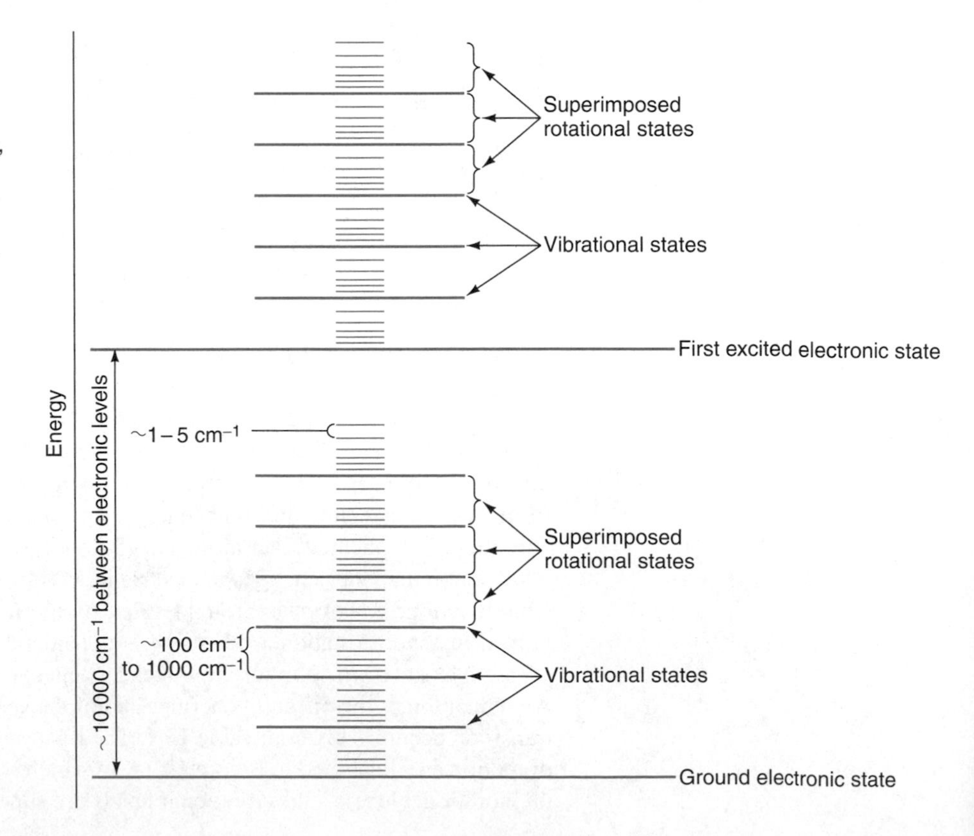

FIGURE 13.14
Typical energy separations for rotational, vibrational, and electronic levels. As permitted by the Born-Oppenheimer approximation, the energy is separated into the three types, so that rotational levels are superimposed on vibrational levels and vibrational levels on electronic levels.

vibrating. Similarly, in dealing with rotations we can consider only the equilibrium internuclear distances when we calculate moments of inertia.

If we make spectroscopic observations in the microwave and far-infrared regions, the only transitions that occur are in the rotational state of the molecule. *Pure rotational spectra* observed in those regions of the spectrum are therefore relatively simple. If we work in the near infrared, vibrational transitions occur and rotational transitions are superimposed since there is ample energy to bring these about; a near-infrared spectrum is therefore a *vibrational-rotational spectrum also called rovibrational spectrum,* and is a good deal more complicated than a pure rotational spectrum. In the visible and ultraviolet regions electronic transitions occur, and at the same time there are rotational and vibrational transitions. *Electronic spectra* are therefore even more complex than the vibrational-rotational spectra.

Condition for Rotational Spectrum

We will deal with pure rotational spectra first. These involve transitions between different rotational states, but there is an important restriction; rotational transitions are only observed in the spectrum if the molecule has a **permanent dipole** moment. The reason for this is that the rotational motion must involve an oscillating dipole (see p. 684), which can interact with an electromagnetic field; otherwise there can be no absorption or emission of radiation. This means, as far as *linear* molecules are concerned, that molecules with a center of symmetry (point group $D_{\infty h}$, see Figure 12.22a), such as N_2 and C_2H_2, do not have a pure rotational spectrum, whereas those without a center of symmetry (point group $C_{\infty v}$, see Figure 12.22b), such as HF and HCN, do have a pure rotational spectrum.

Diatomic Molecules

The rotational spectrum of a diatomic molecule can be treated by considering the molecule to be a rigid rotor. The solution of the Schrödinger equation (Section 11.10; Eq. 11.217) for this problem shows that the magnitude of the angular momentum vector is given by the equation

$$|\boldsymbol{J}| = \sqrt{J(J+1)}\hbar \tag{13.89}$$

where J is the rotational quantum number. The energy in this approximation is given by Eq. 11.218,

$$E_J = J(J+1)\frac{\hbar^2}{2I} \tag{13.90}$$

where I is the moment of inertia and J can have the values 0, 1, 2, ...

Statistical Weighting Factor

The eigenfunctions for the rotation of diatomic molecules are the spherical harmonics $Y_J^M(\theta,\phi)$ (see Eq. 11.189), and therefore the quantum numbers that describe a diatomic molecule are J and M. That is, each rotational level with quantum number J has a $2J + 1$ degeneracy which arises in the same way that the m_l values come from the degeneracy of the l states in the hydrogen atom. That is, the angular momentum vector $\boldsymbol{J}$ along the Z axis has $2J + 1$ different values according to its orientation in space with respect to a fixed axis. Each of the states has the same energy, and all are equally probable. A **statistical weighting factor** of $2J + 1$ is therefore assigned to each rotational level. The relative population of states depends on this statistical factor and also on the Boltzmann factor (Eq. 15.35), according to which the probability decreases as the energy increases. The combination of these

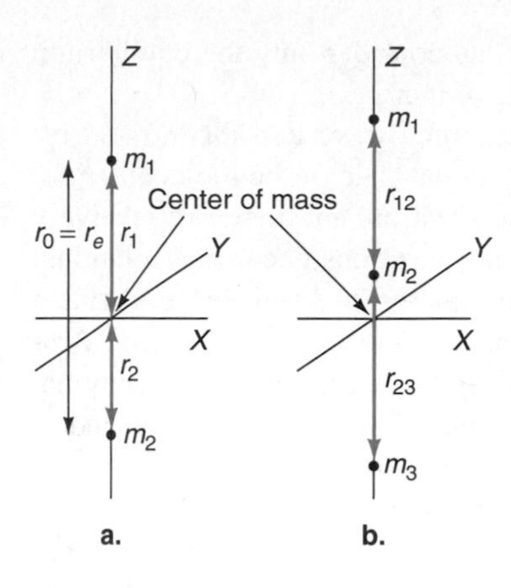

FIGURE 13.15
(a) A diatomic molecule. (b) A linear triatomic molecule. The center of mass depends on the relative masses.

two factors causes the population and hence intensity of the rotational states to increase initially and then decrease as the energy increases.

The moment of inertia of a diatomic molecule is calculated in the following way (see Figure 13.15a). The moment of inertia about the axis of the molecule (the Z axis) is zero. The moments about any axes at right angles to the Z axis are equal and are given by

$$I = m_1 r_1^2 + m_2 r_2^2 \tag{13.91}$$

where r_1 and r_2 represent the distances of the two atomic masses, m_1 and m_2, respectively, from the center of mass of the molecule. If r_0 is the internuclear distance, the values of r_1 and r_2 are given by

$$r_1 = \frac{m_2}{m_1 + m_2} r_0 \qquad r_2 = \frac{m_1}{m_1 + m_2} r_0 \tag{13.92}$$

Substitution of these expressions into Eq. 13.91 gives, after some reduction,

$$I = \frac{m_1 m_2}{m_1 + m_2} r_0^2 = \mu r_0^2 \tag{13.93}$$

Reduced Mass

where μ is known as the **reduced mass** and for a diatomic molecule is given by

$$\mu = \frac{m_1 m_2}{m_1 + m_2} \tag{13.94}$$

As we have seen, by the Born-Oppenheimer approximation (pp. 580 and 664) we take the distance r_0 to be the equilibrium distance r_e.

Pure rotational spectra are simple in appearance, consisting of equally spaced lines. The energy difference for the absorption $J + 1 \leftarrow J$ between two neighboring rotational states is found from Eq. 13.90 to be

$$\Delta E = E_{J+1} - E_J = [(J + 1)(J + 2) - J(J + 1)]\frac{\hbar^2}{2I} \tag{13.95}$$

$$= 2(J + 1)\frac{\hbar^2}{2I} = 2(J + 1)\frac{h^2}{8\pi^2 I} \tag{13.96}$$

The frequency ν_j associated with the transitions is therefore

$$\nu_j = \frac{\Delta E}{h} = 2(J + 1)\frac{h}{8\pi^2 I} \tag{13.97}$$

It is usual to use wavenumbers $\tilde{\nu}$ instead of frequencies ($\tilde{\nu} = \nu/c$), and the corresponding equation is

$$\tilde{\nu}_j = 2(J + 1)\frac{h}{8\pi^2 Ic} \tag{13.98}$$

This equation is conveniently written as

$$\tilde{\nu}_j = 2(J + 1)\,\tilde{B} \tag{13.99}$$

where $\tilde{B}$ is given by

$$\tilde{B} = \frac{h}{8\pi^2 Ic} \tag{13.100}$$

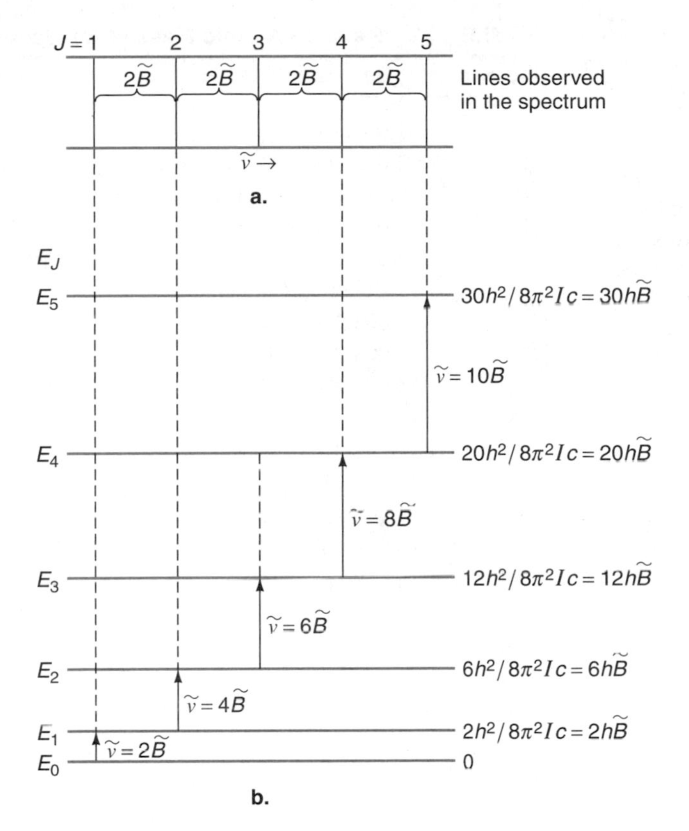

FIGURE 13.16
The constant $2\tilde{B}$ spacing of a pure rotational spectrum is shown in (a). The linearly increasing energy differences between the corresponding transitions are shown in (b).

Rotational Constant

and is known as the **rotational constant.** It follows from Eq. 13.99 that the rotational spectrum will consist of equally spaced lines, the separation between neighboring lines being $2\tilde{B}$. This is shown in Figure 13.16, and it is seen to result from the fact that the spacing between neighboring energy levels increases linearly with J.

A simple interpretation of this selection rule is that photons have unit spin and therefore have a spin angular momentum of $\hbar$; thus, since angular momentum is conserved in a transition, the rotational quantum number must change by $\Delta J = \pm 1$. Rotational transitions can arise only if there is a permanent dipole moment μ_{ro}. Homonuclear diatomic molecules, such as H_2 and Cl_2, do not possess permanent dipoles, whereas heteronuclear diatomics, such as HCl, CO, do. The selection rules can be obtained by evaluating the expressions (see Eq. 13.17) that apply to rotational wave functions:

$$\langle JM \mid \mu_{ro} \mid J'M' \rangle = \int Y_J^M(\theta,\phi)^* \, \mu_{ro} Y_{J'}^{M'}(\theta,\phi) d\tau \qquad (13.101)$$

The important point is that the integral is non-zero, and therefore requires selection rules only for rotational transitions where

$$\Delta J = \pm 1 \qquad \text{and} \qquad \Delta M - 0, \pm 1 \qquad (13.102)$$

Since the rotational levels are degenerate with respect to the M quantum number, only the selection rule for J is relevant.

It follows that if the spacing between rotational lines is measured, the moment of inertia of the molecule may be determined. This is illustrated by the following example.

Atomic weight data were published by R. D. Vocke, Jr., in "Atomic Weights of the Elements 1997"; isotopic compositions data were published by K. J. R. Rosman and P. D. P. Taylor in "Isotopic composition of the elements, 1997." The relative atomic masses of the isotopes were published by G. Andi and A. H. Wapstra in the 1995 Update to the Atomic Mass Evaluation.

The number(s) in parentheses are the uncertainties relating to 1 standard deviation in the last digits(s) quoted.

TABLE 13.2 Relative Atomic Masses for Several Isotopes*

Isotope	Rel. Atomic Mass	% Natural Abundance
^1_1H	1.007 825 0321(4)	99.9885(70)
^2_1H	2.014 101 7780(4)	0.0115(70)
^3_1H	3.016 049 2675(11)	—
^3_2He	3.016 029 3097(9)	$1.37(3) \times 10^{-4}$
^4_2He	4.002 603 2497(10)	99.999863(3)
$^{12}_6C$	12 exactly (by definition)	98.93(8)
$^{13}_6C$	13.003 354 8378(10)	1.07(8)
$^{14}_6C$	14.003 241 988(4)	—
$^{14}_7N$	14.003 074 0052(9)	99.632(7)
$^{15}_7N$	15.000 108 8984(9)	0.368(7)
$^{16}_8O$	15.994 914 6221(15)	99.757(16)
$^{17}_8O$	16.999 131 50(22)	0.038(1)
$^{18}_8O$	17.999 160 4(9)	0.205(14)
$^{19}_9F$	18.998 403 20(7)	100
$^{35}_{17}Cl$	34.968 852 71(4)	75.78(4)
$^{37}_{17}Cl$	36.965 902 60(5)	24.22(4)
$^{79}_{35}Br$	78.918 337 6(20)	50.69(7)
$^{81}_{35}Br$	80.916 291(3)	49.31(7)
$^{127}_{53}I$	126.904 468(4)	100

*The data are from: "Atomic Weights and Isotopic Composition for All Elements," J. S. Coarsey and R. A. Dragoset, NIST (National Institute of Standards and Technology), Physics Laboratory; www.physics.nist.gov.

EXAMPLE 13.9 Absorption by $H^{35}Cl$ occurs in the far infrared near $\tilde{\nu} = 200$ cm^{-1}, and the spacing between the neighboring lines is 20.89 cm^{-1}. Find the moment of inertia and the internuclear distance in $H^{35}Cl$.

Solution From Eq. 13.99, with $J = 0$ (which gives the spacing),

$$\tilde{B} = \frac{\Delta\tilde{\nu}}{2} = \frac{20.89\ (\text{cm}^{-1})}{2} = 10.445\ \text{cm}^{-1}$$

From Eq. 13.100 the moment of inertia is

$$I = \frac{h}{8\pi^2 \tilde{B}c} = \frac{6.626 \times 10^{-34}\ \text{J s}}{8\pi^2 \times 10.445\ (\text{cm}^{-1}) \times 2.998 \times 10^{10}\ (\text{cm s}^{-1})}$$
$$= 2.680 \times 10^{-47}\ \text{kg m}^2$$

The reduced mass μ for a diatomic molecule is given by Eq. 13.94 and the relative atomic masses are given in Table 13.2.

$$\mu = \frac{M_r(\text{Cl})M_r(\text{H})}{M_r(\text{Cl}) + M_r(\text{H})} \times \frac{1}{6.022 \times 10^{23}}$$
$$= \frac{(34.96885)\ (1.007825)}{34.96885 + 1.007825} \times \frac{1}{6.022 \times 10^{23}}$$
$$= 1.627 \times 10^{-24}\ \text{g} = 1.627 \times 10^{-27}\ \text{kg}$$

The equilibrium distance r_e is found from $I = \mu r_e^2$:

$$r_e = \sqrt{\frac{I}{\mu}} = \left(\frac{2.680 \times 10^{-47}}{1.627 \times 10^{-27}}\right)^{1/2} = 1.283 \times 10^{-10} \text{ m}$$

$$= 0.1283 \text{ nm}$$

Such distances are known to still better accuracy.

Linear Triatomic Molecules

The moment of inertia of a linear triatomic molecule is given by

$$I = m_1 r_{12}^2 + m_3 r_{23}^2 - \frac{(m_1 r_{12} - m_3 r_{23})^2}{m} \tag{13.103}$$

where $m = m_1 + m_2 + m_3$ and the individual atomic masses and distances are shown in Figure 13.15b. There are now two distances, r_{12} and r_{23}, to determine and this may be done by studying two molecules that are isotopically different. For example, deuterium may be substituted for hydrogen in HCN, and the pure rotational spectra of HCN and DCN studied. Such a study is possible only if the bond lengths do not change on isotopic substitution. This latter requirement assumes that the electronic potential curves do not change. These curves depend on the electronic structure and nuclear charges but not upon the nuclear masses. This assumption holds to a high degree of accuracy for most atoms, although substituting D for H may give values that differ by 0.0005 nm. The method of determining the internuclear distances is illustrated by the following example.

EXAMPLE 13.10 The $\tilde{B}$ values for HCN and DCN are as follows:

$$\tilde{B}\,(\text{HCN}) = 1.478 \text{ cm}^{-1}$$

$$\tilde{B}\,(\text{DCN}) = 1.209 \text{ cm}^{-1}$$

Calculate the bond lengths H—C (= D—C) and C—N.

Solution From Eq. 13.100 the moments of inertia are equal to $h/8\pi^2\tilde{B}$, and from the $\tilde{B}$ values are found to be

$$I(\text{HCN}) = 1.89 \times 10^{-46} \text{ kg m}^2$$

$$I(\text{DCN}) = 2.32 \times 10^{-46} \text{ kg m}^2$$

The atomic masses are

H: $m_1 = 1.674 \times 10^{-27}$ kg D: $m_1' = 3.344 \times 10^{-27}$ kg

C: $m_2 = 1.993 \times 10^{-26}$ kg N: $m_3 = 2.325 \times 10^{-26}$ kg

For HCN,

$$\text{total mass, } m = 4.485 \times 10^{-26} \text{ kg}$$

For DCN,

$$\text{total mass, } m' = 4.652 \times 10^{-26} \text{ kg}$$

Substitution of these two sets of values into Eq. 13.103 gives two simultaneous quadratic equations. Their solution is

$$r_{12} = r(\text{H—C}) = r(\text{D—C}) = 0.106 \text{ nm}$$

$$r_{23} = r(\text{C—N}) = 0.116 \text{ nm}$$

As the value of J increases, the molecule precesses faster and centrifugal distortions can be observed. These are readily taken into account by the addition of more terms to Eq. 13.90.

Microwave Spectroscopy

The $H^{35}Cl$ molecule has a moment of inertia of 2.680×10^{-47} kg m^2 and this gives a $\tilde{B}$ value of 10.445 cm^{-1}; there are then lines at 20.89 cm^{-1}, 41.78 cm^{-1}, and so on. Measurements in this region of the spectrum can be made by the methods of infrared spectroscopy, in which the radiation is passed through a cell and then a diffraction grating, and the transmitted radiation detected by a heat-sensitive device. For HCl the rotational spectrum falls in the far-infrared region of the spectrum. Heavier and more complicated molecules, however, have larger moments of inertia and consequently the rotational frequencies are much lower. For example, $^{16}O^{12}C^{32}S$ has a moment of inertia of 1.38×10^{-45} kg m^2, which is about 50 times that of HCl, and its $\tilde{B}$ value is 0.203 cm^{-1}. The lines are therefore at 0.406 cm^{-1}, 0.812 cm^{-1}, and so on. It is essentially impossible to make measurements at these *microwave* frequencies by infrared techniques. Larger molecules can, however, be studied by the techniques of microwave spectroscopy.

In microwave spectroscopy one technique is to generate the radiation by an electronically controlled oscillator, called a *klystron,* and it is possible for the radiation to be highly monochromatic, covering an extremely narrow band of wavelengths. The radiation is passed through a hollow metal waveguide of suitable dimensions to direct the waves. The substance under investigation is present in a portion of this waveguide sealed between mica windows. The cell is maintained at about 0.01 to 0.1 Torr because at higher pressures collisional broadening can give a continuous spectrum. However, because of the metal cell, even solids such as NaCl can be heated sufficiently to give adequate vapor pressure for a sample. The microwave beam that emerges from the waveguide absorption cell interacts with a metal antenna linked to a crystal detector, the signal from which is amplified and then recorded on a cathode-ray oscilloscope. The radiation is relatively weak in intensity, but the absorption lines are quite sharp. The sensitivity of the microwave technique is about 10^5 times that of an infrared grating spectrometer. Frequency measurements can in fact be made to more than *seven significant figures.* The precision with which interatomic distances can be obtained by the use of this technique was until recently limited not by the precision of the frequency measurements but by the precision with which the Planck constant had been determined. The Planck constant, however, is now known to eight significant figures [$6.626\,176(36) \times 10^{-34}$ J s^{-1}].

Rotational energies for molecules are in the order of 5×10^{-22} J, or about 0.1 k_BT at room temperature. The rotational energy levels are so close together, and so little energy is required to excite them that a molecule, even at room temperature, is likely to be in an excited rotational level. Thus a transition will involve a jump from one excited state to another. Because so many rotational states are available, rotational absorptions are relatively weak in intensity, and efforts must be made to minimize line broadening resulting from molecular collisions. See Section 14.2. Because large molecules have many vibrational states, the associated rotational states will overlap making it extremely difficult to separate the many lines. However, azulene and β-fluoronaphthalene are two of the larger molecules that have successfully been investigated.

Much of what we know about molecules in outer space comes from an analysis of the microwave spectra. From data within our own atmosphere, early microwave evidence[3] has established that the opening of the ozone hole over the Antarctic, and now over the Arctic, is a result of the presence of chlorofluorocarbons.

The results of microwave experiments are usually quoted not in wavenumbers but in frequencies. For example, for the $^{12}C^{16}O$ spectrum a line has been reported at $1.152\,712\,04 \times 10^{11}\ s^{-1}$, which can be written as 115. 271 204 MHz or as 115.271 204 GHz. This corresponds to $3.845\,033\,49\ cm^{-1}$.

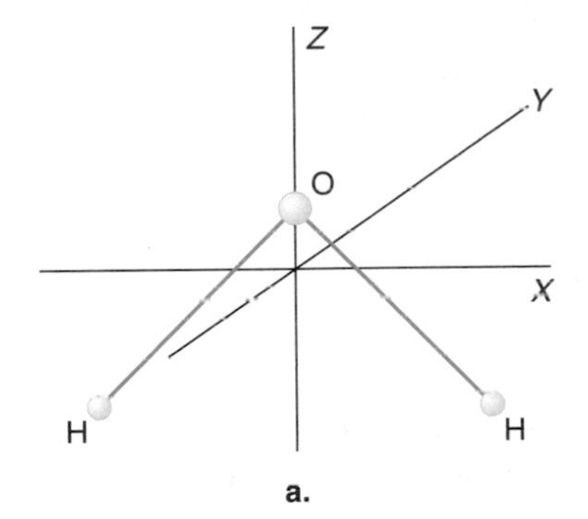

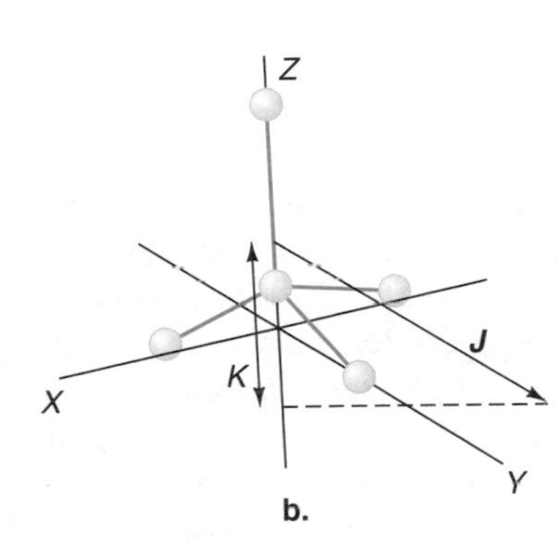

FIGURE 13.17
(a) The water molecule, an example of an asymmetric top. The principal moments of inertia, about the X, Y, and Z axes, are all different. (b) A symmetric top molecule. The moments of inertia about any axis at right angles to the Z axis are equal. Shown in the molecular frame, X, Y, Z where K is the projection of $\mathbf{J}$ on the symmetry axis. The projection of $\mathbf{J}$ on the laboratory axis, X, Y, Z, is J_Z, with values M.

Nonlinear Molecules

Linear molecules have a single moment of inertia. The problem of choice of axes about which to calculate moments of inertia for nonlinear molecules can become complicated because the three Cartesian axes (X, Y, Z) may be drawn arbitrarily through the center of mass and rigidly fixed to the molecule in any orientation. Fortunately, starting with any chosen set of Cartesian axes and the moments of inertia derived therefrom, a surface ellipsoid can be formed which not only is characteristic of the molecule but also has a set of unique axes known as the **principal axes of inertia.** It is generally easiest to calculate moments of inertia about these three principal axes, giving different moments of inertia I_a, I_b, and I_c. The convention is to order these as $I_a \geq I_b \geq I_c$. Table 13.3 classifies nonlinear molecules with respect to their moments of inertia. The rigid rotator Hamiltonian operator of a nonlinear molecule is given by

$$\hat{H} = \frac{\hbar \hat{J}_x^2}{2I_a} + \frac{\hbar \hat{J}_y^2}{2I_b} + \frac{\hbar \hat{J}_z^2}{2I_c} \tag{13.104}$$

The net result of the foregoing can best be expressed in terms of the symmetry of the molecule, since symmetry operations can only exchange identical atoms and cannot change the location of the principal axes. There is at least one principal axis passing through the center of mass about which the moment of inertia is a *maximum.* This must correspond to a rotational symmetry axis even if it is a C_1. Thus, if the molecule has an axis of symmetry, this is bound to be a principal axis. A reflection plane must contain two of the principal axes. It should be clear that a center of symmetry must lie at the center of mass coordinates. The foregoing often allows the principal axes to be determined by inspection. A few important cases follow.

If a molecule has three moments of inertia that are different, it is referred to as an **asymmetric top** (see Table 13.3). An example is the water molecule, which is illustrated in Figure 13.17a. Other examples are acetic acid and glyoxal.

[3]R.A. Kerr, "Halocarbons linked to ozone hole," *Science, 236,* 1182(1987).

TABLE 13.3 Classification of Nonlinear Molecules

Type of Molecule	Moments of Inertia
Spherical top	$I_a = I_b = I_c$
Symmetric top	$I_a = I_b \neq I_c$ or $I_a \neq I_b = I_c$
Asymmetric top	$I_a \neq I_b \neq I_c$

If two of the three moments of inertia are equal (i.e., if there are two different moments of inertia), the molecule is known as a **symmetric top.** The ellipsoid rotates about the third unique axis. Examples are molecules belonging to the point group C_{nv}, where $n \geq 3$. For $n = 3$, examples are ammonia and methyl chloride (Figure 13.17b). Note from this figure that there are two axes about which the molecule can precess. In this case, the Hamiltonian operator for a symmetric top is

$$\hat{H} = \frac{\hbar^2}{2I_a}(\hat{J}_x^2 + \hat{J}_y^2) + \frac{\hbar^2 \hat{J}_z^2}{2I_c} = \frac{\hbar^2 \hat{J}^2}{2I_a} + \frac{\hbar^2 \hat{J}_z^2}{2}\left(\frac{1}{I_c} - \frac{1}{I_a}\right) \tag{13.105}$$

More commonly, the notation A and B is used for the different rotational constants which from Eq. 13.105 can be identified with

$$\hat{H} = \tilde{B}hc\,\hat{J}^2 + (\tilde{A} - \tilde{B})hc\,\hat{J}_z^2 \tag{13.106}$$

The Schrödinger equation can be solved for this case when it is realized that not only is $\hat{J}^2$ an operator of the spherical harmonics, but so is $\hat{J}_z$ in this principal axis frame. In this case, however, the eigenvalue of $\hat{J}_z$ is given a different letter, the quantum number K

$$\hat{H}Y_J^K(\theta,\phi) = hc\left(\tilde{B}J(J+1) + (\tilde{A} - \tilde{B})K^2\right)Y_J^K(\theta,\phi) \tag{13.107}$$

The values of K span all $(2J + 1)$ integers from $-J$ to $+J$.

It should be pointed out that a molecule is not viewed in its principal axis frame but rather in the laboratory frame. In the laboratory frame, the spherical harmonics are not eigenfunctions of the Hamiltonian operator, Eq. 13.106. In general, symmetric-top molecules require three quantum numbers for their complete description. These are J, the rotational quantum number; K, the quantum number which arises from the projection of $\boldsymbol{J}$ on the symmetry axis as shown in Figure 13.17b; and M, the quantum number which arises from the projection of $\boldsymbol{J}$ on the laboratory axis. A symmetric top molecule is said to be *oblate* if $(A < B)$, that is, it is flattened at the poles like a pancake. If the two largest moments of inertia are equal, as they are in methyl chloride, the symmetric top is *prolate* $(A > B)$, like a cigar. Indeed, a linear molecule has one moment equal to zero and the other two moments of inertia are equal, and so is a special case of a prolate top.

If all three moments of inertia are equal, the molecule is known as a **spherical top** (see Table 13.3); an example is methane. The ellipsoid has degenerated into a sphere and so, all molecules that belong to a cubic point group, such as T, T_d, O, and O_h, are spherical tops.

The pure rotational spectra of nonlinear molecules are more complex than those of linear molecules, and their interpretation requires the use of two or three quantum numbers. For a treatment of such spectra, the reader is referred to more advanced texts.

The Stark Effect

J. Stark received the Nobel Prize in physics in 1919 for discovering the Stark effect in 1913.

Since a molecule that has a rotational spectrum also has an electric dipole moment, an electric field will cause an interaction. Known as the *Stark effect,* in honor of the German physicist Johannes Stark (1874–1957), the application of the field causes the $2J + 1$ degenerate rotational energy levels to be split into $2J + 1$ lines, and multiplet structure is observed for all lines with $J > 0$. This allows J values to be assigned to particular observed spectral lines, since the lowest frequency line observed need not be the one for $J = 0$. Since the number of Stark components depends on J, unambiguous assignments can be made.

A second advantage of the Stark effect is that the dipole moment may easily be found because the absorption line shift in an electric field depends both on the electric field and on the dipole moment.

13.4 Vibrational-Rotational Spectra of Molecules

The spectra observed in the infrared region involve vibrational transitions accompanied by rotational transitions. Infrared spectra thus consist of series of *bands,* each band corresponding to changes in the vibrational state of the molecule, and each line in the band corresponding to a superimposed change in the rotational state.

The infrared bands obtained with substances in the gas phase consist of fairly sharp lines, since the molecules can rotate freely. When the substance is in the liquid phase or in solution, however, the molecules cannot rotate freely, and the bands are blurred. It is not possible to make infrared measurements on substances in aqueous solution, since water absorbs intensely in this region and completely masks the spectrum of the solute. Measurements in the infrared must therefore be made with the substance in the gas phase, or in a medium that does not absorb. A common modern technique is to disperse the sample in a suitable inorganic salt, usually potassium bromide. The sample is mixed with the powdered crystalline salt, which is then pressed into a transparent disk; this is mounted in a holder that is supported in the beam of the infrared instrument.

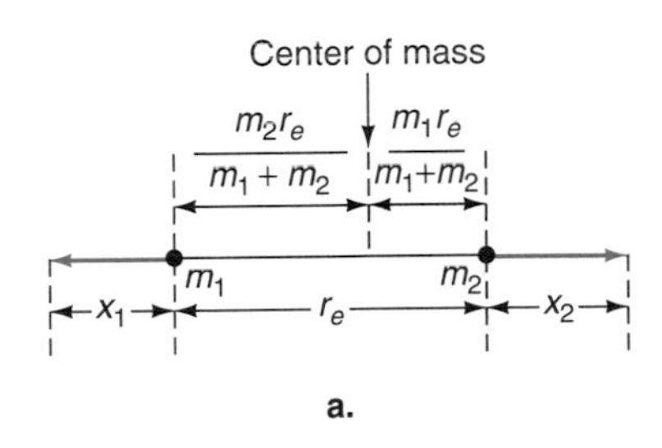

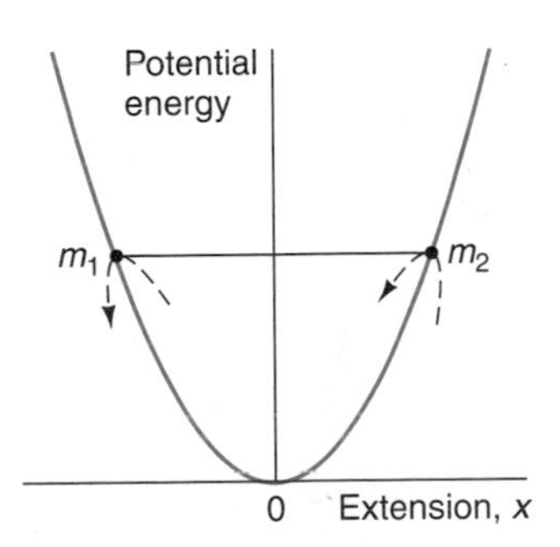

FIGURE 13.18
(a) The displacements of the masses m_1 and m_2 during the vibration of a diatomic molecule. (b) The potential-energy curve (a parabola) for a harmonic oscillator, showing the masses, m_1 and m_2 reaching their maximum extension and then receding along the original path.

Diatomic Molecules

The simplest infrared spectra are those of diatomic molecules, which are normally studied in the gas phase. In Section 11.6 we treated the quantum mechanics of the harmonic oscillator, and the energy (Eq. 11.160) was found to be

$$E_v = \left(v + \frac{1}{2}\right)h\nu_0 \quad v = 0, 1, 2, \ldots \tag{13.108}$$

where v is the *vibrational quantum number.* The frequency of vibration ν_0 can be related to the *force constant k* in the following way. Figure 13.18a shows two masses m_1 and m_2 connected by a spring. The center of mass is taken as the origin, and if the distance between the masses when the system is at rest is r_e, the center of mass is at distances

$$\frac{m_2}{m_1 + m_2} r_e \quad \text{and} \quad \frac{m_1}{m_1 + m_2} r_e$$

from the masses m_1 and m_2 respectively. Suppose that at a particular time during the vibration the masses are displaced by distances x_1 and x_2, as shown in Figure 13.18a. If moments are taken about the center of mass

$$m_1\left(\frac{m_2}{m_1 + m_2}r_e + x_1\right) = m_2\left(\frac{m_1}{m_1 + m_2}r_e + x_2\right) \tag{13.109}$$

from which it follows that

$$m_1x_1 = m_2x_2 \tag{13.110}$$

If x is the extension,

$$x = x_1 + x_2 \tag{13.111}$$

then

$$x_1 = \frac{m_2}{m_1 + m_2}x \tag{13.112}$$

$$x_2 = \frac{m_1}{m_1 + m_2}x \tag{13.113}$$

Force Constant

If the motion is harmonic, the restoring force on each mass is proportional to the extension x; the proportionality constant k is known as the **force constant** for the bond. This force is mass times acceleration, so that the equation of motion for particle 1 is (compare with Eq. 11.11)

$$m_1\ddot{x}_1 = -kx \tag{13.114}$$

Double differentiation with respect to time of Eq. 13.112 gives

$$\ddot{x}_1 = \frac{m_2}{m_1 + m_2}\ddot{x} \tag{13.115}$$

and therefore

$$\frac{m_1m_2}{m_1 + m_2}\ddot{x} = -kx \tag{13.116}$$

This may be written as

$$\mu\ddot{x} = -kx \tag{13.117}$$

where μ, the reduced mass, is given by Eq. 13.94:

$$\mu = \frac{m_1m_2}{m_1 + m_2} \tag{13.118}$$

A solution of Eq. 13.117 is

$$x = A \cos 2\pi\nu_0 t \tag{13.119}$$

double differentiation of which leads to

$$\ddot{x} = -4\pi^2\nu_0^2 A \cos 2\pi\nu_0 t \tag{13.120}$$

$$= -4\pi^2\nu_0^2 x \tag{13.121}$$

Comparison of this with Eq. 13.117 shows that

$$k = 4\pi^2\nu_0^2\mu \tag{13.122}$$

and the frequency of motion is

$$\nu_0 = \frac{1}{2\pi}\sqrt{\frac{k}{\mu}} \tag{13.123}$$

This is of the same form as Eq. 11.15, which was for a single mass m attached by a spring to an infinite mass. The only difference is that the mass is now replaced by the reduced mass μ.

Since the restoring force F is proportional to the extension x,

$$F = -kx \tag{13.124}$$

and the force is the negative of the derivative of the potential energy E_p,

$$F = -\frac{dE_p}{dx} \tag{13.125}$$

it follows that the potential energy is given by

$$E_p = -\int F\,dx = k\int x\,dx \tag{13.126}$$

$$= \tfrac{1}{2}kx^2 \tag{13.127}$$

This is the equation of a parabola (Figure 13.18b). Simple harmonic motion therefore corresponds to the movement of a particle back and forth on the parabola. At the extremities of the motion (points A and B in Figure 13.18b) the system has no kinetic energy and maximum potential energy. As it approaches the equilibrium position, the kinetic energy increases to a maximum and the potential energy decreases; the sum of the two remains constant. The kinetic energy is then reconverted into potential energy as the system continues its motion.

A more realistic shape of a potential-energy curve is shown in Figure 13.19. The lower regions of this curve are satisfactorily represented by a parabola, shown as the dashed curve. The parabolic curve, corresponding to the harmonic oscillator, therefore represents a good starting point for the treatment of molecular vibrations, at any rate at low vibrational energies. We will consider this situation first and later see what modifications are required to give a more realistic representation of the vibration of an actual molecule.

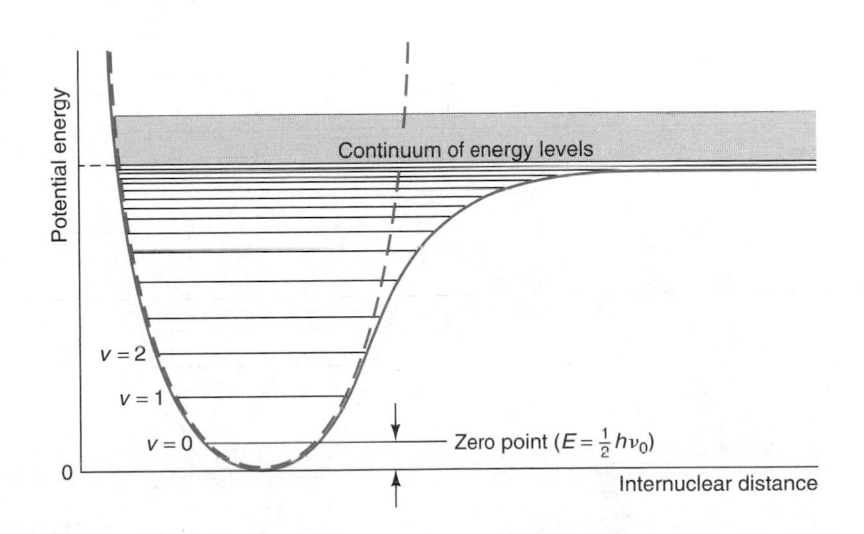

FIGURE 13.19
The actual potential-energy curve for a diatomic molecule (solid curve) with a parabola imposed on it (dashed line). The lowest horizontal lines, representing the quantized vibrational energy levels, are approximately equally spaced. However, as the curve deviates from the parabola, the levels become closer together.

To interpret vibrational-rotational spectra on this model we use the assumption that the vibrational and rotational energies are separable and that the energy can be expressed as the sum of them.

If we combine Eqs. 13.108 and 13.90, with $\tilde{B} = h/(8\pi^2 Ic)$ (Eq. 13.100), we obtain

$$E_{v,J} = \left(v + \frac{1}{2}\right)h\nu_0 + \tilde{B}J(J + 1)hc \tag{13.128}$$

If $\Delta v = 0$, we have the pure rotational spectrum; otherwise we have a vibrational spectrum with rotational changes superimposed. The situation is represented schematically in Figure 13.20, which shows two vibrational states corresponding to $v = 0$ and $v = 1$. The rotational quantum numbers in the $v = 0$ state are designated J'' and those in the $v = 1$ state are J'. To the extent that the behavior is harmonic, the average internuclear distance is r_e for all vibrational states, so that the rotational constant $\tilde{B}$ may be assumed to be the same for all values of v and J.

For strict harmonic motion the selection rules for vibrational-rotational transitions are usually

$$\Delta v = \pm 1 \quad \text{and} \quad \Delta J = \pm 1 \tag{13.129}$$

In other words, the vibrational state can change only to a neighboring one, and at the same time there must be a change to a neighboring rotational level. A transition between an upper level having quantum numbers v' and J' and a lower level with v'' and J'' involves an energy change of

$$\Delta E_{v',J'} = E_{v',J'} - E_{v'',J''} \tag{13.130}$$

$$= (v' - v'')h\nu_0 + \tilde{B}[J'(J' + 1) - J''(J'' + 1)]hc \tag{13.131}$$

Spectroscopists usually work with wavenumbers rather than frequencies, and find it convenient to use energies divided by hc. They define a quantity

$$T = E/hc \tag{13.132}$$

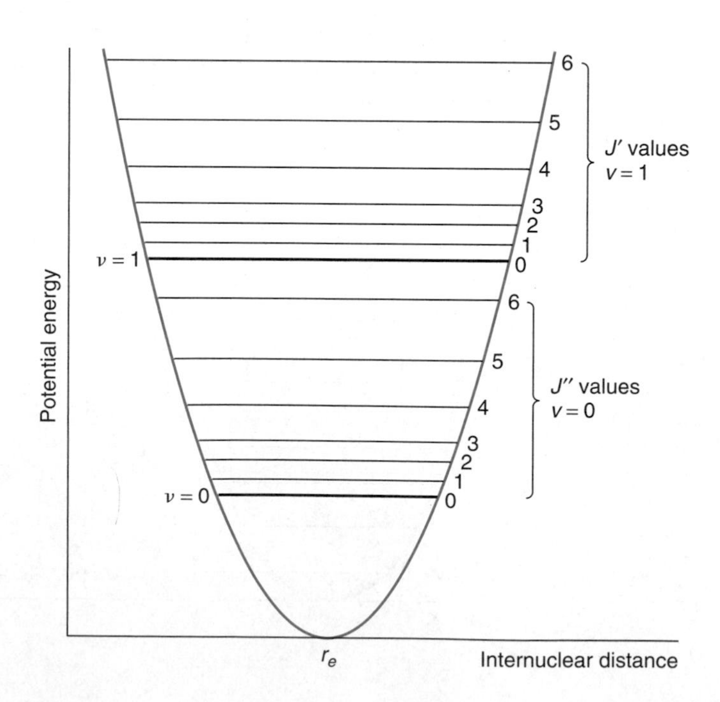

FIGURE 13.20
A potential-energy curve, showing the two lowest vibrational states ($v = 0$ and $v = 1$) with rotational states superimposed.

Spectroscopic Term

known as a **term,**[4] which has the SI unit of m^{-1}, and is usually expressed as cm^{-1} (as is the wavenumber). Thus instead of Eq. 13.131 we can write

$$\Delta T_{v',J'} = (v' - v'')\tilde{\nu}_0 + \tilde{B}\,[(J'(J' + 1) - J''(J'' + 1)] \qquad (13.133)$$

If $v' - v'' = 1$,

$$\Delta T_{v,J} = \tilde{\nu}_0 + \tilde{B}[J'(J' + 1) - J''(J'' + 1)] \qquad (13.134)$$

Associated with this change of $\tilde{\nu}$ there are two allowed changes of J. If $J' = J'' + 1$, the allowed wavenumbers are given by

$$\tilde{\nu} = \tilde{\nu}_0 + 2(J'' + 1)\tilde{B} = \tilde{\nu}_0 + 2J'\tilde{B} \qquad (13.135)$$

P, Q, and R Branches

J'' can have the values 1, 2, . . . , so that there is a series of equally spaced lines. This series is known as the *R branch.* If, on the other hand, $J' = J'' - 1$, the allowed wavenumbers are given by

$$\tilde{\nu} = \tilde{\nu}_0 - 2(J' + 1)\tilde{B} = \tilde{\nu}_0 - 2J''\tilde{B} \qquad (13.136)$$

This corresponds to a series of equally spaced lines on the low-wavenumber side of $\tilde{\nu}_0$, and is known as the *P branch.* In some cases $\Delta J = 0$ is permitted, and the resulting line, of wavenumber $\tilde{\nu}_0$, is known as the *Q branch.*

Figure 13.21a shows a representation of the infrared spectrum of HCl. The absorption lines constituting the P branch are on the low-frequency side, and the R

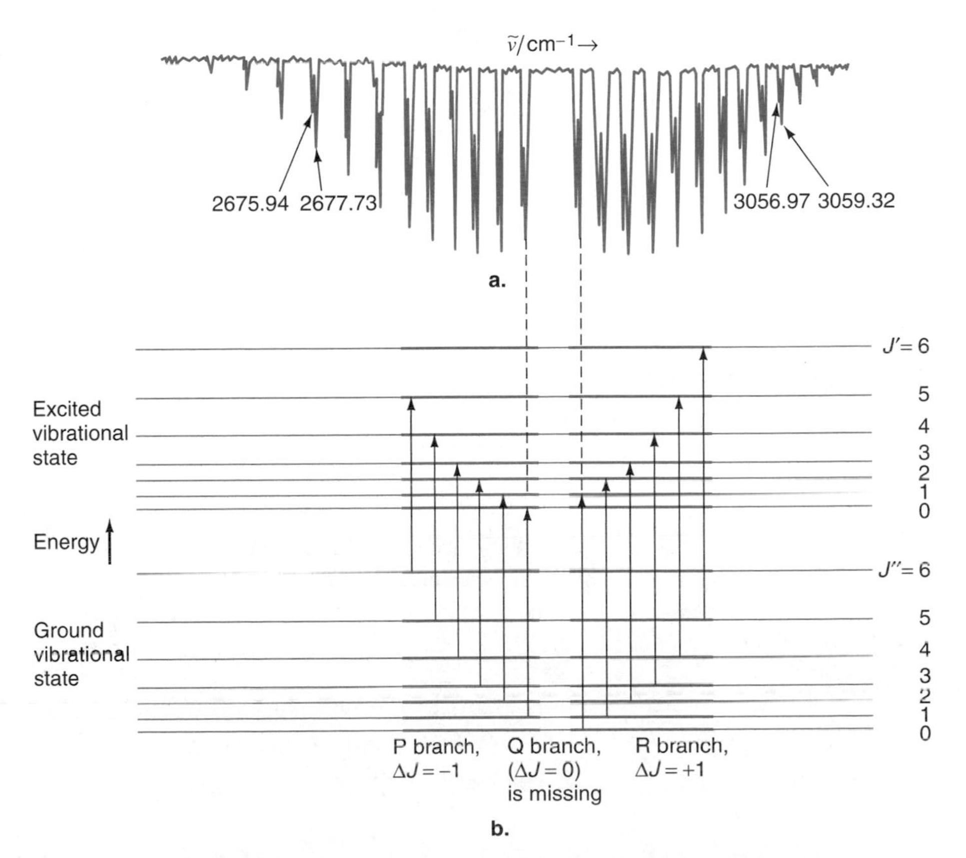

FIGURE 13.21
(a) The vibrational-rotational spectrum of HCl, showing the P and R branches. The absorption band is shown under high resolution where the splitting of the lines is due to the presence of the two isotopes ^{35}Cl and ^{37}Cl. There is a missing line in between the two bands because $\Delta J = 0$ is forbidden. (b) Rotational levels corresponding to two vibrational states, $v = 0$ and $v = 1$, showing the transitions that give rise to the P and R branches.

[4]If only electronic energy is involved the term is known as the *electronic term,* and denoted by the symbol T_e. The *vibrational term* is given the symbol G, and the *rotational term* the symbol F.

branch is on the high-frequency side. In between them is a gap, because the transition $\Delta J = 0$, in this case, is not permitted by the selection rules. Figure 13.21b shows the rotational levels for the two vibrational states and indicates the transitions that give rise to the absorption lines shown in Figure 13.21a. By measuring the separation between lines in the P and R branches, the value of $\tilde{B}$ can be obtained using Eqs. 13.134 and 13.136. It is therefore possible to calculate the intermolecular distance in a diatomic molecule.

Deviations from the simple type of behavior that we have just described arise because of the **anharmonicity** of molecular vibrations. If a potential-energy curve had a true parabolic shape, the bond could never be broken. In reality, as shown in Figure 13.19, at higher internuclear separations the potential energy is lower than represented by the parabola; the deviation becomes greater and greater as the bond length increases. Eventually the curve is horizontal as the molecule dissociates into atoms.

Because at higher internuclear separations the true potential is less confining than the parabolic approximation, the actual quantized vibrational energy levels become more and more closely spaced at higher energies than would be the case for a parabolic energy curve. Eventually, when nothing is holding the atoms together, there is no significant separation between the energy levels (as with a free particle or a particle in a box of large dimensions), and we say that there is a *continuum* of energy levels.

Anharmonic Vibrations

When there are significant deviations from the parabolic curve, the vibrations are no longer harmonic and are said to be **anharmonic.** In addition, overtones and combination bands are observed in the spectrum.

There are various ways of improving parabolic potential-energy curves in order to make them represent the experimental behavior more closely. One procedure is to express the vibrational energy term $G(v)$ $(= E_v/hc)$ as a power series in $v + \frac{1}{2}$:

$$G(v) = \tilde{\nu}_0\left[\left(v + \frac{1}{2}\right) - x_e\left(v + \frac{1}{2}\right)^2 + y_e\left(v + \frac{1}{2}\right)^3 - \cdots\right] \tag{13.137}$$

where $\tilde{\nu}_0, x_e, y_e \cdots$ are constants. It is usually sufficient to include only the first anharmonic term:

$$G(v) = \tilde{\nu}_0\left[\left(v + \frac{1}{2}\right) - x_e\left(v + \frac{1}{2}\right)^2\right] \tag{13.138}$$

Anharmonicity Constant

and x_e is then called the **anharmonicity constant.** A typical value for it is 0.01. According to this expression the higher levels are closer together, as observed experimentally. This expression can be written as

$$G(v) = \tilde{\nu}_0\left[1 - x_e\left(v + \frac{1}{2}\right)\right]\left(v + \frac{1}{2}\right) \tag{13.139}$$

which means that the apparent oscillation wavenumber is $\tilde{\nu}_0[1 - x_e(v + \frac{1}{2})](v + \frac{1}{2})$. The wavenumber therefore decreases as v increases. The zero-point energy term G_0 is given by substituting $v = 0$ into this equation:

$$G_0 = \frac{1}{2}\left(1 - \frac{1}{2}x_e\right)\tilde{\nu}_0 \tag{13.140}$$

If $x_e = 0.01$, this zero-point energy is close to that for the harmonic oscillator.

Whereas for the harmonic oscillator the selection rule is $\Delta v = \pm 1$, that for the anharmonic oscillator is

$$\Delta v = \pm 1, \pm 2, \pm 3, \text{etc.} \tag{13.141}$$

Multiple jumps can thus occur, but the probability is less and the line intensity less, when v changes by more than 1. At ordinary temperatures most molecules are in the lowest vibrational state, and we can see from Eq. 13.139 that the following are the energy changes in a transition from the lowest level to $v = 1$, $v = 2$, and $v = 3$, with no change of J:

$$\Delta E_1 = E_1 - E_0 = hc\tilde{\nu}_0(1 - 2x_e) \tag{13.142}$$

$$\Delta E_2 = E_2 - E_0 = 2hc\tilde{\nu}_0(1 - 3x_e) \tag{13.143}$$

$$\Delta E_3 = E_3 - E_0 = 3hc\tilde{\nu}_0(1 - 4x_e) \tag{13.144}$$

The wavenumber of the transition corresponding to ΔE_1 lies close to $\tilde{\nu}_0$ and is known as the *fundamental* absorption. The transitions corresponding to ΔE_2 and ΔE_3 occur at much reduced intensity and are known as the *first overtone* and the *second overtone,* respectively. These transitions are, of course, accompanied by a rotational fine structure.

At higher temperatures (over 500 K for a typical molecule) there may be enough molecules in the $v = 1$ state to give a weak absorption corresponding to a transition to higher states. For the transition to $v = 2$,

$$E_{2\leftarrow 1} = E_2 - E_1 = hc\tilde{\nu}_0(1 - 4x_e) \tag{13.145}$$

This weak absorption will thus occur at a slightly lower frequency than the fundamental and will increase in intensity as the temperature is raised. Because of this, such bands are known as *hot bands.*

An equation that closely approximates the experimental potential-energy curves for diatomic molecules was suggested in 1929 by the American physicist Philip M. Morse (b. 1903). This equation, known as the **Morse potential function,** is

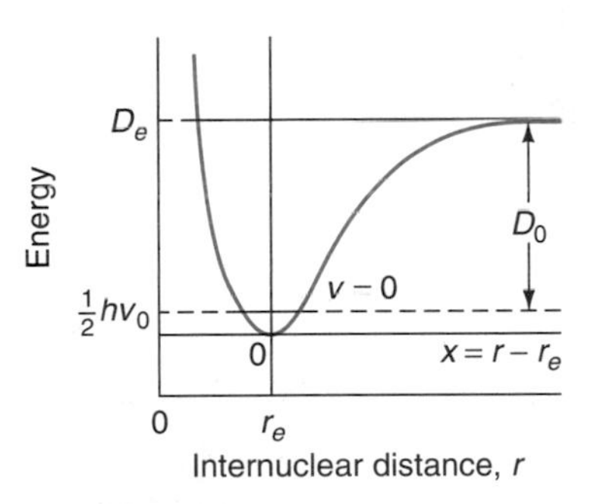

FIGURE 13.22
The potential-energy curve predicted by the Morse equation, Eq. 13.146.

$$E_p = D_e(1 - e^{-ax})^2 \tag{13.146}$$

where $x = r - r_e$ (i.e., the extension of the bond from its equilibrium distance), and D_e and a are constants. The curve represented by this equation is shown in Figure 13.22. When $x = 0$, $E_p = 0$; when x approaches infinity, E_p approaches D_e. The quantity D_e is therefore the *classical* dissociation energy and is equal to the experimental dissociation energy plus the zero-point energy. One advantage of the equation is that if the form of the potential-energy curve can be deduced from spectroscopic data, the curve can be fitted to the Morse function; the energy D_e is then known, and the dissociation energy D_0 can easily be calculated. For further discussion see Example 13.12 and Section 13.6.

Coupling of Rotational and Vibrational Motion: The Separability Assumption

The assumption of separability of the rotational and vibrational motion is justified because the difference between the energies of rotational and vibrational motions is large enough that the two motions are effectively uncoupled (see Figure 13.14). Consider a tightly stretched string and along side a loosely stretched string. It is difficult to pluck one and have the other resonate. For quantum systems, there is no major mechanism in the Hamiltonian that mixes processes which have large energy differences, such as those for rotational and vibrational motions. We note that there is a widely held but incorrect notion, found in many textbooks, that vibronic uncoupling is a result of the Born-Oppenheimer approximation (page 580). This is not the case. The Born-Oppenheimer approximation applies only to the uncoupling of the nuclear and electron motion by virtue of the large mass differences between the two (and not the large energy differences between the two). Hence, the full wave function for atoms becomes, in the Born-Oppenheimer approximation,

$$\psi_{total} = \psi_{nuclear}\psi_{electronic} \tag{13.147}$$

The effect of electronic spin is included in the electronic wave function because of the inclusion of the Slater determinent, thereby simplifying the expression.

We have also seen that electrons have spin so that the spin wave functions must be included. Moreover, nuclei can have nuclear spin. In all, we can usually write

$$\psi_{total} = \psi_{nuclear}\psi_{electronic}\psi_{nuclear\ spin} \tag{13.148}$$

For molecules, in addition to the wave function for the motion of electrons and spins, there are other motions due to molecular rotations and the vibrations. For molecules, then, it is sometimes justified to express the total wave function as

The wave function for nuclear motion is normally incorporated into the rotational-vibrational wave function for further simplification of the expression.

$$\psi_{total} = \psi_{electronic}\psi_{vibrational}\psi_{rotational}\psi_{nuclear\ spin} \tag{13.149}$$

This product of wave functions is convenient to work with because each individual wave function is independently normalized. That means that if we are working with pure rotational motion, with the Hamiltonian operator, $\hat{H}_{rotational}$, then the matrix elements simplify to

$$\langle\psi_{total}|\hat{H}_{rotational}|\psi_{total}\rangle = \langle\psi_{rotational}|\hat{H}_{rotational}|\psi_{rotational}\rangle \tag{13.150}$$

because all the other wave functions commute with the rotational Hamiltonian and are orthogonal. What all this means is that we are assuming that the total Hamiltonian operator for a molecule can be written as a sum

$$\hat{H}_{total} = \hat{H}_{electronic} + \hat{H}_{rotational} + \hat{H}_{vibrational} + \hat{H}_{nuclear\ spin} \tag{13.151}$$

However, this is not always the case. In general we have to verify that no coupling exists between one motion and another to justify the product form of the wave functions. In the case of rotational and vibrational motion, the energies of two are widely separated, and we assume that the two are uncoupled, and thus we are justified in writing

$$\begin{aligned}\langle\psi_{total}|(\hat{H}_{rotational} + \hat{H}_{vibrational})|\psi_{total}\rangle = &\langle\psi_{rotational}|\hat{H}_{rotational}|\psi_{rotational}\rangle\\ &+ \langle\psi_{vibrational}|\hat{H}_{vibrational}|\psi_{vibrational}\rangle\end{aligned} \tag{13.152}$$

If, however, there is coupling between the vibronic motion, as certainly happens in some cases, then the Hamiltonian operator will have a term to describe this, and the rotational and vibrational wave functions are no longer a product. Instead, we write

$$\langle\psi_{total}|\hat{H}_{ro\text{-}vibrational}|\psi_{total}\rangle = \langle\psi_{ro\text{-}vibrational}|\hat{H}_{ro\text{-}vibrational}|\psi_{ro\text{-}vibrational}\rangle \tag{13.153}$$

where *ro* stands for rotational, and the Hamiltonian operator takes the form

$$\hat{H}_{ro\text{-}vibrational} = \hat{H}_{rotational} + \hat{H}_{vibrational} + \hat{H}'_{coupling} \tag{13.154}$$

In this case, by virtue of $\hat{H}'_{coupling}$ the total wave function is

$$\psi_{total} = \psi_{electronic}\psi_{ro\text{-}vibrational}\psi_{nuclear\ spin} \tag{13.155}$$

where

$$\psi_{ro\text{-}vibrational} \neq \psi_{rotational}\psi_{vibrational} \tag{13.156}$$

Only if the rotational and vibrational motions are uncoupled can we express the energy as a sum.

Normal Modes of Vibration

When objects such as a drum or guitar string are struck, the system undergoes a vibration. Depending upon how and where the strike occurs, resonances are set up that persist for some time and gradually fade out. Molecules, as we have seen, also vibrate and their vibrations are sustained by the thermal energy that surrounds them. The process of decomposition of the vibrational motion of a molecule into the minimum number of characteristic modes is called **normal mode analysis.** This analysis is a standard procedure and is useful because it simplifies the analysis of vibrational spectra, revealing in the process information about structure, symmetry, bond lengths, bond strengths, the types of vibrational motion possible, and the potential energy function that describes the molecule.

Normal modes are analyzed by modifying the energy potential which is perturbed by the finite number of vibrational motions available to the N atoms in a molecule. The actual separation of these vibrational motions into the normal modes is well established in classical mechanics, and it is this process that is called normal mode analysis. Once done, the potential energy term in the Hamiltonian is simplified into a sum of vibrational terms, after which the Schrödinger equation is solved.

A diatomic molecule vibrates according to a simple pattern and has only one vibrational frequency. The vibrational motions of molecules having more than two atoms are a superposition of a number of basic vibrations, known as **normal modes of vibration.** In order to bring out this concept, we will work through the dynamics for a linear triatomic molecule.

The linear triatomic system is shown schematically in Figure 13.23a. In this treatment we will ignore bending vibrations and will consider only displacements along the axis of the molecule. The center of mass is again taken as the origin, and if moments are taken about the origin when the molecule is at rest,

Stretching Vibrations

$$m_1(r_{12} - y) = m_2 y + m_3(r_{23} + y) \tag{13.157}$$

where y is the distance between the center of mass and the atom B. If the three masses are displaced to the right by the distances x_1, x_2, and x_3,

$$m_1(r_{12} - y - x_1) = m_2(y + x_2) + m_3(r_{23} + y + x_3) \tag{13.158}$$

and therefore

$$m_1 x_1 + m_2 x_2 + m_3 x_3 = 0 \tag{13.159}$$

The displacements cause the distance between particles A and B to increase by $x_2 - x_1$; the restoring force is therefore $k_{12}(x_2 - x_1)$, where k_{12} is the force constant for the bond. The equation of motion for particle A is therefore

$$k_{12}(x_2 - x_1) = m_1 \ddot{x}_1 \tag{13.160}$$

Similarly for particle C

$$k_{23}(x_3 - x_2) = -m_3 \ddot{x}_3 \tag{13.161}$$

The equations for particle B need not be written since they are simply linear combinations of Eqs. 13.159 to 13.161.

Elimination of x_2 in Eqs. 13.160 and 13.161 by use of Eq. 13.159 leads to

$$m_1 m_2 \ddot{x}_1 + k_{12}[(m_1 + m_2)x_1 + m_3 x_3] = 0 \tag{13.162}$$

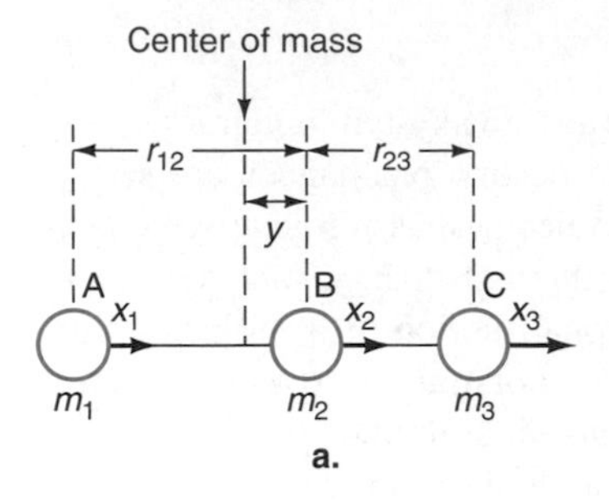

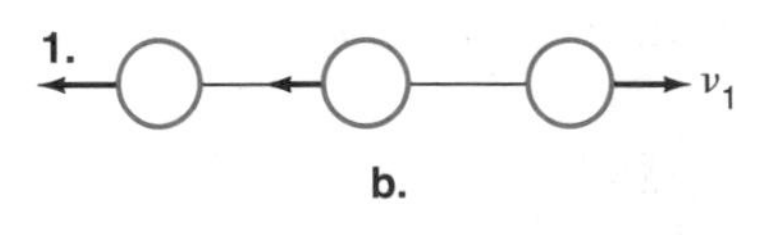

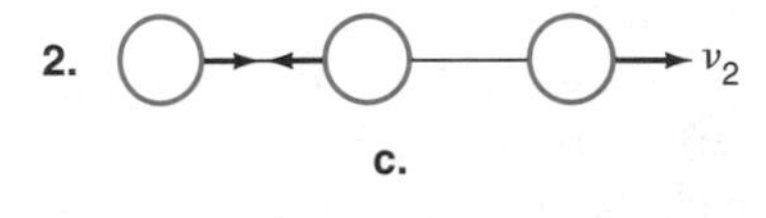

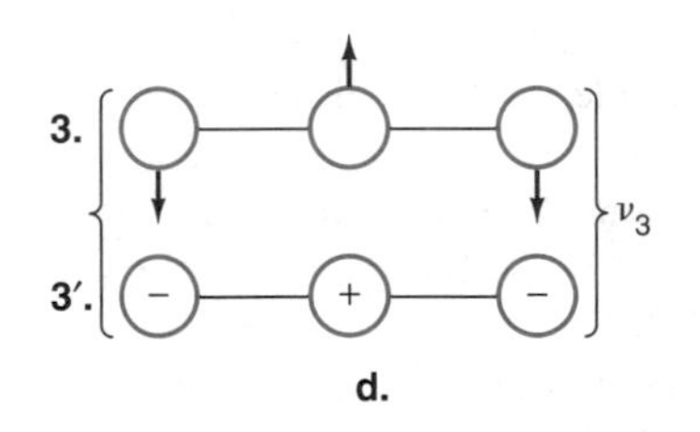

FIGURE 13.23
(a) Stretching vibrations of a linear triatomic molecule. (b) The symmetric stretching (breathing) vibration of a linear triatomic molecule. (c) The antisymmetric stretching vibration. (d) The two degenerate bending vibrations.

and

$$m_2 m_3 \ddot{x}_3 + k_{23}[m_1 x_1 + (m_2 + m_3)x_3] = 0 \tag{13.163}$$

To solve these equations, we look for solutions of the form

$$x_1 = A_1 \cos 2\pi \nu t \tag{13.164}$$

$$x_3 = A_3 \cos 2\pi \nu t \tag{13.165}$$

with the same frequency ν for both displacements. Double differentiation gives

$$\ddot{x}_1 = -4\pi^2\nu^2 x_1 = -\lambda x_1 \tag{13.166}$$

$$\ddot{x}_3 = -4\pi^2\nu^2 x_3 = -\lambda x_3 \tag{13.167}$$

where λ has been written for $4\pi^2\nu^2$. Insertion of these expressions into Eqs. 13.162 and 13.163 leads to two simultaneous equations in x_1 and x_3:

$$[-\lambda m_1 m_2 + k_{12}(m_1 + m_2)]x_1 + k_{12} m_3 x_3 = 0 \tag{13.168}$$

$$m_1 k_{23} x_1 + [-\lambda m_2 m_3 + k_{23}(m_2 + m_3)]x_3 = 0 \tag{13.169}$$

Equations of this type are only consistent with one another provided that

$$\begin{vmatrix} -\lambda m_1 m_2 + k_{12}(m_1 + m_2) & k_{12} m_3 \\ m_1 k_{23} & -\lambda m_2 m_3 + (m_2 + m_3)k_{23} \end{vmatrix} = 0 \tag{13.170}$$

This determinental equation is a quadratic equation in λ:

$$\lambda^2 - \left(\frac{m_1 + m_2}{m_1 m_2} k_{12} + \frac{m_2 + m_3}{m_2 m_3} k_{23}\right)\lambda + \frac{k_{12} k_{23}(m_1 + m_2 + m_3)}{m_1 m_2 m_3} = 0 \tag{13.171}$$

Its solution gives two values of λ and therefore two frequencies ν_1 and ν_2. This equation may be written as

$$\lambda^2 - b\lambda + c = 0 \tag{13.172}$$

and the roots are

$$\lambda_1 = \frac{b - \sqrt{b^2 - 4c}}{2} \tag{13.173}$$

and

$$\lambda_2 = \frac{b + \sqrt{b^2 - 4c}}{2} \tag{13.174}$$

Since $\lambda_2 > \lambda_1$, the frequency ν_2 is the larger of the two frequencies. If Eq. 13.174 for λ is inserted into either Eq. 13.168 or Eq. 13.169, there results an equation from which it is possible to evaluate the ratio x_1/x_3; when this is done, x_1/x_3 is found to have a positive value. If, on the other hand, the lower root λ_1 is inserted into Eq. 13.168 or Eq. 13.169, the ratio x_1/x_3 is negative.

The significance of this result is that the lower frequency ν_1, which relates to λ_1, is the frequency of a symmetric vibration in which the bonds stretch and shorten in unison, as shown in Figure 13.23b. The higher frequency ν_2 is the frequency of an asymmetric vibration in which one bond shortens while the other is stretching, as shown in Figure 13.23c.

Bending Vibrations

A similar treatment of the **bending vibrations,** in terms of a force constant for the bending, leads to the result that there are two *degenerate* (i.e., equivalent) bending motions in planes at right angles to one another. In the upper representation in

Figure 13.23d, the bending is occurring in the plane of the paper, while in the lower representation it is at right angles to the plane of the paper. A bending motion in any plane can be treated as a superposition of the bending motions in these two planes; expressed differently, any bending motion can be resolved into the bending motions in these two planes. In the same way, any stretching motion that the molecule can undergo is the resultant of the two basic motions represented in Figure 13.23b and c.

Normal Modes of Vibration

The stretching and bending motions shown in Figure 13.23b–d are known as **normal modes of vibration,** and they play a similar role in vibrational theory as do motions along the *X*, *Y*, and *Z* axes in translational theory. For any linear triatomic molecule there are four normal modes of vibration; the two bending modes are degenerate. In each of the modes all three atoms are either completely in phase or completely out of phase with each other and move with the same frequency. The motion of each atom, with reference to the center of mass, corresponds to simple harmonic motion.

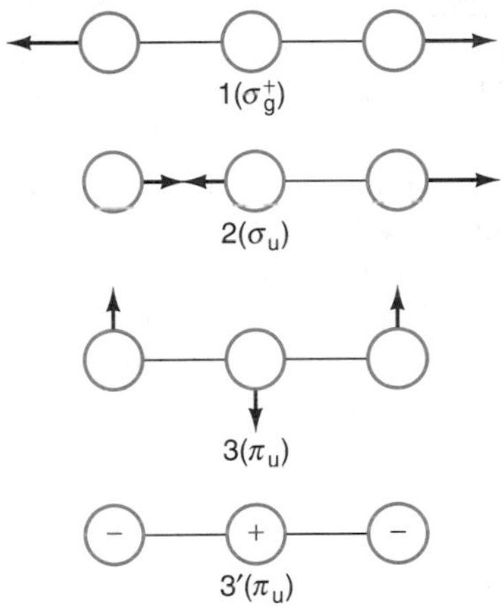

The question of the number of normal modes in a molecule is discussed in Section 15.2. For a linear molecule the number is $3N - 5$, where N is the number of atoms in the molecule; for the linear triatomic molecule that we have just considered the number is $3(3) - 5 = 4$, and these four modes are represented in Figure 13.23. For a nonlinear molecule the number is $3N - 6$; for benzene, for example, there are $3(12) - 6 = 30$ normal modes. Each one of these has a characteristic vibrational frequency, some of them degenerate.

A detailed treatment of molecular vibrations is outside the scope of the present book, and the reader is referred to more specialized texts such as some of those listed in Suggested Reading (pp. 712-713). Mention should, however, be made of the relationship between the vibrational modes and the symmetry of the molecule as indicated by its point group. Each vibrational mode must transform according to one of the symmetry species of the molecule. We may illustrate this for the normal modes of the symmetrical linear carbon dioxide molecule, shown in Figure 13.24a. The point group is $D_{\infty h}$, an abbreviated character table for which is given in Appendix E (p. 1028). The symmetric vibration 1 belongs to the Σ_g^+ symmetry species, since all symmetry operations leave the vibration unchanged. It is usual to denote the vibration as σ_g^+, using the lowercase Greek letter. The antisymmetric vibration 2 is antisymmetric with respect to reflection in the σ_h plane and with respect to inversion, but symmetric with respect to a σ_v reflection; it therefore belongs to the Σ_u^+ symmetry species and is denoted σ_u^+. The reader should confirm that the two bending vibrations are π_u. These assignments are important in connection with the selection rules, as will be discussed later.

FIGURE 13.24
Normal modes of vibration. (a) The carbon dioxide molecule, which belongs to the point group $D_{\infty h}$. (b) The water molecule, which belongs to the point group C_{2v}. The character tables for these point groups are to be found in Appendix E (p. 1028).

Infrared Spectra of Complex Molecules

For a diatomic molecule there is only one set of bands in the infrared, since there is only a single vibrational frequency. With more complicated molecules there are generally a large number of sets of bands corresponding to vibrational-rotational transitions, because of the various vibrational frequencies corresponding to the different normal modes of vibration. Each one of these bands has a P and an R branch corresponding to whether the rotational quantum number in the upper vibrational level is lower or higher, respectively, than in the ground state. In some cases $\Delta J = 0$ is also allowed, and then there is additionally a Q branch.

The presence of infrared bands is restricted by another factor. We saw earlier that a *diatomic* molecule can give a vibrational-rotational spectrum only if it has a permanent dipole moment. This rule is a special case of a more general selection rule for vibrational-rotational transitions, namely that the normal-mode vibration must give rise to an **oscillating dipole moment.** For a diatomic molecule this requires the molecule to have a permanent dipole moment, because if a diatomic molecule has no dipole moment, the moment remains zero throughout the vibration. It is, however, possible for a more complex molecule initially having no dipole moment to produce an oscillating moment during the course of its vibration. Consider, for example, the normal modes of vibration of the carbon dioxide molecule, shown in Figure 13.24a. In the purely symmetric mode the dipole moment remains zero during the course of the vibration, and this vibration is not active in the infrared spectrum. The bending vibrations give rise to a small moment, and therefore they give rise to an infrared spectrum, although it may be of weak intensity. The antisymmetric vibration, on the other hand, produces an oscillating dipole moment, and there is a spectrum arising from this motion. The carbon dioxide spectrum is an example in which the $\Delta J = 0$ transition is allowed, so that there is a Q branch.

Selection Rules for Vibrational-Rotational Spectra

These symmetry restrictions may be formulated in terms of group theory. We have seen that the normal modes of vibration must correspond to one of the symmetry species of the point group to which the molecule belongs. Let us suppose that a molecule is vibrating in a normal mode and that at a particular instant it has a dipole moment $\boldsymbol{\mu}$. This dipole moment, being a vector, has components along the X, Y, and Z axes, and we can write these components as $\boldsymbol{\mu}_x$, $\boldsymbol{\mu}_y$, and $\boldsymbol{\mu}_z$. Suppose that one of these components, $\boldsymbol{\mu}_x$, is affected by all the symmetry operations in the same way as the normal mode of vibration itself. If this is the case, the component will vary in magnitude as the vibration occurs (i.e., will be an oscillating dipole). If, on the other hand, the component $\boldsymbol{\mu}_x$ is of a different symmetry species from the normal mode, its magnitude will remain unchanged as the vibration occurs. It follows that if all the components $\boldsymbol{\mu}_x$, $\boldsymbol{\mu}_y$, and $\boldsymbol{\mu}_z$ are of different symmetry species from the normal mode, there can be no oscillation of the dipole moment and hence no vibrational-rotational spectrum corresponding to that normal mode of vibration. However, if one or more of the three components is of the same symmetry species as the normal mode, the infrared spectrum is allowed, that is, the vibration is infrared active.

Since the components $\boldsymbol{\mu}_x$, $\boldsymbol{\mu}_y$, and $\boldsymbol{\mu}_z$ are along the X, Y, and Z axes, their symmetry properties are the same as those of x, y, and z coordinates, respectively. The symmetry restriction can therefore be stated as follows:

> **There will be a vibrational-rotational spectrum corresponding to a normal mode of vibration only if that mode belongs to the same symmetry species as one or more of three coordinates x, y, and z.**

We may illustrate this rule by some examples. Consider first the normal modes of vibration for carbon dioxide, shown in Figure 13.24a. On p. 707 is an abbreviated character table for $D_{\infty h}$, which shows how the x, y, and z coordinates transform. The purely symmetric vibration is of symmetry species Σ_g^+ and is designated σ_g^+ (the lowercase letters are used); however, none of the coordinates belongs to this species, and this normal mode therefore does not give rise to a spectrum. The antisymmetric vibration, however, is σ_u^+, to which z belongs; there is therefore a spectrum. The two degenerate bending vibrations are π_u, and x and y are also of this species. There can therefore be a spectrum, although it will be weak since the bending vibrations give rise to only a small dipole moment.

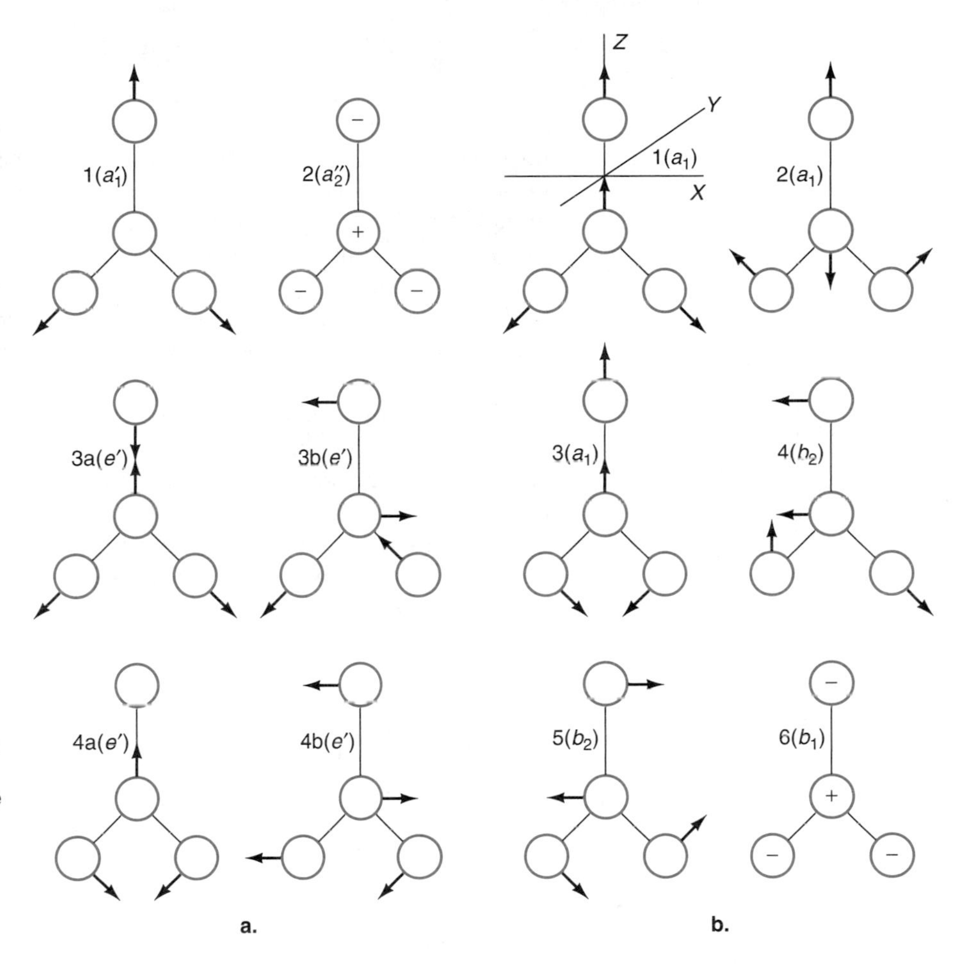

FIGURE 13.25
(a) The normal modes of vibration of a planar symmetrical molecule such as BF_3, which belongs to the point group D_{3h}. (b) The normal modes of vibration of a molecule such as CH_2O, which belongs to the point group C_{2v}. The character tables are in Appendix E (p. 1028).

The case of water is shown in Figure 13.24b, and the reader should verify the assignment of symmetry species with reference to the character table. All three normal modes are either a_1 or b_2, and the coordinate z is A_1 and y is B_2. All three modes therefore give rise to an infrared spectrum.

Two more complicated cases are illustrated in Figure 13.25. In Figure 13.25 are shown the six normal modes of vibration of a planar symmetrical molecule such as BF_3 (point group D_{3h}). The reader should verify the assignment of symmetry species for the vibrations and for the axes. Vibration 2 is of species a_2'', which is the same as z; it therefore gives rise to infrared bands. Vibrations $3a$, $3b$, $4a$, and $4b$ are e', as are x and y, so that these vibrations also give a spectrum. Vibration 1, however, is a_1', which does not correspond to x, y, or z; it therefore gives no spectrum.

Figure 13.25b is for a molecule like CH_2O, which belongs to the point group C_{2v}. The reader should verify that from the character table it can be deduced that all the normal modes give rise to an infrared spectrum.

Some of the selection rules for infrared and Raman spectra (Section 13.5) are conveniently summarized in the appendix to this chapter (p. 707). It should be noted that intuitive arguments, such as we used earlier for the vibrations of CO_2 (Figure 13.24a), are not reliable for more complicated systems.

Characteristic Group Frequencies

The complete analysis of the infrared spectrum of a molecule containing more than a dozen atoms is difficult and sometimes impossible. Useful information can, however, be obtained from the spectra of quite large molecules if one makes use of simple theory and empirical correlations. In particular, individual functional groups in molecules absorb in characteristic regions of the infrared spectrum. For example, a molecule containing the alcohol group, O—H, always shows an absorption in the 3100–3500 cm^{-1} region of the spectrum, while molecules containing the carbonyl group, C=O, characteristically absorb in the 1600–1800 cm^{-1} region. Table 13.4 gives a few wavenumbers and wavelengths for different kinds of bonds.

There are two main reasons why these convenient simplifications occur. If a bond involves a hydrogen atom attached to a much heavier atom such as an oxygen atom, some of the normal modes of vibration involve the light atom moving with respect to the rest of the molecule, which does not move much in relation to the center of gravity. A simple example is provided by the spectrum of HCN. One of the normal-mode wavenumbers is 2089 cm^{-1}, and this relates to the symmetric normal mode of vibration:

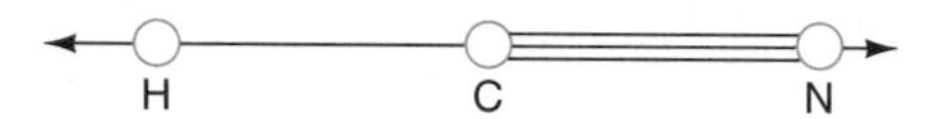

When this vibration occurs, the C—H bond stretches considerably compared with the C≡N bond, because of the relative lightness of the H atom, and the C and N atoms do not move much relative to the center of mass because of the tightness of the C≡N bond. This normal vibration is therefore close to being simply a stretching of the C—H bond. Similarly, with an alcohol, some of the normal modes of vibration are close to being the simple stretching of the O—H bond.

Another factor that tends to favor characteristic frequencies is the presence of multiple bonds. Here the effect is due to the wide disparity between the force constant of the multiple bond and the force constants for other bonds in the molecule. A simple example is again provided by HCN, which has a normal-mode wave number of 3312 cm^{-1} for the antisymmetric vibration:

It happens that the force constant for the C≡N bond is over three times that for the H—C bond, and the normal-mode analysis shows that this antisymmetric vibration is largely the stiff vibration of the C≡N bond.

TABLE 13.4 Characteristic Group Frequencies

Bond	Wavenumber, $\tilde{\nu}$ cm^{-1}	Wavelength, λ μm
O—H, N—H	3100–3500	3.2–2.85
C—H	2800–3100	3.6–3.25
C—O, C—N	800–1300	12.5–7.7
C=C, C=O	1600–1800	6.25–5.55
P—O	1200–1300	8.3–7.7

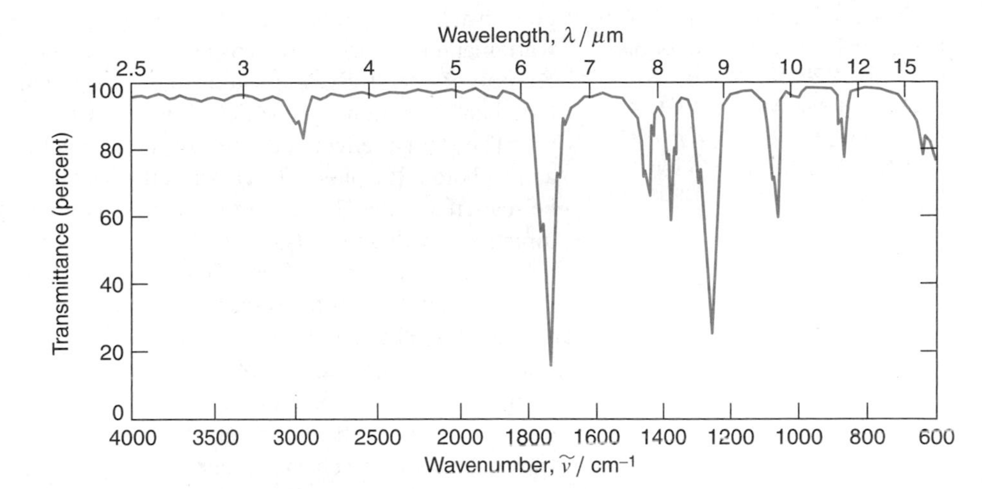

FIGURE 13.26
The infrared spectrum of methyl acetate. This spectrum was obtained with the material in the liquid phase.

Bonds involving hydrogen atoms and multiple bonds therefore tend to have characteristic group frequencies. There are also some effects that work against characteristic group frequencies; the most important is when there is considerable symmetry so that the molecule contains two or more equivalent bonds. A simple example is provided by the water molecule (Figure 13.24b), which has normal mode frequencies of 3756, 3652, and 1595 cm^{-1}. The lowest of these may be attributed to the bending vibration. Neither of the two higher frequencies, however, can be ascribed to the stretching of one bond, since the two O—H bonds are equivalent. The stretching motions of the two bonds will therefore be strongly coupled, and in these two normal modes the two bonds stretch either in phase or out of phase with one another.

These limitations, however, are more prominent with fairly small molecules. With molecules of any complexity the characteristic group frequencies can provide useful supporting evidence for the presence of specific functional groups. Moreover, because of the large number of vibrational modes in a molecule of any size, the infrared spectrum is complicated and is unique to the molecule. Figure 13.26 shows a typical infrared spectrum; since this was obtained with the substance in the liquid phase, the rotational fine structure is blurred, but the features of this spectrum will not be found with any other molecule and will allow a convincing identification of the substance to be made. Because the infrared spectrum of a substance is so uniquely characteristic, it is often referred to as a *fingerprint spectrum.*

13.5 Raman Spectra

When a beam of light passes through a medium, a certain amount of the light is scattered and can be detected by making observations perpendicular to the incident beam. Most of the light is elastically scattered off molecules without a change in wavelength, an effect known as Rayleigh scattering. If the light is scattered inelastically, the result is a change in frequency between the incident and scattered light. This scattered light can have higher or lower wavelengths from the incident light. The effect is small since most of the light is scattered elastically. Spectra resulting from the inelastic scattering of light were first observed in 1928 by the Indian

C. V. Raman received the Nobel Prize in physics in 1930 for his study in 1928 of Raman scattering. He was knighted by King George V in 1929; Krishnan was knighted in 1946 at the last investiture held by the British in India.

physicist Sir Chandrasekhara Venkata Raman (1888–1970) and his coworker Sir Kariamanikkam Srinivasa Krishnan (1898–1961), and this effect is known as the **Raman effect.** It gives an alternate and complementary way of studying the rotational and vibrational properties of molecules.

The Raman effect is a result of the interaction that occurs between a molecule and a photon. If a photon had a perfectly elastic collision with the molecule, it would be scattered with no change of frequency. The radiation would then be emitted in all directions, with an intensity inversely proportional to the fourth power of the wavelength. Because of this dependency, shorter wavelengths are more prominent in scattered light and particles suspended in a gas are rendered visible by a light beam.

On the other hand, in an inelastic collision, energy is exchanged between the molecule and the incident photon, according to quantum selection rules. Raman scattering is a two-photon process, with one photon going in and one coming out, and, as a result of the conservation of angular momentum of the photons (spin of 1 each), the spin of each can remain unchanged or change by two units. If enough energy is available, a vibrational transition may occur, as well as a rotational transition.

It should be apparent that the incident light can be of any wavelength since scattered light does not depend upon the Bohr condition (Eq. 13.18). Common practice is to use monochromatic light and observe the change in frequency due to the inelastic scattering. If the molecule gains energy, the selection rule is then the same as for the infrared spectra of anharmonic oscillators; namely $\Delta v = \pm 1, \pm 2, \ldots$. Generally only the lowest vibrational level is occupied to any extent and the transition $v = 1 \leftarrow v = 0$ with rotational states superimposed gives the strongest Raman band.

Stokes lines are named in honor of George Gabriel Stokes (1819–1903), who discovered fluorescence in 1852. (Fluorescence, however, is not a photon scattering effect.)

If the molecule gains energy ΔE, the photons will be scattered with reduced energy $h\nu_0 - \Delta E$. This results in **Stokes lines** of lower frequency than the frequency ν_0 of the incident beam; their frequency is

$$\nu = \nu_0 - \frac{\Delta E}{h} \tag{13.175}$$

Anti-Stokes lines correspond to a decrease in molecular energy, which can occur if the molecule is initially in an excited state. If the molecule loses energy, the frequency of these lines is

$$\nu = \nu_0 + \frac{\Delta E}{h} \tag{13.176}$$

Molecules are generally not in their excited states, and therefore these lines are much weaker than the Stokes lines.

Figure 13.27a is a schematic representation of a pure rotational Raman spectrum. The line corresponding to the incident beam, of frequency ν_0, is known as the **Rayleigh line.** For pure rotational transitions the energy is given in Eq. 13.96; but since for a Raman line $\Delta J = \pm 2$, the energy difference between two levels is

$$\Delta E = E_{J+2} - E_J = \tilde{B}hc(4J + 6) \tag{13.177}$$

where for anti-Stokes lines J is the rotational quantum number for the lower state. The frequency of the lines is thus

$$\tilde{\nu} = \tilde{B}(4J + 6) \tag{13.178}$$

The first line is thus $6\tilde{B}$ from the Rayleigh line and the subsequent spacing is $4\tilde{B}$ (see Figure 13.27). The anti-Stokes lines are referred to as the S-branch. For Stokes

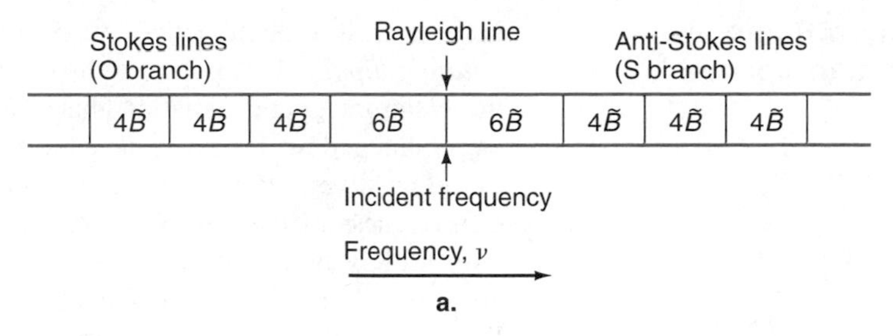

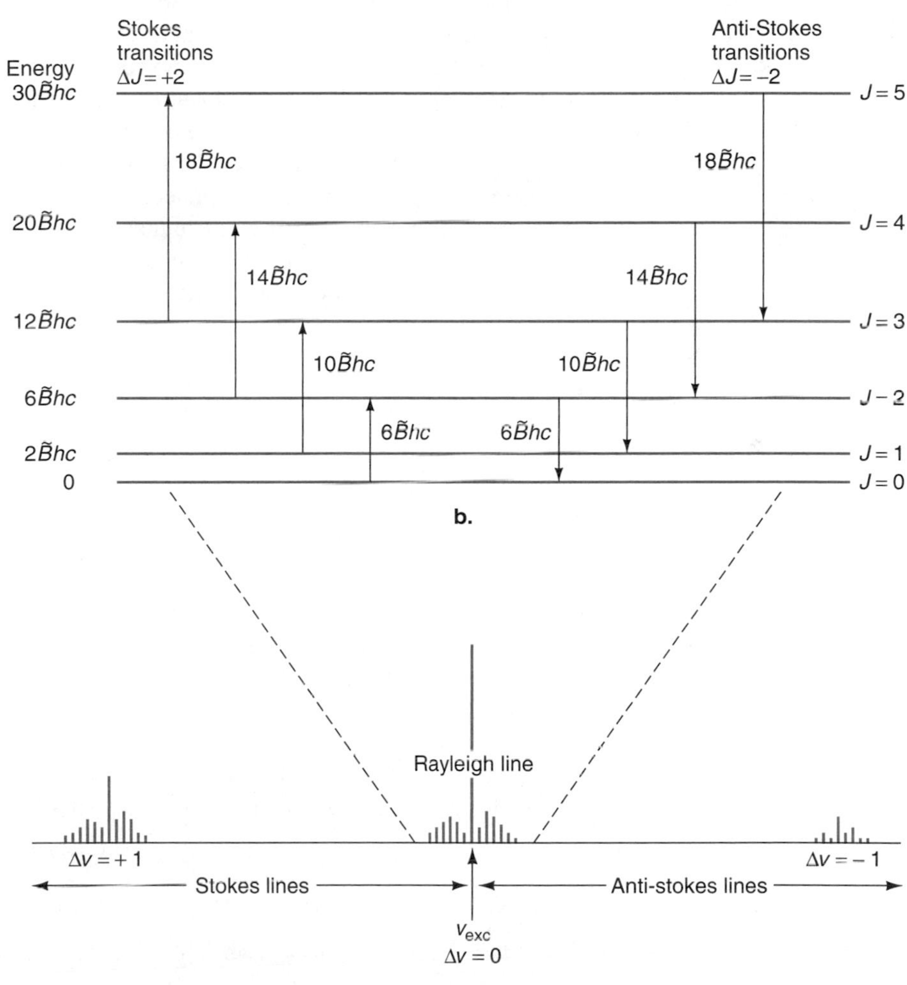

FIGURE 13.27
(a) Schematic representation of a Raman spectrum arising from pure rotational transitions. (b) The rotational transitions giving rise to Raman lines. (c) Stokes and anti-stokes lines for $\Delta v = 0, \pm 1, \ldots$.

lines (the O-branch) the spacings are the same (note the letter assignments O, P, Q, R, and S for ΔJ transitions of $+2$, $+1$, 0, -1, and -2, respectively).

An important advantage of Raman spectroscopy is that it is not necessary to work in the infrared and microwave regions of the spectrum. The region is determined by the choice of frequency of the incident light and can be the visible region. Also, with Raman spectra it is possible to work with aqueous solutions since one can choose an incident frequency where there is no absorption by water.

Another advantage of Raman spectroscopy is that certain lines appear in the Raman spectrum that do not appear in the infrared. The reason is that to be active in the Raman spectrum, a vibration must produce an *oscillating polarizability* α. This

Selection Rules for Raman Spectra

is in contrast to being active in the infrared where a vibration must produce an *oscillating dipole.* This polarizability is the result of the oscillating electric field $\boldsymbol{E}$ in the electromagnetic radiation that causes a distortion in the molecule. The positively charged nuclei are attracted to the negative pole of the field and the electrons to the positive pole. The polarizability of a molecule is therefore a measure of the effectiveness of the field $\boldsymbol{E}$ in disturbing the electron clouds. A molecule or atom with loosely bound electrons is more polarizable than one in which the electrons are tightly bound to the nucleus. For example, xenon is more polarizable than helium. The polarizability of a molecule therefore increases with an increase in the number of electrons, and as a consequence the intensity of the scattered radiation increases with an increase of molecular mass. In general, a bond is more polarizable when it is lengthened than when it is shortened, because in the extended bond the electrons are less under the control of the nuclei.

The selection rules for the Raman effect arise from the fact that a dipole moment can be induced in a molecule. When a molecule is distorted in an electric field, an induced electric dipole moment is produced and the molecule is said to be *polarized.* The induced dipole $\boldsymbol{\mu}$ depends on the strength of the electric field and the polarizability α of the molecule:

$$\boldsymbol{\mu} = \alpha \boldsymbol{E} \tag{13.179}$$

From Eq. 13.1, a rotation or vibration of frequency ν_0 will cause the electric field $\boldsymbol{E}$ to oscillate with this frequency:

$$\boldsymbol{E} = 2\boldsymbol{E}^0 \cos 2\pi\nu_0 t \tag{13.180}$$

and thus the induced dipole moment also oscillates with the same frequency:

$$\boldsymbol{\mu} = \alpha\boldsymbol{E} = 2\alpha\boldsymbol{E}^0 \cos 2\pi\nu_0 t \tag{13.181}$$

Raman lines appear if the molecular rotations or vibrations bring about an oscillating polarizability. For example, if a vibrational frequency ν_{vib} changes the polarizability, the variation of polarizability is given by

$$\alpha = \alpha_0 + R\cos 2\pi\nu_{\text{vib}}t \tag{13.182}$$

where α_0 is the polarizability before vibration and R is a coefficient that relates the change of α to the vibration. From Eq. 13.179 we have

$$\boldsymbol{\mu} = \alpha\boldsymbol{E} = 2(\alpha_0 + R\cos 2\pi\nu_{\text{vib}}t)\boldsymbol{E}^0 \cos 2\pi\nu_0 t \tag{13.183}$$

This can be cast, with the use of a trigonometry product formula, into the form

$$\boldsymbol{\mu} = 2\alpha_0\boldsymbol{E}^0 \cos 2\pi\nu_0 t + R\boldsymbol{E}^0\,[\cos 2\pi(\nu_0 - \nu_{\text{vib}})t + \cos 2\pi(\nu_0 + \nu_{\text{vib}})t] \tag{13.184}$$

The components $\nu_0 \pm \nu_{\text{vib}}$ appear in addition to the incident frequency. The selection rules can be obtained by substituting Eq. 13.184 into Eq. 13.17 and evaluating the integral. Vibrational selection rules are $\Delta v = 0, \pm 1, \pm 2, \pm 3, \ldots$ and $\Delta J = 0, \pm 2$. If $\Delta v = 0$, then only rotational transitions occur and a rotational Raman spectrum is observed. Although the higher vibrational modes can occur, the $\Delta v = \pm 1$ is by far the most intense, giving rise to the Stokes and anti-Stokes lines.

Raman spectra are therefore found with homonuclear diatomic molecules like H_2 and O_2, which have no infrared spectra. The situation with polyatomic molecules is a little more complicated. We saw earlier that the symmetric vibration of carbon dioxide does not give rise to an infrared spectrum because there is no change in dipole moment, although the antisymmetric mode does give an infrared spectrum as a result of the oscillating dipole moment. The situation with Raman spectra is the converse. The symmetric vibration stretches and compresses both bonds, and the polarizability therefore varies. As a result, this vibration gives rise to Raman lines. The antisymmetric vibration, on the other hand, does not lead to a net polarizability variation because as one bond lengthens, the other bond shortens, and the two effects cancel. Thus there are no Raman lines associated with the antisymmetric vibration. To summarize the situation for CO_2: Symmetric vibrations are infrared inactive and Raman active. Antisymmetric vibrations are infrared active and Raman inactive. This can be expressed in a more general way in the form of an exclusion rule: If the molecule has a center of symmetry, vibrations that are infrared inactive are Raman active and *vice versa*. In view of these differences, it is advantageous to make a parallel study of both infrared and Raman spectra.

The selection rules for vibrational Raman spectra of polyatomic molecules are conveniently formulated with reference to the symmetry species of the vibrations. We have seen in our discussion of infrared spectra that an induced dipole moment $\boldsymbol{\mu}$ may be resolved along the axes X, Y, and Z. In considering Raman spectra we must also consider the inducing field $\boldsymbol{E}$, and this also has components along the X, Y, and Z axes. The dipole moment vector $\boldsymbol{\mu}$ is the product of the polarizability α and the inducing field (Eq. 13.181). Suppose that we consider the dipole moment $\boldsymbol{\mu}_x$ in the X-direction induced by a field in the Y-direction:

$$\boldsymbol{\mu}_x = \alpha_{xy}\boldsymbol{E}_y \tag{13.185}$$

The component $\boldsymbol{\mu}_x$ behaves like a translation x and it can either remain unchanged or change sign under a symmetry operation. Similarly $\boldsymbol{E}_y$ behaves like a translation y. Suppose that a particular symmetry operation changes the sign of $\boldsymbol{\mu}_x$ and also changes the sign of $\boldsymbol{E}_y$; by Eq. 13.179 this operation leaves the sign of α_{xy} unchanged. The component of the polarizability α_{xy} thus changes in the same way as x/y or as xy.

This conclusion applies to all the polarizability components, which are α_x^2, α_y^2, α_z^2, α_{xy}, α_{yz}, and α_{xz}. Each one is affected by the symmetry operations of the group in the same way as the corresponding product of coordinates, namely x^2, y^2, z^2, xy, yz, and xz. In order for there to be a Raman spectrum, one or more of these components must belong to the same symmetry species as one of the normal modes of vibration. To obtain the selection rules for Raman spectra we must therefore compare the symmetry species of the vibrations with those of the products x^2, y^2, z^2, xy, yz, and xz (see the appendix to this chapter, p. 707).

As an example, consider the carbon dioxide molecule, shown in Figure 13.24a. The character table (p. 707) shows the species of the previous products, and we see that the symmetric σ_g^+ vibration is of the same symmetry species as z^2; it therefore gives a Raman spectrum. The antisymmetric σ_u^+ vibration does not, nor do the bending vibrations. These conclusions are the same as those arrived at earlier.

Similarly we can see that for water (Figure 13.24b) all three vibrations are Raman active. For boron trifluoride (Figure 13.25a) the vibrations $\nu_1(a_1')$, $\nu_{3a}(e')$, $\nu_{3b}(e')$, $\nu_{4a}(e')$, and $\nu_{4b}(e')$ are Raman active, while $\nu_2(a_2'')$ is Raman inactive. For formaldehyde (Figure 13.25b) all modes are Raman active.

13.6 Electronic Spectra of Molecules

The electronic spectra of particles result from the electronic part of the total wave function (Eq. 13.147). This has already been discussed for atoms (Section 13.2). When treating the electronic spectra of molecules, it is usual to discuss diatomic (linear) molecules first and then treat polyatomic molecules. In this section we survey the description of and results for the electronic spectra of linear molecules.

We have seen (Figure 13.14) that transitions in which there is a change from one electronic state to another generally involve much higher energies than those in which there are vibrational and rotational changes only. Electronic transitions usually give rise to absorption or emission of radiation in the visible and ultraviolet regions of the spectrum. Associated with electronic transitions there are also changes of vibrational energy, which give rise to spectral bands, and changes of rotational energy, which give the fine structure of the bands. If the substance is in the gas phase, the molecular electronic spectrum is sharply defined, but in solution there is considerable blurring. Even in solution, however, substances give characteristic spectra that are useful for their identification or for determinations of their concentration.

From Eq. 13.149, it is seen that the total wave function is written as a product. Although rotational and vibrational transitions accompany electronic transitions, the motions are effectively uncoupled. Therefore, from Eq. 13.151, the energy of an electronic transition is a sum of the three energies,

$$\Delta E = \Delta E_{electronic} + \Delta E_{vibrational} + \Delta E_{rotational} \tag{13.186}$$

with

$$\Delta E_{electronic} > \Delta E_{vibrational} > \Delta E_{rotational} \tag{13.187}$$

Spectroscopic transitions involving combinations of electronic, vibrational, and rotational transitions can occur, and these are governed by selection rules.

Whereas pure rotational and vibrational-rotational spectra are not always exhibited by molecules, every substance can give an electronic spectrum because there are always excited states to which molecules can be raised by absorption of radiation. There are selection rules for electronic transitions, but before considering these we must deal with the classification of electronic states.

In the study of the structure of matter, it should be clear by now that there is a common theme in the way the different angular momenta interact. It is the vector model that embodies the ways in which angular momenta add to give a resultant, which is the vector sum of the components. We have seen this, for example, in Russell-Saunders coupling. The presence of magnetic or electric moments causes one angular momentum to precess about another. The larger moments dominate, so the other angular momenta with smaller moments precess about the larger ones. The origin does not have to be internal motions. External magnetic and electric fields can also be introduced, and the angular momentum in molecules can precess about these (this occurs in NMR and ESR spectroscopy).

Term Symbols for Linear Molecules

The designations used for electronic states of molecules are somewhat analogous to those for atoms, which we considered in Section 13.2, but there are important differences. Whereas for an atom we deal with the resultant total orbital angular momentum $\boldsymbol{L}$, with quantum number L, for a linear molecule we must consider the component of orbital angular momentum along the axis of the molecule (the Z axis; see Figure 13.28). Along the internuclear axis of a linear molecule there can be an electrostatic field produced by the charge separation of the nuclei.

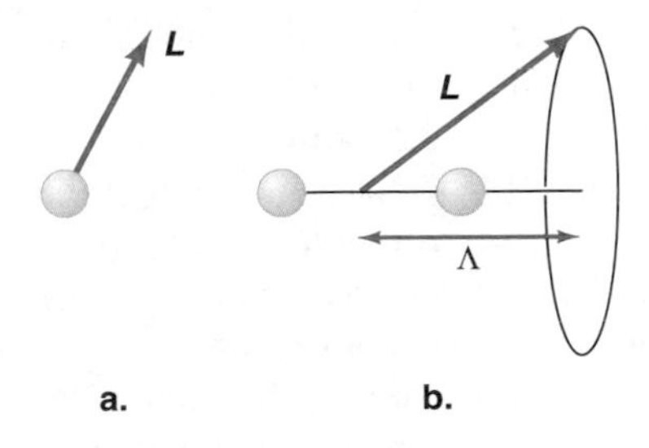

FIGURE 13.28
a. The angular momentum vector ***L*** is in a spherically symmetric field about the atomic nucleus. There is no axis produced about which ***L*** can precess. **b.** An asymmetric field is produced because of the two positive charges on the nuclei. These cause a symmetry axis along the molecular axis of the diatomic molecule about which ***L*** precesses.

To visualize this, consider the case of an atom, Figure 13.28a, and then move from an atom to a diatomic molecule, Figure 13.28b. In an atom, the angular momentum $\boldsymbol{L}$ is in the spherically symmetric field of the atomic nucleus. In a diatomic molecule, the nuclei are separated, producing an asymmetric field due to the different positive nuclear charges, and this produces the axis about which $\boldsymbol{L}$ can precess. As a result, L is no longer a good quantum number, as in the case of atoms. The projection of $\boldsymbol{L}$ along the symmetry axis of a diatomic molecule, however, is well defined. Therefore, we introduce a new angular momentum operator $\hat{\boldsymbol{\Omega}}$, which has quantum numbers of $\Lambda = |\Delta M_L|$ equal to 0, 1, 2, ... See Figure 13.28b.

The quantum number Λ is therefore not the analog of the atomic quantum number L, which relates to the resultant angular momentum of the atom, but is equivalent to the atomic quantum number $|M_L|$ for an atom when there is an electric field along one axis.

The value of the quantum number Λ is related to the symmetry of the electron cloud with respect to the Z axis. When $\Lambda = 0$, the cloud is cylindrically symmetrical about the internuclear axis, and the state is given the term symbol Σ (Greek capital letter sigma), which is the Greek equivalent of the S state ($L = 0$) for atoms. The term symbols for all Λ values are in fact the Greek equivalents of the corresponding capital letters used with respect to the L values for atoms:

Atoms		*Linear Molecules*	
$L = 0$	S	$\Lambda = 0$	Σ (sigma)
$L = 1$	P	$\Lambda = 1$	Π (pi)
$L = 2$	D	$\Lambda = 2$	Δ (delta)
$L = 3$	F	$\Lambda = 3$	Φ (phi)

Since $\Lambda = |\Delta M_L|$ has values of 0, 1, 2, . . . , and M_L has values of 0, ± 1, ± 2, ± 3, . . . , all the Λ states are doubly degenerate except the Σ state, a phenomenon called Λ doubling.

In addition to the orbital angular momentum being quantized along the electric field axis of a linear molecule, there are also electron spins. These sum to a resultant $\boldsymbol{S}$ and in the presence of the magnetic field generated by $\Lambda \neq 0$, the $\boldsymbol{S}$ vector is quantized along the molecular axis, as shown in Figure 13.29, for the same reasons as for $\boldsymbol{L}$. This leads to the vector $\boldsymbol{\Sigma}$ as shown in Fig. 13.29, which has the quantum number Σ, with values $-S$ to $+S$, and therefore a multiplicity of $(2S + 1)$. On the other hand, if $\Lambda = 0$, there is no internal magnetic field generated, $\hat{\Sigma}$ is not defined, and $\hat{S}$ remains as in the case for atoms.

In the absence of the molecular rotation, we are thus led to the existence of three vector operators which describe the orbital and spin angular momentum of linear molecules, $\hat{\boldsymbol{S}}$, $\hat{\boldsymbol{\Sigma}}$, and $\hat{\boldsymbol{\Lambda}}$. When $\hat{\boldsymbol{\Lambda}}$ and $\hat{\boldsymbol{\Sigma}}$ are defined, the total angular moment operator is usually introduced as the sum

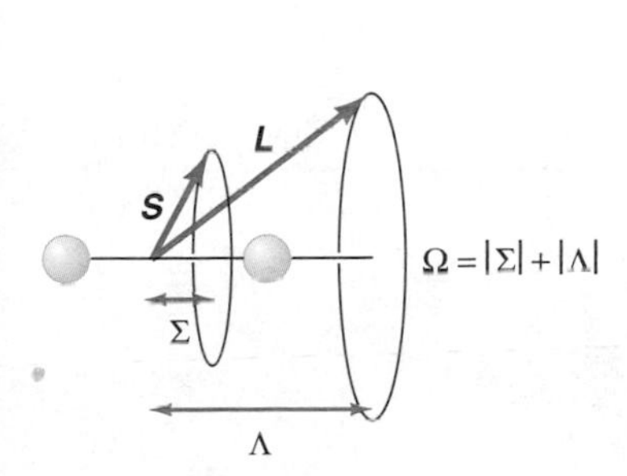

FIGURE 13.29
Quantization of spin and angular momentum vectors in a linear molecule.

$$\hat{\boldsymbol{\Omega}} = \hat{\boldsymbol{\Lambda}} + \hat{\boldsymbol{\Sigma}} \quad (\Lambda \neq 0) \tag{13.188}$$

Since $\hat{\boldsymbol{\Lambda}}$ and $\hat{\boldsymbol{\Sigma}}$ are coaxial (Fig. 13.29), only the magnitudes of $\hat{\boldsymbol{\Omega}}$ are relevant:

$$\Omega = |\Lambda + \Sigma| \text{ for } (\Lambda \neq 0) \tag{13.189}$$

In general the term symbols for diatomic molecules are written as

$$^{2S+1}\Lambda \tag{13.190}$$

A linear molecule belongs either to the point group $C_{\infty v}$ or to the point group $D_{\infty h}$. If it has no center of symmetry, as with LiH, shown in Figure 12.22b, it is $C_{\infty v}$, and in that case a stretch is either Σ^+ or Σ^- depending on whether the state is symmetric or antisymmetric with respect to any of the σ_v planes passing through the axis of the molecule. If the linear molecule has a center of inversion, it is $D_{\infty h}$ (Figure 12.22a) and an additional classification is then possible, according to whether the wave function remains unchanged (g) or changes sign (u) on inversion.

For example, in the ground state of H_2 the spins are paired ($S = 0$), there is no angular momentum along the axis ($\Lambda = 0$), and there is no change of sign on reflection in a σ_v axis (+) or on inversion (g); the term symbol is therefore $^1\Sigma_g^+$. This is readily understood from our discussion of the hydrogen molecule in Section 12.2; in particular, the molecular orbitals are made up of the sum of 1s atomic orbitals, which are spherically symmetrical. The situation with O_2, however, is different because π orbitals are involved and two of the electrons are unpaired. We see from the molecular orbital diagram in Figure 12.26b that these two unpaired electrons are of g symmetry; this is also seen from Figure 12.25b, which in addition shows that the sign changes (−) on reflection in the XY plane. Since these π electrons are in orbitals at right angles to the molecular axis, they do not contribute to the angular momentum along that axis; thus $\Lambda = 0$ and the state is a Σ state. The term symbol for ground-state O_2 is therefore $^3\Sigma_g^-$. A potential energy diagram for O_2 is shown in Figure 13.30, which illustrates still another point. The figure shows that

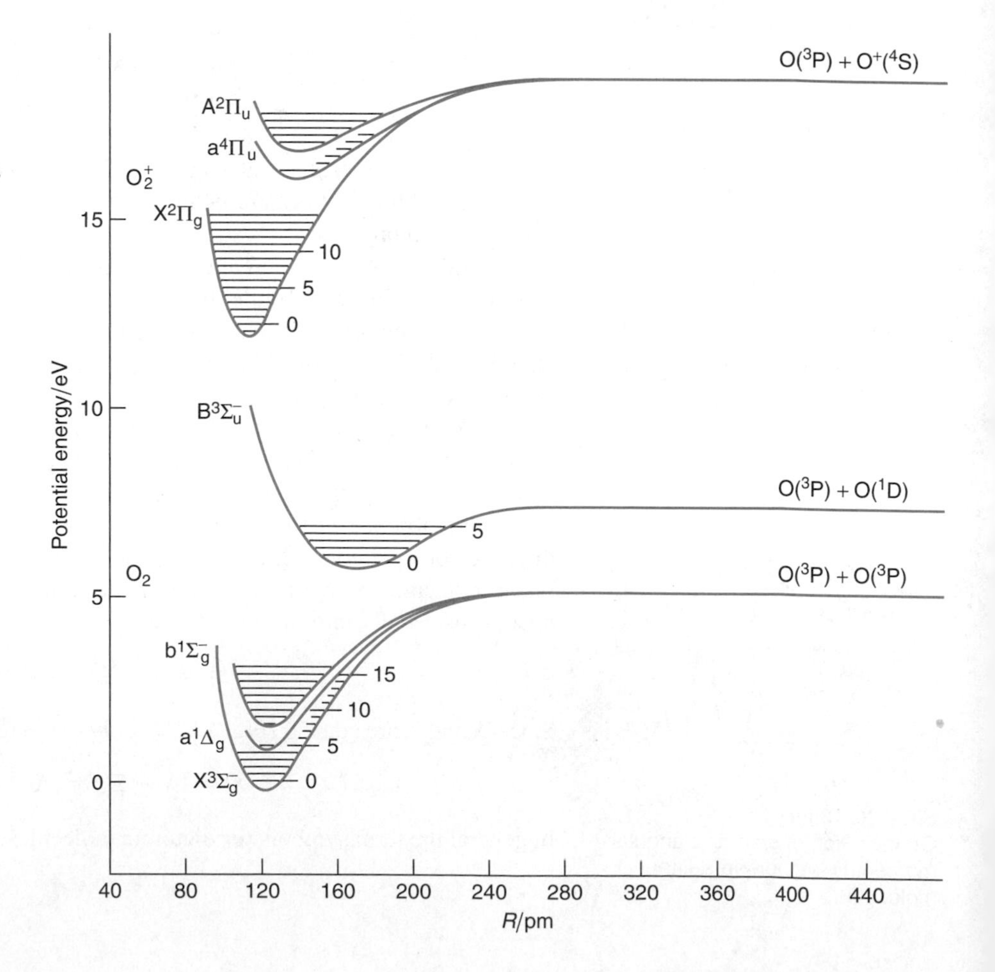

FIGURE 13.30
Potential energy diagram for O_2. Not all vibrational states are shown. (Gilmore, Forrest R., *Potential Energy Curves for* N_2, NO, O_2, *and Corresponding Ions*, RAND, RM-4034-1-PR, April 1966. Used by permission.)

electronic states of molecules are also given letter symbols. The letter X represents the ground electronic state of the linear molecule. Excited states of the same multiplicity are given the letters A, B, C, . . ., in order of increasing energy, whereas excited states with different multiplicity are labeled using lowercase letters a, b, c, ...

The term symbols for *nonlinear* molecules are again based on the symmetry species for the appropriate point groups. For further details the reader should consult the texts listed in Suggested Reading on page 712.

Selection Rules

The spin selection rule for molecules is the same as for atoms. For light molecules the rule

$$\Delta S = 0 \tag{13.191}$$

is obeyed fairly strictly. Thus, singlet to singlet and triplet to triplet transitions may occur, but singlet to triplet transitions are forbidden. For heavier molecules changes of multiplicity are more likely.

Just as for atoms, the selection rule is $\Delta L = 0, \pm 1$, and that for linear molecules is

$$\Delta \Lambda = 0, \pm 1 \tag{13.192}$$

For Σ–Σ transitions the selection rules are

$$\begin{aligned} + &\leftrightarrow + \\ - &\leftrightarrow - \\ + &\not\leftrightarrow - \end{aligned} \tag{13.193}$$

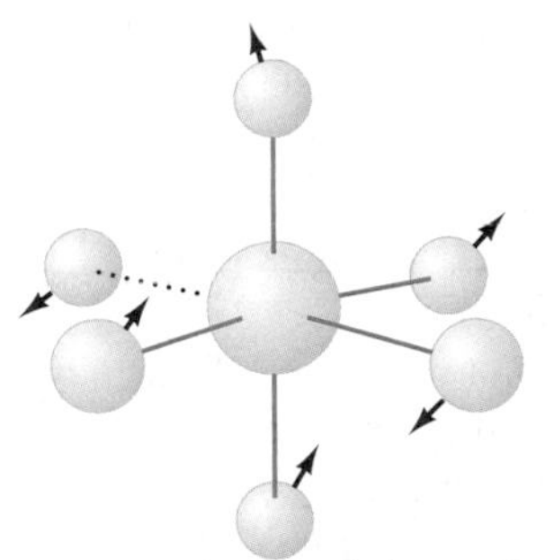

Symmetrical molecule beginning an asymmetrical vibration

a.

The **Laporte selection rule** also applies for centrosymmetric molecules (those with a center of inversion, point group $D_{\infty h}$). Thus an even state tends to transform into an odd state, that is, the only allowed transitions are those that are accompanied by a change of parity (i.e., a change on inversion about a center of symmetry):

$$\begin{aligned} \mathrm{g} &\leftrightarrow \mathrm{u} \\ \mathrm{g} &\not\leftrightarrow \mathrm{g} \\ \mathrm{u} &\not\leftrightarrow \mathrm{u} \end{aligned} \tag{13.194}$$

Exceptions to these restrictions are not uncommon, however, especially with heavier molecules. For example, in Figure 13.31 an asymmetrical vibration in a centrosymmetric complex destroys the center of symmetry, allowing an otherwise forbidden g $\leftrightarrow$ g transition. This distortion allows a d $\leftrightarrow$ d transition in the octahedral complex shown.

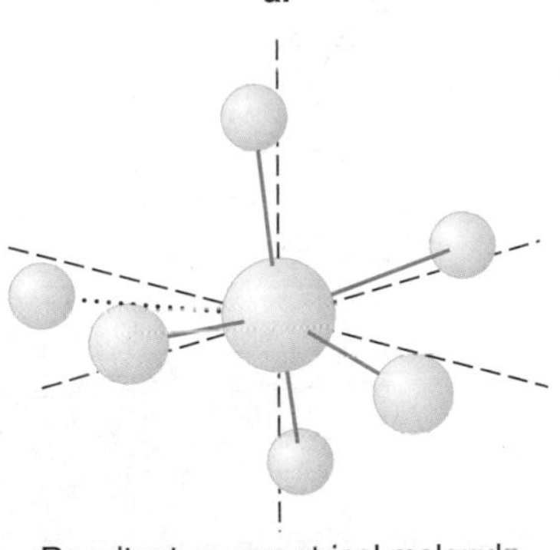

Resultant asymmetrical molecule

b.

FIGURE 13.31
A centrosymmetric complex is parity forbidden in (a) from having a d $\leftrightarrow$ d transition within the molecule. An asymmetric vibration (b) removes the center of inversion and a $e_g \leftarrow t_{2g}$ transition becomes weakly allowed.

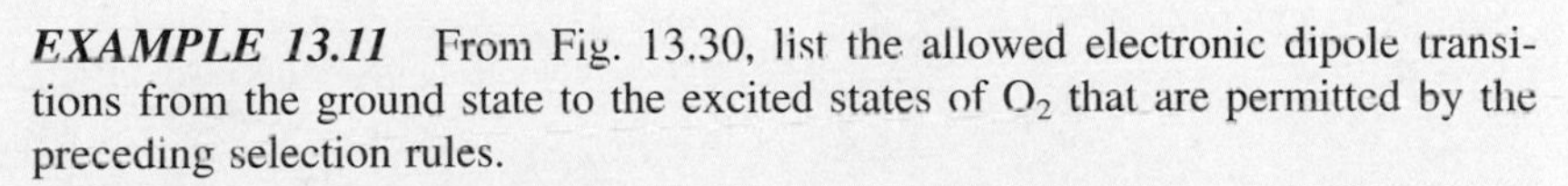

EXAMPLE 13.11 From Fig. 13.30, list the allowed electronic dipole transitions from the ground state to the excited states of O_2 that are permitted by the preceding selection rules.

Solution According to the selection rules, starting with a $^3\Sigma_g^-$ state a transition from the Σ state to a Σ state is permitted as well as to a Π state but no others. However, the excited $^2\Pi_g$ state is of an oxygen molecule ion, O_2^+, and is a doublet. Only a triplet state to a triplet state transition is permitted. These requirements eliminate all transitions shown except the transition to the $^3\Sigma_u^-$ state where g $\leftrightarrow$ u is permitted as well as $- \leftrightarrow -$.

The strongest bands in the absorption spectrum of the oxygen molecule, the so-called *Schumann-Runge* bands, at about 200 nm, arise from the transition $^3\Sigma_u^- \leftarrow {}^3\Sigma_g^-$. This is the transition that is allowed in the example and obeys all the selection rules previously given. The oxygen spectrum also includes some weak bands that violate the selection rules. For example, at about 760 nm there are bands due to a $^1\Sigma_g^+ \leftarrow {}^3\Sigma_g^-$ transition, and there are weak bands at about 1300 nm that arise from a $^1\Delta_g \leftarrow {}^3\Sigma_g^-$ transition. In both of these transitions there is a change of multiplicity, and both violate the g $\nleftrightarrow$ g prohibition.

It is of some interest to note that these two excited states would absorb in the visible and in the infrared range. But in spite of the large amount of oxygen in the atmosphere, only a small fraction of the solar radiation incident on the earth is removed. This shows that the selection rule that transitions occur only between states of like multiplicity is strongly obeyed.

What we have not discussed to this point are the cases in which the actual rotation of the molecule is taken into account. These are described in the rigid rotator section (Section 11.10), where only the rotation of the diatomic molecule is considered. This led to the angular momentum vector operator $\hat{\boldsymbol{J}}$ for which the quantum number J is relevant. In general the angular momenta discussed earlier ($\hat{\boldsymbol{L}}$, $\hat{\boldsymbol{S}}$, $\hat{\boldsymbol{\Lambda}}$, and $\hat{\boldsymbol{\Sigma}}$) can couple with the rotational motion of the actual molecules to form different total angular momenta. There are various way that this can happen, and these are treated by the four Hund's coupling cases.

It is worthwhile to consider first how these angular momenta can add. Consider, with reference to Figure 13.32, three general vectors, $\boldsymbol{r}_1$, $\boldsymbol{r}_2$, and $\boldsymbol{r}_3$, coupled together in different ways to give the same resultant, $\boldsymbol{r}$. Only three different ways exist to obtain $\boldsymbol{r}$ depending on the order of coupling, as shown.

The order in which vectors add is mathematically not important. Physically, the more strongly coupled angular momenta couple first and the weaker ones last. The four **Hund's cases** describe these. The convention is that the final total angular mo-

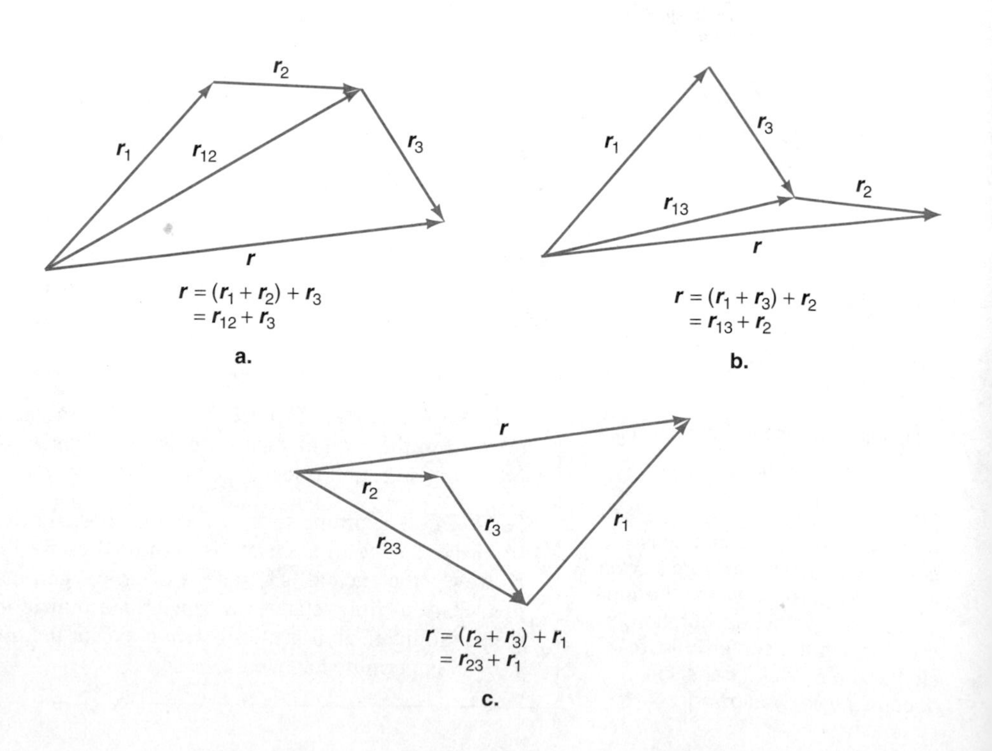

FIGURE 13.32
The importance of the order of coupling vectors. The vectors $\boldsymbol{r}_1$, $\boldsymbol{r}_2$, $\boldsymbol{r}_3$, and $\boldsymbol{r}$ in all three cases are the same, only the order of coupling changes: (a) $\boldsymbol{r}_1$ and $\boldsymbol{r}_2$ couple first (b) $\boldsymbol{r}_1$ and $\boldsymbol{r}_3$ couple first (c) $\boldsymbol{r}_2$ and $\boldsymbol{r}_3$ couple first.

mentum is always designated by $\boldsymbol{J}$. We shall use the convention that the molecular rotation, discussed in Section 11.10, is denoted by $\boldsymbol{N}$ rather than $\boldsymbol{J}$. How these various angular momenta [$\boldsymbol{N}$, $\boldsymbol{L}$ (Λ), and $\boldsymbol{S}$ (Σ)] couple together depends on the relative strengths of magnetic moments created by each, and the strength of the electrostatic field produced by the nuclei along the internuclear axis.

Hund's case a is shown in Figure 13.33a. This is the case we have previously discussed for the quantization of $\boldsymbol{L}$ and $\boldsymbol{S}$ along the molecular axis, first to form $\boldsymbol{\Omega}$, after which they couple to $\boldsymbol{N}$, giving $\boldsymbol{J}$.

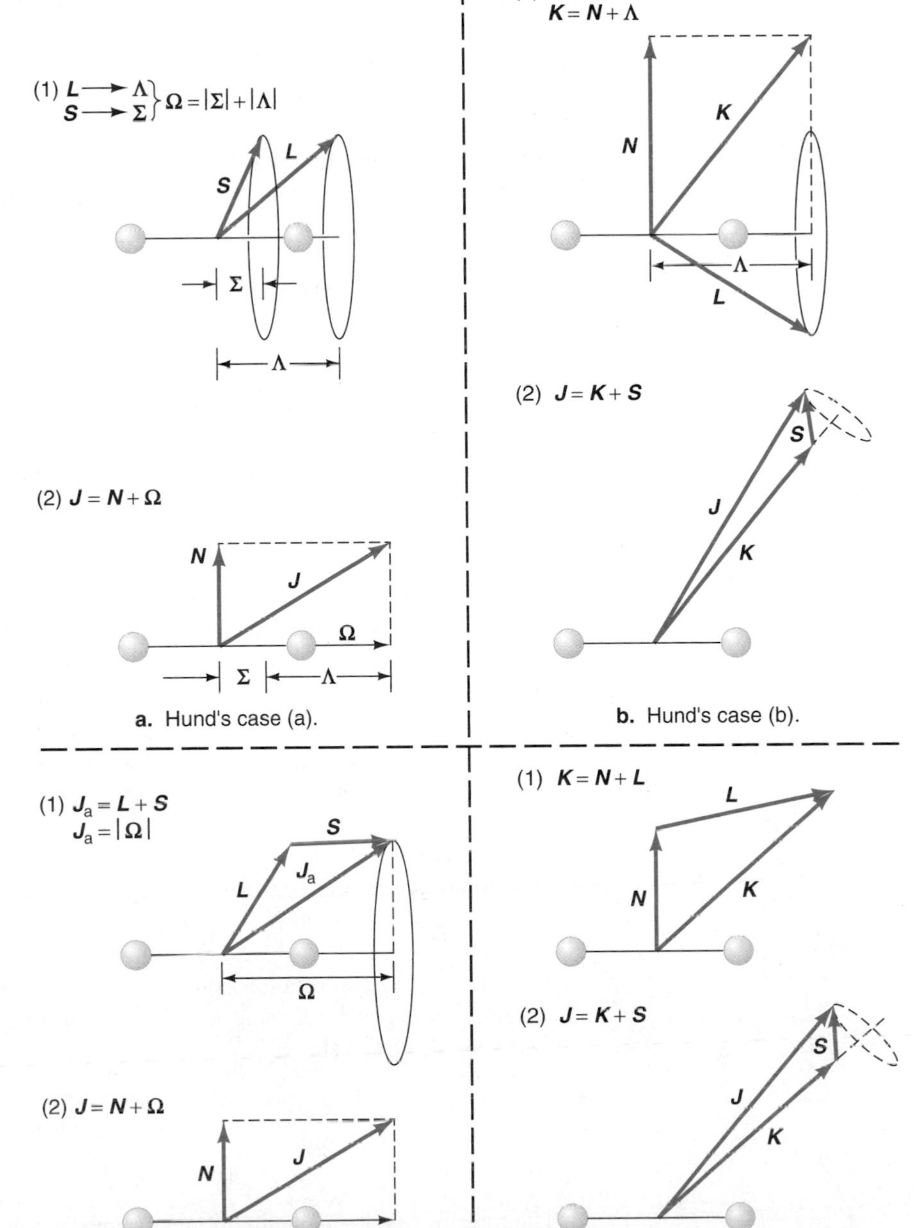

FIGURE 13.33
(a) Hund's coupling case a. (1) $\boldsymbol{L}$ and $\boldsymbol{S}$ precess around the molecular symmetry axis to give projections Λ and Σ that sum to $\boldsymbol{\Omega}$. (2) The molecular rotation $\boldsymbol{N}$ then couples to $\boldsymbol{\Omega}$ to give the total angular momentum $\boldsymbol{J}$.
(b) Hund's coupling case b. (1) $\boldsymbol{L}$ precesses around the molecular symmetry axis to give Λ, but $\boldsymbol{S}$ is weakly coupled to $\boldsymbol{L}$. Hence the molecular rotation $\boldsymbol{N}$ first couples to Λ to give an intermediate angular momentum $\boldsymbol{K}$. (2) $\boldsymbol{S}$ precesses around $\boldsymbol{K}$ to give $\boldsymbol{J}$.
(c) Hund's coupling case c. (1) $\boldsymbol{L}$ and $\boldsymbol{S}$ are strongly coupled to give $\boldsymbol{J}_a$, which precesses around the molecular axis to give $\boldsymbol{\Omega}$. (2) This $\boldsymbol{\Omega}$ couples to $\boldsymbol{N}$ to give $\boldsymbol{J}$.
(d) Hund's coupling case d. (1) The orbital angular momentum $\boldsymbol{L}$ couples to the molecular rotation $\boldsymbol{N}$, rather than the symmetry axis, to give $\boldsymbol{K}$. (2) $\boldsymbol{K}$ can couple to $\boldsymbol{S}$, but usually this is small and can be neglected.

Hund's case b is shown in Figure 13.33b. In this case, the electron spins are weakly coupled to the field along the molecular axis. Hence $\hat{\Sigma}$ is not defined; $\hat{\boldsymbol{\Omega}}$ first couples to $\boldsymbol{N}$, giving an intermediate angular momentum $\boldsymbol{K}$, which then couples to the molecular rotation $\boldsymbol{N}$ to give $\boldsymbol{J}$.

Hund's case c is shown in Figure 13.33c. In this case both $\boldsymbol{L}$ and $\boldsymbol{S}$ are weakly coupled to the molecular axis and so, first, couple together to form a resultant $\boldsymbol{J}_a$. This intermediate angular momentum does couple to the molecular axis to produce $\boldsymbol{\Omega}$, after which $\hat{\boldsymbol{\Omega}}$ and $\boldsymbol{N}$ couple to give $\boldsymbol{J}$.

Hund's case d is shown in Figure 13.33d. Finally, neither $\boldsymbol{L}$ nor $\boldsymbol{S}$ couples to the molecular axis so none of $\hat{\Lambda}$, $\hat{\Sigma}$, or $\hat{\boldsymbol{\Omega}}$ is formed. Rather, in this case $\boldsymbol{L}$ couples to $\boldsymbol{N}$ giving an intermediate denoted again by $\boldsymbol{K}$, and this, in turn as $\boldsymbol{K}$, couples to $\boldsymbol{S}$ to give $\boldsymbol{J}$.

These four cases add to the rich electronic spectra observed for the diatomic molecules. Moreover, situations arise whereby one coupling case changes to another, similar to that found in the Paschen-Back effect (p. 663) discussed previously. More information about the details of electronic structure and spectra of molecules can be found in the references at the end of the chapter.

The Structure of Electronic Band Systems

Since electronic spectra also involve vibrational changes, they are often called *vibronic spectra.* Each band in a vibronic spectrum is associated with a change in vibrational energy and consists of many closely spaced rotational lines on one side that usually have a sharp edge, or *band head.* Beyond this band head the intensity falls sharply to zero, while on the other side of the band the intensity falls off more gradually and is said to be *shaded* or *degraded.*

In the Born-Oppenheimer approximation each electronic state of a diatomic molecule is represented by a curve showing potential energy plotted against internuclear distance. On this curve the various vibrational states can be indicated by horizontal lines. For a molecule containing more than two atoms the energy must in principle be represented by a multidimensional surface in which the potential energy is plotted against a number of distances and angles that specify the configuration of the molecule. Although polyatomic molecules are more difficult to deal with in practice, the basic ideas are the same as for diatomic molecules, to which our discussion will initially be confined.

The equilibrium bond distances and the shapes of potential-energy curves are in general different for different electronic states of a molecule, and the spacing of the vibrational levels is therefore different. Figure 13.34 shows two electronic states of a diatomic molecule. At ordinary temperatures the vast majority of the molecules will be in the ground electronic state and be at the lowest ($v'' = 0$) vibrational level. The absorption spectrum of such a molecule will therefore consist of a series of bands corresponding to the following transitions to vibrational levels of the upper electronic state:

$$v' = 0 \leftarrow v'' = 0$$
$$v' = 1 \leftarrow v'' = 0$$
$$v' = 2 \leftarrow v'' = 0, \text{ etc.}$$

The transitions corresponding to these bands are shown in Figure 13.34. Such a series of bands is called a *progression,* and each band is usually labeled $v' - v''$ with

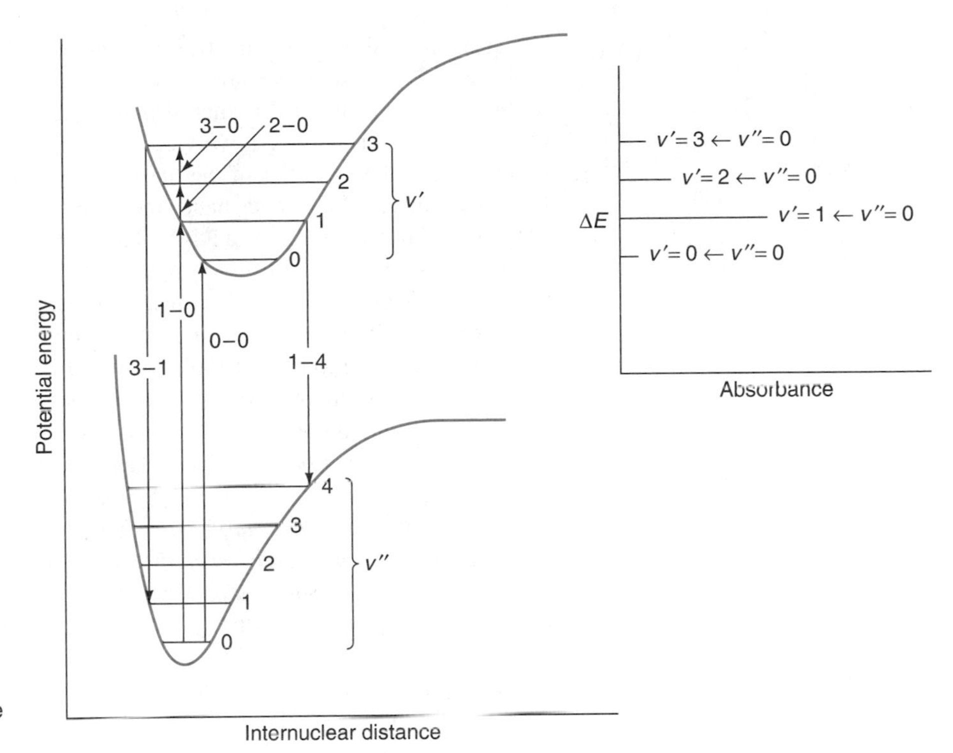

FIGURE 13.34
Two electronic states of a diatomic molecule, showing a number of vibrational levels. Transitions giving rise to absorption bands (↑) and to emission bands (↓) are shown. The relative magnitude of the absorbance at a particular ΔE is represented by the length of the lines in the figure to the right.

the *upper* state given first; the absorption bands indicated here would thus be labeled 0–0, 1–0, 2–0, and 3-0. For an emission spectra to be studied, however, energy must be supplied to the system in some way, with the production of electronically excited species in which higher vibrational levels may be populated. Transitions from these levels can occur to vibrational levels in the electronic ground state, and some of these are represented in Figure 13.34. It should be evident that there may be many more bands in emission than in absorption. Some of the absorption and emission bands will be of the same frequency, but there will be emission bands (for example, the 3–1 and 1–4 bands shown in the figure) that do not appear in absorption. The detailed analysis of electronic spectra of molecules is often a matter of considerable difficulty.

There can be large differences between the intensities of different bands, and the intensity pattern is helpful in the analysis of spectra. An important principle that governs the probabilities of transitions from one vibrational level to another is the **Franck-Condon principle.** This was first expressed qualitatively by the German-American physicist James Franck (1882–1964), who based his principle on the relative speeds of electronic transitions and vibrational motions. An electronic transition may take place in 10^{-15}–10^{-18} s and is so rapid compared with vibrational motion ($\approx 10^{-13}$ s) that immediately after an electronic transition has occurred, the internuclear distance is essentially the same as before. Franck therefore argued that the most probable electronic transitions are those that can be represented as vertical transitions in diagrams such as Figure 13.34.

James Franck shared with Gustav Ludwig Hertz (1887–1975) the 1925 Nobel Prize for physics for their joint work on electron-atom collisions (the Franck-Hertz effect), which clearly demonstrated the quantization of atomic energy levels.

This idea was put on a more rigorous quantum-mechanical basis by the American physicist Edward Uhler Condon (1902–1974). An important point in considering transitions of this kind relates to the most probable internuclear distances in vibrating molecules. According to classical mechanics, this distance is changing more slowly near the turning point of the motion; thus in Figure 13.34 the molecule is vibrating more slowly near the extremities of the vibrations, where the motions are reversing, than when it is passing through the equilibrium position. If classical mechanics applied, we would therefore represent transitions as occurring to and from the extremities of the lines representing the vibrational levels. In quantum mechanics, on the other hand, the situation is somewhat different, as may be seen by reference to Figure 11.14b. There we see that for the *lowest* vibrational state ($v = 0$) the probability is a maximum at the *center* of the vibration. For all other vibrational states, there is a maximum slightly inside each end of the classical vibrational amplitude, and there are smaller maxima in between.

Before a transition, most molecules will be in the ground vibrational state $v'' = 0$ at the most probable position (the center of the line). According to the Franck-Condon principle, the electronic transition is so fast that there is no chance for the interatomic distance to change. The transition must be a vertical transition from the ground state to the positions vertically above it on the curve representing the excited electronic state. *The maximum absorption occurs between the ground state and the excited vibrational level that has its most probable distance vertically above that of the ground state.* Thus in Figure 13.34 the 1–0 transition is more probable than the 0–0 transition. Transitions to other excited states occur, but the intensity of absorption is reduced because of the lower probability of finding the electron at that location within the potential well. This can be expressed quantum mechanically by saying that the transition probability between two energy levels is given by the overlap integral between the two vibrational wave functions. In representing the probabilities of transitions we should therefore draw the connecting arrows between the center of the amplitude for $v'' = 0$ and the v' levels vertically above it. Any offset of the vertical lines from the center of the ground vibrational state is for clarity in the drawing. This is the procedure used in Figure 13.34. The absorbances for electrons at the most probable distance are shown in the figure as individual lines to the right of the transitions. This is, of course, an oversimplification because there are smaller probabilities at other positions in $v'' = 0$, so the weaker transitions occur from those positions also. These transitions have the effect of changing the lines of the absorbance shown in Figure 13.34 into a broader spectrum.

Modern laser femtosecond (1 fs $= 10^{-15}$ s) spectroscopy happens so fast that only electronic transitions occur and all other motions, including vibrations, are frozen. This not only allows experimental study of the Franck-Condon principle, but also gives details of the electronic structure of intermediate transition state complexes in chemical reactions and molecular dissociations.

Excited Electronic States

Various types of electronic transitions may be involved in the formation of electronically excited molecules. The simplest situation arises with diatomic molecules, where an electron in a bonding orbital may be transferred to a higher orbital, which may be an antibonding orbital. As an example, Figure 13.35 shows potential-energy curves for the ground state of the hydrogen molecule and also for a number of excited states. In the ground state of hydrogen both electrons are in the $1s\sigma_g$ orbital (Figure 12.5), and this electronic state is of symmetry $^1\Sigma_g^+$. Some of the higher orbitals available to electrons are shown in Figure 12.26, and one or both of

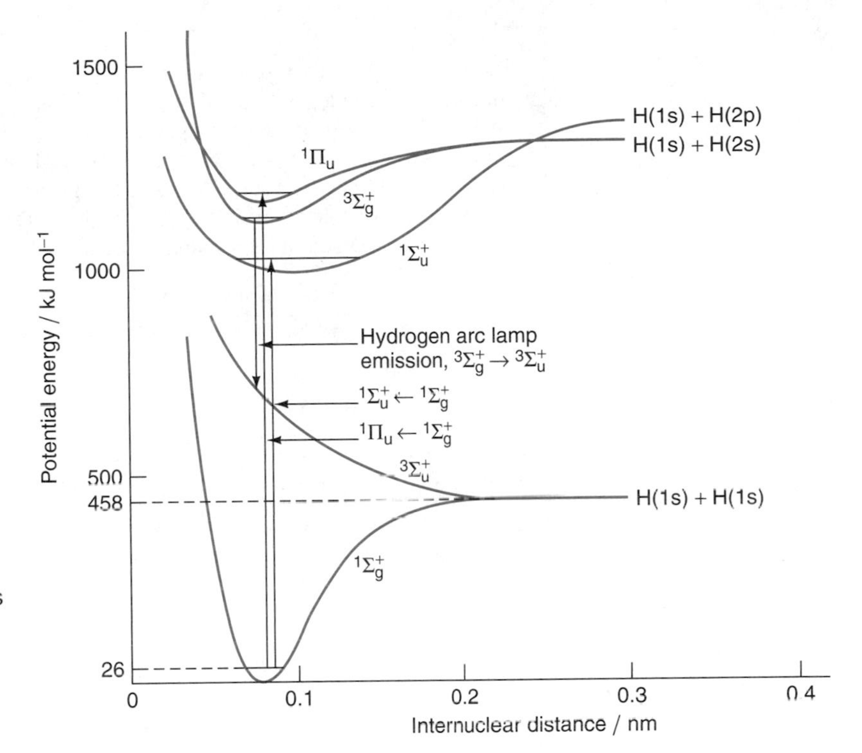

FIGURE 13.35 Schematic potential-energy curves for the ground state and some excited states of the hydrogen molecule. Lines representing absorption from v'' are displaced from the center of the ground state for clarity.

the $1s\sigma_g$ electrons can be promoted to these orbitals. As a result there are a large number of excited states, some of which are shown in Figure 13.35. The lowest excited state is the one in which an electron has been promoted to the $1s\sigma_g^*$ orbital, and since this is antibonding, the state is repulsive; this is the $^1\Sigma_u^+$ state shown in the figure. Promotion to higher orbitals leads to states of higher energy that dissociate not into two ordinary hydrogen atoms but into atoms in excited states. States in which both electrons have opposite spins ($S = 0$) are singlet states, while those in which the spins are the same ($S = 1$) are triplets. Single-triplet transitions are forbidden by the selection rules, which apply rather rigorously to a molecule as light as hydrogen. The absorption spectrum of H_2 shows bands corresponding to the $^1\Sigma_u^+ \leftarrow {}^1\Sigma_g^+$ and $^1\Pi_u \leftarrow {}^1\Sigma_g^+$ transitions. In a hydrogen arc lamp, molecules are produced in the $^3\Sigma_g^+$ state, and the characteristic emission is due to the transition $^3\Sigma_g^+ \rightarrow {}^3\Sigma_u^+$. Information from these states can be used to determine dissociation energies as discussed on page 679.

It is possible to find the dissociation energy when several lines arise in vibronic transitions. The idea is that the energies of each contributing vibrational level sum to give the dissociation energy. However, some of the vibrational levels out to the dissociation limit may not be visible. A useful approximation is to use the *Birge-Sponer extrapolation.* A plot of the energy differences in cm^{-1} is made against the vibrational quantum number, v. (See Problem 13.44.) A straight-line extrapolation of observed values is made to obtain the intercept on the abscissa which corresponds to the vibrational quantum number for dissociation. The area under the curve is the dissociation energy, D_0. This method gives a linear plot when the anharmonicity is small or taken into account. Otherwise a curve should be used to obtain the best approximation since a linear plot tends to overestimate the dissociation energy (see Figure 13.36).

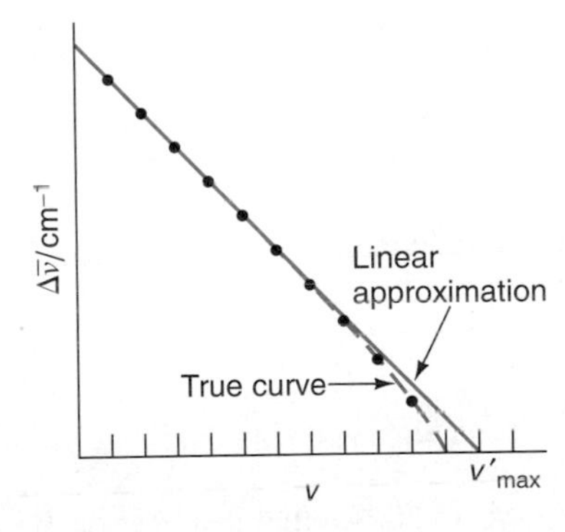

FIGURE 13.36 A Birge-Sponer extrapolation to find v'_{max} and the dissociation energy. Energies in cm^{-1} are plotted against their corresponding vibrational quantum numbers.

EXAMPLE 13.12 In the ultraviolet spectrum of molecular oxygen, the Schumann-Runge bands correspond to the transition ${}^3\Sigma_u^- \leftarrow {}^3\Sigma_g^-$. The ground state dissociates into two ground-state oxygen atoms (^{3}P) and the excited state dissociates into one ^{3}P and one ^{1}D. This ^{1}D level lies 190 kJ mol^{-1} above the ground state. The wavenumbers, $\tilde{\nu}$, for absorption from the ground vibrational level v'' to the v'^{th} vibrational level of the excited state are given. Determine the dissociation energy of molecular oxygen in the ground state and in its excited state.

v'	$\tilde{\nu}$, cm^{-1}	$\Delta\tilde{\nu}$, cm^{-1}	v'	$\tilde{\nu}$, cm^{-1}	$\Delta\tilde{\nu}$, cm^{-1}
1	50 062.6		9	54 641.8	
		662.8			436.4
2	50 725.4		10	55 078.2	
		643.6			381.6
3	51 369.0		11	55 460.0	
		619.6			343.6
4	52 988.6		12	55 803.1	
		590.4			303.7
5	52 579.0		13	56 107.3	
		564.4			253.0
6	53 143.4		14	56 360.3	
		533.5			210.3
7	53 676.9		15	56 570.6	
		500.1			
8	54 177.0				
		464.8			

Solution In Figure 13.22 the changes representing the ground state and excited state dissociation energies (D_0'' and D_0', respectively) are indicated along with the value of E_{ex}, the atomic excitation energy. We will apply Eq. 13.137 for the energy levels of the anharmonic oscillator rewritten in terms of a wavenumber for a molecule in the vth vibrational level.

$$\tilde{\nu} = \tilde{\nu}_e + \tilde{\nu}_0\left[\left(v + \tfrac{1}{2}\right) - x_e\left(v + \tfrac{1}{2}\right)^2 + y_e\left(v + \tfrac{1}{2}\right)^3 \ldots\right]$$

where the added term $\tilde{\nu}_e$ represents the energy in the zero vibrational level and corresponds to the electronic energy; $\tilde{\nu}_0$ is the fundamental vibrational wavenumber; and x_e and y_e are the anharmonicity constants.

For an electronic transition between the zero vibrational level of the ground state and the v'^{th} vibrational level of the excited state, we have

$$\begin{aligned}\tilde{\nu} &= \tilde{\nu}' - \tilde{\nu}'' \\ &= \tilde{\nu}_e + \left[\left(v' + \tfrac{1}{2}\right)\tilde{\nu}_0' - \tfrac{1}{2}\tilde{\nu}_0''\right] - \left[x_e\left(v' + \tfrac{1}{2}\right)^2\tilde{\nu}_0' - \left(\tfrac{1}{2}\right)^2 x_e\tilde{\nu}_0''\right] + \ldots\end{aligned}$$

where $\tilde{\nu}_e$ is the electronic energy change. From Figure 13.30, we find that energy $D_0'' + E_{ex}$ needed for an oxygen molecule in the ground state to dissociate to give a O(${}^1D$) and O*(${}^1D$) requires the energy when the wavenumber, $\tilde{\nu}$, corresponds to $v' = v'_{\text{max}}$. This corresponds to the dissociation limit, $\tilde{\nu}_{\text{max}}$.

Using the Birge-Sponer extrapolation, a plot of $\Delta\tilde{\nu}$ against v' and extrapolation to zero $\Delta\tilde{\nu}$ gives $\tilde{\nu}'_{max}$. See Problem 13.44. Such a plot gives $v'_{max} = 17$, and the corresponding area under the curve which represents $\tilde{\nu}_{max}$ is 56 858.6 cm^{-1}.

The energy is then

$$\begin{aligned} D_0'' + E_{ex} &= Lhc\tilde{\nu}_{max} \\ &= (6.022 \times 10^{23} \text{ mol}^{-1})(6.626 \times 10^{-34} \text{ J s})(2.998 \times 10^{8} \text{ m s}^{-1}) \\ &\quad \times (5.685\ 86 \times 10^{6} \text{ m}^{-1}) \\ &= 6.8017 \times 10^{5} \text{ J mol}^{-1} \end{aligned}$$

The area under the curve when $v' = 0$ is 7115 cm^{-1}. D_0' is found as above to be 8.511×10^{4} J mol^{-1}. The value of E_{ex} is given as 1.90×10^{5} J mol^{-1}, therefore,

$$\begin{aligned} D_0'' &= (6.80 \times 10^{5} - 1.90 \times 10^{5}) \text{ J mol}^{-1} \\ &= 490 \text{ kJ mol}^{-1} \end{aligned}$$

Chromophores

A somewhat different situation may be found with more complex molecules containing groups known as **chromophores** (Greek *chroma,* color; *phoros,* carrying). For example, substances such as acetone that contain the carbonyl group, C=O, all exhibit a weak absorption band having a maximum at about 285 nm and a stronger absorption band near 200 nm. The absorption coefficient for the band at 285 nm is usually around 10 dm^3 mol^{-1} cm^{-1}. At 200 nm the absorption coefficient is usually between 10^3 and 10^4 dm^3 mol^{-1} cm^{-1}. These two absorption bands are easily understood with reference to the nature of the carbonyl bond.

Some of the electrons relating to the carbonyl group are nonbonding and are given the symbol n. Others are in the π bond (valence-bond theory) or π orbital (molecular-orbital theory). Figure 13.37 shows the types of transitions that can occur. A nonbonding (n) electron can be promoted into an empty antibonding π orbital based on carbon, and the transition marked $\pi^* \leftarrow n$ at ~285 nm is known as a $\pi^* \leftarrow n$ *transition,* read "n-to-π-star transition." Since $\pi^* \leftarrow n$ transitions are often symmetry forbidden, the absorption is weak. The more intense absorption at about 200 nm is caused by the promotion of a π electron into the antibonding π^* orbital in the electronically excited state. This is known as a $\pi^* \leftarrow \pi$ transition. This is the same mechanism responsible for the absorption found in other double bonds such as the carbon-carbon double bond, C=C.

The presence of chromophores such as the carbonyl group often causes the molecule to absorb in the visible part of the spectrum and therefore to be colored. The contribution of a particular chromophore to the absorption spectrum is considerably altered by conjugation with other chromophores. Thus, an isolated carbon-carbon double bond exhibits a $\pi^* \leftarrow \pi$ transition at about 180 nm. However, if there is conjugation (i.e., alternating double and single bonds), the absorption is shifted to much higher wavelengths of around 450–500 nm.

Numerous electronic-spectrum assay procedures for determining concentrations are based on the absorption by chromophores. For example, the reaction of ninhydrin with most amino acids yields a purple product having a maximum absorption at 570 nm. Ninhydrin has a molar absorption coefficient $\epsilon_{1\text{ cm}}^{1\,M}$ of $\approx 10^{11}$ dm^3 mol^{-1} cm^{-1}, which is extremely large, and this means that amino acids in amounts of the order of 10 nanomoles (nmol) can be estimated. This reagent is used in forensic science (e.g., to obtain the fingerprints of a person who has handled a piece of paper).

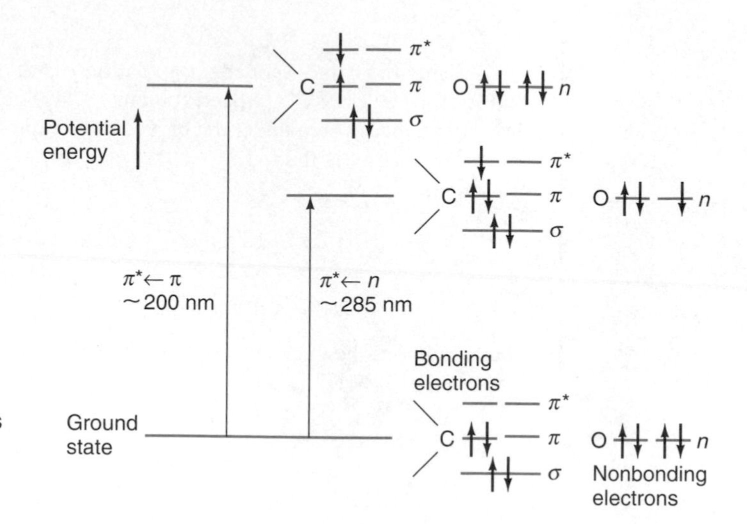

FIGURE 13.37
Two types of electronic transitions that can occur with a chromophore such as the carbonyl group.

The Fate of Electronically Excited Species

After an electronically excited molecule or radical has been produced (e.g., by absorption of radiation or in an electric discharge), a number of things can happen to it.

Fluorescence and Phosphorescence

In the first place, the species can emit radiation and pass into a lower state, which may be the ground state or a lower excited state. This emission of radiation is referred to as **fluorescence.**[5] This term is generally restricted to processes in which there is no change of multiplicity, in contrast to **phosphorescence,** where there is a multiplicity change and in which the emission is slower. Transitions giving rise to fluorescence are included in Figure 13.34. Whereas absorption generally occurs from the ground vibrational state, fluorescent emissions often give excited vibrational states, as shown in Figure 13.34, and therefore correspond to longer wavelengths than observed in absorption. This shift to longer wavelengths is called the **Stokes shift.** Fluorescence often occurs after certain radiationless processes have occurred, and this is considered later.

Stokes Shift

The formation of an excited species can also lead to *molecular dissociation,* which can occur by different mechanisms according to the relative positions of the potential-energy curves for ground and excited states. Figures 13.38a and b relate to a diatomic molecule AB and show two situations that are of particular importance. In Figure 13.38a a *vertical* transition from the lowest vibrational level ($v'' = 0$) of the ground state takes the molecule to point a on the diagram. Here the energy level is higher than for A + B, and as a result the molecule dissociates in its first vibration. The spectrum corresponding to this transition shows no rotational fine structure and appears *diffuse,* because the excited molecule AB* does not live long enough for any rotations to occur. Other transitions from $v'' = 0$ may take the system to lower vibrational levels, such as to point b for which $v' = 1$, and then no dissociation can occur. The band corresponding to this $1 \leftarrow 0$ transition will have a rotational fine structure, as will the $1 \rightarrow 0$ emission band.

Predissociation

Another situation, corresponding to what is known as **predissociation,** is illustrated in Figure 13.38b. A transition from $v'' = 0$ to $v' = 2$, for example, will

[5]This word is derived from the mineral fluorspar (Latin *fluor,* flowing; German *Spat,* crystalline material).

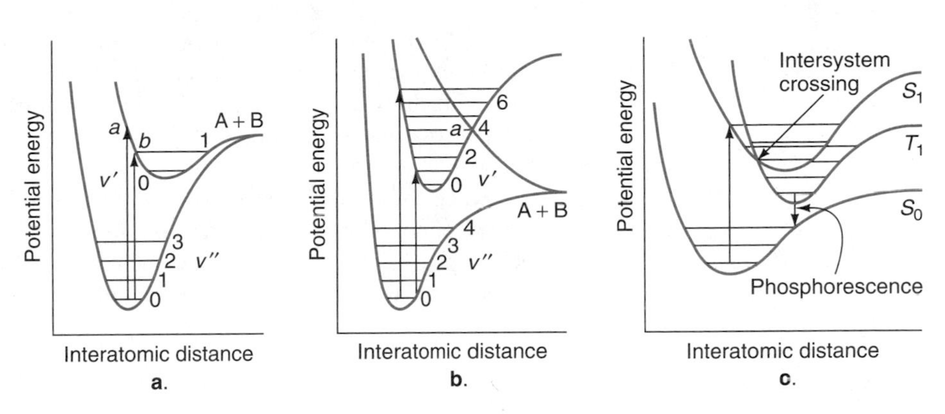

FIGURE 13.38
(a) Potential-energy curves for a diatomic molecule. The transition to point *a* leads to immediate dissociation. (b) Potential-energy curves for the case of predissociation. (c) Potential-energy curves that will lead to phosphorescence. Absorption from the singlet ground state S_0 gives the singlet excited state S_1; after a transition to the triplet state T_1, the molecule phosphoresces.

produce an AB* molecule that has insufficient energy to dissociate, and a rotational fine structure will therefore be observed. However, the diagram shows a *repulsive* state of AB* that crosses the upper potential-energy curve at a level corresponding to $v' = 4$, and $4 \leftarrow 0$ transition therefore produces a species that, in its first vibration, reaches point *a* where it undergoes a transition to the repulsive state and dissociates. The $4 \leftarrow 0$ band in the absorption spectrum therefore appears diffuse, as no rotations can occur. Transitions to a vibrational level such as $v' = 7$ may, on the other hand, produce a species that does not easily dissociate, since when it vibrates, it moves rather rapidly past point *a* and as a result the transition probability is reduced. The $v' = 7$ vibrational state of AB* may therefore have a long enough lifetime for rotations to occur, producing a rotational fine structure corresponding to this $7 \leftarrow 0$ transition.

A characteristic feature of a predissociation spectrum is therefore that there is a diffuse region, corresponding to transitions to the repulsive state followed by dissociation, and that at longer and shorter wavelengths the spectrum shows a rotational fine structure. This is a different situation from that found with ordinary dissociation (Figure 13.38a), where the rotational fine structure can only appear on the lower-energy (longer-wavelength) side of the diffuse bands.

These diagrams (Figure 13.38a and b) relate to a diatomic molecule AB, but the principles are similar for more complex molecules. However, some additional matters must be taken into consideration for molecules having more than two atoms. If a *diatomic* molecule has enough energy to undergo dissociation, it is bound to do so in the period of its first vibration (i.e., within 10^{-13}–10^{-14} s). However, it is possible for a *polyatomic* molecule to have sufficient energy to dissociate without dissociating in the first vibration. This is because the energy may not at first be suitably distributed in the dissociation coordinate for bond rupture to occur. As the molecule vibrates, the energy is constantly redistributed between the normal modes, and eventually a vibration may lead to dissociation. As a result the lifetime of a complex molecule with enough energy to dissociate may be 10^{-8} s or more, which is several orders of magnitude greater than that of a diatomic molecule.

During this period of time a number of alternative processes may occur to a polyatomic molecule, such as the following:

Processes Brought About by Collisions (Processes 1–2)

1. *Vibrational Relaxation.* During this process the molecule in an excited vibrational state undergoes collisions with other molecules and drops to lower vibrational levels within the same electronic state. In the gas phase, infrared emission is observed when collisions are infrequent.

2. *Collisional Quenching* or *External Conversion.* On collision with other molecules the electronically excited molecule may pass into a lower electronic state, which is often the ground state. The electronic energy is converted in this process into translational and internal energy and is eventually dissipated as heat.

Since these two processes *involve collisions,* their rates depend on the frequency of collisions and therefore on the availability of the colliding molecules. In the gas phase these processes may occur within 10^{-8} s at ordinary pressures, but in the liquid phase they are much faster and may occur in 10^{-12}–10^{-14} s.

The following processes that a molecule may undergo *do not involve collisions:*

Processes Not Involving Collisions (Processes 3–5)

3. *Internal Quenching* or *Internal Conversion.* In this process the electronic energy is converted into vibrational energy of the molecule itself, which therefore acts as a self-quencher. In the gas phase, this is the dominant route for excited state deactivation of large molecules with overlapping excited states.

The following processes are *radiative:*

4. *Fluorescence.* We have already mentioned fluorescence, which involves emission of radiation (generally with no change of multiplicity) from an excited singlet state and occurs in 10^{-9}–10^{-8} s. The photon can drop to any of the vibrational levels of the ground state. This is usually a much slower process than vibrational relaxation, quenching, or internal conversion.

Phosphorescence

5. *Phosphorescence.* The term *fluorescence* is usually restricted to processes in which there is no change of multiplicity. The term *phosphorescence,* on the other hand, is applied to processes in which there is a change of multiplicity. Many examples of phosphorescence involve triplet $\rightarrow$ singlet transitions and are preceded by processes in which an excited singlet state is converted into an excited triplet state. Figure 13.38c shows a typical system of potential-energy curves that lead to phosphorescence. The molecule in a singlet ground electronic state S_0 absorbs radiation that excites it to the state S_1, the potential-energy curve for which crosses that for an excited triplet state T. The species S_1 can therefore make an *intersystem crossing* to the triplet state at the point where the two curves cross. This is a *radiationless transition* and it often occurs only with difficulty. However, if the molecule is not too light, there is a considerable amount of spin-orbit coupling, in which states with different spin and orbital angular momenta mix because they have the same total angular momentum. This triplet state may be vibrationally excited and can undergo vibrational relaxation on collision with other molecules. The triplet species is then in its ground vibrational state where it cannot readily lose radiation since the transition involves a change of multiplicity (triplet $\rightarrow$ singlet). However, because of spin-orbit coupling the emission does occur, but slowly, sometimes occurring over periods of seconds or minutes. The long time for emission normally allows nonradiative processes to dominate the deactivation and, consequently, to observe phosphorescence, immobilization techniques are often used. These include using a rigid medium and/or low temperatures.

Figure 13.39, known as a **Jablonski diagram,** summarizes some of the processes that can occur after absorption of radiation. Solid vertical lines represent processes accompanied by the emission of radiation, while radiationless transi-

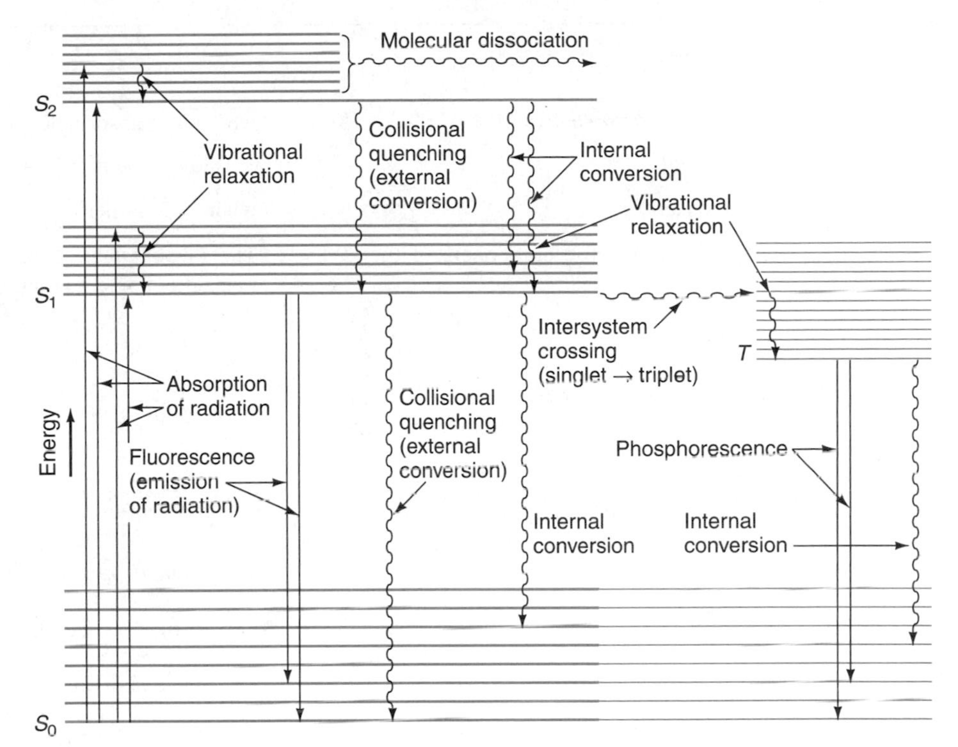

FIGURE 13.39
A Jablonski diagram, which shows the ground singlet state S_0, two excited singlet states S_1 and S_2, and a triplet state, T. Various possible processes are represented. The internal conversions $S_2 \rightarrow S_1$ and $S_1 \rightarrow S_0$ are often at the same energy because the excited states often lie within the potential energy surface of the lower energy state.

tions are indicated by wavy lines. More processes are possible than shown, and further details are to be found in some of the books listed in Suggested Reading (pp. 712-713).

Appendix: Symmetry Species Corresponding to Infrared and Raman Spectra

Symmetry Species	Infrared Spectra			Raman Spectra					
	x	y	z	x^2	y^2	z^2	xy	xz	yz
C_{2v}	b_1	b_2	a_1	a_1	a_1	a_1	a_2	b_1	b_2
C_{3v}	e	e	a_1	a_1, e	a_1, e	a_1	e	e	e
$C_{\infty v}$	π	π	σ^+	σ^+, δ	σ^+, δ	σ^+	δ	π	π
$D_{\infty h}$	π_u	π_u	σ_u^+	σ_g^+, δ_g	σ_g^+, δ_g	σ_g^+	δ_g	π_g	π_g
D_{3h}	e'	e'	a_2''	a_1', e'	a_1', e'	a_1'	e'	e''	e''
C_{2h}	b_u	b_u	a_u	a_g	a_g	a_g	a_g	b_g	b_g
D_{6h}	e_{1u}	e_{1u}	a_{2u}	a_{1g}, e_{2g}	a_{1g}, e_{2g}	a_{1g}	e_{2g}	e_{1g}	e_{1g}

A more complete table is given in G. Herzberg, *Infrared and Raman Spectra,* New York: D. Van Nostrand Co., 1945, p. 252.

KEY EQUATIONS

The *electric and magnetic components* of the EM field:

$$E_x(\omega,t) = 2E_x^0 \cos \omega t$$

$$H_x(\omega,t) = 2H_x^0 \cos \omega t$$

The energy density for the electric component of the EM field:

$$w(\omega) = \frac{3}{2}\epsilon_0 [(E_x^0)^2]_{ave}$$

The *transtition probability* of ending up in eigenstate k after starting from eigenstate i for a molecule with electric dipole moment μ_x:

$$c_k^*(t)\, c_k(t) = \frac{1}{\hbar^2}[\langle\psi_i \mid \mu_x \mid \psi_k\rangle]^2 (E_x^0)^2 t$$

The *Einstein coefficient for absorption*:

$$\frac{dN_i}{dt} = -N_i B_{ki}\, \rho(\omega_{ki});\; B_{ki} = B_{ik}$$

The *Einstein coefficient of spontaneous emission; $k \rightarrow i$:*

$$\frac{dN_k}{dt} = -N_k A_{ki}$$

The *Einstein coefficient of stimulated emission; $k \rightarrow i$:*

$$\frac{dN_k}{dt} = -N_k B_{ik}\, \rho(\omega_{ik})$$

where

$$B_{ki} = \frac{2\pi[<\psi_i\, |\mu_x|\, \psi_k>]^2}{3\hbar^2}$$

$$\frac{A_{ki}}{B_{ki}} = \frac{8\pi h \nu_{ki}^3}{c^3}$$

Lambert-Beer law:

$$A = \log_{10}\frac{I_0}{I} = \epsilon c l$$

where A = absorbance; ϵ = absorption coefficient.

$$\log_{10} T\% = 2 - A$$

where $T\%$ = percentage transmittance.

The *gyromagnetic* (magnetogyric) *ratio* is the ratio of the magnetic moment and the angular momentum:

$$\gamma \equiv \frac{\mu_l}{L} = -\frac{e}{2m_e}$$

for an orbiting electron.

The Bohr magneton:

$$\mu_B = \frac{eh}{4\pi m_e}$$

The *magnetic moment* from both $\hat{\boldsymbol{L}}$ and $\hat{\boldsymbol{S}}$

$$\hat{\boldsymbol{\mu}} = -\frac{e}{2m_e}(g_L \hat{\boldsymbol{L}} + g_S \hat{\boldsymbol{S}})$$

The *magnetic moment* from $\hat{\boldsymbol{J}}$:

$$\hat{\boldsymbol{\mu}} = -\frac{e}{2m_e}(g_L \hat{\boldsymbol{J}})$$

The *Landé-g-factor* is defined as

$$g_J = 1 + \frac{J(J+1) - L(L+1) + S(S+1)}{2J(J+1)}$$

Frequencies of rotational transitions:

$$\tilde{\nu}_j = 2(J+1)\tilde{B}$$

where $\tilde{B} = h/(8\pi^2 c\, I)$ and I = moment of inertia.

Morse potential function for the dissociation energy D_e:

$$E_p = D_e(1 - e^{-ax})^2$$

where $x = r - r_e$.

PROBLEMS

Absorption of Radiation

13.1. The molar absorption coefficient of human hemoglobin (molecular weight 64 000) is 532 $dm^3\ cm^{-1}\ mol^{-1}$ at 430 nm. A solution of hemoglobin in a cuvette having a light path of 1 cm was found at that wavelength to have a transmittance of 76.7%. Calculate the concentration in mol dm^{-3} and in g dm^{-3}.

13.2. A spectrophotometer has a meter that gives a reading directly proportional to the amount of light reaching the detector. When the light source is off, the reading is zero. With pure solvent in the light path, the meter reading is 78; with a 0.1 M solution of a solute in the same solvent, the meter reading is 55. The light path is 0.5 cm. Calculate the absorbance, the transmittance, and the molar absorption coefficient.

13.3. The transmission of a potassium chromate solution was measured at a wavelength of 365 nm using a cell with a 1.0-cm path length. The data are as follows:

Transmission	0.357	0.303	0.194	0.124
Conc. × 10^4/mol dm^{-3}	0.90	1.10	1.50	1.90

Calculate the molar Naperian absorbance and the molar decadic absorbance.

13.4. An aqueous solution containing 0.95 g of oxygenated myoglobin (M_r = 18 800) in 100 cm^3 gave a transmittance of 0.87 at 580 nm, with a path length of 10.0 cm. Calculate the molar absorption coefficient.

13.5. A substance in aqueous solution at a concentration of 0.01 *M* shows an optical transmittance of 28% with a path length of 2 mm. Calculate the molar absorption coefficient of the solute. What would be the transmittance in a cell 1-cm thick?

13.6. The molar absorption coefficient of hemoglobin at 430 nm is 532 dm^3 mol^{-1} cm^{-1}. A solution of hemoglobin was found to have an absorbance of 0.155 at 430 nm, with a light path of 1.00 cm. Calculate the concentration.

13.7. A 10 μM solution of a substance gave an absorbance of 0.1028 with a light path of 1 cm. Calculate the molar absorption coefficient. What would be the percentage light transmittance of a 1-μm solution with the same light path?

13.8. Two substances of biological importance, NAD^+ and NADH, have equal absorption coefficients, 1.8×10^4 dm^3 mol^{-1} cm^{-1}, at 260 nm (a wavelength at which absorption coefficients are equal is known as the *isosbestic point*[6]). At 340 nm, NAD^+ does not absorb at all, but NADH has an absorption coefficient of 6.22×10^3 dm^3 mol^{-1} cm^{-1}. A solution containing both substances had an absorbance of 0.215 at 340 nm and of 0.850 at 260 nm. Calculate the concentration of each substance.

13.9. The transmittance of a 0.01 *M* solution of bromine in carbon tetrachloride, with a path length of 2 mm, is 28%. Calculate the molar absorption coefficient of bromine at that wavelength. What would the percentage transmittance be in a cell 1-cm thick?

13.10. An acid HA ionizes in aqueous solution into H^+ and A^- ions. At a wavelength of 430 nm HA does not absorb light, but A^- does so with an absorption coefficient of 458 dm^3 cm^{-1} mol^{-1}. A solution of the acid at a concentration of 0.1 *M* was found to have a transmittance of 1.47% at 430 nm with a path length of 1 cm and at 25 °C. Calculate the dissociation constant of HA at 25 °C, and $\Delta G°$ for the dissociation process.

Atomic Spectra

13.11. In the Balmer series of the hydrogen atom, the first emission line is observed at 656.3 nm. Calculate the value of the Rydberg constant. What is the energy of the light quanta emitted during the transition?

[6]From the Greek prefix *iso-*, the same, and *sbestos*, quench. This word is sometimes incorrectly written as "isobestic."

13.12. The ground state of the Li atom has the electronic configuration $1s^22s^1$. What is its spectroscopic term? If the 2s electron is excited to the 2p state, what terms are then possible?

13.13. Suppose that an excited state of the carbon atom has the electronic configuration $1s^22s^22p^13p^1$. What are the possible spectroscopic terms?

13.14. What are the terms for the following electronic configurations?

a. Na ($1s^22s^22p^63p^1$)
b. Sc ($1s^22s^22p^63s^23p^64s^23d^1$)

13.15. What values of *J* may arise in the following terms?

$$^1P,\ ^3P,\ ^4P,\ ^1D,\ ^2D,\ ^3D,\ ^4D$$

***13.16.** Calculate the Landé-g-factor for a $^2P_{1/2}$ level. What would be the anomalous Zeeman splitting for this level in a magnetic field of 4.0 T?

***13.17.** Calculate the spacing between the lines for a $^3D_1 \rightarrow {}^3P_0$ transition, in an anomalous Zeeman experiment with a magnetic field of 4.0 T.

Rotational and Microwave Spectra

13.18. The separation between neighboring lines in the pure rotational spectrum of $^{35}Cl^{19}F$ is found to be 1.023 cm^{-1}. Calculate the interatomic distance.

13.19. The lines in the pure rotational spectrum of HF are 41.9 cm^{-1} apart. Calculate the interatomic distance. Predict the separation between the lines for DF and TF.

13.20. In the microwave spectrum of $^{12}C^{16}O$ the separation between lines has been measured to be 115270 MHz. Calculate the interatomic distance.

***13.21.** The $J = 0 \rightarrow J = 1$ line in the microwave absorption spectrum of $^{12}C^{16}O$ and of $^{13}C^{16}O$ was measured by Gillam et al., [*Phys. Rev. 78,* 140(1950)]. In its ground vibrational state, the former has the value 3.842 35 cm^{-1} and the latter, the value 3.673 37 cm^{-1}. Calculate

a. the bond length of the $^{12}C^{16}O$ molecule,
b. the relative atomic mass of ^{13}C,
c. the bond length of the $^{13}C^{16}O$ molecule.

13.22. The microwave spectrum of $^{16}O^{12}C^{32}S$ shows absorption lines separated by 12.163 GHz. That of $^{16}O^{12}C^{34}S$ shows lines separated by 11.865 GHz. The determination of the bond distances involves solving two simultaneous quadratic equations, which is best done by successive approximations. To avoid all that labor, simply confirm that the results are consistent with $r(O—C)$ = 116 pm and $r(C—S)$ = 156 pm.

Vibrational-Rotational and Raman Spectra

13.23. The maximum potential energy that a diatomic molecule can store is $\frac{1}{2}kx^2$, where x is the amplitude of vibration. If the force constant k is 1.86×10^3 N m^{-1}, calculate the maximum amplitude of vibration for the CO molecule in the $v = 0$ vibrational state. Compare this to the bond length obtained in Problem 13.21. Use the value of u_r in that problem.

13.24. Consider the following molecules: H_2, HCl, CO_2, CH_4, H_2O, CH_3Cl, CH_2Cl_2, H_2O_2, NH_3, and SF_6. Which of them will give

a. a pure rotational spectrum,
b. a vibrational-rotational spectrum,
c. a pure rotational Raman spectrum,
d. a vibrational Raman spectrum?

13.25. Analysis of the vibrational-rotational spectrum of the $H^{35}Cl$ molecule shows that its fundamental vibrational frequency ν_0 is 2988.9 cm^{-1}. Calculate the force constant of the H—Cl bond.

13.26. A few transitions in the P and R branches of the infrared spectrum of $H^{35}Cl$ spectrum are identified below.

J''	0	1	2	3	4	5	6
P (cm^{-1})		2865.10	2843.62	2821.56	2798.94	2775.76	2752.04
R (cm^{-1})	2906.24	2925.90	2944.90	2963.29	2981.00	2998.04	

Using Eqs. 13.134 or 13.135 as appropriate, calculate $\tilde{\nu}_0$ and the rotational constant $\tilde{B}$.

13.27. Comparison of the results of Problem 13.26 to experimental values for $H^{35}Cl$ ($\tilde{\nu}_0$ = 2990 cm^{-1}, and $\tilde{B}$ = 10.59 cm^{-1}) shows that Eqs. 13.134 and 13.135 do not accurately relate the observed transitions to the rotational quantum numbers. Much of the error results from not taking the anharmonicity of the potential energy curve into account. Using the definition (see Eq. 13.138) $T_{v,J} = (v + \frac{1}{2})\tilde{\nu}_0 - (v + \frac{1}{2})^2\,\tilde{\nu}_0 x_e + J(J+1)\tilde{B}$, derive a more accurate expression for $\Delta T_{v,J}$ for the $v'' = 0 \rightarrow v' = 1$ transitions of a diatomic molecule. [Even this is an approximate treatment because we are ignoring the coupling of rotations with vibrations.]

13.28. From the results of Problems 13.26 and 13.27 and the experimental value of $\tilde{\nu}_0$ for $H^{35}Cl$ given in Problem 13.27, estimate the value of the anharmonicity constant x_e. Use the average of the P and R branch values for $\tilde{\nu}_0$ from Problem 13.26.

13.29. The vibrational Raman spectrum of $^{35}Cl_2$ shows series of Stokes and anti-Stokes lines; the separation between the lines in each of the two series is 0.9752 cm^{-1}. Estimate the bond length in Cl_2.

13.30. The dissociation energy of H_2 is 432.0 kJ mol^{-1} and the fundamental vibrational frequency of the molecule is 1.257×10^{14} s^{-1}. Calculate the classical dissociation energy. Estimate the zero-point energies of HD and D_2 and their dissociation energies.

13.31. A molecule AB_2 is known to be linear but it is not known whether it is B—A—B or A—B—B. Its infrared spectrum is found to show bands corresponding to three normal modes of vibration. Which is the structure?

13.32. The frequency of the O—H stretching vibration in CH_3OH is 3300 cm^{-1}. Estimate the frequency of the O—D stretching vibration in CH_3OD.

13.33. The spectroscopic constants for the OH radical are $\tilde{\nu}_0$ = 3737.76 cm^{-1}, $\tilde{\nu}_0 x_e$ = 84.8813 cm^{-1}, $\tilde{B}$ = 18.9108 cm^{-1}. Predict the frequencies at which (a) the P branch transitions ending in, and (b) the R branch beginning in, J = 0, 1, 2 will be observed.

13.34. Irradiation of acetylene with mercury radiation at 435.83 nm gives rise to a Raman line at 476.85 nm. Calculate the vibrational frequency that corresponds to this shift.

13.35. The fundamental vibrational frequency of $H^{127}I$ is 2309.5 cm^{-1}. Calculate the force constant of the bond.

***13.36.** The following are some normal modes of vibration for several molecules:

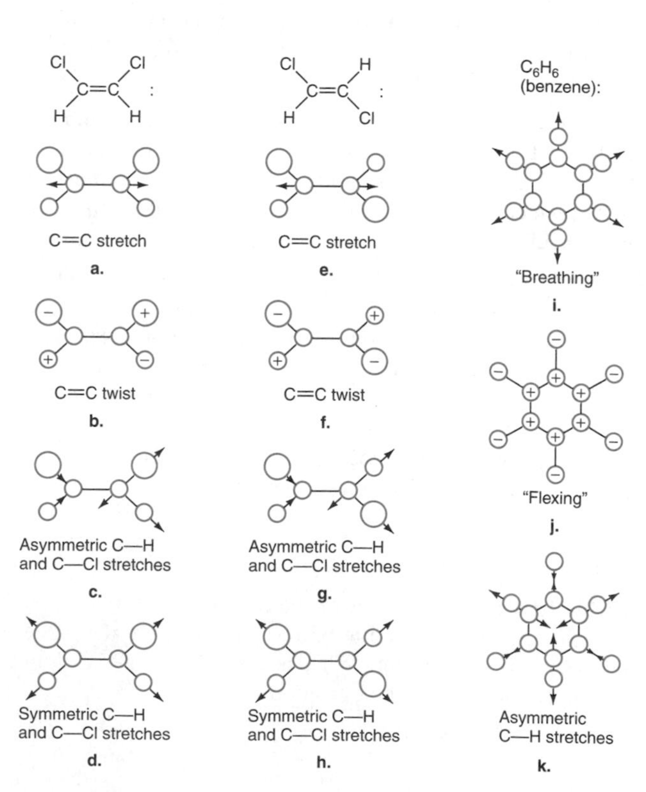

In each case, determine the point group and refer to Appendix E on p. 1028 to determine whether the vibration is active in the infrared and in the Raman spectrum. Then check your conclusions by reference to the appendix to this chapter (p. 707).

***13.37.** Prove that the force constant k corresponding to the Morse potential function (Eq. 13.146) at small bond extensions is

$$k = 2D_e a^2$$

Calculate the vibration frequency ν_0 on the basis of the following Morse parameters for $H^{35}Cl$:

$$D_e = 4.67 \text{ eV}$$

$$a = 1.85 \times 10^8 \text{ cm}^{-1}$$

***13.38.** The Morse function is only one of several models used to describe the behavior of the potential energy of diatomic molecules. A popular and very accurate model introduced by Murrell and Sorbie [*J. Chem. Soc., Faraday Trans. 2, 70,* 1552(1974)] is the so-called "Extended Rydberg function," which is written as

$$E_p(x) = -D_e(1 + a_1 x + a_2 x^2 + a_3 x^3)e^{-a_1 x},$$

where $x = r - r_e$, as in the case of the Morse potential of Eq. 13.146, and the a_i are constants for a given molecule.

a. Derive an expression for the force constant in terms of the parameters D_e and a_i.
b. Show that in order for a function of this form to have a minimum at $r = r_e$, a_1 must be both the coefficient of x and the exponential parameter.

***13.39.** Yet another model for a diatomic potential is the Bond Order function used by Garcia and Laganá [*Mol. Phys. 56,* 621(1985)], which is given as

$$E_p(x) = -D_e \sum_{n=1}^{N} c_n \exp(-n\beta x).$$

Show that for $N = 2$, with appropriate choices for the coefficients c_1 and c_2, this is identical to the Morse potential of Eq. 13.146 expressed as $E_p = De(1 - e^{-\beta x})^2 - De$.

***13.40.** The parameters for the bond order (see Problem 13.39) potential of the $^{35}Cl^{16}O$ radical with $N = 4$ are (in atomic units) $c_1 = 2.691\ 042$, $c_2 = -2.545\ 521$, $c_3 = 1.017\ 916$, $c_4 = -0.163437$, $De = 0.10302$, and $\beta = 1.763\ 768$. What is the vibrational frequency (in cm^{-1}) predicted by this model? [Note that the force constant can be expressed in units of energy area^{-1}.]

***13.41.** A model for the ^{14}N–^{14}N–$^{16}O^+$ ion assigns the following force constants for the two stretching frequencies: $k_{12} = 1092.8$ N m^{-1}, and $k_{23} = 890.68$ N m^{-1}. Use Eqs. 13.171–13.174 to calculate the two stretch frequencies obtained from the model.

***13.42.** The fundamental and a few successive overtones of the H_2^+ molecule lie at 2191, 2064, 1941, 1821, 1705, 1591, 1479 and 1368 cm^{-1}, respectively.

a. Starting from Eq. 13.139, derive an expression that can be used to obtain both $\tilde{\nu}_0$ and x_e by a suitable analysis of the data.
b. Perform the analysis and calculate both $\tilde{\nu}_0$ and x_e.

Electronic Spectra

13.43. Sketch potential energy curves for a diatomic molecule in its ground electronic state and in an excited state, consistent with the following observations:

a. There is a strong $0 \leftarrow 0$ absorption band, and strong $0 \rightarrow 0$, $1 \rightarrow 1$, and $2 \rightarrow 2$ emission bands.
b. The strongest absorption band is $4 \leftarrow 0$, and the strongest emission band is $0 \rightarrow 2$.
c. There is no sharp rotational fine structure in absorption, but there is a sharp emission spectrum.
d. The absorption spectrum shows a well-defined fine structure for the $0 \leftarrow 0$, $1 \leftarrow 0$, $2 \leftarrow 0$, $3 \leftarrow 0$, and $4 \leftarrow 0$ transitions and for the $6 \leftarrow 0$ and $7 \leftarrow 0$, but not in between.

***13.44.** Using the data in Example 13.12 on p. 702, determine the area under the curve in a plot of $\Delta\nu$ against v. Extrapolate to zero $\Delta\nu$ to obtain ν'_{max}, since the Birge-Sponer extrapolation shows that at that point $v' = v'_{max}$. A better value may be obtained by a nonlinear extrapolation. What are the values of v' not given in the table?

13.45. Calculate the dissociation energy of the hydrogen molecule ion from the vibrational energy level separations of H_2^+. The values for the transitions $1 \leftarrow 0$, $2 \leftarrow 1$, . . . , are, respectively, 2191, 2064, 1941, 1821, 1705, 1591, 1479, 1368, 1257, 1145, 1033, 918, 800, 677, 548, 411, with all values given in cm^{-1}. Use both a linear plot and a curve to obtain answers.

13.46. The electronic spectra of diatomic molecules in the gas phase typically show extensive vibrational structure superimposed on the broader electronic transition. Taking the equilibrium geometry of the ground electronic state to be the zero energy, the $\tilde{G}(v'')$ of Eq. 13.138 can be used to express the energies of the vibrational states v'' (in cm^{-1}) of this electronic state. Denoting the minimum energy of the excited electronic state as $\tilde{T}_e$ (in cm^{-1}), the vibrational energy levels v' of the excited state can be expressed as $\tilde{T}_e + \tilde{G}(v')$. Derive an expression for the frequencies $\tilde{\nu}$ of the transitions $v'' \rightarrow v'$ between the vibrational levels of the ground and excited electronic states. (Note that the vibrational frequencies and anharmonicity constants are not the same for the ground and excited electronic states.)

13.47. An easy and reliable way to analyze the electronic spectrum of a diatomic molecule is to use the equation derived in Problem 13.46 as the model for a multiple regression analysis (several plotting packages and mathematics packages such as Mathcad can perform this task) to simultaneously identify the five unknowns, $\tilde{T}_e$, $\tilde{\nu}_0'$, $\tilde{\nu}_0' x_e'$, $\tilde{\nu}_0''$, and $\tilde{\nu}_0'' x_e''$ [McNaught, *J. Chem. Ed.* 57, 101(1980)]. The following data are from the electronic spectrum of iodine.

$0 \rightarrow v'$	λ (nm)	$1 \rightarrow v'$	λ (nm)	$2 \rightarrow v'$	λ (nm)
17	567.2	15	581.0	10	607.3
18	564.2	16	577.8	11	603.1
19	561.5	17	574.2	12	599.1
20	558.5	18	571.3	13	595.5
21	555.8	19	568.3	14	591.8
22	553.0	20	565.2	15	588.1
23	550.1	21	559.6	16	584.8
24	547.8	22	556.9	17	581.2
25	542.7	23	554.2	18	578.1
26	540.7	24	551.8	19	575.1
27	538.5	25	549.0	20	572.4

Perform a multiple regression analysis and identify the spectroscopic parameters of the ground and excited electronic states. Literature values are (in cm^{-1}) $\tilde{T}_e$ = 15730, $\tilde{\nu}_0' = 132.1$, $\nu_0' x_e' = 1.051$, $\tilde{\nu}_0'' = 214.5$, and $\tilde{\nu}_0'' x_e'' =$ 0.614.

13.48. The dissociation energy (from the zero-point level) of the ground state $O_2(^3\Sigma_g^-)$ molecule is 5.09 eV. There exists an electronically excited $^3\Sigma_u^-$ state of O_2, whose zero-point level lies 6.21 eV above the zero-point level of the ground state. The ground-state molecule dissociates into two ground-state $O(^3P)$ atoms, while the $^3\Sigma_u^-$ species dissociates into one ground-state $O(^3P)$ atom and an O* (^1D) atom that lies 1.97 eV above the ground state. Sketch the potential-energy curves and calculate the dissociation energy of $O_2(^3\Sigma_u^-)$ into O + O*(^1D).

13.49. The spectroscopic dissociation energy D_0 is the energy required to dissociate the molecule in its ground vibrational state. This is always slightly smaller than the actual depth of the electronic potential energy because of the zero-point energy of the molecule (see Figure 13.22). Given that for HCl, D_e = 4.6173 eV, $\tilde{\nu}_0$ = 2989 cm^{-1}, and $\tilde{\nu}_0 x_e$ = 52.82 cm^{-1}, calculate the value of D_0 for HCl.

13.50. The dissociation energy (from the zero-point level) of the ground state $NO(X^2\pi)$ molecule is 6.6 eV. There exists an electronically excited $B^2\Pi$ state of NO whose zero-point level lies at 5.7 eV above the zero-point level of the ground state. The ground-state molecule dissociates into ground state $N(^4S)$ + $O(^3(P))$, while the $B^2\Pi$ species dissociates into two ground-state atoms $N(^2D)$ + $O(^3P)$ that lie 3.3 eV above its ground state. Sketch the potential energy curves and calculate the dissociation energy of NO into $N(^2D)$ + $O(^3P)$.

13.51. Sodium vapor, which consists mainly of Na_2 molecules, has a system of absorption bands in the green, the origin of the 0, 0 band being at 20 302.6 cm^{-1}. From the spacing of the vibrational levels it can be deduced that the dissociation energy of the upper state is 0.35 eV. The dissociation of the excited Na_2 gives a normal atom and an atom that emits the yellow sodium D line at 589.3 nm. Calculate the energy of dissociation of Na_2 in its ground state.

For additional problems, see the books listed at the end of the problems in Chapter 12 (pp. 634-635).

Essay Questions

13.52. State the laws of Lambert and Beer, and write an equation comprising the two laws.

13.53. Explain clearly what is meant by absorbance and transmittance, and derive a relationship between them.

13.54. Give an account of the fundamental origins of ultraviolet and infrared spectra.

13.55. Explain the selection rules for infrared spectra, with examples.

SUGGESTED READING

The following five books of Gerhard Herzberg give authoritative and detailed accounts of atomic and molecular spectroscopy. His series of books is often referred to as the "bible" of spectroscopy:

G. Herzberg, *Atomic Spectra and Atomic Structure,* New York: Prentice Hall, 1937; New York: Dover, 2nd edition, 1944.

G. Herzberg, *Molecular Spectra and Molecular Structure, II. Infrared and Raman Spectra of Polyatomic Molecules,* New York: D. Van Nostrand Co., 1945.

G. Herzberg, *Molecular Spectra and Molecular Structures, III. Electronic Spectra and Electronic Structure of Polyatomic Molecules,* New York: D. Van Nostrand Co., 1966.

G. Herzberg, *Molecular Structure, I. Spectra of Diatomic Molecules,* New York: D. Van Nostrand Co., 1950.

G. Herzberg, *The Spectra and Structures of Simple Free Radicals: An Introduction to Molecular Spectroscopy,* New York: Dover, 1988. This small book provides an admirable account of the subject, with excellent diagrams.

The following books also deal with molecular spectroscopy:

C. N. Banwell, *Fundamentals of Molecular Spectroscopy,* 3rd edition, New York: McGraw-Hill, 1966, 1983.

J. Barriol, *Elements of Quantum Mechanics with Chemical Applications,* New York: Barnes & Noble, Inc., 1971. (This is a more advanced text but brings with it much clarity.)

G. M. Barrow, *Introduction to Molecular Spectroscopy,* New York: McGraw-Hill, 1962. This contains an excellent treatment of molecular vibrations with reference to infrared spectroscopy.

C. E. Dykstra, *Quantum Chemistry and Molecular Spectroscopy,* Englewood Cliffs, NJ: Prentice Hall, 1992.

J. D. Graybeal, *Molecular Spectroscopy,* New York: McGraw-Hill, 1988.

J. M. Hollas, *Modern Spectroscopy,* 2nd edition, New York: Wiley, 1992.

M. Orchin and H. H. Jaffe, *Symmetry, Orbitals, and Spectra,* New York: Wiley-Interscience, 1971.

J. I. Steinfeld, *Molecules and Radiation: An Introduction to Modern Molecular Spectroscopy,* 2nd edition, Cambridge, MA: MIT Press, 1985.

W. S. Struve, *Fundamentals of Molecular Spectroscopy,* New York: McGraw-Hill, 1989.

E. B. Wilson, J. C. Decius, and P. C. Cross, *Molecular Vibrations,* New York: McGraw-Hill, 1955; Dover reprint, 1980.

For interesting accounts of the origins of Raman spectroscopy, see

J. C. D. Brand, "The discovery of the Raman effect," *Notes & Records of the Royal Society, 43,* 1–23(1989).

F. A. Miller and G. B. Kauffman, "C. V. Raman and the discovery of the Raman effect," *J. Chem. Ed., 66,* 795–800(1989).

Some Modern Applications of Spectroscopy

14

PREVIEW

Chapter 13 covered many traditional areas of spectroscopy developed before 1950. Since the late 1940s, rapid developments in electronics have allowed the perfection of techniques which were unthought of earlier. Indeed, some laborious techniques, such as obtaining Raman spectra, have become much simplified, allowing for rapid advancement in many areas of spectroscopy. This chapter treats a number of topics which have been developed only in the last 50 years or so, particularly in the last decade.

Several techniques utilize lasers. A *laser beam* is intense and covers a narrow range of wavelengths. *Laser spectroscopy* provides some important information as well as greatly enhancing the applicability of some spectroscopic techniques. The quality of the spectral data is related to the *spectral line widths* which, in turn, are related to the *Doppler effect* and to the *lifetimes* of excited species.

OBJECTIVES

After studying this chapter, the student should be able to:

Lasers

- Understand the basic ideas and requirements of lasers and the different ways in which lasers are produced.
- Know the different types of lasers that require three and four quantum states with a persistent population inversion.
- Know the difference between continuous wave (CW) and pulsed lasers.
- Compare Raman and IR spectroscopy and understand hyper-Raman spectroscopy.
- Appreciate that lasers can be used to study chemical processes down to the femtosecond time scale, thereby allowing the study of short-lived transient molecules.
- Understand some of the special applications of lasers.

Magnetic Spectroscopy

- Appreciate the similarities between electronic spectroscopy and magnetic spectroscopy.
- Understand the lifting of degeneracy of magnetic energy levels in a magnetic field.

Nuclear Magnetic Resonance (NMR) Spectroscopy

- Appreciate NMR in terms of the quantum-mechanical properties of spins, and their ability to couple vectorially.
- Apply the transition probability formula of Chapter 13 to magnetic transitions.
- Recognize that magnetic fields can be time dependent and oscillate in the laboratory coordinate frame.
- Understand that the return to equilibrium is determined by two relaxation times: the spin-lattice relaxation time T_1 and the spin-spin relaxation time T_2.
- Study nuclear spins in a magnetic field and know how Larmor precession occurs.
- Appreciate the advantage of studying spin systems in the rotating frame.
- Appreciate the advantages of improving the signal-to-noise ratio by the use of pulsed NMR.
- Know that the non-equilibrium magnetization precesses and decays in the XY laboratory frame and induces a voltage in a coil as it decays away (the free induction decay, FID).
- Understand the origin of spin echoes and follow the details of the formation of a stimulated echo by the application of two pulses.
- Follow the details of one-dimensional NMR, which arises from the Fourier transform of the FID (FT NMR).
- Know that the three basic pieces of information obtained from the liquid-state NMR experiment are the chemical shift, the signal intensity, and the multiplet structure.
- Understand the basic principles of 2D NMR.
- Recognize how NMR leads to an understanding of the structure and dynamics of solids and liquids.
- Recognize that a number of experimental methods, both pulsed and mechanical, can be used to improve the quality of the spectrum by reducing the line broadening found in solids.

Electron Magnetic Resonance (EMR)

- Understand that the coupling of an external magnetic field with an electron spin is more than a thousand times stronger than with a nuclear spin.
- Appreciate some of the types of EMR: ESR (electron spin resonance); EPR (electron paramagnetic resonance); CW and pulsed EMR; FT ESR; ESEEM, ENDOR, etc.
- Follow the details of the analysis of an ESR experiment with an organometallic compound.

Photoelectron Spectroscopy (PES)

- Recognize that photoelectrons are generated from the photoelectric effect.
- Know that there are two basic forms of PES. One is UPS, which uses ultraviolet light; the other is XPS, which uses X rays.

Photoacoustic Spectroscopy (PAS)

- Appreciate that this technique gives information similar to, and complementary to, IR spectroscopy.

Chiroptical Methods

- Know that these all arise from the use of plane-polarized light.
- Appreciate some aspects of optical activity and polarimetry as well as optical rotatory dispersion (ORD) and circular dichroism (CD).

Mass Spectrometry

- Appreciate the application of this technique to the study of molecular structure.
- Understand that this type of spectroscopy measures the mass-to-charge ratio by ionizing the sample, passing the sample through a mass analyzer, and detecting the charged masses.
- Appreciate how modern techniques of fast atom/ion bombardment (FAB), matrix-assisted laser desorption (MALDI), and electrospray ionization (ESI) have extended the technique to the study of large molecules, including biopolymers.

Three types of spectroscopy involve resonance effects: nuclear magnetic resonance (NMR), electron magnetic resonance (EMR), and the less widely used Mössbauer spectroscopy. The first two techniques, concerned with the behavior of substances in a magnetic field, have played a major role in the understanding of the chemistry of all types of compounds, including simple organic molecules, polymers, organometallic compounds, and biopolymers. In addition, the imaging technique used in medicine (MRI) is based on NMR. The NMR technique involves the alignment of nuclear spins in a magnetic field. Today pulsed Fourier-transform NMR (FT-NMR) is used almost exclusively, and has evolved from one-dimensional spectroscopy to two-, three-, and four-dimensional NMR spectroscopy. In contrast, the majority of EMR experiments are continuous wave (CW) rather than pulsed, although pulsed EMR is sometimes carried out. Aside from the difference in magnitude between the nuclear and electron spins, the descriptions of the two are identical. Pulsed EMR suffers from the difficulty of delivering the short, high-powered pulses that are required.

Mössbauer spectroscopy is quite different from magnetic spectroscopy in that it arises from the absorption of γ (gamma) rays by nuclei. The spectra are made possible by Mössbauer's discovery of a way to generate highly homogeneous beams of γ rays and to control their frequency by use of the Doppler effect.

Other types of spectroscopy do not directly measure absorption or emission between energy levels. In *Raman spectroscopy* (Section 13.5), light undergoes a frequency change by interaction with the material through which is passes. *Photoelectron spectroscopy* measures the difference in energy between the incident photon and the ejected electron and provides a measure of the orbital energies. *Photoacoustic spectroscopy,* while involving absorption, is concerned with the sound waves that are produced. *Circular dichroism* and *optical rotatory dispersion* involve measurements of the rotation of light by a molecule.

Mass spectroscopy is a different type of spectroscopy altogether, involving charged particles rather than electromagnetic radiation. Recent developments in this field have extended the technique to large biopolymers.

In this chapter we look at some of the more recent developments in spectroscopy brought about by advances in electronics and some insightful ideas that have their basis in physics. We begin by investigating the characteristics of different kinds of lasers.

14.1 Laser Spectroscopy

Lasers are high-energy, monochromatic, coherent light sources. They have many applications, ranging from eye surgery to the measuring of distances with extreme accuracy. Here we are interested in the use of lasers in chemistry to provide invaluable information at different frequencies and on different time scales. For example, laser-generated femtosecond (10^{-15} s) pulses in the visible and UV ranges are faster than all molecular motions except electronic transitions. This permits detailed study of molecules and can give valuable structural information not only for short-lived transition-state species in chemical reactions but also for fragments produced in mass spectrometry. To maximize the use of lasers, it is essential, then, to examine how lasers work and discover their limitations.

The idea of harnessing the orbital energy of electrons for amplification occurred independently over 1951–1953 to several workers in the microwave field. These included the American physicist Charles H. Townes (b. 1915) and the Russians Nikolai G. Basov (1922–2001) and Aleksandr M. Prokhorov (b. 1916). In 1954 Townes built the first MASER, which stands for microwave amplification by the stimulated emission of radiation. In 1958 A. L. Schawlow and Townes outlined the theory and proposed a technique for use of a device which would operate at visible frequencies, which became known as light amplification by the stimulated emission of radiation, or its acronym, LASER. For their work, Townes, Basov, and Prokhorov received the 1964 Nobel Prize in physics.

Requirements for Stimulated Emission

The basic conditions for laser action are presented in Section 13.1. There it is shown that the Einstein coefficients describe *absorption* B_{ik} from the lower i to the higher k level; *stimulated emission* B_{ki} from the excited k level back to the i level; and *spontaneous emission* A_{ki} from the excited k level back to the i level. See spontaneous emission in Figures 13.3–13.5. We have shown that $B_{ki} = B_{ik}$; see Eq. 13.21. Moreover, Figures 13.3 and 13.4 show that as a two-level system is excited by the absorption of light, two processes—stimulated emission and spontaneous emission—cause the excited states to relax back to the ground state. Different processes govern these two relaxations. When a molecule absorbs a photon and then relaxes by spontaneous emission, the photon is emitted in all directions with equal probability. In contrast, stimulated emission occurs in the same direction as the incident photon. The effect of this is that the photons emitted will travel in the same direction as the exciting beam and stimulate other emissions in the same direction, thereby causing amplification of the light in that direction. (See Figure 14.1.)

In order to induce a system to lase, several conditions must be met. First, we need an active medium that has the correct optical properties. A second requirement is that the population of the excited state must be greater than that of the lower state (inversion) and be maintained that way if lasing is to persist. Third, since the rate of stimulated emission is quite low because the initial intensity is low, before the amplification process can develop, a way must be found to permit the stimulated photons to stimulate others and so continue to amplify their intensity.

A variety of different active media can be used, and some of these are discussed later. In a two-level system, the competition between optical absorption and emission can never be overcome. As light is absorbed, the lower energy state is depleted and the upper level is populated until the two are equal. This process is called **saturation** and cannot cause a population inversion.

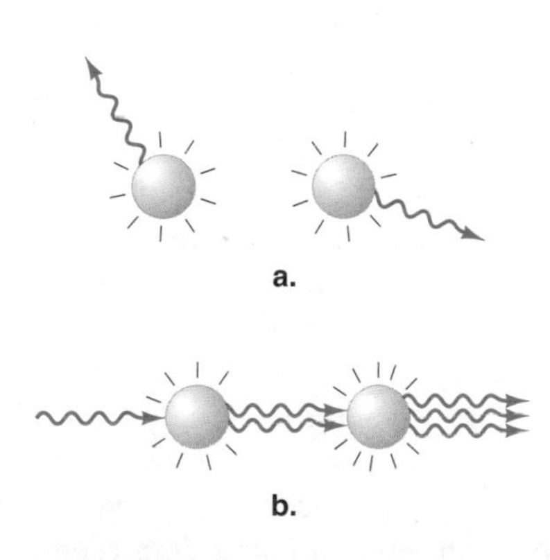

FIGURE 14.1
(a) Spontaneous emission.
(b) Stimulated emission. Reproduced with permission from D. L. Andrews, *Lasers in Chemistry,* New York: Springer, 1997, Figure 1.1.

The first successful lasers used a three-level system, but four-level systems are now widely used. In order to create a population inversion, energy must be supplied to the active medium to populate the excited levels, which causes the system to move away from the equilibrium Boltzmann distribution (Eq. 13.29). This process is called **pumping** and can be accomplished by flashes of light, by electric sparks, and by chemical energy. If it is possible to sustain a population inversion and the laser operates continuously, then a *continuous wave (CW)* laser beam can be produced. Figure 14.2(a) shows a Boltzmann distribution and Figure 14.2(b) an inverted population (see the longer heavy line at the higher energy level). If it is not possible to sustain an inverted population, the amplification builds up and is discharged in pulses and then builds again. This is called **pulsed emission.** Generally, creating continuous emission is more difficult than producing pulsed emission. Another problem, discussed in Section 13.1, Eqs. 13.29 and 13.30, is that unwanted spontaneous emission increases as the third power of the frequency and thus makes it more difficult to produce a laser that operates at higher frequencies than at lower. These difficulties are slowly being overcome.

Since stimulated emission amplifies in one direction, the sample chamber must reflect the beam back onto itself. Mirrors accomplish this, as shown in Figure 14.3. In laser applications, this chamber is referred to as a two-mirror optical **cavity** or **resonator.** In most optical work, it is known as a **Fabry-Perot interferometer** or

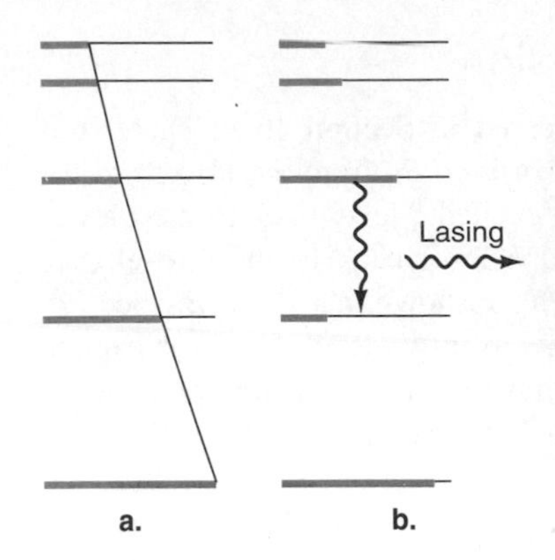

FIGURE 14.2
(a) Boltzmann distribution.
(b) Inverted population necessary for lasing.
The length of the horizontal lines give a sense of the relative populations.

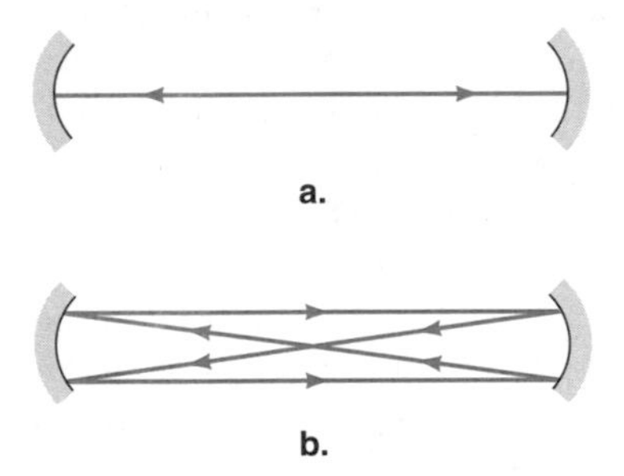

FIGURE 14.3
Use of concave mirrors to keep stimulated emissions within the cavity. (a) TEM_{00} and (b) TEM_{10}. Reproduced with permission from D. L. Andrews, *Lasers in Chemistry,* Springer, 1997, Figures 1.5a and b.

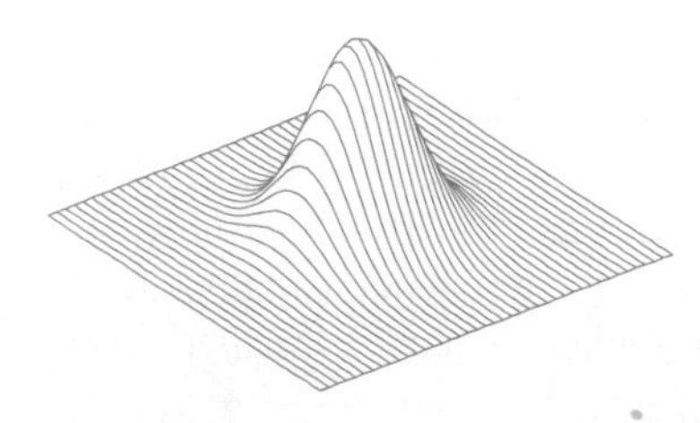

FIGURE 14.4
The TEM_{00} with a Gaussian distribution. Reproduced with permission from D. L. Andrews, *Lasers in Chemistry,* New York: Springer, 1997, Figure 1.11a.

as a **Fabry-Perot etalon** when the spacing l between the mirrors is firmly fixed. As the photons pass back and forth through the cavity and medium, further amplification occurs. After many passes between the mirrors, a portion of the beam reaches a partially reflecting mirror and escapes as a laser beam. The use of concave mirrors helps to keep the reflected light in the cavity and produces various modes called "transverse electromagnetic modes," or TEM_{mn}. Normally, the TEM_{00} known as the *uniphase mode* is retained because it has the desired beam profile, as shown in Figure 14.4, with the peak intensity focused about the center of the beam. Much effort is spent in developmental work to achieve this desired shape.

Figure 14.5 illustrates the process taking place inside the laser cavity. Surrounding the cavity is an external energy source, such as a flash lamp, which pumps the excited states. Although at first stimulated emission occurs in all directions, only those photons that are reflected back into the medium are amplified. The others are lost, as shown in Figure 14.5a. In time, the amplification is increased to the point that the laser beam passes through the partially reflective mirror, as shown in Figure 14.5b.

In Section 11.1, in which standing waves were described, it was shown that constructive interference occurs when an integral number of half-wavelengths fits between two reflecting walls (compare Figure 11.5 with Figure 14.5b). A similar conclusion is reached for a particle-in-a-box (see Section 11.6). This same phenomenon exists for the photons traveling back and forth between the two reflective mirrors separated by a distance l. When light of wavelength λ satisfies the equation

$$\frac{n\lambda}{2} = l \tag{14.1}$$

for a wave with integer n, the cavity is said to resonate and each value of n produces a wave called a mode, as depicted by the beam in Figure 14.5.

Properties of Laser Light

Laser light has a number of features that distinguish it from normal, chaotic white light. First, because stimulated emission occurs between fixed energy levels, laser light is monochromatic, although some broadening does occur. Moreover, by virtue of the standing resonant wave that is set up in the cavity, the emitted laser beam (see Figure 14.5b) has a bandwidth that is coherent both *spatially,* in that all the radiation is in phase across the beam, and *temporally,* which means that the radiation remains in phase as it moves away from the cavity. Bandwidth broadening does occur, however, with the major sources being Doppler broadening and lifetime broadening. These are treated in Section 14.2. If the laser beam were truly monochromatic, the coherence length would be infinite. Coherence length means the distance over which the laser beam stays in phase; that is, the distance over which the wavelengths that make up the beam have phase coherence. A typical laser beam is often seen to bounce around mirrors and through lenses in a room. This is not related to the coherence length. Rather, this is a result of the optics that stop the laser beam from diverging. Because of the broadening, the coherence length is shortened. Although coherence lengths of up to 100 m have been produced, a more typical value is 1 mm. This result may be compared with a coherence length of only a few hundred nanometers for ordinary light coming from an incandescent light bulb.

Laser light is also intense. Typically, on a warm day, the intensity of sunlight on the earth is of the order of 1000 watts per square metre (W m^{-2}). A moderately

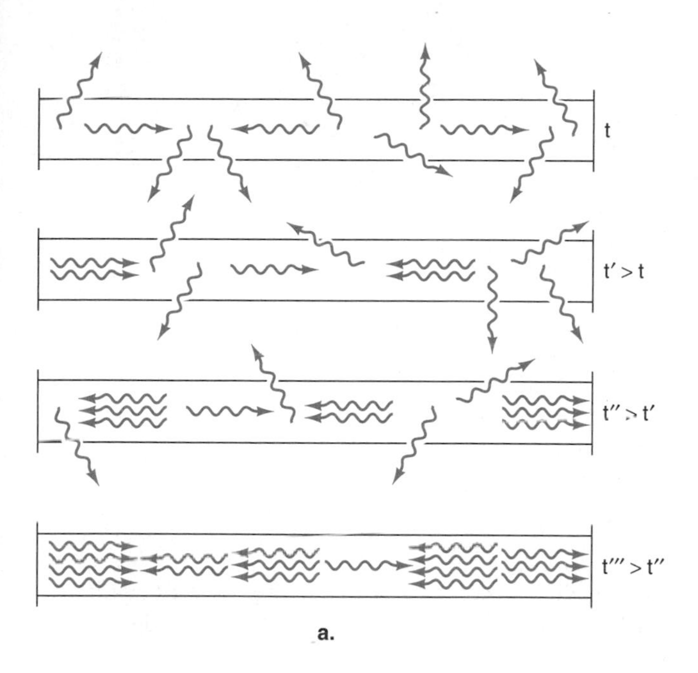

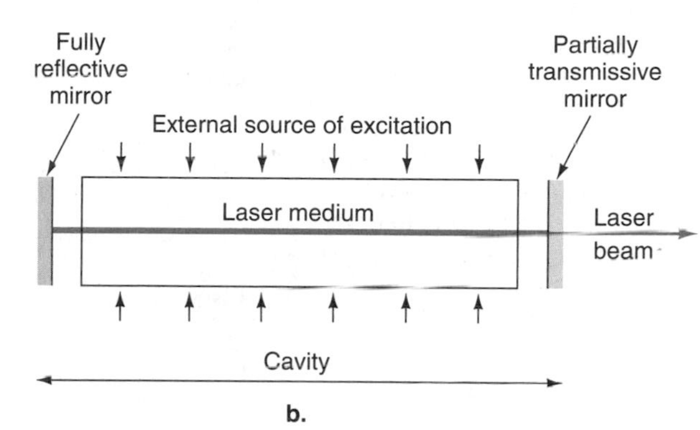

FIGURE 14.5
(a) Stimulated and spontaneous emission occurs in all directions, but amplification of the stimulated emission occurs only between the two reflecting mirrors. (b) The amplification process between two mirrors. The mirror on the right is partially reflecting so that the laser beam can escape. Reproduced with permission from D. L. Andrews, *Lasers in Chemistry,* New York: Springer, 1997, Figure 1.4.

powerful laser can easily be 10 000 times more intense. This is partly due to the fact that a laser beam can be focused onto a small region of space. The smallest region possible has a radius of approximately the wavelength of the laser, and intensities of up to 10^{13} W m^{-2} have been produced.

Three- and Four-Level Lasers

The first successful lasers were three-level systems, exemplified by the ruby laser. Although four-level lasers dominate modern applications in chemistry, it is instructive to review the way in which a three-level system works. Consider Figure 14.6. In order to have a laser emission between levels 2 and 1, it is necessary to maintain a population inversion between these two levels. If it is possible to pump level 3, which then quickly decays to level 2, then population inversion is accomplished. To do this, the transition from 3 to 2 must be fast and be accomplished not by emitting light, but by other fast relaxation mechanisms such as inelastic collisions with other

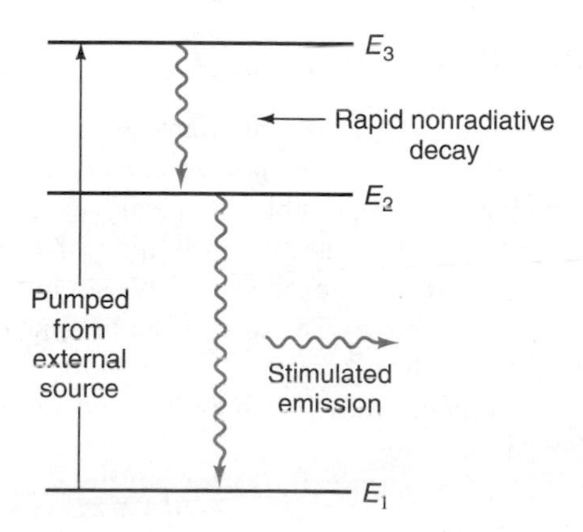

FIGURE 14.6
A three-level system capable of stimulated emission. A population inversion is maintained by pumping level 3, which decays to level 2.

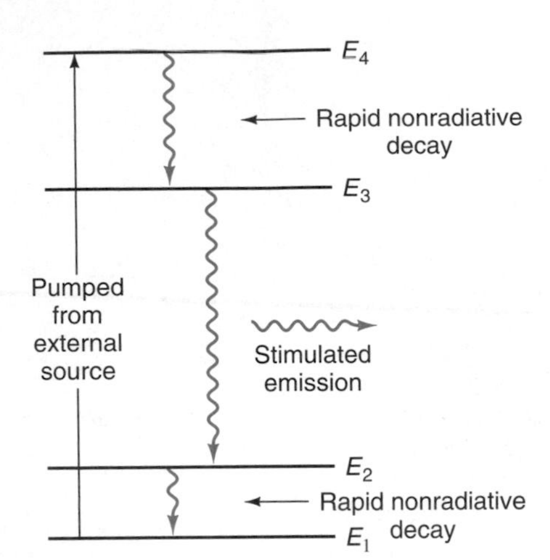

FIGURE 14.7
A four-level system capable of stimulated emission. Improvement over the three-level system is gained by having a rapid nonradiative depletion of level 2 so that a population inversion between levels 3 and 2 can be better maintained.

molecules in the active medium. Note, however, that as the laser process continues, the ground state 1 is replenished in a three-level system, and this works against the maintenance of a population inversion.

A four-level laser system helps to eliminate the problem brought about by replenishing the lower level of the lasing pair. In Figure 14.7, the lasing action results from transitions between levels 3 and 2. The upper level is pumped in the same way as the three-level system, but now, by virtue of the fast collisional decay from the second level to the ground state, level 2 is continually being depleted while level 3 is continually being pumped. This maintains a population inversion more easily than in the three-level case.

14.2 Spectral Line Widths

Transitions between different sets of well-defined energy levels result in spectral lines of different widths. There are several reasons why the lines spread over a range of frequencies. In molecular spectra in the gaseous state, superposition of a number of energy levels that are close together may cause broadening, but there are two more fundamental reasons.

Lifetime Broadening

One mechanism which is important even at low temperatures is a quantum-mechanical effect called *lifetime broadening*. Because of the Heisenberg uncertainty relation it is impossible to specify exactly the energies of the levels in a transition. Only if the state has an infinite lifetime can its energy be specified exactly; in reality, since no excited state has an infinite lifetime, no excited state has a precisely defined energy. The shorter the lifetime of an excited state, the broader the expected spectral lines.

The uncertainty principle may be used to estimate the extent of broadening. From Eq. 11.64,

$$\Delta E \Delta \tau \geqslant \hbar/2 \tag{14.2}$$

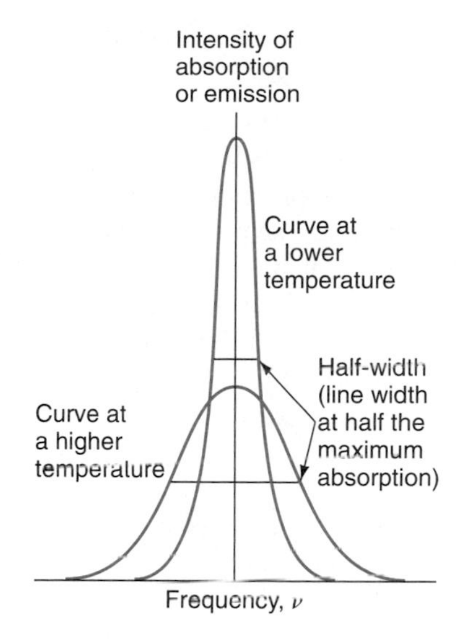

FIGURE 14.8
Line widths at two different temperatures as determined by the Gaussian distribution.

The energy spread ΔE is therefore inversely proportional to the lifetime $\Delta\tau$, and it follows that the corresponding broadening of the line is

$$\Delta\tilde{\nu}/\text{cm}^{-1} \approx \frac{2.7 \times 10^{-12}}{\Delta\tau/\text{s}} \tag{14.3}$$

The radiative lifetime of an electronically excited species may be as long as 10^{-7} s, which leads to a line width of $\approx 2.7 \times 10^{-5}$ cm^{-1}. When there can be molecular dissociation or internal conversion, however, the lifetime may be 10^{-11} s or less; this value gives a broadening of about 0.27 cm^{-1}, which can easily be measured with a high-resolution instrument.

Doppler Broadening

The Doppler effect with sound waves was first recognized by the Austrian physicist Christian Johann Doppler (1803–1853), who developed the theory of it in 1842.

Another mechanism that leads to a broadening of a spectral laser line is the Doppler effect for light. The molecules in the active medium are traveling in all directions, but the light is being amplified and emitted in only one. From special relativity theory, an object that approaches a static observer with a speed u and at the same time emits radiation of frequency ν appears to be emitting radiation at a frequency $\nu' = 1/[1 - (u/c)]\ \nu$. If the object recedes from the observer, its radiation appears to have a frequency $\nu' = 1/[1 + (u/c)]\ \nu$. Since molecules in a gas travel at high speeds in all directions, there is a range of Doppler shifts and the line that is recorded is therefore broadened. The magnitude of this broadening can be predicted from the distribution of velocities as given by the Maxwell distribution (Eq. 1.91). We thus find that the predicted shape of this distribution in a single direction is a bell-shaped Gaussian curve, as shown in Figure 14.8. The distribution is temperature dependent, broadening as the temperature rises. The width $\Delta\nu$ at half-height of any line may be found from

$$\Delta\nu = 2\left(\frac{\nu}{c}\right)(2k_B T/m)^{1/2} \tag{14.4}$$

where m is the mass of the species involved in the transition. To obtain the maximum resolution of spectral lines, the spectra should be taken at the lowest possible temperature. High temperatures (such as the temperature of the sun) may be determined by measuring the broadening of a particular spectral line (see Problem 14.1).

14.3 Types of Lasers

There are a number of different types of laser, which can be classified as follows:

1. **Optically pumped solid-state lasers:** The three-level ruby laser is of this type. The four-level neodymium YAG laser is the most commonly used.
2. **Molecular gas lasers:** The carbon dioxide and nitrogen lasers are examples of these. Chemical lasers also fit into the group and are distinguished by the fact that the energy supplied to excite the upper states is of chemical origin. **Excimer** lasers make use of the excited state of a two-atom system.
3. **Atomic and ionic gas lasers:** These include the helium-neon laser, the argon-ion laser, and the krypton-ion laser.
4. **Other types of lasers:** These include semiconductor diode lasers, dye lasers, and free-electron lasers.

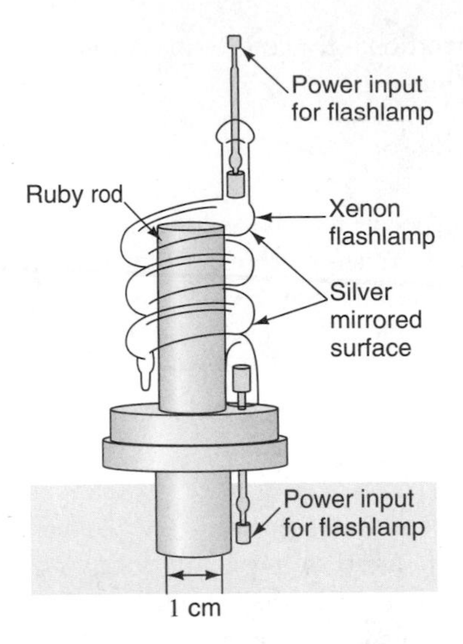

FIGURE 14.9
Drawing of the first synthetic-ruby laser, developed by H. Maiman in 1960. The laser produced red light in pulses and could not operate continuously. (Figure "Picture of the first ruby laser showing the flash tube wrapped around the ruby rod." Courtesy of Hughes Aircraft Company. Used with permission.)

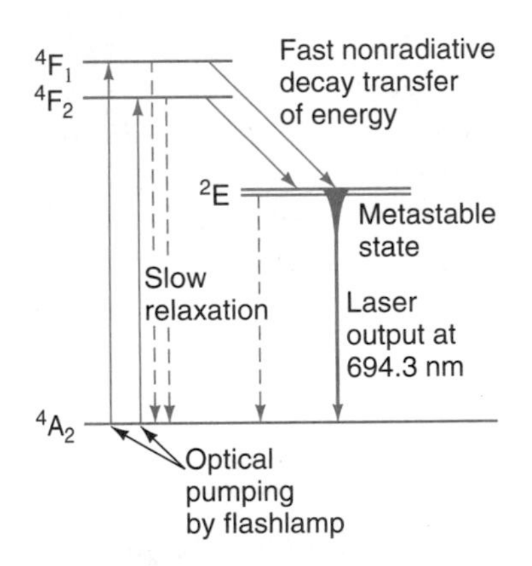

FIGURE 14.10
Energy levels in a ruby laser showing transitions resulting in occupancy of the metastable state ^{2}E with subsequent laser output in a typical three-level laser.

In the following, we summarize the salient features of some of these types of lasers.

Optically pumped solid-state lasers: A prime example of solid-state lasers is the **ruby laser,** in which the medium is a rod of ruby; this is sapphire, Al_2O_3, in which one out of every 10^2 to 10^3 Al^{3+} ions has been replaced by a Cr^{3+} ion; this gives the resulting crystal its characteristic red color. The population inversion is usually brought about by means of xenon flashlamps that are placed around the ruby in a highly reflective housing in order to focus the lamp emission on the laser rod. A line drawing of the first ruby laser is shown in Figure 14.9, and its energy levels are shown in Figure 14.10. The ruby laser is an example of a *three-level laser* (Figure 14.6). The ground state, $E_1(={}^4A_2)$, and the excited state, $E_2(={}^2E)$, constitute two levels. There is a *third set* of levels, two of which—the *F* levels—are shown in Figure 14.10.

Some lasers operate continuously, but a common technique is to pulse the exciting light for short periods of time (e.g., a few nanoseconds), and the power developed in a pulsed laser may be as high as a gigawatt (GW = 10^9 W). The ruby laser is generally pulsed using 2 J pulses for about 10 ns. This corresponds to a power output of about 0.2 GW and produces light of 694.3-nm wavelength, which lies toward the red end of the visible spectrum. Of particular importance is the fact that the light is now highly monochromatic. Also, if the laser medium (e.g., the ruby) has a suitable geometry, the beam divergence can be small, perhaps of the order of a milliradian, and is coherent.

Another solid-state laser is the **neodymium laser.** It normally is formed using Nd^{3+} ions in low concentration in an yttrium aluminum garnet. It is then commonly known as a *YAG laser.* This laser operates in the infrared most efficiently at 1064 nm. The energy scheme for this *four-level laser* is shown in Figure 14.11. In this scheme a more efficient transfer of energy exists in which the higher excited metastable level emits to a lower excited metastable state that is initially empty or only slightly populated. Often *frequency doubling* is used (see page 726), the beam being converted into radiation with double its initial frequency. The wavelength is then 532 nm, in the green region of the visible spectrum.

Molecular gas lasers: Several *gas lasers* are commonly used and can generate high power, because the generated heat produced in the excitation process is easily dissipated by the rapid passage of gas through the cavity. In the **helium-neon laser,** helium atoms are excited by an electric discharge to the metastable $1s^12s^1$ configuration, as shown in Figure 14.12. Transitions by such discharges are not restricted by selection rules. The energy is transferred through collisions to neon, which has excited states of appropriate energy. In effect, upper levels are heavily populated by the transfer, whereas the lower levels have the lower normal number of electrons as determined by the Boltzmann distribution. There is thus a population inversion, and lasing action occurs at 632.8 nm and at several other wavelengths.

The **carbon dioxide laser** also uses an energy exchange. In this device nitrogen is vibrationally excited by an electric discharge and again there are no symmetry restrictions. This energy is transferred through electronic exchanges and ionic collisions to the ν_3 antisymmetric stretching vibrational levels of CO_2, as shown in Figure 14.13. The laser action takes place between the lowest excited level of the ν_3 vibration and the lowest excited level of the sym-

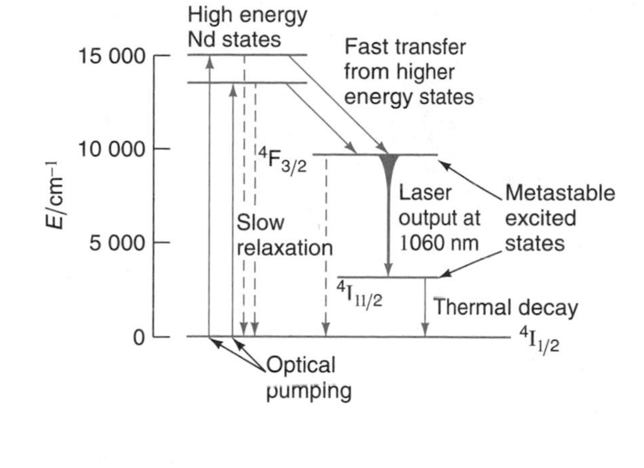

FIGURE 14.11
Energy levels in a neodymium laser showing transitions resulting in occupancy of the metastable state $^4F_{3/2}$. This state has a lifetime of 5×10^{-4} s before dropping to the $^4I_{11/2}$ state, which is populated only slightly by thermal electrons. This is an example of a four-level laser.

metric stretch ν_1. The strongest emission occurs at 10.59 μm, with a second emission possible at 9.6 μm.

The first chemical laser was constructed in 1965 by the American chemists J. V. V. Kasper and George Claude Pimentel (1922–1989). It was based on the reaction $H + Cl_2 \rightarrow HCl^* + H$, in which the HCl is formed with a population inversion of vibrational states. The feasibility of using this reaction in a laser was suggested in 1961 by John Polanyi on the basis of his molecular dynamics studies (Section 9.11).

Atomic and ionic gas lasers: Both **argon-ion** and **krypton-ion lasers** operate on the same principle. An electric discharge is passed through the working gas held at a pressure of about 1 Torr. For argon, both Ar^+ and Ar^{2+} ions are formed in excited states. Laser action drops the electrons to a lower state, where the ions emit ultraviolet radiation at 72 nm and are then neutralized. Because of the large number of lower states, there are many wavelengths in the spectra, but the emission from Ar^+ primarily gives light of 488 nm (in the blue region) and 514 nm (in the green region). The krypton-ion laser gives a wider range of wavelengths, the most intense being at 647 nm (in the red region).

Another type of laser that provides extremely high power in the beam is the **chemical laser,** in which a chemical reaction generates molecules already in an inverted population state. Such lasers have been envisioned by the military for several applications. In the "on again" Strategic Defense Initiative, better known as SDI or "Star Wars," photolysis of F_2 forms F atoms which attack H_2 molecules, with the production of HF and H. The H in turn attacks F_2 to form vibrationally excited or "hot" HF molecules. These occur with inverted populations, and through laser action directly convert chemical energy into coherent radiation in a range between 2.6 μm and 3.0 μm.

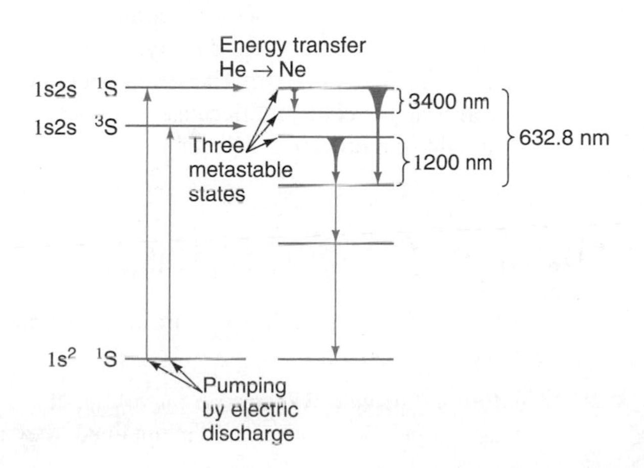

FIGURE 14.12
Energy levels in a helium-neon laser showing transitions resulting in population inversion in three metastable states.

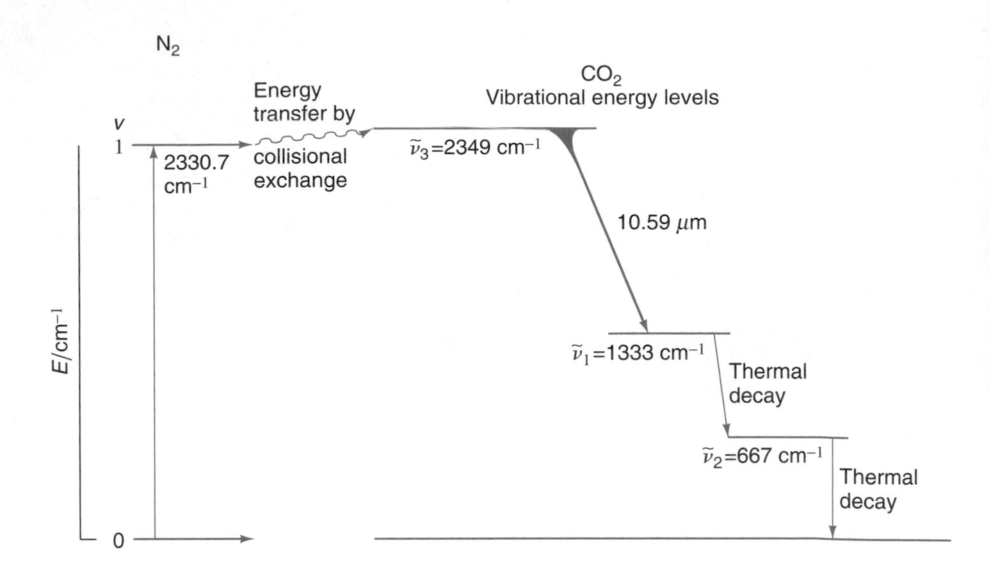

FIGURE 14.13
Energy levels in the CO_2 laser showing transitions leading to energy transfer from N_2 to CO_2 with the resultant laser action of CO_2.

Excimer

Another technique to produce a population inversion in molecular gas lasers occurs in *excimer lasers.* An **excimer** is a combination of two atoms in an excited state. The shedding of the excitation energy is accompanied by the dissociation of the excimer. The population inversion is easy to obtain since there are no populated ground states. In Figure 14.14, for instance, XeCl* or KrF* can be formed by passing an electric discharge through the components Xe and Cl_2 or Kr and F_2 in a neon carrier. The excited Cl or F atom becomes attached to the noble gas atom and remains in an excited state for about 10 ns. This population-inverted state can only give rise to dissociation. Laser action for XeCl* occurs at 308 nm, while the KrF excimer produces radiation at 249 nm.

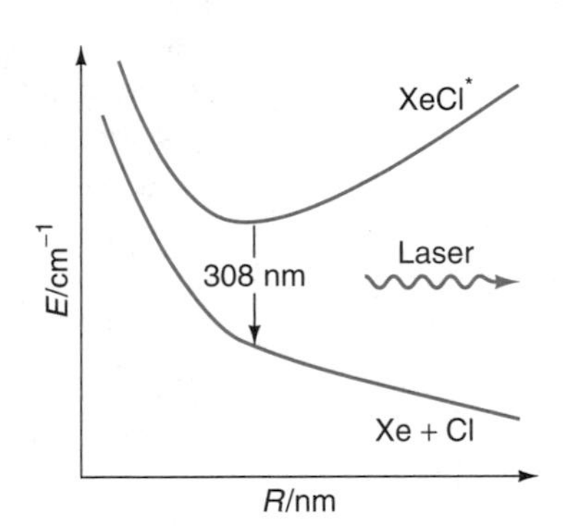

FIGURE 14.14
Diagrammatic representation of an excimer laser. In this case an energetically excited molecule (XeCl*) radiatively decomposes, emitting laser light at 308 nm.

Other types of lasers: Although most lasers have a specific wavelength, tuneable chemical *dye lasers* have been introduced that within limits can be adjusted to a desired wavelength of emission. These are called tuneable lasers.

A recent development is the *gallium-arsenide p-n junction laser.* This laser operates with a typical input of 135 W producing about 9 W of light power at 840 nm. Radiation is generated as approximately 20 A of current is passed through the semiconductor junction, which consists of a *p*-type layer separated by a 2.0-μm spacer from an *n*-type layer, the whole thickness being about 0.1 nm. (See Section 16.6.) The recombination radiation is emitted parallel to the spacer and is coherent. Because there is a direct conversion of electrical energy to light, the process is efficient.

14.4 Laser Techniques for Chemistry

Various special laser techniques are used for chemical applications, and we now consider some of them.

Continuous wave (CW) lasers: Such lasers have already been discussed. As long as a population inversion is maintained, they produce a constant output.

Pulsed emission lasers: For many applications, optical methods are used to change the CW output to a pulsed output. Pulsing has a variety of advantages, one being that it allows high-power laser irradiation to occur for short periods, such as within the time scale of some chemical processes. For chemical applications, pulses from the nanosecond to the femtosecond range are usually appropriate.

Conversion methods can change CW lasers to pulsed lasers, but some lasers are inherently pulsed lasers. For these lasers it is not possible to sustain a continuous population inversion, and as a result these lasers are charged and discharged in pulses.

Tuneable lasers: Since laser action requires the existence of two levels of well-defined energy separation, lasers cannot usually be continuously tuned. Such a feature would be desirable for use in probing specific spectroscopic properties of molecules. When, however, there are many appropriate pairs of states in an active medium, it is possible to tune the laser to excite specific pairs of energy levels. Some dyes offer this option.

Nonlinear optics: A variety of optical methods have also been introduced to permit changing the laser output to a more useful frequency. One technique takes advantage of the fact that the electric-field component of a laser is so large that the polarization produced in a sample crystal is no longer linear. Nonequilibrium processes sometimes lead to a linear response. Ohm's law is an example; the applied electromotive force E_{mf} causes an electrical current I to flow, controlled by the resistance, R, of the material:

$$I = E_{mf}/R \tag{14.5}$$

Many other examples of linear response exist: A temperature difference causes a heat flow; a concentration difference causes a mass flow. Normally when the force F is not large, the flow (flux) ϕ displays a linear response to the force,

$$\phi = LF \tag{14.6}$$

where L is a coefficient that determines the magnitude of the flux. The electrical polarization, P, also normally displays a linear relationship with the electric field component E of the applied electromagnetic radiation.

$$P = \beta E \tag{14.7}$$

where β is the optical polarizability. Since the electric field component of a laser can be so great that nonlinear processes occur, this linear equation must be modified to

$$P = \beta_1 E + \beta_2 E^2 + \beta_3 E^3 + \cdots \tag{14.8}$$

If the light is sinusoidal, Eq. 13.180 applies so that

$$E = E_{\text{o}} \cos 2\pi\nu_{\text{o}}t \tag{14.9}$$

where E_{o} is the amplitude of the radiation field in the crystal. The second term in Eq. 14.8 adds a new term to P:

$$P = \beta_2 E_{\text{o}}^2 \cos^2 2\pi\nu_{\text{o}}t \tag{14.10}$$

or

$$P = 1/2\beta_2 E_{\text{o}}^2 (1 + \cos 4\pi\nu_{\text{o}}t) \tag{14.11}$$

Analysis of these nonlinear terms reveals that they are harmonics of the original laser frequency. For example, the first such overtone results in a frequency doubling of the input frequency. This effect is called **frequency doubling** or **optical second-harmonic generation.** In some cases it is possible to take advantage of harmonic generation to obtain a desired frequency output. Spontaneous emission increases as the third power of the frequency (see Eq. 13.29), and this makes it more difficult to produce high-frequency lasers. Use of the optically generated harmonics can overcome this difficulty.

In controlling and producing pulses, a variety of experimental techniques have been developed to fulfill differing criteria. Some of the more important of these are as follows.

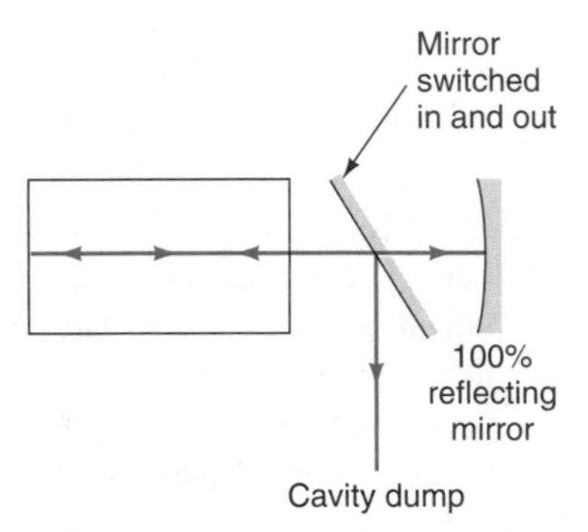

FIGURE 14.15
Cavity dumping produces pulsed lasers. Stimulated emission is allowed to increase between two 100% reflecting mirrors. A third mirror is switched in and deflects the laser beam from the cavity, thereby dumping the energy of the laser beam.

Pulses produced by cavity dumping: See Figure 14.15. This technique can be understood by considering the laser action sustained by two mirrors that are 100% reflective. Suppose that a third mirror is introduced at a different angle to divert the laser light to the target. Since the third mirror replaces one of the end mirrors, the reflective resonance process is terminated and the cavity is emptied, or "dumped." Once the deflecting mirror is removed, the laser action is reestablished. Such pulses are in the nanosecond range.

Pulses produced by Q-switching: Laser cavities allow for the buildup of radiation energy as the stimulated amplification continues. The energy in the cavity increases, but there are energy losses. The ability of a cavity to store energy is usually expressed in terms of a **Q-factor.**

If, during laser action, one of the mirrors is temporarily rendered nonreflective, the Q-factor will decrease. At the same time, the pumping process will continue, allowing an abnormal increase in the population difference between the two levels. If the mirror is then turned on, a sudden high-intensity pulse will result. This technique is known as **Q-switching** and is implemented by introducing a shutter action within the cavity, thereby switching a reflecting mirror on and off. Again, these pulses are in the nanosecond range.

Pulses produced by mode locking: Another technique that has been used to produce optical pulses as short as 6 fs (6×10^{-15} s) is known as mode locking. The wavelength of such a pulse is about 2 μm.

From Eq. 14.1 we see that a standing wave, called a *mode,* is formed for each value of n. There may typically be as many as 10,000 modes resonating in a cavity. The frequency separation between two modes is found, by using $\nu = c/\lambda$, to be

$$\Delta\nu = c/2l \tag{14.12}$$

where l is the distance between the mirrors. To see what can happen with modes of different phase displacement (see Figure 11.3), imagine three randomly distributed (out-of-phase) sinusoidal modes with frequencies ν, 2ν, and 3ν. Addition of these gives the total field amplitude; its square gives the intensity. Both appear as a randomly distributed output, with no repetitive peaks because constructive and destructive interferences are random. If, however, the phases were locked together, the amplitude and intensity would appear as a sharp, repetitive peak. This means that the time for the locked modes to make

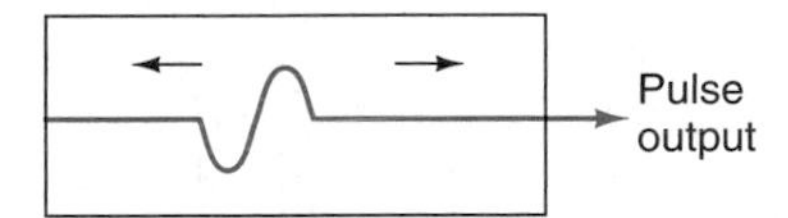

FIGURE 14.16
Mode-locking technique to produce pulsed lasers. By modulating the cavity at the appropriate frequency, one mode survives, which travels back and forth in the cavity and generates a pulsed laser.

one round trip in the cavity would be the same, and rather than a standing wave being set up, a single pulse would travel back and forth (Figure 14.16) in a resonance time of

$$t_r = 2l/c \tag{14.13}$$

Bringing all the phases into coincidence to generate the single pulse is called *modulating the cavity.* This means that the length of the cavity is changed with a frequency $\Delta\nu$. This can be accomplished by an acoustic generator, which is an example of an active mode-locking technique. Mode-locking techniques that do not involve a mechanical modulator are said to be *passive.* Such techniques can involve the use of dyes, which work by absorbing the lower frequency modes, leaving the higher frequency modes to continue the amplification process. In both the active and passive techniques, short pulses of laser light are generated in short intervals. Picosecond to femtosecond pulses are possible, separated by intervals of the order of nanoseconds.

The Use of Lasers in Chemical Spectroscopy

The introduction of lasers into spectroscopy has transformed the field, causing every branch to undergo fundamental changes. The various motions of molecules, from low-energy rotations to high-energy electronic transitions, can be probed by the laser technique. Perhaps the most important area to benefit from lasers is Raman spectroscopy, which in its traditional form suffers from low-intensity scattering. Today Raman spectroscopy usually means laser Raman spectroscopy.

The main properties of lasers just discussed make lasers so successful in spectroscopy. High-intensity pulses of monochromatic collimated beams of coherent light focused onto samples make the difference. The scope of laser spectroscopy is enormous, and we can only begin to cover this area. If there is one drawback to lasers compared with conventional spectroscopic methods, especially in the IR, it is the lack of tuneability. Various techniques have been devised to help overcome this, such as the use of magnetic fields **(laser magnetic resonance)** and electric fields **(laser Stark spectroscopy),** but it is still difficult to lift degeneracies and bring the molecules into resonance with the laser.

The use of lasers in chemistry can be classified into two general areas. The first uses lasers in the traditional areas of spectroscopy. These study the **structure of matter** using the range of rotational to electronic states and provide detailed information on structure, symmetry, and internal motion. Traditional spectroscopy is broadened to include the study of extremely short-lived transition-state species in reactions and the study of fragments in mass spectrometry. This broad area is simply called **laser spectroscopy.**

The second area of application of lasers in chemistry is the use of photons to initiate chemical reactions. Photons can be delivered to molecules that subsequently undergo chemical reactions. Examples of such applications are found in flash photolysis, laser isotope separation, and vibrational photochemistry. This area is generally called **laser photochemistry.**

We will summarize some of the most important applications of lasers to spectroscopy and photochemistry.

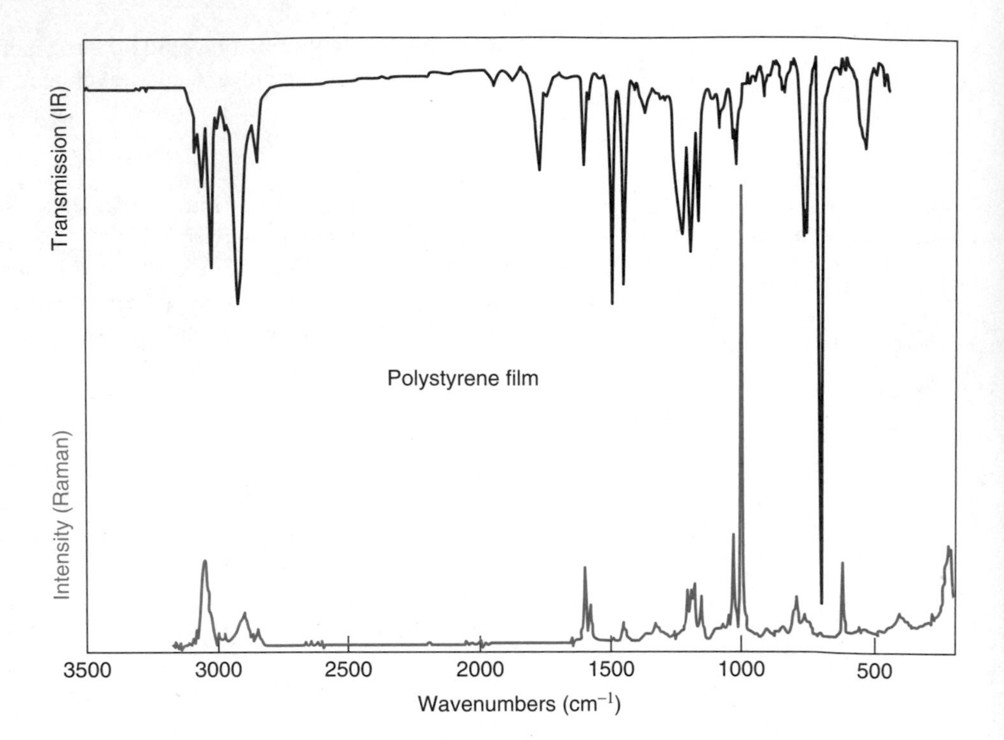

FIGURE 14.17
Comparison between infrared spectroscopy and Raman spectroscopy. Note that different spectral lines appear as well as those in common. The intensities of the common lines differ. Compliments of Inphotonics Inc. (www.inphotonics.com).

Laser Raman Spectroscopy

As seen in Section 13.5, Raman and infrared spectra obey different selection rules so that the position and intensity of their spectral lines are different and complementary. For instance, symmetric stretches produce strong Raman lines that are missing in IR spectroscopy. The asymmetric stretches can appear in both Raman and IR, but the intensities differ. Figure 14.17 shows a comparison between the IR and Raman spectrum of polystyrene film.

Traditional Raman spectroscopy is difficult to carry out because of the low intensity of the *inelastically* scattered light. Raman spectroscopy is, however, not dependent upon the fact that the light source is in resonance with molecular transitions. It is no wonder, then, that Raman spectroscopy underwent a major change with the introduction of lasers and, indeed, was the first spectroscopy to benefit. Besides the greater intensity, which resolved hitherto unseen details, the narrow line width of the laser increased the resolution of the observations. Figure 14.18 displays the Raman spectrum of air, which shows the symmetric stretches of the diatomic molecules of nitrogen and oxygen in air. Such spectra allow for the accurate determination of bond lengths. Raman spectra are often complicated, making analysis difficult. But often, with the use of isotope labeling and by taking differences between spectra, these obstacles can be overcome and thus allow for the determination of hydrogen bonding strengths, bond orders, bond lengths, and geometries.

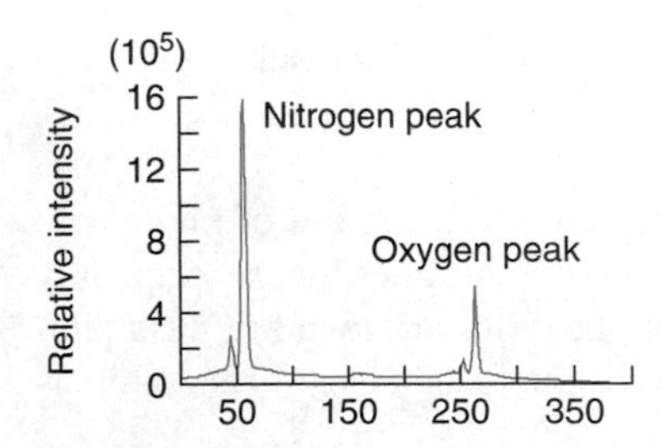

FIGURE 14.18
Laser Raman spectrum of air.

Hyper-Raman Spectroscopy

New types of experiments in Raman spectroscopy became possible by using lasers. Q-switching enables nonlinear effects (see Eq. 14.8) to be investigated for the first time; the technique is called **hyper-Raman spectroscopy.** From Eq. 14.11, note that when the β^2 terms become important, there is a scattering of light at $2\nu_o$ in addition to the usual Raleigh scattering at ν_o. The lines around $2\nu_o$ are caused by what is called hyper-Raleigh scattering. Stokes and anti-Stokes lines again appear. The hyper-Raleigh scattering is well removed from the normal Raleigh scattering, which is always present. In contrast, the normal Raleigh scattering occurs only if the molecule does not have a center of inversion. Analysis of hyper-Raman effects involves more extensive symmetry conditions than normal Raman spectroscopy, and analysis of these spectra reveals structures and lines that are normally undetected.

Resonance Raman Spectroscopy

Traditional Raman spectroscopy depends upon the scattering of light from the source. If the laser is tuned so that the frequency corresponds to a spectroscopic transition, the nature of the spectrum changes and reveals difference features. This is called **resonance Raman spectroscopy.** The intensity of the spectra increases, particularly from a few strongly vibrational modes, giving new structural information as a result of different selection rules. One particular application is found in the study of **chromophores.** These sites are regions within molecules that have localized electronic absorption. Chromophores are particularly common in biological systems. Moreover, use of Q-switching makes it possible to study the kinetics of biological processes.

A wealth of information is obtained from resonance Raman spectroscopy of chromophores. Vibrational frequencies can be determined; metal coordination geometry and the ligand environment can be probed; metal-ligand bond strengths can be found. In addition, there is the ability to assign electronic states. This is possible because the spectra are sensitive to small structural changes at the metal sites. Moreover, kinetics can be monitored over a large temperature range on small amounts of sample.

Figure 14.19 shows a time-resolved resonance Raman spectrum. Techniques for inducing and following chemical reactions via lasers can generally be called **pump-probe techniques.** One or more lasers are used to initiate the **photochemical process** and another is used to monitor the reaction's progress. Figure 14.19 from

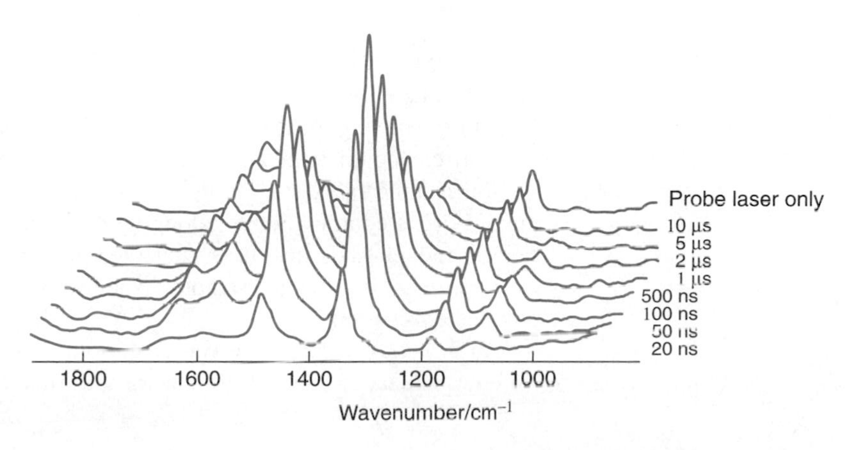

FIGURE 14.19
Time-dependent spectrum observed on photolysis of an anthraquinone derivative in the presence of $NaNO_2$ (0.1 mol dm^{-3}); pump laser 337 nm; probe laser 480 nm. The spectrum corresponds to that of the radical anion. Reproduced from D. Phillips, J. N. Moore, and R. E. Hester, "Time-resolved resonance raman spectroscopy applied to anthraquinone photochemistry," *J. Chem. Soc. Faraday Trans. 2, 82*, 2093(1986), Figure 4. Reproduced by permission of the Royal Society of Chemistry.

the laboratory of Moore and Hester at the University of York, UK, shows the formation and decay of the radical anion on anthraquinone-2,6-disulphonate. The spectra are a series of spectral snapshots of reacting intermediaries. Experiments of this type allow for the investigation of species with lifetimes as short as femtoseconds.

Such photochemical techniques allow the course of the reaction to be followed, and makes it possible for chemical reactions to be induced selectively. An important area for this technique is in **laser-induced isotope separation,** which has considerable economic value to the nuclear energy industry. This technique makes use of the differences in energy levels between isotopes. The laser is tuned to the frequency of the desired isotope and causes it to be ionized, thereby allowing it to be removed by an applied electric field.

Coherent Anti-Stokes Raman Scattering (CARS)

The simultaneous use of two lasers in Raman spectroscopy has led to a different effect, which results from the wave mixing of the laser light. One technique, called **coherent anti-Stokes Raman scattering (CARS)** has found useful industrial applications. Wave mixing of one laser with frequency ν_o with another of frequency ν_1 results in the mixing of components producing one of frequency $2\nu_o - \nu_1$. If ν_1 is chosen to be any one of the Stokes lines, $\nu_1 = \nu_o - \Delta\nu_1$, then the other coherent component of the mixed laser light is equal to the anti-Stokes frequency.

$$2\nu_o - \nu_1 = 2\nu_o - (\nu_o - \Delta\nu_1) = \nu_o + \Delta\nu_1 \tag{14.14}$$

Stimulated emission results, and the normally low-intensity anti-Stokes lines become intense and remain coherent. Different selection rules allow modes that are inactive in Raman and IR spectroscopy to be observed using the CARS technique. Since the region where two laser beams cross is small, as is the sample size in the gas phase, powerful pulsed laser beams are required to obtain spectra. Moreover, there is almost no Doppler broadening and the emission of the coherent light is easily picked out from a fluorescent background. When it is possible to cross two laser beams, CARS is used to study the process of reactions, such as those occurring in flames. This not only allows analysis of the components present but also gives a method of accurately determining the temperature of the process.

Laser-Induced Fluorescence (LIF)

The advantage LIF has over normal fluorescence spectroscopy is the ability to use a laser to selectively excite certain electronic levels so that the specific fluorescence pathway can be followed and the specifically excited fluorescent emission obtained.

Since tuneability is desirable, dye lasers are used, although the non-tuneable argon laser is commonly used along with frequency doubling. LIF is used to study flames and plasmas and uses characteristic fluorescence frequencies to identify trace elements in atomized samples. The technique also allows for accurate temperature determination.

In the study of the fluorescence of molecules, LIF has the ability to monitor the different relaxation rates of the spectra. Information about chemically unstable sys-

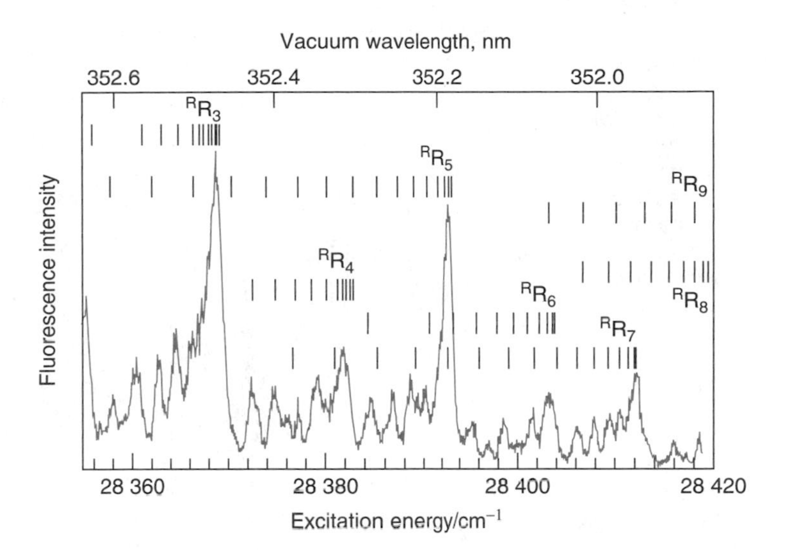

FIGURE 14.20 Partially rotationally resolved fluorescence of the 4_0^1 vibronic band of the $\hat{A}^1A_2 - \hat{X}^1A_1\ 4_0^1$ transition in formaldehyde present in a methane flame obtained at a height $H = 1$ mm above the burner surface using dye laser excitation. Sets of series of rotational lines are assigned known rotational and vibrational constants. Reproduced from Joel E. Harrington and Kermit C. Smyth, "Laser-induced fluorescence measurements of formaldehyde in a methane/air diffusion flame," *Chem. Phys. Lett. 202,* 196(1993), Figure 4, with permission from Elsevier Science.

tems, bond dissociation, transition-state complexes, and general chemical reaction kinetics can be obtained. The use of pulsed lasers in the ps and fs range permits these rapid chemical processes to be monitored.

Figure 14.20 shows the LIF spectrum of formaldehyde. Rotation-vibration structure is observed from formaldehyde in a methane flame. Note the well-resolved rotational series that can be used to assign rotational and vibrational constants.

Zero Kinetic Energy Raman Spectroscopy (ZEKE)

Normally, a number of vibrational levels are populated in the ground electronic state of a molecule. If only the ground state vibrational level were populated, the resulting spectra would improve in resolution. Section 2.7 discussed the Joule-Thomson effect, in which a gas cools as it expands. To accomplish the cooling, supersonic jets have been made to produce molecules in their lowest vibrational energy level. Another benefit of cooling is that the gas molecules move much more slowly, with the result that Doppler broadening is reduced. The result is a dramatic reduction in spectral broadening.

Figure 14.21a shows a highly monochromatic scan of Ag_2 resulting in all possible states as a mixture of direct ions and autoionization peaks. Figure 14.21b shows how ZEKE simplifies the spectrum, eliminating the autoionization and the direct ionization peaks, leaving the pure vibrational progression in the molecular cation.

Application of Lasers to Other Branches of Spectroscopy

Lasers play an important role in conventional electronic spectroscopy in that they provide a source of efficient, fixed wavelength excitation. This is advantageous for use in luminescence (spontaneous emission) spectroscopy. In this area, the most commonly used laser sources are in the visible region (Ar ion, He-Ne, and Kr ion

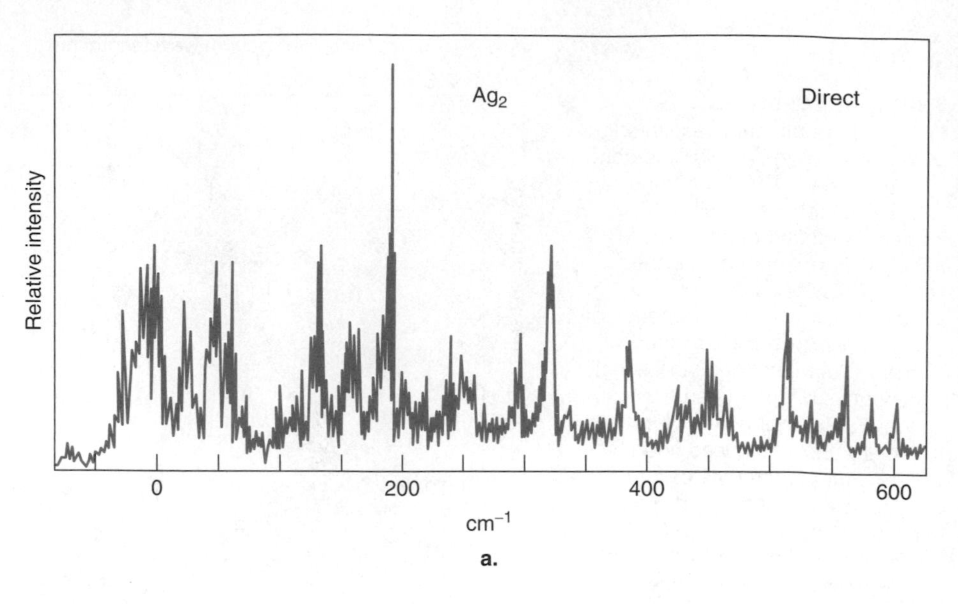

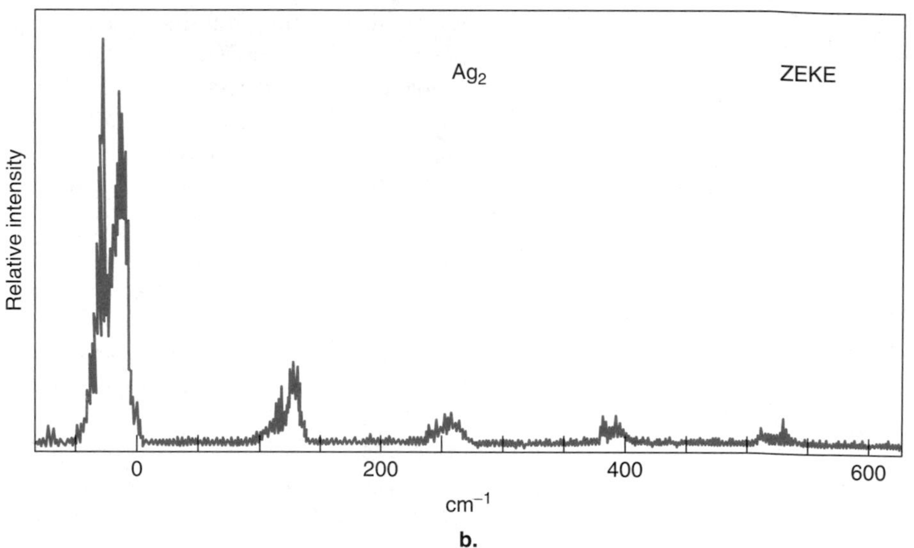

FIGURE 14.21
The effects of ZEKE Raman spectroscopy on Ag_2. (a) Non-ZEKE Raman spectroscopy showing all possible states as a mixture of direct ions and autoionization peaks. (b) The effect of additional selection by the ZEKE process of defining zero kinetic energy. The autoionizing and directly ionizing peaks are eliminated, leaving only the pure vibrational progression in the molecular cation. Reproduced from Chahan Yeretian, R. H. Hermann, H. Ungar, H. L Selzle, E. W. Schlag, and S. H. Lin, "Breakdown of the Born-Oppenheimer approximation in ZEKE states of Ag_2," *Chem. Phys. Lett. 239,* 61(1995).

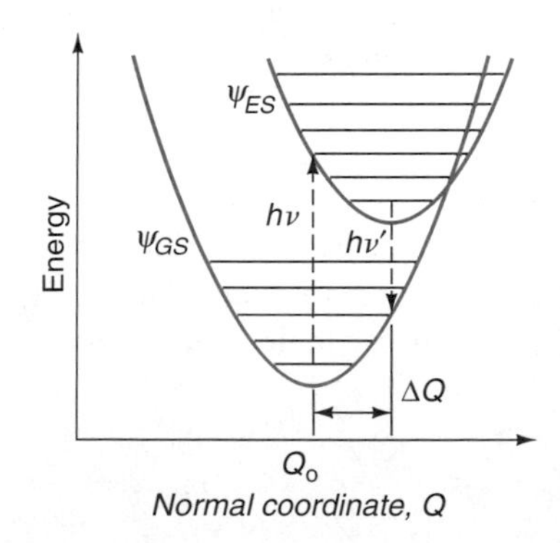

FIGURE 14.22
Schematic showing luminescence transitions generated by laser excitation. Compliments of John Grey, graduate student, Chemistry Department, McGill University.

lasers) as well as diode lasers that are in the near infrared (NIR) and excimer (XeCl) lasers in the UV. Many case studies in luminescence spectroscopy require that the sample be cooled at least to the boiling point of liquid nitrogen and, quite often, to that of liquid helium. This is done because most luminescence spectra at room temperature give broad and unresolved bands. Since the sample is frozen, the quantum yields tend to increase significantly, and more vibrational fine structure appears.

Configuration coordinate diagrams provide a useful model for what is observed spectroscopically and help us to understand the spectral features, including intensity distributions and band profiles. Figure 14.22 shows a simple schematic diagram of a luminescence transition brought about by laser excitation. The spectrum can be understood in terms of how the excited molecular configuration is different from that of the ground state.

Some of the most interesting systems available for study by luminescence spectroscopy are the transition-metal complexes. In these systems, spin-orbit coupling and Jahn-Teller effects can often be observed and studied quantitatively. Transition-metal complexes usually exhibit an electric-dipole forbidden d-d transition in the visible region, giving rise to their broad range of colors. Their spectra also reveal a wealth of information about their electronic structures and the changes that they undergo when electronic transitions occur.

14.5 Magnetic Spectroscopy

So far we have said only a little about the magnetic properties of molecules. We have explained paramagnetism in terms of MO theory, as arising from the presence of unpaired electrons (p. 625). The Zeeman and anomalous Zeeman effects have been explained in a similar fashion (pp. 658–663). Now we must discuss further the magnetic properties of atoms and molecules.

We first deal with some general aspects of magnetic properties, then treat some special applications, namely nuclear magnetic resonance (NMR) and electron magnetic resonance (EMR); the latter includes electron paramagnetic resonance (EPR) and electron spin resonance (ESR). We will not discuss the field of nuclear quadrupole resonance (NQR) beyond mentioning its origins. Magnetic effects are manifest in other areas of science, but their greatest relevance to chemistry results from a widespread use of NMR as a structural tool, rivaled only by X-ray diffraction.

Magnetic Susceptibility

Several different types of magnetism are possible. In the presence of an external magnetic field, and if unpaired electrons are present, **electron paramagnetism** is observed. If the nuclei possess a nuclear spin, **nuclear paramagnetism** is observed, an effect that is about a thousand times smaller than electron paramagnetism. When neither nuclear spins nor unpaired electrons are present, **diamagnetism** is observed; this is the weak repulsion of a substance from regions of high magnetic field. There is also **ferromagnetism,** which gives rise to permanent magnetization, as exhibited by a bar magnet made of iron. Here we are concerned only with nuclear and electronic paramagnetism.

In electronic spectroscopy we observe differences between energy levels when electromagnetic radiation is passed through media. Electromagnetic radiation does not require a medium (we see the light from distant stars as it passes through a vacuum, for example), and we introduce the permittivity of free space ϵ_o to apply to the passage through a vacuum. When the radiation passes through matter, the permittivity is $\epsilon_o\epsilon$, where ϵ is the relative permittivity, usually called the *dielectric constant* (p. 266).

A similar situation exists for the magnetic component of the electromagnetic radiation. We use the symbol H to represent the magnetic field strength. When the radiation passes through matter the magnetic flux density, given the symbol B, is related to H by

$$\boldsymbol{B} = \mu_o (1 + \chi) \boldsymbol{H} \tag{14.15}$$

Here μ_o is the magnetic permittivity of free space ($\mu_o = 1/(c^2\epsilon_o)$) and χ is the magnetic susceptibility. The value of χ changes for different materials and is positive for paramagnetic systems and negative for diamagnetic systems.

Magnetic and Electric Moments

In Section 13.2 we discussed electric and magnetic moments. Electronic spectroscopy arises from the interaction of atoms and molecules with the electric component of electromagnetic radiation. Similarly, magnetic moments exist in atoms and molecules, and these interact with the magnetic component of the radiation. On comparing these two terms for electromagnetic radiation, we see that they have similar forms. One difference found in magnetic spectroscopy is that it is common to apply a large static magnetic field that is time independent, whereas large electric (Stark) fields are seldom used in electronic spectroscopy. The magnetic field is written as

$$\boldsymbol{B} = ZB_o + X2B_x^o \cos(\omega t) \tag{14.16}$$

where $\omega = 2\pi\nu$. It is seen that the magnetic field $\boldsymbol{B}$ is a vector quantity. The static field of strength B_o lies along the laboratory Z axis, and the time-dependent part oscillates in the laboratory X axis with intensity $2B_x^o$.

In Figure 14.23 some arrangements are shown that give rise to various electric and magnetic moments. There is an electric monopole, a dipole, a quadrupole, an octapole, etc., and these are produced by certain distributions of electrons. There is, however, no experimental evidence that a magnetic monopole exists, but magnetic dipoles, quadrupoles, etc., do exist.

In a static magnetic field B_o an electron displays a linear splitting, the Zeeman effect, into two magnetic levels with energies of

$$E_{\pm\frac{1}{2}} = \pm\frac{1}{2} g_s \mu_B B_o \tag{14.17}$$

which we have seen before (Section 13.2) and nuclear spins display a similar lifting of degeneracy as that shown for electron spins in Figure 13.13. The symbol $\hat{\boldsymbol{S}}$ is used for the electron-spin quantum-mechanical operator. This is the basis for ESR, which is concerned with the transitions between these two energy levels of electron spin, or combinations of them arising from spin couplings.

The magnetic dipole moment arises from an intrinsic spin that a *nucleus* might possess. Various ground-state configurations of the nucleus display intrinsic spin properties. Hydrogen ^{1}H has a nuclear spin of $\frac{1}{2}$, deuterium D (= ^{2}H) has a spin of 1. Carbon ^{12}C has no spin, but its isotope ^{13}C has a spin of $\frac{1}{2}$. ^{15}N has a spin of $\frac{1}{2}$, but

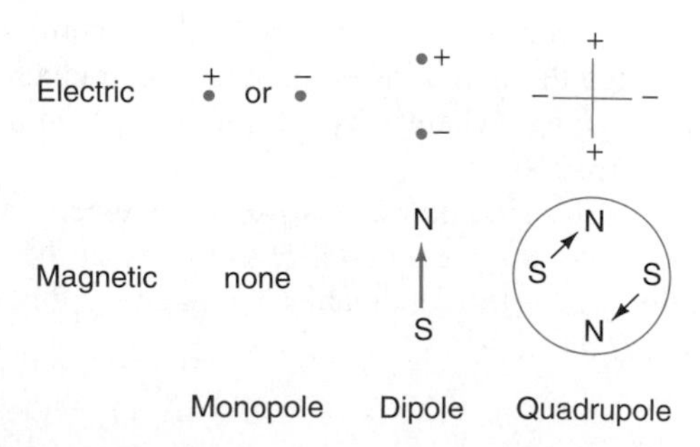

FIGURE 14.23
Electric and magnetic moments showing positive and negative poles for several arrangements.

TABLE 14.1 Some Spin Properties of Nuclei

Isotope	Abundance %	Spin	Gyromagnetic Ratio $\gamma = 10^8$ rad T^{-1} s^{-1}	Relative Sensitivity*	Nuclear g factor g_N	μ/μ_N
^{1}n		$\frac{1}{2}$		0.32139	−3.8263	−1.91304
^{1}H	99.985	$\frac{1}{2}$	2.675	1.00000	5.5856	2.79285
^{2}H	0.015	1	0.411	0.00965	0.857387	0.85744
^{3}H		$\frac{1}{2}$	28.533	1.21354		2.97896
^{7}Li	92.5	$\frac{3}{2}$	10.3964	0.29355		3.25643
^{13}C	1.10	$\frac{1}{2}$	0.673	0.01591	1.404398	0.70241
^{14}N	99.634	1	0.193	0.00101	0.40347	0.40376
^{15}N	0.366	$\frac{1}{2}$	−0.271	0.00104		0.28319
^{17}O	0.038	$\frac{5}{2}$	−0.363	0.02910		−1.89379
^{19}F	100	$\frac{1}{2}$	2.517	0.83400	5.257	2.62887
^{23}Na	100	$\frac{3}{2}$	7.0761	0.09270		2.21752
^{27}Al	100	$\frac{5}{2}$	6.9704	0.20689		3.64151
^{29}Si	4.67	$\frac{1}{2}$	−0.531	0.00786		−0.55529
^{31}P	100	$\frac{1}{2}$	1.083	0.06652		1.13160
^{35}Cl	75.77	$\frac{3}{2}$	0.2621	0.00472	0.54727	0.82187
^{37}Cl	24.23	$\frac{3}{2}$	0.2181	0.00272	0.4555	0.68412

μ_N: Nuclear magneton $= e\hbar/2m_p = 5.050\ 783\ 17(20) \times 10^{-27}$ J T^{-1}
μ/μ_N: Nuclear magnetic moment in units of the nuclear magneton μ_N
*Constant H_o, relative to ^{1}H(= 1), assuming equal number of nuclei and constant T.

Values of Relative Sensitivity are calculated from the expression $0.0076508\ (\mu/\mu_N)^3(I + 1)/I^2$ where μ is the nuclear magnetic moment in units of the nuclear magneton μ_N.

the most common isotope ^{14}N has spin 1. Table 14.1 lists the most common nuclear spins along with their gyromagnetic ratios. The symbol used for spin is usually I, and a nucleus of spin I has $2I + 1$ levels, referred to as its **multiplicity.**

Pauli pointed out in 1924 that all elementary particles (atoms, and molecules with non-zero angular momentum) possess a magnetic moment. For nuclear spins, the moment operator is

$$\hat{\boldsymbol{\mu}}_i = \gamma_i \hbar \hat{\boldsymbol{I}}_i \tag{14.18}$$

where γ_i is the nuclear gyromagnetic ratio for spin i. The nuclear-spin angular momentum operator is $\hat{\boldsymbol{I}}_i$. From Table 14.1 it is seen that the gyromagnetic ratios differ for different nuclei, and we have indicated these differences by the subscript i in Eq. 14.18. Note that, except for tritium, the ^{1}H nucleus has the greatest magnetic sensitivity, a property that is exploited in NMR.

Magnetic moments are not restricted to the intrinsic angular momenta of electrons and nuclei, called **spin angular momenta.** A rotating charge can also produce a magnetic moment; electric motors work in this way. As electrons move around a coil, they produce a magnetic field. In a similar way, as a diatomic molecule rotates, its electrons swirl about and produce a magnetic moment given by

$$\hat{\boldsymbol{\mu}}_J = \gamma_J \hbar \hat{\boldsymbol{J}} \tag{14.19}$$

This is the same $\hat{\boldsymbol{J}}$ that we encountered in the rigid rotor (and which is modified to $\hat{\boldsymbol{N}}$ in the Hund cases; see Section 13.6). The quantity J can have many values, in contrast to the fixed value that the intrinsic nuclear or electron spins possess, but nonetheless a rotational magnetic moment exists, and for each J there are $2J + 1$ magnetic levels in a magnetic field.

Magnetic Interaction Leading to Spectra Consequences

Just as in the case of electronic spin-spin and spin-orbit couplings, etc., so too there are couplings that apply to magnetic systems. In general, there are interactions arising from the angular momentum of the spin of each atom; from the rotational motions of the molecule; and from couplings to electric field gradients. There are many ways that local magnetic fields generated in molecules can arise. As these local magnetic fields couple together, they reveal spectral details different from, but analogous to, the spin-spin and spin-orbit couplings seen for electrons.

It is also possible to follow the effects of varying the external magnetic field from the strongly coupled situation at low or zero magnetic field to the limit of uncoupling in a high external magnetic field. The external magnetic field is time independent and points along the Z axis, which is defined by the orientation of the north and south poles of the magnetic field, or along the axis of the coil if the field is produced by a superconducting solenoid magnet. There is a Hamiltonian operator for each contribution of the interactions leading to the total energy.

We will briefly examine each interaction in turn. The total Hamiltonian operator includes terms for magnetic Zeeman effect, spin-spin, spin-rotation, quadrupole, and diamagnetic interactions.

Magnetic Zeeman interaction and chemical shielding: The interaction between the external magnetic field B_o and the magnetic moments in the system has the form

$$\hat{H}_I^{\text{Zeeman}} = -\hbar\gamma_I (1 - \sigma_I)\,\hat{\boldsymbol{I}}_z B_o \tag{14.20}$$

A quantity of major importance in NMR, the chemical shielding, σ_I appears. It is related to the chemical shift in high-resolution liquid spectra. Note that, depending on its value, it will move the position of the resonance line away from the frequency that is determined solely by the gyromagnetic ratio, γ_I. The chemical shift varies, depending upon the local magnetic fields in which the nucleus finds itself because of differing electron charge densities surrounding the nucleus and shielding it to a greater or lesser extent.

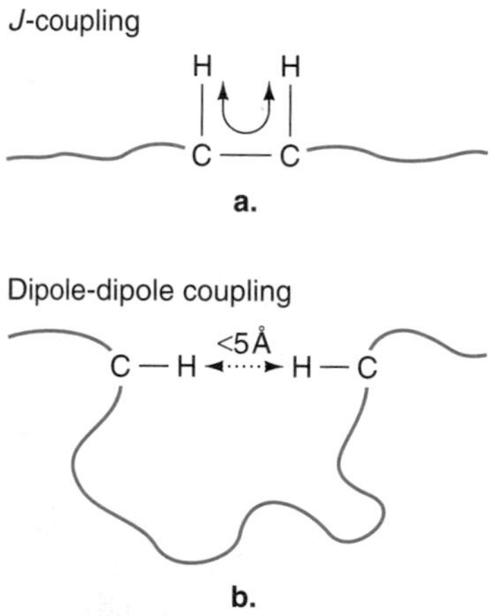

Indirect scalar spin-spin *J*-coupling: The magnetic moment of one spin will set up a small local magnetic field that can couple with nearby spins. There are two types of spin-spin coupling. One is a scalar. The scalar spin-spin coupling has the form

$$\hat{H}_{I_1 I_2}^{J\text{-coupling}} = \hbar J_{12} \hat{\boldsymbol{I}}_1 \cdot \hat{\boldsymbol{I}}_2 \tag{14.21}$$

where J_{12} is the spin-spin coupling constant (not to be confused with the J quantum number). This coupling is of major importance to NMR as it is responsible for the multiplet structures commonly observed in the high-resolution NMR spectra of liquids. Sometimes it is advantageous to remove this coupling, for example, between a carbon-13 nucleus and a proton, and various pulsed NMR techniques have been devised to do this. For protons, J-coupling usually is restricted to protons separated by no more than three bonds. (See Figure 14.24a.)

Direct spin-spin coupling: Another form of spin-spin coupling is through space (as opposed to "through-the-bonds" J-couplings) and depends on the actual sepa-

J-coupling

a.

Dipole-dipole coupling

b.

FIGURE 14.24
Schematics of dipolar coupling showing (a) through-bonds *J*-coupling, which usually extends no more than three chemical bonds and (b) through-space dipole-dipole coupling, which is present when protons are less than about 5 Å apart.

ration, r, of two spins (see Figure 14.24b). It is called direct spin-spin coupling and the Hamiltonian operator has the form

$$\hat{H}_{I_1 I_2}^{\text{direct spin-spin}} = \frac{\hbar^2 \gamma_1 \gamma_2}{r^3} (3\cos^2\theta - 1)\left[\hat{I}_{1z}\hat{I}_{2z} - \frac{1}{2}(\hat{I}_{1x}\hat{I}_{2y} + \hat{I}_{1y}\hat{I}_{2x})\right] \quad (14.22)$$

The strength of this coupling depends on the inverse cubed power of the separation of the spins. Typical distances over which this interaction is detected are less than about 5 Å (10^{-10} m). By performing nuclear Overhauser NMR experiments (NOE), these couplings can be observed. NMR gives an alternative way to measure distances between atoms in molecules compared with such techniques as X-ray crystallography. As a result, NMR, via NOE, gives valuable structural information and has played a vital role in, for example, determining the folding structures of proteins.

Direct spin-spin coupling is also a major cause of spin relaxation in NMR of liquids and solids. In solids it can cause considerable line broadening, which masks the hyperfine multiplet structures due to J-coupling. Removing the spin-spin coupling in solids is therefore frequently desirable. There are several ways of doing this, but the most commonly used is **magic angle spinning (MAS).** That is, the solid sample is spun at high speeds at the angle θ to B_o, where $3\cos^2\theta - 1 = 0$, that is $\theta = 54.74°$, thereby removing this term.

Spin-rotation interaction: The spin-rotation interaction results from a nuclear spin coupling to the magnetic moment caused by molecular rotation. Its form is simply

$$\hat{H}_{IJ}^{spin\text{-}rotation} = c_{sr}\hat{\boldsymbol{I}} \cdot \hat{\boldsymbol{J}} \quad (14.23)$$

This mechanism is usually small in most NMR experiments.

Nuclear quadrupole interactions: Nuclei with $I > \frac{1}{2}$ possess quadrupole moments. When such a nucleus is in a non-uniform electric field, the energy of the nucleus depends on its orientation with respect to the electric field gradient, which in turn determines the orientation of I in B_o. A spin of $\frac{1}{2}$ has only two orientations and these cannot be distinguished in a quadrupole field. Hence the $\pm\frac{1}{2}$ levels are degenerate. For $I = 1$ the $\pm$ 1 levels are indistinguishable in a quadrupole field, but the 0 level has different energy. Generally, for nuclear spins with magnitudes $I > \frac{1}{2}$, different orientations of the nuclear spin are not equivalent and are sensitive to the electron density due to the electric quadrupole moment. The form of the Hamiltonians involves $I > \frac{1}{2}$ and I_J.

The quadrupole interaction leads to additional splittings and characteristic line shapes in solids. In liquids, rapid molecular motions average this anisotropic interaction to zero, but it also provides a slow relaxation mechanism. The Hamiltonian operators include (high magnetic field limit)

$$\hat{\boldsymbol{H}}_I^{quadrupole} = d_I\,[3\hat{I}_z^2 - I(I + 1)] \quad (14.24)$$

and

$$\hat{\boldsymbol{H}}_J^{quadrupole} = d_J\,[3\hat{J}_z^2 - J(J + 1)] \quad (14.25)$$

where the quadrupole strengths are given by d_I and d_J.

Other contributions: Just as internal molecular motions influence electronic spectra, so can they influence magnetic spectra. These effects are generally negligible but should be considered. For example, molecular vibrations can give diamagnetic contributions that can cause small energy shifts, as can centrifugal distortions.

Of more interest, especially as high magnetic fields are reached, are nonlinear magnetic field effects. In complete analogy to its electronic case, where nonlinear optical effects are particularly prevalent in laser spectroscopy due to the presence of intense laser beams (Eq. 14.8), at large magnetic fields nonlinear effects can be manifest. The form of this contribution is in complete analogy to the electric field case. See Eq. 14.10. Its Hamiltonian operator is

$$\hat{H}^{nonlinear} = f\hat{I}^2 B_o^2 \tag{14.26}$$

where f is a small, nonlinear correction.

Finally, not included here are various types of molecular motions that can lead to line narrowing and provide a mechanism for relaxations. These are important and give valuable insight into molecular dynamics. The mechanisms discussed from Eq. 14.20 to Eq. 14.26, when included in the Schrödinger equation, give only the eigenvalues and eigenfunctions.

14.6 Nuclear Magnetic Resonance Spectroscopy

Nuclear magnetic resonance (NMR) spectroscopy has become one of the most versatile types of spectroscopy in current use. NMR was demonstrated in 1946 independently by Edward Mills Purcell (b. 1912) and Felix Bloch (1905–1983). It was originally suggested that just as an electric moment interacts with the electric component of electromagnetic radiation, giving rise to electronic spectroscopy, a magnetic moment will interact with the magnetic component of the radiation, resulting in magnetic resonance. The calculated frequency that a magnetic moment should absorb is in the short-wave and very-high-frequency (VHF) radio-frequency (*rf*) range. Purcell and Bloch succeeded in causing the hydrogen nuclei in molecules to resonate. In the 1950s it was discovered that the magnetic moments show great sensitivity to the electronic environment in which they find themselves. For example, hydrogen atoms have differing electron densities depending upon the nature of their bonding. This effect is by no means limited to hydrogen, but at the time instrument sensitivity restricted probing only to the hydrogen nucleus.

These differences in chemical environment shift the frequencies so that the proton resonances are characteristic of the type of chemical group to which they are bonded. The discovery of such **chemical shifts** was immediately recognized as a tremendous benefit to chemists because it provided a means of identifying structural components and functional groups of molecules.

Another advantage of NMR is that the magnetic moments can couple with other magnetic moments in the molecule. At high magnetic field values this effect is weak but important because it causes, as we shall see, a splitting of the resonances into specific multiplets that are easily related to the number and type of spin to which a proton is coupled. This again is of utility in determining the structures of molecules.

Finally, the intensities of the observed resonances are proportional to the number of protons that are resonating at the same frequency. These intensities are used to measure concentrations of selected species, and are of value in structure determination and in chemical analysis.

One of the characteristics of magnetic resonance is that the effect is weak. In fact, even with hydrogen, which (after tritium, T = ^3H) gives the strongest signal, many spectra must be accumulated and many signals must be averaged before an adequate spectrum can be obtained. Up until the early 1970s, NMR instrumentation was continuous wave (CW) and, in order to obtain a spectrum, the sample had to be scanned many times to be able to reduce the noise. Usually in CW-NMR experiments a constant source of radio-frequency (*rf*) radiation is set up, and the external magnetic field, B_o, is slowly varied. This causes the Zeeman splitting to vary, which means that one by one the frequencies are detected as the Bohr resonance condition is met. CW techniques do not lend themselves to efficient data collection and have been completely replaced by pulse techniques made possible by Fourier-transform data analysis.

We can get some idea of how Fourier-transform methods work as follows. Frequency and time are the inverse of each other, and Fourier transform gives the mathematical relationship between the two. All the information in the frequency domain can be found in the time domain and vice versa. Since the CW technique involves scanning through the transition frequencies one by one, and detecting one resonance after the other in the frequency domain, how would this information appear in the time domain? If all these frequencies are simultaneously excited by a strong *rf* pulse, a train of diminishing-amplitude oscillations would result in the time domain containing all of the information that exists in the frequency domain. The technique used is therefore to perturb the system in the time domain by giving the system a sudden jolt with an *rf* pulse. The response of the system is called the **free induction decay (FID)** and is shown in Figure 14.25a. It oscillates and decays (often in a few milliseconds) and contains all the frequency information that is contained in the frequency domain. Since the relaxation back to the equilibrium state that existed before the pulse occurred is relatively fast compared to CW scans (in the millisecond range rather than in the second range), it is possible to repeat the experiment many times in a much shorter time period relative to the CW experiments. This greatly speeds up the signal averaging process. After accumulating enough data in the time domain, it is usual to Fourier-transform the averaged time-dependent signal to obtain the well-known frequency domain spectrum. This technique, called **Fourier transform NMR (FT-NMR),** is now virtually the only method used. Figure 14.25 shows the FID and the spectrum from the Fourier transform of the FID for ethanol dissolved in deuterated chloroform, $CDCl_3$; this solvent does not disturb the spectrum. This Fourier-transform method revolutionized the NMR technique in that it allowed a great many spectra to be taken quickly and accurately. It opened the door to making NMR, in the hands of the chemist, an accurate and valuable tool for structural investigations of molecules in solution and, more recently, of solids. Moreover, greatly speeding up the acquisition of data made it possible to perform routine NMR experiments on nuclei of much lower sensitivity than protons.

It is evident that the low sensitivity of NMR, although in one sense a disadvantage, is a source of strength. The nuclear magnetic moments do not readily couple to other angular momenta in the system, especially at high magnetic fields. As a result, NMR spectra are sparse, making analysis relatively simple. However, as molecular complexity increases, the spectra also become more complex. Even though the same phenomena are present (chemical shift, spin coupling, and line intensity) when complex and large molecules are studied (in particular, biomolecules such as proteins), interpretation of the spectra becomes intractable using a conventional one-dimensional plot of intensity versus frequency. In 1971 it was proposed

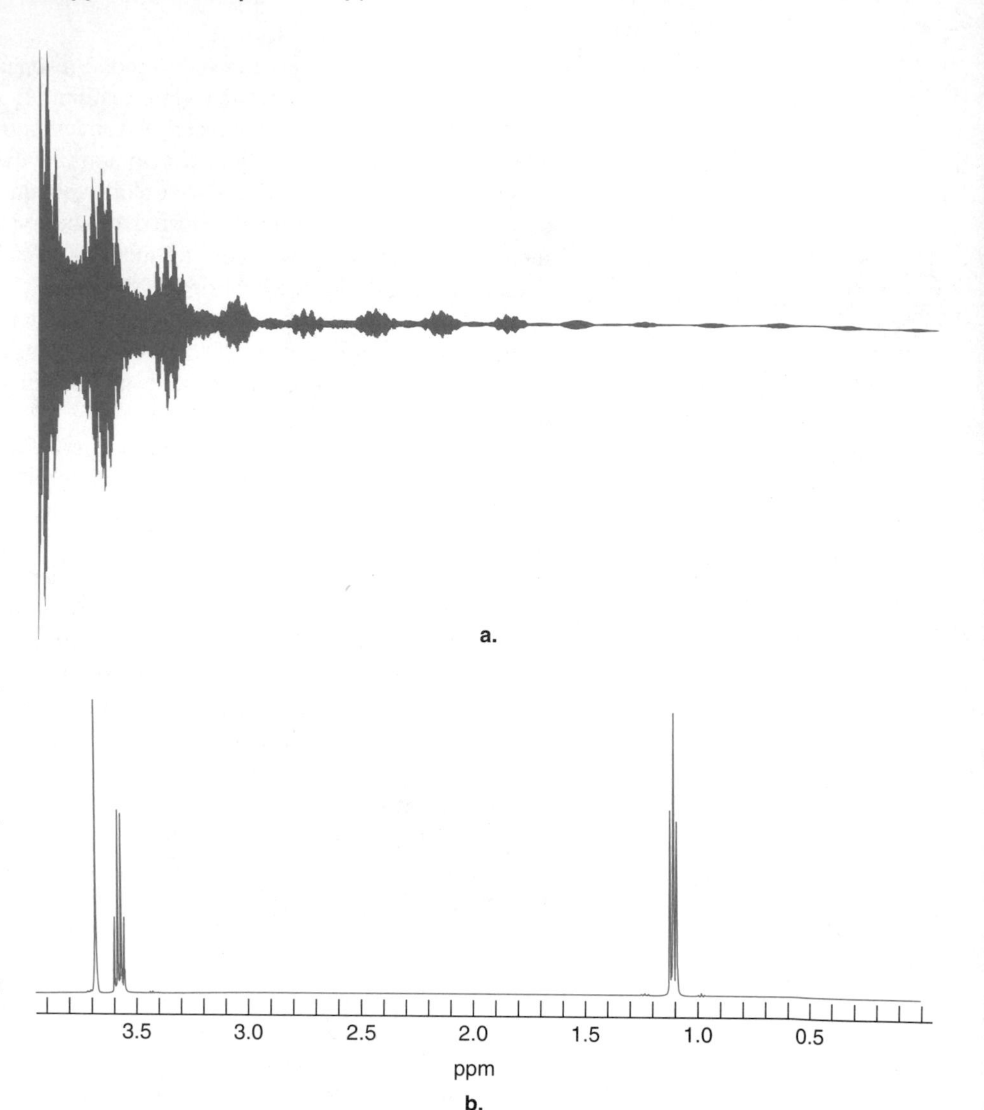

FIGURE 14.25
(a) FID showing the decay of the time-dependent signal that is detected by an *rf* coil. The experiment is repeated many times to signal-average the individual FIDs and the resultant is amplified, processed, and displayed after Fourier transformation. NMR high-resolution spectrum of ethanol in deuterated chloroform $CDCl_3$. The spectrum was collected at a Varian Unity 500 MHz spectrometer with four scans. (b) Spectrum of ethanol after the Fourier transform of the FID in (a). Compliments of Dr. Zhicheng (Paul) Xia, Chemistry Department, McGill University.

Jean Jeener is a physicist at the Free University of Brussels. In 2001, he was awarded the International Society of Magnetic Resonance (ISMAR) prize for his experimental and theoretical contributions to spin thermodynamics in solids and his invention of 2D-FT-NMR.

Richard R. Ernst received the Nobel Prize in Chemistry in 1991, for his contributions to the development of the methodology of high resolution nuclear magnetic resonance (NMR) spectroscopy.

by Jean Jeener (b. 1931) in a lecture that if a system was pulsed twice, and the delay between the first and second pulse was systematically changed, a double Fourier transform could be performed on the data and plotted in two frequency dimensions. At the lecture was Swiss-born Richard R. Ernst (b. 1933); he took up this idea and produced the first 2D-NMR spectrum.

The advantage of 2D NMR is that it enables the dense and complicated 1D spectrum to be spread out from a line into a plane. Displaying the data in this manner enabled complicated biomolecules to be interpreted routinely, allowing NMR to give valuable structural information even for proteins. Since that time, 3D and 4D NMR have been developed, and NMR has moved into the life sciences as a valuable structural tool.

At about the same time that multidimensional NMR was being developed, several researchers found that NMR could be used for the imaging of objects, such as biological materials. This technique was pioneered by Peter Mansfield and Raymond Andrews in England and Paul Lauterbur (b. 1929) in the United States. The imaging of biological material depends on the fact that all living systems contain water, and that as a result of the different environments of the water in various regions of the body, the relaxation times (T_1) of protons change throughout the body. These varying relaxation times can be used as a parameter to construct an image. Computer graphing and imaging technology are used, and the variation in the relaxation times at various locations is used to create an image that is particularly sensitive to soft tissue. This is in contrast to X-ray imaging, which is more sensitive to hard tissue such as cartilage and bone.

The way in which the technique is employed is to place the patient's body in a magnetic field gradient in such a way that the Zeeman splitting, and hence the resonance frequency, varies across the body. Then, when pulsed, the signal is both frequency and spatially encoded. Throughout the 1980s this technique was refined and taken over by the large medical instrumentation companies, and it is now used as a sensitive medical technique for detection of abnormalities, particularly in, but not restricted to, the brain.

The advantage of NMR imaging is that, in contrast to X rays, it is sensitive to soft tissue that contains a high concentration of water. Moreover, NMR imaging machines use low-frequency radio waves, which are not destructive and are much safer than X rays. Since the word "nuclear" has bad connotations to the general public, that word was dropped and the technique of NMR imaging is now simply called **magnetic resonance imaging (MRI).**

Another Area of NMR Research

Over the past few years, work has begun which would use NMR spectroscopy in a totally different role from those described so far. Imagine the $+\frac{1}{2}$ and $-\frac{1}{2}$ spin states of a proton to represent the computer bits 0 and 1. These quantum spins in this role are called qubits. If a series of polarized spins on a molecule could be controlled as in a computer register by using NMR techniques, then superposition of quantum states could occur, thereby leading to algorithms, which could provide a major time savings in computing.

The Machinery of NMR Spectroscopy

There are two major aspects to NMR spectroscopy: recording the spectrum and interpreting it. The latter involves knowing how the magnetic field lifts the degeneracies and how the hyperfine coupling is manifest. This is a time-independent quantum-mechanical problem. It entails setting up a secular determinant for the number of spins involved and taking into account the types of spins (^{1}H, ^{13}C, ^{31}P, etc.), the different chemical shifts, the spin couplings, and the relaxation. The consequences of magnetic equivalences (such as the three H atoms on a methyl group) must also be included. Once done, the energy-level diagram is known so that on applying the selection rule $\Delta M = \pm 1$ for the resonant nuclei, it is straightforward to obtain a "stick" spectrum, or a series of lines, which locate the frequencies. In the area of organic chemistry, a number of rules are developed to recognize common features in the spectrum which are always indicative of certain functional groups, for example, $-CH_3$, $-CH_2$, $-NH_2$, $-OH$, $-C=O$. Determination of the structures of organic molecules is not discussed here, but we consider the main techniques for obtaining the spectra.

The basic NMR experiment involves obtaining a transient time-domain signal, which is averaged to improve the signal-to-noise (S/N) ratio and the Fourier transform. The understanding of pulsed NMR requires a time-dependent treatment, which is based on time-dependent perturbation theory (Section 13.1). The treatment requires that we follow the time evolution of the nuclear spins as they respond to small magnetic fields produced by bursts of radio frequency, called *pulses*.

Pulsed NMR

Nuclear paramagnetism arises from the summation of the magnetic moments from all the molecules in a sample. This magnetization $\boldsymbol{M}$ is a vector and, in a static magnetic field, the magnetization precesses around the magnetic field direction

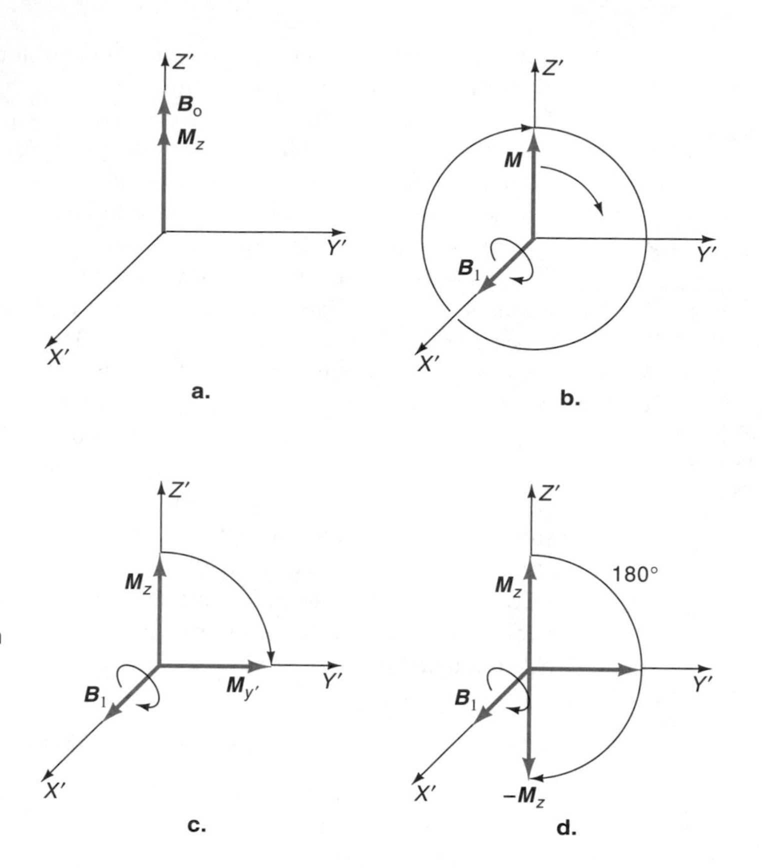

FIGURE 14.26
(a) The Z component of nuclear paramagnetism in the rotating frame at equilibrium in a static magnetic field. (b) The effect of the application of an "on-resonance" time-dependent magnetic field. In the rotating frame, which is rotating at the same *rf* frequency as the applied time-dependent magnetic field, the $X'Y'$ component appears static. This causes the magnetization to precess around the small *rf* field. (c) Same as in (b) but the *rf* field is stopped after the magnetization has undergone a 90-degree rotation. This is called a 90-degree pulse. (d) Same as in (b) but the *rf* field is stopped after the magnetization has undergone a 180-degree rotation. This is called a 180-degree pulse.

(cf. Figure 11.21). It is customary to view the magnetization in a frame that is rotating in the XY plane at the same frequency as the magnetization so that the magnetization vector appears stationary and aligns with the Z axis, as shown in Figure 14.26a. This rotating frame is denoted $X'\ Y'\ Z'$. A small magnetic field, with magnitude B_1, is applied along the X' axis, which is tuned to the Larmor frequency, defined as $\omega_o = \gamma B_o$. This causes the Z' component of the magnetization to be rotated in the $Y'Z'$ plane around the X' axis, as shown in Figure 14.26b. Suppose the *rf* is applied on resonance for a time that rotates the magnetization from the Z' axis to the Y' axis. This defines a 90-degree pulse, shown in Figure 14.26c, that has duration of

$$t_{90} = \frac{\pi}{2\gamma B_1} \tag{14.27}$$

Typically a current of about 5 A is fed into a transmitter coil, which generates a power of about 500 W. For protons, a 90-degree pulse lasts for about 5 μs. A 180-degree resonance pulse has a time duration of (see Figure 14.26d)

$$t_{180} = \frac{\pi}{\gamma B_1} \tag{14.28}$$

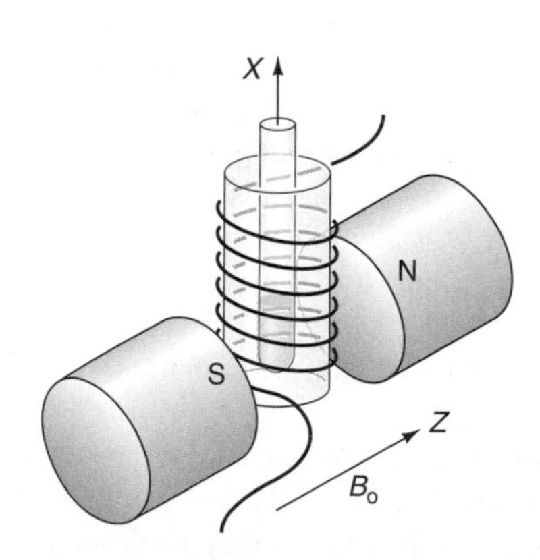

FIGURE 14.27
(a) Schematic of the magnetic field setup in a typical NMR spectrometer. The large external magnetic field defines the laboratory Z axis. A coil surrounds the sample and produces the smaller time-dependent B_1 field.

This analysis gives a remarkably simple picture of the basic NMR experiment. The schematic given in Figure 14.27 shows a sample placed in an NMR tube and surrounded by a coil. The coil is a solenoid that is used to deliver pulses of *rf* electromagnetic radiation in the XY plane. The sample and coil are placed in a large static magnetic field and the following occurs:

1. The system equilibrates in the static field $B_o z$ (Figure 14.26a).
2. A 90-degree pulse is applied, say, along the X' axis; see Figure 14.26b. This rotates the magnetization so that it lies along the Y' axis; see Figure 14.26c.
3. In the laboratory frame, the M_y component precesses with a spectrum of frequencies that are characteristic of the sample. These induce a voltage in the same coil that delivers the pulse, oriented along the X laboratory axis, which is amplified and stored. (See Figure 14.27.)
4. The signal decays, giving an FID. The decay of the signal is due to one or more of several randomly fluctuating interactions and magnetic field inhomogeneities, and is called **T_2 relaxation.** Some spins precess faster than ω_o, while others precess slower than ω_o. Eventually the magnetization loses all its coherence (T_2 relaxation) and the signal vanishes. This is the FID. (See Figure 14.25a, which is the actual FID for ethanol.)

T_2 Relaxation

5. During the FID, the magnetization also relaxes back to lie along the Z' axis in a time known as the **T_1 relaxation,** whence equilibrium is again established (see Figure 14.26a). It is customary to wait for a time period of five times the value of T_1 to ensure that the sample is at equilibrium before repeating the experiment to improve the signal-to-noise (S/N) ratio. This delay ensures that equilibrium has been reestablished so that upon repeating the experiment, the maximum signal is created.

T_1 Relaxation

We have just introduced two important relaxation times: the spin-spin relaxation time T_2 and the spin-lattice relaxation time T_1. They are discussed in more detail in the following treatment.

Spin Echoes

Spin systems can be made to echo. This means that it is possible to allow the individual spins to dephase during the FID until no signal is induced in the detection coil. At a given time (less than the spin-spin relaxation time T_2 but after the FID has decayed), the spins are **time reversed.** They then retrace their paths and refocus at a later time, producing an echo. Spin systems readily display this phenomenon in the solid and liquid phase and are studied in magnetic spectroscopy: NMR, EMR, MRI, and NQR. Spin systems are not the only systems that can produce echoes. In reality, echoes are a result of time reversing a process, thereby allowing the system to return to its original state. There is, however, an irreversible component to the time evolution of systems, and this cannot be reversed and cannot contribute to the echo.

It is not, of course, possible to change the direction in which time proceeds. By time reversal, it is meant that the symmetry of a process is studied when the time variable t is changed to $-t$ in the equations. Some equations do not change (these are even to time reversal) and some change sign (these are odd to time reversal). For example, consider the velocity, $\mathbf{v}(t)$, or angular momentum, $\mathbf{L}(t) = \mathbf{r} \times \mathbf{p}(t)$. Both of these quantities are odd to time reversal:

$$\mathbf{v}(t) = \frac{dx}{dt} \xrightarrow{t \to -t} \frac{dx}{-dt} = -\mathbf{v}(t) \tag{14.29}$$

Likewise, $\boldsymbol{L}(-t) = -\boldsymbol{L}(t)$. For example, if a billiard ball is projected perpendicularly towards a cushioned rail, it will bounce and return along exactly the same path used when originally striking. The velocity has been reversed in the bouncing, which can be viewed as a time-reversal process. The kinetic energy of the ball, however, remains unchanged and so is even to time reversal:

$$E_k(-t) = \frac{1}{2}m\boldsymbol{v}(-t)^2 = E_k(t) = \frac{1}{2}m\boldsymbol{v}(t)^2 \tag{14.30}$$

In the above it is tacitly assumed that the ball bounces back elastically, whereas in fact, irreversible processes take place due to the ball not being perfectly elastic, as well as the effects of heat, noise, air resistance, etc. This means that, in reality, upon bouncing back to the same position from which it started, the ball's velocity and kinetic energy have a magnitude somewhat smaller than the initial velocity and kinetic energy. These are irreversible losses.

In order to understand spin echoes, it is instructive to analyze the process step by step. This is all the more important because most of the NMR experiments currently performed do not consist of a single pulse, but a series of pulses called a **pulse sequence.** In order to understand the effects of pulse sequences on spin systems it is necessary to know how spins respond to pulses and to the time delays between pulses.

Pulse Sequence

A spin echo is actually the refocusing of spin coherence that has already dephased. It was first demonstrated by Erwin Hahn using a pulse sequence

$$\left(\frac{\pi}{2}\right)_x - \tau - \left(\frac{\pi}{2}\right)_x - \tau - \text{Hahn echo} \tag{14.31}$$

The notation $\left(\frac{\pi}{2}\right)_x$ means that a 90-degree pulse is applied along the X axis in the rotating frame. Rather than analyze this pulse sequence, we will use one called the Meiboom-Gill sequence and given by

$$\left(\frac{\pi}{2}\right)_x - \tau - (\pi)_Y - \tau - \text{echo} \tag{14.32}$$

This is shown in Figure 14.28.

Once a $\left(\frac{\pi}{2}\right)_x$ pulse is applied to a spin system at equilibrium, the magnetization is rotated by 90 degrees and ends up along the Y' axis (Figure 14.28a, steps 1 and 2). Figure 14.28b shows the corresponding positions along the time axis during the pulse sequence. At position 2, the FID starts. However, the static magnetic field, in spite of attempts to create it as homogeneous as possible, actually varies across the sample. This means that as the various spin components start to fan out as shown in step 3, similar spins in different locations in the sample experience slightly different field strengths, B_o. The magnetic field inhomogeneity causes the precessional motions to vary and the FID to decay faster than if it were in a totally homogeneous field. The time for the FID to decay is therefore less than T_2 and is denoted by T_2^*. The time between the first and second pulse is τ, and during this time period the phase coherence of the magnetization is lost and no signal can be detected.

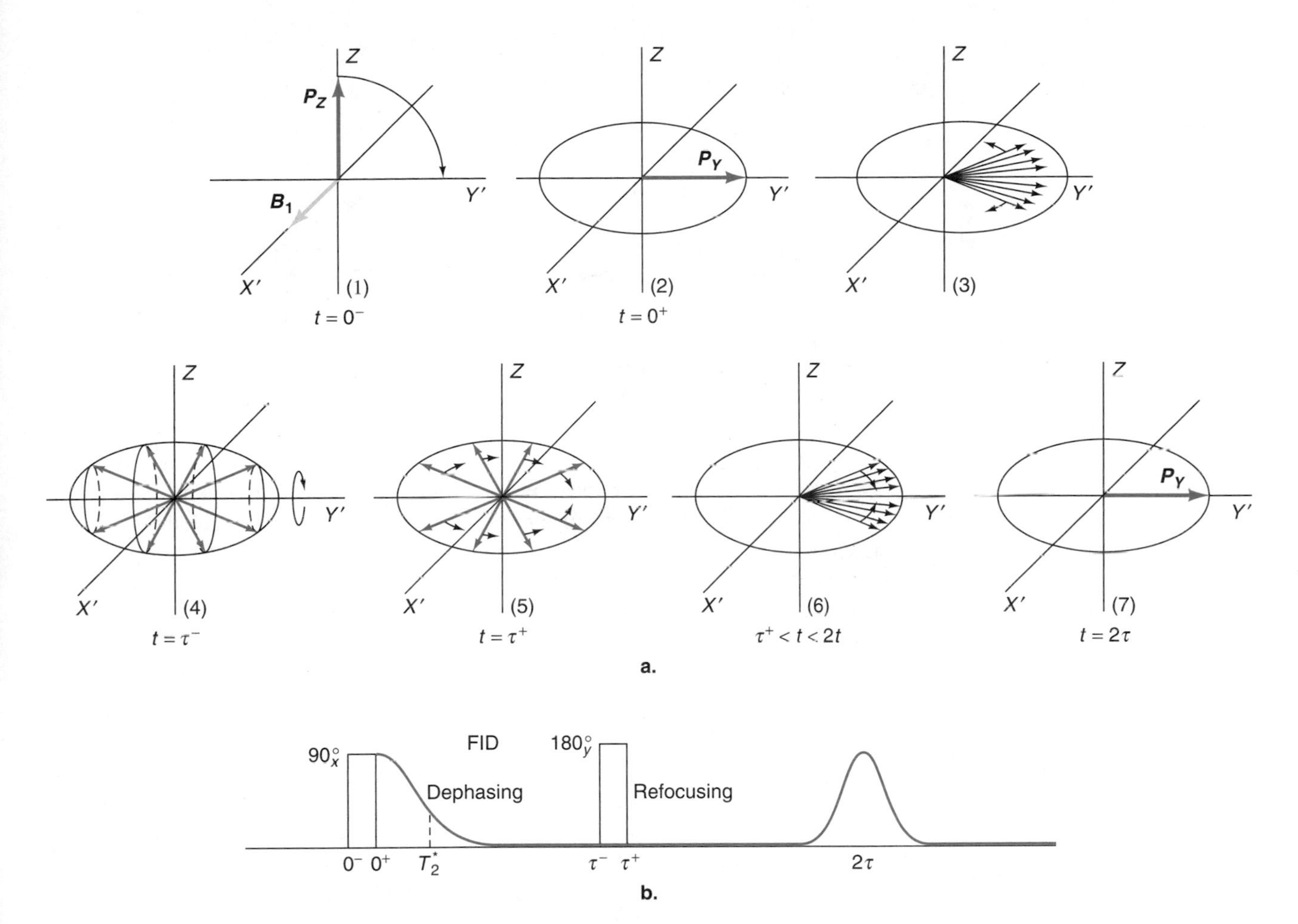

FIGURE 14.28
Stimulated NMR echo. (a) The dynamics of the magnetization in the rotating frame resulting from the pulse sequence. (b) The times in parts (a)1 to (a)7 correspond to the times along the axis in (b). (1) The 90° pulse; (2) the beginning of the FID; (3) the fanning out of *XY* components due to magnetic field inhomogeneities; (4) the flipping and time reversing of the dephased components by the application of a 180° pulse around the *Y* axis in the rotating frame; (5 and 6) the evolution back to the starting position, and (7) the refocusing and formation of the echo.

At time τ after the first pulse, a π pulse is applied around the Y' axis, $(\pi)_Y$ (step 4), and this flips over all the fanned-out spins. The effect is like the billiard ball bouncing off the cushioned rail because now all the precessional motion of the spins is reversed and they start to refocus (steps 5 and 6) until they return back to their original phase coherence at time $t = 2\tau$ and form an echo (step 7). Another way to visualize this is to imagine a pack of runners running around a circular track. Some run faster and others slower, and they start to fan out around the track. If they all suddenly stopped and started running back, the faster runners would catch up to the slower ones and all would arrive at the starting gate simultaneously.

As a result of magnetic field inhomogeneities, it is not possible to measure T_2 from the FID that is governed by $T_2{}^*$. If the Meiboom-Gill pulse sequence (Eq. 14.32) is modified to that shown in Figure 14.29, then each time an echo is produced, the echo amplitude will be smaller than the previous echo due to irreversible T_2 relaxation. Measuring the echo amplitude enables T_2 to be determined. The pulse sequence shown in Figure 14.29 is called the Carr-Purcell-Meiboom-Gill pulse sequence.

NMR Instrumentation

NMR instrumentation is a result of precision engineering that combines magnetic fields of high strength and homogeneity; *rf* amplifiers and receivers to generate

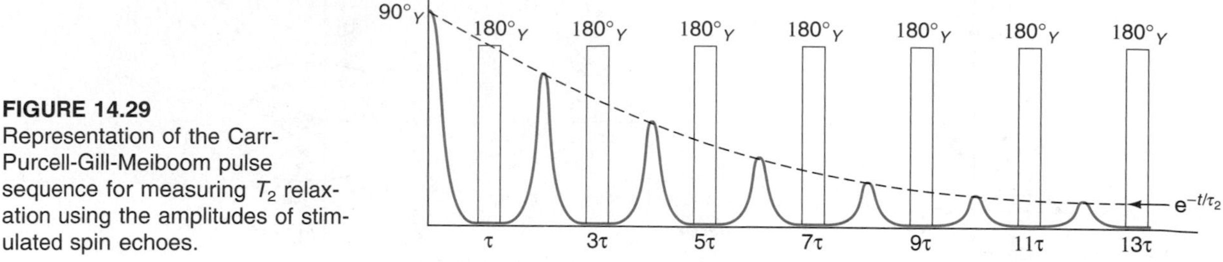

FIGURE 14.29
Representation of the Carr-Purcell-Gill-Meiboom pulse sequence for measuring T_2 relaxation using the amplitudes of stimulated spin echoes.

pulses and detect signals; and computers to control the experiment, store and process data, and display the results.

Figure 14.30 shows a block diagram of a typical NMR spectrometer. The static magnetic field, B_o, is usually produced by a superconducting solenoid. Magnetic field strengths are measured in units of Tesla (T), but it is more common to designate an NMR spectrometer by the magnet's ability to split the two spin states of a proton and quote this as a frequency. Larger fields produce larger splittings in the megahertz region. Nowadays, commonly used instruments are in the 500- to 600-MHz range, although spectrometers operating at 900 MHz are available commercially.

One-Dimensional NMR Experiments

Figure 14.25 shows the basic one-dimensional NMR experiment (1D NMR) for a simple molecule, liquid ethanol, CH_3CH_2OH. Figure 14.25b is the Fourier-transformed 1D frequency spectrum of this molecule obtained by Fourier transformation of the FID shown in Figure 14.25a. There are three basic pieces of information obtainable from high-resolution spectra of molecules, and these are discussed using Figure 14.25b as an example.

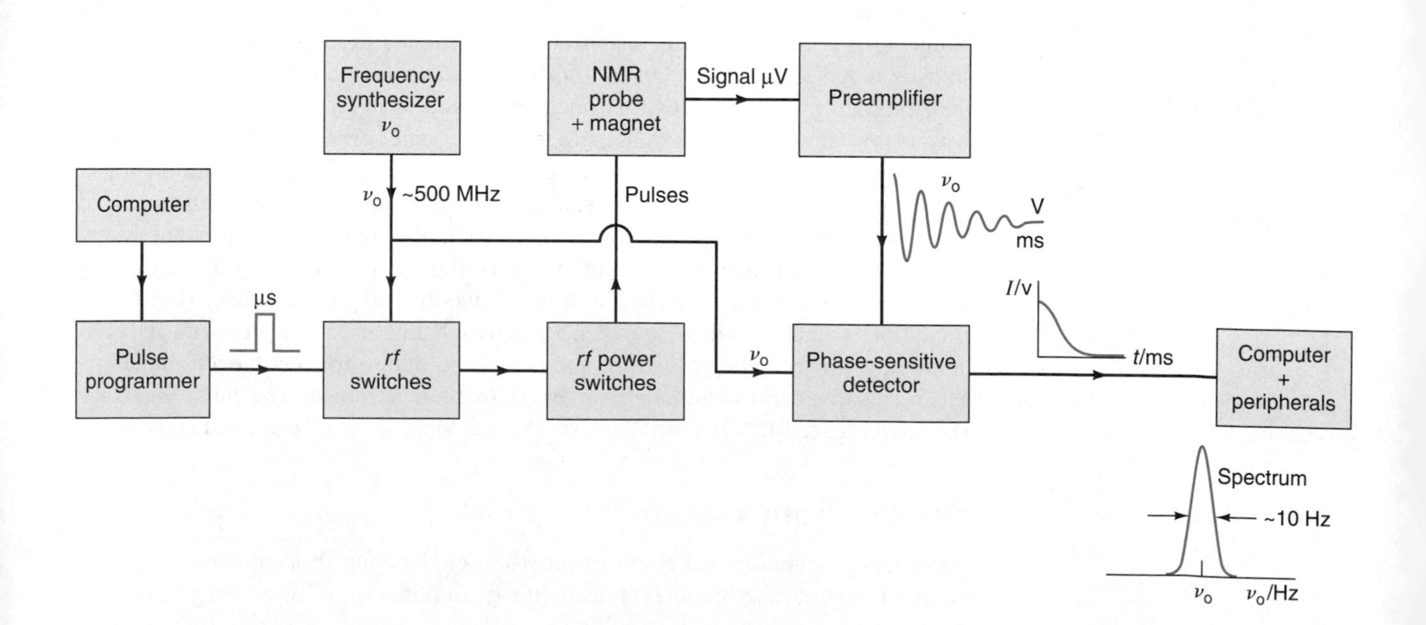

FIGURE 14.30
The main components of an NMR instrument.

The chemical shifts: The methyl protons, the two $-CH_2$ protons and the alcohol $-OH$ proton, all resonate at different positions due to their differing chemical environments (Eq. 14.20). The positions at which a resonance occurs give information that helps to determine the type of functional group to which the protons belong.

The signal intensity: The intensity of the signal is proportional to the number of protons present. Hence, as shown in Figure 14.25b, one-half the intensity is due to the three methyl protons; one-third is due to the $-CH_2$ protons, and the remaining one-sixth is due to the $-OH$ proton. This information helps to determine the number of different types of functional groups.

The multiplet structure: Because of through-bond *J*-coupling, Eq. 14.21, a resonance can split into a multiplet in a well-defined way. This gives information about the number and type of atoms that are close to each other in the molecule.

Of course, NMR experiments are performed on other nuclei. ^{13}C (^{12}C has no spin), ^{15}N, and ^{31}P all have spins of $\frac{1}{2}$. To record spectra of nuclei other than ^1H, the spectrometer frequency must be lowered to be equal to the Larmor frequency of the particular nucleus being studied (see Table 14.1). ^{13}C NMR can be done on samples with a natural abundance (0.1%) of this isotope, but because of the low concentration, many FIDs have to be averaged to obtain a good spectrum and this is time consuming. Therefore, when possible, the samples are often isotope enriched.

Moreover, interactions other than chemical shielding and *J*-coupling could be present, but for spins of $\frac{1}{2}$, these two dominate. Sample spinning at about 10 Hz is performed to reduce the effects of magnetic field inhomogeneities, which can broaden the lines.

Chemical Shifts

Chemical shifts arise due to small changes in electron density around the nuclei. Equation 14.20 shows this through the chemical shielding, σ,

$$\hat{H}_{Zeeman} = -\gamma \hbar B_o (1 - \sigma)\hat{I}_z \tag{14.33}$$

For a range of different environments, the chemical shielding, σ, gives a range of frequencies, $\omega_{o1} = \omega_o(1 - \sigma_1)$; $\omega_{o2} = \omega_o(1 - \sigma_2)$; etc.

For some purposes the value of σ may be considered a contribution from diamagnetic and paramagnetic effects. If the *diamagnetic contribution* σ_d dominates, the applied field causes orbital motion of the electrons in the atoms of the molecule, resulting in circulation of charge and generation of an opposing magnetic field. This was first described in the 1940s by Willis E. Lamb, Jr. (b. 1913). The effect reduces field intensity at the nucleus and the frequency at which the resonance appears and the value of $\sigma > 0$. The effective field at the nucleus is less than the applied field, and we say that the nucleus is *shielded.*

Willis E. Lamb, Jr., received the 1955 Nobel Prize in physics for the discovery, in 1947, of the "Lamb shift" in the hydrogen spectrum.

Calculation of σ_d depends on the knowledge of the electron density about the nucleus and is obtained from the *Lamb formula,*

$$\sigma_d = \frac{e^2 \mu_o}{3m_e} \int_0^\infty r\,\rho(r)dr \tag{14.34}$$

where ρ is the electron density calculated from the square of the wave function of the atom of interest.

If the *paramagnetic contribution* σ_p dominates, the applied field has caused the electrons to circulate throughout the molecule using unoccupied orbitals. This increases the magnetic field intensity at the nucleus. In such cases, $\sigma_p < 0$, so that the effective magnetic field is greater than the applied field, and we say that the nucleus is deshielded. Shielding is favored for electrons localized on a single atom or around a linear double or triple bond. Evidence exists that paramagnetic interaction may be significant in ^{13}C and ^{19}F studies.

Molecules contain different functional groups that have characteristic resonances, which can be used to identify them. Of course, different spectrometers will have slightly different field strengths, B_o, so that the frequencies of these characteristic resonances vary, making them difficult to tabulate. In order to avoid this problem, chemical shifts are measured relative to a reference. Suppose that a reference material resonates at the reference frequency $\omega_R = 2\pi\nu_R$. Then the sample can be measured relative to this reference by specifying how far away it is from the frequency. Chemical shifts are defined as the dimensionless ratio

$$\frac{\omega_s - \omega_R}{\omega_R} = \frac{\sigma_s - \sigma_R}{\sigma_R} \tag{14.35}$$

which is independent of B_o. The usual way of reporting resonance positions is to give them in parts per million (ppm):

$$\delta = \frac{\sigma_s - \sigma_R}{\sigma_R} \times 10^6 \tag{14.36}$$

Previously, internal references were commonly used. These came from placing a chemical inside the sample tube, giving an accurately known resonance. Sample resonances were measured relative to this standard. Today, external references that accurately determine the field strength are commonly used.

For simple molecules, it is usually possible to determine which functional groups are present from the values of the chemical shifts. Tables of such data are widely available, and examples of ^1H and ^{13}C chemical shifts for some common functional groups found in molecules are given in Table 14.2.

TABLE 14.2 Typical Range of δ Values

Molecule	Chemical Shift
	^1H resonances
RCH_2^*R'	1.1–1.5
RCH_3^*	0.8–1.2
RNH_2^*	1–3
ROH^*	2–5
$RCOCH_3^*$	2–3
$-OCH_3^*$	3–4
$ArCH_3^*$	2–3
ArH^*	6–9
$ArOH^*$	5–8
$RC(O)H^*$	9–10
$RCOOH^*$	10–13
	^{13}C resonances
$(R_3C^*)^-$	20–100
RC^*H_3	0–70
$>C{=}C^*<$	100–160
$C^*{-}X$ in ArX	100–160
$R{-}C^*{=}N$	110–150
$R{-}C^*OOH$	160–190
$R{-}C^*HO$	190–210
$(R_3C^*)^+$	200–300

The asterisk marks the ^1H or ^{13}C resonances observed.

J-coupling

Different interactions couple spins. In the presence of large magnetic fields, *J*-coupling, Eq. 14.21, produces perturbations on the various chemically shifted resonances that are split into multiplets. Figure 14.25b shows a spectrum where the methyl proton line is split into a triplet; the $-CH_2$ proton line is split into a quartet and the $-OH$ proton line splitting is too small to be observed.

Two Coupled Spins

The methods of Chapter 11 can be used to describe the coupling of two spins. There are two states for a single spin of $\frac{1}{2}$, $|\alpha\rangle$ and $|\beta\rangle$, but there are four states for two spins of $\frac{1}{2}$:

$$\begin{aligned} &|\alpha_1\alpha_2\rangle \\ &|\alpha_1\beta_2\rangle;\ |\beta_1\alpha_2\rangle \\ &|\beta_1\beta_2\rangle \end{aligned} \tag{14.37}$$

The Hamiltonian operator for two spins with J-coupling is

$$\hat{\boldsymbol{H}}/\hbar = -\omega_{o1}\hat{I}_{z1} - \omega_{o2}\hat{I}_{z2} + J_{12}\hat{\boldsymbol{I}}_1 \cdot \hat{\boldsymbol{I}}_2 \tag{14.38}$$

Using the properties of angular momentum, shown in Eq. 11.212, the secular determinant is obtained:

$$\begin{vmatrix} H_{11}-E & 0 & 0 & 0 \\ 0 & H_{22}-E & H_{23} & 0 \\ 0 & H_{32} & H_{33}-E & 0 \\ 0 & 0 & 0 & H_{44}-E \end{vmatrix} = 0 \tag{14.39}$$

where

$$\begin{aligned} H_{11} &= -\frac{\hbar}{2}(\omega_{o1}+\omega_{o2}) + \frac{\hbar}{4}J_{12} \\ H_{22} &= -\frac{\hbar}{2}(\omega_{o1}-\omega_{o2}) - \frac{\hbar}{4}J_{12} \\ H_{33} &= +\frac{\hbar}{2}(\omega_{o1}-\omega_{o2}) - \frac{\hbar}{4}J_{12} \\ H_{44} &= +\frac{\hbar}{2}(\omega_{o1}+\omega_{o2}) + \frac{\hbar}{4}J_{12} \\ H_{23} &= H_{32} = \frac{\hbar}{2}J_{12} \end{aligned} \tag{14.40}$$

EXAMPLE 14.1 Calculate two matrix elements, H_{22} and H_{23}, to verify Eq. 14.40.

Solution H_{ij} are matrix elements and are found from the four states given in Eq. 14.37,

$$\begin{aligned} H_{22}/\hbar &= \langle\alpha_1\beta_2|\hat{H}|\alpha_1\beta_2\rangle/\hbar \\ &= \langle\alpha_1\beta_2|-\omega_{o1}\hat{I}_{z1} - \omega_{o2}\hat{I}_{z2} + J_{12}(I_{x1}I_{x2} + I_{y1}I_{y2} + I_{z1}I_{z2})|\alpha_1\beta_2\rangle \end{aligned}$$

The operator $\hat{I}_z$ terms are diagonal, while the $\hat{I}_{x1}\hat{I}_{x2}$ and $\hat{I}_{y1}\hat{I}_{y2}$ terms require a change from the $|\alpha_1\beta_2\rangle$ to the $|\beta_1\alpha_2\rangle$ state. Therefore only the $\hat{I}_z$ components remain, giving

$$H_{22} = -\frac{\hbar}{2}(\omega_{o1}-\omega_{o2}) - \frac{\hbar}{4}J_{12}$$

For the off-diagonal matrix elements

$$H_{23} = \langle\alpha_1\beta_2|\hat{H}|\beta_1\alpha_2\rangle$$

and now the states must change from $|\alpha_1\beta_2\rangle$ to $|\beta_1\alpha_2\rangle$. Only the $I_{x1}I_{x2}$ and $I_{y1}I_{y2}$ terms can do this. Again, using the properties of angular momentum,

$$\begin{aligned} H_{23}/\hbar &= \langle\alpha_1\beta_2|\hat{H}|\beta_1\alpha_2\rangle/\hbar \\ &= J_{12}\langle\alpha_1\beta_2|(I_{x1}I_{x2} + I_{y1}I_{y2})|\beta_1\alpha_2\rangle = +J_{12}/2 \end{aligned}$$

Use has been made of the orthogonality of the spin states, Eq. 11.225.

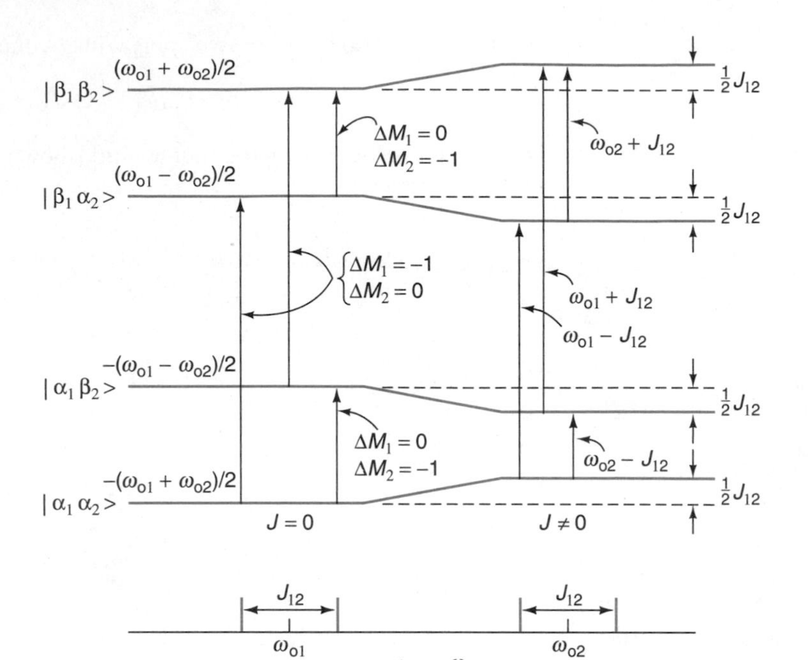

FIGURE 14.31
NMR of two heteronuclear spins (AX) showing the two sets of transitions and the resulting J-splittings produced.

The secular determinant, Eq. 14.39, blocks out into two 1×1 matrices and one 2×2 matrix. Consider first the case where $|\omega_{o1} - \omega_{o2}| >> |J_{12}|/2$. In this case the off-diagonal terms can be dropped and the secular determinant is diagonal. The eigenvalues are given by Equation 14.40 and are shown in Figure 14.31. Notice that the J_{12} coupling splits the lines to produce a doublet. A physical picture of this process is that the J-coupling produces a local magnetic field that adds or subtracts from the applied magnetic field B_o. Hence the neighboring spins see two slightly different local magnetic fields, and an energy splitting results. A generalization of this is the origin of the multiplet structure observed in NMR, such as shown in Figure 14.25b for ethanol.

The case here that ignores the off-diagonal terms in Eq. 14.39 is called the X approximation and is valid when

$$|\omega_{o1} - \omega_{o2}| >> |J_{12}|/2 \tag{14.41}$$

Two spins that obey this condition are denoted AX.

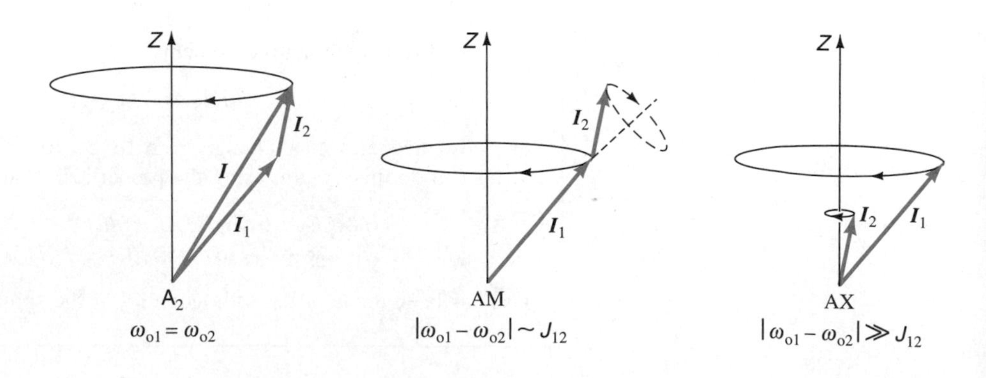

FIGURE 14.32
The progression from strongly coupled homonuclear spins (A_2) through the AM case to the AX case.

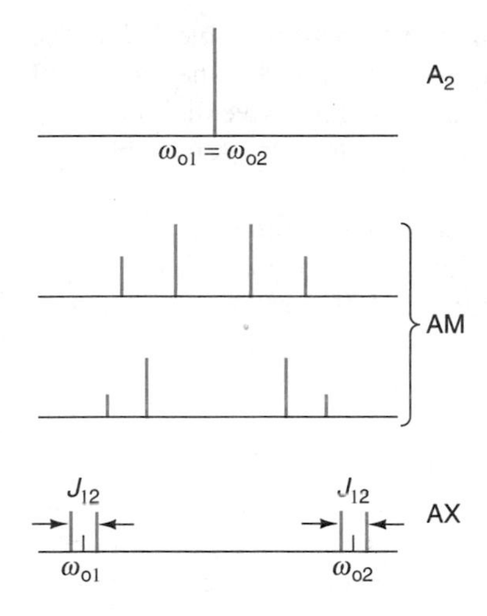

FIGURE 14.33
Stick spectra of two coupled spins of ½ from the A_2 through the AM case to the AX case.

In the case when the two spins have resonance frequencies that do not satisfy Eq. 14.41, the 2 × 2 block in the secular determinant, Eq. 14.39, must be solved. When $\omega_{o1} = \omega_{o2}$, the two spins are strongly coupled and are called an A_2 system. When $\omega_{o1} \approx \omega_{o2}$, they are called an AB or an AM spin system. Figure 14.32 shows these cases in terms of vector coupling and uncoupling.

When the 2 × 2 secular determinant is solved exactly, the four eigenvalues are

$$
\begin{aligned}
E_{11} = E_1 &= -\frac{\hbar}{2}(\omega_{o1} + \omega_{o2}) + \frac{\hbar}{4}J_{12} \\
E_{22} = E_2 &= -\frac{\hbar}{2}\sqrt{(\omega_{o1} - \omega_{o2})^2 + J_{12}^2} - \frac{\hbar}{4}J_{12} \\
E_{33} = E_3 &= +\frac{\hbar}{2}\sqrt{(\omega_{o1} - \omega_{o2})^2 + J_{12}^2} - \frac{\hbar}{4}J_{12} \\
E_{44} = E_4 &= +\frac{\hbar}{2}(\omega_{o1} + \omega_{o2}) + \frac{\hbar}{4}J_{12}
\end{aligned}
\tag{14.42}
$$

Figure 14.33 shows how the spectrum varies as ω_{o1} and ω_{o2} are changed from being different (AX) to equal (A_2). The line intensities are determined by the wave functions, which are not given.

Classification of Spectra

The Hamiltonian operator for two spins can be solved exactly, but as the number of spins increases, the dimension of the secular determinant increases. Historically, perturbation theory was used to classify spectra roughly based upon the order of the perturbation used. (See page 566.) The classifications are as follows:

Zeroth-order spectra (not in current use): These ignore all J-coupling in the calculations. (See, for example, Figure 14.31 on the left side.)

First-order spectra: Only the Z component of the spins is used, which reduces the J-coupling term in the Hamiltonian operator, Eq. 14.38, to $J_{12}\hat{I}_{1z}\hat{I}_{2z}$. This results in a diagonal secular determinant. (See Figure 14.31 on the right side.)

Second-order and higher spectra: These result from keeping the second-order and higher perturbations terms in the Hamiltonian operator. It is usually easier to solve the secular equation exactly for these cases, using computational methods. (See Eq. 14.42) From the first-order cases, however, the main aspects of high-resolution J-coupled NMR spectroscopy can usually be understood. Complications arise due to magnetically equivalent spins and overlap of multiplet splitting. Treatment of these can be found in the NMR texts listed at the end of this chapter.

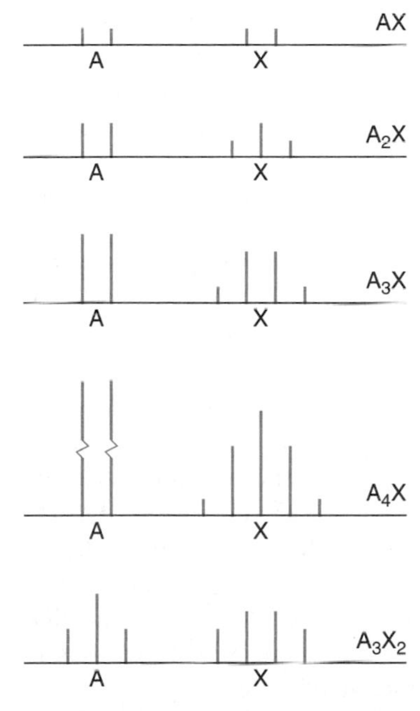

FIGURE 14.34
Various J-coupled spectra for the cases AX, A_2X, A_3X, A_4X, and A_3X_2. See Figure 14.31 for the energy splittings for the AX. Compare the multiplet structure of the A_3X_2 case to the $-CH_3$ and $-CH_2$ multiplet structure shown in Figure 14.25 for ethanol.

It is possible to work out other cases, such as A_2X, A_3X, A_2X_2, AX_3, etc., but these become cumbersome and are better treated by computational methods. What is learned from these cases is apparent from the AX and A_2X cases. In the latter, the single X spin (with two states α and β) splits the two A spin resonance into a doublet, while the two A spins (with four states $\alpha\alpha$, $\alpha\beta$, $\beta\alpha$, and $\beta\beta$) split the X resonance into a triplet. Since the $\alpha\beta$ and $\beta\alpha$ states are degenerate, there are two transitions from these states with double the intensity of the other two lines in the triplet.

Figure 14.34 shows the cases for A_2X, A_3X, A_4X, and A_3X_2. In general, we can state the following results as being valid for first-order spectra: For n magnetically equivalent spins, a resonance is split into a multiplet with $n + 1$ lines. The relative

TABLE 14.3 Binomial Distribution

1
1 1
1 2 1
1 3 3 1
1 4 6 4 1
1 5 10 10 5 1
1 6 15 20 15 6 1
1 7 21 35 35 21 7 1

intensities of the multiplet lines follow the binomial distribution (Table 14.3). The total intensity of all the lines in all the multiplets is proportional to the number of spins present. For example, the two multiplets of an AX system have the same relative intensity, while for the A_3X_2 case, 3/5 of the intensity is in the A multiplet and 2/5 in the X multiplet.

The A_3X_2 system, Figure 14.34, explains most of the features of the ethanol spectrum of ethanol dissolved in CCl_4, Figure 14.25b. The two $-CH_2$ protons split the three methyl protons into a triplet and the three $-CH_3$ protons split the two $-CH_2$ protons into a quartet. The $-OH$ proton is not split by the two $-CH_2$ protons because the *J*-coupling between them is too small for the splitting to be resolved.

Complexities arise from increasing the number of spins and having a range of *J* values and chemical shifts. As such complexities increase, the spectra become more dense and it becomes increasingly difficult, if not impossible, to analyze them. Fortunately, two-dimensional NMR **(2D NMR)** can resolve many of these difficulties.

Two-Dimensional NMR

The idea of two-dimensional NMR is to introduce two time domains and perform two Fourier transformations on the two signals to give two frequency domains. The results are plotted on a two-dimensional surface that displays the spectrum in a plane (2D NMR) rather than on a line (1D NMR). Figure 14.25b for ethanol is a 1D spectrum. Ethanol is too simple a molecule to illustrate the features of 2D NMR, hence the molecule strychnine is used. The chemical formula for strychnine is shown in Figure 14.35 and a 2D spectrum of this molecule is shown in Figure 14.36, displayed as a 2D contour plot. Just as a relief map gives the elevation of mountains and other geographical features of the earth, so the 2D contour spectrum gives the positions of peaks in 2D frequency space. Note that there are peaks along the skew-diagonal (commonly called simply "the diagonal" by NMR spectroscopists) and off-diagonal peaks. The 2D spectrum has expanded the conventional 1D spectrum into a 2D plane, making it easier to identify and assign peaks. The 1D spectrum is also shown in Figure 14.36 as the projection of the 2D spectrum on the top and left of the 2D plot.

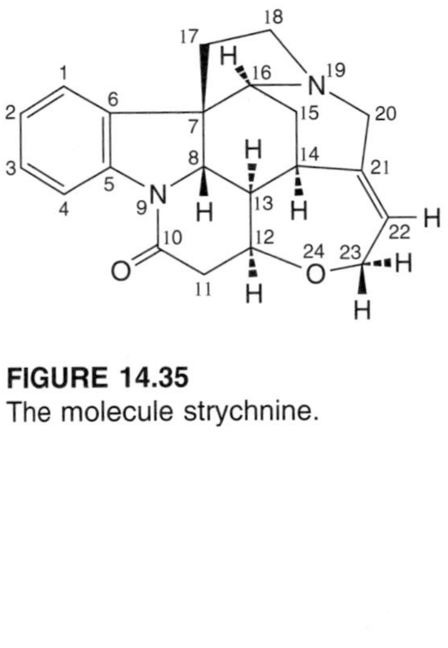

FIGURE 14.35
The molecule strychnine.

Strychnine, shown in Figure 14.35, can be used to demonstrate the main features of 2D NMR. Strychnine has 22 protons and 21 carbon atoms. Five carbons have two protons (11, 15, 17, 18, 20, and 23), while the others have either one proton (1, 2, 3, 4, 8, 12, 13, 14, 16, 22) or no protons (5, 6, 7, 10, 21). The representation of a molecule as a chemical structure such as in Figure 14.35 has limitations; although it gives a faithful representation of the chemical bonds present, it appears planar. Few molecules are planar, and much of their chemical reactivity and many of their properties depend upon their three-dimensional shape. Two-dimensional NMR helps to resolve the 3D structure of molecules.

Two-dimensional NMR is a result of collecting data in two time domains, giving a signal that is a function of two times, $S(t_1, t_2)$. Upon double-Fourier transformation, the signal becomes $S(F_1, F_2)$. In contrast, a 1D-NMR experiment collects the signal $S(t_1)$ and Fourier-transforms it to $S(F_1)$. A 1D pulsed experiment is depicted in Figure 14.37a.

The way the second time domain is introduced to produce a 2D-NMR spectrum makes use of two or more pulses in a sequence. Four regions can be identified; these are shown in Figure 14.37b: preparations, evolution, mixing, and acquisition.

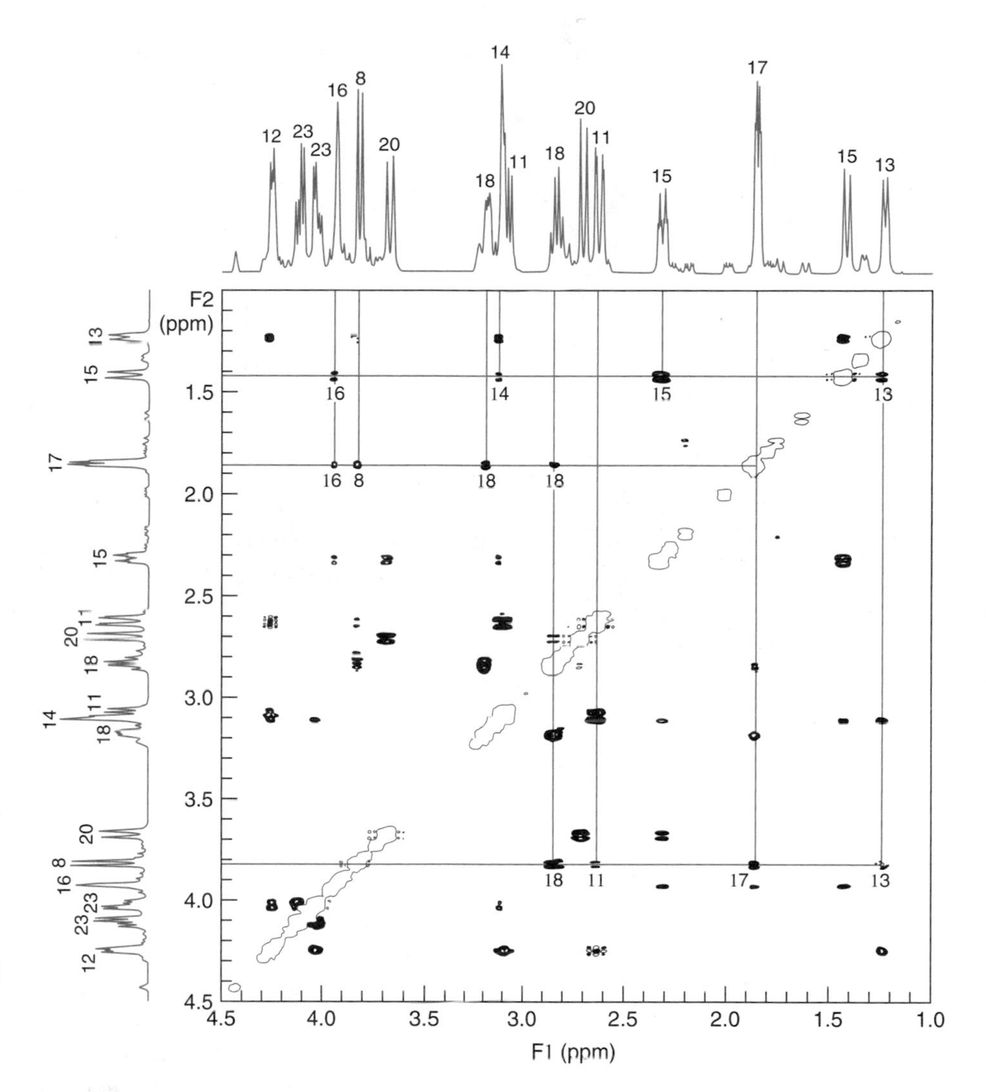

FIGURE 14.36
Two-dimensional NOESY spectrum of strychnine given as a contour plot. Contour plots are useful for the location of peak positions. In this case, every peak that is off the diagonal corresponds to protons that are within 5 Å of each other. See text for interpretation of this spectrum. Compliments of Paul Zia, Chemistry Department, McGill University.

Preparation period: This period allows the spin system to reach equilibrium before the first pulse is applied. Although not always the case, the majority of experiments start with a $(\pi/2)_x$ pulse on the equilibrium state.

Evolution period: During this period, the spins are allowed to evolve under chemical shift and J-coupling to develop characteristic coherences and polarizations. It is during this period, for example, that spin correlations and quantum coherences can develop. The time duration of this period is called t_1 and it is incrementally changed from one pulse sequence to the next. As a result of the incremental change in the length of the t_1 time period, the FID, acquired in the final acquisition step, is modified. In this way the signal is a function of two times, $S(t_1, t_2)$.

Mixing period: During this period, the spins have the opportunity to transfer and mix coherences. The way this is done determines the type of correlations and polarizations that are detected in the final acquisition period. This mixing period may be created by a simple pulse, or by a more complicated series of

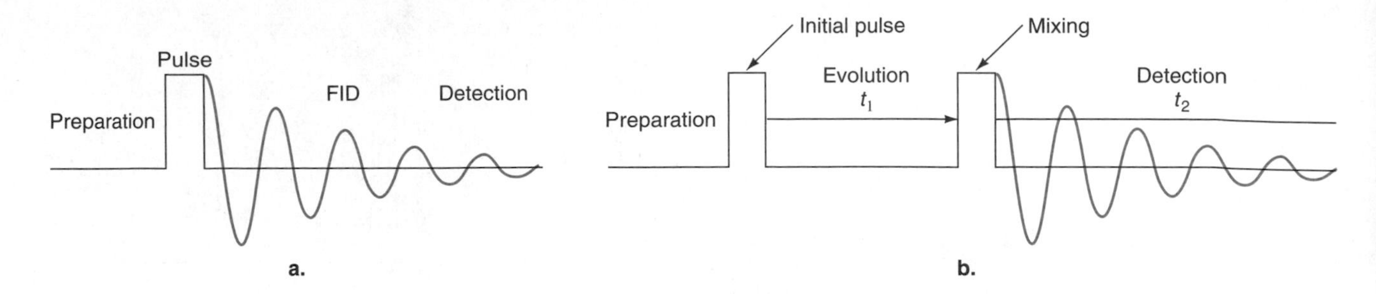

FIGURE 14.37
Comparison of one- and two-dimensional NMR. (a) One-dimensional NMR: A single pulse generates an FID. FT of the FID gives a 1D NMR spectrum. (b) Two-dimensional NMR, which defines two time domains t_1 and t_2 with two (or more) pulses. The preparation, evolution, mixing, and detection (acquisition) phases are indicated.

pulses and delays that the spectroscopist chooses to accomplish different experiments. As we proceed, some of the more important experiments will be discussed.

Acquisition period: There is no difference between the acquisition period for 1D and 2D experiments. The FID is collected in a time period t_2 resulting in the signal $S(t_1, t_2)$. By choosing different ways these four periods are designed, different NMR experiments are now routinely done.

The experiments we will discuss were all done with a Varian Unity-500 NMR spectrometer at 25 °C on a 10% sample of strychnine dissolved in deuterated chloroform, $CDCl_3$.

NOESY (Nuclear Overhauser Effect SpectroscopY): The idea of the basic Overhauser experiment is to excite nuclear spins selectively and observe the (I_z) polarization transfer to neighboring protons via the through-space dipole-dipole interaction, Eq. 14.22. This polarization transfer is called an Overhauser enhancement. The NOESY experiment is useful for structure determination of molecules, polymers, and biopolymers. In fact, it is one of the most useful of all 2D experiments because it gives off-diagonal peaks when protons are close to each other. The through-space direct dipole-dipole interaction that couples spins depends upon the inverse sixth power of their separation. This means that only when spins are less than about 5 Å apart does a NOESY cross peak appear. Notice that the off-diagonal peaks in Figure 14.36 are symmetrical on either side of the diagonal. The information on one side is repeated on the other.

Figure 14.36 is the 2D proton NOESY spectrum for strychnine. The peaks that appear in the two-dimensional plane have been projected to the top and to the left of the plot. These correspond to the 1D proton spectrum of strychnine. The numbers above the peaks refer to the protons on the molecule shown in Figure 14.35. In this case the range of chemical shifts is from 1.0 to 4.5 ppm. Within this range 18 of the 22 protons are observed. Of the remaining five protons, four (1–4) lie on the aromatic ring and the other (22) is attached to a carbon with the double bond. These five are chemically shifted and lie beyond 4.5 ppm.

The process of relating each peak to a specific proton in the molecule is called **spectral assignment.** Spectral assignment is crucial for successful interpretation of 2D NMR, and a number of techniques are used to aid in this important step. Once the assignment is done, it is possible to determine which protons are lying close to one another.

The NOESY data introduce distance constraints to the structure shown in Figure 14.35. To satisfy these constraints, the three-dimensional molecular structure must fold and twist accordingly. Nowadays computer programs are available that enable these constraints to be satisfied by using the techniques of **distance geometry.** To see these constraints, examine the lines drawn on Figure 14.35. One of the two protons on carbon 15 is within 500 pm of protons 13, 14, 16, and the other proton on carbon 15. Another line shows that the two magnetically equivalent protons on carbon 17 are close to protons on carbons 8, 16, and 18. This constrains, at least, the position of the bridge from carbon 7 to nitrogen 19. The final line shows that the proton on carbon 8 is close to those on carbons 18, 11, 17, and 13. Further examination of the figure reveals other distance constraints that the reader can verify.

This is useful in determining how molecules fold and in determining the three-dimensional structures of molecules. In the case of proteins, if two protons at two positions far apart along the molecular peptide chain show a NOESY peak, then that molecule must be folded in such a way as to satisfy that constraint. Knowledge of the three-dimensional structures of a protein is essential to the understanding of its biochemistry. The chemical activity of a molecule is known as its **function.** Knowing the functions of molecules is one of the major goals of chemical research.

DQF-COSY (Double Quantum Filtered-Correlation Spectroscopy): A large number of peaks in the NMR experiment are single-quantum coherences (having selection rule $\Delta M = \pm 1$) that clutter the 2D display along the diagonal. By removing these by double quantum filtered techniques, DQF (having selection rule $\Delta M = \pm 2$), the spectra simplify but the same J-coupling connectivity information is retained as for the single quantum COSY. Were these singlet peaks not filtered out, the spectrum would be considerably more dense and intense along the diagonal and perhaps swamp some cross peaks that lie close to the diagonal.

A COSY spectrum displays a set of diagonal peaks and a pattern of off-diagonal peaks that is symmetrical on either side of the diagonal. The DQF-COSY peaks are also "absorptive," meaning they are all positive (or have the same phase) as opposed to dispersive (which means that the peaks can have both positive and negative phases and positive and negative values). The usefulness of DQF-COSY experiments is that they reveal the effects of J-coupling. Since J-coupling normally can persist only through three chemical bonds, COSY off-diagonal peaks tell us which protons lie adjacent to each other.

Figure 14.38 is the DQF-COSY spectrum of strychnine. The couplings are restricted to spins that are no more than three bonds apart, and the projection of the peaks to the top and to the left is the normal 1D spectrum, just as in the NOESY spectrum. Yet the positions of the off-diagonal peaks are different in the COSY experiment. Again the spectral assignment is displayed by the numbers that refer to the protons attached to carbon atoms in the molecule shown in Figure 14.35. Since the COSY experiment displays off-diagonal peaks no more than three bonds apart, the ordering, or **correlation,** of successive protons can be determined. Drawn on Figure 14.38 are patterns connecting diagonal peaks. There is also a distinctive pattern of the cross peaks from which the J-coupling constant can be measured. It is known that various organic functional groups show COSY peaks in distinct regions of the 2D space. One pattern on Figure 14.38 shows that protons 11 and 12 are on adjacent carbons. Others show correlations between the two protons on carbon 15, between the protons on carbons 8 and 13, between the two protons on carbon 20, and

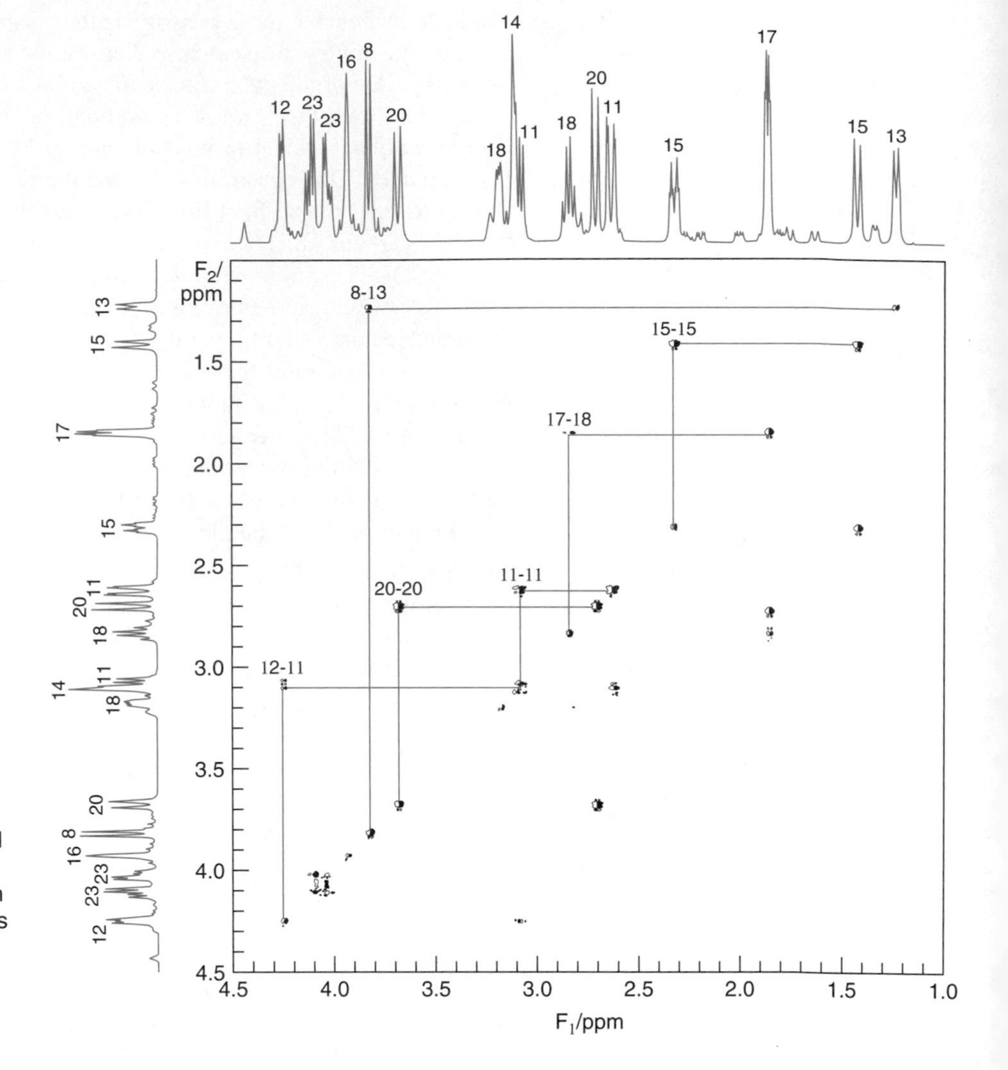

FIGURE 14.38
Two-dimensional DQF-COSY spectrum of strychnine. The proton resonances are numbered and relate to the molecule in Figure 14.35. The patterns drawn on the spectrum correlate protons no more than three bonds apart. Compliments of Paul Zia, Chemistry Department, McGill University.

between those on carbon 17 and 18. With practice, the user can quickly identify the various structural features in a molecule from the various patterns of peaks in a COSY experiment, although this becomes more challenging as the complexity of the molecule increases.

HS(M)QC (Heteronuclear Single (Multi) Quantum Coherence): In this 2D experiment the F_2 axis plots the proton spectrum and the F_1 axis the ^{13}C spectrum. Each peak corresponds to a C-H bond in HSQC. Similar information is available from the HMQC experiment. Hence, once the protons are assigned, it is possible to know which proton is coupled to which carbon. Figure 14.39 shows the HSQC spectrum for strychnine having the same proton range as the NOESY, COSY, and TOCSY, from 1.0 ppm to 4.5. This shows 16 H-C, of the 21 peaks. The remaining five peaks can be resolved and lie in the range greater than 4.5 ppm and are not presented here. Since this is a heteronuclear experiment, the spectrum is not symmetric and no diagonal peaks exist.

Such data are extremely useful for structure determination. It is instructive to analyze this spectrum in more detail. Recall from Figure 14.35 that the 22

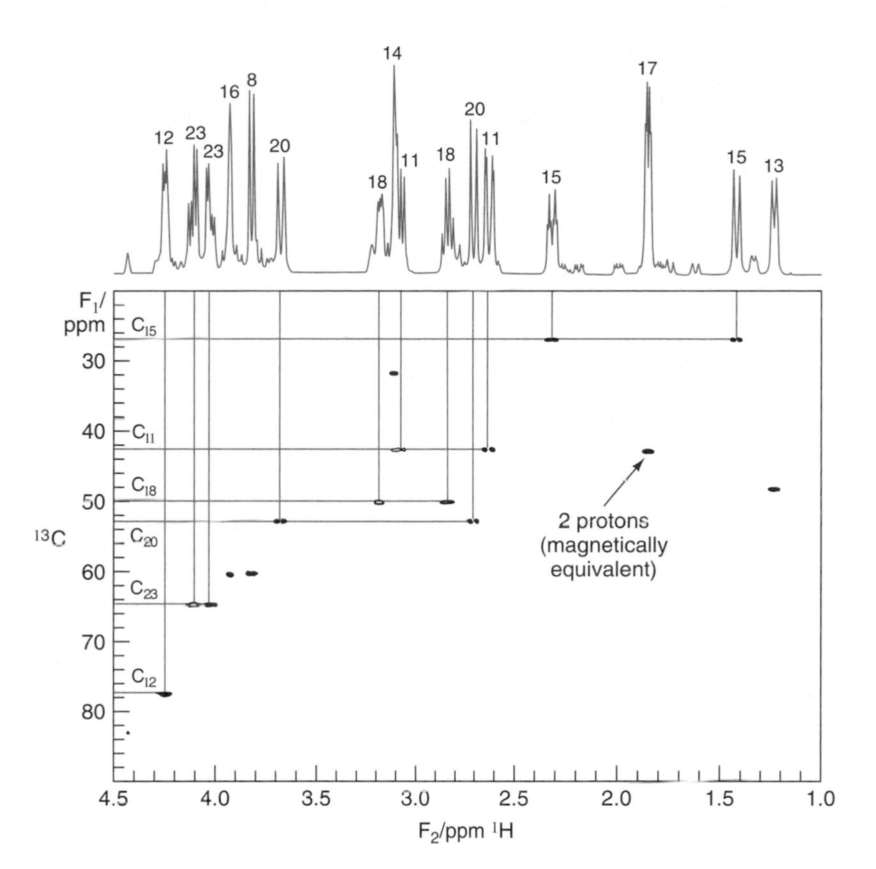

FIGURE 14.39
Two-dimensional HSQC spectrum of strychnine; ^{13}C range 20–90 ppm; ^{1}H range 1–4.5 ppm. The HSQC spectrum of strychnine shows a peak for every ^{1}H-^{13}C bond so that each peak corresponds to a ^{1}H (or magnetically equivalent protons) bonded to one carbon. Compliments of Paul Zia, Chemistry Department, McGill University.

protons are bonded to 21 carbon atoms. Five carbons have two protons (11, 15, 17, 18, 20, and 23), while the others have either one proton (1, 2, 3, 4, 8, 12, 13, 14, 16, 22) or no protons (5, 6, 7, 10, 21). This means there should be 22 peaks in the spectrum. However, one pair of protons (17) is magnetically equivalent and shows only one peak (Figure 14.39), which reduces the total number to 21 peaks. Sixteen of the 21 peaks are clearly visible in the spectrum in Figure 14.39 but, as mentioned earlier, all 21 peaks could be observed if the range were expanded to 10 ppm. When one carbon has two protons attached, then along that carbon resonance (horizontal line), two proton peaks are observed, unless the protons are magnetically equivalent. This can be seen from lines drawn on Figure 14.39. From this spectrum it is possible to determine the carbon resonances from knowing the proton resonances, and vice versa.

Note that the chemical shift range of the carbons is about ten times greater than that for protons. This is due to the greater electron density around carbon atoms, which leads to a larger range of chemical shifts. The information from such a spectrum is useful for the assignment of spectra.

Examples of the Use of NMR in Chemistry

The details of how the experiments just discussed are performed are beyond the scope of this book. It is, however, instructive to give some final examples of liquid-state NMR.

2D NMR is now a routine technique that allows the determination of structures, chemical reactivity, and dynamics of complicated molecules. Of course, by including more than two time periods, it is possible to develop 3D [$S(t_1, t_2, t_3)$] and 4D NMR [$S(t_1, t_2, t_3, t_4)$]. This has been done and has particular applicability to determining the structure of complicated systems such as polymers, proteins, and biopolymers for which 2D spectra become too dense.

A protein is a biopolymer consisting of many amino acids joined together by peptide linkages. Multidimensional NMR can be used to obtain the secondary structure (helix, coil, or beta sheet) and the tertiary structure (folding). There are, however, only 20 naturally occurring amino acids and these repeat numerous times in the primary protein sequence. As such, many resonances overlap, making the assignment of the spectra difficult. Once the spectral assignment is done, however, the NOESY experiment can be used to determine the secondary and tertiary structures by the use of distance geometry. Such techniques are useful for polypeptides up to about 70 amino acid residues. The current technique of 3D heteronuclear correlation spectroscopy limits the size of proteins that can be studied by NMR to less than about 350 amino acid residues.

Rather than describe an application involving biopolymers, it is more instructive to give a more chemical application of NMR that makes use of some of the 2D experiments already described. Consider the reagent Cp_2ZrCl_2/2BuLi (Negishi's reagent), which has been extensively used in organic synthesis. (The Cp group [C_5H_5] is the h5-cyclopentadienyl group. Bu is the standard abbreviation for "butyl.") In the laboratory of John F. Harrod of McGill University the extraordinary complexity and unexpected chemistry that occur with this reagent have been demonstrated using almost exclusively NMR techniques. Figure 14.40a displays the one-dimensional ^{1}H-NMR spectrum of Cp_2ZrCl_2/2BuLi. The ^{1}H resonances can be assigned with the aid of the HMQC experiment shown in Figure 14.40b. Once done, the 1D spectrum, shown in Figure 14.40c, reveals three major species at -20 °C. These are indicated on the spectrum at positions 2a, 3a, and 4a. Following the assignment process, NOESY is used to detect the dynamic exchange processes that occur under different conditions (see Figure 14.39c.) From these and similar NMR experiments, which are performed at different times throughout the organic synthesis procedure and at different temperatures, it is possible to map out various sequences of reactions that occur during the decomposition of the reagent.

Such a study shows the enormous power of NMR because all the assignments and conclusions are drawn from NMR data. In deducing possible reaction sequences, model compounds are proposed as possible candidates. These, too, are then studied by NMR so that various products in the proposed scheme can be confirmed.

Solid-State NMR

So far the discussion of NMR has focused on liquids. Solid-state NMR is a large field with numerous experiments that also allow the structure and dynamics of solids to be studied. The essential difference between liquid-state NMR and solid-state NMR lies in the increased number of interactions that are not detected in liquids because they are averaged out. The parts of the interactions that are averaged out in liquids have different spatial components and are therefore anisotropic in nature. Through sample spinning and thermal random motions, anisotropies tend to average out for liquids, whereas for rigid solids they persist. Of these interactions,

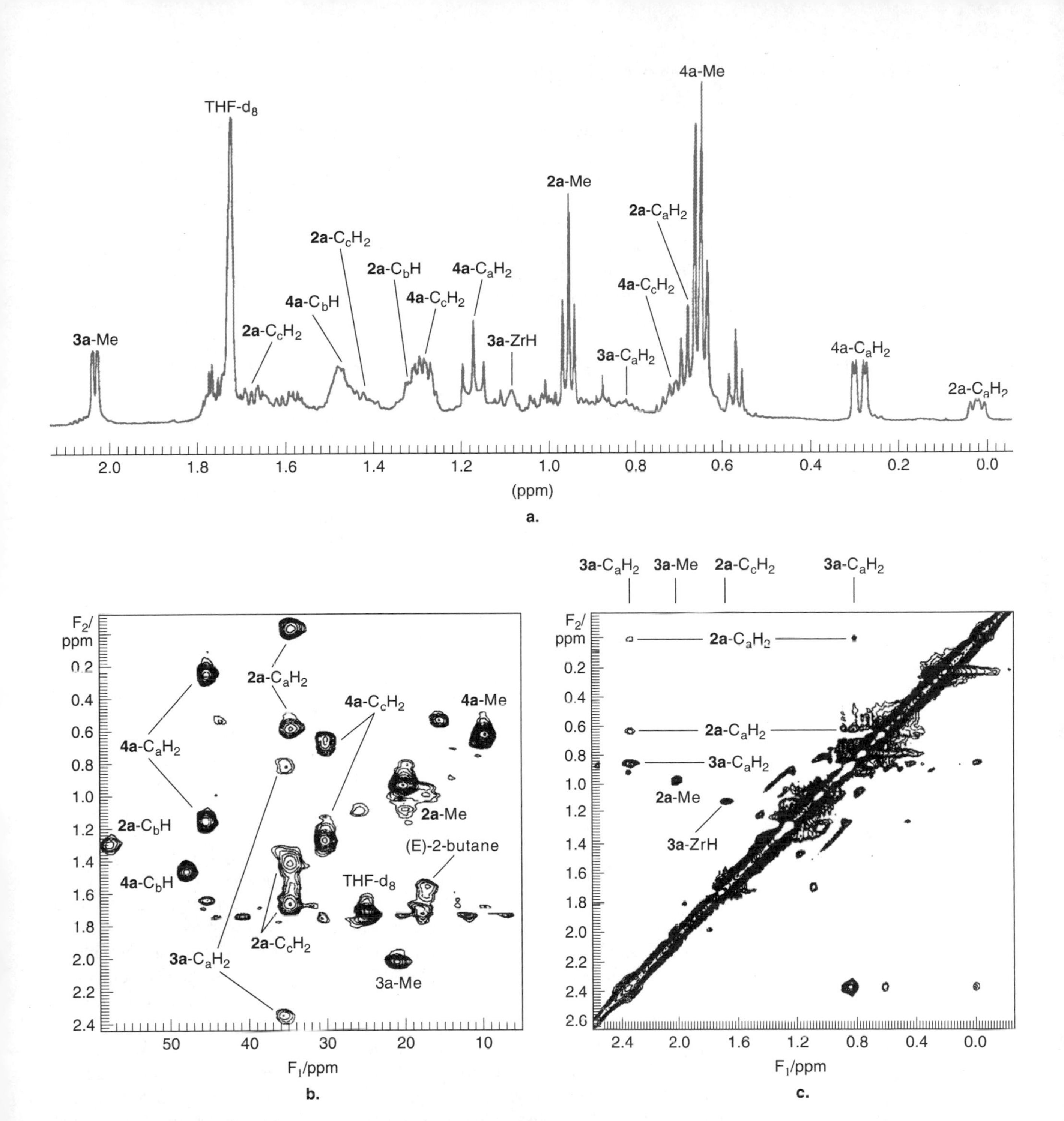

FIGURE 14.40
(a) ^{1}H NMR spectrum of a mixture of **2a, 3a,** and **4a** in THF-d_8 at -20 °C. The multiplicity of the C_aH_2 group in **2a** and **4a** is characteristic of small scale with two diastereotopic geminal protons having considerably different coupling constants, guache and eclipsed, to the same vincinal neighbor. (b) ^{1}H-^{13}C HMQC spectrum of a mixture of **2a, 3a,** and **4a** in THF-d_8 at -40 °C. Pairs of cross peaks having a common ^{13}C chemical shift belong to diastereotopic methylene groups. (c) ^{1}H-^{1}H NOESY spectrum of a mixture of **2a** and **3a** in THF-d_8 at 0 °C. Both C_aH_2 resonances of **3a** are in dynamic exchange with each other and with every C_aH_2 signal of **2a.** There is, however, no exchange with C_aH_2 and C_aH_2 pairs of **2a.** Reproduced with permission from D. Dioumaev and J. F. Harrod, "Species present in a Cp_2ZrCl_2/BuLi mixture," *Organometallics, 16, 7,* 1454 (1997), Figures **a., b., c.,** by permission of The Royal Society of Chemistry.

three have significant impact on the NMR spectra of solids. These are the presence of direct dipole-dipole interactions, Eq. 14.22; chemical shift anisotropy (CSA); and, for spins with $I > \frac{1}{2}$, the nuclear quadrupole interaction, Eq. 14.24. The major effect of these anisotropic interactions is to broaden the spectra to the point that many features are lost. Considerable effort has been devoted to techniques to remove these anisotropies to narrow the spectra. This is accomplished by magic angle spinning, various pulse sequences, and combinations of the two. Further discussion is beyond the scope of this book.

14.7 Electron Magnetic Resonance (EMR)

Whereas NMR arises from intrinsic nuclear spins, **electron magnetic resonance (EMR)** arises from the spin $\frac{1}{2}$ of the electrons. The electron Bohr magneton is more than three orders of magnitude greater than the average nuclear magneton. This difference means that the Zeeman splitting due to an electron is more than 1000 times greater than for nuclei, thereby moving the Larmor frequency from the VHF radio frequency range into the microwave region of the EM spectrum. This accounts for the major differences in experimental technique.

EMR can generally be classified into either **electron spin resonance (ESR),** based on the presence of one unpaired electron, or more generally, **electron paramagnetic resonance (EPR)** for systems with several unpaired electrons. The two terms, ESR and EPR, are, however, used interchangeably. EMR can further be separated into two broad areas based on the method used to obtain a spectrum, **continuous wave (CW)** or **pulsed** techniques. EMR is restricted to the study of molecules and ions that have at least one unpaired electron spin. The majority of ESR experiments are CW, although recently developed pulsed experiments give additional insight and information similar to and complementary to that provided by NMR. Generally, pulsed techniques in NMR were initially motivated by the ability to improve the signal-to-noise ratio. The same improvement is not obtained in ESR for several reasons, one being that the thousandfold increase in the strength of the interaction obviates the need for the same level of signal averaging needed for NMR.

ESR was first observed in 1944 by the Russian physicist Eugeny Konstantinovich Zavoisky (1907–1976).

ESR is an important magnetic resonance technique for studying chemical processes involving ions and free radicals. This excludes most molecules, but nitric oxide (NO), which has 15 electrons, is one example of a stable molecule with an odd number of electrons. Free radicals, such as H and CH_3, also have an odd electron and diradicals have two unpaired electrons as found in certain molecules, such as O_2, which has a triplet ground state. Charged species such as $[C_6H_6]^-$ also show resonance effects because of an unpaired electron. Transition elements with an unpaired electron, such as copper, will also show resonance. Aside from its applications in chemistry, EMR is used in the investigation of materials—the physics of metals, for example—as well as in applications in the life sciences and medicine.

Figure 14.41a shows a schematic diagram of an ESR spectrometer. Microwaves are generated by an oscillator called a *klystron tube* or *Gunn diode* and are directed into the sample under investigation by means of a hollow metal *waveguide* appropriate in size to contain the wavelength of the radiation. The tube is normally kept under vacuum in order to reduce line broadening, which would otherwise occur because of collisions. The sample, containing approximately 10^{11} spins, is maintained in a glass or quartz container supported between the poles of an

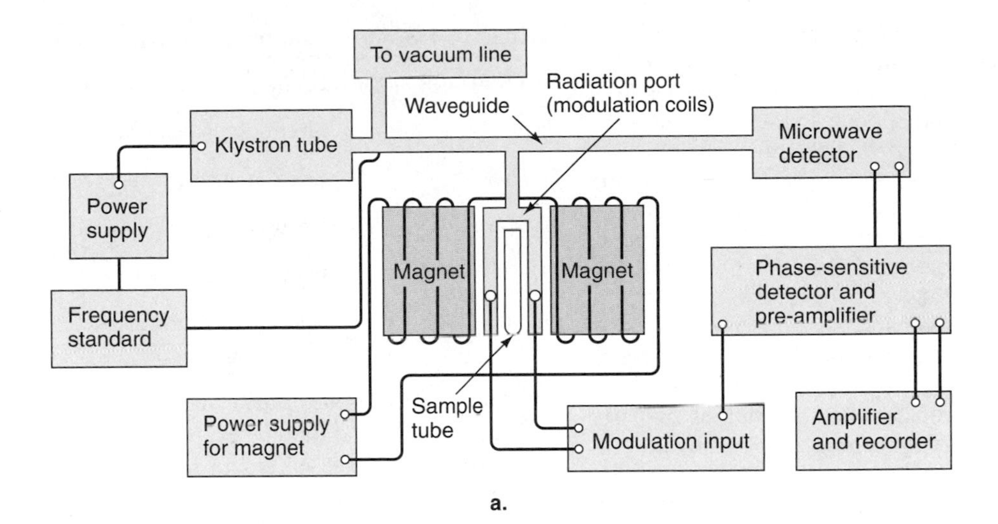

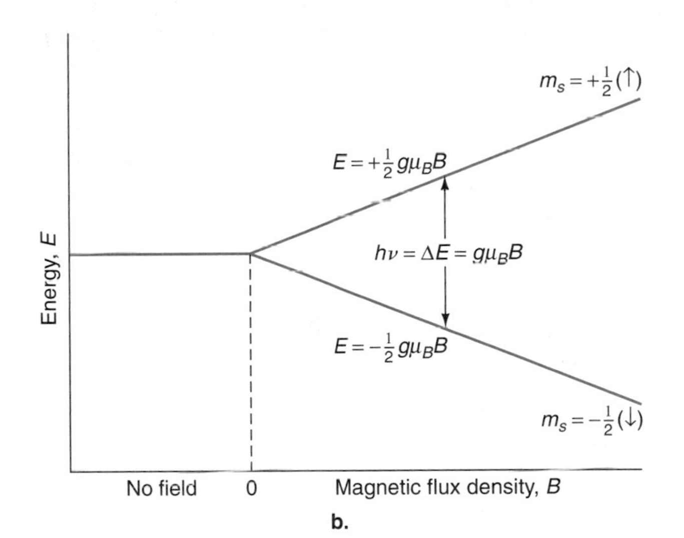

FIGURE 14.41
(a) Schematic diagram of an electron spin resonance spectrometer. (b) The variation of the electron spin energy with the magnetic flux density.

electromagnet operated over a range of field strengths from zero up to about 10 T. It is usual to operate the klystron tube at a fixed frequency and to vary the magnetic field until resonance is achieved. Spectrometers operate over a wide range of frequencies in various bands. Commercial instruments operate in the *X-band* of the microwave region (8 to 10.9 GHz, with 9.5 GHz being most common). Others operate in the *K-band* (27 to 35 GHz), although the complete microwave spectrum has been used (from about 1.5 to several hundred GHz). Typically the modulation frequency may range over 100 kHz, which corresponds to a range of approximately 10^{-2} T, or 100 G in the magnetic field.

The theoretical description of ESR is similar in every way to NMR, once the magnitude of the Zeeman splitting is taken into account. Figure 14.41b shows the linear Zeeman splitting of an electron spin with energy difference given by

$$\Delta E = h\nu = g\mu_B B_o \tag{14.43}$$

where μ_B is the Bohr magneton and g is the Landé-g factor given by Eq. 13.87. Recall that the g factor results from electron spin-orbit coupling. Chemical shielding

The magnetic moment of the nucleus is proportional to the angular momentum, and it is in the opposite direction. In a magnetic field the nuclear magnetic dipole precesses about the direction of the field, in the same manner as for the spinning electron. The component of the magnetic moment in the direction of the field (the *Z*-direction) is $\mu_{I,z} = g_N M_I \mu_N$. The Bohr magneton μ_B has been replaced by the **nuclear magneton** μ_N, defined by $\mu_N = eh/4\pi m_p = e\eta/2m_p$, where m_p is the mass of the proton. The quantum number m_l has been replaced by M_I, which relates to nuclear spin, and Eq. 14.14 also involves the **nuclear *g* factor** g_N. Actual nuclear magnetic moments differ from μ_N and they are normally expressed as $\mu'_N = g_N I$ or in units of μ_N, $g_N = \mu_N/I$. Thus, to obtain g_N for the atomic nucleus of interest, the handbook value of μ_N is divided by the spin. Some values of this factor are given in Table 14.1 (p. 735). An important point to note is that the nuclear magnetic moments are much smaller than those due to orbital motion or to electron spin. The reason is that the nuclear magneton μ_N is much smaller than the Bohr magneton μ_B because the much greater nuclear mass appears in the denominator of the expression. The value of the nuclear magneton is $\mu_N = 1.60219 \times 10^{-19}$(C) 6.6262×10^{-34}(J s)/4 ($1.672\,65 \times 10^{-27}$)(kg) $= 5.0508 \times 10^{-27}$ J T^{-1}, to be compared with 9.274×10^{-24} J T^{-1} for the Bohr magneton.

TABLE 14.4 Some Common EMR Experiments

FT-ES(M)(P)R	Fourier transform electron spin (magnetic)(paramagnetic) resonance
ESEEM	Electron spin-echo envelope modulation
ENDOR	Electron nuclear double resonance
ELDOR	Electron-electron double resonance
FDMR	Fluorescence detected magnetic resonance
ODMR	Optically detected magnetic resonance

influences the positions of the resonances, and hyperfine interactions occur between the electron spin and the nuclear spins. These are entirely equivalent to the NMR case, giving chemical shifts and multiplet structures much like the multiplet patterns found in *J*-coupling.

If we consider pulsed ESR, then the precessional motions are best viewed in the rotating frame. An electron paramagnetic vector is produced, and this can be pulsed by bursts of microwave frequency to rotate the magnetization through various angles. FT methods are employed **(FT ESR);** spin echoes are observed (Hahn echoes, stimulated echoes, as well as an ESR unique experiment called **ESEEM**). The relaxation times T_1 and T_2 arise; 2D experiments equivalent to NMR can be generated (e.g., COSY); and double resonance experiments **(ENDOR, ELDOR)**, as well as a variety of others **(triple resonance ENDOR, FDMR, ODMR)** are performed. Finally, attempts are being made to use the electron spin for medical imaging, much the same way that NMR is the basis for MRI. Table 14.4 list some of the more common EMR experiments.

The relaxation time in EMR is much faster than the NMR relaxation times. In fact, paramagnetic impurities (such as oxygen) in NMR samples dominate all other relaxation mechanisms.

EMR techniques are applied to both liquids and powders and single crystals in the solid state. In solids both the *g* factor and the hyperfine coupling are usually anisotropic tensors and are analyzed along the same lines as for NMR.

EXAMPLE 14.2 An EPR instrument is operating at a frequency of 9.100 GHz, and measurements are made with atomic hydrogen. Resonance is observed at a magnetic flux density of 0.3247 T. Calculate the *g* value for the electron in the hydrogen atom.

Solution From Eq. 14.43, the *g* value is given by

$$g = \frac{h\nu}{\mu_B B}$$

$$= \frac{(6.626 \times 10^{-34}\ \text{J s} \times 9.100 \times 10^{9}\ \text{s}^{-1})}{(9.274 \times 10^{-24}\ \text{J T}^{-1} \times 0.3246\ \text{T})} = 2.0024$$

Hyperfine Structure

The type of ESR spectrum that is commonly observed is shown in Figure 14.42. Usually more than one peak is observed, resulting in hyperfine spectral structure.

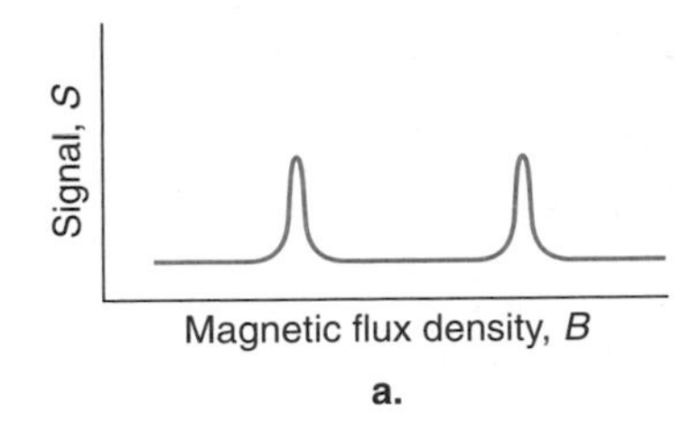

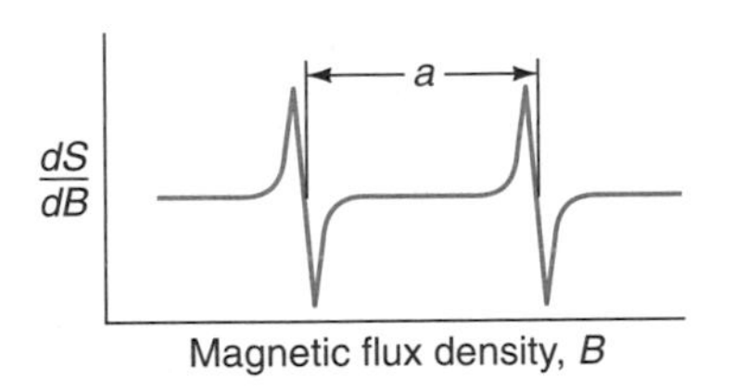

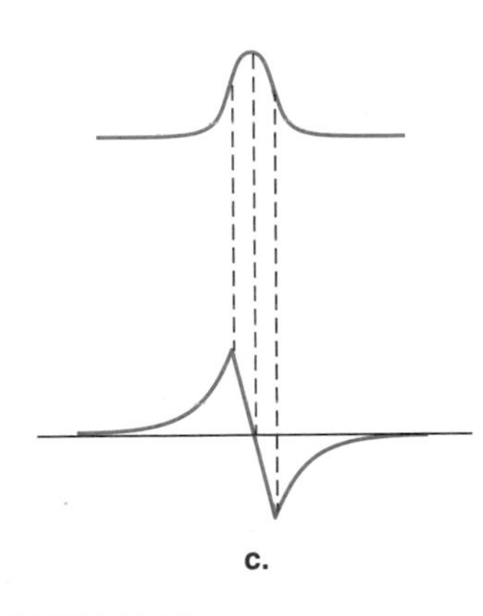

FIGURE 14.42
(a) An ESR absorption curve, in which the signal S is shown as a function of the magnetic flux density B. (b) The first derivative of the signal, dS/dB, is shown as a function of B. Such a signal is observed when a phase-sensitive detector is used. The value of a is the hyperfine splitting constant. (c) Detail of the relationship between the figures in (a) and (b).

The most important contribution to this splitting of lines is the interaction between the electron spin magnetic moment and the nuclear spin magnetic moment. This coupling is of the type AX_n. The form of the interaction is $\hat{\boldsymbol{S}} \cdot \boldsymbol{A} \cdot \hat{\boldsymbol{I}}$ for each nuclear spin coupled to an electron. Here $\hat{\boldsymbol{S}}$ is the electron spin operator and $\boldsymbol{A}$ is the, possibly anisotropic, coupling tensor. The n nuclear spins lead to multiplet splitting of the electron resonance lines in the same way as seen earlier (Figure 14.34). The nuclear spins can have various magnitudes I with $(2I + 1)$ possible orientations.

The electron–nuclear magnetic interaction energy is small in comparison with the electron–spin resonance energy and is independent of the strength of the magnetic field, just as the hyperfine coupling constants are field independent. Figure 14.43 shows the energy splitting obtained with an electron spin coupling to a spin of $\frac{1}{2}$. Measuring the hyperfine coupling constants gives one of the important parameters available from EMR experiments.

> ***EXAMPLE 14.3*** How many ESR lines are expected from a ^{35}Cl atom?
>
> ***Solution*** ^{35}Cl has a spin quantum number $I = \frac{3}{2}$ and there are $2I + 1$ possible M values. Therefore the values are $+\frac{3}{2}$, $+\frac{1}{2}$, $-\frac{1}{2}$, and $-\frac{3}{2}$. The hyperfine structure in the ESR spectrum therefore consists of four lines.

In practical examples there is usually one unpaired electron in an ion or free radical that can couple to many of the different nuclear spins that are present. Consequently the ESR spectrum can be split into many hyperfine lines, often too numerous to interpret. Moreover, the fast relaxation and anisotropic hyperfine interactions lead to broad ESR spectra. A method of circumventing these problems is to perform an **ENDOR** experiment. The basic idea is to do NMR on the nuclei and ESR on the electron simultaneously. Since a nucleus usually couples to only one electron, the spectrum is simpler, allowing the hyperfine coupling constant between the electron spin and the nuclear spin to be measured. ENDOR is a double resonance experiment which applies a constant microwave frequency at the electron-spin Larmor frequency to almost equalize the populations of the two Zeeman levels (causing saturation). At the same time, a second *rf* field is applied to the nuclear spins to induce nuclear Zeeman transitions. Much like cross-polarization experiments in NMR, the hyperfine coupling between the electron spin and the nuclear spin provides a pathway for the two spins to communicate. This modulates the ESR lines, which are being simultaneously monitored by the same resonance frequency that is being applied to the electron spin. ENDOR experiments are done in both the liquid and solid phase and are usually performed at low temperature. Pulsed ENDOR is also performed.

As mentioned previously, pulsed techniques provide a less advantageous signal-to-noise ratio for EMR than for NMR. Moreover, since the electron spin Zeeman splitting is a thousand times greater than in NMR, it is more difficult to produce the pulses and control the flip angles. EMR pulses are in the nanosecond range, compared to the microsecond range for NMR, and this pushes the limits of current electronics, pulse generators, and switching techniques. It is difficult using pulsed techniques to cover the wide frequency range necessary to excite electron spin systems even though the EMR pulses are of much shorter duration than for NMR. Since

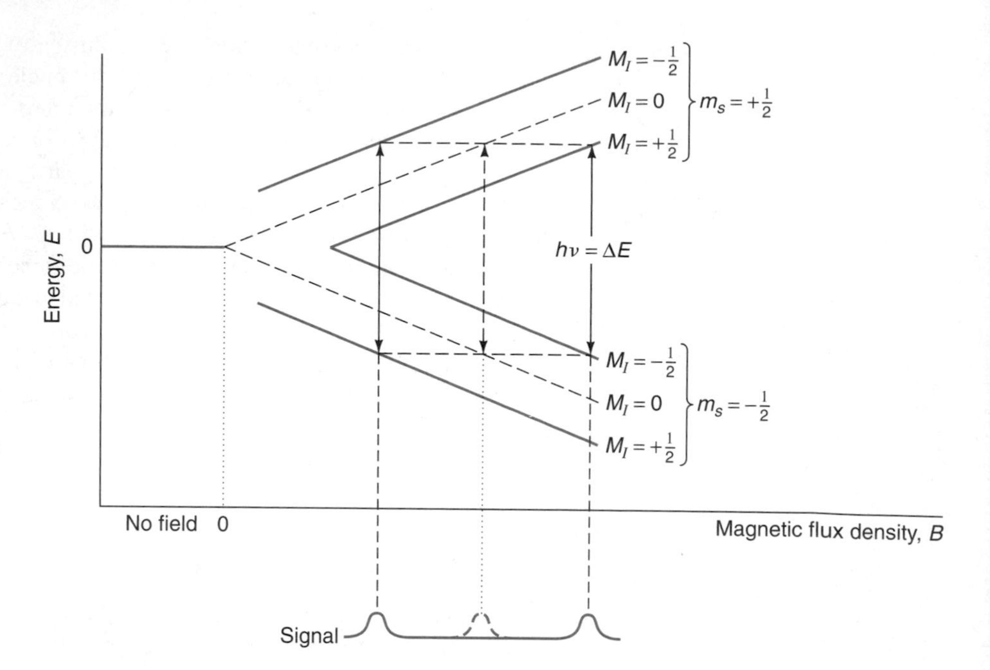

FIGURE 14.43
The energy splitting obtained with a hydrogen atom, for which the permitted nuclear spin quantum numbers are $+\frac{1}{2}$ and $-\frac{1}{2}$. The dashed lines correspond to the situation in Figure 14.41, where there is no interaction with the nucleus.

time and frequency are inverse Fourier parameters, longer pulses cover smaller frequency bands and shorter pulses are needed to cover larger frequency bands.

Nevertheless, pulsed EMR has undergone rapid development in recent years so that, as mentioned, pulsed experiments similar to those in NMR can be performed in EMR. Normal pulsed FT EMR is used to measure relaxation times and to produce a frequency spectrum in one and two dimensions. The experiment gives information about the nuclear hyperfine interactions and complements the ENDOR experiment. The experiment also can be applied to study the structure of proteins (complexed to metals), which may not crystallize for X-ray diffraction studies or may be too large for NMR experiments. In the ESEEM experiment, two microwave pulses are applied to the system and the electron spin echo (ESE) is observed. By repeating the experiment with longer times between pulses, the relaxation of the echo can be monitored. Not only are the electron-spin transitions excited by the pulses, but also some nuclear-spin transitions occur and these are coupled to the electron spin via hyperfine and quadrupole mechanisms. The relaxations result from combined effects of the electrons coupled to the nuclei.

As an example of the application of EMR to obtain structural information and hyperfine coupling constants, consider the spectrum of compound $Cp_2Ti^{III}(NC_5H_5)(CH_3)$, for which the titanium ion has an unpaired electron. In Figure 14.44a the structure of this complex is shown and in Figure 14.44b an ESR spectrum is given which displays well-separated coupling constants. In this case, the unpaired titanium electron couples via a one-bond hyperfine interaction to the spin $\frac{1}{2}$ of the ^{31}P nucleus ($a_p = 28.6$ G) and via a two-bond coupling to the proton nuclear spin of $\frac{1}{2}$ ($a_H = 4.3$ G). Although ^{29}Si has a nuclear spin of $\frac{1}{2}$, the isotropic abundance is too small to detect from the spectrum. Analysis enables the bond lengths and strengths to be determined.

In contrast, Figure 14.44c shows the ESR spectrum for the same complex with two overlapping coupling constants. In this case the hyperfine coupling constant between the nitrogen and the electron spin is $a_N = 1.75$ G, and between the protons

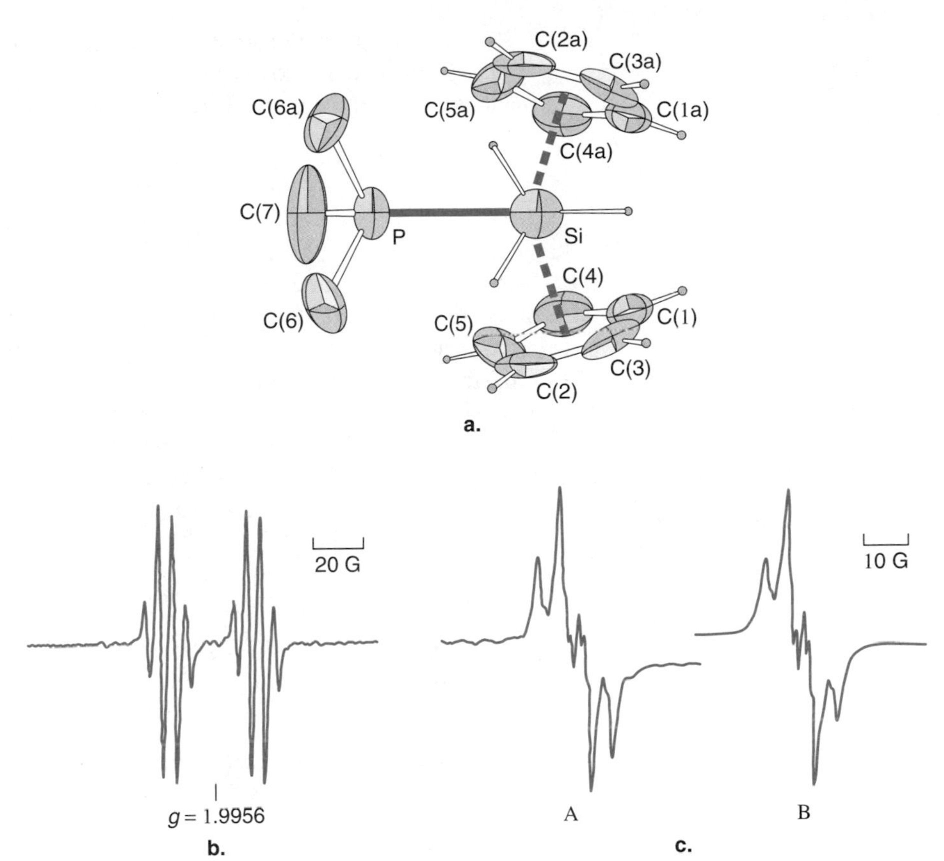

FIGURE 14.44
(a) A view of the structure of $Cp_2Ti(SiH_3)(PMe_3)$ looking along the Si-Ti bond with Ti hidden. (b) EPR spectrum of $Cp_2Ti(SiH_3)(PMe_3)$ in toluene at room temperature for well-separated hyperfine couplings. (c) EPR spectrum of $Cp_2Ti(SiH_3)(PMe_3)$ (A) with overlap of hyperfine couplings, along with a numerical simulation (B). Reproduced with permission from W. Hao, A. Lebuis, and J. F. Harrod, "Synthesis and structural characterization of $Cp_2Ti(SiH_3)(PMe_3)$," *Chem. Commun. 1089* (1998), Figures a., b. and c., by permission of The Royal Society of Chemistry.

and the electron spin it is $a_H = 3.75$ G. These are determined by simulating the spectrum. Such simulations are essential to the interpretation of many ESR spectra. Using computational methods, the values of overlapping hyperfine coupling constants can be deduced. Shown in Figure 14.44c, part B is a simulated spectrum which can be compared to the actual spectrum in Figure 14.44c, part A. The agreement in this case is not perfect, since small and complicated couplings to the protons of the pyridine ring are not included.

14.8 Mössbauer Spectroscopy

R. L. Mössbauer received the Nobel Prize in physics in 1961 for discovering the Mössbauer effect.

In 1958 the German physicist Rudolf Ludwig Mössbauer (b. 1929) discovered what has come to be known as the **Mössbauer effect.** It is concerned with the emission of γ (Greek *gamma*) rays, which under ordinary conditions are emitted with a considerable spread of wavelength. This is because of the recoil of the emitting nuclei. However, in Mössbauer's experiments the emitting nuclei were part of a crystal, the recoil of which is very small, and the γ rays are therefore emitted with an exceedingly narrow spread of wavelength. In some experiments the spread is as low as 1 part in 10^{13}. An important application of these highly homogeneous γ rays was to provide a crucial test of Einstein's theory of relativity. They have also found useful spectroscopic applications.

To take a specific example, the primary source of the γ rays is often the ^{57}Co nucleus. This decays slowly (with a half-life of 270 days) to give an excited ^{57}Fe nucleus, which emits γ rays with a half-life of 2×10^{-7} s:

$$^{57}\text{Co} \xrightarrow{t_{1/2} = 270 \text{ days}} {}^{57}\text{Fe}^* \xrightarrow{t_{1/2}=2\times10^{-7}\text{s}} {}^{57}\text{Fe}$$
$$\searrow \gamma \text{ rays}$$

The frequency of the emitted γ rays is about 3.5×10^{18} Hz, which corresponds to an energy of 2.32×10^{-15} J or 14.5 keV. The half-life of 2×10^{-7} s corresponds to a frequency uncertainty of about 4×10^{5} Hz (see Problem 14.13). If this were the only factor, the ratio of the frequency uncertainty to the frequency would be (4×10^{5} Hz)/(3.5×10^{18} Hz), which is about 10^{-13}. There thus would be a high degree of homogeneity. However, under ordinary conditions the nucleus recoils and the homogeneity is much smaller. Mössbauer's important contribution was to overcome this difficulty by holding the nuclei rigidly in a crystal.

EXAMPLE 14.4 Calculate the frequency range of γ rays of frequency 3.5×10^{18} Hz emitted by (a) a free ^{57}Fe atom and (b) a ^{57}Fe atom held in a crystal of mass 1 g.

Solution The momentum of the photon is $h\nu/c$ (from Eq. 11.55), and if the recoiling entity is of mass m and recoils with a velocity u,

$$\frac{h\nu}{c} = mu$$

The energy $h\nu$ is $6.626 \times 10^{-34} \times 3.5 \times 10^{18}$ J $= 2.32 \times 10^{-15}$ J.

a. The mass of the ^{57}Fe atom is 57×10^{-3} kg/$6.022 \times 10^{23} = 9.47 \times 10^{-26}$ kg. The speed of recoil is therefore

$$u = \frac{h\nu}{cm} = \frac{2.32 \times 10^{-15}(\text{J})}{2.998 \times 10^{8}(\text{m s}^{-1}) \times 9.47 \times 10^{-26}(\text{kg})}$$
$$= 81.7 \text{ m s}^{-1}$$

We saw in Section 14.2 that the change of wavelength due to the Doppler effect is $u\lambda/c$, and the corresponding change of frequency is thus $u\nu/c$. The frequency range is therefore

$$\Delta\nu = \frac{81.7(\text{m s}^{-1}) \times 3.5 \times 10^{18}(\text{Hz})}{2.998 \times 10^{8}(\text{m s}^{-1})}$$
$$= 9.5 \times 10^{11} \text{ Hz}$$

This is much larger than the frequency uncertainty of 4×10^{5} Hz that arises from the half-life of the parent nucleus.

b. If the mass is 1 g, the speed of recoil is

$$u = \frac{2.32 \times 10^{-15}(\text{J})}{2.998 \times 10^{8}(\text{m s}^{-1}) \times 10^{-3}(\text{kg})} = 7.74 \times 10^{-21} \text{ m s}^{-1}$$

The frequency range is now

$$\Delta\nu = \frac{7.74 \times 10^{-21}(\text{m s}^{-1}) \times 3.5 \times 10^{18}(\text{Hz})}{2.998 \times 10^{8}(\text{m s}^{-1})}$$

$$= 9.04 \times 10^{-11} \text{ Hz}$$

Since this is negligible compared with 4×10^5 Hz, the latter is the uncertainty in the frequency.

Mössbauer spectroscopy involves the *resonant absorption of γ rays by atomic nuclei*. For example, the γ rays emitted by a $^{57}Fe^*$ nucleus, of frequency 3.5×10^{18} Hz, may be absorbed by another ^{57}Fe nucleus. The frequency at which resonant absorption occurs is affected to a small extent by the valence state of the atom containing the absorbing nucleus. For example, ^{57}Fe in Fe_2O_3 absorbs at a slightly different frequency from ^{57}Fe in $Fe(CN)_6^{4-}$ or in hemoglobin. Similarly, ^{119}Sn in Sn(II) covalent molecules absorbs at a different frequency from ^{119}Sn in Sn(IV) compounds. The differences may be as large as 100 MHz; this is exceedingly small compared with the frequency of the γ radiation itself, but it is much larger than the uncertainty in the frequency. Consequently, the frequency changes brought about by the different valence states can be measured satisfactorily.

To do this it is necessary to be able to control the frequency of the γ radiation, and this is done by taking advantage of the Doppler effect. The source of radiation is mounted on a sliding support, which is moved at a carefully controlled rate. The following example shows that quite small speeds are sufficient to produce frequency shifts of 100 MHz or less, which are required for resonance to be achieved.

EXAMPLE 14.5 Calculate the frequency shift brought about by moving a ^{57}Co source at a speed of 5 mm s^{-1}.

Solution The Doppler shift is of magnitude $\nu u/c$, where u is the speed of the source. The frequency of the radiation is 3.5×10^{18} Hz, and the shift is therefore

$$\Delta\nu = \frac{3.5 \times 10^{18}(\text{Hz}) \times 5 \times 10^{-3}(\text{m s}^{-1})}{2.998 \times 10^{8}(\text{m s}^{-1})} = 5.84 \times 10^{7} \text{ Hz}$$

$$= 58.4 \text{ MHz}$$

The remarkable feature of Mössbauer spectroscopy is that although the energies of the γ rays are extremely large, much smaller energy differences can be measured. This was only made possible by Mössbauer's device of limiting the range of frequencies by holding the emitting source in a crystal lattice.

14.9 Photoelectron Spectroscopy

The importance of the photoelectric effect in the development of quantum theory was discussed in Section 11.1 (see Eq. 11.37). Photoelectron spectroscopy (PES) measures the kinetic energy of the electrons emitted when molecules of a sample are ionized. The process is

$$M + h\nu = M^+ + e^- \tag{14.44}$$

where M is an atom or molecule. In ultraviolet photoelectron spectroscopy (UPS), the ionizing radiation for a gas in the UV spectrum is normally 58.4 nm from a helium discharge lamp. In X-ray photoelectron spectroscopy (XPS), X-ray photons produce the ionization of the gas or solid sample. The emitted electrons are called *photoelectrons.* UPS is used to study the electronic configuration of valence electrons, and XPS studies the core electrons.

Information is gained from the process in Eq. 14.44 by use of the principle of conservation of energy:

$$E(e^-) = h\nu - [E(M^+) - E(M)] \tag{14.45}$$

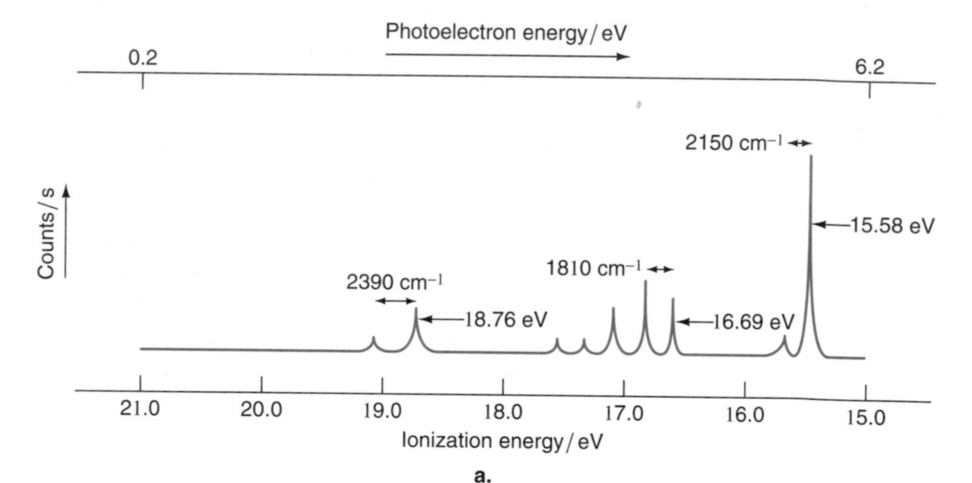

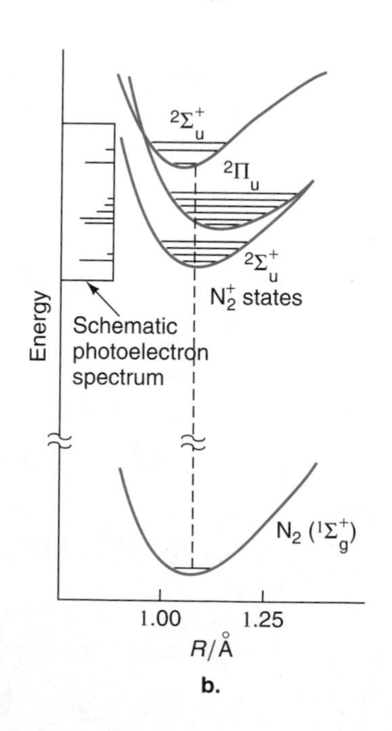

FIGURE 14.45
Photoelectron spectrum of nitrogen (a) showing ejection of an electron from three orbital states. (From *Molecular Spectroscopy* by Ira N. Levine, Copyright © 1975. Reprinted by permission of John Wiley & Sons, Inc.) In (b) are shown the states of the excited nitrogen molecule corresponding to the observed lines in the photoelectron spectrum. (DeKock, Roger, and Gray, Harry B., *Chemical Structure and Bonding,* Second Edition. Copyright © 1989. Used with permission from Roger DeKock.)

where $E(\mathrm{M})$ and $E(\mathrm{M}^+)$ are the energies of the molecule M and the ion M^+ formed by the ionization, and $E(e^-)$ is the kinetic energy of the photoelectron. Since $h\nu$ is known and $E(e^-)$ is measured, the difference between the energies of the ion and the original molecule is obtained. Comparing with Eq. 11.37, it is seen that the energy difference is the same as the work function. This is the binding energy for electrons in different electronic states.

Either of two plots gives the photoelectric spectrum. One is a plot of the rate of electron emission against the electron kinetic energy; the other a plot of the rate of emission against ionization energy. Figure 14.45a shows the spectrum for nitrogen with two energy scales, the photoelectron energy and the ionization energy. Note that since the 58.4-nm radiation gives an energy of 21.2 eV, the sum of the two scales is 21.2 eV at any point. The three groups of lines in the figure correspond to ionization from three different orbitals, with the individual lines in each group being the vibrational levels. Figure 14.45b shows the states of electron occupancy as deduced from the photoelectron spectrum in a. Since this is an electron transition to excited electronic states, the Franck-Condon principle applies. But notice that the occupancy (as indicated by the intensity) of the excited vibrational states, especially in the $^2\Pi_u$ state, corresponds with the high probability of finding the electron near the endmost positions for all but the $v = 1$ vibrational levels of the harmonic oscillator curve.

Because the sample is at room temperature, only the ground vibrational state is significantly occupied. From Section 12.6, the configuration of N_2 in the ground state is

$$\mathrm{N_2:}\quad 1s\sigma_g^2 1s\sigma_u^{*2} 2s\sigma_g^2 2s\sigma_u^{*2} 2p\pi_u^4 2p\sigma_g^2, \quad {}^1\Sigma_g^+$$

The two lines starting at 15.88 eV correspond to the removal of a $2p\sigma_g$ bonding orbital electron. The resulting state is

$$\mathrm{N_2^+:}\quad 1s\sigma_g^2 1s\sigma_u^{*2} 2s\sigma_g^2 2s\sigma_u^{*2} 2p\pi_u^4 2p\sigma_g^1, \quad {}^2\Sigma_g^+$$

Going to the left, the next set of lines beginning at 16.69 eV arises from the removal of a $2p\pi_u$ bonding electron. The final excited state corresponds to removal of a $2s\sigma_u^*$ antibonding electron.

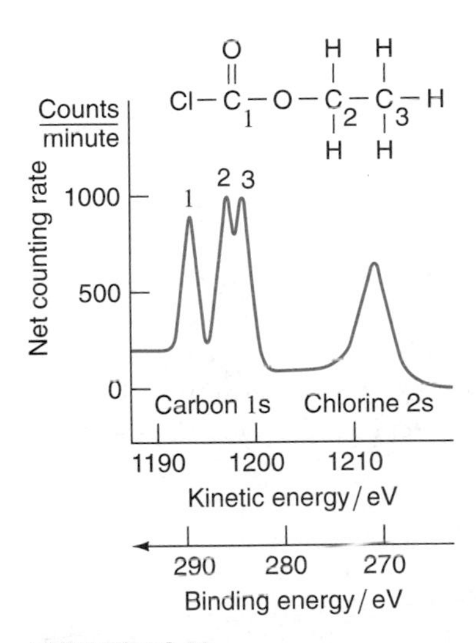

FIGURE 14.46
Photoelectron spectrum of ethyl chloroformate. The three lines in the carbon 1s spectrum appear in the same order from left to right as the carbon atoms drawn in the structure. (From: James T. L., *Journal of Chemical Education*, Vol. 48, pp. 712–718, 1971. Used with permission.)

As indicated earlier, these techniques are not limited to atoms or elements. However, when more energy is required to examine inner levels, an X-ray source is used. X-ray sources are discussed in more detail in Section 16.2. K_α radiation from Mg has energy of 1253.6 eV and from Al of 1486.6 eV. These values are considerably larger than from UV sources, and the photoelectrons so produced can have kinetic energies up to these values. The energy differences between 1s levels for C, O, N, etc., are large, so that individual elements are easily identified by their energy and are referred to as the carbon 1s spectrum, etc. In Figure 14.46 the carbon 1s X-ray photoelectron spectrum is shown for ethyl chloroformate. Although the 1s orbitals are deeply buried in the atom, small differences in electronic environments of the three carbon atoms influence the carbon 1s electron binding energy, resulting in a "chemical shift" leading to a small difference in peak position. Such shifts permit insight into the nature of bonding and of the valence state.

See Section 18.5 for a further discussion of these methods for the study of surfaces.

14.10 Photoacoustic Spectroscopy

A different kind of absorption, originally observed by Alexander Graham Bell (1847–1922), occurs with the energy changes in the form of sound waves. In *photoacoustic spectroscopy* (PAS) a monochromatic beam of radiation is chopped or pulsed so that the sample alternatively has energy or no energy impinging upon it. During photon exposure, the surface may either absorb or not absorb the radiation. If energy is absorbed, the surface is heated and, along with it, the surrounding air. Subsequently, during the time the beam is interrupted, the air cools. Thus the air undergoes expansion because of the heating and contraction because of the cooling process, thereby creating an alternating wave. The chopping of the light beam must be slow enough to allow cooling but fast enough to generate a detectable beat or sound wave using a microphone. If no heat is absorbed, no sound wave is generated.

In practice, a frequency of about 50 Hz is used in photoacoustic spectroscopy. A plot of the intensity is made as the wavelength of the light is varied. The absorption spectrum so produced is particularly useful in examining opaque samples, either solids or strongly absorbing liquids. PAS is used in both the solid and gaseous phases. The information obtained is essentially the IR absorption spectrum, and that is useful for chemical identification. Laser PAS (LPAS) improves sensitivity and is an analytical tool. It can measure heats of formation of metastable reaction intermediates, can probe the kinetics of photochemical products, and can study the thermodynamics of solvated ions and molecules.

14.11 Chiroptical Methods

Chiroptical methods are optical methods that depend on how a plane of polarized light behaves upon interaction with a chiral center. The rotation of plane-polarized light may be followed with a polarimeter, leading to an area of study called *polarimetry*. The variation of the rotation with wavelength is known as *optical rotatory dispersion* (ORD). Also, an optically active substance shows a change in its index of refraction and in its absorption when circularly polarized light is used. The variation of the refractive index with wavelength is known as **circular birefringence.** The variation of the absorption with wavelength is referred to as **circular dichroism** (CD). In the following four subsections, the nature of these interactions is examined.

The Nature of Polarized Light

In Figure 11.2 electromagnetic radiation is shown in terms of sinusoidally varying electric and magnetic fields, perpendicular to each other and to the transverse direction of propagation. Ordinary light consists of electric vectors in all possible directions perpendicular to the direction of propagation. See Figure 14.47a. No particular direction is preferred and we speak of an *unpolarized* light wave. When the electric field at a particular point always vibrates in the same direction, Figure 14.47b, the wave is said to be *plane polarized, linearly polarized,* or just *polarized.* The plane-polarized electric vector E may be thought of as the resultant of two vec-

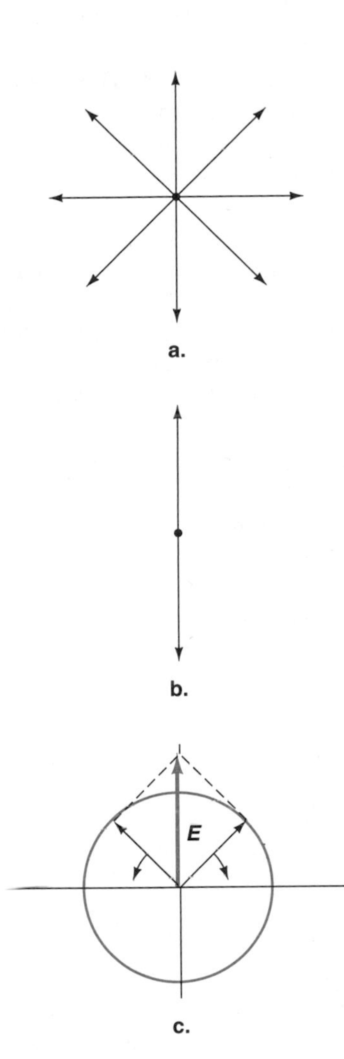

FIGURE 14.47
In each frame the electric field of electromagnetic radiation is shown perpendicular to the direction of propagation coming out of the plane of the paper toward the reader. (a) An unpolarized light beam showing equal probability for the direction of the electric vector. Only one E vector exists at any one time, but each direction is equally probable. (b) A linearly polarized light beam with the electric field vector vibrating in the vertical direction. (c) Plane-polarized light as a special case of circularly polarized light. The vertical vector E is the resultant of two equal vectors rotating in opposite directions.

tors, a right circularly polarized vector and a left circularly polarized vector. Each vector rotates in opposite directions as shown in Figure 14.47c.

Plane polarization may be accomplished in several ways, which will be mentioned before discussing applications in chemistry.

1. *Selective absorption can cause polarization.* Perhaps the most useful technique is to use a **dichroic** material, which transmits waves that have electric field vectors vibrating in only one particular plane and which absorbs all others. Edwin Herbert Land (1909–1991) in 1932 made such a material, known as *polaroid,* by orienting thin sheets of long-chain polyvinyl alcohol impregnated with iodine. The net effect is that the longitudinal direction of the material conducts electrons, so that polaroid absorbs light in the parallel direction to the surface, and transmits light in the transverse direction perpendicular to the molecular chains. In 1937, Land founded the Polaroid Corporation, and he invented his famous Polaroid Land Camera in 1947.

2. *Reflection can cause polarization.* A second way to achieve polarization is by reflection. Imagine an unpolarized light beam reflected from a surface. At an angle of incidence (the angle between the beam of light and a normal to the surface) of 90° (the grazing angle) and at an angle of 0°, no polarization occurs. In between these angles the polarization increases until the reflected beam is completely polarized when the angle of reflection equals the angle of refraction. The angle to the normal at which this happens is called *Brewster's angle* and is a function of the wavelength of light. This type of polarization accounts for the glare from sunlight reflected from bodies of water and snow; the glare originates with the light that is polarized parallel to the surface.

3. *Double refraction causes polarization.* Within an amorphous material, such as glass, the velocity of light does not change, and there is only one index of refraction. However, in certain crystals, such as calcite (calcium carbonate) and sodium nitrate, two speeds of light are found; consequently there are two values for the index of refraction. That this is the case is easy to demonstrate. If a crystal of calcite is held over a single line of print type, as observed through the crystal, the type and its duplicate are seen. Such materials (the two mentioned give the largest differences of refractive index) are referred to as *double-refracting* or *birefringent* materials. The important point is that the two rays formed are polarized in mutually perpendicular directions.

4. *Scattering causes polarization.* We have already seen with Raman spectroscopy (Section 13.5) that light incident on the electrons of the molecules can be absorbed and reradiated as light. Such absorption and reradiation is called *scattering.* When unpolarized light sets the molecules (for instance, those of a gas) vibrating, the horizontal component of the electric field vector causes a horizontal vibration whereas the vertical part causes a vertical vibration. Consequently, horizontal and vertical polarization is effected and can be verified in the atmosphere by looking through a rotating sheet of polaroid held up to the sky. The alternating dark and light regions seen through the polaroid upon rotation verify the polarization.

Optical Activity and Polarimetry

A substance is said to be optically active if it rotates the plane of polarized light transmitted through it. Optical activity may occur in a gas, in a liquid, or in a solid.

For most compounds optical activity arises from the presence of an *asymmetric carbon atom* (a chiral center, where four different groups are attached to a carbon atom). In such a case (except for a *meso*-compound where there is a plane of symmetry between two asymmetric carbons), the compound and its mirror image are not superimposable. This latter fact provides a simple test to determine whether optical activity exists in a compound. From the point of view of symmetry, if an improper symmetry axis, S_n, exists, including a mirror plane, or a center of symmetry, no optical activity is possible.

For some solids optical activity may only occur in the crystalline state. Solid quartz, for example, loses its optical activity when melted. This is because the optical activity can result from a crystalline arrangement of atoms or molecules in a right- or left-handed spiral. In the study of proteins and nucleic acids, there is an intrinsic optical activity arising from the contributions of the asymmetric carbons of the individual amino acids. In addition, where there is α-helical content, there is another contribution to the optical rotation which originates in the asymmetry of the α helix. Indeed, from ORD and CD spectra it is possible to study the conformation of those molecules.

The magnitude of the optical activity is measured using a monochromatic light source and a polarimeter. Figure 14.48 represents such a device. The unpolarized light, often from the sodium D-line, is incident upon a polarizer which has the function of forming a polarized beam. This is often done with a *Nicol prism,* which is formed by cutting a calcite prism in half along a suitable axis and cementing the sections together. At the interface, one component is totally reflected to the side, where it is absorbed by a black surface. The other component is freely transmitted through the medium whose optical activity is to be examined. The light is intercepted by an analyzer, such as another Nicol prism. The relative angular position of the two prisms for maximum transmission of light of a given wavelength is observed with and without the chiral sample. The difference in angle is the angle of rotation α. If the analyzer is turned to the right, or clockwise as viewed looking toward the light source, the sample is said to be *dextrorotatory.* The angle is then given a positive sign. If the optical rotation causes the analyzer

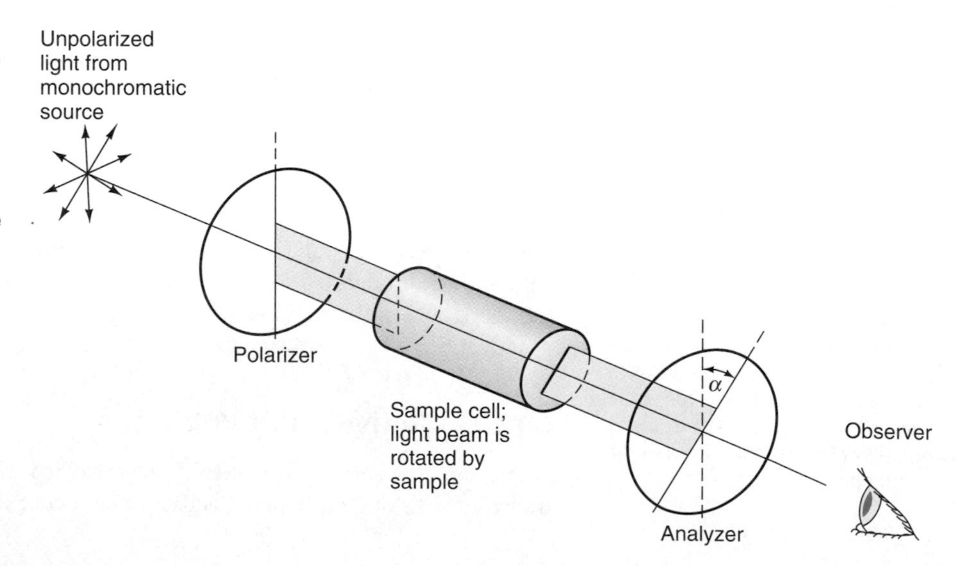

FIGURE 14.48
A typical polarimeter for solution work. A monochromatic light source passes through the polarizer on the left and through the solution. The second polarizer, called an *analyzer,* is initially aligned without the optically active sample in place so that all the polarized light passes through. With the sample in place, the polarized light is rotated and the analyzer must be turned through an angle α to allow the light to pass.

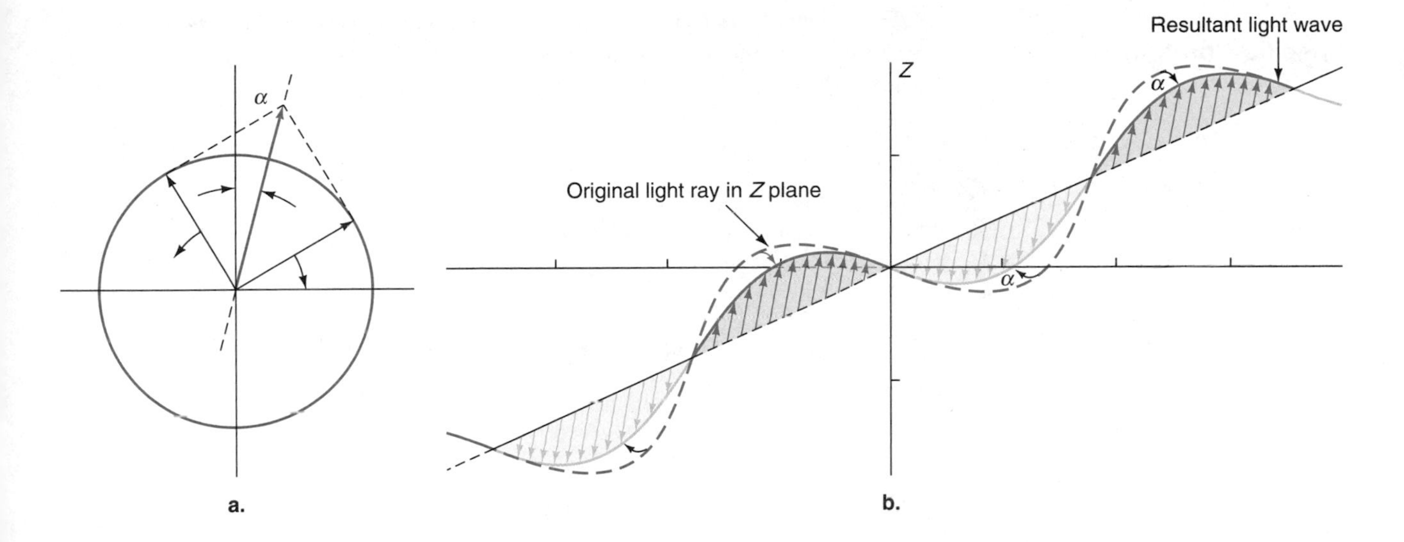

FIGURE 14.49
In (a) the resultant of the electric vector has rotated through an angle α because of the two different refractive indices in the sample. In (b) is shown the circular polarization that arises because of the two waves being out of phase.

to be turned to the left, the sample is said to be *levorotatory* and the angle is given a negative sign.

The reason for the observed rotation is that the chiral centers interact with the right- and the left-circularly polarized light vectors in the polarized light as though the sample has two refractive indices n_L and n_R. The optical rotation α' is related to the difference between these refractive indices by

$$\alpha' = \frac{\pi}{\lambda}(n_L - n_R) \tag{14.46}$$

If, for instance, $n_L > n_R$, then the two electric vectors will have a difference in propagation velocity as shown in Figure 14.49a. This puts the two vectors out of phase and causes a circular helix of propagation to be formed, as in Figure 14.49b, where the rotation of the plane of polarized light is seen. Since the wavelength of visible light is short, large values of α' can be obtained with even a small difference in the refractive index.

EXAMPLE 14.6 A solution of an optically active solute whose concentration is 0.110 g cm^{-3} rotates the sodium 589 nm D-line of plane polarized light by +25.4 deg in a 1.00-dm-long sample tube. What is $n_L - n_R$?

Solution Eq. 14.46 gives the optical rotation α'. We can find the net rotation as the light passes through the sample by multiplying by the length of the sample tube, l. To obtain α in degrees, we multiply by $180°/\pi$, and obtain

$$\alpha = \frac{180°\alpha'}{\pi} l = \frac{180°}{\lambda}(n_L - n_R)l$$

Rearrangement gives

$$n_L - n_R = \frac{\alpha\lambda}{l(180°)} = \frac{(25.4°)(589 \times 10^{-9}\ \text{m})}{(0.100\ \text{m})(180°)} = 8.31 \times 10^{-7}$$

Specific Rotation

Experimentally it is found that the magnitude of rotation is directly proportional to the concentration c and to the length l of the sample tube. The **specific rotation** is defined as

$$[\alpha]_A = \frac{\alpha}{cl} \tag{14.47}$$

and is generally listed as degree $dm^{-1}\ cm^3\ g^{-1}$ (the SI unit is rad $m^2\ kg^{-1}$).

EXAMPLE 14.7 Using the data in the previous example, what is the value of $[\alpha]_A$?

Solution Using Eq. 14.47 we have

$$[\alpha]_A = \frac{+25.4°}{(0.110\ g\ cm^{-3})(10^{-3}\ kg\ g^{-1})(10^2\ cm\ m^{-1})^3(0.100\ m)}$$
$$= 2.31°\ m^2\ kg^{-1}$$

(IUPAC recommends that the older units be retained because of the data in the literature. The value would then be 2.31° $dm^{-1}\ cm^3\ g^{-1}$.)

Specific rotations cover a wide range. For small organic molecules the values using the sodium D-line range up to approximately 70°; +66° for sucrose, and −12° for (−)-tartaric acid. Much larger values may be found for optically active coordination compounds.

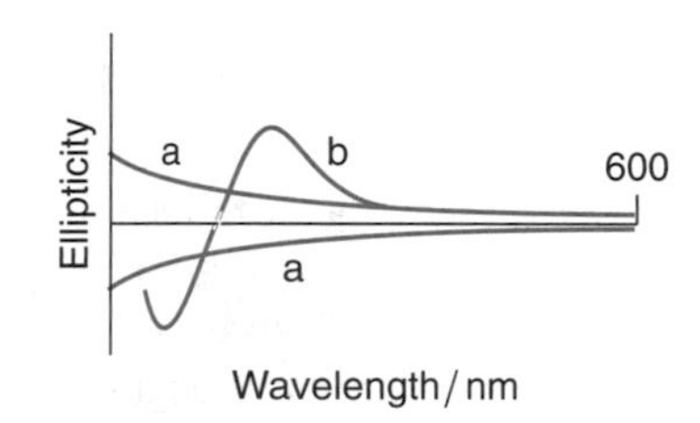

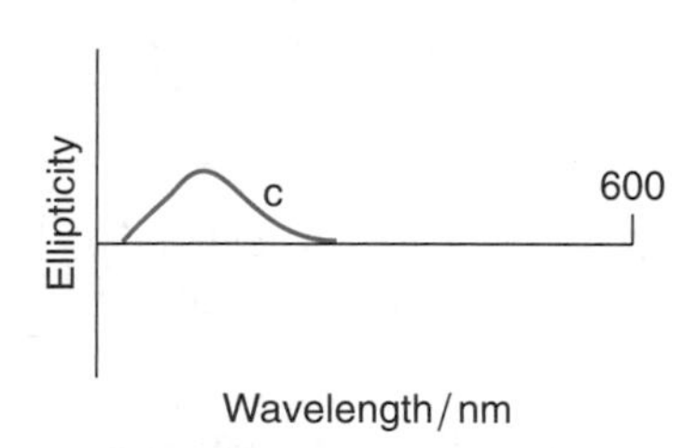

FIGURE 14.50
Types of curves seen in chiroptical techniques. (a) A simple ORD curve showing the monotonically changing angle of rotation with wavelength. (b) An ORD curve with one Cotton effect. Several absorptions may occur in one sample with the Cotton effects overlapping. (c) A single Cotton effect in a CD curve.

Optical Rotatory Dispersion (ORD)

Polarimetry and optical rotatory dispersion both measure the extent to which a beam of linearly polarized light is rotated by a chiral or helical sample. For a nonabsorbing sample, that is, for one without chromophores, both techniques give equivalent spectra with polarimetry limited to the wavelength of the excitation source. In the absence of absorption by a chromophore, the ORD spectrum changes monotonically with wavelength. See Figure 14.50a.

If the optically active center is adjacent to a chromophore that absorbs the polarized beam, anomalous rotations in the ORD spectrum are produced, as shown in Figure 14.50b. This behavior, known as the *Cotton effect* (named after the French physicist A. Cotton), is superimposed on the monotonically changing ORD curve. It is limited to the wavelength range of the absorption band and shows maximum and changed sign minimum.

Circular Dichroism (CD)

Circular dichroism is concerned with the change in the index of refraction, and also with the differential absorption by the chromophore of one of the two circularly polarized waves. Figure 14.50c shows a CD curve with a single Cotton effect. Figure 14.51a shows how elliptically polarized light is generated by the reduction in intensity of the right circularly polarized beam caused by absorption. When the two

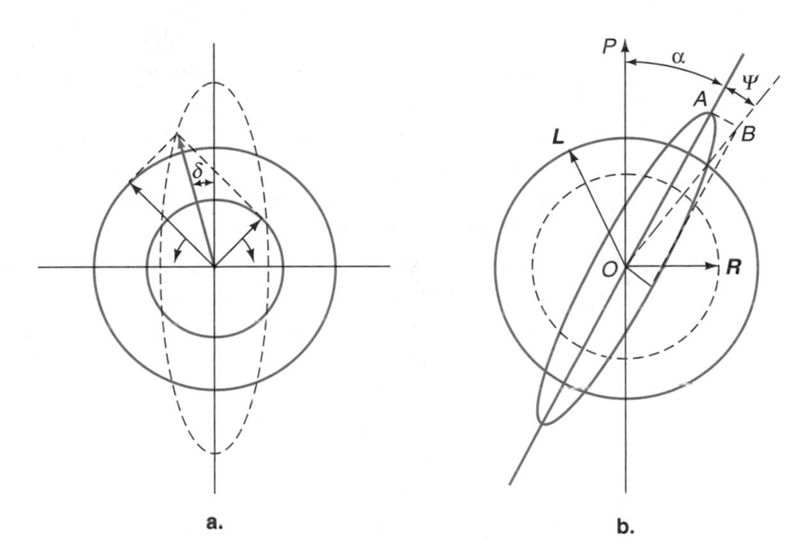

FIGURE 14.51
(a) The representation of elliptically polarized light arising from two circularly polarized beams having different extinction coefficients. (b) The representation of elliptically polarized light combining the effects of a difference in refractive index and in absorption. *OP* is direction of the initial polarization; ***OL*** and ***OR*** are the absorption vectors are left (L) and right (R) circularly polarized light; α is the angle of rotation; Ψ is the ellipticity.

effects are combined, as in Figure 14.51b, the long axis of the ellipse is rotated by α. The ellipticity is defined by $\Psi = \tan^{-1}(\mathrm{OA/OB})$. Ψ is directly proportional to the absorbance difference,

$$\Psi = \pi(\epsilon_L - \epsilon_R)/\lambda \tag{14.48}$$

where ϵ is the molar absorptivity, the L and R standing for the left and right circularly polarized light, and λ is the wavelength at which the ellipticity is measured.

CD can have both positive and negative deviations from the baseline. An advantage that CD has over ORD is that there is no signal at wavelengths where there is no absorption; consequently, the baseline is easy to define. Besides applications to the determination of conformations, CD has applications to analytical determinations which promise to become a major thrust in future years.

14.12 Mass Spectrometry

Mass spectrometry is of a different character from the spectroscopic techniques discussed so far. Ordinary spectroscopy is concerned with the interactions between atoms or molecules and electromagnetic radiation, whereas mass spectrometry is a technique that separates charged molecules and free radicals according to their mass-to-charge ratio.

The origin of mass spectrometry lies in the nineteenth-century investigations of the properties of electric discharges. In 1897 Joseph John Thomson (1856–1940) caused a beam of cathode-ray particles to pass through electric and magnetic fields and measured their deflection, in this way determining the ratio of the mass of an electron to its charge. In 1907 Thomson went on to study positively charged ions, and in 1919 he constructed what can be regarded as a prototype of the mass spectrometer. He pointed out that with further development the technique would be valuable to chemists as an analytical tool.

Mass spectrometry is one of the most widely used of all spectroscopic techniques. For the better part of a century it has provided structural information about small molecules, and recent developments have led to enormous improvement in sensitivity and detection limits. Today mass spectroscopy can study polymers up to relative molecular weights of 300 000. It can be used to study covalent interactions, to sequence proteins and polypeptides, to monitor drug interactions, and to give valuable information about protein folding. Much chemical analysis is now being done by mass spectrometry, and many fully computerized commercial instruments make it easy to make routine analyses.

In addition, mass spectrometers are powerful research instruments; they lead to important information about the structures of molecules and about the kinetics of chemical reactions involving ions. For example, mass spectrometry provides a great deal of thermochemical information about neutral molecules and ions. It is also important in the investigation of the unimolecular decompositions of organic ions in the gas phase.

The basic idea behind mass spectrometry in its original form is that a gaseous sample is introduced into an evacuated chamber in which a beam of electrons converts the molecules into ions. For example, when a water molecule in the gas phase is struck by an electron, an electron may be knocked out of it with the formation of an H_2O^+ radical ion, which is positively charged and has an odd electron:

$$H_2O + e^- \rightarrow H_2O^+ + 2\,e^-$$

Similar processes occur with other molecules. In some mass-spectrometric techniques, the ions so produced are accelerated by an electric field. The beam is passed through a magnetic field, and sweeping the field brings the ions sequentially on to

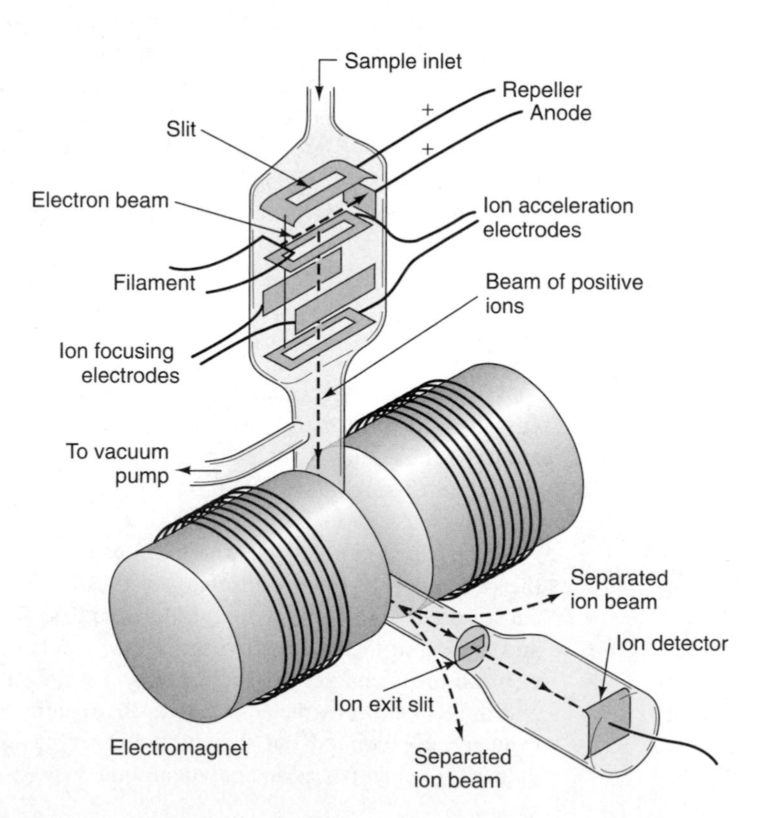

FIGURE 14.52
Schematic diagram of a mass spectrometer.

the detector, producing a spectrum of all the masses present. The deflected beam is focused in such a way that ions of the same mass-to-charge ratio arrive together at the detection device.

Mass spectrometers have three main features, as shown in Figure 14.52: the ion source, the detector, and the mass analyzer. The sample can be introduced into the ionization chamber in a number of ways. If the sample is a volatile liquid, a small amount of it can be placed on a probe and inserted into the ionization chamber. Other methods of introducing the sample include the use of capillary devices.

Several techniques have been used to generate the ions, as follows. The original method is referred to as **electron ionization (EI),** in which picomole amounts of the sample are bombarded with an electron beam. This leads to the formation of both positive and negative ions. This technique is limited by the fact that it can be used only for volatile substances, which means molecules of molecular weight less than about 400. For these, extensive tables and databases have been compiled, and for identification purposes, they can be compared to the fragmentation patterns obtained.

Because of the limitation of volatility, mass spectroscopy was limited to smaller molecules until the early 1980s. For larger molecules, three other ionization techniques are now commonly used, and will be mentioned only briefly. These techniques have proved valuable for analytical purposes but are not so important in research, since the processes occurring are complex and yet to be fully understood.

One technique useful for the analysis of larger molecules is **fast atom/ion bombardment (FAB),** which can be used for molecules with molecular weights up to about 4000. The sample is dissolved in a nonvolatile liquid (the matrix); commonly used solvents are m-nitrobenzyl alcohol and glycerol, to which may be added compounds containing sodium or potassium. The sample is bombarded with a kilovolt-energy beam of atoms or ions, such as argon atoms or cesium ions. This produces positively charged species to which ions such as Na^+ and K^+ have been added. The FAB technique allows the formation of multiply charged ions, which reduces the mass-to-charge ratio.

Another technique for larger molecules is **matrix-assisted laser desorption/ ionization (MALDI),** which is applicable to molecular weights up to about 300 000. The major difference between MALDI and FAB is that MALDI uses a solid rather than a liquid matrix, and that the ionization source is a laser. The matrix absorbs strongly in the ultraviolet spectrum, with the production of only positively charged species. The technique is sensitive to sample concentrations in the femtomole range.

A third technique is **electrospray ionization (ESI),** in which the sample is dissolved in a volatile liquid and sprayed from a region of strong electric field at the tip of a metal nozzle, maintained at a voltage of approximately 4000 V. This produces some negative species but mainly positive polyprotonated molecules. Since multiply charged species are formed, a range of mass-to-charge ratios exists for the same mass. The technique has a sensitivity from picomoles to femtomoles.

A quite different technique has recently come into use. This involves the use of an **ion trap,** developed independently by Wolfgang Paul (b. 1913) and Hans Georg Dehmelt (b. 1922), who shared one-half of the 1989 Nobel Prize for physics for this work. An ion trap is a device of particular shape which uses controlled electromagnetic fields to isolate molecular ions in a restricted space, where they can be studied over long periods of time. By varying the field, the ions can be caused to leave the trap in order of ascending mass/charge ratio. A number of commercial instruments using ion-trap mass spectrometry are now available.

KEY EQUATIONS

Nonlinear effects of the electron polarization at large laser powers

$$P = \beta_1 E + \beta_2 E^2 + \beta_3 E^3 + \cdots$$

Relationship between the magnetic field in a vacuum and in a sample

$$\boldsymbol{B} = \mu_o (1 + \chi)\boldsymbol{H}$$

Time-dependent magnetic field

$$\boldsymbol{B} = ZB_o + X2B_x^o \cos \omega t$$

Energy of an electron spin in a magnetic field

$$E_{\pm 1/2} = \pm\frac{1}{2} g_s \mu_B B_o$$

Nuclear magnetic moment

$$\boldsymbol{\mu}_i = \gamma_i \hbar \hat{\boldsymbol{I}}_i$$

Zeeman Hamiltonian

$$\hat{H}_I^{Zeeman} = -\hbar\gamma_I (1 - \sigma_I)\hat{\boldsymbol{I}} \cdot \boldsymbol{B}$$

Indirect scalar spin-spin J coupling

$$\hat{H}_{I_1 I_2}^{J\text{-coupling}} = \hbar J_{12} \hat{\boldsymbol{I}}_1 \cdot \hat{\boldsymbol{I}}_2$$

Direct dipole-dipole coupling

$$\hat{H}_{I_1 I_2}^{Direct\ spin\text{-}spin} = \frac{\hbar^2 \gamma_1 \gamma_2}{r^3}(3\cos^2\theta - 1)\left[\hat{I}_{1z}\hat{I}_{2z} - \frac{1}{2}(\hat{I}_{1x}\hat{I}_{2y} + \hat{I}_{1y}\hat{I}_{2x})\right]$$

Spin rotation interaction

$$\hat{H}_{IJ}^{Spin\text{-}rotation} = c_{sr}\hat{\boldsymbol{I}} \cdot \hat{\boldsymbol{J}}$$

Quadrupole interaction

$$\hat{H}_I^{quadrupole} = d_I\,[3\hat{\boldsymbol{I}}_z^2 - I(I + 1)]$$

Relationship between chemical shielding and chemical shift

$$\delta = \frac{\sigma_s - \sigma_R}{\sigma_R} \times 10^6$$

Optical rotation

$$\alpha' = \frac{\pi}{\lambda}(n_L - n_R)$$

PROBLEMS

Spectral Line Widths

14.1. The sun emits a spectral line at 677.4 nm and it has been identified as due to an ionized ^{57}Fe atom, which has a molar mass of 56.94 g mol^{-1}. The spectral line has a width of 0.053 nm. Estimate the temperature of the surface of the sun.

14.2. Estimate the lifetime of a state that, because of lifetime broadening, gives rise to a line of width

a. 0.01 cm^{-1},

b. 0.1 cm^{-1},

c. 1.0 cm^{-1},

d. 200 MHz.

14.3. Lasers are commonly used in physical chemistry research laboratories to generate the reactive intermediates required in the study of elementary reactions. For example, highly reactive oxygen atoms in the ground electronic state [O(^{3}P)] can be generated from NO_2 by laser-induced dissociation: $NO_2 \xrightarrow{h\nu} NO(v', J') + O(^3P)$. The velocity distribution of the oxygen atoms thus generated (using a laser with λ = 355 nm) shows two broad peaks centered at 900.00 and 1400.00 m s^{-1}, respectively, and leads to the formation of the NO(v' = 0) and the NO(v' = 1) states [Hradil et al., *J. Chem. Phys. 99,* 4455(1993)].

a. Identify the NO vibrational state that corresponds to each O atom velocity peak, and justify your choice.

b. If a total of 7400 cm^{-1} of energy is available to the fragments, what are the velocities of the NO fragment corresponding to each of the peaks in the oxygen velocity distribution? (These velocities are measured with respect to a fixed laboratory coordinate system and, therefore, are the "absolute" velocities of the fragments. Assume that the NO molecules are the J = 0 rotational state.) For the NO molecule, $\tilde{\nu}_0$ = 1904 cm^{-1}.

Resonance Spectroscopy

14.4. Calculate the magnetic flux density that is required to bring a free electron (g = 2.0023) into resonance in an EPR spectrometer operating at a wavelength of 8.00 mm.

14.5. An EPR spectrometer is operated at a frequency of 10.42 GHz and a study is made of methyl radicals. Resonance is observed at a magnetic flux density of 0.37175 T.

a. Calculate the g value of the methyl radical.

b. Calculate the field required for resonance when the spectrometer is operating at 9.488 GHz.

14.6. How many hyperfine lines would you expect to find in the ESR spectrum of ^2H, ^{19}F, ^{35}Cl, ^{37}Cl? Calculate the nuclear magnetic moment is each case. (Refer to Table 14.1, p. 735.)

14.7. Determine the number of hyperfine lines expected in the ESR spectrum of ^{63}Cu, which has a spin I of 1.

14.8. In a nuclear magnetic resonance instrument operating at a frequency of 60 MHz, at what magnetic fields would you expect to observe resonance with $^1H^{35}Cl$?

14.9. The chemical shift δ of methyl protons in acetaldehyde is 2.20 ppm, and that of the aldehydic proton is 9.80 ppm. What is the difference in the effective magnetic field for the two types of proton when the applied field is 1.5 T? If resonance is observed at 60 MHz, what is the splitting between the methyl and aldehyde proton resonances?

14.10 The nuclear spin quantum number I of the ^{39}K nucleus is $\frac{3}{2}$, and the nuclear g factor is 0.2606. How many orientations does the nucleus have in a magnetic field? At what frequency would there be resonance in a field of 1.0 T?

14.11. The ^{11}B nucleus has a spin I of $\frac{3}{2}$ and a nuclear g factor of 1.7920. At what field would resonance be observed at 60 MHz?

14.12. The Fourier transform is the mathematical foundation for much of modern spectroscopy. The idea behind Fourier transform spectroscopy is to use a pulse of energy, which contains many frequencies, to probe the sample many times each second. The cumulative signal recorded from the pulses, which is a time-dependent oscillatory function $F(t)$, has the general form shown in Figure 14.25a. The Fourier transform of such a function results in a function of frequencies, which we may denote $I(\omega)$, obtained as

$$I(\omega) = A\mathrm{Re}\left[\int_0^{\infty} F(t)e^{i\omega t}\,dt\right]$$

where Re represents the real part of the function. Perform the Fourier transform of $F(t) = [\cos(\omega_1 t) + \cos(\omega_2 t)]\exp(-t/T)$, where T is a relaxation time, and obtain an expression for $I(\omega)$. Analyze the behavior of this function by assigning arbitrary values to the constants and plotting it as a function of ω. Use the fact that $e^{i\omega t} = \cos \omega t + i \sin \omega t$, and

$$\int_0^{\infty} \cos(at)\cos(bt)e^{-t/T}\,dt = T\frac{[1 + (a^2 + b^2)T^2]}{[1 + (a - b)^2T^2][1 + (a + b)^2T^2]}$$

14.13. The lifetime of $^{57}Fe^*$ is 2×10^{-7}s. Calculate the uncertainty in the frequency of the γ radiation emitted and in the wavenumber.

14.14. The free radical CH_3 is found experimentally to be planar. Give an interpretation of this result in terms of orbital hybridization. What microwave spectrum would the radical show? What vibrations would be active in the infrared?

For additional problems, see the books listed at the end of the problem section in Chapter 12 (p. 634).

14.15. A complete photoelectron spectrum of the nitrogen molecule (see Figure 14.45) is given by Bock and Mollère [*J. Chem. Educ. 51,* 506(1974)]. The spectrum extends from about 411 eV (the peak corresponding to the $1s\sigma_g$ is at approximately 410 eV) to about 15.0 eV, with the peak corresponding to the $2p\sigma_g$ electron occurring at 15.88 eV. If we were to use a radiation source of wavelength 58.4 nm, which of the peaks in the spectrum can be studied? What will be the wavelength required to extend the range to 410 eV?

Essay Questions

14.16. Explain the selection rules for Raman spectra, with examples.

14.17. Describe the physical interaction that leads to nuclear magnetic resonance.

14.18. Give an account of the fundamental principles underlying Mössbauer spectroscopy.

SUGGESTED READING

For accounts of lasers and laser spectroscopy, see:

P. W. M. Millonni and J. H. Eberly, *Lasers,* New York: Wiley, 1988.

W. T. Silfrast, *Laser Fundamentals,* New York: Cambridge University Press, 1996. An excellent and readable text.

O. Svelto, *Principles of Lasers* (3rd ed.), New York: Plenum Press, 1989.

For a more advanced treatment of lasers in the field of chemistry, see:

D. L. Andrews, *Lasers in Chemistry* (3rd ed.), New York: Springer-Verlag, 1997.

S. F. Jacobs, Ed., *Laser Photochemistry, Tunable Lasers, and Other Topics,* Reading, MA: Addison-Wesley, Advanced Book Program, 1976.

For electron spin resonance spectroscopy, see:

J. E. Wertz and J. R. Bolton, *Electron Spin Resonance: Elementary Theory and Practical Applications.* New York: Chapman and Hall, 1986.

For more on Mössbauer spectroscopy, see:

G. J. Long, (Ed.), *Mössbauer Spectroscopy Applied to Inorganic Chemistry,* New York: Plenum Press, 1984.

For accounts of magnetic resonance spectroscopy, see:

J. W. Akitt, *NMR and Chemistry: An Introduction to Modern NMR Spectroscopy* (3rd ed.), New York: Chapman & Hall, 1992.

E. Breitmaier and W. Voelter, *Carbon-13 NMR Spectroscopy: High Resolution Methods and Applications to Organic Chemistry and Biochemistry* (3rd ed.), New York: VCH, 1987.

A. E. Derome, *Modern NMR Techniques for Chemical Research,* Oxford: Pergamon, 1987.

C. E. Dykstra, *Quantum Chemistry & Molecular Spectroscopy,* Englewood Cliffs, NJ: Prentice Hall, 1992.

R. K. Harris, *Nuclear Magnetic Spectroscopy,* New York: Pitman Publishing, 1983.

W. Kemp, *NMR in Chemistry: A Multinuclear Introduction,* New York: MacMillan, 1988.

K. A. McLauchlan, *Magnetic Resonance,* Oxford: Clarendon Press, 1972.

D. L. Pavia, G. M. Laupman, and G. S. Kriz, *Introduction to Spectroscopy, A Guide for Students of Spectroscopy,* (3rd ed.), New York: Harcourt College Publishers, 2001.

J. K. M. Sanders and B. K. Hunter, *Modern NMR Spectroscopy: A Guide for Chemists,* Oxford University Press, 1993.

J. K. M. Sanders and B. K. Hunter, *Modern NMR Spectroscopy: A Workbook of Chemical Problems,* Oxford University Press, paperback, 1993.

Y. Takeuchi and A. P. Marchand, *Applications of NMR Spectroscopy to Problems in Stereochemistry and Conformational Analysis,* New York: VCH, 1986.

C. H. Townes and A. L. Schawlar, *Microwave Spectroscopy,* New York: McGraw-Hill, 1955; Dover reprint, 1975.

D. A. R. Williams, *NMR Spectroscopy: Analytical Chemistry by Open Learning,* Chichester and New York: Wiley, 1986.

The early "Bible" of NMR spectroscopy still makes interesting reading:

J. A. Pople, W. G. Schneider, and H. J. Bernstein, *High-Resolution Nuclear Magnetic Resonance,* New York: McGraw-Hill, 1959.

For more advanced treatments of the subject, see:

R. Freeman, *Spin Choreography, Basic Steps in High Resolution NMR,* Oxford: Oxford University Press, 1998.

H. Friebolin (translated by J. F. Becconsall), *Basic One and Two Dimensional NMR Spectroscopy,* New York: VCH, 1991.

H. Günther, *NMR Spectroscopy, Basic Principles, Concepts, and Applications in Chemistry* (2nd ed), West Sussex, England: John Wiley and Sons Ltd., 1998.

G. E. Martin and A. S. Zektzer, *Two-dimensional NMR Methods for Establishing Molecular Connectivity: A Chemist's Guide to Experiment Selection, Performance, and Interpretation,* New York: VCH, 1988.

For photoelectron spectroscopy, see:

T. L. James, "Photoelectric spectroscopy," *Journal of Chemical Education, 48,* 712(1971).

D. W. Turner, C. Baker, A. D. Baker, and C. R. Brundle, *Molecular Photoelectron Spectroscopy,* New York: Wiley Interscience, 1970.

For chiroptical methods, see:

E. Charney, *The Molecular Basis of Optical Activity: Optical Rotatory Dispersion and Circular Dichroism,* New York: Wiley, 1979.

P. Crabbé, *ORD and CD in Chemistry and Biochemistry: An Introduction,* New York: Academic Press, 1972.

N. Purdie and E. A. Swallows, "Analytical applications of polarimetry, optical activity, dispersion, and circular dichroism," *Analytical Chemistry, 61,* 77A–89A(1989).

K. Schmidt-Rohr and H. W. Spiess, *Multidimensional Solid-State NMR and Polymers,* San Diego: Academic Press, 1994.

For mass spectrometry, see:

Gas Phase Ion and Neutral Thermochemistry, *J. Phys. Chem. Reference Data, 17,* Supplement 1, 1988. This book contains many articles by different authors dealing with a range of topics in mass spectrometry.

J. L. Holmes, "Assigning structures to ions in the gas phase," *Organic Mass Spectrometry, 20,* 169(1985).

M. Mann, C. K. Meng, and S. F. Wong, "Electrospray ionization," *Mass Spectrometry Reviews, 9,* 37(1990).

R. E. March and J. F. J. Todd (Eds.), *Practical Aspects of Ion Trap Mass Spectrometry,* 3 vols. CRC Press, Boca Rica, Florida, 1985.

Statistical Mechanics

PREVIEW

This chapter is concerned with interpreting the macroscopic behavior of matter by applying a statistical treatment to large groups of molecules. This subject is known as *statistical mechanics,* and the branch dealing with thermodynamic properties is known as *statistical thermodynamics.*

The energy relating to the motions of atoms and molecules can be classified as translational, rotational, and vibrational energy. A molecule having N_a atoms has $3N_a$ degrees of freedom, and these are divided between the various modes of motion. According to classical mechanics there is *equipartition of energy* between degrees of freedom. However, measurements of heat capacities of gases show that equipartition occurs only in the limit of very high temperatures, and this can be explained on the basis of quantum theory.

To treat these problems more satisfactorily, we need to have a statistical treatment of the way molecules distribute themselves among the various energy levels. The fundamental equation dealing with this is the *Boltzmann distribution law,* according to which the number of molecules in a given state is proportional to $\exp(-\epsilon_i/k_BT)$, where ϵ_i is the energy of the state and k_B is the Boltzmann constant, which is the gas constant divided by the Avogadro constant. The sum of the $\exp(-\epsilon_i/k_BT)$ terms over all energy states is known as the *partition function.* Boltzmann's distribution law was based on an earlier derivation by Clerk Maxwell of the distribution of speeds of gas molecules.

A partition function can relate to an individual molecule or to an assembly of molecules. This partition function can be used to obtain the various thermodynamic quantities, such as entropy, enthalpy, and Gibbs energy. To evaluate partition functions it is simplest to make the assumption that the translational, rotational, vibrational, electronic, and nuclear forms of energy can be treated separately. In that case there are compact expressions for the partition functions, which are easily evaluated if we have the necessary data about the molecules; in particular we need the mass, the moments of inertia, the vibrational frequencies, and the relative electronic energy levels.

An important application of molecular statistics is in the calculation of equilibrium constants. One useful device for doing this is to define certain functions, such as the enthalpy function and the Gibbs energy function. These can be calculated from partition functions and have been tabulated for many molecules. From them the equilibrium constant can be calculated. Equilibrium constants also can be derived directly from partition functions.

OBJECTIVES

After studying this chapter, the student should be able to:

- Understand that thermodynamics gives relationships between macroscopic observables and that these can be evaluated using statistical mechanics.
- Appreciate the difference between an assembly and and ensemble.
- Calculate the number of ways that particles can be distributed among states.
- Explain how Stirling's approximation formula and the method of Lagrangian undetermined multipliers are used in the derivation of the Boltzmann distribution function.
- Appreciate that energy in molecules exists in a number of different modes of motion, and that under some circumstances the energy is distributed equally amongst these levels (the principle of equipartition of energy).
- Realize that the partition function is the most fundamental quantity available in equilibrium statistical mechanics, and that from the partition function all thermodynamic observables can be calculated.
- Appreciate how partition functions can be obtained for the different types of molecular motion: translational, vibrational, rotational, electronic, and nuclear.
- Understand why the magnitude of thermal energy is about $k_B T$.
- Use the molecular partition function to obtain expressions for equilibrium constants.
- Use the ideas of equilibrium statistical mechanics to define a quasi-equilibrium state, and from this develop the equations of transition-state theory.

In the earlier chapters of this book we have been concerned almost exclusively with the macroscopic properties of matter. In Chapters 11–13 we went to the other extreme and dealt with the structure and behavior of individual atoms and molecules. In this chapter we will see how the macroscopic properties can be related to the behavior of the microscopic systems.

In principle, if the positions and velocities of all particles in a given system were known at any one time, the behavior of the system, at that time and at later times, could be determined by applying the laws of classical and quantum mechanics. There are, however, two reasons why this is impossible. In the first place, because of the Uncertainty Principle the actual positions and velocities of individual molecules cannot be known. Secondly, because of the enormous numbers of molecules present in the systems with which chemists and physicists normally deal, the calculation of macroscopic properties from the states of the individual molecules presents insuperable computational difficulties.

Instead of attempting this impossible problem, we make use of statistical methods, with the object of predicting the most probable behavior of a large collection, or *assembly,* of molecules. The assembly will be characterized by certain properties, such as the total volume, the number of molecules, and the total energy, but the individual molecules are distributed over a range of states; for example, the coordinates and velocities of the individual atoms differ from one molecule to another. This way of proceeding is referred to as **statistical mechanics,** an expression first used in 1902 by J. Willard Gibbs in the title of his book *Elementary Principles in Statistical Mechanics.*

Statistical mechanics makes full use of the information derived from quantum mechanics, but it is not concerned with the precise states of individual molecules. Instead it directs its attention to the most probable states and hence deduces the macroscopic properties. Although it only indicates the most probable behavior, the number of molecules in the assembly is usually so large that the most probable behavior is very close to the behavior actually observed. Only if the number of molecules is small will there be observable statistical fluctuations from the most probable behavior.

The subject of statistical mechanics is a very vast one, and here we will deal only with those aspects that are of particular application to chemical problems. We shall be concerned, for example, with the distribution laws for energies and molecular speeds and with the question of average and most probable energies and speeds. We will then see how the thermodynamic properties can be calculated from these distribution laws. This particular aspect is known as statistical thermodynamics, which among other things allows us to calculate equilibrium constants from such basic information as the shapes and sizes of individual molecules, the vibrational frequencies, and the zero-point energies of the molecules.

Maxwell's Demon

The idea of treating physical and chemical problems on the basis of statistics was born in December 1867 in a letter written by James Clerk Maxwell (1831–1879) to a physicist friend, Peter Guthrie Tait. In this letter Maxwell for the first time suggested the correct interpretation of the second law of thermodynamics (Chapter 3), and he did so by discussing a hypothetical way in which the law could be violated. Maxwell considered a vessel divided into two compartments A and B, separated by a partition which had a hole in it, at which he stationed a creature which Kelvin later referred to as a "demon." The gas in A was at a higher temperature than that in B, and the temperatures would normally become equal over a period of time. Maxwell pointed out, however, that if the demon allowed only slow molecules to pass from A to B, and only fast molecules to pass from B to A, the gas in A would become hotter and the gas in B cooler; the second law would have been violated! Maxwell emphasized that his purpose had been merely to show that the second law is true only as a matter of statistics. In a letter to Lord Rayleigh in 1870 Maxwell commented that the second law of thermodynamics

> has the same degree of truth as the statement that if you throw a tumblerful of water into the sea you cannot get the same tumblerful out again.

This point of view, which we all take for granted today, was by no means the popular one at the time. In 1866, for example, Ludwig Boltzmann (1844–1906) published a paper entitled "On the mechanical meaning of the second law of thermodynamics," in which he attempted to derive the law on the basis of mechanics. A similar position was later taken by Clausius, who had himself done so much to establish the second law (Section 3.1); Clausius, in fact, maintained this position to the end of his life. Boltzmann, however, soon became convinced that Maxwell was right to regard the law as a statistical one, and he went on to make contributions of the greatest importance from that point of view, putting the science of statistical mechanics on a firm foundation.

15.1 Forms of Molecular Energy

Statistical mechanics is primarily concerned with the distribution of energies, and we will begin by considering the different forms of molecular energy. One form is *translational energy,* which is the energy arising from the movement of the center of mass of a molecule from one position to another. We have already discussed this form of energy in Section 1.9, where we saw that the average kinetic energy per molecule ϵ_k, resulting from translational motion, is equal to $\frac{3}{2}k_BT$ (Eq. 1.50 on p. 20). This average energy can be split into three equal contributions for the energies corresponding to motion along the three Cartesian axes X, Y, and Z:

$$\epsilon_x = \epsilon_y = \epsilon_z = \frac{1}{2}k_BT \tag{15.1}$$

Equipartition of Energy

This equal division of the total kinetic energy $\frac{3}{2}k_BT$ into three contributions of $\frac{1}{2}k_BT$ is referred to as the **equipartition of energy.** It is obvious that there must be an equal division as far as translational energy is concerned, since there is no reason why any one direction should be favored over the others. We refer to the components of speed and energy along the three Cartesian axes as **three degrees of translational freedom.**

In Section 13.3, the rotational energies of molecules were discussed. For a nonlinear molecule (Figure 13.17) there are three degrees of rotational freedom, corresponding to rotation about the three Cartesian axes. For a linear molecule (Figure 13.15) there are two degrees of rotational freedom, corresponding to rotation about two axes at right angles to the axis of the molecule. The reason that we do not count the rotation about the axis of the linear molecule is that the mass is almost completely concentrated in the atomic nuclei, so that the moment of inertia corresponding to these rotations is very small. It follows from Eq. 11.218 on p. 556, in which the moment of inertia is in the denominator of the energy expression, that the rotational quantum levels are very far apart, so that the corresponding rotational transitions do not occur.[1]

Vibrational energy was discussed in Section 13.4, where we saw that the vibrational motion can be resolved into a certain number of normal modes of vibration, which are the degrees of freedom for vibration.

To describe the position of an atom in space we need to specify three position coordinates x, y, and z. If, therefore, a molecule contains N_a atoms, $3N_a$ position coordinates must be specified. It is much more convenient, however, to work with degrees of freedom, which give us much more insight into the atomic motions. If a molecule contains N_a atoms, it must have $3N_a$ degrees of freedom, and it is easy to see how these are distributed among the different kinds of motion. Let us consider some simple examples of molecules in the gas phase:

1. A *monatomic molecule,* such as He. There are 3 degrees of freedom, all corresponding to translational motion.
2. A *diatomic molecule,* such as H_2. There are now $3 \times 2 = 6$ degrees of freedom, 3 of which are for translation. Since the molecule is linear, there are 2 degrees of rotational freedom. There is therefore 1 degree of vibrational freedom. This is expected since in Section 13.4 (Figure 13.18) it was found that there is just one mode of vibrational motion for a diatomic molecule.
3. A *linear triatomic molecule,* such as CO_2. There are $3 \times 3 = 9$ degrees of freedom, of which 3 are for translation and 2 for rotation. That leaves 4 degrees of vibrational freedom. These four vibrational modes were shown in Figure 13.23b–d.
4. A *nonlinear triatomic molecule,* such as H_2O. Again there are 9 degrees of freedom, of which 3 are for translation. Now, however, 3 are for rotation, and therefore $9 - 6 = 3$ are for vibration. The three vibrational modes were shown in Figure 13.24b.

If we look at Figure 13.23, we can easily see why there is this difference between the numbers of vibrational and rotational degrees of freedom for linear and nonlinear molecules. The vibrations designated 1, 2, and 3 in Figure 13.23 correspond to the 1, 2, and 3 vibrations in Figure 13.24b. However, vibration 3′ for the linear molecule, if applied to the nonlinear molecule, would become a rotation about the X axis.

We can generalize our conclusions and give the following numbers of degrees of freedom for a gaseous polyatomic molecule containing N_a atoms:

	Translational	Rotational	Vibrational	Total
Linear	3	2	$3N_a - 5$	$3N_a$
Nonlinear	3	3	$3N_a - 6$	$3N_a$

Note that the number of vibrational modes can be very large; for benzene, for example, with $N_a = 12$, there are $3 \times 12 - 6 = 30$ vibrational degrees of freedom.

[1]The explanation commonly given in textbooks is that the atoms can be treated as point masses, so that no change occurs when a linear molecule rotates about its own axis. This explanation may satisfy some people, but it did not satisfy Clerk Maxwell, who considered this problem very deeply over many years. He remained convinced that even a linear molecule must have three degrees of rotational freedom.

Molar Heat Capacities of Gases: Classical Interpretations

Important information about energy distributions in molecules is given by the values of the molar heat capacities of gases. It is simplest first to consider the molar heat capacities at constant volume $C_{V,m}$, since the gas then performs no work of expansion and the heat capacity arises solely from the change in internal energy as the temperature is raised (Section 2.4). The situation is most straightforward for a monatomic gas, since its internal energy then resides solely in the translational motion. We have seen that the average internal energy per molecule is $\frac{1}{2}k_BT$ for each degree of freedom or $\frac{3}{2}k_BT$ for the three translational degrees of freedom. For 1 mol the internal energy is therefore

$$U_m = \frac{3}{2}Lk_BT = \frac{3}{2}RT \tag{15.2}$$

where L is the Avogadro constant and R is the gas constant. We therefore predict that $C_{V,m}$ is

$$C_{V,m} = \frac{dU}{dT} = \frac{3}{2}R \tag{15.3}$$

In other words, the value of $C_{V,m}/R$ should be $\frac{3}{2}$ or 1.5. This prediction is confirmed by the experimental values for some monatomic gases, as shown in Table 15.1.

The situation for other than monatomic molecules is, however, considerably more complicated. Obviously some energy will reside in the other degrees of freedom, and the question is: how much? Let us consider first the *rotational motions*. If the molecule is linear, we can express its rotational energy as

$$\epsilon_{rot} = \tfrac{1}{2}I_x\omega_x^2 + \tfrac{1}{2}I_y\omega_y^2 = \tfrac{1}{2}I(\omega_x^2 + \omega_y^2) \tag{15.4}$$

For a linear molecule the two moments of inertia I_x and I_y are the same and can be written simply as I; ω is the angular velocity about the axis indicated. If the molecule is nonlinear, we have

$$\epsilon_{rot} = \tfrac{1}{2}I_x\omega_x^2 + \tfrac{1}{2}I_y\omega_y^2 + \tfrac{1}{2}I_z\omega_z^2 \tag{15.5}$$

There are now three moments of inertia (two or all of which may be the same). In either case, the rotational energy is proportional to a sum of velocity-squared terms, just as in the case of translational energy. We first consider the situation on the basis of classical physics, according to which there will be equipartition of energy between all the degrees of freedom. If that were the case, each of the degrees of rotational freedom would carry $\frac{1}{2}k_BT$ of energy per molecule. Thus, for a linear molecule there is a contribution of $2 \times \frac{1}{2}k_BT = k_BT$ per molecule, or of RT per mole. For a nonlinear molecule the energy is predicted to be $\frac{3}{2}k_BT$ per molecule or $\frac{3}{2}RT$ per mole. The corresponding contributions to $C_{V,m}$ are thus R and $\frac{3}{2}R$, respectively.

In the third column of Table 15.1 are the contributions to $C_{V,m}/R$ that arise from the translational and rotational motions. For a linear molecule this contribution is $\frac{3}{2}$ from translation and 1 from rotation, giving $\frac{5}{2}$ or 2.50. For a nonlinear molecule we have $\frac{3}{2} + \frac{3}{2} = 3$. We see that the prediction is now closer to the experimental value than if we had considered translation alone. With H_2, N_2, HF, etc., the agreement is excellent, but with the other molecules the theoretical values are too low. With

TABLE 15.1 Experimental and Calculated Values of $C_{V,m}/R$ for Gases at 298.15 K

	$C_{V,m}/R$		
		Calculated:	
Molecule	**Experimental**	**trans. + rot.**	**trans. + rot. + vib.**
Monatomics			
He, Ne, Ar, Kr, Xe	1.500	1.500	1.500
Diatomics			
H_2	2.468	2.500	3.500
N_2	2.503	2.500	3.500
HF	2.505	2.500	3.500
HCl	2.504	2.500	3.500
HBr	2.506	2.500	2.500
O_2	2.533	2.500	3.500
NO	2.590	2.500	3.500
Cl_2	3.083	2.500	3.500
Linear triatomics			
CO_2	3.466	2.500	6.500
CS_2	4.447	2.500	6.500
Nonlinear triatomics			
H_2O	3.040	3.000	6.000
NO_2	3.447	3.000	6.000
SO_2	3.796	3.000	6.000
Linear polyatomics			
C_2H_2	4.304	2.500	9.500
Nonlinear polyatomics			
NH_3	3.288	3.000	9.000
P_4	7.078	3.000	9.000

The experimental values in this table are calculated from the $C_{P,m}$ values given in the *JANAF Thermochemical Tables* (Ed. M. W. Chase et al.), New York: American Chemical Society and the American Institute of Physics, 3rd edition, 1985.

these other molecules there must therefore be a contribution from the vibrational degrees of freedom.

The situation with *vibrational motion* is a little different. In a vibration, the atoms have not only kinetic energy but also potential energy, since they are displaced from their equilibrium positions. The total energy in a single vibrational mode must therefore be written as the sum of two terms, one for kinetic energy and one for potential energy. If there is equipartition of energy, *each* of these terms will contribute $\frac{1}{2}k_BT$ to the energy. There is therefore a contribution from each normal mode of k_BT per molecule, or RT per mole. The corresponding contribution to $C_{V,m}/R$ would thus be 1.00 per normal vibrational mode.

Consider the case of CO_2, a linear triatomic molecule. Its predicted $C_{V,m}/R$ value, assuming equipartition of energy, is $\frac{3}{2}$ from translational, 1 from the two rotational degrees of freedom, and 4 from the four vibrational modes; the total is therefore $\frac{3}{2} + 1 + 4 = 6.50$. This value, together with similarly obtained values for other molecules, is included in the last column of Table 15.1.

LUDWIG BOLTZMANN (1844–1906)

Boltzmann was born in Vienna and was educated at the University of Vienna, receiving his doctorate in 1867. Since he had a rather discontented disposition, he held a number of appointments during his career, making seven moves from one university to another. At various times he held positions at the universities of Vienna, Graz, Munich, and Leipzig. Boltzmann's appointment in 1900 to the University of Leipzig, at the invitation of Wilhelm Ostwald who was professor of physical chemistry there, was because he was unhappy at the University of Vienna. However, two years later he decided that he preferred Vienna after all and returned there to succeed himself as professor of theoretical physics. At Vienna he also lectured on the philosophy of science.

In 1866, while still in his early twenties, Boltzmann published a paper in which he attempted to show that the second law of thermodynamics is a purely mechanical law. Two years later, having read Maxwell's 1860 paper on the distribution of molecular speeds, he realized that this approach was incorrect and that the second law must be explained on the basis of statistics. He therefore extended Maxwell's treatment of the distribution of molecular speeds and showed that the probability that a system is in a state having energy E is proportional to $e^{-E/k_B T}$, where k_B is now called the Boltzmann constant.

In 1872 Boltzmann developed the theory of the approach of systems to equilibrium, and of transport processes in gases. His theorem on the approach to equilibrium has become known as Boltzmann's H theorem. In 1871 Boltzmann showed how all the properties of a system can be calculated from his distribution function, and this can be regarded as the birth of the science later called statistical mechanics. In 1877 Boltzmann gave his famous relationship

$$S = k_B \ln W$$

where S is the entropy and W is the number of possible molecular configurations corresponding to a given state of the system (these configurations are now known as *microstates*). W is often simply called the probability. This formula is engraved on Boltzmann's tombstone in Vienna.

Boltzmann also made important contributions to a wide range of topics in mathematics, physics, chemistry, and philosophy. He had a particular interest in Maxwell's electromagnetic theory. Although his achievements are now recognized to be very great, in his time there was much controversy about them, particularly about his statistical work. A number of influential scientists, led by Ostwald and Ernst Mach (1866–1916), were of the opinion that atoms and molecules have no reality, but are mere fictions; Ostwald, for example, believed that everything could be explained in terms of energy, and he even named his house *Energie.*

This point of view struck at the root of much of Boltzmann's life work; he considered that his statistical work would have been a waste of time if atoms did not exist. Although he defended his position against his critics and gained support from a number of influential scientists, he became increasingly despondent and made several unsuccessful suicide attempts. In 1906, accompanied by his wife and daughter, he went on holiday to Duino, a resort on the Adriatic Sea near Trieste, then a part of Austria. There, while his wife and daughter were out swimming, he hanged himself from the window of his hotel room. It is a sad irony that some of the evidence that later convinced even Ostwald of the existence of atoms, that of Jean Perrin (Section 19.2), was greatly influenced by Boltzmann's work on the statistics of molecular motion.

References: S. G. Brush, *Dictionary of Scientific Biography, 2,* 260–268(1970). S. G. Brush, *The Kind of Motion That We Call Heat,* Amsterdam: North-Holland Publishing Co., 1976.

For the monatomic gases, where there is no rotation or vibration, the value in the last column agrees with the experimental value. In all other cases, however, the theoretical values are much too large. The conclusion drawn from these various comparisons is that the vibrational modes are not making their expected contributions to the molar heat capacity; in other words, **the assumption of equipartition of energy is not generally valid.**

The agreement for the monatomic gases indicates that the assumption is valid for the translational motions, and there is no reason to doubt, from the figures in Table 15.1, that the rotations are also behaving as predicted by the assumption of equipartition.

Molar Heat Capacities of Gases: Quantum Restrictions

If we were to look at the heat capacity data at *very high temperatures,* we would find that there was much better agreement with the values in the last column of the table, which means that there is equipartition of energy at high enough temperatures. At lower temperatures, the principle of equipartition of energy breaks down, particularly for the vibrational modes. In Section 13.3 (Figure 13.14) we have seen that the vibrational levels are much more widely spaced than the rotational and translational levels. Because of this wider spacing, energy cannot move as freely into and out of vibrational modes; a transition can occur only if a sufficiently large quantum of energy is involved. These ideas were first put forward in 1906 by Albert Einstein, who applied them quantitatively to the heat capacities of solids, a matter dealt with in Section 16.6. As noted in Section 11.1 on p. 509, Niels Bjerrum dealt successfully with the specific heats of gases on the basis of quantum theory in 1911 to 1914, and in 1913 Einstein and Stern deduced a zero-point energy from an analysis of the specific heat of hydrogen gas at low temperatures.

15.2 Principles of Statistical Mechanics

We now develop the general principles of statistical mechanics, a subject concerned with this problem of how energy is distributed among different modes of motion. These principles were developed for the most part by Ludwig Boltzmann, in a series of papers from 1868 onwards. His work was particularly inspired by treatments by Clerk Maxwell, in 1860 and 1867, of the distribution of the speeds of the molecules of a gas. Maxwell's work, which was considered in Section 1.11, was concerned only with translational energy, and Boltzmann extended it to include all forms of energy.

We start by considering a large number N of identical atoms or molecules in the gas phase; from now on we will refer to them simply as molecules, which includes atoms and ions. Such a collection is known as a **system** or an **assembly.** We assume the molecules to be *independent of each other,* by which we mean that the state of each is unaffected by the state of any other. A number of discrete energy levels are available to the molecules, and in Figure 15.1 we represent these levels as compartments, which are labeled, starting with the lowest energy as ϵ_0 and higher levels as ϵ_1, ϵ_2, etc. The numbers of molecules in each level are represented as n_0, n_1, n_2, etc.

Energy level	Number of molecules in level
ϵ_i	n_i
ϵ_5	n_5
ϵ_4	n_4
ϵ_3	n_3
ϵ_2	n_2
ϵ_1	n_1
ϵ_0	n_0

FIGURE 15.1
A series of energy levels ϵ_0, ϵ_1, ϵ_2, . . . , represented as compartments into which n_0, n_1, n_2, . . . , molecules can go.

In any assembly of a large number N of identical molecules, corresponding to a given total energy, there are very many different ways in which the molecules can be distributed among the energy levels. For a given total energy we have a given *state* of the system, and we may ask how many ways there are of achieving this state. This is equivalent to asking: in how many ways can we put n_0 molecules into ϵ_0, n_1 into ϵ_1, and so on? For example, suppose that for the first three levels we have $n_0 = 3$, $n_1 = 4$, and $n_2 = 2$. If we consider the ϵ_0 level first, with three molecules in it, we have N choices for the first molecule; once that is in the level, we have $N - 1$ choices; then $N - 2$. If the molecules were all distinguishable, there would therefore be $N(N - 1)(N - 2)$ ways of selecting the three molecules to go into the lowest energy level. However, we are considering N identical and therefore indistinguishable molecules, so that the order in which we choose them does not matter. Thus, if the three molecules in $\epsilon_0 = 0$ are labeled a, b, and c, the choices abc, bca, cba, acb, bac, and cab all amount to the same selection. To obtain the correct number of choices we should therefore divide the product $N(N - 1)(N - 2)$ by 6. In general, if n molecules go into a level, we divide by $n!$, which is the number of ways we can select n molecules. Thus, for the ϵ_0 level, with $n_0 = 3$, the number of ways is

$$\frac{N(N-1)(N-2)}{3!}$$

For the second level, ϵ_1, in which there are four molecules, we can choose the first in $N - 3$ ways, the second in $N - 4$ ways, the third in $N - 5$ ways, and the fourth in $N - 6$ ways. Again, we must divide, in this case by 4!, and the number of ways is therefore

$$\frac{(N-3)(N-4)(N-5)(N-6)}{4!}$$

Similarly, for the ϵ_2 level, with $n_2 = 2$, there are

$$\frac{(N-7)(N-8)}{2!}$$

ways of selecting.

Complexions

The product of these three expressions gives the total number of ways of choosing three molecules to go into ϵ_0, four to go into ϵ_1, and two to go into ϵ_2. This number of ways is known as the number of **complexions,** and in our example it is given by

$$\frac{N(N-1)(N-2)(N-3)(N-4)(N-5)(N-6)(N-7)(N-8)}{3!4!2!}$$

So far we have considered only the first three levels; in reality we have to distribute all the N molecules among the levels. This means that the numerator in this expression will continue with $(N - 9)(N - 10) \cdots$ until we have used up all the molecules; the numerator is thus $N!$. The denominator is the product of the $n_i!$ values for all the levels, which we write as $\prod_i n_i!$. The general expression for the number of complexions is therefore

$$\Omega = \frac{N!}{\prod_i n_i!} \tag{15.6}$$

However, there is an important additional factor, arising from the fact that the total energy corresponding to a given state is fixed. We cannot distribute the molecules in any way we like among the energy levels; we are restricted to distributions that are consistent with the total energy of the system. To understand this in a simple way, suppose that we have three molecules to distribute among four energy levels ϵ_0, ϵ_1, ϵ_2, and ϵ_3. We will suppose that the energy levels are equally spaced, with a difference of ϵ between neighboring levels, so that

$$\epsilon_0 = 0, \qquad \epsilon_1 = \epsilon, \qquad \epsilon_2 = 2\epsilon, \qquad \epsilon_3 = 3\epsilon$$

as shown in Figure 15.2a. We will suppose that the total energy of the system is 3ϵ, and the question we ask is: how many complexions are there in which three molecules are in the various energy levels, such that the total energy is 3ϵ?

These complexions are shown in Figure 15.2b. One distribution, designated Ω_1 in the figure, is the one in which each of the three molecules is in level ϵ_1, so that it has an energy ϵ; the total energy is 3ϵ. Since the molecules are indistinguishable, the order of filling is immaterial (i.e., *abc* is the same as *acb*, etc.), so that there is only one complexion corresponding to this distribution. This conclusion is confirmed by use of Eq. 15.6, since

$$\Omega_1 = \frac{3!}{0!3!0!0!} = 1 \tag{15.7}$$

(Remember that $0! = 1$.)

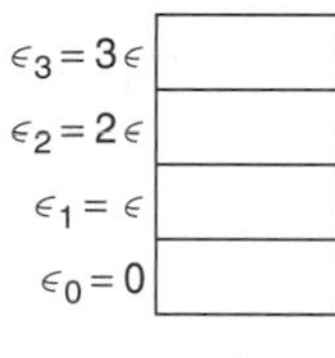

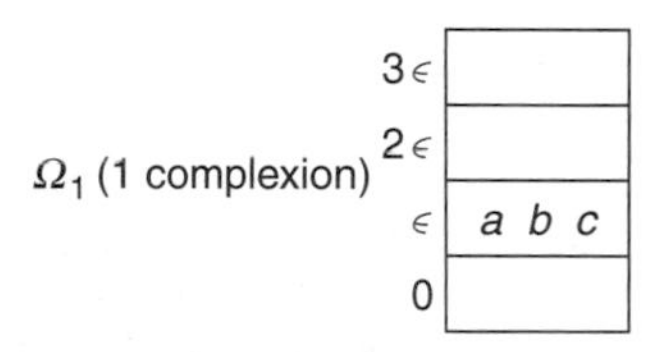

Ω_2 (3 complexions)
3∈ a b c
2∈
∈
0 b c a c a b

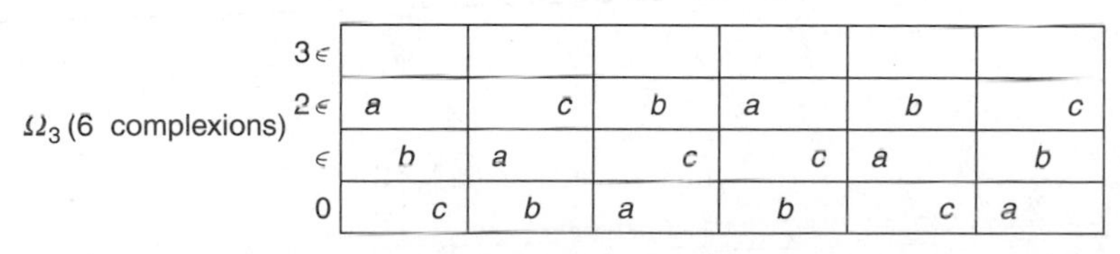

FIGURE 15.2
(a) Four equally spaced energy levels, shown as compartments into which we are going to put three molecules, the total energy of the system being 3ϵ. (b) The various ways (i.e., complexions) in which the three molecules can go into the energy levels. The ways can be grouped into Ω_1, Ω_2, and Ω_3.

A second distribution, designated Ω_2 and shown in the figure, involves having one of the molecules in ϵ_3, with energy 3ϵ, and the other two in ϵ_0, with energy zero. This can be done in three different ways, since any one of the molecules can be in ϵ_3, leaving the other two in ϵ_0. Use of Eq. 15.6 again confirms that there are three complexions:

$$\Omega_2 = \frac{3!}{2!0!0!1!} = 3 \tag{15.8}$$

A third possible arrangement is to have one molecule in $\epsilon_0(\epsilon = 0)$, one in $\epsilon_1(\epsilon = 1)$, and one in $\epsilon_2(\epsilon = 2)$. The figure shows that there are six ways of doing this, as confirmed by use of Eq. 15.6:

$$\Omega_3 = \frac{3!}{1!1!1!0!} = 6 \tag{15.9}$$

In statistical mechanics, each type of distribution, such as Ω_1, is referred to as a **state.**[2] The system of three molecules that we are considering will be constantly shifting from one state to another. Thus, at a given instant we may have the Ω_1 distribution, with all three molecules in ϵ_1, but collisions may bring about a transition to one of the three Ω_2 complexions; two of the molecules drop down to ϵ_0 and one is promoted to ϵ_3, there being no change of total energy. It is easy to see intuitively that *all the complexions are equally probable.* In our example, the Ω_2 distribution is three times as probable as the Ω_1, because there are three ways of forming it as compared with one way for Ω_1. Similarly, Ω_3 is six times as probable as Ω_1 and twice as probable as Ω_2.

We now come to a point of very great importance. In our example of three molecules we see that one state, Ω_3, is more probable than the other two, but the difference is not enormous. However, we are usually dealing with very large numbers of molecules, of the order of 10^{23} and more. We then find that one state has a much larger number of complexions than any of the others.

To illustrate this point, suppose that we had an assembly of N molecules, where N is very large. We can imagine a state in which all the N molecules are in different energy levels. The number of complexions is then, from Eq. 15.6,

$$\Omega = \frac{N!}{1!1!\cdots} = N! \tag{15.10}$$

This is the largest possible value of Ω for this particular state. If two of the molecules were in the lowest level and another were promoted, in order to keep the energy the same, the number of complexions would drop to

$$\Omega = \frac{N!}{2!1!1!\cdots} = \frac{1}{2}N! \tag{15.11}$$

To go to the limit, if all the molecules except one were in the lowest level, the number of complexions would be

$$\Omega = \frac{N!}{(N-1)!1!} = N \tag{15.12}$$

which is vastly smaller than $N!$.

[2]Note that the word *state* is here being used in a more restricted sense than in thermodynamics (Section 2.3).

The important point that we have made is that when we are dealing with very large numbers of molecules, a certain state will greatly predominate over the others. This is the state corresponding to the broadest distribution of the molecules over all the energy levels.

The Boltzmann Distribution Law

This discussion has provided us with the background to proceed in a more formal way, with the objective of finding a general expression for the distribution of molecules among energy levels.

We start by writing the total number W of complexions; this is the sum of the number of complexions for each state:

$$W = \sum \Omega = \sum \frac{N!}{\prod_i n_i!} \tag{15.13}$$

It is understood that the summation is over all the states. There are two additional restrictions. The first is that the sum of all the n_i values is equal to the total number N of molecules in the assembly:

$$n_1 + n_2 + n_3 + \cdots = \sum_i n_i = N \tag{15.14}$$

The second is that the total energy is fixed; if we call the total energy E, this means that

$$n_1\epsilon_1 + n_2\epsilon_2 + \cdots = \sum_i n_i\epsilon_i = E \tag{15.15}$$

We now apply the conclusion that one of the Ω values is predominant, and we take the largest term in Eq. 15.13 to be the value of W. What we therefore do is to refer to Eq. 15.6, which defined Ω, and find what values of n_1, n_2, etc., will make Ω a maximum. Any small change in these values, namely δn_1, δn_2, etc., will produce a small change $\delta\Omega$, and the condition that Ω has its maximum value is

$$\delta\Omega = \frac{\partial\Omega}{\partial n_1}\delta n_1 + \frac{\partial\Omega}{\partial n_2}\delta n_2 + \cdots + \frac{\partial\Omega}{\partial n_i}\delta n_i + \cdots = 0 \tag{15.16}$$

If there were no restrictions on the n_i values, we could obtain the maximum by setting each $\partial\Omega/\partial n_i$ equal to zero. This, however, is not possible because of Eqs. 15.14 and 15.15, which do not allow the n_i values to be varied independently. Instead we make use of the *Lagrange method of undetermined multipliers,* introduced by the astronomer and mathematician Joseph Louis Lagrange (1736–1813). In this method the two restrictive conditions

$$\delta n_1 + \delta n_2 + \cdots + \delta n_i + \cdots = 0 \tag{15.17}$$

and

$$\epsilon_1\delta n_1 + \epsilon_2\delta n_2 + \cdots + \epsilon_i\delta n_i + \cdots = 0 \tag{15.18}$$

(which came from Eqs. 15.14 and 15.15) are first multiplied by constants, which we write as α and $-\beta$, respectively:

$$\alpha(\delta n_1 + \delta n_2 + \cdots + \delta n_i + \cdots) = 0 \tag{15.19}$$

$$-\beta(\epsilon_1\delta n_1 + \epsilon_2\delta n_2 + \cdots + \epsilon_i\delta n_i + \cdots) = 0 \tag{15.20}$$

We then take the sum of Eqs. 15.16, 15.19, and 15.20 and obtain

$$\delta\Omega = \left(\frac{\partial\Omega}{\partial n_1} + \alpha - \beta\epsilon_1\right)\delta n_1 + \left(\frac{\partial\Omega}{\partial n_2} + \alpha - \beta\epsilon_2\right)\delta n_2 + \cdots + \left(\frac{\partial\Omega}{\partial n_i} + \alpha - \beta\epsilon_i\right)\delta n_i + \cdots = 0 \tag{15.21}$$

The only way that this can hold for any values of the independent small quantities δn_1, δn_2, etc., is if every one of the enclosed terms is equal to zero; in other words, for all values of i,

$$\frac{\partial\Omega}{\partial n_i} + \alpha - \beta\epsilon_i = 0 \tag{15.22}$$

Since it is mathematically difficult to determine α and β from the equation in this form, we replace Ω by $\ln \Omega$. This is permissible since $\ln \Omega$ reaches a maximum when Ω does. The equation we need to solve is therefore

$$\frac{\partial \ln \Omega}{\partial n_i} + \alpha - \beta\epsilon_i = 0 \tag{15.23}$$

To solve this equation, we first take the natural logarithms of both sides of Eq. 15.6 and obtain

$$\ln \Omega = \ln N! - \sum_i \ln n_i! \tag{15.24}$$

James Stirling (1692–1770) was a Scottish mathematician whose *Methodus differentialix* (1730) contains many important relationships.

We now make use of Stirling's approximation, which for values of n greater than about 10 reduces satisfactorily to

$$\ln n! = n \ln n - n \quad \text{or} \quad n! = \frac{n^n}{e^n} \tag{15.25}$$

Application of this approximation to Eq. 15.24 gives

$$\ln \Omega = \ln N! - \sum_i (n_i \ln n_i - n_i) \tag{15.26}$$

Then, since $\ln N!$ is constant,

$$\frac{\partial \ln \Omega}{\partial n_i} = -\frac{\partial}{\partial n_i}(n_i \ln n_i - n_i) \tag{15.27}$$

$$= -\ln n_i - 1 + 1 = -\ln n_i \tag{15.28}$$

Insertion of this into Eq. 15.23 gives

$$\ln n_i = \alpha - \beta\epsilon_i \tag{15.29}$$

and therefore

$$n_i = e^{\alpha} e^{-\beta\epsilon_i} \tag{15.30}$$

This is conveniently written as

$$n_i = A e^{-\beta\epsilon_i} \tag{15.31}$$

where A is equal to e^{α}.

In Section 11.1 we proved that β is equal to $1/k_BT$, where T is the absolute temperature and k_B is the Boltzmann constant. This result allows us to write Eq. 15.31 as

$$n_i = Ae^{-\epsilon_i/k_BT} \tag{15.32}$$

The total number of molecules N is

$$N = \sum_i n_i = A \sum_i e^{-\epsilon_i/k_BT} \tag{15.33}$$

Therefore

$$\frac{n_i}{N} = \frac{e^{-\epsilon_i/k_BT}}{\sum_i e^{-\epsilon_i/k_BT}} \tag{15.34}$$

This is the fundamental equation of the **Boltzmann distribution.** It gives us the fraction of the molecules that are in a specified state of energy ϵ_i. If we are interested in two states of energies ϵ_i and ϵ_j, we also have, from Eq. 15.32,

$$\frac{n_i}{n_j} = \frac{e^{-\epsilon_i/k_BT}}{e^{-\epsilon_j/k_BT}} = \exp\left[\frac{-(\epsilon_i - \epsilon_j)}{k_BT}\right] \tag{15.35}$$

That is, the ratio of the populations in the two energy states depends on the energy difference $\epsilon_i - \epsilon_j$.

Some special cases of the Boltzmann principle have already been considered. In Section 1.10, for example, we derived the equation for the distribution of pressure in the atmosphere. Section 1.11 was concerned with the Maxwell distribution law, which deals with the distribution of molecular speeds, i.e., translational energies. Boltzmann's great contribution was to show that a similar relationship applies to any kind of energy.

15.3 The Partition Function

In Section 1.11 we were concerned exclusively with the translational motions of gas molecules. This is an important topic in itself, particularly as it led to a simple way of identifying the quantity β with $1/k_BT$, but the Boltzmann distribution law has a much wider applicability. We will now apply it to a variety of different kinds of energy.

The law in its most general form is expressed in Eq. 15.34. The denominator in this expression has a special significance, and it is given the symbol q or Q and called the **partition function:**

$$q \equiv \sum_i e^{-\epsilon_i/k_BT} \tag{15.36}$$

Its German name *Zustandsumme* (sum over states) indicates its significance more clearly.

The Molecular Partition Function

When the partition function relates to a single molecule, we call it the **molecular partition function** or the *particle partition function* and give it the symbol q. We

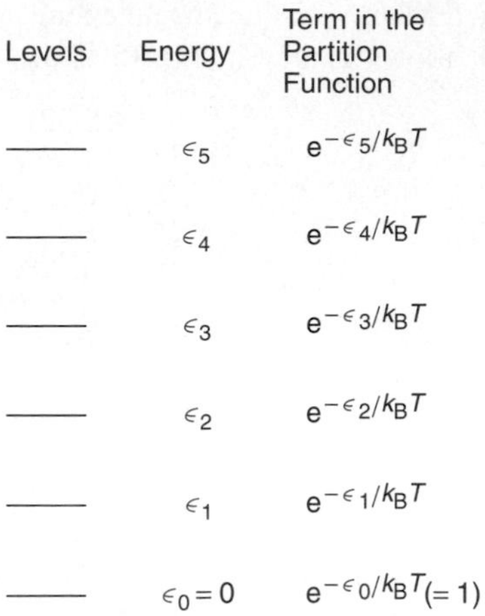

Levels	Energy	Term in the Partition Function
——	ϵ_5	$e^{-\epsilon_5/k_BT}$
——	ϵ_4	$e^{-\epsilon_4/k_BT}$
——	ϵ_3	$e^{-\epsilon_3/k_BT}$
——	ϵ_2	$e^{-\epsilon_2/k_BT}$
——	ϵ_1	$e^{-\epsilon_1/k_BT}$
——	$\epsilon_0 = 0$	$e^{-\epsilon_0/k_BT} (= 1)$

FIGURE 15.3
A series of energy levels and the corresponding terms in the partition function.

will consider this type of partition function first. Figure 15.3 shows a number of molecular energy levels, starting with ϵ_0, which is taken to be zero. The terms for each level are indicated, and their sum is

$$q = e^{-0/k_BT} + e^{-\epsilon_1/k_BT} + e^{-\epsilon_2/k_BT} + \cdots \tag{15.37}$$

The first term is unity, the second is a fraction, the third a smaller fraction, and so on. The partition function is thus unity plus a series of fractions that becomes smaller and smaller. If the energy levels are very close together, the second term is close to unity, and there may be a very large number of terms that are close to unity. This is the situation with translational energy; the levels are very closely spaced, and there are many terms that are close to unity. The partition function is thus very large; we shall see later that for translational motion in a volume of ordinary dimensions the molecular partition function at 300 K is 10^{30} or more.

At the other extreme, if the energy levels are widely spread, even the second term $\exp(-\epsilon_1/k_BT)$ is very small, and subsequent ones are even smaller. The partition function will thus be close to unity (apart from *degeneracy,* which is discussed later). This is often the situation for electronically excited states. For example, the first electronically excited state of the Cl_2 molecule is 18 310.5 cm^{-1} above the ground state. This is 3.64×10^{-19} J, and at 300 K the term is

$$\exp\left(\frac{-3.64 \times 10^{-19}\ \text{J}}{k_BT}\right) = e^{-87.8} = 7.4 \times 10^{-39} \tag{15.38}$$

This is negligible compared to unity, so that the partition function is close to unity.

The situation with rotational and vibrational energy levels is in between these two extremes. Vibrational partition functions are often close to unity because of the fairly wide separation of the energy levels. The separation is much less for the rotational levels, and the partition functions are usually substantially larger than unity.

The significance of the partition function is that it gives an indication of the number of states that are accessible to the system. At the absolute zero only the ground state is accessible, and the partition function is unity. As the temperature is raised, more and more states become accessible, and as T approaches infinity, all states become accessible. A system that has a large number of accessible states is inherently more likely to exist than one with a more limited accessibility. We will see later in this chapter that this concept is of great significance in connection with the evaluation of equilibrium constants.

Degeneracy

It is possible for more than one state to have an energy equal to ϵ_i, and the energy level is then said to be *degenerate.* If the energy level i has g_i states, we can write

$$q = \sum_i g_i e^{-\epsilon_i/k_BT} \tag{15.39}$$

where the factor g_i is known as the **degeneracy** or the **statistical weight.** We can now rewrite Eq. 15.34 in a form that includes the degeneracies:

$$\frac{n_i}{N} = \frac{g_i e^{-\epsilon_i/k_BT}}{\sum_i g_i e^{-\epsilon_i/k_BT}} = \frac{g_i e^{-\epsilon_i/k_BT}}{q} \tag{15.40}$$

EXAMPLE 15.1 In a C—I bond, the iodine atom has a low frequency of vibration because of its large mass. As a result, energy levels higher than ϵ_0 are occupied to a greater extent than for atoms such as hydrogen bound to carbon. If the frequency of vibration of iodine in a C—I bond is 5.0×10^{12} s^{-1} at 298 K, calculate the value of n_8/n_0, the value of q, and the value of n_8/N.

Solution The value of the energy difference $\epsilon_i - \epsilon_0$ corresponds to $n_i h\nu$ since each level is higher than the previous one by $h\nu$. Thus

$$\epsilon_8 - \epsilon_0 = 8h\nu = 8(6.626 \times 10^{-34} \text{ J s})(5.0 \times 10^{12} \text{ s}^{-1}) = 2.65 \times 10^{-20} \text{ J}$$

From Eq. 15.35 we can write

$$\ln \frac{n_8}{n_0} = -\frac{1}{k_B T}(\epsilon_8 - \epsilon_0) = -\frac{2.65 \times 10^{-20}(\text{J})}{1.38 \times 10^{-23} \times 298(\text{J})} = -6.44$$

Thus $n_8/n_0 = 1.59 \times 10^{-3}$.

We can calculate the partition function by adding the $e^{-\epsilon_i/k_B T}$ terms until the value becomes negligible. The first terms are as follows:

i	0	1	2	3	4	5	6	7	8
$e^{-\epsilon_i/k_B T}$	1	0.447	0.200	0.090	0.040	0.018	0.008	0.004	0.002

The sum of these is 1.8, and this is a good approximation to the partition function q. We will later see a simpler way of obtaining such a partition function, by summation of a series. The ratio n_8/N is

$$\frac{n_8}{N} = \frac{e^{-(\epsilon_8 - \epsilon_0)/k_B T}}{q} = \frac{e^{-6.44}}{1.8} = 8.9 \times 10^{-4}$$

The Canonical Partition Function

The molecular partition function q relates to a single molecule, but in thermodynamics we are concerned with assemblies of molecules. We therefore also define a *partition function Q for the system,* which relates to an assembly of the large number N of molecules. If N is the number of molecules in 1 mol (6.022×10^{23}), it is called the *molar partition function.*

The **system partition function,** usually known as the **canonical partition function,**[3] is defined by

$$Q = \sum_i e^{-E_i/k_B T} \tag{15.41}$$

where the summation is now taken over all the energy levels for all the molecules in the assembly. We must now see how the molecular partition function q is related to the partition function Q. An important distinction arises, according to whether the molecules are localized and therefore *distinguishable* or are nonlocalized and *indistinguishable*.

An ideal crystal is a simple example of a system in which the molecules (or atoms or ions) comprising it are localized. If we label the molecules a, b, c, etc., we see that molecule a is *distinguishable* from molecule b simply because it is in a different

Specification of a thermodynamic state requires knowledge of only a few macroscopic properties, such as pressure. Recall from Section 1.9 that on a short time scale the individual values of the momentum change considerably as the particles collide with the wall. A calculation of the pressure by following the trajectories in a gas sample over a period of time sufficient to obtain a result independent of the time interval of measurement is impossible, even with the most modern computers.

In order to make calculations, an ensemble of systems is envisioned. (See Section 15.10, p. 828.) Three

(continued on next page)

[3]The word *canonical* is used to refer to a standard formula.

ensembles are widely used, depending upon the independent variables that are fixed and depending upon the type of contact of the molecules (including atoms and ions) with the surroundings:
- microcanonical ensemble;
  - V, N, E; no contact
- canonical ensemble;
  - V, N, T; thermal contact
- grand canonical ensemble;
  - V, μ, T; thermal and material contact.

position. Each molecule, such as a, will have a set of energy levels ϵ_1^a, ϵ_2^a, ϵ_3^a, etc. The system partition function Q can therefore be expressed as

$$Q = \sum_i \exp\left[\frac{-(\epsilon_i^a + \epsilon_i^b + \epsilon_i^c + \cdots)}{k_B T}\right] \tag{15.42}$$

This summation can be factored into

$$Q = \left(\sum_i e^{-\epsilon_i/k_B T}\right)_a \left(\sum_i e^{-\epsilon_i/k_B T}\right)_b \left(\sum_i e^{-\epsilon_i/k_B T}\right)_c \cdots \tag{15.43}$$

where the first term in parentheses is the summation for molecule a, the second for b, and so on. Since all of these terms are the same, the molecules being identical, we can write

$$Q = \left(\sum_i e^{-\epsilon_i/k_B T}\right)^N \tag{15.44}$$

where N is the number of molecules. We therefore obtain the important result that for this type of system

Distinguishable Molecules

$$Q = q^N \tag{15.45}$$

In a gas, however, the situation is different, because the molecules or atoms composing it are not localized and are therefore *indistinguishable*. Suppose, for example, that molecule a has the energy ϵ_a; molecule b has the energy ϵ_b; and so on. The total energy is therefore

$$E = \epsilon_a + \epsilon_b + \epsilon_c + \cdots \tag{15.46}$$

However, the state corresponding to this energy is indistinguishable from the state that we would obtain by interchanging molecules a and b or any pairs of molecules. In this case we therefore count too many states if we proceed as we did for the crystal. We must divide by the number of ways we can permute N molecules, that number being $N!$. For systems of this kind we therefore have the relationship

Indistinguishable Molecules

$$Q = \frac{q^N}{N!} = \left(\frac{qe}{N}\right)^N \tag{15.47}$$

In the second expression here we have made use of Stirling's approximation (Eq. 15.25).

15.4 Thermodynamic Quantities from Partition Functions

The reason that partition functions are of such great importance is that all the thermodynamic quantities can be calculated from them. This means that the entire macroscopic behavior of matter at equilibrium can be predicted if we have evaluated the partition function. For example, if we know the partition functions for all the substances in a chemical reaction, we can calculate the Gibbs energy of each

The transition from probabilities to thermodynamic quantities is understandable by considering the energy. The probability that any member of the distribution is in a state of given energy is given by Eq. 15.41. This is a rapidly decreasing function of energy. However, this is multiplied by the number of states, g_i, corresponding to each energy. Plotted against energy, this is a rapidly increasing function. Thus the partition function Q (Eq. 15.50) is a sharply peaked function about the mean energy of almost all members.

substance and can therefore calculate the equilibrium constant at any temperature. We will see in Section 15.7 exactly how this is done.

First we will consider the relationship between the partition function and the internal energy U of the system. When all the molecules are in their lowest possible energy states, the internal energy has its minimum value, which we call U_0. In general the internal energy is U and what we can obtain from the partition function Q for the system is the value of $U - U_0$. The way we do this is to consider a very large number N^* of assemblies, having energies E_1^*, E_2^*, etc., with n_1^*, n_2^*, etc., being their respective distribution numbers. The large collection of assemblies is known as an **ensemble,** and we use the starred quantities for the ensemble. The energy of each ensemble is represented by a term $n_i^* E_i^*$, and the average value of $U - U_0$ for the large collection of assemblies is[4]

$$U - U_0 = \frac{\sum_i n_i^* E_i^*}{N^*} \tag{15.48}$$

For an ensemble of N^* assemblies the ratio n_i^*/N^* is given by

$$\frac{n_i^*}{N^*} = \frac{g_i e^{-E_i^*/k_B T}}{Q} \tag{15.49}$$

(compare Eq. 15.40, which is for a single assembly). The partition function Q for the system is given by

$$Q = \sum_i g_i e^{-E_i^*/k_B T} \tag{15.50}$$

where E_i^* is the energy of the ith assembly and g_i is the degeneracy or statistical weight (Eq. 15.39). The value of $U - U_0$ is therefore

$$U - U_0 = \frac{\sum_i g_i E_i^* e^{-E_i^*/k_B T}}{Q} \tag{15.51}$$

The E_i^* values do not depend on temperature and we can differentiate Eq. 15.50 and obtain

$$\left(\frac{\partial Q}{\partial T}\right)_V = \frac{1}{k_B T^2} \sum_i g_i E_i^* e^{-E_i^*/k_B T} \tag{15.52}$$

The expression for $U - U_0$ can therefore be written as

$$U - U_0 = \frac{k_B T^2}{Q} \left(\frac{\partial Q}{\partial T}\right)_V \tag{15.53}$$

or as

$$U - U_0 = k_B T^2 \left(\frac{\partial \ln Q}{\partial T}\right)_V \tag{15.54}$$

[4]More advanced presentations of this topic discuss in detail an assumption that is involved here, namely that the time average of $U - U_0$ is the same as the average over a large number of assemblies. This is known as the *ergodic hypothesis*.

TABLE 15.2 Thermodynamic Quantities in Terms of the Partition Function Q for the System

$U - U_0 = k_B T^2 \left(\frac{\partial \ln Q}{\partial T}\right)_V$
$S = k_B T \left(\frac{\partial \ln Q}{\partial T}\right)_V + k_B \ln Q$
$A - U_0 = -k_B T \ln Q$
$H - U_0 = k_B T^2 \left(\frac{\partial \ln Q}{\partial T}\right)_V + N k_B T$
$G - U_0 = -k_B T \ln Q + N k_B T$

If N is the number of molecules in 1 mol, it can be replaced by the Avogadro constant L (= 6.022×10^{23} mol^{-1}), and the thermodynamic quantity is then the molar quantity.

This expression is given in Table 15.2, along with other expressions for thermodynamic quantities.

To obtain the entropy we can make use of the relationship

$$\int_0^T dS = \int_0^T \frac{C_V}{T}\, dT \tag{15.55}$$

or

$$S - S_0 = \int_0^T \frac{C_V}{T}\, dT \tag{15.56}$$

where S_0 is the entropy at the absolute zero. The heat capacity C_V is

$$C_V = \left(\frac{\partial U}{\partial T}\right)_V = \frac{\partial}{\partial T}\left[k_B T^2 \left(\frac{\partial \ln Q}{\partial T}\right)_V\right] \tag{15.57}$$

$$= k_B \left[T^2 \left(\frac{\partial^2 \ln Q}{\partial T^2}\right)_V + 2T \left(\frac{\partial \ln Q}{\partial T}\right)_V\right] \tag{15.58}$$

Then

$$S - S_0 = k_B \int_0^T \left[T^2 \left(\frac{\partial^2 \ln Q}{\partial T^2}\right)_V + 2T \left(\frac{\partial \ln Q}{\partial T}\right)_V\right] dT \tag{15.59}$$

Integration by parts leads to

$$S - S_0 = k_B T \left(\frac{\partial \ln Q}{\partial T}\right)_V + k_B \ln Q - k_B \ln Q_0 \tag{15.60}$$

where Q_0 is the value of Q at the absolute zero. The values at the absolute zero may be equated to one another,

$$S_0 = k_B \ln Q_0 \tag{15.61}$$

and therefore

$$S = k_B T \left(\frac{\partial \ln Q}{\partial T} \right)_V + k_B \ln Q \tag{15.62}$$

$$= \frac{U - U_0}{T} + k_B \ln Q \tag{15.63}$$

To obtain the Helmholtz energy we use the relationship $A = U - TS$, and with Eq. 15.63 we obtain

$$A - U_0 = -k_B T \ln Q \tag{15.64}$$

(Note that at the absolute zero $A_0 = U_0$.) Since $P = -(\partial A/\partial V)_T$ (Eq. 3.118) we also have

$$P = k_B T \left(\frac{\partial \ln Q}{\partial V} \right)_T \tag{15.65}$$

The definition of the enthalpy, $H = U + PV = U + Nk_BT$, with Eq. 15.54 gives

$$H - U_0 = k_B T^2 \left(\frac{\partial \ln Q}{\partial T} \right)_V + Nk_B T \tag{15.66}$$

The Gibbs energy, $G = A + PV = A + Nk_BT$, is similarly found, with Eq. 15.64, to be

$$G - U_0 = -k_B T \ln Q + Nk_B T \tag{15.67}$$

These various expressions are included in Table 15.2. The corresponding expressions in terms of the molecular partition function q are easily obtained (see Problem 15.7).

If two substances are present, the procedure is as follows. Suppose that, as in a mixture of two gases, the molecules of each type are indistinguishable. The partition function for the system is then

$$Q = \frac{q_a^{N_a} q_b^{N_b}}{N_a! N_b!} \tag{15.68}$$

We can now make use of the definition of the chemical potential (Section 4.1); thus for the substance of type a, of which there are N_a molecules, the chemical potential per molecule of a is

$$\mu_a = \left(\frac{\partial A}{\partial N_a} \right)_{T,V} = -k_B T \left(\frac{\partial \ln Q}{\partial N_a} \right)_{T,V} \tag{15.69}$$

From Eq. 15.68 we obtain

$$\ln Q = N_a \ln q_a + N_b \ln q_b - \ln N_a! - \ln N_b! \tag{15.70}$$

$$= N_a \ln q_a - N_a \ln N_a + N_a + N_b \ln q_b - \ln N_b! \tag{15.71}$$

using Stirling's approximation. Thus

$$\left(\frac{\partial \ln Q}{\partial N_a} \right)_{T,V} = \ln q_a - \ln N_a = \ln \frac{q_a}{N_a} \tag{15.72}$$

The **molecular chemical potential** is therefore

$$\mu_a = -k_B T \ln \frac{q_a}{N_a} \tag{15.73}$$

This type of expression can be used in the treatment of chemical equilibrium.

15.5 The Partition Function for Some Special Cases

We have already seen in Section 15.1 that the energy of a molecule may be translational, ϵ_t; rotational, ϵ_r; and vibrational, ϵ_v. To these must be added electronic energy, ϵ_e, and nuclear energy, ϵ_n, and the total energy is

$$\epsilon_{tot} = \epsilon_t + \epsilon_r + \epsilon_v + \epsilon_e + \epsilon_n \tag{15.74}$$

If the energy of a molecule is the sum of *independent* energy terms, the corresponding partition function is the product of the individual contributions to the partition function.

Because these energies are additive, the partition function is a product of the contributions for these different types of energy. For example, suppose there is a series of vibrational energies ϵ_j^v and another series of rotational energies ϵ_k^r. There is then a set of total energies given by

$$\epsilon_i = \epsilon_j^v + \epsilon_k^r \tag{15.75}$$

Then, from Eq. 15.36, the partition function for these two forms of energy is

$$q = \sum_i e^{-\epsilon_i/k_BT} = \sum_j \sum_k \exp\left[\frac{-(\epsilon_j^v + \epsilon_k^r)}{k_BT}\right] \tag{15.76}$$

$$= \sum_j e^{-\epsilon_j^v/k_BT} \sum_k e^{-\epsilon_k^r/k_BT} = q_v q_r \tag{15.77}$$

In general, for the five forms of energy,

$$q = q_t q_r q_v q_e q_n \tag{15.78}$$

Translational Motion

We will first consider a monatomic gas and will assume, as is usually the case at ordinary temperatures, that only the lowest electronic state has to be considered. There is therefore only translational energy, which consists of the energies in the three independent directions,

$$\epsilon_t = \epsilon_x + \epsilon_y + \epsilon_z \tag{15.79}$$

The translational partition function is also a product, for the reason we have just discussed:

$$q_t = q_x q_y q_z \tag{15.80}$$

Solving for q_t simply involves determining the values of the allowed energies. The translational energies are given by the solution of the Schrödinger equation for a particle of mass m confined in a three-dimensional box with sides of length a, b, c in the X-, Y-, and Z-directions, respectively. The solution, given in Section 11.6 (Eq. 11.150 on p. 533), is

$$\epsilon_t = \epsilon_x + \epsilon_y + \epsilon_z = \frac{h^2}{8m}\left(\frac{n_x^2}{a^2} + \frac{n_y^2}{b^2} + \frac{n_z^2}{c^2}\right) \tag{15.81}$$

where the integers n_x, n_y, n_z are *quantum numbers* and specify the energy states available to the molecule.

The partition function can then be written as

$$q_t = \sum_{n_x}\sum_{n_y}\sum_{n_z}\exp\left[-\frac{(\epsilon_x + \epsilon_y + \epsilon_z)}{k_B T}\right] = q_x q_y q_z \tag{15.82}$$

It has been split into three factors into which the values of the individual energies are inserted. For q_x the expression is

$$q_x = \sum_{n_x} e^{-n_x^2 h^2/8ma^2 k_B T} \tag{15.83}$$

the summation being from 0 to ∞. The spacing between translational levels is extremely small, and the sum in Eq. 15.83 can therefore be replaced by an integral:

$$q_x = \int_0^{\infty} e^{-n_x^2 h^2/8ma^2 k_B T}\, dn_x \tag{15.84}$$

The value of the integral can be obtained from the appendix to Chapter 1 (p. 43) and the result is

$$q_x = \frac{(2\pi m k_B T)^{1/2} a}{h} \tag{15.85}$$

Similar expressions are obtained for q_y and q_z. The result for q_t is

$$q_t = q_x q_y q_z = \frac{(2\pi m k_B T)^{3/2} abc}{h^3} = \frac{(2\pi m k_B T)^{3/2} V}{h^3} \tag{15.86}$$

since *abc* is the volume *V*.

EXAMPLE 15.2 Calculate the translational partition function for a hydrogen atom at 300 K in a container of 1-m^3 volume.

Solution The translational partition function is found from Eq. 15.103. The mass of the hydrogen atom is

$$\frac{1.008 \times 10^{-3}}{6.022 \times 10^{23}} = 1.674 \times 10^{-27}\ \text{kg}$$

With $V = 1$ m^3, the partition function is therefore

$$q_t = \frac{[2\pi \times 1.674 \times 10^{-27}(\text{kg}) \times 1.381 \times 10^{-23}(\text{J K}^{-1}) \times 300(\text{K})]^{3/2} 1(\text{m}^3)}{[6.626 \times 10^{-34}(\text{J s})]^3}$$

$$= 9.89 \times 10^{29}$$

Note that the complete partition function, for a specified volume, is dimensionless. Later, in dealing with equilibrium constants, we will use partition functions per unit volume. Translational partition functions per unit volume typically have the magnitude of ~10^{32} m^{-3} at 300 K, and we will see that the rotational and vibrational functions are much smaller.

An expression for the entropy of an ideal monatomic gas was first derived in 1911 by O. Sackur and H. Tetrode. For a system of N indistinguishable molecules, $\ln Q$ is $N \ln (qe/N)$ (Eq. 15.47), and substitution of the expression for q in Eq. 15.86 gives

$$\ln Q = N \ln\left[\frac{V}{Nh^3}(2\pi m k_B T)^{3/2}\right] + N \tag{15.87}$$

Then from Eq. 15.63, and making use of the fact that $U - U_0 = \frac{3}{2}Nk_BT$, we obtain, for the molar entropy,

Sackur-Tetrode Equation

$$S_m = \frac{U_m - U_{0,m}}{T} + k_B \ln Q = \frac{5}{2}R + R \ln\left[\frac{V_m}{Lh^3}(2\pi m k_B T)^{3/2}\right] \tag{15.88}$$

This is known as the **Sackur-Tetrode equation,** and it gives excellent agreement for monatomic gases. We can replace V_m by Lk_BT/P and then obtain for an ideal gas at 1 bar pressure

$$S_m/\text{J K}^{-1}\,\text{mol}^{-1} = -9.620 + 12.47 \ln M_r + 12.47 \ln T/K \tag{15.89}$$

where M_r is the relative molecular mass ("molecular weight"). At 25 °C

$$S_m/\text{J K}^{-1}\,\text{mol}^{-1} = 108.74 + 12.47 \ln M_r \tag{15.90}$$

Rotational Motion

The rotational energy of a linear molecule, treated as a rigid rotor, was given in Eq. 11.218 on p. 556:

$$\epsilon_r = J(J+1)\frac{\hbar^2}{2I} \tag{15.91}$$

where I is the moment of inertia and J is the rotational quantum number, having values 0, 1, 2, 3, ... The rotational partition function is therefore

$$q_r = \sum_i g_i e^{-\epsilon_i/k_BT} = \sum_J (2J+1) \exp\left[-\frac{J(J+1)\hbar^2}{2Ik_BT}\right] \tag{15.92}$$

As seen in Section 13.3, the degeneracy of a rotational level is $2J + 1$.

In general the summation in this equation cannot be expressed in closed form. However, if the energy levels are sufficiently close together, the summation can be replaced by an integral. This is a satisfactory approximation for most molecules at room temperature and above. Under these conditions, therefore,

$$q_r = \int_0^\infty (2J+1) \exp\left[-\frac{J(J+1)\hbar^2}{2I\,k_BT}\right] dJ \tag{15.93}$$

This integration is easily performed if we recognize that $d[J(J+1)] = (2J+1)\,dJ$. We then have

$$q_r = \int_0^\infty \exp\left[-\frac{J(J+1)\hbar^2}{2I\,k_BT}\right] d[J(J+1)] = \frac{2I\,k_BT}{\hbar^2} \tag{15.94}$$

There is another important feature of the rotational partition function that has to be recognized. In the treatment we have given it was assumed that all the rotational energy levels are accessible to the molecule. This is indeed true for unsymmetrical linear molecules such as HCl, HCN, and OCS, and the rotational partition function for these molecules is correctly given by Eq. 15.94. For symmetrical linear molecules such as H_2, CO_2 and C_2H_2, however, not all the energy levels given by Eq. 15.91 are allowed. Which ones are allowed depends on the nature of the nuclei, but the result for symmetrical linear molecules is that only half of the rotational energy levels can be taken up by the molecule. The partition function in Eq. 15.94 must therefore be divided by 2.

Symmetry Number

This factor of 2 is known as the **symmetry number** and given the symbol σ. The partition function for a linear molecule is therefore given in general by

$$q_r = \frac{2Ik_BT}{\sigma\hbar^2} \tag{15.95}$$

where σ is unity for an unsymmetrical linear molecule and is 2 for a symmetrical one. A simple way of determining the symmetry number, and one that is also applicable to nonlinear molecules, is to imagine the identical atoms to be labeled and to count the number of different but equivalent arrangements that are obtained when the molecule is rotated. Thus, for CO_2 there are two such arrangements:

$$O^1{=}C{=}O^2 \quad \text{and} \quad O^2{=}C{=}O^1$$

An equivalent way of determining the symmetry number is to add all the proper rotations (including E) in the point group of the molecule. The value of Q_r for a system of N molecules is simply q_r^N since the indistinguishability of the molecules has already been taken into account in the translational partition function. Application to thermodynamic quantities is straightforward. For example, for the rotational energy we have, for symmetrical or nonsymmetrical diatomic molecules,

$$U_{m,r} = k_BLT^2\left(\frac{\partial \ln q_r}{\partial T}\right) = RT^2\left(\frac{2Ik_B}{\sigma\hbar^2}\right)\frac{\sigma\hbar^2}{2Ik_BT} = RT \tag{15.96}$$

which is the classical result. Note that the symmetry number does not affect the value of this thermodynamic property, since q appears as the derivative of the logarithm.

EXAMPLE 15.3 The internuclear separation in the hydrogen molecule is 0.074 nm.

a. Calculate the molecular rotational partition function at 300 K.
b. Calculate also $\ln Q_r$ for a system of $N = 6.022 \times 10^{23}$ molecules.

Solution First we calculate the moment of inertia of the hydrogen molecule. The distance between each atom and the center of gravity is $0.074 \times 10^{-9}/2$ m, and the moment of inertia is therefore

$$I = 2 \times 1.674 \times 10^{-27}(\text{kg}) \times \left[\frac{0.074 \times 10^{-9}}{2}(\text{m})\right]^2 = 4.583 \times 10^{-48} \text{ kg m}^2$$

a. The molecular rotational partition function is therefore, from Eq. 15.95, with $\sigma = 2$

$$q_r = \frac{8\pi^2 \times 4.583 \times 10^{-48}(\text{kg m}^2) \times 1.381 \times 10^{-23}(\text{J K}^{-1}) \times 300(\text{K})}{2 \times [6.626 \times 10^{-34}(\text{J s})]^2} = 1.707$$

This is an unusually small value for a rotational partition function; this arises because of the small mass of the H atom and the small internuclear separation. Rotational partition functions are more typically 10–1000. The values are much smaller than for translational partition functions (see Example 15.2 on p. 803) because of the much wider separation of the rotational levels.

b. For a system of $N = 6.022 \times 10^{23}$ molecules,

$$\ln Q_r = N \ln q_r = 6.022 \times 10^{23} \ln 1.707 = 3.22 \times 10^{23}$$

The derivation of the rotational partition function for nonlinear molecules is more difficult since the expressions for the allowed energies are more complicated. The partition function must now contain terms involving the *principal moments of inertia,* which are the moments of inertia I_A, I_B, and I_C about three perpendicular axes (Section 13.3). The final result is

$$q_r = \frac{8\pi^2(8\pi^3 I_A I_B I_C)^{1/2}(k_B T)^{3/2}}{\sigma h^3} \tag{15.97}$$

Again, we obtain the symmetry number σ by labeling identical atoms and seeing how many equivalent arrangements can be formed by rotation. For example, for water

O O
H^1 H^2 H^2 H^1

the symmetry number is two. For ammonia, which is nonplanar, σ is three. The methyl radical is planar, and the possible arrangements are

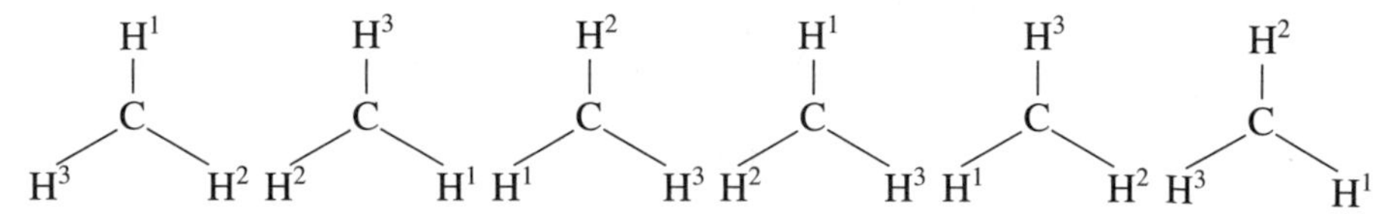

Its symmetry number is therefore six.

Vibrational Motion

The spacings between vibrational levels are appreciable compared to the room-temperature value of $k_B T$. We therefore expect that only a few of the terms in the partition function will contribute significantly to the energy. In Section 11.6 (Eq. 11.160 on p. 535) we found that the energy of the harmonic oscillator is

$$\epsilon_v = \left(v + \tfrac{1}{2}\right)h\nu_0 \qquad v = 0, 1, 2 \tag{15.98}$$

where v is a quantum number, h is the Planck constant, and $\nu(= (1/2\pi)\sqrt{k_h/\mu})$ is the classical frequency of the oscillator. (For simplicity we have dropped the subscript 0 on the ν.) However, in statistical mechanics the zero of energy is most commonly taken to be at $v = 0$ (i.e., at the zero-point level). The energies related to this level are $vh\nu$, and the corresponding partition function is

$$q_v = \sum_{n=0}^{\infty} e^{-vh\nu/k_B T} \tag{15.99}$$

The summation is over all levels, but in practice only a few levels contribute substantially to the sum. Because of the wide spacing of the energy levels compared to $k_B T$, the sum cannot be replaced by an integral. Thus q_v must be written out as a sum:

$$q_v = \sum e^{-vh\nu/k_B T} \tag{15.100}$$

If we substitute x for $e^{-h\nu/k_B T}$ the first few terms are

$$q_v = 1 + x + x^2 + \cdots \tag{15.101}$$

The series $1 + x + x^2 + \cdots$, when x is positive and less than unity (i.e., for all cases of physical interest), is $1/(1 - x)$. Thus

$$q_v = \frac{1}{1-x} = \frac{1}{1 - e^{-h\nu/k_B T}} \tag{15.102}$$

Characteristic Temperature

It is convenient to refer to the quantity $h\nu/k_B$ as the **characteristic temperature,** with the symbol θ_{vib}. Equation 15.102 may then be rewritten as

$$q_v = \frac{1}{1 - e^{-\theta_{vib}/T}} \tag{15.103}$$

When θ_{vib} is large compared to T, the fraction of molecules in *excited states* (levels where $v > 0$) is small. For example, θ_{vib} for HCl is 4140 K, and for the exponential term this gives a value of 1.0×10^{-6} at 300 K. Table 15.3 gives some characteristic temperatures.

Equation 15.103 can be simplified at two extremes of temperature. At very low temperatures, where $\theta_{vib}/T \gg 1$, the exponential term becomes negligible compared to unity and therefore

$$q_v \approx 1 \tag{15.104}$$

At high temperatures, when $\theta_{vib}/T \ll 1$, the exponential in the denominator may be expanded and only the first term accepted. Thus

$$e^{-\theta_{vib}/T} = 1 - \frac{\theta_{vib}}{T} \tag{15.105}$$

and therefore

$$q_v = \frac{T}{\theta_{vib}} \tag{15.106}$$

TABLE 15.3 Characteristic Temperatures,* θ_{vib}

Molecule		θ_{vib}/K	Molecule		θ_{vib}/K
H_2		6210	CO		3070
N_2		3340	NO		2690
O_2		2230	HCl		4140
Cl_2		810	HBr		3700
Br_2		470	HI		3200
CO_2	θ_1	1890	H_2O	θ_1	5410
	θ_2	3360		θ_2	5250
	$\theta_3 = \theta_4$	954		θ_3	2290

*T. L. Hill, *Introduction to Statistical Thermodynamics,* Reading, MA: Addison-Wesley, 1960.

EXAMPLE 15.4 From the data in Table 15.3, calculate, with reference to $v = 0$, the molecular vibrational partition function for H_2 at (a) 300 K and (b) 3000 K.

Solution The value of θ_{vib} is 6210 K, and the molecular vibrational partition function with reference to $v = 0$ is, from Eq. 15.103

$$q_v = \frac{1}{1 - e^{-6210/T}}$$

a. At $T = 300$ K,

$$q_v = \frac{1}{1 - e^{-6210/300}} = \frac{1}{1 - 1.024 \times 10^{-9}} = 1$$

b. At $T = 3000$ K,

$$q_v = \frac{1}{1 - e^{-6210/3000}} = \frac{1}{1 - 0.1262} = 1.144$$

Note that even at high temperatures these molecular vibrational partition functions are not far from unity.

From Eq. 15.103 we find for N independent harmonic oscillators that $Q_v = q_v^N$ (indistinguishability is not a factor because it has been taken into account in the translational term). Thus

$$\ln Q_v = N \ln q_v = -N\left[\frac{\theta_{vib}}{2T} + \ln(1 - e^{-\theta_{vib}/T})\right] \quad (15.107)$$

The energy is given in terms of the partition function as

$$U_v = k_B T^2\left(\frac{\partial \ln Q_v}{\partial T}\right)_V = k_B N T^2\left[\frac{\theta_{vib}}{2T^2} + \frac{\theta_{vib}/T^2}{(e^{\theta_{vib}/T} - 1)}\right] \quad (15.108)$$

If the ground-state energy $U_{0,v}$ is defined as $Nh\nu/2$, then $R\theta_{vib}/2$ is the zero-point energy, and the molar energy may be written as

$$U_v - U_{0,v} = RT\frac{\theta_{vib}/T}{(e^{\theta_{vib}/T} - 1)} \quad (15.109)$$

This is the same result that would have been obtained if we had used the partition function in Eq. 15.102.

When $h\nu_{vib}$ is much less than $k_B T(\theta_{vib}/T \rightarrow 0)$, Eq. 15.109 reduces to the expected classical value. Since under this condition $e^{\theta_{vib}/T} = 1 + \theta_{vib}/T + \cdots$, we may write for the molar energy

$$(U - U_0)_{v,m} = RT\frac{\theta_{vib}/T}{(1 + \theta_{vib}/T) - 1} = RT \quad (15.110)$$

It is also informative to compare the size of the energy terms expected in the classical and quantum-mechanical expressions. At 298 K the average value of the classical kinetic and potential energies is RT, or 2480 J mol^{-1}. If 2×10^{20} J is the energy spacing between typical vibrational levels, then

$$\frac{\theta_{\text{vib}}}{T} = \frac{h\nu}{k_{\text{B}}T} = \frac{2 \times 10^{-20}(\text{J})}{1.38 \times 10^{-23}(\text{J K}^{-1}) \times 298(\text{K})} = 4.9 \tag{15.111}$$

Thus the energy for this one vibrational frequency is

$$(U - U_0)_v = RT\frac{\theta_{\text{vib}}/T}{e^{\theta_{\text{vib}}/T} - 1} = \frac{8.3145(298)(4.9)}{e^{4.9} - 1} = 91 \text{ J mol}^{-1} \tag{15.112}$$

The large difference of approximately 2400 J between the values is the result of the quantum effects. As the temperature increases, the vibrational contribution to the energy increases.

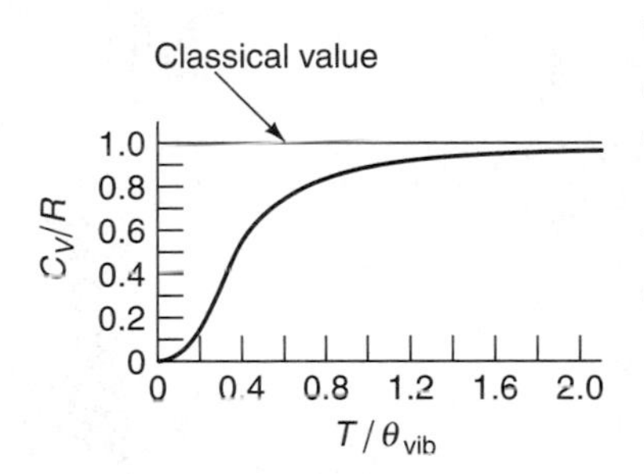

FIGURE 15.4
A plot against T/θ_{vib} of the Einstein function, given by Eq. 15.113.

The heat capacity due to the vibrational modes is an important quantity in many theoretical studies since it can be measured accurately. The form of C_v is given by differentiation of $U - U_0$ with respect to T, or from Eq. 15.109. The result is

$$\frac{\overline{C}_V(\text{vib})}{R} = \left(\frac{\theta_{\text{vib}}}{T}\right)^2 \frac{e^{\theta_{\text{vib}}/T}}{(e^{\theta_{\text{vib}}/T} - 1)^2} \tag{15.113}$$

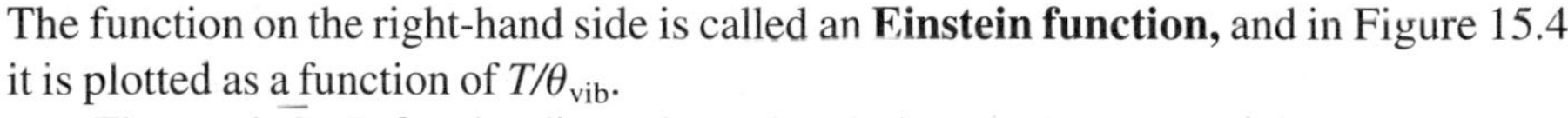

The function on the right-hand side is called an **Einstein function,** and in Figure 15.4 it is plotted as a function of T/θ_{vib}.

The total $\overline{C}_V/R$ for the diatomic molecule is calculated by adding the translational and rotational contributions to Eq. 15.113. Thus

$$\frac{\overline{C}_V}{R} = \frac{3}{2} + \frac{2}{2} + \left(\frac{\theta_{\text{vib}}}{T}\right)^2 \frac{e^{\theta_{\text{vib}}/T}}{(e^{\theta_{\text{vib}}/T} - 1)^2} \tag{15.114}$$

A polyatomic molecule such as CO_2, which has four modes of vibration, two of which are degenerate and thus equal in frequency, has three distinct Einstein functions, one doubly weighted because of the degeneracy. The value $\frac{5}{2}$ arises from the translational and rotational contributions:

$$\frac{\overline{C}_V}{R} = \frac{5}{2} + \left(\frac{\theta_{\text{vib},1}}{T}\right)^2 \frac{e^{\theta_{\text{vib},1}/T}}{(e^{\theta_{\text{vib},1}/T} - 1)^2} + \left(\frac{\theta_{\text{vib},2}}{T}\right)^2 \frac{e^{\theta_{\text{vib},2}/T}}{(e^{\theta_{\text{vib},2}/T} - 1)^2} + 2\left(\frac{\theta_{\text{vib},3}}{T}\right)^2 \frac{e^{\theta_{\text{vib},3}/T}}{(e^{\theta_{\text{vib},3}/T} - 1)^2} \tag{15.115}$$

A listing of partition functions is given in Table 15.4.

The Electronic Partition Function

The electronic levels of an atom or molecule are usually much farther apart than the vibrational levels. For the hydrogen atom, for example, the first stable excited electronic state ($^1\Sigma_u^+$; see Figure 13.35 and the discussion on page 701) is about 1.6×10^{-18} J (10.2 eV) above the lowest state, and at ordinary temperatures this state makes a negligible contribution to the partition function. As a useful rule we can say that if the energy $\Delta\epsilon$ divided by k_B is greater than 5 K, the level can be ignored. At 300 K this means that $\Delta\epsilon/k_B$ is greater than 1500 K (i.e., that $\Delta\epsilon$ is greater than about 2×10^{-20} J or about 0.12 eV); this energy corresponds to about 110 cm^{-1}.

Another factor is the **degeneracy** of an electronic state. The total electronic angular momentum of an atom is determined by the quantum number j, which has positive values of between $l + s$ and $l - s$. For every j value there are $2j + 1$ possible orientations in a magnetic field, corresponding to identical energies, and the

TABLE 15.4 Partition Functions for Different Types of Molecular Motion

Motion	Degrees of Freedom	Partition Function	Order of Magnitude
Translation	3	$\frac{(2\pi m k_B T)^{3/2}}{h^3}$ (per unit volume)	10^{31}–10^{32} m^{-3}
Rotation (linear molecular)	2	$\frac{2Ik_B T}{\sigma \hbar^2}$	10–10^2
Rotation (nonlinear molecular)	3	$\frac{8\pi^2(8\pi^3 I_A I_B I_C)^{1/2}(k_B T)^{3/2}}{\sigma h^3}$	10^2–10^3
Vibration (per normal mode)	1	$\frac{1}{1 - e^{-h\nu/k_B T}}$	1–10
Restricted rotation	1	$\frac{(2I' k_B T)^{1/2}}{\hbar}$	1–10

m = mass of molecule
I = moment of inertia for linear molecule
I_A, I_B, and I_C = moments of inertia for a nonlinear molecule, about three axes at right angles to one another
I' = moment of inertia for restricted rotation
ν = normal-mode vibrational frequency
k_B = Boltzmann constant
h = Planck constant
T = absolute temperature
σ = symmetry number
It is useful to remember that the power to which h appears is equal to the number of degrees of freedom.

electronic degeneracy is therefore $2j + 1$. For example, the ground state of the hydrogen atom is a $^2S_{1/2}$ state, so that j is $\frac{1}{2}$ and $2j + 1$ is equal to 2. The ground state of the iodine atom is $^2P_{3/2}$, and the degeneracy is therefore 4; in both of these cases the excited states can be ignored at ordinary temperatures.

The oxygen atom is a little more complicated. The lowest state is $^3P_2$, but there are two states very close to it: $^3P_1$, which is 157.4 cm^{-1} above the lowest level, and $^3P_0$, which is 226.1 cm^{-1} above the lowest level. Higher states can be neglected at all but the highest temperatures. The degeneracies of the three levels are 5, 3, and 1, respectively. The electronic partition function is, therefore,

$$q_e = 5e^{-0/k_B T} + 3e^{-157.4\ hc/k_B T} + e^{-226.1\ hc/k_B T}$$

At 300 K this gives $5 + (3 \times e^{-0.0075}) + e^{-0.0108} = 8.97$.

Nearly all stable molecules have $^1\Sigma$ ground terms, and their excited states are high enough to be neglected except at very high temperatures; since $J = 0$, there is no degeneracy and their electronic partition function is unity. Important exceptions are O_2, NO, and most free radicals.

The Nuclear Partition Function

There is also a contribution to the partition function arising from nuclear spins. This has to be taken into account in the calculation of absolute values, such as entropies.

For ordinary chemical reactions, however, the nuclear spins do not affect the Gibbs energy change and therefore do not affect the equilibrium constant. The nuclear spin factors will therefore not be considered here, but see the note on page 385 concerning *ortho*- and *para*-hydrogen.

15.6 The Internal Energy, Enthalpy, and Gibbs Energy Functions

Section 15.5 has shown us how partition functions are evaluated, on the basis of spectroscopic and other information. Earlier, in Section 15.4, we saw how the thermodynamic quantities can be calculated from the partition functions. We are thus provided with a fundamental way of obtaining the properties of substances and equilibrium constants of reactions from the basic molecular properties as determined by spectroscopy and in other ways.

To do this it has proved convenient to define thermodynamic functions, as follows:

1. The *internal energy function,* or the *thermal energy,* defined as $U - U_0$, where U is the internal energy at any temperature and U_0 is its value at the absolute zero.
2. The *enthalpy function,* defined as $(H - H_0)/T$ where H is the enthalpy at temperature T and H_0 is its value at the absolute zero.
3. The *Gibbs energy function,* defined as $(G - H_0)/T$ where G is the Gibbs energy at temperature T.

A particular advantage of the last of these functions is that it does not vary strongly with temperature, so that tables can give values at widely separated temperatures; the values at intermediate temperatures can be obtained by interpolation. The internal energy and enthalpy functions are commonly given in tables only at 25 °C.

The **internal energy function** $U - U_0$, also known as the thermal energy, comes at once from the fact that C_V is defined as $(\partial U/\partial T)_V$ (Eq. 2.26 on p. 61); this leads to

Internal Energy Function

$$U_{\text{thermal}} = U - U_0 = \int_0^T C_V \, dT \tag{15.116}$$

The integration constant U_0 is the value of U when $T = 0$ (i.e., when all the molecules are in their lowest energy states). The thermal energy can be obtained from knowledge of C_V over a range of temperatures and also from the partition function, as explained in Section 15.4.

A function similar to the thermal energy, but involving the enthalpy, is obtained as follows:

Enthalpy Function

$$H = U + PV = U_0 + U - U_0 + PV \tag{15.117}$$

and the function $H - U_0$ is therefore

$$H - U_0 = U - U_0 + PV \tag{15.118}$$

For the special case of 1 mol of an ideal gas,

$$H_m - U_{0,m} = U_m - U_{0,m} + RT \tag{15.119}$$

EXAMPLE 15.5 Calculate $U - U_0$ and $H - U_0$ for 1 mol of SO_2 at 298 K, assuming that the rotational levels make their classical contribution. The vibrational frequencies are $\tilde{\nu}_1 = 1151.4$ cm^{-1}, $\tilde{\nu}_2 = 517.7$ cm^{-1}, and $\tilde{\nu}_3 = 1361.8$ cm^{-1}. The electronic contribution may be neglected.

Solution The SO_2 molecule is nonlinear. Its three translational degrees of freedom each contribute $\frac{1}{2}RT = 1243.4$ J mol^{-1} to the internal energy function, as do the three rotational modes. The translational contribution is thus 3730.2 J mol^{-1}, as is the rotational contribution. The vibrational contributions must be calculated separately. Since $h\nu = hc\tilde{\nu}$, where c is the velocity of light, the energies of the levels may be calculated as follows:

$$\epsilon_1 = hc\tilde{\nu}_1$$
$$= 6.626 \times 10^{-34} \text{ J s} \times 2.998 \times 10^{10} \text{ cm s}^{-1} \times 1151.4 \text{ cm}^{-1} = 2.28 \times 10^{-20} \text{ J}$$
$$\epsilon_2 = 6.626 \times 10^{-34} \times 2.998 \times 10^{10} \times 517.7 = 1.03 \times 10^{-20} \text{ J}$$
$$\epsilon_3 = 6.626 \times 10^{-34} \times 2.998 \times 10^{10} \times 1361.8 = 2.70 \times 10^{-20} \text{ J}$$

Values for θ_{vib}/T and $(U - U_0)_{vib}$, obtained by the use of Eqs. 15.111 and 15.112, are

Wavenumber	θ_{vib}/T	$(U - U_0)_{vib}$
1151.4 cm^{-1}	5.54	54.3 J mol^{-1}
517.7 cm^{-1}	2.50	555.9 J mol^{-1}
1361.8 cm^{-1}	6.57	22.9 J mol^{-1}

The internal energy function is thus the sum of these contributions:

$$U - U_0 = 3730.2 + 3730.2 + 54.3 + 555.9 + 22.9 = 8093.5 \text{ J mol}^{-1}$$

The enthalpy function is given by Eq. 15.136:

$$H - U_0 = U - U_0 + RT = 8\,093.5 + 2477.7 = 10\,571.2 \text{ J mol}^{-1}$$

These functions are particularly useful for calculating energies and enthalpies of reaction at elevated temperatures, where direct experimental determinations may be difficult. For example, if $H - U_0$ can be calculated for the reactant and products of a reaction, the enthalpy change in the reaction is given by

$$\Delta H = \Sigma(H - U_0)_{products} - \Sigma(H - U_0)_{reactants} \tag{15.120}$$

Alternatively, ΔH may have been determined experimentally at one temperature T, say 25 °C; then

$$\Delta H = \Delta U_0 + \Delta H - \Delta U_0 = \Delta U_0 + \Delta(H - U_0) \tag{15.121}$$

The value of $\Delta(H - U_0)$ at T_1, calculated from the partition function, therefore allows ΔU_0 to be obtained. From the value of $\Delta(H - U_0)$ at some other temperature T_2, the ΔH at T_2 can therefore be calculated.

For the calculation of equilibrium constants, the Gibbs energy and enthalpy functions are particularly useful. At the absolute zero $U_0 = H_0$ and, therefore, for the standard quantities

$$H^\circ - H_0^\circ = H^\circ - U_0^\circ \tag{15.122}$$
$$= (U^\circ - U_0^\circ) + PV \tag{15.123}$$

TABLE 15.5 Gibbs Energy and Enthalpy Functions for Substances in the Vapor State (Standard Pressure = 1 bar)

	$-(G^\circ - H_0^\circ)/T$ J/K^{-1} mol^{-1}		$\Delta H^\circ_{298.15}$	$H^\circ_{298.15} - H^\circ_0$
Substance	298.15 K	1000 K	kJ mol^{-1}	kJ mol^{-1}
H_2	102.28	137.07	0	8.468
N_2	162.49	198.04	0	8.669
O_2	177.02	212.19	0	8.680
H_2O	155.62	196.83	−241.82	9.902
CO	168.57	204.18	−110.53	8.665
CO_2	182.39	226.54	−393.51	9.360
CH_4	152.63	199.35	−74.81	9.99
CH_3OH	201.17	257.65	−200.66	11.427
NH_3	159.08	203.80	−46.11	10.045

These values are from *JANAF Thermochemical Tables* (Ed. M. W. Chase et al.), New York: American Chemical Society and the American Institute of Physics, 3rd Edition, 1985. This compilation gives data for many other substances, and for many other temperatures.

For 1 mol

$$H_m^\circ - H_{0,m}^\circ = (U_m^\circ - U_{0,m}^\circ) + RT \tag{15.124}$$

The quantity $H_m^\circ - H_{0,m}^\circ$ can be determined from thermochemical data, making use of the relationship

$$H_m^\circ - H_{0,m}^\circ = \int_0^T C_{P,m}\, dT \tag{15.125}$$

Alternatively it can be obtained from the partition function, making use of the expression in Table 15.2; the result is

$$H_m^\circ - H_{0,m}^\circ = kT^2\left(\frac{\partial \ln Q}{\partial T}\right)_V + RT \tag{15.126}$$

Standard Enthalpy Function

It is this function divided by T that is referred to as the **standard enthalpy function.**

The Gibbs energy function is arrived at as follows. Since $G^\circ = H^\circ - TS^\circ$, we can write

Gibbs Energy Function

$$\frac{G^\circ - H_0^\circ}{T} = \frac{H^\circ - H_0^\circ}{T} - S^\circ \tag{15.127}$$

The quantity is the **standard Gibbs energy function,** and it can be evaluated from the partition function by making use of the expressions in Table 15.2.

Table 15.5 shows a few values of the standard Gibbs energy function, standard enthalpies of formation at 25 °C, and $H_{298}^\circ - H_0^\circ$. We shall now see how these values are used to calculate equilibrium constants.

15.7 The Calculation of Equilibrium Constants

For any reaction

$$\Delta H^\circ = \Delta H_0^\circ + \Delta(H^\circ - H_0^\circ) \tag{15.128}$$

where the symbol Δ represents the sum of the values for the products minus the sum for the reactants. Similarly, for the Gibbs energy change,

$$\frac{\Delta G^\circ}{T} = \frac{\Delta H_0^\circ}{T} + \frac{\Delta(G^\circ - H_0^\circ)}{T} \tag{15.129}$$

The quantities $H^\circ - H_0^\circ$ and $G^\circ - H_0^\circ$ can be evaluated for any substance at any temperature, by the methods indicated in Section 15.6. If, therefore, ΔH° or ΔG° is known at any temperature, the value of ΔH_0° can be obtained. This enables ΔG° to be obtained at any other temperature, by the use of Eq. 15.129, and using the value of $\Delta(G^\circ - H_0^\circ)/T$ at that temperature.

The equilibrium constant is then obtained through the equation

$$\Delta G^\circ = -RT \ln K \tag{15.130}$$

Since the standard state for the values given in Table 15.5 is 1 bar, the equilibrium constant obtained is K_P, in units involving bars.

The following example illustrates the application of Gibbs energy and enthalpy functions to equilibrium problems.

EXAMPLE 15.6 Use the data in Table 15.5 and Appendix D to calculate, at 1000 K, ΔG° and K_P for the reaction

$$CO + 2H_2 \rightarrow CH_3OH$$

Solution The standard enthalpy change for the reaction at 298.15 K is, from data in Appendix D,

$$\begin{aligned}\Delta H_{298}^\circ &= \Delta H_{f,298}^\circ(CH_3OH) - \Delta H_{f,298}^\circ(CO) - 2\Delta H_{f,298}^\circ(H_2)\\ &= -200.66 + 110.53 - 0 = -90.13 \text{ kJ mol}^{-1}\end{aligned}$$

Also

$$\begin{aligned}\Delta(H_{298}^\circ - H_0^\circ) &= (H_{298}^\circ - H_0^\circ)(CH_3OH) - (H_{298}^\circ - H_0^\circ)(CO) - 2(H_{298}^\circ - H_0^\circ)(H_2)\\ &= 11.427 - 8.665 - 2(8.468) = -14.18 \text{ kJ mol}^{-1}\end{aligned}$$

From Eq. 15.128 we then have, for ΔH° at the absolute zero,

$$\Delta H_0^\circ = \Delta H_{298}^\circ - \Delta(H_{298}^\circ - H_0^\circ) = -90.13 + 14.18 = -75.95 \text{ kJ mol}^{-1}$$

The Gibbs energy change at 1000 K is then

$$\begin{aligned}\Delta G^\circ &= \Delta H_0^\circ + \Delta(G^\circ - H_0^\circ)\\ &= -75\,950 - 1000(257.65 - 204.05 - 2 \times 136.98)\\ &= 144\,410 \text{ J mol}^{-1} = 144.4 \text{ kJ mol}^{-1}\end{aligned}$$

Then

$$\ln K_p^\circ = -\frac{\Delta G^\circ}{RT} = -17.37$$

and therefore

$$K_p = 2.9 \times 10^{-8} \text{ bar}^{-2}$$

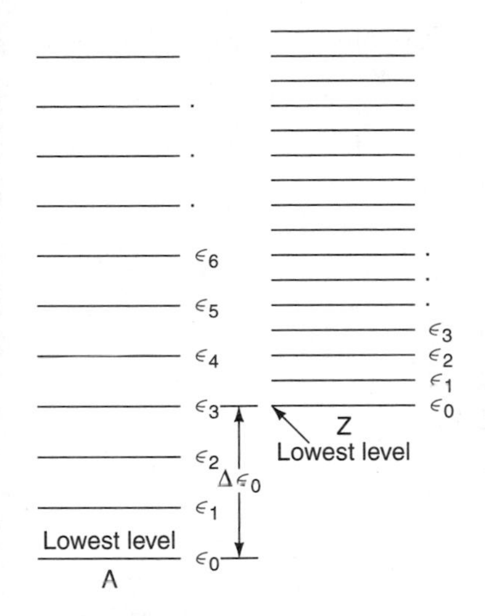

FIGURE 15.5
Two substances A and Z at equilibrium, showing the energy levels and the difference $\Delta\epsilon_0$ between the zero-point levels.

Direct Calculation from Partition Functions

If the thermodynamic functions discussed in Section 15.6 are not available for all the substances in a chemical reaction, the equilibrium constant must be calculated directly from the partition functions.

We will first arrive at the equilibrium equation in an intuitive fashion and then obtain it in a more formal way. Suppose that two substances A and Z are at equilibrium

$$A \rightleftharpoons Z$$

In Figure 15.5 the various energy levels for the two molecules are represented schematically. Although for a given molecule we conventionally take the lowest level as having an energy of zero, we obviously cannot do this when we have molecules present together at equilibrium, since there is, in general, an energy difference between the lowest levels. In the figure we have shown the lowest level of Z as being $\Delta\epsilon_0$ higher than the lowest level of A. If there were no other accessible levels, this would at once give the result as

$$K = \frac{[Z]}{[A]} = e^{-\Delta\epsilon_0/k_BT} \tag{15.131}$$

However, the other energy levels obviously must be taken into account, and this is conveniently done through the partition function. We saw in Section 15.3 that the partition function is related to the accessibility of the various energy levels and is a measure of the inherent probability of the existence of a particular molecule. Thus, if the zero-point levels were the same for reactants and products ($\Delta\epsilon_0 = 0$), the equilibrium constant would be simply the ratio of the molecular partition functions q_Z/q_A. With an energy difference of $\Delta\epsilon_0$ the equilibrium constant is

$$K = \frac{q_Z}{q_A} e^{-\Delta\epsilon_0/k_BT} \tag{15.132}$$

For a general chemical reaction

$$aA + bB \rightleftharpoons yY + zZ$$

the corresponding equation is

$$K = \frac{q_Y^y q_Z^z}{q_A^a q_B^b} e^{-\Delta\epsilon_0/k_BT} \tag{15.133}$$

The standard state to which K is related depends on how the partition functions are formulated. The simplest procedure is to express the q's as related to unit volume (1 m^3), in which case the equilibrium constant is a molecular equilibrium constant corresponding to the standard state 1 m^{-3}; conversion to other units is easily carried out.

The derivation of the expression for the equilibrium constant is based on Eq. 15.67:

$$G° - U_0° = k_BT \ln Q + Nk_BT \tag{15.134}$$

where N is 6.022×10^{23}. The standard molar Gibbs energy change for the reaction we are considering is

$$\Delta G° = yG_Y° + zG_Z° - aG_A° - bG_B° \tag{15.135}$$

(To simplify the notation, we omit the subscripts m.) Use of Eq. 15.134 leads to

$$\Delta G^\circ = yU^\circ_{0,\mathrm{Y}} + zU^\circ_{0,\mathrm{Z}} - aU^\circ_{0,\mathrm{A}} - bU^\circ_{0,\mathrm{B}} - RT\ln\left[\frac{\left(\frac{q_\mathrm{Y}}{N}\right)^y\left(\frac{q_\mathrm{Z}}{N}\right)^z}{\left(\frac{q_\mathrm{A}}{N}\right)^a\left(\frac{q_\mathrm{B}}{N}\right)^b}\right] \tag{15.136}$$

$$= \Delta U^\circ_0 - RT\ln\left[\frac{\left(\frac{q_\mathrm{Y}}{N}\right)^y\left(\frac{q_\mathrm{Z}}{N}\right)^z}{\left(\frac{q_\mathrm{A}}{N}\right)^a\left(\frac{q_\mathrm{B}}{N}\right)^b}\right] \tag{15.137}$$

$$= \Delta U^\circ_0 - RT\ln \frac{q_\mathrm{Y}^y q_\mathrm{Z}^z}{q_\mathrm{A}^a q_\mathrm{B}^b}N^{-\Sigma\nu} \tag{15.138}$$

where $\Sigma\nu$, the stoichiometric sum, is $y + z - a - b$. Then

$$\Delta G^\circ = -RT\ln\left(\frac{q_\mathrm{Y}^y q_\mathrm{Z}^z}{q_\mathrm{A}^a q_\mathrm{B}^b}N^{-\Sigma\nu}\mathrm{e}^{-\Delta U_0/RT}\right) \tag{15.139}$$

It then follows that

$$K^\circ = \frac{q_\mathrm{Y}^y q_\mathrm{Z}^z}{q_\mathrm{A}^a q_\mathrm{B}^b}N^{-\Sigma\nu}\mathrm{e}^{-\Delta U_0/RT} \tag{15.140}$$

This is the thermodynamic equilibrium constant, which is dimensionless. Since we have taken the volume to be m^3, the practical equilibrium constant has the same numerical value but has units involving mol m^{-3}. Conversion to other units is easily carried out, as illustrated in the following example.

EXAMPLE 15.7 Calculate the equilibrium constant K_c for the reaction

$$\mathrm{H_2} \rightleftharpoons 2\mathrm{H}$$

at (a) 300 K, and (b) 3000 K, making use of the following information: ground state of H atom is $^2\mathrm{S}_{1/2}$; higher states can be ignored. H—H dissociation energy = 431.8 kJ mol^{-1}; H—H internuclear distance = 0.074 nm; $\theta_{\mathrm{vib}}(\mathrm{H_2})$ = 6210 K.

Solution According to Eq. 15.140 the equilibrium constant for the reaction is

$$K = \frac{q^2(\mathrm{H})}{q(\mathrm{H_2})N^{\Sigma\nu}}\mathrm{e}^{-\Delta U_0/RT} = \frac{q^2(\mathrm{H})}{q(\mathrm{H_2})N}\mathrm{e}^{-\Delta U_0/RT}$$

since for this reaction $\Sigma\nu = 1$. We have already obtained some of the contributions to the partition functions in earlier examples.

a. At 300 K the translational partition function for H, with $V = 1\ \mathrm{m}^3$, is 9.89×10^{29} (Example 15.2. p. 803); since the ground state of H is a doublet, the electronic partition function is 2. The complete partition function for H at 300 K is therefore, for $V = 1\ \mathrm{m}^3$,

$$q(\mathrm{H},\ 300\ \mathrm{K}) = 2 \times 9.89 \times 10^{29} = 1.978 \times 10^{30}$$

The complete partition function for $\mathrm{H_2}$ is the product of the translational, rotational, vibrational, and electronic factors; the last is unity. Since $m^{3/2}$ enters into q_t, the value for $\mathrm{H_2}$ is

$$q_t(\mathrm{H_2},\ 300\ \mathrm{K}) = 2^{3/2} \times 9.89 \times 10^{29} = 2.797 \times 10^{30}$$

The rotational partition function was found (Example 15.3 on p. 805) to be 1.707, and the vibrational partition function (Example 15.4 on p. 808) to be 1. The electronic partition function is 1. The total partition function is therefore, for $V = 1\ \text{m}^3$,

$$q_{\text{total}}(\text{H}_2, \text{K}) = 2.797 \times 10^{30} \times 1.707 \times 1 \times 1$$
$$= 4.78 \times 10^{30}$$

The thermodynamic (dimensionless) equilibrium constant K_c° is therefore

$$K_c^\circ = \frac{q^2(\text{H})}{q(\text{H}_2)N} e^{-\Delta U_0/RT} = \frac{(1.978 \times 10^{30})^2}{4.78 \times 10^{30} \times 6.022 \times 10^{23}} \exp\left(\frac{-431\ 800}{8.3145 \times 300}\right)$$
$$= 8.95 \times 10^{-70}$$

The practical equilibrium constant (having units) is thus

$$K_c = 8.95 \times 10^{-70}\ \text{mol m}^{-3} = 8.95 \times 10^{-73}\ \text{mol dm}^{-3}$$

b. Since the translational partition function is proportional to $T^{3/2}$, and the rotational partition function is proportional to T, the values at 3000 K, for 1 m^3, are

$$q(\text{H}, \quad 3000\ \text{K}) = 1.978 \times 10^{30} \times 10^{3/2} = 6.255 \times 10^{31}$$
$$q_t(\text{H}_2, \quad 3000\ \text{K}) = 2.797 \times 10^{30} \times 10^{3/2} = 8.845 \times 10^{31}$$
$$q_r(\text{H}_2, \quad 3000\ \text{K}) = 1.707 \times 10 = 17.07$$
$$q_V(\text{H}_2, \quad 3000\ \text{K}) = 1.144 \qquad \text{(see Example 15.4 on p. 808)}$$

The total partition function for H_2 is thus

$$q_{\text{total}}(\text{H}_2, \quad 3000\ \text{K}) = 8.845 \times 10^{31} \times 17.07 \times 1.144 = 1.727 \times 10^{33}$$

The equilibrium constant is therefore

$$K_c^\circ = \frac{(6.255 \times 10^{31})^2}{1.727 \times 10^{33} \times 6.022 \times 10^{23}} \exp\left(-\frac{431\ 800}{8.3145 \times 3000}\right) = 0.114$$

and

$$K_c = 0.114\ \text{mol m}^{-3} = 1.14 \times 10^{-4}\ \text{mol dm}^{-3}$$

The dissociation is negligible at 300 K but is appreciable at 3000 K.

An excellent example of the calculation of an equilibrium constant for a much more complicated reaction, namely

$$C_2H_4 + H_2 \rightleftharpoons C_2H_6$$

has been given by Guggenheim.[5]

The significance of symmetry numbers in calculations of equilibrium constants is illustrated by the following example.

[5] E. A. Guggenheim, *Trans. Faraday Soc., 37,* 97(1941); see also K. S. Pitzer, *J. Chem. Phys., 5,* 469(1937); K. S. Pitzer and W. D. Gwanon, *J. Chem. Phys., 10,* 428(1942); *16,* 303(1948).

EXAMPLE 15.8 Calculate the equilibrium constant at 300 K for the reaction

$$H_2 + D_2 \rightleftharpoons 2HD$$

The relative atomic masses are H = 1.0078 and D = 2.014 and the vibrational wavenumbers are

$$H_2\text{: } 4371 \text{ cm}^{-1}; \qquad HD\text{: } 3786 \text{ cm}^{-1}; \qquad D_2\text{: } 3092 \text{ cm}^{-1}$$

The internuclear separation is 0.074 nm for all three forms.

Solution The value of ΔU_0 for the reaction can be calculated from the zero-point energies:

$$H_2\text{:} \quad \frac{1}{2}h\tilde{\nu}c = \frac{1}{2} \times 6.626 \times 10^{-34} \times 4371 \times 2.998 \times 10^{10} = 4.341 \times 10^{-20}\text{J}$$

$$HD\text{:} \quad \frac{1}{2}h\tilde{\nu}c = \frac{1}{2} \times 6.626 \times 10^{-34} \times 3786 \times 2.998 \times 10^{10} = 3.760 \times 10^{-20}\text{J}$$

$$D_2\text{:} \quad \frac{1}{2}h\tilde{\nu}c = \frac{1}{2} \times 6.626 \times 10^{-34} \times 3092 \times 2.998 \times 10^{10} = 3.071 \times 10^{-20}\text{J}$$

$$\Delta\epsilon_0 = [(2 \times 3.760) - 4.341 - 3.071] \times 10^{-20} = 1.08 \times 10^{-21}\text{J}$$

$$e^{-\Delta U_0/RT} = e^{-\Delta\epsilon_0/k_BT} = e^{-0.2607} = 0.771$$

The masses and moments of inertia of the three species are

	Mass/kg	I/kg m^2
H_2:	3.345×10^{-27}	4.583×10^{-48}
HD:	5.018×10^{-27}	$6.109 \times 10^{-48}\left[= \dfrac{1.674 \times 3.345 \times 10^{-27}}{5.018 \times (0.074 \times 10^{-9})^{-2}}\right]$
D_2:	6.689×10^{-27}	9.157×10^{-48}

The equilibrium constant is obtained from Eq. 15.140 with $\Sigma\nu = 0$:

$$K = \frac{q^2(\text{HD})}{q(\text{H}_2)q(\text{D}_2)}e^{-\Delta U_0/RT}$$

$$= \frac{\left[\dfrac{(2\pi m_{\text{HD}}k_BT)^3}{h^6}\right]\cdot\left(\dfrac{8\pi^2 I_{\text{HD}}k_BT}{\sigma_{\text{HD}}h^2}\right)^2}{\left[\dfrac{(2\pi m_{\text{H}_2}k_BT)^{3/2}}{h^3}\right]\cdot\left(\dfrac{8\pi^2 I_{\text{H}_2}k_BT}{\sigma_{\text{H}_2}h^2}\right)\cdot\left[\dfrac{(2\pi m_{\text{D}_2}k_BT)^{3/2}}{h^3}\right]\cdot\left(\dfrac{8\pi^2 I_{\text{D}_2}k_BT}{\sigma_{\text{D}_2}h^2}\right)}e^{-\Delta U_0/RT}$$

$$= \frac{m^3{}_{\text{HD}}}{(m_{\text{H}_2}m_{\text{D}_2})^{3/2}}\cdot\frac{I^2{}_{\text{HD}_2}}{I_{\text{H}_2}I_{\text{D}_2}}\cdot\frac{\sigma_{\text{H}_2}\sigma_{\text{D}_2}}{(\sigma_{\text{HD}})^2}$$

$$= \frac{(5.018)^3}{(3.348 \times 6.689)^{3/2}} \times \frac{(6.109)^2}{4.583 \times 9.157} \times \frac{2 \times 2}{1} \times 0.771$$

$$= 1.19 \times 0.889 \times 4 \times 0.771 = 3.26$$

The equilibrium constant for this reaction is not far from 4, the ratio of the symmetry numbers. It is easy to see why this should be so. If we were to start with equal amounts of H_2 and D_2 and allowed the reaction to go to equilibrium, each atom in the molecules present would have one-half a chance of being H and one-half a chance of being D. Thus in a molecule AB, the chance that A is H and that also B is H is $\frac{1}{2} \times \frac{1}{2} = \frac{1}{4}$. The chance that both atoms are D is similarly $\frac{1}{2} \times \frac{1}{2} = \frac{1}{4}$. The molecule HD, however, can be arrived at in two ways. The atom A may be H and atom B may be D, the probability of which is $\frac{1}{2} \times \frac{1}{2} = \frac{1}{4}$. Also, atom A may be

D and atom B may be H, with a probability of $\frac{1}{2} \times \frac{1}{2} = \frac{1}{4}$. The probability that a molecule is HD is thus $\frac{1}{4} + \frac{1}{4} = \frac{1}{2}$. The ratio

$$H_2{:}HD{:}D_2$$

is therefore

$$\tfrac{1}{4}{:}\tfrac{1}{2}{:}\tfrac{1}{4} \quad \text{or} \quad 1{:}2{:}1$$

and the equilibrium constant $[HD]^2/[H_2][D_2]$ is thus roughly $2^2/1 \times 1 = 4$. The mass effects just about cancel out ($1.19 \times 0.889 = 1.058$), and the exponential factor (0.771) reduces the constant from 4 to 3.26.

15.8 Transition-State Theory

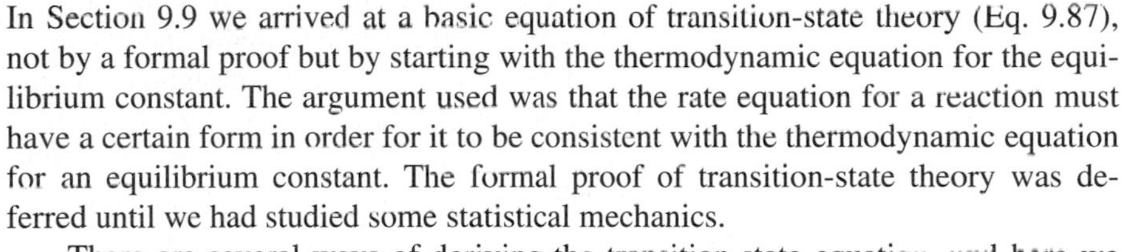
In Section 9.9 we arrived at a basic equation of transition-state theory (Eq. 9.87), not by a formal proof but by starting with the thermodynamic equation for the equilibrium constant. The argument used was that the rate equation for a reaction must have a certain form in order for it to be consistent with the thermodynamic equation for an equilibrium constant. The formal proof of transition-state theory was deferred until we had studied some statistical mechanics.

There are several ways of deriving the transition state equation, and here we give only one of them. As with all scientific theories, certain assumptions have to be made in order to obtain a conveniently simple equation. The most important of these is that the activated complexes are assumed to be formed in a state of equilibrium, referred to as *quasi-equilibrium,* with the reactants. This assumption is one that is often misunderstood, and must first be discussed in a little detail.

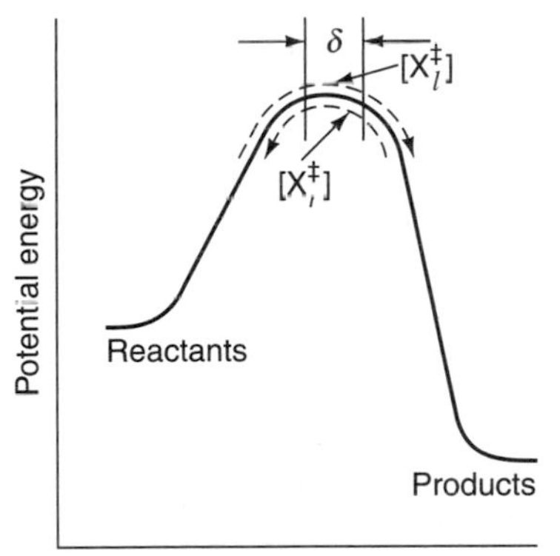

FIGURE 15.6
A schematic potential-energy profile, showing a region of length δ which defines arbitrarily the activated state.

The Assumption of "Quasi-Equilibrium"

Figure 15.6 shows a schematic *potential-energy profile,* which is a section through a potential-energy surface such as shown in Figures 9.14 and 9.15. This diagram shows potential energy plotted against the distance that the system has traveled along the reaction path. The two vertical lines near the maximum, separated by a distance δ, are introduced quite arbitrarily to define the activated state; any molecular complexes lying between the lines are by definition activated complexes, and we will see that this distance δ makes no difference to the final result.

For simplicity we will consider a reaction

$$A + B \rightleftharpoons Y + Z$$

in the gas phase, and will assume that the substances behave ideally (some deviations from ideality have been considered in Section 9.10). Suppose first that the reaction has been allowed to go to complete equilibrium. When this is so, everything in the system, including the activated complexes, is at equilibrium. We can therefore write

$$\left(\frac{[X^{\ddagger}]}{[A][B]}\right)_{eq} = K_c \qquad (15.141)$$

for the equilibrium between the reactant molecules A and B and the activated complexes $X^{\ddagger}$. We can also express this equilibrium constant in terms of partition functions, as was done in Section 15.7, using $q^{\ddagger}$ as the complete partition function for the activated complexes:

$$\left(\frac{[X^{\ddagger}]}{[A][B]}\right)_{eq} = K_c = \frac{q^{\ddagger}}{q_A q_B} e^{-E_0/RT} \tag{15.142}$$

Therefore

$$[X^{\ddagger}] = [A][B]\frac{q^{\ddagger}}{q_A q_B} e^{-E_0/RT} \tag{15.143}$$

Here E_0 is the difference between the molar zero-point energy of the activated complexes and that of the reactants. Since this exponential term has been included, the partition functions must be evaluated with respect to the zero-point levels of the molecules.

The activated complexes are of two types. Some of them in the immediate past were reactant molecules, and these are designated $X_l^{\ddagger}$ since they are moving from left to right in the diagram. In the immediate past the remainder of them were product molecules, and these are designated $X_r^{\ddagger}$ as they are moving from right to left. Since the system has been assumed to be in complete equilibrium the concentrations of $X_l^{\ddagger}$ and $X_r^{\ddagger}$ are equal, so that each concentration is equal to one-half of the total concentration of activated complexes:

$$[X_l^{\ddagger}] = \tfrac{1}{2}[X^{\ddagger}] \quad \text{and} \quad [X_r^{\ddagger}] = \tfrac{1}{2}[X^{\ddagger}] \tag{15.144}$$

The concentration of $X_l^{\ddagger}$ complexes is thus given by

$$[X_l^{\ddagger}] = \frac{1}{2}[A][B]\frac{q^{\ddagger}}{q_A q_B} e^{-E_0/RT} \tag{15.145}$$

So far we have been concerned with the system with everything at complete equilibrium. Suppose now that we imagine in some way sweeping the products Y and Z out of the system; what will happen? Obviously there will be no more $X_r^{\ddagger}$ complexes, since they can be formed only from Y and Z, which are no longer present. The concentration of the $X_l^{\ddagger}$ complexes, however, must remain unchanged, since the formation of these complexes cannot be influenced by whether or not product molecules are present.

It is important to appreciate that in transition-state theory

when one states that the activated complexes are in equilibrium with the reactants one is referring only to those activated complexes $X_l^{\ddagger}$ that were reactant molecules in the immediate past. That is the implication of the term *quasi-equilibrium* that is used to refer to this situation.

Derivation of the Transition-State Theory Equation

We now make use of Eq. 15.143 to derive the equation of transition-state theory. Consider the partition function $q^{\ddagger}$, which relates to the activated complexes. If the molecule A contains N_a atoms, and B contains N_b atoms, the activated complex contains $N_a + N_b$ atoms. The complex has three degrees of translational freedom, and if it is nonlinear it has three degrees of rotational freedom and therefore $3(N_a + N_b) - 6$ degrees of vibrational freedom. We must now focus attention on one of these degrees of vibrational freedom, the one that corresponds to the passage of the complex over the potential-energy barrier. This motion is of a different character from the other vibrational motions of the complex, since there is no restoring force; the complex can pass into products without any restraint. The factor for a vibrational motion is (Eq. 15.103)

$$\frac{1}{1 - \exp(-h\nu/k_B T)}$$

but for the very loose vibration we take this to the limit at which the frequency ν approaches zero; this is obtained by expanding the exponential and accepting only the first two terms:

$$\lim_{\nu=0} = \frac{1}{1 - \exp(-h\nu/k_B T)} = \frac{1}{1 - (1 - h\nu/k_B T)} = \frac{k_B T}{h\nu} \tag{15.146}$$

The complete partition function $q^{\ddagger}$ for the activated complexes may therefore be expressed as

$$q^{\ddagger} = q_{\ddagger}\frac{k_B T}{h\nu} \tag{15.147}$$

where $q_{\ddagger}$ is the remaining product of factors which now for a nonlinear molecule relates to $3(N_a + N_b) - 7$ degrees of vibrational freedom [$3(N_a - N_b) - 6$ if the activated complex is linear]. Equation 15.143 thus becomes

$$[X^{\ddagger}] = [A][B]\frac{k_B T}{h\nu}\frac{q_{\ddagger}}{q_A q_B}e^{-E_0/RT} \tag{15.148}$$

The next stage in the argument is to consider the significance of the frequency ν. If the vibration were an ordinary vibration of a chemical bond, the frequency would be the reciprocal of the time that it takes for the bond to make a complete vibration; in terms of Figure 15.6 it is the reciprocal of the time it takes for the system to traverse the distance δ *twice*. What we need in considering the reaction rate is the frequency of taking the trip of length δ only once, which is 2ν. The rate of reaction v is the concentration of the complexes $X_l^{\ddagger}$ that are moving from left to right divided by the time that it takes them to make the trip, or multiplied by the frequency 2ν corresponding to the trip:

$$v = 2\nu[X_l^{\ddagger}] \tag{15.149}$$

However, since $[X_l^{\ddagger}] = \frac{1}{2}[X^{\ddagger}]$, we have

$$v = \nu[X^{\ddagger}] \tag{15.150}$$

From Eq. 15.148 we thus obtain

$$v = [A][B]\frac{k_B T}{h}\frac{q_{\ddagger}}{q_A q_B}e^{-E_0/RT} \tag{15.151}$$

The rate constant, defined by $v = k[A][B]$, is thus

$$k = \frac{k_B T}{h}\frac{q_{\ddagger}}{q_A q_B}e^{-E_0/RT} \tag{15.152}$$

Eyring Equation

This is the transition-state equation, often known as the Eyring equation. We have presented the proof a little more explicitly than is usual, as there is often confusion about the factors of 2 and $\frac{1}{2}$ that are involved. Every so often people convince themselves that the formula is in error by a factor of 2, but this is because they do not properly understand the points about $[X^{\ddagger}]$ being *twice* the concentration of complexes moving from left to right, and about ν being the frequency of taking the *return* trip.

EXAMPLE 15.9 Calculate, on the basis of transition-state theory, the rate constant at 300 K for the reaction

$$\mathrm{H + HBr \rightarrow H_2 + Br}$$

Use the following data:

Barrier height E_0 (from zero-point level) = 5.0 kJ mol^{-1}

H—Br internuclear distance = 141.4 pm

H—Br vibrational wavenumber = 2650 cm^{-1}.

Take the activated complex to be linear, with the following characteristics derivable from quantum mechanics:

H—H distance = 150 pm

H—Br distance = 142 pm

Wavenumber for the symmetrical stretch = 2340 cm^{-1}

Wavenumber for the two degenerate bending vibrations = 460 cm^{-1}

(The fourth frequency corresponds to passage over the col.)

Solution First we calculate the partition functions for the reactants and the activated complex.

Hydrogen Atom

On p. 803 we calculated the partition function of the hydrogen atom, of mass 1.674×10^{-27} kg, to be, at 300 K,

$$9.89 \times 10^{29}\ \mathrm{m^{-3}}$$

HBr Molecule

$$\text{Mass of Br atom} = \frac{79.90 \times 10^{-3}\ \mathrm{kg\ mol^{-1}}}{6.022 \times 10^{23}\ \mathrm{mol^{-1}}}$$

$$= 1.327 \times 10^{-25}\ \mathrm{kg}$$

$$\text{Moment of inertia, } I = \frac{m_1 m_2}{m_1 + m_2}\, d_{\mathrm{HBr}}^2$$

$$= \frac{1.674 \times 10^{-27} \times 1.327 \times 10^{-25}}{(1.674 + 132.7) \times 10^{-27}} \times (1.414 \times 10^{-10})^2\ \mathrm{kg\ m^2}$$

$$= 3.305 \times 10^{-47}\ \mathrm{kg\ m^2}$$

The translational partition function for the HBr molecule is

$$q_{t,\mathrm{HBr}} = \frac{[2\pi(m_{\mathrm{H}} + m_{\mathrm{Br}})k_{\mathrm{B}}T]^{3/2}}{h^3}$$

and insertion of the values gives

$$q_{t,\mathrm{HBr}} = 7.10 \times 10^{32}\ \mathrm{m^{-3}}$$

The rotational partition function for the HBr molecule is

$$q_{r,\mathrm{HBr}} = \frac{8\pi^2 I k_B T}{h^2}$$
$$= 24.6$$

The vibrational partition function is

$$q_{v,\mathrm{HBr}} = \frac{1}{1 - \exp(-h\nu/k_B T)}$$

The wavenumber 2650 cm^{-1} corresponds to a frequency of 7.94×10^{13}s^{-1}. This leads to

$$\frac{h\nu}{k_B T} = \frac{6.626 \times 10^{-34}\,\mathrm{J\,s} \times 7.94 \times 10^{13}\,\mathrm{s}^{-1}}{1.381 \times 10^{-23}\,\mathrm{J\,K}^{-1} \times 300\,\mathrm{K}}$$
$$= 12.7$$

This is sufficiently large to cause q_v to be close to unity. The total partition function for HBr is thus

$$q_{\mathrm{HBr}} = 7.10 \times 10^{32} \times 24.6\ \mathrm{m}^{-3}$$
$$= 1.75 \times 10^{34}\ \mathrm{m}^{-3}$$

The Activated Complex

The mass of the activated complex is $2m_\mathrm{H} + m_\mathrm{Br} = 1.37 \times 10^{-25}$ kg, and the translational contribution to the partition function is therefore

$$q_t^{\ddagger} = \frac{(2\pi \times 1.37 \times 10^{-25} \times 1.381 \times 10^{-23} \times 300)^{3/2}}{(6.626 \times 10^{-34})^3}\ \mathrm{m}^{-3}$$
$$= 7.32 \times 10^{32}\ \mathrm{m}^{-3}$$

To obtain the rotational contribution we first calculate the moment of inertia. Let x pm be the distance between the center of mass of the complex and the end H atom:

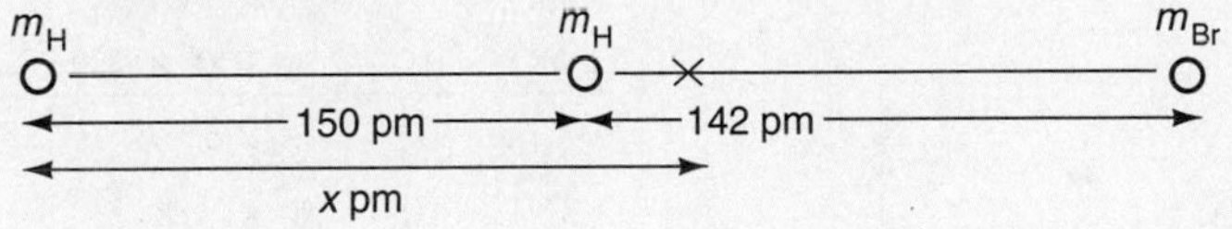

m_H m_H m_Br

150 pm 142 pm

x pm

The following equation can then be set up to obtain the position of the center of mass:

$$m_\mathrm{H}x + m_\mathrm{H}(x - 150) = m_\mathrm{Br}(292 - x)$$

Insertion of the masses leads to

$$x = 283.3$$

Thus, the moment of inertia of the activated complex is $\sum_i m_i r_i^2$ where r_i = distance from the center of mass.

$$I = m_\mathrm{H}(283.3 \times 10^{-12}\ \mathrm{m})^2 + m_\mathrm{H}(133.3 \times 10^{-12}\ \mathrm{m})^2 + m_\mathrm{Br}(8.7 \times 10^{-12}\ \mathrm{m})^2$$

Insertion of the masses leads to

$$I = 1.71 \times 10^{-46}\ \mathrm{kg\ m}^2$$

The rotational partition function for the complex is thus

$$q_r^{\ddagger} = \frac{8\pi^2 \times 1.74 \times 10^{-46} \times 1.381 \times 10^{-23} \times 300}{(6.626 \times 10^{-34})^2}$$
$$= 129.7$$

The vibrational partition function corresponding to the real vibrational frequencies of the activated complex is

$$q_v^{\ddagger} = \frac{1}{(1 - e^{-h\nu_1/k_BT})(1 - e^{-h\nu_2/k_BT})^2}$$

Insertion of

$$\nu_1 = 2340 \text{ cm}^{-1} = 7.013 \times 10^{13} \text{ s}^{-1}$$
$$\nu_2 = 460 \text{ cm}^{-1} = 1.37 \times 10^{13} \text{ s}^{-1}$$

leads to

$$q_v^{\ddagger} = \frac{1}{(1 - e^{-11.2})(1 - e^{-2.19})^2}$$
$$= 1.27$$

Thus, the complete partition function for the activated complex is

$$q_{\ddagger} = 7.32 \times 10^{32} \times 129.7 \times 1.27 \text{ m}^{-3}$$
$$= 1.21 \times 10^{35} \text{ m}^{-3}$$

The rate constant for the reaction is given by Eq. 15.152:

$$k = \frac{k_BT}{h} \frac{q_{\ddagger}}{q_A q_B} e^{-E_0/RT}$$

Insertion of the partition functions, with $E_0 = 5.0$ kJ mol^{-1}, gives

$$k = \frac{1.381 \times 10^{-23} \times 300 \times 1.21 \times 10^{35} \text{ m}^{-3} \text{ s}^{-1}}{6.626 \times 10^{-34} \times 9.89 \times 10^{29} \times 1.75 \times 10^{34}} e^{-5000/8.3145 \times 300}$$
$$= 4.367 \times 10^{-17} \text{ m}^3 \text{ s}^{-1} \times e^{-2.004}$$
$$= 4.367 \times 10^{-17} \text{ m}^3 \text{ s}^{-1} \times 0.1347$$
$$= 5.88 \times 10^{-18} \text{ m}^3 \text{ s}^{-1}$$

These are molecular units; multiplication by L (= 6.022×10^{23} mol^{-1}) and by 1000 dm^3/m^{-3} gives

$$k = 3.54 \times 10^9 \text{ dm}^3 \text{ mol}^{-1} \text{ s}^{-1}$$

The preexponential factor, A, in the expression $k = Ae^{-E_0/RT}$ is

$$A = 2.63 \times 10^{10} \text{ dm}^3 \text{ mol}^{-1} \text{ s}^{-1}$$

Thermodynamic Formulation of Transition-State Theory

In Section 9.9 we presented the equation of transition-state theory, Eq. 9.87, in terms of thermodynamic parameters:

$$k = \frac{k_B T}{h} e^{-\Delta^{\ddagger}G^\circ/RT} = \frac{k_B T}{h} e^{\Delta^{\ddagger}S^\circ/R} e^{-\Delta^{\ddagger}H^\circ/RT}$$

The equation derived in this section, Eq. 15.152, was

$$k = \frac{k_B T}{h} \frac{q_{\ddagger}}{q_A q_B} e^{-E_0/RT}$$

and the factor that appears in it,

$$\frac{q_{\ddagger}}{q_A q_B} e^{-E_0/RT}$$

closely resembles the expression (Eq. 15.142) for the equilibrium constant between the reactants and the activated complexes. The difference is that this expression involves $q_{\ddagger}$, from which the contribution corresponding to crossing the barrier has been omitted, whereas the equilibrium expression (Eq. 15.142) involves the complete partition function $q^{\ddagger}$ for the activated complexes. We can define a modified equilibrium constant $K_c^{\ddagger}$ such that

$$K_c^{\ddagger} = \frac{q_{\ddagger}}{q_A q_B} e^{-E_0/RT} \tag{15.153}$$

so that Eq. 15.152 can be written as

$$k = \frac{k_B T}{h} K_c^{\ddagger} \tag{15.154}$$

If K_c is now expressed in terms of the entropy of activation $\Delta^{\ddagger}S^\circ$ and the enthalpy of activation $\Delta^{\ddagger}H^\circ$, we have

$$k = \frac{k_B T}{h} e^{\Delta^{\ddagger}S^\circ/R} e^{-\Delta^{\ddagger}H^\circ/RT} \tag{15.155}$$

which is the equation we used (Eq. 9.87) in Section 9.9.

Extensions of Transition-State Theory

The version of transition-state theory we have considered, the original version, is now often referred to as *conventional transition-state theory.* Over the years there have been many improvements to the theory, in order to take care of the assumptions and approximations that are involved in the conventional theory. These improvements have led to much better agreement with experiment, but always at the expense of much computational work that is very time-consuming, even on the latest computers. Only two of the modifications can be mentioned here, very briefly.

With some potential-energy surfaces it is possible for the system to pass through the activated state and then be bounced back and return to the reactant state. Allowing for this requires a detailed knowledge of the potential-energy surface. Deviations from conventional transition-state theory arising from this cause are often dealt with in terms of a *transmission coefficient,* usually given the symbol κ (Greek kappa), which represents the probability that an activated complex passes into the product state.

Another modification to conventional transition-state theory is that quantum-mechanical tunneling must sometimes be taken into account. As we saw in Section 9.9, tunneling can be significant only if a light particle such as an electron or a hydrogen atom is transferred in the process. This is a field in which much more research needs to be done. At the present time it seems impossible to make reliable estimates, from potential-energy surfaces, of the probability of tunneling even for the simplest of reactions, and it remains necessary to rely on experimental measurements.

For further information about these modifications of the theory, reference should be made to some of the books and articles listed in the Suggested Reading section at the end of this chapter.

15.9 The Approach to Equilibrium

In 1872 Boltzmann published a paper in which he attempted to deal with the way in which systems approach a state of equilibrium. Approach to equilibrium had previously been considered qualitatively from two different points of view: in terms of the dissipation of energy, following Kelvin, or—better—in terms of an increase of entropy, as was done by Clausius. Boltzmann's intention was to deal with the matter in a quantitative way. By going through a rather lengthy mathematical treatment based on kinetic theory, he was led to a function, now given the symbol H, which involves among many other terms the Boltzmann factor exp $(-E/k_BT)$. A simple way of thinking about the function H is to note that with its sign changed it is proportional to the entropy; it is a measure of the negative of the entropy, which is sometimes called *negentropy*. Clausius had shown in his formulation of the second law that the total entropy is bound to *increase* as time goes on. It follows, since H is proportional to the negative of the entropy, that the value of H for any system plus its environment is bound to *decrease* when any spontaneous process occurs. Boltzmann's theorem was originally referred to as his *minimum theorem*, but is now usually known as **Boltzmann's *H* theorem**.

Boltzmann's *H* theorem

The significance of the function H is that it is supposed to extend the definition of entropy to include states that are not at equilibrium, in contrast to pure thermodynamics, which deals only with systems at equilibrium. In other words, it is a theory of the dynamic approach to equilibrium. An essential feature of the theorem is that Boltzmann considered at first that it was based entirely on mechanical arguments, and that probability theory did not enter into it.

Over the years many objections have been raised to Boltzmann's treatment, particularly to his assertion that probability did not enter into the theorem. One of the objections was based on the so-called *reversibility paradox* suggested by Kelvin in 1874 and again by Josef Loschmidt two years later. The essence of the objection is as follows. Suppose that we first consider the passage of a system such as a gas from a state at which it is not at equilibrium to its state of equilibrium; according to Boltzmann's treatment, H must be decreasing. There could be no objection to the existence of a state in which all the molecular motions are reversed, when the process would be occurring in the opposite direction, away from equilibrium, with H now increasing. How then can Boltzmann insist, as a matter of mechanics, that H must always decrease?

Boltzmann's response to this, in a paper that appeared in 1877, was that there is a vast number of possible equilibrium states, but if we specify a particular initial state, we are confining ourselves to just that one state. As a matter of probability, therefore, motion towards equilibrium is vastly more likely than motion in the opposite direction. He admitted that we could choose states that would move away from equilibrium with an increase in H and a decrease in entropy. However, such a situation would be highly unlikely. This argument of course involves probabilities and contradicts what Boltzmann had implied in his 1872 article, namely that the decrease in H follows from pure kinetic theory and that there is no need to invoke probability theory.

A related problem, referred to as the *recurrence paradox*, arose a good deal later from a theorem in mechanics first propounded in 1893 by the French mathematician and philosopher Jules Henri Poincaré (1854–1912). According to this theorem, any system such as a gas must, in the course of time, eventually reach every configuration possible. Poincaré, and later and more persistently the German mathematician and physicist Friedrich Ferdinand Zermelo (1871–1953), argued that therefore the H theorem cannot always be valid, since a system can pass to any other state. Boltzmann's answer to this challenge, published in 1896, was similar to his answer to the reversibility paradox. It is true that in principle a particular chosen state can recur if we wait long enough, but we would have to wait an immensely long time. Thus, although the H theorem could in principle be violated, such a violation is highly unlikely. Obviously he was again invoking probability to justify the H theorem.

It is of course now clear that we cannot avoid probability theory. Much confusion resulted from the fact that for many years Boltzmann was inconsistent in his statements about the matter, sometimes using the probability argument and sometimes saying that it was not necessary to invoke probability theory. When pressed, Boltzmann invoked the probability argument to justify his argument but sometimes insisted that he had arrived at his function H on the basis of pure mechanics. Boltzmann seems never to have properly understood the apparent paradox, which is a rather subtle one. He had apparently derived the function H from pure dynamics, without explicitly involving probability, and it does decrease with time, but how is this possible? The answer is that in his derivation of the expression for H he had made some hidden assumptions about the characteristics of typical collisions between atoms. He had, as a convenient approximation, neglected unusual types of collisions, and by doing so had unwittingly "loaded the dice"; he had left out the terms that would cause H to lead to the wrong conclusion and had included only the ones that led to the second law. His H theorem had therefore not done what he had originally intended it to do: lead to the second law by a purely dynamical argument. That does not mean that it was a waste of time. It does allow us to gain an understanding of the approach to equilibrium. The kinetic theory he developed was useful in leading ultimately to the understanding that the second law is no more than a matter of chance—but a chance that is for all intents and purposes a certainty.

In connection with his work on the H theorem Boltzmann was led to the important conclusion that the closer any distribution was to the equilibrium one, the more probable it was. By reasoning in this way Boltzmann was led to his famous relationship between entropy and probability, W:

$$S = k_{\mathrm{B}} \log W$$

Entropy and Probability

In this equation, which is engraved on Boltzmann's tombstone in Vienna, W is the number of possible ways of making a given distribution of atoms or molecules, corresponding to a given total energy of the system. This equation allows an expression for the entropy to be obtained from the statistical distribution, and from this expression the other thermodynamic properties can be calculated.

We should note that in this relationship Boltzmann had actually introduced an important extension to the Clausius definition of entropy. We saw in Section 3.4 that when considering an entropy change it is important to include the informational entropy contributions. For example, in dealing with the entropy change when two gases are brought together, we get a different answer if we use the information that the two gases are different from what we get if we think they are identical. If we use Clausius's formulation of entropy we have to add the informational entropy rather arbitrarily. The Boltzmann formulation, on the other hand, is more satisfactory in that it deals with the informational contribution quite naturally.

15.10 The Canonical Ensemble

The methods of statistical mechanics that have been presented so far in this chapter are somewhat simplified. They are adequate for systems in which there are no interactions between molecules, and they therefore provide a useful introduction to the subject. The more complicated methods that must be used when molecular interactions are important will now be mentioned briefly; the student is referred to more advanced texts (see Suggested Reading) for further details.

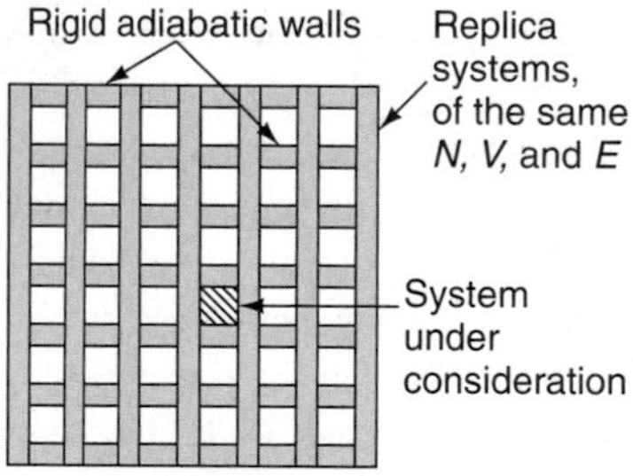

Boltzmann's original system of statistical mechanics, on which the previous discussion of this chapter was based, considered first the molecules of the system and the distribution of energy among them. Each way of distributing the molecules of the system among the energy states was considered to give a *distinguishable state* of the system. Boltzmann later extended those ideas in various ways.

Rigid adiabatic walls

Replica systems, of the same N, V, and E

System under consideration

FIGURE 15.7
Representation of a microcanonical ensemble, consisting of a number of systems each contained within a vessel having adiabatic walls. Each system is a replica of the system under consideration, having the same values of the number of molecules N, the volume V, and the energy E.

In his important book published in 1902, entitled *Elementary Principles in Statistical Mechanics,* J. Willard Gibbs introduces the idea of an **ensemble** of systems. An ensemble is a large number of imaginary replicas of the system under consideration. If the system is a gas maintained adiabatically, the number of molecules N, the volume V, and the energy E, are all fixed. We can then imagine an ensemble of such systems, as shown in Figure 15.7, each one maintained within adiabatic walls. Each system has the same N, V, and E. Such a system, in which the energy E is constant, is known as a **microcanonical ensemble.** The reason for using the idea of such an ensemble is to make it easier to visualize the averaging arguments that are used in statistical mechanics.

There are two postulates which are put forward to allow average properties to be calculated:

Postulate 1. There are properties, such as thermodynamic functions, which relate to the individual systems in the microcanonical ensemble, and the average of these properties can be taken to correspond to the experimentally observed value of the property.

Postulate 2. In a microcanonical ensemble, the individual systems are distributed with equal probability over all the possible quantum states. This is called the *principle of equal a priori probabilities.*

This postulate of equal *a priori probabilities* is fundamental to statistical mechanics, as it is this that allows the thermodynamic properties to be calculated by use of the statistical mechanical equations.

Gibbs also introduced the idea of a *canonical* (as opposed to microcanonical) ensemble, which consists of a large number of systems having the same values of N, V, and T (instead of E). Each system is therefore regarded as thermostatically controlled. The value of E then fluctuates about some average value, which determines the temperature of the entire ensemble. There are also other ensembles, such as the *grand canonical* ensemble, which are less commonly used.

KEY EQUATIONS

Number of complexions, for N molecules distributed among energy levels:

$$\Omega = \frac{N!}{\Pi_i n_i!}$$

Boltzmann distribution law:

$$\frac{n_i}{N} = \frac{g_i e^{-\epsilon_i/k_BT}}{\sum_i g_i e^{-\epsilon_i/k_BT}} = \frac{g_i e^{-\epsilon_i/k_BT}}{q}$$

where q is the *molecular partition function.*

Maxwell distribution law for the distribution of speeds:

$$\frac{dN}{N} = 4\pi\left(\frac{m}{2\pi k_BT}\right)^{3/2} e^{-mu^2/2k_BT}\, u^2\, du$$

(see also Table 1.3).

Maxwell-Boltzmann distribution law for the distribution of energies (Section 1.11):

$$\frac{dN}{N} = \frac{2\pi}{(\pi k_BT)^{3/2}} e^{-\epsilon/k_BT}\epsilon^{1/2}\, d\epsilon$$

System partition function Q:

Distinguishable molecules: $$Q = \left(\sum_i q_i e^{-\epsilon_i/k_BT}\right)^N = q^N$$

Indistinguishable molecules: $$Q = \frac{q^N}{N!} \approx \left(\frac{qe}{N}\right)^N$$

For *thermodynamic quantities* in terms of Q, see Table 15.2.

Partition functions for some simple cases:

Translation: $$q_t = \frac{(2\pi mk_BT)^{3/2}V}{h^3}$$

Rotation: $$q_r = \frac{(8\pi^2 Ik_BT)}{\sigma h^2}$$

(linear molecule; σ = symmetry number)

$$q_r = \frac{(8\pi^2(8\pi^3 I_A I_B I_C)^{1/2}k_BT)^{3/2}}{\sigma h^3}$$

(nonlinear molecule)

Vibration: $$q_v = \frac{1}{1 - e^{-h\nu/k_BT}} = \frac{1}{1 - e^{-\theta_{vib}/T}}$$

Sackur-Tetrode equation for molar entropy:

$$S_m = \frac{5}{2}R + R\ln\left[\frac{V_m(2\pi mk_BT)^{3/2}}{Lh^3}\right]$$

Equilibrium constants in terms of molecular partition functions:

$$K^u = \frac{q_Y^y q_Z^z \cdots}{q_A^a q_B^b \cdots} N^{-\Sigma\nu} e^{-\Delta U_0/RT}$$

($\Sigma\nu$ = stoichiometric sum).

According to transition-state theory

$$k = \frac{k_B T}{h} \frac{q_{\ddagger}}{q_A q_B} e^{-E_0/RT} = \frac{k_B T}{h} e^{\Delta^{\ddagger} S^{\circ}/R} e^{-\Delta^{\ddagger} H^{\circ}/RT}$$

PROBLEMS

Thermodynamic Quantities from Partition Functions

15.1. Obtain an expression for C_P in terms of the partition function Q for the system.

15.2. Obtain expressions for each of the functions $U^\circ - U_0^\circ$, S°, $A^\circ - U_0^\circ$, $H^\circ - U_0^\circ$, and $G^\circ - U_0^\circ$, in terms of the molecular partition function q, for (a) distinguishable molecules and (b) indistinguishable molecules. (Use the Stirling approximation $N! = N^N/e^N$.)

15.3. Obtain an expression for the pressure P in terms of the molecular partition function q, for (a) distinguishable molecules and (b) indistinguishable molecules. Express the result in terms of the number of molecules N and also the amount of substance n.

15.4. Two systems are identical in all respects except that in one the molecules are distinguishable and in the other they are indistinguishable. Calculate the difference between their molar entropies.

15.5. Calculate the molar energy at 25 °C of a monatomic gas absorbed on a surface and forming a completely mobile layer. (Such a system may be regarded as a two-dimensional gas.)

15.6. The partition function for each degree of vibrational freedom is $1/(1 - e^{-h\nu/k_BT})$ (Eq. 15.102). Obtain from this expression the limiting value of the vibrational contribution to C_V as T approaches infinity.

***15.7.** Consider a hypothetical substance for which the molecules can exist in only two states, one of zero energy and one of energy ϵ. Write down its molecular partition function q. Assuming the molecules to be distinguishable, obtain expressions for the following, in the limit of very high temperatures:

a. The molar internal energy $U_m - U_{0,m}$.
b. The molar entropy S_m.
c. The molar enthalpy $H_m - U_{0,m}$.
d. The molar Gibbs energy $G_m - U_{0,m}$.

15.8. Chemical reactions often lead to the formation of products whose energy distributions show significant deviations from the statistical distribution of Eq. 15.40. In a study of an elementary reaction with $^{16}O^1H$ radical as one of the products [Zhang, van der Zande, Bronikowski, and Zare, *J. Chem. Phys. 94,* 2704(1994)], the following rotational distribution was observed for the OH($v = 0$) state (normalized such that $\Sigma n_J/N = 1$). Compare this to the statistical distribution expected from Eq. 15.40 at 298 K. The equilibrium bond distance of OH is 0.96966 Å.

J	7	8	9	10	11	12	13
n_J/N	0.0181	0.0232	0.0356	0.0475	0.0377	0.0762	0.1045
J	14	15	16	17	18	19	20
n_J/N	0.1266	0.1459	0.1466	0.1306	0.0907	0.0167	0.0000

Partition Functions for Some Special Cases

15.9. Starting with Eq. 15.86, obtain an expression for the molar internal energy U_m of an ideal monatomic gas.

15.10. Calculate the molecular translational partition functions q_t for (a) N_2, (b) H_2O, (c) C_6H_6 in a volume of 1 m^3 at 300 K. In each case, calculate also ln $Q_{t,m}$, where $Q_{t,m}$ is the molar translational partition function.

15.11. The internuclear distance for N_2 is 0.1095 nm. Determine the molecular rotational partition function q_r and ln Q for N_2 at 300 K.

15.12. Use the data in Table 15.3 (p. 807) to calculate, with reference to $v = 0$, the molecular vibrational partition function for CO_2 at (a) 300 K and (b) 3000 K.

15.13. Expressions such as the Sackur-Tetrode equation for the entropy contain a term ln (constant × T). At temperatures close to the absolute zero this term has large negative values, and the expression therefore leads to a negative value of the entropy. Comment on this.

15.14. Calculate the entropy of argon gas at 25 °C and 1 bar pressure.

15.15. From the data in Table 15.3, calculate, with reference to $v = 0$, the molecular vibrational partition function for Br_2 at (a) 300 K and (b) 3000 K.

15.16. Give the symmetry numbers of the following molecules: C_3O_2 (carbon suboxide), CH_4, C_2H_4, C_2H_6 in the staggered conformation, C_2H_6 in the eclipsed conformation, $CHCl_3$, C_3H_6 (cyclopropane), C_6H_6 (benzene), NH_2D, CH_2Cl_2.

15.17. Show that the rotational partition function for a linear molecule can be expressed as

$$q_r = k_B T/\sigma Bh$$

where B is the rotational constant defined by Eq. 13.63.

***15.18.** Calculate the molar translational entropy of chlorine gas at 25 °C and 0.1 bar pressure.

***15.19.** The carbon monoxide molecule has a moment of inertia of 1.45×10^{-46} kg m^2 and its vibrational frequency is 6.50×10^{13} s^{-1}. Calculate the translational, rotational, and vibrational contributions to the molar entropy of carbon monoxide at 25 °C and 1 bar pressure.

15.20. Suppose that a system has equally spaced energy levels, the separation between neighboring levels being $\Delta\epsilon$. Prove that the fraction of the molecules in state i, having energy ϵ_i greater than the energy of the lowest level, is

$$1 - e^{(-\Delta\epsilon/k_B T)} e^{-\epsilon_i/k_B T}$$

What is the limiting value of this fraction as $T \rightarrow \infty$? Explain your answer.

***15.21.** Deduce the following from the Sackur-Tetrode equation (Eq. 15.88), which applies to an ideal monatomic gas:

a. The dependence of entropy on relative molecular mass M_r; also, obtain an expression for dS_m/dM_r.
b. The dependence of heat capacity C_P on relative molecular mass.
c. The dependence of entropy on temperature; also obtain an expression for dS_m/dT.

***15.22.** Molecules absorbed on a surface sometimes behave like a two-dimensional gas. Derive an equation, analogous to the Sackur-Tetrode equation 15.88, for the molar entropy of such an adsorbed layer of atoms, in terms of the molecular mass m and the surface area A. What would be the molar entropy if 10^{10} argon atoms were adsorbed on an area of 1 cm^2 at 25 °C?

Calculation of Gibbs-Energy Changes and Equilibrium Constants

15.23. Use the data in Table 15.5 to calculate the molar Gibbs energy of formation of H_2O(g) at 298.15 K.

15.24. From the data in Table 15.5, calculate K_P at 1000 K for the "water-gas" reaction

$$CO_2(g) + H_2(g) \rightleftharpoons CO(g) + H_2O(g)$$

15.25. From the data in Table 15.5, calculate K_P for the reaction

$$N_2(g) + 3H_2(g) \rightleftharpoons 2NH_3(g)$$

at (a) 298.15 K and (b) 1000 K.

15.26. Without making detailed calculations but by using symmetry numbers, estimate the equilibrium constants for the following reactions:

a. $^{35}Cl - ^{35}Cl + ^{37}Cl \rightleftharpoons ^{35}Cl - ^{37}Cl + ^{35}Cl$
b. $^{35}Cl - ^{35}Cl + ^{37}Cl - ^{37}Cl \rightleftharpoons 2^{35}Cl - ^{37}Cl$
c. $C^{35}Cl_4 + ^{37}Cl \rightleftharpoons C^{37}Cl^{35}Cl_3 + ^{35}Cl$
d. $N^{35}Cl_3 + ^{37}Cl \rightleftharpoons N^{37}Cl^{35}Cl_2 + ^{35}Cl$
e. $^{35}Cl_2O + ^{37}Cl \rightleftharpoons ^{37}Cl^{35}ClO + ^{35}Cl$

(Because of the similarity of the masses, these estimates will be quite accurate.)

D. M. Bishop and K. J. Laidler, *J. Chem. Phys.*, *42*, 1688(1965), have defined a *statistical factor* for a reaction as the number of equivalent ways in which a reaction can occur. Thus for reaction (a) from left to right the statistical factor is 2, since the ^{37}Cl atom can abstract either of the two ^{37}Cl atoms. For the reverse reaction the statistical factor r is 1, since the ^{35}Cl atom can only abstract the ^{35}Cl atom in order to give the desired products. If two identical molecules are involved, the statistical factor must be taken as the number of equivalent products divided by 2; thus for reaction (b) from right to left the statistical factor is $\frac{1}{2}$.

Bishop and Laidler proved that the ratio l/r of statistical factors is always equal to the ratio $\sigma_A\sigma_B/\sigma_Y\sigma_Z$ of symmetry numbers. Verify that this is true for the given reactions.

This statistical factor procedure is useful in providing a simple insight into the factors that appear in equilibrium constants.

***15.27.** Calculate the equilibrium constant at 1000 °C for the dissociation

$$I_2 \rightleftharpoons 2I$$

given the following information: moment of inertia of I_2 = 7.426×10^{-45} kg m^2, wavenumber for I_2 vibration = 213.67 cm^{-1}, ΔU_0 = 148.45 kJ mol^{-1}. The I atom is in a $^2P_{3/2}$ state; neglect higher states.

***15.28.** Calculate the equilibrium constant K_P for the dissociation

$$Na_2 \rightleftharpoons 2Na$$

at 1000 K, using the following data: internuclear separation in Na_2 = 0.3716 nm, vibrational wavenumber $\tilde{\nu}$ = 159.2 cm^{-1}, ΔU_0 = 70.4 kJ mol^{-1}. The Na atom is in a $^2S_{1/2}$ state; neglect higher states.

***15.29.** Calculate the equilibrium constant K_P at 1200 K for $Cl_2 \rightleftharpoons 2Cl$, from the following data: internuclear separation in Cl_2 = 199 pm, wavenumber for vibration = 565.0 cm^{-1}, ΔU_0 = 240.0 kJ mol^{-1}. The ground state of Cl is a doublet, $^2P_{\frac{3}{2},\frac{1}{2}}$, the separation between the states being 881 cm^{-1}.

***15.30.** An indication of the spread of energies in the molecules comprising a sample is provided by the *root-*

mean-square deviation of the energy. This quantity, more simply called the *fluctuation* in energy, is defined as

$$\delta\epsilon = \sqrt{\langle\epsilon^2\rangle - \langle\epsilon\rangle^2}$$

where $\langle\epsilon^2\rangle$ is the average value of ϵ^2 and $\langle\epsilon\rangle$ is the average value of ϵ. The probability P_1 that a molecule is in a given state is given by Eq. 15.34:

$$P_i = \frac{n_i}{N} = \frac{\mathrm{e}^{-\epsilon_i/k_BT}}{\sum_i \mathrm{e}^{-\epsilon_i/k_BT}} = \frac{\mathrm{e}^{-\epsilon_i/k_BT}}{q}$$

where q is the molecular partition function. Therefore,

$$\langle\epsilon^2\rangle = \sum_i P_i\,\epsilon_i^2 \quad \text{and} \quad \langle\epsilon\rangle = \sum_i P_i\,\epsilon_i$$

Prove that the square of the fluctuation is given by

$$\delta\epsilon^2 = \frac{1}{q}\left(\frac{\partial^2 q}{\partial\beta^2}\right) - \frac{1}{q^2}\left(\frac{\partial q}{\partial\beta}\right)_V^2$$

where $\beta = 1/k_BT$. Then use this relationship to prove that for a harmonic oscillator the fluctuation of the vibrational energy is

$$\delta\epsilon_{\text{vib}} = \frac{h\nu}{\mathrm{e}^{h\nu/2k_BT} - \mathrm{e}^{-h\nu/2k_BT}}$$

Show from this relationship that as $T \to \infty$,

$$\epsilon_{\text{vib}} = k_BT$$

15.31. Calculate the isotopic ratio K_H/K_D at 300 K for the reactions

$$H_2 \rightleftharpoons 2H \quad \text{and} \quad D_2 \rightleftharpoons 2D$$

Take the zero-point energies of H_2 and D_2 to be 26.1 kJ mol^{-1} and 18.5 kJ mol^{-1}, respectively.

Transition-State Theory

15.32. On the basis of transition-state theory, make rough estimates of the preexponential factors at 300 K for the following types of gas reactions:

a. A bimolecular reaction between an atom and a diatomic molecule, with the formation of a linear activated complex.
b. A bimolecular reaction between two diatomic molecules, the activated complex being nonlinear with one degree of restricted rotation.
c. A bimolecular reaction between two nonlinear molecules, the activated complex being nonlinear with no restricted rotation.
d. A trimolecular reaction between three diatomic molecules, the activated complex being nonlinear with one degree of restricted rotation.

Take the translational partition functions (for three degrees of freedom) to be 10^{33} m^{-3}, the rotational functions for each degree of freedom to be 10, the function for a restricted rotation to be 10, and the vibrational functions to be unity. Express the calculated preexponential factors in molecular units (m^3 s^{-1}) and in molar units (dm^3 mol^{-1} s^{-1}).

15.33. The rate constant for the reaction

$$2NO + O_2 \rightarrow 2NO_2$$

has been found to be proportional to T^{-3}. Suggest an explanation for this behavior.

15.34. Deduce the temperature dependency of the preexponential factor of the following types of reactions, where A represents an atom, L a linear molecule, and N a nonlinear molecule:

$$A + L; \quad A + N; \quad L + L; \quad L + N; \quad N + N$$

15.35. The rate of the reaction

$$O^+ + N_2 \rightarrow NO^+ + N$$

has been found over a certain temperature range to be proportional to $T^{-0.5}$ (M. McFarland et al., *J. Chem. Phys., 59,* 6620(1973). How can this be explained in terms of simple transition-state theory?

15.36. For the case of two atoms giving a product, A + B $\rightleftharpoons$ $[AB]^\ddagger$ $\rightarrow$ Product, show that transition-state theory yields essentially the same expression for the rate constant as the collision theory expression of Eq. 9.76.

15.37. In mass-spectrometric experiments, P. Kebarle and coworkers (*J. Chem. Phys., 52,* 212(1970) have found that under certain conditions the rate constant for the reaction

$$N^+ + N_2 + N_2 \rightarrow N_3^+ + N_2$$

is proportional to $T^{-2.5}$. Suggest an explanation for this behavior.

15.38. On the basis of transition-state theory, and assuming the vibrational partitions to be temperature-independent, deduce the temperature dependence of the preexponential factor for each of the following reactions:

a. $2ClO \rightarrow Cl_2 + O_2$
b. $NO + O_3 \rightarrow NO_2 + O_2$
c. $NO_2 + F_2 \rightarrow NO_2F + F$
d. $2NOCl \rightarrow 2NO + Cl_2$
e. $2NO + Br_2 \rightarrow 2NOBr$

15.39. Benzaldehyde is oxidized by permanganate in aqueous solution. Suppose that the aldehydic hydrogen atom is replaced by a deuterium atom; what can be said on the basis of transition-state theory about the kinetic isotope ratio k_H/k_D at 25 °C? The wavenumber of the aldehydic C—H vibration is 2900 cm^{-1}.

15.40. A transition-state theory study of the reaction O(^{3}P) + HCl generated the following information at T = 600 K:

Reactants

O	q_t	1.767×10^{32} m^{-3}	(for 3 degrees of freedom)
HCl	q_t	6.084×10^{32} m^{-3}	(for 3 degrees of freedom)
	q_r	39.40	(for 2 degrees of freedom)
	$\tilde{\nu}_0$	2991.0 cm^{-1}	

Transition state (bent)

q_t	1.050×10^{33} m^{-3}	(for 3 degrees of freedom)
q_r	1730	(for 3 degrees of freedom)
$\tilde{\nu}_1$	1407.9 cm^{-1}	
$\tilde{\nu}_2$	266.8 cm^{-1}	

The maximum of the reaction path is at 45.97 kJ mol^{-1}. Calculate the rate constant at this temperature.

(*Note:* Some interesting problems on transition-state theory involving the use of a computer are to be found in S. J. Moss and C. J. Coady, *J. Chem. Ed., 60,* 455(1983).)

Essay Questions

15.41. The molar entropy of a gas increases with the temperature and with the molecular weight. Give a physical explanation of these two effects.

15.42. Explain the factors that influence the magnitudes of partition functions, and comment on the magnitudes of the molecular partition functions for translational, rotational, and vibrational energy. What characteristics of a molecule will lead to a high value of (a) q_t, (b) q_r, and (c) q_v?

15.43. Give a general account of the principle of equipartition of energy, emphasizing in particular the circumstances under which the principle fails.

15.44. Discuss the assumptions and limitations of conventional transition-state theory. Mention briefly procedures that have been used to overcome some of the limitations.

15.45. Give an account of quantum-mechanical tunneling in chemical reactions, indicating the type of experimental evidence that is helpful in providing evidence for tunneling.

15.46. On the basis of the treatment of equilibrium constants in terms of partition functions, explain the factors that account for the effects of isotopic substitution.

SUGGESTED READING

E. C. Andrews, *Equilibrium Statistical Mechanics,* New York: Wiley, 1975.

D. Chandler, *Introduction to Modern Statistical Mechanics,* Oxford University Press, 1987. This small book places special emphasis on modern applications.

C. E. Hecht, *Statistical Thermodynamics and Kinetic Theory,* New York: W.H. Freeman and Company, 1990.

T. L. Hill, *Introduction to Statistical Thermodynamics,* Reading, MA: Addison-Wesley, 1960; Dover reprint 1986.

T. L. Hill, *Statistical Mechanics: Principles and Selected Applications,* New York: Dover reprint 1987. This is an advanced book.

H. S. Leff and A. F. Rex (Eds.), *Maxwell's Demon: Entropy, Information, Computing,* Princeton University Press, 1990.

D. Lindley, *Boltzmann's Atom: The Great Debate That Launched a Revolution in Physics,* New York: The Free Press (Simon and Schuster), 2001. This book, written for the general reader, gives an excellent analysis of the significance of Boltzmann's work.

D. A. McQuarrie, *Statistical Mechanics,* New York: Harper and Row, 1976.

L. K. Nash, *Elements of Statistical Thermodynamics,* 2nd edition, Reading, MA: Addison-Wesley, 1972.

O. K. Rice, *Statistical Mechanics, Thermodynamics, and Kinetics,* San Francisco: W.H. Freeman, 1967.

G. S. Rushbrooke, *Introduction to Statistical Mechanics,* Oxford: Clarendon Press, 1962.

For accounts of conventional transition-state theory and of extensions to it, see

S. H. Bauer, A random walk among accessible states, *J. Chem. Ed., 63,* 377(1986).

K. J. Laidler, *Chemical Kinetics,* 3rd edition, New York: Harper & Row, 1984, especially Chapter 4.

K. J. Laidler, A lifetime of transition-state theory, *The Chemical Intelligencer, 4,* 39(1998).

P. Pechukas, Transition-state theory, *Ann. Rev. Phys. Chem., 32,* 159(1981).

P. Pechukas, Recent developments in transition-state theory, *Ber. Bunsenges. Phys. Chem., 86,* 372(1982).

D. G. Truhlar and B. C. Garrett, Variational transition-state theory, *Accounts of Chemical Research, 13,* 440(1980); *Ann. Rev. Phys. Chem., 35,* 159(1984).

G. D. Truhlar, W. L. Hase, and J. T. Hynes, The current status of transition-state theory, *J. Phys. Chem., 87,* 2664(1983).

The Solid State 16

PREVIEW

In the first section of this chapter we examine the form and regularity of crystals, a subject that has been studied for well over a century and is known as *crystallography.* Symmetry is important in this study, but not all of the point groups discussed in Section 12.4 are possible for the whole crystal: Only 32 arrangements of atoms or ions can be found in solids. With the application of translations, screw axes, and glide planes to the different lattice types, 230 different three-dimensional crystal patterns or *space groups* result.

Much information concerning the arrangement of atoms in a crystal is gained by the use of X rays, and this is the subject of Section 16.2. When X rays impinge on a crystalline solid, a *diffraction pattern* is obtained, and it can be related to the interatomic spacings. Analysis of these patterns allows the crystal structure (placement of atoms or molecules and their orientation) to be determined. Section 16.3 is concerned with details of some of the various experimental methods used for studying solids.

Section 16.4 discusses the various theories of solids. The structure of crystals can be considered in terms of two models: the *bond model,* which considers bonds extending over the entire crystal, and the *band model,* which is analogous to molecular-orbital theory. In the development of the bond model, the Madelung constant, concerned with the charge interactions in ionic crystals, is useful. The close packing of spheres is discussed along with several crystal structures normally found in metals.

Another important property of solids is their specific heat, and this is discussed in terms of statistical thermodynamics in Section 16.5. Solids show a wide range of electrical conductivity behavior, which is discussed in Section 16.6.

Metals are good conductors, while many covalent solids are poor conductors, or *insulators.* There are also *semiconductors,* showing intermediate behavior. In addition, metals at temperatures close to the absolute zero behave as *superconductors,* having essentially zero resistivities. Recent work has also shown that certain other solids become superconductors at much higher temperatures than do the metals. The various types of conductivity behavior are discussed in this section in terms of the band theory of solids.

The final section, Section 16.7, deals with optical properties of solids, including color centers and the luminescence of solids.

OBJECTIVES

After studying this chapter, the student should be able to:

- Explain the characteristics of amorphous and crystalline solids and understand that crystalline solids can generally be classified as molecular or ionic.
- Appreciate that many of the properties of crystalline solids are directly related to the periodicity of the crystal structure.
- Understand that symmetry considerations of solids lead to the 14 Bravais lattices and that the unit cell is the basic building block of crystals.
- Appreciate the need for and the properties of the reciprocal lattice; in particular, that the reciprocal lattice vector is of fundamental importance to the theory of crystals.
- Understand that the Miller indices describe a family of planes in real lattice space and a series of points in reciprocal lattice space.
- Understand the basics of X-ray crystallography and the Bragg equation. Appreciate the Bragg's law formulation in both real and reciprocal lattice space.
- Understand experimental methods of X-ray crystallography.
- Recognize that the scattering of X rays, neutrons, and electrons is usually elastic, and that for X rays the scattering intensity depends upon the form factors. Understand that the presence of lattice planes in different crystal structures leads to the presence or absence of some diffraction patterns due to interference.
- Explain the basic ideas behind the bond model and the band model of solids. Understand ionic crystal energy in relation to the Born-Haber cycle.
- Explain how the closest packing of spheres leads to the basic structure of metals, and that the holes produced by different packings allow for other atoms to occupy different interstitial locations.
- Appreciate that the Einstein and Debye models for the heat capacities of solids describe their low-temperature behavior.
- Recognize that electrons make a negligible contribution to the heat capacity of a solid, and this can only be understood by introducing quantum statistics.
- Understand how Fermi-Dirac (Bose-Einstein) statistics lead to the Fermi-Dirac (Bose-Einstein) distribution function. Appreciate the role of the Pauli exclusion principle in the formulation of quantum statistics.
- Appreciate that quantum statistics leads to the concept of the Fermi energy along with its properties.
- From the band theory of metals, know how bands are formed and how they lead to an understanding of the electrical properties of solids. In particular, understand the differences between metals, intrinsic and extrinsic semiconductors, and insulators.
- Explain the effects of *n*- and *p*-type doping of semiconductors and how doping introduces donor and acceptor levels. Understand how the depletion zone arises in *n*-*p*-type junctions and the concepts of forward and reverse bias.
- Discuss the basic ideas of superconductivity and the optical properties of solids.

The characteristic of the solid state is that the substance can maintain itself in a definite shape that is little affected by changes in temperature and pressure. In some solids, such as *glasses* and *amorphous* materials like pitch, there is no long-range order in the arrangement of atoms or molecules at the atomic level. Such materials are really supercooled liquids. Other solids are characterized by a complete regularity of their atomic or molecular structures. Such solids are called *crystals* (Greek *krystallos,* clear ice). Regularity in external form and sharp melting point generally distinguish crystals from amorphous solids.

In the first part of this chapter we focus on the basic feature of crystals, the regularity of their atomic arrangements, and how this is manifested in the periodicity of their structural patterns.

16.1 Crystal Forms and Crystal Lattices

Examine a lattice in 3D and contrast a lattice from its basis; Lattices; Lattice and basis.

Place a basis of atoms in the correct location on the three cubic cells; Crystal forms; Build a lattice.

Crystalline minerals and gems have been admired and studied since the dawn of man. The beauty of crystalline forms took on scientific meaning with the quantitative study of crystalline interfacial angles (the angle between two adjacent faces of a crystal). In 1669 Niels Stensen (Nicolaus Steno) (1638–1686) found that the corresponding interfacial angles in different quartz crystals were always the same. Once the *reflection goniometer* (Greek *gonia metron,* angle measure) was invented in 1809 by the English chemist William Hyde Wollaston (1766–1828), it was rapidly established that other substances also exhibited this same constancy, which is now expressed as the *First Law of Crystallography.* Figure 16.1 shows examples of this first law for three different external shapes or habits.

This constancy of angles may seem surprising because of the large number of shapes a given crystalline material may have. However, the reason for the occurrence of different habits is that atoms or molecules may be deposited on the faces of a crystal in different ways. Suppose, for example, that a perfect cubic crystal is growing in a solution. If one face of the cube preferentially receives more atoms or molecules from the solution than the other sides, the crystal will elongate, and the face receiving more material will not grow as rapidly in area as the other faces. The most rapidly formed rectangular sides are those on which the atoms or molecules deposit most slowly.

In 1678 the Dutch physicist Christiaan Huygens (1629–1695) theorized that the existence of cleavage planes in crystals could be accounted for by the regular

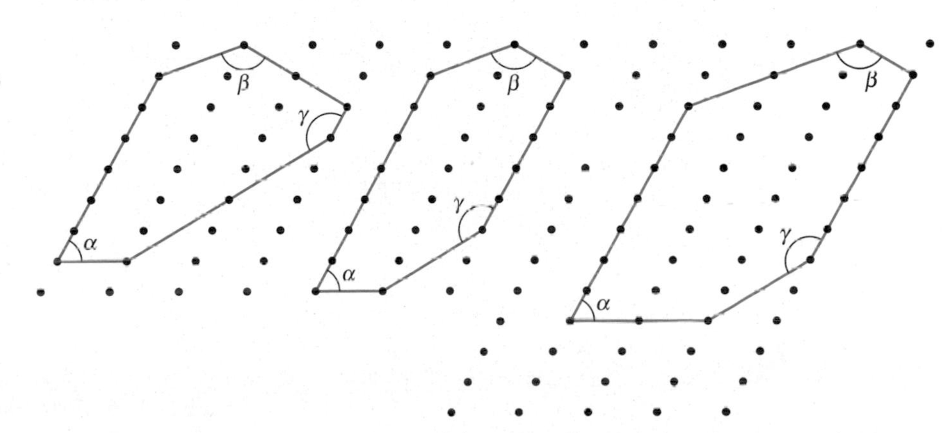

FIGURE 16.1
Constancy of the interfacial angles is shown for three different crystal habits of a material. Each of the angles α, β, and γ has the same value in the three figures.

View a 3D lattice and locate unit cells; Crystal forms; Three dimensions.

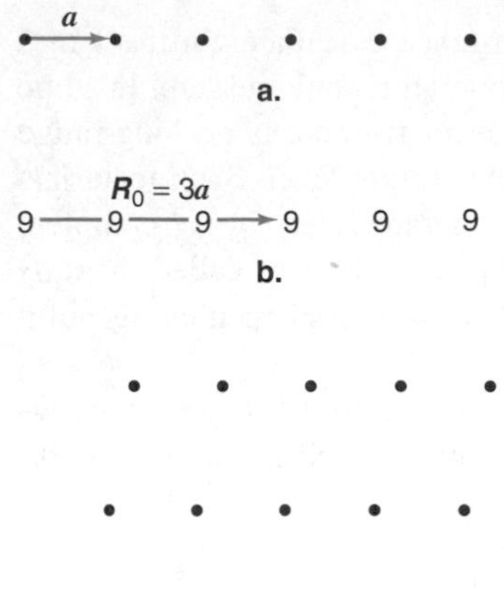

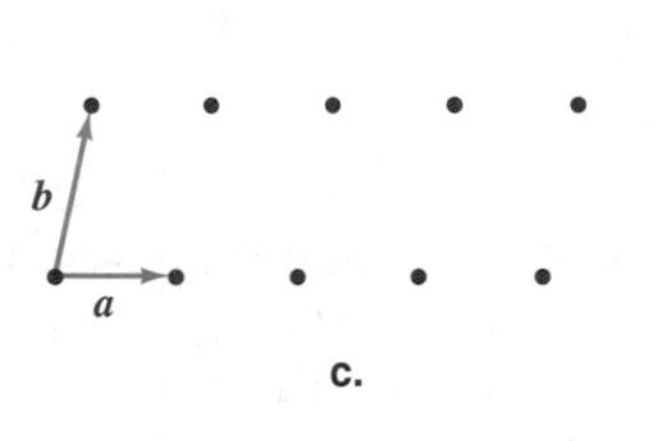

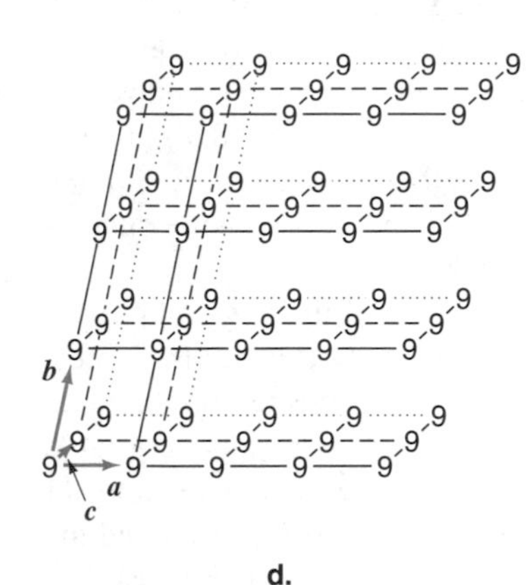

FIGURE 16.2
Patterns representing the formation of a crystal lattice. (a) One-dimensional lattice showing the primitive translation vector $\boldsymbol{a}$. (b) The motif 9 is identically placed on the lattice points of (a). (c) Two-dimensional lattice using the vectors $\boldsymbol{a}$ and $\boldsymbol{b}$. (d) Three-dimensional array showing the repetition of the motif by the vectors $\boldsymbol{a}$, $\boldsymbol{b}$, and $\boldsymbol{c}$.

packing of spheroidal particles in layers. This idea was extended in 1784 by the French mineralogist René Just Haüy (1743–1822). After accidentally dropping a piece of calcareous spar and noting the similarity between the shapes of the fragments, he suggested that the crystal was formed from little cubes or polyhedra, which he called the *molécules intégrantes.* The orderly internal arrangement of these building blocks produced the regular external faces of the crystals. Haüy is recognized as the founder of crystallography for this geometrical *Law of Crystallization,* and for his formulation of the *Law of Rational Intercepts,* which will be explained below.

Crystals are distinguished from amorphous solids by having a representative unit of structure repeated in space at regular intervals on a periodic array of points called a **lattice.** A lattice is merely a uniform arrangement of points in space. The structural unit of a crystal is called a **basis,** which may consist of an atom, an ion, (or ions, such as Na^+ and Cl^-, which form the basis for NaCl) or a molecule (which may be as large as a protein molecule). In a crystal a basis is present at each lattice point in such a way that the environment of each basis is the same throughout the crystal.

A one-dimensional lattice is generated by the continual repetition of a point after translation through a distance a. This is shown in Figure 16.2a. A linear array is obtained in Figure 16.2b, where the motif 9 is placed on each lattice point. (The motif 9 is used to represent a basis since it has no symmetry of its own.) The *translation operation* takes the point or motif into itself and is, therefore, a symmetry operation. It is usual to describe such a translation by the vector $\boldsymbol{a}$.

If the translation $\boldsymbol{a}$ is combined with a translation $\boldsymbol{b}$ in another direction, a two-dimensional or **plane lattice** is obtained by repetition of each $\boldsymbol{a}$ derived point in the direction $\boldsymbol{b}$. A two-dimensional lattice generated from the vectors $\boldsymbol{a}$ and $\boldsymbol{b}$ is shown in Figure 16.2c.

A third translation out of the $\boldsymbol{a}$–$\boldsymbol{b}$ plane repeats the entire plane at intervals $\boldsymbol{c}$. This generates a *space lattice.* Placement of our motif on the space lattice produces the *crystal lattice* or three-dimensional lattice array shown in Figure 16.2d. Each point in the crystal lattice has exactly the same environment as any other lattice point.

The Unit Cell

It is usually convenient to divide the lattice into *primitive cells* that completely generate the three-dimensional lattice under the action of suitable translation operations.

A line joining any two lattice points represents a possible translation vector. Indeed, there exists an indefinite number of translations that can be used to generate the lattice. Figure 16.3 shows a plane lattice and illustrates several possible choices of $\boldsymbol{a}$ and $\boldsymbol{b}$.

We will use the primitive translation vectors $\boldsymbol{a}, \boldsymbol{b}, \boldsymbol{c}$ to define the lattice or crystal axes. These axes $\boldsymbol{a}, \boldsymbol{b}, \boldsymbol{c}$ comprise the three adjacent edges of a parallepiped. A translation using these axes can be described in terms of the lattice translation vector

$$\boldsymbol{R}_o = n_1\boldsymbol{a} + n_2\boldsymbol{b} + n_3\boldsymbol{c} \tag{16.1}$$

where the n's are integers. This vector gives the displacement of the crystal parallel to itself. Note in Figure 16.3 that $\boldsymbol{R}_o$ in Set 5 and in Set 6 cannot be formed from integral combinations of $\boldsymbol{a}_1$ and $\boldsymbol{b}_3$ or $\boldsymbol{a}_2$ and $\boldsymbol{b}_2$. This is because $\boldsymbol{b}_3$ in Set 5 and the combination $\boldsymbol{a}_2$ and $\boldsymbol{b}_2$ in Set 6 are not primitive translation vectors.

Watch a unit cell form from its atoms; Crystal forms; Unit cell.

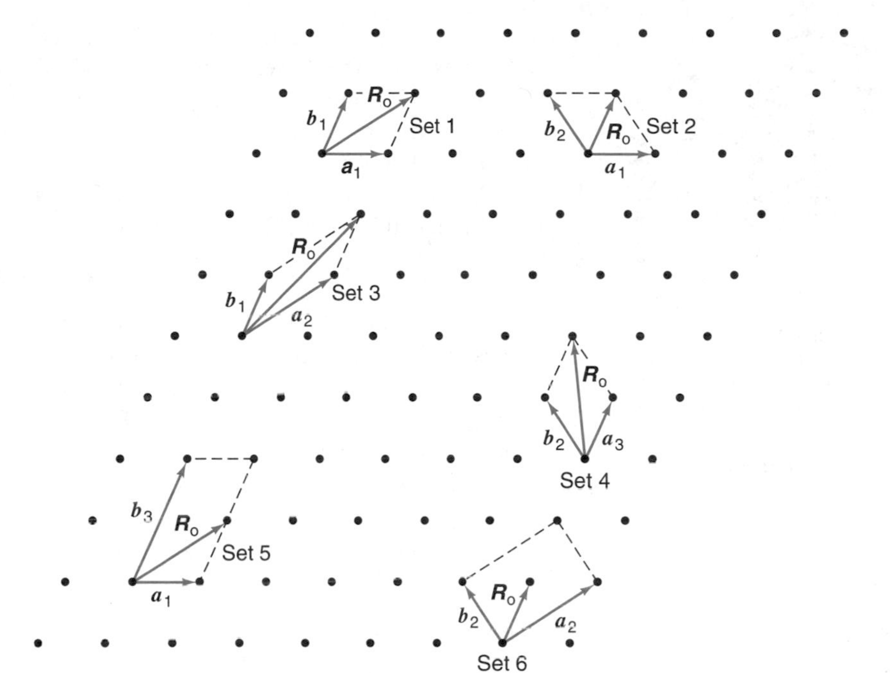

FIGURE 16.3
The translation vectors a and b may be chosen in several ways to generate a plane lattice. The vectors a_1, a_2, a_3, and b_1, b_2 are primitive translation vectors and form primitive lattice cells in the first four sets. The lattices in Sets 5 and 6 are not primitive because the cells formed by vectors and dotted lines contain a lattice point(s) other than at the corners. In Sets 1 to 4, R_o is formed by integral combination of the base vectors. (See Eq. 16.1.)

Examine in three dimensions, and rotate, the three cubic cells; Crystal forms;Types.

The *primitive cell,* therefore, is defined by the primitive crystal axes $\boldsymbol{a}, \boldsymbol{b},$ and $\boldsymbol{c}$. It has a volume V_c from vector analysis of

$$V_c = \boldsymbol{a} \cdot \boldsymbol{b} \times \boldsymbol{c} \tag{16.2}$$

and has a lattice point at each corner of the crystal.

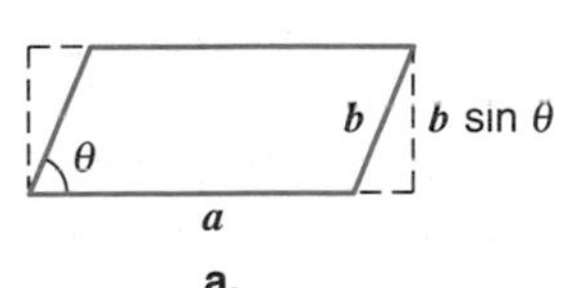

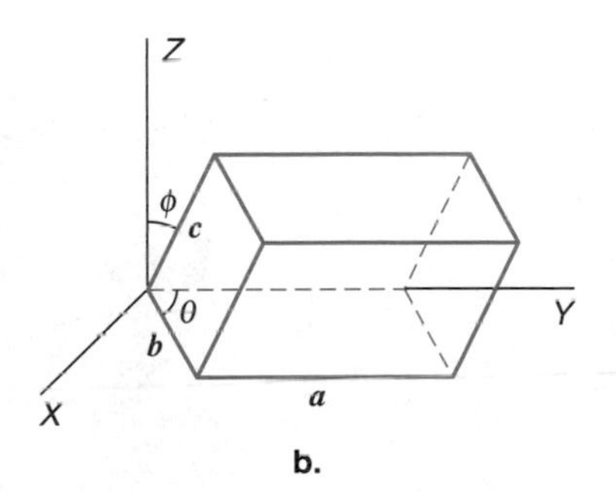

FIGURE 16.4
(a) A parallelepiped. (b) A three-dimensional cell with no orthogonality between the three edges.

EXAMPLE 16.1 Show that the area of a 2D parallelepiped is given by $|\boldsymbol{a}\times\boldsymbol{b}|$.

Solution Refer to Figure 16.4a. Area = $|\boldsymbol{a}\times\boldsymbol{b}| = ab \sin \theta$. This is equivalent to a square of dimensions a and $b \sin \theta$.

EXAMPLE 16.2 Show that the volume of a general primitive cell is $\boldsymbol{a}\cdot\boldsymbol{b}\times\boldsymbol{c}$.

Solution The area of the base is $|\boldsymbol{b}\times\boldsymbol{c}| = bc \sin \theta$, from Example 16.1. From Figure 16.4b, the height is $a \cos \phi$, giving the volume as $abc \sin \theta \cos \phi$, which is just $\boldsymbol{a}\cdot\boldsymbol{b}\times\boldsymbol{c}$.

In Figure 16.3, the choice of pairs of primitive translation vectors in Sets 1 through 4 defines two-dimensional primitive cells. A primitive cell in three dimensions is shown on the left side of Figure 16.5. The *primitive cell is a minimum volume cell that has only one lattice point per cell.* Since there are eight corners and only one-eighth of each point belongs to the cell of interest, there is, therefore, $8 \cdot \frac{1}{8} = 1$ lattice point that belongs to the unit cell. On the other hand, the lattice

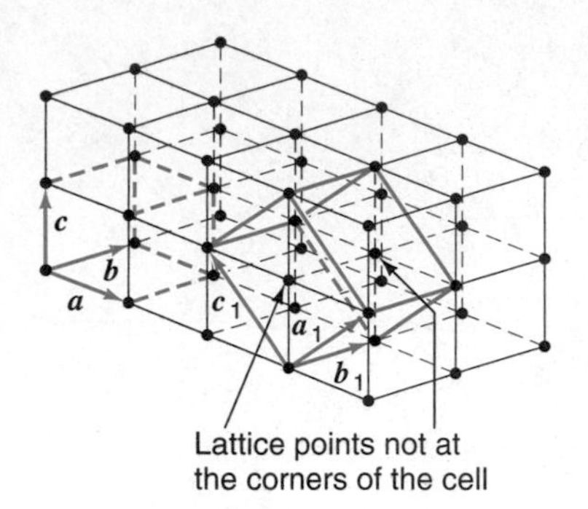

FIGURE 16.5
A unit cell on the left defined by the vectors $\boldsymbol{a}$, $\boldsymbol{b}$, and $\boldsymbol{c}$. The cell to the right is a multiple cell.

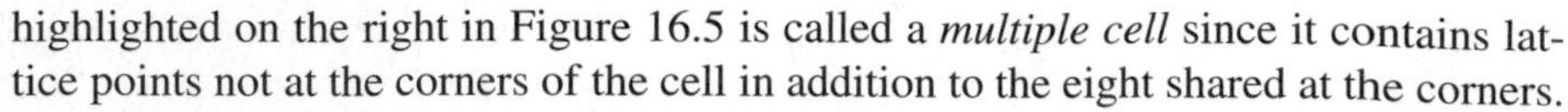
highlighted on the right in Figure 16.5 is called a *multiple cell* since it contains lattice points not at the corners of the cell in addition to the eight shared at the corners.

Either a primitive or a multiple cell may be chosen as a **unit cell** of the lattice. Generally several unit cells are possible. When unit cells larger than the primitive cell are used, it is because they better represent the symmetry of the lattice.

By virtue of the regularity or periodicity of repeated unit cells, a property $f(\boldsymbol{R})$ in one unit cell is identical to those translated by any integer multiple, n, of $\boldsymbol{R}_o$,

$$f(\boldsymbol{R}) = f(\boldsymbol{R} + n\boldsymbol{R}_0) \tag{16.3}$$

This is depicted in Figure 16.6. For example, X-ray and electron scattering techniques depend upon the electron density, $\rho(\boldsymbol{R})$, that exists within the unit cell. Suppose, for example, that the unit cell is given by Figure 16.7 and that there is a certain electron density at position $\boldsymbol{R}$. The contribution from the nth atom to this is $\rho_n(\boldsymbol{R} - \boldsymbol{R}_n)$. In general, the electron density within a unit cell at position R is a result of the contribution from each of the N atoms, and is given by

$$\rho\,(\boldsymbol{R}) = \sum_{n=1}^{N} \rho_n(\boldsymbol{R} - \boldsymbol{R}_n) \tag{16.4}$$

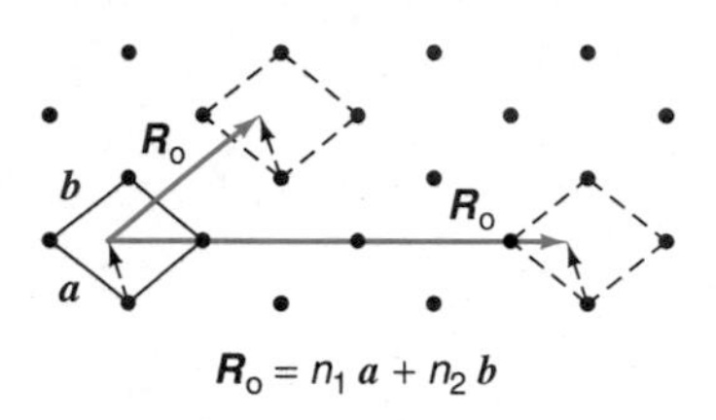

FIGURE 16.6
A two-dimensional lattice showing that a vector $\boldsymbol{r}$ in one unit cell can be located in any other unit cell by use of the lattice vector $\boldsymbol{R}_0 = n_1\boldsymbol{a} + n_2\boldsymbol{b}$ where n_1 and n_2 are integers.

The same electron density can be found by translating $\rho(\boldsymbol{R})$ throughout the crystal according to Eq. 16.3. This unique periodicity of crystals lends itself to Fourier analysis, from which many of the properties of crystals can be understood. As shown later with Bloch functions, quantum-mechanical wave functions are also periodic in a crystal.

Symmetry Properties

In addition to the translations that propagate the crystal, crystals also possess rotational axes. Although all the rotational axes studied in Section 12.5 are possible for individual molecules, the necessity of filling all spaces with repeating units places a restriction on the types of rotational axes that a crystal may have.

This is easily demonstrated if we try to cover a plane surface with regular pentagons; we cannot avoid overlapping the pentagons or leaving spaces between them. The C_5 rotational axis is therefore impossible with repeating units, such as we have in a crystal, and similar considerations lead to the conclusion that only C_2, C_3, C_4, and C_6 rotational axes are possible. (The axis C_1 is also possible but is

Create a cubic unit cell of any dimension and study the charge density at different locations; Lattice types; Charge density.

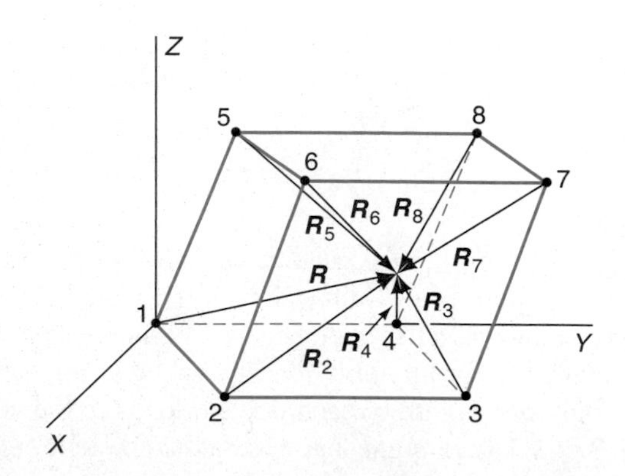

FIGURE 16.7
An example of the contribution to the electron density at a position $\boldsymbol{R}$ from all the atoms in a unit cell. See Eq. 16.4.

This notation is named in honor of Carl Heinrich Hermann (1898–1961) and Charles Victor Mauguin (1878–1958). Their work was done independently.

Fill 2D space with shapes of C_3, C_4, and C_6 symmetry and find that space cannot be filled with shapes of C_5 symmetry; Lattice types; Filling space.

Study the 5 lattice types of Figure 16.8 and understand the periodicity of the lattice; Lattice types; 2D lattice types.

equivalent to doing nothing.) According to the *Hermann-Mauguin* or *international*[1] *crystallographic notation,* a C_n rotational axis is listed simply as n, and therefore we say that only the rotational axes 2, 3, 4, or 6 are possible. Crystals must be constructed of subunits having one of the four axial symmetries just noted. The observed natural crystal forms must reflect this inner regularity in structure.

From considerations like these it is found that only five different types of two-dimensional lattices are possible. The most general, the *oblique lattice* shown in Figure 16.8a, contains a twofold axis normal to the lattice plane but no other symmetry elements. The placement of a *mirror plane*, designated m in the international system, would require lattice points to lie on rows parallel and perpendicular to the mirror plane. Two possibilities exist within the framework of a twofold axis: the *rectangular lattice* (Figure 16.8b) and the *centered-rectangular lattice* or *diamond lattice* (Figure 16.8c). The *hexagonal lattice* is a special case of the centered-rectangular lattice. Here the symmetry increases to a threefold or a sixfold axis. This is shown in Figure 16.8d. Figure 16.8e shows a *square lattice* that can accommodate a fourfold axis on each lattice point in addition to mirror planes.

For the sake of completeness, several other designations in the Hermann-Mauguin system should be mentioned. The *rotary-inversion axes* (n-fold rotation

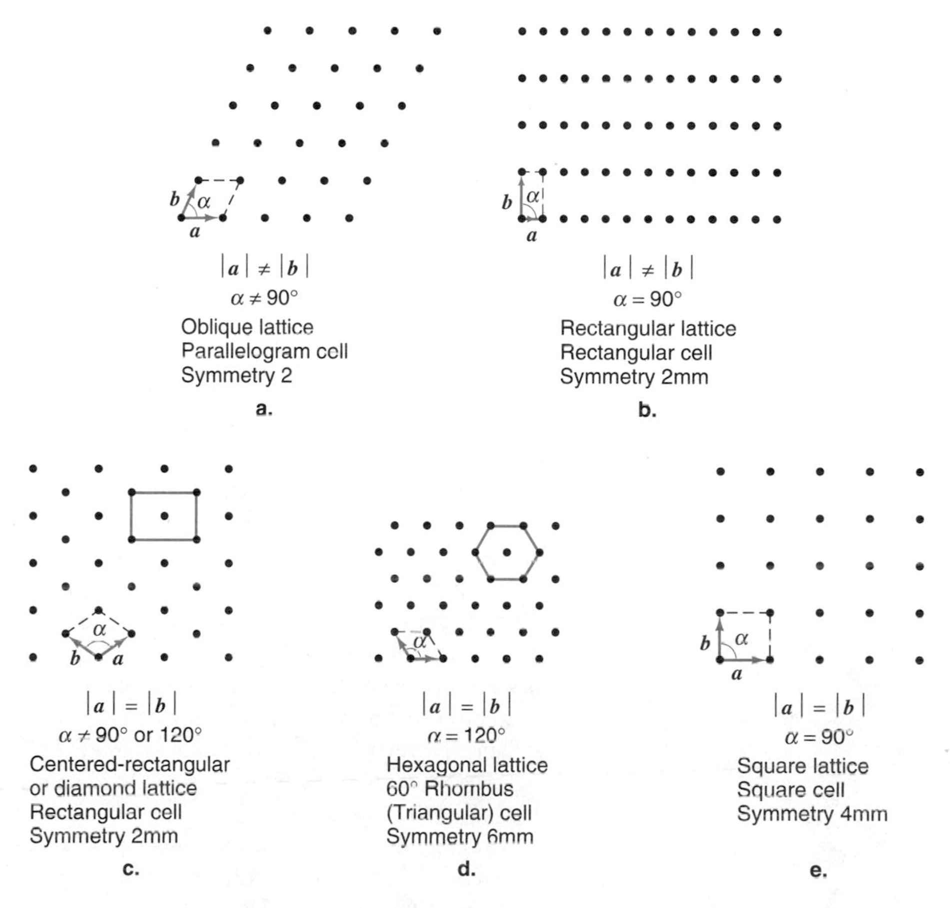

FIGURE 16.8
The five types of plane lattices. The lattice type and its axes are given along with the conventional cell normally chosen. The point group symmetry of the lattice is also given. The notation mm means that two mirror planes are present in addition to the rotational axis.

Vary the lattice spacing and create different lattice types; Lattices; Lattice plane.

Locate neighbors, next-nearest neighbors, and so on, on a lattice plane; Lattices; Nearest neighbors.

Use simple tools to draw and measure cells on a lattice plane that you can adjust to different shapes; Lattices; Lattice spacing.

[1]Norman F. M. Henry and Kathleen Lonsdale (Eds.), *International Tables for X-Ray Crystallography,* (Vol. 1), *Symmetry Groups,* Birmingham, England: Kynoch Press, 1952.

followed by inversion in a plane perpendicular to the axis) are represented by $\bar{n}$ (read n-bar) where n represents the number of equivalent positions in 360°. These are in addition to the four rotations mentioned above. Several equivalences are immediately evident. The inversion i is equivalent to a $\bar{1}$ axis. A $\bar{2}$ axis has the same effect as a mirror plane. A special designation n/m is used for a mirror plane perpendicular to an n-fold axis.

Two other types of symmetry operations are possible in crystals. The first of these is the *screw axis* formed by a proper rotation axis with a translation parallel to the rotation axis. The action is equivalent to the motion of a point on the thread of a screw, hence the name. The second operation is also a combination of two operations: reflection in a plane followed by a translation parallel to the reflection plane. The symmetry element is known as a *glide plane*. The screw axis and glide plane are not point-symmetry operations. For further discussion the reader is directed to the books listed in Suggested Reading (pp. 897–898).

Point Groups and Crystal Systems

Understand the periodicity of the lattice by moving the unit cell from one lattice point to another; Lattice types; Translation vectors.

The original crystallographers had no way to determine the periodicity of the atoms in crystals and had to deduce crystal structures from the external form or *morphology* of the crystal. The symmetry axes are lines, and they and the mirror planes intersect at a common point at the center of the crystal. The sets of symmetry elements found in a particular crystal were therefore designated *point groups* and were used for purposes of classification. Crystals having the same point-group symmetry belong to the same *crystal class*. Because the proper and improper rotational axes are restricted to 1, 2, 3, 4, and 6, only 32 point groups are possible, and these constitute the 32 crystal classes. The effect of screw axes and glide planes cannot be considered here because they are not visible in the external morphology, the translations being too small to be seen.

Space Lattices

We expect that the number of lattices that can fill three-dimensional space is limited, just as in two dimensions only 5 lattice types are permitted. In 1848 the French physicist Auguste Bravais (1811–1863) showed that the point-symmetry groups of the 5 two-dimensional lattices allow only 14 different lattice types in three dimensions. These distinct types of lattices are referred to as the Bravais lattices and are shown in Figure 16.9.

Among the 14 lattices is 1 general lattice type (triclinic) and 13 special types which occur because of symmetry restrictions. It is convenient to divide the 14 lattices into *seven systems* based on the seven unit cells conventionally used: *triclinic, monoclinic, orthorhombic, tetragonal, cubic* or *isometric, trigonal,* and *hexagonal.* The conventional cell axes and angles are summarized in Table 16.1, and the axial and angular relationships are defined in Figure 16.10.

The seven systems are further subdivided into classes based on whether they are *primitive* (P), *face-centered* (F), *side-* or *end-centered* (A, B, or C), *body-centered* (I, German *Innenzentriertes* or inner), or *rhombohedral* (R). These classes are shown in the figure along with the symmetry axes.

The cells in Figure 16.9 are conventional cells, and they are not always primitive cells. The reason for using a nonprimitive cell is that it has a more obvious

Study Figure 16.9 by choosing any of the unit cells and rotating them; lattice types; Bravais lattices.

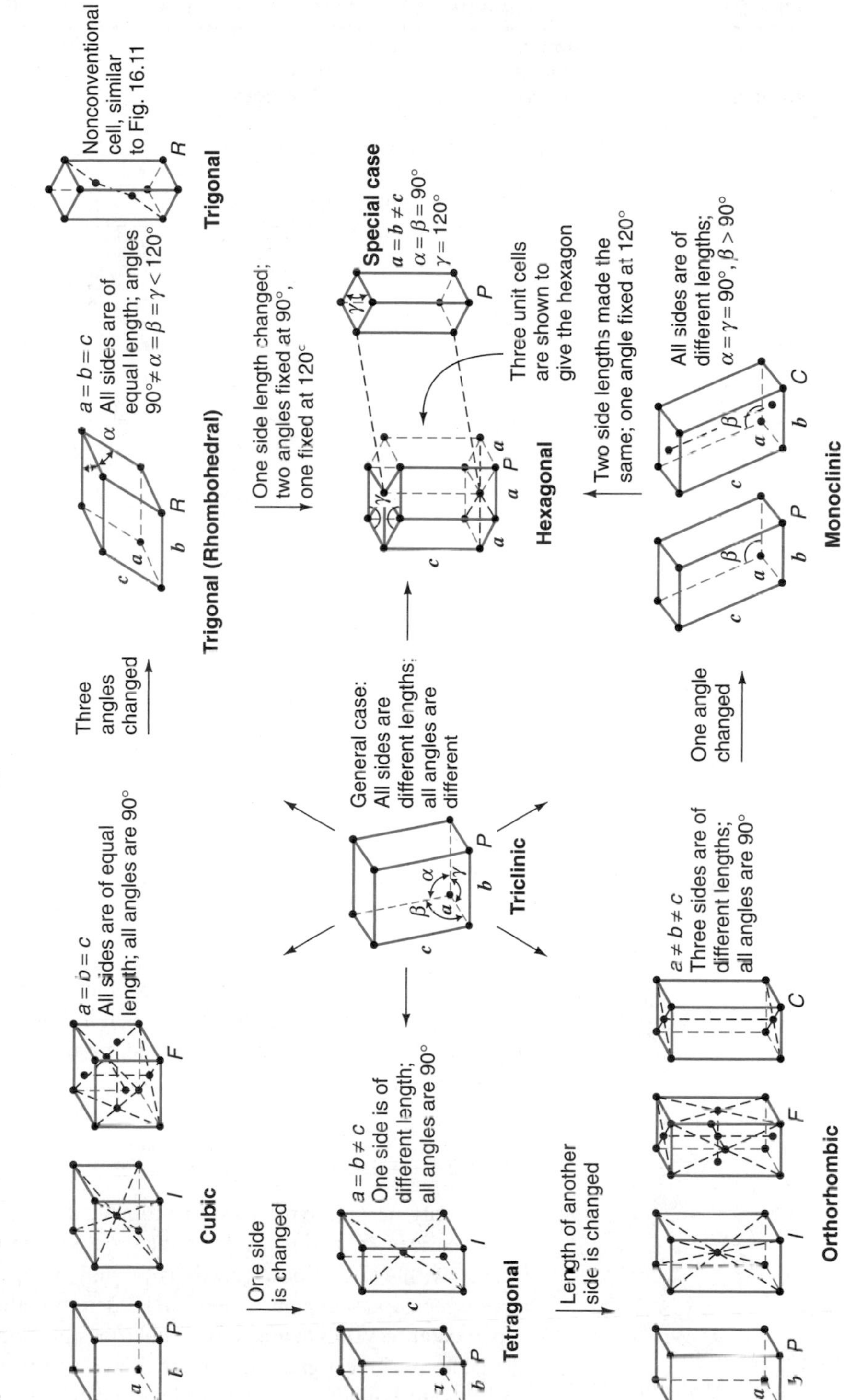

FIGURE 16.9
The 14 Bravais lattices. Conventional cells are shown. For the trigonal case, a nonconventional cell is also shown. $a = |\boldsymbol{a}|$, $b = |\boldsymbol{b}|$, and $c = |\boldsymbol{c}|$

TABLE 16.1 The Seven Crystal Systems and 14 Three-Dimensional Bravais Lattice Types

System/Examples	Number of Lattices in System	Lattice Symbols	Restrictions on Conventional Cell Axes and Angles	Number of Space Groups
Triclinic $CuSO_4 \cdot 5H_2O$ $K_2Cr_2O_7$	1	*P*	$a \neq b \neq c$ $\alpha \neq \beta \neq \gamma$	2
Monoclinic $CaCO_3$ (calcite) $NaNO_3$	2	*P*, *C*	$a \neq b \neq c$ $\alpha = \gamma = 90° \neq \beta$	13
Orthorhombic $CaSO_4$ $NaNO_3$	4	*P*, *C*, *I*, *F*	$a \neq b \neq c$ $\alpha = \beta = \gamma = 90°$	59
Tetragonal SnO_2 $LiNH_2$	2	*P*, *I*	$a = b \neq c$ $\alpha = \beta = \gamma = 90°$	68
Isometric or cubic NaCl CaF_2	3	*P* or sc *I* or bcc *F* or fcc	$a = b = c$ $\alpha = \beta = \gamma = 90°$	36
Trigonal (special case of hexagonal system) $KClO_3$ Au_2Te_3	1	*R* (rhombohedral)	$a = b = c$ $\alpha = \beta = \gamma \neq 90°$	25
Hexagonal CdS SiO_2	1	*P*	$a = b \neq c$ $\alpha = \beta = 90°$ $\gamma = 120°$	27
	14 Bravais Lattices		Total number of space groups = 230	

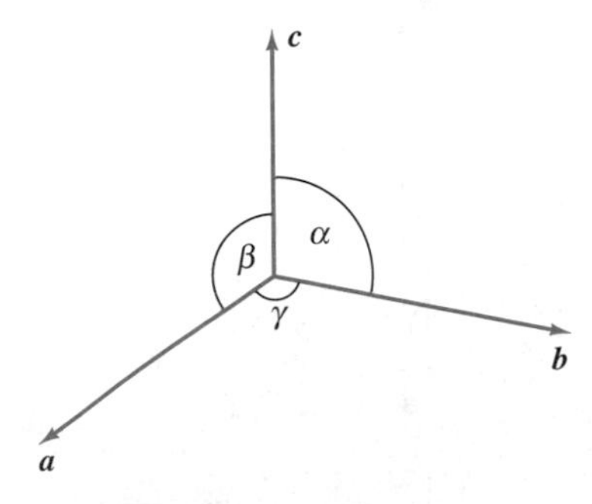

FIGURE 16.10
Crystal axes ***a***, ***b***, ***c***. α is the angle between ***b*** and ***c*** (opposite the *a* axis). β is the angle between ***c*** and ***a*** (opposite the *b* axis). γ is the angle between ***a*** and ***b*** (opposite the *c* axis).

connection with the point-symmetry elements. To illustrate this, the primitive cell for a body-centered cubic lattice is shown in Figure 16.11. Finally we note that it is the points that constitute the lattice, and not the lines, which are inserted as a matter of convenience.

Space Groups

Just as only certain symmetry operations can be combined with the 5 two-dimensional lattices (resulting in 32 point groups), the combination of the 14 Bravais lattices with translations, screw axes, and glide planes results in 230 different three-dimensional crystal patterns. The determination of the space group to which a particular crystal belongs was not possible until diffraction techniques were available to investigate the internal symmetry of crystals. (We have already pointed out that screw axes and glide planes are not observed in a visual observation of the crystal.) The determination of the structure of a crystal is simplified if the space group is known because, if the asymmetric portion of the unit cell is determined experimentally, the rest of the structure may be obtained through symmetry.

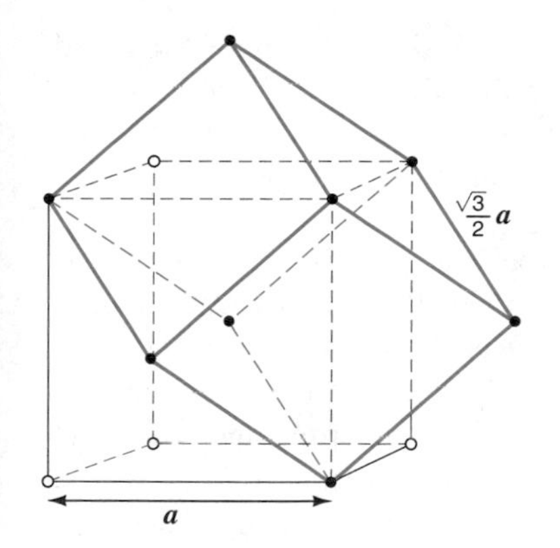

FIGURE 16.11
Demonstration showing how a primitive cell can be formed from a body-centered cubic lattice. The rhombohedron has an edge of $\frac{1}{2}\sqrt{3}a$, where the angle between adjacent edges is the tetrahedral angle 109°28′.

Study the lattice vector and show how points in a unit cell are periodic throughout the lattice; a lattice types; Translational vectors.

Use the lattice vector to generate a plane lattice; lattice types; Generate a lattice.

Periodicity and the Reciprocal Lattice

According to Eq. 16.3, crystal properties are periodic under translation by the lattice vector Eq. 16.1. Since the vectors $\boldsymbol{a}$, $\boldsymbol{b}$, and $\boldsymbol{c}$ have lengths which correspond to the edges of a unit cell, the n's in Eq. 16.1 are integers. The electron density, being a periodic function, can be written as a Fourier series,

$$\rho(\boldsymbol{R}) = \sum_{G} \rho_G \, e^{i\boldsymbol{G}\cdot\boldsymbol{R}} \tag{16.5}$$

where ρ_G are the Fourier coefficients. From this, we are naturally led to defining the reciprocal lattice vector $\boldsymbol{G}$, which is of fundamental significance for solids.

To define the reciprocal lattice, impose the periodic condition, Eq. 16.3, on the electron density $\rho(\boldsymbol{R})$,

$$\rho(\boldsymbol{R} + \boldsymbol{R}_o) = \sum_{G} \rho_G \, e^{i\boldsymbol{G}\cdot(\boldsymbol{R}+\boldsymbol{R}_o)} = \rho(\boldsymbol{R}) \tag{16.6}$$

which can only be true for integral values of n in the expression

$$\boldsymbol{G}\cdot\boldsymbol{R}_o = 2\pi n \tag{16.7}$$

To ensure this, write $\boldsymbol{G}$ as

$$\boldsymbol{G} = h\boldsymbol{a}^* + k\boldsymbol{b}^* + l\boldsymbol{c}^* \tag{16.8}$$

and require

$$\boldsymbol{a}^*\cdot\boldsymbol{a} = 2\pi \qquad \boldsymbol{b}^*\cdot\boldsymbol{b} = 2\pi \qquad \boldsymbol{c}^*\cdot\boldsymbol{c} = 2\pi \tag{16.9}$$

This satisfies Eq. 16.7. The three vectors $\boldsymbol{a}^*$, $\boldsymbol{b}^*$ and $\boldsymbol{c}^*$ are called the fundamental, or primitive, translation vectors for the reciprocal lattice and are defined in terms of the cross product of the primitive translation vectors of the real lattice,

$$\begin{aligned} \boldsymbol{a}^* &= 2\pi\boldsymbol{b}\times\boldsymbol{c} / V_c \\ \boldsymbol{b}^* &= 2\pi\boldsymbol{c}\times\boldsymbol{a} / V_c \\ \boldsymbol{c}^* &= 2\pi\boldsymbol{a}\times\boldsymbol{b} / V_c \end{aligned} \tag{16.10}$$

EXAMPLE 16.3 Show that Eq. 16.9 is satisfied using the definitions of Eq. 16.10.

Solution (See Example 16.2, $V_c = \boldsymbol{a}\cdot\boldsymbol{b}\times\boldsymbol{c}$)

$$\begin{aligned} \boldsymbol{a}^*\cdot\boldsymbol{a} &= 2\pi\boldsymbol{b}\times\boldsymbol{c}\cdot\boldsymbol{a} / V_c = 2\pi\boldsymbol{a}\cdot\boldsymbol{b}\times\boldsymbol{c} / V_c = 2\pi \\ \boldsymbol{b}^*\cdot\boldsymbol{b} &= 2\pi\boldsymbol{c}\times\boldsymbol{a}\cdot\boldsymbol{b} / V_c = 2\pi\boldsymbol{a}\cdot\boldsymbol{b}\times\boldsymbol{c} / V_c = 2\pi \\ \boldsymbol{c}^*\cdot\boldsymbol{c} &= 2\pi\boldsymbol{a}\times\boldsymbol{b}\cdot\boldsymbol{c} / V_c = 2\pi\boldsymbol{a}\cdot\boldsymbol{b}\times\boldsymbol{c} / V_c = 2\pi \end{aligned}$$

where the cyclic property of the triple product has been used

$$\boldsymbol{a}\cdot\boldsymbol{b}\times\boldsymbol{c} = \boldsymbol{b}\cdot\boldsymbol{c}\times\boldsymbol{a} = \boldsymbol{c}\cdot\boldsymbol{a}\times\boldsymbol{b}$$

and the order of the dot product is unimportant since $\boldsymbol{a}\cdot\boldsymbol{b} = \boldsymbol{b}\cdot\boldsymbol{a}$.

Substitution of Eqs. 16.1 and 16.8 into Eq. 16.7 and the use of Eq. 16.10 reveal that

$$\boldsymbol{G}\cdot\boldsymbol{R}_o = 2\pi\,(hn_a + kn_b + ln_c) = 2\pi n \tag{16.11}$$

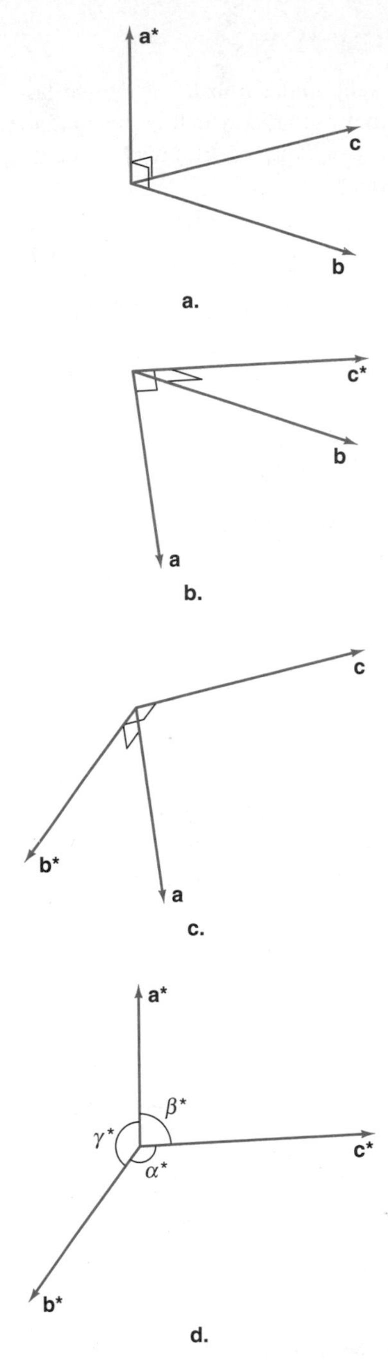

FIGURE 16.12
(a), (b), and (c). The primitive vectors for the reciprocal lattice are perpendicular to the planes produced by the primitive lattice vectors in real space. (d) The reciprocal axes $\boldsymbol{a}^*$, $\boldsymbol{b}^*$, and $\boldsymbol{c}^*$ and the angles; compare with Figure 16.10.

Since all the n's can take only integral values, the quantities hkl also are restricted to integral values. Below it is shown that hkl are the Miller indices. Before discussing them, the reciprocal lattice primitive vectors, $\boldsymbol{a}^*$, $\boldsymbol{b}^*$, and $\boldsymbol{c}^*$, are explained in more detail because reciprocal lattice space is the most convenient way of viewing most crystal properties.

From their definition in Eq. 16.10, the reciprocal lattice primitive vectors $\boldsymbol{a}^*$, $\boldsymbol{b}^*$, and $\boldsymbol{c}^*$ are perpendicular to the three planes formed by the lattice vectors $\boldsymbol{a}$, $\boldsymbol{b}$, and $\boldsymbol{c}$. This is shown in Figure 16.12. The reciprocal lattice vectors form a 3D space, shown in Figure 16.12d, constructed from $\boldsymbol{a}^*$, $\boldsymbol{b}^*$, and $\boldsymbol{c}^*$ and the angles α^*, β^*, and γ^*. Compare Figure 16.12d to Figure 16.10. Lattice space and reciprocal lattice space, although related, are different and cannot be written on one figure. The two spaces are related by the Fourier transform.[2] Whereas the lattice vectors $\boldsymbol{a}$, $\boldsymbol{b}$, and $\boldsymbol{c}$ have dimensions of length, the reciprocal lattice vectors, $\boldsymbol{a}^*$, $\boldsymbol{b}^*$, and $\boldsymbol{c}^*$, have dimensions of reciprocal length. Moreover, from Figure 16.12 it is seen that the planes in real space become points in reciprocal lattice space. This is one of the main uses of reciprocal lattice space, since diffraction techniques scatter from planes of lattice points in real space and are manifest as points in reciprocal lattice space. It is necessary, therefore, to have a way of determining the planes of lattice points in real lattice space and relating these to reciprocal lattice space. Miller indices do this.

Crystal Planes and Miller Indices

From Eq. 16.1, the lattice vector $\boldsymbol{R}_o$ generates the lattice by translating the primitive lattice vectors $\boldsymbol{a}$, $\boldsymbol{b}$, and $\mathbf{c}$ by integral values, n_a, n_b, and n_c. For example, if we decide to translate through values $n_a = 1, 2, 3, \ldots$ while keeping n_b and $n_c = 0$, a series of planes containing lattice points is generated in the $\boldsymbol{a}$ direction, see Figure 16.13. Similarly, translating by values $n_b = 1, 2, 3, \ldots$ while keeping n_a and $n_c = 0$, generates planes in the b direction, and likewise in the c direction by $n_c = 1, 2, 3, \ldots$ while keeping n_a and $n_b = 0$. In general, any integral combination of n_a, n_b, and n_c generates a plane in real lattice space that contains lattice points.

From the definitions of the primitive reciprocal lattice vectors, $\mathbf{G}$, Eq. 16.8, along with Eq. 16.9, we find that Eq. 16.11 must hold for integral values of hkl, or

$$(hn_a + kn_b + ln_c) = n \tag{16.12}$$

Expressed differently, Eq. 16.12 requires that $(n_a n_b n_c)$ have an inverse relationship to (hkl) for each of h, k, l for arbitrary integers n_1, n_2, n_3.

$$n_a = \frac{n_1}{h} \quad n_b = \frac{n_2}{k} \quad n_b = \frac{n_3}{l} \tag{16.13}$$

---

[2]The function $f(x)$ has a Fourier transform defined by

$$g(k) = \frac{1}{\sqrt{2\pi}} \int_{-\infty}^{+\infty} f(x) e^{ikx}\, dx$$

and the inverse transform is

$$f(x) = \frac{1}{\sqrt{2\pi}} \int_{-\infty}^{+\infty} g(k) e^{-ikx}\, dk$$

Such transforms allow us to view a function in two different domains and are used primarily as an aid in analysis. In Fourier transform (FT) NMR spectroscopy, for instance, the waveform and the spectrum are Fourier transforms of each other.

Create different families of planes in a lattice and relate these to the lattice vector; Lattice types; Lattice planes.

Law of Rational Intercepts

Create planes in real lattice space and relate these to the Miller indicies; Miller indicies.

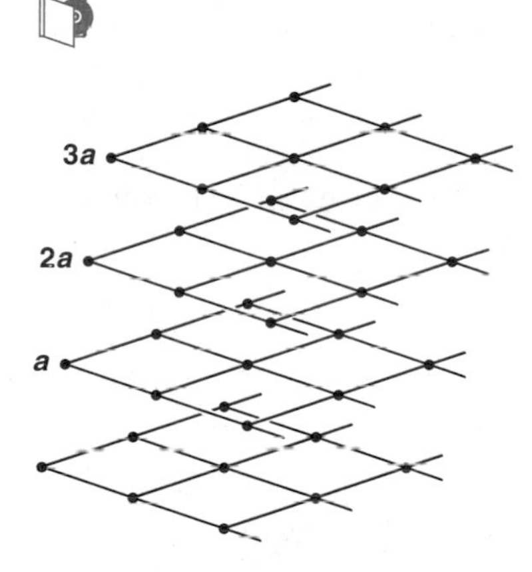

FIGURE 16.13
In real lattice space, planes corresponding to a certain set of Miller indices can be generated by use of the lattice vector $\boldsymbol{R}_o$.

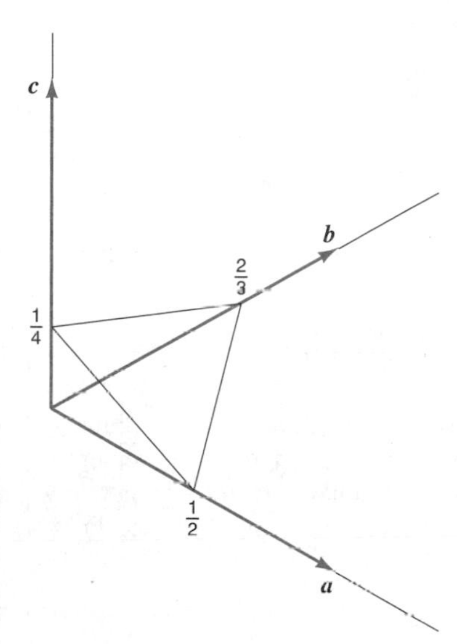

FIGURE 16.14
A plane cutting the axes at $\frac{1}{2}\boldsymbol{a}$, $\frac{2}{3}\boldsymbol{b}$, and $\frac{1}{4}\boldsymbol{c}$. The Miller indices are (438).

That is, if $(n_a n_b n_c)$ correspond to a plane of lattice points in real space, then (hkl) correspond to a point in reciprocal lattice space.

This was first noticed by Haüy on studying cleaved crystals. He noticed that the corresponding faces of the different fragments were equivalent. In order to describe those crystal faces Haüy proposed his **Law of Rational Intercepts.** The law states that it is always possible to find a set of axes that can be used to describe a crystal face in terms of intercepts along the axes. The reciprocals of these intercepts are small whole numbers. These numbers, h, k, and l, are called the **Miller indices** in honor of the British mineralogist William Hallowes Miller (1810–1880), who in 1838 first suggested their use for indexing planes.

Crystal planes pass through lattice points and are parallel to crystal faces. They are also described by the three integers h, k, l, either positive or negative. We can view parallel planes as cutting a unit length of each axis of the unit cell into an integral number of equal parts: the $\boldsymbol{a}$ axis into h equal parts, the $\boldsymbol{b}$ axis into k parts, and the $\boldsymbol{c}$ axis into l parts. Since all parallel planes are exactly alike, we generally consider only the plane nearest to the origin passing through the set of points closest to the origin. The next plane parallel to this will pass through the second nearest set of points to the origin, and so forth. The important point is that each plane passes through the dividing points and through the lattice points of the crystal. Such a set of planes will have the interplanar spacing d equal to the distance of the first plane from the origin.

For example, consider Figure 16.14, in which the noncoplanar axes a, b, c are shown with unit length vectors $\boldsymbol{a}, \boldsymbol{b}, \boldsymbol{c}$. The arrangement of atoms in the lattice is such that a plane can intercept these axes at $\frac{1}{2}$ on the a axis, $\frac{2}{3}$ on the b axis, and $\frac{1}{4}$ on the c axis. The reciprocals of these in the order a, b, c is $\frac{2}{1}, \frac{3}{2}$, and $\frac{4}{1}$. We clear the fractions by multiplying through by 2. The Miller indices are (438) and are placed in parentheses.

The procedure for finding the Miller indices for any set of planes follows:

1. Determine the intercepts.
2. Find the reciprocals of the intercepts.
3. Clear fractions.

If an intercept is at infinity (that is, the plane is parallel to one of the axes), the reciprocal is 0. Several examples of planes that have been indexed are shown in Figure 16.15. If a plane cuts an axis on the negative side of the origin, the corresponding index is negative and is indicated with a minus sign over the index, that is, $(\bar{h}kl)$.

> ***EXAMPLE 16.4*** Determine the indices of a plane (Miller indices) that intersects the axes at $\boldsymbol{a}/2$, $\boldsymbol{b}/2$, and $\boldsymbol{c}/2$.
>
> ***Solution*** From the indices, the reciprocals are 222 and are the values of hkl. This plane is parallel to the 111 plane.

It is important to know the hkl values because a particular face of a substance may have chemical properties different from those presented by another face. For example, the 111 faces of palladium have catalytic activity different from the other faces.

In the hexagonal crystal system the lattice can be described by four hexagonal axes, $\boldsymbol{a}_1$, $\boldsymbol{a}_2$, $\boldsymbol{a}_3$, $\boldsymbol{c}$, and the indices are written $(hkil)$ where $i = -(h + k)$. A symbol that is often used to enclose the Miller indices is the pair of braces, { }; it

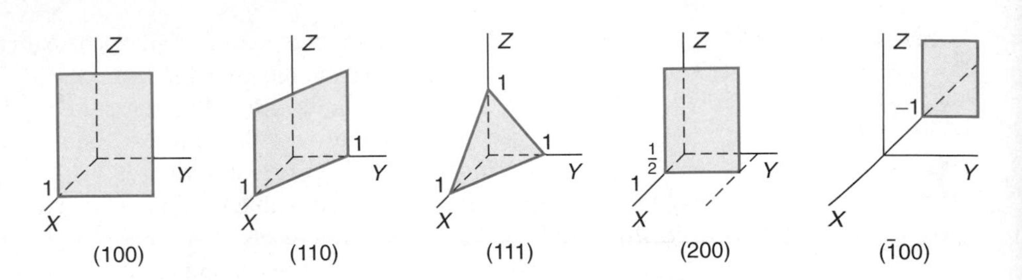

FIGURE 16.15
Miller indices of several planes in a cubic crystal. In this case the primitive vectors are $a = X$, $b = Y$, and $c = Z$.

represents those planes that are equivalent by symmetry. Thus we represent the faces of a cube by {100} and this represents the (100), (010), (001) planes.

Finally, the mapping of planes in real lattice space into points in reciprocal lattice space (*hkl*) is illustrated. The Miller indices in Figure 16.15 are shown in reciprocal lattice space in Figure 16.16. In addition, the plane shown in Figure 16.14 with Miller index (438) is also expressed in Figure 16.16. Of course, there are as many points in reciprocal lattice space as there are sets of planes in real space. Figure 16.16 shows only a representative few.

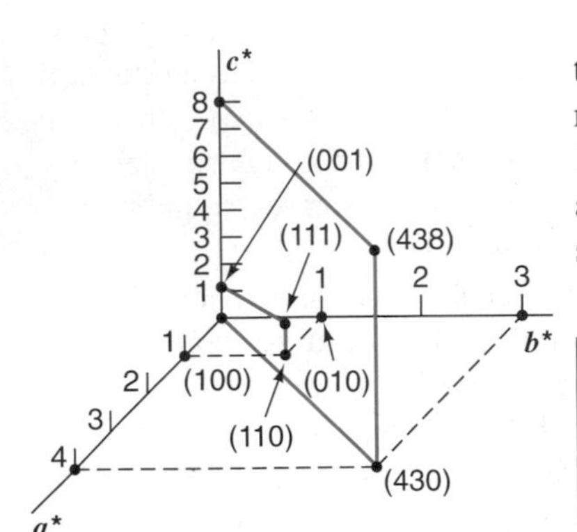

FIGURE 16.16
Reciprocal lattice space showing a few points defined by their Miller indices that correspond to planes in real lattice space. The point (438) corresponds to the plane shown in Figure 16.14.

Compare and contrast a 3D lattice and reciprocal lattice space; Vary the lattice vectors and unit cell to see the simultaneous effects in reciprocal lattice space; Reciprocal lattice.

EXAMPLE 16.5 Show that the volume of the unit cell in reciprocal lattice space is proportional to V_c^{-1}.

Solution The volume of the unit cell in reciprocal lattice space from Example 16.2 is $V_{reciprocal\ lattice} = \boldsymbol{a}^* \cdot \boldsymbol{b}^* \times \boldsymbol{c}^*$ which, by use of Eq. 16.10, gives

$$V_{reciprocal\ lattice} = \frac{(2\pi)^3}{V_c^3}\left[(\boldsymbol{b}\times\boldsymbol{c})\cdot[(\boldsymbol{c}\times\boldsymbol{a})\times(\boldsymbol{a}\times\boldsymbol{b})]\right]$$

In order to work out the vector algebra, the following properties are required:

Cyclic permutation: $\boldsymbol{a}\cdot(\boldsymbol{b}\times\boldsymbol{c}) = \boldsymbol{b}\cdot(\boldsymbol{c}\times\boldsymbol{a}) = \boldsymbol{c}\cdot(\boldsymbol{a}\times\boldsymbol{b})$
Triple vector product: $\boldsymbol{A}\times(\boldsymbol{B}\times\boldsymbol{C}) = \boldsymbol{B}(\boldsymbol{A}\cdot\boldsymbol{C}) - \boldsymbol{C}(\boldsymbol{A}\cdot\boldsymbol{B})$

Making use of the triple vector product with the substitutions: $\boldsymbol{A} = (\boldsymbol{c}\times\boldsymbol{a})$; $\boldsymbol{B} = \boldsymbol{a}$; $\boldsymbol{C} = \boldsymbol{b}$ gives

$$V_{reciprocal\ lattice} = \frac{(2\pi)^3}{V_c^3}\left[(\boldsymbol{b}\times\boldsymbol{c})\cdot\left[\boldsymbol{a}\,[(\boldsymbol{c}\times\boldsymbol{a})\cdot\boldsymbol{b}] - \boldsymbol{b}\,[(\boldsymbol{c}\times\boldsymbol{a})\cdot\boldsymbol{a}]\right]\right]$$

The last term is zero since $\boldsymbol{a}$ is perpendicular to $\boldsymbol{c}\times\boldsymbol{a}$. Then using the cyclic permutation gives

$$V_{reciprocal\ lattice} = \frac{(2\pi)^3}{V_c^3}\,[(\boldsymbol{b}\times\boldsymbol{c})\cdot\boldsymbol{a}]\,[(\boldsymbol{c}\times\boldsymbol{a})\cdot\boldsymbol{b}] = \frac{(2\pi)^3}{V_c^3}\,[\boldsymbol{a}\cdot(\boldsymbol{b}\times\boldsymbol{c})]^2 = \frac{(2\pi)^3}{V_c}$$

Hence the volume in reciprocal space is inversely proportional to the volume in real lattice space. The reason for the factor $(2\pi)^3$ lies in the requirement that Eq. 16.9 applies and is imposed because traditionally reciprocal lattice space treats the wave vector $\boldsymbol{k}$ rather than the inverse wavelength λ^{-1} (see Eq. 16.25 below).

A significant property of the reciprocal lattice vector $\boldsymbol{G}$ is that it is perpendicular to the plane (*hkl*). The Miller indices map out a plane of lattice points in real space which cut the real axes $\boldsymbol{a}$, $\boldsymbol{b}$, and $\boldsymbol{c}$, respectively, at positions $n_a = 1/h$; $n_b = 1/k$;

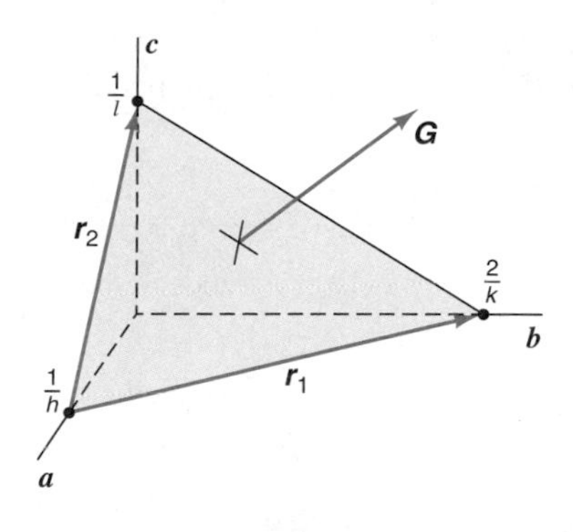

FIGURE 16.17
The crystal lattice coordinate system showing a plane with Miller indices (*hkl*) and the reciprocal lattice vector ***G***. The distance from the origin to the plane is given by d_{hkl}.

Compare and contrast a 2D lattice and reciprocal lattice space; Vary the lattice and unit cell with a simultaneous effect in reciprocal lattice space and find the orthogonality between the two spaces; Reciprocal lattice; 2D reciprocal lattice.

Study the reciprocal lattice vector and generate different reciprocal lattices; Reciprocal lattice; Reciprocal lattice vector.

Study and generate a reciprocal lattice using the reciprocal lattice vector; Reciprocal lattice; General reciprocal lattice.

$n_c = 1/l$. This is shown in Figure 16.17. To show that $\boldsymbol{G}$ is perpendicular to this plane, construct two vectors that lie on the plane,

$$\begin{aligned} \boldsymbol{r}_1 &= n_b\boldsymbol{b} - n_a\boldsymbol{a} \\ \boldsymbol{r}_2 &= n_c\boldsymbol{c} - n_a\boldsymbol{a} \end{aligned} \tag{16.14}$$

From Figure 16.17 the vector cross product, $\boldsymbol{r}_1 \times \boldsymbol{r}_2$, defines a vector perpendicular to this plane, which is given by

$$\begin{aligned} \boldsymbol{r}_1 \times \boldsymbol{r}_2 &= (n_b\boldsymbol{b} - n_a\boldsymbol{a}) \times (n_c\boldsymbol{c} - n_a\boldsymbol{a}) \\ &= n_a\, n_b\, n_c \left(\frac{\boldsymbol{b}\times\boldsymbol{c}}{n_a} + \frac{\boldsymbol{c}\times\boldsymbol{a}}{n_b} + \frac{\boldsymbol{a}\times\boldsymbol{b}}{n_c} \right) \\ &= n_a\, n_b\, n_c \left(h\boldsymbol{a}^* + k\boldsymbol{b}^* + l\boldsymbol{c}^* \right) \frac{V_c}{2\pi} \\ &= \frac{n_a\, n_b\, n_c V_c}{2\pi}\boldsymbol{G} \end{aligned} \tag{16.15}$$

Eq. 16.13 has been used, along with the definitions of the reciprocal primitive vectors in Eq. 16.10. Hence the reciprocal lattice vector $\boldsymbol{G}$ is perpendicular to the plane, as seen in Figure 16.17. The unit vector in this direction is therefore

$$\hat{\boldsymbol{G}}^* = \frac{\boldsymbol{G}}{|\boldsymbol{G}|} \tag{16.16}$$

It is now possible to determine the distance between members of any family of planes in real lattice space. One plane passes through the origin parallel to the next plane defined by the Miller indices (*hkl*). That is, this plane cuts the *a* axis at $\boldsymbol{a}/h$, the *b* axis at $\boldsymbol{b}/k$ and the *c* axis at $\boldsymbol{c}/l$. Any of the possible values of $\boldsymbol{a}$, $\boldsymbol{b}$, or $\boldsymbol{c}$ will do. Using the vector $\boldsymbol{a}$, the projection along $\hat{\boldsymbol{G}}^*$ gives the distance from the origin to the (*hkl*) plane as

$$d_{hkl} = \frac{1}{h}\boldsymbol{a}\cdot\hat{\boldsymbol{G}}^* = \frac{2\pi h}{h|\boldsymbol{G}|} \tag{16.17}$$

or

$$d_{hkl} = \frac{2\pi}{|\boldsymbol{G}|} \tag{16.18}$$

where the perpendicular distance between planes is written as d_{hkl}. It is useful to have a general formula that relates d_{hkl} to the *lattice constants* *a*, *b*, *c*, cell angles, and the Miller indices of the planes. It is

$$\begin{aligned} d_{hkl}^2 = {} & [1 - \cos^2\alpha - \cos^2\beta - \cos^2\gamma + 2\cos\alpha\cos\beta\cos\gamma]/ \\ & [(h^2/a^2)\sin^2\alpha + (k^2/b^2)\sin^2\beta + (l^2/c^2)\sin^2\gamma \\ & + 2(hl/ac)(\cos\alpha\cos\gamma - \cos\beta) \\ & + 2(hk/ab)(\cos\alpha\cos\beta - \cos\gamma) \\ & + 2(kl/bc)(\cos\beta\cos\gamma - \cos\alpha)] \end{aligned} \tag{16.19}$$

Indices of Direction

The direction in a crystal is often important and is also expressed as a set of indices where the integers are written between brackets, [*hkl*]. These are the smallest integers referred to the axes that have the ratio of the components of a vector in the desired direction. For example, in a cubic crystal the X axis is the [100] direction; the Y axis is the [010] direction. In cubic systems the direction $[hkl]$ is perpendicular to a plane (hkl) having the same indices.

EXAMPLE 16.6 Derive an equation for the value of d_{hkl} between planes in the cubic unit cell.

Solution In the cubic cell all angles are 90° so all sines are equal to one and all cosines are equal to zero. Eq. 16.19 reduces to

$$d_{hkl} = a/(h^2 + k^2 + l^2)^{1/2} \qquad \text{where } (a = b = c) \qquad (16.20)$$

Determination of the corresponding expressions for orthorhombic and tetragonal cells is left as Problem 16.18.

Another useful general expression is the volume of a unit cell. It is given by

$$V_c = abc(1 - \cos^2 \alpha - \cos^2 \beta - \cos^2 \gamma + 2 \cos \alpha \cos \beta \cos \gamma)^{1/2} \qquad (16.21)$$

This equation reduces to $V_c = abc$ if the cell is cubic, tetragonal, or orthorhombic.

16.2 X-Ray Crystallography

The Origin of X Rays

Röntgen received the Nobel Prize in physics in 1901 for his discovery of X rays.

In 1895 the German physicist Wilhelm Conrad Röntgen (1845–1923) discovered X rays. X rays are produced when high-energy electron beams (1–100 keV) strike a metal target. The X rays are high-energy photons ($E = h\nu$), the spectrum of which is generated by two different processes within the metal target. A continuous spectrum is formed when the high-speed electrons lose energy as they are slowed down rapidly by multiple collisions with the target. In this process electrons are ejected from different energy levels. These collisions and ejections give rise to *white radiation* or *Bremstrahlung* (German *Bremse,* brake, and *Strahlung,* ray), so called because the radiation is continuous with a wide range of unquantized wavelengths. The second process superimposes on the continuous radiation spectrum discrete peaks (quantized) that are characteristic of the target material. These peaks arise from electron transitions between the lowest-lying quantum levels. These levels have a special nomenclature. Corresponding to the atomic energy levels $n = 1, 2, 3, \ldots$, the electrons are said to be in the $K, L, M, \ldots$ shells. See Figure 16.18, where the more important transitions for X rays are shown.

When one electron is knocked out of the K shell by a high-energy electron, an L- or M-shell electron may drop to replace it within 10^{-14} to 10^{-6} s. An electron in a still higher energy level drops to replace the L or M electron, and so on, until the deficiency is made up with a free electron. The emitted radiation results in spectral lines that are named after the shell where the transition ends. Thus, an electron dropping from the L-shell gives rise to a K spectral line that is characteristic of the atom with an energy equal to the energy difference between the L and the K states. The most probable transition with the highest intensity is named alpha, followed by beta, gamma, . . . , with the indices 1, 2, 3, . . . , defining transitions within the subset. Since two electron states exist from which the electron descends in the L shell, spin-

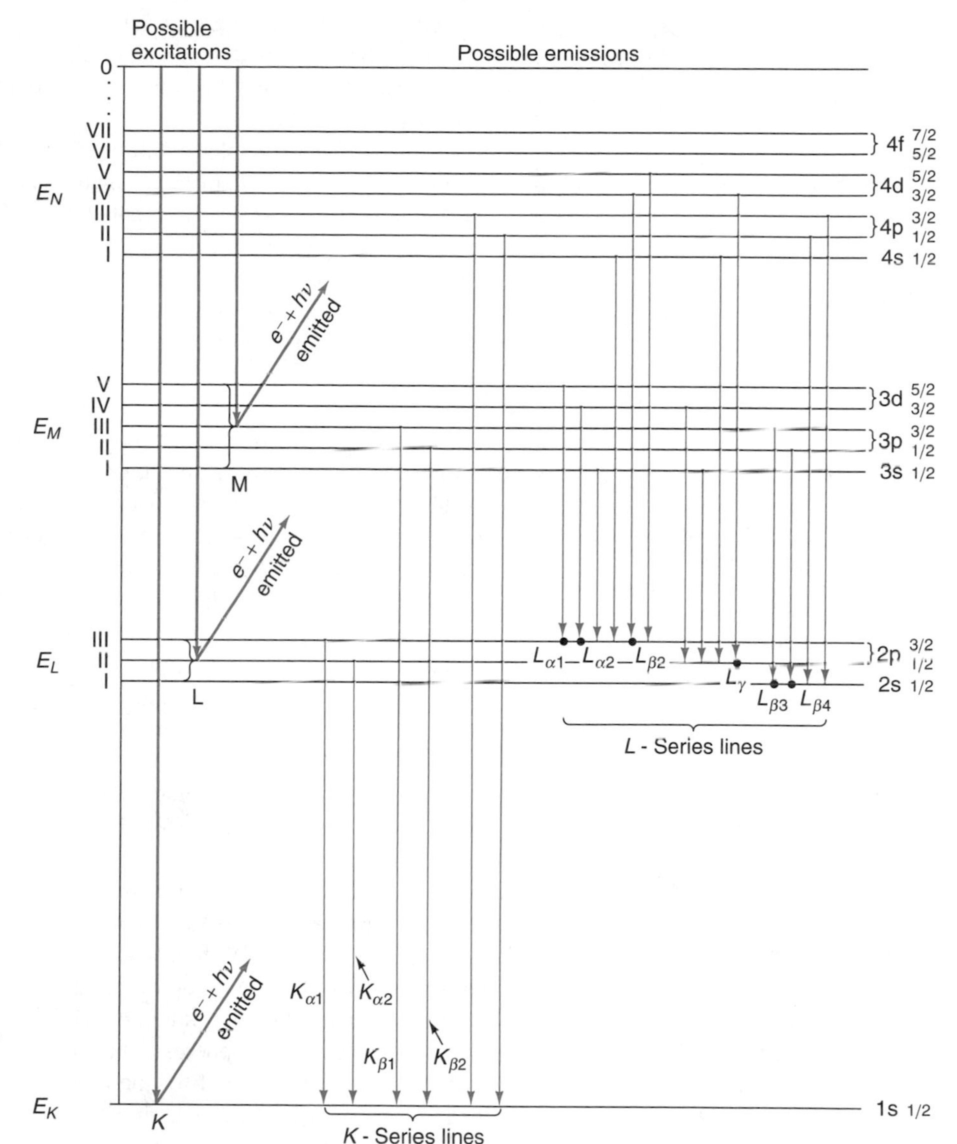

FIGURE 16.18
Some energy levels in the excitation and emission of X rays. *K*, *L*, and *M* excitations are shown. Each excitation results in an excited electron ejected from the excited level. Each exciting electron, shown on the left, produces an energy background which is not quantized and is known as *Bremstrahlung.* The possible emissions, on the right, result from the fall of an electron from the higher state to the lower state, and are accompanied by the emission of a characteristic X ray. The more prominent lines are labeled.

orbit coupling occurs, producing two closely spaced lines, $K_{\alpha1}$, and $K_{\alpha2}$. In a similar manner, if an electron from the *M* level replaces the one removed from the *K* level, the energy change results in two $K_{\beta1}$ lines. These lines, commonly referred to as a K_β line, have greater energy (shorter wavelength) but are of lesser intensity because this transition is less favorable. The transitions are governed by selection rules that the azimuthal quantum number must change by ±1, that is p $\leftrightarrow$ s, d $\leftrightarrow$ p, and so forth. In addition the spin must change by 0 or ±1, as shown. Figure 16.19 shows how the intensity varies with the wavelength in the case of copper, a target material commonly used to produce X rays. Generally, only K_α and K_β radiation are of consequence in structure determination.

For atoms of low atomic number, another effect predominates over emission of a characteristic X-ray photon. In this effect, an electron falls from a higher energy state to a lower energy state but transfers its energy by electrostatic interaction to a

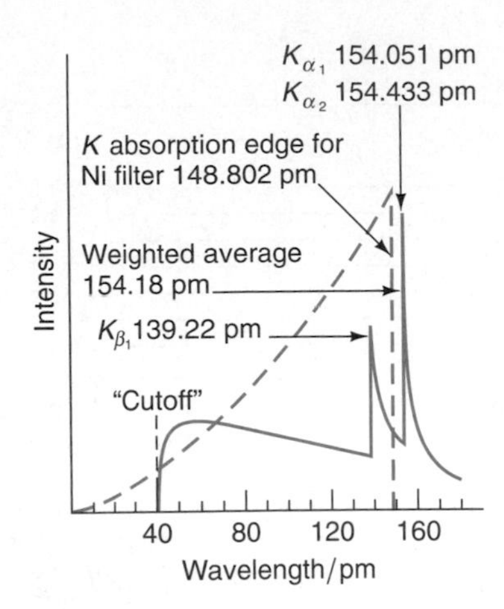

FIGURE 16.19
X-ray spectrum of copper at 35 kV showing the characteristic peaks. As the voltage of the exciting electrons increases, the cutoff and the peak intensity of the white radiation are shifted to shorter wavelengths.

M. von Laue received the Nobel Prize in physics in 1914 for the study of X rays from their diffraction by crystals and for showing that X rays are electromagnetic waves (1912).

The Braggs received the Nobel Prize in physics in 1915 for the study of diffraction of X rays in crystals. Both were later knighted.

bound electron, which then escapes the atom. The ejected electron is known as an **Auger** electron and carries energy that is characteristic of the atom and the transition. The Auger electron has become an important tool of analysis, especially for the physicist (Section 18.6).

In order to simplify the interpretation of X-ray spectra, the beam of X rays must be made monochromatic. Two different methods can be applied: filters or diffraction crystals. The latter method provides monochromatic X rays but at greatly reduced intensity. Specifically for X rays generated from copper, the former method takes advantage of the fact that the *K* absorption edge of nickel metal foil is 1.488 02 Å and is effective for filtering out the K_β radiation from the K_α radiation of copper. (See Figure 16.19, where the absorption edge of nickel is superimposed on the copper X-ray spectrum.) In general, the element immediately preceding the target element in the periodic table has a *K* absorption edge that makes it an effective filter.

An early triumph of the application of X rays rose out of the fact that the X-ray spectra of elements are simple and have characteristic wavelengths similar to one another. In 1913, the English physicist Henry Gwyn Jeffries Moseley (1887–1915) measured, for a series of elements, the characteristic frequencies of the *K* lines. From the data he derived a relationship from which he was able to deduce the charge on the nucleus. This relationship later defined the concept of **atomic number, *Z*,** as the number of protons, and correctly ordered several elements in the periodic table according to atomic number.

It was proposed in the 1890s that crystals were composed of atoms arranged like closely packed spheres. After it was realized that X rays have wavelengths in the order of 100 pm, the German physicist Max Theodor Felix von Laue (1879–1960) suggested in 1912 that crystals could serve as gratings for the diffraction of X rays. His colleagues W. Friedrich and P. Knipping experimentally confirmed his theory using "white" X rays that had a wide range of wavelengths. They were unable to determine the structure of the crystal from the observed results, however. At about the same time the English physicist William Henry Bragg (1862–1942) and his son (William) Lawrence Bragg (1890–1978) were experimenting with X rays having a narrow range of wavelengths. They were prevailed upon to test the theory that the minerals sylvite (KCl) and halite (NaCl) have face-centered cubic packing of spheres. The younger Bragg found that the observed X-ray pattern was that expected for the predicted structure and thereby revolutionized the physical sciences by pioneering X-ray methods for determining structures.

It should be noted that although it is commonly thought that X rays are independent of the ionic state of the atom, Figure 16.18 shows that some transitions originate from the valence shell. In such a case, there will be changes in wavelength or line shape caused by valence and those changes can be measured for multivalent elements at high resolution. Such variations have been shown in a study of sulfide, sulfite, and sulfate pollutants.[3]

The Bragg Equation

The ability to scatter X rays depends on the number of electrons in an atom. Furthermore, since the atoms in a crystal are ordered and lie in planes, each plane can

[3]L. S. Burke and J. V. Gilfrich, *Spectrochim. Acta, 33B*, No. 7, 305(1978).

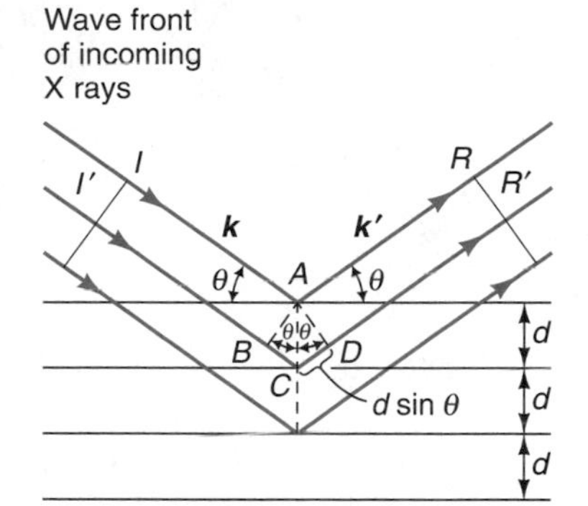

FIGURE 16.20
Derivation of the Bragg equation, $2d \sin \theta = n\lambda$. The diffraction of X rays may be considered as reflections from different planes.

diffract X rays. Since Friedrich used inhomogeneous ("white") X rays, his first diffraction pattern (Laue pattern) was difficult to interpret.

The two Braggs simplified the situation by using near-monochromatic X rays having a narrow wavelength range. They then found that the X rays were diffracted only at certain angles that depended on the wavelength and the interplanar spacings.

W. L. Bragg's explanation for this phenomenon may be followed from Figure 16.20, where several parallel planes of a crystal are shown. X rays impinge on the crystal at an angle θ. The incident X rays of wavelength λ are reflected specularly (as if by a mirror), the angle of incidence being equal to the angle of reflection. A small part of the beam is reflected from surface atoms such as from point A. Some of the beam will penetrate to lower planes and will be reflected. Constructive interference (Section 11.1) of the beam at RR' occurs only if the difference in distances traveled by the two beams is an integral number of wavelengths, that is, only if the path IAR differs from that of $I'CR'$ by a whole number of wavelengths. Otherwise no diffraction will be observed.

The necessary condition for reinforcement is, therefore, that the distance $BC + CD$ be equal to an integral number of wavelengths ($n\lambda$). From Figure 16.20, $\sin \theta = CD/d$ or

$$CD = d \sin \theta \quad \text{and} \quad BC = d \sin \theta \tag{16.22}$$

This condition requires that

$$n\lambda = 2d \sin \theta \tag{16.23}$$

Bragg Equation

which is the **Bragg equation,** often known as **Bragg's law.**

Study Bragg scattering and vary the distance between lattice planes; X-ray crystallography; Bragg scattering.

Homogeneous X rays of fixed λ are usually used and for a given set of lattice planes (hkl) [e.g., (010) planes], d is fixed. Whether a diffraction maximum is found depends on θ, known as the *glancing angle* or *angle of incidence*. As θ increases, a series of maxima is obtained corresponding to n equal to 1, 2, 3, etc. Thus an X-ray diffraction pattern is obtained in which successive maxima are called the first-order, second-order, third-order, etc., reflections, depending on the value of n. As n increases, these reflections decrease in intensity but are observable beyond $n = 20$. Second-order reflections are mathematically equivalent to a first-order reflection ($n = 1$) from planes with half the spacing. For example, if we are dealing with (100) planes, the second-order reflection is equivalent to a first-order reflection from the (200) planes (i.e., planes with half the spacing).

Note that $\sin \theta$ has a maximum value of 1. Thus, for a first-order reflection $\lambda = 2d$. If $2d$ is smaller than λ, no reflection maximum can occur. On the other hand, if $\lambda \ll d$, the X rays are diffracted through extremely small angles.

EXAMPLE 16.7 Consider the three X-ray reflections shown taken from three different crystals. In each an X-ray beam from the left is reflected from a set of dots representing lattice points whose planes have different spacings corresponding to different values of d_{hkl}. These reflections take place only at certain angles as given by the Bragg equation. For Cu $K_\alpha = 0.15418$ nm, calculate the value of $\sin \theta$ and compare it to the value of the angle found by measuring the figure.

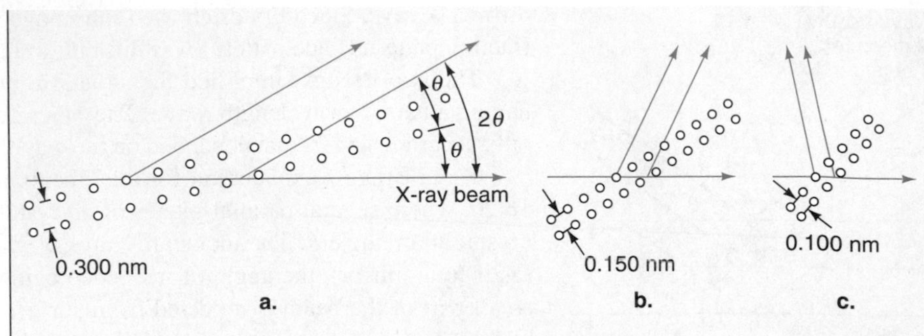

Solution For figure (a)

$$\sin\theta = \frac{0.15418\ \text{nm}}{2(0.300\ \text{nm})} = 0.257 \qquad \theta = 14.9^\circ$$

Similarly for (b) $\theta = 30.9^\circ$

(c) $\theta = 50.4^\circ$

Note that the angle θ between the beam and the planes increases as the spacing d decreases.

From a knowledge of the experimentally observed diffraction angles it is possible to determine the interplanar spacings of a crystal. Then from a knowledge of all the spacings the unit cell size and shape can be determined. Since the intensities depend on the arrangement and nature of the basis within each unit cell, some complications do arise with respect to intensity. This problem will be discussed in Section 16.3.

X-Ray Scattering

The scattering of X rays from crystals is elastic. This means that the energy of the incoming X ray remains unchanged, although the direction changes. From Figure 16.20 the incoming wave is denoted by $\boldsymbol{k}$ and the outgoing wave by $\boldsymbol{k}'$, but with $k = k'$ (see Figure 16.21). The quantity $\boldsymbol{k}$ is called the **circular wave vector.** It is related to the momentum of the wave by

$$\boldsymbol{p} = \hbar\boldsymbol{k} \tag{16.24}$$

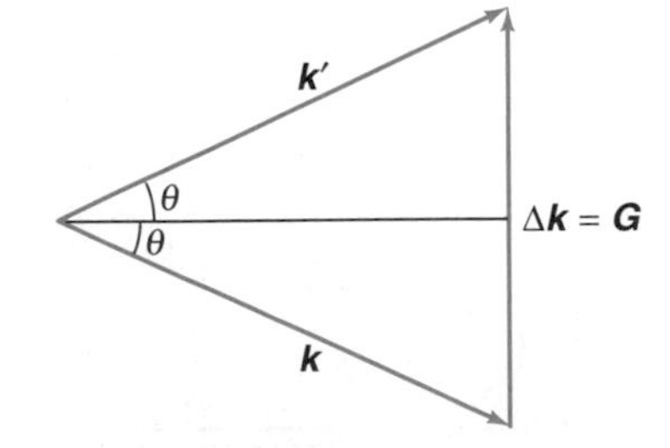

FIGURE 16.21
Since the scattering is elastic, the magnitudes of the incoming wave vector, $\boldsymbol{k}$, and the scattered wave vector, $\boldsymbol{k}'$, are the same. The triangle is therefore isosceles and the change in direction is equal to $\Delta\boldsymbol{k}$. This is Bragg's law in reciprocal lattice space.

Its magnitude is related to the frequency and wavelength of the X rays, which remain unchanged before and after elastic scattering by

$$\frac{hk}{2\pi} = \frac{\hbar\omega}{c} = \frac{h\nu}{c} = \frac{h}{\lambda} \tag{16.25}$$

That is, for elastic scattering, the direction of $\boldsymbol{k}$ changes to $\boldsymbol{k}'$, while the wavelength before and after the scattering is the same,

$$k = k' = \frac{2\pi}{\lambda} \tag{16.26}$$

If the momentum in Eq. 16.24 refers to free electrons or neutrons rather than X rays, Eqs. 16.24 and 16.26 still hold, while Eq. 16.25 involves the mass of the particle,

$$E_k = \hbar\omega = \frac{p^2}{2m} = \frac{\hbar^2k^2}{2m} \tag{16.27}$$

The difference between the incoming X rays, characterized by $\boldsymbol{k}$, and the outgoing wave, characterized by $\boldsymbol{k}'$, defines the scattering vector, $\Delta\boldsymbol{k}$,

$$\Delta\boldsymbol{k} = \boldsymbol{k}' - \boldsymbol{k} \tag{16.28}$$

Then, from Figure 16.21, and using Eq. 16.26, it follows that

$$k' \sin\theta = k \sin\theta = \frac{\Delta k}{2} \tag{16.29}$$

Substitution of the wavelength for k gives an expression similar to the Bragg equation, Eq. 16.23

$$\lambda = \frac{4\pi}{\Delta k} \sin\theta \tag{16.30}$$

To relate this to the Bragg equation, for $n = 1$, it must follow that

$$d_{hkl} = \frac{2\pi}{\Delta k} \tag{16.31}$$

Comparison with Eq. 16.18 gives the important result that the change in direction for the elastic scattering of waves from a latttice plane is equal to the reciprocal lattice vector, $\boldsymbol{G}$,

$$|\boldsymbol{G}| = \Delta k \tag{16.32}$$

and more generally,

$$\boldsymbol{G} = \Delta\boldsymbol{k} = h\boldsymbol{a}^* + k\boldsymbol{b}^* + l\boldsymbol{c}^* \tag{16.33}$$

This result is significant for diffraction because it states that the elastic scattering of waves from different planes (hkl) is in the direction of the reciprocal lattice vector.

Ewald space

A useful way of displaying elastic scattering is to map the planes in real lattice space into reciprocal lattice space and construct the **Ewald sphere,** as shown in Figure 16.22. Some scattering planes are shown in Figure 16.20 and are numbered 1 to 4. These can be considered as a series of semireflecting mirrors. If there were only one mirror, then any angle of the incident X-ray beam would be scattered, but because of the set of mirrors, only those angles that satisfy the Bragg equation are diffracted. If the Miller index for one plane is (100), then the others are (200), (300), (400), ... In fact, there are many such planes in a crystal, making many angles. As the X rays pass through the crystal, the diffraction from these planes leads to diffraction patterns. In order to satisfy the Bragg equation, a crystal is positioned along the beam and rotated. (In contrast, a powder sample contains all the crystallites arranged at all angles, so no rotation of the sample is required.)

In reciprocal lattice space, the Ewald sphere is drawn so that the X-ray beam goes through the center of the sphere (Figure 16.22). The Ewald sphere is placed so that the origin of the reciprocal lattice lies on the sphere's surface at the point where the X-ray beam leaves the sphere. As the crystal rotates in real space, the reciprocal lattice also rotates, but about M its origin on the Ewald sphere. As various points in reciprocal lattice space cut the sphere, the Bragg condition is satisfied. This is true because the radius of the sphere is $k = 2\pi/\lambda$. That is, when the sphere intercepts two or more reciprocal lattice points, the scattering is constructive giving a particular intensity that is detected by a detector positioned along each direction. It is more likely that the second point does not lie on the circle (or sphere in 3D), and the crystal must

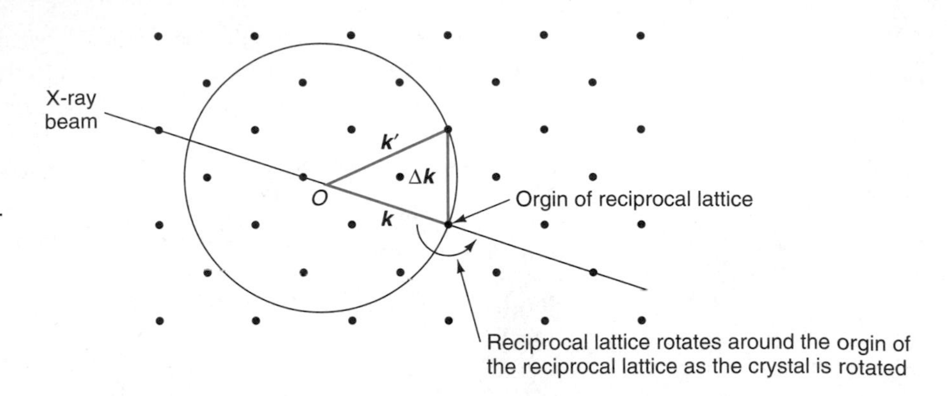

FIGURE 16.22
The Ewald sphere in two dimensions. Many planes exist in a crystal, but not many planes contain lattice points. Planes exist at the Miller indices. Here in reciprocal lattice space the planes are points. The crystal is rotated until two points lie on the circle (sphere), and at that position Bragg's law is obeyed and scattering occurs.

be rotated to satisfy the scattering condition. The Ewald sphere is a statement of Bragg's law in reciprocal lattice space. Any reciprocal lattice points falling on this sphere must obey Bragg's equation.

Vary the wavelength, lattice spacing, and orientation of a reciprocal lattice to see the use of the Ewald sphere; X-ray crystallography; Ewald sphere.

For any arbitrary λ and position of the crystal, no reciprocal lattice point will fall on the sphere. However, rotating the crystal corresponds to rotating the lattice about its origin, and it brings various reciprocal lattice points into coincidence with the surface of the Ewald sphere. Another way of visualizing this focuses on the size of the sphere of reflection. The diameter of the sphere is $2\pi/\lambda$ so that every reciprocal lattice point within a *limiting sphere* with radius $2\pi/\lambda$ represents a potential reflection. If shorter wavelengths are used, the effect is to enlarge the sphere and increase the number of reflections. A corresponding reflection is observed by the detector and can be related to the (hkl) indices of the reflecting plane.

Elastic Scattering, Fourier Analysis, and the Structure Factor

When monochromatic radiation is used, the values of the interplanar spacings of a crystal can be determined from the observed diffraction angles. Thus it is possible to deduce the unit cell dimensions, as shown earlier. For simple crystals the symmetry and unit cell size may be enough information to arrive at the exact structure. More often, the problem remains of determining the arrangement of the atoms or ions within the unit cell. The information needed to establish this is contained in the intensities of the diffraction spots.

The crystal structure can alter the intensities of the various spots that are observed in the following way. Consider a lattice on which is placed a basis that consists of two different atoms, as shown in Figure 16.23. The Bragg equation gives the angle of scattering in terms of the interplanar spacings. At some angle θ, the atoms of each type scatter X rays in phase with other atoms in the same array. However, because of the angular difference between the two arrays, the path length for scattered X rays is longer from one type of atom than it is from the other type. As a result, there is a phase difference between the waves generated from the arrays of the two types of atoms. Interference occurs, therefore, and reduces the intensity of the diffracted beam.

X-ray scattering is caused by the electrons in an atom, and each different type of atom thus has a different ability to scatter X rays. Furthermore, because the electron distribution can be "smeared out" unsymmetrically under the influence of adjoining atoms, the ability of an atom to scatter X rays depends on the space group it is in.

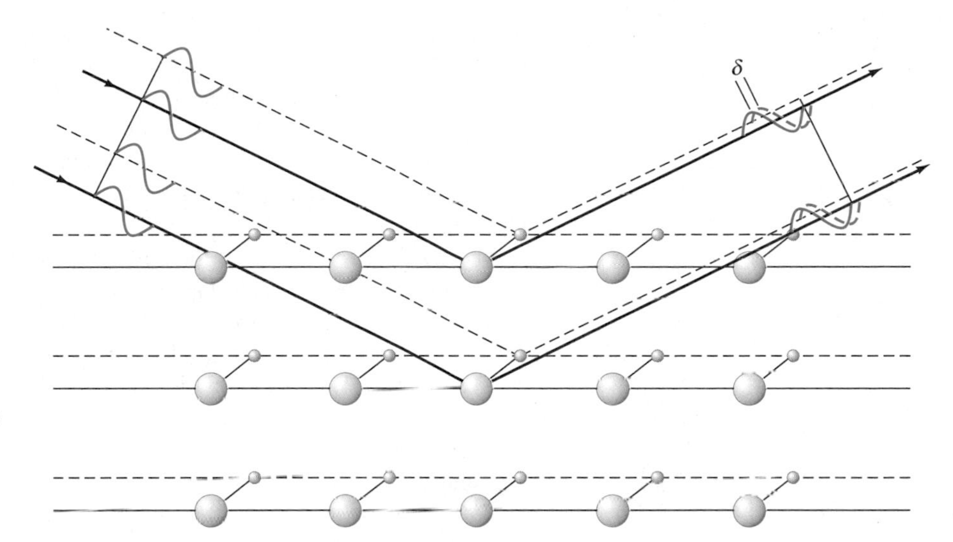

FIGURE 16.23
X rays scattered from a crystal with a basis of two different atoms in mutually displaced lattice arrays. The phase difference between the two waves is represented by δ.

Although the electron density function can be quite complex for large molecules in a crystal, it is periodic and can be represented by a series of sums of sine and cosine functions, known as **Fourier series** after the French mathematician Jean Baptiste Joseph Fourier (1768–1830). In essence the Fourier series representation of a function uses adjustable constants to force a sufficiently large number of sine and cosine waves that may be written as an exponential term to reproduce the original function.

To handle these situations and to put the conclusions arrived at in the previous sections on a more formal footing, the periodic properties of the lattice are exploited by means of Fourier analysis. We have already used Fourier series to define the reciprocal lattice. The elastic scattering of X rays depends upon the electron density, $\rho(\boldsymbol{R})$ (Eq. 16.4), and how the different atoms within the unit cell scatter. An incoming wave has the form $\exp(i\boldsymbol{k}\cdot\boldsymbol{R})$ and the outgoing wave has the form $\exp(i\boldsymbol{k}'\cdot\boldsymbol{R})$. The elastic scattering amplitude, $F(\boldsymbol{k}, \boldsymbol{k}')$ depends on the difference, $\exp[i(\boldsymbol{k}' - \boldsymbol{k})\cdot\boldsymbol{R}]$

$$F(\boldsymbol{k}, \boldsymbol{k}') = \int dV\, \rho(\boldsymbol{R})\, e^{i(\boldsymbol{k}' - \boldsymbol{k})\cdot\boldsymbol{R}} \tag{16.34}$$

Following the same procedure as for Eqs. 16.5 and 16.6, the scattering amplitude depends upon the periodicity of the electron density $\rho(\boldsymbol{R})$, Eq. 16.5, revealing

$$F(\boldsymbol{k}, \boldsymbol{k}') = \sum_{G} \int dV\, \rho_G(\boldsymbol{R})\, e^{i(\Delta\boldsymbol{k} - \boldsymbol{G})\cdot\boldsymbol{R}} \tag{16.35}$$

Mathematically, this integral is negligible unless the phase change, $\Delta\boldsymbol{k}$, between the scattering beams, is equal to $\boldsymbol{G}$, the reciprocal lattice vector, Eq. 16.33. Again, using the scattering amplitude and Fourier analysis, we are immediately led to the reciprocal lattice statement of Bragg's law, as we saw in Eq. 16.33.

The details available from X-ray diffraction, however, depend upon the variation of electron density throughout the unit cell. This electron density is periodically reproduced and repeated throughout the crystal. Equation 16.4 expresses the electron density at each position within a unit cell. It originates from the N atoms contained therein. Substitution of Eq. 16.4 into Eq. 16.35 and rearrangement, gives a Fourier series expression for the scattering amplitude with useful properties for elastic scattering,

$$F(\boldsymbol{G}) = \sum_{n=1}^{N} f_n(\boldsymbol{G})e^{-i\boldsymbol{G}\cdot\boldsymbol{R}_n} \tag{16.36}$$

where now the argument of $F(\boldsymbol{k}, \boldsymbol{k}')$ has been changed to $F(\boldsymbol{G})$. By using the substitution $\boldsymbol{u} = \boldsymbol{R} - \boldsymbol{R}_n$, the **form factors** f_n are identified as,

$$f_n(\boldsymbol{G}) = \int dV\, \rho_G(\boldsymbol{u})\, e^{-i\boldsymbol{G}\cdot\boldsymbol{u}} \tag{16.37}$$

The integral is over the unit cell volume. The significance of Eq. 16.37 lies in the fact that the form factors, f_n, for each of the N atoms which comprise the basis within a unit cell are the scattering centers. The type of atom, with varying number of electrons, leads to different values for f_n. Small atoms like hydrogen, with only one electron, have a small f_{H} value and are often undetected by X-ray techniques, whereas larger atoms with more electrons, like sodium, have larger form factors, f_{Na}.

The structure, therefore, of the unit cell determines the scattering process and it is usual to specify the scattering amplitudes for a particular lattice plane (hkl). Moreover, it is also useful to express the positions of the nth atom in the unit cell by its coordinates $\boldsymbol{r}_n = (x_n, y_n, z_n)$. The values of x_n, y_n, and z_n are only Cartesian for cubic, tetragonal, and orthorhombic unit cells, but the notation is used for all cells. Combining these notions relates the scattering amplitude to the scattering from one plane for a specific type of unit cell. The resulting expression is called the **structure factor**,

$$F(hkl) = \sum_{n=1}^{N} f_n \, \mathrm{e}^{-i2\pi(hx_n + ky_n + lz_n)} \tag{16.38}$$

Structure factors are central to the interpretation of elastic scattering results. The intensity of scattering is proportional to the amplitude squared,

$$I(hkl) \propto F(hkl)^* \, F(hkl) = |F(hkl)|^2 \tag{16.39}$$

Examples of the use of the structure factors are given later in this chapter.

Finally, if the scattering amplitude, $F(\boldsymbol{k}, \boldsymbol{k}') = F(\boldsymbol{G})$ (Eq. 16.34), is recognized as the Fourier transform of the electron density, $\rho(\boldsymbol{R})$, then the electron density is given by the inverse Fourier transform of the scattering amplitude. Using the equations above and the definition of the Fourier transform and its inverse[2] gives

$$\rho(x_n y_n z_n) = \frac{1}{V_c} \sum_{hkl} F(hkl) \, \mathrm{e}^{-i2\pi(hx_n + ky_n + lz_n)} \tag{16.40}$$

Equation 16.40 represents the electron density in direct space, whereas Eq. 16.38 represents the structure factors in terms of electron density in reciprocal space. In other words, when the phase is known, Eq. 16.40 allows us to recover the electron density of the crystal in real space from the observed diffraction amplitudes in reciprocal space.

There is an inherent problem in this method. Although $F(hkl)$ is the amplitude of the wave scattered by the hkl plane, the measured intensity of the beam is proportional to $|F(hkl)|^2$. Because of the way in which the original data are recorded, point by point, the phase relations are lost. Thus the transformation from Fourier space to real space is complicated by the need to determine the phase. In 1953 a solution to this problem was found by Max F. Perutz, who was working on the structure of hemoglobin at Cambridge University. He used the method of *isomorphous replacement,* in which heavy atoms, such as mercury or uranium atoms, were introduced into the molecules without altering the crystal structure. His work, along with that of others on the structures of large molecules of biological importance, is referred to again in Section 16.3 on page 862.

Direct methods, based on statistical techniques, have also been developed to treat the phase problem. In addition, *structure* refinement even gives an estimate of errors in position and bond length.

Also in 1953, Herbert Aaron Hauptman (b. 1917) and Jerome Karle (b. 1918) published a monograph containing a set of probabilistic formulas and measures for addressing the phase problem. This work, for which both men received the 1985 Nobel Prize in chemistry, led to the development of direct methods for determination of crystal structures.

16.3 Experimental Methods and Applications

Several X-ray techniques have been developed for structural studies, and each has its advantages and disadvantages. Photographic methods of recording data are described in order to present the principles in a more pictorial manner. Diffractometers now routinely detect the intensity of an X-ray beam by means of scintillation counters or proportional-counter tubes. Also, sophisticated computer programs control the automatic determination of even complicated crystal structures from the measured intensity and angular relations. It is not uncommon to solve the structure of a well-formed single crystal of a molecule having a hundred atoms in a matter of minutes, and to refine the structure within a day, something outside the realm of possibility only a generation ago.

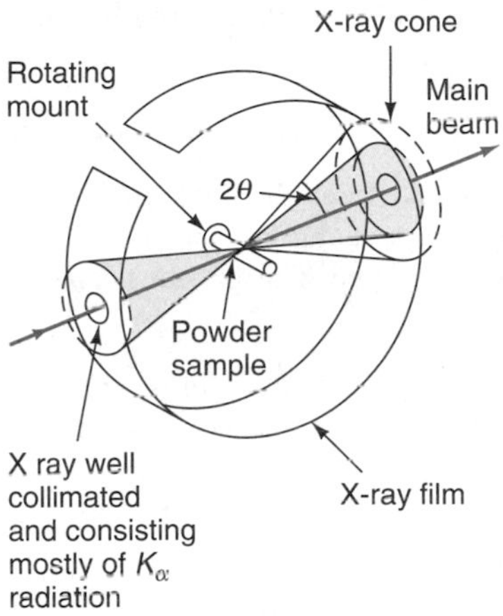

FIGURE 16.24
Schematic representation of the X-ray powder technique. X-ray film is held in the metal camera. Areas on film are segments of X-ray cones diffracted from powder sample.

The Laue Method

In the **Laue method** a single crystal is irradiated by a beam of continuous wavelength X rays. X rays are emitted at angles and wavelengths for which the Laue equations are satisfied. These equations are derived by considering the X-ray beams as reflected from the surface of the crystal as though from a diffraction grating.

A flat film receives the diffracted beams and the diffraction pattern is composed of a series of spots that shows the symmetry of the crystal. This Laue method is used extensively for the rapid determination of crystal *orientation,* such as directions within a crystal, and of *symmetry.* In the former case the crystal is oriented in a goniometer and rotated until a desired direction is found, as indicated by the X-ray pattern. The Laue method is almost never used for crystal *structure* determinations because different orders may be reflected from a single plane on account of the continuous wavelengths employed. This hinders the determination of the reflected intensity, which is needed in crystal structure determinations.

The Powder Method

Peter Debye won the 1936 Nobel Prize for chemistry for his contributions to molecular structure, including dipole moments, electron diffraction, and X-ray diffraction.

When single crystals are not available, or for a variety of purposes including the identification of crystalline materials, a simpler technique may be used. The crystalline material is ground to a powder which then presents all possible orientations to a collimated X-ray beam consisting primarily of K_α radiation. This technique was developed by Peter Joseph William Debye (1884–1966) and Paul Scherrer. The diffracted beam actually forms cones from each scattering plane and in the past has generally been recorded on a strip of X-ray film that has been rolled in a circle centered on the powdered specimen. The experimental apparatus is schematically drawn in Figure 16.24 and a typical X-ray powder pattern is shown in Figure 16.25. If the crystals used are too coarse, the arcs in the powder pattern will appear speckled instead of as firm lines.

We might surmise that with so many crystals there would be utter confusion. However, each set of planes has a certain spacing so its θ is fixed. Any crystal not at the correct angle θ will not reflect the beam. Such an arrangement is all that is required to differentiate between the three crystal possibilities in the cubic system.

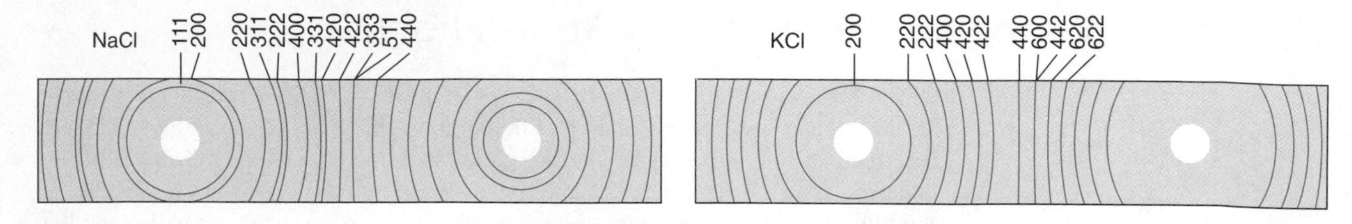

FIGURE 16.25
X-ray diffraction patterns of NaCl and KCl using a powder camera technique shown in Figure 16.24. A crystal of NaCl was used in the first determination of a crystal structure by X-ray crystallography.

The diffracted rays form cones concentric with the original beam and make an angle of 2θ with the direction of the original beam. This is the same result that would occur from mounting a single crystal and turning it through all possible orientations in the X-ray beam.

EXAMPLE 16.8 Determine the lattice parameter a and index the lines (that is, determine each hkl) obtained from a powder pattern of the cubic system β-brass (CuZn). The X-ray wavelength is 229 nm from a $\text{Cr}K_a$ source. The 2θ values are 45.8°, 66.7°, 84.7°, 102°, 120.8°, and 144.5°.

Solution For each value of 2θ calculate the value of $d_{hkl} = \lambda/2 \sin\theta$. Then solve for $a = d_{hkl}\sqrt{h^2 + k^2 + l^2}$ such that small whole number values of h, k, and l give a constant value of a.

2θ	$d_{hkl} = \lambda/2 \sin\theta$	$a = d_{hkl}\sqrt{h^2 + k^2 + l^2}$	
45.8°	0.294 nm	$= 0.294\sqrt{1^2 + 0^2 + 0^2}$	= 0.294 nm
66.7°	0.208 nm	$= 0.208\sqrt{1^2 + 1^2 + 0^2}$	= 0.295 nm
84.7°	0.170 nm	$= 0.170\sqrt{1^2 + 1^2 + 1^2}$	= 0.294 nm
102.0°	0.147 nm	$= 0.147\sqrt{2^2 + 0^2 + 0^2}$	= 0.294 nm
120.8°	0.132 nm	$= 0.132\sqrt{\_^2 + \_^2 + \_^2}$	= 0.295 nm
144.5°	( )	= (	) = 0.294 nm

Filling in the blanks in the last two lines is left to the reader. From the constancy of a, we are confident that the value of a for β-brass is correct. The values of h, k, l chosen to give a may seem arbitrarily determined but can be arrived at more systematically either analytically or graphically. These techniques can be used to determine more complicated systems. A graphical method based on analytical equations is available and is known as the *Hull-Davey Charts.* These are relevant for fiber diffraction and for powers, but not for single crystals.

This technique is particularly important for the determination of lattice types and unit cell dimensions.

Rotating Crystal Methods

In the *rotating crystal methods* a single crystal is oriented so that a monochromatic X-ray beam is perpendicular to a direct-lattice axis. Rotation of the crystal causes different atomic planes to diffract X rays whenever the value of θ satisfies the Bragg equation. Since different levels of reciprocal lattice points are normal to the direct axis, a rotation about the axis causes each level of points to intersect the sphere of reflection in a circle, as shown in Figure 16.26a. The diffracted rays pass from the center of the sphere through these circles, forming cones. The zero level is a flat cone with the other cone axes coincident and parallel to the rotating axis, as shown in Figure 16.26b.

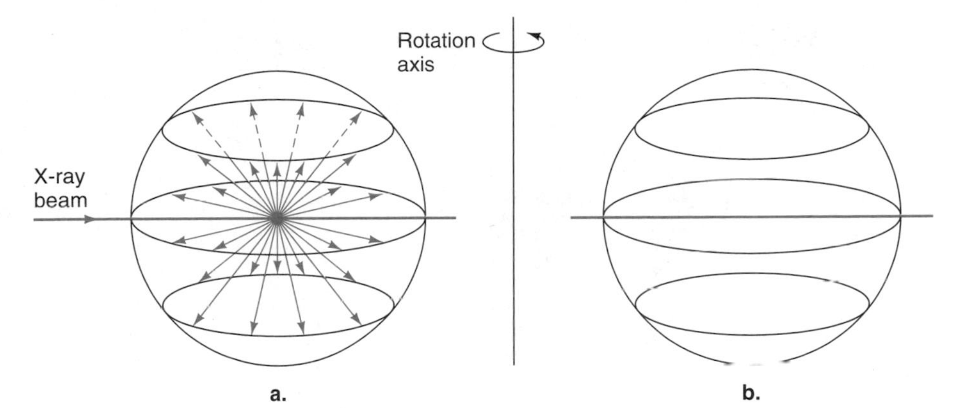

FIGURE 16.26
Relation between diffracted X rays in (a), the direct lattice, and in (b), the reciprocal lattice. The X-ray beam is perpendicular to a direct lattice axis. In (b) the diffracted X rays form circles with the sphere of reflection at three different levels.

In order to record the diffracted X rays, X-ray film is mounted cylindrically around the sample and concentrically with the rotating sample holder. The X-ray beam passes through the crystal at right angles to the film and the crystal holder. X rays diffracted from all planes parallel to the rotation axis occur as reflected spots in a horizontal line on the film. Other planes cause diffraction both above and below this horizontal plane. These reflected spots occurring in lines are called *layer lines*.

Rotating crystal techniques include *oscillating-crystal* methods in which the crystal is oscillated through a limited angular range instead of making a full 360° rotation. This reduces the possibility of overlapping reflections. In general, these techniques are used to align crystals, to measure the cell edge, and to obtain crystal symmetry information.

Perhaps the most familiar device for making rotation and oscillation photographs is the *Weissenberg* camera. Another such device is the *precession* camera, designed in the 1940s by the American crystallographer Martin Julian Buerger. For more detailed discussion of these methods, the reader is referred to the books listed in Suggested Reading (pp. 897–898).

X-Ray Diffraction

We now consider briefly two experimental techniques that are used to study the details of polymer structure. A powerful and widely used method is **X-ray diffraction**. Here we will mention only a few points relating particularly to macromolecules. The analysis of X-ray diffraction patterns is difficult for large molecules because of the many distances involved. It is the electrons that scatter the X rays, and the image calculated from the diffraction pattern thus reveals the distribution of electrons within the molecule. The usual procedure with macromolecules is to use a computer to calculate the electron densities at a regular array of points and to make the image visible by drawing contour lines through points of equal electron density. These contour lines can be drawn on clear plastic sheets, and a three-dimensional image can be obtained by stacking the maps one above the other or by using computer programs capable of producing three-dimensional computer-screen images. With an instrument of high resolving power the atoms appear as peaks on the image map.

Isomorphous Replacement

An important contribution to the analysis of X-ray patterns for macromolecules was made in 1953 by the Austrian-British molecular physicist Max Ferdinand Perutz (1914–2002). His method of **isomorphous replacement** involves the

Dorothy Crowfoot Hodgkin (1910–1994)

Dorothy Crowfoot was born of British parents in Cairo, where her father was working for the Egyptian education service. She received her early education in England, and in 1928 went to Somerville College, Oxford (which is named in honor of another distinguished scientist, Mary Somerville). Before obtaining her degree at Oxford she had become interested in the X-ray crystallography of large molecules. From 1932 to 1934 she carried out research at Cambridge in collaboration with J. D. Bernal and I. Fankuchen; the work was on the X-ray analysis of the digestive enzyme pepsin. In 1934 she returned to Oxford to teach and carry out research. In 1937 she married Thomas L. Hodgkin. They had three children.

At the outbreak of World War II it was realized that penicillin would be of great importance in the treatment of wounds, and much work was begun to establish its structure. Dorothy Hodgkin's X-ray crystallographic study, done in parallel with the chemical studies of Howard Florey and Ernst Chain, played an important role in leading this project to a successful conclusion. Her work was particularly notable in involving, for the first time, an electronic computer in the solution of a structural problem.

Beginning in about 1950 Dorothy Hodgkin began work on the structure of the vitamin B_{12} molecule, which is about four times as large as the penicillin molecule, and is one of the most complicated molecules of a nonprotein nature. The work again involved the use of electronic computers, and much study of a purely chemical kind. It was largely for her success with this extremely difficult problem that Dorothy Hodgkin was awarded the 1964 Nobel Prize for chemistry.

In 1960 Dorothy Hodgkin was appointed Wolfson Research Fellow of the Royal Society, and in the same year became Professorial Fellow of Somerville College. She was elected a Fellow of the Royal Society in 1947, and received its Royal Medal in 1956 and its Copley Medal in 1976. She was also elected a foreign member of various academies, including the U.S. National Academy of Sciences and the Russian Academy of Science. In 1965 she was appointed to the Order of Merit, the most exclusive of British honors.

References: *Biographical Dictionary of Nobel Laureates in Chemistry,* American Chemical Society, Washinton, D.C., 1993.

Sharon Bertseh McGrayne, *Nobel Prize Women in Science,* New York: Birch Lane Press, 1993, pp.225–254.

preparation and study of crystals into which heavy atoms, such as atoms of mercury or uranium, have been introduced without altering the crystal structure. This technique, which overcame the phase problem mentioned on page 858, led rapidly to the detailed analysis of a number of macromolecules. For this and other important work on X-ray diffraction, Perutz shared with Sir John Cowdery Kendrew (b. 1917) the 1962 Nobel Prize in chemistry. Perutz's work was particularly concerned with the structure of the hemoglobin molecule.

No contribution to the determination, by X-ray diffraction, of the structures of large molecules has been greater than that of Dorothy Crowfoot Hodgkin (1910–1994). Following work on penicillin in World War II, Hodgkin decided to tackle the much more difficult problem of vitamin B_{12}, which is a red coordination compound containing a cobalt atom and has a molecular weight of 1355, much larger than that of any molecule previously investigated by the same technique. She obtained the first X-ray photograph of the vitamin in 1948, and during the next six years she and her students obtained photographs of analogous molecules into which various heavy metal substitutions had been made. To analyze the results she first used an IBM calculating machine which employed punch cards; in later work she used one of the new electronic computers. The final paper with the structure of vitamin B_{12}, the largest molecule to have had its structure determined in complete detail, was published in 1957. This achievement gained her the 1964 Nobel Prize in chemistry.

Hodgkin's other great achievement was the determination of the three-dimensional structure of insulin. She took her first X-ray photograph of this molecule in 1935, but it took persistent work over more than 30 years for the structure to be established. Insulin contains 51 amino acids, and by the methods of chromatography, Frederick Sanger (b. 1918) determined the ordering of the chains, for which he was awarded the 1958 Nobel Prize in chemistry. (He received a second Nobel Prize in 1958 for his work on nucleic acids.) By 1969 Hodgkin was able to supplement Sanger's structure of insulin by elucidating the three-dimensional arrangement of the amino acid side groups which, through hydrogen bonding and disulfide bridges, give rise to the overall structure of the molecule. She showed that the long chains were folded into compact molecules that formed hexamers around two zinc atoms.

In spite of important technical developments, the establishment of a macromolecular structure by X-ray diffraction is still an extremely time-consuming process. One difficulty is that the substance must be in pure crystalline form. The technique cannot be applied to macromolecules in aqueous solution, and there is always the question whether the structure of the molecule in solution is the same as that in the crystalline state. In spite of difficulties, the X-ray structures of a large number of macromolecules now have been determined, and the results have made a valuable contribution to the understanding of the behavior of such molecules, particularly in relation to biological systems.

Electron Diffraction

Electrons can also be diffracted by a crystal lattice. This was shown in 1927, simultaneously and independently, by Davisson and Germer and by G.P. Thomson and Reid (Section 11.3). This was the first experimental proof of de Broglie's postulate on the wave nature of electrons.

Electron beams are far more efficiently scattered than are X-ray beams. Although X-ray diffraction patterns are analyzed on the basis that each X-ray photon has undergone just one deflection, this simplification is true for electron diffraction only when the crystalline samples are extremely thin. With solids, electron diffraction studies can be made only on surfaces, films, and thin crystals. The technique is also useful in determining the molecular structure and vibrations of gaseous molecules. The first electron diffraction pattern from a gaseous sample was obtained by R. Wierl in 1931. In gas-phase experiments the sample gas at about 10 Torr is admitted to the scattering chamber maintained at about 10^{-6} Torr. The intensity of the atomic-scattering curve varies in a smooth fashion with the angle of scattering and is largely nucleus dependent. The molecular scattering curve containing the desired information is superimposed on the previous curve and fluctuates fairly rapidly with θ when the energy of the beam is about 40 kV.

The de Broglie wavelength λ of an electron is related to its energy ϵ by the equation $\epsilon = h^2/2m\lambda^2$, where $m = 9.11 \times 10^{-31}$ kg is the mass of the electron.

This expression comes from the kinetic energy of the particle, $\epsilon = p^2/2m$, with the substitution $\lambda = h/p$. A convenient relationship for electron diffraction is

$$\lambda/\text{Å} = \frac{12.25}{(\epsilon/\text{eV})^{1/2}} \quad \text{or} \quad \lambda/\text{pm} = \frac{1225}{(\epsilon/\text{eV})^{1/2}} \tag{16.41}$$

For an electron that has been accelerated through a typical potential difference of 30–40 kV, the wavelength is approximately 6–7 pm (0.06–0.07 Å).

The 1986 Nobel Prize in physics was shared by Ernst Ruska (1906–1988) for his invention of an electron microscope in 1931. The instrument that was the forerunner of modern electron microscopes was constructed by James Hillier (b. 1915) in 1937.

The diffraction of electron beams also provides information about macromolecular structure. An electron accelerated by 10 000 V has a velocity of 5.9×10^7 m s^{-1}, and its de Broglie wavelength (Section 11.3) is 12 pm. This is somewhat less than bond distances, and therefore it is possible to diffract an electron beam by a crystal used as a diffraction grating. Electron beams have the advantage over X rays in that they can be brought into sharp focus by appropriate arrangements of electric and magnetic fields. This has led to the development of electron microscopes, which are capable of resolving images as small as 0.5 nm in diameter. This is not quite fine enough to allow interatomic distances in solids to be measured, but valuable information about macromolecular structures has been obtained by the use of this technique.

Neutron Diffraction

Other nuclear particles also have been diffracted. Neutrons are particularly useful because they interact with unpaired electron magnetic moments and with the nuclear magnetic moments. As a result neutron diffraction is complementary to the other two methods discussed, being used to study magnetic ordering in crystals and ferromagnetism and antiferromagnetism.

A nearly monochromatic beam of neutrons is selected by a crystal monochromator from the range of wavelengths available (Figure 16.27). The wavelength of these neutrons also is governed by the de Broglie equation $\lambda = h/p$. The neutron energy is given by $\epsilon = h^2/(2M_n\lambda^2)$ where $M_n = 1.675 \times 10^{-24}$ kg is the mass of the neutron. This simplifies to

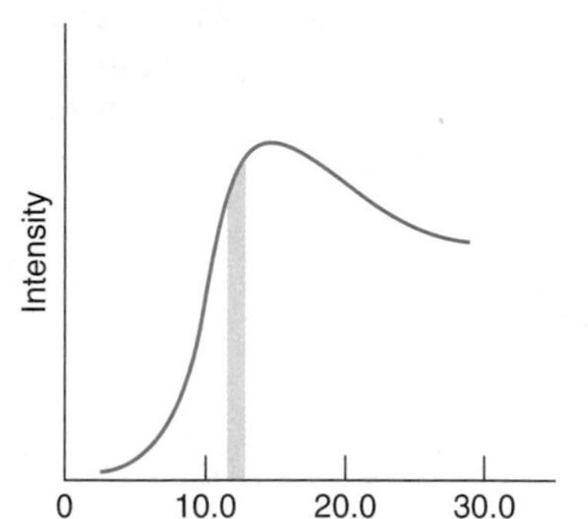

FIGURE 16.27
Intensity plotted against wavelength for a neutron beam. The beam from a reactor can have particular wavelengths selected (hatched area) by a crystal monochromator.

$$\lambda/\text{Å} = \frac{0.28}{(\epsilon/\text{eV})^{1/2}} \quad \text{or} \quad \lambda/pm = \frac{28}{(\epsilon/\text{eV})^{1/2}} \tag{16.42}$$

Neutron diffraction studies have been useful in the location of protons. Deuterium scatters neutrons very efficiently. However, hydrogen atoms, having only one electron, scatter X rays and electrons only weakly and, therefore, are not readily detected by them. In organic chemistry and in intermetallic hydrides, it is difficult, if not impossible, to locate the positions of the hydrogen atoms. However, the location of the proton nucleus can be precisely determined by deuterium substitution for a proton followed by neutron scattering.

Interpretation of X-Ray Diffraction Patterns

The diffraction patterns of cubic systems are the simplest to treat, because reflections are not always observed from all sets of planes that can pass through the

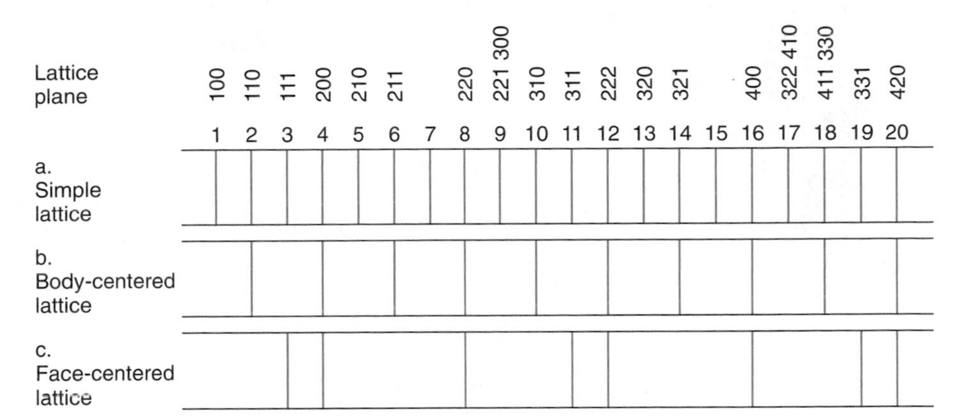

FIGURE 16.28 Powder-pattern lines expected for the three types of cubic lattices. Uniform spacing between actual lines does not occur since the spacings are modulated by the value of sin θ.

crystal. If a set of planes is placed exactly between the planes of another set, interference of the rays reflected from the two sets may cause the absence of the line from the X-ray powder pattern.

Figure 16.28 illustrates this point. Whereas a simple cubic lattice gives observed reflections from all the planes, face-centered and body-centered cubic lattices show reflections from only certain planes. Studying the results of Figure 16.28 in more depth reveals that for fcc lattices the Miller indices, (hkl), are either all even or all odd, while for bcc lattices the sum of $h + k + l$ must be even. These observations have a simple explanation in terms of the structure factors. These can be found by knowing the locations of atoms within the unit cell.

Structure Factor: Simple Cubic (sc) Lattice

In Eq. 16.38, the sum is over all the atoms in the unit cell. The location of atoms in a sc unit cell are at the corners of a cube. One corner is located at the origin, $(x_1y_1z_1) = (000)$, while the others are located at the corners where the planes meet at, $(x_ny_nz_n) = \{100\} = \{110\} = \{111\}$ for $n = 2$ to 8. The curly brackets indicate all permutations. Thus $\{100\} = (100), (010), (001)$. The phase factor in Eq. 16.38 is, therefore,

$$\exp[-i2\pi(hx_n + ky_n + lz_n)] = 1 \tag{16.43}$$

in all cases. Therefore reflection from any plane (hkl) is possible, as depicted in Figure 16.28a. Since each unit cell contains $\frac{1}{8}$ of an atom, and all the atoms are equivalent, the sc structure factor is

$$F_{sc}(hkl) = \frac{1}{8}\sum_{n=1}^{8} f_n = f_{sc} \tag{16.44}$$

Structure Factor: Body-Centered (bcc) Lattice

In addition to the corner atoms, the single body-centered atom, f_2, has coordinates $(x_2y_2z_2) = (\frac{1}{2}, \frac{1}{2}, \frac{1}{2})$, which gives

$$F_{bcc}(hkl) = f_1 + f_2 e^{-i\pi(h+k+l)}$$
$$= f_1 + f_2 \quad \text{for } h+k+l \text{ even}$$
$$= f_1 - f_2 \quad \text{for } h+k+l \text{ odd} \tag{16.45}$$

CsCl is an example of a sc system with CsCl as basis. The powder pattern, however, has strong lines when $h + k + l$ is even and weak lines when $h + k + l$ is odd. In the case of identical atoms, $f_1 = f_2$, the structure is bcc and the weak lines are systematically absent as seen in Figure 16.28b.

Structure Factor: Face-Centered Cubic (fcc) Lattice

In the fcc structure, eight atoms are located at the corners of the unit cell ($\frac{1}{8} f_1$ each) and six atoms are located in the faces of the unit cell ($\frac{1}{2} f_2$ each). The corner atoms have the same coordinates as the sc lattice. The face-centered atoms have coordinates $\{\frac{1}{2}, \frac{1}{2}, 0\}$ and $\{\frac{1}{2}, \frac{1}{2}, 1\}$, giving the fcc structure factor as

$$F_{fcc}(hkl) = [f_1 + f_2(e^{-i\pi(h+k)} + e^{-i\pi(k+l)} + e^{-i\pi(h+l)})] \tag{16.46}$$

Because the atoms in both locations are identical, $f_1 = f_2$, but we will retain the distinction for now. Since the sum of two even integers is even, and the sum of two odd integers is also even, if (hkl) are all even or all odd, the structure factor is

$$F_{fcc}(hkl) = f_1 + 3f_2 \quad \text{with } (hkl) \text{ all even or all odd.} \tag{16.47}$$

On the other hand, if two of *hkl* are even and one is odd, or one of *hkl* is odd and two even, the structure factor is found from Eq. 16.46 to be

$$F_{fcc}(hkl) = f_1 - f_2 \tag{16.48}$$

Clearly the intensity of scattering from such planes is much more reduced than for the case in Eq. 16.47. If the face-centered atoms are identical to the corner atoms, then the structure factors in the two cases are, respectively, $F_{fcc} = 4f_1$ and 0. This is illustrated in Figure 16.28c.

The origins of missing reflection planes are the presence of interleaving planes that cause interference and the relative values of the form factors, f_n. In compounds, the different atoms may cause certain reflections to have diminished intensity, sometimes almost to complete extinction. An example of this is found in a comparison of the X-ray patterns from sodium chloride and potassium chloride. Both have fcc structures and are expected to have (111) planes. See Figure 16.25. In NaCl the scattering factor for Na is much less than for Cl and the intensity of the (111) planes is slightly reduced by interference, but the number of electrons responsible for interference in potassium and chlorine is almost the same and the interference is almost complete. Potassium chloride has no (111) line and thus appears to have a simple cubic lattice unless more careful measurements are made.

The fact that the (111) planes do occur in the X-ray pattern is strong evidence that the structural units for KCl are ions, and not atoms. If potassium and chlorine occurred as neutral atoms in the crystal, they would not be isoelectronic, whereas K^+ and Cl^- are isoelectronic. Therefore the two isoelectronic ions have similar electron densities and, consequently, the form factors f_{K^+} and f_{Cl^-} are approximately equal, whereas $f_K < f_{Cl}$. The presence of the (111) plane is consistent with a sc lattice and confirms the ionic nature of the crystal.

The number of particles per unit cell may also be calculated if the density and size of the cell are known. For example, the unit cell dimension for sodium chloride is found to be 0.564 nm (5.64 Å); its density is 2.163 g cm^{-3} at 25 °C; its molar mass is 58.443 g mol^{-1}. Since the density of the crystal is the density of the unit cell, we may write

$$\text{density of cell} = D = \frac{\text{mass of cell}}{\text{volume of cell}} \tag{16.49}$$

If n is the number of chemical formula units in the cell having the mass M, Eq. 16.49 may be written as

$$D = \frac{nM}{V} = \frac{n \times \text{molar mass}}{L \times \text{volume of unit cell}} \tag{16.50}$$

Rearrangement and substitution of values for NaCl give

$$\begin{aligned} n &= 2.163(\text{g/cm}^{-3})(5.64\ \text{Å})^3 10^{24}(\text{cm}^3\ \text{Å}^{-3}) \\ &\quad \times 6.022 \times 10^{23}(\text{mol}^{-1})/58.443(\text{g mol}^{-1}) \\ &= 3.999 \end{aligned}$$

The number of formula units in a unit cell must be an integer. Therefore, there must be four sodium and four chloride ions in a unit cell.

EXAMPLE 16.9 Copper crystallizes in a cubic unit cell with a lattice constant of a = 3.6153 Å. Determine in which cubic form copper occurs.

Solution From Eq. 16.50 the volume is $a^3 = (3.6153 \times 10^{-10})^3 = 4.7253 \times 10^{-29}$. The atomic mass of copper is 63.546 g mol^{-1} and the density is 8.92×10^6 g m^{-3}. Rearrangement and substitution into Eq. 16.50 gives

$$\begin{aligned} n &= \frac{(8.92 \times 10^{-6}\ \text{g m}^{-3})(4.7253 \times 10^{-29}\ \text{m}^3)(6.022 \times 10^{23}\ \text{mol}^{-1})}{63.546\ \text{g mol}^{-1}} \\ &= 3.99 \approx 4 \end{aligned}$$

This calculation shows that there are 4 copper atoms in the unit cell. In a simple cube only $\frac{1}{8}$ of each corner atom belongs to the unit cell, thus giving a total of 1 copper atom. In the face-centered cubic cell, there are atoms in the centers of each of six faces and $\frac{1}{2}$ of each atom belongs to the unit cell. This is an additional 3 atoms, giving a total of four atoms in the unit cell. Copper is, therefore, face-centered cubic.

16.4 Theories of Solids

We have seen how X-ray diffraction techniques can give dimensional data on atomic, ionic, and molecular systems. Now we investigate how the elementary entities are held together in crystal systems. Two different approaches have been developed. Analogous to the valence-bond theory, the *bond model* of the solid state considers the crystal as an array of atoms, each atom possessing electrons used to form bonds with its neighbors. Such bonds extend over the entire crystal and may be of several types, including ionic and covalent bonds. The second method is known as the *band model* and is analogous to molecular orbital theory for molecules. In this approach the nuclei are fixed in the crystal lattice and then an electron

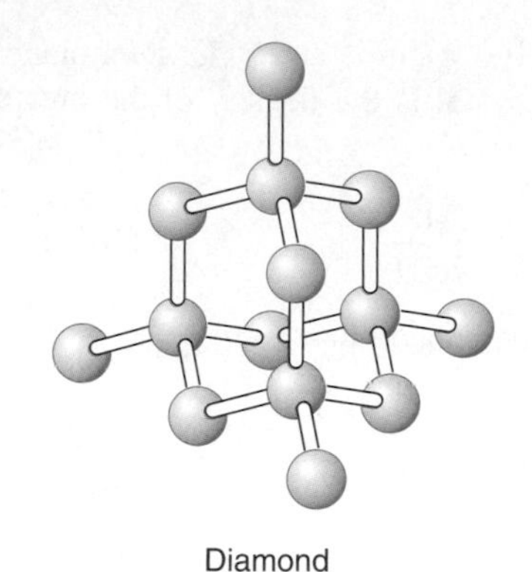

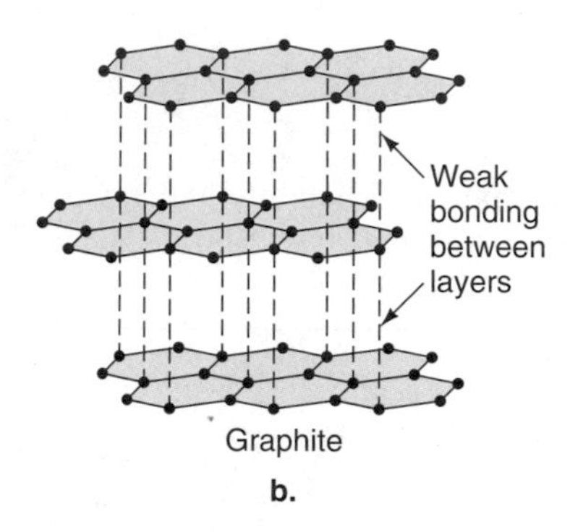

FIGURE 16.29
Two structures of carbons: (a) diamond; (b) graphite. The recently discovered forms called buckminsterfullerene that have the external shape of a soccer ball and consist of 60 or 72 carbon atoms are not shown.

"glue" is added. We will treat this method in Section 16.6. Both methods provide useful and complementary information.

Bonding in Solids

Several bond types may occur in crystalline solids depending on the nature of the entities making up the structure. Although the following classification is not always exact, it provides a useful framework from which to start.

1. *Van der Waals Bonding.* *Molecular crystals* is a name given to substances in the solid state that are held together by forces resulting from the interaction of inert atoms or essentially saturated molecules. Examples are solid nitrogen, carbon tetrachloride, and benzene.

2. *Ionic Bonding.* This type of bond occurs as a result of the interaction of charged ions. The ionic bond is both spherically symmetrical and undirected. There are no distinct ionic molecules but rather each ion is surrounded by as many oppositely charged ions as are spatially possible under the limitation of charge neutrality.

3. *Covalent Bonding.* Many crystals of nonmetals contain covalent bonds that result from the sharing of electrons between adjacent atoms. The crystal is constructed, therefore, to allow every atom to complete its electron octet. A wide variety of structures are possible. For example, carbon may exist as *diamond* shown in Figure 16.29 and may be represented by a face-centered cubic lattice with two carbon atoms at (000) and $(\frac{1}{4}\,\frac{1}{4}\,\frac{1}{4})$. Carbon may also exist as graphite, in which the planes have hexagonal symmetry.

Compounds may also crystallize in forms that allow the best use of space and bonding. For example, in the zinc blende (ZnS) structure each atom is surrounded tetrahedrally by unlike atoms. The *zinc blende structure* shown in Figure 16.30 is related to that of diamond with Zn at (000) and S at $(\frac{1}{4}\,\frac{1}{4}\,\frac{1}{4})$.

4. *Intermediate-Type Bonding.* Bonds of this type occur when a large ion, usually an anion, has its electron distribution distorted, or polarized, by an oppositely charged ion. Usually the effect is greatest for large anions and smallest for small cations since the positive charge tends to hold the electrons in place.

Compare chemical bonding in solids; Theories of solids; Covalent to ionic.

5. *Hydrogen Bonding.* These bonds play an important role in many crystal structures. Ice is an important example and is discussed in Section 17.5.

6. *Metallic Bonding.* In metals, electrons in an energy band are shared amongst the metallic nuclei. The bond is not highly directional and is spread over the entire crystal, a process that increases the stability of the crystal.

Ionic, Covalent, and van der Waals Radii

Several generalizations regarding atomic size may be drawn from the precise data on the dimensions of crystals as provided by X-ray analysis. For example, if we consider oppositely charged ions in a crystal of salt, an equilibrium position will exist between the ions when the attractive and repulsive forces just cancel. This distance may be considered as the sum of the radii **(ionic crystal radii)** of two oppositely charged ions in contact in a stable crystal.

The **covalent radius** is an effective radius for an atom when it is covalently bonded in a compound. Its value is obtained from X-ray determinations by assigning half the homonuclear bond length to each of the atoms. Radii obtained in this manner have been found to be additive to within a few hundredths of a nanometer.

Ionic and covalent radii involve the distance between directly bonded atoms. The **van der Waals radius** is one-half of the distance of closest approach of two like atoms or molecules that are not bonded. Fairly accurate drawings of the shapes of molecules may be made using van der Waals radii to show how molecules fit together.

Binding Energy of Ionic Crystals

Cohesive Energy

The stability and properties of ionic compounds are directly related to the interaction of the ions in the crystalline solid. Two points of reference are possible. If we compare the total energy of the solid with the energy of the same number of free neutral atoms at infinite distance from one another, the difference (free atom energy minus the crystal energy) is known as the **cohesive energy.** The crystal is stable if its total energy is lower in the crystal state than in the free state.

Lattice Energy

The second point of reference compares the energy of the cations and anions in a dilute gas phase, where the interactions are small, to a crystalline state, where the interactions are larger. This energy per mole is termed the **lattice energy** ΔE_c or, sometimes, the *crystal energy* of the compound. This energy represents the total potential energy of interaction of the ions at their specific locations in the given lattice structure. The lattice energy may be calculated from basic principles as well as measured experimentally.

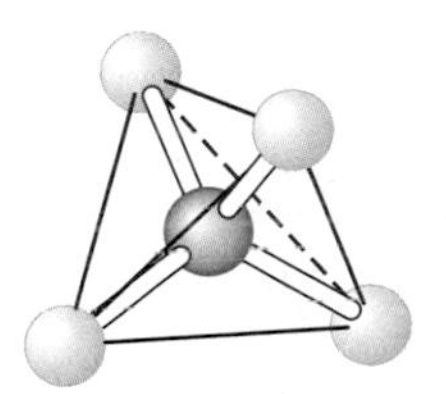

Tetrahedral unit

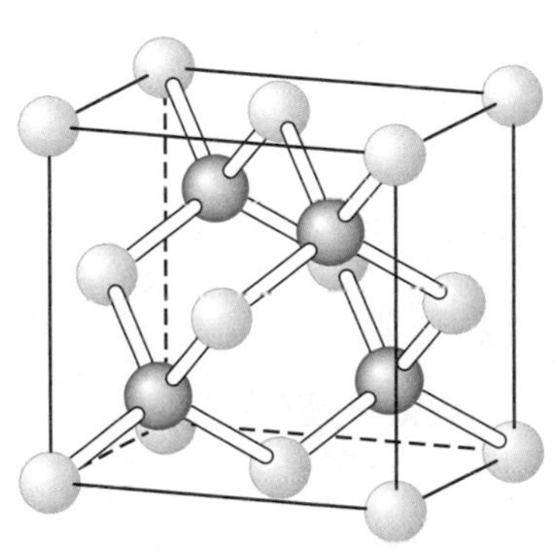

Zinc blende (cubic ZnS)

FIGURE 16.30
The zinc blende (ZnS) structure. Each sulfur atom is tetrahedrally surrounded by four zinc atoms. If all the atoms were the same, such as in carbon, the diamond structure would result.

The theoretical calculation of the lattice energy is based on two major contributions:

1. The *Electrostatic Interactions between Ions.* The attractive potential of two ions with charges $z_i e$ and $z_j e$ has the form $-z_i z_j e^2/4\pi\epsilon_0 \epsilon r$. See Section 17.3 and the discussion leading to Eq. 17.12.

2. The *van der Waals Repulsive Energy.* This is modeled by the form $be^{-r/\rho}$ and operates to fix the equilibrium distance in a stable crystal. The values of b and ρ are determined from the requirement of equilibrium in the crystal. The expression for this repulsive term is often written as B/r^n. See Eq. 17.20.

The lattice energy ΔE_c of a single ion is a sum

$$(\Delta E_c)_i = \sum_{j \neq i} E_{ij} \tag{16.51}$$

which includes all interactions involving the ith ion. The lattice energy is found by summing Eq. 16.51 over all pairs of ions and includes the two types of contribution previously noted. The factor $\frac{1}{2}$ is introduced to ensure that each pair is counted only once. Thus

$$\Delta E_c = \tfrac{1}{2} \sum_{i,j} \frac{z_i z_j e^2}{4\pi\epsilon_0 \epsilon r_{ij}} + \tfrac{1}{2} b \sum_{i,j} e^{-r_{ij}/\rho} \tag{16.52}$$

Born-Mayer Potential

where r_{ij} is the interionic distance. This is known as the *Born-Mayer potential.*

The first sum may be thought of as the interaction of a positive charge, first with all negative charges forming a sphere closest to it and then with all other positive and negative charges in concentric spheres at increasing distances. Thus the interaction consists of alternately positive and negative terms of diminishing magnitude. In effect, this relates all the interionic distances to the nearest-neighbor separation r_0 and the lattice constants d_{ij} of the crystal through the equation

$$r_{ij} = d_{ij} r_0 \tag{16.53}$$

In 1918 E. Madelung calculated sums of the first type appearing in Eq. 16.52. The major electrostatic attraction for 1 mol of the ionic crystal is the *Madelung energy* E_M:

$$E_M = \frac{L\alpha e^2}{4\pi\epsilon\epsilon_0 r_0} \tag{16.54}$$

Madelung Constant

where α is the **Madelung constant,** important in the theory of ionic crystals.[4] This constant is dimensionless and characteristic of the particular crystal structure. Substitution of Eq. 16.54 into Eq. 16.52 and minimization with respect to r_0 allows the result to be written as

$$\Delta E_c = \frac{L\alpha e^2}{4\pi\epsilon_0 r_0}\left(1 - \frac{\rho}{r_0}\right) \tag{16.55}$$

where ρ is a constant that can be evaluated from measurements of the compressibility. Consequently, a theoretical value for ΔE_c can be found.

The Born-Haber Cycle

It is appropriate to test the theoretical model just derived. Although the lattice energy ΔE_c for the reaction

$$M^+(g) + X^-(g) \rightarrow MX(c) \tag{16.56}$$

Compare the ionic bond energy of NaCl and MgO; Theories of solids; Ionic bond energy.

cannot be measured directly, it is possible to use measurable thermochemical data to arrive at a value for ΔE_c using the **Born-Haber cycle.** The symbol (c) refers to the crystalline state. We start the cycle with metallic atom M and nonmetallic molecule X_2 in the gaseous state:

$$\begin{array}{ccc} M^+(g) + X^-(g) & \xrightarrow{-\Delta E_c} & MX(c) \\ \uparrow IE \quad \uparrow -EA & & \downarrow -\Delta_f H \\ M(g) + X(g) & \underset{\frac{1}{2}D_0}{\xleftarrow{\Delta_{sub}H}} & \left\{\begin{array}{c} M(s) \\ + \\ \frac{1}{2}X_2(g) \end{array}\right\} \end{array}$$

The energy terms are $\Delta_f H$ = molar enthalpy of formation of MX(s); $\Delta_{sub}H$ = molar enthalpy of sublimation of M(s); D_0 = molar dissociation energy of Cl_2(g) into atoms; EA = electron affinity; and IE = the ionization energy of Na. (All energies must be in the same units: EA and IE are usually not in molar units.) In this cycle the system returns to its initial state and, consequently, the total energy absorbed must be zero. We may write, therefore,

$$\Delta U = 0 = -\Delta E_c - \Delta_f H + (\Delta_{sub}H + \tfrac{1}{2}D_0) + IE - EA \tag{16.57}$$

[4]See R. P. Grosso Jr., J. T. Fermann and W. J. Vining, *J. Chem. Ed.*, **78** 1198-1202 (2001).

or

$$\Delta E_c = -\Delta_f H + \Delta_{\text{sub}} H + \tfrac{1}{2} D_0 + IE - EA \tag{16.58}$$

Of the quantities on the right side of this equation accurate values for electron affinities have been most difficult to obtain. Values for IE and D_0 are obtainable from spectroscopy. Practically speaking, sufficient data are available only for comparisons within the alkali-halide crystals. Good agreement between the two methods is obtained for some crystals, but where large deviations exist, large non-ionic contributions to the lattice energy occur.

EXAMPLE 16.10 Calculate the ratio of the lattice energies for the ionic crystals MgO and LiF, making use of Figure 16.31 and the following thermodynamic data:

Process	Equation	ΔH/kJ mol^{-1}
Mg heat of sublimation	$Mg(s) \rightarrow Mg(g)$	+155
Ionization energy	$Mg(g) \rightarrow Mg^{2+}(g) + 2e^-$	+2175
Oxygen bond energy	$\frac{1}{2} O_2(g) \rightarrow O(g)$	+247
Electron affinity	$O(g) + 2e^- \rightarrow O^{2-}(g)$	+741
Formation from elements	$Mg(s) + \frac{1}{2} O_2 \rightarrow MgO(s)$	−610
Lattice energy	$Mg^{2+}(g) + O^{2-}(g) \rightarrow MgO(s)$	ΔE_c

Process	Equation	ΔH/kJ mol^{-1}
Li heat of sublimation	$Li(s) \rightarrow Li(g)$	+161
Ionization energy	$Li(g) \rightarrow Li^+(g) + e^-$	+520
Fluorine bond energy	$\frac{1}{2} F_2(g) \rightarrow F(g)$	+77
Electron affinity	$F(g) + e^- \rightarrow F^-(g)$	−328
Formation from elements	$Li(s) + \frac{1}{2} F_2 \rightarrow LiF(s)$	−617
Lattice energy	$Li^+(g) + F^-(g) \rightarrow LiF(s)$	ΔE_c

Solution An easily visualized solution is obtained by inserting the values from the tables into Figure 16.31. Then it is quickly found that the lattice energies are

$$\text{for MgO: } \Delta E_c = -(610 + 155 + 2175 + 741) = -3928 \text{ kJ mol}^{-1}$$

$$\text{for LiF: } \Delta E_c = -(161 + 520 + 77 - 328 - 617) = -1047 \text{ kJ mol}^{-1}$$

The ratio of the two is

$$\frac{\Delta E_c(\text{MgO})}{\Delta E_c(\text{LiF})} = \frac{-3928}{-1047} = 3.75$$

It is easy to understand this ratio in terms of simple electrostatics, according to which the ratio should be

$$\frac{\Delta E_c(\text{MgO})}{\Delta E_c(\text{LiF})} = \frac{\dfrac{(+2)(-2)e^2}{4\pi\epsilon_o\epsilon r_{\text{MgO}}}}{\dfrac{(+1)(-1)e^2}{4\pi\epsilon_o\epsilon r_{\text{LiF}}}} = 4$$

This leads to a value of 4 shown if the radii are the same, in reasonable agreement with experiment.

Study Figure 16.31 in detail; Theories of solids; Born-Haber cycle.

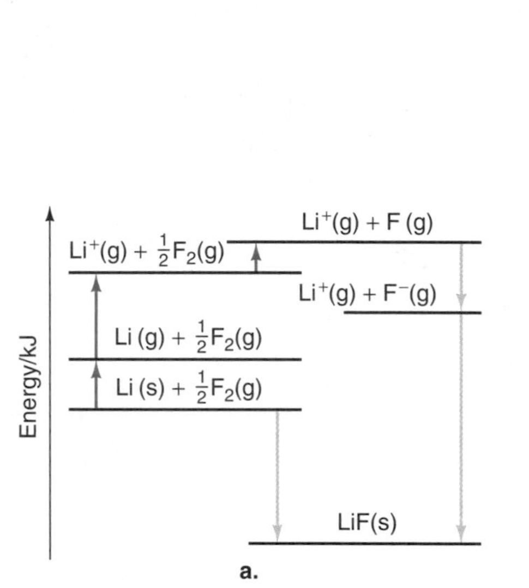

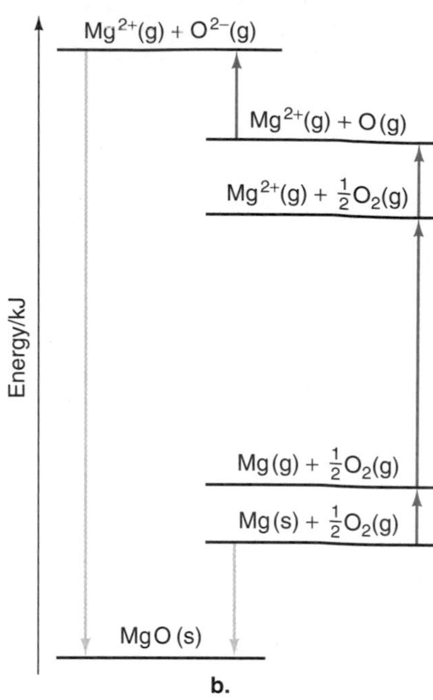

FIGURE 16.31
The Born-Haber cycle for (a) LiF and for (b) MgO.

Watch spheres pack to form a lattice; Theories of solids; Cubic close packing.

The Structure of Metals: The Closest Packing of Spheres

Metals and other crystals that do not have highly directional bonds generally achieve their lowest energy when each entity is surrounded by the greatest possible number of neighbors. The structure of metals may, therefore, be discussed in terms of the way in which the atoms are packed together.

The ordering of atoms in a metal can most easily be seen by considering the atoms as spheres of identical size. *Closest packing* is the most efficient arrangement of spheres to fill an available space. A single layer of spheres occupies space most efficiently when each sphere is arranged to have six spheres around it. (See

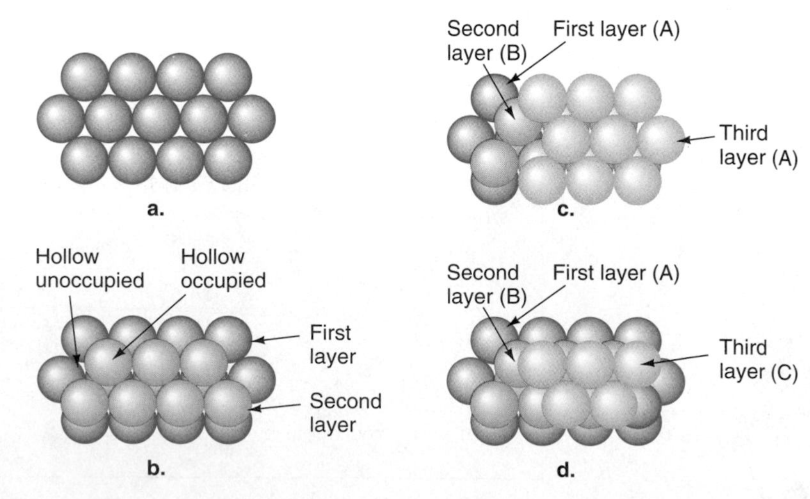

FIGURE 16.32
Formation of two closest-packed arrangements. (a) Base layer showing hexagonal symmetry. (b) Second layer superimposed on top of first showing some hollows of the first layer not occupied. (c) Third layer directly above first layer forming the hexagonal close-packed layers. (d) Third layer is set in alternate hollows forming the cubic close-packed layers.

Drag a lattice plane over another to create octahedral and tetrahedral holes; Theories of solids; Ionic solids; Lattice holes.

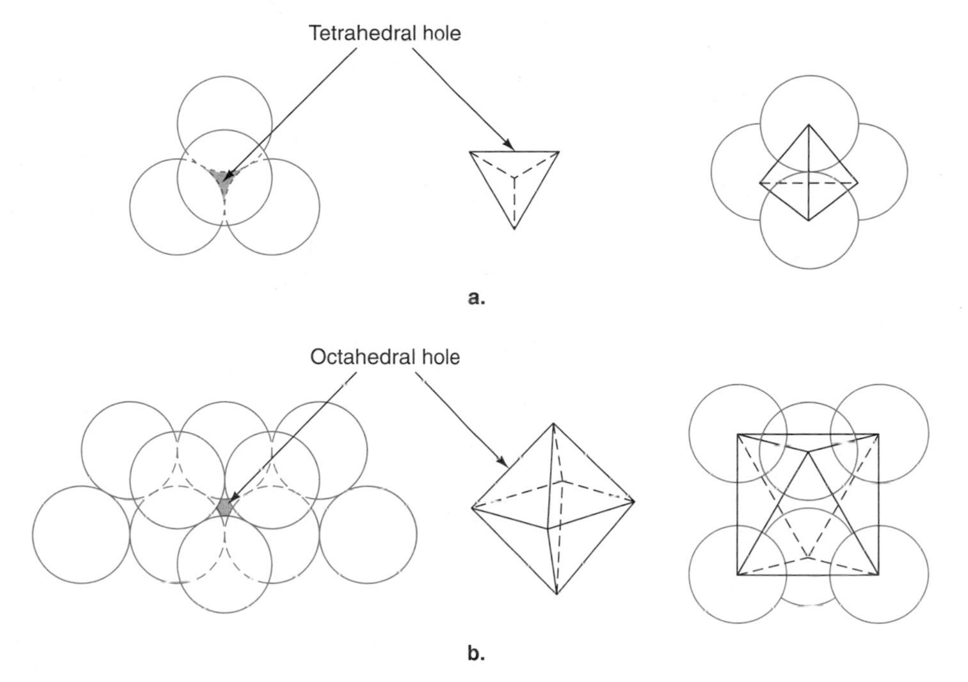

FIGURE 16.33
The formation of (a) tetrahedral and (b) octahedral holes.

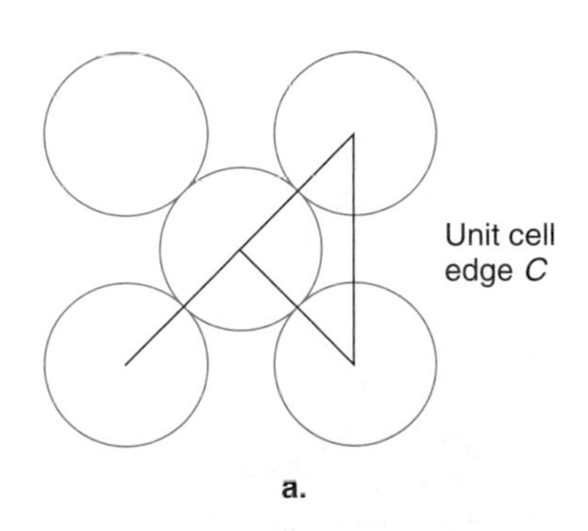

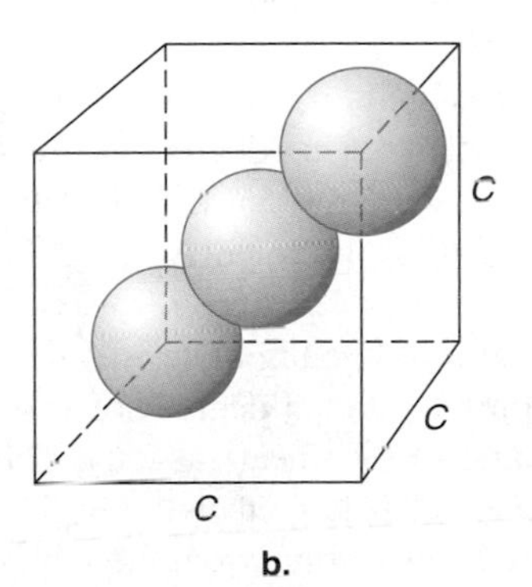

FIGURE 16.34
(a) A face-centered cubic unit cell showing the unit cell edge *C*. (b) A body-centered cubic system showing the unit cell length *C*.

Problem 16.3.) Such an arrangement is shown in Figure 16.32a. The second layer of spheres cannot occupy all the hollows in the first layer, as seen in Figure 16.32b. When the third layer is put in place, two different arrangements are possible. In Figure 16.32c the third layer is placed directly above the first layer. This results in **hexagonal close packing** and, if the order continues in the same fashion, the layer arrangement is designated *AB AB AB* . . . , where letters represent the layers. The second arrangement places the third layer over the unfilled hollows of the first layer. This results in **cubic close packing,** shown in Figure 16.32d. This is described by *ABC ABC* . . . . Both methods of packing[5] result in the same fraction of space occupied and the same number of neighbors, 12.

These two types of closest packings give rise to two types of holes (also called voids or intersticies) between layers. A **tetrahedral hole** is formed when four spheres are arranged at the corners of a tetrahedron. This occurs when a triangular void in one plane has a sphere directly over it, as shown in Figure 16.33a. If two triangular voids are joined at their bases to form a hole surrounded by six spheres, an **octahedral hole** results; the spheres are arranged on the corners of an octahedron. The tetrahedral holes are comparatively small, accommodating a sphere with a radius of 0.23 that of the surrounding layer of spheres. Octahedral holes are large enough to accept spheres with a radius of 0.41 that of the surrounding spheres. The importance of the holes is easily seen. For instance, many oxides have either hexagonal or cubic closest packing of the oxygen atom with the smaller metal atoms in the holes. In general, the properties of materials are greatly influenced by imperfections occurring in crystal lattices. An *interstitial atom* will alter the orderly arrangement of atoms and thereby cause changes in the electrical properties of the crystal.

---

[5]Both types of packing were suggested by W. Barlow, *Nature, 29,* 186, 205, 404(1883), long before the advent of X-ray analysis.

Metallic Radii

Metallic radii are calculated from the unit cell dimensions of the metal using the assumption that the metal ions are in contact in the particular structure. Cubic close-packed layers, shown in Figure 16.32d, are the (111) planes of a face-centered cubic structure. In the fcc unit cell, {100} planes form the faces and have metal-to-metal contact across the diagonal of the face, as shown in Figure 16.34a. From this figure the metal radius is calculated to be $C/(2\sqrt{2})$. Examples of metals that crystallize in the fcc structure are copper, silver, and gold.

Metals often crystallize in the body-centered cubic arrangement shown in Figure 16.34b. This arrangement is not closest packing. Since there is a larger fraction of empty space in such systems, the metals with this structure tend to be more malleable and less brittle. Examples are sodium and the low-temperature form of iron, α-Fe.

16.5 Statistical Thermodynamics of Crystals: Theories of Heat Capacities

According to classical theory the heat capacity of a solid is independent of temperature. In 1819 the French physicists Pierre Louis Dulong (1785–1838) and Alexis Thèrèse Petit (1791–1820) stated their famous law that the atoms of all simple bodies have the same heat capacity. In modern terms this law can be stated as

$$C_V \approx 3Nk_B \approx 3R \approx 25 \text{ J K}^{-1} \text{ mol}^{-1} \tag{16.59}$$

where R is the gas constant and k_B is the Boltzmann constant. Recall that the value of heat capacities, as discussed in Chapter 15, is based upon the classical idea of equipartition of energy, which is valid at high temperature. The energy of the system depends on the partition of energy among the degrees of freedom. For a simple solid, containing N atoms, there are $3N$ degrees of freedom, from which Eq. 16.59 follows. Although the law of Dulong and Petit is in good agreement with experimental values for most solid elements at room temperature, there are several exceptions: boron, silicon, and carbon (diamond) all have values that are too low. A more serious problem, however, is that all substances have heat capacities that approach zero as the temperature approaches absolute zero.

A complete treatment of the heat capacity of solids should include the motions of electrons as well as atoms. Since electrons are particles having degrees of freedom, they might be expected to contribute greatly to the heat capacity. In fact for insulators (See Section 16.6) only a negligible contribution is found; whereas in metals a small contribution is found.

If one were to consider a solid classically and treat it as one big molecule, then the three translational degrees of freedom and the three rotational degrees of freedom of the crystal would make a negligible contribution to the heat capacity. The $3N - 6$ vibrational degrees of freedom are left and since N is large then $3N - 6 = 3N$. The major contribution to the heat capacity of solids must come from the vibrational motion of the crystal. The partition function for this is given by Eqs. 15.102 and 15.103. These equations, however, assume that the vibrations are governed by simple harmonic motion. In his treatment of the heat capacities of solids, Einstein, influenced by Planck's treatment of blackbody radiation, assumed that there was only one vibrational frequency. Although this agreed with the high temperature

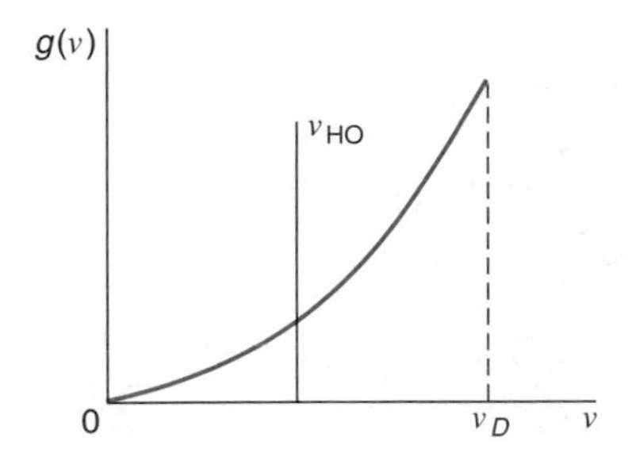

FIGURE 16.35
The distribution of frequencies for solids at low temperatures showing the single harmonic oscillator frequency assumed by Einstein, ν_{HO} and the distribution used by Debye. The cutoff of the Debye frequency at ν_D is a result of the total number of frequencies being equal to $3N$.

limit of the heat capacity of solids, it failed at low temperatures. Debye solved this problem by assuming that a distribution of vibrational frequencies that is valid at low temperatures is valid at all temperatures. We now treat these two cases.

Figure 16.35 shows the distribution of frequencies, $g(\nu)$, for the Einstein and Debye models. The Einstein model assumes one vibrational frequency where ν_{HO} is the harmonic oscillator frequency, while the Debye model assumes a distribution that gives the correct low-temperature result. He then used the same low-temperature assumption for high temperatures, but a cutoff occurs because the total number of oscillators cannot exeed $3N$.

The Einstein Model

In 1906 Einstein introduced the idea that each atom in a solid is an independent harmonic oscillator and that each has the same fundamental frequency ν. Einstein then extended Planck's quantum theory to complete his model.

The average energy of a harmonic oscillator according to Planck's theory is given by Eq. 11.33, namely

$$\bar{\epsilon} = \frac{h\nu}{e^{h\nu/k_BT} - 1} \tag{16.60}$$

For an Einstein solid consisting of N atoms, the average energy is

$$\bar{E} = 3N\frac{h\nu}{e^{h\nu/k_BT} - 1} \tag{16.61}$$

where the factor of 3 comes from the fact that each atom has three degrees of freedom, as previously discussed.

Since the heat capacity at constant volume is defined as $C_V = (\partial E/\partial T)_V$, **Einstein's heat capacity equation** is

$$C_V = 3Nk_B\left(\frac{h\nu}{k_BT}\right)^2 \frac{e^{h\nu/k_BT}}{(e^{h\nu/k_BT} - 1)^2} \tag{16.62}$$

This equation explains the breakdown of classical theory. In the limit of zero temperature, where the heat capacity must approach zero, Eq. 16.62 may be written as

$$C_V \approx 3Nk_B\left(\frac{h\nu}{k_BT}\right)^2 e^{-h\nu/k_BT} \qquad (T \to 0) \tag{16.63}$$

This approaches zero as $T \to 0$. However, it is known that C_V approaches zero as T^3, and Eq. 16.63 does not show this behavior. However, Eq. 16.62 does have a limiting value of $3R$ as $T \to \infty$, in accordance with experiment.

EXAMPLE 16.11 Show that the limiting value of Einstein's heat capacity equation gives the value from Eq. 16.59.

Solution For the classical expression to apply, the temperature must be high, that is $T \gg \frac{h\nu}{k_B}$ in Eq. 16.62. Then $\frac{h\nu}{k_BT} \ll 1$ and the exponential term can be expanded:

$$e^{h\nu/k_BT} = 1 + h\nu/k_BT$$

plus other smaller terms which are neglected. Substitution into Eq. 16.62 for $N = L$ particles gives

$$C_{V,m} \cong 3R(h\nu/k_BT)^2(1 + h\nu/k_BT)/(1 + h\nu/k_BT - 1)^2$$
$$\cong 3R = 3 \times 8.3145 = 24.944 \text{ J mol}^{-1} \text{ K}^{-1}$$

which is the classical value.

The Debye Model

When it was recognized that the heat capacity approached zero too rapidly according to the Einstein model, Peter Debye in 1912 introduced his model of heat capacities. He considered a solid to be a three-dimensional isotropic continuum in which elastic waves could be excited with a continuous range of frequencies. This continuous range of frequencies was cut off at a particular frequency ν_D, which was determined by the number of oscillators present. These vibrations are characterized by a distribution function $g(\nu)$ so defined that $g(\nu)\, d\nu$ gives the number of oscillators in the interval between ν and $\nu + d\nu$.

The logarithm of the partition function for this system may be written as

$$\ln Q = \sum_{i=1}^{i=\nu_D} g(\nu_i) \ln q(\nu_i) \qquad (16.64)$$

where $q(\nu_i)$ is the partition function of the harmonic oscillator defined by Eq. 15.103. Since there is a continuum of frequencies in this case, the summation may be replaced by an integral. Using this latter expression, Debye found the vibrational energy of a solid to be

$$E = \int_0^\infty g(\nu) \frac{h\nu}{e^{h\nu/k_BT} - 1}\, d\nu \qquad (16.65)$$

In the case of the Einstein solid, the distribution of frequencies is simply a Dirac delta function, $g(\nu) = \delta(\nu - \nu_{HO})$, which gives Eq. 16.62. In contrast, Debye used the Rayleigh-Jeans relation (Section 11.1) for the distribution function, $g(\nu) = c\nu^2$, where c is a constant. Since the number of oscillators is fixed at $3N$, there is a cutoff at a maximum frequency called the Debye frequency ν_D; see Figure 16.35. This not only gives the shape and cutoff frequency, but allows the constant to be determined

$$\int_0^\infty g(\nu)\, d\nu = c \int_0^{\nu_D} \nu^2\, d\nu = 3N = \frac{c\nu_D^3}{3} \qquad (16.66)$$

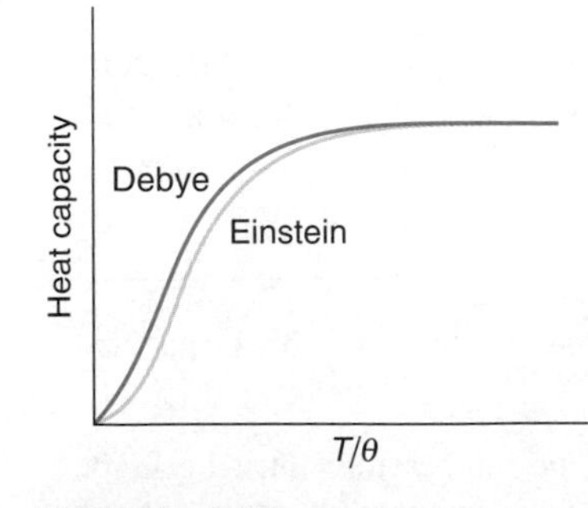

FIGURE 16.36
A comparison of the heat capacity as calculated by the Debye and Einstein theories. Experimental points fit the Debye curve fairly well.

leading to the Debye distribution of oscillators as

$$g(\nu) = \frac{9N\nu^2}{\nu_D^3} \quad \text{for } \nu < \nu_D$$
$$g(\nu) = 0 \qquad \text{for } \nu > \nu_D \qquad (16.67)$$

and it is this function that is plotted in Figure 16.35. Using this directly in Eq. 16.65 leads to

$$E = 9N\left(\frac{k_BT}{h\nu_D}\right)^3 k_BT \int_0^{x_m} \frac{x^3}{e^x - 1}\, dx \qquad (16.68)$$

where $x = h\nu/k_BT$ and $x_m = h\nu_D/k_BT$.

The heat capacity is then

$$C_V = \left(\frac{\partial E}{\partial T}\right)_V = 9Nk_B\left(\frac{T}{\Theta_D}\right)^3 \int_0^{x_m} \frac{x^4 e^x}{(e^x - 1)^2}\,dx \tag{16.69}$$

where Θ_D, known as the **Debye temperature,** is defined by

$$\Theta_D = \frac{h\nu_D}{k_B} \tag{16.70}$$

and is independent of temperature. The Debye temperature is characteristic of a particular substance. A few values of Θ_D are 150 K for Na, 315 K for Cu, and 1860 K for C.

The Debye model fits the experimental low-temperature heat capacities better than the Einstein model. At temperatures much less than the Debye temperature, Eq. 16.69 reduces to

$$C_V = \frac{12}{5}\pi^4 Nk\left(\frac{T}{\Theta_D}\right)^3 = 1.944 \times 10^3\left(\frac{T}{\Theta_D}\right)^3 \text{ J mol}^{-1}\text{ K}^{-1} \tag{16.71}$$

This expression of the **Debye T^3 law** should hold at $T < \Theta_D/10$. For $T > \Theta_D/10$, $C_V/\text{J K}^{-1}\text{ mol}^{-1} = 464.4(T/\Theta_D)^3$. A comparison plot of the heat capacity curves from the Debye and Einstein theories is shown in Figure 16.36.

Fermi-Dirac Statistics

In the electron theory of metals each atom gives one valence electron to the electron gas and these electrons are completely mobile. If there are N electrons, they should contribute $\frac{3}{2}Nk_B$ to the normal heat capacity of $\frac{3}{2}Nk_B$, but the observed electronic contribution at room temperature is generally less than 0.01 or 1% of this value. We have just seen that the Debye theory fairly accurately describes the heat capacity without including electrons. How can the electrons not contribute to the heat capacity? This question posed a difficulty for the early development of the theory and was resolved by application of quantum statistics and its consequences.

Visualization of the Quantum Statistics Function

Quantum statistics not only provides a means to understand why electrons make a negligible contribution to the heat capacity of solids, but also forms the basis for understanding electrical conductivity in solids. In the following subsection, the Fermi-Dirac and Bose-Einstein statistics are explained along with the distribution functions that accompany them. But to understand this we must have a mental picture of the way these different statistics arise.

Figure 16.37 shows how bands are formed in metals. If we assume for now that a metal is one large molecule, the quantum states arise as densely packed energy

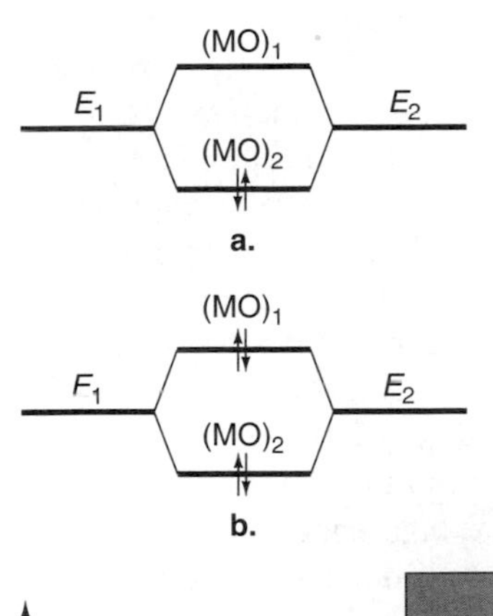

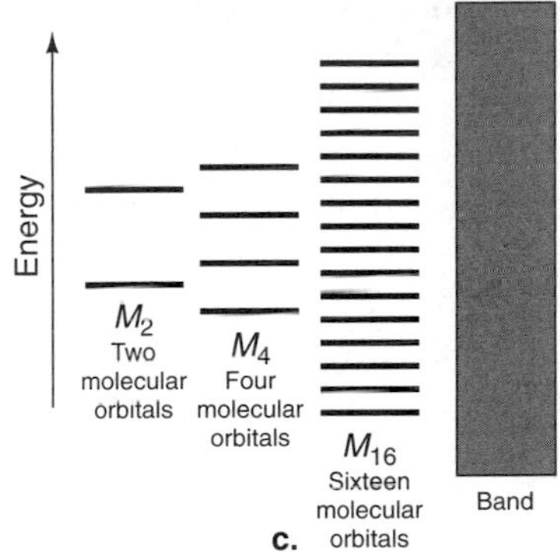

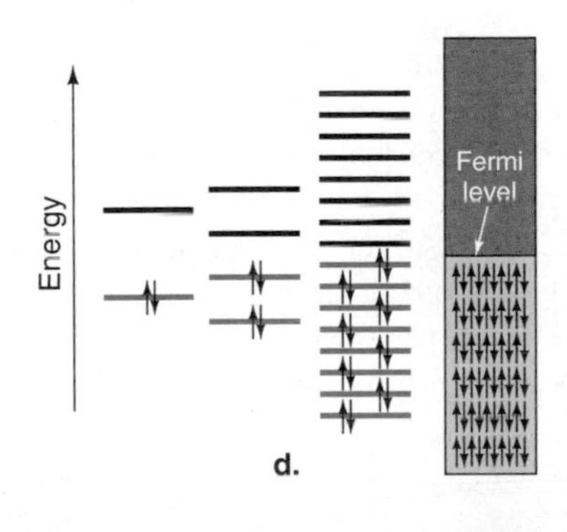

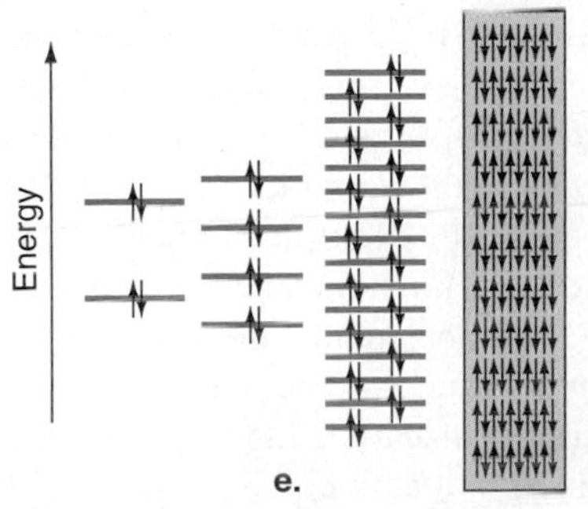

FIGURE 16.37
(a) Simple MO theory of two levels showing the discrete splitting and the half-filling of the levels by two electrons. (b) Same as (a) except now the four electrons fill the two levels. (c) The formation of bands by use of more and more atomic orbitals to produce MO's until the energy splittings have merged into a virtual continuum. (d) Since the number of electrons is only sufficient to half-fill the MO energy levels, only half the band is filled. The level of filling is called the Fermi level, as indicated. (e) The same as (d) but now there are sufficient electrons to fill the band completely. In all cases the Pauli exclusion principle is obeyed.

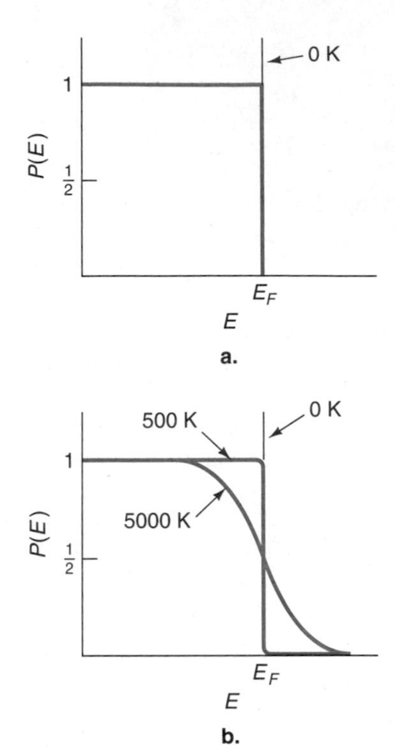

FIGURE 16.38
(a) The Fermi-Dirac distribution function plotted for 0 K. Note the sharp step at E_F. (b) The Fermi-Dirac distribution function plotted for 500 K and 5000 K for the case

$$T_F = \frac{E_F}{k_B} = 50\ 000\ \text{K}$$

Notice that even at 5000 K the probability of occupancy of higher energy states is small.

levels, called *bands*, along with other regions where no states exist, called *gaps*. In spite of the fact that the energy states in the bands are close together, they are still filled, according to the Pauli principle, with two electrons per level. Figure 16.37 shows electrons partially filling a band at absolute zero of temperature. That is, the electrons are in their ground state and, at $T = 0$ K, electrons have no energy to jump into the next highest level. This gives a distribution of the electrons at $T = 0$ K as a step function, shown in Figure 16.38a. In fact, this is the Fermi-Dirac distribution function at $T = 0$ K. The energy of the electrons in the highest energy level at absolute zero is called the Fermi energy and the highest occupied energy level is called the Fermi level. Figure 16.38b shows at E_F that the probability of occupancy of the levels is $P(E) = \frac{1}{2}$, (only $\frac{1}{2}$ of the available states are fully occupied).

Figure 16.39 shows what might be expected at higher temperatures. Only electrons in the highest energy levels can be promoted above the Fermi energy, and the Fermi-Dirac distribution function changes from the step function to a smoothing out of the step as shown in Figure 16.38a. Note that most of the electrons are unaffected by increasing the temperature. Eventually, at high enough temperatures the Fermi-Dirac distribution function does approach a Boltzmann distribution, but long before this happens, the metal would be vaporized. For a metal, the Fermi-Dirac distribution function at room temperature is not much different from that at $T = 0$ K. As far are metals are concerned, a low-temperature approximation at room temperature is better than the high-temperature approximations we have used before.

Quite the opposite is true for bosons. While electrons are fermions, the most relevant boson for us is the photon. For bosons, there is no restriction as to how many can exist in an energy level. At $T = 0$ K, all the bosons occupy the ground state.

Quantum Statistics

Now that we can visualize how electrons fill the quantum states in solids, we will discuss in more detail the quantum statistics and the corresponding distribution functions. This is in preparation for the treatment of electrical conductivity and to explain why electrons make a negligible contribution to the heat capacities of solids. Two concepts, which have fundamental relevance to the theory of solids, especially electrical conductivity, are discussed. One concept is that electrons are fermions and therefore must obey the Pauli exclusion principle. This means that electrons must fill the quantum states of solids according to the same rules that apply to the electronic structure of atoms.

As explained in more detail and in Section 16.6, a solid, considered as one large molecule, does not display a discrete set of energy levels like an atom or molecule, but has many closely spaced energy levels that form bands. Although the energy levels are densely distributed in bands, they are nevertheless quantum states that can accommodate at most two electrons each. As a result, at $T = 0$ K electrons fill the levels according to the rules up to and only up to the Fermi level.

Such considerations force us to reexamine the Pauli exclusion principle (see Section 11.12), leading to a surprising consequence. The Boltzmann distribution function does not hold for electrons (fermions). Nor, incidentally, does it hold for photons (bosons). This leads to two new, different distribution functions. One is for particles that are fermions, called the Fermi-Dirac distribution function, and the other is for bosons, called the Bose-Einstein distribution function.

Plot the density of states, Figure 16.38, for Na, K, and Cu as a function of T; Statistical thermodynamics; Fermi-Dirac distribution.

The origin of both of these functions is based on the symmetry requirements of the wave functions, exactly as discussed in Section 11.12. That is, the wave function for a fermion must be antisymmetric to particle interchange and for a boson it must be antisymmetric. Recall that particles, such as electrons, are indistinguishable from one another; that is, the results of a calculation of the energies must be the same whether we designate the two electrons as either 1 and 2 or 2 and 1. The condition is that if the permutation operator $\hat{P}_{12}$ operates on a wave function $\psi(1,2)$, then

$$\hat{P}_{12}\,\psi(1,2) = \pm\psi(2,1) \tag{16.72}$$

The Fermi-Dirac distribution function was named in honor of Enrico Fermi (1901–1954) and Paul Dirac (1902–1984).

The Bose-Einstein distribution function was named in honor of Einstein and the Indian physicist Satyendra Nath Bose (1894–1974).

where the plus sign holds for bosons and the minus sign for fermions. We have seen that we can make a set of wave functions that describe electrons as antisymmetric by introducing Slater determinants. These symmetry requirements, Eqs. 11.232 to 11.238, lead immediately to the Fermi-Dirac distribution function or to the Bose-Einstein distribution function.

The antisymmetry condition for fermions means that not all wave functions that are available are acceptable. Only those that are antisymmetric to particle interchange are allowed. Fermi-Dirac statistics are used to count these. Similarly, Bose-Einstein statistics are used to count all the allowed symmetric wave functions for bosons. Quantum statistics takes both cases into account.

There are, moreover, two more important aspects of quantum statistics. First, the number of energy states available to a system plays a role. Both the Fermi-Dirac and Bose-Einstein distribution functions approach the Boltzmann distribution function as the energy increases. As this happens, more and more quantum states are available to the particles, and quantum statistics becomes irrelevant. However, the band structure of solids does not have enough energy levels at lower temperatures for there to be a Boltzmann distribution of electrons. At temperatures where there would be sufficient energy levels available, all metals would be vaporized! As a result, at temperatures at which we study solids, for metals below their melting points, say 2000 K or less, the Fermi-Dirac distribution function is similar to that at $T = 0$ K (see Figure 16.38).

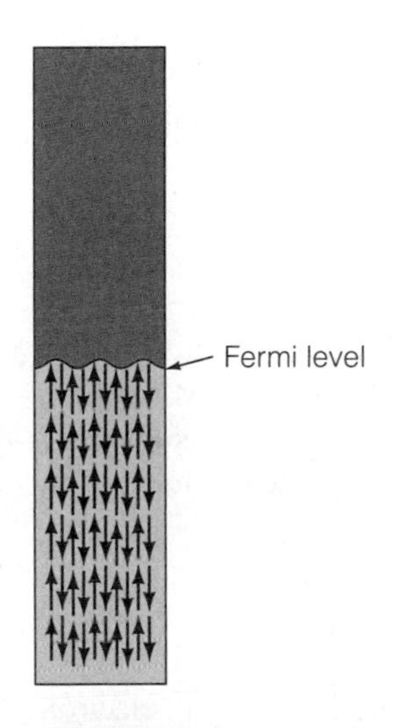

FIGURE 16.39
A half-filled band showing the Fermi energy level. Obeying the Pauli exculsion principle, the electrons fill up to this level at absolute zero. At higher temperatures, some of the electrons in the upper levels can be promoted to levels above the Fermi energy level. The distribution is sometimes referred to as the "sea of electrons" model.

To put the previous ideas into a more mathematical framework, the derivation of the Boltzmann distribution function, Eq. 15.35, is modified. In that case, the sum over states was performed using the weighting factor g_i. This accounts for the degeneracy of each energy state E_i. In order to arrive at the two distribution functions that include quantum statistics, all that is necessary is to replace g_i with a weighting factor that accommodates Fermi-Dirac statistics, or one that accommodates Bose-Einstein statistics. Without presenting the details here,[6] distribution functions for the two types of quantum statistics have a simple, yet important, difference:

$$P(E) = \frac{1}{e^{(E-E_F)/k_BT} \pm 1} \tag{16.73}$$

The plus sign applies to Fermi-Dirac statistics and the minus sign applies to Bose-Einstein statistics. The energy E_F is the Fermi energy.

[6]Terrence L. Hill, *An Introduction to Statistical Thermodynamics*, Reading, MA: Addison-Wesley, 1962; Dover reprint, 1982.

EXAMPLE 16.12 Use the Fermi-Dirac distribution function to find the chance that an energy level E_i is occupied at $T = 0$ K in two cases: first, if the energy is less than that of the Fermi energy, $E_i < E_F$, and second, if the energy is greater than that of the Fermi energy, $E_i > E_F$.

Solution *Case 1:* $E_i < E_F$, as $T \to 0$, for a single energy level, E_i (Eq. 16.73) becomes

$$N_i = \lim_{T \to 0} \frac{1}{e^{(-|E_i - E_F|)/k_B T} + 1} = 1$$

which means that there is a 100% chance that the state is occupied if it lies below the Fermi energy level at $T = 0$.

Case 2: $E_i > E_F$, as $T \to 0$, for a single energy level, E_i (Eq. 16.73) becomes

$$N_i = \lim_{T \to 0} \frac{1}{e^{(+|E_i - E_F|)/k_B T} + 1} = 0$$

which means that there is a 0% chance that the state is occupied if it lies above the Fermi energy level at $T = 0$.

The Fermi-Dirac distribution at $T = 0$ K is a step function, as shown in Figure 16.38a. At higher temperatures, only the electrons lying close to the Fermi energy can be promoted to energy states with $E_i > E_F$. In other words, few of the electrons have any freedom to move into states that are available, as most are restricted to energy levels deeper within the bands. The immediate consequence is that since few electrons are free and able to store energy, they make a negligible contribution to the heat capacity. In the following section, this contribution is calculated.

Determination of the Fermi Energy

Since the Fermi energy is the energy of the highest energy level that is filled by the available electrons in a solid (at $T = 0$ K), to determine this energy we need to know the splittings of the energy levels and then distribute the electrons amongst these levels according to the Pauli principle. The standard approach is to treat an electron as a free particle in a box so that it has energy

$$E = \frac{h^2}{8\,mL^3}(n_x^2 + n_y^2 + n_z^2) \tag{16.74}$$

with the n's being integers with values 1 to ∞. Here m is the mass of the electron and L is the length of the crystal, which therefore has volume of $V = L^3$. As the values of the quantum numbers, n_{x_i}, n_{y_i}, n_{z_i} vary and determine the energy levels available to the electrons, it is found that they are closely spaced. We have encountered this situation earlier in dealing with the translational degrees of freedom and their contribution to the partition function (see Section 15.5). In that case the n's are treated as continuous variables, and sums are replaced by integral. Applying the same idea here, consider that the term in Eq. 16.74 determines a sphere in the n_{x_i}, n_{y_i}, n_{z_i} directions with radius R given by

$$R = (n_x^2 + n_y^2 + n_z^2) \tag{16.75}$$

The sphere is bigger the larger the quantum number, and the states with energy less than E are all contained within the sphere. The volume of the sphere is, of course, $V = \frac{4}{3}\pi R^3$ but the n's are restricted to being positive. Therefore only $\frac{1}{8}$ of the volume of the sphere (one quadrant) is retained, and so the number of states with energy less than E is $\frac{1}{8} \times \frac{4}{3}\pi R^3$. Since only two electrons can occupy a given level, the number of electrons found below energy E is

$$N(E) = 2 \times \frac{1}{8} \times \frac{4}{3}\pi R^3 = \frac{\pi}{3}\left(\frac{8mE}{h^2}\right)^{3/2} V \tag{16.76}$$

where the last equality uses Eqs. 16.74 and 16.75 with $L^3 = V$.

Contained in this sphere, therefore, are many possible quantum states determined by all possible values of the n's with two electrons per level. If we can find the radius that corresponds to just filling all these states with electrons, we will have found the Fermi energy. Rather than knowing the number of states with energy less than E, it is more useful to know the density of states, $g(E)$, between E and dE. This is given by

Plot Figure 16.40 as a function of T for Na, K, and Ca; Statistical thermodynamics; Density of states distribution.

$$g(E)\,dE = \frac{dN(E)}{dE}dE = \frac{8\sqrt{2}\pi m^{3/2}}{h^3}VE^{1/2}\,dE \tag{16.77}$$

The number of electrons $n(E)$ that lie between energy E and dE is found by multiplying the density of states $g(E)$ by the Fermi-Dirac distribution function (Eq. 16.73), giving

$$n(E)\,dE = \frac{DE^{1/2}}{e^{(E - E_F)/k_BT}} + 1\,dE \tag{16.78}$$

where the constant $D = 8\sqrt{2}\pi m^{3/2} / h^3$.

A plot of $n(E)$ against E is shown in Figure 16.40. The Fermi energy can be calculated by requiring that the total number of electrons per unit volume be n. Since the energy of the electrons must be between 0 and ∞,

$$n = \int_0^\infty n(E)\,dE = D\int_0^\infty \frac{E^{1/2}\,dE}{e^{(E - E_F)/k_BT} + 1} \tag{16.79}$$

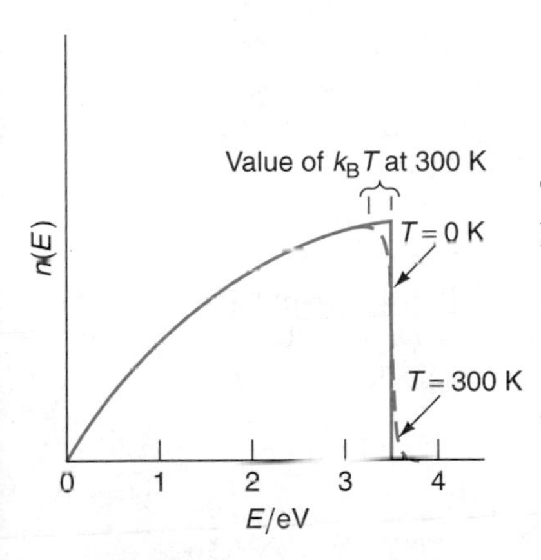

FIGURE 16.40
A plot of $n(E)$, the electron distribution function, against energy E, for a metal at 0 K (firm line) and 300 K (dashed line). In this example the Fermi energy E_F is 3.5 eV.

Then, at $T = 0$ K, as shown in Example 16.12, no states are occupied above the Fermi energy. Eq. 16.79 thus becomes

$$n = D\int_0^\infty E^{1/2}\,dE = D\int_0^{E_F} E^{1/2}\,dE = \frac{2}{3}DE_F^{3/2} \tag{16.80}$$

With the expression $D = 8\sqrt{2}\pi m^{3/2}/h^3$ this equation becomes

$$E_F = \frac{h^2}{2m}\left(\frac{3n}{8\pi}\right)^{2/3} \tag{16.81}$$

The $n^{2/3}$ term shows that the Fermi energy only gradually increases with increasing electron concentration. The $\frac{2}{3}$ power is a consequence of the double occupancy of each level. The Fermi energy for several metals along with several other parameters are listed in Table 16.2.

TABLE 16.2 Fermi Energy and Fermi Temperature for Several Metals

Metal	Electron Concentration $\times 10^{-28}/\text{m}^{-3}$	Fermi Energy, E_F/eV	Fermi Temperature, T_F/K
Na	2.65	3.23	3.75×10^4
K	1.40	2.12	2.46×10^4
Cu	8.49	7.05	8.12×10^4
Ag	5.85	5.48	6.36×10^4
Au	5.90	5.53	6.41×10^4

A similar calculation for the average energy, E_{av}, obtained by modifying Eq. 16.79, gives

$$E_{\text{av}} = D\int_0^{\infty} \frac{E^{3/2}}{e^{(E - E_{\text{F}})/k_{\text{B}}T} + 1}\, dE \tag{16.82}$$

which can be evaluated at $T = 0$ K to give

$$E_{\text{av}} = D\int_0^{E_{\text{F}}} E^{3/2}\, dE = \frac{2}{5}DE_{\text{F}}^{5/2} = \frac{3}{5}\, nE_{\text{F}} \tag{16.83}$$

The last equality makes use of Eq. 16.80. From Table 16.2 a typical value for the Fermi energy is about 5 eV for metals, and since the average energy of an electron is $\frac{3}{5}\,E_{\text{F}}$, the average energy of the electrons is less than the Fermi energy.

EXAMPLE 16.13 Give a rough approximation of the fraction of electrons that are excited from below E_{F} to above E_{F} when Ag is heated from 0 K to 300 K.

Solution Only electrons within a range of $k_{\text{B}}T$ from E_{F} are affected by the temperature change. Therefore, electrons with energies of $k_{\text{B}}T \approx 4 \times 10^{-21}$ J = 0.025 eV, on either side of E_{F}, are affected. The fraction is roughly given by $k_{\text{B}}T/E_{\text{F}}$, although a more careful analysis gives $9k_{\text{B}}T/16E_{\text{F}}$. For silver with E_{F} = 5.48 eV, using the former value gives $k_{\text{B}}T/E_{\text{F}} = 0.025/5.48 = 0.0046$ or 0.46%. Thus, only a small portion of the total electrons contribute to the conduction.

The contribution of electrons to the heat capacity of metals can now be calculated using the previous results and the Fermi-Dirac distribution function. Since we have established that room temperature corresponds to a low temperature for the distribution of electrons in solids, the Fermi-Dirac distribution function can be expanded and the first term is linear in temperature. This adds a term to Eq. 16.69 to give

$$C_V = 9R\left(\frac{T}{\Theta_D}\right)^3 \int_0^{x_m} \frac{x^4 e^x}{(e^x - 1)^2}\, dx + \gamma T \tag{16.84}$$

where the contribution, γ, from electrons to the molar heat capacity is

$$\gamma = R\,\frac{k_{\text{B}}\pi^2}{2E_{\text{F}}} \tag{16.85}$$

and R is the gas constant.

Values of γ range from 0.00138 J mol^{-1} K^{-2} for sodium to 0.0092 J mol^{-1} K^{-2} for the γ form of manganese.

The electronic contribution to heat capacity is important only at very low temperatures. Using Eq. 16.71 with the added γT term of Eq. 16.84 gives

$$C_V = 1.944 \times 10^3 \left(\frac{T}{\Theta_D}\right)^3 + \gamma T \tag{16.86}$$

Debye-Sommerfeld Equation

and is known as the **Debye-Sommerfeld equation.**[7] This equation allows an evaluation of Θ_D and γ if we divide through by T,

$$\frac{C_V}{T} = \frac{1.944 \times 10^3\, T^2}{\Theta_D^3} + \gamma \tag{16.87}$$

where a plot of C_V/T against T^2 gives a straight line with slope = $1.994 \times 10^3/\Theta_D^3$ and intercept γ.

EXAMPLE 16.14 From the data provided in the chapter, calculate the heat capacity of Na at 10 K.

Solution Use Eq. 16.87 and substitute the temperature and values of the constants.

$$C_V = 1.944 \times 10^3 \left(\frac{10}{150}\right)^3 \text{ J mol}^{-1}\text{ K}^{-1} + 0.00138 \text{ J mol}^{-1}\text{ K}^{-2}(10\text{ K})$$

$$= 0.5760 + 0.0138 = 0.5898 \text{ J mol}^{-1}\text{ K}^{-1}$$

In summary, electrons are not the only particles that follow Fermi-Dirac statistics. Electrons, protons, neutrons, and any species with an odd number of electrons, protons, and neutrons are called **fermions,** the half-integral spin being described by Fermi-Dirac statistics. The use of these statistics requires the use of antisymmetric wave functions, which is the same as saying that the Pauli principle applies.

The **Bose-Einstein statistics** requires the use of symmetric wave functions. Particles such as photons and species with an integral spin are called **bosons.**

16.6 Electrical Conductivity in Solids

Important information about the nature of solids is provided by their electrical conductivities. **Metals** are good conductors of electricity, their conductivities usually lying in the range 10^6 to 10^8 S m^{-1}(= Ω^{-1} m^{-1}) where S is the seimens; the values decrease with increasing temperature. Many other solids, particularly those in which there is covalent bonding, have much lower conductivities, ranging from 10^{-8} to 10^{-20} S m^{-1} and are referred to as **insulators.** There are also some materials having rather special and unusual electrical properties.

Semiconductors are solids that have conductivities intermediate between those of the typical metals and typical insulators. The conductivities increase exponentially with increasing temperature, in contrast to metals whose conductivities decrease linearly with increasing T. Many semiconductors are crystalline materials,

[7]Constants for the Debye-Sommerfeld equation, for a variety of elements, are to be found in the *Handbook of Physics and Chemistry* (The CRC Handbook).

but since 1980 there have been many important technical advances, particularly with amorphous materials, many of them involving silicon. Semiconductors have many practical applications as evidenced by their fundamental role in computer technology, diodes, and radiation detectors.

When metals and certain other solids are cooled to very low temperatures, their conductivities become exceedingly high; their resistivities, in fact, become essentially zero. This effect is referred to as **superconductivity.** The value of the critical temperature T_c at which the resistivity becomes zero varies considerably with the metal. For aluminum it is 1.2 K, for lead it is 7.2 K, and for the intermetallic compound Nb_3Sn, $T_c = 18$ K. A few oxides are superconductors, also with low T_c values. Until 1986 the highest known value for T_c was 23 K, found with Nb_3Ge. Materials are now known with T_c value above 138 K, and they are already finding many applications both in research and technology. They have been used, for example, in highly sensitive magnetic field detectors, and in electrical storage devices. Further discussion of superconductors is found under a separate heading on page 890.

The remainder of this section is concerned with theories of solids which provide some explanation for the wide range of conductivity behavior that is observed, and with the other properties of solids.

Metals: The Free-Electron Theory

Metals have smaller cohesive energies than most ionic or covalent compounds. Besides having high electrical conductivities, they also have high thermal conductivities. The first attempt to explain these properties was the *free-electron model* of metals, developed in 1902–1904 independently by the German physicist Paul Karl Ludwig Drude (1863–1906) and the Dutch physicist Hendrik Antoon Lorentz (1853–1928). According to the free-electron theory, the valence electrons (conduction electrons) move freely, like the molecules in an ideal gas, about lattice points at which positive ions are fixed. In an electric field, $\boldsymbol{E}$, the electrons easily move along the potential gradient, and the result is an electric current.

Drude and Lorentz attempted to explain the electrical conductivity from a classical approach using the equations of motion of a particle in an electric field $\boldsymbol{E}$. The drift velocity v_d of the electrons is the average velocity attained between collisions, and the relaxation time τ gives a measure of the time between collisions. At constant electric field the acceleration of an electron is $\boldsymbol{a} = v_d/\tau = -e\boldsymbol{E}/m_e$, and therefore

$$v_d = -\frac{e\boldsymbol{E}\tau}{m_e} \tag{16.88}$$

Since the electric current density J is the charge on N electrons per unit volume, $-Ne$, multiplied by their velocity, v_d, we may write

$$J = -Nev_d \tag{16.89}$$

Combining these equations gives

$$J = \frac{Ne^2}{m_e}\tau\boldsymbol{E} = \sigma\boldsymbol{E} \tag{16.90}$$

where σ is the electrical conductivity. This expression is consistent with Ohm's law. The drift velocity per unit electric field is known as the mobility μ:

TABLE 16.2 Electrical Conductivities of Some Metals at 298 K.

Atom	σ 10^5(ohm cm)$^{-1}$	Atom	σ 10^5(ohm cm)$^{-1}$
Li	1.06	Rb	0.76
Be	2.70	Zr	0.23
Na	2.05	Mo	1.83
Mg	2.23	Rh	2.08
Al	3.69	Pd	0.932
K	1.35	Ag	6.18
Ti	0.23	Cd	1.38
Cr	0.794	Ir	1.96
Mn	0.0694	Pt	0.93
Fe	1.013	Au	4.43
Co	1.72	Hg	0.10
Ni	1.40	Pb	0.474
Cu	5.84	Gd	0.070
Zn	1.66	U	0.39
Ga	0.67	Np	0.085

Values vary considerably with temperature, especially at lower temperatures if phase changes occur. Below 50 k, the values are sensitive to impurities.

$$\mu = \frac{v_d}{E} = \left|-\frac{e\tau}{m_e}\right| \tag{16.91}$$

and is related to the conductivity σ by

$$\sigma = Ne\mu = Ne^2\,\tau/me \tag{16.92}$$

Although this theory explains some aspects of electrical conductivity in metals, it fails when applied to the calculation of heat capacities (Section 16.5). The theory also does not provide a satisfactory explanation for the differences between metals, insulators, and semiconductors. In particular it does not explain the fact that the electrical conductivity of a metal decreases as the temperature is raised, whereas the conductivities of semiconductors and insulators increase. In the following text we outline an extension of the free-electron theory, which is more satisfactory in dealing with the properties of various types of solid.

Metals, Semiconductors, and Insulators: Band Theory

An electron energy level in a solid may be treated in two different ways: by considering the interaction of localized electrons and their energy levels, or in terms of an essentially free-electron approximation. The first approach is similar to the formation of two energy levels in the molecular-orbital description of the hydrogen atom (Section 12.2). For example, if five atoms are placed in a row with equal separations between the atoms, according to a Hückel MO-type treatment the original energy level would split into five energy levels, as shown in Figure 16.41. With each of the N atoms in a crystal contributing N orbitals, the energy may be written in terms of the Coulombic integral J and the exchange integral K:

$$\epsilon_n = J + 2K\cos\frac{2\pi m}{N} \tag{16.93}$$

where $m = 0, \pm1, \pm2, \ldots$. No matter how many atoms are present, the energy levels lie between the energy limits $J + 2K$ and $J - 2K$. As N becomes large (in the order of 10^{22} atoms cm^{-3}), the spacing between levels becomes so small that the energy appears as a continuum. Such a group of levels is called an **energy band.**

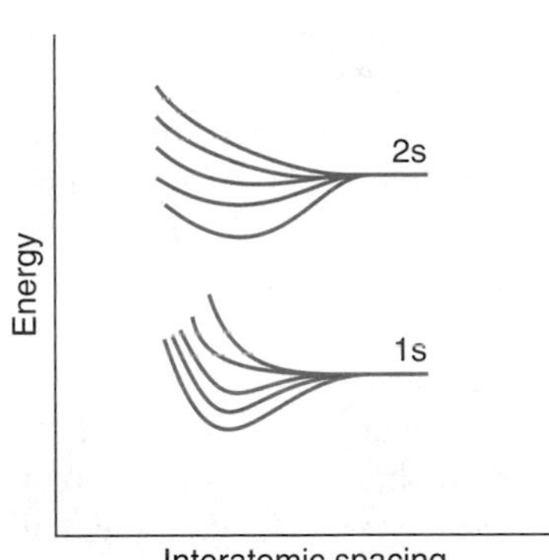

FIGURE 16.41
A potential-energy diagram for a row of five equally spaced atoms showing the formation of bands. Between levels a separation occurs, but there is only a small separation of energy levels within a band.

The second approach considers an assembly of electrons free to move in a crystal and then considers the manner in which they interact with the ionic lattice. This is essentially an application of the "particle-in-the-box" problem (Section 11.6) with periodic boundary conditions, the solutions of which can be written in the form

$$\psi = e^{ikx} \tag{16.94}$$

where k is a wavenumber.

Here p and λ are the momentum and wavelength of the electron. The corresponding energy when the potential energy is assumed to be zero within the box is

$$E = \frac{k^2\hbar^2}{2m} \tag{16.95}$$

and essentially one continuous band of energy levels is formed for different values of k. When the periodicity of the ions in the lattice is considered, we find that the potential energy also varies periodically, $E_p(R) = E_p\,(R + n\,R_o)$. The solution, ϕ_k, to this problem is the product of the wave function of a free electron $\psi(\boldsymbol{R})$ and a function that has the periodicity of the lattice, $u(\boldsymbol{R}) = u(\boldsymbol{R} + n\,\boldsymbol{R}_o)$

$$\phi_k(\boldsymbol{R}) = e(i\boldsymbol{k} \cdot \boldsymbol{R})u_k(\boldsymbol{R}) \tag{16.96}$$

The periodicity property leads to the expression

$$\phi_k(\boldsymbol{R}) = \phi_k(\boldsymbol{R} + n\boldsymbol{R}_o) = e^{(in\boldsymbol{k}\cdot\boldsymbol{R}_o)}\phi_k(\boldsymbol{R}) \tag{16.97}$$

From Eq. 16.1 the lattice vector $\boldsymbol{R}_o$ contains three lattice vectors written here as $n_a\boldsymbol{a}$, $n_b\boldsymbol{b}$, and $n_c\mathbf{c}$. Consequently, in each direction, the exponential term is unchanged by changing the three components of the wave vector $\boldsymbol{k} = (k_a, k_b, k_c)$ by

$$k_a = k_a + m_a(2\pi/a) \tag{16.98}$$

where m_a is an integer. Similar equations exist for k_b and k_c. What this means is that the components of the wave vector are relevant only up to the modulus; $(2\pi/a)$, $(2\pi/b)$, and $(2\pi/c)$ (i.e., the coefficients which express the measure of value of k_a). Restricting the ranges of k_a, k_b, and k_c to m_a, m_b, and m_c, respectively, all equal to 1 defines a region called the first **Brillouin zone**. The values of the wave vectors must, therefore, lie between

Brillouin Zone

$$-\frac{\pi}{a} \leq k_a \leq +\frac{\pi}{a} \quad \text{or} \quad -\frac{1}{2a} \leq \lambda^{-1} \leq +\frac{1}{2a} \tag{16.99}$$

and likewise for k_b and k_c.

Apart from the mathematical consequences of the periodicity on the wave functions of electrons, the Brillouin zone has a geometric interpretation. First, notice that the above equations again lead to the concept of the reciprocal lattice, as seen from Eq. 16.99 and from the exponential in Eq. 16.97 for each component. For example, the primitive lattice vector $\boldsymbol{a}$ and the modulus $(2\pi/a)$ condition on the wave vectors lead to the same definitions as in Eq. 16.10 for the reciprocal lattice. Next, from the restrictions such as Eq. 16.99, the Brillouin zones are regions defined in reciprocal lattice space. That is, Eq. 16.99 means that the first Brillouin zone contains the origin of the reciprocal lattice and extends out to one-half of each of the reciprocal lattice primitive vectors $\boldsymbol{a}^*$, $\boldsymbol{b}^*$, and $\boldsymbol{c}^*$. The boundary, therefore, is determined by the bisector of the distance from the origin to the nearest reciprocal lattice points. The second Brillouin zone is defined by the area enclosed by the bisectors of the distance from the origin to the next-nearest reciprocal lattice points, excluding the first Brillouin zone. This is shown in Figure 16.42a for a cubic lattice. The first Brillouin zone can be mapped into real lattice space. The cell so produced is called the **Wigner-Seitz primitive unit cell**.

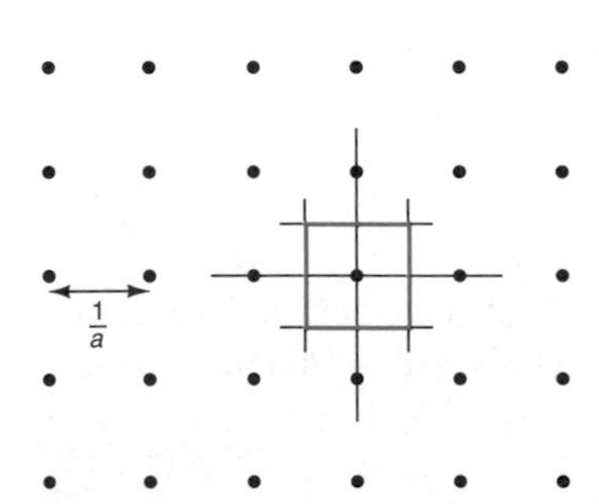

FIGURE 16.42
Reciprocal lattice space showing the first Brillouin zone (see Eq. 16.95).

Equations 16.96 and 16.97, developed in 1928, are known as the Bloch functions, named after Felix Bloch (1905–1983). The great advantage of these functions is that the solution to different periodic potentials need be obtained only in the first Brillouin zone to which the wave vector $\boldsymbol{k}$ is restricted. By varying the potential from weak to strong, the eigenvalues and eigenfunctions also vary from the nearly free electron model, giving electrical conduction, to the case where the electrons are tightly bound to the atoms, which describes insulators.

This situation is similar to the normal quantum mechanical solution we encountered in Chapter 11. There energy was found to be quantized and as such discrete allowed energy levels arise and all others are forbidden. In the case of electrons in solids, rather than single atoms or molecules, the allowed energy regions are called **energy bands** and the forbidden regions are called **energy gaps.** Doing

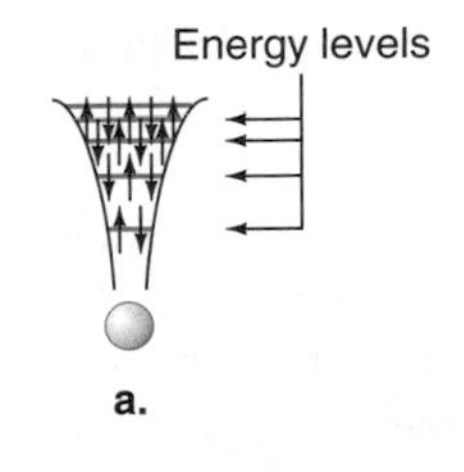

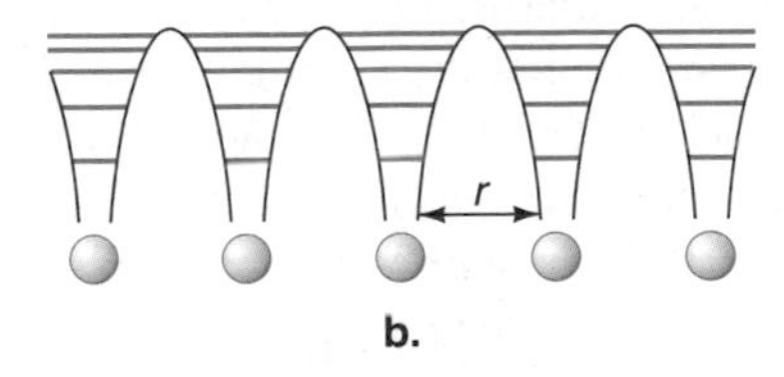

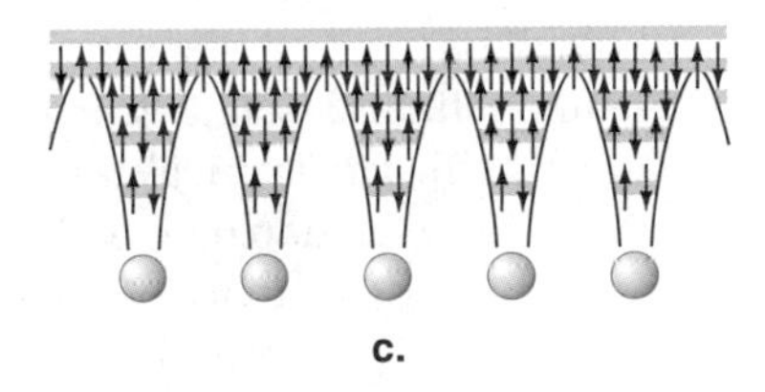

FIGURE 16.43
(a) Schematic of the potential energy around one atom showing the energy levels and the electron configuration. (b) A one-dimensional lattice of atoms brought together but not quite at their equilibrium separations. Some of the energy levels overlap. (c) The result of moving the atoms to their equilibrium positions. Now the single atom wave functions mix and the energy levels turn into bands. Some of the electrons are localized on an atom, while the upper electrons in the higher bands can be delocalized over the whole crystal.

quantum mechanics in the first Brillouin zone reveals this **band structure**. Consider, first, a single atom centered at the origin, shown in Figure 16.43a. That atom has a certain potential, E_p, shown in the figure with a set of energy levels determined by solving the quantum mechanics of a single particle. Figure 16.43b shows a one-dimensional array of atoms along the X axis with the potential $E_p(x + na) = E_p(x)$. Depending on the overlap of the energy levels between neighboring atoms and the height and shape of the potential energy, when quantum mechanics is done on the array, a band structure results. This is shown in Figure 16.43c. If the barrier is low (Figure 16.44a), the electrons are free to move about the bands as in the nearly free electron model. In the other limit of a high barrier (Figure 16.44b), the electrons are tightly bound to the atoms.

As for atoms, the Aufbau process (Section 11.12) also applies to solids. The band structure is a set of closely-spaced energy levels, into which electrons are placed. The Pauli exclusion principle must be obeyed, as discussed in Section 16.5. To visualize the process, consider two atoms coming together. This is the same situation that we encountered in MO theory in the formation of a diatomic molecule. We first determine the MO's, then fill these with the electrons available. Figures 16.37a and 16.37b show two cases of half-filled and filled MO's. Figure 16.37c shows the MO's for several atoms, while Figures 16.37d and 16.37e show the cases for a crystal band structure resulting from N atoms having, in this example, N and $2N$ electrons which must go into the N levels following the Pauli principle. That is, the electrons available may or may not fill a band. When the bands are not filled, there remain unoccupied energy states into which electrons can be easily promoted by the thermal energy available and transported throughout the crystal. This is the conduction case common to all metals. In Figure 16.37c to e the black band has a certain density of states, Eq. 16.77. In Figure 16.37d, electrons fill the band halfway at 0 K, and the highest filled energy level is the Fermi level (Eq. 16.81). In contrast, if all the energy levels are filled, there are no vacant levels for electrons to move into, electron transport is not possible, and the material is an insulator, as shown in Figure 16.44b.

The situation is frequently more complicated than that just discussed; some bands lie close together, others are far apart, and in yet others, gaps occur within bands or discrete energy levels appear between the bands in the gaps. By extending the MO model depicted in Figure 16.37, not only s bands are produced, but also p bands and d bands. As seen in Section 16.5, which discusses heat capacities of solids, many of the unique properties of materials depend upon the details of the

Follow the steps in Figure 16.43; Electrical conductivity in solids; Bonding in metals.

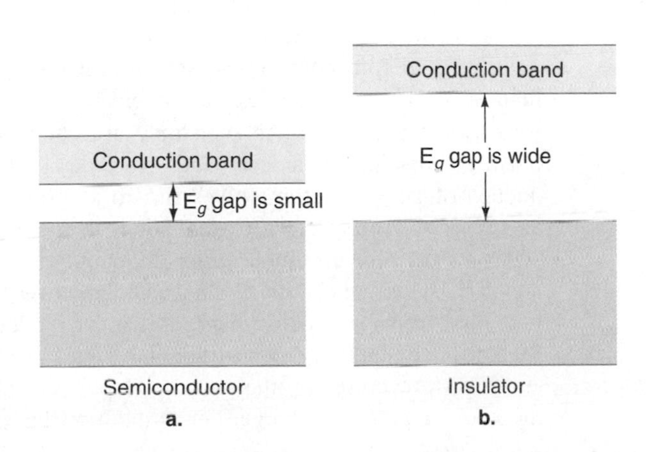

FIGURE 16.44
The relation of energy gaps and conduction bands to the valence band in the two types of solids.

band structure. Generally, however, it is possible to contrast insulators and semiconductors as done in Figure 16.44a and b. In both cases, the electrons fill the lowest bands, which are called the valence bands.

Intrinsic Semiconductor

In **insulators** and **semiconductors,** the valence band is completely filled and an energy gap exists between it and the next higher energy band. If the energy gap is wide compared with k_BT, there is little chance for the electron to be excited into an empty conduction band, and the material is an *insulator.* If there is only a small energy gap, the material is an **intrinsic semiconductor,** the electron being excited easily into the conduction band. If the gap is wide but the solid is "doped" with impurity atoms, it may be possible to establish levels within the gap that facilitate the movement of electrons into the conduction band. In that case we have what are known as **impurity semiconductors,** or **extrinsic semiconductors.** Silicon displays such properties, and is widely used in electronic technology.

Impurity (Extrinsic) Semiconductor

Most technologically important semiconductors are of the extrinsic type, in which the charge carrier production is determined by trace amounts of impurities known as *point defects*. For example, at room temperature gallium arsenide has a band energy gap of about 1.4 eV; it is an insulator since no significant electrical conductivity exists until its melting point of 1511 K is reached. However, doping it with impurities (between about 0.1% to less than 1 ppm) allows the conductivity of the sample to be controlled.

B	C	N	O	F	Ne
Al	Si	P	S	Cl	Ar
Ga	Ge	As	Se	Br	Kr
In	Sn	Sb	Te	I	Xe
Tl	Pb	Bi	Po	At	Rn

FIGURE 16.45
Part of the periodic table showing the main group elements and the metalloids. Silicon and germanium can be made into *n*- or *p*-type extrinsic semiconductors by doping with small amounts of elements to the right (one extra electron) or to the left (one fewer electron) of Si or Ge, respectively.

Figure 16.45 shows a portion of the periodic table with the division of metalloids between metals and nonmetals. (In spite of the name, the metalloids are not metals since the electrical conductivity does not decrease with increasing temperature. However, arsenic can be prepared with a smooth lustrous surface, a characteristic of metals.) Metallic character increases as we move down the periodic table. Consider the metalloids, such as silicon and germanium, characterized by a filled valence band and a low-lying conduction band (Figure 16.46a). Since the electrons follow the Fermi-Dirac distribution function, thermal energy can promote electrons from the valence band to the conduction band, leaving positive regions close to the surface of the valence band (Figure 16.46a). These positive regions are called *holes*, and are best described as empty orbitals. Their presence is the main reason why the electrical conductivity of intrinsic semiconductors increases exponentially with increasing temperature.

The impurities used to change these semiconductors to extrinsic semiconductors are generally of two types. With reference to the portion of the periodic table in Figure 16.45, there are those elements on the left of the group containing Si and Ge, and those lying to the right. The left group, notably B, Al, and Ga, are impurities that enter the silicon lattice structure and occupy sites otherwise occupied by a silicon atom. These substitutional impurities have one electron fewer than Si or Ge, and hence extract an electron from nearby Si-Si bonds in order to form their own bonds. As such they are called **acceptor atoms** and they leave positive holes in the valence band. More detailed calculations show that the introduction of acceptor atoms gives rise to a series of acceptor states that lie in the energy gap, usually close to the filled valence band (see Figure 16.46b). These states result from the impurity, or "dopant," and are fixed because the impurities are fixed within the lattice. They can, however, accept electrons that are easily promoted from the valence band; this leaves holes in the top of the valence band. Electrical conductivity comes about by the movement of the holes through the crystal. More correctly, the motion of positive holes is equivalent to the movement of negatively charged electrons in the opposite direction. Since these

p-Type Semiconductor

Watch n- and p-type semiconductions form; Electrical conductivity of solids, n-type and p-type.

n-Type Semiconductor

impurities leave positive holes, acceptor atom dopants create ***p*-type semiconductors**, the letter p standing for "positive."

In contrast, doping Si and Ge with atoms that lie to the right of them on the periodic table (Figure 16.45), phosphorus and arsenic in particular, produces one electron more than Si or Ge. Since only four electrons are needed for the dopants to bond the semimetal lattice, the extra electron is available for conduction. The dopants create a series of bound states which again lie in the energy gap, but calculations show that these usually lie close to the conduction band (Figure 16.46c), and the extra electrons from the impurity are easily promoted to the conduction band as shown. Electrical conduction of this type, due to electrons with negative charge being promoted to the conduction band, produces ***n*-type semiconductors**, the letter n standing for "negative."

> ***EXAMPLE 16.15*** Suppose that we dope germanium, a semiconductor, with a small amount of arsenic. Determine on the basis of the valence shell electrons whether the dopant produces an n- or p-type semiconductor.
>
> ***Solution*** The valence shell of Ge is $4s^2 4p^2$. Arsenic's valence shell is $4s^2 4p^3$. When put into the lattice of Ge atoms, the extra 4p electron in As is out of place, so to speak, and is free to move through the conduction band. Therefore, As is an electron donor and forms an n-type semiconductor.

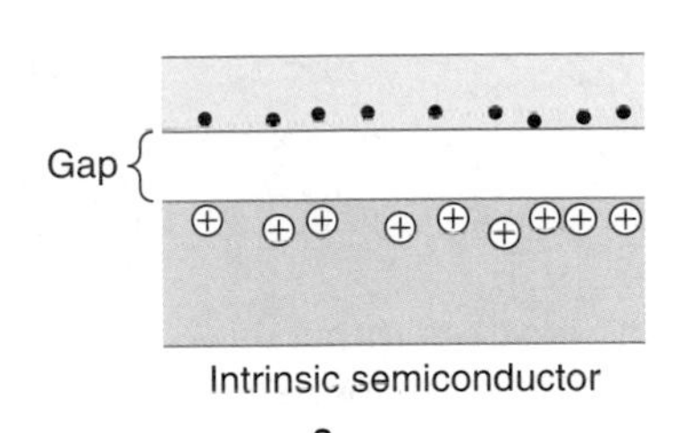

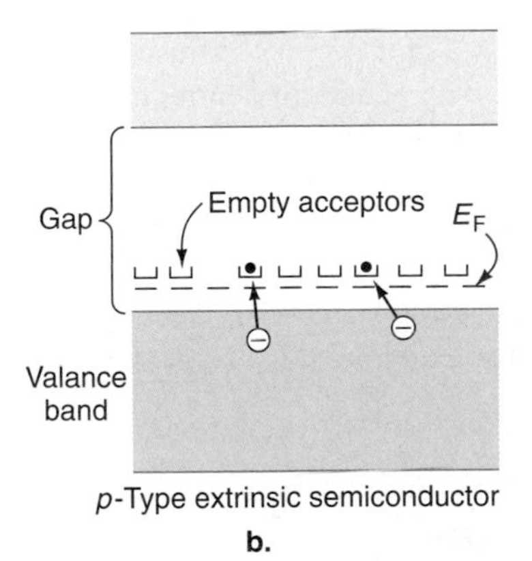

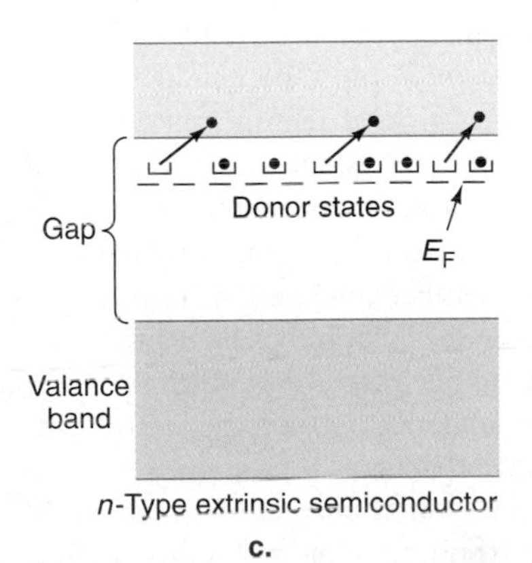

p-n Junction

The essential feature of a transistor and many other semiconductor devices is the p-n junction. A p-n junction is formed at the boundary between one region of a semiconductor with p-type impurities and another region containing n-type impurities. We can understand the behavior of the p-n junction by considering the band diagram shown in Figure 16.47. This shows the relative energies of the bands at equilibrium, which are drawn so that the Fermi level in the p-type region is equal to the Fermi level in the n-type region. A depletion zone is formed on either side of the junction, which contains a few free electrons in the conduction band or holes in the valance band. An electric field exists across the zone. Some electrons from the n-type region have sufficient thermal energy to surmount the potential energy barrier and diffuse into the p-type region, creating a recombination current i_{nr}. Some electron-hole pairs are generated by thermal excitation in the depletion zone and are swept out of the zone by the electric field giving rise to a generation current

FIGURE 16.46
(a) An intrinsic semiconductor has a low-lying conduction band and a filled lower valence band. Since the gap is small, electrons can be excited to the conduction band as the temperature increases. This leaves "holes" in the valence band while the electrons move into the conduction band. (b) A p-type extrinsic semiconductor showing the acceptor level close to the valence band. Electrons can be promoted to these states and in doing so leave holes in the valence band. (c) An n-type extrinsic semiconductor showing the presence of donor states. At absolute zero of temperature, the electrons fill to the donor levels, thereby defining the Fermi energy. At higher temperatures, the electrons are excited to the conduction band.

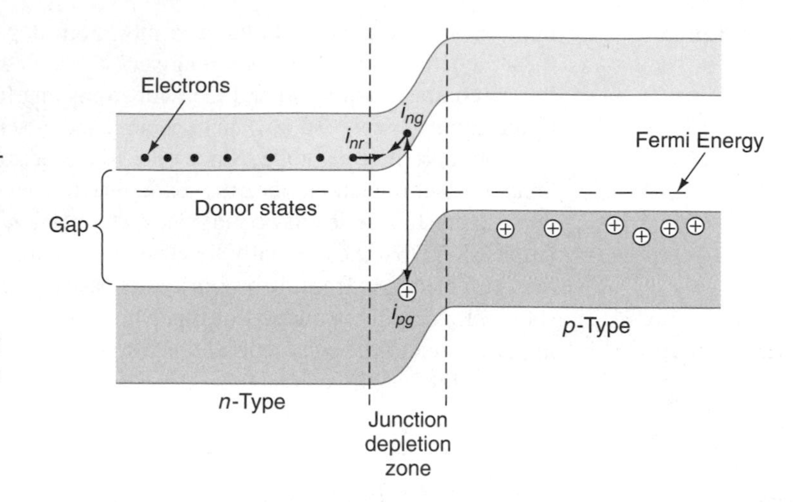

FIGURE 16.47
An *p-n* junction at equilibrium showing the formation of a depletion zone. The generation currents, and i_{pg}, and the recombination currents, i_{rg} and i_{rg}, are shown.

Follow Figure 16.47 for different biases; Electrical conductivity in solids; Biased systems.

The main work on transistors was done in the AT&T Bell research laboratories by Walter Houser Brattain (1902–1987), William Bradford Shockley (1910–1989), and John Bardeen (1908–1991). The first successful transistor was achieved in 1947, and in 1951 a highly effective *n-p-n* germanium transistor was produced. In 1956 the three men shared the Nobel Prize for physics for their achievement.

electron i_{ng}. At equilibrium, these two currents are equal, $|i_{nr}|=|i_{ng}|$, and no net current flows. Similarly for $|i_{nr}|=|i_{pg}|$.

Now we apply a forward bias by connecting the positive terminal of a battery to the *p*-type region and the negative terminal to the *n*-type region. This decreases the electric field in the junction region, and the difference between energy levels on the *p* and *n* sides. It becomes easier for electrons in the *n* region to climb the potential energy hill and diffuse to the *p* region, and likewise for holes to diffuse into the *n* region, that is, i_{nr} and i_{pr} increase exponentially. The generation currents do not change significantly so there is now a net current across the junction. Connecting the battery with reverse polarity gives "reverse bias," and the current in the reverse direction is much smaller than with the same potential difference in the forward direction.

A device that exhibits this behavior is called a diode. Diodes are the basic building blocks of many semiconductor devices, for example, transistors and computer gates.

Superconductivity

In 1908, the Dutch physicist Heike Kamerlingh-Onnes of the University of Leiden was the first person to liquefy helium and subsequently study the conductivity of metals at low temperatures. From that study he found in 1911 that mercury at sufficiently low temperatures ($\approx\ < 4$ K) exhibits zero resistance to the flow of electricity. Such conductors are known as **superconductors**, and the remarkable property that electrons move through the superconductor without resistance is known as **superconductivity**. Many applications for this phenomenon are possible, but for most chemists the most important may be the creation of large magnetic fields for use in NMR. Some superconducting metal alloys such as Nb_3Sn or NbTi with a T_c (**critical temperature** below which the material becomes superconducting) of about 18 to 23 K were found in 1953 to have mechanical properties suitable to form wire. This is important since a coil can be wound with the wire and submerged in an inner jacket containing liquid helium (boiling point = 4 K) to produce

Study the Meissner effect; Electrical conductivity in solids; Superconductivity.

a superconducting coil. An outer jacket containing liquid nitrogen (boiling point = 77 K) is further made to surround this inner jacket. Thus as an electrical current from the line flows through the coil, the moving electrons create a magnetic field. When the maximum design field is attained, the electrical source is disconnected, and the current continues to flow without resistance. To maintain the coil in a superconducting state, liquid helium and liquid nitrogen must be added periodically as heat conduction from the room gradually boils off the liquids.

For the important discovery of this class of superconductors, Bednorz and Muller shared the 1987 Nobel Prize in physics, the time between the discovery and the award being the shortest in history.

Discovering a superconductor that would eliminate the need for liquid He and N_2 cooling and operate at room temperature seemed quite impossible. But as sometimes happens, a totally unexpected breakthrough occurred. J. Georg Bednorz (b. 1950) and Karl Alexander Muller (b. 1927), working at the IBM laboratories in Zürich, Switzerland, discovered in 1986 a *ceramic* material (not expected to conduct electricity at all) with a T_c of approximately 34 K, a critical temperature much higher than any other at that time. The discovery of this material (La_2CuO_4, in which some of the La^{3+} ions had been replaced by Ba^{2+} ions, has a well-known perovskite structure) stimulated much research. Shortly thereafter, in 1987, M. K. Wu and the Chinese-American materials scientist Ching-Wu Chu announced a new superconductor with a T_c of approximately 90 K with a formula $YBa_2Cu_3O_{7-\delta}$, where δ represents a small decrease in oxygen concentration to make the superconducting orthorhombic form of the compound rather than the nonsuperconducting tetrahedral form. Because of the stoichiometry, this material is referred to as a 123-cuprate superconductor; its structure is shown in Figure 16.48. It is important because only liquid nitrogen is needed to make it superconducting. In most of its applications it is known as a coated superconductor because of the metallic sheath used to surround it, but only about 1-meter lengths in wire form can be made even after years of research. This is an indication of the difficulty of making these cuprates. Still other problems, related to its poor ability to carry current in a magnetic field, plague its commercial use. Many similar materials have been made based on the perovskite structure, but the highest T_c achieved has been about 139 K. Among the many cuprates found, a 2223 superconductor ($T_c \approx 110$ K) based on bismuth ($Bi_2Sr_2Ca_2Cu_3O_{10}$, known as BSCCO and pronounced "bisco") has been formed in kilometer lengths, but it cannot commercially compete with copper because of cost, except in special applications.

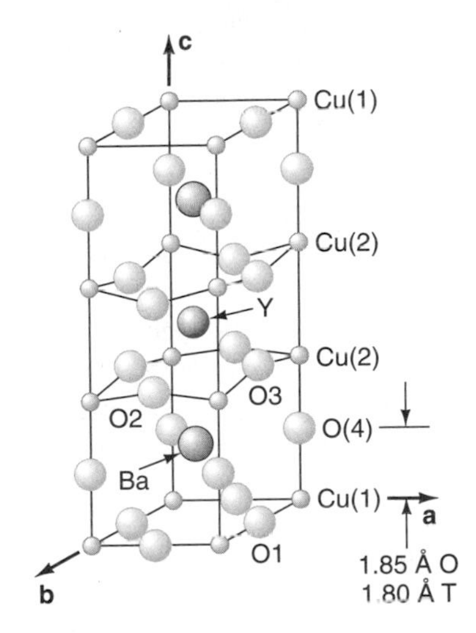

FIGURE 16.48
Crystal structure of $YBa_2Cu_3O_{7-\delta}$. There are O-Cu-O bonds in planes marked Cu(2) on either side of the yttrium atoms, but the O-Cu-O chains in the **b** direction also carry current. The octahedral structure occurs with < 25% of the O1 layer occupied.

Unfortunately, the explanation of superconductivity has proved to be a difficult problem in theoretical physics, especially for the high-T_c superconductors. The first detailed theory of superconductivity was put forward in 1957 by the American physicists John Bardeen (1908–1991), Leon N. Cooper (b. 1930), and John Robert Schrieffer (b. 1931). Their theory, known as BCS theory, was based on an idea, previously put forward by Cooper, that electrons can be bound together in pairs by means of **phonons**. A phonon is a quantum of vibrational energy, sometimes treated as if it were an elementary particle exibiting momentum. See Figure 16.49 for a representation of how this occurs. According to Cooper's theory, the phonon can overcome the electrostatic repulsion between two electrons and bind them together. This effect is known as *phonon-mediated pairing*, or Cooper pairing. Cooper pairs, quantum mechanical tunneling, the Debye temperature (proportional to the characteristic phonon frequency of the lattice), and a number of other theoretical concepts play important roles in BCS theory, which was used as a guide to further research. For example, in conventional metal superconductors, phonons (lattice vibrations) push electrons into pairs that form a superconducting condensate.

For their theory of superconductivity Bardeen, Cooper, and Schrieffer shared the 1972 Nobel Prize for physics. This was Bardeen's second Nobel Prize in physics.

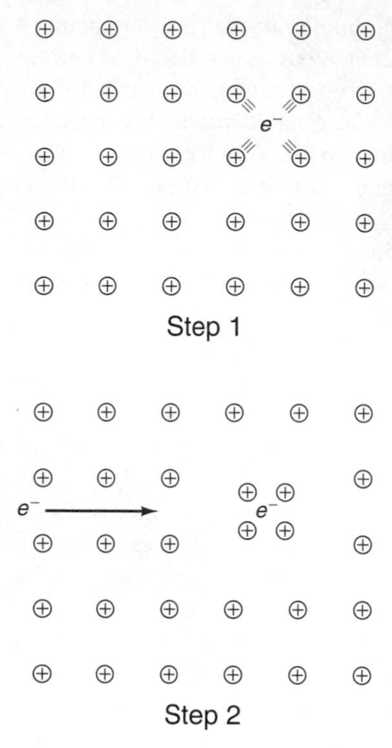

FIGURE 16.49
Mechanism of superconductivity through electron-phonon interaction. The process occurs in two steps. (1) The lattice contracts locally as an electron is attracted to the surrounding ions. (2) The second electron comes by that site and is attracted to the distorted lattice. The net effect is an attraction between two electrons.

In these systems lighter isotopes have higher phonon frequencies, resulting in higher T_c's. This effect is absent in the high-T_c cuprates. In the cuprates, the superconductivity depends on subtle, phonon-free coupling between electrons in the Cu-O-Cu structure, but the theory is still not developed for the cuprates.

Great excitement was generated in this field in January 2001 by the announcement of a totally unexpected new superconductor of pure magnesium diboride, MgB_2. The magnesium atoms form a hexagonal layer with the boron atoms forming a graphite-like honeycomb layer. Its T_c is an unimpressive 39 K, but it is comparatively inexpensive, available off the shelf, and has the important property of fitting the BSC theory. The hope is that through calculations of MgB_2's electronic band structure enough will be learned to advance superconductivity theory so that the way for improvements in T_c in the cuprates will be seen.

Superconductivity is found in another interesting class of compound, T_c's of 40 K being sometimes achieved. The compounds are formed from C_{60}, which is known as fullerene. Its structure resembles that of a soccer ball and is sometimes known as a "Bucky ball" after Buckminster Fuller, who designed the geodesic dome. It consists of only C atoms and is another form of pure carbon. Considering C_{60} as though it were a molecule, Cs_3C_{60} may be thought of as an ionic salt with the $3Cs^+$ cations and a ${C_{60}}^{3-}$ anion, called a fulleride. C_{60} is highly symmetrical and has a threefold degenerate t_{tu} LUMO in this neutral state. HOMO (Highest Occupied Molecular Orbital) and LUMO (Lowest Unoccupied Molecular Orbital) are known as **frontier orbitals** because it is in those orbitals that most of the chemistry of molecules takes place. The h_u HOMO and t_{tu} LUMO levels form bands that are capable of accepting up to six electrons from the intercalation cations. The conductivity arises from this weak overlap of the wave functions and can be understood within the framework of the Bardeen-Cooper-Schrieffer theory. Many other interesting properties of the fullerenes and their derivatives exist, but it is not clear that much higher values of T_c can be achieved. For further information, the reader is directed to the Suggested Readings.

16.7 Optical Properties of Solids

There are 15 recognized, distinct sources of color production in matter. Several are discussed in the realm of physics because the processes involved are from geometrical and physical optics; namely diffraction (color from diffraction gratings), dispersive refraction (color from rainbows), interference (color from soap bubble films), and scattering (color of the blue sky).

We have already treated vibrations and simple excitations from a chemistry perspective. These generate color from incandescence (flames), gas excitations (vapor lamps), and vibrations and rotations (iodine). Transitions between molecular orbitals generally cause color in organic compounds (dyes) and in charge transfer reactions involving changes of oxidation state between two different ions (Prussian blue, a colorant). Transitions that can be explained by ligand field effects involve transition-metal compounds (chrome green) and transition-metal impurities (ruby). Transitions involving energy bands can also produce color. Four separate processes can be treated using energy bands: color in metals (copper), in pure semiconductors (galena), in doped semiconductors (LED's [Light Emitting Diodes]), and in color centers (amethyst). As some of these are well-treated elsewhere, we will focus on just a few of these mechanisms.

Transition Metal Impurities and Charge Transfer

Absorption and emission of electromagnetic radiation by solids occur in the ultraviolet, visible, and near-infrared portions of the spectrum. Often the absorption is caused by impurities or activators that have energy levels lying within the energy gap of the pure material. For example, pure corundum, Al_2O_3 (Chapter 6) is colorless, but the addition of a small amount of Cr^{3+} ($Cr^{3+} / Al^{3+} \approx 1/2000$) introduces a substitutional impurity, replacing Al^{3+}. The result is ruby, which has a deep red color caused by the ligand field splitting of the d orbitals of chromium.

Instead of a Cr^{3+} impurity, substitution of a few hundredths of a percent of Fe^{2+} and Ti^{4+} for two Al ions on adjacent sites allows light absorption resulting in the color of blue sapphire. This is caused by a transfer of one electron, resulting in the absorption of about 2.2 eV. This charge transfer reaction is

$$Fe^{2+} + Ti^{4+} \rightarrow Fe^{3+} + Ti^{3+}.$$

Many variations of this type, such as

$$Fe_A^{2+} + Fe_B^{3+} \rightarrow Fe_A^{3+} + Fe_B^{2+},$$

are possible using different elements and charges in different substrates.

Color and Luster in Metals

From the point of view of band theory, for about 1 mol of n-valent metal, there are over $n \times 10^{23}$ equivalent electrons that form an essentially continuous band of energy states. The electrons fill the energy band from the bottom up to the Fermi energy level. Light falling on the surface is absorbed at any energy. Since the energy of photons of visible light is in the range of the cohesive energies of metals (1–3 eV), light interacts strongly with the free electrons in the metals. It is absorbed so strongly that the light cannot penetrate far, generally less than about one wavelength. Thus the energy is radiated to the ground state quite close to the surface of the metal. The light induces an electrical current on the metal surface, which immediately reemits the light, giving rise to the metallic luster and the reflection characteristic of metals.

In semiconductors, there is absorption of light. This corresponds to the excitation of electrons from the valence band across the energy gap to the conduction band. In insulators, absorption appears to occur at defects in the normally forbidden energy gap.

Color Centers: Nonstoichiometric Compounds

F Centers

When alkali halides are heated in the presence of their alkali metal, atom vapor defects are introduced to make the crystal slightly nonstoichiometric. As a result the crystals take on various colors: LiF, pink; NaCl, yellow; KCl, blue-violet. The most common type of defect is the **F center**, from the German Farbenzentre (color center). An F center is formed in NaCl in the following way. A sodium metal ionizes on the surface of NaCl. The ion attracts a chloride ion from the surface nearby. This forms an anion vacancy on the surface that is attracted by a free electron. This combination is free to migrate into the bulk crystal. In essence the electron in an F center is surrounded by six Na^+ ions with which it is shared in hydrogenlike orbitals. The electron trapped at the vacancy has formed an energy level system inside the band gap of the crystal. The electron produces the observed color by being excited into an absorption band. Thus the observed color corresponds to an electron transition from a ground state to an excited state. For the alkali halides, the E_a of the transition is given by $E_a/eV = 0.257/(d/\text{nm})^{1.83}$, where d is the anion-cation distance.

Many other centers exist in alkali halides. There are **M centers** consisting of two adjacent F centers. The **F′ center** is an F center that has trapped two electrons. A **V_K center** consists of two adjacent F centers, but with only one trapped electron. Also important is an **electron color center**, which is merely an electron trapped in a location where it is not normally found.

Color centers may also be produced by irradiation that produces light-absorbing species that may be either hole centers or electron centers. If, for example, an electron is released from an electron center by heat, then we say that bleaching occurs when it recombines with a hole center to return to the pre-irradiation state.

Exciton

Energy levels formed just below the conduction band are excited states of an electron-hole pair, called an **exciton.** Exciton states generally are localized and do not contribute to the electrical conduction process.

Luminescence in Solids

Aside from fluorescence and phosphorescence emission, radiation from solids is observed in **thermoluminescence** and in lasers. In thermoluminescence the electrons in a solid are excited at low temperatures by the absorption of radiation, and electrons are trapped in the forbidden gap region. When the sample is heated, the electrons receive enough energy to escape the traps, and visible light is sometimes emitted. This property of thermoluminescence has been used to date the time at which pottery was fired.

Some of the characteristics of lasers are described in Section 14.1.

KEY EQUATIONS

Lattice vector:

$$\boldsymbol{R}_0 = n_a\boldsymbol{a} + n_b\boldsymbol{b} + n_c\boldsymbol{c}$$

Reciprocal lattice vector:

$$\boldsymbol{G} = h\boldsymbol{a}^* + k\boldsymbol{b}^* + l\boldsymbol{c}^*$$

Primitive vectors and reciprocal vectors:

$$\boldsymbol{a}^*\cdot\boldsymbol{a} = \boldsymbol{b}^*\cdot\boldsymbol{b} = \boldsymbol{c}^*\cdot\boldsymbol{c} = 2\pi$$

$$\boldsymbol{a}^* = 2\pi\boldsymbol{b}\times\boldsymbol{c}/V_c$$
$$\boldsymbol{b}^* = 2\pi\boldsymbol{c}\times\boldsymbol{a}/V_c$$
$$\boldsymbol{c}^* = 2\pi\boldsymbol{a}\times\boldsymbol{b}/V_c$$

Orthoganality:

$$\boldsymbol{G}\cdot\boldsymbol{R}_0 = 2\pi(hn_a + kn_b + ln_c) = 2\pi n$$

Distance between planes with Miller indices klm:

$$d_{hkl} = \frac{2\pi}{|\boldsymbol{G}|}$$

Scattering of wave vectors:

$$\Delta\boldsymbol{k} = \boldsymbol{k}' - \boldsymbol{k}$$

Bragg's law in lattice and reciprocal lattice space:

$$n\lambda = 2d\sin\theta$$
$$\lambda = \frac{4\pi}{\Delta k}\sin\theta$$
$$\boldsymbol{G} = \Delta\boldsymbol{k} = h\boldsymbol{a}^* + k\boldsymbol{b}^* + l\boldsymbol{c}^*$$

where λ is the wavelength of the X ray, d is the interplanar spacing and θ is the angle of incidence.

Structure factor:

$$F(hkl) = \sum_{n=1}^{N} f_n \exp(-i2\pi(hx_n + ky_n + lz_n))$$

Wavelength from *electron diffraction:*

$$\lambda/\text{Å} = \frac{12.25}{(\epsilon/\text{eV})^{1/2}} \quad \text{or} \quad \lambda/\text{pm} = \frac{1225}{(\epsilon/\text{eV})^{1/2}}$$

Wavelength from *neutron diffraction:*

$$\lambda/\text{Å} = \frac{0.28}{(\epsilon/\text{eV})^{1/2}} \quad \text{or} \quad \lambda/\text{pm} = \frac{28}{(\epsilon/\text{eV})^{1/2}}$$

Einstein's heat capacity equation:

$$C_V = 3Nk_B\left(\frac{h\nu}{k_BT}\right)^2 \frac{e^{h\nu/k_BT}}{(e^{h\nu/k_BT} - 1)^2}$$

Debye's heat capacity equation:

$$C_V = 9Nk_B\left(\frac{T}{\Theta_D}\right)^3 \int_0^{x_m} \frac{x^4e^x}{(e^x - 1)^2}\,dx$$

where $\Theta_D = h\nu_D/k_B$ and is known as the *Debye temperature.*

Debye's T^3 law:

$$C_V = \frac{12}{5}\pi^4 Nk_B\left(\frac{T}{\Theta_D}\right)^3$$

Contribution from electronics to the heat capacity of solids,

$$\gamma = R\frac{k_B\,\Pi^2}{2E_F}$$

Fermi-Dirac (plus sign) and Bose-Einstein (minus sign) distribution functions.

$$P(E) = \frac{1}{e^{(E-E_F)/k_BT} \pm 1}$$

Fermi Energy of a free electron gas

$$E_F = \frac{h^2}{2m}\left(\frac{3n}{8\pi}\right)^{2/3}$$

PROBLEMS

Crystal Lattices, Unit Cells, Density

16.1. How many basis groups are there in

a. an end-centered lattice;
b. a primitive lattice?

16.2. How many lattice points are there in a unit cell of

a. a face-centered lattice;
b. a body-centered lattice?

***16.3. a.** Determine the efficiency of area utilization in packing circles onto the lattice points of a square lattice.
b. Compare that value with the efficiency of packing circles onto a triangular lattice.
c. Which packing uses area more efficiently and by how much?

16.4. a. Calculate the percentage of free space (volume of the cell minus the volume of the atoms in the unit cell) in each of the three cubic lattices if all atoms in each are of equal size and touch their nearest neighbors.
b. Using the calculated values, determine which of the three structures represents the most efficient packing (least amount of used space).

16.5. Derive an equation to relate the density D of a right-angled unit cell to its edge lengths a, b, and c and the number of formula units z per unit cell.

16.6. Silver crystallizes in a face-centered cubic unit cell with a silver atom on each lattice point.

a. If the edge length of the unit cell is 4.0862 Å, what is the atomic radius of silver?

b. Calculate the density of silver.

16.7. Barium crystallizes with an edge length of 5.025 Å in a body-centered cubic unit cell.

a. Calculate the atomic radius of barium using this information.
b. Calculate the density of barium.

16.8. Aluminum crystallizes in a face-centered cubic lattice with an aluminum atom on each lattice point with the edge length of the unit cell equal to 4.0491 Å.

a. Calculate the atomic radius of aluminum.
b. Determine the density of aluminum.

16.9. Crystals of p, p'-dibromo-α, α'-difluorostilbene $[BrC_6H_4C(F)=]_2$ are orthorhombic with edge lengths a = 28.32 Å; b = 7.36 Å; c = 6.08 Å. If there are four molecules in a unit cell, calculate the density of the crystal.

16.10. How many formula units exist in pure crystalline Si, which occurs in a face-centered cubic lattice, if its density is 2.328 99 g cm^{-3} and its cell length is a = 5.431 066 Å? The atomic mass of Si is 28.085 41 g mol^{-1}.

16.11. Sodium chloride crystallizes in a face-centered cubic lattice with four NaCl units per unit cell. If the edge length of the unit cell is 5.629 Å, what is the density of the crystal? Compare your answer to the value given in the *CRC Handbook.*

16.12. LiH crystallizes with a face-centered cubic structure. The edge length of the unit cell of LiH is 4.08 Å. Assume anion-anion contact to calculate the ionic radius of H^-.

Compare your answer to the value using the radius of Li^+ as 0.68 Å.

16.13. KCl is tetramolecular and crystallizes in a face-centered cubic lattice. If the edge length is 6.278 Å, what is the density of KCl? Compare your answer to the value in the *CRC Handbook*.

16.14. Calcium fluoride crystallizes in a face-centered cubic lattice where $a = b = c$, and it has a density of 3.18 g cm^{-3}. Calculate the unit cell length for CaF_2.

Miller Indices and the Bragg Equation

16.15. Calculate the Miller indices of the parallel planes in a cubic lattice that intercepts the unit cell length at $x = a$, $y = \frac{1}{2}a$, and $z = \frac{2}{3}a$.

16.16. Determine the distance (i.e., d value) of the closest plane parallel to the 100, 110, and 111 faces of the cubic lattice.

16.17. What are the Miller indices of the plane that cuts through the crystal axes at

a. $(2a, b, 3c)$;
b. $(2a, -3b, 2c)$;
c. $(a, b, -c)$?

16.18. Determine the value of d_{hkl} in terms of the cell constants and angles for

a. the orthorhombic unit cell,
b. the tetragonal unit cell.

16.19. Calculate the separation between planes in a cubic lattice with unit cell length of 389 pm when the indices are

a. 100;
b. $11\bar{1}$;
c. $12\bar{1}$.

16.20. Copper sulfate single crystals are orthorhombic with unit cells of dimensions a = 488 pm, b = 666 pm, c = 832 pm. Calculate the diffraction angle from Cu K_α X rays (λ = 154.18 pm) for first-order reflections from the (100), (010), and (111) planes.

16.21. Determine the angle of reflection when copper K_α radiation (0.154 18 nm) is incident on a cubic crystal with a lattice constant d_{hkl} of 0.400 nm.

16.22. Single crystals of $FeSO_4$ are orthorhombic with unit cell dimensions a = 482 pm, b = 684 pm, c = 867 pm. Calculate the diffraction angle from Te K_α, X rays (λ = 45.5 pm) from the (100), (010), and (111) planes.

16.23. Single crystals of $Hg(CN)_2$ are tetragonal with unit cell dimensions a = 967 pm and c = 892 pm. Calculate the first-order diffraction angles from the (100) and (111) planes when Cu K_α X rays (λ = 154 pm) are used.

16.24. A two-dimensional lattice is depicted in Figure 16.1 with planes superimposed on it parallel to the third direction. Determine the Miller indices for each set of planes representing the external habit of the left-hand crystal.

16.25. The layers of atoms in a crystal are separated by 325 pm. At what angle in a diffractometer will diffraction occur using

a. molybdenum K_α X rays (λ = 70.8 pm);
b. copper K_α X rays (λ = 154 pm)?

16.26. Calculate the wavelength of an electron that is accelerated through a potential difference of approximately 40 kV.

Interpretation of X-Ray Data

16.27. Find the X-ray wavelength that would give a second-order reflection (n = 2) with a θ angle of 10.40° from planes with a spacing of 4.00 Å.

16.28. A substance forms cubic crystals. A powder pattern shows reflections that have either all even or all odd indices. What type of unit cell does it have?

16.29. A powder pattern of a cubic material has lines that index as (110), (200), (220), (310), (222), (400). What is its type of unit cell?

***16.30.** The successive $\sin^2\theta$ values obtained from a powder pattern for α-Fe are 1, 2, 3, 4, 5, 6, 7, 8, 9, etc.

a. If iron is in the cubic system, which type of unit cell is present?
b. If a copper X-ray tube is used (λ = 154.18 pm), calculate the length of the side of the unit cell and the value of θ from (100) planes. The density of α-Fe is 7.90 g cm^{-3}.
c. What is the radius of the iron atom if the central atom in the cubic cell is assumed to be in contact with the corner atoms?

16.31. Potassium metal has a density of 0.856 g cm^{-3} and has a body-centered cubic lattice. Calculate the length of the unit cell a and the distance between (200), (110), and (222) planes. Potassium has an atomic mass of 39.102 g mol^{-1}.

16.32. Low-angle lines in the Cu K_α powder pattern of KCl are found to be at θ = 14.18°, 20.25°, and 25.10°. Find the crystal type from these data. (For Cu K_α λ = 154.18 pm.) What other information is needed for a definitive determination?

16.33. The smallest observed diffraction angle of silver taken with Cu K_α radiation (λ = 154.18 pm) is 19.076°. This angle is associated with the (111) plane in the cubic close-packed structure of silver.

a. Determine the value of the unit cell length a.
b. If D(Ag) = 10.500 g cm^{-3} and M = 107.87 g mol^{-1}, calculate the number of atoms in the unit cell.

16.34. Sodium fluoride is known to form a cubic closed-packed structure. The smallest angle obtained with Cu K_α radiation (λ = 1.5418 Å) is 16.72° and is derived from the (111) planes. Find the value of a, the unit cell parameter.

16.35. The X-ray powder pattern of NaCl is taken with a chromium tube giving Cr $K_\alpha = 229.1$ pm. The θ values of the lines are: 20°36′, 23°58′, 35°4′, 42°21′, 44°43′, 54°20′, 62°17′, 65°16′. From these data determine the value of each d_{hkl} and index the lines. From the *hkl* values, show that this is a face-centered system.

Bonding in Crystals and Metals

16.36. Cadmium sulfide has been used as a yellow pigment by artists. The sulfide crystallizes with cadmium occupying $\frac{1}{2}$ of the tetrahedral holes in a closest-packed array of sulfide ions. What is the formula of cadmium sulfide?

16.37. Rutile is a mineral that contains titanium and oxygen. The structure of rutile may be described as a closest-packed array of oxygen atoms with titanium in $\frac{1}{2}$ of the octahedral holes. What is the formula of rutile? What is the oxidation number of titanium?

16.38. A tetrahedral hole is shown in Figure 16.33. Determine the largest sphere of radius r that can fit into a tetrahedral hole when the surrounding four spherical atoms of the lattice are in contact. Let the lattice atom have radius R.

16.39. An octahedral hole is surrounded by six spheres of radius R in contact. If one-sixth of each of the six coordinating spheres contributes to the volume of the octahedron surrounding the hole, calculate the maximum radius of the sphere that can be accommodated.

16.40. Calculate the value of ΔE_C of the RbBr from the following information: $\Delta_f H = -414$ kJ mol^{-1}; I (ionization energy, Rb) $= 397$ kJ mol^{-1}; $\Delta_{\text{sub}}H(\text{Rb}) = 84$ kJ mol^{-1}; $D_0(\text{Br}_2) = 192$ kJ mol^{-1}; A (electron affinity, Br) $= 318$ kJ mol^{-1}.

Supplementary Problems

16.41. Some of the d spacings for the mineral canfieldite (Ag_8SnS_6) are 3.23, 3.09, 3.04, 2.81, and 2.74 Å obtained with Cu K_α X rays ($\lambda = 1.5418$ Å).

a. Find the corresponding angles of diffraction.
b. This is a cubic system with $a = 21.54$ Å; determine the *hkl* values for the first 3 d spacings.

16.42. A copper selenide mineral (Cu_5Se_4) called athabascaite is orthorhombic with $a = 8.227$, $b = 11.982$, $c = 6.441$. Strong intensity lines using Cu K_α X rays ($\lambda = 154.18$ pm) are observed at 12.95°, 13.76°, and 14.79°. Determine the d spacings and assign *hkl* values to these lines.

***16.43.** Zinc blende is the face-centered cubic form of ZnS with Zn at 0, 0, 0; $\frac{1}{2}, \frac{1}{2}$, 0; 0, $\frac{1}{2}, 0$; $\frac{1}{2}, \frac{1}{2}$, 0 and with S at $\frac{1}{4}, \frac{1}{4}, \frac{1}{4}$; $\frac{1}{4}, \frac{3}{4}, \frac{3}{4}$; $\frac{3}{4}, \frac{1}{4}, \frac{3}{4}$; $\frac{3}{4}, \frac{3}{4}, \frac{1}{4}$.

a. Determine the structure factor from the (111) planes that gives rise to the lowest angle reflection at $\theta = 14.30°$ using Cu K_α ($\lambda = 154.18$ pm).
b. Calculate the dimension a of the unit cell.

16.44. Calculate the Debye temperature of tungsten that is isotropic (an assumption of the Debye model). The cutoff frequency is given by

$$\nu_D = \left(\frac{9N}{4\pi V}\right)^{1/3}\left(\frac{1}{c_l^3} + \frac{2}{c_t^3}\right)^{-1/3}$$

where

$$c_l = 5.2496 \times 10^5 \text{ cm s}^{-1}$$

and

$$c_t = 2.9092 \times 10^5 \text{ cm s}^{-1}$$

are the longitudinal and transverse elastic wave velocities, respectively, in tungsten.

Essay Questions

16.45. List the 14 Bravais lattices and group them into P, I, F, C, and R cells.

16.46. Explain why the initial X-ray investigation of the two face-centered cubic structures, NaCl and KCl, showed that NaCl was face centered whereas KCl was simple cubic.

16.47. If ΔH_c were required rather than ΔE_c, what modification of the Born-Haber cycle would be needed?

16.48. X-ray diffraction is often used to measure residual stress in metals. Suggest that change in the measured parameters allows this determination.

16.49. Gold diffuses faster in lead at 300 °C than does sodium chloride in water at 15 °C. Point defects based on vacancies can account for such high rates. For an ionic material, suggest ways in which such vacancies can occur without altering the stoichiometry of the crystal.

SUGGESTED READING

General accounts of solid state chemistry and physics are given in the following publications:

S. L. Altmann, *Band Theory of Solids: An Introduction from the Point of View of Symmetry,* Oxford University Press, 1991.

J. S. Blakemore, *Solid State Physics,* 2nd edition, Cambridge University Press, 1985.

W. Borchardt-Ott, *Crystallography, An Introduction,* 2nd edition, New York: Springer Verlag, 1997.

F. C. Brown, *The Physics of Solids,* New York: Benjamin, 1967.

M. J. Buerger, *Contemporary Crystallography,* New York: McGraw-Hill, 1970.

J. K. Burdett, *Chemical Bonding in Solids,* Oxford University Press, paperback, 1995.

G. Burns, *Solid State Physics,* New York: Academic Press, 1987.

G. B. Carpenter, *Principles of Crystal Structure Determination,* New York: W. A. Benjamin, 1969.

A. K. Cheetham and P. Day, *Solid State Chemistry: Techniques,* Oxford University Press, 1988.

J. R. Christman, *Fundamentals of Solid State Physics,* New York: John Wiley & Sons, 1988.

P. A. Cox, *The Electronic Structure and Chemistry of Solids,* Oxford University Press, 1987.

W. A. Harrison, *Electronic Structure and the Properties of Solids: The Physics of the Chemical Bond,* San Francisco: Freeman, 1980 Dover reprint.

A. Kelly, G. W. Groves, and P. Kidd, *Crystallography and Crystal Defects,* New York: John Wiley & Sons, 2000.

C. Kittel, *Introduction to Solid State Physics,* 6th edition, New York: Wiley, 1986.

L. Pauling, *The Nature of the Chemical Bond,* 3rd edition, Ithaca, NY: Cornell University Press, 1960.

D. G. Pettifor, *Bonding and Structure in Molecules and Solids,* Oxford University Press, paperback, 1995.

C. N. R. Rao and J. Gopalkrishnan, *New Directions in Solid State Chemistry: Structure, Synthesis, Properties, Reactivity and Materials Design,* Cambridge University Press, 1986.

L. Smart and E. Moore, *Solid State Chemistry: An Introduction,* London: Chapman and Hall, 2nd edition, 1995.

G. H. Stout and L. H. Jensen, *X-Ray Structure Determination, A practical guide,* New York: John Wiley & Sons, 1989.

D. Tabor, *Gases, Liquids and Solids, and Other States of Matter,* 3rd edition, Cambridge University Press, 1991.

B. K. Tanner, *Introduction to the Physics of Electrons in Solids,* Cambridge University Press, 1995.

A. F. Wells, *Structural Inorganic Chemistry,* 5th edition, Oxford University Press, 1984.

A. R. West, *Solid State Chemistry and its Applications,* New York: John Wiley, 1985.

A. R. West, *Basic Solid State Chemistry,* New York: John Wiley, 1988.

Some special aspects of the subject are treated in

M. J. Buerger, *X-Ray Crystallography: An Introduction to the Investigation of Crystals by their Diffraction of Monochromatic Radiation,* Huntington, NY: R. E. Krieger, 1980.

M. A. Carrondo and G. A. Jeffrey (Eds.), *Chemical Crystallography with Pulsed Neutrons and Synchrotron X-Rays,* Dordrecht: D. Reidel, 1988.

P. P. Ewald (Ed.), *Fifty Years of X-ray Diffraction,* Utrecht, The Netherlands: International Union of Crystallography, 1962, pp. 6–75.

C. Kittel, *Quantum Theory of Solids,* New York: John Wiley, 1987.

For more on semiconductors, see

D. K. Ferry, *Semiconductors,* New York: Macmillan, 1991.

For accounts of superconductivity, see

G. Burns, *High Temperature Superconductivity,* Academic Press, 1992.

R. J. Cava, "Oxide superconductors," *J. Am. Ceram. Soc., 83*[1],5–28(2000).

P. G. de Gennes, *Superconductivity of Metals and Alloys,* Redwood City, CA: Addison-Wesley, 1989.

Y. A. Dubitsky and A. Zaopo, "Fullerene-based superconductors and their electrical applications," *La Chimica e l' Industria, 82,* 299(2000).

A. B. Ellis, "Superconductors: Better levitation through chemistry," *J. Chem. Ed., 84,* 836–841(1987).

D. K. Finnemore, K. E. Gray, D. M. Maley, D. O. Welch, D. K. Christen, and D. M. Kroeger, "Coated conductor development: An assessment," *Physica C, 320,* 1–8(1999).

R. Hawsey, and D. Peterson, "Coated conductors: The next generation of high-T_c Wires," *Superconductor Industry,* Fall, 1996.

Y. Iijima, and K. Matsumoto, "High-temperature-superconductor coated conductors: Technical progress in Japan," *Supercond. Sci. Technol., 13* 68–81(2000).

V. Z. Kresin, H. Morawitz, and S. A. Wolf, *Mechanisms of Conventional and High T_c Superconductivity,* Oxford University Press, 1993.

F. A. Matsen, "Three theories of superconductivity," *J. Chem. Ed., 84,* 842–846(1987).

For crystallographic data, reference should be made to

J. L. C. Daams, P. Villars, and J. H. N. van Vucht, *Atlas of Crystal Structure Types for Intermetallic Phases,* Materials Park, OH 44073: The Materials Information Society, 1991.

N. F. M. Henry and Kathleen Lonsdale (Eds.), *International Tables for X-Ray Crystallography,* Birmingham, England: Kynoch Press, 1952.

P. Villars and L. D. Calvert, *Pearson's Handbook of Crystallographic Data for Intermetallic Phases,* Materials Park, OH 44073: The Materials Information Society, 1991.

The Liquid State

17

PREVIEW

In liquids there is neither complete disorder nor complete order, and as a result, liquids are more difficult to treat theoretically than either gases or solids. This chapter deals with the properties and structure of liquids from several points of view.

Section 17.1 starts from what we know of nonideal gases, and considers what modifications need to be made to deal with liquids. In the van der Waals equation for a nonideal gas, the term a/V^2 is the internal pressure of the gas. To the extent that the van der Waals equation applies to a liquid, the term a/V^2 is the internal pressure of the liquid. The internal pressure varies considerably with the type of liquid.

Section 17.2 discusses liquids in comparison with solids. When a solid melts, there is a decrease in *long-range order.* Liquids exhibit little long-range order but have some *short-range order.* This can be represented by a *radial distribution function,* which specifies the number of molecules that occur, on the average, at a given distance from a particular position. Experimental information about radial distribution functions is provided by X-ray and neutron-diffraction studies, and can be compared with theoretical predictions.

Liquid structure depends on the nature and strength of the *intermolecular forces,* and Section 17.3 discusses these forces. In ordinary liquids the most important forces are the *dipole-dipole, dipole-(induced dipole),* and *dispersion forces.* A particularly important intermolecular force in certain liquids is the *hydrogen bond.*

The theories of liquids, which are outlined in Section 17.4, can be classified as

1. *Lattice theories,*
2. Theories based on *statistical mechanics,* and,
3. Theories based on *computer simulation* of the molecular motions in a liquid.

Today most research on liquids is based on computer simulation. However, the lattice theories provide us with important insights into liquid structure and are therefore discussed in this section. The fundamental statistical-mechanical treatments of liquids are unfortunately too advanced to be covered in this book. However, some account is given of the principles, methods, and results of the computer simulations that have been carried out. A useful simulation method is the *Monte Carlo method,* which involves averaging over different molecular configurations. The *molecular-dynamical method* involves averaging over time. These two methods should converge to the same result.

Section 17.5 is concerned with the properties and structure of *liquid water,* which behaves quite differently from any other liquid. In Section 17.6 the hydrophobic effect is discussed in terms of hydrogen bonding between water molecules.

OBJECTIVES

After studying this chapter, the student should be able to:

- Relate the properties of liquids to those of gases.
- Discuss the concept of the internal pressure of a gas and of a liquid.
- Understand the significance of the radial distribution function and the number density of atoms in liquids.
- Explain how the different types of intermolecular force depend on distances and on electric charge, dipole moment, and polarizability.
- Understand the nature of the hydrogen bond and of the hydrophobic effect.
- Describe the different theories that explain the liquid state.
- Appreciate the special and unusual features of liquid water.

A simple definition of a liquid is that it is a material that assumes the shape of a container without necessarily filling it. A gas, by contrast, both takes the shape of a container and fills it, while a solid neither takes the shape nor fills the container. This definition is satisfactory on the whole, but a few materials present difficulty. Glasses, and some polymers, appear to be solid, but at higher temperatures they behave somewhat like liquids even before they melt (see page 909).

We have seen in Chapter 16 that crystals can conveniently be classified according to the kinds of forces that hold the atoms, ions, or molecules together. A similar classification is useful for liquids. Table 17.1 shows the main different types of solids and the corresponding types of liquids, and indicates some of their more important properties. We see that there are three main classes of liquids, in contrast to the four main classes of solids: liquids corresponding to covalent solids do not exist. The intermolecular forces involved in holding liquids together are considered in Section 17.3.

Liquids are more difficult to treat theoretically than either gases or solids. They do not exhibit the randomness of gases or the ordered arrangement of solids, but lie somewhere in between. As an introduction to the theory of liquids we will first present a comparison of them with gases (Section 17.1), and then with solids (Section 17.2).

17.1 Liquids Compared with Dense Gases

As discussed in Chapter 1, ideal gases obey the relationship $PV = nRT$, and this equation can be modified in various ways to interpret nonideal behavior. The earliest

TABLE 17.1 Characteristics of Different Types of Solids and Liquids

Type of Solid	Properties	Examples	Type of Liquid	Properties	Examples	Constituent Elementary Entity (solid and liquid)	Binding Force (solid and liquid)
Ionic crystals	High melting points	NaCl ZnO	Ionic liquids	High boiling points	Molten salts	Ions	Electrostatic
Covalent crystals	Very high melting points	Diamond Si SiO_2	*	*	*	Atoms	Covalent bonds
Molecular crystals	Low melting points	Ar CH_4 CO_2	Molecular liquids	Low boiling points	H_2O C_2H_5OH	Molecules	Dispersion; dipole-dipole; hydrogen bonds
Metallic crystals	Moderate to high melting points	Li Al Fe	Metallic liquids	Moderate to high boiling points	Molten metals	Positive ions in electron gas	Electrostatic; resonance

*No corresponding liquids; covalent crystals such as diamond sublime and no liquid state exists.

and most famous modification was that of van der Waals, whose equation for 1 mol gas is (Eq. 1.100)

$$\left(P + \frac{a}{V_m^2}\right)(V_m - b) = RT \tag{17.1}$$

The term b allows for the volume occupied by the molecules, and the term a/V_m^2 allows for the intermolecular forces. This equation leads to a P-V relationship such as shown in Figures 1.12 and 1.14. Thc left-hand regions of these curves, where the pressure increases rather strongly as the volume decreases, correspond to the liquid state, where the compressibility is much lower than that of the gas. The van der Waals equation therefore provides an interpretation of the liquid state and can be used to explain many of the interesting properties of liquids.

One of these, which we already noted in Section 1.13, is that liquids can exist in states of *negative* internal pressure. This arises because the horizontal portion, where gas and liquid are in equilibrium, is actually an S-shaped curve determined by solving the van der Waals equation. This curve can actually go below the volume axis corresponding to $P = 0$. Experimentally this has been known to be the case for many years. For example, if the space above the mercury in a simple barometer tube is completely filled with a pure liquid, and the tube brought to a vertical position, it is possible for the mercury to stand at a height greater than that corresponding to the barometric pressure. The liquid is therefore under tension, which means that its pressure is negative, and it can remain indefinitely in this condition. If the liquid is agitated, it breaks up into a vapor-liquid mixture, and the mercury level drops. It has been shown that mercury itself can withstand a negative pressure of about 100 atm. The condition for realizing negative pressures in liquids is that no nuclei (e.g., dust) should be present.

Another property of liquids and gases that can also be interpreted in terms of equations such as the van der Waals equation is the *law of the rectilinear diameter.* This was discussed in Section 1.12 (see Figure 1.19).

Internal Pressure

The term a/V_m^2, which appears in the van der Waals equation, is the **internal pressure** P_i of the gas or liquid. We recall from Section 3.8 (Eq. 3.145) that

$$\left(\frac{\partial U}{\partial V}\right)_T = -P + T\left(\frac{\partial P}{\partial T}\right)_V \tag{17.2}$$

Thermal Pressure

The quantity $(\partial U/\partial V)_T$ is the internal pressure, and $T(\partial P/\partial T)_V$ is sometimes known as the **thermal pressure;** thus

$$\text{external pressure } P = \text{thermal pressure } P_t - \text{internal pressure } P_i \tag{17.3}$$

For an ideal gas the right-hand side of Eq. 17.2 is equal to zero, and the internal pressure is therefore zero. For nonideal gases the internal pressure becomes appreciable, while for liquids it is usually much greater than the external pressure.

The internal pressure of a gas or liquid can be calculated from the P-V-T relationship. We showed (Eq. 3.144) that if the van der Waals equation applies,

$$\left(\frac{\partial U}{\partial V}\right)_T = \frac{a}{V_m^2} \tag{17.4}$$

and this equation is conveniently used to obtain a rough estimate of the internal pressure of a gas or a liquid.

EXAMPLE 17.1 Making use of the data given in Table 1.5, estimate the internal pressure of water vapor at 20 °C and 1 atm pressure and of liquid water at 20 °C; the density of water at 20 °C is 1.00 g cm^{-3}.

Solution From Table 1.5, the van der Waals constant a is 0.5536 Pa m^6 mol^{-1}. The molar volume of water vapor at 20 °C is

$$V = \frac{8.3145(\text{J K}^{-1}\text{ mol}^{-1}) \times 293.15(\text{K})}{1.013\ 25 \times 10^5(\text{Pa})} = 0.02405 \text{ m}^3 \text{ mol}^{-1}$$

The internal pressure is therefore

$$P_i = \left(\frac{\partial U}{\partial V}\right)_T = \frac{0.5536(\text{Pa m}^6 \text{ mol}^{-2})}{(0.02405)^2(\text{m}^6 \text{ mol}^{-2})} = 957 \text{ Pa}$$

This is much less than the external pressure (1 atm = 1.013 25 bar = 1.013 25 × 10^5 Pa).

The molar mass of water is 18.008 g mol^{-1}, and since the density of liquid water is 1.00 g cm^{-3}, the molar volume is 18.008 cm^3 mol^{-1} = 1.8008 × 10^{-5} m^3 mol^{-1}. The internal pressure is therefore

$$P_i = \left(\frac{\partial U}{\partial V}\right)_T = \frac{0.5536(\text{Pa m}^6 \text{ mol}^{-2})}{(1.8008 \times 10^{-5})^2(\text{m}^6 \text{ mol}^{-2})} = 1.707 \times 10^9 \text{ Pa}$$

This pressure, equal to 16 850 atm, is now very much greater than the external pressure.

Water, because of its very strong hydrogen bonds, has an exceptionally high internal pressure. In general, dipolar liquids have larger internal pressures than nonpolar liquids because of the greater attractive forces between the molecules. More about the structure of water is to be found in Section 17.5.

Since the internal pressure involves the intermolecular forces, it varies markedly with the external pressure, which affects the average distances between the molecules. At very high external pressures the repulsive forces predominate, and the internal pressure can have a large negative value.

Internal pressure plays an important role in connection with solubilities, as first realized by Joel Hildebrand. If two liquids have similar internal pressures, they may obey Raoult's law (Section 5.4) fairly closely. Two liquids differing considerably in internal pressure usually show positive deviations from Raoult's law, which means that their mutual solubility is reduced.

Internal Energy

The internal energy of an ideal gas is entirely kinetic and depends on the temperature and not on the volume. For a monatomic gas, for example, the kinetic energy per mole is $\frac{3}{2}RT$. For a nonideal gas or a liquid, there is, in addition, potential energy due to the intermolecular forces, and this depends on the average separation between the molecules. If the volume changes the intermolecular forces are altered and therefore the potential energy depends on the volume as well as on the temperature.

The internal energy E_i of a liquid may be expressed as the sum of the kinetic and potential energies,

$$E_i = E_k + E_p \tag{17.5}$$

We can obtain an expression for U by integrating Eq. 17.2:

$$E_i = U = \int \left\{ T\left(\frac{\partial P}{\partial T}\right)_V - P \right\} dV + I(T) \tag{17.6}$$

where $I(T)$ is the constant of integration; by the rules of integration of partial derivatives the constant of integration is a function of temperature, but at a given temperature it is constant and corresponds to the constant of integration for ordinary derivatives. We can determine the value of $I(T)$ by noting that as the volume becomes infinite, the fluid will behave as an ideal gas. We can denote the internal energy under these conditions by U_∞, which for a monatomic fluid is equal to $\frac{3}{2}RT$. Equation 17.6 may thus be written as

$$U = \int \left[T\left(\frac{dP}{dT}\right)_V - P \right] dV + U_\infty \tag{17.7}$$

Caloric Equation of State

This equation has been known as the **caloric equation of state.** It allows the internal energy to be calculated if P-V-T relationships for the gas are known.

The situation is particularly simple if the system obeys the van der Waals equation. In that case the expression within the integral is equal to a/V_m^2 (Eqs. 3.145 and 3.148). Equation 17.7 therefore gives, for the internal energy E_i

$$E_i = U = \int \frac{a}{V_m^2} dV + U_\infty \tag{17.8}$$

$$= -\frac{a}{V_m} + U_\infty \tag{17.9}$$

EXAMPLE 17.2 Making use of the data in Table 1.5 for argon, estimate the internal energy of the following:

a. Gaseous argon at 273.15 K and 1 atm.
b. Gaseous argon at its normal boiling point at 1 atm pressure, 87.3 K, at which the molar volume is 6.79 dm^3 mol^{-1}.
c. Liquid argon at its boiling point, at which the molar volume is 0.0287 dm^3 mol^{-1}.
d. Then estimate the enthalpy of vaporization of argon at its boiling point.

Solution

a. The kinetic energy E_∞ at 273.15 K is

$$E_k = E_\infty = \tfrac{3}{2}RT = \tfrac{3}{2} \times 8.3145 \times 273.15 \text{ J mol}^{-1}$$
$$= 3406.6 \text{ J mol}^{-1}$$

The molar volume at 273.15 K is 22.4 dm^3 mol^{-1} = 0.0224 m^3 mol^{-1}, and the potential energy is therefore

$$E_p = -\frac{a}{V_m} = -\frac{0.1355(\text{Pa m}^6\ \text{mol}^{-2})}{0.0224(\text{m}^3\ \text{mol}^{-1})}$$
$$= -6.05\ \text{J mol}^{-1}\ (\text{Pa m}^3 \equiv \text{kg m}^2\ \text{s}^{-2} \equiv \text{J})$$

The total internal energy is thus

$$E = E_k + E_p = 3406.6 - 6.05 = 3401\ \text{J mol}^{-1}$$

b. For the gas at 87.3 K,

$$E_k = \frac{3}{2} \times 8.3145 \times 87.3 = 1088.8\ \text{J mol}^{-1}$$
$$E_p = \frac{0.1355}{0.00679} = -20.0\ \text{J mol}^{-1}$$
$$E = 1069\ \text{J mol}^{-1}$$

c. For the liquid at 87.3 K,

$$E_k = 1088.8\ \text{J mol}^{-1}$$
$$E_p = -\frac{0.1355}{2.87 \times 10^{-5}} = -4721.3\ \text{J mol}^{-1}$$
$$E = -3633\ \text{J mol}^{-1}$$

d. $\Delta_{\text{vap}}U = 1069 - (-3633) = 4702\ \text{J mol}^{-1}$

$\Delta_{\text{vap}}H = \Delta_{\text{vap}}U + P(V_{\text{vap}} - V_{\text{liq}})$

With $P = 1\ \text{atm} = 1.013\ 25 \times 10^5$ Pa,

$\Delta_{\text{vap}}H = 4702 + 101\ 325(0.00679 - 0.0000287)$

$= 4702 + 685 = 5387\ \text{J mol}^{-1}$

Different estimates of these quantities will, of course, be obtained if different equations of state are employed. The van der Waals equation gives a useful idea of magnitudes, but considerable improvement can be achieved by the use of some of the other equations.

If the liquid is not monatomic, the kinetic energy is greater than $\frac{3}{2}RT$ because of the contributions from vibrational and rotational energy. Usually, because of quantum effects, the vibrational contributions are unimportant. At ordinary temperatures there is usually a contribution of RT for a linear molecule and of $\frac{3}{2}RT$ for a nonlinear molecule.

EXAMPLE 17.3 Estimate the internal energy of liquid water at 20 °C, assuming the translational and the rotational energies to be $\frac{3}{2}RT$ each and making use of the van der Waals constants given in Table 1.5; molar volume of liquid water at 20 °C = 18.008 cm^3 mol^{-1}.

Solution

$$E_k = \tfrac{3}{2}RT + \tfrac{3}{2}RT$$
$$= 3 \times 8.3145 \times 293.15 = 7312 \text{ J mol}^{-1}$$
$$E_p = -\frac{a}{V} = -\frac{0.5532(\text{Pa m}^6 \text{ mol}^{-2})}{1.8008 \times 10^{-5} (\text{m}^3 \text{ mol}^{-1})}$$
$$= -30\,720 \text{ J mol}^{-1}$$

The total internal energy is therefore

$$E = E_k + E_p = 7312 - 30\,720 = -23\,408 \text{ J mol}^{-1}$$

17.2 Liquids Compared with Solids

We will now consider how the organization in a liquid compares with that in a solid. Properties such as potential energy, which depend on the arrangement of atoms or molecules, are known as **configurational properties.**

The distinction between solids and liquids arises from the different configurations. At absolute zero, according to classical physics, the atoms, molecules, or ions in a perfect crystal are at rest in a regular space lattice, as we have seen in Chapter 16. As the temperature is raised, the particles vibrate about the rest positions, but these positions still remain in a regular geometrical arrangement; the *long-range* order is still maintained. As the temperature is raised still further, at a certain temperature the vibrational motions achieve high amplitudes so that the ordered lattice structure breaks down abruptly. This is the phenomenon of *melting.* At any temperature there is competition between the intermolecular attractions, which tend to produce the orderly arrangement that exists in the perfect crystal, and the kinetic energy, which tends to destroy this arrangement and produce the liquid (and eventually the gaseous) state.

Melting occurs very abruptly at a particular temperature; there is no continuous gradation of properties between crystal and liquid. It is possible for there to be a very limited amount of *short-range disorder* in a crystal. However, quite rigorous geometrical arrangements must be satisfied in a crystal structure, and if too much short-range disorder is introduced, there is a serious disruption of the long-range structure and melting will suddenly occur. If we focus attention on a particular molecule in a liquid, its immediate neighbors will not be grouped around it in a completely random fashion but will be in an arrangement that shows some resemblance to the structure in the solid. The next-nearest neighbors show a much smaller degree of order with respect to the molecule in question, and there is no correlation at all between the positions of the molecules separated by greater distances. Liquids therefore exhibit some short-range order but no long-range order. By contrast, solids exhibit both short-range and long-range order, whereas ideal gases exhibit no order at all. The fluidity of liquids arises as a result of the absence of long-range order.

The distance over which the short-range order exists in a liquid is usually a few atomic or molecular diameters. The distance decreases as the temperature is raised.

Radial Distribution Functions

Short-range order is conveniently represented by means of a **radial distribution function,** which specifies the number of elementary entities to be found, on the average, at a distance r from a particular position.

If we consider first a perfect crystal at absolute zero, the radial distribution function is determined precisely by the geometry of the lattice. Sodium, for example, crystallizes in a body-centered cubic lattice, as shown in Figure 17.1a, in which the dimensions are shown. There are 8 atoms at a distance 367 pm from the central atom, 6 atoms at a distance 424 pm, 12 atoms at a distance 600 pm, and so on. These numbers N_r are represented as a function of distance in Figure 17.1b. This

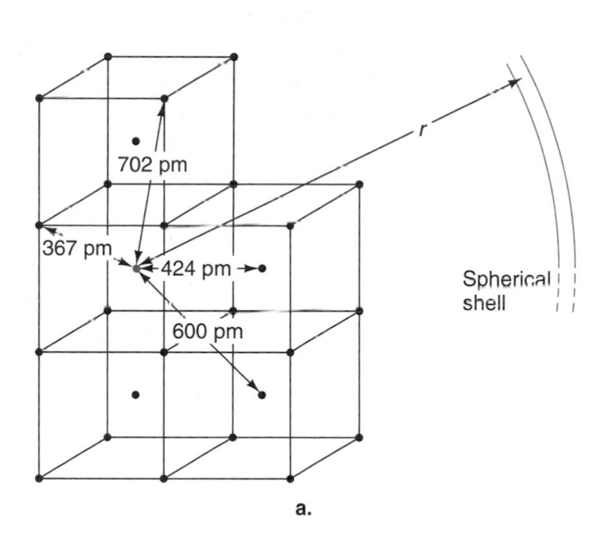

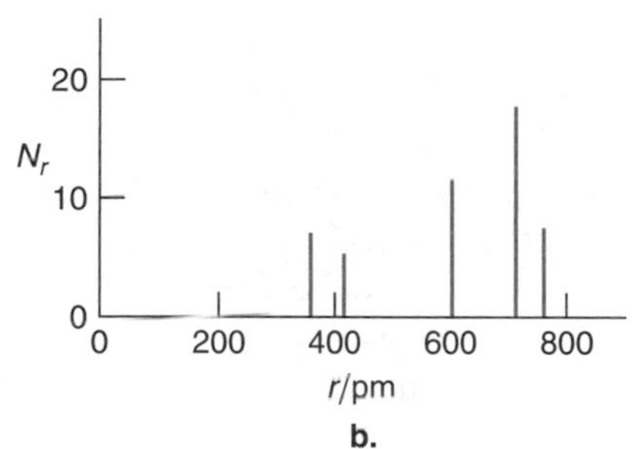

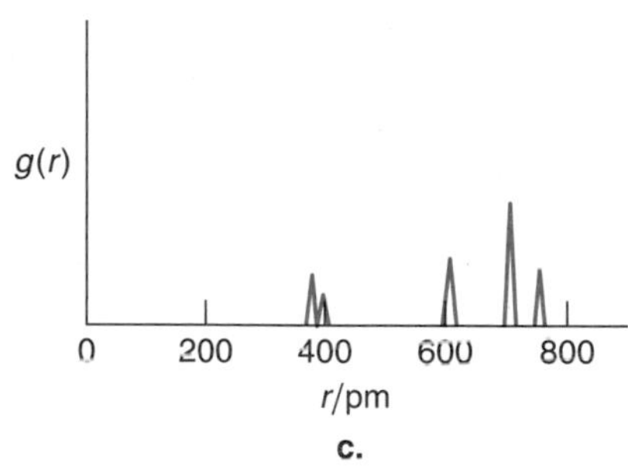

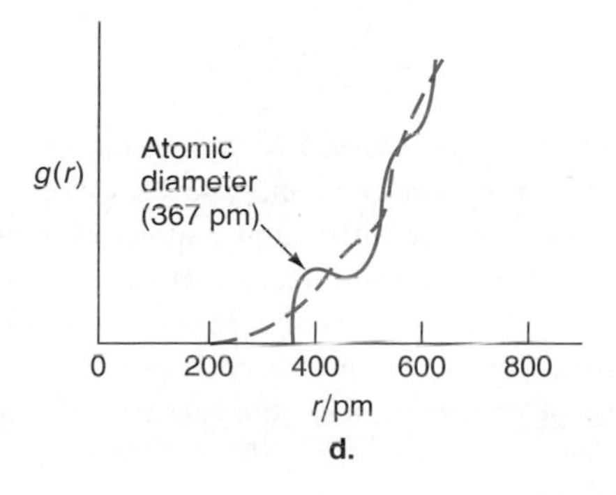

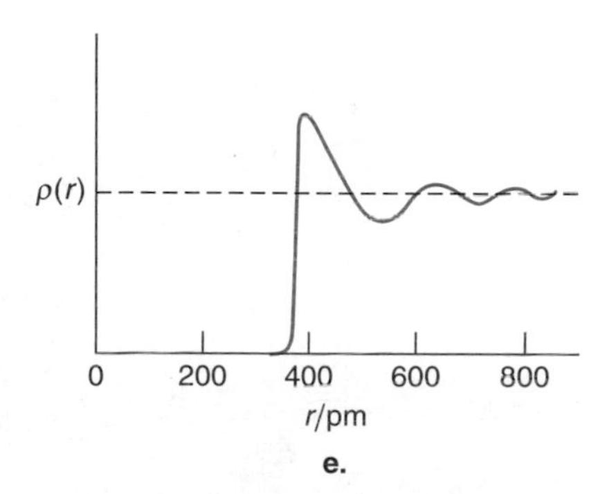

FIGURE 17.1
(a) A body-centered cubic lattice, showing the interatomic distances that apply to solid sodium. (b) The number N_r of sodium atoms at various distances from a particular atom in the solid at the absolute zero. (c) A radial distribution diagram for solid sodium at a higher temperature. (d) Radial distribution diagram for liquid sodium. (e) The number density $\rho(r)$ of sodium atoms plotted against distance r for liquid sodium.

diagram, consisting of vertical lines, shows the radial distribution function in the solid at absolute zero.

At higher temperatures the atoms are not at fixed positions, because of the vibrations. We must now consider the average number $g(r)\,dr$ of atoms that lie within a spherical shell of thickness dr and at a distance r from the rest position of the central atom (Figure 17.1a). At absolute zero, $g(r)$ is equal to N_r, but at higher temperatures the vertical lines are broadened into curves (Figure 17.1c) that increase in width as the temperature rises. The area under any of the peaks corresponds to the number of atoms on the corresponding spherical shell for the lattice at absolute zero. We see from Figure 17.1c that the long-range order is still preserved in the solid.

The radial distribution function for the liquid is shown in Figure 17.1d. If there were complete disorder, the number of atoms whose centers lie between r and $r + dr$ from the central atom would be proportional to the volume of the spherical shell ($4\pi r^2\,dr$) and therefore to r^2. The dashed curve in Figure 17.1d, a parabola, shows such an r^2 dependence. The full curve gives the actual distribution in the liquid. At larger values of r the actual distribution curve is close to the curve for r^2, which means that there is no long-range order. At smaller r values, however, the curve deviates from the r^2 curve, showing that there is some short-range order. There is an initial maximum at about 400 pm, corresponding to a shell of nearest neighbors, and an inflexion at 500–600 pm corresponding to a second group of next-near neighbors. At greater distances the order is completely lost.

An alternative way of representing the situation is to plot the *number density* of the atoms for any distance r. This quantity, $\rho(r)$, is the number of atoms per unit volume and is the ratio of the number in a spherical shell to the volume of the shell:

$$\rho(r) = \frac{g(r)\,dr}{4\pi r^2\,dr} = \frac{g(r)}{4\pi r^2} \tag{17.10}$$

This function has the form shown in Figure 17.1e. At large distances, $\rho(r)$ is equal to the mean number density for the liquid as a whole, ρ_0. At short distances, there are maxima and minima, reflecting the short-range order.

The previous considerations apply strictly only to spherical molecules, in which the direction in space has no effect. Nonspherical molecules, such as benzene, have a tendency to become aligned in a particular orientation, and the distribution function depends on angles as well as on r. The distribution functions obtained by X-ray and neutron diffraction, now to be considered, are averaged over all angles.

X-Ray Diffraction

In Section 16.3 we discussed X-ray diffraction studies on solids. Shortly after von Laue's pioneering work on solids, several workers made similar studies on liquids. It was found that the diffraction pattern obtained from a liquid is not nearly so sharp as that from a crystal. Instead of the pattern of closely spaced spots forming a ring, found with a solid (Figure 16.25), a liquid produces only one or two diffuse rings surrounding an intense central spot. This lack of sharpness is due to the lack of long-range order, but the existence of the rings shows that there is some short-range order.

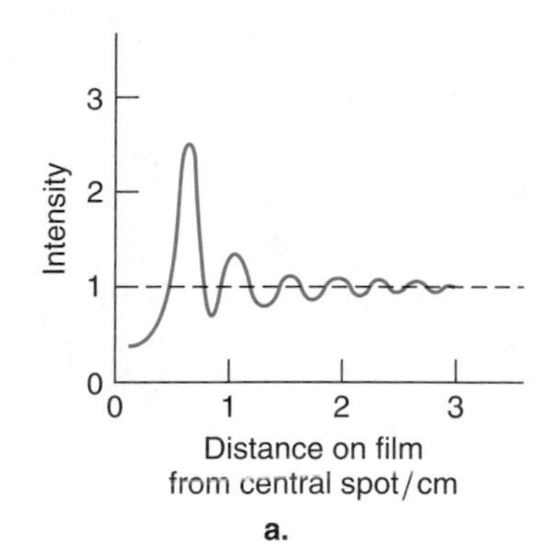

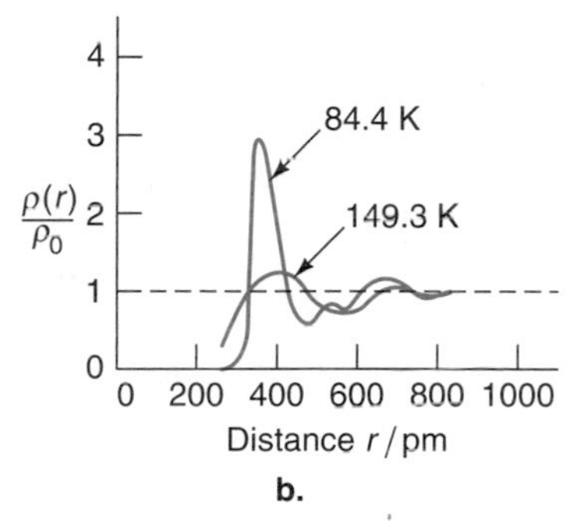

FIGURE 17.2
(a) A photometric tracing of an X-ray diffraction photograph, obtained with liquid mercury.
(b) Radial distribution functions for liquid argon at 84.4 K and 149.3 K.

An X-ray diffraction experiment gives the intensity of the diffracted rays as a function of the angle of diffraction. A typical example is shown in Figure 17.2a, for liquid mercury. The problem is then to derive the radial distribution function, and this is done by a procedure worked out in 1927 by P. Zernicke and J. A. Prins, involving the use of a Fourier inversion. Figure 17.2b shows radial distribution curves for liquid argon at two temperatures, obtained in this way by A. Eisenstein and N. S. Gingrich in 1942. These curves show the number density $\rho(r)$ divided by ρ_0, the value at large distances. At the lower temperature, 84.4 K, there is a well-defined maximum at about 340 pm, which is approximately the diameter of the argon atom. At twice this distance the order has essentially disappeared. At this temperature the number density at 340 pm is slightly more than 3 times the value for the liquid as a whole. At the higher temperature, 149.3 K, the short-range order is much less; the number density is now somewhat less than 1.5 times the value for the bulk of the liquid.

X-ray diffraction studies of liquid water are considered in Section 17.5.

Neutron Diffraction

Radial distribution functions for liquids also may be obtained by the study of the diffraction of neutrons. Like other particles, neutrons show wave properties and have a wavelength λ given by de Broglie's relationship $\lambda = h/mu$. The usual procedure is first to diffract a beam of neutrons by a single crystal of lead or copper so as to produce a monochromatic beam of wavelength about 0.11 nm. This beam then falls on the liquid, and scattered intensities are studied at various positions. Complications arise if the liquid contains an element with more than one isotope, since neutrons are scattered by the nuclei. An advantage of the neutron-scattering technique is that neutrons are scattered by light atoms such as H and He, for which X-ray scattering is negligible.

Glasses

Some materials occur in the *glassy* or *vitreous state,* and they present an interesting problem since they show properties of both solids and liquids. At sufficiently high temperatures they flow like liquids, but when cooled they do not solidify at a fixed temperature; instead their viscosity increases steadily until finally their physical properties are essentially those of solids. Glasses can usefully be described as *supercooled liquids.* Most liquids can be supercooled, but supercooled liquids are generally unstable and will solidify suddenly if they are further cooled or are agitated; glasses, however, can be induced to crystallize only with difficulty, if at all.

X-ray diffraction experiments on glasses have shown that they give the diffuse ring pattern that is characteristic of liquids, and that even at low temperatures they do not show the type of pattern found with solids. It is therefore concluded that, although having the outward appearance of solids, glasses are disordered on the atomic scale. This behavior can be understood from the standpoint of regarding a glass as a supercooled liquid. The viscosity of all liquids increases with decrease of temperature (see Section 18.1). Thus, if a highly viscous liquid is cooled sufficiently rapidly, atomic rearrangements are brought essentially to a standstill, and the disordered configuration of the liquid is retained. At sufficiently low temperatures the viscosity is so high that the material has the appearance of a solid in spite of having the atomic structure of a liquid.

17.3 Intermolecular Forces

When molecules are near enough to influence one another, forces of attraction and repulsion come into play. If there were no forces of attraction, all matter would be gaseous, since there would be nothing to bring the molecules together in the solid and liquid states. The behavior of matter in condensed phases is determined by the balance between the forces of attraction and repulsion.

The force F between two molecules is related to the potential energy E_p by the equation

$$F = -\frac{dE_p}{dr} \tag{17.11}$$

where r is the distance between them. Potential energy is a relative quantity, and it is usual to take it to be zero when the molecules are separated by an infinite distance. It is usually more convenient to deal with energy rather than force, and Figure 17.3a shows the typical variation with distance r of the potential energy for two interacting molecules; Figure 17.3b shows the corresponding variation of the force.

There are a number of types of intermolecular forces, and these are listed in Table 17.2, which indicates how the force and the energy vary with distance and gives some typical values. A useful classification of energies is into **short-range** and **long-range** energies. Short-range energies are those which vary strongly with the distance; the exchange energies involved in covalent bond formation (Section 12.2) and repulsive energies fall in this category. Both of these energies have their origin in the interaction between wave functions of the atoms or molecules concerned.

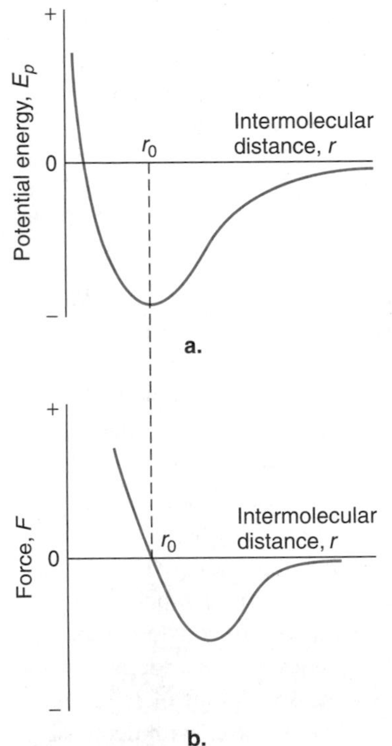

FIGURE 17.3
(a) The variation of potential energy with distance for two interacting molecules. (b) The corresponding variation of force.

Long-range energies vary less strongly with the distance, and they are the energies that can be qualitatively understood in terms of classical electrostatics. The ion-dipole energies, the induction energies [ion-(induced dipole) and dipole-(induced dipole) energies], and the dispersion energies fall into this category. These energies will be considered in a little more detail later in this section.

Another factor of importance is **pairwise additivity.** If three molecules A, B, and C are close to one another, we can consider the energies of the individual pairs A-B, B-C, and A-C and ask whether the total energy is the sum of these three energies. If so, we have pairwise additivity; otherwise not. The electrostatic energy is exactly pair additive, and it can be shown by quantum-mechanical theory that the dispersion energy is almost pair additive. When, however, electrostatic induction is involved, there is no pair additivity, as will be seen later when we consider ion-(induced dipole) and dipole-(induced dipole) interactions.

Ion-Ion Forces

If two charges z_Ae and z_Be are separated by a distance r, the force of attraction or repulsion obeys the inverse-square law, and the force and potential energy obey the equations

$$F = -\frac{z_A z_B e^2}{4\pi\epsilon_0 \epsilon r^2} \quad \text{and} \quad E_p = \frac{z_A z_B e^2}{4\pi\epsilon_0 \epsilon r} \tag{17.12}$$

TABLE 17.2 Types of Intermolecular Forces

Force	Dependence on Distance of: Force	Energy	Range	Pairwise Additive	Typical Potential-Energy Values: E_p per Molecule J	E_p per Mole kJ mol^{-1}
Ion-ion	r^{-2}	r^{-1}	Long	Yes	-1.16×10^{-18}	-680
Ion-dipole	r^{-3}	r^{-2}	Long	Yes	-1.20×10^{-19}	-72
Ion-(induced dipole)	r^{-5}	r^{-4}	Long	No	-1.08×10^{-19}	-65
Dipole-dipole	r^{-7}	r^{-6}	Long	Yes	-2.54×10^{-20}	-15.3
Dipole-(induced dipole)	r^{-7}	r^{-6}	Long	No	-1.56×10^{-21}	-0.9
Dispersion	r^{-7}	r^{-6}	Long	Yes	-7.34×10^{-20}	-44.2
Repulsion	r^{-13}	r^{-12}	Short	Yes	—	—

Calculated for the following values: unit (electronic) charge 1.602×10^{-19} C; dipole moment of 1 D (3.338×10^{-30} C m); polarizability of 1.5×10^{-30} m^3; intermolecular distance 2.00×10^{-10} m. The dispersion energy values are for water molecules (Table 17.3).

Here ϵ_0 is the permittivity of a vacuum and ϵ is the relative permittivity or dielectric constant. For two ions in a vacuum, $\epsilon = 1$, but (as we have seen in the introduction to Chapter 7) if the ions are in solution, an appropriate value for ϵ must be used. The energies in Table 17.2 are calculated for a vacuum.

Ion-Dipole Forces

The concept of dipole moment has been discussed in Section 12.4, p. 601. Figure 17.4a shows a dipolar molecule with an ion at a distance r from its center of charge. The force between them depends on the angle θ; the arrangement in Figure 17.4b gives maximum attraction, and the arrangement in Figure 17.4c gives the maximum repulsion. In general it can be shown that, provided r is much larger than the length of the dipole, the force and potential energy are given by

$$F = \frac{z_A e \mu \cos\theta}{4\pi\epsilon_0 \epsilon r^3} \qquad E_p = -\frac{z_A e \mu \cos\theta}{4\pi\epsilon_0 \epsilon r^2} \tag{17.13}$$

where μ is the dipole moment.

EXAMPLE 17.4 A unit positive charge is situated at a distance of 200 pm from a molecule of dipole moment 3.338×10^{-30} C m (1.00 D). Calculate the maximum force and the corresponding potential energy.

Solution The maximum force from Eq. 17.13, with $\epsilon = 1$, is

$$F = \frac{e\mu}{4\pi\epsilon_0 r^3} = \frac{1.602 \times 10^{-19}(\text{C})\ 3.338 \times 10^{-30}(\text{C m})}{4\pi(8.854 \times 10^{-12})(\text{C}^2\ \text{N}^{-1}\ \text{m}^{-2})[2.00 \times 10^{-10}(\text{m})]^3}$$

$$= 6.01 \times 10^{-10}\ \text{N}$$

The corresponding potential energy, relative to the separated species, is

$$E_p = -\frac{e\mu}{4\pi\epsilon_0 r^2} = -1.20 \times 10^{-19}\ \text{J} = -72.3\ \text{kJ mol}^{-1}$$

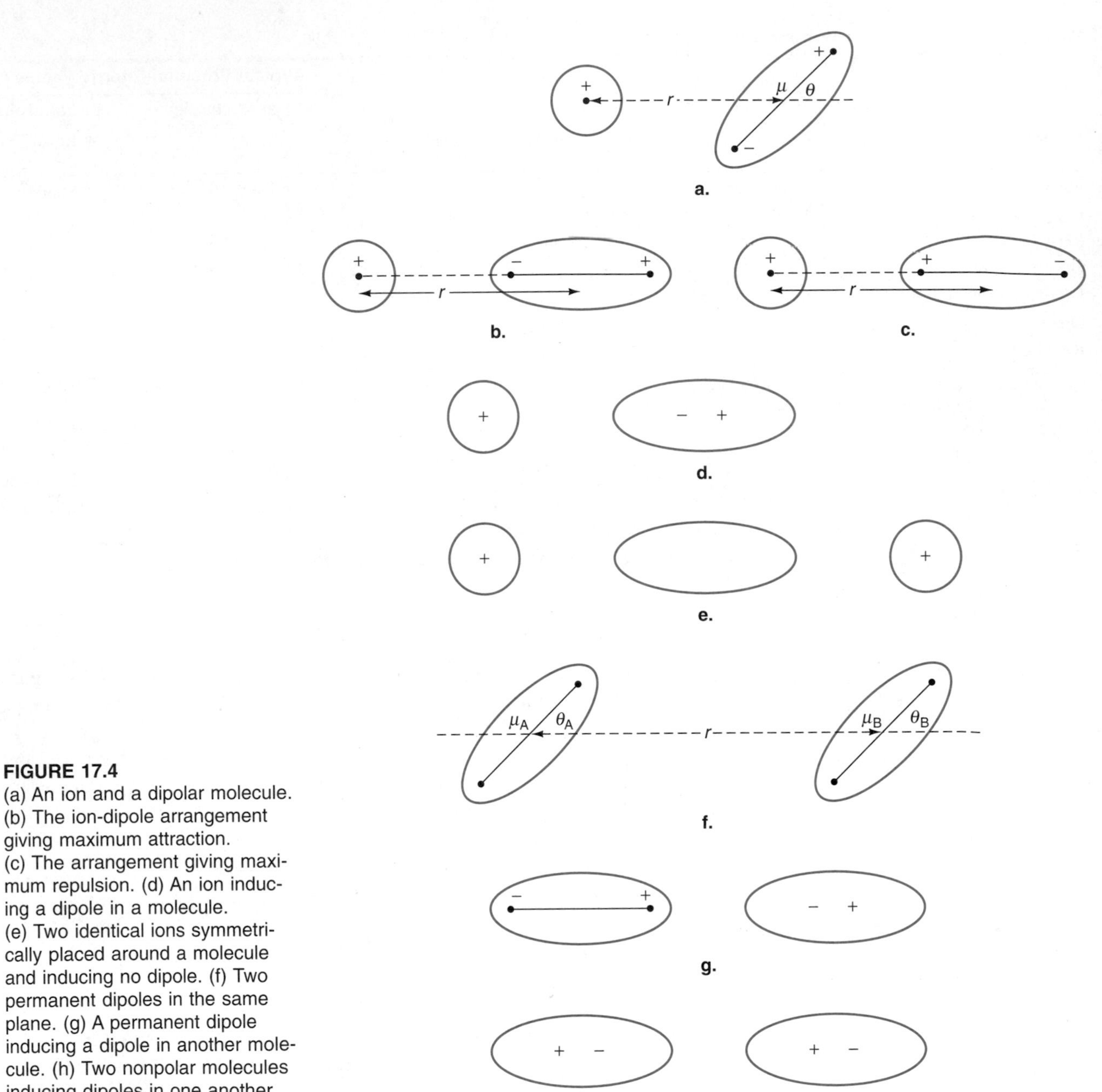

FIGURE 17.4
(a) An ion and a dipolar molecule. (b) The ion-dipole arrangement giving maximum attraction. (c) The arrangement giving maximum repulsion. (d) An ion inducing a dipole in a molecule. (e) Two identical ions symmetrically placed around a molecule and inducing no dipole. (f) Two permanent dipoles in the same plane. (g) A permanent dipole inducing a dipole in another molecule. (h) Two nonpolar molecules inducing dipoles in one another (dispersion forces).

An atom, or a molecule that has no permanent dipole moment, can have a dipole moment induced in it by an electric field. If the field strength is E, the induced dipole moment is αE, where α is the electric polarizability of the molecule. The dipole is formed in the direction of the field, and there is therefore always attraction between an ion and an induced dipole. If an ion of charge $z_A e$ is at a distance r from a molecule of polarizability α, the potential energy is

$$E_p = -\frac{\alpha(z_A e)^2}{8\pi\epsilon_0 \epsilon r^4} \tag{17.14}$$

EXAMPLE 17.5 A unit charge is at a distance of 200 pm from a molecule having no dipole moment but a polarizability of 1.5×10^{-30} m^3. Calculate the potential energy.

Solution From Eq. 17.14,

$$E_p = \frac{\alpha e^2}{8\pi\epsilon_0 r^4} = -\frac{1.5 \times 10^{-30}(\text{m}^3)[1.602 \times 10^{-19}(\text{C})]^2}{8\pi(8.854 \times 10^{-12})(\text{C}^2\ \text{N}^{-1}\ \text{m}^{-2})[2.00 \times 10^{-10}(\text{m})]^4}$$

$$= -1.08 \times 10^{-19}\ \text{J} = -65.0\ \text{kJ mol}^{-1}$$

Note that this is not much less than the energy calculated in the previous example for the ion-(permanent dipole) force.

As noted in Table 17.2, the ion-(induced dipole) forces are not pair additive, and it is easy to understand why this is so. If an ion is inducing a dipole in a molecule (Figure 17.4d) and another ion is brought up to the molecule, it will alter the dipole induced by the first ion. For example, if two identical ions are on two sides of a molecule, there will be no induced dipole (Figure 17.4e), and therefore no ion-(induced dipole) effect at all.

Dipole-Dipole Forces

Figure 17.4f shows two permanent dipoles in the same plane. If the distance between them is much greater than the length of the dipoles, the potential energy is given by

$$E_p = -\frac{\mu_A \mu_B}{4\pi\epsilon_0 \epsilon r^3}(2\cos\theta_A \cos\theta_B - \sin\theta_A \sin\theta_B) \tag{17.15}$$

When the dipoles are aligned in the same direction, the force of attraction is a maximum and the potential energy is

$$E_p = -\frac{\mu_A \mu_B}{2\pi\epsilon_0 \epsilon r^3} \tag{17.16}$$

In a gas or a liquid the potential energy will be an average over the various orientations of the molecules. Orientations giving rise to a lower potential energy are favored over those giving a higher potential energy, in accordance with the Boltzmann distribution. When this is taken into account, with an averaging over all orientations, the potential energy is found to be

$$E_p = -\frac{\mu_A^2 \mu_B^2}{24\pi^2\epsilon_0^2\epsilon^2 k_B T r^6} \tag{17.17}$$

where k_B is the Boltzmann constant and T is the temperature. Note that the dependence is now on the inverse sixth power of the distance.

EXAMPLE 17.6 Calculate the potential energy at 25 °C for two molecules of dipole moments 3.338×10^{-30} C m (1.00 D) separated by 2.00×10^{-10} m.

Solution From Eq. 17.17,

$$E_p = -\frac{[3.338 \times 10^{-30}(\text{C m})]^4}{24\pi^2[8.854 \times 10^{-12}(\text{C}^2\ \text{N}^{-1}\ \text{m}^{-2})]^2 1.381 \times 10^{-23} \times 298.15(\text{J})[2 \times 10^{-10}(\text{m})]^6}$$

$$= -2.54 \times 10^{-20}\ \text{J} = -15.3\ \text{kJ mol}^{-1}$$

A permanent dipole can induce a dipole in a neighboring molecule (Figure 17.4g). If two molecules both have a dipole moment μ and a polarizability α, each will induce a dipole moment in the other, and the resulting potential energy is

$$E_p = -\frac{\alpha\mu^2}{2\pi\epsilon_0\epsilon r^6} \tag{17.18}$$

If only one of the molecules has a permanent dipole moment, the potential energy has just half this value.

These dipole-(induced dipole) energies are not pair additive.

Hydrogen Bonds

A particularly important bond, which has its main origin in dipole-dipole interactions, is the **hydrogen bond.** With this bond the intermolecular energy is exceptionally large, and the bond is particularly important for the liquid state.

In a hydrogen bond a hydrogen atom is bonded to two atoms A and B, the structure A····H····B usually being close to colinear. The atoms A and B are most commonly oxygen, nitrogen, or fluorine. The hydrogen atom is usually more strongly bound to one atom than to the other. For example, in the hydrogen fluoride dimer

H—F····H—F

the intermolecular H····F bond is longer and weaker than the bond in the HF molecule.

Quantum-mechanical calculations have confirmed the early view that the main contribution to the binding in the hydrogen bond is the electrostatic interaction energy between the two dipoles. In the case of stronger hydrogen bonds, however, the theoretical work has shown that there is a significant contribution from a valence-type interaction involving the overlap of orbitals. Thus in the case of a hydrogen bond

A—H····B

the orbitals of A—H overlap those of B, and there is a partial transfer of electrons from B to the A····H bond.

Hydrogen bonds play a particularly important role in the structure of water, as will be discussed in Section 17.5.

Dispersion Forces

The forces considered so far occur only if the molecules are charged or have permanent dipole moments. However, there must be attractive forces even between neutral molecules having zero dipole moments, since liquefaction always occurs at sufficiently low temperatures. The forces in such cases are known as **dispersion forces,** and the theory of them was worked out in 1930 on the basis of quantum mechanics by Fritz London.

The origin of the dispersion forces is as follows. Suppose that two completely nonpolar atoms or molecules are close together, as shown in Figure 17.4h. On the average the electron clouds are arranged symmetrically, but at any given instant the electron distribution in one of the molecules may be unsymmetrical, as shown in the figure. The molecule is therefore momentarily a dipole, and it induces a dipole in the neighboring molecule. Both molecules are hence dipoles at this instant, and the direction of the dipoles is such that they attract one another. Since the electron clouds are in rapid motion, there is a rapid fluctuation of dipoles, but at every instant the dipole in each molecule is inducing one in the other, and there is attraction. The effect is greatest if the molecules are of high polarizability.

London's quantum-mechanical treatment of the problem led to the result that for two identical molecules of polarizability α the potential energy is

$$E_p = -\frac{3h\nu_0\alpha^2}{4r^6} \tag{17.19}$$

where ν_0 is the frequency of oscillation of the dipoles and h is the Planck constant.

Repulsive Forces

If two atoms or molecules are brought close together, the electron clouds will interpenetrate and will no longer be able to shield the nuclei. There is then a repulsive force, and the energy rises as the intermolecular distance is reduced.

Theoretical treatments of this problem have led to the result that the potential energy due to this repulsive effect can be satisfactorily expressed by

$$E_p = \frac{B}{r^n} \tag{17.20}$$

where B and n are constants. For many molecules the best value for n is 12, and the value of B is usually determined by use of the experimental intermolecular distance.

Resultant Intermolecular Energies

Table 17.3 gives some idea of the magnitudes of the different kinds of attractive energies. The values are calculated from data given by F. London and relate to an intermolecular separation of 500 pm. It is to be seen that the dipole-(induced dipole) forces are always quite small. For He and Xe, where there are no dipole forces, the only contribution is from the dispersion forces, which give a substantial attraction for Xe because of its very high polarizability. Carbon monoxide has only a very small dipole moment and the only significant force is the dispersion force. For the molecules with larger dipole moments the dipole forces are more important,

TABLE 17.3 Attractive Energy Contributions for an Intermolecular Separation of 500 pm

Molecule	Dipole Moment $\mu/10^{-30}$ C m	Polarizability $\alpha/10^{-30}$ m^3	E_p/J mol^{-1} Due to Dipole-Dipole Forces	Dipole-(Induced Dipole) Forces	Dispersion Forces
He	0	0.20	0	0	−4.6
Xe	0	4.00	0	0	−850.0
CO	0.40	1.99	−0.012	−0.22	−260.0
HCl	3.43	2.63	−72.0	−2.0	−405.0
NH_3	5.01	2.21	−324.0	−39.0	−360.0
H_2O	6.14	1.48	−732.0	−39.0	−180.0

and for water the hydrogen bonding (mainly dipole-dipole) forces make the largest contribution to the attraction.

For uncharged molecules the attractions all depend on r^{-6}, and if the repulsive contribution depends on r^{-12}, the net energy is

$$E_p = -\frac{A}{r^6} + \frac{B}{r^{12}} \tag{17.21}$$

where A and B are constants. This function is known as the **Lennard-Jones 6–12 function,** after the British theoretical physicist John Lennard-Jones (1894–1954), who first suggested it. It corresponds to a curve of the form shown in Figure 17.3a, and it has been widely used in the study of interactions between uncharged molecules. If r_e is the equilibrium intermolecular separation and E_0 is the classical dissociation energy, $dE_p/dr = 0$ at $r = r_0$ and $E_p = -E_0$ at $r = r_e$; from these boundary conditions we find that

$$E_p = E_0\left\{\left(\frac{r_0}{r}\right)^{12} - 2\left(\frac{r_0}{r}\right)^6\right\} \tag{17.22}$$

This is often a more convenient form of the equation. Because of the theoretical difficulties of calculating the constants A and B, it is usual to make use of experimental values of E_0 and r_0.

17.4 Theories and Models of Liquids

There are three important, different approaches to the understanding of liquid structure:

1. **Lattice theories,** in which the liquid is treated as if it were a somewhat distorted solid.
2. Theories based firmly on the principles of **statistical mechanics.**
3. Treatments based on **computer simulation** of the motions of a large number of molecules of a liquid.

It is undoubtedly the case that the third approach, utilizing the massive computing capacity available today, provides the most accurate representation of the behavior of a liquid. On the other hand, computer simulation generally does not provide the kind of insight into chemical behavior that can often be provided by simpler and less accurate models. We will therefore give brief accounts of some early models of liquids before outlining what has been learned from the more modern treatments.

The lattice theories can be further subdivided as follows:

1. **Free-volume** or **cell** theories, in which the molecules are assumed to be present at lattice sites, all of which are occupied. Whereas in the solid the atoms, ions, or molecules simply vibrate about their lattice positions, in the liquid they have more freedom and move within a "cell" created by the surrounding molecules.
2. **Hole** or **"significant structure"** theories, which resemble the free-volume theories, but in which some of the lattice sites are considered to be unoccupied.

Free-Volume or Cell Theories

A number of fairly simple theoretical treatments have been based on the idea that each molecule in a liquid is confined by its neighbors to a small region of space. According to these theories the local environment of each molecule is much the same as it is in the solid, although there is no long-range order.

The basic ideas in the simplified treatment we will give are similar to those in the theories proposed in 1937 by Lennard-Jones and A. F. Devonshire, and independently by Henry Eyring and Joseph O. Hirschfelder (1911–1990). We assume that all the "cages" in which the molecules are confined are identical and that the molecules move entirely independently of one another in their cages.

Suppose that the molecules in the liquid are arranged in a simple cubic lattice, that the molecules are incompressible spheres of diameter d, and that the distance between lattice positions is a (Figure 17.5a). If we consider motion along the X axis, we see that the distance in which the central molecule is free to move is $2a - 2d$. It can move this distance in three dimensions, so that the volume of the cube in which it can move (i.e., the free volume) is

$$V_f = (2a - 2d)^3 = 8(a - d)^3 \tag{17.23}$$

Alternatively, we might suggest that the molecule can move in a sphere of radius $a - d$, in which case the estimated free volume is

$$V_f = \frac{4\pi}{3}(a - d)^3 \tag{17.24}$$

which is somewhat smaller. The true volume probably lies somewhere in between.

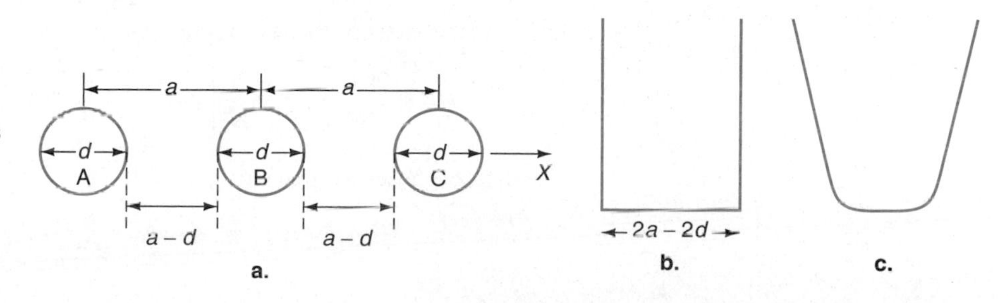

FIGURE 17.5
(a) The relationship between free volume and the molecular diameters and lattice distances. (b) The square-well potential-energy function assumed in the simplest application of free-volume theories of liquids; the molecule has free translational motion. (c) The type of potential-energy variation in the treatments of Lennard-Jones and Devonshire.

Free volumes of liquids can be estimated in a number of ways. One way is to make use of the speed of sound in the liquid. For most liquids this speed, u_l, is 5 to 10 times the speed of sound, u_g, in the gas. The reason for this is illustrated in Figure 17.5a, which shows three molecules A, B, and C in a straight line. If a sound wave travels from left to right, molecule A collides with B and the signal is transmitted almost instantaneously to the opposite side of B. Thus, although molecule A travels only the distance $a - d$ in order to strike B, the sound wave has effectively traveled the distance a. The ratio of sound velocities is thus

$$\frac{u_l}{u_g} = \frac{a}{a - d} \tag{17.25}$$

Thus $a - d$, and hence the free volume, can be estimated from the values of u_l, u_g, and a.

We will first consider the construction of a partition function for the liquid on the basis of the assumption that each molecule undergoes free translational motion in a free volume V_f (Figure 17.5b). The translational partition function per molecule (Eq. 15.86) is then

$$q_t = \frac{(2\pi m k_B T)^{3/2}}{h^3} V_f \tag{17.26}$$

This must be multiplied by an internal partition function q_i that takes account of the rotational and vibrational motions of the molecules. In addition, we must multiply by the factor $e^{-\epsilon/k_B T}$, where ϵ is the energy of the molecule inside its cage, with respect to the molecule in the gas phase. The complete partition function per molecule is therefore

$$q = \frac{(2\pi m k_B T)^{3/2}}{h^3} V_f q_i e^{-\epsilon/k_B T} \tag{17.27}$$

There are alternative procedures for expressing the partition function for an assembly of N molecules in the liquid phase. We have so far developed the argument on the assumption that the molecules are *localized,* being confined to cages at the various lattice sites. In reality, this is more like the behavior in a solid, and if we keep to this assumption, the partition function for the assembly is simply q raised to the Nth power (Eq. 15.45):

$$Q = q^N = \left[\frac{(2\pi m k_B T)^{3/2}}{h^3} V_f q_i e^{-\epsilon/k_B T}\right]^N \tag{17.28}$$

Alternatively, if we now relax this assumption and say that each molecule has access to the *entire* free volume of the liquid, we must replace V_f by $V_f N$. At the same time we must divide the partition function by $N!$, which is the number of ways of distributing the N molecules over all the N sites (Eq. 15.47). The partition function found on the basis of this assumption is

$$Q' = \frac{1}{N!}\left[\frac{(2\pi m k_B T)^{3/2}}{h^3} N V_f q_i e^{-\epsilon/k_B T}\right]^N \tag{17.29}$$

The ratio of these two partition functions is

$$\frac{Q'}{Q} = \frac{N^N}{N!} \tag{17.30}$$

and if we make use of Stirling's approximation (Eq. 15.25), this reduces to

$$\frac{Q'}{Q} = e^N \tag{17.31}$$

To help us to decide which of the two formulations, Eqs. 17.28 or 17.29, is the more satisfactory, it is useful to calculate the entropy difference between them. In Table 15.2 we saw that the entropy is related to the partition function by Eq. 15.62:

$$S = k_B T \frac{d \ln Q}{dT} + k_B \ln Q \tag{17.32}$$

The difference in the entropies calculated from Eqs. 17.28 and 17.29 is therefore

$$S' - S = k_B T \frac{d \ln(Q'/Q)}{dT} + k_B \ln \frac{Q'}{Q} \tag{17.33}$$

Since Q'/Q is e^N and is independent of T, the first term vanishes; the second is Nk_B:

$$S' - S = Nk_B \tag{17.34}$$

The molar entropy is obtained by dividing by the amount of substance n:

$$S'_m - S_m = \frac{Nk_B}{n} = Lk_B = R \tag{17.35}$$

Communal Entropy

Thus, the liquid model in which the molecules have complete access to the *entire* free volume of 1 mol of liquid gives an entropy of R greater than the model in which the molecules are confined to their *individual* free volumes. This additional entropy contribution associated with access to the entire free volume is known as the **communal entropy.** At one time it was thought that the partition function Q applies to the *solid* state and Q' to the *liquid* state and that on melting there is a sudden transition from one to the other with an increase of entropy due largely to the increase in communal entropy. Certain experimental results support this point of view. For example, for metals the constituent units are atoms, and there are therefore no internal degrees of freedom to cause complications. Experimentally, entropies of fusion of metals lie in the range of 7.0 to 9.6 J K^{-1} mol^{-1} and, therefore, are similar to the value of R (8.3145 J K^{-1} mol^{-1}). Moreover, substances for which there are internal degrees of freedom, which may be expected to play a more important role in the liquid state, generally have entropies of fusion in excess of R. It was therefore suggested that when melting occurs, there is an increase of entropy of R due to the communal entropy, and an additional increase if the molecules have more rotational and vibrational freedom in the liquid than in the solid state.

The cell or free-volume theories of liquids have been developed and improved in a number of ways. In the account we have given, the molecules were assumed to be moving in a square potential-energy well (Figure 17.5b). Lennard-Jones and Devonshire improved the treatment considerably by taking into account the intermolecular forces. This means that the potential energy varies in the manner shown in Figure 17.5c. Other refinements that have been introduced allow for the occupancy of cells by more than one molecule; this may be significant at higher temperatures.

Once a partition function has been set up, all the thermodynamic properties can be calculated by making use of the relationships listed in Table 15.2. It is also possible to calculate radial distribution functions and to compare them with experiment.

These improvements to the theory did not, however, lead to the good agreement with experiment that can now be obtained by computer simulation. As a result, the cell approach is no longer used by research workers concerned with making quantitative predictions, although it is still helpful in providing valuable insights.

Hole or "Significant Structure" Theories

One improvement to the cell or free-volume theories involves the assumption that some of the lattice sites may be unoccupied; in other words, the liquid contains *holes.* This idea was first suggested in 1936 by Henry Eyring, who over several decades greatly developed the idea into what he called his theory of *significant structures.* Again, we will discuss the theory only in its simplest form.

When most solids melt, the density decreases, and there is also a decrease in the number of nearest neighbors as determined by diffraction experiments. There is a still further decrease in the density and in the number of neighbors as the temperature of the liquid is raised. These facts suggest that when we employ a lattice model for a liquid, we ought to consider the possibility that some of the lattice sites are empty. The presence of holes, more or less randomly distributed in the lattice, gives an alternative interpretation of the entropy increase on melting. Moreover, there is evidence to suggest that the viscosity of most liquids depends mainly on the volume. This can be explained by the hypothesis that liquids contain holes and that diffusion and flow in liquids can be interpreted in terms of a movement of these holes.

Although Eyring's theories of liquids were fruitful in their time, they are now not much used. However, they still provide useful qualitative insights into the behavior of liquids.

Partition Functions for Liquids

The construction of a satisfactory partition function for a liquid is a matter of considerable difficulty and is beyond the scope of this book. The reader is referred to advanced treatments of statistical mechanics, some of which are listed at the end of this chapter.

Computer Simulation of Liquid Behavior

A mole consists of over 10^{23} molecules, and no computer is (or is likely to be) large enough to simulate the behavior of that number of particles. However, modern computers can deal with the dynamics of a few thousand molecules and can carry out calculations that take into account the intermolecular forces acting on each molecule. The pioneers in this type of investigation were B. J. Alder and T. E. Wainwright, who in 1959 treated a group of 32 molecules as hard spheres. Subsequently, computations have been made for groups of thousands of molecules, and realistic intermolecular forces have been taken into account. The motions of the molecules are calculated on the basis of classical mechanics, positions and velocities often being determined at intervals of 1 fs (10^{-15} s).

Monte Carlo Method

The properties of the liquid can then be calculated by an averaging procedure, and this can be done in two different ways. One method is the **Monte Carlo method,** so named because it is analogous to what happens in a gambling casino, in which dice act as random-number generators. In the Monte Carlo method, many molecular configurations are selected and the potential energy of each one is calculated; the various configurations are then weighted by the Boltzmann factor and the properties calculated on this basis. The Monte Carlo method involves *averaging over different molecular configurations.*

Various computing devices have been used to make the Monte Carlo method more practicable. If molecular configurations are selected entirely at random, a very large majority will involve such interpenetration of orbitals that they are highly improbable. Computations made in this way are very time consuming and inefficient. Much improvement has been brought about by the use of a procedure suggested in 1953 by N. Metropolis and his collaborators. In essence, the molecules move in a random fashion from one configuration to another, but if a molecule comes too close to its neighbors, it is required to bounce back to its original position. In this way the procedure avoids making computations for the very unlikely configurations for which the potential energies are excessively large. Another complication with the Monte Carlo method is that, since the number of molecules has to be limited, a substantial proportion are at the surface so that surface effects are important. This effect can be minimized by taking into account a group of surrounding molecules ("ghosts") that move in unison with the molecules under consideration.

Molecular-Dynamical Method

In the **molecular-dynamical method** the molecules are assigned certain initial positions and velocities, and the equations of motion are set up in terms of the intermolecular forces. The computer then calculates the positions and velocities at various later times. The properties of the liquid are then obtained by *averaging over time.* Equilibrium is established fairly rapidly, after perhaps four collisions per molecule. The properties of the liquid can then be calculated from the positions, velocities, and potential energies at equilibrium. Corrections can again be made for surface effects.

Both the Monte Carlo and molecular-dynamical methods have given very satisfactory agreement with experiment. For example, it has proved possible to reproduce the *P-V-T* relationships for a number of liquids. The Monte Carlo method has been somewhat more useful for liquids at equilibrium, but the molecular-dynamical method has the additional advantage of being able to deal with the nonequilibrium properties of liquids—for example, with properties such as viscosity and diffusion.

17.5 Water, The Incomparable Liquid

Water has probably done more than any other chemical substance to determine our biological environment. It is an essential constituent of all living systems, and aside from this role it is a remarkably useful solvent for many scientific and technical purposes.

Although more investigation has been done on water than on any other liquid, there still remains uncertainty about the details of its structure. As will be seen, many of the experimental results can be interpreted in different ways.

It is of interest to compare and contrast the properties of water with those of the hydrides of some of the neighboring elements in the periodic table. Table 17.4 gives the melting points, boiling points, dielectric constants, dipole moments,

TABLE 17.4 Properties of Water and Related Substances

	NH_3	H_2O	HF	H_2S
M.P./K	195	273	184	187
B.P./K	240	373	293	212
Dielectric constant, ϵ	25(195 K)	79(298 K)	84(273 K)	9(187 K)
Dipole moment, μ/D	1.47	1.85	1.82	0.97
Dimer binding energy/kJ mol^{-1}	16	24	24	16

polarizabilities, and binding energies for the dimers of NH_3, H_2O, HF, and H_2S. Only water is liquid at room temperature, the others being gases, so that the binding energy between water molecules is greater than with the other three. This fact is not correlated in a simple way with the dipole moments or with the binding energies for the dimers. Water and hydrogen fluoride have similar dipole moments and similar dimer binding energies, but in liquid water each molecule can form twice as many hydrogen bonds as can a hydrogen fluoride molecule; that explains why water is a liquid and HF is a gas. The dimer bond strengths in NH_3 and H_2S are weaker, and this can be attributed to the lower dipole moments. The higher boiling point of water compared with the other three is thus due partly to the strength of the hydrogen bond (related to the high dipole moment), and partly to the number of bonds that can be formed by each molecule. Ammonia can form three hydrogen bonds, and hydrogen sulphide two, showing that it is their lower dipole moments that play the vital role in making them gases at room temperature.

The correlation between dielectric constant and dipole moment is again not a simple one, as seen from Table 17.4. High dielectric constants, such as found with HF and H_2O, arise not only from the high dipole moments but from the interaction between neighboring dipoles in the liquid state. A high dielectric constant tends to make a liquid a good ionizing solvent, since it facilitates the separation of ions, but solvating power is also an important factor (Section 7.9). Thus liquid HCN has a much higher dielectric constant (158) than water, but is a poorer ionizing solvent owing to its lower ability to solvate the ions.

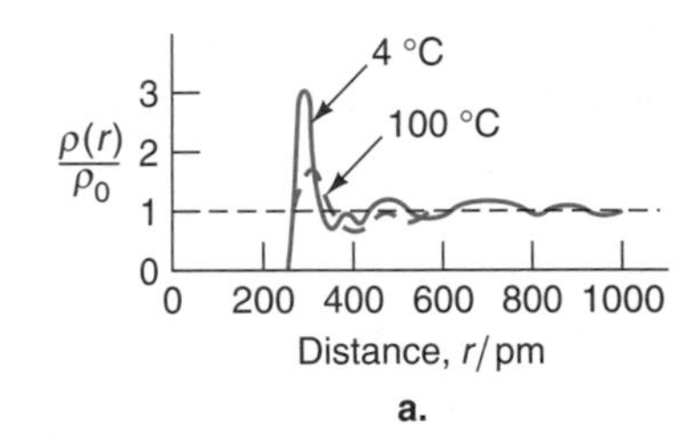

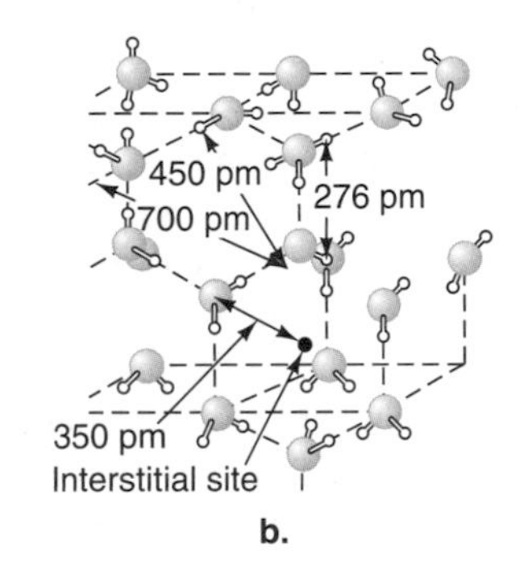

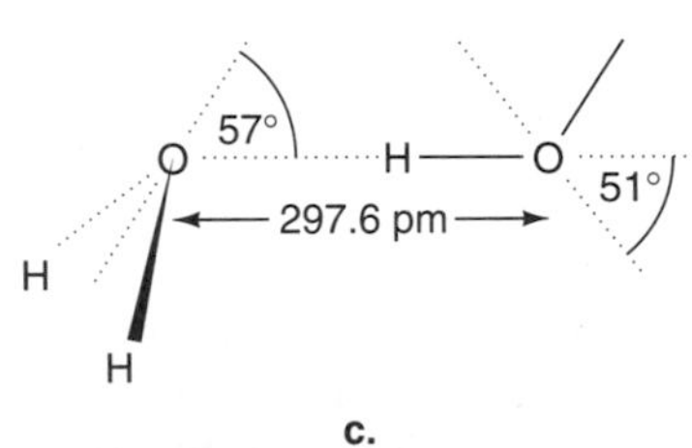

FIGURE 17.6
(a) Experimental radial distribution curves for water at 4 °C and at 100 °C. (b) The arrangement of water molecules in the crystal of ordinary ice (ice-I), showing some of the distances. (c) The structure of the water dimer.

Experimental Investigations of Water Structure

The structure of water has been investigated experimentally in various ways. Diffraction studies, using both X rays and neutrons, have been carried out, and there have also been many studies of infrared and Raman spectra. As will be seen, none of the results lead to a completely clear-cut answer as to the structure of liquid water.

The first detailed X-ray diffraction study of liquid water was made in 1938 by J. Morgan and B. E. Warren. More detailed studies, covering a range of temperature and involving both X-ray and neutron diffraction, were begun in 1966 by A. H. Narten and his coworkers and have been continued more recently. The radial distribution curves at 4 °C and at 100 °C are shown in Figure 17.6a. The function $\rho(r)$ is zero up to about 250 pm, which means that 250 pm is the effective molecular diameter. Above 700 pm at 4 °C and a somewhat greater distance at 100 °C, the function is essentially unity, which means that the local order does not extend beyond this distance. At both temperatures there is a well-defined peak at 290 nm. When $\rho(r)$ is integrated over the volume element $4\pi r^2\, dr$ in this shell, the number of nearest neighbors is calculated to be 4.4 at both temperatures; this value in fact applies from 4 °C to 200 °C. The $\rho(r)$ function also shows weak peaks at 350 pm, 450 pm, and 700 pm.

The structure of ordinary ice (ice-I) has been determined by X-ray diffraction and is shown in Figure 17.6b. Each H_2O molecule is surrounded tetrahedrally by four H_2O molecules, and the O—O distance is 276 pm. The strong peak at 290 pm for liquid water, with a coordination number of 4.4, therefore suggests a very similar tetrahedral arrangement in the liquid. The peak obtained at 350 pm with the liquid does not correspond to any distance in the solid. However, as indicated in Figure 17.6b, ice-I has interstitial sites at a distance of 350 pm from each oxygen atom. It therefore appears that when ice melts, some of the water molecules move from their tetrahedral sites into these interstitial sites, thus giving rise to the peak at 350 pm. There is a contraction in volume of about 9% when ice melts, and this is explained by a partial collapse of the tetrahedral structure, with a movement of some of the molecules into the interstitial sites. The peaks in the $\rho(r)$ function for water at 450 pm and 700 pm are consistent with the tetrahedral arrangement, as shown in Fig. 17.6b.

X-ray and Raman spectra results show that there are frequencies corresponding to the stretching and bending frequencies found with water vapor. In addition there are frequencies corresponding to hindered rotation of the water molecules, and to bending of hydrogen bonds. The results suggest that there are two types of environment for the water molecule.

Intermolecular Energies in Water

A great deal of effort has gone into the study of the intermolecular energies involved in liquid water, as these are essential to an understanding of the structure. In particular, meaningful computer simulations cannot be made unless one has a reliable intermolecular potential.

The forces involved are the dispersion forces, the hydrogen bonding forces (mainly dipole-dipole forces), and the rather weak dipole-(induced dipole) forces (Table 17.3). The latter, since they involve induction, are not pair additive; in other words, it is not completely satisfactory to consider only the forces between neighboring water molecules. Since it is difficult to include many-body terms in the intermolecular potential it is usual to neglect them, and this may be justifiable since they are so weak.

Much experimental work and many quantum-mechanical calculations have been made on the interaction between neighboring water molecules, i.e., on the water dimer. It appears that the binding energy in the dimer is about 23 kJ mol^{-1}, and that the favored structure in the dimer is as shown in Figure 17.6c. The preferred angular arrangement is seen to be consistent with the tetrahedral arrangement found in the ice structure (Figure 17.6b). As will be discussed, a modified version of this structure is believed to exist in the liquid.

Models of Liquid Water

A model for liquid water must be consistent with the X-ray and neutron diffraction patterns, the infrared and Raman spectra, and the dielectric constant. It also must explain the unusual fact that the density of water is greater than that of ice.

Several models have been proposed, all of them bearing a similarity to the structure of ice (Figure 17.6b). The models proposed fall into three main classes:

1. **Mixture models,** in which there are some regions in which the ice structure is maintained, and others in which there is a different liquid structure. These regions are not permanent, but by the breaking and forming of hydrogen bonds they move from place to place. A particular version of this model involves water dimers floating in a liquid structure.
2. **Uniform models,** in which the structure is uniform throughout the liquid, but in which the hydrogen bonds are distorted from the geometry that exists in the ice structure.
3. **Interstitial models,** in which the liquid on the whole has the same structure as in the solid (but with some hydrogen bond distortion), but in which water molecules occupy some of the empty space in the ice lattice. As noted in Figure 17.6b, certain interstitial sites occur in the ice structure, and some of these may be filled when ice melts. This model provides a simple explanation of the greater density of water as compared with ice.

The considerable amount of experimental work that has been done suggests that none of these models are adequate, but that the truth lies somewhere between them. The infrared and Raman spectra seem to favor the interstitial models, although they can be reconciled with the uniform models. The diffraction results, on the other hand, seem on the whole to favor the uniform models rather than the interstitial models, although as already noted they provide some evidence for a small proportion of interstitial water molecules. On the whole there is little evidence in favor of the mixture models. The conclusion seems to be a compromise between a uniform model and an interstitial model, with greater emphasis on the uniform model. In other words, when ice melts the main effect is a modification of the hydrogen-bonded structure, with a small proportion of the interstitial sites becoming occupied by water molecules.

Computer Simulation of Water Structure

A number of computer simulations have been made, usually by the Monte Carlo method, and using different intermolecular potentials. Since it is difficult to include many-body terms a pair-additive potential is usually assumed, and since the inductive forces are weak this is probably satisfactory. On the whole the computer simulations support the uniform models of the structure. They do not support the mixture models, or any model that suggests that there is a sharp distinction between water molecules that are hydrogen bonded and those that are not.

17.6 The Hydrophobic Effect

At the beginning of Chapter 12, in Table 12.1, we made a brief reference to hydrophobic bonds, which occur when nonpolar groups are present in aqueous solution. Such groups only *appear* to be attracted to one another; there is little real attraction. The apparent attraction arises for the most part from the behavior of the surrounding water molecules and is a result of the peculiar structure of liquid water. Thus, instead of speaking of a hydrophobic bond, it is better to speak of the hydrophobic effect.

Since nonpolar groups such as methyl groups have only tiny dipole moments there is little dipole-dipole attraction between them, but the apparent attraction between them when they are in an aqueous environment is much larger. Some light is thrown by the thermodynamic data for processes in which nonpolar groups enter an aqueous environment. Suppose, for example, that ethane in the liquid state dissolves in water; the thermodynamic changes have been measured to be

$$\Delta H^\circ = -10.5 \text{ kJ mol}^{-1} \quad \Delta S^\circ = -87.9 \text{ J K}^{-1}\text{ mol}^{-1} \quad \Delta G^\circ = 16.3 \text{ kJ mol}^{-1}$$

As we expect from the low solubility of ethane in water, the ΔG° value is positive. What is significant is that this positive value of ΔG° arises not from a positive ΔH° value but from a strongly negative entropy change.

This enthalpy decrease when liquid ethane dissolves in water can be explained in terms of increased hydrogen bonding when nonpolar groups are present in aqueous solution. At first sight it might be thought that there would be less hydrogen bonding between water molecules because of the interference of the nonpolar groups. The fact that there is more hydrogen bonding can be explained in terms of the ideas put forward in 1945 by the American chemist Henry S. Frank and his coworkers, which were discussed in Section 7.9. The basic idea is that when a non polar group is present in water, the neighboring water molecules arrange themselves into a structure that has more of a crystalline character than ordinary liquid water. Frank used the word "icebergs" to describe such structures, which are also formed around ions, but emphasized that the structures do not correspond exactly to those in ice. In particular, the icebergs have a higher density than water, whereas ice has a lower density.

The reason for the formation of the icebergs around nonpolar groups is that the neighboring water molecules, being deprived of the opportunity of forming hydrogen bonds with the molecules on the other side of the nonpolar groups, become oriented in a special way with respect to the neighboring water molecules and form hydrogen bonds with them. The result is that the water molecules around the nonpolar groups are held more rigidly than in ordinary water, and as a result there is a lowering of entropy.

When nonpolar groups in water come into close association with each other there is a decrease in the hydrogen bonding in their neighborhood. In other words, the iceberg formation is reduced and there is an increase in entropy. It is this increase in entropy that for the most part causes this association to occur. The hydrophobic effect is thus an apparent, but not real, attraction between nonpolar groups in aqueous solution, resulting from reduction in hydrogen-bond formation with a corresponding entropy increase.

KEY EQUATIONS

Internal pressure of a fluid:

$$\text{internal pressure} \equiv \left(\frac{\partial U}{\partial V}\right)_T$$

If the van der Waals equation is obeyed,

$$\text{internal pressure} = \frac{a}{V_m^2}$$

Caloric equation of state:

$$U = \int \left\{ T\left(\frac{\partial P}{\partial T}\right)_V - P \right\} dV + U_\infty$$

If the van der Waals equation is obeyed, internal pressure is

$$E_i = U = -\frac{a}{V} + U_\infty$$

Intermolecular potential energies:

Ion-ion: $$E_p = \frac{z_A z_B e^2}{4\pi\epsilon_0 \epsilon r}$$

Ion-dipole: $$E_p = -\frac{z_A e\mu \cos\theta}{4\pi\epsilon_0 \epsilon r^2}$$

Ion-(induced dipole): $$E_p = -\frac{\alpha(z_A e)^2}{8\pi\epsilon_0 \epsilon r^4}$$

(α = polarizability)

Dipole-dipole (average): $$E_p = -\frac{\mu_A^2 \mu_B^2}{24\pi^2 \epsilon_0^2 \epsilon^2 k_B T r^6}$$

Dispersion (London's formula): $$E_p = -\frac{3h\nu_0 \alpha^2}{4r^6}$$

Repulsive: $$E_p = \frac{B}{r^{12}}$$

The *Lennard-Jones 6–12 potential:* $$E_p = -\frac{A}{r^6} + \frac{B}{r^{12}} = E_0\left[\left(\frac{r_0}{r}\right)^{12} - 2\left(\frac{r_0}{r}\right)^6\right]$$

(r_0 = equilibrium separation)

PROBLEMS

Thermodynamic Properties of Liquids

17.1. The density of liquid ethanol at 20 °C is 0.790 g cm^{-3}, and the van der Waals constant a is 1.218 Pa m^6 mol^{-2}. Estimate the internal pressure and the potential-energy contribution to the internal energy.

17.2. In Example 17.3 (page 905), we obtained the internal pressure of liquid water from the van der Waals constant a. A more reliable value is obtained by use of Eq. 17.2, from the *thermal pressure coefficient* $(\partial P/\partial T)_V$; this quantity is the ratio α/κ of the coefficient of expansion $\alpha[= (1/V)(\partial V/\partial T)_P]$ to the compressibility $\kappa[= -(1/V)(\partial V/\partial P)_T]$. For water at 1 bar pressure and 298 K the thermal pressure coefficient is 6.60×10^6 Pa K^{-1}. Calculate the internal pressure.

17.3. The density of liquid benzene at 0 °C is 0.899 g cm^{-3}, and the van der Waals constant a is 1.824 m^6 mol^{-2} Pa. Estimate the internal pressure and the potential-energy contribution to the internal energy.

17.4. Make a better estimate of the internal energy of liquid benzene from its thermal pressure coefficient $(\partial P/\partial T)_V$, which at 298 K and 1 bar pressure is 1.24×10^6 Pa K^{-1}.

17.5. Calculate the internal pressures of the following liquids at 298 K and 1 bar pressure from their thermal pressure coefficients, which are as follows:

Hg:	4.49×10^6 Pa K^{-1}
n-Heptane:	8.53×10^5 Pa K^{-1}
n-Octane:	1.01×10^6 Pa K^{-1}
Diethyl ether:	8.06×10^5 Pa K^{-1}

17.6. The thermal pressure coefficient $(\partial P/\partial T)_V$, for CCl_4 vapor at 298 K and 10 Pa pressure, is 115 Pa K^{-1}. That for liquid CCl_4 at 298 K and 1 bar pressure is 1.24×10^6 Pa K^{-1}. Calculate the internal pressures of the vapor and the liquid under these conditions.

***17.7.** The following data apply to liquid acetic acid at 1 atm pressure and 293 K: density, $d = 1.049$ g cm^{-3}; coefficient of expansion, $\alpha = 1.06 \times 10^{-3}$ K^{-1}; compressibility, $\kappa = 9.08 \times 10^{-10}$ Pa^{-1}; van der Waals constant, $a = 1.78$ m^6 Pa mol^{-2}. Make two estimates of the internal pressure P_i, (a) using α and κ and (b) using a.

***17.8. a.** Derive the relationship

$$C_P - C_V = \frac{\alpha^2 VT}{\kappa}$$

where α is the coefficient of expansion and κ is the compressibility.

b. The value of $C_{V,m}$ for liquid CCl_4 at 298 K and at 1 bar pressure is 89.5 J K^{-1} mol^{-1}. Obtain the value of $C_{P,m}$ using the following data: $V_m = 97$ cm^3 mol^{-1}; $\alpha = 1.24 \times 10^{-3}$ K^{-1}; $\kappa = 10.6 \times 10^{-5}$ bar^{-1}.

c. Calculate $C_{P,m} - C_{V,m}$ for liquid acetic acid using the data given in Problem 17.7.

Intermolecular Energies

17.9. A liquid having a molar volume of 50 cm^3 is converted into a vapor having a molar volume of 50 dm^3. By what factor does the average intermolecular energy change?

17.10. Calculate the maximum energy of attraction, in J and in kJ mol^{-1}, when a Ca^{2+} ion is separated from a molecule of dipole moment 6.18×10^{-30} C m (= 1.85 D; this is the dipole moment of water) by a distance of 500 pm in a vacuum.

17.11. Calculate the energy of attraction, in J and in kJ mol^{-1}, when a Ca^{2+} ion is separated in a vacuum from a Cl^- ion by a distance of 500 pm.

17.12. Calculate the energy of attraction, in J and in kJ mol^{-1}, when a Ca^{2+} ion is separated in a vacuum by a distance of 500 pm, from a nonpolar molecule (having zero dipole moment) but a polarizability of 2.0×10^{-30} m^3.

17.13. Calculate the average energy of attraction, in J and in kJ mol^{-1}, for two molecules of dipole moments 6.18×10^{-30} C m separated in a vacuum at 25 °C by a distance of 500 pm.

17.14. The following values for A and B in the Lennard-Jones 6–12 function (Eq. 17.21) have been given for N_2:

$$A = 1.34 \times 10^{-5} \text{ J pm}^6$$
$$B = 3.42 \times 10^{10} \text{ J pm}^{12}$$

Calculate the equilibrium separation r_0 and the classical dissociation energy E_0, in J and in J mol^{-1}.

***17.15.** The following data apply to HBr: dipole moment, $\mu = 2.60 \times 10^{-30}$ C m; polarizability, $\alpha = 3.58 \times 10^{-30}$ m^3; oscillation frequency, $\nu_0 = 3.22 \times 10^{15}$ s^{-1}. Estimate the dipole-dipole, dipole-(induced dipole), and dispersion energies in J and in kJ mol^{-1} for two HBr molecules separated by 500 pm, at 25 °C.

***17.16.** The following are the polarizabilities and oscillation frequencies for Ne, Ar, and Kr:

	Ne	Ar	Kr
Polarizability, $\alpha/10^{-30}$ m^3	0.396	1.63	2.48
Frequency, $\nu_0/10^{15}$ s^{-1}	5.21	3.39	2.94

Calculate the dispersion energies for Ne, Ar, and Kr corresponding to a separation of 500 pm. Related data for He and Xe are given in Table 17.3; plot the five calculated values against the boiling points of the noble gases:

	He	Ne	Ar	Kr	Xe
Boiling point, T_b/K	4.22	27.3	87.3	119.9	165.1

***17.17.** In Table 17.3 and Problem 17.16 the dispersion energies of noble gases were calculated for a constant interatomic distance of 500 pm. More realistic values are:

	He	Ne	Ar	Kr	Xe
Interatomic distances/pm	240	320	380	400	420

Recalculate the dispersion energies for these distances, and again plot the five values against the boiling points, which were given in Problem 17.16.

The experimental value for the enthalpy of vaporization of liquid argon is 6.7 kJ mol^{-1}. Make an estimate of the enthalpy of vaporization from your calculated value of E_p (at 380 pm), assuming the liquid to have a close-packed structure with each atom having 12 nearest neighbors.

***17.18.** Estimate the interaction energy between an argon atom and a water molecule at a separation of 600 pm, which is approximately the distance of closest approach. The necessary data are: H_2O: dipole moment, $\mu = 6.18 \times 10^{-30}$ C m; Ar: polarizability, $\alpha = 1.63 \times 10^{-30}$ m^3.

Argon forms a solid hydrate, $Ar{\cdot}5H_2O$, but the binding energy between Ar and H_2O is about 40 kJ mol^{-1}, which is a good deal larger than the energy calculated from the dipole moment and polarizability. Suggest a reason for this discrepancy.

***17.19. a.** The Lennard-Jones potential

$$E = -\frac{A}{r^6} + \frac{B}{r^n}$$

can be formulated in a different way by expressing A and B in terms of the minimum energy E_{min} and the value r_0 of r at the minimum energy. Obtain the expression for E in terms of E_{min} and r_0.

b. If r^* is the value of r when $E = 0$, obtain the relationship between r^* and r_0.

c. The Lennard-Jones potential is often used with $n = 12$, and the equations are then simpler. Obtain E in terms of E_{min} and r_0 and in terms of E_{min} and r^*, for this special case of $n = 12$.

Essay Questions

17.20. Explain qualitatively how intermolecular forces of attraction are related to the following properties of a liquid:

a. vapor pressure;
b. enthalpy of vaporization;
c. normal boiling point;
d. entropy of vaporization.

17.21. Explain clearly the difference between dipole-dipole and London (dispersion) forces. With reference to a few examples, discuss the magnitudes of attractive energies arising from these forces.

SUGGESTED READING

For general accounts of the liquid state, see

A. F. M. Barton, *The Dynamic Liquid State,* London: Longmans, 1974.

J. D. Bernal, The liquid state, *Scientific American, 203,* 124–134(1960); reproduced in *Readings in the Physical Sciences,* Vol. 2, pp. 549–556, San Francisco: W. H. Freeman, 1969.

P. A. Egelstaff, *An Introduction to the Liquid State,* Oxford University Press, 1992.

J. O. Hirschfelder, C. F. Curtiss, and R. B. Bird, *Molecular Theory of Gases and Liquids,* New York: Wiley, 1954.

J. N. Murrell and E. A. Boucher, *Properties of Liquids and Solutions,* New York: John Wiley, 1982. This book gives a particularly clear and not too advanced account of the subject.

J. A. Pryde, *The Liquid State,* London: Hutchinson Scientific Library, 1966.

J. S. Rowlinson, *Liquids and Liquid Mixtures,* London: Butterworth, 1959.

D. Tabor, *Gases, Liquids and Solids,* 2nd edition, Cambridge University Press, 1979. This fairly small book gives a clear outline of the subject.

H. N. V. Temperley, J. S. Rowlinson, and G. S. Rushbrook (Eds.) *Physics of Simple Liquids,* New York: Wiley, 1968.

Lattice theories of the liquid state are treated in

J. A. Barker, *Lattice Theories of the Liquid State,* New York: Macmillan, 1963.

H. Eyring and M. S. Jhon, *Significant Liquid Structures,* New York: John Wiley, 1969.

For treatments of liquids by fundamental statistical mechanics see

C. A. Croxton, *Liquid State Physics: A Statistical Mechanical Introduction,* Cambridge: University Press, 1974, paperback, 1994. In spite of its title, this book presents the subject at an advanced level.

H. L. Friedman, *A Course in Statistical Mechanics,* Englewood Cliffs, NJ: Prentice Hall, 1985, Chapters 7, 8, and 9.

D. A. McQueen, *Statistical Mechanics,* New York: Harper and Row, 1976, Chapters 13 and 14.

Murrell and Boucher, *op. cit.;* pp. 42–52 give a helpful outline of the methods.

For a discussion of hydrogen bonds, including a quantum-mechanical treatment, see

L. C. Allen. A simple model of hydrogen bonding. *J. Amer. Chem. Soc., 97,* 6921–6940(1975).

General accounts of computer simulation methods, such as are used with liquids, are given in

K. Binder (Ed.), *The Monte Carlo Method in Condensed Matter Physics,* Berlin: Springer-Verlag, 1992. This is a more advanced treatment.

F. Neelamkavil, *Computer Simulation and Modeling,* New York: Wiley, 1987.

D. Nicholson and N. G. Parsonage, *Simulation methods,* Chapter 4 (pp. 135–182) of *Computer Simulation and the Statistical Mechanics of Adsorption,* New York: Academic Press, 1982.

For accounts of the properties and structure of water, see

J. O'M. Bockris and K.-T. Jeng, *Water Structure at Interfaces; the Present Situation,* Amsterdam: Elsevier, 1990.

D. Eisenberg and W. J. Kauzmann, *The Structure and Properties of Water,* Oxford: Clarendon Press, 1969.

F. Franks (Ed.), *Water, A Comprehensive Treatise,* London and New York: Plenum Press, 1972.

F. Franks, *Water,* London: The Royal Society of Chemistry, 1984.

Murrell and Boucher, *op. cit.,* Chapter 8.

Surface Chemistry and Colloids

18

PREVIEW

Substances tend to become *adsorbed* on surfaces, and Section 18.1 of this chapter deals with the basic principles of adsorption. *Adsorption* can be classified as *physisorption* or *van der Waals adsorption,* and as *chemisorption,* which is much stronger since covalent bonding is involved. Adsorption equilibria can be treated on the basis of a kinetic model and also by the methods of statistical mechanics (Sections 18.2 and 18.3). The simplest type of chemisorption is that described by the *Langmuir adsorption isotherm,* which applies when the surface is homogeneous and when the forces between adsorbed molecules can be neglected. This type of adsorption is characterized by the fact that the surface can become *saturated* by the adsorbed molecules. Molecules on surfaces are sometimes *dissociated.*

Surface-catalyzed reactions (Section 18.4) involve the adsorption of the reacting molecules on the catalytic surface, and the Langmuir isotherms are convenient for treating such reactions. When there are two reacting substances there may be a *Langmuir-Hinshelwood mechanism,* in which reaction is between two adsorbed molecules. Alternatively, reaction may occur between an adsorbed molecule and a molecule in the gas phase, in which case we speak of a *Langmuir-Rideal mechanism.* Complications arising from surface heterogeneity are discussed briefly in Section 18.5.

The detailed structures of solid surfaces and adsorbed layers can be investigated by a large number of experimental techniques, the most important of which are described briefly in Section 18.6. Some of the more modern techniques, such as *field-ion microscopy,* and *scanning-tunneling microscopy,* are capable of providing direct images of individual atoms on a surface.

An important property of liquid surfaces (Section 18.7) is the *surface tension,* which results from intermolecular attractions. One result of surface tension is the tendency of certain liquids to rise in a capillary tube, an effect known as *capillarity.*

Liquid films on surfaces (Section 18.8) have been studied particularly by the use of Irving Langmuir's *film balance,* which measures *surface pressure.* Liquid films may be *coherent,* in which case there is a critical area per molecule; such films behave like two-dimensional liquids. There are also *noncoherent* or *gaseous* films, which behave like two-dimensional gases.

Colloidal dispersions (Section 18.9) are intermediate between true solutions and suspensions; the particles are either very large molecules or aggregates, but they are too small to be seen under the microscope. Their properties result from the fact that their ratio of surface area to volume is very large, so that the surface properties are important. There are a number of different types of colloidal systems, including *sols, gels, emulsions,* and *micelles.* An important property of colloids is the *Tyndall effect,* which involves the scattering of light. The ultramicroscope is a useful instrument for studying colloidal systems, and in it *Brownian movement* can be observed.

The electrical properties of colloidal systems are important and arise from the attachment of ions to surfaces.

OBJECTIVES

After studying this chapter, the student should be able to:

- Describe the two types of adsorption, physisorption and chemisorption.
- Explain how a unimolecular layer is formed and describe the magnitude of the energy involved in this process. Relate this to the potential-energy curves for the two types of adsorption.
- Develop the equations necessary to derive the Langmuir isotherm and plot the fraction of surface covered against concentration for adsorption with and without dissociation.
- Describe how competitive adsorption occurs and explain it with equations.
- Explain the conditions under which the BET isotherm occurs.
- Understand the thermodynamics and statistical mechanics of adsorption.
- Describe how unimolecular reactions occur as treated in terms of the Langmuir-Hinshelwood and Langmuir-Rideal mechanisms.
- Describe inherent and induced heterogeneity.
- Understand the techniques used to examine surfaces and what information is obtained from each.
- Understand how surface tension develops and the equations that describe it.
- Understand the Kelvin equation, and how to apply it.
- Describe how a film balance works, and the difference between coherent and noncoherent films.
- Describe the various aspects of colloidal systems.
- Understand the Tyndall effect and the electrical properties of colloidal systems.
- Discuss the various aspects of gels and emulsions.

A boundary that separates two phases is known as a *surface* or an *interface.* Surfaces show special properties that are different from those of the phases themselves. For example, the surface of a solid often shows a strong affinity for molecules that come into contact with it and which are said to be **adsorbed.** A simple way of visualizing adsorption is in terms of additional valence bonds at the surface, which are available for bonding; in reality, however, the situation is more complex.

In this chapter we will deal with the behavior of different kinds of surfaces and, in particular, with the nature of adsorption. We will first consider solid-gas interfaces, which are the easiest to understand and which provide a basis for understanding more complex surfaces. Attention will be given to the way in which molecules are attached to surfaces, to the thermodynamics and statistical mechanics of adsorption, and to the mechanisms of chemical reactions at surfaces. Colloidal systems are also briefly considered in this chapter, since the properties of colloids are determined to a very considerable extent by surface effects.

Some of the earliest work on adsorption and surface catalysis was carried out in 1834 by Michael Faraday, who was particularly concerned with gas reactions on surfaces. He realized that such reactions occur in adsorbed films but thought that the main effect of the solid catalyst was to exert an attractive force on the gas molecules and to cause them to be present at much higher concentrations than in the main body of the gas. However, this idea is shown to be false by the fact that in certain cases different surfaces give rise to different products of reaction. For example, ethanol decomposes mainly into ethylene and water on an alumina catalyst, and mainly into acetaldehyde and hydrogen on copper. This and many other results clearly indicate that specific chemical forces must be involved when molecules become attached to surfaces.

18.1 Adsorption

Physisorption

It is now recognized that there are two main types of adsorption. In the first type the forces are of a physical nature and the adsorption is relatively weak. The forces correspond to those considered by J. H. van der Waals in connection with his equation of state for gases (Section 1.13) and are known as **van der Waals forces.** This type of adsorption is known as **physical adsorption, physisorption,** or **van der Waals adsorption.** The heat evolved when a mole of gas becomes physisorbed is usually small, less than 20 kJ. This type of adsorption plays only an unimportant role in catalysis, except for certain special types of reactions involving free atoms or radicals.

Chemisorption

In the second type of adsorption, first considered in 1916 by the American chemist Irving Langmuir (1881–1957), the adsorbed molecules are held to the surface by covalent forces of the same general type as those occurring between atoms in molecules. The heat evolved per mole for this type of adsorption, known as **chemisorption,** is usually comparable to that evolved in chemical bonding, namely 100 to 500 kJ mol^{-1}.

An important consequence of chemisorption is that after a surface has become covered with a single layer of adsorbed molecules, it is *saturated;* additional adsorption can occur only on the layer already present, and this is generally weak

Langmuir won the 1932 Nobel Prize, mainly for his many contributions to surface chemistry. He also made some important contributions to G. N. Lewis's theory of electronic structure and valency.

adsorption. Langmuir thus emphasized that chemisorption involves the formation of a *unimolecular layer.* Many investigations on surfaces of known area have confirmed that chemisorption ceases after a unimolecular layer is formed, but that physisorption may give rise to additional layers.

Activated Adsorption

It was suggested in 1931 by Hugh Stott Taylor (1890–1974) that chemisorption is frequently associated with an appreciable activation energy and may therefore be a relatively slow process. For this reason chemisorption is often referred to as **activated adsorption.** By contrast, van der Waals adsorption requires no activation energy and therefore occurs more rapidly than chemisorption.

The relationship between physisorption and chemisorption is illustrated by the potential-energy diagram shown in Figure 18.1. It relates to the adsorption of hydrogen on a surface such as that of a metal. Physisorption occurs first, the intact molecule being held loosely to the surface by dispersion forces (Section 17.3). The potential-energy curve for chemisorption, with dissociation of the hydrogen molecule, has a much deeper minimum, corresponding to stronger bonding. The crossing of the two curves shows that there is a small energy of activation for the chemisorption process. This diagram shows that initial physisorption is an important feature of the chemisorption process. If there were no physisorption, there

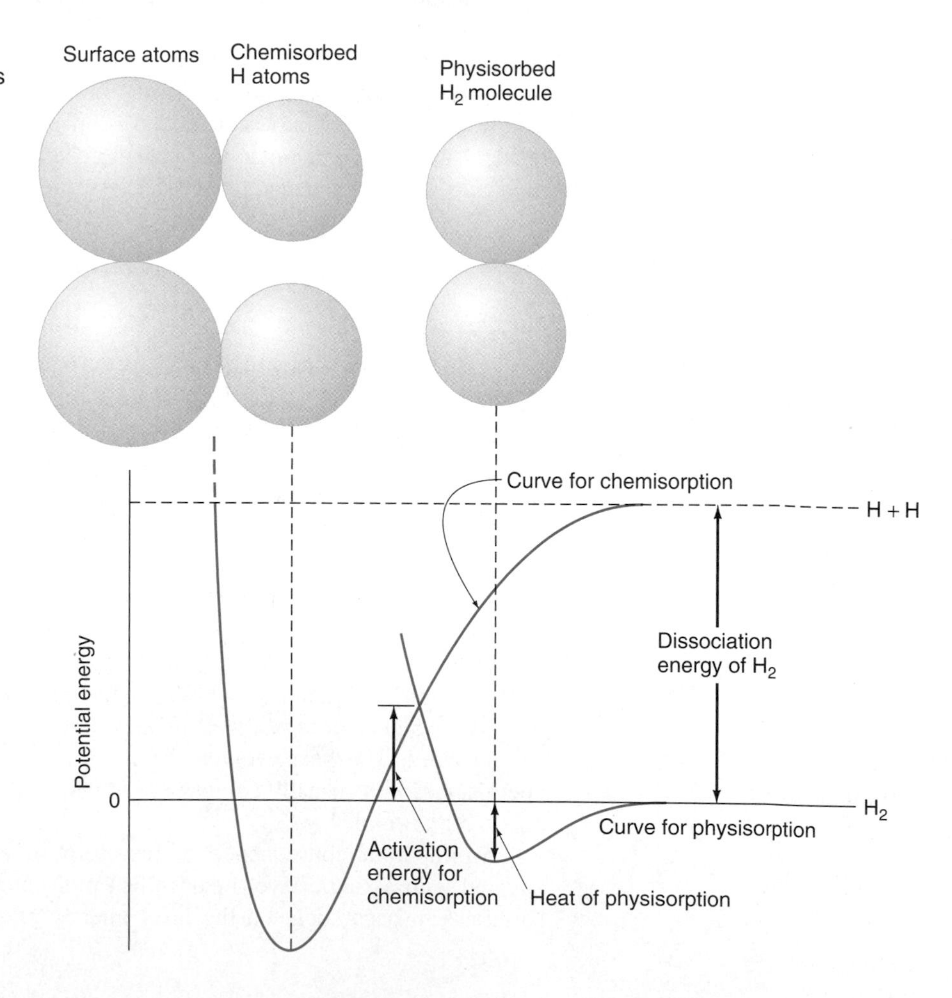

FIGURE 18.1
Schematic potential-energy curves for the physisorption and chemisorption of hydrogen on a surface.

would be a much higher activation energy for the chemisorption; the process via the physisorbed stage is therefore always favored.

Another important concept, suggested in 1925 by H. S. Taylor, is that solid surfaces are never completely smooth and that adsorbed molecules will be attached more strongly to some surface sites than to others. This is particularly important in connection with catalysis, since chemical reaction may occur predominantly on certain sites, which Taylor referred to as *active centers.* Surface heterogeneity is discussed in further detail in Section 18.5.

18.2 Adsorption Isotherms

An equation that relates the amount of a substance attached to a surface to its concentration in the gas phase or in solution, at a fixed temperature, is known as an **adsorption isotherm.**

The Langmuir Isotherm

The simplest isotherm was first obtained in 1916 by Irving Langmuir. The basis of the derivation of the Langmuir isotherm is that all parts of the surface behave in exactly the same way as far as adsorption is concerned. Suppose that, after equilibrium is established, a fraction θ of the surface is covered by adsorbed molecules; a fraction $1 - \theta$ will not be covered. The rate of adsorption will then be proportional to the concentration [A] of the molecules in the gas or liquid phase and also proportional to the fraction of the surface that is bare, because adsorption can only occur when molecules strike the bare surface. The rate of adsorption v_a is thus

$$v_a = k_a[\mathrm{A}](1 - \theta) \tag{18.1}$$

where k_a is a rate constant relating to the adsorption process. The rate of desorption v_d is proportional only to the number of molecules attached to the surface, which in turn is proportional to the fraction of surface covered:

$$v_d = k_d\theta \tag{18.2}$$

where k_d is a rate constant for the desorption process. At equilibrium, the rates of adsorption and desorption are the same; thus

$$k_a[\mathrm{A}](1 - \theta) = k_d\theta \tag{18.3}$$

or

$$\frac{\theta}{1 - \theta} = \frac{k_d}{k_a}[\mathrm{A}] \tag{18.4}$$

The ratio k_a/k_d is an equilibrium constant and can be written as K; then

$$\frac{\theta}{1 - \theta} = K[\mathrm{A}] \tag{18.5}$$

or

$$\theta = \frac{K[\mathrm{A}]}{1 + K[\mathrm{A}]} \tag{18.6}$$

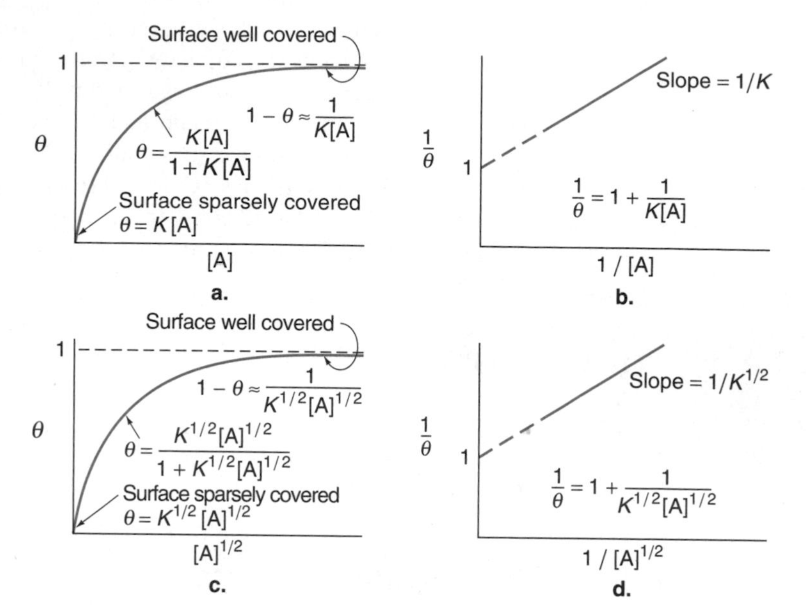

FIGURE 18.2
(a) Schematic plots of θ (fraction of surface covered) against [A] for a system obeying the Langmuir adsorption isotherm, without dissociation. (b) Reciprocal Langmuir plots. (c) Langmuir plots for the case of adsorption with dissociation. (d) Reciprocal Langmuir plots for adsorption with dissociation.

A graph of θ against [A] is shown in Figure 18.2a. At sufficiently low concentrations we can neglect K[A] in comparison with unity, and then θ is proportional to [A]. We can write Eq. 18.6 as

$$1 - \theta = \frac{1}{1 + K[A]} \tag{18.7}$$

so that at very high concentrations

$$1 - \theta \approx \frac{1}{K[A]} \tag{18.8}$$

Note that the **Langmuir isotherm** (Eq. 18.6) is of exactly the same form as the Michaelis equation (Eq. 10.82). A distinctive feature of the isotherm is that the surface becomes *saturated* with molecules at high pressures.

In order to test the Langmuir isotherm it is best to use a reciprocal plot, as shown in Figure 18.2b. The usefulness of the isotherm in obtaining estimates of surface areas is illustrated by the following example.

EXAMPLE 18.1 Benzene adsorbed on graphite is found to obey the Langmuir isotherm to a good approximation. At a pressure of 1.00 Torr the volume of benzene adsorbed on a sample of graphite was found to be 4.2 mm^3 at STP (0 °C and 1 atm pressure); at 3.00 Torr it was 8.5 mm^3. Assume a benzene molecule to occupy 30 $Å^2$ and estimate the surface area of the graphite.

Solution Suppose that the amount adsorbed when the surface is saturated is x mm^3; the fractions of surface covered at the two pressures are thus $4.2/x$ and $8.5/x$. From Eq. 18.6 we can thus set up the two equations

$$\frac{4.2}{x} = \frac{1.0K}{1 + 1.0K} \qquad \frac{8.5}{x} = \frac{3.0K}{1 + 3.0K}$$

The solutions of these simultaneous equations are

$$K = 0.318 \quad \text{and} \quad x = 17.4$$

The maximum amount of benzene adsorbed is thus 17.4 mm^3.

$$1 \text{ mol at STP occupies } 22.4 \text{ L} = 2.24 \times 10^7 \text{ mm}^3$$

$$17.4 \text{ mm}^3 \text{ is thus } 17.4/(2.24 \times 10^7) \text{ mol} = 7.77 \times 10^{-7} \text{ mol}$$
$$= 7.77 \times 10^{-7} \times 6.022 \times 10^{23} = 4.68 \times 10^{17} \text{ molecules.}$$

The estimated area of the surface is thus

$$4.68 \times 10^{17} \times 30 \text{ Å}^2 = 1.40 \times 10^{19} \text{ Å}^2 = 0.140 \text{ m}^2$$

Adsorption with Dissociation

The type of adsorption to which Eq. 18.6 applies may be formulated as

$$\text{A} + \underset{|}{\overset{|}{-\text{S}-}} \rightleftharpoons \overset{\text{A}}{\underset{}{\overset{|}{-\text{S}-}}}$$

where S represents a surface site and A the substance being adsorbed. In certain cases there is evidence that the process of adsorption is accompanied by the dissociation of the molecule when it becomes attached to the surface. For example, when hydrogen gas is adsorbed on the surface of many metals, the molecules are dissociated into atoms each of which occupies a surface site. This type of adsorption may be represented as

$$\text{A}_2 + \overset{|}{-\text{S}}\overset{|}{-\text{S}-} \rightleftharpoons \overset{\text{A}}{\overset{|}{-\text{S}}}\overset{\text{A}}{\overset{|}{-\text{S}-}}$$

The process of adsorption is now a reaction between the gas molecule and *two* adjacent surface sites, and the rate of adsorption is therefore

$$v_a = k_a[\text{A}](1 - \theta)^2 \tag{18.9}$$

The desorption process involves reaction between *two* adsorbed atoms, and the rate is therefore proportional to the square of the fraction of surface covered,

$$v_d = k_d\theta^2 \tag{18.10}$$

At equilibrium the rates are equal, and therefore,

$$\frac{\theta}{1 - \theta} = \left(\frac{k_a}{k_d}[\text{A}]\right)^{1/2} \tag{18.11}$$

$$= K^{1/2}[\text{A}]^{1/2} \tag{18.12}$$

where K is equal to k_a/k_d. This equation can be written as

$$\theta = \frac{K^{1/2}[\text{A}]^{1/2}}{1 + K^{1/2}[\text{A}]^{1/2}} \tag{18.13}$$

A plot of θ against $[A]^{1/2}$ is shown in Figure 18.2c. When the concentration is very small, $K^{1/2}[A]^{1/2}$ is much smaller than unity, and θ is then proportional to $[A]^{1/2}$. Equation 18.13 may be written as

$$1 - \theta = \frac{1}{1 + K^{1/2}[A]^{1/2}} \tag{18.14}$$

so that at high concentrations, when $K^{1/2}[A]^{1/2} \gg 1$,

$$1 - \theta = \frac{1}{K^{1/2}[A]^{1/2}} \tag{18.15}$$

The fraction of the surface that is bare at high concentrations is therefore inversely proportional to the square root of the concentration.

Competitive Adsorption

The isotherm for two substances adsorbed on the same surface is of importance in connection with inhibition and with the kinetics of surface reactions involving two reactants. Suppose that the fraction of surface covered by molecules of type A is θ_A and that the fraction covered by B is θ_B. The fraction bare is $1 - \theta_A - \theta_B$. If both substances are adsorbed without dissociation, the rates of adsorption of A and B are

$$v_a^A = k_a^A[A](1 - \theta_A - \theta_B) \tag{18.16}$$

and

$$v_a^B = k_a^B[B](1 - \theta_A - \theta_B) \tag{18.17}$$

The rates of desorption are

$$v_d^A = k_d^A\theta_A \tag{18.18}$$

$$v_d^B = k_d^B\theta_B \tag{18.19}$$

Equating Eqs. 18.16 and 18.18 leads to

$$\frac{\theta_A}{1 - \theta_A - \theta_B} = K_A[A] \tag{18.20}$$

where K_A is equal to k_a^A/k_d^A. From Eqs. 18.17 and 18.19 we obtain

$$\frac{\theta_B}{1 - \theta_A - \theta_B} = K_B[B] \tag{18.21}$$

where K_B is k_a^B/k_d^B. Equations 18.20 and 18.21 are two simultaneous equations that can be solved to give, for the fractions covered by A and B, respectively,

$$\theta_A = \frac{K_A[A]}{1 + K_A[A] + K_B[B]} \tag{18.22}$$

$$\theta_B = \frac{K_B[B]}{1 + K_A[A] + K_B[B]} \tag{18.23}$$

Equation 18.22 reduces to Eq. 18.6 if [B] = 0 or if $K_B = 0$, which means that substance B is not adsorbed. It follows from Eqs. 18.22 and 18.23 that the fraction of the surface covered by one substance is reduced if the amount of the other substance is increased. This is because the molecules of A and B are competing with one another for a limited number of surface sites, and we speak of **competitive adsorption.** There is evidence that sometimes two substances are adsorbed on two different sets of surface sites, in which case there is no competition between them.

For gaseous systems, the concentration terms in all the equations derived in this section can, of course, be replaced by pressures; the equilibrium constants are then pressure equilibrium constants.

Other Isotherms

The various isotherms of the Langmuir type are based on the simplest of assumptions; all sites on the surface are assumed to be the same, and there are no interactions between adsorbed molecules. Systems that obey these equations are often referred to as showing **ideal adsorption;** the Langmuir equations have the same significance in connection with adsorption as has the ideal gas law $PV = nRT$ in connection with the behavior of gases. Systems frequently deviate significantly from the Langmuir equation. This may be because the surface is not uniform, and also there may be interactions between adsorbed molecules; a molecule attached to a surface may make it more difficult, or less difficult, for another molecule to become attached to a neighboring site, and this will lead to a deviation from the ideal adsorption equation.

Nonideal systems can sometimes be fitted to an empirical adsorption isotherm due to the German physical chemist Herbert Max Finlay Freundlich (1880–1941); according to this equation the amount of a substance adsorbed, x, is related to the concentration c by the equation

Freundlich Isotherm

$$x = kc^n \tag{18.24}$$

where k and n are empirical constants. The value of n is usually less than unity.

This equation does not correspond to saturation of the surface; the amount adsorbed keeps increasing as c increases. If Eq. 18.24 applies, a plot of ln x against ln c will give a straight line, of slope n. This isotherm can be arrived at theoretically in terms of surface heterogeneity and also in terms of repulsive forces between adsorbed molecules.

BET Isotherm

Another isotherm of importance was proposed in 1938 by the American scientists Stephen Brunauer (1903–1986), Paul Emmett (1900–1985), and Edward Teller (b. 1908). Known as the **BET isotherm,** it is an extension of the Langmuir treatment to allow for the physisorption of additional layers of adsorbed molecules. It was derived by balancing the rates of adsorption and condensation for the various layers, and involved the assumption that there is one enthalpy of adsorption for the first layer, and that the enthalpy of liquefaction applies to the second and subsequent layers. The equation has been written in various forms, of which a convenient, simple form is

$$\frac{P P_0}{V(P_0 - P)} = \frac{1}{V_0 K} + \frac{P}{V_0} \tag{18.25}$$

Here V is the volume of gas adsorbed at pressure P, and V_0 the volume that can be adsorbed as a monolayer; P_0 is the saturation vapor pressure, and K is the equilibrium constant for the adsorption. The BET isotherm is particularly useful for determining surface areas, as illustrated by the following example.

EXAMPLE 18.2 The following data relate to the adsorption of nitrogen at 77 K on a 1.00-g sample of silica gel:

P/kPa	15.2	54.8
V/cm^3(STP)	135	247

At 77 K the saturation vapor pressure P_0 of nitrogen is 101.3 kPa. Estimate the surface area of the gel, taking the molecular area of nitrogen to be 1.62×10^{-19} m^2.

Solution Insertion of the data into the BET equation gives the following two simultaneous equations:

$$0.132 = \frac{1}{V_0 K} + \frac{15.2}{V_0} \quad \text{and} \quad 0.261 = \frac{1}{V_0 K} + \frac{54.8}{V_0}$$

These equations can easily be solved to obtain V_0 and K, and the result is

$$V_0 = 303\ (\text{cm}^3) \qquad K = 0.040\ (\text{kPa}^{-1})$$

At STP, 22.4 L = 22 400 cm^3 is the volume occupied by 1 mol. A volume of 303 cm^3 thus contains 303/22 400 = 0.0135 mol; $0.0135 \times 6.022 \times 10^{23} = 8.15 \times 10^{21}$ molecules. Since each molecule occupies 1.62×10^{-19} m^2, the estimated surface area is

$$8.15 \times 10^{21} \times 1.62 \times 10^{-19} = 1320\ \text{m}^2$$

This example illustrates the general method used in estimating areas using the BET isotherm. In practice one always has more than a pair of data, and then graphical methods are used to obtain the constants K and V_0, and hence the surface area.

18.3 Thermodynamics and Statistical Mechanics of Adsorption

A considerable number of measurements have been made of the enthalpy and entropy changes that occur on adsorption and desorption. As we have seen, van der Waals adsorption and chemisorption can be distinguished by the magnitudes of the enthalpy changes. Heat is always liberated on adsorption, and the enthalpy of adsorption is therefore always negative.

The reason that enthalpies of adsorption must be negative is that the adsorption process inevitably involves a decrease in entropy. This is because a molecule in the gas phase or in solution has more freedom of motion than one that is attached to a surface. In view of the thermodynamic relationship

$$\Delta G = \Delta H - T\Delta S \tag{18.26}$$

a process with a negative ΔS must have a negative ΔH if ΔG is to be negative (i.e., if the process is to occur to an appreciable extent).

By the methods of statistical mechanics (Chapter 15) it is possible to derive isotherms of the Langmuir form, and they express the constant K in terms of partition functions. This was first done in 1935 by the British mathematical physicist Sir Ralph Howard Fowler (1889–1944). We will consider only the case of the adsorption of a gas without dissociation, to which Eq. 18.6 applies. Suppose that the volume of the gas is V and that the area of the surface is S. The total number of molecules in the gas phase may be written as N_g, the number of adsorbed molecules as N_a, and the number of bare sites at equilibrium as N_s. The concentrations of these species are as follows:

Concentration in gas phase: $c_g = \dfrac{N_g}{V}$

Concentration of adsorbed molecules: $c_a = \dfrac{N_a}{S}$

Concentration of bare sites: $c_s = \dfrac{N_s}{S}$

The equilibrium constant for the adsorption process is

$$K_c = \frac{c_a}{c_g c_s} = \frac{N_a/S}{(N_g/V)(N_s/S)} = \frac{N_a}{(N_g/V)N_s} \tag{18.27}$$

We saw in Section 15.7 that an equilibrium constant is equal to the ratio of the partition functions multiplied by an exponential term involving the difference between the zero-point levels of reactants and products:

$$K_c = \frac{q_a}{q_g q_s}\mathrm{e}^{-\Delta E_0/RT} \tag{18.28}$$

The partition functions q_a and q_s to be used in this expression are those for unit surface area; q_g is for unit volume. Then

$$\frac{c_a}{c_s} = c_g \frac{q_a}{q_g q_s}\mathrm{e}^{-\Delta E_0/RT} \tag{18.29}$$

If θ is the fraction of the surface that is covered,

$$\frac{c_a}{c_s} = \frac{\theta}{1-\theta} \tag{18.30}$$

and, therefore,

$$\frac{\theta}{1-\theta} = c_g \frac{q_a}{q_g q_s}\mathrm{e}^{-\Delta E_0/RT} \tag{18.31}$$

The partition function q_g per unit volume may be written as

$$q_g = \frac{(2\pi m k_\mathrm{B} T)^{3/2}}{h^3} b_g \tag{18.32}$$

where b_g represents the rotational and vibrational factors in the partition function. The adsorption sites have very little freedom of motion and their partition function q_s may be taken as unity. The partition function q_a for the adsorbed molecules involves only internal factors, which may be written as b_a. The adsorption isotherm thus becomes

$$\frac{\theta}{1-\theta} = c_g \frac{h^3}{(2\pi m k_\mathrm{B} T)^{3/2}} \frac{b_a}{b_g}\mathrm{e}^{-\Delta E_0/RT} \tag{18.33}$$

This equation has the same form as Eq. 18.5, but now the constant K is given in explicit form.

This equation is applicable to the situation where the adsorbed molecules are localized on the surface. This is usually the case with chemisorption, because of the strength of the binding of the adsorbed molecules to the surface. In some systems where there is weak binding and sparse surface coverage, there is evidence that the adsorbed molecules can move fairly freely on the surface; that is to say, they have two degrees of translational freedom. When this is the case, another equation applies (see Problem 18.9).

18.4 Chemical Reactions on Surfaces

An important concept in connection with surface reactions is the *molecularity,* which is the number of reactant molecules that come together during the course of reaction; we do not count the surface sites. The molecularity of a surface reaction is deduced from the kinetics on the basis of the experimental results and of theoretical considerations. Reactions involving a single reacting substance are usually, but not invariably, unimolecular. The mechanism of the surface-catalyzed ammonia decomposition is, for example, usually unimolecular. On the other hand, the kinetics of the decomposition of acetaldehyde on various surfaces can only be interpreted on the hypothesis that two acetaldehyde molecules, adsorbed on neighboring surface sites, undergo a bimolecular reaction. Reactions involving two reacting substances are usually bimolecular. When reactant molecules are dissociated on the surface, the reaction may involve interaction between an atom or radical and a molecule; for example, the exchange reaction between ammonia and deuterium on iron is a bimolecular interaction between a deuterium atom and an ammonia molecule.

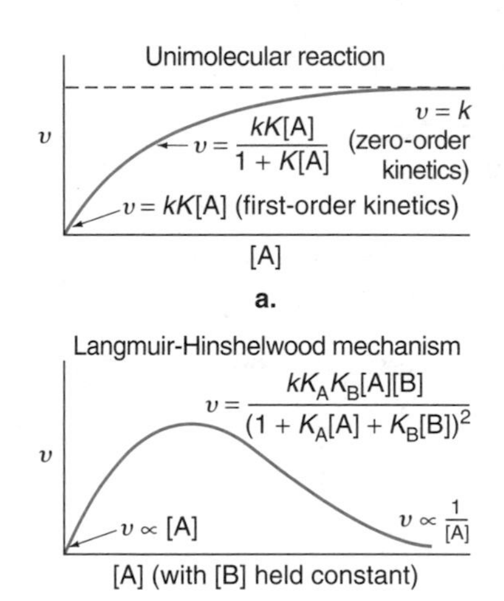

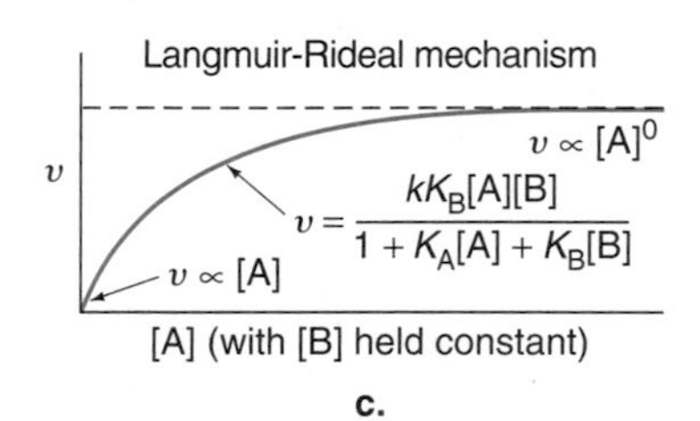

FIGURE 18.3
The variation of rate with concentration for various types of surface reactions. (a) Simple unimolecular processes. (b) A bimolecular reaction occurring by a Langmuir-Hinshelwood mechanism. (c) A bimolecular reaction occurring by a Langmuir-Rideal mechanism.

Unimolecular Reactions

Surface reactions involving one molecule may be treated in terms of the Langmuir adsorption isotherm (Eq. 18.6). In the simplest case the rate of reaction is proportional to θ and is thus

$$v = k\theta = \frac{kK[A]}{1 + K[A]} \tag{18.34}$$

This equation is of the same form as the Michaelis-Menten equation (Eq. 10.82), but we should note that it has been derived on the assumption of a rapid adsorption equilibrium followed by slow chemical reaction.

The dependence of v on [A], shown in Figure 18.3a, is exactly the same as that given by the Langmuir isotherm (Figure 18.2). At sufficiently high concentrations the rate is independent of the concentration, which means that the kinetics are zero order. At low concentrations, when $K[A] \ll 1$, the kinetics are first order. A good example of this type of behavior is the decomposition of ammonia into nitrogen and hydrogen on a tungsten surface.

Sometimes a substance other than the reactant is adsorbed on the surface, with the result that the effective surface area, and therefore the rate, are reduced. Suppose that a substance A is undergoing a unimolecular reaction on a surface and that a nonreacting substance I, known as an *inhibitor* or a *poison,* is also adsorbed. If the fraction of the surface covered by A is θ and that covered by I is θ_i, we have from Eq. 18.22 that

$$\theta = \frac{K[\text{A}]}{1 + K[\text{A}] + K_i[\text{I}]} \tag{18.35}$$

where K and K_i are the adsorption constants for A and I. The rate of reaction, equal to $k\theta$, is thus

$$v = \frac{kK[\text{A}]}{1 + K[\text{A}] + K_i[\text{I}]} \tag{18.36}$$

In the absence of inhibitor this equation reduces to Eq. 18.34.

A case of special interest is when the surface is only sparsely covered by the reactant but is fairly fully covered by the inhibitor. In other words

$$K_i[\text{I}] \gg 1 + K[\text{A}] \tag{18.37}$$

and the rate is then

$$v = \frac{kK[\text{A}]}{K_i[\text{I}]} \tag{18.38}$$

We can easily understand this relationship if we remember that when a surface is sparsely covered (as it now is by A), the coverage is proportional to [A] (Eq. 18.6); whereas if it is almost fully covered (as it now is by I), the fraction that is bare is inversely proportional to [I] (Eq. 18.8). A good example of Eq. 18.38 is provided by the decomposition of ammonia on platinum, the rate law for which is

$$v = \frac{k[\text{NH}_3]}{[\text{H}_2]} \tag{18.39}$$

Since hydrogen is a product of reaction, there is progressive inhibition as reaction proceeds. There is no appreciable inhibition by the other reaction product, nitrogen.

Bimolecular Reactions

When a solid surface catalyzes a bimolecular process, there are two possible mechanisms. One of these, first suggested by Langmuir and further developed by Hinshelwood, involves reaction between two molecules adsorbed on the surface. Such a mechanism may be formulated as follows.

Hinshelwood's 1956 Nobel Prize for chemistry (shared with N. N. Semenov) was for his many pioneering contributions to chemical kinetics, including his work on surface reactions and explosions (Section 10.8).

$$\text{A} + \text{B} + \text{—}\overset{|}{\text{S}}\text{—}\overset{|}{\text{S}}\text{—} \rightleftharpoons \text{—}\overset{\overset{\text{A}}{|}}{\text{S}}\text{—}\overset{\overset{\text{B}}{|}}{\text{S}}\text{—} \rightarrow \underset{\text{activated complex}}{\text{—}\overset{\overset{\text{A}\cdots\text{B}^\ddagger}{\vdots\quad\vdots}}{\text{S—S}}\text{—}} \rightarrow \text{—}\overset{|}{\text{S}}\text{—}\overset{|}{\text{S}}\text{—} + \text{Y} + \text{Z}$$

In the first step the two molecules A and B become adsorbed on neighboring sites on the surface. Reaction then takes place, by way of an activated complex, to give the reaction products.

Langmuir-Hinshelwood Mechanism

When this **Langmuir-Hinshelwood mechanism** applies, the rate is proportional to the probability that A and B are adsorbed on neighboring sites, and this is proportional to the product of fractions of the surface, θ_A and θ_B, covered by A and

B. These fractions were given in Eqs. 18.22 and 18.23, and the rate of reaction is therefore

$$v = k\theta_A\theta_B \tag{18.40}$$

$$= \frac{kK_AK_B[A][B]}{(1 + K_A[A] + K_B[B])^2} \tag{18.41}$$

If [B] is held constant and [A] varied, the variation of rate is as shown in Figure 18.3b. The rate first rises, passes through a maximum, and then decreases toward zero. The physical explanation of the falling off of the rate at high concentrations is that one reactant displaces the other as its concentration is increased. The maximum rate corresponds to the presence of the maximum number of neighboring A-B pairs on the surface.

At sufficiently low concentrations of A and B, Eq. 18.41 predicts second-order kinetics, and this has been observed in a number of systems. Rate maxima have also often been observed. A case of special interest is when one reactant (e.g., A) is weakly adsorbed and the other strongly adsorbed. This means that $K_B[B] \gg 1 + K_A[A]$, and it follows from Eq. 18.41 that

$$v = \frac{kK_A[A]}{K_B[B]} \tag{18.42}$$

The order is thus 1 for A and −1 for B. An example is the reaction between carbon monoxide and oxygen on quartz, where the rate is proportional to the pressure of oxygen and inversely proportional to the pressure of carbon monoxide.

An alternative mechanism for a bimolecular surface process is for the reaction to occur between a molecule that is not adsorbed (e.g., A) and an adsorbed molecule (B). Such a mechanism was also considered by Langmuir and was further developed by the British physical chemist Sir Eric K. Rideal (1890–1974). This mechanism may be represented as

$$\mathrm{A} + \underset{}{-\overset{\displaystyle \mathrm{B}}{\overset{|}{\mathrm{S}}}-} \rightarrow \underset{\text{activated complex}}{-\overset{\displaystyle \mathrm{A}^{\ddagger}\cdots\mathrm{B}}{\overset{\vdots}{\mathrm{S}}}-} \rightarrow -\overset{|}{\mathrm{S}}- + \text{products}$$

The adsorption of A may occur; it is simply postulated, in this mechanism, that an adsorbed A does not react.

Langmuir-Rideal Mechanism

The rate for this **Langmuir-Rideal mechanism** is proportional to the concentration of A and to the fraction of the surface that is covered by B:

$$v = k[A]\theta_B \tag{18.43}$$

$$= \frac{kK_B[A][B]}{1 + K_A[A] + K_B[B]} \tag{18.44}$$

The variation of v with [A] is shown in Figure 18.2c; the same type of curve is found if v is plotted against [B]. There is now no maximum in the rate, and this provides a possible way of distinguishing between this mechanism and the Langmuir-Hinshelwood mechanism.

Not many ordinary chemical reactions occur by a Langmuir-Rideal mechanism. There is evidence, however, that radical combinations on surfaces sometimes occur in this way. The combination of hydrogen atoms, for example, is sometimes a first-order reaction, and it appears to occur by the mechanism

$$\mathrm{H} + -\underset{|}{\overset{\mathrm{H}}{\overset{|}{\mathrm{S}}}}- \rightarrow -\overset{|}{\mathrm{S}}- + \mathrm{H_2}$$

The fraction of the surface covered by hydrogen atoms is

$$\theta = \frac{K[\mathrm{H}]}{1 + K[\mathrm{H}]} \tag{18.45}$$

and the rate of combination is thus

$$v = k[\mathrm{H}]\theta = \frac{kK[\mathrm{H}]^2}{1 + K[\mathrm{H}]} \tag{18.46}$$

At lower temperatures the surface may be fully covered, so that $K[\mathrm{H}] \gg 1$ and

$$v = k[\mathrm{H}] \tag{18.47}$$

and the kinetics are first order. At higher temperatures the coverage decreases and if $1 \gg K[\mathrm{H}]$, the rate is

$$v = kK[\mathrm{H}]^2 \tag{18.48}$$

An increase in order from 1 to 2 has in fact been observed as the temperature is raised.

In 1912, and later, Langmuir made important investigations on the production of hydrogen atoms at hot tungsten surfaces, such as used in tungsten-filament incandescent lamps. The rate is proportional to the square root of the hydrogen pressure, and the mechanism is believed to be the reverse of that just given, namely

$$\mathrm{H_2} + -\overset{|}{\mathrm{S}}- \rightarrow -\overset{\mathrm{H}}{\overset{|}{\mathrm{S}}}- + \mathrm{H}$$

18.5 Surface Heterogeneity

In our treatment of adsorption and of chemical reactions on surfaces we have so far been assuming that all the surface sites are of the same character. The Langmuir adsorption isotherm, for example, is based on this assumption of surface homogeneity. In reality, as H. S. Taylor first pointed out, surfaces are never absolutely smooth, and we have to take account of variations in surface activity.

Various lines of experimental evidence provide evidence for the *inherent heterogeneity* of surfaces. For example, if a metal is heated for a period of time, its capacity for adsorption and catalysis is usually decreased. This is due to sintering of the surface, resulting in a decrease in the number of atoms that constitute the most active centers.

There is also much kinetic evidence for the variability of surfaces. For example, the decomposition of ammonia on molybdenum is retarded by nitrogen, but as the surface becomes saturated by nitrogen, the rate of the decomposition does not fall to zero. This suggests that the reaction can occur on certain surface sites on which the nitrogen cannot be adsorbed.

In view of the fact that surfaces show such variability, it may at first seem surprising that fairly simple kinetic laws often apply to surface-catalyzed reactions. The reason is that because of the exponential term in the rate equation a reaction will occur predominantly on the most active surface sites. For example, suppose that a reaction can occur at ordinary temperatures on 10% of the surface sites with an activation energy of 50 kJ mol^{-1} and on the remaining 90% with activation energies ranging from 80 to 150 kJ mol^{-1}. In this situation the amount of reaction occurring on the 90% of less active sites will be negligible compared with that occurring on the 10% of most active sites. The variability of the 90% of the surface will therefore have little effect on the kinetic behavior.

Aside from an *inherent heterogeneity,* resulting in a true variation in surface sites, there can also be an *induced heterogeneity* resulting from the *interactions* between adsorbed molecules. Suppose, for example, that there is a significant repulsion between molecules that are adsorbed side by side on a surface. The result will be that as the substance is progressively adsorbed, the first molecules will not be close together, but subsequent molecules will necessarily be close to other molecules and will be adsorbed less strongly. The adsorption behavior is thus very similar to that occurring when the surface is inherently heterogeneous; the first molecules will be attached more strongly than the later ones.

The heat evolved on chemisorption usually falls as the surface is progressively covered. This can be due to inherent heterogeneity, to induced heterogeneity, or to a combination of the two.

Promotion

Many surface catalysts are much more efficient if they are *promoted.* A **promoter** is a substance that in itself has little or no catalytic power but which when added to a catalyst greatly improves its performance. A pure iron surface, for example, is not a particularly good catalyst for the synthesis of ammonia from nitrogen and hydrogen; however, addition of small amounts of various materials, such as Al_2O_3, brings about a very considerable improvement in the catalytic performance. Promoters bring about their action in a variety of ways. Some of them act mainly by changing the structure of the surface, with an increase in the surface area. Promoters also act by producing surface regions, such as phase boundaries, which have a greater catalytic activity than other parts of the surface.

Fritz Haber (1868–1934) won the 1918 Nobel Prize for chemistry for the synthesis of ammonia from the elements; he brought hydrogen and nitrogen together at high pressure in the presence of a promoted iron catalyst.

Most catalysts are easily poisoned by substances that are strongly adsorbed at the active centers on the surface. Since the proportion of active centers may be small, minute amounts of poison can cause a very considerable decrease in the effectiveness of a solid surface.

18.6 The Structure of Solid Surfaces and of Adsorbed Layers

A considerable amount of experimental and theoretical work has been directed toward understanding the structures of solid surfaces and adsorbed layers. Only a very brief account can be given here; for further details the reader is directed to the many books and review articles on the subject, some of which are listed in Suggested Reading (pp. 964–965).

Many highly effective experimental techniques have been developed to investigate solid surfaces. This section deals only with a few that are of particular importance and still being used extensively by research workers in this field.

Photoelectron Spectroscopy (XPS and UPS)

As noted in Section 11.1, the photoelectric effect was discovered by Hertz in 1887 and played a part, in 1905, in leading Einstein to his quantum theory of light. Although there were many early investigations of the effect, it was not until the late 1940s that the technique began to be used for studying the electronic structures of various forms of matter.

Photoelectron spectroscopy is carried out with ultraviolet radiation (UPS) and also with X radiation (XPS). Particularly since the 1970s, the techniques have been applied to the study of adsorbed layers on surfaces. For example, comparisons have been made between the photoelectron spectra of gaseous benzene, liquid benzene, and benzene chemisorbed on surfaces and have provided useful, detailed information about the nature of the chemisorption process.

Important investigations using photoelectron spectroscopy have been carried out by M. W. Roberts and coworkers and described by him in an article listed under Suggested Reading (pp. 964–965). Some of the work was concerned with the adsorption of carbon monoxide on transition metals such as molybdenum and tungsten. At room temperature carbon monoxide is adsorbed on such metals with a high heat of adsorption (~ 300 kJ mol^{-1}), and the adsorbed state is described as the β-state. On cooling to 80 K further adsorption occurs with a much lower heat of adsorption (~ 60 kJ mol^{-1}), and this new adsorbed layer is referred to as the γ-state. Photoelectron spectroscopic studies of the two types of adsorbed layers revealed very different behavior. The carbon monoxide in the β-state has a C-O binding energy that is very much less than in the free molecule, while the C-O energy of the γ-state is similar to that in the free molecule. The spectroscopy has therefore revealed a clear distinction between chemisorbed (β-state) carbon monoxide molecules and physisorbed (γ-state) molecules.

The chemisorbed carbon monoxide molecules do not exhibit the type of orbital structure that is found in the free molecules. The conclusion from the spectroscopic studies is that the carbon monoxide is essentially dissociated when it is chemisorbed on these metals. The results indicate that the function of the metal in catalysis is to pump electrons into antibonding orbitals of the chemisorbed molecule, in this way making the molecule more reactive.

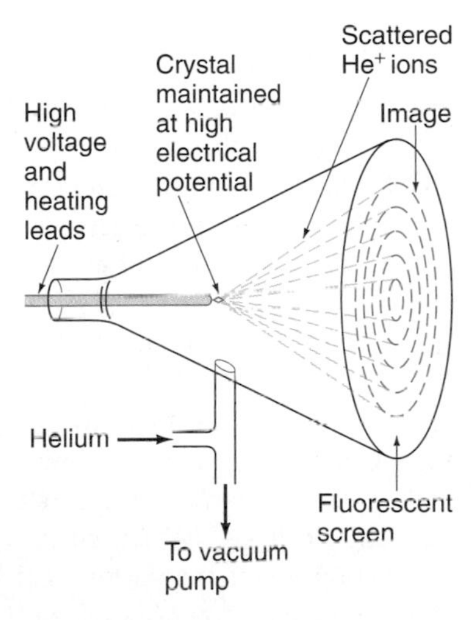

FIGURE 18.4
Schematic diagram showing the principle of the field-ion microscope for the investigation of surfaces.

Field-Ion Microscopy (FIM)

The field-ion microscope was developed in 1951 by Erwin Wilhelm Müller (1911–1977), who was then in Berlin; from 1952 until his death he was at Pennsylvania State University. The instrument was a further development of the **field-emission microscope** that Müller himself had invented in 1936. The FIM was the first instrument capable of providing direct images of individual atoms on a solid surface. From 1952, and for about 20 years, it remained the only instrument able to image individual atoms, and although no longer unique in this regard it still remains a useful and powerful research tool.

A schematic diagram of the apparatus employed in FIM is shown in Figure 18.4. The instrument involves the projection of ions from the tip of a sharply pointed needle to a fluorescent screen. A stream of a gas such as helium enters an evacuated chamber at low pressure (1–100 mPa) and an electric field of about 10^9 V cm^{-1} is applied to the crystal. This is high enough to produce He^+ ions, which are repelled radially from the surface. The ions are accelerated toward a fluorescent screen that displays a highly magnified image of the surface. The ions are produced preferentially by protruding surface atoms, and the image that appears on the screen consists of a pattern of spots, each spot representing an individual surface atom. Analysis of the image thus permits an identification of the positions of the various atoms on the surface. From the symmetry of the patterns it is possible to observe the Miller indices of the observed planes (Section 16.1). A disadvantage of the technique is that the sample must be exceedingly small and must be stable against the high electric fields that have to be used.

The field-ion microscope has been applied to a variety of problems. It has been used to investigate the diffusion and interaction of individual surface atoms, and to follow the course of chemical reactions on surfaces. In the field of metallurgy it has been valuable in the study of the details of defect structures, such as those due to radiation damage.

Auger Electron Spectroscopy (AES)

Auger spectroscopy is also extensively used to investigate the near-surface region of solids. As discussed in Section 16.2, the Auger effect involves the fall of an electron from a higher to a lower energy level, the energy released causing another bound electron (the Auger electron) to be released with a well-defined amount of energy (Figure 16.19).

The application of Auger spectroscopy to surface chemistry involves the use of an electron beam to excite electrons near the surfaces of solids, resulting in the emission of electrons. The energy spectrum of the Auger electrons can then be used to determine the types of atoms present to a depth of about 100 nm (1000 Å) from the surface. At greater depths the Auger electrons have too great a chance of undergoing a collision resulting in a change of energy.

If some kind of erosion process can be used to remove surface atoms layer by layer, AES can provide a depth profile of the composition of a solid at and below its surface.

Low-Energy Electron Diffraction (LEED)

Low-energy electron diffraction is the coherent back-scattering of low-energy electrons from the uppermost few atomic layers of a solid surface. The electrons used are in the energy range from 5 to 250 eV, which penetrate the surface to about 0.5 to 2 nm, which is about 1 to 4 atomic layers. The scattered beam thus reveals the properties of the surface atoms rather than those of the atoms in the bulk of the solid.

The apparatus is shown schematically in Figure 18.5. In order for there to be effective diffraction, the de Broglie wavelength of the electron beam must be similar to the interatomic distances. The incident electrons have a narrow range of energies (i.e., are monochromatic), and the back-scattered electrons are caused to impinge on a fluorescent screen. The structural information about the surface is provided by the elastically scattered electrons, which have not lost energy in the scattering process. If the atoms on the surface are arranged in an ordered pattern, the diffraction spots on the fluorescent screen are small and of high intensity. Analysis of the distribution of the spots and of their intensity provides detailed information about the atomic configuration of the surface.

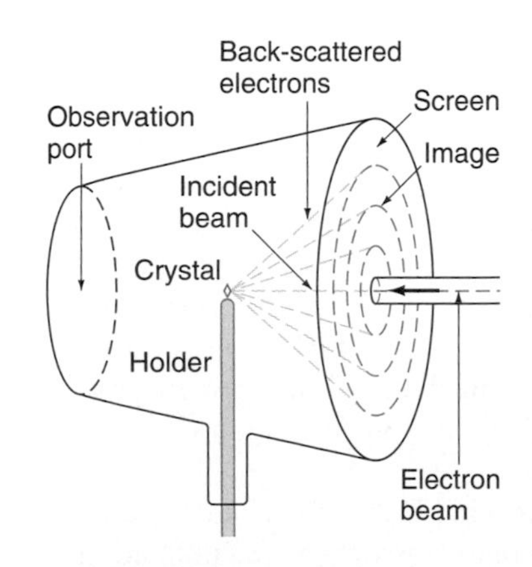

FIGURE 18.5
Schematic diagram showing the principle of the low-energy electron diffraction (LEED) technique for the study of surfaces.

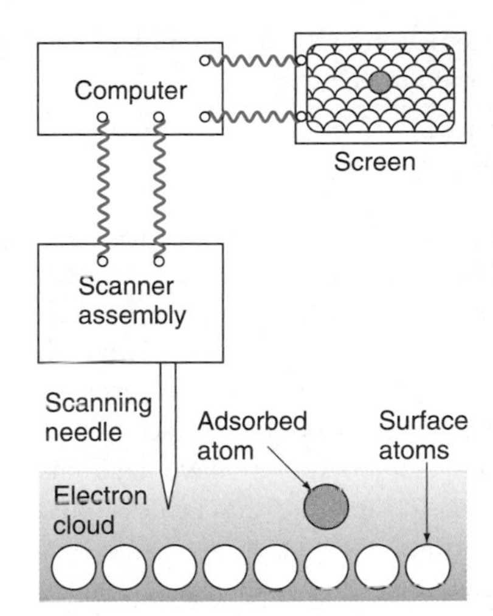

FIGURE 18.6
Schematic diagram of the scanning tunneling microscope.

The original LEED technique developed from the 1927 investigation of Davisson and Germer on electron diffraction (Section 11.3), but it was not until the early 1970s that the method could be successfully applied to solid surfaces. At about the same time a modified technique, **inelastic low-energy electron diffraction (ILEED),** was developed. The original technique involves inelastic scattering, and the analysis of inelastic scattering provides additional information such as the vibrations of substances adsorbed on surfaces.

Scanning Tunneling Microscopy (STM)

The scanning tunneling microscope (STM) is a highly effective device capable of studying surfaces on an atomic distance scale. The technique was invented in the early 1980s by Gerd Karl Binnig (b. 1947) and Heinrich Rohrer (b. 1933) of the IBM Corporation's Zürich Research Division. They shared the 1986 Nobel Prize for physics[1] for this work.

The principle of the method is illustrated schematically in Figure 18.6. An extremely sharp tip (so sharp that it essentially ends as a single atom) scans the entire surface of an electrical conductor, without actually touching the surface, at a height comparable to an interatomic distance. At such short distances the tip is within the electron cloud of the surface atoms, and electrons are able to cross the gap by quantum-mechanical tunneling (Section 15.8, Figure 15.7). There is therefore a flow of electrons, i.e., a current passes, when a potential difference is applied across the gap. The potential difference applied is typically 10 mV for a metal and up to 3 V for a semiconductor.

Tips used in the technique are made by the electrochemical etching of wires of tungsten or other metals. The tip is mounted on a scanner assembly that causes the tip to sweep across the entire surface. The current that passes is extremely sensitive to the distance between the surface and the tip. A computerized feedback system senses the current, and if the tip comes too close to a surface atom, the feedback system causes it to move away from the surface so that the tip never touches the surface. The computer processes the motion of the tip as the surface is scanned and displays the resulting image on a graphics screen or a plotter. One therefore sees a highly resolved three-dimensional image of the surface. A distance of 10 Å on the surface may appear as 10 cm on the screen, so that there is a magnification of 10^9.

STM has already found a wide range of applications in physics, chemistry, materials science, and biology. Large biological molecules such as DNA have been imaged by depositing them onto smooth surfaces, such as a single crystal of graphite. Prodding with the STM tip has even been used to move atoms into desired positions.

In 1993, an even closer look at a surface using STM was achieved by M. F. Crommie, C. P. Lutz, and D. M. Eigler [*Nature, 363,* 524(1993)]. They used a copper surface, and according to band theory (Section 16.6) electrons are regarded as forming a "sea" flowing around and over the atomic nuclei. At the surface of a metal there is therefore a two-dimensional "electron sea" flowing over the various defects and irregularities that occur on the surface. Stationary interference patterns are formed, and the STM experiments were able to reveal the surface defects from

[1]They shared the prize with Ernst August Friedrich Ruska (1906–1988) for his invention of the electron microscope in 1932.

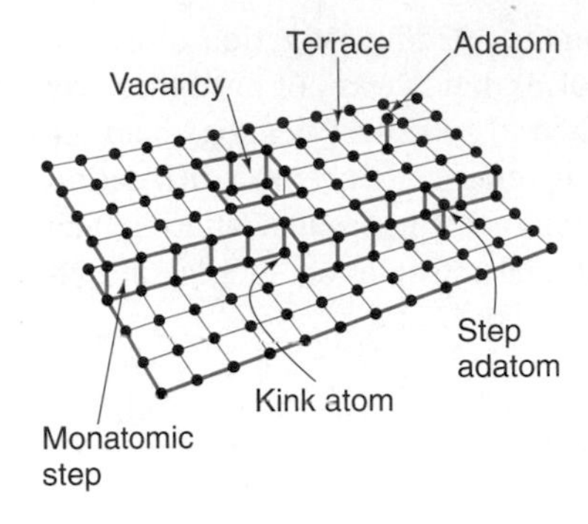

FIGURE 18.7
Schematic diagram of a solid surface, showing some of the features commonly observed.

the interaction of the STM tip with the interference patterns. It was also possible to study the electron interaction with individual adsorbed molecules.

Details of the Solid Surface

The various experimental studies of solid surfaces and adsorbed layers have revealed a number of very important results, which can be mentioned only very briefly. In the first place, surfaces are never smooth, and on the atomic scale they are highly heterogeneous. This is represented schematically in Figure 18.7, which shows that there are various kinds of surface sites; there are atoms in terraces, atoms in steps, atoms at kinks, and adatoms, which project from the surface. These types of surface atoms differ in the number of near neighbors they have; adatoms have few, while atoms in terraces have more. Usually there are few adatoms and many more step, kink, and terrace atoms. The various types of atoms differ very markedly in their chemical behavior; there are large differences in heats of adsorption and in catalytic activity.

Under ordinary experimental conditions, surfaces are covered by a layer of adsorbed molecules, which are removed only by prolonged evacuation. Bare surface atoms are in a state of high Gibbs potential energy, and a lowering results from the formation of an adsorbed layer. Since adsorption is usually associated with only a small activation energy, it occurs rapidly, so that a perfectly clean surface will at once become covered if it is exposed to the atmosphere.

There is a good deal of exchange between atoms and molecules that are adsorbed at the different surface sites. This is because the activation energies for diffusion across a surface are considerably smaller than those for desorption into the gas phase. Under most conditions adsorbed species will soon establish themselves in a state of equilibrium.

An important aspect of an adsorbed layer is that a surface dipole is usually produced, as a result of different electronegativities. Such dipoles are particularly important with ionic solids. Such dipoles constitute an **electric double layer,** which we have already discussed, particularly in connection with electrokinetic effects (Section 10.14).

18.7 Surface Tension and Capillarity

A molecule in the interior of a liquid is, on the average, attracted equally in all directions by its neighbors, and there is therefore no resultant force tending to move it in any direction. On the other hand, at the surface of a liquid that is in contact with vapor there is practically no force attracting the surface molecules away from the liquid, and there is therefore a net inward attraction on the surface molecules. This is represented in Figure 18.8a. If the surface area of a liquid is increased, more molecules are at the surface, and work must be done for this to occur. A surface therefore has an excess Gibbs energy, relative to the interior of the liquid; the SI unit of this energy per unit surface is $\text{J m}^{-2} = \text{kg s}^{-2} = \text{N m}^{-1}$. This relationship shows that the excess surface energy per unit area is force per unit length.

In 1805 the English physicist and physician Thomas Young (1773–1829) showed that surfaces behave as if a membrane were stretched over them. An analogy is provided by a soap bubble, which because of the tension of the membrane

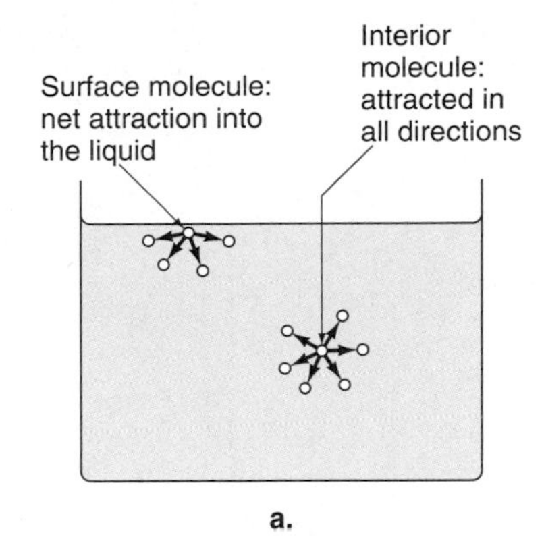

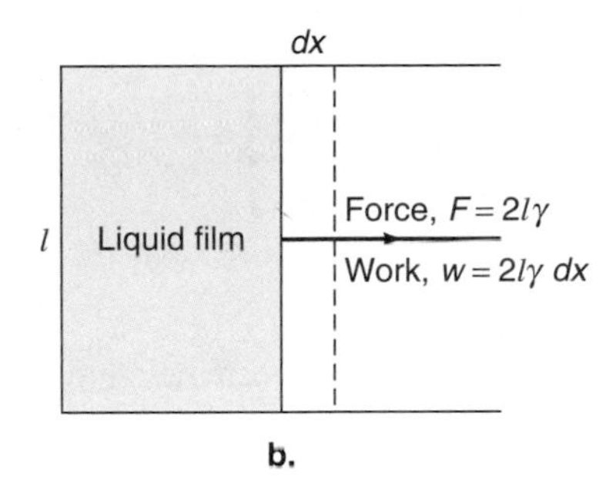

FIGURE 18.8
(a) Attractive forces acting on a surface molecule and on a molecule in the interior of a liquid.
(b) A wire frame supporting a liquid film.

assumes a spherical form. In the same way a liquid, particularly if it is suspended in liquid with which it is immiscible so as to eliminate the effects of gravity, tends to become spherical. Very small drops of any liquid are almost exactly spherical; larger ones are flattened by their weights.

Figure 18.8b shows a thin film, such as a soap film, stretched on a wire frame having a movable side of length l. The film has two sides, and the total length of the side of the film is $2l$. The force F required to stretch the film is proportional to $2l$,

$$F = \gamma(2l) \tag{18.49}$$

and the proportionality constant γ is known as the **surface tension.** Its SI unit is N m^{-1} = J m^{-2}, and it is therefore the surface energy per unit area. Thus if the piston in Figure 18.8b moves a distance dx, the area (on the two sides) increases by $2l\ dx$, and the work done is $2l\ dx$. The ratio of the work done to the increase in surface area is therefore

$$\frac{\text{work done}}{\text{increase in surface area}} = \frac{2l\gamma\ dx}{2l\ dx} = \gamma \tag{18.50}$$

Note that this work done is the increase in Gibbs energy; **the increase in Gibbs energy is thus the increase in surface area multiplied by the surface tension.**

When two liquids are in contact with each other, as in the case of oil floating on water, the interfacial tension arises from differences between the intermolecular forces at the two surfaces in contact.

Several methods are used to measure surface tension. The simplest and most commonly employed is the *capillary-rise method,* illustrated in Figure 18.9a. A tube of small internal radius r is inserted vertically into a liquid, and if the liquid wets the glass, the level of liquid will rise in the tube. The simplest situation, found if there is complete wetting of the tube, is if the angle between the meniscus and the surface is zero. If this is the case, the force F_1 is acting directly upward. This force is equal to the surface tension multiplied by the length of contact between the edge of the liquid and the tube; this length is equal to $2\pi r$ and therefore

$$F_1 = 2\pi r\gamma \tag{18.51}$$

At equilibrium this force is exactly balanced by the force F_2 due to the weight of the liquid in the capillary. The volume of this is $\pi r^2 h$, and if the density is ρ, the force is

$$F_2 = \pi r^2 h\rho g \tag{18.52}$$

where g is the acceleration of gravity. Therefore

$$2\pi r\gamma = \pi r^2 h\rho g \tag{18.53}$$

or

$$\gamma = \frac{rh\rho g}{2} \tag{18.54}$$

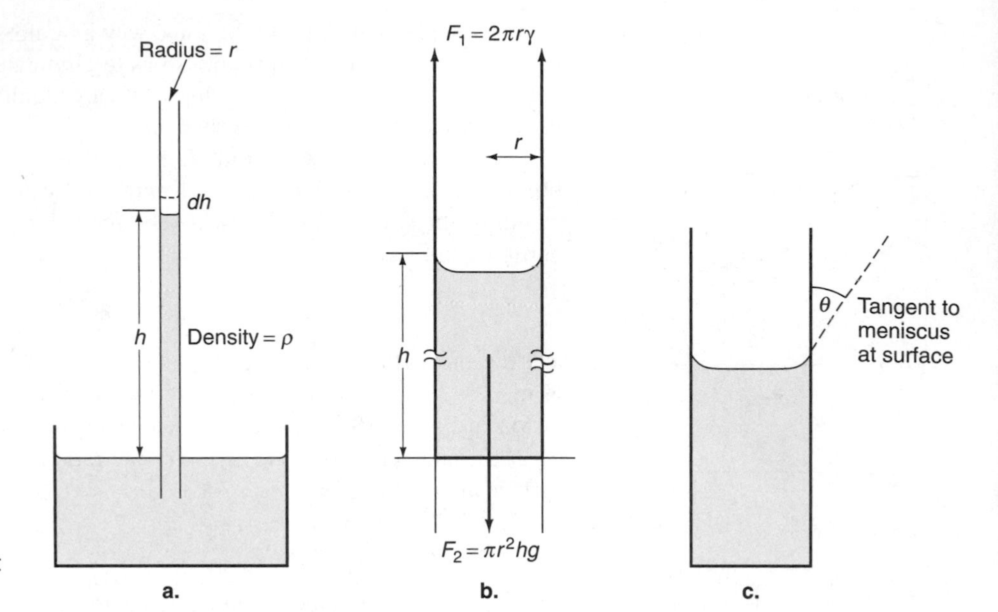

FIGURE 18.9
(a) The capillary-rise method for measuring surface tension.
(b) The forces for the case in which the liquid completely wets the glass. (c) The liquid does not completely wet the surface.

If the liquid does not completely wet the glass, there is an angle θ between the meniscus and the surface, as shown in Figure 18.9c, and when this is the case, Eq. 18.54 becomes

$$\gamma = \frac{rh\rho g}{2 \cos \theta} \tag{18.55}$$

An interesting consequence of the existence of surface tension is that the vapor pressure of a spherical droplet of a liquid may be considerably larger than that of the bulk liquid. This matter was first considered by William Thomson (later Lord Kelvin), and the resulting equation is often known as the **Kelvin equation.** Suppose that the ordinary vapor pressure of a liquid is P_0 and that when it is present in droplets of radius r, the vapor pressure is P. The Gibbs energy change when dn mol of liquid is transferred from a plane surface to a droplet is then

$$dG = dn\, RT \ln \frac{P}{P_0} \tag{18.56}$$

This change can also be calculated from the surface energy change that results from the increase in surface area. The volume of dn mol is $M\, dn/\rho$, where M is the molar mass and ρ is the density. The droplet has a surface area of $4\pi r^2$, and if the radius increases by dr, the increase in volume is $4\pi r^2\, dr$; thus

$$\frac{M\, dn}{\rho} = 4\pi r^2\, dr \tag{18.57}$$

and therefore

$$dr = \frac{M}{4\pi r^2 \rho}\, dn \tag{18.58}$$

The increase in surface area of the droplet is

$$dA = 4\pi(r + dr)^2 - 4\pi r^2 = 8\pi r\, dr \tag{18.59}$$

TABLE 18.1 The Vapor Pressure of Bulk Water and of Water Droplets at 25 °C, Calculated from the Kelvin Equation (Eq. 18.63)

Droplet Radius, r/m	Vapor Pressure, P/kPa	P/P_0
∞ (bulk water)	3.167	1
10^{-6}	3.170	1.001
10^{-7}	3.202	1.011
10^{-8}	3.519	1.111
10^{-9}	9.12	2.88

and the increase in surface Gibbs energy is this quantity multiplied by γ:

$$dG = 8\pi r\gamma\, dr \tag{18.60}$$

Insertion of the expression for dr in Eq. 18.58 leads to

$$dG = \frac{2\gamma M}{\rho r}\, dn \tag{18.61}$$

Equating Eqs. 18.56 and 18.61 gives

$$dn\, RT \ln \frac{P}{P_0} = \frac{2\gamma M}{\rho r}\, dn \tag{18.62}$$

or

Kelvin Equation

$$\ln \frac{P}{P_0} = \frac{2\gamma M}{\rho r RT} \tag{18.63}$$

Some values for water droplets, calculated from this relationship, are shown in Table 18.1.

The fact that the vapor pressure of a tiny droplet can be so much higher than that of the bulk liquid poses an interesting question regarding the process of condensation of a vapor. Suppose, for example, that air saturated with water is chilled. The vapor then becomes supersaturated and is in a metastable state. However, if the condensation first produces tiny droplets, these will have vapor pressures many times that of the bulk liquid, and they should immediately evaporate again. How, then, can condensation ever get started?

There are two possibilities. One is that, as a matter of chance, a large number of molecules may come together to form a droplet large enough that it does not at once reevaporate. This, however, is not likely, and this explanation does not often apply. More often it is dust particles that act as nuclei for supersaturated vapors, or alternatively the condensation may occur at the surface of the vessel. It is well known that supersaturation is much more likely—condensation is much more difficult—in the complete absence of dust particles.

18.8 Liquid Films on Surfaces

The spreading of oil on the surface of water has been of scientific interest for a very long time. In 1757 Benjamin Franklin, while traveling by ship to England, noticed that oil on a stormy sea has the effect of stilling the waves. On arriving in England he made a number of demonstrations of the effect by pouring "a teaspoonful of oil" on various bodies of water, including a pond at Clapham Common and Derwent Water in the Lake District. It is possible from his rough data to calculate the thickness of his layers of oil to be a few Ångstroms, in agreement with later estimates (see Problem 18.25).

Little work along similar lines was done for many years. From 1880 until 1933 Agnes Pockels (1862–1935) carried out a number of important experiments in the kitchen of her home in Brunswick, Lower Saxony. She first studied the relationship between the area occupied by a film and the surface tension and found that the behavior was different above and below a certain critical area. By 1891, when her first paper was published through the intervention of Lord Rayleigh, she had designed a rectangular tin trough that can be regarded as the prototype of the instrument later used extensively by Irving Langmuir (Figure 18.10).

Lord Rayleigh carried out a number of investigations along the same lines, and in 1899 he concluded that the critical area observed by Pockels corresponds to the molecules in the film being closely packed, in a *unimolecular layer.*

Important advances were made from 1917 when Irving Langmuir began to use a **film balance,** which was a technical improvement over the apparatus used by Agnes Pockels. Langmuir's apparatus, shown in Figure 18.10, measures the surface pressure, π_s, of the film, which is the force, F, exerted on it divided by the length of the edge along which the force is exerted; its SI unit is N m^{-1} = kg s^{-2}, which is the same as that of the surface tension. In fact, we see from Figure 18.8b and Eq. 18.49 that for a soap film the surface pressure π (equal to $F/2l$ since the film has two sides) is the same as the surface tension. For a film lying on the surface of water the surface pressure is equal to the *decrease* in surface tension brought about by the film.

The essential features of Langmuir's film balance are shown in Figure 18.10. A trough is filled to the brim with water, and a fixed barrier, which may be a strip of mica, floats on the surface; special devices are used to prevent leakage of the film past the float. The fixed barrier is suspended from a torsion wire by means of which the pressure is measured. A movable barrier rests on the sides of the

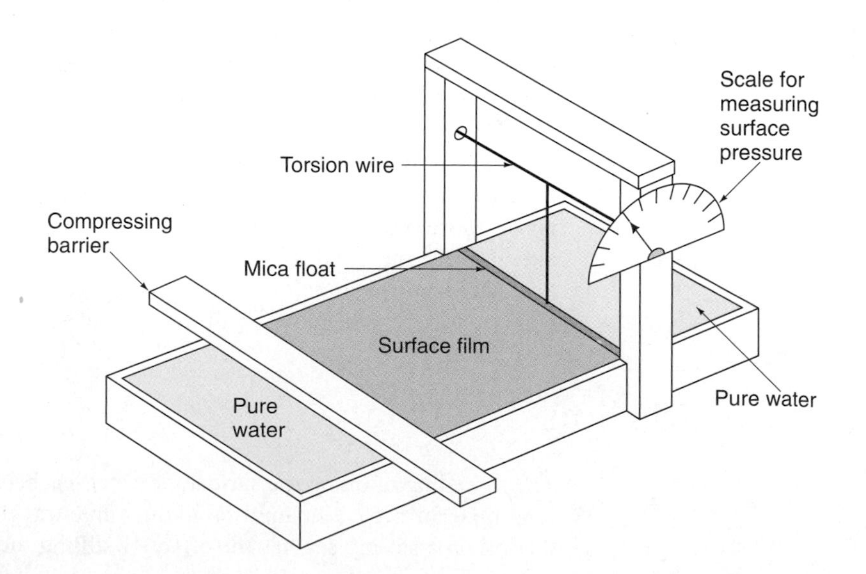

FIGURE 18.10
Diagrammatic representation of Langmuir's film balance.

AGNES POCKELS (1862–1935)

Agnes Pockels was born in Venice, which at the time was part of the Austrian Empire. Her father, Captain Theodor Pockels, was an officer in the Royal Austrian Army, from which he took early retirement because of ill health. After 1871 the family lived in Brunswick, Lower Saxony, where Agnes was educated at the Municipal High School for Girls. When she left school she was required to remain at home since both her mother and father were in poor health. For many years Agnes was responsible for the household management and some of the nursing of her parents.

In 1880, when she was 18, Agnes Pockels began a series of important investigations on surface films which were to have far-reaching consequences. In 1881 she measured surface tension by suspending small buttons, held at an aqueous surface, from a wooden beam balance. In 1882 she devised a rectangular trough with which, by means of a sliding strip of tin, she determined the relationship between surface area and surface tension; this simple apparatus was the prototype of the Langmuir film balance (Figure 18.10) and of the more sophisticated modern instruments. During the course of these investigations she discovered that the behavior of a surface film was different above and below a certain critical area. Above this critical area the area varies to a considerable extent with the tension, while below it "the displacement of the partition makes no impression on the tension."

Almost all of her work was carried out at the kitchen sink of her house. She found little support for her work from German scientists, and there at first seemed little possibility that she would be able to publish her results. However, her younger brother Friedrich (Fritz) (1865–1913), who later became a professor of physics, suggested that she should communicate her results to John William Strutt, 3rd Baron Rayleigh (1842–1919), who was working in the same field. She did so in 1891, and Rayleigh was most helpful; he had his wife translate the article into English and arranged for it to be published in *Nature.* During later years, and until 1933, Pockels made a number of further investigations in the same field.

It was not until she was quite old that her work received any official recognition. On the occasion of her seventieth birthday in 1932 the distinguished colloid chemist Wolfgang Ostwald, the son of Wilhelm Ostwald, published a tribute to her in the *Kolloid Zeitschrift,* and in the same year she was awarded an honorary doctorate by the Carolina-Wilhelmina University of Brunswick.

References: C. H. Giles and S. D. Forrester, The origins of the surface film balance, *Chemistry and Industry,* 2 January, 1971, pp. 43–53. This article contains much biographical information about Agnes Pockels, Lord Rayleigh, and others, including a number of portraits.

M. Elizabeth Derrick, Agnes Pockels, 1862–1935, *J. of Chem. Ed. 59,* 1030–1031(1982).

trough and is in contact with the surface of the water. The area available to the film is varied by moving this barrier. Subsequent to Langmuir's pioneering work many studies of monolayers have been carried out, using essentially the same method, but with surface balances in which there have been considerable technical improvements.

Studies of the relationship between the surface area and the surface pressure have shown that two-dimensional monolayers can exist in states that are analogous to the solid, liquid, and gaseous states of three-dimensional matter. Surface films are found to be of two general kinds:

1. **Coherent Films.** For these there is a critical area per molecule; at larger areas the pressure is quite small, but when the critical area is reached there is little

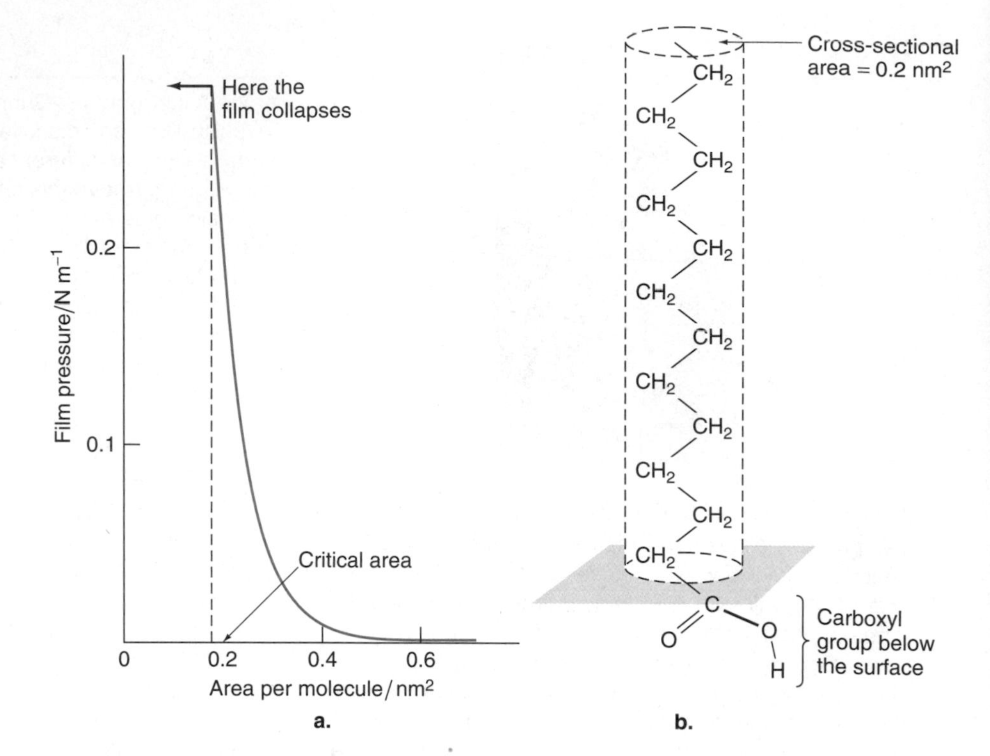

FIGURE 18.11
(a) Relationship between film pressure and molecular area, for a fatty acid such as stearic acid [octadecanoic acid, $CH_3(CH_2)_{16}CO_2H$]. (b) Diagrammatic representation of a fatty acid molecule at the surface of water.

further decrease in area with increase in pressure. An example of this type of behavior is shown in Figure 18.11a, which shows the type of pressure-area curve obtained with a long-chain fatty acid such as stearic acid. The critical area is about 0.2 nm^2. This behavior is analogous to that shown by a three-dimensional gas at a temperature below its critical temperature (compare Figure 1.18); when the gas is compressed sufficiently to form the liquid, the compressibility is then very much smaller than for the gas. The area of 0.2 nm^2 found with fatty acids is the area of cross section of the hydrocarbon chain, and when the critical area is reached, it is concluded that the molecules are standing up with their polar —COOH groups in the water, as shown in Figure 18.11b.

When the surface pressure reaches a very high value, the area suddenly decreases. The reason for this is that when the molecules are pressed very tightly together, further compression causes them to pile on top of one another.

2. **Noncoherent Films.** For noncoherent films there is no critical area. Such films obey, to a good approximation, the equation

$$\pi_s A = k_B T \tag{18.64}$$

where π_s is the surface pressure and A is the area per molecule adsorbed. This equation is the two-dimensional analog of the ideal gas equation, and noncoherent films therefore behave like a two-dimensional gas; they are often called **gaseous films.** There are not many examples of gaseous films, which tend to occur at higher temperatures; the molecular energies are then too high to permit the formation of the condensed phase.

18.9 Colloidal Systems

The word *colloid* (Greek *kolla,* glue; *-oeides,* like) was introduced in 1861 by the Scottish chemist Thomas Graham (1805–1869) to refer to substances that diffused slowly and would not pass through parchment.

In a true solution, such as one of sugar or salt in water, the solute particles consist of individual molecules or ions. At the other extreme there are **suspensions,** in which the particles contain more than one molecule and are large enough to be seen by the eye or at least under a microscope. Between these extremes are to be found the **colloidal dispersions,** in which the particles may contain more than one molecule but are not large enough to be seen in a microscope. It is impossible to draw a distinct line between colloidal dispersions, true solutions, and suspensions. The lower limit of microscopic visibility is about 2×10^{-7} m (0.2 μm or 200 nm), and this can be taken as the upper limit of the size of colloidal particles. The lower limit can be taken to be roughly 5×10^{-9} m (5 nm). This figure is comparable to the diameters of certain macromolecules, such as starch and proteins. Solutions of these substances therefore exhibit colloidal behavior, and although they may involve single molecules, they are conveniently classified as colloidal systems.

The essential properties of colloidal systems are due to the fact that the ratio of surface area to volume is very large. A true solution is a one-phase system, but a colloidal dispersion behaves as a two-phase system, since for each particle there is a definite surface of separation between it and the medium in which the particles are dispersed (the **dispersion medium**). At this surface certain characteristic properties, such as adsorption and electric potential, make themselves evident, since the total surface area in a colloidal dispersion can be very large. For example, the surface area of a cube of 1-cm edge is 6×10^{-4} m^2, but if the cube is subdivided into 10^{18} cubes of 10^{-8}-m edge, typical of colloidal systems, the total area is $10^{18} \times 6 \times 10^{-16} = 600$ m^2; the area has thus increased by a factor of 10^6. The fact that surface effects are important in colloid behavior is therefore not surprising.

The term **disperse phase** is used to refer to the particles that are present in the *dispersion medium.* Both the disperse phase and the dispersion medium may be solid, liquid, or gaseous. Since gases are always completely miscible, we cannot have a gas-in-gas colloidal dispersion, but all the other eight combinations are possible and are listed in Table 18.2, which gives the general name for each colloidal system and some well-known examples. In this section we will be mainly concerned with sols and emulsions.

TABLE 18.2 Types of Colloidal Systems

Dispersion Medium	Disperse Phase	Name of System	Examples
Gas	Liquid	Aerosol	Fog, mist, clouds
Gas	Solid	Aerosol	Smoke
Liquid	Gas	Foam	Whipped cream
Liquid	Liquid	Emulsion	Milk, mayonnaise
Liquid	Solid	Sol	Gold in water
Solid	Liquid	Gel	Jelly
Solid	Solid	Gel	Ruby glass, gold in glass
Solid	Gas	Solid foam	Pumice, styrofoam

Lyophobic and Lyophilic Sols

Colloidal dispersions of solids in liquids (i.e., sols) can be roughly divided into two types:

From the Greek *lysis,* loosening, dissolving; *phobia,* fear of. Lyophobic therefore literally means *fear of dissolving* or *solvent fearing.* Note also the use of the word *hydrophobic* in the expression *hydrophobic effect* (Section 17.6). Hydrophobic bonding results from the fact that when nonpolar groups come together in aqueous solution, there is a decrease in hydrogen bonding between water molecules and a resulting increase in entropy.

1. **Lyophobic Sols.** These can be called *hydrophobic sols* if the dispersion medium is water. The term *hydrophobic* was first used in 1905 by the French physicist Jean Baptiste Perrin (1870–1942) to denote a disperse phase, such as gold or arsenic sulfide, which has a low affinity for water; it is now applied more generally in surface chemistry to refer also to water-repellent surfaces. Since there is low affinity for the solvent, lyophobic sols are relatively unstable. For example, when lyophobic systems are evaporated, solids are obtained that cannot easily be reconverted into sols, and the addition of electrolytes to lyophobic sols frequently causes coagulation and precipitation.

2. **Lyophilic (liquid-loving) Sols.** These are sols in which there is a strong affinity between the disperse phase and the molecules comprising the dispersion medium. These sols are much more stable, and they behave much more like true solutions—which in the case of macromolecules they really are.

Colloidal dispersions are prepared in a variety of ways. Some substances are known as *intrinsic colloids* and readily form sols when they are brought into contact with a suitable dispersion medium. A colloidal solution of starch, for example, is easily prepared by introducing starch into boiling water. Intrinsic colloids are usually lyophilic, and they are usually either macromolecules (such as proteins) or long-chain molecules with polar end groups, which tend to aggregate and form particles of colloidal size, known as **micelles;** soaps and other detergents are of this type. *Extrinsic colloids,* on the other hand, do not form colloidal dispersions readily, and special methods have to be used; lyophobic colloids are frequently of this kind. The methods used fall into two classes:

1. **Condensation Methods.** The materials are initially in true solution. Chemical reactions are used to produce the sol; care is taken—by controlling concentrations, for example—to prevent the growth of the particles and consequent precipitation. For example, sols of various oxides have been prepared by hydrolysis of salts.

2. **Dispersion Methods.** Material originally in massive form is disintegrated into particles of colloidal dimensions. In one procedure, known as **peptization** (Greek *pepticos,* promoting digestion), the disintegration is brought about by the action of a substance known as a *peptizing agent.* For example, cellulose is peptized by the addition of organic solvents, such as ethanol-ether mixtures, leading to the familiar "collodion" sol. In this example the solvent is itself the peptizing agent, but frequently something else must be added; thus, precipitates of certain metal hydroxides in water can be peptized by dilute alkali hydroxides. Physical dispersion methods are also useful for producing colloidal dispersions. For example, a *colloid mill,* through which the dispersion medium and the substance to be dispersed are passed, grinds the material into colloidal particles. Another technique is *electrical disintegration,* in which an arc is struck between metal electrodes under water, with the production of a metal sol; this is often known as *Bredig's method,* after the Polish-German physical chemist Georg Bredig (1868–1944).

Light Scattering by Colloidal Particles

Although by definition colloidal particles are too small to be seen in the microscope, they can be detected by optical means. When light passes through a medium that contains no particles larger than about 10^{-9} m in diameter, the path of light cannot be detected and the medium is said to be *optically clear.* When colloidal particles are present, however, some of the light is scattered, and the incident beam passes through with weakened intensity. The first investigation of this phenomenon was made in 1871 by the British physicist John Tyndall (1820–1893), and the scattering is known as the **Tyndall effect;** the path of light through the medium, made visible as a result of the scattering, is known as the *Tyndall beam.* A sunbeam is a well-known example of a Tyndall beam, the light being scattered by dust particles.

Tyndall Effect

The type of apparatus used in a light-scattering experiment is shown schematically in Figure 18.12. The detector of the scattered light is mounted in such a way that the intensity of scattering can be measured at various scattering angles. Analysis of the scattering as a function of the angle provides valuable information about the sizes and shapes of colloidal particles; when these are single macromolecules, the technique is therefore useful in determining molar masses.

The theory of light scattering is very complicated and here we will merely outline the main ideas. The proportion of incident light that is scattered increases with an increase in the number and size of the particles. If the intensity of the incident radiation is I_0 and l is the length of the light path through the scattering medium, the intensity of the transmitted radiation is given by

$$I = I_0 e^{-\tau l} \tag{18.65}$$

where τ is known as the **turbidity.** In 1871 Lord Rayleigh deduced that for a spherical particle of radius r, which is much smaller than the wavelength λ of the radiation, the intensity of the radiation scattered through an angle θ is given by

$$I_\theta = K_s \frac{r^6}{\lambda^4}(1 + \cos^2 \theta) \tag{18.66}$$

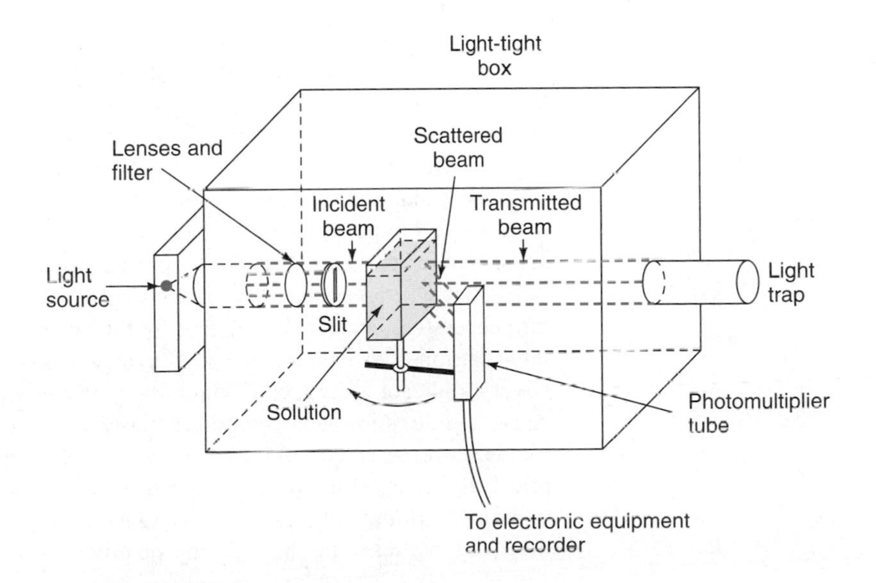

FIGURE 18.12 Schematic diagram of the type of apparatus used in a light-scattering experiment.

The constant K_s involves the refractive indices of the particle and the medium. The inverse fourth-power dependence on the wavelength means that with visible light the blue is the most scattered. It explains the blue of the sky, and the fact that colloidal solutions often appear blue. It also explains why photographs can be obtained under misty conditions by the use of filters that do not allow much blue light to pass through them.

Rayleigh's formula of Eq. 18.66 has been applied to the estimation of particle sizes and therefore to the estimation of molecular weights. When the particle dimensions are not small compared to the wavelength, and the particles are not spherical, the theory is much more complicated. Important contributions to the theory have been made by the German physicist Gustav Mie (1868–1957) in 1908, and in more recent years by Peter Debye, by Bruno Hasbrouck Zimm (b. 1920), and by Paul Doty (b. 1920).

Ultramicroscope

Zsigmondy received the 1925 Nobel Prize for physics for his work in colloid chemistry, including the invention of the ultramicroscope.

Less detailed studies can be carried out with an instrument known as an **ultramicroscope,** invented in 1903 by the Austro-German chemist Richard Adolf Zsigmondy (1865–1929) and his physicist colleague H. Siedentopf. In this instrument a beam is passed through the colloidal system, and individual particles can be seen under the microscope as flashes of scattered light. Particles as small as 5–10 nm in diameter, much too small to be directly visible, can be detected in this way. This is a useful technique for counting particles in a sol; if the amount of material present as disperse phase is known, it is then possible to make an estimate of the size of the particles. Particles detected in an ultramicroscope undergo continuous and rapid motion in all directions, and this is known as **Brownian movement,** named after the Scottish botanist Robert Brown (1773–1858), who first observed it in pollen grains seen under an ordinary microscope (see Section 19.2).

Electrical Properties of Colloidal Systems

When two electrodes are placed in a sol and an electric potential is applied across them, the particles in general move in one direction or another. This migration of colloidal particles in an electric field is called **electrophoresis,** which we considered on p. 486. It is one example of a transport property (see Chapter 19). The motion occurs as a result of the ζ (zeta) potential, the theory of which we met in Section 10.14, where the theory of electrophoresis was also considered. Both lyophobic and lyophilic sols undergo electrophoresis.

The electric double layer and the ζ potential are important for colloids in other ways. The stability of a hydrophobic sol, for example, is very much dependent on the charges on the surface of the particles; repulsion between particles carrying the same charges prevents them from approaching one another and forming larger particles that will precipitate out. The charges at the surface of colloidal particles are influenced to a considerable extent by the adsorption of ions. For example, if a dilute solution of a silver salt is added to an excess of dilute sodium iodide solution, a silver iodide sol is formed in which the particles are negatively charged. This is because iodide ions become preferentially adsorbed on the surface of the silver iodide, as represented in Figure 18.13a; the Na^+ ions remain in solution but form a positively charged atmosphere because of the electrostatic attractions. If, on the other hand, dilute iodide solution is added to an excess of dilute silver nitrate solution, the particles in the sol are positively charged. Now the silver ions are

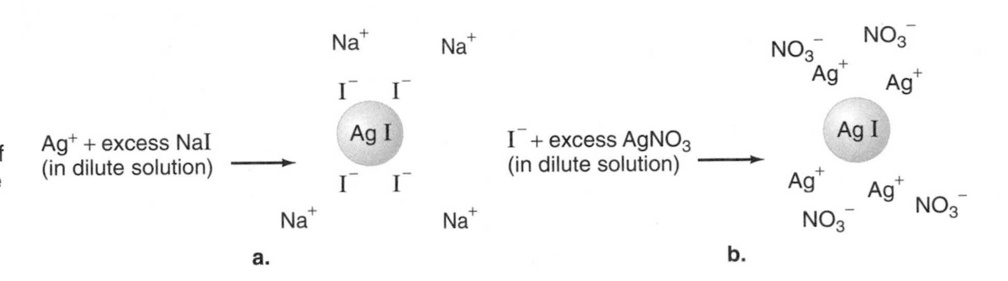

FIGURE 18.13
Silver iodide sols. (a) The particles are formed in the presence of excess iodide ions, which become adsorbed. (b) The particles are formed in the presence of excess silver ions, which are adsorbed.

preferentially adsorbed on the particles, the nitrate ions forming the atmosphere, as shown in Figure 18.13b.

The situation with lyophilic sols is a little different, in that the charges on the particles often result largely from the ionization of the material constituting the disperse phase.

Protein molecules in general bear positive or negative charges because of the ionizations of —COOH, —NH_3^+, —OH, and other groups. At low pH values the protein bears a net positive charge because groups like —NH_3^+ are present but there are no negatively charged groups; at high pH values negative groups like —COO^- and —O^- are present but there are no positively charged groups and the molecule has a net negative charge. At the same time the charge is influenced by the specific adsorption of ions. At some intermediate pH, the **isoionic point,** the protein does not migrate in an electric field.

Isoionic Point

Added electrolytes influence the stability of colloidal particles by affecting their charges. Again the behavior is different for lyophobic and lyophilic sols. Usually lyophobic particles are easily coagulated, with the formation of a visible precipitate, by the addition of electrolytes. The reason is that ions of opposite charge to that on the surface of the particles will easily become adsorbed, because of the electrostatic attraction, and will bring about a reduction in the charge of the particles, which will repel one another less strongly. This is illustrated in Figure 18.14a for a gold sol particle, precipitated by the addition of sodium chloride. Initially the particle has a negative charge, because of the adsorption of anions. When sodium chloride is added, the excess sodium ions replace some of the iodide ions, thereby reducing the surface charge and therefore the repulsion, and precipitation occurs.

The precipitating action of ions of opposite sign to those on the surface of the particle is greater the higher the valence of the ion. These statements regarding the influence of sign and valence on the precipitation of colloidal particles were made in 1882 by H. Schulze and in 1900 by the English biologist Sir William Bate Hardy (1864–1934) and are incorporated as the **Schulze-Hardy rule.**

Lyophilic sols differ from lyophobic sols in that much larger amounts of electrolytes are required to bring about precipitation. The reason is that lyophilic particles are to some extent protected by a layer of bound water molecules. This is illustrated in Figure 18.14b for a protein such as gelatin, which can be regarded as surrounded by a film of water through which ions cannot easily penetrate. Higher concentrations of electrolytes do, however, exert a *salting-out effect,* as we have discussed in Section 7.11; this results from the tendency of the ions to bind water molecules, which are thus less available to the substance being precipitated.

The high sensitivity of lyophobic particles to precipitation by electrolytes can sometimes be reduced by the addition of a lyophobic substance, which is said to be

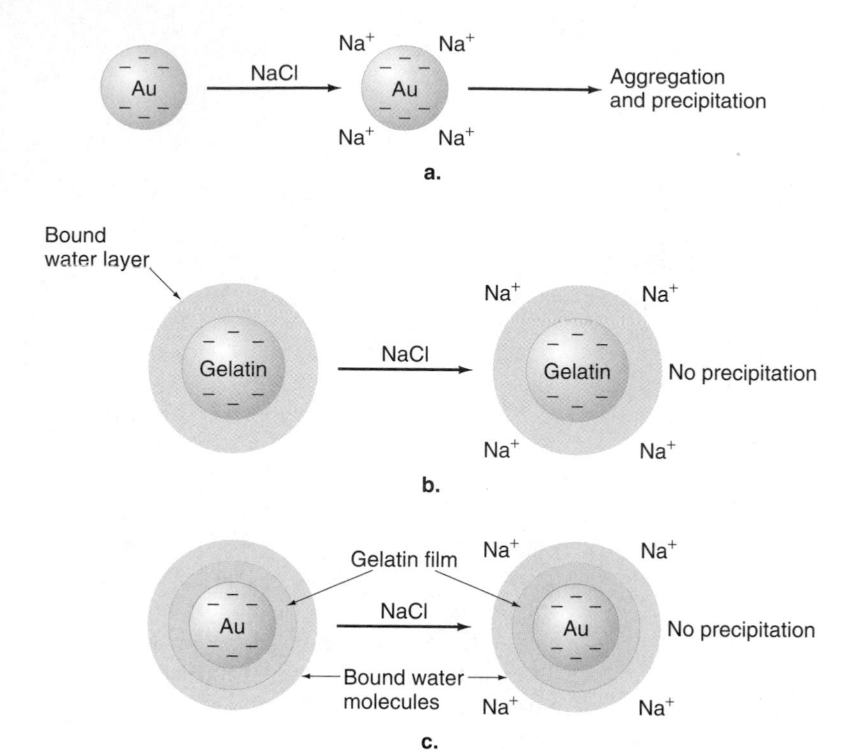

FIGURE 18.14
(a) Precipitation of a hydrophobic sol. The attachment of Na^+ ions to negatively charged gold particles allows them to aggregate. (b) The film of water surrounding a hydrophilic particle inhibits the attachment of ions. (c) Protective action of gelatin attached to a gold particle.

a **protective colloid** and to exert a *protective action.* As illustrated in Figure 18.14c, the protective substance becomes adsorbed on the surface of the lyophobic particles and protects them from the approach of ions of opposite sign.

Gels

Under certain circumstances it is possible to cause a lyophilic sol to coagulate and yield a semirigid jellylike mass that includes the whole of the liquid present in the sol. The product so obtained is known as a **gel,** and there are two types, *elastic gels* and *nonelastic gels.* A number of food products, such as jellies, jams, and cornstarch puddings, are elastic gels. A well-known example of an inelastic gel is that of silicic acid, commonly known as *silica gel.* The essential distinction between elastic and nonelastic gels is their behavior on dehydration. Partial dehydration of an elastic gel leads to the formation of an elastic solid from which the original sol can be regenerated by the addition of water. Dehydration of a nonelastic gel, on the other hand, leads to a glass or powder, which has little elasticity.

Imbibition

Another distinction between elastic and nonelastic gels relates to their ability to take up solvent. If an elastic gel such as gelatin is placed in water, it swells, water having been *imbibed* by the gel; the process is known as **imbibition.** Nonelastic gels, on the other hand, may take up solvent but they do not swell; the liquid enters the pores of the gel but, since the walls are rigid, the volume of the gel does not change.

Gel formation occurs in particular with molecules that can exist as extended chains. As the sol turns into a gel, the chains become interlocked so that the viscos-

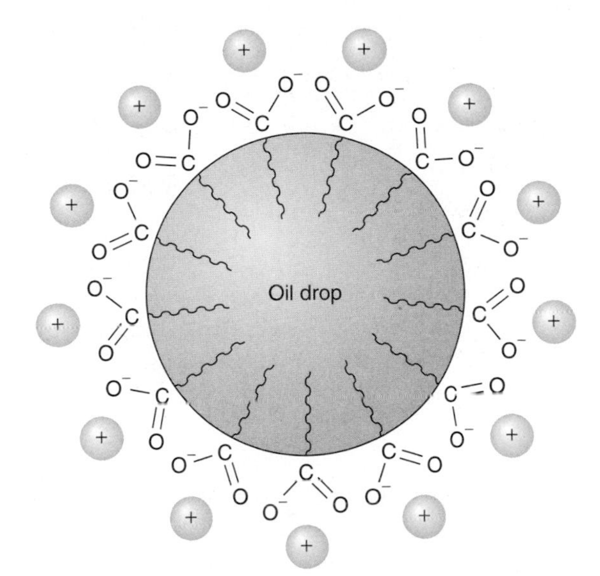

FIGURE 18.15
The structure of a micelle, in which a particle or drop is stabilized by molecules having an anionic group and a long hydrocarbon chain.

ity increases and eventually a semi-solid material is produced. The dispersion medium is held by capillary forces between the chains, some of the molecules exerting a specific solvating effect.

Emulsions

An emulsion consists of droplets of one liquid dispersed in another liquid. The droplets are usually from 0.1 to 1 μm in diameter and hence are larger than sol particles. Emulsions are generally unstable unless a third substance, known as an **emulsifying agent** or a *stabilizing agent,* is present. Soaps and detergents are effective emulsifying agents, particularly for oil-water emulsions. They consist of long-chain hydrocarbon molecules each having at one end a polar group such as a carboxylic acid or sulfonic acid group. These molecules are readily adsorbed at oil-water interfaces; the hydrocarbon chains become attached to the oil and the polar groups to the water, as shown schematically in Figure 18.15. The term **micelle** (Latin *micella,* small crumb, diminutive of the Latin *mica,* crumb) was used by the Canadian-American chemist James William McBain (1882–1953) and the American chemist William Draper Harkins (1873–1951) to refer to particles that are stabilized by emulsifying agents in this manner.

The action of emulsifying agents is to reduce the interfacial tension between the two phases. The effect of surface tension, as we have seen in Section 18.7, is to cause the surface to become as small as possible. A high surface tension between the disperse phase and the dispersion medium will therefore tend to cause an emulsion to separate into two bulk phases (i.e., to *coagulate*). The adsorption of an emulsifying agent on an interface reduces this tension and therefore decreases the tendency of the emulsion to coagulate.

An emulsion with no stabilizing agent has properties similar to those of lyophobic sols; for example, they are easily coagulated by electrolytes. Stabilized emulsions, on the other hand, behave more like lyophilic sols and are only affected by electrolytes at high concentrations.

KEY EQUATIONS

Langmuir isotherms for fraction θ of surface covered:

Simple adsorption: $$\theta = \frac{K[A]}{1 + K[A]}$$

Adsorption with dissociation: $$\theta = \frac{K^{1/2}[A]^{1/2}}{1 + K^{1/2}[A]^{1/2}}$$

Competitive adsorption: $$\theta_A = \frac{K_A[A]}{1 + K_A[A] + K_B[B]}$$

Surface tension γ:

Definition: $$\gamma \equiv \frac{\text{work done}}{\text{increase in surface area}}$$

For a capillary rise h in tube of radius r: $$\gamma = \frac{rh\rho g}{2 \cos \theta}$$

Kelvin equation (pressure of a droplet):

$$\ln \frac{P}{P_0} = \frac{2\gamma M}{\rho r RT}$$

PROBLEMS

Adsorption Isotherms

18.1. A surface is half-covered by a gas when the pressure is 1 bar. If the simple Langmuir isotherm (Eq. 18.6) applies:

a. What is K/bar^{-1}?
b. What pressures give 75%, 90%, 99%, 99.9% coverage?
c. What coverage is given by pressures of 0.1 bar, 0.5 bar, 1000 bar?

18.2. Show that, if V is the volume of gas adsorbed at pressure P, and the Langmuir isotherm is obeyed, a plot of P/V against P is linear. Explain how, from such a plot, the volume V_0 corresponding to complete coverage and the isotherm constant K can be determined.

18.3. The following results were reported by Langmuir for the adsorption of nitrogen on mica at 20 °C:

Pressure/atm	2.8	4.0	6.0	9.4	17.1	33.5
Amount of gas adsorbed/mm^3 at 20 °C and 1 atm	12.0	15.1	19.0	23.9	28.2	33.0

a. Make a linear plot of these values in order to test the Langmuir isotherm, Eq. 18.6. If it applies, evaluate the constant K.
b. Suppose that 10^{15} molecules cover 1 cm^2 of the surface. Make an estimate of the effective surface area in Langmuir's experiment.

18.4. a. Show that for small coverages a system obeying the Langmuir isotherm will give a linear plot of ln (θ/P) against P, with a slope of unity.
b. What is the slope if ln (V_a/P) is plotted against V_a at small coverages? (V_a is the volume of gas adsorbed.)

18.5. The following are the volumes of ammonia, reduced to STP, adsorbed by 1 g of charcoal at 0 °C:

Pressure/kPa	6.8	13.5	26.7	53.1	79.4
Volume/cm^3	74	111	147	177	189

Make a plot to see if the data are consistent with the Langmuir isotherm. If so, evaluate the constants K and V_0, the volume adsorbed when the surface is saturated.

18.6. a. Suggest a method of making a linear plot to test the applicability of the Brunauer, Emmett, and Teller (BET) isotherm (Eq. 18.25) when volumes adsorbed, V, are known at various pressures.
b. Show that the BET equation reduces to the Langmuir isotherm when $P_0 \gg P$.

18.7. The following data were obtained for the adsorption of krypton on a 1.21 g sample of a porous solid:

Pressure/Torr	1.11	3.08
Volume adsorbed/cm^3 (STP)	1.48	1.88

If the saturation vapor pressure is 19.0 Torr, estimate a surface area for the solid, assuming that a molecule of krypton occupies an area of 2.1×10^{-21} m^2.

***18.8.** Derive the equation

$$\frac{\theta}{1-\theta} = c_g^{1/2} \frac{h^{3/2}}{(2\pi m k_B T)^{3/4}} \frac{b_a}{b_g^{1/2}} e^{-\Delta E_0/2RT}$$

for the case of adsorption with dissociation (i.e., $A_2 + 2S \rightleftharpoons 2(S - A)$; ΔE_0 is the energy of adsorption per mole.

***18.9.** Derive the equation

$$c_a = c_g \frac{h}{(2\pi m k_B T)^{1/2}} \frac{b_a}{b_g} e^{-\Delta E_0/RT}$$

for the case of adsorption where the adsorbed molecules are completely mobile on the surface (i.e., have two degrees of translational freedom).

Kinetics of Surface Reactions

18.10. A first-order surface reaction is proceeding at a rate of 1.5×10^{-4} mol dm^{-3} s^{-1} and has a rate constant 2.0×10^{-3} s^{-1}. What will be the rate and the rate constant if

a. the surface area is increased by a factor of 10?
b. the amount of gas is increased tenfold at constant pressure and temperature?
If these values of v and k apply to a reaction occurring on the surface of a spherical vessel of radius 10 cm:
c. What will be the rate and rate constant in a spherical vessel, of the same material, of radius 100 cm, at the same pressure and temperature?
d. Define a new rate constant k' that is independent of the gas volume V and the area S of the catalyst surface.
e. What would be its SI unit?

18.11. A zero-order reaction is proceeding at a rate of 2.5×10^{-3} mol dm^{-3} s^{-1} and a rate constant 2.5×10^{-3} mol dm^{-3} s^{-1}.

a. How will the changes a, b, and c in Problem 18.10 affect the rate and the rate constant in this case?
b. Again, define a rate constant that is independent of S and V.
c. What would be its SI unit?

18.12. The decomposition of ammonia on platinum,

$$2NH_3 = N_2 + 3H_2$$

is first order in NH_3 and the rate is inversely proportional to the hydrogen concentration (Eq. 18.39). Write the differential rate equation for the rate of formation of hydrogen, dx/dt, in terms of the initial concentration of ammonia, a_0, and the concentration x of hydrogen at time t.

***18.13.** On the basis of the mechanism given on p. 943, derive an expression for the rate of formation of hydrogen atoms when hydrogen gas is in contact with hot tungsten. Under what conditions is the order of reaction one-half?

18.14. A unimolecular surface reaction is inhibited by a poison I and obeys Eq. 18.36. If E is the activation energy corresponding to the reaction of the adsorbed substrate molecule (i.e., corresponding to k) and ΔH_A and ΔH_I are the enthalpies of adsorption of A and I, what is the activation energy

a. at very low concentrations of A and I?
b. at a very high concentration of A and a very low concentration of I?
c. at a very low concentration of A and a very high concentration of I?

18.15. Suppose that a reaction,

$$A \rightarrow Y + Z$$

occurs initially as a homogeneous first-order reaction (rate constant k) but that the product Z is adsorbed on the surface and catalyzes the reaction according to a law that is zero order in A and first order in Z (i.e., the term in the rate equation is $k_c[Z]$). Obtain a differential equation for the rate of appearance of Z, and integrate it to give z as a function of time.

18.16. Suggest explanations for the following observations, in each case writing an appropriate rate equation based on a Langmuir isotherm:

a. The decomposition of phosphine (PH_3) on tungsten is first order at low pressures and zero order at higher pressures, the activation energy being higher at the higher pressures.
b. The decomposition of ammonia on molybdenum is retarded by the product nitrogen, but the rate does not approach zero as the nitrogen pressure is increased.
c. On certain surfaces (e.g., Au) the hydrogen-oxygen reaction is first order in hydrogen and zero order in oxygen, with no decrease in rate as the oxygen pressure is greatly increased.
d. The conversion of para-hydrogen into ortho-hydrogen is zero order on several transition metals.

Surface Tension and Capillarity

18.17. The surface tension of water at 20 °C is 7.27×10^{-2} N m^{-1} and its density is 0.998 g cm^{-3}. Assuming a contact angle θ of zero, calculate the rise of water at 20 °C in a capillary tube of radius (a) 1 mm and (b) 10^{-3} cm. Take g = 9.81 m s^{-2}. (Capillaries in a tree have radii of about 10^{-3} cm, but sap can rise in a tree to much greater heights than obtained in this calculation. The reason is that the rise of sap depends to a considerable extent on osmotic flow; because of evaporation the leaves contain solutes of higher concentration than the trunk of the tree, and osmotic flow therefore occurs to the leaves.)

18.18. The density of liquid mercury at 273 K is 13.6 g cm^{-3} and the surface tension is 0.47 N m^{-1}. If the contact angle is 140°, calculate the capillary depression in a tube of 1-mm diameter.

18.19. The density of water at 20 °C is 0.998 g cm^{-3} and the surface tension is 7.27×10^{-2} N m^{-1}. Calculate the ratio between the vapor pressure of a mist droplet having a mass of 10^{-12} g and the vapor pressure of water at a plane surface.

18.20. The two arms of a U-tube have radii of 0.05 cm and 0.10 cm. A liquid of density 0.80 g cm^{-3} is placed in the tube, and the height in the narrower arm is found to be 2.20 cm higher than that in the wider arm. Calculate the surface tension of the liquid, assuming $\theta = 0$.

18.21. A tube is placed in a certain liquid and the capillary rise is 1.5 cm. What would be the rise if the same tube were placed in another liquid that has half the surface tension and half the density of the first liquid? Assume that $\theta = 0$ in both cases.

18.22. When a certain capillary tube is placed in water, the capillary rise is 2.0 cm. Suppose that the tube is placed in the water in such a way that only 1.0 cm is above the surface; will the water flow over the edge? Explain your answer.

***18.23.** A layer of benzene, of density 0.8 g cm^{-3}, is floating on water of density 1.0 g cm^{-3}, and a vertical tube of internal diameter 0.1 mm is inserted at the interface. It is observed that there is a capillary rise of 4.0 cm and that the contact angle is 40°. Calculate the interfacial tension between water and benzene.

18.24. A liter of water at 20 °C is broken up into a spray in which the droplets have an average radius of 10^{-5} cm. If the surface tension of water at 20 °C is 7.27×10^{-2} N m^{-1}, calculate the Gibbs energy change when the droplets are formed.

Surface Films

18.25. Benjamin Franklin demonstrated on a number of occasions that a teaspoonful of oil put on water would produce a layer half an acre in area (1 acre = 4840 square yards; 1 yard = 0.915 m). Assume a teaspoonful to be 1 cm^3, and estimate the thickness of the film.

18.26. A fatty acid was spread on the surface of water in a Langmuir film balance at 15 °C, and the following results obtained:

Area/cm^2 μg^{-1}	5.7	28.2	507	1070	2200	11100
Surface pressure/ 10^{-3} N m^{-1}	30	0.3	0.2	0.1	0.05	0.01

Estimate the molecular weight of the acid and the area per molecule when the film was fully compressed.

18.27. N. K. Adam carried out surface film studies using a Langmuir film balance 14.0 cm in width having a floating barrier 13.8 cm long. In one investigation he introduced 52.0 μg of 1-hexadecanol ($C_{16}H_{33}OH$) onto the surface and measured the force on the float at various lengths of the film, obtaining the following results:

Length/cm	Force on float/10^{-5} N
20.9	4.14
20.3	8.56
20.1	26.2
19.6	69.0
19.1	108.0
18.6	234
18.3	323
18.1	394
17.8	531

Estimate the area per molecule when the film was fully compressed.

Essay Questions

18.28. Describe some of the most important characteristics of a chemisorbed layer. In what ways does a physisorbed (van der Waals) layer differ?

18.29. Derive the Langmuir adsorption isotherms for two substances competitively adsorbed on a surface. Show how these equations interpret the kinetics of bimolecular surface reactions, distinguishing between Langmuir-Hinshelwood and Langmuir-Rideal mechanisms.

18.30. Explain clearly the distinction between inherent and induced heterogeneity of surfaces.

18.31. Explain the difference between lyophilic and lyophobic sols, with reference to some of the properties in which they differ.

18.32. What information can be obtained from light-scattering experiments on colloidal particles in aqueous solution?

A number of problems relating to the transport properties of colloidal systems are to be found at the end of Chapter 19.

SUGGESTED READING

For general treatments of surface chemistry and physics, see

A. W. Adamson, *Physical Chemistry of Surfaces,* New York: Interscience, 1968; 4th edition, 1982.

R. J. Hunter, *Introduction to Modern Colloid Science,* Oxford University Press, 1994.

M. Kerker (Ed.), *Surface Chemistry and Colloids,* London: Butterworth, 1975.

S. R. Morrison, *The Chemical Physics of Surfaces,* New York: Plenum, 1977.

D. J. Shaw, *Introduction to Colloid and Surface Chemistry,* Oxford; Boston: Butterworth-Heinemann, 1966; 4th edition, 1992.

G. A. Somorjai, *Chemistry in Two Dimensions: Surfaces,* Ithaca, NY: Cornell University Press, 1981.

A. Zangwill, *Physics at Surfaces,* Cambridge: University Press, 1988.

For reviews of work on special aspects of the physics of solid surfaces, see the above publications and also

G. K. Binnig and H. Rohrer, The scanning tunneling microscope, *Scientific American, 253,* 50–56(1985). This is a very clear article by the inventors of the method, who shared the 1986 Nobel Prize in physics for the work.

C. J. Chen, *Introduction to Scanning Tunneling Microscopy,* New York: Oxford University Press, 1993.

L. J. Clarke, *Surface Crystallography: An Introduction to Low Energy Electron Diffraction,* New York: Wiley-Interscience, 1985.

J. G. Dash, *Films on Solid Surfaces: The Physical Chemistry of Physical Adsorption,* New York: Academic Press, 1975.

C. B. Duke, Low energy electron diffraction (LEED), in *Encyclopedia of Physics* (Eds. R. G. Lerner and G. L. Trigg), New York: VCH Publishers, 2nd edition, 1990, pp. 651–653.

D. E. Eastman and F. J. Himpsel, Photoelectron spectroscopy, in *Encyclopedia of Physics, op. cit.,* pp. 908–913.

G. Ertl and J. Küppers, *Low Energy Electrons and Surface Chemistry,* Weinheim; Deerfield Beach, FL: VCH Publishers, 2nd edition, 1985.

R. J. Hamers, Scanning tunneling microscopy, in *Encyclopedia of Physics, op. cit.,* pp. 1084–1085.

R. J. Hamers, Atomic resolution surface spectroscopy with the scanning tunneling microscope, *Annual Reviews of Physical Chemistry, 40,* 531–559(1989).

G. L. Kellogg, Field-ion microscopy, in *Encyclopedia of Physics, op. cit.,* pp. 385–388.

E. W. Müller and T. T. Tsong, *Field Ion Microscopy: Principles and Applications,* New York: Elsevier, 1969. Müller was the inventor of the technique.

D. Nicholson and N. G. Parsonage, *Computer Simulation and the Statistical Mechanics of Adsorption,* New York: Academic Press, 1982.

S. L. Sharp, R. J. Warmack, J. P. Gourdonnet, I. Lee, and T. L. Ferrell, Spectroscopy and imaging using the scanning-tunneling microscope, *Accounts of Chemical Research, 26,* 377–382(1993).

C. Tanford, *Ben Franklin Stilled the Waves: An informal history of pouring oil on water, with reflections on the ups and downs of scientific life in general,* Durham: Duke University Press, 1989.

M. Thompson, M. D. Baker, A. Christie, and J. F. Tyson, *Auger Electron Spectroscopy,* New York: Wiley-Interscience. 1985.

T. T. Tsong, Studies of surfaces at atomic resolution: atom-probe and field ion microscopy, *Surface Science Reports, 8,* 127–209(1988).

Surface catalysis is treated in

G. C. Bond, *Heterogeneous Catalysis: Principles and Applications,* Oxford: Clarendon Press, 1987.

C. M. Friend, Catalysis on surfaces, *Scientific American,* April, 1993, pp. 74–79. This article is particularly concerned with the new techniques for observing individual molecules on surfaces.

R. P. H. Gasser, *An Introduction to Chemisorption and Catalysis on Metals,* Oxford: Clarendon Press, 1985.

R. J. Madix, Molecular transformations on single crystal rhodium catalyst, *Science, 233,* 1159–1166(1986).

M. W. Roberts, New perspectives in surface chemistry and catalysis, *Chemical Society Reviews, 6,* 373–391(1977). This article, based on a Tilden Lecture, gives a clear account, with examples, of the use of photoelectron spectroscopy (and to some extent of low-energy electron diffraction) in exploring the detailed nature of adsorbed layers and of catalytic processes.

J. T. Yates, Surface chemistry, *Chemical & Engineering News,* March 30, 1992, pp. 22–35.

Transport Properties

PREVIEW

This chapter deals with properties that relate to the movement of molecules in fluids. One such property is *viscosity,* which is a measure of the frictional resistance in a fluid when there is an applied shearing force. The coefficient of viscosity is defined in terms of Newton's law of viscous flow, according to which the shearing force is proportional both to the area and to the relative motion of two planes. Poiseuille's law is concerned with the rate with which fluid (liquid or gas) flows through a tube.

The theory of the viscosity of a gas is different from that of the viscosity of a liquid. In a gas the reason for the frictional force or drag between two moving parallel planes is that molecules are passing from one plane to another. The expression for the viscosity of a gas, derived on the basis of kinetic theory in terms of the molecular diameters, shows that the viscosity increases with the square root of the absolute temperature. With a liquid, on the other hand, the viscosity decreases with temperature, in proportion to $e^{E_{vis}/RT}$, where E_{vis} is the activation energy for the process in which layers of molecules move past one another. The viscosities of solutions are usually treated empirically.

Diffusion (Section 19.2) is a process in which solute molecules tend to flow from regions of higher concentration to regions of lower concentration. The laws of diffusion are summarized in two laws due to Fick: The first of these is concerned with the rate of flow across a concentration gradient, the second with the rate of change of concentration as a result of diffusion. These laws can be applied to some special situations, such as diffusion across a membrane and Brownian movement. The diffusion coefficient can be related to the frictional force and, in the case of ionic diffusion, to the ionic mobility. The *Stokes-Einstein* equation expresses the diffusion coefficient in terms of the radius of the diffusing particle and the viscosity of the medium.

Molecules in solution can be made to *sediment* by application of a large effective gravitational field in an ultracentrifuge (Section 19.3). On the basis of the theory of sedimentation, measurements in the ultracentrifuge allow molar masses to be determined.

OBJECTIVES

After studying this chapter, the student should be able to:

- Define the various aspects of viscosity, Newtonian or laminar flow, and the coefficient of viscosity.
- Understand the various ways to measure viscosity.
- Appreciate the Poiseuille equation and how to use it to determine the viscosity of liquids and gases.
- Apply the concepts of viscosity in liquids to solutions using the specific and intrinsic viscosity.
- Understand and be able to use Fick's first and second laws of diffusion.
- Recognize Brownian movement as a diffusion effect and its relationship to the random-walk equation.
- Understand the relationship between viscosity and diffusion.
- Derive the Einstein equation, $D = k_B T/f$, and know the Nernst-Einstein equation.
- Understand Stokes's law and the Stokes-Einstein equation.
- Understand facilitated and active transport.
- Explain what sedimentation is and how an ultracentrifuge works.

Hydrodynamics is from the Greek *hydor*, water; *dynamics*, power. The term *fluid dynamics* is more appropriate than *hydrodynamics* when nonaqueous systems are involved.

Properties that depend on rates of movement of matter or energy are known as **transport properties.** When aqueous systems are involved, we can also speak of hydrodynamic properties, although this expression is often used for liquids other than water. Important transport properties are viscosity, diffusion, and sedimentation. A shearing force acting on a liquid produces a relative motion of different planes, and the extent of this motion is measured by a property known as the **fluidity,** the reciprocal of which is the **viscosity.** Diffusion is produced by a force resulting from a Gibbs energy difference between various positions in a solution in which there is a concentration gradient. Sedimentation occurs as a result of a gravitational field or of a centrifugal force, which produces an effective gravitational field.

Transport can also be induced by the application of an electric field. We have already discussed, in Chapter 7, how simple ions move when a potential is applied; this motion contributes to the conductivity of a solution. Certain molecules, such as proteins, contain charged groups and the charges depend on the pH of the solution. Such molecules in solution may move in an electric field; this type of motion is referred to as **electrophoresis** (Greek *pherein,* to carry). A related type of transport occurs when an electric potential is applied across a charged membrane immersed in a solution; there is then movement of the solvent molecules, and we speak of electroendosmosis (Greek *endon,* within; *osmos,* impulse), or **electroosmosis.** These electrical transport properties have already been considered in Section 10.14 (pp. 483-487). Heat conduction is usually treated in physics and chemical engineering textbooks and is not included in this chapter, which is restricted to the transport of matter.

19.1 Viscosity

The viscosity of a fluid is a measure of the frictional resistance it offers to an applied shearing force. If a fluid is flowing past a surface, the layer adjacent to the surface hardly moves; successive layers have increasingly higher velocities. Figure 19.1 shows two parallel planes in a fluid, separated by a distance dx and having velocities of flow differing by dv. According to **Newton's law of viscous flow,** the frictional force F, resisting the relative motion of two adjacent layers in the liquid, is proportional to the area A and to the velocity gradient dv/dx:

Newtonian Flow

$$F = \eta A \frac{dv}{dx} \tag{19.1}$$

Coefficient of Viscosity

The proportionality constant η is known as the **coefficient of viscosity** or simply as the *viscosity.* Its reciprocal, the *fluidity,* is given the symbol ϕ.

The SI unit for the coefficient of viscosity η is kg m^{-1} s^{-1} or N s m^{-2}; a commonly used unit, the *poise* (g cm^{-1} s^{-1}), is one-tenth of the SI unit. The type of flow to which Eq. 19.1 applies is called **laminar,** *streamline,* or **Newtonian flow.** In flow of this kind there is superimposed on the random molecular velocities a net component of velocity in the direction of flow. Streamline flow is observed if the velocity of flow is not too large; with very rapid flow the motion becomes *turbulent* and Eq. 19.1 no longer applies.

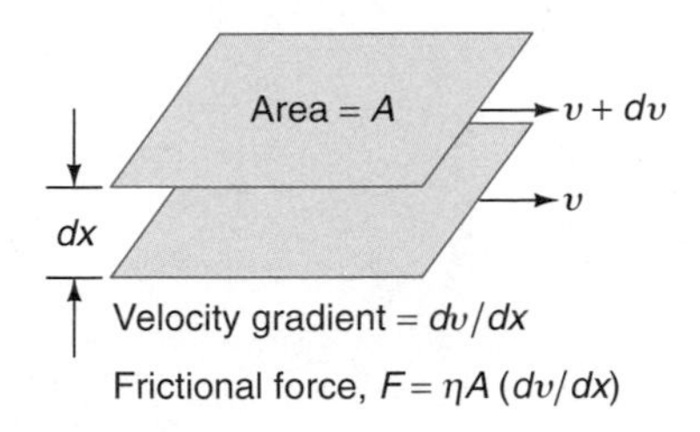

FIGURE 19.1
The definition of the coefficient of viscosity, η. Two parallel layers of fluid, of area A, are separated by a distance dx, and the difference between their velocities is dv.

Measurement of Viscosity

Viscosity is usually studied by allowing the fluid to flow through a tube of circular cross section and measuring the rate of flow. From this rate, and with the knowledge of the pressure acting and the dimensions of the tube, the coefficient of viscosity can be calculated on the basis of a theory developed in 1844 by the French physiologist Jean Leonard Poiseuille (1799–1869). Consider an incompressible fluid flowing through a tube of radius R and length l, with a pressure P_1 at one end and a pressure P_2 at the other (Figure 19.2a and b). The liquid at the walls of the tube is stagnant; the rate of flow increases to a maximum at the center of the tube. A cylinder of length l and radius r has an area of $2\pi rl$, and according to Eq. 19.1 the frictional force is

$$F = -\eta \frac{dv}{dr} 2\pi rl \qquad (19.2)$$

where the velocity gradient dv/dr is a negative quantity. This force is exactly balanced by the force driving the fluid in this cylinder. This force is the pressure difference $P_1 - P_2$ multiplied by the area πr^2 of the cylinder; thus

$$-\eta \frac{dv}{dr} 2\pi rl = \pi r^2 (P_1 - P_2) \qquad (19.3)$$

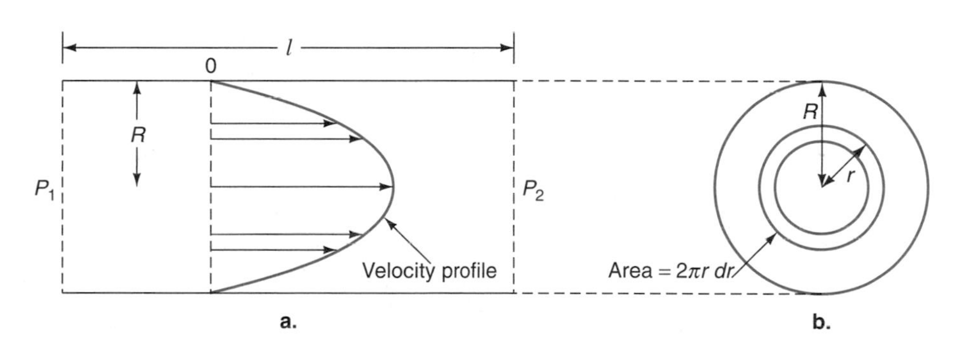

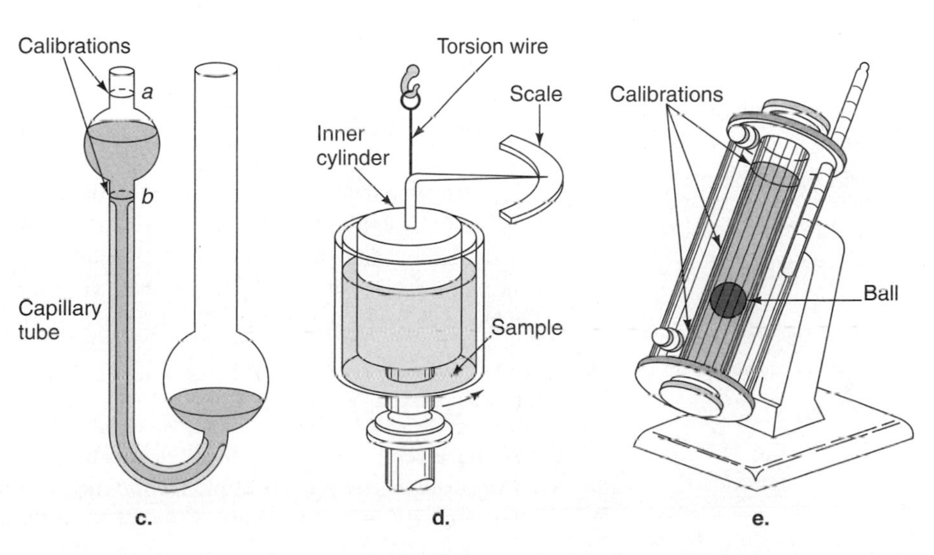

FIGURE 19.2
The measurement of viscosity. (a) Flow through a tube. (b) Cross section of tube. (c) An Ostwald viscometer for measuring the viscosity of a liquid. (d) Couette rotating-cylinder viscometer. (e) Falling-ball viscometer.

or

$$dv = -\frac{r}{2\eta l}(P_1 - P_2)\,dr \tag{19.4}$$

Integration of this gives

$$v = -\frac{(P_1 - P_2)}{4\eta l}r^2 + \text{constant} \tag{19.5}$$

The velocity v is zero when $r = R$; the constant of integration is thus

$$\text{constant} = \frac{(P_1 - P_2)}{4\eta l}R^2 \tag{19.6}$$

and therefore

$$v = \frac{(P_1 - P_2)}{4\eta l}(R^2 - r^2) \tag{19.7}$$

The total volume of liquid flowing through the tube in unit time, dV/dt, is obtained by integrating over each element of cross-sectional area. Each element has an area of $2\pi r\,dr$ (Figure 19.2b) and therefore

$$\frac{dV}{dt} = \int_0^R 2\pi r v\,dr \tag{19.8}$$

$$= \frac{(P_1 - P_2)\pi}{4\eta l}\left\{R^2 \int_0^R r\,dr - \int_0^R r^3\,dr\right\} \tag{19.9}$$

$$= \frac{(P_1 - P_2)\pi R^4}{8\eta l} \tag{19.10}$$

Poiseuille Equation

This is the **Poiseuille equation,** and it enables η to be calculated from measurements of the rate of flow dV/dt in a tube of known dimensions, if the pressure difference $P_1 - P_2$ is known.

A commonly employed instrument for measuring the viscosity of a liquid is the *Ostwald viscosimeter* or *viscometer,* illustrated in Figure 19.2c. One measures the time that it takes for a quantity of liquid to pass through the tube, from one position to another, under the force of its own weight. Usually the instrument is calibrated by the use of a liquid of known viscosity. Another piece of apparatus used for viscosity measurements is the *Couette rotating-cylinder viscometer* (Figure 19.2d). In this instrument a liquid is caused to rotate in an outer cylinder, and it causes a torque to be applied to the torsion wire attached to the inner cylinder. The apparatus is calibrated and the viscosity is calculated from the torque. Another device for measuring viscosity is the falling-ball viscometer (Figure 19.2e); the viscosity is calculated from the time required for the ball to fall from one position to another.

The pressure difference $P_1 - P_2$ varies with the time but is proportional to the density; the densities of the calibrating liquid and of the sample must therefore be known.

Viscosities of Gases

We have seen that viscosity arises because in a fluid there is a *frictional force,* or *drag,* between two parallel planes moving with different velocities. The theory of the viscosity of a gas is very different from the theory of the viscosity of a liquid.

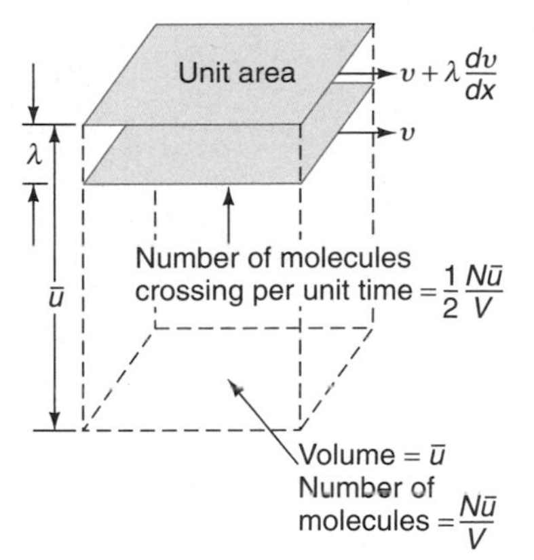

FIGURE 19.3
The kinetic theory of gaseous viscosity.

In the case of gas, the origin of the frictional drag between the two parallel planes is that molecules are passing from one plane to the other. A helpful analogy is provided by two trains traveling in the same direction on parallel tracks but with speeds differing by Δv. Suppose that the passengers are eccentric enough to amuse themselves by jumping from one train to the other. A passenger of mass m jumping from the faster train to the slower one transports momentum $m\Delta v$ to the slower train and therefore tends to increase its speed. Conversely, the passengers in the slower train who jump to the faster one remove $m\Delta v$ of momentum from it and tend to slow it down. As a result, the speeds of the trains tend to equalize, and the net effect is the same as if there were a frictional force acting between the trains.

The analogous situation in a flowing gas is represented in Figure 19.3. The lower plane is moving with velocity v. We consider the separation between the planes to be the mean free path λ of the gas (Section 1.9). The reason for this choice is that λ is the average distance that the gas molecules move between successive collisions, so that a molecule that has experienced a collision in one plane and is moving toward the other plane will, on the average, experience its next collision in that plane. In so doing it transports momentum from one plane to the other. If dv/dx is the velocity gradient, the difference in the velocities of the two planes, separated by λ, is $(dv/dx)\lambda$, and the transfer of momentum when the molecule moves from one plane to the other is

$$m\lambda\frac{dv}{dx}$$

Figure 19.3 shows two parallel planes of unit area, and we now consider how many molecules move from one plane to the other in unit time. The average molecular speed is $\bar{u}$, and in the figure we have constructed a rectangular prism of unit cross-sectional area and of length $\bar{u}$, indicated by dashed lines. The volume of that prism is $\bar{u}$, and if the gas has N molecules in a volume V, the number of molecules in the prism is $N\bar{u}/V$. Half of these molecules are moving in an upward direction and half in a downward direction, and on the average they travel the entire distance $\bar{u}$ in unit time. The number that cross unit area per unit time in one direction is therefore

$$\frac{1}{2}\cdot\frac{N\bar{u}}{V}$$

This is the number that jump from one plane to the other in unit time. Since each molecule transfers momentum of $m\lambda\ (dv/dx)$, momentum transported in unit time is

$$\frac{1}{2}\cdot\frac{N\bar{u}}{V}\cdot m\lambda\cdot\frac{dv}{dx}$$

Force is rate of change of momentum, and this expression is therefore the force F acting between the planes of unit area:

$$F = \frac{1}{2}\cdot\frac{N\bar{u}}{V}\cdot m\lambda\cdot\frac{dv}{dx} \tag{19.11}$$

Insertion of this expression into Eq. 19.1, with $A = 1$, gives

$$\frac{1}{2}\cdot\frac{N\bar{u}}{V}\cdot m\lambda\cdot\frac{dv}{dx} = \eta\frac{dv}{dx} \tag{19.12}$$

and therefore

$$\eta = \frac{Nm\bar{u}\lambda}{2V} \tag{19.13}$$

Since the density ρ of the gas is Nm/V, this equation can be written as

$$\eta = \tfrac{1}{2}\rho\bar{u}\lambda \tag{19.14}$$

In Equation 1.68 we saw that the mean free path λ of a gas is equal to $V/\sqrt{2}\pi d^2 N$, where d is the molecular diameter, and the insertion of this into Eq. 19.13 gives

$$\eta = \frac{m\bar{u}}{2\sqrt{2}\pi d^2} = \frac{m\bar{u}}{2\sqrt{2}\sigma} \tag{19.15}$$

In the second expression πd^2 has been replaced by the collision cross section σ. Table 1.3 (p. 32) gave us the expression $(8k_BT/\pi m)^{1/2}$ for $\bar{u}$, and therefore

$$\eta = \frac{(mk_BT)^{1/2}}{\pi^{3/2}d^2} = \frac{(mk_BT)^{1/2}}{\pi^{1/2}\sigma} \tag{19.16}$$

These equations for the coefficient of viscosity of a gas lead to some interesting predictions, all of which are confirmed by experiment. For example, we see from Eq. 19.16 that η depends on the molecular mass, the molecular diameter, and the temperature. The density or pressure of the gas does not affect the viscosity. This may seem surprising at first sight, but the explanation is not hard to find. At higher densities more molecules jump from one layer to the next, but λ is smaller and each jump involves the transport of less momentum. These two effects just counteract each other.

Second, from Eq. 19.16 we see that the viscosity increases with the square root of the temperature. We shall see in the next subsection that this prediction, supported by experiment, is in marked contrast to the behavior of a liquid, for which the viscosity decreases with increasing temperature. The explanation can be seen either from Eq. 19.14 or from Eq. 19.15. Raising the temperature does not change ρ, λ, m, or d, but it increases $\bar{u}$, which is proportional to the square root of the absolute temperature. More molecules cross from one layer to the next as the temperature goes up, and this increases the drag and therefore the viscosity.

Finally, we can see from Eq. 19.16 that if we had two gases for which the molecules had identical masses but for which the molecular diameters d were different, the gas of higher d would have the lower viscosity. The reason for this is that an increase in d decreases the mean free path, so that each jump involves the transfer of less momentum and therefore produces less drag. Equation 19.16 would lead us to the conclusion that a hypothetical gas consisting of molecules of zero diameter would have infinite viscosity, which is obviously impossible. Molecules of zero size, of course, cannot collide with each other, and there is no change of momentum as a result of collisions; the equations we have derived therefore do not apply to this situation (see Problem 19.9).

Equation 19.16 provides us with a means of obtaining molecular diameters from viscosity measurements, and this method has often been used. Table 19.1

TABLE 19.1 Viscosities of Gases, Molecular Diameters, and Mean Free Paths (at 25 °C and 1 bar)

Gas	Viscosity, η / 10^{-6} kg m^{-1} s^{-1}	Molecular Diameter/nm: From Viscosity	Molecular Diameter/nm: From van der Waals b	Mean Free Path/nm
He	18.6	0.225	0.248	180.6
H_2	8.42	0.281	0.276	115.8
N_2	16.7	0.386	0.314	61.4
O_2	18.09	0.383	0.290	62.7
CO_2	13.8	0.475	0.324	40.6

gives some diameters obtained in this way and compares them with values calculated from the van der Waals constant b (Section 1.13). From the diameters we can calculate mean free paths (Eq. 1.68), and some values are included in Table 19.1.

EXAMPLE 19.1 The viscosity of carbon dioxide at 25 °C and 101.325 kPa is 13.8×10^{-6} kg m^{-1} s^{-1}. Estimate the molecular diameter.

Solution It is most convenient to use Eq. 19.16. The molecular mass is

$$m = \frac{44.01}{6.022 \times 10^{23}} = 7.308 \times 10^{-23} \text{ g} = 7.308 \times 10^{-26} \text{ kg}$$

From Eq. 19.16,

$$d = \frac{(mk_BT)^{1/4}}{\pi^{3/4}\eta^{1/2}}$$

$$= \frac{[7.308 \times 10^{-26} \text{ (kg)} \times 1.381 \times 10^{-23} \text{(J K}^{-1}) \times 298.15\text{(K)}]^{1/4}}{\pi^{3/4}[13.8 \times 10^{-6} \text{(kg m}^{-1} \text{ s}^{-1})]^{1/2}}$$

$$= 4.75 \times 10^{-10} \text{ m} = 0.475 \text{ nm}$$

[The units are kg$^{1/4}$ J$^{1/4}$ kg$^{-1/2}$ m$^{1/2}$ s$^{1/2}$; since J = kg m^2 s^{-2}, this becomes kg$^{1/4}$ (kg m^2 s^{-2})$^{1/4}$ kg$^{-1/2}$ m$^{1/2}$ s$^{1/2}$ = m.]

Viscosities of Liquids

The viscous behavior of liquids is very different from that of gases. For example, whereas gas viscosities increase with rising temperature, the viscosities of liquids decrease. In 1913 it was first shown empirically by the Spanish physical chemist J. de Guzman that viscosity obeys a law of the Arrhenius type, a result that was confirmed in 1916 by Arrhenius himself. We have seen that fluidity is the reciprocal of viscosity, and the variation of fluidity ϕ with temperature can be expressed as

$$\phi = \frac{1}{\eta} = A_{fl}e^{-E_{fl}/RT} \tag{19.17}$$

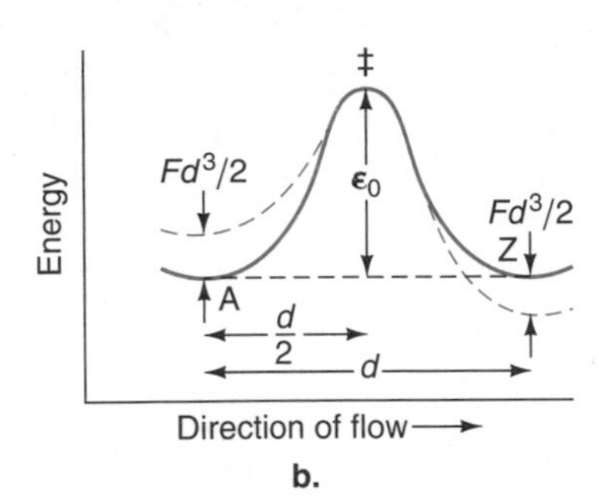

FIGURE 19.4
Liquid viscosity. (a) Two liquid layers moving with respect to each other under the influence of a force F per unit area.
(b) Potential-energy diagram for viscous flow. The solid line shows the potential-energy profile when there is no shearing force and the dashed line shows the profile when there is a shearing force.

where A_{fl} is the preexponential factor for the flow process and E_{fl} is the activation energy. The viscosity itself can be expressed as

$$\eta = A_{vis} e^{E_{vis}/RT} \tag{19.18}$$

where A_{vis} ($= 1/A_{fl}$) and E_{vis} ($= -E_{fl}$) are the preexponential factor and activation energy, respectively. The latter is readily obtained from the slope of a plot of ln η against $1/T$.

The detailed theory of the viscosity of liquids is very complicated, and here we will present only the basic ideas and a simplified version of the treatment. Figure 19.4a shows two layers of molecules in a liquid; the upper layer moves to the right more rapidly than the lower. Additional molecules are present above and below these two layers, and in order for motion to occur the molecules will have to push neighboring molecules aside. As a result, there is an energy barrier to the flow process. Figure 19.4b shows a plot of potential energy against the distance moved by a given molecule. At positions A and Z the molecule is in a state of minimum potential energy, but because of the repulsions of neighboring molecules it has to surmount an energy barrier of ϵ_0. If d is the distance between the initial and final positions of the molecule, the molecule will have to travel a distance of $d/2$ to reach the top of the barrier, which will be symmetrical.

Suppose that a shearing force F per unit area is bringing about the relative displacement of the two planes. As far as a given molecule is concerned, this force acts on the effective area occupied by the molecule. For simplicity we will assume this area to be d^2. The force acting upon a single molecule is thus Fd^2, and in order for the molecule to reach the top of the potential-energy barrier this force will be exerted through a distance $d/2$. The effect of the force is thus to provide energy of $Fd^2 \times d/2 = Fd^3/2$. This means that, as shown by the dashed line in Figure 19.4b, the height of the barrier in the left-to-right direction is reduced by $Fd^3/2$, whereas that in the right-to-left direction is raised by $Fd^3/2$.

In the absence of the shearing force, the rate constant for the passage of a molecule over the potential-energy barrier, of height ϵ_0, can be written

$$k_0 = A'_{vis} e^{-\epsilon_0/k_B T} \tag{19.19}$$

where A'_{vis} is the preexponential factor for the process. When the force is applied, the rate constant for the process from left to right is

$$k_1 = A'_{vis} \exp\left[-\frac{(\epsilon_0 - Fd^3/2)}{k_B T}\right] = k_0 e^{Fd^3/2k_B T} \tag{19.20}$$

while that for the right-to-left process is

$$k_{-1} = A'_{vis} \exp\left[-\frac{(\epsilon_0 + Fd^3/2)}{k_B T}\right] = k_0 e^{Fd^3/2k_B T} \tag{19.21}$$

These rate constants represent the numbers of times a molecule crosses the barrier in the two directions. The rate of crossing is

$$k = k_1 - k_{-1} = k_0 (e^{Fd^3/2k_B T} - e^{-Fd^3/2k_B T}) \tag{19.22}$$

Each time a crossing occurs, the molecule moves a distance d. Multiplication by d therefore gives the distance traveled in unit time, that is, the relative velocity Δv:

$$\Delta v = dk_0 (e^{Fd^3/2k_B T} - e^{-Fd^3/2k_B T}) \tag{19.23}$$

In ordinary viscous flow, F is sufficiently small that $Fd^3 \ll 2k_BT$. We can therefore expand the exponentials and accept only the first terms:

$$\Delta v = dk_0\left[1 + \frac{Fd^3}{2k_BT} - \left(1 - \frac{Fd^3}{2k_BT}\right)\right] = \frac{k_0Fd^4}{k_BT} \tag{19.24}$$

The velocity gradient dv/dx is $\Delta v/d$:

$$\frac{dv}{dx} = \frac{k_0Fd^3}{k_BT} \tag{19.25}$$

Since F is the force per unit area, Eq. 19.1 becomes

$$F = \eta\frac{k_0Fd^3}{k_BT} \tag{19.26}$$

from which it follows that

$$\eta = \frac{k_BT}{k_0d^3} \tag{19.27}$$

Introduction of the expression for k_0 (Eq. 19.19) gives

$$\eta = \frac{k_BT}{A'_{vis}d^3}e^{\epsilon_0/k_BT} \tag{19.28}$$

Since d^3 is the volume occupied by a single molecule in the liquid state, we can replace it by V_m/L, where V_m is the molar volume of the liquid and L is the Avogadro constant. Equation 19.28 can thus be written as

$$\eta = \frac{Lk_BT}{A'_{vis}V_m}e^{\epsilon_0/k_BT} = \frac{RT}{A'_{vis}V_m}e^{E_{vis}/RT} \tag{19.29}$$

This is of the same form as the empirical equation (Eq. 19.18), and it predicts the same temperature dependence. The activation energy per mole, $E_{vis} = \epsilon_0L$, proves to be about one-third to one-quarter of the molar enthalpy of vaporization of the liquid; some values are given in Table 19.2. In other words, for a molecule to push another aside in order to pass to a new equilibrium position, it has to expend one-third to one-quarter of the energy required to remove a molecule entirely from the liquid state.

TABLE 19.2 Activation Energies for Viscous Flow in Liquids, Compared with Enthalpies of Vaporization at the Boiling Point*

Liquid	E_{vis}/kJ mol^{-1}	ΔH_{vap}/kJ mol^{-1}	$\Delta_{vap}H/E_{vis}$
CCl_4	10.5	27.6	2.6
C_6H_6	10.6	27.9	2.6
CH_4	3.01	7.6	2.5
$CHCl_3$	7.4	27.8	3.8
CH_3COCH_3	6.9	26.8	3.9
H_2O†	12.1	37.6	3.2

*Values taken from R. H. Ewell and H. Eyring, *J. Chem. Phys.*, *5*, 726(1937).

†The energy of activation for viscous flow in water varies considerably with temperature, as a result of the changes in hydrogen-bonded structure; E_{vis} is 21.1 kJ mol^{-1} at 0 °C and 8.8 kJ mol^{-1} at 150 °C.

Viscosities of Solutions

It is difficult to work out a satisfactory treatment of the viscosities of solutions based on fundamental principles. As a result, it is usually more satisfactory to proceed empirically. In this way it has proved possible to obtain information about the sizes and shapes of solute molecules, particularly of macromolecules, from the viscosities of their solutions. When a solute is added to a liquid such as water, the viscosity is usually increased. Suppose that the viscosity of a pure liquid is η_0 and that the viscosity of a solution is η. The **specific viscosity,** defined as

Specific Viscosity

$$\text{specific viscosity} = \frac{\eta - \eta_0}{\eta_0} \tag{19.30}$$

is the increase in viscosity $\eta - \eta_0$ relative to the viscosity of the pure solvent. Division of the specific viscosity by the mass concentration of ρ of the solution gives what is known as the **reduced specific viscosity** [SI unit: $m^3\ kg^{-1}$]:

Reduced Specific Viscosity

$$\text{reduced specific viscosity} = \frac{1}{\rho} \cdot \frac{\eta - \eta_0}{\eta_0} \tag{19.31}$$

This quantity, however, has a contribution from the intermolecular interactions between the solute molecules. This contribution can be eliminated by extrapolating the reduced specific viscosity to infinite dilution; then we obtain what is known as the **intrinsic viscosity** [SI unit: $m^3\ kg^{-1}$], given the symbol $[\eta]$:

Intrinsic Viscosity

$$[\eta] = \lim_{\rho \to 0} \left[\frac{1}{\rho} \frac{\eta - \eta_0}{\eta_0} \right] \tag{19.32}$$

This quantity is the fractional change in the viscosity per unit concentration of solute molecules, at infinite dilution.

Various empirical relationships have been proposed to relate the intrinsic viscosity $[\eta]$ to the relative molecular mass M_r. The most successful of these is the equation

$$[\eta] = kM_r^{\alpha} \tag{19.33}$$

where K and α are constants. This equation was proposed independently by Herman Francis Mark (1895–1992) in 1938 and by R. Houwink in 1941 and is usually called the **Mark-Houwink equation.** It applies to series of molecules of similar structure. If it is obeyed, a plot of log $[\eta]$ against log M_r will be a straight line. Figure 19.5 shows some results plotted in this way for various polyisobutenes in diisobutene as solvent. By the use of such calibration curves, the relative molecular masses of unknown samples can be determined.

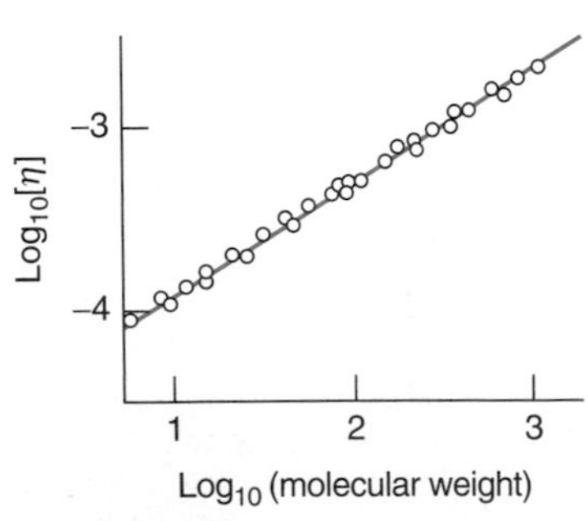

FIGURE 19.5
Logarithm of intrinsic viscosity of polyisobutenes plotted against logarithm of relative molecular mass, for solutions in diisobutene at 20 °C.

Attempts have been made to relate the value of α in Eq. 19.33 to the shape of the molecule. If the molecules are spherical, the intrinsic viscosity $[\eta]$ is independent of the size of the molecules, so that α is equal to zero. In agreement with this, all globular proteins, regardless of their size, have essentially the same $[\eta]$. If a molecule is elongated, its molecules are more effective in increasing the viscosity, and $[\eta]$ is larger; values of 1.3 or higher are frequently obtained for molecules that exist in solution as extended chains. Long-chain molecules that are coiled in solution give intermediate values of α, frequently in the range 0.6 to 0.75.

19.2 Diffusion

If solutions of different concentrations are brought into contact with each other, the solute molecules tend to flow from regions of higher concentration to regions of lower concentration, and there is ultimately an equalization of concentration. The driving force leading to diffusion is the Gibbs energy difference between regions of different concentration.

Fick's Laws

The German physiologist Adolf Eugen Fick (1829–1901) formulated in 1855 two fundamental laws of diffusion. According to **Fick's first law,** the rate of diffusion dn/dt of a solute across an area A, known as the *diffusive flux* and given the symbol J, is

$$J = \frac{dn}{dt} = -DA\frac{\partial c}{\partial x} \tag{19.34}$$

where $\partial c/\partial x$ is the concentration gradient of the solute, and dn is the amount of solute crossing the area A in time dt.

The SI unit of dn/dt is mol s^{-1}, that of A is m^2, and that of $\partial c/\partial x$ is mol m^{-4}; that of the *diffusion coefficient* D is therefore m^2 s^{-1} and that of flux per unit area is mol m^{-2} s^{-1}. In practice, diffusion coefficients are commonly expressed as cm^2 s^{-1}, and in biological work the "fick" is sometimes used as a unit; 1 fick = 10^{-11} m^2 s^{-1}.

Fick also derived an equation for the rate of change of concentration as a result of diffusion. Figure 19.6 shows a system of cross-sectional area A, having a concentration c at position x and a concentration $c + dc$ at position $x + dx$. Because there is an increase in concentration as x increases, the net diffusion occurs from right to left in the diagram. The flux at x can be written as $J(x)$ and that at $x + dx$ as $J(x + dx)$, which is given by

$$J(x + dx) = J(x) + \frac{\partial J}{\partial x}dx \tag{19.35}$$

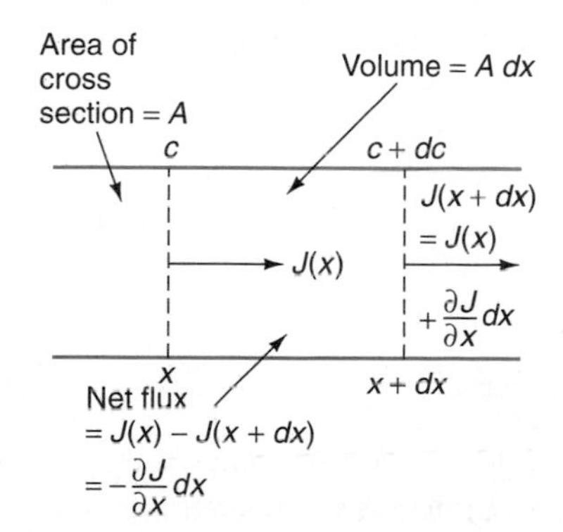

FIGURE 19.6
One-dimensional diffusion; a diagram illustrating Fick's second law.

The net flux into the region between x and $x + dx$ is thus

$$J_{net} = J(x) - J(x + dx) = -\frac{\partial J}{\partial x}dx \tag{19.36}$$

The net rate of increase in concentration in this element of volume is the net flux divided by the volume, which is $A\,dx$:

$$\frac{\partial c}{\partial t} = -\frac{1}{A}\cdot\frac{\partial J}{\partial x} \tag{19.37}$$

From Eq. 19.34, J is $-DA(\partial c/\partial x)$ and therefore

$$\frac{\partial c}{\partial t} = -\frac{1}{A}\cdot\frac{\partial}{\partial x}\left(-DA\frac{\partial c}{\partial x}\right) = \frac{\partial c}{\partial x}\left(D\frac{\partial c}{\partial x}\right) \tag{19.38}$$

If D is independent of the distance x (as is always true to a good approximation), this equation reduces to

$$\frac{\partial c}{\partial t} = D\frac{\partial^2 c}{\partial x^2} \tag{19.39}$$

This is **Fick's second law of diffusion** for the special case of diffusion in one dimension (e.g., along the X axis). In liquids and solutions, which are said to be *isotropic,* D has the same value in all directions. Some solids are *anisotropic,* which means that D is not the same in all directions.

Solutions of Diffusion Equations

Equation 19.39 is a second-order, linear, and homogeneous differential equation. Its solution depends on the nature of the domain through which diffusion is taking place and on the initial conditions (i.e., the concentrations at various positions at some time that may be taken as $t = 0$). Some initial and boundary conditions present a difficult mathematical problem, and often explicit expressions cannot be obtained.

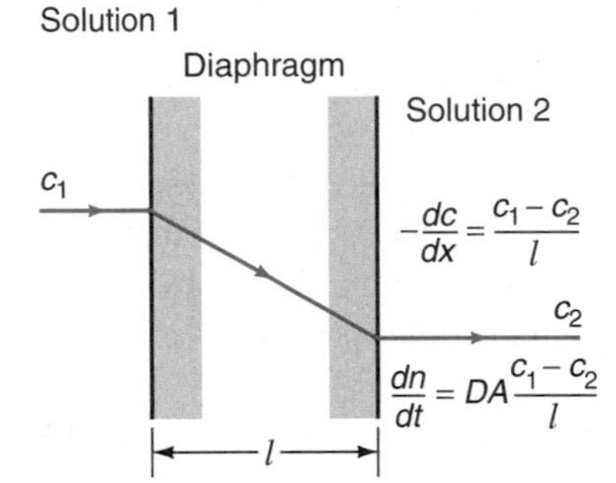

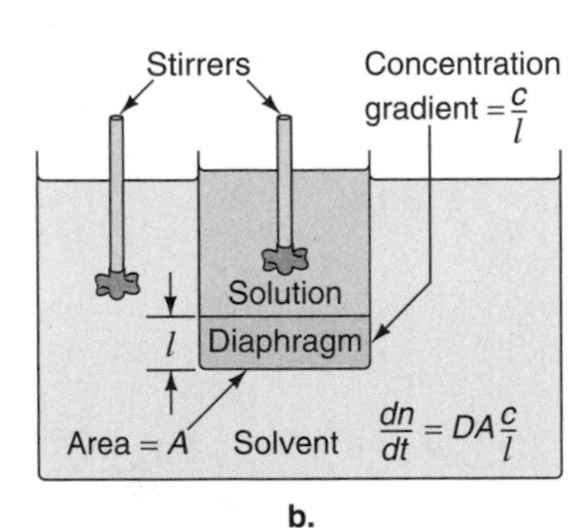

FIGURE 19.7
(a) Stirred solutions of concentrations c_1 and c_2 separated by a porous diaphragm of thickness l. In the steady state the concentration within the diaphragm changes linearly from c_1 to c_2. (b) Schematic diagram of simple apparatus for the measurement of diffusion constants.

A particularly simple situation is when solutions at two different concentrations are separated by a porous diaphragm in such a way that the solutions can be stirred and maintained at uniform concentration. A convenient technique for measuring diffusion rates in water is to separate two solutions by a sintered glass membrane of thickness l, as shown schematically in Figure 19.7b. Such a membrane contains pores filled with water, and it is assumed that the diffusion through the membrane occurs at the same rate as in water (a correction can be made for the thickness of cross section occupied by the glass). When such a system is set up, it is found experimentally that a steady state is soon established; that is, the rate of diffusion does not change with time as long as the concentrations c_1 and c_2 on the two sides remain the same (this will be true to a good approximation if the solution volumes are large). In order for there to be a steady state, the concentration gradient must be uniform across the membrane; otherwise the rates of flow would vary across the membrane and there would be accumulation or depletion of material in certain regions of the membrane and therefore no steady state. In other words, the concentration within the membrane must fall linearly, as shown in Figure 19.7a. The concentration gradient is thus given by

$$-\frac{dc}{dx} = \frac{c_1 - c_2}{l} \tag{19.40}$$

and by Fick's first law the rate of flow through the membrane is

$$\frac{dn}{dt} = DA\frac{c_1 - c_2}{l} \tag{19.41}$$

The diffusion coefficient D is therefore calculated from the measurement of the rate of flow; the area A, thickness l, and concentration difference $c_1 - c_2$ are readily determined. Some diffusion coefficients are given in Table 19.3.

The case just considered is a very simple one in which the concentration has been forced to vary linearly from c_1 to c_2. Another case of interest is when there

TABLE 19.3 Diffusion Coefficients at 20 °C

Diffusing Substance	Relative Molecular Mass	Solvent	Diffusion Coefficient, D/m^2 s^{-1}
H_2	2	(gas)*	1.005×10^{-4}
O_2	32	(gas)*	1.36×10^{-5}
H_2O	18	$H_2O^{\dagger}$	25.4×10^{-10}
CH_3OH	32	H_2O	13.7×10^{-10}
H_2NCONH_2	60	H_2O	11.8×10^{-10}
Glycerol	92	H_2O	8.3×10^{-10}
Sucrose	342	H_2O	5.7×10^{-10}
Insulin	41 000	H_2O	8.2×10^{-11}
Horse hemoglobin	68 000	H_2O	6.3×10^{-11}
Urease	470 000	H_2O	3.5×10^{-11}
Tobacco mosaic virus	31 400 000	H_2O	5.3×10^{-12}
Phenol	94	Benzene	15.8×10^{-10}
Bromoform	353	Benzene	16.9×10^{-10}
I_2	254	Benzene	19.3×10^{-10}

*For self-diffusion at 101.325 kPa, calculated from viscosities (Table 19.1) using Eq. 19.57.
†Self-diffusion.

is an *instantaneous plane source* at a particular plane in a liquid. A simple way of arriving at the equations applicable to this case is to note that a general solution of Eq. 19.39 for Fick's second law is

$$c = \alpha t^{-1/2} e^{-x^2/4Dt} \tag{19.42}$$

where α is a constant. The function on the right-hand side of this equation is commonly known as a **delta** function. That this is a general solution may be verified by substitution into Eq. 19.39.[1] When $t \to 0$, this function corresponds to $c = 0$ everywhere except at $x = 0$, where $c \to \infty$. In other words, this case corresponds to solute present at a plane at the origin $x = 0$; since it is present in zero volume, the concentration is infinite. The constant α is related to the "strength" of the source (i.e., to the amount n_0 of solute initially present at $x = 0$). Since the amount of solute remains the same at all times, n_0 is equal to the integral

$$A \int_{-\infty}^{\infty} c \, dx$$

[1]Thus, from Eq. 19.42,

$$\frac{\partial c}{\partial t} = \alpha\left(-\frac{1}{2} t^{-3/2} e^{-x^2/4Dt} + \frac{x^2}{4Dt^{5/2}} e^{-x^2/4Dt}\right)$$

$$\frac{\partial c}{\partial x} = -\frac{\alpha x}{2D} t^{-3/2} e^{-x^2/4Dt}$$

$$\frac{\partial^2 c}{\partial x^2} = -\frac{\alpha t^{-3/2}}{2D} e^{-x^2/4Dt} + \frac{\alpha x^2}{4D^2 t^{5/2}} e^{-x^2/4Dt}$$

whence

$$\frac{\partial c}{\partial t} = D \frac{\partial^2 c}{\partial x^2}$$

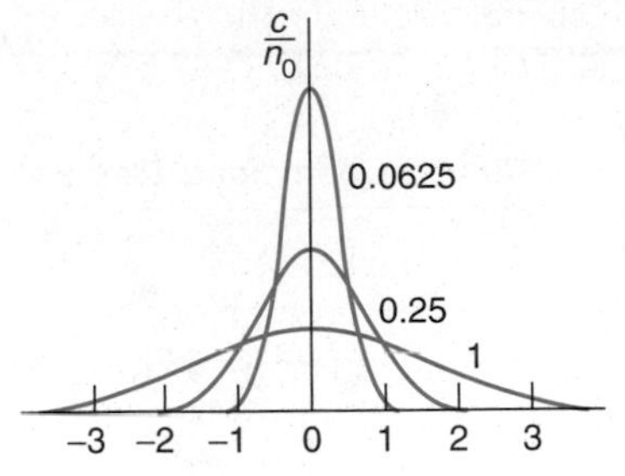

FIGURE 19.8
Plots of c/n_0 against distance from an instantaneous plane source at $x = 0$. At this plane there are initially n_0 molecules of solute, at infinite concentration (volume = 0). The numbers on the curves are values of *Dt*.

at any time t. Introduction of the expression for c (Eq. 19.42) and integration lead to

$$n_0 = \alpha A \int_{-\infty}^{\infty} t^{-1/2} e^{-x^2/4Dt}\, dx = 2\alpha(\pi D)^{1/2} A \tag{19.43}$$

The constant α is thus given by

$$\alpha = \frac{n_0}{2(\pi D)^{1/2} A} \tag{19.44}$$

and Eq. 19.42 becomes

$$c = \frac{n_0}{2(\pi D t)^{1/2} A} e^{-x^2/4Dt} \tag{19.45}$$

The concentration c in this expression is the amount of solute per unit distance; if the cross-sectional area is unity, c is the amount of solute per unit volume.

Figure 19.8 shows plots of c/n_0 against x, for three different values of Dt. This plot shows how the solute molecules spread out from the instantaneous plane source located at $x = 0$.

Brownian Movement

An interesting aspect of diffusion involves focusing attention on individual solute particles. This line of investigation, which turned out to be a very fruitful one, was initiated in 1828 by an observation made by the Scottish botanist Robert Brown (1773–1858). He observed pollen grains under a microscope, and finding them to be in constant and irregular motion he concluded that this behavior was a property of living matter. Later, however, he observed the same motion with dye particles, and no longer could suggest any explanation. The correct answer was not given until 1888 when Louis Georges Gouy (1854–1926) made a careful investigation of the motion, which had come to be called **Brownian movement** or **Brownian motion.** By eliminating a number of other possibilities he concluded that the explanation must lie in the thermal movements of the molecules of the liquid, which cause suspended particles to move by colliding with them.

This point of view was considerably developed by Einstein in a series of papers published from 1905 to 1908. His object was to develop relationships that could be used to confirm the atomic hypothesis (which was still not universally accepted) and to determine the Avogadro constant. In his series of papers he gave a comprehensive treatment of the theory of diffusion, and of the sedimentation of particles in a gravitational field (Section 19.3).

Einstein used the methods of statistical physics, without considering the details of collisions between molecules, and therefore obtained results that could not be easily visualized. At about the same time the Polish physicist Marian von Smoluchowski (1872–1917) also developed a treatment of the Brownian movement but from a different point of view: He explicitly considered the collisions between a suspended particle and the solvent molecules, obtaining results that were consistent with those of Einstein. Later he proposed that Brownian movement of rotation could be studied experimentally by means of a tiny mirror supported in a fluid on a thin quartz fiber. This was later achieved, with results that agreed with Smoluchowski's treatment.

One important result obtained by Einstein was an expression for the probability that a particle will diffuse a distance x, in any direction, in time t. One must, of course, allow a certain spread of distance, and we may call $P(x)\,dx$ the probability that the molecule has diffused a distance between x and $x + dx$. This probability is the number of solute molecules between x and $x + dx$ divided by the total number. Thus

$$P(x)\,dx = \frac{c(x)\,dx}{n_0} = \frac{1}{2(\pi Dt)^{1/2}} e^{-x^2/4Dt}\,dx \tag{19.46}$$

We now ask: What is the mean square distance $\overline{x^2}$ traversed by a solute molecule in time t? (We do not ask what is the mean distance $\bar{x}$, since diffusion is equally probable in both directions, and $\bar{x} = 0$.) The mean square distance is given by

$$\overline{x^2} = \int_{\infty}^{\infty} x^2 P(x)\,dx \tag{19.47}$$

Substitution of Eq. 19.46 into this, and evaluation of the integral, leads to

Random-Walk Equation

$$\overline{x^2} = 2Dt \tag{19.48}$$

An alternative derivation of Eq. 19.48 is also instructive. Figure 19.9a shows three parallel planes of unit area separated by distances x. If x is sufficiently small, the concentration gradient can be taken to be linear, and Figure 19.9b shows how the concentration varies from the coordinate $-x/2$ to $+x/2$. If the average concentration in the left-hand compartment is c, that in the right-hand compartment is

$$c + \frac{\partial c}{\partial x}x$$

and these are also the concentrations at the coordinates $-\frac{1}{2}x$ and $\frac{1}{2}x$, respectively.

Suppose now that x is the average distance that a molecule diffuses in time t. If we consider the flux J_1 from left to right across the cross section at $x = 0$, we see that half of the molecules that are in the left-hand volume (i.e., $\frac{1}{2}cx$ molecules) will cross in time t; thus

$$J_1 = \frac{1}{2}\frac{cx}{t} \tag{19.49}$$

The flux from right to left is one-half of the number of molecules in the right-hand volume, $\frac{1}{2}[c + (\partial c/\partial x)x]x$, divided by t:

$$J_{-1} = \frac{\frac{1}{2}\left(c + \frac{\partial c}{\partial x}x\right)x}{t} \tag{19.50}$$

The net flux from left to right, $J_1 - J_{-1}$, is therefore

$$J = \frac{\frac{1}{2}cx - \frac{1}{2}\left(c + \frac{\partial c}{\partial x}x\right)x}{t} \tag{19.51}$$

$$= -\frac{x^2}{2t}\frac{\partial c}{\partial x} \tag{19.52}$$

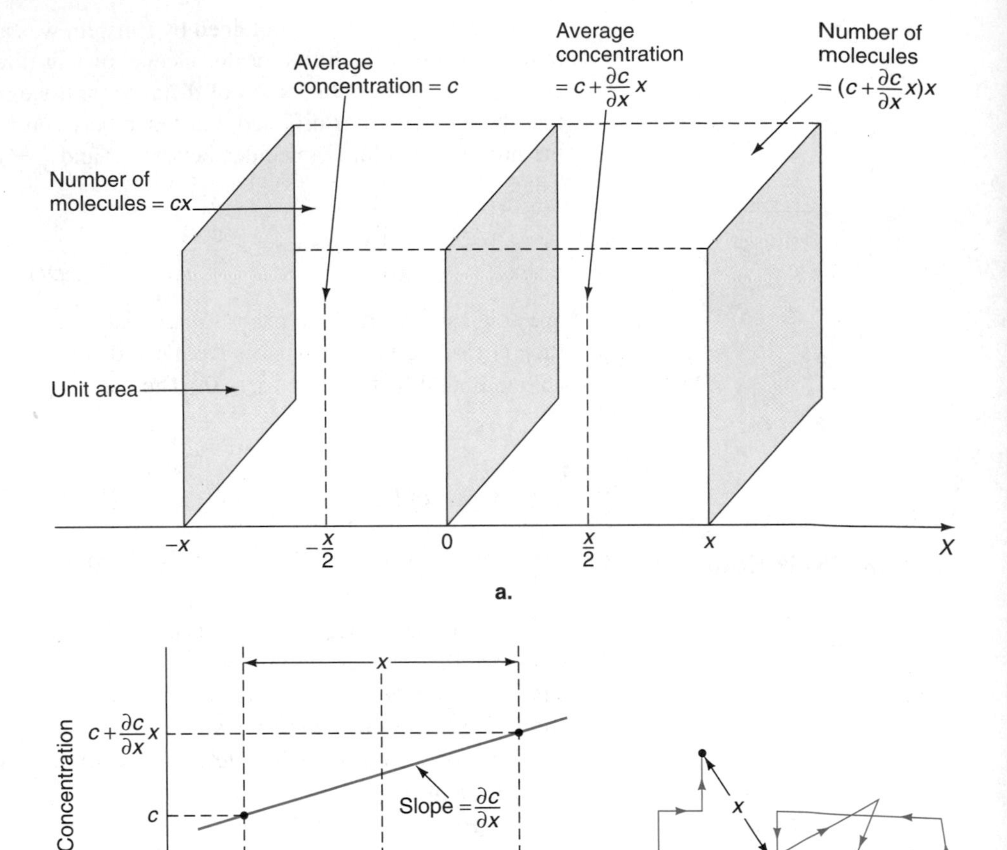

FIGURE 19.9
(a) Three parallel planes separated by distances x. (b) The corresponding concentration gradient. (c) Schematic representation of the random walk, such as is observed in Brownian movement.

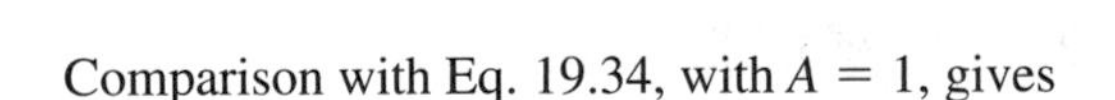

Comparison with Eq. 19.34, with $A = 1$, gives

$$D = \frac{x^2}{2t} \tag{19.53}$$

This is equivalent to Eq. 19.48, since it is the average value of x^2 that is given in this derivation.

Equation 19.48 is often referred to as the **random-walk equation** or, more colloquially, as the equation for the drunkard's walk, since it corresponds to completely random motion, with no sense of direction. Figure 19.9c shows the type of motion to which it corresponds. Experimental work on Brownian movement, based on the Einstein-Smoluchowski treatments and carried out by Jean Perrin, is considered later.

EXAMPLE 19.2 The diffusion coefficient for carbon in α-Fe is 2.9×10^{-8} $cm^2\ s^{-1}$ at 500 °C. How far would a carbon atom be expected to diffuse in 1 year (3.156×10^7 s)?

Solution From Eq. 19.48

$$\overline{x^2} = 2 \times 2.9 \times 10^{-8} \times 3.156 \times 10^7 \text{ cm}^2$$
$$= 1.83 \text{ cm}^2$$

and therefore the expected distance is

$$(\overline{x^2})^{1/2} = 1.35 \text{ cm}$$

Such diffusion has important practical consequences. For example, diffusion of carbon into the area of a mechanical weld might lead to weakening of the weld and possible rupture. Also, a fresh crystal surface becomes contaminated by diffusion of impurities from within the bulk crystal.

Self-Diffusion of Gases

The problem of the diffusion of one gas into another is a somewhat difficult one, because of the different diameters and mean square velocities of the two kinds of molecules. Here we will be content to consider *self-diffusion,* in which all the molecules are identical in diameter and mass. Coefficients for self-diffusion can be measured by isotopically labeling some of the molecules and determining how the labeled gas diffuses into the unlabeled.

For a gas the mean free path λ represents a step that is random with respect to preceding and succeeding steps, and Eq. 19.48 will therefore apply to this step. The time it takes a molecule to travel λ is $\lambda/\bar{u}$, where $\bar{u}$ is the average speed. We therefore replace x^2 by λ^2 and t by $\lambda/\bar{u}$ in Eq. 19.48 and obtain

$$\lambda^2 = \frac{2D\lambda}{\bar{u}} \tag{19.54}$$

or

$$D = \tfrac{1}{2}\lambda\bar{u} \tag{19.55}$$

Introduction of the expressions for λ and $\bar{u}$ (Eq. 1.68 and Table 1.2) gives

$$D = \frac{V}{\pi d^2 N}\left(\frac{k_B T}{\pi m}\right)^{1/2} = \frac{V_m}{\pi d^2 L}\left(\frac{k_B T}{\pi m}\right)^{1/2} = \frac{V_m}{L\sigma}\left(\frac{k_B T}{\pi m}\right)^{1/2} \tag{19.56}$$

where σ ($= \pi d^2$) is the collision cross section.

Comparison of Eq. 19.55 with Eq. 19.14 for the coefficient of viscosity η gives the following relationship between D and η for a gas:

η-*D* Relationship

$$\eta = \rho D \tag{19.57}$$

where ρ is the density (Nm/V).

EXAMPLE 19.3 If the molecular diameter for O_2 is 0.38 nm, estimate the self-diffusion coefficient and the viscosity of O_2 at 25 °C and 101.325 kPa pressure. How far would the molecule diffuse, on the average, in a second?

Solution $PV = nRT = Nk_BT$ and therefore

$$\frac{V}{N} = \frac{k_BT}{P} = \frac{1.381 \times 10^{-23}(\text{J K})^{-1} \times 298.15(\text{K})}{1.013\ 25 \times 10^5(\text{Pa})} = 4.06 \times 10^{-26}\ \text{m}^3$$

The mass of the oxygen molecules is $32.0 \times 10^{-3}/6.022 \times 10^{23} = 5.31 \times 10^{-26}$ kg. Therefore from Eq. 19.56,

$$D = \frac{4.06 \times 10^{-26}(\text{m}^3)}{\pi \times [0.38 \times 10^{-9}(\text{m})]^2}\left[\frac{1.381 \times 10^{-23} \times 298.15(\text{J})}{\pi \times 5.31 \times 10^{-26}(\text{kg})}\right]^{1/2}$$
$$= 1.4 \times 10^{-5}\ \text{m}^2\ \text{s}^{-1} = 0.14\ \text{cm}^2\ \text{s}^{-1}$$

From Eq. 19.57,

$$\eta = \rho D = \frac{NmD}{V} = \frac{5.31 \times 10^{-26}(\text{kg}) \times 1.4 \times 10^{-5}(\text{m}^2\ \text{s}^{-1})}{4.06 \times 10^{-26}(\text{m}^3)}$$
$$= 1.8 \times 10^{-5}\ \text{kg m}^{-1}\ \text{s}^{-1}$$

In 1 second

$$\overline{x^2} = 2 \times 0.14\ \text{cm}^2 = 0.28\ \text{cm}^2$$
$$\sqrt{\overline{x^2}} = 0.53\ \text{cm}$$

Driving Force of Diffusion

Consider the diffusion across a distance dx over which there is a concentration change from c to $c + dc$ (see Figure 19.10). The force that drives the solute molecules to the more dilute region can be calculated from the difference between the molar Gibbs energy at concentration c and that at concentration $c + dc$. This molar Gibbs energy difference is

$$dG = G_{c+dc} - G_c = RT\ln\left(\frac{c + dc}{c}\right) = RT\ln\left(1 + \frac{dc}{c}\right) \tag{19.58}$$

Since dc/c is very small,

$$\ln\left(1 + \frac{dc}{c}\right) \approx \frac{dc}{c} \tag{19.59}$$

and therefore

$$dG = \frac{RT\,dc}{c} \tag{19.60}$$

This Gibbs energy difference is the work w done *on* the system in transferring a mole of solute from concentration c to $c + dc$. The work done *by* the system, $-w$, in transferring a *molecule* of solute from $c + dc$ to c is thus

$$-w = -\frac{RT}{L}\frac{dc}{c} \tag{19.61}$$

$$= -\frac{k_BT\,dc}{c} \tag{19.62}$$

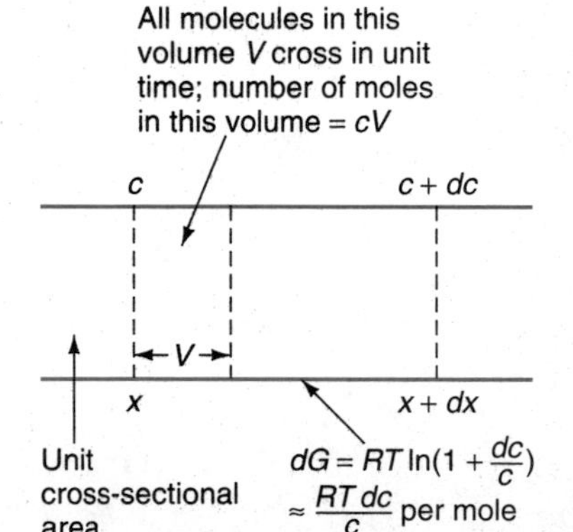

FIGURE 19.10
The driving force of diffusion.

This work $-w$ is done over a distance dx, and if F_d is the driving force leading to diffusion

$$-w = F_d\,dx \tag{19.63}$$

The driving force per molecule is thus given by combining Eqs. 19.62 and 19.63:

$$F_d = -\frac{k_B T}{c}\frac{dc}{dx} \tag{19.64}$$

Frictional Force

When a driving force is applied to a molecule, its speed increases until the **frictional force** F_f acting on it is equal to the driving force. The molecule has then attained a *limiting speed.* Most theories of the frictional force on a molecule, such as that of Stokes (considered later), lead to the conclusion that the force is directly proportional to the speed; thus

$$F_f = fv \tag{19.65}$$

Frictional Coefficient

where f is known as the *frictional coefficient.* The limiting velocity is thus attained when

$$fv = \frac{k_B T}{c}\frac{dc}{dx} \tag{19.66}$$

or

$$cv = -\frac{k_B T}{f}\frac{dc}{dx} \tag{19.67}$$

All molecules within a distance v of a given unit cross-sectional area will cross that area in unit time (Figure 19.10). The number of molecules in that volume is cv, which is therefore the flux J:

$$J = -\frac{k_B T}{f}\frac{dc}{dx} \tag{19.68}$$

Comparison of this molecularly derived diffusion equation with Fick's first law, Eq. 19.34 with $A = 1$, shows that the diffusion coefficient is given by

Einstein Equation

$$D = \frac{k_B T}{f} \tag{19.69}$$

This equation was first derived by Albert Einstein in 1905.

Diffusion and Ionic Mobility

We saw in Section 7.5 that the mobility of an ion in solution is defined as the speed with which the ion moves under a unit potential gradient. Since the diffusion coefficient of a species is related to the speed with which it moves under a unit concentration gradient, the mobilities of ions and, therefore, the molar ionic conductivities are proportional to their diffusion coefficients. We shall now obtain the proportionality factor.

A unit potential gradient is by definition one that exerts a unit force on a unit charge; it will therefore exert a force of Q''N(kg m s^{-2}) on an ion having a charge

Q^uC. The mobility[2] u_i of an ion having a charge Q is thus the speed with which it moves when a force Q acts upon it. Its speed when it is acted upon by a unit force is therefore u_e/Q. When an ion moves in an electric potential gradient, it soon attains a limiting (or equilibrium) speed at which the frictional force F_f is equal but opposite to the electric force, and this speed is

$$v = \frac{u_i}{Q} F_f \tag{19.70}$$

Comparison of this equation with Eq. 19.65 shows that the frictional coefficient f is given by

$$f = \frac{Q}{u_i} \tag{19.71}$$

and insertion of this into the Einstein equation (Eq. 19.69) gives

$$D = \frac{k_B T}{Q} u_i \tag{19.72}$$

If the ion has a charge number of $|z_i|$, its charge Q is $F|z_i|/L$, where F is the Faraday constant and L is the Avogadro constant. By Eq. 7.64 its ionic conductivity $\lambda°$ is equal to Fu_i, so that $u_i = \lambda°/F$. Equation 19.72 therefore becomes

Nernst-Einstein Equation

$$D = \frac{k_B T}{F|z_i|/L} \cdot \frac{\lambda°}{F} = \frac{RT}{F^2|z_i|}\lambda° \tag{19.73}$$

The equation was first derived in 1888 by Nernst. In view of its relationship to Einstein's equation (Eq. 19.69), Eq. 19.73 is often known as the **Nernst-Einstein equation.** In practice, diffusion coefficients have to be measured for more than one type of ion; for electrolytes involving two ions of equal and opposite charges (e.g., NaCl, $ZnSO_4$), Nernst showed that the average diffusion coefficient is

$$D = \frac{2D_+D_-}{D_+ + D_-} \tag{19.74}$$

where D_+ and D_- are the individual diffusion coefficients.

EXAMPLE 19.4 Estimate the diffusion coefficient of NaCl in water at 25 °C from the molar ionic conductivities given in Table 7.3, Section 7.7.

Solution From Eq. 19.73, at 25 °C,

$$D = \frac{8.3145(\text{J K}^{-1}\text{ mol}^{-1}) \times 298.15(\text{K})}{(96\,500)^2(\text{C}^2\text{ mol}^{-2})}\lambda°(\Omega^{-1}\text{ cm}^2\text{ mol}^{-1})$$
$$= 2.66 \times 10^{-7}\lambda°\text{J }\Omega^{-1}\text{ C}^{-2}\text{ cm}^2$$
$$= 2.66 \times 10^{-7}\lambda°\text{ cm}^2\text{ s}^{-1}$$

(since $\Omega = \text{V A}^{-1}$, J = V C, and C = A s).

[2]In this section we use the symbol u_i to avoid confusion with u, the molecular velocity. Again we use the superscript u to indicate the *value* of the electric charge Q, which equals Q^uC.

Then, for Na^+,

$$D_+ = 2.66 \times 10^{-7} \times 50.1 = 1.33 \times 10^{-5}\ \text{cm}^2\ \text{s}^{-1}$$

For Cl^-

$$D_- = 2.66 \times 10^{-7} \times 76.4 = 2.03 \times 10^{-5}\ \text{cm}^2\ \text{s}^{-1}$$

For NaCl, using Eq. 19.74,

$$D = \frac{2 \times 1.33 \times 10^{-5} \times 2.03 \times 10^{-5}}{(1.33 + 2.03) \times 10^{-5}}$$
$$= 1.61 \times 10^{-5}\ \text{cm}^2\ \text{s}^{-1}$$

Stokes's Law

The diffusion coefficient depends on the ease with which the solute molecules can move. In aqueous solution, the diffusion coefficient of a solute is a measure of how readily a solute molecule can push aside its neighboring water molecules and move into another position. An important aspect of the theory of diffusion is how the magnitudes of the frictional coefficients f and, hence, of the diffusion coefficients D, depend on the properties of the solute and solvent molecules.

Examination of the values given in Table 19.3 shows that diffusion coefficients tend to decrease as the molecular size increases. This is easy to understand, since a larger solute molecule has to push aside more solvent molecules during its progress and will therefore move more slowly than a smaller molecule. It is difficult to develop a precise theory of diffusion coefficients, but in 1851 the British physicist George Gabriel Stokes (1819–1903) considered a simple situation in which the solute molecules are so much larger than the solvent molecules that the latter can be regarded as continuous (i.e., as not having molecular character). Imagine a very large solute molecule surrounded by very small solvent molecules, so small that they lose their independent character and appear as a continuum sea to the solute. For such a system Stokes deduced that the frictional force F_f opposing the motion of a large spherical particle of radius r moving at speed v through a solvent of viscosity η is given by

Stokes's Law

$$F_f = 6\pi r \eta v \qquad (19.75)$$

The frictional coefficient is therefore

$$f = 6\pi r \eta \qquad (19.76)$$

It then follows from Eq. 19.69 (the Einstein equation) that when Stokes's law applies, the diffusion coefficient is given by

Stokes-Einstein Equation

$$D = \frac{k_B T}{6\pi r \eta} \qquad (19.77)$$

This is often referred to as the **Stokes-Einstein equation.**

Measurement of D in a solvent of known viscosity therefore permits a value of the radius r to be calculated. Such a calculation is not very satisfactory for solute molecules, however, for several reasons. In the first place, Stokes's law is based on the assumption of very large spherical particles and a continuous solvent, and it involves some error even for approximately spherical macromolecules. Second, the solute molecules may not be spherical, and this introduces an additional error. Furthermore, solute molecules are commonly solvated, and in moving through the solution they transport some of their solvation layer. In spite of these drawbacks, Eq. 19.77 has proved useful in providing approximate values of molecular sizes.

EXAMPLE 19.5 The diffusion coefficient for glucose in water is 6.81×10^{-10} $m^2\ s^{-1}$ at 25 °C. The viscosity of water at 25 °C is 8.937×10^{-4} kg $m^{-1}\ s^{-1}$, and the density of glucose is 1.55 g cm^{-3}. Estimate the molar mass of glucose, assuming that Stokes's law applies and that the molecule is spherical.

Solution Since Stokes's law applies, D and η are related by Eq. 19.77 and the radius of the molecule is given by

$$\begin{aligned} r &= \frac{k_B T}{6\pi\eta D} \\ &= \frac{1.38 \times 10^{-23}(\text{J K}^{-1}) \times 298.15(\text{K})}{6 \times 3.1426 \times 8.937 \times 10^{-4}(\text{kg m}^{-1}\ \text{s}^{-1}) \times 6.81 \times 10^{-10}(\text{m}^2\ \text{s}^{-1})} \\ &= 3.59 \times 10^{-10}\ \text{J kg}^{-1}\ \text{m}^{-1}\ \text{s}^2 \\ &= 3.59 \times 10^{-10}\ \text{m} = 0.359\ \text{nm} \end{aligned}$$

since 1 J = 1 kg $m^2\ s^{-2}$. The estimated volume of the molecule is thus

$$\tfrac{4}{3}\pi(3.59 \times 10^{-10})^3 = 1.94 \times 10^{-28}\ \text{m}^3$$

and its mass is

$$1.94 \times 10^{-28}(\text{m})^3 \times 1.55 \times 10^6(\text{g m}^{-3}) = 3.01 \times 10^{-22}\ \text{g}$$

The molar mass M is this mass multiplied by the Avogadro constant L:

$$M = 3.01 \times 10^{-22} \times 6.022 \times 10^{23} = 181.3\ \text{g mol}^{-1}$$

This agrees very well with the true molar mass of 180.2 g mol^{-1}.

Perrin's Experiments on Brownian Movement

Einstein's and Smoluchowski's treatments of the Brownian movement provided the basis for some of the experiments made by Jean Baptiste Perrin (1860–1942) on suspended particles. His first experiments were on the sedimentation of particles and will be considered in Section 19.3. He also carried out many experiments in which he observed individual gamboge and mastic particles under the microscope and determined their displacements after various periods of time. With the use of Einstein's equation 19.48 for the mean-square displacement, and making use of the Stokes-Einstein equation 19.77, he then estimated the Boltzmann constant k_B and hence, since $k_B = R/L$, estimated the Avogadro constant L. His results confirmed Einstein's conclusion that the mean-square displacement of a particle is proportional to the square root of the time, a conclusion that had come under attack. Perrin also investigated rotational Brownian motion, and again confirmed Einstein's and Smoluchowski's conclusions.

The great importance of Perrin's investigations is that they provided a convincing demonstration of the existence of atoms and molecules. For many years

Wilhelm Ostwald, for example, had maintained that the atomic theory—which he called a "mere hypothesis"—was no more than a convenient fiction, and that atoms did not have a real existence. In 1909, however, Ostwald finally conceded that the experiments of Perrin, together with those of J. J. Thomson on the deflection of electrons and positive particles, provided a convincing demonstration of the real existence of atoms and molecules.

Diffusion through Membranes

The speed with which molecules and ions can pass through membranes is a matter of great importance, particularly in biology. In work with membranes it is convenient to define a quantity known as the **permeability coefficient** P, which is defined as the flux through unit area when there is a unit concentration difference. We have previously considered the case of two stirred solutions, of concentrations c_1 and c_2, separated by a sintered glass diaphragm of thickness l. For such a system the rate of permeation is given by Eq. 19.41, and the flux through unit area is

$$J/A = D\frac{c_1 - c_2}{l} \tag{19.78}$$

The permeability coefficient P is the flux per unit area per unit concentration difference and therefore

$$P \equiv \frac{J/A}{c_1 - c_2} = \frac{D}{l} \tag{19.79}$$

Since the SI unit for D is $m^2\ s^{-1}$, that for P is $m\ s^{-1}$.

It is not necessarily the case that the solute molecules will be just as soluble in the membrane as in the solvent. For example, a membrane might be composed of a certain amount of lipid material, and a solute molecule might have a certain proportion of nonpolar groups. Such a solute might well be more soluble in the membrane than in water. This effect can be represented by use of a partition or distribution coefficient K_p, which is simply an equilibrium constant; it is the ratio of the concentrations of the solute in the membrane and in the solvent. In this situation the concentrations at the surface of the membrane, as shown in Figure 19.11, are K_pc_1 and K_pc_2. The gradient is now given by

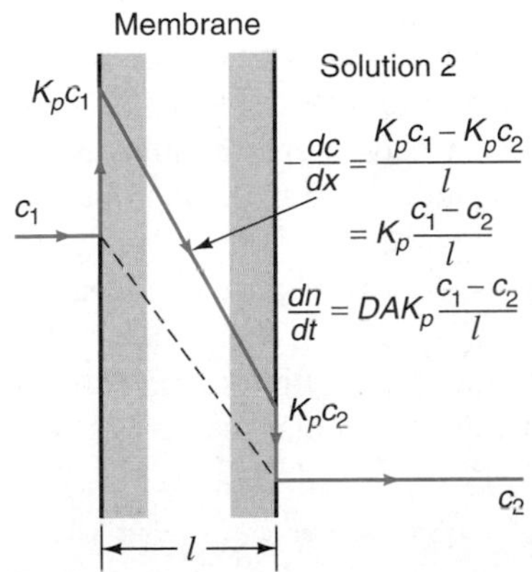

FIGURE 19.11
Stirred solutions of concentrations c_1 and c_2, separated by a membrane of thickness l. The solubility of the solute is different in the membrane and in the solvent (in the diagram the solubility is shown as higher in the membrane).

$$-\frac{dc}{dx} = \frac{K_pc_1 - K_pc_2}{l} = K_p\frac{c_1 - c_2}{l} \tag{19.80}$$

and the rate of diffusion (compare Eq. 19.40 and Eq. 19.41) is

$$\frac{dn}{dt} = DAK_p\frac{c_1 - c_2}{l} \tag{19.81}$$

Thus the rate, the apparent diffusion coefficient (DK_p), and the apparent permeability coefficient (DK_p/l) are all altered by the factor K_p. Since K_p may be greater or less than unity, there may be an enhancement or a diminution in the rate of permeation of the membrane. Sometimes this partitioning effect leads to rates that are several powers of 10 higher than would be obtained in its absence. Conversely, the rate of diffusion through the membrane will be abnormally small if the solute is much less soluble in the membrane than in the solution. For example, a carbohydrate such as sucrose is much less soluble in a fatty membrane than in water, where there is extensive hydrogen bonding; because of this effect it diffuses slowly through a fatty membrane.

Abnormally high rates of transport of solute molecules across membranes can arise from causes other than high solubility in the membrane. Sometimes solute molecules become attached to so-called *carrier* molecules, which remain in the membrane and transport the solute molecule from one side to another. If the solvent molecule is strongly attached to the carrier molecule, the concentration of the solute-carrier complex will be large, and the concentration gradient will be large for the complex. A high rate of permeation can result, and this effect is known as **facilitated transport.**

Facilitated Transport

In facilitated transport the direction of flow is consistent with the concentration gradient. A completely different effect, frequently found in biological systems, is when the flow is contrary to the concentration gradient; the solute molecules move from the low-concentration side to the high-concentration side of the membrane. An example is the formation of the gastric juice in humans, where HCl flows from a solution of pH approximately 7 (i.e., $[H^+] \approx 10^{-7}$ M) into a solution that is approximately 0.1 M in HCl.

Active Transport

This effect is known as **active transport.** It is brought about by a coupling of the diffusion process with an exergonic[3] chemical reaction. Suppose, for example, that in a biological system at 37 °C active transport were occurring against a 10/1 concentration gradient. The Gibbs energy difference per mole is

$$\Delta G = RT \ln 10 = (8.3145 \times 310 \ln 10) \text{ J mol}^{-1} = 5.9 \text{ kJ mol}^{-1}$$

A reaction exergonic by this amount and coupled with the transport process would therefore be able to give transport against the 10/1 gradient. The details of such coupling processes in biological systems are the subject of much investigation.

19.3 Sedimentation

In Section 19.2 we dealt with the movement of a solute as a result of a concentration gradient. Molecules in solution can also be made to move by subjecting them to other forces. For example, a solution or suspension may simply be allowed to stand. If the solute molecules are small, there will be no change in the distribution in space, since the thermal motion will counteract the tendency of the molecules to move in the gravitational field. Very large particles, however, will *sediment* on standing. In order for smaller particles to undergo *sedimentation,* it is necessary to increase the effective gravitational field by subjecting the solution to centrifugal motion. The velocity of sedimentation can lead to values of molar masses, as will now be explained.

Sedimentation Velocity

Suppose that a particle of mass m, having a specific volume (volume per unit mass) of V_2, is in a liquid of density ρ (mass per unit volume). The volume of liquid dis-

[3]An exergonic reaction (p. 126) is one for which ΔG is negative, and which therefore occurs spontaneously.

placed by the particle is V_2m and the mass displaced is $V_2m\rho$. The net force F_g acting on the particle as a result of the gravitational field is therefore

$$F_g = mg - V_2m\rho g = (1 - V_2\rho)mg \tag{19.82}$$

where g is the acceleration of gravity. The particle will reach a limiting speed v when this force is equal to the frictional force F_f, which is equal to the frictional coefficient f multiplied by the speed:

Theodor Svedberg was awarded the 1926 Nobel Prize for his work in colloid chemistry, particularly for studies of Brownian motion. Most of his best-known work, on the ultracentrifuge, was done later.

$$F_f = fv = (1 - V_2\rho)mg \tag{19.83}$$

If Stokes's law applies to the particle, F_f is given by Eq. 19.75, and therefore

$$6\pi r\eta v = (1 - V_2\rho)mg \tag{19.84}$$

The limiting speed is

$$v = \frac{(1 - V_2\rho)mg}{6\pi r\eta} \tag{19.85}$$

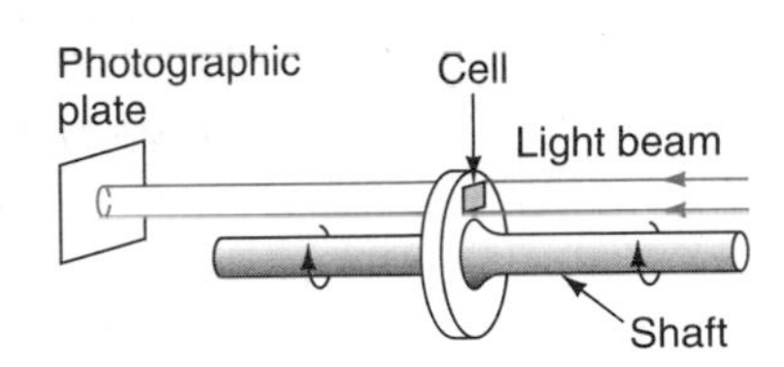

FIGURE 19.12
Schematic diagram of an ultracentrifuge. The shaft is rotated at very high speeds so that the effective gravitational field in the cell is very large and brings about sedimentation. The distribution of solute in the cell is determined from the blackening of the photographic plate at various positions. More satisfactorily, the distribution can be obtained from the refractive index, which varies approximately linearly with concentration. In the *schlieren* method, special methods of illumination are used that lead to darkening of the photographic plate corresponding to regions where the refractive index changes rapidly.

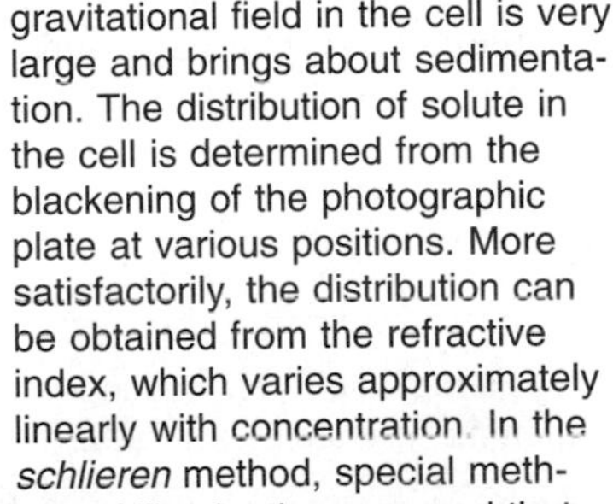

Experiments on sedimentation under the earth's gravitational field were carried out in 1908 by Jean Perrin, who used a microscope to observe particles of the pigment gamboge, and obtained results in good agreement with the theory of sedimentation. Work in the earth's rather weak field is limited to particles of radii greater than about 10^{-5} m (0.01 mm), since otherwise there is little sedimentation and too much interference from the Brownian motion. Even the largest macromolecules have effective radii of not much more than 10^{-8} m (10 nm), so that their sedimentation cannot be observed in the gravitational field of the earth.

In order to observe the sedimentation of such small particles it is necessary to employ much higher effective gravitational fields. These are brought about by means of an **ultracentrifuge,** in which solutions are rotated at speeds of 80 000 and above revolutions per minute, which produces fields of 3×10^5 g and above. The development of such techniques is largely due to the Swedish physical chemist Theodor Svedberg (1884–1971), whose work along these lines was started in 1923 and who devoted much study to the characterization of protein molecules and other macromolecules. A schematic diagram of an ultracentrifuge is shown in Figure 19.12. For a centrifugal field of force we replace g in Eq. 19.83 by ω^2x, where ω is the angular velocity and x is the distance from the center of rotation. Thus

$$v = \frac{(1 - V_2\rho)m\omega^2x}{f} \tag{19.86}$$

The quantity

$$s = \frac{v}{\omega^2x} \tag{19.87}$$

Sedimentation Coefficient

is known as the **sedimentation coefficient;** it is the sedimentation rate v when the centrifugal acceleration is unity. For a given molecular species in a given solvent at a given temperature, s is a characteristic quantity. Its SI unit is the second, but it is often expressed in *Svedberg units,* equal to 10^{-13} s.

If Stokes's law applies, f is $6\pi\eta r$ and therefore, from Eq. 19.86,

$$v = \frac{(1 - V_2\rho)m\omega^2x}{6\pi\eta r} \tag{19.88}$$

The sedimentation coefficient is thus

$$s = \frac{(1 - V_2\rho)m}{6\pi\eta r} \tag{19.89}$$

Use of this equation, however, is unreliable, since Stokes's law is valid only for very large spherical particles. It is more satisfactory to express f by the use of Eq. 19.69, according to which it is k_BT/D. Insertion of this expression into Eq. 19.86 then gives

$$v = \frac{D(1 - V_2\rho)m\omega^2 x}{k_BT} \tag{19.90}$$

or

$$s = \frac{D(1 - V_2\rho)m}{k_BT} \tag{19.91}$$

$$= \frac{D(1 - V_2\rho)M}{RT} \tag{19.92}$$

The molar mass M can therefore be calculated from a measurement of the sedimentation coefficient and the diffusion coefficient:

Svedberg Equation

$$M = \frac{RTs}{D(1 - V_2\rho)} \tag{19.93}$$

This equation, derived by Svedberg in 1929, has been the basis of many measurements of molar masses. For precise determination the values of s, D, and V_2 should be extrapolated to infinite dilution.

EXAMPLE 19.6 Suppose that a protein molecule has a radius of 10^{-8} m and a density of 1.15 g cm^{-3}, and is in water at 20 °C, which has a viscosity of 1.002×10^{-3} kg m^{-1} s^{-1} and a density of 0.997 g cm^{-3}.

a. Use the Stokes-Einstein equation to make an estimate of the diffusion coefficient.
b. Make an estimate of the sedimentation coefficient.
c. Neglecting sedimentation, estimate how far the molecule would diffuse in 1 minute.
d. Neglecting diffusion, estimate how far the molecule would sediment in 1 minute.
e. How far would the molecule sediment in 1 minute if it were in an ultracentrifuge, 10 cm from the axis of revolution, with a revolution speed of 20 000 rpm?

Solution

a. From the Stokes-Einstein equation, Eq. 19.77,

$$D = \frac{k_BT}{6\pi r\eta} = \frac{1.38 \times 10^{-23} \times 293.15 \text{ J}}{6\pi \times 10^{-8} \text{ m} \times 1.002 \times 10^{-3} \text{ kg m}^{-1} \text{ s}^{-1}}$$
$$= 2.14 \times 10^{-11} \text{ m}^2 \text{ s}^{-1}$$

b. The sedimentation coefficient can be calculated from Eq. 19.91. The molecular mass m is

$$m = (4/3)\pi r^3 \rho = (4/3)\pi \times (10^{-8}\ \text{m})^3 \times 1.15 \times 10^3\ \text{kg m}^{-3}$$
$$= 4.82 \times 10^{-21}\ \text{kg}$$

(This corresponds to a molecular weight of 2 900 000.) Then, by Eq. 19.91,

$$s = \frac{D(1 - V_2\rho)m}{k_B T}$$
$$= \frac{2.14 \times 10^{-11}\ \text{m}^2\ \text{s}^{-1}(1 - 0.997/1.15)4.82 \times 10^{-21}\ \text{kg}}{1.38 \times 10^{-23} \times 293.15\ \text{J}}$$
$$= 3.39 \times 10^{-12}\ \text{s}$$

c. The distance of diffusion in 1 minute is estimated from Einstein's equation 19.48:

$$\sqrt{x^2} = \sqrt{2Dt} = \sqrt{2 \times 2.14 \times 10^{-11}\ \text{m}^2\ \text{s}^{-1} \times 60\ \text{s}}$$
$$= 5.07 \times 10^{-5}\ \text{m}$$

d. The sedimentation coefficient is the rate of sedimentation for an acceleration of gravity of unity. In a gravitational field of $g = 9.8$ m s^{-2} the rate of sedimentation is $3.39 \times 10^{-12} \times 9.8$ m s^{-1} = 3.32×10^{-11} m s^{-1}, and the distance moved in 1 minute is 2.0×10^{-9} m. This motion would, of course, be swamped by the diffusion (Brownian movement).

e. The rate of sedimentation in an ultracentrifuge is given by Eq. 19.87

$$v = s\omega^2 x$$

The speed of revolution ω is

$$(20\ 000\ \text{rpm}/60\ \text{s min}^{-1}) \times 2\pi = 2094\ \text{rad s}^{-1}$$

The rate of sedimentation is therefore

$$v = 3.39 \times 10^{-12}\ \text{s} \times (2094\ \text{rad s}^{-1})^2 \times 0.1\ \text{m}$$
$$= 1.49 \times 10^{-6}\ \text{m s}^{-1}$$

The molecule would therefore sediment 8.9×10^{-5} m in one minute.

The estimates made in this example will not be of great reliability because of the use of the Stokes-Einstein equation. If experimental determinations were made of the diffusion coefficient and the sedimentation coefficient, a more reliable molecular weight would be obtained by the use of Eq. 19.93.

Sedimentation Equilibrium

An alternative method of using the ultracentrifuge to measure molar masses is to allow the distribution of particles to reach equilibrium. As sedimentation occurs in the ultracentrifuge, a concentration gradient is established, and this will cause the molecules to diffuse in the opposite direction. Eventually the system reaches a state of equilibrium at which the rate with which the solute is driven outward by the centrifugal force just equals the rate with which it diffuses inward under the influence of the concentration gradient.

The velocity v with which the particles travel as a result of the centrifugal field is given by Eq. 19.86. All particles within a distance v of a given cross-sectional area A will cross that area in unit time, and if the molecular concentration is c, there are vcA particles that cross; the sedimentation flux is thus vcA and, by Eq. 19.86, is

$$J(\text{sedimentation}) = vcA = \frac{(1 - V_2\rho)m\omega^2 xcA}{f} \tag{19.94}$$

The diffusive flux through area A is

$$J(\text{diffusive}) = -AD\frac{dc}{dx} \tag{19.95}$$

$$= -\frac{Ak_BT}{f}\frac{dc}{dx} \tag{19.96}$$

At equilibrium these two rates are equal, and we obtain

$$\frac{dc}{c} = \frac{M(1 - V_2\rho)\omega^2 x\, dx}{RT} \tag{19.97}$$

where M (= mL = mR/k_B) is the molar mass. Integration between two positions x_1 and x_2, at concentrations c_1 and c_2, leads to

$$M = \frac{2RT\ln(c_2/c_1)}{(1 - V_2\rho)\omega^2(x_2^2 - x_1^2)} \tag{19.98}$$

Thus, if measurements of the relative concentrations are made at two positions, after equilibrium has been established, this equation can be used to calculate the molar mass. The value so obtained is the *mass-average molar mass*.

EXAMPLE 19.7 A pure sample of a protein in water is centrifuged to equilibrium at 25 °C and at 15 000 revolutions per minute. The top of the centrifuge tube is 6.77 cm from the axis of revolution, and the concentrations at distances 0.450 cm and 1.25 cm from the top of the tube are measured to be 3.16 g cm^{-3} and 17.84 g cm^{-3}, respectively. The density of water at 25 °C is 0.997 g cm^{-3} and the specific volume of the protein V_2 is 0.750 cm^3 g^{-1}. Calculate the molecular weight.

Solution Use is made of Eq. 19.98. The value of $1 - V_2\rho$ is

$$1 - 0.750 \times 0.997 = 0.252$$

The two distances x_1 and x_2 from the axis of rotation are 7.22×10^{-2} m and 8.02×10^{-2} m. The speed of revolution ω is

$$(15\,000 \text{ rpm}/60 \text{ s min}^{-1}) \times 2\pi = 1570 \text{ rad s}^{-1}$$

The molar mass is thus, by Eq. 19.98,

$$M = \frac{2 \times 8.3145 \times 298.15 \text{ (J mol}^{-1}) \ln(17.84/3.16)}{0.252(1570 \text{ rad s}^{-1})^2[(8.02 \times 10^{-2})^2 - (7.22 \times 10^{-2})^2] \text{ m}^2}$$

$$= 11.3 \text{ kg mol}^{-1}$$

(since J mol^{-1}/s^{-2} m^2 = kg m^2 s^{-2} mol^{-1}/m^2 s^{-2} = kg mol^{-1}).

The molecular weight is thus 11 300.

The sedimentation-equilibrium method does not require an independent measurement of the diffusion coefficient, in contrast to the sedimentation-velocity method. However, the time required for complete equilibrium to be established is so long that the method is often inconvenient to use, especially if the relative molecular mass is greater than 5000.

In order to overcome this drawback, a modification to the sedimentation-equilibrium method was proposed in 1947 by the Canadian physicist William J. Archibald. At the top meniscus of the cell and at the bottom of the cell, there can be no net flux, and the sedimentation-equilibrium equations apply at these sections at all times. Shortly after the ultracentrifuge is brought to its top speed, therefore, concentrations in these special sections can be determined and M calculated from Eq. 19.93. This modification of the sedimentation-equilibrium method greatly increases its applicability.

If a solution of a substance of low molar mass is ultracentrifuged, equilibrium is established within a fairly short period of time, and there will be a density gradient in the solution. If a substance of high molar mass is added, it will float in this solution at the particular position at which its density is equal to the density of the solution. If the macromolecular substance is made up of fractions of different molar masses, it will separate into fractions that will remain at different planes in the cell. This technique, known as *density-gradient ultracentrifugation,* has proved useful in establishing the different kinds of molecules in a macromolecular sample.

KEY EQUATIONS

Viscosity:

Definition (when Newton's law applies):

$$F \equiv \eta A \frac{dv}{dt}$$

Poiseuille equation:

$$\frac{dV}{dt} = \frac{(P_1 - P_2)\pi R^4}{8\eta l}$$

Kinetic theory expressions, for a gas:

$$\eta = \frac{Nm\bar{u}\lambda}{2V}$$

$$= \frac{1}{2}\rho\bar{u}\lambda$$

$$= \frac{(mk_B T)^{1/2}}{\pi^{3/2} d^2}$$

Kinetic theory expression, for a liquid:

$$\eta = A_{vis} e^{E_{vis}/RT}$$

Intrinsic viscosity of a solution:

$$[\eta] = \lim_{\rho \to 0} \left[\frac{1}{\rho} \cdot \frac{\eta - \eta_0}{\eta_0} \right]$$

Mark-Houwink equation:

$$[\eta] = kM_r^{\alpha}$$

(M_r = relative molecular mass)

Diffusion:

Fick's first law:

$$J = \frac{dn}{dt} = -DA\frac{dc}{dx}$$

Fick's second law:

$$\frac{\partial c}{\partial t} = D\frac{\partial^2 c}{\partial x^2}$$

Random-walk equation:

$$\overline{x^2} = 2Dt$$

Einstein equation:

$$D = \frac{k_B T}{f}$$ (f = frictional coefficient)

Nernst-Einstein equation:

$$D = \frac{RT}{F^2|z_i|}\lambda^{\circ}$$

Stokes-Einstein equation:

$$D = \frac{k_B T}{6\pi r \eta}$$

Nernst equation:

$$D = \frac{2D_+ D_-}{D_+ + D_-}$$

Sedimentation:

Definition of sedimentation coefficient:

$$s \equiv \frac{v}{\omega^2 x}$$

Svedberg's equation:

$$M = \frac{RTs}{D(1 - V_2 \rho)}$$

PROBLEMS

Viscosity

19.1. In a normal adult at rest the average speed of flow of blood through the aorta is 0.33 m s^{-1}. The radius of the aorta is 9 mm and the viscosity of blood at body temperature, 37 °C, is about 4.0×10^{-3} kg m^{-1} s^{-1}. Calculate the pressure drop along a 0.5 m length of the aorta.

19.2. A typical human capillary is about 1 mm long and has a radius of 2 μm. If the pressure drop along the capillary is 20 Torr,

a. calculate the average linear speed of flow of blood of viscosity 4.0×10^{-3} kg m^{-1} s^{-1};
b. calculate the volume of blood passing through each capillary per second; and
c. estimate the number of capillaries in the body if they are supplied by the aorta described in Problem 19.1.

***19.3.** The viscosity of ethylene at 25.0 °C and 101.325 kPa is 9.33×10^{-6} kg m^{-1} s^{-1}. Estimate

a. the molecular diameter,
b. the mean free path,
c. the frequency of collisions Z_A experienced by a given molecule, and
d. the collision density Z_{AA}.

***19.4.** For nonassociated liquids the fluidity ϕ (i.e., the reciprocal of the viscosity) obeys to a good approximation an equation of the Arrhenius form

$$\phi = A e^{-E_a/RT}$$

where A and E_a are constants.

a. For liquid CCl_4 the viscosity at 0.0 °C is 1.33×10^{-3} kg m^{-1} s^{-1} and the activation energy E_a is 10.9 kJ mol^{-1}. Estimate the viscosity at 40.0 °C.
b. The Arrhenius equation does not apply well to associated liquids such as water, but it can be used over a limited temperature range. At 20.0 °C the viscosity of water is 1.002×10^{-3} kg m^{-1} s^{-1} and the activation energy for fluidity 18.0 kJ mol^{-1}. Estimate the viscosity at 40.0 °C.

19.5. At 20.0 °C the viscosity of pure toluene is 5.90×10^{-4} kg m^{-1} s^{-1}. Calculate the intrinsic viscosities of solutions containing 0.1 g dm^{-3} of polymer in toluene and having the following viscosities:

a. 5.95×10^{-4} kg m^{-1} s^{-1}
b. 6.05×10^{-4} kg m^{-1} s^{-1}
c. 6.27×10^{-4} kg m^{-1} s^{-1}

These solutions are sufficiently dilute that the reduced specific viscosity can be taken to be the intrinsic viscosity.

***19.6.** Suppose that solutions (a) and (c) in Problem 19.5 correspond to polymers of molecular weights 20 000 and 40 000, respectively. Assuming the Mark-Houwink equation 19.33 to apply, make an estimate of the molecular weight of the polymer in solution (b).

19.7. a. The activation energy for the fluidity of n-octane is 12.6 kJ mol^{-1} and the viscosity at 0 °C is 7.06×10^{-4} kg m^{-1} s^{-1}. Estimate the viscosity at 40.0 °C, assuming the Arrhenius equation to apply.
b. A better temperature law for the viscosity of n-octane has been found to be

$$\eta = A(T/\text{K})^{-1.72} e^{543/(T/\text{K})}$$

where T is expressed in kelvins.

Make another estimate of the viscosity at 40 °C. What is the effective activation energy at 20 °C?

***19.8.** Over its entire liquid range the viscosity of water is represented to within 1% by the following empirical formula:

$$\ln\left(\frac{\eta(20\ °\text{C})}{\eta(t\ °\text{C})}\right) = \frac{3.1556(t - 20.0) + 1.925 \times 10^{-3}(t - 20.0)^2}{109.0 + t}$$

where t is the value of the temperature in degrees Celsius. Make a better estimate of the viscosity of water than obtained by the use of the Arrhenius equation (Problem 19.4b).

To what activation energies does this empirical expression correspond at

a. 20 °C?
b. 100 °C?

Give a qualitative explanation for the difference between the two values.

***19.9.** Consider a hypothetical gas in which the molecules have mass but no size and do not interact with each other.

a. What would be the viscosity of such a gas?
b. Suppose instead that the molecules have zero size but attract one another. What can you then say about the viscosity?
c. If they repel one another, what would the viscosity be?
d. Give a clear explanation of your conclusions in all three cases.

Diffusion

19.10. The molecular diameter of the helium atom is 0.225 nm. Estimate, at 0 °C and 101.325 kPa,

a. the viscosity of the gas,
b. the self-diffusion coefficient,
c. the mean speed of the molecules,
d. the mean free path,
e. the collision frequency Z_A, and
f. the collision density Z_{AA}.

19.11. Calculate the mean square distance traveled by a molecule of H_2 at 20 °C and 101.325 kPa in 10 s ($D = 1.005 \times 10^{-4}$ m^2 s^{-1}).

19.12. Solutions of (a) glucose ($D = 6.8 \times 10^{-10}$ m^2 s^{-1}) and (b) tobacco mosaic virus ($D = 5.3 \times 10^{-12}$ m^2 s^{-1}) were maintained at a constant temperature of 20 °C and without agitation for 100 days. How far would a given molecule of each be expected to diffuse in that time?

19.13. Estimate the diffusion coefficient of cupric sulfate in water at 25 °C from the molar conductivities given in Table 7.3 (p. 291).

19.14. Estimate the diffusion coefficient of sodium acetate in water at 25 °C from the following mobility values:

Na^+: 5.19×10^{-4} cm^2 V^{-1} s^{-1}
CH_3COO^-: 4.24×10^{-4} cm^2 V^{-1} s^{-1}

19.15. The diffusion coefficient for horse hemoglobin in water is 6.3×10^{-11} m^2 s^{-1} at 20 °C. The viscosity of water at 20 °C is 1.002×10^{-3} kg m^{-1} s^{-1} and the specific volume of the protein is 0.75 cm^3 g^{-1}. Assume the hemoglobin molecule to be spherical and to obey Stokes's law, and estimate its radius and the molecular weight.

19.16. If the diffusion coefficient for insulin is 8.2×10^{-11} m^2 s^{-1} at 20 °C, estimate the mean time required for an insulin molecule to diffuse through a distance equal to the diameter of a typical living cell (≈ 10 μm).

***19.17.** A colloidal particle is spherical and has a diameter of 0.3 μm and a density of 1.18 g cm^{-3}. Estimate how long it will take for the particle to diffuse through a distance of 1 mm in water at 20 °C ($\eta = 1.002 \times 10^{-3}$ kg^{-3} m^{-1} s^{-1}; the density of water at 20 °C = 0.998 g cm^{-3}). (See also Problem 19.21.)

Sedimentation and Diffusion

19.18. Diphtheria toxin was found to have, at 20 °C, a sedimentation coefficient of 4.60 Svedbergs and a diffusion coefficient of 5.96×10^{-7} cm^2 s^{-1}. The toxin has a specific volume of 0.736 cm^3 g^{-1}, and the density of water at 20 °C is 0.998 g cm^{-3}. Estimate a value for the molecular weight of the toxin.

19.19. A protein has a sedimentation coefficient of 1.13×10^{-12} s^{-1} at 25 °C and a diffusion coefficient of 4.2×10^{-11} m^2 s^{-1}. The density of the protein is 1.32 g cm^{-3} and that of water at 25 °C is 0.997 g cm^{-3}. Calculate the molecular weight of the protein.

***19.20.** A protein of molecular weight 60 000 has a density of 1.31 g cm^{-3} and in water at 25 °C ($\rho = 0.997$ g cm^{-3}; $\eta = 8.937 \times 10^{-4}$ kg m^{-1} s^{-1}) it has a sedimentation coefficient of 4.1×10^{-13} s^{-1}. Calculate the frictional coefficient f

a. from the sedimentation coefficient, and
b. by the use of Stokes's law.

Suggest a reason why the two values are not quite the same.

***19.21.** How long will it take the particle from Problem 19.17 to sediment a distance of 1 mm in the earth's gravitational field ($g = 9.81$ m s^{-2})?

***19.22.** An aqueous colloidal solution contains spherical particles of uniform size and of density 1.33 g cm^{-3}. The diffusion coefficient at 25 °C is 1.20×10^{-11} m^2 s^{-1}; make an estimate of the sedimentation coefficient ($\rho(H_2O) = 0.997$ g cm^{-3}; $\eta(H_2O) = 8.937 \times 10^{-4}$ kg m^{-1} s^{-1}).

19.23. At 20 °C, γ-globulin has a sedimentation constant of 7.75×10^{-13} s, a diffusion coefficient in water of 4.8×10^{-11} m^2 s^{-1}, and a density of 1.353 g cm^{-3}. The density of water at 20 °C is 0.998 g cm^{-3}.

a. Estimate the molecular weight of γ-globulin.
b. Assuming the Stokes-Einstein equation to apply, estimate the radius of the protein molecule. The viscosity of water at 20 °C is 1.002×10^{-3} kg m^{-1} s^{-1}.

***19.24.** A sample of human hemoglobin had a sedimentation constant of 4.48 Svedbergs in water at 20 °C and a diffusion coefficient of 6.9×10^{-11} m^2 s^{-1}. The specific volume of human hemoglobin is 0.749 cm^3 g^{-1}, and the density of water at 20 °C is 0.998 g cm^{-3}.

a. Estimate the molecular weight of human hemoglobin.
b. How far would a molecule diffuse in 1 minute?

c. Neglect diffusion and estimate how far a molecule would sediment in 1 minute.
d. In a centrifuge rotating at 15 000 rpm, how far would a molecule sediment in 1 minute if it were 20 cm from the center of rotation?
e. Assume the molecule to be spherical and estimate its radius by the use of the Stokes-Einstein equation.
f. Estimate the radius from the molecular weight and the density.

***19.25.** In the first decade of the twentieth century Jean Perrin carried out important investigations on the sedimentation, in the gravitational field, of particles of gamboge, mastic, and other paint pigments. Consider particles of density 1.2 g cm^{-3} and of the following radii:

a. 1 mm
b. 0.1 mm
c. 10 μm
d. 1 μm
e. 10 nm

In each case, assuming Stokes's law to apply, estimate the distance the particle would sediment in 1 hour, in water at 20 °C (viscosity, $\eta = 1.002 \times 10^{-3}$ kg m^{-1} s^{-1} and density 0.998 g cm^{-3}).
f. In the case of the particle having a radius of 10 nm, what speed of rotation in an ultracentrifuge would be required to bring about a sedimentation of 1 mm in 1 hour? Take the distance from the axis of rotation to be 10 cm.

***19.26.** Perrin also carried out, using pigment particles, experiments on Brownian movement in which he determined distances traveled by individual particles in various periods of time. For each of the particles of five different radii mentioned in the previous problem, estimate the diffusion coefficient and the average distance traveled in 1 hour. Take the water temperature again as 20°C and use the data of Problem 19.25.

***19.27.** A pure protein in water is centrifuged to equilibrium at 25 °C and at 25 000 rpm. At distances of 8.34 cm and 9.12 cm from the axis of rotation the concentrations of the protein are measured to be 3.52 g cm^{-3} and 22.49 g cm^{-3}, respectively. The specific volume V_1 of the protein is 0.78 cm^3 g^{-1}, and the density of water at 25 °C is 0.997 g cm^{-3}. Calculate the molecular weight of the protein.

***19.28.** A protein has a molecular weight of 1 000 000 and a specific volume of 0.81 cm^3 g^{-1}. In an ultracentrifuge at 25 °C, what speed of revolution is required to produce at equilibrium a concentration ratio of 20/1 at distances 10.00 cm and 9.00 cm from the axis of revolution? The density of water at 25 °C is 0.997 g cm^{-3}.

Essay Questions

19.29. Explain how the rate of diffusion through a membrane depends on

a. the size of the diffusing substance, and
b. its solubility in the membrane.

19.30. Explain clearly the different mechanisms involved in the viscosity of gases and the viscosity of liquids.

SUGGESTED READING

For general accounts of the kinetic theory and of transport properties, see

R. S. Bradley, *The Phenomena of Fluid Motions,* Reading, MA: Addison-Wesley, 1967.

J. O. Hirschfelder, C. F. Curtiss, and R. B. Bird, *Molecular Theory of Gases and Liquids,* New York: Wiley, 1954.

L. B. Loeb, *Kinetic Theory of Gases: Being a text and reference book whose purpose is to combine the classical deductions with recent experimental advances in a convenient form for student and investigator,* New York: Dover, 3rd edition, 1961. This edition of a book which first appeared in 1927 presents the theory very clearly and includes many applications.

For treatments of diffusion, see

J. Crank, *The Mathematics of Diffusion,* Oxford: Clarendon Press, 1970; 2nd edition, 1979.

W. Jost, *Diffusion in Solids, Liquids, and Gases,* New York: Academic Press, 1960.

P. G. Shewman, *Diffusion in Solids,* New York: McGraw-Hill, 1963.

W. D. Stein, *Transport and Diffusion Across Cell Membranes,* Orlando, FL: Academic Press, 1986.

W. D. Stein, *Channels, Carriers, and Pumps: An Introduction to membrane transport,* San Diego, California: Academic Press, 1990.

For more on Brownian movement, see

M. J. Nye, *Molecular Reality: A Perspective on the Life and Work of Jean Perrin,* London and New York: Watson Publishing International, 1972.

J. B. Perrin, *Brownian Movement and Molecular Reality,* translated by F. Soddy, London: Taylor and Francis, 1910.

J. B. Perrin, *Atoms,* translated by D. L. Hammick, London: Constable, 1916.

Units, Quantities, and Symbols: The SI / IUPAC Recommendations

APPENDIX A

Scientists now use a system of units known as the *Système International d'Unités*, abbreviated SI. In this system, called a *coherent* system, the units for all derived physical quantities are obtained from certain base units by multiplication or division, without the use of any numerical factors. Several articles have been written about SI, which represents an extension of the metric system.[1]

The International Union of Pure and Applied Chemistry (IUPAC) strongly endorses the use of SI and has taken the matter much further by recommending terms, definitions, and symbols to be used in all branches of chemistry. The latest official set of recommendations for physical chemistry is *Quantities, Units, and Symbols in Physical Chemistry,* prepared for publication by a committee under the chairmanship of Ian Mills and published by Blackwell Scientific Publications, Oxford, Boston, etc., in 1987. This publication is usually referred to as "the Green Book," for the not surprising reason that it has a green cover. The student will find this publication extremely useful, as it contains the necessary information about both units and physical quantities, with their precise definitions, in a very compact form. This Appendix is based on the Green Book, and also makes more explicit some of the implications of the IUPAC recommendations.

Another IUPAC publication of considerable value is *Compendium of Chemical Terminology: IUPAC Recommendations,* compiled by the late Victor Gold, K. L. Loening, A. D. McNaught, and P. Sehmi, and published by Blackwell Scientific Publications, Oxford, Boston, etc., 1987. This publication is often referred to as "the Gold Book," partly because of its cover, but partly because of the great contribution to it made by Victor Gold.

[1]The official account of SI is *Le Système International d'Unités* (SI), originally published in 1973 by the *Bureau International des Poids et Mesures* and obtainable from OFF 1L1B, 48 rue Gay-Lussac, F 75005, Paris 5e. Authorized English translations of the official accounts, entitled *SI, The International System of Units,* have been published jointly by the U.S. National Bureau of Standards (Publication 330, available from the Government Printing Office, Washington, D.C. 20402) and by the U.K. National Physical Laboratory (available from Her Majesty's Stationery Office, P.O. Box 276, London SW8 5DT). Another useful publication is *The Metric Guide,* published in 1976 by the Council of Ministers of Education of Canada and available from O1SE Publications Sales, 252 Bloor Street, West, Toronto, Ontario, M5S 1V6. (This publication begins with a very clear and elementary explanation of the metric system and later deals in some detail with SI units.) Also see: B.N. Taylor, *Guide for the Use of the International System of Units (SI), NIST Publication* 811, 1995 Edition, Superintendent of Documents, U.S. Printing Office, Washington, D.C. 20402, 1995.

The SI base units and their IUPAC-recommended symbols are as follows:

Physical Quantity	Symbol for Quantity	SI Unit	Symbol for Unit
Length	l	metre*	m
Mass	m	kilogram	kg
Thermodynamic temperature	T	kelvin	K
Time	t	second	s
Electric current	I	ampere	A
Luminous intensity	I_v	candela	cd
Amount of substance	n	mole	mol

*In the Green Book, published in the U.K., the U.S., and elsewhere, the French spelling *metre* is always used. Some U.S. scientists use this spelling, others prefer the American spelling *meter*. We use the international spelling in this book; it has the advantage of distinguishing between metre, the unit, and meter, the instrument.

In addition there are two supplementary units for angles:

Physical Quantity	Symbol for Quantity	SI Unit	Symbol for Unit
Plane angle	θ, ϕ, etc.	radian	rad
Solid angle	Ω	steradian	sr

Note that in SI the symbol for a physical quantity is always printed in *italic* type, whereas a symbol for a unit is printed in roman type. In this book we use italics for two useful non-SI symbols, namely m (= mol kg^{-1}) and M (= mol dm^{-3}). Luminous intensity is seldom used in physical chemistry, but the other units are of importance and are considered in further detail later.

The following prefixes are used with SI units:

Fraction	Prefix	Symbol	Multiple	Prefix	Symbol
10^{-1}	deci	d	10	deca*	da
10^{-2}	centi	c	10^{2}	hecto	h
10^{-3}	milli	m	10^{3}	kilo	k
10^{-6}	micro	μ	10^{6}	mega	M
10^{-9}	nano	n	10^{9}	giga	G
10^{-12}	pico	p	10^{12}	tera	T
10^{-15}	femto	f	10^{15}	peta	P
10^{-18}	atto	a	10^{18}	exa	E
10^{-21}	zepto	z	10^{21}	zetta	Z
10^{-24}	yocto	y	10^{24}	yotta	Y

*Note: deka is used in the U.S.

Length

The metre was redefined in 1983 as the distance traveled by light in a vacuum during a time interval of 1/299 792 483 of a second.

The following multiple units of length are commonly used:

$$
\begin{aligned}
1 \text{ decimetre} &= 10^{-1} \text{ metre} & (1 \text{ dm} &= 10^{-1} \text{ m})\\
1 \text{ centimetre} &= 10^{-2} \text{ metre} & (1 \text{ cm} &= 10^{-2} \text{ m})\\
1 \text{ millimetre} &= 10^{-3} \text{ metre} & (1 \text{ mm} &= 10^{-3} \text{ m})\\
1 \text{ micrometre} &= 10^{-6} \text{ metre} & (1\ \mu\text{m} &= 10^{-6} \text{ m})\\
1 \text{ nanometre} &= 10^{-9} \text{ metre} & (1 \text{ nm} &= 10^{-9} \text{ m})\\
1 \text{ picometre} &= 10^{-12} \text{ metre} & (1 \text{ pm} &= 10^{-12} \text{ m})\\
1 \text{ femtometre} &= 10^{-15} \text{ metre} & (1 \text{ fm} &= 10^{-15} \text{ m})\\
1 \text{ kilometre} &= 10^{3} \text{ metre} & (1 \text{ km} &= 10^{3} \text{ m})
\end{aligned}
$$

The micrometre was formerly called the micron, and the nanometre was formerly called the millimicron; the use of these older terms is discouraged. The angstrom (Å), a non-SI unit, is still commonly employed:

$$1 \text{ Å} = 10^{-10} \text{ m} = 10^{-8} \text{ cm} = 10^{-1} \text{ nm} = 100 \text{ pm}$$

Some workers are now expressing interatomic distances in picometres, such distances ranging from about 100 to 200 pm.

Volume

In SI the basic unit of volume is the cubic metre (m^3). The liter is now defined as equal to one cubic decimetre (dm^3). The cubic centimetre should be written as cm^3, not as cc.

Mass

The kilogram is the mass of a platinum-iridium block in the custody of the *Bureau International des Poids et Mesures* at Sèvres, France.

Although the kilogram is the base unit of mass, its multiples are expressed with reference to the gram (g); it obviously would be absurd to express a gram as a millikilogram! Thus 10^{-6} kg is written as 1 milligram (1 mg).

Time

The second is the duration of 9 192 631 770 periods of the radiation corresponding to the transition between the two hyperfine levels of the ground state of the cesium-133 atom.

The second should be used as far as possible, but it is sometimes convenient to use larger units:

1 minute (min) = 60 s	1 day = 86 400 s
1 hour = 3600 s	1 year = 3.1558×10^7 s

Thermodynamic Temperature

The kelvin is strictly defined as the fraction 1/273.16 of the temperature interval between the absolute zero and the triple point of water. The temperature θ in degrees Celsius is defined as

$$\frac{\theta}{°\mathrm{C}} = \frac{T}{K} - 273.15$$

The use of the symbols °K and deg is not recommended.

Electric Current

The ampere is that constant current which, if maintained in two straight parallel conductors of infinite length, of negligible cross section, and placed 1 metre apart in vacuum, would produce between these conductors a force equal to 2×10^{-7} newton per metre of length (1 newton = 1 kg m s^{-2}).

Amount of Substance

The mole is the amount of substance that contains as many elementary units as there are atoms in 12 g of carbon-12 ($^{12}_{6}C$). The elementary unit must be specified and may be an atom, a molecule, an ion, a radical, an electron, or any other elementary particle. It may also be any specified group or fraction of such entities such as a polymer or a fraction of a molecule. For example,

1 mol of HgCl has a mass of 236.04 g
1 mol of Hg_2Cl_2 has a mass of 472.08 g
1 mol of Hg has a mass of 200.59 g
1 mol of $\frac{1}{2}CuSO_4$ has a mass of 79.80 g
1 mol of $\frac{1}{2}Cu^{2+}$ has a mass of 31.77 g
1 mol of electrons has a mass of 5.486×10^{-4} g

The amount of substance should not be referred to as the "number of moles"; it is not a number but a quantity having units.

Concentration and Molality

The word *concentration* means the amount of substance divided by the volume of the solution. The SI unit is mol m^{-3}, but the usual unit is moles per liter (mol dm^{-3}), which is conveniently abbreviated as *M*. The use of equivalents and normalities (N) is not recommended. Using the word *molarity* to mean concentration is not recommended because of the possibility of confusion with molality.

The word *molality* means the amount of solute divided by the mass of the solvent. The usual units are mol kg^{-1}, for which the symbol *m* is frequently used; to avoid confusion with the symbol for metre, it is best to use italics.

Molar Quantities

The word *molar* used before the name of a quantity may mean "divided by amount of substance." For example, molar volume is the volume divided by the amount of substance. The word *molar* is also used to mean "divided by concentration," where "concentration" is "amount of substance divided by volume." A useful rule is that molar can be used whenever mol appears in the unit used for the quantity. The subscript *m* can be used to indicate a molar quantity; for example, molar volume can be written as V_m (= V/n). If there is no danger of ambiguity, the subscript *m* can be omitted; this is always the case when a numerical value is given, since mol appears in the unit.

Derived Units

Table A.1 gives some SI-derived units. The convention for writing the abbreviations should be noted. Prefixes are not separated from the unit (e.g., kg, not k g); otherwise units are separated (e.g., m s^{-1}, not ms^{-1}). The solidus can be used, as in m/s for m s^{-1}, but it should not appear more than once in the same expression unless parentheses are used. Thus the unit for entropy must not be written as J/K/mol. Acceptable forms for this unit are

$$\mathrm{J\,K^{-1}\,mol^{-1}} \qquad \mathrm{J(K\,mol)^{-1}} \qquad \mathrm{J/(K\,mol)}$$

and the correct pronunciation of the unit is "joule per kelvin mole," not "joule per kelvin per mole." The solidus is very useful for dividing quantities by units in tables or figures, where the numerical value is needed. For example, a table heading for entropy changes could be ΔS/J K^{-1} mol^{-1}.

TABLE A.1 SI-Derived Units

Physical Quantity	Name of Unit	Symbol	Definition	Alternative Form	Named After
Force	newton	N	kg m s^{-2}		Isaac Newton (1642–1727)
Pressure	pascal	Pa	kg m^{-1} s^{-2}	N m^{-2}	Blaise Pascal (1623–1662)
Energy	joule	J	kg m^2 s^{-2}	N m	James Prescott Joule (1818–1889)
Power	watt	W	kg m^2 s^{-3}	J s^{-1}	James Watt (1737–1819)
Electric charge	coulomb	C	A s		Charles Auguste de Coulomb (1736–1806)
Electric potential difference	volt	V	kg m^2 s^{-3} A^{-1}	J C^{-1}	Allesandro Volta (1745–1827)
Electric resistance	ohm	Ω	kg m^2 s^{-3} A^{-2}	V A^{-1}	Georg Simon Ohm (1787–1854)
Electric conductance	siemens	S	kg^{-1} m^{-2} s^3 A^2	Ω^{-1}	Werner von Siemens (1816–1892)
Electric capacitance	farad	F	kg^{-1} m^{-2} s^4 A^2	C V^{-1}	Michael Faraday (1791–1867)
Magnetic flux	weber	Wb	kg m^2 s^{-2} A^{-1}	V s	Wilhelm Weber (1804–1891)
Magnetic inductance	henry	H	kg m^2 s^{-2} A^{-2}	V A^{-1} s	Joseph Henry (1799–1878)
Magnetic flux density	tesla	T	kg s^{-2} A^{-1}	V s m^{-2}	Nikola Tesla (1856–1943)
Frequency	hertz	Hz	s^{-1}		Heinrich Hertz (1857–1894)
Activity (of radioactive source)	becquerel	Bq	s^{-1}		Antoine Henri Becquerel (1852–1908)
Absorbed dose (of radiation)	gray	Gy	m^2 s^{-2}	J kg^{-1}	Louis Harold Gray (1905–1965)

TABLE A.2 Some Non-SI Units

Physical Quantity	Name of Unit	Symbol	SI Equivalent
Pressure	(standard) atmosphere	atm	101 325 Pa
Pressure	torr	Torr	101 325 Pa/760
Pressure	bar	bar	10^5 Pa
Energy	thermochemical calorie	cal	4.184 J
Energy	electronvolt	eV	1.602×10^{-19} J
Energy	erg	erg	10^{-7} J
Length	angstrom	Å	10^{-10} m
Volume	liter	L	1 dm^3
Electric charge	electrostatic unit	esu	3.336×10^{-10} C
Force	dyne	dyn	10^{-5} N
Viscosity	poise	poise	0.1 kg m^{-1} s^{-1}
Diffusion coefficient	stokes	stokes	10^{-4} m^2 s^{-1}
Magnetic flux density	gauss	G	10^{-4} T
Dipole moment	debye	D	3.336×10^{-30} C m
Magnetic moment	Bohr magneton	μ_B	9.274×10^{-24} J T^{-1}

Some scientists separate units by means of a raised dot, but this is not done in the Green Book, and seems unnecessary in print, but may be useful for handwriting.

A few non-SI units are still in common use, and some of these are listed in Table A.2, together with their SI equivalents. Since 1 bar is a standard state for thermodynamic quantities, it is entirely acceptable to continue to use it. The bar and the angstrom (Å) are, in fact, approved by IUPAC for temporary use with SI units. The other units in Table A.2, however, are being progressively abandoned.

Physicochemical Quantities and Symbols

The most important physicochemical quantities, and most of the ones used in this book, are listed in Table A.3. This table is consistent with the IUPAC Green Book, which gives some alternative symbols; here we give only those used in this book. We have included definitions of less well-known quantities when they can be expressed compactly; in other cases we have referred to the section of this book in which they are explained.

Each physical quantity has an SI unit, which is either an SI base unit or a combination of base units. For example, the SI unit of energy is the joule (J), which is kg m^2 s^{-2}. However, it is quite appropriate, and may be more convenient, to use multiples or submultiples, such as kilojoules (kJ). Sometimes particular multiples are in common use, and when this is the case, they are indicated in the last column of Table A.3 (Customary Multiple). A blank in this column means that there is no strong preference for any multiple.

It is important to note that a physical quantity is the product of a number and a unit. For example, the Avogadro constant L is 6.022×10^{23} mol^{-1} and is the product of the number 6.022×10^{23} and the unit mol^{-1}. This being so, it should not be called the "Avogadro number"; it is not a number (i.e., a dimensionless quantity).

TABLE A.3 Physicochemical Quantities, with their Symbols, Definitions, and Units

A.3.1 Space, Time, and Related Quantities

Quantity	Symbol	Definition	SI Unit	Customary Multiple
Length	l		m	
Height	h		m	
Radius	r		m	
Diameter	d		m	
Path length	s		m	
Wavelength	λ	Section 11.1	m	cm
Wavenumber	$\tilde{\nu}$	$\tilde{\nu} = 1/\lambda$	$\mathrm{m^{-1}}$	$\mathrm{cm^{-1}}$
Area	A		$\mathrm{m^2}$	
Volume	V		$\mathrm{m^3}$	$\mathrm{m^3}$, $\mathrm{dm^3}$(L)
Time	t		s	s
Frequency	ν		$\mathrm{s^{-1}} \equiv \mathrm{Hz}$	$\mathrm{s^{-1}}$, MHz, GHz
Relaxation time	τ		s	s
Velocity	v (or $\mathbf{v}$)	$v = ds/dt$	$\mathrm{m\ s^{-1}}$	$\mathrm{m\ s^{-1}}$
Molecular velocity	u	$u = ds/dt$	$\mathrm{m\ s^{-1}}$	$\mathrm{m\ s^{-1}}$
Speed of light	c		$\mathrm{m\ s^{-1}}$	$\mathrm{m\ s^{-1}}$
Angular velocity	ω		$\mathrm{rad\ s^{-1}}$	$\mathrm{rad\ s^{-1}}$
Acceleration of gravity	g		$\mathrm{m\ s^{-2}}$	$\mathrm{m\ s^{-2}}$
Probability	P		*	*

*In these tables an asterisk indicates that the quantity is dimensionless.

A.3.2 Mechanical and Related Quantities

Quantity	Symbol	Definition	SI Unit	Customary Multiple
Mass	m		kg	kg, g
Reduced mass	μ	Section 11.6	kg	kg, g
Specific volume	v	$v \equiv V/m$	$\mathrm{m^3\ kg^{-1}}$	$\mathrm{dm^3\ g^{-1}}$
Density	ρ	$\rho \equiv m/V$	$\mathrm{kg\ m^{-3}}$	$\mathrm{g\ dm^{-3}}$, $\mathrm{g\ cm^{-3}}$
Relative density	d	$d \equiv \rho/\rho_0$	*	*
Moment of inertia	I	$I = \sum_i m_i r_i^2$	$\mathrm{kg\ m^2}$	$\mathrm{kg\ m^2}$
Momentum	$\mathbf{p}$	$\mathbf{p} = m\mathbf{v}$	$\mathrm{kg\ m\ s^{-1}}$	$\mathrm{kg\ m\ s^{-1}}$
Force	$\mathbf{F}$	$F = dp/dt$	$\mathrm{N = kg\ m\ s^{-2}}$	N
Angular momentum	$\mathbf{L}$	$\mathbf{L} = \mathbf{r} \times \mathbf{p}$	$\mathrm{J\ s \equiv kg\ m^2\ s^{-1}}$	$\mathrm{kg\ m^2\ s^{-1}}$
Pressure	P	$P \equiv F/A$	$\mathrm{Pa \equiv kg\ m^{-1}\ s^{-2}}$	Pa, kPa, bar
Isothermal compressibility	κ	$\kappa \equiv \frac{1}{V}\left(\frac{\partial V}{\partial P}\right)_T$	$\mathrm{Pa^{-1}}$	$\mathrm{Pa^{-1}}$
Coefficient of expansion	α	$\alpha \equiv \frac{1}{V}\left(\frac{\partial V}{\partial T}\right)_P$	$\mathrm{K^{-1}}$	$\mathrm{K^{-1}}$
Work	w	$w \equiv \int F\,ds$	$\mathrm{J \equiv kg\ m^2\ s^{-2}}$	J, kJ
Energy	E	Section 1.2	$\mathrm{J = kg\ m^2\ s^{-2}}$	J, kJ
Potential energy	E_p	$\Delta E_p \equiv \int F\,ds$	$\mathrm{J \equiv kg\ m^2\ s^{-2}}$	J, kJ
Kinetic energy	E_k	$E_k \equiv \frac{1}{2}mv^2$	$\mathrm{J \equiv kg\ m^2\ s^{-2}}$	J, kJ
Hamiltonian	H	$H(p, q) \equiv E_p + E_k$	$\mathrm{J \equiv kg\ m^2\ s^{-2}}$	J

TABLE A.3 Physicochemical Quantities, with their Symbols, Definitions, and Units (Continued)

A.3.2 Mechanical and Related Quantities (Continued)

Quantity	Symbol	Definition	SI Unit	Customary Multiple
Power	P	$P \equiv dw/dt$	$\mathrm{W} \equiv \mathrm{J\ s^{-1}} \equiv \mathrm{kg\ m^2\ s^{-3}}$	W
Viscosity	η	$\mathrm{F} \equiv \eta A(dv/dx)$	$\mathrm{Pa\ s} \equiv \mathrm{kg\ m^{-1}\ s^{-1}}$	Pa s, $\mathrm{g\ cm^{-1}\ s^{-1}}$ (≡ poise)
Intrinsic viscosity	$[\eta]$	Section 19.1	$\mathrm{m^3\ kg^{-1}}$	$\mathrm{dm^3\ g^{-1}}$
Fluidity	ϕ	$\phi \equiv 1/\eta$	$\mathrm{kg^{-1}\ m\ s}$	
Kinematic viscosity	ν	$\nu \equiv \eta/\rho$	$\mathrm{m^2\ s^{-1}}$	
Diffusion coefficient	D	$dn/dt \equiv -DA(dc/dx)$	$\mathrm{m^2\ s^{-1}}$	$\mathrm{cm^2\ s^{-1}}$
Surface tension	γ	dw/dA	$\mathrm{N\ m^{-1}} \equiv \mathrm{kg\ s^{-2}}$	$\mathrm{N\ m^{-1}}$
Angle of contact	θ		rad	rad

A.3.3 Molecular and Related Quantities

Quantity	Symbol	Definition	SI Unit	Customary Multiple
Atomic weight	A_r		*	*
Molecular weight	M_r		*	*
Molar mass	M		$\mathrm{kg\ mol^{-1}}$	$\mathrm{g\ mol^{-1}}$
Amount of substance	n	$n \equiv m/M$	mol	mol
Number of molecules	N		*	*
Avogadro constant	L	$L \equiv N/n$	$\mathrm{mol^{-1}}$	$\mathrm{mol^{-1}}$
Molar volume	V_m	$V_m \equiv V/n$	$\mathrm{m^3\ mol^{-1}}$	$\mathrm{m^3\ mol^{-1}}$
Dipole moment	μ	Section 12.4	C m	
Symmetry number	σ	Section 15.5	*	*
Statistical weight (degeneracy)	g	Section 15.3	*	*
Collision diameter	$d_{\mathrm{AA}}, d_{\mathrm{AB}}$	Section 1.9	m	nm, pm
Mean free path	λ	$\lambda = \sqrt{2}\pi d^2 L$	m	nm, pm
Mole fraction of B	x_{B}	$x_{\mathrm{B}} \equiv n_{\mathrm{B}}/\sum_i n_i$	*	*
Molality of B in solvent A	m_{B}	$m_{\mathrm{B}} \equiv n_{\mathrm{B}}/n_{\mathrm{A}}M_{\mathrm{A}}$	$\mathrm{mol\ kg^{-1}}$	$\mathrm{mol\ kg^{-1}}$
Concentration of B (formerly called "molarity")	c_{B},[B]	$c_{\mathrm{B}} \equiv n_{\mathrm{B}}/V$	$\mathrm{mol\ m^{-3}}$	$\mathrm{mol\ dm^{-3}}$
Mass concentration of B	ρ_{B}	$\rho_{\mathrm{B}} \equiv m_{\mathrm{B}}/V$	$\mathrm{kg\ m^{-3}}$	$\mathrm{g\ dm^{-3}}$
Partition function (assembly)	Q	Section 15.3	*	*
Partition function (molecule)	q	Section 15.3	*	*
Atomic number	Z	number of protons in nucleus	*	*
Neutron number	N	number of neutrons in nucleus	*	*
Mass number	A	$A \equiv Z + N$	*	*
Mass of electron	m_e		kg	kg
Planck constant	h	Section 11.1	J s	J s
Bohr radius	a_0	$a_0 \equiv h^2\epsilon_0/\pi m_e e^2$	m	nm, pm
Magnetic moment	μ_I	Section 11.9, 13.2	$\mathrm{A\ m^2}$	$\mathrm{A\ m^2}$
Bohr magneton	μ_B	$\mu_B \equiv eh/4\pi m_e$	$\mathrm{A\ m^2}$	$\mathrm{A\ m^2}$
Nuclear magneton	μ_N	$\mu_N \equiv eh/4\pi m_p$	$\mathrm{A\ m^2}$	$\mathrm{A\ m^2}$
Gyromagnetic ratio	γ	$\gamma \equiv \mu/L$	$\mathrm{C\ kg^{-1}} \equiv \mathrm{A\ s\ kg^{-1}}$	$\mathrm{C\ kg^{-1}}$
g factor	g	Section 13.2		*

TABLE A.3 Physicochemical Quantities, with their Symbols, Definitions, and Units (Continued)

A.3.4 Thermodynamic and Related Quantities

Quantity	Symbol	Definition	SI Unit	Customary Multiple
Thermodynamic or absolute temperature	T	Section 3.1	K	K
Celsius temperature	θ	$\theta/°C \equiv T/K - 273.15$	°C	°C
(Molar) gas constant	R	$R \equiv \lim_{p\to 0} PV_m/T$	$J\ K^{-1}\ mol^{-1}$	$J\ K^{-1}\ mol^{-1}$
Boltzmann constant	k_B	$k_B \equiv R/L$	$J\ K^{-1}$	$J\ K^{-1}$
Heat	q	Section 2.4	J	J, kJ
Work done on the system	w	$w \equiv \int F\, ds$	J	J, kJ
Internal energy	U	Section 2.4, 2.5	J	J, kJ
Molar internal energy	U_m	$U_m \equiv U/n$	$J\ mol^{-1}$	$kJ\ mol^{-1}$
Enthalpy	H	$H \equiv U + PV$	J	J, kJ
Molar enthalpy	H_m	$H_m = H/n = U_m + PV_m$	$J\ mol^{-1}$	$kJ\ mol^{-1}$
Standard molar enthalpy of formation	$\Delta_f H_m$	Section 2.5	$J\ mol^{-1}$	$kJ\ mol^{-1}$
Heat capacity at constant volume	C_V	$C_V \equiv (\partial U/\partial T)_V$	$J\ K^{-1}$	$J\ K^{-1}$
Molar heat capacity at constant volume	$C_{V,m}$	$C_{V,m} \equiv C_V/n$	$J\ K^{-1}\ mol^{-1}$	$J\ K^{-1}\ mol^{-1}$
Specific heat capacity at constant volume	c_V	$c_V \equiv C_V/m$	$J\ K^{-1}\ kg^{-1}$	$J\ K^{-1}\ g^{-1}$
Heat capacity at constant pressure	C_P	$C_P \equiv (\partial H/\partial T)_P$	$J\ K^{-1}$	$J\ K^{-1}$
Molar heat capacity at constant pressure	$C_{P,m}$	$C_{P,m} \equiv C_P/n$	$J\ K^{-1}\ mol^{-1}$	$J\ K^{-1}\ mol^{-1}$
Specific heat capacity at constant pressure	c_p	$c_p \equiv C_P/m$	$J\ K^{-1}\ kg^{-1}$	$J\ K^{-1}\ g^{-1}$
Ratio of heat capacities	γ	$\gamma = C_P/C_V$	*	*
Compression factor	Z	$Z \equiv PV_m/RT$	*	*
Joule-Thompson coefficient	μ	$\mu \equiv (\partial T/\partial P)_H$	$K\ Pa^{-1}$	$K\ Pa^{-1}$
Stoichiometric coefficient of B (positive for products, negative for reactants)	ν_B	Sections 2.5, 9.2	*	*
Stoichiometric sum	$\Sigma\nu$	$\Sigma\nu \equiv \sum_i \nu_i$	*	*
Extent of reaction	ξ_i	$\xi_i \equiv (n - n_0)/\nu$	mol	mol
Entropy	S	$dS \equiv dq_{rev}/T$	$J\ K^{-1}$	$J\ K^{-1}$
Molar entropy	S_m	$S_m \equiv S/n$	$J\ K^{-1}\ mol^{-1}$	$J\ K^{-1}\ mol^{-1}$
Gibbs energy	G	$G \equiv H - TS$	J	J, kJ
Molar Gibbs energy	G_m	$G_m \equiv G/n$	$J\ mol^{-1}$	$kJ\ mol^{-1}$
Standard molar Gibbs energy change	ΔG_m°	Section 3.7	$J\ mol^{-1}$	$kJ\ mol^{-1}$
Helmholtz energy	A	$A \equiv U - TS$	J	J, kJ
Molar Helmholtz energy	A_m	$A_m \equiv A/n$	$J\ mol^{-1}$	$kJ\ mol^{-1}$
Standard molar Helmholtz energy change	ΔA_m°		$J\ mol^{-1}$	$kJ\ mol^{-1}$
Chemical potential of B	μ_B	$\mu_B \equiv (\partial G/\partial n_B)_{A,\ldots,T,P\ldots}$	$J\ mol^{-1}$	$kJ\ mol^{-1}$
Absolute activity of B	λ_B	$\lambda_B \equiv \exp(\mu_B/RT)$	*	*
Relative activity of B	a_B	$a_B \equiv \lambda_B/\lambda_B^*$	*	*
Fugacity	f	Section 3.8	Pa	Pa
Osmotic pressure	π	Section 5.8	Pa	Pa
Activity coefficient, mole fraction basis	γ_x, f	$\gamma_{x,B} \equiv a_B/x_B$	*	*
Activity coefficient, molality basis	γ_m, y	$\gamma_{m,B} \equiv a_B/m_B$	*	*
Activity coefficient, concentration basis	γ_c, y	$\gamma_{c,B} \equiv a_B/c_B$	*	*
Equilibrium constant with respect to pressure (ideal gases)	K_P	$K_P \equiv (P_Y^y P_Z^z \ldots / P_A^a P_B^b \ldots)_{eq}$	$Pa^{\Sigma\nu}$ ($\Sigma\nu$ = stoichiometric sum)	
Equilibrium constant with respect to concentration (ideal systems)	K_c	$K_c \equiv ([Y]^y[Z]^z \ldots/[A]^a[B]^b \ldots)_{eq}$	$(mol\ m^{-3})^{\Sigma\nu}$	$(mol\ dm^{-3})^{\Sigma\nu}$
Standard equilibrium constant	K	Numerical value of K_P or K_c	*	*

TABLE A.3 Physicochemical Quantities, with their Symbols, Definitions, and Units (Continued)

A.3.5 Chemical Reactions

Quantity	Symbol	Definition	SI Unit	Customary Multiple
Rate of conversion	ξ_i	$\xi_i \equiv di/dt$	mol s^{-1}	mol s^{-1}
Rate of reaction (at constant V)	v	$v \equiv (1/\nu_i)(dc_i/dt)$	mol m^{-3} s^{-1}	mol dm^{-3} s^{-1}
Rate of consumption of A (at constant V)	v_A	$v_A \equiv -dc_A/dt$	mol m^{-3} s^{-1}	mol dm^{-3} s^{-1}
Rate of formation of Z (at constant V)	v_Z	$v_Z \equiv dc_Z/dt$	mol m^{-3} s^{-1}	mol dm^{-3} s^{-1}
Partial order	α, β	Section 9.3	*	*
Overall order	n	$n \equiv \alpha + \beta + \cdots$	*	*
Rate constant, rate coefficient	k	$v \equiv k[A]^\alpha[B]^\beta \ldots$	$(m^3\ mol^{-1})^{n-1}\ s^{-1}$	$(dm^3\ mol^{-1})^{n-1}\ s^{-1}$
Activation energy	E	$E \equiv R[\partial \ln k/d(1/T)]_P$	J mol^{-1}	kJ mol^{-1}
Preexponential factor	A	$k \equiv Ae^{-E/RT}$	$(m^3\ mol^{-1})^{n-1}\ s^{-1}$	$(dm^3\ mol^{-1})^{n-1}\ s^{-1}$
Standard Gibbs energy of activation	$\Delta^\ddagger G°$	$k \equiv \frac{k_B T}{h} e^{-\Delta^\ddagger G°/RT}$	J mol^{-1}	kJ mol^{-1}
Standard enthalpy of activation	$\Delta^\ddagger H°$	$k \equiv \frac{k_B T}{h} e^{\Delta^\ddagger S°/R} e^{-\Delta^\ddagger H/RT}$	J mol^{-1}	k J mol^{-1}
Standard entropy of activation	$\Delta^\ddagger S°$		J K^{-1} mol^{-1}	J K^{-1} mol^{-1}
Collision frequency	Z_A	Section 1.9	s^{-1}	s^{-1}
Collision density	Z_{AA}, Z_{AB}	Section 1.9	m^{-3} s^{-1}	dm^{-3} s^{-1}
Half-life	$t_{1/2}$	Section 9.4	s	s
Relaxation time	τ	Section 9.5	s	s
Collision diameter	d_{AA}, d_{AB}	Section 1.9	m	nm, pm
Reaction probability	P_r		*	*
Reaction cross section	σ	$\sigma \equiv \pi d^2 P_r$	m^2	nm^2, pm^2

A.3.6 Electricity and Magnetism

Quantity	Symbol	Definition	SI Unit	Customary Multiple
Electric current	I		A	A, mA
Quantity of electricity	Q	$Q = \int I\,dt$	C ≡ A s	C
Elementary charge	e		C	C
Electric potential	V	$V \equiv w/Q$	V ≡ m^2 kg s^{-3} A^{-1}	V, mV, kV
Electric field strength	$\boldsymbol{E}$	$\boldsymbol{E} = \boldsymbol{F}/Q$	V m^{-1}	V m^{-1}
Capacitance	C	$C = Q/\Delta V$	F ≡ m^{-2} kg^{-1} s^4 A^2	F
Permittivity of vacuum	ϵ_0	See p. 971	F m^{-1}	F m^{-1}
Permanent dipole moment	μ	Section 12.4	C m	C m
Electrical resistance	R	$R = \Delta V/I$	Ω ≡ V A^{-1}	Ω, mΩ
Electrical conductance	G	$G \equiv 1/R$	S ≡ Ω^{-1} ≡ A V^{-1}	S, Ω^{-1}
Magnetic flux density	$\boldsymbol{B}$	Section 13.2, p. 1012	T ≡ kg s^{-2} A^{-1}	T
Magnetic field strength	$\boldsymbol{H}$	Sections 11.9, 13.2	A m^{-1}	A m^{-1}
Magnetic flux	Φ	See p. 1012	Wb ≡ V s	Wb
Permeability	μ	See p. 1011, 1012	H m^{-1} ≡ kg m s^{-2} A^{-2}	H m^{-1}
Permeability of vacuum	μ_0	See p. 1011, 1012	H m^{-1}	H m^{-1}

TABLE A.3 Physicochemical Quantities, with their Symbols, Definitions, and Units (Continued)

A.3.7 Electrochemistry

Quantity	Symbol	Definition	SI Unit	Customary Multiple
Faraday constant	F	$F \equiv Le$	$\mathrm{C\ mol^{-1}}$	$\mathrm{C\ mol^{-1}}$
Electromotive force	E	Section 8.1	V	V
Electrode potential	E	Section 8.2	V	V
Standard electrode potential	$E°$	Section 8.2	V	V
Charge number of ion B	z_B	$z_B \equiv Q_B/e$	*	*
Charge number of cell reaction	z	Section 8.3	*	*
Electric mobility of B	u_B	Section 7.5	$\mathrm{m^2\ V^{-1}\ s^{-1}}$	$\mathrm{cm^2\ V^{-1}\ s^{-1}}$
Transport number of B	t_B	Section 7.6	*	*
Electrolytic conductivity	κ	Section 7.2	$\mathrm{S\ m^{-1} \equiv \Omega^{-1}\ m^{-1}}$	$\mathrm{S\ cm^{-1} \equiv \Omega^{-1}\ cm^{-1}}$
Molar conductivity of electrolyte	Λ	$\Lambda \equiv \kappa/c$	$\mathrm{S\ m^2\ mol^{-1}}$	$\mathrm{S\ dm^2\ mol^{-1}}$
Molar conductivity of ion B	λ_B	$\lambda_B \equiv Fu_B$	$\mathrm{S\ m^2\ mol^{-1}}$	$\mathrm{S\ m^2\ mol^{-1} \equiv \Omega^{-1}\ m^2\ mol^{-1}}$
Ionic strength	I	$I \equiv \frac{1}{2}\sum_i c_i z_i^2$	$\mathrm{mol\ m^{-3}}$	$\mathrm{mol\ dm^{-3}}$
Overpotential	η	$\eta \equiv E(I) - E(I = 0)$	V	V
Electric potential	ψ	Section 10.14	V	V
Inner electric potential	ϕ	Section 10.14	V	V
Electrokinetic potential	i	Section 10.14	V	V, mV
Thickness of diffusion layer	δ	Section 10.14	m	
Electro-osmotic or electrophoretic mobility	v_0	Section 10.14	$\mathrm{m^2\ V^{-1}\ s^{-1}}$	$\mathrm{cm^2\ V^{-1}\ s^{-1}}$

A.3.8 Electromagnetic Radiation and Spectroscopy

Quantity	Symbol	Definition	SI Unit	Customary Multiple
Planck constant	h	Section 11.1	J s	J s
Radiant energy density	ρ_ν	Section 11.1	$\mathrm{J\ m^{-3}\ Hz^{-1}}$	$\mathrm{J\ m^{-3}\ s}$
Intensity of light	I		cd	cd
Transmittance	T	$T \equiv I(\text{transmitted})/I(\text{incident})$	*	*
Absorbance	A	$A \equiv \log_{10}(1/T)$	*	*
(Linear decadic) absorption coefficient	a	$a \equiv A/l$	$\mathrm{m^{-1}}$	$\mathrm{cm^{-1}}$
Molar (decadic) absorption coefficient	ϵ	$\epsilon \equiv A/lc$	$\mathrm{m^2\ mol^{-1}}$	$\mathrm{cm^2\ mol^{-1}}$
Quantum yield	Φ	Section 10.6	*	*
Turbidity	τ	$I \equiv I_0 e^{-\tau l}$	$\mathrm{m^{-1}}$	$\mathrm{cm^{-1}}$
Frequency	ν	$\nu \equiv c\lambda^{-1}$	$\mathrm{s^{-1}}$	$\mathrm{s^{-1}}$
Rotational constant	$\tilde{B}$	Section 13.3	$\mathrm{m^{-1}}$	$\mathrm{cm^{-1}}$
Rotational quantum number	J	Sections 11.0, 13.3	*	*
Vibrational quantum number	v	Sections 11.6, 13.4	*	*
Wavelength	λ	$\lambda \equiv c\nu^{-1}$ (Section 11.1)	m	m
Wavenumber	$\tilde{\nu}$	$\tilde{\nu} \equiv \nu c^{-1}$	$\mathrm{m^{-1}}$	$\mathrm{cm^{-1}}$

Any quantity can be converted into a dimensionless quantity by dividing it by the unit quantity, and this has to be done before taking a logarithm. A superscript ° is used to indicate the thermodynamic equilibrium constant, $K°$, which is dimensionless. In this book we have sometimes used a superscript u (for "unitless") to indicate that the dimensionless form of the quantity is being used.

The IUPAC Green Book emphasizes that the use of a physical quantity or its symbol should not imply any particular choice of unit. For example, we may have a solution of concentration 1 mol dm^{-3}, which may be alternatively expressed as 10^3 mol m^{-3} or as 1 μmol mm^{-3} or as 10^{-3} mol cm^{-3}. The concentration is the same whichever unit we choose, and if we write an equation involving concentration, using the symbol c, we must not assume that any particular unit is used. This being the case, it is unsatisfactory, for example, to define pH as

$$\text{pH} = \log_{10}[\text{H}^+]$$

This is a very common procedure, but it presupposes that the reader knows that the unit of $[\text{H}^+]$ is to be taken as mol dm^{-3} (which is not even the SI unit) and is then prepared to drop the unit in taking the logarithm. The only satisfactory definition of pH is

$$\text{pH} = -\log_{10}([\text{H}^+]/\text{mol dm}^{-3})$$

Here we are dividing the concentration by the unit, which gives a number, and we are then taking the logarithm of the number.

In tables we list numbers, and in figures we plot numbers. Table headings and axes should therefore show the physical quantity divided by the unit. For example, in a plot of velocity against time, the axes could be labeled v/m s^{-1} and t/s.

Standard Temperature and Pressure

It has been noted in Chapter 2 that the standard pressure for thermodynamic quantities is now the bar (10^5 Pa) instead of the atmosphere (1.01325×10^5 Pa). It should be emphasized that this change in definition does not extend to other properties, such as boiling points; it is still acceptable to quote boiling points at 1 atm pressure.

Chemists have long found it useful to make use of the **standard temperature and pressure** (STP), by which is meant 273.15 K (0 °C) and 1 atm pressure. The molar volume of an ideal gas at STP is 22.4138 dm^3 mol^{-1}, and for most purposes the easily remembered value of 22.4 dm^3 mol^{-1} is adequate. In view of the modern use of 1 bar as the standard pressure it is now convenient to define a **new standard temperature and pressure** (NSTP) as 298.15 K (25 °C) and 1 bar pressure. The molar volume of an ideal gas at NSTP is 24.7894 dm^3 mol^{-1}. The three-digit form of this, 24.8 dm^3 mol^{-1}, is as precise and as easy to remember as 22.4 dm^3 mol^{-1}. Moreover, for many calculations it is adequate to use 25 dm^3 mol^{-1}, which gives an error of less than 1%.

Electrical Units

The units used in electricity and magnetism require some special discussion. The force of attraction or repulsion between charged bodies was first used by the French engineer Charles Auguste de Coulomb (1736–1806) as the basis for the definition of the unit of electric charge. He defined the electrostatic unit (esu) of charge as

> that point charge which, when placed 1 cm from a similar point charge in a vacuum, is repelled by a force of one dyne.

The force is proportional to the product of the charges and inversely proportional to the distance of separation. Thus if two charges Q are separated by a distance r, Coulomb's definition can be expressed by the equation

$$F/\text{dyn} = \frac{(Q/\text{esu})^2}{(r/\text{cm})^2} \tag{A.1}$$

One dyne is the force that gives an acceleration of 1 cm s^{-2} to a mass of 1 gram, and is therefore equal to 10^{-5} newton (N); the newton is the force that produces an acceleration of 1 m s^{-2} in a mass of 1 kg. Thus

$$F/\text{N} = 10^{-5}\ F/\text{dyn} \tag{A.2}$$

The SI unit of charge is the coulomb (C) and the SI unit of distance is the metre. It can be shown that 1 coulomb represents the same physical situation as 1 esu unit of charge multiplied by 2.998×10^9, which is the speed of light divided by 1 dm s^{-1}. Thus 1 C = 2.998×10^9 esu and, therefore,

$$2.998 \times 10^9\ Q/\text{C} = Q/\text{esu} \tag{A.3}$$

Introduction of Eqs. A.2 and A.3 into Eq. A.1 gives

$$10^5\ F/\text{N} = \frac{(2.998 \times 10^9\ Q/\text{C})^2}{(100r/\text{m})^2} \tag{A.4}$$

and therefore

$$F/\text{N} = \frac{8.99 \times 10^9 (Q/\text{C})^2}{(r/\text{m})^2} \tag{A.5}$$

It has proved convenient to write this equation as

$$F = \frac{Q^2}{4\pi\epsilon_0 r^2} \tag{A.6}$$

Permittivity of a Vacuum

where ϵ_0, known as the *permittivity of a vacuum*, is given as

$$\begin{aligned}\epsilon_0 &= (1/4\pi) \times 8.99 \times 10^9\ \text{C}^2\ \text{N}^{-1}\ \text{m}^{-2}\ (\equiv \text{F m}^{-1}):\\ &= 8.854 \times 10^{-12}\ \text{C}^2\ \text{N}^{-1}\ \text{m}^{-2}\ (\equiv \text{F m}^{-1})\end{aligned}$$

The reason for the introduction of 4π is that certain equations, such as the Poisson equation, are simplified when this is done. Also see p. 1012.

Relative Permittivity

In a medium other than a vacuum, ϵ_0 in all the equations is replaced by $\epsilon_0\epsilon$, where ϵ is known as the *relative permittivity* or more usually as the *dielectric constant.* Thus the force between two charges Q_1 and Q_2 is

$$F = \frac{Q_1 Q_2}{4\pi\epsilon_0\epsilon r^2} \tag{A.7}$$

and the corresponding potential energy of interaction is

$$E_p = \frac{Q_1 Q_2}{4\pi\epsilon_0\epsilon r} \tag{A.8}$$

There is an important fundamental difference between the electrostatic units and the SI units for electricity and magnetism. The former are *three-dimensional units,* by which is meant that in the electrostatic system all electrical and magnetic quantities can be expressed in terms of not more than *three* base units. In SI, however, which is based on the electromagnetic system, some quantities have to be expressed in terms of four base units, namely metre, kilogram, second, and ampere. The SI units of electricity and magnetism are therefore said to be *four dimensional.* An example is the ohm, which in terms of base SI units is kg m^2 s^{-3} A^{-2}. Because of this dimensional difference there is not in general a one-to-one correspondence between SI and electrostatic units.

Magnetic Units

As far as magnetic field is concerned, the property that is of most fundamental importance (e.g., from the standpoint of the effects of magnetic fields on magnetic moments arising from electron and nuclear spin) is the *magnetic flux density,* or *magnetic induction,* which is given the symbol B. The SI unit of this quantity is the tesla (T), which is kg s^{-2} A^{-1} or V s m^{-2}. The old unit is the gauss (G), which is 10^{-4} T.

The *magnetic flux* Φ is the magnetic flux density multiplied by the area through which the magnetic field is passing, and its SI unit is therefore

$$\mathrm{T\ m^2 \equiv kg\ m^2\ s^{-1}\ A^{-1} \equiv V\ s}$$

This unit is given a special name, the weber (Wb). The old unit of magnetic flux is the maxwell ($\equiv$ gauss $\times$ cm^2), equal to 10^{-8} Wb.

Another magnetic quantity that is frequently encountered is the *magnetic field strength H.* Its SI unit is A m^{-1}. The magnetic field strength is related to the magnetic flux density B by the equation

$$\boldsymbol{H} = \frac{\boldsymbol{B}}{\mu} \tag{A.9}$$

where μ is the *permeability* of the medium through which the field is passing. The SI unit of permeability is

$$\frac{\mathrm{kg\ s^{-2}\ A^{-1}}}{\mathrm{A\ m^{-1}}} \equiv \mathrm{kg\ m\ s^{-2}\ A^{-2}}$$

A special unit, the henry, is the unit of *inductance* and it is defined as kg m^2 s^{-2} A^{-2}; it is given the symbol H. The unit of permeability can therefore be written as H m^{-1}. If the field is passing through a vacuum, the permeability is given the symbol μ_0. The *permeability of a vacuum* has the value of $4\pi \times 10^{-7}$ H m^{-1} exactly.

Physical Constants

APPENDIX B

For each physical quantity, the one-standard-deviation uncertainty is given by the last two digits in parentheses.

These values are from P. J. Mohr and B. N. Taylor, *J. Phys. Chem.* Ref. Data *28*, 1713(1999) and *Rev. Med. Phys. 72*, 351(2000).

Quantity	Symbol	Value
Avogadro constant	L	$6.022\ 141\ 99\ (47) \times 10^{23}\ \text{mol}^{-1}$
Elementary charge	e	$1.602\ 176\ 462\ (63) \times 10^{-19}\ \text{C}$
Gas constant	R	$8.314\ 472\ (15)\ \text{J K}^{-1}\ \text{mol}^{-1}$
Boltzmann constant	$k_B \equiv R/L$	$1.380\ 650\ 3\ (24) \times 10^{-23}\ \text{J K}^{-1}$
Planck constant	h	$6.626\ 068\ 76\ (52) \times 10^{-34}\ \text{J s}$
	$\hbar = h/2\pi$	$1.054\ 571\ 596\ (82) \times 10^{-34}\ \text{J s}$
Speed of light in vacuum	c	$299\ 792\ 458\ \text{m s}^{-1}$ exactly
Zero of the Celsius scale	T_0	273.15 K exactly
Faraday constant	$F \equiv Le$	$9.648\ 534\ 15\ (39) \times 10^{4}\ \text{C mol}^{-1}$
Permittivity of vacuum	$\epsilon_0 \equiv 1/\mu_0 c^2$	$8.854\ 187\ 817\ldots \times 10^{-12}\ \text{F m}^{-1}$
Permeability of vacuum	μ_0	$4\pi \times 10^{-7}\ \text{H m}^{-1}$ exactly
Rest mass of electron	m_e	$9.109\ 381\ 88\ (72) \times 10^{-31}\ \text{kg}$
Rest mass of proton	m_p	$1.672\ 621\ 58\ (13) \times 10^{-27}\ \text{kg}$
Rest mass of neutron	m_n	$1.674\ 928\ 6\ (10) \times 10^{-27}\ \text{kg}$
Rydberg constant	R	$1.097\ 373\ 156\ 854\ 9\ (83) \times 10^{7}\ \text{m}^{-1}$
Hartree energy	E_h	$4.359\ 743\ 81\ (34) \times 10^{-18}\ \text{J}$
Bohr radius	a_0	$5.291\ 772\ 083\ (19) \times 10^{-11}\ \text{m}$
Bohr magneton	μ_B	$9.274\ 008\ 99\ (37) \times 10^{-24}\ \text{J T}^{-1}$
Nuclear magneton	μ_N	$5.050\ 783\ 17\ (20) \times 10^{-27}\ \text{J T}^{-1}$
Electron magnetic moment	μ_0	$9.284\ 763\ 62\ (37) \times 10^{-24}\ \text{J T}^{-1}$
Landé g factor for free electron	g_e	$2.002\ 319\ 304\ 373\ 7\ (82)$

(*Continued*)

Quantity	Symbol	Value
Proton gyromagnetic ratio	γ_P	$2.675\ 222\ 12\ (11) \times 10^8\ s^{-1}\ T^{-1}$
Standard atmospheric pressure	P_0	$1.013\ 25 \times 10^5$ Pa exactly
Molar volume, ideal gas,		
$P = 1$ bar, $\theta = 25$ °C		$24.789\ 71\ (21)\ L\ mol^{-1}$
$P = 101.325$ kPa, $T = 273.15$ K		$22.413\ 996\ (39) \times 10^{-3}\ m^3\ mol^{-1}$
Standard acceleration of gravity	g_n	$9.806\ 65\ m\ s^{-2}$ exactly
Ratio of circumference to diameter of circle	π	3.141 592 653 6
Base of natural logarithms	e	2.718 281 828

Some Mathematical Relationships

APPENDIX C

In this appendix we list, without proof, some mathematical relationships that are used in this book.[1]

Differentials

$$\frac{d}{dx}(yz) = y\frac{dz}{dx} + z\frac{dy}{dx} \qquad \frac{d}{dx}\left(\frac{y}{z}\right) = \frac{z(dy/dx) - y(dz/dx)}{z^2}$$

$$\frac{df(y)}{dx} = \frac{df(y)}{dy}\cdot\frac{dy}{dx}$$

$$\frac{de^x}{dx} = e^x \qquad \frac{de^y}{dx} = e^y\frac{dy}{dx}$$

$$\frac{d\ln x}{dx} = \frac{1}{x} \qquad \frac{d\ln y}{dx} = \frac{1}{y}\cdot\frac{dy}{dx}$$

Integrals

If an integration is performed, and no additional information is provided, the result involves a constant of integration I. For example,

$$\int x^2\,dx = \frac{x^3}{3} + I \tag{C.1}$$

The constant of integration can be evaluated if the value of the integral is known for a particular value of the variable x. An integral of the kind shown in Eq. C.1 is known as an *indefinite integral*.

Alternatively, the integral may be a *definite integral,* which means that the integration is performed between certain limiting values of the variable. For example, the definite integral

$$\int_b^a x^2\,dx$$

[1]There are many sources of additional mathematical information; for example, CRC *Handbook of Chemistry and Physics* (82nd ed.), CRC Press, Boca Raton, FL, 2001; revised editions appear annually.

means that x can have an upper limit of a and a lower limit of b. The solution is

$$\int_b^a x^2\,dx = \left.\frac{x^3}{3}\right|_b^a = \frac{a^3 - b^3}{3} \tag{C.2}$$

Some useful integrals are listed here:

$$\int x^n\,dx = \frac{x^{n+1}}{n+1} + I \quad \text{if } n \neq -1 \qquad \int \frac{dx}{x} = \ln x + I$$

$$\int e^{ax}\,dx = \frac{1}{a}e^{ax} + I \qquad \int \sin x\,dx = -\cos x + I$$

$$\int \cos x\,dx = \sin x + I \qquad \int x\,dy = xy - \int y\,dx \quad \text{(integration by parts)}$$

$$\int \ln ax\,dx = x \ln ax - x + I$$

Definite integrals useful for statistical problems are to be found in the appendix to Chapter 1 (p. 43).

Integration of Differential Equations

In most of the differential equations in this book, the variables can be separated and the integrations can therefore be performed independently. Suppose, for example, that we have the differential equation

$$a\frac{dx}{x} + by\,dy = 0 \tag{C.3}$$

where x and y are variables and a and b are constants. The variables are already separated and integration proceeds as follows:

$$a\int \frac{dx}{x} + b\int y\,dy = 0 \tag{C.4}$$

$$a \ln x + \frac{by^2}{2} = I \tag{C.5}$$

The constant of integration can be obtained from a known value of y at some known value of x. This is often a *boundary condition* (i.e., the value of one variable is known when the other is zero or infinity or has some other "boundary" value).

For examples of the integration of differential equations, see Eqs. 2.84–2.86 and 9.32–9.35.

Partial Derivatives

If a function z depends on two or more variables, the *partial derivative* relates to the dependence of z on one variable, with all other variables held constant. For example, if $z(x, y)$ is a function of x and y, the partial derivative

$$\left(\frac{\partial z}{\partial x}\right)_y$$

expresses the dependence of z on x when y is held constant.

In such a case, the differential dz can be expressed as

$$dz = \left(\frac{\partial z}{\partial x}\right)_y dx + \left(\frac{\partial z}{\partial y}\right)_x dy \tag{C.6}$$

Figure C.1 gives an interpretation of this relationship. Examples of its use are to be found in Eqs. 2.113–2.114 and 3.112–3.115.

According to *Euler's reciprocity theorem,* in the case of exact differentials (e.g., state functions), the order of differentiation does not matter. Thus, if z is a function of the two variables x and y,

$$\left[\frac{\partial}{\partial x}\left(\frac{\partial z}{\partial y}\right)_x\right]_y \equiv \left[\frac{\partial}{\partial y}\left(\frac{\partial z}{\partial x}\right)_y\right]_x \tag{C.7}$$

These double differentials may be written more compactly as

$$\frac{\partial^2 z}{\partial x\, \partial y} = \frac{\partial^2 z}{\partial y\, \partial x} \tag{C.8}$$

Examples of the use of this theorem are to be found in Eqs. 3.121–3.125. Another important relationship is that

$$\left(\frac{\partial z}{\partial x}\right)_y = \frac{1}{(\partial x/\partial z)_y} \tag{C.9}$$

If z is a function of x and y, Eq. C.6 applies, and we can hold z constant and divide by dx:

$$0 = \left(\frac{\partial z}{\partial x}\right)_y + \left(\frac{\partial z}{\partial y}\right)_x \left(\frac{\partial y}{\partial x}\right)_z \tag{C.10}$$

FIGURE C.1

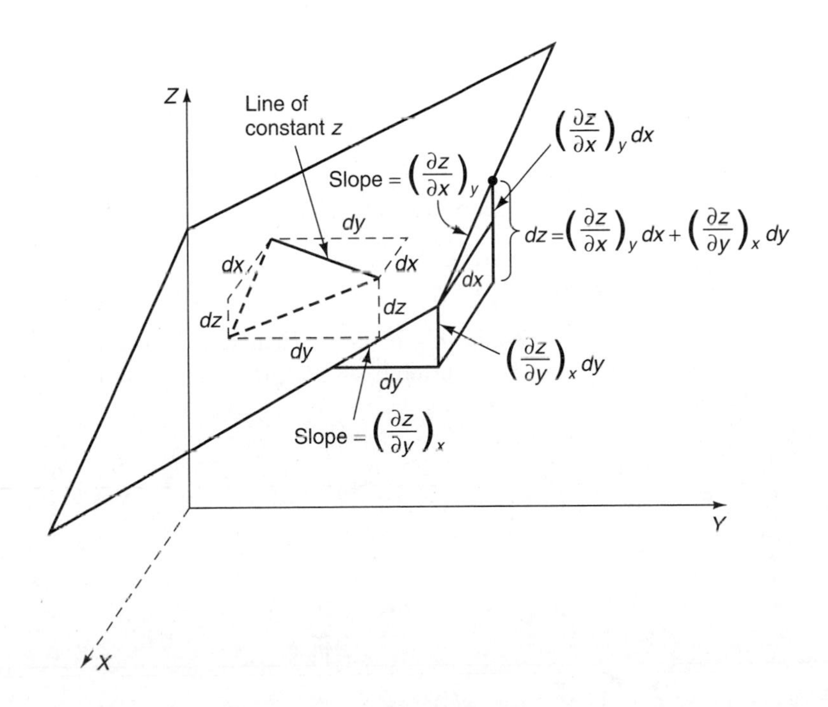

Then applying Eq. C.9,

$$\left(\frac{\partial x}{\partial y}\right)_z\left(\frac{\partial y}{\partial z}\right)_x\left(\frac{\partial z}{\partial x}\right)_y = -1 \tag{C.11}$$

or

$$\left(\frac{\partial z}{\partial x}\right)_y = -\left(\frac{\partial z}{\partial y}\right)_x\left(\frac{\partial y}{\partial x}\right)_z \tag{C.12}$$

This is *Euler's chain relationship.* An interpretation of it is also shown in Figure C.1.

Exact and Inexact Differentials

Some functions such as U and H depend only on the state of the system. The integral of the differential of such a state function is the difference between the values at the two limits. The value of the integration is path independent and the differential is said to be *exact.*

If the integral of a differential depends on the path chosen in going from A to B, the differential is said to be *inexact.* The symbol đ is used to indicate an inexact differential.

Proof that a differential is exact is based on Euler's *criterion for exactness.* The total differential dz of z is related to the differentials dx and dy by the equation

$$dz = M(x, y)\, dx + N(x, y)\, dy \tag{C.13}$$

where M and N are functions of the variables x and y. If z has a definite value at each point in the X-Y plane, it also must be a function of x and y. This allows us to write

$$dz = \left(\frac{\partial z}{\partial x}\right)_y dx + \left(\frac{\partial z}{\partial y}\right)_x dy \tag{C.14}$$

which is the same as Eq. C.6. Comparing Eqs. C.13 and C.14 shows that

$$M(x, y) = \left(\frac{\partial z}{\partial x}\right)_y; \quad N(x, y) = \left(\frac{\partial z}{\partial y}\right)_x \tag{C.15}$$

Since mixed second derivatives are equal from Eq. C.7, then

$$\left(\frac{\partial M}{\partial y}\right)_x = \left(\frac{\partial N}{\partial x}\right)_y \tag{C.16}$$

This equation must be satisfied if dz is an exact differential. Thus, $dz = y\, dx$ is not an exact differential, but the sum of two inexact differentials might be. For example, $dz = y\, dx + x\, dy$ is an exact differential.

Standard Enthalpies, Entropies, and Gibbs Energies of Formation[1]

APPENDIX D

Standard states: 25.00 °C; 1 bar pressure[2] unless otherwise noted; 1 *m* for substances in aqueous solution (aq): g = gas, s = solid, l = liquid. Also see next table for CODATA Thermodynamic Data. Data in this table are not repeated in the next.[3] The data presented here are from the *Handbook of Chemistry and Physics*, 82nd edition, D. Lide, Ed., CPC Press, Boca Raton, FL, 2001. See pp. 1023 and 1027 for additional sources.

INORGANIC SUBSTANCES

	$\Delta_f H°$ kJ mol^{-1}	$\Delta_f G°$ kJ mol^{-1}	$S°$ J K^{-1} mol^{-1}
Aluminum			
$AlCl_3$(s)	−704.2	−628.8	109.3
Al_2Cl_6(g)	−1290.8	−1220.4	490.0
*$Al(OH)_3$(s)	−1276	—	—
*$Al_2(SO_4)_3$(s)	−3441	−3100	239
Barium			
Ba(s)	0	0	62.5
*Ba^{2+}(aq)	−537.6	−560.8	9.6
*$BaCO_3$(s)	−1216	−1138	112
$BaCl_2$(s)	−855.0	−806.7	123.7
BaF_2(s)	−1207.1	−1156.8	96.4
BaO(s)	−548.0	−520.3	72.1
$Ba(OH)_2$(s)	−944.7	−859.4	107
*$Ba(OH)_2 \cdot 8H_2O$(s)	−3342	−2793	427
$BaSO_4$(s)	−1473.2	−1362.2	132.2
Beryllium			
$BeCl_2$(s)	−490.4	−445.6	75.8
BeF_2(s)	−1026.8	−979.4	53.4
Bismuth			
Bi(s)	0	0	56.7
$BiCl_3$(s)	−379.1	−315.0	177.0
Bi_2O_3(s)	−573.9	−493.7	151.5
Boron			
BCl_3(l)	−427.2	−387.4	206.3
*B_2H_6(g)	36.4	86.7	232.1
Bromine			
BrCl(g)	14.6	−0.96	240.1
BrF_3(g)	−255.6	−229.4	292.5
BrF_3(l)	−300.8	−240.5	178.2
Cadmium			
*Cd^{2+}(aq)	−75.90	−77.61	−73.2
$CdCl_2$(s)	−391.5	−343.9	115.3
CdO(s)	−258.4	−228.7	54.8
Calcium			
*$CaBr_2$(s)	−682.8	−663.6	130
*$CaCO_3$(s)	−1207	−1128	88.70
$CaCl_2$(s)	−795.4	−748.8	108.4
CaF_2(s)	−1228.0	−1135.6	68.5
CaI_2(s)	−181.5	−142.5	41.4
$Ca(NO_3)_2$(s)	−938.2	−742.8	193.2
CaO(s)	−634.9	−603.3	38.1
$Ca(OH)_2$(s)	−985.2	−897.5	83.4
$Ca_3(PO_4)_2$(s)	−4120.8	−3884.7	236.0
$CaSO_4$(s)	−1434.5	1322.0	106.5

[1]A listing of absolute entropies, $S°$, at 25 °C and 1 bar pressure, is given in Table 3.2, p. 122.
[2]Before 1982 the standard pressure was 1 atmosphere (1.01325×10^5 Pa); in that year the IUPAC Commission on Thermodynamics recommended that the standard be changed to 1 bar (exactly 10^5 Pa). All data given in this book relate to 1 bar unless otherwise noted. Data relating to 1 atm can be converted to 1 bar by a procedure clearly explained in R. D. Freeman, "Conversion of standard thermodynamic data to the new standard-state pressure," *J. Chem. Ed., 52,* 681–686(1985). The changes are always small, often within the experimental error.
[3]An asterisk on the left of the chemical entry indicates all entries are referenced to 1 atm pressure. Numerical data marked with an asterisk are at 1 atm. All other data are at 1 bar.

INORGANIC SUBSTANCES

	$\Delta_f H°$ kJ mol^{-1}	$\Delta_f G°$ kJ mol^{-1}	$S°$ J K^{-1} mol^{-1}
Carbon (See also the table of organic substances.)			
C (diamond)	1.90	2.90	2.38
C (graphite)	0	0	5.74
$CCl_4(g)$	−95.7	−60.63	309.7
$CCl_4(l)$	−128.2	−65.27	216.2
$C_2N_2(g)$	306.7	297.2	241.9
$C_3O_2(g)$	−93.72	−109.8	276.4
$C_3O_2(l)$	−117.3	−105.0	181.1
$COCl_2(g)$	−219.1	−204.9	283.5
$COS(g)$	−142.0	−169.2	231.5
$CS_2(l)$	89.0	64.6	151.3
Chlorine			
$ClF_3(g)$	−163.2	−123.0	281.6
$ClO_2(g)$	102.5	120.5	256.8
$Cl_2O(g)$	80.3	97.9	266.2
Chromium			
Cr(s)	0	0	23.8
$Cr_2O_3(s)$	−1139.7	−1058.1	81.2
*$CrO_4^{2-}(aq)$	−881.2	−727.8	50.21
*$Cr_2O_7^{2-}(aq)$	−1490	−1301	261.9
Cobalt			
Co(s)	0	0	30.0
CoO(s)	−237.9	−214.2	53.0
$Co(OH)_2$ (pink solid)	−539.7	−454.3	79.0
Copper			
*$CuCO_3 \cdot Cu(OH)_2(s)$	−1051	−893.7	186
$Cu_2O(s)$	−168.6	−146.0	93.1
CuO(s)	−157.3	−129.7	42.6
$Cu(OH)_2(s)$	−449.8	−373	108
*$CuSO_4 \cdot 5H_2O(s)$	−2279.6	−1880.1	300.4
Hydrogen			
$H^+(aq)$	0	0	0
$H_2(g)$	0	0	130.7
HCl(aq)	−167.2	−131.3	56.48
HCN(g)	135	125	201.7
$HNO_3(l)$	−174.1	−80.7	155.6
*$HNO_3(aq)$	−207.4	−113.3	146.4
$H_2O(g)$	−241.826	−228.6	188.8
$H_2O(l)$	−285.830	−237.2	69.95
$H_2O_2(g)$	−136.3	−105.6	232.7
$H_2O_2(l)$	−187.8	−120.4	110
$H_2S(g)$	−20.6	−33.4	205.8
$H_2SO_4(l)$	−814.0	−690.0	156.9
*$H_2SO_4(aq)$	−909.3	−744.6	20.08
Iodine			
IBr(g)	40.8	3.7	258.8
ICl(g)	17.8	−5.5	247.6
ICl(l)	−23.9	−13.6	135.1
Iron			
Fe(s)	0	0	27.3
$Fe^{2+}(aq)$	−89.1	−78.90	−137.7
$Fe^{3+}(aq)$	−48.5	−4.7	−315.9
$FeCO_3(s)$	−740.6	−666.7	92.88
$FeCl_3(s)$	−399.5	−334.0	142.3
FeO(s)	−272.0	−251.5	60.75
$Fe_2O_3(s)$	−824.2	−742.2	87.4
$Fe_3O_4(s)$	−1118.4	−1015.4	146.4
$Fe(OH)_3(s)$	−823.0	−696.6	107
Lead			
$PbO_2(s)$	−277.4	−217.3	68.6
$PbSO_4(s)$	−920.0	−813.0	148.5
Lithium			
LiCl(s)	−408.6	−384.4	59.3
$Li_2O(s)$	−597.9	−561.2	37.6
*LiOH(s)	−484.9	−439.0	42.8
*$LiNO_3(s)$	−483.1	−381.1	90.0
Magnesium			
$MgCl_2(s)$	−641.3	−591.8	89.6
$MgCO_3(s)$	−1096	−1012	65.7
$MgF_2(s)$	−1124.2	−1071.1	57.2
$Mg(OH)_2(s)$	−924.5	−833.5	63.2
$MgSO_4(s)$	−1284.9	−1170.6	91.6
Manganese			
*Mn(s)	0	0	32.0
*$Mn^{2+}(aq)$	−220.8	−228.1	−73.6
$MnO_2(s)$	−520.0	−465.0	53.1
$MnO_4^-(aq)$	−541.4	−447.2	191.2
Nitrogen			
$N_2(g)$	0	0	191.6
$NF_3(g)$	−132.1	−90.6	260.8
$NH_3(aq)$	−80.29	−26.57	111.3
$NH_4Br(s)$	−270.8	−175.2	113.0
$NH_4Cl(s)$	−314.4	−203.0	94.56
$NH_4F(s)$	−464.0	−348.8	71.96
$NH_4HCO_3(s)$	−849.4	−666.1	121
$NH_4I(s)$	−201.4	−112.5	117.0
$NH_4NO_3(s)$	−365.6	−183.9	151.1
*$NH_4NO_3(aq)$	−339.9	−190.7	259.8

	$\Delta_f H°$ kJ mol^{-1}	$\Delta_f G°$ kJ mol^{-1}	$S°$ J K^{-1} mol^{-1}
$(NH_4)_2SO_4$(s)	−1180.9	−901.7	220.1
N_2H_4(g)	95.4	159.4	238.5
N_2H_4(l)	50.6	149.3	121.2
NO(g)	91.3	87.6	210.8
N_2O(g)	82.05	104.2	219.7
NO_2(g)	33.2	51.3	240.1
N_2O_4(g)	11.1	97.82	304.2
N_2O_4(l)	−19.5	97.40	209.2
N_2O_5(g)	11.3	115.1	355.7
NOBr(g)	82.2	82.4	273.7
NOCl(g)	51.71	66.07	261.6
Oxygen			
O_3(g)	142.7	163.2	238.8
OF_2(g)	24.5	41.8	247.3
Phosphorus			
P (red)	−17.6	−12.1	22.8
PCl_3(g)	−287.0	−267.8	311.7
PCl_3(l)	−319.7	−272.3	217.1
PCl_5(g)	−374.9	−305.0	364.5
PCl_5(s)	−443.5	—	—
PH_3(g)	5.4	13.4	210.1
P_4O_{10}(s)	−2984	−2698	228.9
PO_4^{3-}(aq)	−1277	−1019	−222
Potassium			
K(l)	2.28	0.26	71.46
K(s)	0	0	64.18
KBr(s)	−393.8	−380.7	95.9
KCN(s)	−113	−101.9	128.5
KCl(s)	−436.5	408.5	82.6
$KClO_3$(s)	−397.7	−296.3	143.1
*$KClO_4$(s)	−432.8	−303.2	151.0
KF(s)	−567.3	−537.8	66.6
KI(s)	−327.9	−324.9	106.3
KNO_3(s)	−494.6	−394.9	133.1
KOH(s)	−424.6	−378.7	78.89
KOH(aq)	−482.4	−440.5	91.63
K_2SO_4(s)	−1437.8	−1321.4	175.6
Silicon			
SiH_4(g)	34.3	56.9	204.6
Si_2H_6(g)	80.3	127.3	272.7
Silver			
*AgBr(s)	−100.4	−96.90	107
*AgI(s)	−61.84	−66.19	115
*$AgNO_3$(s)	−124.4	−33.5	140.9
*Ag_2O(s)	−31.0	−11.2	121
*Ag_2SO_4(s)	−715.9	−618.5	200.4

	$\Delta_f H°$ kJ mol^{-1}	$\Delta_f G°$ kJ mol^{-1}	$S°$ J K^{-1} mol^{-1}
Sodium			
*Na(l)	2.41	0.50	57.86
Na_2(g)	142.1	103.9	230.2
NaBr(s)	−361.1	−349.0	86.8
Na_2CO_3(s)	−1131	−1044	135.0
$NaHCO_3$(s)	−950.8	−851.0	102
NaCl(s)	−411.2	−384.1	72.1
*NaCl(aq)	−407.3	−393.1	115.5
$NaClO_3$(s)	−365.8	−262.3	123.4
$NaClO_4$(s)	−383.3	−254.9	142.3
NaF(s)	−576.6	−546.3	51.1
NaH(s)	−56.3	−33.5	40.0
NaI(s)	−287.8	−286.1	98.5
$NaNO_3$(s)	−467.9	−367.1	116.5
$NaNO_3$(aq)	−447.4	−373.2	205.4
Na_2O_2(s)	−510.9	447.7	95.0
NaOH(s)	−425.6	−379.5	64.48
NaOH(aq)	−469.2	−419.2	48.1
NaH_2PO_4(s)	−1537	−1386	127.5
Na_2HPO_4(s)	−1748	−1608	150.5
Na_3PO_4(s)	−1917	−1789	173.8
$NaHSO_4$(s)	−1125	−992.9	113
Na_2SO_4(s)	−1387	−1270.2	149.6
Na_2SO_4(aq)	−1387.1	−1268	138.1
$Na_2SO_4 \cdot 10H_2O$(s)	−4327	−3647	592.0
$Na_2S_2O_3$(s)	−1123	−1028	155
Sulfur			
*S_8(g)	102.3	49.16	430.2
*S_2Cl_2(g)	−18.4	−31.8	331.5
SF_6(g)	−1209	−1105	291.7
*SO_2(g)	−296.8	−300.1	248.2
*SO_2(g)	−395.7	−371.1	256.8
*SO_4^{2-}(aq)	−909.3	−744.5	20.1
*$S_2O_3^{2-}$(aq)	−648.5	−522.5	67
*SO_2Cl_2(g)	−364.0	−320.0	311.9
SO_2Cl_2(l)	−394.1	−314	207
Tin			
Sn (gray)	−2.1	0.1	44.1
$SnCl_4$(l)	−511.3	−440.1	258.6
Titanium			
$TiCl_4$(l)	−804.2	−737.2	252.3
Uranium			
UF_6(g)	−2147.4	−2063.7	377.9
UF_6(s)	−2197.0	−2068.5	227.6
Zinc			
$ZnCl_2$(s)	−415.1	−369.4	111.5

SOME ORGANIC SUBSTANCES

Formula	Name	$\Delta_f H°$ kJ mol^{-1}	$\Delta_f G°$ kJ mol^{-1}	$S°$ J K^{-1} mol^{-1}
$CH_4(g)$	Methane(g)	−74.6	−50.5	186.3
$C_2H_2(g)$	Ethyne(g)	227.4	209.9	200.9
$C_2H_4(g)$	Ethene(g)	52.4	68.4	219.3
$C_2H_6(g)$	Ethane(g)	−84.0	−32.0	229.2
$C_3H_8(g)$	Propane(g)	−103.8	−23.4	270.3
$C_4H_{10}(g)$	Butane(g)	−125.7	−17.15*	310.1*
$C_6H_6(g)$	Benzene(g)	82.9	129.7	269.2
$C_6H_6(l)$	Benzene(l)	49.1	124.5	173.4
$C_6H_{12}(g)$	Cyclohexane(g)	−123.4	31.8*	298.2*
$C_6H_{12}(l)$	Cyclohexane(l)	−156.2	26.7*	204.3*
$C_{10}H_8(g)$	Naphthalene(g)	150.6	224.1	333.1
$C_{10}H_8(s)$	Naphthalene(s)	78.5	201.6	167.4
$CH_2O(g)$	Formaldehyde(g)	−108.6	−102.5	218.8
$CH_3OH(g)$	Methanol(g)	−201.0	−162.3	239.9
$CH_3OH(l)$	Methanol(l)	−239.2	−166.6	126.8
$CH_3CHO(g)$	Acetaldehyde(g)	−166.2	−133.0	263.8
$CH_3CHO(l)$	Acetaldehyde(l)	−192.2	−127.6	160.2
$CH_3CH_2OH(g)$	Ethanol(g)	−234.8	−167.9	281.6
$CH_3CH_2OH(l)$	Ethanol(l)	−277.6	−174.8	160.7
$C_6H_5OH(s)$	Phenol(s)	−165.1	−50.42*	144.0
$(CH_3)_2CO(g)$	Acetone(g)	−217.1	−152.7	295.3
$(CH_3)_2CO(l)$	Acetone(l)	−248.4	−155.7*	199.8
$CH_3COOH(g)$	Acetic acid(g)	−432.2	−374.2	283.5
$CH_3COOH(l)$	Acetic acid(l)	−484.3	−389.9	159.8
*$CH_3COOH(aq)$	Acetic acid(aq)	−488.3	−396.6	178.7
$C_6H_5COOH(s)$	Benzoic acid(s)	−385.2	−245.3*	167.6
$CH_3NH_2(g)$	Methylamine(g)	−22.5	32.7	242.9
$C_6H_5NH_2(g)$	Aniline(g)	87.5	166.7*	317.9
$C_6H_5NH_2(l)$	Aniline(l)	31.6	149.1*	191.3*
$C_6H_{12}O_6(s)$	Glucose(s)	−1273.3	−910.4*	212.1*

CODATA Thermodynamic Data at 1 Bar Pressure

This table of thermodynamic data is derived from the work of the Committee on Data for Science and Technology (CODATA) to establish internationally agreed values for properties of chemical substances. A value of 0 in the $\Delta_f H°$ column of an element indicates the reference state for that element. The standard pressure is 100 000 Pa (1 bar). See the original work cited in the next paragraph for such details as the dependence of gas-phase entropy values on the choice of standard-state pressure.

The original work of the Committee appeared in: J. D. Cox, D. D. Wagman, and V. A. Medvedev, *CODATA Key Values for Thermodynamics*, New York: Hemisphere Publishing Corp., 1989.

The data under $\Delta_f G°(298.15)$ are taken from the *Handbook of Chemistry and Physics*, 82nd edition, cited in the previous section, and are at 1 bar pressure. Some data, indicated by an asterisk, are for 1 atm (101.325 kPa). This table does not repeat any entries given in the preceding table.

Substance	$\Delta_f H°$ kJ mol^{-1}	$S°$ J K^{-1} mol^{-1}	$H°-H°(0)$ kJ mol^{-1}	$\Delta_f G°$ kJ mol^{-1}
Ag(s)	0	42.55 ± 0.20	5.745 ± 0.020	0
Ag(g)	284.9 ± 0.8	172.997 ± 0.004	6.197 ± 0.001	
Ag^+(aq)	105.79 ± 0.08	73.45 ± 0.40		
AgCl(s)	−127.01 ± 0.05	96.25 ± 0.20	12.003 ± 0.020	−109.8
Al(s)	0	28.30 ± 0.10	4.540 ± 0.020	0
Al(g)	330.0 ± 4.0	164.554 ± 0.004	6.919 ± 0.001	
Al^{+3}(aq)	−538.4 ± 1.5	−325 ± 10		
AlF_3(s)	−1510.4 ± 1.3	66.5 ± 0.5	11.62 ± 0.04	−1431.1
Al_2O_3 (s, corundum)	−1675.7 ± 1.3	50.92 ± 0.10	10.016 ± 0.020	−1582.3
Ar(g)	0	154.846 ± 0.003	6.197 ± 0.001	0
B (s, rhombic)	0	5.90 ± 0.08	1.222 ± 0.008	0
B(g)	565 ± 5	153.436 ± 0.015	6.316 ± 0.002	521.0
BF_3(g)	−1136.0 ± 0.8	254.42 ± 0.20	11.650 ± 0.020	−1119.4
B_2O_3(s)	−1273.5 ± 1.4	53.97 ± 0.30	9.301 ± 0.040	−1194.3
Be(s)	0	9.50 ± 0.08	1.950 ± 0.020	0
Be(g)	324 ± 5	136.275 ± 0.003	6.197 ± 0.001	286.6
BeO(s)	−609.4 ± 2.5	13.77 ± 0.04	2.837 ± 0.008	0
Br(g)	111.87 ± 0.12	175.018 ± 0.004	6.197 ± 0.001	82.4
Br(aq)	−121.41 ± 0.15	82.55 ± 0.20		
Br_2(l)	0	152.21 ± 0.30	24.52 ± 0.01	0
Br_2(g)	30.91 ± 0.11	245.468 ± 0.005	9.725 ± 0.001	
C (s, graphite)	0	5.74 ± 0.10	1.050 ± 0.020	0
C(g)	716.68 ± 0.45	158.100 ± 0.003	6.536 ± 0.001	
CO(g)	−110.53 ± 0.17	197.660 ± 0.004	8.671 ± 0.001	
CO_2(g)	−393.51 ± 0.13	213.785 ± 0.010	9.365 ± 0.003	
CO_2 (aq, undissoc.)	−413.26 ± 0.20	119.36 ± 0.60		
CO_3^{-2}(aq)	−675.23 ± 0.25	−50.0 ± 1.0		
Ca(s)	0	41.59 ± 0.40	5.736 ± 0.040	0
Ca(g)	177.8 ± 0.8	154.887 ± 0.004	6.197 ± 0.001	144.0

Substance	$\Delta_f H°$ kJ mol^{-1}	$S°$ J K^{-1} mol^{-1}	$H°-H°(0)$ kJ mol^{-1}	$\Delta_f G°$ kJ mol^{-1}
Ca^{+2}(aq)	−543.0 ± 1.0	−56.2 ± 1.0		
CaO(s)	−634.92 ± 0.90	38.1 ± 0.4	6.75 ± 0.06	−603.3
Cd(s)	0	51.80 ± 0.15	6.247 ± 0.015	0
Cd(g)	111.80 ± 0.20	167.749 ± 0.004	6.197 ± 0.001	51.8
Cd^{+2}(aq)	−75.92 ± 0.60	−72.8 ± 1.5		
CdO(s)	−258.35 ± 0.40	54.8 ± 1.5	8.41 ± 0.08	−228.7
$CdSO_4 \cdot 8/3H_2O$(s)	−1729.30 ± 0.80	229.65 ± 0.40	35.56 ± 0.04	
Cl(g)	121.301 ± 0.008	165.190 ± 0.004	6.272 ± 0.001	105.3
Cl^-(aq)	−167.080 ± 0.10	56.60 ± 0.20		
ClO_4^-(aq)	−128.10 ± 0.40	184.0 ± 1.5		
Cl_2(g)	0	223.081 ± 0.010	9.181 ± 0.001	0
Cs(s)	0	85.23 ± 0.40	7.711 ± 0.020	0
Cs(g)	76.5 ± 1.0	175.601 ± 0.003	6.197 ± 0.001	49.6
Cs^+(aq)	−258.00 ± 0.50	132.1 ± 0.5		
Cu(s)	0	33.15 ± 0.08	5.004 ± 0.008	0
Cu(g)	337.4 ± 1.2	166.398 ± 0.004	6.197 ± 0.001	297.7
Cu^{+2}(aq)	64.9 ± 1.0	−98 ± 4		
$CuSO_4$(s)	−771.4 ± 1.2	109.2 ± 0.4	16.86 ± 0.08	−662.2
F(g)	79.38 ± 0.30	158.751 ± 0.004	6.518 ± 0.001	62.3
F^-(aq)	−335.35 ± 0.65	−13.8 ± 0.8		
F_2(g)	0	202.791 ± 0.005	8.825 ± 0.001	0
Ge(s)	0	31.09 ± 0.15	4.636 ± 0.020	0
Ge(g)	372 ± 3	167.904 ± 0.005	7.398 ± 0.001	331.2
GeF_4(g)	−1190.20 ± 0.50	301.9 ± 1.0	17.29 ± 0.10	−1150.0
GeO_2 (s, tetragonal)	−580.0 ± 1.0	39.71 ± 0.15	7.230 ± 0.020	−521.4
H(g)	217.998 ± 0.006	114.717 ± 0.002	6.197 ± 0.001	203.3
H^+(aq)	0	0		0
HBr(g)	−36.29 ± 0.16	198.700 ± 0.004	8.648 ± 0.001	−53.4
HCO_3^-(aq)	−689.93 ± 0.20	98.4 ± 0.5		
HCl(g)	−92.31 ± 0.10	186.902 ± 0.005	8.640 ± 0.001	−95.3
HF(g)	−273.30 ± 0.70	173.779 ± 0.003	8.599 ± 0.001	−275.4
HI(g)	26.50 ± 0.10	206.590 ± 0.004	8.657 ± 0.001	1.7
HPO_4^{-2}(aq)	−1299.0 ± 1.5	−33.5 ± 1.5		
HS^-(aq)	−16.3 ± 1.5	67 ± 5		
HSO_4^-(aq)	−886.9 ± 1.0	131.7 ± 3.0		
H_2(g)	0	130.680 ± 0.003	8.468 ± 0.001	0
H_2O(l)	−285.830 ± 0.040	69.95 ± 0.03	13.273 ± 0.020	−237.1
H_2O(g)	−241.826 ± 0.040	188.835 ± 0.010	9.905 ± 0.005	−228.6
$H_2PO_4^-$(aq)	−1302.6 ± 1.5	92.5 ± 1.5		
H_2S(g)	−20.6 ± 0.5	205.81 ± 0.05	9.957 ± 0.010	−33.4
H_2S (aq, undissoc.)	−38.6 ± 1.5	126 ± 5		
H_3BO_3(s)	−1094.8 ± 0.8	89.95 ± 0.60	13.52 ± 0.04	−968.9
H_3BO_3 (aq, undissoc.)	−1072.8 ± 0.8	162.4 ± 0.6		
He(g)	0	126.153 ± 0.002	6.197 ± 0.001	0
Hg(l)	0	75.90 ± 0.12	9.342 ± 0.008	0
Hg(g)	61.38 ± 0.04	174.971 ± 0.005	6.197 ± 0.001	31.8
Hg^{+2}(aq)	170.21 ± 0.20	−36.19 ± 0.80		

Substance	$\Delta_f H°$ kJ mol^{-1}	$S°$ J K^{-1} mol^{-1}	$H°-H°(0)$ kJ mol^{-1}	$\Delta_f G°$ kJ mol^{-1}
HgO (s, red)	−90.79 ± 0.12	70.25 ± 0.30	9.117 ± 0.025	−58.5
Hg_2^{+2}(aq)	166.87 ± 0.50	65.74 ± 0.80		
Hg_2Cl_2(s)	−265.37 ± 0.40	191.6 ± 0.8	23.35 ± 0.20	−210.7
Hg_2SO_4(s)	−743.09 ± 0.40	200.70 ± 0.20	26.070 ± 0.030	−625.8
I(g)	106.76 ± 0.04	180.787 ± 0.004	6.197 ± 0.001	70.2
I^-(aq)	−56.78 ± 0.05	106.45 ± 0.30		
I_2 (s, rhombic)	0	116.14 ± 0.30	13.196 ± 0.040	0
I_2(g)	62.42 ± 0.08	260.687 ± 0.005	10.116 ± 0.001	19.3
K(s)	0	64.68 ± 0.20	7.088 ± 0.020	0
K(g)	89.0 ± 0.8	160.341 ± 0.003	6.197 ± 0.001	60.5
K^+(aq)	−252.14 ± 0.08	101.20 ± 0.20		
Kr(g)	0	164.085 ± 0.003	6.197 ± 0.001	0
Li(s)	0	29.12 ± 0.20	4.632 ± 0.040	0
Li(g)	159.3 ± 1.0	138.782 ± 0.010	6.197 ± 0.001	126.6
Li^+(aq)	−278.47 ± 0.08	12.24 ± 0.15		
Mg(s)	0	32.67 ± 0.10	4.998 ± 0.030	0
Mg(g)	147.1 ± 0.8	148.648 ± 0.003	6.197 ± 0.001	112.5
Mg^{+2}(aq)	−467.0 ± 0.6	−137 ± 4		
MgF_2(s)	−1124.2 ± 1.2	57.2 ± 0.5	9.91 ± 0.06	−1071.1
MgO(s)	−601.60 ± 0.30	26.95 ± 0.15	5.160 ± 0.020	−569.3
N(g)	472.68 ± 0.40	153.301 ± 0.003	6.197 ± 0.001	455.5
NH_3(g)	−45.94 ± 0.35	192.77 ± 0.05	10.043 ± 0.010	−16.4
NH_4^+(aq)	−133.26 ± 0.25	111.17 ± 0.40		
NO_3^-(aq)	−206.85 ± 0.40	146.70 ± 0.40		
N_2(g)	0	191.609 ± 0.004	8.670 ± 0.001	0
Na(s)	0	51.30 ± 0.20	6.460 ± 0.020	0
Na(g)	107.5 ± 0.7	153.718 ± 0.003	6.197 ± 0.001	77.0
Na^+(aq)	−240.34 ± 0.06	58.45 ± 0.15		
Ne(g)	0	146.328 ± 0.003	6.197 ± 0.001	0
O(g)	249.18 ± 0.10	161.059 ± 0.003	6.725 ± 0.001	231.7
OH^-(aq)	−230.015 ± 0.040	−10.90 ± 0.20		
O_2(g)	0	205.152 ± 0.005	8.680 ± 0.002	0
P (s, white)	0	41.09 ± 0.25	5.360 ± 0.015	0
P(g)	316.5 ± 1.0	163.199 ± 0.003	6.197 ± 0.001	280.1
P_2(g)	144.0 ± 2.0	218.123 ± 0.004	8.904 ± 0.001	103.5
P_4(g)	58.9 ± 0.3	280.01 ± 0.50	14.10 ± 0.20	24.4
Pb(s)	0	64.80 ± 0.30	6.870 ± 0.030	0
Pb(g)	195.2 ± 0.8	175.375 ± 0.005	6.197 ± 0.001	162.2
Pb^{+2}(aq)	0.92 ± 0.25	18.5 ± 1.0		
$PbSO_4$(s)	−919.97 ± 0.40	148.50 ± 0.60	20.050 ± 0.040	−813.0
Rb(s)	0	76.78 ± 0.30	7.489 ± 0.020	0
Rb(g)	80.9 ± 0.8	170.094 ± 0.003	6.197 ± 0.001	53.1
Rb^+(aq)	−251.12 ± 0.10	121.75 ± 0.25		
S (cr, rhombic)	0	32.054 ± 0.050	4.412 ± 0.006	0
S(g)	277.17 ± 0.15	167.829 ± 0.006	6.657 ± 0.001	236.7
SO_2(g)	−296.81 ± 0.20	248.223 ± 0.050	10.549 ± 0.010	−300.1
SO_4^{-2}(aq)	−909.34 ± 0.40	18.50 ± 0.40		

Substance	$\Delta_f H^\circ$ kJ mol^{-1}	S° J K^{-1} mol^{-1}	$H^\circ - H^\circ(0)$ kJ mol^{-1}	$\Delta_f G^\circ$ kJ mol^{-1}
S_2(g)	128.60 ± 0.30	228.167 ± 0.010	9.132 ± 0.002	79.7
Si(s)	0	18.81 ± 0.08	3.217 ± 0.008	0
Si(g)	450 ± 8	167.981 ± 0.004	7.550 ± 0.001	405.5
SiF_4(g)	−1615.0 ± 0.8	282.76 ± 0.50	15.36 ± 0.05	−1572.8
SiO_2(s, α-quartz)	−910.7 ± 1.0	41.46 ± 0.20	6.916 ± 0.020	−856.3
Sn (s, white)	0	51.18 ± 0.08	6.323 ± 0.008	0
Sn(g)	301.2 ± 1.5	168.492 ± 0.004	6.215 ± 0.001	266.2
Sn^{+2}(aq)	−8.9 ± 1.0	−16.7 ± 4.0		
SnO (s, tetragonal)	−280.71 ± 0.20	57.17 ± 0.30	8.736 ± 0.020	−251.9
SnO_2 (s, tetragonal)	−577.63 ± 0.20	49.04 ± 0.10	8.384 ± 0.020	−515.8
Th(s)	0	51.8 ± 0.5	6.35 ± 0.05	0
Th(g)	602 ± 6	190.17 ± 0.05	6.197 ± 0.003	560.7
ThO_2(s)	−1226.4 ± 3.5	65.23 ± 0.20	10.560 ± 0.020	−1169.2
Ti(s)	0	30.72 ± 0.10	4.824 ± 0.015	0
Ti(g)	473 ± 3	180.298 ± 0.010	7.539 ± 0.002	428.4
$TiCl_4$(g)	−763.2 ± 3.0	353.2 ± 4.0	21.5 ± 0.5	
TiO_2 (s, rutile)	−944.0 ± 0.8	50.62 ± 0.30	8.68 ± 0.05	−888.8
U(s)	0	50.20 ± 0.20	6.364 ± 0.020	0
U(g)	533 ± 8	199.79 ± 0.10	6.499 ± 0.020	488.4
UO_2(s)	−1085.0 ± 1.0	77.03 ± 0.20	11.280 ± 0.020	−1031.8
UO_2^{+2}(aq)	−1019.0 ± 1.5	−98.2 ± 3.0		
UO_3 (s, gamma)	−1223.8 ± 1.2	96.11 ± 0.40	14.585 ± 0.050	−1145.7
U_3O_8 (V, VI)(s)	−3574.8 ± 2.5	282.55 ± 0.50	42.74 ± 0.10	−3369.5
Xe(g)	0	169.685 ± 0.003	6.197 ± 0.001	0
Zn(s)	0	41.63 ± 0.15	5.657 ± 0.020	0
Zn(g)	130.40 ± 0.40	160.990 ± 0.004	6.197 ± 0.001	94.8
Zn^{+2}(aq)	−153.39 ± 0.20	−109.8 ± 0.5		
ZnO(s)	−350.46 ± 0.27	43.65 ± 0.40	6.933 ± 0.040	−320.5

SOURCES OF THERMODYNAMIC DATA

Detailed and authoritative sources of thermochemical data are:

The NBS Tables of Chemical Thermodynamic Properties. Selected Values for Inorganic and C_1 and C_2 Organic Substances in SI Units (Ed. D. D. Wagman *et al.*), New York: American Chemical Society and the American Institute of Physics, 1982.

JANAF Thermochemical Tables (Ed. M. W. Chase *et al.*), New York: American Chemical Society and the American Institute of Physics, 3rd edition, 1985. This compilation gives data over a wider range of temperature than the *NBS Tables*.

Both of the following publications were issued as Supplements to the *Journal of Physical and Chemical Reference Data,* which publishes reliable and up-to-date data of all kinds:

CODATA Key Values for Thermodynamics (Ed. J. D. Cox *et al.*), New York: Hemisphere, 1989.

I. Barin, *Thermochemical Data of Pure Substances,* Weinheim: VCH, 2nd edition, 1993.

Very convenient and up-to-date, but less detailed, compilations of thermochemical data (and many other physicochemical data) are:

CRC Handbook of Physics and Chemistry (commonly known as the "Rubber Handbook") Boca Raton, FL: CRC Press. This valuable compilation appears in revised editions annually, 82 editions having been published.

Tables of Physical and Chemical Constants (*"Kaye and Laby"*; originally compiled in 1911 by G. W. C. Kaye and J. H. Laby, and now by a committee, with revised editions appearing from time to time), London and New York: Longmans, 15th edition, 1986.

VNR Index of Chemical and Physical Data, New York: Van Nostrand Reinhold, 1992.

Character Tables for Some Important Symmetry Groups in Chemistry

APPENDIX E

1. Nonaxial Groups

C_1	E
A	1

C_s	E	σ_h		
A'	1	1	x, y, R_z	x^2, y^2, z^2, xy
A''	1	-1	z, R_x, R_y	yz, xz

C_i	E	i			
A_g	1	1		R_x, R_y, R_z	x^2, y^2, z^2 xy, xz, yz
A_u	1	-1	x, y, z		

2. C_n Groups

C_2	E	C_2			
A	1	1	z	R_z	x^2, y^2, z^2, xy
B	1	-1	x, y	R_x, R_y	yz, xz

C_3	E	C_3	C_3^2			$\epsilon = \exp(2\pi i/3)$
A	1	1	1	z	R_z	$x^2 + y^2, z^2$
E	$\begin{Bmatrix}1 \\ 1\end{Bmatrix}$	ϵ ϵ^*	ϵ^* ϵ	x, y	R_x, R_y	$(x^2 - y^2, xy)(yz, xz)$

C_4	E	C_4	C_2	C_4^3			
A	1	1	1	1	z	R_z	$x^2 + y^2, z^2$
B	1	-1	1	-1			$x^2 - y^2, xy$
E	$\begin{Bmatrix}1 \\ 1\end{Bmatrix}$	i $-i$	-1 -1	$-i$ i	x, y	R_x, R_y	(yz, xz)

*Represents the complex conjugate.

C_5	E	C_5	C_5^2	C_5^3	C_5^4			$\epsilon = \exp(2\pi i/5)$
A	1	1	1	1	1	z	R_z	$x^2 + y^2, z^2$
E_1	$\{1$	ϵ	ϵ^2	ϵ^{2*}	$\epsilon^*\}$	x, y	R_x, R_y	(yz, xz)
	$\{1$	ϵ^*	ϵ^{2*}	ϵ^2	$\epsilon\}$			
E_2	$\{1$	ϵ^2	ϵ^*	ϵ	$\epsilon^{2*}\}$			$(x^2 - y^2, xy)$
	$\{1$	ϵ^{2*}	ϵ	ϵ^*	$\epsilon^2\}$			

C_6	E	C_6	C_3	C_2	C_3^2	C_6^5			$\epsilon = \exp(2\pi i/6)$
A	1	1	1	1	1	1	z	R_z	$x^2 + y^2, z^2$
B	1	−1	1	−1	1	−1			
E_1	$\{1$	ϵ	$-\epsilon^*$	−1	$-\epsilon$	$\epsilon^*\}$	x, y		(xz, yz)
	$\{1$	ϵ^*	$-\epsilon$	−1	$-\epsilon^*$	$\epsilon\}$		R_x, R_y	
E_2	$\{1$	$-\epsilon^*$	$-\epsilon$	1	$-\epsilon^*$	$-\epsilon\}$			$(x^2 - y^2, xy)$
	$\{1$	$-\epsilon$	$-\epsilon^*$	1	$-\epsilon$	$-\epsilon^*\}$			

3. D_n Groups

D_2	E	$C_2(z)$	$C_2(y)$	$C_2(x)$			
A	1	1	1	1			x^2, y^2, z^2
B_1	1	1	−1	−1	z	R_z	xy
B_2	1	−1	1	−1	y	R_y	xz
B_3	1	−1	−1	1	x	R_x	yz

D_3	E	$2C_3$	$3C_2$			
A_1	1	1	1			$x^2 + y^2, z^2$
A_2	1	1	−1	z	R_z	
E	2	−1	0	x, y	R_x, R_y	$(x^2 - y^2, xy)(xz, yz)$

D_4	E	$2C_4$	$C_2(= C_4^2)$	$2C_2'$	$2C_2''$			
A_1	1	1	1	1	1			$x^2 + y^2, z^2$
A_2	1	1	1	−1	−1	z	R_z	
B_1	1	−1	1	1	−1			$x^2 - y^2$
B_2	1	−1	1	−1	1			xy
E	2	0	−2	0	0	x, y	R_x, R_y	(xz, yz)

D_6	E	$2C_6$	$2C_3$	C_2	$3C_2'$	$3C_2''$			
A_1	1	1	1	1	1	1			$x^2 + y^2, z^2$
A_2	1	1	1	1	−1	−1	z	R_z	
B_1	1	−1	1	−1	1	−1			
B_2	1	−1	1	−1	−1	1			
E_1	2	1	−1	−2	0	0	x, y	R_x, R_y	(xz, yz)
E_2	2	−1	−1	2	0	0			$(x^2 - y^2, xy)$

4. C_{nv} Groups

C_{2v}	E	C_2	$\sigma_v(xz)$	$\sigma_v'(yz)$			
A_1	1	1	1	1	z		x^2, y^2, z^2
A_2	1	1	−1	−1		R_z	xy
B_1	1	−1	1	−1	x	R_y	xz
B_2	1	−1	−1	1	y	R_x	yz

C_{3v}	E	$2C_3$	$3\sigma_v$			
A_1	1	1	1	z		$x^2 + y^2, z^2$
A_2	1	1	−1		R_z	
E	2	−1	0	x, y	R_x, R_y	$(x^2 - y^2, xy)(xz, yz)$

C_{4v}	E	$2C_4$	C_2	$2\sigma_v$	$2\sigma_d$			
A_1	1	1	1	1	1	z		$x^2 + y^2, z^2$
A_2	1	1	1	−1	−1		R_z	
B_1	1	−1	1	1	−1			$x^2 - y^2$
B_2	1	−1	1	−1	1			xy
E	2	0	−2	0	0	x, y	R_x, R_y	(xz, yz)

C_{6v}	E	$2C_6$	$2C_3$	C_2	$3\sigma_v$	$3\sigma_d$			
A_1	1	1	1	1	1	1	z		$x^2 + y^2, z^2$
A_2	1	1	1	1	−1	−1		R_z	
B_1	1	−1	1	−1	1	−1			
B_2	1	−1	1	−1	−1	1			
E_1	2	1	−1	−2	0	0	x, y	R_x, R_y	(xz, yz)
E_2	2	−1	−1	2	0	0			$(x^2 - y^2, xy)$

5. C_{nh} Groups

C_{2h}	E	C_2	i	σ_h			
A_g	1	1	1	1		R_z	x^2, y^2, z^2, xy
B_g	1	−1	1	−1		R_x, R_y	xz, yz
A_u	1	1	−1	−1	z		
B_u	1	−1	−1	1	x, y		

C_{3h}	E	C_3	C_3^2	σ_h	S_3	S_3^5			$\epsilon = \exp(2\pi i/3)$
A'	1	1	1	1	1	1		R_z	$x^2 + y^2, z^2$
E'	$\begin{cases}1\\1\end{cases}$	ϵ ϵ^*	ϵ^* ϵ	1 1	ϵ ϵ^*	ϵ^* ϵ	x, y		$(x^2 - y^2, xy)$
A''	1	1	1	−1	−1	−1	z		
E''	$\begin{cases}1\\1\end{cases}$	ϵ ϵ^*	ϵ^* ϵ	−1 −1	$-\epsilon$ $-\epsilon^*$	$-\epsilon^*$ $-\epsilon$		R_x, R_y	(xz, yz)

C_{4h}	E	C_4	C_2	C_4^3	i	S_4^3	σ_h	S_4			
A_g	1	1	1	1	1	1	1	1		R_z	x^2+y^2, z^2
B_g	1	−1	1	−1	1	−1	1	−1			x^2-y^2, xy
E_g	1	i	−1	$-i$	1	i	−1	$-i$		(R_x, R_y)	(xz, yz)
	1	$-i$	−1	i	1	$-i$	−1	i			
A_u	1	1	1	1	−1	−1	−1	−1	z		
B_u	1	−1	1	−1	−1	1	−1	1			
E_u	1	i	−1	$-i$	−1	$-i$	1	i	x, y		
	1	$-i$	−1	i	−1	i	1	$-i$			

6. D_{nh} Groups

D_{2h}	E	$C_2(z)$	$C_2(y)$	$C_2(x)$	i	$\sigma(xy)$	$\sigma(xz)$	$\sigma(yz)$			
A_g	1	1	1	1	1	1	1	1			x^2, y^2, z^2
B_{1g}	1	1	−1	−1	1	1	−1	−1		R_z	xy
B_{2g}	1	−1	1	−1	1	−1	1	−1		R_y	xz
B_{3g}	1	−1	−1	1	1	−1	−1	1		R_x	yz
A_u	1	1	1	1	−1	−1	−1	−1			
B_{1u}	1	1	−1	−1	−1	−1	1	1	z		
B_{2u}	1	−1	1	−1	−1	1	−1	1	y		
B_{3u}	1	−1	−1	1	−1	1	1	−1	x		

D_{3h}	E	$2C_3$	$3C_2$	σ_h	$2S_3$	$3\sigma_v$			
A_1'	1	1	1	1	1	1			x^2+y^2, z^2
A_2'	1	1	−1	1	1	−1		R_z	
E'	2	−1	0	2	−1	0	x, y		(x^2-y^2, xy)
A_1''	1	1	1	−1	−1	−1			
A_2''	1	1	−1	−1	−1	1	z		
E''	2	−1	0	−2	1	0		R_x, R_y	(xz, yz)

D_{4h}	E	$2C_4$	C_2	$2C_2'$	$2C_2''$	i	$2S_4$	σ_h	$2\sigma_v$	$2\sigma_d$			
A_{1g}	1	1	1	1	1	1	1	1	1	1			x^2+y^2, z^2
A_{2g}	1	1	1	−1	−1	1	1	1	−1	−1		R_z	
B_{1g}	1	−1	1	1	−1	1	−1	1	1	−1			x^2-y^2
B_{2g}	1	−1	1	−1	1	1	−1	1	−1	1			xy
E_g	2	0	−2	0	0	2	0	−2	0	0		(R_x, R_y)	(xz, yz)
A_{1u}	1	1	1	1	1	−1	−1	−1	−1	−1			
A_{2u}	1	1	1	−1	−1	−1	−1	−1	1	1	z		
B_{1u}	1	−1	1	1	−1	−1	1	−1	−1	1			
B_{2u}	1	−1	1	−1	1	−1	1	−1	1	−1			
E_u	2	0	−2	0	0	−2	0	2	0	0	x, y		

D_{6h}	E	$2C_6$	$2C_3$	C_2	$3C_2'$	$3C_2''$	i	$2S_3$	$2S_6$	σ_h	$3\sigma_d$	$3\sigma_v$			
A_{1g}	1	1	1	1	1	1	1	1	1	1	1	1			$x^2 + y^2, z^2$
A_{2g}	1	1	1	1	−1	−1	1	1	1	1	−1	−1		R_z	
B_{1g}	1	−1	1	−1	1	−1	1	−1	1	−1	1	−1			
B_{2g}	1	−1	1	−1	−1	1	1	−1	1	−1	−1	1			
E_{1g}	2	1	−1	−2	0	0	2	1	−1	−2	0	0		R_x, R_y	(xz, yz)
E_{2g}	2	−1	−1	2	0	0	2	−1	−1	2	0	0			$(x^2 - y^2, xy)$
A_{1u}	1	1	1	1	1	1	−1	−1	−1	−1	−1	−1			
A_{2u}	1	1	1	1	−1	−1	−1	−1	−1	−1	1	1	z		
B_{1u}	1	−1	1	−1	1	−1	−1	1	−1	1	−1	1			
B_{2u}	1	−1	1	−1	−1	1	−1	1	−1	1	1	−1			
E_{1u}	2	1	−1	−2	0	0	−2	−1	1	2	0	0	x, y		
E_{2u}	2	−1	−1	2	0	0	−2	1	1	−2	0	0			

7. D_{nd} Groups

D_{2d}	E	$2S_4$	C_2	$2C_2'$	$2\sigma_d$			
A_1	1	1	1	1	1			$x^2 + y^2, z^2$
A_2	1	1	1	−1	−1		R_z	
B_1	1	−1	1	1	−1			$x^2 - y^2$
B_2	1	−1	1	−1	1	z		xy
E	2	0	−2	0	0	x, y	(R_x, R_y)	xz, yz

D_{3d}	E	$2C_3$	$3C_2$	i	$2S_6$	$3\sigma_d$			
A_{1g}	1	1	1	1	1	1			$x^2 + y^2, z^2$
A_{2g}	1	1	−1	1	1	−1		R_z	
E_g	2	−1	0	2	−1	0		(R_x, R_y)	$(x^2 - y^2, xy)$, (xz, yz)
A_{1u}	1	1	1	−1	−1	−1			
A_{2u}	1	1	−1	−1	−1	1	z		
E_u	2	−1	0	−2	1	0	x, y		

D_{4d}	E	$2S_8$	$2C_4$	$2S_8^3$	C_2	$4C_2'$	$4\sigma_d$			
A_1	1	1	1	1	1	1	1			$x^2 + y^2, z^2$
A_2	1	1	1	1	1	−1	−1		R_z	
B_1	1	−1	1	−1	1	1	−1			
B_2	1	−1	1	−1	1	−1	1	z		
E_1	2	$\sqrt{2}$	0	$-\sqrt{2}$	−2	0	0	x, y		
E_2	2	0	−2	0	2	0	0			$(x^2 - y^2, xy)$
E_3	2	$-\sqrt{2}$	0	$\sqrt{2}$	−2	0	0		R_x, R_y	(xz, yz)

D_{6d}	E	$2S_{12}$	$2C_6$	$2S_4$	$2C_3$	$2S_{12}^5$	C_2	$6C_2'$	$6\sigma_d$			
A_1	1	1	1	1	1	1	1	1	1			$x^2 + y^2,\ z^2$
A_2	1	1	1	1	1	1	1	−1	−1		R_z	
B_1	1	−1	1	−1	1	−1	1	1	−1			
B_2	1	−1	1	−1	1	−1	1	−1	1	z		
E_1	2	$\sqrt{3}$	1	0	−1	$-\sqrt{3}$	−2	0	0	x, y		
E_2	2	1	−1	−2	−1	1	2	0	0			$(x^2 - y^2, xy)$
E_3	2	0	−2	0	2	0	−2	0	0			
E_4	2	−1	−1	2	−1	−1	2	0	0			
E_5	2	$-\sqrt{3}$	1	0	−1	$\sqrt{3}$	−2	0	0		R_x, R_y	(xz, yz)

8. The S_n Groups

S_4	E	S_4	C_2	S_4^3			
A	1	1	1	1		R_z	$x^2 + y^2, z^2$
B	1	−1	1	−1	z		$x^2 - y^2, xy$
E	$\begin{cases}1\\1\end{cases}$	i $-i$	−1 −1	$-i$ i	x, y	R_x, R_y	(xz, yz)

9. Cubic Groups

T	E	$4C_3$	$4C_3^2$	$3C_2$			$\epsilon = \exp(2\pi i/3)$
A	1	1	1	1			$x^2 + y^2 + z^2$
E	1 1	ϵ ϵ^*	ϵ^* ϵ	1 1			$(2z^2 - x^2 - y^2,$ $x^2 - y^2)$
T	3	0	0	−1	x, y, z	R_x, R_y, R_z	(xy, xz, yz)

T_h	E	$4C_3$	$4C_3^2$	$3C_2$	i	$4S_6$	$4S_6^5$	$3\sigma_h$			$\epsilon = \exp(2\pi i/3)$
A_g	1	1	1	1	1	1	1	1			$x^2 + y^2 + z^2$
A_u	1	1	1	1	−1	−1	−1	−1			
E_g	1 1	ϵ ϵ^*	ϵ^* ϵ	1 1	1 1	ϵ ϵ^*	ϵ^* ϵ	1 1			$(2z^2 - x^2 - y^2,$ $x^2 - y^2)$
E_u	1 1	ϵ ϵ^*	ϵ^* ϵ	1 1	−1 −1	$-\epsilon$ $-\epsilon^*$	$-\epsilon^*$ $-\epsilon$	−1 −1			
T_g	3	0	0	−1	−3	0	0	−1		R_x, R_y, R_z	(xz, yz, xy)
T_u	3	0	0	−1	−3	0	0	1	x, y, z		

T_d	E	$8C_3$	$3C_2$	$6S_4$	$6\sigma_d$			
A_1	1	1	1	1	1			$x^2 + y^2 + z^2$
A_2	1	1	1	−1	−1			
E	2	−1	2	0	0			$(2z^2 - x^2 - y^2,$ $x^2 - y^2)$
T_1	3	0	−1	1	−1		R_x, R_y, R_z	
T_2	3	0	−1	−1	1	x, y, z		(xy, xz, yz)

O	E	$6C_4$	$3C_2(= C_4^2)$	$8C_3$	$6C_2$			
A_1	1	1	1	1	1			$x^2 + y^2 + z^2$
A_2	1	−1	1	1	−1			
E	2	0	2	−1	0			$(2z^2 - x^2 - y^2,$ $x^2 - y^2)$
T_1	3	1	−1	0	−1	x, y, z	R_x, R_y, R_z	
T_2	3	−1	−1	0	1			(xy, xz, yz)

O_h	E	$8C_3$	$6C_2$	$6C_4$	$3C_2(= C_4^2)$	i	$6S_4$	$8S_6$	$3\sigma_h$	$6\sigma_d$			
A_{1g}	1	1	1	1	1	1	1	1	1	1			$x^2 + y^2 + z^2$
A_{2g}	1	1	−1	−1	1	1	−1	1	1	−1			
E_g	2	−1	0	0	2	2	0	−1	2	0			$(2z^2 - x^2 - y^2,$ $x^2 - y^2)$
T_{1g}	3	0	−1	1	−1	3	1	0	−1	−1		R_x, R_y, R_z	
T_{2g}	3	0	1	−1	−1	3	−1	0	−1	1			(xz, yz, xy)
A_{1u}	1	1	1	1	1	−1	−1	−1	−1	−1			
A_{2u}	1	1	−1	−1	1	−1	1	−1	−1	1			
E_u	2	−1	0	0	2	−2	0	1	−2	0			
T_{1u}	3	0	−1	1	−1	−3	−1	0	1	1	x, y, z		
T_{2u}	3	0	1	−1	−1	−3	1	0	1	−1			

10. $C_{\infty v}$ and $D_{\infty h}$ for Linear Molecules

$C_{\infty v}$	E	$2C_\infty^\phi$	…	$\infty\sigma_v$			
$A_1 \equiv \Sigma^+$	1	1	…	1	z		$x^2 + y^2, z^2$
$A_2 \equiv \Sigma^-$	1	1	…	−1		R_z	
$E_1 \equiv \Pi$	2	$2 \cos \Phi$	…	0	x, y	(R_x, R_y)	xz, yz
$E_2 \equiv \Delta$	2	$2 \cos 2\Phi$	…	0			$(x^2 - y^2, xy)$
$E_3 \equiv \Phi$	2	$2 \cos 3\Phi$	…	0			
…	…	…	…	…			

$D_{\infty h}$	E	$2C_\infty^\phi$	σ_h	…	$\infty\sigma_v$	i	$2S_\infty^\phi$	…	∞C_2			
Σ_g^+	1	1	1	…	1	1	1	…	1			$x^2 + y^2, z^2$
Σ_g^-	1	1	1	…	−1	1	1	…	−1		R_z	
Π_g	2	$2 \cos \Phi$	−2	…	0	2	$-2 \cos \Phi$	…	0		R_x, R_y	(xz, yz)
Δ_g	2	$2 \cos 2\Phi$	2	…	0	2	$2 \cos 2\Phi$	…	0			$(x^2 - y^2, xy)$
Φ_g	…	…	−2	…	…	…	…	…	…			
Σ_u^+	1	1	−1	…	1	−1	−1	…	−1	z		
Σ_u^-	1	1	−1	…	−1	−1	−1	…	1			
Π_u	2	$2 \cos \Phi$	2	…	0	−2	$2 \cos \Phi$	…	0	x, y		
Δ_u	2	$2 \cos 2\Phi$	−2	…	0	−2	$-2 \cos 2\Phi$	…	0			
Φ_u	…	…	2	…	…	…	…	…	…			

Answers to Problems

Note: In comparing your answers with the values listed, in some cases a difference in the last place may occur if rounding was used in the various steps of the problem. Slight differences in updated thermodynamic data may cause minor changes in a few cases.

Chapter 1

1.1 298 kJ; 476 kJ **1.2** 879 °C **1.3** 4.13×10^{-23} J **1.4** 92.6 W **1.5** (a) extensive (b) intensive (c) intensive (d) intensive **1.6** 977.2 Torr **1.7** (a) 1 Torr = 133.322 3684 Pa (b) 3.24×10^{7} m^{-3} **1.8** 1 atm = 101.325 kPa **1.9** 1 mm = 0.077 Torr **1.10** 0.359 dm^3 **1.11** 0.470 dm^3 **1.12** 0.500 dm^3 **1.13** (a) 0.0409 mol dm^{-3}; 2.46×10^{22} dm^{-3} (b) 4.03×10^{-11} mol dm^{-3}; 2.43×10^{13} dm^{-3} **1.14** 956 Torr **1.15** (a) 0.225 dm^3 (b) 108 g mol^{-1} **1.16** 31.7 g mol^{-1} **1.17** 28.36 g mol^{-1} **1.18** 96.82 kPa **1.19** $x_{N_2} = 0.61106$; $x_{CO_2} = 0.38894$, $P_{N_2} = 35.4$ bar; $P_{CO_2} = 22.5$ bar; $P_t = 57.9$ bar **1.20** 26.6 cm^3 **1.21** 1020 m^3 **1.22** 1240 g; 20 atm **1.23** (a) $\ln P/P_0 = -M/RT(9.807 \text{ m s}^{-2} z - 5 \times 10^{-6} \text{ s}^{-2} z^2)$ (b) 2.73×10^{-5} **1.24** 210 Torr **1.25** $\ln P/P_0 = \frac{Mg}{Ra} \ln(T_0 - az)/T_0$ **1.26** 49.7 °C **1.27** 150 g mol^{-1} **1.28** 2.19 min **1.29** 2400 J **1.30** (a) 0.123 mol (b) 7.41×10^{22} (c) 515.2 m s^{-1} (d) 6.174×10^{-21} J (e) 457 J **1.31** 1.15 **1.32** 6.54×10^{-8} m; 7.26×10^{9} s^{-1}; 8.9×10^{34} m^{-3} s^{-1} **1.33** $\lambda = RT/\sqrt{2}\pi d^2 LP$ **1.34** 2.73×10^{12} s^{-1}; 3.05×10^{37} m^{-3} s^{-1} **1.35** 6.17×10^{-8} m **1.36** (a) 1.044×10^{-4} m (b) 1.37×10^{-7} m (c) 1.39×10^{-10} m **1.37** 3.60×10^{18} m **1.38** 7.44×10^{23} mol^{-1}

1.39 (a) $\sqrt{\frac{3\pi}{8}} = 1.085$ (b) $\frac{2}{\sqrt{\pi}} = 1.128$ **1.40** (a) 10 900 K (b) 173 000 K **1.41** (a) 8.77×10^{-2}; no effect of mass or temperature (b) 0.89; no effect of mass **1.42** They are equal. **1.43** (a) 1.155 (b) 11.55 **1.44** $(2m/2\pi k_B T)^{1/2} \exp(-mu_x^2/2k_B T)$; zero **1.45** $(\pi \epsilon_x k_B T)^{-1/2} \exp(-\epsilon_x/k_B T)\, d\epsilon_x$; $\frac{1}{2}k_B T$ **1.46** $(m/k_B T) \exp(-mu^2/2k_B T) u\, du$; $\exp(-\epsilon^*/k_B T)$ **1.47** Excluded volume for a pair of molecules is $\frac{4}{3}\pi d^3 = 8V$. Excluded volume per molecule, $b = 4$ V.

1.48 Similar to Fig. 1.21 **1.49** (a) 700.7 kPa (b) 650 kPa **1.50** Ideal: 7.46 L; using compression factor: 3.77 L. Error = 97.9% assuming compression factor gives the more accurate result. **1.51** $b = 0$, therefore, the 1st and 2nd derivatives cannot vanish simultaneously ∴ no critical point. **1.52** 0.24 dm^3 mol^{-1}; 0.34 dm^3 mol^{-1}. They are the same. **1.53** $r = 1.62 \times 10^{-10}$ m **1.54** $T_B = 2T_C$ **1.55** (b) $\frac{PV_m}{RT} = 1 + \frac{b}{V_m} + \left(\frac{b}{V_m}\right)^2 + \cdots - \frac{a}{RT}\frac{1}{V_m}$ (c) $B(T) = b - \frac{a}{RT}$; $C(T) = b^2$ (d) $b = a/(RT_B)$, $T_B = a/(bR)$ **1.56** $T_B = a/Rb$ **1.57** 1.15; 0.753; 0.46 dm^3 **1.58** $P = 0.57$ bar; $T = 87.1$ K **1.59** $2V_c RT_c$; $V_c/2 = b$ **1.60** $Z = 1 + \frac{1}{RT}\left(b - \frac{a}{RT}\right) P + \left(\frac{b}{RT}\right)^2 P^2 = 1.01$ $V_m = 4.99$ L **1.61** Let P go to zero, $z = \frac{PV}{RT} = 1$ **1.62** 0.5563 Pa m^6 mol^{-2}; 0.0638×10^{-3} m^3 mol^{-1} **1.63** 9.01 bar; 8.67 bar; 8.57 bar; 8.61 bar **1.64** 0.162×10^{-3} m^3 mol^{-1} **1.65** 2nd $= -(a + RTb)$; 3rd $= a^2/(2RT) + ab + RTb^2$; $RT/V_m - (a + RTb)/V_m^2$

Chapter 2

2.1 1.40 kJ **2.2** 6.025 kJ mol^{-1}; 0.165 J mol^{-1} **2.3** 37.6 kJ mol^{-1}; $-w = 3.06$ kJ mol^{-1} **2.4** -5.65 kJ mol^{-1} **2.5** (a) $\Delta U = -972.7$ kJ mol^{-1}; (b) $\Delta H =$ 975.2 kJ mol^{-1} **2.6** (a) 72.7 °C; (b) 4.34 kg **2.7** 3.79 kJ; zero **2.8** (a) $V = \frac{1}{6}\pi D^3$; (b) 30.6 kJ **2.9** 75.4 J K^{-1} mol^{-1} **2.10** 314 kJ; 314 s **2.11** $(M - Vd)gh$; the same **2.12** Apply Euler's theorem to the total differential dP. **2.13** Differential is exact. $U = \frac{1}{2}x^2y^2$ + constant **2.14** $\Delta H° = -492.55$ kJ **2.15** (a) -3274.9 kJ mol^{-1}; (b) -3278.6 kJ mol^{-1} **2.16** 65.2 kJ mol^{-1} **2.17** (a) -727.7 kJ mol^{-1};

(b) -237.5 kJ mol^{-1} (c) -202.2 kJ mol^{-1}
2.18 -1560.5 kJ mol^{-1} **2.19** $d = 1.7267$ J K^{-1} mol^{-1}; $e = 9.3424 \times 10^{-2}$ J K^{-2} mol^{-1}; $f = -8.714 \times 10^2$ J K mol^{-1}; plot function in range $15 \leq T \leq 275$ and $10 \leq T \leq 25$. **2.20** Measure the heats of combustion of graphite, CO(g), and of CO_2(g) and use Hess's law. **2.21** 53.2 kJ mol^{-1}
2.22 $d = 0.801$ J K^{-1} mol^{-1}; $e = 1.303 \times 10^{-3}$ J K^{-2} mol^{-1}; $f = -2.199 \times 10^4$ J K mol^{-1}
2.23 -44.2 kJ mol^{-1} **2.24** (a) -203.4 kJ mol^{-1}; (b) -291.2 kJ mol^{-1} **2.25** -144.2 kJ mol^{-1}
2.26 -492.5 kJ mol^{-1} **2.27** -2.1 kJ mol^{-1}
2.28 -908.8 kJ mol^{-1} **2.29** 4.88 kJ mol^{-1}
2.30 45.09 kJ mol^{-1} **2.31** 84.8 MJ h^{-1}
2.32 (a) No ice remains: 12.0 °C; (b) 250 g; 0 °C
2.33 -395.03 kJ mol^{-1} **2.34** -105.9 kJ mol^{-1}
2.35 (a) -5635 kJ mol^{-1}; -5635 kJ mol^{-1}; (b) -2231 kJ mol^{-1} **2.36** -280.00 kJ mol^{-1}
2.37 8.244 dm^3 **2.38** (a) 4 bar (b) 4.540 kJ (c) 21.1 J K^{-1} **2.39** (a) zero (b) 4.22 kJ (c) 4.22 kJ (d) 547 kPa (e) 6.21 kJ (f) 5.89 kJ
2.40 (a) 15.5 dm^3 (b) 1.66 kJ (c) 5.88 kJ (d) 5.88 kJ (e) 4.22 kJ **2.41** (a) zero (b) 800 kPa (c) 3.15 kJ (d) 3.15 kJ (e) zero **2.42** (a) 3632 J (b) zero (c) 3632 kJ **2.43** (a) 7310 J (b) zero (c) 7310 J **2.44** (a) 182.9 K (b) -430 J, -604 J
2.45 (a) 17.00 dm^3 (b) 204.6 K (c) -398.4 J K^{-1}; -521.9 J K^{-1} **2.46** (a) $29.83 + 8.2 \times 10^{-3}(T/\text{K})$ (b) zero **2.47** Show that the ratio of the volumes at the intersection points is always the same.
2.48 Show that $(\partial C_V/\partial V)_T = 0$. **2.49** 173.5 K
2.50 226.0 K; -1.5 kJ mol^{-1}; -2.1 kJ mol^{-1}
2.51 (a) 0.62 bar (b) -100 J mol^{-1}
2.52 Use definition of C_V and Euler's chain rule.
2.53 Apply $dP = \left(\frac{\partial P}{\partial T}\right)_V dT + \left(\frac{\partial P}{\partial V}\right)_T dV$ to $P = \frac{nRT}{V}$
2.54 221 K; -8020 J; $-11\,200$ J **2.55** 10.07 kJ
2.56 120.3 K; 6.78 kJ; 9000 J; -2.218 kJ; -3.70 kJ
2.57 $\Delta H = \frac{5}{2} R\left(\frac{PV_2}{nR} - T_1\right)$ **2.58** $q = 0.08$ kJ; $w = -2.30$ kJ; $\Delta U = -2.22$ kJ; $\Delta H = -3.70$ kJ
2.59 (a) 295.8 kg (b) 1.790×10^5 kJ
2.60 (a) -622 J (b) -718 J **2.61** $C_p - C_v = R$
2.62 $\left(\frac{\partial T}{\partial P}\right)_V = \left(\frac{V_m - b}{R}\right)$ **2.63** (a) 9.76 kJ mol^{-1} (b) 10.16 kJ mol^{-1} **2.64** (a) 1.15 kJ mol^{-1} (b) 1.21 kJ mol^{-1} **2.65** $dP = -\frac{P}{V_m - b} dV_m - \frac{a(3V_m - 2b)}{V_m^3 (V_m - b)} dV_m + \frac{PdT}{T} + \frac{adT}{V_m^2 T}$ **2.66** 4930 K
2.67 11.9 kJ; zero; 390 J **2.68** 17.7 kJ; -4.70 kJ; -9.41 kJ **2.69** 2.73 kJ; -18.9 J; -32.4 J
2.70 -731 J mol^{-1}

Chapter 3

3.1 (a) 80% (b) 30 kJ (c) 150 J K^{-1} (d) 150 J K^{-1} (e) zero (f) -150 kJ **3.2** 7.01 kJ **3.3** (b) 40 kJ; -30 kJ; 400 K **3.4** (a) 25% (b) 3.2 kJ (c) 2.4 kJ (d) 8.0 J K^{-1} (e) zero (f) -3.2 kJ (g) 8.0 kJ
3.5 2.29×10^{10} kJ; 265 MW **3.6** From Eq. 3.128 substitute from the ideal gas law to obtain $-P + RT/V_m = 0$. **3.7** 87.3 J K^{-1} mol^{-1}; 88.0 J K^{-1} mol^{-1}; 108.8 J K^{-1} mol^{-1}; 109.7 J K^{-1} mol^{-1}
3.8 (a) -242.9 J K^{-1} mol^{-1} (b) -456.8 J K^{-1} mol^{-1}
3.9 (a) -198.11 J K^{-1} mol^{-1} (b) 176.0 J K^{-1} mol^{-1}
3.10 186.46 J K^{-1} mol^{-1} **3.11** (a) 3.52 J K^{-1} mol^{-1} (b) 2.11 J K^{-1} mol^{-1} **3.12** 21.9 J K^{-1}
3.13 25.99 J K^{-1} **3.14** 4.61 J K^{-1} mol^{-1}
3.15 -345.4 J K^{-1} mol^{-1} **3.16** (a) 19.1 J K^{-1} mol^{-1}; -19.1 J K^{-1} mol^{-1} (b) 64.2 K; 19.1 J K^{-1} mol^{-1}; zero; Net $\Delta S = 19.1$ J K^{-1} mol^{-1}
3.17 $\Delta S° = \int_{273}^{373} \frac{C_{P,m}(l)}{T} dT + \frac{40\,670}{373.0} = 132.59$ J K^{-1} mol^{-1} for heating the liquid and then vaporization; $\Delta S° = \frac{44\,920}{273} + \int_{273}^{373} \left(\frac{30.54}{T} + 10.29 \times 10^{-3}\right) dT - 8.3145 \int_{0.01602}^{1.0} \frac{1}{P} dP = 132.59$ J K^{-1} mol^{-1}
3.18 (a) positive (b) positive (c) negative (d) positive **3.19** $\int_{T_1}^{T_2} \frac{C_P}{T} dT = n \int_{T_1}^{T_2} \frac{C_{P,m}}{T} dT$; if ideal, $\Delta S = C_P \ln T_2/T_1 = n\, C_{P,m} \ln T_2/T_1$ **3.20** 42.4 J K^{-1}
3.21 (a) -6.28 J K^{-1} (b) 7.10 J K^{-1} (c) 0.82 J K^{-1}
3.22 (a) sys, -30.52 J K^{-1} (b) sys, 24.49 J K^{-1} (c) sys, 1.57 J K^{-1} **3.23** 36.8 J K^{-1} mol^{-1} **3.24** irreversible; 0.4 **3.25** -21.63 J K^{-1} mol^{-1}; 21.87 J K^{-1} mol^{-1}; 0.24 J K^{-1} mol^{-1} **3.26** 1.34 J K^{-1} **3.27** 1.51 J K^{-1}
3.28 3.08 J K^{-1} **3.29** 57.6 J K^{-1} **3.30** -23.73 J K^{-1}; 24.78 J K^{-1}; 1.05 J K^{-1} **3.31** -23.45 J K^{-1}; 24.31 J K^{-1}; 0.86 J K^{-1} **3.32** 0.8 J K^{-1} **3.33** All paths give 9.994 J K^{-1} mol^{-1} **3.34** -59.18 J K^{-1}; 64.08 J K^{-1}; 4.90 J K^{-1} **3.35**(a) 24.8 J K^{-1}; (b) 10.8 J K^{-1} **3.36** 217.3 J K^{-1} mo1^{-1}
3.37 -237.46 kJ mol^{-1}; 459.2 J K^{-1} mol^{-1}
3.38 3.06 kJ mol^{-1}; $\Delta U = 37.5$ kJ mol^{-1}; $\Delta G =$ zero; 108.8 J K^{-1} mol^{-1} **3.39** (a) zero; (b) -213 J mol^{-1}
3.40 zero; zero; 19.14 J K^{-1} mol^{-1}; -5.706 kJ mol^{-1}; -5.706 kJ mol^{-1} **3.41** (a) a′ -34.1 kJ mol^{-1} b′ 16.9 kJ mol^{-1} c′ 85.00 kJ mol^{-1} (b) 501 K
3.42 $\Delta_c H° = -894.73$ kJ mol^{-1}
3.43 147.6 J K^{-1} mol^{-1}; 44.0 kJ mol^{-1}; -19.4 kJ mol^{-1}
3.44 (a) ΔH, ΔU (b) ΔS (c) ΔH (d) ΔH (e) none (f) ΔG (g) ΔU (h) none
3.45 1.485 kJ mol^{-1} **3.46** -4.28 kJ mol^{-1}
3.47 247.86 J K^{-1} mol^{-1} **3.48** 500 J mol^{-1}; zero; zero; -1.73 kJ mol^{-1}; 5.76 J K^{-1} mol^{-1}; $q_{rev} = -1.73$ kJ mol^{-1}, $w_{rev} = 1.73$ kJ mol^{-1}

3.49 zero; 30 kJ mol^{-1}; 33.1 kJ mol^{-1}; ΔS 114.6 J K^{-1} mol^{-1}; -9.66 kJ mol^{-1} **3.50** -40.6 kJ mol^{-1}; -103 J K^{-1} mol^{-1}; -2.15 kJ mol^{-1} **3.51** ΔU_m -1.12 kJ mol^{-1}; 1.50 kJ mol^{-1}; 2.22 J K^{-1} mol^{-1} **3.52** -890.4 kJ mol^{-1}; -817.9 kJ mol^{-1}; -243.2 J K^{-1} mol^{-1} **3.53** (a) and (b) True for an ideal gas. (c) True only if the process is reversible. (d) True only for the total entropy. (e) True only for an isothermal process occurring at constant pressure. **3.54** 127.9 J K^{-1} mol^{-1}; -7.13 kJ mol^{-1} **3.55** -100.4 kJ mol^{-1}; -76.5 kJ mol^{-1}; -80.0 J K^{-1} mol^{-1} **3.56** 48.2%; 28.7% **3.57** 37.0 W **3.58** (a) 59.6% (b) 11.9% (c) 6.6% **3.59** 1.82×10^5 kJ **3.60** 22.6 **3.61** (a) 78 kJ (b) 412 kJ **3.62** Use Eq. 3.128 and substitute in $\left(\frac{\partial P}{\partial T}\right)_V = \frac{\alpha}{\kappa}$ **3.63** $\left(\frac{\partial H}{\partial P}\right)_T = V(-\alpha T)$ **3.64** (a) $\alpha = 1/T$ (b) $\kappa = 1/P$ **3.65** Use $dU = T\,dS - P\,dV$; differentiate with respect to V and T. Then use Maxwell's equation and apply the van der Waals equation.

3.66 $\mu = \dfrac{\frac{RT}{P} - V_m}{C_{P,m}}$ **3.71** (a) $RT \ln\left(\frac{f_2}{p_2}\right) = \int_{P_1}^{P_2}\left(V_m - \frac{RT}{P}\right) dP$; rearrange and divide by RT.

(b) $\ln\left(\frac{f}{P_2}\right) = \int_0^{P_2}\left(\frac{Z-1}{P}\right) dP = \left(\frac{b}{RT} - \frac{A}{RT^{5/3}}\right) P_2$

3.72 $f = 268$ bar

Chapter 4

4.1 2.7 mol dm^{-3} **4.2** 42 mol **4.3** 26 mol **4.4** 2.48 bar **4.5** 3.77×10^{-5} mol dm^{-3}; 2.82×10^{-6} bar **4.7** 0.5; 0.0321 mol dm^{-3}; 0.795 bar; 0.795; none **4.8** 0.165 mol dm^{-3}; 4.09 bar; 11.5 **4.9** 7.94×10^{-9} bar; 2.56×10^{-10} mol dm^{-3}; 3.97×10^{-9} **4.10** 1.5; 4 **4.11** 0.065 mol **4.12** $K_P = 1.08$ bar; $K_c = 4.35 \times 10^{-2}$ mol dm^{-3}; $K_x = 1.81$ and $K_P = 1.45$ bar; $K_c = 5.84 \times 10^{-2}$ mol dm^{-3}; $K_x = 0.234$ **4.13** $K_P = \frac{y}{2(1-y)} \times \frac{1}{P_{O_2}^{1/4}}$ **4.14** $K_P = 1.87$ bar$^{-1/4}$ **4.15** $x_{H_2} = x_{I_2} = x/n = 9.94 \times 10^{-2}$; $x_{HI} = (n - 2x)/n = 0.801$ **4.16** (a) 14.42 kJ mol^{-1} (b) 0.01 mol dm^{-3}; 0.045 mol dm^{-3} **4.17** 60.4 kJ mol^{-1} **4.18** (a) 0.321 (b) 408 (c) 130.9; 12.6 kJ mol^{-1} **4.19** (a) -33.26 kJ mol^{-1}; 6.71×10^5 bar^{-2} (b) -242.02 kJ mol^{-1}; 2.5×10^{42} bar^{-2} (c) -100.97 kJ mol^{-1}; 4.9×10^{17} bar^{-1} (d) 60.62 kJ mol^{-1}; 9.5×10^{-13}

4.20

	K_c	K_x
(a)	1092 (mol dm^3)$^{-2}$	6.71×10^5
(b)	4.0682×10^{39} (mol dm^{-3})$^{-2}$	2.5×10^{42}
(c)	1.9766×10^{16} (mol dm^3)$^{-1}$	4.9×10^{17}
(d)	9.5×10^{-13}	9.5×10^{-13}

4.21 (a) 28.5 kJ mol^{-1}; 16.1 kJ mol^{-1} (b) 1.994 mol; 1.994 mol; 6.30×10^{-3} mol; 6.30×10^{-3} mol **4.22** 9.81×10^{-5} dm^3 mol^{-1}; 1.66×10^5 mol dm^{-3}; 16.3 **4.23** 0.0319; 2.138; 0.0682; -6.66 kJ mol^{-1} **4.24** 3.2×10^{-8} mol dm^{-3} **4.25** (a) zero (b) positive (c) 0.0404 mol dm^{-3}; 7.96 kJ mol^{-1} (d) > 1 bar (e) negative **4.26** (a) yes (b) yes (c) positive **4.27** (a) -100.97 kJ mol^{-1}; -136.94 kJ mol^{-1}; -120.6 J K^{-1} mol^{-1}; 1 bar (b) 4.89×10^{17} bar^{-1} (c) 1.21×10^{19} dm^3 mol^{-1} (d) -108.9 kJ mol^{-1} (e) -96.2 J K^{-1} mol^{-1} (f) 7.38×10^{12} bar^{-1} **4.28** (a) -457.14 kJ mol^{-1}; -483.64 kJ mol^{-1}; -88.88 J K^{-1} mol^{-1} (b) 1.222×10^{80} bar^{-1} (c) -281.6 kJ mol^{-1}; 2.96×10^6 bar^{-1} **4.29** 2.371×10^{-43} **4.30** (a) 35.1 J K^{-1} mol^{-1}; (b) 1.66×10^5 mol dm^{-3} (c) -30.6 kJ mol^{-1}; 2.27×10^5 mol dm^{-3} **4.31** (a) -6.84 kJ mol^{-1} (b) 0.060 bar; 0.940 bar **4.32** (a) 56.8 kJ mol^{-1} (b) -56.8 kJ mol^{-1} **4.33** (a) 55.9 kJ mol^{-1}; -80.5 J K^{-1} mol^{-1} (b) 2.38×10^{14} mol^2 dm^{-6} **4.34** (a) 2.259 bar; -4.55 kJ mol^{-1}; 92.75 kJ mol^{-1}; 144.6 J K^{-1} mol^{-1} (b) 0.0404 mol dm^{-3}; 17.96 kJ mol^{-1} (c) 0.0650 M; 0.0650 M; 0.0700 M; 0.0350 M **4.35** -173.13 kJ mol^{-1}; -29.10 kJ mol^{-1}; -483.1 J mol^{-1}; 1.25×10^5 bar^{-3} **4.36** -102.8 kJ mol^{-1}; -74.1 kJ mol^{-1}; -96.3 J K^{-1} mol^{-1} **4.37** (a) -7.30 kJ mol^{-1} (b) -18.7 kJ mol^{-1}; from left to right **4.38** (a) 41.16 kJ mol^{-1}; 28.62 kJ mol^{-1}; 42.06 J K^{-1} mol^{-1} (b) 9.68×10^{-6} (c) ΔH_T°/J mol^{-1} $= 47367 - 12.55(T/\text{K}) + 1.17 \times 10^{-3}(T/\text{K})^2 + 76.6 \times 10^4/(T/\text{K})$ (d) $\ln K_P = 15.60 - 5697/(T/\text{K}) - 1.51 \ln(T/\text{K}) + 1.41 \times 10^{-4}(T/\text{K}) + (4.61 \times 10^4)/(T/\text{K})^2$ (e) 0.71 **4.39** 1.39×10^{-12} bar; 7.92×10^{-12} bar; 5.30×10^{-11} bar; 609 kJ mol^{-1}; 316.8 kJ mol^{-1}; 209 J K^{-1} mol^{-1} **4.40** (a) 0.0638; 0.202; 0.415 (b) 1.36×10^{-4} mol dm^{-3}; 16.0×10^{-4} mol dm^{-3}; 92.6×10^{-4} mol dm^{-3} (c) 1.216 kPa; 17.0 kPa; 113.4 kPa (d) 150 kJ mol^{-1}; 139 kJ mol^{-1} (e) 18.8 kJ mol^{-1}; 103 J K^{-1} mol^{-1} **4.41** 12.9 kJ mol^{-1}; -42.16 kJ mol^{-1}; -183 J K^{-1} mol^{-1} **4.42** 8.8×10^5; 500.6 K **4.43** 35.1 J K^{-1} mol^{-1}; K_c 2.26×10^5 mol dm^{-3} **4.44** 54.02 J K^{-1} mol^{-1} (1 mol dm^{-3}); 740 K **4.45** -44.9 J K^{-1} mol^{-1}; 895 K **4.46** 32.055 kJ mol^{-1}; 96.74 J K^{-1} mol^{-1}; 0.274 bar; 3.212 kJ mol^{-1} **4.47** 1.36×10^{-57} bar^{-1}; $3.37 \times$

10^{-56} dm^3 mol^{-1}; 2.72×10^{-57}; 5.22×10^{-29}
4.48 131.9; 131.9; 131.9; 0.85; no effect
4.49 3.98 kJ mol^{-1}; 0.22; 2.20 kJ mol^{-1}, 0.43; 0.42 kJ mol^{-1}, 0.85; -1.36 kJ mol^{-1}, 1.67; -3.14 kJ mol^{-1}, 3.24; -4.93 kJ mol^{-1}, 6.27; 44.4 °C
4.50 $\bar{\nu} = \frac{[SA]}{[M]} = \frac{n}{\frac{1}{K_S[A]} + 1} = \frac{nK_S[A]}{1 + K_S[A]}$

Chapter 5

5.1 1.51×10^4 bar at 1000 K **5.2** $\Delta P = 3.612 \times 10^3$ J m^{-3} $= 3.612 \times 10^3$ Pa **5.3** 42.7 kJ mol^{-1}
5.4 $P_2 = 0.730$ atm; assumption is not justified.
5.5 984 Pa **5.6** 2nd **5.7** 46.8 kJ mol^{-1}
5.8 55.07 kJ mol^{-1} **5.9** 30.105 kJ mol^{-1}, % error = 5.7% **5.10** 33.8 kJ mol^{-1}, 0.399 atm
5.11 328.48 K **5.12** $T_{trp} = 265.5$ K, $P = 3.411 \times 10^5$ Pa **5.13** 373.13 K **5.14** 27.55 Torr
5.15 $\left(\frac{\partial \Delta G}{\partial P}\right)_T = \Delta V$ **5.16** $\ln P_2/P_1 = \Delta_{vap}H_m/M \ln(T_2/T_1)[(RT_1 + M)/(RT_2 + M)]$
5.17 126 Torr **5.18** $P_2 = 1.53 \times 10^9$ Pa
5.19 $P_{tol} = 0.11$ bar, $P_{ben} = 0.205$ bar, $P_{total} = 0.316$ bar
5.20 375 K **5.21** 0.357 **5.22** $x_2 = (m_2/\text{mol kg}^{-1})(M_1/\text{g mol}^{-1})/1000$ **5.23** 3:5 volume ratio, 105 °C
5.24 $x_2 = C_2M_1/\rho_1 = \frac{(C_2/\text{mol dm}^{-3})(M_1/\text{g mol}^{-1})}{1000\ \rho_1/\text{g cm}^{-3}}$
5.25 $\mu_1 = RT \ln x_1 + \mu_1^0$ **5.26** Temperature change is 3.72 °C, which is 2×1.86 °C, the change for 1 mol of ions. **5.27** $c_2/\text{mol dm}^{-3} = (m_2/\text{mol kg}^{-1})(\rho_1/\text{g cm}^{-3})$
5.28 MW = 118.54 g mol^{-1}, Sn **5.29** $18.068 - 0.015\,744\, m^2 + 0.001\,7016\, m^3$ **5.30** $V_2 = M_2/\rho - (M_1x_1 + M_2x_2)(x_1/\rho^2)\, d\rho/dx_2$ **5.31** 0.037 56 dm^3 mol^{-1}
5.32 $a_c = 0.181$; $f_c = 0.630$ **5.33** 66.1 g mol^{-1}
5.34 positive deviation **5.35** 19.411 kPa, 0.527, 0.473 **5.36** 90 Torr, $k' = 671$ Torr **5.37** (a) 1.06×10^{-5}, 5.15×10^{-6} (b) $5.88 \times 10^{-4}\,M$; $2.86 \times 10^{-4}\,M$
5.38 0.0200 m **5.39** 181.4 g mol^{-1}
5.40 160 g mol^{-1} **5.41** 1.80 **5.42** 97.3%
5.43 $a_2 = 0.3399$; $f_2 = 1.03$ **5.44** 0.966; $a_1 = 0.941$; $f_1 = 0.975$ **5.45** 0 to -1.73 kJ mol^{-1}; $\Delta_{mix}H$ must be negative or only slightly positive.
5.46 0.970; 0.980 **5.47** (a) $\mu_A^* + RT \ln x_A + Cx_B^2$ (b) $\ln \gamma_A = C(x_B^2 - 1)/RT = 0$ when $x_B \to 1$
5.48 $x_1 = 0.302$ **5.49** $-0.003\,66$ K
5.50 249.6 kPa **5.51** 0.9737; 0.995 **5.52** 39.4 g mol^{-1} **5.53** $\alpha = (i - 1)/(\nu - 1)$, $i = 1.94$, $\alpha = 0.94$
5.54 310 g mol^{-1}; 9% error **5.55** 36.2 g mol^{-1}
5.56 from Eq. 5.125, 0.068 K; from Eq. 5.126, 0.0703 K
5.57 171.4 kJ mol^{-1} **5.58** 2.72 MPa

Chapter 6

6.1 2 degrees of freedom; one phase **6.2** water saturated with nicotine and nicotine saturated with water; 2 **6.3** (a) 2 (b) 3 (c) 2 **6.4** 2
6.5 2 **6.6** 3 **6.8** (a) 60 °C (b) 88% B (c) 53% B (d) 79% B in distillate; 88.1 g of B; 23.4 g of A **6.9** 1.2 **6.10** 112.4
6.11 (a) $P = P_2^* + (P_1^* - P_2^*)x$, (b) Eq. 6.11
6.12 $w_A/w_B = n_AM_A/n_BM_B = P_A^*M_A/P_B^*M_B$
6.13 5.51 kg **6.14** mass of chlorobenzene/mass of water = 1.93 **6.15** 0.172; 0.828 **6.17** 2.126×10^{4} K^{-1} **6.18** 6 **6.19** 38.1 atm **6.20** 58% solid and 42% liquid, $X_{Si} = 0.31$ **6.21** 0.28
6.22 (a) 8.9% phenol, 70.0% phenol (b) 1 phase present (c) 2 phases at 63.0 °C; 19.6% phenol and 52.5% phenol **6.23** $Y_3Fe_5O_{12}$ at 37% Y_2O_3 **6.24** eutectic temperature: 128 °C; composition: 26 wt % Au and 74% thallium.
6.25 (b) β first forms with liquid. At 1430 °C, β converts to α and α + L. About 50 °C lower all is solid α. **6.26** 1st eutectic (monotectic) at 680 °C at a composition of 10.5 wt % Mg; 2nd eutectic (monotectic) at 560 °C at a composition of 35 wt % Mg; 3rd eutectic at 360 °C at a composition of 65 wt % Mg. **6.27** Empirical formula of 1st compound is $MgCu_2$. Empirical formula of 2nd compound is Mg_2Cu.
6.31 Vapor pressures are listed in order of decreasing value as hydration decreases. **6.33** (a) Stable triple point R, liquid, and vapor. (b) Metastable triple point W, *L*, *V*. (c) Stable triple point W, R, *L*. If we assume that a solid cannot be superheated, the triple point is unstable. **6.36** (a) At 20% C, solid A disappears. (b) Two solid phases disappear at 50% liquid. (c) Solid salt ceases to exist. **6.37** (a) peritectic (b) eutectic (c) melting point (d) incongruent melting (e) phase transition **6.38** Solid spinel at 1950 K with *L*; at 1875 K all solid spinel; 1400 K Mn_3O_4 + spinel; below 1285 K corundum and Mn_3O_4 + corundum.
6.39 (a) 95.5% toluene, 4% acetic acid, 0.5% H_2O (b) 1% toluene, 37% acetic acid, 62% H_2O; 4*B* to 1*A*
6.40 As B is added, 2 layers are formed, one rich in A and the other rich in C. As B is added, the mutual solubility of A and C increases until at 30 mol % B, the 3 liquids become miscible in all proportions.
6.41 (a) region *AEa*: K_2CO_3 in equilibrium with conjugate liquids *a* and *c*; region *Aac*: K_2CO_3 in equilibrium with water-rich saturated solution; region *abc*: 2 conjugate liquids; region *AcB*: K_2CO_3 in equilibrium with alcohol-rich saturated solution (b) moves along line joining *x* and *A*. (c) separatory funnel (d) state moves along line joining *y* and *D*.

Chapter 7

7.1 33.8 mA **7.2** 2.4 mA **7.3** 14.3 h
7.4 1.51×10^{-3} mol dm^{-3} **7.5** 9.11 μM **7.6** 4.07×10^{-6} mol^2 dm^{-6} **7.7** 180 S cm^2 mol^{-1} **7.8** 1.87 ×

10^{-5} mol dm^{-3} **7.9** $K_b = 1.37 \times 10^{-5}$ mol dm^{-3} **7.10** $\Lambda° = 151.41$ S cm^2 mol^{-1} **7.11** 30 Ω^{-1}cm^2 mol^{-1}; 4.0×10^{-5} mol dm^{-3} **7.12** $\Lambda_c = 2.59 \times 10^{-4}$ S cm^{-1} **7.13** $\kappa = 5.528 \times 10^{-8}$ S cm^{-1}
7.14 (a) 30.5 nm (b) 0.673 nm
7.15 129.9 Ω^{-1}cm^2 mol^{-1} **7.16** 0.0403
7.17 (a) 0.317; 0.683 (b) 36.5 Ω^{-1} cm^2 mol^{-1}; 78.5 Ω^{-1} cm^2 mol^{-1}; 3.78×10^{-4} cm^2 V^{-1} s^{-1}; 8.14×10^{-4} cm^2 V^{-1} s^{-1} **7.18** 0.4428; 0.5572
7.19 3.63×10^{-3} cm^{2}V^{-1}s^{-1}; 7.91×10^{-4} cm^{2}V^{-1}s^{-1}
7.20 5.19×10^{-2} cm^2 s^{-1}; 7.92×10^{-2} cm s^{-1}
7.21 0.070 cm s^{-1}; 0.138 cm s^{-1} **7.22** (a) 2.30×10^{-19} J (b) 2.30×10^{-25} J (c) 2.30×10^{-19} J
7.23 0.358 nm; 4.96 kJ mol^{-1}
7.24 −407.1 kJ mol^{-1}; −877.9 kJ mol^{-1}; −394.1 kJ mol^{-1} **7.25** −1051.4 kJ mol^{-1}; −372.3 kJ mol^{-1}; −1828.7 kJ mol^{-1}; −4500.6 kJ mol^{-1}; −355.7 kJ mol^{-1}; −341.9 kJ mol^{-1} **7.26** 0.1 *M*; 0.3 *M*; 0.4 *M*; 0.3 *M*; 1.0 *M* **7.27** 0.026
7.28 (a) 55.90 kJ mol^{-1}; (b) 1.46×10^{-5} *M*
7.29 For $a = 0.1$

I	0.01	0.10	0.50	1.00	1.50	2.00
$\log_{10} \gamma_\pm$	−0.049	−0.15	−0.29	−0.38	−0.44	−0.49

7.30 124 kJ mol^{-1} **7.31** 0.0148 g dm^{-3}; 2.70×10^{11} mol^3 dm^{-9} **7.32** 0.025 *M*; 0.033 *M*; 0.006 67 *M*; 0.0167 *M* **7.33** (a) 1.14×10^{-3} mol dm^{-3} (b) 8.08×10^{-5} mol dm^{-3} **7.34** 0.090 mol dm^{-3}
7.35 31 J K^{-1} mol^{-1} **7.36** (a) 14.3 kJ mol^{-1} (b) 12.5 kJ mol^{-1} **7.37** 4.88×10^{-4} *M* **7.38** 1.27×10^{-5} *M* **7.39** 43.77 kJ mol^{-1} **7.40** 0.879; 0.598; 0.773 **7.41** palmitate side: [Na^+] = 0.18 *M*, [Cl^-] = 0.08 *M*; other side: [Na^+] = [Cl^-] = 0.12 *M* **7.42** 3.8×10^{-8} **7.43** (a) H_3PO_4 (b) and (c) $H_2PO_4^-$ (d) HPO_4^{2-} (e) PO_4^{3-} **7.44** palmitate side: [K^+] = 0.16 *M*, [Cl^-] = 0.01 *M*; other side: [K^+] = [Cl^-] = 0.04 *M*

Chapter 8

8.1 (a) $H_2 \rightarrow 2H^+ + 2e^-$; $Cl_2 + 2e^- \rightarrow 2Cl^-$; $H_2 + Cl_2 \rightarrow 2H^+ + 2Cl^-$, $z = 2$; $E = E° - \frac{RT}{2F} \ln(a^2_{H^+} a^2_{Cl^-})^u$ (b) $E° + \frac{RT}{2F} \ln(a^2_{H^+} a^2_{Cl^-})^u$
(c) $2Ag(s) + Hg_2Cl_2(s) \rightarrow 2AgCl(s) + 2Hg(s)$
$E = E°$ (no concentration dependence) (d) $E = E° - \frac{RT}{F} \ln(a_{H^+} a_{I^-})^u$ (e) $E = \frac{RT}{F} \ln \frac{a_1}{a_2}$ **8.2** (a) AH_2 is oxidized by B. (b) 0.44 V (c) none **8.3** −0.90 V
8.4 $H_2 \rightarrow 2H^+(1\ m) + 2e^-$; $2e^- + 2H^+ (c) + F^{2-} \rightarrow S^{2-}$; overall reaction: $2H^+ (c) + F^{2-} + H_2 \rightarrow 2H^+(1\ m) + S^{2-}$

$$E = E° - \frac{RT}{2F} \ln \frac{[S^{2-}]}{[F^{2-}]c^2}$$

8.5 (a) $E° = E°_{Ce^{4+}|Ce^{3+}} - E°_{Fe^{3+}|Fe^{2+}} = 0.67$ V
(b) $E° = E°_{Ag^+|Ag} - E°_{AgCl|Ag} = 0.5777$ V
(c) $E° = E°_{HgO|Hg} - E°_{H_2O|H_2} = 0.9254$ V
8.6 9.10×10^7 dm^6 mol^{-2} **8.7** 5.31×10^{-11}
8.8 −24.1 kJ mol^{-1} **8.9** −237.4 kJ mol^{-1}
8.10 1.66×10^6 dm^3 mol^{-1}; half will form Cu^{2+} and half Cu **8.11** 0.1446 V, Sb **8.12** 0.0178 V **8.13** −0.157 V **8.14** (a) −0.0365 V (b) −0.16 V **8.15** 0.5549 V **8.16** 0.934 V **8.17** 0.463 *M*
8.18 10.4 mV **8.19** Au, 1.90×10^{-32} mol dm^{-3}; acceptable **8.20** −11.8 mV **8.21** −29.6 mV
8.22 (a) 2.26×10^{-14} C (b) 1.5×10^{-6}
8.23 (a) −0.152 V (b) −0.147 V **8.24** 0.48
8.25 (a) $2Tl(s) + Cd^{2+}(0.01m) + 2Cl^-(0.02m) \rightarrow 2TlCl(s) + Cd(s)$ (b) −0.054 V; 0.105 V
8.26 right-hand side; 37 mV **8.27** 6.87×10^{14}; $K'' = 6.87 \times 10^{12}$ **8.28** −0.118 V **8.29** (b) 0.839; 0.161 **8.30** 8.2×10^{-8} mol^2 kg^{-2} **8.31** (a) 1.2×10^5; (b) 1.2×10^4 **8.32** (a) 0.216 V; −41.7 kJ mol^{-1}; 4.21 J K^{-1} mol^{-1}; (b) −40.4 kJ mol^{-1}
8.33 (a) $Cd(Hg) + Hg_2SO_4(s) + \frac{8}{3} H_2O(l) \rightarrow CdSO_4 \cdot \frac{8}{3} H_2O(s) + 2Hg(l)$;
(b) −196.5 kJ mol^{-1}; −199.4 kJ mol^{-1}; −9.65 J K^{-1} mol^{-1} **8.34** -2.8834×10^{-4} V K^{-1}, −27.820 J K^{-1} mol^{-1} **8.35** -1.7477×10^{-3} V K^{-1}
8.36 (a) −603.8 kJ mol^{-1} (b) −780.0 kJ mol^{-1}
8.37 (a) 0.58 V (b) −119 kJ mol^{-1} **8.38** 5.03×10^{-13} mol^2 kg^{-2}; 7.09×10^{-7} mol kg^{-1} **8.39** 2.49
8.40 9.24×10^{-9} mol kg^{-1}; 8.45×10^{-17} mol^2 kg^{-2}
8.41 0.8509 **8.42** (a) 2750 kJ mol^{-1} (b) 2740 kJ mol^{-1} **8.43** (a) 0.679 (b) 0.665, good agreement **8.44** 1.010×10^{-14}

Chapter 9

9.1 (a) 3 (b) Both rates are 3.6×10^{-3} mol dm^{-3} s^{-1}. (c) none (d) decreased by factor of 8; none
9.2 1; 1; 6.2×10^{-4} dm^3 mol^{-1} s^{-1} **9.3** 2; 1; 1.64×10^{-3} dm^6 mol^{-2} s^{-1} **9.4** (a) 5.18 hr (b) 0.669
9.5 6.64 times **9.6** (a) 77 ps (b) 77 ns
9.7 (a) 0.540 μg (b) 0.291 μg (c) 0.177 μg
9.8 5.967×10^{-3} min^{-1}, 116 min **9.9** 4.32×10^{-4} mol dm^{-3} s^{-1} **9.10** $\frac{-d\,[NOCl]}{dt} = k[NOCl]^2 = \frac{k}{RT}(3P_0 - 2P_t)^2$ **9.11** 2.0 **9.12** $k = 5.61 \times 10^{-7}$ s^{-1} (a) 61.6% (b) 37.9% (c) 0.78%
9.13 $t_{1/2}$ = 87.0 days; $k = 9.22 \times 10^{-8}$ s^{-1} (a) 2653 (b) 233 **9.14** 290 s

9.15

t	P_A/bar	P_B/bar	P_C/bar	P/bar
0	2.00	0.00	0.00	2.00
$t_{1/2}$	1.00	1.00	0.50	2.50
$2t_{1/2}$	0.67	1.33	0.67	2.67
∞	0.00	2.00	1.00	3.00

9.16 Start with $\frac{dx}{dt} = k(a_0 - x)^n$, integrate and substitute $x = a_0/2$; $t_{1/2} = \frac{2^{(n-1)} - 1}{k a_0^{(n-1)}(n-1)}$.
9.17 2nd order, $k = 2.3182 \times 10^{-5}(M\ s)^{-1}$
9.18 Let c be the concentration produced by 1 dose. When the concentration reaches nc, the concentration will fall to $(n - 1)\,c$ during the interval between successive doses. Then the next dose restores the concentration to a "steady state." Show math calculation.

9.19 $kt = \frac{1}{2(2b_0 - a_0)}\left[\ln \frac{a_0(b_0 - x)}{b_0(a_0 - 2x)}\right]$

9.20 $\frac{2a_0 x - x^2}{a_0^2 (a_0 - x)^2} = 8kt,\ t_{1/2} = \frac{3}{8a_0^2 k}$

9.21 Take ratio of $\frac{d[Y]}{dt}$ to $\frac{d[Z]}{dt}$ and integrate at $t = 0$, $[Y] = [Z]$ and $I = 0$. Q.E.D.
9.22 Take the rate of consumption of A and the rates of formation of B and C. Integrate $\frac{-d\,[A]}{dt}$ and insert into $\frac{d\,[B]}{dt}$. With boundary conditions $t = 0$ and $[B] = 0$, obtain $[A]_0 = [A] + [B] + [C]$. Then substitute.
9.23 (a) $k_1 = \frac{x_e}{(2a_0 - x_e)t} \ln \frac{[a_0 x_e + x(a_0 - x_e)]}{a_0(x_e - x)}$

(b) $K = \frac{a_0 - x_e}{x_e^2}$, solve for x_e in terms of K and substitute into k_1. (c) $k_1 = 1.51 \times 10^{-4}\ s^{-1}$; $k_2 = 3.73 \times 10^{-4}\ dm^3\ mol^{-1}\ s^{-1}$; $7.46 \times 10^{-3}\ dm^6\ mol^{-2}\ s^{-1}$
9.24 $\frac{d\Delta x}{dt} = -(k_1 + 2k_{-1}x_e)\Delta x$. Integrate and apply boundary conditions. The relaxation t^* is the time corresponding to $\frac{(\Delta x)_0}{\Delta x} = e$. Q.E.D.

9.25 51.2 kJ mol^{-1} **9.26** form $\Delta^{\ddagger}G$ = 143.0 kJ mol^{-1} **9.27** $E_a = 9.73$ kJ mol^{-1}, $A = 9.81 \times 10^{13}$ min^{-1} **9.28** (a) 6.68×10^{-3} (b) 6.62 **9.29** $E_a = nRT + E$
9.30 (a) 36.159 kJ mol^{-1}; (b) experimental: 2.37×10^{-13} cm^3 molecule^{-1} s^{-1}; theoretical: 2.17×10^{-13} cm^3 molecule^{-1} s^{-1}
9.31 49.5 kJ mol^{-1} **9.32** 120.7 kJ mol^{-1}
9.33 −1.5 **9.34** 53.5 kJ mol^{-1} **9.35** (a) $k_1 a_e b_e = k_{-1}z_e = 0$ (b) $\frac{dx}{dt} = k_1(a_e - x)(b_e - x) - k_1(z_e - x)$
(c) $\frac{dx}{dt} = [k_1(a_e + b_e) + k_{-1}]\,x$ (d) $\frac{1}{t^*} = 2k_1 a_e + k_{-1}$
(e) $\frac{1}{t^*} = 2\sqrt{k_1 k_{-1} z_e} + k_{-1}$
9.36 $k_{-1} = 79.3\ s^{-1}$, $k_1 = 28.5\ M^{-1}\ s^{-1}$, $K = 0.360\ M^{-1}$ **9.37** $k_{-1} = 1.25 \times 10^{12}\ M^{-1}\ s^{-1}$, $k_1 = 2.25 \times 10^{-4}\ s^{-1}$ **9.38** 409 **9.39** $\Delta^{\ddagger}H^{\circ}$ = 138.8 kJ mol^{-1}; −76.7 J K^{-1} mol^{-1}; 190.4 kJ mol^{-1}; $A = 1.02 \times 10^{10}$ dm^3 mol^{-1} s^{-1} **9.40** 84.1 kJ mol^{-1}; 81.6 kJ mol^{-1}; 100.7 kJ mol^{-1}; 7.48×10^9 s^{-1}; −64.1 J K^{-1} mol^{-1}
9.41 $\Delta^{\ddagger}G^{\circ}$ = 108.8 kJ mol^{-1}; E = 90.0 kJ mol^{-1}; 87.5 kJ mol^{-1}; $A = 4.94 \times 10^9$ s^{-1}; −68.0 J K^{-1} mol^{-1} **9.42** $\Delta^{\ddagger}H^{\circ}$ = 348.6 kJ mol^{-1}; 730 J K^{-1} mol^{-1} **9.43** (a) 3.04×10^{31} m^{-3} s^{-1} (b) 8.60×10^{-7} dm^3 mol^{-1} s^{-1}; −62.1 J K^{-1} mol^{-1}
9.44 3.08×10^{-11} s^{-1}; $\Delta^{\ddagger}H^{\circ}$ = 109.5 kJ mol^{-1}; 119.4 kJ mol^{-1}; −33.3 J K^{-1} mol^{-1} **9.45** 4.18×10^{17} dm^3 mol^{-1} s^{-1}; 117.5 kJ mol^{-1}; $\Delta^{\ddagger}G^{\circ}$ = 92.4 kJ mol^{-1}; 84.1 J K^{-1} mol^{-1} **9.46** 2
9.47 1.92 dm^3 mol^{-1} s^{-1}; −2 **9.48** $\log_{10} k = \log_{10} k_0 - B\,[z_A{}^2 + z_B{}^2]\sqrt{I}$; inconsistent
9.49 7.54×10^7 dm^3 mol^{-1} s^{-1}; zero **9.50** 2
9.51 3.25×10^{-4} dm^3 mol^{-1} s^{-1}
9.52 −16.5 cm^3 mol^{-1} **9.53** −18.7 cm^3 mol^{-1}
9.54 −14.3 cm^3 mol^{-1} **9.55** 10.4 cm^3 mol^{-1}; 10.9 cm^3 mol^{-1}; 20.0 cm^3 mol^{-1}

Chapter 10

10.1 $v = k_1[A][B]$ **10.2** $v = k_2\left(\frac{k_1}{k_{-1}}\right)^{1/2}[A]^{1/2}[B]$

10.3 (a) $v = \frac{k_1 k_2}{k_{-1}}[A][B]$ (b) $v = k_1[A]$

10.4 2A → X (very slow); X + 2B → 2Y + 2Z (very fast) **10.5** 2 simultaneous reactions
10.6 2 consecutive reactions

10.7 $v_{N_2O_5} = \frac{k_1 k_2 [N_2O_5][NO]}{k_{-1}[NO_2] + k_2[NO]}$

10.8 $v = \frac{k_1 k_2}{k_{-1}}[NO]^2[O_2]$ if $k_{-1} >> k_2[O_2]$

10.9 $v = v_{HCl} = k_2\,[Cl][CH_4] = k_2\left(\frac{k_1}{k_{-1}}\right)^{1/2}[Cl_2]^{1/2}\,[CH_4]$

10.10 $v_{O_3} = \frac{2k_1 k_2 [O_3]^2}{k_{-1}[O_2] + k} = 2k_1[O_3]^2$ in the absence of O_2

10.11 $E = \frac{k_2 E_1 + k_{-1}(E_1 + E_2 - E_{-1})}{k_{-1} + k_2}$

10.12 $v = k_2[A^*] = \frac{k_1 k_2 [A]^2}{k_{-1}[A] + k_2}$

10.13 fraction, $f = \frac{c_0 kt}{1 + c_0 kt}$ **10.14** $v = \frac{k_1 k_2 k_3}{k_{-1} k_{-2}}[H_2][I_2]$

10.15 (a) A in excess: $v_z = (k_1 k_3/k_2)[B]^2$ (b) B in excess: $v_z = k_1[A][B]$ **10.16** 306 nm
10.17 2 molecules of reactant are decomposed for each photon absorbed.

$$H + HI \rightarrow H_2 + I$$
$$I + I \rightarrow I_2$$

Net reaction $2HI + h\nu \rightarrow H_2 + I_2$
10.18 47 s **10.19** 0.76 mol
10.20 1.28×10^{15} photons

10.21 $v = k_2\left(\frac{2}{k_4}\right)^{1/2} I_a^{1/2}[CHCl_3]$

10.22 18.1 eV; 7.4 eV

10.23 $k_{obs} = k_2\left(\frac{2k_1}{k_4}\right)^{1/2}$ **10.24** 4.6×10^{-4} mol dm^{-3}

10.25 $v = k_2\left(\frac{k_1}{k_{-1}}\right)^{1/2}[I_2]^{1/2}[CH_3CHO]$

10.26 $k = GK^{\rho/\rho'} = GK^{\alpha}$ **10.27** $t_{1/2} = 2.48 \times 10^6$ s ($\approx$ 29 days) **10.28** $v = k_3[Y] = \frac{k_1k_2[S][OH^-]}{k_{-1} + k_2}$

10.29 Differentiate k with respect to $[H^+]$ and set equal to zero. **10.30** Slope is $0.53 = \beta$.
10.31 Species

$CH_3—C(O^-) = CH_2$ reacts with Br_2, giving rate independent of $[Br_2]$.

10.32 $Cu^{2+} + H_2 \underset{k_{-1}}{\overset{k_1}{\rightleftarrows}} CuH^+ + H^+$; $CuH^+ + Cu^{2+} \xrightarrow{k_2} 2Cu^+ + H^+$ **10.33** $2Ag^+ + H_2 \rightarrow 2AgH^+$ first term; second term explained by

$Ag^+ + H_2 \rightleftharpoons AgH + H^+$
$AgH + Ag^+ \rightarrow 2Ag + H^+$ or $AgH^+ + Ag$

10.34 Initial fast equilibrium: $Ag^+ + Ce^{4+} \rightleftarrows Ce^{3+} + Ag^{2+}$ followed by a slow reaction $Ag^{2+} + Tl^+ \rightarrow Tl^{2+} + Ag^+$. This may be followed by a fast reaction: $Tl^{2+} + Ce^{4+} \rightarrow Tl^{3+} + Ce^{3+}$. Overall rate is: $v = k_2 K_1 [Ce^{4+}][Ag^+][Tl^+]/[Ce^{3+}]$ **10.35** 2.0 mmol dm^{-3}
10.36 16 μmol dm^{-3}; 0.22 μmol dm^{-3} s^{-1}; Curve 1 is nonlinear; Curve 2 is too sensitive to high $\frac{1}{[S]}$ values; Curve 3 gives evenly weighted data.

10.37 (a) $E = 8.2$ kJ mol^{-1}; 5.7 kJ mol^{-1}; $\Delta^{\ddagger}G =$ 42.9 kJ mol^{-1}; -124 J K^{-1} mol^{-1}
(b) $\Delta G = -19.6$ kJ mol^{-1}; $\Delta^{\ddagger}H^{\circ} = 14.2$ kJ mol^{-1}; 113 J K^{-1} mol^{-1} **10.38** 5.8 mmol dm^{-3}; $[E]_o = 4.0 \times 10^{-5}$ mol dm^{-3}; 12.5 s^{-1}
10.39 $E = 90.5$ kJ mol^{-1}; 87.9 kJ mol^{-1}; $\Delta^{\ddagger}G =$ 108.7 kJ mol^{-1}; $\Delta^{\ddagger}S^{\circ} = -66.5$ J K^{-1} mol^{-1}

10.40 $v = \frac{k_1k_2k_3[E]_0[A][B]}{k_{-1}k_3 + k_1k_3[A] + k_2k_3[B] + k_1k_2[A][B]}$

10.41 (a) $v = \frac{k_2[E]_0[S]}{K_m\left(1 + \frac{[I]}{K_i}\right) + [S]}$

(b) $\epsilon = \frac{\frac{K_m}{K_i}[I]}{K_m\left(1 + \frac{[I]}{K_i}\right) + [S]}$

10.42 $k_c = \frac{k_1k_2}{k_{-1} + k_2}$; $K_m = \frac{k_{-1} + k_2}{k_1}\frac{k_3}{k_2 + k_3}$

10.43 Suppose that ΔH_m is positive: K_m increases with increasing T and at low T will be smaller than [S]. Rate increases with increasing T. At higher T, K_m will be larger than [S], and the effective activation energy is $E_c - \Delta H_m$. Rate can then go through a maximum.

10.44

$$v = \frac{k_1k_2k_3[E]^0[A][B]}{k_{-1}(k_{-2} + k_3) + k_1(k_{-2} + k_3)[A] + k_2(k_{-2} + k_3)[B] + k_1k_2[A][B]}$$

10.45 $v = \frac{k_1k_2k_3k_4[E]^0[A][B]}{k_1k_2k_4[A] + k_3k_4(k_{-1} + k_2)[B] + k_1k_3(k_2 + k_4)[A][B]}$

10.46 Chain length is $\frac{k_p[M]^{1/2}}{(k_i k_t)^{1/2}[C]^{1/2}}$

10.47 Write the steady-state equations: sum is $2I - k_t(\Sigma[R_n])^2 = 0$ Rate of removal of monomer is $v = k_p[M]\Sigma[R_n] = k_p\left(\frac{2I}{k_t}\right)^{1/2}[M]$

Chapter 11

11.1 (a) 9.22×10^{14} s^{-1} (b) 3.08×10^4 cm^{-1}
(c) 6.11×10^{-19} J; 3.81 eV; 368 kJ mol^{-1}
(d) 2.04×10^{-25} kg m s^{-1} **11.2** (a) 1.526 m
(b) 1.302×10^{-25} J; 8.127×10^{-7} eV; 0.07841 J mol^{-1} (c) 4.342×10^{-34} kg m s^{-1}
11.3 1.73×10^{16} s^{-1} **11.4** 0.3 s^{-1}
11.5 (a) 0.88 N m^{-1} (b) 0.021 m s^{-1} **11.6** k_BT
11.7 1.39×10^{20}; 1.204×10^{-27} kg m s^{-1}
11.8 1.82 eV; 0.23 eV **11.9** (a) 12.1 pm
(b) 29.3 pm (c) 6.65 pm (d) 2.58 pm
11.10 10^{-10} m s^{-1} **11.11** (a) 1.875×10^6 m s^{-1}; 388 pm (b) 1.875×10^7 m s^{-1}; 38.8 pm
(c) 5.93×10^8 m s^{-1}; 1.23 pm **11.12** u
11.13 (a) 248 nm; (b) 1.078×10^6 m s^{-1}
11.14 (a) 2.410×10^{-21} J (b) 2.410×10^{-23} J
11.15 (a) 1.016×10^{-12} m (b) 1.96×10^{-34} m

11.16 (a) $\frac{1}{\sqrt{2}}(\psi_1 + \psi_2)$

(b) $\frac{1}{\sqrt{2}}(\psi_1 - \psi_2)$ (c) $\frac{1}{\sqrt{3}}(\psi_1 + \psi_2 + \psi_3)$

(d) $\frac{1}{\sqrt{3}}\left(\psi_1 - \frac{1}{\sqrt{2}}\psi_2 + \frac{3}{\sqrt{2}}\psi_3\right)$

11.17 $\frac{d}{dx}(-Aae^{-ax}) = Aa^2e^{-ax}$, eigenvalue $= a^2$
11.18 $\cos 2\pi m_l = 1$ and $\sin 2\pi m_l = 0$. This can occur only if m_l is an integer.

11.19 $\psi^*\psi e^0 = \psi^*\psi$ **11.20** $\frac{h}{2\pi i}\frac{\partial}{\partial x}$; apply to $\phi(x)$ and $\psi(x)$. **11.21** (a) k is eigenfunction with eigenvalue 0 (b) not an eigenfunction
(c) not an eigenfunction (d) an eigenfunction with an eigenvalue k (e) not an eigenvalue

(f) is an eigenfunction with the eigenvalue ik.

11.22

m_l	2	1	0	−1	−2
θ/degrees	35.3	65.9	90.0	114.1	144.7

11.23 The operator corresponds to $x^2 + y^2$ $(= a^2)$. The physical property is the square of the radius of the base of the angular momentum vector as it rotates about the X-axis. **11.24** Both energy and position would be known.
11.25 (a) 1.13×10^4 eV (b) 1.13×10^{12} eV

11.26 (a) $A = \sqrt{\frac{ab}{b-a}}$ (b) $\frac{ab}{b-a} \ln \frac{b}{a}$ **11.27** 5; 11

11.28 no **11.29** (a) $\frac{-h^2}{8\pi^2 m} \nabla^2 \psi + E_p(x, y, z)\psi = E\psi$

(b) $-\frac{1}{X}\frac{\partial^2 X}{\partial x^2} = \frac{8\pi^2 mE_x}{h^2}$ (c) $C = \sqrt{\frac{2}{a}}$

(d) If $a = b = c$, $E = \left(n_x^2 + n_y^2 + n_z^2\right) \frac{h^2}{8ma^2}$

11.30 $P_{QM} = \frac{1}{3} - \frac{1}{n\pi}\left[\sin\left(\frac{4\,n\pi}{3}\right) - \sin\left(\frac{2\,n\pi}{3}\right)\right]$

11.31 (a) $P_{Cl} = \int_{a/3}^{2a/3} \frac{dx}{a} = \frac{1}{3}$

(b) $n \to \infty$, second term vanishes. **11.32** (a) 1.53×10^{-17} J = 95.3 eV (b) 1.53×10^{-9} J = 9.53×10^9 eV **11.33** The appropriate integral is $\frac{2}{\pi}\int_0^{\pi} \sin my \sin ny \, dy$. Its value is zero. Wave functions are orthogonal.

11.34 $1.0132 \frac{h^2}{8ma^2}$; 1.32% **11.35** (a) 1, 2

(b) $r = \frac{3a_0}{2z}\left(3 - \sqrt{3}\right)$ and $= \frac{3a_0}{2z}\left(3 + \sqrt{3}\right)$

11.36 2.34×10^{-20} J; 7.03×10^{-20} J

11.37 $\hat{H} = -\frac{h^2}{8\pi^2 I}\frac{\partial^2}{\partial\phi^2}$ **11.38** 2.18×10^{-18} J = 13.60 eV

11.39 2.19×10^6 m s^{-1}; 3.32×10^{-10} m = 332 pm; n **11.40** $r = a_0/Z$ **11.41** (a) 9.1046×10^{-31} kg; 9.1070×10^{-31} kg (b) 656.30 nm
11.42 4.052×10^{-6} m; 4.903×10^{-20} J
11.43 -10.9×10^{-18} J; −0.25 au **11.44** 1.26
11.45 1.84 **11.46** (a) 6.15 (b) 4.9 (c) 2.2

11.47 $Z_{eff} = 27/16$; $E = -\frac{2.848e^2}{a_0}$; the effective charge is less than the true charge because of screening.

11.48 $\frac{4}{a_0^3} r^2 e^{-2r/a_0} \, dr$

11.49 $\sum_{l,m} [\Theta_{l,m}(\theta)\,\Phi_m(\phi)][\Theta_{l,m}(\theta)\,\Phi_m(\phi)]^* = \frac{3}{4\pi}$

Chapter 12

12.1 336 pm; −341 kJ mol^{-1} **12.2** 55%
12.3 (a) 0.46 D (b) 5.9% **12.4** $BeCl_2$, linear; SF_6, octahedral; H_3O^+, trigonal pyramidal; NH_4^+, tetrahedral; PCl_6^-, octahedral; AlF_6^{3-}, octahedral; PO_4^{3-}, tetrahedral; CO_2, linear; SO_2, bent; NH_3^{2+}, trigonal planar; CO_3^{2-}, trigonal planar; NO_3^-, trigonal planar **12.5** 17.7%; 11.8%; 5.5%; 1.8%
12.6 1, 3, 2, 3, 2.5, 3, 1.5, 1, 2 **12.7** (a) C_2, CN (b) O_2, F_2, NO **12.8** Show that $\int (1s_A + 1s_B) \times (1s_A - 1s_B)\, d\tau = 0$. **12.9** (a) Multiply out $t_3 \times t_3^*$ and take integrals of each term = 1. (b) Multiply out $t_2 \times t_4^*$ and take integrals of each term = 0.

12.10 $a = \frac{1}{\sqrt{3}}$, $b = \sqrt{\frac{2}{3}}$, $c = \sqrt{\frac{2}{3}}$

12.11 $\psi_1 = \frac{1}{\sqrt{2}}(s + p_z)$, $\psi_2 = \frac{1}{\sqrt{2}}(s - p_z)$

12.12 $\langle\psi_1|L^2|\psi_1\rangle = \frac{4}{3}\hbar^2$, same for ψ_2 and ψ_3.
12.13 (a) Based on Figure 12.26 a, the molecule is diamagnetic. (b) Based on Figure 12.26 b, the molecule is paramagnetic. **12.14** Linear: $E = \alpha$, $\alpha \pm \beta\sqrt{2}$; trianglular: $\alpha + 2\beta$, $\alpha - \beta$, $\alpha - \beta$. Triangular arrangement is lower in energy.
12.15 c_n/c_1 = 1.000, 1.6180, 1.6180, 1.0000
c_n = 0.3717, 0.6015, 0.6015, 0.3717
12.16 (a) D_{3h} (b) C_{2v}(c) $D_{\infty h}$
12.17 C_{3v}, C_{2v}, D_{2h}, C_{2v}, C_{2v}, D_{3h}, D_{3h}, $D_{\infty h}$
12.18 Yes, but has not been detected because of interconversion between forms. **12.19** (a) σ_g^+ (b) a_1, b_2, a_1 (c) σ_g^+, σ_u^+ (d) a'_1 (e) a_1 (f) σ^+
12.20 Cyclopropane: C_3 axis is also the S_3 axis, $S_3 = C_3\sigma_h$. Cyclopentane: C_5 axis is also the S_5 axis, $S_5 = C_5\sigma_h$. C_1, C_2, C_{2v}, C_{3v}, and $C_{\infty v}$.
12.21 E_g symmetry (d_{z^2} and $d_{x^2-y^2}$)

Chapter 13

13.1 13.86 g dm^{-3} **13.2** 0.152: 0.705; ϵ = 3.04 dm^3 cm^{-1} mol^{-1} **13.3** A = 0.112, 1117 m^2 mol^{-1} **13.4** 12.0 dm^3 cm^{-1} mol^{-1}
13.5 A = 2.76, 0.17% **13.6** 2.91×10^{-4} mol dm^{-3}
13.7 1.028×10^4 dm^3 mol^{-1} cm^{-1}, 97.7%
13.8 NADH: 3.46×10^{-5} mol dm^{-3}; NAD^+: 1.04×10^{-5} mol dm^{-3}
13.9 276 dm^3 cm^{-1} mol^{-1}, 0.17%
13.10 1.668×10^{-4} M, 21.6 kJ mol^{-1}
13.11 1.097×10^7 m^{-1}, 182.3 kJ mol^{-1}
13.12 $^2S_{1/2}$, $^2P_{3/2}$ and $^2P_{1/2}$
13.13 $^1S_0$ $^1P_1$ $^1D_2$, $^3D_3$ $^3D_2$ $^3D_1$, $^3P_2$ $^3P_1$ $^3P_0$, $^3S_1$
13.14 (a) $^2P_{3/2}$ $^2P_{1/2}$ (b) $^2D_{5/2}$ $^2D_{3/2}$ ^1D
13.15 ^1P: $J = 1$; ^3P: $J = 2, 1, 0$; ^4P: $J = \frac{5}{2}, \frac{3}{2}, \frac{1}{2}$;

^{1}D: $J = 2$; ^{2}D: $J = \frac{5}{2}, \frac{3}{2}$; ^{3}D: $J = 3, 2, 1$; ^{4}D: $J = \frac{7}{2}, \frac{5}{2}, \frac{3}{2}, \frac{1}{2}$ **13.16** 2/3, 1.24 cm^{-1} **13.17** 0.93 cm^{-1} **13.18** 164 pm **13.19** 92 pm, 22.1 cm^{-1}, 15.4 cm^{-1} **13.20** 113 pm **13.21** (a) 0.1131 nm (b) 13.0007 (c) 0.113 12 nm **13.22** Find B values and moments of inertia. Use Eq. 13.103 to find I. Compare values. **13.23** 0.0479 Å, 4.2% change **13.24** (a) HCl, H_2O, CH_3Cl, CH_2Cl_2, H_2O_2, NH_3 (b) all except H_2 (c) all except CH_4 and SF_6 (d) all **13.25** 511.6 kg s^{-2} **13.26** P branch, $\tilde{\nu}_0 = 2888.70$ cm^{-1}, $\tilde{B} = 11.307$ cm^{-1} **13.27** $\Delta T_{0,J'' \rightarrow 1,J'} = \tilde{\nu}_0 - 2\tilde{\nu}_0 x_e + \tilde{B}[J'(J' + 1) - J''(J'' + 1)]$ **13.28** $x_e = 1.540 \times 10^{-2}$ **13.29** 199 pm **13.30** 457.1 kJ mol^{-1}, 21.7 kJ mol^{-1}, 435.4 kJ mol^{-1}, 17.7 kJ mol^{-1}, 439.4 kJ mol^{-1} **13.31** A − B − B **13.32** 2400 cm^{-1} **13.33** (a) 3530.2, 3492.4, 3454.5 (b) 3568.0, 3605.8, 3643.6 **13.34** 1973.7 cm^{-1} **13.35** 314.3 kg s^{-1} **13.36** *cis*-$C_2H_2Cl_2$: The point group is C_{2v}.

(a) a_1; active in infrared and Raman
(b) a_2; inactive in infrared, active in Raman
(c) b_1; active in infrared and Raman
(d) a_1; active in both infrared and Raman
trans-$C_2H_2Cl_2$: The point group is C_{2h}.
(e) a_g; inactive in infrared, active in Raman
(f) a_u; active in infrared, inactive in Raman
(g) b_u; active in infrared, inactive in Raman
(h) a_g; inactive in infrared, active in Raman
Benzene: The point group is D_{6h}.
(i) a_{1g}; inactive in infrared, active in Raman
(j) a_{2u}; active in infrared, inactive in Raman
(k) b_{1u}; inactive in both infrared and Raman

13.37 8.93×10^{13} s^{-1} **13.38** (a) $k = D_e(a_1^2 - 2a_2)$ (b) From $-D_e(a_1 - \alpha) = 0$; $\alpha = a_1$ **13.39** $E_p(x) = -D_e(c_1 e^{-\beta x} + c_2 e^{-2\beta x})$ and $E_p(x) = -D_e(2e^{-\beta x} - e^{-2\beta x}) + D_e$; $c_1 = 2$ and $c_2 = -1$ **13.40** 853.58 cm^{-1} **13.41** 1049.2 cm^{-1}, 1890.2 cm^{-1}

13.42 (a) $\frac{\Delta\tilde{G}_{v'}}{v' - v''} = (\tilde{\nu}_0 - \tilde{\nu}_0 x_e) - \tilde{\nu}_0 x_e (v' + v'')$ (b) $\tilde{\nu}_0 = 2298$ cm^{-1}, $x_e = 2.552 \times 10^{-2}$

13.44 56 858.6 cm^{-1}, $\Delta\tilde{\nu} = 160$, 100, and 9 **13.45** Area under linear extrapolation is 22 550 cm^{-1}; under the curve, it is 22 350 cm^{-1}. **13.46** $\tilde{\nu} = \tilde{T}_e + \left[\left(v' + \frac{1}{2}\right)\tilde{\nu}_0' - \left(v' + \frac{1}{2}\right)^2 \tilde{\nu}_0' x_e'\right] - \left[\left(v'' + \frac{1}{2}\right)\tilde{\nu}_0'' - \left(v'' + \frac{1}{2}\right)^2 \tilde{\nu}_0'' x_e''\right]$ **13.47** $\tilde{\nu}(\text{cm}^{-1}) = 15\,738.5 + \left[\left(v' + \frac{1}{2}\right)131.1 - \left(v' + \frac{1}{2}\right)^2 0.9736\right] - \left[\left(v'' + \frac{1}{2}\right)221.7 - \left(v'' + \frac{1}{2}\right)^2 2.763\right]$ **13.48** 0.85 eV = 82.0 kJ mol^{-1}

13.49 $D_0 = D_e - (\frac{1}{2}\tilde{\nu}_0 - \frac{1}{4}\tilde{\nu}_0 x_e) = 4.4336$ eV

13.50 5.7 eV + 3.3 eV = 9.0 eV.

13.51 2.518 eV, 2.104 eV; dissociation energy of Na_2 = 73.7 kJ mol^{-1}

Chapter 14

14.1 4.7×10^5 K **14.2** (a) 2.7×10^{-10} s (b) 2.7×10^{-11} s (c) 2.7×10^{-12} s (d) 4.05×10^{-10} s **14.3** (a) NO ($v' = 0$) formation corresponds to higher O atom velocity. (b) $v' = 0$ state: $E_0 = \frac{1}{2}hc\tilde{\nu}_0 = 1.855 \times 10^{-20}$ J; $v' = 1$ state: $E_1 = \frac{3}{2}hc\tilde{\nu}_0 = 5.654 \times 10^{-20}$ J **14.4** 1.324 T **14.5** (a) 2.0026 (b) 0.3385 T **14.6** ^{2}H: 3, 4.329; ^{19}F:2, 13.28; ^{35}Cl: 4, 4.144; ^{37}Cl: 4, 3.455 **14.7** 3 **14.8** 1.41 T; 14.4 T **14.9** −11.4 μT; 456 Hz **14.10** 4; 1.99 MHz **14.11** 4.39 T **14.12** $I(\omega) = A\left[\int_0^\infty \cos(\omega_1 t)\cos(\omega t)\, e^{-t/T}\, dt + \int_0^\infty \cos(\omega_2 t)\cos(\omega t)\, e^{-t/T}\, dt\right]$ **14.13** 3.98×10^5 Hz; 1.33×10^{-5} cm^{-1} **14.14** sp^2, none, a_2'' and e′ **14.15** $E = h\frac{c}{\lambda} = 21.2$ eV; $\lambda = 3.02$ nm

Chapter 15

15.1 $C_P = \left\{\frac{\partial}{\partial T}\left[k_B T^2 \left(\frac{\partial \ln Q}{\partial T}\right)_V\right]\right\}_P + Nk_B$

15.2 (a) $U - U_0 = Nk_B T^2 \left(\frac{\partial \ln q}{\partial T}\right)_V$; $S = Nk_B T \left(\frac{\partial \ln q}{\partial T}\right)_V + Nk_B \ln q$;

$A - U_0 = -Nk_B T \ln q$; $H - U_0 = Nk_B T^2 \left(\frac{\partial \ln q}{\partial T}\right)_V + Nk_B T$;

$G - U_0 = -Nk_B T \ln \frac{q}{e}$ (b) $U - U_0 = Nk_B T^2 \left(\frac{\partial \ln q}{\partial T}\right)_V$;

$S = Nk_B T \left(\frac{\partial \ln q}{\partial T}\right)_V + Nk_B \ln \frac{qe}{N}$;

$A - U_0 = -Nk_B T \ln \frac{qe}{N}$; $H - U_0$: $Nk_B T^2 \left(\frac{\partial \ln q}{\partial T}\right)_V + Nk_B T$; $G - U_0 = -Nk_B T \ln \frac{q}{N}$ **15.3** (a) and (b) $P = k_B T \left(\frac{\partial \ln Q}{\partial V}\right)_T = nRT \left(\frac{\partial \ln q}{\partial V}\right)_T$ **15.4** 446.9 J K^{-1} mol^{-1}

15.5 2.48 kJ mol^{-1} **15.6** 8.3145 J K^{-1} mol^{-1}

15.7 (a) $\frac{1}{2}L\epsilon$ when $T \rightarrow \infty$ (b) $R \ln 2$ (c) RT (d) RT

15.8 $n_{J/N} = \frac{(2J + 1)}{q_r} e^{\frac{-J(J+1)\hbar^2}{2Ik_B T}}$ **15.9** $U_m - U_{0,m} = (3/2)RT$

15.10 (a) 1.45×10^{32}; 1.223×10^{25} (b) 7.47×10^{31}; 1.183×10^{25} (c) 6.75×10^{32}; 1.315×10^{25} **15.11** 51.9; 2.38×10^{24} **15.12** (a) 1.09 (b) 42.8 **15.13** Assumption of closely spaced levels is not valid. **15.14** 154.7 J K^{-1} mol^{-1} **15.15** (a) 1.26

(b) 6.90 **15.16** 2, 12, 6, 2, 6, 3, 6, 12, 1, 2 **15.17** $kBT/\sigma Bh$ **15.18** 186.2 J K^{-1} mol^{-1} **15.19** 150.3 J K^{-1} mol^{-1}; 47.2 J K^{-1}; very small **15.20** zero **15.21** (a) S_m = constant + (3/2)R ln M_r (b) no dependence; S_m = constant + (5/2)R ln T; = $5R/2T$; (c) $dS/dT = 5R/2T$ **15.22** 162 J K^{-1} mol^{-1} **15.23** −228.3 kJ mol^{-1} **15.24** 0.695 **15.25** (a) 6.47×10^5 bar^{-1} (b) 3.58×10^{-7} bar^{-2} **15.26** (a) 2 (b) 4 (c) 4 (d) 3 (e) 2 **15.27** 0.180 bar **15.28** 1.68 bar **15.29** 2.14×10^{-5} bar **15.30** $\delta\epsilon_\nu = h\nu/[(1 + h\nu/2k_BT) - (1 - h\nu/2k_BT)]$ **15.31** 14.9 **15.32** (a) 4×10^6 dm^3 mol^{-1} s^{-1} (b) 4×10^6 dm^3 mol^{-1} s^{-1} (c) 4×10^3 dm^3 mol^{-1} s^{-1} (d) 4×10^{-29} dm^3 mol^{-1} s^{-1} **15.34** $T^0, T^{-0.5}, T^{-1}, T^{-1.5}, T^{-2}$ **15.35** For linear complex $T\ T^{2.5}/T^{1.5}\ T^{2.5} = T^{-0.5}$ **15.36** Eq. 15.152 gives $k = \left(\frac{k_BT}{h}\right)\frac{q\ddagger}{q_Aq_B}e^{-E_0/k_BT}$

15.37 $N^+ \sim T^{1.5}$, $N_2 \sim T^{2.5}$, complex nonlinear $\sim T^3$. Therefore, $T\ T^3/T^{1.5}(T^{2.5})^2 = T^{-2.5}$ **15.38** (a) T^{-1}, (b) $T^{-1.5}$ (c) $T^{-1.5}$ (d) T^{-2} (e) $T^{-0.5}$ **15.39** C–H bond cleavage is greater than that of C–D bond. Rate of C–H form is greater. **15.40** 5.659×10^{-21} m^3 s^{-1}

Chapter 16

16.1 (a) 2 basis groups (b) 1 basis group **16.2** (a) 4 lattice points (b) 2 lattice points **16.3** (a) 78.5% (b) 90.7% (c) triangular, 1.16 times **16.4** (a) simple, 47.64%; body-centered, 31.98%; face-centered, 25.95% (b) fcc **16.5** $D = \frac{Mz}{abcL}$ **16.6** (a) 1.4447 Å (b) 10.498 g cm^{-3} **16.7** (a) 2.176 Å (b) 3.595 g cm^{-3} **16.8** (a) 1.432 Å (b) 2.700 g cm^{-3} **16.9** 1.96 g cm^{-3} **16.10** 8 **16.11** 2.176 g cm^{-3} **16.12** d_{H-} = 1.44 Å **16.13** 2.001 g cm^{-3} **16.14** 546 pm **16.15** (243) **16.16** $d_{100} = a$; $d_{110} = a/\sqrt{2}$; $d_{111} = a/\sqrt{3}$ **16.17** (362), $(3\bar{2}3)$, $(11\bar{1})$

16.18 (a) $d_{hkl} = \left(\frac{h^2}{a^2} + \frac{k^2}{b^2} + \frac{l^2}{c^2}\right)^{-1/2}$

(b) $d_{hkl} = \left(\frac{1}{\frac{h^2 + k^2}{a^2} + \frac{l^2}{c^2}}\right)^{1/2}$

16.19 d_{100} = 389 pm, d_{111} = 225 pm, d_{121} = 159 pm **16.20** θ_{100} = 9.08°, θ_{010} = 6.64°, θ_{111} = 12.49° **16.21** θ = 11.1° **16.22** θ_{100} = 2.71°, θ_{010} = 1.91°, θ_{111} = 3.63° **16.23** θ_{100} = 4.57°, θ_{111} = 8.15° **16.24** A, (0, 1, 0); B, (−1, 1, 0); C, (2, 1, 0); D, (1, 1, 0) **16.25** (a) θ = 6.25° (b) θ = 13.70° **16.26** 6.12 pm **16.27** 1.44 Å **16.28** fcc **16.29** bcc **16.30** (a) bcc (b) 286 pm, θ = 32.6° (c) 123.8 pm **16.31** (200), 266.7 pm; (110), 377.1 pm; (222), 154.0 pm **16.32** cubic system **16.33** (a) 408.6 pm (b) N (Ag) = 4.0; fcc **16.34** a = 464.2 pm **16.35** Calculate $d_{hkl} = \lambda/2 \sin\theta$ and index with $a = d_{hkl}\sqrt{(h^2 + k^2 + l^2)}$. For 20°36, hkl = 111, a = 564.0 **16.36** CdS **16.37** +4 **16.38** $r = 0.225\ R$ **16.39** $r = 0.414\ R$ **16.40** 673 kJ mol^{-1} **16.41** (a)

d/pm	323	309	304	284	274
Θ	13.81°	14.45°	14.69°	15.75°	16.33°

(b) d = 323 pm, hkl = (622); d = 309 pm, hkl = (444); d = 304 pm, hkl = (543) **16.42** $d_{12.95°}$ = 344 pm, hkl = (201); $d_{13.76°}$ = 324 pm, hkl = (002); $d_{14.79°}$ = 302 pm, hkl = (040) **16.43** (a) $4f_{Zn} - 4if_s$ (b) 540.5 pm **16.44** 384 K

Chapter 17

17.1 3580 bar; −20.9 kJ mol^{-1} **17.2** 19 660 bar **17.3** 2416 bar; −21.0 kJ mol^{-1} **17.4** 3690 bar **17.5** 13 400 bar; 2540 bar; 3010 bar; 2400 bar **17.6** 0.343 bar; 3690 bar **17.7** (a) 3420 bar (b) 5440 bar **17.8** (b) 132.0 J K^{-1} mol^{-1} (c) 20.7 J K^{-1} mol^{-1} **17.9** 10^{-6} **17.10** −42.9 kJ mol^{-1} **17.11** -9.23×10^{-19} J; −556 kJ mol^{-1} **17.12** -1.48×10^{-20} J; −8.89 kJ mol^{-1} **17.13** -1.22×10^{-21} J; −735 J mol^{-1} **17.14** 415 pm; -1.31×10^{-21} J; −789 J mol^{-1} **17.15** -2.78×10^{-23} J; −16.8 J mol^{-1}; -1.31×10^{-21} J; −790 J mol^{-1} **17.16** He: -7.28×10^{-24} J, −4.38 J mol^{-1}; Ne: -2.60×10^{-23} J; −15.2 J mol^{-1}; Ar: -2.86×10^{-22} J; −172 J mol^{-1}; Kr: -5.66×10^{-22} J; −341 J mol^{-1} Xe: -1.41×10^{-21} J, −850 J mol^{-1} **17.17** 358 J mol^{-1}; 227 J mol^{-1}; 893 J mol^{-1}; 1.30 kJ mol^{-1}; 2.42 kJ mol^{-1}; 5.3 kJ mol^{-1} **17.18** -1.20×10^{-23} J; −7.22 J mol^{-1} **17.19** (a) $E/E_{min} = -[n/(6 - n)](r_0/r)^6 + [6/(6 - n)](r_0/r)^n$ (b) $(r^*/r)^{n-6} = 6/n$ (c) $E = 4E_{min}[(r^*/r)^6 - (r^*/r)^{12}]$

Chapter 18

18.1 (a) 1 bar^{-1} (b) 3 bar, 9 bar, 99 bar, 999 bar (c) 0.091, 0.33, 0.999

18.2 $\frac{P}{V} = \frac{1 + KP}{V_0K} = \frac{1}{V_0K} = \frac{P}{V_0}$

18.3 (a) 3.82 dm^3 mol^{-1} (b) 0.1 m^2 **18.4** (a) A plot of ln (θ/P) against ln θ is linear with a slope of −1. (b) ln (V/P) against V has a slope of $-1/V_0$ **18.5** $K = 7.35 \times 10^{-2}$ kPa^{-1}, 222 cm^3 **18.6** (a) Plot $P/V(P_0 - P)$ against P.

(b) If $P_0 \gg P$, the isotherm is $\frac{P}{V} = \frac{1}{V_0 K} + \frac{P}{V_0}$. $\theta = KP/(1 + KP)$ **18.7** 950 cm^3

18.8 $\frac{\theta}{1-\theta} = C_g^{1/2} \frac{h^{3/2}\, b_a}{(2\pi m k_B T)^{3/4}\, b_g^{1/2}} e^{-\Delta E_0/2RT}$

18.9 $C_a = C_g \frac{h}{(2\pi m k_B T)^{1/2}} \frac{b_a}{b_g} e^{-\Delta E_0/RT}$

18.10 (a) 1.5×10^{-3} mol dm^{-3} s^{-1}, 2.0×10^{-2} s^{-1} (b) 1.5×10^{-5} mol dm^{-3} s^{-1}, 2.0×10^{-4} s^{-1} (c) 1.5×10^{-5} mol dm^{-3} s^{-1}, 2.0×10^{-4} s^{-1} (d) $k' = kV/S$ (e) m s^{-1} **18.11** (a) a′ 2.5×10^{-2} mol dm^{-3} s^{-1}, 2.5×10^{-2} s^{-1} (b′) 2.5×10^{-4} mol dm^{-3} s^{-1}, 2.5×10^{-4} s^{-1} (c′) 2.5×10^{-4} mol dm^{-3} s^{-1}, 2.5×10^{-4} s^{-1} (b) $k' = kV/S$; mol m^{-2} s^{-1} (c) mol m^{-2} s^{-1} **18.12** $ka_0/x - 2k/3$ **18.13** $v = k/K^{1/2}\,[H_2]^{1/2}$ **18.14** (a) $E + \Delta H_A$ (b) E (c) $E + \Delta H_A - \Delta H_I$ **18.15** $z = ka_0/(k_s - k)\,[1 - e^{-(k_s - k)t}]$ **18.16** (a) $v = kK[A]$; $E_{a(\text{observed})} = E_{a(0)} + \Delta H_{ad}$ (b) Eq. 18.36 (c) Eq. 18.44 (d) Eq. 18.15, $v = k/K$ **18.17** (a) 1.49 cm (b) 1.49 m **18.18** −10.8 mm **18.19** 1.0017 **18.20** 0.086 N m^{-1} **18.21** 1.5 cm **18.22** No, meniscus forms when the radius of curvature is half what it is in a longer tube. **18.23** 2.56×10^{-3} N m^{-1} **18.24** 2.182 kJ **18.25** 5 Å **18.26** 216 g mol^{-1}; 0.204 nm^2 **18.27** Compressed layer = 0.19 nm^2

Chapter 19

19.1 0.49 Torr **19.2** (a) 0.333 mm s^{-1} (b) 4.19×10^{-15} m^3 (c) 2.00×10^{10} **19.3** (a) 0.516 nm (b) 34.3 nm (c) 1.38×10^{10} s^{-1} (d) 1.70×10^{35} m^{-3} s^{-1} **19.4** (a) 7.20×10^{-4} kg m^{-1} s^{-1} (b) 6.25×10^{-4} kg m^{-1} s^{-1} **19.5** (a) 0.085 m^3 kg^{-1} (b) 0.254 m^3 kg^{-1} (c) 0.627 m^3 kg^{-1} **19.6** 29 200 **19.7** (a) 3.48×10^{-4} kg m^{-1} s^{-1} (b) 4.33×10^{-4} kg m^{-1} s^{-1}; 8.7 kJ mol^{-1} **19.8** 6.53×10^{-4} kg m^{-1} s^{-1} (a) 17.5 kJ mol^{-1} (b) 12.1 kJ mol^{-1} **19.9** (a) zero (b) and (c) Viscosity decreases with increasing T. **19.10** (a) 1.78×10^{-5} kg m^{-1} s^{-1} (b) 9.97×10^{-5} m^2 s^{-1} (c) 1202 m s^{-1} (d) 1.655×10^{-7} m (e) 7.263×10^9 s^{-1} (f) 9.75×10^{34} m^{-3} s^{-1} **19.11** 4.5 cm **19.12** (a) 10.8 cm (b) 0.96 cm **19.13** 8.67×10^{-6} cm^2 s^{-1} **19.14** 1.20×10^{-5} cm^2 s^{-1} **19.15** 3.4 nm; 132 000 g mol^{-1} **19.16** 0.61 s **19.17** 3.5×10^5 s **19.18** 70 860 **19.19** 272 600 **19.20** (a) 5.80×10^{-11} kg s^{-1} (b) 4.43×10^{-11} kg s^{-1} **19.21** 1.12×10^5 s **19.22** 3.44×10^{-11} s **19.23** (a) 150 000 (b) 4.46 nm **19.24** (a) 62 700 (b) 9.1×10^{-5} m (c) 2.6×10^{-10} m (d) 1.33×10^{-5} m (e) 3.10 nm (f) 2.66 nm **19.25** (a) 1552 mh^{-1} (b) 15.52 m h^{-1} (c) 15.52 cm h^{-1} (d) 0.15 mm h^{-1} (e) 1.5×10^{-7} m h^{-1}; for $r = 10^{-8}$m (f) 7700 rpm **19.26** 2.142×10^{-16} m^2 s^{-1}, 1.24 μm; 2.142×10^{-15} m^2 s^{-1}, 3.9 μm; 2.142×10^{-14} m^2 s^{-1}, 12.4 μm; 2.142×10^{-13} m^2 s^{-1}, 39 μm; 2.142×10^{-12} m^2 s^{-1}, 390 μm **19.27** 252 000 **19.28** 1.9×10^3 rpm

INDEX

A page number in **boldface** indicates that a physical quantity is defined or that an equation or concept is defined and explained. A table is indicated by (T), and (B) means that a biography is given on that page. Attention is called to Appendix A (pp. 999–1012), which includes all of the quantities used in this book, with their units and definitions.

C

D

E

F

The Greek Alphabet

A, α . . . *Alpha*	H, η . . . *Eta*	N, ν . . . *Nu*	T, τ . . . *Tau*
B, β . . . *Beta*	Θ, ϑ, θ . . . *Theta*	Ξ, ξ . . . *Xi*	Υ, υ . . . *Upsilon*
Γ, γ . . . *Gamma*	I, ι . . . *Iota*	O, o . . . *Omicron*	Φ, φ, ϕ . . . *Phi*
Δ, δ . . . *Delta*	K, κ . . . *Kappa*	Π, π . . . *Pi*	X, χ . . . *Chi*
E, ϵ . . . *Epsilon*	Λ, λ . . . *Lambda*	P, ρ . . . *Rho*	Ψ, ψ . . . *Psi*
Z, ζ . . . *Zeta*	M, μ . . . *Mu*	Σ, σ . . . *Sigma*	Ω, ω . . . *Omega*

SI Prefixes

Fraction	Prefix	Symbol	Multiple	Prefix	Symbol
10^{-1}	deci	d	10	deka	da
10^{-2}	centi	c	10^{2}	hecto	h
10^{-3}	milli	m	10^{3}	kilo	k
10^{-6}	micro	μ	10^{6}	mega	M
10^{-9}	nano	n	10^{9}	giga	G
10^{-12}	pico	p	10^{12}	tera	T
10^{-15}	femto	f	10^{15}	peta	P
10^{-18}	atto	a	10^{18}	exa	E
10^{-21}	zepto	z	10^{21}	zetta	Z
10^{-24}	yocto	y	10^{24}	yotta	Y

SI Base Units

Physical Quantity	Symbol for Quantity	SI Unit	Symbol for Unit
Length	l	metre	m
Mass	m	kilogram	kg
Time	t	second	s
Thermodynamic temperature	T	kelvin	K
Electric current	I	ampere	A
Luminous intensity	I_{ν}	candela	cd
Amount of substance	n	mole	mol

Physical Constants*

Speed of light in vacuum	c	2.998×10^{8} m s^{-1}
Elementary charge	e	1.602×10^{-19} C
Avogadro constant	L	6.022×10^{23} mol^{-1}
Gas constant	R	8.3145 J K^{-1} mol^{-1}
		0.083 15 bar dm^{3} K^{-1} mol^{-1}
		0.082 06 atm dm^{3} K^{-1} mol^{-1}
Molar volume at STP (0 °C and 1 atm pressure)	V_m	22.41 dm^{3} mol^{-1}
Molar volume at NSTP** (25 °C and 1 bar pressure)	V_m	24.79 dm^{3} mol^{-1}
Boltzmann constant	k_B	1.381×10^{-23} J K^{-1}
Planck constant	h	6.626×10^{-34} J s
Faraday constant	F	96 485 C mol^{-1}
Electron rest mass	m_e	9.109×10^{-31} kg
Proton rest mass	m_p	1.673×10^{-27} kg
Rydberg constant	R	1.097×10^{7} m^{-1}
Permittivity of vacuum	ϵ_0	8.854×10^{-12} C^{2} N^{-1} m^{-2} (or F m^{-1})
	$1/4\pi\epsilon_0$	0.8988×10^{10} N m^{2} C^{-2}
Standard acceleration of gravity	g	9.807 m s^{-2}

*See Appendix B for additional constants and for values to more significant figures.

**"New Standard Temperature and Pressure." Note the change of both temperature and pressure.